Proceedings of the

International Congress of Mathematicians

August 21–29, 1990, Kyoto, Japan

I

Proceedings of the International Congress of Mathematicians

ICM-90
International Congress of Mathematicians
KYOTO

August 21–29, 1990
Kyoto, Japan

Volume I

The Mathematical Society of Japan

Springer-Verlag

Tokyo Berlin Heidelberg New York
London Paris Hong Kong Barcelona
Budapest

International Congress of Mathematicians
August 21–29, 1990, Kyoto, Japan

Editor:

Ichiro Satake
Mathematical Institute
Faculty of Science
Tohoku University
Sendai 980, Japan

The logo for the ICM-90, designed by Kazuyoshi Aoki and Yuji Komai, symbolizes a Japanese stone lantern, the first character for Kyoto, as well as the character for 10^{16}.

With 96 figures, including 11 halftone illustrations

ISBN 4-431-70047-1 Set (2 volumes)
Springer-Verlag Tokyo Berlin Heidelberg New York

ISBN 3-540-70047-1 in 2 Bänden
Springer-Verlag Berlin Heidelberg New York Tokyo

ISBN 0-387-70047-1 Set (2 volumes)
Springer-Verlag New York Berlin Heidelberg Tokyo

Library of Congress Cataloging-in-Publication Data
International Congress of Mathematicians (1990: Kyoto, Japan)
Proceedings of the International Congress of Mathematicians, August 21-29, 1990, Kyoto / edited by Ichiro Satake.
p. cm. Includes bibliographical references. ISBN 0-387-70047-1
1. Mathematics - Congresses. I. Satake, Ichiro. 1927–. II. Title QA1.I82 1990 510–dc20
91-4972 CIP

Printed in Hong Kong

Typesetting: Asco Typesetting Ltd., Hong Kong; Springer TEX in-house system and typesetting output by Universitätsdruckerei H. Stürtz AG, Würzburg, Fed. Rep. of Germany
Printing and binding: Permanent Typesetting & Printing Co., Ltd., Hong Kong

543210

Preface

The *Proceedings* of the International Congress of Mathematicians 1990, held in Kyoto from August 21 through 29, 1990, are printed in two volumes.

Volume I contains the official record of the Congress, the list of ordinary members, the reports on the work of the Fields Medalists and the Nevanlinna Prize Winner, the plenary addresses, as well as the invited addresses in Sections 1–6. Volume II contains the invited addresses in Sections 7–18. A complete author index is included in both volumes.

The short communications made by members at the Congress are not included in the *Proceedings*, but the names of the communicators are marked in the list of ordinary members. Summaries of these contributions, if received in time, were printed in the *Abstracts*, issued to members during the Congress.

Sendai, April, 1991 The editor

Contents

Volume I

Volume II

Past Congresses

1897 Zürich
1900 Paris
1904 Heidelberg
1908 Rome
1912 Cambridge, UK
1920 Strasbourg
1924 Toronto
1928 Bologna
1932 Zürich
1936 Oslo
1950 Cambridge, USA
1954 Amsterdam
1958 Edinburgh
1962 Stockholm
1966 Moscow
1970 Nice
1974 Vancouver
1978 Helsinki
1982 Warsaw (held in 1983)
1986 Berkeley

Kyoto International Conference Hall

Past Fields Medalists and Rolf Nevanlinna Prize Winners

Recipients of Fields Medals

1936 Lars V. Ahlfors
Jesse Douglas

1950 Laurent Schwartz
Atle Selberg

1954 Kunihiko Kodaira
Jean-Pierre Serre

1958 Klaus F. Roth
René Thom

1962 Lars Hörmander
John W. Milnor

1966 Michael F. Atiyah
Paul J. Cohen
Alexander Grothendieck
Stephen Smale

1970 Alan Baker
Heisuke Hironaka
Sergei P. Novikov
John G. Thompson

1974 Enrico Bombieri
David B. Mumford

1978 Pierre R. Deligne
Charles L. Fefferman
Grigorii A. Margulis
Daniel G. Quillen

1982 Alain Connes
William P. Thurston
Shing-Tung Yau

1986 Simon K. Donaldson
Gerd Faltings
Michael H. Freedman

Rolf Nevanlinna Prize Winners

1982 Robert E. Tarjan

1986 Leslie G. Valiant

Audience in the Main Hall of the Kyoto International Conference Hall

Organization of the Congress

The International Congress of Mathematicians 1990 was held in Kyoto at the invitation of the Science Council of Japan (JSC) and the Mathematical Society of Japan (MSJ) under the cosponsorship of the Japan Society of Mathematical Education, the History of Science Society of Japan, the Institute of Actuaries of Japan, the Japan Society for Software Science and Technology, the Japan Statistical Society, and the Operations Research Society of Japan. The Congress was also supported by the Ministry of Education, Science and Culture, by Kyoto Prefecture, Kyoto City, Kyoto University, and the Information Processing Society of Japan.

The members of the Organizing Committee and its subcommittees are listed on the following pages. However, until the Organizing Committee was established on August 14, 1989, all preparations were carried out by the Committee of ICM-90, which was set up in the MSJ in December, 1986.

The scientific program of the Congress was in the hands of the Program Committee, which was appointed by the International Mathematical Union (IMU) in May, 1987. Its members were Nicolaas H. Kuiper (Chairman), Vladimir I. Arnold, Alain Connes, Ronald L. Graham, Heisuke Hironaka, Masaki Kashiwara, Robert P. Langlands, Sigeru Mizohata, and Daniel G. Quillen.

Recipients of the Fields Medals and the Rolf Nevanlinna Prize were selected by the respective committees appointed by the IMU in April, 1988. The Fields Medal Committee consisted of Ludwig D. Faddeev (Chairman ex officio), Michael F. Atiyah, Jean-Michel Bismut, Enrico Bombieri, Charles L. Fefferman, Kenkichi Iwasawa, Peter D. Lax, and Igor Shafarevich. The Rolf Nevanlinna Prize Committee consisted of László Lovász (Chairman), Alexandre J. Chorin, Michael Rabin and Volker Strassen.

The Organizing Committee was responsible for the other activities of the Congress. The Japan Travel Bureau (JTB) handled accommodations for the participants as the official travel agent.

All sessions of the Congress took place inside the Kyoto International Conference Hall. There were 15 one-hour Plenary Addresses and 138 forty-five-minute Section Lectures in 18 Sections. In addition to these invited lectures, 5 forty-five-minute lectures were invited by the International Commission on Mathematical

Instruction. Participants gave 620 Short Communications in total and conducted more than 40 informal seminars and other meetings.

Various social events were arranged for all participants. Ancient court music and dance accompanied the Opening Ceremony. A reception followed the Presentation of the Works of the Fields Medalists and the Rolf Nevanlinna Prize Winner. A cello recital was held in the evening of August 23, and traditional music performances in the afternoon of August 25. Sunday, August 26 was a holiday, on which several guided tours were arranged by JTB. The Closing Ceremony was followed by a banquet with an entertainment of folk music and dance.

In the evening of August 22, Kyoto Prefecture, Kyoto City, and Kyoto University were hosts to a reception for the Fields Medalists, the Rolf Nevanlinna Prize Winner, invited speakers, and officials of the IMU.

The Organizing Committee of the International Congress of Mathematicians 1990

Honorary Members

Shôkichi Iyanaga
  Member of the Japan Academy, University of Tokyo, Prof. Emer.

Kôsaku Yosida†
  Member of the Japan Academy, University of Tokyo and Kyoto University, Prof. Emer.

Mitio Nagumo
  Osaka University, Prof. Emer.

Saburo Kawai
  Former Chairman of The Institute of Actuaries of Japan, Honorary Chairman of The Kyoei Life Insurance Company, Limited

Masanori Yamanouchi
  Former Chairman of The Institute of Actuaries of Japan

Executive Committee

President
  Kunihiko Kodaira, University of Tokyo, Prof. Emer.

Vice Presidents
  Kiyosi Itô, Kyoto University, Prof. Emer.
  Hikosaburo Komatsu, University of Tokyo

General Secretary
  Huzihiro Araki, RIMS, Kyoto University

Secretary Treasurer
  Hirosi Toda, Kyoto University

Secretaries
  Heisuke Hironaka, Member of the Japan Academy, Harvard University
  Akio Hattori, Meiji University, President of the Mathematical Society of Japan
  Shigeru Iitaka, Gakushuin University
  Sigeru Mizohata, Osaka Electro-Communication University
  Masayoshi Nagata, Okayama University of Science

Kodaira, 1954 Fields Medalist, President of the Organizing Committee

Itô, Honorary President of the Congress

Members

Hiroshi Fujita, Meiji University
Ichiro Satake, Tohoku University
Itiro Tamura, Tokyo Denki University

Organizing Committee

Members representing the Science Council of Japan (JSC)

Hirotugu Akaike, Institute of Statistical Mathematics, Member of JSC, 4th Division
Huzihiro Araki, RIMS, Kyoto University, Liaison Committee for Physics
Hiroshi Fujita, Meiji University, Member of JSC, 4th Division
Nobuyuki Ikeda, Osaka University, Liaison Committee for Mathematics
Seizô Itô, Kyorin University, Liaison Committee for Mathematics
Hikosaburo Komatsu, University of Tokyo, Liaison Committee for Mathematics
Yoshitomo Matsuo, Science University of Tokyo, Liaison Committee for Mathematics
Sigeru Mizohata, Osaka Electro-Communication University, Liaison Committee for Mathematics
Hisasi Morikawa, Nagoya University, Liaison Committee for Mathematics
Masayoshi Nagata, Okayama University of Science, Member of JSC, 4th Division
Sadao Nakajima, Department of Physics, Tokai University, Member of JSC, 4th Division
Ichiro Satake, Tohoku University, Liaison Committee for Mathematics
Kei Takeuchi, University of Tokyo, Member of JSC, 3rd Division
Tadashi Ueno, College of Arts and Sciences, University of Tokyo, Member of JSC, 4th Division
Masaya Yamaguti, Ryukoku University, Liaison Committee for Mathematics

Members representing the Sponsoring Societies

Akio Hattori, Meiji University, President of the Mathematical Society of Japan
Heisuke Hironaka, Harvard University
Shigeru Iitaka, Gakushuin University
Kiyosi Itô, Kyoto University, Prof. Emer.
Tosihusa Kimura, Science University of Tokyo
Kunihiko Kodaira, University of Tokyo, Prof. Emer.
Isamu Mogi, President of Bunkyo University, President of the Japan Society of Mathematical Education
Shingo Murakami, Osaka University
Shuji Naito, Managing Director of Mitsui Mutual Life Insurance Company, President of The Institute of Actuaries of Japan
Takushiro Ochiai, University of Tokyo

Tadao Oda, Tohoku University
Itiro Tamura, Tokyo Denki University
Hirosi Toda, Kyoto University
Toshio Yamazaki, President of The History of Science Society of Japan
Nobuo Yoneda, Department of Information Science, University of Tokyo, President of the Japan Society for Software Science and Technology

Subcommittees of the Organizing Committee

General Committee

Chairman
Hiroshi Fujita, Meiji University

Members
Shōzo Koshi, Hokkaido University
Mutsuhide Matsumura, University of Tsukuba
Yukiyosi Kawada, University of Tokyo, Prof. Emer.
Takuo Fukuda, Tokyo Institute of Technology
Koichi Ogiue, Tokyo Metropolitan University
Mitsuo Morimoto, Sophia University
Yuji Ito, Keio University
Tokitake Kusama, Waseda University
Takeyuki Hida, Nagoya University
Nobuo Shimada, Okayama University of Science
Kenji Ueno, Kyoto University
Shinzo Watanabe, Kyoto University
Mikio Sato, RIMS, Kyoto University
Masaki Kashiwara, RIMS, Kyoto University
Akira Mizuhara, Kyoto Sangyo University
Yuzo Hosono, Kyoto Sangyo University
Shôrô Araki, Osaka City University
Reiko Sakamoto, Nara Women's University
Fujitsugu Hosokawa, Kobe University
Kiyosato Okamoto, Hiroshima University
Akio Kudō, Kyushu University
Shigeko Miyajima

Finance Committee

Chairman
Itiro Tamura, Tokyo Denki University

Members
Takeyuki Hida, Nagoya University
Nobuyuki Ikeda, Osaka University
Hirosi Toda, Kyoto University

Science Committee

Chairman
- Sigeru Mizohata, Osaka Electro-Communication University

Members
- Huzihiro Araki, RIMS, Kyoto University
- Hiroshi Fujita, Meiji University
- Heisuke Hironaka, Harvard University
- Toshihide Ibaraki, Faculty of Engineering, Kyoto University, Appointed by The Operations Research Society of Japan
- Masao Iri, Faculty of Engineering, University of Tokyo
- Masaki Kashiwara, RIMS, Kyoto University
- Hikosaburo Komatsu, University of Tokyo
- Yukio Matsumoto, University of Tokyo
- Yoshitomo Matsuo, Science University of Tokyo, Appointed by the Japan Society of Mathematical Education
- Tamotsu Murata, Momoyama-gakuin University, Appointed by The History of Science Society of Japan
- Akihiro Nozaki, International Christian University, Appointed by the Japan Society for Software Science and Technology
- Takushiro Ochiai, University of Tokyo
- Tadao Oda, Tohoku University
- Yujiro Ohya, Faculty of Engineering, Kyoto University
- Kiyosato Okamoto, Hiroshima University
- Ichiro Satake, Tohoku University
- Ryoichi Shimizu, Institute of Statistical Mathematics, Appointed by The Japan Statistical Society
- Shinzo Watanabe, Kyoto University

Local Committee

Chairman
- Masayoshi Nagata, Okayama University of Science

Members
- Huzihiro Araki, RIMS, Kyoto University
- Hirosi Toda, Kyoto University
- Masayoshi Miyanishi, Osaka University
- Fujitsugu Hosokawa, Kobe University
- Haruo Murakami, Faculty of Engineering, Kobe University
- Akira Kono, Kyoto University
- Yuzo Hosono, Kyoto Sangyo University
- Masaki Kashiwara, RIMS, Kyoto University
- Masaki Maruyama, Kyoto University
- Shinzo Watanabe, Kyoto University
- Kyoji Saito, RIMS, Kyoto University
- Kenji Ueno, Kyoto University

Tetsuji Miwa, RIMS, Kyoto University
Chiaki Tsukamoto, Kyoto Institute of Technology
Takahiro Kawai, RIMS, Kyoto University
Yasuo Yamazaki, RIMS, Kyoto University

Publications Committee

Chairman
Ichiro Satake, Tohoku University

Members
Tadao Oda, Tohoku University
Shige Toshi Kuroda, Gakushuin University
Takushiro Ochiai, University of Tokyo
Kazuo Okamoto, College of Arts and Sciences, University of Tokyo

Foreign Affairs Committee

Chairman
Akio Hattori, Meiji University

Members
Shingo Murakami, Osaka University
Reiji Takahashi, Sophia University
Takushiro Ochiai, University of Tokyo
Shinichi Kotani, University of Tokyo
Shigeko Miyajima

Publicity Committee

Chairman
Shigeru Iitaka, Gakushuin University

Member
Mitsuo Morimoto, Sophia University

Committee of ICM-90

The members of the Organizing Committee and the members of the General Committee except Prof. Nakajima of the former constitute the Committee of ICM-90, which was responsible for the preparations of ICM-90 until August 14, 1989, when the Organizing Committee started. Officers of the Committee of ICM-90 are the following:

President
Kunihiko Kodaira, University of Tokyo, Prof. Emer.

Vice Presidents
Kiyosi Itô, Kyoto University, Prof. Emer.
Heisuke Hironaka, Harvard University

Chairman
Hikosaburo Komatsu, University of Tokyo
Secretary
Huzihiro Araki, Kyoto University
Treasurer
Hirosi Toda, Kyoto University
Executive Members
Shigeru Iitaka, Gakushuin University
Sigeru Mizohata, Osaka Electro-Communication University
Masayoshi Nagata, Okayama University of Science

The Committee of ICM-90 has the same subcommittees as the Organizing Committee and the following Fund Raising Committee:

Fund Raising Committee

Chairman
Kiyosi Itô, Kyoto University, Prof. Emer.
Vice Chairmen
Akio Hattori, Meiji University
Takeyuki Hida, Nagoya University
Hiroaki Hijikata, Kyoto University
Ken Hirose, Waseda University
Tosihusa Kimura, Science University of Tokyo
Haruo Murakami, Faculty of Engineering, Kobe University
Shingo Murakami, Osaka University
Minoru Nakaoka, Okayama University of Science
Masahiko Saito, College of Arts and Sciences, University of Tokyo
Tosiya Saito, Kawaijuku
Members
Shôrô Araki, Osaka City University
Koji Doi, Ritsumeikan University
Yasutaka Ihara, RIMS, Kyoto University
Shigeru Iitaka, Gakushuin University
Mitsuru Ikawa, Osaka University
Hideyuki Matsumura, Nagoya University
Shigeko Miyajima
Masatake Mori, Faculty of Engineering, University of Tokyo
Hisasi Morikawa, Nagoya University
Mitsuo Morimoto, Sophia University
Takushiro Ochiai, University of Tokyo
Akihiro Okuyama, Kobe University
Kayo Otsuka, Faculty of Economics, Teikyo University
Tatuo Simizu, Polytechnic Consultants Inc.
Fumiyuki Terada, Waseda University
Nobuo Yoneda, Department of Information Science, University of Tokyo

List of Donors

International Mathematical Union
Science Council of Japan
Mathematical Society of Japan
Members of the Mathematical Society of Japan

The Life Insurance Association of Japan
The Taniguchi Foundation
Association of Tokyo Stock Exchange Regular Members
The Federation of Bankers Associations of Japan
The Marine & Fire Insurance Association of Japan, Inc.
The Trust Company Association of Japan

Japan Federation of Construction Contractors, Inc.
The Federation of Electric Power Companies
Fujitsu Limited
Hitachi, Ltd.
The Inamori Foundation
Mitsubishi Electric Corporation
NEC Corporation
Incorporated Foundation Oriental Life Insurance Cultural Development Center
Toshiba Corporation
Commemorative Association for the Japan World Exposition (1970)
Japan Automobile Manufacturers Association, Inc.
Japan Association for Mathematical Sciences

IBM Japan, Ltd.
Nippon Telegraph and Telephone Corporation
The Kajima Foundation
Matsushita Electric Industrial Co., Ltd.
Oki Electric Industry Co., Ltd.
Nihon Unisys, Ltd.
The Portopia 81 Foundation

Nippon Steel Corporation
Hattenkai Volunteers
Kyocera Corporation
Murata Machinery, Ltd.
Murata Manufacturing Co., Ltd.
NTT Data Communications Systems Corporation
Omron Corporation
Sanyo Electric Co., Ltd.
Sharp Corporation
Shimadzu Corporation
Suntory Limited
Wacoal Corporation

NKK Corporation
Kawasaki Steel Corporation
Fuji Country Co., Ltd.
Mitsuru Sangyo Co., Ltd.
Tokyo Shoseki Co., Ltd.
Sumitomo Metal Industries, Ltd.

Kobe Steel, Ltd.
Asahi Glass Co., Ltd.
The Norinchukin Bank
Shinko-Shuppansha Keirinkan Publishing Co., Ltd.
Toyobo Co., Ltd.
Kubota Corporation
Dainippon Ink & Chemicals, Inc.
Daiwa House Industry Co., Ltd.
The Ito Foundation
Japan Meat Conference
Japan Tobacco Inc.
Kake Educational Institution, Okayama University of Science
Kao Corporation
Meiji Milk Products Co., Ltd.
Nippon Meat Packers, Inc.
Renown Incorporated
Suken Shuppan Co., Ltd.
Nisshin Steel Co., Ltd.
Anonymous

Ricoh Company, Ltd.
Nakayama Steel Works, Ltd.
Godo Steel, Ltd.
Bridgestone Corporation
Dentsu Inc.
Ebara Corporation
Hakuhodo Inc.
Hitachi Hi Soft, Ltd.
Isetan Company Ltd.
Ito-Yokado Co., Ltd.
Japan Information Processing Service Co., Ltd.
The Japan Research Institute, Limited
K. I. Systems Inc.
Kumon Institute of Education
Mitsukoshi Ltd.
Obunsha Co., Ltd.
Osaka Shoseki Co., Ltd. Publishers
Toppan Printing Co., Ltd.

Daido Steel Co., Ltd.
Topy Industries, Limited
Gendaisugakusha Co., Ltd.
Kyouritsu Shuppan Co., Ltd.
Takaoka Electric Mfg. Co., Ltd.
Yodogawa Steel Works, Ltd.
Mitsubishi Steel Mfg. Co., Ltd.
The Japan Steel Works, Ltd.

Faddeev giving the opening speech to the Congress
(From left to right: *Araki, Komatsu, Itô, Hattori, Kondo, Lovász, Hori, Aramaki, Tanabe, Nishijima, Onozawa*)

Opening Ceremonies

The opening ceremonies of the Congress were held in the Event Hall of the Kyoto International Conference Hall, starting at 9:30 a.m. on August 21, 1990. The Gagaku Club of Tenri University entertained the participants with a *Gagaku* (court music) recital entitled *Etenraku*. Academician Ludwig D. Faddeev, President of the International Mathematical Union (IMU), opened the Congress with the following speech:

> After this symbolic opening let me formally declare the International Congress of Mathematicians (ICM-90) in Kyoto open.
>
> This is the first Congress in the history of the International Mathematical Union to take place outside of Europe and North America. This is consonant with the main goal of the Union – the promotion of mathematical research throughout the world. For this reason, the proposal of the Japanese Committee of Mathematicians to hold the Congress in Kyoto was enthusiastically accepted by the General Assembly of IMU four years ago. I believe that I can already express admiration for the efforts being made by the Organizing Committee to make the Congress work effectively. It appears that the attendance is the highest in the history of ICM.
>
> The scientific program of ICM is traditionally in the hands of the Program Committee appointed by the Executive Committee of IMU. Let me disclose to you the list of its members. Professor Nicolaas H. Kuiper was appointed as chairman. Professors Vladimir I. Arnold, Alain Connes, Ronald L. Graham, Heisuke Hironaka, Masaki Kashiwara, Robert P. Langlands, Sigeru Mizohata, and Daniel G. Quillen served as members. We shall witness the effectiveness of their choice of speakers in the coming days.
>
> Now I come to the important duty of designating the President of the Congress. Let me nominate Professor Hikosaburo Komatsu for this post.
>
> *Acclamation.*
>
> I take this as confirmation of the proposal. Thank you.

Komatsu, President of the Congress, showing the first sheet of the ICM-90 commemorative stamps

Professor Komatsu, as President of the Congress, made the following address:

Thank you Mr. President. I am honored to serve as President of the Congress.

Ladies and Gentlemen,

On behalf of the Organizing Committee I would like to welcome you to the International Congress of Mathematicians 1990 (ICM-90). I very much hope that your stay in the historical city of Kyoto will be a pleasant one.

First of all, let me explain how the Congress has been prepared. It was more than seven years ago that the Liaison Committee for Mathematics in the Science Council of Japan (JSC), which is the Committee for Mathematics in Japan, and the Mathematical Society of Japan (MSJ) started planning to host ICM-90. After the Feasibility Committee chaired by Professor Kiyosi Itô made a careful initial investigation of this issue, MSJ decided in 1985 to invite ICM-90 to Kyoto. Fortunately this invitation was accepted by the Site Committee of IMU and was approved by the General Assembly of IMU at Oakland in 1986.

The actual preparations started when MSJ formed the Committee of ICM-90 in December, 1986. Its Executive Committee consists of the following:

President	Kunihiko Kodaira
Vice Presidents	Kiyosi Itô
	Heisuke Hironaka
Chairman	Hikosaburo Komatsu
Secretary	Huzihiro Araki
Treasurer	Hirosi Toda
Members	Shigeru Iitaka
	Sigeru Mizohata
	Masayoshi Nagata

The full Committee of ICM-90 had an additional 29 members.

By the end of 1987, the Japan Society of Mathematical Education, the History of Science Society of Japan, the Institute of Actuaries of Japan, the Japan Society for Software Science and Technology, the Japan Statistical Society, and the Operations Research Society of Japan decided to sponsor the Congress and elected their representatives to the Committee of ICM-90 and its Science Committee.

In June, 1989, JSC made the final decision of sponsoring ICM-90 with the consent of the Government. Then, according to the general rules of JSC, the Committee of ICM-90 was reorganized into the Organizing Committee in August, 1989.

Second, as the President of IMU reported, the scientific program of the Congress and, in particular, the list of invited speakers were prepared by the Program Committee, appointed by the Executive Committee of IMU. However, a few additional speakers working in Japan, and also substitutes for speakers who declined were invited by the Organizing Committee on its own initiative, in which the latter were selected from the list of alternatives made by the Program Committee.

Finally 15 speakers were invited to give one-hour Plenary Addresses. Among them two declined and were replaced by other choices. Also 144 speakers were invited to give 45-minute Section Lectures, 19 were substituted, and 6 lectures were cancelled. In addition, the International Commission on Mathematical Instruction invited five 45-minute speakers.

We have benefited very much by the recent reconciliation in the world politics and the prosperity of the Japanese economy. As far as I know, no invited speakers declined for political or economic reasons.

Adopting Professor Mary Rudin's proposal, the last General Assembly of IMU in 1986 recommended that Subfields of Mathematics, Women, and Mathematicians in Small Countries should not be overlooked when speakers to the ICM are selected. I believe both the Program Committee and the Organizing Committee have respected this recommendation.

The selection of additional speakers and the scheduling of scientific program were carried out by the Science Committee in the Organizing Committee, chaired by Sigeru Mizohata.

Thirdly, the budget of this Congress amounted to approximately 300,000,000 Yen (approximately US $ 2,000,000). One third of the revenue is the registration fees, one third the donations from private corporations, and the rest consists of subventions from IMU, JSC and MSJ, the donations by individual members of MSJ, and the miscellaneous income (here the donations by 1138 members of MSJ are actually more than the total of all the other subventions). We are very sorry that we had to set a high registration fee of 30,000 Yen but we were forced to do so because the Japanese tax regulations do not allow us to receive tax-exempted donations exceeding the total amount of registration fees.

In addition to the budget of the Organizing Committee, the Japan Association for Mathematical Sciences, MSJ, the Oriental Life Insurance Cultural Development Center, the Commemorative Association for the Japan World Exposition (1970), and The Kajima Foundation allocated a total of 60,000,000 Yen to support 269 foreign participants mostly from developing countries or countries with foreign currency restrictions. This number includes 47 IMU Scholars whose travel expenses are supported by the Special Development Fund of IMU.

Most funds from these foundations also originated from donations solicited from enterprises by the Organizing Committee.

The biggest contributors are insurance and electronic companies. On behalf of all the participants, I would like to thank all these bodies for their generous donations. A list of donors is in the Program and will be published in the Proceedings.

It is not an easy task to raise so much money. But I must confess that it was a pleasant one, too, because every executive I met for this purpose showed a liking for mathematics and appreciated that mathematics had played an important role in the development of the Japanese economy.

The Fund Raising Committee is chaired by Kiyosi Itô and its office was set up in Gakushuin University with Shigeru Iitaka in charge.

Assistance with visa applications was carried out by Shinichi Kotani at the office located in the University of Tokyo.

The other preparations have been planned and implemented by General Secretary Huzihiro Araki, Treasurer Hirosi Toda and the Local Committee chaired by Masayoshi Nagata. The Secretariat was set up in the Research Institute for Mathematical Sciences, Kyoto University.

The Proceedings of the Congress will be edited by the Publications Committee with Ichiro Satake as the chairman. Its office is in Tohoku University.

I would like to thank these institutions and, in particular, Kyoto University for their kind cooperation.

Today, 4,000 mathematicians from 83 countries have assembled here to review our scientific achievements over the last few years and to set goals for the future in all fields of mathematics ranging from pure mathematics through applied mathematics to mathematical education. This seems to be an almost megalomaniac dream at this time of specialization. I do not know any other discipline which attempts to hold this kind of congresses regularly.

I have often wondered why mathematicians do have Congresses and what Congresses mean to them. My answer is that Congresses are to mathematicians what *Bon* and New Year Festivities are to Japanese, in which they abandon their daily life completely.

Japanese are believed to work continuously without vacations, but that is not true. Even in the Edo Period there were two one-week long holidays. One is the New Year Festivities and the other is the *Bon Festivities* which take place a week earlier than this time of the year.

On these holidays people are relieved from labor and go back to their native home. People are not allowed to cook on the first days of the Festivities, so that they have a busy time preparing all meals before the holidays start.

New Year Festivities are associated with the future. We renew everything we can and start again. *Bon Festivities* are for the past. We receive ancestors' ghosts, make conversations with them, and then send them back. In cities like Kyoto people decorate their entrance halls with their treasures and keep the doors open. The whole city becomes a big museum. In the countryside people gather in the village square and dance. There is no distinction between performers and spectators; all dance. That is the way Japanese refresh themselves, inherit their traditions and unite. *Bon* and New Year Festivities also give young people the opportunity to meet together to make a new family.

I would like to ask all speakers at the Congress to make their lectures accessible to a wide audience, and not just to the specialists, at least in the first part. This is certainly not the time to work in a daily manner.

I hope you all enjoy this big event.

Thank you.

Professor Akio Hattori, President of the Mathematical Society of Japan, spoke as follows:

Mr. Chairman, Distinguished Guests, Ladies and Gentlemen,

I should like to extend a warm welcome to you all for having gathered from all parts of the world to attend this Congress. The Japanese mathematical community is proud of a pioneer; I am referring to Professor Kiyosi Itô, a renowned probabilist. He has been instrumental in bringing this Congress to Japan and in its organization. I propose that Professor Itô be elected Honorary President of this Congress.

Professor Itô was elected by acclamation and spoke as follows:

Mr. Chairman, Distinguished Guests, Ladies and Gentlemen,

It is my great honor to speak to you as Honorary President of the Congress.

Before now, some Japanese mathematicians have tried to bring the Congress to Japan. As early as the 1960s, Professors S. Iyanaga and Y. Kawada began campaigning for the Congress to be held in Japan. Now that this Congress is starting, I feel a certain nostalgia for their pioneering efforts.

When we formed the Organizing Committee chaired by Professor Kunihiko Kodaira, the first Japanese Fields medalist, our big problem was how to raise funds. In this respect, I would like to give credit to Professor Kodaira. Impressed by his fame and his fine personality Mr. H. Tanimura, Professor Kodaira's friend, together with Messrs. R. Ishikawa and M. Kitoku set up a fund-raising program for us and appealed to leading companies. We thank them for their tremendous efforts, which enabled us to make concrete plans for the Congress. To our great regret, Professor Kodaira is unable to come here today. I would like to express our sincere gratitude to him on this occasion.

The next problem was that we anticipated there would be less participants than usual because of Japan's geographic location and the worldwide economic upheaval. Professor H. Komatsu proposed to make a grant available to help participants from abroad. This idea was welcomed by several foundations whose support contributed to the increase in the number of participants.

In Japan, people have been more concerned with science and technology than with mathematics. Despite this, we are proud of two exceptional projects which have helped mathematicians: the International Symposia supported by Mr. Toyosaburo Taniguchi and secondly, the Exchange Program of the Japan Association for Mathematical Sciences established by Professor H. Hironaka, the second Japanese Fields medalist. Through these projects many Japanese mathematicians have acquired a sense of international cultural exchange, which is perhaps one of the main aims of the Congress.

A large number of members of the Mathematical Society of Japan helped the Organizing Committee in many respects, academically or nonacademically. At the final stage of preparation a tremendous amount of business had to be carried out within a limited time. We are very grateful to Professors H. Komatsu, H. Araki, H. Toda, S. Iitaka, S. Kotani, and their secretaries for all their hard work.

We are very happy to welcome you here in Kyoto. Please enjoy the lectures and participate as much as possible in the informal discussions.

Thank you.

Professor Jiro Kondo, President of the Science Council of Japan, gave the following welcome address:

Mr. Chairman, Distinguished Guests, Ladies and Gentlemen,

It is a great pleasure to have this opportunity, on behalf of the Science Council of Japan, to speak to all of you who have gathered here from as many as 80 countries, at the opening of the International Congress of Mathematicians 1990.

The Science Council of Japan was established in 1949 as government organization representing qualified Japanese scientists both internally and internationally, covering all scientific fields consisting of Cultural, Social and Natural Sciences. The aim of the Council is to promote scientific development and to improve administration, industry and living standards through science.

Since that time, we have been working to contribute to the progress of science in cooperation with the academic organizations of the world by sponsoring many international congresses here in Japan, and by sending Japanese delegations to international congresses held overseas. We do this because we believe that the promotion of international scientific exchanges is one of our most important duties.

We have opened today the International Congress of Mathematicians in cooperation with the Mathematical Society of Japan, the Japan Society of Mathematical Education, the History of Science Society of Japan, the Institute of Actuaries of Japan, the Japan Society of Software Science and Technology, the Japan Statistical Society, and the Operations Research Society of Japan. It is an extraordinary pleasure for me to have this opportunity to be with so many distinguished scientists from around the world, and to listen to your lectures and presentations on the recent achievement.

The first Meeting of the International Congress of Mathematicians was held in Zürich in 1897. Since then, the Congress has been held nearly every four years somewhere in the world, and this is the first Congress to be held outside Europe and North America. I offer my heartfelt congratulations on the remarkable growth of the Congress in terms of both quality and number of participants and at the same time I express my great pleasure in being able to welcome the Congress to Japan for the first time. Since I graduated from the Department of Mathematics, Imperial University of Kyoto in 1940, I am very happy to hold this Congress here in Kyoto. After my graduation I worked mostly in the field of mathematical applications, such as theoretical aerodynamics, operations research, systems engineering and environmental sciences.

Recent international trends clearly indicate the importance of basic research in science and technology. In particular, mathematical sciences are entrusted with the crucial mission of providing the backbone for all sciences and the theoretical basis for the new era of information science in the twenty-first century, and they must also live up to people's high expectations. Moreover, because of the rapid progress in computer technology, many scientists in other fields, as well as engineers and businessmen, are coming to expect a greater contribution from mathematicians.

As researchers of all branches of mathematics gather from all over the world at this International Congress, it is expected that much interaction among the different branches will be made and that this will result in a new development of mathematics as a unified discipline. While we see recently a strong tendency of specialization in all branches of science, I have learned that essential progress in mathematics has often arisen out of the unexpected connection of different areas. The interaction of researchers with different cultural backgrounds is also expected to produce such an effect. I hear that this Congress is organized with an emphasis on interdisciplinary and international interactions and efforts have been made to provide various opportunities for contact among participants. I hope that through the interaction with neighbouring fields of science which will inject new blood into mathematics and through friendship among participants crossing the boundaries of different branches of mathematics and different countries, this Congress serves as a stepping stone towards active worldwide research cooperation in the future.

I earnestly hope that everyone who participates in this Congress, will gain worthwhile experience in both fundamental and applied aspects of mathematics.

In closing, I sincerely hope for the great success of the Congress. I also hope that all of you from abroad will enjoy their stay in Japan. I believe that this Congress will become truly memorable for you through contact with fellow scientists, and I wish this to be a chance for you to learn more about Kyoto, the ancient capital of Japan, and about Japanese culture.

Thank you.

The following congratulatory telegram from Prime Minister Toshiki Kaifu was read both in Japanese and in English by Professor Huzihiro Araki:

PLEASED TO EXTEND A HEARTY WELCOME TO ALL DELEGATES FROM AROUND THE WORLD AT THE OPENING OF THE 11TH GENERAL ASSEMBLY OF THE INTERNATIONAL MATHEMATICAL UNION AND THE INTERNATIONAL CONGRESS OF MATHEMATICIANS 1990 HELD IN JAPAN UNDER THE COSPONSORSHIP OF THE SCIENCE COUNCIL OF JAPAN, MATHEMATICAL SOCIETY OF JAPAN, JAPAN SOCIETY OF MATHEMATICAL EDUCATION, THE OPERATIONS RESEARCH SOCIETY OF JAPAN, THE HISTORY OF SCIENCE SOCIETY, JAPAN SOCIETY FOR SOFTWARE SCIENCE AND TECHNOLOGY, THE JAPAN STATISTICAL SOCIETY AND THE INSTITUTE OF ACTUARIES OF JAPAN.

I WISH A GREAT SUCCESS IN THIS INTERNATIONAL CONGRESS.

TOSHIKI KAIFU
PRIME MINISTER

Mr. Kosuke Hori, Minister of Education, Science and Culture gave a congratulatory address as follows:

Distinguished Delegates, Ladies and Gentlemen of the Audience,

It is a real pleasure and privilege for me to be afforded this opportunity to address congratulatory greetings to this distinguished audience at the opening of the Twenty-first International Congress of Mathematicians.

First of all, I would like to extend my sincerest welcome to you all, especially to those who have travelled far in order to come to Japan. At the same time, I am deeply grateful and wish to pay respect to all the people who have made great efforts in mathematical research.

Recently, pure mathematics, applied mathematics, mathematical education, and other fields of mathematics have achieved remarkable progress in research. This International Congress has been held every four years since

1897 for the purpose of bringing together mathematicians in all fields from all over the world in order to share the significant progress achieved in each area of mathematics and to help further new progress in mathematics. In these respects, the International Congress of Mathematicians has attained excellent achievements.

I am told that this International Congress is the first to be held in Asia, and distinguished mathematicians from Japan and abroad who participate in it will present many new discoveries in mathematics. I am confident that this Congress will further enrich the knowledge that is common property to all human beings.

I hope that those participants who have travelled from overseas will have the opportunity to acquaint themselves with some aspects of Japanese culture and society and that they will have a pleasant stay in Japan.

Finally, I would like to pay our deepest respects to those people who have already worked so hard in organizing this Congress and I anticipate that it will be highly successful.

Thank you very much for your attention.

Mr. Teiichi Aramaki, Governor of Kyoto Prefecture, gave a congratulatory address as follows:

Distinguished Delegates, Ladies and Gentlemen of the Audience,

As Governor of Kyoto Prefecture, I would like to extend my sincerest congratulations for the Twenty-first International Congress of Mathematicians to be held here in Kyoto from today, with distinguished mathematicians assembled from various countries of the world.

The First International Congress of Mathematicians was held in Zürich, Switzerland in 1897 and since then, the congress has produced excellent results, establishing a long history and tradition. The fact that the Fields Medal, which is said to be the Nobel Prize of mathematics, and the Nevanlinna Prize have been awarded to superb mathematicians has attracted the attention of not only the people concerned but also many others.

It is a great pleasure and privilege for us living in Kyoto Prefecture that this authoritative International Congress is being held in Kyoto which itself is recognized as an academic and scientific center. On behalf of the people of Kyoto Prefecture, I would like to extend a cordial welcome to all of you.

The International Congress of Mathematicians has been held only in western cities so far and so it is a great pleasure and honor that this is the first International Congress of Mathematicians to be held outside Europe and North America. I am sure that this fact will further encourage many Japanese mathematicians.

Finally, I would like to pay our deepest respects to the scientists of the Research Institute for Mathematical Sciences, Kyoto University and to those who have gone to great pains in organizing this International Congress. I anticipate that each of the participating mathematicians will take an active role in the discussions and that the Congress will be highly successful.

Thank you for your attention.

Dr. Tomoyuki Tanabe, Mayor of Kyoto City, gave the following congratulatory address:

Ladies and Gentlemen,

Firstly, please let me welcome you all to Kyoto.

I am delighted that the Twenty-first International Congress of Mathematicians is starting today and on behalf of the citizens of Kyoto, I would like to extend a cordial welcome to you all.

As you may know, Kyoto is an ancient city with a history that looks back on 1,200 years of culture and tradition. This is why Kyoto is said to be the spiritual root of Japan.

Kyoto is rich in cultural heritage and a variety of time-honored customs based on Japanese tradition still exist as part of our everyday way of life. Traditional industries which embody the quintessence of Japanese folk art and technology can also be found.

On the other hand, Kyoto is a scientific city which has produced several superb Nobel Prize winners and Fields Medalists. At the same time, Kyoto is an industrial city which has given birth to world leaders in high-tech industries.

About ten years ago, we declared Kyoto as a City Open to the Free Exchange of World Cultures. In this connection, it is a profound feeling of pleasure and privilege that the International Congress of Mathematicians is being held here in Kyoto.

I am confident that everyone who has travelled from afar to participate in this International Congress, will renew old friendships as well as begin new acqaintances. I also hope that you will enjoy the mixture of traditional Japanese culture and the modern atmosphere of our city.

Let me close my greetings by wishing the International Congress of Mathematicians success and combine this with a wish for your good health and happiness.

Thank you for your attention.

Dr. Yasunori Nishijima, President of Kyoto University, delivered this congratulatory address:

Ladies and Gentlemen,

It is my honor and great pleasure to greet this distinguished gathering at the opening of the International Congress of Mathematicians.

I have learned that this Congress had its first meeting in 1897. That is also the year that Kyoto University was founded as the Kyoto Imperial University and this marked the establishment of the modern academe in the old capital. Ninety-four years have passed since then and today Kyoto welcomes this gathering of the world's notable mathematicians to Japan for the first time. Kyoto is most honored to be the chosen site.

Last week, the General Conference of the International Association of Universities took place in Helsinki, Finland. It was very cool and dry, unlike the hot and wet weather here. The Chancellor, Professor Olli Lehto, and all the members of the University of Helsinki, including the most delightful of students, made the meeting a great success.

The theme of the Conference was "The Mission of the University: Universality, Diversity, and Interdependence," and I had the opportunity to chair the whole-day commission on "Diversity within Universality." In this rapidly and dramatically changing world, Diversity tends to be emphasized. However, at the same time, Universality is the human aspiration to create a higher degree of wisdom for the future and this is the mission of the world's academic community.

After the exciting and fruitful meeting in Helsinki, I returned to Japan a few days ago. In the airport transit lounge, I met Chancellor Lehto, who was on his way to this Congress. He was holding a small bag on his lap with great care. I thought it was the medal for the Rolf Nevanlinna Prize which will be presented at today's Ceremony. I said, "Is that top secret?" Chancellor Lehto simply smiled and nodded. We boarded the same airplane for the last stretch of our long journey from Helsinki to Kyoto. This morning, following the performance of a classic Japanese court dance, *Bugaku*, the presentation of the Fields Medals and the Rolf Nevanlinna Prize will reveal all the secrets.

I really think mathematics is the core of the universality of human wisdom. Mathematics has since ancient times conveyed wisdom throughout the history of human endeavor. At the same time, mathematics covers all branches of knowledge and leads us along the road to Truth and Reason.

David Hilbert said, "Mathematics is an organism for whose vital strength the indissoluble union of the parts is a necessary condition."

I would like to congratulate all those who have worked so hard to prepare and organize this most important meeting. I anticipate it will be a great success.

I wish all participants a most fruitful and pleasant meeting in Kyoto.

On behalf of Takashi Fukaya, Minister of Posts and Telecommunications, Mr. Tomoyuki Onozawa, Director General of Posts, presented a sheet of commemorative stamps with the following letter:

To Professor Hikosaburo Komatsu, President of the International Congress of Mathematicians.

It is of high significance and a matter for congratulations that the Twenty-first International Congress of Mathematicians is being held here in Kyoto, for the first time in Asia.

The Ministry of Posts and Telecommunications has issued a specially designed postage stamp to commemorate this event, and the first sheet of stamps is now presented to you.

From Takashi Fukaya, Minister of Posts
and Telecommunications
On August 21, 1990.

Kyoto Gagaku-Kai then performed a *Bugaku* (court dance) entitled *Gosechi no Mai* accompanied by the Gagaku Club of Tenri University which played a *Gagaku* (court music).

Academician Ludwig D. Faddeev, Chairman of the Fields Medal Committee, announced the recipients of the Fields Medals as follows:

The Fields Medal and Prize Committee is appointed by the Executive Committee of the IMU. For this term the following people were appointed and worked for the Committee: Professors M. Atiyah, J. M. Bismut, E. Bombieri, C. Fefferman, K. Iwasawa, P. D. Lax and I. Shafarevich. Since it is the duty of the President of the IMU to chair the work of this Committee, I served as chairman.

After thorough consideration of the material at our disposal, we decided to award four Medals. The recipients are

Vladimir G. Drinfeld
Vaughan F. R. Jones
Shigefumi Mori
Edward Witten

I believe that these names are well known to the mathematical community throughout the world. Their scientific contributions will be described during the afternoon session.

Let me proceed to the pleasant task of presenting the awards. I ask the Honorable Minister Hori to do this.

The winners came forward and received their medals and prize checks from Mr. Kosuke Hori, Minister of Education, Science and Culture.

Professor László Lovász announced the recipient of the Rolf Nevanlinna Prize as follows:

The Nevanlinna Prize Committee consisted of Alexandre Chorin from Berkeley, Michael Rabin from Jerusalem, Volker Strassen from Konstanz, and Lázló Lovász from Budapest as chairman. After considering a number of outstanding candidates, the Committee decided to award the Rolf Nevanlinna

Prize to Alexander A. Razborov from the Steklov Institute in Moscow, for his groundbreaking work on lower bounds for circuit complexity.

The winner came forward and received his medal and prize check from Mr. Kosuke Hori, Minister of Education, Science and Culture.

The opening ceremonies adjourned at 11:00 a.m.

Closing Ceremonies

The closing ceremonies were held in the Main Hall of the Kyoto International Conference Hall, starting at 1:15 p.m. on August 29, 1990.

Academician Ludwig D. Faddeev, President of the International Mathematical Union began the closing ceremonies with the following words:

> Time runs very fast and we have come to the end of our Congress. I believe that we can now judge very highly the results of the scientific program and congratulate the Program Committee on their success. Personally, I was glad to observe how prominently Mathematical Physics was represented in its connections with other domains of Mathematics.
>
> The work of the Congress has been so smooth that some of you did not notice the efforts of the Organizing Committee to achieve this. We are all highly indebted to our Japanese colleagues for their excellent job.
>
> Let me also add that the general atmosphere of the Congress has been very friendly and this has allowed us to concentrate on purely mathematical problems.
>
> It is my pleasure to inform you that the Emperor and Empress of Japan invited the winners of the Fields Medals and the Nevanlinna Prize to visit them in Tokyo. The Honorary President of the Congress, Professor Itô, as well as Professor Hironaka and I were also present. The interest expressed by the Imperial Family in the work of our Congress is a great honor to us all.
>
> Let me now inform you of the results of the General Assembly which took place in Kobe just before the Congress. It represented 52 members of the IMU most of whom attended the meeting.
>
> During the last few years Saudi Arabia has been accepted as a new member. Spain and Israel have upgraded their membership to Group III.
>
> The General Assembly elected new Committees and Commissions.
>
> The new Executive Committee is composed as follows:

President	J. L. Lions
Vice-Presidents	J. Coates
	D. Mumford

Secretary	J. Palis
Members	J. Arthur
	A. Dold
	H. Komatsu
	L. Lovász
	E. Zehnder

I shall remain on the Executive Committee as the past President.

Half of the Executive Committee has changed. So it is appropriate to cite one of the resolutions adopted at the General Assembly: "The General Assembly gives special thanks to Professor Olli Lehto for his excellent work as Secretary to the IMU during the last eight years, ably assisted by Mrs. Mäkeläinen. It also thanks the University of Helsinki and the Finnish Ministry of Education for their generous support of the IMU secretariat over this period."

The General Assembly also appointed the Commissions.

The ICMI Executive Committee:

President	M. de Guzmán
Vice-Presidents	J. Kilpatrick
	Lee Peng-Yee
Secretary	M. Niss
Members	Yu. L. Ershov
	E. Luna
	A. Sierpinska

The ex-officio members are: the past President of ICMI, the President of IMU, the Secretary of IMU and the IMU representative at CTS (ICSU). The next ICME will take place in 1992 in Quebec.

Commission on Developments and Exchange:

Chairman	M. S. Narasimhan
Members	P. Bérard
	C. Camacho
	A. Grunbaum
	A. O. Kuku
	J. Mawhin
	T. Ochiai
	P. L. Papini
	Wu Wen Tsun

The ex-officio members are: the past Chairman of CDE, the President of IMU, the Secretary of IMU and the IMU representative at COSTED (ICSU).

Besides the administrative business, there was lively and important discussion at the General Assembly on the role of Applied Mathematics and its balance in the program of the ICM, the increasing relevance of Mathematics in Industry, and related problems of mathematical education.

Finally the Site Committee made its proposal to the General Assembly on the time and location of the next Congress, which was adopted after some

Chatterji inviting the Congress to Zürich

discussion. The next Congress, ICM-94 will be held in Zürich, Switzerland. I shall now step down and give the floor to Professor Chatterji.

Professor S. D. Chatterji of the Ecole Polytechnique Fédérale de Lausanne invited the audience to the next International Congress of Mathematicians with these words:

Ladies and Gentlemen,

On behalf of the Swiss Mathematical Society and the entire Swiss mathematical community, it is my great privilege to invite you all to the 1994 International Congress of Mathematicians in Zürich. As you know, the honor and responsibility of organizing the Congress have fallen on Zürich twice before in the past: in 1897 and 1932. Zürich is an important financial, commercial and cultural centre of Switzerland; it is also the seat of two of our important institutions of higher education—the Swiss Federal Institute of Technology (ETH) and the University of Zürich. Situated in beautiful natural surroundings in the heart of Europe, Zürich is easily accessible by rail, road and air.

Our Japanese hosts have set such high levels of hospitality and efficiency at this Congress that it would be difficult to match them. However, we shall do our best to make your participation at the 1994 Congress agreeable and fruitful.

Rendezvous then for August 1994 in Zürich.

Dr. Jose Felipe Voloch of the Instituto de Matemática Pura e Aplicada, Brazil, and Professor Carol Wood of Wesleyan University, USA, President-Elect of the Association of Women in Mathematics, were invited to give comments on ICM-90.

Professor Hikosaburo Komatsu, President of the Congress, closed the meeting and the Congress with these words:

Thank you for your kind remarks. The organizers of the Congress are really rewarded to hear them. However, all praise should go to Professors Kenji Ueno, Masaki Maruyama, and Chiaki Tsukamoto, as well as to Miss Tanaka, Mrs. Ichiki, Miss Ishii to name just a few of the people who have carried out all the plans of the Congress so perfectly.

Our thanks are also due to Professors Gleason and Mesirov for sending us the material they prepared for the last Congess in Berkeley. This was of great help to us.

This Congress has been admired for its smooth organization but this was true only in appearance. In spite of all the devoted works, I often had to remain in the Secretariat to solve many minor problems. Even when I was attending a lecture, the buzzer, you might have heard, called me back there. Therefore, I must admit that my impression of the Congress is very partial. Nevertheless, I felt that we are at another turning point in the history of mathematics. The previous one was marked at the Second Congress in 1900 when Hilbert gave his famous lecture. Since then we have obtained an enormous number of general results by axiomatization and abstract formulation, often at the hands of mathematical giants. This time it is a transition from abstract simplification to more concrete synthesis. We are now in a fortunate time when we can solve many problems which remained open for many years in spite of all the efforts of past generations of mathematicians. We no longer have a single genius but many people work together developing new strong streams. It was only many brooks last time at the Congress in Berkeley. They meet together and now we see a big river or a sea or even an ocean.

On behalf of all participants, I would like to praise Professor Kuiper and the other members of the Program Committee for their outstanding work in selecting the invited speakers. I would also like to thank the speakers for their admirable efforts.

At our time of democracy, international cooperation is indispensable and so is the unity of mathematics. We are proud of the fact that this Congress has helped these goals.

As reported in the Daily Bulletin, the Congress has been attended by 3,954 ordinary members, 452 accompanying members, and 92 child members from 76 countries.

Unfortunately, all the preregistered members from 7 countries were unable to attend because of the crisis in the Persian Gulf. World politics has once again cast its shadow on the Congress. I do hope that the next Congress in 1994 will be a truly universal one.

I declare the Congress closed.

The closing ceremonies adjourned at 1:45 p.m.

Scientific Program

Invited One-Hour Addresses at the Plenary Sessions

Invited Forty-Five Minute Addresses at the Section Meetings

Section 1: Mathematical Logic and Foundations

There were 15 Short Communications in this section.

Section 2: Algebra

There were 59 Short Communications in this section.

Section 3: Number Theory

There were 29 Short Communications in this section.

Section 4: Geometry

There were 38 Short Communications in this section.

Section 5: Topology

There were 43 Short Communications in this section.

Section 6: Algebraic Geometry

There were 25 Short Communications in this section.

Section 7: Lie Groups and Representations

There were 23 Short Communications in this section.

Section 8: Real and Complex Analysis

There were 73 Short Communications in this section.

Section 9: Operator Algebras and Functional Analysis

There were 52 Short Communications in this section.

Section 10: Probability and Mathematical Statistics

There were 37 Short Communications in this section.

Section 11: Partial Differential Equations

There were 48 Short Communications in this section.

Section 12: Ordinary Differential Equations and Dynamical Systems

There were 41 Short Communications in this section.

Section 13: Mathematical Physics

There were 39 Short Communications in this section.

Section 14: Combinatorics

There were 25 Short Communications in this section.

Section 15: Mathematical Aspects of Computer Science

There were 5 Short Communications in this section.

Section 16: Computational Methods

There were 26 Short Communications in this section.

Section 17: Applications of Mathematics to the Sciences

There were 18 Short Communications in this section.

Section 18: History, Teaching and the Nature of Mathematics

There were 22 Short Communications in this section.

In addition there were 2 Short Communications in the Post Deadline Session.

List of Participants

The 4102 participants are listed alphabetically.
The asterisks indicate those 651 participants who gave or coauthored short communications.
Those 151 with § signs were absent, although they were either invited participants or have paid the registration fee.

ABD-ALLA, Abo-el-nour N., EGYPT
ABE, Eiichi, JAPAN
ABE, Hitoshi, JAPAN
ABE, Kojun, JAPAN
ABE, Masanori, JAPAN
ABE, Naoto, JAPAN
ABE, Takatsugu, JAPAN
ABE, Takehisa, JAPAN
ABE, Yukitaka, JAPAN
ABERER, Karl, SWITZERLAND
ABI KHZAM, Farouk F.*, LEBANON
A'CAMPO, Norbert, SWITZERLAND
ACHUTHAN, Paninjukunnath*, INDIA
ADACHI, Kenzo, JAPAN
ADACHI, Masahisa*, JAPAN
ADACHI, Toshiaki, JAPAN
ADACHI, Yuko, JAPAN
ADAMS, David R., U.S.A.
ADAMYAN, Vadim M.*, U.S.S.R.
AGAOKA, Yoshio, JAPAN
AGEMI, Rentaro, JAPAN
AGOU, Simon J.*, FRANCE
AGUILAR, Marcelo A., MEXICO
AHARA, Kazushi*, JAPAN
AHERN, Patrick R., U.S.A.
AHMED, Naseer*, PAKISTAN
AHMEDOU, Haouba, MAURITANIA
AHUJA, Mangho*, U.S.A.
AIBA, Akira, JAPAN
AIDA, Haruko, JAPAN
AIDA, Shigeki, JAPAN
AIKAWA, Hiroaki*, JAPAN
AIKOU, Tadashi, JAPAN
AITCHISON, Iain Roderick*, AUSTRALIA
AKAGI, Misao, JAPAN
AKAHIRA, Masafumi*, JAPAN
AKAHORI, Takao, JAPAN
AKAMATSU, Toyohiro, JAPAN
AKAO, Kazuo, JAPAN
AKASHI, Shigeo§, JAPAN
AKIBA, Shigeo, JAPAN
AKIMOTO, Yoshihisa, JAPAN
AKITA, Yoshihiko, JAPAN
AKIYAMA, Hiroshi, JAPAN
AKIYAMA, Keiichi, JAPAN
AKIYAMA, Kenzi, JAPAN
AKIYAMA, Koichiro, JAPAN
AKUTAGAWA, Kazuo, JAPAN
AKUTSU, Takashi, JAPAN
AKUTSU, Tatsuya, JAPAN
ALAMATSAZ, M. H.§, IRAN
ALAMOLHODAEI, Hassan, IRAN
ALBAR, Muhammad Alawi§, SAUDI ARABIA
ALBER, Solomon J.*, U.S.A.
ALBERTONI, Sergio, ITALY
ALBERTSON, Michael O., U.S.A.
AL-DHAHIR, M. Wassel§, KUWAIT
ALEFELD, Goetz E., F.R.G.
ALEKSEEV, Anatoly Semenovich, U.S.S.R.
ALEXANDROV, Alexander D., U.S.S.R.
ALEXOPOULOS, George, FRANCE
ALIKHANI-KOOPAEI, Aliasghar*, IRAN
ALLAN, Graham R., U.K.
ALMEIDA, Jorge M.G., PORTUGAL
ALON, Noga M., ISRAEL
AL-SHAKHS, Adnan A.§, SAUDI ARABIA
ALSINA, Claudi, SPAIN
ALTMAN, Allen B., U.S.A.
AMANO, Kazuo, JAPAN
AMASAKI, Mutsumi, JAPAN
AMEMIYA, Ichiro, JAPAN
AMFILOKHIEV, Alexandr A., U.S.S.R.

AMIR, Dan, ISRAEL
AMITSUR, Shimshon A., ISRAEL
ANAND, Kailash K.*, CANADA
ANANTHARAMAN, Claire, FRANCE
ANASTASIEI, Mihai*, ROMANIA
ANCONA, Vincenzo, ITALY
ANDERSON, Jack M.*, U.S.A.
ANDERSSON, Karl G., SWEDEN
ANDO, Hideo, JAPAN
ANDO, Shiro*, JAPAN
ANDO, Shoichi, JAPAN
ANDO, Tetsuo, JAPAN
ANDO, Tsuyoshi, JAPAN
ANDO, Yutaka, JAPAN
ANDOU, Yuuki, JAPAN
ANDREATTA, Marco, ITALY
ANDRONIKOF, Emmanuel, FRANCE
ANDROUAIS, Anne M., FRANCE
ANDRUSKIEWITSCH, Nicolas, ARGENTINA
ANICHINI, Giuseppe*, ITALY
ANTOINE, Jean-Pierre, BELGIUM
ANZAI, Kazuo, JAPAN
AOKI, Kazuyoshi, JAPAN
AOKI, Kenji, JAPAN
AOKI, Noboru, JAPAN
AOKI, Nobuo, JAPAN
AOKI, Norihiro, JAPAN
AOKI, Shigeru, JAPAN
AOKI, Takashi, JAPAN
AOKI, Takatoshi, JAPAN
AOMOTO, Kazuhiko, JAPAN
AOYAMA, Hiroshi, JAPAN
AOYAMA, Yoichi, JAPAN
ARAI, Hiroshi, JAPAN
ARAI, Hitoshi*, JAPAN
ARAI, Masaharu, JAPAN
ARAKAWA, Tsuneo*, JAPAN
ARAKELYAN, Norair, U.S.S.R.
ARAKI, Huzihiro, JAPAN
ARAKI, Shoro, JAPAN
ARAKI, Testu, JAPAN
ARAZY, Jonathan§, ISRAEL
ARENS, Richard F., U.S.A.
ARIKI, Susumu, JAPAN
ARIMA, Satoshi*, JAPAN
ARISAKA, Nakaaki, JAPAN
ARISAWA, Mariko, JAPAN
ARISUMI, Tadashi, JAPAN
ARKOWITZ, Martin*, U.S.A.
ARMENDARIZ, Efraim P., U.S.A.
AROCA, José-Manuel*, SPAIN
ARROWSMITH, David K.*, U.K.
ARTEMIADIS, Nicolas K., GREECE
ARUMUGAM, Gurusamy*, INDIA
ARVANITOYEORGOS, Andreas*, U.S.A.
ARYA, Shashi Prabha*, INDIA
ASADA, Akira*, JAPAN
ASADA, Mamoru, JAPAN
ASADA, Teruko, JAPAN
ASAEDA, You*, JAPAN
ASAHI, Masahiro, JAPAN
ASAI, Kazuto, JAPAN
ASAI, Keisui, JAPAN
ASAI, Teruaki, JAPAN
ASAKAWA, Hidekazu, JAPAN
ASAKURA, Fumioki*, JAPAN
ASAKURA, Soichi, JAPAN
ASAMOTO, Noriko, JAPAN
ASANO, Hiroshi, JAPAN
ASANO, Kazuo, JAPAN
ASANO, Kiyoshi, JAPAN
ASANO, Kouhei, JAPAN
ASANO, Shigemoto, JAPAN
ASANO, Tetsuo, JAPAN
ASH, J.Marshall*, U.S.A.
ASHIKAGA, Tadashi, JAPAN
ASHINO, Ryuichi, JAPAN
ASHOUR, Attia Abdel-Salam, EGYPT
ASKEY, Richard A., U.S.A.
ASOH, Shin-ichi, JAPAN
ASOH, Yasuhiro, JAPAN
ASTALA, Kari, FINLAND
ATSUJI, Atsushi, JAPAN
ATSUMI, Tsuyoshi, JAPAN
ATTALLAH, Sabria M. K.*, EGYPT
AUPETIT, Bernard H., CANADA
AUSLANDER, Maurice, U.S.A.
AVDISPAHIĆ, Muharem, YUGOSLAVIA
AWERBUCH, T. E., U.S.A.
AXELSSON, Axel O., NETHERLANDS
AXLER, Sheldon, U.S.A.
AYUPOV, Shavkat A., U.S.S.R.
AZARPANAH, Fariborz, IRAN
AZIMI, Parviz*, IRAN
AZUHATA, Takashi, JAPAN
AZUKAWA, Kazuo, JAPAN
AZUMAYA, Goro, U.S.A.
BAAK, Eric, NETHERLANDS
BAAYEN, P. C., NETHERLANDS
BABAI, László, HUNGARY
BACHELIS, Gregory F.*, U.S.A.
BACHELOT, Agnès, FRANCE
BACHELOT, Alain*, FRANCE
BADE, William George, U.S.A.
BADIOZZAMAN, Abdol J.§, IRAN
BAGCHI, Sitadri N.*, INDIA
BAILEY, Rosemary A., U.K.
BAILY, Walter L. Jr., U.S.A.
BAK, Anthony, F.R.G.
BAKER, Roger C.*, U.K.
BALAKRISHNAN, V. K., U.S.A.
BALOG, Antal, HUNGARY
BANCHOFF, Thomas F., U.S.A.
BANDO, Shigetoshi, JAPAN
BANNAI, Eiichi, JAPAN
BANNAI, Etsuko, JAPAN

BĀNULESCU, Martha C.*, ROMANIA
BARBASCH, Dan M., U.S.A.
BARBOSA, Joao L. M., BRAZIL
BARDOS, Claude W., FRANCE
BARLOTTI, Adriano, ITALY
BARLOW, Martin T., U.K.
BARNES, Frank William*, KENYA
BARRETT, David E.*, U.S.A.
BARTH, Karl F., U.S.A.
BARTLE, Robert G., U.S.A.
BASS, Hyman, U.S.A.
BATTY, Charles J. K., U.K.
BAUER, Heinz, F.R.G.
BAUTISTA-RAMOS, Raymundo, MEXICO
BAXTER, Rodney J., AUSTRALIA
BAYOD, Jose M.*, SPAIN
BEAULIEU, Liliane*, CANADA
BECKER, Helmut, F.R.G.
BECKER, Jerry P., U.S.A.
BECKER, Ronald I., SOUTH AFRICA
BEDFORD, Eric, U.S.A.
BEHBOODIAN, Javad*, IRAN
BEHFOROOZ, G. G.*, U.S.A.
BEILINSON, Alexander, U.S.S.R.
BELAGE, Abel, FRANCE
BELL, Howard E.*, CANADA
BELTRAMETTI, Mauro C.*, ITALY
BENABDALLAH, Khalid, CANADA
BENSON, Clark T., U.S.A.
BENSON, David J., U.K.
BEN-YELLES, Choukri-Bey, ALGERIA
BÉRARD, Pierre H., FRANCE
BERG, Christian*, DENMARK
BERGER, Marcel, FRANCE
BERGER, Ruth I.*, U.S.A.
BERGERON, François*, CANADA
BERNALDEZ, Jose M., PHILIPPINES
BERTHELOT, Pierre R., FRANCE
BERTIN, Emile M. Y., NETHERLANDS
BESCHLER, Edwin F., U.S.A.
BESSIS, Jean-Pierre, FRANCE
BEZARD, Max Jack, FRANCE
BEZDEK, Andras, HUNGARY
BHANDARI, Ashwani K., INDIA
BHATTARAI, Hom Nath, NEPAL
BICHARA, Alessandro, ITALY
BIERSTEDT, Klaus D.*, F.R.G.
BION-NADAL, Jocelyne*, FRANCE
BIRMAN, Joan S., U.S.A.
BIRNIR, Björn*, ICELAND
BISHOP, Christopher J., U.S.A.
BJÖRCK, Göran*, SWEDEN
BJÖRNER, Anders, SWEDEN
BLACKADAR, Bruce E., U.S.A.
BLACKBURN, Norman, U.K.
BLASIUS, Don M., U.S.A.
BLEILER, Steven A.*, U.S.A.
BLOCH, Spencer J., U.S.A.
BLUM, Lenore C., U.S.A.
BODNARESCU, M. V., F.R.G.
BOECHERER, Siegfried, F.R.G.
BOGOLUBOV, N. N. Jr.*, U.S.S.R.
BOGOMOLOV, Fedor A., U.S.S.R.
BOILEAU, Michel§, FRANCE
BOIVIN, André, CANADA
BOJARSKI, Bogdan*, POLAND
BÖKSTEDT, Marcel Anders, F.R.G.
BONAHON, Francis, FRANCE
BONGIORNO, Benedetto, ITALY
BOOTH, Geoffrey L.*, SOUTH AFRICA
BORWEIN, Peter B., CANADA
BOUKRICHA, Abderrahmen, TUNISIA
BOURGAIN, Jean§, FRANCE
BOURGUIGNON, Jean-Pierre, FRANCE
BOUZAR, Chikh, ALGERIA
BOYLE, Ann K., U.S.A.
BOZORGNIA, Abolghasem*, IRAN
BRAMBILA-PAZ, Leticia G.*, MEXICO
BRATTELI, Ola, NORWAY
BRATTSTRÖM, G. B., SWEDEN
BRAVO, Rafael*, SPAIN
BRECKENRIDGE, John Wylie, AUSTRALIA
BREEN, Lawrence S., FRANCE
BREMNER, Andrew, U.S.A.
BRENNAN, Joseph P.*, U.S.A.
BRENNER, Sheila, U.K.
BRENTI, Francesco, U.S.A.
BRIDGES, Douglas S.*, NEW ZEALAND
BRODMANN, Markus Peter*, SWITZERLAND
BRODSKY, Mikhail*, U.S.A.
BROWDER, Felix E., U.S.A.
BROWN, Leon*, U.S.A.
BROWN, Robert F., U.S.A.
BRUCE, James William, U.K.
BRUCK, Ronald E. Jr.§, U.S.A.
BRUNDU, Michela, ITALY
BRUNING, Erwin A. K.*, SOUTH AFRICA
BRÜNING, Jochen Wulf, F.R.G.
BRUNS, Winfried*, F.R.G.
BRUYÈRE, Véronique, BELGIUM
BRZEZINSKI, Juliusz, SWEDEN
BSHOUTY, Daoud H.*, ISRAEL
BUCHSBAUM, David A., U.S.A.
BUCHWEITZ, Ragnar-Olaf, CANADA
BUCKI, Andrzej J.§, U.S.A.
BUI, Huy-Qui, NEW ZEALAND
BUJALANCE, Emilio*, SPAIN
BUJALANCE, José A.*, SPAIN
BULGARELLI, Ulderico*, ITALY
BULTER, Michael C. R., U.K.
BÜNING, Herbert*, F.R.G.
BURCHARD, Paul*, U.S.A.
BURENKOV, Victor Ivanovich*, U.S.S.R.
BURSTALL, Francis E., U.K.
BUTLER, Lynne M.*, U.S.A.
BUTZ, Jeffrey R., U.S.A.

BYLEEN, Karl E., U.S.A.
BYRNE, Catriona M., F.R.G.
BYUN, Chang Ho, KOREA, R. of
CAMACHO, César, BRAZIL
CAMERON, Peter J., U.K.
CANFELL, Michael John, AUSTRALIA
CANO, Felipe*, SPAIN
CAPPELL, Sylvia E., U.S.A.
CARAYOL, Henri L., FRANCE
CARLESON, Lennart, SWEDEN
CARLSON, James Andrew, U.S.A.
CARLSON, Jon F., U.S.A.
CARNEIRO, Mario J. D., BRAZIL
CARROLL, Robert W.*, U.S.A.
CARTER, Jack A., U.S.A.
CARTER, Roger W., U.K.
CASADIO TARABUSI, Enrico, ITALY
CASSELMAN, William A., CANADA
CASSON, Andrew J., U.S.A.
CATANESE, Fabrizio M.E., ITALY
CATERINA, Maniscalco, ITALY
CATTANEO, Uberto*, SWITZERLAND
CATTANI, Eduardo H., U.S.A.
CAYFORD, Afton H., U.S.A.
CECCHERINI, Pier Vittorio, ITALY
CERAMI, Giovanna M., ITALY
CERCIGNANI, Carlo*, ITALY
CHADEMAN, Arsalan*, IRAN
CHALEYAT-MAUREL, Mireille M., FRANCE
CHAN, Clara S., U.S.A.
CHAN, Raymond H., HONG KONG
CHAN, Shih-Ping, SINGAPORE
CHANDNA, Om P.*, CANADA
CHANG, Der-Chen E., U.S.A.
CHANG, Joo Sup, KOREA, R. of
CHANG, Kun Soo*, KOREA, R. of
CHANG, Mei-Chu, CHINA-TAIWAN
CHANG, Sun-Yung A., U.S.A.
CHANSAENWILAI, Petcharat, THAILAND
CHARI, Vyjayanthi, INDIA
CHATTERJI, Srishti D., SWITZERLAND
CHAVENT, Guy*, FRANCE
CHEE, P. S.*, MALAYSIA
CHEMIN, Jean-Yves*, FRANCE
CHEN, Chao-Nien, CHINA-TAIWAN
CHEN, Larry Lung-Kee§, U.S.A.
CHEN, Louis H. Y.*, SINGAPORE
CHEN, Yun-Gang*, CHINA, P. R. of
CHENG, Chong-Qing*, CHINA, P. R. of
CHENG, Jih-Hsin*, CHINA-TAIWAN
CHEREDNIK, Ivan V., U.S.S.R.
CHERUBINI, Alessandra, ITALY
CHEW, Tuan Seng, SINGAPORE
CHIBA, Keiko, JAPAN
CHIBA, Toru, JAPAN
CHIBA, Toshiyuki§, JAPAN
CHIEN, C.-S.*, CHINA-TAIWAN
CHIKATA, Hiroshi, JAPAN
CHIN, Chou-Hsieng Joe*, CHINA-TAIWAN
CHIN, William, U.S.A.
CHISTOV, Alexander L., U.S.S.R.
CHIYONOBU, Taizo, JAPAN
CHO, Su Yon, KOREA, R. of
CHO, Tae Geun, KOREA, R. of
CHODA, Hisashi, JAPAN
CHODA, Marie*, JAPAN
CHOE, Boo Rim*, KOREA, R. of
CHOE, Geon H.*, KOREA, R. of
CHOU, Kai Seng*, HONG KONG
CHOUDHARY, Ram Chandra*, INDIA
CHOW, Bennett, U.S.A.
CHRIST, Carol S., U.S.A.
CHRIST, Lily E., U.S.A.
CHRIST, Michael, U.S.A.
CHRISTODOULOU, Demetrios, U.S.A.
CHU, Sydney C. K., HONG KONG
CHU, Wenchang*, CHINA, P. R. of
CILIBERTO, Carlo, ITALY
CILIBERTO, Ciro, ITALY
CLAESSON, Tomas, SWEDEN
CLARK, W. Edwin§, U.S.A.
CLEMENS, Laura E., U.S.A.
CLOZEL, Laurent Y., FRANCE
CNOP, Ivan J., BELGIUM
COATES, John Henry, U.K.
CODDINGTON, Earl A.§, U.S.A.
COGDELL, James W., U.S.A.
COHEN, Donald S., U.S.A.
COHEN, Maurice, CANADA
COHN, Harvey§, U.S.A.
COHN, Paul M.*, U.K.
COHN, Richard M., U.S.A.
COIFMAN, Ronald R., U.S.A.
COLLINS, Peter J.*, U.K.
COLMEZ, Pierre, FRANCE
CONREY, John Brian, U.S.A.
CONTE, Alberto, ITALY
CONTESSA, Maria, ITALY
CONTRERAS-BARANDIARÁN, Gonzalo, PERU
COOK, Stephen A., CANADA
COOKE, Kenneth L., U.S.A.
COPPEL, William Andrew, AUSTRALIA
CORDUNEANU, Constantin C.*, U.S.A.
CORNEA, Aurel, F.R.G.
CORON, Jean-Michel, FRANCE
COSTA, David G., BRAZIL
COSTE, Michel, FRANCE
COSTES, Constantine N., U.S.A.
CRAPO, Henry*, FRANCE
CRAWLEY-BOEVEY, William W., F.R.G.
CREE, George C., CANADA
CROFT, Hallard Thomas, U.K.
CUBEDDU, Carmen, ITALY
CUEVAS, Félix Humberto, PANAMA
CUI, Chengri, CHINA, P. R. of

CUNTZ, Joachim, F.R.G.
CURTIS, Philip C. Jr., U.S.A.
CURTO, Raul E., U.S.A.
DĂDĂRLAT, Marius Dumitru, ROMANIA
DAHMARDAH, Habib-Olah*, IRAN
DAJCZER, Marcos, BRAZIL
D'AMBROSIO, Ubiratan, BRAZIL
DAMON, James Norman, U.S.A.
DARAFSHEH, Mohammad-Reza*, IRAN
DARMAWIJAYA, Soeparna, INDONESIA
DARMON, Henri René*, CANADA
DATE, Etsuro, JAPAN
DATE, Tamotsu, JAPAN
DAUBEN, Joseph W., U.S.A.
DAVERMAN, Robert J., U.S.A.
DAVID, Sinnou, FRANCE
DAVIES, Roy Osborne§, U.K.
DAVIS, Chandler, CANADA
DEAKIN, Michael A. B.*, AUSTRALIA
DEBNATH, Lokenath*, U.S.A.
DE GRAAF, Jan, NETHERLANDS
DE GUZMÁN, Miguel, SPAIN
DEHGHAN, Mohammad Ali*, IRAN
DEKSTER, Boris V.*, CANADA
DE LA PEÑA, José Antonio M., MEXICO
DE LA PRADELLE, Arnaud, FRANCE
DEL RIEGO, Lilia*, MEXICO
DEMUTH, Michael*, G.D.R.
DENG, Dong-gao*, CHINA, P. R. of
DEWAN, Kum Kum*, INDIA
DIACONIS, Persi, U.S.A.
DIAMOND, Fred I., U.S.A.
DIAS-AGUDO, Fernando R., PORTUGAL
DICKENSTEIN, Alicia M., ARGENTINA
DIEKERT, Volker, F.R.G.
DIMITRIC, Radoslav M.*, YUGOSLAVIA
DINAR, Nathan, ISRAEL
DING, Shisun, CHINA, P. R. of
DING, Tongren*, CHINA, P. R. of
DINH, Dung*, VIETNAM
DINH, Quang Luu, VIETNAM
DIONNE, Philippe-A., FRANCE
DI PIAZZA, Luisa, ITALY
DLAB, Vlastimil*, CANADA
DLOUSSKY, Georges, FRANCE
DOBRUSHIN, Roland L., U.S.S.R.
DO CARMO, Manfredo P., BRAZIL
DODSON, C. T. J.*, CANADA
DOGUWA, Sani Ibrahim, NIGERIA
DOHI, Yutaka, JAPAN
DOI, Masahiro§, JAPAN
DOI, Shinichi, JAPAN
DOI, Yukio, JAPAN
DOKU, Isamu*, JAPAN
DOLCI, Paolo V., ITALY
DÔMAE, Takanobu, JAPAN
DONALDSON, James A., U.S.A.
DOPLICHER, Sergio, ITALY
DOUADY, Adrien, FRANCE
DOUGLAS, Ronald George, U.S.A.
DRANISHNIKOV, Alexander N., U.S.S.R.
DRINFELD, Vladimir G., U.S.S.R.
DUGGAL, Krishan Lal*, CANADA
DUKE, William D.§, U.S.A.
DUMORTIER, Freddy E., BELGIUM
DUOANDIKOETXEA, Javier, SPAIN
DUPONT, Johan L.*, DENMARK
DURFEE, Alan H.*, U.S.A.
DURRETT, Richard Timothy, U.S.A.
DVORETZKY, Aryeh, ISRAEL
DWIVEDI, Brij Narayan*, INDIA
DYMACEK, Wayne M.*, U.S.A.
DYN, Nira*, ISRAEL
DZHUMADIL'DAEV, A. S.*, U.S.S.R.
DZINOTYIWEYI, Heneri A. M.*, ZIMBABWE
DZRBASHIAN, Mhitar M., U.S.S.R.
DZURAEV, Abduhamid D.*, U.S.S.R.
EBERLEIN, Patrick Barry, U.S.A.
EBIHARA, Madoka*, JAPAN
ECALLE, Jean Pascal, FRANCE
EDA, Katsuya, JAPAN
EDAMATSU, Takasi, JAPAN
EDWARDS, Robert D., U.S.A.
EGASHIRA, Shinji, JAPAN
EGGHE, Leo C.*, BELGIUM
EGUCHI, Kazuo, JAPAN
EGUCHI, Masaaki, JAPAN
EGUCHI, Tohru, JAPAN
EHR, Carolyn K., U.S.A.
EHRENPREIS, Leon, U.S.A.
EI, Shinichiro, JAPAN
EIDA, Atsuhiko, JAPAN
EIN, Lawrence, U.S.A.
EJIRI, Norio, JAPAN
EKHAGUERE, Godwin O. S., NIGERIA
ELLERS, Erich W.*, CANADA
ELLERS, Harald E., CANADA
ELLIOTT, George A.*, DENMARK
ELLIS, Alan J., HONG KONG
ELWORTHY, Kenneth David, U.K.
EMARA, Salah Abbas Ahmed*, EGYPT
ENDO, Akira, JAPAN
ENDO, Mikihiko, JAPAN
ENDO, Shizuo, JAPAN
ENGL, Heinz W.*, AUSTRIA
ENOCK, Michel, FRANCE
ENOKI, Ichiro, JAPAN
ENOMOTO, Fumihiko, JAPAN
ENOMOTO, Kazuyuki*, JAPAN
ENOMOTO, Masatoshi, JAPAN
ENRIQUEZ, Benjamin David, FRANCE
ENTA, Yoichi, JAPAN
EPIFANIO, Giuseppina, ITALY
ERBLAND, John Paul, U.S.A.
ERDMANN, Karin, U.K.
ERIKSSON, Folke S.*, SWEDEN

ERIKSSON-BIQUE, Sirkka-L. A.*, FINLAND
ERNÉ, Marcel T.*, F.R.G.
ESCULTURA, Edgar E.*, PHILIPPINES
ESNAULT, Hélène I., FRANCE
ESSÉN, Matts R.*, SWEDEN
ESTRADA, Ricardo*, COSTA RICA
EVANS, David E., U.K.
EVANS, Michael John, CANADA
EZAWA, Hiroshi, JAPAN
EZEILO, James O.C., NIGERIA
FACCHINI, Alberto, ITALY
FADDEEV, Ludwig D., U.S.S.R.
FAGHIH, Nezameddin*, IRAN
FALK, Michael J., U.S.A.
FAN, Peng, CHINA-TAIWAN
FANIA, M.Lucia, ITALY
FARADZEV, Igor A., U.S.S.R.
FARAHI, Mohammad H.*, IRAN
FARMANESH, Mounesali, IRAN
FATHI, Albert*, U.S.A.
FAVIILI, Franco, ITALY
FEE, Greg J., CANADA
FEIG, Ephraim§, U.S.A.
FEIGIN, Boris L., U.S.S.R.
FEIT, Walter, U.S.A.
FELDMAN, Joel S., CANADA
FENG, Keqin*, CHINA, P. R. of
FERENCZI, Sebastien*, FRANCE
FERNANDEZ, Arturo A.*, SPAIN
FIGÀ-TALAMANCA, Alessandro, ITALY
FILLMORE, Peter A., CANADA
FLACH, Matthias, F.R.G.
FLATH, Daniel E.*, U.S.A.
FLATO, Moshé, FRANCE
FLATTO, Leopold, U.S.A.
FLEISCHER, Alexander*, U.S.S.R.
FLICKER, Yuval Z., ISRAEL
FLOER, Andreas, U.S.A.
FOLLAND, Gerald B., U.S.A.
FOROUZANFAR, Abdol M.*, IRAN
FÖRSTER, Karl-Heinz, F.R.G.
FOSSUM, Robert M., U.S.A.
FOSTER, B. L.*, U.S.A.
FOXBY, Hans B., DENMARK
FRANGOS, Nicholas E., U.S.A.
FRANKS, John M., U.S.A.
FREI, Guenther H., SWITZERLAND
FRIBERG, Jöran, SWEDEN
FRIED, David, U.S.A.
FRIEDLAND, Shmuel*, U.S.A.
FRIEDLANDER, John B., CANADA
FRÖBERG, B. Ralf, SWEDEN
FU, Hong Chen*, CHINA, P. R. of
FUCHINO, Sakaé, JAPAN
FUGLEDE, Bent*, DENMARK
FUJII, Akio, JAPAN
FUJII, Junichi, JAPAN
FUJII, Kazuyuki, JAPAN
FUJII, Masatoshi, JAPAN
FUJII, Michihiko, JAPAN
FUJII, Michikazu, JAPAN
FUJII, Seiji, JAPAN
FUJII, Shinichi, JAPAN
FUJII, Toshiakira, JAPAN
FUJIIE, Setsuro, JAPAN
FUJI'I'E, Tatsuo, JAPAN
FUJIKI, Akira, JAPAN
FUJIKOSHI, Yasunori*, JAPAN
FUJIMAGARI, Tetsuo, JAPAN
FUJIMOTO, Hirotaka, JAPAN
FUJIMOTO, Yoshio, JAPAN
FUJIMURA, Shigeyoshi, JAPAN
FUJISAKI, Genjiro, JAPAN
FUJISAKI, Rieko, JAPAN
FUJISAWA, Taro, JAPAN
FUJITA, Hiroshi, JAPAN
FUJITA, Hiroyuki, JAPAN
FUJITA, Hisaaki, JAPAN
FUJITA, Kazunori, JAPAN
FUJITA, Kenji, JAPAN
FUJITA, Kumiko, JAPAN
FUJITA, Osamu, JAPAN
FUJITA, Takahiko, JAPAN
FUJITA, Takao§, JAPAN
FUJIWARA, Daisuke, JAPAN
FUJIWARA, Hidenori, JAPAN
FUJIWARA, Kazuhiro, JAPAN
FUJIWARA, Koji, JAPAN
FUJIWARA, Masahiko, JAPAN
FUJIWARA, Shigeki, JAPAN
FUJIWARA, Tsukasa, JAPAN
FUJIWARA, Tsuyoshi, JAPAN
FUKAGAI, Eiji, JAPAN
FUKAGAI, Nobuyoshi, JAPAN
FUKAISHI, Hiroo, JAPAN
FUKAMIYA, Masanori§, JAPAN
FUKAYA, Kenji, JAPAN
FUKUCHI, Minoru, JAPAN
FUKUDA, Hiromichi, JAPAN
FUKUDA, Nozomu, JAPAN
FUKUDA, Takashi, JAPAN
FUKUDA, Takuo, JAPAN
FUKUHARA, Kenzo, JAPAN
FUKUHARA, Shinji§, JAPAN
FUKUI, Kazuhiko, JAPAN
FUKUI, Seiichi, JAPAN
FUKUI, Toshizumi, JAPAN
FUKUMOTO, Yoshihiro, JAPAN
FUKUSHI, Takeo, JAPAN
FUKUSHIMA, Masatoshi, JAPAN
FUKUTAKE, Takayoshi, JAPAN
FUKUYAMA, Katusi, JAPAN
FUNAHASHI, Ken-ichi, JAPAN
FUNAKI, Tadahisa, JAPAN
FUNAKOSI, Shunsuke, JAPAN
FUNAKURA, Takeo, JAPAN

FUNAYA, Bokurou, JAPAN
FURSTENBERG, Hillel Harry, U.S.A.
FURUKAWA, Yasukuni, JAPAN
FURUSHIMA, Mikio, JAPAN
FURUSHO, Yasuhiro, JAPAN
FURUTA, Kohji, JAPAN
FURUTA, Mikio, JAPAN
FURUTA, Tomonori, JAPAN
FURUTA, Yoshiomi, JAPAN
FURUTO, David Masaru, U.S.A.
FURUTSU, Hirotoshi, JAPAN
FURUYA, Kiyoko, JAPAN
FURUYA, Masako, JAPAN
FUTAKI, Akito§, JAPAN
GABAI, David, U.S.A.
GAMBAUDO, Jean-Marc*, FRANCE
GAMKRELIDZE, Revaz V., U.S.S.R.
GARCÍA-CUERVA, José, SPAIN
GARDNER, Robert B., U.S.A.
GARLING, David J. H., U.K.
GARNER, Cyril W.L., CANADA
GARZON, Max H., U.S.A.
GAUDRY, Garth Ian, AUSTRALIA
GAULD, David B.§, NEW ZEALAND
GEBAUER, Rüdiger, U.S.A.
GEDDES, Keith, CANADA
GELLER, Daryl N., U.S.A.
GEOGHEGAN, Ross, U.S.A.
GERAMI, Nasrollah*, IRAN
GETIMANE, Mário Frengue, MOZAMBIQUE
GETZLER, Ezra, AUSTRALIA
GHOUSSOUB, Nassif, CANADA
GHYS, Etienne, FRANCE
GIACARDI, Livia*, ITALY
GIGA, Mariko, JAPAN
GIGA, Yoshikazu*, JAPAN
GILES, John Robilliard*, AUSTRALIA
GILLARD, Roland D., FRANCE
GILLET, Henri A., U.S.A.
GILPIN, Michael J.*, U.S.A.
GINGOLD, Harry*, U.S.A.
GIORDANO, Thierry*, SWITZERLAND
GIRKO, Vyacheslav L.*, U.S.S.R.
GIUCULESCU, Alexandru*, ROMANIA
GIUSTI, Mark François, FRANCE
GŁAZEK, Kazimierz*, POLAND
GLEASON, Andrew M., U.S.A.
GLOWINSKI, Roland, U.S.A.
GOCHO, Toru, JAPAN
GODA, Hiroshi, JAPAN
GOKA, Yosihiro, JAPAN
GOLAN, Jonathan Samuel, ISRAEL
GOLDBACH, Ronald W., NETHERLANDS
GOLDBLATT, Robert*, NEW ZEALAND
GOLDSTEIN, Jerome A., U.S.A.
GOLDWASSER, Shafi, U.S.A.
GOLODETS, Valentin Y.*, U.S.S.R.
GOMEZ-MONT, Xavier A., MEXICO
GOMI, Kensaku, JAPAN
GOMIKAWA, Taturo, JAPAN
GONCHAR, Andrei A., U.S.S.R.
GONSHOR, Harry*, U.S.A.
GONZALEZ-ACUÑA, Francisco, MEXICO
GONZALEZ-SPRINBERG, G., FRANCE
GOODEARL, Kenneth R., U.S.A.
GOODMAN, Fredrick M., U.S.A.
GOODMAN, Jacob E., U.S.A.
GOODWILLIE, Thomas G., U.S.A.
GORDON, Cameron M., U.S.A.
GOSSEZ, Jean-Pierre*, BELGIUM
GOTO, Midori, JAPAN
GOTO, Morikuni, JAPAN
GOTO, Ryushi, JAPAN
GOTO, Shiro, JAPAN
GOTO, Yukinori, JAPAN
GOWRISANKARAN, Kohur, CANADA
GRAFAKOS, Loukas*, GREECE
GRAHAM, John J., AUSTRALIA
GRAHAM, R. L., U.S.A.
GRANDSARD, Francine L., BELGIUM
GRANIRER, Edmond E., CANADA
GRANVILLE, Andrew J., U.K.
GRATZER, George A., CANADA
GRAY, Jeremy John, U.K.
GREENE, Curtis, U.S.A.
GREENSPOON, Arthur, U.S.A.
GREGUŠ, Michal§, CZECHOSLOVAKIA
GREUEL, Gert-Martin, F.R.G.
GRIGGS, Jerrold R., U.S.A.
GRIGIS, Alain, FRANCE
GRIGORCHUK, Rostislav Ivanovich, U.S.S.R.
GRIGORIEFF, Rolf Dieter, F.R.G.
GRIMALDI-PIRO, Anna, ITALY
GROBBELAAR, Marié*, SOUTH AFRICA
GROENEWALD, Nico J.*, SOUTH AFRICA
GROETSCH, Charles W.*, U.S.A.
GRÓSS, Leonard, U.S.A.
GROVE, Karsten, DENMARK
GRUNDMAN, Helen G., U.S.A.
GRÜTER, Michael, F.R.G.
GUARALDO, Rosalind, U.S.A.
GUBELADZE, Iosif D., U.S.S.R.
GUEST, Martin A.*, U.S.A.
GUIBAS, Leonidas J., U.S.A.
GUICHARDET, Alain, FRANCE
GUILLOPE, Laurent, FRANCE
GULLIVER, Robert D., U.S.A.
GUNDLACH, Karl-Bernhard, F.R.G.
GÜNTHER, Matthias, G.D.R.
GUO, Mao Zheng, CHINA, P. R. of
GURJAR, Rajendra Vasant*, INDIA
GYOJA, Akihiko, JAPAN
GYÖRI, Ervin, HUNGARY
GYŐRY, Kálmán*, HUNGARY
HA, Huy Bang*, VIETNAM
HABGGER, Nathan Bernard, FRANCE

HADANO, Toshihiro, JAPAN
HADIAN DEHKORDI, Masoud*, IRAN
HAG, Kari Jorun, NORWAY
HAG, Per, NORWAY
HAHN, Saul G., U.S.A.
HAIDA, Minoru, JAPAN
HAJARNAVIS, Charudatta R., U.K.
HAKEDA, Josuke, JAPAN
HAKEDA, Nobuhiro§, JAPAN
HAKOSALO, Milja Riitta, U.S.A.
HALISTE, Kersti A., SWEDEN
HAMACHI, Toshihiro, JAPAN
HAMADA, Hozumi*, JAPAN
HAMADA, Yusaku, JAPAN
HAMANA, Masamichi*, JAPAN
HAMANA, Yuji, JAPAN
HAMANAKA, Yojiro, JAPAN
HAMASAKI, Atsumi, JAPAN
HANDA, Kenji, JAPAN
HANO, Jun-ichi, JAPAN
HANOUZET, Bernard, FRANCE
HANSEN, Frank*, DENMARK
HANSEN, Vagn L.*, DENMARK
HANSEN, Wolfhard, F.R.G.
HANS-GILL, Rajinder Jeet, INDIA
HAPPEL, Dieter, F.R.G.
HARA, Masao, JAPAN
HARA, Shin-ichiro, JAPAN
HARA, Tadayuki, JAPAN
HARA, Takuya, JAPAN
HARA, Tamio, JAPAN
HARA, Yoshihito, JAPAN
HARA, Yuko, JAPAN
HARADA, Masana, JAPAN
HARADA, Mikio, JAPAN
HARAGUCHI, Yuri R., CHILE
HARDER, Günter, F.R.G.
HARDY, Kenneth, CANADA
HARIKAE, Toshio, JAPAN
HARRIS, Michael H., U.S.A.
HART, R. Neal, U.S.A.
HARTEN, Amiram, ISRAEL/U.S.A.
HARZHEIM, Egbert R.*, F.R.G.
HASEGAWA, Fumio, JAPAN
HASEGAWA, Izumi, JAPAN
HASEGAWA, Keizo, JAPAN
HASEGAWA, Kenji, JAPAN
HASEGAWA, Kohji, JAPAN
HASEGAWA, Ryo, JAPAN
HASEGAWA, Ryu, JAPAN
HASEGAWA, Takayuki, JAPAN
HASEGAWA, Youjirou, JAPAN
HASHIGUCHI, Kosaburo, JAPAN
HASHIGUCHI, Masao, JAPAN
HASHIGUCHI, Norikazu, JAPAN
HASHIMOTO, Hideya§, JAPAN
HASHIMOTO, Mitsuyasu, JAPAN
HASHIMOTO, Takahiro, JAPAN
HASHIMOTO, Takashi, JAPAN
HASHIMOTO, Yoshiaki, JAPAN
HASHIMOTO, Yoshihiko, JAPAN
HASHIMOTO, Yoshitake, JAPAN
HASHIMOTO, Yuji, JAPAN
HASSANI, Akbar*, IRAN
HASUI, Satoshi, JAPAN
HASUMI, Morisuke, JAPAN
HATA, Masayoshi, JAPAN
HATADA, Kazuyuki, JAPAN
HATORI, Asako, JAPAN
HATORI, Osamu*, JAPAN
HATTORI, Akio, JAPAN
HATTORI, Tetsuya§, JAPAN
HATTORI, Toshiaki, JAPAN
HATTORI, Yasunao, JAPAN
HATTORI, Yoshinori, JAPAN
HATVANI, László*, HUNGARY
HAWKES, John, U.K.
HAWKES, Trevor Ongley, U.K.
HAYAKATA, Takayuki, JAPAN
HAYAKAWA, Kantaro, JAPAN
HAYAKAWA, Keizou, JAPAN
HAYASAKA, Satoshi, JAPAN
HAYASHI, Etsuo, JAPAN
HAYASHI, Heima, JAPAN
HAYASHI, Hiroki, JAPAN
HAYASHI, Hiroshi§, JAPAN
HAYASHI, Kazumichi, JAPAN
HAYASHI, Kiyoshi, JAPAN
HAYASHI, Makoto, JAPAN
HAYASHI, Makoto, JAPAN
HAYASHI, Mikihiro, JAPAN
HAYASHI, Nakao*, JAPAN
HAYASHI, Takahiro, JAPAN
HAYASHI, Yoshiki, F.R.G.
HAYASHIDA, Tsuyoshi§, JAPAN
HAYASIDA, Kazuya, JAPAN
HAYMAN, Walter Kurt, U.K.
HEARST, III, William R., U.S.A.
HEBISCH, Waldemar, POLAND
HECKMAN, Gerrit, NETHERLANDS
HEDBERG, Lars Inge, SWEDEN
HEILIGMAN, Mark Isaac, U.S.A.
HEINONEN, Juha M., U.S.A.
HEINTZ, Joos U., ARGENTINA
HEINZE, Joachim, F.R.G.
HELLSTRÖM, Lennart, SWEDEN
HELMINCK, Aloysius Gerarous, NETHERLANDS
HELMINCK, Gerard G.F., NETHERLANDS
HENRICHS, Rolf Wim*, F.R.G.
HERMAN, Richard H.§, U.S.A.
HERRERA, Rutilio A.*, COSTA RICA
HERRMANN, Manfred H., F.R.G.
HERZ, Prof., CANADA
HERZOG, Jürgen R. G.§, F.R.G.
HIAI, Fumio, JAPAN

HIBI, Takayuki, JAPAN
HIDA, Takeyuki, JAPAN
HIDAKA, Fumio*, JAPAN
HIGA, Tatsuo, JAPAN
HIGASHIYAMA, Teiko, JAPAN
HIGUCHI, Yasunari, JAPAN
HIGUCHI, Yoshiki, JAPAN
HIJIKATA, Hiroaki, JAPAN
HIKIDA, Mizuho, JAPAN
HIKITA, Teruo, JAPAN
HILLEL, Joel*, CANADA
HILSUM, Michel*, FRANCE
HINO, Yosuke, JAPAN
HINOKUMA, Takanori, JAPAN
HINZ, Andreas M.*, F.R.G.
HIRABAYASHI, Mikihito, JAPAN
HIRACHI, Kengo, JAPAN
HIRAGA, Kaoru, JAPAN
HIRAHATA, Hirotoshi*, JAPAN
HIRAI, Etsuko, JAPAN
HIRAI, Takeshi, JAPAN
HIRAI, Yasuhisa, JAPAN
HIRAIDE, Koichi, JAPAN
HIRAKAWA, Fumiko§, JAPAN
HIRAKAWA, Kosaburo, JAPAN
HIRAKAWA, Masataka, JAPAN
HIRAMATSU, Toyokazu, JAPAN
HIRANO, Masaki, JAPAN
HIRASAWA, Yoshikazu, JAPAN
HIRATA, Kazuhiko, JAPAN
HIRATA, Koichi, JAPAN
HIRATA, Masaki, JAPAN
HIRAYAMA, Minoru, JAPAN
HIROHASHI, Masanori, JAPAN
HIROMI, Genkõ, JAPAN
HIROMORI, Katsuhisa, JAPAN
HIRONAKA, Eriko, JAPAN
HIRONAKA, Heisuke, JAPAN
HIRONAKA, Yumiko*, JAPAN
HIROSE, Ken, JAPAN
HIROSE, Susumu, JAPAN
HIROSHIMA, Tsutomu, JAPAN
HISANO, Tamao, JAPAN
HISHIDA, Toshiaki, JAPAN
HITOTUMATU, Sin, JAPAN
HITSUDA, Masuyuki, JAPAN
HOANG, Le Minh, VIETNAM
HODGSON, Bernard R., CANADA
HOFER, Helmut H., F.R.G.
HÖFER, Thomas, F.R.G.
HOFMANN, Bernd, G.D.R.
HOHTI, Aarno*, FINLAND
HOJO, Shun-ichi, JAPAN
HOLLAND, Samuel S. Jr., U.S.A.
HOLME, Audun, NORWAY
HOLMES, Philip J., U.S.A.
HOLOPAINEN, Ilkka O., FINLAND
HOLT, Fred B., U.S.A.
HOLTE, John M.*, U.S.A.
HOMMA, Katsumi, JAPAN
HOMMA, Kiyomi, JAPAN
HOMMA, Masaaki, JAPAN
HOMMA, Tatsuo, JAPAN
HONARY, Bahman K.*, IRAN
HONARY, Taher Ghasemi*, IRAN
HONDA, Ikuji, JAPAN
HONDA, Kin-ya, JAPAN
HONDA, Motohiro, JAPAN
HONDA, Naofumi, JAPAN
HONG, Imsik, JAPAN
HONG, Sungpyo, KOREA, R. of
HONG, Yi, CHINA, P. R. of
HOOPER, Robert Clark, U.S.A.
HORA, Akihito, JAPAN
HORAI, Masako, JAPAN
HORI, Kenta, JAPAN
HORI, Teturô, JAPAN
HORIE, Kuniaki*, JAPAN
HORIE, Mitsuko, JAPAN
HORIGUCHI, Hiroshi, JAPAN
HORIGUCHI, Syunzi, JAPAN
HORIKAWA, Eiji, JAPAN
HORIUCHI, Annick, FRANCE
HORIUCHI, Kiyomitsu, JAPAN
HORIUCHI, Ryutaro, JAPAN
HORIUCHI, Toshio*, JAPAN
HÖRMANDER, Lars V.§, SWEDEN
HOSHINA, Takao, JAPAN
HOSHIRO, Toshihiko, JAPAN
HOSOBUCHI, Masami, JAPAN
HOSOH, Toshio, JAPAN
HOSOKAWA, Fujitsugu, JAPAN
HOSONO, Yuzo, JAPAN
HOSOYA, Naoto, JAPAN
HOSTE, Jim, U.S.A.
HOTTA, Kôsaku, JAPAN
HOTTA, Ryoshi§, JAPAN
HOU, Jinchuan*, CHINA, P. R. of
HOU, Shui-Hung, HONG KONG
HOU, Zixin*, CHINA, P. R. of
HOUH, Chorng-Shi, U.S.A.
HOUSTON, Johnny, U.S.A.
HOUZEL, Christian, FRANCE
HRUSHOVSKI, Ehud, U.S.A.
HSIEH, Po-Fang*, U.S.A.
HSIEH, Ying-Hen*, CHINA-TAIWAN
HU, Bizhong, CHINA, P. R. of
HU, Hesheng, CHINA, P. R. of
HU, Thakyin, CHINA-TAIWAN
HUANG, Cheng-Gui§, CHINA, P. R. of
HUANG, Jing-Song, U.S.A.
HUANG, Qichang*, CHINA, P. R. of
HUANG, Wei-Zhang*, CHINA, P. R. of
HUANG, Xun-Cheng, U.S.A.
HUANG, Zhiyuan, CHINA, P. R. of
HUBBARD, John Hamal, U.S.A.

HUBBUCK, John Reginald, U.K.
HUGHES, Anne C.*, U.S.A.
HUGHES, Kenneth Robert, SOUTH AFRICA
HUKUM, Singh*, INDIA
HULEK, Klaus W.*, F.R.G.
HUNEKE, Craig L.*, U.S.A.
HURUYA, Tadasi, JAPAN
HUXLEY, Martin Neil, U.K.
HUZII, Mituaki, JAPAN
HWANG, Jun Shung, CHINA-TAIWAN
IARROBINO, A. A. Jr.*, U.S.A.
IBARAKI, Toshihide, JAPAN
IBISCH, Horst D., FRANCE
IBUKIYAMA, Tomoyoshi, JAPAN
ICHIDA, Ryosuke, JAPAN
ICHIHARA, Kanji, JAPAN
ICHIJI, Hiroshi, JAPAN
ICHIJYO, Yoshihiro, JAPAN
ICHIKAWA, Akira, JAPAN
ICHIKAWA, Kaoru*, JAPAN
ICHIKAWA, Takashi, JAPAN
ICHIMURA, Humio, JAPAN
ICHINOSE, Takashi, JAPAN
ICHIRAKU, Shigeo, JAPAN
IDA, Tomoya, JAPAN
IGARASHI, Kan, JAPAN
IGARASHI, Masayuki§, JAPAN
IGARI, Katsuju§, JAPAN
IGARI, Satoru, JAPAN
IGUSA, Kiyoshi, U.S.A.
IHARA, Shin-ichiro, JAPAN
IHARA, Shunsuke, JAPAN
IHARA, Yasutaka, JAPAN
II, Kiyotaka, JAPAN
IIDA, Hirokazu, JAPAN
IIDA, Masato, JAPAN
IITAKA, Shigeru*, JAPAN
IKAWA, Mitsuru, JAPAN
IKAWA, Osamu, JAPAN
IKAWA, Toshihiko, JAPAN
IKEBE, Nobunori, JAPAN
IKEDA, Akira, JAPAN
IKEDA, Fumio, JAPAN
IKEDA, Hideo, JAPAN
IKEDA, Kaoru, JAPAN
IKEDA, Kazumasa, JAPAN
IKEDA, Koichiro, JAPAN
IKEDA, Nobuyuki, JAPAN
IKEDA, Noriaki, JAPAN
IKEDA, Shin, JAPAN
IKEDA, Tamotsu, JAPAN
IKEDA, Toru, JAPAN
IKEDA, Toshiharu, JAPAN
IKEDA, Toshio, JAPAN
IKEDA, Tsutomu, JAPAN
IKEDA, Yoshito, JAPAN
IKEGAMI, Giko, JAPAN
IKEGAMI, Teruo, JAPAN
IKEHATA, Masaru, JAPAN
IKESHOJI, Kiyoshi, JAPAN
IKEYAMA, Tamotsu, JAPAN
IKOMA, Yoshio, JAPAN
ILIADIS, Stavros*, GREECE
ILLMAN, Sören A., FINLAND
IL'YASHENKO, Julij S., U.S.S.R.
IM, John J., U.S.A.
IMADA, Naotaka, JAPAN
IMAFUKU, Kentaro, JAPAN
IMAI, Hideo, JAPAN
IMAI, Hideo, JAPAN
IMAI, Hiroshi, JAPAN
IMAI, Keiko, JAPAN
IMAI, Masataka, JAPAN
IMAI, Toshihiro, JAPAN
IMANISHI, Hideki, JAPAN
IMAYOSHI, Yoichi, JAPAN
IMBRIE, John Z., U.S.A.
INABA, Takashi, JAPAN
INAGAKI, Nobuo§, JAPAN
INATOMI, Akira, JAPAN
INNAMI, Nobuhiro, JAPAN
INOUE, Akihiko, JAPAN
INOUE, Junko, JAPAN
INOUE, Jyunji, JAPAN
INOUE, Kazuyuki, JAPAN
INOUE, Masahisa, JAPAN
INOUE, Takayuki, JAPAN
INOUE, Tetsuo*, JAPAN
INOUE, Toru, JAPAN
INOUE, Yoshinari, JAPAN
INOUE, Yoshiyuki, JAPAN
ION, Patrick D. F., U.S.A.
IONESCU, Paltin, ROMANIA
IONESCU, Paul-Cristodor H., ROMANIA
IORIO Jr., Rafael José, BRAZIL
IOZZI, Alessandra, ITALY/U.S.A.
IRAQI MOGHADDAM, G. H., IRAN
IRI, Masao§, JAPAN
IRIE, Shoji, JAPAN
IRITE, Takao, JAPAN
IROKAWA, Susumu, JAPAN
ISBELL, John R., U.S.A.
ISEKI, Kiyoshi, JAPAN
ISHIBASHI, Hiroyuki, JAPAN
ISHIBASHI, Makoto§, JAPAN
ISHIBASHI, Yasunori, JAPAN
ISHIDA, Hisashi, JAPAN
ISHIDA, Masanori, JAPAN
ISHIDA, Mitsuko, JAPAN
ISHIGAMI, Shigeo, JAPAN
ISHIGAMI, Yoshiyasu*, JAPAN
ISHIHARA, Hajime*, JAPAN
ISHIHARA, Ikuo, JAPAN
ISHIHARA, Junichi, JAPAN
ISHIHARA, Kazuo, JAPAN
ISHIHARA, Muneichi, JAPAN

ISHIHARA, Toru, JAPAN
ISHII, Akira, JAPAN
ISHII, Hidenori, JAPAN
ISHII, Hitoshi, JAPAN
ISHII, Ippei, JAPAN
ISHII, Katsuyuki, JAPAN
ISHII, Kazuhiko, JAPAN
ISHII, Noburo, JAPAN
ISHII, Shihoko, JAPAN
ISHII, Yutaka, JAPAN
ISHIKAWA, Goo, JAPAN
ISHIKAWA, Hirofumi, JAPAN
ISHIKAWA, Masao, JAPAN
ISHIKAWA, Masato, JAPAN
ISHIKAWA, Michio, JAPAN
ISHIKAWA, Mituo, JAPAN
ISHIKAWA, Nobuhiro, JAPAN
ISHIKAWA, Saneaki, JAPAN
ISHIKAWA, Shiro, JAPAN
ISHIKAWA, Takeshi, JAPAN
ISHIKAWA, Takeshi, JAPAN
ISHIKAWA, Tsuneo, JAPAN
ISHIKAWA, Yasushi*, JAPAN
ISHIKAWA, Yoichiro, JAPAN
ISHIMOTO, Hiroyasu, JAPAN
ISHIMURA, Naoyuki, JAPAN
ISHIMURA, Ryuichi, JAPAN
ISHITOYA, Kiminao, JAPAN
ISHIZAKI, Katsuya, JAPAN
ISIDRO, José M., SPAIN
ISKOVSKIH, Vasilii A., U.S.S.R.
ISOGAI, Eiichi, JAPAN
ISOZAKI, Hiroshi, JAPAN
ISU, Minoru, JAPAN
ITAI, Masanori*, U.S.A.
ITANO, Mitsuyuki, JAPAN
ITATSU, Seiichi, JAPAN
ITO, Hideji, JAPAN
ITO, Hidekazu, JAPAN
ITO, Hiroshi, JAPAN
ITO, Hiroya, JAPAN
ITO, Hiroyuki, JAPAN
ITO, Ichiro, JAPAN
ITO, Keiichi*, JAPAN
ITO, Kiyosi, JAPAN
ITO, Masami, JAPAN
ITO, Masayuki, JAPAN
ITO, Masayuki, JAPAN
ITO, Ryuichi, JAPAN
ITO, Seizo, JAPAN
ITO, Takashi, JAPAN
ITO, Tatsuro, JAPAN
ITO, Toshikazu, JAPAN
ITO, Yoshifumi*, JAPAN
ITO, Yoshifusa, JAPAN
ITO, Yoshihiko, JAPAN
ITO, Yoshihiko, JAPAN
ITO, Yuji, JAPAN
ITOH, Jin-ichi, JAPAN
ITOH, Masa-aki, JAPAN
ITOH, Shiroh, JAPAN
ITOH, Takashi, JAPAN
ITOH, Tatsuo*, JAPAN
ITOH, Yoshiaki*, JAPAN
ITOKAWA, Yoe, JAPAN
IVANOV, Aleksander A., U.S.S.R.
IVANOV, Andrei V., U.S.S.R.
IVANOV, Kamen Ganchev, BULGARIA
IVANŠIĆ, Ivan*, YUGOSLAVIA
IWAHORI, Nagayoshi, JAPAN
IWAHORI, Nobuko, JAPAN
IWAI, Akira, JAPAN
IWAI, Mieko§, JAPAN
IWAI, Toshihiro, JAPAN
IWAKAMI, Tatsuo, JAPAN
IWAMOTO, Masayuki, JAPAN
IWAMOTO, Takashi, JAPAN
IWANAGA, Yasuo, JAPAN
IWANOWSKI, Peter, F.R.G.
IWASAKI, Chisato, JAPAN
IWASAKI, Ichiro, JAPAN
IWASAKI, Nobuhisa, JAPAN
IWASAKI, Shiro, JAPAN
IWASAWA, Kenkichi§, JAPAN
IWATA, Kazuo, JAPAN
IWATA, Koichi, JAPAN
IWATANI, Teruo, JAPAN
IWATSUKA, Akira, JAPAN
IWAYAMA, Eiko, JAPAN
IYANAGA, Shokichi, JAPAN
IZÉ, Antonio F.*, BRAZIL
IZEKI, Hiroyasu, JAPAN
IZUCHI, Keiji*, JAPAN
IZUMI, Hideo*, JAPAN
IZUMI, Masaki, JAPAN
IZUMI, Shuzo, JAPAN
IZUMISAWA, Masataka, JAPAN
IZUMIYA, Shyuichi, JAPAN
JACO, William H., U.S.A.
JACOB, Niels, F.R.G.
JAEGER, Arno*, F.R.G.
JAROSZ, Krzysztof M., U.S.A.
JAVAME GHAZVINI, M. J., IRAN
JENKINS, James A., U.S.A.
JENSEN, Norbert, F.R.G.
JI, Xinhua*, CHINA, P. R. of
JIJTCHENKO, Aleksei B., U.S.S.R.
JIMBO, Michio, JAPAN
JIMBO, Shuichi, JAPAN
JIMBO, Toshiya, JAPAN
JIN, Gyo Taek, KOREA, R. of
JIN, Naondo, JAPAN
JOHNSON, Dave, U.K.
JOHNSON, Gerald W., U.S.A.
JOHNSON, Roy A., U.S.A.
JOLLY, Michael Summerfield*, U.S.A.

JONES, Lowell Edwin, U.S.A.
JONES, Mark C. W.*, U.K.
JONES, Peter W., U.S.A.
JONES, Vaughan F. R., NEW ZEALAND
JORGENSEN, Palle E. T.*, U.S.A.
JUANG, Jonq*, CHINA-TAIWAN
JURKIEWICZ, Jerzy, POLAND
JUTILA, Matti Ilmari*, FINLAND
KABEYA, Yoshitsugu, JAPAN
KACHI, Hideyuki, JAPAN
KACHI, Yasuyuki, JAPAN
KADISON, Richard V.§, U.S.A.
KADO, Jiro, JAPAN
KADOYAMA, Makoto, JAPAN
KAGA, Toshihiro, JAPAN
KAGESAWA, Masataka, JAPAN
KÅGESTEN, Owe K., SWEDEN
KAGEYAMA, Sanpei, JAPAN
KAHAN, W. M., U.S.A.
KAHANE, Jean-Pierre, FRANCE
KAHN, Bruno*, FRANCE
KAIMANOVICH, Vadim A., U.S.S.R.
KAIZU, Satoshi*, JAPAN
KAIZUKA, Tetsu§, JAPAN
KAJI, Hajime, JAPAN
KAJIKAWA, Yuji*, JAPAN
KAJIKIYA, Ryuji, JAPAN
KAJIMOTO, Hiroshi, JAPAN
KAJITANI, Kunihiko, JAPAN
KAJITANI, Yoji*, JAPAN
KAJIWARA, Joji*, JAPAN
KAJIWARA, Takeshi, JAPAN
KAJIWARA, Tsuyoshi, JAPAN
KAKEHI, Katsuhiko, JAPAN
KAKEHI, Tomoyuki, JAPAN
KAKIE, Kunio, JAPAN
KAKIMIZU, Osamu, JAPAN
KAKITA, Takao, JAPAN
KAKIUCHI, Hideaki, JAPAN
KAKO, Takashi, JAPAN
KAKUBA, Stevens J. K., UGANDA
KAKUDA, Yuzuru, JAPAN
KALLIANPUR, Gopinath, U.S.A.
KALONI, Purna N.§, CANADA
KALYABIN, Gennadiy A.*, U.S.S.R.
KAMADA, Seiichi, JAPAN
KAMAE, Teturo, JAPAN
KAMAL, Ahmed A. M.*, EGYPT
KAMATA, Masayoshi, JAPAN
KAMBAYASHI, Tatsuji, JAPAN
KAMEDA, Masumi, JAPAN
KAMEI, Eizaburo, JAPAN
KAMETAKA, Yoshinori*, JAPAN
KAMETANI, Makoto, JAPAN
KAMETANI, Yukio, JAPAN
KAMEYAMA, Atsushi, JAPAN
KAMIMURA, Yutaka, JAPAN
KAMINKER, Jerry, U.S.A.
KAMIYA, Hisao, JAPAN
KAMIYA, Hisashi, JAPAN
KAMIYA, Noriaki, JAPAN
KAMIYA, Shigeyasu, JAPAN
KAMIYA, Tadashi, JAPAN
KAMO, Shizuo, JAPAN
KAMYAD, A. Vahidian*, IRAN
KANAI, Masahiko, JAPAN
KANAI, Norio, JAPAN
KANAYAMA, Hiroshi, JAPAN
KANDA, Mamoru, JAPAN
KANDA, Shigeo, JAPAN
KANEDA, Eiji, JAPAN
KANEDA, Masaharu, JAPAN
KANEKO, Akira, JAPAN
KANEKO, Hiroshi, JAPAN
KANEKO, Jyoichi, JAPAN
KANEKO, Makoto, JAPAN
KANEKO, Masanobu, JAPAN
KANEMAKI, Shoji, JAPAN
KANEMARU, Tadayoshi, JAPAN
KANEMITSU, Mitsuo, JAPAN
KANEMITSU, Shigeru, JAPAN
KANENOBU, Taizo*, JAPAN
KANETA, Hitoshi, JAPAN
KANETO, Takeshi, JAPAN
KANEYUKI, Soji, JAPAN
KANG, Ming-Chang*, CHINA-TAIWAN
KANG, Seok-Jin, KOREA, R. of
KANG, Tae Ho, KOREA, R. of
KANIUTH, Eberhard*, F.R.G.
KANJIN, Yuichi, JAPAN
KANNO, Kenji*, JAPAN
KANNO, Tsuneo, JAPAN
KANO, Masako, JAPAN
KANO, Tadayoshi, JAPAN
KANTOR, Jean-Michel, FRANCE
KANTOWSKI, Eleanore L., U.S.A.
KANZAKI, Teruo§, JAPAN
KAPLAN, Hadassah, ISRAEL
KAPLANSKY, Irving, U.S.A.
KAPOVICH, Michael Erikovich, U.S.S.R.
KAPUR, Aruna*, INDIA
KARAMATSU, Yoshikazu, JAPAN
KARAMZADEH, Omid*, IRAN
KARANDIKAR, Rajeeva L., INDIA
KARBE, Manfred, F.R.G.
KARHUMÄKI, Juhani E. U.*, FINLAND
KARLSSON, Thomas Nils§, SWEDEN
KARZANOV, Alexander V., U.S.S.R.
KASAHARA, Koji, JAPAN
KASAHARA, Yasushi, JAPAN
KASAHARA, Yuji, JAPAN
KASAI, Shin-ichi, JAPAN
KASAJIMA, Tomomi, JAPAN
KASHANI, S. M. B.*, IRAN
KASHIHARA, Kenji, JAPAN
KASHIMURA, Ryohei, JAPAN

KASHIWABARA, Takuji, JAPAN
KASHIWADA, Toyoko, JAPAN
KASHIWAGI, Toshio§, JAPAN
KASHIWAGI, Yoshimi, JAPAN
KASHIWARA, Hiroko, JAPAN
KASHIWARA, Masaki, JAPAN
KASONGA, Raphael Abel, TANZANIA
KASUE, Atsushi, JAPAN
KATAOKA, Kiyoomi, JAPAN
KATASE, Kiyoshi, JAPAN
KATAYAMA, Shigeru, JAPAN
KATAYAMA, Shin-ichi, JAPAN
KATAYAMA, Yoshikazu, JAPAN
KATO, Akio*, JAPAN
KATO, Fumio*, JAPAN
KATO, Hiroshi, JAPAN
KATO, Hisako, JAPAN
KATO, Hisao§, JAPAN
KATO, Junji, JAPAN
KATO, Kazuhisa, JAPAN
KATÔ, Kazuko, JAPAN
KATO, Kazuya, JAPAN
KATO, Keiichi, JAPAN
KATO, Masahide, JAPAN
KATO, Masakimi, JAPAN
KATO, Minoru, JAPAN
KATO, Mitsuyoshi, JAPAN
KATO, Sadao, JAPAN
KATO, Shin, JAPAN
KATO, Shin-ichi, JAPAN
KATO, Suehiro, JAPAN
KATO, Syohei, JAPAN
KATO, Toyonori, JAPAN
KATO, Yoshifumi, JAPAN
KATO, Yoshio, JAPAN
KATORI, Makoto*, JAPAN
KATSUDA, Atsushi, JAPAN
KATSUMATA, Yasuo*, JAPAN
KATSURA, Shinji, JAPAN
KATSURA, Toshiyuki, JAPAN
KATSURADA, Hidenori, JAPAN
KATSURADA, Masanori, JAPAN
KATSURADA, Masashi, JAPAN
KATSUURA, Hidefumi*, U.S.A./JAPAN
KAUFFMAN, Louis Hirsch, U.S.A.
KAUL, Saroop K., CANADA
KAWABE, Jun, JAPAN
KAWADA, Takayuki, JAPAN
KAWADA, Yukiyosi, JAPAN
KAWADA, Yutaka, JAPAN
KAWAGUCHI, Hiroaki*, JAPAN
KAWAGUCHI, Shinji, JAPAN
KAWAGUCHI, Shokei§, JAPAN
KAWAGUCHI, Tomoaki*, JAPAN
KAWAHARA, Yusaku, JAPAN
KAWAHIGASHI, Yasuyuki, JAPAN
KAWAI, Ei-ichiro, JAPAN
KAWAI, Shigeo, JAPAN
KAWAI, Takahiro, JAPAN
KAWAI, Toru, JAPAN
KAWAI, Tsuneyuki, JAPAN
KAWAKAMI, Masato, JAPAN
KAWAKAMI, Satoshi*, JAPAN
KAWAKAMI, Tomohiro, JAPAN
KAWAKUBO, Katsuo, JAPAN
KAWAMATA, Yujiro, JAPAN
KAWAMOTO, Fuminori, JAPAN
KAWAMOTO, Naoki, JAPAN
KAWAMOTO, Shin-ichi, JAPAN
KAWAMURA, Kazuhiro, JAPAN
KAWAMURA, Michihiko, JAPAN
KAWAMURA, Shinzo, JAPAN
KAWANAGO, Tadashi, JAPAN
KAWANAKA, Noriaki, JAPAN
KAWANO, Haruaki§, JAPAN
KAWANO, Yasuhito, JAPAN
KAWASAKI, Hidefumi, JAPAN
KAWASAKI, Tetsuro, JAPAN
KAWASHIMA, Seiji, JAPAN
KAWASHIMA, Toshio, JAPAN
KAWASHITA, Mishio, JAPAN
KAWASUMI, Shogo, JAPAN
KAWATA, Shigeto, JAPAN
KAWAZOE, Takeshi, JAPAN
KAWAZU, Kiyoshi, JAPAN
KAWAZUMI, Nariya, JAPAN
KAZAMA, Hideaki, JAPAN
KAZAMA, Ken-ichiro, JAPAN
KAZAMAKI, Norihiko, JAPAN
KAZUMI, Tetuya, JAPAN
KEARSLEY, Mary J., U.K.
KEARTON, Cherry, U.K.
KEATING, Kevin Patrick, U.S.A.
KEEN, Linda, U.S.A.
KEGEL, Otto H., F.R.G.
KELLOGG, Franklin R.§, U.S.A.
KEMER, Alexander R., U.S.S.R.
KENDEROV, Peter Stoyanov, BULGARIA
KENIG, Carlos E., U.S.A.
KENMOTSU, Katsuei, JAPAN
KERA, Kazuo, JAPAN
KERAYECHIAN, Asghar*, IRAN
KERMAN, Ron A., CANADA
KERNER, Otto, F.R.G.
KESTENBAND, Barbu C.*, U.S.A.
KEUM, Jonghae, U.S.A./KOREA, R. of
KEUNE, Frans J., NETHERLANDS
KHALIFAH, M.-F. A.-S.*, EGYPT
KHANDUJA, Sudesh K.*, INDIA
KHARAGHANI, Hadi, CANADA
KHELLADI, Abdelkader, ALGERIA
KHRUSHCEV, Sergei V., U.S.S.R.
KIDA, Masanari, JAPAN
KIDA, Yuji, JAPAN
KIGAMI, Jun, JAPAN
KIGURADZE, Ivan T., U.S.S.R.

KIHARA, Hiroshi, JAPAN
KIJIMA, Yoichi, JAPAN
KIKKAWA, Michihiko, JAPAN
KIKUCHI, Akira§, JAPAN
KIKUCHI, Fumio*, JAPAN
KIKUCHI, Hiroko, JAPAN
KIKUCHI, Isamu, JAPAN
KIKUCHI, Kazunori, JAPAN
KIKUCHI, Keiichi, JAPAN
KIKUCHI, Keisuke, JAPAN
KIKUCHI, Koji, JAPAN
KIKUCHI, Masato, JAPAN
KIKUCHI, Norio, JAPAN
KIKUCHI, Shigeki, JAPAN
KIKUCHI, Teppei, JAPAN
KIKUMASA, Isao, JAPAN
KILPELÄINEN, Tero A. V., FINLAND
KIM, Sung Sook, KOREA, R. of
KIM, TongHo, JAPAN
KIM, Yong Woon, KOREA, R. of
KIM, Young Ho, KOREA, R. of
KIMN, Ha-Jine, KOREA, R. of
KIMURA, Hironobu, JAPAN
KIMURA, Hiroshi, JAPAN
KIMURA, Ikuo, JAPAN
KIMURA, Isamu, JAPAN
KIMURA, Kazuhiro, JAPAN
KIMURA, Kyoko, U.S.A.
KIMURA, Makoto, JAPAN
KIMURA, Minako, JAPAN
KIMURA, Morishige, JAPAN
KIMURA, Noriaki, JAPAN
KIMURA, Takahisa, JAPAN
KIMURA, Takashi, JAPAN
KIMURA, Tatsuo, JAPAN
KIMURA, Tetsuzo, JAPAN
KIMURA, Tosihusa, JAPAN
KINOSHITA, Shin'ichi, JAPAN
KINOSHITA, Takafumi, JAPAN
KINOSHITA, Yasushi, JAPAN
KINOSITA, Kiyosi§, JAPAN
KIRILLOV, A. N., U.S.S.R.
KIRK, Paul A., U.S.A.
KISAKA, Masashi, JAPAN
KISHI, Masanori, JAPAN
KISHIMOTO, Kazuo, JAPAN
KISO, Kazuhiro, JAPAN
KISTER, James M., U.S.A.
KISTER, Jane E., U.S.A.
KITA, Hiroo*, JAPAN
KITADA, Akihiko*, JAPAN
KITADA, Toshiyuki, JAPAN
KITADA, Yasuhiko, JAPAN
KITAGAWA, Keiichiro, JAPAN
KITAGAWA, Takashi, JAPAN
KITAGAWA, Tosio§, JAPAN
KITAGAWA, Yoshihisa, JAPAN
KITAHARA, Haruo, JAPAN
KITAHARA, Kiyoshi, JAPAN
KITAKUBO, Shigeru, JAPAN
KITAMI, Tsutomu§, JAPAN
KITAMURA, Ichiziro, JAPAN
KITAMURA, Kazue, JAPAN
KITAMURA, Shin-ichi, JAPAN
KITAZAKI, Kuniaki, JAPAN
KITAZUME, Masaaki, JAPAN
KIYOHARA, Kazuyoshi, JAPAN
KIYOHARA, Mineo, JAPAN
KIYOTA, Masao, JAPAN
KIYOTA, Shoichiroh, JAPAN
KIZUKA, Takashi, JAPAN
KLAWE, Maria M., CANADA
KLEIN, Abraham A.*, ISRAEL
KLEISLI, Heinrich*, SWITZERLAND
KLIN, Mihail Haimovich, U.S.S.R.
KNAPP, Anthony W., U.S.A.
KNOBLOCH, Eberhard H., F.R.G.
KNOPFMACHER, Arnold*, SOUTH AFRICA
KO, Ki Hyoung, KOREA, R. of
KOBAYASHI, Kazuaki, JAPAN
KOBAYASHI, Keijiro§, JAPAN
KOBAYASHI, Keiko, JAPAN
KOBAYASHI, Kusuo, JAPAN
KOBAYASHI, Masako, JAPAN
KOBAYASHI, Masanori*, JAPAN
KOBAYASHI, Mei*, U.S.A.
KOBAYASHI, Minoru, JAPAN
KOBAYASHI, Mitsuko, JAPAN
KOBAYASHI, Osamu, JAPAN
KOBAYASHI, Satoshi§, JAPAN
KOBAYASHI, Shigeru*, JAPAN
KOBAYASHI, Shoji, JAPAN
KOBAYASHI, Shoshichi, U.S.A.
KOBAYASHI, Takao, JAPAN
KOBAYASHI, Teiichi, JAPAN
KOBAYASHI, Toshimasa, JAPAN
KOBAYASHI, Toshiyuki, JAPAN
KOBAYASHI, Tsuyoshi, JAPAN
KOBAYASHI, Yoshikazu, JAPAN
KOBAYASHI, Yuji*, JAPAN
KOBAYASHI, Zenji, JAPAN
KOBAYASI, Kazuo§, JAPAN
KOBORI, Akira§, JAPAN
KOÇAK, Cevdet*, TURKEY
KOCH, Helmut, G.D.R.
KOCHERLAKOTA, Rama Rao*, U.S.A.
KODA, Junji, JAPAN
KODA, Takashi, JAPAN
KODAIRA, Kunihiko§, JAPAN
KODAKA, Kazunori, JAPAN
KODAMA, Akio, JAPAN
KODAMA, Tetsuo, JAPAN
KODAMA, Yukihiro, JAPAN
KODERA, Heiji, JAPAN
KOGISO, Takeyoshi, JAPAN
KOH, Jee Heub*, U.S.A.

KOHDA, Atsuhito, JAPAN
KOHMOTO, Susumu, JAPAN
KOHNO, Mitsuhiko, JAPAN
KOHNO, Toshitake, JAPAN
KOIKE, Kazuhiko, JAPAN
KOIKE, Masao, JAPAN
KOIKE, Minoru§, JAPAN
KOIKE, Naoyuki, JAPAN
KOIKE, Satoshi, JAPAN
KOIKE, Shigeaki, JAPAN
KOISO, Miyuki, JAPAN
KOISO, Norihito, JAPAN
KOITABASHI, Masanori, JAPAN
KOIZUMI, Shoji, JAPAN
KOIZUMI, Sumiyuki, JAPAN
KOJIMA, Hisashi, JAPAN
KOJIMA, Isao, JAPAN
KOJIMA, Jun, JAPAN
KOJIMA, Kiyofumi, JAPAN
KOJIMA, Makoto, JAPAN
KOJIMA, Sadayoshi, JAPAN
KOJYO, Hidemaro§, JAPAN
KOKETSU, Akio, JAPAN
KOKUBU, Hiroshi, JAPAN
KOLLÁR, János, U.S.A.
KOLOSKOV, Valentin Yurievich*, U.S.S.R.
KOLYVAGIN, Victor A., U.S.S.R.
KOMATSU, Gen, JAPAN
KOMATSU, Hikosaburo, JAPAN
KOMATSU, Kazushi, JAPAN
KOMATSU, Keiichi*, JAPAN
KOMATSU, Michiharu, JAPAN
KOMATU, Yusaku, JAPAN
KOMEDA, Jiryo, JAPAN
KOMIYA, Kaname, JAPAN
KOMIYA, Katsuhiro, JAPAN
KOMORI, Tetsushi, JAPAN
KOMORI, Youhei, JAPAN
KOMURA, Mutsumi, JAPAN
KOMURA, Takako, JAPAN
KOMURA, Yukio, JAPAN
KOMURO, Motomasa, JAPAN
KOMURO, Naoto, JAPAN
KONDO, Jiro, JAPAN
KONDO, Seizo, JAPAN
KONDO, Shigeyuki, JAPAN
KONDO, Shoichi, JAPAN
KONDO, Takeshi, JAPAN
KONGSAKORN, Kannika, THAILAND
KONNO, Hiroshi, JAPAN
KONNO, Kazuhiro, JAPAN
KONNO, Norio*, JAPAN
KONNO, Reiji, JAPAN
KONNO, Yasuko, JAPAN
KONNO, Yoshihiko, JAPAN
KONO, Akira, JAPAN
KONO, Hiroaki, JAPAN
KONO, Norio, JAPAN
KONO, Shigeo, JAPAN
KOOHARA, Akira, JAPAN
KOPFERMANN, Klaus, F.R.G.
KOPPELBERG, Sabine, F.R.G.
KORBAŠ, Július*, CZECHOSLOVAKIA
KORENBLUM, Boris*, U.S.A.
KOREVAAR, Jacob*, NETHERLANDS
KORI, Tosiaki, JAPAN
KORIYAMA, Akira§, JAPAN
KOSACHEVSKYA, Elena Aleksandrovna, U.S.S.R.
KOSAKI, Hideki, JAPAN
KOSCHORKE, Ulrich*, F.R.G.
KOSHI, Shozo*, JAPAN
KOSHIBA, Shunichi, JAPAN
KOSHIBA, Yoichi, JAPAN
KOSHIBA, Zenichiro, JAPAN
KOSHIKAWA, Hiroaki, JAPAN
KOSHITANI, Shigeo, JAPAN
KOSUDA, Masashi, JAPAN
KOTA, Osamu, JAPAN
KOTAKE, Takeshi, JAPAN
KOTANI, Motoko, JAPAN
KOTANI, Shinichi, JAPAN
KOTO, Toshiyuki, JAPAN
KOTUS, Janina, POLAND
KOUNO, Akira, JAPAN
KOUNO, Masaharu, JAPAN
KOYAMA, Akira, JAPAN
KOYAMA, Shin-ya, JAPAN
KOYAMA, Toshiko, JAPAN
KOYAMA, Yoshinori§, JAPAN
KOYANAGI, Tsunehira, JAPAN
KOZUKA, Kazuhito, JAPAN
KRANZ, Przemo T.*, U.S.A.
KRASNOSCHEKOV, Pavel, U.S.S.R.
KRASNY, Robert, U.S.A.
KRESS, Rainer, F.R.G.
KRGOVIĆ, Dragica N.*, YUGOSLAVIA
KRICHEVER, Igor M., U.S.S.R.
KRISZTIN, Tibor*, HUNGARY
KRONHEIMER, Peter B., U.K.
KUBO, Fumio*, JAPAN
KUBO, Izumi, JAPAN
KUBO, Kyoko*, JAPAN
KUBO, Masahiro*, JAPAN
KUBO, Yoshiko, JAPAN
KUBO, Yoshiyuki, JAPAN
KUBOTA, Koji, JAPAN
KUBOTA, Mitsuru, JAPAN
KUDO, Aichi, JAPAN
KUDO, Akio*, JAPAN
KUDO, Osamu, JAPAN
KUGA, Kenichi, JAPAN
KÜHNEL, Wolfgang, F.R.G.
KUIPER, Nicolaas H., FRANCE
KUKU, Aderemi Oluyomi*, NIGERIA
KUMAGAI, Donna*, U.S.A.

KUMAGAI, Takashi, JAPAN
KUMAHARA, Keisaku, JAPAN
KUMAKURA, Gakuji, JAPAN
KUMAZAWA, Masaaki, JAPAN
KUNITA, Hiroshi, JAPAN
KUNZ, Ernst A., F.R.G.
KUO, H.-H.*, U.S.A.
KUPERBERG, Gregory J.*, U.S.A.
KUPERBERG, Krystyna M.*, U.S.A.
KUPIAINEN, Antti J., FINLAND
KURAMITSU, Hirofumi, JAPAN
KURAMOTO, Yoshiyuki, JAPAN
KURANO, Kazuhiko, JAPAN
KURATA, Kazuhiro, JAPAN
KURATA, Masahiro, JAPAN
KURATA, Yoshiki, JAPAN
KURATSUBO, Shigehiko, JAPAN
KUREPA, Duro R.*, YUGOSLAVIA
KURIBAYASHI, Akikazu*, JAPAN
KURIBAYASHI, Izumi, JAPAN
KURIBAYASHI, Katsuhiko, JAPAN
KURIBAYASHI, Yukio, JAPAN
KURIHARA, Akira, JAPAN
KURIHARA, Masato, JAPAN
KURIHARA, Mitsunobu, JAPAN
KURIYAMA, Ken, JAPAN
KURKE, Herbert, G.D.R.
KURODA, Koji, JAPAN
KURODA, Shigetoshi, JAPAN
KURODA, Sige-Nobu, JAPAN
KUROGI, Tetsunori, JAPAN
KUROKAWA, Nobushige, JAPAN
KUROKAWA, Takahide, JAPAN
KUROKAWA, Takashi, JAPAN
KUROKAWA, Toshiko, JAPAN
KUROKI, Gen, JAPAN
KUROSAKI, Yasuhiro, JAPAN
KUROSE, Akinari, JAPAN
KUROSE, Hideki, JAPAN
KUROSE, Takashi, JAPAN
KURPITA, Bohdan I., U.K.
KURSHAN, Robert P., U.S.A.
KUSABA, Toshikuni, JAPAN
KUSAMA, Tokitake, JAPAN
KUSANAGI, Hiroyuki, JAPAN
KUSUDA, Masaharu, JAPAN
KUSUNOKI, Yukio, JAPAN
KUSUOKA, Shigeo, JAPAN
KUWABARA, Ruishi, JAPAN
KUWADA, Masahide, JAPAN
KUWAE, Kazuhiro, JAPAN
KUZNETSOV, Nicolai Vasilievičh, U.S.S.R.
KWAK, Jin Ho, KOREA, R. of
KWON, Yonghoon, KOREA, R. of
KYE, Seung-Hyeok, KOREA, R. of
KYUNO, Shoji, JAPAN
LABAHN, George, CANADA
LABARCA, Rafael E., CHILE
LACEY, H. Elton, U.S.A.
LACOMBLEZ, Chantal, FRANCE
LADAS, Gerasimos*, U.S.A.
LADDE, Gangaram S.*, U.S.A.
LADYZENSKAYA, Olga A., U.S.S.R.
LAFFEY, Thomas Joseph, IRELAND
LAHTINEN, Aatos O.*, FINLAND
LAI, Hang-Chin, CHINA-TAIWAN
LAINE, Ilpo E., FINLAND
LAKSHMIBAI, V.*, U.S.A.
LAM, Kee Yuen, CANADA
LAMB, David A., U.S.A.
LAMBERT, William M., COSTA RICA
LAMPRECHT, Erich H. K., F.R.G.
LANDMAN, Alan§, U.S.A.
LANE, Mary C., U.S.A.
LANGLAIS, Michel R.*, FRANCE
LANIN, Anatolii I., U.S.S.R.
LAPIDUS, Michel L.*, U.S.A.
LARIJANI, Mohammad-Javad A.§, IRAN
LAU, Anthony To-Ming, CANADA
LAU, Kee-Wai, HONG KONG
LAUMON, Gerard H., FRANCE
LAURITZEN, Niels, DENMARK
LAVRENTIEV, Mihail M., U.S.S.R.
LAWNICZAK, Anna T.*, CANADA/U.S.A.
LAWRENCE, Harry Robert, U.S.A.
LAWRENCE, Ruth Jayne, U.S.A.
ŁAWRYNOWICZ, Julian, POLAND
LAZARSFELD, Robert K., U.S.A.
LAZARUS, Andrew J.*, U.S.A.
LAZEBNIK, Felix G.*, U.S.A.
LÊ, Dung Trang, VIETNAM
LE, Hai Khoi*, VIETNAM
LE, Hong-Van*, VIETNAM
LEANDRE, Remi, FRANCE
LEBEAU, Gilles, FRANCE
LEBLANC, Emile A.*, CANADA
LEBORGNE, Daniel§, FRANCE
LE CAM, Lucien M.§, FRANCE
LEE, Chung-Nim, U.S.A.
LEE, Jong-Eao John§, CHINA-TAIWAN
LEE, Jyh-Hao*, CHINA-TAIWAN
LEE, Peter Martin, U.K.
LEE, Sa-Ge*, KOREA, R. of
LEE, Yuh-Jia, CHINA-TAIWAN
LEHTO, Olli, FINLAND
LEIBOWITZ, Daniela, ISRAEL
LEICHTWEISS, Kurt*, F.R.G.
LEMAIRE, Jean-Michel, FRANCE
LEMARIÉ, Pierre Gilles*, FRANCE
LEMPP, Steffen, U.S.A.
LEUNG, Ka Hin*, SINGAPORE
LEUNG, Pui-Fai Fred, SINGAPORE
LEVESQUE, Claude, CANADA
LEVIATAN, Dany, ISRAEL
LEVINE, Jerome P., U.S.A.
LEVITIN, Michael R., U.S.S.R.

LEWIS, Adrian S., CANADA
LEWIS, Keith Allen, U.S.A.
LEWIS, Wayne, U.S.A.
LI, Bang-He*, CHINA, P. R. of
LI, Chun Wah*, HONG KONG
LI, Shangzhi, CHINA, P. R. of
LIANG, Wen-Hai, CHINA, P. R. of
LIANG, Xue-Zhang*, CHINA, P. R. of
LIAO, Keren*, CHINA, P. R. of
LICKORISH, W. B. Raymond, U.K.
LIH, Ko-Wei*, CHINA-TAIWAN
LIM, Chong-Keang, MALAYSIA
LIN, En-Bing*, U.S.A.
LIN, Fang Hua, U.S.A.
LIN, Qing, CHINA, P. R. of
LIN, Song-Sun*, CHINA-TAIWAN
LIND, Douglas A., U.S.A.
LINDENSTRAUSS, Joram, ISRAEL
LINDQVIST, Peter L.*, FINLAND
LINTON, Fred E.J., U.S.A.
LIONS, Pierre Louis, FRANCE
LIPMAN, Joseph, U.S.A.
LIPSHUTZ, Seymour, U.S.A.
LIPSCOMB, Stephen Leon, U.S.A.
LIPSMAN, Ronald L., U.S.A.
LIRON, Nadav, ISRAEL
LIU, De-Fu*, CHINA, P. R. of
LIU, Fon-Che*, CHINA-TAIWAN
LIU, Ming-Chit, HONG KONG
LIULEVICIUS, Arunas L.§, U.S.A.
LIVORNI, Elvira Laura, ITALY
LOGAN, George D., U.S.A.
LOI, Phan Hung, U.S.A.
LONG, Yiming*, CHINA, P. R. of
LOPEZ, Luis-Miguel, FRANCE
LORENZ, Dan H., ISRAEL
LORENZINI, Dino J.*, U.S.A.
LORING, Terry A., U.S.A.
LOVÁSZ, László, HUNGARY
LOVE, Eric Russell*, AUSTRALIA
LOZANO, Maria-Teresa*, SPAIN
LOZI, René P.*, FRANCE
LU, Qi-Keng*, CHINA, P. R. of
LU, Shanzhen*, CHINA, P. R. of
LU, Sylvia Chin-Pi*, U.S.A.
LUCKE, John Edwin, U.S.A.
LUDWIG, Garry, CANADA
LUFT, Erhard*, CANADA
LUK, Hing-Sun, HONG KONG
LUKEŠ, Jaroslav*, CZECHOSLOVAKIA
LUNA, Domingo, FRANCE
LUO, Dingjun*, CHINA, P. R. of
LUO, Tie, U.S.A.
LUSE, Dzidra*, U.S.S.R.
LUSZTIG, George, U.S.A.
LÜTZEN, Jesper, DENMARK
LYUBEZNIK, Gennady, U.S.A.
MA, Lawrence Kwan Ho, U.S.A.
MA, Zhiming, CHINA, P. R. of
MABUCHI, Toshiki, JAPAN
MACHIDA, Hajime, JAPAN
MACHIDA, Yoshinori, JAPAN
MACKENZIE, Kirill C. H.*, U.K.
MACKEY, George W., U.S.A.
MADER, Wolfgang K.W., F.R.G.
MAEBASHI, Toshiyuki, JAPAN
MAEDA, Fumi-Yuki, JAPAN
MAEDA, Hidetoshi*, JAPAN
MAEDA, Hironobu, JAPAN
MAEDA, Masao, JAPAN
MAEDA, Michie, JAPAN
MAEDA, Shigeru, JAPAN
MAEDA, Shuichiro, JAPAN
MAEDA, Takashi, JAPAN
MAEDA, Yoshiaki, JAPAN
MAEDA, Yoshitaka, JAPAN
MAEHARA, Kazuhisa*, JAPAN
MAEHARA, Shôji, JAPAN
MAEJIMA, Makoto, JAPAN
MAESONO, Yoshihiko, JAPAN
MAGID, Andy R., U.S.A.
MAHMOODIAN, Ebadollah Sayed*, IRAN
MINAMOTO, Ryuichi, JAPAN
MAITANI, Fumio*, JAPAN
MAJDA, Andrew J., U.S.A.
MAJIMA, Hideyuki, JAPAN
MAKAROV, Vitalii S.*, U.S.S.R.
MÄKELÄINEN, Tuulikki, FINLAND
MAKI, Haruo, JAPAN
MAKIHARA, Hiroshi, JAPAN
MAKINO, Isao, JAPAN
MAKINO, Tetu*, JAPAN
MAKS, Johannes G.*, NETHERLANDS
MAKSOUDOV, Faramaz, U.S.S.R.
MALLIAVIN, Marie-Paule, FRANCE
MALLIAVIN, Paul, FRANCE
MALTSEV, Arkadii A., U.S.S.R.
MANABE, Hiroki, JAPAN
MANABE, Shojiro, JAPAN
MANDAI, Takeshi, JAPAN
MANFREDI, Juan J., U.S.A.
MANIN, Yuri Ivanovich§, U.S.S.R.
MANN, Larry N., U.S.A.
MANSFIELD, Kevin Graham, NEW ZEALAND
MANTINI, Lisa A.*, U.S.A.
MARATHE, Kishore B.*, U.S.A.
MARCELO, Reginaldo M., PHILIPPINES
MARĈENKO, Vladimir Alexandroviĉ, U.S.S.R.
MARGERIN, Christophe M., FRANCE
MARGULIS, Grigorii A., U.S.S.R.
MARINO, Mario, ITALY
MARITZ, Pieter*, SOUTH AFRICA
MARKUSHEVICH, Dmitri G.*, U.S.S.R.
MARSH, Marcus M., U.S.A.
MARSHALL, Donald E., U.S.A.
MARTENS, Henrik H., NORWAY

MARTIN, Christiane C.M., FRANCE
MARTINELLI, Enzo, ITALY
MARTINEZ, André, FRANCE
MARTINEZ, Ernesto*, SPAIN
MARTINEZ-VILLA, Roberto, MEXICO
MARTIO, Olli Tapani, FINLAND
MARTUCCI, Giovanni*, ITALY
MARUBAYASHI, Hidetoshi, JAPAN
MARUMOTO, Yoshihiko, JAPAN
MARUO, Kenji, JAPAN
MARUO, Osamu, JAPAN
MARUYAMA, Akira, JAPAN
MARUYAMA, Fumitsuna, JAPAN
MARUYAMA, Ken-ichi, JAPAN
MARUYAMA, Masaki, JAPAN
MARUYAMA, Naomasa, JAPAN
MARUYAMA, Noriko, JAPAN
MARUYAMA, Toru, JAPAN
MASCARELLO, Maria, ITALY
MASCIONI, Vania*, SWITZERLAND
MASE, Shigeru, JAPAN
MASHIMO, Katsuya, JAPAN
MASHINCHI, Mashaalah*, IRAN
MASTERSON, John T.*, U.S.A.
MASUDA, Kayo, JAPAN
MASUDA, Kazuo, JAPAN
MASUDA, Kazuyoshi, JAPAN
MASUDA, Kyuya, JAPAN
MASUDA, Masao, JAPAN
MASUDA, Takashi, JAPAN
MASUDA, Teruo, JAPAN
MASUDA, Tetsuya, JAPAN
MASUI, Takahiro, JAPAN
MASUOKA, Akira, JAPAN
MATAGA, Yoshiharu, JAPAN
MATANO, Hiroshi, JAPAN
MATET, Pierre*, FRANCE
MATHIEU, Olivier, FRANCE
MATHIEU, Philippe§, BELGIUM
MATHIEU, Yves E.§, FRANCE
MATROSOV, Vladimir M.*, U.S.S.R.
MATSUBARA, Yoshie, JAPAN
MATSUDA, Hiroo, JAPAN
MATSUDA, Shigeki, JAPAN
MATSUDA, Toshimitsu, JAPAN
MATSUE, Hirofumi, JAPAN
MATSUE, Yumiko, JAPAN
MATSUGU, Yasuo, JAPAN
MATSUI, Kiyosi, JAPAN
MATSUI, Taku, JAPAN
MATSUKI, Mihoko*, JAPAN
MATSUKI, Toshihiko, JAPAN
MATSUMOTO, Hiroyuki, JAPAN
MATSUMOTO, Kazuko, JAPAN
MATSUMOTO, Kazuo, JAPAN
MATSUMOTO, Keiji, JAPAN
MATSUMOTO, Kengo, JAPAN
MATSUMOTO, Kohji, JAPAN
MATSUMOTO, Makoto, JAPAN
MATSUMOTO, Makoto, JAPAN
MATSUMOTO, Shigenori, JAPAN
MATSUMOTO, Takanori, JAPAN
MATSUMOTO, Waichiro*, JAPAN
MATSUMOTO, Yukio, JAPAN
MATSUMURA, Akitaka, JAPAN
MATSUMURA, Hideyuki, JAPAN
MATSUMURA, Mutsuhide, JAPAN
MATSUMURA, Toru, JAPAN
MATSUO, Yoshitomo, JAPAN
MATSUOKA, Choichiro, JAPAN
MATSUOKA, Katsuo, JAPAN
MATSUOKA, Sachiko, JAPAN
MATSUOKA, Takashi, JAPAN
MATSUSHITA, Osamu, JAPAN
MATSUSHITA, Yasuo, JAPAN
MATSUURA, Shigetake, JAPAN
MATSUURA, Shozo, JAPAN
MATSUYAMA, Hiroshi, JAPAN
MATSUYAMA, Yoshio*, JAPAN
MATSUZAKI, Katsuhiko*, JAPAN
MATSUZAWA, Jun-ichi, JAPAN
MATSUZAWA, Tadato, JAPAN
MATTILA, Pertti E. J.*, FINLAND
MATUDA, Tizuko, JAPAN
MATUI, Akinori, JAPAN
MATUMOTO, Hisayoshi, JAPAN
MATUMOTO, Takao, JAPAN
MATUNAGA, Saburô, JAPAN
MATUSZEWSKI, Roman, BELGIUM
MATUURA, Takahide*, JAPAN
MAUDE, Ronald, U.K.
MAUDUIT, Christian*, FRANCE
MAULDIN, Richard D., U.S.A.
MAUMARY, Serge, SWITZERLAND
MAWATA, Christopher P., U.S.A.
MAZUR, Barry C., U.S.A.
McCARTHY, Randy*, U.S.A.
McCOY, Barry M., U.S.A.
McCOY, Peter A.*, U.S.A.
MCCRORY, Clint G., U.S.A.
McDONALD, James Wade, U.S.A.
McDUFF, Dusa, U.S.A./U.K.
McKAY, James H.*, U.S.A.
McKEAN, Henry P., U.S.A.
McMINN, Trevor J., U.S.A.
McMULLEN, Curt, U.S.A.
MEGGINSON, Robert E., U.S.A.
MEHRI, Bahman*, IRAN
MEHTA, Madan Lal*, FRANCE
MEHTA, Vikram B., INDIA
MEIER, David, SWITZERLAND
MELROSE, Richard B., AUSTRALIA
MENDOZA, Leonardo, VENEZUELA
MERKLEN, Hector A., BRAZIL
MESHKANI, Ali*, IRAN
MESTRANO, Nicole, FRANCE

MEURMAN, Arne Erik, SWEDEN
MEYER, David A., U.S.A.
MEYER, Jean-Pierre G., U.S.A.
MEYER, Yves F., FRANCE
MIATELLO, Roberto Jorge, ARGENTINA
MICHAEL, Ernest A., U.S.A.
MICHAELIS, Walter J.*, U.S.A.
MICKELSSON, Jouko A., FINLAND
MIDORIKAWA, Hisaichi, JAPAN
MIKAMI, Kentaro, JAPAN
MIKAMI, Shunsuke, JAPAN
MIKAMI, Toshio, JAPAN
MIKI, Hidefumi, JAPAN
MIKI, Hiroo, JAPAN
MIKI, Yoshikazu, JAPAN
MILICIC, Dragan, U.S.A.
MILLER, Arnold W.*, U.S.A.
MILLER, Sanford S.*, U.S.A.
MILLETT, Kenneth C.*, U.S.A.
MILLSON, John J., CANADA
MILMAN, Vitali D., ISRAEL
MILNER, Eric C., CANADA
MILNES, Paul, CANADA
MIMACHI, Katsuhisa, JAPAN
MIMAR, Arman, U.S.A.
MIMURA, Mamoru, JAPAN
MIMURA, Masayasu, JAPAN
MIMURA, Yoshio, JAPAN
MIN, Kyong Jin, KOREA, R. of
MIN, Kyung Chan, KOREA, R. of
MINAMI, Haruo, JAPAN
MINAMI, Nariyuki, JAPAN
MINAMI, Toshiro, JAPAN
MINDA, Carl David*, U.S.A.
MINEMURA, Katsuhiro, JAPAN
MINGO, James A., CANADA
MINN, Jooha, KOREA, R. of
MIRON, Radu*, ROMANIA
MISAGHIAN, Manochehr§, IRAN
MISAKI, Norihiro, JAPAN
MISAWA, Tetsuya, JAPAN
MISCENKO, Evguenii F., U.S.S.R.
MITAMURA, Takashi, JAPAN
MITCHELL, Josephine M., U.S.A.
MITOMA, Itaru, JAPAN
MITROPOLSKY, Yurii A., U.S.S.R.
MITSUI, Takayoshi, JAPAN
MITSUI, Taketomo, JAPAN
MITSUMATSU, Yoshihiko, JAPAN
MIWA, Megumu, JAPAN
MIWA, Tatsuro*, JAPAN
MIWA, Tetsuji, JAPAN
MIYACHI, Akihiko, JAPAN
MIYADA, Itiro, JAPAN
MIYADERA, Isao§, JAPAN
MIYAGAWA, Masami, JAPAN
MIYAGAWA, Yukitaka, JAPAN
MIYAHARA, Yoshio, JAPAN
MIYAJIMA, Kimio, JAPAN
MIYAJIMA, Shizuo, JAPAN
MIYAKE, Masatake, JAPAN
MIYAKE, Toshitsune, JAPAN
MIYAMOTO, Haruo, JAPAN
MIYAMOTO, Masahiko, JAPAN
MIYAMOTO, Munemi, JAPAN
MIYANISHI, Masayoshi, JAPAN
MIYANO, Hiroshi*, JAPAN
MIYANO, Satoru, JAPAN
MIYAOKA, Etsuo, JAPAN
MIYAOKA, Reiko, JAPAN
MIYAOKA, Yoichi, JAPAN
MIYASAKA, Masanori, JAPAN
MIYATA, Yoichiro, JAPAN
MIYATAKE, Sadao, JAPAN
MIYAUCHI, Mutsuo, JAPAN
MIYAWAKI, Isao, JAPAN
MIYAZAKI, Chikashi, JAPAN
MIYAZAKI, Hitoshi, JAPAN
MIYAZAKI, Isao, JAPAN
MIYAZAKI, Ken-ichi, JAPAN
MIYAZAKI, Yoichi*, JAPAN
MIYAZAWA, Yasuyuki, JAPAN
MIYOSHI, Shigeaki, JAPAN
MIZOGUCHI, Minoru, JAPAN
MIZOHATA, Kiyoshi, JAPAN
MIZOHATA, Sigeru, JAPAN
MIZUHARA, Akira, JAPAN
MIZUHARA, Takahiro, JAPAN
MIZUMACHI, Hitoshi, JAPAN
MIZUMACHI, Ryuichi§, JAPAN
MIZUMOTO, Hisao, JAPAN
MIZUNO, Hirobumi, JAPAN
MIZUNO, Toru, JAPAN
MIZUNOYA, Takeshi, JAPAN
MIZUTA, Yoshihiro, JAPAN
MIZUTANI, Akira, JAPAN
MIZUTANI, Tadayoshi, JAPAN
MOCHIZUKI, Kiyoshi, JAPAN
MOEANADDIN, Rahim*, IRAN
MOEGLIN, Colette, FRANCE
MOGHADDAM, M. R. R.*, IRAN
MOGI, Isamu, JAPAN
MOHAMMADI, Hassanabadi A.*, IRAN
MOHEBI, Hossein, IRAN
MOHRI, Masayuki, JAPAN
MOHSENI, Mahmoud M.*, IRAN
MOKOBODZKI, Gabriel, FRANCE
MOLCHANOV, Stanislav A., U.S.S.R.
MOLCHANOV, Vladimir F.*, U.S.S.R.
MØLLER, Jesper M.*, DENMARK
MOMOSE, Fumiyuki, JAPAN
MONTAKHAB, Mohammad Sadegh§, IRAN
MOODY, Robert Vaughan§, CANADA
MOON, John W., CANADA
MOORE, Charles N., U.S.A.
MOORE, John D., U.S.A.

MORA, Leonardo Enrique, BRAZIL
MORAN, Gadi W.*, ISRAEL
MORDUKHOVICH, Boris S.*, U.S.A.
MORENO, Guillermo, MEXICO
MORENO SOCÍAS, Guillermo, SPAIN/FRANCE
MORGENSTERN, Rudolf, F.R.G.
MORI, Eon, JAPAN
MORI, Hiroshi, JAPAN
MORI, Hiroyuki, JAPAN
MORI, Makoto, JAPAN
MORI, Masao*, JAPAN
MORI, Masatake, JAPAN
MORI, Seiki, JAPAN
MORI, Shigefumi, JAPAN
MORI, Shosuke, JAPAN
MORI, Takakazu, JAPAN
MORI, Toshio, JAPAN
MORIKAWA, Hisashi, JAPAN
MORIKAWA, Naoto, JAPAN
MORIKAWA, Ryozo, JAPAN
MORIMOTO, Akira, JAPAN
MORIMOTO, Hiroaki, JAPAN
MORIMOTO, Hiroko, JAPAN
MORIMOTO, Hiroshi, JAPAN
MORIMOTO, Hiroshi, JAPAN
MORIMOTO, Kanji, JAPAN
MORIMOTO, Masaharu, JAPAN
MORIMOTO, Masayosi, JAPAN
MORIMOTO, Mayumi, JAPAN
MORIMOTO, Mitsuo, JAPAN
MORIMOTO, Shoji*, JAPAN
MORIMOTO, Tohru, JAPAN
MORIMOTO, Yoshinori, JAPAN
MORIOKA, Tatsushi, JAPAN
MORISHITA, Kazuhiko, JAPAN
MORISHITA, Mitsuji, JAPAN
MORITA, Jun, JAPAN
MORITA, Shigeyuki, JAPAN
MORITA, Yasuo, JAPAN
MORITA, Yoshihisa, JAPAN
MORITA, Yoshiyuki, JAPAN
MORIWAKI, Atsushi, JAPAN
MOROSAWA, Shunsuke, JAPAN
MOROZ, B. Z., ISRAEL
MORRISON, David R.§, U.S.A.
MORTON, Hugh Reynolds*, U.K.
MOSCOVICI, Henri, U.S.A.
MOSER, Jürgen, SWITZERLAND
MOSER, Lucy, U.S.A.
MOTEGI, Kimihiko*, JAPAN
MOTEGI, Masanori, JAPAN
MOTOHASHI, Nobuyoshi, JAPAN
MOTOSE, Kaoru, JAPAN
MOUNTFORD, Tom*, U.S.A.
MOURAD, Karim Joseph, U.S.A.
MOURRAIN, Bernard, FRANCE
MOUSSAVI, Ahmad*, IRAN
MUKAI, Juno, JAPAN
MUKAI, Shigeru, JAPAN
MUKAWA, Takayuki, JAPAN
MUKOYAMA, Kazuo, JAPAN
MULASE, Motohico, U.S.A.
MULCAHY, Marjorie, U.S.A.
MURA, Toshio§, U.S.A.
MURAI, Jousin, JAPAN
MURAI, Takafumi, JAPAN
MURAKAMI, Haruo, JAPAN
MURAKAMI, Jun, JAPAN
MURAKAMI, Kouichi, JAPAN
MURAKAMI, Maki, JAPAN
MURAKAMI, Shingo, JAPAN
MURAMATU, Tosinobu, JAPAN
MURAMORI, Takao, JAPAN
MURAMOTO, Katsushi, JAPAN
MURASE, Nobuyuki, JAPAN
MURASUGI, Kunio, CANADA
MURATA, Minoru*, JAPAN
MURATA, Tamotsu, JAPAN
MURATA, Yoshihiro, JAPAN
MURAYAMA, Mitutaka, JAPAN
MURAZAWA, Tadashi, JAPAN
MURO, Masakazu*, JAPAN
MURRE, Jacob P., NETHERLANDS
MURTY, Vijayakumar, CANADA
MUTA, Masanori, JAPAN
MUTANGADURA, Simba Arthur*, ZIMBABWE
MUTI, Nobol*, JAPAN
MUTO, Yosio§, JAPAN
MUTOU, Hideo, JAPAN
NAEENI, S. M. K., IRAN
NAGAHARA, Takasi, JAPAN
NAGAI, Hideo, JAPAN
NAGAI, Osamu, JAPAN
NAGAI, Sinpei, JAPAN
NAGAI, Tamao, JAPAN
NAGAMATI, Sigeaki*, JAPAN
NAGAMOCHI, Hiroshi, JAPAN
NAGANO, Azuma§, JAPAN
NAGANUMA, Hidehisa, JAPAN
NAGAO, Hirosi, JAPAN
NAGAO, Hisao, JAPAN
NAGAO, Koichi, JAPAN
NAGAOKA, Katsufumi§, JAPAN
NAGAOKA, Ryosuke, JAPAN
NAGASAKA, Kenji*, JAPAN
NAGASAKA, Yukio, JAPAN
NAGASAKI, Ikumitsu, JAPAN
NAGASAWA, Takeyuki, JAPAN
NAGASE, Haruo, JAPAN
NAGASE, Masayoshi, JAPAN
NAGASE, Michihiro, JAPAN
NAGASE, Noriaki, JAPAN
NAGASE, Teruko, JAPAN
NAGASHIMA, Takashi, JAPAN

NAGATA, Chigusa, JAPAN
NAGATA, Chihiro, JAPAN
NAGATA, Masatsugu, JAPAN
NAGATA, Masayoshi, JAPAN
NAGATOMO, Yasuyuki, JAPAN
NAGAYA, Hitoshi, JAPAN
NAGAYAMA, Haruya*, JAPAN
NAGISA, Masaru, JAPAN
NAGURA, Toshinobu, JAPAN
NÄKKI, Raimo T., FINLAND
NAIM, Linda, FRANCE
NAITO, Hirotada, JAPAN
NAITO, Hisashi, JAPAN
NAITO, Manabu, JAPAN
NAITO, Satoshi, JAPAN
NAITO, Shuji, JAPAN
NAITO, Toshiki, JAPAN
NAKABO, Shigekazu, JAPAN
NAKADA, Hitoshi, JAPAN
NAKADA, Masami, JAPAN
NAKAGAMI, Keiko, JAPAN
NAKAGAMI, Yoshiomi, JAPAN
NAKAGAWA, Hisao, JAPAN
NAKAGAWA, Jin, JAPAN
NAKAGAWA, Kiyokazu, JAPAN
NAKAGAWA, Noriaki, JAPAN
NAKAGAWA, Ryosuke, JAPAN
NAKAGAWA, Yasuhiro, JAPAN
NAKAGAWA, Yoko, JAPAN
NAKAGAWA, Yoko, JAPAN
NAKAGOSHI, Norikata, JAPAN
NAKAHARA, Hayao, JAPAN
NAKAHARA, Toru*, JAPAN
NAKAHATA, Koji, JAPAN
NAKAI, Isao, JAPAN
NAKAI, Mitsuru, JAPAN
NAKAI, Yoshinobu, JAPAN
NAKAJIMA, Atsushi, JAPAN
NAKAJIMA, Fumio, JAPAN
NAKAJIMA, Hiraku, JAPAN
NAKAJIMA, Iwao, JAPAN
NAKAJIMA, Katsuhiko, JAPAN
NAKAJIMA, Kazufumi, JAPAN
NAKAJIMA, Shoichi, JAPAN
NAKAJIMA, Tsugukazu, JAPAN
NAKAKI, Tatsuyuki, JAPAN
NAKAMA, Yutaka, JAPAN
NAKAMITSU, Kuniaki, JAPAN
NAKAMULA, Ken, JAPAN
NAKAMURA, Akira, JAPAN
NAKAMURA, Gen, JAPAN
NAKAMURA, Hiroaki, JAPAN
NAKAMURA, Hiroyuki, JAPAN
NAKAMURA, Iku§, JAPAN
NAKAMURA, Kirio, JAPAN
NAKAMURA, Masataka, JAPAN
NAKAMURA, Masato, JAPAN
NAKAMURA, Masatoshi, JAPAN
NAKAMURA, Riichiro§, JAPAN
NAKAMURA, Tetsuo, JAPAN
NAKAMURA, Tokushi, JAPAN
NAKAMURA, Yoshihiro, JAPAN
NAKAMURA, Yoshimasa, JAPAN
NAKAMURA, Yoshio, JAPAN
NAKAMURA, Yukio, JAPAN
NAKANE, Michiyo*, JAPAN
NAKANE, Shizuo, JAPAN
NAKANE, Takashi, JAPAN
NAKANISHI, Kōki, JAPAN
NAKANISHI, Shizu, JAPAN
NAKANISHI, Tomoki, JAPAN
NAKANISHI, Toshihiro, JAPAN
NAKANISHI, Yasutaka*, JAPAN
NAKANO, Minoru, JAPAN
NAKANO, Minoru, JAPAN
NAKANO, Shigeo, JAPAN
NAKANO, Shin, JAPAN
NAKANO, Tetsuo, JAPAN
NAKANO, Yoshihiro, JAPAN
NAKAO, Masahiro, JAPAN
NAKAO, Mitsuhiro, JAPAN
NAKAO, Mitsuhiro T.*, JAPAN
NAKAO, Shintaro, JAPAN
NAKAOKA, Akira, JAPAN
NAKAOKA, Minoru, JAPAN
NAKASHIMA, Katsuya, JAPAN
NAKASHIMA, Kiichi, JAPAN
NAKASHIMA, Masaharu, JAPAN
NAKASHIMA, Toru, JAPAN
NAKASHIMA, Toshiki, JAPAN
NAKATA, Tomoichi, JAPAN
NAKATA, Yoshimoto, JAPAN
NAKATSUKA, Harunori, JAPAN
NAKATSUKASA, Takuma, JAPAN
NAKAUCHI, Nobumitsu*, JAPAN
NAKAYAMA, Chikara, JAPAN
NAKAYAMA, Hiromichi, JAPAN
NAKAYAMA, Maki*, SOUTH AFRICA
NAKAYAMA, Noboru, JAPAN
NAKAYAMA, Takashi, JAPAN
NAKAYASHIKI, Atsushi, JAPAN
NAKAZATO, Hiroshi, JAPAN
NAKAZAWA, Hideaki, JAPAN
NAKAZAWA, Noriyuki§, JAPAN
NAKAZI, Takahiko, JAPAN
NAMBA, Kanji, JAPAN
NAMBA, Makoto*, JAPAN
NAMIKAWA, Yoshinori, JAPAN
NAMIKAWA, Yukihiko, JAPAN
NAMIOKA, Isaac, U.S.A.
NANBU, Tokumori, JAPAN
NAPALKOV, Valeusive, U.S.S.R.
NAPPI, Chiara R., U.S.A.
NARA, Chie, JAPAN
NARASIMHAN, Madumbai Seshachalu, INDIA

NARAYANASWAMI, Pallasena P., CANADA
NARAZAKI, Takashi, JAPAN
NARITA, Junichiro, JAPAN
NARITA, Masahiro, JAPAN
NARITA, Ryoichi, JAPAN
NARUKAWA, Kimiaki, JAPAN
NARUSE, Hiroshi, JAPAN
NARUSHIMA, Hiroshi, JAPAN
NASATYR, Emile Ben*, U.K.
NASHED, M. ZUHAIR, U.S.A.
NATSUME, Toshikazu, U.S.A.
NAUDÉ, Cornelia G., SOUTH AFRICA
NAUDÉ, Gert, SOUTH AFRICA
NAVARRO, Milagros P., PHILIPPINES
NAVARRO, Vicente, SPAIN
NAWA, Hayato§, JAPAN
NAWATA, Masako, JAPAN
NAYATANI, Shin, JAPAN
NEDELEC, Jean Claude, FRANCE
NEGAMI, Seiya, JAPAN
NEGISHI, Aiko, JAPAN
NEISHTADT, Anatoli I., U.S.S.R.
NEMENZO, Fidel R., PHILIPPINES
NEMOTO, Hiroaki, JAPAN
NEMOTO, Seiji, JAPAN
NESETRIL, Jaroslav, CZECHOSLOVAKIA
NESTERENKO, Yuri, U.S.S.R.
NETO, Orlando M. B., PORTUGAL
NEUBERGER, Barbara O., U.S.A.
NEUBERGER, John W., U.S.A.
NEUEN, Annette I., SWITZERLAND
NEUENSCHWANDER, Daniel, SWITZERLAND
NEVANLINNA, Olavi, FINLAND
NEWHOUSE, Sheldon E., U.S.A.
NEWMAN, Michael F., AUSTRALIA
NGHIEM, Xuan Hai*, FRANCE
NGO, Dac Tan, VIETNAM
NGUIFFO BOYOM, Michel B.*, FRANCE
NGUYEN, Dai Tien, VIETNAM
NGUYEN, Dinh Cong, VIETNAM
NGUYEN, Dinh Tri, VIETNAM
NGUYEN, Duc Tuan*, VIETNAM
NGUYEN, Khoa Son, VIETNAM
NGUYEN, Minh Tri*, VIETNAM
NGUYEN, Quang Do Thong*, FRANCE
NGUYEN, Van Chan*, VIETNAM
NGUYEN, Viet Dung*, VIETNAM
NICOLAS, Jean-Louis*, FRANCE
NIELAND, Hendrik Maarten, NETHERLANDS
NII, Shunsaku, JAPAN
NIIKURA, Yasuo, JAPAN
NIINO, Kiyoshi, JAPAN
NIITSUMA, Hiroshi, JAPAN
NIJENHUIS, Albert*, U.S.A.
NIKNAM, Assadollah*, IRAN
NIKOLSKII, Nikolai K.*, U.S.S.R.
NIKOLTJEVA-HEDBERG, M., SWEDEN
NIKULIN, Vjacheslav V., U.S.S.R.
NINNEMANN, Olaf, F.R.G.
NINOMIYA, Hirokazu, JAPAN
NINOMIYA, Nobuyuki*, JAPAN
NINOMIYA, Syoiti, JAPAN
NINOMIYA, Yasushi, JAPAN
NISHI, Haruko, JAPAN
NISHIDA, Goro, JAPAN
NISHIDA, Hiroshi, JAPAN
NISHIDA, Koji, JAPAN
NISHIDA, Takaaki, JAPAN
NISHIGAKI, Sei-ichi, JAPAN
NISHIHARA, Kenji, JAPAN
NISHIHARA, Masaru*, JAPAN
NISHII, Ryuei, JAPAN
NISHIKAWA, Akira, JAPAN
NISHIKAWA, Masayuki, JAPAN
NISHIKAWA, Nobutaka, JAPAN
NISHIKAWA, Seiki, JAPAN
NISHIMI, Jisho, JAPAN
NISHIMORI, Toshiyuki, JAPAN
NISHIMORI, Yasunori, JAPAN
NISHIMOTO, Katsuyuki*, JAPAN
NISHIMURA, Jun-ichi, JAPAN
NISHIMURA, Takashi, JAPAN
NISHIMURA, Takeshi, JAPAN
NISHIMURA, Toshio, JAPAN
NISHIMURA, Yasuichiro, JAPAN
NISHINO, Tetsuro*, JAPAN
NISHIO, Kazuhiro, JAPAN
NISHIO, Masaharu, JAPAN
NISHIO, Masaru, JAPAN
NISHIOKA, Michio, JAPAN
NISHITANI, Tatsuo, JAPAN
NISHIURA, Yasumasa, JAPAN
NISHIYAMA, Akishige, JAPAN
NISHIYAMA, Kiminao, JAPAN
NISHIYAMA, Kyo, JAPAN
NISHIZAWA, Kiyoko, JAPAN
NISIO, Makiko, JAPAN
NISTOR, Victor, ROMANIA
NITTA, Takashi, JAPAN
NIWA, Masahiko, JAPAN
NIWA, Shinji, JAPAN
NIWANO, Eikazu, JAPAN
NIWASAKI, Takashi, JAPAN
NIZAMUDDIN, Kazi G.*, U.S.A.
NODA, Akio, JAPAN
NODA, Kazunari, JAPAN
NOGAMI, Akihiro, JAPAN
NOGAMI, Yoshiko, JAPAN
NOGI, Tatsuo, JAPAN
NOGUCHI, Junjiro, JAPAN
NOGUCHI, Mitsunori, JAPAN
NOHDA, Nobuhiko*, JAPAN
NOKURA, Tsugunori, JAPAN
NOMA, Atsushi, JAPAN
NOMAKUCHI, Kentaro, JAPAN

NOMOTO, Hisao, JAPAN
NOMURA, Kazumasa*, JAPAN
NOMURA, Yasutoshi, JAPAN
NOMURA, Yuji, JAPAN
NONO, Kiyoharu, JAPAN
NONOGUCHI, Norimichi, JAPAN
NOUMI, Masatoshi, JAPAN
NOZAKI, Akihiro, JAPAN
NOZAWA, Ryohei, JAPAN
NUNZIANTE, Diana§, ITALY
OAKU, Toshinori, JAPAN
OBATA, Morio, JAPAN
OBATA, Nobuaki*, JAPAN
OBAYASHI, Tadao, JAPAN
OBERSCHELP, Arnold*, F.R.G.
OBERSTE-VORTH, Ralph W., U.S.A.
OCHIAI, Hiroyuki, JAPAN
OCHIAI, Shoji, JAPAN
OCHIAI, Takushiro, JAPAN
OCNEANU, Adrian, U.S.A.
ODA, Tadao, JAPAN
ODA, Takayuki*, JAPAN
ODAI, Yoshitaka*, JAPAN
ODANI, Kenzi, JAPAN
ODLYZKO, Andrew M., U.S.A.
OEDA, Kazuo, JAPAN
OEHMKE, Robert H., U.S.A.
OEHMKE, Theresa M., U.S.A.
OELLERMANN, Ortrud Ruth*, SOUTH AFRICA
OETTLI, Werner, F.R.G.
O'FARRELL, Anthony G., IRELAND
OGATA, Shoetsu, JAPAN
OGAWA, Akihiko, JAPAN
OGAWA, Hiroyuki, JAPAN
OGAWA, Shigeyoshi*, JAPAN
OGAWA, Toshiyuki, JAPAN
OGAWA, Tsukane, JAPAN
OGAWA, Yosuke, JAPAN
OGIUE, Koichi, JAPAN
OGOMA, Tetsushi, JAPAN
OGUCHI, Kunio, JAPAN
OGUISO, Keiji, JAPAN
OGURA, Kazunori, JAPAN
OGURA, Shohei, JAPAN
OGURA, Yukio, JAPAN
OH, Jae-Pill, KOREA, R. of
OH, Junghwan, KOREA, R. of
OHARA, Jun*, JAPAN
OHASHI, Masakazu, JAPAN
OHATA, Koichi, JAPAN
OHBUCHI, Akira, JAPAN
OHHAMA, Minako, JAPAN
OHI, Takeo, JAPAN
OHKUBO, Toshio, JAPAN
OHKURO, Shigeru, JAPAN
OHMIYA, Mayumi, JAPAN
OHMORI, Zyuzyu, JAPAN
OHNISHI, Isamu, JAPAN
OHNITA, Yoshihiro, JAPAN
OHNO, Hiloshi*, JAPAN
OHNO, Koji, JAPAN
OHNO, Masahiro, JAPAN
OHNO, Mayumi, JAPAN
OHNO, Shûichi, JAPAN
OHORI, Masayuki, JAPAN
OHSITA, Akihiro, JAPAN
OHTA, Akira, JAPAN
OHTA, Haruto, JAPAN
OHTA, Hiroshi, JAPAN
OHTA, Masami, JAPAN
OHTA, Minolu, JAPAN
OHTA, Takuya§, JAPAN
OHTAKE, Hiromi, JAPAN
OHTSUKA, Fumiko, JAPAN
OHTSUKA, Makoto, JAPAN
OHTSUKI, Tomotada, JAPAN
OHYA, Yujiro, JAPAN
OHYAMA, Yoshiyuki, JAPAN
OHYAMA, Yousuke, JAPAN
OIKAWA, Kotaro, JAPAN
OISHI, Ryuzo, JAPAN
OKA, Hiroe, JAPAN
OKA, Hirokazu, JAPAN
OKA, Masaaki, JAPAN
OKA, Masatoshi, JAPAN
OKA, Mutsuo, JAPAN
OKA, Shinpei§, JAPAN
OKA, Yukimasa, JAPAN
OKABAYASHI, Shigeyoshi, JAPAN
OKABE, Hiroshi§, JAPAN
OKABE, Tsuneharu, JAPAN
OKABE, Yasunori, JAPAN
OKADA, Kanae, JAPAN
OKADA, Mari, JAPAN
OKADA, Masae, JAPAN
OKADA, Masami, JAPAN
OKADA, Masazumi, JAPAN
OKADA, Nolio, JAPAN
OKADA, Norio, JAPAN
OKADA, Soichi, JAPAN
OKADA, Tatsuya, JAPAN
OKADA, Toshinao, JAPAN
OKADA, Tsutomu, JAPAN
OKADA, Tsutomu, JAPAN
OKADA, Yasunori*, JAPAN
OKADO, Mosato, JAPAN
OKAI, Takayuki, JAPAN
OKAJI, Takashi, JAPAN
OKAMOTO, Hisashi, JAPAN
OKAMOTO, Hisashi, JAPAN
OKAMOTO, Kazuo, JAPAN
OKAMOTO, Kiyosato, JAPAN
OKAMOTO, Shigemi, JAPAN
OKAMURA, Hideaki, JAPAN
OKANO, Takashi, JAPAN

OKANO, Takeshi, JAPAN
OKAYASU, Takashi, JAPAN
OKAYASU, Takateru, JAPAN
OKAZAKI, Kimiko, JAPAN
OKAZAKI, Ryotaro, JAPAN
OKIYAMA, Satoru, JAPAN
OKOCHI, Hiroko, JAPAN
OKSENDAL, Bernt*, NORWAY
OKUBO, Katsumi, JAPAN
OKUBO, Kazuyoshi, JAPAN
OKUGAWA, Kotaro, JAPAN
OKUHARA, Hiroshi, JAPAN
OKUMURA, Hirozo, JAPAN
OKUMURA, Masafumi, JAPAN
OKUMURA, Yoshihide, JAPAN
OKUNO, Toshinao, JAPAN
OKURA, Hiroyuki, JAPAN
OKUTSU, Kosaku*, JAPAN
OKUYAMA, Akihiro, JAPAN
OKUYAMA, Tetsuro, JAPAN
OKUYAMA, Yasuo, JAPAN
OKUYAMA, Yukihiko, JAPAN
OLECH, Czesław, POLAND
OLECHE, Paul Odhiambo, KENYA
OLLIVIER, François, FRANCE
OLSEN, Catherine L., U.S.A.
OMACHI, Eriko, JAPAN
OMORI, Hideki, JAPAN
OMOTO, Chikaharu, JAPAN
ONO, Akira, JAPAN
ONO, Hiroakira, JAPAN
ONO, Kaoru, JAPAN
ONO, Kosuke, JAPAN
ONO, Takashi, JAPAN
ONO, Tomoaki, JAPAN
ONO, Yukio§, JAPAN
ONODA, Nobuharu, JAPAN
ONYANGO-OTIENO, Vitalis P., KENYA
OOHASHI, Tsunemichi, JAPAN
OOISHI, Akira, JAPAN
ORAZOV, Mered B., U.S.S.R.
ORIHARA, Akio, JAPAN
ORLOVA, Luidila, U.S.S.R.
O'ROURKE, Francesca S.*, U.K.
ORRO, Patrice, FRANCE
ORSTED, Bent, DENMARK
OSADA, Hirofumi, JAPAN
OSADA, Masayuki, JAPAN
OSAKA, Hiroyuki, JAPAN
OSAWA, Ikuko, JAPAN
OSAWA, Takeo, JAPAN
O'SHEA, Donal B., U.S.A.
OSHIKAWA, Keiichi, JAPAN
OSHIMA, Kazuyuki, JAPAN
OSHIMA, Toshio, JAPAN
OSHIMA, Yoichi, JAPAN
OSHIRO, Kiyoichi, JAPAN
OSIKAWA, Motosige, JAPAN
OSIPOV, Yuri S., U.S.S.R.
OTA, Kaori Imai, JAPAN
OTA, Minoru, JAPAN
OTA, Schoichi, JAPAN
OTAL, Javier*, SPAIN
OTOFUJI, Takashi, JAPAN
OTSUKA, Kayo, JAPAN
OTSUKA, Kenichi, JAPAN
OTSUKI, Nobukazu, JAPAN
OTSUKI, Tominosuke, JAPAN
OTWAY, Thomas H., U.S.A.
OUCHI, Moto, JAPAN
OUCHI, Osamu, JAPAN
OUCHI, Sunao, JAPAN
OWA, Shigeyoshi*, JAPAN
OWADA, Hiromoto, JAPAN
OWUSU-ANSAH, Twum*, GHANA
OYABU, Takashi, JAPAN
OZAWA, Kazuo, JAPAN
OZAWA, Masanao, JAPAN
OZAWA, Tetsuya, JAPAN
OZEKI, Hideki, JAPAN
OZEKI, Ikuzo*, JAPAN
OZEKI, Kiyota, JAPAN
OZEKI, Tomoko, JAPAN
OZONE, Jun, JAPAN
PACELLA, Filomena, ITALY
PACKER, Judith A.*, U.S.A.
PAGANI, Carlo Domenico, ITALY
PAHK, Dae Hyeon*, KOREA, R. of
PAK, Hong Kyung, KOREA, R. of
PAK, Jim Suk, KOREA, R. of
PALEV, Tchavdar Dimitrov, BULGARIA
PALIS, Jacob Jr., BRAZIL
PANDEY, Jagdish N.*, CANADA
PARANJAPE, Kapil H., INDIA
PARDINI, Rita M., ITALY
PARK, Bong-Kyu, KOREA, R. of
PARK, Chan-Young, KOREA, R. of
PARK, Hye Sook, KOREA, R. of
PARK, Kyewon K., U.S.A.
PARK, Sang Ro§, KOREA, R. of
PARK, Yong Moon, KOREA, R. of
PARTANEN, Juha Veikko*, FINLAND
PARTIS, Michael T., AUSTRALIA
PARVATHAM, Rajagopalan*, INDIA
PĀSĀRESCU, Ovidiu-Florin A., ROMANIA
PASCU, Eugen D.*, ROMANIA
PASEMAN, Gerhard Raymond, U.S.A.
PASK, David A., AUSTRALIA
PASSARE, Mikael, SWEDEN
PATHAK, Ram Shankar*, INDIA
PAWŁUCKI, Wiesław J.*, POLAND
PEARS, Alan R., U.K.
PEDERSEN, Gert K., DENMARK
PEDICCHIO, M. Cristina*, ITALY
PEKONEN, Osmo E. T., FINLAND
PELCZYŃSKI, Aleksander, POLAND

PELLER, Vladimir V.*, U.S.S.R.
PELTONEN, Kirsi Hannele, FINLAND
PENA, Juan Manuel*, SPAIN
PERCUS, Ora Engelberg*, U.S.A.
PERSSON, Jan, SWEDEN
PERSSON, Ulf A., SWEDEN
PETROVIĆ, Ljiljana M.*, YUGOSLAVIA
PHAM, Huy Dien*, VIETNAM
PHAM, Loi Vu, VIETNAM
PHAN, Thien Thach*, VIETNAM
PHILIPPIN, Gérard Auguste, CANADA
PHILLIPS, N. C., U.S.A.
PIARD, Alain A.P., FRANCE
PICKERING, Douglas A.§, CANADA
PIER, Jean-Paul, LUXEMBOURG
PIMSNER, Michael Viktor, F.R.G.
PINCHON, Didier M., FRANCE
PIPPENGER, Nicholas J., CANADA
PISIER, Gilles J., FRANCE
PLATONOV, Vladimir P., U.S.S.R.
PLOTKIN, Eugene*, U.S.S.R.
POGORELOV, Aleksei V.*, U.S.S.R.
POINT, Nelly, FRANCE
POKHARIYAL, Ganesh Prasad*, KENYA
POPA, Sorin Teodor, U.S.A.
PORRU, Giovanni*, ITALY
POTYAGAILO, Leonid D.*, U.S.S.R.
POURABDOLLAH, M. A.*, IRAN
POURCIN, Genevieve, FRANCE
POWELL, Wayne B., U.S.A.
PRASAD, Dipendra, INDIA
PRASAD, Gopal, INDIA
PREISS, David, U.K.
PRESSLEY, Andrew N., U.K.
PRICE, David T., U.S.A.
PRIETO, Carlos, MEXICO
PROPPE, Harold§, CANADA
PTÁK, Vlastimil, CZECHOSLOVAKIA
PURZITSKY, Norman, CANADA
PUTINAR, Mihai I., ROMANIA
PYBER, La'szló, HUNGARY
QIAN, Tao*, CHINA, P. R. of
QIU, Peizhang*, CHINA, P. R. of
QIU, Shuxi, CHINA, P. R. of
QUINN, Declan P.F., U.S.A.
QUINN, Frank S., U.S.A.
QURAISHI, Rehana*, INDIA
RABINDRANATHAN, M., U.S.A.
RADHA, Srinivasa Raghavan*, INDIA
RADJABALIPOUR, Mehdi*, IRAN
RĂDULESCU, Florin G., ROMANIA
RAGNEDDA, Francesco§, ITALY
RAGUSA, Alfio, ITALY
RAIMONDO, Mario, ITALY
RAJVANSHI, Niti§, INDIA
RALLIS, Stephen, U.S.A.
RAMANAN, Sundararaman, INDIA
RAMM, Alexander G.*, U.S.A.
RAMSEY, Helen Mary, U.K.
RAN, Ziv, U.S.A.
RANDELL, Richard, U.S.A.
RANSFORD, Thomas J., U.K.
RASOULIAN, Amid§, IRAN
RASSIAS, Themistocles M.§, GREECE
RATANAPRASERT, Chaweewan, THAILAND
RAUBENHEIMER, Heinrich*, SOUTH AFRICA
RAUGEL, Geneviève, FRANCE
RAUTMANN, Reimund K. A.*, F.R.G.
RAWLINGS, Don Paul*, U.S.A.
RAYNER, Francis J.*, U.K.
RAZAVI, Asdollah, IRAN
RAZBOROV, Alexandre A., U.S.S.R.
RECH, Mathias, SWITZERLAND
RECKNAGEL, Winfried, F.R.G.
RECSKI, András*, HUNGARY
REES, S. Mary, U.K.
REID, Alan W., U.K.
REID, Miles, U.K.
REITEN, Idun, NORWAY
REJALI, Ali*, IRAN
REMENYI, Maria, F.R.G.
REN, Jiagang, CHINA, P. R. of
RENAULT, Jean Nicolas, FRANCE
RENEGAR, James M., U.S.A.
REPOVŠ, Dušan*, YUGOSLAVIA
RESHETIKHIN, Nicolai, U.S.A.
REUTENAUER, Christophe*, CANADA
REYES, Bory Juan*, CUBA
RHAI, Tong-Shieng, CHINA-TAIWAN
RHEE, Choon J.*, U.S.A.
RIBES, Luis, CANADA
RICHARDSON, Mike Gerard, U.K.
RICKMAN, Seppo U., FINLAND
RIEDER, Gisèle Ruiz, U.S.A.
RIEMENSCHNEIDER, Oswald, F.R.G.
RIERA, Gonzalo, CHILE
RIGBY, John F.*, U.K.
RINGEL, Claus Michael*, F.R.G.
RIPPON, Philip Jonathan*, U.K.
RISHEL, Thomas W., U.S.A.
ROAN, Shi-Shyr, CHINA-TAIWAN
ROBBIANO, Lorenzo, ITALY
ROBBINS, Neville, U.S.A.
ROBERTS, Anne D., U.S.A.
ROBERTS, Paul C., U.S.A.
ROBINSON, Alan C., U.K.
RODINO, Luigi, ITALY
RÖDL, Vojtech, CZECHOSLOVAKIA/U.S.A.
ROERO, Clara Silvia*, ITALY
ROGGENKAMP, Klaus W., F.R.G.
ROMANOV, Vladimir G., U.S.S.R.
ROMERO M., Guillermo, MEXICO
ROONEY, Paul G., CANADA
ROOS, Guy, FRANCE
ROSATI, Mario, ITALY

ROSENBERGER, Gerhard*, F.R.G.
ROSENKNOP, John Z.*, F.R.G.
ROSENTHAL, Haskell P., U.S.A.
ROTHBERGER, Fritz§, CANADA
ROWE, David E., U.S.A.
ROWLEY, Christopher A., U.K.
ROY, Marie Francoise, FRANCE
ROZANOV, Yurii A., U.S.S.R.
RUAN, Zhong-Jin, U.S.A.
RUBINSTEIN, Joachim H.*, AUSTRALIA
RUDIN, Mary E., U.S.A.
RUDIN, Walter, U.S.A.
RUITENBURG, Wim B.*, U.S.A.
RUIZ, Mari-Jo P., PHILIPPINES
RUMELY, Robert S.*, U.S.A.
RUSHING, Thomas Benny, U.S.A.
RUTTER, John W.*, U.K.
SABURI, Yutaka, JAPAN
SAEKI, Osamu, JAPAN
SAGAN, Bruce E.*, U.S.A.
SAID, Hassan Bin*, MALAYSIA
SAIKAWA, Kazuhiko, JAPAN
SAIKI, Kazuyuki, JAPAN
SAITO, Haruhito, JAPAN
SAITO, Hiroshi, JAPAN
SAITO, Hiroshi, JAPAN
SAITO, Kazuyuki, JAPAN
SAITO, Kichisuke, JAPAN
SAITÔ, Kimiaki, JAPAN
SAITO, Kyoji, JAPAN
SAITO, Masa-Hiko, JAPAN
SAITO, Masahiko, JAPAN
SAITO, Morihiko, JAPAN
SAITO, Mutsumi, JAPAN
SAITO, Satoru, JAPAN
SAITO, Seiji, JAPAN
SAITO, Shiro, JAPAN
SAITO, Shiroshi, JAPAN
SAITO, Shuji, JAPAN
SAITO, Takeshi, JAPAN
SAITO, Tatsuhiko, JAPAN
SAITOH, Hitoshi*, JAPAN
SAITOH, Saburou*, JAPAN
SAITOH, Toshifumi, JAPAN
SAITOU, Noboru, JAPAN
SAKA, Koichi, JAPAN
SAKAGAWA, Hideko, JAPAN
SAKAGUCHI, Koji, JAPAN
SAKAGUCHI, Masanori, JAPAN
SAKAGUCHI, Michinori, JAPAN
SAKAGUCHI, Shigeru§, JAPAN
SAKAI, Akira, JAPAN
SAKAI, Fumio, JAPAN
SAKAI, Koukichi, JAPAN
SAKAI, Makoto, JAPAN
SAKAI, Masami, JAPAN
SAKAI, Sakuko, JAPAN
SAKAI, Shoichiro, JAPAN
SAKAI, Takashi, JAPAN
SAKAI, Yuji, JAPAN
SAKAKI, Makoto, JAPAN
SAKAKIBARA, Kenichi, JAPAN
SAKAKIBARA, Nobuhisa, JAPAN
SAKAMAKI, Kazuhiro, JAPAN
SAKAMOTO, Koichi§, JAPAN
SAKAMOTO, Kunio, JAPAN
SAKAMOTO, Reiko, JAPAN
SAKAMOTO, Takanori, JAPAN
SAKAMOTO, Takeshi, JAPAN
SAKAMOTO, Yoshiteru§, JAPAN
SAKANAGA, Toshiyuki, JAPAN
SAKANE, Yusuke, JAPAN
SAKANO, Kazunori, JAPAN
SAKAROVITCH, Jacques, FRANCE
SAKATA, Hiroshi, JAPAN
SAKATA, Toshio, JAPAN
SAKUMA, Makoto, JAPAN
SAKUMA, Motoyoshi, JAPAN
SAKURADA, Kuninori, JAPAN
SAKURAI, Susumu, JAPAN
SAKURAI, Takatoshi, JAPAN
SAKURAI, Yoshitane, JAPAN
SAKURAMOTO, Atsushi, JAPAN
SALAHITDINOV, Mahmud S., U.S.S.R.
SALBERGER, Per E., FRANCE
SALVATORE, Addolorata, ITALY
SALZER, Herbert E.*, U.S.A.
SAMARSKII, Alexander, U.S.S.R.
SANADA, Katsunori, JAPAN
SANCHEZ-RUIZ, Luis M.*, SPAIN
SANEKATA, Nobuhiro, JAPAN
SANKARAN, Gregory K., U.K.
SANNAMI, Atsuro, JAPAN
SANO, Kimiro, JAPAN
SANO, Shigeru*, JAPAN
SANO, Takashi, JAPAN
SANTHA, Miklos, FRANCE
SANUGI, Bahrom Bin*, MALAYSIA
SAPER, Leslie David, U.S.A.
SARNAK, Peter, U.S.A.
SASABUCHI, Syoichi, JAPAN
SASAJIMA, Yoshiaki, JAPAN
SASAKI, Chikara, JAPAN
SASAKI, Hiroki, JAPAN
SASAKI, Takehiko, JAPAN
SASAKI, Takeshi, JAPAN
SASAKI, Tetsuo, JAPAN
SASAKI, Toru, JAPAN
SASAKURA, Nobuo, JAPAN
SASANO, Kazuhiro, JAPAN
SASAO, Akira, JAPAN
SASAYAMA, Hiroyoshi*, JAPAN
SATAKE, Ichiro, JAPAN
SATAKE, Ikuo, JAPAN
SATO, Atsushi, JAPAN
SATO, Atsushi, JAPAN

SATO, Daihachiro*, JAPAN
SATO, Eiichi, JAPAN
SATO, Enji, JAPAN
SATO, Fumihiro, JAPAN
SATO, Hajime, JAPAN
SATO, Hiroki, JAPAN
SATO, Hiroshi, JAPAN
SATO, Hisashi, JAPAN
SATO, Humitaka, JAPAN
SATO, Iwao, JAPAN
SATO, Kei, JAPAN
SATO, Ken-iti*, JAPAN
SATO, Kenkichi§, JAPAN
SATO, Kunio, JAPAN
SATO, Masahisa, JAPAN
SATO, Masako, JAPAN
SATO, Mikio, JAPAN
SATO, Nobuaki, JAPAN
SATO, Shin, JAPAN
SATO, Shizuko, JAPAN
SATO, Shuichi, JAPAN
SATO, Shun, JAPAN
SATO, Takeshi, JAPAN
SATO, Takeyoshi, JAPAN
SATO, Takuji, JAPAN
SATO, Yoshihisa, JAPAN
SATO, Yoshitaka, JAPAN
SATO, Yumiko, JAPAN
SATOH, Takakazu, JAPAN
SATOH, Yukiti, JAPAN
SAVIN, Gordan, U.S.A.
SAWA, Tatuo, JAPAN
SAWADA, Ken, JAPAN
SAWADA, Toshio, JAPAN
SAWAE, Ryuichi, JAPAN
SAWAHATA, Michimasa, JAPAN
SAWAI, Hitoshi, JAPAN
SAWAKI, Sumio, JAPAN
SAWASHIMA, Ikuko, JAPAN
SAXL, Jan, U.K.
SAYEKI, Hidemitsu§, CANADA
SCEPIN, Evgenii Vitalievich*, U.S.S.R.
SCHAAL, Werner G.H., F.R.G.
SCHAFLITZEL, Reinhard*, F.R.G.
SHAPIRA, Pierre B., FRANCE
SCHEIN, Boris M., U.S.A.
SCIFFMANN, Gerard M., FRANCE
SCHIRESON, Max L.*, U.S.A.
SCHIRMER, Helga H.§, CANADA
SCHMEELK, John*, U.S.A.
SCHMICKLER-HIRZEBRUCH, Ulrike, U.S.A.
SCHMID, Wilfried, U.S.A.
SCHNEIDER, Hans, U.S.A.
SCHNIZER, Ansgar Wolfgang, AUSTRIA
SCHOENFELD, Lowell, U.S.A.
SCHREIBER, Bert M., U.S.A.
SCHWARTZ, Jean-Marie M., FRANCE
SCHWARZ, Albert, U.S.S.R.
SCOPPOLA, Carlo M., ITALY
SCULLY, Catherine N.*, U.K.
SEDDIGHI, Karim§, IRAN
SEGAL, Graeme Bryce, U.K.
SEGAL, Jack, U.S.A.
SEGAWA, Shigeo, JAPAN
SEGERT, Jan*, U.S.A.
SEI, Fumihiro§, JAPAN
SEIBERT, Peter*, MEXICO
SEITOH, Akira, JAPAN
SEKI, Go, JAPAN
SEKIGAWA, Hisao, JAPAN
SEKIGAWA, Kouei, JAPAN
SEKIGUCHI, Hideko, JAPAN
SEKIGUCHI, Katsusuke, JAPAN
SEKIGUCHI, Takeshi, JAPAN
SEKIGUCHI, Tsutomu, JAPAN
SEKINE, Masayuki, JAPAN
SEKINE, Mituhiro, JAPAN
SEKINE, Tadayuki, JAPAN
SEKINO, Kaoru, JAPAN
SEKIZAWA, Masami, JAPAN
SEKO, Yoshiki, JAPAN
SELFRIDGE, John L., U.S.A.
SELIGMAN, George B., U.S.A.
SEMENOV, Evgeny*, U.S.S.R.
SENBA, Takasi, JAPAN
SENDA, Akio, JAPAN
SENDOV, Blagovest Hristov, BULGARIA
SENN, Walter M., SWITZERLAND
SEO, Yuki, JAPAN
SERIZAWA, Hisamitsu, JAPAN
SESAY, Mohamed W.I.§, U.S.A.
SESHADRI, Conjeeveram S., INDIA
SETO, Hisayoshi, JAPAN
SEYDI, Hamet, SENEGAL
SHARFUDDIN, Shaik M.*, BANGLADESH
SHARIF, Habib, IRAN
SHARMA, Banwari Lal, INDIA
SHARP, Rodney Y., U.K.
SHASTRI, Anant R.*, INDIA
SHATZ, Stephen S.§, U.S.A.
SHAW, Sen-Yen*, CHINA-TAIWAN
SHEN, Junpei, CHINA, P. R. of
SHEU, Albert Jeu-Liang*, U.S.A.
SHEU, Shey Shiung*, CHINA-TAIWAN
SHI, Ning-Zhong*, CHINA, P. R. of
SHI, Zhong-Ci*, CHINA, P. R. of
SHIBA, Masakazu*, JAPAN
SHIBAGAKI, Wasao, JAPAN
SHIBANO, Hiroki, JAPAN
SHIBAOKA, Yasumitsu§, JAPAN
SHIBATA, Katsuyuki*, JAPAN
SHIBATA, Keiichi, JAPAN
SHIBATA, Tetsutaro, JAPAN
SHIBATA, Toshio, JAPAN
SHIBAYAMA, Kenshin, JAPAN
SHIBUKAWA, Youichi, JAPAN

SHIBUTANI, Norimasa, JAPAN
SHIEH, Narn R., CHINA-TAIWAN
SHIGA, Hironori, JAPAN
SHIGA, Hiroo, JAPAN
SHIGA, Hiroshige, JAPAN
SHIGA, Kiyoshi, JAPAN
SHIGA, Koji, JAPAN
SHIGA, Tokuzo, JAPAN
SHIGEKAWA, Ichiro, JAPAN
SHIGEMATSU, Keiichi, JAPAN
SHIM, Jae-Ung, KOREA, R. of
SHIMA, Hirohiko, JAPAN
SHIMA, Tadashi, JAPAN
SHIMADA, Hideo, JAPAN
SHIMADA, Ichiro, JAPAN
SHIMADA, Nobuo, JAPAN
SHIMADA, Tsutomu, JAPAN
SHIMAKAWA, Kazuhisa, JAPAN
SHIMAKURA, Norio, JAPAN
SHIMATANI, Kenichiro, JAPAN
SHIMENO, Nobukazu, JAPAN
SHIMIZU, Akinobu, JAPAN
SHIMIZU, Atutoshi, JAPAN
SHIMIZU, Hideo, JAPAN
SHIMIZU, Kenichi, JAPAN
SHIMIZU, Ryoichi*, JAPAN
SHIMIZU, Satoru, JAPAN
SHIMIZU, Yoshiyuki, JAPAN
SHIMIZU, Yuji, JAPAN
SHIMODA, Norio, JAPAN
SHIMODA, Yasuhiro, JAPAN
SHIMOMURA, Hiroaki, JAPAN
SHIMOMURA, Katsunori, JAPAN
SHIMOMURA, Shun, JAPAN
SHIMOMURA, Takashi, JAPAN
SHIMURA, Hiroyuki, JAPAN
SHIMURA, Michio, JAPAN
SHIMURA, Takaaki, JAPAN
SHIN, Kilho, JAPAN
SHINAGAWA, Mitsuo, JAPAN
SHINKAI, Kenzo*, JAPAN
SHINODA, Juichi, JAPAN
SHINODA, Ken-ichi, JAPAN
SHINOHARA, Masahiko, JAPAN
SHINOHARA, Yaichi, JAPAN
SHINOHARA, Yoshitane*, JAPAN
SHINOMIYA, Yoichi§, JAPAN
SHINOZUKA, Shigeo§, JAPAN
SHIODA, Tetsuji, JAPAN
SHIOHAMA, Katsuhiro, JAPAN
SHIOJI, Naoki, JAPAN
SHIOMI, Tatsuyuki, JAPAN
SHIOTA, Jun, JAPAN
SHIOTA, Ken-ichi, JAPAN
SHIOTA, Masahiro, JAPAN
SHIOTA, Takahiro, U.S.A.
SHIOTA, Yasunobu, JAPAN
SHIOYA, Masahiro, JAPAN
SHIOZAKI, Yasutoshi, JAPAN
SHIRAIWA, Kenichi, JAPAN
SHIRAKI, Mitsunobu, JAPAN
SHIRASOU, Takeo, JAPAN
SHIRATANI, Katsumi, JAPAN
SHIROSAKI, Manabu, JAPAN
SHISHIKURA, Mitsuhiro, JAPAN
SHIUE, Wei Kei*, U.S.A.
SHIZUKAWA, Akira, JAPAN
SHIZUTA, Yasushi, JAPAN
SHOHOJI, Takao§, JAPAN
SHOJI, Mayumi, JAPAN
SHOJI, Toshiaki, JAPAN
SHOKUROV, Vyacheslav V., U.S.S.R.
SHPEKTOROV, Sergey Victorovich, U.S.S.R.
SHPILRAIN, Vladimir E.*, U.S.S.R.
SHUKLA, Pradeep K.*, U.S.A.
SHUM, Kar-Ping*, HONG KONG
SHUSTIN, Eugenii I., U.S.S.R.
SI, Si, MYANMAR
SIBONY, Nessim, FRANCE
SIBUYA, Masaaki, JAPAN
SIDARTO, Kuntjoro Adji, INDONESIA
SIDDIQI, Jamil A.*, CANADA
SIDDIQI, Rafat Nabi*, KUWAIT
SIELKIN, Michail V.*, U.S.S.R.
SIEMONS, Johannes, U.K.
SIERSMA, Dirk, NETHERLANDS
SIGAL, Islael Michael, CANADA/U.S.A.
SIKORAV, Jean-Claude, FRANCE
SIM, Chiaw Hock, MALAYSIA
SIMIS, Aron, BRAZIL
SIMIZU, Tatuo*, JAPAN
SIMMS, John C.*, U.S.A.
SIMON, Imre, BRAZIL
SIMON, Jacques C. H., FRANCE
SIMONYI, Gábor*, HUNGARY
SIMPSON, Carlos Tschuodi, U.S.A.
SINAI, Yakov G., U.S.S.R.
SINGH, Neeta*, INDIA
SIRAO, Tunekiti, JAPAN
SIU, Man-Keung, HONG KONG
SJÖGREN, Peter, SWEDEN
SJÖLIN, Per, SWEDEN
SKANDALIS, Georges, FRANCE
SKAU, Christian F., NORWAY
SKORA, Richard K.*, U.S.A.
SKUBACHEVSKII, Alexander L.*, U.S.S.R.
SLAMAN, Theodore A., U.S.A.
SLEMROD, Marshall*, U.S.A.
SLODOWY, Peter J., F.R.G.
SLOVIC, Harold Geoffrey, U.S.A.
SMALØ, Sverre Olaf, NORWAY
SMITH, Martha K., U.S.A.
SMITH, Perry B., U.S.A.
SMITH, Roy Campbell, U.S.A.
SMITH, Stuart H., CANADA
SMITH, Stuart P., U.S.A.

SMITH, Wayne S., U.S.A.
SMOGOR, Louis E., U.S.A.
SOGO, Hideyo, JAPAN
SOLARIN, Adewale R. T.*, NIGERIA
SOLBERG, Øyvind, NORWAY
SOLOMON, David R.*, U.K.
SOLOMON, Ronald M., U.S.A.
SOMA, Sumihiko, JAPAN
SOMEKAWA, Mutsuro, JAPAN
SOMER, Lawrence Eric*, U.S.A.
SOMMER, Manfred, F.R.G.
SOMOLINOS, Alfredo*, SPAIN
SORIA, Fernando, SPAIN
SOTO-ANDRADE, Jorge*, CHILE
SOUČEK, Jiří, CZECHOSLOVAKIA
SOUKUP, Lajos, HUNGARY
SOULÉ, Christophe J., FRANCE
SPAGNOLO, Sergio A., ITALY
SPANIER, Fred, U.S.A.
SPINADEL, Vera W. de*, ARGENTINA
SPINELLI, Giancarlo, ITALY
SRINIVAS, Vasudevan, INDIA
STAFFANS, Olof J., FINLAND
STALLINGS, John R.*, U.S.A.
STAN, Ioan I.*, ROMANIA
STANILOV, Grosio*, BULGARIA
STANLEY, Lee J.*, U.S.A.
STANLEY, Richard P., U.S.A.
STANTON, Robert J., U.S.A.
STEEB, Willi Hans*, SOUTH AFRICA
STEEL, John R., U.S.A.
STEENBRINK, Joseph H. M., NETHERLANDS
STEFĂNESCU, Mirela I.*, ROMANIA
STEFANOV, Plamen D., BULGARIA
STEFÁNSSON, Jón R., ICELAND
STEGEMAN, Jan D., NETHERLANDS
STEGENGA, David A., U.S.A.
STEGER, Tim J.§, U.S.A.
STEINER, Richard J., U.K.
STEPHENSON, Ken, U.S.A.
STERN, Ronald J.§, U.S.A.
STERNHEIMER, Daniel H., FRANCE
STETKAER, Henrik, DENMARK
STETTER, Hans J., AUSTRIA
STEVENS, Glenn H., U.S.A.
STEWART, Ian Nicholas, U.K.
STIEGLER, Karl Drago§, F.R.G.
STOB, Michael J., U.S.A.
STORCH, Uwe, F.R.G.
STORMER, Erling, NORWAY
STORVICK, David Arne*, U.S.A.
STRANO, Rosario, ITALY
STRAY, Arne*, NORWAY
STRÖMBERG, Jan-Olov, SWEDEN
STROOKER, Jan R., NETHERLANDS
STROTH, Gernot*, F.R.G.
STRUWE, Michael, SWITZERLAND
STURM, Karl Theodor*, SWITZERLAND
STURMFELS, Bernd, U.S.A.
SUCCI, Francesco, ITALY
SUCCI CRUCIANI, Rosanna, ITALY
SUDO, Masahiro, JAPAN
SUDO, Masaki, JAPAN
SUEHIRO, Naoki*, JAPAN
SUEYOSHI, Yutaka, JAPAN
SUGA, Shuichi, JAPAN
SUGAHARA, Kunio, JAPAN
SUGANO, Takashi, JAPAN
SUGANO, Tosiaki, JAPAN
SUGAWA, Toshiyuki, JAPAN
SUGAWARA, Masahiro, JAPAN
SUGAWARA, Takeshi, JAPAN
SUGAWARA, Tamio, JAPAN
SUGIE, Jitsuro, JAPAN
SUGIE, Toru, JAPAN
SUGIMOTO, Mitsuru, JAPAN
SUGIMOTO, Shin, JAPAN
SUGINO, Ken, JAPAN
SUGITA, Hiroshi, JAPAN
SUGITA, Kimio, JAPAN
SUGITANI, Sadao, JAPAN
SUGIURA, Mitsuo, JAPAN
SUGIYAMA, Ken-Ichi, JAPAN
SUITA, Nobuyuki, JAPAN
SUKETA, Masaki, JAPAN
SUKLA, Indulata*, INDIA
SULLIVAN, John M., U.S.A.
SULTANGAZIN, Umirzak M., U.S.S.R.
SUMA, Yousuke, JAPAN
SUMI, Makiko, JAPAN
SUMI, Toshio*, JAPAN
SUMIHIRO, Hideyasu, JAPAN
SUMIOKA, Takeshi, JAPAN
SUMITOMO, Takeshi*, JAPAN
SUMMERS, Danny*, CANADA
SUMNERS, De Witt L., U.S.A.
SUN, Wen Xiang*, CHINA, P. R. of
SUN, Ziqi, U.S.A.
SUNADA, Toshikazu, JAPAN
SUNAGA, Junichi, JAPAN
SUNDARAM, Sheila, U.S.A.
SUNDBERG, Gösta, SWEDEN
SUNDER, Viakalathur Shankar, INDIA
SUNLEY, Judith S., U.S.A.
SUNOUCHI, Choichiro, JAPAN
SUNOUCHI, Haruo, JAPAN
SURI, Subhash, U.S.A.
SUTHERLAND, Colin Eric, AUSTRALIA
SUTO, Kiyokazu, JAPAN
SUTTI, Carla Nodari§, ITALY
SUWA, Noriyuki§, JAPAN
SUWA, Tatsuo, JAPAN
SUYAMA, Yoshihiko, JAPAN
SUZUKI, Atsushi, JAPAN
SUZUKI, Chisato, JAPAN
SUZUKI, Haruo, JAPAN

SUZUKI, Hideaki, JAPAN
SUZUKI, Humio, JAPAN
SUZUKI, Kazumasa*, JAPAN
SUZUKI, Kazuo, JAPAN
SUZUKI, Kazuya, JAPAN
SUZUKI, Komei*, JAPAN
SUZUKI, Kunie, JAPAN
SUZUKI, Kyoko, JAPAN
SUZUKI, Makoto, JAPAN
SUZUKI, Masaaki, JAPAN
SUZUKI, Masakazu, JAPAN
SUZUKI, Masashi, JAPAN
SUZUKI, Michio, U.S.A.
SUZUKI, Naoyoshi, JAPAN
SUZUKU, Norio, JAPAN
SUZUKI, Osamu, JAPAN
SUZUKI, Satoshi, JAPAN
SUZUKI, Shinichi, JAPAN
SUZUKI, Takashi, JAPAN
SUZUKI, Takeru, JAPAN
SUZUKI, Toshio, JAPAN
SUZUKI, Toshio, JAPAN
SUZUKI, Toshiyuki, JAPAN
SUZUKI, Yoshiya, JAPAN
SUZUKI, Yuji§, JAPAN
SUZUKI, Yuki, JAPAN
SY, Polly Wee, PHILIPPINES
SYMONDS, Peter, CANADA
SZABO, Zoltan I.§, BOTSWANA
SZŐNYI, Tamás*, HUNGARY
TABATA, Masahisa*, JAPAN
TABATA, Minoru, JAPAN
TACHIKAWA, Atsushi, JAPAN
TACHIKAWA, Hiroyuki, JAPAN
TACHIZAWA, Kazuya, JAPAN
TADA, Minoru, F.R.G.
TADA, Toshimasa, JAPAN
TAFLIN, Erik, FRANCE
TAGUCHI, Yoshiko, JAPAN
TAGUCHI, Yuichiro, JAPAN
TAHARA, Ken-ichi, JAPAN
TAHATA, Rataka, JAPAN
TAHERIZADEH, Abdol-Javad*, IRAN
TAINAKA, Akihiro, JAPAN
TAJIMA, Shinichi, JAPAN
TAKADA, Ichiro, JAPAN
TAKADA, Yoshikazu, JAPAN
TAKAGI, Hiroyuki, JAPAN
TAKAGI, Izumi, JAPAN
TAKAHASHI, Hideo, JAPAN
TAKAHASHI, Hideyuki, JAPAN
TAKAHASHI, Hiroaki, JAPAN
TAKAHASHI, Katsuaki, JAPAN
TAKAHASHI, Katsuo, JAPAN
TAKAHASHI, Katsutoshi, JAPAN
TAKAHASHI, Ken-ichi, JAPAN
TAKAHASHI, Makoto, JAPAN
TAKAHASHI, Masahiro, JAPAN
TAKAHASHI, Moto-o, JAPAN
TAKAHASHI, Reiji, JAPAN
TAKAHASHI, Satoshi, JAPAN
TAKAHASHI, Sechiko, JAPAN
TAKAHASHI, Shuichi, JAPAN
TAKAHASHI, Tadashi, JAPAN
TAKAHASHI, Takeshi, JAPAN
TAKAHASHI, Tetsuya, JAPAN
TAKAHASHI, Toyofumi, JAPAN
TAKAHASHI, Tsunero, JAPAN
TAKAHASHI, Yasuji, JAPAN
TAKAHASHI, Yoichiro, JAPAN
TAKAHASI, Mitiko, JAPAN
TAKAHASI, Sin-Ei, JAPAN
TAKAI, Hiroshi, JAPAN
TAKAKU, Akira, JAPAN
TAKAKURA, Tatsuru, JAPAN
TAKAKUWA, Shoichiro, JAPAN
TAKAMUKI, Takashi, JAPAN
TAKAMURA, Shigeru, JAPAN
TAKANE, Martha Imay, MEXICO
TAKANO, Katsuo, JAPAN
TAKANO, Kazuhiko, JAPAN
TAKANO, Kyoichi, JAPAN
TAKANO, Mitio§, JAPAN
TAKANO, Yoshiko*, JAPAN
TAKANOBU, Satoshi, JAPAN
TAKAO, Hiroshi, JAPAN
TAKAOKA, Masaki§, JAPAN
TAKARAJIMA, Itaru, JAPAN
TAKASAKI, Kanehisa, JAPAN
TAKASE, Koichi, JAPAN
TAKASE, Mitsuo, JAPAN
TAKASHIMA, Keizo, JAPAN
TAKASU, Satoru, JAPAN
TAKATA, Toshie, JAPAN
TAKAYAMA, Nobuki, JAPAN
TAKAYANAGI, Hideshi, JAPAN
TAKEBAYASHI, Tadayoshi, JAPAN
TAKEBE, Takashi, JAPAN
TAKEDA, Masayoshi, JAPAN
TAKEDA, Yoshifumi, JAPAN
TAKEDA, Yuichi, JAPAN
TAKEDA, Yuichiro, JAPAN
TAKEDA, Ziro§, JAPAN
TAKEGOSHI, Kensho, JAPAN
TAKEHANA, Hiroaki, JAPAN
TAKEI, Yoshitsugu, JAPAN
TAKEMOTO, Fumio, JAPAN
TAKENAKA, Shigeo, JAPAN
TAKENOUCHI, Osamu*, JAPAN
TAKEO, Fukiko, JAPAN
TAKESAKI, Masamichi, U.S.A.
TAKESHITA, Akira, JAPAN
TAKEUCHI, Akira, JAPAN
TAKEUCHI, Atsuko, JAPAN
TAKEUCHI, Hiroshi, JAPAN
TAKEUCHI, Jiro, JAPAN

TAKEUCHI, Junji, JAPAN
TAKEUCHI, Kei, JAPAN
TAKEUCHI, Kensuke, JAPAN
TAKEUCHI, Kentaro, JAPAN
TAKEUCHI, Kisao, JAPAN
TAKEUCHI, Kiyohiko, JAPAN
TAKEUCHI, Mamoru, JAPAN
TAKEUCHI, Masahito, JAPAN
TAKEUCHI, Masaru, JAPAN
TAKEUCHI, Nobuko, JAPAN
TAKEUCHI, Shigeru, JAPAN
TAKEUCHI, Teruo, JAPAN
TAKEUCHI, Yasuhiro§, JAPAN
TAKEUCHI, Yasuji, JAPAN
TAKEUCHI, Yoshihiro, JAPAN
TAKEUTI, Gaisi, U.S.A.
TAKEUTI, Yosie, JAPAN
TAKEWAKA, Yoshie, JAPAN
TAKIGAWA, Shinya, JAPAN
TAKIZAWA, Osamu*, JAPAN
TALAGRAND, Michel, FRANCE
TALL, Franklin D.*, CANADA
TAMAKI, Dai, JAPAN
TAMAKI, Kazuhiro, JAPAN
TAMAMURA, Akie, JAPAN
TAMANO, Ken-ichi, JAPAN
TAMANOI, Hirotaka, JAPAN
TAMARI, Fumikazu, JAPAN
TAMBARA, Daisuke, JAPAN
TAMBOUR, Torbjörn, SWEDEN
TAMBURINI, Chiara M., ITALY
TAMURA, Hiroshi, JAPAN
TAMURA, Itiro§, JAPAN
TAMURA, Saburo, JAPAN
TAMURA, Yozo, JAPAN
TAN, Henry K., U.S.A.
TAN, Ki-Seng*, U.S.A.
TANABE, Hiroki, JAPAN
TANABE, Kunio, JAPAN
TANABE, Susumu, JAPAN
TANABÉ, Susumu, JAPAN
TANAHASHI, Katsumi, JAPAN
TANAHASHI, Kotaro, JAPAN
TANAKA, Giichi, JAPAN
TANAKA, Hiroshi, JAPAN
TANAKA, Hiroshi, JAPAN
TANAKA, Junzo, JAPAN
TANAKA, Katsumi, JAPAN
TANAKA, Kazunaga, JAPAN
TANAKA, Kazuyuki, JAPAN
TANAKA, Mahito, JAPAN
TANAKA, Minoru, JAPAN
TANAKA, Naoki, JAPAN
TANAKA, Ryuichi, JAPAN
TANAKA, Shohei, JAPAN
TANAKA, Tadashi, JAPAN
TANAKA, Tamaki*, JAPAN
TANAKA, Tsutomu, JAPAN
TANAKA, Yohei, JAPAN
TANAKA, Yukiyasu, JAPAN
TANDAI, Kwoichi*, JAPAN
TANEMURA, Hideki, JAPAN
TANG, Francis C. Y., CANADA
TANGO, Hiroshi, JAPAN
TANI, Atusi§, JAPAN
TANI, Seiichi, JAPAN
TANI, Tsugio, JAPAN
TANIGAWA, Harumi, JAPAN
TANIGAWA, Yoshio, JAPAN
TANIGUCHI, Hajime, JAPAN
TANIGUCHI, Kazuo*, JAPAN
TANIGUCHI, Masaru, JAPAN
TANIGUCHI, Setsuo, JAPAN
TANIHIRA, Yumiko, JAPAN
TANIKAWA, Akio, JAPAN
TANIKAWA, Masao, JAPAN
TANISAKI, Toshiyuki, JAPAN
TANIYAMA, Kouki, JAPAN
TANNO, Shukichi, JAPAN
TANUMA, Kazumi, JAPAN
TARAMA, Shigeo, JAPAN
TARDOS, Eva, HUNGARY/U.S.A.
TARKALANOV, Krassimir D.§, BULGARIA
TARTAR, Luc C., U.S.A.
TARUI, Jun§, U.S.A.
TARUMI, Tomoyuki, JAPAN
TASAKI, Hiroyuki, JAPAN
TASHIRO, Yoshihiro, JAPAN
TATSUUMA, Nobuhiko, JAPAN
TATUZAWA, Tikao§, JAPAN
TAWARA, Seisui, JAPAN
TAYA, Hisao, JAPAN
TAYLOR, John, U.K.
TAYLOR, John Christopher, CANADA
TAYLOR, Laurence R., U.S.A.
TAYLOR, Michael E., U.S.A.
TAYOSHI, Takao, JAPAN
TAZAWA, Yoshihiko, JAPAN
TELLO, Angela, PANAMA
TENENBLAT, Keti, BRAZIL
TENGSTRAND, Anders Sven, SWEDEN
TERADA, Itaru, JAPAN
TERADA, Junko, JAPAN
TERADA, Toshiaki, JAPAN
TERADA, Toshiji, JAPAN
TERAGAITO, Masakazu, JAPAN
TERAI, Naoki, JAPAN
TERAI, Nobuhiro, JAPAN
TERAO, Taro, JAPAN
TERAOKA, Yoshinobu, JAPAN
TERASAWA, Jun, JAPAN
TERUYA, Tamotsu, JAPAN
THAKUR, Dinesh S., INDIA
THOMASON, Robert W., U.S.A.
THOMASSEN, Carsten, DENMARK
THOMEE, Vidar C., SWEDEN

THOMPSON, Robert J., U.S.A.
THOMSEN, Momme Johs*, F.R.G.
THORNLEY, Gillian M., NEW ZEALAND
TIAN, Gang, CHINA, P. R. of
TIMOURIAN, James G., U.S.A.
TITANI, Satoko, JAPAN
TITI, Edriss S.*, U.S.A.
TOBIN, Sean J., IRELAND
TODA, Hirosi, JAPAN
TODA, Nobushige, JAPAN
TODOROV, Andrey N., BULGARIA
TODOROV, Gordana Glisa, U.S.A.
TOGARI, Yoshio, JAPAN
TOGNOLI, Alberto, ITALY
TOKI, Keisuke, JAPAN
TOKITA, Akira, JAPAN
TOKITA, Takesi, JAPAN
TOKIZAWA, Masamichi, JAPAN
TOKUNAGA, Hideya, JAPAN
TOKUNAGA, Hiroo, JAPAN
TOMABECHI, Satosi, JAPAN
TOMARI, Masataka, JAPAN
TOMARU, Tadashi, JAPAN
TOMEI, Carlos, BRAZIL
TOMIDA, Masamichi, JAPAN
TOMINAGA, Hiroyuki, JAPAN
TOMINAGA, Hisao, JAPAN
TOMINAGA, Tadashi, JAPAN
TOMISAKI, Matsuyo, JAPAN
TOMITA, Koichi, JAPAN
TOMITA, Minoru, JAPAN
TOMITA, Yoshihito, JAPAN
TOMIYAMA, Jun, JAPAN
TOMIZAWA, Akira, JAPAN
TONEGAWA, Takayuki, JAPAN
TON-THAT, Tuong*, U.S.A.
TOPIWALA, Pankaj N., U.S.A.
TORIUMI, Satoshi, JAPAN
TORRES, Rodolfo H., U.S.A./ARGENTINA
TOSE, Nobuyuki*, JAPAN
TOTOKI, Haruo, JAPAN
TOURÉ, Saliou, IVORY COAST
TOUTOUNIAN, Faezeh B.-V.*, IRAN
TOWA, Yuji, JAPAN
TOYODA, Masanori, JAPAN
TOYONARI, Toshitaka, JAPAN
TRAPANI, Camillo, ITALY
TRAUTMANN, Günther*, F.R.G.
TRICERRI, Franco, ITALY
TROTMAN, David, FRANCE
TRUE, Hans*, DENMARK
TSANG, Kai-Man, HONG KONG
TSUBOI, Akito, JAPAN
TSUBOI, Shoji, JAPAN
TSUBOI, Takashi, JAPAN
TSUCHIDA, Kensei, JAPAN
TSUCHIHASHI, Hiroyasu, JAPAN
TSUCHIKAWA, Masao, JAPAN
TSUCHIKURA, Tamotsu, JAPAN
TSUCHIMOTO, Yoshifumi, JAPAN
TSUCHIYA, Akihiro, JAPAN
TSUCHIYA, Masaaki, JAPAN
TSUCHIYA, Morimasa§, JAPAN
TSUCHIYA, Nobuo, JAPAN
TSUCHIYA, Takuya, JAPAN
TSUDA, Ei*, JAPAN
TSUDA, Koichi, JAPAN
TSUDA, Teruko, JAPAN
TSUGAWA, Osamu§, JAPAN
TSUGE, Masayoshi, JAPAN
TSUJI, Hajime, JAPAN
TSUJI, Mikio, JAPAN
TSUJI, Naomitsu, JAPAN
TSUJI, Tadashi, JAPAN
TSUJII, Masato*, JAPAN
TSUJII, Yoshiki, JAPAN
TSUJIOKA, Kunio, JAPAN
TSUJISHITA, Toru, JAPAN
TSUKADA, Kazumi, JAPAN
TSUKADA, Makoto, JAPAN
TSUKAHARA, Kumiko, JAPAN
TSUKAHARA, Shigeo, JAPAN
TSUKAMOTO, Chiaki, JAPAN
TSUKIMOTO, Hiroshi, JAPAN
TSUKUI, Yasuyuki, JAPAN
TSUMURA, Hirofumi, JAPAN
TSUNODA, Shuichiro, JAPAN
TSURUMI, Kazuyuki, JAPAN
TSUSHIMA, Ryuji, JAPAN
TSUSHIMA, Yukio, JAPAN
TSUTSUI, Toru, JAPAN
TSUTSUMI, Akira, JAPAN
TSUTSUMI, Yoshio, JAPAN
TSUYUMINE, Shigeaki, JAPAN
TSUZUKI, Nobuo, JAPAN
TSUZUKU, Tosiro, JAPAN
TUGUÉ, Tosiyuki, JAPAN
TUKADA, Kunio, JAPAN
TUNG, Shih Ping*, CHINA-TAIWAN
TURAEV, Vladimir G., U.S.S.R.
TURQUETTE, Atwell R.§, U.S.A.
TYLLI, Hans-Olav J., FINLAND
TYMCHATYN, Edward Dmytro, CANADA
TYURIN, Andrei N., U.S.S.R.
TZANNES, Vassilis*, GREECE
UCHIBORI, Tomio, JAPAN
UCHIDA, Daiyu, JAPAN
UCHIDA, Koji, JAPAN
UCHIDA, Motoo*, JAPAN
UCHIDA, Yoshiaki, JAPAN
UCHIKOSHI, Keisuke, JAPAN
UCHIMURA, Keisuke, JAPAN
UCHIUMI, Masauki, JAPAN
UCHIYAMA, Jun, JAPAN
UCHIYAMA, Kohei, JAPAN
UCHIYAMA, Koichi, JAPAN

UCHIYAMA, Mitsuru, JAPAN
UDA, Toshio, JAPAN
UDRIŞTE, Constantin N.*, ROMANIA
UE, Masaaki, JAPAN
UEDA, Hideharu, JAPAN
UEDA, Masaru, JAPAN
UEDA, Tetsuo, JAPAN
UEDA, Yoshika, JAPAN
UEHARA, Tsuyoshi, JAPAN
UEKI, Naomasa, JAPAN
UEMURA, Hideaki, JAPAN
UEMURA, Yoshiaki, JAPAN
UENO, Kazuo, JAPAN
UENO, Kazuyoshi, JAPAN
UENO, Kenji, JAPAN
UENO, Kimio, JAPAN
UENO, Machi, JAPAN
UENO, Tadashi, JAPAN
UENO, Yoshiaki, JAPAN
UESAKA, Hiroshi, JAPAN
UESU, Kagumi, JAPAN
UESU, Tadahiro, JAPAN
UETAKE, Tsuneo, JAPAN
UHLENBECK, Karen K., U.S.A.
UJIIE, Katsumi, JAPAN
UKAI, Seiji, JAPAN
UKONN, Michihisa, JAPAN
ULECIA, Teresa*, SPAIN
ULMER, Douglas L., U.S.A.
ULRICH, Bernd, U.S.A.
ULUÇAY, Cengiz§, TURKEY
UMAYA, Noriaki, JAPAN
UMEDA, Tomio, JAPAN
UMEDA, Toru, JAPAN
UMEDA, Yoshio, JAPAN
UMEGAKI, Hisaharu, JAPAN
UMEHARA, Masaaki, JAPAN
UMEMURA, Hiroshi§, JAPAN
UMENO, Takashi, JAPAN
UMEZAWA, Toshio, JAPAN
UMEZU, Yumiko, JAPAN
UNGER, Luise Susanne*, F.R.G.
UNO, Katsuhiro, JAPAN
UNSURANGSIE, Sumalee, THAILAND
URABE, Hironobu, JAPAN
URABE, Tohsuke, JAPAN
URAKAWA, Hajime, JAPAN
URATA, Toshio, JAPAN
USA, Takeshi, JAPAN
USAMI, Hiroyuki, JAPAN
USAMI, Yoko, JAPAN
USHIJIMA, Takeo, JAPAN
USHIJIMA, Teruo*, JAPAN
USHIKI, Shigehiro, JAPAN
USHITAKI, Fumihiro, JAPAN
USUI, Sampei, JAPAN
UTAMURA, Motoaki, JAPAN
UTSUDE, Hisaki, JAPAN
UWANO, Yoshio, JAPAN
UZAWA, Masakatsu, JAPAN
VÄISÄLÄ, Jussi*, FINLAND
VALIBOUZE, Annick Angèle, FRANCE
VALLA, Giuseppe, ITALY
VALLEJO, Ernesto, MEXICO
VALLI, Giorgio, ITALY
VAN DAELE, Alfons, BELGIUM
VAN DEN DRIES, Lou, U.S.A.
VAN DER GEER, Gerard§, NETHERLANDS
VAN DER KALLEN, Wilberd L., NETHERLANDS
VAN DER PUT, Marius, NETHERLANDS
VAN DER WALT, A. P. J., SOUTH AFRICA
VAN ENTER, Aernout C.D.*, NETHERLANDS
VAN GEEL, Jan M. H.*, BELGIUM
VANHECKE, Lieven N.A., BELGIUM
VAN LINT, Jacobus H., NETHERLANDS
VAN WYK, Leon*, SOUTH AFRICA
VARCHENKO, Alexandre, U.S.S.R.
VAROPOULOS, Nicolas Thodore, FRANCE
VASIU, Adrian, ROMANIA
VAUGHAN, Jerry E.*, U.S.A.
VAVILOV, Nikolai A., U.S.S.R.
VEGA, Luis, U.S.A.
VELDSMAN, Stefan*, SOUTH AFRICA
VELOSO, José M. M.§, BRAZIL
VELU, Jacques, FRANCE
VENEMA, Gerard A.*, U.S.A.
VERNIER-PIRO, Stella*, ITALY
VIANA, Marcelo, BRAZIL
VIEHWEG, Eckart E., F.R.G.
VINCENT, Georges A., SWITZERLAND
VIRO, Oleg Ya., U.S.S.R.
VLADIMIROV, Vasilii S., U.S.S.R.
VOEVODIN, Valentin V., U.S.S.R.
VOGEL, Wolfgang*, G.D.R.
VOJTA, Paul A., U.S.A.
VO KHAC, Khoan*, FRANCE
VOLBERG, Alexander L., U.S.S.R.
VOLOCH, José Felipe, BRAZIL
VOLOVICH, Igor V., U.S.S.R.
VOROS, André, FRANCE
VU, Dinh Hoa, G.D.R.
VU, Kim Tuan*, VIETNAM
WADA, Hidekazu, JAPAN
WADA, Hideo, JAPAN
WADA, Junzo, JAPAN
WADA, Kouichi, JAPAN
WADA, Masaaki, U.S.A.
WADA, Ryoko, JAPAN
WADA, Takakazu, JAPAN
WADA, Tomoyuki, JAPAN
WAJNRYB, Bronislaw, ISRAEL
WAKABAYASHI, Isao, JAPAN
WAKABAYASHI, Seiichiro, JAPAN
WAKAE, Masami, JAPAN
WAKATSUKI, Taketo, JAPAN

WAKAYAMA, Masato, JAPAN
WAKIMOTO, Minoru, JAPAN
WALDSPURGER, Jean-Loup, FRANCE
WALKER, Elbert A., U.S.A.
WALL, Charles T. C., U.K.
WALLACE, David A.R.*, U.K.
WALLACH, Nolan R., U.S.A.
WALTER, Wolfgang*, F.R.G.
WAN, Yieh-Hei, U.S.A.
WAN, Zhe-xian, SWEDEN
WANG, Cun-Zheng, CHINA, P. R. of
WANG, Shicheng*, CHINA, P. R. of
WANG, Shikun*, CHINA, P. R. of
WASHIHARA, Masako, JAPAN
WASHINGTON, Lawrence C., U.S.A.
WASHIO, Tadashi, JAPAN
WASSERMANN, Antony J., U.K.
WATABE, Mutsuo, JAPAN
WATABE, Tsuyoshi, JAPAN
WATAMORI, Yoko, JAPAN
WATANABE, Atumi, JAPAN
WATANABE, Hiroshi, JAPAN
WATANABE, Hisako, JAPAN
WATANABE, Hisao, JAPAN
WATANABE, Jiro, JAPAN
WATANABE, Junsei, JAPAN
WATANABE, Kazuhisa, JAPAN
WATANABE, Keiichi, JAPAN
WATANABE, Keiichi, JAPAN
WATANABE, Kimio, JAPAN
WATANABE, Kinji, JAPAN
WATANABE, Kiyoshi, JAPAN
WATANABE, Masayuki, JAPAN
WATANABE, Michiaki§, JAPAN
WATANABE, Nobuya, JAPAN
WATANABE, Osamu, JAPAN
WATANABE, Seiji, JAPAN
WATANABE, Shinzo, JAPAN
WATANABE, Tadashi, JAPAN
WATANABE, Takao, JAPAN
WATANABE, Takesi, JAPAN
WATANABE, Tetsuo, JAPAN
WATANABE, Toshihiro, JAPAN
WATANABE, Toshikazu, JAPAN
WATANABE, Yoshikazu, JAPAN
WATANABE, Yoshiyuki, JAPAN
WATARI, Chinami, JAPAN
WATARI, Satoru, JAPAN
WATATANI, Yasuo, JAPAN
WATKINS, John J., U.S.A.
WEBB, Jeffrey R. L., U.K.
WEBBER, David Brian, U.K.
WEDER, Ricardo*, MEXICO
WEE, In-Suk*, KOREA, R. of
WEHRHAHN, Karl H.*, AUSTRALIA
WEIBEL, Charles A., U.S.A.
WEISFELD, Morris, U.S.A.
WEISS, Asia Ivic, CANADA
WEISS, William, CANADA
WELCH, Philip D.*, U.K.
WELK, Reiner, F.R.G.
WENZEL, Christian*, U.S.A./F.R.G.
WEST, Alan, U.K.
WESTON, Kenneth W.*, U.S.A.
WHITMAN, Nancy C.*, U.S.A.
WIDMAN, Kjell-Owe H., SWITZERLAND
WIEGAND, Sylvia M., U.S.A.
WIGDERSON, Avi, ISRAEL
WIGLEY, Neil M.*, CANADA
WILLIAMS, Daniel A., U.S.A.
WILLIAMS, Robert F., U.S.A.
WILSON, James M., U.S.A.
WILSON, John Stuart, U.K.
WILSON, Pelham M.H., U.K.
WINKLER, Jörg, F.R.G.
WISNER, Robert J., U.S.A.
WITTE, Dave, U.S.A.
WITTEN, Edward, U.S.A.
WITTEN, Louis, U.S.A.
WOGEN, Warren R.*, U.S.A.
WOLPER, James Samuel*, U.S.A.
WONG, Philip Pit Wang, HONG KONG
WONG, Raymond Y., U.S.A.
WONG, Roderick S.-C., CANADA
WONG, Yau-Chuen*, HONG KONG
WOOD, Carol S., U.S.A.
WORONOWICZ, Stanislaw Lech, POLAND
WORSLEY, Keith John, CANADA
WRIGHT, David James*, U.S.A.
WU, Liang Sen*, CHINA, P. R. of
WU, Pei Yuan*, CHINA-TAIWAN
WU, Yihren*, U.S.A.
WULFSOHN, Aubrey*, ISRAEL
XIONG, Jincheng*, CHINA, P. R. of
XU, Dao Yi*, CHINA, P. R. of
XU, Yichao*, CHINA, P. R. of
YABU, Yasuhiko, JAPAN
YABUTA, Kozo, JAPAN
YAGASAKI, Tatsuhiko, JAPAN
YAGI, Akiko, JAPAN
YAGI, Atsushi, JAPAN
YAGI, Hirotomo, JAPAN
YAGI, Katsumi, JAPAN
YAGISHITA, Kimio, JAPAN
YAGITA, Nobuaki, JAPAN
YAGUCHI, Teruo*, JAPAN
YAJIMA, Yukinobu, JAPAN
YAKU, Takeo, JAPAN
YAMADA, Akira, JAPAN
YAMADA, Hiromichi, JAPAN
YAMADA, Kotaro, JAPAN
YAMADA, Naoki, JAPAN
YAMADA, Osanobu, JAPAN
YAMADA, Shuji§, JAPAN
YAMADA, Sumio, JAPAN
YAMADA, Toshihiko, JAPAN

YAMADA, Toshio, JAPAN
YAMADA, Yoshio, JAPAN
YAMADA, Yuji, JAPAN
YAMAGAMI, Shigeru, JAPAN
YAMAGATA, Hideo*, JAPAN
YAMAGATA, Kunio, JAPAN
YAMAGISHI, Kikumichi, JAPAN
YAMAGISHI, Yoshikazu, JAPAN
YAMAGUCHI, Hakuki, JAPAN
YAMAGUCHI, Hiroshi, JAPAN
YAMAGUCHI, Keizo, JAPAN
YAMAGUCHI, Kohhei, JAPAN
YAMAGUCHI, Masaru, JAPAN
YAMAGUCHI, Michinari, JAPAN
YAMAGUCHI, Seiichi, JAPAN
YAMAGUCHI, Tadashi, JAPAN
YAMAGUCHI, Takao, JAPAN
YAMAGUCHI, Toshihiro, JAPAN
YAMAGUTI, Kiyosi*, JAPAN
YAMAGUTI, Masaya, JAPAN
YAMAHARA, Hideo*, JAPAN
YAMAKAWA, Aiko, JAPAN
YAMAKI, Hiroyoshi, JAPAN
YAMAMOTO, Hiro-o, JAPAN
YAMAMOTO, Kazuhiro, JAPAN
YAMAMOTO, Kensho, JAPAN
YAMAMOTO, Makoto, JAPAN
YAMAMOTO, Masahiro, JAPAN
YAMAMOTO, Minoru, JAPAN
YAMAMOTO, Shuichi, JAPAN
YAMAMOTO, Sunao, JAPAN
YAMAMOTO, Takanori, JAPAN
YAMAMOTO, Tetsuro, JAPAN
YAMAMOTO, Tetsuya, JAPAN
YAMAMOTO, Yoshihiko, JAPAN
YAMAMOTO, Yutaka, JAPAN
YAMAMURA, Ken, JAPAN
YAMANAKA, Hajime, JAPAN
YAMANAKA, Takesi, JAPAN
YAMANE, Hiroyuki, JAPAN
YAMANO, Gosuke, JAPAN
YAMANO, Takehisa, JAPAN
YAMANOSHITA, Tsuneyo, JAPAN
YAMANOUCHI, Takehiko, JAPAN
YAMASAKI, Aiichi, JAPAN
YAMASAKI, Hideki, JAPAN
YAMASAKI, Masayuki, JAPAN
YAMASAKI, Motohiro, JAPAN
YAMASAKI, Toshiharu, JAPAN
YAMASAKI, Yasuo, JAPAN
YAMASHIMA, Shigeho, JAPAN
YAMASHITA, Hajime*, JAPAN
YAMASHITA, Satoshi, JAPAN
YAMATO, Hajime, JAPAN
YAMATO, Kazuo, JAPAN
YAMATO, Kenji, JAPAN
YAMATO, Yuiti, JAPAN
YAMAUCHI, Kazunari, JAPAN
YAMAUCHI, Masatoshi, JAPAN
YAMAUCHI, Norio, JAPAN
YAMAWAKI, Noriaki*, JAPAN
YAMAYA, Atsushi, JAPAN
YAMAZAKI, Chikao§, JAPAN
YAMAZAKI, Keijiro, JAPAN
YAMAZAKI, Masao, JAPAN
YAMAZAKI, Mitsuru, JAPAN
YAMAZAKI, Tadashi, JAPAN
YAMAZAKI, Taeko, JAPAN
YAN, Jia-An*, CHINA, P. R. of
YANAGAWA, Minoru*, JAPAN
YANAGAWA, Takaaki, JAPAN
YANAGI, Kenjiro, JAPAN
YANAGIHARA, Hiroshi, JAPAN
YANAGIHARA, Niro, JAPAN
YANAGIMOTO, Shigekazu*, JAPAN
YANAGISAWA, Takashi, JAPAN
YANAGISAWA, Taku, JAPAN
YANAI, Hiromichi, JAPAN
YANG, Jae-Hyun, KOREA, R. of
YANG, Paul C., U.S.A.
YANO, Koichi, JAPAN
YANO, Tamaki*, JAPAN
YAP, Leonard Y. H.*, SINGAPORE
YASUDA, Kumi, JAPAN
YASUGI, Mariko, JAPAN
YASUI, Tsutomu, JAPAN
YASUKURA, Osami, JAPAN
YASUMOTO, Masahiro, JAPAN
YASUNAGA, Hisatoshi, JAPAN
YASUO, Minato, JAPAN
YASURAOKA, Yuzo, JAPAN
YASUTOMI, Sinichi, JAPAN
YASUURA, Hiroto, JAPAN
YATES, Samuel*, U.S.A.
YE, Rugang, U.S.A.
YEH, R. Z., U.S.A.
YEH, Yeong-Nan*, CHINA-TAIWAN
YEUNG, Kit Ming, HONG KONG
YLINEN, Kari E., FINLAND
YOCCOZ, Jean-Christophe, FRANCE
YOKOGAWA, Kouji, JAPAN
YOKOI, Hideo, JAPAN
YOKOI, Yoshitaka, JAPAN
YOKONUMA, Takeo, JAPAN
YOKOTA, Hisashi, JAPAN
YOKOTA, Ichiro*, JAPAN
YOKOTA, Toshiro, JAPAN
YOKOTA, Yoshiyuki, JAPAN
YOKOTE, Ichiro, JAPAN
YOKOYAMA, Kazunori, JAPAN
YOKOYAMA, Kazuo, JAPAN
YOKOYAMA, Misako, JAPAN
YOKOYAMA, Mutsumi, JAPAN
YOKOYAMA, Ryozo, JAPAN
YOKOYAMA, Shin, JAPAN
YOKOYAMA, Takuji, JAPAN

YOKURA, Shoji*, JAPAN
YONEDA, Kaoru, JAPAN
YONEMOTO, Hiroshi, JAPAN
YONEZAWA, Yoshitaka, JAPAN
YOO, Il, KOREA, R. of
YOO, Ki-Jo, KOREA, R. of
YOON, Goog-Joong, KOREA, R. of
YOR, Marc J., FRANCE
YORINAGA, Masataka, JAPAN
YOROZU, Shinsuke§, JAPAN
YOSHIDA, Eiji, JAPAN
YOSHIDA, Haruyo*, JAPAN
YOSHIDA, Hidenobu, JAPAN
YOSHIDA, Hiroaki, JAPAN
YOSHIDA, Hiroyuki, JAPAN
YOSHIDA, Jun-ichi, JAPAN
YOSHIDA, Kenichi, JAPAN
YOSHIDA, Kiyoshi, JAPAN
YOSHIDA, Mamoru, JAPAN
YOSHIDA, Masaaki, JAPAN
YOSHIDA, Masaharu, JAPAN
YOSHIDA, Minoru, JAPAN
YOSHIDA, Nakahiro, JAPAN
YOSHIDA, Nobuo, JAPAN
YOSHIDA, Norio, JAPAN
YOSHIDA, Toshio, JAPAN
YOSHIHARA, Hisao*, JAPAN
YOSHIHARA, Ken-ichi, JAPAN
YOSHIKAWA, Atsushi, JAPAN
YOSHIKAWA, Kazuo, JAPAN
YOSHIKAWA, Kenichi, JAPAN
YOSHIKAWA, Masaki, JAPAN
YOSHIKAWA, Masayuki, JAPAN
YOSHIMATSU, Yumiko, JAPAN
YOSHIMOTO, Akinori, JAPAN
YOSHIMURA, Mitsuhiko, JAPAN
YOSHIMURA, Zenichi, JAPAN
YOSHINAGA, Etsuo, JAPAN
YOSHINAGA, Ken-ichi, JAPAN
YOSHINO, Ken-ichi, JAPAN
YOSHINO, Masafumi, JAPAN
YOSHINO, Takashi, JAPAN
YOSHINO, Yuji, JAPAN
YOSHINOBU, Yasuo, JAPAN
YOSHIOKA, Akiko, JAPAN
YOSHIOKA, Akira, JAPAN
YOSHIOKA, Tsuneo, JAPAN
YOSHITOMI, Kentaro, JAPAN
YOSHIZAWA, Hisaaki, JAPAN
YOSHIZAWA, Taro, JAPAN
YOSIDA, Setuzo, JAPAN
YOSIOKA, Masanori, JAPAN
YOTSUTANI, Shoji, JAPAN
YOZAWA, Takashi, JAPAN
YUAN, Ya-Xiang*, CHINA, P. R. of
YUI, Noriko, CANADA
YUNG, Tin Gun, HONG KONG
YURI, Michiko, JAPAN
ZABCZYK, Jerzy, POLAND
ZAFARANI, Jafar*, IRAN
ZAGIER, Don B., F.R.G.
ZAHAROPOL, Radu*, U.S.A.
ZAHEDANI, Heydar Z.*, IRAN
ZAHEDI, Mohammad Mehdi*, IRAN
ZAICEV, Gennadi Ju.§, U.S.S.R.
ZAINODIN, Haji Jubok, MALAYSIA
ZALCMAN, Lawrence A., ISRAEL
ZARETTI, Anna, ITALY
ZARHIN, Yuri G., U.S.S.R.
ZASLAVSKY, Boris G.*, U.S.S.R.
ZAYED, Ahmed I.*, U.S.A.
ZDRAVKOVSKA, Smilka, U.S.A.
ZEEMAN, E. Christopher, U.K.
ZEINAL-HAMADANI, Ali§, IRAN
ŹELAZKO, Wiesław T.*, POLAND
ZELDITCH, Steven, U.S.A.
ZELEVINSKY, Andrey V., U.S.S.R.
ZELMANOV, Efim I., U.S.S.R.
ZHANG, De-Qi, CHINA, P. R. of
ZHANG, Jin-Hao, CHINA, P. R. of
ZHANG, Li Hua, CHINA, P. R. of
ZHANG, Shunian*, CHINA, P. R. of
ZHANG, Zezeng, CHINA, P. R. of
ZHANG, Zhi-Fen*, CHINA, P. R. of
ZHARN, P. Joseph§, U.S.A.
ZHOU, Xueguang*, CHINA, P. R. of
ZHOU, Zixiang*, CHINA, P. R. of
ZHOU, Zuo Ling*, CHINA, P. R. of
ZHU, Xiao-Wei, U.S.A.
ZIEGLER, Zvi, ISRAEL
ZIEMIAN, Bogdan*, POLAND
ZIMMER, Horst Günter*, F.R.G.
ZIMMERMAN, Grenith J., U.S.A.
ZIMMERMANN, Benno, SWITZERLAND
ZIMMERMANN-HUISGEN, Birge K., U.S.A.
ZOKAEI, Mohammad§, IRAN

Membership by Nationality

Country	Members
Algeria	3
Argentina	10
Australia	21
Austria	4
Bangladesh	2
Belgium	12
Brazil	15
Bulgaria	8
Canada	77
Chile	4
China, People's Republic of	67
China-Taiwan	29
Colombia	1
Costa Rica	2
Cuba	1
Czechoslovakia	8
Denmark	15
Egypt	6
El Salvador	1
Finland	29
France	123
German Democratic Republic	6
Germany, Federal Republic of	98
Ghana	1
Greece	8
Hong Kong	14
Hungary	15
Iceland	2
India	40
Indonesia	2
Iran	57
Iraq	1
Ireland	4
Israel	30
Italy	69
Ivory Coast	1
Japan	2409
Kenya	2
Korea, Republic of	41
Lebanon	1
Luxembourg	1
Malaysia	7
Mauritania	1
Mexico	14
Mozambique	1
Myanmar	1
Nepal	1
Netherlands	25
New Zealand	7
Nigeria	5
Norway	11
Pakistan	1
Panama	2
Peru	2
Philippines	7
Poland	17
Portugal	4
Romania	18
Saudi Arabia	2
Senegal	1
Sierra Leone	1
Singapore	3
South Africa	13
Spain	24
Sweden	32
Switzerland	14
Tanzania	1
Thailand	4
Tunisia	1
Turkey	2
Uganda	2
United Kingdom	94
Uruguay	1
USA	396
USSR	110
Venezuela	2
Vietnam	21
Yugoslavia	9
Zimbabwe	2
Stateless	3
Total	4102

The Work of the Fields Medalists and the Rolf Nevanlinna Prize Winner

The works of the Fields Medalists and the Rolf Nevanlinna Prize Winner were presented as follows:

Fields Medalists

The Rolf Nevanlinna Prize Winner

Fields Medalists: *Witten, Mori, Jones, Drinfeld*

Razborov (left), the Rolf Nevanlinna Prize winner, *Hori* and *Lovász*

On the Mathematical Work of Vladimir Drinfeld

Yuri Ivanovich Manin *

Steklov Mathematical Institute, 42 Vavilova, 117966 GSP-1, Moscow, USSR

I

Drinfeld has written his first published paper when he was a schoolboy. He proved there a nice result in the style of Hardy's classic treatise "Inequalities" and solved a problem to which R. A. Rankin devoted two notes. This paper still makes an interesting reading. It starts a series of Drinfeld's works which can be considered somewhat isolated in the general context of his mathematical production but which contains such worthy results as:

- a proof that the cuspidal degree zero cycles on a modular curve generate a torsion subgroup of the Jacobian;
- classification of instantons (ADHM construction, jointly with Atiyah, Hitchin and Manin);
- reduction theory of completely integrable systems of KdV-type (jointly with Sokolov));
- a sharp asymptotic upper bound for the number of points of a curve defined over a finite field of order p^{2n} (jointly with Vladut);
- a proof that an $SO(n+1)$-invariant finitely additive measure on S^n is Lebesgue for $n = 2$ and 3 (the cases $n = 1$ and $n \geq 4$ were treated earlier by Banach, Margulis and Sullivan).

The limitations of space and time make it impossible for me to review Drinfeld's contributions referred to above, and I shall concentrate upon the two subjects that were Drinfeld's main preoccupation in the last decade. These are Langlands' program and quantum groups. In both domains, Drinfeld's work constituted a decisive breakthrough and prompted a wealth of research.

II

Langlands' program is a series of conjectures, theorems and insights aimed to an understanding of the Galois groups of local and global fields of dimension one, that is $\mathbb{Q}$, $\mathbb{F}_p(t)$, their finite extensions and completions.

* Delivered by Michio Jimbo.

Proceedings of the International Congress of Mathematicians, Kyoto, Japan, 1990

One can convincingly argue that these Galois groups constitute a primary object of number theory, more fundamental than the integers themselves. Anyway, most of the classical themes of number theory, like prime numbers, L-functions, and modular forms reveal a lot of hidden structure when viewed from the Galois-theoretic angle.

The highest achievement of the classical period was the class-field theory which gave a description of $\mathrm{Gal}(k^{\mathrm{ab}}/k)$ in terms of the arithmetics of k. If k is local, $\mathrm{Gal}(k^{\mathrm{ab}}/k)$ is (canonically isomorphic to) the profinite completion of k^*; if k is global, $\mathrm{Gal}(k^{\mathrm{ab}}/k)$ is the profinite completion of the group A_k^*/k^* where A_k denotes the adèle ring of k. Of course, to describe $\mathrm{Gal}(k^{\mathrm{ab}}/k)$ is the same as to describe one-dimensional representations of $\mathrm{Gal}(\overline{k}/k)$. Langlands suggested that the next step should be a description of certain n-dimensional representations of $\mathrm{Gal}(\overline{k}/k)$. The main content of his conjectures, in a very simplified and imprecise form, can be summarized in two statements. First, n-dimensional representations of $\mathrm{Gal}(\overline{k}/k)$ are in a natural bijection with automorphic irreducible infinite-dimensional representations of $GL(n,k)$ (resp. $GL(n,A_k)$) in the local (resp. global) case. Second, this correspondence (the non-commutative reciprocity law) is described via the identification of the L-functions that can be constructed in a natural way for both Galois representations and the adèlic general linear group representations. In the classical (global) cases when the structure of these L-functions is known, they are Mellin's transforms of modular forms which represent, roughly speaking, the de Rham aspect of the cohomology of certain modular spaces, while Galois representations embody their étale cohomology. Thus, Langlands' program is related to Grothendieck's motives, a largely conjectural universal cohomology theory of algebraic varieties.

Drinfeld proved Langlands' conjectures for $GL(2)$ over global fields of finite characteristic. His decisive contribution was the discovery of a new class of modular spaces and their detailed investigation. While the classical modular spaces parametrize elliptic curves, abelian varieties or Hodge structures, Drinfeld had an astonishing idea that in order to treat the finite characteristic case one should parametrize objects of a new kind. The first approximation is now called Drinfeld modules (earlier examples of which were introduced by Carlitz). A more general notion indispensable for the complete theory is known under the equally uninspiring names "shtuka" (meaning approximately "a piece of something" in English) and "F-sheaf" (which Drinfeld uses for the lack of something better; as a mild reproach I must say that in the domain of terminology his imagination produces less brilliant solutions than in the theorem-proving).

For completeness, I shall give below a formal definition of an F-sheaf and the statement of the main theorem of Drinfeld's theory.

Let X be a smooth complete model of the ground field $k = \mathbb{F}_q(X)$. A (left) F-sheaf of rank d over a scheme S is a diagram

$$\begin{array}{ccc} & & \mathscr{L} \\ & {\scriptstyle i}\nearrow & \\ \mathscr{F} & & \\ & {\scriptstyle j}\searrow & \\ & & (\mathrm{id}_X \times \mathrm{Fr}_S)^* \mathscr{L} \end{array}$$

in which $\mathscr{L}$ and $\mathscr{F}$ are locally free sheaves on $X \times S$ of rank d, i and j are injective, Coker(i) is an invertible sheaf on the graph Γ_α of a certain S-point $\alpha : S \to X$, Coker(j) is an invertible sheaf on a similar graph Γ_β.

F-sheaves can be rigidified by imposing a level structure. F-sheaves with a non-trivial level structure admit coarse moduli schemes, which can be conveniently compactified. For $d = 2$, $GL(2, A_k)$ acts upon compactified moduli spaces, much in the same way as $GL(2, A_{\mathbb{Q},f})$ acts upon moduli spaces of elliptic curves via Hecke operators. A Galois-theoretic analysis of this action upon the cohomology furnishes the following results.

Let $W_k \subset \mathrm{Gal}(\overline{k}^s/k)$ be the group of all automorphisms inducing upon $\overline{\mathbb{F}}_q$ an integer power of the Frobenius morphism. Fix a prime $l \neq \mathrm{char}(k)$. Denote by Σ_1^l the set of continuous representations $\varrho : W_k \to GL(n, \overline{\mathbb{Q}}_l)$ with the following properties: i) $\mathrm{Im}(\varrho) \subset GL(n, F)$ for a finite extension F of $\mathbb{Q}_l$; ii) inertia subgroups of almost all points of k are contained in $\mathrm{Ker}(\varrho)$.

Let now E be an algebraically closed field of characteristic 0. A cuspidal representation of $GL(2, A_k)$ is, by definition, an irreducible representation over E belonging to the space of E-valued cusp forms. Denote by Σ_2^l the set of cuspidal representations over $\mathbb{Q}_l$.

The principal theorem of Drinfeld states that Σ_1^l and Σ_2^l are bijective with respect to a map conserving L-functions.

Drinfeld's proof is long and involved, both technically and logically. Although it is contained in a series of journal papers, the most comprehensive treatment of the whole theory is given in his two still unpublished manuscripts (totalling up to about 800 type-written pages). It is to be hoped that after a final brushing up they will be given to a book publisher. The reason it has not been done to this moment seems to be Drinfeld's preoccupation with another new and fascinating subject – quantum groups.

III

Formally speaking, quantum groups constitute a vaguely defined subclass of Hopf algebras. Their first examples were discovered by the mathematical physicists of Leningrad school, students and collaborators of L. D. Faddeev. Drinfeld first summarized the basic definitions and results of this theory, largely conceived or systematized in his own work, in his talk at the Berkeley ICM four years ago. This report and several articles of M. Jimbo played a decisive role in the crystallization of this new domain and drew to it attention of many mathematicians.

Somewhat schematically, one can describe several principal themes of this theory in the following way:

a) For many decades, a common wisdom was that simple Lie groups and Lie algebras are rigid objects: they are classified by discrete data (Dynkin diagrams), they do not vary in a continuous family. A remarkable discovery, due in this generality to Drinfeld and Jimbo, is that this ceases to be true if one considers the deformations in the class of non-commutative and cocommutative Hopf algebras.

More precisely, for any simple Lie algebra $\mathfrak{g}$ (and more generally, for algebras defined by Cartan root data) there exists a one-parametric deformation $U_q(\mathfrak{g})$ of the universal enveloping algebra $U(\mathfrak{g})$, with comultiplication and antipode. More than that, the whole structure and representation theory deforms in this way, with many remarkable new twists.

b) The properties of quantum groups are closely related to certain remarkable non-linear algebraic equations defining the so-called Yang-Baxter operators. The simplest Yang-Baxter equation is a condition on a linear operator $R \in \mathrm{End}(V \otimes V)$ where V is a linear space. The condition is that R satisfies the braid relation: $R_{12}R_{23}R_{12} = R_{23}R_{12}R_{23}$ as operators on $V^{\otimes 3}$. A more general Yang-Baxter equation concerns operators depending on spectral parameters. Many solutions of Yang-Baxter equations were discovered by the specialists working in the two-dimensional statistical physics because such solutions give rise to exactly solvable lattice models generalizing the famous Ising model.

Drinfeld introduced the notion of a universal Yang-Baxter operator which is an invertible element in the (completed) tensor square of a Hopf algebra. He has proved its existence in $U_q(\mathfrak{g})^{\otimes 2}$ and has given a general "double" construction allowing one to generate Yang-Baxter operators in the representation categories of various Hopf algebras.

c) Until recently, the quantum group theory lacked a classification theorem describing precisely what class of objects we want to consider and what is the structure of this class. Such a theorem was recently formulated and proved in the two important papers by Drinfeld. It can be compared with the first theorems of Lie establishing the relations between Lie algebras and local Lie groups.

Indeed, this theorem is local, even formal, in nature, because Drinfeld considers formal deformations of formal groups (or, in the dual language, of their universal enveloping algebras). He also imposes from the very beginning a Yang-Baxter type condition ("quasi-triangularity"). He proves then that the whole deformation is defined by the lowest order non-trivial data.

In the course of proof, he introduces a new notion of quasi-Hopf algebra, weakening in an appropriate way the coassociativity condition on the comultiplication. He connects quasi-Hopf algebras with the Knizhnik-Zamolodchikov differential equation whose monodromy is related to the Drinfeld-Jimbo R-operators as was shown by Kohno. Finally, he shows that in a formal situation the quasi-Hopf algebras can be reduced to the usual Hopf algebras by a kind of gauge transformation. The proof involves killing a lot of cohomological obstructions arising in the complex perturbative scheme.

d) One should mention also the notions of a Poisson-Lie group and a Poisson-Lie action introduced by Drinfeld at an earlier stage of his work on Yang-Baxter equations. They form one of the most basic differential geometric structures connected with Hamiltonian mechanics, and their role in future will certainly grow.

IV

I hope that I conveyed to you some sense of broadness, conceptual richness, technical strength and beauty of Drinfeld's work for which we are now honoring him with the Fields medal. For me, it was a pleasure and a privilege to observe at a close distance the rapid development of this brilliant mind which taught me so much.

The Work of Vaughan F. R. Jones

Joan S. Birman

Department of Mathematics, Columbia University, New York, NY 10027, USA

It gives me great pleasure that I have been asked to describe to you some of the very beautiful mathematics which resulted in the awarding of the Fields Medal to Vaughan F. R. Jones at ICM '90.

In 1984 Jones discovered an astonishing relationship between von Neumann algebras and geometric topology. As a result, he found a new polynomial invariant for knots and links in 3-space. His invariant had been missed completely by topologists, in spite of intense activity in closely related areas during the preceding 60 years, and it was a complete surprise. As time went on, it became clear that his discovery had to do in a bewildering variety of ways with widely separated areas of mathematics and physics, some of which are indicated in Figure 1. These included (in addition to knots and links) that part of statistical mechanics having to do with exactly solvable models, the very new area of quantum groups, and also Dynkin diagrams and the representation theory of simple Lie algebras. The central connecting link in all this mathematics was a tower of nested algebras which Jones had discovered some years earlier in the course of proving a theorem which is known as the "Index Theorem".

My plan is to begin by discussing the Index Theorem, and the tower of algebras which Jones constructed in the course of his proof. After that, I plan to return to the chart in Figure 1 in order to indicate how this tower of algebras served as a bridge between the diverse areas of mathematics which are shown on the chart. I will restrict my attention throughout to one very special example of the tower construction, and so also to one special example of the associated link invariants, in order to make it possible to survey a great deal of mathematics in a very short time. Even with the restriction to a single example, this is a very ambitious plan. On the other hand, it only begins to touch on Vaughan Jones' scholarly contributions.

1. The Index Theorem

Let $\mathbf{M}$ denote a von Neumann algebra. Thus $\mathbf{M}$ is an algebra of bounded operators acting on a Hilbert space $\mathscr{H}$. The algebra $\mathbf{M}$ is called a *factor* if its center consists only of scalar multiples of the identity. The factor is *type* II_1 if it admits a linear functional, called a trace, $\mathrm{tr} : \mathbf{M} \to \mathbb{C}$, which satisfies the following three conditions:

Proceedings of the International Congress
of Mathematicians, Kyoto, Japan, 1990

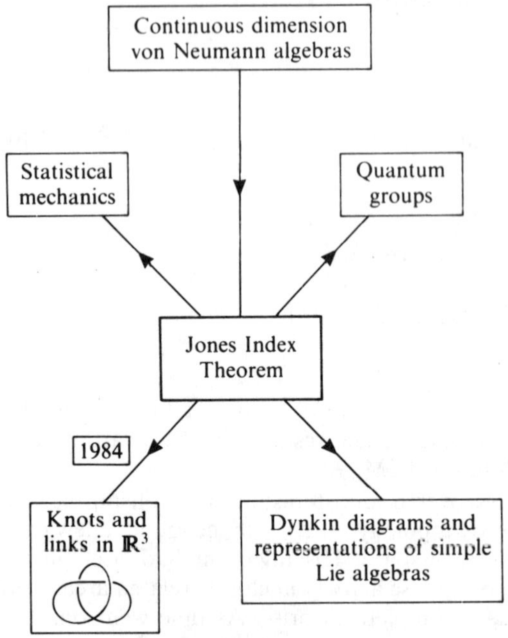

Fig. 1. The Jones Index Theorem

$$\mathrm{tr}(\mathbf{xy}) = \mathrm{tr}(\mathbf{yx}) \text{ for all } \mathbf{x}, \mathbf{y} \in \mathbf{M}$$
$$\mathrm{tr}(\mathbf{1}) = 1\,.$$
$$\mathrm{tr}(\mathbf{xx}^*) > 0 \text{ for all } \mathbf{x} \in \mathbf{M}, \text{ where } \mathbf{x}^* \text{ is the adjoint of } \mathbf{x}.$$

In this situation it is known that the trace is unique, in the sense that it is the only linear form satisfying the first two conditions. An old discovery of Murray and von Neumann was that factors of type II_1 provide a type of "scale" by which one can measure the dimension $\dim_{\mathbf{M}}(\mathscr{H})$ of $\mathscr{H}$. The notion of dimension which occurs here generalizes the familiar notion of integer-valued dimensions, because for appropriate $\mathbf{M}$ and $\mathscr{H}$ it can be any non-negative real number or ∞.

The starting point of Jones' work was the following question: if $\mathbf{M}_1$ is a type II_1 factor and if $\mathbf{M}_0 \subset \mathbf{M}_1$ is a *subfactor*, is there any restriction on the real numbers which occur as the ratio

$$\lambda = \dim_{\mathbf{M}_0}(\mathscr{H}) / \dim_{\mathbf{M}_1}(\mathscr{H}) \;?$$

The question has the flavor of questions one studies in Galois theory. On the face of it, there was no reason to think that λ could not take on any value in $[1, \infty]$, so Jones' answer came as a complete surprise. He called λ the *index* $[\mathbf{M}_1 : \mathbf{M}_0]$ of $\mathbf{M}_0$ in $\mathbf{M}_1$, and proved a type of rigidity theorem about type II_1 factors and their subfactors.

The Jones Index Theorem. *If* $\mathbf{M}_1$ *is a* II_1 *factor and* $\mathbf{M}_0$ *a subfactor, then the possible values of the index* $[\mathbf{M}_1 : \mathbf{M}_0]$ *are restricted to:*

$$[4,\infty] \cup \{4\cos^2(\pi/p)\,, \quad \textit{where } p \in \mathbb{N}, p \geq 3\}\,.$$

Moreover, each real number in the continuous part of the spectrum $[4,\infty]$ *and also in the discrete part* $\{4\cos^2(\pi/p),\ p \in \mathbb{N}, p \geq 3\}$ *is realized.*

We now sketch the idea of the proof, which is to be found in [Jo1]. Jones begins with the type II_1 factor $\mathbf{M}_1$ and the subfactor $\mathbf{M}_0$. There is also a tiny bit of additional structure: In this setting there exists a map $\mathbf{e}_1 : \mathbf{M}_1 \to \mathbf{M}_0$, known as the *conditional expectation* of $\mathbf{M}_1$ on $\mathbf{M}_0$. The map $\mathbf{e}_1$ is a *projection*, i.e. $(\mathbf{e}_1)^2 = \mathbf{e}_1$.

His first step is to prove that the ratio λ is independent of the choice of the Hilbert space $\mathscr{H}$. This allows him to choose an appropriate $\mathscr{H}$ so that the algebra $\mathbf{M}_2 = \langle \mathbf{M}_1, \mathbf{e}_1 \rangle$ generated by $\mathbf{M}_1$ and $\mathbf{e}_1$ makes sense. He then investigates $\mathbf{M}_2$ and proves that it is another type II_1 factor, which contains $\mathbf{M}_1$ as a subfactor, moreover the index $|\mathbf{M}_2 : \mathbf{M}_1|$ is equal to the index $|\mathbf{M}_1 : \mathbf{M}_0|$, i.e. to λ. Having in hand another II_1 factor $\mathbf{M}_2$ and its subfactor $\mathbf{M}_1$, there is also a trace on $\mathbf{M}_2$ which (by the uniqueness of the trace) coincides with the trace on $\mathbf{M}_1$ when it is restricted to $\mathbf{M}_1$, and another conditional expectation $\mathbf{e}_2 : \mathbf{M}_2 \to \mathbf{M}_1$. This allows Jones to iterate the construction, to build algebras $\mathbf{M}_1, \mathbf{M}_2 \ldots$ and from them a family of algebras:

$$\mathbf{J}_n = \{\mathbf{1}, \mathbf{e}_1, \ldots, \mathbf{e}_{n-1}\} \subset \mathbf{M}_n\,, \quad n = 1, 2, 3, \ldots\,.$$

Rewriting history a little bit in order to make the subsequent connection with knots a little more transparent, we now replace the $\mathbf{e}_k$'s by a new set of generators which are units, defining:

$$\mathbf{g}_k = q\mathbf{e}_k - (1 - \mathbf{e}_k)\,,$$

where

$$(1-q)(1-q^{-1}) = 1/\lambda\,.$$

The $\mathbf{g}_k$'s generate $\mathbf{J}_n$, because the $\mathbf{e}_k$'s do, and we can solve for the $\mathbf{e}_k$'s in terms of the $\mathbf{g}_k$'s. So

$$\mathbf{J}_n = \mathbf{J}_n(q) = \{\mathbf{1}, \mathbf{g}_1, \ldots, \mathbf{g}_{n-1}\}\,,$$

and we have a *tower of algebras*, ordered by inclusion:

$$\mathbf{J}_1(q) \subset \mathbf{J}_2(q) \subset \mathbf{J}_3(q) \subset \ldots\,.$$

The parameter q, which replaces the index λ, is the quantity now under investigation.

The parameter q is woven into the construction of the tower. First, defining relations in $\mathbf{J}_n(q)$ depend upon q:

(1) $$\mathbf{g}_i\mathbf{g}_k = \mathbf{g}_k\mathbf{g}_i \quad \text{if } |i-k| \geq 2\,,$$

(2) $$\mathbf{g}_i\mathbf{g}_{i+1}\mathbf{g}_i = \mathbf{g}_{i+1}\mathbf{g}_i\mathbf{g}_{i+1}\,,$$

(3_q) $$\mathbf{g}_i^2 = (q-1)\mathbf{g}_i + q\,,$$

(4) $$\mathbf{1} + \mathbf{g}_i + \mathbf{g}_{i+1} + \mathbf{g}_i\mathbf{g}_{i+1} + \mathbf{g}_{i+1}\mathbf{g}_i + \quad + \mathbf{g}_i\mathbf{g}_{i+1}\mathbf{g}_i = 0\,.$$

A second way in which q enters into the structure involves the trace. Recall that since $\mathbf{M}_n$ is type II_1 it supports a unique trace, and since $\mathbf{J}_n$ is a subalgebra it does too, by restriction. This trace is known as a *Markov trace*, i.e. it satisfies the important property:

$$\mathrm{tr}(\mathbf{w}\mathbf{g}_n) = \tau(q)\,\mathrm{tr}(\mathbf{w}) \quad \text{if } \mathbf{w} \in \mathbf{J}_n, \tag{5_q}$$

where $\tau(q)$ is a fixed function of q. Thus, for each fixed value of q the trace is multiplied by a fixed scalar when one passes from one stage of the tower to the next, by multiplying an arbitrary element of $\mathbf{J}_n$ by the new generator $\mathbf{g}_n$ of $\mathbf{J}_{n+1}$.

Relations (1) and (2) above have an interesting geometric meaning, familiar to topologists. They are defining relations for the *n-string braid group*, $\mathbf{B}_n$, discovered by Emil Artin [Ar] in a foundational paper written in 1923. We pause to discuss braids.

An n-braid may be visualized by a weaving pattern of strings in 3-space which join n points on a horizontal plane to n corresponding points on a parallel plane, as illustrated in the example in Figure 2, where $n = 4$. In the case $n = 3$, the familiar braid in a person's hair gives another example. The strings are allowed to be stretched and deformed, the key features being that strings cannot pass through one-another and always proceed directly downward in their travels from the upper plane to the lower one. The equivalence class of weaving patterns under such deformations is an *n-braid*. One multiplies braids by concatenation and erasure of the middle plane. This multiplication makes them into a group, the n-string braid group $\mathbf{B}_n$. The identity is a braid which, when pulled taut, goes over to n straight lines. Generators are the $n-1$ elementary braids which (by an abuse of notation) we continue to call $\mathbf{g}_1, \ldots, \mathbf{g}_{n-1}$. The pictures in Figure 3 show that relations (1) and (2) hold between the generators of $\mathbf{B}_n$. In fact, Artin proved they are *defining relations* for $\mathbf{B}_n$. Thus for each n there is a homomorphism from the n-string braid group $\mathbf{B}_n$ into the Jones algebra $\mathbf{J}_n(q)$, and from the group algebra $\mathbb{C}\mathbf{B}_n$ onto $\mathbf{J}_n(q)$.

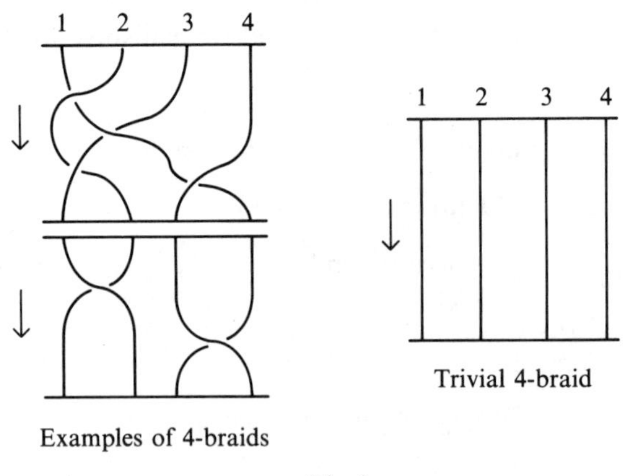

Examples of 4-braids

Trivial 4-braid

Fig. 2

Returning to the business at hand, i.e. the proof of the Index Theorem, Jones next shows that properties (1), (2), (3_q) and (5_q) suffice for the calculation of the

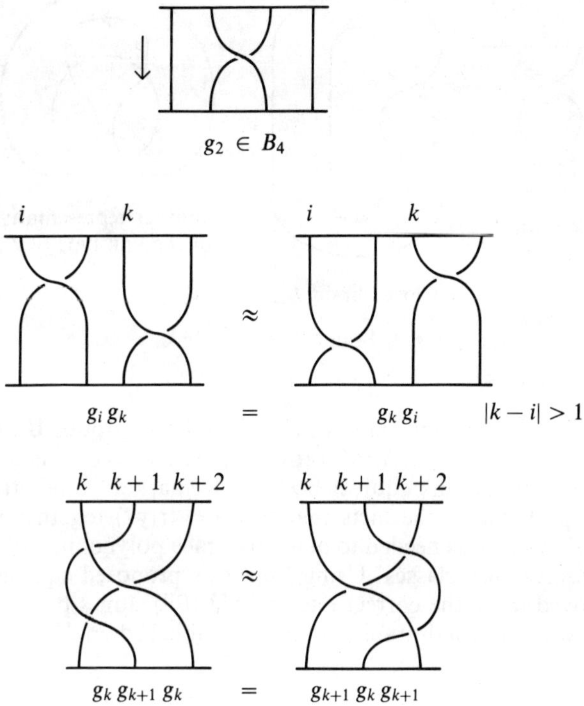

Fig. 3. Generators and defining relations in $\mathbf{B}_n$

trace of an arbitrary element $\mathbf{x} \in \mathbf{J}_n(q)$. It turns out that trace($\mathbf{x}$) is an *integer polynomial* in $(\sqrt{q})^{\pm 1}$. (We will meet it again in a few moments as the Jones polynomial associated to $\mathbf{x}$.) Jones proof of the Index Theorem is concluded when he shows that the infinite sequence of algebras $\mathbf{J}_n(q)$, with the given trace, could not exist if q did not satisfy the restrictions of the Index Theorem.

2. Knots and Links

We have already seen hints of topological meaning in $\mathbf{J}_n(q)$ via braids. There is more to come. Knots and links are obtained from braids by identifying the initial points and end points of a braid in a circle, as illustrated in Figure 4. It was proved by J. W. Alexander in 1928 that every knot or link arises in this way. Earlier we described an equivalence relation on weaving patterns which yields braids, and there is a similar (but less restrictive) equivalence relation on knots, i.e. a knot or link *type* is its equivalence class under isotopy in 3-space. Note that isotopy in 3-space which takes one closed braid representative of a link to another closed braid representative will pass through a sequence of representatives which are not closed braids in an obvious way. For example see the 2-component link which is illustrated in Figure 4. The left picture is an obvious closed braid representative, whereas the right is not.

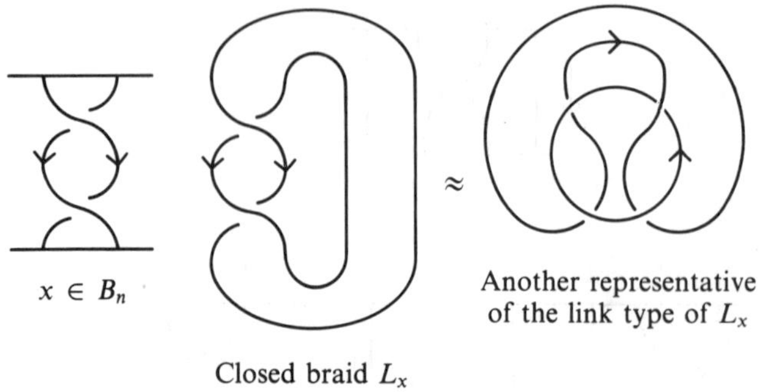

Fig. 4. Braids determine links

Let $\mathbf{B}_\infty$ denote the disjoint union of all of the braid groups $\mathbf{B}_n$, $n = 1, 2, 3, \ldots$. In 1935 the mathematician A. A. Markov proposed the equivalence relation on $\mathbf{B}_\infty$ which corresponds to link equivalence [M]. Remarkably, the properties of the trace, or more particularly the facts that $\mathrm{tr}(\mathbf{xy}) = \mathrm{tr}(\mathbf{yx})$ together with property (5_q), were exactly what was needed to make the trace polynomial into an invariant on Markov's equivalence classes! Using Markov's proposed equivalence relation (which was proved to be the correct one in 1972 [Bi]), Jones proved, with almost no additional work beyond results already established in [Jo1], the following theorem:

Theorem [Jo3]. *If* $\mathbf{w} \in \mathbf{B}_\infty$, *then (after multiplication by an appropriate scalar, which depends upon the braid index n) the trace of the image of* $\mathbf{w}$ *in* $\mathbf{J}_n(q)$ *is a polynomial in* $(\sqrt{q})^{\pm 1}$ *which is an invariant of the link type defined by the closed braid* $\mathbf{L_w}$.

The invariant of Jones' theorem is the one-variable *Jones polynomial* $\mathbf{V}_x(q)$. Notice that the independent "variable" in this polynomial is essentially the index of a type II_1 subfactor in a type II_1 factor! It's discovery opened a new chapter in knot and link theory.

3. Statistical Mechanics

We promised to discuss other ways in which the work of Jones was related to yet other areas of mathematics and physics, and begin to do so now. As it turned out, when Jones did his work the family of algebras $\mathbf{J}_n(q)$ were already known to physicists who were concerned with *exactly solvable models* in Statistical Mechanics. (For an excellent introduction to this topic, see R. Baxter's article in these Proceedings.) One of the simplest examples in this area is known as the *Potts model.* In that model one considers an array of "atoms" arranged at the vertices of a planar lattice with m rows and n columns as in Figure 5. Each "atom" in the system has various possible spins associated to it, and in the simplest case, known as the Ising model, there are two choices, "+" for spin up and "−" for spin down. We have indicated one of the 2^{nm} choices in Figure 5, determining a

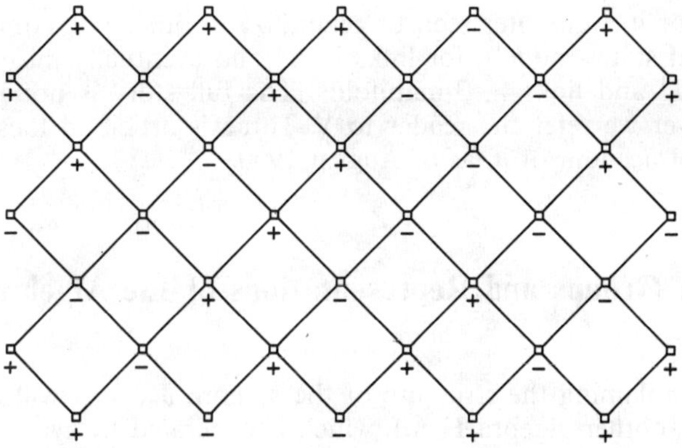

Fig. 5

state of the system. The goal is to compute the free energy of the system, averaged over all possible states.

Letting σ_i denote the spin at site i, we note that an edge e contributes an *energy* $E_e(\sigma_i, \sigma_j)$, where σ_i and σ_j are the states of the endpoints of e. Let E be the collection of lattice edges. Let β be a parameter which depends upon the temperature. Then the Gibbs *partition function* $\mathbf{Z}$ is given by the formula:

$$\mathbf{Z} = \sum_{\sigma_1,\dots,\sigma_{mn}} \prod_{e \in E} \exp(-\beta E_e(\sigma_i, \sigma_j)).$$

All of this is microscopic, nevertheless the major macroscopic thermodynamic quantities are functions of the partition function. In particular, the free energy $\mathbf{F}$, the object of interest to us at this time, is given by $\mathbf{Z} = \exp(-\beta \mathbf{F})$.

To compute the manner in which the atoms in one row of the lattice interact with atoms in the next, physicists set up the *transfer matrix* $\mathbf{T}$, which expresses the row-to-row interactions. It turned out that, in order to be able to calculate the free energy, the transfer matrices must satisfy conditions known as the *Yang-Baxter equations* and (to the great surprise of everyone) they turned out to be the braid relations (1) and (2) in disguise! (Remark: before Jones' work, to the best of our knowledge, it was not known that the Yang-Baxter equation was related to braids or knots.) Even more, the algebra which the transfer matrices generate in the Ising model, known to physicists as the *Temperley-Lieb algebra*, is our algebra $\mathbf{J}_n(q)$. The partition function $\mathbf{Z}$ is related in a very simple way to the transfer matrix:

$$\mathbf{Z} = \operatorname{trace}(\mathbf{T})^m.$$

In fact, it is closely related to the Jones trace.

The initial discovery of a relationship between the Potts model and links was reported on in [Jo3]. It opened a new chapter in the flow of ideas between mathematics and physics. We give an explicit example of a way in which the relationship of Jones' work to physics led to new insight into mathematics. Learning that the partition function was a sum over states of the system, Louis Kauffman was led to seek a decomposition of the Jones polynomial into a related sum over "states" of knot diagram, and arrived in [K1] at an elegant "states model" for the Jones polynomial. The Jones polynomial, and Kauffman's states model for it, were later seen to generalize to other polynomial invariants, with associated states models, for links in S^3 and eventually into invariants for 3-manifolds M^3 and links in 3-manifolds. The full story is not known at this writing, however we refer the reader to V. Turaev's article in these *Proceedings* for an excellent account of it, as of August 1990.

4. Quantum Groups and Representations of Lie Algebras

We begin by explaining the structure of the algebra $\mathbf{J}_n(q)$. It will be convenient to begin with another algebra $\mathbf{H}_n(q)$, which is generated by symbols $\mathbf{g}_1, \ldots, \mathbf{g}_{n-1}$ (which now have a third meaning), with defining relations (1), (2) and (3_q). The algebra $\mathbf{H}_n(q)$ is very well-known to mathematicians. It's the *Iwahori-Hecke algebra*, also known as the *Hecke algebra of the symmetric group* [Bo]. Its relationship with the symmetric group is simple to describe and beautiful. Notice that when $q = 1$, relation (3_q) simplifies to $(\mathbf{g}_k)^2 = 1$. One recognizes (1), (2) and (3_1) as defining relations for the group algebra $\mathbb{C}\mathbf{S}_n$ of the symmetric group $\mathbf{S}_n$. Here $\mathbf{g}_k$ is to be re-interpreted as a transposition which exchanges the symbols $\mathbf{k}$ and $\mathbf{k+1}$. In this way we may view $\mathbf{H}_n(q)$ as a "q-deformation" of the complex group algebra $\mathbb{C}\mathbf{S}_n = \mathbf{H}_n(1)$.

The algebra $\mathbb{C}\mathbf{S}_n$ is *rigid*, that is if one deforms it in this way its irreducible summands continue to be irreducible summands of the same dimension, in fact $\mathbf{H}_n(q)$ is actually algebra-isomorphic to $\mathbb{C}\mathbf{S}_n$ for generic q. Thus $\mathbf{H}_n(q)$ is a direct sum of finite dimensional matrix algebras, its irreducible summands being in one-to-one correspondence with the irreducible representations of the symmetric group $\mathbf{S}_n$. In this setting, Jones showed in [Jo2] that for generic q the algebra $\mathbf{J}_n(q)$ may be interpreted as the algebra associtated to the q-deformations of those irreducible representations of $\mathbf{S}_n$ which have Young diagrams with at most two rows.

We now explain how $\mathbf{H}_n(q)$ is related to quantum groups. It will be helpful to recall the classical picture. The fundamental representation of the Lie group GL_n acts on $\mathbb{C}^n$, and so its k-fold tensor product acts naturally on $(\mathbb{C}^n)^{\otimes k}$. The symmetric group $\mathbf{S}_k$ also acts naturally on $(\mathbb{C}^n)^{\otimes k}$, permuting factors. (Remark: In this latter action, the representations of $\mathbf{S}_k$ which are relevant are those whose Young diagrams have $\leq n$ rows.) As is well known, the actions of GL_n and $\mathbf{S}_k$ are each other's commutants in the full group of linear transformations of $(\mathbb{C}^n)^{\otimes k}$. If one now replaces GL_n and $\mathbb{C}\mathbf{S}_k$ by the quantum group $U_q(GL_n)$ and the Hecke algebra $\mathbf{H}_k(q)$ respectively, then the remarkable fact is that $U_q(GL_n)$ and $\mathbf{H}_k(q)$ are still each other's commutants [Ji]. The corresponding picture for $\mathbf{J}_n(q)$ is obtained by restricting to GL_2 and to representations of $\mathbf{S}_k$ having Young diagrams with at most 2 rows.

We remark that these are not isolated instances of algebraic accidents, but rather special cases of a phenomenon which relates a large part of the mathematics of quantum groups to finite dimensional matrix representations of the group algebra $\mathbb{C}\mathbf{B}_n$ which support a Markov trace (e.g. see [BW]).

5. Dynkin Diagrams

Dynkin diagrams arise in the tower construction which we described in § 1 via the inclusions of the algebras $\mathbf{J}_n(q)$ in the Jones tower. The inclusions for the Jones tower are very simple, and correspond to the Dynkin diagram of type A_n. However, other, more complicated towers may be obtained by replacing the II_1 factor $\mathbf{M}_1$ in the tower construction of § 1 above by $\mathbf{M}_1 \cap (\mathbf{M}_0)'$, where $(\mathbf{M}_0)'$ is the commutant of $\mathbf{M}_0$ in $\mathbf{M}_1$. We refer the reader to [GHJ] for an introduction to this topic and a discussion of the "derived tower" and the Dynkin diagrams which occur. The connections with the representations of simple Lie algebras can be guessed at from our discussion in § 4 above.

6. Concluding Remarks

I hope I have succeeded in showing you some of the ways in which Jones' work created bridges between the areas of mathematics which were illustrated in Figure 1. To conclude, I want to indicate very briefly some of the ways in which those bridges have changed the mathematics which many of us are doing.

There is another link polynomial in the picture, the famous *Alexander polynomial.* It was discovered in 1928, and was of fundamental importance to knot theory, both in the classical case of knots in S^3 and in higher dimensional knotting. Shortly after Jones' 1984 discovery, it was learned that in fact both the Alexander and Jones polynomial were specializations of the 2-variable Jones polynomial. That discovery was made simultaneously by five separate groups of authors: Freyd and Yetter, Hoste, Lickorish and Millett, Ocneanu, and Przytycki and Traczyk, a simple version of the proof of Ocneanu being given in [Jo3]. One of the techniques used in the proof was the combinatorics of link diagrams, and that technique led to the discovery of yet another polynomial, by Louis Kauffman [K2].

From the point of view of algebra, the Jones polynomial comes from a trace function on $\mathbf{J}_n(q)$, and the 2-variable Jones polynomial from a similar trace on the full Hecke algebra $\mathbf{H}_n(q)$. Beyond that, there is another algebra, the so-called Birman-Wenzl algebra [BW], and Kauffman's polynomial is a trace on it. Even more, physicists who had worked with solutions to the Yang Baxter equation, realized that they knew of still other Markov traces, so they began to grind out still other polynomials, in initially bewildering confusion. That picture is fairly well understood at this moment, however the work of Witten [W] indicates there are still other, related, link invariants. The generalizations are vast, with much work to be done.

There is also a different and very direct way in which Jones has had equal influence. His style of working is informal, and one which encourages the free and open interchange of ideas. During the past few years Jones wrote letters to various people which described his important new discoveries at an early stage, when he did not yet feel ready to submit them for journal publication because he had much more work to do. He nevertheless asked that his letters be shared, and so they were widely circulated. It was not surprising that they then served as a rich source of ideas for the work of others. As it has turned out, there has been more than enough credit to go around. His openness and generosity in this regard have been in the best tradition and spirit of mathematics.

References

[Ar] Artin, E.: Theorie der Zöpfe. Hamburg Abh. **4** (1925) 47–72

[Bi] Birman, J.: Braids, links and mapping class groups. Ann. Math. Stud. Princeton Univ. Press, Princeton 1974, pp. 37–69

[Bo] Bourbaki, N.: Groupes et algebres de Lie. Hermann, Paris 1968, Chapitre IV

[BW] Birman, J., Wenzl, H.: Braids, link polynomials and a new algebra. Trans. AMS **313** (1989) 249–273

[GHJ] Goodman, F., de la Harpe, P., Jones, V.: Coxeter graphs and towers of algebras. MSRI Publications, vol. 15. Springer, Berlin Heidelberg New York 1989

[Jo1] Jones, V.F.R.: Index for subfactors. Invent. math. **72** (1983) 1–25

[Jo2] Jones, V.F.R.: Braid groups, Hecke algebras and subfactors. In: Geometric methods in operator algebras. Pitman Research Notes in Mathematics, vol. 123, 1986, pp. 242–273

[Jo3] Jones, V.F.R.: Hecke algebra representations of braid groups and link polynomials. Ann. Math. **126** (1987) 335–388

[Ji] Jimbo, M.: A q-analogue of $U(gl(N+1))$, Hecke algebra, and the Yang Baxter equation. Letters in Math. Physics **II** (1986) 247–252

[K1] Kauffman, L.: A states model for the Jones polynomial. Topology **26** (1987) 385–407

[K2] Kauffman, L.: An invariant of regular isotopy. Trans. AMS (to appear)

[M] Markov, A.A.: Über die freie Äquivalenz geschlossener Zöpfe. Recueil Mathematique Moscou **1** (1935) 73–78

[W] Witten, E.: Quantum field theory and the Jones polynomial. In: Braid group, knot theory and statistical mechanics, eds. Yang and Ge. World Scientific Press, 1989

On the Work of Shigefumi Mori

Heisuke Hironaka

Department of Mathematics, Harvard University, Cambridge, MA 02138, USA

The most profound and exciting development in algebraic geometry during the last decade or so was the *Minimal Model Program* or *Mori's Program* in connection with the classification problems of algebraic varieties of dimension three. Shigefumi Mori initiated the program with a decisively new and powerful technique, guided the general research direction with some good collaborators along the way, and finally finished up the program by himself overcoming the last difficulty. The program was constructive and the end result was more than an existence theorem of minimal models. Even just the existence theorem by itself was the most fundamental result toward the classification of general algebraic varieties in dimension 3 up to birational transformations. The constructive nature of the program, moreover, provided a way of factoring a general birational transformation of threefolds into elementary transformations (divisorial contractions, flips and flops) that could be explicitly describable in principle. Mori's theorems on algebraic threefolds were stunning and beautiful by the totally new features unimaginable by those algebraic geometers who had been working, probably very hard too, only in the traditional world of algebraic or complex-analytic surfaces. Three in dimension was in fact a quantum jump from two in algebraic geometry.

Historically, to classify algebraic varieties has always been a fundamental problem of algebraic geometry and even an ultimate dream of algebraic geometers. During the early decades of this century, many new discoveries were made on the new features in classifying algebraic surfaces, unimaginable from the case of curves. They were mainly done by the so-called Italian school of algebraic geometers, such as Guido Castelnuovo (1865–1952), Federigo Enriques (1871–1946), Francesco Severi (1989–1961) and many others. Since then, there have been several important modernization, precision, reconstruction with rigor, extensions, in the theory of surfaces. The most notable among those were the works by Oscar Zariski during the late 1950s (especially, Castelnuovo's criterion for rational surfaces and minimal models of surfaces) and those by Kunihiko Kodaira during the early 1960s (especially, detailed study on elliptic surfaces and complex-analytic extensions, especially non-algebraic). In particular, Kodaira and his younger colleague S. Iitaka have produced many talented followers and collaborators. Up to and after the "Algebraic Surfaces" by I. R. Shafarevich, Yu. I. Manin, B. G. Moishezon, et al. (1965), the Russian school of algebraic geometers

Proceedings of the International Congress
of Mathematicians, Kyoto, Japan, 1990

advanced the study of some important algebraic surfaces and their deformations, especially the study of Torelli problems. As for the extensions to positive characteristics, outstanding and remarkable were the works of D. Mumford and E. Bombieri, and their followers during the late 1960s and 1970s. There have been so many other important contributors to the theory of surfaces that I would not try here to list all the important names and works in the theory of surfaces. There were also several unique works on classification problems of non-complete surfaces or complete surfaces with specified divisors, whose studies were instigated as being "2.5 dimensional" by S. Iitaka around 1977.

As for the higher dimensional algebraic varieties, the decade of 1970's saw three lines of new progress, just to name a few. One was the discovery, in the early 1970's, of the gap between the *rationality* and *unirationality*. The second was an attempt to classify *Fano* 3-*folds*, of which the birational classification was begun by Yu. I. Manin (1971), V. A. Iskovskih (1971, 1977–78), and followed by V. V. Shokurov (1979). However the biregular classification of Fano 3-folds in case $B_2 > 1$, with existence and number of moduli, had to wait a few more years to be achieved by Mori, jointly with S. Mukai, after a new powerful technique was invented by Mori in 1980. (The biregular classification in case $B_2 = 1$ was recently completed by S. Mukai using moduli of vector bundles on $K3$ surfaces.) The third line of progress was an early version of serious attempts toward *higher-dimensional classification* problems, largely inspired by S. Iitaka's bold conjectures proposed around 1970, and especially pressed hard during the latter half of the 1970's, by the Tokyo school of complex-algebraic geometers. The essence of their results was reported by H. Esnault in her Bourbaki talk, exp. 568 (1981).

Notably, K. Ueno produced some structure theorems in higher dimensions during 1977–79, which appeared at that time to boost the program of Iitaka et al. However, the scope of their achievements were limited toward the original goal of classifying general algebraic 3-folds in a style of extending the beautiful old theory of classifying surfaces. Their crucial limitation was the lack of a good higher-dimensional analog to Zariski's theory of minimal models. Technically, as it later became clear, the drawback was the absence of Mori's stunning success in analyzing birational transformations from the point of view of extremal contractions in dimensions higher than 2. In short, a new insight had to be injected into the classification program and it came from the technique of extremal rays inaugurated by Mori which inspired M. Reid, J. Kollár, Y. Kawamata et al. to begin working in earnest in search of minimal models around the turn of the decade.

Early in 1979, Mori brought to algebraic geometry a completely new excitement, that was his proof of *Hartshorne's conjecture*, proposed in 1970, which said that the projective spaces are the only smooth complete algebraic varieties with ample tangent bundles. It was also exciting news to differential geometers, such as Y.-T. Siu and S.-T. Yau who subsequently found an independent proof for Frankel's conjecture of 1961, which was implied by Hartshorne's. It is not clear that Hartshorne's conjecture can actually be proven by the differential-geometric method of Siu and Yau. In all approaches, the most important step was to show the existence of *rational curves* in the manifold in question. Mori's idea was simple and natural as good ones were always so, while the proof was not. The idea was that, under some numerical conditions, rational curves should be obtained by deforming a given curve inside the manifold and degenerating it into a bunch of curves of lower genera. But the difficulty was proving such plentifulness of deformations. There, Mori's ingenuity was to overcome this difficulty first in the cases of *positive characteristics*, where the Frobenius maps did a miracle, and then deduce the complex case from them.

Mori extended and reformulated his new and powerful technique of finding rational curves, which was referred to as *extremal rays* in the *cone of curves* by himself in another monumental paper of his on "Threefolds whose canonical bundles are not nef" (*nef* = numerically effective). The cone of curves in a projective 3-fold was defined to be the closure of the real convex hull generated by the numerical equivalence classes (or classes in the second homology group over the real numbers) of irreducible algebraic curves, and an extremal ray was to be an edge of the cone within the region where the canonical divisor takes negative intersection number. Mori's discovery was that the cone was locally finitely generated in the canonically negative region and that each extremal ray was generated by a *rational curve*. In the above "not nef" paper, moreover, Mori established an outstanding theorem which, in dimension three, completely classified the geometric structures of the family of rational curves corresponding to an extremal ray. The union of such curves was either an irreducible divisor (either a smooth projective line bundle, or a projective plane, or a quadric in a projective 3-space) or the entire 3-fold. The former led to a birational blowing down while the latter to a fibration map (either conic bundle, or Del Pezzo fiber space or Fano 3-fold). In each case, the description was precise and the target variety (blown-down birational model or the base of fibration) was again *projective*. This result was absolutely stunning to anybody who have had experiences with non-projective or even non-algebraic examples of birational transformations, and it was so beautiful as to encourage many algebraic geometers once again to look into the birational problems in dimension 3. Subsequently, Mori completed the *classification of Fano 3-folds*, jointly with S. Mukai in 1981, and he established a *criterion for uniruledness*, jointly with Y. Miyaoka in 1985. The Fano 3-folds and uniruled 3-folds were those to be investigated separately in all details and there are still left open some important problems about those special 3-folds. Excluding these, the most important general problem was to prove the *existence of minimal models* in which the canonical bundles were nef.

Mori's "not nef" paper showed, for the first time in the history of algebraic geometry, that a general birational transformation between smooth algebraic 3-folds was not completely untouchable and in fact "finitely manageable" in some sense. Naturally the dimension 3 was far more complicated than the dimension 2, and moreover some sort of singularities (seemingly finitely classifiable after Mori) were created in the process of factoring it into elementary transformations. Excited with Mori's discoveries, several algebraic geometers began clustering to him, this time with much clearer vision and better hope than ever, to work on factorization of birational transformations, and more importantly to work on defining what *minimal models* should be and how to obtain them, notably M.

Reid, V. V. Shokurov, Y. Kawamata, J. Kollár and of course S. Mori himself. *Mori's Program*, named by J. Kollár and meant to imply the process of obtaining minimal models, was as follows: Start from any projective smooth variety X, which is not birationally uniruled, and find a finite succession of "elementary" birational transformations by which X is transformed to a "minimal model". Of course, the central questions here were the meanings of "elementary transformation" and "minimal model". Firstly, the inverse of a blowing-up with a smooth center in a smooth 3-fold (even in a 3-fold with a "mildly" singular point) was definitely elementary. It was called a *divisorial contraction*. This certaily decreased the Néron-Severi number and hence it should make the variety closer to its minimal model. Suppose we could not make a divisorial contraction any more and the Néron-Severi number reached its minimum. Is the variety then good enough to be called a minimal model? The answer was clearly *no*, although the singularities created by those divisorial contractions were quite acceptable. Unlike the 2-dimensional cases, the resulting 3-fold can still have *extremal rays* which cause unpleasant behaviors of the canonical and pluricanonical bundles. The importance of these bundles had been clearly recognized even from the time of Castelnuovo and Enriques in the theory of classification problems as well as in the theory of deformations. Thus, to *eliminate an extremal ray*, a completely new type of elementary transformation was needed. This was the one later called *flip*, which was, roughly speaking, to take out extremal (canonically negative, or with negative intersection number with the canonical divisor) rational curves and put some rational curves back in with a new imbedding type so that new curves are canonically positive. Such a surgery type operation is unique if it exists. The *existence*, however, turned out to be extremely delicate and hard to prove.

As early as in 1981, encouraged by Mori's "not nef" paper, Miles Reid made fairly clear the idea of what minimal models should be and what elementary transformations would be, by publishing a paper on "Minimal models of canonical 3-folds", which was an expansion of his lecture "Canonical 3-folds" in the Journées de Géométrie Algébrique d'Angers 1980. It then became absolutely clear (mildly suggested by Mori's work before) that some special kind of singularities must be permitted in the good notion of minimal models. M. Reid introduced the notions of *canonical singularity* and *terminal singularity*, which turned out to be very useful in the minimal model program of which he was the first to conjecture in literature. (See Miles Reid "Decomposition of Toric morphisms" 1983.) The former was, as the name suggests, the kind of singularity that appeared in the canonical models and could be quite horrible algebraically and geometrically. The latter was the kind that behaved much better than the former and which people hoped to be the only kind to appear in the minimal models. In any case, Reid's definition was simple and clear and some basic theorems were proved by himself, in which he showed that terminal singularities were closely related to the deformations of the classic singularities of Du Val.

Historically, P. Du Val in 1934 systematically studied the singularities that did not affect the condition of adjunction, that is, in the language of Reid, the canonical singularities in dimension 2. In 1966, M. Artin extended and modernized the classification of what he called *rational singularities*. Du Val singularities were exactly rational singularities of multiplicity 2, or *rational double points*. After M. Reid's introduction of canonical and terminal singularities, rational singularities and their deformations were studied once again. Extensive and direct studies on terminal singularities followed M. Reid's works and by 1987 a complete classification of terminal singularities in dimension 3 was achieved with technically useful lemmas by a combination of works by several mathematicians, V. I. Danilov, D. Morrison, G. Stevens, S. Mori, J. Kollár, N. Shepherd-Barron, and others. Some experimental works were done about higher dimensional terminal singularities, such as the one in dimension 4 by S. Mori, D. Morrison and I. Morrison which seemed to indicate far more complexity than the case of dimension 3.

Back to the minimal model problem, Mori's technique of *extremal ray contraction* had to be generalized to singular varieties, at least to the varieties with only terminal singularities, which were even "rationally factorial". It does not look completely hopeless that eventually Mori's all characteristics method can be modified and extended to such singular cases. (See J. Kollár's work that is to appear in the proceedings of the Algebraic Geometry Satellite Conference at Tokyo Metropolitan University, 1990.) However it then was not done in Mori's way. Instead, the generalization was obtained first by generalizing Kodaira's Vanishing Theorem and then by making an ingenious use of this generalization. Here, and subsequently, the contributions of Y. Kawamata were big to the minimal model program.

The celebrated *Vanishing Theorem* was proven by K. Kodaira in 1953. Numerous generalizations, in special cases or in general, were done especially during the 1970's and 80's mostly in an effort to investigate the nicety of the structure of the canonical ring of a projective variety. Here the *canonical ring* means the graded algebra generated by sections of pluricanonical bundles (say, on a desingularized model). As for the nicety, they looked for the property of *being finitely generated* of the canonical algebra or "rationality or better" properties of the singularities of the canonical model. Y. Kawamata had a very clear objective in generalizing and using the Vanishing Theorem, that was to generalize Mori's theory of extremal ray contractions and then verify the minimal model program.

The operation called *flip* is clearly directed, i.e., it changes "canonically negative" to "canonically positive". There is a similar operation for a rational curve with zero intersection number with the canonical divisor, which is called a *flop*. Unlike flips, the inverse of flops are again flops. A flop is symmetric in this sense. A flip changes a variety into another, birational to and better than the original, better in singularities, better in terms of pluricanonical bundles, and so on. Historically, there had been known many examples of flops. The simplest flop, between smooth 3-folds, was known and used even by earlier algebraic geometers. Some flops were shown to be useful in studying degeneration of $K3$ surfaces by V. Kulikov in 1977. In contrast, examples of flips are not so easy to find because 3-folds had to be singular in order to have singularities improved. P. Francia gave an explicit example in his paper published in 1980. Other examples were seen in a paper of M. Reid published in 1983, here with prototypes of minimal models. In 1983, V. V. Shokurov published the *Non-vanishing Theorem*, which was proven by using the Vanishing Theorem. His theorem implied that flips cannot be done infinitely many times.

For Mori's Program, therefore, not only the divisorial contractions are finite but also the flips are finite in any sequence. The very final problem was hence to prove the *existence* of flips for given extremal rational curves, after all the previous works by S. Mori, M. Reid, J. Kollár, V. V. Shokurov and Y. Kawamata, just to name some of the most important contributors to the Program. At any rate, Y. Kawamata reduced the problem of the flip to the existence of a "nice" doubly anticanonical divisor *globally* in a neighborhood of a given extremal rational curve, while the existence had been only proven by M. Reid *locally* about each point of the curve. This seemingly small gap between *global* and *local* was actually enormous. Mori finally overcame this gap by checking cases after cases with very delicate and intricate investigations and established the final existence theorem of algebraic flips by reducing them to sequences of simpler analytic flips and smaller contractions. This monumental paper of Mori was published in the very first issue of the new Journal of the American Mathematical Society, establishing a constructive existence theorem of minimal models which had been shown to have many important consequences.

We need much more of Mori's originality to break the stubborn prejudice that it is a Herculean task to extend the classical classification theorems to all dimensions.

References

1. *Algebraic Manifolds with Ample Tangent Bundles*

[1-1] (with H. Sumihiro) On Hartshorne's conjecture. J. Math. Kyoto Univ. **18** (1978) 223–238

[1-2] Projective manifolds with ample tangent bundles. Ann. Math. **110** (1979) 593–606

[1-3] Hartshorne's conjecture and extremal ray (in Japanese). Sugaku **35**, no. 3 (1983) 193–209 [English translation: Hartshorne conjecture and extremal ray. Sugaku Expositions. Amer. Math. Soc. **1**, no. 1 (1988) 15–37]

2. *Threefolds with Non-nef Canonical Bundles*

[2-1] Threefolds whose canonical bundles are not numerically effective. Proc. Nat. Acad. Sci. USA **77** (1980) 3125–3126

[2-2] Threefolds whose canonical bundles are not numerically effective Ann. Math. **116** (1982) 133–176

[2-3] Threefolds whose canonical bundles are not numerically effective. In: Algebraic Threefolds. Proc. of C. I. M. E. Session, Varenna, 1981 (A. Conte, ed.). Lecture Notes in Mathematics, vol. 947. Springer, Berlin Heidelberg New York 1982, pp. 155–189

3. *Classification of Fano 3-folds*

[3-1] (with S. Mukai) Classification of Fano 3-folds with $B_2 \geq 2$. Manuscripta Math. **36** (1981) 147–162

[3-2] (with S. Mukai) On Fano 3-folds with $B_2 \geq 2$. In: Algebraic Varieties and Analytic Varieties (S. Iitaka, ed.). Advanced Stud. in Pure Math., vol. 1. Kinokuniya and North-Holland, 1983, pp. 101–129

[3-3] Cone of curves and Fano 3-folds. Proc. of ICM'82 in Warsaw 1983, pp. 127–132

[3-4] (with S. Mukai) Classification of Fano 3-folds with $B_2 \geq 2$, I. In: Algebraic and Topological Theories – to the memory of Dr. Takehiko Miyata (M. Nagata et al., eds.). Kinokuniya, Tokyo, 1985, pp. 496–545

4. *Characterization of Uniruledness*

[4-1] (with Y. Miyaoka) A numerical criterion of uniruledness. Ann. Math. **124** (1986) 65–69

5. *Completion of the Minimal Model Program, or Mori's Program*

[5-1] On 3-dimensional terminal singularities. Nagoya Math. J. **98** (1985) 43–66

[5-2] Classification of higher dimensional varieties. In: Algebraic Geometry Bowdoin 1985 (S. J. Bloch, ed.), Proc. Symp. in Pure Math., vol. 46, part 1, 1987, pp. 269–331

[5-3] Flip theorem and the existence of minimal models for 3-folds. J. Amer. Math. Soc. **1** (1988) 117–253

[5-4] (with C. Clemens and J. Kollár) Higher Dimensional Complex Geometry, A Summer Seminar at the Univ. of Utah, Salt Lake City, 1987, Astérisque **166**, Soc. Math. France (1988)

[5-5] (announcement of joint work with J. Kollár) Birational classification of algebraic 3-folds. In: Algebraic Analysis, Geometry, and Number Theory, Proc. of JAMI Inaugural Conf. (J.-I. Igusa, ed.), Supplement to Amer. J. Math., Johns Hopkins Univ. Press, 1989, pp. 307–311

On the Work of E. Witten

Ludwig D. Faddeev

Steklov Mathematical Institute, Leningrad 191011, USSR

It is a duty of the chairman of the Fields Medal Committee to appoint the speakers, who describe the work of the winners at this session. Professor M. Atiyah was asked by me to speak about Witten. He told me that he would not be able to come but was ready to prepare a written address. So it was decided that I shall make an exposition of his address adding my own comments. The full text of Atiyah's address is published separately.

Let me begin by the statement that Witten's award is in the field of Mathematical Physics.

Physics was always a source of stimulus and inspiration for Mathematics so that Mathematical Physics is a legitimate part of Mathematics. In classical time its connection with Pure Mathematics was mostly via Analysis, in particular through Partial Differential Equations. However quantum era gradually brought a new life. Now Algebra, Geometry and Topology, Complex Analysis and Algebraic Geometry enter naturally into Mathematical Physics and get new insights from it. And[1]

In all this large and exciting field, which involves many of the leading physicists and mathematicians in the world, Edward Witten stands out clearly as the most influential and dominating figure. Although he is definitely a physicist (as his list of publications clearly shows) his command of mathematics is rivalled by few mathematicians, and his ability to interpret physical ideas in mathematical form is quite unique. Time and again he has surprised the mathematical community by a brilliant application of physical insight leading to new and deep mathematical theorems.

Now I come to description of the main achievements of Witten. In Atiyah's text many references are given to Feynman Integral, so that I begin with a short and rather schematic reminding of this object.

In quantum physics the exact answers for dynamical problems can be expressed in a formal way as follows:

$$Z = \int e^{iA} \prod_x d\mu$$

where A is an action functional of local fields – functions of time and space variables x, running through some manifold M. The integration measure is a

[1] Small print type here and after refers to Atiyah's text.

Proceedings of the International Congress
of Mathematicians, Kyoto, Japan, 1990

product of local measures for values of fields in a point x over all M. The result of integration Z could be a number or function of parameters defining the problem – coupling constants, boundary or asymptotical conditions, etc.

In spite of being an ill-defined object from the point of view of rigorous mathematics, Feynman functional integral proved to be a powerful tool in quantum physics. It was gradually realized that it is also a very convenient mathematical means. Indeed the geometrical objects such as loops, connections, metrics are natural candidates for local fields and geometry produces for them interesting action functionals. The Feynman integral then leads to important geometrical or topological invariants.

Although this point of view was expressed and exemplified by several people (e.g., A. Schvarz used 1-forms ω on a three dimensional manifold with action $A = \int \omega d\omega$ to describe the Ray-Singer torsion) it was Witten who elaborated this idea to a full extent and showed the flexibility and universality of Feynman integral.

Now let me follow Atiyah in description of the main achievements of Witten in this direction.

1. Morse Theory

His paper [2] on supersymmetry and Morse theory is obligatory reading for geometers interested in understanding modern quantum field theory. It also contains a brilliant proof of the classic Morse inequalities, relating critical points to homology. The main point is that homology is defined via Hodge's harmonic forms and critical points enter via stationary phase approximation to quantum mechanics. Witten explains that "supersymmetric quantum mechanis" is just Hodge-de Rham theory. The real aim of the paper is however to prepare the ground for supersymmetric quantum field theory as the Hodge-de Rham theory of infinite-dimensional manifolds. It is a measure of Witten's mastery of the field that he has been able to make intelligent and skilful use of this difficult point of view in much of his subsequent work.

Even the purely classical part of this paper has been very influential and has led to new results in parallel fields, such as complex analysis and number theory.

2. Index Theorem

One of Witten's best known ideas is that the index theorem for the Dirac operator on compact manifolds should emerge by a formally exact functional integral on the loop space. This idea (very much in the spirit of his Morse theory paper) stimulated an extensive development by Alvarez-Gaumé, Getzler, Bismut and others which amply justified Witten's view-point.

3. Rigidity Theorems

Witten [7] produced an infinite sequence of such equations which arise naturally in the physics of string theories, for which the Feynman path integral provides a heuristic explanation of rigidity. As usual Witten's work, which was very precise and detailed in its formal aspects, stimulated great activity in this area, culminating in rigorous proofs of these new rigidity theorems by Bott and Taubes [1]. A noteworthy aspect of these proofs is that they involve elliptic function theory and deal with the infinite sequence of operators simultaneously rather than term by term. This is entirely natural from Witten's view-point, based on the Feynman integral.

4. Knots

Witten has shown that the Jones invariants of knots can be interpreted as Feynman integrals for a 3-dimensional gauge theory [11]. As Lagrangian, Witten uses the Chern-Simons function, which is well-known in this subject but had previously been used as an addition to the standard Yang-Mills Lagrangian. Witten's theory is a major breakthrough, since it is the only intrinsically 3-dimensional interpretation of the Jones invariants: all previous definitions employ a presentation of a knot by a plane diagram or by a braid.

Although the Feynman integral is at present only a heuristic tool it does lead, in this case, to a rigorous development from the Hamiltonian point of view. Moreover, Witten's approach immediately shows how to extend the Jones theory from knots in the 3-sphere to knots in arbitrary 3-manifolds. This generalization (which includes as a specially interesting case the empty knot) had previously eluded all other efforts, and Witten's formulas have now been taken as a basis for a rigorous algorithmic definition, on general 3-manifolds, by Reshetikin and Turaev.

Now I turn to another beautiful result of Witten – proof of positivity of energy in Einstein's Theory of Gravitation.

Hamiltonian approach to this theory proposed by Dirac in the beginning of the fifties and developed further by many people has led to a natural definition of energy. In this approach a metric γ and external curvature h on a space-like initial surface $S^{(3)}$ embedded in space-time $M^{(4)}$ are used as parameters in the corresponding phase space. These data are not independent. They satisfy Gauss-Codazzi constraints – highly nonlinear PDE. The energy H in the asymptotically flat case is given as an integral of indefinite quadratic form of $\nabla\gamma$ and h. Thus it is not manifestly positive. The important statement that it is nevertheless positive may be proved only by taking into account the constraints – a formidable problem solved by Yau and Schoen in the late seventies and as Atiyah mentions, "leading in part to Yau's Fields Medal at the Warsaw Congress".

Witten proposed an alternative expression for energy in terms of solution of a linear PDE with the coefficients expressed through γ and h. This equation is

$$\mathscr{D}^{(3)}\psi = 0$$

where $\mathscr{D}^{(3)}$ is the Dirac operator induced on $S^{(3)}$ by the full Dirac operator on $M^{(4)}$. Witten's formula somewhat schematically can be written as follows:

$$H(\psi_0, \psi_0) = \int (|\nabla\psi|^2 + \psi^* G\psi) dS$$

where ψ_0 is the asymptotic boundary value for ψ and G is proportional to the Einstein tensor $R_{ik} - \frac{1}{2}g_{ik}R$. Due to the equation of motion $G = T$, where T is the energy-momentum tensor of matter and thus manifestly positive. So the positivity of H follows.

This unexpected and simple proof shows another ability of Witten – to solve a concrete difficult problem by specific elegant means.

On the Work of Edward Witten

Michael Atiyah

Trinity College, Cambridge CB2 1TQ, England

1. General

The past decade has seen a remarkable renaissance in the interaction between mathematics and physics. This has been mainly due to the increasingly sophisticated mathematical models employed by elementary particle physicists, and the consequent need to use the appropriate mathematical machinery. In particular, because of the strongly non-linear nature of the theories involved, topological ideas and methods have played a prominent part.

The mathematical community has benefited from this interaction in two ways. First, and more conventionally, mathematicians have been spurred into learning some of the relevant physics and collaborating with colleagues in theoretical physics. Second, and more surprisingly, many of the ideas emanating from physics have led to significant new insights in purely mathematical problems, and remarkable discoveries have been made in consequence. The main input from physics has come from quantum field theory. While the analytical foundations of quantum field theory have been intensively studied by mathematicians for many years the new stimulus has involved the more formal (algebraic, geometric, topological) aspects.

In all this large and exciting field, which involves many of the leading physicists and mathematicians in the world, Edward Witten stands out clearly as the most influential and dominating figure. Although he is definitely a physicist (as his list of publications clearly shows) his command of mathematics is rivalled by few mathematicians, and his ability to interpret physical ideas in mathematical form is quite unique. Time and again he has surprised the mathematical community by a brilliant application of physical insight leading to new and deep mathematical theorems.

Witten's output is remarkable both for its quantity and quality. His list of over 120 publications indicates the scope of his research and it should be noted that many of these papers are substantial works indeed.

In what follows I shall ignore the bulk of his publications, which deal with specifically physical topics. This will give a very one-sided view of his contribution, but it is the side which is relevant for the Fields Medal. Witten's standing as a physicist is for others to assess.

Proceedings of the International Congress
of Mathematicians, Kyoto, Japan, 1990

Let me begin by trying to describe some of Witten's more influential ideas and papers before moving on to describe three specific mathematical achievements.

2. Influential Papers

His paper [2] on supersymmetry and Morse theory is obligatory reading for geometers interested in understanding modern quantum field theory. It also contains a brilliant proof of the classic Morse inequalities, relating critical points to homology. The main point is that homology is defined via Hodge's harmonic forms and critical points enter via stationary phase approximation to quantum mechanics. Witten explains that "supersymmetric quantum mechanics" is just Hodge-de Rham theory. The real aim of the paper is however to prepare the ground for supersymmetric quantum field theory as the Hodge-de Rham theory of infinite-dimensional manifolds. It is a measure of Witten's mastery of the field that he has been able to make intelligent and skilful use of this difficult point of view in much of his subsequent work.

Even the purely classical part of this paper has been very influential and has led to new results in parallel fields, such as complex analysis and number theory.

Many of Witten's papers deal with the topic of "Anomalies". This refers to classical symmetries or conservation laws which are violated at the quantum level. Their investigation is of fundamental importance for physical models and the mathematical aspects are also extremely interesting. The topic has been extensively written about (mainly by physicists) but Witten's contributions have been deep and incisive. For example, he pointed out and investigated "global" anomalies [3], which cannot be studied in the traditional perturbative manner. He also made the important observation that the η-invariant of Dirac operators (introduced by Atiyah, Patodi and Singer) is related to the adiabatic limit of a certain anomaly [4]. This was subsequently given a rigorous proof by Bismut and Freed.

One of Witten's best known ideas is that the index theorem for the Dirac operator on compact manifolds should emerge by a formally exact functional integral on the loop space. This idea (very much in the spirit of his Morse theory paper) stimulated an extensive development by Alvarez-Gaumé, Getzler, Bismut and others which amply justified Witten's view-point.

Also concerned with the Direc operator is a beautiful joint paper with Vafa [5] which is remarkable for the fact that it produces sharp uniform bounds for eigenvalues by an essentially topological argument. For the Dirac operator on an odd-dimensional compact manifold, coupled to a background gauge potential, Witten and Vafa prove that there is a constant C (depending on the metric, but independent of the potential) such that *every interval of length C contains an eigenvalue.* This is not true for Laplace operators or in even dimensions, and is a very refined and unusual result.

3. The Positive Mass Conjecture

In General Relativity the positive mass conjecture asserts that (under appropriate hypotheses) the total energy of a gravitating system is positive and can only be zero for flat Minkowski space. It implies that Minkowski space is a stable ground state. The conjecture has attracted much attention over the years and was established in various special cases before being finally proved by Schoen and Yau in 1979. The proof involved non-linear P. D. E. through the use of minimal surfaces and was a major achievement (leading in part to Yau's Fields Medal at the Warsaw Congress). It was therefore a considerable surprise when Witten outlined in [6] a much simpler proof of the positive mass conjecture based on linear P. D. E. Specifically Witten introduced spinors and studied the Dirac operator. His approach had its origin in some earlier ideas of supergravity and it is typical of Witten's insight and technical skill that he eventually emerged with a simple and quite classical proof. Witten's paper stimulated both mathematicians and physicists in various directions, demonstrating the fruitfulness of his ideas.

4. Rigidity Theorems

The space of solutions of an elliptic differential equation on a compact manifold is naturally acted on by any group of symmetries of the equation. All representations of compact connected Lie groups occur this way. However, for very special equations, these representations are trivial. Notably this happens for the spaces of harmonic forms, since these represent cohomology (which is homotopy invariant). A less obvious case arises from harmonic spinors (solutions of the Dirac equation), although the relevant space here is the "index" (virtual difference of solutions of D and D^*). This was proved by Atiyah and Hirzebruch in 1970. Witten raised the question whether such "rigidity theorems" might be true for other equations of interest in mathematical physics, notably the Rarita-Schwinger equation. This stimulated Landweber and Stong to investigate the question topologically and eventually Witten [7] produced an infinite sequence of such equations which arise naturally in the physics of string theories, for which the Feynman path integral provides a heuristic explanation of rigidity. As usual Witten's work, which was very precise and detailed in its formal aspects, stimulated great activity in this area, culminating in rigorous proofs of these new rigidity theorems by Bott and Taubes [1]. A noteworthy aspect of these proofs is that they involve elliptic function theory and deal with the infinite sequence of operators simultaneously rather than term by term. This is entirely natural from Witten's view-point, based on the Feynman integral.

5. Topological Quantum Field Theories

One of the remarkable aspects of the Geometry/Physics interaction of recent years has been the impact of quantum field theory on low-dimensional geometry (of 2, 3 and 4 dimensions). Witten has systematized this whole area by showing that there are, in these dimensions, interesting *topological* quantum field theories [8], [9], [10]. These theories have all the formal structure of quantum field theories but they are purely topological and have no dynamics (i.e. the Hamiltonian is zero). Typically the Hilbert spaces are finite-dimensional and various traces give well-defined invariants. For example, the Donaldson theory in 4 dimensions fits into this framework, showing how rich such structures can be.

A more recent example, and in some ways a more surprising one, is the theory of Vaughan Jones related to knot invariants, which has just been reported on by Joan Birman. Witten has shown that the Jones invariants of knots can be interpreted as Feynman integrals for a 3-dimensional gauge theory [11]. As Lagrangian, Witten uses the Chern-Simons function, which is well-known in this subject but had previously been used as an addition to the standard Yang-Mills Lagrangian. Witten's theory is a major breakthrough, since it is the only intrinsically 3-dimensional interpretation of the Jones invariants: all previous definitions employ a presentation of a knot by a plane diagram or by a braid.

Although the Feynman integral is at present only a heuristic tool it does lead, in this case, to a rigorous development from the Hamiltonian point of view. Moreover, Witten's approach immediately shows how to extend the Jones theory from knots in the 3-sphere to knots in arbitrary 3-manifolds. This generalization (which includes as a specially interesting case the empty knot) had previously eluded all other efforts, and Witten's formulas have now been taken as a basis for a rigorous algorithmic definition, on general 3-manifolds, by Reshetikin and Turaev.

Moreover, Witten's approach is extremely powerful and flexible, suggesting a number of important generalizations of the theory which are currently being studied and may prove to be important.

One of the most exciting recent developments in theoretical physics in the past year has been the theory of 2-dimensional quantum gravity. Remarkably this theory appears to have close relations with the topological quantum field theories that have been developed by Witten [12]. Detailed reports on these recent ideas will probably be presented by various speakers at this congress.

6. Conclusion

From this very brief summary of Witten's achievements it should be clear that he has made a profound impact on contemporary mathematics. In his hands physics is once again providing a rich source of inspiration and insight in mathematics. Of course physical insight does not always lead to immediately rigorous mathematical proofs but it frequently leads one in the right direction, and technically correct proofs can then hopefully be found. This is the case with Witten's work. So far his insight has never let him down and rigorous proofs, of the standard we mathematicians rightly expect, have always been forthcoming. There is therefore no doubt that contributions to mathematics of this order are fully worthy of a Fields Medal.

References

1. R. Bott and C. H. Taubes: On the rigidity theorems of Witten. J. Amer. Math. Soc. **2** (1989) 137
2. E. Witten: Supersymmetry and Morse theory. J. Diff. Geom. **17** (1984) 661
3. E. Witten: An $SU(2)$ anomaly. Phys. Lett. **117 B** (1982) 324
4. E. Witten: Global anomalies in string theory. Proc. Argonne-Chicago Symposium on Geometry, Anomalies and Topology (1985)
5. E. Witten and C. Vafa: Eigenvalue inequalities for Fermions in gauge theories. Comm. Math. Phys. **95** (1984) 257
6. E. Witten: A new proof of the positive energy theorem. Comm. Math. Phys. **80** (1981) 381
7. E. Witten: Elliptic genera and quantum field theory. Comm. Math. Phys. **109** (1987) 525
8. E. Witten: Topological quantum field theory. Comm. Math. Phys. **117** (1988) 353
9. E. Witten: Topological gravity. Phys. Lett. B **206** (1988) 601
10. E. Witten: Topological sigma models. Comm. Math. Phys. **118** (1988) 411
11. E. Witten: Quantum field theory and the Jones polynomial. Comm. Math. Phys. **121** (1989) 351
12. E. Witten: On the structure of the topological phase of two dimensional gravity. Nuclear Phys. B **340** (1990) 281

The Work of A. A. Razborov

László Lovász

Department of Computer Science, Eötvös University, H-1088 Budapest, Hungary
and
Department of Computer Science, Princeton University, Princeton, NJ 08544, USA

Perhaps the most difficult and deepest field in computer science is the derivation of lower bounds for the computational complexity of various problems (i.e., proving that no algorithm whatsoever can solve the given problem within a certain time or space). While these questions are very natural extensions of classical problems on undecidability, they turn out much more difficult. Up to some years ago, very few results were known, and these gave very poor bounds and concerned very restricted models of computation. At the same time, some of the most exciting unsolved problems in theoretical computer science (like $\mathscr{P} = \mathscr{N}\mathscr{P}$) are in this area; to prove those, one would need lower bounds on the running time which is exponential in the size of the input.

One conclusion drawn from the situation was that the methods of mathematical logic used in decidability theory are not powerful enough to yield negative results in complexity theory, and that combinatorial methods would be needed to obtain such results.

The idea of using combinatorial methods is suggested by the *Boolean circuit* model of computation. In this model, the computation is described by an acyclic directed graph, whose nodes (also called *gates*) correspond to elementary steps in the computation. Nodes with no edge entering them are called *input gates*, and the input of the computation is an assignment of a Boolean value (0 or 1) to each input gate. Every other gate computes some simple (Boolean) function of the values computed by the the tails of edges entering it; this recursively defines a value of each gate. For simplicity, we shall assume that there is only one gate with no edge leaving; this serves as the *output gate.*

Every algorithm in other possible models (Turing machine, RAM machine etc.) can be transformed into a Boolean circuit (more exactly, into a series of Boolean circuits, one for each input size). The number of gates in the circuit corresponds to the running time of the algorithm. Another important parameter of the circuit is its *depth*, i.e., the maximum length of a path from an input node to an output node. This corresponds to the running time of the algorithm if parallel processing is allowed; in this case, the *width* (the maximum number of gates mutually inaccessible by paths) corresponds to the number of processors needed for this degree of parallelization.

The two simplest kinds of gates are AND gates and OR gates, computing the logical conjunction and disjunction of Boolean values, respectively. A Boolean circuit consisting of such gates can compute only monotone functions. If we

Proceedings of the International Congress
of Mathematicians, Kyoto, Japan, 1990

allow, however, that each input node receives either a variable or its negation, then every Boolean function can be computed using only AND and OR gates.

The famous $\mathscr{P} \neq \mathscr{NP}$ problem would follow if one could show the following. For every $n \geq 2$, let D_n be a Boolean circuit with AND and OR gates, whose input gates correspond to pairs (i, j) with $1 \leq i < j \leq n$. Every input to such a circuit, i.e., every assignment of values 0 or 1 to the input gates, corresponds to a graph on $\{1, \ldots, n\}$. Assume that D_n computes the value 1 if and only if this graph has a clique larger than $\sqrt{n}$ (or has a Hamiltonian circuit, or is 3-colorable; we could use here any $\mathscr{NP}$-complete property). Conjecture: *the number of gates in D_n is necessarily superpolynomial (perhaps exponential) in n.*

While we are still far from being able to prove this, in the last few years combinatorial methods for proving good (exponential) lower bounds on the circuit complexity of various Boolean functions have been developed. These methods still impose restrictions on the computation model, but these restrictions are quite natural and much weaker than those considered previously. One class of results is for bounded-depth (or, more generally, small-depth) circuits. This development is the result of contributions from many authors (Furst–Saxe–Sipser 1981, Ajtai (1983), Yao (1985), Hastad (1986), Razborov (1987)). The other (mathematically perhaps even more difficult) class of results concerns monotone circuits; this restriction of the model was introduced by Valiant, and the breakthrough is due to Razborov (1985a, 1986). We start with describing this fundamental result.

While every monotone Boolean function can be computed by a monotone Boolean circuit, it can be expected that monotone circuits are less powerful than general circuits. So (as suggested by Valiant), exponential (or at least superpolynomial) lower bounds on the monotone circuit complexity of monotone problems in $\mathscr{NP}$ might be easier to prove, and at the same time they may eventually lead to exponential lower bounds for general circuits.

Razborov (1985a) justified part of this expectation by proving the following fundamental result.

Theorem. *For deciding the existence of a clique with given size in a graph, no polynomial size monotone circuits exist.*

This basic result and its powerful proof method have immediately inspired much further work. Andreev (1985) used similar techniques to obtain an exponential lower bound on a less natural $\mathscr{NP}$-complete problem. Alon and Boppana (1987), by strengthening the combinatorial arguments of Razborov, proved an exponential (not merely superpolynomial) lower bound on the monotone circuit complexity of the clique problem.

Razborov (1985b) proved that even the bipartite matching problem (decide whether a bipartite graph has a perfect matching, which is a special case of the clique problem) needs superpolynomial-size monotone circuits. This problem is solvable in polynomial time if we do not require monotonicity of the circuit; therefore this result separates monotone and non-monotone polynomial time. Such separation results are extremely rare. Tardos [T] used Razborov's methods to prove the existence of a polynomial time computable monotone Boolean function which takes an exponential number of gates if we want to compute it by a monotone Boolean circuit.

The proof technique of Razborov is also very important: he introduces a very powerful new method, namely approximation. The method applies in many situations, but to fill in the details, difficult special considerations are needed. We shall sketch the method below; it is, however, best described through another of its applications to the complexity of bounded-depth circuits.

Every Boolean function can be computed by a circuit with depth 2 (this follows from the existence of the "conjunctive normal form"), but circuits with small depth will typically have exponential size even for quite simple Boolean functions. The parity function (or XOR function) of n bits is defined as the sum of these bits modulo 2. It was proved by Hastad (1986) that *a circuit with* AND *and* OR *gates, with n input bits and depth d, computing the parity function of these bits, has at least* $\exp\left(0.1n^{1/(d-1)}\right)$ *gates.*

Razborov (1985b) showed that the approximation method can also be applied to prove lower bounds on the size of non-monotone but small-depth circuits. Among the really simple functions of n bits, besides AND, OR and XOR, the majority function comes to mind: the value of this function is 1 if at least half of the input bits are 1, and 0 otherwise. Razborov proved an exponential lower bound for the size of small-depth circuits computing the majority function. An interesting feature of the result is that he can even allow XOR gates, computing the mod 2 sum of arbitrarily many bits in a single step. (This shows the complementary nature of this result and Hastad's.) To be exact, Razborov's important result is the following:

Theorem. *A circuit with* AND, OR *and* XOR *gates, with n input bits and depth d, computing the majority function of these bits, has at least* $\exp\left(0.1n^{(1/2d)}\right)$ *gates.*

This application of the approximation method is in fact much simpler, and it is a "textbook" example of a lower-bound proof combining algebraic, probabilistic, and combinatorial ideas.

To sketch the general idea, assume that we want to prove an exponential lower bound on the number of gates in a Boolean circuit computing some Boolean function f. One introduces a notion of "distance" between Boolean functions and specifies a class of "simple" Boolean functions so that the distance of the function f from any "simple" function is exponentially large. Furthermore, one defines gates which "approximate" the original gates in the sense that if we replace an original gate by the approximating gate, the new circuit computes a Boolean function which is at a distance at most 1 from the original. These approximating gates must have the property that if all the original gates are replaced by their approximations, the new circuit will compute a "simple" function. So this new function is exponentially far from the original function f, and hence, we must have changed exponentially many gates. In particular, the circuit must have exponentially many gates.

In the proof of the lower bound for small-depth circuits, "simple" functions will be those expressible as polynomials over GF(2) with small degree. We measure the distance of two functions in terms of the number of inputs on which they differ. An elegant argument shows that the majority function is "far" from every polynomial with low degree. Approximations are only needed for the AND gates; this is achieved by a nice randomized construction.

For proving his lower bound for the monotone complexity of the k-clique problem, Razborov has to use much more involved constructions, and the proofs are based on difficult combinatorial methods, in particular on the powerful method of "sunflowers" from hypergraph theory.

In a recent paper (1989), he analyses the approximation method and in a rather novel manner, is able to establish its boundaries. (This to me shows that this is indeed a "method" and not just a language to present various arguments.) He shows among others that for Boolean circuits without any restriction, this method can yield at most quadratic lower bounds (while, as we have seen, both for monotone and bounded-depth circuits it yields exponential lower bounds).

Razborov's work (of which here we discussed only one cluster of results) was received with great interest and discussed enthusiastically at seminars all around the world. In an area where any step forward seemed almost hopeless (but which was at the same time a central area of theoretical computer science) his results meant that deep methods could be developed and to obtain strong lower bounds for algorithms was not impossible.

References

Ajtai, M.: Σ_1^1-formulae on finite structure. Ann. Pure Appl. Logic **24** (1983) 1–48

Alon, N., Boppana, R.B.: The monotone circuit complexity of Boolean functions. Combinatorica **7** (1987) 1–22

Andreev, A.E.: On a method for obtaining lower bounds for the complexity of individual monotone functions. Dokl. Akad. Nauk SSSR **282** (1985) 1033–1037 (Russian) [English transl.: Soviet Math. Dokl. **31**, 530–534]

Furst, M., Saxe, J., Sipser, M.: Parity, circuits, and the polynomial time hierarchy. Proc. 22nd IEEE Symp. on Found. of Comp. Sci., 1981, pp. 260–270

Hastad, J.: Almost optimal lower bounds for small depth circuits. Proc. 18th ACM Symp. on Theory of Comp., 1986, pp. 6–20

Razborov, A.A.: Lower bounds on the monotone circuit complexity of some Boolean functions. Dokl. Akad. Nauk SSSR **281** (1985a) 798–801 (Russian) [English transl.: Soviet Math. Dokl. **31**, 354–357]

Razborov, A.A.: A lower bound on the monotone network complexity of the logical permanent. Math. Zametki **37**, (1985b) 887–900 (Russian) [English transl.: Math. Notes of the Acad. Sci. USSR **37**(6) 485–493]

Razborov, A.A.: Lower bounds for the monotone complexity of Boolean functions. Proc. of the Internatl. Congress of Math., Berkeley, Calif., 1986, pp. 1478–1487

Razborov, A.A.: Lower bounds on the size of bounded depth network over a complete basis with logical addition. Math. Zametki **41**, (1987) 598–607 (Russian) [English transl.: Math. Notes of the Acad. Sci. USSR **41**(4) 333–338]

Razborov, A.A.: On the method of approximations. Proc. 21st Annual ACM Symp. on Theory of Computing, 1989, pp. 167–176

Tardos, É.: The gap between monotone and non-monotone circuit complexity is exponential. Combinatorica **8** (1988) 141–142

Yao, A.C.-C.: Separating the polynomial-time hierarchy by oracles. Proc. 26th IEEE Symp. on Found. of Comp. Sci., 1985, pp. 1–10

Invited One-Hour Addresses at the Plenary Sessions

Algebraic K-Theory, Motives, and Algebraic Cycles

Spencer Bloch

Department of Mathematics, University of Chicago, Chicago, IL 60637, USA

The concept of motive was introduced by A. Grothendieck. Roughly, one wants to think of smooth projective varieties over a field k as objects in an additive category $\mathscr{V}(k)$ with morphisms $X \to Y$ being correspondences (i.e., suitable equivalence classes of algebraic cycles on $X \times Y$.) The category of motives, $\mathscr{M}(k)$, is a $\mathbb{Q}$-linear abelian (or weaker, Karoubian) category equipped with a tensor product, an invertible Tate twist functor $? \otimes \mathbb{Q}(1) : \mathscr{M}(k) \to \mathscr{M}(k)$, and a cohomology functor $H^* : \mathscr{V}(k) \to \mathscr{M}(k)$ which is universal for cohomology functors like Betti, Hodge, or étale cohomology. Standard conjectures about algebraic cycles imply $\mathscr{M}(\bar{k})$ is semi-simple, but these conjectures have proved quite intractable.

Inspired by work of Deligne on mixed Hodge structures, Beilinson started to study a more general abelian category of mixed motives $\mathscr{M}\mathscr{M}(k)$. The smallest full abelian subcategory of $\mathscr{M}\mathscr{M}(k)$ containing the Tate motives $\mathbb{Q}(n)$, $n \in \mathbb{Z}$, and closed under extensions, would be the category $\mathscr{M}\mathscr{T}\mathscr{M}(k)$ of mixed Tate motives. Objects in this category would have a canonical weight filtration with $\mathbb{Q}(n)$ of weight $-2n$, and the functor associating to an object M the corresponding $\bigoplus(\mathrm{gr}^W_{-2n}M)(-n)$ should be a fibre functor making $\mathscr{M}\mathscr{T}\mathscr{M}(k)$ a neutral Tannakian category. In other words, automorphisms of the fibre functor compatible with tensor product form an algebraic group $\mathrm{Gal}(\mathscr{M}\mathscr{T}\mathscr{M}(k))$ which, because it must preserve the weight filtration, is of the form $\mathscr{N} \rtimes \mathbb{G}_m$ with $\mathscr{N}$ nilpotent. The category $\mathscr{M}\mathscr{T}\mathscr{M}(k)$ is equivalent to the category of representations over $\mathbb{Q}$ of $\mathscr{N} \times \mathbb{G}_m$. The graded Lie algebra

$$\mathscr{L}(k) \overset{\text{def}}{=} \mathrm{Lie}(\mathscr{N}),\ \mathscr{L}(k) = \mathscr{L}_{-1} \oplus \mathscr{L}_{-2} \oplus \cdots \tag{0.1}$$

is a fundamental invariant of k. Beilinson further stipulated that $\mathscr{M}\mathscr{T}\mathscr{M}(k)$ should be related to the Quillen K-theory of the field k via the basic formula

$$\mathrm{Ext}^i_{\mathscr{M}\mathscr{T}\mathscr{M}(k)}(\mathbb{Q}, \mathbb{Q}(n)) \cong \mathrm{gr}^n_\gamma K_{2n-i}(k) \otimes \mathbb{Q} \tag{0.2}$$

where the grading is with respect to the γ-filtration on $K_*(k)$.

It seems likely this expected relation between motives and K-theory has a natural interpretation in terms of algebraic cycles. Although technical problems involving moving lemmas currently block any definitive results, I want to show that cycles can be used to define reasonable candidates for motivic cohomology and also for the Lie algebra associated to $\mathscr{M}\mathscr{T}\mathscr{M}(k)$. The basic cycle-theoretic objects, the higher Chow groups $\mathrm{CH}^r(X, n)$ associated to a variety X and non-negative integers

Proceedings of the International Congress
of Mathematicians, Kyoto, Japan, 1990

n and r, are integrally defined (i.e. not $\mathbb{Q}$-vector spaces). In § 3 I exploit this to show how, in an arithmetic situation, these groups give natural generalizations of the global points of an abelian variety over a number field and of the corresponding group Ш. One is led to a general Birch-Swinnerton-Dyer conjecture for motives. This conjecture has been proved by K. Kato [BK] for the case of the Riemann zeta function.

Everything I know about mixed motives I learned from Beilinson and Deligne. I am grateful to M. Hanamura for many helpful conversations about these ideas. A. Suslin pointed out an error in the proof of the moving lemma in [Bl1]. I have another proof, but it is too complicated to inspire confidence at this point, so I prefer to formulate the relation between K-theory and higher Chow groups conditionally. Results about a spectral sequence relating higher Chow groups and K-theory are joint work with S. Lichtenbaum. Finally, the arithmetic conjectures in the last section were inspired by joint work with K. Kato on Tamagawa numbers of motives [BK]. Over the years, he has graciously shared credit with me for any number of his ideas, including a number of those discussed in § 3.

1. Higher Chow Groups

Let X be a variety over a field k. A cosimplicial scheme is built from affine spaces $\Delta_n = \mathrm{Spec}(k[t_0, \ldots, t_n]/(\sum t_i - 1))$ with face maps $\Delta_{n-1} \hookrightarrow \Delta_n$ defined by $t_i = 0$ and degeneracies $t_i \mapsto t_i + t_{i+1}$. For any $r \geq 0$ and any algebraic k-scheme X, a simplicial abelian group $\mathscr{Z}^r(X, \cdot)$ is defined by taking $\mathscr{Z}^r(X, n)$ to be the free abelian group generated by irreducible subvarieties $V \subset X \times \Delta_n$ of codimension r meeting all faces of codimension ≥ 1 properly. The boundary maps $\partial_i : \mathscr{Z}^r(X, n) \to \mathscr{Z}^r(X, n-1)$ are given by restriction to the i-th face. One defines the higher Chow groups

$$\mathrm{CH}^r(X, n) \stackrel{\mathrm{def}}{=} \pi_n(\mathscr{Z}^r(X, \cdot)) \cong H_n(\mathscr{Z}^r(X, \cdot)). \tag{1.1}$$

In particular, one checks easily $\mathrm{CH}^r(X, 0)$ is the classical Chow group, and $\mathrm{CH}^1(X, 1) \cong \Gamma(X, \mathscr{O}_X^*)$ for X regular.

Conjecture (1.2) (Moving Lemma for Cycles). *Let $U \subset X$ be Zariski open. Write $\mathscr{Z}^r(U, \cdot)' = \mathrm{Image}[\mathscr{Z}^r(X, \cdot) \to \mathscr{Z}^r(U, \cdot)]$. Then for all r, X, and U, the inclusion $\mathscr{Z}^r(U, \cdot)' \hookrightarrow \mathscr{Z}^r(U, \cdot)$ is a quasi-isomorphism.*

Conjecture (1.2) would immediately yield a long exact localization sequence for higher Chow groups. With more work, one gets:

Theorem (1.3) [Bl1]. *Assume the moving lemma for cycles holds. Then there are isomorphisms*

$$\mathrm{CH}^r(X, n) \otimes \mathbb{Q} \cong \mathrm{gr}^r_\gamma K'_n(X) \otimes \mathbb{Q}, \tag{1.3.1}$$

where K' is the K-theory of coherent sheaves, and gr_γ is the γ-grading.

Let T be a smooth k-variety. Let $D_1, \ldots, D_m$ be smooth divisors on T, and assume all intersections of the D_i are transversal. The multi-relative K-theory $K_*(T; D_1, \ldots, D_m)$ is defined via a homotopy fibre construction. One has long exact sequences

$$K_*(T; D_1, \ldots, D_m) \to K_*(T; D_1, \ldots, D_{m-1}) \to K_*(D_m; D_1 \cap D_m, \ldots, D_{m-1} \cap D_m)$$
$$\to K_{*-1}(T; D_1, \ldots, D_m) \to \cdots$$
$$\to K_0(D_m; D_1 \cap D_m, \ldots, D_{m-1} \cap D_m).$$

For $W \subset T$ closed, one defines $K_{*W}(T; D_1, \ldots, D_m)$ again by a homotopy fibre so there are long exact sequences

$$K_{*W}(T; D_1, \ldots, D_m) \to K_*(T; D_1, \ldots, D_m) \to K_*(T - W; D_1, \ldots, D_m)$$
$$\to K_{*-1W}(T; D_1, \ldots, D_m) \to \cdots \to K_0(T - W; D_1, \ldots, D_m).$$

Conjecture (1.4) (Moving Lemma for Relative K_0). *Let X be a smooth k-variety. Write $\delta_j = X \times \Delta_{n-1} \hookrightarrow X \times \Delta_n$ for the codimension 1 faces. Let $\mathscr{V}_i \subset X \times \Delta_n$ be the union of all algebraic subsets of dimension i meeting all faces $\bigcap \delta_j$ of codimension ≥ 1 properly. Let $W \subset \mathscr{V}_i$ be an algebraic set of dimension $\leq i - 1$. Then the map*

$$K_{0, W - \mathscr{V}_{i-1} \cap W}(X \times \Delta_n - \mathscr{V}_{i-1}; \delta_0, \ldots, \delta_n) \to K_{0, \mathscr{V}_i - \mathscr{V}_{i-1}}(X \times \Delta_n - \mathscr{V}_{i-1}; \delta_0, \ldots, \delta_n)$$

is zero.

Intuitively, this conjecture says a class in K_0 supported on $W - W \cap \mathscr{V}_{i-1}$ can be moved on $\mathscr{V}_i$ to a class supported on $\mathscr{V}_{i-1}$.

Theorem 1.5 (with S. Lichtenbaum). *Let X be a smooth k-variety, and assume the moving lemma for relative K_0 holds for X and all $n, i \geq 0$. Then there is a spectral sequence*

$$E_2^{p,q} = \mathrm{CH}^{-q}(X, -p - q) \Rightarrow K_{-p-q}(X). \tag{1.5.1}$$

This should be compared with the Atiyah-Hirzebruch spectral sequence in topological K-theory. Unlike the localization spectral sequence in algebraic K-theory, (1.5.1) may be non-trivial for $X = \mathrm{Spec}(k)$. Viewing $\mathscr{Z}^r(X, \cdot)$ as a complex in negative degrees, one may conjecture the complex

$$\Gamma(r)^{\cdot} \stackrel{\text{def}}{=} \mathscr{Z}^r(X, *)[-2r] \tag{1.6}$$

satisfies the axioms formulated by Beilinson and Lichtenbaum. The spectral sequence can be rewritten

$$E_2^{p,-q} = H^{p+q}(\Gamma(q)) \Rightarrow K_{q-p}(X). \tag{1.7}$$

2. Motives

The philosophy is something like this. Associated to a smooth variety X there should be a Tannakian category $\mathscr{MTM}(X)$ of mixed Tate motives, with a fibre functor $M \mapsto H^0(\bigoplus \mathrm{gr}_{2n}^W R_{\mathrm{Betti}}(M)(n))$, where $R_{\mathrm{Betti}}(M)$ is a Tate variation of Hodge structure on X, the Betti realization of M. There should be a space $\mathrm{Mot}(X)$ with $\pi_1(\mathrm{Mot}(X))^\wedge \cong \mathrm{Gal}(\mathscr{MTM}(X))$ (where "$\wedge$" means nilpotent completion) so mixed

Tate motives (e.g. the $\mathbb{Q}(n)$) can be thought of as $\mathbb{Q}$-local systems on $\mathrm{Mot}(X)$. Indeed, one might hope for natural $\mathbb{Z}$-local systems $\mathbb{Z}(n)$. Moreover one expects isomorphisms in the derived category

$$\Gamma(r)^{\cdot} \cong R\Gamma(\mathrm{Mot}(X), \mathbb{Z}(r)), \tag{2.1}$$

and hence

$$\mathrm{CH}^r(X, p) \cong H^{2r-p}(\mathrm{Mot}(X), \mathbb{Z}(r)). \tag{2.2}$$

Let $\mathrm{Grass}_{X,n+p,n}$ denote the Grassmannian of rank n quotients of $\mathcal{O}_X^{\oplus n+p}$. By analogy with the work of Dwyer and Friedlander [DF], one might hope to identify the K-theory $K_*(X)$ with the homotopy of the space of sections

$$\pi_*\left[\varinjlim_n \varinjlim_p \mathrm{Hom}(\mathrm{Mot}(X), \mathrm{Mot}(\mathrm{Grass}_{X,n+p,n}))\right].$$

In [op. cit.] the l-adic étale homotopy groups of the fibre are calculated:

$$\pi_i\left[\varinjlim \varinjlim \mathrm{fibre}(\mathrm{Grass}_{X,n+p,n,\text{ét}} \to X_{\text{ét}})\right] \simeq \begin{cases} 0 & i = 2j+1 \\ \mathbb{Z}_l(j) & i = 2j. \end{cases} \tag{2.3}$$

The homotopy groups of the fibre $\varinjlim \varinjlim \mathrm{Mot}(\mathrm{Grass}_{X,n+p,n}) \to \mathrm{Mot}(X)$, viewed as local systems on the base should, by the same token, be 0 in odd degrees and $\mathbb{Z}(j)$ in degree $2j$. The usual spectral sequence in topology converging to the homotopy groups of the space of sections of a fibration

$$E_2^{p,-q} = H^p(\text{Base}, \pi_q(\text{Fibre})) \Rightarrow \pi_{q-p}(\text{Space of Sections}) \tag{2.4}$$

would give

$$E_2^{p,-q} = H^p(\mathrm{Mot}(X), \mathbb{Z}(q/2))) \Rightarrow K_{q-p}(X) \tag{2.5}$$

where by convention $\mathbb{Z}(q/2) = (0)$ for q odd. Note that all even differentials d_{2n} are necessarily 0. Using this it is straightforward to check that the spectral sequences (1.7) and (2.5) are at least potentially in agreement up to renumbering (décalage).

Of course, all this is completely speculative. One can, however, make a start at understanding the space $\mathrm{Mot}(X)$. For $X = \mathrm{Spec}(F)$ a field, one expects $\mathrm{Mot}(X) = \mathrm{Mot}(F)$ to be a $K(\pi, 1)$ with $\pi^\wedge = \mathrm{Gal}(\mathcal{MTM}(F))$. (Note this should not be the general case. Indeed, for X a flag variety over F, $\mathcal{MTM}(X)$ should equal $\mathcal{MTM}(F)$, but the motivic cohomology groups

$$\mathrm{gr}_\gamma^n K_{2n-i}(X) \otimes \mathbb{Q} \neq \mathrm{gr}_\gamma^n K_{2n-i}(F) \otimes \mathbb{Q}.) \tag{2.6}$$

Beilinson and Deligne have made progress in writing down what should be the Lie algebra $\mathcal{L}(F)$ (0.1). I want to propose a candidate $\mathcal{L}_{\text{cycle}}(F)$ for $\mathcal{L}(F)$ constructed using algebraic cycles.

Let F be an algebraically closed field. Write $X = \mathbb{P}_F^1 - \{1\}$, $X^s = X \times \cdots \times X$ (s times). Faces on X^s are obtained by setting various coordinate functions $= 0$ or ∞. The symmetric group $\mathcal{S}_s$ acts on X^s. Let $\mathcal{V}^r(s)$ denote the group of codimension r algebraic cycles with $\mathbb{Q}$-coefficients on X^s which meet all faces properly and which are alternating with respect to the action of $\mathcal{S}_s$. Define

$$\partial = \sum (-1)^{i-1}(\partial_i^0 - \partial_i^\infty), \tag{2.7}$$

where ∂_i^j is the restriction map on cycles obtained by setting the i-th coordinate function $= j$. The map ∂ preserves alternating representations so we have $\partial : \mathscr{V}^r(s) \to \mathscr{V}^r(s-1)$.

Define a graded group $\mathscr{N}^i = \bigoplus_r \mathscr{V}^r(2r-i)$ and a complex $\mathscr{N}^{\cdot}$ which is $\mathscr{N}^i$ in degree i with boundary $\partial : \mathscr{N}^i \to \mathscr{N}^{i+1}$ induced by ∂ on $\mathscr{V}^r(s)$. One has a product structure $\mathscr{V}^r(s) \times \mathscr{V}^t(u) \to \mathscr{V}^{r+t}(s+u)$ given by obvious (external) product of cycles composed with projection on the alternating representation. This induces a graded product structure $\mathscr{N}^i \times \mathscr{N}^j \to \mathscr{N}^{i+j}$. With this product, $\mathscr{N}^{\cdot}$ is a DGA (i.e. associative and graded commutative with the usual identities for ∂. Note $\mathscr{N}^i \neq (0)$ for $i < 0$.)

Define $\mathscr{N}^{0+} = \mathrm{Ker}(\mathscr{N}^0 \to \mathscr{V}^0(0))$. Let $J = [\bigoplus_{i \le -1} \mathscr{N}^i \oplus \mathscr{N}^{0+}] \cdot \mathscr{N}^{\cdot}$. The ideal J is graded but not stable under ∂. Let $\mathscr{J}$ be the differential graded ideal generated by J. $\mathscr{J}$ is generated as an ideal by $\bigoplus_{i \le -1} \mathscr{N}^i \oplus \mathscr{N}^{0+} \oplus \partial \mathscr{N}^{0+}$ because $\mathscr{V}^0(p) = (0)$ for $p \neq 0$ so $\partial \mathscr{N}^{-1} \subset \mathscr{N}^{0+}$. One shows for F algebraically closed and $i > 0$,

$$\textstyle\bigwedge^i (\mathscr{N}/\mathscr{J})^1 \subset (\mathscr{N}/\mathscr{J})^i. \tag{2.8}$$

Writing

$$\mathscr{R} = \{x \in (\mathscr{N}/\mathscr{J})^1 | \partial x \in \textstyle\bigwedge^2 (\mathscr{N}/\mathscr{J})^1\}, \qquad \mathscr{R} = \mathscr{R}_1 \oplus \mathscr{R}_2 \oplus \cdots \tag{2.9}$$

one proves

$$\partial \mathscr{R} \subset \textstyle\bigwedge^2 \mathscr{R} \subset \bigwedge^2 (\mathscr{N}/\mathscr{J})^1. \tag{2.10}$$

As a consequence, one has a complex

$$\mathscr{R} \xrightarrow{\delta} \textstyle\bigwedge^2 \mathscr{R} \xrightarrow{\delta} \bigwedge^3 \mathscr{R} \to \cdots. \tag{2.11}$$

Define $\mathscr{L}_{\text{cycle}}(F)$ to be the pro-vector space dual to $\mathscr{R}$, i.e. write $\mathscr{R} = \varinjlim \mathscr{R}_\alpha$ with $\mathscr{R}_\alpha$ finite dimensional and take $\mathscr{L}_{\text{cycle},\alpha} = \mathrm{Hom}(\mathscr{R}_{\text{cycle},\alpha}, \mathbb{Q})$, $\mathscr{L}_{\text{cycle}}(F) = \varprojlim \mathscr{L}_{\text{cycle},\alpha}$. The graded pro-Lie algebra $\mathscr{L}_{\text{cycle}}(F)$ is zero in non-negative degrees.

A (mixed Tate) motive on $\mathrm{Spec}(F)$ is a finite dimensional graded $\mathbb{Q}$-vector space V with a representation (compatible with gradings) $\mathscr{L}(F) \to \mathrm{End}(V)$. This leads to

Conjecture (2.12). *A (mixed Tate) motive on* $\mathrm{Spec}(F)$ *is an algebraic cycle with coefficients in* $\mathrm{End}(V)$, $Z \in \mathscr{R} \otimes \mathrm{End}(V)$ *which is of degree* 0 *for the tensor product grading, and which satisfies* $\delta Z = [Z, Z]$.

By way of example, I want to describe the polylogarithm motives as cycles. (This discussion is modeled on Deligne's interpretation of a conjecture of Zagier on the higher K-theory of number fields [D1].) Write $\mathscr{R} = \bigoplus \mathscr{R}_r$ with $\mathscr{R}_r \subset \mathscr{V}^r(2r-1)/$ (relations). One checks that

$$\mathscr{R}_1 = \mathscr{V}^1(1)/(\text{relations}) \cong F^* \otimes \mathbb{Q}. \tag{2.13}$$

Define a map

$$\varrho_r : \mathbb{Z}[F - \{0, 1\}] \to \mathscr{R}_r \tag{2.14}$$

as follows: take

$$\varrho_1(a) = (1 - a) \in \mathscr{R}_1 \cong F^* \otimes \mathbb{Q}. \tag{2.15}$$

For $r \geq 2$, let $T_0, \ldots, T_{r-1}$ be homogeneous coordinates on $\mathbb{P}^{r-1}$. For $a \in F - \{0, 1\}$, let $V_a \subset X^{2r-1}$ be the subvariety described parametrically by

$$\begin{aligned} V_a = \{(aT_{r-2}/T_{r-1}, \ldots, aT_0/T_{r-1}, (T_0 - T_1)/T_0, (T_1 - T_2)/T_1, \ldots, \\ (T_{r-2} - T_{r-1})/T_{r-2}, (T_{r-1} - aT_0)/T_{r-1})\}. \end{aligned} \tag{2.16}$$

One checks that V_a meets faces properly. Define

$$\varrho_r(a) = \text{alt}(V_a) \tag{2.17}$$

where alt is the projection on the alternating representation of $\mathscr{S}_{2r-1}$. (The cycle $\varrho_2(a)$ was found by Totaro [T].) One verifies for $r \geq 2$

$$\partial\varrho_r(a) = -(a)\cdot\varrho_{r-1}(a) = \varrho_{r-1}(a)\cdot(a). \tag{2.18}$$

Define $\mathscr{R}_{\log,r} = \text{Image}(\varrho_r)$. It follows that $\mathscr{R}_{\log} \subset \mathscr{R}$ is a graded sub-co-Lie algebra, with dual graded Lie algebra $\mathscr{L}_{\log}$ a quotient of $\mathscr{L}_{\text{cycle}}$.

Let $N \geq 2$ be an integer. The N-polylog motive, M_N, should be viewed as an object in $\mathscr{MTM}(\mathbb{P}^1 - \{0, 1, \infty\})$. We consider its specialization $M_{N,z}$ at some F-point $z \in \mathbb{P}^1 - \{0, 1, \infty\}$. Define (following Deligne [op. cit.]) $(N + 1) \times (N + 1)$ matrices (viewed as acting on $\mathbb{Q}^{[0,N]}$):

$$e_0 = \begin{pmatrix} 0 & & & & & 0 \\ 0 & 0 & & & & \\ 0 & 1 & 0 & & & \\ 0 & 0 & 1 & 0 & & \\ & & & & \ddots & \\ & 0 & & & 1 & 0 \end{pmatrix} \qquad e_1 = \begin{pmatrix} 0 & & 0 \\ 1 & 0 & \\ & \ddots & \\ & 0 & 0 \end{pmatrix}$$

Then $M_{N,z} \in \mathscr{R}_{\log} \otimes \text{End}(\mathbb{Q}^{[0,N]})$ should be associated to an algebraic cycle Z_z. Taking

$$Z_z = -(z) \otimes e_0 - \sum_{r=1}^{N} \varrho_r(z) \otimes \text{ad}(e_0)^{r-1}(e_1) \tag{2.19}$$

yields

$$\partial Z_z = [Z_z, Z_z] \tag{2.20}$$

so the resulting map $\mathscr{L}_{\text{cycle}} \twoheadrightarrow \mathscr{L}_{\log} \to \text{End}(\mathbb{Q}^{[0,N]})$ is a representation.

Of course much work remains to be done, for example, to spread out the construction to define $\mathscr{MTM}(T)$ for a variety T over a field. It seems reasonable to expect a construction analogous to (2.12), applied to sums of cycles of codimension r_i on $T \times X^{2r_i-1}$ which are equidimensional over T, should yield objects in $\mathscr{MTM}(T)$.

3. Arithmetic

As suggested above, assuming the moving Conjecture (1.2), the higher Chow groups of a variety over a field can be thought of as giving an integral structure for motivic cohomology. In [BK], we define a Tamagawa number for motives, and we use this to give an integral version of the Beilinson conjectures relating values of a regulator map on rational motivic cohomology with values of L-functions. Our formulation only involved motivic cohomology $\otimes \mathbb{Q}$, but availability of integral motivic cohomology suggests a natural interpretation for the group Ш appearing in the Tamagawa volume formula. I will outline these ideas briefly. In [op. cit.], the local constructions are completely proven and do not depend on any conjectures. This forces us to work with an integral model of our variety. In what follows I will indicate (following the brief discussion at the end of § 5 of op. cit.) what is needed to formulate a conjecture depending only on the variety over $\mathbb{Q}$. For a general variety X, this formulation involves a number of other "standard" conjectures about étale cohomology, L-functions, and K-theory. However, for varieties like curves and rational surfaces the necessary conditions can be verified directly.

Let X be a smooth projective variety over $\mathbb{Q}$. We think of $\mathrm{CH}^r(X, 2r-n)$ "=" $H^n(\mathrm{Mot}(X), \mathbb{Z}(r))$. It is suggestive (although I do not know how precise) to think of the motive $H^n(X, \mathbb{Z}(r))$ as the "local system" on $\mathrm{Mot}(\mathrm{Spec}(\mathbb{Q}))$ given by $H^n(\mathrm{Fibre}[\mathrm{Mot}(X) \to \mathrm{Mot}(\mathrm{Spec}(\mathbb{Q}))], \mathbb{Z})(r)$. Assuming $2r > n$, this local system would have "weights" < 0 and hence no non-zero sections, so one would get

$$\mathrm{CH}^r(X, 2r-n) \to \mathrm{Ext}^1_{\mathrm{Mot}(\mathrm{Spec}(\mathbb{Q}))}(\mathbb{Z}, H^{n-1}(X, \mathbb{Z}(r))). \tag{3.1}$$

In the case $2r = n$,

$$\begin{aligned}\mathrm{CH}^r(X)_0 &\overset{\mathrm{def}}{=} \mathrm{Ker}[\mathrm{CH}^r(X) \to \mathrm{Hom}_{\mathrm{Mot}(\mathrm{Spec}(\mathbb{Q}))}(\mathbb{Z}, H^{2r}(X, \mathbb{Z}(r)))] \\ &\to \mathrm{Ext}^1_{\mathrm{Mot}(\mathrm{Spec}(\mathbb{Q}))}(\mathbb{Z}, H^{2r-1}(X, \mathbb{Z}(r))).\end{aligned} \tag{3.2}$$

One hopes that (3.1) and (3.2) are isomorphisms.

Next one wants to understand $\mathrm{Ext}^1_{\mathrm{Mot}(\mathrm{Spec}(\mathbb{Q}_p))}(\mathbb{Z}, H^{n-1}(X, \mathbb{Z}(r)))$ for $p \leq \infty$. We assume henceforth that $n - 2r \leq -1$, so the motive

$$M \overset{\mathrm{def}}{=} H^{n-1}(X, \mathbb{Z}(r))$$

has weight $n - 1 - 2r \leq -2$.

Let MHS/$\mathbb{R}$ denote the category of mixed Hodge structures defined over $\mathbb{R}$, i.e. endowed with an $\mathbb{R}$-linear conjugation F_∞. Following [D3], one expects

$$\begin{aligned}\mathrm{Ext}^1_{\mathrm{Mot}(\mathrm{Spec}(\mathbb{R}))}(\mathbb{Z}, M) &\cong \mathrm{Ext}^1_{\mathrm{MHS}/\mathbb{R}}(\mathbb{Z}, H^{n-1}(X, \mathbb{Z}(r))) \\ &\cong [H_B^{n-1}(X_{\mathbb{C}}, \mathbb{C})/\{H_B^{n-1}(X_{\mathbb{C}}, \mathbb{Z}(r)) + F^r H_B^{n-1}(X_{\mathbb{C}}, \mathbb{C})\}]^+.\end{aligned} \tag{3.3}$$

Note that this is a Lie group.

The situation at the non-archimedean primes is more delicate. One wants

$$\mathrm{Ext}^1_{\mathrm{Mot}(\mathrm{Spec}(\mathbb{Q}_p))}(\mathbb{Z}, M) \subset H^1(\mathrm{Gal}(\bar{\mathbb{Q}}_p/\mathbb{Q}_p), M(\hat{\mathbb{Z}})). \tag{3.4}$$

where $M(\hat{\mathbb{Z}}) = H_{\text{ét}}^{n-1}(X_{\bar{\mathbb{Q}}}, \hat{\mathbb{Z}}(r))$. There are defined [op. cit.] subgroups

$$H_e^1 \subset H_f^1 \subset H_g^1 \subset H^1(\mathrm{Gal}(\bar{\mathbb{Q}}_p/\mathbb{Q}_p), M(\hat{\mathbb{Z}})). \tag{3.5}$$

If elements in H^1 are viewed as extensions, the $\mathbb{Z}_l$ part ($l \neq p$) of H_f^1 (resp. H_g^1) coincides with extensions which split after tensoring with $\mathbb{Q}_l$ and restricting to the maximal unramified extension of $\mathbb{Q}_p$ (resp. all extensions). The $\mathbb{Z}_p$ part of H_f^1 (resp. H_g^1) are crystalline (resp. de Rham) extensions, i.e. those which split when tensored with the Fontaine rings B_{cris} (resp. B_{DR}). In [op. cit.] the main object of study is H_f^1, which should be thought of as $\mathrm{Ext}^1_{\mathrm{Mot}(\mathrm{Spec}(\mathbb{Z}_p))}$. One knows the $\mathbb{Z}_l$ part of H_g^1/H_f^1 is torsion free of rank equal to the rank of the space of coinvariants for frobenius acting on $\mathrm{Hom}_I(M(\mathbb{Q}_l), \mathbb{Q}_l(1))$, where $I \subset \mathrm{Gal}(\bar{\mathbb{Q}}_p/\mathbb{Q}_p)$ is the inertia group (and there is a similar interpretation of the p-part in terms of $\mathrm{Crys}(M^*(1))/(1-f)$, cf. op. cit.) Standard conjectures say that these ranks are all equal and coincide with the order of vanishing of the local factor of the dual L-function $L_p(M^*(1), s)^{-1}$ at $s = 0$, i.e. with the rank of $H_{2d-n+1,\text{ét}}(X_{\mathbb{F}_p}, \mathbb{Q}_l(r-1-d))$, where $d = \dim X_{\mathbb{F}_p}$. What one expects is that the cycle class map [Bl2]

$$\mathrm{CH}^r(X_{\mathbb{Q}_p}, 2r-n) \to H^1(\mathrm{Gal}(\bar{\mathbb{Q}}_p/\mathbb{Q}_p), M(\hat{\mathbb{Z}})) \tag{3.6}$$

has image $\subset H_g^1$ and maps $\mathrm{CH}^r(X_{\mathbb{Z}_p}, 2r-n)$ into H_f^1. Moreover, one has a chern class map $\mathrm{gr}_\gamma^{r-1} K'_{2r-n-1}(X_{\mathbb{F}_p}) \otimes \mathbb{Q}_l \to H_{2d-n+1,\text{ét}}(X_{\mathbb{F}_p}, \mathbb{Q}_l(r-1-d))$ which by Jannsen's generalization of Tate's conjecture [J] should be an isomorphism. Putting these ideas together with localization yields (conjecturally)

$$\begin{aligned} \mathbb{Q} \otimes H_g^1/H_f^1 &\overset{i}{\hookleftarrow} [\mathrm{CH}^r(X_{\mathbb{Q}_p}, 2r-n)/\mathrm{CH}^r(X_{\mathbb{Z}_p}, 2r-n)] \otimes \mathbb{Q} \\ &\cong [\mathrm{gr}_\gamma^r K_{2r-n}(X_{\mathbb{Q}_p})/\mathrm{gr}_\gamma^r K_{2r-n}(X_{\mathbb{Z}_p})] \otimes \mathbb{Q} \cong \mathrm{gr}_\gamma^{r-1} K'_{2r-n-1}(X_{\mathbb{F}_p}) \otimes \mathbb{Q}, \end{aligned} \tag{3.7}$$

the inclusion i being an isomorphism after tensoring with $\hat{\mathbb{Q}}$. Let $N = (H_g^1/H_f^1) \cap \mathrm{image}(i)$, and "define" $\mathrm{Ext}^1_{\mathrm{Mot}(\mathrm{Spec}(\mathbb{Q}_p))}(\mathbb{Z}, M) \subset H_g^1$ to be the pre-image of N. One has

$$0 \to \mathrm{Ext}^1_{\mathrm{Mot}(\mathrm{Spec}(\mathbb{Z}_p))}(\mathbb{Z}, M) \to \mathrm{Ext}^1_{\mathrm{Mot}(\mathrm{Spec}(\mathbb{Q}_p))}(\mathbb{Z}, M) \to N \to 0 \tag{3.8}$$

so $\mathrm{Ext}^1_{\mathrm{Mot}(\mathrm{Spec}(\mathbb{Q}_p))}(\mathbb{Z}, M)$ can be thought of as a topological group containing $\mathrm{Ext}^1_{\mathrm{Mot}(\mathrm{Spec}(\mathbb{Z}_p))}(\mathbb{Z}, M)$ as an open compact subgroup. In the special case $X = \mathrm{Spec}(\mathbb{Q})$, $n = r = 1$, one sees from op. cit. Example 3.9 that $H_f^1 \cong \mathbb{Z}_p^\times$, $H_g^1 \cong \mathbb{Q}_p^\times \otimes \hat{\mathbb{Q}}$, and $N = \mathbb{Z}$ embedded in H_g^1/H_f^1 in the natural way, so $\mathrm{Ext}^1_{\mathrm{Mot}(\mathrm{Spec}(\mathbb{Q}_p))}(\mathbb{Z}, \mathbb{Z}(1)) \cong \mathbb{Q}_p^\times$.

It will be convenient to write

$$\begin{aligned} B(\mathbb{Z}_p) = H_f^1 = \mathrm{Ext}^1_{\mathrm{Mot}(\mathrm{Spec}(\mathbb{Z}_p))}(\mathbb{Z}, M);\ B(\mathbb{Q}_p) = \mathrm{Ext}^1_{\mathrm{Mot}(\mathrm{Spec}(\mathbb{Q}_p))}(\mathbb{Z}, M); \\ B(\mathbb{R}) = \mathrm{Ext}^1_{\mathrm{Mot}(\mathrm{Spec}(\mathbb{R}))}(\mathbb{Z}, M). \end{aligned} \tag{3.9}$$

One of the main results in op. cit. was to endow the groups (compact for $p < \infty$) $B(\mathbb{Z}_p)$ and $B(\mathbb{R})$ with Haar measures ω_p (depending on a choice of trivialization of $\det(H^n_{\mathrm{DR}}(X/\mathbb{Q})/F^r)$). We extend the ω_p to measures on the locally compact groups $B(\mathbb{Q}_p)$, $p \leq \infty$. Writing $L(M, s) = \prod L_p(M, s)$ for the L-function associated to M, one has for almost all p,

$$\omega_p(B(\mathbb{Z}_p)) = L_p(M, 0)^{-1}. \tag{3.10}$$

The resulting measure μ on

$$B(\mathbb{A}_{\mathbb{Q}}) = \prod_{p \leq \infty} B(\mathbb{Q}_p) \text{ (restricted direct product with respect to } B(\mathbb{Z}_p)) \tag{3.11}$$

is independent of the choice of trivialization. Note that this measure converges (i.e. gives finite non-zero measure to the compact set $\prod B(\mathbb{Z}_p)$) when $n - 2r \leq -2$. In general, it is necessary to assume meromorphic continuation of the L-function, define $\omega_p' = \omega_p \cdot L_p(M, 0)$ for all p outside a finite set S of bad primes ($\omega_p' = \omega_p$ for $p \in S$), and then define

$$\mu = \mu'/L_S(M, 0) \tag{3.12}$$

where L_S is the partial L-function with $p \in S$ omitted.

Assume $n - 1 - 2r \leq -3$. The Tamagawa number conjecture, as formulated in op. cit., reads

$$\mu(B(\mathbb{A}_{\mathbb{Q}})/B(\mathbb{Q})) = \# H^0(\mathrm{Gal}(\bar{\mathbb{Q}}/\mathbb{Q}), M^*(1)(\mathbb{Q}/\mathbb{Z}))/\# \tilde{Ш}(M) \tag{3.13}$$

where

$$\begin{aligned} B(\mathbb{Q}) = \{x \in H^1(\mathrm{Gal}(\bar{\mathbb{Q}}/\mathbb{Q}), M(\hat{\mathbb{Z}})) | x \otimes \mathbb{Q} \in \mathrm{image}[\mathrm{gr}^r_\gamma K_{2r-n}(X) \otimes \mathbb{Q} \\ \to H^1(\mathrm{Gal}(\bar{\mathbb{Q}}/\mathbb{Q}), M(\hat{\mathbb{Q}}))]\}; \end{aligned}$$

$$\tilde{Ш} = \mathrm{Ker}\left[\frac{H^1(\mathrm{Gal}(\bar{\mathbb{Q}}/\mathbb{Q}), M(\mathbb{Q}/\mathbb{Z}))}{B(\mathbb{Q}) \otimes \mathbb{Q}/\mathbb{Z}} \to \bigoplus_{p \leq \infty} \frac{H^1(\mathrm{Gal}(\bar{\mathbb{Q}}_p/\mathbb{Q}_p), M(\mathbb{Q}/\mathbb{Z}))}{B(\mathbb{Q}_p) \otimes \mathbb{Q}/\mathbb{Z}}\right]. \tag{3.14}$$

Construction (3.15). Assume $n - 1 - 2r \leq -2$.

(i) There is a natural map

$$\mathrm{CH}^r(X_{\mathbb{Q}}, 2r - n) \to B(\mathbb{Q}). \tag{3.15.1}$$

(ii) Assume moreover $H^{n-1}_{\text{ét}}(X_{\bar{\mathbb{Q}}}, \hat{\mathbb{Z}})$ has no torsion. Then there is a natural map

$$Ш(\mathrm{CH}^r(X_{\mathbb{Q}}, 2r - n - 1)) \to \tilde{Ш}(M). \tag{3.15.2}$$

Here, of course,

$$\begin{aligned} Ш(\mathrm{CH}^r(X_{\mathbb{Q}}, 2r - n - 1)) \stackrel{\text{def}}{=} \mathrm{Ker}(&\mathrm{CH}^r(X_{\mathbb{Q}}, 2r - n - 1) \\ &\to \prod_{p \leq \infty} \mathrm{CH}^r(X_{\mathbb{Q}_p}, 2r - n - 1)). \end{aligned} \tag{3.15.3}$$

The arrow in (3.15.1) is the cycle class [Bl2]. The first step in constructing the map (3.15.2) is to consider the exact sequence (with $\mathscr{Z}^r(X, m)$ as in § 1)

$$0 \to Z\mathscr{Z}^r(X_{\bar{\mathbb{Q}}}, 2r - n) \to \mathscr{Z}^r(X_{\bar{\mathbb{Q}}}, 2r - n) \stackrel{\partial}{\to} B\mathscr{Z}^r(X_{\bar{\mathbb{Q}}}, 2r - n - 1) \to 0. \tag{3.15.4}$$

Taking galois cohomology yields a map

$$\begin{aligned} \mathrm{Ker}[\mathrm{CH}^r(X_{\mathbb{Q}}, 2r - n - 1) &\to \mathrm{CH}^r(X_{\bar{\mathbb{Q}}}, 2r - n - 1)] \\ &\to H^1(\mathrm{Gal}(\bar{\mathbb{Q}}/\mathbb{Q}), Z\mathscr{Z}^r(X_{\bar{\mathbb{Q}}}, 2r - n)) \\ &\to H^1(\mathrm{Gal}(\bar{\mathbb{Q}}/\mathbb{Q}), \mathrm{CH}^r(X_{\bar{\mathbb{Q}}}, 2r - n)) \end{aligned} \tag{3.15.5}$$

and hence a map

$$Ш(\mathrm{CH}^r(X_{\mathbb{Q}}, 2r - n - 1)) \to Ш(H^1(\mathrm{Gal}(\bar{\mathbb{Q}}/\mathbb{Q}), \mathrm{CH}^r(X_{\bar{\mathbb{Q}}}, 2r - n))). \tag{3.15.6}$$

The rest of the argument is more complicated, and will not be given in detail. Note, however, that in many cases of interest (e.g. $r = n = 2$ and X a curve; or $r = 2, n = 3$, and X a surface over $\mathbb{Q}$ whch is $\bar{\mathbb{Q}}$-rational; or $r = 2$, $n = 1$, and $X = \mathrm{Spec}(K)$ for K a number field) one has an exact sequence (3.15.7)

$$0 \to H^{n-1}(X_{\bar{\mathbb{Q}}}, \mathbb{Q}/\mathbb{Z}(r)) \to \mathrm{CH}^r(X_{\bar{\mathbb{Q}}}, 2r - n) \to \mathrm{CH}^r(X_{\bar{\mathbb{Q}}}, 2r - n) \otimes \mathbb{Q} \to 0. \quad (3.15.7)$$

In these cases, a straightforward diagram chase gives

$$Ш(H^1(\mathrm{Gal}(\bar{\mathbb{Q}}/\mathbb{Q}), \mathrm{CH}^r(X_{\bar{\mathbb{Q}}}, 2r - n))) \to \tilde{Ш}(M). \quad (3.15.8) \quad \square$$

Conjecture (3.16). *The map* (3.15.1) *is surjective, and identifies* $B(\mathbb{Q})$ *with the image of* $\mathrm{CH}^r(X_{\mathbb{Q}}, 2r - n) \to \mathrm{CH}^r(X_{\bar{\mathbb{Q}}}, 2r - n)$. *The map* (3.15.2) *is an isomorphism* $Ш\mathrm{CH}^r(X_{\mathbb{Q}}, 2r - n - 1) \cong \tilde{Ш}(M)$.

The author attaches particular importance to the possibility of giving a geometric interpretation of $\tilde{Ш}(M)$ via the higher Chow groups.

Example (3.17). Using an important result of Suslin [Su]

$$\mathrm{CH}^r(K, r) \cong K_r^{\mathrm{Milnor}}(K) \quad (3.17.1)$$

for any field K, and assuming the moving Conjecture (1.2), one gets a localization sequence for any curve

$$K_2(k(C)) \to \coprod_{P \in C} k(P)^* \to \mathrm{CH}^2(C, 1) \to 0, \quad (3.17.2)$$

whence $\mathrm{CH}^2(C, 1) \cong SK_1'(C)$. Let X be a smooth surface, and let $\mathscr{S} \subset X$ be the union of all curves in X. Repeating the above localization argument yields

$$K_2(k(X)) \to SK_1'(\mathscr{S}) \to \mathrm{CH}^2(X, 1) \to 0, \quad (3.17.3)$$

so $\mathrm{CH}^2(X, 1) \cong H^1(X, \mathscr{K}_2)$ (Zariski cohomology of the sheaf $K_2[Q]$).

Assume now that X is a surface over $\mathbb{Q}$ which is $\bar{\mathbb{Q}}$-rational, it follows [CT] that for any number field K,

$$\mathrm{CH}^2(X_K, 1) \cong T(K), \quad (3.17.4)$$

where T is the algebraic torus with character group the $\mathrm{Gal}(\bar{\mathbb{Q}}/\mathbb{Q})$-module $\mathrm{NS}(X_{\bar{\mathbb{Q}}})$ (= Néron-Severi group of divisors on $X_{\bar{\mathbb{Q}}}$.) In this case

$$M(\hat{\mathbb{Z}}) = H^2_{\text{ét}}(X_{\bar{\mathbb{Q}}}, \hat{\mathbb{Z}}(2)) \cong NS(X_{\bar{\mathbb{Q}}}) \otimes \mathbb{Z}(1) \quad (3.17.4)$$

and it is not hard to show

$$\tilde{Ш}(M) \cong Ш(T) \stackrel{\text{def}}{=} \mathrm{Ker}[H^1(\mathrm{Gal}(\bar{\mathbb{Q}}/\mathbb{Q}), T(\bar{\mathbb{Q}})) \to \prod H^1(\mathrm{Gal}(\bar{\mathbb{Q}}_p/\mathbb{Q}_p), T(\bar{\mathbb{Q}}_p))]. \quad (3.17.5)$$

The Tamagawa number conjecture (3.13) (suitably modified because here $n - 1 - 2r = -2$) follows in this case from work of Ono [O].

One knows in general for a rational surface that there is an inclusion

$$\mathrm{Ker}(\mathrm{CH}^2(X_{\mathbb{Q}}) \to \mathrm{CH}^2(X_{\bar{\mathbb{Q}}})) \hookrightarrow H^1(\mathrm{Gal}(\bar{\mathbb{Q}}/\mathbb{Q}), T(\bar{\mathbb{Q}})) \quad [\mathrm{CT}]. \tag{3.17.6}$$

If we assume in addition that X admits a rational pencil of genus 0 curves, a beautiful theorem of Salberger [Sa] says that this inclusion induces an isomorphism

$$Ш\mathrm{CH}^2(X_{\mathbb{Q}}, 0) = Ш\mathrm{CH}^2(X_{\mathbb{Q}}) \cong Ш(T). \tag{3.17.7}$$

Exercise (3.18). Show that for F a number field, $Ш\mathrm{CH}^2(\mathrm{Spec}(F), 2)$ coincides with the "wild kernel"

$$\mathrm{Ker}(K_2(F) \to \prod K_2(F_v)), \tag{3.18.1}$$

where v runs through all places of F, and F_v denotes the completion of F at v. Show the Tamagawa number conjecture for $\zeta_F(2)$ is compatible with Lichtenbaum's conjecture.

Bibliography

[Be1] Beilinson, A.A.: Polylogarithm and cyclotomic elements. Preprint

[Be2] Beilinson, A.A.: Notes on absolute Hodge cohomology. Contemporary Mathematics, vol. 55, part 1 (1986)

[Be3] Beilinson, A.A., Goncharov, A.B., Schechtman, V.V., Varchenko, A.N.: Aomoto dilogarithms, mixed Hodge structures, and motivic cohomology of pairs of triangles on the plane. In: The Grothendieck Festschrift, vol. 1. (Progress in Mathematics Series, vol. 86) Birkhäuser, Boston 1990

[Be4] Beilinson, A.A., MacPherson, R., Schechtman, V.V.: Notes on motivic cohomology. Duke Math. J. **54**, no. 2 (1987) 679–710

[Bl1] Bloch, S.: Algebraic cycles and higher K-theory. Adv. Math. (1986)

[Bl2] Bloch, S.: Algebraic cycles and the Beilinson conjectures. Contemporary Mathematics, vol. 58, part 1 (1986)

[BK] Bloch, S., Kato, K.: L-functions and Tamagawa numbers of motives. In: The Grothendieck Festschrift, vol. 1 (Progress in Mathematics Series, vol. 86.) Birkhäuser, Boston 1990

[CT] Colliot-Thélène, J.-L.: Hilbert's theorem 90 for K_2, with application to the Chow groups of rational surfaces. Invent. math. **71** (1983) 1–20

[D1] Deligne, P., Interprétation motivique de la conjecture de Zagier reliant polylogarithmes et régulateurs. Preprint

[D2] Deligne, P.: Le groupe fondamental de la droite projective moins trois points. In: Galois groups over $\mathbb{Q}$. Math. Sci. Res. Inst. Publ. 16. Springer, Berlin Heidelberg New York 1989

[D3] Deligne, P.: Valeurs de fonctions L et périodes d'intégrales. In: AMS Proc. Symp. Pure Math., vol. 33, part 2, pp. 313–346 (1979).

[DM] Deligne, P., Milne, J.S.: Tannakian categories, in Hodge cycles, motives, and Shimura varieties. (Lecture Notes in Mathematics, vol. 900.) Springer, Berlin Heidelberg New York 1982

[DF] Dwyer, W., Friedlander, E., Algebraic and étale K-theory. Preprint

[J] Jannsen, U.: Mixed motives and algebraic K-theory. (Lecture Notes in Mathematics, vol. 1400.) Springer, Berlin Heidelberg New York 1990

[Kl] Kleiman, S.: Motives. In: Proc. 5th Nordic Summer School, Oslo 1970. Wolters-Noordhoff, Holland 1972

[L1] Lichtenbaum, S.: New results on weight two motivic cohomology. In: The Grothendieck Festschrift, vol. III. (Progress in Mathematics Series, vol. 86.) Birkhäuser, Boston 1990

[L2] Lichtenbaum, S.: The construction of weight two arithmetic cohomology. Invent. math. **88** (1987) 183–215

[M] Manin, Yu.: Correspondences, motives, and monoidal transformations. Mat. Sbornik **119** (1968) 475–507

[O] Ono, T.: On the Tamagawa number of algebraic tori. Ann. Math. **78**, (1963) 47–73

[Q] Quillen, D.: Higher algebraic K-theory I. In: Algebraic K-theory I. (Lecture Notes in Mathematics, vol. 341.) Springer, Berlin Heidelberg New York 1973

[Sa] Salberger, P.: Zero cycles on rational surfaces over number fields. Invent. math. **91**, no. 3, (1988) 505–524

[Su] Suslin, A.: Algebraic K-theory of fields. ICM 1986, pp. 222–244

[T] Totaro, B.: Thesis, Berkeley 1990

[Z1] Zagier, D.: Polylogarithms, Dedekind zeta functions, and the algebraic K-theory of fields. Preprint

[Z2] Zagier, D.: The Bloch-Wigner-Ramakrishnan polylogarithm function. Math. Ann. **286** (1990) 613–624

Computational Complexity of Higher Type Functions

Stephen A. Cook *

University of Toronto, Toronto, Ontario M5S 1A4, Canada

Abstract. The customary identification of feasible with polytime is discussed. Next, higher type functions are presented as a way of giving computational meaning to theorems. In particular, if a theorem has a "feasibly constructive" proof, the associated functions should be polytime. However examples are given to illustrate the difficulty of capturing the notion of polytime for higher type functions. Finally, König's Lemma is used to illustrate a theorem whose computational meaning is naturally expressed by functions of type level 3.

1. Traditional Complexity and Polytime

The title of Alan Cobham's 1964 lecture [Cob] "The intrinsic computational difficulty of functions" expresses well the subject matter of complexity theory. A major subgoal of the theory is to identify and study the so-called *feasible* functions. It has become customary to identify the notion of feasible with that of polynomial time computable (polytime, for short). In this section we will examine this identification.

In defining polytime, it is important to emphasize that the parameter used is always the length $|x|$ of the input number x, where $|x| = \lceil \log_2(x+1) \rceil$ = the number of bits in the binary notation for x.

Definition 1. $f : \mathbb{N} \to \mathbb{N}$ is *polytime* iff there is a Turing machine M and a number k such that for all $x \in \mathbb{N}$, M on input x computes $f(x)$ within $O(|x|^k)$ steps.

The class of polytime functions is very robust. Thus, for example, "Turing machine" in the above definition can be replaced by any reasonable abstraction of a computer, and the class remains unchanged.

The equation feasible = polytime needs interpretation before it can be justified. It is not literally true, since the function $f(x) = 2^{|x|^{1000}}$ is polytime but is not computationally feasible in any sense. (We assume that radix notation is used for

* Supported by the Natural Sciences and Engineering Research Council of Canada.

Proceedings of the International Congress
of Mathematicians, Kyoto, Japan, 1990

numbers.) Thus we must restrict attention to functions arising from "natural" computational problems.

Convention 2. NAT is the class of functions arising from natural computational problems.

A problem whose description requires mention of a large exponent such as 1000 is considered contrived, and not natural.

For the sake of argument here, we identify "feasible" with "practically solvable" in the following sense:

Convention 3. *PracSolv* is the class of functions f which can be computed by some actual computer on all inputs of 10,000 bits or less.

Note that a practical program for computing f need not be known. We only require that one exists. The Convention 3 is not as technology dependent as it may appear. There are physical limits on the power of any conceivable actual computer, as illustrated by the examples below.

The equation feasible = polytime can now be replaced by a somewhat more precise statement.

Thesis 4. PracSolv $\cap$ NAT = polytime $\cap$ NAT.

There are countless examples arising from computing theory and practice which tend to support this thesis. To give just a few, the functions $f(x) = x^2$, $f(x, y) = \gcd(x, y)$, and $f(x) = \lceil \sqrt{x} \rceil$ all lie clearly in both the left side and right side of 4. On the other hand, the function $f(x) = 2^x$ lies clearly outside both the left side and right side of 4. This function is not in PracSolv, because when x has length 1000 bits, $f(x)$ has length about 2^{1000} bits, which cannot be physically written using all the atoms on Earth to express bits.

There are also 0-1 valued functions in NAT which are clearly outside both sides of Thesis 4. For instance, let f represent truth in WS1S (the weak second order theory of one successor). That is, $f(x) = 1$ if x codes (say using ASCII conventions) a true formula in WS1S, and $f(x) = 0$ otherwise. Then f is computable, but provably not polytime [Mey]. Furthermore Stockmeyer [Sto] proved that any Boolean circuit that correctly decides the truth of an arbitrary WS1S formula of length 616 symbols must have more than 10^{123} gates. Thus any conceivable actual computer that is correctly programmed to solve an arbitrary such instance would take until the sun burns out to solve at least one of them.

There are many natural problems whose status with respect to both sides of Thesis 4 is questionable. Among these are the 300 or so "NP complete" problems listed in the book by Garey and Johnson [GJ]. None of these is known to be polytime and none is known to be in PracSolv. If any one of these problems is polytime, then provably they all are (i.e. $P = NP$). It is conceivable that one (and therefore all) are polytime, but nevertheless the satisfiability problem has no practical algorithm. This would refute Thesis 4. Unless and until this unlikely situation is demonstrated, however, the assumption $P \neq NP$ together with the thesis 4 provides a very useful orientation for programmers in the field. Anyone who is trying to write a program for a problem and discovers the problem is NP complete would be well advised to change goals and work on a more feasible version of the problem.

There are other potential candidates for refuting Thesis 4, and it is instructive to consider some of these.

CANDIDATES FOR (polytime ∩ NAT) – (PracSolv ∩ NAT)

A) Factoring in $\mathbb{Q}[x]$

Lenstra, Lenstra and Lovasz [LLL] were the first to prove that factoring univariate polynomials over the rationals can be performed in polytime. The original runtime proved for their algorithm was $O(n^{12})$, and it did not appear to be practical. However more recent versions [Scho] have smaller runtimes and do appear to work on large inputs in a reasonable amount of time. Before [LLL] there were programs which effectively solved the problem in most cases, although these programs were known to "blow up" on a small set of inputs. These "bad"inputs can arise in algebraic number field computations.

B) Minor-Closed Graph Families

Robertson and Seymour [RS] proved the remarkable result that every minor-closed family of finite graphs has a finite obstruction set. Here an undirected graph H is a *minor* of G iff H can be obtained (up to isomorphism) from G by contracting certain edges and deleting certain edges and vertices. A family $\mathcal{T}$ of graphs is *minor-closed* iff whenever $G \in \mathcal{T}$ and H is a minor of G, then $H \in \mathcal{T}$. A graph G_0 is an *obstruction* of $\mathcal{T}$ iff $G_0 \notin \mathcal{T}$, but if H is any proper minor of G_0, then $H \in \mathcal{T}$. Thus to determine whether a given graph G is in $\mathcal{T}$ it suffices to check, for each obstruction G_0 of $\mathcal{T}$, whether G_0 is a minor of G.

Robertson and Seymour also proved that for each fixed H there is a polytime algorithm which takes an arbitrary graph G as input and determines whether H is a minor of G. It follows that every minor-closed graph family has a polytime recognition algorithm. Later the minor checking algorithm was improved to run in time $O(n^3)$, but in either case the runtime increases by an extremely large constant factor with increasing H. None of these algorithms is practical when H has four or more vertices.

The most famous minor-closed family is the class of planar graphs, which have the Kuratowski graphs K_5 and $K_{3,3}$ as the obstruction set. More generally, for any fixed surface, the set of graphs embeddable on that surface forms a minor-closed family and hence is in polytime. When the surface is a torus, there are some 800 known obstructions, and there may be more. Another interesting minor-closed family consists of graphs embeddable in $\mathbb{R}^3$ in such a way that no two cycles are linked. See [FL] for other examples.

Although the work of Robertson and Seymour does not yield practical algorithms for the recognition of any of these minor-closed families, such algorithms are known in some cases (e.g. planar graphs). Also the knowledge springing from [RS] that these problems are polytime has spurred research to find practical algorithms, with success in some cases [FL]. Time will tell whether some natural minor-closed family resists long and concerted efforts to find a practical recognition algorithm.

CANDIDATES FOR (PracSolv ∩ NAT) – (polytime ∩ NAT)

A) Linear Programming

The simplex method has long been an extremely useful practical method for solving linear programming problems, even though the method, or at least some versions of it, are known to be exponential in the worst case. Thus linear programming used to be a candidate for the above difference set. However, in 1979 Khachian [Kha] proved that the problem is in polytime, using a very different algorithm.

B) Probabilistic Polytime (Verified Answer)

There are some natural problems not known to have (deterministic) polytime algorithms but which do have verified probabilistic polytime algorithms. These "Las Vegas" algorithms require a random source of input bits (e.g. coin tosses). For every problem instance the algorithms terminate in polytime with high probability, and when they do terminate the answer given is always correct. An early example of such a problem is factoring univariate polynomials over large finite fields [Ber]. A more recent example is the set of prime numbers. Probabilistic algorithms for compositeness have been known for some time [SS], [Rab], but these have one-sided error, as in C) below. Recently Goldwasser and Kilian [GK] found a probabilistic polytime algorithm for prime testing which is always correct when it gives an output, but it fails to give an output for certain prime inputs. This was improved [AH] to yield a provably correct but impractical Las Vegas algorithm for prime recognition. A practical version has also been developed [AM], but this algorithm lacks a proof that the runtime is polynomial.

It turns out that both the polynomial factoring algorithm [Ber] and the practical prime recognition algorithm [AM], although probabilistic in principle, work perfectly well in practice when the random source of bits is replaced by some simple deterministically generated sequence of bits. Thus there exist plausible deterministic polytime algorithms for these problems, and it may be only a question of time until someone proves them correct. In fact, Miller [Mil] proved that prime recognition is deterministic polytime, assuming the Extended Riemann Hypothesis.

C) Probabilistic Polytime (One-Sided Error)

Some probabilistic set recognition algorithms always terminate in polytime; if they accept the input then that answer is always correct, but if they reject the input then there is a small error probability. This error probability can be made arbitrarily small simply by repeating the computation a number of times, assuming that the random input bits are truely random and independent.

One example [Schw] of a problem with this kind of probabilistic algorithm, but not known to be in (deterministic) polytime, is the set of nonsingular $n \times n$ matrices whose entries are linear polynomials over $\mathbb{Z}$ in n variables. The probabilistic algorithm consists of choosing small random integer values for the variables and evaluating the resulting determinant. If the value is nonzero, then the original matrix is certainly nonsingular; if the value is 0, that is evidence that the original matrix was singular. The experiment can be repeated any number of times. Assuming we have access to a truely random source of bits, this is a practical algorithm. However in practice truely random sources are scarce or non-existent, so pseudo random number generators with weakly random seeds are used. These *seem* to work well, but if the algorithm repeatedly outputs "singular" for a given input matrix it is hard to justify having total confidence in the result. Hence on the one hand the problem may not really be in PracSolv, while on the other hand it may be deterministic polytime for the reasons outlined in B) above.

A second example in this category is testing irreducibility of polynomials in many variables over $\mathbb{Q}$, using "concise" representation [vzG]. Here a "yes" answer from the probabilistic algorithm means the input polynomial is certainly irreducible, while a "no" leaves some room for doubt. The algorithm has been extended [KalT] to find likely factorization in case of a "no" answer. For each input polynomial of size n (in concise representation), the probability of the factorization algorithm producing a false factor can be made much less than 2^{-n}. Nevertheless, the possibility of error persists, because there is no known deterministic polytime algorithm for testing divisibility of polynomials in many variables.

In sum, there are no very convincing candidate counterexamples to thesis 4. However some natural polytime problems, especially certain minor-closed graph families, certainly do not have known practical algorithms. But the tendency is for practical algorithms for a problem to be discovered, once it is known that the problem is polytime and a concerted effort is made.

2. Realizing Constructive Proofs

A constructive proof of a theorem provides computational information. For example, if the theorem has the form $\forall x \exists y A(x, y)$ the proof should (at least implicitly) provide an algorithm for computing a function f such that $\forall x A(x, f(x))$ holds. We say that f *realizes* the assertion $\forall x \exists y A(x, y)$. We are interested in studying the complexity of functions f needed to realize such assertions; for example, determining whether there is a polytime such f. The notion of a *feasibly constructive* proof is discussed in [CU]. A necessary condition for a proof of a theorem to be feasibly constructive is that it provide polytime algorithms for any functions needed to realize the theorem.

Here are some theorems with feasibly constructive proofs:

Kuratowski's Theorem. Every finite graph G can either be embedded into the plane, or one of the graphs K_5 and $K_{3,3}$ can be homeomorphically embedded in G. If f is a realizing function, then $f(G)$ is either a planar embedding of G or identifies an embedding of either K_5 or $K_{3,3}$ in G. One can in fact infer a polytime such f from the usual proof of Kuratowski's theorem.

Many theorems of graph theory and finite combinatorics can be feasibly realized in a similar manner, including Hall's theorem and Menger's theorem.

The Extended Euclidean Algorithm. For all integers a and b there exist integers x, y, and d such that $d = ax + by$ and $d|a$ and $d|b$. Again Euclid's proof shows how x, y, and d can be computed from a and b in time at worst $O((\log a + \log b)^2)$.

An interesting example whose feasibility status is unknown is Fermat's "Little Theorem", which we state in contrapositive form: For all integers n and a there exists an integer d such that if $0 < a < n$ and $a^{n-1} \not\equiv 1 \pmod{n}$ then $1 < d < n$ and d divides n. A feasibly constructive proof of this theorem would provide a polytime f which produces a divisor $d = f(a, n)$ for n whenever a and n satisfy the hypotheses above. Such an f would provide a major breakthrough in the quest for practical algorithms for large integer factorization. This is because if $n > 1$ is composite and not one of the rare "Carmichael numbers" [HW] then at least half of the integers a such that $0 < a < n$ satisfy $a^{n-1} \not\equiv 1 \pmod{n}$. (The a's in the multiplicative group $\mathbb{Z}_n^*$ which satisfy $a^{n-1} \equiv 1 \pmod{n}$ form a proper subgroup of $\mathbb{Z}_n^*$, assuming n is not a Carmichael number.)

3. Type 2 Functions

Higher type functions take functions (as well as numbers) as arguments. We may as well suppose that such functions take numbers (as opposed to functions) as values, because of the following device. If $F : \mathbb{N}^{\mathbb{N}} \to \mathbb{N}^{\mathbb{N}}$, then we define $F_1 : \mathbb{N}^{\mathbb{N}} \times \mathbb{N} \to \mathbb{N}$ by

$$F_1(g, x) = F(g)(x).$$

Thus F_1 is for practical purposes the same as F.

Higher type functions may be needed to realize theorems for two reasons. First, the theorem may make an assertion about functions, as in $\forall g \forall x \exists y A(g, x, y)$. Second, even if the subject matter is numbers rather than functions, the usual way of realizing an implication $A \to B$ is by a function F such that for all objects g realizing A, $F(g)$ realizes B.

In order to discuss the complexity of such functions F we must explain how to compute them. For now, we restrict attention to "type 2" functions $F(g, x)$ which take as arguments a numerical function $g : \mathbb{N} \to \mathbb{N}$ and a number $x \in \mathbb{N}$. An algorithm for computing F takes x as input in the ordinary way, but it accesses the function g via an "oracle". Thus a machine executing the algorithm has a special query register (or tape). When the machine assumes a query state with a number y in this register, the "oracle" replaces y by $g(y)$.

This is the only mechanism by which the machine can acquire knowledge of the input g. It follows that for fixed g and x, the value $F(g,x)$ can only depend on $g(y)$ for finitely many values y.

In measuring the runtime of the machine which computes $F(g,x)$, we shall make the convention that each oracle call $g(y)$ takes just $|g(y)|$ steps. This pins down the definition of runtime, but does not solve the problem of defining a complexity class, such as polytime, for type 2 functions. For type 1 functions, polytime means time polynomial in the length of the input. To generalize this to type 2, we must decide what is the "length" of an input function $g : \mathbb{N} \to \mathbb{N}$. The definition below is a first step in that direction.

Definition 5. A type 2 function F is in the class OPT (oracle polytime) iff there is an algorithm which computes F whose runtime on input (g,x) is bounded by a polynomial in m and $|x|$, where m is the length of the maximum query answer $g(y)$ returned during the computation.

If we restrict attention to inputs g which are 0-1 valued, then OPT provides a perfectly reasonable way to formalize polytime for type 2 functions. This was the original method used to define a polytime version of Turing reducibility between sets [Coo1].

However, for arbitrary input functions g, OPT is too broad to be identified with polytime. For example, if

$$F(g,x) = g^{|x|}(x)$$

(g composed with itself $|x|$ times and evaluated at x) then F is in OPT. But if we define g_0 by $g_0(y) = y^2$, and let $f(x) = F(g_0,x)$, then $f(x) = x^{2^{|x|}}$, which is not polytime. Thus there are type 2 functions in OPT which, when interpreted as operators on the set of type 1 functions, do not preserve the set of polytime functions.

Nevertheless, we will take $F \in$ OPT to be a necessary condition for F to be intuitively polytime. This is enough to show that the pigeon hole principle below has no feasibly constructive proof.

Proposition 6 (Pigeon Hole Principle). $\forall g \forall x \exists a \exists b[a,b \leq x$ *and* $(g(a) \geq x$ *or* $(a \neq b$ *and* $g(a) = g(b)))]$

A functional F realizing the above proposition would satisfy $F(g,x) = \langle a,b \rangle$, where $\langle a,b \rangle$ is a natural number coding the pair (a,b) using a standard pairing function, and a and b satisfy the condition in Proposition 6.

Theorem 7. *No function F in* OPT *realizes the Pigeon Hole Principle.*

The proof is an easy consequence of the definition of OPT. Consider an algorithm which computes an F which realizes the Pigeon Hole Principle and fix some input $x > 0$. Each time the algorithm poses a distinct new query $y \leq x$ we choose the answer $g(y)$ to be the least natural number not yet given as a query answer for this computation. In this way we can force the algorithm to query $g(y)$ for each y in $\{0,1,\ldots,x\}$ before it finds a and b satisfying Proposition 6.

Thus the algorithm must make at least $x + 1$ different queries, a number which grows exponentially in $|x|$.

4. The Basic Polytime Functions

As mentioned above, it seems that some definition of length for functions must be given in order to give a proper generalization for polytime to type 2 functions. One possibility is to define the length $|g|$ of g to be a function, as follows:

Definition 8. If $g : \mathbb{N} \to \mathbb{N}$, then the length $|g|$ of g is the type one function defined by

$$|g|(n) = \max_{|y| \leq n} |g(y)|.$$

This definition and the development below appears in [Kap] and [KC].

We define a type 2 polynomial to be any type 2 function $P(L, n)$ represented by a "polynomial expression" in the type 1 variable L and the numerical variable n. More formally, we give a recursive definition of *polynomial expression* as follows:

a) The variable n and the constants 0 and 1 are each polynomial expressions.
b) If P and Q are polynomial expressions, so are $P + Q$, $P \cdot Q$, and $L(P)$.

Each polynomial expression represents a type 2 function (which by definition is a type 2 polynomial) in the obvious way. An example of a type 2 polynomial is

$$P(L, n) = L(L(n + 3) \cdot L(n)).$$

Definition 9. The type 2 functional $F(g, x)$ is *basic polytime* iff for some type 2 polynomial $P(L, n)$ some oracle algorithm computes $F(g, x)$ within $P(|g|, |x|)$ steps, for all g and x.

It can be shown that every basic polytime function is in OPT. Thus, for example, the bounded maximum function

$$M(g, x) = |g|(|x|) = \max_{|y| \leq |x|} |g(y)|$$

is not in OPT and hence not basic polytime, since $2^{|x|}$ oracle calls are necessary to evaluate $M(g, x)$ in case g is identically 0. This shows that the permissible runtime bounds for basic polytime functions are not necessarily themselves basic polytime, because $M(g, x) = P(L, n) = L(n)$ is a type 2 polynomial, where $L = |g|$ and $n = |x|$. On the other hand the length maximum function

$$LM(g, x) = \max_{y \leq |x|} |g(y)|$$

is basic polytime, with runtime bound

$$P(L, n) = O(n \cdot L(n)),$$

where $n = |x|$ and $L = |g|$.

The basic polytime functions have nice closure properties, and (unlike OPT functions) they preserve the set of type 1 polytime functions. What is more remarkable they turn out essentially to coincide with the polynomial operators defined by Mehlhorn [Meh] in 1976. The latter functions were defined using Cobham's [Cob] limited recursion on notation, and the definition does not mention machines or runtimes. Polynomial operators were introduced to generalize the notion of polytime reducibility from sets to that of functions.

Theorem 10 [Kap, KC]. *A type 2 function $F(g,x)$ is basic polytime iff it is a polynomial operator.*

It is straightforward to prove the "if" direction in the above result. The converse requires considerable effort, however, because of the necessity of pre-bounding the growth rate each time the operation of limited recursion on notation is applied to a function.

Theorem 10 reinforces the conclusion that the basic polytime functionals form a natural robust class. However we will give an example in the next section of a type 2 function outside the class that still seems to merit the label "polytime".

5. Realizing Well-Quasi-Orders

Definition 11. Let Q be a set and $\preceq$ be a binary relation in Q. Then $(Q,\preceq)$ is a *quasi-order* iff $\preceq$ is reflexive and transitive. $(Q,\preceq)$ is a well-quasi-order (wqo) if in addition, for every infinite sequence $q_0, q_1, \ldots$, from Q, there exists indices i, j with $i < j$ and $q_i \preceq q_j$.

Note that $(\mathbb{N},\leq)$ is a wqo; and in fact every well-order is a wqo. A more interesting example is $(\mathcal{G},\leq_m)$, where $\mathcal{G}$ is the set of finite graphs and $H \leq_m G$ iff H is a minor of G. Robertson and Seymour's monumental work [RS], which will cover more than 1500 pages, shows that $(\mathcal{G},\leq_m)$ is a wqo. An immediate corollary is the fact that each minor-closed graph family has a finite obstruction set, as mentioned in Sect. 1.

Robertson and Seymour's proof is not only long, it is nonconstructive. Yet the assertion that some pair $(Q,\preceq)$ is a wqo should have computational significance, as suggested by the following definition.

Definition 12. A function F *realizes* a wqo $(Q,\preceq)$ iff for all $g : \mathbb{N} \to Q$, $F(g) = \langle i,j \rangle$, where $i < j$ and $g(i) \preceq g(j)$.

We are interested in the complexity of the functions F required to realize a given wqo $(Q,\preceq)$. This gives an indication of what techniques are required to prove that $(Q,\preceq)$ is a wqo. In the case of $(\mathbb{N},\leq)$, it turns out that there is a realizing basic feasible functional. We first note however that the "brute force" algorithm is not polytime. By brute force we mean: query $g(0), g(1), g(2), \ldots g(j)$ until $g(j-1) \leq g(j)$. This algorithm could require up to $g(0)+1$ queries, and is not in OPT.

A basic polytime realizing function $F(g)$ for $(\mathbb{N}, \leq)$ can be computed using binary search. We initialize $a = 0$ and $b = g(0) + 1$. First compare $g(a)$ and $g(b)$; if $g(a) \leq g(b)$ then output $< a, b >$ and halt. Otherwise we note that

$$(*) \qquad 0 < g(a) - g(b) < b - a.$$

Let $c = \lfloor \frac{1}{2}(a + b) \rfloor$. Assign to the pair (a, b) either (a, c) or (c, b) as follows. If after one of these assignments $a < b$ and $g(a) \leq g(b)$ then make that assignment, output $< a, b >$, and halt. Otherwise choose the assignment which maintains the invariant (*) and repeat the above starting with the computation of c.

This algorithm terminates after at most $|g(0)|$ iterations. The largest query argument is $g(0) + 1$, so the time charged for each query is at most $|g|(|g(0) + 1|)$. Hence the runtime is bounded by $P(L) = O(L(0) \cdot (L(L(0) + 1) + L(0)))$, where $L = |g|$. Hence the functional it computes is basic polytime.

As a second example, consider the lexicographical order $(\mathbb{N} \times \mathbb{N}, \preceq)$, where $(a, b) \preceq (a', b')$ iff either $a < a'$ or ($a = a'$ and $b \leq b'$). Then no basic polytime functional realizes this wqo. In fact no polytime function realizes this wqo under any reasonable definition of polytime. To see this, we define for each parameter $c \in \mathbb{N}$ a sequence $g_c(0), g_c(1), g_c(2), \ldots$ of pairs as follows:

$$\begin{array}{l}
\langle c, c \rangle, \langle c, c-1 \rangle, \ldots, \langle c, 0 \rangle \\
\langle c-1, c^{|c|} \rangle, \langle c-1, c^{|c|} - 1 \rangle, \ldots, \langle c-1, 0 \rangle \\
\langle c-2, c^{|c|^2} \rangle, \ldots, \langle c-2, 0 \rangle \\
\quad \vdots \\
\langle 0, c^{|c|^c} \rangle, \ldots, \langle 0, 0 \rangle \\
\langle 0, 0 \rangle, \ldots \ .
\end{array}$$

We argue below that the function f defined by $f(c, i) = g_c(i)$ is polytime computable. Suppose however that we assume that F realizes the wqo $(\mathbb{N} \times \mathbb{N}, \preceq)$ and define h by

$$(**) \qquad h(c) = F(\lambda i f(c, i)) \ .$$

Since the sequence $g_c(0), g_c(1), g_c(2), \ldots$ is strictly decreasing with respect to $\preceq$ for at least the first $c^{|c|^c}$ values, it follows that $j > c^{|c|^c}$, where $h(c) = \langle i, j \rangle$. Hence h is not polytime. Thus (**) indicates a sense in which F does not preserve polytime functions.

To see that f is polytime computable, note that if $g_c(i) = \langle c-k, c^{|c|^k} \rangle$ for some k, $1 \leq k \leq c$, then $i > c^{|c|^{k-1}}$. Hence $|c|^k = O(|i|)$, so in general $|g_c(i)| = O(|i|^2 + |c|)$.

As a third example, consider the length lexicographical order $(\mathbb{N} \times \mathbb{N}, \preceq')$, defined by $(a, b) \preceq' (a', b')$ iff either $|a| < |a'|$ or ($|a| = |a'|$ and $|b| \leq |b'|$). We then have

Theorem 13 [Coo2]. *No basic polytime F realizes $(\mathbb{N} \times \mathbb{N}, \preceq')$.*

Nevertheless, the brute force algorithm defines a realizing function F which seems intuitively to be polytime. In particular, F is in OPT, and F preserves polytime functions in the sense of (**). This suggests that the class of basic polytime functions should be enlarged if it is to include all intuitively polytime type 2 functions (see [Coo2]).

As a fourth and final example we return to graph minors.

Theorem 14 [FRS]. *Suppose F realizes $(\mathcal{G}, \leq_m)$. Then there is $f : \mathbb{N} \times \mathbb{N} \to \mathcal{G}$ such that $f(c, i)$ has at most $c + i$ nodes, and if*

$$h(c) = F(\lambda i f(c, i)),$$

then h grows faster than any function provably recursive in Peano Arithmetic.

Peano Arithmetic refers to the standard first order theory of $\mathbb{N}$ under $+$ and $\times$. Roughly speaking, it can formalize any argument not involving infinite sets. In particular, Ackermann's function is provably recursive in Peano Arithmetic. Thus any function realizing $(\mathcal{G}, \leq_m)$ is infeasible in a very strong sense.

6. Type 3 Functions

The most natural way to realize the Weak König's Lemma (König's Lemma for binary trees) is via a type 3 function. We state the lemma in contrapositive form, to give it computational content.

Proposition 15 (Weak König's Lemma (WKL)). *If every path is finite in a binary tree, then there is a uniform bound on path length in the tree.*

A realizing functional B should take a function K, which bounds paths, to a uniform bound $B(K)$. To make sense of the path-bounding function K, its argument g should code a potentially infinite branch. This branch can be specified by a function $g : \mathbb{N} \to \{0, 1\}$, where the binary sequence $g(0), g(1), g(2), \ldots$ indicates successive choices (left or right) of children of nodes in the tree, starting with the root. Then $K(g)$ is an upper bound on the longest path in the tree which is specified by an initial segment of the sequence $g(0), g(1), \ldots$. Thus two different functions g and g' might both be extensions of the same finite maximal path in the tree, and we do not require that the two upper bounds $K(g)$ and $K(g')$ for this path be the same. Any total function $K \in \mathbb{N}^{\{0,1\}^{\mathbb{N}}}$ provides enough information to specify a uniform bound $B(K)$ on path length, even though K does not uniquely determine the tree.

Definition 16. The type 3 function B *realizes* WKL iff for all $K \in \mathbb{N}^{\{0,1\}^{\mathbb{N}}}$ and all $g \in \{0, 1\}^{\mathbb{N}}$ there exists $g' \in \{0, 1\}^{\mathbb{N}}$ such that $g(i) = g'(i)$ for all $i \leq K(g')$, and $K(g') \leq B(K)$.

Notice that we do not require $K(g) \leq B(K)$, but only $K(g') \leq B(K)$ for some g' which extends the same finite maximal path as g. (For an alternative definition of realizing WKL see [Fef].)

It turns out that not just the complexity of B is in question, but there is even doubt about whether a computable B exists which realizes WKL in the above sense. In fact computability theory for type 3 functions suffers because there are at least two different plausible but inconsistent definitions of computable function in the literature. Fortunately these definitions are essentially equivalent for type 2 (and type 1) functions; namely *computable* means computable by a Turing machine with oracles. The two definitions we have in mind are Kleene's general recursive functionals and the so-called recursively countable functionals (see [GH] for an excellent presentation of both).

It turns out that no Kleene general recursive functional B realizes WKL. Roughly speaking, this is because an algorithm for computing B can access its input function K only by posing query arguments g to an oracle, which returns $K(g)$. Further, the only possible arguments g that can be posed are recursive. But it turns out (see [GH]) that there exists an infinite binary tree with no infinite recursive branch. Of course this tree has arbitrarily long finite branches. Thus one can define K such that $K(g)$ is an upper bound for the longest path which g extends, when g is recursive, and $K(g)$ is arbitrary if g is not recursive. When presented with such an input K, no Kleene algorithm can compute a sensible value for $B(K)$.

On the other hand, there is a recursively countable function B which realizes WKL. This is partly because recursively countable type 3 functions have as their domain only the set of "continuous" type 2 functions K. Here K is *continuous* iff for each $g \in \{0,1\}^{\mathbb{N}}$, there exists $m \in \mathbb{N}$ such that $K(g)$ is determined by the restriction of g to $\{0, 1, \ldots, m-1\}$. (In other words, $K(g') = K(g)$ for all g' such that $g'(i) = g(i)$, $0 \le i < m$.) A recursively countable oracle only queries its input at finite restrictions $g_0 = g \mid \{0, 1, \ldots, m-1\}$ of a function g. The oracle may return $K(g)$, or the symbol $\perp$, denoting "unknown". But it can only return $\perp$ for a finite number of restriction sizes m. By systematically querying K at all possible finite 0-1 sequences g_0 of length m, where $m = 1, 2, \ldots$ the algorithm will (by König's Lemma) eventually reach a value of m for which the oracle can give a value $K(g)$ for each g_0 of length m. The algorithm then outputs the maximum of all these values of $K(g)$ as the value of $B(K)$.

7. Conclusion

One general phenomenon we have seen is that the higher the type level, the harder it is to pin down suitable definitions for complexity classes. Thus for type 3 functions we are not sure what computable means. For type 2 functions we are not sure what polytime means. We might add that for type 1 functions we are not sure what linear time means.

We have talked about the notion of realizing theorems informally, but logicians have given various general formal definitions of realizability [Tro]. This development includes specification of a formal system whose theorems are to be realized. Buss's system IS_2^1 [Bus] is especially interesting here, since in some sense it captures first order polynomial time reasoning. In particular, the functions provably recursive in IS_2^1 are precisely the polytime functions. The underlying logic is the intuitionistic predicate calculus, and hence the proofs are "feasibly constructive" (see [CU]).

In [CU] a class of functionals of all finite types is introduced using the typed lambda calculus and a type 2 "recursor" for limited recursion on notation. This class, called the *basic feasible functionals* in [CK], turns out at type level 2 to coincide with the basic polytime functionals discussed in Sect. 4. In [CU] it is shown that the basic feasible functionals provide realizability interpretations for the system IS_2^1 in two different senses: Kreisel's "modified realizability" and Gödel's "*Dialectica* interpretation". These interpretations are used to study the proving power of IS_2^1 and related systems.

A major goal of studying formal systems such as IS_2^1 is proving independence results and conservative extension results. An example of the latter, proved in [Fer] and [Fef], is that when a version of Weak König's Lemma is added as an axiom to IS_2^1, there are no new resulting theorems of the form $\forall x \exists y A(x, y)$, where $A(x, y)$ is a polytime relation.

A nice (but so far elusive) result would be to show that a standard theorem, such as Fermat's Little Theorem, is independent of IS_2^1. This would not show that the factor finding function $f(a, n)$, discussed in Sect. 2, is not polytime computable. But it would show that no algorithm for such a function has a feasibly constructive correctness proof. The significance of this lies in the following general observation: Natural problems solvable by polytime functions can be solved by polytime functions with algorithms which have correctness proofs that involve only polytime notions.

Acknowledgements. I am grateful to Erich Kaltofen, Stephen Bellantoni, Toni Pitassi, and other colleagues for pointing out errors and suggesting improvements in the original manuscript.

References

[AH] Adleman, L.M., Huang, M.-D.A.: Recognizing primes in random polynomial time. Proc. 19th ACM Symp. on Theory of Computing 1987, pp. 462–469

[AM] Atkin, O.L., Morain, F.: Elliptic curves and primality proving. Math. Comput. (September 1990) (submitted)

[Ber] Berlekamp, E.R.: Factoring polynomials over large finite fields. Math. Comp. **24** (1970) 713–735

[Bus] Samuel R. Buss: The polynomial hierarchy and intuitionistic bounded arithmetic. In: Selman, A.L. (ed) Structure in Complexity Theory. Lecture Notes in Computer Science, vol. 223. Springer, Berlin Heidelberg New York 1986, pp. 77–103

[Cob] Cobham, A.: The intrinsic computational difficulty of functions. Proc. of the 1964 International Congress for Logic, Methodology, and the Philosophy of Science (Y. Bar-Hillel, ed.). North-Holland, Amsterdam 1964, pp. 24–30

[Coo1] Cook, S.A.: The complexity of theorem-proving procedures. Proc. 3rd ACM Symp. on Theory of Computing 1971, pp. 151–158

[Coo2] Cook, S.A.: Computability and complexity of higher type functionals. Proc. MSRI Workshop on Logic from Computer Science, Y. Moschovakis, ed. (1990) (to appear)

[CK] Cook, S.A., Kapron, B.M.: Characterizations of the basic feasible functionals of finite type. Proc. MSI Workshop on Feasible Mathematics, S. Buss and P. Scott, eds. Birkhäuser 1990, pp. 71–96

[CU] Cook, S., Urquhart, A.: Functional interpretations of feasibly constructive arithmetic. Technical Report 210/88, University of Toronto (1988). Extended Abstract in Proc. 21st ACM Symp. on Theory of Computing 1989, pp. 107–112 (to be submitted for publication)

[Edm] Edmonds, J.: Paths, trees, flowers. Canad. J. Math. **17** (1965) 449–467

[Fef] Feferman, S.: Milking the "Dialectica" interpretation. Manuscript (1990)

[FL] Fellows, M., Langston, M.: Nonconstructive tools for proving polynomial time decidability. JACM **35**(3) (1988) 727–739

[Fer] Ferreira, F.: Polynomial time computable arithmetic and conservative extensions. Ph.D. thesis, Dept. of Mathematics, Pennsylvania State University 1988

[FRS] Friedman, H., Robertson, N., Seymour, P.: The metamathematics of the graph minor theorem. Contemp. Math. **65** (1987) 229–261

[GH] Gandy, R.O., Hyland, J.M.E.: Computable and recursively countable functions of higher type. Logic Colloquium 76, North-Holland 1977, pp. 407–438

[GJ] Garey, M.R., Johnson, D.S.: Computers and intractability: A guide to the theory of NP-completensss. Freeman, San Francisco 1979

[vzG] von zur Gathen, J.: Irreducibility of multivariate polynomials. J. Comp. System Sci. **31** (1985) 225-264

[GK] Goldwasser, S., Kilian, J.: A provable correct and probably fast primality test. Proc. 18th ACM Symp. Theory Comp. 1986, pp. 316–329

[HW] Hardy, G.H., Wright, E.M.: An introduction to the theory of numbers, fifth edn. Oxford 1979

[KalT] Kaltofen, E., Trager, B.: Computing with polynomials given by black boxes for their evaluations: Greatest common divisors, factorization, separation of numerators and denominators J. Symb. Comp. **9**(3) (1990) 301–320

[Kap] Kapron B.: Feasible computation in higher types. Ph.D. thesis, University of Toronto 1990

[KC] Kapron, B., Cook, S.: A new characterization of Mehlhorn's polynomial time functionals. Manuscript (1990) (submitted)

[Kha] Khachian, L.G.: A polynomial time algorithm for linear programming. Dokl. Akad. Nauk SSSR **244**(5) (1979) 1093–1096. Translated in Sov. Math. Dokl. **20**, 191–194

[LLL] Lenstra, A.K., Lenstra, H.W., Lovász, L.: Factoring polynomials with rational coefficients. Math. Ann. **261** (1982) 515–534

[Meh] Mehlhorn, K.: Polynomial and abstract subrecursive classes. JCSS **12** (1976) 147–148

[Mey] Meyer, A.R.: Weak monadic second-order theory of successor is not elementary-recursive. Lecture Notes in Mathematics, vol. 453. Springer, Berlin Heidelberg New York 1975, pp. 132–154

[Mil] Miller, G.L.: Riemann's hypothesis and tests for primality. J. Comp. System Sci. **13** (1976) 300–317

[Rab] Rabin, M.O.: Probabilistic algorithms for testing primality. J. Number Theory **12** (1980) 128–138

[RS] Robertson, N., Seymour, P.: Graph minors $I, \ldots, XV$. J. Combinatorial Theory (Ser. B), beginning 1983. (Not all papers are completed.)

[Scho] Schönhage, A.: Factorization of univariate integer polynomials by diophantine approximation and an improved basis reduction algorithm. Proc. ICALP '84. Lecture Notes in Computer Science, vol. 172. Springer, Berlin Heidelberg New York 1984, pp. 436–447

[Schw] Schwartz, J.T.: Fast probabilistic algorithms for verification of polynomial identities. J. ACM **27** (1980) 701–717

[Sto] Stockmeyer, L.J.: The complexity of decision problems in automata theory and logic. Ph.D. thesis, MIT (1974). Report TR-133, MIT Laboratory for Computer Science

[SS] Solovay, R.M., Strassen, V.: A fast Monte-Carlo test for primality. SIAM J. Comp. **6** (1977) 84–85. Correction: **7** (1978) 118

[Tro] Troelstra, A.S.: Metamathematical investigation of intuitionistic arithmetic and analysis. Lecture Notes in Mathematics, vol. 344. Springer, Berlin Heidelberg New York 1973

Conformal Field Theory and Cohomologies of the Lie Algebra of Holomorphic Vector Fields on a Complex Curve

Boris L. Feigin

Institute of Solid State Physics, USSR Academy of Sciences, Chernogolovka
142432 Moscow Region, USSR

Introduction

In this text we shall deal with only one aspect of the conformal field theory – the concept of the modular functor. Recall the main structure of the conformal field theory. Let M be a complex curve. First of all the theory associates to M a finite dimensional vector space $H(M)$. The construction of $H(M)$ depends on the choice of a complex structure on M, but really the correspondence $M \to H(M)$ is topological by nature. Our approach to the definition of $H(M)$ is the following. Let $\mathrm{Dif}(M)$ be the group of all C^∞-diffeomorphisms of M. The group $\mathrm{Dif}(M)$ acts on the space of all complex structures on M.

If a group G acts on a manifold S and n is a point of S we can define G_n – the stability group of n. If G acts transitively on S, then $S \cong G/G_n$. It is possible to define the object "G_n" for a non-transitive action so that we have always $S \cong G/$"G_n". Assume that a group H acts transitively on S and H_n – the stability group of n; define G_n as the following diagram in the category of groups: $G \to H - H_n$. In other words, G_n is the "intersection" G and H_n in H.

So we can assign to a complex structure "com" on M the "subgroup" $\mathrm{Dif}_{\mathrm{com}}(M) \subset \mathrm{Dif}(M)$. It can be shown, that for different complex structures com_1 and com_2 the "subgroups" $\mathrm{Dif}_{\mathrm{com}_1}(M)$ and $\mathrm{Dif}_{\mathrm{com}_2}(M)$ are "conjugate" in some "derived" sense.

We want to think about the space $H(M)$ as the space of cohomologies of the group $\mathrm{Dif}_{\mathrm{com}}(M)$ with coefficients in some one-dimensional representation. This representation is related to the value of the central charge of the theory. In this text we shall give some arguments in favour of this fact. Really, we shall construct the Lie algebra of the group $\mathrm{Dif}_{\mathrm{com}}(M)$, and realize $H(M)$ as the space of "integrable" homologies of this Lie algebra with coefficients in some distinguished one-dimensional representation.

So, from this point of view, a minimal conformal field theory is the investigation of the group of symmetries of a complex structure on a surface. By the same method we can explore the symmetries of other geometrical structures. For example, the Wess-Zumino theory is attached to the group of holomorphic transformations of a vector bundle on M.

Proceedings of the International Congress
of Mathematicians, Kyoto, Japan, 1990

In more general cases we have a sheaf of associative algebras on M (e.g. conformal field theory associated with a W-algebra). The modular functor is the homologies of this object.

The contents of the text are clear from the titles of the sections. The main result is a theorem of the Section VI.

Now a few words about the references. Information about continuous cohomologies of the Lie algebras of vector fields on smooth manifolds is contained in the book: D.B. Fuchs, *Cohomologies of infinite dimensional Lie algebras*, Plenum Publishing Corporation, New York 1986.

Conformal field theories, associated with Kac-Moody algebras are discussed by Akihiro Tsuchiya, Kenji Ueno, and Yasuhiko Yamada in the article *Conformal field theory on universal family of stable curves with gauge symmetries* (Advanced Studies in Pure Math. **19** (1989) 459–565).

The Segal's article in these Proceedings is devoted to the notion of modular functor. Determinant bundles on the moduli space of curves are explored in: A.A. Beilinson and V.A. Schechtman, *Determinant bundles and Virasoro algebras*, Comm. Math. Phys. **118** (1988) 651.

Minimal conformal field theories were constructed in: A.A. Belavin, A.M. Polyakov and A.B. Zamolodchikov, Nucl. Phys. **B241** (1984) 333.

The theory of singular support of representations of a Lie algebra is developped in O. Gabber, *The integrability of the characteristic variety*, Amer. J. Math. **103** (1981) 445.

In the article: B.L. Feigin and B.L. Tsygan, *Riemann-Roch theorem and Lie algebra cohomology I*, Proceedings of the winter school on geometry and physics, Srni, 9–16 January, 1988, pp. 15–52, you can find the calculations that may be useful for generalization of the results of this text on the manifolds of dimension greater than one.

Ideas of deformation theory are in the works of Gerstenhaber, Quillen, Halperin, Stasheff, and others.

The construction of the modular functor in the Virasoro case is contained in: B.L. Feigin and D.B. Fuchs, *Cohomology of some nilpotent subalgebras of the Virasoro and Kac-Moody Lie algebras*, IGP, vol. 5, no. 2 (1988).

I. Definition of the Lie Algebra of Vector Fields on a Complex Manifold

Let M be a complex manifold of dimension n. What is the Lie algebra of holomorphic vector fields on M?

Let T be the tangent bundle of M, $T = T^{0,-1} \oplus T^{-1,0}$ – the Hodge decomposition of the complexification of T. The space of C^∞ sections $\Gamma(T^{-1,0})$ of the bundle $T^{-1,0}$ is closed with respect to the bracket. Locally, an element of $\Gamma(T^{-1,0})$ has the form

$$\sum_i f_i(z_1, \dots, z_n, \bar{z}_1, \dots, \bar{z}_n) \frac{\partial}{\partial z_i}.$$

Denote by $T^{i,j}$ the bundle $(\bigwedge^{-i}T^{-1,0})\otimes(\bigwedge^{-j}T^{0,-1})$, where i,j are integers. In particular $T^{1,1}$ is the bundle of volume forms. Consider the Dolbeault resolution of the sheaf of holomorphic vector fields on M:

$$0 \to T^{-1,0} \xrightarrow{\bar\partial} T^{-1,1} \xrightarrow{\bar\partial} T^{-1,2} \to \cdots T^{-1,n} \to 0.$$

The bracket on the space $\Gamma(T^{-1,0})$ is extended in a natural way on the space $\Gamma(T^{-1,0})\oplus\Gamma(T^{-1,1})\oplus\cdots\oplus\Gamma(T^{-1,n})$. (Since elements of $\Gamma(T^{-1,i})$ are (0, 1)-forms with values in the bundle $T^{-1,0}$). It is easy to see that this extended bracket defines a structure of differential. Lie superalgebra on the following complex

$$0 \to \Gamma(T^{-1,0}) \xrightarrow{\bar\partial} \Gamma(T^{-1,1}) \xrightarrow{\partial} \Gamma(T^{-1,2}) \to \cdots \Gamma(T^{-1,n}) \to 0.$$

We shall call this differential superalgebra the Lie algebra of holomorphic vector fields on M since it is the space of global sections of the resolution of the sheaf of holomorphic vector fields on M. We shall denote this complex by $\Gamma(\mathrm{Lie}(M))$. It is clear that for affine M the algebra $\Gamma(\mathrm{Lie}(M))$ is equivalent to $\mathrm{Lie}(M)$, where $\mathrm{Lie}(M)$ is the Lie algebra of holomorphic vector fields on M. Another approach to the definition of the Lie algebra of vector fields on a manifold is the following. Choose a covering $\{U_i, i\in I\}$ of M, where each U_i is affine. Let us associate to any subset $i_1 < i_2 < \cdots < i_l$ of I the Lie algebra $\mathrm{Lie}(U_{i_1}\cap U_{i_2}\cap\cdots\cap U_{i_l})$. Using this set of data we can define by standard method a cosimplicial object in the category of Lie algebras. We omit details of the construction, but it is possible to show that such cosimplicial Lie algebras for different coverings are equivalent in some sense. In particular, we can use the covering consisting of all open sets. So, we shall call this cosimplicial object the Lie algebra of vector fields on M and denote it by $\mathrm{Lie}_\Delta(M)$. The differential Lie superalgebra $\Gamma(\mathrm{Lie}(M))$ defines in the standard way a cosimplicial Lie algebra which is equivalent to $\mathrm{Lie}_\Delta(M)$.

Remark. Let A be a group, ξ an A-bundle on M, $\mathrm{End}(\xi)$ the bundle of endomorphisms of ξ. Denote by $\mathrm{End}(\xi)\otimes T^{0,i}$ the bundle of $(0,i)$-forms with values in $\mathrm{End}(\xi)$. The Dolbeault complex of $\Gamma\,\mathrm{End}(\xi)$:

$$0 \to \Gamma(\mathrm{End}(\xi)\otimes T^{0,0}) \xrightarrow{\bar\partial} \Gamma(\mathrm{End}(\xi)\otimes T^{0,1}) \xrightarrow{\bar\partial} \cdots \to \Gamma(\mathrm{End}(\xi)\otimes T^{0,n}) \to 0$$

has an obvious structure of differential Lie superalgebra. This object is the natural candidate for the Lie algebra of the endomorphisms of the bundle. We can also give the cosimplicial version of this definition, using coverings of M. Note that all statements in this text have their counterpart for the algebra $\Gamma\,\mathrm{End}(\xi)$, or for the cosimplicial version $\mathrm{End}_\Delta(\xi)$.

II. Continuous Cohomologies of Lie(M)

Using the methods of the Gel'fand-Fuchs theory of the continuous cohomologies of the Lie algebra of vector fields on a smooth manifold it is possible to define and to calculate the continuous cohomologies of the differential Lie superalgebra $\Gamma\,\mathrm{Lie}(M)$. First we recall main definitions and results from the Gel'fand-Fuchs theory.

Let N be a C^∞-manifold and $L(N)$ be the Lie algebra of C^∞-vector fields on N; $L(N)$ is a topological Lie algebra with respect to the C^∞-topology. Continuous cohomologies of $L(N)$ are defined as the cohomologies of the subcomplex of the standard cohomological complex of $L(N)$ which consists of continuous cochains. The procedure of calculation of the continuous cohomologies $H_c^*(L(N))$ is divided into two parts – "local" and "global". The aim of the "local" part is the calculation of the continuous cohomologies of the Lie algebra W_n of formal vector fields on $\mathbb{R}^n$, where $n = \dim N$. An element of W_n is a vector field on the formal neighbourhood of the origin in $\mathbb{R}^n$. Gel'fand and Fuchs proved, that $H_c^*(W_n)$ is isomorphic to the algebra of cohomologies of some topological space X_n. The space X_n may be described as a preimage of $2n$-skeleton of the base space of the universal $\mathrm{GL}(n, \mathbb{C})$-bundle in the total space of this bundle. It is important for us that X_n is a $\mathrm{GL}(n, \mathbb{C})$-space and therefore a $\mathrm{GL}(n, \mathbb{R})$-space, since $\mathrm{GL}(n, \mathbb{R}) \subset \mathrm{GL}(n, \mathbb{C})$.

The aim of the "global" part of the calculation is to "sew" the local calculations at all points of N. To do this let us associate with the tangent bundle of N the bundle $\varrho : U \to N$, of which each fiber is isomorphic to X_n. The main theorem on continuous cohomologies asserts that the algebra $H_c^*(L(N))$ is isomorphic to the cohomologies of the topological space of all sections of the bundle ϱ. Note that in the most important case (for us) where $n = 1$ the space X_n is homotopic to the three-dimensional sphere S^3.

The differential Lie superalgebra $\Gamma(\mathrm{Lie}(M))$ is also equipped with C^∞-topology. So we can define the continuous cohomologies and calculate them using the same technique. Let us formulate the result in the case $\dim M = 1$.

Proposition. *In the case* $\dim M = 1$, *the algebra* $H_c^* \Gamma(\mathrm{Lie}(M))$ *is isomorphic to the algebra of cohomologies of the space of all continuous maps* $\mathrm{Hom}(M, S^3)$ *from M (as a topological space) into* S^3.

It is not hard to find out the structure of the algebra $H^*(\mathrm{Hom}(M, S^3), \mathbb{C})$. Consider the natural map $\theta : M \times \mathrm{Hom}(M, S^3) \to S^3$; $\theta(m, f) = f(m)$, where $m \in M$, $f \in \mathrm{Hom}(M, S^3)$. The corresponding map between cohomologies gives us a map $\mathbb{C} \cong H^3(S^3, \mathbb{C}) \to H^*(M, \mathbb{C}) \otimes H^*(\mathrm{Hom}(M, S^3), \mathbb{C})$ which we can rewrite as a set of three maps $\theta_i : H_i(M, \mathbb{C}) \to H^{3-i}(\mathrm{Hom}(M, S^3), \mathbb{C})$, $i = 0, 1, 2$. It can be proved that the algebra $H^*(\mathrm{Hom}(M, S^3), \mathbb{C})$ is a free skew-commutative algebra and the images of θ_1, θ_2, θ_3 are the space of generators. Now we shall write down the explicit formulas for the cochains in the standard complex of $\Gamma(\mathrm{Lie}(M))$ which represent the classes corresponding to the images of $\{\theta_i\}$.

We need some preparations. First we recall the definition of holomorphic projective connections on a one-dimensional manifold. A projective connection defines the following holomorphic differential operator of order three:

$$D : \Gamma(T^{-1,0}) \to \Gamma(T^{2,0})$$

such that in a neighbourhood of an arbitrary point three independent solutions of the equation $D(\varphi) = 0$ constitute the Lie algebra which is isomorphic to $\mathrm{sl}_2(\mathbb{C})$. This operator is defined by the projective connection up to a constant. We can fix this constant if we suppose that in the local coordinate Z the highest symbol of D is equal to $\left(\frac{\partial}{\partial Z}\right)^3$. An operator D with these properties defines a projective connection.

Remark. Fix a closed curve $\gamma \in M$ without self-intersections. After restriction on γ the operator D defines the operator of the second hamiltonian structure in the KdV theory. Note that a projective connection provides the family of operators:

$$D_i : \Gamma(T^{-1,i}) \to \Gamma(T^{2,i}).$$

Locally, the kernel of the operator D_i consists of antiholomorphic tensors of type (0, 1) with values in the kernel of $D_0 = D$. We suppose that all operators D_i have the same symbol. It is easy to see that D_i is adjoint to D_{1-i}. (Recall that for any differential operator $U : E \to F$, where E, F are vector bundles on a manifold, the adjoint operator acts from the bundle $F^* \otimes \mathrm{Vol}$ into $E^* \otimes \mathrm{Vol}$, where Vol is the bundle of volume forms, and * is a sign for the dual bundle.)

Let us consider the following maps:

$$\Psi_0 : \Lambda^3\Gamma(T^{-1,0}) \to \Gamma(T^{0,0}),$$
$$\Psi_1 : \Lambda^2\Gamma(T^{-1,0}) \to \Gamma(T^{1,0}).$$

Here the operator Ψ_1 is given by the formula: $\Psi_1(a, b) = (Da)b - a(Db)$, a, $b \in \Gamma(T^{-1,0})$. The map Ψ_0 is characterized up to a constant by the property: Ψ_0 is a $\Gamma(T^{-1,0})$-homomorphism. (Note that the spaces $\Gamma(T^{i,j})$ are $\Gamma(T^{-1,0})$-modules.) The explicit formula for Ψ_0:

$$\Psi_0(U_1, U_2, U_3) = \mathrm{Det}\begin{vmatrix} f_1 & f_2 & f_3 \\ f_1' & f_2' & f_3' \\ f_1'' & f_2'' & f_3'' \end{vmatrix},$$

$$U_i \in \Gamma(T^{-1,0}), \qquad U_i = f_i(z, \bar{z})\frac{\partial}{\partial z},$$

where z is a local coordinate on M, and "$'$" means the derivation $\frac{\partial}{\partial z}$. The value of the determinant does not depend on the choice of a coordinate. We need also the maps Ψ_2, Ψ_3.

$$\Psi_2 : \Lambda^2\Gamma(T^{-1,0}) \otimes \Gamma(T^{-1,1}) \to \Gamma(T^{0,1}),$$
$$\Psi_3 : \Gamma(T^{-1,0}) \otimes \Gamma(T^{-1,1}) \to \Gamma(T^{1,1}).$$

The formula for Ψ_3 is similar to the formula for Ψ_1, namely $\Psi_3(a, b) = D_0(a)\cdot b - a\cdot D_1(b)$, $a \in \Gamma(T^{-1,0})$, $b \in \Gamma(T^{-1,1})$. The map Ψ_2 is a $\Gamma(T^{-1,0})$-homomorphism and is given by the same determinant as Ψ_0, where $U_1 = f_1\frac{\partial}{\partial z}$, $U_2 = f_2\frac{\partial}{\partial z}$ and $U_3 = f_3\frac{\partial}{\partial z}d\bar{z}$.

Proposition. *There is a map φ from the standard homological complex of the Lie algebra $\Gamma(\mathrm{Lie}(M))$ into the de Rham complex of M. The construction of φ is clear from the following diagram. In the top of the diagram is a fragment of the homological complex and the de Rham complex is on the bottom.*

$$\begin{array}{l|l|l|l}
\Lambda^0\Gamma(T^{-1,0}) & \Gamma(T^{-1,0}) & \Lambda^2\Gamma(T^{-1,0}) & \Lambda^3\Gamma(T^{-1,0}) \\
 & \oplus & \oplus & \\
 & \Gamma(T^{-1,0})\otimes\Gamma(T^{-1,1}) & \Lambda^2\Gamma(T^{-1,0})\otimes\Gamma(T^{-1,1}) & \quad\Psi_0 \\
\varphi & \quad\Psi_3 & \Psi_2 & \\
 & & \Gamma(T^{1,0}) \leftarrow \quad \Psi_1 & \\
 & & \oplus & \quad\downarrow \\
 & \hookrightarrow \Gamma(T^{1,1}) & \hookrightarrow \Gamma(T^{0,1}) & \Gamma(T^{0,0})
\end{array}$$

Note that the de Rham complex here is reflected and shifted.

Now let β be an element of $H_i(M)$. We can assign a continuous cochain of the Lie algebra $\Gamma(\mathrm{Lie}(M))$ by the formula $u \to \langle \beta, \varphi(u)\rangle$, where u is a chain, and $\langle\ ,\ \rangle$ is the integration of $\varphi(u)$ over the cycle β. It is easy to see that this construction gives us the desired map $H_i(M) \to H_c^{3-i}(\Gamma(\mathrm{Lie}(M)))$.

Remark 1. The following part of our diagram

$$\begin{array}{ccccccc}
 & \leftarrow & \Lambda^2\Gamma(T^{-1,0}) & \leftarrow & \Lambda^3\Gamma(T^{-1,0}) & \leftarrow & \\
 & & \downarrow & & \downarrow & & \\
0 & \leftarrow & \Gamma(T^{1,0}) & \leftarrow & \Gamma(T^{0,0}) & \leftarrow & 0
\end{array}$$

is a morphism from the standard complex of the Lie algebra $\Gamma(T^{-1,0})$ into the Dolbeault complex of M. For non-compact M the continuous cohomologies of $\Gamma(T^{-1,0})$ is infinite-dimensional at least in dimension three. This map of complexes induces the map $V^* \to H_c^3(\Gamma(T^{-1,0}))$ where V is the space of holomorphic functions on M. But for compact M we have $\dim H_c^i(\Gamma(T^{-1,0})) < \infty$ for arbitrary i. Let us formulate the important corollary from the results of this section.

Corollary. *The dimension of the space $H_c^1(\Gamma(\mathrm{Lie}(M)))$ is equal to one if M is compact. It means that the algebra $\Gamma(\mathrm{Lie}(M))$ has a non-trivial one-dimensional representation.*

Remark 2. If $\mathfrak{U}$ is a finite dimensional Lie algebra and $v: \mathfrak{U} \to \mathrm{End}(V)$ is a finite dimensional representation of $\mathfrak{U}$, then the map $a \mapsto \mathrm{tr}\, v(a)$, $a \in \mathfrak{U}$ is a 1-cocycle of $\mathfrak{U}$. We can interpret this 1-cocycle as an element of the space $H^0(\mathfrak{U}, \mathfrak{U}^*)$. This construction can be easily generalized. The function $a^n \mapsto \mathrm{tr}(v(a^n))/n!$ is an element $\mathrm{tr}_v^{(n)}$ of the space $H^0(\mathfrak{U}, S^n(\mathfrak{U}^*))$. In other word, the representation v defines a functional $\mathrm{tr}_v \in H^0(\mathfrak{U}, S^*(\mathfrak{U}^*))$, $\mathrm{tr}_v = \sum \mathrm{tr}_v^{(n)}$. Let us consider the Lie superalgebra $\mathfrak{U} \oplus \xi\mathfrak{U} = \mathfrak{U} \otimes \mathbb{C}[\xi]$, where $\mathbb{C}[\xi]$ is a Grassmann algebra in one variable; $\mathfrak{U} \oplus \xi\mathfrak{U}$ is a graded

Lie superalgebra with $\deg(\mathfrak{U}) = 0$, $\deg(\xi\mathfrak{U}) = -1$. It is evident that $H^*(\mathfrak{U} \oplus \xi\mathfrak{U}) \cong H^*(\mathfrak{U}, S^*(\mathfrak{U}^*))$. So we have $\mathrm{tr}_v \in H^0(\mathfrak{U} \oplus \xi\mathfrak{U})$. This idea can be applied to the Lie algebra $\Gamma\,\mathrm{Lie}(M)$. Namely, using Gel'fand-Fuchs technique it is possible to find out the continuous cohomologies of the algebra $\Gamma\,\mathrm{Lie}(M) \oplus \xi\Gamma\,\mathrm{Lie}(M)$. This problem is also divided into two parts: "local" and "global". The local part consists of the calculation of the algebra $H_c^*(W_1 \oplus \xi W_1)$ where W_1 is the algebra of formal vector fields on a line. The answer is the following:

$$H_c^*(W_1 \oplus \xi W_1) \cong H_c^*(W_1) \oplus H_c^*(W_1, W_1^*) \oplus H_c^*(W_1, S^2 W_1^*) \oplus \cdots;$$

$$H_c^i(W_1) \cong \mathbb{C} \quad \text{if } i = 0, 3 \quad \text{and} \quad H_c^i(W_1) \cong 0 \quad \text{if } i \neq 0, 3;$$

$$H_c^2(W_1, S^i W_1^*) \cong H_c^3(W_1, S^i W_1^*) \cong \mathbb{C}$$

and in other dimensions the cohomologies are equal to zero ($i \neq 0$). The multiplication in the space $H_c^*(W_1 \oplus \xi W_1)$ is trivial. The cohomologies of the algebra $\Gamma\,\mathrm{Lie}(M) \oplus \xi\Gamma\,\mathrm{Lie}(M)$ are the same as those of the space of all continuous maps of M (as a topological space) into $\mathbb{Y}$, where $\mathbb{Y}$ is the space satisfying $H^*(\mathbb{Y}, \mathbb{C}) \cong H_c^*(W_1 \oplus \xi W_1)$. In particular, we can conclude that $H_c^0(\Gamma\,\mathrm{Lie}(M) \oplus \xi\Gamma\,\mathrm{Lie}(M))$ is an algebra of polynomials in an infinite set of variables $\sigma_1, \sigma_2, \ldots, \in H_c^0(\Gamma\,\mathrm{Lie}(M) \oplus \xi\Gamma\,\mathrm{Lie}(M))$. Therefore we can associate with a representation v of $\Gamma\,\mathrm{Lie}(M)$ an analog of trace – an element of the space $\mathbb{C}[\sigma_1, \sigma_2, \ldots]$.

III. Description of a 1-Cocycle of the Cosimplicial Lie Algebra $\mathrm{Lie}_\Delta(M)$

In this section we shall suppose that M is compact and one-dimensional.

Choose the following simple Zarisky covering of M. Let p be a point of M, $U_1 = M\{p\}$, U_2 is a small neighborhood of p; $\{U_1, U_2\}$ is a covering and let $\mathrm{Lie}_0(M)$ be the following diagram of Lie algebras:

$$\begin{array}{ccc} \mathrm{Lie}(U_1) & & \mathrm{Lie}(U_2) \\ & \searrow^{\pi_1} \quad {}^{\pi_2}\swarrow & \\ & \mathrm{Lic}(U_1 \cap U_2) & \end{array} \qquad (*)$$

We shall work with this diagram in the category of Lie algebras instead of the full object $\mathrm{Lie}_\Delta(M)$. A 1-cocycle of $\mathrm{Lie}_0(M)$ is the following set of data: $\{v_1, v_2, \varrho\}$, here ϱ is a 2-cocycle of the Lie algebra $\mathrm{Lie}(U_1 \cap U_2)$, v_i is a 1-cochain of $\mathrm{Lie}(U_i)$ and $\pi_i^*\varrho = dv_i$, where d is the differential in the standard cohomological complexe. Let z be a local coordinate at p, and ϱ is a standard Virasoro-type cocycle:

$$\varrho\left(f_1\frac{\partial}{\partial z}, f_2\frac{\partial}{\partial z}\right) = \mathrm{Res}_{z=0}\left(\frac{\partial^2 f_1}{\partial^2 z}\frac{\partial f_2}{\partial z} - \frac{\partial^2 f_2}{\partial^2 z}\frac{\partial f_1}{\partial z}\right) dz.$$

It can be proved that the restrictions of ϱ on $\mathrm{Lie}(U_1)$ and $\mathrm{Lie}(U_2)$ are homologous to zero. It means that we can find v_1 and v_2 such that $dv_i = \varrho$; these equations

$(i = 1, 2)$ have only one solution since $H^1(\mathrm{Lie}(U_i)) = 0$. Let us write down the formula for ϱ without using a local coordinate.

Let us fix a projective connection on M and the corresponding operator D. The expression for the cochain: $\tilde{\varrho}(u_1, u_2) = \mathrm{Res}_p(Du_1 \cdot u_2 - u_1 \cdot Du_2)$, $u_i \in \mathrm{Lie}(U_1 \cap U_2)$. It is evident that $\tilde{\varrho}$ is homologous to ϱ. The restriction of $\tilde{\varrho}$ onto $\mathrm{Lie}(U_i)$ is zero since in this case $(Du_1)u_2 - u_1(Du_2)$ is holomorphic on U_i, if $u_1, u_2 \in \mathrm{Lie}(U_i)$.

So, we have constructed the class in the space $H^1(\mathrm{Lie}_0(M))$. Really this is the same class as we constructed earlier for $\Gamma(\mathrm{Lie}(M))$. Note that our construction of the cochain of the algebra $\Gamma(\mathrm{Lie}(M))$ depends on the choice of projective connections, but the corresponding cohomology classes are equal. The same is true for the algebras $\mathrm{Lie}_0(M)$ and $\mathrm{Lie}_\Delta(M)$. In this section we have constructed a map $H^2(\mathrm{Lie}(U_1 \cap U_2)) \to H^1(\mathrm{Lie}_0(M)) \cong H^1(\mathrm{Lie}_\Delta(M))$. It enables us to associate a one-dimensional representation of the algebra $\mathrm{Lie}_\Delta(M)$ to any value of the central charge of the Virasoro algebra.

It is not hard to calculate the whole space of cohomologies of $\mathrm{Lie}_\Delta(M)$. The result is: $H^*(\mathrm{Lie}_0(M)) \cong H^*(\mathrm{Lie}_\Delta(M)) \cong H_c^*(\Gamma(\mathrm{Lie}(M)))$. This is a consequence of the following fact. Let N be an affine one-dimensional manifold and $\mathrm{Lie}(N)$ is a Lie algebra of algebraic vector fields on N.

Proposition. *The natural pairing between the spaces* $H_i(\mathrm{Lie}(N))$ *and* $H_c^i(\Gamma(\mathrm{Lie}(N)))$ *is non-degenerate for arbitrary* i. *The algebra* $H_c^*(\Gamma(\mathrm{Lie}(N)))$ *is free and skew-commutative with one generator in degree three and* r *generators in degree 2, where* $r = \dim H^1(N, \mathbb{C})$.

IV. Homologies of the Lie Algebra of Holomorphic Vector Fields on a Compact One-dimensional Manifold M

As in the previous section we choose the diagram $(*)$ as a model for the Lie algebra of holomorphic vector fields $\mathrm{Lie}_\Delta(M)$. Let us fix a one-dimensional representation of the diagram $\mathrm{Lie}_0(M)$ which we denote by $\square_c$, $c \in \mathbb{C}$. It means that the corresponding element $H^1(\mathrm{Lie}_0(M))$ is the image of the class $a \in H^2(\mathrm{Lie}(U_1 \cap U_2))$ which is represented by the cocycle:

$$\left(f\frac{\partial}{\partial z}, g\frac{\partial}{\partial z}\right) \mapsto \frac{c}{24}\mathrm{Res}_0(f''\cdot g' - f'\cdot g'').$$

We shall denote by $\mathrm{Vir}(U_1 \cap U_2) = \mathrm{Lie}(U_1 \cap U_2) \oplus \mathbb{C}C$ the central extension of $\mathrm{Lie}(U_1 \cap U_2)$, where C is the central element, and the cocycle a corresponds to the value $c = 1$.

The diagram $(*)$ is the same as the set of mappings:

$$\begin{array}{ccccc} & & \mathrm{Lie}(U_1) & & \\ & \nearrow & & \searrow & \\ \mathrm{Lie}_0(M) & & & & \mathrm{Lie}(U_1 \cap U_2) \\ & \searrow & & \nearrow & \\ & & \mathrm{Lie}(U_2) & & \end{array} \qquad (**)$$

Informally speaking, $\mathrm{Lie}_0(M)$ is an intersection of the subalgebras $\mathrm{Lie}(U_1)$ and $\mathrm{Lie}(U_2)$ in $\mathrm{Lie}(U_1 \cap U_2)$. By the standard arguments in the spirit of the Frobenius reciprocity law the homologies $H_i(\mathrm{Lie}_0(M), \square_c)$ are isomorphic to

$$H_i(\mathrm{Lie}(U_1), \mathrm{Ind}(\mathrm{Lie}_0(M), \square_c; \mathrm{Lie}(U_1))).$$

Here we denote by the symbol $\mathrm{Ind}(A, M; B)$ the induced representation of the Lie algebra B from the representation M of subalgebra $A \subset B$. It is evident from the diagram (**) that the representation $\mathrm{Ind}(\mathrm{Lie}_0(M), \square_c; \mathrm{Lie}(U_1))$ is isomorphic to the restriction onto $\mathrm{Lie}(U_1)$ of the $\mathrm{Vir}(U_1 \cap U_2)$-representation $\mathrm{Ind}(\mathrm{Lie}(U_2) \oplus \mathbb{C}C, 1_c; \mathrm{Vir}(U_1 \cap U_2))$. Here 1_c is the one-dimensional module of the algebra $\mathrm{Lie}(U_2) \oplus \mathbb{C}C$, where $\mathrm{Lie}(U_2)$ acts in trivially on 1_c and C acts by multiplication with $c \in \mathbb{C}$. It is easy to see that if the genus of M is greater than one, then the representation $\mathrm{Ind}(\mathrm{Lie}_0(M), \square_c; \mathrm{Lie}(U_1))$ is free over the universal enveloping algebra of $\mathrm{Lie}(U_1)$. Now we are ready to formulate the result about the homologies but first we recall the standard notations about the representations of the Virasoro algebra. The algebra Vir is the central extension of the Lie algebra of vector fields on a punctured disk around the origin in $\mathbb{C}$. Let C be the central element of Vir. There are two important subalgebras Vir^+ and $\overline{\mathrm{Vir}}^+$ in $\mathrm{Vir} : \mathrm{Vir}^+ = \left\{ f(z)\frac{\partial}{\partial z} \oplus \mathbb{C}C \,|\, f(0) = 0 \right\}$ and $\overline{\mathrm{Vir}}^+ = \left\{ f(z)\frac{\partial}{\partial z} \oplus \mathbb{C}C \right\}$, where f is regular at $z = 0$. Denote by 1_c the one-dimensional $\overline{\mathrm{Vir}}^+$-module, where C acts by multiplication with $c \in \mathbb{C}$; $1_{h,c}$ denotes the one-dimensional Vir^+-module, whose corresponding character is $f(z)\frac{\partial}{\partial z} \mapsto hf'(0)$, $C \mapsto c$, for $h, c \in \mathbb{C}$. We also put $M_c = \mathrm{Ind}(\overline{\mathrm{Vir}}^+, 1_c; \mathrm{Vir})$ and $M_{h,c} = \mathrm{Ind}(\overline{\mathrm{Vir}}^+, 1_{h,c}; \mathrm{Vir})$. Usually $M_{h,c}$ and M_c are called a Verma module and a parabolic Verma module respectively.

We can attach to an arbitrary point $p \in M$ the Virasoro algebra $\mathrm{Vir}(p)$ which is the central extension of the Lie algebra of holomorphic vector fields in a punctured neighborhood $U(p)$ of p. The algebras $\overline{\mathrm{Vir}}^+(p)$ and $\mathrm{Vir}^+(p)$ consist of regular vector fields and regular vector fields which are zero at p respectively. So, for any $p \in M$ we have the $\mathrm{Vir}(p)$-modules $M_{h,c}(p)$ and $M_c(p)$. It is easy to see that the representation $M_c(p)$ is isomorphic to $\mathrm{Ind}(\mathrm{Lie}(U_2) \oplus \mathbb{C}C, 1_c; \mathrm{Vir}(U_1 \cap U_2))$.

Proposition. *Suppose that the genus of* $M > 1$. *Then for any* $i \neq 0$ $H_i(\mathrm{Lie}_0(M), \square_c) \cong H_i(\mathrm{Lie}_\Delta(M), \square_c) = 0$; $H_0(\mathrm{Lie}_0(M), \square_c) \cong H_0(\mathrm{Lie}_\Delta(M), \square_c) \cong M_c(p)/(\mathrm{Lie}(U_1)M_c(p))$; *the last term is the space of* $\mathrm{Lie}(U_1)$ – *coinvariants in the module* $M_c(p)$.

Now we fix a set of distinct points $p_1, p_2, \ldots, p_n$ of M and let $W(p_1, p_2, \ldots, p_n) = M \backslash \{p_1, p_2, \ldots, p_n\}$. We want to describe the induced representation $\mathrm{Ind}(\mathrm{Lie}_\Delta(M), \square_c: \mathrm{Lie}\ W(p_1, p_2, \ldots, p_n))$. Let Z_n be the sum $\mathrm{Lie}\ U(p_1) \oplus \mathrm{Lie}\ U(p_2) \oplus \cdots \mathrm{Lie}\ U(p_n)$ and let $\hat{Z}_n$ be the central extension of the Lie algebra Z_n which corresponds to the cocycle:

$$\left(f_1(z_1)\frac{\partial}{\partial z_1} + f_2(z_2)\frac{\partial}{\partial z_2} + \cdots + f_n(z_n)\frac{\partial}{\partial z_n},\right.$$

$$\left. g_1(z_1)\frac{\partial}{\partial z_1} + g_2(z_2)\frac{\partial}{\partial z_2} + \cdots + g_n(z_n)\frac{\partial}{\partial z_n}\right)$$

$$\mapsto \frac{c}{24}\sum (f_i''\cdot g_i' - f_i'\cdot g_i''), \quad c \in \mathbb{C},$$

where z_i is a local coordinate at p_i. It is clear that the algebra $\hat{Z}_n$ acts on the space $M_c(p_1) \otimes M_c(p_2) \otimes \cdots \otimes M_c(p_n)$. The Lie algebra Lie $W(p_1, p_2, \ldots, p_n)$ is a subalgebra of Z_n, and the imbedding θ : Lie $W(p_1, p_2, \ldots, p_n) \to Z_n$ sends a vector field $a \in \mathrm{Lie}(W(p_1, p_2, \ldots, p_n))$ to the set of the Laurent expansions of a at the points p_1, $p_2, \ldots, p_n$. It may be shown that the imbedding θ has a lift to an imbedding $\hat{\theta}: \mathrm{Lie}(W(p_1, p_2, \ldots, p_n)) \to \hat{Z}_n$. Using the same arguments as in the beginning of this section we can see that the module $\mathrm{Ind}(\mathrm{Lie}_\Delta(M), \square_c; \mathrm{Lie}(W(p_1, p_2, \ldots, p_n)))$ is isomorphic to the restriction of the $\hat{Z}_n$-module $M_c(p_1) \otimes M_c(p_2) \otimes \cdots \otimes M_c(p_n)$ on $\mathrm{Lie}(W(p_1, p_2, \ldots, p_n))$.

It is easy to define the (cosimplicial) Lie subalgebra $\mathrm{Lie}_\Delta(M, p_1, p_2, \ldots, p_n) \subset \mathrm{Lie}_\Delta(M)$ which consists of holomorphic vector fields on M with zeros at the points $p_1, p_2, \ldots, p_n$.

Each point p_i defines the homomorphism $\varphi(h_i): \mathrm{Lie}_\Delta(M, p_1, p_2, \ldots, p_n) \to \mathbb{C}$, $h_i \in \mathbb{C}$ which assigns to a vector field in a neighbourhood of p the number by the rule: $\varphi(h_i)\left(f_1(z_1)\frac{\partial}{\partial z_1}\right) = h_i f(0)$. We also have a map $\mu: \mathrm{Lie}_\Delta(M, p_1, p_2, \ldots, p_n) \to \mathrm{Lie}_\Delta(M) \to \mathbb{C}$, where the last arrow corresponds to the representation $\square_c$. The tensor product of all these 1-dimensional modules gives us the representation $\square_{h_1, \ldots, h_n; c}$ of the algebra $\mathrm{Lie}_\Delta(M, p_1, p_2, \ldots, p_n)$. Consider the imbedding

$$\mathrm{Lie}_\Delta(M, p_1, p_2, \ldots, p_n) \to \mathrm{Lie}(W(p_1, p_2, \ldots, p_n))$$

and the corresponding induced representation.

Proposition. 1) *The module* $\mathrm{Ind}(\mathrm{Lie}_\Delta(M, p_1, p_2, \ldots, p_n), \square_{h_1, \ldots, h_n; c}; \mathrm{Lie}(W(p_1, p_2, \ldots, p_n)))$ *is isomorphic to the restriction of the representation of the Lie algebra* $\hat{Z}_n$ *in the space* $M_{h_1, c}(p_1) \otimes M_{h_2, c}(p_2) \otimes \cdots \otimes M_{h_n, c}(p_n)$ on $\mathrm{Lie}(W(p_1, p_2, \ldots, p_n))$.

2) For $i \neq 0$, $H_i(\mathrm{Lie}_\Delta(M, p_1, p_2, \ldots, p_n), \square_{h_1, \ldots, h_n; c}) = 0$ *if the genus* $g(M) > 1$ *or* $g(M) = 1$ *and* $n \geq 1$ *or* $g(M) = 0$ *and* $n \geq 3$.

3) $H_0(\mathrm{Lie}_\Delta(M, p_1, p_2, \ldots, p_n), \square_{h_1, \ldots, h_n; c}) \cong M_{h_1, c}(p_1) \otimes M_{h_2, c}(p_2) \otimes \cdots \otimes M_{h_n, c}(p_n)/\mathrm{Lie}(W(p_1, p_2, \ldots, p_n))(M_{h_1, c}(p_1) \otimes M_{h_2, c}(p_2) \otimes \cdots \otimes M_{h_n, c}(p_n))$.

Remark. Let γ_i be a small circle around the point $p_i \in M$ and $D(\gamma_i)$ is the group of diffeomorphisms of γ_i with itself; $D(\gamma_i)$ is a real infinite-dimensional Lie group, and let $\mathrm{Lie}^{\mathbb{R}}(\gamma_i)$ be the Lie algebra of the group $D(\gamma_i)$. Denote by $\mathrm{Lie}(\gamma_i) = \mathbb{C} \otimes_{\mathbb{R}} \mathrm{Lie}^{\mathbb{R}}(\gamma_i)$

the complexification of $\mathrm{Lie}^{\mathbb{R}}(\gamma_i)$. It is impossible to construct the complexification of the group $D(\gamma_i)$, but it is possible to define a local complex group with the Lie algebra $\mathrm{Lie}(\gamma_i)$. Fix two subalgebras L^+ and L^- in $\mathrm{Lie}(\gamma_1) \oplus \mathrm{Lie}(\gamma_2) \oplus \cdots \oplus \mathrm{Lie}(\gamma_n)$. The algebra Lie^+ consists of vector fields which admit extensions on the small disks with the bounderies $\gamma_1, \ldots, \gamma_n$. The Lie algebra L^- is the image of $\mathrm{Lie}(W) \subset \oplus \mathrm{Lie}(\gamma_i)$, where W is the manifold obtained by deleting the disks from M. Let D, D^+, D^- be the local groups with the Lie algebras $\oplus \mathrm{Lie}(\gamma_i)$, L^+, L^- respectively. Define also the local group $D_0^+ \subset D^+$, where an element of the Lie algebra of D_0^+ is a set of vector fields on the disks $(f_1, \ldots, f_n)$ such that $f_i(p_i) = 0$.

Denote by $M(g, n)$ the moduli space of curves of genus g with n marked points. Our object $(M, p_1, \ldots, p_n)$ is a point of $M(g, n)$. It is easy to see that the double coset space $D_0^+ \backslash D/D^-$ is isomorphic to the formal neighbourhood of $(M, p_1, \ldots, p_n)$ in the $M(g, n)$. The double coset space $D^+ \backslash D/D^-$ is the formal neighbourhood of M in the moduli space $M(g) = M(g, 0)$. As in the Section III a value $c \in \mathbb{C}$ of the central charge defines a central extension of the Lie algebra $\oplus$ $\mathrm{Lie}(\gamma_i)$ and therefore a central extension of the local group D^∞. This extension is the map $\hat{D}_c \xrightarrow{\lambda} D$, whose the kernel is isomorphic to $\mathbb{C}$. Let us fix liftings $\mu_0 : D_0^+ \to \hat{D}_c$, $\mu_1 : D^- \to \hat{D}_c$. The local group D_0^+ has the family of one-dimensional representations $\theta(h_1, \ldots, h_n) : (\exp(u_1), \ldots, \exp(u_n)) \mapsto \exp(\sum h_i u_i'(P_i))$, where $h_i \in \mathbb{C}$, and u_i is a little vector field on the disk with boundary γ_i.

Using these data we can construct a line bundle $\xi_{h_1, \ldots, h_n; c}$ over the formal neighbourhood of $(M, p_1, \ldots, p_n) \in M(g, n)$. First define a line bundle ξ_c by the map $\xi_c : \mu_0(D_0^+) \backslash \hat{D}_c/\mu_1(D^-) \to D_0^+ \backslash D/D^-$. The representation $\theta(h_1, \ldots, h_n)$ defines a homogeneous line bundle on $D_0^+ \backslash D$, hence a line bundle $\xi_{h_1, \ldots, h_n}$ over $D_0^+ \backslash D/D^-$. Now put $\xi_{h_1, \ldots, h_n; c} = \xi_{h_1, \ldots, h_n} \otimes \xi_c$.

Denote by ξ_i the following line bundle on $M(g, n)$; the fiber of ξ_i is the tangent vector space to M at p_i. Let Δ be the pull-back on $M(g, n)$ of the bundle of volume forms on $M(g, 0)$ (it is given by the natural map from $M(g, n)$ onto $M(g, 0)$). Then, the bundle $\xi_{h_1, \ldots, h_n; c}$ is isomorphic to the tensor product $\xi_1^{h_1} \otimes \xi_2^{h_2} \otimes \cdots \otimes \xi_n^{h_n} \otimes \Delta^{c/26}$. (If the numbers h_1, ..., h_n and $c/26$ are integers, then $\xi_{h_1, \ldots, h_n; c}$ is a bundle defined over the whole $M(g, n)$, otherwise $\xi_{h_1, \ldots, h_n; c}$ exists only in a small neighbourhood.

The space of coinvariants $M_{h_1, c}(p_1) \otimes M_{h_2, c}(p_2) \otimes \cdots \otimes M_{h_n, c}(p_n)/\mathrm{Lie}(W(p_1, p_2, \ldots, p_n))(M_{h_1, c}(p_1) \otimes M_{h_2, c}(p_2) \otimes \cdots \otimes M_{h_n, c}(p_n))$ can be identified with the space of generalized sections of the bundle $\xi_{h_1, \ldots, h_n; c}$ with support in the point $(M, p_1, \ldots, p_n)$. More precisely; the space of coinvariants is:

$$H^N_{(M, p_1, \ldots, p_n)}(\xi_{h_1, \ldots, h_n; c}),$$

where $H^N_X(F)$ is the local cohomology of a sheaf F with support in a submanifold X, in our case $N = n + 3g - 3$.

V. An Analytic Version of the Homology of the Lie Algebra of Holomorphic Vector Fields on M with Coefficients in the One-dimensional Representation $\square_c$

In this section we shall calculate the homology of the differential Lie superalgebra $\Gamma(\mathrm{Lie}(M))$ with coefficients in a one-dimensional representation. Let us write down the standard homological complex:

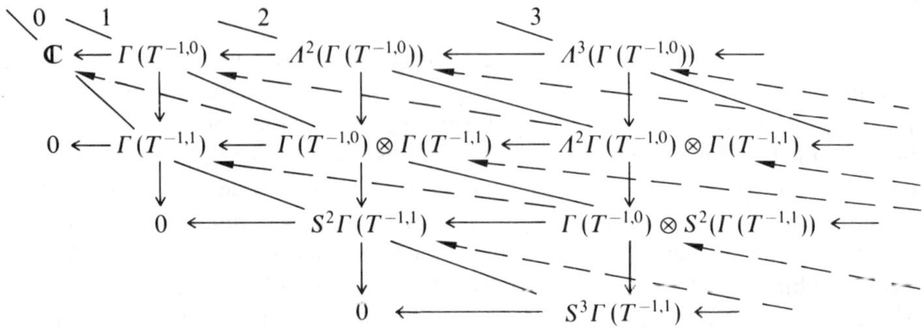

Here the horizontal and vertical arrows are derived from the bracket in $\Gamma(\mathrm{Lie}(M))$. They constitute the differential in the standard complex of $\Gamma(\mathrm{Lie}(M))$ with trivial coefficients. The dashed arrows are the following maps:

$$\Lambda^i\Gamma(T^{-1,0})\otimes S^j\Gamma(T^{-1,1})\to\Lambda^{i-1}\Gamma(T^{-1,0})\otimes S^{j-1}\Gamma(T^{-1,1}).$$

They arise as a contraction with the multiple of the tensor corresponding to the map $\Gamma(T^{-1,0})\otimes\Gamma(T^{-1,1})\overset{\Psi_3}{\to}\Gamma(T^{1,1})\to\mathbb{C}$, where the last map is the integration of forms over the manifold. (Recall that the map Ψ_3 is related to the operator D with the highest symbol $(\partial/\partial z)^3$. If we multiply Ψ_3 with $c/24$ we get the one-dimensional representation of $\Gamma(\mathrm{Lie}(M))$ corresponding to the value c of the central charge.)

Our complex has a natural filtration $\{K_n\}$, the subcomplex K_n consisting of the first n columns. First let us calculate the first term of the spectral sequence, i.e., the homologies of the columns. If the genus of M is greater than one, then the first term is concentrated in degree zero and is equal to the sum $\bigoplus S^iV$, where V is the space of homologies of the complex $\Gamma(T^{-1,0})\to\Gamma(T^{-1,1})$, of $\dim V=3g-3$. The first term of the spectral sequence coincides with the limit term.

Proposition. $H_*(\Gamma\,\mathrm{Lie}(M),\square_c)\cong H_*(\mathrm{Lie}_\Delta(M),\square_c)$.

The previous calculation illustrates this fact. We saw in the Section III, that $H_i(\mathrm{Lie}_\Delta(M),\square_c)=0$, for $i\neq 0$; and $H_0(\mathrm{Lie}_\Delta(M),\square_c)$ is the space of distributions on the space of complex structures on M.

Remark 1. Recall the idea of the deformation theory (in a very rough form). Let A be an algebraic object (algebra, Lie algebra, algebraic manifold), RA be a resolution of A. For example, if A is a ring, then RA is a simplicial free ring; if A is a commutative algebra, then RA is the Tyurin resolution. Let G be the group of transformations of RA and $H^*(G)$ be the algebra of cohomologies of G. Then, the spectrum of the skew-commutative algebra $H^*(G)$ is the space of versal deformations of A. The model for deformation is the standard complex of G with coefficients in RA. The results of the Sections V and IV are realizations of this idea in the case where A is a one-dimensional manifold M. We associated to M the cosimplicial Lie algebra of transformations of M (or the differential Lie algebra $\Gamma(\mathrm{Lie}(M))$). In terms of this cosimplicial Lie algebra we calculated the small neighbourhood of M in the moduli space. A representation π of the group (or Lie algebra) of transformations defines a sheaf over the moduli space (versal deformations), of which the space of sections is $H^*(G, \pi)$.

Remark 2. There is another (very informal) way of thinking about all these facts. Let M be a compact complex manifold and $\mathrm{Dif}(M)$ be a group of all C^∞-diffeomorphisms of M and $\mathrm{Dif}_c(M) \subset\subset \mathrm{Dif}(M)$ be the subgroup of diffeomorphisms preserving the complex structure. If $\dim M = 1$ and the genus of M is greater than one, then $\mathrm{Dif}_c(M)$ is very little (finite). But we want to think about the group $\mathrm{Dif}_c(M)$ in a "derived" sense. The Lie algebra of $\mathrm{Dif}_c(M)$ is the differential Lie algebra $\Gamma(\mathrm{Lie}(M))$ and the Euler characteristic of the complex $\Gamma(\mathrm{Lie}(M))$ is the formal dimension of the group $\mathrm{Dif}_c(M)$. So, if M is a curve, then $\dim(\mathrm{Dif}_c(M))$ is negative and equal to $3 - 3g$, where g is the genus of M. The double coset space $\mathrm{Dif}_c(M)\backslash \mathrm{Dif}(M)/\mathrm{Dif}(M)$ is exactly the space of all complex structures on M. In this remark we suppose $\dim M = 1$. The group of $\mathrm{Dif}_c(M)$ has a negative dimension, so the dimension of the double coset space is equal to $3g - 3 = \dim M(g, 0)$. A one-dimensional representation of the group $\mathrm{Dif}_c(M)$ defines a homogeneous line bundle over the factor space $\mathrm{Dif}_c(M)\backslash \mathrm{Dif}(M)$ and, therefore, over the moduli space. By this construction we get powers of the determinant bundle. From this point of view it is very interesting to investigate all representations of the group $\mathrm{Dif}_c(M)$.

Example. Consider the complex $L_n = \{\Gamma(T^{n,0}) \to \Gamma(T^{n,1})\}$, $n \in \mathbb{Z}$. It is clear that L_n is a $\Gamma(\mathrm{Lie}(M))$-module. It means that we have a representation $v_n : \Gamma(\mathrm{Lie}(M)) \to \mathrm{End}(L_n)$. The trace of this representation is an element of $H^1(\Gamma(\mathrm{Lie}(M)))$. The corresponding value of the central charge is equal to $-2(6n^2 - 6n + 1)$.

From the presentation of $M(g, 0)$ as a double coset space $\mathrm{Dif}_c(M)\backslash \mathrm{Dif}(M)/\mathrm{Dif}(M)$ we see that there is a $\mathrm{Dif}_c(M)$-bundle $* = \mathrm{Dif}(M)/\mathrm{Dif}(M) \to M(g)$.

This is a principle $\mathrm{Dif}_c(M)$-bundle. We can apply the Weil construction of the characteristic classes to this bundle. Let G be a Lie group and $\mathscr{G}$ be the Lie algebra of G. The usual Weil homomorphism associates with a principal G-bundle $E \to S$ and an invariant polynomial p on the space $\mathscr{G}$, $p \in S^n(\mathscr{G}^*)$ a cohomology class of the space S of dimension $2n$. In our case we have also a map: $H_c^0(\Gamma(\mathrm{Lie}(M)), S^j(\Gamma(\mathrm{Lie}(M)^*)) \to H^{2j}(M(g, 0), \mathbb{C})$.

We saw that the algebra $H_c^0(\Gamma(\mathrm{Lie}(M)), S^j(\Gamma(\mathrm{Lie}(M)^*))$ is an algebra of polynomials in an infinite number of variables $\mathbb{C}[\sigma_1, \sigma_2, \ldots]$. So we get the (well-known) map $\mathbb{C}[\sigma_1, \sigma_2, \ldots] \to H^*(M(g, 0), \mathbb{C})$. The calculation of this map gives us the Riemann-Roch theorem for families of curves.

VI. Integrable Homologies of the Lie Algebra $\mathrm{Lie}_\Delta(M)$

For any open subset (in the Zarisky topology) $U \subset M$ we have an imbedding $\mathrm{Lie}_\Delta(M) \subset \mathrm{Lie}(U)$. A representation v of the algebra $\mathrm{Lie}_\Delta(M)$ is actually the set of $\mathrm{Lie}(U)$-modules $H(U) \cong \mathrm{Ind}(\mathrm{Lie}_\Delta(M), v; \mathrm{Lie}(U))$ and isomorphisms $\varphi(U_1, U_2)$: $\mathrm{Ind}(\mathrm{Lie}(U_2), H(U_2); \mathrm{Lie}(U_1)) \to H(U_1)$, where U_1 is a subset of U_2. Such morphisms $\varphi(U_1, U_2)$ should satisfy the evident compatibility condition: for a couple of imbeddings $U_3 \to U_2 \to U_1$ the composition of the maps

$$\mathrm{Ind}(\mathrm{Lie}(U_2), \mathrm{Ind}(\mathrm{Lie}(U_1), H(U_1); \mathrm{Lie}(U_2)); \mathrm{Lie}(U_3))$$

$$\xrightarrow{\mathrm{Ind}\,\varphi(U_2, U_1)} \mathrm{Ind}(\mathrm{Lie}(U_2), H(U_2); \mathrm{Lie}(U_3)) \xrightarrow{\varphi(U_3, U_2)} H(U_3)$$

is equal to $\varphi(U_3, U_1)$.

In a more general context it is reasonable to call such a set of data as a representation of the algebra $\mathrm{Lie}_\Delta(M)$ also in the case when the maps $\varphi(U_1, U_2)$ are not necessarily isomorphisms. It is easy to define the homologies of the algebra $\mathrm{Lie}_\Delta(M)$ with coefficients in such a representation.

Let $\mathfrak{U}$ be a Lie algebra and π be a representation of $\mathfrak{U}$. Let us denote by $ss(\pi) \subset \mathfrak{U}^*$ the singular support of π. We shall call π integrable (or lisse) if $ss(\pi) = 0$. Note that for finite dimensional $\mathfrak{U}$ the class of integrable representations coincides with the class of finite dimensional ones. We shall call a representation of the algebra $\mathrm{Lie}_\Delta(M)$ lisse if for any open set $U \subset M$ the singular support of the corresponding $\mathrm{Lie}(U)$-module $H(U)$ is zero. (We also suppose that $H(U)$ has a finite set of generators.)

Proposition. *Let π be a representation of the Virasoro algebra from the category of representations with highest weight. We suppose that the central element $C \in$ Vir acts on π by multiplication with a constant $c \in \mathbb{C}$. The following conditions are equivalent:*

(1) *π is a lisse representation of* Vir.

(2) $c = c_{p,q} = 1 - \dfrac{6\cdot(p-q)^2}{p\cdot q}$, *where p, q are two natural relatively prime numbers. The representation π decomposes as $\pi_1 \oplus \pi_2 \oplus \cdots \oplus \pi_N$, where π_i is an irreducible representation of the corresponding minimal conformal field theory.*

Let us denote by $L_{h,c}$ the irreducible quotient of the Verma representation $M_{h,c}$. Fix two natural relatively prime numbers (p, q). The irreducible representations of the minimal (p, q)-theory are $L_{h,c_{p,q}}$, where $h = h_{r,s}$ is equal to $\dfrac{(rp - sq)^2 - 1}{4pq}$, $1 \le r \le p - 1$, $1 \le s \le r$; $L_{h_{1,1}}$ is called the vacuum representation.

Now we attach to a minimal (p, q)-theory a lisse representation of the algebra $\mathrm{Lie}_\Delta(M)$. A $\mathrm{Lie}_\Delta(M)$-module $\square_{c_{p,q}}$ is defined in the following way. Consier the open set $W(p_1, \ldots, p_n) = M \backslash \{p_1, \ldots, p_n\}$. We previously attached to each point p_i a copy of the Virasoro algebra and let $L(p_i)$ be its vacuum representation. The

Lie algebra Lie $W(p_1, \dots, p_n)$ acts on the space $L(p_1) \otimes L(p_2) \otimes \cdots \otimes L(p_n) = \overline{\square}_{c_{p,q}}(W(p_1, \dots, p_n))$. The induced representation from the Lie $W(p_1, \dots, p_n)$-module $\overline{\square}_{c_{p,q}}(W(p_1, \dots, p_n))$ to the algebra Lie $W(p_1, \dots, p_{n+1})$ is isomorphic to $L(p_1) \otimes L(p_2) \otimes \cdots \otimes L(p_n) \otimes M_{0,c_{p,q}}(p_{n+1})$. So it is clear that there is a natural map

$$\operatorname{Ind}(\text{Lie } W(p_1, \dots, p_n); \overline{\square}_{c_{p,q}}; \text{Lie } W(p_1, \dots, p_{n+1})) \to \overline{\square}_{c_{p,q}}(W(p_1, \dots, p_{n+1})).$$

We see that $\overline{\square}_{c_{p,q}}$ is the lisse quotient of the $\text{Lie}_\Delta(M)$-representation $\square_{c_{p,q}}$. If $c \neq c_{p,q}$ for any p, q, then the module $\square_c$ has no lisse quotient.

Fix a set of distinct points $\{a_1, \dots, a_l\}$ of M. We can define in a similar way the notion of lisse representations of $\text{Lie}_\Delta(M, a_1, \dots, a_l)$. An example of a lisse representation of $\text{Lie}_\Delta(M, a_1, \dots, a_l)$ is the following. Let us attach to each point a_i the number $h(a_i) = h_{r_i, s_i}$. If a point a is not in $\{a_1, \dots, a_l\}$ put $h(a) = 0$. The representation $\overline{\square}(h(a_1), \dots, h(a_l); c_{p,q})$ associates to the open set $W(p_1, \dots, p_n)$ the space $L_{h(p_1);c_{p,q}} \otimes L_{h(p_2);c_{p,q}} \otimes \cdots \otimes L_{h(p_n);c_{p,q}}$.

The conformal field theory defines the so-called modular functor. This functor associates with a Riemann surface a finite dimensional vector space (another name of this space is the conformal block for the holomorphic statistical sum). The definition of the modular functor for the minimal conformal field theory is the following: $H(M, p, q) = L_{0,c_{p,q}}/\mathfrak{U} \cdot L_{0,c_{p,q}}$. Here $L_{0,c_{p,q}}$ is the vacuum representation of Vir attached to an arbitrary point of M, and $\mathfrak{U}$ is the Lie algebra of vector fields on M-{point}.

Theorem. 1) *The algebra* $H_c^*(\Gamma \text{ Lie}(M))$ *acts on the space* $H_*(\text{Lie}_\Delta(M), \overline{\square}_{c_{p,g}})$ *and the dual space is a free* $H_c^*(\Gamma \text{ Lie}(M))$*-module with generators in degree zero.*

2) $H_0(\text{Lie}_\Delta(M), \overline{\square}_{c_{p,q}}) \cong H(M, p, q)$.

A similar statement is true for the algebra $\text{Lie}_\Delta(M, a_1, a_2, \dots, a_n)$.

Elliptic Methods in Variational Problems

Andreas Floer *

Ruhr-Universität Bochum, Postbox 102148, W-4630 Bochum, Fed. Rep. of Germany

1. Introduction

Morse theory, invented in the 1930s, has had remarkable applications both in Topology and Analysis. Applied to infinite dimensional manifolds, it leads to existence proofs for solutions of certain differential equations which entirely circumvent the problem of calculation. One of the earliest examples is Schnirelman's existence proof for closed geodesics on spheres obtained by considering the energy function on loop space. Conversely, Morse theory on the same function led in 1956 to Bott's periodity theorem [5] for classical Lie groups.

In the 70ies the hypotheses for infinite dimensional Morse theory were formalized into a condition of "essential positivity" of the function [1], and an analytic compactness condition [23], the much discussed Palais–Smale condition. In the following decade, attention focussed on certain essentially positive functions which just fail to satisfy this second condition, called "borderline cases". The most prominent examples are the minimal-surface problem [25] and Yang-Mills theory in dimension 4. These problems were not only these very "natural " problems from the geometric and physical point of view, but also revealed similar special properties.

In this contribution, we consider functions which are also of "Palais Smale borderline" type, but which are not "essentially positive". The point is that while the usual approach to Morse theory does not apply to such cases, a different approach yields applications to existence of solutions to differential equations [2], as well as to topology. Instead of developing an abstract setting for this method, we present the two main applications. First, however, we discuss the two different approaches to Morse theory in the finite dimensional case.

2. Morse Theory in Finite Dimensions

Consider a smooth real valued function $f : M \to \mathbb{R}$ on a smooth (finite dimensional) manifold M. Morse theory is concerned with the relation between the

* † May 15, 1991.

[1] Meaning boundedness from below and finiteness of the Morse indices.

[2] i.e. Hamilton's equation.

Proceedings of the International Congress
of Mathematicians, Kyoto, Japan, 1990

critical set

$$C(f) = \{x \in M \mid df(x) = 0\} \tag{1}$$

of f and the topology of M. Here, we assume that $C(f)$ is "nondegenerate", i.e. that for every $x \in C(f)$, the second derivative of f (called the Hessian of f at x) is a nondegenerate quadratic form. That such relation should exist is seen most easily in the "waterlevel picture": Consider as a standard example of a manifold the two-torus T^2, embedded in 3-space $\mathbb{R}^3$ in the usual way as the surface of a tire. Let f be given by the height [3], such that $C(f)$ consists of four points. Now, imagine $\mathbb{R}^3$ to be flooded, and consider the portion

$$M_a = \{x \in M \mid f(x) \leq a\} \tag{2}$$

of M below the "waterlevel" $a \in \mathbb{R}$. The crucial observation is that the topology of M_a changes precisely at those levels which are the values of a critical point of f. Moreover, up to homotopy, this change is described by gluing to $M_{a-\varepsilon}$ the "unstable manifold" of x, which is a cell whose dimension is called the Morse index $\mu(x)$, and equals the dimension of the negative "eigenspace" of the Hessian at x. This successive attaching of cells implies the well-known "Morse inequalities" in homology, which can be stated as follows:

> If M is closed, then $H_*(M)$ is obtained from a chain complex, which is given by a boundary operator ∂ on the free abelian group over the critical set.

We will denote this "chain complex" by

$$C_*(f) \simeq \mathbb{Z}^d \qquad \text{with} \qquad d = \mid C(f) \mid \; .$$

To make this procedure more precise, we have to introduce an auxiliary metric γ on M, and consider the corresponding "gradient flow trajectories" of f, i.e. smooth paths $u : \mathbb{R} \to M$ satisfying the equation

$$\frac{du(t)}{dt} + \nabla_\gamma f(u(t)) = 0 \, . \tag{3}$$

With the exception of the constant solutions $u(t) \equiv x$ for $x \in \mathscr{C}(f)$, the function decreases along gradient flow trajectories. The stable and unstable manifolds of $x \in \mathscr{C}(f)$ are the sets $W_\gamma^\pm(x)$ of all trajectories u with $\lim_{t\to\pm\infty} = x$.[4] The metric on $TM(x)$ also converts the Hessian into a symmetric operator on $TM(x)$ whose negative eigenspace $E_\gamma^-(x)$ is the tangent space at x of $W_\gamma(x)$.[5]

There is a different, in some sense dual approach to Morse theory, which instead of the unstable manifolds considers only the spaces

$$\mathscr{M}_\gamma(x, y) = \{u : \mathbb{R} \to M \mid \quad u \text{ satisfies (5).} \quad \} = W_\gamma^-(x) \cap W_\gamma^+(y) \, . \tag{4}$$

[3] The tire assumed to be in upright position.

[4] We can always identify trajectories and points in M through the correspondence $u \hat{=} u(0)$.

[5] In fact, it is globally diffeomorphic to $W_\gamma^-(x)$.

with

$$\begin{cases} \dfrac{du}{dt} &= -\nabla_\gamma f(u(t)) \ , \\ \lim\limits_{t\to-\infty} u(t) = & x \ , \\ \lim\limits_{t\to\infty} u(t) \ = & y \ . \end{cases} \tag{5}$$

These are spaces of gradient flow lines connecting two given critical points $x, y \in C(f)$.

For "generic" γ, these intersections are transversal, resulting in smooth manifolds $\mathcal{M}_\gamma(x, y)$ of dimension $\mu(x) - \mu(y)$. Such gradient flows are called "Morse Smale flows". Note that if we assume an orientation on $E_\gamma^-(x)$ at all critical points $x \in C(f)$, then $\mathcal{M}_\gamma(x, y)$ is naturally oriented as the intersections of the oriented $W_\gamma^-(x)$ with the co-oriented $W_\gamma^+(y)$. Finally the translational symmetry implies a splitting

$$\mathcal{M}_\gamma(x, y) = \hat{\mathcal{M}}_\gamma(x, y) \times \mathbb{R} \cup \{ \text{ constant trajectories } \} \ . \tag{6}$$

where $\mathcal{M}_\gamma(x, y)$ will be called the reduced trajectory space. To formulate the Morse inequalities, we consider for $x, y \in \mathscr{C}(f)$ the integer

$$< \partial x \mid y >= \# \hat{\mathcal{M}}_\gamma(x, y) \tag{7}$$

where # applied to an oriented manifold denotes its algebraic number, counted with sign, if it is discrete, and is zero otherwise. In other words, $< \partial x \mid y >$ is the number of "isolated trajectories" of the $\nabla_\gamma f$-flow, i.e. of components of $\mathcal{M}_\gamma(x, y)$ of type $\mathbb{R}$, between x and y, counted positively, if its "intrinsic" orientation points in the flow direction and negatively otherwise.

We interprete (7) as the matrix elements of an endomorphism on $C_*(f)$. That is, if $x \in \mathscr{C}(f)$, then ∂x is the formal sum over the endpoints of all isolated trajectories originating at x, counted by sign. Then we have the following

Theorem 1 (Morse Inequalities). *If M is closed and (f, γ) defines a Morse Smale flow on M, then*

1. *the discrete parts of the sets $\hat{\mathcal{M}}_\gamma(x, y)$ are finite, and* $\partial\partial = 0$
2. $\ker / \operatorname{im} \partial \cong H_*(M)$

In the example of the torus, we may have to perturb the metric given by standard embedding slightly to obtain a Morse Smale flow. For example, there should be no trajectories connecting the two critical points of Morse index 1. For any Morse Smale flow, however, we will find that isolated trajectories come in pairs of opposite sign: the boundary operator is zero, since $H_*(M) = C_*(f) = \mathbb{Z}^4$. (This is called a perfect Morse function.) An example with nontrivial ∂ is easily constructed e.g. by creating an additional local maximum of f.

Let us have just a glimpse of the proof of the boundary property of ∂. The matrix element $< \partial\partial x \mid y >$ of the homomorphism $\partial\partial$ for $x, y \in \mathscr{C}(f)$ is given by the algebraic number of "double trajectories", i.e. pairs of isolated flow lines connecting x and z with a stopover at a third critical point y. In this situation, it can be shown that if we change the outgoing direction at x slightly in a particular way, we generate a 1-parameter family in $\hat{\mathscr{M}}_\gamma(x,z)$ for any Morse Smale flow. In fact, there exist local diffeomorphisms

$$\hat{\mathscr{M}}_\gamma(x,y) \times (0,1] \times \hat{\mathscr{M}}_\gamma(y,z) \hookrightarrow \hat{\mathscr{M}}_\gamma(x,z)\,. \tag{8}$$

Now, every non-compact component of a 1-dimensional manifold must have two ends, and compactness of M implies that such ends always correspond to "double trajectories". Analysing the orientation involved, we find that each of these pairs of double trajectories cancel in $< \partial\partial x \mid y >$. We see that compactness properties of the spaces $\mathscr{M}_\gamma(x,y)$ are crucial in the definition of ∂ as well as in the proof of its boundary property. These compactness properties can be derived from compactness of M, but they would follow in the same way from compactness of

$$S = \{\, u(0) \mid u \in \mathscr{M}_\gamma(x,y) \quad \text{for some} \quad x,y \in \mathscr{C}(f)\,\} \tag{9}$$

as subset of M. Hence, if this set, the union of all bounded flow trajectories, is compact in M, then we can associate to it the homology group

$$I_*(S,f,\gamma) = \ker / \mathrm{im}\ \partial\,. \tag{10}$$

If M is noncompact, then this group will in general differ from $H_*(M)$; it is a new invariant of the flow and of its compact invariant set S. One can show that $I_*(S,f,\gamma)$ is invariant under the deformation of f and γ (through non–Morse–Smale flows) as long as S remains compact throughout the deformation. In fact, $I_*(S,f,g)$ can be shown to be the homology of the "Conley Index" of S; which is the homology of a neighbourhood U of S relative to a suitably defined "exit set" $A \subset U$. For details, see [10] and [17].

3. The Symplectic Action

We want to extend the idea of the Conley index to infinite dimensional variational problems. Consider the "phase space" $\mathbb{R}^{2n}$, of positions and moments of a linear mechanical system, parameterized by $Z = \{p_i, q_i\}_{i=1}^n$. For every loop $z : \mathbb{S}^1 \to \mathbb{R}^{2n}$, we define the "action function"

$$a_h(z) = \int pdq + \int h_s(z(s))ds \tag{11}$$

where $h : \mathbb{S}^1 \times \mathbb{R}^{2n} \to \mathbb{R}$ is a smooth function, considered as a "time dependent Hamiltonian". In the simplest case, it is the sum of the kinetic term $\frac{m}{2}p^2$ and a potential term $V_s(q)$. Hamilton's principle states that the critical points of a_h in $\Omega(\mathbb{R}^{2n}) = C^\infty(\mathbb{S}^1, \mathbb{R}^{2n})$ are the solutions of Hamilton's equation. If $\mathbb{R}^{2n}$ is identified with $\mathbb{C}^n$ by means of a complex structure J and a metric g, then these equations take the form[6]

$$J\frac{dz}{ds} + \nabla_g h_s(z(s)) = 0\,. \tag{12}$$

[6] Here, $\omega := g(J\cdot,\cdot)$ is assumed to be skew–symmetric, and the equation (12) is essentially determined by this "symplectic form" alone.

The left hand side of (12) can be considered as the gradient of a_h on $\Omega(\mathbb{R}^{2n})$ with respect to the "L^2–metric" on this infinite dimensional space defined by g.

To apply Morse theory to Hamilton's principle, one might extend $\Omega(\mathbb{R}^{2n})$ to some Banach space $\bar{\Omega}$, and modify the gradient to a smooth vector field on $\bar{\Omega}$. Then the gradient flow would be available as a homotopy on $\bar{\Omega}$ decreasing the action. Note, however, that the Hessian of a_h at a critical point is a zero–order perturbation of the selfadjoint operator $J\frac{d}{dz}$, which has infinite dimensional positive and negative eigenspaces. Hence, the unstable manifolds are infinite dimensional cells, the addition of which does not change the homotopy type. Therefore, one cannot relate critical points of a_h to the topology of subsets of $\Omega(\mathbb{R}^{2n})$. Nevertheless, the problem was attacked in 1978 by Rabinowitz [24]: He restricted the function to finite dimensional Fourier-subspaces of loops, and used a method of Morse theory which was "stabilized" with an increase of dimension. Thus the topological invariants used by Rabinowitz were trivial in the total loopspace, but existed nevertheless in some sort of a "stable limit". In the following decade, one came to understand these invariants in the framework of a special homology theory in $\Omega(\mathbb{R}^{2n})$ which pays special respect to the decomposition into positive and negative Fourier components of loops, and in which the unstable and stable manifolds would define cycles and cocycles, respectively. Though this idea was never made precise, it inspired various existence proofs for closed solutions of Hamiltonian systems in $\mathbb{R}^{2n}$.

The dual approach to Morse theory can be used to define such "half–infinite" dimensional homology groups without having to consider infinite dimensional subsets. Since we are interested in individual trajectories rather than in the entire gradient flow, we do not have to "regularize" the gradient, but can consider maps $u : \mathbb{R} \times \mathbb{S}^1 \to \mathbb{R}^{2n}$ satisfying the equation

$$\frac{\partial u(s,t)}{\partial t} + J\frac{\partial u(s,t)}{\partial s} + \left(\nabla_g h_s\right)(u\,(s,t)) = \left(\bar{\partial}_J u + D_g h(u)\right)(s,t) = 0\,. \tag{13}$$

Here, we have combined the derivative in the flow time and the differential part of the gradient of a_h to the Cauchy Riemann operator, whose zero set is the set of holomorphic maps from $\mathbb{R} \times \mathbb{S}^1 \cong \mathbb{C}/i\mathbb{Z}$ to $\mathbb{C}^n$. Thus the "gradient flow equation" for a_h is a zero order perturbation of the equation for "holomorphic curves" in $\mathbb{C}^n$, which Gromov [21] introduced to symplectic geometry in 1985. Ellipticity of $\bar{\partial}_J$ and nondegeneracy of $\mathscr{C}(a_h)$ implies that the spaces

$$\mathscr{M}_h(x) = \left\{\, u : \mathbb{R} \times \mathbb{S}^1 \to \mathbb{R}^{2n} \mid u \text{ satisfies (15) } \right\} \tag{14}$$

$$\begin{cases} \bar{\partial}_J u + D_g h(u) = 0\ , \\ \lim\limits_{t\to-\infty} u(t,s) = x(s)\ , \\ \lim\limits_{t\to\infty} u(t,s) = y(s) \end{cases} \tag{15}$$

are, for generic h and J, finite dimensional manifolds, with an orientation defined through a "determinant bundle". Moreover, Gromov's compactness theory of (pseudo)-holomorphic curves implies that the integers (7) are still defined for suitable h.

The trajectory spaces (14) can be defined equally well if $\mathbb{R}^{2n}$ is replaced by any symplectic manifold (P, ω) with symplectic (i.e. nondegenerate closed) form $\omega \in \Omega^2(P)$. In fact, let us assume that P is compact. In this case, Gromov's compactness theory predicts that the spaces $\mathcal{M}_\gamma(x, y)$ are compact (in the topology of local convergence) up to the phenomenon of "bubbles": A sequence u_k in $\mathcal{M}_\gamma(x, y)$ may converge to u outside a point $(s, t) \in \mathbb{S}^1 \times \mathbb{R}$, but may "blow up" to form a holomorphic map $\left(S^2, \infty\right) \longrightarrow (P, u\,(s,t))$ in arbitrarily small neighbourhoods of (s, t). Such "holomorphic spheres" always define nontrivial elements in $\pi_2(P)$. This leads to the following chain complex:

Theorem 2. *Let* (P, ω) *be a compact symplectic manifold with* $\pi_2(P) = 0$. *Then for generic* h *and* J, *the trajectory spaces* (14) *define a boundary operator* $\partial = \partial_{h,J}$ *on* $C_*(a_h)$.

Its homology is independent of the choice of h and J. For the proof of the boundary property $\partial\partial = 0$, the "gluing maps" (8) are now defined by a version of Taubes' grafting constructions, which was first defined for instantons, but applies to other elliptic equations as well. The crucial point is invariance property, which implies that the "new homology group" depends only on P, and that it may be calculated e. g. for "small" Hamiltonians. Since $\mathcal{C}(a_0)$ is the set of constant loops, it turns out that it coincides with $H_*(P)$. The result [18] is a "Morse inequality" between the set of closed solutions of (12) and the homology of P; in particular, the number of such solutions must be greater than or equal to the dimension of $H_*(P)$. Recall that all solutions were assumed to be non-degenerated. In the general case, the method can be extended to yield "cuplenght estimates". This fact, conjectured by Arnold in [3,2], had been established by Conley and Zehnder in [8] for the torus T^{2n}, applying the Conley index to a "reduced" finite dimensional problem. The construction of the complex in Theorem 2 was in fact an attempt to define the Conley index directly in an infinte dimensional situation. It can be defined in various other situations involving the symplectic action, e.g. to Lagrangian intersections [15] and symplectic manifolds with contact boundaries[7]. It is not always as readily calculable, in fact, it may yield new invariants in symplectic geometry which are yet to be explored.

4. Gauge Theory

The method described in the preceeding section uses strongly the special geometric and analytic features of the symplectic action function. There exists, however, another natural variational problem with remarkably similar properties. To describe it, we have to replace the loop space $\Omega(P)$ by the space of SU_2-connections on a closed oriented three-manifold M. Recall that all SU_2-bundles of M are trivial, and that with respect to such a trivialization, the space of connections on $M \times SU_2$ can be identified with

$$\mathcal{A}_M := \Omega^1(M) \otimes \mathrm{su}_2 \tag{16}$$

[7] Work in progress.

where su_2 is the Lie algebra of SU_2. A connection a defines a covariant derivative $\nabla_a\xi = \nabla\xi + a\xi$ of "sections" $\xi.\ M \to \mathbb{C}^2$, where ∇ is the usual derivative, and su_2 acts on $\mathbb{C}^2$ in the usual way. The gauge group

$$\mathscr{G}_M = C^\infty(M, SU_2) \tag{17}$$

of bundle transformations of $M \times \mathbb{C}^2$ also acts on such sections, and hence on $\mathscr{A}_M$ through

$$g^{-1}\nabla_a g = \nabla_{g_* a}\ ;\ g_* a = g^{-1}\nabla g + g^{-1}ag\,. \tag{18}$$

The curvature-2-form F_a of $a \in \mathscr{A}_M$ is $\mathscr{G}_M$-equivariant, where $\mathscr{G}_M$ acts linearly on $\Omega^2(M) \otimes su_2$, i.e. we have $F_{g_* a} = g^{-1}F_a g^{-1}$. Such a $\mathscr{G}_M$-equivariant map can also be considered as a section of an infinite dimensional vector bundle[8] over the quotient space $\mathscr{B}_M = \mathscr{A}_M/\mathscr{G}_M$. The latter is known to contain as an open subset the infinite dimensional manifold $\mathscr{B}^*_M$ of irreducible connection classes.

So far, we have not made use of the dimension of M. However, the particular position of $\Omega^2(M)$ in the De Rham complex of M lends a special aspect to gauge theory in dimensions 2, 3 and 4. In the present case, note that curvature defines a 1-form on $\mathscr{A}_M$, which turns out to be exact. More precisely, it defines up to an additive constant a function $\boldsymbol{S} : \mathscr{A}_M \to \mathbb{R}$ by

$$\int_M \operatorname{tr}(F_a \wedge b) = d\boldsymbol{S}(a)b\,. \tag{19}$$

This function is almost gauge invariant: It defines a function $\boldsymbol{S} : \mathscr{B}_M \to \mathbb{R}/\mathbb{Z}$ first considered by Chern and Simons. Thus the critical point set $\mathscr{C}(\boldsymbol{S})$ of this Chern Simons functions is the set of flat connections, which can be described algebraically through their holonomy:

$$\begin{aligned}\mathscr{C}(\boldsymbol{S}) &= \{a \in \mathscr{A}_M \mid F_a = 0\}\,/\,\mathscr{G}_M \\ &= \operatorname{Hom}(\pi_1(M), SU_2)\,/\operatorname{ad}(SU_2)\,.\end{aligned} \tag{20}$$

Though the actual description of (17) may be a rather complicated problem, we cannot really expect infinite dimensional Morse theory to be of much help. In fact, the objective in this case is entirely reversed: If R_M is related by such a Morse theory to some homology group, then this group could define a new invariant of M. Since the Hessian of $\boldsymbol{s}$ at $\alpha \in R_M$, which is given by the symmetric bilinear form

$$D^2\boldsymbol{S}(\alpha)(\xi,\zeta) = \int_M \operatorname{tr}(\xi \wedge d_\alpha\zeta) \tag{21}$$

on $\Omega^1(M) \otimes su_2$, has again an infinite dimensional negative and positive subspaces, we cannot expect the topology of $\mathscr{B}_M$ itself to come in. Just as solutions of the Cauchy-Riemann equation on $\mathbb{R} \times S^1$ are the gradient flow trajectories for the symplectic action, the gradient flow trajectories for $\mathbf{S}$ are given by instantons on $\mathbb{R} \times M$. To see this, note that every connection on $\mathbb{R} \times M$ is gauge equivalent

[8] With fibre $\Omega^2(M) \otimes su_2$.

to a connection a with component in $\mathbb{R}$-direction[9] vanishing. Such a connection describes a family in $\mathscr{B}_M$ by $a(s) = a_{\{s\times M\}}$. If γ_M is a metric on M, and a metric on $\mathbb{R} \times M$ is defined by $(ds)^2 + \gamma_M$, then the antiself duality equation on $\mathbb{R} \times M$ takes on the form

$$0 = \frac{d}{ds}a(s) + *_M F_{a(s)} \tag{22}$$

where $*_M$ denotes Hodge duality in $\Omega^*(M)$ with respect to γ_M. By (17), we can consider $*F_a$ as the gradient of the Chern-Simons function with respect to the L^2-metric on $\mathscr{A}_M$ defined by γ_M. We arrive at the conclusion that instantons on $\mathbb{R} \times M$ represent "trajectories of the L^2-gradient flow" of the Chern-Simons function $\boldsymbol{S}$, even though such a flow is not defined everywhere as a homotopy on $\mathscr{B}_M$. The analogue of (4) and (14) is now the family of instanton spaces

$$\mathscr{M}(\alpha \mid \mathbb{R} \times M \mid \beta) = \{a \in \mathscr{A}_{\mathbb{R}\times M} \mid \quad A \text{ satisfies (24)} \quad \}/\mathscr{G}_{\mathbb{R}\times M} \tag{23}$$

with

$$\begin{cases} F_a^+ &= 0 \ , \\ \lim\limits_{t\to-\infty} a(t) = \alpha \ , \\ \lim\limits_{t\to\infty} a(t) &= \beta \ . \end{cases} \tag{24}$$

The connection $\alpha \in \mathscr{C}(\boldsymbol{S})$ is nondegenerate as a critical point of $\mathbf{S}$ in $\mathscr{B}_M$ iff the first cohomology group $H^1_\alpha(M, \mathrm{su}_2)$ with coefficients twisted by α vanishes. Under this condition, it follows from Taubes' analysis that "generically" the spaces (18) are smooth finite dimensional manifolds. (Here, we may need perturbations not only of the metric but also of $\boldsymbol{S}$ itself, and such perturbations may also be used to make $\mathscr{C}(\boldsymbol{S})$ nondegenerate whenever this is not a priori the case. However, for the sake of brevity we will have to ignore this point here.) Orientations can be derived from Donaldson's work [11, 12] and a "weak compactness theory" of instanton spaces was developed by Uhlenbeck [28,29]. With these three ingredients, we can again set up the dual "Morse inequalities". Assume first that $H_1(M) = 0$, to avoid "abelian" elements of $C(\boldsymbol{S})$. Denote by $C_*(M)$ the free abelian group over $\mathscr{C}(\boldsymbol{S}) \setminus \{\theta\}$ (which is finite when nondegenerate), and define

$$\partial : C_*(M) \to C_*(M) \qquad < \partial\alpha, \beta >= \#\hat{\mathscr{M}}(\alpha \mid \mathbb{R} \times M \mid \beta) \tag{25}$$

with # and $\hat{\mathscr{M}}$ defined as before. The proof of the boundary property $\partial\partial = 0$ now relies on an analogue of (8) which is almost the original Taubes grafting procedure. The result is the following new topological invariant:

Theorem 3. *We have $\partial\partial = 0$, and the homology groups*

$$F_*(M) = \ker\partial/\mathrm{im}\partial$$

[9] Which we will consider as the zero–component a_0.

do not depend on the choice of the metric γ_M. Moreover, for every oriented homology 3-sphere there are perturbations of **S** whose critical set is nondegenerate, and which give rise to a chain complex $(C_*(M,h), \partial_{\gamma,h})$, whose homology depends only on M.

The groups $F_*(M)$ were calculated by Fintushel and Stern for Brieskorn spheres $M(p,q,r)$. Here $\boldsymbol{S}$ truns out to be "perfect", and $F_*(M) = C_*(M)$. For more complex Brieskorn spheres $M(p_1, \ldots, p_n)$, $\boldsymbol{S}$ is still perfect in the sense that although $C(\boldsymbol{S})\backslash\{\theta\}$ is not discrete, its homology coincides with $F_*(M)$, see [4, 22][10]. There is, however, no reason to believe that $\boldsymbol{S}$ will be perfect for any large class of 3–manifolds, and other methods of calculation are needed. These rely very much on the functorial property of F_* with respect to (oriented) cobordism, and on certain exact sequences which define a system of "axioms" for the "homology theory" F_*. To explain functoriality, consider a smooth compact oriented cobordism W between M and N, which we denote as $W : M \to N$. Add half cylinders $\mathbb{R}_- \times M$ and $\mathbb{R}_+ \times N$ to W to obtain an open 4–manifold W_α, and consider a metric on W_α which is cylindrical on the ends. Then we can define finite dimensional, oriented, and weakly compact manifolds $\mathcal{M}_\gamma(\alpha \mid W \mid \beta)$ for $\alpha \in R_M$ and $\beta \in R_N$ in much the same way as in the case $W = \mathbb{R} \times M$. (Of course, the translational symmetry is not defined anymore.) It turns out that, counting the discrete part of the (unreduced) moduli spaces, we obtain a chain homomorphism

$$W_* : C_*(M) \to C_*(N) \qquad < \alpha \mid W_* \mid \beta >= \#\mathcal{M}(\alpha \mid W \mid \beta)\,. \tag{26}$$

The proof of the chain property is similar to the proof of $\partial\partial = 0$. Hence, cobordisms $W : M \to N$ of homology 3–spheres induce homomorphisms $W_* : F_*(M) \to F_*(W)$. This construction turns out to be functorial with respect to the obvious composition law for cobordism.

In the category of 3–dimensional cobordisms between surfaces, such a functor (taking values in finite dimensional Hilbert spaces) has been recently discovered, the Witten–Jones "topological quantum field theory". Its initial definition uses the path integral of the Quantum–Field–theory on 3-dimensions, in which the Chern–Simons function $\boldsymbol{S}$ plays the role of the Lagrangian.[11] The path integral itself does not have an entirely rigorous mathematical meaning, but physical reasoning implies certain axiomatic properties the resulting functor which allow its construction by combinatorial techniques. To a certain degree the functor F_* may also be considered as a topological quantum field theory, though its "axioms", besides functoriality itself, differ strongly from those of Witten–Jones

[10] This is similar to the determination of $I_*(P)$ in the symplectic case.

[11] It is quite remarkable that this function thus leads to two apparently quite independent sets of topological invariants for three manifolds, but no direct relation can be derived from the definition.

theory: the central ingredient seems to be an exact sequence

$$\begin{array}{ccc} & F_*(M) & \\ \swarrow & & \nwarrow \\ F_*(M_0) & \longrightarrow & F_*(M_1) \end{array} \tag{27}$$

where M_0 and M_1 are given by two different kinds of surgery on the same knot in M, see [19]. Moreover, even though such exact sequences can be shown to define a "complete" axiomatic system, i.e. to "characterize" the functor F_* in the same way as the Eilenberg–Steenrod axioms characterize usual homology, they do not yield a straight forward calculation for the groups F_* due to the ambiguities inherent in exact sequences. Exact sequences are nevertheless the most useful tools for calculations in ordinary homology, and we expect the same to be true for F_*. The functorial structure also suggests that F_* should find applications to special (e.g. integral) cobordism groups of 3-manifolds. For example, parts of the work of Fintushel and Stern on the integral cobordism classes of Bricskorn spheres can be put into a "homological" framework, and can thus be extended to more general classes of manifolds. Moreover, F_* comes up in connection with Donaldson's applications of instantons on 4-manifolds. For a manifold W with boundary $\partial W = M$, the Donaldson polynomials can be extended to $\mathbb{Z}$–linear maps

$$S_* H_2(W) \to F_*(M)\,. \tag{28}$$

This allows us to formulate "obstructions" to closing a 4-manifold. (In particular, we have $F_*(M) \neq 0$ whenever a homology 3-sphere M bounds a 4-manifold whose intersection form is "forbidden" by one of Donaldson's theorems.) Moreover, with luck, one may be able to analyse the change of the Donaldson polynomial (22) under handle addition. The exact triangle for surgery on M may be considered as one step in this direction.

References

1. S. Akbulut, J. McCarthy: Casson's invariant for oriented homolgy 3-spheres – an exposition. Mathematical Notes. Princeton University Press, 1987
2. V.I. Arnold: Mathematical methods of classical mechanics. Chapter Appendix 9. Springer, 1978
3. V.I. Arnold: Sur une propriété topologique des applications globalement canoniques de la méchanique classique. C. R. Acad. Paris (1965) **261**, 3719–3722
4. S. Bauer, C. Okonek: The algebraic geometry of representation spaces associated to Seifert fibred homology 3-spheres. Preprint.
5. R. Bott: An application on the Morse theory to the topology of Lie-groups. Bull. Soc. Math. France **84** (1956) 251–281
6. R. Bott: Lectures on Morse theory, old and new. Bull. AMS **7** (1982) 331–358
7. A. Casson: An invariant for homology 3-spheres. 1985. Lectures at MSRI Berkeley

8. C.C. Conley, E. Zehnder: The Birkhoff-Lewis fixed point theorem and a conjecture of V. I. Arnold. Invent. math. **73** (1983) 33–49
9. S.S Chern, J. Simons: Characteristic forms and geometric invariants. Ann. Math. **99** (1974) 48–69
10. C.C. Conley: Isolated invariant sets and Morse index. CBMS Reg. Conf. Series in Math. **38**. AMS, 1978
11. S.K. Donaldson: An application of gauge–theory to the topology of 4-manifolds. J. Diff. Geom. **18** (1983) 269–316
12. S.K Donaldson: The orientation of Yang–Mills moduli spaces and 4-manifold topology. J. Diff. Geom. **26** (1987) 397–428
13. R. Fintushel, R.J Stern: Instanton homology groups of Seifert fibred homology 3-spheres. Preprint
14. R.Fintushel, R.J. Stern: SO(3)-connections and the topology of 4-manifolds. J. Diff. Geom. **20** (1984) 523–539
15. A. Floer: Morse thoery for Lagrangian intersection theory. J. Diff. Geom. **18** (1988) 513–517
16. A. Floer: Symplectic fixed points and holomorphic spheres. Comm. Math. Physics**120** (1989) 576–611
17. A. Floer: The unregularised gradient flow of the symplectic action. Comm. Pure Appl. Math. **41** (1988) 775–813
18. A. Floer: Witten's complex and infinite dimensional Morse theory. J. Diff. Geom. **30** (1989) 207–221
19. A. Floer: Instanton homology and dehn surgery. In preparation.
20. W.M. Goldman: Invariant functions on lie groups and Hamiltonian flows of surface group representation. MSRI preprint, 1985
21. M. Gromov: Pseuodoholomorphic curves in symplectic manifolds. Inv. math. **82** (1985) 307–347
22. P.A. Kirk, P.E. Klassen: Representation spaces of Seifert fibred homology spheres. To appear in Topology
23. R.S. Palais: Morse theory on Hilbert manifolds. Topology **2** (1963) 229–340
24. P. Rabinowitz: Periodic solutions of Hamiltonian systems. Comm. Pure Appl. Math. **31** (1978) 157–184
25. J. Sachs, K.K. Uhlenbeck: The existence of minimal 2-spheres. Ann. Math. **113** (1981) 1–24
26. C.H. Taubes: A framework for Morse theory for the Yang-Mills functionals. preprint Havard University, 1986
27. C.H. Taubes: Self-dual Yang-Mills connections on non-self-dual 4-manifolds. J. Diff. Geom. **17** (1982) 139–170
28. K.K. Uhlenbeck: Connections with L_p-bounds on curvature. Commun. Math. Phys. (1982) 32–32
29. K.K. Uhlenbeck: Removable singularities in Yang-Mills fields. Commun. Math. Phys. (1982) 11–29
30. E. Witten: Supersymmetry and Morse theory. J. Diff. Geom. **17** (1982) 661–692
31. E. Witten: Topological quantum field theory. Preprint IAS, Feb. 1988

Braids, Galois Groups, and Some Arithmetic Functions

Yasutaka Ihara

Research Institute for Mathematical Sciences, Kyoto University, Kyoto 606, Japan

This lecture is about some new relations among the classical objects of the title. The study of such relations was started by [B_1, G, De, Ih_1] from independent motivations, and was developed in [A_3, C_3, A-I, IKY, Dr_2, O, N], etc. It is still a very young subject, and there are several different approaches, each partly blocked by its own fundamental conjectures! But it is already allowing one to glimpse some new features of the classical "monster" $\mathrm{Gal}(\bar{\mathbb{Q}}/\mathbb{Q})$, and providing a bridge connecting $\mathrm{Gal}(\bar{\mathbb{Q}}/\mathbb{Q})$ even with such "modern" objects as the quantum groups [Dr_2]. I will not try to "explain" any general philosophies that are still in the air, but to draw a few lines sketching the concretely visible features of the subject.

§1. Introduction

The absolute Galois group over the rational number field $\mathbb{Q}$, denoted by $G_{\mathbb{Q}} = \mathrm{Gal}(\bar{\mathbb{Q}}/\mathbb{Q})$, is one of the classical mathematical objects which we really need to understand better. It is primarily a (huge) topological group; the automorphism group of the field $\bar{\mathbb{Q}}$ of all algebraic numbers in $\mathbb{C}$, equipped with the Krull topology. It is moreover equipped with arithmetic structure, i.e., the system of conjugacy classes of embeddings $G_{\mathbb{Q}_p} \to G_{\mathbb{Q}}$ of all local absolute Galois groups into $G_{\mathbb{Q}}$. Through the arithmetic structure of $G_{\mathbb{Q}}$, each "natural" representation of $G_{\mathbb{Q}}$ provides the set of all prime numbers with a "natural" additional structure. For example, the character $\chi_N : G_{\mathbb{Q}} \to (\mathbb{Z}/N)^{\times}$ defining the action of $G_{\mathbb{Q}}$ on the group of N-th roots of unity gives rise to the classification of prime numbers modulo N. Study of the group $G_{\mathbb{Q}}$ and its natural representations has an *ultimate* goal to understand the "total structure" of the set of all prime numbers.

Now if X is an algebraic variety over $\mathbb{Q}$, the fundamental group $\pi_1(X(\mathbb{C}), b)$, when suitably completed, is equipped with a natural action of $G_{\mathbb{Q}}$. Here, b is a $\mathbb{Q}$-rational base point of X. Already in some basic cases, this action gives rise to a very big and interesting representation of $G_{\mathbb{Q}}$. What this action amounts to is the following. (We may, and will, assume X to be geometrically connected.) As is well known, all finite (topological) coverings of $X(\mathbb{C})$ and morphisms between two such coverings are algebraic and defined over $\bar{\mathbb{Q}}$. The above action contains all information related to the "field of definition" and the $G_{\mathbb{Q}}$-conjugations of

Proceedings of the International Congress
of Mathematicians, Kyoto, Japan, 1990

these coverings, morphisms and the points above b. In particular, if they are all defined over some Galois extension $\Omega/\mathbb{Q}$ ($\Omega \subset \bar{\mathbb{Q}}$), then $G_{\mathbb{Q}}$ acts via its factor group $\mathrm{Gal}(\Omega/\mathbb{Q})$. (The converse is also valid if the completion of π_1 has trivial center.) Choice of completions of π_1 depends on the family of finite coverings one wants to consider. If this is all the finite coverings, the choice should be the profinite completion $\hat{\pi}_1 = \varprojlim(\pi_1/N)$, the projective limit of all finite factor groups of π_1. It is also important to consider the pronilpotent completion π_1^{nil}, where one allows only those π_1/N (the covering groups) that are nilpotent.

If there is no family of finite coverings of X having a remarkable common field of definition, then the representation of $G_{\mathbb{Q}}$ on $\hat{\pi}_1(X(\mathbb{C}), b)$ would not be so interesting. But already when $X = \mathbb{P}^1 - \{0, 1, \infty\}$, there are several natural families with various remarkable arithmetic features in their fields of definition (see (4) below). One of the reasons why this type of representations is interesting lies in that these different arithmetic features can be viewed simultaneously from a certain height, the representation of $G_{\mathbb{Q}}$ on $\hat{\pi}_1$. Other reasons and motivations will also be explained below, with names of the main contributors. Before this we shall give a sequence of basic examples of X starting with $\mathbb{P}^1 - \{0, 1, \infty\}$. For $n \geq 4$, X_n is the moduli space of ordered n-tuples $(x_1, \ldots, x_n)$ of distinct points on the projective line $\mathbb{P}^1$. The corresponding fundamental group $P_n = \pi_1(X_n(\mathbb{C}), b)$ is the quotient modulo center ($\cong \mathbb{Z}/2$) of the pure sphere braid group on n strings. Note that $X_4 \simeq \mathbb{P}^1 - \{0, 1, \infty\}$, $P_4 \simeq F_2$ (free, rank 2).

Recently, several substantial works have been done in connection with these "big Galois representations." They include the following (1)~(6):

(1) Already for more than a decade, this type of Galois representations has been *used* effectively to construct finite Galois extensions over $\mathbb{Q}$ with given Galois groups, for various cases of finite simple groups (G. V. Belyĭ, M. Fried, B. Matzat, K. Y. Shih, J. G. Thompson, ...). By the Hilbert irreducibility theorem, it suffices to construct a Galois extension over some rational function field $\mathbb{Q}(t_1, \ldots, t_m)$ ($m \geq 1$) having a given Galois group. Many such extensions have been detected among the (function fields of) coverings of X_{m+3} (etc.) by combining (i) only the basic knowledge on how $G_{\mathbb{Q}}$ acts on $\hat{\pi}_1(X(\mathbb{C}), b)$, with (ii) a deep knowledge on specific finite groups and their characters. As there is a distinguished report on this topic in the last ICM [B$_2$] (see also [Ma] for the later development), we shall only recall and stress the following:

Belyĭ proved, among other things in [B$_1$] that the canonical representation

$$\varphi_X : G_{\mathbb{Q}} \to \mathrm{Out}\,\hat{\pi}_1(X(\mathbb{C}))$$

for $X = \mathbb{P}^1 - \{0, 1, \infty\}$ is *injective*. This sets in evidence the importance of the problem to characterize the image. Here, $\mathrm{Out}\,\hat{\pi}_1$ is the outer automorphism group, and φ_X is induced from the $G_{\mathbb{Q}}$-action on $\hat{\pi}_1(X(\mathbb{C}), b)$. φ_X is "independent" of b in the obvious sense.

(2) A. Grothendieck made some basic proposals related to the study of the $G_{\mathbb{Q}}$-action on $\hat{\pi}_1(X(\mathbb{C}), b)$ [G]. One of them is as follows. Take $X = X_4 = \mathbb{P}^1 - \{0, 1, \infty\}$. Then it has the obvious S_3-symmetry but there is no S_3-invariant choice of a

base point b. This already indicates that the use of the fundamental *groupoid* on a suitable S_3-stable set B of base points would be better. (The natural action of S_4 on X_4 factors through its quotient $\simeq S_3$.) He also suggests using all possible combinatorial relations among the $X_n(\mathbb{C})$ $(n = 4, 5, \ldots)$, and that, in a certain sense, the two cases of n with $\dim X_n = 1, 2$ (i.e., $n = 4, 5$) would be basic. In particular, understanding and using full relationship between $X_4(\mathbb{C})$ and $X_5(\mathbb{C})$ would give, presumably, all the crucial non-obvious information on the image of φ_X for $X = X_4$.

We shall describe more about these in the main text; the action of $G_{\mathbb{Q}}$ on the completed fundamental groupoids (§2), Deligne's tangential base points (which will serve as B) (§2.3), Drinfeld's new information on the image of φ_X (§3), and the Lie version of the study of the $G_{\mathbb{Q}}$-action on $\{\pi_1^{\mathrm{nil}}(X_n(\mathbb{C}))\}_n$ (§5).

(3) π_1^{nil} *as a new test case for motivic philosophy.*

In [De], P. Deligne develops a motivic theory of nilpotent quotients of fundamental groups in algebraic geometry. The first main point is that π_1^{nil} is not just a topological group with a $G_{\mathbb{Q}}$-action but is a limit of objects with more structures –: To each $\pi_1 = \pi_1(X(\mathbb{C}), b)$, one can associate, via Malcev, some projective system $\{U^m\pi_1\}_{m\geq 1}$ of linear unipotent algebraic groups $U^m\pi_1$ over $\mathbb{Q}$. One may assume that the group of $\mathbb{Z}$-valued points $(U^m\pi_1)(\mathbb{Z})$ is the quotient modulo torsion of $\pi_1/\pi_1(m+1)$, where $\pi_1(m+1)$ is the $(m+1)$-th member of the lower central series of π_1. The Galois group $G_{\mathbb{Q}}$ acts on the profinite groups $(U^m\pi_1)(\widehat{\mathbb{Z}})$ $(m \geq 1)$, from which the $G_{\mathbb{Q}}$-action on π_1^{nil} can be almost recovered. Under some assumptions on X, the underlying vector space over $\mathbb{Q}$ of the Lie algebra $\mathrm{Lie}(U^m\pi_1)$ has mixed Hodge structure (J. Morgan, D. Sullivan, K. T. Chen, R. Hain, …). Deligne adds more structures to $\mathrm{Lie}(U^m\pi_1)$ (and accordingly, on $U^m\pi_1$), the mixed motif structure, in terms of various *realizations.*

When $X = \mathbb{P}^1 - \{0, 1, \infty\}$, the motivic $\mathrm{Lie}(U^m\pi_1)$ is a successive extension of Tate's motives "$\mathbb{Q}(k)$" $(1 \leq k \leq m)$. Knowledge on the motivic extension of "$\mathbb{Q}$" by "$\mathbb{Q}(k)$" turns out to be crucial in order to understand a certain portion of the Galois action on $(U^m\pi_1)(\widehat{\mathbb{Z}})$. Deligne constructs a basic extension of "$\mathbb{Q}$" by "$\mathbb{Q}(k)$" $(k \geq 3$, odd), describes a certain portion of the Galois action on the double commutator quotient of π_1^{nil}, and under the hypothesis "$\mathrm{Ext}^1(\mathbb{Q}, \mathbb{Q}(k))$ *for such k is one-dimensional*", derives further desirable consequences related to the size and the "$\mathbb{Q}$-rationality" of the Galois image. (See §5.4 below.)

(4) Some new objects in number theory, such as adelic analogues of beta and gamma functions, and higher circular l-units, have been constructed and used to give explicit comparisons of the $G_{\mathbb{Q}}$-actions, on π_1^{nil} of $\mathbb{P}^1 - \{0, 1, \infty\}$, on torsion points of Fermat Jacobians, and on higher circular l-units (G. Anderson, R. Coleman, the author, …). This series of work was started in [Ih$_1$] and was developed in [A$_3$, C$_3$, IKY, Ih$_i$, A-I$_{1,2}$], etc., by combining with other ideas [A$_1$, C$_1$, C$_2$], ….

The following special towers of coverings of $\mathbb{P}^1 - \{0, 1, \infty\}$ are relevant; (i) *the meta-abelian tower,* (ii) *the nilpotent tower,* and (iii) *the genus* 0 *tower.* Here, (i) corresponds to the double commutator quotient $\hat{\pi}_1/\hat{\pi}_1''$ of $\hat{\pi}_1$, (ii) to π_1^{nil}, and (iii) is also quite big – what it generates is bigger than (ii). As for the common field of definition, (i) is related to abelian extensions over the cyclotomic field $\mathbb{Q}(\mu_\infty)$, (ii) to a very natural sequence of Galois extensions over $\mathbb{Q}$ that are nilpotent over $\mathbb{Q}(\mu_\infty)$, and (iii) to (the field generated by) higher circular l-units.

As for (i), the conclusion is that the Galois actions on $\hat{\pi}_1/\hat{\pi}_1''$, on torsion points of Fermat Jacobians, and on roots of circular units, can be compared with each other in terms of *explicit universal formulas*. The size of the Galois image is also measured explicitly. The theory developed to that of Anderson's hyperadelic gamma function, which also plays the role of a bridge connecting Gauss sums with circular units.

As for (ii) and (iii), the main conclusion is that the action of $\sigma \in G_{\mathbb{Q}}$ on π_1^{nil} can be expressed explicitly $\mod l^n$, for any $n \geq 1$, in terms of its action on the group of higher circular l-units.

For more of these, see §6.

(5) V. G. Drinfeld more recently discovered a striking connection between (what is expected to be) the automorphism group of the tower $\{\pi_1(X_n(\mathbb{C}))\}_{n\geq 4}$ and a "universal braid transformation group"

$$(A, \Delta, \varepsilon, \phi, R) \to (A, \Delta, \varepsilon, \phi', R')$$

acting on the structures of quasi-triangular quasi-Hopf algebras [Dr$_2$]. These groups become non-trivial after suitable *completions* of π_1, etc. The Galois image in $\operatorname{Aut}\hat{\pi}_1(X_4(\mathbb{C}))$ is contained in these (essentially the same) groups. (See §4.)

(6) There are several other important works on this subject including the following.

(i) Grothendieck raises a question as to whether the intertwiners of two Galois representations in $\hat{\pi}_1$ arising from two algebraic varieties always correspond to algebraic morphisms in the "non-abelian" situation. In the special case of $\mathbb{P}^1$ minus finitely many points, H. Nakamura [N] gives an affirmative answer.

(ii) T. Oda and Y. Matsumoto each gives, from different viewpoints, a non-abelian analogue of the Néron-Ogg-Šafarevič criterion for good reduction of curves, using π_1^{nil} instead of H_1^{et}; cf. [O].

Due to the lack of time, these will not be included in this lecture.

We shall start by defining the Galois action on the completed fundamental groupoids.

§2. The Basic Definitions

2.1 Let X be an algebraic variety defined over $\mathbb{Q}$, and $X(\mathbb{C})$ be the set of $\mathbb{C}$-rational points of X equipped with the usual (complex analytic) topology. We assume X to be geometrically connected, which is equivalent with $X(\mathbb{C})$ being connected. For $a, b \in X(\mathbb{C})$, call $\pi_1(X(\mathbb{C}); a, b)$ the set of homotopy classes (rel. to a, b) of paths from a to b on $X(\mathbb{C})$. When $a = b$, it is denoted by $\pi_1(X(\mathbb{C}), b)$. The composition rule

$$\pi_1(X(\mathbb{C}); b, c) \times \pi_1(X(\mathbb{C}); a, b) \longrightarrow \pi_1(X(\mathbb{C}); a, c) \tag{2.1.1}$$

gives the system $\{\pi_1(X(\mathbb{C});a,b)\}_{a,b\in B}$ structure of groupoid, the fundamental groupoid of $X(\mathbb{C})$ with base point set B ($B \subset X(\mathbb{C})$). When $B = \{b\}$, $\pi_1(X(\mathbb{C}),b)$ is the fundamental group.

We shall recall the definitions of the profinite completion $\hat{\pi}_1(X(\mathbb{C});a,b)$ and the $G_{\mathbb{Q}}$-action on $\hat{\pi}_1(X(\mathbb{C});a,b)$ when $a,b \in X(\mathbb{Q})$ (the $\mathbb{Q}$-rational points of X).

First, $\hat{\pi}_1(X(\mathbb{C});a,b)$ is the completion of $\pi_1(X(\mathbb{C});a,b)$ with respect to the following topology. Two elements p, p' of the latter set are sufficiently close to each other if the associated "round trip" $p'^{-1} \cdot p$ belongs to a sufficiently small subgroup of $\pi_1(X(\mathbb{C}),a)$ with finite index. The completed set $\hat{\pi}_1(X(\mathbb{C});a,b)$ is then profinite, i.e., compact and totally disconnected topological space. The composition (2.1.1) is continuous and carries over to the completion. Each path class $p \in \pi_1(X(\mathbb{C});a,b)$ gives rise to a "compatible" system of fiber bijections $p_f : f^{-1}(a) \xrightarrow{\sim} f^{-1}(b)$, where f runs over all finite coverings of $X(\mathbb{C})$, and p_f is induced by tracing above a path representing p. Here, "compatible" means that for any f, f', p_f and $p_{f'}$ are compatible with the fiber projections induced from each element of $\mathrm{Hom}(f, f')$. This procedure $p \to \{p_f\}_f$ induces the bijection

$$\hat{\pi}_1(X(\mathbb{C});a,b) \approx \left\{ \begin{array}{l} \text{compatible systems of fiber} \\ \text{bijections} \quad \hat{p}_f : f^{-1}(a) \xrightarrow{\sim} f^{-1}(b) \end{array} \right\}_f .$$

This is because $X(\mathbb{C})$ is locally arcwise connected and locally simply connected, being an underlying space of an analytic space.

Since all finite coverings of $X(\mathbb{C})$ are algebraic and defined over $\overline{\mathbb{Q}}$ (the generalized Riemann existence theorem), we may assume that f runs over all finite etale coverings of $X \otimes \overline{\mathbb{Q}}$. If $a,b \in X(\mathbb{Q})$, each $\sigma \in G_{\mathbb{Q}}$ induces the fiber bijections $f^{-1}(a) \xrightarrow{\sim} (\sigma f)^{-1}(a)$, $f^{-1}(b) \xrightarrow{\sim} (\sigma f)^{-1}(b)$; hence σ acts on $\hat{\pi}_1(X(\mathbb{C});a,b)$ by

$$\{\hat{p}_f\} \longrightarrow \{\sigma \circ \hat{p}_{\sigma^{-1}f} \circ \sigma^{-1}\} .$$

It is easy to see that this action is compatible with the composition induced from (2.1.1) by completion. Therefore, $G_{\mathbb{Q}}$ acts as an automorphism group of the completed fundamental groupoid of $X(\mathbb{C})$ with base point set $B \subset X(\mathbb{Q})$; in particular, as that of the completed fundamental group with base point $b \in X(\mathbb{Q})$. These actions are compatible with the groupoid homomorphisms induced from any algebraic morphisms $X' \to X$ over $\mathbb{Q}$.

Example. Take $X = \mathbb{P}^1 - \{0,\infty\}$ and $b = 1$. Then $\pi_1(X(\mathbb{C}),1) \simeq \mathbb{Z}$, being generated by the class p of the loop $\tau \to \exp(2\pi i\tau)$ $(0 \le \tau \le 1)$, and $\hat{\pi}_1(X(\mathbb{C}),1) = \widehat{\mathbb{Z}} = \varprojlim(\mathbb{Z}/N)$. If f_N is the cyclic covering $t \to t^N$ of X $(N = 1,2,\dots)$, then

$$f_N^{-1}(1) = \{1, \zeta_N, \dots, \zeta_N^{N-1}\}, \quad \zeta_N = \exp(2\pi i/N) .$$

Write $\sigma(\zeta_N) = \zeta_N^{\chi(\sigma)_N}$ $(\sigma \in G_{\mathbb{Q}}, \chi(\sigma)_N \in \mathbb{Z}/N)$. Then p_{f_N} acts on $f_N^{-1}(1)$ by $\theta \longrightarrow \zeta_N\theta$, and σ acts on $f_N^{-1}(1)$ by $\theta \longrightarrow \theta^{\chi(\sigma)_N}$ $(\theta \in f_N^{-1}(1))$. Therefore,

$$\sigma \circ p_{f_N} \circ \sigma^{-1} = p_{f_N}^{\chi(\sigma)_N} . \tag{2.1.2}$$

Call $\chi(\sigma) = \varprojlim \chi(\sigma)_N \in \widehat{\mathbb{Z}}^\times$. Then χ is a (continuous) homomorphism

$$\chi : G_{\mathbb{Q}} \longrightarrow \widehat{\mathbb{Z}}^\times , \tag{2.1.3}$$

called the cyclotomic character. By (2.1.2), σ acts on $\hat{\pi}_1 = \widehat{\mathbb{Z}}$ via $\chi(\sigma)$-multiplication. (Here and in the following, for any associative ring A with unit element, $A^\times$ denotes the group of invertible elements of A.)

2.2 We return to the general case. The Galois group $G_{\mathbb{Q}}$ thus acts on the completed fundamental groupoid $\{\hat{\pi}_1(X(\mathbb{C}); a, b)\}_{a,b\in B}$ $(B \subset X(\mathbb{Q}))$ and in particular on $\hat{\pi}_1(X(\mathbb{C}), b)$ $(b \in X(\mathbb{Q}))$ as an automorphism group:

$$\varphi_{X,b} : G_{\mathbb{Q}} \longrightarrow \operatorname{Aut} \hat{\pi}_1(X(\mathbb{C}), b) \ .$$

The induced homomorphism obtained by forgetting the role of the base point is also of interest:

$$\varphi_X : G_{\mathbb{Q}} \longrightarrow \operatorname{Out} \hat{\pi}_1(X(\mathbb{C}))$$

$(\operatorname{Out}(\hat{\pi}_1) = \operatorname{Aut}(\hat{\pi}_1)/\operatorname{Int}(\hat{\pi}_1)$, where $\operatorname{Int}(\hat{\pi}_1)$ is the inner automorphism group). Replacing $\hat{\pi}_1$ by the pronilpotent completion π_1^{nil}, we obtain $\varphi_{X,b}^{\mathrm{nil}}$, etc.

As for X, we shall mainly consider the varieties

$$X = X_n = \{(x_1, \dots, x_n) \in (\mathbb{P}^1)^n; \quad x_i \neq x_j \quad \text{for} \quad i \neq j\}/PGL(2)$$

$(n \geq 4)$, where $PGL(2) = \operatorname{Aut} \mathbb{P}^1$ acts on X diagonally. Note that

$$X_4 \simeq \mathbb{P}^1 - \{0, 1, \infty\},$$
$$X_5 \simeq (X_4)^2 - \Delta \quad (\Delta : \text{the diagonal}),$$

$P_n = \pi_1(X_n(\mathbb{C}), *)$ is the quotient modulo center $(= \pi_1(PGL(2, \mathbb{C})) \simeq \mathbb{Z}/2)$ of the pure sphere braid group on n strings, and that

$$P_4 \simeq F_2 \quad (\text{free, rank } 2)$$
$$P_5 = F_2 \ltimes F_3 \quad (\text{semi-direct, } F_2 \text{ acts on } F_3) \ .$$

Now let $X = X_4 = \mathbb{P}^1 - \{0, 1, \infty\}$, so that $\pi_1(X(\mathbb{C})) \simeq F_2$. As reviewed in §1, Belyǐ proved that φ_X is then injective. This focuses light on the following

Question 1. What can one say about the images of $\varphi_X, \varphi_X^{\mathrm{nil}}$, etc.?

If one can characterize the image of φ_X explicitly, then one is led to a completely different description of $G_{\mathbb{Q}}$.

Question 2. What can one say about the "elementwise description" of $\varphi_X, \varphi_X^{\mathrm{nil}}$, etc.?

Actually, this *Q 2* is not well posed at this stage of the development of mathematics, as we know no good system of NAMES for elements of $G_{\mathbb{Q}}$. As far as the author knows, no element of $G_{\mathbb{Q}}$ other than the identity element and the complex conjugation (remember that $\overline{\mathbb{Q}} \subset \mathbb{C}$ in our formulation) has an explicit name to identify itself. Nor is it so for $\widehat{F}_2$. Before asking to give an explicit description of the homomorphism φ_X, we must ask ourselves the possibility of giving good names to elements of $G_{\mathbb{Q}}, \widehat{F}_r (r \geq 2)$. If, however, we replace $\widehat{F}_r$ by the nilpotent completion F_r^{nil}, each element of F_r^{nil} then has a good name to identify itself which is obtained from the non-commutative Taylor expansion in r variables with coefficients in $\widehat{\mathbb{Z}} (\cong \prod_l \mathbb{Z}_l)$. Thus, at present, the sense of *Q 2* is to ask to *give* a new system of names for elements of $G_{\mathbb{Q}}$ using φ_X^{nil} *and describe* any interplay in terms of these names. (φ_X^{nil} is no longer injective; hence it concerns with a certain quotient of $G_{\mathbb{Q}}$ which is "big and small".)

We shall mainly discuss:

1) Works of Deligne, Drinfeld, and the author on *Q 1* (§§3, 4, 5),

and

2) Works of Anderson, Coleman, and the author on *Q 2* (§6).

Among them, 1) is closely related to $\varphi_{X,B}$ for $X = X_5$.

2.3 Let $X = \mathbb{P}^1 - \{0, 1, \infty\}$, and F_2 be the free group of rank 2 on two letters x, y. Our first goal is to show that for each $\sigma \in G_{\mathbb{Q}}$, $\varphi_X(\sigma)$ is determined by two "coordinates", $\chi(\sigma)$ and f_σ, where $\chi : G_{\mathbb{Q}} \to \widehat{\mathbb{Z}}^\times$ is the cyclotomic character and f_σ is an element of $\widehat{F}'_2 = (\widehat{F}_2, \widehat{F}_2)$, the commutator subgroup of $\widehat{F}_2$. This can also be done relying more on group-theoretic normalization as in [B_1], [Ih_1], but we proceed more "conceptually" using Deligne's tangential base points [De].

Let $\mathbb{B}$ be the set of "arrows" $\overrightarrow{ij}$ with $i, j \in \{0, 1, \infty\}, i \neq j$. Thus, $\mathbb{B}$ has six elements and the symmetric group S_3 acts simply transitively on $\mathbb{B}$. For $a, b \in \mathbb{B}$, Deligne defines $\pi_1 = \pi_1(X(\mathbb{C}); a, b)$ and the $G_{\mathbb{Q}}$-action on its profinite completion. Topologically, it is clear what π_1 should mean when, in general, a, b are *simply connected subspaces* of $X(\mathbb{C})$. The base point $\overrightarrow{ij}$ plays the same role as the open interval I_{ij} on $\mathbb{R}$ bounded by i, j *and* not containing the third point k from $\{0, 1, \infty\}$. For a finite etale covering $Y \to X$ over $\overline{\mathbb{Q}}$, the fiber above $\overrightarrow{ij}$ consists of points $P \in \overline{Y}(\overline{\mathbb{Q}})$ above i given together with a "topological branch" (i.e., a lifting of I_{ij}) at each P. Here $\bar{f} : \overline{Y} \to \mathbb{P}^1$ is the compactification of f. In order to define the $G_{\mathbb{Q}}$-action on $\hat{\pi}_1(X(\mathbb{C}); a, b)$ for $a, b \in \mathbb{B}$, it suffices to give an *algebraic* interpretation of the branches at P. One way to put it is as follows. (This device proved to be useful [A–I_1].) Let t_{ij} be the linear fractional function $\mathbb{P}^1 \to \mathbb{P}^1$ which maps i, j, k to $0, 1, \infty$, respectively. Then a branch at P is a local embedding of the local ring of $\overline{Y}$ at P into the ring of *Puiseux series* in t_{ij} which extends: (i) the obvious embedding of the local ring of $\mathbb{P}^1$ at i into the ring of power series in t_{ij}, and (ii) the residue field embedding determined by the geometric point P. The corresponding topological branch is obtained by the principle to choose "the positive real root for $t_{ij}^{1/e}$, on I_{ij}". The group $G_{\mathbb{Q}}$ acts on the fibers above $\overrightarrow{ij}$ via its action on the Puiseux coefficients $\in \overline{\mathbb{Q}}$. One may prefer to reinterpret this in terms of the normalization of the fiber product of f with $\operatorname{Spec} \mathbb{C}[t_{ij}^{1/e}]$ (e: the ramification index).

Now we consider $\pi_1(X(\mathbb{C}); \overrightarrow{01}, \overrightarrow{10})$, $\pi_1(X(\mathbb{C}), \overrightarrow{01})$ and the Galois action on their completions. The first set contains an obvious element defined from the interval (0,1). Call it p. The second group contains a small positive loop around 0, called x, and $y = p^{-1} \circ x' \circ p$, where x' is the transform of x by $t \to 1-t$ ($t = t_{01}$). This group is free on x, y.

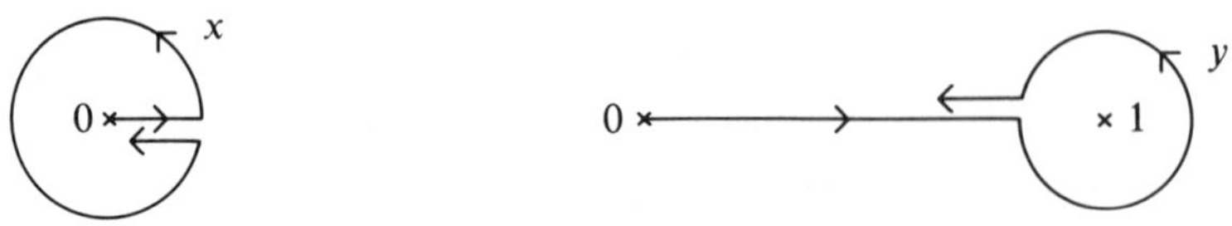

Now for each $\sigma \in G_{\mathbb{Q}}$, put

$$f_\sigma = p^{-1} \circ \sigma(p) \in \hat{\pi}_1(X(\mathbb{C}), \overrightarrow{01}) \ . \tag{2.3.1}$$

Then σ acts on the generators of $\hat{\pi}_1(X(\mathbb{C}), \overrightarrow{01}) = \widehat{F}_2$ as

$$x \longrightarrow x^{\chi(\sigma)}, \quad y \longrightarrow f_\sigma^{-1} \cdot y^{\chi(\sigma)} \cdot f_\sigma \ . \tag{2.3.2}$$

It follows easily that $f_\sigma \in \widehat{F}_2'$, and that (2.3.2) with this requirement characterizes f_σ. When σ is the complex conjugation, $\chi(\sigma) = -1$, $f_\sigma = 1$.

Remark. By (3.1.1)(I),(II) below, it follows also that $z = (xy)^{-1}$ (a loop around ∞) is mapped to $g_\sigma^{-1} z^{\chi(\sigma)} g_\sigma$, where $g_\sigma = f_\sigma(x,z) x^{\frac{1}{2}(1-\chi(\sigma))}$.

Although $\widehat{F}_2$ contains much more than the free words on x, y, we shall express an element of this group conveniently as $f(x,y)$, because it will then make sense to speak of $f(\xi,\eta)$ for any elements $\xi, \eta \in G$ of any profinite group G; the image of f under the unique homomorphism $\widehat{F}_2 \to G$ mapping x, y to ξ, η respectively.

§3. The Galois Action (Profinite)

3.1 So what is the image of the mapping $G_{\mathbb{Q}} \to \widehat{\mathbb{Z}}^\times \times \widehat{F}_2'$ defined by $\sigma \mapsto (\chi(\sigma), f_\sigma)$? The known equations satisfied by $\lambda = \chi(\sigma)$, $f = f_\sigma$ are as follows.

$$\begin{aligned} &\text{(I)} && f(x,y)f(y,x) = 1; \\ &\text{(II)} && f(z,x)z^m f(y,z)y^m f(x,y)x^m = 1, \\ &&& \qquad \text{if} \quad xyz = 1, \quad m = \tfrac{1}{2}(\lambda - 1); \end{aligned} \tag{3.1.1}$$

(III) (Drinfeld) Let $P_5 = \pi_1(X_5(\mathbb{C}), \mathscr{B}_5)$ and $x_{ij} \in P_5$

$(1 \le i, j \le 5)$ be as defined below. Then in $\widehat{P}_5$,

$$f(x_{12}, x_{23}) f(x_{34}, x_{45}) f(x_{51}, x_{12}) f(x_{23}, x_{34}) f(x_{45}, x_{51}) = 1 \ .$$

Remark. Drinfeld's formula given in [Dr$_2$] is in terms of *plane* braid group on 4 strings and is non-cyclic. The above formula is equivalent to his. The author previously wrote down more complicated formulas as 4 transposition relations (w.r.t. (1 2),(2 3),(3 4),(4 5) in S_5 instead of (12345)) in [Ih$_6$]. Drinfeld thinks this type of formula may very well be known to Grothendieck.

As for the more obvious question: "use of X_n $(n \geq 6)\ldots$?", see §3.3.

The definitions of $\mathscr{B}_n$ *and* x_{ij}. Let $\widetilde{\mathscr{B}}_n$ be the space of all n-tuples $(b_1, \ldots, b_n)$ of distinct points of $\mathbb{R}^{\cup}(\infty)$ satisfying the condition: b_{i+1} *is next to* b_i *in the positive direction for all* i $(1 \leq i \leq n-1)$ *including the case of passing through* ∞. Then $PGL_2^+(\mathbb{R})$, the real projective linear group of degree two with positive determinant, acts on $\widetilde{\mathscr{B}}_n$ diagonally, and the quotient space $\mathscr{B}_n = \widetilde{\mathscr{B}}_n / PGL_2^+(\mathbb{R})$ is naturally embedded into $X_n(\mathbb{C})$. The space $\mathscr{B}_n$ is simply connected and hence it makes sense to speak of the fundamental group $P_n = \pi_1(X_n(\mathbb{C}), \mathscr{B}_n)$. It is generated by the elements x_{ij} $(1 \leq i, j \leq n)$ shown below:

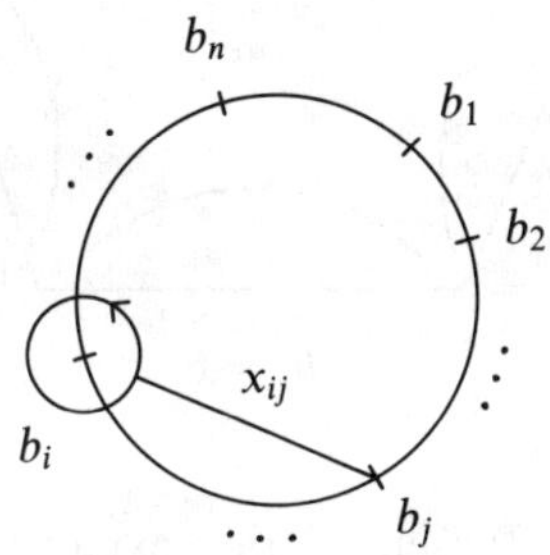

3.2 *Sketch of proof of (3.1.1).* (I) Apply the automorphism $\theta : t \to 1-t$ to both sides of (2.3.1).

(II) Let r be the element of $\pi_1(X(\mathbb{C}); \overrightarrow{10}, \overrightarrow{1\infty})$ corresponding to the rotation of argument π at the point 1:

Then it is easy to see that

$$r^{-1} \cdot \sigma(r) = \theta(x)^{\frac{1}{2}(\chi(\sigma)-1)} \in \hat{\pi}_1(X(\mathbb{C}), \overrightarrow{10}) \ .$$

Therefore, if $q = r \circ p$, we have $\sigma(q) = q y^{\frac{1}{2}(\chi(\sigma)-1)} f_\sigma(x, y)$. Let ω be the automorphism $t \to (1-t)^{-1}$ of X. Then $\omega^2(q)\omega(q)q = 1$. Apply σ on this to obtain (II).

(III) The symmetric group S_5 acts on X_5 from the left by substitution of coordinates, as $(x_i) \to (x_{s^{-1}i})$ $(s \in S_5)$. Let $s = (13524)$. Then for a certain "tangential base point"β and a path class $q \in \pi_1(X_5(\mathbb{C}); \beta, s(\beta))$, one can naturally identify $\pi_1(X_5(\mathbb{C}), \beta)$ with $\pi_1(X_5(\mathbb{C}), \mathscr{B}_5) = P_5$ and show that

$$\sigma(q) = q \cdot f_\sigma(x_{45}, x_{34}) \ , \tag{3.2.1}$$

$$s^4(q) \cdot s^3(q) \cdot s^2(q) \cdot s(q) \cdot q = 1 \ . \tag{3.2.2}$$

The equation (III) follows directly from these. A topological illustration:

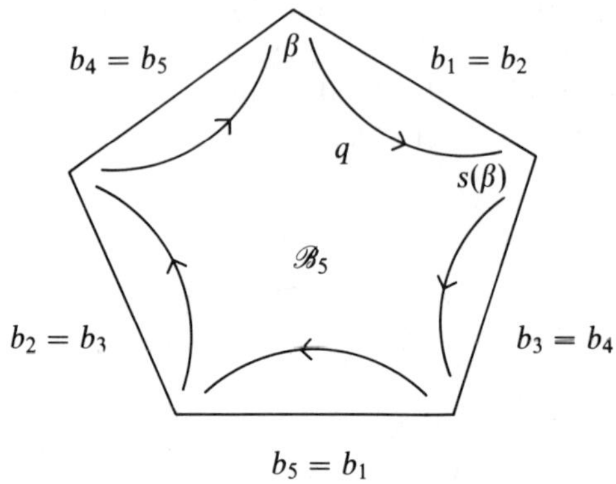

(3.2.3)

The author used a double Puiseux series expansion to define the algebraic interpretation of β.

3.3 As in [Dr$_2$], define $\widehat{GT}$ (the "Grothendieck-Teichmüller group") to be the subgroup of Aut $\widehat{F}_2$ consisting of all automorphisms of the form

$$x \longrightarrow x^\lambda, \quad y \longrightarrow f^{-1} \cdot y^\lambda \cdot f \qquad (\lambda \in \widehat{\mathbb{Z}}^\times, \ f \in \widehat{F}_2') \ ,$$

where λ and f satisfy (3.1.1)(I),(II),(III). (It forms a subgroup!) Then by definition, the image of $\varphi_{X,b}$ ($b = \overrightarrow{01}$) is contained in $\widehat{GT}$. As φ_X is injective (Belyĭ), so is $\varphi_{X,b}$. Therefore, $\varphi_{X,b}$ induces an inclusion

$$\varphi_{X,b} : G_{\mathbb{Q}} \hookrightarrow \widehat{GT} \ . \tag{3.3.1}$$

Two questions arise:

Question 3.3.2 Can one still obtain new relations using X_n for higher n's ?

Question 3.3.3 Can one characterize the image of $G_{\mathbb{Q}}$ using these types of relations (or to ask more strongly, is (3.3.1) already a bijection) ?

If one believes a strict analogy with conformal field theory, then no more essentially new relations would be obtained by using X_n for higher n ([M-S, G,

T-K, Dr_2]). Drinfeld's new quantum group theoretic interpretation of $\widehat{GT}$ (see §4 below) gives a very impressive version of this philosophical reasoning.

See §5.3 for the Lie version of these questions.

§4. Connection with Transformation of Structure of "Complete" Quasi-Triangular Quasi-Hopf Algebras

4.1 Drinfeld introduced the concept of quasi-triangular quasi-Hopf algebra (abbrev. *qtqH* algebra) [Dr_1, Dr_2]. It is more general than Hopf algebra in that the coassociativity and the cocommutativity assumptions are weakened (equalities replaced by conjugations). Thus, a *qtqH* algebra over a field k is a quintuple $(A, \Delta, \varepsilon, \phi, R)$, where (i) A is an associative k-algebra with unit element 1, (ii) $\Delta : A \to A \otimes A$ (comultiplication) and $\varepsilon : A \to k$ (counit) are k-algebra homomorphisms mapping 1 to 1 ($\otimes$: the tensor product over k), such that

$$(\varepsilon \otimes \mathrm{id}) \circ \Delta = (\mathrm{id} \otimes \varepsilon) \circ \Delta = \mathrm{id} \ , \tag{4.1.1}$$

(iii) $\phi \in (A \otimes A \otimes A)^\times$ and $R \in (A \otimes A)^\times$ satisfy:

(A) The basic conjugacy relations:

$$(\mathrm{id} \otimes \Delta)(\Delta(a)) = \phi \cdot (\Delta \otimes \mathrm{id})(\Delta(a)) \cdot \phi^{-1} \ , \tag{4.1.2}$$

$${}^t\Delta(a) = R \cdot \Delta(a) \cdot R^{-1} \ , \tag{4.1.3}$$

for all $a \in A$, where $b \to {}^tb$ ($b \in A \otimes A$) denotes the transposition of two factors. It is *not* assumed that ${}^tR \cdot R = 1$.

(B) The compatibility equalities:

$$(\mathrm{id} \otimes \varepsilon \otimes \mathrm{id})\phi = 1 \ , \tag{4.1.4}$$

(*the pentagon relation*) (4.1.5)

$$(\mathrm{id} \otimes \mathrm{id} \otimes \Delta)\phi \cdot (\Delta \otimes \mathrm{id} \otimes \mathrm{id})\phi = (1 \otimes \phi) \cdot (\mathrm{id} \otimes \Delta \otimes \mathrm{id})\phi \cdot (\phi \otimes 1) \ ,$$

(*two hexagon relations*) omitted (cf. [Dr_2](I·6, ab)). (4.1.6)

Finally, the existence of an antipode is assumed.

As is explained in [Dr_2] the assumptions in (iii) can be understood more conceptually in terms of isomorphisms between tensor products of A-modules. If V, W are (left) A-modules (w.r.t. the k-algebra structure of A), then their tensor product $V \otimes W$ over k is natually an $(A \otimes A)$-module, and via Δ, again an A-module. If V_1, V_2, V_3 are three A-modules, then the $(A \otimes A \otimes A)$-module $V_1 \otimes V_2 \otimes V_3$ can be regarded as an A-module in two different ways, and (4.1.2) imposes that the ϕ-multiplication induce an A-isomorphism

$$(V_1 \otimes V_2) \otimes V_3 \xrightarrow{\sim} V_1 \otimes (V_2 \otimes V_3) \ .$$

The pentagon relation requires the diagram

$$((V_1 \otimes V_2) \otimes V_3) \otimes V_4$$

$$(V_1 \otimes (V_2 \otimes V_3)) \otimes V_4 \qquad\qquad (V_1 \otimes V_2) \otimes (V_3 \otimes V_4) \tag{4.1.7}$$

$$V_1 \otimes ((V_2 \otimes V_3) \otimes V_4) \xrightarrow{\sim} V_1 \otimes (V_2 \otimes (V_3 \otimes V_4))$$

to be commutative. In the diagrams for hexagon relations, the choice of directions of arrows related to "R" is delicate. The two diagrams (3.2.3),(4.1.7) are related to each other via Knizhnik–Zamolodchikov equations ([Dr$_2$]§2)

4.2 Now take $m \in \mathbb{Z}$ and $f(x, y) \in F_2$. If we replace ϕ, R by

$$\begin{cases} \phi' &= \phi \cdot f(R^{21}R^{12},\ \phi^{-1}R^{32}R^{23}\phi)\ , \\ R' &= R \cdot ({}^tRR)^m\ , \end{cases} \tag{4.2.1}$$

respectively, then the 2 equalities in (A) are still satisfied. Here, if $R = \sum a_i \otimes b_i$, $R^{23} = \sum 1 \otimes a_i \otimes b_i$, $R^{32} = \sum 1 \otimes b_i \otimes a_i$, etc. Drinfeld shows that $(A, \Delta, \varepsilon, \phi', R')$ *satisfies* (B) *if and only if* (λ, f) *satisfies the* 3 *equalities of* (3.1.1), *where* $m = \frac{1}{2}(\lambda - 1)$. Actually, the only pairs of λ and f satisfying (3.1.1) are $(\lambda, f) = (\pm 1, 1)$. But if $\mathbb{Z}$ and F_2 are replaced by suitable *completions*, then there are many solutions. It suffices to recall that for $\lambda \in \widehat{\mathbb{Z}}^\times$ and $f \in \widehat{F}_2'$, (3.1.1) was the system of defining equations for the group $\widehat{GT}$, and that $\widehat{GT}$ contains $G_{\mathbb{Q}}$(!). The above two explicit solutions correspond to the identity element and the complex conjugation of $G_{\mathbb{Q}}$. Drinfeld considers some types of *complete* $qtqH$ algebras over the ring of formal power series $k[\![h]\!]$ over a field k of characteristic 0, called *quantized universal enveloping algebras* (*abbrev*. QUE-algebras). In this case, $\otimes$ is replaced by the completed tensor product $\widehat{\otimes}$. For such $qtqH$ algebras, the transformation (4.2.1) will make sense if λ, f are elements of $k^\times$, $F_2^{\mathrm{nil}}(k)$ (the "k-nilpotent completion" of F_2) respectively, and if R is sufficiently close to 1. Thus, one may define the group $GT(k)$ which is still "big" and has a meaning as the group of transformations of structures of QUE-algebras over $k[\![h]\!]$. (The group law for $\widehat{GT}$ corresponds to the reverse of the composition of transformations (4.2.1).) Roughly speaking, one may consider $\widehat{GT}$ as the universal group of transformations of structures of $qtqH$ algebras.

Remark. There is another type of transformations of structures of $qtqH$ algebras called *twists* [Dr$_2$]. The two types of transformations commute with each other, and the twists do not change the structure (as quasi-tensor category) of the category of A-modules.

Question 4.2.2. *Is there any "profinite" $(A, \Delta, \varepsilon, \phi, R)$ over some ring related to $\overline{\mathbb{Q}}$ on which $G_{\mathbb{Q}}$ acts "naturally" and for which the natural action of σ^{-1} coincides with the transformation (4.2.1) obtained by $(\chi(\sigma), f_\sigma)$?*

§5. The Galois Action (Pronilpotent)

5.1 For $\pi_1 = \pi_1(X(\mathbb{C}), *)$, instead of $\hat{\pi}_1$ we may also consider the pronilpotent completion π_1^{nil}, the projective limit of all finite nilpotent factor groups of π_1. It is the direct product of its l-Sylow subgroups $\pi_1^{(l)}$, the pro-l completion of π_1, where l runs over all prime numbers. For a topological group G, we denote by

$$G = G(1) \supseteq G(2) \supseteq \cdots \supseteq G(m+1) \supseteq \cdots$$

its lower central series defined by $G(m+1) = (G, G(m))$ $(m \geq 1)$. Here, (,) is the closure of the algebraic commutator. Note that

$$\bigcap_m G(m) = \{1\}, \quad \text{for} \quad G = \pi_1^{\text{nil}},\ \pi_1^{(l)}\ .$$

The representations $\varphi_X, \varphi_{X,b}$ of $G_{\mathbb{Q}}$ induce the quotient representations $\varphi_X^{\text{nil}}, \varphi_X^{(l)}$, etc. There are various advantages of passage to these quotients: (i) For a "canonical" choice of X such as $\mathbb{P}^1 - \{0, 1, \infty\}$, φ_X^{nil} gives rise to a natural filtration of some quotient of $G_{\mathbb{Q}}$ which is of arithmetic interest. (ii) Each element of π_1^{nil}, for $\mathbb{P}^1 - \{0, 1, \infty\}$ etc., allows an explicit presentation as formal non-commutative power series over $\hat{\mathbb{Z}}$. So, we may ask more explicit questions about the projection of $f_\sigma(x, y)$ on π_1^{nil} (than about $f_\sigma(x, y)$ itself). (iii) For π_1^{nil}, we may use Lie algebra techniques. (iv) The nilpotent quotients of the fundamental group π_1 are, in some sense, close to cohomology groups and are known to have additional structures. Under some assumptions on X, the Lie algebra of $\pi_1/\pi_1(m+1)$ has *mixed Hodge structure* (J. Morgan, D. Sullivan, K. T. Chen, R. Hain), and further, *mixed motif structure* (Deligne).

We shall start by explaining (i).

5.2 For each prime number l, there is a canonical sequence

$$\mathbb{Q} \subseteq \mathbb{Q}^{(l)}(1) \subseteq \cdots \subseteq \mathbb{Q}^{(l)}(m) \subseteq \cdots \subseteq \mathbb{Q}^{(l)}(\infty) = \bigcup \mathbb{Q}^{(l)}(m) \qquad (5.2.1)$$

of (infinite) Galois extensions over $\mathbb{Q}$, starting with $\mathbb{Q}^{(l)}(1) = \mathbb{Q}(\mu_{l^\infty})$ (μ_{l^∞} : the group of roots of unity of l-power order), and an associated graded Lie algebra $\mathfrak{g}^{(l)}$, defined as follows. For each $m \geq 1, \mathbb{Q}^{(l)}(m)$ is the field corresponding to the kernel of the representation

$$G_{\mathbb{Q}} \longrightarrow \text{Out}(F_2^{(l)}/F_2^{(l)}(m+1))$$

induced from $\varphi_X^{(l)}$ for $X = X_4 = \mathbb{P}^1 - \{0, 1, \infty\}$. This kernel will not change if X_4 is replaced by X_n $(n \geq 5)$ [Ih5]. The union field $\mathbb{Q}^{(l)}(\infty)$ corresponds to the kernel of $\varphi_X^{(l)}$ for $X = X_n$, for any $n \geq 4$. It is a pro-l (non-abelian) extension over $\mathbb{Q}(\mu_{l^\infty})$ unramified outside l. For each $m \geq 1$, the Galois group $\mathrm{Gal}(\mathbb{Q}^{(l)}(m+1)/\mathbb{Q}^{(l)}(m))$ is a free $\mathbb{Z}_l$-module of finite rank (call it $r^{(l)}(m)$). It is centralized by $\mathrm{Gal}(\mathbb{Q}^{(l)}(m+1)/\mathbb{Q}^{(l)}(1))$, and as $\mathrm{Gal}(\mathbb{Q}^{(l)}(1)/\mathbb{Q})$-module, has the Tate twist m. The graded Lie algebra $\mathfrak{g}^{(l)}$ is the direct sum of its m-th graded pieces

$$\mathrm{Gal}(\mathbb{Q}^{(l)}(m+1)/\mathbb{Q}^{(l)}(m)) \otimes_{\mathbb{Z}_l} \mathbb{Q}_l \quad (m = 1, 2, 3, \ldots),$$

each of which is a $\mathbb{Q}_l$-module ($\mathbb{Q}_l$: the l-adic number field); cf. [Ih4]. This graded Lie algebra $\mathfrak{g}^{(l)}$ over $\mathbb{Q}_l$ is a standard "approximation" of the filtered Galois group $\mathrm{Gal}(\mathbb{Q}^{(l)}(\infty)/\mathbb{Q})$.

Here is a set of very basic open questions.

Question 5.2.2 (i) What is the structure of $\mathfrak{g}^{(l)}$ *? (ii) Does there exist a natural graded Lie algebra* $\mathfrak{g}$ *over* $\mathbb{Q}$ *such that* $\mathfrak{g}^{(l)} \simeq \mathfrak{g} \otimes \mathbb{Q}_l$ *for all* l *? (iii) In particular, is the rank* $r^{(l)}(m)$ *for each* m *independent of* l *?*

We shall discuss two approaches to *Q 5.2.2 (ii)*.

5.3 We shall define a candidate for the Lie algebra $\mathfrak{g}$ in question (cf. [Ih5]). As before, let $P_n = \pi_1(X_n(\mathbb{C}), \mathscr{B}_n)$ $(n \geq 4)$ (see §3.1), and let $\mathfrak{P}_n$ be the graded Lie algebra over $\mathbb{Q}$ associated with the lower central series of P_n. Thus, the m-th graded piece $\mathrm{gr}^m\mathfrak{P}_n$ of $\mathfrak{P}_n$ is the $\mathbb{Q}$-module

$$(P_n(m)/P_n(m+1)) \otimes \mathbb{Q} \quad (m \geq 1) \ .$$

If X_{ij} $(1 \leq i, j \leq n)$ denotes the element of $\mathfrak{P}_n$ of degree 1 represented by x_{ij} (§3.1), then $\mathfrak{P}_n$ is the graded Lie algebra over $\mathbb{Q}$ generated by the X_{ij}'s which satisfy the fundamental relations

$$X_{ii} = 0, \quad X_{ij} = X_{ji}, \quad \sum_{k=1}^{n} X_{ik} = 0 \qquad (1 \leq i, j \leq n)$$
$$[X_{ij}, X_{kl}] = 0 \qquad \text{if } \{i, j\} \cap \{k, l\} = \phi$$

([K, Ih5]). A *special derivation* of $\mathfrak{P}_n$ of degree m (≥ 1) is a derivation D of $\mathfrak{P}_n$ into itself such that $D(X_{ij}) = [T_{ij}, X_{ij}]$ with some $T_{ij} \in \mathrm{gr}^m\mathfrak{P}_n$ $(1 \leq i, j \leq n)$. Let $\mathscr{D}_n$ be the graded Lie algebra over $\mathbb{Q}$ whose m-th graded piece is the $\mathbb{Q}$-module of all *symmetric special outer* derivations of $\mathfrak{P}_n$ of degree m $(m \geq 1)$. Here, "symmetric" refers to the invariance with respect to the obvious S_n-action on $\mathfrak{P}_n$, and "outer" refers to considering modulo inner derivations.

The sense of considering such a Lie algebra $\mathscr{D}_n$ is that the $G_{\mathbb{Q}}$-action on $P_n^{(l)}$ gives rise to a degree-preserving Lie algebra embedding

$$\mathfrak{g}^{(l)} \hookrightarrow \mathscr{D}_n \otimes \mathbb{Q}_l \quad (n \geq 4) \tag{5.3.1}$$

for each prime number l. Now, when n increases, $\mathscr{D}_n$ gets *smaller*. More precisely, the projection $\mathfrak{P}_n \to \mathfrak{P}_{n-1}$, defined by letting $X_{ij} \to 0$ for i or $j = n$, induces a Lie homomorphism $\mathscr{D}_n \to \mathscr{D}_{n-1}$. We know that this is *injective* for $n \geq 5$. Thus, these embeddings give rise to an infinite chain

$$\mathscr{D}_\infty = \bigcap_{n\geq 4} \mathscr{D}_n \subseteq \cdots \subseteq \mathscr{D}_n \subseteq \cdots \subseteq \mathscr{D}_5 \subseteq \mathscr{D}_4 \ , \tag{5.3.2}$$

of graded Lie algebras over $\mathbb{Q}$. By (5.3.1) and some compatibility,

$$\mathfrak{g}^{(l)} \subseteq \mathscr{D}_\infty \otimes \mathbb{Q}_l \ . \tag{5.3.3}$$

Now we ask the following Lie versions of *Q 3.3.2, Q 3.3.3*:

Question 5.3.4 (i) $\mathscr{D}_\infty = \mathscr{D}_5$? (ii) $\mathfrak{g}^{(l)} = \mathscr{D}_\infty \otimes \mathbb{Q}_l$ *for all* l ?

Remark. From (5.3.2) (5.3.3), it follows that

$$r^{(l)}(m) \leq \dim \mathrm{gr}^m \mathscr{D}_\infty \leq \cdots \leq \dim \mathrm{gr}^m \mathscr{D}_5 \leq \dim \mathrm{gr}^m \mathscr{D}_4$$

for each $m \geq 1$. It is interesting to see which of the $\leq$ are equalities. For $m < 7$, $r^{(l)}(m) = \dim \mathrm{gr}^m \mathscr{D}_4$. But $r^{(l)}(7) = 1$ and $\dim \mathrm{gr}^7 \mathscr{D}_4 = 2$. Terada-Ihara and Drinfeld independently verified that $\dim \mathrm{gr}^7 \mathscr{D}_5 = 1$ ([Ih$_7$, Dr$_2$]), which is in favor of the affirmative aspect of *Q 5.3.4*. See [Ih$_3$, Ih$_4$] for more about these ranks.

Problem 5.3.5. *Construct elements of* $\mathscr{D}_\infty$ *by algebraic or topological means.*

5.4 Deligne constructs a basic motivic extension of the Tate motives, "$\mathbb{Q}$" by "$\mathbb{Q}(m)$" ($m \geq 3$, odd) [De]. Moreover, assuming his conjecture (*loc. cit.* (8.1)) which asserts that such extensions form a 1-dimensional space generated by his extension, and using [So$_{1,2}$], he proves the following.

The Galois representation in $\pi_1^{(l)}/\pi_1^{(l)}(m+1)$ *is "$\mathbb{Q}$-rational" in the following sense (for* $X = \mathbb{P}^1 - \{0, 1, \infty\}$, $\pi_1 = \pi_1(X(\mathbb{C}), \overrightarrow{01})$*): There exists a linear algebraic group* Del_m *over* $\mathbb{Q}$ *and a short exact sequence*

$$1 \longrightarrow U\mathrm{Del}_m \longrightarrow \mathrm{Del}_m \longrightarrow GL(1) \longrightarrow 1 \ ,$$

with $U\mathrm{Del}_m$ *unipotent, all independent of* l*, such that for each* l *the Galois representation in* $\pi_1^{(l)}/\pi_1^{(l)}(m+1)$ *factors through a representation* $G_\mathbb{Q} \to \mathrm{Del}_m(\mathbb{Q}_l)$ *which has an open image at least if* $l > 2$*. Moreover, the abelianization of* $U\mathrm{Del}_m$*, with the* $GL(1)$*-action, decomposes as*

$$(U\mathrm{Del}_m)^{\mathrm{ab}} \simeq \bigoplus_{3\leq k\leq m, odd} \mathbb{A}^1(k)$$

($\mathbb{A}^1$ *: the affine line,* k *: the Tate twist). Finally,* $\{\mathrm{Del}_m\}_{m\geq 1}$ *forms a* $G_\mathbb{Q}$*-compatible projective system.*

In particular, if one assumes the above conjecture on the extension of "$\mathbb{Q}$" by "$\mathbb{Q}(m)$", then *Q 5.2.2 (ii)* will have an affirmative answer with an additional information: the graded Lie algebra $\mathfrak{g}$ is generated by some subset of the form $\{s_m\}_m$, where m runs over *odd* integers ≥ 3 and s_m is of degree m.

§6. Arithmetic Aspects

6.1 In §6, we shall review some works of Anderson, Coleman, the author, ..., on the "elementwise description" and the arithmetic study of the representation φ_X^{nil} for $X = \mathbb{P}^1 - \{0, 1, \infty\}$.

Let $X = \mathbb{P}^1 - \{0, 1, \infty\}$ and, as in §2.3, identify the fundamental group $\pi_1(X(\mathbb{C}), \overrightarrow{01})$ with the free group F_2 on x, y. The Galois group $G_{\mathbb{Q}}$ acts on $\hat{\pi}_1(X(\mathbb{C}), \overrightarrow{01})$, and hence on $\widehat{F}_2$ and $F_2^{\text{nil}} = \prod_l F_2^{(l)}$. Take any $\sigma \in G_{\mathbb{Q}}$. Then the action of σ on $\widehat{F}_2$ can be expressed by two coordinates $\chi(\sigma)$ and f_σ ($\chi(\sigma) \in \widehat{\mathbb{Z}}^\times, f_\sigma \in \widehat{F}_2'$) (see §2.3). Therefore, its action on F_2^{nil} can be expressed by $\chi(\sigma)$ and the projection f_σ^{nil} of f_σ on F_2^{nil}. Call $\mathscr{A} = \widehat{\mathbb{Z}} \ll \xi, \eta \gg$ the *non-commutative* power series algebra in two variables over $\widehat{\mathbb{Z}}$. Then F_2^{nil} can be embedded into $\mathscr{A}^\times$ via $x \to 1 + \xi$, $y \to 1 + \eta$, and f_σ^{nil} can be regarded as an element of $\mathscr{A}^\times$. Each coefficient of the power series f_σ^{nil} may be regarded as *a new invariant* of σ. One is motivated to compare it with other invariants of σ, the old ones (§§6.2, 6.3) or the newly constructed ones (§6.5).

Now as a $\widehat{\mathbb{Z}}$-module, $\mathscr{A}$ is the direct sum

$$\mathscr{A} = \widehat{\mathbb{Z}} \cdot 1 \oplus \mathscr{A} \cdot \xi \oplus \mathscr{A} \cdot \eta \ , \tag{6.1.1}$$

and $(f_\sigma^{\text{nil}})^{-1}$ decomposes as

$$(f_\sigma^{\text{nil}})^{-1} = 1 + A_1\xi + A_2\eta \quad (A_1, A_2 \in \mathscr{A}) \ .$$

Put $\psi_\sigma(\xi, \eta) = 1 + A_1\xi$. Then it follows easily from (3.1.1)(I) that

$$f_\sigma^{\text{nil}} = \psi_\sigma(\eta, \xi) \cdot \psi_\sigma(\xi, \eta)^{-1} \ .$$

Thus, *knowing ψ_σ is equivalent to knowing f_σ^{nil}*. Moreover, ψ_σ is an anti 1-cocycle

$$\psi_{\sigma\tau} = \sigma(\psi_\tau) \cdot \psi_\sigma \qquad (\sigma, \tau \in G_{\mathbb{Q}})$$

with respect to the action of $G_{\mathbb{Q}}$ on $\mathscr{A}$ extending that on F_2^{nil}, and is more convenient for describing the σ-action on abelian subquotients of $\hat{\pi}_1$ ([Ih$_2$]; cf. also [A-I$_2$§2]).[1]

Our first subject now is an explicit formula for the *commutative* power series ψ_σ^{ab} obtained from ψ_σ by letting ξ and η commute. Let $\mathscr{A}^{\text{ab}} = \widehat{\mathbb{Z}}[[\xi, \eta]]$ be the *commutative* formal power series algebra, with the induced $G_{\mathbb{Q}}$-action $1 + \xi \to (1 + \xi)^{\chi(\sigma)}$, $1 + \eta \to (1 + \eta)^{\chi(\sigma)}$ ($\sigma \in G_{\mathbb{Q}}$). Then $G_{\mathbb{Q}} \to (\mathscr{A}^{\text{ab}})^\times$ ($\sigma \mapsto \psi_\sigma^{\text{ab}}$) is a

---

[1] In these papers, the base point is $\overrightarrow{\infty 1}$ and x, y are loops around 0, 1, respectively. So, the definitions are slightly different. See *Remark* at the end of §6.5.

1-cocycle, and it turns out that each coefficient of $\psi_\sigma^{\rm ab}$ can be expressed in terms of "old invariants" of σ which we now recall.

6.2 The Old Invariants. Fix a prime number l.

(i) *The l-adic cyclotomic character* $\chi^{(l)}$ is the l-component of χ, i.e., $\chi(\sigma) = (\chi^{(l)}(\sigma))$, $\chi^{(l)}(\sigma) \in \mathbb{Z}_l^\times$.

(ii) *The cyclotomic elements* (Soulé, Deligne). These are certain continuous mappings

$$\kappa_m^{(l)} : G_{\mathbb{Q}} \longrightarrow \mathbb{Z}_l \quad (m \geq 1, odd) \ ,$$

satisfying the 1-cocycle relation

$$\kappa_m^{(l)}(\sigma\tau) = \kappa_m^{(l)}(\sigma) + \chi^{(l)}(\sigma)^m \kappa_m^{(l)}(\tau) \quad (\sigma, \tau \in G_{\mathbb{Q}}) \ . \tag{6.2.1}$$

[Construction] Let $n \geq 1$ (but $n \geq 2$ if $l = 2$). Put $\zeta_n = \exp(\frac{2\pi i}{l^n})$ and

$$\varepsilon_{m,n} = \prod_a (\zeta_n^a - 1)^{<a^{m-1}>} \ ,$$

where the product is over all integers a such that $0 < a < l^n$ and $(a, l) = 1$; $\langle a^{m-1} \rangle$ is the smallest positive integer congruent to a^{m-1} mod l^n. Note that $\varepsilon_{m,n}$ is totally real and totally positive (because m is *odd*). It is easy to see that each of $\varepsilon_{m,n+1}/\varepsilon_{m,n}$ and $\sigma(\varepsilon_{m,n})/\varepsilon_{m,n}^b$ is an l^n-th power of a totally positive element of $\mathbb{Q}(\mu_{l^\infty})$, where $\sigma \in G_{\mathbb{Q}}$, $b \in \mathbb{Z}$, $b \equiv \chi^{(l)}(\sigma)^{1-m} (\mathrm{mod}\, l^n)$. Hence there is a unique $\kappa_m^{(l)}(\sigma) \in \mathbb{Z}_l$ such that

$$\sigma((\varepsilon_{m,n})^{1/l^n}) = (\sigma(\varepsilon_{m,n}))^{1/l^n} \cdot \zeta_n^{\chi^{(l)}(\sigma)^{1-m} \cdot \kappa_m^{(l)}(\sigma)}$$

holds for all $n \geq 2$. Moreover, $\kappa_m^{(l)}$ satisfies (6.2.1). Here, for any positive real number c, c^{1/l^n} denotes its *positive real* root. By (6.2.1), $\kappa_m^{(l)}$ factors through $\mathrm{Gal}(\mathbb{Q}(\mu_{l^\infty})^{\rm ab}/\mathbb{Q})$, $\mathbb{Q}(\mu_{l^\infty})^{\rm ab}$ being the maximal *abelian* extension of $\mathbb{Q}(\mu_{l^\infty})$. Moreover, by Soulé [So$_{1,2}$], these 1-cocycles $\kappa_m^{(l)}$ do not vanish at least if $l > 2$.

6.3 The following *explicit formula* for the coefficients of $\psi_\sigma^{\rm ab}$ is due to the contributions of Anderson, Coleman, Deligne, the author, Kaneko and Yukinari; cf. [A$_3$, C$_3$, IKY]. (See also [Ich] for a simplification of [IKY].)

For each $\sigma \in G_{\mathbb{Q}}$, define $\kappa_m^*(\sigma) \in \widehat{\mathbb{Z}}^\times = \prod_l \mathbb{Z}_l^\times$ by

$$\kappa_m^*(\sigma) = ((l^{m-1} - 1)^{-1} \kappa_m^{(l)}(\sigma))_l \ .$$

Theorem [A$_3$, C$_3$, IKY]. *The commutative power series $\psi_\sigma^{\rm ab}(\xi, \eta)$ can be expressed explicitly as follows.*

$$\psi_\sigma^{\rm ab}(\xi,\eta) = \exp\left\{ \sum_{m \geq 3, odd} \frac{\kappa_m^*(\sigma)}{m!} ((X+Y)^m - X^m - Y^m) \right\}$$
$$\times \exp\left\{ -\frac{1}{2} \sum_{m \geq 2, even} \frac{b_m(1 - \chi(\sigma)^m)}{m!} ((X+Y)^m - X^m - Y^m) \right\} \ ,$$

where $X = \log(1+\xi)$, $Y = \log(1+\eta)$, *and the constants* b_m *are defined by*

$$\log\left(\frac{1-e^{-t}}{t}\right) = \sum_{m\geq 1} \frac{b_m}{m!} t^m \ .$$

(mb_m *is the m-th Bernoulli number.)*

Note that $\psi_\sigma^{\mathrm{ab}}(\xi,\eta)$ is of the form

$$\gamma_\sigma(\xi)\gamma_\sigma(\eta)\gamma_\sigma((1+\xi)(1+\eta)-1)^{-1} \ .$$

This power series γ_σ is closely related to Anderson's hyperadelic gamma function Γ_σ.

The image of $\sigma \to \psi_\sigma^{\mathrm{ab}}(\xi,\eta)$ was studied closely by Coleman [C3], Ichimura and Kaneko [IK]. The expected image can be figured out via Coleman theory [C1, C2], and the difference from the expected image can be measured in terms of the "Vandiver gap".

6.4 We now explain another aspect of $\psi_\sigma^{\mathrm{ab}}$. It is a connection with the action of σ on the *double* commutator quotient $\widehat{F}_2/\widehat{F}_2''$ of $\widehat{F}_2$, or equivalently, on torsion points of Fermat Jacobians. Put $\mathscr{F} = \widehat{F}_2$, and consider the abelianizations $\mathscr{F}^{\mathrm{ab}} = \mathscr{F}/\mathscr{F}'$ and $\mathscr{F}'^{\mathrm{ab}} = \mathscr{F}'/\mathscr{F}''$ of $\mathscr{F}$ and $\mathscr{F}'$ first as (additive) $\widehat{\mathbb{Z}}$-modules. Then $\mathscr{F}^{\mathrm{ab}} = \widehat{\mathbb{Z}}\underline{x} \oplus \widehat{\mathbb{Z}}\underline{y}$ on which σ acts via $\chi(\sigma)$-multiplication (because of (2.3.2)). Here $\underline{x}, \underline{y}$ are the classes of x, y. Now $\mathscr{F}^{\mathrm{ab}}$ acts on $\mathscr{F}'^{\mathrm{ab}}$ by conjugation. Therefore, $\mathscr{F}'^{\mathrm{ab}}$ may be regarded as a module over the completed group algebra $\widehat{\mathbb{Z}}[[\mathscr{F}^{\mathrm{ab}}]]$. But one can show that *this module is free of rank* 1 *generated by the class* θ' *of* $(x,y) = xyx^{-1}y^{-1}$. As σ acts semi-linearly on $\mathscr{F}'^{\mathrm{ab}}$, this action is presented by the unique element B'_σ of $\widehat{\mathbb{Z}}[[\mathscr{F}^{\mathrm{ab}}]]$ such that

$$\sigma(\theta') = B'_\sigma \cdot \theta' \ . \tag{6.4.1}$$

Now define $B_\sigma \in \widehat{\mathbb{Z}}[[\mathscr{F}^{\mathrm{ab}}]]$ by the formula

$$B'_\sigma = \left(\frac{\underline{x}^{\chi(\sigma)}-1}{\underline{x}-1} \cdot \frac{\underline{y}^{\chi(\sigma)}-1}{\underline{y}-1}\right) B_\sigma \ . \tag{6.4.2}$$

This B_σ is connected with $\psi_\sigma^{\mathrm{ab}}$ as follows. Consider the projection

$$\mathrm{pr} : \widehat{\mathbb{Z}}[[\mathscr{F}^{\mathrm{ab}}]] \longrightarrow \widehat{\mathbb{Z}}[[\xi,\eta]] = \mathscr{A}^{\mathrm{ab}} \quad \begin{cases} \underline{x} \mapsto 1+\xi, \\ \underline{y} \mapsto 1+\eta. \end{cases} \tag{6.4.3}$$

Then

$$\mathrm{pr}\,(B_\sigma) = \psi_\sigma^{\mathrm{ab}} \ . \tag{6.4.4}$$

The projection pr has a big kernel $\mathscr{K}$. In fact, $\mathscr{K} \cdot \theta'$ ($\subset \mathscr{F}'/\mathscr{F}''$) is the kernel of $\mathscr{F}/\mathscr{F}'' \to F_2^{\mathrm{nil}}/(F_2^{\mathrm{nil}})''$. *Thus,* $\psi_\sigma^{\mathrm{ab}}$ *is the power series which describes the* σ*-action on* $(F_2^{\mathrm{nil}})'/(F_2^{\mathrm{nil}})''$ *universally.*

This power series was first treated in [Ih_1], including its connections with the l-power torsion points of Fermat Jacobians of l-power degree and Jacobi sums. We shall explain this briefly keeping in sight its generalized and refined version due to Anderson [A_3]. First, we make the following identification

$$\mathscr{F}^{\mathrm{ab}} = \mathrm{Hom}((\mathbb{Q}/\mathbb{Z})^2, \mu_\infty) \tag{6.4.5}$$

(μ_∞ : the group of roots of unity in $\mathbb{C}$), with $\underline{x}$ (resp. $\underline{y}$) corresponding to $(s,t) \to \exp(2\pi i s)$ (resp. $\exp(2\pi i t)$); $s,t \in \mathbb{Q}/\mathbb{Z}$. Note that the $G_{\mathbb{Q}}$-action on $\mathscr{F}^{\mathrm{ab}}$ is recovered from its action on μ_∞ via (6.4.5). Through (6.4.5) we may regard each element of $\widehat{\mathbb{Z}}[\![\mathscr{F}^{\mathrm{ab}}]\!]$ as a function

$$(\mathbb{Q}/\mathbb{Z})^2 \mapsto \hat{\mathbb{Z}} \otimes \mathbb{Q}(\mu_\infty) \ . \tag{6.4.6}$$

The above B_σ, considered as a function (6.4.6), is the *adelic beta function* $B_\sigma(s,t)$ ($s,t \in \mathbb{Q}/\mathbb{Z}$). It is strikingly analogous to the classical beta function [A_3]. Very roughly speaking, B_σ plays the role of (the classical beta)$^{\sigma-1}$.

Now, the abelian covering of $X = \mathbb{P}^1 - \{0,1,\infty\}$ corresponding to $\mathscr{F}^{\mathrm{ab}}/(N)$ ($N = 1,2,\ldots$) is the Fermat curve

$$Y_N : \{u^N + v^N = 1; \quad uv \neq 0\} \ .$$

The covering map is given by $(u,v) \to u^N$. The abelian coverings over Y_N are controlled by the group of torsion points of the Jacobian of $\overline{Y}_N$ (the compactification of Y_N). Thus the action of σ on $\mathscr{F}/\mathscr{F}''$ is directly tied to that on the group of these torsion points ([Ih_1, $\mathrm{A}_{2,3}$]). As the Frobenius elements act on the latter group by multiplication of Jacobi sums (etc.), these collected together in terms of $\psi_\sigma^{\mathrm{ab}}$ (or B_σ) give a universal expression of Jacobi sums. For $\psi_\sigma^{\mathrm{ab}}$, it is:

Theorem [$\mathbf{Ih}_1$]. *Let l be a prime number, $n \geq 1$, $\mathfrak{p}$ be a prime ideal of $\mathbb{Q}(\mu_{l^n})$ not lying above l, and $\sigma = \sigma_{\mathfrak{p}}$ be a Frobenius element of $\mathfrak{p}$. Then for any l^n-torsion points s,t of $\mathbb{Q}/\mathbb{Z}$ with $s, t, s+t \neq 0$, the special value of the l-component of $\psi_\sigma^{\mathrm{ab}}$ ($\in \mathbb{Z}_l[\![\xi,\eta]\!]$) at $\xi = \exp(2\pi i s) - 1$, $\eta = \exp(2\pi i t) - 1$, is essentially the Jacobi sum (w.r.t. $\mathfrak{p}, l^n, s, t$).*

More generally, the values of $B_\sigma(s,t)$ ($s,t \in \mathbb{Q}/\mathbb{Z}$) are related to Jacobi sums and also Gauss sums [A_3]. Note that the two theorems of §6.3, §6.4, combined, give a direct connection between the circular units and the Jacobi sums.

Anderson [A_3, A_4] defined the hyperadelic gamma function

$$\Gamma_\sigma : \mathbb{Q}/\mathbb{Z} \longrightarrow \text{(some arithmetic ring)}$$

which factors B_σ just as γ_σ factors $\psi_\sigma^{\mathrm{ab}}$. It interpolates Gauss sums, and its "logarithmic derivative" can be given explicitly in terms of circular units so that it forms a bridge connecting Gauss sums and circular units. The last connection was partly established independently by a different method by Miki [Mi]. See Coleman [C_3] for connections with and applications to other aspects of cyclotomy.

6.5 $\psi_\sigma(\xi,\eta)$ and Higher Circular l-Units (Anderson-Ihara [AI$_{1,2}$]). This connection arises from a comparison of the tower of nilpotent coverings and of genus 0 coverings.

Call a finite subset $S \subset \mathbb{P}^1(\mathbb{C})$ *l-elementary*, if S is obtained from $S_0 = \{0, 1, \infty\}$ by finite number of operations of the form $S \to S^{1/l}$ (all l-th roots), $S \to T_{a,b,c}(S)$. Here, $a, b, c \in S$ (distinct), and $T_{a,b,c}$ is the projective linear transformation of $\mathbb{P}^1$ that maps a, b, c to $0, 1, \infty$ respectively.

Definition 6.5.1. $E^{(l)}$ is the subgroup of $\mathbb{C}^\times$ generated by the constituents of $S - \{0, \infty\}$, where S runs over all l-elementary subsets of $\mathbb{P}^1(\mathbb{C})$.

It is easy to see that elements of $E^{(l)}$ are l-units, i.e., element of $\overline{\mathbb{Q}}$ which, together with its reciprocal, is integral over $\mathbb{Z}[1/l]$. They are called *higher circular l-units*. The group $E^{(l)}$ contains such elements as

$$1 - \zeta_n, \quad (1 - (1 - \zeta_n)^{1/l^n})^{1/l^n}, \ldots,$$

where $\zeta_n = \exp(2\pi i/l^n)$ $(n = 1, 2, \ldots)$.

Theorem [A-I$_2$]. *Each coefficient of $\psi_\sigma(\xi,\eta)$ (mod l^n) can be expressed explicitly in terms of the σ-action on $E^{(l)}$.*

A key dialogue : "How to distinguish different l-th roots in $E^{(l)}$ intelligibly?" : "In terms of a natural structure of "forest" with vertices in $E^{(l)}$."

Corollary [A-I$_1$]

$$\mathbb{Q}^{(l)}(\infty) = \mathbb{Q}(E^{(l)}) \ .$$

We shall conclude this lecture with two additional open questions.

Question 6.5.2. i) *Is $\mathbb{Q}^{(l)}(\infty)$ the maximal pro-l extension over $\mathbb{Q}(\mu_{l^\infty})$ unramified outside l ?*

ii) *How big is $E^{(l)}$ and $E^{(l)} \cap \mathbb{Q}(\mu_{l^\infty})$?*

Remark. The non-commutative (resp. commutative) l-adic power series $\psi(\sigma) \in \mathbb{Z}_l \ll u, v \gg$ of [Ih$_2$] §1 (D) Ex. 1 (resp. $F_\sigma \in \mathbb{Z}_l[\![u, v]\!]$ of [Ih$_1$] §2) are related to the above ψ_σ (resp. $\psi_\sigma^{\mathrm{ab}}$) as follows. Write ξ, η instead of u, v respectively. Then $\psi(\sigma)$ and F_σ are the l-components of

$$f_\sigma(y,z) y^{\frac{1}{2}(\chi(\sigma)-1)} \cdot \psi_\sigma(\eta, \xi)$$

and

$$\frac{\underline{x}^{\chi(\sigma)} - 1}{\underline{x} - 1} \cdot \frac{\underline{y}^{\chi(\sigma)} - 1}{\underline{y} - 1} \cdot \underline{y}^{\frac{1}{2}(\chi(\sigma)-1)} \cdot \psi_\sigma^{\mathrm{ab}}(\xi, \eta) \ ,$$

respectively, where $z = (xy)^{-1}$. Note that $f_\sigma^{\mathrm{ab}} = 1$ and that $\psi_\sigma^{\mathrm{ab}}$ is symmetric in ξ and η.

Added in proof. The author found later that Question 5.3.4 (i) has an affirmative answer. The proof is based on [Dr_2, Ih_5].

References

[A_1] Anderson, G.: Cyclotomy and an extension of the Taniyama group. Compositio Math. **57** (1986) 153–217

[A_2] Anderson,G.: Torsion points on Fermat Jacobians, roots of circular units and relative singular homology. Duke Math. J. **54** (1987) 501–561

[A_3] Anderson, G.: (a) The hyperadelic gamma function. Invent. math. **95** (1989) 63–131. (b) ibid (a précis). Adv. Stud. Pure Math. **12** (1987) 1–19

[A_4] Anderson, G.: Normalization of the hyperadelic gamma function. In: Galois groups over $\mathbb{Q}$. Publ. MSRI, no. 16 (1989) 1–31. Springer, Berlin Heidelberg New York

[A-I_1] Anderson, G., Ihara, Y.: Pro-l branched covering of $\mathbb{P}^1$ and higher circular l-units. Ann. Math. **128** (1988) 271–293

[A-I_2] Anderson, G., Ihara, Y.: ibid Part 2. Int'l J. Math. **1** (1990) 119–148

[B_1] Belyi, G. V.: On Galois extensions of a maximal cyclotomic field. Izv. Akad. Nauk USSR **43** (1979) 267–276; transl. Math. USSR Izv. **14** (1980) 247–256

[B_2] Belyi, G. V.: On the commutator of the absolute Galois group. Proc. ICM 1986. American Mathematical Society, Berkeley 1987, pp. 346–349 (in Russian)

[C_1] Coleman, R.: The dilogarithm and the norm residue symbol. Bull. Soc. Math. France **109** (1981) 373–402

[C_2] Coleman, R.: Local units modulo circular units. Proc. Amer. Math. Soc. **89** (1983) 1–7

[C_3] Coleman, R.: Anderson-Ihara theory: Gauss sums and circular units. Adv. Stud. Pure Math. **17** (1989) 55–72

[De] Deligne, P.: Le groupe fondamental de la droite projective moins trois points. In: Galois groups over $\mathbb{Q}$. Publ. MSRI, no. 16 (1989) 79–298. Springer, Berlin Heidelberg New York

[Dr_1] Drinfeld, V.G.: Quasi-Hopf algebras (Russian). Algebra and Analysis **1**, no. 6 (1989) 114–148.

[Dr_2] Drinfeld, V.G.: On quasi-triangular quasi-Hopf algebras and some group closely associated with $Gal(\overline{\mathbb{Q}}/\mathbb{Q})$. Preprint 1990 (in Russian)

[G] Grothendieck, A.: Esquisse d'un programme. Mimeographed note (1984)

[Ich] Ichimura, H.: On the coefficients of the universal power series for Jacobi sums. J. Fac. Sci. Univ. Tokyo **IA 36** (1989) 1–7

[IK] Ichimura, H., Kaneko, M.: On the universal power series for Jacobi sums and the Vandiver conjecture. J. Number Theory **31** (1989) 312–334

[Ih_1] Ihara, Y.: Profinite braid groups. Galois representations and complex multiplications. Ann. Math. **123** (1986) 43–106

[Ih_2] Ihara, Y.: On Galois representations arising from towers of coverings of $\mathbb{P}^1\backslash\{0,1,\infty\}$. Invent. math. **86** (1986) 427–459

[Ih_3] Ihara, Y.: Some problems on three point ramifications and associated large Galois representations. Adv. Stud. Pure Math. **12** (1987) 173–188

[Ih_4] Ihara, Y.: The Galois representation arising from $\mathbb{P}^1 - \{0,1,\infty\}$ and Tate twists of even degree. In: Galois groups over $\mathbb{Q}$. Publ. MSRI, no. 16 (1989) 299–313. Springer, Berlin Heidelberg New York

[Ih_5] Ihara, Y.: Automorphisms of pure sphere braid groups and Galois representations. In: The Grothendieck Festschrift, vol. 2. Progress in Mathematics, vol. 87. Birkhäuser, Basel 1991, pp. 353–373

[Ih_6] Ihara, Y.: Galois groups over $\mathbb{Q}$ and monodromy (in Japanese). In: Prospects of Algebraic Analysis, RIMS report **675** (1988) 23–34

[Ih_7] Ihara, Y.: Derivations of the Lie algebra associated with $F_{0,n}\mathbb{P}^1$ and the image of $\mathrm{Gal}(\overline{\mathbb{Q}}/\mathbb{Q})$ (in Japanese). In: Algebraic number theory, RIMS report **721** (1990) 1–8

[IKY] Ihara, Y., Kaneko, M., Yukinari, A.: On some properties of the universal power series for Jacobi sums. Adv. Stud. Pure Math. **12** (1987) 65–86

[K] Kohno, T.: On the holonomy Lie Algebra and the nilpotent completion of the fundamental group of the complement of hypersurfaces. Nagoya Math. J. **92** (1983) 21–37

[Ma] Matzat, B. H.: Braids and Galois groups. Doğa Mat. **14** (1990) no. 2, 57–69

[M-S] Moore, G., Seiberg, N.: Classical and quantum conformal field theory. Commun. Math. Phys. **123** (1989) 177–254

[Mi] Miki, H.: On the l-adic expansion of certain Gauss sums and its applications. Adv. Stud. Pure Math. **12** (1987) 87–118

[N] Nakamura, H.: Galois rigidity of the etale fundamental groups of punctured projective lines. J. Reine Angew. Math. **411** (1990) 205–216

[O] Oda, T.: A note on ramification of the Galois representation on the fundamental group of an algebraic curve. J. Number Theory **34** (1990) 225–228; Part 2 in preparation

[So_1] Soulé, C.: On higher p-adic regulators. Lecture Notes in Mathematics, vol. 854. Springer, Berlin Heidelberg New York 1981, pp. 372–401

[So_2] Soulé, C.: Éléments cyclotomiques en K-théorie. Astérisque **147–148** (1987) 225–257

[T-K] Tsuchiya, A., Kanie, Y.: Vertex operators in conformal field theory on $\mathbb{P}^1$ and monodromy representations of braid group. Adv. Stud. Pure Math. **16** (1987) 297–372

Von Neumann Algebras in Mathematics and Physics

Vaughan F. R. Jones

Department of Mathematics, University of California, Berkeley, CA 94720, USA

Motivation

The last ten years or so have seen a considerable synthesis in mathematics and mathematical physics. In this talk I will be concerned only with those topics appearing in Fig. 1, all connected by having something to do with braids.

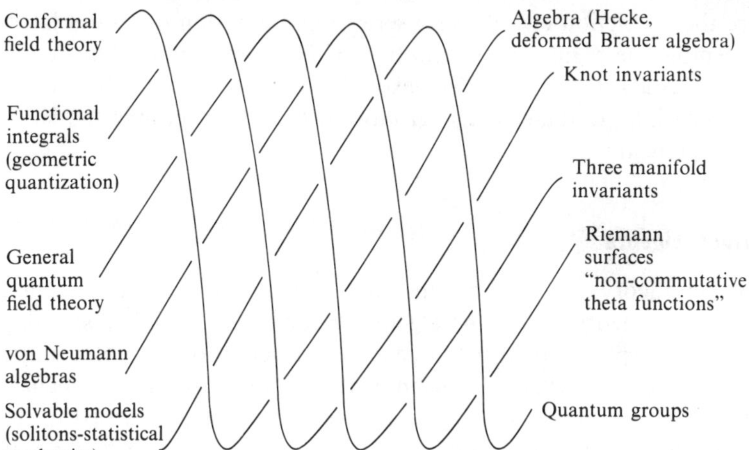

Fig. 1

Many very different themes could be used for a talk such as this one, but I have chosen von Neumann algebras because they are what led me into this circle of ideas. Thus the presentation will be historical rather than logical.

A *von Neumann algebra* M is a $*$-algebra of bounded operators on a Hilbert space $(\mathscr{H}, \langle\ ,\ \rangle)$ which contains the identity and is closed in the weak operator topology, i.e. if a_n is a net of operators in M and $\langle a_n\xi, \eta\rangle \to \langle a\xi, \eta\rangle$ for some a and all ξ and η in $\mathscr{H}$, then a is in M. Most of the interest is when M is infinite dimensional so it should be pointed out at the outset that a finite dimensional von Neumann algebra is just a direct sum of matrix algebras, each acting with a certain multiplicity on $\mathscr{H}$.

Proceedings of the International Congress of Mathematicians, Kyoto, Japan, 1990

These algebras were first introduced by Murray and von Neumann in [MvN1]. Their motivations were many, not the least being a beautiful result of von Neumann in [vN] which showed that von Neumann algebras could equally be defined as *commutants* of self-adjoint sets (notation: if S is a set of bounded operators, S', the commutant of S, is by definition the set of all bounded operators x such that $xs = sx$ for all s in S). This means that von Neumann algebras are the algebras of *symmetries* for any structure belonging to Hilbert space, such as geometric configurations of subspaces or unitary group representations. Let me discuss some other motivations for looking at von Neumann algebras.

a) Unitary Group Representations

The abstract theory of representations of a compact group is very complete but as soon as the group is not compact, many different phenomena occur. A unitary representation no longer decomposes simply as a direct sum of irreducibles, which is reflected in a possibly exotic structure of the commutant of the group. This already happens for, say, the regular representation of the free group on two generators. Of course a semisimple Lie group G is an example of a "type I" group where there is a satisfactory reduction to irreducible representations, but even here the restriction of a discrete series representation of G to a lattice Γ in G will resemble the regular representation of Γ. So even in arithmetic questions (say $G = PSL_2(\mathbb{R})$, $\Gamma = PSL_2(\mathbb{Z})$), "exotic" von Neumann algebras occur quite naturally. For infinite dimensional groups such as loop groups the situation is even more interesting.

b) Abstract Algebra

The abstract theory of finite dimensional semisimple algebras over a field is also a complete theory. When the algebra becomes infinite dimensional, general theory must be replaced by a heterogeneous collection of examples unless some analysis is introduced. Over $\mathbb{C}$ semisimplicity is implied (in finite dimensions) by the existence of a $*$ operation and to say that an algebra is closed in the weak topology is the strongest reasonable condition on it. So von Neumann algebras should be the best behaved abstract family of infinite dimensional algebras. Although the theory is considerably richer than in finite dimensions there do exist non-trivial general results and there is a significant class of von Neumann algebras (those approximable by finite dimensional algebras) for which a complete classification exists.

c) Unbounded Operators

At first sight the important operator d/dx appears unnatural on Hilbert space since it is not defined on all vectors (in $L^2(\mathbb{R})$). However the opposite is true. Provided d/dx is given the right *domain* its graph is a closed subspace and hence natural to Hilbert space. In general an operator is called pre-closed if it is densely defined and the closure of its graph is the graph of an operator. All interesting linear operators seem to be pre-closed.[1]

The trouble with domains comes when one tries to add and multiply unbounded operators. For this reason one would hope to replace unbounded operators by bounded ones. Thus the relations $[P, Q] = \text{id}$ can be handled in the Weyl form with two unitary groups $U(t)$ and $V(t)$ with $U(t)V(s) = e^{2\pi ist}V(s)U(t)$. In general one could use the von Neumann algebra of all bounded operators having the same symmetries as the unbounded ones.

d) Quantum Theory

The very language of quantum mechanics suggests von Neumann algebras. States are vectors in a Hilbert space. Observables are self-adjoint operators and numerical information about observables for systems in states is given by scalar products. Thus it is quite natural to consider a von Neumann algebra of observables associated with any subsystem of a quantum system. Certainly von Neumann was thinking along these lines. Such an approach was indeed adopted as an "algebraic" approach to quantum field theory by Haag and Kastler [HK] who postulated von Neumann algebras associated with regions of space time and satisfying certain causality, positivity and Lorentz covariance conditions. Although it is very abstract and difficult to produce mathematical examples, some things can be deduced from such a general theory and new results in von Neumann algebras have recently added to the possibilities.

1. Factors and Their Types

The spectral theorem shows that abelian von Neumann algebras are abstractly of the form $L^\infty(X, \mu)$ acting with some multiplicity on a Hilbert space. Attention turns immediately to *factors* which are by definition von Neumann algebras with trivial centre. The simplest example of such a factor is $\mathscr{B}(\mathscr{H})$, the algebra of all bounded operators. One of the fundamental discoveries of Murray and von Neumann was that of factors not abstractly isomorphic to any $\mathscr{B}(\mathscr{H})$.

The first example was that of the commutant M of the left regular representation of any discrete group Γ all of whose (non-identity) conjugacy classes are infinite (such as free groups F_n or $PSL_n(\mathbb{Z})$ for $n \geq 2$). One can show that this von Neumann algebra is generated by the right regular representation and that it is a factor. Also if $\xi \in l^2(\Gamma)$ is the characteristic function of the identity then the linear functional $\text{tr}(x) = \langle x\xi, \xi\rangle$ defines a *trace* on M (i.e. $\text{tr}(xy) = \text{tr}(yx)$). There is no such functional on $\mathscr{B}(\mathscr{H})$ if $\dim \mathscr{H} = \infty$.

An infinite dimensional factor admitting such a trace is called a *type* II_1 *factor* and a factor of the form $\mathscr{B}(\mathscr{H}) \otimes$ (a type II_1 factor) is called a type II_∞ factor if $\dim \mathscr{H} = \infty$. A factor which is neither of type I ($\cong \mathscr{B}(\mathscr{H})$) nor of type II is called a type III factor.

[1] I can only think of one exception – the "derivations" in Fox's free differential calculus are not pre-closed on l^2 of the free group.

A II_1 factor shares some of the nice features of a finite dimensional matrix algebra (e.g. it is simple). Its most seductive feature is continuous dimensionality. If one looks at the numbers trace(p) (trace normalized so that trace(1) = 1), where $p \in M_n(\mathbb{C})$ are projections, one obtains the numbers m/n for $m = 0, 1, \ldots, n$, where m is of course the rank of p. In a II_1 factor one obtains the whole unit interval $[0, 1]$.

2. GNS Construction

An important construction is the Gelfand-Naimark-Segal construction. One begins with a $*$-algebra A and a linear functional (state) $\varphi : A \to \mathbb{C}$ with $\varphi(a^*a) \geq 0$. Define $\langle\ ,\ \rangle$ on A by $\langle a, b\rangle = \varphi(b^*a)$. Quotienting if necessary A becomes a pre-Hilbert space and its completion is written $\mathscr{H}_\varphi$. Under mild conditions A will act on $\mathscr{H}_\varphi$ by left multiplication and the von Neumann algebra it generates is said to result from the GNS construction from φ. It should be thought of as the completion of A with respect to φ. Examples of all types of factors can now be obtained, using $A = \otimes_{i=1}^{\infty} M_2(\mathbb{C})$. If $h_i \in M_2(\mathbb{C})$ are positive matrices of trace 1 then the formula $\varphi(\otimes_{i=1}^{\infty} x_i) = \prod_{i=1}^{\infty} \mathrm{trace}(h_i x_i)$ defines a state, said to be a product state. The result of the GNS construction is then always a factor. It is of type I if $h_i = \begin{pmatrix} 1 & 0 \\ 0 & 0 \end{pmatrix}$, of type II_1 if $h_i = (1/2)\begin{pmatrix} 1 & 0 \\ 0 & 1 \end{pmatrix}$ and of type III if $h_i = (1+\lambda)^{-1}\begin{pmatrix} 1 & 0 \\ 0 & \lambda \end{pmatrix}$ for $0 < \lambda < 1$. In the last case the state is known as the Powers state after R. Powers [Pow] who proved that the factors for different λ are mutually non-isomorphic.

3. Modular Theory

The Tomita-Takesaki theory shows that to every weakly continuous state φ on a von Neumann algebra M there is a one parameter group σ_t of automorphisms of M characterized by the KMS condition $\varphi(xy) = \varphi(\sigma_i(y)x)$. It allows one to subdivide the type III factors into type III_λ, $\lambda \in [0, 1]$ where the Powers factors are of type III_λ, $\lambda \in (0, 1)$. Types III_0 and III_1 can be obtained from product states by suitable choices of the h_i's. Generically a factor is of type III_1. For details we refer to Connes' Helsinki congress talk [Co1], or [Ta].

4. Hyperfiniteness

A von Neumann algebra M is *hyperfinite* if there is an increasing sequence of finite dimensional $*$-subalgebras whose union is weakly dense in M. Thus our infinite tensor product factors are hyperfinite, by construction, but Murray and von Neumann showed that the free group II_1 factor is not hyperfinite, nor are the examples coming from lattices in semisimple Lie groups.

All hyperfinite factors are known. Here is a table of them.

Table 1. Hyperfinite factors

Type I_n $n = 1, 2, \ldots, \infty$	One for each n Proof: elementary
Type II_1	A unique factor, denoted R. Uniqueness proved by Murray and von Neumann in [MvN2]
Type II_∞	A unique factor. Uniqueness proved by Connes in [Co2]
Type III_0	One for each ergodic non-transitive flow. Proved by Krieger [Kr] and Connes [Co2]
Type III_λ $0 < \lambda < 1$	Powers factors are the unique examples. Uniqueness proved by Connes [Co2]
Type III_1	A unique example, first analysed by Araki and Woods in [AW]. Uniqueness proved by Haagerup [Ha] and Connes [Co3]

In quantum field theory it is expected that the von Neumann algebra of observables localized in a nice region is a hyperfinite type III_1 factor.

The classification of hyperfinite (also called "injective") factors is a great achievement. Among other things it paves the way for the study of subfactors, to which I now turn.

5. Index for Subfactors

The representation theory of a II_1 factor is very simple. There is a single parameter – a positive real number (or ∞), $\dim_M(\mathscr{H})$, which measures the size of $\mathscr{H}$ compared to the Hilbert space $L^2(M)$ obtained from M by the GNS construction using the trace (for which $\dim_M(L^2(M)) = 1$). Thus if $N \subset M$ are II_1 factors we define the index of N in M to be the real number (≥ 1):

$$[M : N] = \dim_N(L^2(M)) .$$

Examples:

(i) $[N \otimes M_n(\mathbb{C}) : N] = n^2$
(ii) $[M : M^G] = |G|$ if G is a group of outer automorphisms of M.

The next theorem shows that this is an interesting notion.

Theorem [Jo1]. a) *If* $[M : N] < 4$ *then there is an* $n \in \mathbb{Z}$, $n \geq 3$ *for which* $[M : N] = 4\cos^2 \pi/n$.

b) *All values of the index of part* a) *are realized, as is any real number* ≥ 4, *by subfactors of the hyperfinite* II_1 *factor.*

This theorem may be proved by iterating a certain *basic construction* which associates to $N \subset M$ an extension $\langle M, e_N \rangle$ of M where e_N projects from $L^2(M)$ to $L^2(N)$. One obtains a tower M_i with $N = M_0$, $M = M_1$ and $M_n = \langle M, e_1, \ldots, e_{n-1} \rangle$, e_j being orthogonal projection from $L^2(M_j)$ to $L^2(M_{j-1})$.

The e_i's satisfy the following properties

1) $e_i^2 = e_i = e_i^*$
2) $e_i e_j = e_j e_i$ if $|i - j| \geq 2$
3) $e_i e_{i\pm1} e_i = \tau e_i \qquad (\tau = [M : N]^{-1})$
4) $\mathrm{tr}(we_{n+1}) = \tau\,\mathrm{tr}(w)$ if w is a word on $1, e_1, \ldots, e_n$.

The hermitian form $\mathrm{tr}(x^*y)$ on the algebra generated by the e_i's must be positive definite and this forces $[M : N]$ to be $4\cos^2 \pi/n$. We will see later how the same form arises in connection with surgery on three-manifolds and this result about degeneracies of the form is precisely what allows a simple explicit formula for some new three-manifold invariants!

The construction of examples of subfactors proceeds by the explicit construction of a sequence of e_i's and a trace satisfying 1)–4) above. The II_1 factor is then obtained by the GNS construction and the subfactor is that generated by $e_2, e_3, \ldots$. In the case of index > 4 these subfactors have non-trivial centralizer. For more details on what happens in index > 4 see Popa's talk in this volume.

6. Commuting Squares

The tower M_i arising from $N \subset M$ defines two towers of finite dimensional algebras which we will call the centralizer towers. They are the commutants $A_i = M' \cap M_i$ and $B_i = N' \cap M_i$. Clearly $A_i \subset B_i$ and they satisfy the "commuting square" condition (first used by Popa in a different context) that the orthogonal projections from B_{i+1} to A_{i+1} and B_i commute and $B_i \cap A_{i+1} = A_i$.

This condition allows one to control the von Neumann algebra inclusion of the GNS closures of $\cup A_i$ and $\cup B_i$, which would be impossible without the condition.

In fact commuting squares give subfactors automatically by iterating the basic construction and the study of commuting squares constitutes a new and intriguing problem in finite dimensional linear algebra. A machine for producing examples from quantum groups has been developed by Wenzl. Other examples abound – see [Su], [HJ] and [HS], but the general structure of commuting squares is quite unclear.

7. Finite Depth, Classification Results

Actions of finite groups are completely classified on the hyperfinite type II_1 factor ([Jo7]) and one might hope for an extension of these results to finite index subfactors. This is unlikely since one may take any finite set of automorphisms $\alpha_1, \alpha_2, \ldots, \alpha_n$ and form the subfactor

$$\left\{ \left. \begin{pmatrix} x & & & \\ & \alpha_1(x) & & \mathrm{O} \\ \mathrm{O} & & \ddots & \\ & & & \alpha_n(x) \end{pmatrix} \right| x \in R \right\}$$

of $R \otimes M_n(\mathbb{C})$. This subfactor remembers too much about the group generated by $\alpha_1, \alpha_2, \ldots, \alpha_n$ and although if this group is amenable we are in a good situation ([O1]), if it is not there will be no nice classification.

On the other hand there is a class of subfactors, first stressed by Ocneanu, of R for which classification is a well-posed problem. A subfactor $N \subset M$ is said to be of *finite depth* if the dimensions of the centres of the centralizer towers A_i and B_i are bounded. It is then easy to see that the centralizer towers exhibit periodicity beyond a certain point. The index of a finite depth subfactor is given by the square of the norm of the matrix, with integer entries, describing the stabilized inclusion for $A_i \subset A_{i+1}$. See [GHJ].

Popa has shown in [Po] that the stabilized commuting square of centralizer towers is a complete invariant for subfactors of R of finite depth. Another version of the result is claimed by Ocneanu who has a more elaborate and computable version of the invariant.

If the index is less than 4 a complete classification is possible. Coxeter-Dynkin diagrams arise out of the combinatorics of the centralizer towers and according to Ocneanu subfactors of index < 4 are classified by Dynkin diagrams of types A_n, D_{2n}, E_6 and E_8, there being one for each A_n and D_{2n} and two for each of E_6 and E_8. The index of the subfactor is $4\cos^2 \pi/n$, n being the Coxeter number (see [O2, GHJ]). Popa has extended this to index 4, where infinite depth and extended Dynkin diagrams occur (see [GHJ] and Popa's talk in these proceedings).

Popa has given a deep generalization of his result to cases of infinite depth provided certain asymptotic behavior of the combinatorics can be controlled. Combined with ideas of Wasserman in [Was] this gives new results about compact group actions. See Popa's paper in these proceedings.

8. Statistical Mechanical Models

The abstract algebra presented by the e_i relations 1), 2) and 3) was used by Temperley and Lieb [TL, Ba] to show the equivalence of the ice-type and self-dual Potts models of statistical mechanics. The relations are satisfied by certain matrices which combine to give the row-to-row transfer matrices of the models. The same abstract algebra occurs in the models of Andrews, Baxter and Forrester [ABF] but the parameter is now in the discrete series $4\cos^2 \pi/n$. Pasquier used the ADE Coxeter graphs to get more models ([Pa]).

The ice-type model is a vertex model where the interactions between the elementary components of a system take place at the vertices of a graph. The Potts model is a spin model where interactions are on the edges and the ABF and Pasquier models are IRF models with many-spin interactions around faces of a planar graph.

Many more elaborate models can be obtained from quantum groups and there are corresponding algebraic relations generalizing the e_i ones. One may use these models to construct subfactors including the Wenzl ones but apparently others as well ([Jo2, D+]).

The relation of subfactors to the solvability of the model, if any, remains unclear.

9. Bimodules, Hypergroups, Paragroups, Quantized Groups ...

The combinatorics of the centralizer towers are very rich and attempts are being made to extract the data in them. The most ambitious project is Ocneanu's. He uses bimodules and intertwiners and obtains a structure with many properties of an IRF model of statistical mechanics, where the "faces" become closed paths around induction-restriction diagrams. See his notes in these proceedings.

Sunder has used a less detailed structure called a hypergroup which is only supposed to contain the combinatorial structure of tensor products of bimodules ([Su]).

The idea of using bimodules was first stressed by Connes. See [Co4, Jo3].

10. Braid Groups

The braid group B_n on n strings may be presented on $n-1$ generators $\sigma_1, \sigma_2, \ldots, \sigma_{n-1}$ with relations $\sigma_i\sigma_{i+1}\sigma_i = \sigma_{i+1}\sigma_i\sigma_{i+1}$ and $\sigma_i\sigma_j = \sigma_j\sigma_i$ if $|i-j| \geq 2$ (see [Bi]). These relations bear a resemblance to the e_i ones and one may represent the braid group by sending σ_i to $te_i - (1-e_i)$ if $2+t+t^{-1} = \tau^{-1}$. The discrete series corresponds to $t = e^{2\pi i/n}$, σ_i unitary, and the continuous part to $t > 0$, σ_i self-adjoint.

The presence of such braid group representations is a pervasive feature of all the generalizations to do with quantum groups and solvable models. The e_i representations correspond to the 2-dimensional representation of $U_q(sl_2)$. It is not known whether these representations, either collectively or individually, are faithful for $n > 3$. Lawrence has found these representations as natural actions of braid groups on homology groups – see [La]. See also Varchenko's talk in these proceedings.

11. Hecke Algebras

If $H < G$ are groups, the Hecke algebra is the commutant of the representation of G on the vector space of functions on G/H. In the case where G is $GL_n(\mathbb{F}_q)$ and H is the upper triangular matrices, this algebra admits a presentation on generators $g_1, g_2, \ldots, g_{n-1}$ with relations

$$g_i^2 = (q-1)g_i + q, \quad g_ig_{i+1}g_i = g_{i+1}g_ig_{i+1}, \quad g_ig_j = g_jg_i \text{ if } |i-j| \geq 2.$$

For $|q| \neq 1$ or 0 this algebra is isomorphic to the group algebra of the symmetric group S_n (see [Bo]).

Clearly the braid group can be represented in the Hecke algebra in the obvious way. In fact these representations contain the e_i ones (with $q = t$) as direct summands. Under the isomorphism with $\mathbb{C}S_n$, the e_i representations correspond to Young diagrams with at most 2 rows. (See [Jo4].)

12. Knot Polynomials

Braids may be closed by tying the top to the bottom to form links (see below).

The braid $\sigma_1\sigma_2^{-1}\sigma_1\sigma_2^{-1}$:  Its closure: 

It is clear that the closure of a braid $\alpha \in B_n$ does not change if α is conjugated in B_n, nor if α is embedded in B_{n+1} and multiplied by $\sigma_n^{\pm 1}$. These two "Markov moves" generate the equivalence relation of having the same closure. It follows from this and relation 4) in the e_i algebra that a normalized version of the trace of the algebra element representing a braid will be an invariant of the closure of the braid. For a link L this invariant turns out to be a polynomial $V_L(t)$ in the variable t (or $\sqrt{t}$) and can be calculated, though not always rapidly, from the following "skein relation".

Skein Relation ([Cnw]). If L_+, L_- and L_0 are links identical except near one crossing where they are as below

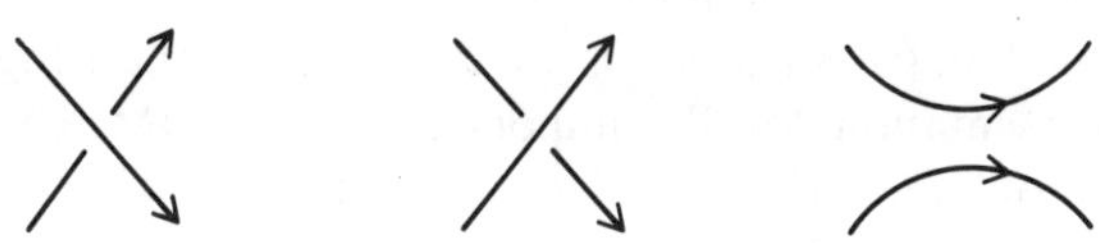

L_+ L_- L_0

then

$$\frac{1}{t}V_{L_+} - tV_{L_-} = (\sqrt{t} - 1/\sqrt{t})V_{L_0}\,.$$

The right-handed trefoil has $V(t) = t + t^3 - t^4$ and the left-handed one has $V(t) = 1/t + 1/t^3 - 1/t^4$. The Alexander polynomial has a similar relation and this prompted many people to develop a 2-variable polynomial called the HOMFLY polynomial $P_L(l,m)$ having arbitrary coefficients in its skein relation ([Jo5, F+, Li1]).

Kauffman found an explicit formula for $V_L(t)$ in [Ka1] from an arbitrary knot diagram which was used by him, Murasugi and Thistlethwaite to prove some old conjectures of Tait about alternating knots ([Ka1, Mu, Th]). He also found another two-variable polynomial generalization of $V(t)$, called the Kauffman polynomial $F(a,x)$. It does not contain the Alexander polynomial.

The polynomials are very useful in calculating the minimal number of crossings for a diagram of a knot and the minimal number of strings for a closed braid representation of a knot ([LT, Mo, FW]).

13. The R-Matrix, Powers State Picture

A special representation of the relations 1)–4) of §5 was discovered by Pimsner and Popa in [PP]. If $e \in \mathrm{End}(\mathbb{C}^2 \otimes \mathbb{C}^2)$ is defined by $e = \tau e_{11} \otimes e_{22} + \sqrt{\tau(1-\tau)}(e_{12} \otimes e_{21} + e_{21} \otimes e_{12}) + (1-\tau)e_{22} \otimes e_{11}$, then if we let e_i on $(\mathbb{C}^2)^{\otimes n}$ be defined between the ith and $(i+1)^{th}$ tensor components as e and the identity on the others, one finds that 1), 2) and 3) are satisfied. Moreover if φ_λ is the Powers state (giving a type III_λ factor), then its restriction to the algebra generated by $1, e_1, \ldots, e_n$ defines a trace satisfying 4). This gives a useful way of calculating $V_L(t)$, with $t = \lambda$. This representation, without φ_λ, was used also in [TL] where the e_i's give the transfer matrix for an ice-type model.

At this point a remarkably rapid development in the understanding of the polynomials was made possible by the existence of the theory of quantum groups. Fadeev's Leningrad group, particularly Sklyanin [Sk] and Kulish and Reshetikhin [KR] had uncovered a new structure related to the ice-type model for which the relevant Lie algebra was $sl_2(\mathbb{C})$. For background see [Fa]. This had been generalized by Jimbo [Ji1] and Drinfeld [Dr] to produce analogues of the braiding matrices obtained from the e_i representation I have just described, one for each finite dimensional representation of every simple Lie algebra. The analogue of the Powers state was soon found and the following picture was established by Reshetikhin [Re] and Rosso [Ro]:

Let $\mathcal{G}$ be a simple finite dimensional complex Lie algebra and let $\mathcal{L}$ be the class of links with distinguished components $C_1, C_2, \ldots, C_n$. Then to each way of assigning finite dimensional representations of $\mathcal{G}$ to $C_1, \ldots, C_n$ there is a polynomial invariant of isotopy for links in $\mathcal{L}$.

The polynomial $V_L(t)$ corresponds to $\mathcal{G} = sl_2$ and the assignment of its 2-dimensional representation to all components. The HOMFLY and Kauffman polynomials are obtained in a similar way from sl_n and the symplectic (or orthogonal) algebras respectively.

In this picture the geometric operations of cabling (and satellites in general) can be understood in terms of tensor products of representations ([MSt]).

Explicit formulae on any (not necessarily braided) picture of a link may be given. This was first done for HOMFLY in [Jo6] and generalized to the Kauffman polynomial in [Tu1].

14. Positivity of the Markov Trace

Ocneanu's approach to the HOMFLY polynomial was a direct generalization of my construction of $V_L(t)$, by defining a trace on the Hecke algebra by the property $\mathrm{tr}(wg_{n+1}) = z\,\mathrm{tr}(w)$ if w is a word on $g_1, g_2, \ldots, g_n$, and where z is a new variable. Subfactors occur for the values of (q, z) allowing a $*$-algebra structure on the Hecke algebra for which the trace is positive. This set of values was determined by Ocneanu, and Wenzl constructed the subfactors and calculated their indices ([F+, We1]). It is convenient to use the variables $\tau = q/(1+q)^2$ and $\eta = (1+z)/(1+q)$. Then the "positivity spectrum" (Fig. 2) is as follows:

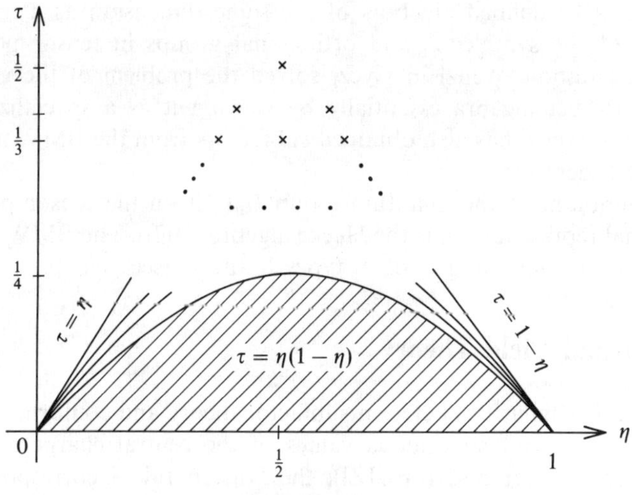

Fig. 2

The discrete points are the intersections in $[0,1] \times [0,1]$ of the curves $\eta = P_k(\tau)/P_{k-1}(\tau)$ with the curves $\eta = \tau P_{k'-2}(\tau)/P_{k'-1}(\tau)$ as indicated, where $P_k(\eta)$ are the polynomials given by $P_{k+1} = P_k - \eta P_{k-1}$, $P_0 = 1$, $P_1 = 1$, and $k + k' = l + 3$, $k = 2, 3, \ldots, l+1$, l indicating the horizontal row on which the point is situated. The index of the subfactor at the point labelled (l, k) is $\sin^2 k\pi/(l+3)/\sin^2 \pi/(l+3)$ ([We1]).

15. Brauer and BMW Algebras

A new algebra was developed in [BW] and [Muk] to play the role for the Kauffman polynomial that the Hecke algebra plays for the HOMFLY polynomial. The idea, inspired by Kauffman, is to add objects like $|\ldots| \asymp |\ldots|$ to the braid group generators. Calling these generators E_i and the usual braid group generators G_i, the BMW algebra with parameters a, x, has presentation:

$$G_iG_{i+1}G_i = G_{i+1}G_iG_{i+1}, \quad G_iG_j = G_jG_i \text{ if } |i-j| \geq 2,$$
$$G_i + G_i^{-1} = x(1+E_i), \quad E_iG_i = G_iE_i = aE_i, \quad E_i^2 = (a + a^{-1} - x)x^{-1}E_i,$$
$$E_iG_{i\pm 1}^{\pm 1}E_i = a^{\mp 1}E_i, \quad E_iG_{i\pm 1}G_i = E_iE_{i\pm 1}.$$

These relations are best understood in terms of diagrams. The third is used because of the Kauffman polynomial. Using Markov traces and the basic construction of subfactor analysis Wenzl determined the structure of the BMW algebra for generic values of the parameters and a lot about it for special values. Its dimension is $1 \cdot 3 \cdots (2n+1)$, n being the number of G_i's as above. It has the Hecke algebra as a natural quotient and the e_i algebra as a natural subalgebra.

Brauer in [Br] defined algebras of the same dimension as a model for the commutant of the symplectic and orthogonal groups in tensor powers of the vector representation. Wenzl in [Wez] solved the problem of the generic structure of the Brauer algebra essentially by viewing it as a specialization of the BMW algebra. Wenzl has also obtained subfactors from the BMW algebra using positivity considerations.

The commutant of the quantum group $U_q(sl_n)$ on the tensor powers of the n-dimensional representation is the Hecke algebra ([Ji2]). The BMW algebra does the same job for quantum groups of types B and C (see [We3]).

16. Conformal Field Theory

There are many similarities between subfactor theory and conformal field theory. I cite the discrete and continuous values of the central charge c ([FQS]), the ADE classification for $c < 1$ ([CIZ]), the "fusion rules" corresponding to the combinatorics of the centralizer towers ([GW]) and the existence of a tractable subclass called rational CFT's corresponding to finite depth subfactors. But the most compelling evidence for a connection is that the braid group representations which occur canonically in the subfactor picture for index < 4 occur also for the $SU(2)$ case of one of the main models of CFT – the Wess-Zumino-Witten model. Following [KZ], Tsuchiya and Kanie calculated the monodromy of the n-point functions in the holomorphic sector of this model and found the e_i-braid group representations (for spin 1/2). Thus the subfactors themselves could be constructed out of the WZW theory but in a somewhat indirect way.

Moore and Seiberg showed in [MSe] that the braid group situation was just the genus zero case of a theory that works in arbitrary genus for any conformal field theory. To do this they checked that the defining relations ([Waj]) for the mapping class group do indeed follow from their axioms for CFT.

Conformal field theories are often obtained as continuum limits of critical 2-dimensional classical statistical mechanical systems. It is not clear that there is any deep relationship between the appearances of the braid group in solvable models and in the CFT of the continuum limit.

17. Algebraic Quantum Field Theory, Superselection Sectors

Several people ([Fr, FRS, Lo]) have noticed that the braiding and Markov trace structure is inherent in any (not necessarily conformal) low dimensional QFT. In the framework of algebraic QFT ([HK]), Doplicher, Haag and Roberts in [DHR] introduced superselection sectors as representations of the observable algebra which are equivalent to the vacuum representation when restricted to the algebra of the causal complement of some bounded region. Haag duality is the property that the algebras of such bounded regions and their causal complements are each other's commutants. This duality is supposed to hold in the vacuum sector. A simple calculation shows that superselection sectors give rise to endomorphisms of the local observable algebra. Using geometric properties of space-time, implementers of the endomorphism give rise to unitary braid group representations where the endomorphism takes σ_i to σ_{i+1}. The Markov trace may be obtained as a simple weak limit.

The square root of the index of the subfactor which is the image of the endomorphism was called the "statistical dimension" of the sector in [DHR], but because they were considering 4 dimensions this dimension was necessarily an integer in their theory.

One of the problems with all this work is that there do not seem to be any concrete examples where the subfactors and braid group representations have been calcualted to the satisfaction of an expert in von Neumann algebras.

18. Loop Groups and Subfactors

The loop group $LSU(2)$ has a discrete series of "positive energy" projective unitary reprseentations labelled by a level l and a spin j, $0 \leq 2j \leq l$. Inspired by [TK] and in an attempt to understand and implement the ideas of superselection sectors, and to provide naturally occuring examples of subfactors, A. Wassermann and I have been working on the best known example, WZW theory, especially for $LSU(2)$. We interpret the superselection sectors as being the discrete series of the loop group for a fixed level. Given an interval I in the circle S^1 (I^c will denote the complementary interval), let $L_I G$ be the group of loops supported in I. The von Neumann algebra $(L_I G)''$ corresponds to the local algebra. Haag duality (which we have proved) says that for the vacuum sector (spin $= 0$), $(L_I G)' = (L_{I^c} G)''$. These von Neumann algebras are type III_1 factors in any sector and we have shown that, if the level is fixed, the representations of $L_I G$ for any spin are unitarily equivalent.

We conjecture that $[(L_I G)' : (L_{I^c} G)''] = \sin^2\{(2j+1)\pi/(l+2)\}/\sin^2(\pi/(l+2))$ and in general that the subfactor is the tensor product of a Wenzl subfactor and the hyperfinite type III_1 factor. The vertex operators of Tsuchiya and Kanie, if made to become unbounded operator valued distributions on S^1, should provide explicit intertwiners between the representations of $L_I G$ for fixed level and different spin.

19. Witten's Interpretation of $V_L(t)$ and Its Generalizations

Motivated by his own work, ideas of Atiyah and Segal and the relations between subfactors, knots and CFT, Witten proposed the following formula for a link L with components $L_1, L_2, \ldots, L_n$ ([Wi1]).

$$V_L(e^{2\pi i/k+2}) = \int_A [\mathscr{D}A] e^{ik\int_{S^3} \operatorname{tr}(A\wedge dA + \frac{2}{3}A\wedge A\wedge A)} \prod_{i=1}^{n} W(L_i)$$

where A runs over all $su(2)$-valued 1-forms on S^3 (modulo the gauge group) and $W(L_i)$ is the trace of the parallel transport using A (in the 2-dimensional representation) along the closed curve L_i. A measure $[\mathscr{D}A]$ with the appropriate properties has not been shown to exist so this formula must be taken in the context of Witten's topological QFT ([A]), which predicts enough formal properties of such an integral for it to be calculated (i.e. identified with $V_L(t)$). This can be thought of as the solvability of this particular topological QFT.

The generalizations are now apparent. $SU(2)$ can be replaced by any compact Lie group G and one may choose any finite dimensional representation of G per component of L. This reproduces the ingredients of the picture that emerged from quantum groups in §13.

More significantly S^3 may be replaced by any closed 3-manifold so that Witten's theory predicts the existence of invariants for links in 3-manifolds and gives explicit formulae for calculating them from a surgery description of the link in the 3-manifold. These formulae have been checked using the Kirby calculus ([FR]) at least for $SU(2)$ in [RT] and [KM] using quantum groups. In [KM] certain explicit evaluations are given in terms of classical invariants. In [Li2] an elementary formula for these invariants occurs using no more than cabling and $V_L(t)$. The key ingredient in the proof is the degeneracy of the trace on the e_i algebra of §5. One wonders whether the 3-manifold invariants may be obtained directly from subfactors.

20. Topological Quantum Field Theory

Witten has developed a formalism in which a quantum field theory in $d+1$ dimensions assigns a "Hilbert space" to any d-dimensional manifold Σ with extra structure and every time this manifold is the boundary of a $d+1$ dimensional manifold M with compatible extra structure there is a vector in the Hilbert space of the boundary. In the simplest case $M = \Sigma \times [0,1]$ the vector is supposed to define an operator giving the time evolution of a system from time 0 to time 1. The Hilbert spaces and vectors are supposed to satisfy certain important and powerful axioms. If the "extra structure" is little more or less than an orientation one talks of *topological* quantum field theory. See [A] where examples are given connected with Donaldson and Floer theory.

In the case of the theory given by the Chern-Simons action $\mathrm{tr}(\mathrm{Ad}\,A + (2/3)A^3)$, $d = 2$ and Witten identifies the Hilbert space corresponding to a surface Σ as being the (finite dimensional) vector space of conformal blocks for the corresponding Wess-Zumino-Witten theory with the same gauge group. This is crucial for his calculations as it allows him to assert that, for the $SU(n)$ theory the vector space corresponding to the sphere with 4 marked points is 2-dimensional. This, together with the "gluing" axioms of topological QFT allows the formal calculation of the functional integral via skein theory.

The Hilbert space for a surface Σ is also deduced via a "geometric quantization" approach using a complex structure on Σ. That the Hilbert space should be independent of the complex structure is interpreted as implying the existence of a flat connection on certain natural bundles over Teichmüller space. See the talk by Tsuchiya in these proceedings.

Witten's surgery formula comes from identifying a basis for the Hilbert space of a torus with the vectors obtained by realizing the torus as the boundary of a solid torus containing one simple homotopically non-trivial closed curve to which an irreducible representation of the compact group is assigned. The action of the diffeomorphisms in this basis (all that is required for surgery formulae given the gluing axioms) is precisely that of $SL(2,\mathbb{Z})$ on the characters of the relevant affine Lie algebra at the given level.

References

[AW] Araki, H., Woods, E.: A classification of factors. Publ. RIMS, Kyoto University (1968) 51–130

[A] Atiyah, M.: Topological quantum field theory. Publ. Math. I.H.E.S. **68** (1989) 175–186

[Ba] Baxter, R.: Exactly solved models in statistical mechanics. Academic Press, New York 1982

[BW] Birman, J., Wenzl, H.: Braids, link polynomials and a new algebra. Trans. AMS **313** (1989) 269–273

[Bi] Birman, J.: Braids, links and mapping class groups. Ann. Math. Studies **82** (1974)

[Bo] Bourbaki, N.: Groupes et algèbres de Lie, IV, V, VI. Masson, Paris 1981

[Br] Brauer, R.: On algebras which are connected with the semisimple continuous groups. Ann. Math. **38** (1937) 856–872

[CIZ] Cappelli, A., Itzykson, C., Zuber, J. B.: Nucl. Phys. B **280** (1987) 445–

[Co1] Connes, A.: von Neumann algebras. Proceedings ICM Helsinki, 1978, vol. 1, pp. 97–109

[Co2] Connes, A.: Classification of injective factors. Ann. Math. **104** (1976) 73–115

[Co3] Connes, A.: Factors of type III_1, property L'_λ and the closure of inner automorphisms. J. Operator Theory **14** (1985) 189–211

[Co4] Connes, A.: Classification des facteurs. In: Operator Algebras and Applications, Proc. Symp. Pure Appl. Math. **38** (1982) part II, 43–109

[Cnw] Conway, J.: An enumeration of knots and links. Computational Problems in Abstract Algebra (ed. J. Leech). Pergamon Press 1969, pp. 329–358

[D+] Date, E., Jimbo, M., Miki, K., Miwa, T.: Cyclic representations of $U_q(sl(n+1,\mathbb{C}))$ at $q^N = 1$. RIMS Preprint 1990

[DHR] Doplicher, S., Haag, R., Roberts, J.: Local observables and particle statistics I, II. Comm. Math. Phys. **23** (1971) 199–230 and **35** (1974) 49–85

[Dr] Drinfeld, V.: Quantum groups. Proc. ICM 1986, vol. 1, pp. 798–820

[Fa] Faddeev, L.: Integrable models in (1 + 1)-dimensional quantum field theory. (Lectures in les Houches, 1982) Elsevier Science Publishers, 1984, pp. 563–608

[FR] Fenn, R., Rourke, C.: On Kirby's calculus of links. Topology **18** (1979) 1–15

[FW] Franks, J., Williams, R.: Braids and the Jones-Conway polynomial. Trans. AMS **303** (1987) 97–108

[FRS] Fredenhagen, K., Rehren, Schroer, B.: Superselection sectors with braid group statistics and exchange algebras. Comm. Math. Phys. **125** (1989) 201–226

[F+] Freyd, P., Yetter, D., Hoste, J., Lickorish, W., Millett, K., Ocneanu, A.: A new polynomial invariant of knots and links. Bull. AMS **12** (1985) 183–190

[FQS] Friedan, D., Qiu, Z., Shenker, S.: Conformal invariance, unitarity and critical exponents in two dimensions. Phys. Rev. Lett. **52** (1984) 1575–1578

[Fr] Fröhlich, J.: Statistics of fields, the Yang-Baxter equation and the theory of knots and links. Proceedings Cargise, ed. G. 't Hooft et al. (1987)

[GHJ] Goodman, F., delaHarpe, P., Jones, V.: Coxeter graphs and towers of algebras. MSRI Publications (Springer) **14** (1989)

[GW] Goodman, F., Wenzl, H.: Littlewood Richardson coefficients for Hecke algebras at roots of unity. Adv. Math.

[HK] Haag, R., Kastler, D.: An algebraic approach to quantum field theory. J. Math. Phys. **5** (1964) 848–861

[Ha] Haagerup, U.: Connes' bicentralizer problem and the uniqueness of the injective factor of type III_1. Acta. Math. **158** (1987) 95–148

[HS] Haagerup, U., Schou, J.: Some new subfactors of the hyperfinite II_1 factor. Preprint 1989

[HJ] delaHarpe, P., Jones, V.: Paires de sous-algèbres semisimples et graphes fortement réguliers. C.R. Acad. Sci. Paris **311** (1990) 147–150

[Ji1] Jimbo, M.: A q-difference analogue of $U(\mathfrak{g})$ and the Yang-Baxter equation. Lett. Math. Phys. **102** (1986) 537–567

[Ji2] Jimbo, M.: A q-analogue of $U(sl(N+1))$, Hecke algebra and the Yang-Baxter equation. Lett. Math. Phys. **11** (1986) 247–252

[Jo1] Jones, V.: Index for subfactors. Invent. math. **72** (1983) 1–25

[Jo2] Jones, V.: Notes on subfactors and statistical mechanics. In: Braid Group, Knot Theory and Statistical Mechanics (ed. Yang and Ge). World Scientific, 1989, pp. 1–25

[Jo3] Jones, V.: Index for subrings of rings. Contemp. Math. **43** (1985) 181–190

[Jo4] Jones, V.: Braid groups, Hecke algeras and type II_1 factors. In: Geometric Methods in Operator Algebras (ed. Araki and Effros). Pitman Res. Notes in Math. (1983) 242–273

[Jo5] Jones, V.: A polynomial invariant for knots via von Neumann algebras. Bull. AMS **12** (1985) 103–111

[Jo6] Jones, V.: On knot invariants related to some statistical mechanical models. Pacific J. Math. **137** (1989) 311–334

[Jo7] Jones, V.: Actions of finite groups on the hyperfinite type II_1 factor. Memoirs AMS **237** (1980)

[Ka] Kauffman, L.: State models and the Jones polynomial. Topology **26** (1987) 395–401

[KM] Kirby, R., Melvin, P.: The three manifold invariants of Witten and Reshetikhin-Turaev. Preprint 1990

[KZ] Knizhnik, V., Zamolodchikov, A.: Current algebra and Wess-Zumino models in two dimensions. Nuc. Phys. B **247** (1984) 83–103

[Kr] Krieger, W.: On ergodic flows and the isomorphism of factors. Math. Ann. **223** (1976) 19–70

[KR] Kulish, P., Reshetikhin, N.: Quantum linear problem for the sine-Gordon equation and higher representations. J. Sov. Math. **23** (1983) 2435–2441

[La] Lawrence, R.: Homological representations of the Hecke algebra. Comm. Math. Phys. (to appear)

[Li1] Lickorish, W.: Polynomials for links. Bull. LMS **20** (1988) 558–588

[Li2] Lickorish, W.: Three-manifold invariants from the combinatorics of the Jones polynomial. Pacific Math. J. (to appear)

[LT] Lickorish, W., Thistlethwaite, M.: Some links with non-trivial polynomials and their crossing-numbers. Comment. Math. Helv. **63** (1988) 527–539

[Lo] Longo, R.: Index of subfactors and statistics of quantum fields I. Comm. Math. Phys. **126** (1989) 217–247

[MSe] Moore, G., Seiberg, N.: Classical and quantum conformal field theory. Comm. Math. Phys. **123** (1989) 177–254

[Mo] Morton, H.: Closed braid representations for a link and its 2-variable polynomial. Preprint Liverpool, 1985

[MSt] Morton, H., Strickland, P.: Jones polynomial invariants for knots and satellites. Univ. of Liverpool Preprint 1989

[Mu] Murasugi, K.: Jones polynomials and classical conjectures in knot theory. Topology **26** (1987) 187–194

[Muk] Murakami, J.: The Kauffman polynomial of links and representation theory. Osaka J. Math. **24** (1987) 745–758

[MvN1] Murray, F., von Neumann, J.: On rings of operators. Ann. Math. **37** (1936) 116–229

[MvN2] Murray, F., von Neumann, J.: On rings of operators IV. Ann. Math. **44** (1943) 716–808

[O1] Ocneanu, A.: Actions of discrete amenable groups on von Neumann algebras. Lecture Notes in Mathematics, vol. 1138. Spinger, Berlin Heidelberg New York 1985

[O2] Ocneanu, A.: Quantized groups, string algebras and Galois theory for algebras. In: Operator Algebras and Applications (eds. Evans and Takesaki), 1988, pp. 119–172

[Pa] Pasquier, V.: Two-dimensional critical systems labelled by Dynkin diagrams. Nucl. Phys. B **285** (1987) 162–172

[Po] Popa, S.: Classification of subfactors: the reduction to community squares. Invent. math. **101** (1990) 19–43

[PP] Pimsner, M., Popa, S.: Entropy and index for subfactors. Ann. Sci. Ec. Norm. Sup. **19** (1986) 57–106

[Pow] Powers, R.: Representations of uniformly hyperfinite algebras and their associated von Neumann rings. Ann. Math. **86** (1967) 138–171

[Re] Reshetikhin, N.: Quantized universal enveloping algebras, the Yang-Baxter equation and invariants of links I, II. LOMI Preprints 1988

[RT] Reshetikhin, N., Turaev, V.: Invariants of 3-manifolds via link polynomials and quantum groups. Invent. math. (to appear)

[Ro] Rosso, M.: Groupes quantiques et modèles à vertex de V. Jones en théorie des noeuds. C. R. Acad. Sci. Paris **307** (1988) 207–210

[Sk] Sklyanin, E.: Some algebraic structures connected with the Yang-Baxter equation. J. Sov. Math. **19** (1982) 1546–1596

[Su] Sunder, V. S.: Π_1-factors, their bimodules and hypergroups. Preprint 1989

[Ta] Takesaki, M.: Tomita's theory of modular Hilbert algebras and its applications. (Lecture Notes in Mathematics, vol. 128.) Springer, Berlin Heidelberg New York 1970

[TL] Temperley, H., Lieb, E.: Relations between the 'percolation' and… . Proc. Roy. Soc. Ser A **322** (1971) 251–280

[Th] Thistlethwaite, M.: A spanning tree expansion of the Jones polynomial. Topology **26** (1987) 297–309

[TK] Tsuchiya, A., Kanie, Y.: Vertex operators in conformal field theory on $\mathbb{P}^1$ and monodromy representations of braid group. Adv. Stud. Pure Math. **16** (1988) 297–372

[Tu] Turaev, V.: The Yang-Baxter equation and invariants of links. Invent. math. **92** (1988) 527–553

[vN] von Neumann, J.: Zur Algebra der Funktionaloperatoren. Math. Ann. **102** (1929) 370–427

[Waj] Wajnryb, B.: A simple presentation for the mapping class group of an orientable surface. Israel J. Math. **45** (1983) 157–174

[Wass] Wassermann, A.: Coactions and Yang-Baxter equations for ergodic actions and subfactors. In: Operator Algebras and Applications (ed. Evans and Takesaki). LMS Lecture Notes **136**, vol. 2 (1988) 203–236

[We1] Wenzl, H.: Hecke algebras of type A_n and subfactors. Invent. math. **92** (1988) 349–383
[We2] Wenzl, H.: On the structure of Brauer's centralizer algebras. Ann. Math. **128** (1988) 173–193
[We3] Wenzl, H.: Quantum groups and subfactors of type B, C and D. Comm. Math. Phys. **133** (1990) 383–432
[Wi1] Witten, E.: Quantum field theory and the Jones polynomial. Comm. Math. Phys. **121** (1989) 351–399

Geometric Algorithms and Algorithmic Geometry

László Lovász

Department of Computer Science, Eötvös University, H-1088 Budapest, Hungary and Department of Computer Science, Princeton University, Princeton, NJ 08544, USA

0. Introduction

In this paper I would like to illustrate two facts. First, that ideas borrowed from convex, projective, and other classical branches of geometry play an important role in the design of algorithms for problems that do not seem to have anything to do with geometry: for problems in optimization, combinatorics, algebra, and number theory. Second, that such applications of geometry suggest some very elementary algorithmic questions concerning geometric notions, whose solution is far from complete. How to compute the volume? How to decide whether or not a convex body contains the other? How to present a convex body as an input to an algorithm? Even partial answers to these problems require a wealth of mathematical ideas, often (again) quite unrelated to the original question. On the other hand, answers to these questions have a very wide range of applicability.

In the first chapter some typical constructions are surveyed which lead from non-geometric problems to geometric ones. Such is, of course, the classical field of the "Geometry of Numbers". This is the application of lattice geometry and convexity to number theory, and was initiated by Minkowski around the turn of the century. We only touch upon this area to point out that the recent shift of interest from structural problems to algorithmic ones has induced a lot of activity, and this new approach has even fertilized classical types of investigations.

We also give a brief introduction to polyhedral combinatorics, developed in the 60s by Ford, Fulkerson, Hoffman, Edmonds, and others. Here polyhedral theory and linear programming are applied to combinatorial optimization problems. This approach yields surprisingly successful algorithms both from a theoretical and practical point of view. We also show how some enumeration problems are related to the computation of the volume of certain polytopes.

The second chapter treats verious forms of presentation of a convex body to an algorithm. This seemingly technical issue leads to a powerful equivalence principle between different ways of presentation. A combination of this principle with polyhedral combinatorics provides (at least theoretically) efficient algorithms for most combinatorial optimization problems that can be solved efficiently by any other means.

Chapter 3 describes a basic construction in algorithmic geometry, namely the Löwner-John ellipsoids. These are used in the ellipsoid method of Shor, Yudin, Nemirovskii and Khachiyan (in particular in establishing the equivalence principle formulated in the previous chapter) and in many other geometric

Proceedings of the International Congress
of Mathematicians, Kyoto, Japan, 1990

algorithms. We also sketch the geometric background of Karmarkar's celebrated linear programming algorithm.

In the last chapter we discuss recent developments concerning the problem of computing the volume of a convex body. After some very discouraging negative results, Dyer, Frieze and Kannan designed a polynomial time randomized algorithm which approximates the volume with an arbitrarily small error. It turns out that the crucial issue is to generate a random point uniformly distributed over a general convex body. The solution of this problem leads to Markov chains, eigenvalues of matrices, differential geometry, and even to some algebraic topology.

There is an extremely important branch of algorithmic geometry which is not treated here; this is usually called computational geometry. Polyhedral combinatorics leads to high-dimensional problems; other – more immediate – applications in image processing and robotics lead to two- and three-dimensional questions. In such cases, a different notion of efficiency is the crucial one. Our concern will be polynomial time; in computational geometry, usually linear or almost-linear time is the target. The interested reader may consult the monographs by Preparata and Shamos (1985) and by Edelsbrunner (1987).

1. Number Theory, Combinatorics, and Convex Sets

In this chapter we illustrate some of the most important constructions which transform algorithmic problems in various branches of mathematics into geometric questions, and in fact into very simple ones. These examples will also serve to help us put the corresponding general algorithmic problems into the right framework.

1.1 Geometry of Numbers

This is the classical area where the application of geometric ideas to non-geometric problems has been very successful ever since Minkowski's work. In the last decade, algorithmic questions arising from numerical methods, primality testing, computational algebra, cryptography, and other areas have revigorated the field; it turns out that to solve these algorithmic problems, often new structural insight is needed.

The classical problem in number theory which lead Minkowski to the "Geometry of Numbers" is the problem of *simultaneous diophantine approximation*: given n real numbers $\alpha_1, \ldots, \alpha_n$, and an "error bound" $\varepsilon > 0$, find integers $p_1, \ldots, p_n$ and q such that $q > 0$ and

$$\left|\alpha_i - \frac{p_i}{q}\right| \leq \frac{\varepsilon}{q} \qquad (i = 1, \ldots, n).$$

The answer is trivial for $\varepsilon = 1/2$ (we can choose q arbitrarily); it was proved by Dirichlet that such integers exist for every $\varepsilon > 0$ and in fact we can require that $q \leq \varepsilon^{-n}$.

While the proof of Dirichlet's theorem is quite easy (using "Dirichlet's Principle"), no efficient algorithm is known to *find* such an approximation. A geometric translation is useful, among others, in finding a solution with a worse bound on q.

Consider the following vectors in $\mathbb{R}^{n+1}$: $e_1, \ldots, e_n$ (the first n basis vectors) and $(\alpha_1, \ldots, \alpha_n, 1)^T$. The linear combinations of these vectors with integer coefficients form a lattice L, whose typical point looks like $(q\alpha_1 - p_1, \ldots, q\alpha_n - p_n, q)^T$ with integral $p_1, \ldots, p_n$ and q. Dirichlet's Theorem is equivalent to asserting that the lattice L has a non-zero lattice point in the brick

$$K = [-\varepsilon, \varepsilon] \times \ldots \times [-\varepsilon, \varepsilon] \times [-\varepsilon^{-n}, \varepsilon^{-n}].$$

Minkowski's famous "First Theorem" shows that the fact that K is a brick is irrelevant: all that matters is that K is a convex body centrally symmetric with respect to the origin with volume at least 2^{n+1} times the determinant of the lattice.

The convex body K can be viewed as the unit ball of a norm, and so Minkowski's theorem guarantees the existence of a "short" non-zero lattice vector in every lattice, measured in an arbitrary norm.

Many other problems turn out equivalent to the existence of a short non-zero lattice vector in appropriately defined lattices. For example, let $f(x) = \sum_{i=1}^n a_i x^i$ be a polynomial with integral coefficients; we want to know whether f is irreducible. Let α be a root of f; for simplicity, assume that α is real. Let K be a sufficiently large real number (computable from f) and let $L \subseteq \mathbb{R}^{n+1}$ be the lattice generated by the vectors $e_i + K\alpha^{i-1} e_{n+1}$, $1 \leq i \leq n$. If f is reducible over the rational field, then L contains a non-zero vector with (euclidean) length at most $2^n \sqrt{\sum_i a_i^2}$. On the other hand, if f is irreducible then every non-zero vector in the lattice is at least 2^n times this length. This fact is the basis of the efficient (polynomial time) algorithm for factoring polynomials (Lenstra, Lenstra and Lovász 1982).

We shall not go into geometric algorithms involving lattices in this survey; instead, we refer to (Lovász 1989) for a survey.

1.2 Polyhedral Combinatorics

Polyhedral combinatorics provides perhaps the most successful general approach to various combinatorial problems. To illustrate the idea, consider the following simple graph-theoretic problem. Let G be a (finite) graph with node set V and edge set E. A set of nodes of G is called *stable* if no two elements of it are connected by an edge. Let $\alpha(G)$ denote the maximum cardinality of a stable set in G. To determine $\alpha(G)$ is difficult (NP-hard) in general; but the approach of polyhedral combinatorics suggests efficiently solvable special cases, as well as efficiently computable estimates of $\alpha(G)$.

Let us construct the following convex polytope: for every stable set S of nodes, let χ^S denote the incidence vector of S (in the space $\mathbb{R}^V$ of vectors indexed by the nodes of G), and let STAB(G) be the convex hull of such incidence vectors. Then

$$\begin{aligned}\alpha(G) &= \max\{|S| : S \text{ stable}\} = \max\{1 \cdot \chi^S : S \text{ stable}\} \\ &= \max\{\textstyle\sum_i x_i : x \in \mathrm{STAB}(G)\}.\end{aligned}$$

(since the maximum of a linear objective function over a polytope is automatically assumed at a vertex).

The hope is to apply the powerful methods of linear programming to find this maximum. To this end, however, we have to find a representation of STAB(G) as the solution set of a system of linear inequalities. Such a representation exists of course; but how to find it? Note that the arguments presented so far would work for the problem of finding a maximum cardinality member in any collection of subsets of a finite set V; where we have to be problem-specific is in finding the linear representation of the corresponding polytope.

There is a very large number of results presenting linear descriptions of various combinatorial polyhedra; we shall restrict ourselves to the stable set polytope and be content with giving a couple of illustrations. A natural starting point is the following set of linear inequalities:

$$0 \le x_i \le 1 \qquad (i \in V), \tag{1}$$

$$x_i + x_j \le 1 \qquad (ij \in E). \tag{2}$$

It is clear that the incidence vector of any stable set, and therefore every vector in STAB(G), satisfies inequalities (1) and (2). The solution set of this system is, however, a larger polytope than STAB(G) in general. In fact, *inequalities* (1) *and* (2) *suffice to describe* STAB(G) *if and only if G is a bipartite graph.* So at least for bipartite graphs, $\alpha(G)$ can be determined using linear programming algorithms. Or, applying the Duality Theorem of linear programming, one can obtain a min-max formula for $\alpha(G)$ (which, after some transformations, turns out to be equivalent to the König Theorem).

If the graph is non-bipartite then it contains a circuit of odd length and we can use such a circuit to add further constraints to (1) and (2): for every odd circuit C we write up the inequality

$$\sum_{i \in V(C)} x_i \le \frac{|C|-1}{2}. \tag{3}$$

Graphs for which (1), (2) and (3) suffice to describe STAB(G) are called *t-perfect*; these graphs are less well understood than bipartite graphs, but several important classes of them are known. We mention *series-parallel graphs*, i.e., graphs which can be obtained by the repeated application of series and parallel extensions (Chvátal 1975, Boulala and Uhry 1979)

Assume that G is t-perfect; then $\alpha(G)$ can be expressed as the optimum value of a linear program with constraints (1)–(2)–(3). Note, however, that (3) includes possibly exponentially many constraints (in $n = |V|$), and so writing up this program and calling a linear program solver would be inefficient. We shall see that general geometric considerations provide efficient solution methods for such systems which do not need the whole system explicitly.

Further natural inequalities valid for STAB(G), but not implied by the previous ones, can be written up. Let B be a clique (a set of mutually adjacent nodes in G). Then every stable set meets B in at most one node and hence the inequality

$$\sum_{i \in B} x_i \le 1 \tag{4}$$

is valid for STAB(G). Those graphs for which (1) and (4) suffice to describe STAB(G) are called *perfect*. This rich class has been defined by Berge (1961) before these polyhedral methods were introduced, motivated by many classes of examples. Let us mention one: Let $(V, \le)$ be a (finite) partially ordered set.

Define a graph G by connecting two elements of V iff they are comparable. *Comparability graphs* obtained this way are perfect.

It is beyond the scope of this paper to treat perfect graphs; we refer to Berge and Chvátal (1984) and Grötschel, Lovász and Schrijver (1988). It should be mentioned, however, that polyhedral methods play a central role in their study, even in proving seemingly elementary properties.

Polyhedral combinatorics is closely related to *integer linear programming*. The basic problem here is to solve a system of linear inequalities *in integers*. For example, the integer solutions of the system (1)–(2) are exactly the incidence vectors of stable sets. The problem of simultaneous diophantine approximation can also be viewed as an integer programming problem: Given α_i, ε, and Q, find a solution of

$$-\varepsilon \le q\alpha_i - p_i \le \varepsilon \qquad (i = 1, \dots, n),$$
$$1 \le q \le Q.$$

For a long while, integer programming and lattice geometry have developed independently, and used rather different methods. A first substantial connection was established by Lenstra (1983), who designed a polynomial time algorithm to solve integer linear programs with a bounded number of variables, using methods borrowed from the geometry of numbers. This approach to integer programming seems to gain further momentum in recent years.

1.3 Enumeration and Volume

Some enumeration problems also have useful translations into geometry. This connection is not so well understood and we only give one example. Let $P = (E, \le)$ be a partially ordered set. A linear order of E which is compatible with the given partial order is called a *linear extension* of the partial order. The number of linear extensions is a measure of how incomplete the partial order is, and it plays an important role in several algorithmic and other questions. There is no reasonable formula or efficient algorithm known to find this number. In fact, no such algorithm can be expected by the recent important result of Brightwell and Winkler (1990), which asserts that to determine the number of linear extensions of a poset is #P-complete. But an efficient approximation algorithm can be obtained, which is based on the following construction (Stanley 1986).

Consider the linear space $\mathbb{R}^E$ and the incidence vectors of filters in P. The convex hull of these incidence vectors is a polytope FILT(P). By the methods of polyhedral combinatorics mentioned in the previous section, it can be shown that FILT(P) is defined by the inequalities

$$0 \le x_i \le 1 \qquad (i \in E),$$
$$x_i \le x_j \qquad (i \le j).$$

This fact can be used, as sketched above, to solve optimization problems involving filters. Right now, however, the following fact is important: *the number of linear extensions of P equals $n!$ times the volume of* FILT(P). This observation reduces the problem of enumerating linear extensions to the problem of determining the volume of a polytope, which is described as the solution set of a small number of simple linear inequalities (in n-space we have $O(n^2)$ inequalities). We shall return to this general geometric problem in Chapter 4.

2. What is a Convex Body?

We have seen that various number theoretic, algebraic, and combinatorial questions can be reduced to quite fundamental problems in geometry, such as computing the volume or finding the maximum of a linear objective function over a convex body. Before treating geometric algorithms to solve such problems, we have to introduce the right framework.

A convex body is a closed, bounded, full-dimensional convex set in $\mathbb{R}^n$. This simple definition becomes insufficient, however, if we are interested in algorithmic questions. In this chapter we discuss our aspects for algorithms and the algorithmic notion of a convex body.

2.1 Basic Algorithmic Problems for a Convex Body

Our condition for the "efficiency" of an algorithm is worst-case polynomial time: this means that there exists a constant $c > 0$ such that for every input of length n, the algorithm makes $O(n^c)$ bit-operations. Here the length of the input is the total number of bits needed to describe the input. We assume that input numbers are always rational, and their contribution to the input length is the total number of digits in the binary representation of the numerator and of the denominator.

For the theory of polynomial time algorithms, and for related notions in complexity theory like NP and NP-hard, see e.g. (Garey and Johnson 1979). To follow this paper, it should suffice to translate "NP-hard" or "#P-complete" as "it is hopeless to find a polynomial-time algorithm for this problem".

We want to study algorithms whose input is a convex body, and with this we run into trouble. How to present a convex body? Various forms occur in various problems. In linear programming, one always considers convex polyhedra, presented as the solution set of a system of linear inequalities. An equally natural form in which convex bodies come up is a convex polytope, presented as the convex hull of an explicitly given set of vectors. In Banach space theory, convex bodies arise as unit balls of norms, where the norm may be given by some formula; e.g. $\sum_i |x_i|^3 \le 1$ defines the unit ball of the ℓ_3-norm. In geometry, a convex body $K \subseteq \mathbb{R}^n$ is sometimes described by its support function: this is essentially the function $\varrho_K : \mathbb{R}^n \to \mathbb{R}$, defined by $\varrho_K(u) = \max\{u \cdot x : x \in K\}$.

The polytope STAB(G) is defined as the convex hull of a set of vectors (incidence vectors of stable sets), but to describe it, it would be very inefficient to list these vectors; it suffices to specify the graph G. So STAB(G) is presented by *implicitly* specifying its vertices.

We are interested in algorithms that are as independent of the specifics of the presentation of the body as possible. It turns out that most geometric algorithms depend on the possibility to carry out one or more of the following tasks:

- *Membership Test:* Given a (rational) vector x, decide whether or not $x \in K$.
- *Separation:* Given a (rational) vector x, decide whether or not $x \in K$, and if not, find a hyperplane separating x from K.
- *Validity Test:* Given a linear inequality $a \cdot x \le \alpha$ (with rational coefficients), decide whether or not the inequality is valid for all $x \in K$.
- *Violation:* Given a linear inequality $a \cdot x \le \alpha$ (with rational coefficients), decide whether or not the inequality is valid for all $x \in K$, and if not, find an $x \in K$ violating it.

It is clear that finding a separating hyperplane is a more difficult task than testing for membership, and finding a violating point is more difficult than testing for validity. Scanning through the examples above, we see that depending on the presentation of K, one or the other of these tasks can be carried out easily while others appear non-trivial. For example, if the body is given as the solution set of linear inequalities, then it is trivial to test membership by substituting x in each of the defining inequalities; and if it violates one, this also yields a separating hyperplane. On the other hand, there is no obvious way to test the validity of an inequality. Similarly, if the body is presented as the convex hull of vectors, then it is trivial to test validity of a linear inequality, but testing membership is non-trivial.

For STAB(G), neither one of the above tasks is easy, at least if we allow only polynomial time in n (note that the input size in $O(n^2)$). In fact, each of the above tasks is NP-hard.

To get presentation-independent results, we define a *membership oracle* as a black box which works as follows: if we plug in a vector $x \in \mathbb{R}^n$, it returns "YES" or "NO". Its answers must be consistent with the interpretation that "YES" means $x \in K$ for some convex body K. We can call such an oracle in an algorithm; one call counts as a single step. Of course, if we have an algorithm to test membership in polynomial time, then we can "put this inside the black box", and this increases the running time by a polynomial factor only.

One can introduce separation, validity, and violation oracles in a similar way.

The following is a surprising and powerful principle (not a theorem!):

Equivalence Principle. *The four oracles above are equivalent from the point of view of polynomial time algorithms.*

In other words, if for a class of convex bodies we have a polynomial time algorithm to solve either one of them, then all the others can be solved in polynomial time. This principle is true only under some technical assumptions and restrictions which, however, seldom cause any problem in its applications. There are various ways to formulate technical conditions making the Equivalence Principle valid; we only sketch some, and refer for a complete discussion to the monograph (Grötschel, Lovász and Schrijver 1988).

One set of these technicalities consists of making the "boundedness" and "full-dimensionality" properties of convex bodies effective: we need to know a number $R > 0$ such that K is contained in the ball with radius R about the origin, and another number $r > 0$ such that K contains a ball with radius r. These numbers must be considered part of the input to any algorithm, so the number of bits needed to write them down must be included in the input size. If the body is given by a membership oracle, and we want to solve the other tasks, then in addition the center of an inscribed ball with radius r must be given in advance. It is easy to argue that without this kind of information, the Equivalence Principle would not be valid.

Another set of limitations comes from numerical errors. We have to re-define the oracles so that an "error bound" $\varepsilon > 0$ is also part of the input, and then allow a small error in the answer. For example, in a membership oracle we should allow either a "YES" or a "NO" answer if the distance of x from the boundary of K is less than ε. One can formulate "weak" versions of all the above oracles in an analogous way. A precise form of the Equivalence Principle holds for these weak versions. (If the convex body K is a polytope with (say) 0-1 vertices, then we do not have to restrict ourselves to weak versions.)

The main ingredient in the proof of the Equivalence Principle is the *Ellipsoid Method*, to be sketched in the next chapter.

2.2 Applications of the Equivalence Principle

Some consequences of the Equivalence Principle are immediate. Since testing membership in a polyhedron presented by an (explicit) system of linear inequalities is trivial, we can test validity of a linear inequality for such polyhedra in polynomial time. This result implies a polynomial time test for the solvability of a system of linear inequalities, which in turn implies a polynomial time algorithm to solve linear programs.

In a similar way it follows that quadratic programs with positive definite constraints can be solved in polynomial time.

Several basic algorithmic issues concerning convex bodies can be solved using just the Equivalence Principle; let us mention two of these. The issue of validity of a linear constraint can be viewed as a special case of either of the following questions:

- Given two convex bodies K_1 and K_2, are they disjoint?
- Given two convex bodies K_1 and K_2, is $K_1 \subseteq K_2$?

Due to the necessary uncertainty around the boundary, in both cases we can only expect an approximate answer; we specify an $\varepsilon > 0$ and if $\mathrm{vol}(K_1 \cap K_2) < \varepsilon$, or if $\mathrm{vol}(K_1 \setminus K_2) < \varepsilon$, then the "YES" answer is acceptable.

Having separation oracles for K_1 and K_2, we can design a separation algorithm for $K_1 \cap K_2$ trivially. Suppressing technical details (the number r), the Equivalence Principle yields a polynomial time solution for the violation problem for $K_1 \cap K_2$. Let $c \cdot x \leq \gamma$ be any inequality which is invalid for all points in K_1, and solve the violation problem for $K_1 \cap K_2$. If the inequality is valid for $K_1 \cap K_2$ then this intersection must be empty; else, the violation algorithm returns a common point.

The second problem is more difficult; in fact, to solve it (even with the ε-tolerance) would yield polynomial-time solutions for NP-complete problems. But using the randomized methods discussed in Chapter 4 below, we can test $K_1 \subseteq K_2$ (with the ε-tolerance) in time polynomial in n and $1/\varepsilon$. (Note that a "truely" polynomial algorithm ought to be polynomial in $\log(1/\varepsilon)$.)

The most interesting applications of the Equivalence Principle are in the field of combinatorial optimization. We illustrate this by two results due to Grötschel, Lovász and Schrijver (see 1988). Consider the problem of determining $\alpha(G)$ for a t-perfect graph G. As discussed earlier, this is equivalent to maximizing the linear objective function $\sum_i x_i$ over STAB(G), which could be easily solved by binary search if we could test validity for STAB(G). By the Equivalence Principle, it would suffice to find a polynomial time membership test for STAB(G). In other words, given a vector $x \in \mathbb{R}^V$, is $x \in$ STAB(G)?

At this point we want to use the linear description of STAB(G), which we have determined for t-perfect graphs: STAB(G) is the solution set of inequalities (1)–(2)–(3) above. So all we have to do is to test whether x satisfies these inequalities.

For (1) and (2), this is trivially checked by simply substituting x in them. But for (3) this does not work: there are (typically) exponentially many inequalities in (3), and we do not want to generate them all.

However, there is a way to check in polynomial time whether or not x satisfies all these inequalities. We may assume that we have checked (1) and (2) and that x satisfies these. For every edge $ij \in E$, let $y_{ij} = 1 - x_i - x_j$. Since x satisfies (2), these numbers are non-negative, and we may consider them as "lengths" of the edges. Let C be an odd circuit in G, then

$$\sum_{i \in E(C)} y_{ij} = |C| - \sum_{i \in V(C)} x_i ,$$

and hence x satisfies the constraint of (3) belonging to this circuit C if and only if the "length" of C is at least 1. So x satisfies (3) iff the length of every odd circuit is at least 1.

Now there is a rather simple version of breadth-first-search that finds the shortest odd circuit in any graph with non-negative edge-lengths (the details of this do not belong here). Using this, all we have to do is to check whether this minimum length is less than 1 or not.

The stability number of a perfect graph can also be determined in polynomial time. This algorithm is also based on the Equivalence Principle, but it is substantially more complicated. If we want to copy the above argument, we run into the following difficulty: the membership test for STAB(G) is equivalent to the validity test for the stable set polytope of the complementary graph, and so the equivalence principle reduces the problem of determining a maximum stable set to (esssentially) the same problem for the complementary graph. One has to go up to the $|V|^2$ dimensional space to apply the Equivalence Principle successfully.

This type of application of the Equivalence Principle is not rare; in fact, it is shown by Grötschel, Lovász and Schrijver (see 1988) that most of those combinatorial optimization problems whose polynomial time solvability was known by (often quite involved) ad hoc algorithms, can be solved in polynomial time using a combination of the Equivalence Principle with very elementary algorithmic ideas. There are also several combinatorial optimization problems which can be solved in polynomial time, but for which no polynomial time algorithm avoiding the use of the Equivalence Principle is known (two very important such problems are finding the independence number of a perfect graph and minimizing a submodular setfunction).

Unfortunately, algorithms derived from the Equivalence Principle are polynomial, but very slow (their running time is a polynomial with a very high degree), and therefore practically useless. This is quite natural, considering how general this method is. These results should be interpreted as "existence proofs" for polynomial time solvability of the problem; having established this much, one can try to design problem-specific algorithms which are more efficient. The remarks above show that this is often very difficult.

3. Convex Bodies and Ellipsoids

3.1 The Löwner-John Ellipsoid

It was proved by Löwner that *for each convex body K, there exists a unique ellipsoid E with minimum volume containing it.* John proved that *if we shrink this ellipsoid of a convex body K from its center by a factor of n, we obtain an ellipsoid that is contained in K.* We call this ellipsoid the *Löwner-John ellipsoid* of the body (see (Grötschel, Lovász and Schrijver 1988) for details). Analogues of the theorems above hold if we consider an inscribed ellipsoid with largest volume.

If we restrict ourselves to centrally symmetric convex bodies, then of course the center will be the center of the Löwner-John ellipsoid as well, and in John's theorem it suffices to shrink by a factor of $\sqrt{n}$.

The Löwner-John ellipsoid itself may be difficult to compute. However, an ellipsoid E with somewhat weaker properties can be computed in polynomial time, using a version of the shallow cut ellipsoid method due to Yudin and Nemirovskii (1976). We call an ellipsoid E a *weak Löwner-John ellipsoid* for K, if E contains K and if we shrink E from its center by a factor of $2n^{3/2}$, we obtain an ellipsoid that is contained in K.

Theorem 3.1 *Given a separation oracle for a convex body K, a weak Löwner-John ellipsoid for K can be computed in polynomial time.*

We remark that if K is given in a more explicit manner (e.g., as the solution set of a system of linear inequalities or the convex hull of a set of vertices) then the factor $n^{3/2}$ can be improved to $2n$.

The algorithm proving Theorem 3.1 is based on a natural proof of John's theorem. Let E be the ellipsoid with smallest volume containing K. Applying an affine transformation, we may assume that E is the unit ball about the origin. Let E' denote the ball with radius $1/n$ about the origin, and assume that E' is not contained in K. Let $x \in E' \setminus K$, and let H be a hyperplane separating x from K. H cuts E into two parts, one of which, say E_1, includes K. Now E_1 is smaller or only slightly larger than a half-ball, since $x \notin E_1$. Therefore it is a routine computation in linear algebra to verify that E_1 can be included in an ellipsoid which has smaller volume than E.

To turn this proof into an algorithm, two ideas have to be added. We want to start out with some ellipsoid containing K (say, with the ball with radius R about the origin), and replace it as in the above proof with ellipsoids of smaller and smaller volume until a weak Löwner-John ellipsoid is obtained. In order to guarantee a good running time, we have to reduce the volume by a substantial factor like $1-(1/n^2)$; this will be achieved if the radius of E' is chosen, say, $1/2n$ instead of $1/n$. The other problem is to find the vector $x \in E' \setminus K$. This is easy if K is the solution set of a system of linear inequalities, but hard in general. What we can do is to test the $2n$ intersection points of the axes with the surface of E' for membership in K; if one of them is not in K, then the separation algorithm yields the hyperplane H. If each of them is in K then K contains the smaller ball obtained from E' by shrinking by a factor of $\sqrt{n}$. Thus E is a weak Löwner-John ellipsoid.

Recently some interesting new results concerning the computation of the "true" Löwner-John ellipsoid have been obtained. If K is a convex polytope presented by an explicit list of its vertices, then for any $\varepsilon > 0$ an approximation of the Löwner-John ellipsoid with error at most ε can be found in polynomial time. The "polar" problem of finding an inscribed ellipsoid with (approximately) maximum volume can also be solved in polynomial time, provided the polytope is presented as the solution set of an explicit system of linear inequalities. Such an algorithm was given by Nesterov and Nemirovskii (1989), and improved by Khachiyan and Todd (1990). These results are very presentation-dependent; Khachiyan and Todd conjecture that to compute an approximate Löwner-John ellipsoid for a polytope presented by linear inequalities is NP-hard.

The diameter and width of the weak Löwner-John ellipsoid can be used to estimate the diameter and width of K with relative error at most $n^{3/2}$. This estimate is not as bad as it looks at a first glance; it can be shown (Bárány and Füredi 1986) that if K is given by any of the oracles above, then any estimate on the width of a body K computable in polynomial time is off by a factor which grows as a power of n.

3.2 The Ellipsoid Method

The algorithm sketched above is the heart of various versions of the algorithm called the Ellipsoid Method, developed by Shor (1970) and Yudin and Nemirovkii (1976). The method became widely known when Khachiyan (1979) applied it to obtain the first polynomial time linear programming algorithm. To show the idea, assume that we want to find the optimum of a linear objective function $c \cdot x$ over a convex body $K \subseteq \mathbb{R}^n$ for which we have a separation oracle. Design a black box which does the following: we plug in a vector $x_0 \in \mathbb{R}^n$. If $x_0 \notin K$, it returns a hyperplane separating x from K. If $x_0 \in K$, it returns the hyperplane $c \cdot x = c \cdot x_0$. This is essentially a separation oracle for the set of vectors in K optimizing the linear objective function $c \cdot x$. Even though this convex set is not full-dimensional, an appropriate modification of the argument above gives that we can use this black box as a separation oracle in the ellipsoid method and obtain an "approximately optimal" point in K.

Once we can solve the optimization problem over K, the validity and violation problems are easily settled.

It is more difficult to handle these problems when K is given by a membership test rather than by a separation oracle. It takes a nice but difficult argument by Yudin and Nemirovskii (1976) to settle this case.

3.3 Karmarkar's Method

The ellipsoid method yields a polynomial time algorithm to solve linear programs, but it is too slow in practice to be useful in actual computations. The classical method to solve linear programs, namely Dantzig's simplex method, takes exponential time in the worst case but works very efficiently for the majority of real-life problems. A linear programming algorithm which is both theoretically efficient (polynomial-time) and competitive with the simplex method in practice was given by Karmarkar (1984). Since this method also has a very geometric background, it is worth sketching here.

Assume that we want to maximize a linear objective function $c \cdot x$ over a convex body K. Also assume that we have already found an interior point x_0 and we want to find a sequence of interior points $x_1, x_2, \ldots$ converging to the optimum point. When stepping from x_i to x_{i+1}, it would be natural to move in the direction of c, or, in other words, orthogonal to the hyperplane $c \cdot x = \text{const}$. This leads us, however, to points close to the boundary and (possibly) far from the optimum, and then we have to move in some other direction.

However, it is not really natural to move orthogonal to c; this is not invariant under affine transformations while the hole task is. The idea of many optimization algorithms is to change the metric of the space so that the notion of "orthogonality" should be tailored to the particular body (such are the variable metric methods, but the ellipsoid method can also be viewed this way).

There is a very natural metric inside every convex body, introduced by Hilbert. The distance of any two points $u, v \in \text{int}(K)$ is defined as the logarithm of the cross ratio of u, v, and the two points of intersection of the line through u and v with the boundary of K. (This construction may also be familiar from the Caley-Klein model of hyperbolic geometries.) It is not difficult to see that if we move in the direction of steepest descent with respect to this metric to, say, half way to the boundary, and then repeat this, then we get very close to the optimum very fast. For example, if K is centrally symmetric with respect to x_0 then we approach the optimum on a straight line and the distance from the optimum is halved at each step.

Unfortunately, there is no easy way to compute the Hilbert metric in a convex body, but we can approximate it in the neighborhood of an interior point v by finding a projective transformation which maps K onto a convex body K' so that the unit ball includes K, K includes the ball with a reasonably large radius ϱ about the origin, and v is mapped onto the origin. Then the euclidean metric will approximate the projective metric. Except for the condition on v, this sounds like the ellipsoid method, but we have more freedom because we are allowed to use projective transformations. (The optimization problem itself is not projective invariant, but one can transform it into projective invariant problems easily; for example, the Violation Problem above is projective invariant.)

The crucial part of Karmarkar's method is to construct such a projective transformation very fast in the case of polytopes presented by an explicit system of m linear inequalities. The inner radius ϱ he achieves is $1/m$; this shows that (unlike the ellipsoid method) this method is sensitive to the number of constraints. This is of course only a very rough sketch; many ingenious details must be added to make this procedure work for really large linear programs. For details, extensions, and related algorithms see the survey of Todd (1989).

4. The Volume of a Convex Body

Now we turn to the fundamental problem of determining, or at least estimating, the volume of a convex body. This question has recently brough exciting developments. For a while, a number of negative results were obtained (Section 4.1), which showed that even to compute an estimate in polynomial time with decent relative error is hopeless. But recently Dyer, Frieze and Kannan (1989) designed a *randomized* polynomial time algorithm (i.e., an algorithm making use of a random number generator) which computes an estimate of the volume such that the probability that the relative error is larger than a prescribed $\varepsilon > 0$ is arbitrarily small. This outstanding result uses a number of tools from probability and geometry, and we sketch it in Section 4.2.

4.1 The Difficulty of Computing the Volume

One thing we can do is to find a weak Löwner-John ellipsoid of the body. Since it is easy to follow how an affine transformation modifies the volume (it multiplies by the determinant of the corresponding matrix), we may assume that the body is contained in the unit ball and contains the ball with radius $1/2n^{3/2}$ about the origin. In this case, we have the trivial bounds on the volume:

$$\frac{\pi^{n/2}}{2^n n^{3n/2}\Gamma(1+n/2)} \leq \mathrm{vol}(K) \leq \frac{\pi^{n/2}}{\Gamma(1+n/2)}.$$

So for an arbitrary convex body (given by, say, a separation oracle), we can compute an upper bound on its volume with relative error at most $2^n n^{3n/2}$.

The following surprising result of Bárány and Füredi (1986), improving a somewhat weaker bound given by Elekes (1986), shows that no substantially better estimate can be given on the volume, at least if the convex body is given by one of the four equivalent oracles discussed in Chapter 2.

Theorem 4.1. *Consider any polynomial time algorithm which assigns to every convex body K, given by (say) a membership oracle, an upper bound $w(K)$ on $\mathrm{vol}(K)$. Then there is a constant $c > 0$ such that in every dimension n there exists a body K for which $w(K) > n^{cn}\mathrm{vol}(K)$.*

Let us sketch the proof of this result. It depends on the following geometric lemma:

Lemma 4.2. *There exists a constant $c > 0$ such that the volume of the convex hull of any $p > 0$ points in the unit ball is less than $p \cdot n^{-cn}$.*

To prove Theorem 4.1, first apply the algorithm to the unit ball B. The algorithm runs and asks the membership of a polynomial number p of points $v_1, \dots, v_p$ from the oracle, and computes an upper bound $w(B)$ of the volume. Now apply the algorithm to the convex hull K of $\{e_1, \dots, e_n, v_1, \dots, v_p\} \cap B$. It follows that the algorithm runs exactly as in the previous case: it asks the same questions from the oracle and therefore it gets the same answers. So the algorithm finds the same estimate $w(K) = w(B)$. But since the volume of K is much smaller than the volume of B, this estimate must have a large relative error.

This result does not say anything about computing the volume of convex bodies given in any specific way, say as the solution set of a system of linear inequalities. However, in both cases the exact computation of the volume is NP-hard (Dyer and Frieze 1988, Khachiyan 1988) and even #P-hard (Khachiyan 1989). (This latter fact also follows from the result of Brightwell and Winkler (1990) mentioned in Section 1.3.)

The approximate computation of the volume of explicitly described polyhedra is an open problem.

4.2 Markov Chains, Isoperimetric Inequalities, and Approximating the Volume

In this section we sketch the randomized algorithm of Dyer, Frieze and Kannan (and variants) for estimating the volume of a convex body $K \subseteq \mathbb{R}^n$. About this body, we only need to assume that it is given by a separation oracle.

At this point, many readers may ask: what's wrong with the straightforward Monte-Carlo algorithm? We have already made the assumption that K is included in the unit ball B. Let us generate many random points in B, and count how often we hit K. This gives us an estimate on the ratio of the volumes of K and B.

The problem is that the volume of K may be smaller than the volume of B by an exponential factor (in n). Hence the first exponentially many random points will miss the body K. This method can be applied to estimate the ratio of the volumes of two convex bodies (one including the other) only if this ratio is not too small.

This suggests the first trick: "connect" K and B by a sequence of convex bodies $K = K_0 \subseteq K_1 \subseteq \ldots \subseteq K_m = B$, so that $\text{vol}(K_i)/\text{vol}(K_{i+1}) \geq 1/2$ (and m is polynomial in n). Then these ratios can be estimated by the Monte-Carlo method, and their product gives an estimate on the ratio $\text{vol}(K)/\text{vol}(B)$. Such a sequence is easily constructed: we can take e.g. $K_i = B \cap (1 + \frac{1}{2n})^i K$. Since K contains the ball $(1/(2n^{3/2}))K$, it follows that $K_i = B$ for $i \geq 4n \log n$.

However, estimating $\text{vol}(K_i)/\text{vol}(K_{i+1})$ by the Monte-Carlo method is not so easy; the key question in this algorithm (and in all versions of it) is:

How to generate a random point (with uniform distribution) in a convex body?

The solution by Dyer, Frieze and Kannan is the following. Consider the lattice of vectors whose coordinates are integer multiples of a sufficiently small number δ. Call two lattice points *adjacent* if their distance is exactly δ. Starting from the origin, take a random walk on the lattice points in K. If we are at a lattice point v, select an adjacent lattice point w at random. If $w \in K$, then move to w; else, stay at v. (For technical reasons, we flip a coin before the move and if it falls on head, we stay where we are anyway.) After an appropriate number of steps, we stop; our current position can be considered as a random point in K.

It is easy to see, using the theory of Markov chains, that if w_t denotes the (random) lattice point obtained after t steps, then the distribution of w_t tends to the uniform distribution over the set V of lattice points accessible from the origin along legal walks. (Note that V is essentially $L \cap K$, except possibly for some lattice points near the boundary of K.)

The problem is to find a good bound on the rate of this convergence (on the *mixing rate* of the Markov chain $(w_0, w_1, \ldots)$). By general results on the mixing rate of Markov chains (Sinclair and Jerrum 1989), we know that this depends on the *conductance* of the Markov chain. This (in our case) can be defined as the largest $\Phi > 0$ such that for every $S \subseteq V$, the number of pairs (u, v) of adjacent lattice points with $u \in S$ and $v \in V \setminus S$ is at least $\Phi \cdot (2n) \cdot \min\{|S|, |V \setminus S|\}$. The least t for which the distribution w_t is essentially uniformly distributed is about $1/\Phi^2$.

So the question is to find a lower bound on the conductance of this Markov chain. Let $S \subseteq V$ and let K_1 be the set of points in K nearer to S than to $V \setminus S$. Then we expect that

(a) the volume of K_1 is about $\delta^n|S|$,
(b) the volume of $K \setminus K_1$ is about $\delta^n|V \setminus S|$,
(c) the surface area of the K_1 inside K is about δ^{n-1} times the number of adjacent pairs of lattice points (u, v) with $u \in S$ and $v \in V \setminus S$.

If we accept these approximations (which is a technically quite difficult part of the argument), then the problem reduces to the following isoperimetric inequality:

Theorem 4.3. *Let K be a convex body in $\mathbb{R}^n$ with diameter d. Let F be a surface with $(n-1)$-dimensional measure f, cutting K into two parts with volumes v_1 and v_2. Then*

$$f \geq \frac{1}{d} \min\{v_1, v_2\}.$$

A weaker inequality (but still sufficient to prove the polynomiality of the volume algorithm) was proved by Dyer, Frieze and Kannan. This form is due to Karzanov and Khachiyan (1990) (who use methods from differential geometry) and to Lovász and Simonovits (1990) (who use the so-called Ham-Sandwich Theorem; to justify the remark that algebraic topology also.plays some role, let us mention that the Ham-Sandwich Theorem is derived from the Brouwer Fixed Point Theorem).

Unfortunately, even with these improvements the algorithm is practically useless; its running time grows with the 16th power of n. The main problem is that random walks are slow in getting to the distant part of K. It is easy to come up with other Markov chains of points in K which "jump around" faster (and also have uniform limit distribution). Unfortunately, to prove the corresponding analogues of the isoperimetric inequality in Theorem 4.3 seems to take new methods. It is clear the algorithm of Dyer, Frieze and Kannan, representing a theoretical breakthrough, is not the last word in the area of volume algorithms.

An algorithm to efficiently generate random points in a convex body has many further applications. Numerical integration by Monte-Carlo methods is an obvious one. We have also mentioned in Section 2.2 the problem to test whether a convex body K_1 is included in another convex body K_2. This can be done by generating many random points in K_1 and test whether these are contained in K_2. (To be convinced that at most a fraction of ε of the volume of K_1 is not contained in K_2, we have to generate about $1/\varepsilon$ points. Thus this algorithm is not "truely" polynomial if ε is part of the input.)

References

Bárány, I., Füredi, Z. (1986): Computing the volume is difficult. Proc. of the 18th Annual ACM Symp. on Theory of Computing. Assoc. Comput. Machin., New York, pp. 442–447

Berge, C. (1961): Färbung von Graphen, deren sämtliche bzw. deren ungerade Kreise starr sind. Wiss. Zeitung, Martin Luther Univ. Halle-Wittenberg, 1961, p. 114

Boulala, M., Uhry, P. (1979): Polytope des indépendants dans un graph série-parallèle. Discrete Math. **27**, 225–243

Brightwell, G., Winkler, P. (1990): Counting linear extensions is #P-complete. Bellcore preprint

Chvátal, V. (1975): On certain polytopes associated with graphs. J. Combin. Theory B **18**, 138–154

Dyer, M., Frieze, A. (1988): On the complexity of computing the volume of a polytope. SIAM J. Comp. **17**, 967–974

Dyer, M., Frieze, A., Kannan, R. (1989): A random polynomial time algorithm for approximating the volume of convex bodies. Proc. of the 21st Annual ACM Symposium on Theory of Computing. Assoc. Comput. Machin., New York, pp. 375–381

Edelsbrunner, H. (1987): Algorithms in combinatorial geometry. Springer, Berlin Heidelberg New York

Elekes, G. (1986): A geometric inequality and the complexity of computing volume. Discrete and Computational Geometry **1**, 289–292

Garey, M.R., Johnson, D.S. (1979): Computers and intractability: A guide to the theory of NP-completeness. Freeman, San Francisco

Grötschel, M., Lovász, L., Schrijver, A. (1986): Relaxations of vertex packing. J. Comb. Theory B **40**, 330–343

Grötschel, M., Lovász, L., Schrijver, A. (1988): Geometric algorithms and combinatorial optimization. Springer, Berlin Heidelberg New York

Khachiyan, L.G. (1988): On the complexity of computing the volume of a polytope. Izv. Akad. Nauk SSSR, Engineering Cypernetics **3**, 216–217

Khachiyan, L.G. (1989): The problem of computing the volume of polytopes is NP-hard. Uspekhi Mat. Nauk **44**, 199–200

Khachiyan, L.G., Karzanov, A. (1990): On the conductance of order Markov chains. Rutgers Univ. Tech. Report DCS TR 268

Khachiyan, L.G., Todd, M.J. (1990): On the complexity of estimating the maximal inscribed ellipsoid for a polytope. Cornell Univ. SORIE Tech. Report 893

Lenstra, H.W., Jr. (1983): Integer programming with a fixed number of variables. Oper. Res. **8**, 538–548

Lenstra, A.K., Lenstra, H.W., Jr., Lovász, L. (1982): Factoring polynomials with integral coefficients. Math. Ann. **261**, 515–534

Lovász, L. (1989): Geometry of numbers and integer programming. In: Mathematical programming, recent developments and applications (eds. M. Iri, K. Tanabe). Kluwer Academic Publishers, Dordrecht Boston London, pp. 177–201

Lovász, L., Simonovits, M. (1990): Mixing rate of Markov chains, an isoperimetric inequality, and computing the volume. Proc. 31st Annual Symp. on Found. of Computer Science. IEEE Computer Soc., New York (to appear)

Preparata, F.P., Shamos, M.I. (1985): Computational geometry, an introduction. Springer, New York Berlin Heidelberg

Todd, M.J. (1989): Recent developments and new directions in linear programming. In: Mathematical programming, recent developments and applications (eds. M. Iri, K. Tanabe). Kluwer Academic Publishers, Dordrecht Boston London, pp. 109–157.

Yudin, B.D., Nemirovskii, A.S. (1976): Informational complexity and efficient methods for the solution of convex extremal problems. Ekonomika i Mat. Metody **12**, 357–369 (Russian) [English transl.: Matekon **13**, 25–45]

Intersection Cohomology Methods in Representation Theory*

George Lusztig

Department of Mathematics, Room 2-276
Massachusetts Institute of Technology, Cambridge, MA 02139, USA

1. Introduction

In recent years, the theory of group representations has greatly benefited from a new approach provided by the topology of singular spaces, namely intersection cohomology (IC) theory.

Let G be a connected reductive algebraic group over an algebraically closed field (which will be assumed to be **C** unless otherwise specified); although G itself is clearly a non-singular variety, the study of the representations of G (or its Lie algebra, or its forms over various fields) leads to singular varieties connected with G.

The main theme of this article is that the invariants of IC theory, applied to the singular varieties arising from G appear again and again in a broad range of problems of representation theory: construction of representations, computation of their character, construction of nice bases for representations.

Chronologically, both the earliest [22, 23] and the most recent [45] such problem was that of constructing canonical bases in certain representations of Iwahori-Hecke algebras (resp. in irreducible, finite dimensional representations of G). The relevant singularities were Schubert singularities (resp. quiver singularities); their local IC provided the necessary ingredients.

Certain misterious and apparently intractable quantities in representation theory, like the multiplicities in the Jordan-Hölder series of Verma modules, were recognized to be expressible in terms of certain local IC spaces and this has led to understanding and computing them.

IC methods are needed for the classification of complex irreducible representations of a reductive group over a finite field; furthermore, the character values of such representations appear to be intimately related to certain IC complexes, the character sheaves.

IC methods have been used to give a new construction of the Springer representations of Weyl groups; an elaboration of this idea has been used to construct representations of affine Iwahori-Hecke algebras and hence of p-adic groups.

* Supported in part by National Science Foundation grant DMS 8702842.

Proceedings of the International Congress of Mathematicians, Kyoto, Japan, 1990

Unfortunately, several important topics had to be omitted from this account. Among them are the use of $\mathscr{D}$-modules, the role of the Fourier-Deligne transform and the geometric theory of automorphic forms (see Laumon [26]).

2. *IC* Theory

This theory has been originally introduced by Goresky and MacPherson [15]. (See also MacPherson's survey [48].) Subsequently, it has been developed into a very powerful tool by Beilinson, Bernstein, Deligne and Gabber [4], who blended *IC* theory into Deligne's theory of weights.

Consider an irreducible complex algebraic variety Y. An *IC datum* for Y consists of an open dense smooth subvariety Y_0 of Y together with a local system $\mathscr{L}$ on Y_0. ($\mathscr{L}$ corresponds to a finite dimensional complex representation of the fundamental group of Y_0.)

In general, it is not possible to extend $\mathscr{L}$ to a local system on the whole of Y; even if such extension existed, local Poincaré duality with coefficients in it would not necessarily hold, when Y has singularities. Now *IC* theory provides a canonical extension $\mathscr{L}^\sharp$ of $\mathscr{L}$ from Y_0 to Y, not necessarily as a local system, but as a complex of sheaves (well defined up to quasi-isomorphism) which has constructible cohomology sheaves $\mathscr{H}^i\mathscr{L}^\sharp$, does satisfy local Poincaré duality and is the most economical possible with these properties.

We say that $\mathscr{L}^\sharp$ is the *IC extension* of $\mathscr{L}$ to Y or that $\mathscr{L}^\sharp$ is an *IC complex with support* Y. The stalks $\mathscr{H}^i_x\mathscr{L}^\sharp$ at points $x \in Y$ are the *local IC spaces* of Y with coefficients in $\mathscr{L}$. The *global IC* of Y with coefficients in $\mathscr{L}$ is by definition the cohomology of Y with coefficients in $\mathscr{L}^\sharp$.

These concepts are also well defined in the case of algebraic varieties over an algebraic closure $\bar{F}_q$ of a finite field F_q; one uses l-adic local systems and l-adic (étale) cohomology.

We shall give three examples, which will be amplified in the following sections.

In the first example (the prototype of a Schubert variety singularity, see §3), we take Y to be the variety of all flags of subspaces $V_1 \subset V_2 \subset V_3$ in $\mathbf{C}^4$ ($\dim V_i = i$) such that V_2 has non-zero intersection with a fixed two dimensional subspace $E_2 \subset \mathbf{C}^4$. We take Y_0 to be the open subset of Y defined by the condition $V_2 \neq E_2$ and we take $\mathscr{L} = \mathbf{C}$. The stalks $\mathscr{H}^i_x\mathscr{L}^\sharp$ of the *IC* extension of $\mathscr{L}$ to Y can be described as follows. If $x \in Y_0$ we have $\dim \mathscr{H}^0_x\mathscr{L}^\sharp = 1$ and $\mathscr{H}^i_x\mathscr{L}^\sharp = 0$ for $i \neq 0$. If $x \in Y - Y_0$ we have $\dim \mathscr{H}^0_x\mathscr{L}^\sharp = 1$, $\dim \mathscr{H}^2_x\mathscr{L}^\sharp = 1$ and $\mathscr{H}^i_x\mathscr{L}^\sharp = 0$ for $i \neq 0, 2$.

In the second example (the prototype of a character sheaf, see §14) we take

$$Y = \{\begin{pmatrix} a & b \\ c & d \end{pmatrix} | a, b, c, d \in \mathbf{C}, ad - bc = 1\}$$

and we take Y_0 to be the open set of Y defined by the condition $a + d \neq \pm 2$ (matrices with distinct eigenvalues). Let $\pi : \tilde{Y}_0 \to Y_0$ be the two-fold covering defined by the equation $z^2 = (a+d)^2 - 4$. We define $\mathscr{L}$ to be the local system on Y_0 whose stalk $\mathscr{L}_y$ at $y \in Y_0$ is the one dimensional vector space of functions

$\pi^{-1}(y) \to \mathbf{C}$ with sum of values equal to 0. The stalks $\mathscr{H}^i_x \mathscr{L}^\sharp$ of the IC extension of $\mathscr{L}$ to Y can be described as follows. If $x \in Y_0$ we have $\mathscr{H}^0_x \mathscr{L}^\sharp = \mathscr{L}_x$ and $\mathscr{H}^i_x \mathscr{L}^\sharp = 0$ for $i \neq 0$. If x is $\pm$ the identity matrix, we have $\dim \mathscr{H}^2_x \mathscr{L}^\sharp = 1$ and $\mathscr{H}^i_x \mathscr{L}^\sharp = 0$ for $i \neq 2$. If x has eigenvalues ± 1 but it is not $\pm$ the identity matrix, then $\mathscr{H}^i_x \mathscr{L}^\sharp = 0$ for all i.

In the third example (the prototype of a cuspidal character sheaf, see §§13, 14) we take

$$Y = \{\begin{pmatrix} a & b \\ c & d \end{pmatrix} | a, b, c, d \in \mathbf{C}, ad - bc = 1, a + d = 2\}$$

and we take Y_0 to be the open set defined by the condition $(b, c) \neq (0, 0)$. We define a two-fold covering $\pi : \tilde{Y}_0 \to Y_0$ by

$$\tilde{Y}_0 = \{(z_1, z_2) \in \mathbf{C}^2 | (z_1, z_2) \neq (0, 0)\}, \quad \pi(z_1, z_2) = \begin{pmatrix} z_1 z_2 + 1 & -z_1^2 \\ z_2^2 & -z_1 z_2 + 1 \end{pmatrix}$$

and we associate to it a local system $\mathscr{L}$ on Y_0 as in the second example. The stalks $\mathscr{H}^i_x \mathscr{L}^\sharp$ of the IC extension of $\mathscr{L}$ to Y can be described as follows. If $x \in Y_0$ we have $\mathscr{H}^0_x \mathscr{L}^\sharp = \mathscr{L}_x$ and $\mathscr{H}^i_x \mathscr{L}^\sharp = 0$ for $i \neq 0$. If x is the identity matrix, we have $\mathscr{H}^i_x \mathscr{L}^\sharp = 0$ for all i.

3. Left Cells in Weyl Groups and Schubert Varieties

In this section we describe the first instance when IC methods made their appearance in representation theory.

The theory of primitive ideals in enveloping algebras leads naturally to a remarkable partition of the Weyl group W of G into subsets called *left cells* (see Joseph [18]); the left cells can be used to parametrize primitive ideals. (This generalizes a partition of the symmetric group S_n which was found earlier by combinatorists.) It is known that each left cell carries a natural representation of W with a canonical basis (in 1-1 correspondence with the elements in the left cell). These representations provide a decomposition of the regular representation of W into representations which are close to being irreducible.

Let H be the Iwahori-Hecke algebra over $\mathbf{Z}[v, v^{-1}]$ corresponding to W; here, v is an indeterminate. H has a standard basis $\tilde{T}_w$ (as a $\mathbf{Z}[v, v^{-1}]$-module) indexed by elements $w \in W$. Its multiplication rule is a deformation of the multiplication rule in the group algebra of W:

$$\tilde{T}_w \tilde{T}_{w'} = \tilde{T}_{ww'} \quad \text{if } l(ww') = l(w) + l(w'),$$

$$(\tilde{T}_s + v^{-1})(\tilde{T}_s - v) = 0 \quad \text{if } l(s) = 1;$$

here, $l : W \to \mathbf{N}$ is the length function.

In [22], Kazhdan and the author came across a new basis C_w ($w \in W$) of H which had some very remarkable properties with respect to multiplication. It can be defined as follows.

Let $^- : H \to H$ be the ring homomorphism which takes $\tilde{T}_w$ to $\tilde{T}_{w^{-1}}^{-1}$ for all w and v to v^{-1}. Let $\mathscr{L}$ be the $\mathbf{Z}[v^{-1}]$-submodule of H generated by the standard basis and let $\pi : \mathscr{L} \to \mathscr{L}/v^{-1}\mathscr{L}$ be the canonical projection. It turns out that for each $w \in W$, there is a unique element $C_w \in \mathscr{L}$ such that $\bar{C}_w = C_w$ and $\pi(C_w) = \pi(\tilde{T}_w)$.

This definition is equivalent to (but simpler than) the one given in [22]; unlike the definition in [22], it does not use the partial order on W.

It turns out that

$$C_w = \sum_{y \in W} (-1)^{l(w)-l(y)} v^{l(w)-l(y)} P_{y,w}(v^{-2}) \tilde{T}_y$$

where the $P_{y,w}$ are polynomials with integral coefficients. Moreover this basis is, in principle, computable.

In terms of this basis, we found a new definition of left cells (which was later shown to be the same as the one of Joseph). Our definition had the advantage of being elementary, of carrying representations of H (not only of W) endowed with canonical bases, and of making sense for an arbitrary Coxeter group.

Let $\mathscr{F}$ be the collection of all subsets $K \subset W$ with the following property: the submodule of H spanned by the elements C_w ($w \in K$) is a left ideal. The definition [22] of left cells can be given as follows. Two elements $w, w' \in W$ are in the same left cell if the following condition is satisfied: for any $K \in \mathscr{F}$ we have $w \in K$ if and only if $w' \in K$.

In [23], Kazhdan and the author found a geometric interpretation of the polynomials $P_{y,w}$ above. To explain the result, we need some notation.

Assume that we are given a Borel subgroup B of G, with unipotent radical U_B and a maximal torus T of B. The orbits of B acting by conjugation on the variety X of all Borel subgroups (the *flag manifold*) of G are naturally parametrized by the elements of W. Let $\mathcal{O}_w$ be the orbit corresponding to $w \in W$ and let $\bar{\mathcal{O}}_w$ be its closure in X (a *Schubert variety*). Let $\mathscr{L}$ be the local system $\mathbf{C}$ on $\mathcal{O}_w$ and let $\mathscr{L}^\sharp$ be its IC extension to $\bar{\mathcal{O}}_w$. If $y, w \in W$, we say that $y \le w$ if $\mathcal{O}_y \subset \bar{\mathcal{O}}_w$. We then showed that, for such y, w, we have $P_{y,w}(v^2) = \sum_i n_{y,w,i} v^i$, where $n_{y,w,i} = \dim \mathscr{H}_x^i \mathscr{L}^\sharp$ (notation of §2), for any $x \in \mathcal{O}_y$. (If $y \not\le w$, we have $P_{y,w} = 0$.) Thus, the polynomials $P_{y,w}$ contain encoded all information about the local IC of Schubert varieties.

Apart from their relevance to the theory of primitive ideals, the theory of left cells in W has been crucial in the study [33] of representations of groups over F_q (and also in [35]); in the opposite direction, the results of [33] gave new information on the structure of left cells, in particular they have made possible the explicit determination [37] of all representations of W which are carried by left cells. (For example, if G is a classical group, these representations of W are multiplicity free, with a number of irreducible components equal to a power of 2.)

Another definition of left cells has been found in [36] in terms of leading coefficients of the structure constants $h_{x,y,z} \in \mathbf{Z}[v, v^{-1}]$ ($x, y, z \in W$) of the algebra H with respect to the basis C_w. Namely, for each $z \in W$, there is a well defined integer $a(z) \ge 0$ such that $v^{a(z)} h_{x,y,z} \in \mathbf{Z}[v]$ for all x, y and $v^{a(z)-1} h_{x,y,z} \notin \mathbf{Z}[v]$ for some x, y. Let $\gamma_{x,y,z^{-1}}$ be the constant term of $(-v)^{a(z)} h_{x,y,z^{-1}}$. Quite remarkably, it turns out that the rule $t_x t_y = \sum_{z \in W} \gamma_{x,y,z^{-1}} t_z$ defines a structure of associative ring with 1 on the free abelian group J with basis $(t_x)_{x \in W}$. Then the left cells of W can be characterized as follows: we have $t_x t_y \neq 0$ if and only if x, y^{-1} are in the same left cell.

The ring J has several interesting features. First, there is a explicit homomorphism of algebras $H \to J \otimes \mathbf{Z}[v, v^{-1}]$ (see [36]) which becomes an isomorphism whenever v is specialized to any non-zero complex number which is either 1 or is not a root of 1. Second, the multiplication in J can be (conjecturally, see [39]) described in very simple terms, in terms of convolution of equivariant vector bundles on some finite sets with some finite group action. (The conjecture of [39] gives a precise description of the apparently complicated pattern in which left cells meet right cells, i.e. images of left cells under $w \to w^{-1}$.)

4. Highest Weight Modules

Let $\mathbf{g}, \mathbf{b}, \mathbf{n}, \mathbf{t}$ be the Lie algebras of G, B, U_B, T. In [22], Kazhdan and the author formulated a conjecture relating the polynomials $P_{y,w}$ (see §3) to multiplicities in the Verma modules of $\mathbf{g}$. More precisely, assuming that λ is a linear form on $\mathbf{t}$, we denote by M_λ the Verma module of $\mathbf{g}$ with highest weight λ and by L_λ its unique irreducible quotient. Let $\varrho : \mathbf{t} \to \mathbf{C}$ be half the sum of the positive roots. Our conjecture was that for all $w \in W$ we have

$$L_{-w(\varrho)-\varrho} = \sum_{y \leq w} (-1)^{l(w)-l(y)} P_{y,w}(1) M_{-y(\varrho)-\varrho} \tag{a}$$

in the appropriate Grothendieck group. By an inversion formula for the $P_{y,w}$ proved in [22], one sees that (a) admits the following equivalent formulation

$$M_{y(\varrho)-\varrho} = \sum_{w \geq y} P_{y,w}(1) L_{w(\varrho)-\varrho} \,. \tag{b}$$

In view of [23], this relates multiplicities in Verma modules to local IC of Schubert varieties. Our conjecture has been proved by Beilinson, Bernstein [2] and by Brylinski, Kashiwara [8] using the theory of $\mathscr{D}$-modules.

Not only the value of $P_{y,w}$ at 1 but also the coefficients of the various powers of v^2 in it have representation theoretic meaning (actually two different ones): they can be interpreted as dimensions of higher Ext groups between a Verma module and a simple module (Vogan [59]) or as multiplicities of a simple module in the successive quotients of the Jantzen filtration of a Verma module (Beilinson, Bernstein [3]).

5. Symmetric Spaces

Let $\theta : G \to G$ be an involution and let K be the identity component of the fixed point set G^θ. Let Y_0 be a K-orbit on X and let $\mathscr{L}$ be a one dimensional K-equivariant local system on Y_0.

By [2, 59], the local IC of the closure $\bar{Y}_0$ with coefficients in $\mathscr{L}$ enters in an essential way in the representation theory of the real reductive group attached to G, θ.

An algorithm for computing this local IC (generalizing that in [22]) has been proposed by Vogan and has been established in [47].

6. Groups over F_q

In this section we take G,X as before, but we assume that the ground field is $\bar{F}_q$ rather than $\mathbf{C}$ and that we are given an F_q-rational structure on G with corresponding Frobenius map $F : G \to G$. This induces a Frobenius map $F : X \to X$. Following [12], we consider for any $w \in W$ the set X_w of all Borel subgroups B' of G such that $B', F(B')$ are in relative position w. The finite group $G(F_q)$ acts on X_w by conjugation; this induces an action on the l-adic cohomology groups of X_w with compact support. Although one has from [12] good information about the alternating sum of these cohomology groups (as a virtual representation) one knows very little about the representations on the individual cohomology groups, due to the fact that X_w is non-proper. The situation is much better if we replace X_w by its closure $\bar{X}_w$ and consider the natural representations of $G(F_q)$ in the global IC groups of $\bar{X}_w$ with constant coefficients. Although $\bar{X}_w$ is singular in general, it is locally isomorphic to a Schubert variety hence its local IC is as in §3. Hence we still have good information on the alternating sum of these IC spaces (as a virtual representation); but this time we can also get good information on the individual IC spaces using especially the Weil conjectures in IC (see [4]). We can in this way compute explicitly [33] the way in which individual IC spaces decompose in irreducible representations and at the same time classify the irreducible representations occuring there (these are the *unipotent representations* of $G(F_q)$).

This method extends to non-unipotent representations (see [33]).

7. Affine Schubert Varieties

In this section we take G to be simply connected, almost simple.

Let $\check{Q}$ be the the lattice of coroots of G and let $\tilde{W}$ be the the semidirect product of W with $\check{Q}$; thus, $\tilde{W}$ is the *affine Weyl group*.

Let $G((\varepsilon))$ (resp. $G[[\varepsilon]]$) be the group of points of G over $\mathbf{C}((\varepsilon))$ (resp. $\mathbf{C}[[\varepsilon]]$) where ε is an indeterminate. The inverse image of B under the canonical homomorphism $G[[\varepsilon]] \to G$ induced by $\varepsilon \to 0$ is denoted by I. Let $\tilde{X}$ be the set of all subgroups of $G((\varepsilon))$ which are conjugate to I. It is called the *affine flag manifold*; a closely related space has been first studied in Bott's work [7]. Now $G((\varepsilon))$ acts transitively on $\tilde{X}$ and the stabilizer of I is known to be I itself. The restriction of this action to I has countably many orbits on $\tilde{X}$ which may be naturally put in 1-1 correspondence (as in Iwahori and Matsumoto [17]) with the elements of $\tilde{W}$. Let $\mathcal{O}_w$ be the I-orbit corresponding to $w \in \tilde{W}$. Now $\mathcal{O}_w$ may be naturally regarded as an affine space over $\mathbf{C}$ of dimension $l(w)$ where $l : \tilde{W} \to \mathbf{N}$ is the length function (see [17]).

One can regard naturally $\tilde{X}$ as an increasing union of projective algebraic varieties over $\mathbf{C}$, each of the varieties in this union being I-stable. (See [23, 31, 58].) Then the closure $\bar{\mathcal{O}}_w$ of $\mathcal{O}_w$ in $\tilde{X}$ is a well defined projective variety over $\mathbf{C}$, called an *affine Schubert variety*. We can then define the relation $y \leq w$ on $\tilde{W}$, as in §3. Now the polynomials $P_{y,w}$ in §3 are well defined for $y, w \in \tilde{W}$ (see [22]). They again record the dimensions of the local IC of $\bar{\mathcal{O}}_w$ at a point in $\mathcal{O}_y \subset \bar{\mathcal{O}}_w$, for $y \leq w$, as in §3 (see [23]).

(The elementary definition of the polynomials $P_{y,w}$ given in [22] makes sense for arbitrary Coxeter groups; it is expected [23] that they always have ≥ 0 coefficients, although it is not clear what the geometrical interpretation of these coefficients should be in the non-crystallographic case. This positivity property has been verified by Alvis, for the finite Coxeter group of type H_4, using a computer.)

8. Highest Weight Modules for Affine Lie Algebras

In this section we assume that G is almost simple, simply connected and of simply laced type. We shall recall some basic facts on affine Lie algebras. Consider the Lie algebras obtained by extending scalars: $\mathbf{g}[\varepsilon, \varepsilon^{-1}] = \mathbf{g} \otimes \mathbf{C}[\varepsilon, \varepsilon^{-1}]$, $\mathbf{g}[\varepsilon] = \mathbf{g} \otimes \mathbf{C}[\varepsilon]$. We regard $\mathbf{g}$ as a subalgebra of $\mathbf{g}[\varepsilon]$ by $x \to x \otimes 1$. It is known that there is a non-trivial central extension $\pi : \tilde{\mathbf{g}} \to \mathbf{g}[\varepsilon, \varepsilon^{-1}]$ of Lie algebras over $\mathbf{C}$ with one dimensional kernel D; moreover, it is unique up to isomorphism. The Lie algebra $\tilde{\mathbf{g}}$ is said to be an *affine Lie algebra*. Let $\tilde{U}$ be its enveloping algebra.

Let $\tau \in \mathbf{t}$ be a coroot and let $\tau', \tau'' \in \tilde{\mathbf{g}}$ be such that $\pi(\tau') = \varepsilon\tau$, $\pi(\tau'') = \varepsilon^{-1}\tau$. Then $[\tau', \tau'']$ is a non-zero element of D, independent of the choice of τ, τ', τ''; we denote it c.

Let $k \in \mathbf{C}$. A $\tilde{U}$-module is said to have *central charge* k if c acts on it as k times identity.

Let $\tilde{\mathbf{p}}$ be the inverse image of $\mathbf{g}[\varepsilon]$ under π. It is known that $\pi : \tilde{\mathbf{p}} \to \mathbf{g}[\varepsilon]$ has a unique cross-section $\mathbf{g}[\varepsilon] \to \tilde{\mathbf{p}}$ which is a homomorphism of Lie algebras (over $\mathbf{C}$); using this we may identify $\tilde{\mathbf{p}} = D \oplus \mathbf{g}[\varepsilon]$ as Lie algebras.

Let $\mathbf{i}$ (resp. $\mathbf{i}_0$) be the inverse image of $\mathbf{b}$ (resp. of $\mathbf{n}$) under the canonical homomorphism $\mathbf{g}[\varepsilon] \to \mathbf{g}$ induced by $\varepsilon \to 0$. We have a direct sum decomposition $\mathbf{i} = \mathbf{t} \oplus \mathbf{i}_0$ as vector spaces. Let $\tilde{\mathbf{t}} = D \oplus \mathbf{t}$, $\tilde{\mathbf{i}} = D \oplus \mathbf{i} = \tilde{\mathbf{t}} \oplus \mathbf{i}_0$ and let $\tilde{\lambda} : \tilde{\mathbf{t}} \to \mathbf{C}$ be a linear form. Consider the $\mathbf{C}$-linear map $\tilde{\mathbf{i}} \to \mathbf{C}$ given by $\tilde{\lambda}$ on $\tilde{\mathbf{t}}$ and by 0 on $\mathbf{i}_0$. This is a Lie algebra homomorphism hence it extends to a homomorphism of the enveloping algebra $\tilde{U}_1$ of $\tilde{\mathbf{i}}$ to $\mathbf{C}$. We induce this to a $\tilde{U}$-module; the resulting $\tilde{U}$-module $M_{\tilde{\lambda}} = \tilde{U} \otimes_{\tilde{U}_1} \mathbf{C}$ is called a Verma module; it has central charge $\tilde{\lambda}(c)$.

Let h be the Coxeter number of G. The existence of Segal-Sugawara operators implies that $M_{\tilde{\lambda}}$ has a unique simple quotient module (denoted $L_{\tilde{\lambda}}$) provided that $\tilde{\lambda}(c)$ is not equal to $-h$.

Let $\tilde{\varrho} : \tilde{\mathbf{t}} \to \mathbf{C}$ be the linear form which is equal to ϱ on $\mathbf{t}$ (see §4) and takes the value h on c. Let Q be the lattice of roots of $\mathbf{g}$ (regarded as a group of linear forms $\mathbf{t} \to \mathbf{C}$). Let $\tilde{W}$ be the semidirect product of W and Q (the affine Weyl group); this may be identified with the affine Weyl group in §7 since we are in the simply laced case. We have an action of $\tilde{W}$ on the space of linear forms on $\tilde{\mathbf{t}}$ as follows: an element $\xi \in Q$ acts by $\tilde{\lambda} \to \tilde{\lambda} + \tilde{\lambda}(c)\xi$; the action of W is the contragredient of the action on $\tilde{\mathbf{t}}$ which is the usual one on $\mathbf{t}$ and the identity on D.

The statements (a) and (b) below (which are entirely analogous to §4(a),(b)) have been recently proved by Casian [11] and by Casian [10], Kashiwara[19]:

$$L_{-w(\tilde{\varrho})-\tilde{\varrho}} = \sum_{y \leq w} (-1)^{l(w)-l(y)} P_{y,w}(1) M_{-y(\tilde{\varrho})-\tilde{\varrho}} \tag{a}$$

$$M_{y(\tilde{\varrho})-\tilde{\varrho}} = \sum_{w \geq y} P_{y,w}(1) L_{w(\tilde{\varrho})-\tilde{\varrho}} \,. \tag{b}$$

Here $w, y \in \tilde{W}$; the sums are in the appropriate Grothendieck groups (but the sum in (b) is infinite) and $P_{y,w}$ are the polynomials in §7 describing the local IC of affine Schubert varieties. Note that $L_{-w(\tilde{\varrho})-\tilde{\varrho}}$ and $L_{w(\tilde{\varrho})-\tilde{\varrho}}$ are well defined since $-w(\tilde{\varrho}) - \tilde{\varrho}$ and $w(\tilde{\varrho}) - \tilde{\varrho}$ take at c the values $-2h$ and 0, which are $\neq -h$. While the identities §4(a),(b) are equivalent to each other, the identities (a),(b) above are quite different: (a) is typical for integral highest weights with central charge $< -h$, while (b) is typical for integral highest weights with central charge $> -h$. Note that in [19] the geometry used is not the one in §7, but that of "affine Schubert varieties of finite codimension" (see [23, 20]). Note also that, in [10, 19], a formula like (b) is proved for any symmetrizable Kac-Moody Lie algebra.

9. Modular Representations

Assume that G is as in §8. Let $\tilde{W}^0$ be the set of all elements in the affine Weyl group $\tilde{W}$ which have maximal length in their left W-coset.

Let p be a (not too small) prime number. Let $\mathscr{C}$ be the category of finite dimensional rational representations (over $\bar{F}_p$) of the algebraic group over $\bar{F}_p$ of the same type as G.

The simple objects of $\mathscr{C}$ are indexed by the set of dominant integral weights $\mathbf{t} \to \mathbf{C}$ but their structure is unknown. They appear as the unique simple quotients of certain "Weyl modules" whose weight structure is well understood (it is given by Weyl's character formula). By Steinberg's tensor product theorem, the structure of simple objects of $\mathscr{C}$ would be known in general if it were known for simple modules with *restricted* highest weight (i.e. with coordinates in $[0, p-1]$). In [29] a conjecture was formulated expressing the character of certain simple modules in $\mathscr{C}$ (including those with restricted highest weight) as linear combinations of Weyl modules with explicit coefficients involving the values at 1 of the polynomials $P_{y,w}(y, w \in \tilde{W}^0)$ as in §7. This conjecture has been tested in a variety of ways and seems to be extremely plausible. However, the connection between local IC of affine Schubert varieties and modular representations is quite misterious. It seems likely that to establish this connection one should use two intermediate steps, which we now explain.

Besides $\mathscr{C}$ we consider two other categories $\mathscr{C}', \mathscr{C}''$ of representations.

$\mathscr{C}'$ has as objects the finite dimensional representations of the quantized enveloping algebra corresponding to G with parameter $\exp(2\pi i/p)$ (in the sense of [43]) in which the p-th powers of the canonical generators K_i (see [43]) act as identity.

$\mathscr{C}''$ has as objects the $\tilde{U}$-modules with central charge $-p-h$ (see §8), of finite length, with all composition factors of form $L_{\tilde{\lambda}}$ (as in §8) where the restriction of $\tilde{\lambda}$ to $\mathbf{t}$ is a dominant integral weight.

In both these categories, the simple objects are naturally indexed by the same set: the set of dominant integral weights $\mathbf{t} \to \mathbf{C}$. The simple modules appear again as the unique simple quotients of certain "Weyl modules" (also contained in our category). For $\mathscr{C}'$, the Weyl modules have the same weight structure as the corresponding simple module of G (over $\mathbf{C}$) hence are described by Weyl's character formula; for $\mathscr{C}''$ a Weyl module is defined by inducing from the enveloping algebra of $\tilde{\mathbf{p}}$ (see §8) to $\tilde{U}$ the representation of $\tilde{\mathbf{p}}$ on which c acts as $-p-h$ times identity and $\mathbf{g}[\varepsilon]$ acts through a finite dimensional irreducible representation of its quotient $\mathbf{g}$.

In [44] it is conjectured that

(a) *the categories $\mathscr{C}'$, $\mathscr{C}''$ are equivalent; the equivalence should preserve the indexing of simple modules and takes Weyl modules to Weyl modules.*

(For this, p need not be a prime.)

Although $\mathscr{C}, \mathscr{C}'$ are not equivalent, one can expect a very close connection between them; in particular, one can expect [43] that

(b) *two simple modules in $\mathscr{C}, \mathscr{C}'$ which have the same restricted highest weight should have the same weight structure.*

Now, in $\mathscr{C}''$, the simple modules can be expressed in terms of Weyl modules with coefficients involving the values at 1 of the polynomials $P_{y,w} (y, w \in \tilde{W}^0)$ as in §7. (This can be deduced by a "translation principle" from §8(a), see [25].) In view of (a), (b), the same should hold for $\mathscr{C}'$ and, in the restricted case, for $\mathscr{C}$. In this way, our conjecture on modular representations would follow from (a) and (b).

In the non-simply laced case one should use twisted affine Lie algebras.

10. Affine Schubert Varieties and Weight Multiplicities

A number of features of the theory of finite dimensional representations of G can be recovered in a purely geometric way from the IC of affine Schubert varieties. We preserve the assumptions of §7, 8. If $\lambda : \mathbf{t} \to \mathbf{C}$ is a dominant weight contained in Q, we define $n_\lambda \in \tilde{W}$ to be the unique element of maximal length in the $W-W$ double coset of $\tilde{W}$ containing λ. Consider the finite dimensional representation L_λ of G with highest weight λ and consider another dominant weight $\lambda' \in Q$. The following result is proved in [31].

In the weight decomposition of L_λ, the weight λ' appears with multiplicity equal to $P_{n_{\lambda'}, n_\lambda}(1)$ if $n_{\lambda'} \leq n_\lambda$ and zero, otherwise.

Taking the polynomial itself instead of its value at 1 we get a q-analog of weight multiplicities. An interpretation of this in representation theoretic terms has been found by R. K. Brylinski [9] (modulo some restrictions which were removed by Ginsburg [16]).

11. Periodic Affine Schubert Varieties

We preserve the setup of §7. We want to discuss (in a somewhat non-rigurous way) some variants of affine Schubert varieties. We need some further notation.

Let $B((\varepsilon))$ (resp. $U_B((\varepsilon))$) be the group of $\mathbf{C}((\varepsilon))$-points of B (resp. U_B). Let $T[[\varepsilon]]$ be the group of $\mathbf{C}[[\varepsilon]]$-points of T. Let $R = T[[\varepsilon]]U_B((\varepsilon))$ be the "identity component" of $B((\varepsilon))$, so that $B((\varepsilon))/R \cong \check{Q}$.

Now I acts by left translation on $G((\varepsilon))/B((\varepsilon))$ with finitely many orbits, which may be put naturally in 1-1 correspondence with the elements of W.

We have a "principal fibration" $\pi : G((\varepsilon))/R \to G((\varepsilon))/B((\varepsilon))$ with group $\check{Q}$ and any I-orbit on $G((\varepsilon))/R$ (for the left translation action) is mapped by π bijectively onto an I-orbit on $G((\varepsilon))/B((\varepsilon))$. Thus, $\check{Q}$ acts freely on the set of I-orbits on $G((\varepsilon))/R$ with quotient space W. More precisely, we see that the set of I-orbits on $G((\varepsilon))/R$ may be put in 1-1 correspondence with $\tilde{W}$; however, this is canonical only up to composition with a translation in $\check{Q}$.

Let F be the fixed point set of T acting on $G((\varepsilon))/R$ by left translation. Each I-orbit $\mathcal{O}$ on $G((\varepsilon))/R$ contains a unique point $x_{\mathcal{O}} \in F$.

The orbits of I on $G((\varepsilon))/R$ are seen to be both of infinite dimension and of infinite codimension. The "closure" $\bar{\mathcal{O}}$ of such an orbit $\mathcal{O}$ can be defined as follows.

Let $G[\varepsilon^{-1}]$ be the group of points of G over $\mathbf{C}[\varepsilon^{-1}]$; let B^- be the Borel subgroup of G such that $B \cap B^- = T$ and let I^- be the inverse image of B^- under the canonical homomorphism $G[\varepsilon^{-1}] \to G$ induced by $\varepsilon^{-1} \to 0$. By definition, $\bar{\mathcal{O}}$ is the union of all I-orbits $\mathcal{O}'$ such that $\mathcal{O}$ has non-empty intersection with the the I^--orbit $I^-x_{\mathcal{O}'}$ of $x_{\mathcal{O}'}$. For such $\mathcal{O}'$, the intersection $I^-x_{\mathcal{O}'} \cap \bar{\mathcal{O}}$ is a (finite dimensional) irreducible variety and $I^-x_{\mathcal{O}'} \cap \mathcal{O}$ is an open dense smooth subvariety of it.

Then $\bar{\mathcal{O}}$ is a union of countably many I-orbits of larger and larger (finite) codimension; the orbits and their closure relation form a periodic pattern, with $\check{Q}$ as the group of periods. The orbit closures $\bar{\mathcal{O}}$ may be called *periodic affine Schubert varieties*.

We can define the local IC of the closure $\bar{\mathcal{O}}$ of one I-orbit along another I-orbit $\mathcal{O}' \subset \bar{\mathcal{O}}$ to be the local IC of $I^-x_{\mathcal{O}'} \cap \bar{\mathcal{O}}$ at $x_{\mathcal{O}'}$ with respect to the local system $\mathbf{C}$ on $I^-x_{\mathcal{O}'} \cap \mathcal{O}$.

I was led to consider these periodic affine Schubert varieties about ten years ago while trying to find support for the conjecture on modular representations mentioned in §9. This has led to a periodicity property of the polynomials $P_{y,w}$, for the affine Weyl group, with $y, w \in \tilde{W}^0$ as in §9, which I could prove directly [28]. The periodicity property means that these polynomials (in the stable range) may be regarded as being indexed by a pair of I-orbits on $G((\varepsilon))/R$; it turns out that they can be interpreted geometrically as the polynomials which record the local IC spaces of the periodic affine Schubert varieties.

The previous discussion suggests that the singularities of periodic affine Schubert varieties might be isomorphic to singularities of (ordinary) affine Schubert varieties. This is false; there is equality only for local IC spaces, which are not sensitive to torsion phenomena. In fact, periodic affine Schubert varieties are less singular than the corresponding ordinary ones; for example, if $G = SL_2$, the periodic affine Schubert varieties are non-singular, while the ordinary ones are rational homology manifolds with singularities.

The periodic affine Schubert varieties have been considered independently by B. Feigin and E. Frenkel in their work [14] on Wakimoto modules. It would be very interesting to find a connection between their representation theoretic considerations and the local IC spaces considered above.

12. Local IC of Closures of Unipotent Classes

Let $\mathscr{U}$ be the variety of unipotent elements of G. We denote by $\mathscr{I}$ the set of all pairs $(C, \mathscr{L})$ where C is a unipotent class in G and $\mathscr{L}$ is an irreducible G-equivariant local system on C defined up to isomorphism. (Thus, $\mathscr{L}$ corresponds to an irreducible representation of the group of components of the centralizer of an element of C; this group is naturally a quotient of the fundamental group of C.) Let $(C, \mathscr{L}) \in \mathscr{I}$. Then the cohomology sheaf $\mathscr{H}^i\mathscr{L}^\sharp$ of the IC complex on the closure $\bar{C}$ of C can be restricted to a unipotent class $C' \subset \bar{C}$ and is a G-equivariant local system there; a given irreducible G-equivariant local system $\mathscr{L}'$ on C' appears with a certain multiplicity, say $m_{C',\mathscr{L}';C,\mathscr{L};i}$ in this restriction. The polynomials

$$\Pi_{C',\mathscr{L}';C,\mathscr{L}} = \sum_i m_{C',\mathscr{L}';C,\mathscr{L};i} v^i \tag{a}$$

describe the local IC of $\bar{C}$ with coefficients in $\mathscr{L}$. We want to describe a natural partition of $\mathscr{I}$ into *blocks*. We define a relation $\leq$ on $\mathscr{I}$ as follows: $(C', \mathscr{L}') \leq (C, \mathscr{L})$ means that $C' \subset \bar{C}$ and $\Pi_{C',\mathscr{L}';C,\mathscr{L}} \neq 0$. A pair $(C, \mathscr{L}) \in \mathscr{I}$ is said to be *minimal* if $(C', \mathscr{L}') \leq (C, \mathscr{L})$ implies $(C', \mathscr{L}') = (C, \mathscr{L})$. For a minimal pair $(C, \mathscr{L})$, we define the block of $(C, \mathscr{L})$ to be the set of all $(C'', \mathscr{L}'') \in \mathscr{I}$ such that $(C, \mathscr{L}) \leq (C'', \mathscr{L}'')$. This block contains a unique minimal pair: $(C, \mathscr{L})$. Any element of $\mathscr{I}$ is contained in a unique block, i.e. the blocks form a partition of $\mathscr{I}$.

Let us fix a minimal pair $(C, \mathscr{L})$ and choose $u \in C$. Let $\mathscr{T}$ be a maximal torus of the centralizer of u in G, let L be the centralizer of $\mathscr{T}$ in G and let $\mathscr{W}$ be the normalizer of $\mathscr{T}$ modulo L; one can identify naturally $\mathscr{W}$ with the Weyl group of a maximal reductive connected subgroup of the centralizer of u.

(b) *The pairs $(C', \mathscr{L}')$ in the block of $(C, \mathscr{L})$ are in natural* 1-1 *correspondence with the irreducible representations of $\mathscr{W}$.*

Originally, a statement of this kind has been established by Springer [57] in the case where $C = \{1\}, \mathscr{L} = \mathbf{C}$; in that case, we have $\mathscr{W} = W$. Springer's construction has been reformulated in IC terms in [30] (see also [6]) and, in [34], the "generalized Springer correspondence" (b) has been established.

The construction of [30, 34] is based on the following idea. We can start with a very simple minded action of $\mathscr{W}$ on a local system on an open dense smooth subset of some irreducible variety (the action being the identity on the base). By functoriality of the IC complex, this extends automatically to an action of $\mathscr{W}$ on the corresponding local IC at any point of our variety and this provides many interesting representations of $\mathscr{W}$.

The Springer correspondence (resp. the generalized one) has been explicitly described in each case, see [1, 51, 52, 55] (resp. [34, 46, 56]).

The polynomials (a) have been described in all cases in the form of tables or algorithms. In the first case which has been considered, type A (see [30]), the computation of these polynomials was based on a connection between unipotent classes and affine Schubert varieties. The case of the block containing $(\{1\}, \mathbf{C})$ for types $\neq A$, was done in [5, 53, 54], using results of [6, 21]. The general case is considered in [35] in the framework of the theory of character sheaves.

These results (in l-adic version) remain valid in characteristic p, at least if p is a good prime for G. In this form they are of direct significance for the representation theory of reductive groups over F_q; indeed, the polynomials (a) for the block of $(\{1\}, \bar{\mathbf{Q}}_l)$ are very closely related with Green functions (character values of generic representations at unipotent elements), while in the case of the other blocks, they are closely related with character values of more degenerate representations at unipotent elements. (See [42].)

13. Cuspidal Pairs

A pair $(C, \mathscr{L})$ is said to be *cuspidal* (cf. [34, 35]) if for any proper parabolic subgroup P of G with unipotent radical U and any unipotent element $g \in P$, the cohomology with compact support of $gU \cap C$ with coefficients in the restriction of $\mathscr{L}$ is zero. This implies that $(C, \mathscr{L})$ forms a block by itself; conversely, if a block of $\mathscr{I}$ consists of a single element, then that element is cuspidal. The cuspidal pairs are quite rare. For example, if G is of type E_8, there is a single cuspidal pair; its C is the unique unipotent class whose fundamental group is the symmetric group S_5 and $\mathscr{L}$ corresponds to the sign representations of S_5. If $G = Sp_{2n}$, then G has exactly one cuspidal pair if n is a triangular number and none, otherwise.

14. Character Sheaves

It is possible to immitate very closely the representation theory of a reductive group over F_q, using certain IC complexes (called *character sheaves*) on G or subvarieties of G as substitutes for the irreducible representations. One obtains in this way a purely geometric character theory which makes sense over any algebraically closed field (in particular, over $\mathbf{C}$).

We refer to [34, 35] for the definition of character sheaves. Here we give two (extreme) examples. Let Y_0 be the set of regular semisimple elements in G. The fundamental group of Y_0 has as a canonical quotient the semidirect product of W with the lattice of coweights of G; consider an irreducible representation of the fundamental group which factors through a finite quotient of this semidirect product and let $\mathscr{L}$ be the corresponding local system on Y_0. The IC complex on G corresponding to $(Y_0, \mathscr{L})$ is an example of a character sheaf. If $G = GL_n$ this gives all character sheaves; for other types, there may be in addition character sheaves supported by a proper closed subset of G; in particular, if G is semisimple, the IC complex on $\bar{C}$ corresponding to a cuspidal pair $(C, \mathscr{L})$ (see §13) is a character sheaf (said to be a cuspidal character sheaf). The existence of these character sheaves with thin support is the main reason why representation theory for groups other than GL_n is so much more complicated than for GL_n itself.

The character sheaves have been classified in [34, 35] in two different ways: in terms of their support and also in terms of objects connected with the dual group. This last classification is extremely similar to that of the irreducible representations of a reductive group over a finite field [33]. In both cases the essential ingredients are: a semisimple element in the dual group, a "special" unipotent element in its centralizer, and a certain (small) finite group.

There are at least two, quite different, applications of character sheaves to representation theory. The first one is to characters of groups over a finite field (see §15). The second one is to the construction of representations of p-adic groups (see §§16, 17).

15. Character Sheaves and Frobenius

In this section we assume that G is as in §6.

The theory of character sheaves remains valid (in the l-adic version) for G, at least in good characteristic (and probably in bad characteristic as well). Consider a character sheaf of A of G such that A and its inverse image F^*A under the Frobenius map $F : G \to G$ have the same support and are isomorphic (as objects in the derived category of that support). Choose a specific isomorphism and, for x a fixed point of F in the support, take the alternating sum of traces of that isomorphism on the stalks of the cohomology sheaves of A at x. We obtain a class function $\chi_A : G^F \to \bar{\mathbf{Q}}_l$ on the finite group $G^F = G(F_q)$ (for $x \in G^F$, not in the support of A, we set $\chi_A(x) = 0$). It is independent, up to a non-zero factor, of the choice of isomorphism. One of the main results of [35] is that, when A runs over the character sheaves as above, the functions χ_A (suitably normalized) have values in the cyclotomic integers and they form an orthonormal basis of the space of all class functions on G^F. Thus, this space has two natural bases; one, of geometric origin, is provided by the functions χ_A; the other one is formed by the characters of the irreducible representations of G^F. The elements in the first basis are essentially computable, hence the character table of G^F would be known if the transition matrix between these two bases would be determined. For GL_n, this transition matrix is the identity, but in general it is not so. However, there is considerable evidence that this transition matrix is quite simple, almost diagonal; more precisely this matrix should consist of small diagonal blocks, and each such diagonal block should be the matrix of a *non-abelian Fourier transform* (as in [33, 39]) over a small finite group. (This has been verified in a number of cases [38].)

16. Some Algebras Attached to Blocks

We fix $r_0 \in \mathbf{C}^*$; let

$$\mathbf{V} = \{(N,\sigma) \in \mathbf{g} \times \mathbf{g} | N \text{ nilpotent, } \sigma \text{ semisimple, } [\sigma, N] = 2r_0 N\}.$$

We have a natural action of G on $\mathbf{V}$ by conjugation on both factors. Let $\tilde{\mathscr{I}}$ be the set of pairs $(\mathscr{O},\mathscr{F})$ where $\mathscr{O}$ is a G-orbit on $\mathbf{V}$ and $\mathscr{F}$ is an irreducible G-equivariant local system on the variety $\mathscr{O}$. We consider some $(\mathscr{O},\mathscr{F}) \in \tilde{\mathscr{I}}$. The first projection defines a map $\mathscr{O} \to \mathbf{c}$ for a well defined nilpotent G-orbit $\mathbf{c}$ on $\mathbf{g}$. We identify $\mathbf{c}$ with a unipotent class C in G using the exponential map; we obtain a map $\mathscr{O} \to C$. We can find a G-equivariant local system $\mathscr{L}$ on C whose inverse image under the last map contains $\mathscr{F}$ as a direct summand. Let $\mathscr{B}$ be the block of $\mathscr{I}$ containing $(C,\mathscr{L})$. One can show that $\mathscr{B}$ is independent of the choice of $\mathscr{L}$; it depends only on $(\mathscr{O},\mathscr{F})$. We thus have a map $(\mathscr{O},\mathscr{F}) \to \mathscr{B}$ from $\tilde{\mathscr{I}}$ to the set of blocks of $\mathscr{I}$; its fibres define a partition of $\tilde{\mathscr{I}}$ into subsets in 1-1 correspondence with the blocks of $\mathscr{I}$. These subsets are called the *blocks of* $\tilde{\mathscr{I}}$.

It turns out that the elements of $\tilde{\mathscr{I}}$ in a fixed block (corresponding, say, to $\mathscr{B}$) index in a natural way the isomorphism classes of simple modules over a certain algebra $\mathbf{H}$.

To define this algebra, we need some notation. Let $(C,\mathscr{L})$ be the unique minimal pair in $\mathscr{B}$; choose $u \in C$ and associate to it $\mathscr{T}, L, \mathscr{W}$ as in §11. We choose a parabolic subgroup P of G having L as Levi subgroup; let U_P be its unipotent radical. Let $P_1,\dots,P_m$ be the parabolic subgroups which contain strictly P and are minimal with this property. Let L_i be the Levi subgroup of P_i which contains L. Let $\mathbf{l}, \mathbf{l}_i, \mathbf{n}'$ be the Lie algebras of L, L_i, U_P. Let $\mathbf{h}, \mathbf{h}_i$ be the centres of $\mathbf{l}, \mathbf{l}_i$. For any linear form $\alpha : \mathbf{h} \to \mathbf{C}$ we set $\mathbf{g}_\alpha = \{x \in \mathbf{g} | [y,x] = \alpha(y)x, \forall y \in \mathbf{h}\}$. For each i, there is a unique non-zero linear form $\alpha_i : \mathbf{h} \to \mathbf{C}$ such that $0 \neq \mathbf{g}_{\alpha_i} \subset \mathbf{l}_i \cap \mathbf{n}'$ and $\mathbf{g}_{\alpha_i/2} = 0$. Let $\mathbf{S}$ be the symmetric algebra of $\mathbf{h}^*$.

Let s_i be the unique non-trivial element of $\mathscr{W}$ which can be represented by an element of L_i. Then $s_1,\dots,s_m$ are Coxeter generators of $\mathscr{W}$. We have a natural action $\xi \to^w \xi$ of $\mathscr{W}$ on $\mathbf{h}$ hence on $\mathbf{S}$. Let $x_0 \in \mathbf{l}$ be the nilpotent element such that $\exp(x_0) = u$. For each i, let c_i be the largest integer ≥ 2 such that $\mathrm{ad}(x_0)^{c_i-2} : \mathbf{l}_i \cap \mathbf{n}' \to \mathbf{l}_i \cap \mathbf{n}'$ is $\neq 0$. We define a $\mathbf{C}$-vector space $\mathbf{H} = \mathbf{S} \otimes \mathbf{C}[\mathscr{W}]$. There is a unique structure of associative $\mathbf{C}$-algebra with unit $1 \otimes e$ on $\mathbf{H}$ such that the following properties hold. First, the maps $\mathbf{S} \to \mathbf{H}$ ($\xi \to \xi \otimes e$) and $\mathbf{C}[\mathscr{W}] \to \mathbf{H}$ ($w \to 1 \otimes w$) are algebra homomorphisms. Second, we have $(\xi \otimes e)(1 \otimes w) = \xi \otimes w$ for any $\xi \in \mathbf{S}, w \in \mathscr{W}$. Finally, we have

(a) $(1 \otimes s_i)(\xi \otimes e) - ({}^{s_i}\xi \otimes e)(1 \otimes s_i) = c_i r_0 (\xi - {}^{s_i}\xi)\alpha_i^{-1} \otimes e, \quad (\xi \in \mathbf{S}, 1 \leq i \leq m).$

One can show that

(b) *the isomorphism classes of simple* $\mathbf{H}$*-modules are in natural* 1-1 *correspondence with the elements in the given block of* $\tilde{\mathscr{I}}$.

The connection between these kind of objects is given in [40]. In that paper an **H**-module is constructed for any element of $\tilde{\mathcal{I}}$. The idea is to apply the *IC* construction [34] of generalized Springer representations using fully its symmetry group (i.e. in equivariant homology); thus we do that construction with parameters in the classifying space of that symmetry group, or a smooth, finite dimensional approximation of it.

In the case where $\mathcal{B}$ is the block of $(\{1\}, \mathbf{C})$, the algebra **H** is closely related to an affine Iwahori-Hecke algebra whose simple modules were classified in [24]. In fact, the proof of (b) shares a number of features with the proof of the main result in [24]; one important difference is that in [24] the main technique is equivariant K-theory where *IC* methods are not available.

In the case where $G = SL_n$, $\mathcal{B}$ is the block of $(\{1\}, \mathbf{C})$ and $r_0 = 1$, the algebra **H** appears also in Drinfeld's work [13] on Yangians of type A. Can one construct finite dimensional representations of Yangians of general type using *IC* methods parallel to those just described ?

17. Unramified Representations of Simple p-Adic Groups

In this section we take G as in §7. Let $\mathcal{G}$ be the group of rational points of a simple split adjoint algebraic group of type dual to G over a nonarchimedean local field with residue field F_q. An irreducible admissible representation V of $\mathcal{G}$ is said to be *unramified* if there exists a parahoric subgroup $\mathcal{P}$ of $\mathcal{G}$ with "unipotent radical" $U_{\mathcal{P}}$ and "reductive part" $\bar{\mathcal{P}} = \mathcal{P}/U_{\mathcal{P}}$ (a reductive group over F_q) such that the $U_{\mathcal{P}}$-invariant part of V contains some unipotent cuspidal representation Ψ of the finite group $\bar{\mathcal{P}}$. (We get the same concept if we omit the word "cuspidal".) The following refinement of the Deligne-Langlands conjecture has been formulated in [32].

There should be a natural 1-1 *correspondence between the set of isomorphism classes of unramified representations of $\mathcal{G}$ and the set of triples (N, s, ϕ) up to G-conjugacy, where N is a nilpotent element of* **g**, *s is a semisimple element of G such that $Ad(s)N = qN$ and ϕ is an irreducible representation of the (finite) group of components of the simultaneous centralizer $Z(N, s)$ of N and s on which the centre of G acts trivially.*

This is rather similar to the classification of unipotent representations for a reductive group over F_q given in [33].

(One can allow $\mathcal{G}$ to be not necessarily split, but an inner form of a split group; this gives rise to a character χ of the centre of G and one should take the same set of parameters except that the centre should be required to act on ϕ according to χ.)

We would like to suggest a way prove this. Let us fix $\mathcal{P}, \Psi$ as above. To these one can associate an Iwahori-Hecke algebra $\mathcal{H}$ corresponding to a certain (extended) Dynkin diagram Δ and some parameters $q^{m_i}, (m_i \geq 1)$, one for each vertex t_i of Δ. Let $\tilde{\Gamma}$ be the (extended) Dynkin diagram corresponding to $\mathcal{G}$ and let Γ be the its subdiagram (of finite type) corresponding to $\mathcal{P}$. By definition, the vertices of Δ are the vertices of $\tilde{\Gamma}$ which are not in Γ, assuming that there are at least two such vertices; otherwise, Δ is defined to be empty. Assume that Δ is non-empty and let t_i be one of its vertices. Let $\mathcal{P}_i \supset \mathcal{P}$ be the parahoric subgroup corresponding to the subgraph whose vertices are those of Γ together with t_i. The representation of $\bar{\mathcal{P}}_i$ induced from $\bar{\mathcal{P}}$ by Ψ decomposes into a sum of two irreducible representations; the quotient of their dimension is known to be of form q^{m_i} for some integer $m_i \geq 1$. (This integer is explicitly known by [27, p.35].) Now let t_i, t_j be two different vertices of Δ; we want to define the multiplicity of the bond joining them in Δ. If Δ has exactly two vertices, this multiplicity is defined to be infinite. If Δ has at least three vertices, we denote by $\mathcal{P}_{i,j} \supset \mathcal{P}$ the parahoric subgroup corresponding to the subgraph whose vertices are those of Γ together with t_i, t_j. The representation of $\bar{\mathcal{P}}_{i,j}$ induced from $\bar{\mathcal{P}}$ by Ψ decomposes according to the representations of an Iwahori-Hecke algebra (with possibly unequal parameters) of type $A_1 \times A_1, A_2, B_2$ or G_2; accordingly, the sought for multiplicity is by definition 0,1,2 or 3. (See again [27, p.35].) One can check that the Δ is an (extended) Dynkin diagram. This completes the definition of $\mathcal{H}$ (except that there is an indeterminacy due to isogeny, but we will skip over this difficulty).

It seems likely (and probably not difficult to prove) that the unramified representations of $\mathcal{G}$ which correspond to our $\mathcal{P}, \Psi$ are in natural 1-1 correspondence with the simple $\mathcal{H}$-modules.

This is well known in the case where $\mathcal{P}$ is an Iwahori subgroup. In that case, $\mathcal{H}$ is an Iwahori-Hecke algebras with equal parameters and one can use the classification [24] of its simple modules.

But it seems to be difficult to extend the K-theoretic method of [24] to the general case. Instead, in the general case we can argue as follows. We can replace $\mathcal{H}$ by its generic version (over $\mathbf{C}[v, v^{-1}]$) and take (as in [41]) certain associated graded algebras (over $\mathbf{C}[r]$) of it and then specialize r to $r_0 = \log(q)/2$. The resulting algebras turn out to be of the type considered in §16 (hence their simple modules can be classified as in §16(b)); on the other hand, from the representation theory of these algebras one can completely recover the representation theory of $\mathcal{H}$, by the results of [41]. This should give the desired result.

18. Cells in Affine Weyl Groups

We preserve the assumptions of §7. The definition of left cells and right cells of W given in [22] (see also §3) extends without change to the case of $\tilde{W}$. (But in this case, this is the only known definition; it is not known how to extend the definition from the theory of primitive ideals).

One can also define a partition of $\tilde{W}$ into *two-sided cells*: the two sided cells are the smallest subsets of $\tilde{W}$ which are unions of left cells and also unions of right cells.

It is known (see [36]) that there are only finitely many left cells in $\tilde{W}$ but their structure is not well understood except in type A (see Shi [50]); however, the two-sided cells of $\tilde{W}$ have been shown in [36] to be in natural 1-1 correspondence with the unipotent classes in the corresponding group over $\mathbf{C}$. The proof in [36] is quite complicated. One ingredient is the use of the function $a : \tilde{W} \to \mathbf{N}$ defined as in §3; another one is the use of the results of [24] on representations of affine Iwahori-Hecke algebras; in addition, the theory of character sheaves also plays a role in the proof.

It is interesting that many invariants of a unipotent class can be reconstructed from the corresponding two-sided cell, in purely combinatorial terms. For example, if a two-sided cell corresponds to the class of the unipotent element u, then the variety of Borel subgroups containing u has dimension $a(z)$ where z is any element in the two-sided cell.

19. Quiver Singularities and Canonical Bases

In this section we assume that G is as in §8. In [45], methods of IC have been used to construct some particularly nice bases of the finite dimensional irreducible representations of $\mathbf{g}$. We will explain how this is done. (Another approach to these questions has been recently found by Kashiwara; see his lecture at this Congress.)

We assume that we are given a basis element in each root space in $\mathbf{n}$ corresponding to a simple root; we denote these elements by $e_1, \ldots, e_n$. (They are in 1-1 correspondence with the vertices of the Dynkin graph, denoted $1, \ldots, n$.)

Let $\mathbf{U}$ (resp. $\mathbf{U}^+$) be the enveloping algebra of $\mathbf{g}$ (resp. of $\mathbf{n}$). Recently, Ringel [49] gave a very elegant description of $\mathbf{U}^+$ in terms of representations of a *quiver* (the Dynkin graph of G with a fixed orientation). All representations of the quiver are assumed to be finite dimensional over $\mathbf{C}$. The dimension of a representation is naturally a vector $\mathbf{d} = (d_1, \ldots, d_n) \in \mathbf{N}^n$.

Given three representations V, V', V'' of the quiver, we can consider the set of all subrepresentations of V'' which are isomorphic to V and such that the corresponding quotient is isomorphic to V'. This set is naturally an algebraic variety and we denote by $c_{V,V',V''}$ its Euler characteristic in cohomology with compact support. Let $\mathscr{R}$ be the $\mathbf{C}$-vector space with basis elements V indexed by the isomorphism classes of representations of the quiver. Then the $c_{V,V',V''}$ can be regarded as the structure constants of an algebra structure on $\mathscr{R}$ and Ringel shows that there is a unique algebra isomorphism $\mathbf{U}^+ \cong \mathscr{R}$ which takes each $e_i \in \mathbf{U}^+$ to the irreducible representation of our quiver corresponding to the vertex i of the Dynkin graph. Via this isomorphism, $\mathbf{U}^+$ aquires from $\mathscr{R}$ a $\mathbf{C}$-basis. Note that this basis depends on the chosen orientation of the Dynkin graph.

We want to look at ways in which a representation of the quiver can be deformed to a more degenerate one. It is known that the isomorphism classes of representations of our quiver correpond to certain orbits of algebraic group actions and we are lead to consider closures of such orbits. One of the ideas of [45] was to exploit the fact that these closures of orbits may be singular hence have invariants coming from IC theory. (These singularities may be called *quiver singularities.*)

We shall define such invariants $j_{V,V'} \in \mathbf{N}$ for any two representations V, V' of the quiver. In the case where V, V' have different dimensions we set $j_{V,V'} = 0$. Assume now that they have the same dimension $\mathbf{d} = (d_1, \ldots, d_n)$. Let

$$\mathbf{E} = \oplus_{i\to j}\mathrm{Hom}(\mathbf{C}^{d_i}, \mathbf{C}^{d_j})$$

sum over all arrows $i \to j$ in the given orientation, and let $G_{\mathbf{d}} = \prod_i \mathrm{GL}_{d_i}(\mathbf{C})$. The group $G_{\mathbf{d}}$ acts naturally on the $\mathbf{C}$-vector space $\mathbf{E}$ by $(g_i) : (f_{ij}) \to (g_j f_{ij} g_i^{-1})$. Any point in $\mathbf{E}_{\mathbf{d}}$ may be regarded as a representation of dimension $\mathbf{d}$ of our quiver and all representations of dimension $\mathbf{d}$ arise in this way. Moreover two points in $\mathbf{E}$ define isomorphic representations if and only if they are in the same $G_{\mathbf{d}}$-orbit. Thus, V, V' correspond to two $G_{\mathbf{d}}$-orbits $\mathscr{O}, \mathscr{O}'$ on $\mathbf{E}$. If $\mathscr{O}$ is contained in the closure of $\mathscr{O}'$, we define $j_{V,V'}$ to be the Euler characteristic of the local IC of that closure (with coefficients $\mathbf{C}$) at a point in $\mathscr{O}$. Otherwise, we set $j_{V,V'} = 0$.

The elements $\sum_V j_{V,V'} V$ (for various representations V' of the quiver) form a new basis of $\mathscr{R}$; upon transfering it to $\mathbf{U}^+$, we obtain a basis of $\mathbf{U}^+$ which, by one of the main results of [45], is independent of the chosen orientation. For this reason, we call it the *canonical basis.*

Assume that we are given an irreducible, finite dimensional representation of $\mathbf{U}$ with a specified lowest weight vector. If we apply the elements of the canonical basis to this lowest weight vector, the resulting elements less 0 will miraculously form a basis for the representation space. This is a canonical basis of the given representation of $\mathbf{U}$. It is a basis with extremely favorable properties (see [45]).

All these bases have natural q-analogs in the framework of quantized enveloping algebras.

References

1. D. Alvis and G. Lusztig: On Springer's correspondence for simple groups of type E_n, $(n = 6, 7, 8)$. Math. Proc. Camb. Phil. Soc. **92** (1982) 65–72
2. A. A. Beilinson and J. Bernstein: Localisation des $\mathfrak{g}$-modules. C.R. Acad. Sci. Paris **292** (1981) 15–18
3. A. A. Beilinson and J. Bernstein: A proof of Jantzen conjecture. Preprint
4. A. A. Beilinson, J. Bernstein and P. Deligne: Faisceaux pervers. Astérisque **100** (1982)
5. W. Beynon and N. Spaltenstein: Green functions of finite Chevalley groups of type E_n $(n = 6, 7, 8)$. J. Algebra **88** (1984) 584–614
6. W. Borho and R. MacPherson: Représentations des groupes de Weyl et homologie d'intersection pour les variétés nilpotentes. C. R. Acad. Sci. Paris **292** (1981) 707–710
7. R. Bott: An application of the Morse theory to the topology of Lie groups. Bull. Soc. Math. France **84** (1956) 251–282
8. J.-L. Brylinski and M. Kashiwara: Kazhdan-Lusztig conjecture and holonomic systems. Invent. math. **64** (1981) 387–410
9. R. K. Brylinski: Limits of weight spaces, Lusztig's q-analogs and fiberings of adjoint orbits. J. Amer. Math. Soc. **2** (1989) 517–533
10. L. Casian: Kazhdan-Lusztig multiplicity formulas for Kac-Moody algebras. C. R. Acad. Sci. Paris **310** (1990) 333–337
11. L. Casian: Kazhdan-Lusztig conjecture in the negative level case (Kac-Moody algebras of affine type). Preprint
12. P. Deligne and G. Lusztig: Representations of reductive groups over a finite field. Ann. Math. **103** (1976) 103–161
13. V. G. Drinfeld: Degenerate affine Hecke algebras and Yangians. Funkt. Anal. Prilozh. **20** (1986) 69–70 (Russian)

14. B. L. Feigin and E. V. Frenkel: Affine Kac-Moody algebras and semi-infinite flag manifolds. Commun. Math. Phys. **128** (1990) 161–189
15. M. Goresky and R. MacPherson: Intersection homology theory. Topology **19** (1980) 135–162
16. V. Ginsburg: Perverse sheaves on loop groups and Langlands duality. Preprint
17. N. Iwahori and H. Matsumoto: On some Bruhat decomposition and the structure of the Hecke ring of p-adic Chevalley groups. Publ. Math. IHES **25** (1965) 237–280
18. A. Joseph: W-module structure in the primitive spectrum of the enveloping algebra of a semisimple Lie algebra. (Lecture Notes in Mathematics, vol. 728.) Springer, Berlin Heidelberg New York 1979, pp. 116–135
19. M. Kashiwara: Kazhdan-Lusztig conjecture for symmetrizable Kac-Moody Lie algebra. Preprint
20. M. Kashiwara and T. Tanisaki: Kazhdan-Lusztig conjecture for symmetrizable Kac-Moody Lie algebra; *II*, Intersection cohomologies of Schubert varieties. Preprint
21. D. Kazhdan: Proof of Springer's hypothesis. Isr. J. Math. **28** (1977) 272–286
22. D. Kazhdan and G. Lusztig: Representations of Coxeter groups and Hecke algebras. Invent. math. **53** (1979) 165–184
23. D. Kazhdan and G. Lusztig: Schubert varieties and Poincaré duality. Proc. Symp. Pure Math. **36** (1980) 185–203, Amer. Math. Soc.
24. D. Kazhdan and G. Lusztig: Proof of the Deligne-Langlands conjecture for Hecke algebras. Invent. math. **87** (1987) 153–215
25. S. Kumar: Proof of Lusztig's conjecture concerning negative level representations of affine Lie algebras. preprint
26. G. Laumon: Faisceaux automorphes liés aux séries d'Eisenstein. Automorphic forms, Shimura varieties and L-functions, vol. I (L. Clozel, J. S. Milne, ed.). Academic Press, New York 1990, pp. 227–281
27. G. Lusztig: Representations of finite Chevalley groups. Regional Conf. Series in Math. 39. Amer. Math. Soc., 1978
28. G. Lusztig: Hecke algebras and Jantzen's generic decomposition patterns. Adv. Math. **37** (1980) 121–164
29. G. Lusztig: Some problems in the representation theory of finite Chevalley groups. Proc. Symp. Pure Math., vol. 37. Amer. Math. Soc. 1981, pp. 313–317
30. G. Lusztig: Green polynomials and singularities of unipotent classes. Adv. Math. **42** (1981) 169–178
31. G. Lusztig: Singularities, character formulas and a q-analog of weight multiplicities. Astérisque **101–102** (1983) 208–229
32. G. Lusztig: Some examples of square integrable representations of semisimple p-adic groups. Trans. Amer. Math. Soc. **227** (1983) 623–653
33. G. Lusztig: Characters of reductive groups over a finite field. Ann. Math. Studies, vol. 107. Princeton University Press, 1984
34. G. Lusztig: Intersection cohomology complexes on a reductive group. Invent. math. **75** (1984) 205–272
35. G. Lusztig: Character sheaves. Adv. Math. **56** (1985) 193–237; *II*, **57** (1985) 226–265; *III*, **57** (1985) 266–315; *IV*, **59** (1986) 1–63; *V*, **61** (1986) 103–155
36. G. Lusztig: Cells in affine Weyl groups. Algebraic groups and related topics. Advanced Studies in Pure Math., vol. 6. Kinokuniya and North-Holland, Tokyo and Amsterdam, 1985; *II*, J. Alg. **109** (1987) 536–548; *III*, J. Fac. Sci. Tokyo University, IA **34** (1987) 223–243; *IV*, J. Fac. Sci. Tokyo University, IA **36** (1989) 297–328
37. G. Lusztig: Sur les cellules gauches des groupes de Weyl. C. R. Acad. Sci. Paris **302** (1986) 5–8
38. G. Lusztig: On the character values of finite Chevalley groups at unipotent elements. J. Algebra **104** (1986) 146–194

39. G. Lusztig: Leading coefficients of character values of Hecke algebras. Proc. Symp. Pure Math. **47** (1987) 253–262. Amer. Math. Soc.
40. G. Lusztig: Cuspidal local systems and graded Hecke algebras. *I*, Publ. Math. I.H.E.S. **67** (1988) 145–202; *II*, in preparation
41. G. Lusztig: Affine Hecke algebras and their graded version. J. Amer. Math. Soc. **2** (1989) 599–635
42. G. Lusztig: Green functions and character sheaves. Ann. Math. **131** (1990) 355–408
43. G. Lusztig: Finite dimensional Hopf algebras arising from quantized universal enveloping algebras. J. Amer. Math. Soc. **3** (1990)
44. G. Lusztig: On quantum groups. J. Algebra **131** (1990) 466–475
45. G. Lusztig: Canonical bases arising from quantized enveloping algebras. J. Amer. Math. Soc. **3** (1990) 447–498; *II*, Progress of Theor. Physics (to appear)
46. G. Lusztig and N. Spaltenstein: On the generalized Springer correspondence for classical groups. Algebraic groups and related topics. Advanced Studies in Pure Math., vol. 6. Kinokuniya and North-Holland, Tokyo and Amsterdam 1985, pp. 289–316
47. G. Lusztig and D. Vogan: Singularities of closures of K-orbits on a flag manifold. Invent. math. **71** (1983) 365–379
48. R. MacPherson: Global questions in the topology of singular spaces. Proc. I.C.M., Warszawa 1983, North-Holland, 1984, pp. 213–235
49. C. M. Ringel: Hall algebras and quantum groups. Invent. math. (1990)
50. J.-Y. Shi: The Kazhdan-Lusztig cells in certain affine Weyl groups. (Lecture Notes in Mathematics, vol. 1179.) Springer, Berlin Heidelberg New York 1986
51. T. Shoji: On the Springer representations of the Weyl groups of classical algebraic groups. Comm. in Alg. **7** (1979) 1713–1745; 2027–2033
52. T. Shoji: On the Springer representations of the Chevalley groups of type $F4$. Comm. Alg. **8** (1980) 409–440
53. T. Shoji: On the Green polynomials of Chevalley groups of type $F4$. Comm. Alg. **10** (1982) 505–543
54. T. Shoji: On the Green polynomials of classical groups. Invent. math. **74** (1983) 239–267
55. N. Spaltenstein: Appendix. Math. Proc. Camb. Phil. Soc. **92** (1982) 73–78
56. T. Shoji: On the generalized Springer correspondence for exceptional groups. Algebraic groups and related topics. Advanced Studies in Pure Math., vol. 6. Kinokuniya and North-Holland, Tokyo and Amsterdam 1985, pp. 317–338
57. T. A. Springer: Trigonometric sums, Green functions of finite groups and representations of Weyl groups. Invent. math. **36** (1976) 173–207
58. J. Tits: Résumé du cours. Annuaire du Collège de France (1981–82)
59. D. Vogan: Irreducible characters of semisimple Lie groups. *I*, Duke Math. J. **46** (1979) 61–108; *II*, The Kazhdan-Lusztig conjectures. Duke Math. J. **46** (1979) 805–859; *III*, Proof of the Kazhdan-Lusztig conjecture in the integral case. Invent. math. **71** (1983) 381–418; *IV*, Character-multiplicity duality. Duke Math. J. **49** (1982) 943–1073

The Interaction of Nonlinear Analysis and Modern Applied Mathematics

Andrew J. Majda *

Department of Mathematics and Program in Applied and Computational Mathematics
Fine Hall, Washington Road, Princeton University, Princeton, NJ 08544, USA

Dedicated to Ron Diperna (1947–1989)

Introduction

Many of the physical phenomena that are currently important research topics in applied mathematics involve strong, irregular, and typically unstable fluctuations with intense activity occurring on extremely small length scales. Examples include fully developed turbulence and the instability of thin layers in fluid flows, phase transitions and homogenization in materials, dendrite formation in crystal growth, the focusing of laser beams, and various problems in classical statistical physics such as the kinetic theory of gases. For a discussion of the mathematical progress in this last topic, see the report of P.L. Lions at this Congress. The quantitative physical phenomena are usually described by solutions of nonlinear partial differential equations in various singular limits involving either small spatial/temporal scales or coefficients of the equation. A prototypical example of such a process is the high Reynolds number limit of the Navier-Stokes equations,

$$\begin{gathered}\frac{\partial v^\varepsilon}{\partial t} + \operatorname{div}\,(v^\varepsilon \otimes v^\varepsilon) = -\nabla p^\varepsilon + \varepsilon \Delta v^\varepsilon\,, \\ \operatorname{div}\, v^\varepsilon = 0\,.\end{gathered} \tag{0.1}$$

Here $v = {}^t(v_1, v_2, v_3)$ is the fluid velocity, p is the scalar pressure, $v \otimes v = (v_i v_j)$, and ε is the reciprocal of the Reynolds number. Fully developed turbulence is characterized by the fact that the parameter ε in (0.1) is extremely small, i.e. $\varepsilon \ll 1$; the formal limiting equations with $\varepsilon = 0$ in (0.1) are called the Euler equations.

The recent development of powerful, user friendly, supercomputers and workstations combined with high resolution graphics and sophisticated numerical algorithms has enabled researchers to discover interesting and unexpected new phenomena in the solutions of nonlinear P.D.E.'s with a resolution often exceeding the capabilities of contemporary experiments. Understanding these phenomena has created the need for a highly interdisciplinary interaction involving ideas from nonlinear analysis, numerical computation, and sophisticated asymptotic methods. The reports by Mimura [51] and Krasny [33] at this Congress illustrate aspects of this type of interaction in modern applied mathematics; some recent

* Partially supported by research grants NSF DMS-9001805, ARO DAAL03-89-K-0013, ONR N00014-89-J-1044.

Proceedings of the International Congress
of Mathematicians, Kyoto, Japan, 1990

papers of the author [46, 36, 37] describe this modern research mode in the context of various problems in fluid dynamics. Except for brief remarks and some references, it will not be possible due to limitations on length to describe examples of such exciting interdisciplinary developments in any detail in this article and the main emphasis will involve the nonlinear analysis of the new phenomena.

The problems which I discuss below have a common theme. There is a sequence of solutions of a nonlinear P.D.E. which satisfies the bound,

$$\int_{\Omega} |v^{\varepsilon}|^2 dy \leq C \tag{0.2}$$

as $\varepsilon \downarrow 0$ so that there is a function v and a subsequence with

$$v^{\varepsilon} \rightharpoonup v \text{ i.e. weakly in } L^2(\Omega)\,. \tag{0.3}$$

In applications, the bound in (0.2) is usually guaranteed by a physical principle such as an energy estimate and the small parameter $\varepsilon \downarrow 0$ represents the limit of increasingly smaller scale fluctuations. The use of the L^2-norm in (0.2) is for simplicity in exposition. Various kinds of defects can happen in such a limiting process. These are illustrated by the heuristic decomposition

$$v^{\varepsilon} = v^{\varepsilon}_{\text{osc}} + v^{\varepsilon}_{\text{conc}} + \bar{v}^{\varepsilon} \tag{0.4}$$

of v^{ε}. Here $\bar{v}^{\varepsilon}$ converges strongly in L^2 so that there is a $\bar{v}$ with

$$\int_{\Omega} |\bar{v}^{\varepsilon} - \bar{v}|^2 dy \to 0 \text{ as } \varepsilon \downarrow 0\,. \tag{0.5}$$

The functions $v^{\varepsilon}_{\text{osc}}$ measure the small scale oscillations in the limit in the sense that

$$|v^{\varepsilon}_{\text{osc}}(y)|_{L^{\infty}} \leq C\,, \tag{0.6A}$$

and there is a $v_{\text{osc}} \in L^{\infty}$ so that

$$\int_{\Omega} \phi v^{\varepsilon}_{\text{osc}} dy \to \int_{\Omega} \phi v_{\text{osc}} dy \qquad \text{for all functions } \phi \in C_0(\Omega)\,, \tag{0.6B}$$

but $v^{\varepsilon}_{\text{osc}}$ does not converge strongly in L^2 to v_{osc}. A prototypical example of a sequence with small scale oscillations in R^1 is given by $v^{\varepsilon}_{\text{osc}} = v\left(\frac{y}{\varepsilon}\right)$ where $v(y)$ is a fixed smooth periodic function. The functions $v^{\varepsilon}_{\text{conc}}$ in the heuristic decomposition in (0.4) measure the development of intense small scale concentrations in the limit in the sense that $v^{\varepsilon}_{\text{conc}}$ satisfies the uniform L^2-bound in (0.2) and there is a function v_{conc} so that

$$v^{\varepsilon}_{\text{conc}} \to v_{\text{conc}} \text{ a.e. in } \Omega\,, \tag{0.7}$$

but nevertheless, $v^{\varepsilon}_{\text{conc}}$ does not converge strongly to v_{conc} in L^2. A prototypical example illustrating concentration in $L^2(R^N)$ is given by the sequence $v^{\varepsilon}_{\text{conc}} = \varepsilon^{\frac{-N}{2}} \varrho\left(\frac{x}{\varepsilon}\right)$ with $\varrho \in C^{\infty}_0$, $\varrho \geq 0$, $\int \varrho = 1$.

Most of the research in nonlinear P.D.E. prior to the late 1970's involved problems where suitable apriori estimates automatically guaranteed strong convergence so that $v^{\varepsilon}_{\text{osc}}$ and $v^{\varepsilon}_{\text{conc}}$ in (0.4) were zero and the relevant nonlinear maps $g(v^{\varepsilon})$ are continuous in the limit. With the strong motivation from understanding many of the new phenomena described earlier, an important current research direction in nonlinear P.D.E.'s involves situations where both fine scale oscillations and/or concentrations occur in various physical problems so that $v^\varepsilon_{\text{osc}}$ and/or $v^\varepsilon_{\text{conc}}$ are non-zero in the heuristic decomposition in (0.4). In the next section some of the mathematical tools and physical problems that have been studied recently with oscillations and/or concentrations are described briefly. Of course, this article is not intended to be a detailed survey and the discussion is limited by both the taste and expertise of the author. The final sections of the article contain a more detailed report on problems with oscillations and concentrations in incompressible fluid flow – this is a rich and important class of problems where both effects occur simultaneously and many interesting problems remain unsolved.

1. The Development of Oscillations and Concentrations in Problems from Applied Mathematics

1A. Some Mathematical Tools

An effective tool for nonlinear analysis associated with a sequence $v^\varepsilon_{\text{osc}}$ satisfying the L^∞-bound in (0.6A) is the Young measure introduced by Tartar in this context. If the vector field v is a mapping from $\Omega \subseteq R^N$ to R^M, the Young measure theorem asserts that there is a family of probability measures on R^M, $\{\nu_y\}$ indexed by $y \in \Omega$ so that after passing to subsequence, weak limits of composite non-linear maps $g(v^\varepsilon)$ are given by

$$\lim_{\varepsilon\to 0} \int_\Omega \phi(y) g(v^\varepsilon_{\text{osc}}) dy = \int_\Omega \phi(y) \int_{R^M} g(\lambda) d\nu_y(\lambda) dy . \tag{1.1}$$

There is strong convergence of $v^\varepsilon_{\text{osc}}$ to v_{osc} on Ω if and only if the probability measure ν_y is given by the Dirac mass, $\delta_{v_{\text{osc}}(y)}$, for each $y \in \Omega$. An elementary probabilistic interpretation of the Young measure utilizing the Lebesque differentiation theorem has been developed recently by Ball [6]. Diperna [18, 19] introduced the concept of measure-valued solution of a system of conservation laws associated with an L^∞ bounded sequence so that only oscillations occur in the limit. The effectiveness of this tool for interpreting the limit of fine scale oscillations in model problems such as the zero dispersion limit of solutions of the KdV equation has been surveyed earlier by Diperna (see [18, 19]). A very general extension of the Young measure theorem for general sequences with only the uniform $L^2(L^p)$ bound in (0.2) has been developed by Diperna and the author [20] for the specific applications of constructing suitable measure-valued solutions of problems from mathematical physics where both fine scale oscillations and concentrations develop in the limit. Strong motivation for this work was provided by the problems involving the high Reynolds number limit of solutions of the Navier-Stokes equations with finite kinetic energy described in (0.1); the limiting solution is a suitable measure-valued solution of the Euler equations with both oscillations and concentrations (see [20]). Unlike (1.1), the generalized Young measures in this context are often singular continuous with respect to Lebesque measure, reflecting the development of concentrations – more will be said about this topic later in this paper.

In physical problems where the phenomena of concentration from (0.7) occurs without the development of oscillations (so that $v^\varepsilon_{\text{osc}} = 0$ in (0.4)) it is very natural to measure the size of the set where concentration occurs; two types of defect measures have been introduced recently for this purpose, the weak* defect measure [35] and the reduced defect measure [22]. With the uniform L^2-bound in (0.2) and $v^\varepsilon \rightharpoonup v$, there is a non-negative Radon measure σ so that

$$\int \varphi d\sigma = \lim_{\varepsilon\to 0} \int_\Omega \varphi |v^\varepsilon - v|^2 dy \tag{1.2}$$

for all $\varphi \in C_0(\Omega)$. The measure σ is the weak* defect measure; if σ is zero on an open set, there is strong convergence on that open set. Motivated by problems in the calculus of variations with critical Sobolev exponents, P.L. Lions [35] has proved a beautiful simple theorem guaranteeing that the weak* defect measure is always concentrated on a rather small set in this context. An illustration of this theorem is provided by the following example: if v^ε is a sequence of functions in R^2 satisfying the bound

$$\int_{R^2} |v^\varepsilon|^2 + \int_{R^2} |\nabla v^\varepsilon| \le C\,, \tag{1.3A}$$

then there is a countable number of points $\{y_j\}_{j=1}^\infty$ and weights $\alpha_j \ge 0$ with $\sum \alpha_j^{1/2} < \infty$, so that necessarily

$$\sigma = \sum_j \alpha_j \delta_{y_j}\,, \tag{1.3B}$$

with δ_{y_j} the Dirac mass at y_j.

The reduced defect measure [22] is the finitely sub-additive outer measure θ defined by

$$\theta(E) = \limsup \int_E |v^\varepsilon - v|^2 dx\,. \tag{1.4}$$

Obviously, the reduced defect measure has the property that $\theta(E) = 0$ for any set E if and only if there is strong convergence on E. Furthermore, for any closed set F, it is easy to see that

$$\theta(F) \le \sigma(F)\,, \tag{1.5}$$

so that $\theta(F)$ can be very small even though $\sigma(F)$ is large. Examples of the use of estimates for the size of the reduced defect measure occur in the important physical problem involving vortex sheets for incompressible fluid flow and will be discussed in the later sections of this paper [22]. In these examples the support of σ can include all of R^3 while θ concentrates on a set of small measure with Hausdorff dimension one and this is crucial for the applications.

The discussion just presented of some tools for measuring phenomena in weak convergence is necessarily sketchy and incomplete. The interested reader should consult the recent lecture notes of L.C. Evans [25] for a more complete discussion of these matters. There is a need to develop other theoretical tools to measure additional structure in weak convergence with defects; one interesting new concept with potential applications is the H-measure introduces independently by Tartar [57] and Gerard [26]. Tartar's lecture at this Congress [58] provides an introduction to these topics.

1B. The Development of Fine Scale "Twinning" and Defects in Materials

Very recently an elegant mathematical theory has been developed to explain at the macroscopic level certain phase transitions in solids as the temperature, θ, varies [7, 8, 13, 16, 32]. In such austenitic-martensitic phase transitions there is a decrease in crystal symmetry at a transition temperature; for example, from a high temperature solid phase with cubic symmetry (austenite) to a low temperature solid phase with tetragonal symmetry (martensite). At the transition temperature, it is observed in experiments that the martensitic phase develops small scale regular fluctuations and is often "finely twinned" along planes related to the crystal lattice and also there is an austensite/finely twinned martensite interface. The key point in the mathematical analysis is the fact that the symmetries in a macroscopic continuum theory of crystals for both the austensite and martensite phases at the critical temperature θ_0 leads to a variational problem to minimize the free energy over deformations $y(x)\colon \Omega \supseteq R^3 \to R^3$ with a structure so that the finite infimum *is not achieved.* In other words, for the appropriate free energy density, $W(A,\theta)$ with $A \in M_+^{3\times 3}$, the set of 3×3 matrices with positive determinant, I is finite with

$$I = \min_y \int_\Omega W\left(\frac{dy}{dx}, \theta_0\right) dx, \tag{1.6}$$

but there is no weak deformation y so that the minimum, I, is achieved. In particular, the appropriate free energy functionals $W(A,\theta)$ at the critical temperature, θ_0, are necessarily not rank-one convex and thus are not quasi-convex [5]. The rigorous theory predicts that in the process of attempting to achieve the minimum, the sequence of deformation gradients, $\frac{dy^\varepsilon}{dx}$, necessarily develops apriori fine scale but regular oscillations in the martensitic phase which admit an elegant characterization via classical Young measures (see (1.1)) as combinations of weighted Dirac masses located on an appropriate set of matrices. Numerical computations [16] confirm the continuum theory and specific tests for detailed experiments based on the mathematical theory have recently been proposed [8]. An extremely readable mathematical account of this interdisciplinary research is presented in the paper by James and Kinderlehrer [32].

In the previous paragraph, an application in materials science with fine scale oscillations was described. Next the solution of a model problem motivated by the issue of prescribing strengths and locations of defects in the theory of liquid crystals [24] will be described where concentrations develop in a certain fashion so that the weak defect measure from (1.2) is nontrivial. The model problem analyzed by Brezis, Coron, and Lieb [10] concerns harmonic maps with values in S^2 and defects at prescribed locations a_i with specified (Brouwer) degree, deg (ϕ, a_i), at the location a_i for $1 \le i \le N$. Thus, the admissible maps are given by

$$\mathscr{E} = \left\{\phi \in C^1(R^3 \sim \cup_{i=1}^N \{a_i\}; S^2) \middle| \deg\, (\phi, a_i) = d_i \text{ and } \int_{R^3} |\nabla\phi|^2 < \infty\right\}$$

The model problem consists of finding

$$E = \inf_{\phi\in\mathscr{E}} \int |\nabla\phi|^2. \tag{1.7}$$

The number E is finite and there is a beautiful explicit formula for this value [10]; as in (1.6), the infimum in (1.7) is not achieved but concentrations develop in the minimizing deformation gradients rather than the fine scale oscillations from (1.6). Specifically, in attempting to achieve the minimum, $v^{\varepsilon} = \nabla\phi^{\varepsilon}$ concentrates so that $v^{\varepsilon} \rightharpoonup 0$ but $|v^{\varepsilon}|^2 \rightharpoonup \sigma$ in the sense of Radon measures where the weak* defect measure is given by $\sigma = 8\pi\delta_C$; the set C consists of a finite union of straight line segments in R^3 defined by some minimal connection, where δ_C is the uniform one-dimensional Lebesque measure on these lines (see [10]).

1C. The Development of Concentrations in Focusing Solutions of Nonlinear Schrödinger Equations

This is a very rich class of problems providing simplified models for focusing laser beams where concentrations develop spontaneously in time in a problem with critical Sobolev exponents. These equations have the form

$$iv_t = -\Delta v - |v|^{4/N} v, \qquad v(x,0) = v_0(x), \tag{1.8}$$

for $x \in R^N$. The equations in (1.8) have a Hamiltonian structure and a special conformal invariance as a consequence of the special exponent in (1.8); furthermore, the L^2-norm is conserved in time on intervals of existence. Weinstein [59] has proved that solutions of (1.8) exist for all time provided that $||v_0||_{L^2} < ||Q||_{L^2}$ where $Q(x)$ is a suitable ground state. For suitable initial data with critical mass so that $||v_0||_{L^2} = ||Q||_{L^2}$, he exploited the conformal invariance to construct explicit focusing solutions which develop dynamic concentrations as $t \nearrow T_* < \infty$ and studied the stability of this process through methods of concentration compactness [60]. In very recent work, Merle [50] has attacked the intriguing question of continuation of these special solutions past the blow up time. He considers sequences of smooth initial data v_0^{ε} regularizing the data $v_0(x)$ which develops singularities so that the regularized data v_0^{ε} satisfy $||v_0^{\varepsilon}|| < ||Q||$ and $||v_0^{\varepsilon} - v_0|| \to 0$. As mentioned earlier, the corresponding solutions $v^{\varepsilon}(x,t)$ are defined globally in time and the limiting behavior as $\varepsilon \downarrow 0$ is analyzed in [50] for $t > T_*$; the singular solution is unstable in time in the sense that the singularity disappears in this limit for $t > T_*$ and suitable special subsequences of v^{ε} converge to explicit defocusing solutions of (1.8) characterized by arbitrary phase shift parameters. There is "loss of information" in time once the dynamic concentration develops because these phase shifts are not uniquely determined and depend on the special subsequence. This author conjectures that appropriate measure-valued solutions [20] of (1.8) arise for $t > T_*$ in the general case in order to reflect the loss of information. The nature of singularity formation for solutions of (1.8) with $||v_0||_{L^2} > ||Q||_{L^2}$ is an active research topic in applied mathematics [34] with rich additional phenomena in the structure of singular solutions beyond those discussed briefly here.

1D. Nonlinear Geometric Optics and Oscillations for Hyperbolic Systems of Conservation Laws

Recently, there has been a great deal of activity in developing the theory of nonlinear geometric optics for general hyperbolic systems of conservation laws in multi-D [29, 42, 30, 28, 49]. As in linear geometric optics, these formal asymptotic theories utilize short wave length approximations so they involve the propagation of oscillations in hyperbolic systems. In nonlinear geometric optics, approximate solutions of general quasi-linear hyperbolic systems are constructed with the general form,

$$v^{\varepsilon} = \bar{v}(x,t) + \varepsilon v_1 \left(\frac{\boldsymbol{\phi}}{\varepsilon}, x, t \right) + 0(\varepsilon), \qquad \varepsilon \ll 1 \tag{1.9}$$

where $\boldsymbol{\phi} = (\phi_1(x,t), ..., \phi_d(x,t))$ is a vector of phase functions. Through self-consistent asymptotic approximations, simplified equations for the phase functions and amplitudes in v_1 are developed and in many instances these approximations can be solved exactly; thus, a great deal of genuine insight can be gained into complex physical problems. One of the theoretical achievements is a general theory for resonantly interacting small amplitude periodic wave trains with many new phenomena that are completely absent in geometric optics for linear problems [42, 30, 49, 12, 43]. In applications, the general theory of nonlinear geometric optics has been used recently to explain a variety of complex phenomena in physical problems involving compressible fluid flow [43, 3, 4], wave propagation in reacting materials [41, 2, 45] and dynamic nonlinear elasticity [49]. For example, a completely new theory of nonlinear instability for supersonic vortex sheets has been developed recently by the author and his recent Ph.D. student, Artola which combines the theory of hyperbolic mixed problems with nonlinear geometric optics [3, 4, 48] in a sophisticated fashion; this theory has been motivated and confirmed by large scale computations of Woodward (see the discussion in [36]. Another important use of nonlinear geometric optics is to extract mathematically tractable model equations which retain some of the essential features of much more complex physical phenomena [52, 44, 45].

This general topic has been mentioned briefly here because it involves interdisciplinary interactions in applied mathematics with the propagation of oscillations, and there is a definite need for more work on the rigorous theory of nonlinear geometric optics beyond the contribution of Diperna and the author [23]. Some important and accessible open theoretical problems are discussed in two review articles [37, 40]. Unlike the linear case, the effects of diffraction are not understood even at the formal level for small amplitude discontinuities. Current surveys of the formal aspects of nonlinear geometric optics including a discussion of problems with diffraction have been given by Hunter [31] and Rosales [53]. In interesting recent work, D. Serre [55, 54] has used the classical Young measure to describe the propagation of large amplitude oscillations in special quasi-linear hyperbolic systems with linearly degenerate wave fields.

1E. Oscillations and Concentrations for Incompressible Fluid Flow

Ingenious numerical computations [15, 14] and physical experiments reveal bewildering complexity at finite times in solutions of the incompressible fluid equations from (0.1) both at high Reynold's number for the Navier-Stokes equations with $\varepsilon \ll 1$ in (0.1) and as various numerical regularization parameters tend to zero for the incompressible Euler equations (the equations in (0.1) with $\varepsilon = 0$). The work of Diperna and the author [20, 22, 21] involving measure-valued solutions of the three-dimensional Euler equations is a first attempt to quantify such observed catastrophic instabilities in a mathematical fashion. Due to the rapid amplification of vorticity in three-dimensional fluid flows [36], very little apriori information is available and the only generally bounded positive physical quantity of an approximating sequence, v^ε, is the local kinetic energy. Thus, it is reasonable to assume the uniform bounds

$$\max_{0\le t\le T} \int_{|x|\le R} |v^\varepsilon|^2(x,t)dx \le C_{R,T} \tag{1.10}$$

for any $R > 0, T > 0$. In [20], measure-valued solutions of the three-dimensional Euler equations are constructed which arise from the weak limit of suitable approximate solution sequences satisfying the uniform bound in (1.10); a prototypical example from [20] is the fact that in the limit as $\varepsilon \downarrow 0$, a family of Leray-Hopf weak solutions of (0.1) converges to a measure-valued solution of the 3-D Euler equations. In general, these measure-valued solutions allow for both oscillations and concentrations and several examples of both of these phenomena are discussed in [20, 21].

The simplest way to generate examples of approximate solution sequences and measure-valued solutions for 3-D Euler is to take a sequence of smooth exact solutions and examine the generalized Young measure in the limit. It is an easy exercise for the reader to check that any sequence of smooth velocity fields

$$v^\varepsilon = (0, 0, v_3^\varepsilon(x_1, x_2)), \tag{1.11A}$$

with the kinetic energy bound in (1.10), defines a sequence of exact smooth solutions of the 3-D Euler equations converging to a measure-valued solution of 3-D Euler; clearly all of the phenomena of development of simultaneous oscillations and concentrations in solutions of the fluid equations can be displayed in such solution sequences by mimicking the constructions below (0.6) and (0.7). If $v_3^\varepsilon \rightharpoonup v_3(x_1, x_2)$ in L^2, then

$$v = (0, 0, v_3(x_1, x_2)) \tag{1.11B}$$

defines a weak solution of the 3-D Euler equations despite the fact that typically $(v_3^\varepsilon)^2$ does not converge weakly to $(v_3)^2$. A more interesting example of an exact solution sequence for the 3-D Euler equations displaying the development of fine scale oscillations [20] is given by

$$v^\varepsilon = \left(v\left(\frac{x_2}{\varepsilon}, x_2\right), 0, w\left(x_1 - v\left(\frac{x_2}{\varepsilon}, x_2\right)t, x_2, \frac{x_2}{\varepsilon}\right)\right), \tag{1.12}$$

where v and w are fixed smooth bounded functions with $v(y, x_2)$ periodic in y; in particular, this sequence satisfies the uniform L^∞-bound, $|v^\varepsilon|_{L^\infty} \le C$. Unlike the example in (1.11), *the weak limit of* (1.12) *is an explicit smooth function but* this smooth function is *no longer an exact solution of the 3-D Euler equations* –

new phenomena occur through the persistence of oscillations and the limiting process is an explicit example of a nontrivial measure-valued solution. It is worth remarking that both of the exact solution sequences for the Euler equations in (1.11) and (1.12) inherently involve three space dimensions; furthermore, these examples are constructed in a systematic fashion beginning with suitable exact two-dimensional solutions, $(0,0)$ and $\left(v\left(\frac{x_2}{\varepsilon}, x_2\right), 0\right)$ respectively (see p. 445 of [17] for a description of the general recipe). Steady exact solutions of the 2-D Euler equations (see [47]) are given by the radial eddies,

$$v = \left(\frac{-x_2}{r^2}, \frac{x_1}{r^2}\right) \int_0^r s\omega(s)ds, \tag{1.13}$$

with $r = (x_1^2 + x_2^2)^{1/2}$. Elementary examples of exact solution sequences for the 2-D Euler equations exhibiting concentrations can be generated in two different ways:

A) Pick a *positive* function $\omega(r) \geq 0$ with bounded support and define the v^ε by

$$v^\varepsilon = \left(\log \frac{1}{\varepsilon}\right)^{-1/2} \varepsilon^{-1} v\left(\frac{x}{\varepsilon}\right) \tag{1.14A}$$

with v given from ω by (1.13).

B) Pick a function $\omega(r)$ with compact support but $\int_0^\infty s\omega(s)ds = 0$ so that there is zero circulation and define v^ε by

$$v^\varepsilon = \varepsilon^{-1} v\left(\frac{x}{\varepsilon}\right) \tag{1.14B}$$

with v given from ω by (1.13).

The scalings in both examples have been chosen to guarantee the uniform local kinetic energy bounds in (1.10). In both examples, $v^\varepsilon \rightharpoonup 0$ but the nonlinear terms in the Euler equations, $v \otimes v$ from (0.1), satisfy

$$v^\varepsilon \otimes v^\varepsilon \rightharpoonup C \begin{pmatrix} 1 & 0 \\ 0 & 1 \end{pmatrix} \delta(x) \tag{1.15}$$

with weak convergence in the sense of measures. Here $\delta(x)$ is the Dirac mass at the origin and the constant C varies in the two examples and depends on the core (see [20] and [21] for the details). Such examples illustrate the phenomena of concentration in solutions of the 2-D Euler equations; these two exact solution sequences generate measure-valued solutions of the 2-D Euler equations where the associated Young measure is singular continuous with respect to Lebesque measure due to the appearance of concentrations. It is useful to note here that in example (1.14A)

$$\int_{R^2} |\text{curl } v^\varepsilon| \leq C, \quad \text{but} \quad \overline{\lim}_{\varepsilon \downarrow 0} \int_{R^2} |\nabla v^\varepsilon| = +\infty \tag{1.16}$$

while $||\nabla v^\varepsilon||_{L^1} \leq C$ in example (1.14B). Other amusing examples of measure-valued solutions for the 3-D Euler equations with fine scale oscillations have been developed in interesting recent work by Brenier [9].

The following general problem is of interest in the present context:

Problem. Characterize the closure in the L^2-weak topology of smooth exact solutions of the 3-D Euler equations.

The above explicit examples all provide insight into this question. There is a need to find other interesting examples like the one in (1.12) which illustrate that the weak closure is larger than merely the weak solutions of 3-D Euler; such solutions involving interacting oscillations and concentrations would be particularly useful.

2. Vortex Sheets, Potential Theory, and Concentration-Cancellation for 2-D Incompressible Flow

Incompressible fluid flow in two-space dimensions is generally much simpler than that in three space dimensions because given the velocity $v = (v_1, v_2)$, the scalar vorticity, $\omega = (v_2)_{x_1} - (v_1)_{x_2}$, is conserved along fluid particle paths for the 2-D Euler equations. Thus, it is a classical fact that smooth solutions of the 2-D Euler equations exist for all time while the analogous question for 3-D Euler is an outstanding and very important open problem due to vorticity amplification [36, 47]. Nevertheless, there are important practical problems involving weak solutions of the 2-D Euler equations, and here recent progress [22, 21, 27, 1, 61] is described in understanding the new mathematical phenomena and structure that occurs for vortex sheet initial data.

A two dimensional incompressible velocity field, v_0, defines *vortex sheet initial data* provided that curl $v_0 = \omega_0$ is a finite Radon measure and v_0 has locally finite kinetic energy. In applications, ω_0 is typically a surface Dirac measure supported on a smooth curve. Recent ingenious large scale numerical computations reveal tremendous complexity in the solutions of the 2-D fluid equations with vortex sheet initial data as various regularization parameters converge to zero; there is also much more complexity in the computed solutions when ω_0 changes sign than if ω_0 has a distinguished sign i.e. $\omega_0 \geq 0$ (see Krasny's report at this Congress [33] and also [46] for a detailed description and references for the numerical results). Furthermore, classical linearized stability analysis of the simplest vortex sheets reveals that they define a classical Hadamard ill-posed initial value problem like that for the Laplace equation, so both the computational and theoretical issues are extremely subtle. All of this complexity combines with the fact that vortex sheets are ubiquitous as approximations to velocity fields in engineering applications [11] to make the study of conceivable weak solutions for the 2-D Euler equations a fascinating contemporary research topic. One large research effort (see [11] for references) restricts the study of vortex sheets to analytic initial curves and attempts to use complex analysis to study the first relatively faint singularity formation in such vortex sheets. In typical computational and practical applications, vortex sheet initial data are usually not defined by real analytic curves [46, 38]. The methods and results described here [22, 21, 27, 1, 61] are completely different and address the global complexity of approximations for vortex sheets for all times with general (non-analytic) vortex sheet initial data through ideas naturally involving the weak topology.

Vortex sheets are regularized [21] by a sequence of globally smooth solutions of the 2-D Euler equations, $v^\varepsilon(x,t)$, satisfying the bounds,

$$\max_{0\leq t\leq T}\left(\int_{|x|\leq R}|v^\varepsilon(x,t)|^2dx+\int_{R^2}|\omega^\varepsilon(x,t)|dx\right)\leq C_{R,T}\,, \tag{2.1}$$

for any $R,T>0$ with $\omega^\varepsilon=\text{curl } v^\varepsilon$; a very general mathematical framework for these and other important regularizations satisfying (2.1) such as the high Reynolds number limit of 2-D Navier-Stokes and computational vortex methods is developed in [21]. If $v^\varepsilon(x,0)\rightharpoonup v_0$ where v_0 defines vortex sheet initial data, does $v^\varepsilon(x,t)$ converge to a weak solution of the 2-D Euler equations with this vortex sheet initial data? Are there new phenomena in the limiting process? There has been partial progress on the first question which is described here along with the fact that striking new mathematical phenomena do occur. In particular, one result in [21] is the existence of a measure-valued solution of 2-D Euler defined for all time with a special structure so that there are only concentrations and not oscillations (such as in (1.12)) in the limit as $\varepsilon\downarrow 0$ with the bounds in (2.1). The key fact proved in [21] is that with (2.1) there is a subsequence v^ε so that $v^\varepsilon\rightharpoonup v$ in L^2_{loc} and

$$\int_0^T\int_{|x|\leq R}|v^\varepsilon-v|dx\,dt\to 0 \tag{2.2}$$

for any $T>0, R$ so that in particular $v^\varepsilon\rightharpoonup v$ a.e. and only concentrations can occur (see (0.7)). The elementary examples for 2-D Euler described in (1.14)–(1.16) above satisfy the bounds in (2.1) and illustrate that the phenomena of concentration can occur with the natural bounds in (2.1) for vortex sheets. An important issue in deciding whether the function v defined in (2.2) is a suitable weak solution of 2-D Euler involves various estimates on the size of the concentration set [22, 21, 27, 1, 61]. These topics are discussed next.

2A. Size of the Weak Star and Reduced Defect Measures for the Problem of Vortex Sheets

The problem of existence for vortex sheet initial data with the bounds in (2.1) has some of the flavor of a problem involving critical Sobolev exponents. This fact is supported by the result proved in [21] that if the approximating sequence v^ε satisfies the L^p-bound on the initial vorticity, $\|\omega_0^\varepsilon\|_{L^p}\leq C$ for some $p>1$, in addition to (2.1), then v^ε converges strongly to v in L^2 and v defines a classical weak solution for 2-D Euler. Also recall that the bounds already presented in (1.3) in discussing concentration compactness resemble those in (2.1) except that in (2.1) only div $v^\varepsilon=0$ and curl v^ε are bounded in L^1 rather than the complete gradient. The elementary example of concentration with positive vorticity in (1.14A) and (1.16) directly illustrates that the bounds in (2.1) for exact solutions of 2-D Euler are not sufficient to guarantee the stronger bound in (1.3A). Thus, one suspects that sequences of smooth solutions of 2-D Euler satisfying (2.1) will not have a weak * defect measure with the simple structure from (1.3B) involving a countable number of Dirac masses. C. Greengard and Thomann [27] have constructed a beautiful family of explicit examples which illustrate this point in dramatic fashion.

Theorem [27]. *Consider an arbitrary positive finite Radon measure σ on R^2, then there is a sequence of exact steady solutions of the 2-D Euler equations, v^ε satisfying the bounds in (2.1) with $v^\varepsilon \rightharpoonup 0$ in L^2 so that the weak*defect measure associated with this sequence is σ.*

Actually, the example in [27] produces uniform Lebesque measure for the unit square as weak* defect but it is an elementary exercise for the reader involving simple functions to deduce the more general result in the theorem from the construction in [27]; the construction superimposes "clouds of vortices" like the ones in (1.14B) but utilizes two scales in such elementary solutions in a clever fashion. In agreement with some of the trends of numerical solutions when the vorticity changes sign [46], the theorem indicates that almost arbitrary complexity can occur in the weak topology in concentrations when the bound in (2.1) is satisfied.

One of the main results in [22] is that despite this almost arbitrary complexity as regards the weak* defect measure for sequences v^ε with the bounds in (2.1), the reduced defect measure, θ, defined in (1.4) nevertheless concentrates on small sets. The following operational definition is useful and implicit in the constructions in [22].

Definition. The reduced defect measure, θ, concentrates on a *set* with *Hausdorff dimension p* provided that there is a constant C depending only on δ and closed sets F_r so that $\theta(F_r) = 0$ with the complement of F_r, F_r^c, satisfying $H_r^{p+\delta}(F_r^c) \leq C$ for any $\delta > 0$ and $r \leq r_0$. Here H_r^d denotes Hausdorff premeasure of dimensional d and level r.

In particular, $H_r^{p+2\delta}\left(F_r^c\right)$ vanishes as $r \downarrow 0$ for any $\delta > 0$. Recall from (1.5) that θ satisfies $\theta(F) \leq \sigma(F)$ for closed sets; thus, it is possible that even though σ has very large support and charges a closed set F_r, θ satisfies $\theta(F_r) = 0$. In fact, the following is true:

Theorem [22]. *A) Assume that v^ε is a sequence of* steady time-independent velocity fields *with div $v^\varepsilon = 0$ and* satisfying the bounds *in* (2.1)*, then the reduced defect measure associated with a subsequence* concentrates on a set with Hausdorff dimension zero.

B) Assume that v^ε is a time-dependent *approximate* solution sequence *for 2-D Euler (see* [21, 22]) satisfying the bounds in (2.1)*, then the reduced defect measure* concentrates *on a set with* space-time cylindrical *Hausdorff dimension one.*

The examples in (1.14) show that the theorem is sharp in general. Here are some brief comments on the proof of the theorem. There is a stream function ψ^ε so that $\Delta\psi^\varepsilon = \omega^\varepsilon$ and $v^\varepsilon = \nabla^\perp\psi^\varepsilon = \left(-\psi^\varepsilon_{x_2}, \psi^\varepsilon_{x_1}\right)$ and thus

$$v^\varepsilon(x,t) = \int_{R^2} K(x-y)\omega^\varepsilon(y,t)dy\,, \tag{2.3}$$

with kernel $K(x) = (2\pi|x|)^{-2}(-x_2, x_1)$. There are two main steps in the proof from [22] for time independent velocity fields. In the first step, (see Theorem 4.1 in [22], the radial distribution function of the vorticity at an arbitrary point is introduced,

$$\omega^\varepsilon(B(s,x)) = \int_{B(s,x)} |\omega^\varepsilon(y)| dy, \tag{2.4}$$

where $B(s,x)$ denotes the open ball of radius s centered at x. The key point is to prove a uniform decay estimate for the vorticity distribution function for an appropriate subsequence, i.e. for any $\delta > 0$ there exists a closed set F_R so that

$$F_R \subset \{x | \omega^\varepsilon(B(s,x)) \leq Ks^\delta, \quad 0 \leq s \leq 1\}, \tag{2.5}$$

and $H_R^{3\delta}(F_R^c) \leq C$. Here the constants K and C depend on δ. For a fixed measure with finite total mass, the conclusion in (2.5) follows from standard covering arguments (see Stein [56]); however, it is *important* here *that the set* F_R in (2.5) *is uniform for the entire sequence* so a different proof is needed – Riesz potentials and a refined Chebyshev inequality for Hausforff premeasure are utilized to achieve this. In the second step (see thm 6.1 in [22], the uniform decay estimate in (2.5) is used with the convolution representation in (2.3) yielding the improved velocity estimate

$$\int_{F_R} |v^\varepsilon|^{p'} \leq C(R) \qquad \text{for some } p' > 2. \tag{2.6}$$

The facts in (2.6) and (2.2) are then combined with simple interpolation to establish that $\theta(F_R) = 0$ and this completes the proof of the theorem for steady velocity fields. For the time-dependent case of the theorem, these estimates are combined with uniform Hölder estimates in time

$$\left(\int_{B(R,0)} |v^\varepsilon(x,t_1) - v^\varepsilon(x,t_2)|^2\right)^{1/2} \leq C_{q,\alpha}|t_1 - t_2|^\alpha \tag{2.7}$$

for $1 < q < 2$, $0 < \alpha < 1$ which are valid for approximate solution sequences for 2-D Euler. One subtlety is that the estimates in (2.7) are also critical in the sense that $C_{q,\alpha} \nearrow \infty$ as $q \nearrow 2$ or $\alpha \nearrow 1$. Recently, L.C. Evans [25] has found another proof of part A) of the theorem for steady velocity fields by utilizing capacity theory.

2B. Concentration-Cancellation

The terminology "concentration-cancellation" was introduced in [21, 22] and describes the phenomena that even though $v^\varepsilon \otimes v^\varepsilon$ does not converge weakly to $v \otimes v$ due to the development of concentrations while $||v^\varepsilon - v||_{L^1} \to 0$, the velocity field v nevertheless defines a weak solution for 2-D Euler – in simple exact solutions such as those in (1.14), the concentrations in (1.15) actually cancel against test functions in the weak form in the limit (see [21]). At the present time, it is an important open problem to determine whether concentration-cancellation always occurs in the limit for solution sequences satisfying the bounds in (2.1) which are appropriate for vortex sheet initial data. Nevertheless, there are some interesting partial results [22, 27, 1, 61]. The first results are given in the following

Theorem [22]. *A) For steady solution sequences for 2-D Euler satisfying the bounds in* (2.1)*, concentration-cancellation always occurs.*

B) For time-dependent solution sequences for 2-D Euler satisfying the bounds in (2.1) *assume additionally that the reduced defect measure,* θ*, concentrates on a*

set with Hausdorff dimension strictly less than one, then concentration-cancellation occurs.

Lack of space prevents a detailed discussion of the proof of the theorem but a few comments are appropriate. The proofs crucially exploit the special nonlinear covariant structure in the weak form of the Euler equations which is a direct consequence of the rotational invariance of solutions of (0.1), i.e. if $v^{\varepsilon}(x,t)$ satisfies 2-D Euler, so does ${}^{t}\sigma v^{\varepsilon}(\sigma x, t)$ for any rotation matrix σ. The proofs also crucially use the fact that the reduced defect measure, θ, concentrates on a small set (see the theorem in the previous subsection) to build a "shielding sequence" of test functions which shields the limit process from the concentration defects (see Section 3 of [22]). A simpler proof of part B) of the theorem was presented in [27]. Recently, in interesting work, Alinhac [1] and Zheng [61] have proved that concentration-cancellation occurs provided it is assumed that the weak star defect measure (not the reduced defect measure!) has support which is a set of roughly one-dimensional Hausdorff measure. These proofs use the "shielding sequence" strategy from [22] in an essential fashion. All of these partial results still leave unanswered the following basic issue:

Problem. Does concentration-cancellation always occur for time-dependent solution sequences with the bound in (2.1) when θ concentrates on a set with space-time cylindrical Hausdorff measure exactly one?

There is probably an affirmative answer to this question when the vorticity has a distinguished sign, i.e. $\omega^{\varepsilon}(x,t) \geq 0$. It should be mentioned here that for smooth solutions of 2-D Euler a non-negative vorticity at later times is automatically guaranteed by a non-negative initial vorticity, $\omega_0^{\varepsilon}(x)$ because the vorticity is transported along fluid particle trajectories. In particular, Theorem 3.1 from [21] guarantees that for $\omega^{\varepsilon}(x,t) \geq 0$ the solution sequence v^{ε} converges strongly to v in L^2 and v defines a weak solution of 2-D Euler provided that the vorticity maximal function using (2.4) decays at the extremely slow rate,

$$\max_{\substack{x\varepsilon R^2 \\ 0\leq t\leq T}} \omega^{\varepsilon}(B(r,x),t) \leq C\left(\log\left(\frac{1}{r}\right)\right)^{-(1+\beta)}, \quad r \leq 1 \tag{2.8}$$

for some $\beta > 0$ and any $T > 0$. On the other hand, it is not difficult to prove [38] the following apriori decay estimate for this maximal function assuming positive vorticity distributions, $\omega^{\varepsilon}(x,t) \geq 0$ with bounded support:

$$\max_{\substack{x\varepsilon R^2 \\ 0\leq t<+\infty}} \omega^{\varepsilon}(B(r,x),t) \leq C\left(\log\left(\frac{1}{r}\right)\right)^{-1/2}, \quad r \leq 1\,. \tag{2.9}$$

Thus, for positive signed vorticity, (2.8) and (2.9) indicate that there is an extremely small discrepancy between conditions guaranteeing strong convergence to a weak solution and the general case. Of course, the elementary examples in (1.14A) can be used to show that concentration does occur with positive vorticity distributions and also that the estimate in (2.9) is sharp so that some assessment of concentration-cancellation is still needed. On the other hand, the elementary examples in (1.14B) illustrate that there is no apriori decay rate for the vorticity maximal function when the vorticity changes sign. The interested reader can consult [46] for a report on the striking computational evidence indicating vastly increased complexity in numerical solutions when the vorticity changes sign.

References

[1] Alinhac, S.: Un phenomene de concentration evanescente pour des flots non-stationnaires incompressibles en dimension deux. Comm. Math. Phys. **127** (1990) pp. 585–596

[2] Almgren, R., Majda, A., Rosales R.: Rapid initiation through high frequency resonant nonlinear acoustics. Phys. Fluids A **2** (1990) 1014–1029

[3] Artola, M., Majda, A.: Nonlinear development of instabilities in supersonic vortex sheets I: the basic kink modes. Physica **28D** (1988) 253–281

[4] Artola, M., Majda, A.: Nonlinear development of instabilities in supersonic vortex sheets II: resonant interaction among kink modes. S.I.A.M. J. Appl. Math. **49** (1989) 1310–1349

[5] Ball, J.: Energy-minimizing configurations in nonlinear elasticity. Proceedings of I.C.M. Warsaw, 1983, pp. 1309–1314

[6] Ball, J.: A version of the fundamental theorem for Young measures. (Lecture Notes in Physics, vol. 344, eds. M. Rascle, D. Serre, M. Slemrod). Springer, Berlin Heidelberg New York 1981, pp. 207–215

[7] Ball, J., James, R.D.: Fine phase mixtures as minimizers of energy. Arch. Rat. Mech. Anal. **100** (1987) 13–52

[8] Ball, J., James, R.D.: Proposed experimental tests of a theory of fine microstructure and the two-well problem. Preprint 1990

[9] Brenier, Y.: The least action principle and the related concept of generalized flows for incompressible perfect fluids. J. Am. Math. Soc. **2** (1989) 225–255

[10] Brezis, H., Coron, J.M., Lieb, E.: Harmonic maps with defects. Comm. Math. Phys. **107** (1987) 649–705

[11] Caflisch, R.: Mathematical aspects of vortex dynamics. S.I.A.M. Publications, Philadelphia 1989

[12] Cehelsky, P., Rosales, R.: Resonantly interacting weakly nonlinear hyperbolic waves in the presence of shocks: A single space variable in a homogeneous time independent medium. Stud. Appl. Math. **74** (1986) 117–138

[13] Chipot, M., Kinderlehrer, D.: Equilibrium configurations of crystals. Arch. Rat. Mech. Anal. **103** (1988) 237–277

[14] Chorin, A.J.: Estimates of intermittency, spectra and blow-up in developed turbulence. Comm. Pure Appl. Math. **34** (1981) 853–866

[15] Chorin, A.J.: The evolution of a turbulent vortex. Comm. Math. Phys. **83** (1982) 517–535

[16] Collins, C., Luskin, M.: The computation of the austenitic-martensitic phase transition. Lecture Notes in Physics, vol. 344 (eds. M. Rascle, D. Serre, M. Slemrod). Springer, Berlin Heidelberg New York 1989, pp. 34–50

[17] Constantin, P., Majda, A.: The Beltrami spectrum for incompressible fluid flows. Comm. Math. Phys. **115** (1988) 435–456

[18] Diperna, R.: Measure-valued solutions to conservation laws. Arch. Rat. Mech. Anal. **88** (1985) 223–270

[19] Diperna, R.: Compactness of solutions to nonlinear P.D.E. Proceedings of I.C.M. Berkeley, 1986, pp. 1057–1063

[20] Diperna, R., Majda, A.: Oscillations and concentrations in weak solutions of the incompresible fluid equations. Comm. Math. Phys. **108** (1987) 667–689

[21] Diperna, R., Majda, A.: Concentrations in regularizations for 2-D incompressible flow. Comm. Pure Appl. Math. **40** (1987) 301–345
[22] Diperna, R., Majda, A.: Reduced Hausdorff dimension and concentration-cancellation for 2-D incompressible flow. J. Am. Math. Soc. **1** (1988) 59–95
[23] Diperna, R., Majda, A.: The validity of nonlinear geometric optics for weak solutions of conservation laws. Comm. Math. Phys. **98** (1985) 313–347
[24] Ericksen, J.L., Kinderlehrer, D. (eds.): Theory and application of liquid crystals, vol. 5. I.M.A. Volumes in Math. and Appl. Springer, New York
[25] Evans, L.C.: Weak Convergence Methods for Nonlinear Partial Differential Equations. Regional Conf. Series in Mathematics. Am. Math. Soc., Providence, RI, 1990
[26] Gerard, P.: Compacite par compensation et regularite 2-microlocale. Seminaire d'EDP (Ecole Polytechnique, Paris), et Article a paraitre 1988–1989
[27] Greengard, C., Thomann, E.: On Diperna-Majda concentration sets for two-dimensional incompressible flow. Comm. Pure Appl. Math. **41** (1988) 295–303
[28] Hunter, J., Keller, J.B.: Weak shock diffraction. Wave Motion **6** (1984) 79–89
[29] Hunter, J.K., Keller, J.B.: Weakly nonlinear high frequency waves. Comm. Pure Appl. Math. **36** (1983) 543–569
[30] Hunter, J.K., Majda, A., Rosales, R.: Resonantly interacting weakly nonlinear hyperbolic waves, II: several space variables. Stud. Appl. Math. **75** (1986) 187–226
[31] Hunter, J.: Nonlinear geometrical optics. To appear in I.M.A. volume in Math. and Appl. titled on *Multi-dimensional Hyperbolic Waves* edited by J. Glimm and A. Majda
[32] James, R.D., Kinderlehrer, D.: Theory of diffusionless phase transformations. (Lecture Notes in Physics, vol. 344, eds. M. Rascle, D. Serre, M. Slemrod). Springer, Berlin Heidelberg New York 1989, pp. 51–84
[33] Krasny, R.: Proceedings of this International Congress, p. 1573
[34] Landman, M., Papanicolaou, G.C., Sulem, C. Sulem, P.L.: Rate of blow-up for solutions of the nonlinear Schrödinger equation in critical dimension. Phys. Rev. **A 38** (1988) 3834–3843
[35] Lions, P.L.: The concentration-compactness principle in the calculus of variations: The limit case, Part I and Part II. Rev. Mat. Iberoamericana, vol. 1, no. 1, pp. 145–201 et no. 2, pp. 45–121 (1985)
[36] Majda, A.: Vorticity, turbulence, and acoustics in fluid flow. The 1990 John von Neumann lecture of the Society for Industrial and Applied Mathematics. To appear in S.I.A.M. Review in 1991
[37] Majda, A.: One perspective on open problems in multi-dimensional conservation laws. To appear in I.M.A. volume in Math. and Appl. titled *Multi-dimensional Hyperbolic Waves* edited by J. Glimm and A. Majda
[38] Majda, A.: Lecture notes on oscillations and concentrations for incompressible fluid flow. Princeton University (1988–1989 academic year)
[39] Majda, A.: Compressible fluid flow and systems of conservation laws in several space variables. (Applied Mathematical Sciences, vol. 53.) Springer, Berlin Heidelberg New York 1984
[40] Majda, A.: Nonlinear geometric optics for hyperbolic systems of conservation laws. In: Oscillation theory, computation, and methods of compensated compactness, IMA vol. 2. Springer, New York 1986 pp. 115–165
[41] Majda, A., Rosales, R.: Nonlinear mean field-high frequency wave interactions in the induction zone. S.I.A.M.J. Appl. Math. **47** (1987) 1017–1039
[42] Majda, A., Rosales, R.: Resonantly interacting weakly nonlinear hyperbolic waves, I: A single space variable. Stud. Appl. Math. **71** (1984) 149–179
[43] Majda, A., Rosales, R., Schonbek, M.: A canonical system of integro-differential equations arising in resonant nonlinear acoustics. Stud. Appl. Math. **79** (1988) 205–261

[44] Majda, A.: High mach number combustion. In: Reacting flows: Combustion and chemical reactors. AMS Lectures in Applied Mathematics **24** (1986) 109–184
[45] Majda, A., Rosales, R.: A Theory for spontaneous mach stem formation in reacting shock fronts, I: The basic perturbation analysis. S.I.A.M. J. Appl. Math **43** (1983) pp. 1310–1334
[46] Majda, A.: Vortex dynamics: Numerical analysis, scientific computing, and mathematical theory. Proceedings First International Congress for Industrial and Applied Mathematics. S.I.A.M. Publications, 1988, pp. 153–182
[47] Majda, A.: Vorticity and the mathematical theory of incompressible fluid flow. Comm. Pure Appl. Math., vol 39, 1986, pp. S187–S220
[48] Majda, A., Artola, M.: Nonlinear geometric optics for hyperbolic mixed problems. In: Analyse mathematique et applications. In honor of J.L. Lions, Gauthier-Villars, Paris 1988, pp. 319–356
[49] Maslov, V.P.: Mathematical aspects of integral optics. Moscow VINITI 1983
[50] Merle, F.: On uniqueness and continuation properties after blow-up time of self-similar solutions of nonlinear Schrödinger equation with critical exponent and critical mass. Preprint 1990
[51] Mimura, M.: Proceedings of this International Congress, p. 1627
[52] Rosales, R., Majda, A.: Weakly nonlinear detonation waves. S.I.A.M. J. Appl. Math. **43** (1983) 1086–1118
[53] Rosales, R.: A introduction to weakly nonlinear geometrical optics. To appear in I.M.A. volume on *Multi-dimensional Hyperbolic Waves* edited by J. Glimm and A. Majda
[54] Serre, D.: Richness and the classification of quasilinear hyperbolic systems. Preprint 1989
[55] Serre, D.: Oscillations non lineaires des systemes hyperboliques: Methods et resultats qualitatifs. Preprint 1989
[56] Stein, E.M.: Singular integrals and the differentiability properties of functions. Princeton University Press, NJ, 1970
[57] Tartar, L.: H-measures, a new approach for studying homogenization, oscillations and concentration effects in partial differential equations. To appear in Proc. Roy. Soc. Edinburgh
[58] Tartar, L.: Proceedings of this International Congress
[59] Weinstein, M.I.: Nonlinear Schrödinger equations and sharp interpolation estimates. Comm. Math. Phys. **87** (1983) 567–576
[60] Weinstein, M.I.: On the structure and formation of singularities in solutions to the nonlinear dispersive evolution equations. Comm. Partial Differential Equations **11** (1986) 545–565
[61] Zheng, Y.: Concentration-cancellation phenomena for weak solutions to certain nonlinear partial differential equations. Ph.D. thesis, Univ. of Calif. at Berkeley (April 1990)

Dynamical and Ergodic Properties of Subgroup Actions on Homogeneous Spaces with Applications to Number Theory

Grigorii A. Margulis

Institute for Problems in Information Transmission, Academy of Sciences of USSR
ul. Ermolovoi 19, Moscow 101447, USSR

A subgroup action on a homogeneous space is a classical object of study in ergodic theory. Some old and new results about ergodic properties of this section will be stated. We shall also formulate results on orbit closures and give applications of these results to number theory and, in particular, to the study of sets of values of indefinite quadratic forms at integral points. In the last section, formulations of some new results about actions on general manifolds and measure spaces will be given.

1. Ergodicity Theorems

Let G be a Lie group, F a subgroup of G and D a closed subgroup of G. Let us consider the natural action of F by left translations on the homogeneous space G/D. This action is called *ergodic* if, for every F-invariant Borel subset A of G/D, either $\mu(A) = 0$ or $\mu((G/D) - A) = 0$ where μ is a G-quasiinvariant measure on G/D. One can easily prove the following two assertions.

Proposition 1. *The action of F on G/D is ergodic if and only if the closure of F acts ergodically on G/D.*

Proposition 2 (Duality Principle). *Let us assume that F is closed. Then the action of F on G/D is ergodic if and only if the action of D on G/F is ergodic.*

We say that the homogeneous space G/D has finite volume if there exists a finite G-invariant measure on G/D. If G/D *has finite volume* and D is discrete then D is called a *lattice.*

1.1 Mautner Property

If G/D carries a G-invariant measure μ then one can define a continuous unitary representation ϱ of the group G on the Hilbert space $L_2(G/D, \mu)$ of square μ-integrable functions on G/D by the formula $(\varrho(g)f)(x) = f(g^{-1}x)$ where $f \in L_2(G/D, \mu)$, $g \in G$ and $x \in G/D$. It turns out that many ergodic properties of the action of F on G/D

Proceedings of the International Congress
of Mathematicians, Kyoto, Japan, 1990

can be reformulated in terms of the representation ϱ. This observation gives an approach to the study of ergodic properties based on representation theory; this approach was first used by I.M. Gelfand and S.V. Fomin (see [28]).

One can easily see that, in the case where G/D has finite volume, the action of F on G/D is ergodic if and only if each $\varrho(F)$-invariant function $f \in L_2(G/D, \mu)$ is constant or, in other words, if f is $\varrho(G)$-invariant. Therefore Lemma 1 stated below is true. Before the formulation of this lemma we give

Definition 1. We say that a subgroup H of G has *Mautner property* with respect to F if, for any continuous unitary representation φ of G on a Hilbert space W and any $w \in W$ such that $\varphi(F)w = w$, we have that $\varphi(H)w = w$.

Lemma 1. *Let the homogeneous space G/D have finite volume, and let H be a subgroup of G. Let us assume that H has Mautner property with respect to F and that the closure of the subgroup generated by F and H acts ergodically on G/D. Then the action of F on G/D is ergodic.*

Mautner property was first used by F.J. Mautner to prove the ergodicity of geodesic flows on symmetric Riemannian spaces (see [43]). C.C. Moore proved Theorem 1 formulated below about Mautner property of Ad-compact normal subgroups. Let us give at first the definition of these subgroups.

Definition 2. The Ad-*compact normal subgroup* for F is the smallest subgroup in the class of all connected normal subgroups H of G satisfying the following condition

(A) for each g from the closure of F, the linear transformation $\mathrm{Ad}(\pi(g))$ is diagonalizable and its eigenvalues have absolute value 1 where Ad is the adjoint representation of G/H and $\pi : G \to G/H$ is the canonical epimorphism.

Remark. It is not difficult to prove that the condition (A) is equivalent to the condition

(A′) the subgroup $\mathrm{Ad}(\pi(F))$ is relatively compact in the group $\mathbf{GL}(\mathfrak{G}/\mathfrak{H})$ of linear transformations of the Lie algebra $\mathfrak{G}/\mathfrak{H}$ of the group G/H.

Let us denote the Ad-compact normal subgroup for F by H_F. If G is connected and semisimple then H_F is the product of all almost simple factors G_i of G such that the restriction of $\mathrm{Ad}(F)$ to the Lie algebra $\mathfrak{G}_i$ of G_i is not relatively compact in $\mathbf{GL}(\mathfrak{G}_i)$.

Theorem 1 (see [45]). *The subgroup H_F has Mautner property with respect to F.*

Remark 1. It is not difficult to show that the closure of the subgroup generated by F and H_F is the maximal subgroup of G having Mautner property with respect to F.

Remark 2. An analogue of Theorem 1 for semisimple groups over locally compact fields of characteristic zero was proved by S.G. Dani in [10].

Roughly speaking, the proof of Theorem 1 is based on the following simple

Lemma 2. *Let G be a topological group, F a subgroup of G, φ a continuous unitary representation of G on a Hilbert space W, and let $g_0 \in G$ and $w \in W$. Assume that, for any neighbourhood $V \subset G$ of the identity $e \in G$, g_0 is contained in he closure of the subset FVF. Then $\varphi(g_0)w = w$ whenever $\varphi(F)w = w$.*

To prove Lemma 2, it is enough to consider the function $f(g) = \langle \varphi(g)w, w \rangle$, $g \in G$, and remark that (a) since φ is unitary and $\varphi(F)w = w$, the function f is constant on a double coset modulo F; (b) the function f is continuous; (c) $f(e) = \|w\|^2$ and $f(g) = \|w\|^2$ if and only if $\varphi(g)w = w$. A special case of Lemma 2 is the following

Lemma 3 (Generalized Mautner Lemma). *Let G be a topological group and let x, $y \in G$ be the elements such that the sequence $\{x^n y x^{-n}\}$ converges to e as $n \to +\infty$. If φ is a continuous unitary representation of the group G on a Hilbert space W, $w \in W$ and $\varphi(x)w = w$ then $\varphi(y)w = w$.*

Example 1. Let $G = \mathbf{SL}_2(\mathbb{R})$ be the group of real unimodular 2×2 matrices and let $g = \begin{pmatrix} a & 0 \\ 0 & a^{-1} \end{pmatrix} \in \mathbf{SL}_2(\mathbb{R})$ where $|a| \neq 1$. Let F denote the cyclic subgroup generated by g, and let

$$U = \left\{ \begin{pmatrix} 1 & t \\ 0 & 1 \end{pmatrix} \middle| t \in \mathbb{R} \right\} \quad \text{and} \quad U^- = \left\{ \begin{pmatrix} 1 & 0 \\ t & 1 \end{pmatrix} \middle| t \in \mathbb{R} \right\}.$$

One can easily check that if $y \in U \cup U^-$ then one of the sequences $\{g^n y g^{-n}\}$ and $\{g^{-n} y g^n\}$ converges to e as $n \to +\infty$. In view of Lemma 3, it implies that both U and U^- have Mautner property with respect to F. But these subgroups generate G. Hence G has Mautner property with respect to F.

Example 2. Let G, g, a, F and U denote the same as in Example 1. If $c \neq 0$ we have that

$$\begin{pmatrix} 1 & c^{-1}(1-a) \\ 0 & 1 \end{pmatrix} \begin{pmatrix} a & 0 \\ c & a^{-1} \end{pmatrix} \begin{pmatrix} 1 & c^{-1}(1-a^{-1}) \\ 0 & 1 \end{pmatrix} = \begin{pmatrix} 1 & 0 \\ c & 1 \end{pmatrix}$$

and hence $\begin{pmatrix} a & 0 \\ c & a^{-1} \end{pmatrix} \in U \begin{pmatrix} 1 & 0 \\ c & 1 \end{pmatrix} U$. It implies that, for any neighbourhood $V \subset G$ of e, g is contained in the closure of the subset UVU. Now applying Lemma 2, we get that F has Mautner property with respect to U. On the other hand according to Example 1, G has Mautner property with respect to F. Hence G has Mautner property with respect to U.

1.2 Criteria of Ergodicity for Semisimple G

Theorem 1 and Lemma 1 imply

Theorem 2 (see [45]). *Let G be a Lie group, F a subgroup of G, and D a closed subgroup of G such that G/D has finite volume. Let H_F denote the* Ad*-compact normal subgroup for F. Then the action of F on G/D is ergodic if and only if the closure of the subgroup FH_F acts ergodically on G/D.*

Using Theorem 2 and Propositions 1 and 2, it is not difficult to prove

Theorem 3 (see [44], [45] and [46]). *Let G, F, D and H_F be the same as in Theorem* 2.

(a) *If the subgroup DH_F is dense in G then the action of F on G/D is ergodic.*
(b) *Let us assume that G is semisimple and that the action of F on G/D is ergodic. Then DH_F is dense in G.*

Definition 2. A lattice $\Gamma \subset G$ in a connected semisimple Lie group without compact factors is *reducible* if G admits connected normal subgroups H, H' such that $HH' = G$, $H \cap H'$ is discrete and $\Gamma/(\Gamma \cap H)(\Gamma \cap H')$ is finite. A lattice is *irreducible* if it is not reducible.

It is known (see [54, Corollary 5.21]) that if a lattice $\Gamma \subset G$ in a connected semisimple Lie group without compact factors is irreducible then, for any nontrivial connected normal subgroup H of G, the subgroup ΓH is dense in G. Therefore Theorem 3 implies

Theorem 4 (see [44] or [80, Theorem 2.2.6]). *Let G be a connected semisimple Lie group without compact factors, Γ an irreducible lattice in G, and $F \subset G$ a subgroup which is not relatively compact in G. Then the action of F on G/Γ is ergodic.*

Let us note that Theorem 4 is used in an essential way in the proofs of the strong rigidity theorem and the superrigidity theorem (see [40] and [47]).

Example 3. Let G, g, F and U denote the same as in Example 1 and let Γ be a lattice in G. Then in view of Theorem 4, both F and U act ergodically on G/Γ. This assertion follows also from Lemma 1 and Examples 1 and 2. Let us note that the ergodicity of actions of F and U on G/Γ implies (in fact, is equivalent to) the classical results of E. Hopf and G.A. Hedlund on ergodicity of geodesic and horocycle flows on a surface of constant negative curvature and finite area (see [29]).

1.3 Criteria of Ergodicity for Nilpotent and Solvable G

For the case of nilpotent G, there is a good ergodicity criterion due to L. Green. Before we formulate this criterion, let us recall that (a) G/D is compact if and only if G/D has finite volume; (b) whenever G/D is compact, the subgroup $D \cdot [G, G]$ is closed and, consequently, $G/D \cdot [G, G]$ is a torus where $[G, G]$ denotes the commutator subgroup of G (see [54, Chap. II]).

Theorem 5 (see [1, Chap. V]). *Let G be a connected nilpotent Lie group, F a subgroup of G, and D a closed subgroup of G such that G/D is compact. The action of F on G/D is ergodic if and only if $\pi(F)$ is dense in the torus $G/D\cdot[G, G]$ where $\pi: G \to G/D\cdot[G, G]$ denotes the canonical epimorphism.*

Remark 1. In [1], Theorem 5 is formulated only in the case of one-parameter F. But the proof given there is applicable in general case.

Remark 2. It is possible to deduce Theorem 5 from Theorem 3(a).

If G is solvable but not nilpotent then the ergodicity criterion is rather complicated (see Appendix 1 to the Russian translation of [1], [5] or [67]). This criterion is based on the reduction to the case of nilpotent G. The reduction consists of two steps: (1) the reduction to the case of solvable groups of type (R) (by definition, a group G is of type (R) if, for each $g \in G$, all eigenvalues of Ad g have absolute value 1); (2) the reduction from the case of solvable groups of type (R) to the case of nilpotent groups. The first reduction is based on Mautner lemma (Lemma 3), and the second one is based on Malcev's construction of semisimple splitting.

1.4 Ergodicity Criterion for Arbitrary G

Let G be a connected Lie group and D a closed subgroup of G. By a quotient of G/D we mean a homogeneous space G/B where $B \subset G$ is a closed subgroup which contains D. If B contains a closed normal subgroup L of G such that the factor group G/L is Euclidean (resp. semisimple) then the quotient G/B is called *Euclidean* (resp. *semisimple*); a connected solvable Lie group is called *Euclidean* if it is locally isomorphic to an extension of a vector group by a compact commutative Lie group. It is clear that there exists the maximal Euclidean (resp. semisimple) quotient G/B of G/D, i.e. if G/B' is another Euclidean (resp. semisimple) quotient of G/D then $B' \supset B$ or, in other words, G/B' is a quotient of G/B. Let us note that if G/D has finite volume then arbitrary (resp. arbitrary Euclidean) quotient of G/D has finite volume (resp. is compact).

Theorem 6 (see [5] and [12]). *Let G be a connected Lie group, $F \subset G$ a subgroup which is either one-parameter or cyclic, and $D \subset G$ a closed subgroup such that G/D has finite volume. The action of F on G/D is ergodic if and only if the actions of F on the maximal Euclidean and semisimple quotients of G/D are both ergodic.*

Remark. It is assumed in [5] and [12] that G/D is so called admissable homogeneous space. But according to [66], this condition is satisfied whenever G/D has finite volume.

1.5 Ergodic Decomposition of Nonergodic Flows

A.N. Starkov obtained a number of results on the ergodic decomposition of nonergodic flows (i.e. nonergodic actions of one-parameter subgroups). Let us formulate some of these results in the form of the following

Theorem 7 (see [65], [67] and [69]). *Let G be a connected Lie group, $F = \{f_t\}$ a one-parameter subgroup of G, and D a closed subgroup of G such that G/D has finite volume. Then there exists two partitions E and $\tilde{E}$ of G into closed F-invariant submanifolds $E(x)$ and $\tilde{E}(x)$ with the following properties*

(i) *$x \in E(x) \subset \tilde{E}(x)$ for each $x \in G/D$;*

(ii) *$E(x) = \tilde{E}(x)$ for almost all $x \in G/D$ (i.e. for all $x \in (G/D) - A$ where $\mu(A) = 0$ and μ is a G-invariant measure on G/D);*

(iii) *for each $x \in G/D$, there exist smooth finite F-invariant measures μ_x on $E(x)$ and $\tilde{\mu}_x$ on $\tilde{E}(x)$;*

(iv) *for each $x \in G/D$, the action of F on $E(x)$ is ergodic (relative to the measure μ_x);*

(v) *$\tilde{E}(x) = \overline{FH_F x}$ and $E(x) = \overline{FI_F x}$ for each $x \in G/D$ where $\bar{A}$ denotes the closure of a subset A, H_F denotes the* Ad*-compact normal subgroup for F, and $I_F \subset H_F$ is a normal subgroup of G. (The subgroup I_F is called the Dani subgroup and is characterized as the smallest connected normal subgroup of G such that, for each $g \in F$, the operator* Ad *g have on $\mathfrak{G}/\mathfrak{I}$ only eigenvalues with absolute value 1 and this operator is diagonalizable on $\mathfrak{G}/\mathfrak{R} + \mathfrak{I}$ where $\mathfrak{G}$ (resp. $\mathfrak{I}$) is the Lie algebra of G (resp. I_F) and $\mathfrak{R}$ is the radical of $\mathfrak{G}$).*

(vi) *for each $x \in G/D$, there exist a connected Lie group G_x, a one-parameter subgroup $F_x = \{f_{x,t}\} \subset G_x$, a closed subgroup $D_x \subset G_x$, and a finite covering π: $G_x/D_x \to E(x)$* such that *G_x/D_x has finite volume, the action of F_x on G_x/D_x is ergodic, and $\pi(f_{x,t}y) = f_t\pi(y)$ for each $y \in G_x/D_x$.*

2. Spectrum and Mixing

We refer to [9] and [63] about the notions of pure point spectrum, Lebesgue spectrum, K-flow, Bernoulli shift etc. and about results from ergodic theory.

2.1 Spectrum of Flows on Homogeneous Spaces

Let G be a connected Lie group, $F = \{f_t\}$ a one-parameter subgroup of G, and D a closed subgroup of G such that G/D has finite volume. Let μ denote the normalized ($\mu(G/D) = 1$) Borel G-invariant measure on G/D. As in Sect. 1, let us define the unitary representation ϱ on $L_2(G/D, \mu)$. For any quotient $E = G/B$ of G/D, let

$$L_2(E) = \{f \in L_2(G/D, \mu) | f(g_1 D) = f(g_2 D) \text{ if } g_1 \in g_2 B\}.$$

It is clear that $L_2(E)$ is $\varrho(F)$-invariant.

Theorem 8 (see [5]). *Let E denote the maximal Euclidean quotient of $M = G/D$. Suppose that the action of F on M is ergodic. Then the restriction of $\varrho(F)$ to $L_2(E)$ has pure point spectrum (the set of eigenvalues being a finitely generated subgroup of the real line of rank equal to the dimension of E). If $L_2(E) \neq L_2(M)$ i.e. if M is not Euclidean, the restriction of $\varrho(F)$ to the orthogonal complement $L_2(M) \ominus L_2(E)$ has Lebesgue spectrum of infinite multiplicity.*

Remark 1. It follows from Theorems 7 and 8 that, also in the case where the action of F on G/D is not ergodic, the space $L_2(G/D, \mu)$ can be decomposed into the direct sum of two $\varrho(F)$-invariant subspaces H_1 and H_2 such that the restriction of $\varrho(F)$ to H_1 has pure point spectrum and the restriction of $\varrho(F)$ to H_2 has Lebesgue spectrum of infinite multiplicity.

Remark 2. The proof of Theorem 8 is based on some results about unitary representations of semi-direct products and about the spectrum of K-flows.

Remark 3. Theorem 8 generalizes some earlier results about the spectrum of flows on homogeneous spaces (see [1, 10, 12, 28, 43, 44, 45, 53, 61, 70, 71]).

2.2 Mixing Properties

Theorem 8 implies the following

Theorem 9 (see [5]). *Suppose that the action of F on G/D is ergodic and that there are no nontrivial Euclidean quotients of G/D. Then the section of $F = \{f_t\}$ on G/D is mixing, i.e.*

$$\lim_{t\to\infty} \mu(f_t A \cap B) = \mu(A)\mu(B)$$

for any measurable susets $A, B \subset G/D$.

Analogues of Theorems 8 and 9 for cyclic F are proved in [12]. In [10, 11] and [12], S.G. Dani obtained criteria when the translation $T_g : G/D \to G/D$, $T_g(x) = gx$, $x \in G/D$, is a K-automorphism, when T_g is a Bernoulli shift, and when T_g has positive entropy. In particular, he proved that T_g has positive entropy if and only if $|\lambda| \neq 1$ for at least one eigenvalue λ of Ad g (the same is true for algebraic groups over nonarchimedean locally compact fields).

B. Marcus proved that, in many cases, the action of F on G/D is mixing of all degrees. In particular, he proved

Theorem 10 (see [34]). *Suppose that the group G is semisimple and has no compact factors and that the action of $F = \{f_t\}$ on G/D is ergodic. Then this action is mixing of all degrees, i.e.*

$$\lim_{t_1,\dots,t\to+\infty} \mu(A_0 \cap f_{t_1} A_1 \cap f_{t_1+t_2} A_2 \cap \cdots \cap f_{t_1+t_2+\dots+t_r} A_r) = \mu(A_0)\cdot\ldots\cdot\mu(A_r)$$

for any integer $r \geq 1$ and any measurable subsets $A_0, A_1, \dots, A_r \subset G/D$.

Combining Theorem 4 and Theorem 10, we get

Theorem 11. *Let G be a connected semisimple Lie group without compact factors, Γ an irreducible lattice in G, and $F \subset G$ a one-parameter subgroup which is not relatively compact in G. Then the action of F on G/Γ is mixing of all degrees.*

A special case of Theorem 11 is

Corollary (see [34]). *The horocycle flow on a surface of constant negative curvature and finite area is mixing of all degrees.*

This corollary proves the conjecture which was posed by Ya.G. Sinai on the International congress of mathematicians in Stockholm (1962). In [34], B. Marcus stated the conjecture that "mixing $\Rightarrow$ mixing of all degrees" for the action of F on G/D.

3. Invariant Measures for Actions of Unipotent Subgroups

Let G be a Lie group. An element u of G will be called *unipotent* if the transformation Ad u of the Lie algebra of G is unipotent, i.e. if all eigenvalues of Ad u are equal to 1. A subgroup U of G will be called *unipotent* if it consists of unipotent elements.

3.1 Description of Finite Ergodic Measures

Let F be a subgroup of G and D a closed subgroup of G. A F-invariant Borel measure μ on G/D is called *F-ergodic* if the action of F on G/D is ergodic relative to μ, i.e., if for every F-invariant Borel subset A of G/D, either $\mu(A) = 0$ or $\mu((G/D) - A) = 0$. Let us note that each F-invariant Borel measure on G/D can be decomposed into a (continuous) sum of F-ergodic measures. It seems impossible to describe in the general case all F-invariant Borel probability measures on G/D. But in view of the following recent fundamental theorem of M. Ratner, this is possible in the case where F is unipotent and D is discrete.

Theorem 12 (see [58, 59] and [60]). *Let G be a Lie group, Γ a discrete subgroup of G (not necessarily a lattice) and U a unipotent subgroup of G. Then for any U-invariant U-ergodic Borel probability measure σ on G/Γ, there exist $x \in G/\Gamma$ and a closed subgroup $H \subset G$ containing U such that the set Hx is closed in G/Γ and σ is a finite H-invariant measure the support of which is Hx. (In other words, H_x is a lattice in H and the measure σ is the image of a H-invariant measure on H/H_x under the canonical map $H/H_x \to Hx$ where $H_x = \{h \in H | hx = x\}$ is the stabilizer of x in H.)*

Roughly speaking, the difference between the case of unipotent subgroups and the general case is explained by the fact that, for an action of unipotent subgroups, the divergence of trajectories is "polynomial" in contrast to an action of general F where the divergence of trajectories is usually exponential.

Theorem 12 proves a conjecture which M. Ratner suggests to call Raghunathan's measure conjecture. It was first posed in [13]. The conjecture was suggested by the results of H. Furstenberg and W. Parry for nilpotent G and of S.G. Dani for the case where G is reductive and U is horosprerical (see [13], if G/Γ is compact see also [72]; a subgroup $U \to G$ is called *horospherical* if there exists $g \in G$ such that

$$U = \{u \in G | g^j u g^{-j} \to e \text{ as } j \to +\infty).$$

Generalizing a result of H. Furstenberg from [27], S.G. Dani proved the conjecture for the case where $G = \mathbf{SL}_2(\mathbb{R})$ (see [14]).

Theorem 12 provides many important ergodic theoretic consequences. In particular, it implies the joinings theorem and the factor theorem (see [59]). The joinings theorem describes invariant ergodic measures for a product of two actions of cyclic unipotent groups, and the factor theorem describes U-invariant measurable partitions of G/Γ where U is a cyclic unipotent subgroup which acts ergodically on G/Γ. These theorems generalize analogous results for $G = \mathbf{SL}_2(\mathbb{R})$ obtained by M. Ratner (see [56] and [57]). Let us formulate a corollary of the joinings theorem.

Theorem 13 (The Rigidity Theorem; see [59]). *Let G_i be a connected semisimple Lie group, Γ_i a lattice in G_i containing no non-trivial normal subgroups of G_i and u_i a unipotent element of G_i, $i = 1, 2$. Let U_i denote the cyclic subgroup generated by u_i. Suppose that the action of U_1 on G_1/Γ_1 is ergodic and there is a measure preserving map $\psi : G_1/\Gamma_1 \to G_2/\Gamma_2$ such that $\psi(xu_1) = \psi(x)u_2$ for μ_1-almost every $x \in G_1/\Gamma_1$, where μ_1 denotes a G_1-invariant measure on G_1/Γ_1. Then one can find $h \in G_1$ and a surjective homomorphism $\alpha : G_1 \to G_2$ such that $\alpha(\Gamma_1) \subset h\Gamma_2 h^{-1}$ and $\psi(g\Gamma_1) = \alpha(g)h\Gamma_2$ for μ_1-almost every $g\Gamma_1 \in G_1/\Gamma_1$. Also α is a local isomorphism whenever ψ is finite to one or G_1 is simple and it is an isomorphism whenever ψ is one-to-one or G_1 is simple with trivial center.*

Theorem 13 generalizes Ratner's rigidity theorem for $G_i = \mathbf{SL}_2(\mathbb{R})$ (see [55]). This theorem had been previously obtained by D. Witte (see [78] and [79]) with the help of methods from [55] and [57].

3.2 Finiteness of Ergodic Measures

Theorem 12 gives the description of finite ergodic measures for actions of unipotent groups. Now let us formulate a theorem about the finiteness of ergodic measures for these actions.

Theorem 14 (see [15]). *Let G be a connected Lie group, Γ a lattice in G, and U a unipotent subgroup of G. Let ν be a locally finite U-invariant measure on G/Γ. Then there exist Borel U-invariant subsets X_i, $i \in \mathbb{N}$, such that $\nu(X_i) < \infty$ for all i and $G/\Gamma = \bigcup_{i=1}^{\infty} X_i$. In particular every locally finite U-ergodic U-invariant measure is finite.*

For the case where $G = \mathbf{SL}_n(\mathbb{R})$ and $\Gamma = \mathbf{SL}_n(\mathbb{Z})$, Theorem 14 is essentially equivalent to the following assertion: for any lattice Λ in R^n and any unipotent one-parameter subgroup $\{u_t\}$ of $\mathbf{SL}_n(\mathbb{R})$, one can find $\delta > 0$ such that the set

$$\Omega(\Lambda, \delta) = \{t \geq 0 \mid \|u_t z\| > \delta \text{ for all } z \in \Lambda - \{0\}\}$$

has positive density. A strengthened version of this assertion is proved in [17] and, in a weak form, it is proved in [35] (namely it is proved in [35] that $\Omega(\Lambda, \delta)$ is unbounded for some $\delta = \delta(\Lambda) > 0$).

Using Theorem 1, one can deduce from Theorem 14

Theorem 15. *Let G be a connected Lie group, Γ a lattice in G, and H a connected subgroup of G such that the quotient of H by its unipotent radical is semisimple (by the unipotent radical of H we mean the maximal connected normal unipotent subgroup of H). Let ν be a locally finite H-invariant measure on G/Γ. Then there exist Borel H-invariant subsets X_i, $i \in \mathbb{N}$, such that $\nu(X_i) < \infty$ for all i and $G/\Gamma = \bigcup_{i=1}^{\infty} X_i$. In particular every locally finite H-ergodic H-invariant measure is finite.*

One can see that Theorem 15 differs from Theorem 14 only in the replacement of a unipotent subgroup U by a connected subgroup H such that the quotient of H by its unipotent radical is semisimple.

The following is a special case of Theorem 15.

Theorem 16. *Let G, Γ and H be the same as in Theorem 15, and let $x \in G/\Gamma$. Suppose that the orbit Hx is closed in G/Γ. Then $H \cap G_x$ is a lattice in G where $G_x = \{g \in G | gx = x\}$ is the stabilizer of x.*

Applying Theorem 16 in the case where $G = \mathbf{SL}_n(\mathbb{R})$, $\Gamma = \mathbf{SL}_n(\mathbb{Z})$, $x = e\Gamma$, and $H \subset G$ is the set of $\mathbb{R}$-rational points in a $\mathbb{Q}$-subgroup of G, we obtain thc following theorem of Borel and Harish-Chandra.

Theorem 17. *Let H be a connected $\mathbb{Q}$-group. Suppose that the quotient of H by its unipotent radical is semisimple. Then $H(\mathbb{Z})$ is a lattice in $H(\mathbb{R})$.*

Remark. In the most general form of Borel-Harish-Chandra theorem, the condition "the quotient of H by its unipotent radical is semisimple" is replaced by a weaker condition "H has no non-trivial $\mathbb{Q}$-rational characters".

4. Orbit Closures

4.1 Raghunathan's Conjecture

Roughly speaking, ergodicity theorems of Sect. 1 claim that "almost all" orbits of subgroup actions on a homogeneous space are dense. But these theorems say nothing about the behaviour of "individual" orbits. M.S. Raghunathan stated a conecture about closures of orbits of unipotent groups and also noted the connection of his conjecture with Oppenheim-Davenport conjecture.

Conjecture 1 (Raghunathan's Conjecture). *Let G be a connected Lie group, Γ a lattice in G and U a unipotent subgroup of G. Then for any $x \in G/\Gamma$, there exists a closed subgroup $L = L(x) \subset G$ containing U such that the closure of the orbit Ux coincides with Lx.*

Conjecture 1 can be generalized in the following way:

Conjecture 2 (Generalized Raghunathan's Conjecture). *Let G be a connected Lie group, Γ a lattice in G and H a subgroup of G. Suppose that H is generated by unipotent* *elements. Then for any $x \in G/\Gamma$ there exists a closed subgroup $L = L(x) \subset G$ containing H such that the closure of the orbit Hx coincides with Lx.*

Raghunathan's conjecture has been proved in the following cases: (a) G is reductive and U is horospherical (see [18]); (b) $G = \mathbf{SL}_3(\mathbb{R})$ and $U = \{u(t)\}$ is a one-parameter unipotent subgroup of G such that $u(t) - I$ has rank 2 for all $t \neq 0$ where I is the identity matrix (see [21]); (c) G is solvable (see [68] and [69]).

Let G, Γ and U be the same as in Conjecture 1. A closed U-invariant subset $Y \subset G/\Gamma$ is called *U-minimal* if it does not contain any proper closed U-invariant subset or, equivalently, if Uy is dense in Y for any $y \subset Y$. It is proved in [38] that any closed U-invariant U-minimal subset of G/Γ is compact. On the other hand since the unipotent subgroup U is nilpotent and, consequently, amenable, we have that, for any compact U-invaraint subset $Y \subset G/\Gamma$, there exists a U-invariant Borel probability measure on Y. In view of these observations, Theorem 12 implies that any closed U-invariant U-minimal subset of G/Γ is an orbit of a subgroup $L \subset G$. This proves Conjecture 1 in the case where the closure of Ux is U-minimal.

Remark 1. It is not possible to assume in Conjecture 2 that H is an arbitrary subgroup of G. For example, if

$$G = \mathbf{SL}_2(\mathbb{R}), \qquad \Gamma = \mathbf{SL}_2(\mathbb{Z}), \qquad D = \left\{ \begin{pmatrix} \lambda & 0 \\ 0 & \lambda^{-1} \end{pmatrix} \middle| \lambda \in \mathbb{R}, \lambda \neq 0 \right\}.$$

then there exists $x \in G/\Gamma$ such that the closure of Dx is not a manifold. On the other hand in Conjecture 2, the condition "H is generated by unipotent elements" is not the most general of all possible conditions. Let $G = \mathbf{SL}_2(\mathbb{R})$, let Γ be a lattice in G, and let $P = \left\{ \begin{pmatrix} a & b \\ 0 & a^{-1} \end{pmatrix} \middle| a, b \in \mathbb{R}, a \neq 0 \right\}$. Then for any point $x \in G/\Gamma$, the orbit Px is dense in G/Γ. Also it seems plausible that if G is a semisimple Lie group without compact factors, $D \subset G$ is a maximal $\mathbb{R}$-split torus in G, $\Gamma \subset G$ is an irreducible lattice, $x \in G/\Gamma$, Dx is relatively compact in G/Γ and $\mathbb{R}$-rank of G is greater than 1, then the closure of Dx is an orbit of a subgroup $L \subset G$.

Remark 2. Let $L \subset G$ be a closed subgroup, $x \in G/\Gamma$ and $G_x = \{g \in G | gx = x\}$. Then the orbit Lx is closed in G/Γ if and only if the canonical map $L/L \cap G_x \to G/\Gamma$ is proper. Using this observation and Theorem 15, it is not difficult to prove that, in Conjectures 1 and 2, Lx admits a finite L-invariant measure or, equivalently, $L \cap G_x$ is a lattice in L.

Remark 3. Using Remark 2 and Borel's density theorem (see [3]), it is not difficult to prove that if $G = \mathbf{SL}_n(\mathbb{R})$, $\Gamma = \mathbf{SL}_n(\mathbb{Z})$ and $x = g\Gamma \in G/\Gamma$ then, in Conjectures 1 and 2, the subgroup $g^{-1}L(x)g$ is algebraic and defined over $\mathbb{Q}$ (more precisely, $g^{-1}L(x)g$ is commensurable with the set of $\mathbb{R}$-rational points of a $\mathbb{Q}$-subgroup of $\mathbf{SL}_n$).

Remark 4. It seems that Conjecture 2 should follows from Conjecture 1. This is so if Γ is an arithmetic subgroup of G. In that case as it was noted by M.S. Raghunathan, the mentioned reduction follows from Remark 3 and the countability of the set of $\mathbb{Q}$-subgroups.

Let us say that an element g of a Lie group is *quasiunipotent* if all eigenvalues of the transformation Ad g have absolute value 1. If Conjecture 2 is proved then it will be possible to prove the following conjecture; for one-parameter F, this conjecture was stated by A.N. Starkov in [69].

($*$) *Let G be a connected Lie group, F a subgroup of G, and D a closed subgroup of G such that G/D has finite volume. Suppose that F is generated by quasiunipotent elements. Then for any $x \in G/D$, the closure of the orbit Fx is a manifold.*

4.2 Closures of Orbits of Non-Quasiunipotent One-Parameter Subgroups

It is not difficult to prove (see for example [68]), that whenever D is a lattice in G, F is one-parameter and F does not consist of quasiunipotent elements, there exists $x \in G/D$ such the closure of Fx is not a manifold.

Conjecture (A) stated below has been proved by S.G. Dani in the following cases: (i) $G = \mathbf{SL}_n(\mathbb{R})$ and $g_t = \operatorname{diag}(e^{-t}, \ldots, e^{-t}, e^{\lambda t}, \ldots, e^{\lambda t})$ where λ is such that the determinant is 1 (see [16] where some results of W.M. Schmidt on badly approximable systems of linear forms are used); (ii) G is a connected semisimple Lie group of $\mathbb{R}$-rank 1 (see [19]).

Conjecture (A). *Let G be a Lie group and Γ a lattice in G. Let $\{g_t\}$ be a one-parameter subgroup of G such that g_1 is not quasiunipotent. Then for any nonempty open subset Ω of G/Γ*

$$\{x \in \Omega \mid \textit{the } \{g_t\}\textit{-oribt of } x \textit{ is bounded (relatively compact)}\}$$

is of Hausdorff dimension equal to the dimension of G.

It is plausible that the following strengthening of Conjecture (A) is true.

Conjecture (B). *Let G, Γ and $\{g_t\}$ be the same as in Conjecture* (A), *and let Y be a finite subset of G/Γ. Then for any nonempty open subset Ω of G/Γ*

$$\{x \in \Omega \mid \textit{the } \{g_t\}\textit{-orbit of } x \textit{ is bounded, the closure of this orbit is of Hausdorff dimension less than the dimension of } G\textit{, and the intersection of this closure with } Y \textit{ is empty}\}$$

is of Hausdorff dimension equal to the dimension of G.

5. Values of Indefinite Quadratic Forms at Integral Points

In this section, we give some applications of results on orbit closures to the study of sets of values of indefinite quadratic forms at integral points.

5.1 Proof of Oppenheim-Davenport Conjecture

According to Meyer s theorem (see [6] or [62]), if B is a rational nondegenerate indefinite quadratic form in n variables and $n \geq 5$ then B represents zero over $\mathbb{Z}$ nontrivially i.e. there exist integers $x_1, \ldots, x_n$ not all equal to 0 such that $B(x_1, \ldots, x_n) = 0$. Theorem 18 stated below can be considered as an analogue of this assertion in the case where B is not a multiple of a rational form. Note that in Theorem 18, the condition "$n \geq 5$" is replaced by a weaker condition "$n \geq 3$".

Theorem 18 (see [36] and [37]; a simplified proof is given in [22] and [39]). *Let B be a real nondegenerate indefinite quadratic form in n variables. Suppose that $n \geq 3$ and that B is not a multiple of a rational form. Then for any $\varepsilon > 0$, there exists $x \in \mathbb{Z}^n - \{0\}$ such that $|B(x)| < \varepsilon$.*

Let us formulate a stronger version of Theorem 18.

Theorem 18′ (see [37]). *Let B and n be the same as in Theorem 18. Then for any $\varepsilon > 0$, there exists $x \in \mathbb{Z}^n - \{0\}$ such that $0 < |B(x)| < \varepsilon$.*

Theorems 18 and 18′ are of course equivalent for forms that do not represent zero over $\mathbb{Z}$ nontrivially. One can easily see that if Theorems 18 and 18′ are proved for some n_0 then they are proved for all $n \geq n_0$. So it is enough to prove these theorems for $n = 3$. We note that, in Theorems 18 and 18′, the condition "$n = 3$" can not be replaced by the condition "$n = 2$"; to see this, consider the form $x_1^2 - \lambda x_2^2$ where λ is an irrational positive number such that $\sqrt{\lambda}$ has a continued fraction development with bounded partial quotients.

Theorem 18 was conjectured by A. Oppenheim in 1929 for $n \geq 5$ (see [48] and [49]) and by H. Davenport in 1946 for all $n \geq 3$ (see [25]). Theorem 18′ was conjectured in 1953 by A. Oppenheim (see [51]).

Theorem 18′ had been proved earlier by B.J. Birch, H. Davenport and H. Ridout for $n \geq 21$ (see [26] or [31, Sect. 42.4]) and by H. Davenport and H. Heilbronn for diagonal forms in 5 variables i.e. for forms B of the type $B(x_1, \ldots, x_5) = \lambda_1 x_1^2 + \cdots + \lambda_5 x_5^2$ (see [25]). G.L. Watson [77] extended the result of Davenport and Heilbronn to forms which include a single cross-product term (say $\lambda_6 x_4 x_5$). R.C. Baker and H.P. Schlickewey [2] proved Theorem 18′ for n equal 20, 19 or 18 but under some restrictions on the signature of B. Theorem 18′ was proved by A. Oppenheim for forms in $n \geq 4$ variables representing zero over Z nontrivially (see [51] and [52]) and by H. Iwaniec [30] and G.L. Watson [76] for some types of forms in 3 and 4 variables. The proofs mentioned above are in the context of analytic number theory and reduction theory.

Theorems 18 and 18′ are deduced from the following

Theorem 19. *Let $G = \mathbf{SL}_3(\mathbb{R})$ and $\Gamma = \mathbf{SL}_3(\mathbb{Z})$. Let B be a real nondegenerate indefinite quadratic form in 3 variables. Let us denoted by H_B the group of elements of G preserving the form B and by Ω the space of lattices in $\mathbb{R}^3$ having determinant 1. (The quotient space G/Γ can be canonically identified with Ω. Under this identification*

$g\Gamma$ goes to $g\mathbb{Z}^3$.) Let G_y denote the stabilizer $\{g \in | gy = y\}$ of $y \in \Omega$. If $z \in \Omega = G/\Gamma$ and the orbit $H_B z$ is relatively compact in Ω then the quotient space $H_B/H_B \cap G_z$ is compact.

Theorem 19 proves generalized Raghunathan's conjecture in a very special case. Now we explain how Theorem 18 is deduced from Theorem 19. As mentioned above, it is enough to prove Theorem 18 for $n = 3$. Suppose that the assertion of Theorem 18 is not true. Then one can easily show using Mahler's compactness criterion that the set $H_B\mathbb{Z}^3$ is relatively compact in Ω. Now we apply Theorem 19 for $z = \mathbb{Z}^3$ and get that the quotient space $H_B/H_B \cap \Gamma$ is compact. Then in view of Borel's density theorem, $H_B \cap \Gamma$ is Zariski dense in H_B. But $\Gamma = \mathbf{SL}_3(\mathbb{Z})$. Therefore the algebraic subgroup $H_B \subset G$ is defined over $\mathbb{Q}$, and hence B is a multiple of a rational form. Contradiction.

5.2 Values of Quadratic Forms at Primitive Integral Points

Let B be a real nondegenerate indefinite quadratic form in n variables. In [51], it is shown for $n \geq 3$ that if the set $(0, \varepsilon) \cap B(\mathbb{Z}^n)$ is not empty for any $\varepsilon > 0$, then the samc is true for the form $-B$. On the other hand, it is clear that $B(\mathbb{Z}^n)$ is invariant under multiplication by square of integers. Thus Theorem 18′ implies that if $n \geq 3$ and B is not a multiple of a rational form then $B(\mathbb{Z}^n)$ is dense in $\mathbb{R}$. Let us consider now the set of values of B at primitive integral points, namely on $P(\mathbb{Z}^n) = \{x \in \mathbb{Z}^n | x \neq ky \text{ for any } y \in \mathbb{Z}^n \text{ and } k \in \mathbb{Z} \text{ with } |k| \geq 2\}$. Then Theorem 18′ doesn't imply, under the condition of that theorem, that $B(\mathfrak{P}(\mathbb{Z}^n))$ is dense in $\mathbb{R}$. Nevertheless it is possible to prove the following

Theorem 20 (see [20]). *Let B be a real nondegenerate indefinite quadratic form on $\mathbb{R}^n$, where $n \geq 3$, which is not a multiple of a rational form. Let B_2 be the corresponding bilinear form defined by $4B_2(v, w) = B(v + w) - B(v - w)$ for all $v, w \in \mathbb{R}^n$. Let a, b, c be such that there exist $v, w \in \mathbb{R}^n$ for which $B(v) = a$, $B(w) = b$, $B_2(v, w) = c$. Then for any $\varepsilon > 0$ there exist $x, y \in \mathfrak{P}(\mathbb{Z}^n)$ such that $|B(x) - a| < \varepsilon$, $|B(y) - b| < \varepsilon$ and $|B_2(x, y) - c| < \varepsilon$. In particular $B(\mathfrak{P}(\mathbb{Z}^n))$ is dense in $\mathbb{R}$.*

Theorem 20 is easily deduced from Theorem 16 and the following strengthening of Theorem 19.

Theorem 21 (see [20]). *Let G, Γ, Ω and H_B be the same as in Theorem 19. Let $x \in \Omega = G/\Gamma$. Then the orbit $H_B x$ is either closed or dense in G/Γ.*

Remark. Generalized Raghunathan's conjecture entails that if $n \geq 3$, B is a nondegenerate indefinite quadratic form on $\mathbb{R}^n$ and $H_B = \{g \in \mathbf{SL}_n(\mathbb{R}) | gB = B\}$ then, for any $x \in \mathbf{SL}_n(\mathbb{R})/\mathbf{SL}_n(\mathbb{Z})$, the orbit $H_B x$ is either closed or dense in $\mathbf{SL}_n(\mathbb{R})/\mathbf{SL}_n(\mathbb{Z})$. This and Theorem 16 easily imply that if B as above is not a multiple of a rational form and B_2 is the corresponding bilinear form then for any $\{a_{ij} | i, j = 1, 2, \ldots, n-1\} \subset \mathbb{R}$ for which there exist $v_1, \ldots, v_{n-1} \in \mathbb{R}^n$ such that $B_2(v_i, v_j) = a_{ij}$ and for any $\varepsilon > 0$ there also exist $x_1, \ldots, x_{n-1} \in \mathfrak{P}(\mathbb{Z}^n)$ such that $|B_2(x_i, x_j) - a_{ij}| < \varepsilon$ for all $i, j = 1, \ldots, n-1$.

As mentioned in Sect. 4, Raghunathan's conjecture is proved in [21] in the case where $G = \mathbf{SL}_3(\mathbb{R})$ and $U = \{u(t)\}$ is a one-parameter unipotent subgroup of G such that $u(t) - I$ has rank 2 for all $t \neq 0$. This and Theorem 16 easily imply the following

Theorem 22 (see [21]). *Let B_1 and B_2 be two real quadratic forms on $\mathbb{R}^3$ for which the following conditions are satisfied:*

(i) *there exists a basis of $\mathbb{R}^3$ such that*

$$B_1(x) = 2x_1x_3 - x_2^2 \qquad \text{and} \qquad B_2(x) = x_3^2$$

where x_1, x_2, x_3 are coordinates of x with respect to the basis;

(ii) *no nonzero linear combination of B_1 and B_2 is a multiple of a rational form.*

Then for any $a, b \in \mathbb{R}$, $b > 0$, and any $\varepsilon > 0$ there exists a primitive integral vector p such that

$$|B_1(p) - a| < \qquad \text{and} \qquad |B_2(p) - b| < \varepsilon.$$

Digressing from quadratic forms, let us note at the end of the section that there a connection between results on closures of orbits of horospherical subgroups and some results on orbits of frames under discrete linear groups (see [13], [17] and [23]). These results on orbits of frames have the flavour of results on Diophantine approximation with matrix argument.

6. Markov Spectrum

Let $B(x) = \sum_{1 \leq i,j \leq n} b_{ij}x_ix_j$, $b_{ij} = b_{ji}$, be a real nondegenerate indefinite quadratic form in n variables and let us denote by Φ_n the set of all such forms. Let $d(B)$ denote the determinant of the matrix (b_{ij}). Let us set

$$m(B) = \inf_{x \in \mathbb{Z}^n - \{0\}} |B(x)| \qquad \text{and} \qquad \mu(B) = \frac{m(B)^n}{|d(B)|}.$$

It is clear that $\mu(\lambda B) = \mu(B)$ for every $\lambda \in \mathbb{R}$, $\lambda \neq 0$. Let M_n denote $\mu(\Phi_n)$. The set M_n is called *Markov spectrum.* It easily follows from Mahler's compactness criterion that M_n is bounded and closed.

Two forms $B, B' \in \Phi_n$ will be called *equivalent* if there exist $g \in \mathbf{SL}_n(\mathbb{Z})$ and $\lambda \in \mathbb{R}$ such that $gB = \lambda B'$. One can easily see that $\mu(B) = \mu(B')$ if B and B' are equivalent.

6.1 Discreteness of Markov Spectrum for $n \geq 3$

In 1880, A.A. Markov described the intersection of M_2 with the segment $(4/9, \infty)$ and described corresponding quadratic forms (see [7] and [41]). It follows from this description that $M_2 \cap (4/9, \infty)$ is a discrete countable subset of $(4/9, 4/5]$ and that, for any $a > 4/9$, there are only finitely many equivalence classes of forms $B \in \Phi_2$ with $\mu(B) > a$. On the other hand, the intersection $M_2 \cap [0, 4/9]$ is not countable and moreover has quite a complicated topological structure (see [7]).

It follows from Meyer's theorem and Theorem 18 that $M_n = \{0\}$ whenever $n \geq 5$. In 1955 (see [8]), J.W.S. Cassels and H.F.P. Swinnerton-Dyer proved the conditional result that Theorem 18 implies the following

Theorem 23. *Let $n \geq 3$ and $\varepsilon > 0$. Then the set $M_n \cap (\varepsilon, \infty)$ is finite and moreover there are only finitely many equivalence classes of forms $B \in \Phi_n$ with $\mu(B) > \varepsilon$.*

For rational forms B, Theorem 23 had been proved earlier by L.Ya. Vulakh (see [75]). Really he proved a stronger assertion that there are only finitely many equivalence classes of forms B with $\mu_0(B) > \varepsilon$ where

$$\mu_0(B) = \frac{m_0(B)^n}{|d(B)|} \quad \text{and} \quad m_0(B) = \inf_{x \in \mathbb{Z}^n - \{0\},\, B(x) \neq 0} |B(x)|.$$

As noted in [74], Vulakh has obtained the complete description of spectra of nonzero minima of rational Hermitian forms.

Both the proof of Theorem 23 based on Theorem 18 and the proof given in [75] are non-effective. In particular these proofs do not give an explicit upper bound for the cardinality of $M_n \cap (\varepsilon, \infty)$ for $n = 3, 4$. It would be very interesting to get such a bound.

For $n = 3$, A.A. Markov determined first 4 values of $\mu(B)^{-1}$ and B.A. Venkov determined next 7 values (see [42] and [73]). These values are

$$3/2,\ 5/2,\ 3,\ 25/8,\ 1200/7^3 = 3{,}498\ldots,\ 7/2,\ 15/4,\ 3 \cdot 13^2/5^3 = 4{,}056,$$
$$112/3^3 = 4{,}148\ldots,\ 135/2^5 = 4{,}218\ldots,\ 9/2.$$

For $n = 4$ and forms of signature 0, A. Oppenheim determined first 7 values of $\mu(B)^{-1}$ (see [50]). These values are

$$9/4,\ 17/4,\ 25/4,\ 117/16,\ 33/4,\ 9,\ 625/64.$$

It should be noted that there are two inequivalent forms which correspond to the value 9. For $n = 4$ and forms of signature -2 or 2, A. Oppenheim determined also the first value of $\mu(B)^{-1}$ which is 7/4.

6.2 A Strengthened Version of the Discreteness Theorem

Let us formulate now a strengthened version of Theorem 23.

Theorem 24 (see [39]). *Let $a, b, c \in \mathbb{R}$ and let*

$$\Phi_n(a, b, c) = \Bigg\{ B \in \Phi_n \,\Bigg|\, \text{there exist } v, w \in \mathbb{R}^n \text{ such that } \frac{B(v)^n}{d(B)} = a,\ \frac{B(w)^n}{d(B)} = b \text{ and } \frac{B_2(v, w)^n}{d(B)} = c \text{ where } B_2 \text{ is the bilinear form corresponding to } B \Bigg\}.$$

Suppose that $n \geq 3$. Then for any $\varepsilon > 0$, there are only finitely many equivalence classes of forms $B \in \Phi_n(a, b, c)$ for which it is not possible to find primitive integral vectors $x, y \in \mathfrak{P}(\mathbb{Z}^n)$ such that

$$\left|\frac{B(x)^n}{d(B)} - a\right| < \varepsilon, \qquad \left|\frac{B(y)^n}{d(B)} - b\right| < \varepsilon \qquad \text{and} \qquad \left|\frac{B_2(x, y)^n}{d(B)} - c\right| < \varepsilon.$$

The proof of Theorem 24 is based on the following slight generalization of Theorem 21.

Theorem 21′ (see [39]). *Let G, Γ, Ω and H_B be the same as in Theorem 21. Let A be a closed H_B-invariant subset of $\Omega = G/\Gamma$. Suppose that $A \neq \Omega$. Then A is the union of finite number of closed H_B-orbits.*

A. Borel and G. Prasad proved in [4] a generalization of Theorem 18′ for a family $\{B_s\}$ where $s \in S$, S is a finite set of places of a number field k containing the set S_∞ of archimedean places, B_s is a quadratic form on k_s^n, and k_s is the completion of k at s. It would be interesting to prove analogues of Theorems 20, 22, 23 and 24 for families $\{B_s\}$.

7. Generic Points

Let X be a locally compact σ-compact topological space and φ a homeomorphism of X. We say that a point $x \in X$ is *generic* (with respect to φ) if there exists a Borel probability measure μ_x on X such that for any bounded continuous function f on X

$$\lim_{n\to\infty} \frac{1}{n} \sum_{i=0}^{n-1} f(\varphi^i x) = \int_X f \, d\mu_x.$$

According to Birkhoff individual ergodic theorem for any φ-invariant Borel probability measure μ on X, one can find a subset $A \subset X$ such that $\mu(A) = 0$ and each point $x \in X - A$ is generic.

The following conjecture was stated in [14] (for the case where $G = \mathbf{SL}_n(\mathbb{R})$ and $\Gamma = \mathbf{SL}_n(\mathbb{Z})$).

Conjecture 3. *Let G be a connected Lie group, Γ a lattice in G and u a unipotent element of G. Then every point $x \in G/\Gamma$ is generic with respect to the homeomorphism $y \to uy$, $y \in G/\Gamma$.*

Conjecture 3 can be strengthened in the following way:

Conjecture 4. *Let G, Γ and u be the same as in Conjecture 3. Then for any $x \in G/\Gamma$, there exist a closed subgroup $L \subset G$ containing u and a L-invariant Borel probability measure μ on G/Γ such that the orbit Lx is closed, the support of the measure μ is Lx, and for any bounded continuous function f on G/Γ*

$$\lim_{n\to\infty} \frac{1}{n} \sum_{i=0}^{n-1} f(u^i x) = \int_{G/\Gamma} f \, d\mu.$$

Remark. Considering appropriate increasing sequences $\{S_n\} \subset U$ and replacing $\frac{1}{n}\sum_{i=0}^{n-1} f(u^i x)$ by $\frac{1}{\sigma(S_n)}\int_{S_n} f(ux)\,d\sigma(u)$ where σ is Haar measure on U, one can formulate an analogue of Conjecture 4 for an arbitrary closed unipotent subgroup $U \subset G$. Let us also note the connection of Conjecture 4 with Theorems 12 and 14.

It is clear that for a cyclic U, Raghunathan's conjecture follows from Conjecture 4. Conjecture 4 was proved for nilpotent G by E. Lesigne (see [32] and [33]) and for $G = \mathbf{SL}_2(\mathbb{R})$ by S.G. Dani and J. Smillie (see [24]; for $\Gamma = \mathbf{SL}_2(\mathbb{Z})$ see also [14]). For $G = \mathbf{SL}_2(\mathbb{R})$, we have the following formulation

Theorem 25 (see [24]). *Let $G = \mathbf{SL}_2(\mathbb{R})$, let Γ be a lattice in G, let $x \in G/\Gamma$, let $\{u_t\}$ be a one-parameter unipotent subgroup of G, and let u denote u_1. Suppose that the orbit $\{u_t x \mid t \in \mathbb{R}\}$ is not periodic i.e. that $u_t x \neq x$ for every $t \neq 0$. Then the orbit $\{u_t x \mid t \in \mathbb{R}\}$ and the sequence $\{u^n x \mid n \in \mathbb{Z}\}$ are uniformly distributed with respect to the G-invariant Borel probability measure μ on G/Γ; i.e. for any bounded continuous function f on G/Γ*

$$\lim_{T\to\infty}\frac{1}{T}\int_0^T f(u_t x)\,dt = \lim_{N\to\infty}\frac{1}{N}\sum_{n=0}^{N-1} f(u^n x) = \int_{G/\Gamma} f\,d\mu.$$

The following number theoretic assertion is deduced from Theorem 25.

Theorem 26 (see [14]). *For $t \in \mathbb{R}$, let $[t]$ denote the largest integer x with $x \leq t$ and let $\{t\} = t - [t]$. For $m, n \in \mathbb{N}^+$, let (m, n) denote the* g.c.d. *of m and n. Then for any irrational θ*

$$\lim_{T\to\infty}\frac{1}{T}\sum_{\substack{0\leq m\leq T\{m\theta\}\\ (m,[m\theta])=1}} \{m\theta\}^{-1} = \frac{1}{\xi(2)} = \frac{6}{\pi^2}$$

where ξ denotes Riemann zeta-function.

Conjecture 4 entails quantative versions of results from Sect. 5. In particular, if this conjecture is proved it will be possible to prove the following assertions.

(i) *Let n, B, B_2, a, b and c be the same as in Theorem 20. Then for any $\varepsilon > 0$, one can find $d = d(\varepsilon, a, b, c, B) > 0$ such that for all sufficiently large S there exist at least dS pairs of primitive integral vectors $x, y \in \mathfrak{P}(\mathbb{Z}^n)$ for which*

$$\|x\| < S, \quad \|y\| < S, \quad |B(x) - a| < \varepsilon, |B(y) - b| < \varepsilon \quad \textit{and} \quad |B_2(x, y) - c| < \varepsilon$$

where $\| \ \|$ denotes the Euclidean norm on $\mathbb{R}^n$. Moreover we can choose $d(\varepsilon, a, b, c, B)$, so that it is a rational function of ε, a, b, c and coefficients of B.

(ii) *Let B_1 and B_2 be the same as in Theorem 22. Then for any $a, b \in \mathbb{R}$, $b > 0$, and any $\varepsilon > 0$, one can find $c = c(\varepsilon, a, b, B_1, B_2)$ such that for all sufficiently large S there exist at least $cS^{1/2}$ primitive integral vectors p for which*

$$\|p\| < S, \qquad |B_1(p) - a| < \varepsilon \qquad \textit{and} \qquad |B_2(p) - b| < \varepsilon.$$

Moreover of $c(\varepsilon, a, b, B_1, B_2)$, can be chosen to be a rational function of ε, a, b and coefficients of B_1 and B_2.

8. Actions of Semisimple Groups and Discrete Subgroups on Manifolds and Measure Spaces

Let us give formulations of some recent results about actions on general manifolds and measure spaces. We refer to [80] and [81] for the definition of notions which are used in these formulations.

Theorem 27 (see [64]). *Let G be a connected simple Lie group with finite center, finite fundamental group, and $\mathbb{R}$-rank$(G) \geq 2$. Let M be a compact manifold and suppose there is a nontrivial real analytic action of G on M preserving a (real analytic) connection and a finite measure. Then $\pi_1(M)$ is not isomorphic to the fundamental group of any complete Riemannian manifold N with negative curvature bounded away from* 0 *and* $-\infty$.

Theorem 28 (see [82]). *Let a locally compact group G act continuously on a compact metrizable space M. Suppose G preserves a finite measure on M with respect to which the G-action is ergodic. Suppose further that G is a semisimple group of higher rank (i.e. G is a finite product $\prod G_i$ where each G_i is the set of k_i-points of a k_i-simple connected k_i-group of k_i-rank at least* 2). *If $P \to M$ is a principal H-bundle on which G acts by principle bundle automorphisms, where H is a real algebraic group, then the algebraic hull of the action is a reductive group with finite center.*

Theorem 29 (see [83]). *Let G be a semisimple group of higher rank and $\Gamma \subset G$ a lattice. Suppose Γ acts smoothly on a compact manifold M preserving a volume density. Then the following are equivalent*:

(a) *The Γ-action has zero entropy, i.e. the entropy $h(\gamma)$ equals* 0 *for all $\gamma \in \Gamma$.*
(b) *The Γ-action is measurably isometric, i.e. it is measurably conjugate to an action which obtained by a homomorphism of Γ into a compact group K which acts continuously on M.*
(c) *There is a Γ-invariant measurable Riemannian metric on M.*

Concluding Remarks Made During the Congress

(i) I learnt at the beginning of the Congress that very recently (about two weeks before the Congress) M. Ratner proved Conjectures 1, 2, 3 and 4. Also recently, N.A. Shah proved some partial results in the direction of the proof of Conjecture 4 (even these results are enough to prove assertions (i) and (ii) from Sect. 7 and the assertion mentioned in the remark after Theorem 21). Proofs of M. Ratner and N.A. Shah are based on the reduction to Theorems 12 and 14.

(ii) Let $G = \mathbf{SL}_2(\mathbb{R})$, let $\Gamma = \mathbf{SL}_2(\mathbb{Z})$ and let $U = \{u_t\}$ be a one-parameter unipotent subgroup of G. For each $t > 0$ there is a unique closed orbit of U on G/Γ of period t, call it C_t ($u_t x = x$ and $u_{t'} x \neq x$ for $x \in C_t$ and $0 < t' < t$). P. Sarnak informed me about the following observation which was essentially due to D. Zagier. The famous Riemann hypothesis about zeroes of Riemann zeta-function is equivalent to the following statement. For any function $f \in C_0^\infty(G/\Gamma)$ (i.e. f is smooth and has a compact support) and any $\varepsilon > 0$

$$\frac{1}{t}\int_{C_t} f\,dl = \int_{G/\Gamma} f\,d\mu + 0_{\varepsilon,f}(t^{-(3/4)+\varepsilon})$$

as $t \to \infty$ where l is the Lebesgue measure on C_t and μ is the G-invariant probability Borel measure on G/Γ. It should be noted that (a) the power $-\frac{3}{4} + \varepsilon$ is different from the usual power $-\frac{1}{2} + \varepsilon$ which appears in ergodic theory; (b) the equivalence remains true if one considers only K-invariant functions f where K is a maximal compact subgroup of G.

References

1. Auslander, L., Green, L., Hahn, F.: Flows on homogeneous spaces (Annals of Mathematics Studies, no 53). Princeton University Press, 1963
2. Baker, R.C., Schlickewey, H.P.: Indefinite quadratic forms. Proc. London Math. Soc. **54** (1987) 383–411
3. Borel, A.: Density properties of certain subgroups of semisimple groups. Ann. Math. **72** (1960) 179–188
4. Borel, A., Prasad, G.: valeurs de formes quadratiques aux points entiers. C. R. Acad. Sci. Paris **307**, Serie 1 (1988) 217–220
5. Brezin, J., Moore, C.C.: Flows on homogeneous spaces: a new look. Amer. J. Math **103** (1981) 571–613
6. Cassels, J.W.S.: Rational quadratic forms. Academic Press, London New York, 1978
7. Cassels, J.W.S.: An introduction to Diophantine approximation. Cambridge Univ. Press, Cambridge, 1957
8. Cassels, J.W.S., Swinnerton-Dyer, H.P.F.: On the product of three homogeneous linear forms and indefinite ternary quadratic forms. Philos. Trans. Roy. Soc. London **248**, Ser. A (1955) 73–96
9. Cornfeld, I.P., Sinai, Ya.G., Fomin, S.V.: Ergodic theory. Gosud. Izd. Fiz. Mat. Lit., Moscow, 1980 (Russian) [English transl.: Springer, Berlin Heidelberg New York 1982].
10. Dani, S.G.: Kolmogorov automorphisms on homogeneous spaces. Amer. J. Math. **98** (1976) 119–163
11. Dani, S.G.: Bernoullian translations and minimal horospheres on homogeneous spaces. J. Ind. Math. Soc. **39** (1976) 245–294
12. Dani, S.G.: Spectrum of an affine transformation. Duke Math. J. **44** (1977) 129–155
13. Dani, S.G.: Invariant measures and minimal sets of horospherical flows. Invent. math. **64** (1981) 357–385
14. Dani, S.G.: On uniformly distributed orbits of certain horocycle flows. Ergod. Theor. Dynam. Syst. **2** (1982) 139–158
15. Dani, S.G.: On orbits of unipotent flows on homogeneous spaces. Ergod. Theor. Dynam. Syst. **4** (1984) 25–34
16. Dani, S.G.: Divergent trajectories of flows on homogeneous spaces and Diophantine approximation. J. Reine Angew. Math. **359** (1985) 55–89
17. Dani, S.G.: On orbits of unipotent flows on homogeneous spaces II. Ergod. Theor. Dynam. Syst. **6** (1986) 167–182
18. Dani, S.G.: Orbits of horospherical flows. Duke Math. J. **53** (1986) 177–188
19. Dani, S.G.; Bounded orbits of flows on homogeneous spaces. Comment. Math. Helv. **61** (1986) 636–660
20. Dani, S.G., Margulis. G.A.: Values of quadratic forms at primitive integral points. Invent. math. **98** (1989) 405–424

21. Dani, S.G., Margulis, G.A.: Orbit closures of generic unipotent flows on homogeneous spaces of SL(3, ℝ). Math. Ann. **286** (1990) 101–128
22. Dani, S.G., Margulis, G.A.: Values of quadratic forms at primitive integral points. l'Enseignement Math. (to appear)
23. Dani, S.G., Raghavan, S.: Orbits of Euclidean frames under discrete linear groups. Isr. J. Math. **36** (1980) 300–320
24. Dani, S.G., Smillie, J.: Uniform distribution of horocycle flows for Fuchsian groups. Duke Math. J. **51** (1984) 185–194
25. Davenport, H., Heilbronn, H.: On indefinite quadratic forms in five variables. J. London Math. Soc. **21** (1946) 185–193
26. Davenport, H., Ridout, H.: Undefinite quadratic forms. Proc. London Math. Soc. **9** (1959) 544–555
27. Furstenberg, H.: The unqiue ergodicity of the horocycle flow. In: Recent Advances in Topological Dynamics (ed. A. Beck). Springer, Berlin Heidelberg New York 1972, pp. 95–115
28. Gelfand, I.M., Fomin, S.V.: Geodesic flows on manifolds of constant negative curvature. Usp. Mat. Nauk **7** (1952) 118–137 (Russian)
29. Hopf, E.: Statistik der geodätischen Linienin Mannigfaltigkeiten negativer Krümmung. Ber. Ver. Sächs. Akad. Wiss. Leipzig **91** (1939) 261–304
30. Iwaniec, H.: On indefinite quadratic forms in four variables. Acta Arith. **33**, 209–229 (1977)
31. Lekkerkerker, C.G.: Geometry of numbers. Walters-Noordhoff, Groningen; North-Holland, Amsterdam London, 1969
32. Lesigne, E.: Théorèmes ergodiques pour une translation sur nilvariété. Ergod. Theor. Dynam. Syst. **9** (1989) 115–126
33. Lesigne, E.: Sur une nilvariété, les parties minimales associées à une translation sont uniquement ergodiques. Preprint.
34. Marcus, B.: The horocycle flow is mixing of all degrees. Invent. math. **46** (1978) 201–209
35. Margulis, G.A.: On the action of unipotent groups in the space of lattices. In: Proc. of the Summer school on group representations (bolyai Janos Math. Soc., Budapest, 1971). Akadémiai Kiado, Budapest 1975, pp. 365–370
36. Margulis, G.A.: Formes quadratiques indéfinies et flots unipotents sur les espaces homogènes. C. R. Acad. Sci. Paris **304**, Serie 1 (1987) 249–253
37. Margulis, G.A.: Discrete subgroups and ergodic theory. In: Proc. of the conference "Number theory, trace formulas and discrete groups" in honour of A. Selberg (Oslo, 1987). Academic Press, London New York, 1988, pp. 377–398
38. Margulis, G.A.: Compactness of minimal closed invariant sets of actions of unipotent groups. Geometriae Dedicata (to appear)
39. Margulis, G.A.: Orbits of group actions and values of quadratic forms at integral points. Preprint IHES, 1990
40. Margulis, G.A.: Discrete subgroups of semisimple groups. Springer, Berlin Heidelberg New York 1990
41. Markov, A.A.: On binary quadratic forms of positive determinant. SPb, 1880; Usp. Mat. Nauk **3**(5) (1948) 7–51 (Russian)
42. Markov, A.A.: Sur les formes quadratiques ternaires indéfinies. Math. Ann. **56** (1903) 233–251
43. Mautner, F.J.: Geodesic flows on symmetric Riemannian spaces. Ann. Math. **65** (1957) 416–431
44. Moore, C.C.: Ergodicity of flows on homogeneous spaces. Amer. J. Math. **88** (1966) 154–178
45. Moore, C.C.: The Mautner phenomenon for general unitary representations. Pacific J. Math. **86** (1980) 155–169

46. Mostow, G.D.: Intersections of discrete subgroups with Cartan subgroups. Indian J. Math. **34** (1970) 203–214
47. Mostow, G.D.: Strong rigidity of locally symmetric spaces. (Annals of Mathematics Studies, no 78). Princeton University Press, 1973
48. Oppenheim, A.: The minima of indefinite quaternary quadratic forms. Proc. Acad. Sci. USA **15** (1929) 724–727
49. Oppenheim, A.: The minima of indefinite quaternary quadratic forms. Ann. Math. **32** (1931) 271–298
50. Oppenheim, A.: Minima of quaternary quadratic forms of signature O. Proc. London Math. Soc. **37** (1934) 63–81
51. Oppenheim, A.: Values of quadratic forms I, II. Quart. J. Math. Oxford Ser. (2), **4** (1953) 54–59, 60–66
52. Oppenheim, A.: Values of quadratic forms III. Monatsh. Math. Phys. **57** (1953) 97–101
53. Parasyuk, O.S.: Horocycle flows on surfaces of constant negative curvature. Usp. Mat. Nauk **8**(3) (1953) 125–126 (Russian)
54. Raghunathan, M.S.: Discrete subgroups of Lie groups. Springer, Berlin Heidelberg New York, 1972
55. Ratner, M.: Rigidity of horocycle flows. Ann. Math. **115** (1982) 597–614
56. Ratner, M.: Factors of horocycle flows. Ergod. Theor. Dynam. Syst. **2** (1982) 465–489
57. Ratner, M.: Horocycle flows: joinings and rigidity of products. Ann. Math. **118** (1983) 277–313
58. Ratner, M.: Strict measure rigidity for unipotent subgroups of solvable groups. Invent. math. **101** (1990) 449–482
59. Ratner, M.: On measure rigidity for unipotent subgroups of semisimple groups. Acta Math. (to appear)
60. Ratner, M.: On Raghunathan's measure conjecture. Preprint.
61. Safonov, A.V.: on the spectral type of ergodic G-induced flows. Funkt. Anal. i Priložen **14**(4) (1980) 81–82 (Russian)
62. Serre, J.P.: A course in arithmetic. Springer, Berlin Heidelberg New York 1973
63. Sinai, Ya.G.: Dynamical systems with countably-multiple Lebesgue spectrum II. Izv. Akad. Nauk SSSR Ser. Mat. **30**, 15–68 (Russian) [English transl.: Amer. Math. Soc. Transl. (2) **68** (1988) 34–88]
64. Spatzier, R.J., Zimmer, R.J.: Fundamental groups of negatively curved manifolds and actions of semisimple groups. Topology (to appear)
65. Starkov, A.N.: The ergodic behaviour of flows on homogeneous spaces. Dokl. Akad. Nauk SSSR **273**, 538–540 (Russian) [English transl.: Soviet Math. Dokl. **28** (1983) 675–676]
66. Starkov, A.N.: On spaces of finite volume. Vestnik MGU **5** (1986) 64–65 (Russian)
67. Starkov, A.N.: The ergodic decomposition for homogeneous flows. Izv. Akad. Nauk SSSR **51** (1987) 1191–1213 (Russian)
68. Starkov, A.N.: Solvable homogeneous flows. Mat. Sbornik **176** (1987) 242–259 (Russian)
69. Szarkov, A.N.: The ergodic decomposition of flows on homogeneous spaces of finite volume. Mat. Sbornik **180** (1989) 1614–1633 (Russian)
70. Stepin, A.M.: Flows on solvable manifolds. Usp. Mat. Nauk **24**(5) (1989) 241–242 (Russian)
71. Stepin, A.M.: Dynamical systems on homogeneous spaces of semisimple groups. Izv. Akad. Nauk SSSR Ser. Mat. **37**, 1091–1107 (Russian) [English transl. Math. USSR Izv. **7** (1973) 1089–1104]
72. Veech, W.A.: Unique ergodicity of horospherical flows. Amer. J. Math. **99** (1977) 827–859
73. Venkov, B.A.: On Markov's extremal problem for ternary quadratic forms. Izv. Akad. Nauk SSSR Ser. Mat. **9** (1945) 429–434 (Russian)
74. Vulakh, L.Ya.: On minima of rational indefinite Hermitian forms. Ann. N.Y. Acad. Sci. **70** (1983) 99–106

75. Vulakh, L.Ya.: On minima of rational indefinite quadratic forms. J. Number Theory **21** (1985) 275–285
76. Watson, G.L.: On indefinite quadratic forms in three or four variables. J. London Math. Soc. **28** (1953) 239–242
77. Watson, G.L.: On indefinite forms in five variables. Proc. London Math. Soc. (3) **3** (1953) 170–181
78. Witte, D.: Rigidity of some translations on homogeneous spaces. Invent. math. **85** (1985) 1–27
79. Witte, D.: Zero entropy affine maps on homogeneous spaces. Amer. J. Math. **109** (1987) 927–961
80. Zimmer, R.J.: Ergodic theory and semisimple groups. Birkhäuser, Boston, 1984
81. Zimmer, R.J.: Actions of semisimple groups and discrete subgroups. Proc. Int. Math. Cong., Berkeley, California, USA, 1986, pp. 1247–1258
82. Zimmer, R.J.: On the algebraic hull of an automorphism group of a principal bundle. Comment. Math. Helv. (to appear)
83. Zimmer, R.J.: Spectrum, entropy, and geometric structures for smooth actions of Kazhdan groups. Preprint

Pseudodifferential Operators, Corners and Singular Limits

Richard B. Melrose

Department of Mathematics, Room 2-243, Massachusetts Institute of Technology
Cambridge, MA 02139, USA

In the first part of my talk I shall describe some of the properties one should expect of a calculus of pseudodifferential operators which corresponds to the microlocalization of a Lie algebra of vector fields. This is not intended to be a formal axiomatic program but it leads one to consider conditions on the Lie algebra for such microlocalization to be possible. The symbolic structure of the calculus also shows how it can be applied in the solution of analytic questions related to the Lie algebra, especially to the inversion of elliptic elements of the enveloping algebra.

In the second part I shall consider several such analytic question which arise in various differential-geometric settings and are, or appear to be, amenable to the application of these pseudodifferential techniques. For those examples which have already been analyzed the Lie algebra is identified and then a specific question is discussed using the calculus of pseudodifferential operators which arises from it.

Finally in the third part of the talk I shall briefly outline a general strategy for the construction of 'the' calculus of pseudodifferential operators which microlocalizes a given Lie algebra satisfying conditions which make it a boundary-fibration structure. This construction applies to most of the situations described in the second part of the talk and, I conjecture, can be extended to apply to the others.

I. Pseudodifferential Operators

Even though many of the examples of interest here arise on singular or non-compact spaces it is very convenient from an analytic point of view to work always in a fixed category of spaces. The simplest class of spaces which seems to be sufficiently wide to allow many different types of problems to be attacked is that of compact manifolds with corners; the morphisms are the b-maps, discussed below. Such a space is a compact topological manifold with boundary, X, with a $\mathscr{C}^\infty$ structure locally modeled on the spaces

$$\mathbb{R}^n_k = [0, \infty)^k \times \mathbb{R}^{n-k}. \tag{1}$$

This fixes the space $\mathscr{C}^\infty(X)$ of smooth functions on X; locally these are just the functions on $\mathbb{R}^n_k$ which are the restrictions of $\mathscr{C}^\infty$ functions on $\mathbb{R}^n$. For

Proceedings of the International Congress
of Mathematicians, Kyoto, Japan, 1990

simplicity we also insist that all the boundary hypersurfaces of X are embedded. If we let $M_1(X)$ be the set of boundary hypersurfaces then this means that each $H \in M_1(X)$ has a defining function

$$\varrho_H \in \mathscr{C}^\infty(X) \text{ s.t. } H = \{\varrho_H = 0\},\ \varrho_H \geq 0,\ d\varrho_H \neq 0 \text{ at } H\,. \tag{2}$$

We denote by $M_{(1)}(X)$ the set of true boundary faces, the connected closed submanifolds locally given by intersection of the boundary hypersurfaces and set $M(X) = M_{(1)}(X) \cup \{X\}$; all these submanifolds are embedded. Limiting the analysis to compact manifolds with corners implies that in many cases a preliminary geometric step involving a compactification or blow-up procedure is required.

It is always possible to embed a compact manifold with corners as a subset of a compact manifold, $\widetilde{X}$, without boundary:

$$\iota_X : X \hookrightarrow \widetilde{X},\ \mathscr{C}^\infty(X) = \iota_X^* \left(\mathscr{C}^\infty(\widetilde{X})\right)\,. \tag{3}$$

This allows one to readily discuss any of the usual structures on a $\mathscr{C}^\infty$ manifold, by restriction from $\widetilde{X}$, provided only that such structures are invariant under diffeomorphisms of $\widetilde{X}$ which fix each point of ∂X. In particular $\mathscr{C}^\infty$ vector fields, the tangent bundle, cotangent bundle, form bundles etc. can all be defined in this way; of course they also have more intrinsic definitions.

In most cases the structure we are seeking is most readily expressed in terms of a space of real $\mathscr{C}^\infty$ vector fields:

$$\mathscr{V}(X) \subset \mathscr{C}^\infty(X; TX)\,. \tag{4}$$

The two most basic properties we insist upon are:

$$\mathscr{V}(X) \text{ is a Lie algebra} \tag{5}$$

under the commutation bracket and

$$\mathscr{V}(X) \text{ is a } \mathscr{C}^\infty(X)\text{-module}\,. \tag{6}$$

Of course the space of all smooth vector fields, $\mathscr{C}^\infty(X; TX)$, has these two properties. However, in order that the vector fields in $\mathscr{V}(X)$ correspond to infinitesmal diffeomorphisms of X we also require that

$$\text{each } V \in \mathscr{V}(X) \text{ is tangent to each } H \in M_1(X)\,. \tag{7}$$

The simplest example is just the space $\mathscr{V}_b(X)$ of all $\mathscr{C}^\infty$ vector fields on X satisfying (7). Thus we can summarize these properties as requiring

$$\mathscr{V}(X) \subset \mathscr{V}_b(X) \text{ is a Lie subalgebra and } \mathscr{C}^\infty(X)\text{-submodule}\,. \tag{8}$$

Near any point $\bar{x} \in X$ there are local coordinates $x_1, \ldots, x_n$ in which $\bar{x}$ is mapped to the origin of $\mathbb{R}^n_k$, for some k. Then $\mathscr{V}_b(X)$ is locally the span, as a $\mathscr{C}^\infty$-module, of the vector fields

$$x_1\partial_{x_1}, \dots, x_k\partial_{x_k}, \partial_{x_{k+1}}, \dots, \partial_{x_n}. \tag{9}$$

There are no relations between these vector fields so there is a natural vector bundle, which we denote ${}^bTX$, of which $\mathcal{V}_b(X)$ forms the space of all sections. Since these sections are vector fields there is a natural vector bundle map

$${}^bTX \longrightarrow TX \tag{10}$$

which is an isomorphism over the interior, but has corank k at a boundary face of codimension k, since the first k elements in (9) are mapped to zero. In general we insist that the Lie algebra $\mathcal{V}(X)$ have a similar structure. Thus we require that there be a vector bundle ${}^{\mathcal{V}}TX$ and vector bundle maps

$${}^{\mathcal{V}}TX \longrightarrow {}^bTX \longrightarrow TX \tag{11}$$

where the overall map is $\iota_{\mathcal{V}}$ and

$$\mathcal{V}(X) = \iota_{\mathcal{V}} \circ \mathcal{C}^\infty(X; {}^{\mathcal{V}}TX). \tag{12}$$

We denote by ${}^{\mathcal{V}}T^*X$ the dual bundle to ${}^{\mathcal{V}}TX$, and by ${}^{\mathcal{V}}\Lambda^kX$ the associated form bundles.

Since $\mathcal{V}$ is a Lie algebra of operators acting on $\mathcal{C}^\infty(X)$ the enveloping algebra is naturally a filtered ring of operators, denoted $\mathrm{Diff}^m_{\mathcal{V}}(X)$, acting on $\mathcal{C}^\infty(X)$. In fact, since the elements of $\mathcal{V}$ are local diffeomorphisms, twisted versions, $\mathrm{Diff}^m_{\mathcal{V}}(X; E, F)$, of these spaces of operators are defined between sections of any two vector bundles, E and F, over X.

One easy way to see why we might be interested in these structures is the following elementary result.

Proposition (13). *Exterior differentiation defines a complex*

$$\mathcal{C}^\infty(X) \xrightarrow{d} \mathcal{C}^\infty(X; {}^{\mathcal{V}}\Lambda^1) \xrightarrow{d} \dots \xrightarrow{d} \mathcal{C}^\infty(X; {}^{\mathcal{V}}\Lambda^q) \tag{14}$$

where q is the fibre dimension of ${}^{\mathcal{V}}TX$. If g is a fibre metric on ${}^{\mathcal{V}}TX$ then the associated Laplacian is an element of $\mathrm{Diff}^2_{\mathcal{V}}(X; {}^{\mathcal{V}}\Lambda^k)$ *for each k, $0 \le k \le q$.*

The assumption (12) means that the filtration of the ring $\mathrm{Diff}^*_{\mathcal{V}}(X)$ is given by symbol maps, $\sigma_{k,\mathcal{V}}$, giving a short exact sequence

$$0 \longrightarrow \mathrm{Diff}^{k-1}_{\mathcal{V}}(X; E, F) \hookrightarrow \mathrm{Diff}^k_{\mathcal{V}}(X; E, F) \xrightarrow{\sigma_{k,\mathcal{V}}} P^k({}^{\mathcal{V}}T^*X; E, F) \longrightarrow 0, \tag{15}$$

where $P^k({}^{\mathcal{V}}T^*X; E, F)$ is the space of $\mathcal{C}^\infty$ homomorphisms of the lifts to ${}^{\mathcal{V}}T^*X$ of the vector bundles E, F which are homogeneous polynomials of degree k on the fibres. An operator is ($\mathcal{V}$-) elliptic if its symbol in this sense is invertible off the zero section of ${}^{\mathcal{V}}T^*X$. The symbol of the Laplacian is always given by multiplication by the square of the length, so it is always elliptic as an element of $\mathrm{Diff}^k_{\mathcal{V}}(X; {}^{\mathcal{V}}\Lambda^k)$.

Now by a *small* calculus of $\mathcal{V}$-pseudodifferential operators we mean linear spaces of operators

$$\Psi^m_{\mathcal{V}}(X;E,F) \ni A : \mathscr{C}^\infty(X;E) \longrightarrow \mathscr{C}^\infty(X;F),\ m \in \mathbb{R}, \tag{16}$$

with (at least) the following additional properties. First these spaces should extend the ring of $\mathcal{V}$-differential operators:

$$\mathrm{Diff}^k\,\mathcal{V}(X;E,E') \subset \Psi^k_{\mathcal{V}}(X;E,E') \tag{17}$$

$$\Psi^m_{\mathcal{V}}(X;E,E') \circ \Psi^{m'}_{\mathcal{V}}(X;E',E'') \subset \Psi^{m+m'}_{\mathcal{V}}(X;E,E''). \tag{18}$$

Moreover we require that the filtration be fixed by symbol maps giving short exact sequences analogous to (15):

$$0 \longrightarrow \Psi^{m-1}_{\mathcal{V}}(X;E,F) \hookrightarrow \Psi^m_{\mathcal{V}}(X;E,F) \overset{\sigma_{m,\mathcal{V}}}{\longrightarrow} S^{[m]}({}^{\mathcal{V}}T^*X;E,F) \longrightarrow 0, \tag{19}$$

where now $S^{[m]}({}^{\mathcal{V}}T^*X;E,F)$ is the quotient of the space of homomorphisms, as before, but now only required to be symbols of order m, (see (79)) by the subspace of symbols of order $m-1$. In fact we require that (19) should split (continuously), i.e. there should be a $\mathcal{V}$-quantization map:

$$q_{\mathcal{V}} : S^m({}^{\mathcal{V}}T^*X;E,F) \longrightarrow \Psi^m_{\mathcal{V}}(X;E,F) \tag{20}$$

which splits (19) and which is surjective modulo the residual algebra

$$\Psi^{-\infty}_{\mathcal{V}}(X;E,F) = \bigcap_m \Psi^m_{\mathcal{V}}(X;E,F). \tag{21}$$

Naturally the map (19) is to be consistent with (15) as is the product formula:

$$\sigma_{m+m',\mathcal{V}}(A \circ B) = \sigma_{m,\mathcal{V}}(A) \circ \sigma_{m',\mathcal{V}}(B). \tag{22}$$

Finally we require the algebra to be asymptotically complete with respect to the filtration. Thus if $A_j \in \Psi^{m-j}_{\mathcal{V}}(X;E,F)$ then there must exist $A \in \Psi^m_{\mathcal{V}}(X;E,F)$ such that

$$A \sim \sum_j A_j, \text{ i.e. } A - \sum_{j<N} A_j \in \Psi^{m-N}_{\mathcal{V}}(X;E,F) \quad \forall\ N \in \mathbb{N}. \tag{23}$$

Then A is determined up to the addition of an element of the residual algebra.

These conditions hold for the calculus of pseudodifferential operators on a compact manifold without boundary. Various of the standard theorems in that case can be extended to any algebra with these properties. For example

Proposition (24). *If $A \in \Psi^m_{\mathcal{V}}(X;E,F)$ is elliptic, in the sense that its symbol has a representative which is invertible outside a compact set, and the conditions* (16)-(21) *hold then there exists $B \in \Psi^{-m}_{\mathcal{V}}(X;F,E)$ such that*

$$A \circ B - \mathrm{Id} \in \Psi^{-\infty}_{\mathcal{V}}(X;F,F),\ B \circ A - \mathrm{Id} \in \Psi^{-\infty}_{\mathcal{V}}(X;E,E). \tag{25}$$

In the familiar case, that is $\partial X = \emptyset$ and $\mathcal{V} = \mathscr{C}^\infty(X;TX)$, the residual calculus consists of smoothing operators, which are compact operators on any Sobolev space. Then (25) is a rather strong conclusion. In other cases, such as those discussed below, the residual calculus does not consist of compact operators. Then for most applications (25) must be improved. The proof of (25) itself is straightforward. Using the surjectivity of the symbol map in (19), the composition formula (22) and the exactness in (19) one can make an initial choice B_1 for B which satisfies the weaker form of (25):

$$A \circ B_1 - \mathrm{Id} = R_1 \in \Psi_{\mathcal{V}}^{-1}(X;F). \tag{26}$$

The space $\Psi_{\mathcal{V}}^{-1}(X;E)$ is an ideal in $\Psi_{\mathcal{V}}^{0}(X;E)$ so, again using (18) the Neumann series for $(\mathrm{Id}+R_1)^{-1}$ converges asymptotically in the sense of (23), allowing $B \sim B_1(\mathrm{Id}+R_1)^{-1}$ to be constructed. To improve (25) we should therefore look for other ideals which allow finer approximations to an inverse to be constructed.

To find such ideals in the calculus we need to assume more about the algebra $\mathcal{V}$. Through any point $x \in X$ let $Q = Q(x)$ be the closure of the set of endpoints of integral curves of $\mathcal{V}$ starting at x. We shall demand that this be always a p-submanifold. This is the strongest condition on a submanifold in the category of manifolds with corners, it just means that near each point of Q there is a coordinate system reducing the point to the origin in $\mathbb{R}^n_k$, and such that Q is locally given by the intersection of coordinate planes; the first k of these being the local boundary hypersurfaces. Restriction to Q of the elements of $\mathcal{V}$ defines a Lie algebra of vector fields on Q :

$$\mathcal{V}(X) \longrightarrow \mathcal{W}_{\mathcal{V}}(Q) \subset \mathcal{V}_b(Q). \tag{27}$$

We demand that

$$\mathcal{W}_{\mathcal{V}}(Q) = \mathcal{W}(Q) \text{ span } \mathcal{V}_b(Q) \text{ over the interior of } Q. \tag{28}$$

It follows that the null space of the restriction (27), $\mathcal{N}(\mathcal{V};Q)$, defines a subbundle

$${}^{\mathcal{V}}NQ \subset {}^{\mathcal{V}}T_QX. \tag{29}$$

The fibres of this bundle are finite dimensional Lie algebras and

$$\mathcal{W}(Q) = \mathcal{C}^\infty(Q; {}^{\mathcal{V}}T_QX/{}^{\mathcal{V}}NQ)$$

i.e. the structure bundle, ${}^{\mathcal{W}}TQ$, of $\mathcal{W}$ is just the quotient bundle ${}^{\mathcal{V}}T_QX/{}^{\mathcal{V}}NQ$.

We actually require that these conditions on the integral leaves of $\mathcal{V}$ hold locally uniformly in an appropriate sense. To explain this we first consider the notion of a b-map between compact $\mathcal{C}^\infty$ manifolds with corners. A $\mathcal{C}^\infty$ map

$$f : X \longrightarrow X' \tag{30}$$

is a *b-map* if, for each boundary hypersurface $H' \in M_1(X')$ of X' the lift of the ideal of functions, $\mathcal{I}(H') \subset \mathcal{C}^\infty(X')$, vanishing at H' is the product of such ideals in X :

$$f^*\mathcal{I}(H') = \prod_{H \in M_1(X)} \mathcal{I}(H)^{e(H,H')}. \tag{31}$$

In terms of defining functions of the boundary hypersurfaces this just means

$$f^*\varrho_{H'} = a_{H'} \prod_{H\in M_1(X)} \varrho_H^{e(H,H')},\ 0 < a_{H'} \in \mathscr{C}^\infty(X)\,. \tag{32}$$

The powers $e(H,H')$ are non-negative integers. One important consequence of (32) is that the differential of f extends to the b-tangent space ${}^bTX$, introduced above:

$${}^bf_* : {}^bT_xX \longrightarrow {}^bT_{f(x)}X' \quad \forall\, x \in X\,. \tag{33}$$

We introduce three refinements of the notion of a b-map related to the properties if this b-differential. Namely a b-map is said to be a *b-submersion* if the b-differential, (33), is everywhere surjective. It is said to be a *b-normal* map if

$$\forall\ H \in M_1(X)\ e(H,H') \neq 0 \text{ for at most one } H' \in M_1(X')\,. \tag{34}$$

Finally a b-map is said to be a *b-fibration* if it is both b-normal and a b-submersion. A fibration of compact manifolds with corners is easily seen to be a b-fibration. The converse statement is not true. For any b-fibration $\phi : X \longrightarrow X'$ the subspace $\mathscr{V}_\phi(X) \subset \mathscr{V}_b(X)$, just the sections of the null space of ${}^b\phi_*$, is a Lie algebra of vector fields satisfying all the conditions imposed so far, as is the weighted version

$$\begin{gathered}\varrho^k\mathscr{V}_\phi(X) = \left\{V \in \mathscr{V}_b(X); V = \varrho^k W,\ W \in \mathscr{V}_\phi(X)\right\}\\ \varrho^k = \prod_{h\in M_1(X)} \varrho_H^{k(h)}\end{gathered} \tag{35}$$

for any map $k : M_1(X) \longrightarrow \mathbb{N}_0 = \{0,1,\ldots\}$.

The final condition we wish to impose on $\mathscr{V}$ is that each boundary face $F \in M(X)$ (including X itself) should have a b-fibration

$$\phi_F : F \longrightarrow L(F) \tag{36}$$

such that

$$Q(x) = \phi_F^{-1}(\phi_F(x))\ \forall\ x \text{ in the interior of } F\,. \tag{37}$$

Moreover we require that the Lie algebra is not too far from being that of the b-fibration corresponding to the interior:

$$\mathscr{V}(X) \supset \varrho^k\mathscr{V}_\phi(X) \text{ for some } k : M_1(X) \longrightarrow \mathbb{N}_0 \tag{38}$$

where $\phi = \phi_X$. Although the integral leaves through the interior points of F are just the leaves of ϕ_F through those points this still allows considerable freedom for the Q to change, in dimension, from face to face.

Definition (39). A boundary-fibration structure on a compact manifold with corners is a subspace $\mathscr{V}(X) \subset \mathscr{V}_b(X)$ satisfying (8), (11), (12), (28), (37) and (38).

With these additional conditions it follows that for each integral submanifold $Q = Q(x)$

$$\text{each fibre } {}^{\mathscr{V}}N_pQ,\ p \in Q, \text{ is a solvable Lie algebra}\,. \tag{40}$$

Notice that we are immediately forced to consider a more general notion than that of a boundary-fibration structure, namely the structure induced by a boundary-fibration structure on each of the leaves. In this extended type of structure there are hidden variables, namely those in the bundle, ${}^{\mathcal{V}}NQ$ of solvable Lie algebras. A small calculus of pseudodifferential operators, still denoted $\Psi^m_{\mathcal{V}}(X)$, corresponding to such a structure should behave as convolution operators on the fibres of the associated bundle of Lie groups.

The ideals needed to obtain finer approximate inverses of elliptic operators are associated with the quotients

$$0 \longrightarrow \mathcal{I}_Q(X)\cdot\mathcal{V}(X) \longrightarrow \mathcal{V}(X) \longrightarrow \mathcal{U}(Q) = \mathcal{C}^\infty(Q;{}^{\mathcal{V}}T_QX) \longrightarrow 0\,. \tag{41}$$

Since $\mathcal{I}(Q)\cdot\mathcal{V}$ is an ideal this gives rise to a homomorphism of the enveloping algebras, which we call the normal operator at Q:

$$N_Q : \mathrm{Diff}^k_{\mathcal{V}}(X;E,F) \longrightarrow\!\!\!\!\rightarrow \mathrm{Diff}^k_{\mathcal{U}}(Q;E,F)\,. \tag{42}$$

We require (as a condition on the $\mathcal{V}$-pseudodifferential operators) that there be an extension of this map

$$N_Q : \Psi^m_{\mathcal{V}}(X;E,F) \longrightarrow\!\!\!\!\rightarrow \Psi^m_{\mathcal{U}}(Q;E,F)\,. \tag{43}$$

The ideal

$$\Psi^{m,1}_{\mathcal{V}}(X;E,F) = \{A \in \Psi^m_{\mathcal{V}}(X;E,F); N_Q(A) = 0\ \forall\ Q \subset \partial X\} \tag{44}$$

and its powers

$$\Psi^{m,k}_{\mathcal{V}}(X;E,F) = \Psi^{m,1}_{\mathcal{V}}(X;E,F)\cdot\Psi^{0,k-1}_{\mathcal{V}}(X;E) \tag{45}$$

are then of considerable interest and in particular we would expect that

$$\Psi^{-\infty,\infty}_{\mathcal{V}}(X;E,F) \ni A : \mathcal{C}^\infty(X;E) \longrightarrow \dot{\mathcal{C}}^\infty(X;F)\,, \tag{46}$$

the latter space being the subspace of $\mathcal{C}^\infty(X;F)$ consisting of the sections vanishing to all order at ∂X.

Error terms of the type in (46) would be extremely satisfactory, but in most practical cases it is not possible to obtain them directly. This can be understood from the discussion of (26). In order to get the error in $\Psi^{-1,1}_{\mathcal{V}}(X;F)$, which could be improved by iteration, as in the proof of (25), we need to be able to solve the *model problem*

$$N_Q(A)\cdot N_Q(B) = \mathrm{Id} \tag{47}$$

exactly, for all integral submanifolds $Q \subset \partial X$. This can often be done, and this is precisely the method used in the problems discussed below. However the inverse, $N_Q(B)$, is seldom in the small calculus associated to $\mathcal{U}$. In practice it is therefore necessary to consider *full* calculi which have more off-diagonal singularities. This is discussed in III below.

II. Applications and Examples

Next I shall consider various examples of Lie algebras of vector fields, $\mathcal{V}$, and give some representative results concerning $\mathcal{V}$-differential operators which can be obtained by use of the calculus of $\mathcal{V}$-pseudodifferential operators.

(i) *b*-Structure

The first example is the one already mentioned, namely $\mathcal{V}_b(X)$. This is naturally defined on any compact manifold with corners. For a compact manifold with boundary the associated b-calculus was described in [22], see also [15; Chapter 18], and called the totally characteristic calculus. For the general case of a compact manifold with corners the small calculus is described in [25] and the full calculus in [23]. Certainly $\mathcal{V}_b(X)$ is a boundary-fibration structure and the calculus has all the properties described above. In the case of a compact manifold with boundary the only integral submanifolds are X and the components of the boundary $H \in M_1(X)$. The b-fibrations required in (37) are all the single-leaf fibrations. Consider the normal operator for the calculus, as indicated in (43). Since this is such a natural calculus the normal operator can be realized as an operator on an associated geometric space. Namely the normal bundle to a component of the boundary, $N\{X;H\}$, $H \in M_1(X)$, can be compactified to a bundle of intervals, denoted X_H over H. This has a natural action of $[0,\infty)$. The normal operator takes the form

$$N_H : \Psi_b^m(X) \twoheadrightarrow \Psi_{b,I}^m(X_H) \tag{48}$$

where the suffix I denotes the $[0,\infty)$-invariant elements. The invertibility properties of these operators can be discussed using the Mellin transform. In fact the Mellin transform of the normal operator gives an entire family of pseudodifferential operators on H, called the indicial family:

$$\mathbb{C} \ni \lambda \longmapsto I(A,\lambda) \in \Psi^m(H)\,. \tag{49}$$

If A is elliptic as a b-pseudodifferential operators this family is elliptic for each λ (in fact the principal symbol is independent of λ) and $I(A,\lambda)$ is invertible except for a discrete set of points:

$$\operatorname{spec}_b(A) = \{\lambda \in \mathbb{C} : I(A,\lambda) \text{ is not invertible}\}\,. \tag{50}$$

Theorem (51) (Melrose and Mendoza [24]). *An elliptic element of $\Psi_b^m(X)$ is Fredholm as a map on the space $\varrho^r\mathcal{C}^\infty(X)$ if and only if*

$$\operatorname{Im}\lambda \neq -r \quad \forall\, \lambda \in \operatorname{spec}_b(A)\,. \tag{52}$$

The formula for the index of such an elliptic b-pseudodifferential operators involves extensions of the work of Atiyah, Patodi and Singer [1] on index theory and the η-invariant.

Metrics on the bundle ${}^bTX$ are related to conic geometry. If X is a compact manifold with boundary then near ∂X it decomposes as a product, since ∂X has a (global) defining function $x \in \mathscr{C}^\infty(X)$. By a *conic metric* (of weight s) on X we mean a Riemann metric on the interior which, near ∂X, takes the form

$$g_b \sim x^{2s}\left[\left(\frac{dx}{x}\right)^2 + h(x,y,dx,dy)\right] \tag{53}$$

for some defining function x, and with $h(0,y,0,dy)$ a metric on ∂X. This is slightly more special than simply a fibre metric on ${}^bTX$. Cheeger considered such questions as the identity of the L^2 cohomology groups of X with respect to g_b (see [6, 7, 8]). These questions can be treated rather directly by constructing a good parametrix for the Laplacian using the b-calculus, see [23].

Another differential-geometric question involving the b-calculus is suggested by the work of Baum, Douglas and Taylor [3]). Namely to find an analytic construction of Poincaré duality for K-theory (Kasparov [16, 17]) for compact manifolds with boundary, extending the original motivation for K-homology of Atiyah.

Theorem (54) (Melrose and Piazza [25]). *A b-quantization map induces isomorphisms realizing Poincaré duality:*

$$K^i(T^*X) \longleftrightarrow K_i(X,\partial X),\ K^i(T^*X,\partial T^*X) \longleftrightarrow K_i(X),\ i=0,1\,.$$

This means that the K_0-homology groups can be identified with equivalence classes of elliptic b-pseudodifferential operators on X. The K_1 groups can be identified with such classes on $X \times [0,1]$, so the calculus on manifolds with corners is useful even if X is a manifold with boundary.

The b-calculus also has other more analytic applications, particularly as related to spaces of conormal distributions.

(ii) 0-Structure

A second natural boundary-fibration structure on any compact manifold with boundary is the 0-structure. The Lie algebra of vector fields is

$$\mathscr{V}_0(X) = \{V \in \mathscr{V}_b(X); V = 0 \text{ at } \partial X\}\,. \tag{55}$$

The associated calculus described in [19] has all the properties introduced above. The integral submanifolds are simply X and the individual points of the boundary. The b-fibrations are therefore again trivial, the single-fibre fibration of X and the point-fibre fibration of the boundary. Since each boundary point is a leaf the fibre of the structure bundle ${}^0T_pX$ is a solvable Lie algebra for each $p \in \partial X$; it is just the homogeneous extension of an abelian algebra. The normal operators at each boundary point are convolution operators on the associated solvable Lie group. Again the naturality of the 0-structure means that the normal operators actually act on a geometric space, in this case on the inward-pointing half of the tangent space to X at p.

Fibre metrics on the structure bundle ${}^0TX$ are always of the form

$$g_0 = \frac{h}{x^2} \tag{56}$$

where h is a metric in the ordinary sense on X and x is a defining function for the boundary. The invariant metric on hyperbolic n-space, represented as the ball $\{|x| < 1\}$ in $\mathbb{R}^n$,

$$g_H = \frac{|dx|^2}{(1 - |x|^2)}$$

is clearly an example of such a metric. In fact the normal operator of the Laplacian of a metric (56) is always reducible to the hyperbolic Laplacian. This can be used to show that the spectral theory (indeed the scattering theory) of a 0-metric is similar to that of the hyperbolic Laplacian. For example

Theorem (57) (Mazzeo and Melrose [19]). *If* $0 < \kappa \in \mathscr{C}^\infty(X)$ *and* $\kappa|dx|_h^2 = 1$ *at* ∂X *then the kernel of the normalized resolvent of the metric* (56),

$$[\Delta - \kappa s(n-s)]^{-1} , \tag{58}$$

extends to a meromorphic family of 0*-pseudodifferential operators,* $s \in \mathbb{C}$.

Mazzeo in [18] used the construction of a parametrix for the Laplacian to find the Hodge cohomology of a 0-metric as in (56). Metrics of this form arise on cuspless, infinite volume, quotients of hyperbolic space by a geometrically finite discrete group (e.g. Schottky groups).

Other problems can be tackled using these operators, in particular the solution operators to elliptic boundary problems lie in this calculus.

(iii) Bergman Geometry

Any $\mathscr{C}^\infty$ bounded strictly pseudoconvex domain $\Omega \subset \mathbb{C}^n$ carries Kähler metrics with Kähler form

$$\omega_\varrho = -i\partial\bar{\partial}\log\varrho \tag{59}$$

where $-\varrho$ is a plurisubharmonic defining function for $\partial\Omega$. The Laplacian of this metric corresponds to a boundary-fibration structure on $X = \Omega_{\frac{1}{2}}$, the manifold obtained from (and diffeomorphic to) Ω by adjoining $\sqrt{\varrho}$ to $\mathscr{C}^\infty(\Omega)$, to give $\mathscr{C}^\infty(\Omega_{\frac{1}{2}})$. Thus there is a $\mathscr{C}^\infty$ map (with fold singularity at the boundary) $\iota_{\frac{1}{2}}$: $X \longrightarrow \Omega$. On X the boundary-fibration structure is determined by the contact form θ on $\partial\Omega$. This form fixes a class of 1-forms on X such that for any representative, Θ,

$$\mathscr{V}_\Theta(X) = \{V \in \mathscr{V}_0(X); \Theta(V) = O(r^2)\} \tag{60}$$

where $r = (\iota_{\frac{1}{2}})^* \sqrt{\varrho} \in \mathscr{C}^\infty(X)$.

Theorem (61) (Epstein, Melrose and Mendoza [11]). *The resolvent of the Bergman-type Laplacian, of the metric corresponding to* (59), *has an analytic extension similar to that of the Laplacian on the ball:*

$$[\Delta - s(1-s)]^{-1}$$

extends from $\operatorname{Re} s > \frac{1}{2}$ *to be a meromorphic family of* Θ*-pseudodifferential operators,* $s \in \mathbb{C}$.

Using this same approach, i.e. by studying the Bergman-type Laplacian, but now on $(n, 0)$ and $(n, 1)$-forms one can provide an alternate solution of Kohn's $\bar{\partial}$-Neumann problem:

$$\begin{aligned} &\bar{\partial} u = f,\ f \in \mathscr{C}^\infty(\Omega; \Lambda^{0,1}),\ \bar{\partial} f = 0 \\ &u \perp \mathscr{H}(\Omega) = \left\{u \in L^2(\Omega),\ \bar{\partial} u = 0\right\}. \end{aligned} \tag{62}$$

Corollary (63). *On any* $\mathscr{C}^\infty$ *strictly pseudoconvex domain Kohn's* $\bar{\partial}$*-Neumann problem* (62) *is solved,* $u = Nf$, *by a* Θ*-pseudodifferential operator of order* -1.

The standard regularity properties, see [13], then follow readily.

(iv) Adiabatic Limit of a Fibration

Consider a fibration of compact manifolds

$$\begin{array}{ccc} F & \text{———} & M \\ & & \downarrow \phi \\ & & Y. \end{array}$$

On the total space, M, consider a 1-parameter family of metrics which becomes singular as the parameter, $x \downarrow 0$:

$$g_x = g_\infty + \frac{\phi^* h}{x^2}. \tag{64}$$

Here g_∞ is a metric on M (or just a symmetric cotensor which is non-degenerate on the fibres) and h is a metric on Y.

This is an example of a singular limit. It was considered by Witten ([28]) in a special case to study the η-invariant. In general the behaviour of the adiabatic limit of the η-invariant was discussed by Bismut and Cheeger [2] and by Dai [9]. The analytic torsion or Ray and Singer can be examined by similar methods (see Dai, Epstein and Melrose [10]).

The associated boundary-fibration structure provides a simple way to describe the adiabatic limit of the Hodge cohomology. The manifold with corners on which we work is $X = M \times [0, 1]$ where the extra factor is the parameter x. Then set

$$\begin{aligned} \mathscr{V}_a(X) = \{V \in \mathscr{V}_b(X); Vx \equiv 0 \text{ and} \\ V \text{ is tangent to the fibres of } \phi \text{ at } x = 0\}. \end{aligned} \tag{65}$$

If local coordinates are taken for the fibration, y in the base and z in the fibres then locally

$$\mathcal{V}_a(X) = \mathrm{sp}\,\{x\partial y, \partial z\} = \mathcal{C}^\infty(X; {}^a T X)\,. \tag{66}$$

The b-fibrations required, in (37), to show that $\mathcal{V}_a(X)$ is a boundary-fibration structure are given by the projection $X \longrightarrow M$ over the interior and $\{x = 1\}$ and by the fibration ϕ over $\{x = 0\}$. The metrics (64) fix a fibre metric on ${}^a T X$ and so, from (13) the Laplacian $\Delta_a \in \mathrm{Diff}^2_a(X; {}^a\Lambda^k)$. Then

$$\mathcal{H}^k = \left\{u \in \mathcal{C}^\infty(X; {}^a\Lambda^k); \Delta_a u = 0\right\}$$

is the space of all $\mathcal{C}^\infty$ sections of a vector bundle over $[0,1]$ with fibre $H^k_x \subset \mathcal{C}^\infty(X_x; {}^a\Lambda^k)$. The behaviour of these fibres at 0 brings out the Leray spectral sequence for the fibration. Namely

Proposition (67) (Mazzeo and Melrose [20]). *For each k the spaces, defined for each $\ell \in \mathbb{N}_0$,*

$$\begin{aligned} E^{k,\ell} = \big\{u_0 \in \mathcal{C}^\infty(X_0; {}^a\Lambda^k); \exists\ \tilde{u} \in \mathcal{C}^\infty(X; {}^a\Lambda^k) \\ \text{with } \tilde{u}\big|_{x=0} = u_0,\ \Delta_a \tilde{u} = 0(x^{2\ell})\big\} \end{aligned} \tag{68}$$

form a decreasing sequence of vector spaces stabilizing to H^k_0, which has the dimension of the cohomology of M :

$$E^{k,0} \supset E^{k,1} \supset E^{k,2} \supset \cdots \supset E^{k,N} = \cdots = E^{k,\infty} = H^k_0\,. \tag{69}$$

Here $E^{k,0} = \mathcal{C}^\infty(X_0; {}^a\Lambda^k)$, $E^{1,k}$ is the space of sections of a vector bundle over Y and $E^{k,2}$ is finite dimensional.

(v) Töplitz Correspondence

Any compact $\mathcal{C}^\infty$ manifold, Y, can be embedded as a totally real submanifold of a complex manifold, Ω, with $\dim_{\mathbb{C}} \Omega = \dim_{\mathbb{R}} Y$, (Bruhat and Whitney [5]). There exists $\varrho \in \mathcal{C}^\infty(\Omega)$ with

$$\varrho \geq 0,\ Y = \{\varrho = 0\}$$

and a non-degenerate minimum at, i.e. non-degenerate Hessian normal to, Y. Then the tubes around Y

$$\Omega_\varepsilon = \left\{z \in \Omega; \varrho(z) \leq \varepsilon^2\right\}$$

are, for $\varepsilon > 0$ small enough, strictly pseudoconvex neighbourhoods of Y. A fibration of Ω near Y and transversal to it can be chosen so that $\mathrm{Im}\,\partial\varrho$ vanishes on the fibres. The Töplitz correspondence is the map obtained by fibre-integration of holomorphic $(n,0)$ forms:

$$T_\varepsilon : \left\{u \in \mathcal{C}^\infty(\Omega_\varepsilon; \Lambda^{n,0});\ \bar{\partial} u = 0\right\} \xrightarrow{\int_{\text{fibre}}} \mathcal{C}^\infty(Y)\,. \tag{70}$$

Boutet de Monvel and Guillemin [4] showed that this map is Fredholm and conjectured the following

Theorem (71) **(Epstein and Melrose [12]).** *There exists* $\varepsilon_0 > 0$ *such that for* $0 < \varepsilon < \varepsilon_0$ *the Töplitz correspondence* (70) *is an isomorphism.*

The main step in the proof is the uniform solution of the $\bar{\partial}$-Neumann problem in Ω_ε as $\varepsilon \downarrow 0$. For this we start with the manifold $M \simeq \Omega_{\varepsilon_0} \times [0, \varepsilon_0]$, obtained by blowing up the singular set in the cone formed by the shrinking tubes,

$$Y \times \{0\} \subset \left\{(z, \varepsilon); \varrho(z) \le \varepsilon^2,\ 0 \le \varepsilon \le \varepsilon_0\right\}.$$

Thus M is a manifold with corners, it has three boundary hypersurfaces. The blow-up procedure introduces an 'adiabatic' boundary H_a, and this has a fibration, $\phi_a : H_a \longrightarrow Y$. There is a trivial boundary at $\{\varepsilon = \varepsilon_0\}$ and the θ-boundary, H_θ. The θ-boundary corresponds to the boundary of Ω and has a contact bundle defined on it. Let X be the manifold with the square-root $\mathcal{C}^\infty$ structure introduced at H_θ, there is a conformal class of 1-forms, with representative Θ, defined at the new H_Θ. On X the α-structure is obtained by combining the Θ-structure of (iii) with the adiabatic structure of (iv):

$$\begin{aligned}\mathcal{V}_\alpha(X) = \big\{V \in \mathcal{V}_b(X); Vx \equiv 0,\\ &V \text{ is tangent to the fibres of } \phi_a \text{ and} \\ &V \text{ vanishes at } H_\Theta \text{ with } \Theta(V) = O(\varrho_\Theta^2)\big\}.\end{aligned} \tag{72}$$

Here ϱ_Θ is a defining function for H_Θ; $\mathcal{V}_\alpha(X)$ is a boundary-fibration structure. The adiabatic Bergman-type metric with Kähler form

$$-i\partial\bar{\partial}\log\left(\frac{\varrho}{\varepsilon^2} - 1\right)$$

is a fibre metric on the structure bundle ${}^{\alpha}TX$ of $\mathcal{V}_\alpha$. Thus, by Proposition (13) the Laplacian is an elliptic element of $\mathrm{Diff}^2_\alpha(X; {}^{\alpha}\Lambda^{p,q})$. One of the crucial steps in proving Theorem (71) is to show that the Bergman projection, onto holomorphic functions on the tube of radius ε, is an α-pseudodifferential operator.

(vi) Analytic Surgery

Suppose M is a compact manifold without boundary and $H \subset M$ is an embedded, oriented closed hypersurface. If ϱ is a defining function for H and x is a parameter consider the 1-parameter family of metrics

$$g_x = a\,d\varrho^2 + (\varrho^2 + x^2)h.$$

Here $a > 0$ is a $\mathcal{C}^\infty$ function in (ϱ, x)-polar coordinates, i.e. on the manifold, $X = \left[M \times [0,1]; H \times \{0\}\right]$, obtained by blowing up $H \times \{0\}$ in $M \times [0,1]$ and h is a metric on M. As $x \downarrow 0$ the metric g_x degenerates to a conic metric (incomplete) on the manifold with boundary obtained by closing $M \backslash H$ with two copies of H. The boundary-fibration structure on X is

$$\mathcal{V}_v = \{V \in \mathcal{V}_b(X); Vx \equiv 0\} . \tag{73}$$

McDonald in his PHD thesis [21] has constructed the v-calculus which contains the inverse of the Laplacian of g_x. This allows one to describe, somewhat as in (iv), the behaviour of the Hodge cohomology as $x \downarrow 0$. Partial results of this type have been obtained by Seeley and Singer [27] and more recently extended by Seeley [26].

(vii) Projective Algebraic Varieties

One rather intriguing question is whether the notion of boundary-fibration in Definition (39) applies to a singular projective algebraic variety M. More precisely let X be a compact manifold with corners obtained from M in two steps. First let $F : \widetilde{M} \longrightarrow M$ be a resolution of M, so the preimage of the singular locus of M is a finite union of hypersurfaces in $\widetilde{M}$ with only normal intersections. From the real point of view these are embedded submanifolds of codimension two, let X be obtained by real blow-up of these submanifolds. Now we ask:

Question (74). Is there are boundary-fibration structure on X (for an appropriate choice of $\widetilde{M}$) such that the induced (Fubini-Study) metric on the regular part of M is a weighted fibre metric on the structure bundle?

A weighted metric is as in (53), i.e. a fibre metric with a conformal factor which is a product of powers of defining functions for boundary hypersurfaces. If such a boundary-fibration structure exists then the construction of the corresponding $\mathcal{V}$-pseudodifferential calculus, as outlined next, should allow a rather direct treatment of the Hodge cohomology of the Laplacian.

III. Microlocalization

In the final part of my talk I wish to give a brief description of a general procedure to construct a calculus of $\mathcal{V}$-pseudodifferential operators. For simplicity I shall only consider the cases involving no parameters, i.e. when $\mathcal{V}$ spans the tangent space over the interior of X. This description therefore only applies to cases (i)–(iii) although the general case is quite similar, simply more complicated. Also for simplicity of presentation I shall suppress various line bundles, typically weighted density bundles, which occur during the construction by describing their sections as 'distributional densities'.

Schwartz' kernel theorem asserts that the space of continuous linear maps

$$A : \dot{\mathcal{C}}^{\infty}(X) \longrightarrow \mathcal{C}^{-\infty}(X) , \tag{75}$$

where the range space is the space of extendible distributions on a compact manifold with corners, can be identified with a space distributional densities over the product $X^2 = X \times X$; namely the sections of the 'kernel density bundle' KD. Thus k_A is the kernel if A if

$$A\psi(x) = \int_X k_A(x, x')\psi(x') \,. \tag{76}$$

If $\partial X = \emptyset$ then under this identification

$$\Psi^m(X) \longleftrightarrow I^m(X^2, \Delta; \mathrm{KD}) \,. \tag{77}$$

The notation for the space of conormal distributions on the right, on X^2 associated to the diagonal, $\Delta = \{(x, x') \in X^2; x = x'\}$, is that of Hörmander ([14]). The differential operators on X are precisely the local pseudodifferential operators so:

$$\mathrm{Diff}^k(X) \longleftrightarrow \left\{k \in I^k(X^2, \Delta; \mathrm{KD}); \mathrm{supp}(k) \subset \Delta\right\} \,. \tag{78}$$

This means that the kernels of differential operators are Dirac sections of KD.

The conormal distributions in (77) are those distributions which are singular only at Δ and in any local coordinates have Fourier transform in directions transversal to Δ a symbol:

$$\begin{gathered} a(x, \xi) = \int k(x, x') e^{i(x-x')\cdot\xi} dx' \in S^m \text{ i.e. satisfies} \\ |\partial_x^\alpha \partial_\xi^\beta a(x, \xi)| \le C_{\alpha,\beta}(1 + |\xi|)^{m-|\beta|} \quad \forall\, \alpha, \beta \,. \end{gathered} \tag{79}$$

A function which is smooth in x and a polynomial in ξ certainly satisfies these estimates. The differential operators are pseudodifferential operators for which a is a polynomial. Thus if we think of the algebra extension

$$\mathrm{Diff}^*(X) \hookrightarrow \Psi^*(X) \tag{80}$$

as microlocalization of $\mathscr{C}^\infty(X; TX)$ it consists in the replacement of all polynomials by all symbols in (79), or all Dirac sections smooth along Δ in (76) by all conormal sections. Then one can think of pseudodifferential operators as non-commutative symbolic functions of vector fields, just as differential operators are non-commutative polynomials in vector fields.

Algebraically it is clear that this construction should be extendible to boundary-fibration structures. Indeed the assumption (12) means that the elements of $\mathscr{V}$ are just linear functions on the fibres of the bundle ${}^{\mathscr{V}}T^*X$. Thus one can think of the elements of $\mathrm{Diff}^*_{\mathscr{V}}(X)$ as corresponding to all polynomials on ${}^{\mathscr{V}}T^*X$. However we need to realize this in a more geometric fashion to generalize (77). Note that $\mathrm{Diff}^k_{\mathscr{V}}(X)$ cannot correspond to all the Dirac sections at the diagonal, since the full algebra of differential operators does!

So we look for a new manifold with corners, the $\mathscr{V}$-stretched product $X^2_{\mathscr{V}}$, which resolves the diagonal from this point of view. More precisely there should be a b-map

$$\beta_{\mathscr{V}} : X^2_{\mathscr{V}} \twoheadrightarrow X^2 \tag{81}$$

which is a diffeomorphism of the interior of $X^2_{\mathscr{V}}$ to the interior of X^2 and a p-submanifold $\Delta_{\mathscr{V}} \subset X^2_{\mathscr{V}}$ such that

$$\beta_{\mathscr{V}} : \Delta_{\mathscr{V}} \longleftrightarrow \Delta \,. \tag{82}$$

Moreover $X^2_{\mathcal{V}}$ should be constructed in such a way that each element of $\mathcal{V}$, lifted to X^2 acting on the left factor of X, is $\beta_{\mathcal{V}}$-related to a vector field on $X^2_{\mathcal{V}}$, i.e. the Lie algebra lifts from the left factor, and so that the lifted algebra is transversal to $\Delta_{\mathcal{V}}$, meaning it spans the normal bundle. It is also natural to demand symmetry of the construction, so that the reflection giving interchange of the factors,

$$\tau : X^2 \ni (x, x') \longmapsto (x', x) \in X^2$$

should lift to diffeomorphism $\tau_{\mathcal{V}}$ of $X^2_{\mathcal{V}}$.

Once we find such a resolution we can define a corresponding class of operators. The fact the $\beta_{\mathcal{V}}$ is a b-map which is an isomorphism over the interior means that

$$\beta^*_{\mathcal{V}} : \mathcal{C}^{-\infty}(X^2) \longleftrightarrow \mathcal{C}^{-\infty}(X^2_{\mathcal{V}}) . \tag{83}$$

Thus the kernel of an operator can just as well be considered as a distributional density on $X^2_{\mathcal{V}}$ as on X^2. Next we need to find a replacement for the space on the right in (77). This is also easy since the assumption that $\Delta_{\mathcal{V}}$, the lifted diagonal, is a p-submanifold means that it is natural just to take $I^m(X^2_{\mathcal{V}}, \Delta_{\mathcal{V}})$ to be the set of restrictions to $X^2_{\mathcal{V}}$ of the conormal distributions with respect to an extension of $\Delta_{\mathcal{V}}$ in an extension of X^2 as in (3). Away from $\Delta_{\mathcal{V}}$ these distributions are $\mathcal{C}^\infty$ so we can impose additional conditions to get

$$\begin{aligned} I^m_{\mathrm{sm}}(X^2_{\mathcal{V}}, \Delta_{\mathcal{V}}) = \{\kappa \in I^m(X^2_{\mathcal{V}}, \Delta_{\mathcal{V}}) ; \kappa \equiv 0 \text{ at all} \\ H \in M_1(X^2_{\mathcal{V}}) \text{ s.t. } H \cap \Delta_{\mathcal{V}} = \emptyset\} . \end{aligned} \tag{84}$$

This allows us to define the small calculus by analogy with (77):

$$\Psi^m_{\mathcal{V}}(X) \longleftrightarrow I^m_{\mathrm{sm}}(X^2_{\mathcal{V}}, \Delta_{\mathcal{V}} ; \mathrm{KD}) . \tag{85}$$

In all the cases (i)–(vi) (and various others) which have been analyzed the stretched product can be defined by a process of blowing up the integral submanifolds Q of $\mathcal{V}$ in ∂X as submanifolds of the boundary of the diagonal in X^2.

This definition automatically gives the inclusion (17) and a symbol map as in (19). The product formula (18) is not so immediate. Before discussing this briefly let me note where the 'full' calculus, alluded to above, comes from. The kernels in (85) are singular just at the diagonal. However it is rather natural to expect similar conormal singularities to arise at the other naturally defined submanifolds of $X^2_{\mathcal{V}}$, namely the boundary hypersurfaces. Thus we can consider more general spaces of conormal distributions, denoted $I^{m,\mathfrak{m}}(X^2_{\mathcal{V}}, \Delta)$, where $\mathfrak{m} : M_1(X) \longrightarrow \mathbb{R}$ associates an order to each boundary face. Then

$$\Psi^{m,\mathfrak{m}}_{\mathcal{V}}(X) \longleftrightarrow I^{m,\mathfrak{m}}(X^2_{\mathcal{V}}, \Delta_{\mathcal{V}} ; \mathrm{KD}) . \tag{86}$$

In practice one should consider polyhomogeneous spaces, with elements having expansions in terms of powers of the defining functions at the boundary hypersurfaces and the diagonal. The small calculus then just corresponds to kernels of order $+\infty$ (meaning rapidly vanishing) at hypersurfaces not meeting $\Delta_{\mathcal{V}}$ and smooth up to the other hypersurfaces (having expansions with non-negative integral powers only).

The most important outstanding point is the proof of a product formula, (18), or more generally for the full calculus (although this is not usually an algebra so the orders must satisfy appropriate bounds for operators to compose). This proof is also carried out geometrically in these cases. The crucial construction is that of the $\mathcal{V}$-triple product, $X^3_{\mathcal{V}}$. This is related to $X^2_{\mathcal{V}}$ in essentially the same way that X^3 is related to X^2. Namely there are three maps

$$\pi^3_{o,\mathcal{V}} : X^2_{\mathcal{V}} \longrightarrow\!\!\!\!\!\rightarrow X^2_{\mathcal{V}} \tag{87}$$

for $o = f, c, s$ corresponding to projection off the left, the central and the right factor of X. The notation corresponds to the fact that in the product formula

$$C = A \cdot B \tag{88}$$

the kernel of C, the composite operator, can be computed from those of A, the second operator, and B, the first operator, by pull-back, product and push-forward operations:

$$\kappa_C = (\pi^3_{c,\mathcal{V}})_* \left[(\pi^3_{s,\mathcal{V}})^* \kappa_A \cdot (\pi^3_{f,\mathcal{V}})^* \kappa_B \right] . \tag{89}$$

Here of course one needs to interpret the operations appropriately in terms of distributional densities. Thus the functorial properties of conormal distributions are of primary importance. In particular it is crucial that the stretched projections (87) be b-fibrations since the push-forward and pull-back of polyhomogeneous conormal distributions under b-fibrations are polyhomogeneous conormal, see [23].

References

1. Atiyah, M.F., Patodi, V.K., Singer, I.M: Spectral asymmetry and Riemannian geometry. Math. Proc. Camb. Phil. Soc. **77** (1975) 43–69
2. Bismut, J.-M., Cheeger, J.: Invariants êta et indices des familles pour les variétés à bord. C.R. Acad. Sci. Paris **305** (1987) 127–130
3. Baum, P., Douglas, R.G., Taylor, M.E.: Cycles and relative cycles in analytic K-theory. J. Diff. Geom. **30** (1989) 761–804
4. Boutet de Monvel, L., Guillemin, V.W.: The spectral theory of Toeplitz operators. Princeton Univ. Press, Princeton 1981
5. Bruhat, F., Whitney, H.: Quelques propriétés fondamentales des ensembles analytiques-réels. Comm. Math. Helv. **33** (1959) 132–160
6. Cheeger, J.: On the Hodge theory of Riemannian pseudomanifolds. Proc. Symp. Pure Math. **36** (1980) 91–146
7. Cheeger, J.: Spectral geometry of singular Riemannian spaces. J. Diff. Geom. **18** (1983) 575–657
8. Cheeger, J.: Eta invariants, the adiabatic approximation and conical singularities. J. Diff. Geom. **26** (1987) 175–221
9. Dai, X.: Adiabatic limits, Non-multiplicativity of the signature and the Leray spectral sequence. PhD Thesis, S.U.N.Y. Stony Brook, 1989
10. Dai, X., Epstein, C.L., Melrose, R.B.: Adiabatic limit of the Ray-Singer analytic torsion. (In preparation)

11. Epstein, C.L., Melrose, R.B., Mendoza, G.: Resolvent of the Laplacian on strictly pseudoconvex domains. Acta math. (To appear)
12. Epstein, C.L., Melrose, R.B.: Shrinking tubes and the $\bar{\partial}$-Neumann problem. Preprint (1990)
13. Folland, G.B., Kohn, J.J.: The Neumann problem for the Cauchy-Riemann complex. Ann. Math. Studies, No. 75, Princeton Univ. Press, Princeton 1972
14. Hörmander, L.: Fourier integral operators I. Acta Math. **127** (1971) 79–183
15. Hörmander, L.: The analysis of linear partial differential operators III. Springer, Berlin Heidelberg New York 1985
16. Kasparov, G.G.: Topological invariants of elliptic operators, I: K-homology. Math. USSR Izv. **9** (1975) 751–792
17. Kasparov, G.G.: Equivariant KK-theory and the Novikov conjecture. Invent. math. **91** (1988) 147–201
18. Mazzeo, R.: Hodge cohomology of a conformally compact metric. J. Diff. Geom. **28** (1988) 309–339
19. Mazzeo, R., Melrose, R.B.: Meromorphic extension of the resolvent on complete spaces with asymptotically negative curvature. J. Funct. Anal. **75** (1987) 260–310
20. Mazzeo, R., Melrose, R.B.: The adiabatic limit, Hodge cohomology and Leray's spectral sequence for a fibration. J. Diff. Geom. **31** (1990) 185–213
21. McDonald, P.T.: The Laplacian for spaces with cone-like singularities. PhD Thesis, M.I.T. (1990)
22. Melrose, R.B.: Transformation of boundary problems. Acta Math. **147** (1981) 149–236
23. Melrose, R.B.: Differential analysis on manifolds with corners. (In preparation)
24. Melrose, R.B., Mendoza, G.: Elliptic operators of totally characteristic type. MSRI Preprint (1983)
25. Melrose, R.B., Piazza, P.: Analytic K-theory on manifolds with corners. Adv. Math. (To appear)
26. Seeley, R.: Conic degeneration of the Dirac operator. Preprint
27. Seeley, R., Singer, I.M.: Extending $\bar{\partial}$ to singular Riemann surfaces. J. Geom. Phys. **5** (1988) 121–136
28. Witten, E.: Global gravitational anomalies. Comm. Math. Phys. **100** (1985) 197–229

Birational Classification of Algebraic Threefolds

Shigefumi Mori

Research Institute for Mathematical Sciences, Kyoto University, Kyoto 606, Japan

§1. Introduction

Let us begin by explaining the background of the birational classification. We will work over the field $\mathbb{C}$ of complex numbers unless otherwise mentioned.

Let X be a non-singular projective variety of dimension r. The canonical divisor class K_X is the only divisor class (up to multiples) naturally defined on an arbitrary X. Its sheaf $\mathcal{O}_X(K_X)$ is the sheaf of holomorphic r-forms. An alternative description is $K_X = -c_1(X)$, where $c_1(X)$ is the first Chern class of X. Therefore it is natural to expect some role of K_X in the classification of algebraic varieties.

The classification of non-singular projective curves C is classical, and summarized in the following table, where $g(C)$ is the genus (the number of holes) of C, $H = \{z \in \mathbb{C} \mid \text{Im} z > 0\}$ and Γ is a subgroup of $SL_2(\mathbb{R})$:

(1.1)

$g(C)$	0	1	≥ 2
$\deg K_C$	-2	0	$2g(C)-2$
C	$\mathbb{P}^1$	$\mathbb{C}/(\text{lattice})$	H/Γ

Here we see three different situations. For instance, everything is explicit if $g(C) = 0$; the moduli (to parametrize curves) is the main interest if $g(C) \geq 2$.

Our interest is in generalizing this to higher dimensions. The first difficulty which arises in the surface case is that there are too many varieties for genuine classification (*biregular classification*).

(1.2) For a non-singular projective surface X and an arbitrary point $x \in X$, there is a birational morphism $\pi : B_xX \to X$ from a non-singular projective surface B_xX such that $E = \pi^{-1}(x)$ is isomorphic to $\mathbb{P}^1$ (E is called a (-1)-*curve*) and π induces an isomorphism $B_xX - E \simeq X - x$.

In view of (1.2), it is impractical to distinguish X from B_xX, $B_yB_xX, \ldots$ if we want a reasonable classification list. More generally, we say that two algebraic varieties X and Y are *birationally equivalent* and we write $X \sim Y$ if there is a birational mapping $X \cdots\to Y$ or equivalently if their rational function fields $\mathbb{C}(X)$ and $\mathbb{C}(Y)$ are isomorphic function fields over $\mathbb{C}$. We did not face this phenomenon in the curve case, since $X \simeq Y$ iff $X \sim Y$ for curves X and Y.

Proceedings of the International Congress
of Mathematicians, Kyoto, Japan, 1990

In view of the list (1.1) for curves, we need to divide the varieties into several classes to formulate more precise problems. This is why the *Kodaira dimension* $\kappa(X)$ of a non-singular projective variety X was introduced by [Iitaka1] and [Moishezon].

(1.3) Let $H^0(X,\mathcal{O}(\nu K_X))$ be the space of global ν-ple holomorphic r-forms ($\nu \geq 0, r = \dim X$), and $\phi_0,\dots,\phi_N$ be its basis. If $N \geq 0$, then

$$\Phi_{\nu K_X} : X \cdots \to \mathbb{P}^N \text{ given by } \Phi_{\nu K}(x) = (\phi_0(x) : \cdots : \phi_N(x))$$

is a rational map. We set $P_\nu(X) = N+1$. It is important that $H^0(X,\mathcal{O}(\nu K_X))$ and $\Phi_{\nu K}$ are birational invariants, that is $X \sim Y$ induces $H^0(X,\mathcal{O}(\nu K_X)) = H^0(Y,\mathcal{O}(\nu K_Y))$ for $\nu > 0$. We set $\kappa(X) = -\infty$ if $P_\nu(X) = 0$ for all $\nu > 0$. If $P_e(X) > 0$ for some $e > 0$, then

$$\kappa(X) := \mathrm{Max}\{\dim \Phi_{\nu K_X}(X) \mid \nu > 0\}.$$

In particular, $P_\nu(X)$ and $\kappa(X)$ are birational invariants of X.

We remark that $\kappa(X) \in \{-\infty, 0, 1, \dots, \dim X\}$, and that X with $\kappa(X) = \dim X$ is said to be *of general type*. We have the following table for curves.

(1.4)

$g(C)$	0	1	≥ 2
$\kappa(C)$	$-\infty$	0	1

To have some idea on higher dimensions, we can use the easy result $\kappa(X\times Y) = \kappa(X) + \kappa(Y)$. In particular,

(1.5) case ($\kappa(X) = -\infty$) $\quad \kappa(\mathbb{P}^1 \times Y) = -\infty$,

(1.6) case ($0 < \kappa(X) < \dim X$)

$$\kappa(\underbrace{E \times \cdots \times E}_{a \text{ times}} \times \underbrace{C \times \cdots \times C}_{b \text{ times}}) = b \text{ if } g(E) = 1 \text{ and } g(C) \geq 2.$$

The case $0 < \kappa(X) < \dim X$ is studied by the Iitaka fibration.

(1.7) **Iitaka Fibering Theorem** [Iitaka2]. *Let X be a non-singular projective variety with $0 < \kappa(X) < \dim X$. Then there is a morphism $f : X' \to Y'$ of non-singular projective varieties with connected fibers such that $X' \sim X$, $\dim Y' = \kappa(X)$ and $\kappa(f^{-1}(y)) = 0$ for a sufficiently general point $y \in Y'$.*

In (1.7), we cannot expect $\kappa(Y') = \dim Y'$ or even $\kappa(Y') \geq 0$. Therefore X' is not so simple as (1.6). Nevertheless (1.7) reduces the case $0 < \kappa(X) < \dim X$ to the cases $\kappa(X) = -\infty, 0, \dim X$. Thus we can explain the birational classification as in (1.1) for higher dimensions.

§2. Birational Classification

For a non-singular projective variety X, we define a graded ring (called the *canonical ring*)

$$R(X) = \oplus_{\nu \geq 0} H^0(X, \mathcal{O}(\nu K_X)).$$

If $\kappa(X) \geq 0$, the ν-canonical image $\Phi_{\nu K}(X)$ is a birational invariant of X. The existence of stable canonical image is interpreted in terms of $R(X)$ by the following easy proposition.

(2.1) **Proposition.** *Let X be a non-singular projective variety of $\kappa(X) \geq 0$. Then $\Phi_{\nu K}(X)$ for sufficiently divisible $\nu > 0$ are all naturally isomorphic iff $R(X)$ is a finitely generated $\mathbb{C}$-algebra.*

(2.2) For X of general type, constructing moduli spaces is one of our main interests. One standard way is to try to find a uniform ν such that $\Phi_{\nu K} : X \cdots\to \Phi_{\nu K}(X)$ is birational for all X and classify the image. One can expect nice properties of the image (canonical model to be explained later) if there is a stable canonical image. Therefore we would like to ask whether the canonical ring is finitely generated for X of general type (2.1).

(2.3) The case $\kappa(X) = \dim X$ suggests to reduce the birational classification of all varieties to the biregular classification of standard models (like $\Phi_{\nu K}(X)$ for sufficiently divisible ν). However, when $\kappa(X) < \dim X$, there are no obvious candidates for the standard models. For $0 \leq \kappa(X)$, we can ask to find some "standard" models.

We only say the following for $\kappa(X) \leq 0$ at this point.

(2.4) For X with $\kappa(X) = 0$, we would like to find some "standard" model $Y \sim X$ and to classify all such Y.

(2.5) For many X with $\kappa(X) = -\infty$, there exist infinitely many "standard" models $\sim X$. To study the relation among these models is a role of birational geometry. We would like to have a structure theorem of such models. One general problem is to see if all such X are *uniruled*, i.e. there exists a rational curve through an arbitrary point of X, or equivalently there is a dominating rational map $\mathbb{P}^1 \times Y \cdots\to X$ for some Y of dimension $n-1$. (It is easy to see that uniruled varieties have $\kappa = -\infty$ as in (1.5).)

Since we use the formulation by Iitaka and Moishezon, one basic problem will be the deformation invariance of κ.

(2.6) **Conjecture** [Iitaka1, Moishezon]. *Let $f : X \to Y$ be a smooth projective morphism with connected fibers and connected Y. Then $\kappa(f^{-1}(y))$ and $P_\nu(f^{-1}(y))$ $(\nu \geq 1)$ are independent of $y \in Y$.*

§3. Surface Case

We review a few classical results on surfaces which may help the reader to understand the results for 3-folds.

The basic result is the inverse process of (1.2).

(3.1) **Castelnuovo-Enriques.** *Let E be a curve on a non-singular projective surface X'. Then E is a (-1)-curve (i.e. $X' = B_x X$ and E is the inverse image of x for some non-singular projective surface X and $x \in X$) iff $E \simeq \mathbb{P}^1$ and $(E \cdot K_{X'}) = -1$. We write* $\operatorname{cont}_E : X' \to X$ *and call it the contraction of the (-1)-curve E.*

Finding a (-1)-curve in every exceptional set, we have the following:

(3.2) **Factorization of Birational Morphisms.** *Let $f : X \to Y$ be a birational morphism of non-singular projective surfaces. Then f is a composition of a finite number of contractions of (-1)-curves.*

Starting with a non-singular projective surface X, we can keep contracting (-1)-curves if there are any. After a finite number of contractions, we get a non-singular projective surface $Y(\sim X)$ with no (-1)-curves. Depending on whether K_Y is *nef* ($(K_Y \cdot C) \geq 0$ for all curves C), $\kappa(X)$ takes different values.

(3.3) Case where K_Y is nef. Then Y is the only non-singular projective surface $\sim X$ with no (-1)-curves. To be precise, if Y' is a such surface, then the composite $Y \cdots\to X \cdots\to Y'$ is an isomorphism. This Y is called the *minimal model* of X and denoted by $X_{\min}$. In this case, we have $\kappa(X) \geq 0$.

(3.4) Case where K_Y is not nef. Then an arbitrary Y' (including Y) which is birational to X and has no (-1)-curves is isomorphic to either $\mathbb{P}^2$ or a $\mathbb{P}^1$-bundle over some non-singular curve. In this case, X has no minimal models and we have $\kappa(X) = -\infty$ by (1.5).

The above (3.3) together with (3.4) says that the birational classification of X with $\kappa \geq 0$ is equivalent to the biregular classification of minimal models.

Based on (3.3) and (3.4), the canonical model is defined.

(3.5) Let X be a non-singular projective surface of general type. Then there exists exactly one normal projective surface $Z(\sim X)$ such that Z has only Du Val (rational double) points and K_Z is ample, where Du Val points are defined by one of the following list.

$$\begin{aligned}
A_n &: xy + z^{n+1} = 0 \ (n \geq 0),\\
D_n &: x^2 + y^2 z + z^{n-1} = 0 \ (n \geq 4),\\
E_6 &: x^2 + y^3 + z^4 = 0,\\
E_7 &: x^2 + y^3 + yz^3 = 0,\\
E_8 &: x^2 + y^3 + z^5 = 0.
\end{aligned}$$

Such Z is called the *canonical model* of X and denoted by X_{can}. The natural map $X_{\min} \cdots \to X_{\text{can}}$ is a morphism which contracts all the rational curves C with $(C \cdot K_{X_{\min}}) = 0$ into Du Val points and is isomorphic elsewhere.

(3.5.1) **Remark.** This X_{can} can also be obtained as $\Phi_{\nu K}(X_{\min}) = \Phi_{\nu K}(X)$ for an arbitrary $\nu \geq 5$ (Bombieri).

(3.6) Let X be a non-singular projective minimal surface with $\kappa = 0$. Thus X has torsion K_X, i.e. some non-zero multiple of it is trivial. There is a precise classification of all such X.

(3.7) The deformation invariance of $\kappa(x)$ and $P_\nu(X)$ was done by [Iitaka3] using the classification of surfaces. [Levine] gave a simple proof without using classification.

§4. The Extremal Ray Theory (The Minimal Model Theory)

The first problem in generalizing the results in §3 to higher dimensions is to find some class of varieties in which there is a reasonable contraction theorem because there is no immediate generalization of (3.1) to 3-folds, since the contraction process inevitably introduces singularities [Mori1]. To define the necessary class of singularities, the first important step was taken by Reid [Reid1,3].

(4.1) **Definition** [Reid3]. Let (X, P) be a normal germ of an algebraic variety (or an analytic space) which is normal. We say that (X, P) has *terminal singularities* (resp. *canonical singularities*) iff

(i) K_X is a $\mathbb{Q}$-*Cartier* divisor, i.e. rK_X is Cartier for some positive integer r (minimal such r is called the *index* of (X, P)), and

(ii) for some (or equivalently, every) resolution $\pi : Y \to (X, P)$, we have $a_i > 0$ (resp. $a_i \geq 0$) for all i in the expression:

$$rK_Y = \pi^*(rK_Y) + \sum a_i E_i,$$

where E_i are all the exceptional divisors and $a_i \in \mathbb{Z}$.

For surfaces, a terminal (resp. canonical) singularity is smooth (resp. a Du Val point). We note that, for projective varieties X with only canonical singularities, the same definitions of $P_\nu(X)$, $\Phi_{\nu K}$ and $\kappa(X)$ work and these are still birational invariants. We can also talk about the ampleness of K_X and the intersection number $(K_X \cdot C) \in \mathbb{Q}$ for such X.

The idea of the cone of curves which is the core of the extremal ray theory was first introduced in Hironaka's thesis [Hironaka].

(4.2) **Definition.** Let X be a projective n-fold. A 1-cycle $\sum a_C C$ is a formal finite sum of irreducible curves C on X with coefficients $a_C \in \mathbb{Z}$. For a 1-cycle Z and a $\mathbb{Q}$-Cartier divisor D, the intersection number $(Z \cdot D) \in \mathbb{Q}$ is defined. Then

$$N_1(X)_{\mathbb{Z}} = \{\text{1-cycles}\}/\{\text{1-cycles } Z \mid (Z \cdot D) = 0 \text{ for all } D\}$$

is a free abelian group of finite rank $\varrho(X) < \infty$. Thus $N_1(X) = N_1(X)_{\mathbb{Z}} \otimes_{\mathbb{Z}} \mathbb{R}$ is a finite dimensional Euclidean space. The classes $[C]$ of all the irreducible curves C span a convex cone $NE(X)$ in $N_1(X)$. Taking the closure for the metric topology, we have a closed convex cone $\overline{NE}(X)$. Then

(4.3) **Cone Theorem.** *If X has only canonical singularities, then there exist countably many half lines $R_i \subset NE(X)$ such that*

(i) $\overline{NE}(X) = \sum_i R_i + \{z \in \overline{NE}(X) \mid (z \cdot K_X) \geq 0\}$,

(ii) for an arbitrary ample divisor H of X and arbitrary $\varepsilon > 0$, there are only finitely many R_i's contained in

$$\{z \in \overline{NE}(X) \mid (z \cdot K_X) \leq -\varepsilon(z \cdot H)\}.$$

Such an R_i is called an *extremal ray* of X if it cannot be omitted in (i) of (4.3). We note that an extremal ray exists on X iff K_X is not nef. Each extremal ray R_i defines a contraction of X.

(4.4) **Contraction Theorem.** *Let R be an extremal ray of a projective n-fold X with only canonical singularities. Then there exists a morphism $f : X \to Y$ to a projective variety Y (unique up to isomorphism) such that $f_*\mathcal{O}_X = \mathcal{O}_Y$ and an irreducible curve $C \subset X$ is sent to a point by f iff $[C] \in R$. Furthermore* $\operatorname{Pic} Y = \operatorname{Ker}[(C\cdot) : \operatorname{Pic} X \to \mathbb{Z}]$ *for such a contracted curve C. This f is called the contraction of R and denoted by* cont_R.

The contraction of an extremal ray is not always birational.

(4.5) *Let X be a smooth projective surface with an extremal ray R. Then* cont_R *is one of the following.*

(i) the contraction of a (-1)*-curve,*

(ii) a $\mathbb{P}^1$*-bundle structure $X \to C$ over a non-singular curve,*

(iii) a morphism to one point, when $X \simeq \mathbb{P}^2$.

The description of all the possible contractions for a nonsingular projective 3-fold X is given in [Mori1]. Here we only remark that $\operatorname{cont}_R X$ can have a terminal singularity $\mathbb{C}^3/<\sigma>$ of index 2, where σ is the involution $\sigma(x, y, z) = (-x, -y, -z)$.

(4.6) The category of varieties in which we play the game of the minimal model program is the category $\mathscr{C}$ of projective varieties with only terminal singularities which are $\mathbb{Q}$-*factorial* (i.e. every Weil divisor is $\mathbb{Q}$-Cartier). The goal of the game is to get a *minimal* (resp. *canonical*) model, i.e. a projective n-fold X with only terminal (resp. canonical) singularities such that K_X is nef (resp. ample). Let us first state the minimal model program which involves two conjectures.

(4.7) Let X be an n-fold $\in \mathscr{C}$. If K_X is nef, then X is a minimal model and we are done. Otherwise, X has an extremal ray R. Then $\operatorname{cont}_R : X \to X'$ satisfies one of the following.

(4.7.1) Case where $\dim X' < \dim X$. Then cont_R is a surjective morphism with connected fibers of dimension > 0 and relatively ample $-K_X$ (like $\mathbb{P}^1$-bundle), and X is uniruled ([Miyaoka-Mori]). This is the case where we can never get a minimal model, and we stop the game since we have the global structure of X, $\mathrm{cont}_R : X \to X'$.

(4.7.2) Case where $\mathrm{cont}_R : X \to X'$ is birational and contracts a divisor. This cont_R is called a *divisorial contraction*. In this case $X' \in \mathscr{C}$ and $\varrho(X') < \varrho(X)$. Therefore we can work on X' instead of X.

(4.7.3) Case where $\mathrm{cont}_R : X \to X'$ is birational and contracts no divisors. In this case, $K_{X'}$ is not $\mathbb{Q}$-Cartier and $X' \notin \mathscr{C}$. So we cannot continue the game with X'. This is the new phenomenon in dimension ≥ 3.

To get around the trouble in (4.7.3) and to continue the game, Reid proposed the following.

(4.8) **Conjecture (Existence of Flips).** *In the situation of (4.7.3), there is an n-fold $X^+ \in \mathscr{C}$ with a birational morphism $f^+ : X^+ \to X'$ which contracts no divisors and such that K_{X^+} is f^+-ample. The map $X \cdots\to X^+$ is called a flip.*

Since $\varrho(X^+) = \varrho(X)$ in (4.8), the divisorial contraction will not occur for infinitely many times. Therefore the following will guarantee that the game will be over after finitely many steps.

(4.9) **Conjecture (Termination of Flips).** *There does not exist an infinite sequence of flips $X_1 \cdots\to X_2 \cdots\to \cdots$.*

Therefore the minimal model program is completed only when the conjectures (4.8) and (4.9) are settled affirmatively.

The conjecture (4.9) was settled affirmatively by [Shokurov1] for 3-folds and by Kawamata-Matsuda-Matsuki [KMM] for 4-folds. (4.8) was first done by [Tsunoda], [Shokurov2], [Mori3] and [Kawamata6] in a special but important case. Finally (4.8) was done for 3-folds by [Mori5] using the work of [Kawamata6] mentioned above.

(4.10) Thus for 3-folds, we can operate divisorial contractions and flips for a finite number of times and get either a minimal model $\in \mathscr{C}$ or an $X \in \mathscr{C}$ which has an extremal ray R of type (4.7.1). Thus we can get 3-fold analogues of results in §3.

(4.11) For simplicity of the exposition, we did not state the results in the strongest form and we even omitted various results. Therefore we would like to mention names and give a quick review.

After the prototype of the extremal ray theory was given in [Mori1], the theory has been generalized to the relative setting with a larger class of singularities (toward the conjectures of Reid [Reid3,4]) by Kawamata, Benveniste, Reid, Shokurov and Kollár (in the historical order) and perhaps some others.

First through the works of [Benveniste] and [Kawamata2], Kawamata introduced a technique [Kawamata3] which was an ingenious application of the Kawamata-Viehweg vanishing ([Kawamata1] and [Viehweg2]). Based on the works by [Shokurov1] (Non-vanishing theorem) and [Reid2] (Rationality theorem), [Kawamata4] developped the technique to prove the Base point freeness theorem (and others) in arbitrary dimensions. The discreteness of the extremal rays was later done by [Kollár1]. As for this section, we refer the reader to the talk of Kawamata.

§5. Applications of the Minimal Model Program (MMP) to 3-Folds

Considering MMP in relative setting, one has the factorization generalizing (3.2):

(5.1) **Theorem.** *Let $f : X \to Y$ be a birational morphism of projective 3-folds with only $\mathbb{Q}$-factorial terminal singularities. Then f is a composition of divisorial contractions and flips.*

Since minimal 3-folds have $\kappa \geq 0$ by the hard result of Miyaoka [Miyaoka1–3], one has the following (cf. (3.3) and (3.4)).

(5.2) **Theorem.** *A 3-fold X has a minimal model iff $\kappa(X) \geq 0$.*

Unlike the surface case, the minimal model of a 3-fold X is not unique; it is unique only in codimension 1. If we are given a $\mathbb{Q}$-factorial minimal model $X_{\min}$, every other $\mathbb{Q}$-factorial minimal model of X is obtained from $X_{\min}$ by operating a simple operation called a *flop* for a finite number of times ([Kawamata6], [Kollár4]). Many important invariants computed by minimal models do not depend on the choice of the minimal model. We refer the reader to the talk of Kollár.

(5.3) **Theorem.** *For a 3-fold X, the following are equivalent.*

(i) $\kappa(X) = -\infty$,

(ii) X is uniruled,

(iii) X is birational to a projective 3-fold Y with only $\mathbb{Q}$-factorial terminal sigularities which has an extremal ray of type (4.7.1).

It will be an important but difficult problem to classify all the possible Y in (iii) of (5.3). There are only finitely many families of such Y with $\varrho(Y) = 1$ ([Kawamata7]).

Since a canonical model exists if a minimal model does ([Benveniste] and [Kawamata2]), one has the following (cf. (3.5)).

(5.4) **Theorem.** *If X is a 3-fold of general type, then X has a canonical model and the canonical ring $R(X)$ is a finitely generated $\mathbb{C}$-algebra.*

The argument for (5.4) can be considered as a generalization of the argument for (3.5.1). However the effective part "$v \geq 5$" of (3.5.1) has not yet been generalized to dimension ≥ 3.

To study varieties X with $\kappa \geq 0$, [Kawamata4] posed the following.

(5.5) **Conjecture (Abundance Conjecture).** *If X is a minimal variety, then rK_X is base point free for some $r > 0$.*

For 3-folds, there are works by [Kawamata4] and [Miyaoka4] (cf. [KMM]). However the torsionness of K for minimal 3-folds with $\kappa = 0$ is unsolved, and it remains to prove:

(5.6) **Problem.** *Let X be a minimal 3-fold with H an ample divisor such that $(K_X^3) = 0$ and $(K_X^2 \cdot H) > 0$. Then prove that $\kappa(X) = 2$.*

(5.7) **Remark** ($\kappa = 0$). The 3-folds X with $\kappa(X) = 0$ and $H^1(X, \mathcal{O}_X) \neq 0$ were classified by [Viehweg1] and (5.5) holds for these. This was based on Viehweg's solution of the addition conjecture for 3-folds, and we refer the reader to [Iitaka4]. However not much is known about the 3-folds X with $\kappa(X) = 0$ (or even K_X torsion) and $H^1(X, \mathcal{O}_X) = 0$: so far many examples have been constructed and it is not known if there are only finitely many families. There is a conjecture of [Reid6] in this direction.

By studying the flips more closely, [Kollár-Mori] proved the deformation invariance of κ and P_v (cf. (3.7)):

(5.8) **Theorem.** *Let $f : X \to \Delta$ (unit disk) be a projective morphism whose fibers are connected 3-folds with only $\mathbb{Q}$-factorial terminal singularities. Then*
(i) $\kappa(X_t)$ is independent of $t \in \Delta$, where $X_t = f^{-1}(t)$,
(ii) $P_v(X_t)$ is independent of $t \in \Delta$ for all $v \geq 0$ if $\kappa(X_0) \neq 0$.

Indeed for such a family X/Δ, the simultaneous minimal model program is proved and the (modified) work of [Levine] is used to prove (5.8). We cannot drop the condition "$\kappa(X_0) \neq 0$" at present since the abundance conjecture is not completely solved for 3-folds.

As for other applications (e.g. addition conjecture, deformation space of quotient surface singularities, birational moduli), we refer the reder to [KMM] and [Kollár6].

§6. Comments on the Proofs for 3-Folds

Many results on 3-folds are proved by using only the formal definitions of terminal singularities. However some results on 3-folds rely on the classification of 3-fold terminal singularities [Reid3], [Danilov], [Morrison-Stevens], [Mori2] and [KSB] (cf. Reid's survey [Reid5] and [Stevens].) The existence of flips and flops heavily rely on it. Thus generalizing their proofs to higher dimension seems hopeless. At present, there is no evidence for the existence of flips in higher dimensions except that they fit in the MMP beautifully. I myself would accept them as working hypotheses. A more practical problem will be to complete the log-version of the minimal model program for 3-folds [KMM]. This is related to the birational classification of open 3-folds and n-folds with $\kappa = 3$. Since log-terminal singularities have no explicit classification, this might be a good place to get some idea on higher dimension. Shokurov made some progress in this direction [Shokurov3].

There are two other results relying on the classification.

(6.1) **Theorem** [Mori4]. *Every 3-dimensional termal singularity deforms to a finite sum of cyclic quotient terminal singularities (i.e. points of the form $\mathbb{C}^3/\mathbb{Z}_r(1,-1,a)$ for some relatively prime positive integers a and r).*

This was used in the Barlow-Fletcher-Reid plurigenus formula for 3-folds [Fletcher] and [Reid5] (cf. also [Kawamata5]). Given a 3-fold X with only terminal singularities, each singularity of X can be deformed to a sum of cyclic quotient singularities $\mathbb{C}^3/\mathbb{Z}_r(1,-1,a)$. Let $S(X)$ be the set of all such (counted with multiplicity). For each $P = \mathbb{C}^3/\mathbb{Z}_r(1,-1,a) \in S(X)$, we let

$$\phi_P(m) = (m - \{m\}_r)\frac{r^2-1}{12r} + \sum_{j=0}^{\{m\}_r-1} \frac{\{aj\}_r(r-\{aj\}_r)}{2r},$$

where $\{m\}_r$ is the integer $s \in [0, r-1]$ such that $s \equiv m(\text{mod } r)$. For a line bundle L on X, let $\chi(L) = \sum_j (-1)^j \dim H^j(X, L)$ and let $c_2(X)$ be the second Chern class of X, which is well-defined since X has only isolated singularities. Then the formula is stated as the following.

(6.2) **The Barlow-Fletcher-Reid Plurigenus Formula.**

$$\chi(\mathcal{O}_X(mK_X)) = \frac{m(m-1)(2m-1)}{12}(K_X^3) + (1-2m)\chi(\mathcal{O}_X) + \sum_{P \in S(X)} \phi_P(m),$$

$$\chi(\mathcal{O}_X) = -\frac{1}{24}(K_X \cdot c_2(X)) + \sum_{P \in S(X)} \frac{r^2-1}{24r}.$$

This is important for effective results on 3-folds (cf. §7).

(6.3) **Theorem** ([KSB]). *A small deformation of a 3-dimensional terminal singularity is terminal.*

This is indispensable in the construction of birational moduli. An open problem in this direction is

(6.4) **Problem.** *Is every small deformation of a 3-dimensional canonical singularity canonical?*

Since this remains unsolved, we cannot put an algebraic structure on

$$\{\text{canonical 3-folds}\}/\text{isomorphisms}.$$

§7. Related Results

I would like to list some of the directions, which I could not mention in the previous sections. This is by no means exhaustive. For instance, I could not mention the birational automorphism groups (cf. [Iskovskih] for the works before 1983) due to the lack of my knowledge.

(7.1) *Effective Classification.* The Kodaira dimension κ is not a simple invariant. For instance, we know that $\kappa(X) = -\infty$ iff $P_\nu(X) = 0 (\forall \nu > 0)$. Therefore $P_{12}(X) = 0$ was an effecitve criterion for a surface X to be ruled, while $\kappa(X) = -\infty$ was not. The 3-dimensional analogue is not known yet.

There are results by Kollár [Kollár2] in the case $\dim H^1(X, \mathcal{O}_X) \geq 3$ (cf. [Mori4]). The Barlow-Fletcher-Reid plurigenus formula (6.2) is applied for instance to get $aK_X \sim 0$ with some effectively given $a > 0$ for 3-folds X with numerically trivial K_X by [Kawamata5] and [Morrison], and to get $P_{12}(X) > 0$ for canonical 3-folds X with $\chi(\mathcal{O}_X) \leq 1$ by [Fletcher].

(7.2) *Differential Geometry.* As shown by [Yau], there are differential geometric results (especially when K is positive) which seem out of reach of algebraic geometry. Therefore we welcome differential geometric approaches. In this direction is Tsuji's construction of Kähler-Einstein metrics on canonical 3-folds [Tsuji].

(7.3) *Characteristic p.* [Kollár5] generalized [Mori1] (extremal rays of smooth projective 3-folds over $\mathbb{C}$) to char p. This suggests the possibility of little use of vanishing theorems in MMP for 3-folds. A goal will be the MMP for 3-folds in char p. However even the classification of terminal singularities is open.

(7.4) *Mixed Characteristic Case.* One can ask about the extremal rays (and so on) for arithmetic 3-folds X/S. The methods of [Shokurov2] and [Tsunoda] might work, if X/S is semistable. In the general case, I do not know any results in this direction.

(7.5) *Analytic or Non-projective 3-Folds.* Studying analytic or non-projective 3-folds will require a substitute for the cone of curves modulo numerical equivalence. However analytic or non-projective minimal 3-folds can be handled by the flop [Kollár4]. There is a work of [Kollár6].

References

[Benveniste] X.Benveniste: Sur l'anneau canonique de certaines variétés de dimension 3. Invent. math. **73** (1983) 157–164

[Danilov] V.I.Danilov: Birational geometry of toric 3-folds. Math. USSR Izv. **21** (1983) 269–279

[Fletcher] A.Fletcher: Contributions to Riemann-Roch on projective 3-folds with only canonical singularities and applications, In: Algebraic Geometry, Bowdoin 1985. Proc. Symp. Pure Math. **46** (1987) 221–232

[Hironaka] H.Hironaka: On the theory of birational blowing-up. Thesis, Harvard University 1960

[Iitaka1] S.Iitaka: Genera and classification of algebraic varieties. 1 (in Japanese). Sugaku **24** (1972) 14–27

[Iitaka2] S.Iitaka: On D-dimensions of algebraic varieties. J. Math. Soc. Japan **23** (1971) 356–373

[Iitaka3] S.Iitaka: Deformations of complex surfaces. II. J. Math. Soc. Japan **22** (1971) 274–261

[Iitaka4] S.Iitaka: Birational geometry of algebraic varieties. Proc. ICM83, Warsaw 1984, pp. 727–732

[Iskovskih] V.A.Iskovskih: Algebraic threefolds with special regard to the problem of rationality. Proc. ICM83, Warsaw 1984, pp. 733–746

[Kawamata1] Y.Kawamata: A generalisation of Kodaira-Ramanujam's vanishing theorem. Math. Ann. **261** (1982) 43–46

[Kawamata2] Y.Kawamata: On the finiteness of generators of the pluri-canonical ring for a threefold of general type. Amer. J. Math. **106** (1984) 1503–1512

[Kawamata3] Y.Kawamata: Elementary contractions of algebraic 3-folds. Ann. Math. **119** (1984) 95–110

[Kawamata4] Y.Kawamata: The cone of curves of algebraic varieties. Ann. Math. **119** (1984) 603–633

[Kawamata5] Y.Kawamata: On the plurigenera of minimal algebraic 3-folds with $K \approx 0$. Math. Ann. **275** (1986) 539–546

[Kawamata6] Y.Kawamata: The crepant blowing-up of 3-dimensional canonical singularities and its applications to the degeneration of surfaces. Ann. Math. **127** (1988) 93–163

[Kawamata7] Y.Kawamata: Boundedness of $\mathbb{Q}$-Fano threefolds. Preprint (1990)

[KMM] Y.Kawamata, K.Matsuda, K.Matsuki: Introduction to the minimal model problem. In: Algebraic Geometry, Sendai 1985. Adv. Stud. Pure Math. **10** (1987) 283–360

[Kollár1] J.Kollár: The cone theorem. Ann. Math. **120** (1984) 1–5

[Kollár2] J.Kollár: Higher direct images of dualizing sheaves. Ann. Math. (2) **123** (1986) 11–42

[Kollár3] J.Kollár: Higher direct images of dualizing sheaves. II. Ann. Math. (2) **124** (1986) 171–202

[Kollár4] J.Kollár: Flops. Nagoya Math. J. **113** (1989) 14–36

[Kollár5] J.Kollár: Extremal rays on smooth treefolds. Ann. Sci. ENS (to appear)

[Kollár6] J.Kollár: Flips, flops, minimal models etc. Proc. Diff. Geom. Symposium at Harvard, May 1990

[Kollár-Mori] J.Kollár, S.Mori: To be written up. (1991)

[KSB] J.Kollár, N.Shepherd-Barron: Threefolds and deformations of surface singularities. Invent. math. **91** (1988) 299–338

[Levine] M.Levine: Pluri-canonical divisors on Kähler manifolds. Invent. math. **74** (1983) 293–303

[Miyaoka1] Y.Miyaoka: Deformations of a morphism along a foliation. In: Algebraic Geometry, Bowdoin 1985. Proc. Symp. Pure Math. **46** (1987) 245–268

[Miyaoka2] Y.Miyaoka: The Chern classes and Kodaira dimension of a minimal variety. In: Algebraic Geometry, Sendai 1985. Adv. Stud. Pure Math. **10** (1987) 449–476

[Miyaoka3] Y.Miyaoka: On the Kodaira dimension of minimal threefolds. Math. Ann. **281** (1988) 325–332

[Miyaoka4] Y.Miyaoka: Abundance conjecture for threefolds: $\nu = 1$ case. Comp. Math. **68** (1988) 203–220

[Miyaoka-Mori] A numerical criterion for uniruledness. Ann. Math. **124** (1986) 65–69

[Moishezon] B.G.Moishezon: Algebraic varieties and compact complex spaces. ICM70, Nice 1971, pp. 643–648

[Mori1] S.Mori: Threefolds whose canonical bundles are not numerically effective. Ann. Math. **116** (1982) 133–176

[Mori2] S.Mori: On 3-dimensional terminal singualrities. Nagoya Math. J. **98** (1985) 43–66

[Mori3] S.Mori: Minimal models for semistable degenerations of surfaces. Lectures at Columbia University 1985, unpublished

[Mori4] S.Mori: Classification of higher-dimensional varieties. In: Algebraic Geometry, Bowdoin 1985. Proc. Symp. Pure Math. **46** (1987) 269–332

[Mori5] S.Mori: Flip theorem and the existence of minimal models for 3-folds. J. Amer. Math. Soc. **1** (1988) 117–253

[Morrison] D.Morrison: A Remark on Kawamata's paper "On the plurigenera of minimal algebraic 3-folds with $K \approx 0$". Math. Ann. **275** (1986) 547–553

[Morrison-Stevens] D.Morrison, G.Stevens: Terminal quotient singularities in dimensions three and four. Proc. AMS **90** (1984) 15–20

[Nakamura-Ueno] I.Nakamura, K.Ueno: An addition formula for Kodaira dimensions of analytic fibre bundles whose fibres are Moishezon manifolds. J. Math. Soc. Japan **25** (1973) 363–371

[Reid1] M.Reid: Canonical threefolds. In: Géometrie Algébrique Angers. Sijthoff & Noordhoff (1980) 273–310

[Reid2] M.Reid: Projective morphisms according to Kawamata. Preprint, Univ. of Warwick (1983)

[Reid3] M.Reid: Minimal models of canonical threefolds. In: Algebraic and Analytic Varieties. Adv. Stud. Pure Math. **1** (1983) 131–180

[Reid4] M.Reid: Decomposition of toric morphisms. Progress in Math. **36** (1983) 395–418

[Reid5] M.Reid: Young person's guide to canonical singularities. In: Algebraic Geometry, Bowdoin 1985. Proc. Symp. Pure Math. **46** (1987) 345–416

[Reid6] M.Reid: The moduli space of 3-folds with $K = 0$ may nevertheless be irreducible. Math. Ann. **278** (1987) 329–334

[Shokurov1] V.Shokurov: The nonvanishing theorem. Izv. Akad. Nauk SSSR Ser. Mat. **49** (1985) 635–651

[Shokurov2] V.Shokurov: Letter to M.Reid, 1985

[Shokurov3] V.Shokurov: Special 3-dimensional flips. Preprint, MPI 1989

[Stevens] J.Stevens: On canonical singularities as total spaces of deformations. Abh. Math. Sem. Univ. Hamburg **58** (1988) 275–283

[Tsuji] H.Tsuji: Existence and degeneration of Kähler-Einstein metric for minimal algebraic varieties of general type. Math. Ann. **281** (1988) 123–133

[Tsunoda] S.Tsunoda: Degenerations of Surfaces. In: Algebraic Geometry, Sendai 1985. Adv. Stud. Pure Math. **10** (1987) 755–764

[Viehweg1] E.Viehweg: Klassifikationstheorie algebraischer Varietäten der Dimension drei. Comp. Math. **41** (1981) 361–400

[Viehweg2] E.Viehweg: Vanishing theorems. J. Reine Angew. Math. **335** (1982) 1–8

[Yau] S.-T.Yau: On the Ricci curvature of a compact Kähler manifold and the complex Monge-Ampère equation I. Comm. Pure Appl. Math. **31** (1978) 339–411

Hyperbolic Billiards

Yakov G. Sinai

Landau Institute of Theoretical Physics, Academy of Sciences of USSR
Moscow, 117334 USSR

1. Definition of Billiards

Billiards are dynamical systems which correspond to the uniform motion of a material point inside a domain on a Riemannian manifold with elastic reflections off the boundary. Thus billiards are geodesic flows on manifolds with boundaries. Specific features of billiards arise when the role of the boundary is much more important than the role of the underlying manifold. Ergodic properties of billiards have been discussed already in the works of Hadamard and Birkhoff. An important contribution was made by the Soviet physicists N.S. Krylov [K1]. During the last two decades the theory of billiards has been developed enormously in various directions and some part of this development will be the main content of this lecture. Billiards constitute an important class of dynamical systems due to the following reasons:

1. some classes of billiards demonstrate a strong chaotic behavior and can be considered among the best examples of deterministic chaos; the theory of such billiards is discussed below in detail;
2. many interesting examples of dynamical systems of physical or mechanical origin can be reduced to billiards; especially it is true for systems where the interaction involves elastic collisions between the particles;
3. a deep analysis of properties of billiards is essential for some problems in the theory of quantum chaos and asymptotical problems in partial differential equations;
4. the theory of billiards suggests many beautiful geometrical problems.

A rich theory of billiards relates to cases where the table of the billiard is a closed compact subset of Euclidean space R^d or d-dimensional flat torus Tor^d, d is the dimension. The tables which we shall consider here can be described as follows. Let there be given r C^∞-functions $f_i(q), 1 \le i \le r$, defined on R^d or Tor^d and such that $f_i^{-1}(0)$ do not contain critical points of f_i. Then $Q = \{q \in R^d \,|\, f_i(q) \ge 0,\ 1 \le i \le r\}$ or $Q = \{q \in Tor^d \,|\, f_i(q) \ge 0,\ 1 \le i \le r\}$. In order to avoid some trivial complications it is natural to assume also that if $q \in Q$ and $f_{i_1}(q) = f_{i_2}(q) = 0$ then grad $f_{i_1}(q)$, grad $f_{i_2}(q)$ are non-collinear. The boundary ∂Q is the union $\partial Q = \bigcup_{i=1}^r \partial Q_i$ where $\partial Q_i = \{q \in \partial Q | f_i(q) = 0\}$. The points of a

Proceedings of the International Congress
of Mathematicians, Kyoto, Japan, 1990

component ∂Q_i which do not belong to other components ∂Q_j, $j \neq i$, are called regular points of the boundary. At regular points $q \in \partial Q$ the unit normal vector $n(q)$ directed inside Q as well as the operator $K(q)$ of the second fundamental form of the boundary are defined. Recall that $n(q+dq) = n(q) + K(q)dq$ where $dq \in \mathcal{T}_q(\partial Q)$ and $\mathcal{T}_q(\partial Q)$ is the tangent space to ∂Q at q. The geometrical properties of the boundary which we shall use below will be expressed in terms of $K(q)$.

As was already mentioned billiards correspond to the uniform motion of a point (billiard ball) inside Q with elastic reflections off ∂Q. This means that at the reflection the normal component of the velocity changes its sign while the tangential component remains unaltered. The phase space M of the billiard is the unit tangent bundle over Q. The points of M are denoted as $x = (q, v)$ where $q \in Q, \quad v \in \mathcal{T}_q(Q), \quad \|v\| = 1$. The canonical projection $x \to q$ is denoted by π. The space M is also a manifold with a piece-wise smooth boundary, the boundary $\partial M = \bigcup_{i=1}^r \partial M_i, \partial M_i = \pi^{-1}(\partial Q_i)$. Regular points of ∂M are such points $x \in \partial M_i$ that $\pi(x)$ are regular points of ∂Q.

Introduce the set of incoming vectors

$$\partial M^- = \{x = (q, v) \in \partial M | q \in \partial Q \text{ is regular and } (v, n(q)) \leq 0\}$$

and the set of outgoing vectors

$$\partial M^+ = \{x = (q, v) \in \partial M | q \in \partial Q \text{ is regular and } (v, n(q)) \geq 0\}.$$

The law of reflection corresponds to the transformation $R : \partial M^- \to \partial M^+$ where

$$Rx = R(q, v) = (q, v - 2(n(q), v)n(q)) \in \partial M^+ \quad \text{if} \quad (q, v) \in \partial M^-.$$

A point moves inside M and when it reaches the boundary ∂M^- it jumps instantly to a new point of ∂M^+ under the action of R. In what follows we consider only a subset $M^1 \subset M$ consisting of such x whose trajectories pass only through regular points of the boundary and have finitely many reflections during any finite interval of time. If $d\mu = \text{const}\, dq d\omega_q(v)$ where $d\omega_q(v)$ is the Lebesgue measure on the unit sphere $S_q^{d-1} \subset \mathcal{T}_q(Q)$ is the normed Liouville measure on M then it is easy to show that $\mu(M^1) = 1$. Thus we may define the one-parameter group $\{S^t\}$ of transformations of M^1 onto itself which corresponds to the billiard motion and preserves μ (see [CFS1]). The flow $\{S^t\}$ is discontinuous because the dynamics is discontinuous. It is an often situation in ergodic theory that discontinuous flows are easier for an analysis than their smooth approximations.

Sometimes it is useful also to consider the induced map or the so-called billiard ball map $T : \partial M^+ \cap M^1 \to \partial M^+ \cap M^1$ which transforms a point $x \in \partial M^+ \cap M^1$ to the point $x^1 \in \partial M^+ \cap M^1$ corresponding to the next reflection. The map T preserves the measure $d\mu_1 = \text{const}\,(n(q), v) \cdot d\sigma(q)\omega_q(v)$ where $d\sigma(q)$ is the Riemannian volume on ∂Q.

The first examples of billiards are billiards inside polygons or polyhedra. Some of them appear in connection with the dynamics of one-dimensional particles on a segment interacting through elastic collisions. The billiards in triangles correspond to the dynamics of two such particles. It turns out that even this simple case is very difficult for an analysis and still many problems remain open. Deep results here were obtained by Kerchkoff, Mazur and Smillie (see [KMS1]) using the theory of quadratic differentials and some methods of algebraic geometry. In particular they showed that for billiards inside polygons with commensurate angles the system has only obvious first integrals. The first applications of these methods appeared in the papers by Veech (see [V1]).

2. Hyperbolic Billiards

The main content of this lecture will concern the theory of hyperbolic billiards, i.e., billiards where the behavior of trajectories has many common features with the behavior of geodesics in manifolds of negative curvature. Take a point $x = (q, v) \in M^1$ and a $(d-1)$-dimensional open C^∞-submanifold $\widetilde{\gamma} \subset Q$ passing through q, orthogonal to v at q and homeomorphic to the $(d-1)$-dimensional disk. Denote by γ its continuous framing by unit normal vectors such that $x \in \gamma$.

Definition 1. A submanifold γ is called a *local stable manifold* (*lsm*) of x if during the dynamics $S^t x$ and $S^t y$ for any $y \in \gamma$ have reflections from the same components of the boundary ∂M for all $t > 0$ and one can find positive numbers $C(x), \lambda(x)$ such that for all $t > 0$

$$\operatorname{dist}(\pi(S^t x), \pi(S^t y)) \leq C(x) \exp\{-\lambda(x)|t|\}.$$

If in this definition we take $t < 0$ we shall get the definition of a *local unstable manifold* (lum) of x. Lsm and lum will be denoted by $\gamma^{(s)}(x)$ and $\gamma^{(u)}(x)$, respectively.

Definition 2. A global stable (unstable) manifold of $x \in M$ consists of all y for which one can find $\tau \geq 0$ $(\tau \leq 0)$ such that $S^\tau y$ belongs to $\gamma^{(s)}(S^\tau x)$ $(\gamma^{(u)}(S^\tau x))$.

We shall denote them by $\Gamma^{(s)}(x)$ and $\Gamma^{(u)}(x)$ and use the abbreviations gsm and gum. Stable and unstable manifolds are the main objects in the theory of smooth hyperbolic dynamical systems like Axiom-A systems of Smale [Sm1] and Anosov systems [An1]. Contrary to the case of smooth hyperbolic systems gsm and gum in billiards have singularities (see examples below) which arise from trajectories tangent to the boundary. Definitions 1 and 2 can be introduced in the same way for the induced map T.

We shall now introduce the main definition.

Definition 3. Billiard in a domain Q is called hyperbolic if μ-almost every point has gsm and gum.

We shall discuss now various conditions under which billiard is hyperbolic. As it will be seen the corresponding problems are particular cases of a more general problem of finding sufficient conditions for a dynamical system to display some kind of a chaotic behavior.

Definition 4. Billiard is called dispersing if at all regular points of the boundary the operator $K(q) > 0$.

This definition implies that at any regular point the boundary is locally strictly concave.

Theorem 1. *Every dispersing billiard is hyperbolic.*

This theorem is in fact some version of the well-known Hadamard-Perron theorem (see [An1, G1, S2] and [KS1]), which gives sufficient conditions for the existence of lsm and lum in the case of smooth dynamical systems. Theorem 1 gives an easy way to construct examples of hyperbolic billiards.

Dispersing billiards are in many respects similar to geodesic flows on compact manifolds of negative curvature. For many applications it is important to consider billiards which correspond to geodesic flows on manifolds of non-positive curvature.

Definition 5. Billiard is called semi-dispersing if at all regular points of the boundary $K(q) \geq 0$.

Semi-dispersing billiards arise naturally when we study the dynamics of hard disks or spheres interacting via elastic collisions (see for example [CFS1]). In those cases the boundary is the union of subsets of cylinders having many flat directions. It is a more difficult problem to find conditions under which a semi-dispersing billiard is hyperbolic. A natural approach is the following (see [CS1]). Let us write down the sequence of positive times of reflections from the boundary, $0 < t_1 < t_2 < \cdots < t_n, \ldots$, and let $\tau_j = t_j - t_{j-1}, \tau_1 = t_1$. Introduce also the points of the boundary where these reflections take place: $q_1, q_2 \ldots$ and $K_j = K(q_j)$. We shall need also the operators U_i mapping $\mathcal{T}_i^+$ onto $\mathcal{T}_i^-$ in parallel to $n(q_i)$ where the subspace $\mathcal{T}_i^+(\mathcal{T}_i^-)$ is orthogonal to $v_i^+(v_i^-), S^{t_i \pm 0}x = (q, v_i^{\pm})$, and the operators V_i mapping $\mathcal{T}_i^+$ onto $\mathcal{T}_{q_i}(\partial Q)$ in parallel to v_i^+. The adjoint operator V_i^* maps $\mathcal{T}_{q_i}(\partial Q)$ onto $\mathcal{T}_i^+$ in parallel to $n(q_i)$. Now write down the following operator-valued continued fraction

$$B^+(x) = \cfrac{I}{\tau_1 I + U_1 \cfrac{I}{2\cos\phi_1 V_1^* K_1 V_1 + U_2 \cfrac{I}{\tau_2 I + \ldots\ldots} U_2^{-1}} U_1^{-1}}. \tag{1}$$

Here $\cos\phi_j = (v_i^+, n(q_i)), \phi_j$ is the reflection angle. For semi-dispersing billiards all operators $V_j^* K_j V_j$ are non-negative. It is easy to show that this continued fraction is converging as soon as $t_n \to \infty$ and $B^+(x) \geq 0$. The following theorem is a particular case of Chernov's theorem (see [C1]).

Theorem 2. *If* $B^+(x) > 0$ *a.e. then the semi-dispersing billiard is hyperbolic.*

This theorem can be derived also from the results in the book by Katok and Strelcyn (see [KS1]).

The operator $B^+(x)$ for $x = (q, v)$ is a self-adjoint operator acting on the $(d-1)$-dimensional space passing through q orthogonal to v. It has a simple geometrical meaning. Namely, if $\gamma^{(u)}(x)$ is a lum of x, $\tilde{\gamma}^{(u)}(x) = \pi(\gamma^{(u)}(x))$ then $\tilde{\gamma}^{(u)}(x)$ is a $(d-1)$-dimensional submanifold lying inside Q and $B^+(x)$ is the operator of the second fundamental form of $\tilde{\gamma}^{(u)}(x)$ at $q = \pi(x)$. The condition $B^+(x) > 0$ means that the submanifold $\tilde{\gamma}^{(u)}(x)$ is strictly concave. The appearance of continued fractions in (1) is quite natural. If we write down the Jacobi equations corresponding to our flow then they will be second order linear differential equations with piecewise constant coefficients having discontinuites at the moments of reflections. It is well-known that the solutions to such equations can be written as continued fractions. The operators $B^+(x)$ are exactly these solutions.

An interesting non-trivial example of the semi-dispersing billiard where one can already see the main difficulties of the theory was proposed by Kramli, Simanyi and Szasz (see [KSS1]). The table consists of the three-dimensional torus from which two non-parallel cylinders are removed. As was already mentioned the system of r hard disks or balls moving uniformly inside a cube and interacting via elastic collisions can be reduced to semi-dispersing billiards. Apparently these billiards are hyperbolic but now it is known only for small values of r.

It is very surprising that hyperbolic billiards are encountered also among two-dimensional convex domains. The first example was proposed in the famous paper by Bunimovich [B1]. The table is the Bunimovich's stadium where the boundary consists of two straight segments and two circular arcs. The boundary is a C^1-curve and it is essential that the curvature is a discontinuous function on the boundary. If it were sufficiently smooth then the corresponding billiard cannot be hyperbolic because a theorem of Lazutkin (see [L1]) says that in such cases there exist so many caustics that the set of their tangent vectors has positive measure. Recently the result by Bunimovich was extended by Bunimovich (see [B2]) and Donnay (see [D1]). Namely the form of the stadium can be perturbed in such a way that two segments remain but the circular arcs can be replaced by arbitrary C^2-arcs whose curvature is close to a constant.

In connection with [B1, B2, D1] a general problem emerges of describing two-dimensional tables for which the billiard is hyperbolic. Wojtkowski in [W1] proposed a general approach to this problem based on the existence of invariant families of cones (see also [S3]) and found several new examples of hyperbolic billiards. Some other examples were constructed in the paper by Markarian (see [M1]).

3. Fundamental Theorem in the Theory of Hyperbolic Billiards

In this section we discuss the problems related to ergodicity and mixing of hyperbolic billiards. We remark first that the existence of lsm and lum for the flow $\{S^t\}$ implies the existence of these manifolds for the induced map T. Their dimensions are equal to $d-1$ while the dimension of ∂M is $2(d-1)$. It is an easy part of the theory of piecewise smooth hyperbolic dynamical systems which says that ergodic components of T and $\{S^t\}$ have positive measures (see, for example, [P1] and [KS1]). The proof is based upon an idea of Hopf [H1] according to which the ergodic component of a.e. x contains also $\Gamma^{(u)}(x)$ and $\Gamma^{(s)}(x)$. Thus we can take a lum $\gamma^{(u)}(x)$ and construct lsm $\gamma^{(s)}(y)$ for every $y \in \gamma^{(u)}(x)$. It needs some efforts to show that starting with an a.e. x we get a set of positive measure. The main ingredient of the proof is the so-called property of absolute continuity of foliations generated by $\Gamma^{(s)}, \Gamma^{(u)}$ (see details in [KS1]). However it is a much more difficult problem to prove that the ergodic component of T or $\{S^t\}$ is unique. The difficulty is due to the fact that because of singularities the sizes of lsm or lum are everywhere discontinuous and can be arbitrarily small. Therefore apriori it is not clear whether one can connect two arbitrary points by a chain of sets of the described form.

In [S2] a theorem was proven which showed how to overcome the difficulty in the case of two-dimensional dispersing billiards. Later it was called the fundamental theorem in the theory of dispersing billiards. The idea was to show that in a neighborhood of a typical point for a given lum $\gamma^{(u)}$ the probability to find $\gamma^{(s)}(y)$, $y \in \gamma^{(u)}$, whose length is greater than $Cl(\gamma^{(u)})$, l is the length, C is an arbitrary constant, tends to 1 as the radius of the neighborhood tends to zero. It follows from this statement that a small neighborhood of this point lies within one ergodic component. The ergodicity follows from an easy statement that the set of typical points is connected. There are two other publications [BS1] and [S4] which contain some modifications of the original proof and a proof of ergodicity based on a slightly different idea.

In our joint paper with Chernov [CS1] the ideas of the fundamental theorem were extended further. Let $R \subset \partial M$ be the union of singular points of ∂M, of tangent points of $\partial M^+ \cup \partial M^-$, i.e., points $x = (q, v)$ where $(n(q), v) = 0$ and the points of discontinuity of T. We imposed the following condition: for a.e. point of R the image under dynamics of a plane framed by unit normals locally extends in all directions. Under this condition we proved that typical points have neighborhoods belonging to one ergodic component (local ergodicity). An immediate consequence of this theorem is

Theorem 3. *Every dispersing billiard is ergodic.*

Kramli, Simanyi and Szasz in [KSS2] made an important improvement in the conditions of the fundamental theorem which is essential for semi-dispersing billiards and gave a very detailed proof of their version of the fundamental theorem which they called “A transversal fundamental theorem”.

Using their improvement they proved the following remarkable result (see [KSS3]).

Theorem 4. *Consider the system of three balls on $Tor^d, d \geq 2$. Then it is ergodic on the subspace of the phase space where the total momentum is zero and the centrum of mass is fixed.*

In the paper by Bunimovich [B3] it was shown that the fundamental theorem is applied to the billiards in stadium and some other convex domains. Thus these billiards are also ergodic.

In systems with hyperbolic behavior ergodicity usually implies much stronger statistical properties. Namely it is easy to show under the conditions of Theorems 3 and 4 that the billiard is a K-flow. It means in particular that the adjoint group of unitary operators has countable Lebesgue spectrum and strong properties of mixing. It is a more difficult problem to prove the Bernoulli property which says that the induced map T is isomorphic to a Bernoulli shift. This was done by Gallavotti and Ornstein in [GO1] for two-dimensional dispersing billiards.

4. Markov Partitions and Strong Statistical Properties of Two-Dimensional Hyperbolic Billiards

If we intend to study more deeply statistical properties of hyperbolic billiards like central limit theorem and decay of time-correlation functions we must use some special symbolic representations of these systems. The most convenient one is the representation which can be obtained with the help of a Markov partition. Let us recall the corresponding definitions (see [Bo1, BoR1, R1, R2, S5, S7]). We shall deal with the induced map T.

Take a point $x \in \partial M^+$ and lum $\gamma^{(u)}(x)$, lsm $\gamma^{(s)}(x)$, choose subsets $\Pi^{(u)} \subset \gamma^{(u)}(x), \Pi^{(s)} \subset \gamma^{(s)}(x)$.

Definition 6. A subset $\Pi \subset \partial M^+$ is called a parallelogram with basic sets $\Pi^{(u)}, \Pi^{(s)}$ if for any $y_1 \in \Pi^{(u)}, y_2 \in \Pi^{(s)}$ the lsm $\gamma^{(s)}(y_1)$ and the lum $\gamma^{(u)}(y_2)$ intersect each other and the intersection $\gamma^{(s)}(y_1) \cap \gamma^{(u)}(y_2)$ consists of one point also belonging to Π.

If Π is a parallelogram then for any $x' \in \Pi$ the sets $\gamma^{(u)}(x') \cap \Pi, \gamma^{(s)}(x') \cap \Pi$ can be taken as basic sets. If Π', Π'' are parallelograms then $\Pi' \cap \Pi''$ is also a parallelogram. Each parallelogram has two natural partitions $\xi^{(u)}(\Pi), \xi^{(s)}(\Pi)$. The elements of these partitions are $C_{\xi^{(u)}(\Pi)}(y) = \Pi \cap \gamma^{(u)}(y)$, $C_{\xi^{(s)}(\Pi)}(y) = \Pi \cap \gamma^{(s)}(y)$ respectively. Assume that we are given a partition η whose elements are parallelograms $\Pi_1, \Pi_2, \ldots, \Pi_r, \ldots$. This means that $\mu(\Pi_{i_1} \cap \Pi_{i_2}) = 0$ for $i_1 \neq i_2$ and $\mu(\cup \Pi_j) = 1$. We shall denote by $\xi^{(u)}, \xi^{(s)}$ the partitions of ∂M^+ where for a.e.x we have $C_{\xi^{(u)}}(x) = C_{\xi^{(u)}(\Pi_j)}(x), C_{\xi^{(s)}}(x) = C_{\xi^{(s)}(\Pi_j)}(x)$ when $x \in \Pi_j$.

Definition 7 (see [Bo1, S5, S7]). The partition η is called Markov partition if for a.e. x

$$T^{-1} C_{\xi^{(u)}}(x) \subseteq C_{\xi^{(u)}}(T^{-1}x), \quad T C_{\xi^{(s)}}(x) \subseteq C_{\xi^{(s)}}(T_x) .$$

In the case of smooth hyperbolic systems natural Markov partitions are usually finite. For billiards and other hyperbolic discontinuous systems such partitions are by necessity countable. If $\mu(\Pi) > 0$ then the basic sets $\Pi^{(u)}, \Pi^{(s)}$

are subsets of $\gamma^{(u)}(x), \gamma^{(s)}(x)$ which do not contain open components and usually they are Cantor-like sets of positive measure. The reason is again due to the fact that in any neighborhood of a typical point lum and lsm can be arbitrarily small. Apparently similar Markov partitions can appear in other dynamical systems with a non-uniform hyperbolic behavior like some classes of one-dimensional maps, Henon maps etc.

Returning to billiards suppose that we have an arbitrary partition η of ∂M^+ which is the phase space for T. Having a partition η we construct the symbolic representation for T as follows. Assume that all Π_j are closed and write down the inclusions $T^n x \in \Pi_{j_n}, -\infty < n < \infty$.

Definition 8. The sequence $j = \{j_n\}$ is called the symbolic representation of x.

It is clear that if $j \Leftrightarrow x$ then $T\Pi_{j_{n-1}} \cap \Pi_{j_n} \neq \phi$. The usefulness of Markov partitions is due to the fact that the inverse statement is in a sense also true: if j is any sequence for which $\mu(T\Pi_{j_{n-1}} \cap \Pi_{j_n}) > 0$ for all $n, -\infty < n < \infty$, then there exists x for which j is a symbolic representation. In other words the phase space ∂M^+ of T is coded by realizations of a topological Markov chain up to a subset of measure zero. In the paper [BS2] we constructed Markov partitions for two-dimensional dispersing billiards. In the recent paper [BCS1] using the same ideas we proved under some technical assumptions the following theorem.

Theorem 5. *For two-dimensional hyperbolic billiards there exist countable Markov partitions. Moreover, the diameters of parallelograms can be made less than any given* $\varepsilon > 0$.

In particular such Markov partitions exist for dispersing billiards, Bunimovich's stadium and its perturbations and some other cases.

Describe now some corollaries which follow directly from Theorem 5.

Corollary 1. *The set of periodic orbits of a hyperbolic billiard is everywhere dense.*

It was shown to us by Pesin how this statement can be proven by "standard arguments from hyperbolic theory". L. Stojanov in [St1] proved that for any multi-dimensional semi-dispersing billiard in R^d the number $N(R)$ of periodic orbits whose period is not more than R is not bigger than $C_1 \exp\{C_1 R\}$ for some $C_1 > 0$. Theorem 5 gives a possibility to obtain an estimation of $N(R)$ from below.

Corollary 2. *The number* $N(R)$ *is bigger than* $\exp\{C_2 R\}$ *for some* $C_2 > 0$ *and all sufficiently large* R.

Apparently as in the case of smooth hyperbolic system there exists the asymptotics $\lim_{R\to\infty} \frac{\ln N(R)}{R}$ connected with the topological entropy of billiard but it still remains as an open problem.

In our paper [BS3] such statistical properties as the central limit theorem, time decay of correlation functions for two-dimensional dispersing billiards were studied. These results can be extended to a wider class of two-dimensional hyperbolic billiards (Bunimovich, Chernov, Sinai – in preparation). Using Markov partitions one can prove for these billiards some version of H-theorem in statistical mechanics.

5. Some Other Results and Concluding Remarks

As one can see from what was said above the theory of dispersing billiards is similar to some extent to the theory of hyperbolic dynamical systems. The theory of semi-dispersing billiards is in the same sense similar to the theory of partially hyperbolic dynamical system. Unfortunately it is not still developed in a sufficiently general manner. Only some particular results have been obtained. Chernov in [Ch1] constructed lsm and lum under rather general conditions. Their dimensions may be less than $d-1$. He gave also (see [Ch2]) a detailed proof of the formula for the measure-theoretic entropy

$$h(\{S^t\}) = \int tr(B^+(x))d\mu(x)$$

which is positive unless $K(q) \equiv 0$ like in the case of flat boundaries. Wojtkowski in [W2] studied the asymptotical behavior of entropy when the mean free path tends to zero.

Some examples of multi-dimensional hyperbolic billiards analogous in sense to billiard in stadium are described in [B3] (see, however, the remarks in [W3]).

Several Hamiltonian systems behave in many respects as hyperbolic billiards. In the papers by Kubo [Ku1] (see also [S6]) the motion of a particle on the two-dimensional torus under the action of some repelling forces was studied. It was shown that it is hyperbolic (in the same sense as in the definition 3). A surprising result was obtained in a recent paper by Donnay and Liverani [DL1]. The authors found examples of attractive potentials for which the dynamics is also hyperbolic. In the paper by Wojtkowski [W4] (see there some other references) a system of particles with different masses moving along a semi-line with a constant acceleration and undergoing elastic collisions was studied. It was shown that it has several positive Lyapunov exponents which implies through the Pesin's formula [P1] that this system has positive entropy and a partially hyperbolic behavior.

Arnold in [Ar1] gave an heuristic explanation why two-dimensional dispersing billiards should behave like geodesic flows on surfaces of negative curvature. He proposed to consider doubles of the corresponding tables and represent them as limits of smooth surfaces of negative (more precisely, non-positive) curvature. However, this idea being very beautiful has a defect. The flow which appears under this limiting transition contains "extra" trajectories which can have common parts with the boundary. In this connection some questions may be asked. For example, one can take the double of the billiard table, its universal covering and ask the following:

1. will the lift of a billiard trajectory to the universal covering go to infinity;
2. if so, what is the structure of points at infinity which are the end-points of the lifted trajectories, going out of some points;
3. will a lifted trajectory of the billiard lie to a finite distance from some Lobachevsky geodesics.

Some answers to these questions were obtained recently by Babenko [Ba1].

In conclusion I thank the Organizational Committee of ICM-90 for the honorable invitation to present this talk. I thank V.I. Arnold and J. Lebowitz who read the text and made many useful remarks. Also I am very glad to express my sincere gratitude to Bunimovich, Chernov, Kramli, Szasz with whom we worked together for many years on various problems concerning billiards.

References

[An1] Anosov D.V.: Geodesic flows on closed Riemannian manifolds of negative curvature. Proc. Steklov Institute **90** (1967) 1–235

[Ar1] Arnold V.I.: Small denominators and problems of stability of motion in classical and celestical mechanics. Uspehi Math. Sci. **18:6** (1963) 85–191

[Ba1] Babenko I.K.: Behavior of trajectories of dispersing billiards on the flat torus. Math. Sbornik. (In press)

[Bo1] Bowen R.: Equilibrium states and ergodic theory of Anosov diffeomorphisms. (Lecture Notes in Mathematics, vol. 470.) Springer, Berlin Heidelberg New York 1975

[BR1] Bowen R., Ruelle D.: The ergodic theory of Axiom A flows. Invent. math. **29** (1975) 181–202

[B1] Bunimovich L.A.: On the ergodic properties of certain billiards. Funct. Anal. Appl. **8** (1974) 254–255

[B2] Bunimovich L.A.: On stochastic dynamics of rays in resonators. Radiofizika **28** (1985) 1601–1602

[B3] Bunimovich L.A.: A theorem on ergodicity of two-dimensional hyperbolic billiards. Comm. Math. Phys. **130** (1990) 599–621

[B4] Bunimovich L.A.: Many-dimensional nowhere dispersing billiards with chaotic behavior. Physica **D33** (1988) 58–63

[BS1] Bunimovich L.A., Sinai Ya.G.: On a fundamental theorem in the theory of dispersing billiards. Math. Sbornik **19** (1973) 407–423

[BS2] Bunimovich L.A., Sinai Ya.G.: Construction of Markov partitions for dispersing billiards. Comm. Math. Phys. **73** (1980) 247–280

[BS3] Bunimovich L.A., Sinai Ya.G.: Statistical properties of Lorentz gas with periodic configuration of scatterers. Comm. Math. Phys. **78** (1981) 479–497

[BCS1] Bunimovich L.A., Chernov N.I., Sinai Ya.G.: Markov partitions for two-dimensional hyperbolic billiards. Uspehi Math. Sci. **45:3** (1990) 97–134

[C1] Chernov N.I.: Construction of transversal fibers for multi-dimensional semi-dispersing billiards. Funct. Anal. Appl. **16** (1982) 35–46

[C2] Chernov N.I.: Entropy of semidispersing billiards. (In preparation)

[CS2] Chernov N.I., Sinai Ya.G.: Ergodic properties of some systems of 2-D disks and 3-D spheres. Uspehi Math. Sci. **42:3** (1987) 153–174

[CFS1] Cornfeld I.P., Fomin S.V., Sinai Ya. G.: Ergodic theory. Springer, Berlin Heidelberg New York 1981

[D1] Donnay V.J.: Convex billiards with positive entropy. (In preparation)

[DL1] Donnay V.J., Liverani C.: Ergodic properties of particle motion in potential fields. Comm. Math. Phys. (In press)

[G1] Gallavotti G.: Lectures on the billiard. (Lecture Notes in Physics, vol. 38. Moser, Ju., ed.) Springer, Berlin Heidelberg New York 1975, pp. 236–295

[GO1] Gallavotti G., Ornstein D.: Billiards and Bernoulli schemes. Comm. Math. Phys. **38** (1974) 83–101

[H1] Hopf E.: Statistik der geodätischen Linien in Mannigfaltigkeiten negativer Krümmung. Ber. Verhandl. Sächs. Akad. Wiss. **91** (1939) 261–304

[K1] Krylov N.S.: Works on the foundations of statistical physics. Princeton University Press, Princeton 1979

[Ku1] Kubo I.: Perturbed billiard systems. Nagoya Math. J. **61** (1976) 1–57

[KMS1] Kerchhoff S., Masur H., Smillie J.: Ergodicity of Billiard flows and quadratic differentials. Ann. Math. **124** (1986) 293–311

[KS1] Katok A.B., Strelcyn J.-M.: Invariant manifolds, entropy and billiards, smooth maps with singularities. (Lecture Notes in Mathematics, vol. 1222.) Springer, Berlin Heidelberg New York 1986

[KSS1] Kramli A., Simany N., Szasz D.: Ergodic properties of semi-dispersing billiards. Two cylindric scatterers in the 3-D torus. Nonlinearity (1989) 311–326

[KSS2] Kramli A., Simanyi N., Szasz D.: A "transversal" fundamental theorem for semi-dispersing billiards. Comm. Math. Phys. **129** (1990) 535–560

[KSS3] Kramli A., Simanyi N., Szasz D.: Three billiard balls on the ν-dimensional torus is a K-flow. Ann. Math. (In press)

[L1] Lazutkin V.F.: The existence of caustics for the billiard problem in a convex domain. Izv. Acad. of Sci. Math. Series **37** (1973) 188–223

[M1] Markarian R.: Billiards with Pesin region of measure one. Comm. Math. Phys. **118** (1988) 87–97

[P1] Pesin Ya.B.: Lyapunov characteristic exponents and smooth ergodic theory. Uspehi Math. Sci. **32:4** (1977) 55–112

[R1] Ruelle D.: A measure associated with Axiom A attractors. Amer. J. Math. **98** (1976) 619–654

[R2] Ruelle D.: Thermodynamic formalism. (Encyclopedia of Math. and its Appl. 5) Addison-Wesley, 1978

[S1] Sinai Ya.G.: Introduction to Ergodic Theory. Princeton University Press, Princeton 1977

[S2] Sinai Ya.G.: Dynamical systems with elastic collisions. Uspehi Math. Sci. **25:2** (1970) 141–192

[S3] Sinai Ya.G.: Appendix to the first edition of G.M.Zaslavsky's book "Statistical irreversibility in non-linear systems" ("Nauka" Press, Moscow) 114–139, 1970

[S4] Sinai Ya.G.: Ergodic properties of Lorentz gas. Funct. Anal. Appl. **13** (1979) 46–59

[S5] Sinai Ya.G.: Markov partitions and U-diffeomorphisms. Funct. Anal. Appl. **2** (1968) 61–82

[S6] Sinai Ya.G.: A "physical" system with positive entropy. Vestnik Moscow State University **5** (1963) 6–12

[S7] Sinai Ya.G.: Gibbs measures in ergodic theory. Uspehi Math. Sci. **27:4** (1972) 21–64

[Sm1] Smale S.: Differentiable dynamical systems. Bull. Amer. Math. Soc. **73** (1967) 747 817

[St1] Stojanov L.: An estimate from above of the number of periodic orbits for semi-dispersed billiards. Comm. Math. Phys. **124** (1989) 217–227

[V1] Veech W.A.: Interval exchange transformations. J. Anal. Math. **33** (1978) 222–272

[W1] Wojtkowski M.P.: Principles for the design of billiards with nonvanishing Lyapunov exponents. Comm. Math. Phys. **105** (1986) 391–414

[W2] Wojtkowski M.P.: Measure theoretic entropy of the system of hard spheres. Erg. Theor. Dyn. Syst. **8** (1988) 133–153

[W3] Wojtkowski M.P.: Linearly stable orbits in 3 dimensional billiards. Comm. Math. Phys. **129** (1990) 319–328

[W4] Wojtkowski M.P.: A system of one dimensional balls with gravity. Comm. Math. Phys. **126** (1990) 507–534

Applications of Non-Linear Analysis in Topology*

Karen Uhlenbeck

Sid Richardson Foundation Regents' Chair in Mathematics, Department of Mathematics, The University of Texas at Austin, Austin, TX 78712, USA

If there is any single characterizing feature of the mathematics of the last few years, it is the interactions among subdisciplines. This activity is well-documented by the talks at ICM90. This lecture contains a small part of the background for applications of non-linear analysis in the field of topology. My discussion covers roughly the last twelve years, with emphasis on the earlier period of this time. There are many other articles in this volume which describe in more detail current areas of research. We refer particularly to Floer's Plenary address, and section lectures by McDuff, Simpson, Tian, Kronheimer, and many talks related to topological quantum field theory.

The specific mathematical tools I am considering in this paper are those of "hard" analysis, by which we mean two things. Hard analysis refers in graduate student slang to the use of estimates. But here it refers as well to what Gromov [Gr] calls "hard" — the realization of "soft" or "flabby" topological concepts via the solution of specific rigid partial differential equations. These techniques are naturally not to every mathematician's taste. Topologists have spent a great deal of effort reproving theorems such as the Bott periodicity theorem in ways more related to the general constructive methods of algebraic topology. For mathematicians like myself these geometric methods provide a concrete geometric realization of what is otherwise very much algebraic abstraction.

1. Background Discussion

Our first step is to illustrate the content of this talk with an example which uses only advanced calculus. We define a real-valued function

$$f : \mathbb{C}^{n+1} - \{0\} \to \mathbb{R}$$

by the formula

$$f(\mathbf{v}) = \frac{(A\mathbf{v} \cdot \bar{\mathbf{v}})}{(\mathbf{v} \cdot \bar{\mathbf{v}})}$$

for $\mathbf{v} = (z^1, \dots, z^n) \in \mathbb{C}^{n+1} - \{0\}$. Here A is any Hermitian complex $(n+1) \times (n+1)$ matrix. It is easy to see that if $\alpha \in \mathbb{C} - \{0\}$, $f(\alpha\mathbf{v}) = f(\mathbf{v})$. So f induces a map

* Partially supported by grants from the American Mathematical Society and the National Science Foundation.

Proceedings of the International Congress of Mathematicians, Kyoto, Japan, 1990

$$[f] = (\mathbb{C}^{n+1} - \{0\})/(\mathbb{C} - \{0\}) = \mathbb{C}P^n \to \mathbb{R}$$

by $[f]([\mathbf{v}]) = f(\mathbf{v})$. The critical points of f (and $[f]$) are points where the derivative vanishes. It is a standard exercise in advanced calculus to show that the equations for critical points of $[f]$ are the lines $[\mathbf{v}]$ for

$$A\mathbf{v} - \lambda \mathbf{v} = 0$$

where $\lambda = f(\mathbf{v})$.

Now the connection with topology comes from the relation between the critical points of $[f]$ and the topology of $\mathbb{C}P^n$. If the eigenvalues of A are distinct, one can compute that the smallest is the minimum, the largest the maximum, and the $n-1$ other critical points occur with index (number of negative directions in the second derivative) $2, 4, \ldots, 2(n-1)$. Standard Morse theory tells us that $\mathbb{C}P^n$ is built from one handle in each even dimension. The Betti numbers $b_{2i} = 1$, $0 \le i \le n$ and $b_{2i+1} = 0$. Of course, this is not the usual way of computing the Betti numbers of $\mathbb{C}P^n$.

However, much closer in spirit to many of our examples is the case when A is a Hermitian projection operator of rank $n-k$. Then the minimum of $[f]$ on $\mathbb{C}P^n$ is zero and it occurs on a $\mathbb{C}P^k$ sitting in $\mathbb{C}P^n$. The other critical points consist of a $\mathbb{C}P^{n-k-1}$ on which the maximum occurs. We obtain a very simple topological result by using the gradient flow for $[f]$ to retract $\mathbb{C}P^n - \mathbb{C}P^{n-k-1}$ into $\mathbb{C}P^k$. The result is that the embedding $\mathbb{C}P^k \subseteq \mathbb{C}P^n$ is a homotopy equivalence up to dimension k.

The prototype theorem for this talk is the Bott Periodicity theorem for the unitary group, as originally proved by Bott in 1959 [B]. Here the theory of ordinary differential equations replaces the advanced calculus of our first example. The space which replaces $\mathbb{C}P^n$ is $C^{\infty}_{-I,I}([0,1], SU(2m))$, or the parametrized smooth curves between $-I$ and $+I$ in the special unitary group $SU(2m)$. The function $[f]$ is replaced by energy

$$E(s) = \frac{1}{2}\int_0^1 |\dot{s}(t)|^2\, dt\,.$$

The equation for critical curves is the equation for geodesics

$$D_t \dot{s}(t) = 0\,.$$

The minimum for this functional occurs on the set of great circles between $(-I, +I)$. This can be checked to be the complex Grassmannian $G(m, 2m)$ of m planes in $2m$ space. All the other critical points have index at least $2m+2$. The gradient flow provides a homotopy equivalence between the loop space and the Grassmannian up to dimension $2m$. Namely for $0 \le i \le 2m$, we have the result that

$$\pi_{i+1} SU(2m) = \pi_i\big(\Omega SU(2)\big) = \pi_i G(m, 2m)\,.$$

There are methods from algebraic topology which show that

$$\pi_{i-1} SU(m) = \pi_{i-1} SU(k)$$

for $i \le 2m \le 2k$ and

$$\pi_i G(m, 2m) = \pi_{i-1} SU(m) \quad \text{for } i \le 2m\,.$$

Bott Periodicity Theorem [B].

$$\pi_{i-1}SU(k) = \pi_{i+1}SU(k), \qquad i \leq k.$$

Other older results which use the theory of geodesics have more to do with differential geometry.

Theorem (Hadamard). *Suppose M is a compact, connected manifold with negative sectional curvature, then $\pi_{i+1}(M) = 0$ for $i \geq 1$.*

One version of this proof is obtained by showing that all geodesics are of shortest length. As a consequence, every connected component of the loop space is topologically trivial. Hence $\pi_i(\Omega(M)) = \pi_{i+1}(M) = 0$ for $i \geq 1$.

Theorem (Myers). *If M is compact with positive Ricci curvature, then $\pi_1(M)$ is finite.*

This can be proved by showing all the minimizing geodesics are short. A general reference is Milnor's book on Morse theory [Mi].

The infinite dimensional loop spaces $\Omega(M)$ were originally handled by retraction onto finite dimensional spaces using piecewise solutions as approximations. This does not work in more than one variable because of the difficulties involved in gluing small solution pieces together. It may have been that mathematicians hoped that the multivariable problems could be easily handled once the proper tools for treating global problems were developed. This turned out to be not quite true.

The modern developments *do* rest entirely on the foundations of functional analysis and elliptic operator theory. The analytic tools are Hölder spaces, Sobolev spaces, embedding theorems, interpolation theorems and the fundamental estimates for elliptic and parabolic systems.

In the 1960s, an ambitious subject called "global analysis" developed with the explicit goal of solving non-linear problems via methods from infinite dimensional differential topology. During this period, a different set of tools was developed. A short list of these tools includes: the notion of Fredholm operator and the the Atiyah-Singer index theorems (1963) [A-S]; the definition of infinite dimensional manifolds [L]; metric structures and refinements such as layer structures and Fredholm structures; the definition of a non-linear Fredholm operator and Smale's extension of the Sard theorem [Sm] (1965); the Palais-Smale conditions and applications in the calculus of variations (1964) [P-S]; in addition, several variants of infinite dimensional degree theorems, K-theory and transversality. A good sense of the spirit of this development can be obtained by browsing through the three volumes of the proceedings of the Berkeley 1968 AMS conference organized by S.S. Chern and S. Smale [C-S].

The optimism of the era of global analysis has ultimately been justified, but this did not happen immediately. The problem is essentially as follows: In order to discover properties of solutions of ordinary or partial differential equations which have global significance, it is essential to make estimates. Now, it usually happens that certain estimates are natural to the problem. Sometimes it may be an estimate on a maximum of a norm, or more usually an integral estimate on solutions is available. Typically the estimate is on the L^2 norm of the first derivative of a solution. In other words, an estimate in the Banach space L_1^2 is natural. However, in order to obtain the results which have topological meaning, the estimates have to imply something about the space of continuous solutions, or at the very least, some information about continuity. This occurs when the Banach space L_1^2 lies in C^0. However, $L_1^2 \subset C^0$ is a dimensionally dependent inclusion. It holds for $n = 1$, or for the case of ordinary differential equations, but not for $n \geq 2$, the dimensions of partial differential equations.

Hence the explanation for the success of the cited examples on loop space is not exactly what it was expected to be. The problems work at least partly because the Sobolev embedding theorem $L_1^2 \subset C^0$ is true in dimension 1, and naive attempts to apply the theories to partial differential equations do not work except in restrictive cases. Many of the applications which were ultimately found are extremely deep. Especially significant are the dimensional differences. In retrospect, one could not expect one single aspect of non-linear analysis to magically provide for a variety of deep applications.

2. Results

S.T. Yau's proof of the Calabi conjecture, published in 1978, showed for the first time the effect which modern methods of solving partial differential equations could have on other fields. The analytic theorem Yau proved was for an arbitrary complex Kähler manifold with non-positive first Chern class. Yau proved that there is a Kähler metric which solves Einstein's equation.

Theorem [Y-1]. *Let M be a complex Kähler manifold with $c_1(M) \leq 0$. Then there is a Kähler metric g in the same Kähler class as the given one with*

$$\mathrm{Ricci}(g) - Rg = 0\,.$$

The topological conditions imply that $R \leq 0$, where the constant R is a multiple the constant scalar curvature of the Einstein metric.

This result had been conjectured by Calabi and partially proved by Aubin. However, its importance in algebraic geometry lies in the following application.

Corollary. *If M is a complex Kähler manifold with $c_1(M) \leq 0$, then*

$$(-1)^n c_2(M) c_1(M)^{n-2}[M] \geq \frac{(-1)^n n}{2(n+1)} c_1(M)^n [M]\,.$$

This theorem is cited in every theorem on the classification of higher dimensional algebraic varieties. It is essentially the only topological restriction known for algebraic manifolds. The theorem is not true for positive Chern class. One of the satisfying results of Yau's proof is that the importance of the topological condition $c_1(M) \leq 0$ is apparent. Yau's original theorem (in the case of complex 3-folds with $c_1 = 0$) is fundamental in the current model for fundamental physics (string theory).

In the ensuing years, the applications of partial differential equations have been extensive, and we give a very brief survey of the initial results in each field. It would not be possible to list all the latest results and their fine points in a general survey article.

Minimal Surfaces

Yau and his coworkers obtained a number of interesting results. Many but not all of these results have alternate proofs. The first theorem is in the spirit of Myer's theorem, and was published in 1980 by Schoen and Yau. I call it the topological positive mass theorem, as it is the topological version of the well-known positive mass theorem of general relativity. (Schoen and Yau used the same techniques to prove the theorem in general relativity.) Scalar curvature is local mass. An alternate proof using linear analysis (Dirac operators) now exists. The analytical basis for the Schoen-Yau proof is a theorem on minimal surfaces.

Theorem [S-Y, S-U]. *If Σ_g is a surface of genus $g \geq 0$, and M any compact manifold with $\pi_1(\Sigma) \subseteq \pi_1(M)$, then there exists an area minimizing branched immersion of Σ in M.*

Application (Topological Positive-Mass) [S-Y]. *If $M = M^3$ has non-negative scalar curvature, then the group $\pi_1(M)$ does not contain $\pi_1(\Sigma_g)$ as a subgroup unless $\Sigma_g = S^1 \times S^1$ and $M^3 = S^1 \times S^1 \times S^1/\Gamma$ is a quotient of the flat torus.*

To prove this, look at the second variation of the minimal surface. The positive scalar curvature forces it to have a negative direction.

Meeks and Yau proved a series of results which show that minimal surfaces in 3-manifolds are embedded. In many cases, this provides alternate more rigid proofs of basic theorems in 3-manifold topology, such as Dehn's lemma, the loop theorem, and the sphere theorem. As part of this program they gave the first proof of the equivariant loop theorem.

Equivariant Loop Theorem (Meeks-Yau). *Let M^3 be a handle-body with boundary Σ_g, and $K \subseteq \mathrm{Diff}(M^3, \Sigma_g)$ be a finite group. Assume M^3 is given a metric in which K acts as isometries, and in which Σ_g has positive scalar curvature (outward like $S^2 = \partial D^3$). Then there exists an embedded minimal disk $(D^2, S^1) \subseteq (M^3, \Sigma_g)$ such that the elements $k \in K$ either leave D^2 invariant, or map D^2 to a disk $k(D^2)$ with $k(D^2) \cap D^2 = \emptyset$.*

The first step in the analysis consists in showing that there is indeed a smallest disc in M^3 with boundary on Σ_g (this leads to a boundary value problem which is a combination of Dirichlet and Neumann conditions). If the solution is not embedded, or if it intersects an iterate, it turns out that it cannot really be of smallest area. Meeks and Yau pioneered the use of 3-manifold techniques to show that minimal surfaces in 3-manifolds are often embedded rather than immersed [M-Y].

The proof due to Siu and Yau of the Frankel conjecture dates from the same year as the proof due to Mori.

Frankel Conjecture. *If M^n is a complex Kähler manifold with positive bisectional curvature, then M^n is biholomorphically equivalent to $\mathbb{C}P^n$.*

Siu and Yau use the minimal 2-spheres shown to exist by Sacks and Uhlenbeck [S-U]. Positive curvature tends to place restrictions on what can be minimal (as in Myer's theorem and in Schoen and Yau's proof of the positive mass conjecture). In this case, Siu and Yau show that the minimal sphere is actually a holomorphic curve [Si-Y].

Finally, we mention a much more recent result. The method due to Sacks and myself for finding minimal 2-spheres in manifolds had been used by Meeks and Yau to handle embedding problems for spheres in 3-manifolds, as well as by Siu and Yau in the Frankel conjecture. However, these proofs use the area minimizing spheres, whereas Micallef and Moore [M-M] later found a use for the non-minimizing critical points. Their isotropic curvature condition is satisfied if the Riemannian curvature is pinched between K and $4K$.

Theorem [S-U]. *If M is a compact manifold and $\pi_i(M) \neq 0$ for some $i \neq 1$, then there is a 2-sphere which is a stationary point of the area functional.*

Sphere Theorem of Micallef and Moore. *If M^n is simply connected and has positive curvature on isotropic 2-planes, then M^n has the homotopy type of a n-sphere.*

Proof. Show that $\pi_i(M) = 0$ for $i \leq [n/2]$. If this is true, the Hurewicz homomorphism and Poincaré duality complete the proof. If $\pi_i(M)$ is the first non-zero homotopy group, a difficult minimax argument leads to the construction of a minimal 2-sphere of index at most $(i - 2)$. However, the curvature condition forces the existence of at least $[n/2] - 1$ directions in which the second variation is negative. This leads to a contradiction.

Gauge Theory and 4-Manifolds

Donaldson's announcement of the restrictions on the topology of 4-manifolds with differentiable structures is a more recent mathematical event. Donaldson's startling use of gauge theory in four dimensions followed almost immediately the successful use of minimal surfaces, and development of these gauge theory techniques is still an active field.

The program instigated and to a great extent carried out by Donaldson consists of encoding the properties of differential structures on 4-manifolds by studying the self-dual Yang-Mills equations on the manifold. The difficulty is that the mathematician must introduce a metric onto the smooth manifold. The first step is to understand the analysis, then the dependence on the metric must still be analyzed.

The non-linear analysis part of Donaldson's theory consists in construction of the space of self-dual solutions of Yang-Mills equations over a conformal manifold M in a bundle with structure group $SU(2)$ and second Chern class $-k$. (In the more recent literature, the orientation is reversed to study the anti-self-dual equations in a bundle of second Chern class $+k$. This fits in better with complex analysis). The basic ingredients is the list of theorems developed by global analysts. The Atiyah-Singer theorem determines the dimension of the moduli space and the Sard-Smale theorem can be used to show it is generically a manifold. Taubes' implicit function theorem developed to construct solutions by gluing instantons on a manifold was later modified by Donaldson to include the construction of solutions on the connected sum $M_1 \# M_2$ from solutions on M_1 and M_2. The solution spaces are not compact. However, the boundary is well-understood via exactly the arguments developed to understand the convergence of minimal surfaces.

Theorem (Donaldson) [D-2]. *If M^4 is a simply connected 4-manifold with positive definite self-intersection form, then the moduli space of solutions to the self-dual Yang-Mills equations with $k = -1$ and group $SU(2)$ is generically an oriented manifold with isolated singularities whose boundary can be identified with M.*

The isolated singularities correspond to solutions where $E = L \oplus L^{-1}$ splits into line bundles. This is a theorem an analyst might well have proved, although it would certainly not be obvious to include the orientability. However, the topological use of this theorem appears a a corollary.

Corollary. *M^4 is topologically the connected sum of $\mathbb{C}P^2$'s.*

Each singularity looks like a $\mathbb{C}P^2$ with a positive orientation. There isn't anything else with definite form oriented cobordant to this sum of $\mathbb{C}P^2$'s.

The theory has been developed further, and more elaborate properties of the solution space of the Yang-Mills equation are used in later results. We refer the readers to a forthcoming survey by Freed [Fr] and the article by Donaldson [D-1].

Some of the properties of the moduli spaces in gauge theory are quite similar to properties of Gromov's pseudoholomorphic curves. McDuff has used these to distinguish different symplectic forms [Gr, McD].

Complex Moduli Spaces

The results about 4-manifolds obtained from the Yang-Mills equation are obtained by looking at the topology of the "moduli spaces of solutions." Here by moduli space we refer to the actual solutions to Yang-Mills divided out by the natural geometric equivalence. This is in many ways similar to older examples of moduli spaces, such as the Riemann moduli space of conformal structures on surfaces. In a development which is related to the moduli space of self-dual Yang-Mills equations, a whole class of problems in algebraic geometry can be put in a general framework which we might call infinite dimensional geometric invariant theory.

The foundational paper is the paper by Atiyah and Bott on Yang-Mills equations over Riemann surfaces [A-B]. An older result of Narasimhan and Sheshadri [N-S] proves that the moduli space of stable holomorphic bundles M_g over a complex curve Σ_g of genus $g \geq 2$ can be identified with the moduli space of projectively flat connections. These connections can be identified with the minima of the Yang-Mills functional on connections in Σ_g. Atiyah and Bott conjectured that there is a very beautiful analytic picture which fits this functional. The symmetry group of Yang-Mills is the real group of gauge transformations $\mathcal{G}$. However, its complexification $\mathcal{G}_{\mathbb{C}}$ acts and stratifies the sets of connections $\mathfrak{A}$ into classes of holomorphic structures $\mathfrak{A}/\mathcal{G}_{\mathbb{C}}$ on the bundle. The Yang-Mills functional is the L^2 norm of moment map for this action. Stable orbits are (essentially) the orbits on which the action is free and the quotient Hausdorff. By a very general principle, the equivariant topology of the stable moduli space can be computed from the topology of $\mathfrak{A}/\mathcal{G}$ by examining the Morse theory.

Atiyah and Bott were not able to carry through the analysis, but obtained their results from algebraic geometry. Frances Kirwan [Ki] carried out this very general program in a finite dimensional setting. This pattern of development is strikingly similar to the original construction of the Morse theory of geodesics, which inspired the very useful finite dimensional development. Donaldson [Do-3,4], Hitchin [Hi], Daskalapoulos [Da], Bradlow [Br], Corlette [Co] and Simpson [Si] have carried out the analysis and extended the picture to cover coupled equations and complex manifolds of higher dimension. Naturally, the topological results are much better in complex dimension 1. This is because of the Sobolev inequalities. We discuss this in more detail in the next section.

We finish this section by stating one of the basic results of these computations. It corresponds to the bundle version of Yau's solution of the Calabi conjecture. The topological consequences are similar, although as with the Calabi and Frankel conjectures, there is also an algebraic proof.

Theorem (Donaldson, Uhlenbeck-Yau) [D-5, U-Y]. *The moduli space of stable bundles on a complex Kähler manifold is isomorphic to the space of irreducible solutions to the holomorphic Yang-Mills equations (an extension of the anti-self dual equations to arbitrary dimension).*

Corollary. *If E is a stable holomorphic bundle of rank r on a complex Kähler manifold of dimension n, then*

$$c_1(E)^2 \wedge \omega^{n-2}[M] \leq \frac{2r}{r+1} c_2(E) \wedge \omega^{n-2}[M] .$$

This is a purely topological statement about the cohomology class of the Kähler form and the characteristic classes of the complex tangent bundle. This inequality is easily seen to be true by applying Chern-Weil theory to the holomorphic Yang-Mills connections.

The Poincaré Conjecture

The Poincaré conjecture refers to one of the best known problems in topology. If a manifold M^n has the homotopy type of S^n, is it S^n (differentiably or continuously)? I don't think that analysts have gotten very close to proving or disproving the Poincaré conjecture. It is perhaps not so well-known that this conjecture has been the inspiration for a number of fundamental developments in analysis.

Most of the ideas have focussed on Einstein's equation. This equation is the Euler-Lagrange equation for a critical metric for the variational integral $\int_M K_g \, d\mu_g$. Here $\operatorname{vol} M = \int_M d\mu_g$ is kept fixed. Here K_g is the scalar curvature of the metric g. Einstein's equation reads

$$\operatorname{Ricci}(g) - \lambda_n K g = 0 .$$

It goes without saying that we know more about a manifold if it has a solution of Einstein's equation on it. In particular, for a 3-manifold Einstein's equation is equivalent to the metric having constant Riemannian curvature. Clearly not every 3-manifold supports a solution to Einstein's equation because there is no metric of constant Riemannian curvature on most 3-manifolds. Nevertheless, this fact may not have been widely recognized by analysts in the past.

Solving Einstein's equation is very difficult. Yamabe [Yam] proposed as the first step to fix the conformal structure and vary only the function describing lengths. This leads to a much studied conformally invariant problem usually called the Yamabe Problem [L-P].

Palais was motivated to construct a very general theory of the calculus of variations [P]. He claims that he hoped to apply it to the Einstein functional. However, he very quickly realized that the critical points of the Einstein functional have infinite index and coindex. Hence they cannot be detected by the topological methods he developed. There is still some hope of using a minimax argument and the known solution to the Yamabe problem. I am not sure how much faith any of us have in this project, though.

Hamilton's results on the Einstein equation (1982) were very surprising and promising. He showed via a heat flow argument that a 3-manifold with positive Ricci curvature supports a solution to Einstein's equation. This makes it the quotient of S^3 by a finite group. This result has been very influential on analysis in general without providing a solution to the Poincaré conjecture.

Finally, I would like to comment that the inspiration of the Poincaré conjecture is still very much with analysts. Thurston's results on 3-manifolds go a long way towards describing the geometry of 3-manifolds. Present thinking is that it may be possible to use some other variant of a curvature integral. An example might be

$$\int_M |\mathrm{Riem}(g)|^p \, d\mu(g)$$

for $p \geq 3/2$ in 3 dimensions. The hope is that, by transposing gauge theoretic techniques over to manifolds, the obstructions to minimizing such integrals can be better understood.

Conjecture (Due to Deane Yang). *If M^3 is aspherical and atoroidal, then the minimum of the integral*

$$\int_M |\mathrm{Riem}(g)|^{3/2} \, d\mu(g)$$

under the constraint $\mathrm{vol}\, M = 1$ *is either zero, or is taken on by a metric of constant negative curvature.*

This result is in fact nearly implied by Thurston's conjectures on 3-manifolds. Yang's conjecture should lead to interesting analysis, even if it doesn't touch the Poincaré conjecture.

3. Analytical Technique

The variety of topological results cited in the previous section is matched by the variety of different analytical methods which were used. There is no neat classification which matches a set of results with a set of techniques. I roughly classify the methods under the subheadings:

Continuity method,
Borderline dimension,
Gauge theory,
Heat equation methods.

This is not an exhaustive list of methods. A theory such as that constructed by Andreas Floer uses nearly all the ideas mentioned in this section and more. Many of the latest developments involve topological index theory constructions on the solution spaces as are needed for Donaldson's invariants of 4-manifolds and applications of Gromov's pseudo-holomorphic curves.

Continuity Method

The continuity method is on the surface naive. The goal is to solve an equation

$$F(g) = 0$$

where F is a non-linear elliptic system. To use the continuity method, start with a trial g_0, and compute $F(g_0) = F_0$. Put in a parameter $\varepsilon \in [0, 1]$ and solve

$$F(g_\varepsilon) = (1 - \varepsilon)F_0 .$$

This is done by showing that $dF(g_\varepsilon)$ is invertible as a map between appropriately introduced Banach spaces, and that the solution g_t stays bounded in the tangent norm. The invertibility of dF implies via the implicit function theorem that the set of $\varepsilon \in [0, 1]$ for which we have a solution to our equation is open. The estimates (and some weak convergence) show this set is closed. The sophistication comes from the usual necessity of dealing with a number of Banach spaces, and from the further necessity of estimating the inverse of dF.

This is the method used by S.T. Yau to solve the Calabi conjecture. While the equation is Einstein's equation

$$\mathrm{Ricci}(g) - \lambda(g) = 0 ,$$

in the Kähler case an equation can be written for a Kähler potential $\varphi(\varepsilon)$ where $g(t) = g_0 + \partial\bar{\partial}\varphi(\varepsilon)$ is a new metric. The equation becomes one for the potential function φ. The first stage of the estimates follow easily from the maximum principle. However, essentially all derivatives of φ have to be estimated, although the estimates become iterative after the third derivatives. No method of solution has been found which avoids estimates (nor would we expect this to happen). However the estimates themselves have been given a more geometric foundation and have found a wider application in the general study of Monge-Ampere equations.

Uhlenbeck and Yau used the same method to solve the holomorphic Yang-Mills equations on a Kähler manifold. Here, the metric is in a bundle, and there is no potential, but there is a maximum principle. By adding an ε, we obtain an invertible equation

$$F_g + \varepsilon \ln g = 0.$$

For $\varepsilon > 0$, it is easy to solve this equation. As $\varepsilon \to 0$, the solution does blow up unless an extra geometric condition of stability is satisfied. We show that, if blow-up does occur, that the normalized solutions $\frac{g(\varepsilon)}{\mu(\varepsilon)}$ converge to a degenerate metric π which violates the stability condition. The sophisticated analysis occurs in higher order estimates, not in the initial outline.

Finally, Taubes' construction of instantons on four manifolds is via an implicit function theorem proved in the same style [T-1]. His original iterative proof can be rewritten using the continuity method. He glues in the standard instantons on $\mathbb{R}^4$, localized to lie in a ball of radius λ into a ball centered at p on an arbitrary compact four manifold. The resulting approximate solution can be modified by a small amount to give a solution to the instanton equation. Donaldson further needs to obtain a moduli space with the parameters (λ, p). Again, it is the estimates and the invertibility of an operator very close to the derivative operator which are essential [D-2].

The continuity method is best suited to cases where invertibility of the derivative is built into the situation. In all the cases just cited, there is a moduli space of solutions whose dimension and structure can be understood by using the appropriate choice of coordinates (or gauge) when looking for the solution. A maximum principle is available for the Kähler-Einstein and Holomorphic Yang-Mills examples. Invertibility in Taubes' construction comes from topological constraints and knowledge of S^4. It remains a question when Taubes' gluing of point localized solutions applies to other partial differential equations. Schoen has applied a somewhat similar idea in constructing solutions of the Yamabe problem with point singularities [S] and Kapouleas [Ka] has shown the existence of many complicated surfaces of constant curvature in $\mathbb{R}^3$ by a not unrelated technique. These problems all exhibit conformal invariance, which allows the scaling of solutions. The resulting approximate solutions have a number of parameters. In the simple cases there is a projection onto a moduli space; in complicated examples the parameters must be chosen carefully.

Borderline Dimension

The critical ingredient in geometric problems is the Sobolev embedding theorems. For the usual mapping examples, $L_1^2 \subset C^0$ is true in dimension 1, false in higher dimensions. Dimension 2 is the borderline or scale invariant dimension, because the integral $\iint |ds|^2 (dx)^2$ scales the same way as $\max_{x \in M} |s(x)|$. Both are scale invariant. A similar phenomena is observed for the Yamabe problem. In dimension n, $L_1^2 \subset L^p$ for $p < \frac{2n}{n-2}$. The relevant power of p in the Yamabe problem is $p = \frac{2n}{n-2}$ [L-P].

For borderline problems, solutions can easily be found using minimization and weak convergence methods. The difficulty is that the limiting functions may not satisfy the constraint satisfied by the approximating functions. A typical example would be to fix a domain bounded by a simple closed curve Γ in the plane. Consider maps $s : D^2 \to \mathbb{R}^2$ such that $s|S^1 : S^1 \to \Gamma$ has degree one. Now minimize

$$\iint_{D^2} \left(\left| \frac{\partial s}{\partial x} \right|^2 + \left| \frac{\partial s}{\partial y} \right|^2 \right) dx\, dy\,.$$

The minimum occurs on a conformal map, and will provide a solution of the Riemann mapping problem. The difficulty is to keep the weak limit of a minimizing sequence from being trivial (i.e., to preserve the condition of $s|S^1 : S^1 \to \Gamma$ is of degree one). One solution is to use the conformal invariance to fix the map on three points of S^1. C.B. Morrey [Mo] used these ideas (which originate with Douglas' solution of the Plateau problem) to find minimal surfaces in arbitrary Riemannian manifolds.

The fundamental observation is that in conformally invariant problems energy estimates plus scale invariance imply that the estimates and convergence are valid except possibly at an isolated set of points. At these points, small neighborhoods may conformally expand to cover large pieces of the geometric solution. The mathematician can recover the geometric solution by blow-up. For $\varepsilon_i \to 0$, set

$$s_i \left(\frac{x - x_0}{\varepsilon_i} \right) = \hat{s}_i(x)\,.$$

Then $\hat{s}_i(x) \to s : \mathbb{R}^n = S^n - \{p\} \to M$ will be a solution on S^n [S-U1].

This principle allows the construction of a large number of two-dimensional minimal surfaces, and underlines our present understanding of the Yamabe problem. Later it came to be fundamental in the analysis of solutions of the self-dual Yang-Mills equation on 4-manifolds. The borderline dimension for Yang-Mills turns out to be 4 instead of 2, and many results for harmonic maps and minimal surfaces have counterparts in theorems on Yang-Mills in dimension four [Ma, Se]. However, the topological results have come from Donaldson's study of the solution spaces of the self-dual Yang-Mills equations. Here existence theorems are proved by the implicit function theorem mentioned in the previous section. Whatever compactness of the solution space has is shown by the same techniques used in the construction of minimal surfaces [Do-2, F-U]. Gromov later applied these same ideas back in two dimensions to study pseudo-holomorphic 2-spheres in symplectic manifolds [Gr, McD].

It is an interesting question whether mathematicians will discover new and useful scale-invariant geometric partial differential equations. These techniques are just waiting to be used again!

Gauge Theory (New Sobolev-Inequalities)

Until gauge field theory appeared in mathematics, well-posed, natural, topological calculus of variations problems seemed confined to one dimension. It is possible to artificially construct variational problems of all sorts. However, the geometrically natural variational problems seem to be the only ones which are useful in topology. These are (a) first order in derivatives and (b) quadratic. Conditions (a) and (b) imply that the natural estimates are L_1^2. Topological results come from C^0, so we are stuck in the classical case of dimension one with geodesics for good problems and in the case of dimension 2 with minimal surfaces for the borderline case.

However, in gauge field theory on a manifold M, the unknown is the connection A which is a one-form. The natural function is the L^2 norm of curvature

$F_A = dA+[A, A]$, which is the first derivative of A. However, topology comes from the overlap functions, which relate A's via their derivatives. Intuitively speaking, curvature is the second derivative of the structure functions of a bundle. The relevant Sobolev embedding is $L_2^2 \subset C^0$. This embedding is true in dimensions 2 and 3, and borderline in dimension 4.

Some basic estimates need to be obtained before analysis can be done. Gauge theory problems have an infinite dimensional symmetry group, and the curvature $F_A = dA + [A, A]$ contains only part of the full derivative of A. These problems are related and are taken care of in the same analytical lemma. Because of the nonlinearity in the problem, the estimates are stated globally in terms of convergence.

Theorem [U-1]. *Let D_i be a sequence of connections on a compact manifold M of dimension n. If F_i is the curvature of D_i, and $F_i \in L^p$ forms a bounded sequence for $p > n/2$, then there exists a subsequence $D_{i'}$ and a sequence of gauge transformations $s_{i'}$ such that $s_{i'}^* D_{i'} \rightharpoonup D$ in L_1^p.*

A weaker version of this lemma applies in the borderline case $p = n/2$.

Yang-Mills in dimensions 2 and 3 gives us new examples of variational problems in which the topology and analysis match. The 2-dimensional example has been extensively studied [A-B] and gives us the topological results on the moduli space of stable bundles over Riemann surfaces. Recently the analytical details were completed by Daskalapoulos [Da]. The 3-dimensional problem is not very well understood geometrically. We do not as yet know how to use the many solutions to Yang-Mills which exist on a 3-manifold.

Heat Equation Methods

One of the earliest global non-linear results was that of Eells and Sampson [E-S]. Recall that the energy functional is defined on maps between two compact Riemannian manifolds $s : M \to N$

$$E(s) = \int_M |ds|^2 \, d\mu .$$

Critical maps are called harmonic maps. Eells and Sampson constructed a harmonic map in every homotopy class of maps $s \in [M, N]$ when N has non-positive sectional curvature. They did this by following L^2 gradient curves. Thus L^2 flow is a non-linear system of parabolic equations

$$\frac{\partial s}{\partial t} = \nabla_s^* \, ds .$$

In fact, early in his career S.T. Yau wrote a number of papers which apply this existence theory [Y-2].

A potential method of solving a non-linear elliptic equation is to solve the associated non-linear parabolic equations and follow the solutions as $t \to \infty$. This at first seems unduly cumbersome, and at least one mathematician expended some effort in finding "better" ways to find the Eells-Sampson harmonic maps [U-2]. In 1983 Richard Hamilton [H] was able to solve Einstein's equation on some 3-manifolds by solving the associated parabolic equation and following the time dependent solution to a solution of the elliptic equation. This result has already been noted in the section on the Poincaré conjecture in section 2. Hamilton's result has had a remarkable effect on geometric PDE, since this result surely cannot be obtained by variational methods. In the wake of Hamilton's work, both the Einstein and the Kähler-Einstein equation have been reexamined and solved in certain contexts using the heat equation method.

The non-linear heat flow was the geometric gradient used by Atiyah and Bott to study the Yang-Mills equations. Their analysis was shown to be rigorous by Donaldson, and was used by Donaldson to solve the holomorphic version of Yang-Mills in arbitrary complex dimension [Do-3,Do-4]. Carlos Simpson [Si] has shown that this approach fits in very nicely with geometric invariant theory. The method can be used to obtain the same results as obtained by Yau and myself [U-Y] and is more philosophically satisfying than the perturbation method.

Strüwe has gone a long way towards showing that many of the useful properties of harmonic maps (or minimal surfaces) can be obtained via the parabolic equation [St]. In many cases, the rigid method of looking for solutions of a geometric elliptic equation via the parabolic equation will give the most delicate results. This fits in best with the topology when there are moduli spaces of solutions, the equation is geometrically natural, and fewer choices are involved in following a parabolic equation compared to picking a minimizing sequence.

4. Failures

There are a number of theorems and conjectures which seem to be suitable for attack by the methods discussed in the previous section. Some mathematicians might regard these as "open problems." This is a matter of perspective. Perhaps, since I confess to having spent considerable time on them myself, the failure is personal.

Mostow-Rigidity Theorem. *If $s : M^n \to N^n$, $n > 2$, is a diffeomorphism between hyperbolic manifolds, then s is homotopic to an isometry.*

The proof of this theorem ought to go somewhat as follows. Let $\hat{s}$ be the harmonic map homotopic to s. The existence of this map dates back to [E-S] in 1963. Because M^n and N^n have constant curvature -1, the map $\hat{s}$ is an isometry.

Unfortunately, the only way we know that $\hat{s}$ is an isometry is by applying Mostow rigidity in a circular argument. This is, of course, only the simplest version of the Mostow rigidity theorem. There are some extensions due to Gromov which ought to be accessible by these means. Ultimately we would generally like to know better the relationship between curvature and volume. So far partial differential equations have not been helpful.

A related theorem by Siu uses harmonic maps to establish rigidity of complex manifolds with negative bisectional curvature [Si]. Other work has been done by Corlette, Gromov and Schoen, but a partial differential equation's proof of Mostow's theorem remains elusive.

Theorem [Se]. *Let $C_d^\infty(S^2, S^2)$ be the space of maps of degree d between S^2 and S^2. Let $M_d^\infty(S^2, S^2) \subseteq C_d^\infty(S^2, S^2)$ be the meromorphic maps of degree d. Then the inclusion of M_d in C_d^∞ is a homotopy equivalence up to dimension d.*

The proof ought to proceed as follows: Consider the energy integral

$$E(s) = \frac{1}{2} \iint_{S^2} |ds|^2 \, d\mu$$

on maps $s : S^2 \to S^2$. The meromorphic functions are the set on which E takes on its minimum. After removing a set of codimension $d + 2$, the gradient flow retracts the remainder of the space of C^∞ functions onto the minimum.

This is a conformally invariant problem and we have not learned to handle the finer points of the topology of the flow! Segal's theorem is actually very extensive, applying to surfaces of arbitrary genus as the domain and a whole class of (positively curved) complex manifolds as the image [Se]. It is related to the Atiyah-Jones conjecture on the topology of instantons embedded in the space of connections [A-J].

There is actually an analytic proof of Segal's theorem. It can pieced together by applying a theorem of Donaldson [Do-6] on monopoles and Taubes' proof of a Morse theory for the monopole equation [T-2]. This seems too unwieldy to be a model proof, however.

Thurston's Techniques for 3-Manifolds. As part of his theory of 3-manifold topology, Thurston develops techniques for analyzing hyperbolic 3-manifolds which depend heavily on rigidly embedded surfaces in the manifolds. These surfaces are geodesically embedded with constant Gaussian curvature, and are broken along geodesics. It is more natural from the point of view of analysis to examine the properties of constant mean curvature surfaces in 3-manifolds. (Zero mean curvature characterizes locally minimizing surfaces.) Up until this date, attempts to replace Thurston's broken surfaces with smooth surfaces have been strikingly unsuccessful. While some results in 3-manifolds are obtained by partial differential equations techniques, Thurston's program has been unaffected by these methods.

Jones-Witten Invariants. Two of the four fields medals at this congress were given out at least in part for a theory designed to produce 3-manifold invariants. There are many approaches to this theory, but the unifying approach of Witten is to start with a classical geometrical integral similar to those which have been used already in the applications of analysis to topology. Witten in fact takes the integral used by Floer. The classical Chern-Simons integral is defined for an $su(N)$ valued one-form A on a 3-manifold.

$$CS(A) = \frac{1}{4\pi} \int_{M^3} \operatorname{tr} \left(dA \wedge A + \frac{2}{3} A \wedge A \wedge A \right).$$

The Jones-Witten invariant for the manifold has the form

$$W(k, N) = \not\!\!\int_{\text{all } A} e^{ikCS(A)} \, dA.$$

The symbol $\not\!\!\int$ indicates the Feynman path integral used heavily in quantum field theory. Unfortunately, this is not a situation which has been made completely rigorous. However, many tools exist as input into calculations in quantum field theory. We list a few, in the hopes of giving some flavor of the subject.

Perturbation Theory. Calculations in quantum field theory which relate to physical experiment are meant to obtain asymptotic formulas as $1/k \to 0$. They are done via Feynman diagrams, which represent power series expansion around a vacuum (usually $A = 0$). In all physical cases I know of, the domain manifold is $\mathbb{R}^n$, $n \leq 4$. These methods would fail here, due to the lack of a proper Green's function.

Stationary Phaze (and Ghosts). There are two separate problems which can be formally dealt with in more geometric problems. One is that of several important classical solutions and the other is the lack of ellipticity. Witten has carried out these asymptotic expansions and obtained the first order asymptotic in k [W-2]. There are clearly severe problems in the theory since further calculations cannot even be guessed at. No one has yet obtained lower order terms for manifolds, although some calculations exist for knots [B-N].

Finite Approximation. The approximation of the infinite dimensional integral $⨎$ by finite integrals is known as lattice gauge theory. A large amount of supercomputer time is spent on more down-to-physics calculations than those of Chern-Simons theory. The value of these type of calculations is not clear. They certainly shed no light on arbitrary 3-manifolds.

Axiomatic Approach The quantum field theory involved in the Chern-Simons theory is particularly simple, and the formulation of the correct axioms seems to have been the greatest success of the theory so far. Atiyah has formuated the axioms for topological quamtum field theory (TQFT) [A-1].

Geometric Quantization. This formulation of quantum theory is probably the best understood by mathematicians. Witten and his students have a successful formulation of the Chern-Simons theory in terms of calculations on the moduli space of flat bundles [A-PD-W].

Canonical Quantization. It is usually necessary to use both the path integral and the Hamiltonian approach in computing the ingredients of quantum field theories. In many contexts, the Hamiltonian approach makes contact with group representations. We know that the input from 2-*d* conformal field theory is very useful in setting up the building blocks for the 3-*d* Chern-Simons theory. We can hence include on our list of useful mathematics:

Group representation theory,
Integrable lattice models of statistical mechanics,
Quantization of completely integrable systems.

Quantum Groups. Finally, some topologists have found that the direct quantum group (Hopf algebra) approach leads to the most direct construction of the Jones-Witten invariants [Q,Wa].

In classical mechanics, Lagrangian and Hamiltonian formulations are equivalent. In quantum field theory, both are necessary and complementary. What remains in question is the consistency (not equivalence) of the facts gained from the different approaches. To someone like myself, who has worked in topological applications of partial differential equations, the situation is analogous to the plurality of approaches which can be made to understanding the Laplace operator on a Lie group. It is possible to get input from a wide variety of mathematical directions: separation of variables, special functions, finite-element approximation schemes, group representations, asymptotic heat kernel methods and a variety of different types of geometric constructions. However, we know what the Laplace operator itself is. This holds the different ideas together. The study of Jones-Witten invariants is similar, but we are missing the central ingredient which corresponds to the Laplace operator — a proper definition of $\not\!\!\int$.

Recent computer calculations of D. Freed and R. Gompf indicate that two approaches (both due to Witten), from conformal field theory and from stationary phase approximation actually agree for lens spaces and some homology spheres [F-G]. Any mathematical demonstrations of this agreement would of necessity contain the proof tht analytic torsion agrees with combinatorial (Reidemeister) torsion. Perhaps we should look forward to a construction of $\not\!\!\int$ which will pull this theory together.

References

[A-1] Atiyah, M.: Topological quantum field theories. Publ. Math. Inst. Hautes Etudes Sci. Paris **68** (1989) 175–186

[A-2] Atiyah, M.: The geometry and physics of knots. Lezioni Lincee, Cambridge University Press, Cambridge, 1990

[A-B] Atiyah, M.F., Bott, R.: The Yang-Mills equations on Riemann surfaces. Phil. Trans. R. Soc. London A**308** (1982) 523–615

[A-J] Atiyah, M.F., Jones, J.: Topological aspects of Yang-Mills theory. Comm. Math. Phys. **61** (1978) 97–118

[A-S] Atiyah, M.F., Singer, I.: The index of elliptic operators on compact manifolds. Bull. Amer. Math. Soc. **69** (1963) 422–433

[A-DP-W] Axelrod, S., Della Pietra, S., Witten, E.: Geometric quantization of Chern-Simons gauge theory. To appear

[B-N] Bar-Natan, D.: Perturbative Chern-Simons theory. Preprint. Princeton University

[B] Bott, R.: The stable topology of the classical groups. Ann. Math. **70** (1965) 313–337

[Br] Bradlow, S.: Special metrics and stability for holomorphic bundles with global sections. J. Diff. Geom. (to appear)

[C-S] Chern, S.S., Smale, S. (eds.): Global Analysis. Proc. of Sym. in Pure Math XIV-XVI, 1968

[Co] Corlette, K.: Flat *G*-bundles with canonical metrics. J. Diff. Geom. **28** (1988) 361–382

[D-1] Donaldson, S.K.: The geometry of 4-manifolds. ICM 86, vol. 1 (1987) pp. 43–54

[D-2] Donaldson, S.K.: A application of gauge theory to 4-dimensional topology. J. Diff. Geom. **18** (1983) 81–83

[D-3] Donaldson, S.K.: A new proof of a theorem of Narasimhan and Seshadri. J. Diff. Geom. **18** (1983) 269–277

[D-4] Donaldson, S.K.: Anti-self-dual connections on complex algebraic surfaces and stable vector bundles. Proc. Lond. Math. Soc. **50** (1985) 1–26

[D-5] Donaldson, S.K.: Infinite determinants, stable bundles and curvature. Proc. London Math. Soc. **55**(3) (1987)

[D-6] Donaldson, S.K.: Nahm's equation and the classification of monopoles. Comm. Math. Phys. **96** (1984) 387–407

[Da] Daskalopoulos, G.: The topology of the space of stable bundles on a Riemann surface. Thesis. University of Chicago, 1989

[E-S] Eells, J., Sampson, J.: Harmonic mapping of Riemannian manifolds. Amer. J. Math. **86** (1964) 109–140

[F] Freed, D.: A survey of the application of gauge theory methods to 4-manifolds. In progress

[F-G] Freed, D., Gompf, R.: Computer calculation of Witten's 3-manifold invariant. Preprint. University of Texas

[F-U] Freed, D., Uhlenbeck, K.: Instantons and 4-manifolds. MSRI Publications #1, (Second edition) 1990

[Gr] Gromov, M.: Soft and hard symplectic geometry. ICM 86, **1** (1987) 81–89

[H] Hamilton, R.: Three manifolds with positive Ricci curvature. J. Diff. Geom. **17** (1982) 255–306

[Hi] Hitchin, N.: The self-duality equations on a Riemann surface. Proc. London Math. Soc. **55** (1987) 59–126

[Ka] Kapouleas, N.: Thesis. Stanford, 1988

[K] Kirwan, F.: Cohomology of quotients in symplectic and algebraic geometry. Mathematical Notes #31. Princeton University Press, 1987

[L] Lang, S.: Introduction to differentiable manifolds. Interscience, New York, 1962

[L-P] Lee, J., Parker, T.: The Yamabe problem. Bull. Amer. Math. Soc. **17** (1987) 37–81

[Ma] Marini, A.: Dirichlet and Neumann boundary value problems for Yang-Mills connections. Thesis. University of Chicago, 1990

[McD] McDuff, D.: Elliptic methods in symplectic geometry. Bull. Amer. Math. Soc. **23** (1990) 311–359

[M-Y] Meeks, III, W., Yau, S.T.: The equivariant loop theorem. In: Morgan and Bass (eds.), The Smith Conjecture. Academic Press, New York 1984

[M-M] Micallif, M., Moore, J.: Minimal 2-spheres and the topology of manifolds with positive curvature on totally isotropic 2-planes. Ann. Math. **127** (1988) 196–228

[Mi] Milnor, J.: Morse theory. Ann. Math. Studies, vol. 51. Princeton University Press, 1968

[Mo] Morrey, C.B.: The problem of Plateau on a Riemannian manifold. Ann. Math. **49** (1948) 807–851

[N-S] Narasimhan, M.S., Sheshadri, S.: Stable unitary vector bundles on a compact Riemann surface. Ann. Math. **89** (1965) 540–567

[P] Palais, R.: Foundations of global non-linear analysis. Benjamin, New York 1968

[P-S] Palais, R., Smale, S.: A generalized Morse theory. Bull. Amer. Math. Soc. **70** (1964) 165–171

[Q] Quinn, F.: Private communication

[S-U1] Sacks, J., Uhlenbeck, K.: The existence of minimal immersions of 2-spheres. Ann. Math. **113** (1981) 1–24

[S-U2] Sacks, J., Uhlenbeck, K.: Minimal immersions of closed Riemann surfaces. Trans. Amer. Math. Soc. **271** (1982) 639–652

[S] Schoen, R.: The existence of weak solutions with prescribed singular behavior for a conformally invariant scalar equation. Comm. Pure Appl. Math. **41** (1988) 317–392

[S-Y] Schoen, R., Yau, S.T.: Existence of incompressible minimal surfaces and the topology of manifolds with non-negative scalar curvature. Ann. Math. **110** (1970) 127–142

[Sed] Sedlacek, S.: A direct method for minimizing the Yang-Mills functional in 4-manifolds. Comm. Math. Phys. **86** (1982) 515–527

[Se] Segal, G.: Topology of spaces of rational functions. Acta Math. **143** (1979) 39–51

[Sim] Simpson, C.: Constructing variations of Hodge structure using Yang-Mills theory and applications to uniformization. J. Amer. Math. Soc. **1** (1988) 867–918

[Si] Siu, Y.T.: The complex analyticity of harmonic maps and the strong rigidity of Kähler manifolds. Ann. Math. **112** (1980) 73–111

[Si-Y] Siu, Y.T., Yau. S.T.: Compact Kähler manifolds of positive bisectional curvature. Inventiones Math. **59** (1980) 189–204

[Sm] Smale, S.: An infinite dimensional version of Sard's theorem. Amer. J. Math. **87** (1965) 861–866

[St] Struwe, M.: On the evolution of harmonic mappings of Riemann surfaces. Comment. Math. Helv. **60** (1985) 558–581

[T-1] Taubes, C.: Self-dual connections on non-self-dual 4-manifolds. J. Diff. Geom. **17** (1982) 139–170

[T-2] Taubes, C.: Min-Max Theory for the Yang-Mills-Higgs Equations. Comm. Math. Phys. **97** (1985) 473–540

[U-1] Uhlenbeck, K.: Connections with L^p bounds on curvature. Comm. Math. Phys. **88** (1982) 11–30

[U-2] Uhlenbeck, K.: Morse theory by perturbation methods with applications to harmonic maps. Trans. Amer. Math. Soc. **267** (1981) 569–583

[U-Y] Uhlenbeck, K., Yau, S.T.: On the existence of Hermitian-Yang-Mills Connections in Stable Vector Bundles. Comm. Pure Appl. Math. (1986) 257–293

[W-1] Witten, E.: Topological quantum field theory. Comm. Math. Phys. **117** (1988) 353–386

[W-2] Witten, E.: Quantum field theory and the Jones polynomial. Comm. Math. Phys. **131** (1989) 351–399

[Ya] Yamabe, H.: On a deformation of Riemannian structures on compact manifolds. Osaka Math. J. **12** (1960) 21–37

[Y-1] Yau, S.T.: On the Ricci curvature of a compact Kähler manifold and the complex Monge-Ampere equation. Comm. Math. Appl. Math. **31** (1978) 339–411

[Y-2] Yau, S.T.: On the fundamental group of compact manifolds of non-positive curvature. Ann. Math. **93** (1971) 579–585

Multidimensional Hypergeometric Functions in Conformal Field Theory, Algebraic *K*-Theory, Algebraic Geometry

Alexandre Varchenko

Moscow Institute of Gas and Oil, Leninski Prospekt 65, 117917 Moscow, USSR

Rudolf Arnheim in the book *Visual Thinking* (L.A. 1969) writes that usually concepts tend to crystallize into simple, well-shaped forms. They are tempted by Platonic rigidity. This creates troubles when the range they are intended to cover includes relevant qualitative differences. The variations can be so different from each other that to see them as belonging to one family of phenomena requires mature understanding. To the young mind, they look as different from each other as did the morning star from the evening star to the ancients.

The notion of a general hypergeometric function was introduced by I.M. Gelfand in the mid 80s. Now it is clear that general hypergeometric functions play a major role in interesting parts of mathematics such as Conformal Field Theory, Representation Theory, Algebraic *K*-Theory, Algebraic Geometry and provide new connections among them.

The general hypergeometric functions are generalizations of the Euler beta-function. The beta-function is the integral of a product of powers of linear functions over the segment. In the generalization the segment is replaced by a polytope and the integral

$$I(\Delta, f, \alpha) = \int_{\Delta} f_1^{\alpha_1} \dots f_N^{\alpha_N}\, dx_1 \dots dx_n$$

is considered as a function of the polytope $\Delta \subset \mathbb{R}^n$, the linear functions $\{f_j\}$, and the exponents $\{\alpha_j\} \subset \mathbb{C}$. The simplest examples are the classical hypergeometric function, the Euler dilogarithm, the volume of a polytope. The systematic study of the general hypergeometric functions was begun only recently in works of I.M. Gelfand's school and K. Aomoto.

There are three basic reasons for the appearance of general hypergeometric functions: the general hypergeometric functions satisfy remarkable differential equations, the general hypergeometric functions satisfy remarkable functional equations, the general hypergeometric functions, as analytic functions of their arguments, have remarkable monodromy groups.

Proceedings of the International Congress
of Mathematicians, Kyoto, Japan, 1990

1. Functional Equations

1.1 Volume of Polytope

The volume of a convex polytope in $\mathbb{R}^3$ has the following properties. The volume does not change under movements of the polytope. If a polytope is divided into two parts by a plane, then the volume of the polytope is equal to the sum of the volumes of the parts. These properties are functional equations of the volume considered as a function on the space of all convex polytopes. The properties of the volume suggest the following definition. The *group of polytopes* in $\mathbb{R}^3$ is the abelian group generated by the symbols (Δ), where $\Delta \subset \mathbb{R}^3$ is any convex polytope, subject to the relations

(a) $(\Delta) = (g\Delta)$ for any motion g.
(b) $(\Delta) = (\Delta_1) + (\Delta_2)$, if Δ is divided by a plane into parts Δ_1, Δ_2.

Is an element of this group uniquely determined by its volume? Is a regular tetrahedron equivalent to a cube of the same volume? These questions form the content of the third Hilbert problem. The third Hilbert problem was solved in Dehn's articles in 1900–1902, before it was published. It turned out that the regular tetrahedron cannot be composed from a cube because the two have different Dehn invariants.

Consider an edge of a polytope. An edge has two characteristics: the length l and the angle θ at the edge. The angle is defined up to a multiple of π and lies in $\mathbb{R}/\pi\mathbb{Z}$. The *Dehn invariant* of a polytope is the expression $D(\Delta) = \sum l_i \otimes \theta_i$, where the sum is taken over all edges of the polytope. The Dehn invariant is an element of the group $\mathbb{R} \otimes_{\mathbb{Z}} (\mathbb{R}/\pi\mathbb{Z})$. It is obvious that the invariant does not change under movements of the polytope and is additive under cuttings. It is easy to see that the Dehn invariant of a cube is equal to zero (as is that of any prism) and that the Dehn invariant of a regular tetrahedron is not equal to zero. So a regular tetrahedron cannot be composed from a cube. According to Sydler, 1965, the equivalence class of a polytope is uniquely determined by its volume and Dehn invariant, see also [C1, Du, DuPS, DuS, S].

This example demonstrates the scheme leading from a hypergeometric function to interesting algebraic concepts. Given a function on some space satisfying some functional equations one considers the abelian group generated by the points of the space subject to relations given by the functional equations of the initial function. If the initial function satisfies some differential equations then the group has additional structure.

1.2 Example

The *Euler dilogarithm* is the function defined by the power series

$$\mathrm{Li}_2(z) = \sum_{m=1}^{\infty} z^m/m^2 \qquad \text{for } |z| < 1.$$

The dilogarithm is a hypergeometric function. It has the integral representation

$$\mathrm{Li}_2(z) = \int_{\triangle} \frac{dx}{x} \wedge \frac{dy}{y}$$

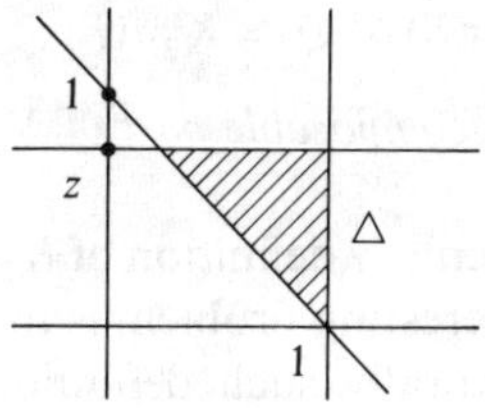

Fig. 1

where the triangle Δ is shown in Fig. 1. The dilogarithm satisfies the differential equation

$$d\,\mathrm{Li}_2(t) = \ln(1 - t)d\ln(t)$$

and the functional equation

$$\mathrm{Li}_2(x) - \mathrm{Li}_2(y) + \mathrm{Li}_2(y/x) - \mathrm{Li}_2\left(\frac{y(1-x)}{x(1-y)}\right) + \mathrm{Li}_2\left(\frac{1-x}{1-y}\right)$$
$$= n^2/6 - \ln(x)\ln\left(\frac{1-x}{1-y}\right).$$

for $0 < y < x < 1$.

The algebraic construction. The *Bloch group* of a field F is the abelian group B_2, generated by the symbols (t), where $t \in F\backslash\{0, 1\}$, subject to the relations

$$(x) - (y) + (y/x) - \left(\frac{y(1-x)}{x(1-y)}\right) + \left(\frac{1-x}{1-y}\right) = 0 \qquad (1)$$

for any $x, y \in F\backslash\{0, 1\}$.

Consider the multiplicative group F^* of a field F and its exterior square $F^* \wedge_{\mathbb{Z}} F^*$. The map

$$t \mapsto (t) \wedge (1 - t),$$

sending the symbol (t), $t \in F\backslash\{0, 1\}$, to the element $(t) \wedge (1 - t)$ of the group $F^* \wedge F^*$, has a remarkable property. It sends the alternating sum of the elements on the left-hand side of (1) to the zero element of the group $F^* \wedge F^*$. This gives a well-defined homomorphism

$$S: B_2 \to F^* \wedge F^*,$$

called the *Bloch complex*.

The homology groups of the Bloch complex are connected with the Quillen K-groups of the field F. In algebraic K-theory with any field F there is associated the sequence of groups $K_n(F)$, $n \geq 0$. For these groups a multiplication $K_p \otimes K_q \to K_{p+q}$ is defined.

Theorem (Matsumoto, Suslin).

1) $$\mathrm{Coker}\, S \simeq K_2(F);$$

2) $$(\mathrm{Ker}\, S) \otimes \mathbb{Q} \cong K_3^{\mathrm{ind}}(F) \otimes \mathbb{Q},$$

where $K_3^{\mathrm{ind}} = K_3/K_1^3$ *is the indecomposable part of* K_3.

This theorem gives an elementary definition of K_2 and the indecomposable part of K_3 (modulo torsion). An interesting problem is to find elementary definitions of all the groups $K_n(F)$. Conjecturally such definitions arise from functional and differential equations of polylogarithmic functions.

There are several possible generalizations of the logarithm and the Euler dilogarithm [A3, Le, GM, HaM]. One of them is the Aomoto polylogarithms.

1.3 Aomoto Polylogarithms

A simplex in $P^n(\mathbb{C})$ is an ordered set $L = (L_0, L_1, \dots, L_n)$ of hyperplanes. A simplex defines the differential n-form

$$\omega_L = d\ln(z_1/z_0) \wedge \cdots \wedge d\ln(z_n/z_0),$$

where $z_i = 0$ is a homogeneous equation of L_i. With a second simplex $M = (M_0, \dots, M_n)$ n-chain Δ_M in $P^n(\mathbb{C})\backslash L$ is connected. Δ_M is a curved oriented

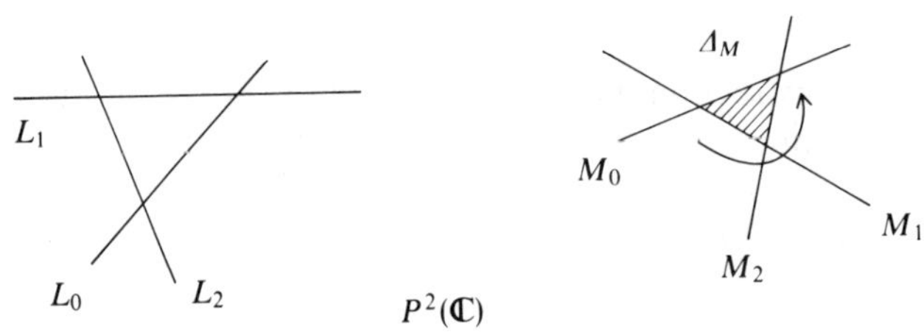

Fig. 2

n-simplex with boundary in M, see Fig. 2. With a pair of simplices an integral

$$a_n(L; M) = \int_{\Delta_M} \omega_L$$

is associated called the *Aomoto polylogarithm* of order n. The integral depends on the choice of the chain Δ_M, but does not change under its deformation. The Aomoto polylogarithm has the following properties.

(2) *Antisymmetry.* The integral is antisymmetric with respect to renumbering of the hyperplanes of the first or the second simplices.

(3) *Additivity with respect to a form.* If $L_0, \dots, L_{n+1}$ are $(n+2)$ hyperplanes and $L^i := (L_0, \dots, \hat{L}_i, \dots, L_{n+1})$, then

$$\sum_{j=0}^{n+1} (-1)^j a_n(L^j; M) = 0.$$

(4) *Additivity with respect to a chain.* If $M_0, \dots, M_{n+1}$ are $(n+2)$ hyperplanes, then

$$\sum_{j=0}^{n+1} (-1)^j a_n(L; M^j) = 0$$

for a suitable choice of Δ_{M^j}.

(5) *Projective invariance.* For every $g \in PGL(n+1, \mathbb{C})$ if $\Delta_{gM} = g\Delta_M$, then

$$a_n(gL; gM) = a_n(L; M).$$

For example, an Aomoto polylogarithm of order 1 is defined by two pairs of points $L = (L_0, L_1)$, $M = (M_0, M_1)$ on $P^1(\mathbb{C})$ and is equal to an integral of the form $\omega_L = d \ln(z_1/z_0)$ over a path going from M_1 to M_0. $a_1(L; M)$ is equal to the logarithm of the cross ratio of the four points (L_0, L_1, M_0, M_1). The Euler dilogarithm is a special case of the Aomoto dilogarithm, see Fig. 1.

There are two connections of polylogarithms of different orders: the multiplication and the differential equation.

(6) The product of Aomoto polylogarithms of orders p, q may be expressed as a sum of Aomoto polylogarithms of order $p+q$, see [BGSV].
(7) The differential of the polylogarithm of order n, considered as a function on the space of all configurations (L, M) may be expressed through suitable polylogarithms of orders $n-1$ and 1 [A3].

Aomoto polylogarithms describe parameters of the mixed Hodge structure of cohomology groups of a pair $P^n(\mathbb{C})\backslash L$, $M\backslash M \cap L$, where L, M are configurations of hyperplanes, see [BMS, BGSV]. For example if two points $L_0, L_1 \in P^1(\mathbb{C})$ are removed and two points $M_0, M_1 \in P^1(\mathbb{C})$ are identified then the mixed Hodge structure of the first cohomology group of this space is defined by the number $\exp(a_1(L_0, L_1; M_0, M_1))$.

1.4 Hopf Algebra of Pairs of Simplices

Taking properties of Aomoto polylogarithms as a starting point, it is possible to suggest a definition of a graded Hopf algebra $A(F) = A_0 \oplus A_1 \oplus A_2 \oplus \cdots$ for any field F [BMS, BGSV]. Here $A_0 := \mathbb{Z}$, A_n is the abelian group, generated by the symbols $(L; M)$, where $L = (L_0, \ldots, L_n)$, $M = (M_0, \ldots, M_n)$ are ordered sets of hyperplanes in $P^n(F)$, subject to relations similar to (2)–(5). The multiplication $\mu_{p,q}: A_p \otimes A_q \to A_{p+q}$ and the comultiplication $\nu_n = \bigoplus_{p=0}^n \nu_{p,n-p}: A_n \to \bigoplus_{p=0}^n A_p \otimes A_{n-p}$ of the Hopf algebra are modeled by properties (6), (7).

There is the sequence of complexes $A[n]$, $n \in \mathbb{Z}_+$, associated with any graded Hopf algebra A. The simplest of these complexes are

$$A[1]: 0 \to A_1 \to 0,$$

$$A[2]: 0 \to A_2 \xrightarrow{\nu_{1,1}} A_1 \otimes A_1 \to 0,$$

$$A[3]: 0 \to A_3 \xrightarrow{\nu_{2,1}+\nu_{1,2}} A_2 \otimes A_1 \oplus A_1 \otimes A_2 \xrightarrow{\nu_{1,1}\otimes 1 - 1 \otimes \nu_{1,1}} A_1 \otimes A_1 \otimes A_1 \to 0.$$

The complex $A[n]$ is concentrated in degrees from 1 to n. Conjecturally, the cohomology groups of these complexes give the K-groups of the field F modulo torsion.

Conjecture (Beilinson) [BGSV].

$$K_n(F) \otimes \mathbb{Q} \simeq \bigoplus_{j=0}^{[(n-1)/2]} H^{n-2j}(A[n-j] \otimes \mathbb{Q}).$$

Theorem [BGSV]. *The conjecture holds for* $n \leq 3$:

$$K_1 \simeq H^1(A[1]), \quad K_2 \simeq H^2(A[2]), \quad K_{3,\mathbb{Q}} \simeq H^3(A[3]_{\mathbb{Q}}) \oplus H^1(A[2]_{\mathbb{Q}}).$$

In the last years efforts have been undertaken to construct a motivic cohomology theory of algebraic manifolds which would be an arithmetical variant of the singular cohomology theory. According to Beilinson a category of motivic sheaves over the spectrum of a field F is a category of graded modules over a suitable algebra and a possible candidate for it could be the algebra $(A(F) \otimes \mathbb{Q})^*$ dual to the algebra $A(F) \otimes \mathbb{Q}$.

1.5 Bloch-Wigner Function

The Euler dilogarithm may be continued to a multivalued analytic function on $\mathbb{C} \setminus \{0, 1\}$. The *Bloch-Wigner* function is its imaginary part

$$D(z) = \mathrm{Im}(\mathrm{Li}_2(z)) + \arg(1 - z) \ln|z|,$$

see [B2, Z]. The Bloch-Wigner function has the following properties.

(8) $D(z)$ is single-valued real analytic on $\mathbb{C}$ except at the points 0 and 1, where it is only continuous.

(9) $$D(x) - D(y) + D(y/x) - D\left(\frac{y(1-x)}{x(1-y)}\right) + D\left(\frac{1-x}{1-y}\right) = 0$$

for any $x, y \in \mathbb{C} \setminus \{0, 1\}$.

In particular, for any field $F \subset \mathbb{C}$ the Bloch-Wigner function defines a homomorphism $B_2(F) \to \mathbb{R}$ of the Bloch group to real numbers.

1.6 Polylogarithms and Zeta-Function

Let F be an algebraic number field of degree n over $\mathbb{Q}$ with r_1 real and r_2 complex places, $r_1 + 2r_2 = n$. Let $\zeta_F(s)$ be the Dedekind zeta-function of F. The value $\zeta_F(2)$ is expressed in terms of the values of the Bloch-Wigner function at points of the field F.

Theorem [Bo, B, Su, Z]. *In the Bloch complex of F the group* Ker S *is isomorphic (modulo torsion) to* $\mathbb{Z}^{r_2}$. *The co-volume of the image of the map* Ker $S \to \mathbb{R}^{r_2}$, *defined by the composition of complex imbeddings and the Bloch-Wigner function, is a non-zero rational multiple of* $\pi^{-2(r_1+r_2)}|d_F|^{1/2}\zeta_F(2)$, *where* d_F *is the discriminant of* F.

D. Zagier conjectured that for any natural number m the value $\zeta_F(m)$ has similar expression in terms of values of the classical polylogarithms of order at most m at points of F, [Z]. Recently A. Goncharov proved this conjecture for $m = 3$.

Let $\mathbb{Z}[F\backslash\{0,1\}]$ be the free abelian group generated by the symbols (t), $t \in F\backslash\{0,1\}$. $\mathbb{Q}[F\backslash\{0,1\}] := \mathbb{Z}[F\backslash\{0,1\}] \otimes \mathbb{Q}$. For any function $D: \mathbb{C}\backslash\{0,1\} \to \mathbb{R}$ there is a homomorphism $D: \mathbb{Q}[\mathbb{C}\backslash\{0,1\}] \to \mathbb{R}$, $D: \sum n_i(t_i) \to \sum n_i D(t_i)$.

Let $\{a_j\}$, $j = 1, \ldots, r_1 + 2r_2$, be all possible imbeddings $F \to \mathbb{C}$, $\overline{a_{r_1+k}} = a_{r_1+r_2+k}$.

Theorem [Go]. *There exist* $z_1, \ldots, z_{r_1+r_2} \in \mathbb{Q}[F\backslash\{0,1\}]$, *such that*

$$\zeta_F(3) = \pi^{3r_2}|d_F|^{-1/2} \det(D_3(\sigma_j(z_i))),$$

where $j = 1, \ldots, r_1 + r_2$,

$$D_3(z) = \mathrm{Re}(\mathrm{Li}_3(z) - \ln|z|\mathrm{Li}_2(z) + \tfrac{1}{3}\ln^2|z|\mathrm{Li}_1(z)),$$

$\mathrm{Li}_n(z)$, $n \geq 1$, *is the classical polylogarithm of order n, defined by the power series*

$$\mathrm{Li}_n(z) = \sum_{m=1}^{\infty} z^m/m^n \qquad \textit{for } |z| < 1.$$

2. Hypergeometric Integrals

2.1 Classical Hypergeometric Function

The hypergeometric series

$$F(a,b;c;z) = 1 + \frac{ab}{1c}z + \frac{a(a+1)b(b+1)}{1\cdot 2c(c+1)}z^2 + \frac{a(a+1)(a+2)b(b+1)(b+2)}{1\cdot 2\cdot 3c(c+1)(c+2)}z^3 + \cdots$$

satisfies the differential equation

$$z(1-z)F'' + (c - (a+b+1)z)F' - abF = 0$$

and has the integral representation

$$\frac{\Gamma(b)\Gamma(c-b)}{\Gamma(c)}F(a,b;c;z) = \int_1^{\infty} t^{a-c}(t-1)^{c-b-1}(t-z)^{-a}\,dt.$$

Thus the classical hypergeometric function has three definitions: as a power series, as the solution of a differential equation, as an integral depending on a parameter.

These objects are associated with the family of the configurations of the triples of the points 0, 1, z of the complex line and are generalized naturally to the case of a family of configurations of hyperplanes in an affine space [A, G, GGZ, GKZ].

2.2 Cohomology of Complement of Configuration of Hyperplanes

Let $\mathscr{C}$ be a finite set of hyperplanes in $\mathbb{C}^n$. Choose a linear equation $f_H = 0$ for any hyperplane. Define the closed differential form df_H/f_H. The *Orlik-Solomon algebra* is the exterior algebra generated by 1 and the forms df_H/f_H, $H \in \mathscr{C}$.

Theorem [Ar1, Bri]. *The Orlik-Solomon algebra is naturally isomorphic to the complex cohomology ring of the complement of a configuration.*

In other words every cohomology class can be represented as a polynomial in the forms df_H/f_H, such a polynomial defines the zero class only if it is equal to zero.

Example [Ar1]. The Orlik-Solomon algebra of the configuration of all diagonal hyperplanes $t_i - t_j = 0$ in $\mathbb{C}^n$ is isomorphic to the exterior algebra generated by 1 and the symbols w_{ij} subject to the relations $w_{ij} = w_{ji}$, $w_{ij} \wedge w_{jk} + w_{jk} \wedge w_{ki} + w_{ki} \wedge w_{ij} = 0$ for pairwise different i, j, k. $P(t) = (1 + t)(1 + 2t)\dots(1 + (n - 1)t)$ is the Poincaré polynomial of the algebra, see also [Or, OS].

V.I. Arnold computed this example in connection with study of superpositions of algebraic functions and Hilbert's 13th problem; see [Ar2].

A configuration is called weighted if a complex number $\alpha(H)$ is assigned to each hyperplane H. The weights define the function

$$l_\alpha = \prod_{H \in \mathscr{C}} f_H^{\alpha(H)}.$$

This function is a multivalued function on the complement of a configuration. The differential of this function has the form

$$dl_\alpha = l_\alpha \sum_{H \in \mathscr{C}} \alpha(H)\, df_H/f_H.$$

A *hypergeometric differential form* of a weighted configuration is any form $l_\alpha \omega$, where ω is a differential form of the Orlik-Solomon algebra.

Hypergeometric forms form a finite-dimensional complex, as the differential of a hypergeometric form is a hypergeometric form:

$$d(l_\alpha \omega) = l_\alpha \sum_{H \in \mathscr{C}} \alpha(H)\, df_H/f_H \wedge \omega.$$

The *weight local system* $\mathscr{S}(\alpha)$ on the complement of a configuration is the complex one-dimensional local system of coefficients with the monodromy around a hyperplane H equal to the multiplication by $\exp(-2\pi i \alpha(H))$.

The cohomology of the complement of a configuration with coefficients in the weight local system is computed by the complex of hypergeometric forms. More precisely for any number t denote by $t\alpha$ the weights $H \mapsto t\alpha(H)$, $H \in \mathscr{C}$, homothetic to the initial ones.

Theorem [SV3]. *For almost all $t \in \mathbb{C}$ the cohomology of the finite dimensional complex of hypergeometric forms with weights $t\alpha$ is naturally isomorphic to the cohomology of the complement of a configuration with coefficients in $\mathscr{S}(t\alpha)$.*

For example, this is true for $t = 0$ according to the Arnold-Brieskorn theorem. Exceptional values of t form a descrete set. Conjecturally the exceptions form explicitly given arithmetical progressions [Ao1].

2.3 Determinant Formulas

The Euler beta-function is the alternating product of Euler gamma-functions:

$$B(\alpha, \beta) = \Gamma(\alpha)\Gamma(\beta)/\Gamma(\alpha + \beta).$$

There is a generalization of this formula to the case of a configuration of hyperplanes in an affine space [V, Se, A6].

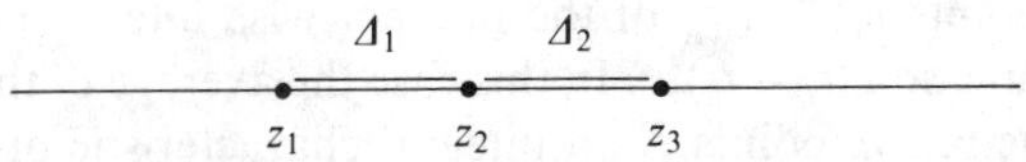

Fig. 3

Example. Consider the configuration of three points z_1, z_2, z_3 on a line, see Fig. 3. The point z_j is the zero of the function $f_j = t - z_j$. Put

$$l_\alpha = (t - z_1)^{\alpha_1}(t - z_2)^{\alpha_2}(t - z_3)^{\alpha_2},$$

$$\omega_1 = \alpha_1 l_\alpha d(t - z_1)/(t - z_1), \qquad \omega_2 = \alpha_2 l_\alpha d(t - z_2)/(t - z_2),$$

then

$$\det\left(\int_{\Delta_i} \omega_j\right) = \frac{\Gamma(\alpha_1 + 1)\Gamma(\alpha_2 + 1)\Gamma(\alpha_3 + 1)}{\Gamma(\alpha_1 + \alpha_2 + \alpha_3 + 1)} \prod_{i \neq j} f_i^{\alpha_i}(z_j).$$

Thus the determinant of integrals of basic hypergeometric forms of a configuration over all bounded components of the complement of the configuration is equal to the product of values of powers of linear functions at the vertices of the configuration up to a multiplicative constant equal to an alternating product of values of the gamma function [V].

The formula has an arithmetical analog. F. Loeser extended it to the case of a configuration of hyperplanes in an affine space over a finite field [Lo]. In this case the gamma-functions are replaced by Gauss sums, the determinant of the hypergeometric integrals is replaced by an alternating product of the determinants of the Frobenius operator in suitable cohomology groups.

3. Hypergeometric Functions and Representation Theory of Lie Algebras

The appearance of hypergeometric functions in the representation theory of Kac-Moody algebras, their quantum deformations, and in the Conformal Field Theory (CFT) is connected with integrals of the form

$$\underbrace{I(t_1, \ldots, t_n)}_{n} = \int \prod_{1 \leq j < k \leq N} (t_j - t_k)^{\lambda_{jk}} \underbrace{dt_{n+1} \wedge \cdots \wedge dt_N}_{m},$$

$$N = m + n.$$

Such integrals correspond to special configurations. The characteristics of these configurations, in particular the homology groups with twisted coefficients of the complement, the complex of hypergeometric forms are interpreted as objects of the representation theory of Kac-Moody algebras. Such integrals satisfy the differential equation which is described in terms of representation theory and is known in CFT as the Knizhnik-Zamolodchikov equation. The branching of such integrals is described in terms of quantum groups corresponding to Kac-Moody algebras.

The appearance of these integrals in physical models isn't surprising. Imagine a model in which points $t_1, \ldots, t_N$ of the line pairwise interact. The interaction is described by the function $(t_j - t_k)^{\lambda_{jk}}$. In this case the average of the interaction over all positions of the last m points is an integral characteristic of the first n points, described by the hypergeometric integral.

In applications the constants of the interaction λ_{jk} have the form

$$\lambda_{jk} = B(v_j, v_k)/\kappa,$$

where $v_1, \ldots, v_N$ are vectors of some complex linear space V, B is a symmetric bilinear form on V, and κ is a complex parameter of the model.

In applications to the representation theory of Kac-Moody algebras, V is the dual space to the Cartan subalgebra of the Kac-Moody algebra, B is the Killing form on it. The vectors $v_{n+1}, \ldots, v_N$ corresponding to the averaged points belong to the set of simple negative roots. The vectors $v_1, \ldots, v_n$ corresponding to the parameters of the integral are the highest weights of the representations of the Kac-Moody algebra.

3.1 Hypergeometric Construction

Assume given

(a) natural numbers $n \leq N$, $N = m + n$;

(b) a complex linear space V, symmetric bilinear form B on V and an ordered set of (weight) vectors $v_1, \ldots, v_N \in V$, not necessarily different.

From these data the construction builds a complex linear space W and a differential equation on a W-valued function $\phi(z_1, \ldots, z_n)$:

$$d\phi = \kappa^{-1} \sum_{1 \leq j < k \leq n} \Omega_{jk} \phi d(z_j - z_k)/(z_j - z_k),$$

where $\Omega_{jk} : W \to W$ are suitable linear operators, κ is a complex parameter.

The Construction. Consider the configuration of all the diagonal hyperplanes $t_j = t_k$ in $\mathbb{C}^N$. Define the weights of the diagonals: $\lambda_{jk} = B(v_j, v_k)/\kappa$.

The weights define the complex of hypergeometric forms, the one-dimensional complex local system $\mathcal{S}$ on the complement of the diagonals with the monodromy around the diagonal $t_j = t_k$ equal to the multiplication by $\exp(2\pi i \lambda_{jk})$. Hypergeometric forms have well-defined integrals over chains with coefficients in $\mathcal{S}$.

Let $\mathbb{C}^N \to \mathbb{C}^n$ be the projection on the first n coordinates. A fiber over a point z is the space $\mathbb{C}^m$ in which the diagonals cut a configuration depending on z. Denote it by $\mathcal{C}(z)$.

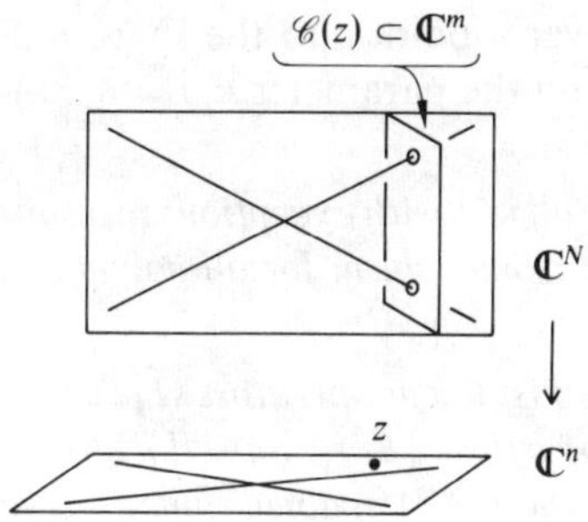

Fig. 4

A point of the base is called *discriminantal* if the configuration in the fiber over it is degenerate. The discriminantal points form the discriminant, the configuration of all diagonal hyperplanes. Over the complement of the discriminant, fibers with distinguished configurations form a locally trivial bundle. The fundamental group of the complement of the discriminant is the pure braid group on n strings.

Consider the restriction $\mathscr{S}(z)$ to the fiber $\mathbb{C}^m \backslash \mathscr{C}(z)$ over a point z of the local system $\mathscr{S}$, and the top homology group $H_m(\mathbb{C}^m \backslash \mathscr{C}(z), \mathscr{S}(z))$ of the fiber with coefficients in this restriction. The *homology bundle* is the complex vector bundle over the complement of the discriminant with fiber $H_m(\mathbb{C}^m \backslash \mathscr{C}(z), \mathscr{S}(z))$ over the point z.

The homology group $H_m(\mathbb{C}^m \backslash \mathscr{C}(z), \mathscr{S}(z))$ depends on z and is uniquely translated along paths in the base. The *Gauss-Manin connection* is this integrable connection on the homology bundle.

The *monodromy representation* of the Gauss-Manin connection is the representation of the pure braid group in the automorphism group of the homology group, induced by translations of the homology group along loops in the complement of the discriminant.

The monodromy representation gives the well-known Burau representation of the braid group in the special case of one-dimensional fiber and equal weights [K2, GiS, Gi]. As will be explained the representations of the braid groups appearing in the theory of quantum groups are closely connected with the monodromy representations of the constructed Gauss-Manin connections.

An integral of a hypergeometric form on $\mathbb{C}^N$ over any cycle in a fiber is equal to zero certainly if this form is the differential of some form or the restriction of this form on any fiber equals zero.

Accordingly we define the *hypergeometric cohomology group* as

$$\mathscr{H}^m = \Omega^m/(Z^m + d\Omega^{m-1}),$$

where Ω^m is the space of hypergeometric m-forms on $\mathbb{C}^N$, $Z^m \subset \Omega^m$ is the subspace of forms with zero restriction on any fiber, $d\Omega^{m-1} \subset \Omega^m$ is the subspace of differentials of hypergeometric $(m-1)$-forms.

The integration defines the homomorphism

$$\imath : H_m(\mathbb{C}^m \backslash \mathscr{C}(z), \mathscr{S}(z)) \to (\mathscr{H}^m)^*$$

of the homology group over a point z to the space independent of z. The homomorphism depends on z and the parameter κ, $\imath = \imath(z, \kappa)$.

Theorem [SV]. *For almost all κ^{-1} with exceptions in a suitable discrete subset of $\mathbb{C}$ the homomorphism $\imath(z, \kappa)$ is an isomorphism for all points z outside the discriminant.*

Theorem [Ao, SV]. *There exist linear operators $\Omega_{jk} : (\mathscr{H}^m)^* \to (\mathscr{H}^m)^*$, $1 \le j < k \le n$, with the following properties. For any locally constant homology class $\Delta(z) \in H_m(\mathbb{C}^m \backslash \mathscr{C}(z), \mathscr{S}(z))$ the $(\mathscr{H}^m)^*$-valued function $\phi(z) := \imath(z, \kappa)[\Delta(z)]$ satisfies the differential equation*

$$d\phi = \kappa^{-1} \sum_{1 \le j < k \le n} \Omega_{jk} \phi d(z_j - z_k)/(z_j - z_k). \tag{10}$$

In other words hypergeometric integrals satisfy the differential equation (10) and for a general κ all solutions of the equation are given by the integrals.

This picture has a symmetry group. Any permutation of the coordinates in $\mathbb{C}^N$ accordingly enumerates the weight vectors $v_1, \ldots, v_N \in V$. Any permutation of the last m coordinates preserving the ordered set of the weight vectors preserves the fibers of the projection, the weights of the diagonals, acts on the complex of hypergeometric forms, on homology groups of fibers, on solutions of the differential equation (10).

The *twisted homology group* $H_m(\mathbb{C}^m \backslash \mathscr{C}(z), \mathscr{S}(z))_{\mathrm{ant}}$ is the antisymmetric part of the group $H_n(\mathbb{C}^m \backslash \mathscr{C}(z), \mathscr{S}(z))$ with respect to the action of the permutation group of the last m coordinates preserving the ordered set of weight vectors. The *twisted hypergeometric cohomology group* $\mathscr{H}^m_{\mathrm{ant}}$ is the antisymmetric part of the hypergeometric cohomology group with respect to the action of the same group.

The result of the construction is the complex linear space $(\mathscr{H}^m_{\mathrm{ant}})^*$ and a $(\mathscr{H}^m_{\mathrm{ant}})^*$-valued differential equation (10) on $\mathbb{C}^n$.

3.2 Representations of Lie Algebras and Knizhnik-Zamolodchikov Equation

As an example consider the Lie algebra $\mathfrak{g} = \mathfrak{sl}_2(\mathbb{C})$ of complex 2×2-matrices with the zero trace. $\mathfrak{g}$ is generated by the standard generators e, f, h subject to the relations $[e, f] = h$, $[h, f] = -2f$, $[h, e] = 2e$.

Fix an invariant scalar product on $\mathfrak{g}$ (the Killing form). Let $\Omega \in \mathfrak{g} \otimes \mathfrak{g}$ be the tensor corresponding to the invariant scalar product (the Casimir operator).

Let $L_1, \ldots, L_n$ be representations of $\mathfrak{g}$, $L = L_1 \otimes \cdots \otimes L_n$. Let Ω_{jk} be the linear operator on $L_1 \otimes \cdots \otimes L_n$, acting as the Casimir operator on $L_j \otimes L_k$ and as the identity operator on the other factors. The *Knizhnik-Zamolodchikov* (KZ) *equation* on the L-valued function $\phi(z_1, \ldots, z_n)$ is the system of the differential equations

$$d\phi = \kappa^{-1} \sum_{1 \le j < k \le n} \Omega_{jk} \phi d(z_j - z_k)/(z_j - z_k),$$

where κ is a complex parameter.

The KZ equation defines the integrable connection on the trivial bundle $L \times \mathbb{C}^n$ with singularities over the diagonals. This connection has a remarkable property:

parallel translations of this connection commute with the action of $\mathfrak{g}$ on fibers. Thus the eigenspaces of the operators e, f, h are invariant under parallel translations.

It turns out that the KZ equation restricted on suitable invariant subspaces coincides with suitable hypergeometric equations constructed above.

More precisely, let $\mathfrak{h} \subset \mathfrak{g}$ be the Cartan subalgebra generated by h. The *Verma module* of $\mathfrak{g}$ with the highest weight $\Lambda \in \mathfrak{h}^*$ is the infinite dimensional representation of $\mathfrak{g}$ generated by one (vacuum) vector v with the properties $ev = 0$, $hv = \langle h, \Lambda \rangle v$. Verma modules are the simplest representations from which all finite-dimensional representations may be constructed. Let $M_1, \ldots, M_n$ be Verma modules with the highest weights $\Lambda_1, \ldots, \Lambda_n \in \mathfrak{h}^*$. Put $M = M_1 \otimes \cdots \otimes M_n$, $\Lambda = \Lambda_1 + \cdots + \Lambda_n$. M is the direct sum of the eigenspaces of the operator h, $M = \bigoplus_{m \geq 0} M_{\Lambda - m\alpha}$, where $M_{\Lambda - m\alpha}$ is the eigenspace with the eigenvalue $\langle h, \Lambda - m\alpha \rangle$, $\alpha \in \mathfrak{h}^*$ is the single positive root of $\mathfrak{sl}_2(\mathbb{C})$. The *vacuum subspace* $\mathrm{Vac}_{\Lambda - m\alpha} \subset M_{\Lambda - m\alpha}$ is the subspace of all vectors annihilated by the operator e.

Theorem [SV, DJMM]. *The vacuum subspace* $\mathrm{Vac}_{\Lambda - m\alpha}$ *and the* KZ *equation restricted on it are canonically isomorphic, respectively, to the space* $(\mathscr{H}^m_{\mathrm{ant}})^*$ *dual to the twisted hypergeometric cohomology group, and to the differential equation with values in* $(\mathscr{H}^m_{\mathrm{ant}})^*$ *constructed by the hypergeometric construction from the projection* $\mathbb{C}^{m+n} \to \mathbb{C}^n$, *the linear space* $\mathfrak{h}^*$, *the Killing form on* $\mathfrak{h}^*$ *and the ordered set of the vectors* $\Lambda_1, \ldots, \Lambda_n, \underbrace{-\alpha, \ldots, -\alpha}_{m} \in \mathfrak{h}^*$.

Corollary. *For a general* κ *all solutions of the* KZ *equation with values in a tensor product of Verma modules are given by the hypergeometric functions.*

An analogous picture has a place if the algebra $\mathfrak{sl}_2(\mathbb{C})$ is replaced by any Kac-Moody algebra, see [SV, DJMM, Ma, Ch, CF, L].

This theorem shows that the KZ equation has topological nature, it is just the Gauss-Manin connection of the simple fiber bundle.

3.3 Hypergeometric Functions in the Conformal Field Theory

The KZ equation was invented in the CFT. Its solutions describe $(n + 1)$-point correlation functions on the Riemann sphere in the Wess-Zumino-Witten model [BPZ, KZ]. In the minimal models of the CFT correlation functions on the sphere have also integral representations in terms of the same configurations [DF]. For the first time integral representations of correlation functions in the CFT appeared in the works by Dotsenko and Fateev.

Any model of the CFT have a certain set of primary fields $\{\phi_n(z, \bar{z})\}$ and an operator algebra

$$\phi_n(z, \bar{z})\phi_m(0, 0) = \sum_p C^p_{nm} z^{\Delta_p - \Delta_m - \Delta_n} \bar{z}^{\bar{\Delta}_p - \bar{\Delta}_m - \bar{\Delta}_n} \phi_p(0, 0) + \cdots,$$

where the numbers $\Delta_n, \bar{\Delta}_n$ are the conformal dimensions of the field ϕ_n, the numbers C^p_{nm} are the structure constants of the model, in this expression the terms containing non-primary fields are omitted, see [BPZ]. The integral representations of the correlation functions allowed one to calculate the structure constants of the minimal models of the CFT and of the Wess-Zumino-Witten model corresponding to $\mathfrak{sl}_2(\mathbb{C})$ [DF, Do]. It turns out that in these cases the structure constants are equal to certain alternating products of values of the gamma-function similar to that appearing in the determinant formula of Sect. 2.3. In these cases any structure constant was represented by a certain hypergeometric integral. Probably in other models of the CFT the structure constants are connected with suitable determinants of hypergeometric integrals which in their turn may be expressed as alternating products of values of the gamma-function.

3.4 Hypergeometric Functions and Quantum Groups

The hypergeometric integrals as functions of the parameters satisfy the KZ equation. The monodromy of the KZ equation is described in terms of quantum groups.

The *monodromy representation* of the KZ equation is the representation of the pure braid group on n strings in the group $\operatorname{Aut}(L_1 \otimes \cdots \otimes L_n)$ generated by analytic continuation of solutions along loops in the base. Denote it by τ_κ. Algebraically we define another representation of the pure braid group.

According to Faddeev, Kulish, Reshetikhin, Sklynin, Jimbo, Drinfeld [Dr1, J] the universal enveloping algebra $U\mathfrak{g}$ of the algebra $\mathfrak{g} = \mathfrak{sl}_2(\mathbb{C})$ is deformed to the quantum universal enveloping algebra $U_q\mathfrak{g}$ (the quantum group) depending on the complex parameter q.

For the quantum group a comultiplication $\Delta : U_q\mathfrak{g} \to U_q\mathfrak{g} \otimes U_q\mathfrak{g}$ is defined. If V_1, V_2 are representations of the quantum group then the comultiplication induces a representation structure on their tensor product. The representations $V_1 \otimes V_2$ and $V_2 \otimes V_1$ are isomorphic. The isomorphism is defined by the formula

$$V_1 \otimes V_2 \xrightarrow{R} V_1 \otimes V_2 \xrightarrow{P} V_2 \otimes V_1,$$

where P is the transposition of the factors and $R \in U_q\mathfrak{g} \otimes U_q\mathfrak{g}$ is the distinguished element called the *universal R-matrix* of the quantum group.

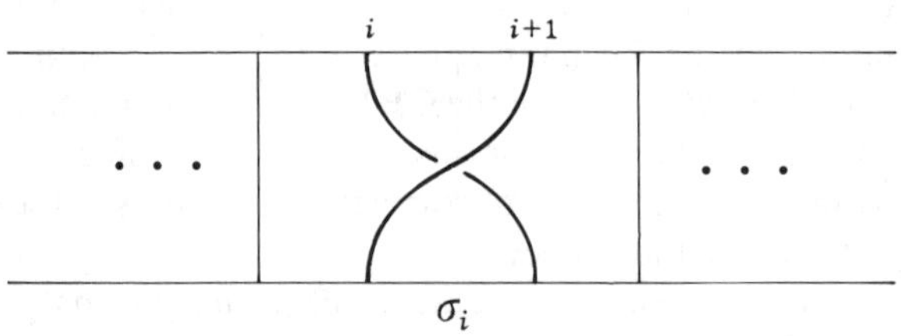

Fig. 5

Let $V_1, \ldots, V_n$ be representations of the quantum group. Then the pure braid group acts on their tensor product. Namely let $\sigma_1, \ldots, \sigma_{n-1}$ be the elementary braids shown in Fig. 5. To any braid σ_i assign the linear operator

$$V_1 \otimes \cdots \otimes V_i \otimes V_{i+1} \otimes \cdots \otimes V_n \to V_1 \otimes \cdots \otimes V_{i+1} \otimes V_i \otimes \cdots \otimes V_n$$

acting as PR on the i-th and $(i+1)$-th factors and as the identity operator on the other factors. These operators define the representation ϱ_q of the pure braid group on n strings in $\mathrm{Aut}(V_1 \otimes \cdots \otimes V_n)$.

For general values of q the representation theory of the quantum group $U_q\mathfrak{g}$ is the same as the representation theory of the algebra $\mathfrak{g}$, any representation of the algebra $\mathfrak{g}$ is deformed canonically to the representation of the quantum group. Let $L_1, \ldots, L_n$ be representations of $\mathfrak{g}$ and $L_{1q}, \ldots, L_{n,q}$ be their quantum deformation.

Theorem [K, Dr2]. *The monodromy representation τ_κ of the KZ equation with values in $\mathrm{Aut}(L_1 \otimes \cdots \otimes L_n)$ is equivalent to the R-matrix representation ϱ_q, $q = \exp(2\pi i/\kappa)$, with values in $\mathrm{Aut}(L_{1,q} \otimes \cdots \otimes L_{n,q})$, if κ is not a rational number.*

An analogous statement holds if $\mathfrak{sl}_2(\mathbb{C})$ is replaced by any Kac-Moody algebra.

This theorem is a wonderful statement. Given the differential equation of the CFT described in terms of the representation theory of the Lie algebra one considers its global characteristic, the monodromy representation defined by analytic continuation of solutions and one gets the representation defined by R-matrix of the quantum group which in its turn is defined as the deformation of the universal enveloping algebra of the initial Lie algebra.

It is interesting that new invariants of knots defined by Jones and others are constructed in terms of the same representations of the braid groups, see [C2, Jo, L, RT].

The realization of the KZ equation as the Gauss-Manin connection allows one to make explicit the equivalence of the representations τ_κ and ϱ_q, $q = \exp(2\pi i/\kappa)$. It turns out that for real $z_1 < \cdots < z_n$ there is a canonical isomorphism of the quantum deformation of the vacuum subspace to the twisted homology group:

$$\mathrm{Vac}_{\Lambda - m\alpha, q} \to H_m(\mathbb{C}^m \setminus \mathscr{C}(z), \mathscr{S}(z))_{\mathrm{ant}},$$

which sends the R-matrix representation to the monodromy representation of the Gauss-Manin connection. The equivalence of the Kohno-Drinfeld theorem is the composition

$$\mathrm{Vac}_{\Lambda - n\alpha, q} \to H_m(\mathbb{C}^m \setminus \mathscr{C}(z), \mathscr{S}(z))_{\mathrm{ant}} \xrightarrow{\iota} (\mathscr{H}^m_{\mathrm{ant}})^* \to \mathrm{Vac}_{\Lambda - n\alpha}.$$

Thus a correspondence between a representation of Lie algebra and its quantum deformation is the correspondence between the twisted homology group and the dual space to the twisted hypergeometric cohomology group given by the integration of the hypergeometric forms.

The details of the proof of this statement and the analogous statement for arbitrary Kac-Moody algebra are being checked in [SV4]. On this subject see also [L].

4. Concluding Remarks

Many interesting works on hypergeometric functions are not mentioned in this talk. Among them there are works by Deligne and Mostow, Beukers and Heckman on monodromy of hypergeometric functions, works by Gelfand, Kapranov, Zelevinsky on combinatorial description of differential equations on hypergeometric functions, works on q-analogs of hypergeometric functions, on difference equations on hypergeometric functions, works by Heckman and Opdam on hypergeometric functions and root systems.

The main problem is to unify numerous and remote investigations on hypergeometric functions into a united theory in which it might be possible to pass from Conformal Field Theory to Algebraic K-Theory by analytic continuation on parameters.

References

[A1] Aomoto, K.: Un théorème du type de M-M concernant l'integrale des fonctions multiformes. J. Math. Pures Appl. **52** (1973) 1–11

[A2] Aomoto, K.: On vanishing of cohomology attached to certain many valued meromorphic functions. J. Math. Soc. Japan **27** (1975) 248–255

[A3] Aomoto, K.: On the structure of integrals of power product of linear functions. Sc. Papers of the College of General Education. Tokyo Univ. **27** (1977) 49–61

[A4] Aomoto, K.: Addition theorem of Abel type for hyper-logarithms. Nagoya Math. J. **88** (1982) 55–71

[A5] Aomoto, K.: Configurations and invariant theory of Gauss-Manin systems. Adv. Stud. Pure Math. **4** (1984) 165–179

[A6] Aomoto, K.: Jacobi polynomials associated with Selberg integrals. SIAM J. Math. Anal. **18** (1987) 545–549

[A7] Aomoto, K.: On the complex Selberg integral. Quart. J. Math. Oxford **38** (1987) 385–399

[A8] Aomoto, K.: Special value of the hypergeometric function ${}_3F_2$ and connection formulae among asymptotic expansions. J. Indian Math. Soc. **51** (1987) 161–221

[A9] Aomoto, K.: Gauss-Manin connection of integral of difference products. J. Math. Soc. Japan **39** (1987) 191–208

[A10] Aomoto, K.: Special values of hyperlogarithms and linear difference schemes. Preprint 1988, pp. 1–47

[A11] Aomoto, K.: Finiteness of a cohomology associated with certain Jackson integrals. Preprint 1989, pp. 1–48

[Ar1] Arnold, V.I.: The cohomology ring of the colored braid group. Math. Notes **5** (1969) 138–140

[Ar2] Arnold, V.I.: Topological invariants of algebraic functions. II. Funct. analys i ego pril. **4** (1970) 1–9

[As1] Askey, R.S.: The q-gamma and q-beta functions. Appl. Anal. **8** (1978) 125–141

[As2] Askey, R.S.: Some basic hypergeometric extensions of integrals of Selberg and Andrews. SIAM J. Math. **11** (1980) 938–951

[Bei] Beilinson, A.: Polylogarithms and cyclotomic elements. Preprint MIT 1990, pp. 1–30

[BGSV1] Beilinson, A., Goncharov, A., Schechtman, V., Varchenko, A.: Aomoto dilogarithms, mixed Hodge structures and motivic cohomology of pairs of triangles on the plane. The Grothendieck Festschrift, vol. 1. Birkhäuser, 1990, pp. 135–172

[BGSV2] Beilinson, A., Goncharov, A., Schechtman, V., Varchenko, A.: Projective Geometry and K-theory. Algebra Anal. **2** (1990) n°3, 78–131

[BMS] Beilinson, A., MacPherson, R., Schechtman, V.: Notes on motivic cohomology. Duke Math. J. **54** (1987) 679–710

[BPZ] Belavin, A., Polyakov, A., Zamolodchokov, A.: Infinite conformal symmetry in two-dimensional quantum field theory. Nucl. Phys. B241 (1984) 333

[BeH] Beukers, F., Heckman, G.: Monodromy for hypergeometric functions ${}_nF_{n-1}$. Invent. math. **95** (1989) 325–354

[B1] Bloch, S.: Higher regulators, algebraic K-theory, and zeta functions of elliptic curves. Lecture Notes. U.C. Irvine, 1977

[B2] Bloch, S.: Applications of the dilogarithm function in algebraic K-theory and algebraic geometry. Proc. of the Intern. Symp. Alg. Geometry. Tokyo, 1978

[Bo] Borel, A.: Commensurability classes and volumes of hyperbolic 3-manifolds. Ann. Sc. Norm. Sup. Pisa **8** (1981) 1–33

[BrGM] Brasselet, J., Goresky, M., MacPherson, R.: Simplicial differential forms with poles. Preprint 1988, pp. 1–40

[Bri] Brieskorn, E.: Sur les groupes de tresses. Séminaire Bourbaki (Lecture Notes in Mathematics, vol. 317.) Springer, Berlin Heidelberg New York, pp. 21–44

[C1] Cartier, P.: Decomposition des polyhèdres: le point sur le troisieme problème de Hilbert. Séminaire Bourbaki, 1985–86, n. 646

[C2] Cartier, P.: Developpements recents sur les groupes de tresses. Applications à la topologie et a l'algèbre. Séminaire Bourbaki, 1989–90, n. 716

[Ch1] Cherednik, V.: Monodromy representations for generalized Knizhnik-Zamolodchikov equation. Preprint ITP-89-74E, Kiev 1990, pp. 1–21

[Ch2] Cherednik, V.: Integral solutions of trigonometric Knizhnik-Zamolodchikov equations and Kac-Moody algebras. Preprint RIMS 1990, pp. 1–31

[CF] Christe, P., Flume, R.: The four-point correlations of all primary operators of the $d = 2$ conformally invariant SU(2) σ-model with Wess-Zumino term. Nucl. Phys. B282 (1987) 466–494

[CW1] Cohen, P., Wolfart, J.: Monodromie et plongement modulaire. Preprint 1989, pp. 1–10

[CW2] Cohen, P., Wolfart, J.: Monodromie des fonctions d'Appell, varietes abéliennes et plongement modulaire. Preprint 1989, pp. 1–6

[DJMM] Date, E., Jimbo, M., Matsuo, A., Miwa, T.: Hypergeometric-type integrals and the sl(2, $\mathbb{C}$) Knizhnik-Zamolodchikov equation. Preprint RIMS (1989) 1–12

[D1] Dehn, M.: Ueber raumgleiche Polyeder. Nachr. Akad. Wiss., Göttingen, Math.-Phys. Kl. 1900, pp. 345–354

[D2] Dehn, M.: Uber den Rauminhalt. Math. Ann. **55** (1902) 465–478

[De1] Deligne, P.: Théorie de Hodge. I. Proc. Int. Cong. Nice 1970, vol. 1, 425–430; II. – Publ. Math. IHES **40** (1971) 5–58; III. – Publ. Math. IHES **44** (1972) 5–77

[De2] Deligne, P.: Equations différentielles à points singuliers réguliers. (Lecture Notes in Mathematics, vol. 163.) Springer, Berlin Heidelberg New York 1970

[DeM] Deligne, P., Mostow, G.D.: Monodromy of hypergeometric functions and non-lattice integral monodromy. Publ. Math. IHES **63** (1986) 5–89

[Do] Dotsenko, V.: Solving the SU(2) conformal field theory with the Wakimoto free field representation. Moscow (1990) 1–31

[DF1] Dotsenko, V., Fateev, V.: Conformal algebra and multipoint correlation functions in 2D statistical models. Nucl. Phys. B240 [FS12] (1984) 312–348

[DF2] Dotsenko, V., Fateev, V.: Four-point correlation functions and the operator algebra in 2D conformal invariant theories with central charge $C \leq 1$. Nucl. Phys. **B251** [FS13] (1985) 691–734

[Dr1] Drinfeld, V.: Quantum groups. Proc. Int. Cong. Math. Berkeley 1986, vol. 1, pp. 798–820

[Dr2] Drinfeld, V.: Quasi-Hopf algebras. Algebra Anal. **1** (1989) 30–46

[Du] Dupont, J.: Algebra of polytopes and homology of flag complexes. Osaka J. Math. **19** (1982) 599–641

[Du2] Dupont, J.: On polylogarithms. Preprint (1987) 1–29

[DuPS] Dupont, J., Parry, W., Sah, C.-H.: Homology of classical Lie groups made discrete, II. J. Algebra **113** (1988) 215–260

[DuS1] Dupont, J., Sah, C.-H.: Scissors congruences, II. J. Pure and Appl. Alg. **25** (1982) 159–195

[DuS1] Dupont, J., Sah, C.-H.: Homology of Euclidean Motion groups made discrete and Euclidean scissors congruences. Preprint 1989, pp. 1–39

[FSV] Feigin, B., Schechtman, V., Varchenko, A.: On algebraic equations satisfied by correlators in WZW model. Letters in Math. Phys. **20**, 291–297 (1990)

[G] Gelfand, I.M.: General theory of hypergeometric functions. Dokl. Akad. Nauk. SSSR **288** (1986) 573–576

[GG] Gelfand, I.M., Graev, M.I.: Duality theorem for general hypergeometric functions. Dokl. Akad. Nauk SSSR **289** (1986) 19–23

[GGZ] Gelfand, I.M., Graev, M.I., Zelevinsky, A.V.: Holonomic systems of equations and series of hypergeometric type. Dokl. Akad. Nauk SSSR **295** (1988) 14–19

[GKZ1] Gelfand, I.M., Kapranov M.M., Zelevinsky, A.V.: Projective dual varieties and hyperdeterminants. Dokl. Akad. Nauk SSSR **305** (1989) 1294–1298

[GKZ2] Gelfand, I.M., Kapranov M.M., Zelevinsky, A.V.: A-discriminants and Cayley-Koszul complexes. Dokl. Akad. Nauk SSSR **307** (1989) 1307–1311

[GKZ3] Gelfand, I.M., Kapranov M.M., Zelevinsky, A.V.: Newton polyhedra of principal A-determinants. Dokl. Akad. Nauk SSSR **308** (1989) 20–23

[GKZ4] Gelfand, I.M., Kapranov M.M., Zelevinsky, A.V.: Hypergeometric functions and toric varieties. Funct. analys i ego pril. **23** (1989) 12–26

[GM] Gelfand, I.M., MacPherson, R.: Geometry in grassmanians and a generalization of the dilogarithm. Adv. Math. **44** (1982) 279–312

[GVZ] Gelfand, I.M., Vasiliev, V.A., Zelevinsky, A.V.: General hypergeometric functions on complex grassmanians. Funct. analys i ego pril. **21** (1987) 23–38

[GV] Gelfand, I.M., Varchenko, A.N.: On Heaviside functions of configuration of hyperplanes. Funct. analys i ego pril. **21** (1987) 1–18

[GZ] Gelfand, I.M., Zelevinsky, A.V.: Algebraic and combinatoric aspects of general theory of hypergeometric functions. Funct. analys i ego pril. **20** (1986) 17–34

[Gi] Givental, A.: Twisted Picard-Lefschetz formulas. Funct. analys i ego pril. **22** (1988), 1, 12–22

[GiS] Givental, A., Schechtman, V.: Monodromy groups and Hecke algebras. Usp. Mat. Nauk **42** (1987), 6, 138

[Go] Goncharov, A.: The classical threelogarithm, algebraic theory of fields and Dedekind zeta functions. Moscow 1990, pp. 1–9

[GorM] Goresky, M., MacPherson, R.: Stratified morse theory. Springer, Berlin Heidelberg New York 1988

[H] Heckman, G.: Root systems and hypergeometric functions II. Comp. Math. **64** (1987) 353–373

[HO] Heckman, G., Opdam, E.: Root systems and hypergeometric functions I. Comp. Math. **64** (1987) 329–352

[HaM] Hain, R., MacPherson, R.: Higher logarithms. Seattle 1990, pp. 1–84

[J] Jessen, B.: The algebra of polyhedra and the Dehn-Sydler theorem. Math. Scand. **22** (1968) 241–256

[Ji] Jimbo, M.: Lett. Math. Phys. **10** (1985) 63–69

[Jo] Jones, V.F.R.: Hecke algebra representations of braid groups and link polynomials. Ann. Math. **126** (1987) 335–388

[KZ] Knizhnik, V., Zamolodchikov, A.: Current algebras and Wess-Zumino models in two dimensions. Nucl. Phys. **B247** (1984) 83–103

[K1] Kohno, T.: Monodromy representations of braid groups and Yang-Baxter equations. Ann. Inst. Fourier **37** (1987) 139–160

[K2] Kohno, T.: Linear representations of braid groups and classical Yang-Baxter equations. Contemp. Math. **78** (1988) 339–363

[K3] Kohno, T.: Integrable connections related to Manin and Schechtman's higher braid groups. Preprint (1989) 1–13

[L1] Lawrence, R.: A topological approach to the representations of the Iwahori-Hecke algebra. International J. Modern Physics A **5** (1990) No 16

[L2] Lawrence, R.: The Homological Approach to Higher Representations. Preprint 1990, pp. 1–9

[L3] Lawrence, R.: A Functorial Approach to the One-Variable Jones Polynomial. Preprint 1990, pp. 1–15

[Le] Lewin, L.: Dilogarithms and associated functions. MacDonald, London, 1958

[Lo] Loeser, F.: Arrangements d'hyperplans et sommes de Gauss Preprint 1989, pp. 1–31

[M] Manin, Yu. I.: Correspondences, motives and monoidal transformations. Mat. Sb. (N.S.) **119** (1968) 475–507

[MaSY] Matsumoto, K., Sasaki, T., Yoshida, M.: The period map of a 4-parameter family of K3 surfaces and the Aomoto-Gelfand hypergeometric function of type (3, 6). Proc. Japan Acad. **64** (1988) 307–310

[Ma] Matsuo, A.: An application of Aomoto-Gelfand hypergeometric functions to the SU(n) Knizhnik-Zamolodchikov equation. Preprint RIMS 1990, pp. 1–18

[Mi1] Milne, S.: A q-analog of the Gauss summation theorem for hypergeometric series on $U(n)$. Adv. Math. **72** (1988) 59–131

[Mil] Milne, S.: A q-analog of hypergeometric series well-poised in SU(n) and invariant G-functions. Adv. Math. **58** (1985) 1–60

[Mo] Mostow, G.D.: Generalized Picard lattices arising from half-integral conditions. Publ. Math. IHES **63** (1986) 91–106

[Mo] Mostow, G.D.: Braids, hypergeometric functions, and lattices. Bull. Am. Math. Soc. **16** (1987) 2, 225–246

[N] Nishiyama, S.: Appell's hypergeometric function F_2 and periods of certain elliptic K3 surfaces. Tokyo J. Math. **10** (1987) 33–67

[O1] Opdam, E.: Root systems and hypergeometric functions III. Comp. Math. **67** (1988) 21–49

[O2] Opdam, E.: Root systems and hypergeometric functions IV. Comp. Math. **67** (1988) 191–209

[O3] Opdam, E.: Some applications of hypergeometric shift operators. Invent. math. **98** (1988) 1–18

[Or] Orlik, P.: Introduction to arrangements. Preprint 1988, pp. 1–127

[OS] Orlik, P., Solomon, L.: Combinatorics and topology of complements of hyperplanes. Invent. math. **56** (1980) 167–189

[Q] Quillen, D.: Higher K-theory I. (Lecture Notes in Mathematics, vol. 341) Springer, Berlin Heidelberg New York, pp. 85–147

[RT] Reshetikhin, N., Turaev, V.: Invariants of 3-manifolds via link polynomial and quantum groups. Invent. math. **103** (1991)

[R] Rogers, L.: On function sum theorems connected with the series $\sum_1 (x^n/n^2)$. Proc. London Math. Soc. **4** (1907) 169–189

[S1] Sah, C.-H.: Homology of classical Lie groups made discrete, I. Comment. Math. Helv. **61** (1986) 308–347

[S2] Sah, C.-H.: Homology of classical Lie groups made discrete, III. J. Pure Appl. Algebra (to appear)

[Se] Selberg, A.: Bemerkninger om et multiplet integral. Norsk Mat. Tidsskr. **26** (1944) 71–78

[SV1] Schechtman, V., Varchenko, A.: Integral representations of N-point conformal correlators in the WZW model. Bonn 1989, pp. 1–21

[SV2] Schechtman, V., Varchenko, A.: Hypergeometric solutions of Knizhnik-Zamolodchikov equations. Letters in Math. Phys. **20** (1990) 279–283

[SV3] Schechtman, V., Varchenko, A.: Arrangements of hyperplanes and Lie algebras homology. Moscow 1990, pp. 1–130

[SV4] Schechtman, V., Varchenko, A.: Quantum groups and homology local systems. Preprint 1990, pp. 1–30

[Su] Suslin, A.: Algebraic K-theory of fields. Proc. of the Int. Congr. Math., Berkeley, 1986, pp. 222–244

[Sy] Sydler, J.-M.: Conditions necessaires et suffissantes pour l'équivalence des polyèdres de l'espace euclidien à trois dimensions. Comment. Math. Helv. **40** (1965) 43–80

[T] Terada, T.: Fonctions hypergeometriques F_1 et fonctions automorphes. – I. – J. Math. Soc. Japan **35** (1983) 451–475. II. – Math. Soc. Japan **37** (1985) 173–185

[TK] Tsuchiya, A., Kanie, Y.: Vertex operators in conformal field theory on $\mathbb{P}^1$ and monodromy representations of braid groups. Adv. Stud. Pure Math. **16** (1988) 297–372

[V1] Varchenko, A.: Euler beta-function, Vandermond determinant, Legendre equation and critical values of linear functions on configuration of hyperplanes. I. Izv. Akad. Nauk SSSR (1989) 1206–1235. II. Izv. Akad. Nauk SSSR (1990) 146–158

[V2] Varchenko, A.: Critical values and determinant of periods. Uspehi Mat. Nauk (1989) 235–236

[V3] Varchenko, A.: Hodge filtration of hypergeometric integrals associated with an affine configuration of general position and a local Torelli theorem. Preprint (1989) 1–13

[V4] Varchenko, A.: Determinant formula for Selberg type integrals. Preprint (1990) 1–3

[Z1] Zagier, D.: The remarkable dilogarithm. Bonn 1987, pp. 1–15

[Z1] Zagier, D.: The Bloch-Wigner-Ramakrishnan polylogarithm function. Bonn 1989, pp. 1–10

Invited Forty-Five Minute Addresses at the Section Meetings

Degree Structures

Theodore A. Slaman

Mathematical Sciences Research Institute
1000 Centennial Drive, Berkeley, CA 94720, USA

The Turing degrees $\mathscr{D}$ were introduced by Kleene and Post ([9], 1954) to isolate and study those properties of the subsets of the natural numbers $\mathbb{N}$ which are expressed purely in terms of relative computability. Intuitively, we form $\mathscr{D}$ by identifying any pair of subsets of $\mathbb{N}$ which are mutually computable and ordering the resulting equivalence classes by relative computability. The natural hierarchies of definability within arithmetic, analysis and higher fragments of set theory all have sharply focused images in $\mathscr{D}$.

We will focus our attention on the second order properties of the Turing degrees. We will phrase our discussion of the known results and especially of the techniques of their proofs within as large a context as possible, so to apply to degree structures based on other forms of relative definability, as well as to $\mathscr{D}$.

Formally, suppose that A and B are subsets of $\mathbb{N}$, henceforth called *reals*. We say that A is *Turing reducible* to B ($A \leq_T B$) if there is a computational procedure which takes an input n from $\mathbb{N}$; over the course of its execution on input n, asks whether various numbers are in B; and, if it receives the correct responses to those questions, after finitely many computational steps returns the answer as to whether n is an element of A. In other words, if we were given B then we would be able to compute A. We say that A and B are Turing equivalent ($A \equiv_T B$) if $A \leq_T B$ and $B \leq_T A$. The Turing degrees D are the $\equiv_T$-equivalence classes. The degree structure $\mathscr{D}$ associated with Turing reducibility is the partial ordering $\langle D, \leq_T \rangle$, the Turing degrees with the ordering inherited from $\leq_T$.

$\mathscr{D}$ is the most intensively studied degree structure but not the only one. Although we cannot introduce them all in detail, we will point out a few other examples. Among the many degree structures on the real numbers obtained by varying the notion of relative definability, we might mention the arithmetic degrees, the hyperarithmetic degrees, the Σ_k-degrees ($A \leq_{\Sigma_k} B$ if A is in the least set which is Σ_k-admissible relative to B), the many-one degrees, the enumeration degrees, the Δ^1_2-degrees and the degrees of constructibility. Alternately, we can retain relative computability and vary the class of reals on which it acts. We can form the Turing degrees of the recursively enumerable sets, the Turing degrees

The author was partially supported by N.S.F. grant DMS-8902437 and Presidential Young Investigator Award DMS-8451748.

Proceedings of the International Congress
of Mathematicians, Kyoto, Japan, 1990

of the Δ_2^0-sets (i.e. the structure $\mathscr{D}(\leq_T 0')$ on the degrees below $0'$) or the Turing degrees of the arithmetically definable sets. We will refer to these examples as *local substructures* of $\mathscr{D}$. For any degree x, we can also look at $\mathscr{D}({}_T\!\geq x)$, the partial order of the degrees greater than or equal to x.

Extending our scope to include degree notions on other than sets of integers, we can also mention the Kleene degrees of sets of reals, the degrees of sets of reals modulo ${}^3E$, and the α-degrees of subsets of an admissible ordinal α. Similarly, we can form their local substructures such as α-recursively enumerable α-degrees or the Kleene degrees of the Π_1^1-subsets of the continuum.

To some extent, in our discussion of these structures, we will follow the historical order in which their properties where discovered. We will focus our discussion on the Turing degrees since historically $\mathscr{D}$ has been the proving ground for the analysis of degree structures. To begin with the obvious: there is a least degree consisting of the computable reals; since there are only countably many computational procedures, each Turing degree is countable and has only countably many predecessors, any two degrees x and y have a least upper bound $x \vee y$ (so $\mathscr{D}$ is an upper-semi-lattice with a natural operation of join) and the cardinality of the whole structure is the same as the continuum. To continue beyond these immediate observations we must directly analyze the notion of relative computability.

In the first section, we will discuss the early results about the Turing degrees. For the most part, these were to the effect that $\mathscr{D}$ has a rich existential theory, showing that $\mathscr{D}$ plays various roles as a universal object. In the second section, we will describe the subsequent results which simultaneously limit $\mathscr{D}$'s universal role and reveal its detailed structure. In the final sections, we will discuss some recent results and conjectures of Slaman and Woodin. If true, the conjectured structural theorem would provide a complete logical characterization of $\mathscr{D}$. We prove the conjecture for the hyperarithmetic degrees and for $\mathscr{D}$ with finitely many additional parameters.

§1. The Embedding Theorems of the Late 1950s, 1960s and Early 1970s

Partial Order Embeddings

In their 1954 paper [9], Kleene and Post used a Baire category argument, one involving a primitive form of Cohen forcing, to show that there are sets of incomparable Turing degree. In fact, Kleene and Post proved that every finite partial order can be embedded in $\mathscr{D}$. Sacks ([18], 1963) extended the Kleene-Post embedding theorem to show that every countable partial ordering, and even every one of size $\aleph_1$ which is locally countable, could be embedded in $\mathscr{D}$.

Sacks conjectured:

(1) (Sacks ibid., 1963) **Conjecture:** A partially ordered set P is imbeddable in the (Turing) degrees if and only if P has cardinality at most that of the continuum and each member of P has at most countably many predecessors.

The Kleene-Post and Sacks partial order embedding results indicated a universal quality of $\mathscr{D}$. Sacks expressed a belief in this quality by conjecturing that the strongest possible purely existential property would be satisfied by $\mathscr{D}$. (Surprisingly, this conjecture is still open.) Furthermore, the algebraic properties of $\mathscr{D}$ seemed to be based more in the nature of relative definability than in specific properties of computability. The proofs of the embedding theorems apply to any of the above mentioned degree notions on the reals which are countably based, i.e. excluding only the Δ^1_2-degrees and the degrees of constructibility.

Initial Segments

Say that a subset I of $\mathscr{D}$ is an *ideal* if it is an initial segment and closed under join. I is a *principal ideal* if in addition it has a greatest element. Spector ([23], 1956) answered a question of Kleene-Post [9] by constructing a two element ideal, i.e. a minimal nontrivial degree. Spector's theorem rules out the potential classification of the principal ideals in $\mathscr{D}$ (as the countable model of an $\aleph_0$-categorical theory). For example, there are two nontrivial ideals which are not isomorphic. Spector's method could be extended to build many examples of isomorphism types of initial segments in $\mathscr{D}$. Prompted by Spector's results, Sacks made the following conjecture.

(2) (Sacks [19], 1966) **Conjecture:** S is a finite, initial segment of the degrees if and only if S is order isomorphic to a finite, initial segment of some upper-semi-lattice with a least member.

Sacks's conjecture postulated that another universal property would hold of $\mathscr{D}$: the initial segments of $\mathscr{D}$ would realize all the finite possibilities. Extending work of Thomason and Lachlan, Lerman ([12], 1971) confirmed this conjecture. Lachlan and Lebeuf ([11], 1976) showed that every initial segment of a countable upper-semi-lattice with least element is isomorphic to an initial segment of $\mathscr{D}$. Ultimately, Abraham and Shore ([1], 1986) showed that every initial segment of an upper-semi-lattice which is locally countable and of cardinality $\aleph_1$ is isomorphic to an initial segment of $\mathscr{D}$. Groszek and Slaman ([4], 1983) showed that the Abraham-Shore theorem is best possible; it is independent of $ZFC + 2^\omega > \aleph_1$ whether every initial segment of an upper-semi-lattice which is locally countable and of cardinality $\aleph_2$ can be embedded in $\mathscr{D}$ as an upper-semi-lattice.

The proofs of these theorems, while more technically demanding than the proofs of the embedding theorems, could still be applied to a wide range of countably based degree structures. The enumeration degrees are the only notable exception.

From the progress on the Sacks conjectures, degree theorists speculated whether $\mathscr{D}$ might have an algebraic characterization, or at least occupy a distinguished position among the upper-semi-lattices.

Questions from the 1960s

Workers in the field faced several fundamental questions. We pose them here for $\mathscr{D}$, but they are completely general. Is there a global structure theory for $\mathscr{D}$. Is the theory of $\mathscr{D}$ specific to Turing reducibility or is it applicable to a general class of degree structures? Is the structure of $\mathscr{D}$ tied to the continuum or is it reflected in the local substructures of $\mathscr{D}$?

We recall some specific questions from that time.

(3) (Sacks [19], 1966) Is the theory of $\mathscr{D}$ decidable?
(4) (Sacks ibid., 1966) Are the Turing degrees and the Turing degrees of the arithmetic sets ($\mathscr{A}$) elementarily equivalent?
(5) (Rogers [17], 1967) For a degree a, let $\mathscr{D}(\geq_T a)$ denote the restriction of $\mathscr{D}$ to those degrees above a. Is it the case that for all a and b, $\mathscr{D}(\geq_T a) \xrightarrow{\sim} \mathscr{D}(\geq_T b)$?
(6) (Rogers ibid., 1967) Is there a nontrivial automorphism of $\mathscr{D}$? (If not then we say that $\mathscr{D}$ is *rigid*.)

Rogers defined a relation R on degrees to be *absolutely definable in* $\mathscr{D}$ if R is invariant under all automorphisms of $\mathscr{D}$.

(7) (Rogers ibid., 1967) Are the Turing jump and the relation *recursively enumerable in* absolutely definable in $\mathscr{D}$? In general, which relations are absolutely definable in $\mathscr{D}$?

§2. Coding and Definability Theorems of the 1970s and Early 1980s

The results of the 1970s to the mid 1980s ruled out any reasonable understanding of the Turing degrees in algebraic terms. But then, the exact properties of $\mathscr{D}$ that make it algebraically intractable were used during this period to settle almost all of the Sacks and Rogers questions. Their solutions illustrate the complexity of $\mathscr{D}$: $\mathscr{D}$ is not decidable; the theory of $\mathscr{D}$ is not equal to the theory of the Turing degrees of the arithmetic sets; there are a and b such that $\mathscr{D}({}_T\!\geq a)$ and $\mathscr{D}({}_T\!\geq b)$ are not isomorphic. Rogers's question whether the jump is definable was only recently settled (the jump is definable), but the techniques in its solution were steadily developed through this period. The only questions that remain open are whether $\mathscr{D}$ is rigid and to give a classification of which relations are definable in $\mathscr{D}$.

The trend is for the global properties of degree structures to follow those of the Turing degrees. The notable exception has been the many-one degrees. There, Ershov ([3], 1975) and Paliutin ([16], 1975) succeeded in obtaining an algebraic characterization for the partial ordering of the many-one degrees. The reader might also see Odifreddi ([15], 1989).

A Cone of Minimal Covers

One of the earlier questions to fall was whether $\mathscr{D}$ is elementarily equivalent to $\mathscr{A}$. Jockusch and Soare ([8], 1970) showed that there is no arithmetic degree x such that every y above x is a minimal cover of some z less than y. Jockusch ([5], 1973) showed that there is a nonarithmetic degree with the above property. Thus, the Turing degrees of the arithmetic sets are not elementarily equivalent to $\mathscr{D}$.

Coding in $\mathscr{D}$ and Undecidability

A primary ingredient in the work during this period was the proving and exploiting of *coding theorems.*

Definition. Suppose that $\mathfrak{A}$ is a model of the finite language $\mathscr{L}$ and $\vec{p}$ is a finite sequence of parameters from $\mathscr{D}$. $\mathfrak{A}$ is *coded by* $\vec{p}$ *in* $\mathscr{D}$ if there is an isomorphic image of $\mathfrak{A}$ whose universe, relations, functions, constants and quantifiers are all first order definable in the language of $\mathscr{D}$ with additional symbols for the parameters $\vec{p}$.

A disparate sequence of coding schemes preceded the one which we have isolated below, see Simpson ([20], 1977) or Nerode-Shore ([13], 1979). Say that *a relation R is countable* if there is a countable subset of the degrees such that all of the solutions to R come from that set.

Coding Lemma (Slaman-Woodin [21], 1986). *For any countable relation R on degrees there are parameters* $\vec{p}$ *such that R is definable in* $\mathscr{D}$ *from* $\vec{p}$.

The coding lemma is uniform in the following sense. For each n, there is a fixed first order formula φ such that for every countable n-ary relation R on $\mathscr{D}$ there is a sequence $\vec{p}$ so that R is defined from $\vec{p}$ using the formula φ in $\mathscr{D}$. The proof of the coding lemma uses finite conditions to construct $\vec{p}$ from R; consequently, it does not use any machinery that is special to the Turing degrees. It applies or can be modified to apply to a very broad class of degree structures, even to some without minimal degrees such as the enumeration degrees, see Slaman-Woodin ([22], to appear).

We can use the coding lemma to present the solution to Sacks's question whether $\mathscr{D}$ is decidable. By the coding lemma, we can both code the standard model of arithmetic and also define the collection of codes of standard models. In addition, we can interpret second order quantifiers over a coded countable model by quantifiers over sequences in $\mathscr{D}$ which define unary relations. Thus, (Simpson [20], 1977) there is an interpretation of second order arithmetic in the first order theory of $\mathscr{D}$. Lachlan ([10], 1968) gave the original solution to Sacks's question but only showed that the theory of $\mathscr{D}$ is not recursive. Simpson calculated the degree of the first order theory of $\mathscr{D}$.)

Shore's Program

The next developments which we will discuss were initiated by Nerode and Shore ([14], 1980) and pursued extensively by Shore and his collaborators.

Suppose that R is a relation on degrees. We note that the degrees which are produced by the coding lemma to define R in $\mathscr{D}$ are recursion theoretically close to R. By this we mean the following. Let X be a real of degree x. If R is recursively presented relative to X then there is a sequence of degrees $\vec{p}$ which codes R in $\mathscr{D}$ and is arithmetic in x. As a special case, there is a sequence $\vec{P_X}$ of sets which is arithmetically definable from X and whose degrees code (in $\mathscr{D}$) an isomorphic copy of the standard model of arithmetic with a unary predicate for X. Thus, the set X is coded in $\mathscr{D}$ by parameters which are near its degree. Furthermore, if X is sufficiently complicated, say above $0'$, then a sequence of parameters whose degrees code X in $\mathscr{D}$ can be found recursively in X, that is below x in $\mathscr{D}$.

The next step was to find a notion of *neighborhood* which would be first order definable in $\mathscr{D}$ and link an arbitrary degree x to the reals coded in its neighborhood. The Jockusch-Soare theorems pointed in the correct direction. Jockusch and Soare defined a filter in $\mathscr{D}$ and proved that it is disjoint from the degrees of the arithmetic sets. Their theorem suggested that there might be a related definable filter whose complement was exactly the arithmetic degrees or even the degrees below $0'$.

The search for such a filter lead to the development of the theory of *REA*-operators, see Jockusch-Shore ([7], 1984). Recently, Cooper ([2], 1990) combined the analysis of *REA*-operators with his study of the degrees of differences of recursively enumerable sets to prove that the Turing jump is definable in the Turing degrees. In fact, Cooper proved the stronger theorem that the relation x *is recursively enumerable in and above* y is definable in $\mathscr{D}$ (solving Roger's question). We remark that this approach is specific to the Turing degrees. The proofs directly exploit the way that Turing reducibility can be recursively approximated.

We list some applications, in the strongest form obtained by these techniques. (Unfortunately, we do not have the space to describe the historical development and give credit where it is due.)

Theorem.

- (Jockusch-Shore [7], 1984) *For all* x *and* y, *if* $\mathscr{D}({}_T\geq x) \stackrel{\sim}{\rightarrow} \mathscr{D}({}_T\geq y)$ *then* x *and* y *have the same arithmetic degree.*
- (Cooper [2], 1990) *If* π *is an automorphism of* $\mathscr{D}$ *then* π *is the identity on all degrees greater than or equal to* $0'''$.
- (Cooper ibid.) *If* R *is a relation on the reals which are Turing above* $0'''$, R *is invariant under Turing degree and* R *is definable in second order arithmetic then the relation induced by* R *is definable in* $\mathscr{D}$.

The last two examples were known to follow from the definability of the jump, see Nerode-Shore ([14], 1980).

A Shortcoming of Method

The three structural properties of $\mathscr{D}$ we isolated above were proven above by following some practical advice: examine the neighborhood of x and recover some information about which reals belong to x. In fact, we will exactly recover the set of representatives of x provided that the set of representatives of x is coded in the neighborhood of x and that the neighborhood of x and the method employed decode the set of its representatives are uniformly recursive in x. Since the statement of the Turing order between reals involves Σ_3^0 and Π_3^0 predicates, x has to provide $0'''$, the information necessary to evaluate these predicates. Consequently, our conclusions for the Turing degrees have all been confined to the realms of sufficiently large degrees or arithmetic equivalence between Turing degrees. Even using Cooper's theorem to get the strongest possible bound on the orbit of $0'$, these methods have (so far) been confined by this three quantifier limit.

We should remark that this shortcoming does not apply to the degrees of constructibility. In that notion of degree, the partial ordering of the degrees of constructibility is below x is just another set constructed from x. In fact, with reasonable set theoretic hypotheses the techniques we have touched upon are sufficient to give a complete analysis of the global theory of the constructibility degrees, in the sense of the next section.

§3. Recent Results

In this section, we will discuss some recent work of Slaman and Woodin ([22], to appear). Unless we specifically indicate otherwise, all of the following results, conjectures and even remarks are drawn from that source.

Assignment of Representatives

Suppose that $\mathscr{E}$ is a degree structure on a set of reals, possibly the set of all reals. Say that $\mathscr{E}$ is determined by the equivalence relation $\equiv_E$ and the ordering $\leq_E$.

Definition. $\mathscr{E}$ has a *first order assignment of representatives* if

(1) for every $\mathscr{E}$-degree x there is a sequence $\vec{p}$ from $\mathscr{E}$ which uniformly codes a representative of x (say using the formulas $\varphi_1, \dots \varphi_k$);

(2) the relation $\vec{p}$ *codes a representative of* x is an $\mathscr{E}$-definable relation (say by the formula ψ).

We say that $\mathscr{E}$ *has a first order assignment of representatives in parameters* if the same conditions hold as above except that we allow $\varphi_1, \dots \varphi_k$ and ψ to mention parameters from $\mathscr{E}$.

Suppose that $\mathscr{E}$ has a first order assignment of representatives. If $\mathscr{E}$ is definable in second order arithmetic and is a degree structure on all of the reals then we say that $\mathscr{E}$ is *biinterpretable with second order arithmetic*. If $\mathscr{E}$ is definable in first order arithmetic and is a degree structure on a uniformly arithmetic set of reals then we say that $\mathscr{E}$ is *biinterpretable with first order arithmetic*. Similarly, we can define biinterpretability with parameters.

If $\mathscr{E}$ is biinterpretable with the standard model second order arithmetic then all of the logical properties of $\mathscr{E}$ can be reduced to the reals. We will list some examples. Others will undoubtedly occur to the reader.

Assuming $\mathscr{E}$ is biinterpretable with second order arithmetic:

- A relation is definable in $\mathscr{E}$ if and only if it is induced by an $\equiv_E$-invariant relation that is definable in second order arithmetic.
- $\mathscr{E}$ has no nontrivial automorphism, i.e. $\mathscr{E}$ is *rigid*.

Assuming that $\mathscr{E}$ is biinterpretable with second order arithmetic, we sketch an example proof. The standard model of first order arithmetic is rigid because each number is definable. Consequently, the standard model of second order arithmetic is also rigid since each subset of $\mathbb{N}$ is determined by its elements. The fact that no representative of a degree can be moved by an automorphism of the standard model of second order arithmetic implies that an automorphism of $\mathscr{E}$ can only move a degree to another one with the same set of representatives. Of course, this last condition is another way of saying that the only automorphism of $\mathscr{E}$ is the identity.

If we allow parameters we obtain the boldface versions of the above conclusions.

Assuming $\mathscr{E}$ is biinterpretable with second order arithmetic using the parameters $\vec{e}$:

- A relation is definable from finitely many parameters in $\mathscr{E}$ if and only if it is induced by an $\equiv_E$-invariant relation on reals that definable from finitely many parameters in second order arithmetic, i.e. is induced by a *projective* relation.
- Any automorphism of $\mathscr{E}$ is determined by its action on $\vec{e}$. That is, $\vec{e}$ is an *automorphism base.*

Just knowing that $\mathscr{E}$ has a finite automorphism base gives us some information about its group of automorphisms. For example, if $\mathscr{E}$ is a degree structure on the reals then the cardinality of the orbit of any finite sequence from $\mathscr{E}$ is bounded by the cardinality of the continuum. Thus, if $\mathscr{E}$ has a finite automorphism base then it has at most continuum many automorphisms. Similarly, if $\mathscr{E}$ is a degree structure on a countable set of reals and has a finite automorphism base then the automorphism group for $\mathscr{E}$ is countable.

A Finite Automorphism Base

We will now sketch a heuristic approach to proving that a degree structure on the reals has a finite automorphism base. The method directly applies to the specific structures of the Turing degrees, the arithmetic degrees, the hyperarithmetic degrees, the Σ_k-admissible degrees, the PTIME Turing degrees and the enumeration degrees.

Let $\mathcal{D}$ denote any of the above degree structures on the reals. In the next few paragraphs we will be measuring definability in terms of a number of jumps. The interpretation of *jump* depends on the interpretation of $\mathcal{D}$. By *arithmetic in,* we mean below some finite number of jumps. The reader may comfortably imagine that we are discussing the Turing degrees with the Turing jump.

For the first step, we prove the coding lemma with its full uniformity. In particular, there is a degree z_0 such that if x is above z_0 then all the representatives of x are coded below x. Some examples for z_0 are $0''$ in the Turing degrees, 0^ω in the arithmetic degrees and the degree of Kleene's $\mathcal{O}$ in the hyperarithmetic degrees. Let z_0 be fixed.

We continue with some countable algebra for $\mathcal{D}$.

Definition. Let I be a countable ideal in $\mathcal{D}$ and let ϱ be an automorphism of I. We say that ϱ is *persistent* if for every x in $\mathcal{D}$ there are an ideal J and an automorphism ϱ^* of J such that $x \in J$, $I \subseteq J$ and ϱ^* agrees with ϱ on I.

Next we prove that every persistent countable automorphism extends to an automorphism of $\mathcal{D}$. The proof employs a generalization of an insight due to Odifreddi and Shore: the coding lemma can be used to show that the restriction of an automorphism of $\mathcal{D}$ to a countable ideal is recursion theoretically close to any uniform upper bound on that ideal. Specifically, we show that if ϱ is a persistent automorphism of the ideal I then ϱ is arithmetic in any upper bound of I.

We now introduce some metamathematical methods. We know that there is a nontrivial automorphism of $\mathcal{D}$ if and only if there is a nontrivial countable automorphism that is persistent. The latter condition is upwards absolute between well-founded models of ZFC.

Applying results of Slaman-Woodin ([21], 1986), we obtain the following theorem.

Theorem. *$\mathcal{D}$ is rigid if and only if $\mathcal{D}$ is biinterpretable with second order arithmetic.*

Continuing the metamathematical discussion, let V denote the universe of sets. Let $V[\mathcal{G}]$ be a generic extension of V. Using the absoluteness theorem, we show that if π is an automorphism of $\mathcal{D}^V$, the degrees in V, then π lifts to an automorphism π^* of $\mathcal{D}^{V[\mathcal{G}]}$, the degrees in the generic extension. By moving to a generic extension of V, we can use the definition of forcing to analyze π^*. In particular, if $\mathcal{G}$ is generic with respect to the partial order to add ω_1 Cohen reals to V then π^* is represented as a continuous function on the set of generic reals. In fact, we can use our proof that $\pi(x)$ is close to x to show the following. There is a recursive functional $\{e\}$ and an integer n such that if G is a Cohen generic real over V then the degree of $\{e\}(G \oplus (\pi^{-1}(Z_0))^{(n)})$ is π^* of the degree of G. Here we let Z_0 denote a representative of z_0, $\pi^{-1}(Z_0)$ denote a representative of $\pi^{-1}(z_0)$ and $(\pi^{-1}(Z_0))^{(n)}$ denote the nth jump of $\pi^{-1}(z_0)$.

Our next step is to extract a representation of π on a comeager set of reals in V from the representation of π^* on the comeager set of generic reals in $V[\mathcal{G}]$.

We prove that the same representation of π (using the functional $\{e\}$ relative to $\pi^{-1}(Z_0)^{(n)}$) holds on the set C of reals which are sufficiently generic relative to $\pi^{-1}(z_0)$. The level of genericity required is only finitely many jumps, in the sense of the reducibility determining $\mathscr{D}$. Furthermore, if there is one G in C such that G and $\{e\}(G \oplus (\pi^{-1}(Z_0))^{(n)})$ have the same degree then π is the identity on the degrees represented in C.

In ([6], 1981), Jockusch and Posner show that the Turing degrees represented by any comeager set of reals generate the Turing degrees under the operations of meet and join. Their argument is completely general, so we may conclude that C is an automorphism base.

Now we can make two observations.

First, suppose that $\pi(z_0) = z_0$. Then the set C consists of those reals which are arithmetically generic relative to z_0. Let g_0 be any degree of an element of C. If π also maps g_0 to g_0 then π must be the identity on C and therefore π must be the identity on all of $\mathscr{D}$. Thus, $\{z_0, g_0\}$ is a finite automorphism base for $\mathscr{D}$.

Second, from the arithmetic representation of π on C relative to $\pi^{-1}(Z_0)$ we can find a function arithmetic in $\pi^{-1}(Z_0)$ which represents π on all the reals. Consequently, the fact that the rigidity of $\mathscr{D}$ is absolute may be explained by the fact that every automorphism of $\mathscr{D}$ is arithmetically definable in a real parameter.

Biinterpretability with Parameters

The fact that $\mathscr{D}$ has a finite automorphism base can be combined with the analysis of persistent automorphisms to show that $\mathscr{D}$ is biinterpretable with second order arithmetic in parameters.

Let Z_0 and G_0 denote representatives of the degrees z_0 and g_0 above. Suppose that ψ is a map from the reals onto $\mathscr{D}$ which induces an automorphism on $\mathscr{D}$, i.e. ψ is degree invariant, preserves order in the sense of $\mathscr{D}$-reducibility and has distinct values on reals of distinct degree. If ψ maps Z_0 to z_0 and G_0 to g_0 then ψ must induce the identity automorphism. In other words, ψ must be an assignment of representatives to $\mathscr{D}$. Let the parameters $\vec{p}_{Z_0}$ code Z_0 and $\vec{p}_{G_0}$ code G_0. With some finesse, we can use the coding lemma to express the following condition in D: *There is a persistent countable assignment of representatives sending Z_0 to z_0, G_0 to g_0 and X to x.* By the characterization of persistent countable automorphisms as restrictions of global automorphisms, this statement is equivalent to one saying that there is a map from the reals to $\mathscr{D}$ with the values as above that induces an automorphism. Of course, this automorphism is the identity. Thus, the statement above defines *X is a representative of x* as expressed in the codes for the real X. Consequently, $\mathscr{D}$ is biinterpretable with second order arithmetic using the parameters $\vec{p}_{Z_0}, z_0, \vec{p}_{G_0}$ and g_0.

Theorem. *The structures of the Turing degrees, the arithmetic degrees, the hyperarithmetic degrees, the Σ_k-admissible degrees, the PTIME Turing degrees and the enumeration degrees are all biinterpretable with second order arithmetic in parameters.*

Special Arguments for the Turing, Arithmetic, Hyperarithmetic and Σ_k-Admissible Degrees

In this section, we will discuss results obtained using special properties common to the Turing, Arithmetic, Hyperarithmetic and Σ_k-admissible reducibilities. We should note that these arguments do not intersect with those of Jockusch-Shore and Cooper and provide new proofs of the three structural properties of the Turing degrees mentioned above.

Since the proofs here are more technically involved than the ones that we have discussed so far, we have to resign ourselves to merely stating results. We can only say that the proofs of these results involve a direct analysis how a continuous function must behave to represent the restriction of a degree structure automorphism to a comeager set.

By the remark of the previous section, for all of our degree structures of interest, rigidity is equivalent to biinterpretability with second order arithmetic. Thus, the central problem for any of these structures is whether it has a nontrivial automorphism.

Theorem.

(1) *The hyperdegrees are rigid (and thus are biinterpretable with second order arithmetic.) Similarly, all of the Σ_k-admissible degrees structures are biinterpretable with second order arithmetic.*

(2) *Any automorphism of the arithmetic degrees is the identity above* 0^{ω}.

(3) *Any automorphism of the Turing degrees is the identity above* $0''$.

Corollary. *There are only countably many automorphisms of the Turing degrees. Any automorphism of the Turing degrees is represented by an arithmetically definable function on reals.*

A similar corollary holds for the arithmetic degrees; any automorphism is hyperarithmetically representable.

Special Arguments for the Turing Degrees

We now restrict $\mathscr{D}$ to denote the Turing degrees. We also drop any restraint against using very special properties of $\mathscr{D}$. In particular, we will make full use of Cooper's theorem that the relation x is recursively enumerable in and above y is definable in $\mathscr{D}$.

Earlier, we sketched a metamathematical proof that any automorphism of $\mathscr{D}$ is arithmetically definable. We can also give more traditional, purely recursion theoretic proof of this fact. Of course, this proof is a more difficult local version of its metamathematical progenitor. With the sharper argument, we can replace the full structure of $\mathscr{D}$ by any ideal in $\mathscr{D}$ which has $0^{(7)}$ as element.

Theorem. *Suppose the I is an ideal in $\mathscr{D}$ and $0^{(7)} \in I$.*

(1) *If $\pi : I \xrightarrow{\sim} I$ then π is represented by an arithmetically definable function on reals.*

(2) *I is biinterpretable in parameters with the fragment of second order arithmetic in which the second order quantifiers range over the reals whose degrees lie in I.*

As a corollary to the theorem, we can demonstrate a connection between the existence of a local automorphism and a global one. By the Kleene basis theorem, if an arithmetic function does not represent an automorphism of $\mathscr{D}$ then there is a counter-example which is recursive in Kleene's $\mathcal{O}$. Thus, we can conclude that any automorphism of $\mathscr{D}(\leq_T \mathcal{O})$ extends to an automorphism of $\mathscr{D}$.

A Concrete Automorphism Base for the Turing Degrees

We already know that there is a finite automorphism base for $\mathscr{D}$. The question arises as to how concrete can we make the base elements. Outside of proving that the Turing degrees are rigid, we can give the best possible result. We show that there is a finite set of recursively enumerable degrees which is an automorphism base for $\mathscr{D}$.

Our approach is to provide a generating family of first order formulas such that the smallest set including the recursively enumerable degrees and closed under definition by these formulas includes the degrees of $0''$ (i.e. z_0) and of a real G which is sufficiently generic for the pair of degrees to form a base. We isolate an result which we prove along the way which is of independent interest.

Theorem.

(1) *$\mathscr{D}(\leq_T 0')$ is biinterpretable with first order arithmetic in parameters. In fact, we may take the parameters to be recursively enumerable degrees.*

(2) *Any automorphism of $\mathscr{D}(\leq_T 0')$ is arithmetically definable.*

Ultimately, we prove that there is a finite set F of recursively enumerable degrees such that $\mathscr{D}$ is biinterpretable with second order arithmetic in the parameters from F. In fact, the analogous theorem holds of any ideal I such that $0^{(7)} \in I$. Thus, we have the following theorem.

Theorem.

(1) *The recursively enumerable degrees are an automorphism base for $\mathscr{D}$. Further, $\mathscr{D}$ is biinterpretable with second order arithmetic using recursively enumerable parameters.*

(2) *If the recursively enumerable degrees are rigid then $\mathscr{D}$ is rigid.*

(3) *If $\mathscr{D}(\leq_T 0')$ is rigid then $\mathscr{D}$ is rigid.*

It is open whether every automorphism of the recursively enumerable degrees extends to one of $\mathscr{D}(\leq_T 0')$ or even to $\mathscr{D}$. Any result along this line would be valuable.

§4. Conjectures

We make the following conjectures jointly with W. H. Woodin.

We say that a real number G is *1-generic* if for every recursively enumerable set S of finite Cohen conditions either there is an element of S which is satisfied by G or there is a neighborhood condition satisfied by G which is incompatible with all the elements of S.

Conjecture I. If I is an ideal in the Turing degrees such that there is the degree of a 1-generic real in I then I has a first order assignment of parameters.

In particular, we believe that the partial ordering of the Turing degrees is biinterpretable with second order arithmetic and that $\mathscr{D}(\leq_T 0')$ is biinterpretable with first order arithmetic.

Conjecture II. The partial ordering of recursively enumerable degrees is biinterpretable with the standard model of first order arithmetic.

We end with a question. Is there a general proof of rigidity and equivalently of biinterpretability with second order arithmetic that is based on simple properties of Turing reducibility and applies to a wide range of degree structures?

References

1. Abraham, U., Shore, R. A.: Initial segments of the degrees of size $\aleph_1$. Israel J. Math. **53** (1986) 1–51
2. Cooper, S. B.: The jump is definable within the structure of the Turing degrees. Bull. A.M.S. **23** (1990) 151–158
3. Ershov, Y. L.: The uppersemilattice of enumerations of a finite set. Alg. Log. **14** (1975) 258–284 ; transl. **14** (1975) 159–175
4. Groszek, M. J., Slaman, T. A.: Independence results on the global structure of the Turing degrees. Trans. Am. Math. Soc. **277** (1983) 579–588
5. Jockusch, C. A., Jr.: An application of Σ_4^0 determinacy to the degrees of unsolvability. J. Sym. Logic **38** (1973) 293–294
6. Jockusch, C. A., Jr., Posner, D.: Automorphism bases for degrees of unsolvability. Israel J. Math. **40** (1981) 150–164
7. Jockusch, C. A., Jr., Shore, R. A.: Pseudo-jump operators II: Transfinite iterations, hierarchies and minimal covers. J. Sym. Logic **49** (1984) 1205–1236
8. Jockusch, C. A., Jr., Soare, R. I.: Minimal covers and arithmetical sets. Proc. Amer. Math. Soc. **25** (1970) 856–859
9. Kleene, S. C., Post, E. L.: The upper semi-lattice of degrees of unsolvability. Ann. Math. **59** (1954) 379–407
10. Lachlan, A. H.: Distributive initial segments of the degrees of unsolvability. Z. Math. Logik Grundlagen Math. **14** (1968) 457–472
11. Lachlan, A. H., Lebeuf, R.: Countable initial segments of the degrees of unsolvability. J. Sym. Logic **41** (1976) 289–300
12. Lerman, M.: Initial segments of the degrees of unsolvability. Ann. Math. **93** (1971) 365–389

13. Nerode, A., Shore, R. A.: Second order logic and first order theories of reducibility. orderings. In: The Kleene Symposium. Barwise, J., Keisler, J., Kunen, K., eds. North-Holland, Amsterdam 1979, pp. 181–200
14. Nerode, A., Shore, R. A.: Reducibility orderings: theories, definability and automorphism. Ann. Math. Logic **18** (1980) 61–89
15. Odifreddi, P. G.: Classical Recursion Theory. North-Holland, Amsterdam 1989
16. Paliutin, E.: Addendum to the paper of Ershov [1975]. Alg. Log. **14** (1975) 284–287; transl. **14** (1975) 176–178
17. Rogers, H., Jr.: Some problems of definability in recursive function theory. In: Sets Models and Recursion Theory, Proc. Summer School in Math. Logic and Tenth Logic Colloq., Lercester, August–September 1965. J.N. Crossley, ed. North-Holland, Amsterdam 1967
18. Sacks, G. E.: Degrees of Unsolvability. Princeton University Press, Princeton, New Jersey 1963
19. Sacks, G. E.: Degrees of Unsolvability, Second Edition. Princeton University Press, Princeton, New Jersey 1966
20. Simpson, S. G.: First order theory of the degrees of recursive unsolvability. Ann. Math. **105** (1977) 121–139
21. Slaman, T. A., Woodin, W. H.: Definability in the Turing degrees. Illinois J. Math. **30** (1986) 320–334
22. Slaman, T. A., Woodin, W. H.: Definability in degree structures. (To appear)
23. Spector, C.: On degrees of recursive unsolvability. Ann. Math. **64** (1956) 581–592

Cohomology and Modules over Group Algebras

Jon F. Carlson

Department of Mathematics, University of Georgia, Athens, GA 30602, USA

Introduction

The last several years have witnessed an increasing interaction between homological algebra and the representation theory of finite groups. The aim of the effort has primarily been the development of a module theory for group rings and algebras. Of course, it is not surprising that homological methods would play a role in such an investigation. After all, cohomology theory really is a theory of extensions and, in small degrees, its relevance to the structure of modules is apparent. However the connection has proven to be deeper. The module theory is directly related to some of the larger features of the cohomology of groups. By this we mean such items as the ring theoretic properties of cohomology – the maximal ideals, Krull dimension, and modules for cohomology rings. In this lecture I will survey some of the developments in this area. For the most part, we will stick to the case in which the group G is finite and the coefficient ring K is a field of characteristic $p > 0$. Nevertheless, many of the results hold true for integral coefficients and for more general groups.

The work which I wish to survey is based on several finiteness conditions. First, we consider only finitely generated modules. But more than that, the foundation rests on two fundamental results of group cohomology.

Theorem (Evens [14]). *$H^*(G, K) = \mathrm{Ext}^*_{KG}(K, K)$ is a finitely generated K-algebra. If M is a KG-module, then $H^*(G, M)$ is a finitely generated module over $H^*(G, K)$.*

Even's Theorem permits us to consider the maximal ideal spectrum, $V_G(K)$, of $H^*(G, K)$. We need not be concerned here that $H^*(G, K)$ is not commutative. The only noncommutativity occurs among elements of odd degree. These, however, are all in the radical and hence are in every maximal ideal. Note that $V_G(K)$ is a homogeneous affine variety. For a KG-module M, we can define $V_G(M)$ to be the subvariety of $V_G(K)$ associated to the annihilator of $\mathrm{Ext}^*_{KG}(M, M) \cong H^*(G, \mathrm{Hom}_K(M, M))$. It is the set of all maximal ideals containing the annihilator.

The other foundational result is the "Dimension Theorem" of Quillen.

Proceedings of the International Congress
of Mathematicians, Kyoto, Japan, 1990

Theorem [19]. *The components of $V_G(K)$ are in one-to-one correspondence with the conjugacy classes of maximal elementary abelian p-subgroups $E(\cong (\mathbb{Z}/p)^r \hookrightarrow G)$ via the restriction maps.*

For algebraists, the best proof of the theorem is given in [21]. All proofs have required some variation on a theorem of Serre [22] which says that for p-groups which are not elementary abelian, the product of all Bocksteins of elements of degree 1 is nilpotent. One of the inspirations in the development of the subject was the extension of Quillen's theorem to the varieties of modules. The first attempt considered only the complexity or dimension of the varieties [3]. The full result was obtained independently by Alperin and Evens and by George Avrunin [4, 5]. The importance of this step is obvious. It says that in this context we can restrict our attention to the elementary abelian p-groups.

Elementary Abelian Groups

For this section assume that the field K is algebraically closed. Suppose that $E = \langle x_1, \cdots, x_n \rangle \cong (\mathbb{Z}/p)^n$ is an elementary abelian p-group. It is well known that

$$H^*(E, K)/\mathrm{Rad} = K[\zeta_1, \cdots, \zeta_n]$$

is a polynomial ring in $n = \mathrm{rank}(E)$ generators. Here Rad denotes the radical of $H^*(G, K)$. If $p = 2$ then $\mathrm{Rad} = 0$ and the generators $\zeta_1, \cdots, \zeta_n$ have degree 1. If $p > 2$ then Rad is generated by the elements of degree 1 and each ζ_i has degree 2. In either case, the maximal ideal spectrum $V_E(K) = K^n$, is affine n-space.

For $\alpha = (\alpha_1, \cdots, \alpha_n) \in K^n$, let

$$u_\alpha = 1 + \sum_{i=1}^{n} \alpha_i(x_i - 1) \in KE\,.$$

It is easy to see that u_α is a unit of order p in KE and that the inclusion map $K\langle u_\alpha \rangle \hookrightarrow KE$ is an injection of group algebras. Moreover KE is a free $K\langle u_\alpha \rangle$-module. Hence, it makes sense to speak of a restriction map on cohomology from E to $\langle u_\alpha \rangle$. Thus, for a KE-module M, the set of all u_α which will measure nonzero cohomology of M should be special. This leads to the following definition.

Definition. Let M be a KE-module with K, E as above. The set

$$V_E^r(M) = \{\alpha \in K^n | M \text{ is not free over } K\langle u_\alpha \rangle\} \cup \{0\}$$

is a homogenous affine variety called the *rank variety* of M.

The fact that it is a variety is fairly easy to show. The proof shows further that $V_E^r(M)$ can be computed directly from elementary information such as the representation $\varrho : E \to GL(m, K)$ which defines M as a module [9,10]. But more importantly we have

Theorem. $V_E(M) \cong V_E^r(M)$.

This is only an isomorphism of algebraic sets, subsets of K^n. There seem to be no natural, corresponding maps on the coordinate rings. For $p = 2$, it can be viewed as an equality. However for $p > 2$, the restriction map to the subgroups $\langle u_\alpha \rangle$ introduces a Frobenius twist on cohomology. The existence of a relationship between the varieties was clear from the beginning. Avrunin and Leonard Scott[6] proved the more difficult containment by altering the Hopf-algebra structure required for cup products.

For more general groups G, the variety $V_G(M)$ can be pieced together from the $V_E(M)$'s, $E \leq G$. This is true in theory. In practice it may be very difficult. Avrunin and Scott also showed that $V_G(M)$ is stratified as in [20].

The variety $V_G(M)$ has several useful properties. Some can be derived easily by considering the rank variety.

Theorem. (a) $V_G(M) = \{0\}$ *if and only if* M *is projective.*

(b) $V_G(M \otimes_K N) = V_G(M) \cap V_G(N)$.

(c) *If* $0 \to A_1 \to A_2 \to A_3 \to 0$ *is exact then* $V_G(A_i) \subseteq V_G(A_j) \cup V_G(A_k)$ *for* $\{i, j, k\} = \{1, 2, 3\}$.

(d) *If* $W \subseteq V_G(K)$ *is a (Zariski) closed subvariety, then there exists a module* M *with* $V_G(M) = W$.

(e) *If* M *is indecomposable then the corresponding projective variety* $\overline{V}_G(M)$ *is connected* [11].

Property (d) has been particulary useful. Its proof is completely constructive. Suppose that we are given $\zeta \in H^m(G, K), m \geq 2$. Let $P_0, \cdots, P_{m-1}$ be minimal projective modules so that

$$0 \to \Omega^m(K) \to P_{m-1} \to \cdots \to P_1 \to P_0 \to K \to 0$$

is exact and $\Omega^m(K)$ has no projective submodules. Then ζ is represented by a cocycle $\hat{\zeta} : \Omega^m(K) \to K$. Assuming that $\zeta \neq 0$, we have an exact sequence

$$0 \to L_\zeta \to \Omega^m(K) \xrightarrow{\hat{\zeta}} K \to 0.$$

By using the rank variety we may show that $V_G(L_\zeta) = V_G(\zeta)$, the subvariety corresponding to the ideal generated by ζ. Thus (d) can be proved by choosing a set of generators for the ideal of W and using (b).

Two Asides

I want to mention two related areas which have been the subject of some recent study. The first involves the structure of the cohomology rings $E^*(M) = \mathrm{Ext}^*_{KG}(M, M) = H^*(G, \mathrm{Hom}_K(M, M))$ for M a KG-module. The variety of M is certainly associated to this ring but not as its maximal ideal spectrum. The problem is that $E^*(M)$ is not commutative. We do not simply mean that it is graded commutative, but rather, it can be noncommutative in an essential way[12]. There exist examples, which are not exotic, where the ring $E^*(M)$ has simple modules of K-dimension greater than one. That is, there exists a surjective homomorphism of $E^*(M)$ onto the ring of $n \times n$ matrices for $n > 1$. However the news is not all bad. $E^*(M)$ is a PI-ring and hence all of its simple modules have finite dimension. Recently Niwasaki has settled a question concerning the nature of the simple modules.

Theorem [17]. *If $U \subset \mathrm{Ext}^*_{KG}(M,M)$ is a maximal ideal then there exists an elementary abelian p-subgroup $E = \langle x_1, \cdots, x_n \rangle \subset G$ and a unit $u_\alpha = 1 + \sum_{i=1}^{n} \alpha_i(x_i - 1) \in KE$ such that U contains the kernel of the composition of the restrictions of $\mathrm{Ext}^*_{KG}(M,M)$ to E and to $\langle u_\alpha \rangle$.*

In another direction, a very similar sort of theory has been developed for the restricted p-Lie algebras associated to an algebraic group in characteristic p (see [18] for a survey). Suppose that G is a connected linear algebraic group defined over the prime field $\mathbb{F}_p$. Let G_1 denote the kernel of the Frobenius map on G. Here G_1 should be regarded as a group scheme or as a functor, since as a group it has only a single element. The Lie algebra $\underline{g}$ of G may be equipped with a p-power operation $x \to x^{[p]}$, thus making it a restricted p-Lie algebra. The representation theory of rational G-modules is related to that of rational G_1-modules via spectral sequences. Moreover the category of rational G_1-modules is equivalent to the category of restricted $\underline{g}$-modules.

For restricted $\underline{g}$-modules it is possible to define cohomological support varieties and rank varieties as in the finite group case. That is, $H^*(\underline{g}, K)$ is a finitely generated ring and $H^*(\underline{g}, M)$ is a finitely generated module over it, provided M is finitely generated. The analogue of the rank variety is the null cone

$$\mathcal{N} = \{x \in \underline{g} \mid x^{[p]} = 0\}$$

which is isomorphic to the maximal ideal spectrum of the cohomology ring [16]. For M, the variety is the set of points $x \in \mathcal{N}$ such that M is not injective as a Kx-module. Many of the same properties for the variety have been shown to hold in this context. A major part of this work was done jointly by Friedlander and Parshall (e.g. see [15]).

It should be noted that our friends in the field of algebraic groups have left us with some sticky problems. What happens with G_2 or G_r for $r \geq 2$? Is $H^*(G_r, K)$ finitely generated? The answers are not known except in some special cases.

Projective Resolutions and Homogeneous Parameters

In recent years, some interesting results have been obtained on systems of parameters for cohomology rings and for the cohomology of a module. Suppose that $\zeta_1, \cdots, \zeta_n$ are homogeneous elements in $H^*(G, K)$ which have the property that $H^*(G, K)$ is a finitely generated module over $K[\zeta_1, \cdots, \zeta_n]$, the polynomial ring in the symbols $\zeta_1, \cdots, \zeta_n$. Another way of stating the condition is that $\cap V_G(\zeta_i) = \{0\}$. For each i, let $m_i = \deg \zeta_i$ and let $\hat{\zeta}_i$ be a cocycle representing ζ_i. Then we have the following commutative diagram.

$$\begin{array}{ccccccccccccc}
 & \cdots & \longrightarrow P_{m_i+1} & \longrightarrow P_{m_i} & \longrightarrow P_{m_i-1} & \longrightarrow P_{m_i-2} & \longrightarrow \cdots \longrightarrow & P_0 & \longrightarrow K & \longrightarrow 0 \\
 & & \downarrow & \hat{\zeta}_i \downarrow & \downarrow & \downarrow & & \| & \| & \\
E^{(i)}: & & 0 & \longrightarrow K & \stackrel{j}{\longrightarrow} L_i & \longrightarrow P_{m_i-2} & \longrightarrow \cdots \longrightarrow & P_0 & \stackrel{\varepsilon}{\longrightarrow} K & \longrightarrow 0 \\
 & & & & \| & \| & & \| & & \\
C^{(i)}: & & & 0 & \longrightarrow C^{(i)}_{m_i-1} & \longrightarrow C^{(i)}_{m_i-2} & \longrightarrow \cdots \longrightarrow & C^{(i)}_0 & \longrightarrow 0 &
\end{array}$$

Here (P_*, ε) is a minimal projective resolution of K. The middle row $E^{(i)}$ is the pushout along $\hat{\zeta}_i$ of the upper row and is exact. The complex $C^{(i)}$ has homology $H_j(C^{(i)}) = K$ for $j = m_i - 1, 0$ and $H_j(C^{(i)}) = 0$ otherwise. Let $X^{(i)}$ be the complex which is obtained by splicing copies of $E^{(i)}$ as

$$(X^{(i)}) \qquad \cdots \to P_0 \stackrel{j\varepsilon}{\longrightarrow} L_i \to \cdots \to P_0 \stackrel{j\varepsilon}{\longrightarrow} L_i \to \cdots \to P_0 \to 0.$$

The first result on systems of parameters for cohomology was the following theorem proven in joint work with Dave Benson some 5 years ago.

Theorem [7]. *$X = X^{(1)} \otimes_K \cdots \otimes_K X^{(n)}$ is a projective resolution of K.*

The proof is very easy. The exactness except in degree zero follows from the Künneth formula, while the projectivity comes mostly from the properties of varieties. Actually the theorem can be stated more generally. If $\mathrm{Ext}^*_{KG}(M, M)$ is finitely generated as a module over $K[\zeta_1, \cdots, \zeta_r]$, $\zeta_i \in H^*(G, K)$, then $X = X^{(1)} \otimes_K \cdots \otimes_K X^{(r)} \otimes_K M$ is a projective resolution of M. Hence every module has a projective resolution which is a tensor product of periodic complexes. Moreover the number of such complexes can be taken to be the complexity of M or the dimension of $V_G(M)$.

With appropriate modifications a similar theorem can be proved for a $\mathbb{Z}G$-lattice. Adem [2] has adapted some of these ideas to get a similar result for groups with finite virtual cohomological dimension.

Remark. *Similar techniques will also prove facts about the exponents of cohomology with integral coefficients. For example, if $H^*(G, \mathbb{Z})$ is finitely generated over $\mathbb{Z}[\zeta_1, \cdots, \zeta_n]$ then the order of G divides the product $\prod \exp(\zeta_i)$, where $\exp(\zeta_i)$ is the additive order of $\zeta_i \in H^*(G, \mathbb{Z})$* [13]. *The method has also been used to deal with some questions concerning equivariant cohomology of topological spaces with an action of a finite group* [1]. *The theorem can be extended to exponents of $\mathbb{Z}G$-lattices and to a similar result on the Farrell cohomology of a group with finite virtual cohomological dimension.*

As an example we wish to consider the case in which $n = 3$. That is, we assume that $H^*(G, M)$ is finitely generated over $K[\alpha, \beta, \gamma]$. Then we can form the Kozsul complex:

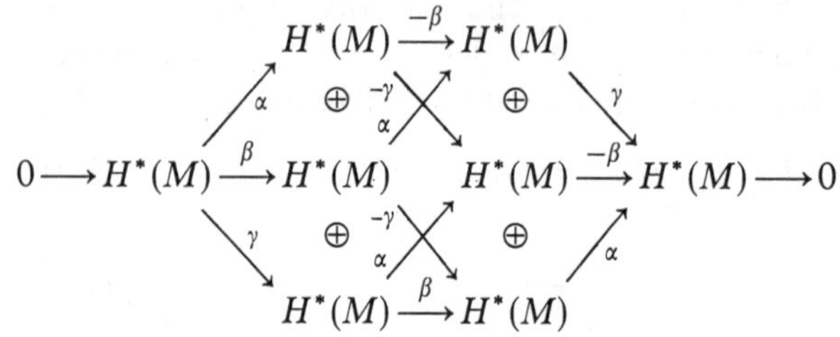

where $H^*(M) = H^*(G,M)$. The maps are cup products with the indicated elements. Let l,m,n be the degree of α, β and γ respectively. We need to assume that the degrees are even if p is odd. The complex has a graded version

$$(*_s) \qquad 0 \longrightarrow H^s \longrightarrow \begin{array}{c} H^{s+l} \\ \oplus \\ H^{s+m} \\ \oplus \\ H^{s+n} \end{array} \longrightarrow \begin{array}{c} H^{s+l+m} \\ \oplus \\ H^{s+l+n} \\ \oplus \\ H^{s+m+n} \end{array} \longrightarrow H^{s+l+m+n} \longrightarrow 0$$

where $H^r = H^r(G,M)$. Of course for any set of parameters $\zeta_1, \cdots, \zeta_t$ we get similar complexes $(*_s)$ which have length t. In any case it is not difficult to prove the following.

Proposition. *If s is sufficiently large then $(*_s)$ is an exact sequence.*

The proof is based on the facts that polynomial rings have finite global dimension and that the Koszul complex with $H^*(G,M)$ replaced by the polynomial ring is a projective resolution of K. The proposition allows us to define a new invariant for modules. Notice first that if we use Tate cohomology then s can be negative. Moreover the sequence $(*_s)$ for M is the same as the complex $(*_{s+1})$ for $\Omega(M)$. Here $\Omega(M)$ is the kernel of the epimorphism $P \to M$ where P is a projective cover for M.

Definition. The *index* of a KG-module M is the least integer r such that $(*_s)$ is exact for all $s \geq r$.

The index of a module is a benchmark for the degree shifting process in cohomology. Clearly $\text{Index}(\Omega(M)) = \text{Index}(M) + 1$. Also we have the following.

Proposition. *The index of M does not depend on the choice of parameters. That is, the index is the same for any set $\zeta_1, \cdots, \zeta_m$ such that $H^*(G,M)$ is finitely generated over $K[\zeta_1, \cdots, \zeta_m]$.*

Some interesting questions are raised by the construction. For example, is the index of trivial module zero? The answer is yes when the p-rank of G is two and in several other situations [8]. What is the index of an irreducible module? Is it zero?

One of the connections with the representation theory is contained in the following. We state this only in the context of the example ($n = 3, \alpha, \beta, \gamma$ as above). The proof follows from results in [8].

Theorem. *If* $s \geq \mathrm{Index}(U \otimes_K M^*)$ *for all simple modules* U, *then there is an exact sequence*

$$\begin{array}{ccccccccc} & & & & \Omega^{s+l+m}(M) & & \Omega^{s+l}(M) & & \\ & & & & \oplus & & \oplus & & \\ 0 & \longrightarrow & \Omega^{s+l+m+n}(M) & \longrightarrow & \Omega^{s+l+n}(M) & \longrightarrow & \Omega^{s+m}(M) & \longrightarrow & \Omega(M) \longrightarrow 0 \\ & & & & \oplus & & \oplus & & \\ & & & & \Omega^{s+m+n}(m) & & \Omega^{s+n}(M) & & \end{array}$$

where the maps are induced by (± 1) *times* α, β, γ *as appropriate.*

There is another view which we may take of this theory. Consider the complex $C^{(i)}$ associated to ζ_i. Recall that $H_j(C^{(i)})$ is K if $j = 0, m_i - 1$ and zero otherwise. Of course, $\zeta_1, \cdots, \zeta_n$ is a homogeneous set of parameters for $H^*(G, K)$. Let

$$C = C^{(1)} \otimes_K \cdots \otimes_K C^{(n)}.$$

Theorem [8]. *The complex* C *is a Poincaré duality complex of projective* KG-*modules. That is, there is a duality* $C^* \to C$ *which is a homotopy equivalence.*

Among other things the duality implies that when $H^*(G, K)$ is a Cohen-Macaulay ring then its Poincaré series, $f(t)$, satisfies the functional equation $f(\frac{1}{t}) = (-t)^r f(t)$, where $r = p\text{-rank}(G)$. It is conjectured that the condition is equivalent to $H^*(G, K)$ being Cohen-Macaulay.

It can be seen that $H^*(C) = \Lambda(\hat{\zeta}_1, \cdots, \hat{\zeta}_n)$, an exterior algebra, where $\hat{\zeta}_i$ is a cohomology element (in degree $\deg(\zeta_i) - 1 = m_i - 1$) coming from the cohomology of $C^{(i)}$. In some cases the cup product here is G-equivariant. We can form the hypercohomology spectral sequence whose E_0-term is

$$E_0^{r,s} = \mathrm{Hom}_{KG}(P_r \otimes C_s, K)$$

and whose E_2-term is

$$E_2^{r,s} = H^r(G, H^s(C)) = H^r(G, K) \otimes_K H^s(C),$$

and which converges to the hypercohomology

$$H_G^{r+s}(C, K) = H^{r+s}(\mathrm{Hom}_{KG}(C, K)).$$

The early differentials in the spectral sequence are simply cup products with the corresponding group cohomology elements. That is,

$$d_{m_i}(\alpha \otimes \hat{\zeta}_i) = \alpha\zeta_i \otimes 1.$$

The higher differentials are a sort of twisted Massey product – a matric Massey product. The results seem to indicate that some of these Massey products are zero in high enough degrees. The questions about the index of the trivial module K are closely related to the nature of the differentials in this spectral sequence. A smiliar construction can be made for the cohomology of modules.

References

1. Adem, A.: Torsion in equivariant cohomology. Comm. Math. Helv. **64** (1989) 401–411
2. Adem, A.: On the exponent of cohomology of discrete groups. Bull. London Math. Soc. **21** (1989) 585–590
3. Alperin, J. L., Evens, L.: Representations, resolutions and Quillen's dimension theorem. J. Pure Appl. Algebra **22** (1981) 1–9
4. Alperin, J. L. and Evens, L.: Varieties and elementary abelian subgroups. J. Pure Appl. Algebra **26** (1982) 221–227
5. Avrunin, G. S.: Annihilators of cohomology modules. J. Algebra **69** (1981) 150–154
6. Avrunin, G. S., Scott, L.: Quillen stratification for modules. Invent. math. **66** (1982) 277–286
7. Benson, D. J., Carlson, J. F.: Complexity and multiple complexes. Math. Zeit. **195** (1987) 221–238
8. Benson, D. J., Carlson, J. F.: Projective resolutions and Poincaré duality complexes. (To appear)
9. Carlson, J. F.: The complexity and varieties of modules. (Lecture Notes in Mathematics, vol. 882.) Springer, Berlin Heidelberg New York, 1981, pp. 415–422
10. Carlson, J. F.: The varieties and the cohomology ring of a module. J. Algebra **85** (1983) 104–143
11. Carlson, J. F.: The variety of an indecomposable module is connected. Invent. math. **77** (1984) 291–299
12. Carlson, J. F.: Cohomology rings of induced modules. J. Pure Appl. Algebra **44** (1987) 85–97
13. Carlson, J. F.: Exponents of modules and maps. Invent. math. **95** (1989) 13–24
14. Evens, L.: The cohomology ring of a finite group. Trans. Amer. Math. Soc. **101** (1961) 224–239
15. Friedlander, E. M., Parshall, B. J.: Support varieties for restricted Lie algebras. Invent. math. **86** (1986) 553–562
16. Jantzen, J. C.: Kohomologie von p-Lie-Algebren und nilpontente Elemente. Abh. Math. Sem. Univ. Hamburg **76** (1986) 191–219
17. Niwasaki, T.: On Carlson's conjecture for cohomology rings of modules. (To appear)
18. Parshall, B. J.: Cohomology of algebraic groups. Proc. Symp. Pure Math. **47** (1987) 233–248
19. Quillen, D.: The spectrum of an equivariant cohomology ring, I, II. Ann. Math. (2) **94** (1971) 549–602
20. Quillen, D.: A cohomological criterion for p-nilpotency. J. Pure Appl. Algebra **1** (1971) 361–372
21. Quillen, D., Venkov, B. B.: Cohomology of finite groups and elementary abelian subgroups. Topology **11** (1972) 317–318
22. Serre, J. P.: Sur la dimension cohomologique des groupes profinis. Topology **3** (1965) 413–420

On Growth in Group Theory

Rostislav I. Grigorchuk

Moscow Institute of Railway Transportation Engineers
ul. Obraszowa 15, Moscow, USSR

1. General

The concept of growth appeared in group theory in the mid-fifties and now it plays an increasing role in this theory. Among different notions concerning the idea of growth in group theory the most important and successful is the notion of growth of a finitely generated (f.g.) group.

We start with the main definition, then enumerate the results concerning this definition and at the end we shall touch some other aspects of growth in group theory.

1.1 The Main Definition

Let G be a finitely generated (f.g.) group with generator system $A = \{a_1, \ldots, a_m\}$, and let $\partial(g)$ be the length of the element $g \in G$ with respect to the system A, in other words the minimal number k such that g can be represented in the form

$$g = a_{i_1}^{\varepsilon_1} \ldots a_{i_k}^{\varepsilon_k},$$

$$\varepsilon_j = \pm 1, \qquad j = 1, \ldots, k.$$

The growth function of the group G with respect to the system A is the function

$$\gamma(n) = \operatorname{card}\{g \in G; \partial(g) \leq n\}.$$

As $\gamma(n)$ depends on the generator system it is convenient to introduce an equivalence relation on the set of growth functions:

$$\gamma_1(n) \sim \gamma_2(n) \Leftrightarrow \exists C(\gamma_1(n) \leq \gamma_2(Cn) \& \gamma_2(n) \leq \gamma_1(Cn))$$

and the preordering relation:

$$\gamma_1(n) \preceq \gamma_2(n) \Leftrightarrow \exists C(\gamma_1(n) \leq \gamma_2(Cn)).$$

The equivalence class $[\gamma_g{}^A(n)]$ is an invariant of the group G. We shall call it the growth degree of the group G.

Obviously the growth degrees of a group and of any of its subgroup of finite index coincide.

The partially ordered (with respect to $\preceq$) set $\mathfrak{W}$ of the growth degrees of f.g. groups will be the main object of our consideration.

Proceedings of the International Congress
of Mathematicians, Kyoto, Japan, 1990

1.2 Examples

If $G = \mathbb{Z}^d$ is the free abelian group of rank d, then $\gamma(n) \sim n^d$. If $G = F_m$ is an (absolute) free group of rank $m \geq 2$ then $\gamma(n) \backsim e^n$. Thus the growth of a group can be polynomial of any degree $d \in \mathbb{Z}_+$ and can be exponential.

1.3 Connection with Geometrical Growth

The notion of the growth function was introduced by A.S. Schwarzc [1] and independently by J. Milnor [2]. This notion is a combinatorial form of the geometrical growth notion explored by V.A. Efremovich [3].

Let M be a Riemannian manifold. Its growth at infinity is characterized by the growth when $r \to \infty$ of the function

$$v(r) = \mathrm{Vol}(B_x(r)),$$

expressing the volume of the geodesic ball of radius r with center at the point $x \in M$. The examples of the Euclidean space $\mathbb{R}^d$ and hyperbolic space $\mathbb{H}^d$ show that this function in the first case grows as the polynomial r^d and in the second case as the exponential function e^r.

On the other hand the growth function $\gamma_G(n)$ of the f.g. group G expresses the volume of the ball $B_1(n)$ of radius n with center at the unit element $1 \in G$ if the group G is supplied with the metric $d(g, h) = \partial(g^{-1}h)$ and Haar measure.

The connection between geometric and algebraic growths can be also illustrated by A.S. Schwarzc's theorem [1]: if the manifold $\widetilde{M}$ is a universal covering manifold of the compact Riemannian manifold M then

$$v_{\widetilde{M}}(r) \sim \gamma_{\pi_1(M)}(r).$$

In other words the growth of the covering manifold $\widetilde{M}$ is the same as the growth of the fundamental group $\pi_1(M)$ of the manifold M.

2. Milnor's Problems

2.1 Some Results

Intensive investigation of the growth functions of f.g. groups began after J. Milnor's paper [2] was published. A series of important results was obtained within a short time interval. Here are some of them.

2.1.1 A f.g. nilpotent group G has polynomial growth (Wolf [4]) like n^d, where

$$d = \sum_k k \operatorname{rank}_Q(G_k/G_{k+1})$$

and $\{G_k\}$ is the lower central series of the group G (Bass [5]).

2.1.2 A f.g. solvable group has exponential growth if it is not virtually nilpotent (Wolf [4], Milnor [6]). (Virtually nilpotent means: contains a nilpotent subgroup of finite index).

2.1.3 A f.g. linear group has exponential growth if it is not virtually nilpotent (Tits [7]).

2.1.4 A free periodic Burnside group

$$B(m, n) = \langle a_1, \ldots, a_m | g^n = 1, g \in G \rangle$$

has exponential growth when $m \geq 2$ and $n \geq 665$ is odd (Adyan [8]).

In 1968 J. Milnor [9] proposed a number of problems concerning growth of groups.

1) "*Is the growth function $\gamma(n)$ necessarily equivalent either to a power of n or to the exponential function 2^n?*" 2) *In particular, is the growth exponent*

$$d = \lim_{n \to \infty} \frac{\log \gamma(n)}{\log n} \tag{1}$$

always either a well-defined integer or infinitely? For which groups is $d < \infty$? (A possible conjecture would be that $d < \infty$ if and only if G contains a nilpotent subgroup of finite index)".

(Division of Milnor's problem into Parts 1) and 2) has been made by me).

As a conjecture Problem 1) was formulated by Wolf [4] and Bass [5].

Conjecture: "*A* f.g. *group of non-exponential growth is virtually nilpotent*".

2.2 Two Definitions

A group whose growth function is not majorized by any function n^d and is not equivalent to the function e^n is called an intermediate growth group. And a group G is called a group of subexponential growth if $\gamma(n) \prec e^n$ (strong inequality), in other words, if

$$\lim_{n \to \infty} \sqrt[n]{\gamma(n)} = 1$$

(the limit exists due to the semi-multiplicativity of the function $\gamma(n)$).

3. Answer to Milnor's Problem 1)

Milnor's problem 1) was solved negatively in [10, 11, 12].

3.1 On the Construction of Intermediate Growth Groups

The simplest example of such a group is the group Γ from the paper [13] which was constructed as a simple example of a f.g. infinite periodic group. Γ is defined as the group of transformations of the interval [0, 1] from which rational points of the form $k/2^n$, $n = 1, 2, \ldots, 0 \leq k \leq 2^n$ are removed. This group is generated by four transformations a, b, c, d, where a is the permutation of the halves of the interval [0, 1] and b, c, d, are defined with the help of infinite periodic words

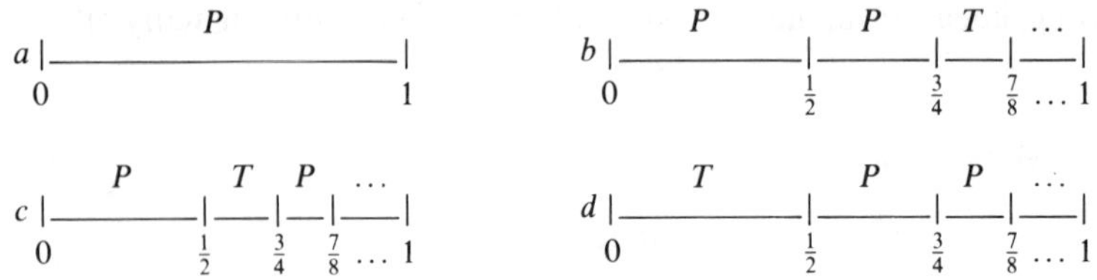

Fig. 1. (P denotes the permutation of the halves of the interval, T is the identity transformation).

$$\begin{array}{l} PPT\ PPT\ldots \\ PTP\ PTP\ldots \\ TPP\ TPP\ldots \end{array}$$

as in Figure 1.

We constructed uncountably many intermediate growth groups on the basis of this example (see [10–12]).

3.2 The Main Idea of the Growth Function $\gamma_\Gamma(n)$ Upper Estimation

The group Γ contains a subgroup H of finite index, which allows an embedding Ψ in the direct product of eight copies of the group Γ:

$$H \overset{\Psi}{\simeq} \hat{H} \leq \Gamma \times \Gamma \times \Gamma \times \Gamma \times \Gamma \times \Gamma \times \Gamma \times \Gamma,$$

and $\hat{H}$ is a subgroup of finite index in Γ^8.

This embedding has the following property: if $g \in H$ and

$$\Psi(g) = (g_1, g_2, \ldots, g_8) \in \hat{H}$$

then

$$\sum_{i=1}^{8} \partial(g_i) \leq k\partial(g),$$

where k is some constant less than 1. Hence

$$\gamma(n) \leq C \sum \gamma(n_1) \ldots \gamma(n_8) \tag{2}$$

(C is some constant), where the summation in (2) is taken over those sets of non-negative integers $n_1, \ldots, n_8$ for which the inequality

$$\sum_{i=1}^{8} n_i \leq kn$$

is valid.

If we assume that

$$\lim_{n\to\infty} \sqrt[n]{\gamma(n)} = \lambda > 1$$

then we shall obtain a contradiction $\lambda < \lambda^k$ from (2). So Γ has sub-exponential growth.

The fact that the function $\gamma(n)$ grows faster than any polynomial results, for example, from the following consideration.

The group Γ is commensurable with its square, in other words the groups Γ and $\Gamma \times \Gamma$ contain isomorphic subgroups of finite index. Hence $\gamma(n)$ satisfies the so-called Δ^2-condition $\gamma(n) \sim \gamma^2(n)$, which results in the lower estimate by some function of the type e^{n^β}, where $\beta > 0$.

However, the lower estimate of the function $\gamma_\Gamma(n)$ can be obtained directly. It is proved more precisely in [11] that

$$e^{\sqrt{n}} \preceq \gamma_\Gamma(n) \preceq e^{n^\alpha},$$

where $\alpha = \log_{32} 31$. This obviously gives a negative answer to Milnor's problem 1).

On the basis of a construction which generalizes this example we proved the following theorem (see [11, 14]).

Theorem 1. *The set of growth degrees of* f.g. *groups has the cardinality of the continuum. It contains a chain of the cardinality of the continuum, and an antichain of the cardinality of the continuum.*

I would like to stress that we are speaking precisely about intermediate degrees of groups in this theorem.

Another example of a group of intermediate growth was constructed by J. Fabrykowski and N. Gupta [15].

3.3 Some Properties of Intermediate Growth Groups

Intermediate growth groups can be constructed both in the class of p-groups (p is any given prime) and in the class of torsion free groups [11, 12].

The intermediate growth groups known up to this moment belong to the class of residually finite groups. Moreover, the groups from the papers [10–12] are residually-p-finite.

The periodic groups constructed in [11, 12] have the property that all their proper factor groups are finite. At the same time in [16] for any prime p, a f.g. p-group of intermediate growth having continuum nonisomorphic factor groups was constructed.

The torsion free group of intermediate growth constructed in [12] admits an invariant linear ordering.

The intermediate growth groups from [11, 12] are not finitely presentable.

There exist recursively presentable (by generating elements and defining relations) intermediate growth groups having solvable word problem. At the same time there are analogous groups with unsolvable word problem [11, 14].

4. On Groups with Polynomial Growth

An exhaustive answer to Part 2) of Milnor's problem was given in the papers [17, 18] and Milnor's conjecture suggested in this part was confirmed. Note that a more general result was obtained in [18], although this paper is mainly devoted to proving Gromov's theorem by methods of nonstandard analysis.

4.1 On Gromov's Theorem

Gromov's theorem proved in [17] gives the description of f.g. groups whose growth functions admit polynomial estimate

$$\gamma(n) \preceq n^d, \tag{3}$$

(d is some constant). This theorem states that a group G is virtually nilpotent if the estimate (3) is valid for its growth function.

Hence the estimate (3) results in an equivalence $\gamma(n) \sim n^d$ for a suitable $d \in \mathbb{Z}_+$.

The proof of this theorem is based on geometric considerations. The left invariant metric $d(g, h) = \partial(g^{-1}h), g, h \in G$ is built on the group G and the sequence

$$X_n = (G, \tfrac{1}{n}d), \qquad n \in \mathbb{N}$$

of metric spaces is considered.

Gromov proved that from this sequence one can extract some subsequence which converges in an exactly defined way to some metric space X_∞ if the condition (3) is valid. A homomorphism $G \to \mathrm{Isom}(X_\infty)$ from the group G into the isometry group of X_∞ arises and $\mathrm{Isom}(X_\infty)$ is a Lie group with a finite number of connected components. Some simple algebraic considerations complete the proof of this theorem.

4.2 On L. van den Dries and A. Wilkie's Theorem

The main result of [18] is: if the inequality

$$\gamma(n) \leq Cn^d$$

is valid on some infinite subset $\mathbb{N}_0 \subseteq \mathbb{N}$ of the set of natural numbers then the group G is virtually nilpotent.

From this statement if follows that the limit (1) (finite or infinite) always exists, and also $d < \infty$ if and only if the group G is virtually nilpotent.

5. On the Growth of Cancellation Semigroups

The notion of the growth function of a f.g. semigroup is defined as in the case of a group. The growth of a semigroup can be very strange even in the case when the condition (3) holds true. But if we consider the class of semigroups with left and right cancellation laws then in this case the statement similar to Gromov's theorem is valid.

5.1 The Nilpotency of Semigroups Due to A.I. Malcev

Let x, y, ξ_1, ξ_2, $\ldots$, ξ_n, $\ldots$ be symbols denoting variables running through a semigroup S. Let us denote $X_0 = x$, $Y_0 = y$ and after that by induction

$$X_{n+1} = X_n \xi_n Y_n, \qquad Y_{n+1} = Y_n \xi_n X_n.$$

The semigroup S whose elements satisfy the identity $X_n = Y_n$ for some n is called nilpotent.

A.I. Malcev proved [19] that the classical nilpotency identity for groups is equivalent to the semigroup identity $X_n = Y_n$ for corresponding n.

5.2 On Cancellation Semigroup of Polynomial Growth

Let $S_0 \subseteq S$ be a subsemigroup. We shall say that S_0 has finite index in S if there exists a finite subset $K \subseteq S$ such that for any $s \in S$ there exists $k \in K$ for which $sk \in S_0$.

Theorem 2. *A* f.g. *cancellation semigroup S has polynomial growth* (*Condition* (3)) *if and only if S contains a nilpotent subsemigroup of finite index.*

The proof of this theorem is based on the remark that for a cancellation semigroup of subexponential growth the Ore condition is satisfied and this semigroup possesses a group $G_S = S^{-1}S$ of left quotients.

The estimate (3) on the growth function of the semigroup S makes it possible to give a polynomial type estimate on the growth function of the group G_S after detailed analysis. Then Gromov's theorem can be applied.

5.3 Some Remarks

In [20] it was proved that if a semigroup S with cancellation has polynomial growth, then the group G_S of left quotients also has polynomial growth of the same degree.

It would be interesting to construct examples of semigroups with cancellation of subexponential growth for which there is a jump of growth degree in the diagram $S \to G_S$.

If one could construct a semigroup S with cancellations of subexponential growth such that G_S has exponential growth, then Problem N12 from [21] will be positively solved.

6. On Lacunae in the Set of Growth Degrees of Residually Nilpotent Groups

Recall that a group G is residually-p if for any nonunit element $g \in G$ there exists a finite p-group K and a homomorphism $\varphi : G \to K$ such that $\varphi(g) \neq 1$.

Theorem 3 [22]. *Let p be any prime, let the* f.g. *group G be residually-p and its growth function satisfy the estimate* (3). *Then G is virtually nilpotent and so has polynomial growth.*

6.1 Proof of Theorem 3

Let F_p be a finite prime field, $F_p[G]$ the group algebra, $\Delta \leq F_p[G]$ the augmentation ideal, in other words the ideal generated by elements of the form $g - 1$, $g \in G$, $\mathrm{gr}(G)$

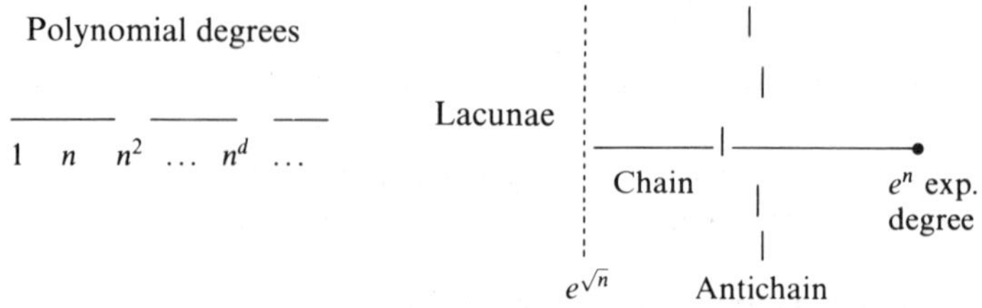

Fig. 2

the associative graded algebra defined by means of the powers of the augmentation ideal

$$\operatorname{gr}(G) = \bigoplus_{n=0}^{\infty} A_n = \bigoplus_{n=0}^{\infty} \Delta^n/\Delta^{n+1},$$

let $f_{G,P}(t)$ be the Hilbert-Poincaré series of the algebra $\operatorname{gr}(G)$:

$$f_{G,P}(t) = \sum_{n=0}^{\infty} a_n(G)t^n,$$

where $a_n(G) = \dim_{F_p} A_n$.

It follows from Lazard's Theorem 3.11 [23] that either the sequence $a_n(G)$ has polynomial growth and then the p-completion $\hat{G}$ is p-analytic or $a_n(G) \succeq e^{\sqrt{n}}$. Besides if $\gamma(n)$ is the growth function of the group G with respect to any system of generators then the inequality $a_n(G) \leq \gamma(n)$, $n = 1, 2, \ldots$ is valid.

Using Lazard's result and Condition (3) we obtain that the growth of coefficients $a_n(G)$ when $n \to \infty$ is polynomial and so the p-completion $\hat{G}$ is analytic. It follows from Tit's theorem [7] that either G contains a free subgroup with two generators or G is a virtually solvable group. But the first case is impossible due to limitation on the growth, and in the second case, we can conclude that G is virtually nilpotent due to results of [4, 6].

Recently A. Lubotzky and A. Mann [24] have pointed out that in Theorem 3 the assumption that G is residually-p can be changed by the assumption that G is residually nilpotent.

6.2 On the Scale of Growth Degrees of Residually-p groups

Let p be any prime and denote by $\mathfrak{W}_p$ the set of growth degrees of f.g. residually-p groups. Theorem 1 formulated above for the whole class of groups is also true for every set $\mathfrak{W}_p$. So, due to Theorem 3 the structure of the partially ordered set $\mathfrak{W}_p$ can be presented in the form as in Figure 2.

7. On the Generating Series of Growth Function

Many questions of a theoretical and applied character require more detailed investigation of the asymptotic behavior of the growth function $\gamma(n)$ than up to the equivalence $\sim$, defined above. For this it is sometimes useful to connect the generating series

$$\Gamma(t) = \sum_{n=0}^{\infty} \gamma(n) t^n$$

to the function $\gamma(n)$.

7.1 Some Cases When the Function $\Gamma(t)$ is Rational for any System of Generators

i) The group is virtually abelian (Benson [25])
ii) The group is a cocompact group of isometries of the hyperbolic space $\mathbb{H}^d$ (Cannon [26])
iii) The group is hyperbolic (Gromov [27]).

For many nilpotent groups the series the $\Gamma(t)$ also represents a rational function (see for example [28]). At the same time the conjecture that for any f.g. nilpotent group and any of its finite generating set the function $\Gamma(t)$ is rational proved to be false. Namely F. Grunewald proved that for the nilpotent group

$$G = \langle a_1, a_2, b_1, b_2, z | [a_1, b_1] = [a_2, b_2] = z, z \in \mathbb{Z}(G),$$
$$[a_1, a_2] = [b_1, b_2] = [a_1, b_2] = [a_2, b_1] = 1 \rangle$$

($\mathbb{Z}(G)$ is the centre of the group G and the function $\Gamma(t)$ defined using the system of generators $\{a_1, a_2, b_1, b_2, z\}$), the generating series $\Gamma(t)$ is not a rational function.

7.2 Information on the Growth Functions of Nilpotent Groups

Let G be a nilpotent group, $\gamma(n)$ the growth function of G with respect to any system of generators, d the power of polynomial growth of G. Then

a) there exists the limit

$$\lim_{n \to \infty} \frac{\gamma(n)}{n^d} = C$$

(Pancu [29]),

b) the estimate

$$\gamma(n) = Cn^d + O(n^{d-1/2})$$

is valid (Grunewald, unpublished).

It is interesting to find new examples of rationality of the growth function $\Gamma(t)$ and also to obtain another information about the generating series of growth functions.

8. Some Problems and Conjectures

The theory of growth degrees of groups has been developed for more than three decades and has already accumulated a lot of unsolved problems. Let me formulate some of them.

8.1 (A well known problem.) *Is there a* f.g. *group of intermediate growth with a finite set of defining relations?*

8.2 *Is it true that if the growth function of a* f.g. *group G grows slower than the function* $e^{\sqrt{n}}$, *then G is virtually nilpotent?*

8.3 *Does there exist a* f.g. *group whose growth function is equivalent to the function* $e^{\sqrt{n}}$?

8.4 *Is it true that any group of subexponential growth is residually finite?*

8.5 *To find the asymptotic behaviour of the growth function of the group Γ from* [13] *(conjecture is that $\gamma(n) \sim e^{\sqrt{n}}$ in this case).*

8.6 *Is it true that for any* f.g. *nilpotent group the function $\Gamma(t)$ is meromorphic or even algebraic?* (F. Grunewald).

Let

$$G = \langle A | r = 1 \ (r \in R) \rangle$$

be f.g. and $\mathfrak{J}$ be the Cayley graph of G with respect to the system of generators A. A spanning tree $T \subseteq \mathfrak{J}$ is regular if it is defined by a finite automaton. A spanning tree T is minimal if every word in T is the shortest.

8.7 **Conjecture.** *G has a rational growth function if and only if $\mathfrak{J}$ has a minimal regular spanning tree* (Machi, Schupp).

9. Other Aspects of Growth in Groups

9.1 Cogrowth

Let a f.g. group G be realized as a factor group F_m/H of the free group F_m of rank m, $H \leq F_m$ a normal subgroup, the elements of which will be further considered as reduced words over the basis of F_m. Denote by $h(n)$ the number of words of length $\leq n$ in H. Obviously $h(n) \leq 2m(2m-1)^{n-1}$. Let

$$\alpha_H = \limsup_{n \to \infty} \sqrt{h(n)}, \qquad \mathscr{H}(t) = \sum_{n=0}^{\infty} h(n) t^n.$$

The value α_H which appeared in [30] is called the growth exponent of the normal subgroup H and belongs to the interval $(\sqrt{2m-1}, 2m-1]$.

In [31] the following amenability criterion was proved.

Theorem 4. *The group $G = F_m/H$ is amenable if and only if $\alpha_H = 2m - 1$.*

The function $\mathscr{H}(t)$ and growth exponent α_H can be defined also for any subgroup $H \leq F_m$. In [31] it was proved that if H is a f.g. group then $\mathscr{H}(t)$ is a rational function which can be effectively calculated if the system of generators of H is known.

9.2 Subgroup Growth

Let G be a f.g. group, and let $\varrho_n(G)$ be the (finite) number of its subgroups of index n. The problem of investigation of the asymptotic behavior of the sequence $\varrho(n)$ when $n \to \infty$ has been drawing more and more interest. For example in [32] F. Grunewald, D. Segal and G. Smith investigated in detail the Dirichlet series

$$\xi_G(s) = \sum^{\infty} \varrho_n(G) n^{-s}$$

of the sequence $\varrho_n(G)$ for the case when G is nilpotent.

On the other hand A. Lubotzky and A. Mann [24] showed that if G is a residually nilpotent f.g. group then $\varrho_n(G)$ has polynomial growth if and only if G is solvable of finite rank.

The general case of a residually finite group is still open. See [24] for other references and comments.

9.3 On the Growth of Graded Algebras Associated to Groups

During investigations of groups different graded algebras associated to these groups arise. For example the associative graded algebra gr G was defined above. Studying the growth of algebras and in particular graded algebras associated to groups is important from different points of view and is connected with different applied aspects of algebra [33–37].

9.4 On the Asymptotic Behavior of Random Walks on Groups

Investigation of random walks on groups was started by H. Kesten in [38]. It is interesting to investigate the asymptotic behavior when $n \to \infty$ of the probability $P_{1,1}{}^{(n)}$ of returning to the unit element after n steps [38, 39], to calculate the entropy of random walks [40], to describe Martin's and Poisson's boundaries [41, 42] and so on.

10. Applications

10.1 To the Theory of Invariant Means

The definition of an invariant means and the corresponding notion of amenable groups belong to J. von Neumann [43].

Finite and abelian groups are amenable. The class of amenable groups is closed under operations of extension and inductive limit.

Let us denote the class of amenable groups by AG, the class of groups without free subgroups with two generators by NF, and the class of elementary amenable groups constructed by J. von Neumann by EG, i.e. the minimal class of groups containing finite groups, abelian groups and closed under the operations of extension and inductive limit. The following imbeddings EG $\subseteq$ AG $\subseteq$ NF hold true.

In 1957 M. Day [44] proposed the following two problems:

(1) *Is it true that* AG = NF?
(2) *Is it true that* AG = EG?

Some mathematicians ascribe Problem (1) to J. von Neumann but there is no written confirmation of the fact that it was he who suggested this problem.

In 1979 to disprove the conjecture AG = NF I suggested A.Yu. Olshanskii to apply the amenability criterion (Theorem 4) to the groups constructed by him at that moment for the solution of some famous problems of group theory. This suggestion was realized in the short paper [45].

A little later S.I. Adyan [46] also applied the amenability criterion to prove that free periodic Burnside groups $B(m, n)$ are nonamenable when $m \geq 2$ and $n \geq 665$ is odd (the conjecture that these groups are nonamenable was expressed by S.I. Adyan in 1975).

The negative answer to the problem (2) was given in [11].

Theorem 5. *There exist uncountably many amenable groups not belonging to the class* EG.

Hence the class AG of amenable groups is much more than the class EG of elementary amenable groups.

Theorem 5 is an easy corollary to the existence of a continuum of nonisomorphic intermediate growth groups.

Using intermediate growth groups we managed [48] to disprove one of Rosenblatt's conjectures [47] (see also [21], Problem 11) about the so-called superamenable groups.

10.2 To Riemannian Geometry

The results concerning the growth functions of groups are widely used in the classification theory of Riemannian manifolds up to quasi-isometries, in the theory of foliations, in the investigation of Laplace operator on Riemannian manifolds (see [49, 50]), ...

10.3 Other Applications

Information concerning the growth functions of groups is used in many, sometimes unexpected branches of mathematics: in the theory of random walks [38, 39], ergodic theory [50, 52], the theory of finite automata [20] and so on.

In my lecture I did not touch upon many other aspects of applications of the notion of growth in group theory, many of which at this moment are only in their initial stage of development.

References

1. Schwarzc, A.S.: A volume invariant of coverings. Dokl. Ak. Nauk USSR **105** (1955) 32–34
2. Milnor, J.: A note on curvature and fundamental group. J. Diff. Geom. **2** (1968) 1–7

3. Efremovic, V.A.: The proximity geometry of Riemannian manifolds, Uspekhi Math. Nauk **8** (1953) 189 (Russian)
4. Wolf, J.: Growth of finitely generated solvable groups and curvature of Riemannian manifolds. J. Diff. Geom. **2** (1968) 424–446
5. Bass, H.: The degree of polynomial growth of finitely generated nilpotent groups. Proc. London Math. Soc. **25** (1972) 603–614
6. Milnor, J.: Growth in finitely generated solvable groups. J. Diff. Geom. **2** (1968) 447–449
7. Tits, J.: Free subgroups in linear groups. J. Algebra **20** (1972) 250–270
8. Adyan S.I.: The Burnside problem and identities in groups. Nauka, Moscow 1975 (Russian) [English transl.: Proc. Steklov Inst. Math. (1970) 142]
9. Milnor, J.: Problem 5603. Amer. Math. Monthly **75** (1968) 685–686
10. Grigorchuk, R.I.: On Milnor's problem of group growth. Dokl. Ak. Nauk SSSR **271** (1983) 31–33 (Russian) [English transl.: Soviet Math. Dokl. **28** (1983) 23–26]
11. Grigorchuk, R.I.: The growth degrees of finitely generated groups and the theory of invariant means. Izv. Akad. Nauk SSSR. Ser. Math. **48** (1984) 939–985 (Russian) [English transl.: Math. SSSR Izv. **25** (1985)]
12. Grigorchuk, R.I.: On the growth degrees of *p*-groups and torsion-free groups. Math. Sbornik **126** (1985) 194–214 (Russian) [English transl.: Math. USSR Sbornik **54** (1986) 185–205]
13. Grigorchuk, R.I.: On the Burnside problem for periodic groups. Funct. Anal. Prilozen. **14** (1980) 53–54 [English transl.: Funct. Anal. Appl. **14**, 41–43]
14. Grigorchuk, R.I.: A groups with intermediate growth function and its applications. Second degree doctoral thesis, Moscow 1985
15. Fabrykowski, J., Gupta, N.: On groups with subexponential growth functions. J. Ind. Math. Soc. **49** (1985) 249–256
16. Grigorchuk, R.I.: The construction of intermediate growth groups having continuum factor groups. Algebra and Logic **23** (1984) 383–394
17. Gromov, M.: Groups of polynomial growth and expanding maps. Publ. Math. IHES **53** (1981) 53–73
18. Van den Dries, L., Wilkie, A.J.: Gromov's theorem on groups of polynomial growth and elementary logic. J. Alg. **89** (1984) 349–374
19. Malcev, A.I.: Nilpotent semigroups. Uchen. Zap. Ivanovsk. Ped. In-ta **4** (1953) 107–111 (Russian)
20. Grigorchuk R.I.: On the cancellation semigroup of polynomial growth. Matem. Zametki **43** (1988) 305–319 (Russian)
21. Wagon, S.: The Banach-Tarski Paradox. Encyclopedia of mathematics and its applications 24. Cambridge University Press 1985
22. Grigorchuk, R.I.: On the Hilbert-Pouncare series of the graded algebras associated to groups. Mat. Sbornik **180** (1989) 307–225
23. Lazard, M.: Groupes analytiques *p*-adiques. Publ. Math. IHES **26** (1965) 389–603
24. Lubotzky, A., Mann, A.: On groups of polynomial subgroup Growth. Preprint 1990, pp. 1–22
25. Benson, M.: Growth series of finite extentions of $\mathbb{Z}^d$ are rational. Invent. math. **73** (1983) 251–269
26. Cannon, J.: The combinatorial structure of cocompact discrete hyperbolic groups. Geometricae Dedicata **16** (1984) 123–148
27. Gromov, M.: Hyperbolic groups. Essays in group theory. MSRI Publications 8 (ed. S.M. Gersten). Springer, Berlin Heidelberg New York 1987, pp. 75–263
28. Shapiro, M.: A geometrical approach to the almost convexity and growth of some nilpotent groups. Preprint 1989
29. Pancu, P.: Croessance des boules et des geodeseques fermees dans les nilvarietes. Ergodic theory and dynamic systems, 1983, pp. 415–445

30. Grigorchuk, R.I.: Symmetrical random walks on discrete groups. Uspekhi Math. Nauk **XXXII** (1977) 217–218 (Russian)
31. Grigorchuk, R.I.: Symmetrical random walks on discrete groups. In: Multicomponent random systems (ed. R.L. Dobrushin, Ya.G. Sinai). Nauka, Moscow 1978, pp. 132–152. [English transl.: Advances in probability and related topics, vol. 6. Marcel Dekker 1980, pp. 285–325]
32. Grunewald, F.J., Segal, D., Smith, G.C.: Subgroups of finite index in nilpotent groups. Invent. math. **93** (1988) 185–233
33. Gelfand, I.M., Kirillov, A.A.: Sur les corps lie's aux algebres enveloppantes des algebres de Lie. Publ. Math. IHES **31** (1966) 5–19
34. Krause, G.R., Lenagan, T.H.: Growth of algebras and Gelfand-Kirillov dimention. Pitman Advanced Publishing Program, Boston London Melbourne 1985
35. Babenco, I.C.: The problems of growth and rationality in Algebra and Topology. Uspekhi Math. Nauk **41** (1989) 95–142
36. Grigorchuk, R.I.: On the topological and metrical types of a regular covering surfaces. Izv. Akad. Nauk SSSR. **54** (1990) 498–536 (Russian)
37. Bereznyi, A.E.: Discrete sub-exponential groups Zap. Nauch Sem. LOMI, **123** (1983) 155–166
38. Kesten, H.: Symmetric random walks on groups. Trans. Amer. Math. Soc. **92** (1959) 336–354
39. Kesten, H.: Full Banach mean values on countable groups. Math. Scand. **7** (1959) 146–156
40. Kaimanovich, V.A., Vershik, A.M.: Random walks on discrete groups: boundary and entropy. Ann. Probab. **11** (1983) 457–490
41. Dynkin, E.B., Maljutov, M.B.: Random walks on groups with a finite number of generators. Soviet Math. Dokl. **2** (1961) 339–402
42. Furstenberg, H.: Random walks and Discrete subgroup of Lie groups. Adv. Probab. Related Topics **1**, 3–63. Dekker, New York (1971)
43. Von Neumann, J.: Zur allgemeinen Theorie des Masses. Fund. Math. **13** (1929) 73–116
44. Day, M.: Amenable semigroups. Ill. J. Math. **1** (1957) 509–544
45. Olshanskii, A.Yu.: On the problem of the existence of an invariant mean on a group. Uspekhi Mat. Nauk **35** (1980) 1165–1166
46. Adyan, S.I.: Random walks on free periodic groups. Izv. Akad. Nauk SSSR, Ser. Mat. **46** (1982) 1139–1149 (Russian) [English transl.: Math. USSR Izv. **21** (1983)]
47. Rosenblatt, J.: Invariant measures and growth conditions. Trans. Amer. Math. Soc. **193** (1974) 33–53
48. Grigorchuk, R.I.: Supramenability and problem of embedding a free semigroups. Funkt. Anal. Priloz. **21** (1987) 74–75 (Russian)
49. Brooks, R.: The fundamental group and the spectrum of the Laplacian. Comm. Math. Helv. **56** (1981) 581–598
50. Varopoulos, N.: Brownian motion and transient groups. Ann. Inst. Fourier (Grenoble) **34** (1984) 243–269

Absolute Integral Closure and Big Cohen-Macaulay Algebras

*Craig Huneke**

Department of Mathematics, Purdue University, West Lafayette, IN 47907, USA

Introduction

Our purpose in this paper is to discuss a recent result by this author jointly with Melvin Hochster which gives the existence of big Cohen-Macaulay (C-M) algebras in equicharacteristic. In general proofs will be omitted, though sometimes sketchy arguments will be given. Before stating the main theorem, we give a review of the basic definitions and concepts used in this paper, particularly the property of a ring's being C-M. Throughout this paper, "ring" means commutative noetherian ring with identity, with one notable exception: the integral closure of a domain R in an algebraic closure of its quotient field will come into play. However, as these rings are never noetherian if $\dim(R) > 0$, no confusion will arise.

Definition 1.1. *Let R be a ring, let M be an R-module, and let $x_1, \ldots, x_n \in R$. The elements $x_1, \ldots, x_n$ are said to be a regular sequence on M (or simply an M-sequence) if* i) *x_i is not a zerodivisor on the module $M/(x_1, \ldots, x_{i-1})M$ for each i between* 1 *and n, and* ii) *$M \neq (x_1, \ldots, x_n)M$.*

Condition ii) is a nondegeneracy restriction, which is especially relevant when the module M is not finitely generated. For example if R is a domain and M is its quotient field, then i) is trivially satisfied, but ii) fails.

The first condition can be rephrased in several ways of which we list two:

i)′ whenever $\sum_{i=1}^{k} y_i x_i = 0$ with $y_i \in M$, $1 \le k \le n$, then $y_k \in (x_1, \ldots, x_{k-1})M$.

If M is finitely generated and the x_i are in the Jacobson radical of R, then i) is equivalent to,

i)″ The first Koszul homology $H_1(\mathbf{x}, M) = 0$, where here and below $\mathbf{x}$ stands for all the x_i. (The Koszul complex $K.(\mathbf{x}, M)$ is the tensor product of the n complexes $0 \to R \xrightarrow{x_i} R \to 0$ with M.)

* Partially supported by the NSF.

Proceedings of the International Congress
of Mathematicians, Kyoto, Japan, 1990

In any case, if the $\mathbf{x}$ form a regular sequence on M, then the first Koszul homology is 0; it is the converse that requires some restrictions.

The (Krull) *dimension* of a ring R, $\dim(R)$, is $\sup\{n \in \mathbf{N} \mid P_0 \subsetneq P_1 \subsetneq \cdots \subsetneq P_n : P_i$ are prime ideals of $R.\}$

A ring $(R, \mathfrak{m})$ is *local* if there is a unique maximal ideal $\mathfrak{m}$ of R. The dimension of a local ring $(R, \mathfrak{m})$ can also be characterized as the least integer n for which there exist n elements $x_1, \ldots, x_n$ with the property that the nilradical of the ideal they generate is exactly $\mathfrak{m}$. Such a system of elements is called a *system of parameters* (s.o.p.). A sequence of elements which can be extended to an s.o.p. will be called *parameters* in this paper.

Definition 1.2. *A local ring R (necessarily noetherian, recall) is* Cohen-Macaulay, *if there is a regular sequence $x_1, \ldots, x_n$ in $\mathfrak{m}$ whose length $n =$* $\dim(R)$. *A noetherian ring R is* C-M *if $R_\mathfrak{p}$ is* C-M *for every prime ideal $\mathfrak{p}$ of R (equivalently, every maximal ideal $\mathfrak{p}$ of R).*

Any sequence as in (1.2) will have to be an s.o.p. Moreover if R is C-M then any s.o.p. will form a regular sequence. In general the length of a maximal regular sequence of elements in the maximal ideal $\mathfrak{m}$ of a local ring $(R, \mathfrak{m})$ is called the *depth of R*, denoted $\operatorname{depth}(R)$. It always holds that $\operatorname{depth}(R) \leq \dim(R)$; R is C-M precisely when equality occurs. Some examples of C-M rings and how they occur are given below.

Example 1.3. Any 0-dimensional ring is C-M vacuously. Any 1-dimensional domain R is C-M, since any non-zero element is a nonzerodivisor.

Example 1.4. Not all 2-dimensional rings, even domains, are C-M (cf. 1.9 below). However, integrally closed 2-dimensional domains are C-M. As integral closure and finite maps play an important role in this paper, we review these concepts here. If $R \subseteq S$, an element $s \in S$ is said to be *integral* over R if s satisfies a monic polynomial equation with coefficients in R. The set of all elements of S integral over R form a ring T called the *integral closure of R in S*. If R is a domain and S is its quotient field, then T is called the integral closure of R. If further $R = T$, R is said to be integrally closed. A homomorphism from R into S is said to be *finite* if S is a finite R-module. This is equivalent to saying S is a finitely generated R-algebra in which every element is integral over R.

Example 1.5. Any regular ring is C-M. A local ring $(R, \mathfrak{m})$ is *regular* if the maximal ideal $\mathfrak{m}$ can be generated by an s.o.p. A noetherian ring is regular if all its local rings at prime (equivalently maximal) ideas are regular local rings. Regular local rings are exactly those local rings in which every finitely generated module has finite projective dimension. Examples of regular rings include polynomial rings and power series rings over fields. Regularity is the analogue of the geometric notion of being non-singular.

Example 1.6. If $(R, \mathfrak{m})$ is C-M local and $x_1, \ldots, x_k$ are parameters (hence also a regular sequence) then the quotient ring $R/(x_1, \ldots, x_k)$ is also C-M. Combining this example with (1.5) gives that the so-called complete intersections are C-M. A *complete intersection* is by definition a ring which is a regular ring modulo an ideal generated by a regular sequence. More generally a local Gorenstein ring is C-M. Gorenstein local rings are characterized by the property that they have finite injective dimension over themselves, or alternatively that every s.o.p. generates an irreducible ideal (*irreducible* means not the intersection of two strictly larger ideals).

Example 1.7. If G is a linearly reductive algebraic group over k acting k-linearly on a polynomial ring $S = k[x_1, \ldots, x_n]$, then the ring of invariants $R = S^G$ is C-M. This is due to Hochster and Roberts [H-R], and includes an enormous number of examples. A more general result was recently obtained by myself and Hochster [H-H1]: if S is any regular local ring containing a field, and R is a subring of S such that the inclusion of R into S splits R-linearly, then R is C-M (in positive characteristic this was known [H-R]; see also [Bo]). Our proof uses the theory of tight closure which is discussed below, cf. (2.3).

Example 1.8. If $X \subseteq \mathbb{P}_k^n$ is a projective variety of dimension d which is arithmetically normal (i.e., its homogeneous coordinate ring R is integrally closed) then X is arithmetically C-M (equivalently, R is C-M) iff for all integers t and all i, $1 \leq i \leq d-1$, $H^i(X, \mathcal{O}_X(t)) = 0$. X is said to be C-M itself if its local rings are C-M. Geometrically C-M varieties are quite an important and good class; for instance Serre duality holds if X is C-M. Furthermore the computation of the degree of X (equivalently the multiplicity of R) is made easy if X is arithmetically C-M – one only needs to take a general linear space L of complementary dimension such that $X \cap L$ is empty, and then $\deg(X) = \dim_k(S/I(X) + I(L))$, where $I(Y)$ is the homogeneous ideal of functions vanishing on Y and $S = k[X_0, \ldots, X_n]$. In fact, this property characterizes when X is arithmetically C-M.

Example 1.9. To see an explicit ring which is *not* C-M take R to be the subring $k[t^4, t^3s, ts^3, s^4]$ of the polynomial ring $k[t, s]$ in two variables over a field k. It is easy to see that t^4 and s^4 form a homogeneous s.o.p. for R but are not a regular sequence, since $t^4(ts^3)^2 = s^4(t^3s)^2$, but t^4 does not divide t^6s^2 *in* R, as it would have to do if R is C-M. The element t^2s^2 is missing. If we adjoin this element to R, we actually arrive at the integral closure of R, which is then C-M as it is a 2-dimensional integrally closed domain (see (1.4)).

Example 1.10. While C-M rings can be highly singular, nonetheless they are in a partial sense homologically trivial over regular rings – this is made precise below for two important classes of rings, complete local rings and nonnegatively graded rings over a field. To this end, let $(R, \mathfrak{m}, k)$ denote either a complete local ring with maximal ideal $\mathfrak{m}$ (*complete* means complete in the topology determined by the powers of the maximal ideal) and coefficient field k (a *coefficient field* is a field k contained in R such that the map from k to $R/\mathfrak{m}$ is an isomorphism; if R contains a field and is complete, then such fields always exist), or let $R = \bigoplus_{i \geq 0} R_i$ be a graded ring with $R_0 = k$, and $\mathfrak{m} = \bigoplus_{i \geq 1} R_i$. In this case $R/\mathfrak{m} \simeq R_0 = k$. We will say we are in case 1 if $(R, \mathfrak{m}, k)$ is complete local, and in case 2 if $(R, \mathfrak{m}, k)$ is graded. In both of these fundamental cases, R can be represented as finite over a regular ring A. In case 2, the Noether normalization theorem shows that if $x_1, \ldots, x_n$ is any homogeneous s.o.p. for R, then the subring $k[x_1, \ldots, x_n]$ of R has the properties that R is a finite A-module, and A is isomorphic to a polynomial ring in n-variables over k, i.e., the elements $x_1, \ldots, x_n$ are algebraically independent. In case 1, the Cohen structure theorem for complete local rings shows that if $x_1, \ldots, x_n$ is any s.o.p. for R and we take A to be the complete subring $k[[x_1, \ldots, x_n]]$ of R generated by $x_1, \ldots, x_n$, then R is a finite A-module, and A is isomorphic to a power series ring in n-variables over k, i.e., $x_1, \ldots, x_n$ are analytically independent. We may thus study the properties of R through thinking of it as an A-module in both cases. The homological properties of A are much nicer in general – regular rings are characterized by the fact that modules over them have finite free resolutions.

The property that R is C-M can now be easily expressed in terms of the A-module structure of R. In both cases, R *is* C-M *iff* R *is a free* A*-module*, i.e., is "homologically trivial" over A. This is a useful point of view.

2. Trivializing Relations on Parameters

Let me return for a moment to (1.9). There we had a 'bad' relation on the parameters t^4 and s^4, namely the relation $t^4(ts^3)^2 = s^4(t^3s)^2$. This relation is nontrivial in the sense that it does not come from the trivial relation on t^4 and s^4, i.e., the relation $t^4(s^4) = s^4(t^4)$. To be precise, if $x_1, \ldots, x_n$ are elements in R, we say a relation $(r_1, \ldots, r_n)$ on these elements (i.e., $\sum r_i x_i = 0$) is *trivial* if it is an R-linear combination of the Koszul relations $(0, \ldots, 0, x_j, 0, \ldots, 0, -x_i, 0, \ldots, 0)$ where the x_j is in the ith spot and the $-x_i$ is in the jth spot. In a C-M local ring all relations on parameters are trivial; this is the fundamental property of C-M rings. However, even if R is not C-M we can ask if a given relation on parameters $x_1, \ldots, x_k$ can be made trivial in some nice extension of R. This phenomenon occurred in (1.9), where by passing to the integral closure of R, we add the element t^2s^2, and then write our nontrivial relation (over R) $(t^2s^6, -t^6s^2)$ as $t^2s^2(s^4, -t^4)$ (over S).

Unfortunately, or perhaps fortunately, else life would be too simple, it is not enough to pass to the integral closure to ensure relations on parameters become trivial, since not all integrally closed rings are C-M. The homogeneous coordinate ring of an elliptic curve cross $\mathbb{P}^1$ is a 3-dimensional integrally closed non C-M ring. Nonetheless one might naively hope that it is possible to trivialize relations on parameters of R in some finite extension S of R. There has been no known natural way to increase depth past 2 (one can obtain depth 2 by taking integral closure). A similar situation prevails for modules over normal rings, where by taking double duals one can pass to modules having depth 2, but no construction is known to 'make' a module have depth bigger than 2 if it does not already possess this property.

In characteristic 0 it is not in general possible to trivialize relations in finite extensions. Indeed, if R is an integrally closed local domain containing the rationals, and $\sum_{i=1}^{k} r_i x_i = 0$ is a nontrivial relation on parameters $x_1, \ldots, x_k$ of R, then this relation can *never* be made trivial by passing to a finite extension S of R: if such an S existed, then $r_k \in (x_1, \ldots, x_{k-1})S$. Let β denote the map $(1/[L:K])\mathrm{tr}_{L/K}$, where K

is the quotient field of R, L the quotient field of S, and tr( ) is field trace. Then β is an R-linear map from S to R, splitting the inclusion of R into S. In particular, $r_k = \beta(r_k) \in (\beta(x_1), \ldots, \beta(x_{k-1}))R = (x_1, \ldots, x_{k-1})R$, a contradiction. This impossibility in characteristic 0 makes the following main theorem (in characteristic p) somewhat miraculous.

Main Theorem 2.1 [H-H5]. *Let $(R, \mathfrak{m})$ be an excellent local domain of positive characteristic p, and let R^+ denote the integral closure of R in an algebraic closure of the quotient field K of R (called an* absolute integral closure *in* [Ar]). *If $x_1, \ldots, x_n$ is an s.o.p. of R, then $x_1, \ldots, x_n$ is a regular sequence in R^+, i.e., R^+ is a big C-M algebra for R.*

Here 'big' refers to the fact that R^+ is not a finitely generated R-module. In fact, since the quotient field of R^+ is algebraically closed, R^+ cannot even be noetherian (unless $R = K$ is a field). We will not give an explicit definition of "excellent" except to say all algebras essentially of finite type over a field and all complete local rings are excellent.

The meaning of (2.1) is that given *any* nontrivial relation on the parameters $x_1, \ldots, x_n$ of R, there will be a finite extension domain S of R in which the relation becomes trivial! Of course, passing to S may reap a harvest of new bad relations, but each one can again be trivialized in a further finite extension. While no one finite extension S may suffice to trivialize all relations (in such a case S would be C-M), by passing to the union of all finite extensions, namely R^+, we do obtain such trivialization.

The Frobenius map $F: R \to R$ sending r to r^p plays an important role in the proof of (2.1), as one might expect. To explain how the Frobenius is used I need to digress to talk about the unmixed part of an ideal.

Every ideal in a noetherian ring R has a primary decomposition. An ideal J is *unmixed* if all the primes associated to primary components of J are minimal over J, i.e., there are no embedded components. (One note for experts; one usually also demands that minimal components of J have the same dimension to call J unmixed, but in the context in which we will use this concept, this latter condition will be automatic.) C-M rings got their name from theorems of Macaulay (for polynomial rings) and Cohen (for general regular rings) which state that in a regular local ring R ideals generated by parameters are unmixed. Nagata showed that this unmixedness property characterizes C-M rings.

If J is an ideal of R, we let J^u be the unmixed part of J, i.e., J^u equals the intersection of the minimal primary components of J. If J is generated by parameters in a local ring R, and R is C-M, then $J = J^u$, so it is no surprise that (2.1) is equivalent to the following claim:

(2.2) If R, R^+, are as in (2.1) and J is an ideal of R generated by parameters $x_1, \ldots, x_k$, then $J^uR^+ = JR^+$. In fact there is a finite extension domain S of R such that $J^uS = JS$. Note however that $(JS)^u$ need not be equal to JS.

The last part of (2.2) follows at once from the first, since J^u is finitely generated, and hence by collecting coefficients in R^+, one can get the extension S.

The first general problem confronting one in attacking this claim, without which one cannot even leave the ground, is this: given two ideals I, J in a ring R, how can one force J to be contained in I in some finite extension S of R? We want this only in the special case where I is generated by parameters and J is I^u, but this case is no easier than the general problem. Of course, one cannot arbitrarily force an ideal J into I in some finite extension; there are obstructions. In fact in characteristic p, if this occurs then J is in the *tight closure* of I. The tight closure of an ideal was introduced in [H-H1], and the study of tight closure was a key in proving (2.1), although the proof is independent of the theory of tight closure. The definition of tight closure is given below.

Definition 2.3. *Let R be a ring of characteristic $p > 0$, let I be an ideal of R, and let $x \in R$. We say that x is in* the tight closure *of I, denoted by I^*, if there exists an element c of R, not in any minimal prime of R, such that $cx^q \in I^{[q]} = \langle i^q : i \in R \rangle$ for all $q = p^e \gg 0$. Note that $I^{[q]}$ is the image of I under the e-th iteration of the Frobenius. (We also note there is a similar definition for the tight closure of a submodule N of a module M, though we do not give the definition here.)*

While this definition is technical, it has proved to be a powerful tool, and has been used by Hochster and myself (cf. [H-H1-10]) to give strong improvements and new proofs in equicharacteristic (especially simple in characteristic p) of many theorems, which are not apparently related: the intersection conjecture [P-S, Hol, R1-2], the criterion for acyclicity of Buchsbaum and Eisenbud [B-E], the syzygy theorem [E-G], the monomial conjecture [Ho1], the Briançon-Skoda theorem [B-S], [L-S], and the Hochster-Roberts theorem mentioned in (1.7) [H-R, K, Bo]. Many other results also ensue from the theory.

We return to the general problem of finite extensions. Although it is true that if J is forced into I in a finite extension then $J \subseteq I^*$, the starting point for us was that we knew this to be true in the case where I is generated by parameters and $J = I^u$ [H-H1, 4.7]. Being in the tight closure gives strong equational constraints upon the elements of J in terms of the elements of I. However, as far as we know it is not necessarily true that if I is generated by parameters in a complete local domain R, then $I^*R^+ = IR^+$, although this is true for parameters of length at most three and is probably true. A careful analysis of the equations for tight closure does lead, nonetheless, to the following lemma which gives a sufficient criterion for pushing J into the extension of I in a finite extension of R:

Equational Lemma 2.4. *Let R be a domain of* char $p > 0$, *and let I, J be ideals of R. Write $I = (x_1, \ldots, x_k)$. If*

$$J^{[p]} \subseteq I^{[p]}R^+ + (x_1 \ldots x_k)^{p-1}JR^+, \tag{\#}$$

then there is a finite extension domain S of R such that $JS \subseteq IS$.

We would like to apply (2.4) in the case where I is generated by parameters and where $J = I^u$. Unfortunately, the condition (#) is quite a strong one and certainly will not be true for general parameters $x_1, \ldots, x_k$. The difficulty in proving (2.2) by using (2.4) is finding appropriate parameters (which we call *extraordinary*, cf. (2.7) below) which are special enough to satisfy (#), with $J = I^u$, but general enough so that they exist, and in enough profusion to allow us to pass from arbitrary s.o.p.'s to extraordinary ones in the proof of (2.2). One thing that makes this hard is that whatever the definition of extraordinary might be, it will surely be lost under finite extensions, so one cannot carry through the entire proof using only special types of parameters. There is a delicate balance between the two necessary properties needed by these special parameters. However, *there is* an appropriate definition: we want s.o.p.'s which kill as much cohomology as possible. To describe this precisely, we need a couple more definitions.

Definition 2.5. *A parameter x of R is said to be a* CM multiplier *if for all parameters $x_1, \ldots, x_{k+1}$, x annihilates the quotient module $N = (x_1, \ldots, x_k) : x_{k+1}/(x_1, \ldots, x_k)$, where, in general, $I : J = \{r \in R : rJ \subseteq I\}$.*

The quotient module N of (2.5) can often be interpreted as some cohomology group of the ring. Notice that $N = 0$ if R is C-M. It is not clear that such x exist; the quantification over all parameters makes the existence of such x a problem. (For instance if we set $x_1 = x^n$ for large n then it is immediate that x^n annihilates N, but not at all clear that x will. Unless x is chosen quite carefully, $xN \neq 0$.) The next lemma, found in [H-H1, 11.5] gives the existence of many such elements in excellent rings.

Lemma 2.6. *Let $x \in R$ be such that R_x is* C-M. *Then there is a fixed power of x which is a CM multiplier.*

In an excellent domain R, the locus of the non Cohen-Macaulay primes of R is closed, and the defining radical ideal has height at least 2, since 1-dimensional domains are C-M. (2.6) guarantees that we can always find two parameters x, y in such a ring, each of which is a CM multiplier. This allows a flexibility which is crucial to the proof.

Definition 2.7. *Parameters $x_1, \ldots, x_k$ in a local ring R are said to be* extraordinary *if x_k is a CM multiplier, and the image of x_i is a CM multiplier in the ring $R_i = R/(x_1, \ldots, x_{i-1})^u$.*

One quick note – in the paper [H-H5], a slightly different requirement is given for x_k but our change here does not affect the proof in any essential way.

Sequences of extraordinary parameters do exist, in any length up to the dimension of the ring, in excellent local domains. Furthermore *any* s.o.p. in such a ring can be moved to an extraordinary s.o.p. by a sequence of the following three types of operations:

i) Switching two of the parameters.
ii) Replacing x_i by $x_i + rx_j$ for any j $\neq$ i,
iii) Replacing parameters $x_1, \ldots, x_k$ with $yx_1, x_2, \ldots, x_k$, where $y, x_2, \ldots, x_k$ are also parameters.

The third condition allows enormous change.

Suppose that we know by induction on k, in the context of (2.2), that any $k-1$ parameters form a regular sequence in R^+. Then it turns out that any extraordinary parameters $x_1, \ldots, x_k$ satisfy the equational lemma (2.4) with $I = (x_1, \ldots, x_k)$ and $J = I^u$. In particular we can trivialize relations on the parameters $x_1, \ldots, x_k$; said otherwise, the map on Koszul homology, $H_1(\mathbf{x}; R) \to H_1(\mathbf{x}; R^+)$ is zero. If this latter property holds for $\mathbf{x}$, which is derived from $\mathbf{y}$ by repeated applications of i)–iii) above, then it also holds for $\mathbf{y}$. This gives us that relations on parameters can be trivialized in R^+, which finishes the sketch of the proof of (2.2) and thus (2.1).

3. Remarks

In this section we give a little background as well as some indications of how the main theorem can be applied. The existence of big C-M modules was shown by Hochster [Ho1] in equicharacteristic around 1975. He went on to show that the existence of such a module, even though not finitely generated, implied many of the so-called "local homological conjectures" in equicharacteristic. Most remain open in mixed characteristic, although the recent work of Paul Roberts [Ro1-2], using the intersection theory of Fulton, MacPherson, and Baum, has resulted in the solution of many of the conjectures, even in mixed characteristic.

Hochster's construction of a big C-M module was universal in the sense that he showed if any Cohen-Macaulay module exists for a given local ring, then his construction must also necessarily give one. This very universality makes it difficult to understand the structure of the modules he constructs. Moreover although the module is canonically constructed from the ring the actual structure of the module is mysterious. It is difficult to interpret its structure in terms of the original ring. Hochster also gave a construction of a big C-M algebra, without identity, [Ho2]. Again it is somewhat universal in nature and hard to understand its structure. On the other hand the ring R^+ of (2.1) has a great deal of structure and provides a wealth of insight into why such modules even exist. Moreover its structure is attached to that of R: for instance the going up and going down theorems hold between R and R^+ in the case where R is integrally closed. Another example is provided in the remarks below concerning the structure of R^+: if R, a char. p excellent local domain, maps to S, also an excellent local domain of char. p, then the map extends (not naturally, however) to a map from R^+ to S^+. If the original map is injective (surjective) then the map from R^+ to S^+ can be chosen to be injective (surjective). In other words, the + construction is enough like a functor to be helpful.

Moreover although (2.2) is false in characteristic 0 one can show, by standard methods of reduction to characteristic p, there exist big Cohen-Macaulay algebras (with identity) in equicharacteristic 0. The existence of big C-M algebras (as opposed to big C-M modules) has recently been used by Ian Aberbach [Ab].

We next make some remarks, which we collect in a lemma, concerning some of what we know about the structure of R^+.

Proposition 3.1. *Let R be an excellent local domain, and let R^+ be as in Section 2. Then,*

i) *the sum of any two prime ideals of* R^+ *is again a prime ideal, or is the whole ring* (*cf.* [Ar]);
ii) *if* $\mathfrak{p} \in \operatorname{Spec}(R)$, *and* $P \in \operatorname{Spec}(R^+)$ *lies over* $\mathfrak{p}$, *then* $(R/\mathfrak{p})^+ \simeq R^+/P$;
iii) *If* W *is a multiplicative system in* R, *then* $(W^{-1}R)^+ \simeq W^{-1}(R^+)$.

Thus R^+ is, both locally and globally, a C-M algebra, even modulo any prime ideal! Part ii) also explains the comments above concerning R^+ mapping onto S^+ when S is a domain which is a homomorphic image of R; simply write $S = R/\mathfrak{p}$, and use ii).

It is interesting to see what (2.1) says when one restricts to case 1 of Example 1.10. In this case R is a complete local domain, finite over a power series subring A generated (as a complete subring of R) by any s.o.p. $x_1, \ldots, x_n$. Since R is finite over A, $R^+ = A^+$. Any given s.o.p. in R can be used to define such an A, so (2.1) says simply that when A is a power series ring over a field of characteristic p, A^+ is a C-M A-algebra. Then it is not difficult to see that Theorem 2.1 is equivalent to the following theorem:

Theorem 3.2. *Let* A *be a power series ring over a field of characteristic* p, *and let* A^+ *be the integral closure of* A *in an algebraic closure of its quotient field. Then* A^+ *is flat over* A.

Lest this theorem sound too simple, recall that it is wildly false for dimension ≥ 3 in characteristic 0.

The flatness of A^+ over A has strong consequences concerning the syzygies of certain finitely generated modules M over complete local non C-M domains R. Suppose that the R-module M comes via base change from a module N over a power series ring A contained in R over which R is finite (i.e., $M \cong N \otimes_A R$). The module N has a finite resolution of the form,

$$\mathbb{F}: 0 \to F_d \to F_{d-1} \to \cdots \to F_1 \to F_0 \to N \to 0,$$

where F_i is a finitely generated free A-module. When we tensor with R to obtain M, the sequence is no longer exact; the homology comes exactly from the failure of R to be C-M. (Recall from 1.10 that R is C-M iff R is free as an A-module.) However since $R^+ = A^+$ is flat over A, when we tensor $\mathbb{F} \otimes_A R$ with R^+ the sequence becomes exact. Thus, e.g., the relations on the generators of M over R, while not coming from the original relations of N over A, *are* linear combinations of the original relations over some finite extension of R. This has profound consequences, especially in the graded case, which we discuss below.

4. The Graded Case

We close this paper by making a few remarks on the graded case. Let R be a domain of char. $p > 0$ and as in case 2 of 1.10. We are able to prove a graded version of

Theorem 2.1 by replacing R^+ by a subring R^{+gr} of R^+ which contains R, and is $\mathbb{N}$-graded extending the grading on R. In this case (2.1) becomes the following theorem.

Theorem 4.1. *Any s.o.p. in R consisting of homogeneous elements is a regular sequence in* R^{+gr}.

If A is a graded polynomial subring of R as in case 2 of 1.10 then $A^{+gr} = R^{+gr}$ is free over A, and the closing remarks of Section 3 apply in this case as well for graded modules and resolutions. Theorem 4.1 can be translated into a vanishing theorem on maps of cohomology of projective varieties as follows.

Theorem 4.2. *Let X be a closed, irreducible, reduced, d-dimensional subscheme of* $\mathbb{P}_k^n$, *where k is a field of characteristic* $p > 0$. *Then for all i,* $1 \leq i \leq d-1$, *and all integers t, there is a finite surjective morphism* $f_{i,t} = f\colon Y \to X$ *from a projective scheme Y, also reduced and irreducible, over a finite extension k′ of k such that the induced map* $H^i(X, O_X(t)) \to H^i(Y, f^*O_X(t))$ *is zero.*

A similar result holds without assuming X is irreducible and reduced; one only needs to assume that X is unmixed, and in this case the corresponding projective scheme Y can be chosen to be unmixed.

One very special case of this result, which already has many consequences is that in any reduced graded ring R, any s.o.p. $x_1, \ldots, x_d$ generated by one forms is "C-M" up to degree 2, i.e., $\sum r_i x_i = 0$, with r_i homogeneous, then up to a trivial syzygy, $\deg(r_i) \geq 2$.

For many problems, where a ring R is assumed to be C-M, what is really important is not that R is C-M, but rather that a given finite set of relations can be trivialized in the ring. Moreover proof analysis often shows that it suffices that the relations be trivialized only is some finite extension of the ring; a good example of this is a connectedness theorem of Hartshorne which says that if R is C-M (or even has Serre's property S_2) then R is connected in codimension 1. An analysis of this proof, in light of (2.1), shows that in characteristic p (and hence also equicharacteristic 0 by reduction to equicharacteristic p) that the assumptions on depth are not really needed, and allow one to prove the following generalization: if R is complete local domain of dimension d and $x_1, \ldots, x_k$ are elements of R with $k \leq d-2$, then the punctured spectrum of $R/(x_1, \ldots, x_k)$ is connected. This was already known and is a result of Faltings. The point is that (2.1) provides considerable insight into what might be true and provides the ammunition to carry proofs through. It seems that significant insight can be gained if one follows the intuition that "all" domains in characteristic p are C-M; while this is false, passing to R^+ will recapture the C-M property.

Acknowledgement. I'd like to thank Mel Hochster and Bill Heinzer for reading a preliminary version of this manuscript and making valuable suggestions.

References

[Ab] Aberbach, I.: *Finite phantom projective dimension and a phantom analogue of the Auslander-Buchsbaum theorem.* Thesis, Univ. of Michigan, 1990

[Ar] Artin, M.: *On the joins of Hensel rings.* Adv. Math. **7** (1971) 282–296

[Bo] Boutot, J.-F.: *Singularitiés rationelles et quotients par les groupes reductifs.* Invent. Math. **88** (1987) 65–68

[B-R] Briançon, J., Skoda, H.: *Sur la cloture integrale d'un ideal de germes de fonctions holomorphes en un point de* $\mathbb{C}^n$. C. R. Acad. Sci. Paris Ser. A **278** (1974) pp. 949–951

[E-G] Evans, G., Griffith, P.: *The syzygy problem.* Annals of Math. **114** (1981) 323–333

[F] Faltings, G.: *A contribution to the theory of meromorphic functions.* Nagoya Math. J. **77** (1980) 99–106

[Ha] Hartshorne, R.: *Complete intersections and connectedness.* Amer. J. Math. **84** (1962) 497–508

[Ho1] Hochster, M.: Topics in the homological theory of modules over commutative rings. C.B.M.S. Regional Conf. Ser. in Math. No. 24 A.M.S., Providence, RI 1975

[Ho2] Hochster, M.: *Big Cohen-Macaulay modules and algebras and embeddability in rings of Witt vectors.* Proceedings of the Queen's Univ. Commutative Algebra Conference, Queen's Papers in Pure Appl. Math. **42** (1975) 106–195

[H-H1] Hochster, M., Huneke, C.: *Tight closure, invariant theory, and the Briançon-Skoda theorem.* Journal of the Amer. Math. Soc. **3** (1990) 31–116

[H-H2] Hochster, M., Huneke, C.: *Phantom homology.* Preprint 1990

[H-H3] Hochster, M., Huneke, C.: *F-regularity, test elements and smooth base change.* In preparation

[H-H4] Hochster, M., Huneke, C.: *Tight closure of parameter ideals and splitting in module-finite extensions.* In preparation

[H-H5] Hochster, M., Huneke, C.: *Infinite integral extensions and big Cohen-Macaulay algebras.* Ann. Math. (to appear)

[H-H6] Hochster, M., Huneke, C.: *Applications of the Cohen-Macaulay property for absolute integral closures.* In preparation

[H-H7] Hochster, M., Huneke, C.: *Tightly closed ideals.* Bull. Amer. Math. Soc. **18** (1988) 45–48

[H-H8] Hochster, M., Huneke, C.: *Tight Closure.* Commutative Algebra, Math. Sci. Res. Inst. Publ., no. 15, Springer, New York, Berlin, Heidelberg, London, Paris, Tokyo 1989, pp. 305–324

[H-H9] Hochster, M., Huneke, C.: *Tight closure and strong F-regularity.* Soc. Math. France Memoire **38** (1989) 119–133

[H-H10] Hochster, M., Huneke, C.: *Absolute integral closures are big Cohen-Macaulay algebras in characteristic* p. Bull. Amer. Math. Soc. **24** (1991) 137–143

[H-R] Hochster, M., Roberts, J.: *Rings of invariants of reductive groups acting on regular rings are Cohen-Macaulay.* Adv. Math. **13** (1974) 115–175

[K] Kempf, G., *The Hochster-Roberts theorem of invariant theory.* Michigan Math. J. **26** (1979) 19–32

[L-S] Lipman, J., Sathaye, A.: *Jacobian ideals and a theorem of Briançon-Skoda*, Michigan Math. J. 28 (1981) 199–222

[P-S] Peskine, C., Szpiro, L.: *Dimension projective finie et cohomologie locale.* I.H.E.S. Publ. Math. **42** (1973) 323–395

[Ro1] Roberts, P.: *The vanishing of intersection multiplicities of perfect complexes.* Bull. Amer. Math. Soc. **13** (1985) 127–130

[Ro2] Roberts, P.: *Le théorème d'intersection.* C. R. Acad. Sc. Paris Ser. I, **304** (1987) 177–180

Identities of Associative Algebras

Alexander R. Kemer

Ulyanovsk Branch of Moscow State University, Str. Tolstogo 42, Ulyanovsk 432700, USSR

Preliminaries

Let F be field, $F\langle X\rangle$ the free associative algebra over F generated by a countable set of variables X. One may consider an element from $F\langle X\rangle$ as a polynomial in non-commutative variables from X.

Definition 1. We shall say that an associative algebra A over F satisfies an identity $f(x_1, \ldots, x_n) = 0$, where $f = f(x_1, \ldots, x_n) \in F\langle X\rangle$, if for arbitrary elements $a_i \in A$ the equality $f(a_1, \ldots, a_n) = 0$ is always valid in A.

Definition 2. The set of all polynomials $f \in F\langle X\rangle$ such that A satisfies the identity $f = 0$ is said to be the ideal of identities of the algebra A. We shall denote this ideal by $T[A]$. An ideal of $F\langle X\rangle$ which is an ideal of identities of some algebra is called a T-ideal.

Definition 3. An Algebra A is said to be PI-algebra if it satisfies a non-trivial identity (i.e. $T[A] \neq 0$).

The structure theory for PI-algebras is well developed. Some results of this theory are classic now. One of them is Kaplansky's theorem which asserts that a primitive PI-algebra is finite dimensional over its centre. Another example is the theorem of Nagata-Higman which asserts that any algebra over a field of zero characteristic satisfying identity $x^n = 0$ is nilpotent.

In 1957 A.I. Shirshov proved his famous Height Theorem:

Theorem (A.I. Shirshov [1]). *For any finitely generated PI-algebra A there exist a number h and elements $a_1, \ldots, a_n \in A$ such that elements*

$$a_{i_1}^{\alpha_1} \cdot \ldots \cdot a_{i_k}^{\alpha_k}, \qquad k < h$$

generate A as a space.

The Height Theorem implies the positive solution of Kurosh's problem for PI-algebras: A finitely generated algebraic PI-algebra is finite dimensional.

Proceedings of the International Congress
of Mathematicians, Kyoto, Japan, 1990

Although Kurosh's problem for PI-algebras was solved earlier in 1948 by I. Kaplansky, Shirshov's approach shows clearly the contributions of both conditions (PI and algebraicity) in order that algebra be finite dimensional. The Height Theorem shows also that finitely generated PI-algebras are close to finite dimensional algebras.

Shirshov's Height Theorem was the first result which gave some information about identities of PI-algebras.

This paper is also devoted to the study of the structure of identities of associative algebras.

1. Finitely Generated PI-Algebras

First of all we recall the famous result in PI-theory which was proved by A. Braun in 1982.

Theorem (A. Braun [2]). *The radical of a finitely generated PI-algebra is nilpotent.*

This theorem has the following important corollary:

Corollary. *A finitely generated PI-algebra satisfies for some n the identity*

$$\sum_{\sigma \in S(n)} (-1)^{\sigma} x_{\sigma(1)} y_1 x_{\sigma(2)} y_2 \cdot \ldots \cdot y_{n-1} x_{\sigma(n)} = 0 \tag{1}$$

where $S(n)$ is the symmetric group of degree n; $(-1)^{\sigma} = 1$ if σ is an even permutation, $(-1)^{\sigma} = -1$ if σ is odd.

The identity (1) is called the Capelli identity of n-th order.

It is well known that if char $F = 0$ then the Capelli identities "distinguish" finitely generated PI-algebras from infinitely generated algebras.

Theorem. *Let an associative PI-algebra A over a field of zero characteristic satisfy the Capelli identity of some order. Then there exists a finitely generated PI-algebra B such that $T[A] = T[B]$.*

If char $F \neq 0$ then the conclusion of this theorem is not true. We shall discuss this case later.

The main result about identities of finitely generated PI-algebras is the following theorem which was proved by the author in 1988.

Theorem 1 ([3, 4]). *For any finitely generated PI-algebra A over an infinite field F there exists a finite dimensional F-algebra C such that $T[A] = T[C]$.*

In other words finitely generated PI-algebras cannot be distinguished from finite dimensional algebras in the language of identities.

If F is a finite field then the conclusion of Theorem 1 is not valid, but Theorem 1 gives a full information about identities which are homogeneous with respect to every variables.

One may consider Theorem 1 as a theorem on classification of finitely generated PI-algebras in the language of identities.

Definition 4. Let Γ be a T-ideal in the free algebra $F\langle X\rangle$. The factor algebra $F\langle X\rangle/\Gamma$ is said to be a relatively free algebra generated by the set of variables X.

Definition 5. An algebra over a field F is called representable if it can be embedded into a matrix algebra of finite order over some extension of the basic field F.

Theorem 1 has the following important corollary:

Corollary. *Any relatively free PI-algebra of finite rank is representable (the basic field is infinite).*

Since the proof of Theorem 1 has been published we give the main ideas of this proof.

Let A be a finitely generated PI-algebra. Using the theorem of Braun and Levin's theorem it is easy to prove that

$$T[A] \supseteq T[C]$$

for some finite dimensional algebra C. That is the initial situation. Then we want to find an intermediary finite dimensional algebra C_1 such that

$$T[A] \supseteq T[C_1] \supsetneqq T[C].$$

Then we want to find a finite dimensional algebra C_2 such that

$$T[A] \supseteq T[C_2] \supsetneqq T[C_1]$$

and so on.

If we shall follow this way then two problems arise.

1) How can one find those intermediary algebras?
2) Does every ascending chain of ideals of identities of finite dimensional algebras terminate?

We shall not discuss the first problem because it is the most technical part of the proof. We construct those algebras using the so-called trace identities.

The solution of the second problem is also technical. The main idea is to translate structure properties of a finite dimensional algebra into the language of identities.

We assume for simplicity that the basic field is algebraically closed. Consider a finite dimensional algebra C with a unit. The algebra C may be represented in the form

$$C = P + \operatorname{Rad} C$$

where P is the semisimple part of C. The algebra P is a direct sum of full matrix algebras $M_{n_i}(F)$. We define the following parametres of the algebra C:

$$\alpha(C) = \sum n_i$$

$$\beta(C) = \dim_F P$$

$$\gamma(C) = \text{the index of nilpotency of Rad } C.$$

The triple

$$t(C) = (\alpha(C), \beta(C), \gamma(C))$$

is called the type of the algebra C. We order the types lexicographically.

How do we translate the properties of the algebra C into the language of identities? For example, the property $\alpha(C) = n$ is translated into the following identity

$$\sum_{\sigma_1,\ldots,\sigma_k \in S(n+1)} (-1)^{\sigma_1} \cdot \ldots \cdot (-1)^{\sigma_k} y^{\sigma_1(1)+\cdots+\sigma_k(1)-k} x_1 y^{\sigma_1(2)+\cdots+\sigma_k(2)-k} \cdot \ldots \cdot x_n y^{\sigma_1(n+1)+\cdots+\sigma_k(n+1)-k} = 0 \qquad (2)$$

where $k = \gamma(C)$.

If $T[C'] \supseteq T[C]$ for another finite dimensional algebra C', then it is proved that there exist finite dimensional algebras $C_1, \ldots, C_m$ such that

$$\alpha(C_i) \leq \alpha(C), \qquad T[C'] = T\left[\bigoplus_{i=1}^{m} C_i\right].$$

Other identities translate some other properties. Finally we can prove the following:

Proposition. *Let C, C' be finite dimensional algebras over an infinite field, $T[C'] \supseteq T[C]$. Then there exist finite dimensional algebras $C_1, \ldots, C_m$ such that*

$$T\left[\bigoplus_{i=1}^{m} C_i\right] = T[C'].$$

This proposition is principal for the solution of the second problem.

2. Infinitely Generated PI-Algebras

The methods of studying the identities of infinitely generated algebras highly depend on the characteristic of the basic field.

2.1 Algebras over Field of Zero Characteristic

Definition 6. Let A be an associative algebra, A_0, A_1 subspaces of A such that

$$A = A_0 \oplus A_1$$

$$A_0A_0,\ A_1A_1 \subseteq A_0; \qquad A_0A_1,\ A_1A_0 \subseteq A_1.$$

Then the algebra $A = A_0 \oplus A_1$ is said to be the $\mathbf{Z}_2$-graded algebra or superalgebra graded by A_0, A_1.

Consider the Grassmann algebra G generated by elements $e_1, e_2, e_3, \ldots$ satisfying the relations $e_i e_j = -e_j e_i$ for every i, j. Let G_0 be the subspace of G generated by monomials of even degree, G_1 the subspace generated by monomials of odd degree. Then $G = G_0 \oplus G_1$ is the Grassmann superalgebra.

Definition 7. The subalgebra $G(A) = G_0 \otimes_F A_0 + G_1 \otimes A_1$ of the algebra $G \otimes_F A$ is called the Grassman Hull of the superalgebra $A = A_0 \oplus A_1$.

In 1981 the author proved that any T-ideal equals the ideal of identities of the Grassmann Hull of some finitely generated PI-superalgebra. This theorem has reduced the studying of identities of infinitely generated PI-algebras to the studying of graded identities of finitely generated PI-superalgebras. We recall that the characteristic of the basic field equals zero. Theorem 1 formulated above is valid also in the graded case ([3]). Hence we obtain the following main result about the identities of PI-algebras over the field of zero characteristic:

Theorem 2 [3]. *For any PI-algebra A there exists a finite dimensional superalgebra C such that*

$$T[A] = T[G(C)].$$

Theorem 2 has an interesting corollary:

Corollary. *A relatively free PI-algebra can be embedded into the matrix algebra of finite order over an algebra satisfying the identity* $[[x, y], z] = 0$ $([x, y] = xy - yx)$.

2.2 Specht's Problem

We want to classify associative PI-algebras in the language of identities. Theorem 2 is a satisfactory theorem of classification for algebras of zero characteristic. In the general case we do not have any hypothesis about the classification.

Definition 8. The minimal set of identities of an algebra A which implies all the other identities of A is said to be the base of identities of the algebra A.

In 1950 W. Specht formulated the following problem [5]: Has any associative algebra of zero characteristic a finite base of identities?

The problem of finite base may be formulated not only for algebras of zero characteristic but also for arbitrary algebras, rings, groups and so on. The problem of finite base may be considered as a strict formulation for problem of classification (if there is no more satisfactory hypothesis).

In 1986 the author [6] solved Specht's Problem positively:

Theorem 3. *Any associative algebra over a field of zero characteristic has a finite base of identities.*

Theorem 3 has other formulations:

1) Any T-ideal is finitely generated as a T-ideal.
2) The set of T-ideals satisfies the ascending chain condition.

In the case of non-zero characteristic the problem of finite base is open, but Theorem 1 yields the positive solution of local Specht's problem for algebras over an infinite field.

Theorem 4. *Any* T-*ideal of a finitely generated free algebra is generated by a finite set of polynomials* (*as a* T-*ideal*).

2.3 Identities of Algebras over a Field of Characteristic *p*

In the case of characteristic p Theorem 2 is not true. Moreover the author has no classification hypothesis. The structure of identities of algebras over fields of characteristic p is not clear and there are no strong results on this theme.

In 1981 I.B. Volichenko formulated the strange problem: Does any PI-algebra over a field of non-zero characteristic satisfy the standard identity

$$\sum_{\sigma \in S(n)} (-1)^{\sigma} x_{\sigma(1)} \cdot \ldots \cdot x_{\sigma(n)} = 0$$

for some n?

It is not difficult to prove that Volichenko's hypothesis is true if and only if any PI-algebra of non-zero characteristic satisfies the symmetrized standard identity

$$\sum_{\sigma \in S(n)} x_{\sigma(1)} \cdot \ldots \cdot x_{\sigma(n)} = 0$$

for some n.

Indeed, let A be any PI-algebra over a field F and G a Grassmann algebra over F. It is well-known that the algebra $A \otimes_F G$ is a PI-algebra. If $A \otimes_F G$ satisfies the standard identity (symmetrized standard identity) of some degree then it is easy to see that the algebra A satisfies the symmetrized standard identity (standard identity) of the same degree.

Note that the symmetrized standard identity of n-th degree is a full linearization of the identity $x^n = 0$.

The author has solved these problems recently.

Theorem 5. *Any PI-algebra over a field of non-zero characteristic satisfies the standard identity and the symmetrized standard identity of some degree.*

Proof. First of all we remark that it is enough to prove the theorem for algebras with unit over algebraically closed field.

Let Γ be the T-ideal of identities of the given algebra with unit $\Gamma \subseteq F\langle X \rangle$, $X = \{x_1, x_2, \ldots\}$. Consider the finitely generated PI-algebras $F_k/\Gamma \cap F_k$, where $F_k = F\langle x_1, \ldots, x_k \rangle$. By Theorem 1

$$T[F_k/\Gamma \cap F_k] = T[A_k]$$

for some finite dimensional algebra A_k.

The primitive images of the algebra $F\langle X\rangle/\Gamma$ are full matrix algebras over the basic field F. The maximal order of these matrix algebras is called the complexity of the T-ideal Γ.

We represent the algebra A_k in the form

$$A_k = P_k + \text{Rad } A_k,$$

where P_k is the semisimple part of A_k. The algebra P_k may be represented in the form

$$P_k = e_0 P_k + e_1 P_k + \cdots + e_s P_k, \qquad s = s(k),$$

where $e_0, e_1, \ldots, e_s$ are orthogonal idempotents, $e_i P_k = M_q(F)$, $i > 0$ (q the complexity of Γ); $e_0 P_k$ a direct sum of other matrix algebras $M_n(F)$, $n < q$.

We prove the theorem by induction. The base of induction ($q = 1$) and the inductive step will be proved simultaneously. Consider the ideal I_k of the algebra A_k generated by all mixed elements $e_i a e_j$, $i \neq j$, $a \in A_k$. Note that it is sufficient to prove that the algebras I_k satisfy the symmetrized standard identity of some degree $n = n(\Gamma)$.

Indeed, if $q = 1$ we put $k = 4n$. The algebra A_k/I_k satisfies the identity

$$\underbrace{[x, y, \ldots, y]}_{p^m} = 0 \tag{3}$$

for some m. Therefore the algebra A_k satisfies the identities

$$S_n^+(z_1[x_1, y_1, \ldots, y_1]t_1, \ldots, z_n[x_n, y_n, \ldots, y_n]t_n) = 0 \tag{4}$$

(S_n^+ is the symmetrized standard polynomial), where z_i, $t_j \in X \cup \{1\}$. Since the identities (4) contain at most k variables, these identities are valid modulo Γ. Hence we obtain that the algebra $F\langle X\rangle/\Gamma$ is an extension of some algebra satisfying (3) by an algebra satisfying the symmetrized standard identity of degree n. It is easy to verify that the full linearization of (3) is the symmetrized standard identity of degree p^m, hence the algebra $F\langle X\rangle/\Gamma$ satisfies the symmetrized standard identity of degree $p^m \cdot n$.

If $q > 1$ then we may assume that the theorem is proved in the case when the complexity is less than q.

Let B be an algebra such that $T[M_q(B)] \supseteq \Gamma$. Since q is the complexity of Γ the complexity of the T-ideal $T[B]$ is equal to 1. Hence by the inductive hypothesis the algebra B must satisfy the symmetrized standard identity of some degree. It is easy to prove that then the algebra $M_q(B)$ must satisfy the symmetrized standard identity of some degree $m = m(\Gamma)$.

Let $h(x_1, \ldots, x_r)$ be a polynomial such that for all s the algebra $M_q(F)$ does not satisfy the identity $h^s = 0$, but the algebra $M_{q-1}(F)$ satisfies the identity $h^s = 0$ for some s.

We put $k = (m + r + 3)n$. Consider the algebra A_k/I_k. This algebra is a direct sum of algebras of the type $M_d(B)$, where B is a local algebra, $d \leq q$. Therefore the algebra A_k/I_k must satisfy the identities

$$f(x_1, \ldots, x_m, y_1, \ldots, y_r, u, v, w) = uS_m^+(x_1, \ldots, x_m)v(h(y_1, \ldots, y_r))^t w = 0 \tag{5}$$

for some m, t, where $u, v, w \in X \cup \{1\}$. Hence we obtain that the algebra A_k satisfies the identities

$$S_n^+(f^{(1)}, \ldots, f^{(n)}) = 0, \tag{6}$$

where $f^{(i)} = f(x_1^{(i)}, \ldots, x_m^{(i)}, y_1^{(i)}, \ldots, y_r^{(i)}, u^{(i)}, v^{(i)}, w^{(i)})$, $u^{(i)}, v^{(i)}, w^{(i)} \in X \cup \{1\}$. Since the identities (6) contain at most k variables these identities are valid modulo Γ. Therefore the algebra $F\langle X\rangle/\Gamma$ is an extension of some algebra satisfying (5) by an algebra satisfying the symmetrized standard identity of degree n. By the inductive hypothesis the identity $h^t = 0$ implies the symmetrized standard identity of some degree n_0. Hence we obtain that the algebra $F\langle X\rangle/\Gamma$ satisfies the symmetrized standard identity of degree $(n_0 + m)\cdot n$.

Let us prove that the algebra I_k satisfies the symmetrized standard identity of some degree $n = n(\Gamma)$.

Let $h(x_1, \ldots, x_r)$ be a central polynomial for the algebra $M_q(F)$. Then the algebra $F\langle X\rangle/\Gamma$ satisfies all identities of the type

$$[h(u_1, \ldots, u_r), v]x^{i_1}[h(u_1, \ldots, u_r), v]x^{i_2}\cdot\ldots\cdot x^{i_{m-1}}[h(u_1, \ldots, u_r), v] = 0. \tag{7}$$

Since $T[A_k] \supseteq \Gamma$, the algebra A_k also satisfies the identities (7). If $m > s = s(k)$ then we substitute into the identity (7) the following elements of A_k: $x = 1$, $v = \sum_{i=1}^m e_i a_i e_{i+1}$, $u_i = b_i$, where b_i are such elements that $h(b_1, \ldots, b_r) = \sum \alpha_i e_i$, $\alpha_i \neq \alpha_j$, $\alpha_i \in F$ (these elements exist because h is a central polynomial for $M_q(F)$). As a result we get the equality $v^m = 0$. Hence we obtain

$$0 = e_1 v^m = e_1 a_1 e_2 a_2 \ldots e_m a_m e_{m+1}.$$

It means that we may assume that $s \leq m$.

If we substitute into (7) $x = b \in I_k$, $v = e_i a e_j$, $i \neq j$, $u_t = b_t$, where b_t are such elements that $h(b_1, \ldots, b_r) = e_j$, then we get

$$va^{i_1}va^{i_2}\ldots a^{i_{m-1}}v = 0.$$

Linearizing this equality we can see that if we substitute into the polynomial $S_n^+(x_1, \ldots, x_n)$ a lot of elements of mixed type $e_i a e_j$, $i \neq j$, then we annihilate this polynomial. Therefore it is sufficient to prove that for all i, j, $i \neq j$, the algebra $B_k = e_i A_k e_j A_k e_i$ satisfies the symmetrized standard identity of some degree $n = n(\Gamma)$.

Put $i_1 = i_3 = i_5 = \cdots = 0$ in (7). Substituting into (7) $v = y + z$, $y \in e_i A_k e_j$, $z \in e_j A_k e_i$; $x \in e_i A_k e_i$, $h(v_1, \ldots, v_r) = e_i$, we get the equality

$$(z - y)^2 x^{j_1}(z - y)^2 x^{j_2}\ldots x^{j_{t-1}}(z - y)^2 = 0$$

where $j_d = i_{2d}$, $t = [(m + 1)/2]$. Multiplying this equality by e_i from the left side we get the identity

$$x^{j_0}yzx^{j_1}yz\ldots x^{j_{t-1}}yzx^{j_t} = 0 \tag{8}$$

for all $y \in e_i A_k e_j$, $z \in e_j A_k e_i$, $x \in e_i A_k e_i$.

Substituting $y = y + y_1$ into (8) and taking the component of degree 1 by y_1, we get the identity

$$x^{j_0}y_1zx^{j_1}yz\cdots + x^{j_0}yzx^{j_1}y_1z\cdots + \cdots = 0. \tag{9}$$

Substitute $y_1 = xy_1$ into (9) and subtract the result from (9) where j_0 is replaced by $j_0 + 1$. We get the identity of the type

$$x^{j_0}yzx^{j_1}(\alpha_1 y_1 zx^{j_2}yz\cdots + \alpha_2 yzx^{j_2}y_1 z\cdots + \cdots) = 0, \tag{10}$$

$\alpha_i \in F$. Then we again substitute $y_1 = xy_1$ into (10) and subtract the result from (10) where j_1 is replaced by $j_1 + 1$. Continue the described process. Finally we get a non-trivial identity of the type

$$\sum_{(j)=(j_0,j_1,\ldots,j_t)} \alpha_{(j)} x^{j_0}yzx^{j_1}\cdot\ldots\cdot yzx^{j_{t-1}}y_1 zx^{j_t} = 0. \tag{11}$$

We repeat this process again t times and then repeat this process with respect to z. Finally we get the identity of the type

$$\sum_{(j)=(j_0,\ldots,j_t)} \beta_{(j)} x^{j_0}y_t z_t x^{j_1}\cdot\ldots\cdot x^{j_{t-1}}y_1 z_1 x^{j_t} = 0$$

for all $x \in e_i A_k e_i$, $y \in e_i A_k e_j$, $z \in e_j A_k e_i$. Hence we obtain that the algebra B_k satisfies the non-trivial identity of the type

$$\sum_{(j)} \beta_{(j)} x^{j_0}x_1 x^{j_1}x_2\cdot\ldots\cdot x_t x^{j_t} = 0 \tag{12}$$

for all $x, x_i \in B_k$.

So we may assume that the algebra $F\langle X\rangle/\Gamma$ satisfies the identity (12). Therefore the algebra A_k satisfies the identity (12). Substituting $x = a + e_j$, $a \in B_k$, $x_1 = e_i b_1 e_j$, $x_2 = e_j b_2 e_i$, $x_i \in B_k$, $i \geq 3$, we see that the algebra B_k satisfies the identity of the type (12) where t is replaced by $t - 1$. □

References

1. Shirshov, A.I.: On rings with identical relations. Mat. Sbornik **43** (1957) 277–283
2. Braun, A.: The radical in finitely generated PI-algebra. Bull. Amer. Math. Soc. **7** (1982) 385–386
3. Kemer, A.R.: Representability of relatively free algebras. Algebra i Logika **27** (1988) 274–294
4. Kemer, A.R.: Identities of finitely generated algebras over infinite field. Izv. Acad. Nauk SSSR. Ser. Mat. **54** (1990) 726–753
5. Specht, W.: Gesetze in Ringen. 1. Math. Z. **52** (1951) 557–589
6. Kemer, A.R.: Finite base of identities of associative algebras. Algebra i Logika **26** (1987) 597–641

Intersection Theory and the Homological Conjectures in Commutative Algebra

Paul C. Roberts

Department of Mathematics, University of Utah, Salt Lake City, UT 84112, USA

The subject of this article is a set of conjectures which have been a central topic of investigation in Commutative Algebra for several years. Many of these conjectures are closely related to questions in Intersection Theory, and in some cases they arose directly from attempts to define intersection multiplicities in an algebraic setting. We begin with a discussion of some of these questions in Intersection Theory and their influence in Commutative Algebra. Next we discuss a few of the main conjectures related to this topic. Finally, we show how recent developments in Intersection Theory have made it possible to settle some of these problems.

1. Introduction: Serre's Definition of Intersection Multiplicities

We begin this section with some background on the problem of defining intersection multiplicities. The general question is as follows: given a variety V, a point p of V, and subvarieties X and Y of V such that p is an isolated point of the intersection $X \cap Y$, we wish to to define the multiplicity of the intersection of X and Y at the point p. We are interested here in an algebraic definition (which does not involve topology, for instance), and in one which is defined locally at the point.

As an illustration, we consider the case in which X and Y are curves in the plane. Let A be the local ring at the point p; that is, the ring of rational functions defined in a neighborhood of p. Let the curves X and Y be defined locally near p as the sets of zeros of polynomials $f(x, y)$ and $g(x, y)$. Then the correct definition for the multiplicity is the length of the quotient ring $(A/(f, g))$, where (f, g) is the ideal generated by f and g (or, in the complex case, the dimension of $A/(f, g)$ as a vector space over $\mathbb{C}$). This definition agrees with intuition; if the curves are smooth and not tangent at p it gives multiplicity one, and if they are tangent it is greater than one. In addition, it satisfies Bézout's Theorem, which states that if V is the projective plane, and if X and Y are subvarieties of V of degrees m and n respectively, then the total number of intersections, counted with multiplicities, is mn. This theorem is very important in enumerative geometry.

Before proceeding, we recall some basic definitions and notation. Let A be a commutative Noetherian local ring with maximal ideal $\mathfrak{m}$ and residue field $A/\mathfrak{m} = k$. Let $\dim(A) = d$ denote the Krull dimension of A; this can be defined

Proceedings of the International Congress of Mathematicians, Kyoto, Japan, 1990

either as the maximum length of a chain of prime ideals or as the minimum number d such that there are elements $x_1, \dots x_d \in \mathfrak{m}$ which generate an ideal primary to the maximal ideal. Such a sequence of elements is called a *system of parameters.* The ring A is *regular* if the maximal ideal can be generated by d elements; in the geometric situation, the local ring at a point is regular when the variety is smooth at the point. The ring is *Cohen-Macaulay* when the system of parameters form a regular sequence; that is, when $ax_i \in (x_1, \dots, x_{i-1})$ implies $a \in (x_1, \dots, x_{i-1})$ for each i. We remark that the above definition of multiplicities for curves works because A is Cohen-Macaulay and (f, g) is a system of parameters in that case.

If A is an integral domain, we say that A is *equicharacteristic* if the characteristic of A is the same as the characteristic of k; otherwise A has *mixed characteristic.*

We now return to the question of defining intersection multiplicities. Let A be a local ring, and let I and J be ideals of A; in the geometric situation, I and J will be the ideals of functions vanishing on subvarieties X and Y. The straightforward generalization of the definition for curves would be to define the multiplicity to be the length of the quotient $A/(I+J)$. However, this will not satisfy Bézout's Theorem (among other things).

The idea of Serre was to add extra correction terms. In addition, it is more convenient to generalize from the quotients A/I and A/J of the ring A to more general finitely generated modules M and N.

Definition (Serre [28]). Let A be a regular local ring and let M and N be finitely generated A-modules such that $M \otimes_A N$ is a module of finite length. Then the intersection multiplicity of M and N is defined to be

$$\chi(M, N) = \sum_{i \geq 0} (-1)^i \operatorname{length}(\operatorname{Tor}_i(M, N)).$$

This definition makes sense since all modules over regular local rings have finite projective dimension. Furthermore, it satisfies Bézout's Theorem, and it agrees with the definition given above in the case of curves. Serre stated three additional properties and proved them in the equicharacteristic case.

1. $\dim(M) + \dim(N) \leq \dim(A)$.
2. (Vanishing) If $\dim(M) + \dim(N) < \dim(A)$, then $\chi(M, N) = 0$.
3. (Positivity) If $\dim(M) + \dim(N) = \dim(A)$, then $\chi(M, N) > 0$.

The first of these Serre proved in general; we note that it can be interpreted as a statement on the topology of intersections in smooth varieties. The second and third were proven in the equicharacteristic case, and were conjectured in the general regular case.

2. The Peskine-Szpiro Intersection Theorem

Attempts to prove Serre's conjectures in the ensuing years included various generalizations of them to non-regular rings. The three statements listed above make sense for any local ring as long as M is assumed to have finite projective dimension (and, of course, $M \otimes_A N$ has finite length). In fact, the first statement makes sense even without the assumption of finite projective dimension, but it is false in this generality; whether it holds with this assumption is not known.

The next major advance was the proof of the "Intersection Theorem" in a quite general situation by Peskine and Szpiro [16, 17]. This theorem states:

Theorem (Intersection Theorem). *Let M be a module of finite projective dimension and let N be a module such that $M \otimes_A N$ has finite length as a module. Then*

$$\dim N \leq \operatorname{proj}\dim M\,.$$

If one replaces projective dimension in this theorem by codimension (i.e. $\dim(A) - \dim(M)$), it becomes the first statement of Serre's conjectures, and, like that statement, it can be considered to describe the topology of the intersection of a variety which is the support of a module of finite projective dimension. The importance of the theorem was demonstrated by the fact that it implies several other conjectures (Peskine-Szpiro [17]):

Corollary (Bass's Conjecture). *If A has a non-zero finitely generated module of finite injective dimension, then A is Cohen-Macaulay.*

Corollary (Zero-Divisor Conjecture). *If M is an A-module of finite projective dimension and x in A is a zero-divisor on A, then x is a zero-divisor on M.*

Shortly thereafter, a stronger version of this theorem was stated (Peskine-Szpiro [18], Roberts [20]):

Theorem (New Intersection Theorem). *Let*

$$0 \to F_k \to \ldots \to F_0 \to 0$$

be a complex of finitely generated free modules with homology of finite length. Then if $F_\bullet$ is not exact, we have $\dim(A) \leq k$.

This theorem implies the original Intersection Theorem.

In the paper cited above, the Intersection Theorem was proven for rings of positive characteristic and for rings whose completion is the completion of a ring of finite type over a ring of characteristic zero. The method, which has since been applied successfully to a wide variety of questions, was to reduce (using the Artin Approximation Theorem) to the case of a ring of positive characteristic p, and then to use iterations of the Frobenius map (the map which sends a to a^p) to prove the theorem in the positive characteristic case. The method of reducing using the Artin Approximation Theorem was extended by Hochster [11] to the general equicharacteristic case. These techniques did not extend to the case of mixed characteristic.

3. The Homological Conjectures

The conjectures of Serre and the Intersection Conjecture in mixed characteristic, together with a number of others, became known as the Homological Conjectures. These conjectures dealt with intersection multiplicities, modules of finite projective dimension, the existence of Cohen-Macaulay modules, and properties of systems of parameters, together with relations between these topics. A summary of these conjectures in 1975 can be found in Hochster [12]; in this monograph Hochster introduced big Cohen-Macaulay modules and proved their existence (thus implying several other conjectures, including the Intersection Conjecture) in the equicharacteristic case. The proof made use of reduction to positive characteristic.

In a somewhat different approach, in Roberts [21] the notion of Cohen-Macaulay complex is defined and proven to exist in the complex algebraic case via the Grauert-Riemenschneider vanishing theorem. The connection between the technique of reduction to positive characteristic and vanishing theorems in cohomology coming from Hodge theory was further shown by a proof of the Kodaira vanishing theorem by reduction to positive characteristic, conjectured by Szpiro and proven recently by Deligne and Illusie [2].

Among these conjectures, as mentioned above, was the question of whether Serre's vanishing and positivity conjectures held for arbitrary rings, when M was assumed to have finite projective dimension. This conjecture was proved for graded modules over a graded ring by Peskine and Szpiro [18] using computations with Hilbert polynomials in a free resolution, and they showed that several other conjectures followed from this one. These results led to specuation that an appropriate theory of Chern classes and Riemann-Roch theorem over general local rings could be used to prove the multiplicity conjectures, and hence many others, in general. The idea was that a bounded complex of free modules $F_\bullet$ would define a sequence of Chern characters $\mathrm{ch}_i(F_\bullet)$, and that these would vanish for i less than the codimension of the support of $F_\bullet$. In codimension one this program was carried out by Foxby [7], using a construction of MacRae [15], leading to proofs of some of the conjectures in low dimension.

However, Dutta, Hochster and McLaughlin [4] constructed an example which showed that this generalization of Serre vanishing was false in dimension three. They construct a module M of finite projective dimension and finite length over the ring $A = k[[X,Y,Z,W]]/(XY - ZW)$, which has dimension three, such that if $N = A/(X,Z)$, then $\chi(M,N) = -1$. Thus the theory of local Chern characters, even if it existed, could not have the hoped for vanishing properties. In the next section we describe such a theory and then show how in spite of these setbacks, it could solve some of the conjectures.

4. Localized Chern Characters

The Chern characters alluded to in the previous section were defined by Baum, Fulton, and MacPherson [1]. There were two major problems in defining such a theory for complexes over a local ring. The first was that there is no cohomology theory. This problem was answered by defining Chern characters as operators on the Chow group. The second problem was that there are no non-trivial vector bundles over a local ring, so that straightforward definitions of Chern characters tend to give zero. The solution here was to give a "localized" definition, with values not in the Chow group of the entire ring, but in the Chow group of the support of the complex. We give a very brief summary of the part of the theory we need; we refer to Fulton [8] for a more complete description.

Let $V = \mathrm{Spec}(A)$, and let Y be a closed subset of V. For each integer i, let $Z_i(Y)$ denote the rational group of cycles of dimension i; that is, $Z_i(Y)$ consists of the free $\mathbb{Q}$-module with generators all reduced and irreducible subschemes of Y of dimension i, or, equivalently, all prime ideals P with $\dim(A/P) = i$. Then the i-th graded piece of the Chow group, denoted $A_i(Y)$, is $Z_i(Y)$ modulo rational equivalence, where rational equivalence is defined by setting the divisor of a rational function on a reduced and irreducible suscheme of dimension $i+1$ to zero.

Let $F_\bullet$ be a bounded complex of free modules with support Z. Then there are Chern characters $\mathrm{ch}_k(F_\bullet)$ for each $k \geq 0$, which map $A_i(X)$ to $A_{i-k}(X \cap Z)$.

The connection with Euler characteristics was given by the local Riemann-Roch theorem (Fulton [8], Example 18.3.12). We state a very special case of this theorem here. Let $F_\bullet$ be a complex of free modules with homology of finite length, or, equivalently, with support at the closed point of $\mathrm{Spec}(A)$. Assume that A is an integral domain, and let $[A]$ denote the class of $\mathrm{Spec}(A)$ in the highest dimensional graded piece of the Chow group $A_d(\mathrm{Spec}(A))$. Then there is a class $\tau(A)$, with $[A]$ as its highest component, which satisfies

$$\chi(F_\bullet) = \sum_{i \geq 0} \mathrm{ch}_i(F_\bullet)(\tau_i(A)) .$$

In this equation, the left hand side is an integer, and the right hand side is an element of the Chow group of the support of $F_\bullet$. In this case, the support of $F_\bullet$ is a point p, so its Chow group reduces to $A_0(p)$, which we identify with $\mathbb{Q}$.

Furthermore, if A is regular (or, more generally, a complete intersection), then $\tau(A) = [A]$.

5. The Proofs of Serre's Vanishing Conjecture and the Intersection Conjecture

Using the theory described in the last section, it is possible to prove Serre's vanishing conjecture without regard to characteristic and to prove the Peskine-Szpiro Intersection Theorem in mixed characteristic. We briefly outline these proofs here and attempt to show how the theory of localized Chern characters is used in their solution.

Let M and N be modules over a regular local ring A such that $M \otimes_A N$ has finite length and $\dim(M) + \dim(N) < \dim(A)$. Let $F_\bullet$ and $G_\bullet$ be free resolutions of M and N respectively. The problem is to show that $\chi(F_\bullet \otimes G_\bullet) = 0$, since the homology of $F_\bullet \otimes G_\bullet$ is precisely $\mathrm{Tor}(M, N)$. By the local Riemann-Roch formula we have

$$\chi(F_\bullet \otimes G_\bullet) = \mathrm{ch}(F_\bullet \otimes G_\bullet)([A]) .$$

We must now use some properties of local Chern characters. First, they are multiplicative, which means that we can write

$$\mathrm{ch}(F_\bullet \otimes G_\bullet)([A]) = \sum_{i+j=d} \mathrm{ch}_i(F_\bullet)(\mathrm{ch}_j(G_\bullet))([A]).$$

Let X and Y denote the supports of M and N respectively; these are also the supports of $F_\bullet$ and $G_\bullet$. Now suppose that $d - j > \dim(Y)$. Then the element $\mathrm{ch}_j(G_\bullet)([A])$, which is in the group $A_{d-j}(Y)$, must be zero, since there can be no subvarieties of Y of dimension larger than the dimension of Y. Thus all terms for which $d - j > \dim(Y)$ are zero. Using the commutativity of Chern characters (proven in Roberts [23]), one shows also that terms with $d - i > \dim(X)$ are zero. The hypothesis that $\dim(X) + \dim(Y) < d$ implies that all terms fall into one of these two sets, and therefore the entire sum is zero, as was to be shown.

This theorem also holds if A is a complete intersection. It was proven independently using Adams operations in K-theory by Gillet and Soulé [9, 10]. It was proven when the singular locus has dimension at most one in Roberts [23]. We note that this argument gives a vanishing theorem for intersection multiplicities defined by local Chern characters for modules of finite projective dimension in general, and this approach appears to work better than the approach based on Euler characteristics for singular points.

We next describe how localized Chern characters are used in the proof of the New Intersection Theorem in mixed characteristic. Let A be a domain of mixed characteristic of dimension d, and let

$$0 \to F_{d-1} \to \ldots \to F_0 \to 0$$

be a complex of free modules with homology of finite length which is not exact. We wish to prove that such a complex cannot exist. Let $\overline{F}_\bullet$ denote $F_\bullet \otimes A/pA$, where p is the characteristic of the residue field of A. Then $\overline{F}_\bullet$ is a complex of free modules over A/pA.

The main idea of the proof is to compute $\mathrm{ch}_{d-1}([A/pA])$ in two different ways. First, using the properties of localized Chern characters and the fact that $\overline{F}_\bullet$ comes from restriction to a divisor of a complex supported at one point, we deduce that $\mathrm{ch}_{d-1}([A/pA]) = 0$. Second, we use that $\overline{F}_\bullet$ is a non-trivial complex of length $d-1$ over a ring of positive characteristic of dimension $d-1$. By means of an asymptotic Euler characteristic introduced by Dutta [3] and put on a firm foundation by Seibert [27], together with a local cohomology argument, we can then show that it must be positive (details can be found in Roberts [24, 25]). This contradiction proves the theorem.

6. Open Questions

We conclude with some remarks on three questions which are still open.

1. Serre's Positivity Conjecture. The method of local Chern characters provides a geometric approach to the question of positivity, but as of yet it has not led to a solution. A result of Tennison [30] states that if one computes $\chi(A/P, A/Q)$, where $\dim(A/P) + \dim(A/Q) = \dim(A)$, and if the subschemes defined by the prime ideals P and Q are not tangent at the closed point of Spec(R) (this condition can be stated precisely in terms of associated graded rings), then $\chi(R/P, R/Q)$ is the product of the multiplicities of the A/P and A/Q and is therefore positive. We remark that an example of Roberts [26] shows that positivity is unlikely for two modules of finite projective dimension in the non-regular case, even using Chern characters rather than Euler characteristics, although this question is also open.

2. The Improved New Intersection Conjecture (Evans-Griffith [5, 6]). This conjecture states the following: Let $F_\bullet = 0 \to F_k \to \ldots \to F_0 \to 0$ be a complex of free modules with finite length homology except possibly for $H_0(F_\bullet)$. Assume that there is a minimal generator of $H_0(F_\bullet)$ which is annihilated by a power of the maximal ideal of A. Then $k \geq \dim(A)$.

It seems reasonable to try to prove this theorem using the methods which worked for the "unimproved" version, but this version appears to be more difficult. This conjecture is equivalent to several others, such as Hochster's Monomial Conjecture (see [13]), and is open in dimension three in mixed characteristic.

3. Another Conjecture on Bounded Complexes. The New Intersection Theorem states that the minimum length of a bounded complex of free modules is d. Suppose that $F_\bullet$ has length exactly d. In this case, the question is whether the cycles Z_i are integral over the boundaries B_i in degrees larger than zero. This result would imply the Monomial Conjecture, it can be proven in positive characteristic by the theory of tight closure of Hochster and Huneke [14], and it has been proven by Rees [19] for the Koszul complex in any characteristic.

References

1. Baum, P., Fulton, W., MacPherson, R.: Riemann-Roch for singular varieties. Publ. Math. IHES **45** (1975) 101–145
2. Deligne, P., Illusie, L.: Relèvements modulo p^2 et décomposition du complexe de de Rham. Invent. math. **89** (1987) 247–270
3. Dutta, S.P.: Frobenius and multiplicities. J. Algebra **85** (1983) 424–448
4. Dutta, S.P., Hochster, M., McLaughlin, J.E.: Modules of finite projective dimension with negative intersection multiplicities. Invent. math. **79** (1985) 253–291
5. Evans, E.G., Griffith, P.: The syzygy problem. Ann. Math. **114** (1981) 323–333
6. Evans, E.G., Griffith, P.: The syzygy problem: a new proof and historical perspective. Commutative Algebra (Durham 1981), London Math Soc. Lecture Note Series **72** (1982) 2–11
7. Foxby, H.-B.: The MacRae invariant. Commutative Algebra (Durham 1981), London Math Soc. Lecture Note Series **72** (1982) 121–128
8. Fulton, W.: Intersection theory. Springer, Berlin Heidelberg New York 1984
9. Gillet, H., Soulé, C.: K-théorie et nullité des multiplicités d'intersection. C. R. Acad. Sc. Paris Série I, no. 3, **300** (1985) 71–74
10. Gillet, H., Soulé, C.: Intersection theory using Adams operations. Invent. math. **90** (1987) 243–277
11. Hochster, M.: The equicharacteristic case of some homological conjectures on local rings. Bull. Amer. Math. Soc. **80** (1974) 683–686
12. Hochster, M.: Topics in the homological theory of modules over commutative rings. Regional Conference Series in Mathematics **24**, 1975
13. Hochster, M.: Canonical elements in local cohomology modules and the direct summand conjecture. J. Algebra **84** (1983) 503–553
14. Hochster, M., Huneke, C.: Tight closure, invariant theory, and the Briançon-Skoda Theorem. J. Amer. Math. Soc. **1** (1990) 31–116
15. MacRae, R.E.: On an application of the Fitting invariants. J. Algebra **2** (1965) 153–169

16. Peskine, C., Szpiro, L.: Sur la topologie des sous-schémas fermés d'un schéma localement noethérien, définis comme support d'un faisceau cohérent localement de dimension projective finie. C. R. Acad. Sci. Paris Sér. A **269** (1969) 49–51
17. Peskine, C., Szpiro, L.: Dimension projective finie et cohomologie locale. Publ. Math. IHES **42** (1973) 47–119
18. Peskine, C., Szpiro, L.: Syzygies et Multiplicités. C. R. Acad. Sci. Paris Sér. A **278** (1974) 1421–1424
19. Rees, D.: Reduction of Modules. Math. Proc. Camb. Philos. Soc. **101** (1987) 431–449
20. Roberts, P.: Two applications of dualizing complexes over local rings. Ann. Sci. Éc. Norm. Sup. **9** (1976) 103–106
21. Roberts, P.: Cohen-Macaulay complexes and an analytic proof of the New Intersection Conjecture. J. Algebra **66** (1980) 220–225
22. Roberts, P.: The vanishing of intersection multiplicities of perfect complexes. Bull. Amer. Math. Soc. **13** (1985) 127–130
23. Roberts, P.: Local Chern characters and intersection multiplicities. Proc. Sympos. Pure Math. **46** 2, Amer. Math. Soc., Providence, R.I. 1987, pp. 389-400
24. Roberts, P.: Le théorème d'intersection. C. R. Acad. Sc. Paris Sér. I, no. 7, **304** (1987) 177–180
25. Roberts, P.: Intersection theorems. Commutative Algebra, Proceedings of an MSRI Microprogram. Springer, Berlin Heidelberg New York 1989, pp. 417–436
26. Roberts, P.: Negative intersection multiplicities on singular varieties. To appear in the Proceedings of the Zeuthen Conference, Copenhagen 1989
27. Seibert, G.: Complexes with homology of finite length and Frobenius functors. J. Algebra **125** (1989) 278–287
28. Serre, J.-P.: Algèbre locale – multiplicités. (Lecture Notes in Mathematics, vol. 11.) Springer, Berlin Heidelberg New York 1961
29. Szpiro, L.: Sur la théorie des complexes parfaits. Commutative Algebra (Durham 1981), London Math. Soc. Lecture Note Series **72** (1982) 83–90
30. Tennison, B.R.: Intersection multiplicities and tangent cones. Math. Proc. Camb. Philos. Soc. **85** (1979) 33–42

The Isomorphism Problem for Integral Group Rings of Finite Groups

Klaus W. Roggenkamp

Report on joint work with L.L. Scott 1980–1990

Mathematisches Institut B, Universität Stuttgart, Pfaffenwaldring 57
W-7000 Stuttgart 80, Fed. Rep. of Germany

1. The History of the Isomorphism Problem

The modern theory of groups originated with the treatments of *Galois* (1811–1832), *Cauchy* (1789–1857) and *Serret* (1819–1885) on finite discontinuous substitution groups. That is permutation groups, or group actions on sets, to use modern language. At that time the theory of abstract groups was only little developed. So it was natural next to study abstract groups by letting them act on sets or linearly on vector spaces. Time has shown that studying linear actions of groups on vector spaces is a much richer theory than letting groups act on sets, since one can invoke the arithmetic of the general linear groups.

In this spirit *W. Burnside*, *F. G. Frobenius* and later *I. Schur* developed the ordinary (complex) representation theory of finite groups in the years 1896–1910.

In modern language, the idea of Frobenius, Burnside and Schur was to study homomorphisms

$$\varphi \; : \; G \to GL(n, R) \; , \tag{1}$$

of G into $(n \times n)$-matrices over the commutative ring R and use the informations on the matrix groups $\mathrm{Im}(\varphi)$ to obtain informations on the abstract finite group G ; originally R was the ring of complex numbers. (Strictly speaking Burnside and Frobenius, especially, only worked with traces of these matrices.)

An important demonstration of the power of complex representation theory, where one did use the arithmetical properties of $\mathrm{Im}(\varphi) \subset GL(n, \mathbb{C})$, was *Burnside's Theorem* (1911): A finite group, whose order involves at most two different primes is solvable.

In modern times a purely group theoretical proof was found, though a theorem of Frobenius of the same spirit has never been proved without representation theory.

Modular representation theory – i.e. representations in $GL(n, K)$, where K is a field such that $\mathrm{char}(K)$ divides the order of the group – lay dormant until *R. Brauer* in 1935 – at the suggestion of I. Schur – developed the theory further, using ring theory, which enters as follows: Given a representation

$$\begin{aligned} \varphi \; : \; & G \to GL(n, R) \subset \mathrm{Mat}(n, R) \; , \\ & g \mapsto \varphi(g) \; . \end{aligned}$$

Proceedings of the International Congress
of Mathematicians, Kyoto, Japan, 1990

Since $\mathrm{Mat}(n,R)$, the ring of $n \times n$ matrices over R, is an R-algebra, we may associate with φ the R-algebra

$$\Lambda_\varphi = \{\Sigma_{g \in G}\, r_g \cdot \varphi(g) | r_g \in R\}\ , \tag{2}$$

generated by $\mathrm{Im}(\varphi)$ over R. Studying the representation φ is tantamount to studying the R-algebra Λ_φ. The universal such R-algebra is the *group ring*, consisting of formal linear combinations,

$$RG = \{\Sigma_{g \in G}\, r_g \cdot g | r_g \in G\}\ ; \tag{3}$$

addition is componentwise, and the R-linear multiplication is induced from the multiplication in G. This group ring maps onto each of the various Λ_φ. The original motivation of Burnside, Frobenius, Schur and Brauer can thus be rephrased as the question:

Which properties of the finite group G are reflected in RG?

Already in early time it was known, that $\mathbb{C}G$ does not reflect all properties of G, since finite abelian groups G and H have isomorphic complex group algebras if and only if $|G| = |H|$.

However, if one considers the group rings KG for all fields K, then the various isomorphism types of abelian groups can be distinguished by their group algebras. That this is not so in general was shown only in 1971 by E. Dade:

Theorem 1 (Dade [D]). *There are two non-isomorphic finite metabelian groups G and H of order $p^6 \cdot q^6$, p and q different primes, such that $KG \simeq KH$ for* **every** *field K.*

It is still an open

Problem 2. Let $\mathrm{char}(K) = p > 0$, and let G and H be p-groups, such that $KG \simeq KH$. Is then $G \simeq H$?

Very little is known here, except, that – by computer-analysis – for 2-groups of order $\leq 2^6$, the problem has a positive answer [Pas] and [RS4].

Now the ring of integers $\mathbb{Z}$ is the universal commutative ring, and if for two finite groups the integral group rings $\mathbb{Z}G$ and $\mathbb{Z}H$ are isomorphic, then $RG \simeq RH$ for **every** commutative ring. Thus we come to

Problem 3 (Isomorphism Problem). Assume that for two finite groups the integral group rings $\mathbb{Z}G$ and $\mathbb{Z}H$ are isomorphic. Is then $G \simeq H$?

A positive answer would imply that representation theory – especially, the integral representation theory of a finite group would determine the group up to isomorphism. This issue has a fundamental appeal in and of itself, aside from the original motivation by Burnside and Frobenius. There is a huge literature [Sa].

The rational group algebra, say for a p-group G of odd order, has the form:

$$\mathbb{Q}G = \mathbb{Q}^+ \prod \left(\prod_{i=1}^{\nu} \mathrm{Mat}(n_i, K_i) \right),$$

where K_i are algebraic number fields, $1 \leq i \leq v$, and $\mathbb{Q}^+$ comes from the trivial representation.

The integral group ring $\mathbb{Z}G$ is then in general a subring of finite index in

$$\mathbb{Z}G \subset \mathbb{Z}^+ \prod \left(\prod_{i=1}^{v} \mathrm{Mat}(n_i, R_i) \right), \tag{4}$$

where R_i are the algebraic integers in the number fields K_i, $1 \leq i \leq v$.

2. Major Results Until 1980

The isomorphism problem was first considered by *Graham Higman*, a student of G. Whitehead, in his thesis in 1939. One of his most spectacular results was:

Theorem 4 (G. Higman [Hi]). *Let u be a unit of finite order in $\mathbb{Z}A$ for a finite abelian group A. Then $u = \pm a$ for some group element $a \in A$.*

The *proof* uses the fact, that the group ring of a finite group G over the commutative ring R is not only an augmented R-algebra, with augmentation

$$\varepsilon_G \ : \ \Sigma_{g \in G}\, r_g \cdot g \to \Sigma_{g \in G}\, r_g\,, \tag{5}$$

but also a Hopf-algebra with antipode

$$*_G \ : \ \Sigma_{g \in G}\, r_g \cdot g \to \Sigma_{g \in G}\, r_g \cdot g^{-1}\,. \tag{6}$$

Let now — more generally — G be a finite group and $u \in \mathbb{Z}G$ a *central* unit of finite order n. (We shall here only deal with the case, where n is odd.) Then $u^* = *_G(u)$ is also a central unit of order n, as well as $v = u \cdot u^*$ (note that u is central), which is fixed under $*_G$, and hence must be equal to 1. Now the coefficient of 1 in v is $1 = \Sigma\, z_g^2$. Since $z_g \in \mathbb{Z}$, we have $u = \pm g_0$ for some $g_0 \in G$.

Though this argument has heavily used the fact that the ring of coefficients is $\mathbb{Z}$, the result is still valid for any integral domain R of characteristic zero, in which no rational prime divisor of $|G|$ is invertible. We shall call such a ring *G-adapted.*

Definition 5. $U(RG)$ stands for the units in RG and

$V(RG) = \{u \in U(RG) \mid \varepsilon_G(u) = 1\}$ are the *units of augmentation* 1.

Then $U(RG) = V(RG) \times U(R)$.

Let us turn to our aim, the isomorphism Problem 4:
If $RG = RH$, then $H \subset U(RG)$, in general however, it will not be contained in $V(RG)$. But this can easily be remedied: Replace H by $H' = \{h \cdot \varepsilon_G(h)^{-1} \mid h \in H\}$. Then $H \simeq H'$ and $RH = RH'$; moreover, $H' \subset V(RG)$.

Thus from now on we will always assume that an equality of group rings $RG = RH$ is always augmented*; i.e. $H \subset V(RG)$.*

Higman's result then says that the abelian group A is unique in $\mathbb{Z}G$ as augmented group of units. This result is a very elegant positive answer to the isomorphism problem.

Actually, Higman, in his 1939 thesis "Units in group rings", was the first to speculate about the isomorphism problem: "Whether it is possible for two non isomorphic groups to have isomorphic group rings I do not know; but the results ... (here) ... suggest that it is unlikely". So, Higman was a bit in favor of a positive answer to the isomorphism problem.

In this connection I should also mention *Richard Brauer*, who in his "Harvard Lecture on Modern Mathematics, Representations of finite groups" in 1963 [Br] listed the isomorphism problem as one of the important open questions in representation theory; however, he did not commit himself in either direction. Contrary to *Hans Zassenhaus*, who firmly believes in a positive answer to the isomorphism problem and even made a much stronger conjecture in 1974 [Za] (cf. below).

The isomorphism problem, $\mathbb{Z}G = \mathbb{Z}H$ as augmented algebras, appears in a new light, if we assume that $*_G = *_H$ (cf. (5)).

Proposition 6 (Banachevski [Ba]). *Let $\mathbb{Z}G = \mathbb{Z}H$ as augmented algebras and assume that both rings have the same antipode i.e. $*_G = *_H$. Then $G = H$ in $\mathbb{Z}G = \mathbb{Z}H$.*

The *proof* is in the same spirit as the proof of Higman's result.

Given an equality of augmented algebras $RG = RH$. Then the elements $\{g \mid g \in G\}$ form a finite subgroup in $V(RH)$, consisting of R-linearly independent elements. *S. D. Berman* [Ber, Sak] has observed, that the linear independence is often automatic for G-adapted rings. This result allows to phrase the isomorphism problem differently:

Problem 7. Let U be a finite subgroup in $V(\mathbb{Z}G)$, with $|G| = |U|$. Is then $U \simeq G$?

In order to gain more insight into the structure of $V(\mathbb{Z}G)$, one should even ask: How is U embedded in $V(\mathbb{Z}G)$? G. Higman gave the answer for abelian G. However, in general, one can not expect that $U = G$, since $V(\mathbb{Z}G)$ need not be abelian and G need not be normal in $V(\mathbb{Z}G)$. Moreover, even for the dihedral group D of order 8, there exists a unit $u \in \mathbb{Q}D \setminus V(\mathbb{Z}D)$, such that $D \neq u \cdot D \cdot u^{-1} \in \mathbb{Z}D$, and this conjugation is not inner in $\mathbb{Z}D$.

In this connection, *H. Zassenhaus*[Za] made a far reaching conjecture:

Zassenhaus Conjecture. *Let $U \leq V(\mathbb{Z}G)$ be a finite subgroup with $|V| = |G|$. Then there exists a unit $a \in \mathbb{Q}G$ with*

$$a \cdot V \cdot a^{-1} = G. \tag{7}$$

A direct extension of Higman's result for G a finite group and R G-adapted is given by

Theorem 8 (Berman [Ber], Glauberman (Passman [Pa]), Saksonov [Sak]). *Let* $\mathbb{Z}G = \mathbb{Z}H$ *as augmented algebras and let*

$$K_g = \sum_{x \in G/C_G(g)} {}^x g$$

be a class sum in G — $C_G(g)$ denotes the centralizer of g in G. Then $K_g = K_h$ is a class sum in RH.

An immediate consequence is:

Corollary 9. *Let N be a normal subgroup of G, and let RG = RH as augmented algebras. Then in RG = RH we have $\sum_{n \in N} n = \sum_{m \in M} m$ for a normal subgroup M in H.*

Using the above result, we can *reformulate the Zassenhaus conjecture:*

Isomorphism Problem over the Class Sums: $\mathbb{Z}G = \mathbb{Z}H$ *implies that there exists an* **isomorphism** $\chi : G \to H$ *inducing the class sum correspondence; i.e. for* $g \in G$,

$$K_g = K_{\chi(g)} \,. \tag{8}$$

The next result pushes the isomorphism problem further to metabelian groups:

Theorem 10 (Whitcomb [Wh], Jackson [J] 1968). *Assume that G is metabelian with a normal abelian subgroup A and G/A abelian.*

$$\text{If} \quad RG = RH, \quad \text{then} \quad G \simeq H$$

for R a G-adapted ring.

This follows easily from Higman's result and by considering the "*small group ring*" $RG/(I(A) \cdot I(G) \cdot RG)$. Here we have denoted for a normal subgroup N of G by $I(N) \cdot RG$, the *augmentation ideal of N*, the kernel of the natural map $RG \to RG/N$. We note that the proof does not give any information of *how G is embedded into V(RH)*.

The following result was in 1989 obtained by Kimmerle, Lyons and Sandling, using the classification of finite simple groups:

Theorem 11 [KLS]. *Let* $\mathbb{Z}G = \mathbb{Z}H$ *as augmented algebras, and let*

$$1 = N_0 < \ldots < N_t = G$$

be a chief series of G. Then H has a chief series of the same length, and both have isomorphic chief factors.

This shows in particular that finite simple groups are determined by their integral group rings – most of them though are already determined by their order.

3. Recent Progress

When, in 1980, we started to consider the isomorphism problem, the Zassenhaus conjecture was at first a tempting target for a counterexample. At that time we were not successful, and eventually started to believe, that the Zassenhaus conjecture might be true for certain classes of groups; in fact, it was a guide for our work on p-groups.

Roughly speaking, the Zassenhaus conjecture states, that there are only few automorphisms of the integral group ring $\mathbb{Z}G$ of the finite group G. After having done plenty of calculations, we were looking for some heuristic evidence for the Zassenhaus conjecture. Borrowing ideas from Lie theory we found it in the fact (first noted by us, though the proof is easy), that every derivation $\delta : \mathbb{Z}G \to \mathbb{Z}G$ is inner.

One might even go one step further than Zassenhaus and ask – this problem was briefly touched upon by *Berman and Rossa* [BR] for complete rings, like the p-adic integers $\hat{\mathbb{Z}}_p$:

Problem 12 (Conjugacy Problem). If $V(\hat{\mathbb{Z}}_pG) = V(\hat{\mathbb{Z}}_pH)$, are G and H conjugate in $V(\hat{\mathbb{Z}}_pG)$, provided G is a p-group ?

3.1 Progress on the Isomorphism Problem

The next results were obtained with Leonard Scott in 1985 [RS1] and 1987 [RS3]:

Theorem 13. *Let G be a finite group such that there exists an exact sequence*

$$1 \to A \to G \to N \to 1,$$

where A is abelian and N is nilpotent. If $\mathbb{Z}G \simeq \mathbb{Z}H$, then G is isomorphic to H. (More generally, we can replace N by a product of p_i-constrained groups G_i with $O_{p_i'}(G_i) = 1$ of relatively prime order, with slight restrictions on A.)

Remark. For a solvable group G the above implies:

a) For every prime p, the group $G/O_{p'}(G)$ is determined by $\mathbb{Z}G$, where $O_{p'}(G)$ is the largest normal subgroup of G of order prime to p.

b) The Sylow subgroups of G are determined – up to isomorphism – by $\mathbb{Z}G$. It was shown in [KR], that for $\mathbb{Z}G = \mathbb{Z}H$ there exists for each prime p and a Sylow p-subgroup P of G a unit $a(p) \in \mathbb{Q}G$ such that $a(p) \cdot P \cdot a(p)^{-1}$ is a Sylow p-subgroup of H.

3.2 Drawbacks and Progress on the Zassenhaus Conjecture

Let us recall, that the Zassenhaus conjecture (7) implies that the action of every automorphisms α of $\mathbb{Z}G$ on the center of $\mathbb{Z}G$ can be compensated there by a group automorphism.

Theorem 14. *The Zassenhaus conjecture is true for the following classes of groups:*

(i) G is nilpotent. (It has been reported to us, that in this case A.Weiss has shown that for every finite subgroup U of $V(\mathbb{Z}G)$, there exists a unit $a \in \mathbb{Q}G$ such that $a \cdot U \cdot a^{-1} \subset G$ [We].)

(ii) There exists a prime $p \in \mathbb{Z}$, such that G has a normal p-subgroup P with $C_G(P)$ a p-group.

On the other hand, the Zassenhaus conjecture is false [RS2, Ro]:

Theorem 15. *There exists a finite metabelian – even supersolvable – group G, and an automorphism α of $\mathbb{Z}G$, commuting with the augmentation ε_G, such that $\alpha \cdot \varrho$ is not a central automorphism for any group automorphism ϱ of G.*

Note that the groups in Dade's example (Theorem 1) are metabelian!

3.3 Progress on the Conjugacy Problem

In connection with the possibility, that group bases may be conjugate (cf. conjugacy Problem 12), we have obtained (1985, 1986) the following result [RS3]:

Theorem 16. *Let G be a finite group with normal Sylow p-subgroup, such that $O_{p'}(G) = 1$. Let α be an augmented automorphism of $\hat{\mathbb{Z}}_pG$. Then there exists a unit $u \in V(\hat{\mathbb{Z}}_pG)$, such that $u \cdot G \cdot u^{-1} = \alpha(G)$.*

The theorem we have proved is actually *much more general:* Let G be a finite p-constrained group with $O_{p'}(G) = 1$ and let $N = O_p(G)$. If α is an augmented automorphism of $\hat{\mathbb{Z}}_pG$, stabilizing $I_{\hat{\mathbb{Z}}_p}(N)G$, then G and $\alpha(G)$ are conjugate in $V(\hat{\mathbb{Z}}_pG)$. This stability always occurs for $N = O_p(G)$, if α arises from an augmented automorphism of $\mathbb{Z}G$.

The result applies in particular to p-groups. For the groups in Theorem 16, the Zassenhaus conjecture is true. Our result is so strong, that it allows to compute the Picard group of the group ring $\mathbb{Z}N$ of a nilpotent group N semilocally [RS, R1]. However, the conjugacy problem has a negative answer both for $V(\mathbb{Z}N)$ and for $V(\hat{\mathbb{Z}}_pN)$. Our results, and the evidence we have gathered up to now, let it appear reasonable to ask whether $V(\hat{\mathbb{Z}}_pG)$ has the

Sylow Property. *Let G be a finite p-group and U a finite subgroup of $V(\hat{\mathbb{Z}}_pG)$.*

$$\textit{Is } U \textit{ conjugate in } V(\hat{\mathbb{Z}}_pG) \textit{ to a subgroup of } G\ ? \tag{9}$$

Even more generally, one might formulate some version of this for the principal p-block, in the spirit of the defect group question of [S3].

Because of the results in [RS] we know, this is true in case $|U| = |G|$. Theorem 16 can be interpreted as a first step to prove Sylow's theorems for finite subgroups of $V(RG)$. In an attempt to prove (9) we recalled that a fancy way of proving Sylow's theorems for the finite group H is to show that the spectrum of the cohomology ring $H^*(H, \mathbb{F}_p)$ is connected; this is also true for the unit group.

More precisely, let V be a profinite p-group, and denote by $H^*(V, \mathbb{F}_p)$ the continuous (even-dimensional for p odd) cohomology ring of V with coefficients in $\mathbb{F}_p$, the field with p elements. If the spectrum of $H^*(V, \mathbb{F}_p)$ is connected, we say that the variety of V is connected. We shall write $VC(V, \mathbb{F}_p)$ for the variety of $H^*(V, \mathbb{F}_p)$.

Though we could not reach our original goal (in the spirit of [Ca]), we were able to prove for R a complete Dedekind domain of characteristic zero with residue field of characteristic p:

Theorem 17. *The following conditions are equivalent for a p-group G:*

(i) Every finite p-subgroup U of $V(RG)$ is conjugate in $V(RG)$ to a subgroup of G.

(ii) For every p-subgroup P of G, the natural inclusion

$$N_G(P)/P \to N_{V(RG)}(P)/P$$

induces a continuous map

$$VC(N_G(P)/P, \mathbb{F}_p) \to VC(N_{V(RG)}(P)/P, \mathbb{F}_p),$$

which is a bijection.

(iii) The variety of $N_{V(RG)}(P)/P$ is connected for every p-subgroup P of G.

(In the statement of the result we have used $N_A(B)$ to denote the normalizer of B in A.)

We want to point out, that the statements (i) and (iii) are not true in general for profinite p-groups; in fact we have examples of unit groups of orders, where (i) is false. It would be interesting to have a *group theoretical criterion* for when (iii) is true for a profinite p-group V and $P = 1$.

Let us return to a discussion of (9). Some years ago we found a proof of (9) for G a 2-group [R1]. Later on we were able to handle groups of order p^3, and we developed a sketch of the proof in the general p-group case in Nov. 1985 [S1]. Since in this proof we had not worked out all the details, we only made at the Arcata-meeting in 1986 the conjecture, that (9) is true for p-groups. (In the meantime, Gary Thompson, a student of Leonard Scott, has worked out the details in [S1, Th].) In October 1986 we learnt that Al Weiss from the University of Alberta [W] had a different proof of:

Theorem 18 (Subgroup Rigidity Theorem). *Let G be a finite p-group, and U a finite subgroup of $V(\hat{\mathbb{Z}}_pG)$. Then U is conjugate in $V(\hat{\mathbb{Z}}_pG)$ to a subgroup of G.*

4. Some Remarks to the Proof of Theorem 16

Questions about isomorphisms of group rings reduce to questions of automorphisms by Kimmerle's trick [Ki, RS]. The main result, Theorem 16, is a consequence of the following statement: Let G be a finite group and R an unramified extension of $\hat{\mathbb{Z}}_p$. Put $N = O_p(G)$.

Theorem 19. *Assume that $C_G(N) \subset N$ and let α be an augmented automorphism of RG with $\alpha(N) \subset I_R(N) \cdot G$ – the induced augmentation ideal of N. Then $\alpha(G)$ is conjugate to G in the units of RG.*

Remark. 1. It turns out, that in the proof of Theorem 19 we do not really require that α stabilizes $I_R(N)\cdot G$, but only that it stabilizes $I_{\mathbb{F}}(N)\cdot G$, where $\mathbb{F} = R/\text{rad}\, R$.

2. We do not know, whether the stabilization of either ideal in 1.) is automatic for an augmented automorphism α of RG. This is indeed the case when N is a Sylow p-subgroup, since then RG/N is the largest R-algebra homomorphic image of RG, which is torsionfree and whose reduction over the residue field $\mathbb{F}$ of R is semi-simple. The obstruction lies in the fact, that we do not know, whether a class sum correspondence holds for classes of p-power elements in RG.

3. The Theorem 19 is likely to be true for any complete discrete valuation domain of characteristic zero with residue field of characteristic p, assuming appropriate generalizations of the results of Weiss [W], which have been obtained by Gary Thompson [Th] and in [R].

Corollary 20. *Let G be a p-constrained group such that $O_{p'}(G) = 1$ for some prime p. Then the isomorphism problem and the Zassenhaus conjecture have a positive answer for RG. (Any two normalized group bases are even p-adically conjugate.)*

Corollary 21. *Let G be a finite group with a normal Sylow p-subgroup for some prime p. Then the defect group of the principal R-block of G — R is an unramified extension of $\hat{\mathbb{Z}}_p$ — is uniquely determined, up to conjugacy in the block, by the principal block, (i.e., if $\alpha : RG \to RG$ is an augmented automorphism, then $\alpha(P)$ and P are conjugate in the principal block).*

This result was extended by L. L. Scott in [S3], who proved it for a finite group with a cyclic Sylow p-subgroup, which is a T.I. set.

We shall now state the main ingredients of the proof of Theorem 19. There are three main ingredients in the proof. The *first one*, which we already proved in April 1986 is:

Theorem 22. *G is a p-constrained group with P a Sylow p-subgroup. R is a complete Dedekind domain of characteristic zero with $pR \neq R$. Let α be a normalized automorphism of the principal block $B_0 = R[G/O_{p'}(G)]$ of RG – note that B_0 is an augmented algebra, the augmentation being induced from that of RG. Assume that α stabilizes the image of P in B_0; i.e. $\alpha(P) = P$ in B_0. Then $\alpha|_P$ — the restriction of α to P — is induced from an automorphism β of $G/O_{p'}(G)$, such that the automorphism induced from β on B_0 agrees with α up to inner automorphisms centralizing P.*

The proof is based on the following two results:

Lemma 23 (Generalized Coleman-Ward Result [C,Wa]). *Let G be a finite group, P a p-subgroup of G, and S an integral domain, in which p is not invertible. Let $V = V(SG)$. Then we have for the normalizers*

$$N_V(P) = N_G(P) \cdot C_V(P) .$$

We shall be using the following *notation*: Given an automorphism α of RG, and an RG-bimodule M, we denote by ${}_{\alpha}M_1$ the bimodule, which has the original action on the right, but the left action is twisted by α, i.e. $x \cdot m \cdot y = \alpha(x)my$, $x, y \in RG$, $m \in M$.

Lemma 24. *R is a complete Dedekind domain of characteristic zero with $pR \neq R$, G is a finite group and B a block of RG with defect group D. Let α be an automorphism of B, which stabilizes the image of D in B, b is the Brauer correspondent in $N = N_G(D)$ to B. Then the Green correspondent to the twisted module ${}_{\alpha}B_1$ on $G \times G$, is ${}_{\beta}b_1$ on $N \times N$ for some automorphism β of RN. Moreover, $\alpha|_D = \beta|_D$, where we regard $D \subset b \subset B$.*

The *second main ingredient* is the following: Let R be an unramified extension of $\hat{\mathbb{Z}}_p$ with residue field $\mathbb{F}$. Let N be a normal p-subgroup of a finite group G, and assume $C_G(N) \subset N$.

Theorem 25. *Let α be an augmented automorphism of RG, with N and G as above, and assume that α stabilizes the ideal $I_{\mathbb{F}}(N) \cdot G$ in $\mathbb{F}G$. Then $\alpha(N)$ is conjugate to N by a unit of RG.*

The *third main ingredient* is an extension of results of Weiss [W] combined with a criterion on the indecomposability of permutation modules, which is the main tool to show that group bases are conjugate. We let R be an unramified extension of $\hat{\mathbb{Z}}_p$, the p-adic integers. $\mathbb{F} = R/pR$ is the residue field of R, and we denote by $x \to \bar{x}$ the natural projection from R to $\mathbb{F}$, and also for any R-module X we put $\overline{X} = X/pX$.

Theorem 26. *Assume that P and Q are p-groups. Let M be an $R(P \times Q)$-module (with the action $(x, y) \cdot m = x \cdot m \cdot y^{-1}$, $x \in P$, $y \in Q$, $m \in M$), which is free of finite rank as RQ-module. If $\overline{M/(M \cdot I_R(Q))}$ is a permutation module for $\mathbb{F}P$, then $\overline{M}$ is a permutation module for $\mathbb{F}(P \oplus Q)$.*

The next result is the place where we use that $C_G(N) \subset N$.

Lemma 27. *Suppose $C_G(N) \subset N$. Then the $\mathbb{F}(N \times G)$-module $\mathbb{F}G$ — with the action $(n, g) \cdot x = n \cdot x \cdot g^{-1}$, $n \in N$, $g \in G$, $x \in \mathbb{F}G$ — is absolutely indecomposable. Moreover, the same is true if the action of N on the left is twisted by an augmented $\mathbb{F}$-algebra automorphism α of $\mathbb{F}G$, or by a group automorphism ϱ of N – i.e., $(n, g) \cdot x = \varrho(n) \cdot x \cdot g^{-1}$ for $n \in N$, $g \in G$ and $x \in \mathbb{F}G$.*

We shall next come to the proof of the "Subgroup Rigidity Theorem" 18 as it was generalized from [W] in [R]. R will be a complete Dedekind domain of finite rank over $\hat{\mathbb{Z}}_p$, the p-adic integers, K is the field of fractions of R and $\mathrm{rad}(R) = \pi \cdot R$. G is a finite p-group. Theorem 18 is an immediate consequence of the following result, which generalizes and was inspired by the work of Al Weiss [W]:

Theorem 28. *Let M be an RG-lattice, and let N be a normal subgroup of G. Assume that*

1. $M{\downarrow}_N$ is a free RN-module,
2. $M/(I(N)\cdot M)$ is a permutation module for G/N.
Then M is an RG-permutation module.

The next result plays a major role in the proof of Theorem 19, though it is also of interest for its own sake. A generalized permutation module is a direct sum of modules induced from one dimensional representations of subgroups.

Theorem 29 [R]. *1. Assume that K is a splitting field for G and all of its subgroups. If ζ be a primitive p-th root of unity in R, then we require $\pi^t \cdot R \subset (1-\zeta)\cdot R$. An RG-lattice M is a generalized permutation module if and only if M/π^t is a generalized permutation module.*

2. If $\pi^t \cdot R \not\subseteq (1-\zeta)\cdot R$, then there is an RG-lattice M, which is not a generalized permutation module, but M/π^t is a generalized permutation module.

Let us point out, where the restrictive condition on R/π^t becomes apparent:

Lemma 30. *Let R be a complete Dedekind domain of finite rank over $\hat{\mathbb{Z}}_p$. The induced map*

$$\mathrm{Ext}^1_{RG}(M,N) \to \mathrm{Ext}^1_{R/\pi^t G}(M/\pi^t, N/\pi^t)$$

is injective for all generalized permutation lattices M and N if and only if either R does not contain a primitive p-th root of unity or $\pi^t \cdot R \subset (1-\zeta)\cdot R$, where ζ is a primitive p-th root of unity.

References

[Ba] Banachevski, B.: Integral group rings of finite groups. Can. Math. Bull. **10** (1967) 635–642

[Ber] Berman, S. D.: On a necessary condition for isomorphisms of integral group rings. Dopovidi Akad. Nauk Ukrain. RSR 1953, 313–316

[BR] Berman, S. D., A.R. Rossa: Integral group rings of finite and periodic groups. Algebra and Math. Logic: Studies in Algebra, pp. 44–53. Izdat., Kiew 1966

[Br] Brauer, R.: Representations of finite groups. In: Harvard Lectures on Modern Mathematics, 1963

[Ca] Carlson, J.: The variety of a module. (Lecture Notes in Mathematics. vol. 1142.) Springer, Berlin Heidelberg New York 1985, pp. 88–95

[C] Coleman, D. B.: On the modular group ring of a *p*-group. Proc. AMS. **5** (1964) 511–514

[D] Dade, E.: Deux groupes finis distinctes ayant la même algèbre de groupes sur tout corps. Math. Z. **119** (1971) 345–348

[Hi] Higman, G.: Units in group rings. D. Phil. Thesis. Oxford University, 1940

[J] Jackson, D.A.: The groups of units of the integral group rings of finite metabelian and finite nilpotent groups. Quat. J. Math. Oxford Ser. **2,** 20 (1969) 319–331

[Ki] Kimmerle, W.: Personal communication 1985

[KLS] Kimmerle, W., R. Lyons, R. Sandling: Composition factors for group rings and Artin's theorem on orders of simple groups. Preprint 1987

[KR] Kimmerle, W, K.W. Roggenkamp: A Sylow like theorem for integral group rings of finite solvable groups. To appear in Archiv der Mathematik (1990)

[Pa] Passman, D.S.: Isomorphic groups and group rings. Pacific J. Math. **15** (1965) 561–583

[Pas] Passman, D.: Group algebras of groups of order p^4 over a modular field. Mich. Math. J. **12** (1965) 405–415

[P] Puig, L.: Pointed groups and construction of characters. Math. Z. **176** (1981) 265–292

[Rey] Reynolds, *W. F.*: Blocks and normal subgroups of finite groups. Nagoya Math. J. **22** (1963) 15–32

[RiS] Ritter, J., S.K. Sehgal: On a conjecture of Zassenhaus on torsion units in integral group rings. Preprint, Univ. Augsburg 1983

[RS] Roggenkamp, K.W., L.L. Scott: Isomorphisms of p-adic group rings. MS Sept. 85, Ann. Math. **126** (1987) 593–647

[RS1] Roggenkamp, K.W., L.L. Scott: The isomorphism theorem for integral group rings of nilpotent by abelian groups. MS, May 86, pp. 1–14 (to be published)

[RS2] Roggenkamp, K.W., L.L. Scott: On a conjecture on group rings by H. Zassenhaus. MS, March 87, pp. 1–39 (to be published)

[RS3] Roggenkamp, K.W., L.L. Scott: A strong answer to the isomorphism problem for finite p-solvable groups with a normal p-subgroup containing its centralizer. MS, 87 (to be published)

[RS4] Roggenkamp, K.W., L.L. Scott: Automorphisms and non abelian cohomology. MS, 89

[R1] Roggenkamp, K.W.: Picard groups of integral group rings of nilpotent groups. Proc. Symp. Pure Math. **47** (1987) 477–485

[R2] Roggenkamp, K.W.: Units in integral metabelian group rings I. Jackson's unit theorem revisited. Quat. J. Math. **23** (1981) 209–224

[R] Roggenkamp, K.W.: Subgroup rigidity of p-adic group rings (Weiss arguments revisited). MS 1989

[Ro] Roggenkamp, K.W.: Observation to a conjecture of H. Zassenhaus. Proceedings of Groups, St. Andrews 1989 (to appear)

[Sak] Saksonov, A.I.: Group rings of finite groups I. Publ. Math. Debrecen **18** (1971) 187–209

[Sa] Sandling, R.: The isomorphism problem for group rings: A survey. (Lecture Notes in Mathematics, vol. 1142.) Springer, Berlin Heidelberg New York 1985, pp. 239–255

[S1] Scott, L.L.: Report on the isomorphism problem. Proc. Symp. Pure Math. **47** (1987) 259–273

[S2] Scott, L.L.: The modular theory of permutation representations. Representation theory of finite groups and related topics. AMS Meeting at Madison, April 1970, pp. 137–144

[S3] Scott, L.L.: Defect groups and the Isomorphism Problem. Preprint 1988

[Th] Thompson, G.: Conjugacy in p-adic group rings. Ph. D. thesis, Univerity of Virginia 1989

[Wa] Ward, H.N.: Some results on the group algebra of a group over a prime field. Seminar on Finite Groups and Related Topics. Harvard University 1960–1961

[W] Weiss, A.: p-adic rigidity of p-torsion. Ann. Math. **127** (1987) 317–332

[We] Weiss, A.: Torsion units in integral group rings. MS, 89

[Wh] Whitcomb, A.: The group ring problem. Ph. D. thesis, Chicago 1968

[Za] Zassenhaus, H.: On the torsion units of finite group rings. Studies in Mathematics, Instituto de Alta Cultura, Lisboa 1974, pp. 119–126.

The Local to Global Principle in Algebraic K-Theory

*Robert W. Thomason**

CNRS, U.F.R. de Mathématiques, Université de Paris 7, F-75251 Paris Cedex 05, France

In the early 70s, Quillen [Q] succeeded in finding a good definition of the higher algebraic K-groups, generalizing the usual Grothendieck group K_0. To a noetherian ring or scheme X, his theory associates two distinct series of K-groups: $K_*(X)$, the theory associated to the exact category of finitely generated projective modules or algebraic vector bundles on X; and $G_*(X)$ or $K'_*(X)$, the theory associated to the abelian category of all finitely generated modules or coherent sheaves on X. Quillen's localization theorem for the K-theory of abelian categories applies to $G_*(X)$, and enables him to prove many fundamental results for this theory, including invariance of G_* under replacing the ring R by the ring of polynomials $R[T]$ or the scheme X by $X[T]$, a Mayer-Vietoris exact sequence for a cover of X by two open subschemes, and a Brown-Gersten local-to-global spectral sequence which reduces many G-theory problems to the case of local rings. For regular rings or schemes X, the existence of finite projective resolutions for all finitely generated modules shows that $K_*(X) = G_*(X)$, so all these good results apply to $K_*(X)$ in the regular case. However if X has singularities, Quillen's methods give many fewer results about $K_*(X)$. Unfortunately, this case arises often: for example the integral group ring of a non-zero finite group is always singular. The essential difficulty was the lack of the key tool: a good localization theorem for K-theory.

The author and Thomas F. Trobaugh have remedied this lack by finding the good localization result, using the techniques of Waldhausen's "algebraic K-theory of spaces" [Wa2] and returning to some ideas of Grothendieck about defining K-theory using "perfect complexes" in place of single vector bundles [SGA6]. Using a characterization of perfect complexes, essentially defined as finite complexes of algebraic vector bundles, as being the finitely presented objects in the derived category of modules, we obtain the result that the K_0 class is the only obstruction to extension up to quasi-isomorphism of a perfect complex on an open subscheme of X to a perfect complex on all of X. This extension result and Waldhausen's theory give the localization theorem for K-theory, Theorem 2.1 below. This in turn unleashes a pack of new fundamental results for K-theory, including the Mayer-

* Author partially supported by NSF, The Johns Hopkins University and Université de Paris 7.

Proceedings of the International Congress
of Mathematicians, Kyoto, Japan, 1990

Vietoris exact sequence, the Brown-Gersten local-to-global spectral sequences for both the Zariski and Nisnevich topologies, the generalizations to schemes from rings of the results of Weibel, Ogle, and Goodwillie controlling the failure of K_* to be invariant under polynomial or nilpotent extension or to satisfy Mayer-Vietoris for closed covers, and the isomorphism between mod l algebraic K-theory localized by inverting the Bott element and etale topological K-theory. Recently Dongyuan Yao has obtained some similar localization results for the K-theory of non-commutative rings. Hiding in the background of all this is the fact that the Waldhausen K-theory of a category of complexes depends essentially only on the derived category; this idea has been much advanced by work in progress of Giffen and Neeman, which unfortunately I will not discuss below.

§ 1. Perfect Complexes and Waldhausen K-Theory

We consider schemes X which are quasi-compact and quasi-separated. *Quasi-compact* means that X is covered by a finite number of open subschemes which are affine, i.e. isomorphic to Spec(R) for some commutative rings R. *Quasi-separated* means that the intersection of any two such affines is again quasi-compact. These conditions are indeed very mild, and are met by all algebraic varieties over a field, by all schemes of finite type over a noetherian ring, and by Spec(R) for any commutative ring R.

Recall that a map between two chain complexes in an abelian category is a *quasi-isomorphism* if it induces an isomorphism on the homology of the complexes. Two complexes E and F are quasi-isomorphic if they are connected by a chain of quasi-isomorphisms

$$E \overset{\sim}{\leftarrow} A \overset{\sim}{\rightarrow} \cdots \overset{\sim}{\rightarrow} F$$

Definition (Grothendieck [SGA6]). *A perfect complex on a scheme X is a complex E of $\mathcal{O}_X$-modules, such that for every affine open $U =$ Spec(R) in X, the restriction $E|U$ is quasi-isomorphic to a finite complex of finitely generated projective R-modules.*

For E to be perfect, it suffices that $E|U$ be quasi-isomorphic on U to a finite complex of such projective modules for all the U in some affine open cover of X. In particular, for $X =$ Spec(R), the perfect complexes are those quasi-isomorphic to such complexes of projective R-modules.

The category of perfect complexes on X is a "category with cofibrations and weak equivalences" in the sense of [Wa2] 1.2, i.e., it is a Waldhausen category. The cofibrations are the maps of complexes which are degree-wise split monomorphisms; the weak equivalences are the quasi-isomorphisms. For Y a closed subspace of X, there is a full Waldhausen subcategory of those perfect complexes which are cohomologically supported on Y; i.e., those which are acyclic on $X - Y$.

The theory of Waldhausen [Wa2] associates higher algebraic K-groups to Waldhausen categories: these are the homotopy groups of a space, or even better, of an infinite loop space or spectrum in the sense of algebraic topology. Let $K_*(X)$ denote the Waldhausen algebraic K-groups associated to the category of perfect complexes on X, and $K(X)$ denote the associated K-theory spectrum. Let $K_*(X$ on $Y)$ and $K(X$ on $Y)$ denote the K-groups and spectrum associated to the Waldhausen subcategory of those perfect complexes on X which are acyclic on $X - Y$.

$K_0(X)$ is the Grothendieck group of [SGA6], generated by quasi-isomorphism classes $[E]$ of perfect complexes on X, modulo an Euler characteristic relation that $[E] = [E'] + [E'']$ whenever one can choose representatives of the quasi-isomorphism classes that fit in a short exact sequence of complexes $0 \to E' \to E \to E'' \to 0$, that is, whenever they are the vertices of an exact triangle in the derived category $D(X)$ of $\mathcal{O}_X$-modules.

Suppose X satisfies the mild hypothesis of having an "ample family of line bundles" in the sense of [SGA6] II.2.2.3. This condition is inherited by all subschemes of X and is satisfied when X is a commutative ring, or is quasi-projective over a commutative ring, or is a separated regular noetherian scheme. In particular, this condition holds for all classical algebraic varieties, since they are quasi-projective over a field. For a scheme X with an ample family of line bundles, a global resolution theorem of Illusie ([SGA6] II.2.2.8) says that every perfect complex on X is globally quasi-isomorphic to a finite complex of algebraic vector bundles defined on all of X. Then the $K_0(X)$ above is isomorphic to the "naive" Grothendieck group of algebraic vector bundles on X, and the Waldhausen $K_*(X)$ are isomorphic to the Quillen K-groups of X ([TT2] 3.10, 3.9).

We have an easy to verify but very useful characterization of perfect complexes in terms of the derived category $D(X)$. Recall that this category is obtained from the category of all chain complexes of $\mathcal{O}_X$-modules by inverting the quasi-isomorphisms (see Verdier [Ve]). $D(X)_{qc}$ denotes the full subcategory of those complexes whose homology sheaves are *quasi-coherent*, i.e., which locally on affine subschemes $\mathrm{Spec}(R)$ are isomorphic to R-modules. $D^+(D)_{qc}$ is the further subcategory of complexes E such that the homology sheaves $H_k(E)$ are 0 for all k sufficiently large.

Proposition 1.1. *Let X be a quasi-compact and quasi-separated scheme. Let E be an object of $D^+(X)_{qc}$. Then the following are equivalent:*

a) *E is a perfect complex*

b) *For any direct system of complexes F_α with quasi-coherent homologies, the canonical map is an isomorphism of morphism sets:*

$$\mathrm{colim}_\alpha\, D(X)(E, F_\alpha) \xrightarrow{\sim} D(X)(E, \mathrm{colim}_\alpha\, F_\alpha)$$

c) *The functor $D(X)(E, \ )$ sends infinite direct sums in $D^+(X)_{qc}$ to direct sums in the category of abelian groups.*

Proof. [TT2] 2.4.3 yields that a) iff b). That b) implies c) is clear, and that c) implies b) follows from a demonstration dual to that of [T3] 2.6. □

Comparing condition b) with Grothendieck's characterization of finitely presented modules or algebras A over a ring or scheme as those such that $\mathrm{Mor}(A, \ )$

preserves direct colimits ([EGA] IV 8.14), we see this proposition characterizes perfect complexes as finitely presented objects in the derived category. This suggests interesting generalizations. Thus, the "perfect complexes" in the stable homotopy category turn out to be the homotopy finite cell spectra: they have an associated K-theory, which turns out to be Waldhausen's $A(\mathrm{pt})$ ([Wa1] § 1). The proposition also suggests that we could apply the yoga for extending finitely presented objects from an open subscheme to all of X ([EGA] I.6.9). This suggestion leads to:

Key Proposition 1.2. *Let X be a quasi-compact and quasi-separated scheme, $j: U \to X$ the inclusion of a quasi-compact open subscheme, and F a perfect complex on U. Then F is quasi-isomorphic to the restriction j^*E of a perfect complex on X if and only if the class $[F]$ of F in $K_0(U)$ is in the image of $j^*: K_0(X) \to K_0(U)$.*

Proof. [TT2] 5.2.2 The basic idea is to write the complex with quasi-coherent homology $Rj_*(F)$ on X as quasi-isomorphic to a direct colimit of perfect complexes, colim $E_\alpha \approx Rj_*F$. Then on U, F is quasi-isomorphic to colim j^*E_α. As F is perfect, Proposition 1.1 shows that this quasi-isomorphism factors through some j^*E_α. Thus F is quasi-isomorphic to a summand of j^*E_α, the restriction of a perfect complex E_α on X. This is the miracle for general U, although it would be trivial for affine U. The next step, more ordinary, is to show that given a morphism on U between restrictions of perfect complexes, $j^*E_1 \to j^*E_2$, one can extend this to a morphism $E_1 \to E_2$ on X after replacing the old E_1 by another perfect complex whose restriction to U is quasi-isomorphic to that of the old one. An induction and excision reduces this to the case where X has an ample family of line bundles. If $X - U$ is a divisor, one succeeds after replacing E_1 by its tensor product with a power of the line bundle of the divisor. For $X - U$ general, one succeeds after replacing E_1 by its tensor power with the positive degree part of a Koszul complex of line bundles for divisors whose intersection is $X - U$. From this extension result on morphisms, it follows that if 2 out of 3 terms in an exact triangle of perfect complexes in $D(U)$ extend, so does the third. We now adapt an idea of Grayson from cofinality theory: consider the abelian monoid of quasi-isomorphism classes of perfect complexes on U, and take the quotient monoid by setting equal to 0 the classes that are restrictions of perfect complexes on X; using the preceding sentence one sees that the class of a perfect complex on U is zero in this quotient if and only if it extends to a perfect complex on X. The miracle above shows this monoid has inverses, and so is an abelian group. We see from the above that this obstruction group to extension satisfies the Euler characteristic relation for exact triangles, and thus in fact that it is the quotient of $K_0(U)$ by the image of $K_0(X)$. □

In particular, this key Proposition gives a K_0-criterion for an extending up to quasi-isomorphism an algebraic vector bundle on U to a perfect complex on X, that is, to a finite complex of algebraic vector bundles on X in the usual case where X has an ample family. No similar criterion is possible for the question of extension to a single vector bundle on X: it is for this reason that it is necessary to go to the framework of perfect complexes and leave the framework of algebraic vector bundles to prove a good localization theorem for algebraic K-theory. The impossibility of such a criterion was noted by Serre ([Se] 5a), who observed that as the tangent vector bundle to the projective plane $\mathbb{P}^2$ is not a sum of line bundles, its preimage on $\mathbb{A}^3 - 0$ is a vector bundle which is not free, and which does not extend to a vector bundle on $\mathbb{A}^3$, even though $K_0(\mathbb{A}^3) \to K_0(\mathbb{A}^3 - 0) = \mathbb{Z}$ is an isomorphism, and though this vector bundle on $\mathbb{A}^3 - 0$ is the quotient of a monomorphism of free vector bundles on $\mathbb{A}^3 - 0$, which of course do extend. The vast literature on the Serre conjecture contains many negative and a few special positive results on the problem of extending to vector bundles on X for X regular: one may consult the articles of Horrocks, Swan, Murthy, Vasserstein, Bass, M. Kumar, Suslin, and Quillen on this conjecture. Very few have dared to say anything about the case where X is singular.

§ 2. The Localization Theorem and Its Basic Consequences

Theorem 2.1 (Localization Theorem) *Let X be a quasi-compact and quasi-separated scheme, and let Y be a closed subspace of X such that $X - Y$ is quasi-compact. Then there is a natural homotopy fibre sequence, and an associated long exact sequence of homotopy groups*:

$$K(X \text{ on } Y) \to K(X) \to K(X - Y)$$

$$\cdots \to K_n(X \text{ on } Y) \to K_n(X) \to K_n(X - Y) \to K_{n-1}(X \text{ on } Y) \to$$

Proof. [TT2] 7.4. For $n \geq 0$, this results easily from the Key Proposition above and Waldhausen's fibration and approximation theorems [Wa2] 1.6.4 and 1.6.7. As the map $K_0(X) \to K_0(X - Y)$ is not always surjective in the singular case, it is necessary to construct new non-zero negative K-groups K_n for $n < 0$ to continue the exact sequence to the right, and to replace the spectrum $K(X)$ by a new non-connective spectrum of which the Waldhausen K-theory spectrum is a covering space. This is done by following an inductive procedure due to Bass [B], which depends on the part of the localization theorem already proved for higher n, and which uses the exact sequence of 2.2 to inductively define K_{n-1} for $n \leq 0$ given K_n. □

The special case of 2.1 where $X - Y$ is affine and Y is a Cartier divisor is due to Quillen [Gr] (at least in light of [TT2] 5.7 which identifies $K(X \text{ on } Y)$ with Quillen's third term $K(H_Y X)$ in this case). Painful efforts to squeeze a bit more out of Quillen's method have been made by Levine [Le3]. The counterexample of Deligne given in Gersten's paper ([Ge] § 7) shows the limits of an exact category approach to localization.

Corollary 2.2 (Bass Fundamental Theorem). *Let X be a quasi-compact and quasi-separated scheme. Denote by $X[T]$ the scheme $X \times \mathbb{A}^1 = X \otimes_{\mathbb{Z}} \mathbb{Z}[T]$, and similarly by $X[T^{-1}]$ another polynomial extension of X, and by $X[T, T^{-1}]$ the Laurent polynomial extension. Then there is a natural exact sequence for all integers $n \in \mathbb{Z}$*:

$$0 \to K_n(X) \to K_n(X[T]) \oplus K_n(X[T^{-1}]) \to K_n(X[T, T^{-1}]) \to K_{n-1}(X) \to 0$$

Similarly for $K_*(X \text{ on } Y)$.

Proof. [TT2] 7.5. This is proved by descending induction on n using the formula of Quillen ([Q] § 8) that $K_n(\mathbb{P}^1_X) \cong K_n(X) \oplus K_n(X)$, and the Mayer-Vietoris Corollary 2.3 to the localization theorem applied to the cover of $\mathbb{P}^1_X$ by $X[T]$ and $X[T^{-1}]$ with intersection $X[T, T^{-1}]$. For X affine and $n \leq 1$, the result is indeed due to Bass [B]; and for X affine and $n \geq 2$, it is due to Quillen [Gr]. □

Corollary 2.3 (Mayer-Vietoris Theorem). *Let U, V be two quasi-compact open subschemes of a quasi-separated scheme X. Then there is a homotopy cartesian square*:

$$\begin{array}{ccc} K(U \cup V) & \longrightarrow & K(U) \\ \downarrow & & \downarrow \\ K(V) & \longrightarrow & K(U \cap V) \end{array}$$

and an associated long exact Mayer-Vietoris sequence:

$$\cdots \to K_n(U \cup V) \to K_n(U) \oplus K_n(V) \to K_n(U \cap V) \to K_{n-1}(U \cup V) \to \cdots$$

Proof. [TT2] 8.1. The localization theorem calculates the homotopy fibres of the horizontal maps in the square, and an easy excision result shows they are equivalent, as U is an open neighborhood of $Y = (U \cup V) - V = U - (U \cap V)$ in $U \cup V$. □

Theorem 2.4 (Brown-Gersten or Local-to-Global Spectral Sequence, Spectral Sequence of Cohomological Descent). *Let X be a noetherian scheme of finite dimension. Then there is a strongly converging spectral sequence whose E^2 term is the cohomology of X for the Zariski topology with coefficients in the sheaf associated to the presheaf of K-groups*:

$$E_2^{p,-q} = H^p(X_{\text{Zar}}; K_q) \Rightarrow K_{q-p}(X)$$

In fact, the augmentation map to the Zariski hypercohomology spectrum ([T1] § 1) *is a homotopy equivalence identifying this spectral sequence to the canonical hypercohomology spectral sequence.*

$$K(X) \xrightarrow{\sim} \mathbb{H}(X_{\text{Zar}}; K)$$

One has the same results after replacing the Zariski topology by the slightly finer Nisnevich topology ([N] § 1, [TT2] Appendix E).

Proof. [TT2] 10.3, 10.8. In the Zariski case, Brown and Gersten showed this would follow from Mayer-Vietoris [BG], and proved the case for regular X or for G-theory. The argument for the Nisnevich topology is similar, but needs a slightly stronger excision result that $K(X \text{ on } Y)$ is unchanged by replacing X by its henselization along Y. Nisnevich [N] proved the Nisnevich case for regular X or for G-theory. Weibel [We4] proved the theorem in the case X has isolated singular points, using ideas of Collino and Pedrini. We note that the Nisnevich topology plays a big role in the work of Kato and Saito [KS], who call it the henselian topology. □

2.5 This spectral sequence reduces many problems to the case of local rings for the appropriate topology, since the stalks of the coefficient sheaves are the values of K_q at these local rings. For the Zariski topology, these are the local rings of X in the usual sense. For the Nisnevich topology, the local rings are the hensel local rings of X, the henselizations of the usual local rings. It is a theorem of Gabber [Ga], completing ideas of Suslin, that for $R_{\mathfrak{m}}$ a hensel local ring and n an integer such that $1/n \in R_{\mathfrak{m}}$, then the reduction map to the residue field, $R_{\mathfrak{m}} \to R/\mathfrak{m}$, induces an isomorphism on mod n K-theory, at least in positive degrees, $K/n_*(R_{\mathfrak{m}}) \cong K/n_*(R/\mathfrak{m})$. Thus many problems for mod n K-theory reduce to the case of fields. For example, this reduction plays an essential role in the proof of the étale cohomological descent Theorem 4.1 below.

2.6 In the previously known case where X is a regular scheme of finite type over a field k, this spectral sequence has interesting relations with classical intersection theory because of the formula for $A^p(X)$, the Chow group of codimension p algebraic cycles on X, inspired by Gersten, proved by Quillen ([Q] § 7.5), and ascribed to Bloch (who in fact did prove the case $p = 2$, the case $p = 1$ being due to Cartier): $A^p(X) = H^p(X; K_p)$. As Nisnevich observed, this works either for the Zariski or the Nisnevich topology. It would be interesting to have a geometric interpretation of $H^p(X; K_p)$ and a relation to some kind of intersection theory in the singular case. Some work on this has been done by Collino [Co], Gillet [Gi], Levine [Le1, Le2, LeW], Pedrini [PW1, PW2], Weibel, and Barbieri Viale [Vi]. In the singular case, the Zariski and Nisnevich cohomologies are different, and there is some evidence that the Nisnevich cohomology works better.

2.7 The hard-core algebraist may have been wondering if all this scheme business gives anything new for the K-theory of ordinary commutative rings. The answer is that it does. For the proof of spectral sequence 2.4 for the case of a noetherian ring A, $X = \mathrm{Spec}(A)$, and thus the reduction of problems about $K(A)$ to the case of local rings, depends on an induction which requires considering arbitrary open subschemes U of $\mathrm{Spec}(A)$, not all of which will be affine when the dimension of A is greater than one. This is one of the reasons why Quillen's localization result with its hypothesis that $X - Y$ be affine did not suffice to prove Theorem 2.4 for noetherian rings. I also offer the following two new results:

Proposition 2.8. *Let R be a commutative ring, and let $r, s \in R$ be two elements such that the ideal generated by r and s is all R. Then there is a long exact Mayer-Vietoris sequence*:

$$\cdots \to K_n(R) \to K_n(R[1/r]) \oplus K_n(R[1/s]) \to K_n(R[1/rs]) \to K_{n-1}(R) \to \cdots.$$

Proof. This is a special case of 2.3 with $U = \mathrm{Spec}(R[1/r])$ and $V = \mathrm{Spec}(R[1/s])$. Since the ideal (r, s) is all of R, $\mathrm{Spec}(R) = U \cup V$. The result would follow from

Quillen's localization theorem [Gr] under the additional hypothesis that either r or s is not a zero-divisor in R, but this hypothesis is often embarrassing, e.g. in the case of group rings. □

Proposition 2.9 (Weibel). *Let R be a commutative ring, and $r \in R$. Denote by $R_r^\wedge$ the completion* $\lim_k R/r^kR$. *Then there is an exact Mayer-Vietoris sequence*:

$$\cdots \to K_n(R) \to K_n(R_r^\wedge) \oplus K_n(R[1/r]) \to K_n(R_r^\wedge[1/r]) \to K_{n-1}(R) \to \cdots$$

Proof. This follows from 2.1 and an excision result that $K(R \text{ on } R/r) \approx K(R_r^\wedge \text{ on } R_r^\wedge/r)$, which Weibel observed could be proved by the method of [TT2] 2.6.3 and 3.19. □

§ 3. Generalizations to Schemes of Some Results Known in the Affine Case

The Mayer-Vietoris Theorem 2.3 often permits one to generalize results known for affine schemes to general quasi-compact separated schemes X, by inducting on the number of affines in an open cover of X. (A second induction on the number of quasi-affines needed to cover X then relaxes the hypothesis of separation to that of quasi-separation). Work of Weibel [We1, We2, We3, We5], Ogle [Og, OW], Geller, Goodwillie [Go], Soulé [So], Staffeldt [Sta], van der Kallen [vdK], and Vorst [Vo] have produced many results controlling the failure of K-theory of rings to be invariant under polynomial or nilpotent extensions, and the failure of Mayer-Vietoris for closed covers of affine schemes. Their demonstrations depend on the use of projective objects and other ring-theoretic techniques, and do not readily generalize to schemes, where the locally projective algebraic vector bundles are not globally projective. However, our method generalizes their conclusions to the case of schemes. As a sample, I give the following results. The first two concern the mod n K-theory defined by Karoubi and Browder; these groups fit in a short exact universal coefficient sequence:

$$0 \to K_q(X) \otimes \mathbb{Z}/n \to K/n_q(X) \to \mathrm{Tor}_{\mathbb{Z}}(K_{q-1}(X), \mathbb{Z}/n) \to 0$$

Theorem 3.1 (Weibel) *Let X be a quasi-compact and quasi-separated scheme, and let $W \to X$ be the total space of a vector bundle over X (or even a torsor under a vector bundle). Let n be an integer such that $1/n \in \mathcal{O}_X$. Then $W \to X$ induces an isomorphism on* mod *n K-groups*:

$$K/n_*(X) \xrightarrow{\sim} K/n_*(W)$$

In particular, this is true when $W = X[T]$.

Proof. [TT2] 9.5. This is deduced by the method above, using the 5-lemma and the Mayer-Vietoris sequence 2.3, from the affine case due to Weibel [We1] 3.3, [We3]. The critical case is that of the polynomial extension $R[T]$ of a ring R. In general, the induced map $K_*(R) \to K_*(R[T])$ has a non-trivial cokernel, although it is obviously injective, split by the map induced by the ring homomorphism sending T to 0. Stienstra [Sti], following work of Almkvist, Grayson, and Bloch, showed that the cokernel was a module over the ring of Witt vectors of R. Weibel observed this shows the cokernel is uniquely n-divisible if $1/n \in R$, yielding the result in the critical case. □

Theorem 3.2 (Weibel). *Let X be a quasi-compact and quasi-separated scheme, with two closed subschemes Y, Z, such that $X = Y \cup Z$ as spaces. Give the intersection $Y \cap Z$ the scheme structure of the fibre product of Y and Z over X. Let n be an integer such that $1/n \in \mathcal{O}_X$. Then there is a homotopy cartesian square:*

$$\begin{array}{ccc} K/n(X) & \longrightarrow & K/n(Y) \\ \downarrow & & \downarrow \\ K/n(Z) & \longrightarrow & K/n(Y \cap Z) \end{array}$$

and an associated long exact Mayer-Vietoris sequence:

$$\cdots \to K/n_q(X) \to K/n_q(Y) \oplus K/n_q(Z) \to K/n_q(Y \cap Z) \to K/n_{q-1}(X) \to \cdots.$$

Proof. [TT2] 9.8. Again one reduces to the affine case, which is due to Weibel, essentially [We2] 1.3. The corresponding statement for K_* in place of K/n_* would not be true, but its failure can be analyzed locally using cyclic homology [OW], which result can be globalized by the method of 3.3 below. □

3.3 Much recent work has focused on cyclic homology HC_*, which serves as a sort of "linear approximation" to algebraic K-theory, and is more readily calculable. Discovered for operator algebras by Connes, cyclic homology has been extended to algebras over a field k by Loday-Quillen and Feigin-Tsygan, and has been extensively developed by J. Block, Brylinksi, Burghelea, Carlsson, Cathelineau, R. Cohen, Goodwillie, J. Jones, Karoubi, Kassel, Ogle, Staffeldt, Vigué-Poirier, Weibel, Wodzicki, and others in a torrent of articles [Lo].

Independently, several people, including J. Block, Loday, and Weibel have proved that HC_* has a Cech-Mayer-Vietoris spectral sequence for a cover of Spec(R) by affine open subschemes. This allows one to extend the definition of $HC_*(X)$ to schemes X over a field k, by taking homology of the total complex of the Cech complex of an affine cover of X, with coefficients in the cyclic homology complex HC. We deduce, as suggested by Weibel:

Theorem 3.4 (Goodwillie). *Let X' be a quasi-compact and quasi-separated scheme over* Spec($\mathbb{Q}$). *Let $i: X \to X'$ be a closed immersion defined by a sheaf of nil ideals. Then there is a natural isomorphism between the relative cyclic homology over $\mathbb{Q}$ and the relative algebraic K-theory:*

$$K_n(X \to X') \cong HC_{n-1}(X \to X').$$

Proof. Here $K(X \to X')$ is the homotopy fibre of the map $i^*: K(X') \to K(X)$, so that the relative K-groups $K_*(X \to X')$ fit in a long exact sequence with $K_*(X)$ and

$K_*(X')$; and similarly for the relative cyclic homology complex $HC(X \to X')$. For X' affine, the main theorem of [Go] gives a natural homotopy equivalence between $K(X \to X')$ and the complex $HC(X \to X')$ shifted one degree and considered as a generalized Eilenberg-MacLane spectrum. The result for general X then follows by Mayer-Vietoris, cf. [TT2] 9.10. □

Levine has remarked that this result is interesting when X is smooth over a field k of characteristic 0, and X' is the singular infinitesimal thickening of X to $X[\varepsilon]/\varepsilon^2$. Then $K_*(X \to X')$ is by Grothendieck's definition the "tangent space" to K_* at X, and the result shows that it indeed equals the tangent space to the "linear approximation" HC_*. One has the spectral sequence of tangent spaces given by applying 2.4 to the fibre of $K(X') \to K(X)$; using cyclic homology to calculate the E_2 term gives:

$$E_2^{p,-q} = H^p(X; \Omega_X^{q-1} \oplus \Omega_X^{q-3} \oplus \Omega_X^{q-5} \oplus \cdots)$$

where Ω_X are the absolute Kähler differentials over $\mathbb{Q}$, not the relative ones over k. This does suggest somewhat obscure relations between this tangent space, Hodge theory, and infinitesimal deformations of algebraic cycles.

§ 4. Comparison of Algebraic and Etale Topological *K*-Theory

4.0 Suppose now that X is a scheme of finite type over a ring k, where k is one of the following: an algebraically closed field, a separably closed field, a number field, a ring of integers in a number field, $\mathbb{Z}_p^\wedge$, $\mathbb{Q}_p^\wedge$, $\mathbb{F}_q[[t]]$, $\mathbb{F}_q((t))$, or $\mathbb{F}_q$. Fix a prime power $l^\nu \geq 3$. Suppose that $1/l \in \mathcal{O}_X$, and that if $l = 2$ then $\mathcal{O}_X$ contains a square root of -1.

If k contains all the l^ν-th roots of unity, this gives a torsion subgroup $\mathbb{Z}/l^\nu$ in the group of units k^*, and hence in $K_1(k)$. By the universal coefficient sequence, this corresponds to an element β in $K/l^\nu{}_2(k)$. One may then localize the ring $K/l^\nu{}_*(X)$ by inverting the image of this Bott element β. If k doesn't contain all the l^ν-th roots of unity, one can still form this $K/l^\nu(X)[\beta^{-1}]$: as Dwyer showed, essentially a power of β exists in $K/l^\nu{}_*(k)$, although β in degree 2 does not ([T1], Appendix A).

Theorem 4.1. *For X as in* 4.0, *there is a strongly converging spectral sequence from etale cohomolgy*:

$$E_2^{p,-q}\begin{Bmatrix} H^p(X_{\text{et}}; \mathbb{Z}/l^\nu(i)) & q = 2i \\ 0 & q \text{ odd} \end{Bmatrix} \Rightarrow K/l^\nu{}_{q-p}(X)[\beta^{-1}]$$

The canonical map which forgets the algebraic structure on an algebraic vector bundle and remembers only the underlying topological vector bundle induces an equivalence with the Dwyer-Friedlander etale topological K-theory [DF]:

$$\varrho : K/l^\nu(X)[\beta^{-1}] \xrightarrow{\sim} K/l^{\nu\text{Top}}(X)$$

Proof. [TT2] 11.5. The method of 2.5 reduces this to the case where X is replaced by its various residue fields, which case was done in [T1], as was the case where X is regular. □

One notes that the results of Suslin-Gabber-Gillet-Thomason [Ga, Su1, Su2, GiT] on the K-theory of strict hensel local rings, the local rings for the etale topology, show that the values of the sheaves $K/l^{\nu}{}_q$ in the etale topology are the i-times Tate-twisted cyclic groups $\mathbb{Z}/l^{\nu}(i)$ for $q = 2i$ even and non-negative, and are 0 for q odd and positive. Thus this spectral sequence is the spectral sequence of etale cohomological descent for $K/l^{\nu}{}_*$, at least in positive degrees. However, it converges to $K/l^{\nu}{}_*(X)[\beta^{-1}]$, and not to $K/l^{\nu}{}_*(X)$, which is in general different. I believe that $K/l^{\nu}{}_q(X)[\beta^{-1}]$ equals $K/l^{\nu}{}_q(X)$ for q sufficiently large (cf. the results of [T2]), but this is definitely false for q small, e.g. for $q = 0$ and X a $K3$ surface over $\mathbb{C}$ or Spec$(R[1/l])$, where R is a ring of integers in a number field which has more than one prime lying over l ([T1] Example 4.5).

The relation between algebraic K-theory and etale cohomology remains a very active area. Leaving aside the infinity of baseless conjectures, there remain some other actual theorems in special cases: the surjectivity results of Soulé in the case of rings of integers, as perfected by Dwyer-Friedlander [DF]; complete results on K_2 of fields due to Merkurjev and Suslin [MS1]; less complete results about K_3 of fields due to Merkurjev-Suslin ([MS2], [MS3]), Levine [Le4], and Rost [R]; and the descent result of Carlsson for $\pi_* BGL_N{}^+$ with respect to finite Galois extensions of certain localized rings of integers. The K_2 result of Merkurjev-Suslin has been extensively employed to study algebraic cycles in codimension 2, notably in the hands of Colliot-Thélène, Coombes, M. Gros, Raskind, Salberger, Sansuc, and Soulé. M. Harada has proved a Riemann-Roch theorem for singular varieties without a quasi-projectivity hypothesis by using the homology version of 4.1 for G-theory to reduce Riemann-Roch to the projection formula ([H], [T4] § 15).

§ 5. The Case of Non-commutative Rings

D. Yao has begun the study of the localization theorem for non-commutative rings [Y]. This involves many new technical problems, and the construction of a sort of algebraic geometry of Grothendieck abelian categories. As a sample of his results, one has:

Theorem 5.1 (D. Yao) *Let R be a ring, not necessarily commutative. Suppose that s_1, $s_2, \ldots, s_n$ are elements of R such that the ideal they generate is all of R. Suppose there exist ring automorphisms ϕ_i for $i = 1, 2, \ldots, n$ such that:*

1) *for all $r \in R$, $\phi_i(r)s_i = s_i r$*
2) $\phi_i(s_j) = s_j$
3) $\phi_i\phi_j = \phi_j\phi_i$

(For example, the s_i could be central, and the $\phi_i =$ id.) Then there is a strongly converging Cech cohomology spectral sequence computing $K_*(R)$ from the K-groups of the localizations:

$$E_2^{p,-q} = H^p\left(\cdots \to \bigoplus_{i_0 < i_1 < \cdots < i_n} K_q(R[1/s_{i_0} \cdot 1/s_{i_1} \cdot \ldots \cdot 1/s_{i_n}]) \to \cdots\right) \Rightarrow K_{q-p}(R).$$

References

[B] H. Bass: Algebraic K-theory. Benjamin 1968

[BG] K. Brown, S. Gersten: Algebraic K-theory as generalized sheaf homology. In: Higher K-theories. (Lecture Notes in Mathematics, vol. 341) Springer, Berlin Heidelberg New York 1973, pp. 266–292

[Co] A. Collino: Quillen's K-theory and algebraic cycles on almost nonsingular varieties Ill. J. Math. **25** (1981) 654–666

[DF] W. Dwyer, E. Friedlander: Etale K-theory and arithmetic. Trans. Amer. Math. Soc. **292** (1985) 247–280

[Ga] O. Gabber: K-theory of henselian local rings and henselian pairs Preprint 1985

[Ge] S. Gersten: The localization theorem for projective modules. Comm. Alg. **2** (1974) 307–350

[Gi] H. Gillet: On the K-theory of surfaces with multiple curves and a conjecture of Bloch. Duke Math. J. **51** (1984) 195–233

[GiT] H. Gillet, R. Thomason: The K-theory of strict hensel local rings and a theorem of Suslin. J. Pure Appl. Alg. **34** (1984) 241–254

[Go] T. Goodwillie: Relative algebraic K-theory and cyclic homology. Ann. Math. **124** (1986) 347–402

[Gr] D. Grayson: Higher algebraic K-theory II (after D. Quillen). In: Algebraic K-theory: Evanston 1976. (Lecture Notes in Mathematics, vol. 551) Springer, Berlin Heidelberg New York 1976, pp. 217–240

[H] M. Harada: A proof of the Riemann-Roch theorem. Preprint 1988

[KS] K. Kato, S. Saito: Global class field theory of arithmetic surfaces. In: Applications of algebraic K-theory to algebraic geometry and number theory, part I. (Contemporary Mathematics, vol. 55) AMS, 1986, pp. 255–331

[vdK] W. van der Kallen: Descent for K-theory of polynomial rings. Math. Z. **191** (1986) 405–415

[Lec] F. Lecomte: Rigidité des groupes de Chow. Duke J. Math. **53** (1986) 405–426

[Le1] M. Levine: Bloch's formula for singular surfaces. Topology **24** (1985) 165–174

[Le2] M. Levine: Zero cycles and K-theory on singular varieties. In: Algebraic geometry: Bowdoin 1985. (Proc. Symp. Pure Mathematics, vol. 46, part 2) AMS 1987, pp. 451–462

[Le3] M. Levine: Localization on singular varieties. Invent. math. **91** (1988) 423–464

[Le4] M. Levine: The indecomposable K_3 of fields. Ann. Sci. Ec. Norm. Sup. (4) **22** (1989) 255–344

[LeW] M. Levine, C. Weibel: Zero cycles and complete intersections on singular varieties. J. Reine Angew. Math. **359** (1985) 106–120

[Lo] J.-L. Loday: Cyclic homology: a survey. In: Geometric and algebraic topology. (Banach Center Publication, vol. 18) PWN 1986, pp. 285–307

[MS1] А.С. Меркурьев, А.А. Суслин: К-когомологий многооЬразий Севери-Брауэра и гомоморфизм норенного вычета. In: Изв. Акал. Наук С.С.С.Р. Сер. Мат. **46** (1982) 1011–1046 [English translation in Math. USSR Izv. **21** (1983) 307–340]

[MS2] A.S. Merkurjev, A.A. Suslin: On the norm residue homomorphism of degree three. LOMI preprint 1986

[MS3] A.S. Merkurjev, A.A. Suslin: On K_3 of a field. LOMI preprint 1987

[N] Y. Nisnevich: The completely decomposed topology on schemes and associated descent spectral sequences in algebraic K-theory. In: Algebraic K-theory: Connections with geometry and topology. Kluwer 1989, pp. 241–342

[Og] C. Ogle: On the K-theory and cyclic homology of a square-zero ideal. J. Pure Appl. Alg. **46** (1987) 233–248

[OW] C. Ogle, C. Weibel: Relative K-theory and cyclic homology (to appear)

[PW1] C. Perdrini, C. Weibel: K-theory and Chow groups on singular varieties. In: Applications of algebraic K-theory to algebraic geometry and number theory, part I. (Contemporary Mathematics, vol. 55) AMS 1986, pp. 339–370

[PW2] C. Pedrini, C. Weibel: Bloch's formula for varieties with isolated singularities. Comm. Alg. **14** (1986) 1895–1907

[Q] D. Quillen: Higher algebraic K-theory I. In: Higher K-theories. (Lecture Notes in Mathematics, vol. 341) Springer, Berlin Heidelberg New York 1973, pp. 85–147

[R] M. Rost: Hilbert's theorem 90 for $K^M{}_3$. Preprint 1986

[Se] J.-P. Serre: Prolongement des faisceaux analytiques cohérents. Ann. Inst. Fourier **16** (1966) 363–374

[So] C. Soulé: Rational K-theory of the dual numbers of a ring of algebraic integers. In: Algebraic K-theory, Evanston 1980. (Lecture Notes in Mathematics, vol. 854) Springer, Berlin Heidelberg New York 1981, pp. 402–408

[Sta] R. Staffeldt: Rational algebraic K-theory of certain truncated polynomial rings. Proc. Amer. Math. Soc. **95** (1985) 191–198

[Sti] J. Stienstra: Operations in the higher K-theory of endomorphisms. In: Current trends in algebraic topology, part 2. AMS 1982, pp. 59–115

[Su1] A.A. Suslin: On the K-theory of algebraically closed fields. Invent. math. **73** (1983) 241–245

[Su2] A.A. Suslin: On the K-theory of local fields. J. Pure Appl. Alg. **34** (1984) 301–318

[Su3] A.A. Suslin: Algebraic K-theory of fields. Proceedings of the International Congress of Mathematicians; Berkeley 1986, vol. 1. AMS, pp. 222–244

[T1] R.W. Thomason: Algebraic K-theory and etale cohomology. Ann. Sci. Ec. Norm. Sup. (4) **18** (1985) 437–552; erratum (4) **22** (1989) 675–677

[T2] R.W. Thomason: Bott stability in algebraic K-theory. In: Applications of algebraic K-theory to algebraic geometry and number theory, part I. (Contemporary Mathematics, vol. 55) AMS 1986, pp. 389–406

[T3] R.W. Thomason: The finite stable homotopy type of some topoi. J. Pure Appl. Alg. **47** (1987) 89–104

[T4] R.W. Thomason: Survey of algebraic vs. etale topological K-theory. In: Algebraic K-theory and algebraic number theory. (Contemporary Mathematics, vol. 83) AMS 1989, pp. 393–443

[TT1] R.W. Thomason, T.F. Trobaugh: Le théorème de localisation en K-théorie algébrique. C.R. Acad. Sci. Paris Ser. I **307** (1988) 829–831

[TT2] R.W. Thomason, T.F. Trobaugh: Higher algebraic K-theory of schemes and of derived categories. To appear in the Festschrift for Grothendieck (Progress in Math.) Birkhäuser, Basel

[Ve] J.-L. Verdier: Catégories dérivées. In: SGA4&1/2: Cohomologie étale. (Lecture Notes in Mathematics, vol. 569) Springer, Berlin Heidelberg New York 1977, pp. 262–311

[Vi] L. Barbieri Viale: Cohomology theories and algebraic cycles on singular varieties. Preprint 1989

[Vo] T. Vorst: Localization of the K-theory of polynomial extensions. Math. Ann. **244** (1979) 333–353

[Wa1] F. Waldhausen: Algebraic K-theory of spaces, localization, and the chromatic filtration of stable homotopy. In: Algebraic topology, Aarhus 1982. (Lecture Notes in Mathematics, vol. 1051) Springer, Berlin Heidelberg New York 1984, pp. 173–195

[Wa2] F. Waldhausen: Algebraic K-theory of spaces. In: Algebraic and geometric topology. (Lecture Notes in Mathematics, vol. 1126) Springer, Berlin Heidelberg New York 1985, pp. 318–419

[We1] C. Weibel: Mayer-Vietoris sequences and module structures on NK. In: Algebraic

K-theory, Evanston 1980 (Lecture Notes in Mathematics, vol. 854) Springer, Berlin Heidelberg New York 1981, pp. 466–493

[We2] C. Weibel: Mayer-Vietoris sequences and mod p K-theory. In: Algebraic K-theory, Oberwolfach 1980. (Lecture Notes in Mathematics, vol. 966) Springer, Berlin Heidelberg New York 1982, pp. 390–407

[We3] C. Weibel: Module structures on the K-theory of graded rings. J. Alg. **105** (1987) 465–483

[We4] C. Weibel: A. Brown-Gersten spectral sequence for the K-theory of varieties with isolated singularities. Adv. Math. **73** (1989) 192–203

[We5] C. Weibel: "Homotopy algebraic K-theory. In: Algebraic K-theory and number theory. (Contemporary Mathematics, vol. 83) AMS 1989, pp. 461–488

[Y] D. Yao: Thesis, The Johns Hopkins University 1990

[SGA6] P. Berthelot, A. Grothendieck, L. Illusie: SGA6: Théories des intersections et théorème de Riemann-Roch. (Lecture Notes in Mathematics, vol. 225) Springer, Berlin Heidelberg New York 1971

[EGA] A. Grothendieck, J. Dieudonné: Eléments de géométrie algébrique. (Publ. Math. I.H.E.S. Nos. 8, 11, 17, 20, 24, 28, 32) Presse Univ. France (1961–1967). Chapitre I refondu: (Grundlehren der mathematischen Wissenschaften, Bd. 166) Springer, Berlin Heidelberg New York 1971

On the Restricted Burnside Problem

Efim I. Zelmanov

Institute of Mathematics, Universitetskii pr. 4, Novosibirsk-90, 630 090 USSR

In 1902, W. Burnside formulated his famous problems for periodic groups [6]:

The Burnside Problem (also known as the Ordinary Burnside Problem): Is it true that every finitely generated group of bounded exponent is finite?

The General Burnside Problem: Is it true that every finitely generated periodic group is finite?

After many unsuccessful attempts to obtain a proof in the late 30s–early 40s the following weaker version of The Burnside Problem was studied: Is it true that there are only finitely many m-generated finite groups of exponent n? In other words the question is whether there exists a universal finite m-generated group of exponent n having all other finite m-generated groups of exponent n as homomorphic images. Later (thanks to W. Magnus [35]) this question became known as **The Restricted Burnside Problem**.

In 1964 E. S. Golod gave a negative answer to The General Burnside Problem (cf. [9]). Since then a considerable array of infinitely generated periodic groups was constructed by other authors (cf. Alyoshin [2], Suschansky [44], Grigorchuk [11], Gupta-Sidki [54]).

In 1968 P. S. Novikov and S. I. Adian [39] constructed counter-examples to The Burnside Problem for groups of odd exponents $n \geq 4381$ (now for odd exponents $n \geq 115$, cf. I. Lysenok [33]). Olshansky's Monsters (cf. [40]) shows how wildly periodic groups may behave.

At the same time there were two major reasons to believe that The Restricted Burnside Problem would have a positive solution. One of these reasons was the reduction theorem obtained by Ph. Hall and G. Higman [14]. Let $n = p_1^{k_1} \dots p_r^{k_r}$, where p_i are distinct prime numbers, $k_i \geq 1$, and assume that (a) The Restricted Burnside Problem for groups of exponents $p_i^{k_i}$ has a positive solution, (b) there are only a finite number of finite simple groups of exponent n, (c) the factor group $\mathrm{Out}(G) = \mathrm{Aut}(G)/\mathrm{Inn}(G)$ is solvable for any finite simple group of exponent n. Then The Restricted Burnside Problem for groups of exponent n also has positive solution.

Another reason was the close relation of The Problem to Lie algebras. Suppose $n = p^k$, where p is a prime number. Then the finite group G of exponent p^k is clearly nilpotent. It is easy to see that it is sufficient to find an upper bound

Proceedings of the International Congress
of Mathematicians, Kyoto, Japan, 1990

$f(m, p^k)$ for the class of nilpotency of all m-generated groups G of exponent p^k. Consider the lower central series $G = \gamma_1(G) > \gamma_2(G) > \cdots > \gamma_s(G) = \langle 1 \rangle, \gamma_{i+1}(G) = (\gamma_i(G), G), 1 \leq i < S$, and the direct sum of abelian groups

$$L_\gamma(G) = \bigoplus_{i=1}^{s-1} \gamma_i(G)/\gamma_{i+1}(G).$$

Brackets $[a_i\gamma_{i+1}(G), b_j\gamma_{j+1}(G)] = (a_i, b_j)\gamma_{i+j+1}(G)$, where $a_i \in \gamma_i(G), b_j \in \gamma_j(G), (\ ,\ )$ is the group commutator, define the structure of a Lie ring on $L_\gamma(G)$ (cf. [4], [12, 34, 48]). It is obvious that the Lie ring $L_\gamma(G)$ has the same class of nilpotency as the group G. If G is generated by elements $x_1, \ldots, x_m$ then $L_\gamma(G)$ is generated by $x_i\gamma_2(G), 1 \leq i \leq m$.

If $n = p$ is a prime number then $L_\gamma(G)$ is an algebra over the field $Z_p, |Z_p| = p$, which satisfies Engel's identity

$$[\ldots[y, \underbrace{x], x], \ldots, x}_{p-1}] = 0 \qquad E_{p-1}$$

(cf. [35]). Thus The Problem for groups of prime exponent has been reduced to the following problem in Lie algebras: is it true that a Lie algebra over Z_p, which satisfies Engel's identity E_{p-1}, is locally nilpotent?

The last problem was successfully solved by A. P. Kostrikin [27, 28] wo solved in this way The Restricted Burnside Problem for groups of prime exponent.

If G is a finite group of prime power exponent p^k then $L_\gamma(G)$ is no longer an algebra over the field Z_p, it is an algebra over the ring $Z(p^k)$ of residues modulo p^k. That's why along with the lower central series $\gamma_i(G)$ we shall consider the lower central p-series of G (cf. [21], [31, 48, 46]):

$$G = G_1 \supset G_2 \supset \ldots,$$

where G_i is the subgroup of G generated by commutators $((\ldots(x_1, x_2), x_3), \ldots, x_r)$, $r \geq i$, and powers $((\ldots(x_1, x_2), x_3), \ldots, x_r)^{p^\ell}, r \cdot p^\ell \geq i$. It is easy to see that $G_i \supseteq \gamma_i(G) \supseteq G_{i \cdot p^k}$,

$$L(G) = \bigoplus_{i \geq 1} G_i/G_{i+1}$$

is an algebra over Z_p. Neither the Lie ring $L_\gamma(G)$ nor $L(G)$ need necessarily satisfy Engel's identity E_{p^k-1} (cf. [13, 16]) but

(1) the Lie algebra $L(G)$ satisfies the linearized Engel's identity E_{p^k-1}, that is, for arbitrary elements $a_1, \ldots, a_{p^k-1} \in L(G)$ we have

$$\Sigma \operatorname{ad}(a_{\sigma(1)}) \ldots \operatorname{ad}(a_{\sigma(p^k-1)}) = 0, \quad \sigma \in S_{p^k-1},$$

(G. Higman, [17]).

(2) for an arbitrary commutator ϱ on the generators $x_iG_2, 1 \leq i \leq r$, we have

$$\operatorname{ad}(\varrho)^{p^k} = 0$$

(I. N. Sanov, [42]).

Now let us turn to what was happening in associative and Lie nil algebras.

A. G. Kurosch [30]* and independently J. Levitzky (cf. [3]) formulated two problems for nil algebras which were similar to Burnside's problems.

The General Kurosh-Levitzky Problem: Is every finitely generated nil algebra nilpotent?

The (Ordinary) Kurosh-Levitzky Problem: Is every finitely generated nil algebra of bounded degree nilpotent?

Actually it was a counterexample to the General Kurosh-Levitzky Problem which was constructed in the paper, [9] of E. S. Golod, then this counterexample was used to construct the first counterexample to the General Burnside Problem. Remark that so far it remains the only counterexample to the General Kurosh-Levitzky Problem.

For the (Ordinary) Kurosh-Levitzky Problem we have a quite different situation. Unlike Group Theory it has only positive solutions in all important classes of algebras. To appreciate the impact this problem had on the Ring Theory let us mention that N. Jacobson's Structure Theory of Algebras was stimulated by the Kurosh Problem and I. Kaplansky introduced the concept of a PI-algebra in search of the most general conditions which ensure the positive solution of the Problem. The following result (in its final form) was due to I. Kaplansky [23].

Theorem. *A finitely generated nil algebra which satisfies a polynomial identity is nilpotent.*

In 1956 A. I. Shirshov suggested another purely combinatorial direct approach to the Kurosh-Levitzky Problem.

Theorem (A. I. Shirshov [43]). *Suppose that an associative algebra A is generated by elements $x_1, \ldots, x_r$ and assume that (1) A satisfies a polynomial identity of degree n, (2) every monomial in $\{x_i\}$ of degree $\leq n$ is nilpotent. Then A is nilpotent.*

It is very important that the nilpotency assumption here is imposed not on every element of A, but only on the monomials in the generators (even on a finite collection of them).

Now let us turn to Lie algebras. It is natural to call an element $a \in L$ nilpotent if the operator $\mathrm{ad}(a)$ is nilpotent. With this definition both Kurosh-Levitzky problems become meaningful for Lie algebras. Moreover, by the results of G. Higman and I. N. Sanov (cf. above) the Lie algebra $L(G)$ of a finite group G of exponent p^n satisfies the assumptions of the Kurosh-Levitzky Problem in the form of A. I. Shirshov (the role of monomials is played by commutators).

In [52, 43] we solved this problem for those Lie algebras which satisfy the linearized Engel's identity E_n.

---

* Actually Kurosh's Problem concerned algebraic algebras but we consider only the important case of nil algebras.

Theorem 1. *Suppose that a Lie algebra L is generated by elements* $x_1, \ldots, x_r$ *and assume that there exist integers* $n \geq 1, m \geq 1$ *such that (1) L satisfies the linearized Engel's identity* E_n, *(2) for an arbitrary commutator* ϱ *on the generators* x_i *we have* $ad(\varrho)^m = 0$. *Then L is nilpotent.*

Corollary. *A Lie ring which satisfies Engel's identity is locally nilpotent.*

From Theorem 1 we derive

Theorem 2. *The Restricted Burnside Problem has a positive solution for groups of exponent* p^k.

By the Reduction Theorem of Ph. Hall and G. Higman, the Restricted Burnside Problem has a positive solution in the class of soluble groups, in particular it has a positive solution for groups of odd exponent (by the celebrated Theorem of W. Feit and J. Thompson [8]) and for groups of exponent $n = p^\alpha q^\beta, p, q$ are prime numbers (cf. W. Burnside, [7]).

The announced classification of finite simple groups (cf. [10]) implies that the Restricted Burnside Problem has a positive solution for groups of arbitrary exponent. Now we shall try to explain briefly the idea of the proof of Theorem 1.

In [50] we proved that to prove Theorems 1, 2 it suffices to prove that a restricted (in the sense of N. Jacobson [18]) Lie algebra over an infinite field, which satisfies an Engel's identity, is locally nilpotent. An element a of a Lie algebra L is called sandwich if

$$[[L,a],a] = 0, \quad [[[L,a],L],a] = 0$$

(cf. A. I. Kostrikin, [27]). In case of odd characteristics the second equality easily follows from the first one, however if char $= 2$ then both conditions are necessary. We call a Lie algebra a sandwich algebra if it is generated by a finite collection of sandwiches. The following theorem is due to A. I. Kostrikin and the author.

Theorem About Sandwich Algebras [49]. *A sandwich Lie algebra is nilpotent.*

This theorem suggests the following plan of attack on Theorem 1 (which has been outlined in [50]). Assume that there exists a nonzero Lie algebra L over an infinite field K which satisfies an Engel's identity but isn't locally nilpotent. Then taking the factor-algebra of L modulo its locally nilpotent radical (cf. [26, 41]) we may assume L doesn't contain any nonzero locally nilpotent ideals. Suppose we manage to construct a Lie polynomial $f(x_1, \ldots, x_r)$ such that f is not identically zero on L and for arbitrary elements $a_1, \ldots, a_r \in L$ the value $f(a_1, \ldots, a_r)$ is a sandwich of L. The K-linear span of $f(L) = \{f(a_1, \ldots, a_r) | a_1, \ldots, a_r \in L\}$ is an ideal in L. By the theorem about sandwich algebras the ideal $Kf(L)$ is locally nilpotent which contradicts our assumption.

However one year of effort didn't bring us a desired sandwich-valued polynomial (its existence a posteriori follows from Theorem 1). Instead in November of 1988 we constructed an even sandwich-valued superpolynomial f, which means that for a Lie superalgebra $L = L_0 + L_1$ satisfying the superization of E_n every value of f is a sandwich of L_0. It turned out to be a good substitute of sandwich-valued polynomials. The sketch of this rather complicated construction appeared in [51]. Unfortunately it worked only for characteristics $\neq 2, 3$.

In January of 1989 we constructed another "generalized" nonzero sandwich-valued polynomials (this time involving "divided powers" of ad-operators), their full linearizations being ordinary polynomials. Every value of such a full linearization is a linear combination of sandwiches. This approach worked for any p (cf. [52, 53]).

Some lengthy computations from the proof (which are really hard to read) may be explained within the framework of Jordan Algebra Theory (cf. [19, 20]). We shall demonstrate the idea for the simpler case $p \neq 2, 3$. The first (less computational) part of the construction of a sandwich-valued polynomial is a construction of a polynomial f such that $f(L) \neq 0$ and for an arbitrary element $a \in f(L)$ we have $\operatorname{ad}(a)^3 = 0$. Choose arbitrary elements $a, b \in f(L)$ and consider the subspaces $\mathcal{Q}^+ = L \operatorname{ad}(a)^2$, $\mathcal{Q}^- = L \operatorname{ad}(b)^2$. Then for an arbitrary element $c \in \mathcal{Q}^-$ the operation $x \circ y = [x, c, y]; x, y \in \mathcal{Q}^+$, defines the structure of a Jordan algebra on $\mathcal{Q}^+$ (cf. [5, 19, 22, 25]). The pair of subspaces $(\mathcal{Q}^-, \mathcal{Q}^+)$ is a so-called Jordan pair (cf. [32, 38]).

For $p = 2$ or $p = 3$ we define $\mathcal{Q}^-$ and $\mathcal{Q}^+$ with the divided powers of adjoint-operators and apply Kevin McCrimmon's Theory of Quadratic Jordan Algebras ([20, 36, 37]).

For odd p we managed to translate Jordan arguments into the language of elementary computations in [52]. For $p = 2$ this substitute didn't work so Jordan Pairs and Algebras played an important role in our paper [53]. However, recently M. Vaughan-Lee succeeded in getting rid of Jordan Algebra Theory even in the case $p = 2$.

Not much is known about the upper bound for classes of nilpotency of r-generated finite groups of exponent p (let alone the exponent $p^k, k > 1$). S. I. Adian and N. N. Repin [1] proved that it grows at least exponentially with respect to p. For comparison let us mention the recent result of A. Belov which asserts that there exists a constant α such that an arbitrary r-generated associative ring which satisfies the identity $x^n = 0$ is nilpotent of degree $\leq r^{\alpha n}$.

Conjecture. *There exists a constant α such that an arbitrary r-generated finite group of exponent p is nilpotent of class $\leq r \cdot \alpha^P$.*

Residually Finite Groups and Compact Groups

The following generalization of Theorem 1 solves the Kurosh-Levitzky Problem (in Shirshov's form) in the class of Lie PI-algebras.

Theorem 3. *Suppose that a Lie algebra L is generated by a finite subset $X \subseteq L$, $|X| = m$ and assume that*

(1) *L satisfies a polynomial identity of degree n,*

(2) *for an arbitrary commutator ϱ on X of weight $\leq h(m,n)$ the operator $ad(\varrho)$ is nilpotent.*

Then L is nilpotent.

This theorem has some applications to the General Burnside Problem in group varieties and to compact groups. It is well known that there are counterexamples to the General Burnside Problem even among residually finite p-groups (such are the counterexamples of E. S. Golod and R. I. Grigorchuk). However Theorem 3 implies

Theorem 4. *A residually finite p-group which satisfies a nontrivial group identity is locally finite.*

Apparently this theorem can be generalized from p-groups to periodic groups in the spirit of the theorem of P. Hall and G. Higman [14].

The assumption of a nontrivial identity can be further weakened to the "infinitesimal" assumption that the adjoint Lie algebra $L(G)$ is PI. The last assertion in its turn is related to the General Burnside Problem for compact groups. V. P. Platonov conjectured that periodic compact (Hausdorff) groups are locally finite. J. S. Wilson [47] proved that (under the assumption that there are finitely many simple sporadic groups) it suffices to prove the conjecture for pro-p-groups. That's what is done in the following theorem.

Theorem 5. *Every periodic pro-p-group is locally finite.*

Indeed, let G be a periodic pro-p-group. Consider the closed subsets $G_{(n)} = \{g \in G | g^{p^n} = 1\}, G = UG_{(n)}$. By Baire's Category theorem one of the subsets $G_{(n)}$ contains some neighborhood, that is $G_{(n)} \supset gH$, where H is a normal subgroup of G of finite index. Then we show that every finitely generated subgroup of H which is invariant under conjugation by g, satisfies an "infinitesimal" identity.

From Theorem 5 combined with [47] and with what is known about locally finite groups [15, 24] there follows

Theorem 6. *Every infinite compact group contains an infinite abelian subgroup.*

Remark as far as Theorem 6 is concerned the reduction to pro-p-groups in [47] didn't use the classification of finite simple groups.

References

1. Adian, S. I., Repin, N. N.: On exponential lower bound for class of nilpotency of Engel Lie algebras. Matem. Zametky **39**, no. 3 (1986) 444–452
2. Alyoshin, S. V.: Finite automata and the Burnside problem on periodic groups. Matem. Zametky **11**, no. 3 (1972) 319–328
3. Amitsur, S. A.: Jacob Levitzki 1904–1956. Israel J. Math. **19** (1974) 1–2

4. Baer, R.: The higher commutator subgroups of a group. Bull. Amer. Math. Soc. **50** (1944) 143–160
5. Benkart, G.: Inner Ideals and the Structure of Lie algebras. Dissertation. Yale University 1974
6. Burnside, W.: On an unsettled question in the theory of discontinuous groups. Quart. J. Pure Appl. Math. **33** (1902) 230–238
7. Burnside, W.: Theory of groups of finite order. Cambridge Univ. Press, 1911; Reprint, Dover, New York 1955
8. Feit, W., Thompson, G.: Solvability of groups of odd order. Pacific J. Math. **13** (1963) 755–1029
9. Golod, E. S.: On nil algebras and residually finite *p*-groups. Izv. Akad. Nauk SSSR **28**, no. 2 (1964) 273–276
10. Gorenstein, D.: Finite simple groups. New York 1982
11. Grigorchuk, R. I.: On the Burnside Problem for periodic groups. Funct. Anal. Appl. **14**, no. 1 (1980) 53–54
12. Grün, O.: Zusammenhang zwischen Potenzbildung und Kommutatorbildung. J. Reine Angew. Math. **182** (1940) 158–177
13. Grunewald, F. J., Havas, G., Mennicke, J. L., Neumann, M. F.: Groups of exponent eight. Bull. Austral. Math. Soc. **20** (1979) 7–16
14. Hall, P., Higman, G.: On the *p*-length of *p*-soluble groups and reduction theorems for Burnside's problem. Proc. London Math. Soc. GN3 (1956) 1–42
15. Hall, P., Kulatilaka, C. R.: A property of locally finite groups. J. London Math. Soc. **39** (1964) 235–239
16. Havas, G., Newmann, M. F.: Application of computers to questions like those of Burnside. (Lecture Notes in Mathematics, vol. 806.) Springer, Berlin Heidelberg New York 1980, pp. 211–230
17. Higman, G.: Lie ring methods in the theory of finite nilpotent groups. Proc. Intern. Congr. Math. Edinburgh 1958, pp. 307–312
18. Jacobson, N.: Restricted Lie algebras of characteristic *p*. Trans. AMS **50** (1941) 15–25
19. Jacobson, N.: Structure and representations of Jordan algebras. AMS, Providence, RI 1969
20. Jacobson, N.: Lectures on quadratic Jordan algebras. Bombay. Tata Institute of Fundamental Research, 1969
21. Jennings, S. A.: The structure of the group ring of a *p*-group over a modular field. Trans. AMS **50** (1941) 175–185
22. Kantor, I. L.: Classification of irreducible transitively differential groups. Sov. Math. Dokl. SN SSSR **5** (1964) 1404–1407
23. Kaplansky, I.: Rings with a polynomial identity. Bull. AMS **54** (1948) 575–580
24. Kargapolov, M. I.: On a problem of O. Ju. Smidt. Siber. Math. Ž. **4** (1963) 232–235
25. Koecher, M.: Imbedding of Jordan algebras into Lie algebras. Amer. J. Math. **89** (1967) 787–816
26. Kostrikin, A. I.: On Lie rings with Engel's condition. Dokl. Akad. Nauk SSSR **108**, no. 4 (1956) 580–582
27. Kostrikin, A. I.: On the Burnside Problem. Izv. Akad. Nauk SSSR **23**, no. 1 (1959) 3–34
28. Kostrikin, A. I.: Sandwiches in Lie algebras. Matem. Sb. **110** (1979) 3–12
29. Kostrikin, A. I.: Around Burnside. Nauka, Moscow 1986
30. Kurosh, A. G.: Problems in ring theory which are related to the Burnside Problem for periodic groups. Izv. Akad. Nauk SSSR **5**, no. 3 (1941) 233–240
31. Lazard, M.: Sur les groupes nilpotents et les anneaux de Lie. Ann. Sci. Ecole Norm. Sup. **71**, no. 3 (1954) 101–190
32. Loos, O.: Jordan pairs. Springer, Berlin Heidelberg New York 1975

33. Lysenok, I.: On the Burnside problem for odd exponents $n \geq 115$. Materials of the International Conference in Algebra, Novosibirsk, 1989
34. Magnus, W.: Über Gruppen und zugeordnete Liesche Ringe. J. Reine Angew. Math. **182** (1940) 142–159
35. Magnus, W.: A connection between the Baker-Hausdorff formula and a problem of Burnside. Ann. Math. **52** (1950) 11–26; Errata Ann. Math. **57** (1953) 606
36. McCrimmon, K.: A general theory of Jordan rings. Proc. Nat. Acad. Sci. USA **56** (1966) 1072–1079
37. McCrimmon, K., Zelmanov, E.: The structure of strongly prime quadratic Jordan algebras. Adv. Math. **69** (1988) 113–222
38. Meyberg, K.: Lectures on algebras and triple systems. Lecture Notes, The University of Virginia, Charlottesville, 1972
39. Novikov, P. S., Adian, S. I.: On infinite periodic groups. I, II, III. Izv. Akad. Nauk SSSR **32**, no. 1 (1968) 212–244; no. 2, 251–254; no. 3, 709–731
40. Olshansky, A. Yu.: Geometry of defining relations in groups. Nauka, Moscow 1989
41. Plotkin, B. I.: Algebraic sets of elements in groups and Lie algebras. Uspekhi Mat. Nauk **13**, no. 6 (1958) 133–138
42. Sanov, I. N.: On a certain system of relations in periodic groups of prime power exponent. Izv. Akad. Nauk SSSR **15** (1951) 477–502
43. Shirshov, A. I.: On rings with identical relations. Mat. Sb. **43** (1957) 277–283
44. Suschansky, V. I.: Periodic p-groups of permutations and the General Burnside Problem. Dokl. Akad. Nauk SSSR **247**, no. 3 (1979) 447–461
45. Vaughan-Lee, M.: The Restricted Burnside's Problem. Oxford University Press, 1990
46. Wall, G. E.: On the Lie ring of a group of prime exponent. Proc. 2nd Intern. Conf. Theory of Groups, Canberra, 1973, pp. 667–690
47. Wilson, J. S.: On the structure of compact torsion groups. Monatsh. Math. **96** (1983) 404–410
48. Zassenhaus, H.: Ein Verfahren, jeder endlichen p-Gruppe einem Lie-Ring mit der Charakteristik p zuzuordnen. Abh. Math. Sem. Univ. Hamburg **13** (1940) 200–207
49. Zelmanov, E. I., Kostrikin, A. I.: A theorem on sandwich algebras. Proceedings of the V. A. Steklov Math. Institute of Akad. Nauk SSSR **183** (1988) 142–149
50. Zelmanov, E. I.: On some problems in the theory of groups and Lie algebras. Mat. Sb. **180**, no. 2 (1989) 159–167
51. Zelmanov, E. I.: On the restricted Burnside problem. Siberian Math. J. **30**, no. 6 (1989) 68–74
52. Zelmanov, E. I.: The solution of the restricted Burnside problem for groups of odd exponent. Izv. Akad. Nauk SSSR **54** (1990)
53. Zelmanov, E. I.: The solution of the restricted Burnside problem for 2-groups. To appear in Mat. Sb.
54. Gupta, N., Sidki, S.: On the Burnside problem for periodic groups. Math. Z. **182** (1983) 385–386

A Riemann-Roch Theorem in Arithmetic Geometry

Henri Gillet *

Department of Mathematics, Statistics, and Computer Science, (m/c 249)
University of Illinois at Chicago, Box 4348, Chicago, IL 60680, USA

1. Introduction

This talk describes joint work with Christophe Soulé; for full details the reader can consult the papers [22, 24].

Given an arithmetic variety X, together with a Hermitian metric on its set of complex points, the arithmetic Riemann-Roch theorem computes, given a Hermitian vector bundle $\bar{E}$ on X, the degree (in the sense of Arakelov [1,2]) of the determinant of its cohomology equipped with the Quillen metric. The Quillen metric is an invariant of the Laplace operator on $X(\mathbb{C})$ with coefficients in E. The computation is in terms of characteristic classes, which are invariants of the Hermitian bundle and the arithmetic variety itself, in the 'arithmetic Chow ring' of the variety. The arithmetic Chow ring is inspired by the ideas of Arakelov (ibid.) and Deligne [15] in the case of arithmetic surfaces, and the arithmetic Riemann-Roch theorem itself generalizes the work of Faltings [16] and Deligne (ibid.) in the relative dimension one case. One application of the theorem, to the existence of sections with small norm, will be described.

2. Preliminaries

2.1 Basic Definitions

Definition 1. In this note by an *arithmetic variety* over $\mathbb{Z}$, we mean a regular scheme X, flat and quasi-projective over $\mathbb{Z}$. However in general we need only assume that the generic fibre $X_{\mathbb{Q}}$ is smooth, and for much of what follows we can also replace the integers $\mathbb{Z}$ by a more general 'arithmetic ring'; see [18]. We shall write F_∞ for the antiholomorphic involution of the complex manifold $X(\mathbb{C})$ induced by complex conjugation.

Definition 2. A *Hermitian bundle* $\bar{E} = (E, h)$ over X consists of a vector bundle E over X, and a choice of C^∞ Hermitian metric (i.e. positive definite Hermitian inner product) h on E over $X(\mathbb{C})$, which is invariant under F_∞.

---

* Supported by N.S.F. grant DMS-8901784.

Proceedings of the International Congress
of Mathematicians, Kyoto, Japan, 1990

Example 1. If $X = \mathrm{Spec}(\mathbb{Z})$, then a Hermitian bundle over X is the same thing as a lattice in $\mathbb{R}^n$ for some n.

Example 2. If $X = \mathrm{Spec}(\mathcal{O}_K)$ for $\mathcal{O}_K$ the ring of integers in a number field K, then a Hermitian bundle $\bar{E}$ consists of a finitely generated projective $\mathcal{O}_K$-module E together with a Hermitian inner product on $E \otimes_{\mathcal{O}_K} K_v$ for each archimedean completion K_v of K. The set of isomorphism classes of Hermitian bundles over X is classified by the double coset space

$$GL_n(K)\backslash GL_n(\mathbf{A}_K)/U \ ,$$

where $\mathbf{A}_K$ is the ring of adeles of K, and U is the maximal compact subgroup of $GL_n(\mathbf{A}_K)$.

These examples show that the if one views $\mathrm{Spec}(\mathcal{O}_K)$ as being compactified by adding the archimedean places of K as points at infinity, then Hermitian vector bundles over $\mathrm{Spec}(\mathcal{O}_K)$ have a similar adelic description to vector bundles over a smooth projective curve. In this case a Hermitian bundle has a well-defined degree. Let $\det(\bar{E})$ be the top exterior power of E, with the induced Hermitian form, and let $a \in \det(E)$ be a non-zero element. Then set

$$\widehat{\deg}(\bar{E}) = \log(\#(\det(E)/\mathcal{O}_K a)) - \sum_v \log ||a||_v$$

where the sum is over the complex embeddings $v : K \hookrightarrow \mathbb{C}$, and $|| \ ||_v$ denotes the absolute value corresponding to the choice of metric on $E \otimes_v \mathbb{C}$. Equivalently the degree is equal to minus the logarithm of the covolume (i.e. the volume of a fundamental domain), with respect to the euclidean measure determined by the Hermitian inner product, of the lattice $E \subset E \otimes \mathbb{R}$. See [1,2].

2.2 Riemann-Roch for Curves, and Its Analog for Lattices

Let X be a smooth projective curve over a field k, and let E be a rank n vector bundle over X. Recall that the degree of E equals the degree of $\det E$, which is the degree of the divisor of a meromorphic section of $\det E$. We write g_X for the genus of X. Then the classical Riemann-Roch theorem for curves is:

$$\chi(X,E) = n(1 - g_X) + \deg(E) \ .$$

If k is a finite field of order q, and V is a finite dimensional k-vector space,then $\dim_k V = \log_q \#(V)$. This leads, given a Hermitian bundle over $\mathrm{Spec}(\mathbb{Z})$, *i.e.* a lattice $\Lambda \subset \mathbb{R}^n$, to the folowing definition of a numerical analog $h^0(\Lambda)$ of the rank of $H^0(E)$ of a vector bundle E over a curve.

Definition 3. Let $\Lambda \subset \mathbb{R}^n$ be a lattice. We set

$$h^0(\Lambda) = \log \#\{x \in \Lambda \mid ||x|| \leq 1\} \ .$$

We also write Λ^* for the dual or 'polar' lattice.

It is an immediate consequence of Minkowski's theorem that there exists a constant a such that:

$$h^0(\Lambda) \geq \widehat{\deg}(\Lambda) + an \ . \tag{1}$$

This is analogous to Riemann's theorem:

$$h^0(E) \geq \deg(E) + (1 - g_X)n \ .$$

See [28] for a discussion of this analogy. The following result is more analogous to the Riemann-Roch theorem, in that it estimates the difference between the two sides of (1). See [23] for a proof.

Theorem 4. *With the above notation,*

$$|h^0(\Lambda) - h^0(\Lambda^*) - \widehat{\deg}(\Lambda)| \leq n \log(6/\sqrt{\pi}) + \log\left(2^{-1}\Gamma\left(\frac{n}{2}+1\right)\right) \ .$$

2.3 Grothendieck-Riemann-Roch for Varieties over a Curve

Let X be a smooth projective variety over a field k, and let E be a vector bundle over X. Suppose that $f : X \to C$ is a map of X onto a smooth projective curve, and let us write $\mathrm{Td}(X/C) = \mathrm{Td}(X)f^*\mathrm{Td}(C)^{-1}$ for the relative Todd class of X over C in any suitable cohomology theory, such as the Chow ring $CH^*(X)_{\mathbb{Q}}$.

On C there is a line bundle $\lambda(E) = \lambda(X/C, E) = \det(Rf_*E)$. The fiber of $\lambda(E)$ at a point $y \in C$ is naturally isomorphic to $\bigotimes_{i\geq 0} \det[H^i(f^{-1}(y), E)]^{(-1)^i}$. (Here, if L is a line bundle, L^{-1} denotes its dual). It follows from the Grothendieck-Riemann-Roch theorem that

$$\deg \lambda(E) = \deg(\mathrm{ch}(E)\mathrm{Td}(X/C)) \ .$$

Here $\deg : CH^*(X)_{\mathbb{Q}} \to CH^0(\mathrm{Spec}(k))_{\mathbb{Q}} \simeq \mathbb{Q}$ is the map which, if X has dimension n, is zero on $CH^i(X)_{\mathbb{Q}}$ for $i \neq n$, and on $CH^n(X)_{\mathbb{Q}}$ takes a dimension zero cycle to its degree. The Riemann-Roch theorem for a curve tells us that:

$$\chi(E) = \chi(f^{-1}(y), E)(1 - g_C) + \deg(\lambda(E)) \ . \tag{2}$$

Here $y \in C$ is arbitrary. Notice that this formula allows one to deduce the Hirzebruch-Riemann-Roch formula for E over X from the corresponding formula for the generic fiber of X over C, together with the computation of $\deg(\lambda(E))$.

3. The Arithmetic Riemann-Roch Theorem

3.1 Motivation

Suppose now that $f : X \to \mathrm{Spec}(\mathbb{Z})$ is a projective arithmetic variety, and we fix a Kähler metric ω on $X(\mathbb{C})$ which is invariant under F_∞. What might a Riemann-Roch theorem for a Hermitian bundle $\bar{E}$ on X look like? If we view f as being analogous to a map to a curve, then (2) suggests that such a theorem combines three ingredients: Theorem 4, the Hirzebruch-Riemann-Roch theorem for E on the generic fibre $X_{\mathbb{Q}}$, and the computation of the degree of a Hermitian line bundle $\lambda(X, \bar{E})$. We need therefore to define a metric on $\lambda(E) = Rf_*(E)$. Note that such a metric on $\lambda(E) = Rf_*(E)$ can be interpreted as measuring the torsion in Rf_*E supported on the 'closed points at infinity.'

3.2 The Quillen Metric

Procedures have been proposed for putting a metric on $\lambda(E) = Rf_*(E)$ by Faltings [16] (for X an arithmetic surface) and Quillen [25] . We shall follow the method of Quillen.

Let $f : M \to B$ be a proper smooth map of complex manifolds. (This of course includes the case of $X(\mathbb{C})$ mapping to a point, for X an arithmetic variety). Suppose that ω is a closed $(1,1)$-form on M which restricts to a Kähler form on each fiber of f. If $\bar{E}$ is a Hermitian holomorphic vector bundle on M, *i.e.* a locally free coherent analytic sheaf E equipped with a C^∞ Hermitian metric, then $\lambda(E) = \lambda(M/B, E) = \det(Rf_*E)$ is a holomorphic line bundle on B. Here Rf_* is the direct image in the derived category of perfect complexes of analytic sheaves. If $b \in B$ is a point, the Hermitian metric on E, together with the induced Kähler metric on $M_b = f^{-1}(b)$ and the $\bar{\partial}$ operator on M_b determine an inner product and a laplace operator Δ_i^b on $A^{0,i}(M_b, E)$. By Hodge theory we may identify canonically the cohomology $H^i(M_b, E)$ of E over M_b, with the subspace $\mathrm{Ker}(\Delta_i^b) \subset A^{0,i}(M_b, E)$. The inner product on $A^{0,i}(M_b, E)$ then induces an inner product, the L^2 inner product h_{L^2}, on $H^i(M_b, E)$ and hence on $\lambda(M_b, E) = \bigotimes_{i \geq 0}[H^i(M_b, E)]^{(-1)^i}$. There is a canonical isomorphism $\lambda(M_b, E) \simeq \lambda(M/B, E)_b$; however in general the L^2 inner product on this complex line will not vary in a C^∞ fashion with $b \in B$. Following Quillen, we modify the L^2 inner product as follows. For each $i \geq 0$ and $b \in B$ we consider the zeta function $\zeta_i^b(s) = \sum 1/\lambda_n^s$,where the sum is over the non zero eigenvalues of Δ_i^b. The sum defining $\zeta_i^b(s)$ converges absolutely for the real part of s large enough, and $\zeta_i^b(s)$ has a meromorphic continuation to the whole complex plane, with a regular point at zero. The Ray-Singer analytic torsion, [26], is then defined as

$$\tau(M_b, \bar{E}) = \sum_{q \geq 0}(-1)^{q+1} q \zeta_q'(0) \ .$$

Finally we define $\lambda(\bar{E}) = (\lambda(E), h_Q = h_{L^2} \exp(\tau(\bar{E})))$. The following theorem was joint work with J.-M. Bismut, and C. Soulé, see [8] for details.

Theorem 5. *Let $f : M \to B$, ω, and $\bar{E}$, be as above. Then:*

1. *The Quillen metric h_Q on the holomorphic line bundle $\lambda(E)$ on B is C^∞.*
2. *There is an equality of $(1,1)$-forms on B:*

$$\mathrm{ch}_1(\lambda(E), h_Q) = \int_f \mathrm{ch}(\bar{E}) \mathrm{Td}(\bar{T}_{M/B}) \ .$$

Here $\mathrm{ch}(\bar{E})$ is the representative of the Chern class in deRham cohomology obtained by applying the Chern-Weil homomorphism to the unique unitary connection of type $(0,1)$ determined by the metric and holomorphic structure on E.

Remark. The Grothendieck Riemann-Roch theorem predicts the equality of cohomology classes which corresponds to the above equation. This result is built on some earlier work of Bismut and Freed on C^∞ determinant bundles for families of Dirac operators.

3.3 Arithmetic Chow Groups

If $f : X \to Y$ is a proper morphism between arithmetic varieties which restricts to a (proper) smooth map $X(\mathbb{C}) \to Y(\mathbb{C})$, and $\bar{E}$ is a Hermitian bundle on X, $\lambda(X/Y, \bar{E})$ is a Hermitian line bundle on Y. It is the isomorphism class, up to torsion, of this Hermitian line bundle which we would like to determine. Notice that the "classical" Grothendieck-Riemann-Roch theorem computes the Chern character $\mathrm{ch}_1(\lambda(X/Y, E)) \in CH^1(Y)_{\mathbb{Q}}$ while by Theorem 5 the Quillen metric determines a form representing the de Rham representative of this class.

If X is an arithmetic variety, and $\bar{E} = (E, h)$ is a Hermitian bundle on X, then we have the Chern character $\mathrm{ch}(E) \in CH^*(X)_{\mathbb{Q}}$ in the Chow group of cycles modulo rational equivalence, and forms $\mathrm{ch}(\bar{E}) \in Z^*(X) \stackrel{\mathrm{def}}{=} \bigoplus_{i \geq 0} Z^{i,i}(X(\mathbb{C}))$ determined by the Chern-Weil homomorphism. (Here $Z^{i,i}(X(\mathbb{C}))$ denotes closed forms of type (i, i)). We shall combine these together into a single theory carrying the characteristic classes we need to compute $\lambda(X/Y, \bar{E})$.

Definition 6. Let X be an arithmetic variety as above. A codimension i *arithmetic cycle* on X is a pair (Z, g_Z) consisting of a codimension i algebraic cycle Z on X, and a current (*i.e.* a form with distribution coefficients) of type $(i-1, i-1)$, such that $dd^c g_Z + \delta_Z$ is C^∞, where δ_Z is the current of type (i, i) given by integration over the complex points of Z. We also require that $F_\infty^* g_Z = (-1)^{i-1} g_Z$. We call g_Z a *Green current* for Z, and we view two Green currents g_Z and g'_Z as equivalent if their difference lies in the image of $\partial + \bar{\partial}$.

If $W \subset X$ is an integral subscheme of codimension $i-1$, and $f \in k(W)$ is a rational function on W, we define $\widehat{\mathrm{div}}(f) = (\mathrm{div}(f), -\log|f|^2)$. Here $\log|f|^2$ is the current of type $(i-1, i-1)$ obtained by integration against the function $\log|f|^2$ on the complex points of W: by the Poincaré-Lelong lemma $\log|f|^2$ is a Green current for $\mathrm{div}(f)$. We say that an arithmetic cycle is rationally equivalent to zero if it is the sum of cycles of the form $\widehat{\mathrm{div}}(f)$, and we define the codimension i *arithmetic Chow group* $\widehat{CH}^i(X)$ of X to be the quotient of the group of codimension i arithmetic cycles by the subgroup of cycles rationally equivalent to zero.

An element of $\widehat{CH}^i(X)$ determines a rational equivalence class of algebraic cycles on X, a C^∞ form representing their de Rham cohomology class on $X(\mathbb{C})$, together with secondary data. In the paper [18] we make this relationship precise and show how to lift the product and pull back structures from cycles and forms to the arithmetic Chow groups:

Theorem 7. *Suppose that X is an arithmetic variety, with $X_{\mathbb{Q}}$ projective. Then there is an exact sequence*

$$CH^{i-1,i}(X) \stackrel{\varrho}{\to} H^{i-1,i-1}(X) \stackrel{a}{\to} \widehat{CH}^i(X) \stackrel{\zeta \oplus \omega}{\to} CH^i(X) \oplus Z^{i,i}(X) \stackrel{c-h}{\to} H^{i,i}(X) .$$

Here $CH^{i-1,i}(X)$ is a group which is up to torsion the weight i part of $K_1(X)$, ϱ is the Beilinson regulator ([4]), $H^{i-1,i-1}(X)$ is the subspace of $H^{i-1,i-1}(X(\mathbb{C}))$ on which F_∞ acts by $(-1)^{i-1}$, a is the map $\alpha \mapsto (0, \alpha)$, ζ is the map forgetting the Green current, $\omega(Z, g_Z) = dd^c g_Z + \delta_Z$, c maps a cycle to its cohomology class and h maps a closed form to its cohomology class. If we had not assumed that $X_{\mathbb{Q}}$ was projective we would have to modify the definition of the Dolbeault cohomology groups in the sequence; cf. [18].

Example 3. Let K be a number field and $X = \mathrm{Spec}(\mathcal{O}_K)$. Then

$$\widehat{CH}^1(X) \simeq K^* \backslash \mathbf{I}_K / U \ ,$$

where $\mathbf{I}_K$ is the idele group of K, and $U \subset \mathbf{I}_K$ is its maximal compact subgroup. If $\mathrm{Cl}(\mathcal{O}_K)$ is the ideal class group of K, the exact sequence of Theorem 7 becomes:

$$\mathcal{O}_K^* \to \mathbb{R}^{r_1+r_2} \to \widehat{CH}^1(X) \to \mathrm{Cl}(\mathcal{O}_K) \to 0 \ .$$

Example 4. For a general arithmetic variety X, let $\widehat{\mathrm{Pic}}(X)$ be the group of isomorphism classes of Hermitian line bundles on X, with the group operation given by tensor product. Then there is an isomorphism

$$\hat{C}_1 : \widehat{\mathrm{Pic}}(X) \to \widehat{CH}^1(X) \ ,$$

given by $\bar{L} \mapsto (\mathrm{div}(s), -\log ||s||^2)$, for s an arbitrary rational section of L.

Example 5. On projective space over the integers $\mathbb{P}^n_{\mathbb{Z}}$, viewed as an arithmetic variety, we have a tautological Hermitian line bundle $\widehat{\mathcal{O}(1)}$, with the metric induced by orthogonal projection from the trivial bundle. If $x : \mathrm{Spec}\mathbb{Z} \to \mathbb{P}^n_{\mathbb{Z}}$ is an integral point, then $\widehat{\deg}\, x^*(\widehat{\mathcal{O}(1)})$ is the Arakelov height of x; cf. [27].

Theorem 8. *1. If $f : X \to Y$ is a map between (regular) arithmetic varieties, there is a pull back map $f^* : \widehat{CH}^*(Y) \to \widehat{CH}^*(X)$. If $f : X \to Y$ and $g : Y \to Z$, then $f^* g^* = (gf)^*$.*

2. $\widehat{CH}^*(X)_{\mathbb{Q}}$ is an associative ring with unit, and the product is compatible, via the maps ζ and ω, with the intersection product on cycles and the wedge product on forms.

3. If $f : X \to Y$ is a proper map of relative dimension d, smooth over $\mathbb{Q}$, there is a direct image map $f_* : \widehat{CH}^*(X) \to \widehat{CH}^{*-d}(Y)$, with the obvious projection formula holding.

The proof of the theorem, and in particular the construction of the product, involves showing that if Y and Z are two cycles on a projective variety over $\mathbb{C}$ which intersect properly, choices of Green currents g_Y and g_Z determine a Green current $g_Y * g_Z$, with good properties, for the product cycle $Y \cdot Z$. This product of Green currents must then be combined with the intersection product of cycles on the regular scheme underlying the arithmetic variety. However, because there is as yet no intersection theory with integral coefficients on a general regular scheme, we use our construction in [17], via Adams operations on K_0, of an intersection theory with rational coefficients. Observe that if $f : X \to \mathrm{Spec}(\mathbb{Z})$ is projective, the pushforward f_* induces a map $\widehat{\deg} : \widehat{CH}^{d+1}(X) \to \mathbb{R}$, where

d is the relative dimension of X over $\mathbb{Z}$; explicitly, $(Z, g_Z) \mapsto \log \#\Gamma(Z, \mathcal{O}_Z) + \frac{1}{2}\int_{X(\mathbb{C})} g_Z$. Composing the product with the degree map gives a pairing $\widehat{CH}^p(X) \otimes \widehat{CH}^{d+1-p}(X) \to \mathbb{R}$. If we restrict this pairing to the subgroup $\widehat{CH}^*(X)_0 = \ker(\omega)$, it factors, at least if X has no bad fibers, through the subgroup of $CH^*(X_{\mathbb{Q}})$ consisting of cycles which are homologically equivalent to zero, where it becomes the height pairing defined by Beilinson and Bloch [5], [13] generalizing the Neron height pairing.

3.4 Characteristic Classes for Hermitian Bundles

In [19] we show that any characteristic class defined by invariant polynomials has a common lifting from the Chow ring and the ring of differential forms to the arithmetic Chow ring. The following theorem summarizes this result in the case of the Chern character, which will be sufficient for our purposes, since the Chern character determines the Todd genus.

Theorem 9. *There is one and only one way of assigning to each Hermitian bundle $\bar{E}$ on an arithmetic variety X a class $\widehat{\mathrm{ch}}(\bar{E}) \in \widehat{CH}^*(X)_{\mathbb{Q}}$, the "arithmetic Chern character", with the following properties:*

1. Under the maps ζ and ω it maps to the classes $\mathrm{ch}(E) \in CH^*(X)_{\mathbb{Q}}$ *and* $\mathrm{ch}(\bar{E}) \in Z^*(X)$.

2. If $\bar{L}$ is a Hermitian line bundle, then $\widehat{\mathrm{ch}}(\bar{L}) = \exp(\widehat{C}_1(\bar{L}))$.

3. Given two Hermitian bundles $\bar{E}$ and $\bar{F}$ we have:

$$\widehat{\mathrm{ch}}(\bar{E} \otimes \bar{F}) = \widehat{\mathrm{ch}}(\bar{E}) \cdot \widehat{\mathrm{ch}}(\bar{F}) ,$$

and

$$\widehat{\mathrm{ch}}(\bar{E} \oplus \bar{F}) = \widehat{\mathrm{ch}}(\bar{E}) + \widehat{\mathrm{ch}}(\bar{F}) .$$

Here $\bar{E} \oplus \bar{F}$ is the orthogonal direct sum.

4. $\widehat{\mathrm{ch}}$ commutes with pullbacks.

Remark. If $\mathcal{E} : 0 \to E' \to E \to E'' \to 0$ is an exact sequence of vector bundles on an arithmetic variety X, and we choose metrics h', h, h'' on the three bundles, then

$$\widehat{\mathrm{ch}}(E', h') + \widehat{\mathrm{ch}}(E'', h'')) - \widehat{\mathrm{ch}}(E, h) = (0, \widetilde{\mathrm{ch}}(\mathcal{E}, h', h, h'')) ,$$

which is the secondary chern character defined by Bott and Chern [14] satisfying $dd^c\widetilde{\mathrm{ch}}(\mathcal{E}, h', h, h'') = \mathrm{ch}(E', h') + \mathrm{ch}(E'', h'') - \mathrm{ch}(E, h)$, *i.e.* it measures the failure of the Whitney sum formula to be true at the level of forms.

3.5 The Main Theorem

Let $f : X \to Y$ be a proper map between nonsingular arithmetic varieties, which restricts to a smooth map $X_{\mathbb{Q}} \to Y_{\mathbb{Q}}$; fix a Kähler metric on $X(\mathbb{C})$ which is invariant under F_∞. Let $\bar{E}$ be Hermitian bundle on X, and let $\lambda(\bar{E})$ be the corresponding determinant line bundle on Y, equipped with the Quillen metric. Let $\bar{T}_{X/Y}$ denote the relative tangent bundle with the metric induced by the

Kähler metric on $X(\mathbb{C})$; note that strictly speaking, unless f is smooth, this relative tangent bundle is a virtual object. Consider the difference

$$\widehat{\mathrm{ch}}_1(\lambda(\bar{E})) - [f_*(\widehat{\mathrm{ch}}(\bar{E})\widehat{\mathrm{Td}}(\bar{T}_{X/Y}))]^1 ,$$

where $[\ ]^1$ denotes the component in $\widehat{CH}^1_{\mathbb{Q}}$. By the 'classical' Grothendieck-Riemann-Roch theorem, together with Theorem 5 we know that this difference has image zero under both ζ and ω, and so by Theorem 7 must lie in the image of the map $a : H^{0,0}(Y) \to \widehat{CH}^1(Y)_{\mathbb{Q}}$. If one computes this difference for $X = \mathbb{P}^1_{\mathbb{Z}}$ and $Y = \mathrm{Spec}(\mathbb{Z})$, one does not get zero, showing that a naive translation of the Grothendieck-Riemann-Roch theorem is false. By computing, with the help of D. Zagier, the difference for the trivial bundle on $X = \mathbb{P}^n_{\mathbb{Z}}$ for all n, we were led in [20] to define an "arithmetic Todd genus":

Definition 10. Consider the power series

$$R(x) = \sum_{m\ \mathrm{odd}} [2\zeta'(-m) + \zeta(-m)(1 + 1/2 + \ldots + 1/m)] \left(\frac{x^m}{m!}\right) .$$

Using the splitting principle, we have an additive characteristic class $R(E) \in H^*(X)$ for complex vector bundles E over topological spaces X, such that $R(L) = R(C_1(L))$ for a line bundle L. Now given $f : X \to Y$ as above, define the *arithmetic Todd genus*:

$$\mathrm{Td}^A(X/Y) = \widehat{\mathrm{Td}}(\bar{T}_{X/Y})(1 + a(R(T_{X(\mathbb{C})/Y(\mathbb{C})}))) .$$

Remark. Notice that power series $R(x)$ used to define the correction term in Td^A has a mysterious relationship to the power series used to define the Todd genus itself:

$$\mathrm{Td}(x) = 1 - x\sum_{m\geq 0} \zeta(-m)\frac{x^m}{m!} .$$

Theorem 11. *Let $f : X \to Y$, $\bar{E}$, be as above. Then we have*

$$\widehat{\mathrm{ch}}_1(\lambda(\bar{E})) = [f_*(\widehat{\mathrm{ch}}(\bar{E})\mathrm{Td}^A(\bar{T}_{X/Y}))]^1 .$$

The strategy of the proof is based on the method used to prove the "classical" version of the theorem. That is, to factor the map f as the composition of a closed immersion $i : X \to \mathbb{P}^n_Y$ followed by the projection $g : \mathbb{P}^n_Y \to Y$. The main ingredients are:

1. The computation, with Zagier, of the analytic torsion of the trivial bundle on $\mathbb{P}^n_Y$, [20].

2. Following the series of papers [6] (by J.-M. Bismut), and [9] and [10] (with J.-M. Bismut and C. Soulé), given a resolution $\eta : \bar{\xi}. \to i_*E$, by a complex of Hermitian bundles, the 'torsion' of $\bar{\xi}.$ is defined. This is a current on $\mathbb{P}^n(\mathbb{C})$, and in [9] it is shown that it plays for closed immersions between complex manifolds a role similar to that played by the Ray-Singer torsion for submersions. In [10] we then prove a Riemann-Roch theorem for i.

3. The papers [7], by J.-M. Bismut, and [11], by J.-M. Bismut and G. Lebeau, compute the ratio of the Quillen metrics on $\lambda(\bar{E})$ and $\lambda(\bar{\xi}.)$ in terms of the the torsion of $\bar{\xi}$. This can be viewed as showing that $f_! = g_! i_!$ in the K-theory of Hermitian bundles, up to a correction involving the exotic class $R(N_{X/\mathbb{P}^n_Y})$.

Remark. 1. If $Y = \mathrm{Spec}(\mathbb{Z})$, then the theorem gives:

$$\widehat{\deg}\lambda(\bar{E}) = \widehat{\deg}[\widehat{\mathrm{ch}}(\bar{E})\mathrm{Td}^A(X/\mathbb{Z})] \ ,$$

which is the analogue of the Riemann-Roch theorem for a map to curve that we were looking for.

2. We have proved a more general result, in the style of [3], in which X is allowed to have singularities away from its generic fibre. This requires a careful analysis of the behaviour of complexes under the Grassmannian-graph construction of [3].

3. G. Faltings has recently given a more direct proof of the result, without appealing to the results of [6, 7, 9, 10, 11].

3.6 An Application

Let X be an arithmetic variety of dimension $d+1$, and fix a Kähler metric on $X(\mathbb{C})$. Let $\bar{E}$ be a Hermitian vector bundle, and $\bar{L}$ a Hermitian line bundle, on X. Suppose that L is relatively ample over X and that $C_1(\bar{L})$ is strictly positive. Write $h^0(X, \bar{E})$ for $\log \#\{s \in H^0(X, E) \mid \|s\|_{L^2} \leq 1\}$. The proof of the following theorem uses an earlier, weaker, version of the arithmetic Riemann-Roch theorem, Minkowski's theorem, and a result of J.-M. Bismut and E. Vasserot, [12]. See [21] for further details.

Theorem 12. *If* $\widehat{\deg}(\hat{C}_1(\bar{L})^{d+1}) > 0$*, then:*

$$h^0(\bar{E} \otimes \bar{L}^n) \geq n^{d+1} \frac{r}{(d+1)!} \widehat{\deg}(\hat{C}_1(\bar{L})^{d+1}) + O(n^d \log(n)) \ .$$

In particular, if n is sufficiently large, $E \otimes L^n$ *has a non-zero section with* L^2 *norm less than one.*

References

1. Arakelov, S.J.: Intersection theory for divisors on an arithmetic surface. Math. USSR Izv. **8** (1974) 1167–1180
2. Arakelov, S.J.: Theory of intersections on an arithmetic surface. Proc. Int. Cong. of Math., Vancouver, vol. 1, 1978, pp. 405–408
3. Baum, P., Fulton, W., Macpherson, R.: Riemann-Roch for singular varieties. Publ. Math. I.H.E.S. **45** (1975) 101–146
4. Beilinson, A.A.: Higher regulators and values of L-functions. J. Sov. Math. **30** (1985) 2036–2070
5. Beilinson, A.A.: Height pairings between algebraic cycles. Contemp. Math. **67** (1987) 1–24

6. Bismut, J.-M.: Superconnection currents and complex immersions. Invent. math. **99** (1990) 59–113
7. Bismut, J.-M.: Koszul complexes, harmonic oscillators, and the Todd class. Prépublication Orsay 1989, pp. 89–96
8. Bismut, J.-M., Gillet, H., Soulé, C.: Analytic torsion and holomorphic determinant bundles, I, II, III. Comm. Math. Phys. **115** (1988) 49–78, 79–126, 301–351
9. Bismut, J.-M., Gillet, H., Soulé, C.: Bott–Chern currents and complex immersions. Duke Math. J. **60** (1990) 255–284
10. Bismut, J.-M., Gillet, H., Soulé, C.: Complex immersions and Arakelov geometry. In: The Grothendieck Festschrift, vol. I (P. Cartier et al., ed.). Birkhäuser, Boston 1990, pp. 249–231
11. Bismut, J.-M., Lebeau, G.: Immersions complexes et métriques de Quillen (to appear).
12. Bismut, J.-M., Vasserot, E.: Comportement asymptotique de la torsion analytique associée aux puissance d'un fibré en droites positif. C.R. Acad. Sci. Paris **307** (1988) 779–781
13. Bloch, S.: Height pairings for algebraic cycles. In: Proc. Luminy conference on algebraic K-theory. J. Pure App. Alg. **34** (1984) 119–145
14. Bott, R., Chern S.S.: Hermitian vector bundles and the equidisitribution of the zeroes of their holomorphic sections. Acta math. **114** (1968) 71–112
15. Deligne, P.: Le déterminant de la cohomologie. In: Current Trends in Arithmetical Algebraic Geometry (K.Ribet, ed.). Contemp. Math. **67** (1987) 93–177
16. Faltings, G.: Calculus on arithmetic surfaces. Ann. Math. **119** (1984) 387–424
17. Gillet, H., Soulé, C.:Intersection theory using Adams operations. Invent. math. **90** (1987) 243–277
18. Gillet, H., Soulé, C.: Arithmetic intersection theory. Publ. Math. I.H.E.S. **72** (1990) 94–174
19. Gillet, H., Soulé, C.: Characteristic classes for algebraic vector bundles with Hermitian metric. Ann. Math. **131** (1990) 163–238
20. Gillet, H., Soulé, C.: Analytic torsion and the arithmetic Todd genus. Topology **30** (1991) 21–54
21. Gillet, H., Soulé, C.: Amplitude arithmétique. C.R. Acad. Sci. Paris **307** (1988) 887–890
22. Gillet, H., Soulé, C.: Un théorème de Riemann-Roch-Grothendique arithmétique. C.R. Acad. Sci. Paris **309** (1989) 929–932
23. Gillet, H., Soulé, C.: On the number of lattice points in convex symmetric bodies and their duals. Prépublication I.H.E.S. 1990. Israel Math. (to appear)
24. Gillet, H., Soulé, C.: An arithmetic Riemann-Roch theorem. Preprint.
25. Quillen, D.: Determinants of Cauchy-Riemann operators over a Riemann surface. Funct. Anal. Appl. **14** (1985) 31–34
26. Ray, D.B., Singer, I.M.: Analytic torsion for complex manifolds. Ann. Math. **98** (1973) 154–177
27. Silverman, J.H.: A survey of the theory of height functions, In: Current Trends in Arithmetical Algebraic Geometry (K.Ribet, ed.). Contemp. Math. **67** (1987) 269–278
28. Weil, A.: Sur l'analogie entre les corps de nombres algébriques, et les corps de fonctions algébriques. Collected Papers, vol. 1. Springer, Berlin Heidelberg New York 1980, pp. 236–240

Area, Lattice Points and Exponential Sums

Martin N. Huxley

School of Mathematics, University of Wales, College of Cardiff, Senghenydd Road
Cardiff CF2 4AG. Wales, UK

Suppose you have a closed curve. How do you find the area inside? While I was writing my first paper on exponential sums and lattice points, my seven year old daughter came home from school and said. "I know how you find the area of a curve. You count the squares". In other words, copy the curve onto squared paper and count how many squares lie inside the curve. If you want the area to greater accuracy, use paper with smaller squares. If the side of the squares is $1/M$, and you count N squares, then the area A is approximately N/M^2. As a number theorist I prefer to take the squares as unit squares, the curve as enlarged by a factor M, and the relation as being that N is approximately AM^2.

Some squares are inconveniently cut by the curve. When do you count them?

Rule 1. Count a square if its centre lies inside.
Rule 2. Count a square if its lower left corner lies inside.
Rule 3. Count all incomplete squares as half a square.

Rules 1 and 2 are really the same rule: if we shift the squared paper by half a unit in the x and y directions, then the corners of the squares are now where the centres were before. Rule 3 is locally like Rule 2, but it counts extra squares where the curve has maxima and minima, and where x takes its extreme values. Take Rule 2 as the basic rule. The corners of squares are the points in two dimensional space with integer coordinates x, y, the lattice points.

How accurate is the rule? If the curve fits into a rectangular box C high, B broad (in terms of unit squares), then the number of squares cut is at most $2B + 2C + 4$. As M tends to infinity, this discrepancy has order of magnitude M, whilst the area has order M^2.

Lattice points on the curve limit the accuracy. Whether by convention one counts them in or out, changing M to $M + \varepsilon$ makes little change to the area, and counts them all in, whilst changing M to $M - \varepsilon$ counts them all out. If the curve is a polygon, especially a rectangle with sides parallel to the axes, there can be M lattice points on the sides of the rectangle, so that order of magnitude M is then best possible.

So take a smooth curve, convex for convenience. What is the mathematics available? The curve is described in a coordinate-free way by an equation connecting the arc length s with the tangent angle ψ, the radius of curvature being $ds/d\psi$. The

Proceedings of the International Congress
of Mathematicians, Kyoto, Japan, 1990

lattice of integers has an algebraic isomorphism group SL(2, Z), the two by two integer matrices of determinant one. The lattice is also periodic in two dimensions, so that you can take Fourier transforms. These tell you that for a smooth curve the discrepancy $N - AM^2$ has root mean square lying between bounded multiples of $\sqrt{M}$ (bounded in terms of the shape of the curve). The mean is taken over translations of the curve. As a consequence, for some constant c,

$$|N - AM^2| < c\sqrt{M}$$

for at least half the positions of the curve on the squared paper. The two dimensional Fourier transform does not help much with pointwise upper bounds, even for a circle centred on a lattice point, the Gauss circle problem.

The upper bound so far corresponds to replacing the curve by a step function. Voronoi and Sierpinski used Archimedes' idea of replacing the curve by a polygon. The polygon is chosen so that the gradients of its sides are small rational numbers. Small is measured by the height norm: the height of a/q in lowest terms is $\max(|a|, |q|)$. As far as I know, they did not treat the general curve, but only the hyperbola (for the Dirichlet divisor problem) and the circle. If you assume merely that the curve has a radius of curvature bounded away from zero and infinity, then you can show that the discrepancy $|N - AM^2|$ has order of magnitude at most $(M \log^2 M)^{2/3}$ by using the estimate for the discrepancy of a right-angled triangle in terms of the continued fraction expansion of the gradient a/q. Sierpinski had this result in the Gauss circle problem without the logarithmic factor, so being a perfect circle is a little help. The sides of the polygon have rational gradients, so that they may contain many lattice points.

The next step is obvious with hindsight: instead of approximating the curve by a piecewise constant or piecewise linear function, approximate by a piecewise quadratic function. Here the argument becomes highly analytic. Define $e(t)$ to be $\exp 2\pi it$, and $\varrho(t)$ to be the row-of-teeth function, with $\varrho(t) = 1/2 - t$ for $0 \le t < 1$, and with period one. Then $\varrho(t)$ has the Fourier series

$$\varrho(t) \approx \sum_{\substack{h=-\infty \\ h \neq 0}}^{\infty} \frac{e(ht)}{2\pi ih}.$$

The number of integers n in an interval $a < n \le b$ is $b - a + \varrho(b) - \varrho(a)$. I can now explain Iwaniec and Mozzochi's attack on the circle problem (1987). It is closely related to Bombieri and Iwaniec's great paper (1986) on the size of the Riemann zeta function. Bombieri and Iwaniec's paper turns out to be related to Jutila's work on the Dirichlet series of modular forms. The connection is that modular forms are functions of two dimensional lattices.

The first step is to divide the curve into arcs corresponding to the sides of the Voronoi-Sierpinksi polygon, and to take a new basis for the integer lattice so that (q, a) is a basis vector. This brings in a 2×2 integer matrix. The curve is then approximated as a quadratic in a coordinate n in the direction of the vector (q, a) tangential to the curve.

The sum involving $\varrho(t)$ gives an exponential sum in two summands n and h when the Fourier series for $\varrho(t)$ is truncated. The next two steps are Poisson summation in each variable. There is a serious danger of error terms adding up. It is lucky that the integral to compute is a Bessel function of order one half which has an exact expression of the form

$$e(t)/\sqrt{t}.$$

After approximating, the main terms involve an inner product:

$$e(x^{(k,l)} \,.\, y^{(j)}),$$

where $x^{(k,l)}$ is a four dimensional vector

$$(kl, l, l\sqrt{k}, l/\sqrt{k})$$

indexed by two integers k and l, and $y^{(j)}$ is a four dimensional vector constructed from the quadratic approximation. The index j refers to the arc of the curve.

If a/q has small height (a major arc), then k and l take a bounded number of values, and you estimate trivially (in fact one need not go so deep). If a/q has large height (a minor arc), then k and l run through complicated ranges. When these have been simplified, the next step is a form of the large sieve.

The modern large sieve is best described as a sort of Sobolev inequality relating the discrete L_2 norm of a function to the ordinary L_2 norms of the function and its derivatives. The discrete L_2 norm in one dimension is

$$\sup_S \sum_S |f(z_i)|^2$$

taken over sets $S = \{z_1, \ldots, z_R\}$ with $|z_i - z_j| \geq \delta$ for $i \neq j$. In number-theoretic applications we have a fixed set S, usually corresponding to a subset of the rational numbers, and the function $f(z)$ is an exponential sum, which can be regarded as an integral with respect to a discrete measure supported on the integers. In this case there is a duality principle and an underlying bilinear form, in which the vectors x are on the same footing as the vectors y. A further generalisation replaces the condition $|z_i - z_j| \geq \delta$ by a factor in the upper bound which counts the number of pairs of points differing by less than δ (the convolution of the discrete measure supported on the points of S with itself, against a kernel function supported on a δ-neighbourhood of the diagonal).

The next two steps are counting the number of coincidences among sums of pairs of x vectors, which is number theory, and the number of coincidences among pairs of y vectors, which is number theory mixed with analysis. A coincidence between two y vectors means that on two different arcs, the curve weaves between lattice points in the same way. A large set of mutually coincident vectors will add up to a systematic error in the discrepancy $N - AM^2$.

Rather than write down the conditions for entries of the y vectors to coincide, I try to describe them. A y vector corresponds to a minor arc and a two by two integer matrix. Coincidence in the first entries means that the two integer matrices are close in the sense that PQ^{-1} is a matrix with small integer entries, and so PQ^{-1} is a short word in the generators $\begin{pmatrix} 1 & 1 \\ 0 & 1 \end{pmatrix}$ and $\begin{pmatrix} 1 & 0 \\ 1 & 1 \end{pmatrix}$. Coincidence in the second entry involves the matrix, the constant term and the gradient. Coincidence in the third entry is a relation between the denominators q of the gradients a/q and the coefficients of n^2 in the quadratic approximations. Coincidence in the fourth entry involves all the coefficients of the approximating polynomial, but if the other three entries coincide, then the condition simplifies to one involving the constant terms and the denominators q of the gradients a/q.

What arguments can one use in counting the number of pairs of coincident vectors? The first is compactness, or in its discrete version, Dirichlet's box principle. If there are many points in a bounded region, then there is a small set containing many of them. The second principle is approximation, or the mean value theorem. A smooth function on a small set can be approximated by a polynomial, or still better, a linear function. The third principle is the boon and bane of analytic number theory. The only arbitrarily small integer is zero. An inequality which is strengthened too much turns into an equation. When these three arguments are used in turn, we pass from discrete to continuous and back to discrete again.

At present we are overestimating the number of coincidences because we cannot handle the constant term in the quadratic polynomial. So the second and fourth entries of the y vectors are not used. Iwaniec and Mozzochi (1987) could count coincidence only for the rectangular hyperbola. Huxley and Watt (1988) generalised the relevant lemma in Bombieri and Iwaniec (1986), getting bounds for one dimensional exponential sums, and Huxley (1990) adapted the Iwaniec-Mozzochi method to general smooth closed curves to get

$$N = AM^2 + O(M^{7/11}(\log M)^{47/22}).$$

Can this be improved? There is some hope of using the second and fourth entries of the y vectors, which would give a small improvement on the exponent 7/11. To reach one half one must avoid taking moduli. Bombieri and Iwaniec (1986) suggest using the Fourier theory of the modular group, which acts on the vectors (q, a) corresponding to the sides of the Voronoi-Sierpinski polygon. This Fourier theory was suggested by Selberg (1956) and worked into a usable form by Kuznietsov (1980) and Deshouillers and Iwaniec (1982).

There are many related questions. Bombieri and Pila (1989) have shown by algebraic geometry that if the curve is not algebraic, then the number of integer points on it is $O(M^\varepsilon)$ for any positive ε, improving the earlier result of Swinnerton-Dyer (1974). Huxley (1988, 1989) has considered the number of integer points within a small distance δ of the curve, using a mixture of elementary and exponential sum techniques. The lattice point problem is a special case of rounding error in numerical approximation, and one can give estimates there too (Huxley 1991).

This method seems to be purely two dimensional, because a quadratic approximation to a surface contains too many terms. Even the continued fraction rule does not generalise. But I hope that someone will take up the challenge of finding an arithmetic method for counting lattice points in three or more dimensions.

References

Bombieri, E. (1987): Le grand crible dans le théorie analytique des nombres, 2de edn. Astérisque, Paris

Bombieri, E., Iwaniec, H. (1986): On the order of $\zeta(1/2 + it)$. Ann. Sc. Norm. Sup. Pisa Cl. Sci. (4) **13**, 449–472

Bombieri, E., Pila, J. (1989): The number of integral points on arcs and ovals. Duke Math. J. **59**, 337–357

Deshouillers, J.-M., Iwaniec, H. (1982): Kloosterman sums and Fourier coefficients of cusp forms, Invent. math. **70**, 219–288

Fouvry, E., Iwaniec, H. (1989): Exponential sums for monomials, J. Number Theory **33**, 311–333

Heath-Brown, D.R., Huxley, M.N. (1990): Exponential sums with a difference. Proc. London Math. Soc. (3) **61**, 227–250

Huxley M.N. (1988): The fractional parts of a smooth sequence. Mathematika **35**, 292–296

Huxley M.N. (1989): The integer points close to a curve. Mathematika **36**, 198–215

Huxley M.N. (1990): Exponential sums and lattice points. Proc. London Math. Soc. (3) **60**, 471–502

Huxley M.N. (1991): Exponential sums and rounding error. J. London Math. Soc. (to appear)

Huxley, M.N., Kolesnik, G. (1991): Exponential sums and the Riemann zeta function III. Proc. London Math. Soc. (3) **62** (to appear)

Huxley, M.N., Watt, N. (1988): Exponential sums and the Riemann zeta function. Proc. London Math. Soc. (3) **57**, 1–24

Huxley, M.N., Watt, N. (1989): Exponential sums with a parameter. Proc. London Math. Soc. (3) **59**, 233–252

Iwaniec, H., Mozzochi, C.J. (1988): On the divisor and circle problems. J. Number Theory **29**, 60–93

Jutila, M. (1987): Lectures on a method in the theory of exponential sums. Tata Institute Lectures in Mathematics and Physics, vol. 80. Springer, Bombay

Krätzel. E., (1988): Lattice points. D.V.W., Berlin

Kuznietsov, N.V. (1980): Peterson's conjecture for cusp forms of weight zero and Linnik's conjecture on sums of Kloosterman sums. Mat. Sbornik **111**. 334–383

Selberg, A. (1956): Harmonic analysis and discontinuous groups in weakly symmetric Riemannian spaces with applications to Dirichlet series. J. Indian Math. Soc. **20**, 47–87

Sierpinski, W. (1906): Sur un problème du calcul des fonctions asymptotiques. Prace Mat.-Fiz **17**, 77–118

Swinnerton-Dyer, H.P.F. (1974): The number of lattice points on a convex curve. J. Number Theory **6**, 128–135

Voronoi, G. (1906): Sur un problème du calcul des fonctions asymptotiques. J. Reine Angew. Math. **126**

Watt, N. (1989a): A problem on semicubical powers. Acta Arith. **52**, 119–140

Watt, N. (1989b): Exponential sums and the Riemann zeta function II, J. London Math. Soc. **39**, 385–404

Watt, N. (1990): A problem on square roots of integers. Periodica Math. Hung. **21**, 55–64

Watt, N. (1991): Exponential sums with a character. (To appear)

Generalized Class Field Theory

Kazuya Kato

Department of Mathematics, Faculty of Science
University of Tokyo, Hongo, Bunkyo-ku, Tokyo 113, Japan

This paper is a survey of the K-theoretic generalization of class field theory.

For a field K, let K^{ab} be a maximal abelian extension of K, that is, the union of all finite abelian extensions of K in a fixed algebraic closure of K. The classical local (resp. global) class field theory says that if K is a finite extension of the p-adic (resp. rational) number field $\mathbf{Q}_p$ (resp. $\mathbf{Q}$), the Galois group $\mathrm{Gal}(K^{ab}/K)$ is approximated by the multiplicative group $K^\times$ (resp. the idele class group C_K), and via this approximation, we can obtain knowledge on abelian extensions of K.

In §1 (resp. §2), we give a K-theoretic generalization of the classical local (resp. global) class field theory. There finite extensions of $\mathbf{Q}_p$ (resp. $\mathbf{Q}$) are replaced by "higher dimensional local fields" (resp. finitely generated fields over prime fields), and the group $K^\times$ (resp. C_K) is replaced by Milnor's K-group $K_n^M(K)$ (resp. by the K_n^M-idele class group), where n is the "dimension" of K.

In §3, we discuss some other aspects of generalizations of local class field theory.

In §4, we discuss generalizations of the classical ramification theory to higher dimensional schemes.

1. Local Class Field Theory

An n-dimensional local field is defined inductively as follows. A 0-dimensional local field is a finite field. For $n \geq 1$, an n-dimensional local field is a complete discrete valuation field whose residue field is an $(n-1)$-dimensional local field.

For example, a finite extension of $\mathbf{Q}_p$ is a one dimensional local field.

For a field k, let $K_*^M(k)$ be Milnor's K-group of k defined by

$$K_q^M(k) = ((k^\times)^{\otimes q})/J$$

where J is the subgroup of the q-fold tensor product of $k^\times$ (as a $\mathbf{Z}$-module) generated by elements of the form $a_1 \otimes \ldots \otimes a_q$ such that $a_i + a_j = 1$ for some $i \neq j$. (Cf. Milnor [Mi].) The main result of the local class field theory of an n-dimensional local field is the following (Parsin [Pa$_2$], Kato [Ka$_1$] II).

Proceedings of the International Congress
of Mathematicians, Kyoto, Japan, 1990

Theorem 1. *Let K be an n-dimensional local field. Then, there exists a canonical homomorphism*

$$\varrho : K_n^M(K) \longrightarrow \mathrm{Gal}(K^{\mathrm{ab}}/K)$$

which induces an isomorphism $K_n^M(K)/N_{L/K}K_n^M(L) \xrightarrow{\cong} \mathrm{Gal}(L/K)$ for each finite abelian extension L of K. Here $N_{L/K} : K_n^M(L) \to K_n^M(K)$ is the norm homomorphism ([Ka$_1$] II 1.7). *The correspondence $L \mapsto N_{L/K}K_n^M(L)$ is a bijection from the set of all finite abelian extensions of K in a fixed algebraic closure of K onto the set of all open subgroups of $K_n^M(K)$ of finite indices. (For the definition of the openness of a subgroup of $K_n^M(K)$, see* [Ka$_2$]*).*

2. Global Class Field Theory

Let X be a proper integral scheme over the ring of rational integers $\mathbf{Z}$ and let K be the function field of X. For simplicity we assume here $\mathrm{char}(K) = 0$ and that K has no ordered field structure.

For a non-zero coherent ideal I of $\mathcal{O}_X$ and for $q \geq 1$, define the sheaf of abelian groups $K_q^M(\mathcal{O}_X, I)$ on X_{zar} by

$$K_q^M(\mathcal{O}_X, I) = \mathrm{Ker}(K_q^M(\mathcal{O}_X) \to K_q^M(\mathcal{O}_X/I))$$

where

$$K_q^M(\mathcal{O}_X) = ((\mathcal{O}_X^\times)^{\otimes q})/J$$

with J the subgroup sheaf of the tensor product generated by local sections of the form $a_1 \otimes \ldots \otimes a_q$ such that $a_i + a_j = 1$ for some $i \neq j$, and $K_q^M(\mathcal{O}_X/I)$ is defined similarly. Define

$$C_I(X) = H^n(X_{\mathrm{zar}}, K_n^M(\mathcal{O}_X, I)), \quad \text{where} \quad n = \dim(X)\,.$$

If I and I' are non-zero coherent ideals of $\mathcal{O}_X$, the inclusion $I \subset I'$ induces a surjection $C_I(X) \to C_{I'}(X)$. The main result of the class field theory of K is the following.

Theorem 2.1. *(1) $C_I(X)$ is a finite group for any I.*

(2) We have a canonical isomorphism of profinite abelian groups

$$\varprojlim_I C_I(X) \cong \mathrm{Gal}(K^{\mathrm{ab}}/K)\,,$$

where I ranges over all non-zero coherent ideals of $\mathcal{O}_X$.

(3) For a non-empty regular open subscheme U of X, there exists a canonical isomorphism of profinite abelian groups

$$\varprojlim_I C_I(X) \cong \pi_1^{\mathrm{ab}}(U)\,,$$

where I ranges over all non-zero coherent ideals of $\mathcal{O}_X$ *such that* $U \cap \mathrm{Spec}(\mathcal{O}_X/I) = \phi$, *and* $\pi_1^{\mathrm{ab}}(U)$ *is the quotient of the algebraic fundamental group* $\pi_1(U)$ *of U modulo the closure of its commutator subgroup.*

(4) If X is regular, then $C_{\mathcal{O}_X}(X)$ *is isomorphic to the group* $CH_0(X)$ *of the classes of zero cycles on X modulo rational equivalence, and we have a canonical isomorphism of finite abelian groups*

$$CH_0(X) \cong \pi_1^{\mathrm{ab}}(X)\,.$$

An essential part of this theorem was proved by Bloch [Bl$_1$] in the case $\dim(X) = 2$. Cf. also [Pa$_1$] for the two dimensional case. The general case was proved in [KS$_2$] (see also S. Saito [SS$_1$]) by using the method in Bloch [Bl$_1$], except that we adopted in [KS$_2$] another definition of $C_I(X)$ which uses Nisnevich topology [Ni], a Grothendieck topology defined by Nisnevich [Ni] (called the henselian topology in [KS$_2$]), instead of Zariski topology. ($C_I(X) \overset{\mathrm{def}}{=} H^n(X_{\mathrm{Nis}}, K_n^M(\mathcal{O}_X, I))$ in [KS$_2$].). It was found later that Nisnevich topology and Zariski topology give the same $C_I(X)$ [KS$_3$].

The relation of the above theorem with the classical global class field theory is that if $X = \mathrm{Spec}(O_K)$ for a finite extension K of $\mathbf{Q}$, then $C_I(X)$ coincides with the ideal class group of conductor I of O_K.

The positive characteristic version of (4) of the above theorem is that if X is a proper smooth variety over a finite field k, $CH_0(X)^0 = \mathrm{Ker}(\deg : CH_0(X) \to \mathbf{Z})$ is finite and canonically isomorphic to $\mathrm{Ker}(\pi_1^{\mathrm{ab}}(X) \to \mathrm{Gal}(k^{\mathrm{ab}}/k))$. This fact was proved in [KS$_1$], and another proof was given in Colliot-Thélène, Sansuc and Soulé [CSS] and Gros [Gr]. That the former group surjects onto the latter was proved long ago by Lang whose paper [Lan] is the first work on the higher dimensional class field theory.

As an application of the generalized global class field theory, we have

Theorem 2.2. *For any scheme S of finite type over* $\mathbf{Z}$, *the abelian group* $CH_0(S)$ *is of finite type.*

The case $\dim(S) = 2$ of this theorem was proved by Bloch [Bl$_1$], and the general case was proved in [KS$_2$].

Some finiteness theorems on $\pi_1^{\mathrm{ab}}(S)$ for arithmetic schemes S were proved in Katz-Lang [KL].

3. Some Aspects of the Generalized Local Class Field Theory

I give rough indications on some aspects of generalizations of local class field theory, which are not included in §1.

3.1 Explicit Reciprocity Law

The explicit reciprocity law for a finite extension of $\mathbf{Q}_p$ has been generalized to higher dimensional local fields of mixed characteristic by Vostokov-Kirrilov [VK] and Vostokov [V]. A generalization to complete discrete valuation fields of mixed characteristic $(0, p)$ with residue field F such that $[F : F^p] < \infty$ is given in [Ka$_4$], as an application of the p-adic cohomology theory of Fontaine-Messing [FM].

Let K be "the most important two dimensional local field"

$$(\varprojlim_n (\mathbf{Z}/p^n\mathbf{Z})[[q]][q^{-1}]) \otimes_{\mathbf{Z}} \mathbf{Q}$$

where q is the q-invariant in the theory of moduli of elliptic curves. This field appears as a certain p-adic completion at infinity of a modular curve. As is shown in [Ka$_5$], the explicit reciprocity laws of the two dimensional local fields $K(\zeta_{p^n})$, where ζ_{p^n} is a primitive p^n-th root of 1, are related to the special values of L-functions of elliptic modular forms and to an Iwasawa theory of elliptic modular forms, just as the explicit reciprocity laws of $\mathbf{Q}_p(\zeta_{p^n})$ are related to the special values of Riemann zeta function and to the classical Iwasawa theory.

3.2 Semi-Global Theories

The class field theory of curves over (usual) local fields was studied by Coombes [Co] and S. Saito [SS$_2$], and that of surfaces over local fields was studied recently by S. Saito and Salberger.

The class field theory of complete discrete valuation fields whose residue fields are (usual) global fields of positive characteristic was sought for first by Ihara [Ih], and studied by [Ka$_1$] III.

The class field theory of two dimensional complete noetherian local rings with finite residue fields (for example $\mathbf{Z}_p[[T]]$) was studied by S. Saito [SS$_3$].

3.3 Serre's Local Class Field Theory

Serre's local class field theory of a complete discrete valuation field with algebraically closed residue field [Se$_2$], and its generalization by Hazewinkel to the perfect residue field case [Ha], are generalized to the imperfect residue field case as a duality theorem of the following form: If K is a complete discrete valuation field with residue field F such that $\mathrm{char}(F) = p > 0$ and $[F : F^p] = p^r < \infty$, then the Galois cohomologies

$$(*) \qquad R\Gamma(K, \mathbf{Z}/p^n\mathbf{Z}(s)) \quad \text{and} \quad R\Gamma(K, \mathbf{Z}/p^n\mathbf{Z}(t)), \qquad s+t=r+1$$

are in perfect duality via the dualizing functor $R\,\mathrm{Hom}_{\mathbf{Z}/p^n\mathbf{Z}}(\ , W_n\Omega^r_{F,\log})[r+1]$. Here the objects $(*)$ and the logarithmic part of the de Rham-Witt sheaf $W_n\Omega^r_{F,\log}$ ([Il]) are not regarded just as a complex of abelian groups or as objects on the small etale site on $\mathrm{Spec}(F)_{\mathrm{et}}$, but regarded as objects on a much bigger site. (Precisely speaking, they are regarded as objects on the site of schemes S over

F which are locally isomorphic to "relative perfections" ([Ka$_3$] I, II) of smooth schemes over F, endowed with the etale topology.) The details of this duality theorem are given in [Ka$_3$] III.

For example, if F is an algebraically closed field, this duality and the inclusion

$$U_K/(U_K)^{p^n} \subset K^\times/(K^\times)^{p^n} = H^1(K, \mathbf{Z}/p^n\mathbf{Z}(1))$$

induces $H^1(K, \mathbf{Z}/p^n\mathbf{Z}) \cong \mathrm{Ext}^1(U_K, \mathbf{Z}/p^n\mathbf{Z})$, where U_K is the unit group of K which is not regarded just as a group, but regarded as a pro-algebraic group over F. This reproduces the p-primary part (the essential part) of Serre's local class field theory

$$\pi_1(U_K) \cong \mathrm{Gal}(K^{\mathrm{ab}}/K).$$

4. Ramification Theory

In the classical ramification theory, we mainly consider a finite extension B/A of discrete valuation rings with perfect residue fields. We have the following three kinds of important invariants of ramification: The different $\delta(B/A) \in \mathbf{Z}$; in the case of Galois extension with Galois group G, the Lefschetz numbers

$$i(\sigma) = \mathrm{length}(B/I_\sigma) \in \mathbf{Z} \quad \text{for} \quad \sigma \in G - \{1\},$$

with I_σ the ideal of B generated by $\{\sigma(b) - b\,;\, b \in B\}$; also in the Galois case, the Artin (or Swan) conductors of representations of G. Is it possible to generalize these invariants and relations between them such as

$$(*) \qquad \delta(L/K) = \sum_{\sigma \in G - \{1\}} i(\sigma)$$

to higher dimensional schemes?

Concerning $\delta(B/A)$ and $i(\sigma)$, Bloch obtained in [Bl$_2$] a nice ramification theory in higher dimensions, generalizing $\delta(B/A)$ and $i(\sigma)$ to zero cycle classes on the ramification locus (i.e. to elements of $CH_0(E)$ where E is the ramification locus). His projection formula [Bl$_2$] (7.1) is an extension of (*) to two dimensional schemes.

Concerning conductors, there are many different attempts of generalizations (Deligne [De], Laumon [La], S. Saito [SS$_4$], Berthelot [Be$_1$, Be$_2$]). In the following, I only introduce my method on the conductors of one dimensional Galois representations [Ka$_7$] which is closely related to the generalized local class field theory. I generalize, under the influence of Bloch's theory [Bl$_2$], the Swan conductors to zero cycle classes on the wild ramification locus in the two dimensional case (the plan exists even for dimension ≥ 3), and give applications Theorems (4.1)–(4.3).

Let X be an excellent connected normal scheme, U a regular dense open subscheme of X, and let $\chi : \pi_1^{\mathrm{ab}}(U) \to \Lambda^\times$ be a continuous homomorphism where Λ is a discrete (commutative) field. We are interested in the wild ramification of χ

on X. We assume that the wild ramification locus E of χ on X (with the reduced scheme structure) is a disjoint union of schemes of finite type over perfect fields. This condition is satisfied for example if X is of finite type over $\mathbf{Z}$ or over a perfect field.

First, we have a divisor called "Swan conductor divisor" ([Ka$_6$]) $\mathrm{sw}(\chi) = \sum_{\mathfrak{p}} \mathrm{sw}_{\mathfrak{p}}(\chi)\, \overline{\{\mathfrak{p}\}}$ on X ($\overline{\{\mathfrak{p}\}}$ denotes the closure of $\{\mathfrak{p}\}$), where $\mathfrak{p}$ ranges over all points of codimension one. The integer $\mathrm{sw}_{\mathfrak{p}}(\chi)$ has the following properties: (i) If $\dim(X) = 1$, it coincides with the classical Swan conductor of χ at $\mathfrak{p}$. (ii) $\mathrm{sw}_{\mathfrak{p}}(\chi) > 0$ if and only if χ is wildly ramified at $\mathfrak{p}$. Furthermore, if the following condition

(C) X is regular and $D = (X - U)_{\mathrm{red}}$ is a divisor with normal crossings on X,

we have a canonical global section

$$\mathrm{rsw}(\chi) \in \Gamma(E, \mathcal{O}_E \otimes_{\mathcal{O}_X} \Omega^1_X(\log(D)) \otimes_{\mathcal{O}_X} \mathcal{O}_X(\mathrm{sw}(\chi)))$$

called the refined Swan conductor ([Ka$_7$] I). Here $\Omega^1_X(\log(D))$ is the $\mathcal{O}_X$-module on X_{et} defined by generators $\mathrm{dlog}(a)$ $(a \in j_*(\mathcal{O}_U^\times)$ with j the inclusion map $U \to X)$ subject to natural relations such as "$a\mathrm{dlog}(a)$ is additive in a", and $\mathcal{O}_X(\mathrm{sw}(\chi))$ denotes the invertible $\mathcal{O}_X$-module corresponding to the divisor $\mathrm{sw}(\chi)$. In this case, E coincides with the support of the divisor $\mathrm{sw}(\chi)$. It can be shown that the $\mathcal{O}_E$-module $\mathcal{O}_E \otimes_{\mathcal{O}_X} \Omega^1_X(\log(D))$ is locally free of rank (locally) $\dim(E)+1$. If X is of finite type over $\mathbf{Z}$, these $\mathrm{sw}(\chi)$ and $\mathrm{rsw}(\chi)$ have explicit descriptions in terms of K-theoretic class field theory ([Ka$_7$] I). We say (X, U, χ) is clean if at any point x of E, the stalk of $\mathrm{rsw}(\chi)$ at x (which is always non-zero) is a part of a basis of $\mathcal{O}_E \otimes_{\mathcal{O}_X} \Omega^1_X(\log(D)) \otimes_{\mathcal{O}_X} \mathcal{O}_X(\mathrm{sw}(\chi))$ at x. If (X, U, χ) is clean, we define a cycle $\mathrm{Char}(X, U, \chi)$ on the vector bundle $\mathcal{O}_E \otimes_{\mathcal{O}_X} \Omega^1_X(\log(D))$ over E by

$$\mathrm{Char}(X, U, \chi) = \sum_{\mathfrak{p}} \mathrm{sw}_{\mathfrak{p}}(\chi)\, \mathrm{Image}(\varphi_{\mathfrak{p}}),$$

where $\varphi_{\mathfrak{p}}$ is the map $\mathcal{O}_E \otimes_{\mathcal{O}_X} \mathcal{O}_X(-\mathrm{sw}(\chi)) \to \mathcal{O}_E \otimes_{\mathcal{O}_X} \Omega^1_X(\log(D))$ induced by $\mathrm{rsw}(\chi)$, and $\mathrm{Image}(\varphi_{\mathfrak{p}})$ is regarded as a subbundle of $\mathcal{O}_E \otimes_{\mathcal{O}_X} \Omega^1_X(\log(D))$. In the clean case, we define our generalization of the Swan conductor

$$c(X, U, \chi) \in CH_0(E)$$

to be $(-1)^{\dim(E)}$ times the intersection class of the two cycles $\mathrm{Char}(X, U, \chi)$ and the zero section of the vector bundle $\mathcal{O}_E \otimes_{\mathcal{O}_X} \Omega^1_X(\log(D))$. This $\mathrm{Char}(X, U, \chi)$ is an analogue of the charateristic cycle in the theory of $\mathcal{D}$-modules.

Now without the condition (C), I conjecture that there exists a proper birational morphism $f : X' \to X$ such that $f^{-1}(U) \xrightarrow{\cong} U$ and $(X', f^{-1}(U), \chi)$ is clean. It can be proved ([Ka$_7$] I) that if $\dim(X) \le 2$, this conjecture is true and

$$c(X, U, \chi) \stackrel{\mathrm{def}}{=} f_* c(X', f^{-1}(U), \chi) \in CH_0(E)$$

is independent of the choice of such X'.

Theorem 4.1. *Let X be a proper normal surface over an algebraically closed field k, U a regular dense open subscheme of X, ℓ a prime number different from* char(k), $\chi : \pi_1^{\mathrm{ab}}(U) \to \overline{\mathbf{F}}_\ell^\times$ *a continuous homomorphism, and $\mathscr{F}$ the $\overline{\mathbf{F}}_\ell$-sheaf on U_{et} of rank one corresponding to χ. Then,*

$$\chi(U,\mathscr{F}) = \chi(U,\overline{\mathbf{F}}_\ell) - \deg(c(X,U,\chi)).$$

Here $\chi(\;)$ denotes the Euler-Poincaré characteristic $\sum_i (-1)^i \dim_{\overline{\mathbf{F}}_l} H^i_{\mathrm{et}}(\;)$.

This theorem was generalized by T. Saito [ST$_2$] to the case $\dim(X)$ is arbitrary under a certain assumption on χ .

The Riemann-Roch formula for the Euler-Poincaré characteristics of ℓ-adic sheaves ($\ell \neq$ characteristic), which should generalize the Grothendieck-Ogg-Shafarevich formula on ℓ-adic sheaves on curves to higher dimensional varieties, was sought for first by Grothendieck. A formula in the characteristic zero case was obtained by MacPherson [Ma]. In the positive characteristic case, results for surfaces, of types different from (4.1), have been obtained by Deligne [De], Laumon [Lau] (using ideas of Deligne), S. Saito [SS$_4$]. Deligne is the first person who had the idea of the characteristic cycle of an ℓ-adic sheaf. Recently, Berthelot obtained a Riemann-Roch formula for $\mathscr{D}$-modules with Frobenius in characteristic p by defining the characteristic cycle of a $\mathscr{D}$-module with Frobenius ([Be$_2$] basing on his theory [Be$_1$]). There should be close relations between his characteristic cycle, the characteristic cycle of Deligne, and the characteristic cycle $\mathrm{Char}(X,U,\chi)$ discussed above.

Theorem 4.2. *Let A be a complete discrete valuation ring with field of fractions k and with perfect residue field F. Let X be a regular connected A-scheme which is proper flat of relative dimension one over A, U a dense open subscheme of $X_k = X \otimes_A k$, let ℓ be a prime different from* char(F), *let* $\chi : \pi_1^{\mathrm{ab}}(U) \to \overline{\mathbf{F}}_\ell^\times$ *be a continuous homomorphism, and let $\mathscr{F}$ be the $\overline{\mathbf{F}}_\ell$-sheaf of rank one on U corresponding to χ . Assume χ is at worst tamely ramified at points in $X_k - U$. (This last condition is satisfied automatically if* char$(k) = 0$*). Then*

$$\mathrm{sw}(R\Gamma((U_{\overline{k}})_{\mathrm{et}},\mathscr{F})) = \mathrm{sw}(R\Gamma((U_{\overline{k}})_{\mathrm{et}},\overline{\mathbf{F}}_\ell)) + \deg(c(X,U,\chi)).$$

Here $\mathrm{sw}(R\Gamma(\;\;)) = \sum_i (-1)^i \mathrm{sw}(H^i(\;\;))$, with sw the Swan conductor of a representation of $\mathrm{Gal}(k^{\mathrm{sep}}/k)$.

In [Bl$_2$], Bloch obtained a formula which expresses $\mathrm{sw}(R\Gamma((U_{\overline{k}})_{\mathrm{et}},\overline{\mathbf{F}}_\ell))$ in terms of the differential module $\Omega^1_{X/A}$.

Finally we discuss a conjecture of Serre. Let B be a regular local ring and let G be a finite subgroup of $\mathrm{Aut}(B)$. Assume that the following conditions (i)(ii) are satisfied.

(i) For any $\sigma \in G - \{1\}$, B/I_σ is of finite length where I_σ denotes the ideal of B generated by $\{\sigma(b) - b \,;\, b \in B\}$.

(ii) Let $A = \{b \in B \,;\, \sigma(b) = b \quad \text{for any} \quad \sigma \in G\}$. Then, A is noetherian and $A/m_A \to B/m_B$ is an isomorphism (m_* denote the maximal ideals).

Define the function $a_G : G \to \mathbf{Z}$ by

$$a_G(\sigma) = -\operatorname{length}(B/I_\sigma) \quad \text{for} \quad \sigma \in G - \{1\},$$

$$a_G(1) = -\sum_{\sigma \in G - \{1\}} a_G(\sigma).$$

If $\dim(B) = 1$, a_G is a character of a representation of G called Artin representation. Serre conjectures that a_G is a character of a representation of G even when $\dim(B) > 1$ ([Se$_1$]).

Theorem 4.3. *The conjecture is true if* $\dim(B) = 2$.

This theorem was proved in Kato, S. Saito, T. Saito [KSS] in the equal characteristic case, and in T. Saito [ST$_1$] in the mixed characteristic case under a certain assumption.

The outline of the proof of (4.3) given in [Ka$_7$] II is very similar to the proof for the one dimensional case. It is sufficient to prove that if χ is the character of a representation of G, then

$$\operatorname{Card}(G)^{-1} \sum_{\sigma \in G} a_G(\sigma)\chi(\sigma) \in \mathbf{Z}.$$

By using the two dimensional version in Bloch [Bl$_2$] of the formula (*) at the bigining of §4 and by the theory of Brauer, just as in the one dimensional case, we are reduced to the case χ is of degree one. We may assume that A is complete, A/m_A is perfect, and χ is wildly ramified. Let $X = \operatorname{Spec}(A)$, $E = \operatorname{Spec}(A/m_A) \subset X$ and $U = X - E$. Then, we can prove

$$\operatorname{Card}(G)^{-1} \sum_{\sigma \in G} a_G(\sigma)\chi(\sigma) = 1 + \deg(c(X, U, \chi)) \in \mathbf{Z}$$

where deg is the canonical isomorphism $CH_0(E) \xrightarrow{\cong} \mathbf{Z}$.

References

[Be$_1$] Berthelot, P.: Cohomologie rigid et théorie des $\mathscr{D}$-modules. Preprint

[Be$_2$] Berthelot, P.: In preparation

[Bl$_1$] Bloch, S.: Algebraic K-theory and class field theory for arithmetic surfaces. Ann. Math. **114** (1981) 229–266

[Bl$_2$] Bloch, S.: Cycles on arithmetic schemes and Euler characteristics of curves. Proc. Symp. Pure Math. AMS **46** Part 2 (1987) 421–450

[CSS] Colliot-Thélène, J.-L., Sansuc, J.-J., Soulé, C.: Torsion dans le groupe de Chow de codimension deux. Duke Math. J. **50** (1983) 763–801

[Co] Coombes, K. R.: Local class field theory for curves. Contemp. Math. **55** I (1983) 117–134

[De] Deligne, P.: Letter to Illusie, Nov. 4, 1976

[FM] Fontaine, J. M., Messing, W.: p-adic periods and p-adic etale cohomology. Contemp. Math. **67** (1987) 179–207

[Fu] Fulton, W.: Intersection theory. Springer, Berlin Heidelberg New York 1984

[Gr] Gros, M.: Sur la partie p-primaire du groupe de Chow de codimension deux. Comm. Alg. **13** (1985) 2407–2420

[Ha] Hazewinkel, M.: Corps de classes local, Appendix to Demazure, M., Gabriel, P.: Groupes algébriques, North-Holland Publ. 1970, pp. 648-674

[Ih] Ihara, Y.: On a problem on some complete p-adic function fields (in Japanese). Kokuroku 41, RIMS, Kyoto Univ. (1968) 7–17

[Il] Illusie, L.: Complexe de de Rham-Witt et cohomologie cristalline. Ann. Sci. Ec. Norm. Sup. **12** (1976) 501–661

[Ka_1] Kato, K.: A generalization of local class field theory by using K-groups, I., J. Fac. Sci. Univ. of Tokyo, Sec. IA, **26** (1979) 303–376; II, ibid. **27** (1980) 603–683; III, ibid. **29** (1982) 31–43

[Ka_2] Kato, K.: Class field theory and algebraic K-theory. (Lecture Notes in Mathematics, vol. 1016.) Springer, Berlin Heidelberg New York, pp. 109–127

[Ka_3] Kato, K.: Duality theories for p-primary etale cohomology. Algebraic and topological theories. I, Kinonuniya (1985) 127–148; II, Comp. Math. **63** (1987) 259–270; III. In preparation

[Ka_4] Kato, K.: The explicit reciprocity law and the cohomology of Fontaine-Messing. Preprint

[Ka_5] Kato, K.: p-adic Hodge theory and values of zeta functions of elliptic cusp forms. In preparation

[Ka_6] Kato, K.: Swan conductors for characters of degree one in the imperfect residue field case. Contemp. Math. **83** (1989) 101–132

[Ka_7] Kato, K.: Class field theory, $\mathscr{D}$-modules, and ramification on higher dimensional schemes. I, to appear in Amer. J. Math.; II. In preparation

[KL] Katz, N., Lang, S.: Finiteness theorems in geometric class field theory. l'Enseignement Mathématique **27** (1981) 285–319

[KS_1] Kato, K., Saito, S.: Unramified class field theory of arithmetical surfaces. Ann. Math. **118** (1983) 241–275

[KS_2] Kato, K., Saito, S.: Global class field theory for arithmetic schemes. Contemp. Math. **5** Part I (1986) 255–330

[KS_3] Kato, K., Saito, S.: In preparation

[KSS] Kato, K., Saito, S., Saito, T.: Artin characters for algebraic surfaces. Amer. J. Math. **110** (1988) 49–76

[Lan] Lang, S.: Unramified class field theory over function fields in several variables. Ann. Math. **64** (1956) 286–325

[Lau] Laumon, G.: Caractéristique d'Euler Poincaré des faisceaux constructibles sur une surface. Asterisque **101–102** (1983) 193–207

[Ma] MacPherson, R.: Chern classes of singular varieties. Ann. Math **100** (1974) 423–432

[Mi] Milnor, J.: Algebraic K-theory and quadratic forms. Inv. math. **9** (1970) 318–344

[Ni] Nisnevich, A.: Arithmetic and cohomology invariants of semi-simple group schemes and compactifications of locally symmetric spaces. Funct. Anal. Appl. **14**, no. 1 (1980) 61–62

[Pa_1] Parsin, A. N.: Abelian coverings of arithmetic schemes. Sov. Math. Dokl. **19**, no. 6 (1978) 1438–1442

[Pa_2] Parsin, A. N.: Local class field theory. Proc. Steklov Inst. Math. **165** (1985) 157–185

[Se_1] Serre, J.-P.: Sur la rationalité des représentations d'Artin. Ann. Math. **72** (1960) 406–420

[Se_2] Serre, J.-P.: Sur les corps locaux à corps résiduel algébriquement clos. Bull. Soc. Math. France **89** (1961) 105–154

[SS_1] Saito, S.: Unramified class field theory for arithmetic schemes. Ann. Math. **121** (1985) 251–281

[SS$_2$] Saito, S.: Class field theory for curves over local fields. J. Number Theory **21** (1985) 44–80

[SS$_3$] Saito, S.: Class field theory for two dimensional local rings. Adv. Stud. Pure Math. **12** (1987) 343–373

[SS$_4$] Saito, S.: General fixed point formula for an algebraic surfaces and the theory of Swan represenations for two dimensional local rings. Amer. J. Math. **109** (1987) 1009–1042

[ST$_1$] Saito, T.: Bloch's 0-cycles and Artin characters of arithmetic surfaces. Preprint (1986)

[ST$_2$] Saito, T.: The Euler numbers of ℓ-adic sheaves of rank 1 in positive characteristic. Preprint

[Vo] Vostokov, S. V.: Explicit construction of class field theory for a multidimensional local field. Math. USSR Izv. **26** (1986) 263–287

[VK] Vostokov, S. V., Kirillov A. N.: Normed pairing in a two dimensional local field. J. Sov. Math. **30**, no. 1 (1985) 1847–1853

On the Mordell-Weil Group and the Shafarevich-Tate Group of Modular Elliptic Curves

*Victor Alecsandrovich Kolyvagin**

Steclov Mathematical Institute, u. Vavilova 42, 117966 Moscow GSP-1, USSR

The main purpose of this paper is to describe some recent results pertaining to the diophantine analysis of elliptic curves. A new element is an extension of the set of explicit cohomology classes, see Section 2.

1. The Conjecture of Birch and Swinnerton-Dyer and the Hypothesis of Finiteness of the Shafarevich-Tate Group

Let E be an elliptic curve defined over the field of rational numbers $\mathbb{Q}$, for example, by its Weierstrass equation $y^2 = 4x^3 - g_2x - g_3$. Let R be a finite extension of $\mathbb{Q}$. We are interested in the group $E(R)$ called the Mordell-Weil group of E over R and the Shafarevich-Tate group $Ш(R, E)$. The group $Ш(R, E)$ is, by definition, $\ker(H^1(R, E) \to \prod_v H^1(R(v), E))$, where v runs through the set of all places (equivalence classes of valuations) of R, $R(v)$ is the v-adic completion of R. For an arbitrary extension L of $\mathbb{Q}$, we let $\bar{L}$ denote an algebraic closure of L. If V/L is a Galois extension, then $G(V/L)$ denotes its Galois group, and $H^1(L, E) = H^1(G(\bar{L}/L), E(\bar{L}))$.

Let Y be some set of algebraic curves over R. By definition, the Hasse principle holds for Y, if for all $X \in Y$ one has: $X(R)$ is nonempty $\Leftrightarrow X(R(v))$ is nonempty for each v. The group $Ш(R, E)$ is the obstacle to the Hasse principle for the set $Y(R, E)$ of main principal homogeneous spaces over E defined over R. In particular, the Hasse principle holds for $Y(R, E)$ if and only if the group $Ш(R, E)$ is trivial.

According to the Mordell-Weil theorem, $E(R) \simeq F \times \mathbb{Z}^{r(R,E)}$, where $F \simeq E(R)_{\text{tor}}$ is a finite group, and $r(R, E)$ is a nonnegative integer called the rank of E over R. Concerning the group $Ш(R, E)$, it is conjectured that it is finite. In general, it is known that $Ш(R, E)$ is a torsion group (being a subgroup of the torsion group $H^1(R, E)$) and for a natural number M its subgroup $Ш(R, E)_M$ is finite. If A is an abelian group, we let A_M denote its subgroup of all elements of exponents M. Only recently in works of Rubin and the author, the finiteness of $Ш(R, E)$ was proved for some E and R. We shall discuss these results later.

---

* This work was partly prepared during a visit of the author at the Max-Planck-Institute für Mathematik at Bonn. He wishes to express his gratitude for the support and hospitality provided by this institute.

Proceedings of the International Congress of Mathematicians, Kyoto, Japan, 1990

The elements of $E(R)_{\mathrm{tor}}$ can be effectively calculated. For example, let R be $\mathbb{Q}$ and let E be defined by an equation $u^2 = w^3 + \alpha w + \beta$, where $\alpha, \beta \in \mathbb{Z}$, $\delta = 4\alpha^3 + 27\beta^2 \neq 0$ (this is always possible). According to the Nagell-Lutz theorem, if $P \in E(\mathbb{Q})_{\mathrm{tor}}$ is nonzero, then $u(P) = 0$ or $u(P)^2 | \delta$. Mazur determined all possible types of $E(\mathbb{Q})_{\mathrm{tor}}$, in particular, $[E(\mathbb{Q})_{\mathrm{tor}}] \leq 16$.

We are interested here in the case $R = \mathbb{Q}$. No algorithm is known in general for calculating $r(\mathbb{Q}, E)$ and generators of $E(\mathbb{Q})/E(\mathbb{Q})_{\mathrm{tor}}$. But recently here and in the study of $\text{Ш}(R, E)$ essential progress was made.

More specifically, it is connected to advances towards proving the Birch-Swinnerton-Dyer conjecture (BSD) which predicts a connection between the arithmetic of E and its L-function.

We let $L(E, s)$ denote the L-function of E over $\mathbb{Q}$, defined for $\mathrm{Re}(s) > 3/2$ as

$$\prod_q L_q(E, s) = \sum_{n=1}^{\infty} a_n n^{-s}, \qquad a_n \in \mathbb{Z}.$$

Here q runs through the set of rational primes. Let $N \in \mathbb{N}$ be the conductor of E. If $(q, N) = 1$, then $L_q(E, s) = (1 - a_q q^{-s} + q^{1-2s})^{-1}$, where $a_q = q + 1 - [\tilde{E}(\mathbb{Z}/q\mathbb{Z})]$, $\tilde{E}$ being the reduction of E modulo q (E has the good reduction at q). If $q | N$, then $L_q(E, s) = 1, (1 \pm q^{-s})^{-1}$ depending on the type of bad reduction of E at q.

Assume that E is modular, that is there exists a weak Weil parametrization $\gamma : X_0(N) \to E$ [12]. Here $X_0(N)$ is the modular algebraic curve over $\mathbb{Q}$ parametrizing classes of isogenies of elliptic curves with cyclic kernel of order N. According to the Taniyama-Shimura-Weil conjecture, every elliptic curve over $\mathbb{Q}$ is modular. Then $L(E, s)$ has an analytic continuation to an entire function on the complex plane which satisfies a functional equation

$$Z(E, 2 - s) = \varepsilon Z(E, s) \tag{1}$$

where $Z(E, s) = (2\pi)^{-s} N^{s/2} \Gamma(s) L(E, s)$ and $\varepsilon = \pm 1$ depends on E.

An analogeous L-function $L(R, E, s)$ of E over R can be defined (its definition is essential for us only up to a finite product of Euler factors), having analogous properties. We let $ar(R, E)$ denote the oder of vanishing $L(R, E, s)$ at $s = 1$. According to BSD, one conjectures the identity:

$$r(R, E) = ar(R, E). \tag{2}$$

Moreover BSD connects the first nonzero coefficient of the expansion of $L(R, E, s)$ around $s = 1$ with the order of $\text{Ш}(R, E)$ (using the hypothesis that $\text{Ш}(R, E)$ is finite) and other parameters of E, but we do not go into this here.

In the sequel we will omit the letter $\mathbb{Q}$ in the notations $\text{Ш}(\mathbb{Q}, E)$, $r(\mathbb{Q}, E)$, $ar(\mathbb{Q}, E)$. It follows from (1) that $ar(E)$ is even when $\varepsilon = 1$, $ar(E)$ is odd when $\varepsilon = -1$. E is called even or odd, respectively.

For $R = \mathbb{Q}$ the current state of conjecture (2) and of the hypothesis of finiteness of $\text{Ш}(E)$ is expressed by the result:

Theorem 1. *The equality $r(E) = ar(E)$ holds and $\text{Ш}(E)$ is finite if $ar(E) \leq 1$.*

We remark that empirical material shows that curves with $ar(E) > 1$ compose a relatively small part in the set of all curves. Apparently (taking into account the Taniyama-Shimura-Weil conjecture), Theorem 1 covers a substantial part of all elliptic curves over $\mathbb{Q}$.

Further we discuss a scheme of the proof of Theorem 1, formulate earlier results and give some examples.

Let D be a fundamental discriminant of the imaginary-quadratic field $K = \mathbb{Q}(\sqrt{D})$ such that $D \equiv \square (\text{mod } 4N)$, $D \neq -3, -4$. As E is modular, there exists the Heegner point $P_D \in E(K)$ (which will be defined later), it satisfies the condition:

$$\sigma e P_D = -\varepsilon e P_D \tag{3}$$

where $e =$ exponent of $E(\mathbb{Q})_{\text{tor}}$, σ is the generator of $G(K/\mathbb{Q})$. The author proved [6]–[8]:

Theorem 2. *The equality* $r(E) = ar(E)$ *holds and* Ш(E) *is finite if* 1) $ar(E) \leq 1$, 2) $\exists D | P_D$ *has infinite order.*

From the Gross and Zagier results [5] it follows

Theorem 3. *If* $(D, 2N) = 1$, *then* $ar(K, E) \geq 1$, $ar(K, E) = 1 \Leftrightarrow P_D$ *has infinite order.*

Waldspurger [21] for $ar(E) = 1$ and, independently, Bump, Friedberg, Hoffstein [2] and M. Murty, V.K. Murty [14] for $ar(E) = 0$ proved

Theorem 4. *If* $ar(E) \leq 1$, *then* $(D, 2N) = 1$ *and* $ar(K, E) = 1$ *for an infinite set of values of* D.

So from Theorems 3, 4 it then follows that condition 2) in Theorem 2 follows from condition 1), that is Theorem 2 is equivalent to Theorem 1.

From (1) we have that $ar(E) = 0 \Rightarrow \varepsilon = 1$, $ar(E) = 1 \Rightarrow \varepsilon = -1$. Using (3), we deduce from the conditions: P_D has infinite order, $r(K, E) = 1$, and $ar(E) \leq 1$, that $r(E) = ar(E)$. The kernel of the natural homomorphism Ш$(E) \to$ Ш(K, E) is Ш$(E) \cap H^1(G(K/\mathbb{Q}), E(K)) \subset$ Ш$(E)_2$ which is a finite group.

Thus Theorem 2 is a consequence of the author's result [8]:

Theorem 5. *The equality* $r(K, E) = 1$ *holds, and* Ш(K, E) *is finite, if* P_D *has infinite order.*

We note that Theorems 5, 3 give (1) for $R = K$ when $ar(K, E) = 1$. The inequality $r(E) \geq 1$ when $ar(E) = 1$ follows already from Theorem 3 and Waldspurger's result.

A subclass in the class of modular elliptic curves is formed by elliptic curves with complex multiplication: $\text{End}(E) \neq \mathbb{Z}$ and then $\text{End}(E)$ is an order with class number one of an imaginary-quadratic extension k of $\mathbb{Q}$. We let W' denote this subclass. The modular invariant $j = g_2^3/(g_2^3 - 27g_3^2)$, which runs through all rational numbers on the set of elliptic curves over $\mathbb{Q}$, takes on 13 values on the set W'.

The specific property of a curve from W' is the possibility to use, in studying it, the theory of abelian extensions of k because $E(\mathbb{Q})_{\text{tor}} \subset E(k^{ab})$ for $E \in W'$. In particular, by using so called elliptic units, Coates and Wiles [3] proved (2) for $E \in W'$, $ar(E) \neq 0$. Recently Rubin [17], also using elliptic units (we will come back to this later), proved under the same condition that Ш(E) is finite. This gave the first examples of finite groups Ш(E). Moreover he proved that, for $E \in W'$, $ar(E) = 1 \Rightarrow r(E) \leq 1$.

2. Explicit Cohomology Classes

Now we discuss briefly the method of proof of Theorem 5.

For an arbitrary extension L of $\mathbb{Q}$ the exact sequence $0 \to \mathrm{E}_M \to \mathrm{E}(\bar{L}) \to E(\bar{L}) \to 0$ $(E_M = E(\bar{\mathbb{Q}})_M)$ induces the exact sequence

$$0 \to E(L)/ME(L) \to H^1(L, E_M) \to H^1(L, E)_M \to 0. \tag{4}$$

The Selmer group $S_M(R, E)$, by definition, is the subgroup of $H^1(R, E_M)$ consisting of elements whose image in $H^1(R(v), E_M)$ lies in $E(R(v))/ME(R(v))$ for all places v of R. In particular, (4) induces the exact sequence

$$0 \to E(R)/ME(R) \to S_M(R, E) \to \text{Ш}\,(R, E)_M \to 0. \tag{5}$$

It is known (the weak Mordell-Weil theorem) that $S_M(R, E)$ is a finite M-torsion group. In particular, Ш$(R, E)_M$ is a finite group as we remarked before.

Let $R = K$. If $P = P_D$ has infinite order, then we define $C = C_D$ to be the maximal natural number dividing the image of P in $E(K)/E(K)_{\text{tor}} \simeq \mathbb{Z}^{r(K,E)}$. We let $C = 0$ if $P \in E(K)_{\text{tor}}$. Thus P has infinite order $\Leftrightarrow C \neq 0$. We let S'_M denote the factor group of $S_M(K, E)$ modulo the subgroup generated by P. Taking into account (5) and the Mordell-Weil theorem: $E(K) \simeq F \times \mathbb{Z}^{r(K,E)}$, with F finite, Theorem 5 will follow from the existence of $C' \in \mathbb{N}$ such that $C'S'_M = 0\, \forall\, M \in \mathbb{N}$.

The non-degenerate alternating Weil pairing $[\ ,\ ]_M : E_M \times E_M \to \mu_M = \bar{\mathbb{Q}}^*_M$ induces a pairing

$$\langle\ ,\ \rangle_{M,v} : H^1(K(v), E_M) \times H^1(K(v), E_M) \to H^2(K(v), \mu_M).$$

For $v = \infty$ the field $K(\infty) \simeq \mathbb{C}$ and the corresponding cohomology groups are trivial. For $v \neq \infty$ the group $H^2(K(v), \mu_M)$ is identified canonically with $\mathbb{Z}/M\mathbb{Z}$ by local class field theory. If $a, b \in H^1(K, E_M)$, then $\langle a, b\rangle_{M,v} \stackrel{\text{def}}{=} \langle a(v), b(v)\rangle_{M,v}$, where $a(v)$, $b(v)$ are the localizations of a, b. According to global class field theory (the reciprocity law) $\langle a, b\rangle_{M,v} \neq 0$ only for a finite set of places v and the following relation holds:

$$\sum_{v \neq 0} \langle a, b\rangle_{M,v} = 0. \tag{6}$$

Relation (6) can be considered as a condition on a if an element b is fixed. To use (6) for the study of $S_M(K, E)$ it is necessary to find explicit elements b. This was my strategy. Thus I constructed a set T of explicit elements of $H^1(K, E_M)$ by using Heegner points over ring class fields of K. The special properties of these elements allowed to deduce from (6) with $a \in S_M(K, E)$ and $b \in T$ the relation $C'S'_M = 0$ for some $C' \in \mathbb{N}$, the divisor and main component of which is C.

Now we describe the construction of an element from T. First we define the Heegner points. Fix an ideal i in the ring of integers O of K such that $O/i \simeq \mathbb{Z}/N\mathbb{Z}$ (i exists in view of the assumptions on D). If $\lambda \in \mathbb{N}$, then K_λ denotes the ring class field of K of conductor λ. It is a finite abelian extension of K. Let O_λ be $\mathbb{Z} + \lambda O$, $i_\lambda = i \cap O_\lambda$. If $(\lambda, N) = 1$, we define the point $z_\lambda \in X_0(N)(K_\lambda)$ as corresponding to the class of the isogeny $\mathbb{C}/O_\lambda \to \mathbb{C}/i_\lambda^{-1}$, where i_λ^{-1} is the inverse of i_λ in the group of proper O_λ-ideals. We let $y_\lambda = \gamma(z_\lambda) \in E(K_\lambda)$, $P = P_D =$ the norm of y_1 from K_1 to K. The points y_λ, P are called Heegner points (corresponding to the parametrization $\gamma: X_o(N) \to E$, $K = \mathbb{Q}(\sqrt{D})$ and i).

We use the notation p (or p with (a subscript) for rational primes which do not divide N and remain prime in K. We let Λ^r denote the set of all products $p_1 \ldots p_r$ with distinct p_m, $\Lambda = \bigcup_{n=1}^{\infty} \Lambda^r$.

Let $\lambda \in \Lambda$, $G_\lambda = G(K_\lambda/K_1)$. The group G_λ is the direct product of the subgroups $G_{\lambda,p} = G(K_\lambda/K_{\lambda/p})$ for $p|\lambda$. The natural homomorphism $G_{\lambda,p} \to G_p$ is an isomorphism. The group G_p is isomorphic to the group $\mathbb{Z}/(p+1)\mathbb{Z}$. For each p, we fix a generator $t_p \in G_p$; $t_p \in G_{\lambda,p}$ denotes the corresponding generator of $G_{\lambda,p}$. We let $\mathrm{Tr}_p = \sum_{j=0}^{p} t_p^j$. Recall that $\sum_{n=1}^{\infty} a_n n^{-s} = L(E, s)$ for $\mathrm{Re}(s) > 3/2$. For $p|\lambda$ one finds the relation:

$$\mathrm{Tr}_p y_\lambda = a_p y_{\lambda/p}. \tag{7}$$

The relations (7) are the basis for the definition of explicit cohomology classes.

Let Δ_λ denote the ring $\mathbb{Z}[G_\lambda]$. We define a Δ_λ-module B_λ in the following way. Let F_λ be the direct sum $\sum_{\eta|\lambda} \Delta_q$, where G_λ acts on Δ_η by the natural homomorphism $\Delta_\lambda \to \Delta_\eta$. Let 1_η denote the unit of Δ_η, H_λ be the Δ_λ-submodule of F_λ generated by the elements $\mathrm{Tr}_p 1_\eta - a_p 1_{\eta/p}$ for all $p|\eta|\lambda$. Then $B_\lambda = F_\lambda/H_\lambda$.

It is not difficult to prove that $(B_\lambda)_{\mathrm{tor}} = 0$. Let $1'_\eta$ be the image of 1_η in B_λ, then $\{1'_\eta, \eta|\lambda\}$ is a system of generators of B_λ over Δ_λ. By (7) $\exists!$ homomorphism $\varphi: B_\lambda \to E(K_\lambda)$ such that $1'_\eta \to y_\eta$. We let $I_p = -\sum_{j=1}^{p} j t_p^j \in \Delta_\lambda$, $I_\lambda = \prod_{p|\lambda} I_p$. Let Q_λ be the element $I_\lambda 1'_\lambda$.

For $M \in \mathbb{N}$ we define $\Lambda(M)$ as the subset of Λ consisting of elements λ such that $M|(p+1)$, $M|a_p \, \forall\, p|\lambda$. Further, $\Lambda^r(M) = \Lambda^r \cap \Lambda(M)$. We claim that $(1-g)Q_\lambda \in MB_\lambda$ for $\lambda \in \Lambda(M)$ and $g \in G_\lambda$. It is enough to verify this for $g = t_p$, where $p|\lambda$. It is clear that

$$(1 - t_p)I_p = \mathrm{Tr}_p - (p + 1). \tag{8}$$

Thus, we have $(1 - t_p)Q_\lambda = I_{\lambda/p}(1 - t_p)I_p 1'_\lambda = I_{\lambda/p}(\mathrm{Tr}_p - (p+1))1'_\lambda = I_{\lambda/p}(a_p 1'_{\lambda/p} - (p+1)1'_\lambda) \in MB_\lambda$.

As $(B_\lambda)_{\mathrm{tor}} = 0$, there exists a unique element $((1-g)Q_\lambda)/M \in B_\lambda$. We define the element $\tau'_\lambda(M) \in H^1(K_1, E_M)$ to be the class of the cocycle:

$$\psi: g \mapsto (g - 1)(\varphi(Q_\lambda)/M) + \varphi(((1 - g)Q_\lambda)/M),$$

where $g \in G(\bar{K}_1/K_1)$. The element $\tau_\lambda(M) \in H^1(K, E_M)$ we define as the corestriction of $\tau'_\lambda(M)$. We call T the set $\{\tau_\lambda(M), M \in \mathbb{N}, \lambda \in \Lambda(M)\}$.

Let (b) denote the image of $b \in H^1(K, E_M)$ in $H^1(K, E)_M$, $c_\lambda(M) = (\tau_\lambda(M))$. That is, $c_\lambda(M)$ is the corestriction of the element of $H^1(K_1, E)_M$ defined by the cocycle, $g \mapsto \varphi((1-g)Q_\lambda)/M)$. If $\lambda \in \Lambda^r(M)$, then the automorphism $\sigma \in G(K/\mathbb{Q})$ acts on $c_\lambda(M)$ by multiplication by $(-1)^{r+1}\varepsilon$. The symbol $\langle a, b\rangle_{M,v}$ depends only on (b), if $a \in S_M(K, E)$.

The elements $c_p(M)$ were defined first, see [6]. This allowed to prove the relation $C'(\sigma + \varepsilon)S_M(K, E) = 0$, which is equivalent to the finiteness of $E(\mathbb{Q})$ and Ш(E) when $\varepsilon = 1$, and to the finiteness of $E_{(D)}(\mathbb{Q})$ and Ш$(E_{(D)})$ when $\varepsilon = -1$. Here $E_{(D)}$ is the elliptic curve (the form of E over K) defined by the equation $Dy^2 = 4x^3 - g_2x - g_3$.

In [8] there were defined elements $\tau_\lambda(M)$ for some subset of the set $\{M \in \mathbb{N}, \lambda \in \Lambda(M)\}$ containing the set $\{M | (M, d) = 1, \lambda \in \Lambda(M)\}$, where $d =$ exponent of $E(\mathbb{K})_{\mathrm{tor}}$, $\mathbb{K}$ is the composite of the $K_{\lambda'}$ for $\lambda' \in \Lambda$. By using here the modules B_λ and the property $(B_\lambda)_{\mathrm{tor}} = 0$ we shake off the additional restrictions on (M, λ) when $(M, d) > 1$. The relation (6) with $(b) = c_\lambda(M)$ when $\lambda \in \Lambda^r(M)$, $r \leq 2$, allowed to prove the relation $C'S'_M = 0$.

We note that an application of the elements $\tau_\lambda(M)$ when $\lambda \in \Lambda^r$ with arbitrary $r \geq 0$ allowed in [8] to pass from a relation of the type C Ш$(K, E) = 0$ to a relation of the type $[$Ш$(K, E)] | C^2$. Because of the existence on Ш(K, E) of a non-degenerate (as Ш(K, E) is finite) alternate Cassels pairing with values in $\mathbb{Q}/\mathbb{Z}$, it then follows that the second relation implies the first relation.

In [20] Thaine used the cyclotomic units for a new proof of annihilating relations in the ideal class groups of real abelian extensions of $\mathbb{Q}$. Rubin [16] adapted Thaine's approach, using elliptic units instead of cyclotomic units, for proving annihilating relations in the ideal class groups of abelian extensions of the imaginary-quadratic field $k = \mathrm{End}(E) \otimes \mathbb{Q}$ when $E \in W'$. By using the natural connection between ideal class groups and the Selmer group $S_M(\mathbb{Q}, E)$ Rubin proved an universal annihilating relation for $S_M(\mathbb{Q}, E)$ by the condition that $ar(E) = 0$.

A comparison of the approaches of Thaine [20] and of the author [6] for proving annihilating relations in the ideal class groups and in the Selmer groups, respectively, suggested the possibility in [7] of combining them into a single general framework. A further step was a construction and use in [8] of sets of cohomology classes of the type T, both in the theory of modular elliptic curves and in the theory of ideal class groups of abelian extensions of $\mathbb{Q}$ or an imaginary-quadratic extension of $\mathbb{Q}$. For information on this theory and some further applications we refer to the papers [8, 1, 4, 9, 10, 11, 13, 15, 18, 19].

3. Examples

Example 1 (Rubin [17]). For the curves with complex multiplication ($k = \mathbb{Q}(\sqrt{-1})$) $y^2 = x^3 - x$, $y^2 = x^3 + 17x$ we have: $r(E) = ar(E) = 0$, Ш$(E) = 0$, $\mathbb{Z}/2\mathbb{Z} + \mathbb{Z}/2\mathbb{Z}$, respectively.

Example 2 (Kolyvagin [7]). Let $E : y^2 = 4x^3 - 4x + 1$. It is an odd modular curve without complex multiplication, of conductor $N = 37$. Let $(D, 2N) = 1$. The curves $E_{(D)}$:

$$Dy^2 = 4x^3 - 4x + 1 \tag{9}$$

are even and have no complex multiplication. For computation of $L(E_{(D)}, 1)$ and C_D the following identity can be used:

$$L(E_{(D)}, 1) = 2\sum_{n=1}^{\infty} \frac{a_n}{n}\left(\frac{D}{n}\right)\exp(-2\pi n/(|D|\sqrt{37})) = (2\Omega_-/\sqrt{D})C_D^2 \tag{10}$$

where Ω_--the imaginary period of E, $\left(\frac{D}{n}\right)$-the Legendre symbol. See [22] for (10); the connection between $L(E_{(D)}, 1)$ and C_D is a consequence of the results of Gross and Zagier [5].

Let $L(E_{(D)}, 1) \neq 0$ or, equivalently, $C_D \neq 0$. Then $E_{(D)}(\mathbb{Q})$ is finite and, moreover, is trivial because always $E_{(D)}(\mathbb{Q})_{\text{tor}} = 0$. That is equation (9) has no solutions in rational numbers. Further, $Ш(E_{(D)})$ is finite and $C_D Ш(E_{(D)}) = 0$. For example, if $D = -7, -11$ then $C_D = 1$, so $Ш(E_{(D)}) = 0$. See [7] for further information on this example.

We recall that $C_D \neq 0$ for an infinite set of values of D according to a result of Waldspurger.

It is a classical fact tht $E(\mathbb{Q}) \simeq \mathbb{Z}$ is generated by the point ($y = 1, x = 0$). Of course, $ar(E) = 1$, see [22], for example. The author proved [8] that $Ш(E) = 0$.

References

1. Bertolini, M., Darmon, H.: Kolyvagin's descent and Mordell-Weil groups over ring class fields. J. Reine Angew. Math. **412** (1990) 63–74
2. Bump, D., Friedberg, S., Hoffstein, J.: A non vanishing theorem for derivatives of automorphic L-functions with applications to elliptic curves. Bull. Amer. Math. Soc. **21** (1989) 89–93
3. Coates, J., Wiles, A.: On the conjecture of Birch and Swinnerton-Dyer. Invent. math. **39** (1977) 223–251
4. Gross, B.H.: Kolyvagin's work on modular elliptic curves. Proceedings of Durham Conference on L-functions and Arithmetic, 1989. Cambridge University Press (to appear)
5. Gross, B.H., Zagier, D.B.: Heegner points and derivatives of L-series. Invent. math. **84** (1986) 225–320
6. Kolyvagin, V.A.: Finiteness of $E(\mathbb{Q})$ and $Ш(E, \mathbb{Q})$ for a subclass of Weil curves. Izv. Akad. Naùk SSSR, Ser. Mat. **52** (1988) 522–540 [English transl.: Math USSR Izv. **32** (1989) 523–542]
7. Kolyvagin, V.A.: On the Mordell-Weil and Shafarevich-Tate groups for Weil elliptic curves. Izv. SSSR, Ser. Mat. **52** (1988) 1154–1180 [English transl.: Math USSR Izv. **33** (1989) 474–499]
8. Kolyvagin, V.A.: Euler systems. The Grothendieck Festschrift, vol. 2. Progr. in Math., vol. 87. Boston: Birkhäuser 1991, pp. 435–483
9. Kolyvagin, V.A., Logachev, D.Y.: Finiteness of the Shafarevich-Tate group and the group of rational points for some modular abelian varieties. Algebra and Analysis 1, no. 5 (1989)
10. Kolyvagin, V.A.: On the structure of Shafarevich-Tate groups. Proceedings of USA-USSR Symposium on Algebraic Geometry, Chicago, 1989. (Lecture Notes in Mathematics.) Springer, Berlin Heidelberg New York (to appear)
11. Kolyvagin, V.A.: On the structure of Selmer groups. (1990). Math. Ann. (to appear)

12. Mazur, B., Swinnerton-Dyer, H.P.F.: Arithmetic of Weil curves. Invent. math. **25** (1974) 1–61
13. McCallum, W.G.: Kolyvagin's work on Shafarevich-Tate groups. Proceedings of Durham Conference on *L*-functions and Arithmetic, 1989. Cambridge University Press (to appear)
14. Murty, M.R., Murty, V.K.: Mean values of derivatives of modular *L*-series. (1989). Ann. Math. (to appear)
15. Perrin-Riou, B.: Travaux de Kolyvagin et Rubin. Séminaire Bourbaki 717 (1989/1990)
16. Rubin, K.: Global units and ideal class groups. Invent. math. **89** (1987) 511–526
17. Rubin, K.: Tate-Shafarevich group and *L*-functions of elliptic curves with complex multiplications. Invent. math. **89** (1987) 527–560
18. Rubin, K.: The Main Conjecture. Appendix to: Cyclotomic fields I–II by S. Lang. (Graduate Texts in Mathematics, vol. 121.) Springer, New York Berlin Heidelberg 1990, 2nd edn., pp. 397–419
19. Rubin, K.: The "main conjectures" of Iwasawa theory for imaginary quadratic fields. Invent. math. **103** (1991) 25–68
20. Thaine, F.: On the ideal class groups of real abelian extensions of $\mathbb{Q}$. Ann. Math. **128** (1988) 1–18
21. Waldspurger, J.-L.: Sur les valeurs de certaines fonctions *L* automorphes en leur centre de symmetrie. Comp. Math. **54** (1985) 173–242
22. Zagier, D.B.: Modular points, modular curves, modular surfaces and modular forms. (Lecture Notes in Mathematics, vol. 1111.) Springer, Berlin Heidelberg New York 1985, pp. 225–246

La Transformation de Fourier Géométrique et ses Applications

Gérard Laumon

Université Paris-Sud, URA D 0752, Bâtiment 425, F-91405 Orsay Cedex, France

1. La transformation de Fourier géométrique

[Br, De1, Il1, Ka1, Ka-La, La1]

Dans tout cet exposé, on fixe deux nombres premiers distincts p et ℓ, une clôture algébrique $\overline{\mathbb{Q}}_\ell$ du corps $\mathbb{Q}_\ell$ des nombres ℓ-adiques et un caractère additif non trivial $\psi : \mathbb{F}_p \longrightarrow \overline{\mathbb{Q}}_\ell^\times$. On fixe aussi un corps algébriquement clos k de caractéristique p. On désignera par q une puissance de p ($q = p^f$, $f \in \mathbb{N}^\times$), par $\mathbb{F}_q$ l'unique sous-corps de k à q éléments et par $\mathrm{Frob}_q \in \mathrm{Gal}(k/\mathbb{F}_q)$ l'inverse de l'élément de Frobenius ($\mathrm{Frob}_q(\alpha)^q = \alpha$, $\forall \alpha \in k$). On utilisera librement le formalisme des 6 opérations de Grothendieck entre les catégories dérivées $D_c^b(X, \overline{\mathbb{Q}}_\ell)$ (X un k-schéma de type fini) ainsi que le formalisme des t-structures (cf. [SGA4, SGA5, SGA7, B-B-D, De2 et Ek]).

Soient S un k-schéma de type fini, $\pi : E \longrightarrow S$ un fibré vectoriel de rang constant d, $\pi' : E' \longrightarrow S$ le fibré dual, $\mathrm{pr} : E' \times_S E \longrightarrow E$, $\mathrm{pr}' : E' \times_S E \longrightarrow E'$ les projections canoniques et $\langle , \rangle : E' \times_S E \longrightarrow \mathbb{A}_S^1$ l'accouplement de dualité. Sur $\mathbb{A}_S^1 = S[u]$, on a le revêtement d'Artin-Schreier

$$S[v] \longrightarrow S[u], \; v \longmapsto u = v^p - v$$

qui est étale, galoisien, de groupe de Galois $\mathbb{F}_p$. Si l'on pousse ce $\mathbb{F}_p$-torseur par le caractère ψ^{-1} on obtient un $\overline{\mathbb{Q}}_\ell$-faisceau lisse de rang 1 (un "$\overline{\mathbb{Q}}_\ell^\times$-torseur") sur $\mathbb{A}_S^1$, noté $\mathscr{L}_\psi$. La *transformation de Fourier géométrique*, inventée par Deligne, est l'opération

$$\mathscr{F}_\psi : D_c^b(E, \overline{\mathbb{Q}}_\ell) \longrightarrow D_c^b(E', \overline{\mathbb{Q}}_\ell),$$

$$\mathscr{F}_\psi(-) = R\,\mathrm{pr}'_!(\mathrm{pr}^*(-) \overset{L}{\otimes} \langle , \rangle^* \mathscr{L}_\psi)[d].$$

Théorème 1.1 (Deligne). *(i) $\mathscr{F}_\psi$ est une équivalence de catégories triangulées, de quasi-inverse $\mathscr{F}'_\psi(-)(d)$, où $\mathscr{F}'_\psi$ est la transformation de Fourier géométrique pour $\pi' : E' \longrightarrow S$ (on a identifié E'' à E par $e \longmapsto (e' \longmapsto -\langle e', e \rangle)$).*

(ii) Si $E = E_1 \times_S E_2$ pour deux fibrés vectoriels E_1 et E_2 de rang constant sur S, on a

$$\mathscr{F}_\psi\Big(\mathrm{pr}_1^*(-) \overset{L}{\otimes} \mathrm{pr}_2^*(-)\Big) \simeq \mathrm{pr}_1'^* \mathscr{F}_{\psi,1}(-) \overset{L}{\otimes} \mathrm{pr}_2'^* \mathscr{F}_{\psi,2}(-),$$

Proceedings of the International Congress
of Mathematicians, Kyoto, Japan, 1990

où $\mathscr{F}_{\psi,i}(-)$ est la transformation de Fourier géométrique pour E_i et $\mathrm{pr}_i : E_1 \times_S E_2 \longrightarrow E_i$ (resp. $\mathrm{pr}'_i : E'_1 \times_S E'_2 \to E'_i$) la projection canonique.

(iii) Si $f : E_1 \longrightarrow E_2$ est un morphisme de fibrés vectoriels de rang constant sur S, on a

$$f'^{*}\mathscr{F}_{\psi,1}(-)[d_2 - d_1] \simeq \mathscr{F}_{\psi,2}Rf_!(-),$$

où f' est le transposé de f, où d_i est le rang de E_i et où $\mathscr{F}_{\psi,i}$ est la transformation de Fourier géométrique pour E_i.

Théorème 1.2 (Verdier). *Pour tout $K \in \mathrm{ob}\, D_c^b(E, \overline{\mathbb{Q}}_\ell)$, la flèche d'oubli des supports*

$$R\,\mathrm{pr}'_!(\mathrm{pr}^* K \overset{L}{\otimes} \langle,\rangle^* \mathscr{L}_\psi) \longrightarrow R\,\mathrm{pr}'_*(\mathrm{pr}^* K \overset{L}{\otimes} \langle,\rangle^* \mathscr{L}_\psi)$$

est un isomorphisme dans $D_c^b(E', \overline{\mathbb{Q}}_\ell)$.

Compte-tenu de la dualité de Verdier, de la t-exactitude à droite de f_* pour f affine (prouvée par Artin) et de la forme forte des conjectures de Weil (prouvée par Deligne), le Théorème 1.2 admet les corollaires suivants:

Corollaire 1.3. *On a un isomorphisme canonique*

$$R\mathscr{H}om(\mathscr{F}_\psi(K), \pi'^! L) \simeq \mathscr{F}_{\psi^{-1}}(R\mathscr{H}om(K, \pi^! L))(d)$$

bi-fonctoriel en $(K, L) \in \mathrm{ob}(D_c^b(E, \overline{\mathbb{Q}}_\ell)^{\mathrm{opp}} \times D_c^b(S, \overline{\mathbb{Q}}_\ell))$.

Corollaire 1.4. *$\mathscr{F}_\psi$ est t-exact et induit une équivalence de catégories abéliennes*

$$\mathscr{F}_\psi : \mathrm{Perv}(E, \overline{\mathbb{Q}}_\ell) \longrightarrow \mathrm{Perv}(E', \overline{\mathbb{Q}}_\ell)$$

de quasi-inverse $\mathscr{F}'_{\psi}(-)(d)$.

Corollaire 1.5. *Si S et $E \xrightarrow{\pi} S$ sont définis sur $\mathbb{F}_q$, pour tout $K \in \mathrm{ob}\, D_c^b(E, \overline{\mathbb{Q}}_\ell)$ défini sur $\mathbb{F}_q$ et pur de poids w, $\mathscr{F}_\psi(K) \in \mathrm{ob}\, D_c^b(E', \overline{\mathbb{Q}}_\ell)$ est défini sur $\mathbb{F}_q$ et pur de poids $w + d$.*

2. Application aux sommes trigonométriques

[Br, Ka2, Ka3, Ka-La]

Dans ce paragraphe $S = \mathrm{Spec}(k)$ et $E = \mathrm{Spec}(k[x_1, \ldots, x_d])$ (avec sa $\mathbb{F}_q$-structure naturelle). On identifie alors E' à $\mathrm{Spec}(k[x'_1, \ldots, x'_d])$ de sorte que

$$\langle x', x \rangle = \sum_{\delta=1}^{d} x'_\delta x_\delta \,.$$

Pour tout schéma X de type fini sur k, défini sur $\mathbb{F}_q$, pour tout $x \in X(\mathbb{F}_q) = \{x \in X(k) \mid \mathrm{Frob}_q(x) = x\}$ et pour tout $M \in \mathrm{ob}\, D_c^b(X, \overline{\mathbb{Q}}_\ell)$, défini sur $\mathbb{F}_q$, Frob_q agit sur la fibre de M en x et on note $t_M(x)$ la trace de cet endomorphisme de M_x. La formule des traces de Grothendieck pour Frob_q entraîne immédiatement:

Proposition 2.1. *Pour tout* $K \in \operatorname{ob} D_c^b(E, \overline{\mathbb{Q}}_\ell)$*, défini sur* $\mathbb{F}_q$*, et tout* $x' \in E'(\mathbb{F}_q) = (\mathbb{F}_q)^d$*, on a*

$$t_{\mathscr{F}_\psi(K)}(x') = (-1)^d \sum_{x \in (\mathbb{F}_q)^d} t_K(x) \psi \circ \operatorname{tr}_{\mathbb{F}_q/\mathbb{F}_p}(\langle x', x\rangle)$$

(en d'autres termes, $t_{\mathscr{F}_\psi(K)}$ *est, au signe près, la transformée de Fourier de la fonction* t_K*).*

Soit R une sous-$\mathbb{Z}$-algèbre de type fini de $\mathbb{C}$, soit X un R-schéma affine, lisse, purement de dimension relative m, et soit $f = (f_1, \ldots, f_d) : X \longrightarrow \mathbb{A}_R^d$ un R-morphisme fini. Pour tout homomorphisme d'anneaux $v : R \longrightarrow k$ avec $v(1) = 1$, on note $(-)_v$ le changement de base par v. Alors, $K_v = (f_v)_* \overline{\mathbb{Q}}_\ell[m]$ est un $\overline{\mathbb{Q}}_\ell$-faisceau pervers sur $\mathbb{A}_k^d = E$, K_v est défini sur $\mathbb{F}_q \subset k$ dès que $\mathbb{F}_q \supset R/\operatorname{Ker}(v)$ et K_v est pur de poids m. Par suite, $\mathscr{F}_\psi(K_v)$ est aussi sur $\overline{\mathbb{Q}}_\ell$-faisceau pervers sur $\mathbb{A}_k^d = E'$, $\mathscr{F}_\psi(K_v)$ est aussi défini sur $\mathbb{F}_q \subset k$ dès que $\mathbb{F}_q \supset R/\operatorname{Ker}(v)$ et $\mathscr{F}_\psi(K_v)$ est aussi pur de poids $m + d$ (cf. (1.4), (1.5)). De plus, d'après (2.1), on a

$$t_{\mathscr{F}_\psi(K_v)}(x') = (-1)^{d+m} \sum_{x \in X_v(\mathbb{F}_q)} \psi \circ \operatorname{tr}_{\mathbb{F}_q/\mathbb{F}_p}(\langle x', f_v(x)\rangle_v)$$

pour tout $\mathbb{F}_q$ avec $k \supset \mathbb{F}_q \supset R/\operatorname{Ker}(v)$.

Comme l'a remarqué le premier Brylinski, il résulte du théorème de structure des $\overline{\mathbb{Q}}_\ell$-faisceaux pervers simples et purs qu'il existe un entier $\chi_v \geq 0$ et un polynôme non nul $\varphi_v(x')$ dans $(R/\operatorname{Ker}(v))[x'_1, \ldots, x'_d]$ tels que

$$\left| \sum_{x \in X_v(\mathbb{F}_q)} e^{\frac{2\pi i}{p} \operatorname{tr}_{\mathbb{F}_q/\mathbb{F}_p}(\langle x', f_v(x)\rangle_v)} \right| \leq \chi_v q^{m/2} \tag{2.2}$$

dès que $k \supset \mathbb{F}_q \supset R/\operatorname{Ker}(v)$ et que $x' \in (\mathbb{F}_q)^d$ satisfait $\varphi_v(x') \neq 0$.

Théorème 2.3 (Katz et Laumon). *Il existe* $\chi \in \mathbb{N}$ *et* $\varphi(x') \in R[x'_1, \ldots, x'_d]$ *non nul tels que, pour tout* k *et tout* v, $\chi_v = \chi$ *et* $\varphi_v = v(\varphi)$ *fassent marcher les estimations* (2.2) *ci-dessus.*

Exemple 2.4 (Adolphson et Sperber [Ad-Sp], Denef et Lœser [De-Lo]). Soient $I \subset \mathbb{Z}^m$ avec $|I| = d$, $X = (\mathbb{G}_{m,R})^m$ et $f = (f_i)_{i \in I} : X \longrightarrow \mathbb{A}_R^I = \mathbb{A}_R^d$ avec $f_i(x) = x^i$. Alors f est fini (et par suite (2.3) s'applique) si et seulement si l'enveloppe convexe de I dans $\mathbb{R}^m$ est de dimension m et contient 0 dans son intérieur. Dans ce cas, χ et φ sont déterminés explicitement dans loc. cit.

3. Localisation de caractéristiques d'Euler-Poincaré et de déterminants [Ka1, Il1, La1]

Soient X un k-schéma de type fini, $A = \mathrm{Spec}(k[x])$, $D = A \cup \{\infty\}$ la droite projective correspondante, $f : X \longrightarrow D$ un k-morphisme propre, $A' = \mathrm{Spec}(k[x'])$ la droite affine duale de A et $D' = A' \cup \{\infty'\}$ la droite projective correspondante. On pose $Y = f^{-1}(\infty)$, $U = X - Y = f^{-1}(A)$. On note $j : A \hookrightarrow D$, $j' : A' \hookrightarrow D'$ les inclusions, $\mathrm{pr}_X : D' \times_k X \longrightarrow X$, $\mathrm{pr}_{D'} : D' \times_k X \longrightarrow D'$ les projections canoniques, $\mathscr{L}_\psi(x'x)$ le $\overline{\mathbb{Q}}_\ell$-faisceau lisse de rang 1 sur $A' \times_k A$ associé au revêtement d'Artin-Schreier $v^p - v = x'x$ et au caractère ψ et prolongé par 0 à $D' \times_k D$ tout entier et $\mathscr{L}_\psi(x'f)$ le $\overline{\mathbb{Q}}_\ell$-faisceau sur $D' \times_k X$ image réciproque de $\mathscr{L}_\psi(x'x)$ par $D' \times_k f$. Enfin, soit $K \in \mathrm{ob}\, D_c^b(X, \overline{\mathbb{Q}}_\ell)$.

Posons

$$K' = j'_! \mathscr{F}_\psi (j^* Rf_* K)[-1] \in \mathrm{ob}\, D_c^b(D', \overline{\mathbb{Q}}_\ell)\,.$$

On a encore

$$K' = R\,\mathrm{pr}_{D'*}(\mathrm{pr}_X^* K \overset{L}{\otimes} \mathscr{L}_\psi(x'f))\,.$$

Si l'on note o' l'origine de A', $\eta_{o'}$ (resp. $\eta_{\infty'}$) le point générique de l'hensélisé de D' en o' (resp. ∞') et $\overline{\eta}_{o'}$ (resp. $\overline{\eta}_{\infty'}$) un point géométrique localisé en $\eta_{o'}$ (resp. $\eta_{\infty'}$), on a le triangle distingué

$$R\Gamma_c(U,K) \longrightarrow K'_{\overline{\eta}_{o'}} \longrightarrow R\Gamma(X, R\Phi_{\overline{\eta}_{o'}}) \longrightarrow \tag{3.1}$$

($K'_{o'} = R\Gamma_c(U,K)$) et l'isomorphisme

$$K'_{\overline{\eta}_{\infty'}} \xrightarrow{\sim} R\Gamma(X, R\Phi_{\overline{\eta}_{\infty'}}) \tag{3.2}$$

($K'_{\infty'} = 0$), où $R\Phi_{\overline{\eta}_{o'}} \in \mathrm{ob}\, D_c^b(X, \overline{\mathbb{Q}}_\ell[\mathrm{Gal}(\overline{\eta}_{o'}/\eta_{o'})])$ (resp. $R\Phi_{\overline{\eta}_{\infty'}} \in \mathrm{ob}\, D_c^b(X, \overline{\mathbb{Q}}_\ell[\mathrm{Gal}(\overline{\eta}_{\infty'}/\eta_{\infty'})])$) sont les cycles évanescents pour $\mathrm{pr}_{D'}$ relativement à $\mathrm{pr}_X^* K \overset{L}{\otimes} \mathscr{L}_\psi(x'f)$ en o' (resp. ∞').

Le support de $R\Phi_{\overline{\eta}_{o'}}$ est clairement contenu dans Y. Soit $V \subset U$ l'ouvert formé des points $x \in U$ tels que f et les $\overline{\mathbb{Q}}_\ell$-faisceaux de cohomologie $\mathscr{H}^i(K)$ ($i \in \mathbb{Z}$) soient lisses en x et soit $Z = U - V$ le fermé complémentaire.

Théorème 3.3 (Laumon). *Le support de $R\Phi_{\overline{\eta}_{\infty'}}$ est contenu dans $Y \cup Z$.*

Ce théorème est l'analogue en cohomologie ℓ-adique du principe de la phase stationnaire: si U est une variété différentielle $\mathscr{C}^\infty$, si $f : U \longrightarrow \mathbb{R}$ est une fonction $\mathscr{C}^\infty$ et si ω est une densité $\mathscr{C}^\infty$ à support compact sur U telle que

$$\mathrm{Supp}(\omega) \cap \{x \in U | df(x) = 0\} = \emptyset\,,$$

on a

$$\int_U e^{ix'f}\omega = O(x'^{-n})$$

pour tout entier $n \geq 0$ quand $x' \in \mathbb{R}$ tend vers ∞.

On ne peut pas comparer directement $K_{\overline{\eta}_{o'}}$ et $K_{\overline{\eta}_{\infty'}}$. Par contre, $K_{\overline{\eta}_{o'}}$ et $K_{\overline{\eta}_{\infty'}}$ ont même caractéristique d'Euler-Poincaré. On déduit donc de (3.1) , (3.2) et (3.3) que :

$$(3.4) \qquad \chi(X,K) = \chi(Y,K) - \chi(Y, R\Phi_{\overline{\eta}_{o'}}) + \chi(Y \cup Z, R\Phi_{\overline{\eta}_{\infty'}}).$$

Application 3.5. Si X est une variété projective sur k et si $K_1, K_2 \in \mathrm{ob}\, D_c^b(X, \overline{\mathbb{Q}}_\ell)$ sont localement isomorphes pour la topologie étale sur X, Deligne a démontré que $\chi(X,K_1) = \chi(X,K_2)$ (cf. [Il2]). On peut en donner une autre démonstration par récurrence sur la dimension de X basée sur (3.4).

On peut aussi comparer les déterminants de $K_{\overline{\eta}_{o'}}$ et $K_{\overline{\eta}_{\infty'}}$. En particulier, on peut déduire de (3.1), (3.2) et (3.3) la conjecture de Deligne suivante :

Théorème 3.6 (Laumon). *Soient X une courbe projective, lisse et connexe sur k, ω une 1-forme méromorphe non identiquement nulle sur X et $K \in \mathrm{ob}\, D_c^b(X, \overline{\mathbb{Q}}_\ell)$. On suppose X, ω et K définis sur $\mathbb{F}_q$. Alors la "constante globale" de Grothendieck*

$$\varepsilon(X,K) = \det(-\mathrm{Frob}_q, R\Gamma(X,K))^{-1}$$

admet la formule du produit

$$\varepsilon(X,K) = q^{(1-g)r(K)} \prod_{x \in |X|} \varepsilon(X_{(x)}, K|X_{(x)}, \omega|X_{(x)})$$

où g est le genre de X, $r(K)$ est le rang générique de K et, pour tout point fermé x de X, $X_{(x)}$ est l'hensélisé de X en x et $\varepsilon(X_{(x)}, K|X_{(x)}, \omega|X_{(x)})$ est la "constante locale" de Deligne, Dwork et Langlands.

Récemment, Loeser a obtenu par la même méthode une formule du produit pour le déterminant de la cohomologie étale du complémentaire d'un arrangement d'hyperplans dans $\mathbb{P}_k^n$ à valeurs dans un $\overline{\mathbb{Q}}_\ell$-faisceau lisse de rang 1 de type Kummer (cf. [Lo]).

4. Transformations de Fourier locales [La1]

Il y en a de trois types $\mathscr{F}_\psi^{(o,\infty')}$, $\mathscr{F}_\psi^{(\infty,o')}$ et $\mathscr{F}_\psi^{(\infty,\infty')}$. On ne s'intéressera ici qu'au premier type. Soient $T = \mathrm{Spec}(k[[\varpi]])$ (resp. $T' = \mathrm{Spec}(k[[\varpi']])$) avec sa $\mathbb{F}_q$-structure naturelle (ϖ (resp. ϖ') est une indéterminée sur k), η (resp. η') le point générique et s (resp. s') le point fermé de T (resp. T'), $\overline{\eta}$ (resp. $\overline{\eta}'$) un point géométrique localisé en η (resp. η') et $\mathscr{G}$ (resp. $\mathscr{G}'$) la catégorie des $\mathrm{Gal}(\overline{\eta}/\eta)$ (resp. $\mathrm{Gal}(\overline{\eta}'/\eta')$)-modules sur $\overline{\mathbb{Q}}_\ell$. Si $V \in \mathrm{ob}\,\mathscr{G}$, on note encore V le $\overline{\mathbb{Q}}_\ell$-faisceau lisse sur η correspondant et $V_!$ sur prolongement par zéro à T tout entier. Alors par définition

$$\mathscr{F}_\psi^{(o,\infty')}(V) = R^1\Phi_{\overline{\eta}'}(\mathrm{pr}^*(V_!) \overset{L}{\otimes} \mathscr{L}_\psi(\varpi/\varpi'))_{(s',s)} \in \mathrm{ob}\,\mathscr{G}'$$

où $\mathrm{pr} : T' \times_k T \longrightarrow T$ et $\mathrm{pr}' : T' \times_k T \longrightarrow T'$ sont les projections canoniques et où $\mathscr{L}_\psi(\varpi/\varpi')$ est le $\overline{\mathbb{Q}}_\ell$-faisceau lisse de rang 1 sur $\eta' \times_k T$ associé au revêtement d'Artin-Schreier $v^p - v = \varpi/\varpi'$ et au caractère ψ et prolongé par zéro à $T' \times_k T$ tout entier.

Théorème 4.1 (Laumon).

(i) $\mathscr{F}_\psi^{(o,\infty')}$ *est un foncteur exact de* $\mathscr{G}$ *vers* $\mathscr{G}'$.

(ii) Pour tout $V \in \mathrm{ob}\,\mathscr{G}$, *on a*

$$\begin{aligned} r(\mathscr{F}_\psi^{(o,\infty')}(V)) &= r(V) + s(V) \\ s(\mathscr{F}_\psi^{(o,\infty')}(V)) &= s(V) \end{aligned}$$

où $r(-)$ *est le rang et* $s(-)$ *le conducteur de Swan, de sorte que* $\mathscr{F}_\psi^{(o,\infty')}(\mathscr{G}) \subset \mathscr{G}'_{[0,1[}$, *où* $\mathscr{G}'_{[0,1[}$ *est la sous-catégorie pleine de* $\mathscr{G}'$ *formée des objets* V' *dont tous les sous-quotients irréductibles* W' *vérifient* $s(W') < r(W')$.

(iii) Le foncteur $\mathscr{F}_\psi^{(o,\infty')} : \mathscr{G} \longrightarrow \mathscr{G}'_{[0,1[}$ *est une équivalence de catégories abéliennes.*

(iv) $\mathscr{F}_\psi^{(o,\infty')}(V)^\vee \simeq \mathscr{F}_{\psi^{-1}}^{(o,\infty')}(V^\vee)(1)$.

(v) Si V *est défini sur* $\mathbb{F}_q$, *il en est de même de* $\mathscr{F}_\psi^{(o,\infty')}(V)$ *et la "constante locale"* $\varepsilon_0(T, V, d\varpi)$ *de Deligne, Dwork et Langlands est égale à*

$$\varepsilon_0(T, V, d\varpi) = (-1)^{r(V')} \det(V')(\varpi')$$

où $V' = \mathscr{F}_\psi^{(o,\infty')}(V)$ *(on a identifié* $\mathrm{Gal}(\overline{\mathbb{F}_q((\varpi'))}/\mathbb{F}_q((\varpi')))^{\mathrm{ab}}$ *à* $(\mathbb{F}_q((\varpi'))^\times)^\wedge$ *par la théorie du corps de classe abélien).*

Cette transformation de Fourier locale donne une construction cohomologique locale de la représentation d'Artin et de la "constante locale". Elle a aussi permis à Henniart de démontrer la conjecture de Langlands locale numérique, i.e. de dénombrer les objets simples de $\mathscr{G}$ de conducteur borné (cf. [He]).

5. Une nouvelle preuve d'une conjecture de Weil [La1]

Si K est un $\overline{\mathbb{Q}}_\ell$-faisceau pervers, irréductible, défini sur $\mathbb{F}_q$ et pur de poids 0, sur une droite projective D définie sur $\mathbb{F}_q$, Deligne a montré que $R\Gamma(D, K)$ est aussi pur de poids 0. La transformation de Fourier géométrique permet d'en donner une nouvelle démonstration, inspirée par la preuve de Witten des inégalités de Morse. Supposons que $D = A \cup \{\infty\}$, où $\infty \in D(\mathbb{F}_q)$ et $A = \mathrm{Spec}(k[x])$ est défini sur $\mathbb{F}_q$, et que les $\overline{\mathbb{Q}}_\ell$-faisceaux de cohomologie $\mathscr{H}^i(K)$ ($i \in \mathbb{Z}$) sont lisses en ∞. Alors la restriction de $\mathscr{F}_\psi(K|A)$ à $A' - \{o'\}$ ($A' = \mathrm{Spec}(k[x'])$) est de la forme $\mathscr{J}'[1]$ pour un $\overline{\mathbb{Q}}_\ell$-faisceau lisse irréductible $\mathscr{J}'$. La remarque essentielle est que $\mathscr{J}'$ est un constituant du $\overline{\mathbb{Q}}_\ell$-faisceau lisse **réel**

$$\mathscr{H}^{-1}(\mathscr{F}_\psi((K \oplus K^\vee)|A) \oplus \mathscr{F}_{\psi^{-1}}((K \oplus K^\vee)|A))|A' - \{0'\}$$

où $K^\vee = R\mathscr{H}om(K, \overline{\mathbb{Q}}_\ell[2](1))$ est le dual de Verdier de K. Par suite, d'après un théorème clé de Deligne, $\mathscr{J}'$ est lui même pur d'un certain poids (on a fixé un isomorphisme $\iota : \overline{\mathbb{Q}}_\ell \xrightarrow{\sim} \mathbb{C}$, de sorte que $\iota \circ \psi^{-1} = \overline{\iota \circ \psi}$).

6. La transformation de Radon géométrique [Br, Il1]

Il s'agit d'une variante homogène de la transformation de Fourier. Soient $\pi : P = \mathbb{P}(E) \longrightarrow S$ et $\pi' : P' = \mathbb{P}(E') \longrightarrow S$ les fibrés projectifs associés aux fibrés vectoriels du paragraphe 1. On note $Z \subset P' \times_S P$ la variété d'incidence entre ces deux fibrés projectifs en dualité et $\varrho : Z \longrightarrow P$, $\varrho' : Z \longrightarrow P'$ les projections canoniques. La *transformation de Radon géométrique*, inventée par Brylinski, est l'opération

$$\mathscr{R} : D_c^b(P, \overline{\mathbb{Q}}_\ell) \longrightarrow D_c^b(P', \overline{\mathbb{Q}}_\ell)$$

$$\mathscr{R}(-) = R\varrho'_* \varrho^*(-)[d-2].$$

On note $\mathscr{S} \subset \mathrm{Perv}(P, \overline{\mathbb{Q}}_\ell)$ (resp. $\mathscr{S}' \subset \mathrm{Perv}(P', \overline{\mathbb{Q}}_\ell)$) la sous-catégorie strictement pleine formée des $\overline{\mathbb{Q}}_\ell$-faisceaux pervers sur P (resp. P') isomorphes à $\pi^* L[d-1]$ (resp. $\pi'^* L'[d-1]$) pour un $L \in \mathrm{ob}\,\mathrm{Perv}(S, \overline{\mathbb{Q}}_\ell)$ (resp. $L' \in \mathrm{ob}\,\mathrm{Perv}(S, \overline{\mathbb{Q}}_\ell)$) ; c'est une sous-catégorie de Serre.

Théorème 6.1 (Brylinski). *(i) Pour tout* $K \in \mathrm{ob}\,\mathrm{Perv}(P, \overline{\mathbb{Q}}_\ell)$ *et tout entier* $i \neq 0$, ${}^p\mathscr{H}^i(\mathscr{R}(K)) \in \mathrm{ob}\,\mathscr{S}'$.

(ii) Le foncteur $K \longmapsto {}^p\mathscr{H}^0(\mathscr{R}(K))$ *induit une équivalence de catégories abéliennes de la catégorie quotient* $\mathrm{Perv}(P, \overline{\mathbb{Q}}_\ell)/\mathscr{S}$ *sur la catégorie quotient* $\mathrm{Perv}(P', \overline{\mathbb{Q}}_\ell)/\mathscr{S}'$ *; un quasi-inverse est induit par* $K' \longmapsto {}^p\mathscr{H}^0(\mathscr{R}'(K'))(d-2)$, *où* $\mathscr{R}'$ *est la transformation de Radon géométrique pour* P'.

Brylinski applique ce théorème à l'étude, dans le cas singulier, de la monodromie des pinceaux de Lefschetz.

Dans [La2], nous utilisons cette transformation de Radon pour l'étude des faisceaux automorphes cuspidaux de Drinfeld.

7. Une application à la théorie des représentations des groupes réductifs sur les corps finis [Kaz-La]

Soient V un espace vectoriel de dimension 2 sur k et $\omega \in \Lambda^2 V - \{0\}$ une forme volume. On suppose que V et ω sont définis sur $\mathbb{F}_q$. On identifie le dual V' de V à V à l'aide de la forme symplectique ω. On a donc une involution

$$\mathscr{F}_\psi(-)(1) : \mathrm{Perv}(V, \overline{\mathbb{Q}}_\ell) \longrightarrow \mathrm{Perv}(V, \overline{\mathbb{Q}}_\ell)$$

si l'on considère V comme un fibré vectoriel sur $\mathrm{Spec}(k)$ de manière évidente. Soit $\mathscr{A}$ la catégorie abélienne des paires (A, φ) où $A \in \mathrm{ob}\,\mathrm{Perv}(V, \overline{\mathbb{Q}}_\ell)$ et $\varphi : \mathscr{F}_\psi(\mathrm{Frob}_q^* A)(1) \xrightarrow{\sim} A$ est un isomorphisme dans $\mathrm{Perv}(V, \overline{\mathbb{Q}}_\ell)$. Soit $K(\mathscr{A})$ le groupe de Grothendieck de $\mathscr{A}$. On considère l'application

$$\mathrm{ob}\,\mathscr{A} \times \mathrm{ob}\,\mathscr{A} \longrightarrow \overline{\mathbb{Q}}_\ell$$

$$((A_1, \varphi_1), (A_2, \varphi_2)) \longmapsto \sum_{i \geq 0} (-1)^i \,\mathrm{tr}(\varphi^i, \mathrm{Ext}^i_{\mathrm{Perv}(V, \overline{\mathbb{Q}}_\ell)}(A_1, DA_2))$$

où $DA_2 = R\mathcal{H}om(A_2, \overline{\mathbb{Q}}_\ell[2](1))$, où les Ext^i $(i \geq 0)$ sont les Ext^i de Yoneda dans la catégorie abélienne $\mathrm{Perv}(V, \overline{\mathbb{Q}}_\ell)$ (ce sont des $\overline{\mathbb{Q}}_\ell$-espaces vectoriels de dimension finie, nuls pour $i \gg 0$, d'après [Be]) et où φ^i est induit par φ_1 et φ_2. Elle induit un accouplement

$$(,) : K(\mathscr{A}) \times K(\mathscr{A}) \longrightarrow \overline{\mathbb{Q}}_\ell$$

dont on notera $N(\mathscr{A})$ le noyau.

Le groupe algébrique sur k, $\mathrm{Aut}(V,\omega) \times \mathbb{G}_{m,k}$, agit sur V par $(g,t)\cdot v = g(tv) = tg(v)$ d'une part et par $(g,t)\cdot v = g(t^{-1}v) = t^{-1}g(v)$ d'autre part, d'où deux actions de $\mathrm{Aut}(V,\omega) \times \mathbb{G}_{m,k}$ sur $\mathrm{Perv}(V, \overline{\mathbb{Q}}_\ell)$ échangées par $\mathscr{F}_\psi(-)(1)$. On en déduit une action de $\mathrm{Aut}(V,\omega)(\mathbb{F}_q) \times \{t \in k | t^{q+1} = 1\}$ sur $K(\mathscr{A})$ qui respecte l'accouplement $(,)$ et donc son noyau.

Théorème 7.1 (Kazhdan). (*i*) *$K(\mathscr{A})/N(\mathscr{A})$ a une structure naturelle de $\overline{\mathbb{Q}}_\ell$-espace vectoriel de dimension finie sur lequel $\mathrm{Aut}(V,\omega)(\mathbb{F}_q) \times \{t \in k | t^{q+1} = 1\}$ agit.*

(ii) Si $\chi : \{t \in k | t^{q+1} = 1\} \longrightarrow \overline{\mathbb{Q}}_\ell^\times$ est un caractère régulier ($\chi^2 \neq 1$) la composante isotypique σ_χ de χ dans $K(\mathscr{A})/N(\mathscr{A})$ est la représentation irréductible de la série discrète de $\mathrm{Aut}(V,\omega)(\mathbb{F}_q) \simeq SL_2(\mathbb{F}_q)$ associée à χ par la construction de Weil.

Dans [Kaz-La], nous construisons un analogue de $K(\mathscr{A})/N(\mathscr{A})$ pour un groupe semi-simple et simplement connexe arbitraire (à la place de $\mathrm{Aut}(V,\omega)$) et conjecturons une généralisation de (7.1).

Remarque 7.2. Pour d'autres applications à la théorie des représentations des groupes réductifs sur les corps finis, voir [Br, De3, Lu].

Remerciement. Je remercie Mme Le Bronnec pour la magnifique composition de ce manuscrit.

Bibliographie

[Ad-Sp] Adolphson, A., Sperber, S.: Exponential sums and Newton polyhedra, cohomology and estimates. Ann. Math. **130** (1989) 367–406

[Be] Beilinson, A.: On the derived category of perverse sheaves. (Lecture Notes in Mathematics, vol. 1289.) Springer, Berlin Heidelberg New York 1987, pp. 27–41

[B-B-D] Beilinson, A., Bernstein, J., Deligne, P.: Faisceaux pervers. Astérisque **100** (1983)

[Br] Brylinski, J.-L.: Transformations canoniques, Dualité projective, Théorie de Lefschetz, Transformations de Fourier et sommes trigonométriques. Astérisque **140–141** (1986) 3–134

[De1] Deligne, P.: Lettre à D. Kazhdan du 29 novembre 1976

[De2] Deligne, P.: La conjecture de Weil II. Publ. Math. IHES **52** (1980) 137–252

[De3] Deligne, P.: Métaplectique. Communication personnelle

[De-Lo] Denef, J., Lœser, F.: Weights of exponential sums, intersection cohomology, and Newton polyhedra. Preprint 1990

[Ek] Ekedahl, T.: On the ℓ-adic formalism in the “Grothendieck Festschrift”, vol. 2. Birkhäuser, Basel (to appear)

[He] Henniart, G.: La conjecture de Langlands locale numérique pour $GL(n)$. Ann. Sci. Ec. Norm. Sup. **21** (1988) 497–544

[Il1] Illusie, L.: Deligne's ℓ-adic Fourier transform. Proceedings of Symposia in Pure Math., vol. 46, part 2. A.M.S. 1987, pp. 151–163

[Il2] Illusie, L.: Théorie de Brauer et caractéristique d'Euler-Poincaré d'après P. Deligne. Astérisque **82–83** (1981) 161–172

[Ka1] Katz, N.M.: Travaux de Laumon, Séminaire Bourbaki. Astérisque **161-162** (1988) 105–132

[Ka2] Katz, N.M.: Gauss sums, Kloosterman sums, and monodromy. Ann. Math. Stud., vol. 116. Princeton Univ. Press, 1988

[Ka3] Katz, N.M.: Perversity and exponential sums. Tohoku Math. J. (to appear)

[Ka-La] Katz, N.M., Laumon, G.: Transformation de Fourier et majoration de sommes exponentielles. Publ. Math. IHES **62** (1985) 361–418

[Kaz-La] Kazhdan, D., Laumon, G.: Gluing of perverse sheaves and discrete series representations. J. Geom. Phys. **5** (1988) 63–120

[La1] Laumon, G.: Transformation de Fourier, constantes d'équations fonctionnelles et conjecture de Weil. Publ. Math. IHES **65** (1987) 131–210

[La2] Laumon, G.: Correspondance de Langlands géométrique pour les corps de fonctions. Duke Math. J. **54** (1987) 309–359

[Lo] Lœser, F.: Arrangements d'hyperplans et sommes de Gauss. Prépublicaton École Polytechnique, Palaiseau 1989

[Lu] Lusztig, G.: Fourier transforms on a semi-simple Lie algebra over $\mathbb{F}_q$. (Lecture Notes in Mathematics, vol. 1271.) Springer, Berlin Heidelberg New York 1987, pp. 117–188

[SGA] Grothendieck, A., et al.: Séminaire de Géométrie Algébrique du Bois-Marie, SGA4, SGA5, SGA7. (Lecture Notes in Mathematics, vols. 269, 305, 589, 288, 340.) Springer, Berlin Heidelberg New York 1972 à 1977

Algebraic Independence of Values of Analytic Functions

Yuri Nesterenko

Department of Mathematics, Moscow State University, 119899, Moscow, USSR

The first result on the algebraic independence of values of analytic functions was obtained by F. Lindemann in 1882:

If $a_1, \ldots, a_m$ are algebraic numbers linearly independent over $\mathbb{Q}$, then numbers

$$e^{a_1}, \ldots, e^{a_m}$$

are algebraically independent over $\mathbb{Q}$.

This means that for any polynomial $P \in \mathbb{Z}[x_1, \ldots, x_m]$, $P \not\equiv 0$, the inequality $|P(e^{a_1}, \ldots, e^{a_m})| > 0$ holds. The algebraic independence of any numbers $\omega_1, \ldots, \omega_m$ over $\mathbb{Q}$ implies, in particular, that all the numbers ω_i are transcendental. Side by side with problems of the algebraic independence of numbers, an object of study in the theory of transcendental numbers is to obtain quantitative characteristics of independence. By a measure of algebraic independence of complex numbers $\omega_1, \ldots, \omega_m$ we mean a function of two variables $f(d, H)$ such that for any polynomial $P \in \mathbb{Z}[X_1, \ldots, X_m]$, $P \not\equiv 0$, whose degree does not exceed d and whose coefficients have modulus not greater than H, the estimate

$$|P(\omega_1, \ldots, \omega_m)| > f(d, H)$$

holds.

It is easy to prove for all complex numbers $\omega_1, \ldots, \omega_m$ the inequality

$$\phi(\bar{\omega}; d, H) = \min |P(\omega_1, \ldots, \omega_m)| < H^{-c_1 d^m},$$

where $c_1 = 1/(2m!)$ and minimum is taken over all polynomials

$$P \in \mathbb{Z}[x_1, \ldots, x_m], \qquad P \not\equiv 0, \qquad \deg P \leq d, \qquad H(P) \leq H$$

($H(P)$ is the maximum of the moduli of the coefficients of P). But any lower estimate of this minimum $\phi(\bar{\omega}; d, H)$ depends on the individual properties of numbers ω_1, $\ldots$, ω_m and estimates are obtained only for special kinds of numbers.

In 1932 K. Mahler [16] proved that, under the conditions of Lindemann's theorem,

$$\phi(e^{a_1}, \ldots, e^{a_m}; d, H) > H^{-c_2 d^m},$$

Proceedings of the International Congress
of Mathematicians, Kyoto, Japan, 1990

where $c_2 > 0$ depends only on numbers $\alpha_1, \ldots, \alpha_m$, and H exceeds some bound, which depends on $\alpha_1, \ldots, \alpha_m$ and d.

1. *E*-Functions

Function $e^{\alpha z}$ with algebraic number α, $\alpha \neq 0$, belong to a certain class of entire functions, introduced in 1929 by C. Siegel [39] and called by him *E*-functions. This class of functions contains hypergeometric functions

$$pFq\left(\begin{matrix} a_1, \ldots, a_p \\ b_1, \ldots, b_q \end{matrix}\middle| z^{q-p+1}\right), \qquad p \leq q,$$

with rational parameters a_i, b_j and is a ring, closed with respect to differentiation, integration between 0 and z, and replacing z by λz, where λ is algebraic. After the basic articles of Siegel the complete result was proved in 1955 by A.B. Shidlovskij [38].

Theorem 1. *Suppose that the E-functions $f_1(z), \ldots, f_m(z)$ form a solution of the system of linear differential equations*

$$y_k' = q_{k0}(z) + \sum_{i=1}^{m} q_{ki}(z) y_i, \qquad k = 1, 2, \ldots, m, \tag{1}$$

$q_{ki}(z) \in \mathbb{C}(z)$; *suppose that α is an algebraic number not equal to zero and the singular points of the system* (1). *Then*

$$\operatorname{tr} \deg_{\mathbb{Q}} \mathbb{Q}(f_1(\alpha), \ldots, f_m(\alpha)) = \operatorname{tr} \deg_{\mathbb{C}(z)} \mathbb{C}(z)(f_1(z), \ldots, f_m(z)).$$

Lindemann's theorem follows from the theorem of Shidlovskij, if we take $f_i(z) = e^{\alpha_i z}$ and $\alpha = 1$.

Theorem 1 gave rise to a great number of results concerning specific *E*-functions (see [38]). For example, the function

$$\Psi_k(z) = \sum_{n=1}^{\infty} \frac{z^{kn}}{(n!)^k},$$

satisfies a linear differential equation of order k with coefficients in $\mathbb{C}(z)$; it can be proved that, for any algebraic number $\alpha \neq 0$, the $r(r+1)/2$ numbers $\Psi_k^{(i)}(\alpha)$, $0 \leq i < k$, $1 \leq k \leq r$, are algebraically independent over $\mathbb{Q}$.

There exists an important problem, connected with applications of Theorem 1: *proving the algebraic independence of solutions of linear differential equations over* $\mathbb{C}(z)$. In different specific cases it was the subject of many papers. We mention here only recent papers by W. Salichov [34, 35] and F. Beukers, D. Brownawell and G. Heckmann [2]. This subject is related to the Picard-Vessiot theory of linear differential equations and the calculation of Galois groups of such equations.

In 1962, using the Shidlovskij results, S. Lang [13] obtained a quantitative variant of the above theorem in a form similar to the Mahler's theorem (see also the article by Galochkin [8]). In the next assertion, which was proved in 1977 [20], the dependence on d of the lower bound for H is effective.

Theorem 2. *Suppose that the E-functions $f_1(z), \ldots, f_m(z)$ and the number α satisfy the conditions of Theorem 1. There exists constant $\tau > 0$, which depends only on the system (1), the functions $f_1(z), \ldots, f_m(z)$ and the number α, such that for*

$$\ln \ln H > \tau d^{2m} \ln(d + 1)$$

the following inequality holds:

$$\phi(f_1(\alpha), \ldots, f_m(\alpha); d, H) > H^{-c_3 d^m},$$

where $c_3 = 4^m \varkappa^m (m\varkappa^2 + \varkappa + 1)$, and $\varkappa$ is the degree of the algebraic field, which contains α and the coefficients of Taylor expansions of $f_i(z)$.

In many cases the constant τ may be computed effectively, [4] and [25].

2. Mahler's Functions

Here we briefly discuss the values of power series (in general, in several variables) which satisfy certain functional equations. Mahler was the first to study such series in 1929 (see [17]). Besides the works of Mahler, we mention here the results of J. Loxton and A.J. van der Poorten, K.K. Kubota, D.W. Masser. Recently interesting relations between Mahler's functions and automata theory were found [14].

For an integer $n \geq 1$ let $T = (t_{ij})$ be a nonsingular matrix of order n with nonnegative integer entries such that none of the eigenvalues of T is a root of unity. We define a transformation $T\colon \mathbb{C}^n \to \mathbb{C}^n$ by the rule

$$T(z_1, \ldots, z_n) = \left(\prod_{i=1}^{n} z_i^{t_{1i}}, \ldots, \prod_{i=1}^{n} z_i^{t_{ni}} \right).$$

The next theorem was proved by Loxton and van der Poorten [15].

Theorem 3. *Let $A(\bar{z})$ be an $m \times m$ matrix whose entries are rational functions of the variable $\bar{z}$ with algebraic coefficients such that* det $A(0) \neq 0$. *Let $\bar{f}(\bar{z}) = (f_1(\bar{z}), \ldots, f_m(\bar{z}))$ be a vector of power series in $\bar{z} = (z_1, \ldots, z_n)$ with coefficients in an algebraic number field K, $[K : \mathbb{Q}] < \infty$, which converge in some neighbourhood U of the origin in $\mathbb{C}^n$ and satisfy the equations*

$$\bar{f}(\bar{z}) = A(\bar{z}) * \bar{f}(T\bar{z}).$$

If the matrix A and vector $\bar{\alpha}$ with algebraic components satisfy some technical conditions, then

$$\operatorname{tr\,deg}_{\mathbb{Q}} \mathbb{Q}(f_1(\bar{\alpha}), \ldots, f_m(\bar{\alpha})) = \operatorname{tr\,deg}_{\mathbb{C}(\bar{z})} \mathbb{C}(\bar{z})(f_1(\bar{z}), \ldots, f_m(\bar{z})).$$

Recently K. Nishioka [27], using a method introduced in [23] and based on the general theory of elimination (see § 5), in the case $n = 1$ proved the same assertion under weaker conditions: α is an algebraic number such that $0 < |\alpha| < 1$, $A(\alpha^{d^l})$ is nonsingular and $a(\alpha^{d^l}) \neq 0$ for all $l \geq 0$, where $a(z)$ is a common denominator of entries of $A(z)$.

The first estimate for the measure of algebraic independence of values of Mahler's functions were proved in 1985 in [23]. But this estimate was not effective in the degrees of polynomials. Using the zero estimate for Mahler functions, which was proved by Nishioka, Becker found effective algebraic independence measures of those numbers (see [1, 28]).

Theorem 4. *Let $A(z)$ be an $m \times m$ matrix and $B(z)$ be an m-dimensional vector whose entries are rational functions of one variable z with algebraic coefficients. Let $\bar{f} = (f_1(z), \ldots, f_m(z))$ be a vector of power series with algebraic coefficients which converge in some neighbourhood U of the point $z = 0$ and let $\bar{f}$ satisfy*

$$\bar{f}(z^t) = A(z) * \bar{f}(z) + B(z),$$

where $t \geq 2$ is an integer, and which are algebraically independent over $\mathbb{C}(z)$. Suppose that α is an algebraic number, $\alpha \in U$, $0 < |\alpha| < 1$, and the numbers α^{t^l}, $l \geq 0$, are distinct from the poles of $A(z)$ or $B(z)$. Then

$$\phi(f_1(\alpha), \ldots, f_m(\alpha); d, H) > \exp(-c_4 d^m(\ln H + d^{m+2})).$$

3. Exponential Functions

One of the problems suggested by D. Hilbert in 1900 was the conjectural transcendence of values of the function α^z with algebraic base $\alpha \neq 0, 1$ at algebraic irrational points β. This assertion was proved in 1934 by A.O. Gelfond and independently by Th. Schneider. The natural analog of the Lindemann's theorem is the next

Gelfond's Conjecture. *If $\alpha \neq 0, 1$ is an algebraic number and algebraic numbers $\beta_1, \ldots, \beta_m$ and 1 are linearly independent over $\mathbb{Q}$, then the numbers*

$$\alpha^{\beta_1}, \ldots, \alpha^{\beta_m}$$

are algebraically independent over $\mathbb{Q}$.

This problem is equivalent to *the algebraic independence over $\mathbb{Q}$ of numbers α^{β}, $\alpha^{\beta^2}, \ldots, \alpha^{\beta^{d-1}}$*, where β is an algebraic number of degree $d \geq 3$. It was proved for $d = 3$ by Gelfond [9] in 1948. Now the best result in general case is the lower estimate $\operatorname{tr\,deg} K \geq \left[\frac{d+1}{2}\right]$ for transcendence degree over $\mathbb{Q}$ of the field $K = \mathbb{Q}(\alpha^{\beta}, \alpha^{\beta^2}, \ldots, \alpha^{\beta^{d-1}})$, where [ ] denotes the integer part.

Let us consider a more general situation. Suppose $a_1, \ldots, a_p$ and $b_1, \ldots, b_q$ be complex numbers which for any nonzero vectors $\bar{k} = (k_1, \ldots, k_p) \in \mathbb{Z}^p$ and $\bar{l} = (l_1, \ldots, l_q) \in \mathbb{Z}^q$ satisfy the inequalities

$$|k_1 a_1 + \cdots + k_p a_p| > \exp(-\gamma |\bar{k}| \ln |\bar{k}|), \qquad |\bar{k}| = \max_{1 \leq i \leq p} |k_i|,$$

$$|l_1 b_1 + \cdots + l_q b_q| > \exp(-\gamma |\bar{l}| \ln |\bar{l}|), \qquad |\bar{l}| = \max_{1 \leq i \leq q} |l_i|,$$

where γ is a positive constant. In particular, it is well known that this condition is satisfied by algebraic numbers and by the logarithms of algebraic numbers which are linearly independent over $\mathbb{Q}$. Traditionally the three sets of numbers have been studied

1) $e^{a_i b_j}$, 2) $a_i, e^{a_i b_j}$, 3) $a_i, b_j, e^{a_i b_j}$, $1 \le i \le p, 1 \le j \le q$.

Many estimates for transcendence degree of these sets of numbers were proved after appearance of Gelfond's articles [9, 10]. We mention here papers by R. Tijdeman, A. Shmelev, W.D. Brownawell, G. Chudnovsky, M. Waldschmidt, E. Reyssat. P. Philippon, and others (see [40]). The best result now is the next assertion.

Theorem 5. *Among the numbers in the sets* 1)–3) *there are at least*

$$\left[\frac{pq}{p+q}\right], \quad \left[\frac{pq+p}{p+q}\right], \quad \frac{pq}{p+q},$$

respectively which are algebraically independent over $\mathbb{Q}$.

The lower estimate for tr deg K follows immediately from this theorem if in the set 2) we take $p = q = d$, $a_i = \beta^{i-1}$, $b_i = \beta^{i-1} \ln \alpha$, $i = 1, \ldots, d$.

P. Philippon in 1984 [31] proved a little weaker estimate for the sets 1), 2) and the estimate for the set 3) in Theorem 5. The main tools for obtaining these results were introduced by the author [20–23] and P. Philippon [32], and are connected with application of some classical ideas of "constructive" commutative algebra, developed in the works of K. Hentzelt, E. Noether, B.L. van der Waerden, W.L. Chow (see Section 5) to the theory of transcendental numbers. The estimates for the sets 1) and 2) in the Theorem 5 were established in 1987 by G. Diaz [6] who used a new construction of auxiliary function in the analytical part of the proof (for another variant see [24]).

For $d = 3$ the estimate for integral polynomials at the point $(\alpha^{\beta}, \alpha^{\beta^2}, \ldots, \alpha^{\beta^{d-1}})$ was established by Gelfond and Feldman in 1950. It was later improved by W.D. Brownawell [3] and G. Diaz [7]. In the general case $d \ge 4$ such an estimate is not proved yet. But it is possible to prove a lower estimate for the maximum of absolute values at this point of any set of polynomials which generate an ideal in $\mathbb{Z}[x_1, \ldots, x_{d-1}]$ with rank higher than $\dfrac{d-1}{2}$ (see [23]).

4. Elliptic Functions

Let $\wp(z)$ be the Weierstrass elliptic function with algebraic invariants g_2, g_3. This function satisfies an algebraic differential equation with algebraic coefficients and an addition theorem. That is why its transcendence behaviour is similar to the behaviour of e^z. First results about the transcendence of values of this function were proved by Schneider in 1934 [36]. The next assertion, the elliptic analogue of the Lindemann's theorem, was proved in 1983 by G. Wustholz [42] and P. Philippon [30] after the previous paper of G. Chudnovsky [5].

Theorem 6. *If* $\wp(z)$ *has the complex multiplication over the field* $k = \mathbb{Q}(\tau)$ and $\alpha_1, \ldots, \alpha_m$ *are algebraic numbers, linearly independent over field k, then the numbers*

$$\wp(\alpha_1), \ldots, \wp(\alpha_m)$$

are algebraically independent over $\mathbb{Q}$.

Now it is possible to estimate a measure of the algebraic independence of these numbers (see [26]).

Theorem 7. *Let* $\wp(z)$, $\alpha_1, \ldots, \alpha_m$ *be the same as in Theorem* 6, *H and d be positive numbers such that*

$$\ln \ln H > c_5 * d^m * \ln(d + 1), \qquad d \geq 1.$$

Then the following inequality holds

$$\phi(\wp(\alpha_1), \ldots, \wp(\alpha_m); d, H) > H^{-c_6 d^m}. \tag{2}$$

Here c_5 *and* c_6 *are a positive constants, depending only on* $\alpha_1, \ldots, \alpha_m$, g_2 *and* g_3.

This inequality was established for $m = 1$ by G. Chudnovsky [5] in 1980. In the general form with exponent $m + \varepsilon$ instead of m in the right hand side of the inequality (2) it was proved by E.M. Jabbouri [11] as a consequence of a result for algebraic groups.

Now let $\wp(z)$ have algebraic invariants g_2, g_3 and possibly be without complex multiplication. Let $\alpha_1, \ldots, \alpha_m$ be algebraic numbers linearly independent over $\mathbb{Q}$. One can prove (see [26]) that, among the numbers $\wp(\alpha_1), \ldots, \wp(\alpha_m)$ at least there are $\left[\frac{m}{2}\right]$ algebraically independent over $\mathbb{Q}$ and there exists a quantitative result like in the exponential case.

In the general case (g_2 and g_3 may be transcendental numbers) if $a_1, \ldots, a_p$ and $b_1, \ldots, b_q$ are complex numbers with almost the same condition on linear forms in a_i and b_j there are results about the lower estimates for the transcendence degree of the fields $L_1 = \mathbb{Q}(g_2, g_3, \wp(a_1 b_1), \ldots, \wp(a_p b_q))$, $L_2 = \mathbb{Q}(g_2, g_3, a_1, \ldots, a_p, \wp(a_1 b_1), \ldots, \wp(a_p b_q))$, $L_3 = \mathbb{Q}(g_2, g_3, a_1, \ldots, a_p, b_1, \ldots, b_q, \wp(a_1 b_1), \ldots, \wp(a_p b_q))$.

Small estimates were proved by W.D. Brownawell and K.K. Kubota, D.W. Masser and G. Wüstholz, R. Tubbs. In 1983 D.W. Masser and G. Wüstholz [19] under the assumption that g_2 and g_3 are algebraic, proved for $t = \operatorname{tr}\deg L_1$ the estimate

$$2^{t+2}(t + 8) \geq pq/(p + 2q).$$

Later in the papers of M. Waldschmidt [41] was published a sketch of the proof of the inequalities

$$\operatorname{tr}\deg L_1 \geq \frac{pq}{p + 2q} - 1, \qquad \operatorname{tr}\deg L_2 \geq \frac{pq + p}{p + 2q} - 1, \qquad \operatorname{tr}\deg L_3 \geq \frac{pq + 2q}{p + 2q} - 1.$$

The paper [37] of S. Shestakov contains the proof of the inequality $\operatorname{tr}\deg L_2 \geq \left[\frac{pq + p}{p + 2q}\right]$ from which follows "one half" of the elliptic variant of Gelfond's conjecture: *if* g_2, g_3 *are algebraic,* $\mathfrak{p}(z)$ *has complex multiplication over a quadratic field k,* β *is algebraic of degree d over k and* $u \neq 0$ *is such that* $\mathfrak{p}(u)$ *is algebraic, then among the numbers* $\mathfrak{p}(\beta u), \mathfrak{p}(\beta^2 u), \ldots, \mathfrak{p}(\beta^{d-1} u)$ *at least* $\left[\frac{d+1}{2}\right]$ *are algebraically independent over* $\mathbb{Q}$. In [29] a weaker estimate was proved.

The paper [41] contains estimates for fields, generated by the values of both exponential and elliptic function.

We should note here that recent articles of D.W. Masser and G. Wüstholz [18], P. Philippon [33] contain estimates for the number of zeros of polynomial defined on group varieties, which are very useful in the proofs of algebraic independence of numbers, especially of the values of elliptic and abelian functions.

5. Algebraic Base of Methods

Any homogeneous unmixed ideal $I \subset \mathbb{Z}[x_0, \ldots, x_m]$ may be characterized by numbers $\deg I$, $H(I)$, which are analogous to $\deg P$, $H(P)$ for polynomials $P \in \mathbb{Z}[x_0, \ldots, x_m]$. One may also define $|I(\bar{\omega})|$ for the point $\bar{\omega} = (\omega_0, \ldots, \omega_m) \in \mathbb{C}^{m+1}$, analogous to $|P(\bar{\omega})|$.

For exact definitions we need the notion of rank for ideals. The rank of the prime ideal $\mathfrak{p} \subset \mathbb{Z}[x_0, \ldots, x_m]$ is the maximal length of any increasing chain of prime ideals terminating with $\mathfrak{p}$. The rank of any ideal I is the minimal rank of prime ideals $\mathfrak{p}$, containing I. The rank of an ideal I will be denoted by $h(I)$. Recall that an ideal $I \subset \mathbb{Z}[x_0, \ldots, x_m]$ is called unmixed if all of the primary components of I have rank $h(I)$.

Let I be a homogeneous unmixed ideal of the ring $\mathbb{Z}[x_0, \ldots, x_m]$ and $r = m + 1 - h(I) \geq 1$. We introduce the linear forms

$$L_i(x) = \sum_{j=0}^{m} u_{ij} x_j, \qquad i = 1, 2, \ldots, r,$$

where u_{ij} are variables, $\bar{u}_i = (u_{i0}, \ldots, u_{im})$, and denote $\bar{I}(r)$ the ideal of $\mathbb{Z}[\bar{u}_1, \ldots, \bar{u}_m]$, consisting of the polynomials G for each of which there exists a natural number M such that Gx_i^M for all $i = 0, \ldots, m$ belong to the ideal generated in $\mathbb{Z}[x_0, \ldots, x_m, \bar{u}_1, \ldots, \bar{u}_r]$ by the linear forms $L_1, \ldots, L_m$ and the elements of I. One can prove that $\bar{I}(r) = (F)$ is a principal ideal. The polynomial F is a Chow form of I. We define (see [21]) $\deg I$ as the degree of F in the variables $\bar{u}_1$ and $H(I) = H(F)$. Now consider r skew-symmetric matrices $S^{(i)} = \|s^i_{jk}\|$, $0 \leq j, k \leq m$, $1 \leq i \leq r$, where we suppose that, except for the skew-symmetry $s^i_{jk} + s^i_{kj} = 0$, the variables s^i_{jk} are not connected by any algebraic relation over the ring $\mathbb{Z}[x_0, \ldots, x_m]$. If $æ(F)$ is the polynomial in the variables s^i_{jk}, $j < k$, which is obtained from F by substituting the vectors $S^{(i)}\bar{\omega}$, $i = 1, \ldots, r$, for the variables $\bar{u}_i$, we define

$$|I(\bar{\omega})| = |\bar{\omega}|^{-r * \deg I} * H(æ(F)).$$

Now the problem may be put as follows. For a given point $\bar{\omega} = (\omega_0, \ldots, \omega_m) \in \mathbb{C}^{m+1}$ we are searching for a lower estimate for $|I(\bar{\omega})|$ in terms of the $H(I)$ and $\deg I$. For the principal ideal $I = (P)$, its rank equals 1, and the quantities $|I(\bar{\omega})|$ and

$|P(\bar{\omega})|$, $\deg I$ and $\deg P$, $H(I)$ and $H(P)$ are closely related, which gives us possibility to obtain the lower estimate for $|P(\bar{\omega})|$ in terms of the properties of P.

For example, the proof of the Theorem 7 is based on the next assertion.

Theorem 8. *Let D and H be positive numbers satisfying the inequalities*

$$\ln \ln H > c_7 * D^m * \ln(D+1), \qquad D \geq 1.$$

For every integer r, $1 \leq r \leq m$ there exist constants $\lambda_r > 0$ and $\mu_r \geq 0$ such that for all homogeneous unmixed ideals $I \subset \mathbb{Z}[x_0, \ldots, x_m]$ with conditions $h(I) = m + 1 - r$,

$$\deg I \leq \lambda_r D^{m+1-r}, \qquad \ln H(I) \leq \lambda_r D^{m-r} \ln H$$

the inequality holds ($\bar{\omega} = (1, \mathfrak{p}(\alpha_1), \ldots, \mathfrak{p}(\alpha_m))$)

$$\ln|I(\bar{\omega})| \geq -\mu_r(D * \ln H(I) + \deg I * \ln H)D^{r-1}.$$

The Theorem 8 may be proved by induction on r. The step of induction contains two stages.

i) *Reduction of the estimate of $|I(\bar{\omega})|$ to the analogous estimate for prime ideals.* It is based upon the fact that numbers $\deg I$, $H(I)$ and $|I(\bar{\omega})|$ have almost linear behaviour under decomposition of the ideal into primary ideals.

Proposition 1 (see [22]). *Suppose that I is an unmixed homogeneous ideal in $\mathbb{Z}[x_0, \ldots, x_m]$, $h(I) \leq m$; there exist prime ideals $\mathfrak{p}_1, \ldots, \mathfrak{p}_s$ associated with I and natural numbers $k_1, \ldots, k_s$ such that*

1) $\sum_{l=1}^{s} k_l * \deg \mathfrak{p}_l \sim \deg I$,

2) $\sum_{l=1}^{s} k_l * \ln H(\mathfrak{p}_l) \leq \ln H(I) + m^2 \deg I$,

3) $\sum_{l=1}^{s} k_l * \ln |\mathfrak{p}_l(\bar{\omega})| \leq \ln |I(\bar{\omega})| + m^3 \deg I$.

ii) *Increasing of the rank of ideal.* The algebraic base of this stage is the next assertion.

Proposition 2 (see [24]). *Suppose that $\mathfrak{p}$ is a homogeneous prime ideal of $\mathbb{Z}[x_0, \ldots, x_m]$, $\mathfrak{p} \cap \mathbb{Z} = (0)$, $h(\mathfrak{p}) \leq m$ and Q is a homogeneous polynomial in $\mathbb{Z}[x_0, \ldots, x_m]$, $Q \notin \mathfrak{p}$. If $r = m + 1 - h(I) \geq 2$, then there exists an unmixed homogeneous ideal $J \subset \mathbb{Z}[x_0, \ldots, x_m]$ whose zeros coincide with the zeros of the ideal $(\mathfrak{p}, Q)$, for which $h(J) = m - r + 2$, and such that*

1) $\deg J \leq \deg \mathfrak{p} * \deg Q$,

2) $\ln H(J) \leq \deg Q * \ln H(\mathfrak{p}) + \deg \mathfrak{p} * \ln H(Q) + m(r+1) * \deg \mathfrak{p} * \deg Q$,

3) *If $\bar{\omega} \in \mathbb{C}^{m+1}$, $|Q(\bar{\omega})| < |\bar{\omega}|^{\deg Q}$ and ϱ is the distance between $\bar{\omega}$ and the set of zeros of the ideal $\mathfrak{p}$, then the following inequality holds*

$$\ln |J(\bar{\omega})| \leq \ln A + \deg Q * \ln H(\mathfrak{p}) + \deg \mathfrak{p} * \ln H(Q) + 11m^2 * \deg \mathfrak{p} * \deg Q,$$

where $A = |Q(\bar{\omega})| * |\bar{\omega}|^{-\deg Q}$, *if* $\varrho < |Q(\bar{\omega})| * |\bar{\omega}|^{-\deg Q}$, *and* $A = |\mathfrak{p}(\bar{\omega})|$ *in opposite case.*

If $r = 1$ *the right side of the inequality in* 3) *is nonnegative.*

If we can construct some polynomial Q, which is sufficiently small at the point $\bar{\omega}$ and does not belong to the ideal $\mathfrak{p}$, then the inequality from 3) and the induction assumption about the $|J(\bar{\omega})|$ give us a lower estimate for $|\mathfrak{p}(\bar{\omega})|$. The inequalities from 1) and 2) enable us to recompute this estimate in terms of $\deg \mathfrak{p}$ and $H(\mathfrak{p})$.

The constructions of the desired polynomials Q are based on the analytic properties of the functions considered and of course are different in different problems (for example in Theorems 4, 5, 7).

It is possible to axiomatize the situation and to prove a general assertion in which from the existence of a set of polynomials, sufficiently small at $\bar{\omega}$ and independent in some sense, follows a lower estimate for the number of algebraically independent coordinates of the vector $\bar{\omega}$. The first assertion of this kind for polynomials in one variable was proved by Gelfond [10] (see [40] for history). Recently P. Philippon [32] by means of the above ideas established a general algebraic independence criterion. One of the forms of it is

Theorem 9. *Suppose* $\omega_1, \ldots, \omega_m$ *are complex numbers algebraically independent over* $\mathbb{Q}$, $\bar{\omega} = (\omega_1, \ldots, \omega_m)$; γ *and* η *are positive numbers. Suppose that for any* $N \geq N_0$ *there exist polynomials* $P_1, \ldots, P_M \in \mathbb{Z}[x_1, \ldots, x_m]$ *with finite set of common zeros in the ball* $B(\bar{\omega}, \exp(-3\gamma N^\eta))$ *and with conditions*

$$\deg P_i + \ln H(P_i) \leq N, \qquad i = 1, \ldots, M,$$

$$\max_{1 \leq i \leq M} \{|P_i(\omega_1, \ldots, \omega_m)|\} < \exp(-\gamma N^\eta).$$

If γ *is sufficiently large with respect of* m, *then the inequality* $m > \eta - 1$ *holds.*

Jabbouri [12] obtained a quantitative form of this criterion which made it possible to estimate in this way the measure of algebraic independence of numbers.

References

1. Becker, P.G., Nishioka, K.: Measures for the algebraic independence of the values of Mahler type functions. C.R. Math. Rep. Sci. Canada **XI**, no. 3 (1989) 89–93
2. Beukers, F., Brownawell, W.D., Heckman, G.: Siegel normality. Ann. Math. **127** (1988) 279–308
3. Brownawell, W.D.: On the Gelfond-Feldman measure of algebraic independence. Comp. Math. **38**, no. 3 (1979) 355–368
4. Brownawell, W.D.: The effectivity of certain measures of algebraic independence. In: Seminaire de Theorie des Nombres, Paris 1984–1985. Birkhäuser, Basel 1986, pp. 41–50
5. Chudnovsky, G.V.: Algebraic independence of the values of elliptic functions at algebraic points; Elliptic analogue of the Lindemann-Weierstrass theorem. Invent. math. **61** (1980) 267–290
6. Diaz, G.: Grands degres de transcendance pour des familles d'exponentielles. C.R. Acad. Sci. Paris, Ser. 1 **305**, no. 5 (1987) 159–162

7. Diaz, G.: Une nouvelle mesure d'independance algebrique pour $(\alpha^{\beta}, \alpha^{\beta^2})$. Acta Arithmetica. **LVI** (1990) 25–32
8. Galochkin, A.I.: Estimate for the relative transcendence measure of the values of *E*-functions. Mat. Zametki **3**, no. 4 (1968) (Russian) 377–386
9. Gelfond, A.O.: On algebraic independence of algebraic powers of algebraic numbers. Dokl. Akad. Nauk USSR **64**, no. 3 (1949) (Russian) 277–280
10. Gelfond, A.O.: On the algebraic independence of transcendental numbers of certain classes. Usp. Mat. Nauk **4**, no. 5 (1949) (Russian) 14–48. [English transl.: Amer. Math. Soc. Transl. **2** (1962)]
11. Jabbouri, E.M.: Mesures d'independance algebrique de valeurs de fonctions elliptiques et abeliennes. C.R. Acad. Sci. Paris, Ser. I **303**, no. 9 (1986).
12. Jabbouri, E.M.: Sur un critere d'independance algebrique de P. Philippon. In: Seminaire d'Arithmetique de Saint-Etienne 1986–1987, IV.
13. Lang, S.: A transcendence measure for *E*-functions. Mathematika **9** (1962) 157–161
14. Loxton, J.H.: Automata and transcendence. In: Baker, A. (ed.) New advances in transcendence theory. Cambridge University Press, 1988, pp. 215–228
15. Loxton, J.H., van der Poorten, A.J.: Arithmetic properties of the solutions of a class of functional equations. J. Reine Angew. Math. **330** (1982) 159–172
16. Mahler, K.: Zur Approximation der Exponentialfunktion und des Logarithmus I. J. Reine Angew. Math. **166** (1932) 118–136
17. Mahler, K.: Arithmetische Eigenschaften einer Klasse transzendental-transcendenter Functionen. Math. Z. **32** (1930) 545–585
18. Masser, D.W., Wüstholz, G.: Zero estimates on group varieties I. Invent. math. **64** (1981) 489–516; II, Invent. math. **80** (1985) 233–267
19. Masser, D.W., Wüstholz, G.: Fields of large transcendence degree generated by values of elliptic functions. Invent. math. **72** (1983) 407–464
20. Nesterenko, Yu. V.: Estimates for the orders of zeros of functions of a certain class and applications in the theory of transcendental numbers. Izv. Akad. Nauk SSSR, Ser. Mat. **41**, no. 2 (1977) (Russian) 253–284. [English transl.: Math. USSR Izv. **11**, no. 2 (1977)]
21. Nesterenko, Yu. V.: On the algebraical independence of algebraic numbers to algebraic powers. In: Bertrand, D., Waldschmidt, M. (eds.) Approximations diophantiennes et nombres transcendants, Luminy 1982. (Progress in Mathematics, vol. 31). Birkhäuser, Basel 1983, pp. 199–220
22. Nesterenko, Yu. V.: On algebraic independence of algebraic powers of algebraic numbers. Mat. Sb. **123**(165), no. 4 (1984) (Russian) 435–459. [English. transl.: Math. USSR Sb. **51**, no. 2 (1985)]
23. Nesterenko, Yu. V.: On a measure of the algebraic independence of the values of certain functions. Mat. Sb. **128**(170), no. 4 (1985) (Russian) 545–568. [English transl.: Math. USSR Sb. **56**, no. 2 (1987)]
24. Nesterenko, Yu. V.: On a transcendence degree of certain fields generated by values of exponential function. Matem. Zametki **46**, no. 3 (1989) (Russian) 40–49
25. Nesterenko, Yu. V.: Effective bounds for the algebraic independence measure of the values of *E*-functions. Vestnik Moskov. Univ., Ser. I, no. 4 (1988) (Russian) 85–88
26. Nesterenko, Yu. V.: On a measure of the algebraic independence of the values of elliptic functions. In: Philippon, P. (ed) Approximations diophantiennes et nombres transcendants, Luminy 1990 (to appear)
27. Nishioka, K.: New approach in Mahler's method. J. Reine Angew. Math. **407** (1990) 202–219
28. Nishioka, K.: Algebraic independence measures of the values of Mahler functions. (To appear)
29. Philippon, P.: Varieties abeliennes et independance algebrique I. Invent. math. **70** (1983) 289–318

30. Philippon, P.: Varietes abeliennes et independance algebrique II. Invent. math. **72** (1983) 389–405
31. Philippon, P.: Criteres pour l'independance algebrique de familles de nombres. Math. Repts. Acad. Sci. Can. **6**, no. 5 (1984) 285–290
32. Philippon, P.: Criteres pour l'independance algebrique. Inst. Hautes Etudes Sci. Publ. Math. **64** (1986) 5–52
33. Philippon, P.: Lemmes de zeros dans les groupes algebriques commutatifs. Bull. Soc. Math. France **114** (1986) 355–383
34. Salichov, W.: Formal solutions of linear differential equations and their application in transcendental number theory. Trudy Moskov. Mat. Obshch. **51** (1988)
35. Salichov, W.: On algebraic independence of the values of hypergeometric *E*-functions. Dokl. Akad. Nauk SSSR **307**, no. 2 (1988) 284–287
36. Schneider, Th.: Transzendenzuntersuchungen periodischer Funktionen II; Transzendenzeigenschaften elliptischer Funktionen. J. Reine Angew. Math. **172** (1934) 70–74
37. Shestakov, S.: On transcendence degree of the fields generated by the values of elliptic function. Vestn. Moscow State Univ. (To appear)
38. Shidlovskij, A.B.: Transcendental Numbers. Nauka, Moskow, 1987 (Russian). [English transl.: Transcendental numbers. W. de Gruyter, Berlin New York 1989]
39. Siegel, C.L.: Über einige Anwendungen diophantischer Approximationen. Abh. Preuss. Akad. Wiss. Berlin, Phil.-Math. Kl., no. 1 (1929–1930) 1–70
40. Waldschmidt, M.: Algebraic independence of transcendental numbers. Gelfond's method and its developments. In: Jager, W., Moser, J., Remmert, R. (eds.) Perspectives in Mathematics, Anniversary of Oberwolfach. Birkhäuser, Basel 1984, pp. 551–571
41. Waldschmidt, M.: Algebraic independence of values of exponential and elliptic functions. J. Indian Math. Soc. **48** (1985) 215–228
42. Wüstholz, G.: Über das abelsche Analogon des Lindemannschen Satzes. Invent. math. **72** (1983) 363–388
43. Wüstholz, G.: Multiplicity estimates on group varieties. Ann. Math. **129** (1989) 501–517

Diophantine Problems and Linear Groups

Peter C. Sarnak

Department of Mathematics, Stanford University, Stanford, CA 94305, USA

1. Group Varieties

In this lecture we describe and exploit the relation between analytic Diophantine problems on homogeneous varieties and harmonic analysis on the corresponding groups. For the case $G = SL(2)$ this relation has been well studied and striking applications to analytic number theory have been found, especially by Iwaniec and his collaborators [I1]. Our focus here will be on general G where this aspect of the theory is still in a primitive state.

Let $V \subset A^n$ be a variety in affine n space and we assume that V is defined over $\mathbf{Q}$ (or more generally a number field $E/\mathbf{Q}$). Let $V(\mathbf{Z})$ denote the integral points in V and define the counting functions N by

$$N(T,V) = |\{m \in V(\mathbf{Z}) \mid \|m\| \leq T\}| \tag{1.1}$$

$$N(T,V,\xi,q) = |\{m \in V(\mathbf{Z}) \mid \|m\| \leq T,\ m \equiv \xi \pmod q\}|, \tag{1.2}$$

where $T \geq 1, q \in \mathbf{N}$, $\xi \in V(\mathbf{Z})$, and $\|\cdot\|$ is some suitably chosen Euclidean norm on A^n.

The basic problem is to determine the asymptotic behavior of N as $T \to \infty$. In certain special cases where there are sufficiently many additive terms in the equations defining V, the Hardy–Littlewood method may be applied, see Schmidt [Sc]. One obtains an asymptotic in the form

$$N(T,v) \sim c\,T^{\beta}, \qquad \beta \in \mathbf{N}. \tag{1.3}$$

The constant c is given by a product of local densities [Sc]. Another case, due to Moroz [M], that can be successfully handled by algebraic number theory is algebraic varieties of the form $f_1(x_1) = f_2(x_2) = \ldots = f_r(x_r)$, where $f_j(x_j)$ is a norm form of an order in a number field. Besides these cases[1] very little is known even in terms of upper bounds for N, see S.D. Cohen's bound described in Serre [SER]. However, if we assume that V is also a homogeneous variety for a linear algebraic group action, then much more can be said. In particular if $V \simeq H\backslash G$ with $H(\mathbf{R})\backslash G(\mathbf{R})$ an affine symmetric space then a reasonably complete

[1] Also see [F-M-T] for rational points on homogeneous projective varieties.

Proceedings of the International Congress
of Mathematicians, Kyoto, Japan, 1990

theory can be developed, see [D-R-S]. We give some concrete examples of the latter which go beyond the well studied case of quadratic forms.

Example 1. (a) For $k \neq 0$ let

$$V_k = \{x_{ij},\ i = 1,2,3,\ j = 1,2,3 \mid \det(x_{ij}) = k\} \subset A^9 . \tag{1.4}$$

V_1 is the group variety $G = SL(3)$, while for the general k, G acts on V_k by multiplication. In this way V_k decomposes into finitely many homogeneous spaces for G. We have [D-R-S]

$$N(T, V_k) = c_k\, T^6 + O(T^{23/4}) \qquad \text{as } T \to \infty .$$

Here and elsewhere the constant c_k is explicitly computable and is given in terms of products of local densities.

(b) Let

$$\begin{aligned}\varphi(x_1, x_2, x_3, x_4, x_5, x_6) &= x_1x_4x_6 + 2x_2x_3x_5 - x_1x_5^2 - x_6x_2^2 - x_4x_3^2 \\ &= \det \begin{pmatrix} x_1 & x_2 & x_3 \\ x_2 & x_4 & x_5 \\ x_3 & x_5 & x_6 \end{pmatrix} = \det X .\end{aligned} \tag{1.5}$$

Let

$$W_k = \{x \in A^6 \mid \varphi(x) = k\} , \qquad k \neq 0 .$$

$G = SL(3)$ acts on W_k by

$$X \mapsto {}^t gXg .$$

Again W_k decomposes into finitely many G orbits. We have

$$N(T, W_k) \sim d_k\, T^3 \qquad \text{as } T \to \infty .$$

(c) Let

$$D(a, b, c, d) = 18abcd + b^2c^2 - 4ac^3 - 4b^3d - 27a^2d^2$$

be the homogeneous form of degree four in four variables which is the discriminant of binary cubics. Let

$$U_k = \{(a, b, c, d) \mid D(a, b, c, d) = k\} .$$

$SL(2)$ acts on binary cubics and hence on U, which is again a homogeneous space. Relative to the norm

$$\|(a, b, c, d)\|^2 = a^2 + \frac{b^2}{3} + \frac{c^2}{3} + d^2$$

we have for $k \neq 0$

$$N(T, U_k) \sim C_k\, T^{2/3} \qquad \text{as } T \to \infty .$$

For the purposes of this report we specialize now to the case that V is a group variety G. We assume that $G(\mathbf{R})$ is semisimple and that we have a realization

$$G \subset GL(m) \subset A^{m^2}. \tag{1.6}$$

Without loss of generality we assume that

$$K = G(\mathbf{R}) \cap O(m, \mathbf{R}) \tag{1.7}$$

is a maximal compact subgroup of $G(\mathbf{R})$. Using (1.6) we get from the Euclidean norm on A^{m^2} a norm, $\|\cdot\|$, on $G(\mathbf{R})$ and hence on V. Let

$$\alpha(G) = \lim_{T\to\infty} \frac{\log \int_{\|g\|\le T} dg}{\log T}. \tag{1.8}$$

$\alpha(G)$ depends on the realization (1.6) but does not depend on the Haar measure used in definition (1.8). As usual let $\Gamma(q)$ be the principal congruence subgroup of $\Gamma(1) = G(\mathbf{Z}) = V(\mathbf{Z})$, that is

$$\Gamma(q) = \{\gamma \in G(\mathbf{Z}) \mid \gamma \equiv I \pmod q\}. \tag{1.9}$$

Denote by $W(q)$ the index $[\Gamma(1) : \Gamma(q)]$.

Theorem 1.1 [D-R-S]. *For* $\xi \in G(\mathbf{Z})$

$$N(T, V, \xi, q) \sim \frac{c_V}{W(q)} T^\alpha \qquad \text{as } T \to \infty.$$

Remarks 1.1. (a) A special case of this is well known. That is in hyperbolic spaces ($G(\mathbf{R}) = SO(n,1)$) where the problem reduces to counting lattice points in a non Euclidean ball. This goes back to Delsarte [DE], Selberg [SEL2], Huber [HUB], Patterson [P], and Lax-Phillips [L-P]. Also in this case, Elstrodt-Mennicke-Grunewald [E-G-M] have investigated the arithmetic aspects of the asymptotics. Bartels [BA] obtains the asymptotics of lattice points in a non-Euclidean ball (which in general is quite different from the above counting problem)[2] for general Riemannian symmetric spaces, when $\Gamma\backslash G(\mathbf{R})$ is compact.

(b) The constant c_V is essentially $\mathrm{Vol}\,(G(\mathbf{Z})\backslash G(\mathbf{R}))^{-1}$ which is non-zero by the Borel-Harish Chandra theorem [B-HC]. Moreover, by the theory of Tamagawa numbers, see [KNE], c_V is given by a product of local densities which at almost all places is the Hardy-Littlewood local density.

(c) If G is not semisimple some care must be taken. For example, if $V = \{(x,y) \mid x^2 - y^2 = 1\}$ then $V(\mathbf{Z})$ is finite, while for norm forms of number fields the asymptotics is of the form $c\,(\log T)^r$.

Theorem 1.1 can be improved to give remainder terms as well but these depend on the spectrum of $L^2(\Gamma(q)\backslash G(\mathbf{R})/K)$. More precisely, let $G(\mathbf{R}) = NAK$ be an

[2] See also Terras [TE], p. 248.

Iwasawa factorization of $G(\mathbf{R})$ and $W = \text{Weyl}(G(\mathbf{R}), A)$ the corresponding Weyl group. $S = G(\mathbf{R})/K$ is a globally Riemannian symmetric space. The algebra of invariant differential operators on S is isomorphic to the algebra of polynomials on $\mathfrak{a}^*_{\mathbf{C}}$ (where $\mathfrak{a} = \text{Lie}(A)$) which are W invariant [V]. Thus the spectrum of this ring of invariant operators acting on $L^2(\Gamma(q)\backslash G(\mathbf{R})/K)$ may be identified with a subset of $\mathfrak{a}^*_{\mathbf{C}}/W$ [V]. In fact it is a subset of $\hat{G}^1(\mathbf{R})$, the set of $\lambda \in \mathfrak{a}^*_{\mathbf{C}}/W$ which correspond to positive spherical functions. Note too that this spectral parameter space comes with a natural topology. We denote the above spectrum by $\sigma^1(\Gamma(q)\backslash G(\mathbf{R}))$, the super 1 denoting "class-1" representations [KN]. It consists of discrete eigenvalues as well as unitary Eisenstein series when $\Gamma(q)\backslash G$ is not compact [LS]. In our normalization of $\mathfrak{a}^*_{\mathbf{C}}$, $\varrho = 1/2$ sum of positive roots, corresponds to the spherical function 1 and $i\mathfrak{a}^*_{\mathbf{R}}$ to the tempered spectrum denoted $\hat{G}^1_{\text{temp}}(\mathbf{R})$. Note that the notion of spectrum $\sigma^1(\Gamma\backslash G(\mathbf{R}))$ makes sense for any discrete subgroup $\Gamma \leq G(\mathbf{R})$.

Definition 1.2. $\Gamma \leq G(\mathbf{R})$ (Γ discrete and not necessarily of finite volume) is called *tempered* if $\sigma^1(\Gamma\backslash G(\mathbf{R})) \subset \hat{G}^1_{\text{temp}}(\mathbf{R})$.

Theorem 1.3 [B-L-S]. *If $\Gamma < G(\mathbf{R})$ is tempered then*

$$N(T,\Gamma) = \sum_{\|\gamma\| \leq T} 1 = O_\varepsilon(T^{\alpha/2+\varepsilon}), \quad \textit{for all } \varepsilon > 0.$$

One might (optimistically) hope that the "Ramanujan conjectures" of $G = SL(2)$ (see Selberg [SEL], Satake [SA]) would be true in this general setting. That is for all $q \geq 1$

$$\sigma^1(\Gamma(q)\backslash G(\mathbf{R})) \subset \hat{G}^1_{\text{temp}}(\mathbf{R}) \cup \{\varrho\}. \tag{1.13}$$

Such a bound would lead to optimally small remainder terms in Theorem 1.1, at least for smoothed versions thereof. Unfortunately (1.13) is not true in general. The first counterexamples to (1.13) (at least to its analogue at the finite places) were found in $SP(4)$, numerically by Kurokawa [KU] and using θ-liftings by Howe and Piatetski-Shapiro [H-PS]. Actually, we can see that (1.13) is not true in general for certain structural reasons that come from an examination of $N(T,\Gamma)$. Indeed the truth of (1.13) would imply the same for any $\Gamma(q) \subset \Gamma \subset \Gamma(1)$, i.e., any congruence subgroup. Now suppose that $\Delta = \cap_{i=1}^{\infty}\Gamma_i$ is an intersection of congruence groups. Δ need no longer be of finite index in $G(\mathbf{Z})$. It is easy to see [B-L-S2] that

$$\sigma^1(\Delta\backslash G(\mathbf{R})) \subset \bigcap_{j=1}^{\infty} \overline{\bigcup_{i=j}^{\infty} \sigma^1(\Gamma_i\backslash G(\mathbf{R}))}\,. \tag{1.14}$$

In particular if $\sigma^1(\Delta\backslash G(\mathbf{R}))$ is nontempered (and does not contain ϱ) then the same must be true of $\sigma^1(\Gamma\backslash G(\mathbf{R}))$ for $\Gamma = \cap_{i=1}^{\ell}\Gamma_i$ where ℓ is sufficiently large. That is (1.13) fails for suitable large q. To find Δ as above with $\sigma^1(\Delta\backslash G(\mathbf{R}))$ nontempered we note that if $H < G$ is a Q-subgroup then $\Delta = H(\mathbf{Z})$ is the intersection of congruence groups [B-L-S2]. Moreover we can compute $N(T, H(\mathbf{Z}))$ by virtue of

$H(\mathbf{Z})$ being (essentially) a lattice in $H(\mathbf{R})$. Hence using Theorem 1.3 we can show $H(\mathbf{Z}) = \Delta$ is nontemprered by showing it has too many elements. We illustrate this with an example.

Let $G = SO(n+1,1) \subset GL(n+2, \mathbf{R})$ be the orthogonal group of the form

$$F(x_1, \ldots, x_{n+2}) = x_1^2 + x_2^2 + \ldots + x_{n+1}^2 - x_{n+2}^2 . \tag{1.15}$$

Then

$$\alpha(G) = n \qquad \text{for } n \geq 1 . \tag{1.16}$$

If $H < G$ corresponds to the subgroup fixing $x_1, \ldots, x_m$ then $H \simeq SO(n+1-m, 1)$ and

$$\alpha(H(\mathbf{Z})) = n - m . \tag{1.17}$$

Hence H is nontempered if

$$1 \leq m < \frac{n}{2} . \tag{1.18}$$

Thus for $n \geq 3$, $G(\mathbf{Z})$ has congruence subgroups with exceptional spectrum, that is for which (1.13) fails. In fact, a more refined analysis shows that

$$\left\{ \varrho = \frac{n}{2}, \frac{n}{2} - 1, \frac{n}{2} - 2, \ldots, \frac{n}{2} - \left[\frac{n}{2}\right] \right\} \tag{1.19}$$

are limit points of exceptional eigenvalues of congruence subgroups of $G(\mathbf{Z})$. An explicit lower bound for the automorphic spectrum $\overline{\cup_q \sigma(\Gamma(q) \backslash G(\mathbf{R}))}$ is developed in [B-L-S2] along these and related lines. The results show the important role played by all Q-subgroups in the general Ramanujan conjectures. In fact one obtains a powerful and general method to construct automorphic forms and spectrum. This new method complements the only other general method for constructing automorphic forms viz. theta liftings [Ho].[3] In particular it leads to new results on nonvanishing of Betti numbers of arithmetic lattices [B-L-S2].

To end this section a word about the proofs of Theorem 1.1 and its generalizations. The problem reduces to one of counting points of $\Gamma = G(\mathbf{Z})$ (or subgroups thereof) lying in an expanding family of regions $R_T \subset G(\mathbf{R})$. One forms the Γ invariant function

$$F_T(g) = \sum_{\gamma \in \Gamma} \chi_{R_T}(\gamma g) ,$$

where χ_R is the characteristic function of R. Under suitable assumptions about R_T, F_T can be asymptotically spectrally analyzed. When successful, one deduces from the asymptotic behavior of certain eigenfunctions on $G(\mathbf{R})/K$, such as the spherical functions, that the main term in the asymptotics of $F_T(1)$ comes from the component of F_T along constants. The latter is $\mathrm{Vol}(R_T)/\mathrm{Vol}(G(\mathbf{Z}) \backslash G(\mathbf{R}))$. Various L^2 averaging arguments reduce the spectral analysis to a minimum so that a result like Theorem 1.3 may be deduced without any detailed knowledge of the nature of the spectrum (here $\mathrm{Vol}(\Gamma \backslash G(\mathbf{R})) = \infty$).

---

[3] We add, however, that when lifting by the trace formula can be carried out successfully it typically yields the most complete picture known – see for example Arthur-Clozel [A-C].

2. Density Hypothesis

We have noted that the naive Ramanujan conjecture is false in most cases. Borrowing from the analytic theory of the zeta function we formulate certain weaker but still very useful hypotheses. If $\nu \in \hat{G}^1(\mathbf{R})$ is an eigenvalue occuring in $L^2(\Gamma(p)\backslash S)$, p a large prime, then we get a representation of the finite group $\Gamma(1)/\Gamma(p)$ on the eigenspace corresponding to ν. Under suitable assumptions $\Gamma(1)/\Gamma(p) \simeq G(\mathbf{Z}/p\mathbf{Z})$ and we may use the lower bounds on the dimensions of the smallest nontrivial irreducible representations of $G(\mathbf{Z}/p\mathbf{Z})$ [H-H, LUS] to deduce a lower bound on the multiplicity $m(\nu, \Gamma(p))$ of ν in $L^2(\Gamma(p)\backslash S)$

$$m(\nu, \Gamma(p)) \gg p^{\delta(G)}, \quad \delta(G) \geq 1. \tag{2.1}$$

In particular in such a case if an exceptional eigenvalue occurs it must do so with high multiplicity. The substitute for Ramanujan is a sharp upper bound for this multiplicity. Notice that we are sticking to the principal congruence subgroups $\Gamma(q)$. In fact this is forced on us in view of the Q-subgroup phenomenon discussed in the previous Section. We begin with a basic conjecture concerning the integer points on the group variety V_G. For what follows we assume that V_G is as in (1.6) and the norm is the Euclidean one on $\mathbf{R}^{m^2}$.

Main Conjecture 2.1 [S-X]. *For* $T, q \geq 1$

$$N(T, V, I, q) \ll_\varepsilon \frac{T^{\alpha+\varepsilon}}{W(q)} + T^{\alpha/2}, \quad \varepsilon > 0. \tag{2.2}$$

Remarks 2.1. (a) The point of this conjecture is the uniformity in T and q. The value of T for which it is most useful (and hardest to prove) is when the terms on the right hand side are equal. Theorem 1.1 implies (2.2) is true for fixed q.

(b) The Main Conjecture (MC) follows from the naive Ramanujan conjecture 1.13.

(c) MC has been established in two important cases:

(i) If Γ is any arithmetic lattice in $G(\mathbf{R}) = SL(2,\mathbf{R})$ or $SL(2,\mathbf{C})$ (see [BO] for a classification) and $\Gamma(q)$ are corresponding congruence subgroups then MC is true [S-X].

(ii) J. Katznelson [KA] has recently established MC for $G = SL(m)$, $m \geq 2$.

We consider the implications of MC. We begin with the density hypothesis and consider only the case $G(\mathbf{R})$ of (real) rank 1. The reason is that in this case there is a natural ordering of the spectrum. In the general case one must order by regions. If $G(\mathbf{R})$ is rank 1 then $\hat{G}^1(\mathbf{R})$ is contained in

$$i\mathfrak{a}^*_{\mathbf{R}} \cup (0, \varrho] \subset \mathfrak{a}^*_{\mathbf{C}}. \tag{2.3}$$

For $0 \leq \nu \leq \varrho$ let

$$M(\nu, q) = \sum_{\nu \leq \mu \leq \varrho} m(\mu, \Gamma(q)). \tag{2.4}$$

Thus $M(\nu, q)$ counts the number of exceptional eigenvalues which are more nontempered than ν. Clearly

$$M(\varrho, q) = 1$$
$$M(0, q) = O(W(q)).$$

The Density Hypothesis is the linear interpolation of these

Density Hypothesis 2.2 [S-X]. *Assume* $G(\mathbf{R})$ *is rank 1, then*

$$M(\nu, q) = O_\varepsilon(W(q)^{1-\nu/\varrho+\varepsilon}), \quad \textit{for all } \varepsilon > 0.$$

Proposition 2.3 [S-X]. MC $\Rightarrow$ *Density Hypothesis.*

Remarks 2.3. (a) Combining (2.3) with Remark 2.1 (c) we find that the Density Hypothesis is true for any arithmetic lattice Γ (relative to its principal congruence subgroups) in $SL(2, \mathbf{R})$ or $SL(2, \mathbf{C})$. For $\Gamma = SL(2, \mathbf{Z})$ this was first proved by H. Iwaniec and J. Szmidt [I-S] and M. Huxley [HU] using trace formulae. Iwaniec [I2] has obtained an even stronger bound for $\Gamma_0(q)$ in $SL(2, \mathbf{Z})$.

The Main Conjecture also implies bounds on the multiplicities of the $\pi \in \hat{G}(\mathbf{R})$, the unitary dual of $G(\mathbf{R})$, which appear discretely in the decomposition of the regular representation of $G(\mathbf{R})$ on $L^2(\Gamma(q)\backslash G(\mathbf{R}))$. Denote this multiplicity by $m(\pi, \Gamma(q))$. For π class 1, π corresponding to ν as in the previous Section, we have $m(\nu, \Gamma(q)) = m(\pi, \Gamma(q))$ by the duality theorem [G-G-PS]. If $\pi \in \hat{G}(\mathbf{R})$ we let $p(\pi)$ be the infimum over all $p \geq 2$ such that all matrix coefficients of π are in $L^p(G)$ [KN]. For $\pi = \varrho$ the trivial representation, $p(\pi) = \infty$ and $m(\pi, \Gamma(q)) = 1$. For $\pi \in L^2(G)$, i.e., π discrete series, $p(\pi) = 2$ and $m(\pi, \Gamma(q))$ is of order $W(q)$, see deGeorge-Wallach [G-W]. Linearly interpolating leads to

Conjecture 2.3 [S-X]. *Let* $\pi \in \hat{G}(\mathbf{R})$

$$m(\pi, \Gamma(q)) = O_\varepsilon(W(q)^{2/p(\pi)+\varepsilon}).$$

We have

Proposition 2.4 [S-X]. *If* $G(\mathbf{R})$ *is of rank 1 then* MC $\Rightarrow$ *Conjecture* 2.3.

Remarks 2.4. (a) If we drop the assumption that $G(\mathbf{R})$ is of rank 1 then we can prove (in the case $G(\mathcal{O}_E)\backslash G(\mathbf{R})$ is compact) that MC implies a somewhat weaker bound for multiplicities.

(b) Conjecture (2.3) as well as the others are typically false for noncongruence families of $G(\mathbf{Z})$. See for example the Abelian covers in Section 3.

(c) Combining the bound in Conjecture 2.3 with Matsushima's formula [B-W] in the case $\Gamma\backslash G(\mathbf{R})$ is compact, gives rather sharp bounds for the Betti numbers of principal congruence subgroups of Γ.

Finally Conjecture 2.3 implies certain "arithmetic vanishing theorems" and in particular lower bounds on small eigenvalues. The lower bound (2.1) applies to any representation $\pi \in \hat{G}(\mathbf{R})$ and hence we conclude:

Under Conjecture 2.3, if $p(\pi) > 2/\delta(G)$ then $m(\pi, \Gamma(q)) = 0$ for all q. (2.5)

As an application of this where MC has been established we have the following [S-X]:

Let $\Gamma \leq SL(2,\mathbf{R})$ (or $SL(2,\mathbf{C})$) be arithmetic. Thus Γ is commensurable with $G(\mathcal{O}_E)$ where E is a number field over $\mathbf{Q}$ and G is the unit group of a quaternion algebra A over E which is ramified at all infinite places of E except one, call it v, and $A \otimes v \simeq M(2,\mathbf{R})$. Let $\Gamma(P) = G(P) \cap \Gamma$, where P is a prime ideal of $\mathcal{O}_E$. Denote by $\lambda_1(P)$ the smallest non zero eigenvalue of the Laplacian on $\Gamma(P)\backslash\mathbf{H}$, $\mathbf{H}$ being the hyperbolic plane. Clearly $\lambda_1(P) \leq \lambda_1(\Gamma)$ and the last may be arbitrarily small, see Randol [R].

Proposition 2.5 [S-X]. *For* Norm (P) *large*

$$\min\left(\lambda_1(\Gamma), \frac{5}{36}\right) \leq \lambda_1(\Gamma(P)) \leq \lambda_1(\Gamma).$$

The number 5/36 also appears in Huxley [HU] for essentially the same reason; viz. (2.5). This kind of vanishing theorem in (2.5) can also be derived for the graphs constructed in [L-P-S] to prove, in an entirely elementary fashion, that they are expanding families with good expansion coefficients.

Arithmetic vanishing results of the type of (2.5) have been established unconditionally for certain (in particular noncompact) congruence lattices in $SO(n,1)$ by Elstrodt-Grunewald-Mennicke [E-G-M2] and Li-Piatetski Shapiro-Sarnak [L-PS-S]. The method used in these papers is quite different (and special) and goes via Kloosterman sums as in Selberg [SEL].

3. Variations

We have restricted our discussion to $G(\mathbf{R})$ and its spectrum. There are numerous Diophantine problems which are related to $G(\mathbf{Q}_p)$ for finite p. We illustrate this with certain Diophantine problems studied by Linnik [LIN, LIN2]. Let $F(x_1,\ldots,x_r)$ be a homogeneous form in r variables and of degree d. For $m \in \mathbf{N}$ and $x \in \mathbf{Z}^r$ for which

$$F(x) = m \tag{3.1}$$

we associate to x,

$$x' = \frac{x}{m^{1/d}}. \tag{3.2}$$

Thus $F(x') = 1$. With the use of his ergodic method Linnik was able, in certain cases, to study the distribution of these points x' on the surface $V_1 = \{F = 1\}$ as $m \to \infty$.

For example, let

$$F(x_1,x_2,\ldots,x_r) = x_1^2 + x_2^2 + \ldots + x_r^2, \qquad r \geq 3. \tag{3.3}$$

The basic problem is the equidistribution with respect to Lebesgue measure of the x' on the unit sphere S^{r-1}. Linnik [LIN] proved this for $r \geq 4$ and conditionally for $r = 3$ [LIN]. By the use of classical θ–functions [S1] this problem is immediately reduced to one of estimating Fourier coefficients of classical modular forms (of half integer weight if r is odd). For $r \geq 4$ standard estimates give the result. For $r = 3$ the estimates needed lie much deeper and were only proven recently by Iwaniec [I2]. The general case of this problem for quadratic forms in three variables requires similar estimates for Fourier coefficients of Maass forms as well. These were established by W. Duke [D] whose work resolves this instance of Linnik's problem completely.

Concerning forms of higher degree Linnik and Skubenko [L-S] considered the form

$$F(x_{ij}) = \det(x_{ij}), \quad i, j = 1, \ldots, n. \tag{3.4}$$

Using the ergodic method they show that if $\Omega \subset \mathbf{R}^{n^2}$ is a (nice) compact set and

$$R_\Omega(m) = |\{x \in \mathbf{Z}^{n^2} \mid F(x') \in \Omega\}| \tag{3.5}$$

then

$$R_\Omega(m) \sim \frac{\text{meas}\,(\Omega')}{\zeta(2)\zeta(3)\ldots\zeta(n)} \prod_{p_i^{k_i}} \frac{(p_i^{k_i+1}-1)(p_i^{k_i+2}-1)\ldots(p_i^{k_i+n-1}-1)}{(p_i-1)(p_i^2-1)\ldots(p_i^{n-1}-1)}, \tag{3.6}$$

where $\Omega' = \Omega \cap SL(n, \mathbf{R})$, meas (Ω') its Euclidean content, and $m = p_1^{k_1} p_2^{k_2} \ldots p_\nu^{k_\nu} \to \infty$.

We outline a proof of (3.6) using harmonic analysis on $\hat{G}(\mathbf{Q}_p)$. Let $G(\mathbf{R}) = SL(n, \mathbf{R})$, $\Gamma = SL(n, \mathbf{Z})$. The usual Hecke operators [TE], T_p (or more generally T_m) are defined by

$$\left.\begin{aligned} &T_p : L^2(\Gamma \backslash G(\mathbf{R})) \to L^2(\Gamma \backslash G(\mathbf{R})) \\ &T_p H(g) = \sum_{\substack{\det M_j = p \\ M_j \text{ integral} \\ M_j \not\equiv M_k \ (\text{mod}\ \Gamma)}} H\left(\frac{1}{p^{1/n}} M_j g\right) \end{aligned}\right\} \tag{3.7}$$

Let $f \in C_0^\infty(G(\mathbf{R})$, then

$$\sum_{\substack{\det X = p \\ X,\, n\times n \text{ integral}}} f(x') = (T_p H)(I), \tag{3.8}$$

where

$$H(g) = \sum_{\gamma \in \Gamma} f(\gamma g). \tag{3.9}$$

Now

$$\int_{\Gamma \backslash G(\mathbf{R})} H(g) dg = \int_{G(\mathbf{R})} f(g) dg \tag{3.10}$$

(for a fixed Haar measure dg) from which it is easy to see that (3.6) follows from

$$\| T_m |_B \|_2 \ll m^{n-1-\delta} \tag{3.11}$$

for $\delta > 0$ ($\delta = \delta(n)$), where

$$B = \left\{ h \in L^2(\Gamma\backslash G(\mathbf{R})) \mid \int_{\Gamma\backslash G(\mathbf{R})} h(g)dg = 0 \right\}.$$

For $n \geq 3$, (3.11) follows from the classification of the spherical dual $\hat{G}(\mathbf{Q}_p)$ [TA], together with the standard relation between Hecke operators and representations of $G(\mathbf{Q}_p)$. For $n = 2$ we need well known, nontrivial bounds on the Fourier coefficients of holomorphic and Maass forms on $SL(2,\mathbf{Z})\backslash\mathbf{H}$ to establish (3.11).

By examining more carefully the best δ that can be achieved in (3.11) (again this is the "Ramanujan Conjecture") or the best estimates for Fourier coefficients in the other cases, one can go well beyond the equidistribution of the points x'. In fact one can examine how well the points x' with $F(x) = m$ cover the set V_1. The harmonic analysis method above leads to sharp results [D-R-S]. For an elementary approach concerning the last problem, in the case $SL(n)$, see Harman [HA]. Determining the optimal covering exponent for these points (as $m \to \infty$) is an interesting and apparently difficult problem [D-R-S].

Finally we make comments related to Conjecture 2.3 but in connection with other families of coverings of $G(\mathbf{Z})\backslash S$. Assume that $H_1(\Gamma,\mathbf{Z}) \simeq \mathbf{Z}^r$ with $r \geq 1$. Then there are uniquely defined $(\mathbf{Z}/q\,\mathbf{Z})^r$ Galois coverings of $\Gamma\backslash S$ given by $\Phi(q)\backslash S$ where

$$\Phi(q) = \{\gamma \in \Gamma \mid p(\gamma) \equiv 0 \pmod{q}\}, \tag{3.12}$$

$p : \Gamma \to H_1(\Gamma,\mathbf{Z})$ being the canonical projection. An interesting example of this is the family $F_q : x^q + y^q = 1$ of Fermat curves. These are $(\mathbf{Z}/q\,\mathbf{Z})^2$ Galois coverings of $\Gamma(2)\backslash\mathbf{H}$ where $\mathbf{H} = \{z \mid \operatorname{Im} z > 0\}$ and $\Gamma(2) < SL(2,\mathbf{Z})$ is the principal congruence subgroup of level 2, see [LA1]. Another example is the family of Hirzebruch surfaces [HI].

For such families the behavior of $m(\pi,\Phi(q))$, for $\pi \in \hat{G}(\mathbf{R})$ fixed, is closely tied to another Diophantine problem; one of counting the number of q–division points in an r dimensional torus which lie on a real analytic subvariety, see [S2]. For $\pi \in \hat{G}(\mathbf{R})$ which correspond to cohomology, the subvariety V in question is algebraic in the sense that it is a zero set (or intersection of such) of a trigonometric polynomial. For these there is a definite structure to the set of such torsion points lying on V. This was conjectured by Lang [LA2] and proved in Laurent [LAU]. See also [S2] for an effective solution. As a consequence one has

Theorem 3.1. [S2] *Let $M(q) = \Phi(q)\backslash S$ as above, then the j–th Betti numbers $\beta^{(j)}(M(q))$ are polynomial periodic in q, that is*

$$\beta^{(j)}(M(q)) = \sum_{\nu=0}^{r} a_\nu^{(j)}(q)q^\nu$$

with $a_\nu^{(j)}(q)$ periodic in q.

This is in sharp contrast to the congruence family. For the general $\pi \in \hat{G}(\mathbf{R})$ we cannot conclude such strong results. However, using some recent work of Bombieri-Pila [B-P] one can get asymptotic results for $m(\pi,\Phi(q))$ for the Fermat and related families [P-S]. Actually for the Fermat family we prove in [P-S] an exact formula for the number of exceptional eigenvectors coming from Eisenstein series residues. In the notation of (2.4) with obvious modifications

$$M^{\mathrm{Eis}}(0,q) = 6\left[\frac{q}{8}\right]. \tag{3.13}$$

What is striking about (3.13) is that the constant term of the Eisenstein series for the noncongruence groups $\Phi(q)$ corresponding to F_q, are not (as far as is known) computable in terms of classical zeta like functions.

We have touched on some aspects of the analysis of $G(\mathbf{Z})\backslash G(\mathbf{R})$ which are directly related to analytic Diophantine problems. The aspects of the theory relating to L-functions associated to automorphic forms on general G, have been well developed at the hands of Langlands, Piatetski-Shapiro, and their coworkers.

Acknowledgements. I would like to thank M. Burger, W. Duke, Jian Shu Li, Z. Rudnick, and X. Xue. My various collaborations with them form the core of this report.

References

[A-C] J. Arthur, L. Clozel: Simple algebras, base change, and the advanced theory of the trace formula. Ann. Studies, vol. 120, Princeton 1989

[BA] H.J. Bartels: Nichteuklidische Gitterpunktprobleme und Gleichverteilung in linearen algebraischen Gruppen. Comment. Math. Helv. **57** (1982) 158–172

[BO] A. Borel: Commensurability classes of Hyperbolic 3 manifolds. Ann. Sci. Norm. Sup. Pisa, Cl. Sci. **8** (1981) 1–33

[B-HC] A. Borel, Harish–Chandra: Arithmetic subgroups of algebraic groups. Ann. Math. **75** (1962) 485–535

[B-W] A. Borel, N. Wallach: Continuous cohomology, discrete subgroups, and representations of reductive groups. Ann. Math. Studies, vol. 94, Princeton 1980

[B-P] E. Bombieri, J. Pila: The number of points on curves and ovals. Duke Math. J. **59** (1989) 337–358

[B-L-S1] M. Burger, Li Juan Shu, P. Sarnak: Ramanujan duals and automorphic spectrum. Preprint 1990

[B-L-S2] M. Burger, Li Juan Shu, P. Sarnak. In preparation

[DE] J. Delsarte: Sur le Gitter Fuchsien. C.R. Acad. Sci. Paris **214** (1942) 147–149

[D] W. Duke: Hyperbolic distribution problems and half integral weight Maass forms. Inv. math. **92** (1988) 73–90

[D-R-S] W. Duke, Z. Rudnick, P. Sarnak: The density of integral points on affine homogeneous varieties. Preprint 1990

[E-G-M1] J. Elstrodt, F. Grunewald, J. Mennicke: Arithmetic applications of the hyperbolic lattice point theorem. Proc. London Math. Soc. (3) **57** (1988) 239–283

[E-G-M2] J. Elstrodt, F. Grunewald, J. Mennicke: Poincaré series, Kloosterman sums and eigenvalues of the Laplacian for congruence groups acting on hyperbolic spaces. C.R. Acad. Sci. Paris **305** I (1987) 537–581

[F-M-T] J. Franke, Yu. Manin, Yu Tschinkel, Rational points of bounded height on Fano varieties. Inv. math. **95** (1989) 421–435

[G-G-PS] I. Gelfand, M. Graev, I. Piatetski-Shapiro: Generalized function, vol. 6. Saunders, Philadelphia 1969

[GW] de George, N. Wallach: Limit formulas for multiplicities in $L^2(\Gamma\backslash G)$. Ann. Math. **107** (1978) 133–150

[HA] G. Harman: Approximation of real matrices by integral matrices. J. Number Theory **34** (1990) 63–81

[H-H] M. Harris, C. Hering: On the smallest degrees of projective representations of the group $PSL(n, q)$. Can. J. Math. **23** (1971) 90–102

[HO] R. Howe: θ-series and invariant theory. Proc. Symp. A.M.S. XXXIII (1979) 315–322

[H-PS] R. Howe, I. Piatetski–Shapiro: A counter example to the generalized Ramanujan conjecture for quasi split groups. Proc. Symp. A.M.S. XXXIII (1979) 315–322

[HUB] H. Huber: Über eine neue Klasse automorpher Funktionen und Gitterpunkt Problem in der hyperbolischen Ebene. Comment. Mat. Helv. **30** (1956) 20–62

[Hu] M. Huxley: Exceptional eigenvalues and congruence groups. AMS Contemp. Math. **53** (1985) 341–351

[I1] H. Iwaniec: Spectral theory of automorphic functions and recent developments in analytic number theory. Proc. Int. Cong. 1986, Berkeley, pp. 444–456

[I2] H. Iwaniec: Fourier coefficients of modular forms of half integral weight. Inv. math. **87** (1987) 385–401

[I3] H. Iwaniec: Small eigenvalues for congruence groups. Acta Arithmetica, to appear.

[I-S] H. Iwaniec, J. Szmidt: Density theorems for exceptional eigenvalues of the Laplacian for congruence groups. Banach Center Publ. **17** (1985) 317–331

[KN] A. Knapp: Representation theory of semisimple groups. Princeton U. Press 1986

[KA] J. Katznelson, Ph.D. Thesis, Stanford (to appear)

[KNE] M. Knesser: Semisimple algebraic groups. In: Algebraic number theory, eds. Cassels and Frölich. Academic Press 1986, pp. 250–265

[KU] N. Kurokawa: Examples of eigenvalues of Hecke operators on Siegel modular forms of degree two. Inv. math. **49** (1978) 149–165

[LA1] S. Lang: Introduction to algebraic and Abelian functions. Springer, Berlin Heidelberg New York 1982

[LA2] S. Lang: Fundamentals of Diophantine Geometry. Springer, Berlin Heidelberg New York 1983

[LAU] M. Laurent: Equations Diophantiennes exponentielles. Inv. math. **78** (1984) 299–327

[L-P] P. Lax, R. Phillips: The asymptotic distribution of lattice points in Euclidean and non Euclidean spaces. J. Funct. Anal. **46** (1982) 280–350

[LI] Li Jian Shu: Kloosterman zeta functions on complex hyperbolic spaces. Preprint.

[L-PS-S] Li Jian Shu, I. Piatetski–Shapiro, P. Sarnak: Poincaré series for $SO(n, 1)$. Proc. Ind. Acad. Sc. **97** (1987) 231–237

[LIN] Y.V. Linnik: Ergodic properties of algebraic fields. Ergebnisse der Mathematik, vol. 45. Springer, Berlin Heidelberg New York 1967

[LIN2] Y.V. Linnik: Additive problems and eigenvalues of the modular operators. Proc. Int. Cong. Stockholm 1962, pp. 270–284

[L-S] Y.V. Linnik, B.F. Skubenko: Asymptotic distribution of integral matrices of third order. Vest. Leniner. Univ. Ser. Math. **13** (1964) 25–36

[L-P-S] A. Lubotzky, R. Phillips, P. Sarnak: Ramanujan graphs. Combinatorica **8** (1988) 261–277

[LUS] G. Lustig: Irreducible representations of finite classical groups. Inv. math. **43** (1977) 125–177

[M] B. Moroz: On the numbe of integral points on a norm form variety in a cube like domain. J. Number Theory **27** (1987) 106–110

[P] S.J. Patterson: A lattice point problem in hyperbolic space. Mathematica **22** (1975) 81–88

[P-S] R. Phillips, P. Sarnak: The spectrum of Fermat curves. GAFA **2** (1991) 80–146

[R] B. Randol: Small eigenvalues of the Laplace operator on compact Riemann surfaces. Bull. A.M.S. **80** (1974) 996–1000

[S1] P. Sarnak: Some applications of modular forms. Cambridge University Press (to appear)

[S2] P. Sarnak: Betti numbers of congruence subgroups. To appear in Israel Math. J.

[S-X] P. Sarnak, X. Xue: Bounds for multiplicities of representations. To appear in Duke Math. J.

[SA] Satake: Spherical functions and the Ramanujan conjecture. Proc. A.M.S. **9** (1967) 258–364

[SEL] A. Selberg: On the estimation of Fourier coefficients of modular forms. Proc. Symp. Pure Math. **8** (1965) 1–15

[SEL2] A. Selberg: Harmonic analysis and discontinuous groups in weakly symmetric Riemannian spaces with applications to Dirichlet series. J. Ind. Math. Soc. **20** (1956) 47–87

[SER] J.P. Serre: Lectures on the Mordell–Weil theorem. Vieweg, Aspects Math. 1989

[SC] W. Schmidt: The density of integer points on homogeneous varieties. Acta Math. **154** (1985) 243–296

[TA] M. Tadic: Spherical unitary dual of the general linear group over a non Archimedian local field. Ann. Inst. Fourier **36** (1986) no. 2, 47–55

[TE] A. Terras: Harmonic analysis on symmetric spaces and applications II. Springer, Berlin Heidelberg New York 1987

[V] V. S. Varadarajan: The eigenvalue problem on negatively curved compact locally symmetric manifolds. A.M.S. Contemp. Math. **53** (1985) 449–463

Theory of Mordell-Weil Lattices

Tetsuji Shioda

Department of Mathematics, Faculty of Science, Rikkyo University
Nishi-Ikebukuro 3-chome, Tokyo 171, Japan

0. Introduction

The basic idea of Mordell-Weil lattices is to view the Mordell-Weil group (= the group of rational points on an elliptic curve or an abelian variety defined over some "global" field) as a Euclidean lattice by means of suitable inner product.

The necessary setup for this has been known for some time. First of all, the finite generation of such a group was established by Mordell and Weil in 1920s (thus named after them) in the case of elliptic curves over $\mathbb{Q}$ [Mo] or Jacobian varieties over a number field [W1]. After the general theory of abelian varieties was founded by Weil [W2], it was extended by Néron and Lang in 50s to a more general situation [L]. Second the notion of the canonical height on abelian varieties was developed by Néron, Tate and Manin in 60s [N2, T1, T3, M1; cf. L, Se2]. A more geometric method, based on the theory of elliptic surfaces [K], was used by the author [S1] and Cox-Zucker [CZ] in 70s.

Certainly the lattice-theoretic feature of Mordell-Weil groups has appeared in various works, e.g. in the statement of the Birch-Swinnerton-Dyer conjecture. But the idea to view them as lattices seems relatively new, and a systematic study of Mordell-Weil lattices has been done only very recently by the author [S3, S4, S5, S6, S7] and independently by N. Elkies [E], in the case of elliptic curves over function fields, or equivalently, in the case of elliptic surfaces. Actually the general scope of the theory of Mordell-Weil lattices should cover also the case of algebraic surfaces with higher genus fibration and the case of arithmetic surfaces as well (cf. [F, H] for the latter). But, in this talk, we shall restrict our attention to the case of elliptic surfaces.

In Part I, we review the definition and the basic results on Mordell-Weil lattices, and in Part II, we consider the Galois representations and algebraic equations arising from them, and discuss some applications. In the case of rational elliptic surfaces, we obtain some new insight into various problems related to E_6, E_7, E_8—a mysteriously rich subject in Mathematics. In particular, we can answer some questions raised by Weil [W3] and Manin [M2] concerning the Galois representation arising from the 27 lines on a cubic surface.

Proceedings of the International Congress
of Mathematicians, Kyoto, Japan, 1990

Acknowledgement. I would like to thank Professor J-P. Serre for many valuable comments on my work. I have had helpful discussions or correspondences with T. Ekedahl, N. Elkies, D. Gross, E. Horikawa, K. Oguiso, K. Saito, F. Sato, I. Shimada and P. Slodowy, and I thank them all.

Part I

1. Definition of Mordell-Weil Lattices

Let $K = k(C)$ be the function field of a smooth projective curve C over an algebraically closed field k. Let E be an elliptic curve defined over K, and let $E(K)$ denote the group of K-rational points of E, with the origin $O \in E(K)$. For what follows, the main reference is [S7].

Our basic tool is the associated elliptic surface $f: S \to C$ (the Kodaira-Néron model of E/K) and the intersection theory on S. Recall [K, N1, T2] that S is a smooth projective surface over k and f is a relatively minimal fibration with the generic fibre E. We can naturally identify the global sections of $f: S \to C$ with the K-rational points of E, and so $E(K)$ denotes the group of sections. For $P \in E(K)$, (P) will denote the image curve of $P: C \to S$.

We assume throughout that $(*)$ f is not smooth, i.e., there is at least one singular fibre. Then $E(K)$ is finitely generated by the Mordell-Weil theorem. On the other hand, let $N = NS(S)$ be the Néron-Severi group of S, i.e. the group of divisors modulo algebraic equivalence. It is a free module of finite rank under the assumption $(*)$, and it becomes an indefinite integral lattice with respect to the intersection pairing $(D \cdot D')$. Let T be the sublattice of N generated by the zero section (O), a fibre F and the irreducible components of fibres. Then T is a direct sum:

$$T = \langle (O), F \rangle_{\mathbb{Z}} \oplus V, \qquad V = \bigoplus_{v \in R} T_v \tag{1.1}$$

where $R = \{v \in C | F_v = f^{-1}(v) \text{ reducible}\}$ and T_v is generated by the irreducible components of F_v other than the identity component; each T_v is a root lattice of type A, D, E up to the sign. Let $L = T^{\perp}$ be the orthogonal complement of T in N. Then L is a negative-definite even integral lattice, as follows from the Hodge index theorem, the adjunction formula and the canonical bundle formula; see [S7, § 7]. We call T or L the *trivial* or *essential* sublattice of $NS(S)$.

Now the map $P \to (P)$ mod T induces an isomorphism:

$$E(K) \simeq NS(S)/T \tag{1.2}$$

(cf. [S7, Th.1.3]). By splitting this isomorphism, we get a unique homomorphism ([S7, § 8])

$$\varphi: E(K) \to NS(S) \otimes \mathbb{Q} \tag{1.3}$$

such that

$$\varphi(P) \equiv (P) \bmod T \otimes \mathbb{Q}, \qquad \operatorname{Im}(\varphi) \perp T \quad \text{and} \quad \operatorname{Ker}(\varphi) = E(K)_{\text{tor}}. \tag{1.4}$$

Theorem 1.1. *For $P, P' \in E(K)$, let*

$$\langle P, P' \rangle = -(\varphi(P) \cdot \varphi(P')). \tag{1.5}$$

Then it defines the structure of a positive-definite lattice on $E(K)/E(K)_{\text{tor}}$, called the Mordell-Weil lattice *of E/K or of $f: S \to C$.*

The explicit formula of the *height pairing* (1.5) is as follows:

$$\langle P, P' \rangle = \chi + (PO) + (P'O) - (PP') - \sum_{v \in R} \operatorname{contr}_v(P, P'). \tag{1.6}$$

Here χ is the arithmetic genus of S (a positive integer under $(*)$). $(PQ) = ((P)\cdot(Q))$ is the intersection number of the curves (P) and (Q). The local contribution term $\operatorname{contr}_v(P, P')$ is a non-negative rational number, which is non-zero only if both P and P' pass through non-identity components of F_v; it is then determined by the type of F_v and the position of the components hit by P, P' ([S7, § 8]).

On the other hand, let $E(K)^0$ be the subgroup of $E(K)$ consisting of those sections which pass through the identity component of every fibre. It is a torsion-free subgroup of finite index in $E(K)$.

Theorem 1.2. *With respect the height pairing, $E(K)^0$ is a positive-definite even integral lattice, which is isomorphic via the map φ to the opposite lattice L^- of the essential sublattice L. It will be called the* narrow Mordell-Weil lattice *of E/K or of $f: S \to C$.*

Observe that

$$\langle P, P' \rangle = \chi + (PO) + (P'O) - (PP') \in \mathbb{Z} \qquad \text{if } P \text{ or } P' \in E(K)^0, \tag{1.7}$$

$$\langle P, P \rangle = 2\chi + 2(PO) \geq 2\chi \qquad \text{for any } P \in E(K)^0, P \neq 0. \tag{1.8}$$

The former shows that $E(K)/E(K)_{\text{tor}}$ is contained in the dual lattice M^* of $M = E(K)^0$.

Theorem 1.3. *Assume further that $NS(S)$ is unimodular. Then $E(K)/E(K)_{\text{tor}}$ is equal to the dual lattice M^* of $M = E(K)^0$. Moreover we have $[M^* : M] = \det M = (\det T)/|E(K)_{\text{tor}}|^2$ (see* [S7, Th.9.1]).

2. Basic Invariants of MWL

By the very definition, the Mordell-Weil lattices (abbreviated henceforth as MWL) are of Diophantine nature and have quite rich structure, as will be seen later. But, apart from that, they provide a purely algebraic method for constructing lattices of some interest.

In general, given a lattice L, one would like to know its basic invariants such as the rank $\operatorname{rk}(L)$, the determinant $\det(L)$ and, if L is positive-definite (which we assume now), the minimal norm $\mu(L)$; recall (cf. [CS] for what follows) that $\det(L) = |\det(\langle x_i, x_j \rangle)|$ for any $\mathbb{Z}$-basis $\{x_i\}$ of L, and $\mu(L) = \mathrm{Min}(\langle x, x\rangle | x \in L - \{0\})$. Now a lattice L of rank r gives rise to a sphere packing in $L \otimes \mathbb{R} = \mathbb{R}^r$ by spheres of radius $\varrho_L = \frac{1}{2}\sqrt{\mu(L)}$. The density $\Delta(L)$ is the ratio of the volume of one sphere to that of the fundamental domain of L, i.e. $\sqrt{\det(L)}$. The quantity $\delta(L) = \Delta(L)/\mathrm{vol}(\text{unit sphere})$ is called the center density: we have $\delta(L) = (\sqrt{\mu(L)}/2)^r/\sqrt{\det(L)}$.

As for Mordell-Weil lattices, it follows from the results of § 1 that their invariants can be expressed in terms of geometric data on the associated elliptic surface:

Theorem 2.1. *Let $M = E(K)^0$ be the narrow MWL of E/K. Then*

$$r = \mathrm{rk}(M) = \varrho(S) - 2 - \sum_{v \in R} (m_v - 1), \tag{2.1}$$

where $\varrho(S) = \mathrm{rk}\, NS(S)$ is the Picard number of S and m_v is the number of irreducible components of the singular fibre F_v. Further we have

$$\det M = \det NS(S) \cdot v^2/\det T, \qquad v = [E(K) : E(K)^0]. \tag{2.2}$$

$$\mu(M) = 2\chi + \mathrm{Min}\{(PO) | P \in E(K)^0, P \neq 0\} \geq 2\chi. \tag{2.3}$$

It is not so easy in general to evaluate the rank and det, though we have a good lower bound 2χ for μ. In particular, it is a nontrivial problem to find E/K with large rank. A notable exception to this is the results of Shafarevich and Tate[ShT] and of the author [S2] that the rank can be arbitrarily large in case $p = \mathrm{char}(K) > 0$; the proof uses supersingular Fermat curves [ShT] or supersingular Fermat surfaces [S2]. From the viewpoint of MWL, Elkies and the author have independently studied these cases and obtained, among others, the following (see [E] and [S4], cf. [G]):

Theorem 2.2. *Let E be the elliptic curve $y^2 = x^3 + t^{q+1} + 1$ over $K = k_1(t)$, k_1 any extension of the finite field of q^2 elements with $q = p^e \equiv -1 \pmod 6$. Then $E(K) = E(K)^0$ is an even integral lattice with the invariants:*

$$r = 2q - 2, \det = q^{2(\chi - 1)}/b, \mu = 2\chi \quad (\chi = (q+1)/6), \tag{2.4}$$

where b is a positive integer (an even power of p) such that

$$b = 1\ (e = 1), \quad b \geq p^{2(e-1)((p-5)/6)^e}\ (e > 1) \tag{2.5}$$

in which equality holds if $e = 3$. The integer b is equal to the order of the Shafarevich-Tate group of E/K for $k_1 = \mathbb{F}_{q^2}$ or of the Brauer group of S/k_1 (cf. [T1]). In particular, we have

$$\delta = \sqrt{b} \cdot ((q+1)/12)^{q-1}/q^{(q-5)/6}. \tag{2.6}$$

N.B. This gives denser sphere packings than previously known ones in certain dimensions; for instance, for $q = 41$, we have $r = 80$ and $\log_2 \delta = 40.14$, the previous record being 36 (cf. [CS, Table 1.3], [E], [S4, Ex.1.3]).

3. Rational Elliptic Surfaces

With the notation of § 1, suppose that S is a *rational* elliptic surface, i.e. birational to $\mathbb{P}^2$. Then we have $C = \mathbb{P}^1$, $K = k(t)$, $\chi = 1$, and $N = NS(S)$ is unimodular of rank $\varrho = 10$. Hence $r \leq 8$ by (2.1).

The structure of the MWL and MW group of a rational elliptic surface is completely determined by the reducible singular fibres. Namely, look at the embedding of lattices, deduced from (1.1):

$$V = \bigoplus_{v \in R} T_v \subset H = \langle (O), F \rangle^\perp \subset N. \tag{3.1}$$

Note that H is a negative-definite even unimodular lattice of rank 8, i.e. the root lattice E_8 up to the sign, and V is a direct sum of root sublattices (cf. [B, CS] for root lattices). Theorem 1.3 gives:

Theorem 3.1. *For a rational elliptic surface, the narrow MWL is isomorphic to the orthogonal complement of V in E_8, and the MWL is its dual lattice. The torsion subgroup of $E(K)$ is V'/V, where V' is the primitive closure of V in E_8, i.e. $V' = V \otimes \mathbb{Q} \cap E_8$. Thus we have*

$$E(K) \simeq (V^\perp)^* \oplus (V'/V), \qquad E(K)^0 \simeq V^\perp, \tag{3.2}$$

where the isomorphism preserves the pairing. (See [S7, Th.10.3]).

The structure theorem in the case of rank $r = 8$, 7 or 6 is:

Theorem 3.2. *The narrow MWL $E(K)^0$ is the root lattice E_8, E_7 or E_6, according to whether $f: S \to \mathbb{P}^1$ has (i) no reducible fibres, (ii) only one reducible fibre of type* I_2 *or* III ([*K*]), *or (iii) only one reducible fibre of type* I_3 *or* IV. *Further $E(K)$ is torsion-free and isomorphic to E_8, E_7^* or E_6^* accordingly. (Cf.* [S7, Th.10.4]).

More generally, we have a complete structure theorem by using Dynkin's classification of $V \subset E_8$ (see [OS]). As a consequence, we obtain an effective result on the generators:

Theorem 3.3. *The Mordell-Weil group of any rational elliptic surface is generated by the sections P with $(PO) = 0$, hence by those P with $\langle P, P \rangle \leq 2$. If E/K is defined by the Weierstrass equation*

$$y^2 + a_1(t)xy + a_3(t)y = x^3 + a_2(t)x^2 + a_4(t)x + a_6(t), \tag{3.3}$$

with $a_i(t) \in k[t]$, $\deg a_i(t) \leq i$, then $E(K)$ is generated by rational points $P = (x, y)$ of the form:

$$x = gt^2 + at + b, \qquad y = ht^3 + ct^2 + dt + e \qquad (g, a, \ldots, e \in k). \tag{3.4}$$

There are at most 240 *such rational points. (Cf.* [S7, Th.10.10]).

Part II

4. Galois Representations Arising from MWL

From now on, we consider the following situation. Let k_0 be a perfect field and k its algebraic closure; let $G = \mathrm{Gal}(k/k_0)$. Let C be an absolutely irreducible smooth projective curve defined over k_0 and $K = k(C)$ or $K_0 = k_0(C)$ the function field of C over k or k_0. Now let E denote an elliptic curve defined over K_0 and consider E/K. The associated elliptic surface $f: S \to C$ is now defined over k_0.

Obviously the Galois group G acts on $E(K)$, and $E(K_0)$ coincides with $E(K)^G$, the subgroup of G-invariants. First we note:

Lemma 4.1. (*i*) *The map* φ *defined by* (1.3) *is* G*-equivariant.* (*ii*) *The height pairing* (1.6) *on* $E(K)$ *is stable under* G. (*Cf.* [S7, Prop.8.13]).

Therefore we get a Galois representation on the MW group

$$\varrho: G = \mathrm{Gal}(k/k_0) \to \mathrm{Aut}(E(K), \langle\ ,\ \rangle) = \text{a finite group} \tag{4.1}$$

and its variant on the MWL or the narrow MWL of E/K, say M,

$$\varrho': G \to \mathrm{Aut}(M) \subset \mathrm{GL}_r(\mathbb{Z}) \qquad (r = \mathrm{rk}(M)). \tag{4.2}$$

Let $\mathcal{K}/k_0$ be the extension corresponding to $\mathrm{Ker}(\varrho)$; equivalently, $\mathcal{K}$ is the smallest extension of k_0 such that $E(\mathcal{K}(C)) = E(K)$. By definition, $\mathcal{K}/k_0$ is a finite Galois extension such that

$$\mathrm{Gal}(\mathcal{K}/k_0) = \mathrm{Im}(\varrho). \tag{4.3}$$

The basic problem on the Galois representation (4.1) is this:

Problem 4.2. Determine the image of ϱ. In particular, we ask: (i) How big or (ii) how small can $\mathrm{Im}(\varrho)$ be?

To study this will be a main theme in the subsequent sections.

Remark 4.3. The Galois representations ϱ or ϱ' arising from MWL are quite different from those arising from the torsion points of an elliptic curve or an abelian variety (e.g. the Tate modules), because we are dealing with points of *infinite order* here!

Next we note the connection to the Hasse zeta functions (cf. [M2, Se1]). We state the result in the simplest situation, which is related to the questions of Weil [W3] and Manin [M2]; see Theorem 8.4.

Proposition 4.4. *Let* $k_0 = \mathbb{Q}$. *Assume that* S *is a rational elliptic surface* (*over* $\bar{\mathbb{Q}}$) *and that the trivial lattice* T *is a trivial* G*-module. Then the Hasse zeta function of* S *over* $\mathbb{Q}$ *is given by*

$$\zeta(S/\mathbb{Q}, s) = \zeta(s)\zeta(s-1)^{10-r}\zeta(s-2)\cdot L(s-1, \varrho, \mathcal{K}/\mathbb{Q}), \tag{4.4}$$

where $\zeta(s)$ is the Riemann zeta function and $L(s, \varrho, \mathcal{K}/\mathbb{Q})$ is the Artin L-function attached to the representation ϱ of $\mathrm{Gal}(\mathcal{K}/\mathbb{Q})$ *on the* MWL.

Proof. By [Se1], the Hasse zeta function is determined by the l-adic representation of G on $H^i_{\text{ét}}(S, \mathbb{Q}_l)$ $(0 \le i \le 4)$. The other cases being trivial, consider the case $i = 2$. For a rational surface, the cycle map induces a G-isomorphism

$$NS(S) \otimes \mathbb{Q}_l \simeq H^i_{\text{ét}}(S, \mathbb{Q}_l(1)). \tag{4.5}$$

On the other hand, the results in § 1 imply

$$NS(S) \otimes \mathbb{Q} = L \otimes \mathbb{Q} \oplus T \otimes \mathbb{Q} \quad \text{and} \quad L \otimes \mathbb{Q} \simeq E(K) \otimes \mathbb{Q}, \tag{4.6}$$

where the latter is a G-isomorphism by Theorem 1.2 and Lemma 4.1. □

(N.B. We have *exact* equality in (4.4), as pointed out by J-P. Serre.)

5. Algebraic Equations Arising from MWL

Given an elliptic curve E/K_0, take a singular fibre $F_v = f^{-1}(v)$ and a G-stable finite subset I of $E(K)$. Then we define

$$\Phi(X) = \Phi(E/K_0, v, I; X) = \prod_{P \in I} (X - \mathrm{sp}'_v(P)) \in k[X]. \tag{5.1}$$

Here, sp'_v is the *specialization map* at v, defined as follows. For any $P \in E(K)$, the section (P) intersects the fibre F_v at a unique smooth point of F_v; call it $\mathrm{sp}_v(P)$. The smooth part $F_v^{\#}$ of F_v is an algebraic group over k, which is the product of $\mathbb{G}_a$ or $\mathbb{G}_m$ by a finite group (at least if char $k \neq 2, 3$; cf. [K, N1, T2]), and the map

$$\mathrm{sp}_v : E(K) \to F_v^{\#}(k) \tag{5.2}$$

is a G-equivariant homomorphism. We denote by $\mathrm{sp}'_v(P)$ the projection of $\mathrm{sp}_v(P)$ to the factor $\mathbb{G}_a$ or $\mathbb{G}_m$.

As a typical example of the set I, we have (by Lemma 4.1)

$$I_n = \{P \in E(K) | \langle P, P \rangle = n\} \qquad (n \ge 0). \tag{5.3}$$

Proposition 5.1. *Assume $v \in C(k_0)$. Then $\Phi(X)$ has coefficients in k_0, i.e. $\Phi(X) \in k_0[X]$. The splitting field of the algebraic equation $\Phi(X) = 0$ over k_0, say $\mathcal{K}'$, is contained in $\mathcal{K}$ defined in §4, and we have $\mathcal{K}' = \mathcal{K}$ if* 1) sp'_v *is injective and* 2) *I contains generators of $E(K)$.*

Proof. Immediate by Galois theory. □

Remark 5.2. The algebraic equations (5.1) look somewhat analogous to the classical "division equations" arising from torsion points on an elliptic curve, whose study goes back to Abel and Galois ([ST, Ch.1]). As noted in Remark 4.3, we are concerned here with points of *infinite order* on an elliptic curve E, and in general the algebraic equations will lead to highly *non-abelian* extensions (see Theorems 6.1, 7.1).

6. Generic Galois Representation and Algebraic Equation of Type E_r

To illustrate the nature of the Galois representations and the algebraic equations arising from MWL, we consider some special cases. Let $E = E_\lambda$ be the elliptic curve, defined by one of the following equations, over $K_0 = k_0(t)$ where $k_0 = \mathbb{Q}(\lambda) = \mathbb{Q}(p_i, q_j)$.

$$
\begin{aligned}
&(\mathrm{E}_8) \quad && y^2 = x^3 + x\left(\sum_{i=0}^{3} p_i t^i\right) + \left(\sum_{i=0}^{3} q_i t^i + t^5\right) \\
& && \lambda = (p_0, p_1, p_2, p_3, q_0, q_1, q_2, q_3) \in \mathbb{A}^8 \\
&(\mathrm{E}_7) \quad && y^2 = x^3 + x(p_0 + p_1 t + t^3) + \left(\sum_{i=0}^{4} q_i t^i\right) \\
& && \lambda = (p_0, p_1, q_0, q_1, q_2, q_3, q_4) \in \mathbb{A}^7 \\
&(\mathrm{E}_6) \quad && y^2 = x^3 + x\left(\sum_{i=0}^{2} p_i t^i\right) + \left(\sum_{i=0}^{2} q_i t^i + t^4\right) \\
& && \lambda = (p_0, p_1, p_2, q_0, q_1, q_2) \in \mathbb{A}^6.
\end{aligned}
\tag{6.1}
$$

Let $f: S_\lambda \to \mathbb{P}^1$ be the associated elliptic surface. The fibre $f^{-1}(\infty)$ is an additive singular fibre of type II, III or IV according to the case (E_r) for $r = 8$, 7 or 6 and $f^{-1}(\infty)^\# \simeq \mathbb{G}_a \times \mathbb{Z}/(9 - r)$ [K, N1, T2]. Assume that λ satisfies the condition:

$$\text{every fibre of } f \text{ over } t \neq \infty \text{ is irreducible.} \tag{6.2}$$

Then the MWL $E(k(t))$ is isomorphic to E_r^* by Theorem 3.2, and we get the Galois representation

$$\varrho = \varrho_\lambda : \mathrm{Gal}(k/k_0) \to \mathrm{Aut}(E_r^*) = \mathrm{Aut}(E_r).$$

Recall that $\mathrm{Aut}(E_r) = W(E_r)$ for $r = 8$ or 7, and $W(E_r)\{\pm 1\}$ for $r = 6$, where $W(E_r)$ denotes the Weyl group of type E_r (cf. [B, CS]). In any case, we have $\mathrm{Im}(\varrho_\lambda) \subset W(E_r)$. It should be noticed that $W(E_r)$ is almost a *simple* group: with the notation of simple groups in Atlas [C], $W(E_6)$ contains $U_4(2)$ (order $2^6 3^4 5$) as a subgroup of index 2, $W(E_7)/\{\pm 1\} \simeq S_6(2)$ ($=\mathrm{Sp}_6(\mathbb{F}_2)$; order $2^9 3^4 5 \cdot 7$), and $W(E_8)$ has a subgroup H of index 2 such that $H/\{\pm 1\} \simeq O_8^+(2)$ (order $2^{12} 3^5 5^2 7$).

Now we assume that λ is *generic*, i.e. p_i, q_j are algebraically independent over $\mathbb{Q}$. Then (6.2) obviously holds.

Theorem 6.1. *Let λ be generic over $\mathbb{Q}$. Then* (i) *the image of the Galois representation ϱ_λ is the full Weyl group $W(E_r)$:*

$$\mathrm{Im}(\varrho_\lambda) = W(E_r). \tag{6.3}$$

(ii) *Let $\mathcal{K}_\lambda/k_0$ be the Galois extension corresponding to* $\mathrm{Ker}(\varrho_\lambda)$, *and let $\{P_1, \ldots, P_r\}$ be a basis of $E(k(t)) \simeq E_r^*$ consisting of minimal vectors. Further let $u_i = \mathrm{sp}'_\infty(P_i) \in \mathcal{K}_\lambda \subset k$. Then $u_1, \ldots, u_r$ are algebraically independent over $\mathbb{Q}$, and we have*

$$\mathcal{K}_\lambda = k_0(u_1, \ldots, u_r) = \mathbb{Q}(u_1, \ldots, u_r) \tag{6.4}$$

$$\mathrm{Gal}(\mathbb{Q}(u_1, \ldots, u_r)/\mathbb{Q}(p_i, q_j)) = W(E_r). \tag{6.5}$$

(iii) *$W(E_r)$ acts on the vector space $\sum \mathbb{Q}u_i$ and hence on the polynomial ring $\mathbb{Q}[u_1, \ldots, u_r]$, and the ring of the invariants is:*

$$\mathbb{Q}[u_1, \ldots, u_r]^{W(E_r)} = \mathbb{Q}[p_i, q_j]. \tag{6.6}$$

In particular, p_i and q_j form the fundamental invariants of the Weyl group $W(E_r)$ and we can explicitly write

$$p_i \text{ or } q_j = J_d(u_1, \ldots, u_r), \tag{6.7}$$

where J_d denotes a $W(E_r)$-invariant of degree d, $d \in \{2, 5, 6, 8, 9, 12\}$, $\{2, 6, 8, 10, 12, 14, 18\}$ or $\{2, 8, 12, 14, 18, 20, 24, 30\}$ for $r = 6, 7$ or 8.

The *universal polynomial of type E_r* is defined as a special case of (5.1), and it has degree $N = 27$, 56 or 240 for $r = 6$, 7 or 8:

$$\Phi_{E_r}(X, \lambda) = \Phi(E_\lambda/\mathbb{Q}(\lambda)(t), \infty, I; X) = \prod_{P \in I} (X - \mathrm{sp}'_\infty(P)) \in \mathbb{Q}(\lambda)[X]. \tag{6.8}$$

Here $I = I_\mu$ is the set of minimal vectors for $r = 8$ or 7 (thus $\mu = 2$ or $3/2$, and $\#I = 240$ or 56 accordingly). For $r = 6$, we take $I = I_\mu^\pm$, either one of 2 orbits of $W(E_6)$ in I_μ ($\mu = 4/3$, $\#I = 27$), and modify the right hand side of (6.8) by $\prod(X - (-2)\,\mathrm{sp}'_\infty(P))$. Then we have

Theorem 6.2. *For λ generic, $\Phi_{E_r}(X, \lambda)$ is a monic irreducible polynomial in X with coefficients in the polynomial ring $\mathbb{Z}[\lambda] = \mathbb{Z}[p_i, q_j]$, and its splitting field over $\mathbb{Q}(\lambda)$ is equal to $\mathcal{K}_\lambda$ in Theorem 6.1.*

Considering the elliptic curve (6.1) over the field $\mathcal{K}_\lambda(t) = \mathbb{Q}(u_1, \ldots, u_r)(t)$, we have the following:

Theorem 6.3. *Suppose $u_1, \ldots, u_r$ are algebraically independent over $\mathbb{Q}$, and define $\lambda = (p_i, q_j)$ by (6.7) and the elliptic curve E_λ by (6.1). Then the Mordell-Weil group $E(\mathbb{Q}(u_1, \ldots, u_r)(t)) \simeq E_r^*$ is of rank r, and has a basis $\{P_1, \ldots, P_r\}$ such that $\mathrm{sp}'_\infty(P_i) = u_i$. More explicitly, we have $P_i = (x, y)$ where x, y are polynomials in t of the form* (3.4) *which are given as follows:*

(i) *If $r = 8$, then $g = u_i^{-2}$, $h = u_i^{-3}$ and*

$$a, \ldots, e \in \mathbb{Q}[u_1, \ldots, u_8][u_i^{-1}] \cap \mathbb{Q}(p_0, \ldots, q_3)(u_i).$$

(ii) *If $r = 7$, then $c = u_i$ and*

$$a, b, d, e \in \mathbb{Q}[u_1, \ldots, u_7] \cap \mathbb{Q}(p_0, \ldots, q_4)(u_i), \quad g = h = 0.$$

(iii) *If $r = 6$, then $a = -2 \cdot u_i$, $c = 1$, and*

$$b, d, e \in \mathbb{Q}[u_1, \ldots, u_6] \cap \mathbb{Q}(p_0, \ldots, q_2)(u_i), \quad g = h = 0.$$

Let us briefly sketch the proof of the above theorems (for details, we refer to [S6, § 8]). By Theorem 3.2, we know the structure of the MWL $E_\lambda(k(t))$. For instance, if $r = 8$, it is E_8 and there exist 240 minimal vectors $P = (x, y)$ of the form (3.4). Substitute (3.4) into the equation (E_8), and we obtain polynomial relations among the coefficients $a, \ldots, g, h$. Then the successive elimination leads to a monic relation of $u = g/h = \mathrm{sp}_\infty(P)$ with coefficients in $\mathbb{Z}[\lambda]$, of degree 240, which must coincide with the universal polynomial of type E_8. This implies (6.6) and (6.7) by comparison of the coefficients of u^d, d being the degree of fundamental invariants. The rest of Theorems 6.1 and 6.2 follow immediately from this, and Theorem 6.3 from a closer look at the elimination process. □

Remark 6.4. Actually the proof shows that, in the above theorems, the ground field $\mathbb{Q}$ can be replaced by any field F, provided that its characteristic is different from a few primes which come into the denominators of the expression (6.7); they are $\{2, \ldots, 7\}$, $\{2, \ldots, 11, 29, 1229\}$ or $\{2, \ldots, 19, 41, 61, 199\}$ for $r = 6, 7$ or 8.

7. Galois Representations of Type E_r over $\mathbb{Q}$

We can compare the results of § 6 with the classical theory of the generic algebraic equation of degree n. Let

$$F(X, \varepsilon) = X^n + \varepsilon_2 X^{n-2} + \cdots + (-1)^n \varepsilon_n \qquad (\varepsilon_i : \text{alg.indep.}/\mathbb{Q})$$

be such an equation, normalized so that the sum of the roots x_i is 0. Let $\mathcal{K}$ be the splitting field of $F(X, \varepsilon)$ over $k_0 = \mathbb{Q}(\varepsilon_2, \ldots, \varepsilon_n)$. Then

(i) $\mathcal{K} = k_0(x_1, \ldots, x_n) = \mathbb{Q}(x_2, \ldots, x_n)$.
(ii) $\mathrm{Gal}(\mathbb{Q}(x_2, \ldots, x_n)/\mathbb{Q}(\varepsilon_2, \ldots, \varepsilon_n)) = \mathfrak{S}_n = W(A_{n-1})$.
(iii) $\mathbb{Q}[x_2, \ldots, x_n]^{W(A_{n-1})} = \mathbb{Q}[\varepsilon_2, \ldots, \varepsilon_n]$.
(iv) ε_i = the fundamental invariants of $\mathfrak{S}_n$.

Thus Theorems 6.1 and 6.2 (based on the theory of MWL) give a complete analogy for the exceptional type E_r ($r = 6, 7, 8$) of what the theory of generic equation does to the classical type A_{n-1}.

The latter theory has a standard application to number theory: (a) construction of Galois extensions of $\mathbb{Q}$ with Galois group $\mathfrak{S}_n$ (via specialization "downstairs" $(\varepsilon_i) \to (a_i) \in \mathbb{Q}^{n-1}$ and Hilbert's irreducibility theorem (cf. [L, Se2]), and (b) construction of an algebraic equation with the prescribed roots (via specialization "upstairs" $(x_i) \to (b_i) \in \mathbb{Q}^n$), which is indeed trivial.

In what follows, we describe similar application of our theory.

First consider the specialization "downstairs": $\lambda \to \lambda^0 \in \mathbb{Q}^r$. We keep the same notation as in § 6, except that we write λ for λ^0 (to simplify printing) so that $\mathcal{K}_\lambda/\mathbb{Q}$ is now a Galois extension.

Theorem 7.1. *Fix $r = 6, 7$ or 8, and let*

$$\Lambda = \Lambda_r = \{\lambda = (p_i, q_j) \in \mathbb{Q}^r \mid \mathrm{Gal}(\mathcal{K}_\lambda/\mathbb{Q}) = W(E_r)\}. \tag{7.1}$$

Then Λ is a Zariski dense subset of $\mathbb{Q}^r$ (more precisely, it is the complement of a "thin set" in $\mathbb{Q}^r$ in the sense of [Se2]). There exist infinitely many $\{\lambda\} \subset \Lambda$ such that $\{\mathscr{K}_\lambda\}$ are linearly disjoint over $\mathbb{Q}$.

Proof. Immediate from Theorem 6.1 (esp. (6.5)) by applying Hilbert's irreducibility theorem [Se2, Ch.10]. □

We note that, for each $\lambda \in \Lambda$, we have an elliptic curve $E_\lambda/\mathbb{Q}(t)$ with the MWL $E_\lambda(\bar{\mathbb{Q}}(t)) = E_\lambda(\mathscr{K}_\lambda(t)) \simeq E_r^*$ and $E_\lambda(\mathbb{Q}(t)) = \{0\}$, and a rational elliptic surface S_λ defined over $\mathbb{Q}$, having the Hasse zeta function (4.4) with non-abelian Artin L-function $L(s, \varrho_\lambda, \mathscr{K}_\lambda/\mathbb{Q})$.

Theorem 7.2. *Every Galois extension of $\mathbb{Q}$ with Galois group $W(E_r)$ is obtained as $\mathscr{K}_\lambda$ for some $\lambda \in \Lambda$. In other words, every such extension arises from the* MWL *of the elliptic curve $E_\lambda/\mathbb{Q}(t)$ for some $\lambda \in \mathbb{Q}^r$.*

Proof. Let $K/\mathbb{Q}$ be any extension with Galois group $W(E_r)$. Viewed as a $W(E_r)$-module, K is equivalent to the regular representation, and hence it contains a subspace, say U, isomorphic to $E_r \otimes \mathbb{Q}$ (a natural irreducible representation of $W(E_r)$). Fixing such an isomorphism, choose a $\mathbb{Z}$-basis $\{u_i\}$ of $E_r^* \subset U$ and define $\lambda = (p_i, q_j)$ by (6.7). Then we have $\lambda \in \mathbb{Q}^r$, since p_i, q_j are $W(E_r)$-invariant. The splitting field $\mathscr{K}_\lambda$ of $\Phi_{E_r}(X, \lambda) \in \mathbb{Q}[X]$ is contained in K and contains U. We claim: $\mathscr{K}_\lambda = K$. Indeed, any $\sigma \in \mathrm{Gal}(K/\mathscr{K}_\lambda)$ acts trivially on $U \subset \mathscr{K}_\lambda$. But U is a faithful representation of $W(E_r)$. Hence $\sigma = 1$, which proves the claim. (Our original proof of the claim used the structure of $W(E_r)$ (cf. § 6). We owe the above simplification to T. Ekedahl). □

We can also prove a formal analogue of Tate's conjecture for abelian varieties Let $\lambda = (p_i, q_j)$, $\lambda' = (p_i', q_j') \in \Lambda$, for $r = 6$ or 7.

Theorem 7.3. *The following conditions are equivalent*:

(i) *The Galois representations ϱ_λ and $\varrho_{\lambda'}$ are equivalent.*

(ii) *$\mathscr{K}_\lambda = \mathscr{K}_{\lambda'}$ and the roots $\{u_i\}$ and $\{u_i'\}$ of the polynomials $\Phi(X, \lambda)$ and $\Phi(X, \lambda')$ defined by* (6.8) *are related (up to ordering) by*

$$u_i' = A(u_i) \text{ (all } i\text{) for some } A(X) = \sum a_\nu X^\nu \in \mathbb{Q}[X].$$

(iii) *$\mathscr{K}_\lambda = \mathscr{K}_{\lambda'}$ and p_i', q_j' can be expressed by certain polynomials in p_i, q_j with $\mathbb{Q}$-coefficients.*

Proof. We consider the case $r = 6$. First assume (i). Then clearly $\mathscr{K}_\lambda = \mathscr{K}_{\lambda'}$, which we denote by K. There are exactly $n = 27$ minimal subfields M of K such that $[M : \mathbb{Q}] = n$, because $K/\mathbb{Q}$ has Galois group $W(E_r)$, which has so many maximal subgroups of index n (cf. [C]). Thus the n subfields $\mathbb{Q}(u_i)$ $(1 \le i \le n)$ of K give all such, and hence $\mathbb{Q}(u_i') = \mathbb{Q}(u_i)$ (all i) up to reordering. Let $u_1' = \sum_{\nu=0}^{n-1} a_\nu u_1^\nu$ $(a_\nu \in \mathbb{Q})$. Then by the action of $\sigma \in \mathrm{Gal}(K/\mathbb{Q})$ on both sides, we see (i) implies (ii). Conversely, if (ii) holds, then the permutations induced by σ on the sets $\{u_i\}$ and $\{u_i'\}$ are identical. Hence we have $\varrho_\lambda \sim \varrho_{\lambda'}$, i.e. (i). The equivalence of (ii) and (iii) follows from (6.6), (6.7).

The case $r = 7$ can be treated in the same way. (We omit the case $r = 8$ here, since more work is needed in addition to the information on the maximal subgroups in [C].) □

Our method allows us to find explicit examples.

Example 7.4 ($r = 6$). Take $\lambda = (1, 0, 1; 1, 1, 1)$. The elliptic curve is:

$$E = E_\lambda : y^2 = x^3 + x(1 + t^2) + (1 + t + t^2 + t^4). \tag{7.2}$$

Then the MWL $E(\bar{\mathbb{Q}}(t)) \simeq E_6^*$. We claim: $\mathrm{Gal}(\mathscr{K}_\lambda/\mathbb{Q}) = W(E_6)$, i.e. $\lambda \in \Lambda_6$. Further $\mathscr{K}_\lambda$ is the splitting field of $F(X) = \Phi(X, \lambda)$ below:

$$\begin{aligned} F(X) = X^{27} &+ 12X^{25} + 60X^{23} + 264X^{21} + 1302X^{19} - 1344X^{18} \\ &+ 3792X^{17} - 5568X^{16} + 22252X^{15} - 8832X^{14} + 57560X^{13} \\ &+ 4224X^{12} + 39025X^{11} - 49728X^{10} + 88516X^9 - 50880X^8 \\ &- 95024X^7 - 150016X^6 - 35840X^5 - 16384X^4 - 104192X^3 \\ &+ 5888X - 4096. \end{aligned}$$

This can be verified as follows. Look at the decomposition of $F(X) \bmod p$ into irreducible factors. For $p = 3$, it has 3 irreducible factors of degree 9 each, denoted symbolically by $(9)^3$. Similarly, for $p = 19$, it has $(2)(5)^3(10)$. Then the claim follows from:

Lemma 7.5 (Serre). *Let C_1 or C_2 be the conjugacy classes in $W(E_6)$ of the elements of order 9 or 10 with cycle type $(9)^3$ or $(2)(5)^3(10)$ in $\mathfrak{S}_{27}$ (cf. [Sw, p.57]). Suppose a subgroup H of $W(E_6)$ has the property that $H \cap C_i \neq \emptyset$ for $i = 1, 2$. Then $H = W(E_6)$.*

Proof. If not, take a maximal subgroup containing H. The possible orders of such are given in [C], but none of them are divisible by both 9 and 10 except one. But this last group does not contain any element of order 9. Hence a contradiction. □

(We can show further that the ramification occurs only at $p = 2, 137, 15784603$.)

Example 7.6 ($r = 7$). Let $\lambda = (1, 1; 1, 1, 1, 0, 1)$ and

$$E = E_\lambda : y^2 = x^3 + x(1 + t + t^3) + (1 + t + t^2 + t^4). \tag{7.4}$$

Then the MWL $E(\bar{\mathbb{Q}}(t)) \simeq E_7^*$, and $\mathrm{Gal}(\mathscr{K}_\lambda/\mathbb{Q}) = W(E_7)$, i.e. $\lambda \in \Lambda_7$. The universal polynomial $F(X) = \Phi(X, \lambda)$ equals $f(Y)$ below ($Y = X^2$), and the splitting field of $f(Y)$ over $\mathbb{Q}$ has the Galois group isomorphic to $W(E_7)/\{\pm 1\} \simeq S_6(2)$ (a simple group of order 1451520).

$$\begin{aligned} f(Y) = Y^{28} &- 36Y^{27} + 594Y^{26} - 6084Y^{25} + 43935Y^{24} - 240192Y^{23} \\ &+ 1039392Y^{22} - 3661764Y^{21} + 10681839Y^{20} - 26088660Y^{19} \\ &+ 53894394Y^{18} - 95282532Y^{17} + 145821463Y^{16} - 194265660Y^{15} \end{aligned}$$

$$+ 223728462Y^{14} - 216948108Y^{13} + 168475770Y^{12} - 98229852Y^{11}$$
$$+ 42796234Y^{10} - 19590492Y^{9} + 14262444Y^{8} - 9949084Y^{7}$$
$$+ 4609696Y^{6} - 1118808Y^{5} + 53521Y^{4} + 160Y^{3} + 4288Y^{2}$$
$$- 5312Y + 1024. \tag{7.5}$$

The proof is similar to Example 7.4. The decomposition type of $F(X)$ mod p is $(7)^8$ for $p = 47$, and $(3)^2(5)^4(15)^2$ for $p = 131$. This suffices to conclude $\mathrm{Gal}(\mathcal{K}_\lambda/\mathbb{Q}) = W(E_7)$ in view of [C]. □

In the above, we have taken $\mathbb{Q}$ as the ground field, but the same idea works in more general situation. Here is one such example.

Example 7.7 ($r = 6$). Let $k_0 = \mathbb{F}_3(s)$ be the rational function field over the finite field $\mathbb{F}_3$. Then the following simple equation

$$F(X) = X^{27} + X^{12} - X^7 - X^2 - s^3 \in \mathbb{F}_3[s][X] \tag{7.6}$$

has the Galois group $W(E_6)$! It arises from the elliptic curve $E/k_0(t)$

$$y^2 = x^3 + xt + st + t^4 \tag{7.7}$$

having the MWL $E(\bar{k}_0(t)) = E_6^*$. This is proven by checking that $F(X)$ mod $\mathfrak{p}$ has the conjugacy type C_1 or C_2 for the prime ideal $\mathfrak{p} = (s - 1)$ or $(s + 1)$ of $\mathbb{F}_3[s]$ and by applying Lemma 7.5.

8. Further Applications

We briefly mention other applications of Theorems 6.1, etc.

1) Construction of Elliptic Curves over $\mathbb{Q}(t)$ with Rank $r = 6, 7, 8$

We consider the analogy for E_r, mentioned before, of (b) writing down an algebraic equation with prescribed roots. By specializing the parameters "upstairs" (u_i) to some $(u_i^0) \in \mathbb{Q}^r$ in Theorems 6.1 or 6.3 in such a way that the condition (6.2) is satisfied, we obtain an elliptic curve $E = E_\lambda$ over $\mathbb{Q}(t)$, together with r rational points P_i. Observe that each P_i has the prescribed value u_i^0 as its essential parameter; e.g. $P_i = (-2u_i^0 t + \cdots, t^2 + \cdots)$ for $r = 6$. The condition (6.2) is equivalent to certain polynomial $\delta_0(u^0) \neq 0$ (see [S6]).

Theorem 8.1. *Assume $\delta_0(u^0) \neq 0$. Then the Mordell-Weil group $E(\mathbb{Q}(t)) \simeq E_r^*$ has rank r $(= 6, 7, 8)$, and $\{P_i\}$ forms a set of generators.*

See [S6] for the proof, as well as for explicit examples.

Corollary 8.2. *Given $E/\mathbb{Q}(t)$ as above, we have an infinite family $\{E^{(\tau)} | \tau \in \mathbb{Q} - \sum\}$ $(\#\sum < \infty)$ of elliptic curves over $\mathbb{Q}$ of rank $\geq r$, with r independent points $P_i^{(\tau)}$, by specializing $t \to \tau \in \mathbb{Q}$. Further*

$$\det(\langle P_i^{(\tau)}, P_j^{(\tau)} \rangle_{\text{can}}/h(\tau)) \to 1/2^r d \qquad \text{as } h(\tau) \to \infty, \tag{8.1}$$

where $\langle\ ,\ \rangle_{\text{can}}$ *is the canonical height and* $h(\tau)$ *the standard height (esp.* $h(\tau) = \log|\tau|$ *for* $\tau \in \mathbb{Z}$*), and* $d = 1, 2$ *or* 3 *for* $r = 8, 7$ *or* 6.

This follows from a result of Néron, Silverman [Si], Tate [T3].

2) The 27 Lines on a Smooth Cubic Surface

In the equation (E_6) of (6.1), let $y = y' \pm t^2$ and $(x : y' : t : 1) = (X : Y : Z : W)$. Then we have a cubic surface $V = V_\lambda^\pm$ in $\mathbb{P}^3$:

$$Y^2W \pm 2YZ^2 = X^3 + X(p_0W^2 + p_1ZW + p_2Z^2) + q_0W^3 + q_1ZW^2 + q_2Z^2W. \tag{8.2}$$

It is smooth if and only if (6.2) holds. Under this assumption, the narrow MWL $E_\lambda(k(t))^0 \simeq E_6$ is isomorphic to the primitive part of $NS(V)$ $(= \langle \omega_V \rangle^\perp$ in the notation of [M2, Ch.4]). The 27 minimal sections of the form $P = (at + b, t^2 + dt + e)$ in the MWL $E_\lambda(k(t))$ are transformed into the 27 lines on $V = V_\lambda^+$ defined by the equation

$$X = aZ + bW, \qquad Y = dZ + eW. \tag{8.3}$$

(Similarly, the 27 sections $-P$ are mapped to the 27 lines on V_λ^-.)

Hence the universal polynomial $\Phi(X, \lambda)$ of type E_6 becomes the "algebraic equation of the 27 lines" on V_λ. As a consequence, we can easily deduce the following from our previous results.

Theorem 8.3. *For* $\lambda = (p_i, q_j)$ *generic over any field* F *of char.* $p \neq 2$, *the Galois group of the 27 lines on the cubic surface* V_λ *over* $F(\lambda)$ *is equal to the full Weyl group* $W(E_6)$.

Proof. This is immediate from Theorem 6.1 and Remark 6.4 if p is different from 3, 5, 7, in which case the assertion can be verified as in Example 7.7. (A similar result holds also for $p = 2$.) □

Theorem 8.4. *For* $\lambda \in \Lambda_6 \subset \mathbb{Q}^6$, *the Galois group of the 27 lines on* V_λ *over* $\mathbb{Q}$ *is* $W(E_6)$, *and the Hasse zeta function of* $V_\lambda/\mathbb{Q}$ *is given by*

$$\zeta(V_\lambda/\mathbb{Q}, s) = \zeta(s)\zeta(s-1)\zeta(s-2)L(s-1, \varrho_\lambda, \mathcal{K}_\lambda/\mathbb{Q}). \tag{8.4}$$

Proof. Immediate from Theorem 7.1. □

Example 8.5. We easily obtain an explicit example of a cubic surface over $\mathbb{Q}$ with the Galois group $W(E_6)$. For instance, Example 7.4 gives:

$$Y^2W + 2YZ^2 = X^3 + XW^2 + XZ^2 + W^3 + ZW^2 + Z^2W. \tag{8.5}$$

The 27 lines are defined over $\mathcal{K}_\lambda$ given by (7.3). Compare [Ek].

We believe that these results clarify some of Weil's statements at the end of his 1954 Congress address [W3]. Also we can essentially answer the problems raised by Manin [M2, Ch.4, 23.13] by translating Theorems 7.2 and 7.3 into the language of cubic surfaces.

We also have a systematic construction of cubic surfaces over $\mathbb{Q}$ such that all of the 27 lines (8.3) are defined over $\mathbb{Q}$.

Example 8.6. *Let V be the cubic surface*

$$Y^2W + 2YZ^2 = X^3 - X(78Z^2 + 59475W^2) + 18226Z^2W + 2848750W^3. \quad (8.6)$$

The 27 lines are given by

$$\begin{aligned}(X, Y) = (411W, -6916W), &(-2Z + 575W, 74Z - 12600W),\\ (-4Z + 275W, 124Z - 2700W), &(-6Z + 15W, 126Z + 1400W),\\ (-8Z - 85W, 56Z + 2700W), &(-10Z - 49W, -110Z - 2376W), \quad (8.7)\end{aligned}$$

etc.; the above 6 correspond to the generators of $E(\mathbb{Q}(t)) \simeq E_6^*$.

3) The 28 Double Tangents to a Plane Quartic Curve

This classical topic is closely connected with the case (E_7). In the same way as in 2) above, we can prove the following:

(i) The Galois group of 28 double tangents on a generic quartic curve over any prime field is $W(E_7)/\{\pm 1\} \simeq S_6(2)$.

(ii) Construction of a quartic over $\mathbb{Q}$ with the Galois group $\simeq S_6(2)$.

(iii) Construction of such with all double tangents defined over $\mathbb{Q}$. As a by-product, we can also prove

(iv) the rationality of the moduli space of plane quartic curves (genus 3) with a flex (over any base field of char $\neq 3$).

4) Deformation of E_r-Singularities

The equation (E_r) in (6.1) defines the universal deformation of the rational double point of type E_r; at $\lambda = 0$, we have $y^2 = x^3 + t^4$, $y^2 = x^3 + xt^3$ and $y^2 = x^3 + t^5$ for $r = 6, 7, 8$. By Theorem 6.1, we can easily reprove some results due to Brieskorn, Tjurina and others (cf. [Br, DP, Sl]). Indeed (6.6) implies immediately that the map

$$\pi : \mathbb{A}^r \to \mathbb{A}^r/W(E_r) \simeq \mathbb{A}^r, \qquad (u_i) \to \lambda = (p_i, q_j) \quad (8.8)$$

is a finite covering, ramified along the discriminant locus D ([B]), so that (in case $k = \mathbb{C}$) the monodromy map below is surjective:

$$\pi_1(\mathbb{C}^r - D) \to W(E_r). \quad (8.9)$$

Further we can describe the stratification of D according to the type of singularities (see [S5]); also it will be evident that our method applies to the case of char $p > 0$ as well.

On the other hand, let $\mathfrak{g}$ denote the simple Lie algebra of type E_r and $\mathfrak{h}$ the Cartan subalgebra, $\Delta = \{\alpha\}$ the root system in $\mathfrak{h}^*$, then the characteristic polynomial of the adjoint representation ad(H) on $\mathfrak{g}$ ($H \in \mathfrak{h} = \mathbb{A}^r$) is given by

$$\begin{aligned} \det(X - \mathrm{ad}(H)) &= X^r \cdot \prod_{\alpha \in \Delta} (X - \alpha(H)) \\ &= X^r \cdot \Phi^*(X, \lambda) \qquad (\lambda = \pi(H)), \end{aligned} \tag{8.10}$$

where $\Phi^*(X, \lambda) = \Phi(E_\lambda/k_0(t), \infty, I_2; X)$ is the special case of our algebraic equation (5.1) (cf. (6.8)), of degree 72, 126 or 240 for $r = 6, 7, 8$. According to K. Saito and Slodowy, the study of the roots of (8.10) led Killing [Ki] (about 100 years ago) to introduce the "root" systems in classifying Lie algebras.

Thus the theory of Mordell-Weil lattices has unexpectedly rich applications even in the simplest case of rational elliptic surfaces. We close this paper with the hope to see much more fruits in future coming from the study of the Mordell-Weil lattices.

References

[B] Bourbaki, N.: Groupes et algèbres de Lie, Chap. 4, 5 et 6. Hermann, Paris 1968

[Br] Brieskorn, E.: Singular elements of semisimple algebraic groups, Proc. ICM Nice 1970, II. Gauthier-Villars, Paris 1971, pp. 279–284

[C] Conway, J. et al: Atlas of finite groups. Clarendon Press, Oxford 1985

[CS] Conway, J., Sloane, N.: Sphere packings, lattices and groups. (Grundlehren der mathematischen Wissenchaften, vol. 290.) Springer, Berlin Heidelberg New York 1988

[CZ] Cox, D., Zucker, S.: Intersection numbers of sections of elliptic surfaces. Invent. math. **53** (1979) 1–44

[DP] Demazure, M., Pinkham, H., Teissier, B.: Séminaire sur les singularités des surfaces. (Lecture Notes in Mathematics, vol. 777.) Springer, Berlin Heidelberg New York 1980

[Ek] Ekedahl, T.: An effective version of Hilbert's irreducibility theorem. In: Sém. théorie des nombres, Paris 1988–89, pp. 241–249, Birkhäuser, Boston Basel Berlin 1990

[E] Elkies, N.: On Mordell-Weil lattices. Arbeitstagung Bonn, 1990

[F] Faltings, G.: Calculus on arithmetic surfaces. Ann. Math. **119** (1984) 387–424

[G] Gross, B.: Group representations and lattices. J. Amer. Math. Soc. **3** (1990) 929–960

[H] Hriljac, P.: Heights and Arakelov's intersection theory. Amer. J. Math. **107** (1985) 23–38

[Ki] Killing, W.: Die Zusammensetzung der stetigen endlichen Transformationsgruppen, II. Math. Ann. **33** (1889) 1–48

[K] Kodaira, K.: On compact analytic surfaces II–III. Ann. Math. **77** (1963) 563–626; **78** (1963) 1–40; Collected Works, vol. III. Iwanami and Princeton Univ. Press, 1975, pp. 1269–1372

[L] Lang, S.: Fundamentals of diophantine geometry. Springer, Berlin Heidelberg New York 1983

[M1] Manin, Ju.: The Tate height of points on an Abelian variety, its variants and applications. Izv. Akad. Nauk SSSR, Ser. Mat. **28** (1964) 1363–1390; A.M.S. Transl. (2) **59** (1966) 82–110

[M2] Manin, Ju.: Cubic forms, 2nd edn. North-Holland 1986

[Mo] Mordell, L.: On the rational solutions of the indeterminate equation of the third and fourth degree. Proc. Camb. Phil. Soc. **21** (1922) 179–192

[N1] Néron, A.: Modèles minimaux des variétés abéliennes sur les corps locaux et globaux. Publ. Math. I.H.E.S. **21** (1964)

[N2] Néron, A.: Quasi-fonctions et hauteurs sur les variétés abéliennes. Ann. Math. **82** (1965) 249–331

[OS] Oguiso, K., Shioda, T.: The Mordell-Weil lattice of a rational elliptic surface. Comment. Math. Univ. St. Pauli **40** (1991) 83–99

[Se1] Serre, J-P.: Facteurs locaux des fonctions zeta des variétés algébriques. Sém. DPP (1969/70); Oeuvres II, pp. 581–592. Springer, Berlin Heidelberg New York

[Se2] Serre, J-P.: Lectures on the Mordell-Weil theorem. Vieweg 1989.

[ShT] Shafarevich, I., Tate, J.: The rank of elliptic curves. Dokl. Akad. Nauk SSSR **175** 770–773; Sov. Math. Dokl. **8** (1967) 917–920

[ST] Shimura, G., Taniyama, Y.: Modern number theory (in Japanese). Kyoritsu Publ., Tokyo 1957

[S1] Shioda, T.: On elliptic modular surfaces. J. Math. Soc. Japan **24** (1972) 1–59

[S2] Shioda, T.: An explicit algorithm for computing the Picard number of certain algebraic surfaces. Amer. J. Math. **108** (1986) 415–432

[S3] Shioda, T.: Mordell-Weil lattices and Galois representation, I, II, III. Proc. Japan Acad. **65A** (1989) 267–271, 296–299, 300–303

[S4] Shioda, T.: Mordell-Weil lattices and sphere packings. (To appear in Amer. J. Math.)

[S5] Shioda, T.: Mordell-Weil lattices of type E_8 and deformation of singularities. (To appear)

[S6] Shioda, T.: Construction of elliptic curves with high rank via the invariants of the Weyl groups. (To appear in J. Math. Soc. Japan)

[S7] Shioda, T.: On the Mordell-Weil lattices. Comment. Math. Univ. St. Pauli **39** (1990) 211–240

[Si] Silverman, J.: Heights and the specialization map for families of abelian varieties. J. Reine Angew. Math. **342** (1983) 197–211

[Sl] Slodowy, P.: Simple singularities and simple algebraic groups. (Lecture Notes in Mathematics, vol. 815.) Springer, Berlin Heidelberg New York 1980

[Sw] Swinnerton-Dyer, H.P.F.: The zeta function of a cubic surface over a finite field. Proc. Camb. Phil. Soc. **63** (1967) 55–71

[T1] Tate, J.: On the conjectures of Birch and Swinnerton-Dyer and a geometric analog. Sém. Bourbaki 1965/66, no. 306

[T2] Tate, J.: Algorithm for determining the type of a singular fiber in an elliptic pencil. (Lecture Notes in Mathematics, vol. 476.) Springer, Berlin Heidelberg New York 1975, pp. 33–52

[T3] Tate, J.: Variation of the canonical height of a point depending on a parameter. Amer. J. Math. **105** (1983) 287–294

[W1] Weil, A.: L'arithmetique sur les courbes algébriques. Acta Math. **52** (1928) 281–315

[W2] Weil, A.: Variétés abéliennes et courbes algébriques, Hermann, Paris 1948/1973

[W3] Weil, A.: Abstract versus classical algebraic geometry. Proc. ICM Amsterdam, vol. III (1954), pp. 550–558; Collected Works, vol. II, pp. 180–188. Springer, Berlin Heidelberg New York 1980

Collapsing Riemannian Manifolds and Its Applications

Kenji Fukaya

Department of Mathematics, Faculty of Science
University of Tokyo, Hongo, Bunkyo-ku, Tokyo 113, Japan

§1

One of the typical problems to which Hausdorff convergence and collapsing of Riemannian manifolds can be applied is the study of local structure of Riemannian manifolds.

Problem 1.1. Let $\mathscr{C}$ be a class of Riemannian manifolds. Can we find a positive number $\varepsilon_{\mathscr{C}}$ such that for each $p \in M \in \mathscr{C}$ there exists a neighborhood U_p of p with the following properties ?

(1.1.1) U_p contains $B_p(\varepsilon_{\mathscr{C}}, M)$, the metric ball of radius $\varepsilon_{\mathscr{C}}$ and centered at p.
(1.1.2) We can control the topology of U_p.

The essential point here is that $\varepsilon_{\mathscr{C}}$ is independent of p and M.

In the case when $\mathscr{C}$ is the class of homogeneous spaces, this problem is solved in Zassenhauss [Z] and Kazdan-Margulis [KM]. In that case, U_p is controled by nilpotent groups. In the case of negatively curved manifolds, this problem is studied by Margulis [M], Heintze [H] and Gromov [G1].

We consider the following two classes.

$$\mathscr{M}_n = \{M \mid \dim M = n, |K_M| \le 1\},$$
$$\mathscr{N}_n = \{M \mid \dim M = n, K_M \ge -1\}.$$

Here K_M stands for the sectional curvature. Problem 1.1 for class $\mathscr{M}_n$ is first studied by Gromov [G2].

§2

In §§2–6, we deal with the class $\mathscr{M}_n$. First let us take the following two extremal cases.

(1) When we consider the subclass $\mathscr{M}_n(i_0)$ of $\mathscr{M}_n$ consisting of manifolds with injectivity radius $> i_0$, we can choose $\varepsilon_{\mathscr{C}} = i_0$, since i_0-ball is diffeomorphic to Euclidean space.

Proceedings of the International Congress
of Mathematicians, Kyoto, Japan, 1990

(2) When we consider the subclass consisting of manifolds with small diameter, we can apply the following theorem. (In this case we choose $U_p = M$.)

Theorem 2.1 (Gromov [G2], Ruh [R]). *There exists $\varepsilon_n > 0$ such that if*

$$|K_M| \cdot \text{Diam } M^2 < \varepsilon_n,$$

then M is diffeomorphic to N/Γ, where

(2.1.1) *N is a nilpotent Lie group,*
(2.1.2) *Γ is a discrete subgroup of the semidirect product $N \tilde{\times} \text{Aut } N$, the group of affine diffeomorphisms of N,*
(2.1.3) *The index, $[\Gamma : \Gamma \cap N]$ is finite.*

(1) is the case when no direction of M is degenerate. (2) is the case when every direction of M is degenerate. In general there are both degenerate and nondegenerate directions. Hence we have to separate those two directions. For this purpose we study the following:

Problem 2.2. For a small neighborhood U_p of p in a Riemannian manifold M^n, find a Riemannian manifold V^m such that

(2.2.1) the injectivity radius of V is larger than a constant depending only on the class $\mathscr{C}$ containing M,
(2.2.2) V is Hausdorff close to U_p.

The Hausdorff closeness between spaces is defined as follows.

Definition 2.3 (Gromov [G3]). Let X and Y be compact metric spaces. A (not necessary continuous) map $f : X \to Y$ is said to be an *ε-Hausdorff approximation* if

(2.3.1) the ε-neighborhood of $f(X)$ is Y,
(2.3.2) $|d(f(x), f(y)) - d(x, y)| < \varepsilon$, for each $x, y \in X$.

We say that the *Hausdorff distance*, $d_H(X, Y)$, between X and Y is smaller than ε if there exist ε-Hausdorff approximations from X to Y and from Y to X.

If V is as in Problem 2.2, a nondegenerate direction in U_p is one which is "parallel" to some direction in V.

§3

In fact the answer to Problem 2.2 itself is negative. But we can prove a bit modifided statement. The construction of the space V in 2.2 is based on a limit argument. In other words, the space V is a limit of Riemannian manifolds contained in class $\mathscr{M}_n$. Hence to study Problem 2.2, we need to study the limit behavior of the sequence of Riemannian manifolds.

First we need to show that there exist sufficiently many convergent sequences. The following result ensures it.

Theorem 3.1 (Gromov [G3]). *Let $\mathscr{S}(D)$ be a set of all Riemannian manifolds with diameter $\le D$ and Ricci curvature ≥ -1. Then $\mathscr{S}(D)$ is precompact in the set of all compact metric spaces, with respect to the Hausdorff distance.*

Next we need to study the following:

Problem 3.2. Let $M_i \in \mathscr{M}_n$ and X be a compact metric space. Suppose that M_i converges to X with respect to the Hausdorff distance. What kind of singularity can X have ?

We are mainly interested in the case when the dimension of X is smaller than that of M. In this case, we say that M_i collapses to X.

The first important example of collapsing Riemannian manifolds is discovered by M. Berger. In his example, a family of metrics on S^3 collapses to S^2. Here S^2 is regarded as the quotient of S^3 by an action of S^1.

In general, one can construct a collapsing family of metrics on manifolds M with torus action, provided each orbit is of positive dimension. The Hausdorff limit of this family is the quotient space M/T, which, in general, is singular. The other important example is a nilmanifold, which collapses to a point. (See [G2, BK].)

The systematic study of the collapsing phenomena for Riemannian manifolds was initiated by Cheeger-Gromov [CG1]. We discuss their results in §5.

Now we present results about Problem 2.2. In case when M_i is a $K(\pi, 1)$-space, we can prove the following:

Theorem 3.3 ([F4]). *Let M_i and X be as in Problem 3.2. Assume $\pi_k(M_i) = 1$, for $k > 1$. Then $X = Y/\Gamma$, where Y is a contractible manifold and Γ a discrete group of isometries.*

In fact we can prove the following:

(3.4.1) $\widetilde{M}_i$ converges to Z. Z is a manifold of the same dimension as M_i. The convergence is compact $C^{1,\alpha}$-convergence of metric tensors.

(3.4.2) Γ_i converges to a group G of isometries of Z.

(3.4.3) The connected component G_0 of G is a simply connected nilpotent Lie group.

(3.4.4) $Y = Z/G_0$, $\Gamma = G/G_0$.

Thus, roughly speaking, the Hausdorff convergence of $K(\pi; 1)$-manifolds is a generalization of the convergence of discrete subgroups of Lie group. Collapsing phenomenon occurs when discrete groups converge to a continuous group.

Before discussing the general case, we give applications of Theorem 3.3.

Theorem 3.5 ([F4]). *There exists a positive number* ε_D *such that if an n-dimensional Riemannian manifold* M $(n > 1)$ *satisfies*

(3.5.1) $\pi_k(M) = 1$, *for* $k > 1$,
(3.5.2) $|K_M| \leq 1$, Diam $M \leq D$,
(3.5.3) $\pi_1(M)$ *does not contain* $\mathbf{Z}^2$,

then

$$\text{Vol}\, M > \varepsilon_D\,.$$

Conjecture 3.6. *The assumption* Diam $M \leq D$ *can be removed.*

In case $K_M \leq 0$, this was proved by Heintze [H], Gromov [G1], Buyalo [Bu].

The following result is not a direct application of Theorem 3.3 but its proof uses an idea of the proof of Theorem 3.3.

Theorem 3.7 (Fukaya-Yamaguchi [FY1]). *There exists a positive number* ε_D *such that if*

$$\varepsilon_D > K_M > -1\,,$$
$$\text{Diam}\, M < D\,,$$

then the universal covering space of M *is diffeomorphic to* $\mathbf{R}^n$.

Theorem 3.7 is an affirmative answer to a conjecture by Gromov [G4].

As we remarked before, the limit of manifolds in class $\mathcal{M}_n$ is not necessary a manifold. The reason is that a singular space can arise as the quotient space M/T of a manifold M by its torus action. Note, however, that the quotient space FM/T is always nonsingular. Here FM is the frame bundle of M and the action of T on FM is a lift of one on M. This observation leads us to the following:

Theorem 3.8 ([F3]). *Let* $M_i \in \mathcal{M}$, FM_i *be their frame bundles equipped with metrics induced by ones on* M. *Let* X *be a compact metric space. Suppose that* FM_i *converges to* X *with respect to Hausdorff convergence.*

Then X *is a smooth manifold with* $C^{1,\alpha}$*-metric.*

Furthermore, there exists a smooth $O(n)$ *action on* X *such that* M_i *converges to* $X/O(n)$ *and that the connected components of isotropy groups are abelian.*

Using Theorems 3.1 and 3.8, we can prove the following:
Let $M \in \mathcal{M}_n$, $p \in M$. Then there exists a manifold V such that

(3.9.1) injectivity radius of $V >$ const > 0,
(3.9.2) $d_H(FB_p(\varepsilon, M), V)$ is small.

§4

To apply Problem 2.2 to Problem 1.1, we need to study the relation between the topological structures of U_p and V_p in 2.2. Namely:

Problem 4.1. Let M_i and X be as in Problem 3.2. Find the relation between the topological structures of M_i and X.

We have the following two results:

Theorem 4.2 (Fukaya [F1,2], Yamaguchi [Y2]). *Let $M_i \in \mathcal{N}_n$ and N be a compact Riemannian manifold. Suppose that M_i converges to N with respect to the Hausdorff distance. Then for each suffciently large i, there exists a map $f_i : M_i \to N$, such that*

(4.2.1) *f_i is a locally trivial fibre bundle,*
(4.2.2) *f_i is an almost Riemannian submersion, namely*

$$e^{\varepsilon_i} > \frac{|f_*(V)|}{|V|} > e^{-\varepsilon_i}$$

holds for vectors $V \in TM$ perpendicular to the fibre,
(4.2.3) *the fibre of f_i is of almost nonnegative curvature. (See §7).*

If we assume $M_i \in \mathcal{M}_n$ in addition, then

(4.2.4) *the fibre is almost flat, namely it satisfies* (2.1.1)–(2.1.3),
(4.2.5) *the structure group of fibration is reduced to the group of affine diffeomorphisms.*

Theorem 4.3 ([F3]). *Let M_i and X be as in Theorem* 3.8. *Then, for each sufficiently large i, there exist $O(n)$ maps $f_i : FM_i \to X$ satisfying* (4.2.1), (4.2.2), (4.2.4), (4.2.5).

It follows from Theorem 4.3 that there exists a singular fibration: $M_i \to X/O(n)$. The singular fibre corresponds to the singular point of $X/O(n)$.
Using the results of §§2, 3, 4, we can prove the following:

Theorem 4.4 ([F5]). *There exists a positive number ε_n such that if $p \in M \in \mathcal{M}_n$ then there exists a neighborhood U_p of p with the following properties.*

(4.4.1) *U_p is diffeomorphic to a vector bundle over N_p/Γ_p, where N_p and Γ_p is as in* (2.1.1)–(2.1.3).
(4.4.2) *U_p contains $B_p(\varepsilon_n, M)$.*

Ghanaat, Min-no, Ruh [GMR] proved a closely related result.

§5

The nilpotent Lie group N_p in Theorem 4.4 depends on the point p. In other words the dimension of the collapsing directions changes from point to point. In fact, consider the metric

$$dt^2 + e^{t-2C}dx^2 + e^{-t-2C}dy^2,$$

on $[-C, C] \times T^2$. (Here x and y are coordinates of T^2.) Then, in a neighborhood of $\{\pm C\} \times T^2$, the dimension of collapsing direction is one, and in a neighborhood of $\{0\} \times T^2$ the dimension of collapsing direction is two.

In order to study the global behavior of the collapsing part of the manifolds, we need to study the above phenomenon. Cheeger and Gromov introduced the notion of the local action of the groups for this purpose.

Definition 5.1 (Cheeger-Gromov [CG1]). An F-structure on a manifold M is a subsheaf $\mathscr{F}$ of the sheaf of vector fields such that

(5.1.1) $\mathscr{F}$ is a sheaf of abelian Lie algebra,
(5.1.2) for each $p \in M$ there exists a neighborhood U_p, its finite covering space $\widetilde{U}_p$, an action of torus, T_p, on $\widetilde{U}_p$, such that the lift of $\mathscr{F}$ to $\widetilde{U}_p$ is induced by the action of T_p.

They proved the following two results:

Theorem 5.2 ([CG1]). *Let M be a manifold. Assume that there exists an F-structure on M such that the dimension of each orbit is nonzero. Then there exists a family of metrics g_ε on M such that*

(5.2.1) $(M, g_\varepsilon) \in \mathscr{M}_n$,
(5.2.2) *the injectivity radius of (M, g_ε) is smaller than ε.*

Theorem 5.3 ([CG2]). *There exists a positive number ε_n with the following properties. Let $M \in \mathscr{M}_n$. Put*

$$M_{\text{thin}} = \{p \in M \mid \text{injectivity radius at } p < \varepsilon_n\}.$$

Then there exists an F-structure $\mathscr{F}$ on M_{thin} such that every orbit of it is of positive dimension.

Cheeger-Gromov applied these results to study Gauss-Bonnet type formula for noncompact manifolds. ([CG 3,4] etc.)

§6

In Theorem 4.4, we obtained a nilpotent structure but the result is local. In Sect. 5, we obtain more global structure but the structure is abelian and not nilpotent. To combine them, we need to introduce the action of sheaf of nilpotent Lie algebra.

Definition 6.1 (Cheeger-Fukaya-Gromov [CFG]). A *nilpotent structure* on a manifold M is a sheaf $\mathcal{N}$ of vector fields such that for each p there exists a neighborhood U_p and an action of group G_p on a Galois covering $\widetilde{U}_p$ of U_p with the following properties.

(6.1.1) The connected component N_p of G_p is nilpotent.
(6.1.2) The deck transformation group Γ_p of the covering: $\widetilde{U}_p \to U_p$ is contained in G_p.
(6.1.3) G_p is generated by N_p and Γ_p.
(6.1.4) $[\Gamma_p : \Gamma_p \cap N_p] = [G_p : N_p]$ is finite.
(6.1.5) The lift of $\mathcal{N}$ to $\widetilde{U}_p$ is generated by the action N_p.

Theorem 6.2 ([CFG]). *There exist ε_n and k_n with the property that, for each $M \in \mathcal{M}_n$ there exists a nilpotent structure $\mathcal{N}$ such that the following holds in addition.*

(6.2.1) *U_p contains $B_{\varepsilon_n}(p, M)$.*
(6.2.2) *The injectivity radius of $\widetilde{U}_p$ is larger than ε_n.*
(6.2.3) *$[G_p : N_p] < k_n$.*

Definition 6.3 ([CFG]). A Riemannian metric g on M is said to be (ε, k)*-round*, if there exists a nilpotent structure $\mathcal{N}$ on M such that it satisfies (6.2.1)–(6.2.3) in addition and that the section of $\mathcal{N}$ is a Killing vector field of the metric g. (In other words the action of N_p is isometry for induced metric.)

Theorem 6.4 ([CFG]). *For each δ there exist $\varepsilon = \varepsilon(\delta, n)$ and $k = k(\delta, n)$ such that for each Riemannian manifold (M, g) in the class $\mathcal{M}_n$ we can find a metric g_δ on M with the following properties.*

(6.4.1) *(M, g_δ) is (ε, k)-round.*
(6.4.2) $|g - g_\delta|_{C^{1,\alpha}} < \delta$.

We expect that there is a result similar to Theorem 5.2 for nilpotent structure. The proof of it is not yet complete.

§7

In this section we consider the class $\mathcal{N}_n$. To study the collapsing phenomena of Riemannian manifolds in class $\mathcal{M}_n$, the starting point was Theorem 2.1, which characterize the manifolds collapsing to a point. The corresponding problem for class $\mathcal{N}_n$ is the study of almost nonnegatively curved manifolds.

Problem 7.1. Find a characterization of manifolds M admitting a metric with

$$K_M \cdot \text{Diam } M^2 > -\varepsilon_n ,$$

where ε_n is a positive small number depending only on dimension.

This problem should be very difficult, since it includes the characterization of nonnegatively curved manifolds. We describe some results concerning the topology of almost nonnegatively curved manifolds. (Some of them hold for manifolds of almost nonnegative Ricci curvature.)

Theorem 7.2 (Gromov-Gallot [G3, Ga]). *There exists a positive number C_n such that if M is an n-dimensional Riemannian manifold with*

$$\text{Ricci}_M \cdot \text{Diam } M^2 > -D ,$$

then

$$b_1(M;\mathbf{Q}) = \text{rank } H_1(M;\mathbf{Q}) \le n - 1 + C_n^D .$$

In particular, if M is of almost nonnegative Ricci curvautre then the first Betti number does not exceed the dimension. This is a generalization of a classical result by Bochner [B].

Theorem 7.3 (Yamaguchi [Y2]). *There exists a positive number ε_n depending only on dimension such that if an n-dimensional Riemannian manifold M satisfies*

$$K_M \cdot \text{Diam } M^2 > -\varepsilon_n ,$$

then a finite cover of M is a fibre bundle over $b_1(M;\mathbf{Q})$-dimensional torus.

If the first Betti number is equal to the dimension then M is diffeomorphic to a torus.

(Compare also [Y1].)

Theorem 7.4 (Fukaya-Yamaguchi [FY2]). *If M satisfies the assumption of Theorem 7.3 then the fundamental group of M contains a nilpotent subgroup of finite index.*

Theorem 7.4 is an affirmative answer to another conjecture in [G4].

In fact we can prove a bit more than Theorem 7.4. (The precise statement is omitted. See [FY2].) As a corollary we have:

Theorem 7.5 ([FY2]). *There exists c_n such that if M satisfies the assumption of Theorem 7.3 then we have*

$$b_1(M;\mathbf{Z}_p) \le n ,$$

for each prime $p > c_n$.

In case when the equality holds, M is diffeomorphic to a torus.

Using Theorem 7.4 we can control the fundamental group of the small neighborhood. Namely we have the following:

Theorem 7.6 ([FY2]). *There exists a positive number ε_n depending only on dimension such that if $M \in \mathcal{N}_n$ and $p \in M$ then the image*

$$\mathrm{Im}\,[\pi_1(B_p(\varepsilon_n, M)) \to \pi_1(B_p(1, M))]$$

contains a nilpotent subgroup of finite index.

Conjecture 7.7. *Theorems 7.3–7.6 still hold if we replace sectional curvature by Ricci curvature.*

In the case when $M \in \mathcal{M}_n$ one can choose a neighborhood U_p which is $K(\pi;1)$ and whose fundamental group is almost nilpotent. (See Theorem 4.4.) But we can not choose such U_p of uniform size for M in $\mathcal{N}_n$, since the nonnegatively curved manifold is not necessary a $K(\pi;1)$-space. Hence Theorem 7.6 does not give enough information to determine the local topological structure of manifolds in the class $\mathcal{N}_n$. So far, our knowledge about higher homology or homotopy groups of (almost) nonnegatively curved manifolds is quite restricted. The best result seems to be Gromov's Betti number estimate, [G5].

As for Hausdorff convergence of manifolds in the class $\mathcal{N}_n$, we have Theorem 4.2 due to Yamaguchi. But we do not know so much about Problem 3.2 for this class. (Recently Burago and Gromov proved that the limit space is almost everywhere a manifold.)

On the other hand, Yamaguchi [Y2] constructed examples of collapsing family of metrics using an action of compact groups. Namely, on a manifold M on which a compact Lie group K acts, there exists a sequence of metrics g_ε such that $(M, g_\varepsilon) \in \mathcal{N}_n$ and that (M, g_ε) converges to M/K for Hausdorff distance. This construction is closely related to the fact that compact Lie group admits a metric of positive sectional curvature.

In noncollapsing situation, (namely the case we assume Vol $M >$ constant > 0, in addition,) Grove-Petersen-Wu [GPW] proved a strong result. They conjectured that the limit space is a topological manifold in this case.

References

[B] S.Bochner: Vector fields and Ricci curvature. Bull. AMS **52** (1946) 776–797

[BK] P.Buser, H.Karcher: Gromov's almost flat manifolds. Asterisque **81** (1981) 1–148

[Bu] V.Buyalo: Volume and fundamental group of a manifold of nonpositive curvature. Math. USSR Sb. **50** (1985) 137–150

[CFG] J.Cheeger, K.Fukaya, M.Gromov: Nilpotent structure and invariant round metric on collapsed Riemannian manifolds. Preprint

[CG1] J.Cheeger, M.Gromov: Collpasing Riemannian manifolds while keeping their curvature bounded I. J. Diff. Geom. **23** (1986) 309–346

[CG2] J.Cheeger, M.Gromov: Collpasing Riemannian manifolds while keeping their curvature bounded II. J. Diff. Geom. **32** (1990) 269–298

[CG3] J.Cheeger, M.Gromov: On the characteristic numbers of complete manifolds of bounded curvature and finite volume. In: Differential Geometry and Complex Analysis. Springer, Berlin Heidelberg New York 1985

[CG4] J.Cheeger: Chopping Riemannian manifolds. Preprint

[F1] K.Fukaya: Collapsing Riemannian manifolds to ones of lower dimension. J. Diff. Geom. **25** (1987) 139–156

[F2] K.Fukaya: Collapsing Riemannian manifolds to ones of lower dimension II. J. Math. Soc. Japan **41** (1989) 333–356

[F3] K.Fukaya: A boundary of the set of the Riemannian manifolds with bounded curvatures and diameters. J. Diff. Geom. **28** (1988) 1–21

[F4] K.Fukaya: A compactness of a set of aspherical Riemannain orbifolds. In: A Fete of Topology, ed. by Matsumoto, Mizutani, Morita. Academic press, Boston 1988

[F5] K.Fukaya: Hausdorff convergence of Riemannian manifolds and its applications. In: Recent topics in differential and analytic geometry, ed. by T.Ochiai. Advanced Studies in Pure Math. 18-I. Kinokuniya, Academic Press, Tokyo Boston 1990

[FY1] K.Fukaya, T.Yamaguchi: Almost nonpositively curved manifolds. J. Diff. Geom. **33** (1991) 67–90

[FY2] K.Fukaya, T.Yamaguchi: The fundamental group of almost nonnegatively curved manifolds. Preprint

[Ga] S.Gallot: A Sobolev inequality and some geometric applications. In: Spectra of Riemannian manifolds. Kaigai, Tokyo 1983

[GMR] P.Ghanaat, Min-Oo, E.Ruh: Local structure of Riemannian manifolds. Preprint

[G1] M.Gromov: Manifolds of negative curvature. J. Diff. Geom. **13** (1978) 223–230

[G2] M.Gromov: Almost flat manifolds. J. Diff. Geom. **13** (1978) 231–241

[G3] M.Gromov (J.Lafontaine and P. Pansu): Structure metrique pour les varietes riemanniennes. Cedic/Fernad Nathan, Paris 1981

[G4] M.Gromov: Synthetic Geometry in Riemannian manifolds. Proceedings of ICM Helsinki 1978, vol. 1, pp. 415–419

[G5] M.Gromov: Curvature, Diameter and Betti numbers. Comment. Math. Helv. **56** (1981) 179–195

[GPW] K.Grove, P.Petersen, J.Wu: Geometry finiteness theorems via controlled topology. Invent. math. **99** (1990) 205–213

[H] E.Heintze: Mannigfaltigkeiten negativer Krümmung. Habilitationsschrift, Universität Bonn

[KM] Kazdan, G.Margulis: A proof of Selberg's hypothesis. Math. USSR Sb. **75** (1968) 162–168

[M] G.Margulis: Discrete group of motions of manifolds of nonpositive curvature. AMS Trans. **109** (1977) 33–45

[R] E.Ruh: Almost flat manifolds. J. Diff. Geom. **17** (1982) 1–14

[Y1] T.Yamaguchi: Manifolds of almost nonnegative Ricci curvature. J. Diff. Geom. **28** (1988) 157–167

[Y2] T.Yamaguchi: Collapsing and Pinching in lower curvature bound. To appear in Ann. Math.

[Z] H.Zassenhaus: Beweis eines Satzes über diskrete Gruppen. Abh. Math. Sem. Hansischen Univ. 12

Le Cercle à l'Infini des Surfaces à Courbure Négative

Etienne Ghys

Ecole Normale Supérieure de Lyon, 46, allée d'Italie F-69007 Lyon, France

1. Introduction

Soit S une surface compacte orientée munie d'une métrique riemannienne g de classe C^∞ à courbure variable strictement négative. Il est bien connu que le revêtement universel $\tilde{S}$ de S peut naturellement être compactifié par l'adjonction d'un bord à l'infini, noté $\partial\tilde{S}$ (voir par exemple [1]). Par définition, un point de $\partial\tilde{S}$ est représenté par un rayon géodésique, c'est-à-dire un plongement isométrique $r : [0, +\infty[\to \tilde{S}$ et deux tels rayons r_1 et r_2 définissent le même point du bord si la distance entre $r_1(t)$ et $r_2(t)$ est uniformément bornée. Si p est un point de $\tilde{S}$, on peut représenter chaque point de $\partial\tilde{S}$ par un unique rayon issu de p. Ainsi, le bord $\partial\tilde{S}$ s'identifie à l'espace des vecteurs unitaires de l'espace tangent à $\tilde{S}$ en p; il est donc homéomorphe à un cercle. Bien sûr, le groupe fondamental Γ de S opère sur $\tilde{S}$ et donc sur le cercle $\partial\tilde{S}$. Nous nous proposons ici de décrire quelques résultats relatifs à cette action. Notre but n'est pas de les démontrer mais d'essayer de les motiver et d'indiquer les liens qui les unissent.

2. L'aspect topologique et la cohomologie bornée

D'un point de vue topologique, il n'y a qu'une action à étudier. C'est un fait bien connu depuis longtemps.

Théorème. *Soient g_1 et g_2 deux métriques à courbure négative sur la même surface compacte S. Alors les actions du groupe fondamental de S sur les bords pour g_1 et g_2 du revêtement universel $\tilde{S}$ sont topologiquement conjuguées.*

Pour montrer ce théorème, on constate qu'un rayon r_1 dans $\tilde{S}$ pour la métrique g_1 n'est pas nécessairement un rayon pour g_2 mais que c'est un "quasi-rayon": la g_2-distance entre $r_1(t)$ et $r_1(t')$ est comprise entre $c^{-1}|t - t'|$ et $c|t - t'|$ pour une certaine constante $c > 0$. Ceci permet de montrer l'existence d'un rayon r_2 pour g_2 qui est à distance bornée de r_1. Les bords de $\tilde{S}$ pour g_1 et g_2 sont alors naturellement identifiés de manière Γ équivariante.

De manière plus conceptuelle, on peut associer un bord à un groupe de type fini G (voir [7] et [19]). Pour cela, on se fixe une partie génératrice finie et on considère

Proceedings of the International Congress
of Mathematicians, Kyoto, Japan, 1990

le graphe de Cayley de G correspondant à cette partie génératrice. Si une arête relie les éléments γ_1 et γ_2 de G, on équipe cette arête d'une métrique qui la rend isométrique à un intervalle de longueur $(n+1)^{-2}$ où $n = \inf(\|\gamma_1\|, \|\gamma_2\|)$ et $\|\gamma_i\|$ désigne la longueur de γ_i par rapport à la partie génératrice choisie. Le graphe de Cayley de G devient ainsi un espace métrique non complet (si G est infini). Le bord ∂G de G est alors défini comme l'espace qu'il faut ajouter à ce graphe pour le compléter. Il ne dépend pas du choix de la partie génératrice. Le groupe $\mathrm{Aut}(G)$ des automorphismes de G opère naturellement sur ∂G. En particulier, G opère sur ∂G (via les automorphismes internes). Ce bord est particulièrement utile lorsque G est un groupe hyperbolique au sens de M. Gromov [19]. Par exemple, si G est le groupe fondamental Γ d'une surface compacte S de genre supérieur ou égal à 2, le bord ∂G s'identifie à $\partial\tilde{S}$. Ainsi, *l'action que nous étudions n'est autre que celle du groupe Γ sur son bord.*

Examinons rapidement le cas particulier où la courbure de la métrique g est -1. Le revêtement universel $\tilde{S}$ s'identifie alors au disque de Poincaré D^2 et son groupe d'isométries directes à $\mathrm{PSL}(2, \mathbb{R})$. Le bord de D^2 est le cercle, ici identifié à la droite projective réelle P^1. Dans ce cas, l'action étudiée de Γ sur P^1 est projective et provient d'un plongement de Γ dans $\mathrm{PSL}(2, \mathbb{R})$ comme sous-groupe discret co-compact. L'espace de ces plongements (à conjugaison près) est l'espace de Teichmüller; il est de dimension finie. Bien que toutes ces actions projectives soient topologiquement conjuguées, il est facile de s'assurer qu'elles ne sont C^1-conjuguées que si elles sont projectivement conjuguées. D. Sullivan [31] montre même que si deux de ces actions sont conjuguées par une application mesurable qui respecte les ensembles négligeables au sens de Lebesgue, alors ces actions sont projectivement conjuguées.

Comment caractériser le type topologique de cette action? Si un groupe discret G opère sur le cercle en respectant l'orientation, on peut construire un fibré en cercles au dessus de l'espace d'Eilenberg-MacLane $K(G, 1)$. La classe d'Euler de ce fibré, élément de $H^2(G, \mathbb{Z})$, est évidemment un invariant de conjugaison topologique (respectant l'orientation). Cet invariant est cependant insuffisant. Si $G = \mathbb{Z}$ par exemple, ce second groupe de cohomologie est trivial. Dans [12], nous introduisons un invariant plus fin, élément de $H_b^2(G, \mathbb{Z})$, second groupe de cohomologie bornée à coefficients entiers [18]. Cette cohomologie est celle du sous-complexe du complexe d'Eilenberg-MacLane formé des cochaînes bornées (comme fonctions de G^n vers $\mathbb{Z}$). Voici comment on procède pour définir l'invariant en question.

Soient f_1, f_2, f_3 trois homéomorphismes directs du cercle S^1 et x un point base sur S^1. Posons $c(f_1, f_2, f_3) = 1$ si les points $f_1(x), f_2(x), f_3(x)$ sont distincts et placés dans un ordre cyclique rétrograde sur le cercle ou si $f_1(x) = f_3(x) \neq f_2(x)$ et posons $c(f_1, f_2, f_3) = 0$ dans les autres cas. Il est aisé de s'assurer que c est un cocycle. Si ϕ est une action du groupe discret G sur le cercle par homéomorphismes directs, on obtient ainsi, par image réciproque, une classe $\phi^*(c)$ dans $H_b^2(G, \mathbb{Z})$.

On vérifie facilement que l'image de cette classe dans la cohomologie usuelle est la classe d'Euler (voir aussi [25]). Nous appellerons cette classe la *classe d'Euler bornée* de l'action ϕ. Si $G = \mathbb{Z}$, on a $H_b^2(G, \mathbb{Z}) \simeq \mathbb{R}/\mathbb{Z}$ et la classe d'Euler bornée n'est autre que le nombre de rotation du générateur de l'action.

Théorème [12]. *Soient ϕ_1 et ϕ_2 deux actions du même groupe G sur le cercle, par homéomorphismes directs. On suppose que toutes les orbites de ϕ_1 et ϕ_2 sont denses dans le cercle. Alors, ϕ_1 et ϕ_2 sont topologiquement conjuguées, par un homéomorphisme direct, si et seulement si les classes d'Euler bornées $\phi_1^*(c)$ et $\phi_2^*(c)$ sont égales dans $H_b^2(G, \mathbb{Z})$.*

Lorsque les orbites ne sont pas supposées denses, on a un énoncé plus faible qui utilise une notion de semi-conjugaison (voir [12] et aussi [32]).

Revenons au cas où G est le groupe fondamental Γ de S. La classe d'Euler d'une action de Γ sur le cercle est un élément de $H^2(\Gamma, \mathbb{Z})$, c'est-à-dire un entier que nous noterons *eu*. Un théorème de Milnor-Wood [35], qui est d'ailleurs l'une des origines de la notion de cohomologie bornée, affirme que cet entier vérifie l'inégalité $|eu| \leq |\chi(S)|$ où $\chi(S)$ désigne la caractéristique d'Euler-Poincaré de S. Dans le cas de l'action de Γ sur le cercle $\partial\tilde{S}$ qui correspond à une métrique à courbure négative, on a en fait l'égalité $eu = \pm\chi(S)$.

Dans un joli article [28], S. Matsumoto montre que si Γ opère sur le cercle et si le nombre d'Euler est maximal, i.e. si $eu = \chi(S)$, alors la classe d'Euler *bornée* ne peut prendre qu'une valeur, à savoir celle correspondant à l'action de Γ sur son bord. Il peut alors en déduire la caractérisation topologique suivante qui répond positivement à une conjecture de W. Goldman [17].

Théorème [28]. *Considérons une surface compacte orientée S de genre supérieur ou égal à 2 et une action du groupe fondamental de S sur le cercle, par homéomorphismes directs. On suppose que toutes les orbites sont denses sur le cercle et que le nombre d'Euler vérifie $eu = \pm\chi(S)$. Alors, cette action est topologiquement conjuguée à l'action naturelle de ce même groupe sur le bord $\partial\tilde{S}$ du revêtement universel de S, pour n'importe quelle métrique à courbure négative sur S.*

L'hypothèse de densité des orbites est en fait inutile si l'on se restreint à des actions suffisamment différentiables.

Théorème [13]. *Considérons une surface compacte orientée S de genre supérieur ou égal à 2 et une action du groupe fondamental de S sur le cercle, par difféomorphismes directs de classe C^2 (resp. analytiques réels). On suppose que le nombre d'Euler du fibré associé vérifie $eu = \pm\chi(S)$ (resp. $eu \neq 0$). Alors, toutes les orbites sont denses dans le cercle.*

La démonstration se fonde sur les théorèmes de structure des feuilletages de codimension 1, à la Denjoy-Sacksteder.

Citons encore un résultat, obtenu avec J. Barge, et dont la démonstration utilise aussi la cohomologie bornée. Soit x un point base sur le bord $\partial\tilde{S}$ correspondant à une métrique g. Si $\gamma_1, \gamma_2, \gamma_3$ sont trois éléments de Γ, on peut considérer l'aire $a(\gamma_1, \gamma_2, \gamma_3)$ du triangle idéal de $\tilde{S}$ de sommets $\gamma_1(x), \gamma_2(x), \gamma_3(x)$. C'est un 2-cocycle borné sur Γ à valeurs réelles. Dans le cas de courbure -1, l'aire de ces triangles vaut $\pm\pi$ et la classe de cohomologie bornée de a est égale à 2π fois la classe d'Euler bornée (considérée comme classe réelle). La réciproque est plus délicate.

Théorème [2]. *Soit g une métrique riemannienne à courbure négative sur la surface compacte S. On suppose que tous les triangles idéaux du revêtement universel $\tilde{S}$ sont d'aire π. Alors, la courbure de g est constante, égale à -1.*

3. La différentiabilité du bord

Soient p et q deux points de $\tilde{S}$. Nous avons déjà observé que $\partial\tilde{S}$ est naturellement identifié aux cercles unités S_p^1 et S_q^1 dans les espaces tangents en p et q à $\tilde{S}$. Ainsi, il existe un homéomorphisme naturel π_{pq} entre S_p^1 et S_q^1. Quel est le degré de régularité de ces homéomorphismes? Un résultat classique de E. Hopf affirme qu'ils sont de classe C^1 [23] (voir aussi [22]). Un résultat plus fort a été obtenu par S. Hurder et A. Katok.

Théorème [24]. *Les homéomorphismes π_{pq} sont de classe $C^{2-\varepsilon}$ pour tout $\varepsilon > 0$.*

Ce résultat est à comparer au suivant sur lequel nous allons nous attarder un peu plus.

Théorème [15]. *Si les homéomorphismes π_{pq} sont de classe C^2, alors la courbure de la métrique considérée est constante.*

Ce théorème fait suite à un théorème local de S. Hurder et A. Katok [24] qui se fondait sur un résultat de [11] que nous mentionnerons plus loin. Il a donné lieu par la suite à un certain nombre de développements, tout spécialement en dimension supérieure. Puisque nous ne traitons ici que du cas des surfaces, nous ne décrirons pas ces développements qui aboutissent à une caractérisation analogue des espaces localement symétriques à courbure négative mais nous renvoyons à [3, 5, 6, 8, 20, 26]. Grâce à ces travaux, et tout particulièrement à ceux de M. Kanai, la démonstration du théorème précédent s'est significativement simplifiée. Nous allons esquisser ici une preuve assez élémentaire.

Soit T_1S le fibré unitaire tangent à S et ϕ_t le flot géodésique de S, agissant sur T_1S. Soient $\mathscr{F}^s$ et $\mathscr{F}^u$ les feuilletages stables et instables faibles de ϕ_t. Les feuilles du relevé $\tilde{\mathscr{F}}^s$ de $\mathscr{F}^s$ au fibré unitaire $T_1\tilde{S}$ sont constituées de tous les vecteurs unitaires qui définissent des rayons géodésiques asymptotes. L'espace des feuilles de $\tilde{\mathscr{F}}^s$ est donc homéomorphe au bord $\partial\tilde{S}$. Les fibres de la fibration en cercles de $T_1\tilde{S}$ sont transverses à $\tilde{\mathscr{F}}^s$ et les applications d'holonomie induites sur ces fibres sont précisément les homéomorphismes π_{pq}. Ainsi, la différentiabilité des π_{pq} est équivalente à celle du feuilletage $\mathscr{F}^s$ (ou, d'ailleurs, de $\mathscr{F}^u$ puisque ces deux feuilletages sont conjugués par l'involution de T_1S envoyant un vecteur sur son opposé).

Plaçons nous donc dans l'hypothèse où ces feuilletages sont de classe C^2. La première étape consiste à montrer qu'ils sont en fait transversalement projectifs. Un disque D transverse à ϕ_t est muni d'une forme d'aire Ω (provenant de la forme de Liouville) et de deux champs de directions L^s et L^u, traces de $\mathscr{F}^s$ et $\mathscr{F}^u$. Ces structures permettent de définir une métrique pseudo-riemannienne q sur D: si v est un vecteur tangent à D et si v_u et v_s sont ses composantes sur L^s et L^u, on pose $q(v) = \Omega(v_u, v_s)$. C'est une métrique de classe C^2 si $\mathscr{F}^s$ est de classe C^2 et L^s et L^u sont les directions isotropes de q. On a ainsi construit une métrique pseudo-riemannienne sur le fibré normal à ϕ_t, évidemment invariante par ϕ_t. La courbure de q est donc une fonction continue constante sur les orbites de ϕ_t; elle est donc constante car ϕ_t est transitif.

En d'autres termes, tous les disques transverses sont localement isométriques à un même plan lorentzien de courbure constante. Il est très facile d'en conclure que $\mathscr{F}^s$ est muni d'une structure transverse projective (ou affine si la courbure de q était nulle). Chaque fibre de $T_1 S$, étant transverse à $\mathscr{F}^s$, est ainsi équipée d'une structure projective ou affine et les applications d'holonomie préservent cette structure. Puisque nous connaissons a priori la topologie de l'action de Γ sur le cercle, on montre que c'est en fait le cas projectif qui se présente et que l'action de Γ sur le cercle est C^2-conjuguée à celle d'un sous-groupe discret co-compact de $\mathrm{PSL}(2, \mathbb{R})$.

Ainsi, si le feuilletage $\mathscr{F}^s$ est de classe C^2, il est C^2-conjugué au feuilletage stable du flot géodésique d'une métrique à courbure -1. Il faut encore en conclure que la métrique considérée est elle même à courbure constante. On dispose actuellement de deux méthodes pour y parvenir. La première consiste à développer encore des arguments analogues à ceux utilisés plus haut et à montrer qu'en fait ϕ_t est conjugué au flot géodésique d'une métrique à courbure constante. On peut alors conclure en appliquant le théorème de rigidité de A. Katok [27] qui caractérise la courbure constante en termes d'entropie. La seconde méthode est due à Y. Mitsumatsu et elle est reliée à l'invariant de Godbillon-Vey. Nous avons choisi de décrire ici cette méthode car cet invariant interviendra encore par la suite.

Rappelons la définition de cet invariant. Soit $\mathscr{F}$ un feuilletage de codimension 1, transversalement orientable et de classe C^2 sur une variété M et soit ω une 1-forme différentielle qui le définit. Il existe alors, d'après le théorème de Frobenius, une forme ω_1 telle que $d\omega = \omega \wedge \omega_1$. Il se trouve que la 3-forme $\omega_1 \wedge d\omega_1$ est fermée et que sa classe de cohomologie ne dépend que de $\mathscr{F}$; c'est la classe de Godbillon-Vey de $\mathscr{F}$, notée $\mathrm{GV}(\mathscr{F})$. Lorsque M est une variété fermée orientée de dimension 3, l'évaluation de $\mathrm{GV}(\mathscr{F})$ sur la classe fondamentale de M est un nombre réel; c'est l'invariant de Godbillon-Vey que nous noterons $gv(\mathscr{F})$. Par exemple, considérons le cas du feuilletage stable $\mathscr{F}^s$ du flot géodésique d'une métrique à courbure constante -1. Nous savons que le fibré $T_1 S$ s'identifie à un espace homogène $\mathrm{PSL}(2, \mathbb{R})/\Gamma$. Sur $\mathrm{PSL}(2, \mathbb{R})$, il existe une base de formes invariantes à droite ω, ω_1, ω_2 telle que:

$$d\omega = \omega \wedge \omega_1; \qquad d\omega_1 = \omega \wedge \omega_2; \qquad d\omega_2 = \omega_1 \wedge \omega_2.$$

La forme ω passe au quotient sur $T_1 S$ et définit le feuilletage $\mathscr{F}^s$. L'invariant de Godbillon-Vey est alors facile à calculer: c'est le volume de $T_1 S$. Le théorème de Gauss-Bonnet donne:

$$gv(\mathscr{F}^s) = 4\pi^2 \chi(S).$$

Dans [29], Y. Mitsumatsu eut l'idée de calculer l'invariant de Godbillon-Vey du feuilletage stable du flot géodésique d'une métrique à courbure négative variable, en supposant celui-ci de classe C^2. Voici le principe de son calcul. Sur $T_1 S$, la théorie du repère mobile permet de construire trois champs de vecteurs V, X_1, X_2 qui forment en chaque point un repère orthonormé et sont tels que:

i) V est le champ associé à l'action de SO(2) sur $T_1 S$,

ii) X_1 engendre le flot géodésique,

iii) $[V, X_1] = X_2$; $[V, X_2] = -X_1$; $[X_1, X_2] = -(k \circ p)V$ où k est la courbure de S et p est la projection de $T_1 S$ sur S.

Il existe une fonction F définie sur $T_1 S$ telle que le feuilletage stable $\mathscr{F}^s$ soit engendré par X_1 et $X_2 + FV$. La condition d'intégrabilité est facile à écrire; elle est équivalente à l'équation de Ricatti:

$$X_1(F) + F^2 + k \circ p = 0$$

où $X_1(F)$ désigne la dérivée de la fonction F dans la direction X_1. On dispose alors de toutes les données nécessaires à l'évaluation de l'invariant $gv(\mathscr{F}^s)$. Tous calculs faits, on trouve:

$$gv(\mathscr{F}^s) = 4\pi^2\chi(S) - 3 \int V(F)^2 \, d\text{vol}.$$

Revenons à notre situation. Si le feuilletage $\mathscr{F}^s$ est de classe C^2, nous avons vu qu'il est C^2-conjugué au feuilletage stable du flot géodésique d'une métrique à courbure -1. L'invariant $gv(\mathscr{F}^s)$ est donc égal à $4\pi^2\chi(S)$. La formule précédente montre alors que $V(F)$ est identiquement nul, c'est-à-dire que F est constant sur les fibres de p. Il est facile d'en conclure que la courbure est constante.

4. Le problème de l'invariance topologique de la classe de Godbillon-Vey

Soit $\mathscr{F}^s$ le feuilletage stable du flot géodésique d'une métrique de S à courbure -1. Nous avons déjà vu que l'invariant de Godbillon-Vey de $\mathscr{F}^s$ ne dépend pas du choix de cette métrique. Est-il possible de perturber $\mathscr{F}^s$ parmi les feuilletages de classe C^∞, hors de l'espace des feuilletages stables de flots géodésiques, de façon à faire varier la classe de Godbillon-Vey?

Nous savons que $\mathscr{F}^s$ peut être défini par une forme ω telle que $d\omega = \omega \wedge \omega_1$ avec $\omega_1 \wedge d\omega_1$ non singulière. Il est clair que cette même propriété est satisfaite pour tous les feuilletages $\mathscr{F}$, disons de classe C^∞, qui sont C^3-proches de $\mathscr{F}^s$. En dérivant la relation $d\omega = \omega \wedge \omega_1$, on constate qu'il existe une forme ω_2 telle que $d\omega_1 = \omega \wedge \omega_2$. Les formes $\omega, \omega_1, \omega_2$ sont alors linéairement indépendantes partout et on peut donc définir deux champs de vecteurs X et Y, tangents à $\mathscr{F}$ par:

$$\omega(X) = 0; \qquad \omega_1(X) = 1; \qquad \omega_2(X) = 0$$
$$\omega(Y) = 0; \qquad \omega_1(Y) = 0; \qquad \omega_2(Y) = 1$$

Des calculs extrêmement simples montrent alors que les champs X et Y préservent le volume $\omega \wedge \omega_1 \wedge \omega_2$ et que $[X, Y] = -Y$. Cette dernière relation décrit l'algèbre de Lie du groupe affine GA de la droite, i.e. du groupe des transformations $x \mapsto ax + b$ $(a > 0)$. On conclut donc que le feuilletage $\mathscr{F}$ est défini par une action localement libre de GA sur $T_1 S$, préservant le volume $\omega \wedge \omega_1 \wedge \omega_2$. Ces actions sont complètement classées dans [11], ce qui permet alors d'obtenir le résultat de rigidité suivant:

Théorème [11]. *Tout feuilletage de classe C^∞ qui est suffisamment C^3-proche du feuilletage stable du flot géodésique d'une métrique à courbure -1, est C^∞-conjugué au feuilletage stable du flot géodésique d'une (autre) métrique à courbure -1.*

Résumons la situation. Toutes les actions de Γ de classe C^2 sur le cercle, de classe d'Euler maximale, sont topologiquement conjuguées. Deux actions projectives provenant de métriques à courbure -1 ne sont C^1-conjuguées que si les métriques correspondantes sont isométriques. Une action de classe C^∞, qui est C^3-proche d'une de ces actions projectives est C^∞-conjuguée à une autre action projective. Peut-on globaliser ce dernier énoncé?

Problème. *Soit Γ le groupe fondamental d'une surface compacte orientée S de genre supérieur ou égal à 2. Considérons une action de Γ sur le cercle par difféomorphismes directs de classe C^∞. Si le nombre d'Euler de l'action est maximal (i. e. égal à $\pm\chi(S)$), peut-on affirmer que l'action est C^∞-conjuguée à une action projective (associée à une métrique à courbure -1)?*

Ainsi, les déformations des feuilletages stables de flots géodésiques, dans l'espace des feuilletages de classe C^∞, gardent un type topologique constant mais ont aussi un invariant de Godbillon-Vey constant. C'est l'une des motivations de la question suivante.

Problème. *Soient $(M_1, \mathcal{F}_1)$ et $(M_2, \mathcal{F}_2)$ deux variétés compactes feuilletées, de codimension 1, transversalement orientables, et de classe C^2. On suppose qu'il existe un homéomorphisme h entre M_1 et M_2 qui envoie $\mathcal{F}_1$ sur $\mathcal{F}_2$. L'isomorphisme h^* entre les cohomologies de M_1 et M_2 envoie-t-il la classe de Godbillon-Vey de $\mathcal{F}_2$ sur celle de $\mathcal{F}_1$?*

Si l'on consent à un élargissement du domaine de définition de la classe de Godbillon-Vey, la réponse à la question précédente est négative. Nous allons mentionner rapidement deux tels élargissements.

Il n'est pas possible de définir un invariant pour les feuilletages de classe C^1. Ceci résulte de la contractibilité du classifiant des feuilletages transversalement orientables, de codimension 1 et de classe C^1, démontrée par T. Tsuboi [33]. La classe de Godbillon-Vey est cependant invariant par conjugaison de classe C^1 (voir [16] et [30]). D'autre part, dans [24], S. Hurder et A. Katok définissent un invariant pour les feuilletages de classe $C^{1+\varepsilon}$, avec $\varepsilon > 1/2$. En particulier, on peut calculer l'invariant du feuilletage stable du flot géodésique d'une métrique à courbure négative variable. Dans ce cas, la formule de Y. Mitsumatsu est encore valable et montre que l'invariant de Godbillon-Vey varie effectivement si la courbure varie. Cette même formule montre d'ailleurs que l'invariant prend sa valeur maximale précisément sur les métriques à courbure constante.

Dans une autre direction, on peut aussi définir l'invariant de Godbillon-Vey pour les feuilletages de classe C^2 par morceaux [10] et on peut aussi se poser le problème précédent dans ce contexte élargi. Nous y répondons négativement à travers le résultat suivant:

Théorème [14]. *L'action du groupe fondamental d'une surface à courbure négative sur le bord de son revêtement universel est topologiquement conjuguée à une action affine par morceaux.*

Ceci donne effectivement une réponse négative au problème car un feuilletage transversalement affine par morceaux a un invariant de Godbillon-Vey nul.

Pour démontrer ce dernier théorème, on utilise le fait que le flot géodésique ϕ_t possède une section de Birkhoff, c'est-à-dire qu'il existe une surface à bord Σ dans $T_1 S$ dont l'intérieur est transverse à ϕ_t et dont le bord est constitué d'orbites périodiques de ϕ_t. De plus, toute orbite de ϕ_t coupe Σ une infinité de fois et on peut donc définir une application de premier retour $r: \Sigma \to \Sigma$ (voir [4, 9]). Les feuilletages $\mathscr{F}^s$ et $\mathscr{F}^u$ tracent sur Σ deux feuilletages transverses ayant un comportement bien contrôlé au voisinage du bord. On montre alors que r est topologiquement conjugué à un difféomorphisme pseudo-Anosov qui agit de manière affine sur l'espace de ses feuilles stables. Par conséquent, $\mathscr{F}^s$ est topologiquement conjugué à un feuilletage transversalement affine par morceaux, les cassures correspondant aux feuilles contenant les courbes du bord de Σ. Une description explicite des homéomorphismes affines par morceaux ainsi obtenus a été récemment donnée par N. Hashiguchi [21].

D'une certaine façon, ce contre-exemple à l'invariance topologique de la classe de Godbillon-Vey est peu satisfaisant. On est tenté de penser que par cette conjugaison topologique, la classe de Godbillon-Vey s'est concentrée dans les singularités du feuilletage affine par morceaux. Il est effectivement possible de définir un "invariant discret" qui tient compte de ces singularités mais on montre que, même en ajoutant cette contribution, la classe n'est pas un invariant topologique.

Les deux extensions du domaine de définition de la classe de Godbillon-Vey que nous venons d'évoquer ont été récemment unifiées par T. Tsuboi [34]. Il définit un pseudogroupe d'homéomorphismes de la droite qui contient à la fois les difféomorphismes de classe $C^{1+\varepsilon}$ ($\varepsilon > 1/2$) et les homéomorphismes qui sont de classe C^2-par morceaux. Il montre ensuite comment définir une classe de Godbillon-Vey pour les feuilletages de codimension 1 dont la structure transverse est modelée sur ce pseudogroupe. Il semble bien que ceci soit "le" domaine maximal d'existence de l'invariant.

Le problème de l'invariance topologique dans le cadre des feuilletages de classe C^2 reste cependant ouvert.

References

1. Ballman, W., Gromov, M., Schroeder, V.: Manifolds of non-positive curvature. Progress in Mathematics **61** (1985)
2. Barge, J., Ghys, E.: Surfaces et cohomologie bornée. Invent. math. **92** (1988) 509–526
3. Benoist, Y., Foulon, P., Labourie, F.: Flots d'Anosov à distributions stables et instables différentiables. C.R. Acad. Sci. Paris 1990
4. Birkhoff, G.D.: Dynamical systems with two degrees of freedom. Trans. Amer. Math. Soc. **18** (1917) 199–300
5. Feres, R.: Geodesic flows on manifolds of negative curvature with smooth horospheric foliations. PhD Caltech. 1989

6. Feres, R., Katok, A.: Invariant tensor fields of dynamical systems with pinched Lyapunov exponents and rigidity of geodesic flows. Ergod. Theor. Dynam. Sys. **3** (1989) 427–433
7. Floyd, W.: Group completions and limit sets of Kleinian groups. Invent. math. **57** (1980) 205–218
8. Foulon, P., Labourie, F.: Flots d'Anosov à distributions de Liapounov différentiables. C.R. Acad. Sci. Paris **309** (1989) 255–260
9. Fried, D.: Transitive Anosov flows and pseudo-Anosov maps. Topology **22** (1983) 299–303
10. Fuchs, D.B., Gabrielov, A.M., Gelfand, I.M.: The Gauss-Bonnet theorem and the Atiyah-Patodi-Singer functionals for the characteristic classes of foliations. Topology **15** (1976) 165–188
11. Ghys, E.: Actions localement libres du groupe affine. Invent. math. **82** (1985) 479–526
12. Ghys, E.: Groupes d'homéomorphismes du cercle et cohomologie bornée. Contemp. Math. **58**, part III (1987) 81–105
13. Ghys, E.: Classe d'Euler et minimal exceptionnel. Topology **26** (1987) 93–105
14. Ghys, E.: Sur l'invariance topologique de la classe de Godbillon-Vey. Ann. Inst. Fourier **37** (1987) 59–76
15. Ghys, E.: Flots d'Anosov dont les feuilletages stables sont différentiables. Ann. Sci. Ec. Norm. Sup. **20** (1987) 251–270
16. Ghys, E., Tsuboi, T.: Différentiabilité des conjugaisons entre systèmes dynamiques de dimension 1. Ann. Inst. Fourier **38** (1988) 215–244
17. Goldman, W.: Discontinuous groups and the Euler class. PhD thesis, Dept. of Mathematics, University of California at Berkeley, 1980
18. Gromov, M.: Volume and bounded cohomology. Publ. Math. I.H.E.S. **56** (1982) 5–100
19. Gromov, M.: Hyperbolic groups. In: Essays in group theory (ed. S.M. Gersten). M.S.R.I. Pub. **8** (1987) 7–263
20. Hamenstadt, U.: A geometric characterization of negatively curved locally symmetric spaces. Preprint 1990
21. Hashiguchi, N.: PL-representations of the geodesic flows on closed surfaces. Preprint 1990
22. Hirsch, M., Pugh, C.: Smoothness of horocycle foliations. J. Diff. Geom. **10** (1975) 225–238
23. Hopf, E.: Statistik der geodätischen Linen in Mannigfaltigkeiten negativer Krümmung. Ber. Verh. Süchs. Akad. Wiss. Leipzig **91** (1939) 261–304
24. Hurder, S., Katok, A.: Differentiability, rigidity and Godbillon-Vey classes for Anosov flows. Preprint IHES, 1987
25. Jekel, S.: An elementary formula and a bound for the Euler class. Preprint
26. Kanai, M.: Geodesic flows of negatively curved manifolds with smooth stable and unstable foliations. Ergod. Theor. Dynam. Sys. **8** (1988) 215–239
27. Katok, A.: Entropy and closed geodesics. Ergod. Theor. Dynam. Sys. **2** (1982) 339–366
28. Matsumoto, S.: Some remarks on foliated S^1 bundles. Invent. math. **90** (1987) 343–358
29. Mitsumatsu, Y.: A relation between the topological invariance of the Godbillon-Vey invariant and the differentiability of Anosov foliations. Adv. Stud. Pure math. **5** (1985) 159–167
30. Raby, G.: Invariance des classes de Godbillon-Vey par C^1-difféomorphismes. Ann. Inst. Fourier **38** (1988) 205–213
31. Sullivan, D.: Discrete conformal groups and measurable dynamics. Bull. Amer. Math. Soc. **6** (1982) 53–73
32. Takamura, M.: Semi-conjugacy and a theorem of Ghys. Preprint
33. Tsuboi, T.: On the foliated products of class C^1. Ann. Math. **130** (1989) 227–271
34. Tsuboi, T.: Area functional and Godbillon-Vey cocycles. Preprint
35. Wood, J.: Bundles with totally disconnected structure group. Comm. Math. Helv. **46** (1971) 257–273

Metric and Topological Measurements of Manifolds

Karsten Grove

Department of Mathematics, University of Maryland, College Park, MD 20742, USA

A basic question in riemannian geometry asks how properties detected by metric measurements of a riemannian manifold are reflected in properties detected by topological measurements. If (M, g) is riemannian manifold, a *metric measurement* can by definition be given in terms of its distance function. A *topological measurement* on the other hand is given in terms of algebraic topological invariants of M and/or of objects associated with M. Examples of invariants resulting from such measurements are e.g. *diameter*, *volume*, and *sectional curvature bounds* on the metric side, and *Betti numbers*, *characteristic numbers* etc. on the topological side.

Here we consider only metric invariants that assign to any (closed) riemannian manifold a number. These invariants can therefore be thought of as functions on the collection $\mathcal{M}$ of all isometry classes of (closed) riemannian manifolds, and as such they divide this class into smaller natural subclasses. Since topological properties are not affected by scaling of the metrics, one needs to normalize, or at least bound the *size* of (M, g). Quantities commonly chosen for this purpose are diameter or volume.

Ultimately one would like to give a complete topological description of a riemannian manifold making as few metric measurements as possible. As a first step one tries to bound the topology in terms of its geometry. Results of that type are usually referred to as *finiteness theorems*. In metrically extreme cases the goal is to determine all the possible topological types. Results of this kind are called *pinching theorems*.

As an illustration, suppose we normalize all closed, connected riemannian n-manifolds, M, $n \geq 2$, so as to have the same diameter as the unit n-sphere, i.e. $\operatorname{diam} M = \pi$. The function *minsec*, that assigns to any such manifold the minimum of its sectional curvatures, is then bounded above by 1, i.e. $\operatorname{minsec} M \leq 1$, according to the classical Bonnet-Myers theorem. For the superlevel sets, $\operatorname{minsec} \geq k$ one has the following finiteness theorem.

0.1 Betti Number Theorem [G2]. *Let $n \geq 2$ be an integer and $k \leq 1$ a real number. There is a constant $C = C(n, k)$, such that for any field, F and any closed riemannian n-manifold, M with* $\operatorname{diam} M = \pi$ *and* $\sec M \geq k$ *one has* $\dim H_*(M; F) \leq C$.

Proceedings of the International Congress
of Mathematicians, Kyoto, Japan, 1990

Near the extreme value 1 of minsec there is only one manifold topologically as stated in the following pinching theorem.

0.2 Diameter Sphere Theorem [GS]. *Any closed riemannian n-manifold M, $n \geq 2$, with* diam $M = \pi$ *and* sec $M > 1/4$ *is homeomorphic to the n-sphere.*

The bound sec $M > 1/4$ is optimal in this theorem as e.g. the real projective space of constant curvature $1/4$ shows. A corresponding *rigidity theorem* for manifolds with diam $M = \pi$ and sec $M \geq 1/4$ was proved in [GG1, 2].

In the general setting of (0.1), one cannot expect much more information (other than possibly improving the constant C [A]). To get more refined information it seems therefore natural to subdivide the superlevel sets minsec $\geq k$ in terms of other metric measurements.

The purpose here is to report on recent progress in this general area. This progress involves on the metric side the use of critical point theory for distance functions as conceived in [GS] (cf. [G2] and for a survey [Gr]), and the use of Hausdorff convergence of metric spaces as initiated in [G1] (cf. [P2] for a survey). On the topological side, powerful results from controlled topology as developed in [CF, F] and [Q1, 2, 3] play a crucial rôle.

1. Distance and Volume Comparison

Throughout we let M denote a closed, connected riemannian n-manifold with $n \geq 2$. Following Rinow [R], S^n_k is the complete, simply connected n-dimensional space form of constant curvature k. The distance function on M, S^n_k, or any other metric space will be denoted by d. To distinguish point in S^n_k from points in other metric spaces we use the notation $\bar{p}, \bar{q}, \ldots$, etc., rather than $p, q, \ldots$, etc.

For each $p \in M$ and $r > 0$, let $B(p, r)$, resp. $D(p, r)$ be the open, resp. closed r-ball in M centered at p. Moreover, let $\mathscr{D}(p) \subset T_pM$ be the region bounded by the tangent cut locus $\mathscr{C}(p)$, i.e. $v \in \mathscr{D}(p)$ if and only if the exponential map $\exp_p : T_pM \to M$ maps the segment, $[0, v]$ in T_pM to a segment, i.e. minimal geodesic in M.

The *injectivity radius* of M, inj M, is by definition the largest $r > 0$ so that $\exp_p | B(0_p, r)$ is injective for all $p \in M$. The smallest $R > 0$ so that some $\exp_p | D(0_p, R)$ is surjective is easily seen to be the number $\min_p \max_q d(p, q)$, which makes sense in any compact metric space. This number is called the *radius* of M and is denoted by rad M.

Obviously rad $M \leq$ diam $M \leq$ 2rad M. In particular, imposing a bound on the diameter is equivalent to imposing a bound on the radius.

Now suppose sec $M \geq k$. This condition can be viewed in the following way: For each $p \in M$ replace the euclidean metric on $B(0_p, r) \subset T_pM$ by a constant curvature k metric via a radial conformal change ($r \leq \pi/\sqrt{k}$ if $k > 0$). Then $\mathscr{D}(p)$ is a proper compact subset of S^n_k for every $p \in M$, except when $k > 0$ and M isometric to S^n_k. In the latter case we interpret $\mathscr{D}(p)$ as S^n_k. The Toponogov triangle comparison theorem is then equivalent to

1.1 Distance Comparison Theorem. *The exponential map* $\exp_p : \mathscr{D}(p) \to M$ *is distance nonincreasing, when* $\mathscr{D}(p) \subset S^n_k$ *is given the induced metric from* S^n_k.

This is the basis also for volume comparison of metrically defined subsets of M with corresponding subsets in S^n_k. For clarity, we list three situations of obvious interest although they are all special cases of one general result [GP2] (cf. also [Du]).

1.2 Half Spaces. Fix $p \in M$ and a closed $Q \subset M$. Consider $\bar{p} = \exp_p^{-1}(p)$ and $\bar{Q} = \exp_p^{-1}(Q)$ in $\mathscr{D}(p) \subset S^n_k$. For the half spaces $H(p, Q) = \{x \in M \,|\, d(x, p) \leq d(x, Q)\}$ and $H(\bar{p}, \bar{Q})$ in M and S^n_k respectively we have

$$\operatorname{vol} H(p, Q) \leq \operatorname{vol} H(\bar{p}, \bar{Q}).$$

1.3. Swiss Cheeses. Define the swiss cheese K relative to $D(p, R)$ and $r : Q \to \mathbb{R}_+$ as the complement $K((Q, r); (p, R)) = D(p, R) - \bigcup_{q \in Q} B(q, r(q))$. Then

$$\operatorname{vol} K((Q, r); (p, R) \leq \operatorname{vol} K((\bar{Q}, \bar{r}); (\bar{p}, R))$$

where $\bar{r} = r \circ \exp_p : \bar{Q} \to \mathbb{R}_+$.

1.4 Ball Collection. For the union $D(Q, r) = \bigcup_{q \in Q} D(q, r(q))$ one has $\operatorname{vol} D(Q, r) \leq \operatorname{vol} D(I(Q), r \circ I^{-1})$, provided $I : Q \to I(Q) \subset S^n_k$ is an isometry.

The estimates (1.2)–(1.4) and (1.7), (1.8) below do not hold under the weaker curvature assumption $\operatorname{Ric} M \geq (n-1)k$. This, however, is sufficient for the following simple extension of the so-called Bishop-Gromov volume comparison theorem.

1.5 Relative Volume Comparison. *For finite* $Q \subset M$, *the function* $\operatorname{vol} D(Q, R)/v^n_k(R)$ *is non-increasing in* R. *Here* $v^n_k(R) = \operatorname{vol} D(\bar{p}, R)$ *in* S^n_k.

This theorem is the basis for various *packing arguments*. A subset $Q \subset M$ is called a *weak R-net* in M if $D(Q, R) = M$. Examples of such nets are provided by maximal sets Q for which $B(q, R/2)$, $q \in Q$ are mutually disjoint. For fixed $n \geq 2$, $k \in \mathbb{R}$ and $D > 0$ this together with (1.5) yields a *covering function*.

1.6 $$N(\varepsilon) = C_1 \varepsilon^{-n}, \quad \varepsilon \in (0, D], \quad C_1 = C(n, k, D) > 0.$$

For any closed riemannian n-manifold M with $\operatorname{diam} M \leq D$ and $\operatorname{Ric} M \geq (n-1)k$, i.e. for every $\varepsilon \in (0, D]$, any such M can be covered by $\leq N(\varepsilon)$ closed ε-balls.

The relative volume comparison (1.5) can also be used to estimate the right hand sides of (1.2) and (1.3) in an important special case: (i) there is an $\bar{r} > 0$ such that $d(\bar{p}, \bar{q}) = \bar{r}$, $\bar{q} \in \bar{Q}$, and (ii) the set of directions of segments $\overline{pq}$, $\bar{q} \in \bar{Q}$ form a weak θ-net, $\theta \geq \pi/2$ in the unit sphere $S^{n-1}_1 \subset T_{\bar{p}} S^n_k$. If furthermore $\bar{q}_1, \bar{q}_2 \in S^n_k$ are chosen so that $d(\bar{p}, \bar{q}_1) = d(\bar{p}, \bar{q}_2) = \bar{r}$ and $\operatorname{ang}(\overline{pq_1}, \overline{pq_2}) = 2\pi - 2\theta$, then

1.7 $$\operatorname{vol} H(\bar{p}, \bar{Q}) \leq \operatorname{vol} H(\bar{p}, \{\bar{q}_1, \bar{q}_2\}), \qquad \text{and}$$

1.8 $$\operatorname{vol} K((\bar{Q}, r); (\bar{p}, R)) \leq \operatorname{vol} K(\{\bar{q}_1, \bar{q}_2\}, r); (\bar{p}, R)), \qquad r \text{ constant}.$$

Now fix $v > 0$ in addition to N, κ and D as above. Combining (1.2) and (1.7) with a *critical point theory argument* for $d: M \times M \to \mathbb{R}$ considered as the distance function in $M \times M$ from the diagonal $\Delta \subset M \times M$ gives constants $C_2 = C_2(n, \kappa, D, v) \geq 1$ and $R = R(n, \kappa, D, v) > 0$ such that

1.9 $$\varrho(\varepsilon) = C_2 \varepsilon, \qquad \varepsilon \in (0, R]$$

is a *contractibility function* for any closed riemannian n-manifold M satisfying $\sec M \geq \kappa$, $\operatorname{diam} M \leq D$ and $\operatorname{vol} M \geq v$, i.e. every ε-ball in such M is contractible in the concentric $C_2 \cdot \varepsilon$-ball.

The existence of covering and contractibility functions as in (1.6) and (1.9) are key ingredients in the finiteness and pinching theorems reported on here.

2. The Gromov-Hausdorff Topology

In this section we outline topological and geometric properties of Gromov-Hausdorff limit spaces (cf. [G1]) suited for our purposes.

Let $X, Y, Z, X_i, i = 1, 2, \ldots$ be compact metric spaces. If X and Y are isometrically embedded in Z, the classical *Hausdorff distance*, d_{H}^Z satisfies $d_{\mathrm{H}}^Z(X, Y) < \varepsilon$ if and only if $Y \subset B(X, \varepsilon)$ and $X \subset B(Y, \varepsilon)$. For the *Gromov-Hausdorff distance* d_{GH} one has $d_{\mathrm{GH}}(X, Y) < \varepsilon$ if and only if $d_{\mathrm{H}}^Z(X, Y) < \varepsilon$ from some metric on $Z = X \amalg Y$ extending the ones on X and Y. Similarly $X = \lim X_i$ in the Gromov-Hausdorff topology if and only if $d_H^Z(X, X_i) \to 0$, $i \to \infty$ in some extended metric on $Z = X \amalg_i X_i$.

2.1 Precompactness Theorem [G1]. *A class χ of compact metric spaces is precompact if and only if there is a common covering function for all $X \in \chi$.*

Now let $\chi(N, \varrho)$ be the class of compact metric spaces which have N in (1.6) as a covering function and ϱ in (1.9) as a contractibility function. Since all spaces having (1.6) as covering function have dimension $\leq n$ by [PS], a result of Borsuk [Bo] combined with (2.1) implies (cf. also [P1]).

2.2 Compactness Theorem [GPW]. *The class* $\chi(N, \varrho)$ is compact and $\dim X \leq n$ *for any* $X \in \chi(N, \varrho)$.

Using the contractibility properties of $X \in \chi(N, \varrho)$ one can construct explicit homotopy equivalences between nearby spaces as in [GP1] (see [P1]). In particular

2.3 Homotopy Finiteness Theorem [P1]. *The class $\chi(N, \varrho)$ contains only finitely many (simple) homotopy types.*

The application of controlled topology is based on the important observation that the homotopy equivalences and corresponding homotopies alluded to above are controlled in size in terms of $d_{\mathrm{GH}}(X, Y)$. In the context of manifolds, this together with the results of Begle [B] and Quinn [Q3] proves

2.4 Almost Compactness Theorem [GPW]. *Let $\mathcal{M}(N, \varrho)$ be the subset of closed topological n-manifolds in $\chi(N, \varrho)$. Any $X \in \overline{\mathcal{M}(N, \varrho)}$ is a generalized n-manifold and if $n \neq 3$ it admits a manifold resolution.*

Since $X \times S^1 \times S^1$ satisfies the disjoint disc property (cf. [D]) it follows from Edwards approximation theorem [E] that it is a manifold. Consequently, by results of Chapman and Ferry [CF, F] there are only finitely many homeomorphism types among $M \times S^1 \times S^1$, $M \in \mathcal{M}(N, \varrho)$. The controlled h-cobordism theorem [Q1, 2] is then used to complete the proof of

2.5 Topological Finiteness Theorem [GPW]. *The class of closed n-manifolds, $n \neq 3$, with metrics having* (1.6) *as covering functions and* (1.9) *as contractibility function contains at most finitely many homeomorphism types.*

We conclude this section with a brief discussion of geometric properties of limit spaces $X = \lim M_i$, where sec $M_i \geq k$ and diam $M \leq D$. First of all, X is an inner metric space, i.e. the distance between points in X is the infimum of lengths of curves joining them. In particular, one has the notion of geodesics in X. Moreover,

2.6 Exponential Maps [GP2]. Let $p \in X = \lim M_i$, sec $M_i \geq k$ and $\bar{p} \in S^n_k$. There is a compact subset $\mathcal{D}(p) \subset S^n_k$ and a distance nonincreasing surjective map $\exp_p : \mathcal{D}(p) \to X$. Moreover, $\exp_p$ maps segments from $\bar{p} \in \mathcal{D}(p) \subset S^n_k$ to segments in X from p, and any segment from p is the image of a segment from $\bar{p}$.

It should be pointed out that the pairs $(\mathcal{D}(p), \exp_p)$ in (2.6) are by no means unique! Using the nontrivial fact that geodesics in X are limits of geodesics in M_i one gets

2.7 Curvature for Limits [GP2]. $X = \lim M_i$, sec $M_i \geq k$ has Toponogov curvature, sec $X \geq k$, i.e. standard distance comparison holds for geodesic triangles in X.

As an interesting consequence, X has everywhere constant dimension. This is of interest in situations of collapse (cf. also [Fu]).

3. The Volume Function

In this and the next section we consider sub- or superlevel sets of functions restricted to the classes of $\mathcal{M}^D_k(n)$, and $\mathcal{M}^R_k(n)$ of closed connected riemannian n-manifolds M, $n \geq 2$ with sec $M \geq k$ and diam $M \leq D$, rad $M \leq R$ respectively.

The following finiteness theorem is an immediate corollary of (1.6), (1.9), (2.3), (2.5) and the fact that every closed topological n-manifold carries at most finitely many inequivalent smooth structures, when $n \neq 4$ (cf. [KS] for $n \geq 5$).

3.1 Diffeo/Top-Theorem [GP1, GPW]. *Fix $n \geq 2$, $k \in \mathbb{R}$ and D, $v > 0$. The class of closed riemannian n-manifolds M with* $\sec M \geq k$, $\operatorname{diam} M \leq D$ *and* $\operatorname{vol} M \geq v$ *contains at most finitely many diffeomorphism-, homeomorphism-, and homotopy types when $n \neq 3, 4$, $n = 4$, and $n = 3$ respectively.*

We now turn to the corresponding metrically extreme cases of small and large volume.

Let $\{M_i\}$ be a sequence in the $\mathcal{M}_k^D(n)$ where $\operatorname{vol} M_i \to 0$ as $i \to \infty$. Then a subsequence will converge in the Gromov-Hausdorff topology (cf. (1.6) and (2.1)) to a *lower dimensional* inner metric space X with $\operatorname{diam} X \leq D$ and Toponogov curvature $\sec X \geq k$. In the exceptional case where X is a riemannian manifold, a thorough investigation has been carried out in [Y2]: For large i, M_i fibers almost riemannian over X, and a finite cover of the fibers, F fiber almost riemannian over the $b_1(F)$-dimensional flat torus.

In the (weakly) extreme case of manifolds M with $\sec M \geq k$, $\operatorname{diam} M \leq D$ and $\operatorname{vol} M \leq \varepsilon$, ε small, one would expect some similar kind of decomposition of M. This of course is related to the structure of general inner metric spaces X with $\sec X \geq k$. For progress in this direction we refer to [BGP].

To treat the other extreme case of manifolds with large volume one first needs to have optimal bounds. Such bounds are not known for the class $\mathcal{M}_k^D(n)$. If on the other hand $\sec M \geq k$ and $\operatorname{rad} M \leq R$ we get from (1.4), (1.3) and (1.8) (cf. [GP4]) that

3.2
$$\operatorname{vol} M \leq \begin{cases} v_k^n(R); & R \leq \pi/2\sqrt{k} \text{ if } k > 0 \\ \dfrac{R}{\pi/\sqrt{k}} \operatorname{vol} S_k^n & \text{if } k > 0 \text{ and } R \geq \pi/2\sqrt{k} \end{cases}$$

There are riemannian metrics on S^n (and $\mathbb{R}P^n$) showing that (3.2) is optimal (cf [GP4]). Moreover, using (1.3), (1.4), (1.8), (1.9), (2.6), (2.7) and controlled topology as in Sect. 2 one can prove the following pinching theorem in this (strongly) extreme case.

3.3 Large Volume Theorem [GP4]. *Given $n \geq 2$, $k \in \mathbb{R}$ and $R \in \mathbb{R}_+$ ($\leq \pi/2\sqrt{k}$ if $k > 0$). There is an $\varepsilon = \varepsilon(n, k, R) > 0$ such that any riemannian n-manifold M with* $\sec M \geq k$, $\operatorname{rad} M \leq R$ *and* $\operatorname{vol} M \geq v_k^n(R) - \varepsilon$ *is topologically either S^n or $\mathbb{R}P^n$.*

In the case $\sec M \geq k > 0$ and $\operatorname{rad} M > \pi/2\sqrt{k}$, of course M is S^n topologically by (0.2). In all cases, however, there are (*singular*) *constant curvature k* models with equality in (3.2), and manifolds with volume close to this are metrically similar to these models (cf. [GP4]). In the special cases $k > 0$ and $R = \pi/2\sqrt{k}$, $\pi/\sqrt{k}$ corresponding to the only nonsingular models, it is possible to conclude diffeomorphism [OSY].

These pinching theorems may also be viewed as a solution to a generalized analog of Aleksandrov's area problem for convex surfaces in $\mathbb{R}^3$ [Al].

4. The Excess and Other Shape Invariants

The Theorems 0.1, 0.2, 3.1 and 3.3 are in essence exhaustive. Clearly therefore, in order to gain more information about individual manifolds $M \in \mathcal{M}^D_{kv}(n)$ where vol $M \geq v$ it is necessary to make more measurements.

Other than *maxsec*, *inj*, and to some extent, *FillRad* [G3], no such invariants seem to have been studied very extensively. This includes the *intermediate diameters*, diam $M = \text{diam}_0\, M \geq \text{diam}_1\, M \geq \cdots \geq \text{diam}_{n-1} M > \text{diam}_n\, M = 0$ considered by Urysohn [U] and more recently in [G3] (cf. also [P2]).

Here we discuss briefly a measurement based on the excess in the triangle inequality. Following [GP3] we define the *excess* of M by exc $M = \min_{(p,q)} \max_x (d(p, x) + d(x, q) - d(p, q))$. Obviously $0 \leq \text{exc}\, M \leq \text{diam}\, M$.

A smooth manifold M admits a Riemannian metric with exc $M = 0$ if and only if M is a twisted sphere (cf. [GP3]). On the other hand, any M admits a riemannian metric with exc M/diam M arbitrarily small.

When restricted to $\mathcal{M}^D_{kv}(n)$, however, the sublevel sets exc $\leq \varepsilon$, ε small, contain only one topological type:

4.2 Exotic Sphere Theorem [GP3]. *Fix $n \geq 2$, $k \in \mathbb{R}$ and D, $v > 0$. There is an $\varepsilon = \varepsilon(n, k, D, v)$ such that any closed riemannian n-manifold M, with* sec $M \geq k$, diam $M \leq D$, vol $M \geq v$ *and* exc $M \leq \varepsilon$, *is a homotopy sphere.*

This result is also optimal in its context. Observe that in contrast to all other known sphere theorems in riemannian geometry, the metrically extreme case is represented not only by the unit sphere S^n_1 but by any exotic sphere!

The other extreme exc M = diam M occurs when M contains three points at maximal distance. This is related to the *triameter* of M, which is the second in an infinite sequence of invariants beginning with the diameter. That and another related infinite sequence beginning with twice the radius and the excess is currently under investigation in [GM].

5. Curvature Free Problems

We conclude in the generality we started, now scaling all closed riemannian n-manifolds M so that vol M = vol $S^n_1 = v^n_1(\pi)$. In this generality we consider the superlevel sets inj $\geq i$.

5.1 Isoembolic Rigidity Theorem [Be]. *The injectivity radius of any closed riemannian n-manifold M so that* vol $M = v^n_1(\pi)$ *satisfies* inj $M \leq \pi$, *with equality if and only if M is isometric to S^n_1.*

5.2 Isoembolic Pinching Theorem [C2]. *For any $n = \mathbb{N}$ there is an $\varepsilon = \varepsilon(n)$, such that any closed riemannian n-manifold M with* vol $M = v^n_1(\pi)$ *and* inj $M \geq \pi - \varepsilon$ *is homeomorphic to S^n.*

5.3 Isoembolic Finiteness Theorem [GPW, Y1]. *Fix $n \in \mathbb{N}$ and $\varrho \in (0, \pi]$. There are at most finitely diffeomorphism-, homeomorphism-, or homotopy types of closed riemannian n-manifolds M satisfying* vol $M = v_1^n(\pi)$, inj $M \geq \varrho$ *and $n \neq 3, 4$, $n = 4$, or $n = 3$ respectively.*

The geodesic flow plays an essential role in the proof of these results. Here we only point out that a basic volume estimate from [C1] implies that the class of closed riemannian n-manifolds M with vol $M \leq v$ and inj $M \geq \varrho$ has a covering function as in (1.6). Since it has a contractibility function as in (1.9) with $C_2 = 1$ by assumption, (5.3) is a corollary of (2.3) and (2.5) just like (3.1) was.

The subclass consisting of closed riemannian n-manifolds M with Ric $M \geq (n-1)k$, diam $M \leq D$ and inj $M \geq \varrho$ has recently been investigated in [AC]. In particular it is shown there, that this class contains at most finitely many diffeomorphism types also in dimensions 3 and 4.

References

[A] U. Abresch: Lower curvature bounds, Toponogov's theorem, and bounded topology, I; II. Ann. Sci. Éc. Norm. Sup. **18**; **20** (1985/87) 563–633, 475–502

[Al] A.D. Aleksandrov: Die innere Geometrie der konvexen Flächen. Academie-Verlag, Berlin 1955

[AC] M.T. Anderson, J. Cheeger: C^α-Compactness for manifolds with Ricci curvature and injectivity radius bounded below. J. Diff. Geom. (to appear)

[B] E.B. Begle: Regular convergence. Duke Math. J. **11** (1944) 441–450

[Be] M. Berger: Une borne inférieure pour le volume d'une variétés riemannienne en function du rayon d'injectivité. Ann. Inst. Fourier **30** (1980) 259–265

[Bo] K. Borsuk: On some metrizations of the hyper space of compact sets. Fund. Math. **41** (1955) 168–201

[BGP] Y. Burago, M. Gromov, G. Perelman: A.D. Alexandrov's spaces with curvatures bounded from below I. Preprint

[CF] T.A. Chapman, S. Ferry: Approximating homotopy equivalences by homeomorphisms. Amer. J. Math. **101** (1979) 583–607

[C1] C.B. Croke: Some isoparimetric inequalities and eigenvalue estimates. Ann. Sci. Éc. Norm. Sup. **13** (1980) 419–435

[C2] C.B. Croke: An isoembolic pinching theorem. Invent. math. **92** (1988) 385–387

[D] R.J. Davermann: Detecting the disjoint disc property. Pac. J. Math. **93** (1981) 277–298

[Du] O. Durumeric: Manifolds of almost half of the maximal volume. Proc. Amer. Math. Soc. **104** (1988) 277–283

[E] R.D. Edwards: The topology of manifolds and cell-like maps. Proc. Int. Congress of Math. 1978. Acad. Sci. Fennica, Helsinki 1980, pp. 111–127

[F] S. Ferry: Homotoping ε-maps to homeomorphisms. Amer. J. Math. **101** (1979) 567–582

[Fu] K. Fukaya: These proceedings

[G1] M. Gromov: Groups of polynomial growth and expanding maps. Publ. Math. IHES **53** (1981) 53–73

[G2] M. Gromov: Curvature, diameter and Betti numbers. Comment. Math. Helv. **56** (1982) 179–195

[G3] M. Gromov: Filling Riemannian manifolds. J. Diff. Geom. **18** (1983) 1–148

[GG1] D. Gromoll, K. Grove: A generalization of Berger's rigidity theorem for positively curved manifolds. Ann. Sci. Éc. Norm. Sup. **20** (1987) 227–239

[GG2] D. Gromoll, K. Grove: The low-dimensional metric foliations of euclidean spheres. J. Diff. Geom. **28** (1988) 143–156

[Gr] K. Grove: Critical point theory for distance functions. Proc. A. M. S. Summer Inst., Los Angeles 1990

[GM] K. Grove, S. Markvorsen: In preparation

[GP1] K. Grove, P. Petersen. V: Bounding homotopy types by geometry. Ann. Math. **128** (1988) 195–206

[GP2] K. Grove, P. Petersen, V: Manifolds near the boundary of existence. J. Diff. Geom. **33** (1991) 379–394

[GP3] K. Grove, P. Petersen, V: A pinching theorem for homotopy spheres. J. Amer. Math. Soc. **3** (1990) 671–677

[GP4] K. Grove, P. Petersen, V: Volume comparison à la Aleksandrov. Preprint

[GPW] K. Grove, P. Petersen, V, J.-Y. Wu: Geometric finiteness theorems via controlled topology. Invent. math. **99** (1990) 205–213

[GS] K. Grove, S. Shiohama: A generalized sphere theorem. Ann. Math. **106** (1977) 201–211

[KS] P. Kirby, L. Siebenmann: Foundational essays on topological manifolds, smoothings and triangulations. Ann. Math. Stud., vol. 88. Princeton Univ. Press 1977

[OSY] Y. Otsu, K. Shiohama, T. Yamaguchi: A new version of differentiable sphere theorem. Invent. math. **98** (1989) 219–228

[P1] P. Petersen, V: A finiteness theorem for metric spaces. J. Diff. Geom. **31** (1990) 387–395

[P2] P. Petersen, V: Gromov-Hausdorff convergence of metric spaces. Proc. A. M. S. Summer Inst. Los Angeles, 1990

[PS] L. Pontrjagin, L. Schnirelmann: Sur une propriété métrique de la dimension. Ann. Math. **33** (1932) 156–162

[Q1] F. Quinn: Ends of maps. J. Ann. Math. **110** (1979) 275–213

[Q2] F. Quinn: Ends of maps, III. Dimensions 4 and 5. J. Diff. Geom. **17** (1982) 503–521

[Q3] F. Quinn: An obstruction to the resolution of homology manifolds. Michigan Math. J. **34** (1987) 285–291

[R] W. Rinow: Die innere Geometrie der metrischen Räume. Springer, Berlin Heidelberg 1961

[U] P. Urysohn: Notes supplémentaires. Fund. Math. **8** (1926) 352–356

[Y1] T. Yamaguchi: Homotopy type finiteness theorems for certain precompact families of Riemannian manifolds. Proc. Amer. Math. Soc. **102** (1988) 660–666

[Y2] T. Yamaguchi: Collapsing and pinching under a lower curvature bound. Ann. Math. **133** (1991) 317–357

Symplectic Invariants

Helmut Hofer

Institut für Mathematik, Ruhr-Universität Bochum, Postbox 102148
W-4630 Bochum 1, Fed. Rep. of Germany

1. Introduction

In 1985 M. Gromov proved in his seminar paper *Pseudoholomorphic Curves in Symplectic Manifolds* [17] among other things a striking rigidity result, the so-called *squeezing theorem.*

Consider the vectorspace $\mathbb{C}^n$ equipped with its usual Hermitian inner product $(\cdot,\cdot)$. Denote by $\langle\cdot,\cdot\rangle = \text{Re}(\cdot,\cdot)$ the associated real inner product and by $\sigma = -Im\,(\cdot,\cdot)$ the usual induced symplectic form. If $\mathbf{B}^{2n}(r)$ denotes the Euclidean r-ball and $\mathbf{Z}^{2n}(R) = \mathbf{B}^2(R) \times \mathbb{C}^{n-1}$ the symplectic cylinder of radius R, M. Gromov proved that $\mathbf{B}^{2n}(r)$ admits a symplectic embedding into $\mathbf{Z}^{2n}(R)$ iff $r \leq R$. This result exhibits a striking C°-rigidity phenomenon in the symplectic category. Hence already on the C°-level, volume-preserving and symplectic maps can be distinguished.

Coming from the variational theory of Hamiltonian dynamics, I. Ekeland and the author observed in [5, 6] that the study of periodic solutions of Hamiltonian systems can be effectively used to prove the squeezing theorem and – more important – gives even infinitely many new symplectic invariants.

Finally combining the key construction in [5, 6] with Floer's Instanton Homology [12–15] one is able to construct a new theory called Symplectic Homology [16].

2. Symplectic Capacities

Let us first motivate why periodic solutions of Hamiltonian systems occur as obstructions in a *symplectic rigidity theory*.

Given any hypersurface S in a symplectic manifold (M, ω) we see that S carries a distinguished 1-dimensional distribution $\mathscr{L}_S \to S$ which is defined by $\mathscr{L}_S = \ker(\omega\,|_S)$. We denote by $L_S(x)$ the leaf through $x \in S$. If S is a compact hypersurface, of particular interest are the closed characteristics, i.e. closed integral curves $L_S(x) \approx S^1$. We shall write $\mathscr{P}(S)$ for the set of periodic characteristics.

Now, suppose, coming back to Gromov's Theorem that we have an optimal symplectic embedding of some open subset $\mathscr{U}$ of $\mathbb{C}^n$ into $\mathbf{Z}^{2n}(r)$. Optimal here

Proceedings of the International Congress
of Mathematicians, Kyoto, Japan, 1990

means that there is no symplectic embedding for any $r' < r$. Of course, it is in general not clear, if there exists an optimal embedding. Identifying $\mathcal{U}$ with its image, we may assume without loss of generality that $\mathcal{U} \subset \mathbf{Z}^{2n}(r)$. Let us even suppose that $\partial\mathcal{U}$ is smooth. If we try to do, what is impossible by our assumption, namely to make $\mathcal{U}$ smaller, one likes to proceed as follows. It seems reasonable to try to push the set $(\partial\mathcal{U}) \cap (\partial\mathbf{Z}^{2n}(r))$ into $\mathbf{Z}^{2n}(r)$. Then the new set $\mathcal{U}'$, at least if $\bar{\mathcal{U}}'$ is compact, would strictly stay away from $\partial\mathbf{Z}^{2n}(r)$ and we would be able to replace $\mathbf{Z}^{2n}(r)$ by a smaller symplectic cylinder. The optimal way to carry out this procedure is, at least locally, to take a Hamiltonian $H : \mathbb{C}^n \to \mathbb{R}$, which is increasing on the parts of the characteristics of $\partial\mathcal{U}$ which touch $\partial\mathbf{Z}^{2n}(r)$. The associated Hamiltonian flow pushes (locally) $\partial\mathcal{U}$ into $\mathcal{U}$. However, there are obvious global obstructions against the existence of Hamiltonians increasing on leaves, namely characteristics in $\mathcal{P}(\partial\mathcal{U})$, i.e. periodic characteristics. There is another more diffuse obstruction, namely Poincare's recurrence: The characteristics properly parameterized define a volume preserving flow on $\partial\mathcal{U}$. So, there is a set of full measure consisting of points which return to every neighbourhood of themselves under the flow. Obviously this is a second obstruction.

The surprising fact is, that the above can be turned into a qualitative and quantitative theory of symplectic rigiditiy.

We start with the theory of symplectic capacities initiated in [5, 6], see also [20, 21, 23, 35]. Motivated by a break through results of C. Viterbo [34] concerning the existence of periodic solutions of Hamiltonian systems on a prescribed energy surface, a very simple, refined approach was given by E. Zehnder and the author in [24]. This is best described as follows. If E denotes the loop space of $\mathbb{C}^n$ and $\mathcal{U}$ is any bounded open set in $\mathbb{C}^n$ let $H : \mathbb{C}^n \to [0, +\infty)$ be a smooth Hamiltonian vanishing on $\mathcal{U}$ but growing sufficiently fast to infinity outside from $\mathcal{U}$. We define a map $\Phi_H : E \to \mathbb{R}$ by

$$\Phi_H(x) = \int x^*\lambda - \int H(x),$$

where the integrals are taken over $S^1 = \mathbb{R}/\mathbb{Z}$, and λ is a primitive of σ. Now, if H is in some sense "sharp" enough at $\mathcal{U}$ it turns out that every critical point of Φ_H with $\Phi_H(x) > 0$ is a 1-periodic solution of $\dot{x} = X_H(x)$, which is close to $\partial\mathcal{U}$. How close the periodic solutions are to $\partial\mathcal{U}$ depends solely on the growth of H outside of $\mathcal{U}$. This observation leads to an existence result of periodic solutions on almost all energy surfaces [24]. Precisely this abundance of periodic solutions is used in [5, 6]. One constructs a universal subset $\mathcal{F}$ of the power set of E and defines for every $H : \mathbb{C}^n \to [0, +\infty)$, which vanishes on some nonempty open set $\mathcal{U}$ a min-max-characterisation defining a number $c(H) \in [0, +\infty)$

$$c(H) := \sup_{F \in \mathcal{F}} \inf \Phi_H(F).$$

Obviously if $H \geq K$ then $c(K) \geq c(H)$. Finally one puts for a bounded subset $\mathcal{U}$ of $\mathbb{C}^n$

$$c(\mathcal{U}) := \inf c(H),$$

where the infimum is taken over all such H which vanish on $\mathcal{U}$. One can even show that for good sets $\mathcal{U}$ the number $c(\mathcal{U})$ is the action of a closed (perhaps iterated) characteristic on $\partial\mathcal{U}$. This number turns out to be a symplectic invariant. Moreover, different choices of $\mathcal{F}$ lead to infinitely many invariants.

These invariants satisfy a set of axioms, and are called symplectic capacities. To give the axioms, let us denote by $\Gamma = (0, +\infty]$ the extended strictly positive half line equipped with the obvious ordering "$\leq$". We view $(\Gamma, \leq)$ as a category. Let $\mathcal{S}$ denote the category of $(2n)$-dimensional symplectic manifolds together with the symplectic embeddings as the morphisms.

Definition 1. A symplectic capacity is a covariant functor $c : \mathcal{S} \to \Gamma$ such that
- $c(M, \alpha\omega) = \mid \alpha \mid c(M, \omega) \quad$ for $\alpha \neq 0$,
- $c(\mathbf{Z}^{2n}(1), \sigma) < +\infty$,

where $\mathbf{Z}^{2n}(1)$ is equipped with the symplectic structure induced from $\mathbb{C}^n$.

We note that a convex combination of symplectic capacities as well as positive multiples are again capacities. So the collection of all capacities build a cone. One can show that this cone is infinite dimensional.

The first capacity constructed in [5] satisfies in addition the normalisation $c(\mathbf{B}^{2n}(1)) = c(\mathbf{Z}^{2n}(1)) = \pi$ and is therefore similar to Gromov's width [19]. But there are many other capacities [6], allowing for example to prove that the polydisk $\mathbf{B}^2(1) \times \ldots \times \mathbf{B}^2(1)$ can only be symplectically embedded into the round ball $\mathbf{B}^{2n}(R)$ iff $R \geq \sqrt{n}$.

Next we present some results which emphasize the importance of symplectic capacities.

Theorem 2. *A smooth embedding $\Psi : \mathbf{B}^{2n}(\varepsilon) \to \mathbb{C}^n$ preserving the capacity of open subsets of $\mathbf{B}^{2n}(\varepsilon)$ for some capacity c, is either symplectic or anti-symplectic.*

Even stronger, a homeomorphism $h : \mathbb{C}^n \to \mathbb{C}^n$ preserving locally capacity, has at every point x, where it is differentiable, either a symplectic or anti-symplectic linearization. If h and $h \times Id_{\mathbb{C}}$ preserve some capacities and h is smooth it is even symplectic. This motivates the following definition

Definition 3. A homeomorphism $h : \mathcal{U} \to V$ between two open sets $\mathcal{U}$ and $\mathcal{V}$ of $\mathbb{C}^n$ is said to be C°-symplectic if h and $h \times Id_{\mathbb{C}} : \mathcal{U} \times \mathbb{C} \to \mathcal{V} \times \mathbb{C}$ preserve every capacity of open subsets of their domains.

In view of the above results this is a natural C°-definition of "symplectic". It is not very practical yet since our knowledge of the cone of capacities is very limited. However, it is by now clear that the symplectic notion is a C°- rather than a C^1-concept. There are many known constructions for a symplectic capacity [5, 6, 16, 20, 21, 23, 35, 36]. A very interesting one is the following, which is only defined for subsets of $\mathbb{C}^n$ [20].

Let $\mathcal{C}$ be the vectorspace of all compactly supported smooth maps $H : [0,1] \times \mathbb{C}^n \to \mathbb{R}$. We define a norm $\|H\|$ by

$$\|H\| := \int_0^1 \left[\sup_{x \in \mathbb{C}^n} H(t,x) - \inf_{x \in \mathbb{C}^n} H(t,x) \right] dt$$

Associated to $H \in \mathscr{C}$ is the time-1-map Ψ_H obtained by integrating the Hamiltonian system associated to H. The collection of all time-1-maps forms a group $\mathscr{D}$ under composition. We define a map $E : \mathscr{D} \to [0, +\infty)$, called the "Energy", by

$$E(\Psi) = \inf\{\|H\| \mid \Psi_H = \Psi\}.$$

One verifies easily that

$$E(\Psi^{-1}) = E(\Psi) = E(\Phi^{-1} \circ \Psi \circ \Phi) \qquad \text{and} \qquad E(\Psi \circ \Phi) \leq E(\Psi) + E(\Phi).$$

Of course the above would be true if $E \equiv 0$ (in fact that was everyone's favorite guess). However,

Theorem 4. *$E(\Psi) = 0$ iff $\Psi = Id$. In particular $d : \mathscr{D} \times \mathscr{D} \to [0, +\infty)$ defined by $d(\Psi, \Phi) = E(\Psi^{-1}\Phi)$ defines a bi-invariant metric on $\mathscr{D}$. Moreover the map $\mathscr{C} \to \mathscr{D} : H \to \Psi_H$ is globally Lipschitz continuous with constant one.*

Thus the infinite dimensional group $\mathscr{D}$ admits a bi-invariant metric. This distinguishes $\mathscr{D}$ from many other diffeomorphism groups. The d-topology on $\mathscr{D}$ is very strange. So, one has for example the following fact: Given any volume preserving diffeomorphism $\Psi : \mathbb{C}^n \to \mathbb{C}^n$ there exists a sequence $(\Psi_k) \subset \mathscr{D}$ such that $\Psi_k \to Id$ in $(\mathscr{D}, d)$ but

$$\int_K |\Psi_k(x) - \Psi(x)|^p \, dx \to 0 \text{ as } k \to +\infty$$

for every compact set K in $\mathbb{C}^n$ and every $1 \leq p < \infty$. Hence there seems to be no relation between the d-topology and the L^p_{loc}-topology, $1 \leq p < \infty$. However, [20], we have

Theorem 5. *The inclusion map $j : (\mathscr{D}, d) \to C^\circ$ (which of course is not continuous) has closed graph. Here C° stands for the Frechet space of continuous maps $\mathbb{C}^n \to \mathbb{C}^n$, equipped with the metrizable compact open topology.*

See also [36] for related results. The d-topology had been introduced in [20] in order to study symplectic fixed point problems.

As Y. Eliashberg pointed out recently [9] the d-topology occurs naturally in the study of the question to what extend does the interior of a symplectic manifold determine the symplectic topology on the boundary, see also [11].

Another important question, which one should mention is the following: *Does there exist a good model for the complection $\tilde{\mathscr{D}}$ of $\mathscr{D}$?* Note that $\tilde{\mathscr{D}}$ turns out to be a topological group! It is interesting to note that for every continuous Hamiltonian H which vanishes at infinity there exists a unique "time-1-map" in $\tilde{\mathscr{D}}$.

The d-topology on $\mathscr{D}$ and the symplectic phase space geometry in $\mathbb{C}^n$ are closely related. Let S be a compact, connected, smooth hypersurface in $\mathbb{C}^n$. Denote by B_S the bounded component of $\mathbb{C}^n \setminus S$. We write c_{EH} for the particular symplectic capacity introduced in [5]. We say the hypersuface S is of contact type if there exists a 1-form λ on S such that $d\lambda = \omega \mid S$ and λ does not vanish on the fibres of $\mathscr{L}_S \to S$. We say S is of restricted contact type if λ can be extended to

$\mathbb{C}^n$ satisfying globally $d\lambda = \omega$ (see [1] for examples). To get a quantitative idea about c_{EH} we just mention the following estimate

$$c_{EH}(B_S) \geq \sup\{\pi r^2 \mid \text{There exists a symplectic embedding } \mathbf{B}^{2n}(r) \to B_S\}.$$

Theorem 6. *Let S be a smooth, compact connected hypersurface of restricted contact type in $\mathbb{C}^n$. If $\Psi \in \mathscr{D}$ with $E(\Psi) \leq c_{EH}(S)$ then there exists $x \in S$ with $\Psi(x) \in L_S(x)$.*

This result has nice applications in a global Poincaré perturbation theory [28, 20]. As a consequence of 2.6 we can define a very intuitive capacity, called the displacement energy.

Definition 7. For a bounded subset $\mathscr{U}$ of $\mathbb{C}^n$ define $c(\mathscr{U}) = \inf\{E(\Psi) \mid \Psi(\bar{\mathscr{U}}) \cap \bar{\mathscr{U}} = \emptyset\}$. For an unbounded subset $T \subset \mathbb{C}^n$ put $c(T) = \sup\{c(S) \mid S \subset T,\ S \text{ bounded}\}$. We call c the displacement energy.

One can show that $c(\mathbf{B}^{2n}(1)) = c(Z^{2n}(1)) = \pi$. We recall the fact that a $\Psi \in \mathscr{D}$ with small energy could be close to a translation in L^p_{loc}, $1 \leq p < \infty$. Nevertheless 2.6 shows that small energy implies a lot of local recurrence.

3. Symplectic Homology

We have already explained the importance of closed characteristics in the theory of symplectic capacities. For the following they might be even more important. The starting point is similar to the construction of symplectic capacities. Namely we associate to suitable Hamiltonians H some invariant. Whereas it were numerical invariants in the contruction of capacities, we associate now to a Hamiltonian H a Floer-Homology group [16]. The contruction is well behaved with respect to the pointwise ordering on the Hamiltonians, in the sense that $H \leq K$ induces a unique morphism between the Floer groups, [16]. Thus instead of taking the infimum of numbers $c(H)$ over the set of Hamiltonians H vanishing on a set S in order to obtain some invariant for S, we just take the direct limit of the Floer groups. This construction explained now in more detail leads to symplectic homology.

The domain of the theory consists of all pairs (M, A), where M is a symplectic manifold of dimension $2n$ and A an open subset of $M \setminus \partial M$, whose closure $\bar{A}$ is contained in some open subset $\mathscr{U}$ of M, with $\bar{\mathscr{U}}$ being compact, such that $\partial\mathscr{U}$ is smooth and of contact type. Moreover $\pi_2(M) = 0$ (however, weaker hypotheses will do as well). A morphism $\Psi : (M, A) \to (N, B)$ is a symplectic embedding $\Psi : M \to N$ satisfying $\Psi(A) \subset B$. One studies now for each pair (M, A) the Hamiltonian systems associated to Hamiltonians $H : M \to \mathbb{R}$ which satisfy $H \mid \bar{A} < 0$, and $H(x) \equiv m(H) \in (0, +\infty)$, if $x \in M \setminus V$ for some open neighbourhood V of $\bar{A}$ contained in $M \setminus \partial M$. Moreover $H(x) \leq m(H)$ for all $x \in M$. Let us denote this class of Hamiltonians by $\mathscr{H}(M, A)$. One is interested in the 1-periodic solutions of the Hamiltonian system $\dot{x} = X_H(x)$. This problem is variational, i.e. the 1-periodic solutions are critical points of some functional Φ_H. Using an equivariant version of Floer's construction of a homology theory one can associate to Φ_H a cochain complex $C(H)$, which is essentially spanned by the 1-periodic solutions of $\dot{x} = X_H(x)$. As a new ingredient we have a real filtration of this cochain complex since the basis vectors carry a numerical invariant, namely their Φ_H-value. There is also a $\mathbb{Z}$-grading through some kind of Maslov-index. All homological algebra constructions like chain homotopy, chain map etc., are supposed to be compatible with the $\mathbb{R}$-filtration. We write

$$C = \bigcup_{\lambda \in \mathbb{R}} \left(\bigoplus_{k \in \mathbb{Z}} C_\lambda^k \right), \qquad C_\mu \subset C_\lambda \text{ for } \quad \lambda \leq \mu .$$

Given two Hamiltonian $H \geq K$ one obtains a natural chain homotopy class (preserving the extra structure) $C(H) \to C(K)$, and for $\lambda \leq \mu$ induced maps $C_\lambda(H)/C_\mu(H) \to C_\lambda(K)/C_\mu(K)$. Applying the functor $\mathrm{Hom}(\cdot, R)$ for some commutative ring we obtain for every pair (M, A) a directed system

$$(\mathrm{Hom}(C_\lambda(H)/C_\mu(H)), R)_{H \in \mathscr{H}(M,A)} .$$

We pass to Homology and take the direct limit denoted by $I_\lambda^\mu(M, A)$. Symplectic embeddings $\Psi : (M, A) \to (N, B)$ induce group morphisms $\Psi^* : I_\mu^\lambda(N, B) \to I_\lambda^\mu(M, A)$. One has long exact sequences as well as isotopy invariance. For the details we refer the reader to [16] or [22]. We list now some applications. The first is a solution of Gromov's polydisk conjecture [22]. Let $a = (a_1, \ldots, a_n)$ be a n-tuplet of positive real numbers. We shall write $o(a)$ for the associated ordered n-tuplet (the a_i's written in increasing order). We write $\mathbf{D}^{2n}(a)$ for the open polydisk $\mathbf{D}^{2n}(a) = \mathbf{B}^2(a_1) \times \ldots \times \mathbf{B}^2(a_n)$.

Theorem 8. *Let a, b be ordered n-tuplets. Suppose $a_j \leq b_j$ for $j = 1, \ldots, n$ and $b_j^2 < a_1^2 + a_j^2$ for $j = 2, \ldots, n$. Assume $\mathbf{D}(r)$ is a third polydisk such that there exist symplectic embeddings $\Psi : \mathbf{D}(a) \to \mathbf{D}(r)$ and $\Phi : \mathbf{D}(r) \to \mathbf{D}(b)$ with $\Phi \circ \Psi$ being isotopic to the standard inclusion. Then we have for some permutation $\sigma(\tau)$ of τ*

$$a \leq \sigma(\tau) \leq b .$$

In particular if $a = b$ we have $a = o(r)$.

Hence, if $\mathbf{D}(a) \simeq \mathbf{D}(b)$ we must have $o(a) = o(b)$, which is Gromov's conjecture.

It follows from the H-principle [18], that the space of symplectic embeddings $\bar{\mathbf{B}}^2(2) \to \mathbf{D}^4(1,1)$ is connected for the compact open topology. Take any such embedding, say $\hat{\Psi}$. Then $\hat{\Psi}(\bar{\mathbf{B}}^2(2))$ has a symplectic tubular neighbourhood. Hence, we might slightly thicken $\mathbf{B}^2(2)$ to obtain a symplectic embedding $\Psi : \mathbf{D}^4(\varepsilon, 2) \to \mathbf{D}^4(1,1)$ for some $0 < \varepsilon < 1$. If now $\Phi : \mathbf{D}^4(\varepsilon, \delta) \to \mathbf{D}^4(1,1)$ is the composition of the standard inclusion of $\mathbf{D}^4(\varepsilon, \delta) \to \mathbf{D}^4(\varepsilon, 2)$ with Ψ for some $0 < \delta < 1$, we obtain the following result

Theorem 9. *Assuming $\varepsilon^2 + \delta^2 > 1$ the symplectic embedding Φ is not isotopic to the standard inclusion $j : \mathbf{D}^4(\varepsilon, \delta) \to \mathbf{D}^4(1,1)$.*

This result is obtained by studying the maps induced by Φ and j in symplectic homology.

There is another application concerned with the question to which extent is the symplectic topology on the boundary determined by the symplectic topology of the interior [9, 11]. Let (M,ω) be a compact symplectic manifold with $\omega = d\lambda$ being an exact symplectic form. Suppose ∂M is of restricted contact type. We define the action spectrum $A(M)$ by

$$A(M) = \left\{ m \int (\lambda \mid P) \mid P \in \mathscr{P}(\partial M), m \in \mathbf{N}^* \right\} .$$

Theorem 10. *Let (M,ω) and (N,σ) be of the type described above. Suppose $M \setminus \partial M$ and $N \setminus \partial N$ are symplectomorphic by an exact symplectic map. Then $A(M) = A(N)$, provided the periodic trajectories on ∂M and ∂N are nondegenerate.*

One can even strengthen the statements by incorporating the Maslov-type-index giving the grading of the symplectic homology groups. The symplectic homology is so far not complete, but already it allows to prove some not so obvious results.

References

1. Benci, V., Hofer H., Rabinowitz, P.: A priori bounds for periodic solutions on hypersurfaces. In: Periodic solutions of Hamiltonian systems and related topics (eds. Rabinowitz et al.). Nato ASI, Ser. C 209. Reidel 1987
2. Ekeland, I.: Convexity methods in Hamiltonian mechanics. (Ergebnisse der Mathematik, 3. Folge, Band 19.) Springer, Berlin Heidelberg New York 1989
3. Ekeland, I.: Une theorie de Morse pour les systemes hamiltoniens convexes. Ann. IHP "Analyse non linéare" **1** (1984) 19–78
4. Ekeland, I., Hofer, H.: Convex Hamiltonian energy surfaces and their closed trajectories. Comm. Math. Phys. **113** (1987) 419–467
5. Ekeland, I., Hofer, H.: Symplectic topology and Hamiltonian dynamics. Math. Z. **200** (1990) 355–378
6. Ekeland, I., Hofer, H.: Symplectic topology and Hamiltonian dynamics II. Math. Z. **203** (1990) 553–567
7. Eliashberg, Y.: Three lectures in symplectic topology. Proceedings of the Conference on Differential Geometry in Cala Conone 1988, to appear
8. Eliashberg, Y.: Filling by holomorphic disks and its applications. Preprint 1989.
9. Eliashberg, Y.: Lecture at a Conference on Dynamical Systems, Lyon, July 1990.
10. Eliashberg, Y.: A theorem on the structure of wave fronts. Funct. Anal. Appl. **21** (1987) 65–72
11. Eliashberg, Y., Hofer, H.: Towards the definition of a symplectic boundary. In preparation
12. Floer, A.: Morse theory for Lagrangian intersections theory. J. Diff. Geom. **18** (1988) 513–517
13. Floer, A.: The unregularised gradient flow of the symplectic action. Comm. Pure Appl. Math. **41** (1988) 775–813
14. Floer, A.: Witten's complex and infinite dimensional Morse theory. J. Diff. Geom. **30** (1989) 207–221

15. Floer, A.: Symplectic fixed points and holomorphic spheres. Comm. Math. Physics **120** (1989) 576–611
16. Floer, A., Hofer, H.: Symplectic homology. In preparation.
17. Gromov, M.: Pseudoholomorphic curves in symplectic manifolds. Invent. math. **82** (1985) 307–347
18. Gromov, M.: Partial differential relations. Ergebnisse der Mathematik 1986.
19. Gromov, M.: Soft and hard symplectic geometry. Proceedings of the ICM at Berkeley 1986. AMS, 1987, pp. 81–89
20. Hofer, H.: On the topological properties of symplectic maps. Proc. Roy. Soc. Edinburgh **115A** (1990) 25–83
21. Hofer, H.: Symplectic capacities. Geometry of Low-dimensional Manifolds: 2. Edited by S. K. Donaldson and C. B. Thomas. London Mathematical Society Lecture Notes 151, pp. 15–34
22. Hofer, H., Wysocki, K.: Applications of symplectic homology. In preparation
23. Hofer, H., Zehnder, E.: A new capacity for symplectic manifolds. Analysis et cetera, edited by P. Rabinowitz and E. Zehnder. Academic Press 1990, pp. 405–428
24. Hofer, H., Zehnder, E.: Periodic solutions on hypersurfaces and a result by C. Viterbo. Inv. math. **90** (1987) 1–7
25. McDuff, D.: Examples of symplectic structures. Invent. math. **89** (1987) 13–36
26. McDuff, D.: Rational and ruled symplectic 4-manifolds. Preprint.
27. McDuff, D.: Elliptic methods in symplectic geometry. Progress in Math. Lecture, Boulder 1989
28. Moser, J.: A fixed point theorem in symplectic geometry. Acta mathematica **141** (1978) 17–34
29. Rabinowitz, P.: Periodic solutions of Hamiltonian systems. Comm. Pure Appl. Math. **31** (1978) 157–184
30. Rabinowitz, P.: Periodic solutions of a Hamiltonian system on a prescribed enery surface. J. Diff. Eq. **33** (1979) 336–352
31. Sikorav, J.C.: Rigidité symplectique dans le cotangent de T^n. Duke Math. J. **59** (1989) 227–231
32. Sikorav, J.C.: Systemes Hamiltoniens et topologie symplectique. Lecture notes Dipartimento Di Matematica Dell‘ Università Di Pisa, ETS Editrice
33. Viterbo, C.: New obstructions to embedding Lagrangian tori. Invent. math. **100** (1990) 301–320
34. Viterbo, C.: A proof of the Weinstein conjecture in $\mathbb{R}^{2n}$. Ann. Inst. Henri Poincare. Analyse non lineare **4** (1987) 337–356
35. Viterbo, C.: Capacités symplectiques et applications. Seminaire Bourbaki, June 1989, Asterisque, no. 695
36. Viterbo, C.: Symplectic topology as the geometry of generating functions. Preprint
37. Weinstein, A.: Lectures on symplectic manifolds. CBMS Reg. Conf. Series in Math. no 29, AMS, Providence, R.I. 1979
38. Weinstein, A.: On the hypotheses of Rabinowitz' periodic orbit theorems. J. Diff. Eq. **33** (1979) 353–388

Embedded Surfaces in 4-Manifolds

Peter B. Kronheimer

Merton College, Oxford OX1 4JD, UK

1. Introduction

The adjunction formula expresses the genus of a smooth curve C in a complex surface X in terms only of the homology class of C and the canonical class K of the surface:

$$2g - 2 = C \cdot C + K \cdot C .$$

An old and still elusive conjecture [Ki] asserts that, if Σ is any smoothly embedded 2-manifold in the same homology class as C, then the genus of Σ is at least as big as the genus of C, so

$$2g - 2 \geq \Sigma \cdot \Sigma + K \cdot \Sigma . \tag{1}$$

With Donaldson's introduction of gauge theory into 4-dimensional differential topology, there seems reason to hope that such a conjecture might well be approachable. However, almost the best result to date on questions of this sort is now twenty years old, proved independently by Rohlin [R] and Hsiang and Szczarba [HS]. Their result is not special to complex surfaces, but exploits a branched covering to prove a non-trivial lower bound on the genus whenever the homology class of Σ is divisible. Results due to Donaldson do allow an occasional improvement of this lower bound, but this still falls short of (1) by a factor of about 2, even in favourable cases.

A somewhat looser version of the above conjecture has been formulated by Morgan; we adapt it slightly here:

Conjecture A. *If X is a simply-connected, oriented 4-manifold for which Donaldson's polynomial invariants are defined and non-zero, and if Σ is any smoothly embedded, oriented 2-manifold with positive self-intersection, then the genus of Σ satisfies the inequality*

$$2g - 2 \geq \Sigma \cdot \Sigma .$$

To explain this conjecture we need to recall a few facts about the polynomial invariants [D2]. These are invariants of X which are defined whenever $b^+(X)$ (the dimension of a maximal positive subspace of the intersection form on the second homology) is odd and not less than 3. This is the case, for example, if X is a complex surface whose canonical bundle admits a non-zero holomorphic

Proceedings of the International Congress of Mathematicians, Kyoto, Japan, 1990

section, and it is one of the two key results of [D2] that the invariants are indeed non-zero for such a surface. Thus simply-connected complex surfaces with $p_g > 0$ satisfy the hypotheses of the conjecture. (Here p_g, the geometric genus, is the dimension of the space of sections of K).

For a complex surface, the condition that K has a section is easily seen to imply that $K \cdot C$ is non-negative whenever C is a curve with positive self-intersection. So Conjecture A is consistent with the adjunction formula and the conjecture (1). It is weaker, in that the delicate term involving the canonical class is missing, and it says nothing about rational surfaces; on the other hand, it is a conjecture which is not restricted to algebraic cycles. An important case in which the two conjectures essentially coincide is when X is a $K3$ surface (a simply-connected surface with trivial canonical class). On account of this case, Conjecture A implies various local versions of (1). Thus it implies that if B^4 is a standard ball in $\mathbb{C}^2$ and C is a smooth, plane algebraic curve meeting ∂B^4 transversely, then the genus of $C \cap B^4$ is minimal among all 2-manifolds in B^4 with the same oriented boundary. This in turn implies, for example, that the unknotting number of the (p, q) torus knot is $\frac{1}{2}(p-1)(q-1)$; see [Mi].

Conjecture A is known to hold if the genus of Σ is 0 or 1; in other words, an embedded sphere or torus cannot have positive self-intersection if X has non-zero polynomial invariants. The fact that a sphere cannot have self-intersection 1 is a special case of the results of [D2] concerning connected sum decompositions; the other cases are due to Morgan, Mrowka and Ruberman [MMR].

The purpose of this article is to outline a proof of Conjecture A, but a proof with a hole. The argument uses gauge theory, but with a modification which makes it a little different from the usual set-up. We consider connections in some auxiliary $SU(2)$ or $SO(3)$ bundle over the manifold, but connections which are defined on the complement of the surface Σ and have non-trivial holonomy around the small linking circles of Σ. Such a set-up has been considered elsewhere, and the similar situation of a punctured 2-manifold is quite well-understood [Se, Si]. For the 4-dimensional case, some important analysis has been done by Sibner and Sibner [SS], but still some things are missing.

The basic properties of the moduli spaces of anti-self-dual connections associated to a compact 4-manifold are deduced from a simple package of results. This is the Kuranishi-type deformation theory, built on a particular Fredholm complex combined with the implicit function theorem. This package is missing for the "branched instantons" which we wish to exploit. In Section 2 we shall explain what one would expect of such a package were it to exist in our case. This will be our black box conjecture. Assuming the existence of a black box with the expected properties, we shall deduce Conjecture A.

This is not the place for a very detailed exposition. If the author, or anyone else, succeeds in making the necessary box, then full details will appear. If it turns out that such a box does not exist, then the reader will be glad that he or she did not read a longer paper.

The author wishes to thank Simon Donaldson for his generous help and advice. The key tunnelling trick described in Section 4 was suggested by him. The author must also thank Ed Witten for clearing up some of the author's early misconceptions and confusion concerning these branched instantons. Finally, thanks are due to Tom Mrowka and Carlos Simpson for their help at various times.

2. Connections with Holonomy

For some basic definitions and properties of anti-self-dual connections over 4-manifolds, we refer to [D2] and the references therein. We consider here an extension of the usual constructions which seems naturally suited to the problem at hand. Let Σ be a connected, oriented 2-manifold embedded in the oriented, Riemannian 4-manifold X, and let E be a hermitian vector bundle over $X \setminus \Sigma$. For the moment we shall take it that E has structure group $SU(2)$. Let $\mathcal{N}$ be the set of all connections A in E (compatible with the $SU(2)$ structure) which are anti-self-dual and have finite action; that is, the curvature $F(A)$ should satisfy

$$*F(A) = -F(A),$$

and

$$\int_{X\setminus\Sigma} |F(A)|^2 < \infty,$$

where $*$ is the Hodge star operator on 2-forms. Let M be the quotient of $\mathcal{N}$ by the action of the automorphisms of the bundle E. This M is the moduli space we would like to study. Note that the connections A are defined only on the complement of the surface Σ and are anti-self-dual with respect to an incomplete metric.

A local study of such connections was made by Sibner and Sibner in [SS]. Their first result is that the finite action condition alone implies that a connection on $X \setminus \Sigma$ has a well-defined holonomy on the small linking circles. To be precise, choose polar coordinates on a small element of plane normal to Σ at a point σ. Let h_r be the holonomy of the connection around the circle of radius r in this plane. Then the result is that, for almost all σ, the conjugacy class of h_r has a well-defined limit as r approaches zero, and this limit is independent of the choice of σ. Thus we can write

$$h_r \sim \exp 2\pi i \begin{pmatrix} \alpha & 0 \\ 0 & -\alpha \end{pmatrix},$$

for some real α in the interval $0 \leq \alpha \leq 1/2$. Let us write M_α for the subset of the moduli space M where the holonomy is α in this parametrization.

The first simple point is that the values 0 and 1/2 are very special values for the holonomy. When $\alpha = 0$, the holonomy matrix is the identity; so there is no geometrical obstruction to extending the connection. It follows from the technical result of [SS] that in this case, an anti-self dual connection A will extend smoothly to all of X, or to be more precise, there is a bundle E_0 over X with an anti-self-dual connection whose restriction to $X \setminus \Sigma$ is isomorphic to A. When α is 1/2, the holonomy is -1. In this case, the connection will not necessarily extend as an $SU(2)$ connection, but in the associated $SO(3)$ bundle the holonomy is again trivial; so A extends to a smooth anti-self-dual connection in some $SO(3)$ bundle $E_{1/2}$ over X. In this sense then, M_0 and $M_{1/2}$ are both ordinary moduli spaces of anti-self-dual connections associated to X. (We shall generally write M_E for the usual moduli space of anti-self-dual connections on a bundle E.)

When $0 < \alpha < 1/2$, there are new phenomena. First, whereas an $SU(2)$ connection on a closed 4-manifold has only one characteristic number, namely the second Chern class k, a connection A with holonomy such as we are considering has two invariants. To defines these, first take a small tubular neighbourhood

N of Σ, and let Y be its boundary, a circle bundle over Σ. Since the holonomy around the circle fibres is non-trivial, its eigenspaces give a decomposition of $E|_Y$ as a sum of line bundles $L \oplus L^*$. We take L to be the eigen-bundle corresponding to the eigenvalue $e^{-2\pi i\alpha}$. On each circle fibre, we can choose a trivialization of E for which the θ component of the connection is approximately

$$A_\theta = \begin{pmatrix} \alpha & 0 \\ 0 & -\alpha \end{pmatrix} d\theta .$$

This preferred trivialization on the fibres allows us to extend $L \oplus L^*$, in an essentially unique way, as a topological bundle on N. Thus we get an extension $\bar{E}$ of E to all of X, and a decomposition of $\bar{E}|_\Sigma$ as a sum of line bundles. We define

$$\begin{aligned} k &= c_2(\bar{E})[X] , \\ l &= -c_1(L)[\Sigma] , \end{aligned}$$

and call these the *instanton* and *monopole* number (the latter by analogy with the topology of monopole solutions on 3-space; see [A]).

Given suitable regularity of the connection near the surface, we can also define k and l by a formula of Chern-Weil type, which involves some correction terms due to the non-trivial topology of the neighbourhood. Near Σ, we should expect the curvature $F(A)$ to be approximately abelian, since the subbundle L on Y should be asymptotically parallel. If this is true to a sufficient degree, we can write

$$F(A) = \begin{pmatrix} \omega & 0 \\ 0 & -\omega \end{pmatrix},$$

with ω having a well-defined limit on Σ, in which case

$$k = -2\alpha l + \alpha^2 \Sigma\cdot\Sigma + \frac{1}{8\pi^2}\int_{X\setminus\Sigma} \mathrm{Tr}(F \wedge F) , \tag{2}$$

and

$$l = \alpha\Sigma\cdot\Sigma - \frac{1}{2\pi i}\int_\Sigma \omega .$$

We shall write $M_{\alpha,k,l}$ for the moduli space of anti-self-dual connections with holonomy α, instanton number k and monopole number l. Our definition makes it, at present, just a set. We can topologize it, mimicking [T2] for example, but it is not clear what sort of a space would result. The rest of this article will be largely based on the assumption that, for a generic choice of Riemannian metric on X, the moduli space $M_{\alpha,k,l}$ will be (except at the reducible or flat connections) a smooth manifold whose dimension is given by the formula

$$d = 8k - 3(b^+ - b^1 + 1) + 4l - (2g - 2) . \tag{3}$$

The first two terms here, involving the 4-dimensional characteristic number of the bundle and the topology of X, are familiar as the dimension of the ordinary moduli spaces associated to a closed 4-manifold. The remaining terms, $4l - (2g - 2)$, involve the surface. The formula is motivated by consideration of parabolic bundles on a complex surface, as is explained in the next section. For the moment, it might be helpful to note that in the case that Σ is S^2 contained in S^4 and k is zero, the formula gives $4l - 1$, in accordance with the dimension of the moduli space of hyperbolic monopoles; see [A] again.

The arguments we shall use later will assume a little more. We wish to consider the moduli spaces $M_{\alpha,k,l}$ as a family as α varies. Let us write $M_{*,k,l}$ for the union of $M_{\alpha,k,l}$ as α runs through $(0, \frac{1}{2})$. We cannot expect that, for a given metric, all the $M_{\alpha,k,l}$ will be smooth manifolds, but we can hope that the union $M_{*,k,l}$ is a manifold with a smooth map to to the half-interval:

Conjecture (Black Box). *For a generic choice of Riemannian metric on X, the space $M_{*,k,l}$ is, except at the reducible or flat connections, a smooth manifold of dimension $d+1$, with d as shown in* (3), *and the map $\alpha : M_* \to (0, \frac{1}{2})$ is smooth. The usual transversality and weak compactness arguments, as applied to the usual moduli spaces of anti-self-dual connections, carry over to this case.*

It is on the basis of this conjecture, and one other described in Section 3, that we shall deduce Conjecture A. The conjecture is, necessarily, loosely phrased. Indeed, it would not be sufficient for our application if the conjecture were correct only by some accident. We have in mind that M_α should be generically a manifold *for the same reasons* that the usual moduli spaces M_E are, and that as α varies, M_α should behave rather as the ordinary moduli spaces do when the Riemannian metric varies in a 1-parameter family [D2]. The principal thing is that there should be a deformation theory describing the local structure of M_α, based on a Fredholm complex. Smoothness and compactness are secondary matters which should follow from standard techniques if the primary machinery is in place. In particular, the results of [SS] are what is needed for a weak compactness theorem along the lines of the essential theorem of Uhlenbeck [U]. Here there will be differences, though, since there will be a bubbling off phenomenon in which both the instanton and monopole number change in the weak limit. Detailed information concerning the possibilities is not important to our argument, but we do need to know that the limiting connection lives in a moduli space of smaller dimension.

3. Parabolic Bundles

Suppose X is a Kähler surface and $\Sigma \subset X$ is a smooth curve. By a *parabolic $SL(2,\mathbb{C})$-bundle* on X we shall mean a rank-2 holomorphic bundle $\mathscr{E} \to X$ with $\Lambda^2\mathscr{E}$ trivial, together with a holomorphic line sub-bundle $\mathscr{L} \subset \mathscr{E}|_\Sigma$. Pick a positive real number α and define the *α-degree* of a rank-1 subsheaf $\mathscr{F} \subset \mathscr{E}$ by the formula

$$\deg_\alpha(\mathscr{F}) = \langle c_1(\mathscr{F}), [\omega]\rangle \pm \alpha\langle[\Sigma], [\omega]\rangle,$$

where ω is the Kähler form. Here the sign is to be $+$ if $\mathscr{F}|_\Sigma$ is contained in $\mathscr{L}$ and is to be $-$ otherwise. Define $(\mathscr{E}, \mathscr{L})$ to be *α-stable* if $\deg_\alpha(\mathscr{F})$ is strictly negative for every such subsheaf.

Conjecture (Parabolic Bundles). *If Σ is a smooth curve in a Kähler surface X and $0 < \alpha < \frac{1}{2}$, then the set of irreducible connections $M^{\mathrm{irr}}_{\alpha,k,l} \subset M_{\alpha,k,l}$ is in one-to-one correspondence with the set of α-stable parabolic bundles $(\mathscr{E}, \mathscr{L})$ with $c_2(\mathscr{E}) = k$ and $c_1(\mathscr{L})[\Sigma] = -l$.*

This result is the expected analogue of the theorem of [D1] relating anti-self-dual connections to stable bundles on a complex surface. The terminology 'parabolic' is taken from [Se] which contains a corresponding result in complex dimension 1. An analytic proof of the result of [Se], together with a large part of what is necessary for the higher-dimensional case, is contained in [Si].

The above conjecture motivates everything else in this paper, but our later arguments are not logically dependent on this general result and will appeal to only one particular case, that of a ruled surface. To be precise, let $N \to \Sigma$ be a holomorphic line bundle of positive degree, and let X be the $\mathbb{C}P^1$ bundle obtained by adding a section at infinity. Let Σ be embedded in X as the zero section. We shall assume the truth of the parabolic bundles conjecture for this pair (X, Σ) in the case $k = 0$.

There are actually a couple of points which make this case technically a little easier. One is the circle symmetry in X, another is the local product structure near Σ. Exploiting these and the results of [Si], the author has no difficulty in constructing the map from parabolic bundles to $M_{\alpha,0,l}$. Surjectivity is still in question.

4. Outline of the Proof

Assuming the package of results in the black box, and assuming also the parabolic bundles conjecture in the case of the ruled surface described above, let us deduce Conjecture A.

For ease of exposition, we shall first treat a slightly simplified situation; the proof will occupy this section and the next. The necessary modifications for the general case are described in Section 6. For the simplified argument, we shall suppose that there is an $SO(3)$ bundle E over X, with $w_2(E)$ non-zero, such that the (ordinary) moduli space of anti-self-dual connections M_E is a finite set of points, cut out transversely by the anti-self-duality equations. The number of points in this moduli space, counted with sign according to their orientation, is a simple example of one of the polynomial invariants of X. We shall denote this number by q and we shall suppose it is non-zero.

Concerning Σ, we shall assume that its genus is *even* and that $w_2(E)$ is non-zero on Σ. This will simplify the excision argument in the following section. We assume also that there is an integer homology class η in $X \setminus \Sigma$ on which $w_2(E)$ is again non-zero. Finally, we shall suppose that $\Sigma \cdot \Sigma$ is square-free. The point of this condition is that it ensures that $\pi_1(X \setminus \Sigma)$ has no non-trivial representations in $SO(3)$, as can be deduced from the Chern-Weil formulae for k and l in the case that the curvature terms vanish.

The assumptions on Σ which we have just listed do not involve a loss of generality. If Σ_0 is any surface which satisfies the hypotheses of Conjecture A but violates the inequality which is the conclusion of that conjecture, then we can soon construct many other surfaces with the same property, and in particular, we can find surfaces Σ which satisfy the additional hypotheses of the previous paragraph. The trick is to take n nearby copies of Σ_0, intersecting transversely with positively-oriented intersection points, and then replace the intersection points with small handles. This gives a surface representing the class $n[\Sigma_0]$ which will violate the inequality of Conjecture A by a margin which is n times larger. We can then form a connected sum of this new surface with any other surface T disjoint from Σ_0 in X, to obtain a representative of $[T] + n[\Sigma_0]$ which still violates Conjecture A once n is large.

We shall take for granted the slight modification of the contents of the previous sections which is necessary to include the case of $SO(3)$ bundles. The difference from the $SU(2)$ case is the presence of the Stieffel Whitney class; we shall not build $w_2(E)$ into our notation for the moduli spaces, but shall take it that $M_{\alpha,k,l}$ denotes now the moduli spaces of connections with the same w_2 as E on $X \setminus \Sigma$. The instanton number now lives in $\mathbb{Z}/4$, and the monopole number l will be a half-integer if $w_2(\bar{E})$ is non-zero on Σ.

After these preliminaries, we can begin the proof. Let k be the instanton number of E and consider the corresponding moduli spaces $M_{\alpha,k,l}$. We shall consider a special value for the monopole number: we take

$$l = \frac{1}{4}(2g-2). \tag{4}$$

The significance of this choice is that the dimension of $M_{\alpha,k,l}$ is now the same as the dimension of the ordinary moduli space M_E: the terms from Σ in the dimension formula (3) cancel out. Note that since g is even, l is a half-integer, as it should be when $w_2(E)$ is non-zero on Σ.

Since, the dimension of M_E is zero, so too is the dimension of $M_{\alpha,k,l}$. The latter will be generically an oriented 0-manifold to which we can attach an integer by counting the points with their signs (here we appeal to the compactness assertion in the black box). Further, the family $M_{*,k,l}$ gives a compact (again), oriented, 1-dimensional cobordism between any two of these 0-manifolds, so the integer we obtain is independent of α. Let us call this integer p. (It is important here that $b^+ \geq 3$, to avoid reducible connections in this one-dimensional family). To prove Conjecture A, we will calculate p in two different ways, using the limiting behaviour as α approaches first 0 and then 1/2.

The first is technically the harder of the two in our case. We shall argue, using small values of α, that $p = 2^g q$. So, roughly speaking, every point of the original moduli space M_E gives rise to 2^g points in M_α, counted algebraically, when α is small. The origin of the number 2^g is in the following result.

Proposition. *Let $\mathscr{E}$ be a stable rank-2 holomorphic bundle on a curve Σ. Assume that the degree of $\mathscr{E}$ is even if the genus of Σ is odd, and that the degree is odd otherwise. Let $\lambda = \frac{1}{2}\deg(\mathscr{E}) - \frac{1}{4}(2g-2)$. Then generically the bundle has precisely 2^g holomorphic line sub-bundles of degree λ. Here the word generic refers to a Zariski open set in the set of pairs $(\mathscr{E}, \Sigma)$.*

For the proof, the Riemann-Roch theorem for families shows that 2^g is the right number: one applies the theorem to the family of bundles $\mathrm{Hom}(\mathscr{L}, \mathscr{E})$ as $\mathscr{L}$ runs through the Jacobian of line bundles of the given degree. It remains to show that the generically expected behaviour does occur, and for this one can consider a particularly degenerate curve, such as a sphere intersecting g tori, for which the problem is easy to analyse.

The idea here is that the above proposition tells us that $p = 2^g q$ in the case that Σ is a complex curve in a complex surface, provided we accept the conjecture of Section 3 concerning parabolic bundles. The proposition tells us that, when l is given by (4), the moduli space of parabolic bundles $(\mathscr{E}, \mathscr{L})$ generically covers the ordinary moduli space of bundles 2^g times (though there are some provisos here, to do with stability). Of course, the point is precisely that our 2-manifold Σ is not holomorphic; nevertheless, we can use an excision argument, based on the fact that a neighbourhood of Σ can always be given a complex structure, to show that the conclusion $p = 2^g q$ is valid generally. Some details of the excision argument are given in the next section.

We have not yet referred to the inequality $2g - 2 \geq \Sigma\cdot\Sigma$, but consider now what happens as α approaches $1/2$. As we said in Section 2, when α is $1/2$ we can identify the moduli space $M_{1/2,k,l}$ with an ordinary moduli space $M_{E_{1/2}}$ for some $SO(3)$ bundle $E_{1/2}$. This bundle is different from the E; it differs by a sort of twist along Σ, and its topology depends on l. The essential piece of information is the formula which relates the instanton number of $E_{1/2}$ to the old k and l: a short calculation gives

$$k(E_{\frac{1}{2}}) = k(E) + l - \tfrac{1}{4}\Sigma\cdot\Sigma\,.$$

(Compare this with (2) when $\alpha = 1/2$). If we recall that l is given by (4), we see that in our case the difference in the instanton numbers is a quarter of $(2g - 2) - \Sigma\cdot\Sigma$, which is precisely the number whose sign we are interested in.

Thus if the inequality of Conjecture A is violated, then $k(E_{1/2})$ is less than $k(E)$. Since the dimension of M_E is zero, this means that $M_{E_{1/2}}$ will have negative dimension and will therefore be empty. (Here again we use the fact that w_2 is non-zero, to rule out the possibility that the moduli space contains the trivial flat connection.) This would mean, however, that p must be zero. For otherwise the moduli spaces $M_{(1/2-\varepsilon),k,l}$ are non-empty for all $\varepsilon > 0$, and as ε goes to zero, Uhlenbeck's compactness theorem gives a limiting anti-self-dual connection which is either in $M_{E_{1/2}}$ (if there is no bubbling off) or in some moduli space of still more negative dimension (if some action is lost in the limit). Thus we have obtained a contradiction, since $p = 2^g q$, and q was supposed to be non-zero.

5. Excision

Excision arguments have become a flexible tool to analyse moduli spaces of anti-self-dual connections. Necessary ingredients are described and used in [T1, T2, Mr] and [MMR].

Consider deforming the Riemannian metric on X so that a collar of the boundary, Y, of the tubular neighbourhood $N \supset \Sigma$ is stretched out into a long cylinder $Y \times [-T, T]$. Let $M(T)$ be the corresponding moduli space of anti-self-dual connections in E. This contains q points, counted with sign. As T goes to infinity, connections in $M(T)$ converge weakly to give us connections on the disjoint union $X' \cup N'$, where these two are non-compact manifolds, with cylindrical ends isometric to $Y \times \mathbb{R}^+$. In our case, a simple dimension count shows that the limiting connection on N' must be flat, and the limiting connection on X' must have total action k (no action is lost). On the end of X', the connection will approach a flat connection on Y, and if we recall that $w_2(E)$ is non-zero on N, we see that this flat connection ϱ must be the pull-back of a flat $SO(3)$ connection on Σ with odd w_2. Write $\mathcal{R}(\Sigma)_{\text{odd}}$ for the space of these flat connections; it is a smooth manifold since all these connections are irreducible.

In the other direction, given an anti-self-dual connection in E over X' with action k, asymptotic to a flat connection from $\mathcal{R}(\Sigma)_{\text{odd}}$, we can "glue" it to the corresponding flat connection on N' to obtain a connection on X. The analysis of this gluing is simplified by the fact $\mathcal{R}(\Sigma)_{\text{odd}}$ is smooth. It can thus be shown that, when counted with signs, the number of such connections on X' is the same integer q. Let us write these connections as $A_1, \dots, A_t$, and let $\varrho_1, \dots, \varrho_t$ be the limiting elements of $\mathscr{R}(\Sigma)_{\mathrm{odd}}$.

So far, this has been for the ordinary connections, without holonomy around Σ. Consider now the moduli space of connections on $N' \setminus \Sigma$ which have holonomy α and monopole number given by (4). There is a component of this moduli space which has the the same dimension as $\mathscr{R}(\Sigma)_{\mathrm{odd}}$ (which can be regarded as a moduli space of flat connections on N') and in which every connection is asymptotic to some element of $\mathscr{R}(\Sigma)_{\mathrm{odd}}$ on the end. Calling this moduli space $M_\alpha(N')$, we have a smooth map $r : M_\alpha(N') \to \mathscr{R}(\Sigma)_{\mathrm{odd}}$, given by this limiting connection [T2]. Although the former space is not compact due to possible bubbling off of monopoles, the weak compactness theorem will guarantee that r has a well-defined *degree*, say D, provided α is small. (For larger α, the degree may jump, on account of reducible connections.) Using Proposition 6.4 from [T2], we can ensure that the connections ϱ_i are regular values for r; so on N' there will be exactly D anti-self-dual connections with holonomy α asymptotic to each ϱ_i.

For small α, the analysis of the ordinary case goes through, and shows that the anti-self-dual connections with holonomy on E over X can be regarded as being obtained by gluing one of the connections A_i on X' to one of the connections of $M_\alpha(N')$. Whereas in the original set-up each A_i could be glued to just one flat connection on N', now there are D possibilities for each (counted algebraically). Thus we see that the total number of instantons with holonomy α (the number we called p) is equal to D times q.

The important thing we have gained now is the information that D depends only on the neighbourhood of Σ, not on the geometry of the rest of X; this is the excision principle. We can therefore calculate this covering degree using any convenient model. The simplest thing is the ruled surface described in Section 3, where the answer comes out as 2^g, for small α, by virtue of the proposition from Section 4 and the conjecture on parabolic bundles.

6. The General Case and Further Comments

For the general case we must start with an $SU(2)$ bundle, not an $SO(3)$ bundle. We therefore arrange that g is odd. Our original moduli space M_E will not be zero-dimensional, but as in [D2] we can cut down the moduli space by divisors V_j corresponding to 2-dimensional homology classes β_j in X. We choose the β_j so that they are orthogonal to $[\Sigma]$ in $H_2(X)$ and are therefore supported in $X \setminus \Sigma$. Provided we are in the stable range [D2], we can cut down the moduli space to a finite set of points whose number (counted algebraically) is an invariant of X and the chosen homology classes: this is the definition of the polynomial invariants.

Everything can be carried through for these cut-down moduli spaces. Being in the stable range means that simple dimension-counting arguments ensure the compactness which the non-zero Stieffel-Whitney class previously provided (see [D2] again). The only technical difference is in the excision step, where we are now forced to deal with $\mathscr{R}(\Sigma)_{\mathrm{even}}$. Since this space contains reducible connections, we need to be sure that, after deformation, we can arrange for the flat connections ϱ_i involved in the gluing to be in the irreducible part. For this step, the positive sign of the self-intersection of Σ puts us in a good situation (see [T2], Proposition 12.2).

At the end of the day, the argument shows that the polynomial invariants vanish on the orthogonal complement of $[\Sigma]$ if the surface contradicts the inequality of the conjecture. But since infinitely many pairwise-independent homology classes can be represented by surfaces contradicting the inequality (see the previous section), it follows that the invariant must vanish identically.

In a more general setting, it seems that the ideas of this paper, if they can be carried through, would be useful for establishing relations between the polynomial invariant associated to a bundle E_0 and that associated to a corresponding twisted bundle $E_{1/2}$; each choice of l would give some relation. To calculate these, one first needs to know the analogue of the proposition of Section 4, for different values of the monopole number. For example, if μ is greater than the value λ given in that proposition, then only special rank-2 bundles $\mathscr{E}$ admit line subbundles of degree μ. This condition defines a subvariety of complex codimension $2(\mu - \lambda)$ in the moduli space of stable bundles over Σ, and one would need to know its dual cohomology class in terms of standard generators. Via the excision argument, this sort of information establishes the relationship between M_0 and M_α for small α, with similar considerations at $\alpha = 1/2$.

References

[A] Atiyah, M.F.: Magnetic monopoles in hyperbolic space. In: Vector Bundles on Algebraic Varieties. Tata Institute of Fundamental Research, Bombay, 1968, pp. 1–33

[D1] Donaldson, S.K.: Anti-self-dual Yang-Mills connections on complex algebraic surfaces and stable vector bundles. Proc. London Math. Soc. **50** (1985) 1–26

[D2] Donaldson, S.K.: Polynomial invariants for smooth four-manifolds. Topology **29** (1990) 257–315

[HS] Hsiang, W.C., Szczarba, R.H.: On embedding surfaces in four-manifolds. In: Algebraic Topology. (Proc. Sympos. Pure Math., vol. 22.) Amer. Math. Soc., Providence, RI, 1971, pp. 97–103

[Ki] Kirby, R.C.: Problems in low-dimensional topology. In: Algebraic and Geometric Topology. (Proc. Sympos. Pure Math., vol. 22, pt. 2.) Amer. Math. Soc., Providence, RI, 1978, pp. 273–312

[Mi] Milnor, J.: Singular Points of Complex Hypersurfaces. (Ann. Math. Studies, vol. 61.) Princeton University Press, 1968

[MMR] Morgan, J.W., Mrowka, T., Ruberman, D.: Preprint (in preparation)

[Mr] Mrowka, T.: A Mayer-Vietoris principle for Yang-Mills moduli spaces. Ph.D. thesis, Berkeley, 1989

[R] Rohlin, V.A.: Two-dimensional submanifolds of four-dimensional manifolds. Funkt. Annal. Prilozen. **5** (1971) 48–60 (Russian). [English transl.: Funct. Anal. Appl. **5** (1971) 39–48]

[Se] Seshadri, C.S.: Moduli of vector bundles on curves with parabolic structures. Bull. Amer. Math. Soc. **83** (1977) 124–126

[SS] Sibner, L.M., Sibner, R.J.: Classification of singular Sobolev connections by their holonomy. Preprint, 1989

[Si] Simpson, C.T.: Constructing variations of Hodge structure using Yang-Mills theory and applications to uniformization. J. Amer. Math. Soc. **1** (1988) 867–918

[T1] Taubes, C.H.: Gauge theory on asymptotically periodic 4-manifolds. J. Diff. Geom. **25** (1987) 363–430

[T2] Taubes, C.H.: L^2-moduli spaces on 4-manifolds with cylindrical ends, I. Preprint, 1990

[U] Uhlenbeck, K.K.: Connections with L^p bounds on curvature. Commun. Math. Phys. **83** (1982) 31–42

Symplectic 4-Manifolds

Dusa McDuff

Department of Mathematics, State University of New York at Stony Brook
Stony Brook, NY 11794, USA

§ 1. Introduction

A symplectic structure on a $2n$-dimensional manifold V is a 2-form ω which is closed and non-degenerate, that is, $d\omega = 0$ and ω^n does not vanish. The main example is the standard form

$$\omega_0 = dx_1 \wedge dx_2 + dx_3 \wedge dx_4 + \cdots + dx_{2n-1} \wedge dx_{2n}$$

on $\mathbb{R}^{2n}$. Darboux showed that every symplectic form is locally diffeomorphic to ω_0. Thus symplectic manifolds, in contrast to Riemannian manifolds, are all locally isomorphic. We are concerned here with questions of their global structure.

When $n = 1$ a symplectic form is just an area form, and the corresponding global topology is well-understood. For example, the existence problem is trivial: clearly, if V is compact, orientable and connected, it has a symplectic structure in each non-zero class in $H^2(V; \mathbb{R})$. The uniqueness problem is almost as easy, since Moser [M] showed that an area form (or, in higher dimensions, a volume form) is determined up to isotopy by its cohomology class. (Two forms ω_0 and ω_1 on V are said to be *isotopic* if there is a family g_t of diffeomorphisms of V such that $g_0 = \text{id.}$ and $g_1^*(\omega_0) = \omega_1$. They are *symplectomorphic* if $g_1^*(\omega_0) = \omega_1$ for some diffeomorphism g_1.) However, in dimensions >2, these questions are far from being understood.

This paper will describe selected topics in the 4-dimensional case, concentrating on my own work. As we shall see, there are situations in which a symplectic form is determined up to symplectomorphism by its cohomology class. For example, Gromov's celebrated uniqueness theorem says that this holds for symplectic forms on $\mathbb{C}P^2$ which are non-degenerate on some embedded 2-sphere in the homology class of $\mathbb{C}P^1$. However, this does not solve the problem of describing all symplectic structures on $\mathbb{C}P^2$ since it is unknown whether this condition holds for every symplectic form on $\mathbb{C}P^2$.

Almost all the results mentioned below use Gromov's technique of J-holomorphic spheres. They exploit the almost complex structures associated to a symplectic structure, and so emphasise the similarities between Kähler, complex and symplectic geometry. However, symplectic geometry cannot always be derived from Kähler (or even complex) geometry. For example, there are many non-Kähler

Proceedings of the International Congress
of Mathematicians, Kyoto, Japan, 1990

symplectic 4-manifolds. (There are even symplectic 4-manifolds which admit no complex structure: see [FGG].) There is also an example in [McD7] of a symplectic 4-manifold which has a disconnected "symplectically convex" boundary. By contrast, any complex manifold with pseudo-convex boundary must have connected boundary. So far, the geometric significance of a symplectic manifold being Kähler is not well understood. It would be very interesting to have some answers to the following questions:

1.1 *Is every simply-connected symplectic 4-manifold Kähler?*

This is not true in dimensions ≥ 10 by [McD1].

1.2 *Is every symplectic form on a 4-manifold determined up to symplectomorphism (or even isotopy) by its cohomology class?*

This is not true in dimension 8, and in dimension 6 there are examples of symplectomorphic but non-isotopic forms: see [McD2].

1.3 *Which 4-manifolds admit a symplectic structure?*

Very little is known about this unless V is a fibration or the form ω is invariant under a non-trivial S^1 action (see [Au, AH]).

§ 2. *J*-Holomorphic Curves in Almost Complex 4-Manifolds

Because the linear symplectic group deformation retracts onto the unitary group, there is a homotopy class of almost complex structures J associated to a symplectic form ω. (Recall that an almost complex structure J on V is an automorphism of the tangent bundle TV of V such that $J^2 = -Id$. Thus TV may be considered as a complex n-dimensional vector bundle, with J corresponding to multiplication by i.) When $n = 1$, every almost complex structure is integrable, that is, it comes from an underlying complex structure on V. This fact is one of the reasons why J-holomorphic curves have such nice properties.

Gromov realised that one could get a handle on the geometry of a symplectic manifold by considering properties of a more restricted family of almost complex structures on V, namely the set $\mathcal{J}(\omega)$ of all J such that $\omega(v, Jv) > 0$ for all non-zero $v \in TV$. Such J are said to be *ω-tame*. Recall from [G1] that the set $\mathcal{J}(\omega)$ is non-empty and contractible.

A map f from a Riemann surface (S, J_0) to (V, J) is said to be *J-holomorphic* if f satisfies the generalized Cauchy-Riemann equation $df \circ J_0 = J \circ df$. Because this equation is elliptic, the space $M_p(J, A)$, of all J-holomorphic maps f which have a fixed compact domain S and represent a fixed homology class A, is a finite dimensional manifold for generic J. Moreover, if J is ω-tame, the quotient of this space by the automorphism group $G = \mathrm{Aut}(S)$ of (S, J_0) is either compact, or has a nice compactification. In fact, as the following proposition shows, homological conditions are often enough to guarantee compactness. For simplicity, we state a result about rational curves, i.e. curves $C = f(S)$ which are J-holomorphic images of spheres. The first Chern class of the complex vector bundle (TV, J) will be denoted by c. A proof may be found in [G1] or [McD4].

Theorem 2.1 [G1]. *Suppose that ω is integral (i.e. $[\omega] \in H^2(V, \mathbb{Z})$) and that $\omega(A) = 1$. Then, for generic ω-tame J, the moduli space $M_p(J, A)/G$ of rational A-curves is a compact manifold of dimension $2(c(A) + n - 3)$. Moreover, the bordism class of the evaluation map*

$$e(J) : M_p(J, A) \times_G S^2 \to V$$

given by $(f, z) \mapsto f(z)$ is independent of $J \in \mathcal{J}(\omega)$.

This result already has many interesting consequences: see [G1, 2] for example. In dimension 4 the theory is much sharper because, as the following results show, the geometric behaviour of the curves is governed by their homology classes.

Theorem 2.2. Positivity of Intersections ([G1]). *Let C and C' be distinct closed J-holomorphic curves in the almost complex 4-manifold V. Then C and C' have only a finite number of intersection points. Each such point x contributes a number $k_x \geq 1$ to the algebraic intersection number $C \cdot C'$. Moreover, $k_x = 1$ iff the curves C and C' intersect transversally at x.*

In particular, the curves C and C' are disjoint iff $C \cdot C' = 0$. This result is almost obvious if neither curve is singular at the point of intersection. The best way to prove the result in general is to perturb the curves so that their intersections avoid the singular points: see [McD8]. The next result gives a homological criterion for a curve to be embedded. By analogy with the integrable case, we define the virtual genus $g(C)$ of a closed curve C in an almost complex 4-manifold to be the number $g(C) = 1 + \frac{1}{2}(C \cdot C - c(C))$. If C is an embedded copy of a closed Riemann surface S, it is easy to check that the virtual genus $g(C)$ equals the genus g_0 of S. Conversely:

Theorem 2.3 [McD5]. *Let $C \subset V^4$ be the J-holomorphic image of a closed Riemann surface S of genus g_0. Then $g(C)$ is an integer which is greater than or equal to g_0, with equality if and only if C is embedded.*

The crucial point here is that one can define for each point $x \in C = \operatorname{Im} f$ a local self-intersection number L. Int.(f, x) of $\operatorname{Im} f$ at x. One proves that this is always ≥ 0, and is strictly > 0 iff x is a singular point of f. In fact, it is not hard to see that one can choose local coordinates near $x \in V$ so that the lowest order terms in the Taylor expansion of f when regarded as a polynomial in z and $\bar{z}$ involve only z. More precisely, one can suppose that $f : D \to \mathbb{C}^2$ has the form

$$z \mapsto (z^k, z^m) + \text{terms in } z, \bar{z} \text{ of order} > m,$$

where $m > k \geq 1$ and k does not divide m. (Here, D is the unit disc in $\mathbb{C}$, and we identify a neighbourhood of x in V with a neighbourhood of $\{0\}$ in $\mathbb{C}^2$ in such a way that J corresponds to an almost complex structure on $\mathbb{C}^2$, which equals the standard almost complex structure J_0 at $\{0\}$.) If m and k are mutually prime, it is easy to see that L. Int.$(f, x) =$ L. Int.$(f_0, x) = (m-1)(k-1)$, where $f_0(z) = (z^k, z^m)$. In general, the proof of Theorem 2.3 involves perturbing f, using the techniques of Nijenhuis and Woolf. A more detailed analysis of the generalized Cauchy-Riemann equation shows that the jet of the J-holomorphic map at a critical point is dominated by its J_0-holomorphic part. More precisely,

Theorem 2.4 [McD8]. *Let $f:(D, 0) \to (\mathbb{C}^2, 0)$ be a J-holomorphic map, where J is an almost complex structure on $\mathbb{C}^2$ which equals the standard structure J_0 at $\{0\}$. Then there is a J_0-holomorphic map f_0 such that for sufficiently small $\varepsilon > 0$ the knots formed by intersecting $f(D)$ and $f_0(D)$ with the sphere $S^3(\varepsilon)$ of radius ε about $\{0\}$ are isotopic.*

§ 3. Applications to Symplectic 4-Manifolds; Uniqueness Results

In order to apply this theory one needs to know that the bordism class of the evaluation map $e(J)$ of Theorem 2.1 is non-trivial. For example, suppose that there is a symplectically embedded 2-sphere C in V^4. It is not hard to see that C can be parametrized in such a way that it is J-holomorphic for some ω-tame J. If in addition the self-intersection number $p = C \cdot C$ is ≥ -1 one can show that that the corresponding moduli space is generically non-empty. Thus V contains a $2(p+1)$-dimensional family of (unparametrized) curves which may be used to analyse the structure of V.

The case $p = -1$ is rather special since then, by Theorem 2.2, there is at most one J-holomorphic curve in each homology class. Such an embedded 2-sphere is said to be *an exceptional sphere*. By analogy with the complex case, a symplectic 4-manifold which contains no exceptional spheres is said to be *minimal.*

The following results describe all minimal symplectic 4-manifolds which contain a symplectic 2-sphere with non-negative self-intersection.

Theorem 3.1 [McD6]. *Let (V, ω) be a closed symplectic 4-manifold (V, ω) which contains a symplectically embedded 2-sphere C with $C \cdot C = 0$. Suppose further that $(V - C, \omega)$ is minimal. Then (V, ω) is ruled, that is, there is a fibration $\pi: V \to M$ whose fibers are symplectically embedded 2-spheres, one of which we may assume to be C. Moreover the symplectic form ω is determined up to symplectomorphism by its cohomology class.*

Sketch of proof. Let A be the homology class of C. One first shows that, for any ω-tame J, the set of A-curves may be compactified by adding "A-cusp-curves". These must consist of two exceptional curves joined at one point, and, because $A \cdot A = 0$, must be disjoint from all J-holomorphic A-curves by Theorem 2.2. Therefore, if J is chosen so that C is J-holomorphic, the minimality of $V - C$ implies that the moduli space of A-curves is compact. Using Theorem 2.2 again, one sees that there is at most one A-curve through each point. It follows that the evaluation map $e(J)$ has degree 1, and that there is exactly one curve through each point. Now observe that these curves are embedded by Theorem 2.3. With a little more work, one can show that they form the fibers of a fibration.

To prove the uniqueness statement, one first reduces to the case $V = S^2 \times S^2$ by some cutting and pasting. Next one constructs a symplectic section of V in the class $B = [\text{pt.} \times S^2]$. We are now in a situation considered by Gromov. Since $B \cdot B = 0$, the B-curves also fiber V. Thus there is a symplectomorphism from (V, ω) to $(S^2 \times S^2, \varrho)$ where ϱ is non-degenerate on all the slices pt. $\times S^2$ and $S^2 \times$ pt. The result now follows from:

Lemma 3.2 [K]. *Suppose that ω is a symplectic form on the product $M' \times M''$ of two compact 2-manifolds which is non-degenerate on all the slices $M' \times$ pt. and pt. $\times M''$. Then ω is isotopic to a product form.*

Proof. This is a straightforward calculation.

An explicit description of all possible symplectic ruled 4-manifolds is given in [McD6] §4. Basically, one gets all the Kähler ruled surfaces considered as symplectic manifolds, i.e. one forgets the complex structure, since this is not relevant. However, note that when V is a non-trivial bundle the cohomology classes $a = [\omega]$ which can be realised in this way satisfy the condition $(a(C))^2 < a^2(V)$, and so do not include every class a with $a^2(V) > 0$. Note also that (V, ω) is itself minimal unless V is the non-trivial bundle over S^2, in which case (V, ω) is $\mathbb{C}P^2$ blown up at one point with a standard Kähler form.

There is a corresponding and even simpler result when the sphere has positive self-intersection. It is the symplectic analogue of the fact that the only minimal complex surfaces which contain a rational curve of positive self-intersection are $\mathbb{C}P^2$ and $\mathbb{C}P^1 \times \mathbb{C}P^1$.

Theorem 3.3 [McD6]. *Let (V, ω) be a closed minimal symplectic 4-manifold (V, ω) which contains a symplectically embedded 2-sphere C with $p = C \cdot C > 0$. Then (V, ω) is symplectomorphic either to $\mathbb{C}P^2$ with its standard Kähler form or to $S^2 \times S^2$ with a product form. In the former case p must equal 1 or 4 and C is either $\mathbb{C}P^1$ or a quadric, and in the latter case p is even, and C is the graph of a holomorphic self-map of S^2.*

We will see in §4 below that, just as in the complex case, one can get rid of exceptional spheres by blowing them down. Thus these two theorems classify all symplectic 4-manifolds which contain a symplectically embedded 2-sphere C with $C \cdot C \geq 0$, modulo the question of the uniqueness of blow ups. They generalize and sharpen uniqueness results which Gromov obtained in [G1] for the manifolds $S^2 \times S^2$ and $\mathbb{C}P^2$. One expects that there are corresponding results for manifolds which contain a symplectically embedded 2-manifold M of higher genus, provided that $M \cdot M$ (or equivalently, $c(M)$) is sufficiently large to guarantee that the index (i.e. formal dimension of the corresponding moduli space) is positive. But it is unknown what happens if $M \cdot M$ is small. For example, if ω is a symplectic form on

$T^2 \times S^2$ which is non-degenerate on one slice $T^2 \times$ pt., it is not known whether ω must be symplectomorphic to a product form.

Note. These uniqueness results all derive from considering families of J-holomorphic spheres, and so apply only to 4-manifolds which contain such spheres. In some circumstances it is better to consider J-holomorphic discs whose boundary is constrained to lie on a 2-manifold M which is totally real, i.e. $J(T_xM) \cap T_xM = \{0\}$, except at isolated points. This approach yields the sharpest results when M is a 2-sphere which is contained in a J-convex hypersurface: see [E, B]. Note that M need not be symplectic.

§ 4. Blowing Up and Down, and Embeddings of Balls

In the category of complex manifolds, there is a well-known operation of blowing up a point in which one replaces a point x by the space of all complex lines through x. More formally, one cuts out a neighbourhood of x which is biholomorphic to $\mathbb{C}^n$ and glues back in a copy of the total space L of the canonical line bundle over $\mathbb{C}P^{n-1}$ by identifying the complement of the zero section in L with the deleted neighbourhood $\mathbb{C}^n - \{0\}$ of x. Thus the point x is replaced by an "exceptional divisor" which is a copy of $\mathbb{C}P^{n-1}$ with normal bundle L. When $n = 2$, this divisor is simply an embedded rational curve C with $C \cdot C = -1$. The converse process of "blowing down" replaces an exceptional divisor by a point.

In the symplectic case, when one blows down an exceptional divisor P one gets not a point but a ball $B(\lambda)$ whose radius λ is related to the cohomology class of ω by the formula $\pi\lambda^2 = \omega(A)$, where A is the positive generator of $H_2(P; \mathbb{Z})$. Conversely, in order to "blow up a point" of (V, ω) one chooses a symplectic embedding g of $B = B(\lambda)$ into V, cuts out the interior of the ball $g(B)$ and then forms the blow up $\tilde{V}$ by identifying the boundary sphere $g(\partial B)$ with $\mathbb{C}P^{n-1}$ via the Hopf map and smoothing the result: see [McD3]. It is not hard to check that ω induces a symplectic form $\tilde{\omega}$ on $\tilde{V}$. Note that $\tilde{V}$ is diffeomorphic to the manifold obtained by putting any integrable tame almost complex structure on a neighbourhood of some point $x \in V$, and then blowing V up at x. The point of the above description is that it gives the construction of $\tilde{\omega}$.

Clearly, by successive blowing down of a finite sequence of exceptional divisors one can make any manifold minimal, i.e. such that it does not contain any exceptional divisors. When $n = 2$, one can use the theory of J-holomorphic curves to control the exceptional divisors and hence can show:

Theorem 4.1 [McD3]. *Every symplectic 4-manifold (V, ω) covers a minimal symplectic manifold (V', ω') which may be obtained from V by blowing down a finite collection of disjoint exceptional curves. Moreover, (V', ω') is determined up to symplectomorphism by the homology classes of the blown down curves.*

Another important question is that of the uniqueness of blow ups. Not much is known about this except when one blows up one point of the standard $\mathbb{C}P^2$. In this case, the resulting manifold (X, ω) is the non-trivial S^2 bundle over S^2. Moreover, it is not hard to show that X contains a symplectically embedded 2-sphere C with $C \cdot C = 0$. Hence the theory of § 3 applies, and one can show that all cohomologous forms are symplectomorphic. As a corollary, one finds:

Theorem 4.2 [McD3]. *For each $\lambda < 1$, the space of symplectic embeddings of the ball $B(\lambda)$ into the open unit ball in $\mathbb{C}^2$ is connected.*

The analogous statement is true if the target space is the standard $\mathbb{C}P^2$. However, this result is surprisingly delicate, and it is unclear if this holds for any other target space, even those as simple as $S^2 \times S^2$ or a convex subset of $\mathbb{C}^2$. This question is of interest because of the following result.

Proposition 4.3 [McD3]. *Let X_k be the complex surface obtained by blowing $\mathbb{C}P^2$ up at k points, and let S be a fixed copy of $\mathbb{C}P^1$ in X_k (disjoint from the blown up points). Then, there is a non-Kähler structure on X_k which is non-degenerate on S iff there are $\lambda_1, \ldots, \lambda_k$ such that the space of symplectic embeddings of the disjoint union $\coprod_i B(\lambda_i)$ into $B(1)$ is disconnected.*

References

[Au] Audin, M.: Hamiltoniens périodiques sur les variétés symplectiques compactes de dimension 4. (Lecture Notes in Mathematics, vol. 1416.) Springer, Berlin Heidelberg New York 1990

[AH] Ahara, K., Hattori, A.: 4-dimensional symplectic S^1-manifolds admitting a moment map. Preprint, Tokyo 1990

[B] Bennequin, D.: Topologie symplectique, convexité holomorphe et structure de contacte. Séminaire Bourbaki, June 1990

[E] Eliashberg, Ya.: Filling by holomorphic discs and its applications. In: Geometry of low dimensional manifolds, vol 2, ed. Donaldson and Thomas. CUP, Cambridge 1990

[FGG] Fernandez, M., Gotay, M., Gray, A.: Four-dimensional parallelizable symplectic and complex manifolds. Proc. Amer. Math. Soc. **103** (1988) 1209–1212

[G1] Gromov, M.: Pseudo-holomorphic curves in almost complex manifolds. Invent. math. **82** (1985) 307–347

[G2] Gromov, M.: Soft and hard symplectic geometry. In: Proceedings of the ICM at Berkeley, 1986. Amer. Math. Soc., 1987, pp. 81–98

[K] Kasper, B.: Examples of symplectic structures on fiber bundles. Ph.D. thesis, Stony Brook 1990

[McD1] McDuff, D.: Examples of simply connected non-Kählerian symplectic manifolds. J. Diff. Geom. **20** (1984) 267–277

[McD2] McDuff, D.: Examples of symplectic structures. Invent. math **89** (1987) 13–36

[McD3] McDuff, D.: Blowing up and symplectic embeddings in dimension 4. To appear in Topology 1991

[McD4] McDuff, D.: Elliptic methods in symplectic geometry. Bull. Amer. Math. Soc. **23** (1990) 311–358

[McD5] McDuff, D.: The local behaviour of holomorphic curves in almost complex 4-manifolds. To appear in J. Diff. Geom. 1991

[McD6] McDuff, D.: The structure of rational and ruled symplectic 4-manifolds. J. Amer. Math. Soc. **3** (1990) 679–712

[McD7] McDuff, D.: Symplectic manifolds with contact-type boundaries. Invent. math. **103** (1991) 651–671

[McD8] McDuff, D.: Singularities of J-holomorphic curves in almost complex 4-manifolds. Preprint, 1991

[M] Moser, J.: On the volume elements on manifolds. Trans. Amer. Math. Soc. **120** (1965) 280–296

Rational Homotopy Theory and Deformation Problems from Algebraic Geometry

John J. Millson

Department of Mathematics, University of Maryland, College Park, MD 20742, USA

This paper is a description of research I have been doing over the last four years, applying some of the methods and ideas of rational homotopy theory as developed by Chen, Quillen and Sullivan, to deformations of flat and holomorphic bundles, complex manifolds and isolated singularities. My work is based on the fundamental observation of Pierre Deligne [D] that "in characteristic zero, a deformation problem is controlled by a differential graded Lie algebra, with quasi-isomorphic differential graded Lie algebras giving the same deformation theory." I would like to thank Pierre Deligne for providing me with this insight and Bill Goldman who was my collaborator in developing much of what follows. I would also like to thank Ragnar Buchweitz, Kevin Corlette, Steve Halperin, Jack Lee, Madhav Nori and Mike Schlessinger for helpful conversations. The interested reader will find details in [GM1, GM2, BM] and [M].

We begin by recalling that Sullivan showed how to recover *explicitly* the rational homotopy type of a simply-connected manifold M by replacing the de Rham algebra $\mathscr{A}^*(M)$ on M by a minimal free differential graded algebra quasi-isomorphic to it having finite dimensional cochain groups and decomposable differential. We recall that two differential graded algebras A and B are quasi-isomorphic if there is a chain of homomorphisms

$$A = A_0 \to A_1 \leftarrow A_2 \cdots \to A_n = B$$

all of which induce isomorphisms of cohomology. The rational homotopy type of M can be calculated from any differential graded algebra quasi-isomorphic to $\mathscr{A}^*(M)$.

We will make use of one concept from the now well-developed homotopy theory of differential graded algebras.

Definition. *A differential graded algebra A is* formal *if it is quasi-isomorphic to a differential graded algebra B with zero differential. The underlying algebra B is then necessarily isomorphic to the cohomology algebra of A.*

We now recall the celebrated theorem of [DGMS].

This work was supported in part by NSF grant DMS-85-01742.

Proceedings of the International Congress
of Mathematicians, Kyoto, Japan, 1990

Theorem. *The de Rham algebra $\mathscr{A}^*(M)$ of a compact Kähler manifold M is formal.*

The theorem is a consequence of the following considerations. We use the complex structure on M to decompose the exterior differential d on M in the usual way, $d = \partial + \bar{\partial}$. Then we have the following quasi-isomorphism

$$(\mathscr{A}^*(M), d) \leftarrow (\ker \partial, \bar{\partial}) \rightarrow \left(\frac{\ker \partial}{\operatorname{im} \partial}, 0\right)$$

The fact that $\bar{\partial}$ induces the zero differential on $\frac{\ker \partial}{\operatorname{im} \partial}$ is an immediate consequence of the "$\partial\bar{\partial}$-lemma" of Kähler geometry, [DGMS].

To carry Sullivan's ideas over to deformation theory, we start with a differential graded Lie algebra over a field $\mathbf{k}$ (we will consider only those algebras with finite dimensional first cohomology groups). Given such an algebra L we choose a complement $C^1(L)$ to the 1-coboundaries $B^1(L) \subset L^1$. We define a functor $A \rightarrow Y_L(A)$ on the category of Artin local $\mathbf{k}$-algebras by

$$Y_L(A) = \{\eta \in C^1(L) \otimes \mathfrak{m} : d\eta + \tfrac{1}{2}[\eta, \eta] = 0\}.$$

Here $\mathfrak{m}$ is the maximal ideal of the Artin local $\mathbf{k}$-algebra A. It is evident that the functor Y_L satisfies the hypotheses of Theorem 2.11 of [Sc1] and is consequently pro-representable by a complete local $\mathbf{k}$-algebra R_L (we will see later as a consequence of our main theorem that the isomorphism class of R_L does not depend on the complement $C^1(L)$).

One can apply the above construction to the following geometric situations:

(i) The twisted de Rham algebra with coefficients in the flat Lie algebra bundle ad P associated to a flat principal G-bundle P over a compact manifold M;
(ii) the Kodaira-Spencer algebra $(\oplus \mathscr{A}^{0,*}(M, T^{1,0}(M)), \bar{\partial})$ associated to a complex manifold M;
(iii) the tangent complex T associated to the germ (V, x) of an isolated singularity in $\mathbb{C}^n$.

It is then reasonably clear (and proved in [GM2]) from the above construction that in cases (i) and (ii) the algebra R_L is the completion of the analytic local ring $\mathfrak{O}_{\mathscr{K}}$ associated to the versal deformation space as constructed by Kuranishi of the given flat connection on P (resp. complex structure on M). In case (iii), it is proved in [BM] following ideas in [SS] that the ring R_T is isomorphic to the completion of the analytic local ring of the versal deformation space of the isolated singularity (V, x). We note two other important cases of differential graded Lie algebras L over a field $\mathbf{k}$ in which R_L is the completion of an analytic local $\mathbf{k}$-algebra.

In case L has zero differential, then R_L is the completion of the analytic local $\mathbf{k}$-algebra associated to the germ $(\mathscr{Q}_L, 0)$ where

$$\mathscr{Q}_L = \{\eta \in L^1 : [\eta, \eta] = 0\}.$$

We note that $\mathscr{Q}_L$ is a quadratic cone canonically associated to L. In case L^1 is finite dimensional, then R_L is the completion of the analytic local $\mathbf{k}$-algebra of the germ $(Y, 0)$ where

$$Y = \{\eta \in C^1 : d\eta + \tfrac{1}{2}[\eta, \eta] = 0\}.$$

Henceforth we will use $\mathscr{K}_L$ to denote the corresponding analytic germ.

Definition. *If (X, x) is an analytic germ which parameterizes a versal family for a deformation theory and L is a differential graded Lie algebra such that $R_L \approx \widehat{\mathfrak{O}}_{X,x}$ then we will say that L is a* controlling differential graded Lie algebra *for that deformation theory.*

We have the following theorem (Theorem 4.1 of [GM2]).

The Comparison Theorem. *Suppose $f : L_1 \to L_2$ is a homomorphism of differential graded Lie algebras such that f induces an isomorphism on first cohomology and an injection on second cohomology. Then R_{L_1} and R_{L_2} are isomorphic.*

Corollary. *The isomorphism class of R_L does not depend on the choice of $C^1(L)$.*

Corollary. *If L_1 and L_2 are quasi-isomorphic, then R_{L_1} and R_{L_2} are isomorphic.*

Remarks. We do not require f to carry the complement $C^1(L_1)$ into $C^1(L_2)$. In case R_{L_1} and R_{L_2} are the complete local **k**-algebras associated to analytic germs (X_1, x_1) and (X_2, x_2), it follows from [A] that (X_1, x_1) and (X_2, x_2) are isomorphic.

If L is quasi-isomorphic to a differential graded Lie algebra with zero differential (one says L is formal), then it follows that

$$\mathscr{K}_L = \{\eta \in H^1(L) : [\eta, \eta] = 0\}.$$

Somewhat surprisingly this frequently happens. Carlos Simpson [S] has shown that the twisted de Rham algebra is formal if M is Kähler, the structure group G of the underlying principal bundle is linear and the monodromy representation $\varrho : \pi_1(M) \to G$ is completely reducible. Using Simpson's Theorem, the above general results and some standard results relating representations of $\pi_1(M)$ and flat connections we obtain the following theorem (recall that if G is a linear algebraic group over $\mathbb{R}$ (or $\mathbb{C}$) then the space of representations $\mathrm{Hom}(\pi_1(M), G)$ is an affine scheme over $\mathbb{R}$ (or $\mathbb{C}$)). The proof of this theorem is in [GM1].

Theorem. *Let G be a linear algebraic group over $\mathbb{R}$ (or $\mathbb{C}$) and M be a compact Kähler manifold. Let $\varrho : \pi_1(M) \to G$ be a completely reducible representation. Then the analytic local ring of $\mathrm{Hom}(\pi_1(M), G)$ at ϱ is isomorphic to that of the quadratic germ $(\mathscr{Q}, 0)$ where*

$$\mathscr{Q} = \{u \in Z^1(\pi_1(M), \mathfrak{g}) : [u, u] = 0 \quad in \quad H^2(\pi_1(M), \mathfrak{g})\}.$$

Here $Z^1(\pi_1(M), \mathfrak{g})$ is the space of Eilenberg-MacLane 1-cocycles with values in the Lie algebra $\mathfrak{g}$ of G and $[u, u]$ denotes the product obtained by combining the cup-product on group cochains with the bracket on $\mathfrak{g}$.

The proofs of the above results are too long to be given here; however, to emphasize the analogy with the theorem of [DGMS] referred to earlier, we give the original proof of [GM1] of the formality of the twisted de Rham algebra in case G is compact. We may decompose the exterior covariant differential d_∇ by type as $d_\nabla = \partial_\nabla + \overline{\partial}_\nabla$ (here ∇ is the covariant derivative operator on sections of ad $P \otimes \mathbb{C}$ associated to the flat connection on P). We then have the quasi-isomorphism

$$(\mathscr{A}^*(M, \operatorname{ad} P_{\mathbb{C}}), d_\nabla) \leftarrow (\ker \partial_\nabla, \overline{\partial}_\nabla) \to \left(\frac{\ker \partial_\nabla}{\operatorname{im} \partial_\nabla}, 0\right).$$

Once again the induced differential on $\frac{\ker \partial_\nabla}{\operatorname{im} \partial_\nabla}$ is zero by a $\partial_\nabla\overline{\partial}_\nabla$-lemma, see [GM1].

There is also an interesting formality result for the Kodaira-Spencer algebra.

Theorem. *Let M^n be a compact Kähler manifold admitting a nowhere zero top degree holomorphic form. Then the Kodaira-Spencer algebra of M is formal. Moreover, the cup-square from $H^1(M, T^{1,0}(M))$ to $H^2(M, T^{1,0}(M))$ is zero.*

Corollary. *The Kuranishi space (versal deformation space) of M is $H^1(M, T^{1,0}(M))$.*

The corollary is Bogomolov's Theorem. Our proof of the above theorem is a reinterpretation of proofs of Tian [Ti] and Todorov [To]. We let ω be the nowhere zero top degree holomorphic form on M. We then obtain an isomorphism of complexes $\Phi : (\mathscr{A}^{0,\bullet}(M, T^{1,0}(M)), \overline{\partial}) \to (\mathscr{A}^{n-1,\bullet}(M), \overline{\partial})$ given by

$$\Phi(\eta) = \iota_\eta \omega .$$

Here $\iota_\eta\omega$ denotes the contraction [FN] of the scalar form ω by the vector form η. We use Φ to transfer the graded Lie bracket from the Kodaira-Spencer algebra to $\bigoplus_{q=0}^n \mathscr{A}^{n-1,q}(M)$.

The Tian-Todorov Lemma. *Suppose $\eta_1 \in \mathscr{A}^{n-1,q_1}(M)$ and $\eta_2 \in \mathscr{A}^{n-1,q_2}(M)$ are both ∂-closed differential forms. Then the (transported) bracket $[\eta_1, \eta_2]$ is ∂-exact.*

Remark. This lemma is the analogue of the lemma in symplectic geometry that the bracket of two symplectic vector fields corresponds to an *exact* 1-form. It may be proved in the same way using the formalism of vector-valued forms of [*FN*].

Once the Tian-Todorov Lemma is proved, the formality of the Kodaira-Spencer algebra follows from the now familiar diagram

$$(\mathscr{A}^{n-1,\bullet}(M), \overline{\partial}) \leftarrow (\ker \partial, \overline{\partial}) \to \left(\frac{\ker \partial}{\operatorname{im} \partial}, 0\right).$$

Here $\ker \partial$ denotes $\ker (\partial : \mathscr{A}^{n-1,\bullet}(M) \to \mathscr{A}^{n,\bullet}(M))$ and $\operatorname{im} \partial$ denotes $\operatorname{im} (\partial : \mathscr{A}^{n-2,\bullet}(M) \to \mathscr{A}^{n-1,\bullet}(M))$. For details and the vanishing of the cup square the reader is referred to [GM2].

A class of examples which are easily analyzed is the class of compact complex parallelizable nilmanifolds. Let $M = \Gamma \backslash N$, N nilpotent complex with Lie algebra $\mathfrak{n}$ defined over $\mathbb{R}$ and Γ a cocompact lattice. Let L be the Kodaira-Spencer algebra and $\overline{L} \subset L$ the image of the left N-invariants. The inclusion $\overline{L} \to L$ is a quasi-isomorphism so $\mathscr{K}_L \cong \mathscr{K}_{\overline{L}}$. It is easy to see that

$$\mathscr{K}_{\overline{L}} = \mathrm{End}_{\mathrm{alg}}(\mathfrak{n})$$

the (germ at 0 of the) affine variety of Lie algebra endomorphisms of $\mathfrak{n}$. If we describe $\mathfrak{n}$ by generators and relations we can produce a very large number of germs that are Kuranishi spaces of complex manifolds. For example, let $\mathfrak{n}$ be the free Lie algebra on two generators X and Y subject to the relations

(i) all $(n+1)$-fold commutators $= 0$
(ii) $\mathrm{ad}^{n-1} X(Y) = \mathrm{ad}^{n-1} Y(X)$.

Then

$$\mathscr{K}_{\overline{L}} = \{(X', Y') \in \mathfrak{n}^2 : X', Y' \text{ satisfy (ii) }\}$$

so the Kuranishi space of M is a homogeneous cone of degree n, see [GM2] for more details.

Remark. These examples provide realizations of all the obstructions to integrating an infinitesimal deformation of the complex structure of a complex manifold M.

We observe that $\overline{L}$ controls the deformation theory of locally left-invariant (i.e., descended from left-invariant complex structure on N) structures on M. Since $\overline{L} \to L$ is a quasi-isomorphism it follows that the two deformation theories are the same, that is, every complex structure on M sufficiently close to the locally bi-invariant one is locally left-invariant. The rest of this paper outlines deeper examples of such a comparison of two deformation theories.

We now describe applications of the above ideas to deformation of isolated singularities. We first summarize the fundamental results of Kuranishi in [K2]. Let V be an analytic subvariety of $\mathbb{C}^N$ with a normal isolated singularity at the origin. Let M be a link of V (the intersection of V with a small sphere centered at the origin). Then M has an induced CR-structure. Let $T^{1,0}(M)$ be the (1,0) subspace of the complexified horizontal subspace $H \otimes \mathbb{C} \subset T(M) \otimes \mathbb{C}$. Choose a complement F to H in $T(M)$. The map τ of [K2], (8), gives

$$E = T^{1,0}(M) \oplus (F \otimes \mathbb{C})$$

the structure of a holomorphic vector bundle over M. Then $(\oplus \mathscr{A}^{0,*}(M, E), \overline{\partial}_b)$ is a complex, and Kuranishi used it to construct a finite-dimensional family $(\mathscr{K}_M, 0)$ of integrable CR-structures on M which is a versal deformation of the given CR-structure modulo a relation (coarser than isomorphism) designed to account for the above choice of sphere. However, he did not give $\mathscr{K}_M$ a complex analytic structure nor did he relate $\mathscr{K}_M$ to the versal deformation space of $(V, 0)$. The first problem was solved for the case dim $V \geq 4$ by Miyajima in [Mi1] completing earlier work of Akahori [Ak1]. In the rest of this paper we will show how the Comparison Theorem can be used to identify $\mathcal{K}_M$ with the versal deformation space of $(V,0)$. The following result is proved in [BM]. It was proved independently by Miyajima in [Mi2] using results of Fujiki [F].

Theorem. *Suppose* $(V,0)$ *is normal and satisfies*

(1) $\dim V \geq 4$;
(2) $\operatorname{depth}_{\{0\}} V \geq 3$.

Then the base space of the versal deformation of $(V,0)$ *is isomorphic to* $\mathcal{K}_M$.

Remarks. The assumption (2) is equivalent to the assumption that the Kohn-Rossi cohomology group $H^1(M,\mathfrak{O})$ vanishes by [Y]. If we do not assume (2), it can be shown that the base space for the versal deformation of $(V,0)$ is isomorphic to a closed subgerm of $\mathcal{K}_M$. In [BM] we give a family of examples such that $(V,0)$ is normal but the deformation space of $(V,0)$ is a *proper* subgerm of $\mathcal{K}_M$.

We will now prove the above theorem by applying the Comparison Theorem many times.

Let T be the tangent complex associated to $(V,0)$, see [B], [Sc3] or [P]. Then as stated above, T is a differential graded Lie algebra that controls the deformation theory of $(V,0)$. We recall that this means that there is an isomorphism

$$R_T \approx \widehat{\mathfrak{O}}_{X,0}$$

where $(X,0)$ is the analytic germ parametrizing the versal deformation of $(V,0)$. Choose any Stein representative V of $(V,0)$ and let L be the Kodaira-Spencer algebra of $U = V - \{0\}$. In [BM] we prove, following [Sc2], that under the assumption $\operatorname{depth}_{\{0\}} V \geq 3$ we have

$$R_T \approx R_L\,.$$

It remains to compare the deformation theory of U with that of M.

We now consider the image of L under the restriction map from U to M. Unfortunately brackets in L do not behave well under this map and consequently it is necessary to replace L by a subalgebra $L_{\tan}$ of forms whose restrictions to M take values tangent to M. The definition of the algebra $L_{\tan}$ is somewhat involved. We may assume that r has no critical points on U and note that $\omega = \frac{i}{2}\partial\overline{\partial}(r^2)$ is the restriction of the Kähler form of $\mathbb{C}^n$. We consider the sub-graded vector space $D^\bullet$ of L defined by

$$D^0 = \{Z \in L^0 : \iota_{\overline{\partial} Z}\partial r|_M = 0\}$$

and for $i \geq 1$

$$D^i = \{\mu \in L^i : \iota_\mu \partial r|_M = 0, \quad \iota_\mu \omega|_M = 0\}\,.$$

It is proved in [Ak2] that $(D^\bullet, \overline{\partial})$ is a complex and it is easily seen that

$$D^+ = \bigoplus_{i\geq 1} D^i$$

is closed under the Frölicher-Nijenhuis bracket [FN]. Furthermore, it is a result of Akahori [Ak2] that the inclusion $D^\bullet \to L^\bullet$ is a quasi-isomorphism of complexes. Thus if we choose a complement C^1 to $\overline{\partial} D^0$ in D^1 and define

$$L_{\tan} = C^1 \oplus \bigoplus_{i \geq 2} D^i ,$$

we find that $L_{\tan}$ is a differential graded Lie algebra and that the inclusion $L_{\tan} \to L$ induces an isomorphism of cohomology in degree greater than or equal to one. Thus by the Comparison Theorem we have (for any choice of C^1)

$$R_{L_{\tan}} \approx R_L .$$

We now identify the image of j^*. To do this we need the CR-analogue of $L_{\tan}$ which was constructed by Akahori in [Ak1] and was the basis of the proof of the Akahori-Miyajima Theorem referred to above. We define $\overline{D}^\bullet$ to be the following sub-graded vector space of Kuranishi's complex $(\oplus \mathscr{A}^{0,*}(M,E), \overline{\partial}_b)$. Let θ be a contact form on M compatible with H. Then we define

$$\overline{D}^0 = \{ Z \in \mathscr{A}^0(M,E) : \iota_{\overline{\partial}_b Z} \theta = 0 \}$$

and for $i \geq 1$,

$$\overline{D}^i = \{ \mu \in \mathscr{A}^{0,i}(M,E) : i_\mu \theta = 0, \quad \iota_\mu d\theta = 0 \} .$$

We then choose a complement $\overline{C}^1$ to $\overline{\partial}_b \overline{D}^0$ in $\overline{D}^1$ and define

$$\overline{L} = \overline{C}^1 \oplus \bigoplus_{i \geq 2} \overline{D}^i .$$

We also define

$$\overline{D}^+ = \bigoplus_{i \geq 1} \overline{D}^i .$$

Now the restriction map j^* followed by the canonical map $\tau : T^{1,0}(U)|M \to E$ of [K2] gives a homomorphism of complexes

$$j^+ : D^+ \to \overline{D}^+ .$$

Let $I^+ = \ker j^+$.

Lemma. *I^+ is an ideal in D^+.*

As a consequence of this lemma we find that $\overline{D}^+$ carries the structure of a differential graded Lie algebra such that j^+ is a homomorphism. Since j^* is surjective we may choose the complement C^1 so that it is carried into $\overline{C}^1$ by j^+. We obtain a homomorphism of differential graded Lie algebras

$$j^+ : L_{\tan} \to \overline{L} .$$

In [Ak1], Akahori proved that the inclusion of $\overline{L}$ into Kuranishi's complex induced an isomorphism of cohomology of degree one or greater. In [Ak1] and

[Mi1], Akahori and Miyajima proved that Kuranishi's family $\mathscr{K}_M$ was isomorphic to the analytic subvariety $\mathscr{K}_{\overline{L}} \subset H^1(\overline{L})$ obtained by applying the well-known construction of [K1] to $\overline{L}$ with a suitable choice of complement $\overline{C}^1$, provided $\dim V \geq 4$. In particular, it is immediate that

$$R_{\overline{L}} \approx \widehat{\mathfrak{O}}_{\mathscr{K}_M,0} .$$

Our theorem then follows by the Comparison Theorem from the fact that j^+ induces isomorphisms on cohomology in degrees 1 and 2 provided $\dim V \geq 4$. To prove this latter fact we consider the following diagram of complexes:

$$\begin{array}{ccc} L & \longrightarrow & \oplus\mathscr{A}^{0,*}(M,E) \\ \uparrow & & \uparrow \\ L_{\tan} & \longrightarrow & \overline{L} \end{array}$$

We have seen that the two vertical arrows are quasi-isomorphisms. But it follows from the discussion in [Y], pp. 81-82, that the top horizontal arrow induces an isomorphism on H^1 and H^2 in case $\dim V \geq 4$.

The theorem of Buchweitz-Millson and Fujiki-Miyajima can be used to compute $\mathscr{K}_M$ if one has enough information about the ideal of the corresponding singularity. In [M] we use differential geometry to compute $\mathscr{K}_M$ directly for a certain class of CR-manifolds. Let $\mathscr{L}$ be a negative line bundle over a compact Kähler manifold N and let $U(\mathscr{L})$ be its unit circle bundle. In [M] we compute $\mathscr{K}_M$ for $M = U(\mathscr{L})$ in case $\mathscr{L}$ is sufficiently negative. Recall that if $\mathscr{L}$ is sufficiently negative, then $\mathscr{L}^{-1}$ gives a projective embedding of N into $\mathbb{CP}^m$ for some m. We then have the following theorem, [M], the CR-analogue of Theorem 2 of [Sc2].

Theorem. *Let $M = U(\mathscr{L})$ as above. If $\mathscr{L}$ is sufficiently negative then $\mathscr{K}_M$ is isomorphic to the parameter space for the versal projective deformation of N in $\mathbb{CP}^m$. Moreover, by Schlessinger's Theorem, this latter space is isomorphic to the deformation space of the singularity $(C_N, 0)$, where C_N is the affine cone over N.*

Example. Let N be a complex torus of dimension n and $\mathscr{L}$ *any* negative line bundle over N. Then by a direct computation one can show

$$\mathscr{K}_{U(\mathscr{L})} \approx \mathbb{C}^{\frac{n(n+1)}{2}} .$$

In case $\mathscr{L}$ is sufficiently negative, the above Theorem produces an isomorphism between $\mathscr{K}_{U(\mathscr{L})}$ and the parameter space of the versal deformation of $(C_N, 0)$ even though

$$\operatorname{depth}_{\{0\}} C_N = 2 .$$

Remark. For any $\mathscr{L}$ as above one can obtain an analytic germ $(C_N, 0)$ by collapsing the zero section. It is reasonable to expect that the parameter space of the versal deformation of $(C_N, 0)$ is again $\mathbb{C}^{\frac{n(n+1)}{2}}$.

References

[A] M. Artin: On solutions to analytic equations. Invent. math. **6** (1968) 277–291

[Ak1] T. Akahori: The new estimate for the sub-bundles E_j and its application to the deformation of the boundaries of strongly pseudo-convex domains. Invent. math. **63** (1981) 311–344

[Ak2] T. Akahori: The new Neumann operator associated with deformation of strongly pseudo-convex domains and its application to deformation theory. Invent. math. **68** (1982) 317–352

[AM] T. Akahori, K. Miyajima: Complex analytic construction of the Kuranishi family on a normal strongly pseudo-convex manifold II. Publ. RIMS, Kyoto University **16** (1980) 811–834

[B] R.O. Buchweitz: Contributions à la théorie des singularités. Ph.D. thesis, University of Paris

[BM] R.O. Buchweitz, J.J. Millson: *CR* geometry and deformations of isolated singularities. In preparation

[D] P. Deligne: Letter to J.J.Millson, April 24, 1986

[DGMS] P. Deligne, P.A. Griffiths, J.W. Morgan, D. Sullivan: Rational homotopy type of compact Kähler manifolds. Invent. math. **29** (1975) 245–274

[F] A. Fujiki: Flat Stein completion of a flat (1,1)-convex concave map. Preprint

[FN] A. Frölicher, A. Nijenhuis: Theory of vector-valued differential forms, Part I. Derivations in the graded ring of differential forms. Proc. Kon. Ned. Akad. Wet. Amsterdam **59** (1956) 338–359

[GM1] W.M. Goldman, J.J. Millson: The deformation theory of representations of fundamental groups of compact Kähler manifolds. Publ. Math. IHES **67** (1988) 43–96

[GM2] W.M. Goldman, J.J. Millson: The homotopy invariance of the Kuranishi space. Ill. J. Math. **34** (1990) 337–367

[K1] M. Kuranishi: Deformations of compact complex manifolds. Les Presses de l'Université de Montréal 1971

[K2] M. Kuranishi: Application of $\bar{\partial}_b$ to deformation of isolated singularities. Proceedings of Symposia in Pure Mathematics **30** (1977) 97–106

[M] J.J. Millson: $\bar{\partial}_b$ and deformations of cones. Preprint

[Mi1] K. Miyajima: Completion of Akahori's construction of the versal family of strongly pseudo-convex *CR* structures. Trans. A.M.S. **277** (1983) 163–172

[Mi2] K. Miyajima: Deformations of a complex manifold near a strongly pseudo-convex real hypersurface and a realization of the Kuranishi family of strongly pseudo-convex CR-structures. Preprint

[P] V.P. Palamodov: Deformations of complex spaces. In: Several Complex Variables IV, Encyclopedia of Mathematical Sciences, vol. 10 (S.G. Gindikin and G.M. Khenkin, eds.) Springer, Berlin Heidelberg New York, pp. 105–194

[S] C. Simpson: Higgs bundles and local systems. Preprint

[Sc1] M. Schlessinger: Functors of Artin rings. Trans. A.M.S. **130** (1968) 208–222

[Sc2] M. Schlessinger: On rigid singularities. Conference on Complex Analysis, Rice University Studies 59 No. 1 (1972) 147–162

[Sc3] M. Schlessinger: Infinitesimal deformation of singularities. Ph.D. thesis, Harvard University

[SS] M. Schlessinger, J. Stasheff: Deformation theory and rational homotopy type. Preprint

[Ti] G. Tian: Smoothness of the universal deformation space of compact Calabi-Yau manifolds and its Peterson-Weil metric. In: S.T. Tau (ed.) Math. Aspects of string theory. World Scientific Publ. 1987

[To] A.N. Todorov: The Weil-Peterson geometry of the moduli space SU ($n \geq 3$) (Calabi-Yau) manifolds I. Preprint

[Y] Stephen S.-T. Yau: Kohn-Rossi cohomology and its application to the complex Plateau problem I. Ann. Math. **113** (1981) 67–110

Geometry of Discriminant and Topology of Algebraic Curves

Eugenii I. Shustin

Department of Mathematics, Kuibyshev State University, ul. Acad. Pavlova 1
Kuibyshev 443011, USSR

Introduction

The subject of this lecture is some aspects of the theory of deformations of plane real and complex algebraic singular curves and its applications. Namely, in this large classical problem we shall consider (i) the independence of singular points deformations, (ii) connections between deformations of singularities and discriminant hypersurface in moduli space of plane curves of a given degree, (iii) applications of these results to classification of algebraic real and complex curves.

First results in this direction concerned the curves with nodes (in complex case – Severi [17], in real case – Brusotti [2]): any reduced curve with nodes can be deformed in the space of curves of a given degree so that arbitrary set of nodes is smoothed, and all other singularities are preserved.

The proof idea is that the germ of locus of curves with a given number of nodes is a transversal intersection of smooth germs of hypersurfaces corresponding to each singularity. The last assertion follows from nonspeciality of some linear system on the curve by means of Riemann-Roch theorem. It should be noted that Harris theorem on irreducibility of variety of given degree curves with a given genus [10] was proved, in fact, by the same ideas. We also point out that many methods of construction of real algebraic curves [8] are based on Brusotti theorem.

In this lecture we shall develop above results and approaches.

1. Notions and Notations

Further on we shall consider only plane algebraic projective real and complex curves (including irreduced and reducible curves). The polynomial defining a curve and the locus of curve points in $\mathbb{C}P^2$ we denote identically. The set of curves of degree n is parametrized by $\mathbb{P}^N$, $N = n(n + 3)/2$ (from now on $\mathbb{P}^N$ means $\mathbb{C}P^N$ and $\mathbb{R}P^N$ at the same time).

Let $F \in \mathbb{P}^N$ and $p \in F$ be an isolated singular point. *The singular point type*, or *singularity*, means the list of characteristic Puissex exponents of all local curve branches and the list of intersection numbers of branches. This is complete topo-

Proceedings of the International Congress
of Mathematicians, Kyoto, Japan, 1990

logical invariant of singular point. Symbols $\mu(p, F)$, $\delta(p, F)$ mean the Milnor number and δ-invariant of the point $p \in F$ respectively. By U_p we denote an open sufficiently small ball in $\mathbb{C}P^2$ centred at point p.

Remark 1. If $F \in \mathbb{R}P^N$, $p \in \mathbb{C}P^2 \setminus \mathbb{R}P^2$ then there is a conjugated singular point $\operatorname{conj} p \in F$. In this case we call the pair $(p, \operatorname{conj} p)$ by "singular point" understanding correspondingly all other terms.

The perturbation of $F \in \mathbb{P}^N$ is a sufficiently close to F curve $G \in \mathbb{P}^N \setminus \{F\}$. By *perturbation of the point* $p \in F$ we call any set $G \cap U_p$, where G is a perturbation of curve FL^m for some integer $m \geq 0$ and some fixed line $L \in (\mathbb{P}^2)^*$ non-crossing U_p. Nonsingular perturbations of the real point are called *smoothings* [25].

Two perturbations of $p \in F$ are called *equivalent*, if they can be transformed one into another by (equivariant in real case) diffeomorphism of U_p close to identity. Two perturbations of p are called *algebraic-topologically* (*AT*) *equivalent*, if they have the same sets of singularities, and (in real case) some neighbourhoods of their real parts can be transformed one into another by equivariant homeomorphism of U_p, close to identity.

By $V(n, S_1, \ldots, S_r)$ we'll denote the locus of degree n irreducible curves with r singular points of types $S_1, \ldots, S_r$ respectively. The germ $M(F) \subset \mathbb{P}^N$ of equisingular (ES) stratum (this is $\mu = \text{const}$ stratum) centered at reduced curve $F \in \mathbb{P}^N$ is called regular, if it is smooth and has codimension $\sum_{z \in F} c(z)$, where $c(z)$ is codimension of $\mu = \text{const}$ stratum in miniversal deformation base of singular point z.

2. Independence of Versal Deformations

Let p be an isolated singular point of curve $F \in \mathbb{P}^N$. Let Π, Π' denote two generic polar curves of F. For any branch Q of curve Π at point p we assign $\tau(Q) = s/r$, where s is the sum of orders of Q at point p and at the infinitely close to p points belonging to Π'. For a reduced curve F we assign $\tau = \max \tau(Q)$. It depends on different singularities of F, and is ever less than 1. Put

$$\lambda(F) = \begin{cases} 25/729, & \tau < 5\sqrt{2}/18 \\ 2\tau^2/9, & 5\sqrt{2}/18 \leq \tau < 5/9 \\ (1 - \tau + \tau^2 + (1 - 2\tau)\sqrt{1 + 2\tau - 2\tau^2})/9, & 5/9 \leq \tau \end{cases}$$

Theorem 1 [23, 24]. *If* (i) *F is reduced and satisfies*

$$\sum_{z \in F} \mu(z, F) \leq 4n - 5 \tag{1}$$

or (ii) *F is reduced and irreducible and satisfies*

$$\sum_{z \in F} (\mu(z, F) + \delta(z, \Pi)) \leq \lambda(F) n^2, \tag{2}$$

then some neighbourhood of F in $\mathbb{P}^N$ *represents a joint versal deformation of all singular points of F.*

This theorem fully describes properties of the discriminant hypersurface germ at $F \in \mathbb{P}^N$ under above conditions. The proof is based on the following: the intersection transversality of $\mathbb{P}^N$ and tangent spaces to orbits of Diff U_p-action for all $p \in \text{Sing } F$ can be concluded from nonspeciality of some linear system on suitable curve by means of Riemann-Roch theorem (in particular, polar curve Π can be taken as such a curve, and then one obtains condition (1)).

Corollary 1. *Under conditions of Theorem* 1 *a suitable perturbation of F realizes perturbations of any singular points, equivalent to a priori given perturbations of these points, while other singularities are preserved.*

This statement does not cover Severi and Brusotti theorems, but it does not contain restrictions on singular points types and their perturbations; also the right hand side of (2) is sharp with respect to exponent at n. For example, if F has only ordinary cusps as their singularities then (2) ensures the independent deformation of all cusps when

$$\#(\text{Sing } F) \leq (7 - \sqrt{13})n^2/162,$$

since the other known estimates [4, 7] give

$$\#(\text{Sing } F) \leq 3n - 1.$$

Corollary 2. *Under conditions of Theorem* 1 $M(F)$ *is regular.*

The above assertions do not contain restrictions on a position of singular points. Adding such restrictions one can weaken inequalities (1), (2) (in [3] there are those estimates for curves with ordinary singular points).

Compare (1) with known counterexamples. According to [14] varieties $V(9, A_{35})$, $V(11, A_{50})$ are not smooth. The Corollary 2 implies the smoothness of $V(9, A_k)$, $V(11, A_m)$ only for $k \leq 31$, $m \leq 39$.

3. Independence of Certain Perturbations Classes

Corollary 1 embraces every perturbation of arbitrary singularities, but it does not show what types of perturbations of each singular point exist. Here we'll use certain other approaches to our problem.

3.1 Perturbations of Quasiordinary Singular Points

Consider the following approach to problem of regularity of $M(F)$. Using Cremona transformations we can obtain that F had only ordinary singularities. Then regularity of $M(F)$ follows (see [3, 7]) from nonspeciality of some linear system on F. The exact assertions are as follows.

Definition 1. Let us consider the full resolution of singular point $p \in F$. Let $p_0 = p$, $p_1, \ldots, p_k$ be consequent infinitely close to p points of this resolution, which belong to branch Q of F centered at p. Define $\tilde{\beta}(Q, F)$ as sum of orders of branch Q at p_0, $\ldots, p_{k-1}$ (i.e. at those points, where proper inverse image of F has singularity, or intersection number of F with exceptional divisor exceeds 1). Let $\beta(Q, F)$ be equal to $\tilde{\beta}(Q, F) - 1$ (resp. $\tilde{\beta}(Q, F)$), if one of $p_1, \ldots, p_{k-1}$ is touch node, i.e. point of two smooth branches contact (resp. no touch node) of full inverse image of F. If order of F at p is 2, then we must subtract the order of Q at p from above defined value $\beta(Q, F)$. Put also

$$\tilde{\beta}(p, F) = \sum \tilde{\beta}(Q, F), \qquad \beta(p, F) = \sum \beta(Q, F).$$

It is easy to show that

$$c(p, F) - \delta(p, F) \leq \beta(p, F) \leq \mu(p, F) - 1$$

and that $\beta(p, F) = r$ for ordinary singular point of order r (this is an intersection point of r smooth branches with different tangents).

Definition 2. For any reduced curve F with irreducible components $F_1, \ldots, F_k$ and for a set $S \subset \operatorname{Sing} F$ we define vectors $d(F), b(F, S) \in \mathbb{Z}^k$ as follows:

$$[d(F)]_i = 3 \deg F_i - 1,$$
$$[b[F, S)]_i = \sum \beta(Q_1, F) + \sum \tilde{\beta}(Q_2, F),$$

where Q_2 (resp. Q_1) runs all branches of F_i centred at points $z \in S$ (resp. $z \in \operatorname{Sing} F \setminus S$). Put $\tilde{b}(F) = b(F, \operatorname{Sing} F)$, $b(F) = b(F, \varnothing)$.

Every inequality for vectors below means inequalities for each pair of corresponding coordinates of vectors.

Theorem 2 [18, 22, 24]. *If F is reduced and satisfying*

$$b(F) \leq d(F) \tag{3}$$

then $M(F)$ is regular. If

$$\tilde{b}(F) \leq d(F) \tag{4}$$

then the variety $V \subset \mathbb{C}P^N$ of curves, having singularities as F and disintegrating into components with the same singularities sets, is irreducible.

This theorem covers the Severi and Brusotti theorems and analogous results on curves with cusps and ordinary singularities [6, 7, 13, 26]. According to [27] varieties $V(6, 6A_2)$, $V(6, 3E_6)$ are reducible. Theorem 2 claims the irreducibility of $V(6, kA_2)$, $V(6, mE_6)$ for $k \leq 4$, $m \leq 2$.

Remark 2. The point $p \in F \setminus \operatorname{Sing} F$ is called *quasisingular* of order $k \geq 1$ with respect to fixed line f, smooth at p, if intersection number $(F \cdot f)_p = k$. We can take into account quasisingularities in Theorem 2 by adding to left hand sides of (3), (4) the values $\beta(p, F) = k - 1$, $\tilde{\beta}(p, F) = k$ respectively. Thus, using the supplemented

Theorem 2 one can prove Harris theorem [10] on irreducibility of $V(n, kA_1)$, $k \geq 1$, Ran theorem [16] on irreducibility of $V(n, kA_1, S)$, where singularity S is transformed into set of quasisingularities and nodes by one blowing-up. Analogous approach gives

Theorem 3. *The variety* $V(n, kA_1, S_1, \dots, S_r)$, *where* $k \geq 0$, $0 \leq r \leq 9$, $S_1, \dots, S_r$ *are ordinary singularities with* $\sum \beta(S_i) \leq 2n$, *is irreducible.*

In particular, this implies that the variety of curves on quadric $\mathbb{C}P^1 \times \mathbb{C}P^1$ with a given bidegree and a given genus is irreducible (see also [16]).

Remark 3. From proof of Theorem 2 it is possible to deduce that coefficients at monomials lieing under Newton diagrams of all singular points can be varied independently. Therefore perturbations determined by these monomials are independent too. Effective construction of those perturbations was suggested by Viro [25].

Definition 3. Singular point $p \in F \in \mathbb{C}P^N$ is called *quasiordinary* (*QO-point*), if (i) for some affine coordinates $p = (0; 0)$, and all sections of polynomial F on edges of the point p Newton diagram are nondegenerated (normal QO-point), or (ii) some embedding $\varphi : U_p \to \mathbb{C}^2$ transforms p into normal QO-point (in the case $p \in \mathbb{R}P^2$, $F \in \mathbb{R}P^N$ we assume $\varphi \circ \text{conj} = \text{conj} \circ \varphi$). Quasisingular points are QO-points in a natural sense.

This notion includes, for instance, zero-modal points, quasihomogeneous singular points. For normal QO-point $\tilde{\beta}(p, F) = k + m - r + 1$, where k, m are maximal absciss and ordinate of points on the Newton diagram, and r is a number of integer points on the Newton diagram.

Modifying the Viro construction [25] we'll define a set of *V-perturbations* of QO-point $p \in F$. Normalize the point p, and denote the Newton diagram of p by Γ and the Newton polygon of normalized curve F_0 by Δ_0. Take a set of polynomials $F_i(x, y) = \phi_i(x, y)x^t y^u$, where x, $y \nmid \phi_i$, with Newton polygons Δ_i, $1 \leq i \leq k$, lieing under Γ. Assume that (i) $\Delta = \bigcup \Delta_i$ is convex, any nonempty intersection $\Delta_i \cap \Delta_j$ is a common edge of Δ_i, Δ_j or a common apex, there is also a continuous convex function $v : \Delta \to \mathbb{R}$, linear on each Δ_i and nonlinear on each union $\Delta_i \cup \Delta_j$, (ii) if $\Delta_i \cap \Delta_j = \gamma$ is a common edge, then sections of F_i, F_j on γ are identical square-free (except x, y) polynomials. Now V-perturbation is constructed by means of set $\{F_i\}$ according to [25]: (i) lift each curve φ_i upon toric manifold determined by Δ_i, (ii) glue these toric manifolds according to arrangement of $\Delta_0, \dots, \Delta_k$, (iii) project the correspondingly glued curves $F_0, \varphi_1, \dots, \varphi_k$ on the plane.

For a given V-perturbation π we'll define the set $T(\pi)$ of oriented graphs without oriented circles. Vertexes of graph $\Lambda \in T(\pi)$ correspond to polygons Δ_i, arcs of Λ correspond to common edges of Δ_i, $0 \leq i \leq k$. To curve F_i, $1 \leq i \leq k$, and to graph $\Lambda \in T(\pi)$ we assign the vector $v(F_i, \Lambda)$. Its coordinates correspond to irreducible components of φ_i, and are equal to numbers of local branches of each component, defining by those edges of Δ_i, which correspond to arcs of Λ coming in Δ_i.

Theorem 4 [19, 22]. *Let $F \in \mathbb{P}^N$ be a reduced curve, $\sigma \subset \operatorname{Sing} F$ be some set of zero-modal singular points, $\tau \subset \operatorname{Sing} F|\sigma$ be some set of QO-points, $\{\pi_z | z \in \tau\}$ be a set of V-perturbations, defined by curves F_{zi}, $z \in \tau$, $1 \leq i \leq j_z$. If there exist graphs $\Lambda_z \in T(\pi_z)$, $z \in \tau$, for which*

$$b(F, S) + \sum_z v(F, \Lambda_z) \leq d(F), \tag{5}$$

$$b(F_{zi}, S_{zi}) + v(F_{zi}, \Lambda_z) \leq d(F_{zi}), \tag{6}$$

where $z \in \tau$, $1 \leq i \leq j_z$, $S = \{z \in \tau | v(F, \Lambda_z) \neq 0\}$, $S_{zi} = \operatorname{Sing} \phi_{zi} \cap \mathbb{C}^2 \backslash \{xy = 0\}$, then a suitable perturbation of F realizes perturbations of $z \in \tau$, which are AT-equivalent to π_z, $z \in \tau$, and perturbations of points $z \in \sigma$, which are equivalent to their a priori given perturbations, while other singularities are preserved.

Remark 4. If all curves F_{zi}, $z \in \tau$, $1 \leq i \leq j_z$, are smooth in $\mathbb{C}^2 \backslash \{xy = 0\}$, or have there only nodes, then (5), (6) can be replaced by (3).

Remark 5. V-perturbations can be defined for arbitrary singular point as follows: the singular point should be resoluted and all infinitely close singular and quasi-singular points should be consequently perturbed.

3.2 Equigeneric and Equiclassical Perturbations

The independence of above perturbations is connected with properties of ES-stratum. Here we consider perturbations, which are connected with equigeneric (EG) stratum $g = \text{const}$ and equiclassical (EC) stratum $g = \text{const}$ and $c = \text{const}$; c is class of curve, i.e. degree of dual curve.

Theorem 5 (see [1, 15]). *For any reduced curve $F \in \mathbb{C}P^N$ there exists a perturbation that turns independently each singular point $z \in F$ into an arbitrary number of nodes from 0 to $\delta(z, F)$.*

This follows from that the tangent cone to EG-stratum at $F \in \mathbb{C}P^N$ is a linear system of adjoining curves of degree n, which is evidently nonspecial. It implies that curves with nodes form a dense subset of EG-stratum in $\mathbb{C}P^N$.

A perturbation from EC-stratum turns (see [5]) a singular point into $3\delta - \varkappa$ nodes and $\varkappa - 2\delta$ cusps, where $\varkappa$ is an intersection number of curve and its polar curve at this point. Namely (see [5]),

Theorem 6. *If a reduced irreducible curve $F \in \mathbb{C}P^N$ satisfies $c > 2g - n$, then the suitable perturbation of F turns independently each singular point p into d nodes and k cusps, where d, k are arbitrary satisfying*

$$0 \leq d \leq \delta, \qquad 0 \leq k \leq \min\{\varkappa - 2\delta, \delta - d\}.$$

According to [1, 4] EG-germ and EC-germ in a versal deformation base are irreducible. Therefore

Theorem 7. *If $F \in \mathbb{C}P^N$ satisfies* (1) *or* (2), *then EG-germ and EC-germ at F in $\mathbb{C}P^N$ are irreducible.*

3.3 Independence of Minimal Smoothings

Give one more example of unconditionaly independent perturbations – minimal smoothings (see [21]). *Minimal smoothing* of singular point $p \in \mathbb{R}P^2$ of order k, at which there are r real local branches of curve $F \in \mathbb{R}P^N$, is (i) a set of r disjunct intervals connecting the points of $F \cap U_p \cap \mathbb{R}P^2$ for $r > 0$, or (ii) a set of $s \leq k/2$ concentred circumferences in $U_p \cap \mathbb{R}P^2$ for $r = 0$.

Theorem 8 [21]. *A suitable perturbation of any reduced curve $F \in \mathbb{R}P^N$ realizes arbitrary a priori given minimal smoothings of all real singular points modulo AT-equivalence, while nodes can be preserved.*

Remark 6. Perturbations from germs of almost all linear families $F + tG$, $t \in \mathbb{R}$, $G \in \mathbb{R}P^N$, realize only minimal smoothings of all singular points.

4. Perturbations of Singular Points and Algebraic Curves

The approach used in Section 3.1 raises the following questions: 1) what part of all perturbations form V-perturbations? 2) what V-perturbations can be realized for a given singular point?

Consider semiquasihomogeneous singular point (that means Newton diagram lies on straight line). Then V-perturbations are constructed by gluing affine curves with complementary Newton triangle. On the other hand (see [23]) any perturbation can be obtained by formal gluing of those affine curves and replacing their singular points by suitable perturbations. If singular points of above affine curves have independent perturbations, then we have (i) *property A*: each perturbation of initial singular point can be obtained from affine curve with a given Newton triangle by means of (equivariant in real case) homeomorphism $\mathbb{C}^2 \to U_p$, or (ii) *property A'*: the same modulo AT-equivalence. If we can vary coefficients of a curve with a given Newton triangle at monomials on the edge of gluing, while a curve topology is preserved, then singular points of the same type have the same sets of V-perturbations modulo homeomorphisms of singular points neighbourhoods (*property B*), or modulo AT-equivalence (*property B'*).

Zero-modal singular points, ordinary 4th order point, point of quadratic contact of 3 smooth branches hold properties A and B. The ordinary 5th order point holds property A [23], and if all local branches are real then property B′ is fulfilled for its smoothings [20]. Properties A′, B′ are fulfilled for smoothings of point of quadratic contact of 4 real branches [20]. Property A′ is fulfilled for smoothings of ordinary 6th order point [12].

5. Applications to Study of Topology of Algebraic Curves

Firstly, above results on irreducibility mean that curves of a given kind are rigidly isotopic, i.e. they are connected by isotopy consisting of the same curves.

It might be interesting to construct curves with a given singularities. For example (see [19]), Theorem 4 implies existance of degree n complex curves with arbitrary cusps number from 0 to $7n^2/36$, and existence of degree n curves over $\mathbb{R}$ with arbitrary number of real cusps from 0 to $3n^2/16$.

All known types of nonsingular real curves were realized in the way of perturbation of singular curves. The majority of latest results about real curves [11, 12, 22, 25] were obtained by Viro method [25] through perturbation of singular points mentioned in Section 4. It should also be noted that there is M-curve of degree 8, whose construction does not satisfy conditions of Viro method and is based on Theorem 4 [22].

Another example of using above results is the complete description of discriminant in the space of plane real quartic curves and complete classification of inflexion points arrangements on these curves [9].

At last we'll mention the Hilbert-Rohn method (see [8, 20]), which allows to construct or to prohibit certain classes of real curves.

References

1. Arbarello, E., Cornalba, M.: A few remarks. Ann. Sci. Ec. Norm. Sup. **16** (1983) 467–488
2. Brusotti, L.: Sulla "piccola variazione" di una curva piana algebrica reali. Rend. Rom. Ac. Lincei (5) **30** (1921) 375–379
3. Catalisano, M.V.: Linear systems of plane curves through fixed "fat" points of $\mathbb{P}^2$. Queen's Pap. Pure Appl. Math. **83** (1989) D1–D34
4. Diaz, S.: Irreducibility of equiclassical stratum. J. Diff. Geom. **29** (1989) 489–498
5. Diaz, S., Harris, J.: Ideals associated to deformations of singular plane curves. Trans. Amer. Math. Soc. **309**, no. 2 (1988) 443–468
6. Gradolato, M.A., Mezzetti, E.: Curves with nodes, cusps and ordinary triple points. Ann. Univ. Ferrara, sez. 7 **31** (1985) 23–47
7. Gudkov, D.A.: On certain questions in topology of plane algebraic curves. Mat. Sb. **58**, no. 1 (1962) 95–127 (Russian)
8. Gudkov, D.A.: Topology of real projective algebraic varieties. Sov. Math. Surv. **29**, no. 4 (1974) 3–79
9. Gudkov, D.A.: Plane real projective quartic curves. (Lecture Notes in Mathematics, vol. 1346.) Springer, Berlin Heidelberg New York 1988, pp. 341–347
10. Harris J.: On the Severi problem. Inv. math. **84**, (1986) 445–461
11. Korchagin, A.B.: New M-curves of degree 8 and 9. Sov. Math. Dokl. **306**, no. 5 (1989) 1038–1041 (Russian)
12. Korchagin, A.B., Shustin, E.I.: Affine curves of degree 6 and smoothings of nondegenerated 6th order singular point. Math. USSR Izvestiya **33**, no. 3 (1989) 501–520
13. Lindner, M.: Über Mannigfaltigkeiten ebener Kurven mit Singularitäten. Arch. Math. **28** (1977) 603–610
14. Luengo, I.: The μ-constant stratum is not smooth. Inv. math. **90** (1987) 139–152
15. Nobile, A.: On specialisation of curves. Trans. Amer. Math. Soc. **282**, no. 2 (1984) 739–748
16. Ran, Z.: Families of plane curves and their limits: Enriques' conjecture and beyond. Ann. Math. **130**, no. 1 (1989) 121–157
17. Severi, F.: Vorlesungen über algebraische Geometrie. Anhang F. Teubner, Leipzig 1921
18. Shustin, E.I.: On varieties of singular algebraic curves. In: Methods of qualitative theory. Gorky Univ. Press, Gorky 1983, pp. 148–163 (Russian). [English translation: Shustin,

E.I.: On Manifolds of Singular Algebraic Curves. Selecta Math. Sov. **10**, no. 1 (1991) 27–37]
19. Shustin, E.I.: Gluing of singular algebraic curves. In: Methods of qualitative theory. Gorky Univ. Press, Gorky 1985, pp. 116–128 (Russian)
20. Shustin, E.I.: The Hilbert-Rohn method and smoothings of algebraic curves singular points. Sov. Math. Dokl. **31**, no. 2 (1985) 282–286
21. Shustin, E.I.: Hyperbolic and minimal smoothings of singular points. In: Methods of qualitative theory. Gorky Univ. Press, Gorky 1986, pp. 165–174 (Russian)
22. Shustin, E.I.: New M-curve of 8th degree. Math. Notices **42**, no. 6 (1987) 180–186 (Russian)
23. Shustin, E.I.: Versal deformations in the space of a fixed degree curves. Funct. Anal. Appl. **21**, no. 1 (1987) 90–91
24. Shustin, E.I.: Smoothness and irreducibility of varieties of singular algebraic curves. In: Arithmetic and geometry of algebraic varieties. Saratov Univ. Press/Kuibyshev branch, Kuibyshev 1989, pp. 102–117 (Russian)
25. Viro, O.J.: Gluing of algebraic hypersurfaces, smoothing of singularities and construction of curves. In: Proc. Leningrad Topological Internat. Conf., Nauka, Leningrad 1983, pp. 147–195 (Russian)
26. Wahl, J.: Deformations of plane curves with nodes and cusps. Amer. J. Math. **96** (1974) 529–577
27. Zariski, O.: Algebraic surfaces. Springer, New York Berlin Heidelberg 1971.

Applications of Hodge Theory to Singularities

Joseph H.M. Steenbrink

Mathematical Institute, University of Nijmegen, Toernooiveld
NL-6525 ED Nijmegen, The Netherlands

1. Introduction

We will describe some old and recent results in the theory of complex hypersurface singularities which have been obtained using methods from mixed Hodge theory. In the last 15 years, there has been a fruitful interaction between singularity theory and Hodge theory. The common interest is the study of degenerations. In the work of W. Schmid [21] the limit mixed Hodge structure of a one parameter polarized variation of Hodge structure was constructed. This has consequences for projective degenerating families, but does not apply to the case of a singularity. (However, Scherk's trick [19] made it apply to the isolated singularity case.) As P. Deligne observed in Metz 1974, the author's geometric approach [23] (filling in a divisor with normal crossings as a special fibre and analyzing the relative logarithmic de Rham complex) works also for the case of a singularity [24, 11].

In cases where a resolution of singularities is readily available, e.g. the Newton non-degenerate singularities, the resulting mixed Hodge structure on the vanishing cohomology has very computable discrete invariants (Hodge numbers, spectrum) which are closely related to interesting topological invariants like monodromy and intersection form. In the isolated singularity case, a description of the Hodge structure without resolution of singularities was first given by A. Varchenko [29] who used asymptotics of integrals over vanishing cycles. This work was the starting point of a description in the language of $\mathscr{D}$-modules, inspired by Pham's Gauss-Manin system [12] and leading to M. Saito's theory of mixed Hodge modules [13, 14, 15]. This formalism enables one to generalize many results to the case of non-isolated singularities.

2. Vanishing Cycles

Let X be a complex manifold of dimension $n+1$ and let $f : X \to \mathbf{C}$ be a holomorphic function. Let $X_t = f^{-1}(t)$ for $t \in \mathbf{C}$ and let $i : X_0 \to X$ be the inclusion mapping. Define

$$X_\infty = \{(x, u) \in X \times \mathbf{C} \mid f(x) = \exp(2\pi i u)\}$$

and let $k : X_\infty \to X$ be given by $k(x, u) = x$. Choose an injective resolution $I^\bullet$ of $\mathbf{C}_{X_\infty}$ and define the *sheaf of nearby cycles* of f as

Proceedings of the International Congress
of Mathematicians, Kyoto, Japan, 1990

$$\psi_f(\mathbf{C}_X) = i^* k_* I^\bullet .$$

Let $x \in X_0$ and choose $\varepsilon, \eta > 0$ with $\eta \ll \varepsilon \ll 1$. By [10], the restriction of f to $\{z \in X \mid |z - x| < \varepsilon, 0 < |f(z)| < \eta\}$ is a C^∞ fibre bundle (the *Milnor fibration*). Let $X_{f,x}$ denote a Milnor fibre, i.e. a typical fibre of the Milnor fibration. Then

$$H^k(\psi_f(\mathbf{C}_X)_x) \simeq H^k(X_{f,x}, \mathbf{C}) .$$

There is a natural morphism of complexes $\mathbf{C}_{X_0} \to \psi_f(\mathbf{C}_X)$ and one obtains the *sheaf of vanishing cycles* $\phi_f(\mathbf{C}_X)$ as the cone over this morphism. This gives a distinguished triangle

$$\mathbf{C}_{X_0} \longrightarrow \psi_f(\mathbf{C}_X) \xrightarrow{can} \phi_f(\mathbf{C}_X) \xrightarrow{+1}$$

and we have

$$H^k(\phi_f(\mathbf{C}_X)_x) \simeq \tilde{H}^k(X_{f,x}, \mathbf{C}) .$$

So $\phi_f(\mathbf{C}_X)$ measures the difference in the local topology of the zero fibre and the nearby fibres of f. In general, we have $\tilde{H}^k(X_{f,x}, \mathbf{C}) = 0$ for $k > n$ and in the isolated singularity case, $\tilde{H}^k(X_{f,x}, \mathbf{C}) = 0$ for $k \neq n$.

Since [10], the study of the topology of the Milnor fibre of a holomorphic function germ has been an important issue in singularity theory. A first topic is the *monodromy*. Define $h : X_\infty \to X_\infty$ by $h(x, u) = (x, u + 1)$. Then we have an induced action $T = (h^*)^{-1}$ on $\psi_f(\mathbf{C}_X)$ and $\phi_f(\mathbf{C}_X)$ and their cohomology sheaves. The latter action is quasi-unipotent by the monodromy theorem [1, 7] and the Jordan blocks of T on $H^k(\phi_f(\mathbf{C}_X)_x)$ are of size at most $k + 1$.

3. Hodge Structure via Resolution

Let f be a holomorphic function on the complex manifold X and let $x \in f^{-1}(0)$. Then there exists a projective bimeromorphic mapping $\pi : \tilde{X} \to X$ such that

- $\tilde{X}$ is smooth;
- $E = (f \circ \pi)^{-1}(0)$ is a divisor with normal crossings on $\tilde{X}$;
- $E_x = \pi^{-1}(x)$ is a union of irreducible components of E.

Let e be a common multiple of all multiplicities of components of E. We map $\mathbf{C}$ to itself by $t \mapsto t^e$ and let $\tilde{Y}$ be the normalization of $\{(z, t) \in \tilde{X} \times \mathbf{C} \mid f(\pi(z)) = t^e\}$. The natural projections from $\tilde{Y}$ to $\tilde{X}$ and $\mathbf{C}$ are denoted by $\tilde{f}$ and ϱ respectively. By the Semi-stable Reduction Theorem [5], π and e can be chosen in such a way that $\tilde{Y}$ is smooth. Then $D = \tilde{F}^{-1}(0)$ is a *reduced* divisor with normal crossings and ϱ is a cyclic covering of degree e branched only along E. We put $D_x = \varrho^{-1}(E_x)$.

By [24] we have isomorphisms

$$\begin{aligned} H^k(\psi_f(\mathbf{C}_X)_x) &\simeq \mathbf{H}^k(D_x, \Omega^\bullet_{\tilde{Y}/\mathbf{C}}(\log D) \otimes \mathcal{O}_{D_x}) \\ &\simeq \mathbf{H}^k(E_x, \varrho_*(\Omega^\bullet_{\tilde{Y}/\mathbf{C}}(\log D) \otimes \mathcal{O}_{D_x})) \end{aligned}$$

where $\mathbf{H}$ denotes hypercohomology.

In [25], a quasi-isomorphism between $\Omega^\bullet_{\tilde{Y}/\mathbf{C}}(\log D) \otimes \mathcal{O}_{D_x}$ and a certain complex $K^\bullet$ has been constructed. The complex $K^\bullet$ carries an increasing *weight filtration*

$L.$ and a decreasing *Hodge filtration* $F^\bullet$ which make it into a *cohomological mixed Hodge complex* in Deligne's sense (after a suitable rational structure has been added). In this way, $H^k(\psi_f(\mathbf{C}_X)_x)$ carries a mixed Hodge structure for each k.

The monodromy action T on $H^k(\psi_f(\mathbf{C}_X)_x)$ can be described in the following way. The semisimple part T_s is induced by the automorphism $(x,t) \mapsto (x, t\exp(2\pi i/e))$ of $\tilde{Y}$. Its nilpotent logarithm $N = -\log(T_u)/2\pi i$ is induced by an endomorphism v of the complex $K^\bullet$ which maps F^p to F^{p-1} and L_w to L_{w-2}. Hence N is a morphism of mixed Hodge structures of type $(-1,-1)$.

In the isolated singularity case, there is a nice description of the space $F^nH^n(X_{f,x}, \mathbf{C})$ due to Varchenko [29], in terms of the *geometrical weight* of holomorphic $(n+1)$-forms on X. As a consequence of this description one concludes that, if $Gr^{2n}_W H^n(X_{f,x}) \neq 0$, then T_s has an eigenvalue 1 on it. This implies the following surprising supplement to the monodromy theorem in the isolated singularity case [2]: if T has a Jordan block of size $n+1$, then for the eigenvalue 1 it has a Jordan block of size n.

4. Hodge Structure via $\mathcal{D}$-Modules

For each complex manifold X we have the coherent sheaf of rings $\mathcal{D}_X$ of germs of holomorphic differential operators on X. It is equipped with the increasing filtration $F.$ by the order of differential operators, such that $Gr^F_\cdot \mathcal{D}_X \simeq \mathrm{Sym}(\Theta_X)$ where Θ_X is the tangent sheaf of X.

Each $\mathcal{D}_X$-module M has its de Rham complex

$$DR(M) = \Omega^\bullet_X \otimes_{\mathcal{O}_X} M[\dim X].$$

This de Rham functor defines an equivalence of categories

$$DR : D^b_{rh}(\mathcal{D}_X) \xrightarrow{\sim} D^b_c(\mathbf{C}_X)$$

between the derived categories of bounded complexes of $\mathcal{D}_X$-modules with regular holonomic cohomology sheaves and of bounded complexes of $\mathbf{C}_X$-modules with constructible cohomology sheaves. The single regular holonomic $\mathcal{D}_X$-modules correspond in this way to the socalled *perverse* $\mathbf{C}_X$-complexes. These are characterized by the fact that for $K \in \mathrm{Perv}(\mathbf{C}_X)$

$$\dim\mathrm{supp}\, \mathcal{H}^i(K),\ \dim\mathrm{supp}\, \mathcal{H}^i(\mathbf{D}K) \leq -i$$

where $\mathbf{D}$ denotes Verdier duality. See [9, 3].

It appears that for X a complex manifold of dimension $n+1$ and $f : X \to \mathbf{C}$ a holomorphic function, $\psi_f(\mathbf{C}_X)[n]$ and $\phi_f(\mathbf{C}_X)[n]$ are perverse [8]. Hence they correspond via the de Rham functor to regular holonomic $\mathcal{D}_X$-modules $\psi_f(\mathcal{O}_X)$ and $\phi_f(\mathcal{O}_X)$. These are defined as follows. Consider the $\mathcal{D}_X[t, \partial_t]$-modules $\mathcal{O}_X[t, \partial_t]$ and $\mathcal{O}_X[t, \partial_t, \partial_t^{-1}]$ where the action is defined by

$$g(h\partial_t^k) = gh\partial_t^k$$

$$v(h\partial_t^k) = v(h)\partial_t^k - v(f)h\partial_t^{k+1}$$

$$t(h\partial_t^k) = fh\partial_t^k - kh\partial_t^{k-1}$$

$$\partial_t(h\partial_t^k) = h\partial_t^{k+1}$$

for g, h sections of $\mathcal{O}_X$ and v a section of Θ_X.

By [8], on $\mathcal{O}_X[\partial_t]$ and $\mathcal{O}_X[\partial_t, \partial_t^{-1}]$ there exist unique decreasing filtrations $V^\bullet$, discretely indexed by $\mathbf{Q}$, such that

1. $tV^\alpha \subset V^{\alpha+1}$ with equality for $\alpha > 0$
2. $\partial_t V^\alpha \subset V^{\alpha-1}$
3. each V^α is a finitely generated $\mathcal{D}_X$-module
4. $\partial_t t - \alpha$ is nilpotent on Gr_V^α.

One has

$$\psi_f(\mathcal{O}_X) = \bigoplus_{0<\alpha\leq 1} Gr_V^\alpha \mathcal{O}_X[\partial_t] = \bigoplus_{0<\alpha\leq 1} \psi_f^\alpha$$

$$\phi_f(\mathcal{O}_X) = \bigoplus_{0<\alpha\leq 1} Gr_V^\alpha \mathcal{O}_X[\partial_t, \partial_t^{-1}] = \bigoplus_{0<\alpha\leq 1} \phi_f^\alpha .$$

Observe that $can : \psi_f^\alpha \tilde{\to} \phi_f^\alpha$ for $\alpha \neq 1$.

The modules $\mathcal{O}_X[\partial_t]$ and $\mathcal{O}_X[\partial_t, \partial_t^{-1}]$ carry the increasing filtration $F.$ given by

$$F_p\mathcal{O}_X[\partial_t] = \bigoplus_{0\leq i\leq p-1} \mathcal{O}_X\partial_t^i$$

$$F_p\mathcal{O}_X[\partial_t, \partial_t^{-1}] = \bigoplus_{i\leq p-1} \mathcal{O}_X\partial_t^i .$$

These induce filtrations $F.$ on $\psi_f(\mathcal{O}_X)$, $\phi_f(\mathcal{O}_X)$ and their de Rham complexes. Moreover we have

$$T_s = \exp(-2\pi i\alpha),\ N = -\partial_t t + \alpha \text{ on } Gr_V^\alpha .$$

The nilpotent endomorphism N of $\psi_f(\mathcal{O}_X)$ determines uniquely an increasing weight filtration L on it by $\mathcal{D}_X$-submodules in such a way that $N(L_j) \subset L_{j-2}$ and $N^k : Gr_L^{n+k} \tilde{\to} Gr_L^{n-k}$. The data of the bifiltered $\mathcal{D}_X$-module

$$(\psi_f(\mathcal{O}_X), F, L)$$

together with the rational structure

$$DR\,\psi_f(\mathcal{O}_X) \simeq \psi_f(\mathbf{C}_X)[n] \simeq \psi_f(\mathbf{Q}_X)[n] \otimes_{\mathbf{Q}_X} \mathbf{C}_X$$

constitute the ingredients of a *mixed Hodge module* on X [15]. We similarly have a mixed Hodge module structure on ϕ_f, with the same L on $\phi_f^\alpha \simeq \psi_f^\alpha$ for $\alpha \neq 1$ but with $N^k : Gr_L^{n+k+1} \tilde{\to} Gr_L^{n-k+1}$ on ϕ_f^1. The natural map $can : \psi_f \to \phi_f$ is a morphism of mixed Hodge modules.

If $f : (X, x) \to (\mathbf{C}, 0)$ has an isolated singularity, ϕ_f has support at x. Because mixed Hodge modules with support at a point correspond to mixed Hodge structures, we have a mixed Hodge structure on the vanishing cohomology again. This case has been studied in detail in [20]. In general, ϕ_f has support along the critical locus Σ of f, and one has to take the restriction to $\{x\}$ in the sense of mixed Hodge modules. This is a very delicate affair, even if Σ is a curve. One has to take iterated restrictions to divisors, and each of these steps involves taking the cone over the morphism *can* between nearby and vanishing cycles (for a defining function for the divisor). The resulting object $i_x^*\phi_f$ is no longer a single mixed Hodge structure, but a complex of mixed Hodge structures, satisfying $H^k(i_x^*\phi_f) \simeq H^{k+n}(X_{f,x}) = 0$ for $k \notin [-\dim \Sigma, 0]$.

If Σ has dimension one, there is a description of ϕ_f in terms of certain topological mappings and its restriction H to $\Sigma \setminus \{x\}$. Choose a holomorphic function germ g on (X,x) such that $\Sigma \cap g^{-1}(0) = \{x\}$. Observe that H is a variation of mixed Hodge structure (up to a shift) and that we have mixed Hodge structures $\psi_g(H)$ and $\phi_g(\phi_f)$. These have the following topological interpretation. At each point of $\Sigma \cap g^{-1}(\delta)$ for some small non-zero δ, take a slice transverse to Σ and let $F^\flat$ denote the union of the Milnor fibres of the restriction of f to all of these transverse slices. There is a natural way to embed $F^\flat$ into the Milnor fibre F of f. We have (taking $n \geq 2$ for convenience)

$$\psi_g(H) \simeq H^{n-1}(F^\flat)$$

$$\phi_g(\phi_f) \simeq H^n(F, F^\flat).$$

The structure of ϕ_f is now determined by H and a pair of mappings

$$\tilde{can} : \psi_g(H) \longrightarrow \phi_g(\phi_f)$$

$$\tilde{var} : \phi_g(\phi_f) \longrightarrow \psi_g(H).$$

Here $\tilde{can}$ is interpreted as the natural map $u : H^{n-1}(F^\flat) \to H^n(F, F^\flat)$ whereas $\tilde{var}$ is related to the map v on homology defined in the following way. Represent $\gamma \in H_{n-1}(F^\flat)$ by a cycle on $F^\flat$ and sweep it around x once following a characteristic homeomorphism of the monodromy of g the local system H. This gives a chain on F with boundary on $F^\flat$, representing the element $v(\gamma)$ of $H_n(F, F^\flat)$. One has $\tilde{var} = \sum_{i \geq 0} (-1)^i v^t (uv^t)^i / (i+1)$. One has the exact sequence of mixed Hodge structures

$$0 \longrightarrow H^{n-1}(F) \longrightarrow H^{n-1}(F, F^\flat) \xrightarrow{\tilde{can}} H^n(F) \longrightarrow 0.$$

Let $f : (X,x) \to (\mathbf{C}, 0)$ and $g : (Y,y) \to (\mathbf{C}, 0)$ be holomorphic function germs. Put

$$Z = X \times Y;\ z = (x,y);\ h = f + g : (Z,z) \longrightarrow (\mathbf{C}, 0).$$

(We will also denote h as $f \oplus g$). A result of Thom and Sebastiani [22] expresses the vanishing cohomology and monodromy of h in terms of the monodromies for f and g. Using the $\mathcal{D}$-module description above, one can show that

$$\phi_h(\mathcal{O}_Z) \simeq p_X^* \phi_f(\mathcal{O}_X) \otimes p_Y^* \phi_g(\mathcal{O}_Y)$$

and relate the Hodge filtrations at both sides. This is a weakened version of a general Thom-Sebastiani formula for mixed Hodge modules due to M. Saito and P. Deligne (unpublished).

5. Discrete Invariants

The main discrete invariants of a mixed Hodge structure (V, F, W) are its *Hodge numbers*

$$h^{pq}(V) = \dim Gr_F^p Gr_{p+q}^W V_{\mathbf{C}}\,.$$

If moreover an automorphism γ of finite order of V is given (like T_s in the case of the vanishing cohomology) it is convenient to split into eigenspaces of γ and to put

$$h_\lambda^{pq}(V,\gamma) = \dim\ker(\gamma - \lambda I \mid Gr_F^p Gr_{p+q}^W V_{\mathbf{C}})\,.$$

These invariants involve the weight filtration, which is in general quite hard to handle. Forgetting about the weight filtration leads one to the weaker invariant of the *spectrum*. For $f : (X,x) \to (\mathbf{C}, 0)$ a hypersurface singularity one defines

$$h_\lambda^p(f,x) = \sum_{j,q}(-1)^j h_\lambda^{pq}(H^j i_x^* \phi_f, T_s)\,.$$

For each $\alpha \in \mathbf{Q}$ let $n_\alpha = h_\lambda^p(f,x)$ where $\lambda = \exp(-2\pi i\alpha)$ and $p \in \mathbf{Z}$ is determined by $n - p < \alpha \le n - p + 1$. Then

$$\mathrm{Sp}(f,x) := \sum_\alpha n_\alpha t^\alpha\,.$$

If $e \in \mathbf{N}$ is such that $T_s^e = I$, then $\mathrm{Sp}(f,x) \in \mathbf{Z}[t^{1/e}]$.

There are formulas for spectra of isolated quasi-homogeneous singularities in terms of the weights, and for Newton-nondegenerate isolated singularities in terms of the Newton diagram [16, 6]. A formula describing the behaviour of the spectrum in series of singularities was proven by M. Saito [17].

For spectra the Thom-Sebastiani formula takes the elegant form

$$\mathrm{Sp}(f \oplus g, (x,y)) = \mathrm{Sp}(f,x)\,\mathrm{Sp}(g,y)$$

(the right hand side is the product in the ring of fractional polynomials). This was conjectured in [24, 27], proved in the isolated singularity case in [29, 30] and in [18] in general.

For isolated singularities, the spectrum is semicontinuous under deformation of the singularity in a certain sense [30, 26]. The proof is based on a semicontinuity result for Hodge numbers and the Thom-Sebastiani formula. It appears that a similar semicontinuity result also holds for certain deformations of non-isolated singularities [28]. This gives e.g. necessary conditions for the existence of (globalizable) deformations between arbitrary normal surface singularities. These correspond to admissible (in the sense of [4]) deformations of weakly normal surface singularities, of which one considers the spectra. The proof uses the full strength of the formalism of mixed Hodge modules.

References

1. E. Brieskorn: Die Monodromie der isolierten Singularitäten von Hyperflächen. Man. Math. **2** (1970) 103–161
2. M.G.M. van Doorn, J.H.M. Steenbrink: A supplement to the monodromy theorem. Abh. Math. Sem. Univ. Hamburg **59** (1989) 225–233
3. M. Goresky, R. MacPherson: Intersection homology II. Invent. math. **71** (1983) 77–129
4. T. de Jong, D. van Straten: Deformations of non-isolated singularities. Chapter 3 of de Jong's thesis, University of Nijmegen 1988
5. G. Kempf, F. Knudsen, D. Mumford, B. Saint-Donat: Toroidal embeddings I. Lecture Notes in Mathematics, vol. 339. Springer, Berlin Heidelberg New York 1973
6. A.G. Khovanskii, A.N. Varchenko: Asymptotics of integrals over vanishing cycles and the Newton polyhedron. Soviet Math. Dokl. **32** (1985) 122–127
7. A. Landman: On the Picard-Lefschetz formula for algebraic manifolds acquiring general singularities. Thesis Berkeley 1967
8. B. Malgrange: Polynômes de Bernstein-Sato et cohomologie évanescente. Astérisque **101–102** (1983) 243–267
9. Z. Mebkhout: Une équivalence de catégories. Compos. Math. **51** (1984) 51–62; Une autre équivalence de catégories. Ibid, pp. 63–88
10. J. Milnor: Singular points on complex hypersurfaces. Ann. Math. Stud. 61. Princeton 1968
11. V. Navarro Aznar: Sur la théorie de Hodge-Deligne. Invent. math. **90** (1987) 11–76
12. F. Pham: Structures de Hodge mixtes associées à un germe de fonction à point critique isolé. Astérisque **101–102** (1983) 268–285
13. M. Saito: Gauss-Manin system and mixed Hodge structure. Proc. Japan Acad. **58**, Ser. A (1982) 29–32
14. M. Saito: Modules de Hodge polarisables. Publ. RIMS Kyoto Univ. **24** (1988) 849–995
15. M. Saito: Mixed Hodge modules. Publ. RIMS Kyoto Univ. 1990
16. M. Saito: Exponents and Newton polyhedra of isolated hypersurface singularities. Math. Ann. **281** (1988) 411–417
17. M. Saito: Preuve d'une conjecture de Steenbrink. Manuscript in preparation.
18. M. Saito, J.H.M. Steenbrink: Thom-Sebastiani formula for spectra of hypersurface singularities. Manuscript in preparation.
19. J. Scherk: On the monodromy theorem for isolated hypersurface singularities. Invent. math. **58** (1980) 289–301
20. J. Scherk, J.H.M. Steenbrink: On the mixed Hodge structure on the cohomology of the Milnor fibre. Math. Ann. **271** (1985) 641–665
21. W. Schmid: Variation of Hodge structure: the singularities of the period mapping. Invent. math. **22** (1973) 211–320
22. M. Sebastiani, R. Thom: Un résultat sur la monodromie. Invent. math. **13** (1971) 90–96
23. J.H.M. Steenbrink: Limits of Hodge structures. Invent. math. **31** (1976) 229–257
24. J.H.M. Steenbrink: Mixed Hodge structure on the vanishing cohomology. In: P. Holm (ed.): Real and complex Singularities. Oslo 1976. Sijthoff-Noordhoff, Alphen a/d Rijn 1977, pp. 525–563
25. J.H.M. Steenbrink: Mixed Hodge structures associated with isolated singularities. Proc. Symp. Pure Math. 40, Part **2** (1983) 513–536
26. J.H.M. Steenbrink: Semicontinuity of the singularity spectrum. Invent. math. **79** (1985) 557–565
27. J.H.M. Steenbrink: The spectrum of hypersurface singularities. Astérisque **179–180** (1989) 163–184

28. J.H.M. Steenbrink: Semicontinuity of the spectrum under certain deformations of non-isolated singularities. Manuscript in preparation
29. A.N. Varchenko: Asymptotic Hodge structure on the vanishing cohomology. Izv. Akad. Nauk SSSR, Ser. Mat. **45** (1981) 540–591 (in Russian) [English transl.: Math. USSR Izv. **18** (1982) 469–512]
30. A.N. Varchenko: On semicontinuity of the spectrum and an upper bound for the number of singular plints of projective hypersurfaces. Dokl. Akad. Nauk. **270** (1983) 1294–1297 (in Russian) [English transl.: Sov. Math. Dokl. **27** (1983) 735–739]

Trace Formulae in Spectral Geometry

Toshikazu Sunada and Manabu Nishio

Department of Mathematics, Faculty of Science, Nagoya University
Nagoya 464-01, Japan

A geometric analogue of the celebrated "Riemann hypothesis" is stated as

$$\lambda_1(M) \geq \lambda_0(\tilde{M}), \tag{1}$$

where $\lambda_1(M)$ is the first *positive* eigenvalue of the Laplacian Δ_M on a closed Riemannian manifold M, and $\lambda_0(\tilde{M})$ is the *bottom* of the spectrum of the Laplacian on the *universal* covering manifold $\tilde{M}$ (see [21]). For instance, if M is a Riemann surface with constant negative curvature, the inequality (1) turns out to be equivalent to the Riemann hypothesis for the Selberg zeta function defined by an Euler product over prime closed geodesics in M (A. Selberg [17]).

The view that we regard (1) as an analogue of the Riemann hypothesis is supported also by the fact that, under finite-fold covering maps, the eigenvalues behave like the non-trivial zeros of the zeta functions of number fields [19]. It is worthwhile, on the other hand, to note that prime closed geodesics play the same role as prime ideals, though, in general, there does not exist a well-suited zeta function connecting the eigenvalues directly with closed geodesics. We may establish, however, a weak relationship between those objects, which is embodied as an analogue of the Weil's explicit formula (cf. S. Lang [13]). This illustrates more intimate kinship between spectral geometry and number theory. To explain it, let $\varrho : \pi_1(M) \to U(N)$ be a unitary representation, and let $\lambda_0(\varrho) \leq \lambda_1(\varrho) \leq \cdots$ be the eigenvalues of the Laplacian Δ_ϱ acting on sections of the *flat* vector bundle associated with ϱ (which may be regarded as an analogue of the non-trivial zeros of an L-function). We define the (even) distribution $\Theta(\varrho) \in \mathscr{D}'(\mathbb{R})$ by

$$\langle \Theta(\varrho), f \rangle = \sum_{k=0}^{\infty} \hat{f}(\sqrt{\lambda_k(\varrho)}), \qquad f \in \mathscr{D}(\mathbb{R}),$$

where $\hat{f}(s) = \int_{-\infty}^{\infty} f(t) \cos(st)\, dt$. We also associate an (even) distribution Θ_α with each *free* homotopy class $\alpha \in [S^1, M]$ of closed curves by setting

$$\langle \Theta_\alpha, f \rangle = \int_{\Gamma_\sigma \backslash X} U_f(\sigma x, x)\, dx, \tag{2}$$

where $X = \tilde{M}$, $\Gamma = \pi_1(M)$, and we identify $[S^1, M]$ with the set $[\Gamma] = \{[\sigma]\}$ of

Proceedings of the International Congress
of Mathematicians, Kyoto, Japan, 1990

conjugacy classes in Γ, and write Γ_σ for the centralizer of σ. The function $U_f(x, y) \in C^\infty(X \times X)$ is the kernel function of the smooth operator

$$\hat{f}(\sqrt{\Delta_X}) = \int f(t) \cos(t\sqrt{\Delta_X})\, dt.$$

We then obtain the following theorem which gives a refinement of the result by J. Chazarain [5] (see also [7, 9] and [25]).

Theorem I. (a) (trace formula) $\Theta(\varrho) = \sum_{[\sigma]\in[\Gamma]} \operatorname{tr} \varrho(\sigma)\Theta_{[\sigma]}$.

(b) *The support of the distribution* $\Theta_{[\sigma]}$ *is contained in the set* $\{t \in \mathbb{R}; |t| \geq l_{[\sigma]}\}$, *where* $l_{[\sigma]}$ *is the length of the shortest closed geodesics in* M *whose homotopy class is* $[\sigma]$, *so that, when the right hand side of* (a) *is applied to a test function in* $\mathscr{D}(\mathbb{R})$, *the sum over* $[\Gamma]$ *reduces to a finite one.*

(c) *The singular support of* $\Theta_{[\sigma]}$ *is contained in the set* $\{\pm$*length of closed geodesics in* M *with the homotopy class* $[\sigma]\}$.

(d) *For the unit element* 1 *in* Γ, *the distribution* $\Theta_{[1]}$ *is temperate.*

(e) *The distribution* $\Theta_{[\sigma]}$ $(\sigma \neq 1)$ *is extended to a functional on the space of test functions* f *satisfying*

$$|f^{(k)}(t)| \leq C \exp(-(h + \delta)|t|), \qquad 0 \leq k < 1 + n/2$$

for some $\delta > 0$ *and* $C > 0$. *Here* h *is the exponential growth rate of volume of geodesic balls in* X. *Moreover, the sum* $\sum \operatorname{tr} \varrho(\sigma) \langle \Theta_{[\sigma]}, f\rangle$ *converges absolutely and equals* $\langle \Theta(\varrho), f\rangle$. *If* Γ *is of polynomial growth, then* $\Theta_{[\sigma]}$ *is also a temperate distribution.*

(f) *For* $f_\tau(t) = (4\pi\tau)^{-1/2} \exp(-t^2/4\tau)$, *we have*

$$\langle \Theta(\mathbb{1}), f_\tau\rangle = \mu_\tau(\Omega),$$
$$\langle \Theta_{[\sigma]}, f_\tau\rangle = \mu_\tau(\Omega_{[\sigma]}),$$

where $\mathbb{1}$ *denotes the trivial representation,* $\Omega_{[\sigma]}$ *is the connected component of the free loop space* $\Omega = C^0(S^1, M)$ *corresponding to the homotopy class* $[\sigma]$, *and* μ_τ *is the Wiener measure on* Ω.

We should point out that, by a slight modification, all the statements above are generalized to the case of normal covering manifolds. We shall give an outline of the proof.

Let X be a normal covering space of M with covering transformation group Γ. From the *finite propagation property* of the wave equation, it follows that, if $d(x, y) > R$ and $\operatorname{supp}(f) \subset \{t \in \mathbb{R}; |t| < R)$, then $U_f(x, y) = 0$. For each $R > 0$, if we put

$$H_R = \{x \in \Gamma_\sigma \backslash X; d(x, \sigma x) \leq R\},$$

then H_R is compact, so that $U_f(\sigma x, x)$ has compact support in $\Gamma_\sigma\backslash X$. Hence the distribution $\Theta_{[\sigma]}$ is well-defined (see (2)) and satisfies (b). Since $\lim_{R\to\infty} \sup R^{-1} \times$

$\log(\mathrm{vol}(H_R)) \leq h$, applying the estimate established by J. Cheeger, M. Gromov and M. Taylor [6], we obtain (e).

Let $\mathbb{U}_{\varrho,f}(p, q) \in C^\infty(M \times M)$ be the kernel function of the operator

$$\hat{f}(\sqrt{\Delta_\varrho}) = \int f(t) \cos(t\sqrt{\Delta_\varrho})\, dt.$$

If we denote by ω the coveing map of X onto M, then

$$\mathbb{U}_{\varrho,f}(\omega(x), \omega(y)) = \sum_{\sigma \in \Gamma} \operatorname{tr} \varrho(\sigma) U_f(\sigma x, y).$$

Moreover, we find

$$\langle \Theta(\varrho), f\rangle = \operatorname{tr} \hat{f}(\sqrt{\Delta_\varrho}) = \int_M \mathbb{U}_{\varrho,f}(p, p)\, dp.$$

The proof of (a) is now carried out in the same way as that of the Selberg trace formula (see [17, 18]). Actually, (a) is specialized down to the Selberg's formula when M is a constant negatively curved surface.

The assertion (c) is shown in a similar manner as [5], by analyzing the wave front set of $\Theta_{[\sigma]}$. In fact, one has the following expression for $\Theta_{[\sigma]}$ in terms of the distribution kernel $U_X(t, x, y)$ of the operator $\cos(t\sqrt{\Delta_X})$:

$$\Theta_{[\sigma]} = (\pi_\sigma)_*(p_\sigma{}^*)^{-1}(D_\sigma{}^*)U_X,$$

where $D_\sigma : \mathbb{R} \times X \to \mathbb{R} \times X \times X$ is the map defined by $D_\sigma(t, x) = (t, \sigma x, x)$, $p_\sigma : \mathbb{R} \times X \to \mathbb{R} \times (\Gamma_\sigma \backslash X)$ is the map induced by the covering projection $X \to \Gamma_\sigma \backslash X$, and $\pi_\sigma : \mathbb{R} \times (\Gamma_\sigma \backslash X) \to \mathbb{R}$ is the projection onto the first factor. We should note that $(D_\sigma{}^*)U_X$ is a Γ_σ-invariant distribution on $\mathbb{R} \times X$, and that the pull-back $p_\sigma{}^*$ yields an isomorphism of $\mathscr{D}'(\mathbb{R} \times (\Gamma_\sigma \backslash X))$ onto the space of Γ_σ-invariant distributions on $\mathbb{R} \times X$. We then make use of the fact that the map π_σ restricted to the support of $(p_\sigma{}^*)^{-1}(D_\sigma{}^*)U_X$ is *proper* (recall that H_R is compact), so that the push-forward $(\pi_\sigma)_*$ can be applied to this distribution. The rest of the proof is carried out by using the wave-front calculus.

The assertion (f) is a consequence of the observation that, for $f = f_\tau$, U_f is the fundamental solution $k(\tau, x, y)$ of the heat equation on X ([18]).

We now proceed to the proof of (d). We first recall that $\Theta(\varrho)$ is a temperate distribution. Indeed, this is deduced from the fact that the counting function $\varphi_\varrho(\lambda) = \#\{\lambda_k(\varrho) \leq \lambda\}$ satisfies $\varphi_p(\lambda) \sim C_n \,\mathrm{vol}(M)\lambda^{n/2}$ as $\lambda \uparrow \infty$, and the equality $\langle \Theta(\varrho), f\rangle = \int \hat{f}(\sqrt{\lambda})\, d\varphi_\varrho(\lambda)$. We wish to establish a similar expression for $\Theta_{[1]}$, say

$$\langle \Theta_{[1]}, f\rangle = \int \hat{f}(\sqrt{\lambda})\, d\varphi(\lambda).$$

To define the increasing function $\varphi(\lambda)$, we first note that $\hat{f}(\sqrt{\Delta_X})$ lies in the von Neumann algebra $\mathrm{End}_\Gamma(L^2(X))$ of Γ-equivariant endomorphisms of $L^2(X)$ and $\langle \Theta_{[1]}, f\rangle = \operatorname{tr}_\Gamma \hat{f}(\sqrt{\Delta_X})$, where tr_Γ is a *trace* (called Γ-trace) in the sense of von Neumann algebras,and is equal, in this case, to

$$\int_{\mathscr{D}} U_f(x, x)\, dx,$$

$\mathscr{D}$ being a *fundamental domain* for the Γ-action (see M.F. Atiyah [1]). Let $\Delta_X = \int \lambda \, dE_\lambda$ be the *spectral resolution* of Δ_X, so that

$$\hat{f}(\sqrt{\Delta_X}) = \int \hat{f}(\sqrt{\lambda}) \, dE_\lambda.$$

We observe that E_λ is of Γ-trace class, and

$$\mathrm{tr}_\Gamma \, \hat{f}(\sqrt{\Delta_X}) = \int \hat{f}(\sqrt{\lambda}) \, d(\mathrm{tr}_\Gamma \, E_\lambda).$$

Hence we have only to set $\varphi(\lambda) = \mathrm{tr}_\Gamma \, E_\lambda$.

Consider the special case $f = f_\tau$. Then $\hat{f}_\tau(s) = e^{-\tau s^2}$, and

$$\int_{\mathscr{D}} k(\tau, x, x) \, dx = \mathrm{tr}_\Gamma \, \hat{f}_\tau(\sqrt{\Delta_X}) = \int e^{-\lambda\tau} \, d\varphi(\lambda).$$

The function $k(\tau, x, x)$ has the same asymptotic behavior as that of the heat kernel on M, and hence $k(\tau, x, x) \sim (4\pi\tau)^{-n/2}$ as $\tau \downarrow 0$, so that, applying the Tauberian theorem to φ, we get

$$\varphi(\lambda) \sim (4\pi)^{n/2} \, \mathrm{vol}(M) \lambda^{n/2} / \Gamma\left(\frac{n}{2} + 1\right). \tag{3}$$

This completes the proof of (d). □

The function $\varphi(\lambda)$ includes much information of the spectrum of Δ_X. For instance, $\varphi(\lambda) = 0$ if $\lambda < \lambda_0(X)$, and $\varphi(\lambda) > 0$ if $\lambda > \lambda_0(X)$. The spectrum $\sigma(\Delta_X)$ coincides with the set of increasing points of φ. Let $\varphi(\lambda) = \varphi_{\mathrm{ac}}(\lambda) + \varphi_{\mathrm{s}}(\lambda) + \varphi_{\mathrm{p}}(\lambda)$ be the (unique) decomposition into an *absolutely continuous* function φ_{ac}, a *singular continuous* function φ_{s}, and a *step* function φ_{p}. If

$$(\Delta_X, L^2(X)) = (\Delta_X, H_{\mathrm{ac}}) \oplus (\Delta_X, H_{\mathrm{s}}) \oplus (\Delta_X, H_{\mathrm{p}})$$

is the decomposition into the absolutely continuous, singular, and pure point parts of Δ_X, then $\varphi_{\mathrm{ac}}(\lambda) = \mathrm{tr}_\Gamma \, E_\lambda P_{\mathrm{ac}}$, $\varphi_{\mathrm{s}}(\lambda) = \mathrm{tr}_\Gamma \, E_\lambda P_{\mathrm{s}}$, and $\varphi_{\mathrm{p}}(\lambda) = \mathrm{tr}_\Gamma \, E_\lambda P_{\mathrm{p}}$, where $P_{\mathrm{ac}}, P_{\mathrm{s}}$, and P_{p} are the orthogonal projections onto H_{ac}, H_{s}, and H_{p} respectively. Therefore the sets of increasing points of the functions φ_{ac} and φ_s are just the absolutely continuous spectrum and the singular spectrum of Δ_X respectively. The point spectrum coincides with the set of jumping points of φ_{p}.

When $X = \tilde{M}$ and M has no *null*-homotopic closed geodesics except for point geodesics, the assertion (c) says that the singular support of $\Theta_{[1]}$ is $\{0\}$. It is an interesting problem to examine the decay of the function $\Theta_{[1]}$ at infinity. For instance, if $\Theta_{[1]}$ is in L^2 at infinity, it is shown that Δ_X has purely absolutely continuous spectrum. In fact, from the above argument, we have

$$\langle \Theta_{[1]}, f \rangle = \int \hat{f}(\lambda) \, d\psi(\lambda) = \langle \dot{\psi}, \hat{f} \rangle \qquad (\psi(\lambda) = \varphi(\lambda^2)),$$

so that $\hat{\dot{\psi}} = \Theta_{[1]} = T_1 + T_2$, where T_1 is a distribution with compact support and $T_2 \in L^2(\mathbb{R})$. In view of the Fourier inversion formula and the Parseval equality, we may write $\dot{\psi} = S_1 + S_2$, where S_1 is analytic and $S_2 \in L^2(\mathbb{R})$. Hence $\dot{\psi}$ is in L^2_{loc}, and φ is absolutely continuous. We conjecture that, if M is nonpositively curved, the function $\Theta_{[1]}$ decays with the order of $|t|^{-\alpha}$ at $t = \pm\infty$ with some $\alpha > 0$.

Example 1. When M is a flat torus and $X = \mathbb{R}^n$,

$$\varphi(\lambda) = (4\pi)^{-n/2} \operatorname{vol}(M)\lambda^{n/2}/\Gamma\left(\frac{n}{2} + 1\right) \qquad (\lambda \geq 0).$$

This follows from the Jacobi's inversion formula for the θ-series (a special case of (a) in Theorem I).

Example 2. Let M be a closed Riemann surface with constant negative curvature (-1), and let X be the universal covering. The Selberg trace formula leads to the identity

$$\operatorname{tr}_\Gamma \exp(-t\Delta_X) = (4\pi)^{-1} \operatorname{vol}(M) \int_0^\infty \exp\left(-t\left(\lambda + \frac{1}{4}\right)\right) \tanh \pi\sqrt{\lambda}\, d\lambda,$$

from which we easily deduce

$$\varphi(\lambda) = \begin{cases} (4\pi)^{-1} \operatorname{vol}(M) \displaystyle\int_0^{\lambda - 1/4} \tanh \pi\sqrt{\lambda}\, d\lambda & \text{for } \lambda \geq \dfrac{1}{4} \\ 0 & \text{for } \lambda < \dfrac{1}{4}. \end{cases}$$

Note that $\varphi(\lambda) \sim \frac{1}{6} \operatorname{vol}(M)(\lambda - \frac{1}{4})^{3/2}$ as $\lambda \downarrow \frac{1}{4}$.

Example 3. Let $X \to M$ be an infinite-fold covering with an *abelian* covering transformation group of rank b. Then $\varphi(\lambda)$ is smooth on an interval $(0, \delta)$ for some $\delta > 0$, and $\varphi(\lambda) \sim C\lambda^{b/2}$ as $\lambda \downarrow 0$. To see this, we make use of (e) in Theorem I:

$$\operatorname{tr} \exp(-\tau\Delta_\chi) = \sum_{[\sigma]} \operatorname{tr} \chi(\sigma)\langle\Theta_{[\sigma]}, f_\tau\rangle,$$

χ being a character of Γ. Integrating the both sides over the character group $\hat{\Gamma}$ with respect to the normalized Haar measure, we have

$$\operatorname{tr}_\Gamma \exp(-t\Delta_X) = \int_{\hat{\Gamma}} \operatorname{tr} \exp(-t\Delta_\chi)\, d\chi,$$

where we have applied the orthogonal relations of characters. Thus we obtain the equation

$$\int_{\hat{\Gamma}} d\chi \int e^{-t\lambda}\, d\varphi_\chi(\lambda) = \int e^{-t\lambda}\, d\varphi(\lambda),$$

which implies $\varphi(\lambda) = \int_{\hat{\Gamma}} \varphi_\chi(\lambda)\, d\chi$. We now recall the fact that $\lambda_0(\mathbb{1}) = 0$ and $\lambda_0(\chi)$ is simple for χ in a small neighborhood of $\mathbb{1}$, from which it follows that when $\lambda \geq 0$ is small enough, $\varphi_\chi(\lambda) = 1$ for χ with $\lambda_0(\chi) \leq \lambda$, and $\varphi_\chi(\lambda) = 0$ otherwise. Hence we have

$$\varphi(\lambda) = \text{vol}\{\chi \in \hat{\Gamma};\ \lambda_0(\chi) \le \lambda\}.$$

Since $\lambda_0(\chi)$ smoothly depends on χ around a neighborhood of $\mathbb{1}$, and the Hessian of $\lambda_0(\chi)$ at $\chi = \mathbb{1}$ is positive definite [11], the assertion follows easily.

For a normal *infinite-fold* covering $X \to M$, the bottom $\lambda_0(X)$ lies in the continuous spectrum of Δ_X [21], that is, $\varphi(\lambda) \to 0$ as $\lambda \downarrow \lambda_0$. In view of the above exmaples, we conjecture that there exist positive constants c and C such that

$$\varphi(\lambda) \sim C(\lambda - \lambda_0)^c \qquad \text{as} \qquad \lambda \downarrow \lambda_0. \tag{4}$$

We now go back to the analogue of "Riemann hypothesis". The inequality (1) does *not* hold in general. Even for constant negatively curved surfaces, we have many counter examples (B. Randol [15], P. Buser [4], and [19]). H. Donnelly [8] established, however, a weak version of (1). To be exact, consider a set $\{M_i\}_{i=1}^{\infty}$ of finite-fold subcoverings of a normal covering $X \xrightarrow{\Gamma} M$, and let Γ_i be the subgroup of finite index in Γ corresponding to the covering map $X \to M_i$. Let us introduce the counting function $\varphi_{M_i}(\lambda) = \max\{k;\ \lambda_k(M_i) \le \lambda\}$, where $\lambda_1(M_i) \le \lambda_2(M_i) \le \cdots$ are the positive eigenvalues of the Laplacian on M_i (so that $\varphi_M = \varphi_{\mathbb{1}} - 1$). The "Riemann hypothesis" (1) for M_i is stated as $\varphi_{M_i}(\lambda) = 0$ for $\lambda < \lambda_0(\tilde{M})$. Suppose that $d(M_i) \to \infty$, where

$$d(M_i) = \inf_{\sigma \in \Gamma_i} \inf_{x \in X} d(x, \sigma x),$$

which, in the case that $X = \tilde{M}$, coincides with the infimum of length of non-null homotopic closed geodesics in M_i. Then, by [8], we have

$$\limsup_{i \to \infty} \text{vol}(M_i)^{-1} \varphi_{M_i}(\lambda) = 0 \qquad \text{if } \lambda < \lambda_0(X),$$

$$\liminf_{i \to \infty} \text{vol}(M_i)^{-1} \varphi_{M_i}(\lambda) > 0 \qquad \text{if } \lambda > \lambda_0(X).$$

In other words, the set of eigenvalues in the interval $[0, \lambda_0(X))$ is "thin" in the totality of eigenvalues on covering manifolds M_i.

We shall give a more precise statement.

Theorem II. *Let $X \xrightarrow{\Gamma} M$ be a normal covering, and let $\{\varrho_i\}_{i=1}^{\infty}$ be a sequence of finite-dimensional unitary representations of Γ. Suppose that*

$$\lim_{i \to \infty} (\dim \varrho_i)^{-1} \operatorname{Re} \operatorname{tr} \varrho_i(\sigma) = 0$$

for every $\sigma \neq 1$ in Γ. Then

$$\lim_{i \to \infty} (\dim \varrho_i)^{-1} \varphi_{\varrho_i}(\lambda) = \varphi(\lambda)$$

at every continuity point λ of φ.

Let $\{M_i\}$ be as above. As an example of $\{\varrho_i\}$, let us take the regular representation ϱ_i of Γ on $L^2(\Gamma/\Gamma_i)$. Then

$$\operatorname{tr} \varrho_i(\sigma) = \#\{\mu\Gamma \in \Gamma\backslash\Gamma_i;\ \mu^{-1}\sigma\mu \in \Gamma_i\}.$$

It is easily checked that $d(M_i) \to \infty$ if and only if, for $\sigma \neq 1$, $\operatorname{tr} \varrho_i(\sigma) = 0$ for sufficiently large i. Furthermore we observe that Δ_{M_i} is unitarily equivalent to Δ_{ϱ_i}. Thus we obtain

Corollary 1. *If $d(M_i) \to \infty$, then*

$$\lim_{i\to\infty} \operatorname{vol}(M_i)^{-1}\varphi_{M_i}(\lambda) = \operatorname{vol}(M)^{-1}\varphi(\lambda)$$

at all the continuity points of φ, and hence for continuity points $\alpha < \beta$ of φ, we have

$$\lim_{i\to\infty} \operatorname{vol}(M_i)^{-1}\#\{\lambda_k(M_i);\ \alpha < \lambda_k(M_i) \leq \beta\} = \operatorname{vol}(M)^{-1}(\varphi(\beta) - \varphi(\alpha)).$$

Especially, $\sigma(\Delta_X) \subset$ the closure of $\bigcup_i \sigma(\Delta_{M_i})$.

This corollary gives a partial generalization of a result by H. Huber [10] (see also [26]).

Corollary 2. *Under the same condition as above, if Γ is amenable, then $\sigma(\Delta_X) =$ the closure of $\bigcup_i \sigma(\Delta_{M_i})$.*

In fact, from amenability, it follows that $\sigma(\Delta_{M_i}) \subset \sigma(\Delta_X)$ for every M_i (see R. Brooks [2], and [19]).

It is interesting to observe a similarity of Corollary 1 to the *thermodynamical-limit-results* in statistical physics. The conjecture (4) on φ bears a resemblance to that on critical exponents for phase transitions.

To prove Theorem II, it suffices to show

$$\lim_{i\to\infty} (\dim \varrho_i)^{-1} \int_0^\infty e^{-\lambda t}\, d\varphi_{\varrho_i}(\lambda) = \int_0^\infty e^{-\lambda t}\, d\varphi(\lambda) \tag{5}$$

(cf. M.A. Shubin [16]). The left hand side of (5) is equal to

$$\begin{aligned}
&\lim_{i\to\infty} (\dim \varrho_i)^{-1} \operatorname{tr} \exp(-t\Delta_{\varrho_i})\\
&\quad = \operatorname{tr}_\Gamma \exp(-t\Delta_X) + \lim_{i\to\infty} \sum_{\substack{[\sigma]\\ \sigma\neq 1}} (\dim \varrho_i)^{-1}(\operatorname{Re} \operatorname{tr} \varrho_i(\sigma))\langle\Theta_{[\sigma]}, f_t\rangle\\
&\quad = \operatorname{tr}_\Gamma \exp(-t\Delta_X) = \int_0^\infty e^{-\lambda t}\, d\varphi(\lambda).
\end{aligned}$$

This completes the proof. □

We are now concerned with a fine structure of the spectrum of Δ_X on a normal infinite-fold covering manifold X. If the covering transformation group Γ is abelian, the spectrum has *band structure*, namely $\sigma(\Delta_X)$ consists of a series of intervals without accumulations ([12]). We conjecture that this is true for general cases. An evidence supporting this conjecture will be explained below.

We identify $L^2(X)$ with $L^2(\Gamma, V) = L^2(\Gamma) \otimes V$ in a natural manner, where $V = L^2(\mathscr{D})$. We define the subalgebra $C_r^*(\Gamma, \mathscr{K})$ of $\mathrm{End}_\Gamma(L^2(X))$ to be the completion of $C_0(\Gamma, \mathscr{K}) = C_0(\Gamma) \otimes \mathscr{K}$ with respect to the operator norm, where $\mathscr{K}$ is the algebra of compact operators of $L^2(\mathscr{D})$, and $C_0(\Gamma) \otimes \mathscr{K}$ acts on $L^2(X)$ by

$$f \in L^2(\Gamma, V) \mapsto Af \in L^2(\Gamma, V)$$

$$(Af)(\sigma) = \sum_{\mu \in \Gamma} A(\mu^{-1}\sigma) f(\mu), \qquad A \in C_0(\Gamma, \mathscr{K}).$$

The algebra $C_r^*(\Gamma, \mathscr{K})$ is what we call the *reduced* group C^*-algebra with value in $\mathscr{K}$. From the finite propagation property for the wave equation, we deduce that $\hat{f}(\sqrt{\Delta_X}) \in C_0(\Gamma, \mathscr{K})$ for $f \in \mathscr{D}(\mathbb{R})$. Given a rapidly decreasing function $f \in \mathscr{S}(\mathbb{R})$, by approximating f by functions f_n in $\mathscr{D}(\mathbb{R})$ in the L^1-sense, we find that $\hat{f}(\sqrt{\Delta_X}) \in C_r^*(\Gamma, \mathscr{K})$. In fact,

$$\|\hat{f}(\sqrt{\Delta_X}) - \hat{f}_n(\sqrt{\Delta_X})\| \leq \sup |\hat{f} - \hat{f}_n| \leq \|f - f_n\|_{L^1}.$$

Especially $\exp(-t\Delta_X) \in C_r^*(\Gamma, \mathscr{K})$. Applying again an approximation-argument, we observe that if two real numbers a and b are *not* in the spectrum of Δ_X ($b > a$), then the projection $E_b - E_a$ is in $C_r^*(\Gamma, \mathscr{K})$. We thus have

Theorem III [22]. *Suppose that there exists a positive constant C such that* $\mathrm{tr}_\Gamma P \geq C$ *for every non-trivial orthogonal projection P in $C_r^*(\Gamma, \mathscr{K})$. Then the spectrum of Δ_X has band structure (possibly with degenerate intervals, say $[\lambda, \mu]$ with $\lambda = \mu$, corresponding to isolated eigenvalues). Moreover, the number of components of the spectrum which intersect with $(-\infty, \lambda]$ has the following asymptotic estimate*

$$\limsup_{\lambda \to \infty} \frac{C\Gamma(1 + n/2)N(\lambda)}{(4\pi)^{-n}\,\mathrm{vol}(M)\lambda^{n/2}} \leq 1.$$

The asymptotic estimate is a consequence of the asymptotic formula (3) for $\varphi(\lambda)$.

A large class of discrete groups seems to satisfy the condition in the above theorem (cf. M. Pimsner [14]). For example, a free product of finite number of finite groups and infinite cyclic groups satisfies the condition with $C =$ the reciprocal of the least common multiple of the orders of the finite groups. The Kadison conjecture says that one may take 1 as C if Γ is torsion free.

We conclude the discussion with pointing out that the method employed throughout may be applied to a Schrödinger operator on X with a Γ-invariant potential, thereby leading to *non-abelian* Bloch's theory.

Acknowledgement. The authors are grateful to H. Donnelly for drawing their attention to the paper [8].

References

1. Atiyah, M.F.: Elliptic operators, discrete groups and von Neumann algebra. Astérisque **32–33** (1976) 43–72

2. Brooks, R.: The fundamental groups and the spectrum of the Laplacian. Comment. Math. Helv. **56** (1981) 581–598
3. Brooks, R.: The spectral geometry of tower of coverings. J. Diff. Geom. **23** (1986) 97–107
4. Buser, P: On Cheeger's inequality: $\lambda_1 \geq \frac{1}{4}h^2$. Am. Math. Soc. Proc. Symp. Pure Math. **36** (1980) 29–77
5. Chazarain, J.: Formule de Poisson pour les variétés riemanniennes. Invent. math. **24** (1974) 65–82
6. Cheeger, J. Gromov, M., Taylor, M.: Finite propagation speed, kernel estimates for functions of the Laplace operator, and the geometry of complete Riemannian manifolds. J. Diff. Geom. **17** (1982) 15–53
7. Donnelly, H.: On the wave equation asymptotics of a compact negatively curved surface. Invent. math. **45** (1978) 115–137
8. Donnelly, H.: On the spectrum of towers. Proc. A.M.S. **87** (1983) 322–329
9. Duistermaat, J.J., Guillemin, V.W.: The spectrum of positive elliptic operators and periodic bicharacteristics. Invent. math. **29** (1975) 39–79
10. Huber, H.: Ueber das spectrum des Laplace-operators auf kompakten Riemannschen flächen, Comment. Math. Helv. **57** (1982) 627–647
11. Katsuda, A., Sunada, T.: Homology and closed geodesics in a compact Riemann surface. Amer. J. Math. **109** (1987) 145–156
12. Kobayashi, T., Ono, K., Sunada, T.: Periodic Schrödinger operators on a manifold. Forum Math. **1** (1989) 69–79
13. Lang, S.: Algebraic number theory. Addison-Wesley, Massachusetts 1970
14. Pimsner, M.: KK-groups of crossed products by groups acting on trees. Invent. math. **86** (1986) 603–634
15. Randol, B.: Small eigenvalues of the Laplace operator on compact Riemann surfaces. Bull. AMS. **130** (1974) 996–1000
16. Shubin, M.A.: The spectral theory and the index of elliptic operators with almost periodic coefficients. Usp. Mat. Nauk **34** (1979) 95–135
17. Selberg, A.: Harmonic analysis and discontinuous groups in weakly symmetric spaces with application to Dirichlet series. J. Indian Math. Soc. **20** (1956) 47–87
18. Sunada, T.: Trace formula, Wiener integrals and asymptotics. Proc. Japan-France Seminor, "Spectra of Riemannian Manifolds", Kaigai Publ. Tokyo 1983, pp. 159–169
19. Sunada, T.: Riemannian coverings and isospectral manifolds, Ann. Math. **121** (1985) 169–186
20. Sunada, T.: Unitary representations of fundamental groups and the spectrum of twisted Laplacians. Topology **28** (1989) 125–132
21. Sunada, T.: Fundamental groups and Laplacians. Proc. Taniguchi Symp. "Geometry and Analysis on Manifolds" 1987. Lecture Notes in Mathematics, vol. 1339. Springer, Berlin Heidelberg New York 1988, pp. 248–277
22. Sunada, T.: Group C^*-algebras and the spectrum of a periodic Schrödinger operator on a manifold. Preprint 1989
23. Sunada, T.: Fundamental groups and Laplacians. Monograph in Japanese. Kinokuniya, Tokyo 1988
24. Sunada, T.: Trace formula for Hill's operators. Duke Math. J. **47** (1980) 529–546
25. Sunada, T.: Trace formula and heat equation asymptotics for a non positively curved manifold. Amer. J. Math. **104** (1982) 795–812
26. DeGeorge, D.L., Wallach, N.R.: Limit formulas for multiplicities in $L^2(\Gamma \backslash G)$. Ann. of Math. **107** (1978) 133–150
27. Efremov, D.V., Shubin, M.A.: Spectrum distribution function and variational principle for automorphic operators on hyperbolic space. Équations aux dérivées partielles, Séminaire 1988–1989 Ecole Polytechnique, Exposé n° VIII.

Kähler-Einstein Metrics on Algebraic Manifolds

Gang Tian *

Department of Mathematics, State University of New York
Stony Brook, NY 11794, USA

1. Introduction

It is one of fundamental problems in differential geometry to find a distinguished metric on a smooth manifold. H. Poincaré's Uniformization theorem settles this problem for Riemann surfaces. That is, there is a unique metric with constant curvature in each Kähler class on a Riemann surface. Trying to generalize it to higher dimensions, E. Calabi conjectured in the 50s the existence of Kähler-Einstein metrics on a compact Kähler manifold with its first Chern class definite. A Kähler-Einstein metric is a Kähler metric with constant Ricci curvature.

In the middle of the 70s, this conjecture was solved by S.T. Yau in case the first Chern class is vanishing and Aubin and Yau, independently, in case the Chern class is negative (cf. [Y1]). The uniqueness in these two cases was done by E. Calabi himself in the 50s. Such Kähler-Einstein metrics were then applied to studying projective manifolds. For instance, Yau used these metrics to show the Miyaoka-Yau inequality on surface of general type, its generalized version in higher dimensions and the characterization of the quotients of the complex hyperbolic spaces (cf. [Y2]). We also refer readers to [CY, Ko, Ts, TY1] for the generalizations of these to quasi-projective manifolds.

However, this conjecture of Calabi still remains open in general in case the first Chern class is positive. In this paper, we will survey the recent progress on this part of Calabi's conjecture, including the uniqueness and the existence of Kähler-Einstein metrics with positive scalar curvature, the outline of the complete solution for Calabi's conjecture in case of complex dimension two, etc. Some related problems will also be discussed.

From now on, we always denote by M a compact Kähler manifold with positive first Chern class $C_1(M)$, that is, M is a smooth Fano variety. Then we can choose a Kähler metric g with its Kähler class ω_g representing $C_1(M)$. In local coordinates $(z_1, \cdots, z_n)$ of M with $\dim_C M = n$, if g is represented by positive hermitian metrices $\{g_{i\bar{j}}(z)\}_{1\le i,j,\le n}$,

$$\omega_g = \frac{\sqrt{-1}}{2\pi} \sum_{i,j=1}^{n} g_{i\bar{j}} dz_i \wedge d\bar{z}_j .$$

* This work is partially supported by a grant from NSF.

Proceedings of the International Congress
of Mathematicians, Kyoto, Japan, 1990

It is well-known that the Ricci curvature form $Ric(g)$ also represents the first Chern class, therefore, there is a smooth function f in $C^\infty(M,R)$ such that

$$Ric(g) = \omega_g + \frac{\sqrt{-1}}{2\pi}\partial\bar{\partial}f$$

and

$$\int_M (e^f - 1)\omega_g^n = 0$$

where $\omega_g^n = \omega_g \wedge \cdots \wedge \omega_g$. In local coordinates, the Ricci curvature has the following expression

$$R_{i\bar{j}} = \partial_i\partial_{\bar{j}} \log \det(g_{k\bar{l}})$$

$$Ric(g) = \frac{\sqrt{-1}}{2\pi}\sum_{i,j=1}^{n} R_{i\bar{j}} dz_i \wedge d\bar{z}_j .$$

It follows from this that the existence of a Kähler-Einstein metric on M is equivalent to the solvability of the following complex Monge-Amperé equations,

$$\begin{cases} \left(\omega_g + \frac{\sqrt{-1}}{2\pi}\partial\bar{\partial}\varphi\right)^n = e^{f-t\varphi}\,\omega_g^n & \text{on } M \\ \left(\omega_g + \frac{\sqrt{-1}}{2\pi}\partial\bar{\partial}\varphi\right) > 0 & \text{on } M \end{cases} \qquad (1.1)_t$$

where $n = \dim_C M$.

In case $(1.1)_1$ has a solution φ, then $\omega_g + \frac{\sqrt{-1}}{2\pi}\partial\bar{\partial}\varphi$ gives the Kähler form of a Kähler-Einstein metric.

Theorem 1.1 (Bando and Mabuchi [BM]). *The solution of* $(1.1)_1$ *is unique modulo the connected component* $\mathrm{Aut}_0(M)$ *of* $\mathrm{Aut}(M)$ *containing the identity if it exists, where* $\mathrm{Aut}(M)$ *denotes the group of all holomorphic automorphisms of M. In particular, it implies the uniqueness of the Kähler-Einstein metric on M if it exists.*

Since Yau's solution of Calabi's conjecture in case of vanishing first Chern class more than fifteen years ago, it has been known that the solvability of $(1.1)_t$ follows from an a priori C^0-estimate for the solutions, namely, there is a uniform constant C depending only on M and g such that for any solution φ of $(1.1)_t$,

$$\sup_M |\varphi| \leq C . \qquad (1.2)$$

However, such an estimate (1.2) does not exist in general due to the analytic obstructions discussed in the next section.

2. Some Obstructions

Let $\eta(M)$ be the Lie algebra of the automorphism group Aut(M), that is, $\eta(M)$ consists of all holomorphic vector fileds on M. In [Fu], Futaki defined an analytic invariant $F : \eta(M) \to C$ as follows,

$$F(X) = \int_M X(f)\omega_g^n, \qquad X \in \eta(M).$$

Theorem 2.1 *Suppose that (M, g) admits a Kähler-Einstein metric with positive scalar curvature. Then*

1. *(Matsushima, 1957 [Ma]) The Lie algebra $\eta(M)$ is reductive;*
2. *(Futaki, 1983, [Fu]) The Futaki invariant F is identically zero on $\eta(M)$.*

Blowing up a one or two points in CP^2, we obtain a complex surface with positive first Chern class and non-reductive Lie algebra of holomorphic vector fields. Therefore, by the result of Matsushima, such a surface does not admit a Kähler-Einstein metric. In [Fu], Futaki constructed a Fano 3-fold with reductive Lie algebra of holomorphic vector fields, but nonvanishing Futaki invariant. These obstructions indicate that there is no a priori C^0-estimate for the solution of $(1.1)_t$ in general.

Both obstructions of Matsushima and Futaki come from holomorphic vector fields. One might expect that the absence of holomorphic vector field, as is the case with most Fano varieties, would exclude these obstructions. However, there is a very disturbing example constructed in [T1]. Before we give this example, we first remark that Calabi's conjecture can be generalized to compact Kähler orbifolds with positive Chern class and the previous discussions are still effective. Using Volume Comparison Theorem [Bi], one can show

Theorem 2.2 [T1]. *Let (M, g) be a n-dimensional Kähler-Einstein orbifold with positive first Chern class. Then for any singular point x in M,*

$$\sharp(G_x) < \frac{(2n-1)^n \operatorname{Vol}(S^{2n})}{C_1(M)^n} \tag{2.1}$$

where G_x is the local uniformization group of x in M, $C_1(M)^n$ is the volume of M with respect to any Kähler metric with the associated Kähler form in $C_1(M)$ and Vol(S^{2n}) *is the volume of S^{2n} with respect to the standard metric. In particular, in case $n = 2$, (2.1) becomes*

$$\sharp(G_x) < \frac{48}{C_1(M)^2}. \tag{2.2}$$

Example 2.1 [T1]. By blowing up CP^2 successively at [0,0,1], [0,1,0], [1,0,0], [0,1,1] seven times, one can obtain a rational surface $\tilde{M}$ without holomorphic vector fields and containing a Hirzebruch-Jung string $D = E_1 + E_2 + E_3 + E_4$ with E_i smooth rational curves and $E_1^2 = E_4^2 = -3$, $E_2^2 = E_3^2 = -2$, $E_i \cdot E_j = 1$ if $i - j = \pm 1$; $= 0$ otherwise.

Let M be the 2-dimensional Kähler orbifold obtained from $\tilde{M}$ by collapsing the Hirzebruch string D. Then M has positive first Chern class $C_1(M)$ with $C_1(M)^2 = 3$ and an orbifold singular point x of type C^2/σ, where $\sigma : (z_1, z_2) \to$

$(e(\frac{1}{16})z_1, e_2(\frac{7}{16})z_2)$ for the Euclidean coordinates z_1, z_2 on C^2, $e(\frac{l}{m}) = \exp(\frac{2\pi l\sqrt{-1}}{m})$. In particular, $\sharp(G_x) = 16$ in equal to $48/C_1(M)^2$. Therefore, by Theorem 2.2, M does not admit any Kähler-Einstein orbifold metric. On the other hand, M has no holomorphic vector filed since $\tilde{M}$ does not.

This example indicates that holomorphic vector fields may not be only obstructions in solving Calabi's conjecture. The author believes that the existence of Kähler-Einstein metric with positive scalar curvature should be closely related to the geometry of pluri-anti-canonical divisors.

3. Some Existence Theorems

In this section, let (M, g) be a compact Kähler manifold with the Kähler form ω_g in $C_1(M)$ as given in the first section. We will give some existence results. First of all, let us mention two known results among others, for special Kähler manifolds with positive first Chern class, in particular, those manifolds having a lot of holomorphic automorphisms. In [Ma2], Matsushima showed that every simply-connected homogeneous Kähler manifold admits a Kähler-Einstein metric. In case M is a P^1-bundle of certain type with C^*-actions which restrict to C^*-actions on the fibers P^1, Koiso and Sakane [KS] showed that M admits a Kähler-Einstein metric iff the Futaki invariant is zero. The manifold M in all these examples has very high homogeneity. In fact, until 1986, Siu [Si] and the author [T2] provided the first examples of compact Kähler-Einstein manifolds with positive first Chern class and without holomorphic vector fields. Precisely, Siu showed the existence of Kähler-Einstein metrics on the 2-dimensional Fermat surface and the surface obtained by blowing up CP^2 at three generic points, while the author showed the existence of Kähler-Einstein metrics on n-dimensional Fermat hypersurfaces of degree n or $n+1$. Their proofs are completely different. In fact, the examples of the author are just corollaries of a general existence theorem proved in the same paper [T2]. Let us first describe this result.

Let G be any compact subgroup in $\mathrm{Aut}(M)$. This group G preserves the first Chern class $C_1(M)$, therefore we may take g, f to be G-invariant. Define

$$P_G(M,g) = \Big\{ \phi \in C^\infty(M,g) \mid \phi \text{ is } G\text{-invariant on } M, \omega_g + \frac{\sqrt{-1}}{2\pi}\partial\bar{\partial}\phi > 0,\ \sup_M \phi = 0 \Big\}$$

$$\alpha(M) = \sup\{\alpha \mid \exists C_\alpha > 0 \text{ s.t. } \int_M e^{-\alpha\phi} dV_g \le C_\alpha \text{ for all } \phi \in P_G(M,g)\}.$$

It is not hard to check that $\alpha(M)$ is independent of the particular choice of G, g, that is, $\alpha(M)$ is a holomorphic invariant.

Theorem 3.1 [T2]. *Let M be a compact Kähler manifold with $C_1(M) > 0$ as above. Then M admits a Kähler-Einstein metric whenever $\alpha(M) > n/n+1$.*

Theorem 3.1 is proved by deriving an a priori C^0-estimate for the solutions of $(1.1)_t$.

Remark. There is another proof of Theorem 3.1 given by W. Ding later in [Di]. His approach is interesting by itself and was based on an inequality proposed by T. Aubin.

Example 3.1. Let M be a smooth complete intersection in CP^{pq} defined by Fermat polynomials $\sum_{i=0}^{pq} a_{ji} z_i^p = 0$ for $1 \leq j \leq q$, where $[z_0, \cdots, z_{pq}]$ is the homogeneous coordinate of CP^{pq} and $p \geq 2, q \geq 1$ are positive integers. Then by Adjunction Formula (cf. [GH]), one can easily see that $C_1(M)$ is just the restriction of the hyperplane section on CP^{pq}. and the maximal compact subgroup G in $\mathrm{Aut}(M)$ contains all transformations of form $\sigma_j : [z_0, \cdots, z_{pq}] \to [z_0, \cdots, e(\frac{1}{p})z_j, \cdots, z_{pq}]$ for $1 \leq j \leq pq$. The dimension n of M is $(p-1)q$. One can show (cf. [T3])

$$\alpha(M) \geq 1 .$$

In particular, these complete intersections have Kähler-Einstein metrics with positive scalar curvature.

Besides those mentioned above, one can find in [TY2, T3, Na] etc. more examples of Kähler-Einstein manifolds with positive scalar curvature by using this theorem.

The same proof as that for Theorem 3.1 in [T2] also shows

Theorem 3.2 (cf. [TY2]). *Any compact Kähler manifold M with positive first Chern class admits a Kähler metric g_t with Kähler form ω_{g_t} in $C_1(M)$ and Ricci curvature $\geq t$ for any t between* $\min\{1, \frac{n+1}{n}\alpha(M)\}$ *and* 1.

By Bishop's Volume Comparison Theorem, one can derive from it

Corollary 3.1. *For any compact Kähler manifold M with $C_1(M) > 0$, we have*

$$C_1(M)^n \leq C_n \alpha(M)^{-n}$$

where C_n is a constant depending only on n.

In case $n = 2$, we have the following complete solution of Calabi conjecture.

Theorem 3.3 [T4]. *Let M be a complex surface with positive first Chern class, i.e., a Del-Pezzo surface. Then M admits a Kähler-Einstein metric iff $\eta(M)$ is reductive.*

We will outline the proof of this theorem in the next section. Next we briefly discuss the application of L^2-estimate for $\bar{\partial}$-operators to estimating $\alpha(M)$ from below.

Theorem 3.4 (Theorem 3.1 in [T2]). *Let $\{\phi_i\}$ be a sequence of functions in $P(M,g)$, λ be a positive number. Then there exists a subsequence $\{i_k\}$ of $\{i\}$ and a subvariety S_λ of M with $\dim_C S_\lambda \leq \dim_C M - 1$ such that*

1. $\forall\, z \in M - S_\lambda$, $\exists\, r > 0$, $C > 0$, s.t.

$$\int_{B_r(z)} e^{-\lambda \phi_{i_k}(w)} dV_g(w) \leq C$$

2. $\forall\, z \in S_\lambda$,

$$\lim_{k\to\infty}\int_{B_r(z)} e^{-\lambda\phi_{i_k}(w)} dV_g(w) = +\infty \quad \text{for all } r.$$

This theorem can be proved with help of Theorem 5.2.4 in Hörmander's book [Ho]. It implies that the solution φ_t of $(1.1)_t$ should behave well outside a subvariety. Note that the degree of this subvariety can be controlled in terms of the Chern numbers of M. Obviously, $\alpha(M) \geq \lambda_0$ is equivalent to $S_\lambda = \emptyset$ for $\lambda < \lambda_0$. Therefore, it is important to understand S_λ for $\lambda \leq 1$. Since there is no nonconstant holomorphic function on M, one can easily show that each S_λ is connected by the same arguments as in the proof of the Theorem 3.4 (cf. [T2]). In fact, recently, A.Nadel has obtained more information on such a subvariety S_λ for $\lambda < 1$ by Bochner identities for $(0,q)$-forms with $0 < q \leq n$. It can be explained as follows. Denote by I_λ the ideal sheaf of S_λ counting multiplicity, then for each open subset $U \subset M$, $I_\lambda(U)$ consists of those local holomorphic functions f such that $\int_U |f|^2 e^{-\lambda\varphi_{i_k}(w)} dV_g$ are uniformly bounded for all k. On the other hand, one can easily choose the metric g such that $Ric(g) + \frac{\sqrt{-1}}{2\pi}\partial\bar\partial\varphi_{i_k} > 0$ on M. Therefore $h^i(M, I_\lambda) = 0$ for all $i \geq 1$, $\lambda < 1$ by same argument as that in the proof of Kodaira's Vanishing Theorem (cf. [KM, Na]).

For more applications of L^2-estimates to bounding $\alpha(M)$ from below, the readers may refer [T5, TY2].

4. Solution of Calabi's Conjecture for Surfaces

Theorem 3.3 gives the complete solution to Calabi's conjecture for complex surfaces. In this section, we describe briefly the three major steps in the proof. Let M be a complex surface with positive first Chern class. In the first step, we deform the complex structure on M to obtain a smooth family of Kähler surfaces $\{M_t\}_{0\leq t\leq 1}$ with positive first Chern class such that $M_1 = M$ and M_0 admits a Kähler-Einstein metric g_0. The existence of such a family follows from the main theorem in [TY2] and the classification theory of complex surfaces. Now we define $E = \{t \in [0,1] \mid M_{t'}$ has a Kähler-Einstein metric for any $t' \leq t\}$. Then E is nonempty. It is not hard to prove by the Implicit Function Theorem that E is open in [0,1]. Therefore, in order to show that $M = M_1$, has a Kähler-Einstein metric, it suffices to prove that E is closed. We should point out that the closedness is the most difficult part in the solution of Calabi's conjecture. We accomplish this in the following two steps. Without losing generality, we may assume that $E = [0,1)$. Let g_t $(0 \leq t < 1)$ be the Kähler-Einstein metrics on M_t. We need to show that g_t converges to a metric g_1 on M in C^2-topology. Since $\{M_t\}_{0\leq t\leq 1}$ is a smooth family of Kähler surfaces with positive first Chern class, there are Kähler metrics $\tilde g_t$ on M_t $(0 \leq t \leq 1)$ such that their Kähler forms $\omega_{\tilde g_t}$ are in $C_1(M_t) = C_1(M)$. Therefore, for each $t < 1$, $\omega_{g_t} = \omega_{\tilde g_t} + \frac{\sqrt{-1}}{2\pi}\partial\bar\partial\varphi_t$ for a unique function φ_t such that

$$\omega_{g_t}^2 = e^{\tilde f_t - \varphi_t}\,\omega_{\tilde g_t}^2 \quad \text{on } M_t \tag{4.1$_t$}$$

where $\tilde f_t$ are smooth functions determined by $\tilde g_t$ $(0 \leq t \leq 1)$, namely,

$$Ric(\tilde{g}_t) - \omega_{\tilde{g}_t} = \frac{\sqrt{-1}}{2\pi}\partial\bar{\partial}\tilde{f}_t \quad \text{on } M_t$$

and

$$\int_{M_t} (e^{\tilde{f}_t} - 1)\, \omega_{\tilde{g}_t}^2 = 0.$$

Now, in order to have g_t converge to a metric on M, it suffices to derive an a priori C^0-estimate of φ_t $(0 \le t \le 1)$.

In the second step in the solution of Calabi's conjecture, we give a partial C^0-estimate for φ_t.

Theorem 4.1. *For any positive integer* $m \equiv 0 \mod 6$, *there is a constant* C_m *independent of* φ_t *such that there are an orthonormal basis* $\{S_i(t)\}_{0\le i\le N_m}$ *of* $H^0(M, K_M^{-m})$ *with respect to the inner product induced by* $\tilde{g}_t$ *and a sequence of positive numbers* $0 < \lambda_0(t) \le \lambda_1(t) \le \cdots \le \lambda_{N_m}(t) = 1$ *satisfying:*

$$\left\| \varphi_t - \sup_M \varphi_t - \frac{1}{m}\log\left(\sum_{i=0}^{N_m} \lambda_i(t)\|S_i(t)\|_{\tilde{g}_t}^2\right)\right\|_{C^0(M_t)} \le C_m \qquad (4.2)_t$$

where $\|\cdot\|_{\tilde{g}_t}$ *are the norms of* $K_{M_t}^{-m}$ *induced by* $\tilde{g}_t$.

Remark. It is interesting to know whether or not m has to be the multiple of 6 in $(4.1)_t$.

The estimate in $(4.1)_t$ in particular implies that each $\varphi_t - \sup_{M_t}\varphi_t$ is uniformly bounded in any compact subset outside the zero locus of $S_{N_m}(t)$. That is why $(4.2)_t$ is called a partial C^0-estimate of φ_t. Since there is no known local estimate for the solutions of complex Monge-Amperé equations, such a partial C^0-estimate is very precious. The key observation in the proof of $(4.2)_t$ is: by some computations and the maximum principle, one can show that $(4.2)_t$ is equivalent to

$$\log\left(\sum_{i=1}^{N_m} \|S_i'(t)\|_{g_t}^2\right) \ge C_m' \qquad (4.3)_t$$

where C_m' is a constant independent of t, $\|\cdot\|_{g_t}$ is the hermitian metric on $K_{M_t}^{-m}$ induced by g_t and $\{S_i'(t)\}_{0\le t\le N_m}$ is an orthonormal basis of $H^0(M, K_M^{-m})$ with respect to the inner product induced by g_t. In [T4], $(4.3)_t$ is proved by using Cheeger-Gromov's compactness Theorem, Uhlenbeck's Yang-Mills estimates and L^2-estimates for $\bar{\partial}$-operators.

In the last step of our solution for Calabi's conjecture, we first prove a numerical criterion for the existence of a Kähler-Einstein metric on M. To describe it, let us denote by $P_{G,m,k}(M, \tilde{g}_1)$ the collection of all G-invariant functions of the form $\frac{1}{m}\log(\sum_{i=1}^k \|S_i\|_{\tilde{g}_1}^2)$, where $\{S_i\}_{0\le t\le N_m}$ is an orthonormal basis of $H^0(M, K_M^{-m})$ with respect to the metric $\tilde{g}_1$, $0 \le k \le N_m$, $N_m = h^0(M, K_M^{-m}) - 1$. Define

$$\alpha_{m,k}(M) = \sup\left\{\alpha \,|\, \exists C_\alpha > 0, \text{ s.t. } \int_M e^{-\alpha\varphi} dV_g \le C_\alpha, \text{ for any } \varphi \text{ in } P_{G,m,k}(M, \tilde{g}_1)\right\}.$$

Theorem 4.2. *Fix an integer m such that* $(4.1)_t$ *hold for all* $t < 1$. *If either* $\alpha_{m,1}(M) > 2/3$ *or* $\alpha_{m,2}(M) > 2/3$ *and* $\alpha_{m,1}(M) > l_0/(l_0+1)$, *where* $l_0 = \max\{0, 2 - \frac{3\alpha_{m,2}(M)-2}{\alpha_{m,2}(M)}\}$, *then M admits a Kähler-Einstein metric.*

Remark. In general, if we replace 3 by $n+1$ and 2 by n in Theorem 4.2 and the partial C^0-estimate in Theorem 4.1 is valid for solutions of $(1.1)_t$, then the above conclusion is still true. It is just Theorem 6.1 in [T5].

In [T4], by direct computation, it is proved that $\alpha_{6,1}(M) \geq 2/3$, $\alpha_{6,2}(M) > 2/3$ for any complex surface M with positive first Chern class. Therefore, the proof for Theorem 3.3 is finished.

The lower bounds of these $\alpha_{m,k}(M)$ seem to be closly related to Mumford's stability of M in CP^{N_m} in the Chow variety. Therefore, it is quite possible that there is a connection between the existence of Kähler-Einstein metrics on M and Mumford's stability of M in the Chow variety.

5. Some Problems Related to Calabi's Conjecture in Higher Dimensions

According to the author's opinion, in order to solve Calabi's conjecture in higher dimensions, one has to understand further the behavior of the solutions of $(1.1)_t$ (cf. § 1). The difficulty is the lack of some fundamental estimates for complex Monge-Amperé equations. To get around it, we posed and verified a partial C^0-estimate for the solutions of $(1.1)_1$, on complex surfaces. Such an estimate is crucial in the solution of Calabi's conjecture for complex surfaces. The auther believes such an estimate holds on n-dimensional complex Kähler manifold (M, g) with positive first Chern class. For reader's convenience, we state its stronger version.

Conjecture. *Given* $t > 0$ *and* $\mu > 0$. *There are constants* $m_0 < m_1$, C, *depending only on* n, t, μ, *satisfying: For any n-dimensional compact Kähler manifold* (M, g) *with* $Ric(g) \geq t\omega_g$ *and* $\sum_{i=1}^{2n} |b_i(M)| \leq \mu$, *where* $b_i(M)$ *is the ith betti number of* M, *there is an integer m in* $[m_0, m_1]$ *such that*

$$\inf_M \log\left(\sum_{i=0}^{N} \|S_i\|_g^2\right) \geq C \tag{5.1}$$

where $\|\cdot\|_g$ *is the hermitian metric on* K_M^{-m} *with* $m\omega_g$ *as its curvature form,* $\{S_i\}_{1\leq i\leq N}$ *is any orthonormal basis of* $H^0(M, K_M^{-m})$ *with respect to the inner product induced by g and* $\|\cdot\|_g$ *and* $N = \dim H^0(M, K_M^{-m}) - 1$.

The author is able to affirm this conjecture under two more assumptions: Ricci $(g) \leq \mu\omega_g$ and $\int_M |\,\mathrm{Rm}\,|_g^n dV_g \leq \mu$, where Rm denotes the riemannian curvature tensor of g.

One may try to establish (5.2) by using Hörmander's L^2-estimate for $\bar{\partial}$-operator and constructing local nonvanishing sections of K_M^{-m} for m large, as we did in [T4] for complex surfaces. It boils down to understanding the local structure of a compact Kähler manifold with positive Ricci curvature. Naturally, it leads to the following problem.

Problem. Classify all complete Ricci-flat Kähler manifolds X with the following properties:

1) There is a point $x_0 \in X$ such that the geodesic balls $B_r(x_0)$ have Euclidean volume growth as r goes to infinity.

2) $\lim_{r\to\infty} r^{-2n+4} \int_{B_r(x_0)} \|R(g)\|^2 dV_g = 0$ where $R(g)$ denotes the curvature tensor of g.

In case $n = 2$, (X, g) is clearly the Euclidean space C^2. In case $n \geq 3$, we can obtain such (X, g) as follows. Define $X = X_1 \times \cdots \times X_l$. Each X_i is a resolution of C^{n_i}/Γ_i with $n_i \geq 3$ and $\Gamma_i \subset SU(n_i)$ such that the push-down to $(C^{n_i}\backslash\{0\})/\Gamma_i$ of the standard holomorphic n_i form on C^{n_i} can be nonvanishingly extended across X_i. In [T6], by using the method in [TY3], it is proved that there exists a complete Ricci flat Kähler metric g_i on X_i with Euclidean volume growth and $\int_{X_i} \|R(g_i)\|^{n_i} dV_{g_i} < +\infty$. Take g to be the product metric of g_i $(1 \leq i \leq l)$. Then (X, g) satisfies 1) and 2) above.

Are such (X, g) the only ones with 1) and 2)?

6. Degeneration of Kähler-Einstein Manifolds

In order to compactify the moduli space of Kähler-Einstein metrics with positive scalar curvature, we are bound to study Kähler-Einstein orbifold metrics as indicated by the work of M.Anderson [An] and Nakajima [Nk]. Such a metric is defined to be a Kähler orbifold metric on a normal Kähler orbifold such that its Ricci curvature form is a constant multiple of its Kähler form. On one hand, there are some analytic obstructions from holomorphic vector fields to the existence of Kähler-Einstein orbifold metrics analogous to those by Matsushima and Futaki in case of Kähler manifolds, on the other hand, the absence of holomorphic vector field does not assure the existence of a Kähler-Einstein orbifold metric as indicated by the example in § 2 (cf. § 2).

Problem 6.1. Determine when a Kähler orbifold $\tilde{M}$ with $C_1(\tilde{M}) > 0$ has a Kähler-Einstein metric.

It can be regarded as a generalized form of Calabi's conjecture. Of course, The discussions in previous sections, as well as those methods in [T2, TY2], etc., can be applied to this general case with slight modification. However, there are some additional difficulties due to the presence of singularities. For instance, the singularity may cause higher multiplicity of the anticanonical divisor, consequently, the invariant $\alpha(\tilde{M})$ gets smaller (cf. Section 3). Nevertheless, one can produce a lot of Kähler-Einstein orbifolds by means of the methods in [T2], etc. For example, we can prove that there is a Kähler-Einstein orbifold metric on the minimal model of the surface obtained by blowing up CP^2 at 7 or 8 points such that no six of them are on a common quadratic curve and exact three of them are collinear.

The simplest Kähler orbifold with positive first Chern class is the 2-dimensional projective variety with only rational double points as singularities. The minimal resolution of it is an almost Del-Pezzo surface, i.e., the rational surface with numerically positive anticanonical line bundle K^{-1} and $K^2 > 0$. It seems to be even nontrivial to solve Problem 6.1 in this particular case. The methods in [T2, T4] can be used for this purpose.

Not every Kähler-Einstein orbifold can be in the boundary of the moduli space of Kähler-Einstein metrics. Here the distance function on the moduli is defined in terms of some suitable limit, such as Hausdorff limit (cf. [Gr, T4]). In [T4], it is proved that the dimensions of all plurianticanonical divisors are preserved under Hausdorff limit for compact complex surfaces. In fact, the same proof also yields

Theorem 6.1. *Let $\{(M_i, g_i)\}$ be a sequence of Kähler-Einstein manifolds of complex dimension n. Suppose that (M_i, g_i) converge to a Kähler-Einstein orbifold (M_∞, g_∞) in the sense (cf. [Gr]): there are subsets V_i in M_i and V_∞ in M_∞ of real Hausdorff dimension less than $2n-2$ with respect to g_i and g_∞, respectively, such that for any $\varepsilon > 0$, there are diffeomorphisms ϕ_i from $M_\infty - B_{2\varepsilon}(V_\infty, g_\infty)$ into $M_i - B_\varepsilon(V_i, g_i)$ satisfying: (1) The image of ϕ_i contains the exterior of $B_{4\varepsilon}(V_i, g_i)$ for each i; (2) The pull-backs $\phi^* g_i$ converge to g_∞ in C^∞-topology. Then $H^0(M_i, K_{M_i}^{-m})$ converge to $H_0(M_\infty, K_{M_\infty}^{-m})$ for all m in Z. In particular, the dimensions of pluri-anticanonical or canonical sections stay same under the above convergence.*

Problem 6.2. Characterize the Kähler-Einstein orbifold in the boundary of the moduli space of Kähler-Einstein manifolds.

In fact, the author does not know an example of Kähler-Einstein orbifold with positive scalar curvature in the boundary of the moduli space such that its anticanonical ring is different from that of those Kähler-Einstein manifolds in the moduli. It is certainly an interesting question to be explored. We end this paper by a thereom on the degeneration of Kähler-Einstein surfaces (cf. Theorem 7.1 in [T4]).

Theorem 6.2. *Let $\{(M_i, g_i)\}$ be a sequence of Kähler-Einstein surfaces with $C_1(M_i)^2$ equal to $9-n$ $(5 \le n \le 8)$ and positive first Chern class. By taking a subsequence, we may assume that they converge to a Kähler-Einstein orbifold (M_∞, g_∞) (cf. [An], [Nk]). Then*

1. *if $n = 8$, then M_∞ has either only rational double points or two singular points of type $C^2/Z_{l,2}$ besides rational points$(2 \le l \le 7)$;*
2. *if $n = 7$, then M_∞ has either only rational double points or two singular points of type $C^2/Z_{1,2}$ besides rational points;*
3. *if $n = 5, 6$, M_∞ has at most two singular points of type $C^2/Z_{l,2}$ or $C^2/Z_{1,3}$ $(1 \le l \le 3)$ besides rational double points. Moreover, in case M_∞ has two of such singular points, one of them must be of type $C^2/Z_{l,2}$, while the other is of type $C^2/Z_{1,3}$.*

Here $Z_{p,q}$ denotes the finite cyclic group in $U(2)$ generated by diag $\left(e(\frac{1}{pq^2}), e(-\frac{1}{pq^2} + \frac{1}{q})\right)$.

We in fact also have the explicit description of those M_∞ with singular points other than rational double points (cf. § 6, § 7 in [T4]). In fact, there are only a few of them.

References

[An] Anderson, M.: Ricci curvature bounds and Einstein metrics on compact manifolds. J. Amer. Math. (1989)

[Au] Aubin, T.: Réduction du cas positif de l'equation de Monge-Ampére sur les varietés Kählerinnes compactes á la démonstration dún intégalité. J. Funct. Anal. **57** (1984) 143–153

[Bi] Bishop, R.T., Crittenden, R.J.: Geometry of manifolds. Pure and Applied Math., vol. XV. Academic Press, New York London 1964

[BM] Bando, S., Mabuchi, T.: Uniqueness of Einstein Kähler metrics modulo connected group actions. (Adv. Stud. Pure Math., no. 10.) Kinokuniya, Tokyo 1987

[Ca] Calabi, E.: The space of Kähler metrics. Proc. Int. Congress Math., Amsterdam, vol. 2 (1954)

[CY] Cheng, S.Y., Yau, S.T.: Inequality between Chern numbers of singular Kähler surfaces and characterization of orbit space of discrete group of $SU(2,1)$. Contemp. Math. **49** (1986) 31–43

[Di] Ding, W.: Remarks on the existence problem of positive Kähler-Einstein metrics. Math. Ann. **282** (1988) 463–471

[Fu] Futaki, S.: An obstruction to the existence of Einstein-Kähler metrics. Invent. math. **73** (1983) 437–443

[GH] Griffiths, P., Harris, J.: Principles of algebraic geometry. Wiley, New York 1978

[Gr] Gromov, M., Lafontaine, J., Pansu, P.: Structure metrique pour les varietes Riemanniennes. Cedic/Fernand, Nathen 1981

[Ho] Hörmander, L.: An intrudction to complex analysis in several variables. Van Nostrand, Princeton 1973

[Ko] Kobayashi, R.: Einstein-Kähler metrics on open algebraic surfaces of general type. Tohoku Math. J. **37** (1985) 43–77

[KS] Koiso, N., Sakane, Y.: Non-homogeneous Kähler-Einstein metrics on compact complex manifolds. (Lecture Notes in Mathematics, vol. 1201.) Springer, Berlin Heidelberg New York 1986

[Ma1] Matsushima, Y.: Sur la structure du group d'homeomorphismes analytiques d'une certaine varietie Kaehlerinne. Nagoya Math. J. **11** (1957) 145–150

[Ma2] Matsushima, Y.: Sur les espaces homogériens Kählériens d'un group réductif. Nagoya Math. J. **11** (1957) 53–60

[Na] Nadel, A.: Multiplier ideal sheaves and existence of Kähler-Einstein metrics of positive scalar curvature. Proc. Natl. Acad. Sci. USA, vol. 86, no. 19 (1989)

[Nk] Nakajima, H.: Hausdorff convergence of Einstein 4-manifolds. J. Fac. Sci. Univ. Tokyo **35** (1988) 411–424

[Si] Siu, Y.T.: The existence of Kähler-Einstein metrics on manifolds with positive anticanonical line bundle and a suitable finite symmetry group. Ann. Math. (1988)

[T1] Tian, G.: Some Notes on Kähler-Einstein metrics with positive scalar curvature. Preprint

[T2] Tian, G.: On Kähler-Einstein metrics on certain Kähler manifolds with $C_1(M) > 0$. Invent. math. **89** (1987) 225–246

[T3] Tian, G.: Kähler metrics on algebraic manifolds. Thesis, Harvard University

[T4] Tian, G.: On Calabi's conjecture for complex surfaces with positive first Chern class. Invent. math. **101**, no. 1 (1990) 101–172

[T5] Tian, G.: On one of Calabi's problem. Proc. of AMS Summer Institute in Santa Cruz, 1989 (to appear)

[T6] Tian, G.: Complete Ricci-flat Kähler metrics on the resolution of the quotients of C^n by finite groups. In preparation

[Ts] Tsuji, H.: An inequality of Chern numbers for open varieties. Math. Ann. **277**, no. 3 (1987) 483–487

[TY1] Tian, G., Yau, S.T.: Existence of Kähler-Einstein metrics on complete Kähler manifolds and their applications to algebraic geometry. Math aspects of string theory (San Diego, California, 1986), pp. 574–628, Adv. Ser. Math. Phys., 1. World Sci. Publishing, Singapore 1987

[TY2] Tian, G., Yau, S.T.: Kähler-Einstein metrics on complex surfaces with $C_1(M)$ positive. Comm. Math. Phys. **112** (1987)

[TY3] Tian, G., Yau, S.T.: Complete Kähler manifolds with zero Ricci curvature, II. Submitted to Invent. math.

[Y1] Yau, S.T.: On the Ricci curvature of a compact Kähler manifold and the complex Monge-Ampére equation, I^*. Comm. Pure Appl. Math. **31** (1978) 339–441

[Y2] Yau, S.T.: On Calabi's conjecture and some new results in algebraic geometry. Proc. Nat. Acad. Sci. USA **74** (1977) 1798–1799

Ensembles Limites et Applications

Francis Bonahon

Department of Mathematics, University of Southern California
Los Angeles, CA 90089-1113, USA

Nous nous proposons de faire le point sur quelques progrès récents accomplis dans l'étude des ensembles limites des variétés à courbure négative, et surtout de montrer comment on peut utiliser les ensembles limites comme moyen d'attaque pour d'autres problèmes de géométrie et de topologie.

Les travaux de l'auteur ont été partiellement subventionnés par la N.S.F. et la Fondation A.P. Sloan. Cet article a été écrit en grande partie pendant que l'auteur visitait l'Université de Californie à Davis, qu'il souhaite vivement remercier de son hospitalité.

1. Ensembles Limites des Variétés de Courbure Négative

Considérons une variété riemannienne complète M à courbure majorée par une constante strictement négative. Dans son revêtement universel $\widetilde{M}$, les rayons géodésiques issus d'un point base $\tilde{x}_0$ balaient $\widetilde{M}$ et, puisque la courbure est négative, ne se recoupent pas deux à deux. On peut ainsi compactifier $\widetilde{M}$ par sa *sphère à l'infini* $\widetilde{M}_\infty$, en adjoignant un point au bout de chacun de ces rayons géodésiques. L'espace $\widetilde{M} \cup \widetilde{M}_\infty$ ainsi défini est homéomorphe à une boule fermée.

L'hypothèse sur la courbure permet de rendre cette construction indépendante du choix du point base $\tilde{x}_0 \in \widetilde{M}$ (voir par exemple [EbON]). En effet, pour chaque rayon géodésique issu d'un autre point base $\tilde{x}'_0$, il existe un unique rayon géodésique issu de $\tilde{x}_0$ qui lui est asymptote. Ceci fournit une identification naturelle entre les rayons géodésiques issus de $\tilde{x}_0$ et ceux issus de $\tilde{x}'_0$, et montre que l'espace $\widetilde{M} \cup \widetilde{M}_\infty$ est en fait indépendant du choix du point base $\tilde{x}_0$. En particulier, l'action du groupe fondamental $\pi_1(M)$ sur $\widetilde{M}$ s'étend continûment à $\widetilde{M} \cup \widetilde{M}_\infty$.

Un cas important est celui où la métrique de M a courbure constante -1. Le revêtement universel $\widetilde{M}$ est alors isométrique à l'espace hyperbolique $\mathbb{H}^n$, défini comme la boule ouverte unité de $\mathbb{R}^n$ munie de la métrique qui est $2/(1-\|x\|^2)$ fois la métrique euclidienne au point x. La compactification de $\mathbb{H}^n$ par sa sphère à l'infini $\mathbb{H}^n_\infty$ est simplement la compactification usuelle de la boule unité par la sphère unité dans $\mathbb{R}^n$. Dans ce cas, $\widetilde{M} \cup \widetilde{M}_\infty$ a une structure particulièrement riche. Par exemple, $\widetilde{M} \cup \widetilde{M}_\infty$ admet une structure différentiable qui est respectée

Proceedings of the International Congress
of Mathematicians, Kyoto, Japan, 1990

par les isométries de $\widetilde{M}$. Ceci n'est en général plus vrai si la courbure n'est pas constante (voir [Gh] et les références qui y sont mentionnées).

Par un argument simple, toutes les orbites de l'action de $\pi_1(M)$ sur $\widetilde{M}$ ont le même ensemble Λ_M de points d'accumulation dans $\widetilde{M} \cup \widetilde{M}_\infty$. Ce sous-ensemble fermé de $\widetilde{M}_\infty$ est l'*ensemble limite* de M. Pour une définition plus "physique", imaginons un observateur situé en un point x_0 de M. La sphère visuelle de cet observateur, formée de tous les rayons géodésiques issus de x_0, s'identifie à l'ensemble de tous les rayons géodésiques issus d'un point $\widetilde{x}_0 \in \widetilde{M}$ de la préimage de x_0, et donc à la sphère à l'infini $\widetilde{M}_\infty$. Si l'on suppose que la lumière se propage le long de géodésiques, notre observateur (ponctuel) voit de nombreuses copies de lui-même, correspondant aux arcs géodésiques de M joignant x_0 à lui-même. L'ensemble limite est alors formé des points d'accumulation de ces images de l'observateur dans sa sphère de directions.

Quand M est compacte, l'ensemble limite Λ_M est la sphère à l'infini $\widetilde{M}_\infty$ toute entière. Mais sinon, Λ_M peut avoir une structure particulièrement complexe dans $\widetilde{M}_\infty$. Par exemple, considérons le cas classique où S est une surface fermée et où $M = S \times \mathbb{R}$ est munie d'une métrique de courbure constante -1 pour laquelle $S \times [-1, +1]$ est à bord convexe. Alors, l'ensemble Λ_M dans $\widetilde{M}_\infty = \mathbb{H}^3_\infty = S^2 \subset \mathbb{R}^3$ est une courbe de Jordan qui, ou bien est un cercle, ou bien n'est nulle part différentiable et est de dimension de Hausdorff strictement comprise entre 1 et 2 (voir [Bow, Su1]). A ce propos, on peut signaler une caractéristique non négligeable des ensembles limites: ils donnent lieu à de jolis dessins (voir par exemple [Th2] pour quelques specimens). En effet, l'action de $\pi_1(M)$ sur la sphère à l'infini $\widetilde{M}_\infty$ respecte l'ensemble limite Λ_M, et lui confère ainsi de multiples symétries fractales.

2. Propriétés de Mesure des Ensembles Limites

Pour simplifier, limitons-nous au cas où la métrique de M a courbure constante, par exemple égale à -1. Nous avons dejà indiqué que son revêtement universel $\widetilde{M}$ est alors isométrique à l'espace hyperbolique $\mathbb{H}^n$, ce qui identifie la sphère à l'infini $\widetilde{M}_\infty$ à la sphère unité $S^{n-1} = \mathbb{H}^n_\infty$ de $\mathbb{R}^n$. Cette identification $\widetilde{M}_\infty \cong S^{n-1}$ est bien définie modulo composition par isométrie de $\mathbb{H}^n$, c'est à dire modulo composition par un élément du groupe de transformations de $\mathbb{R}^n \cup \infty$ engendré par les inversions par rapport aux sphères (rondes) de dimension $n-1$ qui sont orthogonales à S^{n-1}. Il s'ensuit facilement dans ce cas que l'on peut alors parler de propriétés de négligeabilité pour la mesure de Lebesgue, ainsi que de dimension de Hausdorff, pour l'ensemble limite Λ_M dans $\widetilde{M}_\infty$.

L'un des premiers résultats dans cette direction est dû à L.V. Ahlfors [Ah2] qui a montré en dimension inférieure ou égale à 3 que, sous l'hypothèse que l'action de $\pi_1(M)$ sur $\widetilde{M}$ admet un polyhèdre fondamental borné par un nombre fini de faces totalement géodésiques (on dit alors que M est *géométriquement finie*), l'ensemble limite Λ_M est ou bien de mesure de Lebesgue nulle dans la sphère à l'infini $\widetilde{M}_\infty$ ou bien égal à $\widetilde{M}_\infty$ tout entière. Ahlfors émit aussi la conjecture que cette propriété est vraie sous une hypothèse beaucoup plus faible [Ah1]:

Conjecture. Si M est une variété de dimension 3 à courbure négative constante dont le groupe fondamental $\pi_1(M)$ est de type fini, alors l'ensemble limite Λ_M est ou bien de mesure de Lebesgue nulle dans la sphère à l'infini $\widetilde{M}_\infty$ ou bien égal à $\widetilde{M}_\infty$ tout entière.

En dimension 2, une surface à courbure négative constante dont le groupe fondamental est de type fini est toujours géométriquement finie.

Ces résultats d'Ahlfors ont depuis été améliorés par D. Sullivan dans [Su3], où il montre que la dimension de Hausdorff de l'ensemble limite Λ_M d'une variété géométriquement finie M est strictement inférieure à la dimension de la sphère à l'infini $\widetilde{M}_\infty$ quand Λ_M n'est pas $\widetilde{M}_\infty$ tout entière. On pourra également considérer [Su1] pour une extension de ces résultats en dimension supérieure, ainsi que pour une relation entre la dimension de Hausdorff de Λ_M et le spectre du Laplacien sur M. D'un autre côté, Sullivan a aussi exhibé dans [Su2] des variétés (géométriquement infinies) M de dimension 3 pour lesquelles Λ_M est de dimension de Hausdorff 2 sans être égal à $\widetilde{M}_\infty$, et pour lesquelles $\pi_1(M)$ est de type fini.

De gros progrès vers une solution complète de la conjecture d'Ahlfors ont été accomplis depuis une quinzaine d'années, grâce à un programme lancé vers le milieu des années 1970 par W.P. Thurston. En effet, Thurston a alors proposé un modèle conjecturel pour la géométrie des variétés de dimension 3 à courbure négative constante et dont le groupe fondamental est de type fini. Selon cette conjecture, une telle variété M est nécessairement difféomorphe à l'intérieur d'une variété compacte, de sorte que ses bouts sont topologiquement de la forme $S \times [0, \infty[$ où S est une surface compacte. De plus, pour un tel bout difféomorphe à $S \times [0, \infty[$, la surface S peut être décomposée en un nombre fini de sous-surfaces S_i telles que la géométrie de la métrique de M sur chacun de ces $S_i \times [0, \infty[$ est de deux types possibles: Le premier type est analogue à celui que l'on observe pour tous les bouts d'une variété géométriquement finie, et est tel que la métrique sur les sections $S_i \times t$ croît exponentiellement avec t. Le second type se comporte plutôt comme un tube de section bornée, et les tranches $S_i \times t$ avec t entier ont une aire bornée.

Par une généralisation d'un argument de L.V. Ahlfors pour les variétés géométriquement finies [Ah2], Thurston avait observé que toute variété satisfaisant cette conjecture satisfait aussi la conjecture d'Ahlfors, en ce sens que l'ensemble limite Λ_M dans la sphère à l'infini $\widetilde{M}_\infty$ est ou bien de mesure de Lebesgue nulle ou bien égal à $\widetilde{M}_\infty$.

Cette conjecture de Thurston est maintenant démontrée par les travaux successifs de Thurston [Th1], l'auteur [Bo1], et R. Canary [Ca] si l'on sait déjà que M est topologiquement sage, c'est à dire que M est difféomorphe à l'intérieur d'une variété compacte. Les résultats de [Bo1], améliorant ceux de [Th1], ne nécessitent pas l'hypothèse de sagesse topologique sur M (et l'entraînent en fait comme corollaire), mais réclament une condition supplémentaire sur le groupe fondamental $\pi_1(M)$, par exemple que celui-ci ne se décompose pas en un produit libre non-trivial. Les résultats récents de [Ca], qui requièrent la sagesse topologique mais pas d'hypothèse sur le groupe fondamental, utilisent [Bo1] et [Th1] et sont basés sur un astucieux argument de revêtement ramifié.

3. Topologie des Ensembles Limites

Que peut-on dire de l'espace topologique Λ_M, muni de son action de $\pi_1(M)$?

Un point de départ est le cas où M a un cœur convexe compact. Un *cœur convexe* pour M est une sous-variété M_c à bord convexe telle que $M - \operatorname{int}(M_c)$ est difféomorphe à $\partial M_c \times [0, \infty[$. En particulier, toute variété compacte M admet un cœur convexe compact $M_c = M$.

Quand M admet un cœur convexe compact, il est alors connu depuis longtemps que l'espace topologique Λ_M et son action de $\pi_1(M)$ peuvent être décrits uniquement en termes de la structure algébrique du groupe $\pi_1(M)$ (voir [Mo, Ni, Fl, Gr]). En effet, identifions $\pi_1(M)$ à l'orbite d'un point base $\tilde{x}_0 \in \tilde{M}$ situé au-dessus du cœur convexe M_c, ainsi qu'à l'ensemble des sommets du graphe de Cayley de $\pi_1(M)$ associé à un choix de générateurs. Par convexité et compacité de M_c, la fonction distance induite par la métrique de M sur cette orbite est équivalente à celle provenant du graphe de Cayley. Par ailleurs, l'ensemble limite Λ_M est l'espace des suites de l'orbite de $\tilde{x}_0$, où l'on identifie deux telles suites quand elles sont asymptotes, c'est à dire quand l'angle sous lequel elles sont vues depuis $\tilde{x}_0$ converge vers 0. Les références mentionnées plus haut développent plusieurs méthodes pour traduire cette relation d'asymptoticité en termes de la métrique induite sur l'orbite par la métrique de M. Ceci permet de décrire Λ_M de manière purement combinatoire, en termes d'une certaine relation d'asymptoticité sur les suites de sommets du graphe de Cayley de $\pi_1(M)$.

Par exemple, comme on l'a vu au Chap. 1 dans le cas d'une surface, si M est une variété fermée à courbure négative, l'ensemble limite de n'importe quelle métrique de courbure négative à cœur convexe compact sur $M \times \mathbb{R}^n$ est homéomorphe à l'ensemble limite de M, et donc à la sphère unité de l'espace euclidien de même dimension que M.

Quand la métrique de M n'admet pas de cœur convexe compact, on conjecture que l'ensemble limite Λ_M est un certain quotient de l'*ensemble limite algébrique* Λ_M^a associé au groupe $\pi_1(M)$ par la recette qui produirait l'ensemble limite si M avait un cœur convexe compact. Cette conjecture est notamment vérifiée quand la métrique de M est géométriquement finie (voir [Fl, Tu]), auquel cas Λ_M est obtenu à partir de Λ_M^a en pinçant certaines paires de points associées aux sous-groupes de $\pi_1(M)$ correspondant aux cusps de M.

Quand M n'est pas géométriquement finie, cette conjecture n'est guère vérifiée que dans des cas très particuliers, dûs à J.W. Cannon et W.P. Thurston [CaTh]. Dans certains de leurs exemples, l'ensemble limite algébrique Λ_M^a est homéomorphe au cercle tandis que l'ensemble limite Λ_M est une sphère de dimension 2, ce qui fait que la projection $\Lambda_M^a \to \Lambda_M$ fournit une courbe de Peano recouvrant toute la sphère!

4. Utilisations des Ensembles Limites

Après avoir parlé de la structure des ensembles limites, nous nous proposons maintenant de montrer comment on peut utiliser ceux-ci pour attaquer d'autres problèmes de toplogie et de géométrie. L'idée générale est, pour analyser des objets qui ne sont définis que modulo déformation, de considérer des structures sur l'ensemble limite qui sont associées à ces objets et qui sont définies sans aucune relation d'équivalence particulière.

Pour éviter trop de généralité, concentrons-nous sur un exemple. Dans [Bo1], on est amené à considérer l'espace Γ des classes d'homotopie libre de courbes fermées sur une surface fermée M de courbure négative, et on se trouve confronté au problème suivant: définir et étudier des directions asymptotiques dans Γ vis-à-vis de la forme d'intersection géométrique $i : \Gamma \times \Gamma \to \mathbb{N}$, où le *nombre d'intersection géométrique* $i(\alpha, \beta)$ de α et $\beta \in \Gamma$ est le minimum du nombre de points d'intersection (avec multiplicités) de courbes a et b représentant respectivement α et β.

Pour une approche formelle de ce problème, on peut considérer l'espace $\mathbb{R}_+^\Gamma$ des applications de Γ dans $\mathbb{R}_+$, muni de la topologie produit, et plonger Γ dans $\mathbb{R}_+^\Gamma$ en envoyant $\gamma \in \Gamma$ sur l'application $\alpha \mapsto i(\alpha, \gamma)$. Les directions asymptotiques que l'on veut étudier correspondent alors aux demi-droites de $\mathbb{R}_+^\Gamma$ qui sont dans l'adhérence des demi-droites $\mathbb{R}_+\gamma \subset \mathbb{R}_+^\Gamma$, $\gamma \in \Gamma$.

Cette approche formelle n'est malheureusement pas commode à manier. On peut toutefois raisonner de manière plus géométrique, de la façon suivante.

Le point de départ est que, à cause de l'hypothèse de courbure négative, chaque courbe fermée de M est homotope à un unique multiple d'une géodésique fermée. (Par convention, une géodésique fermée ne tourne pas plusieurs fois autour d'une autre géodésique, et représente donc un élément indivisible du groupe fondamental). Ceci établit une bijection entre les éléments de Γ et les géodésiques fermées de M munies d'un poids entier positif.

Relevons la situation dans le revêtement universel $\widetilde{M}$. La préimage d'une géodésique fermée de M donne une famille de géodésiques de $\widetilde{M}$, invariante par l'action du groupe fondamental $\pi_1(M)$. A chaque élément de Γ, on a ainsi associé un poids entier positif et un sous-ensemble $\pi_1(M)$-invariant de l'espace $G(\widetilde{M})$ des géodésiques de $\widetilde{M}$. On vérifie que ce sous-ensemble est fermé et discret. Maintenant, pour analyser un sous-ensemble fermé discret A muni d'un poids positif, un moyen naturel est de considérer la mesure de Dirac ainsi définie, pour laquelle la masse d'un sous-ensemble est le produit du poids par le nombre d'éléments de A contenus dans ce sous-ensemble. On associe ainsi à chaque élément de Γ une mesure de Dirac $\pi_1(M)$-invariante sur $G(\widetilde{M})$.

Nous venons ainsi d'interpréter les éléments de Γ comme éléments de l'espace $\mathscr{C}(M)$ des mesures (de Radon positives) sur $G(\widetilde{M})$ qui sont invariantes par l'action de $\pi_1(M)$. Pour des raisons expliquées dans [Bo1], les éléments de $\mathscr{C}(M)$ sont appelés *courants géodésiques*.

Munissons $\mathscr{C}(M)$ de la *topologie vague*, pour laquelle une suite de mesures α_n converge vers α si et seulement si les intégrales $\alpha_n(f)$ convergent vers $\alpha(f)$ pour toute fonction $f : G(M) \to \mathbb{R}$ à support compact. Notons que l'on peut additionner deux éléments de $\mathscr{C}(M)$, et multiplier un élément de $\mathscr{C}(M)$ par un nombre réel positif. Le résultat suivant est alors démontré dans [Bo3].

Proposition. *Le sous-espace $\mathbb{R}_+\Gamma \subset \mathscr{C}(M)$, formé des éléments de la forme $\lambda\gamma$ où $\lambda \in \mathbb{R}_+$ et $\gamma \in \Gamma \subset \mathscr{C}(M)$, est dense dans $\mathscr{C}(M)$.*

On peut donc considérer les courants géodésiques sur M comme des "classes d'homotopie diffusées".

Pour le moment, cette construction semble dépendre du choix d'une métrique à courbure négative sur M. Toutefois, chaque géodésique de $\widetilde{M}$ est caractérisée par ses deux points limites distincts dans la sphère à l'infini $\widetilde{M}_\infty$, laquelle est aussi égale à l'ensemble limite Λ_M puisque M est compacte. Par conséquent, $G\left(\widetilde{M}\right)$ s'identifie à $\Lambda_M \times \Lambda_M - \Delta$, où Δ est la diagonale du produit. Ainsi qu'on l'a vu au Chap. 3, l'ensemble limite Λ_M peut être décrit en termes de la structure algébrique du groupe fondamental $\pi_1(M)$, et l'espace $\mathscr{C}(M)$ est par conséquent indépendant de la métrique de M et dépend uniquement du groupe $\pi_1(M)$. En fait, toute cette construction peut être developpée dans le cadre plus général des groupes à courbure négative (aussi appelés hyperboliques) au sens de Gromov [Gr]; ce point de vue est exposé dans [Bo3].

Notons que, pour la surface compacte M, l'espace $G\left(\widetilde{M}\right)$ est particulièrement simple puisqu'il est homéomorphe à un anneau ouvert. Toutefois, l'action de $\pi_1(M)$ sur cet anneau est nettement plus complexe.

Revenons maintenant à la forme d'intersection géométrique $i : \Gamma \times \Gamma \to \mathbb{N}$. Dans [Bo1], on montre:

Proposition. *La forme d'intersection géométrique $i : \Gamma \times \Gamma \to \mathbb{N}$ s'étend en une (unique) application bilinéaire continue $i : \mathscr{C}(M) \times \mathscr{C}(M) \to \mathbb{R}_+$.*

En particulier, il existe une application continue $\mathscr{C}(M) \to \mathbb{R}_+^\Gamma$ qui associe $\gamma \mapsto i(\alpha,\gamma)$ à $\alpha \in \mathscr{C}(M)$, et on vérifie que cette application est propre. Un résultat de J.-P. Otal [Ot] affirme que, réciproquement, un courant géodésique α est complètement déterminé par ses nombres d'intersection avec les éléments de Γ (c'est non-trivial même dans le cas apparemment plus simple où α est un élément de Γ). Par densité des multiples réels d'éléments de Γ dans $\mathscr{C}(M)$, on en déduit:

Théorème. *L'application $\mathscr{C}(M) \to \mathbb{R}_+^\Gamma$ qui associe $\gamma \mapsto i(\alpha,\gamma)$ à $\alpha \in \mathscr{C}(M)$ induit un homéomorphisme de $\mathscr{C}(M)$ sur l'adhérence de $\mathbb{R}_+\Gamma$ dans $\mathbb{R}_+^\Gamma$.*

Ainsi, les directions asymptotiques de Γ que l'on voulait étudier au début de cette section correspondent exactement au demi-droites issues de l'origine dans $\mathscr{C}(M)$.

La définition de $\mathscr{C}(M)$ a été inspirée par une construction antérieure, due à W.P. Thurston [Th1, Th3], de "classes d'homotopie de courbes simples diffusées" sur les surfaces appelées *laminations géodésiques mesurées*. En fait, l'espace $\mathscr{ML}(M)$ des laminations géodésiques mesurées se plonge de manière naturelle dans $\mathscr{C}(M)$. De plus, de la même façon qu'un élément $\gamma \in \Gamma$ est représenté par une courbe simple si et seulement si son nombre d'auto-intersection géométrique $i(\gamma, \gamma)$ est égal à 0, $\mathcal{ML}(M)$ est exactement formé des $\alpha \in \mathcal{C}(M)$ tels que $i(\alpha, \alpha) = 0$ (voir [Bo1, Bo2]).

Les courants géodésiques peuvent également être utilisés pour étudier l'espace $\mathcal{M}(M)$ de toutes les classes d'isotopie de métriques à courbure négative sur une surface compacte M. Une difficulté ici est que ces métriques ne sont définies que modulo isotopie, ce qui fournit trop de degrés de liberté. La considération de l'ensemble limite Λ_M permet de "geler" ces isotopies

En effet, si m est une métrique à courbure négative sur la surface M, nous avons vu plus haut que l'espace topologique $G(\widetilde{M}) = \Lambda_M \times \Lambda_M - \Delta$ muni de son action de $\pi_1(M)$ est indépendant de la métrique m. Toutefois, m dépose une structure suplémentaire sur cet espace. En effet, $G(\widetilde{M})$ peut aussi être considéré comme l'espace des orbites du flot géodésique de la métrique m sur $\widetilde{M}$, agissant sur le fibré tangent unitaire $T_1(\widetilde{M})$. Ce flot géodésique respecte la métrique induite par m sur les orbites du flot et, par un résultat classique de Liouville, respecte aussi la forme volume de $T_1(\widetilde{M})$. Il s'ensuit que l'on peut factoriser la mesure volume sur $T_1(\widetilde{M})$ en le produit d'une mesure L_m sur $G(\widetilde{M})$ et de la mesure induite par m le long des orbites du flot. Cette mesure L_m est la *mesure de Liouville* associée à la métrique m. Elle est clairement invariante par l'action de $\pi_1(M)$ et fournit donc un élément de $\mathcal{C}(M)$ associé à la métrique m.

Le nombre d'intersection géométrique des courbes avec cette mesure de Liouville a une interprétation intéressante:

Proposition. *Le nombre d'intersection géométrique $i(\gamma, L_m)$ de $\gamma \in \Gamma$ avec L_m est égal à 4 fois la longueur du multiple de m-géodésique fermée représentant la classe d'homotopie γ.*

On vérifie que la mesure L_m sur $G(\widetilde{M})$ dépend uniquement de la classe d'isotopie de m. Un résultat récent de J.-P. Otal [Ot] affirme que, réciproquement, on peut reconstruire la métrique m à partir de sa mesure de Liouville L_m, modulo isotopie, en ce sens que:

Théorème. *L'application $\mathcal{M}(M) \to \mathcal{C}(M)$ qui à la métrique m associe sa mesure de Liouville L_m est injective, et est un homéomorphisme sur son image.*

Il serait intéressant d'avoir une caractérisation de cette image (voir [Bo2] pour une caractérisation des mesures de Liouville associées aux métriques de courbure constante).

On a ainsi transformé le problème de l'étude de classes d'isotopies de métriques en l'étude de mesures sur un espace donné.

Un sous-espace important de $\mathcal{M}(M)$ est l'*espace de Teichmüller* $\mathcal{T}(M)$ de la surface M, formé des métriques de courbure constante -1. Cet espace est aussi l'espace des classes d'isotopie de structures conformes sur M, et joue par conséquent un important rôle en analyse complexe.

Dans [Th1, Th3], Thurston a introduit une compactification de l'espace de Teichmüller $\mathcal{T}(M)$ en lui adjoignant à l'infini le projectivisé de l'espace $\mathcal{ML}(M)$ des laminations mesurées. L'espace $\mathcal{C}(M)$, qui contient les deux espaces $\mathcal{T}(M)$ et $\mathcal{ML}(M)$, fournit un cadre pour cette construction qui est sans doute plus naturel que celui originellement utilisé par Thurston. En effet, on démontre dans [Bo2] le résultat suivant:

Théorème. *L'union des demi-droites issues de l'origine dans $\mathcal{C}(M)$ qui sont asymptotes à $\mathcal{T}(M) \subset \mathcal{C}(M)$ est exactement égale à $\mathcal{ML}(M)$. La compactification de $\mathcal{T}(M)$ par l'espace de ces demi-droites que l'on obtient de cette façon est exactement la compactification de Thurston.*

De plus, par des travaux de W.P. Thurston et S. Wolpert [Wo], la restriction à $\mathcal{T}(M) \subset \mathcal{C}(M)$ de la forme d'intersection géométrique i se trouve reliée à la forme de Weil-Petersson sur $\mathcal{T}(M)$; voir [Bo2].

L'idée motivant ce programme est que, si l'on convertit des espaces d'objets définis seulement modulo isotopie en des espaces de mesures, on peut utiliser des méthodes d'analyse pour démontrer des propriétés de continuité et développer un calcul différentiel sur ces espaces. Par example, la continuité de la forme d'intersection sur $\mathcal{C}(M)$ fournit immédiatement des propriétés de continuité intéressantes sur la compactification de Thurston de $\mathcal{T}(M)$.

Le développement d'un calcul différentiel se heurte toutefois à une difficulté: En analyse classique, un vecteur tangent à un espace de mesures est une distribution. Mais, comme on l'a indiqué plus haut, la sphère à l'infini $\widetilde{M}_\infty = \Lambda_M$ et donc l'espace $G\left(\widetilde{M}\right)$ ne possèdent pas de structure différentiable privilégiée. On ne peut donc pas parler de distributions sur $G\left(\widetilde{M}\right)$ de manière intrinsèque.

On résoud cette difficulté en observant que Λ_M possède une structure Hölder privilégiée, c'est à dire une fonction distance bien définie modulo la relation d'équivalence qui identifie deux distances d_1 et d_2 quand il existe des constantes $A > 0$ et $\nu > 0$ telles que $A^{-1} d_1(x, y)^{1/\nu} \leq d_2(x, y) \leq A d_1(x, y)^\nu$ pour tous x, y (voir par exemple [Gr]). On peut ainsi parler de fonctions Hölder $G\left(\widetilde{M}\right) \to \mathbb{R}$. Définissons une *distribution Hölder* sur $G\left(\widetilde{M}\right)$ comme une forme linéaire continue sur l'espace des fonctions Hölder $G\left(\widetilde{M}\right) \to \mathbb{R}$ à support compact. Ainsi, si l'on choisit sur $G\left(\widetilde{M}\right)$ une structure différentiable compatible avec la structure Hölder, une distribution Hölder est une distribution au sens usuel avec une forte propriété de régularité. L'espace $\mathcal{H}(M)$ des distributions Hölder sur $G\left(\widetilde{M}\right)$ qui sont invariantes par l'action de $\pi_1(M)$ semble être le bon cadre pour developper un calcul différentiel du type cherché.

Par exemple, considérons l'espace des laminations géodésiques mesurées $\mathcal{ML}(M) \subset \mathcal{C}(M) \subset \mathcal{H}(M)$. Thurston a construit une structure linéaire par morceaux naturelle sur $\mathcal{ML}(M)$, ce qui fournit une notion combinatoire de vecteur tangent à $\mathcal{ML}(M)$. Dans [Bo4], on établit une correspondance entre ces vecteurs tangents et les vecteurs de $\mathcal{H}(M)$ qui sont tangents à son sous-espace $\mathcal{ML}(M)$. Ceci fournit une description beaucoup plus géométrique des vecteurs tangents de $\mathcal{ML}(M)$, interprétant ceux-ci comme laminations géodésiques munies de distributions Hölder transverses invariantes. Réciproquement, il existe une classification combinatoire simple des distributions Hölder transverses invariantes dont on peut munir une lamination géodésique donnée, ainsi qu'un critère simple pour déterminer celles qui proviennent de vecteurs tangents à $\mathcal{ML}(M)$. Par ailleurs, un grand nombre d'objets définis sur $\mathcal{ML}(M)$ ont une extension naturelle au cadre Hölder. En particulier, ceci permet de calculer les dérivées de certaines fonctions longueur sur $\mathcal{ML}(M)$.

De manière similaire, l'espace de Teichmüller $\mathcal{T}(M) \subset \mathcal{C}(M) \subset \mathcal{H}(M)$ admet en chacun de ses points un espace linéaire tangent dans $\mathcal{H}(M)$, isomorphe à l'espacc tangent défini par la structure différentiable naturelle de $\mathcal{T}(M)$.

Références

[Ah1] Ahlfors, L.V.: Finetely generated Kleinian groups. Amer. J. Math. **86** (1964) 412–423

[Ah2] Ahlfors, L.V.: Fundamental polyhedrons and limit sets of Kleinian groups. Proc. Nat. Acad. Sci. USA **55** (1966) 251–254

[Bo1] Bonahon, F.: Bouts des variétés hyperboliques de dimension 3. Ann. Math. **124** (1986) 71–158

[Bo2] Bonahon, F.: The geometry of Teichmüller space via geodesic currents. Invent. math. **92** (1988) 139–162

[Bo3] Bonahon, F.: Geodesic currents on hyperbolic groups. In: Alperin, R. (ed.) Arboreal group theory. (Mathematical Sciences Research Institute Publications, vol. 19). Springer, Berlin Heidelberg New York 1991

[Bo4] Bonahon, F.: Geodesic laminations with transverse distributions. Prépublication, 1990

[Bow] Bowen, R.: Hausdorff dimension of quasi-circles. Publ. Math. IHES **50** (1979) 11–25

[Ca] Canary, R.: Ends of Hyperbolic 3-manifolds. Prépublication. Stanford University, 1990

[CaTh] Cannon, J., Thurston, W.P.: Group invariant Peano curves. Prépublication, Brigham Young University and Princeton University, 1984–90

[EbON] Eberlein, P., O'Neill, B.: Visibility manifolds. Pac. J. Math. **46** (1973) 45–109

[Fl] Floyd, W.J.: Group completions and limit sets of Kleinian groups. Invent. math. **57** (1980) 205–218

[Gh] Ghys, E.: Le cercle à l'infini des surfaces de courbure négative. Ces Comptes-Rendus, p. 501

[Gr] Gromov, M.: Hyperbolic groups. In: Gersten, S.M. (ed.) Essays in group theory. Springer, Berlin Heidelberg New York, 1987, pp. 75–263

[Mo] Morse, H.M.: A one-to-one representation of geodesics on a surface of negative curvature. Amer. J. Math. **43** (1921) 33–51

[Ni] Nielsen, J.: Untersuchungen zur Topologie der geschlossenen zweiseitigen Flächen I. Acta Math. **50** (1927) 189–358; II. Acta Math. **53** (1929) 1–76; III. Acta Math. **58** (1932) 87–167

[Ot] Otal, J.-P.: Le spectre marqué des longueurs des surfaces à courbure négative. Ann. Math. **131** (1990) 151–162

[Su1] Sullivan, D.: The density at infinity of a discrete group of hyperbolic motions. Publ. Math. IHES **50** (1979) 171–202

[Su2] Sullivan, D.: Growth of positive harmonic functions and Kleinian group limit sets of zero planar measure and Hausdorff dimension 2. In: Looijenga, E., Siersma, D., Takens, F. (eds.) Geometry Symposium, Utrecht 1980 (Lecture Notes in Mathematics, vol. 894). Springer, New York Heidelberg Berlin 1981, pp. 127–144

[Su3] Sullivan, D.: Hausdorff measures old and new, and limit sets of geometrically finite Kleinian groups. Acta Math. **153** (1984) 259–277

[Th1] Thurston, W.P.: The topology and geometry of 3-manifolds. Notes de cours, Princeton University, 1976–79

[Th2] Thurston, W.P.: Three-dimensional manifolds, Kleinian groups and hyperbolic geometry. Bull. Amer. Math. Soc. **6** (1982) 357–381

[Th3] Thurston, W.P.: On the geometry and dynamics of diffeomorphisms of surfaces. Bull. Amer. Math. Soc. **19** (1988) 417–431

[Tu] Tukia, P.: A remark on a paper of Floyd. In: Drasin, D., et al. (eds.) Holomorphic functions and moduli II. (Mathematical Sciences Research Institute Publications, vol. 11). Springer, Berlin Heidelberg New York 1988

[Wo] Wolpert, S.: Thurston's Riemannian metric for Teichmüller space. J. Diff. Geom. **23** (1986) 143–174

Foliations and 3-Manifolds*

David Gabai

California Institute of Technology, Pasadena, CA 91125, USA

The goal of this talk is two fold. First in Sections 1 and 2 I survey some key results relating foliations and the topology of 3-manifolds. In particular in Section 2 I will follow the evolution of one enormously important idea "foliated height functions" as it has evolved over the last century. Second in Section 3 I discuss a recent development, the essential lamination.

1. 1-Dimensional Foliations

Since $\chi(M) = 0$ for all closed 3-manifolds M, it follows that M has a smooth vector field, hence a smooth codimension-2 foliation. By requiring a bit of structure on the foliation one imparts great structure to the 3-manifold. I briefly discuss one such structure.

A Seifert fibred space is a compact 3-manifold M which is almost a S^1 bundle over a compact surface, i.e. there exists a projection $\pi : M \to N$ such that for each $x \in N$ there exists a D^2 neighborhood of x such that $\pi^{-1}(D^2) = D^2 \times S^1$ and $\pi((r, \theta_1), (1, \theta_2)) = (r, p\theta_1 + q\theta_2)$ where $p \neq 0$ and p, q are relatively prime and depend on x and $\theta \in \mathbb{R} \bmod 2\pi$.

Remark. Seifert fibred spaces were classified up to fibre preserving homeomorphism in 1928 by Seifert [S]. Their topological classification was obtained by Orlik, Vogt, and Zieschang [OVZ] in 1967. Seifert fibred spaces have one of six geometric structures as discussed in Scott's survey article [Sc1].

In 1972 Epstein [E] showed that M has a C^1 foliation by circles if and only if M is a Seifert fibred space.

Seifert fibred spaces play a central role in 3-dimensional topology. In fact Thurston's geometrization conjecture [Th3] for 3-manifolds asserts that a closed oriented 3-manifold which has no torus [JS], [Jo] or sphere decomposition [M], [K] is either a Seifert fibred space or a hyperbolic 3-manifold.

Very recently (fall 1990) Casson and the author [G1] have independently announced proofs of the conjecture that S^1 convergence groups [GM] are Fuchsian groups. This result implies the Seifert fibred space conjecture, i.e. if M is a closed

* Partially Supported by NSF grant DMS-8902343 and a Sloan foundation research fellowship.

Proceedings of the International Congress
of Mathematicians, Kyoto, Japan, 1990

orientable irreducible 3-manifold, then M is a Seifert fibred space if and only if $\pi_1(M)$ contains an infinite cyclic normal subgroup. [Me] using the work of [Sc2] and others reduced the Seifert fibred space conjecture to the convergence group conjecture.

Regarding the qualitative types of codimension-2 foliations on 3-manifolds I cite the following two remarkable results.

Theorem (Schweitzer [Sw] 1976). *S^3 has a C^1 foliation by lines.*

Theorem (Vogt [V] 1989). *$\mathbb{R}^3$ has a C^1 foliation by circles.*

2. 2-Dimensional Foliations

I will try to give a chronological account of codimension-1 foliations in 3-manifolds. For simplicity, unless otherwise stated all manifolds will be closed and orientable.

In 1863 Möbius proved the following result (see [St]).

Theorem. *If S is a smooth closed surface in $\mathbb{R}^3$, then S is diffeomorphic to a surface of genus g.*

Idea of Proof. Find enough parallel planes to chop S up into discs, annuli, and spheres with 3-holes. Manipulate the pieces to get a standard form. □

This is a foliations proof of a topological theorem. Möbius uses a foliation of $\mathbb{R}^3$ by parallel planes to decompose S, then rebuilds S in a recognizable way. This "idea" will be generalized many times over for increasingly sophisticated applications. Here is one such generalization.

Theorem (Alexander 1923 [A]). *A PL embedded S^2 in $\mathbb{R}^3$ bounds a 3-Ball. (The PL Shoenflies theorem).*

Idea of Proof. Consider a foliation of $\mathbb{R}^3$ by parallel planes. A generic such foliation induces a foliation on S^2 which has a finite number of critical points at isolated levels. Analyzing the foliation and the embedding one observes that saddles can be cancelled with centers after an isotopy of S^2, thus, after isotopy, one obtains an S^2 with exactly one maximum and one minimum. This S^2 evidently bounds a 3-cell, hence so does the original one. □

The study of foliations did not begin as a recognized subject in its own right until the late 1940s with the work of Ehresmann and Reeb. (Perhaps as an offshoot of the study of fibre bundles which was a hot new topic in the 1940s.) The story goes that Ehresmann suggested the following thesis problem to his student Reeb. Show that there exists no codimension-1 foliation on S^3.

Ehresmann-1940s

- Introduced the idea of holonomy [Eh]. Advisor to Reeb and influenced Haefliger's thesis work (whose advisor was DeRham).

Reeb-1948 [R]
- Reeb foliation of S^3.
- Reeb Stability theorem - If F is a codimension-1 foliation on M and L is a compact leaf with trivial holonomy group, then L has a neighborhood in M homeomorphic to $L \times I$ and $F \mid L \times I$ is the product foliation.
- If F is transversely orientable and F contains a S^2 leaf, then $M = S^2 \times S^1$. (He more generally showed that the set of compact leaves with finite π_1 is closed.)

Remark. Later Lickorisch [L], Zieschang [N], and Wood [W] showed that all closed 3-manifolds have codimension-1 foliations.

Haefliger 1959 [H]
- Showed that if γ is a closed curve transverse to F, an analytic foliation of M, then $0 \neq \gamma \in \pi_1(M)$. In particular this implied that M has infinite fundamental group.
- Showed that the union of compact leaves of a codimension-1 foliation is closed in M.

Novikov 1962 [N]. If $M \neq S^2 \times S^1$, F is a C^2 foliation and has no Reeb components then
- $\pi_2(M) = 0$.
- π_1(leaves) inject into $\pi_1(M)$.
- γ transverse to F implies that $0 \neq \gamma \in \pi_1(M)$.

Remark. It was known to experts that C^2 was not crucial, being used only to push curves normally off of surfaces. Thus a transverse line field provides the needed structure in the C^0 case. See also [So] for a C^0 proof.

Sacksteder 1962 [Sa]. If F is a C^2 foliation on M then
- F has no holonomy implies that F is defined by a closed 1-form. In particular the lifted foliation on the universal cover $\tilde{M}$ is the product foliation $\mathbb{R}^2 \times \mathbb{R}$.
- exceptional minimal sets have resiliant leaves, i.e. leaves with holonomy elements which are contractions.

Remark. A C^0 "version" of Sacksteder's first theorem can be found in [I]. Tischler [Ti] showed that a foliation defined by a closed 1-form can be perturbed slightly to obtain a fibration over S^1.

Stallings 1962 [St]
- If M is irreducible, $\partial M \neq \emptyset$ and the inclusion map of $T \to M$ induces an isomorphism on π_1 where T is a component of boundary M, then $M = T \times I$.

Rosenberg 1968 [Ro]
- If F is a foliation without Reeb components, then M is irreducible.
- (with Sondow). If M is C^2 foliated by planes, then M is the 3-torus.

Remark. The Rosenberg-Sondow theorem uses C^2 in an essential way for it invokes Sacksteder's theorem. Although unstated there (see Sect. 3) a C^0 proof follows by [I].

The main result of Novikov's paper is that a Reeb component is the obstruction to isotoping a disc rel boundary into a leaf, where the induced foliation of the D^2 is the standard foliation by circles with one singular point at the origin. (Consider a meridianal disc transverse to a Reeb foliated $D^2 \times S^1$.)

Rosenberg's key observation in the proof of his first theorem is that Novikov's result implies that there is no obstruction to carrying out Alexander's argument to show that a smoothly imbedded S^2 in a foliated manifold without Reeb components bounds a B^3.

A generalization of the Möbius, Alexander, Rosenberg line is the following result of

Rousserie and Thurston 1972 [Rou, Th1]

- If T is a compact embedded π_1 injective surface in M and F is a foliation on M which has no Reeb components, then T can be isotoped to be transverse to F except at a finite number of saddle and circle tangencies.

Remark. A circle tangency looks exactly like the rim of a volcano. Rousserie further shows that the obstruction to isotoping away the circle tangencies in the case of T a torus is the existence of a cylindrical component, i.e. an annulus bundle over S^1 whose boundary components are leaves and whose interior leaves are annuli whose ends spiral "in the same direction" about the ends. More generally it is well known to experts that the obstruction to eliminating the circle tangencies is the existence of generalized Reeb components, i.e. of bundles over S^1 with fibre a compact surface S with boundary. The boundary tori are leaves and the interior leaves are homeomorphic to $\mathring{S}$, and nearly tangent to S except near the ends which spiral in the same direction about the boundary tori.

Thurston 1976 [Th2]

- Defined the Thurston norm on $H_2(M)$. $\|z\| = \{\min -\chi(S') \mid [S] = z \in H_2(M)$ where $S' = S - S^2$ components$\}$.
- compact leaves of foliations without Reeb components are Thurston norm minimizing, i.e. compact leaves are topologically minimal in their homology class.

Remark. Corresponding definitions and results were given for manifolds with boundary. Thurston's Pseudo-Anosov theory [FLP] is also of central importance.

Palmeira 1978 [P]

- If F is a foliation on M without Reeb components, then the universal covering of M is $\mathbb{R}^3$. Furthermore the induced foliation on $\mathbb{R}^3$ is topologically equivalent to a product of a foliation on $\mathbb{R}^2$ and $\mathbb{R}$.

Sullivan 1979 [Su]

- If F is C^2 and taut, (i.e. there exists a closed curve transverse to F which hits all the leaves), then there exists a Riemannian metric on M such that all the leaves are minimal, i.e. locally area minimizing.

Remark. Sullivan's result was inspired by a letter from Herman Gluck. This result was generalized by Harvey and Lawson [HL] who showed that F was in fact calibrated, i.e. there exists a metric so that compact portions of leaves are minimal area in their homology class. A corresponding topological statement can be made regarding the Thurston norm.

Gabai [G2, G3]

- 1981 – If T is a Thurston norm minimizing surface in an oriented irreducible manifold M, (∂M possibly nonempty) then there exists a finite depth taut transversely oriented foliation F on M such that T is a leaf.
- 1985 – If T is a minimal genus surface for a knot in S^3, then there exists a finite depth taut transversely oriented foliation F of $S^3 - \overset{\circ}{N}(k)$ such that T is a leaf of F and $F \mid \partial N(k)$ is a foliation by circles.

Remark. The difference between the 1985 and 1981 theorems is that the foliation $F \mid \partial N(k)$ given by the former would be a suspension of a homeomorphism of the circle with some fixed points. See [G3] for other constructions.

Applications

- If $z \in H_2(M, \partial M)$, then $g(z) = 2t(z)$ where g denotes the gromov norm and t denotes the Thurston norm. Viewed another way we obtain a positive solution to a conjecture of Thurston [Th2] that the norm on homology based on singular or immersed surfaces is equal to the "classical" norm based on embedded surfaces. In particular the immersed genus of a knot equals the embedded genus of a knot, which is the higher genus analogue to Dehn's lemma. [G2]
- Positive solutions to Property R and Poenaru conjectures. More generally we show that a homology $S^2 \times S^1$ manifold N obtained by surgery on a knot k in S^3 is irreducible or $S^2 \times S^1$. Furthermore the extension of a minimal genus surface for k becomes a Thurston norm minimizing surface representing the generator of N. One can view Property R as asserting that $\{S^3 \text{ knot theory}\} \cap \{S^2 \times S^1 \text{ knot theory}\} = 0$. [G3]
- Knots in $S^2 \times S^1$ or torus bundles over S^1, which are not contained in 3-cells, are determined by their complements. [G4]
- If T is a Thurston norm minimizing surface in M disjoint from a boundary component P, then T remains norm minimizing (and hence π_1 injective) in all but at most one of the manifolds obtained by Dehn filling on P. [G4]
- Sela showed by example that the bad surgeries corresponding to distinct norm minimizing surfaces may be distinct. [Se]
- Computing the genus of an oriented link in S^3. [G5], [G6]
- Property P for satellite knots. [G7]
- Superadditivity of knot genus under band connnect sum. [G8], [Sch]
- The classification of knots in solid tori such that non trivial surgery yields a solid torus. In particular there exists a unique knot in $D^2 \times S^1$ such that exactly two surgeries yield $D^2 \times S^1$. [G7] reduced the problem to 1-bridge braids in solid tori and Berge [B] solved the problem for 1-bridge case. See also [G9].

Key elements of the proofs of the theorems.

- Sutured manifold theory. Just as a Haken 3-manifold can be decomposed to a 3-cell via splitting along incompressible surfaces, the content of sutured manifold theory is that a 3-manifold M with nontrivial $H_2(M, \partial M)$ can be decomposed, with control, along taut sufaces to obtain a 3-cell. Dual to this decomposition is a finite depth foliation such that the cores of the finite depth leaves correspond to the splitting taut surfaces.
- Finite depth foliations "are like" compact surfaces, i.e. technically finite depth foliations can be treated more like incompressible surfaces rather than "thick" foliations.
- Height function arguments. A crucial moment in [G3] was the defining the notion of thin presentation of a knot in S^3. A thin presentation is essentially a bridge presentation where the local maxima are required to be as low as possible and the minima as high as possible, thus from the point of view of the horizontal planes in S^3 the knot is as thin as possible. With respect to a thin presentation a lamination can be put into a normal form. In particular there will be level planes where the induced lamination will have no inessential curves. These arguments should be viewed as sophisticated versions of Möbius, Alexander, Rosenberg, ...
- Combinatorial arguments.

I close this section with the following results, whose proofs push variants of the foliated height function arguments to high technical levels.

Gabai 1984 [G10]

- (Simple Loop Conjecture) $f : S \to T$ is a map of closed surfaces and $\emptyset \neq \ker f_\# : \pi_1(S) \to \pi_1(T)$, then there exists a simple loop in $\ker f_\#$.

Gabai Kazez 1985 [GK1]

- (Homotopy Classification of Maps of Surfaces) If $f, g : S \to T$ are maps of positive degree of closed oriented surfaces, then there exists a homeomorphism $h : S \to S$ such that f is homotopic to $g \circ h$ if and only if $f_\#(\pi_1(S)) = g_\#(\pi_1(S))$ and $\deg f = \deg g$.

Remark. The idea behind these results is to view f (or g) as a branched immersion in $T \times I$, then put the double curve locus into normal form. See [GK 2] for the version for maps of closed surfaces.

Gordon Luecke 1988 [GL]

- Knots in S^3 are determined by their complements.

Remark. The proof involved two steps.

i) If there exists two knots with distinct complements, then each knot complement has a foliation by planar surfaces coming from the foliation of S^3 by level S^2's. When these foliations are viewed in a single knot complement one finds two planar surfaces which intersect each other in an essential way. This step was done independently by myself and [GL] using thin presentations.
ii) There do not exist such planar surfaces in S^3. This step required a deep combinatorial topology argument.

3. Essential Laminations

Example. If $f : T \to T$ is a pseudo anosov homeomorphism of a closed surface T and Λ is the f-invariant (stable measured) foliation, then λ, the suspension of Λ in M, the T bundle over S^1 defined by f, is a singular foliation on M. One can get singular foliations in different manifolds by doing Dehn surgery to the singular circles. Fried was aware of this construction in the late 70's. Independently (at the same time) Ghys and Thurston knew how to create nonsingular foliations when the singular locus consisted of $2n$ prongs.

Remark.

i) ([G11] 1981) λ is an example of a singular foliation with a finite number of singular circles. λ restricted to cross sections of neighborhoods of the circles look like index $n/2$ singularities of line fields where $n = 1$ or $n \leq -1$. The author observed that Novikov's theorem extends to such singular foliations provided $n < 0$. (There exists a singular foliation of S^3 with a single singular circle of type 1/2 and no compact leaves. Consider the foliation transverse to the fibration of the (-2,3,7) Pretzel knot.)

ii) ([G11] 1986) Good singular foliations can be constructed transverse to finite depth foliations. In some sense most manifolds contain such foliations.

iii) The notion of essential lamination [GO] developed with Oertel generalizes the above notion of "good" singular foliation. Recently Hatcher and Oertel have shown that essential laminations in non Haken manifolds can be viewed as singular foliations (of a possibly more general type).

Definition. A codimension-1 lamination λ of M^3 is a decomposition of a closed subset of M by surfaces (called leaves) such that M is covered by charts of the form $\mathbb{R}^2 \times \mathbb{R}$ and leaves of λ pass through in $\mathbb{R}^2 \times$ pts. λ is an essential lamination [GO] of M if

i) $M - \lambda$ is irreducible.

ii) No leaf of λ is a torus bounding a solid torus.

iii) If V is the closure (in the path metric) of a complementary region of λ in M, then ∂V is incompressible and end incompressible.

Theorem [GO] 1987. *Let M be a closed oriented 3-manifold with an essential lamination λ, then*

i) *M is irreducible.*

ii) *π_1 (leaves) inject into $\pi_1(M)$.*

iii) *π_1 (transverse efficient loops) inject into $\pi_1(M)$, transverse efficient arcs cannot be homotoped rel boundary into a leaf.*

iv) *the universal cover of M is $\mathbb{R}^3$.*

Remark. Essential laminations generalize the notions of foliations without Reeb components and incompressible surfaces. Like finite depth foliations essential laminations can be manipulated as compact surfaces. While in some sense most 3-manifolds do not contain incompressible surfaces, most 3-manifolds do contain essential laminations.

Example. Essential laminations can be obtained by "blowing air" into the singular loci of the above good singular laminations. See [GO, Ha, O, Na, D] for other constructions.

Conjecture. *If M is a closed aspherical 3-manifold, then M is finitely covered by a manifold with an essential lamination.*

Conjecture. *If M is closed oriented and aspherical, then M has an essential lamination or is Seifert fibred.*

Remark. The second optimistic conjecture implies the first. The first conjecture is implied by Waldhausen's conjecture that every aspherical 3-manifold is finitely covered by a Haken manifold. There is also Thurston's conjecture that every hyperbolic manifold is finitely covered by a surface bundle over S^1. The following result combined with [M2, W2] (see [JN]) shows that there exist examples of aspherical Seifert fibred spaces with out essential laminations.

Theorem (Brittenham [B], Claus [C] 1990). *If λ is an essential lamination of a Seifert Fibred Space, then there exists a sublamination which is either horizontal or vertical (after isotopy).*

Remark. Vertical means that it is a union of Seifert fibres and horizontal means that it is transverse to the Seifert fibration.

The case for λ a compact surface was due to [Wa]. Thurston [Th1] obtained the result when λ was a C^2 foliation in a S^1 bundle. [EHN] generalized that result to C^2 foliations in Haken Seifert fibred spaces.

Theorem ([GK3] 1990). *If M has an essential lamination and $0 \neq \gamma \in \pi_1(M)$, then $\mathring{D}^2 \times S^1$ is the covering space of M with π_1 generated by γ.*

Theorem (Gabai-Kazez 1990). *If M has a pseudoanosov flow and $f, g : M \to M$ are homotopic homeomorphisms, then f is isotopic to g.*

Remark. It suffices to consider the case that $f = \text{id}$. The hypotheses imply the existence of two transverse (stable and unstable) essential laminations. The proof proceeds by first showing that f is isotopic to f' which fixes a simple closed curve common to both laminations.

Homotopy implies isotopy for irreducible 3-manifolds is an old conjecture. Waldhausen [Wa] established the case for M Haken, a number of authors established the case for M a spherical space form or more generally a Seifert fibred space, see [Sc3] and [BO]. See also [HS].

Theorem. *If λ is an essential lamination of the closed 3-manifold M by planes, then M is the 3-torus.*

Proof. The closure of each complementary region of λ, with the path metric, is $\mathring{D}^2 \times I$, therefore λ extends to a foliation also called λ of M by planes. It follows from Theorem 3.1 [I], the lift $\tilde{\lambda}$ of λ to the universal covering of M has space of leaves $\mathbb{R}$. Since each leaf of λ is a plane, $\pi_1(M)$ acts freely and order preserving on $\mathbb{R}$, thus by a theorem of Holder $\pi_1(M)$ is archemedian and hence free abelian. This last argument is essentially Proposition 4.1 of [I]. Now argue as in [Ro] to conclude that $M = S^1 \times S^1 \times S^1$. □

Bibliography

[A] Alexander, J. W.: On the subdivision of 3-space by a polyhedron. Proc. Nat. Acad. Sci. **10** (1924) 6–8

[B] Berge, J.: The knots in $D^2 \times S^1$ with nontrivial Dehn surgeries yielding $D^2 \times S^1$. Topology Appl. (to appear)

[BO] Boileau, M., Otal, J. P.: Groupe des diffeotopies de certains varietes de Seifert. Compt. Rend. Acad. Sci. **303** (1986) 19–22

[Br] Brittenham, M.: Essential laminations in Seifert fibred spaces. Cornell University Thesis, 1990

[C] Claus, E.: University of Texas Thesis. In preparation

[D] Delman, C.: Cornell University thesis, 1991. In preparation

[E] Epstein, D.: Periodic flows on 3-manifolds. Ann. Math. **95** (1972) 66–82

[Eh] Ehresmann, C.: Sur la theorie des varietes feuilletees. Rend. Mat. Appl. Ser. 5, vol. 10, Rome, 1951

[EHN] Eisenbud, D., Hirsch, U., Neumann, W.: Transverse foliations of Seifert bundles and self homeomorphism of the circle. Comm. Math. Helv. **56** (1981) 638–660

[FLP] Fathi, A. Laudenbach, F. Poenaru V.: Travaux de Thurston sur les surfaces. Asterisque **66-67** (1979)

[G1] Gabai, D.: Convergence groups are Fuchsian groups. Preprint

[G2] Gabai, D.: Foliations and the topology of 3-manifolds. J. Diff. Geom. **18** (1983) 445–503

[G3] Gabai, D.: Foliations and the topology of 3-manifolds III. J. Diff. Geom. **26** (1987) 479–536

[G4] Gabai, D.: Foliations and the topology of 3-manifolds II. J. Diff. Geom. **26** (1987) 461–478

[G5] Gabai, D.: Genera of the arborescent links. AMS Memoirs **339** (1986) 1–98

[G6] Gabai, D.: Foliations and genera of links. Topology **23** (1984) 381–394

[G7] Gabai, D.: Surgery on knots in solid tori. Topology **28** (1989) 1–6

[G8] Gabai, D.: Genus is superadditive under band connected sum. Topology **26** (1987) 209–210

[G9] Gabai, D.: 1-Bridge braids in solid tori. Topology Appl. (to appear)

[G10] Gabai, D.: The simple loop conjecture. J. Diff. Geom. **22** (1985) 143–149

[G11] Gabai, D.: Personal communication

[GK1] Gabai, D., Kazez, W.H.: The classification of maps of surfaces. Invent. math. **90** (1987) 219–242

[GK2] Gabai, D., Kazez, W.H.: The classification of maps of nonorientable surfaces. Math. Ann. **281** (1988) 687–702

[GK3] Gabai, D., Kazez, W.H.: Manifolds with essential laminations are covered by solid tori. Preprint

[GL] Gordon, C., Luecke, J.: Knots are determined by their complements. JAMS **2** (1989) 371–415

[GM] Gehring F.W., Martin G.J.: Discrete quasiconformal groups I. Proc. London Math. Soc. (3) **55** (1987) 331–358

[GO] Gabai, D., Oertel, U.: Essential laminations in 3-manifolds. Ann. Math. **130** (1989) 41–73

[H] Haefliger, A.: Varietes feuilletes. Ann. Scuola Norm. Sup. Pisa (3) **16** (1962) 367–397

[Ha] Hatcher, A.: Examples of essential laminations. Preprint

[HL] Harvey, R., Lawson Jr., H. Blaine: Calibrated foliations. Amer. J. Math. **104** (1980) 607–633

[HS] Hass, J., Scott, P.: Homotopy and isotopy in non Haken 3-manifolds. In preparation

[I] Imanishi, H.: On the theorem of Denjoy-Sacksteder for codimension one foliations without holonomy. J. Math. Kyoto U. **14** (1974) 607–634

[J] Johannson, K.: Homotopy equivalences of 3-manifolds with boundary. (Lecture Notes of Mathematics, vol. 761). Springer, Berlin Heidelberg New York 1979

[JN] Jankins, M., Neumann, W.D.: Rotation numbers of products of circle homeomorphisms. Math. Ann. **271** (1985) 381–400

[JS] Jaco, W., Shalen, P.: Seifert fibred spaces in 3-manifolds. Mem. Amer. Math. Soc. **220** 1979

[K] Kneser, H.: Geschlossene Flächen in dreidimensionalen Mannigfaltigkeiten. Jahresber. Deutsch. Math. Verein. **38** (1929) 248–260

[L] Lickorish, W.B.R.: A foliation for 3-manifolds. Ann. Math. (2) **82** (1965) 414–420

[M1] Milnor, J.: A unique factorization theorem for 3-manifolds. Amer J. Math. **84** (1962) 1–7

[M2] Milnor, J.: On the existence of a connection with curvature zero. Comm. Math. Helv. **32** (1957) 215–223

[Me] Mess, G.: The Seifert conjecture and groups which are coarse quasiisometric to planes. Preprint

[N] Novikov, S.P.: Topology of foliations. Trans. Moscow Math. Soc. (1965) 268–304

[Na] Naimi, R.: Personal communication

[O] Oertel, U.: Example of an affine lamination. Personal communication

[OVZ] Orlik, P., Vogt, E., Zieschang, H.: Zur Topologie gefaserter dreidimensionaler Mannigfaltigkeiten. Topology **6** (1967) 49–64

[P] Palmeira, C.F.B.: Open manifolds foliated by planes. Ann. Math. **107** (1978) 109–131

[R] Reeb, G.: Sur certaines proprietes topologiques des varietes feuilletees. Actualites Sci. Indust. no 1183. Hermann, Paris 1952, pp. 91–158

[Ro] Rosenberg, H.: Foliations by planes. Topology **6** (1967) 131–138

[Rou] Rousserie, R.: Plongements dans les varietes feuilletees et classification de feuilletages sans holonomie. IHES **43** (1973) 101–142

[S] Seifert, H.: Topologie dreidimensionaler gefaserter Raum. Acta Math. **60** (1933) 147–288. [English transl.: Topology of 3-dimensional fibred spaces. Pure Appl. Math. Ser. **89** (1980) 360–423

[Sa] Sacksteder, R.: Foliations and pseudogroups. Amer. J. Math. **87** (1965) 79–102

[Sc1] Scott, P.: The geometries of 3-manifolds. Bull. London Math. Soc. **15** (1983) 401–487

[Sc2] Scott, P.: There are no fake Seifert fibred spaces with infinite π_1. Ann. Math. **117** (1983) 35–70

[Sc3] Scott, P.: Homotopy implies isotopy for some Seifert fibred spaces. Topology **24** (1985) 341–351

[Sch] Scharlemann, M.: Sutured manifolds and generalized Thurston norms. J. Diff. Geom. (to appear)

[Se] Sela, Z.: Torus fillings that reduce Thurston norm. Preprint

[St] Stillwell, J.: Classical topology and combinatorial group theory. (Graduate Texts in Mathematics, vol. 72). Springer, Berlin Heidelberg New York 1980

[Sta] Stallings, J.: On fibering certain 3-manifolds. Topology of 3-manifolds. Prentice Hall, 1962, pp. 95–100

[Su] Sullivan, D.: A homological characterization of foliations consisting of minimal surfaces. Comm. Math. Helv. **54** (1979) 218–223

[Sw] Schweitzer, P.: Counterexample to the Seifert conjecture and opening closed leaves of foliations Ann. Math. **100** (1974) 386–400

[Th1] Thurston, W.P.: Foliations of three-manifolds which are circle bundles. Berkeley Thesis 1972

[Th2] Thurston, W.P.: A norm for the homology of 3-manifolds. AMS Memoirs **339** (1986) 99–130

[Th3] Thurston, W.P.: Three dimensional manifolds, Kleinian groups and hyperbolic geometry. BAMS (N.S.) **6** (1982) 357–381

[Ti] Tischler, D.: On fibering certain foliated manifolds over S^1. Topology **9** (1970) 153–154

[V] Vogt, E.: Preprint 1989

[W1] Wood, J.: Foliations on 3-manifolds. Ann. Math. (2) **89** (1969) 336–358

[W2] Wood, J.: Bundles with totally disconnected structure group. Comm. Math. Helv. **46** (1971) 257–273

[Wa] Waldhausen, F.: On 3-manifolds which are sufficiently large. Ann. Math. **87** (1968) 56–88

The Differential Calculus of Homotopy Functors

Thomas G. Goodwillie

Department of Mathematics, Brown University, Providence, R1 02912, USA

0. Introduction

My purpose here is to explain a method in homotopy theory. The following result is perhaps the best example to date of a statement that can be proved by this method:

Theorem 1. *For any 2-connected map of topological spaces $Y \to X$ the fiber of $A(Y) \to A(X)$ and the fiber of $TC(Y) \to TC(X)$ are weakly homotopy equivalent.*

Here A is Waldhausen's algebraic K-theory functor from spaces to spectra, and TC is another functor which I will discuss below. "Fiber" means homotopy fiber.

If we write $\tilde{A}$ for the reduced functor $\tilde{A}(Y) = \text{fiber}(A(Y) \to A(*))$ and similarly for TC, then in the case when X is a point we have the statement:

Corollary 2. *For any 1-connected topological space Y the spectra $\tilde{A}(Y)$ and $\widetilde{TC}(Y)$ are weakly homotopy equivalent.*

I will not say much now about the other functor TC, except that $TC(X)$ is closely related to the free loop space ΛX (the space of all unbased maps from the circle to X) and is easier to study than $A(X)$ from the point of view of algebraic topology.

The theorem stated above represents the work of several people. In particular, the definition of the functor TC, and of a map $A \to TC$ which is crucial to the proof, uses work of Bökstedt-Hsiang-Madsen. A p-completed version of Theorem 1 (proved by the method outlined below) was the main result of [BCCGHM]. The theorem stated above is only a marginal advance over this, since a rational version ([G1], Corollary on p. 349) has been known for some time. (The final steps in the proof of Theorem 1 will appear in [G6].)

1. Summary of the Method

The proof of Theorem 1 uses a kind of deformation theory. The goal is to describe the change in $A(X)$ produced by a given (small) change in X. It turns out that to

Proceedings of the International Congress
of Mathematicians, Kyoto, Japan, 1990

achieve this it is enough to describe the *infinitesimal* change in $A(X)$ produced by an *infinitesimal* change in X. By this I mean: to give an approximate description of the change in $A(X)$ produced by a very small change in X. (A small change in X is a highly connected map $Y \to X$.)

In a little more detail, the method is this:

There is a natural map of spectra from $A(X)$ to $TC(X)$, called the cyclotomic trace map. Denote its homotopy fiber by $F(X)$. There is a constant c such that for any k-connected map of spaces $Y \to X$ the map of homotopy fibers

$$\text{fiber}(A(Y) \to A(X)) \to \text{fiber}(TC(Y) \to TC(X))$$

is $(c + 2k)$-connected. In other words, the map $F(Y) \to F(X)$ is about twice as highly connected as the map $Y \to X$ upon which it depends. By a certain general principle (Proposition 5 below), it follows that the map $F(Y) \to F(X)$ is in fact ∞-connected when the map $Y \to X$ is at least 2-connected. (In other words, up to weak equivalence $F(X)$ depends only on $\pi_1(X)$ if X is connected.) This yields the conclusion of Theorem 1.

The general principle used above is analogous to the following fact from differential calculus: If a function f (in a suitable domain, and satisfying suitable differentiability hypotheses) is such that $|f(x) - f(y)| < C|x - y|^2$, then f is locally constant. A more familiar statement of this fact is that if the derivative of f is identically zero then f is locally constant.

Section 2 explains the idea "derivative of a homotopy functor". Section 3 states the general principle mentioned above. Section 4 discusses what one needs to know about Waldhausen's functor A in order to apply the principle here. Section 5 describes the other functor TC. Section 6 discusses that part of the proof which involves the map from A to TC. Details may be found in [BCCGHM, G2, G3, G4, and G6].

2. Differentiation of a Functor

For a more detailed account of the ideas below, see [G2].

2.1 The Definition

The idea can be made quite general, but for concreteness let us suppose that F is a functor from spaces to spectra. We always assume that it is a *homotopy functor*, meaning that it takes equivalences to equivalences. (Throughout, an *equivalence* of spaces or spectra means a weak homotopy equivalence.)

In calculus the concept of derivative, or differential, of a function f at a point x is a way of systematically describing the quantity $f(y) - f(x)$ with an accuracy like $|y - x|^2$. In a similar way the next two definitions serve to describe the $2k$-homotopy type of the fiber of $F(Y) \to F(X)$ when the map $Y \to X$ is k-connected.

Definition 3. The *derivative* $\partial_x F(X)$ of F at the based space (X, x) is the homotopy colimit (as $k \to \infty$) of the spectra Ω^k fiber$(F(X \vee S^k) \to F(X))$.

The maps in the limit system are (loosely speaking) induced by the diagrams

$$\begin{array}{ccccc} & & F(X \vee S^{k-1}) & \longrightarrow & F(X \vee D^k_+) \sim F(X) \\ & & \downarrow & & \downarrow \\ F(X) & \sim & F(X \vee D^k_-) & \longrightarrow & F(X \vee S^k) \end{array}$$

(Note that $F(X \vee D^k)$ is equivalent to $F(X)$.) Up to equivalence the derivative is determined by X and by knowing which component of X contains the point x. The spectrum $\partial_x F(X)$ is a functor of the based space (X, x), and any based map $(X, x) \to (Y, y)$ which is an equivalence induces an equivalence $\partial_x F(X) \to \partial_y F(Y)$.

There is a more general construction. If $f: Y \to X$ is a map of spaces, think of Y as a space over X, think of the mapping cylinder of f as the *fiberwise cone* of Y over X (another space over X), and denote it by $C_X Y$. Let $\Sigma_X Y$, the *fiberwise suspension* of Y over X, be the union along Y of two copies of $C_X Y$.

Definition 4. The *differential* of $(D_X F)(Y)$, defined for any map $Y \to X$, is the homotopy colimit of the spectra Ω^k fiber$(F(\Sigma^k_X Y) \to F(X))$,

The maps in the limit system are defined using diagrams

$$\begin{array}{ccccc} & & F(\Sigma^{k-1}_X Y) & \longrightarrow & F(C_X \Sigma^{k-1}_X Y) \sim F(X) \\ & & \downarrow & & \downarrow \\ F(X) & \sim & F(C_X \Sigma^{k-1}_X Y) & \longrightarrow & F(\Sigma^k_X Y) \end{array}$$

For fixed X the differential $D_X F$ is a functor from spaces over X to spectra. It is a homotopy functor in the sense that it preserves equivalences, where a map of spaces over X is called an equivalence if as a map of spaces it is a (weak homotopy) equivalence. We have $(D_X F)(X) \sim *$ and $(D_X F)(X \vee S^0) \sim \partial_x F(X)$.

Note that there is a natural map

$$\text{fiber}(F(Y) \to F(X)) \to (D_X F)(Y)$$

The functor $D_X F$ is intended to be an excisive functor that approximates $Y \mapsto$ fiber$(F(Y) \to F(X))$, much as in calculus the differential of a function f at a point x is a linear function that approximates $f(y) - f(x)$. To explain this I need some language.

2.2 Excision

A commutative diagram $\mathscr{X}$ of spaces (or spectra)

$$\begin{array}{ccc} \mathscr{X}(\varnothing) & \longrightarrow & \mathscr{X}(1) \\ \downarrow & & \downarrow \\ \mathscr{X}(2) & \longrightarrow & \mathscr{X}(1, 2) \end{array}$$

is a *cofiber square* if the canonical map to $\mathscr{X}(1, 2)$ from the homotopy pushout (union along $\mathscr{X}(\varnothing)$ of the mapping cylinders of $\mathscr{X}(\varnothing) \to \mathscr{X}(1)$ and $\mathscr{X}(\varnothing) \to \mathscr{X}(2)$) is an equivalence. It is a *fiber square* (resp. k-connected) if the canonical map from $\mathscr{X}(\varnothing)$ to the homotopy pullback (fiber product over $\mathscr{X}(1, 2)$ of the path fibrations of $\mathscr{X}(1) \to \mathscr{X}(1, 2)$ and $\mathscr{X}(2) \to \mathscr{X}(1, 2)$) is an equivalence (resp. k-connected). Equivalently, a square diagram $\mathscr{X}$ of spectra is k-connected if the *iterated fiber*, the homotopy fiber of the map

$$\text{fiber}(\mathscr{X}(\varnothing) \to \mathscr{X}(1) \to \text{fiber}(\mathscr{X}(2) \to \mathscr{X}(1, 2))$$

of homotopy fibers, is $(k - 1)$-connected.

A functor F (say from spaces, or spaces over X, to spectra) is *excisive* if it takes cofiber squares to fiber squares. This is a very strong condition. Homotopy functors occurring in nature usually satisfy a much weaker, but useful, condition, called *stable excision*: there is a constant c_2 such that if the maps $\mathscr{X}(\varnothing) \to \mathscr{X}(1)$ and $\mathscr{X}(\varnothing) \to \mathscr{X}(2)$ in a cofiber square are respectively k_1- and k_2-connected, then the diagram $F(\mathscr{X})$:

$$\begin{array}{ccc} F(\mathscr{X}(\varnothing)) & \longrightarrow & F(\mathscr{X}(1)) \\ \downarrow & & \downarrow \\ F(\mathscr{X}(2)) & \longrightarrow & F(\mathscr{X}(1,2)) \end{array}$$

is $(k_1 + k_2 + c_2)$-connected.

If F satisfies stable excision then, for each X, $D_X F$ satisfies excision; we may think of $D_X F$ as a (reduced) homology theory on the category of spaces over X. Moreover, stable excision for F implies that the map from $\text{fiber}(F(Y) \to F(X))$ to $(D_X F)(Y)$ is approximately $2k$-connected for any k-connected map $Y \to X$.

2.3 The Principle

Theorem 1 is proved by applying the following principle with $F = \text{fiber}(A \to TC)$ and $\varrho = 1$. The term "ϱ-analytic" will be explained in Section 3.

Proposition 5. *If F is a ϱ-analytic functor from spaces to spectra such that $(D_X F)(Y)$ is trivial for all X and all $Y \to X$, then for every $(\varrho + 1)$-connected map $Y \to X$ of spaces the map $F(Y) \to F(X)$ is an equivalence.*

"Trivial" means equivalent to a point (all homotopy groups are trivial). If F satisfies a suitable limit axiom, so that up to equivalence it is determined by its behavior on finite CW complexes, then it is enough to assume that $\partial_x F(X)$ rather than $D_X F$ is trivial.

3. Analytic Functors

"Analyticity" of a homotopy functor F has to do with the behavior of F with respect to cubical diagrams of spaces. By an n-cubical diagram we mean a functor $\mathscr{X}$ from the partially ordered set of all subsets of $\{1, \ldots, n\}$ to the category of spaces. Analyticity of F involves one condition, stable $(n-1)$st order excision, for each n. Stable first-order excision is stable excision as defined in Sect. 2.2.

Stable $(n-1)$st order excision concerns certain n-cubical diagrams $\mathscr{X}$, namely the *strong cofiber cubes*. Call $\mathscr{X}$ a strong cofiber cube if, for each $1 \leq i < j \leq n$ and $S \subset \{1, \ldots, n\} - \{i, j\}$, the diagram

$$\begin{array}{ccc} \mathscr{X}(S) & \longrightarrow & \mathscr{X}(S \cup \{i\}) \\ \downarrow & & \downarrow \\ \mathscr{X}(S \cup \{j\}) & \longrightarrow & \mathscr{X}(S \cup \{i, j\}) \end{array}$$

is a cofiber square. The condition is that there is a constant c_n such that, whenever $\mathscr{X}$ is a strong cofiber cube in which the map $\mathscr{X}(\emptyset) \to \mathscr{X}(i)$ is k_i-connected for all i, with $k_i > \varrho$, then $F(\mathscr{X})$ is $(c_n + \Sigma k_i)$-connected. (An n-cubical diagram of spectra is called k-connected if its iterated fiber – the spectrum obtained by taking homotopy fibers in each of the n directions in turn – is $(k-1)$-connected.) Note that c_n is allowed to be negative.

For $n = 1$ this simply says that there is a constant c_1 such that for any k-connected map $\mathscr{X}(\emptyset) \to \mathscr{X}(1)$ of spaces the map $F(\mathscr{X}(\emptyset)) \to F(\mathscr{X}(1))$ is $(k + c_1)$-connected, at least if $k > \varrho$.

Definition 6. The functor F is ϱ-analytic if it satisfies $(n-1)$st order as above for all $n \geq 1$, and if the numbers c_n are bounded below by $c - \varrho n$ for some constant c.

Most homotopy functors occurring in nature are ϱ-analytic for some ϱ, and in many cases it is a routine matter to verify this. The identity functor from spaces to spaces is 1-analytic, as is Waldhausen's functor A.

The proof of Proposition 5 uses an unusual inductive argument. It is not difficult, but I will not take time to explain it here; see [G3].

Proposition 5 expresses one of two main consequences of analyticity. The other, the existence of a "Taylor tower" for a functor in analogy with the Taylor series of a function, is not used in the proof of Theorem 1. It is explained in [G5].

4. The Derivative of *K*-Theory

In order to use Proposition 5 for proving Theorem 1, it is necessary first of all to know (up to natural equivalence) what the derivative of the functor A is. The answer turns out to be this:

Theorem 7. *For a based space* (X, x) *the spectrum* $\partial_x A(X)$ *is related by a chain of natural equivalences to* $\Sigma^\infty(\Omega(X, x)_+)$.

This is the unreduced suspension spectrum of the based loopspace of X. (The subscript "+" adds a disjoint basepoint.)

Theorem 7 (3.3 of [G2]) is proved indirectly; it is reduced to a corresponding statement (Theorem 7′ below) about smooth manifolds, using a major theorem of Waldhausen:

Theorem 8 (Waldhausen [W1]). *There is a natural weak equivalence of spectra between* $A(X)$ *and the product* $\Sigma^\infty(X_+) \times Wh^{\mathrm{Diff}}(X)$, *where* $Wh^{\mathrm{Diff}}(X)$ *is a natural double delooping of the differentiable pseudoisotopy spectrum* $\mathscr{P}^{\mathrm{Diff}}(X)$.

In view of Theorem 8, Theorem 7 may be rewritten:

Theorem 7′. *For a based space* (X, x) *the spectrum* $\partial_x \mathscr{P}^{\mathrm{Diff}}(X)$ *is related by a chain of natural equivalences to* $\Omega^2\Sigma^\infty(\Omega(X, x)/X)$.

It is notable that, while the relationship between K-theory and pseudoisotopy theory expressed in Theorem 8 is usually viewed as a way of reducing geometry to algebra, in this instance the flow of information is in the other direction. In this connection see also Sect. 5.3.

Recall that the underlying space of the spectrum $\mathscr{P}^{\mathrm{Diff}}(X)$ is essentially defined as a limit of spaces $P^{\mathrm{Diff}}(M)$ for manifolds M (compact, with boundary, of arbitrarily large dimension) of the homotopy type of X. The space $P^{\mathrm{Diff}}(M)$ is the simplicial group of all diffeomorphisms of $M \times I$ which are the identity along $(M \times 0) \cup (\partial M \times I)$.

Therefore, to "compute" $\partial_x \mathscr{P}^{\mathrm{Diff}}(X)$ is essentially to solve the following problem: For a smooth manifold M with an attached handle h of index $k \geq 3$, determine the $2k$-homotopy type of the fiber of $P^{\mathrm{Diff}}(M) \to P^{\mathrm{Diff}}(M \cup h)$. This is done in [G2] using Morlet's "disjunction lemma" and an old-fashioned differentiable general-position argument.

5. The Functor *TC*

I will now say something about the functor TC which occurs in the statement of Theorem 1. There are really two questions to address: How is it defined, and what does it turn out to be?

5.1 Definition of *TC*

I will not be very specific about this. TC is related to Bökstedt's "topological Hochschild homology" (THH). For details see [BCCGHM], [BHM], or [G4].

Recall that, according to one way of thinking about the K-theory of (based, connected) spaces, $A(BG)$ is the K-theory spectrum of the "ring up to homotopy" $\Omega^\infty\Sigma^\infty(|G|_+)$. The latter is to be thought of as the "group ring" $k[G]$ of the simplicial group G over the ground "ring" $k = QS^0$. Heuristically,

$$\begin{aligned}\text{connective spectrum} &= \text{infinite loop space}\\ &= \text{abelian group up to homotopy}\\ &= k\text{-module}\end{aligned}$$

and the group structure of G gives $k[G]$ a multiplication compatible with its additive structure. These ideas can be made precise by using a suitable notion of "ring up to homotopy", for example Bökstedt's notion of FSP (functor with smash product).

For such a "ring" R, Bökstedt defines a K-theory spectrum $K(R)$. Both the Quillen K-theory of rings and the Waldhausen K-theory of spaces are included as special cases (the cases of a discrete ring R and a group ring $k[G]$ respectively). He also defines a spectrum $THH(R)$; heuristically it is the simplicial object

$$\begin{gathered} R \\ \uparrow \downarrow \uparrow \\ R \otimes R \\ \uparrow \downarrow \uparrow \downarrow \uparrow \\ R \otimes R \otimes R \\ \cdots \end{gathered}$$

with face and degeneracy maps given by the product and unit of R, respectively, as in the definition of the standard chain complex for Hochschild homology. The "tensor products" are meant to be over k and are really smash products of spectra.

Bökstedt defines a map of spectra $K(R) \to THH(R)$; it is modeled on the "trace map" from K-theory to Hochschild homology defined by Dennis for an ordinary ring R.

Very roughly speaking, TC is related to THH as cyclic homology is related to Hochschild homology. For any FSP there is a spectrum $TC(R)$ with a map $TC(R) \to THH(R)$. The trace $K(R) \to THH(R)$ lifts to a map $K(R) \to TC(R)$, called the *cyclotomic trace.* (After p-completion this is the same as the map of that same name constructed in [BHM]).

Let the simplicial group G be a loop group for the space X, and let R be $k[G]$. In this case we sometimes write $TC(X)$ instead of $TC(R)$. Thus in this case the cyclotomic trace is a map $A(X) \to TC(X)$. It is this which is used in the proof of the theorem.

5.2 Description of *TC*

From a computational point of view the main thing to know about $TC(X)$ is that it is related in a certain way to the free loop space $\Lambda X = \mathrm{Map}(S^1, X)$. Again let G be a simplicial loop group for X.

First of all, it is fairly easy to see that $THH(k[G])$ is equivalent to $\Sigma^\infty(\Lambda X_+)$. This is essentially because ΛX is equivalent to the realization of the simplicial space

$$\begin{gathered} G \\ \uparrow \downarrow \uparrow \\ G \times G \\ \uparrow \downarrow \uparrow \downarrow \uparrow \\ G \times G \times G \\ \cdots \end{gathered}$$

(the "cyclic bar construction" or "cyclic nerve" of G).

To describe $TC(X)$ we must consider some additional structure that the space ΛX has. Let the circle group S^1 act on ΛX in the usual way, and let $\Delta_p: \Lambda X \to \Lambda X$ be the pth power map (composition with the standard map $S^1 \to S^1$ of degree p).

It turns out that the functor $X \mapsto TC(X)$ is very closely related to the functor $X \mapsto B(X) = \Sigma^\infty \Sigma((ES^1 \times_{S^1} \Lambda X)_+)$, although to say exactly how they are related it is apparently necessary to consider separately the profinite homotopy type and the rational homotopy type.

Concerning the profinite type, the statement is that after p-completion (p a prime) the spectrum $TC(X)$ becomes part of a fiber square

$$\begin{array}{ccc} TC(X) & \longrightarrow & B(X) \\ \downarrow & & \downarrow \scriptstyle\text{Trf} \\ \Sigma^\infty \Lambda X_+ & \xrightarrow[1-\Delta_p]{} & \Sigma^\infty \Lambda X_+ \end{array}$$

Here Trf is the S^1-transfer associated to the bundle

$$(\Lambda X \sim)\ ES^1 \times \Lambda X \to ES^1 \times_{S^1} \Lambda X$$

and $1 - \Delta_p$ is the difference between two stable maps, the identity and the map induced by Δ_p.

This, it turns out, has the consequence that for 1-connected spaces X there is a natural equivalence, after p-completion and passage to reduced functors, between $TC(X)$ and

$$\Sigma^\infty(X_+) \times \text{fiber}(e \circ \text{Trf}: B(X) \to \Sigma^\infty(X_+))$$

where the map is the composition of the transfer and the map induced by evaluation $\Lambda X \to X$ at a point in the circle.

Concerning the rational type, the statement is that for 2-connected maps $Y \to X$ there is a natural equivalence, after rationalization, between the fiber of $TC(Y) \to TC(X)$ and the fiber of $B(Y) \to B(X)$. (This is not, however, induced by a natural map $TC \to B$ or $B \to TC$.)

5.3 Generalizations

Theorem 1 can be generalized so as to apply to more than the K-theory of spaces. There is considerable evidence for the following:

Conjecture 9. *For any* 1*-connected map* $R \to S$ *of FSP's the resulting map of spectra from* fiber$(K(R) \to K(S))$ *to* fiber $(TC(R) \to TC(S)$ *is an equivalence.*

This can be deduced from Theorem 1 in some cases, namely those in which $\pi_0(R)$ $(= \pi_0(S))$ is an integral group ring $\mathbb{Z}[\pi]$. In particular, it is true for the map $QS^0 = k = R \to S = \mathbb{Z}$. Unfortunately, this does not yet amount to a computation of the fiber of $A(*) \to K(\mathbb{Z})$ in any real sense, because $TC(\mathbb{Z})$ is still a fairly mysterious object.

6. The Derivative of *TC*

After producing a map from A to TC, it remains to show that it induces an equivalence $\partial_x A(X) \to \partial_x TC(X)$. This is done in two steps.

The first step is to show that $\partial_x A(X)$ and $\partial_x TC(X)$ are abstractly equivalent, in the sense that these two functors from based spaces (X, x) to spectra are related by a chain of natural equivalences. I have already said that $\partial_x A(X)$ is abstractly equivalent to $\Sigma^\infty \Omega(X, x)_+$. The same is true of $\partial_x TC(X)$. Of course I cannot begin to explain why, since I have not even defined TC here, but to get the idea I invite the reader to work out the equivalences (see Section 2 of [G2]):

$$\partial_x \Sigma^\infty \Lambda(X)_+ \sim \mathrm{Map}(S^1, \Sigma^\infty \Omega(X, x)_+)$$
$$\partial_x B(X)_+ \sim \Sigma^\infty \Omega(X, x)_+.$$

The second step is to prove:

Lemma 10. *The cyclotomic trace $A \to TC$ induces an equivalence $\partial_x A(X) \to \partial_x TC(X)$.*

The trick in proving this is to begin with the case when X is the suspension ΣY of a connected space Y.

To see that this special case is enough, one classifies all the natural maps $\Sigma^\infty \Omega(X, x)_+ \to \Sigma^\infty \Omega(X, x)_+$ in the homotopy category of homotopy functors from based spaces to spectra. It turns out that the only maps which are equivalences when X is a simply-connected suspension are those which are equivalences for all X.

The argument which proves the lemma in the case $X = \Sigma Y$ is essentially the main argument of [CCGH]. It relies on a tool which is only available in the suspension case: the cyclotomic trace can be composed with another natural map as follows:

$$\coprod_{n \geq 1} D_n(Y) \to \Omega \tilde{A}(\Sigma Y) \to \Omega \tilde{T}C(\Sigma Y).$$

Here $D_n(Y)$ is the divided power $\Sigma^\infty(E(\mathbb{Z}/n)_+ \wedge_{\mathbb{Z}/n} Y^{[n]})$. (I am writing $Y^{[n]}$ for the smash product of n copies of Y.) The composed map above induces a map of derivatives

$$\partial_y \left(\coprod_{n \geq 1} D_n(Y) \right) \to \partial_y \Omega \tilde{T}C(\Sigma Y) \sim \Omega \Sigma^\infty \Omega \Sigma Y_+$$

which, more or less by direct examination, is seen to be an equivalence. It follows that the map

$$\Omega \Sigma^\infty \Omega \Sigma Y_+ \sim \partial_y \Omega \tilde{A}(\Sigma Y) \to \partial_y \Omega \tilde{T}C(\Sigma Y) \sim \Omega \Sigma^\infty \Omega \Sigma Y_+$$

induced by the cyclotomic trace is a split surjection, and from this one concludes without much trouble that it is an equivalence.

As a by-product this yields the main result of [CCGH], which can now be viewed as a special case of Corollary 2:

Theorem 11. *For connected spaces Y there is a natural equivalence of spectra*

$$\Omega A(\Sigma Y) \sim \Omega A(*) \times \coprod_{n\geq 1} D_n(Y).$$

References

[B] Bökstedt, M.: Topological Hochschild homology. Preprint, Bielefeld

[BCCGHM] Bökstedt, M., Carlsson, G., Cohen, R., Goodwillie, T., Hsiang, W.-c., Madsen, I.: The algebraic K-theory of simply-connected spaces (p-complete case). Preprint

[BHM] Bokstedt, M., Hsiang, W.-c., Madsen, I.: The cyclotomic trace and algebraic K-theory of spaces. Preprint, Aarhus

[CCGH] Carlsson, G., Cohen, R., Goodwillie, T., Hsiang, W.-c.: The free loop space and the algebraic K-theory of spaces. K-theory **1** (1987) 53–82

[G1] Goodwillie, T.: Relative algebraic K-theory and cyclic homology. Ann. Math. **124** (1986) 347–402

[G2] Goodwillie, T.: Calculus I, The derivative of a homotopy functor. K-Theory **4** (1990) 1–27

[G3] Goodwillie, T.: Calculus II, analytic functors. To appear in K-theory

[G4] Goodwillie, T.: Notes on the cyclotomic trace. In preparation

[G5] Goodwillie, T.: Calculus III, The taylor series of functor. In preparation

[G6] Goodwillie, T.: Calculus IV, Applications to K-theory. In preparation

[W1] Waldhausen, F.: Algebraic K-theory, concordance, and stable homotopy. In: Ann. Math. Studies **113** (1987) 392–417

Dehn Surgery on Knots

Cameron McA. Gordon

Department of Mathematics, The University of Texas at Austin
Austin, TX 78712, USA

1. Introduction

In [D], Dehn considered the following method for constructing 3-manifolds: remove a solid torus neighborhood $N(K)$ of some knot K in the 3-sphere S^3 and sew it back differently. In particular, he showed that, taking K to be the trefoil, one could obtain infinitely many non-simply-connected homology spheres in this way. Let $M_K = S^3 - \overset{\circ}{N}(K)$. Then the different resewings are parametrized by the isotopy class r of the simple closed curve on the torus ∂M_K that bounds a meridional disk in the re-attached solid torus. We denote the resulting closed oriented 3-manifold by $M_K(r)$, and say that it is obtained by *r-Dehn surgery on K*.

More generally, one can consider the manifolds $M_L(\mathbf{r})$ obtained by $\mathbf{r}$-Dehn surgery on a k-component link $L = K_1 \cup \cdots \cup K_k$ in S^3, where $\mathbf{r} = (r_1, \ldots, r_k)$. It turns out that *every* closed oriented 3-manifold can be constructed in this way [Wal, Lic]. Thus a good understanding of Dehn surgery might lead to progress on general questions about the structure of 3-manifolds.

Starting with the case of knots, it is natural to extend the context a little and consider the manifolds $M(r)$ obtained by attaching a solid torus V to an arbitrary compact, oriented, irreducible (every 2-sphere bounds a 3-ball) 3-manifold M with ∂M an incompressible torus, where r is the isotopy class (*slope*) on ∂M of the boundary of a meridional disk of V. We say that $M(r)$ is the result of *r-Dehn filling on M*.

An observed feature of this construction is that

generically, the topology of M persists in $M(r)$.

We shall illustrate this slogan by stating some results that give restrictions on the "exceptional" slopes r for which $M(r)$ represents some sort of degeneration of M. Specifically, we shall consider the following questions:

(1) When is an essential surface destroyed by Dehn filling?
(2) When is an essential surface created by Dehn filling?
(3) When is $M(r)$ "small"?

The results we shall state show that the present state of knowledge on these questions is quite good.

Proceedings of the International Congress
of Mathematicians, Kyoto, Japan, 1990

First notice that, roughly speaking, anything can happen under a *single* Dehn filling. For if we are interested in when $M(r)$ has a certain property, then we can simply start with a closed 3-manifold Q with that property, and let $M = Q - \overset{\circ}{N}(K)$ for some suitable knot K in Q. Then there is a Dehn filling on M that yields Q. This puts a limitation on the kind of results one expects to obtain. However, the theorems we shall state will usually assert that (for any M) the set of exceptional slopes (for the particular kind of degeneration in question) is small. In fact, if $\Delta(r,s)$ denotes the minimal geometric intersection number of two slopes r, s on ∂M, then the theorems will often be of the form that give an upper bound on $\Delta(r,s)$ for any pair of exceptional slopes r and s.

We conclude this introduction with some conventions that we shall adopt throughout the paper. All 3-manifolds and surfaces will be assumed to be compact and orientable, and M will always denote (as above) an irreducible 3-manifold with ∂M an incompressible torus. A knot K will always be assumed to be a (non-trivial) knot in S^3 unless otherwise stated, and we reserve the notation M_K $(= S^3 - \overset{\circ}{N}(K))$ for this case. Slopes on ∂M_K will be parametrized by $\mathbb{Q} \cup \{\infty\}$ in the usual way, using a meridian-longitude basis $\{\mu, \lambda\}$ for $H_1(\partial M_K)$. Thus $r \leftrightarrow a/b$ if and only if $[r] = a\mu + b\lambda$ in $H_1(\partial M_K)$. Here, ∞ denotes $1/0$ (so $M_K(\infty) \cong S^3$).

Finally, since the condition $\Delta(r,s) = 1$ occurs frequently in the sequel, we remark that any set of slopes, such that $\Delta(r,s) = 1$ for every pair of distinct slopes in the set, has at most 3 elements. Also, for slopes on ∂M_K, $\Delta(r,\infty) = 1$ if and only if r is an integer.

2. When Is an Essential Surface Destroyed by Dehn Filling?

A properly embedded surface F in a 3-manifold Q is *essential* if either F is incompressible, not parallel into ∂Q, and not a 2-sphere, or F is a 2-sphere which does not bound a 3-ball in Q.

Essential surfaces play an important role in the theory of 3-manifolds. For instance there is the Kneser-Milnor prime factorization theorem, which involves cutting the manifold up into canonical pieces along essential 2-spheres. Essential tori feature in a similar way in the torus decomposition theory of Jaco-Shalen and Johannson. And of course there is the Haken-Waldhausen theory of irreducible 3-manifolds that contain essential surfaces (*Haken manifolds*), based on the idea of successively cutting the manifold along essential surfaces to eventually obtain a disjoint union of 3-balls.

Coming to the title of this section, let us take the case of a *closed* essential surface in M.

As examples, first consider the case of a (p,q)-*cable* K in a solid torus $S^1 \times D^2$. In other words, take a simple closed curve on $\partial(S^1 \times D^2)$ that winds around p times meridionally and q times longitudinally ($q \geq 2$) and push it into $\operatorname{int}(S^1 \times D^2)$. Then it turns out that there are infinitely many Dehn surgeries on K that yield a solid torus again. In particular, there are infinitely many Dehn surgeries on K under which $\partial(S^1 \times D^2)$ compresses. To put this in the setting of our question, we simply take a knot K_0 in a closed 3-manifold Q (such that $M_0 = Q - \overset{\circ}{N}(K_0)$ is irreducible and ∂M_0 is incompressible), and identify $N(K_0)$ with $S^1 \times D^2$ by some homeomorphism, so that K becomes a knot in Q. (We say that K is a (p,q)-*cable of* K_0.) Then the torus $S = \partial N(K_0)$ is essential in $M = Q - \overset{\circ}{N}(K)$, but compresses in $M(r)$ for infinitely many slopes r.

Analogous examples with surfaces S of higher genus can be constructed by starting with a curve C in the boundary of a handlebody X of genus ≥ 2 such that $\partial X - C$ is incompressible in X, pushing C into $\overset{\circ}{X}$ to get K, say, taking a suitable embedding of X in some closed 3-manifold Q, and letting $M = Q - \overset{\circ}{N}(K)$ and $S = \partial X$.

An example of a different kind is the knot K in $S^1 \times D^2$ illustrated in Fig. 1.

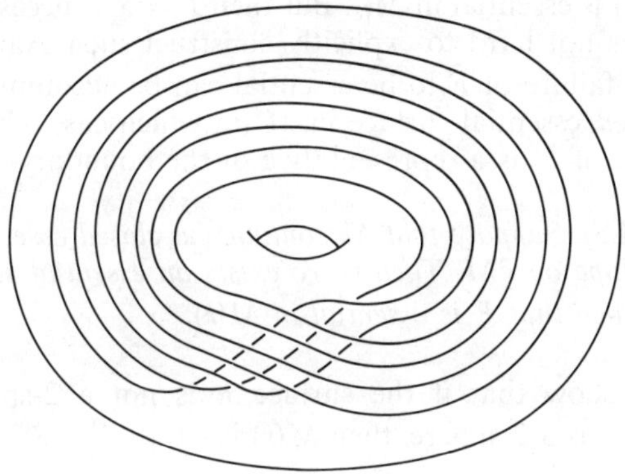

Fig. 1

Let $W = S^1 \times D^2 - \overset{\circ}{N}(K)$. Then, parametrizing slopes on $\partial N(K)$ using the meridian and longitude that come from the embedding of K in S^3 defined by Fig. 1, we have

$$W(\infty) \cong W(18) \cong W(19) \cong S^1 \times D^2,$$

so $\partial(S^1 \times D^2)$ compresses in $W(r)$ for 3 distinct slopes r. (In fact K is the unique non-cable curve in $S^1 \times D^2$ (which is not a core and does not lie in a 3-ball) with this property [Ga2, B1, Sc].) Note that

$$\Delta(18, 19) = \Delta(18, \infty) = \Delta(19, \infty) = 1.$$

Similar examples in handlebodies of genus ≥ 2 have been constructed by Berge (unpublished). Again, as above, these can be used to construct essential closed surfaces S in irreducible 3-manifolds M with torus boundary such that S compresses under 3 distinct Dehn fillings on M.

With these examples in mind we have the following result of Wu.

Theorem 2.1 [Wu2]. *Let S be a closed essential surface in M which compresses in $M(r)$ and $M(s)$. Then either $\Delta(r,s) = 1$, or the core K of the solid torus V in $M(s) = M \cup V$ can be isotoped onto S and S compresses in $M(r)$ for infinitely many r.*

In particular, if S is a torus then either $\Delta(r,s)=1$ or K is a cable. In this case the theorem was originally proved in [CGLS].

Let us now consider when an essential surface *with boundary* in M is destroyed by Dehn filling. By this we mean the following. Let F be an essential surface in M with non-empty boundary, so each boundary component of F has slope r, say. (We say that r is a *boundary-slope*.) Then F gives rise to a closed surface $\widehat{F}$ in $M(r)=M\cup V$ by capping off the boundary components of F with meridian disks of V, and we are interested in the question: is $\widehat{F}$ essential in $M(r)$?

It is clear that the answer is not always "yes." For instance, we can start with a closed surface S in S^3, and take a knot K which punctures S in such a way that $F=S-\mathring{N}(K)$ is essential in M_K. But then $\widehat{F}=S$ is necessarily inessential in $M_K(\infty)=S^3$. (It is not hard to explicitly construct such examples.) However, it turns out that the failure of $\widehat{F}$ to be essential can be accounted for either by the presence of a *closed* essential surface in M (in which case Theorem 2.1 applies), or by a bad choice of F as a representative of the boundary-slope r.

Theorem 2.2 [CGLS]. *Suppose that M contains no closed essential surface, and let r be a boundary-slope on ∂M. Then there exists an essential surface F in M with boundary-slope r such that $\widehat{F}$ is essential in $M(r)$.*

One can also show that if the surface $\widehat{F}$ is not a 2-sphere, then $M(r)$ is irreducible, and if $\widehat{F}$ is a 2-sphere, then $M(r)$ is either $S^1\times S^2$ or a connected sum of two lens spaces.

3. When Is an Essential Surface Created by Dehn Filling?

Suppose that $M(r)=M\cup V$ contains an essential surface S. It is straightforward to show that if S is moved (by an isotopy if $M(r)$ is irreducible, or in general, by a sequence of disk-swappings) so as to minimize the number of components of $S\cap V$, then either S lies in M or $F=S-\mathring{V}$ is an essential surface in M with boundary-slope r. The title of this section refers to the second possibility. In this context we have the following basic result of Hatcher.

Theorem 3.1 [Ha]. *M has only finitely many boundary-slopes.*

Corollary 3.2. *If M does not contain a closed essential surface then $M(r)$ contains a closed essential surface for only finitely many r.*

Since essential spheres and tori play a special role in 3-manifold theory, it is of interest to consider when they are created by Dehn filling.

In the case of essential spheres, the cabling construction described in Sect. 2 provides some examples. For if K is a (p,q)-cable of a knot K_0 in a 3-manifold Q, then $M=Q-\mathring{N}(K)$ contains an essential annulus with boundary-slope r where $\Delta(r,s)=1$, s being the meridian of K. This annulus becomes a 2-sphere in $M(r)$, and in fact it splits $M(r)$ as a connected sum $M_0(r_0)\#L(q,p)$, where $M_0=Q-\mathring{N}(K_0)$ and r_0 is some slope on ∂M_0. If we now choose Q to contain an essential sphere (and K_0 to be any knot in Q such that M_0 is irreducible), then

M is irreducible, whilst $M(r)$ and $M(s) = Q$ contain essential spheres. (It is not obvious, but in fact follows from [GLu1], that, here, $M_0(r_0)$ cannot be S^3.)

Another example where $M(r)$ and $M(s)$ contain essential spheres with $\Delta(r,s) = 1$, which does not come from cabling, is given in [GLi].

The next theorem says that this is the worst that can happen.

Theorem 3.3 [GLu3]. *If $M(r)$ and $M(s)$ each contain an essential sphere then $\Delta(r,s) = 1$.*

Turning to essential tori, a key example here is the exterior M_K of the figure eight knot K (see Fig. 2). M_K does not contain an essential torus, but $M_K(4)$, which is homeomorphic to $M_K(-4)$ since K is amphicheiral, does. Note that $\Delta(4,-4) = 8$.

Fig. 2

Theorem 3.4 [Go]. *If M does not contain an essential torus, but $M(r)$ and $M(s)$ do, then $\Delta(r,s) \leq 8$.*

In fact one can do a little better. The figure eight knot exterior belongs to the family of manifolds obtained by Dehn surgery on one component of the Whitehead link L (see Fig. 3). More precisely, let $W = S^3 - \mathring{N}(L)$ and let $W(p/q)$ be the manifold with torus boundary obtained by p/q-Dehn filling on one boundary component of W (where we use the obvious meridian-longitude parametrization). Then $W(1)$ is homeomorphic to the exterior of the figure eight knot. It also turns out that for $M = W(-5)$, $W(-5/2)$, and $W(2)$, M is irreducible and contains no essential torus, whilst there are r, s such that $M(r)$ and $M(s)$ contain essential tori with $\Delta(r,s) = 8, 7$, and 6 respectively. One can show that these examples are the only ones with $\Delta(r,s) > 5$.

Addendum 3.5 [Go]. *In the setting of Theorem 3.4,*

if $\Delta(r,s) = 8$, then M is homeomorphic to $W(1)$ or $W(-5)$;
if $\Delta(r,s) = 7$, then M is homeomorphic to $W(-5/2)$;
if $\Delta(r,s) = 6$, then M is homeomorphic to $W(2)$.

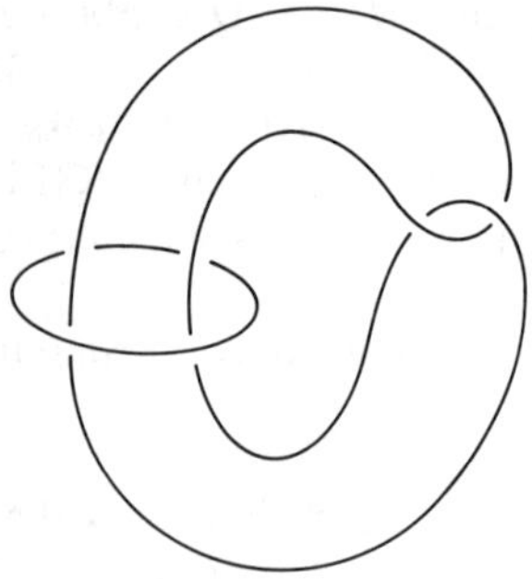

Fig. 3

4. When Is $M(r)$ Small?

Examples of what we might mean by "small" are:

When is $\pi_1(M(r))$ trivial? finite? cyclic?

When is $M(r)$ homeomorphic to S^3? a lens space? $S^1 \times S^2$? a Seifert fibred space?

This is a convenient point at which to recall that by [T], M is either hyperbolic, or Seifert fibred, or contains an essential torus. (For instance, if $M = M_K$ then the last two possibilities correspond respectively to K being a torus knot or a satellite knot.) As it is straightforward to describe $M(r)$ when M is Seifert fibred [He], and as the case when M contains an essential torus is to a large extent covered by Theorem 2.1, the hyperbolic case is the most important to understand. Here there is the following result of Thurston.

Theorem 4.1 [T]. *If M is hyperbolic then $M(r)$ is hyperbolic for all but finitely many r.*

A hyperbolic manifold is one whose interior has a complete Riemannian metric of constant negative (sectional) curvature. If we are willing to sacrifice "constant" in the conclusion of Theorem 4.1, then we have the following theorem, due to Gromov and Thurston, and improved in [BH] using a result of Adams.

Theorem 4.2. *If M is hyperbolic then $M(r)$ has a Riemannian metric of negative curvature for all but at most 24 values of r.*

It is conjectured that a 3-manifold with negative curvature is in fact hyperbolic. Even in the absence of a proof of this conjecture it is known that such a manifold must have infinite fundamental group, cannot contain an essential sphere or torus, cannot be Seifert fibred, etc.

Let us turn to the more specific question: When is $\pi_1(M(r))$ cyclic?

As a first set of examples, let K be a torus knot. Then $M_K(r)$ is a lens space for infinitely many r [Mo]. Note that in this case M_K is Seifert fibred.

A set of examples of a different flavor comes from the knot K in $S^1 \times D^2$ depicted in Fig. 1 and discussed in Sect. 2. Embedding $S^1 \times D^2$ in S^3 by applying k meridional twists to the embedding shown in Fig. 1, we obtain an infinite family of knots K_k, $-\infty < k < \infty$, (which are in fact hyperbolic). The two non-trivial Dehn surgeries on K that yield $S^1 \times D^2$ determine two Dehn surgeries on K_k, with slopes r_k and s_k, say, that yield lens spaces, where $\Delta(r_k, s_k) = \Delta(r_k, \infty) = \Delta(s_k, \infty) = 1$. For instance K_0 is the $(-2, 3, 7)$ pretzel knot shown in Fig. 4 (this example was discovered earlier by Fintushel and Stern [FS]), for which

$$M_{K_0}(18) \cong L(18, 5)\,, \quad M_{K_0}(19) \cong L(19, 7)\,, \quad M_{K_0}(\infty) \cong S^3\,.$$

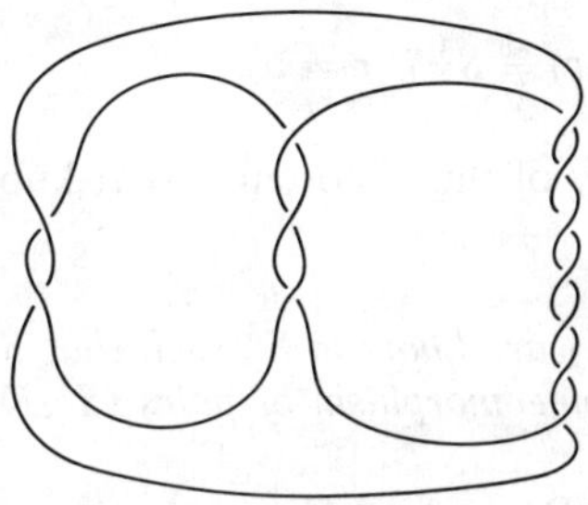

Fig. 4

The following Cyclic Surgery Theorem shows that these examples are extremal.

Theorem 4.3 [CGLS]. *Suppose that M is not Seifert fibred. If $\pi_1(M(r))$ and $\pi_1(M(s))$ are cyclic then $\Delta(r, s) = 1$.*

Less is known about when $\pi_1(M(r))$ is finite. An interesting example here is the manifold $W(-5)$ defined in Sect. 3. It is shown in [We] that there exist r, s with $\Delta(r, s) = 3$ such that $\pi_1(W(-5)(r))$ and $\pi_1(W(-5)(s))$ are finite.

Question 4.4. *If M is hyperbolic and $\pi_1(M(r))$ and $\pi_1(M(s))$ are finite, is $\Delta(r, s) \leq 3$?*

The case where $M = M_K$ for a satellite knot K (so M contains an essential torus) is discussed in [BH], where it is shown that $\Delta(r, s) \leq 5$, and that this bound is attained.

Regarding the question of when $\pi_1(M(r))$ is trivial, the following conjecture is still open.

Conjecture 4.5. *$\pi_1(M(r))$ is trivial for at most one slope r.*

5. Dehn Surgery on Knots in S^3

Continuing the theme of the previous section, let us ask: when can we have two Dehn fillings on M that give small manifolds, one of which is S^3? Clearly there is a Dehn filling on M that gives S^3 if and only if M is homeomorphic to M_K for some knot K in S^3, with the slope that determines the filling corresponding to the meridian of K. So our question becomes: given a knot K in S^3, when does non-trivial Dehn surgery on K yield a small manifold?

First we have the result that S^3 can never be obtained by non-trivial Dehn surgery.

Theorem 5.1 [GLu2]. *$M_K(r) \not\cong S^3$ if $r \neq \infty$.*

An immediate corollary of this theorem is that knots are determined by their complements.

Corollary 5.2. *If K_1 and K_2 are knots in S^3 such that $S^3 - K_1$ is homeomorphic to $S^3 - K_2$, then there is a homeomorphism of pairs $(S^3, K_1) \cong (S^3, K_2)$.*

The following so-called Property P Conjecture, however, is still open, although it is known to be true for many classes of knots.

Conjecture 5.3. *$\pi_1(M_K(r)) \neq 1$ if $r \neq \infty$.*

Of course, Conjectures 4.5, 5.3, and Theorem 5.1 are closely related, and are equivalent if the Poincaré Conjecture is true.

After S^3 (the unique 3-manifold of Heegaard genus zero), we might consider when Dehn surgery on a knot K yields a manifold of Heegaard genus one, *i.e.*, $S^1 \times S^2$ or a lens space.

For homological reasons, the only Dehn surgery that could possibly give $S^1 \times S^2$ is 0-Dehn surgery. However, a result of Gabai implies that this never happens.

Theorem 5.4 [Ga1]. *$M_K(0)$ is irreducible.*

Regarding lens spaces, Berge [B2] gives an explicit construction which yields several infinite classes of knots with a Dehn surgery yielding a lens space. (These include the infinite family described in Sect. 4 with *two* such Dehn surgeries.) It appears to be not entirely outside the bounds of possibility that every such knot can be constructed in this way.

Question 5.5. *Does every knot K such that $M_K(r)$ is a lens space for some r appear in Berge's list?*

An affirmative answer to this question would imply, in particular, the truth of the following conjecture.

Conjecture 5.6. *$M_K(r) \not\cong RP^3$ $(= L(2,1))$, $L(3,1)$, or $L(4,1)$.*

Finally, we mention that for satellite knots, the question of when $\pi_1(M_K(r))$ is cyclic is completely solved.

Theorem 5.7 [Wan, Wu1, BL]. *Let K be a satellite knot. Then $\pi_1(M_K(r))$ is cyclic if and only if K is the $(2pq \pm 1, 2)$-cable of a (p,q)-torus knot and $r = 4pq \pm 1$, in which case $M_K(r) \cong L(4pq \pm 1, 4q^2)$.*

We now turn to the creation of essential spheres and tori by Dehn surgery on a knot K.

As far as spheres are concerned, recall the discussion of cables in Sect. 3. In our present setting of knots in S^3, if K is the (p,q)-cable of K_0 then the boundary-slope of the essential annulus in M_K is pq, and

$$M_K(pq) \cong M_{K_0}(p/q) \# L(q,p) .$$

(It is convenient here to allow K_0 to be unknotted, in which case K is the (p,q)-torus knot and $M_K(pq) \cong L(p,q) \# L(q,p)$. Also, if K_0 is non-trivial then $\pi_1(M_{K_0}(p/q)) \neq 1$ by [CGLS, Corollary 2].) The following Cabling Conjecture of González-Acuña and Short asserts that these are the only examples where essential spheres arise.

Conjecture 5.8 [G-AS]. *$M_K(r)$ contains an essential sphere if and only if K is a (p,q)-cable and $r = pq$.*

Although this conjecture is still open, several partial results exist. For instance, there is Theorem 5.4 above. Also, it is known to be true if K is a satellite knot [Sc], an alternating knot [MT], a strongly invertible knot [E-M], and others [G-AS]. It is also known that if $M_K(r)$ contains an essential sphere then r is an integer [GLu1] and $M_K(r)$ has a lens space summand [GLu2]. (Incidentally, it is an immediate consequence of this last result that no non-prime homology sphere can be obtained by Dehn surgery on a knot in S^3. There remains the interesting question of whether there are *prime* homology spheres that cannot be so obtained.)

Regarding essential tori, some examples are known, but the general picture is not yet clear. However, the following conjecture seems reasonable. (Recall that M_K contains an essential torus if and only if K is a satellite knot.)

Conjecture 5.9. *If K is not a satellite knot and $M_K(r)$ contains an essential torus then $\Delta(r,\infty) \leq 2$.*

(There are examples with $\Delta(r,\infty) = 2$.)

6. The General Knot Complement Problem

Theorem 5.1 above asserts that if $M(r) \cong M(s) \cong S^3$, then $r = s$. More generally, one can ask: when is $M(r) \cong M(s)$? One has to be a little careful with orientations here, for M may have an orientation-reversing automorphism. For instance, if K is an amphicheiral knot (such as the figure eight knot), then $M_K(r) \cong M_K(-r)$. However, if we orient M, then this determines an orientation of $M(r)$, and we can ask: when is $M(r) \cong M(s)$ by an orientation-preserving homeomorphism? No counterexamples are known to the following conjecture.

Conjecture 6.1. *If $M(r) \cong M(s)$ by an orientation-preserving homeomorphism then $r = s$.*

This would imply the truth of the following Oriented Knot Complement Conjecture.

Conjecture 6.2. *If K_1 and K_2 are knots in a closed, oriented 3-manifold Q such that $Q-K_1$ and $Q-K_2$ are homeomorphic by an orientation-preserving homeomorphism, then there exists an orientation-preserving homeomorphism $h : Q \to Q$ such that $h(K_1) = K_2$.*

Mathieu [Ma] has given examples where $Q-K_1$ and $Q-K_2$ are homeomorphic (by an orientation-reversing homeomorphism), but there is *no* homeomorphism of pairs $(Q, K_1) \cong (Q, K_2)$.

7. Intersections of Surfaces

The proofs of the theorems stated above use several techniques from modern 3-dimensional topology. See [Sh] for instance, for a nice account of some of the ideas that went into the proof of the Cyclic Surgery Theorem (Theorem 4.3). However, here we shall focus on just one technique, namely, the analysis of intersections of properly embedded surfaces. The point is that many of the particular forms of what we have called the "degeneration" of the topology of M in $M(r)$ imply the existence of a (useful) surface F in M with non-empty boundary such that each boundary component has slope r on ∂M. For instance, as was pointed out in Sect. 3, if $M(r)$ contains an essential surface S which cannot be moved into M, then M contains an essential surface F with boundary-slope r such that $\widehat{F} \cong S$. Hence if $M(r_1)$ and $M(r_2)$ contain such surfaces S_1, S_2 respectively, then we get corresponding essential surfaces F_1, F_2 in M with boundary-slopes r_1, r_2. Putting F_1 and F_2 in mutual general position, $F_1 \cap F_2$ will consist of circles and arcs properly embedded in F_1 and F_2. Note that the pattern of intersections on the boundary, that is, the triple $(\partial M; \partial F_1, \partial F_2)$ is completely standard: all the components of ∂F_i are parallel, $i = 1, 2$, and each component of ∂F_1 meets each component of ∂F_2 in $\Delta(r_1, r_2)$ points. A final important point is that we may assume that no arc of $F_1 \cap F_2$ is isotopic in F_i (keeping its endpoints fixed) into ∂F_i, $i = 1, 2$. This easily follows from the fact that F_1 and F_2 are essential.

The idea is to then analyze the pattern of arcs of $F_1 \cap F_2$ as they lie on F_1 and F_2. We record the intersections on the boundary by numbering the boundary components of F_i in order as they appear on ∂M, $i = 1, 2$, and labelling the points of $\partial F_1 \cap \partial F_2$ on F_1 with the number of the corresponding boundary component of F_2 (and vice versa). It is convenient to regard the disks $\widehat{F}_i - \mathring{F}_i$ as "fat" vertices, and the arcs of $F_1 \cap F_2$ as edges, so that we get graphs Γ_1, Γ_2 in $\widehat{F}_1, \widehat{F}_2$ respectively, whose vertices and edge-endpoints are labelled as just described. Typically, one now proceeds to show that if $\Delta(r_1, r_2)$ is greater than some Δ_0, then the pair of graphs Γ_1, Γ_2 must contain certain configurations of faces which have topological implications for M (or $M(r_1)$ or $M(r_2)$) inconsistent with the hypotheses. This type of argument was first used by Litherland [Lit].

Theorems 3.3 and 3.4 are proved in this way, the surfaces S_i being spheres and tori respectively. The proof of Theorem 2.1 is similar in spirit, only here the surfaces S_i are disks, whose boundaries lie on the closed surface S, realizing the compression of S in $M(r_i)$, $i = 1, 2$. (This set-up is also an ingredient of the proof of Theorem 4.3.) In Theorem 5.1, the hypothesis is that $M(r_1) \cong M(r_2) \cong S^3$, which of course does not contain any essential surface. Nevertheless, using an idea of Gabai [Ga1], 2-spheres S_i in $M(r_i)$, $i = 1, 2$, can be chosen so that the corresponding punctured spheres F_i in M still satisfy the key condition that no arc of $F_1 \cap F_2$ can be isotoped into the boundary in either F_1 or F_2. This allows one to apply the same philosophy.

8. Conclusion

As mentioned in the Introduction, the importance of Dehn surgery lies in the fact that every closed oriented 3-manifold can be obtained by Dehn surgery on some link in S^3. Thus one might regard a Dehn surgery presentation of a 3-manifold as being analogous to a Heegaard splitting, with the study of Dehn surgery on knots corresponding to the study of Heegaard splittings of genus one. However, I think this analogy is misleading, because, unlike Heegaard splittings of genus one, Dehn surgery on knots already seems to contain much of the complexity of the general case. This, together with the progress that has been made in the case of a single component, makes it hopeful that Dehn surgery might be a useful way to approach general problems about 3-manifolds. At any rate, the next step should be to find appropriate generalizations of the results discussed above to the case where solid tori are attached along the boundary components of an irreducible 3-manifold M whose boundary is a disjoint union of an arbitrary number of tori.

References

[B1] Berge, J.: The knots in $D^2 \times S^1$ which have nontrivial Dehn surgeries that yield $D^2 \times S^1$. Topology and its Applications **38** (1991) 1–19

[B2] Berge, J.: Obtaining lens spaces by surgery on knots. To appear in Proceedings of International Conference on Knots 90. de Gruyter

[BH] Bleiler, S.A., Hodgson, C.G.: Spherical space forms and Dehn surgery. To appear

[BL] Bleiler, S.A, Litherland, R.A.: Lens spaces and Dehn surgery. Proc. Amer. Math. Soc. **107** (1989) 1127–1131

[CGLS] Culler, M., Gordon, C.McA., Luecke, J., Shalen, P.B.: Dehn surgery on knots. Ann. Math. (2) **125** (1987) 237–300

[D] Dehn, M.: Über die Topologie des dreidimensionalen Raumes. Math. Ann. **69** (1910) 137–168

[E-M] Eudave-Muñoz, M.: Band sum of links which yield composite links. To appear

[FS] Fintushel, R., Stern, R.: Constructing lens spaces by surgery on knots. Math. Z. **175** (1980) 33–51

[Ga1] Gabai, D.: Foliations and the topology of 3-manifolds. III. J. Diff. Geom. **26** (1987) 479–536

[Ga2] Gabai, D.: Surgery on knots in solid tori. Topology **28** (1989) 1–6

[G-AS] González-Acuña, F., Short, H.: Knot surgery and primeness. Math. Proc. Camb. Phil. Soc. **99** (1986) 89–102

[Go] Gordon, C.McA.: Boundary slopes of punctured tori in 3-manifolds. To appear

[GLi] Gordon, C.McA., Litherland, R.A.: Incompressible planar surfaces in 3-manifolds. Topology and its Applications **18** (1984) 121–144

[GLu1] Gordon, C.McA., Luecke, J.: Only integral Dehn surgeries can yield reducible manifolds. Math. Proc. Camb. Phil. Soc. **102** (1987) 97–101

[GLu2] Gordon, C.McA., Luecke, J.: Knots are determined by their complements. J. Amer. Math. Soc. **2** (1989) 371–415

[GLu3] Gordon, C.McA., Luecke, J.: Reducible manifolds and Dehn surgery. To appear

[Ha] Hatcher, A.E.: On the boundary curves of incompressible surfaces. Pacific J. Math. **99** (1982) 373–377

[He] Heil, W.: Elementary surgery on Seifert fiber spaces. Yokohama Math. J. **22** (1974) 135–139

[Lic] Lickorish, W.B.R.: A representation of orientable combinatorial 3-manifolds. Ann. Math. **76** (1962) 531–540

[Lit] Litherland, R.A.: Surgery on knots in solid tori, II. J. London Math. Soc. (2) **22** (1980) 559–569

[Ma] Mathieu, Y.: Sur des noeuds qui ne sont pas déterminés par leur complément et problèmes de chirurgie dans les variétés de dimension 3. Thèse, L'Université de Provence, 1990

[MT] Menasco, W.W., Thistlethwaite, M.B.: Surfaces with boundary in alternating knot exteriors. To appear

[Mo] Moser, L.: Elementary surgery along a torus knot. Pacific J. Math. **38** (1971) 734–745

[Sc] Scharlemann, M.: Producing reducible 3-manifolds by surgery on a knot. Topology **29** (1990) 481–500

[Sh] Shalen, P.B.: Representations of 3-manifold groups and applications in topology. Proceedings ICM Berkeley, 1986, pp. 607–614

[T] Thurston, W.: Three dimensional manifolds, Kleinian groups and hyperbolic geometry. Bull. Amer. Math. Soc. **6** (1982) 357–381

[Wal] Wallace, A.H.: Modifications and cobounding manifolds. Can. J. Math. **12** (1960) 503–528

[Wan] Wang, S.: Cyclic Surgery on knots. Proc. Amer. Math. Soc. **107** (1989) 1091–1094

[We] Weeks, J.R.: Hyberbolic structures on three-manifolds. Ph.D. thesis, Princeton University, 1985

[Wu1] Wu, Y-Q.: Cyclic surgery and satellite knots. Topology and its Applications **36** (1990) 205–208

[Wu2] Wu, Y-Q.: Incompressibility of surfaces in surgered 3-manifolds. To appear in Topology

Parametrized Morse Theory and Its Applications

*Kiyoshi Igusa**

Department of Mathematics, Brandeis University, Waltham, MA 02254, USA

We will discuss the basic concepts of parametrized Morse theory and show how they are related to the stability theorem, Waldhausen's $A(X)$ and higher Franz-Reidemeister torsion.

§ 1. Standard Morse Theory

In standard Morse theory one usually takes a compact smooth (C^∞) manifold M with boundary the union of three manifolds $\partial_0 M \cup \partial_1 M \cup D \times I$ where $I = [0, 1]$, $D \times 0 = \partial\partial_0 M$ and $D \times 1 = \partial\partial_1 M$. However for the purpose of these notes we will stick mainly to the case when D is empty. We will also write $X = \partial_0 M$. Thus $\partial M = X \coprod \partial_1 M$.

Any generic smooth function $f: M \to I$ with $f^{-1}(i) = \partial_i M$ for $i = 1, 2$ will be a Morse function with a finite number of nondegenerate singularities $x_1, \ldots, x_m$. Each critical point x_i has an index $\operatorname{ind}(x_i) \geq 0$. We can construct a finite relative CW complex

$$Y = X \cup e_1 \cup e_2 \cup \cdots \cup e_m$$

with one cell e_i for each critical point x_i and $\dim(e_i) = \operatorname{ind}(x_i)$. From this relative cell complex we can construct the cellular chain complex $C_*(Y, X)$ with coefficients in a ring R or with twisted coefficients given by a locally constant sheaf $\mathscr{F}$ over M. If this chain complex is acyclic then we get a torsion invariant $\tau(Y) = \tau(M)$ which is an element of some quotient group of $K_1 R$ independent of the choice of the function f (assuming $K_0 Z \to K_0 R$ is a monomorphism).

Example 1.1. $M = S^1 \times I, X = \partial_0 M = \varnothing, \partial_1 M = S^1 \times \{0, 1\}$ and f is the "height" in the following drawing.

In Example 1.1 there are two critical points x_0, x_1 of indices 0, 1 and $Y = e^0 \cup e^1$ is a cricle. There is no torsion unless we take twisted coefficients in which case $\tau(Y) = \tau(M) = \pm(1 - h^{\pm 1})$ where h is the holonomy of the coefficient sheaf around the circle $S^1 \times 1/2$ in M and the sign ambiguities are due partially to the fact that the 1-cell e^1 is unoriented.

---

* Supported by NSF Grant No. DMS9002512.

Proceedings of the International Congress
of Mathematicians, Kyoto, Japan, 1990

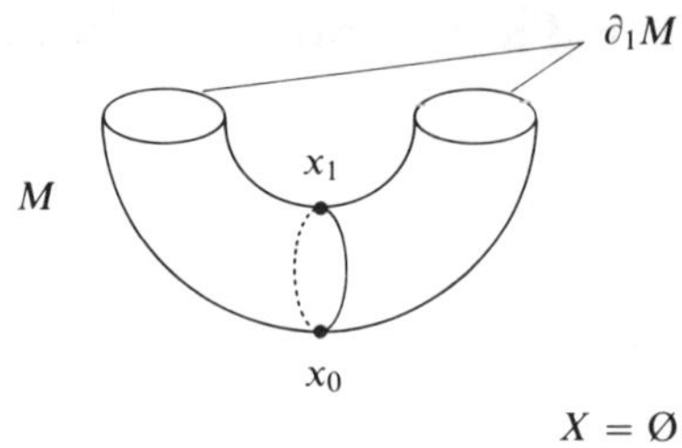

§ 2. Parametrized Morse Theory

The study of parametrized Morse theory began with J. Cerf who used 1 and 2 parameter families of functions on $M = X \times I$ to prove the following theorem with definitions given below.

Theorem 2.1 [C2]. *If X is a simply connected with dimension ≥ 5 then $\mathscr{C}(X)$ is connected.*

Definition 2.2. If X is a compact smooth manifold then the *concordance space* of X is defined to be the space $\mathscr{C}(X) = \mathrm{Diff}(X \times I \text{ rel } X \times 0 \cup \partial X \times I)$ where $\mathrm{Diff}(M \text{ rel } A)$ denotes the space of all self diffeomorphisms of M which are the identity on A with the strong C^∞ topology [C1].

Instead of taking families of functions on a single manifold we will consider functions on "families of manifolds." This is an equivalent point of view. We take a smooth bundle (i.e. submersion) $p: E \to B$ with fiber M so that E contains the trivial subbundle $X \times B$. If $B = S^{k+1}$, the isomorphism classes of such bundles are in one-to-one correspondence with the elements of $\pi_i \mathrm{Diff}(M \text{ rel } X)$. We are interested in calculating these groups and of finding computable invariants.

Given a smooth bundle $M \to E \to B$ as above we will construct maps:

$$\theta: B \to |\mathscr{E}.(X)| = \{\text{cell complexes and characteristic maps}\}$$

$$\lambda\theta: B \to |\mathscr{K}.(Z, F)| = \{\text{filtered chain complexes}\}$$

The second map is constructed from the first by composition with a map $\lambda: |\mathscr{E}.(X)| \to |\mathscr{K}.(Z, F)|$. Taking $B = S^{k+1}$ we will get homomorphisms

$$\theta_+ : \pi_k \mathrm{Diff}(M \text{ rel } X) \to \pi_{k+1} |\mathscr{E}.(X)|$$

$$\lambda_* \theta_* : \pi_k \mathrm{Diff}(M \text{ rel } X) \to \pi_{k+1} |\mathscr{K}.(Z, F)|.$$

It turns out that θ_* is an isomorphism in many cases. Its purpose it to compute the group $\pi_k \mathrm{Diff}(M \text{ rel } X)$. The map $\lambda_* \theta_*$ is used to obtain algebraic invariants to detect particular elements of $\pi_k \mathrm{Diff}(M \text{ rel } X)$.

The idea behind θ and λ is as follows. We take a "good" function $f: E \to I$ and consider it as a family of functions on the fibers $M_t = p^{-1}(t)$ of $p: E \to B$. This will give us a "family of cell complexes" Y_t which we interpret as a map from B to the "space of all cell complexes." By a "linearization" process we get a map to the "space of all filtered chain complexes." We will explain what these terms mean.

§ 3. Waldhausen's Expansion Category

We use a model for the "space of all cell complexes" due to Waldhausen [W1]. Our nomenclature is however different: what we call the "expansion space" is the loop space of Waldhausen's expansion space. Our *expansion space* $|\mathscr{E}.(X)|$ is the geometric realization of a simplicial category $\mathscr{E}.(X)$. The objects of $\mathscr{E}_0(X)$, which form the vertices of $|\mathscr{E}.(X)|$, are given by pairs $(Y, \{\psi_e\})$ where Y is a finite relative cell complex $Y = X \cup e_1 \cup \cdots \cup e_m$ and $\psi_{e_i}: D^{n_i} \to Y$ are *characteristic maps* (i.e. parametrizations) for the cells. The objects of $\mathscr{E}_k(X)$, which form the k-simplices of $|\mathscr{E}.(X)|$, are given by pairs $(Y, \{\psi_e\})$ where Y is a k parameter family of finite relative cell complex $Y = X \times \Delta^k \cup e_1 \times \Delta^k \cup \cdots \cup e_m \times \Delta^k$ and each $\psi_{e_i}: D^{n_i} \times \Delta^k \to Y$ is a k parameter family of characteristic maps. To avoid set theoretic difficulties we assume that Y is a subspace of a fixed infinite dimensional contractible space, e.g., $\mathbb{R}^\infty$. This also clarifies our assertion that morphism are inclusion maps. Figure 3.1 gives an example for $k = 1$.

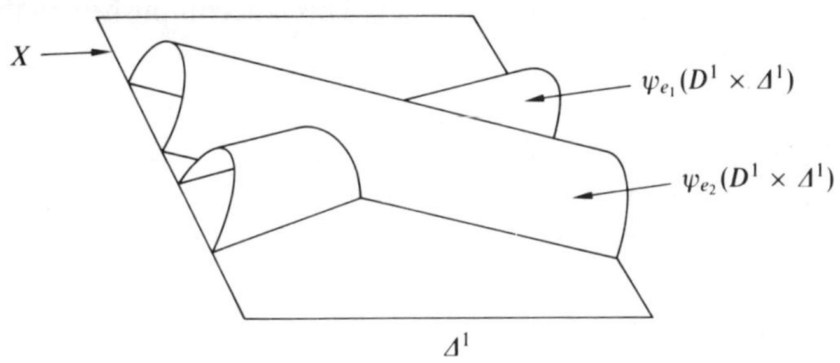

Fig. 3.1

The morphisms of $\mathscr{E}_k(X)$, which may be viewed as gluing maps for the k-simplices of $|\mathscr{E}.(X)|$, are given by expansions. *Expansions* are compositions of *elementary expansions* which are given by inclusion maps: $Y \to Y \cup D^n \times \Delta^k$ where the k parameter family of disks $D^n \times \Delta^k$ is attached to the k parameter family of cell complexes Y along the southem hemisphere $S^{n-1} \times \Delta^k$. Figure 3.2 shows a path in $|\mathscr{E}.(X)|$ which is obtained by gluing two edges $A, B \in \mathscr{E}_k(X)$ along a morphism $f: \partial_1 A \to \partial_0 B$ from an endpoint of A to an endpoint of B.

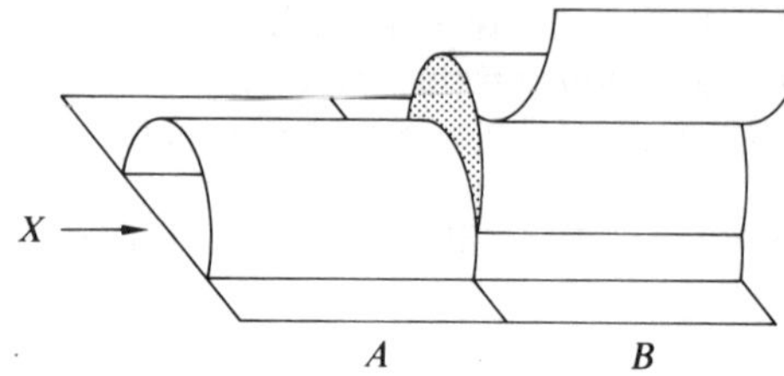

Fig. 3.2

An old unpublished result of Waldhausen gives a computation of the homotopy type of some of the components of the expansion space in terms of the Waldhausen K-theory of X. A proof of this is being prepared for publication in [IW].

Theorem 3.3 (Waldhausen). *$A(X) \simeq Q(X_+) \times B|\mathscr{E}.^h(X)|$ where $A(X)$ = Waldhausen K-theory of X [W2], $Q(X_+) = \Omega^\infty\Sigma^\infty(X \coprod \text{pt.})$ $B|\mathscr{E}.^h(X)|$ is a nonconnected delooping of $|\mathscr{E}.^h(X)|$, $\mathscr{E}.^h(X)$ is the simplicial full subcategory of $\mathscr{E}.(X)$ consisting of pairs (Y, ψ) where $Y \simeq X$.*

§ 4. The Framed Function Theorem

Suppose we have a smooth bundle $M \to E \to B$ as in Section 2 above. Then we would like to construct a map $\theta : B \to |\mathscr{E}.(X)|$. This is accomplished by the following two theorems.

Theorem 4.1 [I1]. *If* $\dim B \leq \dim M$ *there exists a smooth function $f: E \to I$ so that for each $t \in B$, the map $f_t: M_t \to I$ given by restricting f to the fiber over t is a generalized Morse function (GMF), i.e. it satisfies the following conditions.*

a) $f_t^{-1}(0) = X, f_t^{-1}(1) = \partial_1 M_t$,
b) *f_t is nonsingular on ∂M_t,*
c) *f_t has only A_1 and A_2 singularities.*

Since $f: E \to I$ is being considered as a family of functions $f_t: M_t \to I$ we consider the *fiberwise singular set* $\sum(f)$ of f which is defined to be the set of all x in E so that x is a singularity of $f_t: M_t \to I$ where $t = p(x)$ *is the image of* x in B.

Definition 4.2. $y \in M$ is an A_k *singularity* of $g: M \to \mathscr{R}$ if g can be written in local coordinates as

$$g = \pm x_0^{k+1} + \sum_{i=1}^{n} \pm x_i^2 + C.$$

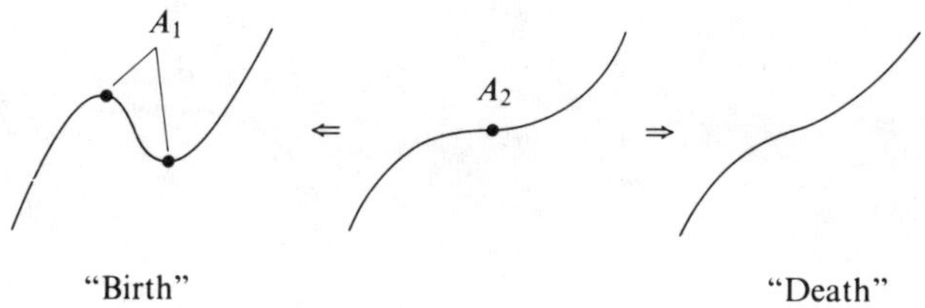

Fig. 4.3

By the Morse lemma A_1 singularities are the same as nondegenerate singularities. A_2 singularities are also called *birth-death* singularities. This is because an A_2 singularity can be perturbed to give either two A_1 singularities in "cancelling position" or to give no singularities. This is illustrated in Fig. 4.3. (As before f is the "height" function given by the distance from the bottom of the page.)

Theorem 4.1 is not sufficient for our purposes. Given a smooth bundle $M \to E \to B$ and a smooth function $f: E \to I$ as described in Theorem 4.1 we can do Morse theory on the family of generalized Morse functions $f_t: M_t \to I$ and we get a family of cell complexes parametrized by B. However we do not get a family of characteristic maps $\{\psi_e\}$ as required in the definition of the expansion category. Therefore we do not get a map from B into its geometric realization. To remedy this situation we have another theorem.

Theorem 4.4 [I2] (Framed Function Theorem). *In Theorem* 4.1 *the smooth map $f: E \to I$ can be chosen so that it admits a "framing." This consists of a Riemannian metric for E and a function ξ on the fiberwise singular set of f of the following form.*

a) *For each critical x of $f_t: M_t \to I$, $\xi(x) = \xi_t(x) = (\xi^1, \ldots, \xi^i)$ is an orthonormal framing for the nonpositive eigenspace of $D^2 f_t(x)$, (i.e. the sum of all eigenspaces of $D^2 f_t(x)$ corresponding to nonpositive eigenvalues).*
b) *For each j, $\xi_t^j(x)$ is a continuous function on its domain.*
c) *At each birth-death singularity the last vector ξ^i lies in* ker $D^2 f_t(x)$ *and the intrinsically defined third derivative $D^3 f_t(x)(\xi^i, \xi^i, \xi^i)$ is positive.*

Furthermore, if dim $B <$ dim M *then (f, ξ) is unique up to framed homotopy.*

We will call the pair (f, ξ) a *fiberwise framed function* on E.

Figure 4.5 illustrates the definition of a framing. In Fig. 4.5.1 there are two A_1 singularities of indices 0, 1. Thus a framing associates a unit tangent vector ξ^1 to the A_1 point of index 1. This goes continuously to the indicated framed A_2 singularity. The unit vector ξ^1 must point to the right at the A_2 singularity by condition (c) in Theorem 4.4. In Fig. 4.5.2 the same A_1 singularities with the opposite framing cannot be cancelled in the same way, although there is a more complicated deformation which eliminates these two framed singularities. In Fig. 4.5.3 the framing vector ξ^2 plays the role of ξ^1 in the first two examples.

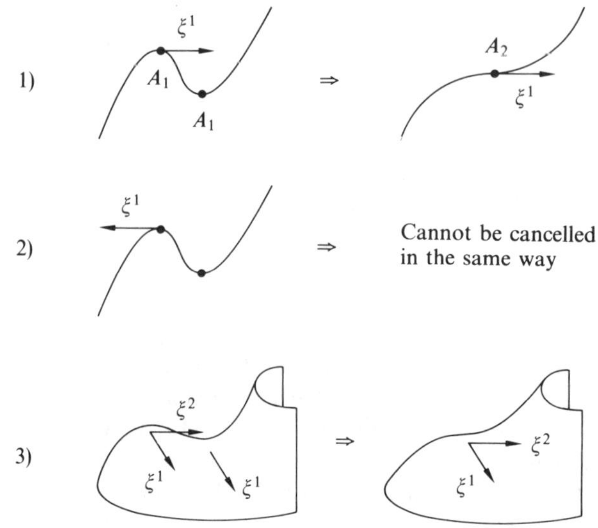

Fig. 4.5

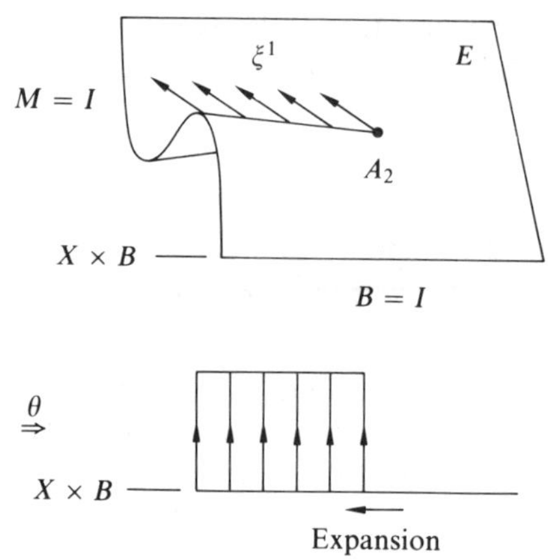

Fig. 4.6

Now suppose that $M \to E \to B$ is a smooth bundle and $(f, \xi): E \to I$ is a fiberwise framed function. Then for each t in B we get a GMF $f_t: E \to I$ and a framing ξ_t. The GMF f_t tells us now to construct a finite relative cell complex Y_t and the framing ξ_t is exactly what is needed to construct the characteristic maps $\{\psi_t\}$ for Y_t. Consequently we get a map $\theta: B \to |\mathscr{E}.(X)|$. The uniqueness of (f, ξ) up to framed homotopy implies that the map θ is unique up to homotopy. We note that the dimension condition dim $B <$ dim M can be circumvented by "stabilization," i.e. by taking the product with a large disk D^N to get $M \times D^N \to E \times D^N \to B$.

Figure 4.6 gives an example of this construction.

§ 5. Application to Pseudoisotopy

Suppose that $M = X \times 1$. Then smooth bundles $M \to E \to S^{k+1}$ are classified by the homotopy group $\pi_k \mathscr{C}(X)$ and the framed function theorem gives us a homomorphism

$$\theta_* : \pi_k \mathscr{C}(X) \to \pi_{k+1} |\mathscr{E}.(X)|$$

This map is an isomorphism if $3k \leq \dim X - 9$ basically because a smooth bundle can be reconstructed from a family of cell complexes by embedding it in Euclidean space and taking a neighborhood. Another way to say this is that we get a highly connected map

$$\theta : B\mathscr{C}(X) \to |\mathscr{E}.^h(X)|$$

where $B\mathscr{C}(X)$ is a nonconnected delooping of $\mathscr{C}(X)$. This leads to the following stability theorem. The details can be found in [I3].

Theorem 5.1 (Stability Theorem). *The suspension map* $\sigma : \mathscr{C}(X) \to \mathscr{C}(X \times I)$ *is* $\sim \dim X/3$*-connected.*

Using Theorem 3.1 we also get another proof of the following famous theorem of Waldhausen [W2]:

Theorem 5.2 (Waldhausen). $A(X) \simeq Q(X_+) \times B^2\mathscr{P}(X)$ *where* $\mathscr{P}(X)$ *is the direct limit of concordance spaces* $\mathscr{C}(X) \to \mathscr{C}(X \times I) \to \cdots \to \mathscr{C}(X \times I^n)$ *with respect to suspension and* $B^2\mathscr{P}(X)$ *denotes a two-fold nonsimply-connected delooping of* $\mathscr{P}(X)$.

The space $\mathscr{P}(X)$ is called the *stable pseudoisotopy* (or *concordance*) space of X.

§ 6. The Space of Filtered Chain Complexes

This section is a report on recent joint work with John Klein.

One way to obtain K-theory invariants for smooth bundles is to triangulate the base space B and associate to each simplex the total singular complex of its inverse image. This gives a functor from the category of simplices of B to a category of chain complexes with additional structure. If this category is carefully constructed it will have homotopy groups closely related to algebraic K-groups. One suitable construction is the simplicial category of "filtered chain complexes."

Suppose that Z is a space, R is an associative ring with 1 and $\mathscr{F}$ is a locally constant sheaf of nonzero finitely generated (f.g.) free R-modules over Z. Then we construct the "space of filtered chain complexes." This is the geometric realization of a simplicial category $\mathscr{K}.(Z, \mathscr{F})$ whose objects in every degree are quadruples $(S, C, \{S^A\}, \{l_{A,e}\})$ where S is a free R-complex, C is a finite poset over Z. $\{S^A\}$ is a family of subcomplexes of S indexed by the order idcals in C and $\{l_{A,e}\}$ is a family of cohomology classes indexed by pairs (A, e) where A is an order ideal in C and e is a minimal element of the complement $C \backslash A$ of A in C. The precise definition is rather technical so we will explain it with two examples. We recall that an *order ideal* in a poset C is a subset A of C so that A contains any element of C which is less than any element of A.

Example 6.1. Let $f: M \to I$ be a Morse function as explained in Section 1, let ξ be a framing for f and let $\mathscr{F}$ be any locally constant sheaf of f.g. free R-modules over M. Then an object $(S, C, \{S^A\}, \{l_{A,e}\})$ of $\mathscr{K}_0(M, \mathscr{F})$ is given as follows.

a) $S = C_*(M, X; \mathscr{F})$ is the relative singular complex of (M, X) with coefficients in $\mathscr{F}$.

b) $C = \sum(f)$ is the singular set of f ordered by critical value. Thus $a < b$ iff $f(a) < f(b)$.
   i) $\text{ind}: C \to N$ is given by $\text{ind}(x) =$ *index of f at x*.
   ii) $p: C \to M$ is the inclusion map.

c) $S^A = C_*(f^{-1}[0, f(e)] \backslash (C \backslash A), X; \mathscr{F})$ for any order ideal A in C where e is a minimal element in $C \backslash A$. (If no such e exists then $A = C$ and $S^C = S$.)

d) If A is an order ideal in C and $e \in A$ is maximal with $\text{ind}(e) = i$ then let

$$l_{A,e}: H_i(S^{A \cup e}, S^A) \simeq \mathscr{F}_{p(e)}$$

be the isomorphism specified by the framing $\xi(e)$ where $\mathscr{F}_{p(e)}$ is the stalk of $\mathscr{F}$ over $p(e)$.

Example 6.2. Let $\pi = \pi_1 X$ and let $B\pi$ be an open $K(\pi, 1)$ neighborhood of X in $\mathbb{R}^\infty$. Let $\mathscr{F}$ be any locally constant sheaf of f.g. free R-modules over $B\pi$. Let $\mathscr{E}.^\pi(X)$ denote the simplicial full subcategory of $\mathscr{E}.(X)$ whose objects are pairs (Y, ψ) so that Y is a subspace of $B\pi$. Then there is a simplicial functor $\lambda: \mathscr{E}.^\pi(X) \to \mathscr{K}.(B\pi, \mathscr{F})$ given in degree zero by $\lambda(Y, \psi) = (S, C, \{S^A\}, \{l_{A,e}\})$ where S, C, $\{S^A\}$, $(l_{A,e}\}$ are given as follows.

a) $S = C_*(Y, X; \mathscr{F}_Y)$ where $\mathscr{F}_Y$ is the restriction of $\mathscr{F}$ to Y.

b) C is the set of cells of Y. (The indexed poset C is actually part of the structure of (Y, ψ).)
   i) $\text{ind}: C \to N$ is given by $\text{ind}(x) = \dim(x)$.
   ii) $p: C \to B\pi$ is given by $p(x) = \psi_x(0)$.

c) $S^A = C_*(Y^A, X; \mathscr{F}_{Y^A})$ where Y^A is the closed subcomplex of Y corresponding to A.

d) If A is an order ideal in C and $e \in A$ is maximal with $\text{ind}(e) = i$ then let

$$l_{A,e}: H_i(S^{A \cup e}, S^A) \simeq \mathscr{F}_{p(e)}$$

be the isomorphism specified by the characteristic map ψ_e.

Let $\mathscr{K}.^h(Z, \mathscr{F})$ denote the simplicial full subcategory of $\mathscr{K}.(Z, \mathscr{F})$ whose objects are those quadruples $(S, C, \{S^A\}, \{l_{A,e}\})$ so that S is acyclic. Assume that Z is connected and that the natural map $K_0 Z \to K_0 R$ is a monomorphism. Then using ideas from the proof of Theorem 3.3 we obtain the following.

Theorem 6.3 [IK]. *There is a homotopy fiber sequence*:

$$|\mathscr{K}.^h(Z, \mathscr{F})| \to Q(Z_+) \to \mathbb{Z} \times B\,\mathrm{Gl}(R)^+.$$

Corollary 6.4 (Higher Franz-Reidemeister Torsion). *Suppose that* $\pi_1 M = \pi_1 X = \pi$ *is a finite group,* $R = \mathbb{C}$ *and* $\mathscr{F}$ *has unitary holonomy, i.e.* $\mathscr{F}$ *is given by a unitary representation of* π. *Let* $M \to E \to S^{2k}$ *be a smooth bundle as in Section 2 so that the map in homology* $H_*(M, X; \mathscr{F}) \to H_*(E, X \times S^{2k}; \mathscr{F})$ *induced by the inclusion of* M *in* E *is a monomorphism. Then there is a naturally defined torsion invariant* $\tau(E) \in \mathbb{R}$. *Furthermore these invariants detect the rational homotopy groups of* $\mathscr{P}(*)$ *and are thus nontrivial for all even* k.

Remark. Higher Franz-Reidemeister torsion was first constructed by Wagoner [W] but only for bundles over circles. Klein [K] was the first to construct higher Franz-Reidemeister torsion invariants for bundles over higher dimensional spheres but he assumed that $H_*(M, X; \mathscr{F}) = 0$.

References

[C1] Cerf, J.: Topologie de certain espaces de plongements. Bull. Soc. Math. France **89** (1961) 227–380

[C2] Cerf, J.: La stratification naturelle des espaces de fonctions différentiables réelles et le théorème de la pseudo-isotopie. I.H.E.S. Publ. Math. **39** (1970) 5–173

[I1] Igusa, K.: On the homotopy type of the space of generalized Morse functions. Topology **23** (1984) 245–256

[I2] Igusa, K.: The space of framed functions. Transactions of the A.M.S. **301** (1987) 431–477

[I3] Igusa, K.: The stability theorem for smooth pseudoisotopies. *K*-Theory **2** (1988) 1–355

[IK] Igusa, K., Klein, J.: Filtered chain complexes and higher Franz-Reidemeister torsion. In preparation

[IW] Igusa, K., Waldhausen, F.: The expansion space and its application to pseudoisotopy theory. In preparation

[K] Klein, J.: The cell complex construction and higher *R*-torsion for bundles with framed Morse function. Ph.D. Thesis, Brandeis University, 1989

[W] Wagoner, J.: Diffeomorphisms, K_2, and analytic torsion. Proc. Symp. Pure Math., vol. 32, part I, AMS (1978) 23–33

[W1] Waldhausen, F.: Algebraic *K*-theory of topological spaces I. Proc. Symp. Pure Math. vol. 32, part I, AMS (1978) 35–60

[W2] Waldhausen, F.: Algebraic *K*-theory of spaces, Lecture Notes in Mathematics, vol. 1126. Springer, Berlin Heidelberg New York 1985, pp. 318–419

Rigidity in Geometry and Topology

F. Thomas Farrell[1] *and Lowell E. Jones*[2]

[1] SUNY, Binghamton, NY 13902, USA
[2] SUNY, Stony Brook, NY 11794, USA

Here is an outline of the paper.

In § 1 we consider whether the class of closed Riemannian manifolds (closed = connected, compact, with no boundary) with sectional curvature satisfying $K \leq 0$ everywhere is smoothly rigid or is topologically rigid. In other words if M, N are two such manifolds with isomorphic fundamental groups, then are M, N diffeomorphic to one another or homeomorphic to one another? We review some of the history of the problem beginning with the rigidity results of Bieberbach and Mostow. Our two main results state that this class of manifolds is not smoothly rigid, but it is topologically rigid in dimensions greater than four.

In § 2 we analyze the space $\mathrm{TOP}(M)$ of all self-homeomorphisms of a closed Riemannian manifold M whose sectional curvature satisfies $K \leq 0$ everywhere. In this setting rigidity expresses itself through the simplicity of the homotopy type of $\mathrm{TOP}(M)$, e.g. the rational homotopy groups of $\mathrm{TOP}(M)$ vanish through a stable range of dimensions greater than one. We review results which reduce this problem to the study of the space $\mathscr{P}(M)$ of stable topological pseudoisotopies of M. Our main result in this section states that up to homotopy the space $\mathscr{P}(M)$ can be constructed in a simple way from the stable pseudoisotopy spectrum $\mathscr{P}_*(S^1)$ of the circle.

In § 3 we formulate a general conjecture whose truth would certainly imply the main topological results of § 1, § 2, and would also imply many other well known conjectures in algebraic K-theory and algebraic topology (e.g. the Novikov Conjectures, the Borel Conjecture in dimensions ≥ 5, and Conjectures 1.8, 1.9, 2.8 of this paper). Roughly speaking our conjecture states that for any connected CW complex X the stable topological pseudoisotopy Ω-spectrum $\mathscr{P}_*(X)$ can be constructed in a simple way from the Ω-spectra $\{\mathscr{P}_*(X_H) : H \in \mathscr{G}(X)\}$, where $\mathscr{G}(X)$ denotes the collection of all cyclic by finite subgroups of $\pi_1 X$ and where $X_H \to X$ denotes the covering projection corresponding to any $H \in \mathscr{G}(X)$. This same conjecture is also made for the Ω-spectra valued functors associated to surgery L-theory, algebraic K-theory, and to smooth pseudoisotopy theory. Based on our recent computations of the surgery L-groups and the stable pseudoisotopy spectra of cocompact discrete subgroups of virtually connected Lie groups (cf. 3.4), we believe it is only a matter of time before these conjectures are verified when $\pi_1 X$ is a discrete subgroup of a virtually connected Lie group. For more general classes of fundamental groups the authors regard these conjectures as only estimates which best fit the known data at

Proceedings of the International Congress
of Mathematicians, Kyoto, Japan, 1990

this time. If these estimates prove incorrect for a larger class of groups, then one should look for the minimal class of subgroups $\mathscr{G}(X)$ of each $\pi_1 X$ which makes the preceeding conjecture true ($\mathscr{G}(X)$ is now forced to be larger then the class of cyclic by finite subgroups of $\pi_1 X$).

§ 1. Rigidity

We begin with a statement of the rigidity results of Bieberbach [5] and of Mostow [26, 27].

1.1 Bieberbach's Rigidity Theorem. *Let $f: N \to M$ be a homotopy equivalence between closed flat (i.e. sectional curvature satisfies $K = 0$ everywhere) Riemannian manifolds. Then f is homotopic to an affine diffeomorphism.*

1.2 Mostow's Rigidity Theorem. *Let $f: N \to M$ be a homotopy equivalence between closed locally symmetric spaces with sectional curvature satisfying $K \leq 0$ everywhere. Suppose that M has no closed one or two-dimensional totally geodesic subspaces which are local factors of M. Then f is homotopic to a diffeomorphism, which becomes an isometry after adjusting the normalizing factors of M.*

These two results suggest the following problem.

1.3 Problem. Let $f: N \to M$ denote a homotopy equivalence between two closed Riemannian manifolds which both have sectional curvature satisfying $K \leq 0$ everywhere. Is f homotopic to a diffeomorphism or to a homeomorphism?

1.4 History of Problem. In addition to the Rigidity Theorems of Bieberbach and Mostow, there are the following results which give partial answers to Problem 1.3. Eells and Sampson [9] showed that f is homotopic to a harmonic map $\bar{f}: N \to M$, and Al'ber [1] and Hartmann [18] showed that $\bar{f}$ is uniquely determined by f if $K < 0$ holds everywhere for M. Lawson and Yau then suggested that $\bar{f}: N \to M$ should be a diffeomorphism whenever $K < 0$ holds everywhere for both N, M. This was proven by Schoen and Yau [33] when dim $M = 2$, and follows from Mostow's Rigidity Theorem when N, M are locally symmetric spaces. The next theorem (cf. 1.5) gives counterexamples to the Lawson-Yau problem in dimensions greater than six, but the problem remains open in dimensions three through six. If $K = 0$ holds everywhere for M in 1.3 then it follows from work of Gromoll and Wolf [17], and from work of Yau [39], that $K = 0$ also holds everywhere for N; thus f must be homotopic to an affine diffeomorphism by Bieberbach's Rigidity Theorem. If M is an irreducible locally symmetric space of rank ≥ 2 Gromov [3] has shown that after rescaling the metric on M f will be homotopic to an isometry. Eberlein [7, 8] independently proved the same result under the hypothesis that the universal cover of M is reducible. Cheeger showed in the mid-1970s that the bundles of orthonormal two-frames $V_2(N)$, $V_2(M)$ are homeomorphic provided $K < 0$ holds everywhere for both N, M; and then, under the same hypothesis, Gromov showed that the unit sphere bundles $S(N)$, $S(M)$ are homeomorphic, via a homeomorphism which preserves the orbits of the geodesic flows. Mishchenko [25] showed that f pulls the rational Pointrjagin classes of M back to those of N; and Farrell and Hsiang [10] showed that $f \times \mathrm{id} : N \times \mathbf{R}^3 \to M \times \mathbf{R}^3$ is properly homotopic to a homeomorphism, where $\mathrm{id} : \mathbf{R}^3 \to \mathbf{R}^3$ denotes the identity map.

The authors [11, 15] have proven the following two theorems which more or less settle Problem 1.3. In the first of these theorems we let M denote any closed real hyperbolic manifold with sectional curvature satisfying $K = -1$ everywhere, and we let $\Sigma_1, \Sigma_2, \ldots, \Sigma_n$ denote a complete list of nondiffeomorphic exotic spheres of dimension equal $\dim M > 4$.

1.5 Theorem. *Given any number $\delta > 0$ there is a finite sheeted covering space projection $\hat{M} \to M$ which satisfies the following properties.*

(a) *No two of the manifolds $\hat{M}$, $\hat{M} \# \Sigma_1, \ldots, \hat{M} \# \Sigma_n$ are diffeomorphic, but they are all homeomorphic.*

(b) *Each $\hat{M} \# \Sigma_i$ supports a Riemannian metric all of whose sectional curvature values lie in the interval $(-1 - \delta, -1 + \delta)$.*

1.6 Theorem. *Let $f : N \to M$ denote a homotopy equivalence between two closed manifolds, where M is a closed Riemannian manifold with sectional curvature satisfying $K \leq 0$ everywhere. Then f is homotopic to a homeomorphism, provided* $\dim M \geq 5$.

1.6.1 Remark. There is the following more general version of 1.6: any homotopy equivalence $h : (N, \partial N) \to (M \times I^k, \partial(M \times I^k))$ of compact manifold pairs, which is a homeomorphism of the boundaries, is homotopic rel ∂ to a homeomorphism, provided $k + \dim M \geq 5$, where $k \geq 0$ is an integer and I^k denotes the k-fold Cartesian product of the unit interval with itself.

That the conclusion to Theorem 1.6 should remain true for any closed aspherical manifold M is known as the *Borel Conjecture*. In the remainder of this section we formulate two conjectures in terms of generalized homology theory and the Whitehead group, which taken together are even more general than Borel's conjecture.

1.7 The Assembly Map and Generalized Homology Theory. We remind the reader that Ω-spectra are the "coefficients" of generalized homology (and cohomology) theories. Let S_* denote an Ω-spectrum, that is S_* is a collection of spaces $\{S_i : i \in \mathbf{Z}\}$ together with given homotopy equivalences $S_i \cong \Omega S_{i+1}$ and base points $s_i \in S_i$ for all $i \in \mathbf{Z}$. For any topological space X and for each integer j define a space $\mathbb{H}_j(X, S_*)$ to be the direct limit space $\lim_{i \to \infty} \Omega^i(X \times S_{j+i}/X \times s_{j+i})$. Note that the collection of spaces $\{\mathbb{H}_j(X, S_*) : j \in \mathbf{Z}\}$ is a Ω-spectrum, which we denote by $\mathbb{H}_*(X, S_*)$ and call the *homology spectrum* for X with coefficients in S_*. Recall that the *homology groups* $H_j(X, S_*)$ for X with coefficients in S_* are defined to be the homotopy groups $\pi_j(\mathbb{H}_*(X, S_*))$ of the Ω-spectrum $\mathbb{H}_*(X, S_*)$.

Let $\mathscr{L}_*^h(\ )$ denote the Ω-spectra valued homotopy functor whose value on a connected CW complex X is the equal to the Ω-spectrum of surgery classifying spaces with four-fold periodicity (cf. [30]) for the group $\pi_1 X$ with orientation data given by the zero homomorphism $w: \pi_1 X \to \mathbf{Z}_2$, where the homotopy groups of $\mathscr{L}_*^h(X)$ are the surgery groups $L_*^h(\pi_1 X, w)$ defined by Wall [37]. Let "pt." denote the space with one point. Note that for each integer j we get a map $\phi_j : X \times \mathscr{L}_j^h(\text{pt.}) \to \mathscr{L}_j^h(X)$ by identifying each $y \times \mathscr{L}_j^h(\text{pt.})$, $y \in X$, with the subset $\mathscr{L}_j^h(y) \subset \mathscr{L}_j^h(X)$. Define $A_j : \mathbb{H}_j(X, \mathscr{L}_*^h(\text{pt.})) \to \mathscr{L}_j^h(X)$ to be the direct limit of the composition of maps

$$\Omega^i(X \times \mathscr{L}_{j+i}^h(\text{pt.})/X \times s_{j+i}) \xrightarrow{\Omega^i(\phi_{j+i})} \Omega^i(\mathscr{L}_{j+i}^h(X)) \cong \mathscr{L}_j^h(X),$$

where s_{j+i} is the base point of $\mathscr{L}_{j+i}^h(\text{pt.})$. Note that the collection of all such maps $A_* : \mathbb{H}_*(X, \mathscr{L}_*^h(\text{pt.})) \to \mathscr{L}_*^h(X)$ is a mapping of Ω-spectra, which is called the *assembly map* (cf. [29, 38]).

1.8 Conjecture. *Let X denote any connected aspherical CW complex with torsion free $\pi_1 X$. Then the assembly map $A_* : \mathbb{H}_*(X, \mathscr{L}_*^h(\text{pt.})) \to \mathscr{L}_*^h(X)$ is an equivalence of Ω-spectra.*

1.9 Conjecture. *The Whitehead group $Wh(\Gamma)$ of any torsion free group Γ is zero.*

1.10 Remark. There is a more general version of Conjecture 1.8 (which we do not write down because of space limitations) that covers the case of surgery when the surgery orientation data $w: \pi_1 X \to \mathbf{Z}_2$ is not assumed to be the zero homomorphism. Conjecture 1.8 (in its dual cohomology form) and Conjecture 1.9 are possibilities that many experts have been aware of for some time (cf. [20, 28, 31]). It has also been known for some time (cf. [24, 37]) that if X is a connected aspherical compact manifold (possibly with boundary) satisfying $\dim(X) \geq 5$ then Conjectures 1.8, 1.9 (with $\Gamma = \pi_1 X$ in 1.9) are together equivalent to the following strong topological rigidity property for X: any homotopy equivalence $h: (N, \partial N) \to (X \times I^k, \partial(X \times I^k))$ of compact manifold pairs, which is a homeomorphism of the boundaries is homotopic rel ∂ to a homeomorphism, where $k \geq 0$ is an integer and I^k is the k-fold Cartesian product of the unit interval with itself. Thus it follows from 1.6.1 that Conjectures 1.8, 1.9 are true for X equal any closed Riemannian manifold with sectional curvature satisfying $K \leq 0$ everywhere. Note that $X \times T^n$ also is a closed Riemannian manifold having $K \leq 0$ everywhere, where T^n is the flat n-torus. So by 1.9 we have that $Wh(\pi_1 X \oplus \mathbf{Z}^n) = 0$ holds for all integers $n \geq 0$. Now this last equality together with the Bass-Heller-Swan formula [4] implies that $\tilde{K}_0(Z\pi_1 X) = 0$ and $K_{-i}(Z\pi_1 X) = 0$ for all integers $i \geq 1$. We have just derived the following theorem.

1.11 Theorem. *Let X denote a closed Riemannian manifold with sectional curvature satisfying $K \leq 0$ everywhere. Then we have that $Wh(\pi_1 X) = 0$, $\tilde{K}_0(Z\pi_1 X) = 0$, and $K_{-i}(Z\pi_1 X) = 0$ for all integers $i \geq 1$. Moreover Conjecture 1.8 is also true for X.*

§ 2. Spaces of Self-homeomorphisms

Let TOP(M) denote the space of all self-homeomorphisms of the closed manifold M. Topologists study TOP(M) by analyzing the three associated spaces $G(M)$,

$\overline{\text{TOP}}(M)$, $P(M)$ of self-homotopy equivalences of M, blocked homeomorphisms of M, and pseudoisotopies of M. We have the inclusions $\text{TOP}(M) \subset \overline{\text{TOP}}(M) \subset G(M)$ since we may think of all of these spaces as being semi-simplicial spaces in which a typical k-simplex Δ consists of a selfmap $f: \Delta^k \times M \to \Delta^k \times M$ of the Cartesian product of the standard k-simplex with M which satisfies the following properties: if $\Delta \in \text{TOP}(M)$ then for each $p \in \Delta^k$ the restricted map $f: p \times M \to p \times M$ is a homeomorphism; if $\Delta \in \overline{\text{TOP}}(M)$ then for each face $\Delta \in \Delta^k$ the restricted map $f: \Delta \times M \to \Delta \times M$ is a homeomorphism; if $\Delta \in G(M)$ then for each face $\Delta \in \Delta^k$ the restricted map $f: \Delta \times M \to \Delta \times M$ is a homotopy equivalence. Recall that a pseudoisotopy of M is a homeomorphism $h: M \times [0, 1] \to M \times [0, 1]$ such that $h|M \times 0$ is equal the inclusion $M \times 0 \subset M \times [0, 1]$.

The following lemma, which is due to Hatcher [19], shows the role of $\overline{\text{TOP}}(M)$ and $P(M)$ in analyzing $\text{TOP}(M)$.

2.1 Lemma. *There is a spectral sequence with* $E^1_{p,q} = \pi_q(P(M \times I^p))$ *which converges to* $\pi_{q+p+1}(\overline{\text{TOP}}(M)/\text{TOP}(M))$.

For any aspherical closed manifold M it is well known that there is a homotopy equivalence $G(M) \cong \text{Out}(\pi_1 M) \times K(\text{center}(\pi_1 M), 1)$, where $\text{Out}(\pi_1 M)$ denotes the outer automorphism group and where $K(\Gamma, 1)$ denotes the aspherical space having $\pi_1(K(\Gamma, 1)) = \Gamma$. If M is a closed aspherical manifold for which Conjectures 1.8, 1.9 hold then we have (by 1.10) that the inclusion $\overline{\text{TOP}}(M) \subset G(M)$ is a homotopy equivalence provided $\dim M \geq 5$. Thus we have the following lemma.

2.2 Lemma. *Let M denote a closed aspherical manifold with* $\dim M \geq 5$ *for which Conjectures* 1.8, 1.9 *hold (e.g. M is a Riemannian manifold with $K \leq 0$ everywhere). Then there is a homotopy equivalence* $\overline{\text{TOP}}(M) \cong \text{Out}(\pi_1 X) \times K(\text{center}(\pi_1 X), 1)$.

The effect of Lemmas 2.1, 2.2 is to reduce the study of $\text{TOP}(M)$ to the study of the spaces of pseudoisotopies $P(M \times I^p)$, $p = 0, 1, 2, 3, \ldots$. Note that there is an "inclusion" map $P(M \times I^p) \to P(M \times I^{p+1})$ gotten by forming the product of any pseudoisotopy $h: (M \times I^p) \times [0, 1] \to (M \times I^p) \times [0, 1]$ with the identity map $\text{id}: I \to I$. Thus we may form the direct limit space $\mathscr{P}(M) = \lim_{p \to \infty} P(M \times I^p)$, which is more amenable to homotopy analysis than its approximating spaces $P(M \times I^p)$. One of the benefits of this stabilization process is that $\mathscr{P}(M)$ now fits in as the zero'th space of an Ω-spectrum $\mathscr{P}_*(M)$ (cf. [19]). Moreover $\mathscr{P}_*(\ )$ is an Ω-spectra valued continuous functor defined on the category of all topological spaces; the value of $\mathscr{P}_i(\ )$, $i \geq 0$, on the space X is the semisimplicial space of all stable pseudoisotopies over $X \times \mathbf{R}^i$ which have compact support in the factor X and are bounded in the factor $\mathbf{R}^i$. Note that any homotopy information obtained about $\mathscr{P}(M)$ can be translated into information about the spaces $P(M \times I^p)$ by using the following lemma.

2.3 Lemma. *For any closed smooth manifold M the inclusion map* $P(M \times I^p) \subset \mathscr{P}(M)$ *induces an isomorphism on homotopy groups through dimension* $d = (\dim M + p - 7)/3$ *provided* $(\dim M + p) > 10$ *holds.*

This last lemma is proven by Igusa [22] for spaces of smooth psdeudoisotopies. Results of Burghelea and Lashof [6], and of Goodwillie (unpublished), then imply that the lemma also holds for spaces of topological pseudoisotopies.

We will now precisely formulate a result which states roughly that $\mathscr{P}_*(M)$ can be constructed in a simple way from $\mathscr{P}_*(S^1)$, where S^1 is the circle. Towards this end we need the terminology and notation of the following two subsections.

2.4 Homology Theory with Coefficients in Stratified and Twisted Ω-Spectra and the Assembly Map. A mapping $f: E \to X$ is called a *generalized Siefert fibration* if there is a triangulation T for X, and for each simplex $\Delta \in T$ there is a continuous map g_Δ: $Y_\Delta \times \Delta \to f^{-1}(\Delta)$, such that the following hold: $f \circ g_\Delta$ is equal projection onto the second factor of $Y_\Delta \times \Delta$, and Y_Δ is connected; the restricted map $g_\Delta | Y_\Delta \times (\Delta - \partial\Delta)$ is a homeomorphism onto $f^{-1}(\Delta - \partial\Delta)$; and the restricted map $g_\Delta | Y_\Delta \times (\sigma - \partial\sigma)$ is a covering space projection onto $f^{-1}(\sigma - \partial\sigma)$ for each face $\sigma \in \Delta$.

Let $\mathscr{S}_*(\ )$ denote a given Ω-spectra valued continuous functor from the category of path connected topological spaces, and let $f: E \to X$ denote a given generalized Siefert fibration. Note that the collection of Ω-spectra $\{\mathscr{S}_*(f^{-1}(p)) : p \in X\}$ is a system of stratified and twisted coefficients over X with respect to which we can (roughly) define the *homology spectrum* for X, denoted by $\mathbb{H}_*(X, \mathscr{S}_*(f))$, by

$$\mathbb{H}_j(X, \mathscr{S}_*(f)) = \lim_{i \to \infty} \Omega^i\left(\left(\bigcup_{p \in X} \mathscr{S}_{i+j}(f^{-1}(p))\right) \Big/ \left(\bigcup_{p \in X} s_{i+j,p}\right)\right)$$

where $s_{i+j,p} \in \mathscr{S}_{i+j}(f^{-1}(p))$ are the base points. The *k-th homology group* of X with respect to the system of coefficients $\{\mathscr{S}_*(f^{-1}(p)) : p \in X\}$, denoted by $H_k(X, \mathscr{S}_*(f))$, is defined to be $\pi_k(\mathbb{H}_*(X, \mathscr{S}_*(f)))$. There is also an *assembly map* $A_* : \mathbb{H}_*(X, \mathscr{S}_*(f)) \to \mathscr{S}_*(E)$ induced as in 1.7 from the inclusions $\mathscr{S}_*(f^{-1}(p)) \to \mathscr{S}_*(E)$. For more details about the homology spectrum $\mathbb{H}_*(X, \mathscr{S}_*(f))$ and the assembly map A_* the reader is referred to Quinn [29], Yamasaki [38], and to Farrell and Jones [14].

2.5 The Generalized Siefert Fibration $p: E \to G$ and the Map $f: E \to M$. Let S denote the space of all maps $\{S^1 \to S^1\}$ which have constant non-zero speed. Note that S is a semi-group with respect to the composition operation. Let F denote the space of all essential closed geodesics $\{g : S^1 \to M\}$ in the closed Riemannian manifold M with sectional curvature satisfying $K \leq 0$. Note that S "acts" on the right of F by $(g, h) \to g \circ h$ for $g \in F, h \in S$, and S "acts" on the right of $F \times S^1$ by $(g, x, h) \to (g \circ h, h^{-1}(x))$ for $g \in F, x \in S^1, h \in S$. We define $p : E \to G$ to be the quotient of the first factor projection $F \times S^1 \to F$ under the right actions of S. Note that the map $F \times S^1 \to M$ defined by $(g, x) \to g(x)$ induces a map $f: E \to M$. We note that G parametrizes the collection of (unparametrized) essential t-simple closed geodesics $g : S^1 \to M$ (where "t-simple" means that there is a no rotation $r : S^1 \to S^1$ other than the identity which satisfies $g = g \circ r$), in fact any such geodesic is equal to the restricted map $f: p^{-1}(y) \to M$ for some $y \in G$. The authors have proven that $p : E \to G$ is a generalized Siefert fibration.

The following two results have been proven by the authors [15].

2.6. Theorem. *Let M be a closed Riemannian manifold with sectional curvature satisfying $K \leq 0$ everywhere. Then the composite map*

$$\mathbb{H}_*(G, \mathscr{P}_*(p)) \xrightarrow{\mathscr{P}_*(f) \circ A_*} \mathscr{P}_*(M)$$

is an equivalence of Ω-spectra, where A_* is the assembly map for the generalized Siefert fibration $p: E \to G$ (cf. 2.4, 2.5) and where $\mathscr{P}_*(f)$ is the image of the map $f: E \to M$ under the functor $\mathscr{P}_*(\ )$.

2.6.1 Remark. Theorem 2.6 reduces the study of $\mathscr{P}_*(M)$ to the Ω-spectrum $\mathscr{P}_*(S^1)$ about which there is a good deal known. For example Waldhausen [35, 36] shows that $\pi_i(\mathscr{P}_*(S^1)) \otimes \mathbf{Q} = 0$ for $i \geq 0$, and Anderson and Hsiang [2] show that $\pi_i(\mathscr{P}_*(S^1)) = 0$ for $i < 0$; results of Igusa [23] imply that $\pi_0(\mathscr{P}_*(S^1)) = \mathbf{Z}_2^\infty$, where $\mathbf{Z}_2^\infty$ denotes the direct sum of a countably infinite number of copies of $\mathbf{Z}_2$. By using Theorem 2.6 and a spectral sequence argument we get these same equalities for $\pi_i(\mathscr{P}_*(M))$.

2.7 Corollary. *Let M be as in 2.6 with* $\dim M > 10$ *and let i be any integer satisfying* $1 \leq i \leq (\dim M - 7)/3$. *Then*

$$\pi_i(\mathrm{TOP}(M)) \otimes \boldsymbol{Q} = \begin{Bmatrix} \mathrm{center}(\pi_1 M) \otimes \boldsymbol{Q}, \textit{if } i = 1 \\ 0, \textit{if } i \neq 1 \end{Bmatrix}.$$

Moreover the forgetful map $\pi_0(\mathrm{TOP}(M)) \to \mathrm{Out}(\pi_1 X)$ is surjective with kernal equal $\boldsymbol{Z}_2^\infty$.

The proof of 2.7 consists of combining Lemmas 2.1, 2.2, 2.3 with Remarks 1.6.1, 2.6.1.

We end this section by recalling a conjecture made by the authors in [14]. To precisely formulate this conjecture we associate to any space X a generalized Siefert fibration $\bar{p}: \bar{E} \to \bar{G}$ as follows. Let $\bar{F}$ denote the space of all essential continuous maps $\{g: S^1 \to X\}$ and let S be the semi-group of 2.5. Note S acts on the right of $\bar{F} \times S^1$ and of $\bar{F}$ as described in 2.5. Our first candidate for $\bar{p}: \bar{E} \to \bar{G}$ is gotten by imitating the construction of $p: E \to G$ in 2.5, i.e. taking the quotient of the projection map $\bar{F} \times S^1 \to \bar{F}$ under the right actions of S. However with this definition it is not clear that $\bar{p}: \bar{E} \to \bar{G}$ is a generalized Siefert fibration. So instead we define $\bar{p}: \bar{E} \to \bar{G}$ to be a semisimplicial version of our first candidate: replace $\bar{F}$ and S by their associated semisimplicial objects and proceed as before. By imitating the construction of $f: E \to M$ in 2.5 we get a map $\bar{f}: \bar{E} \to X$. (The reader is referred to [14; §4] for more details).

2.8. Conjecture. *Let X denote any path connected aspherical CW complex with torsion free $\pi_1 X$. Then the composite map*

$$\mathbb{H}_*(\bar{G}, \mathscr{P}_*(\bar{p})) \xrightarrow{\mathscr{P}_*(\bar{f}) \circ A_*} \mathscr{P}_*(X)$$

is an equivalence of Ω-spectra, where A_* is the assembly map for the generalized Siefert fibration $\bar{p}: \bar{E} \to \bar{G}$ and where $\mathscr{P}_*(\bar{f})$ is the image of the map $\bar{f}: \bar{E} \to X$ under the functor $\mathscr{P}_*(\ )$.

2.8.1 Remark. Unlike Conjectures 1.8, 1.9, which have been around for some time, Conjecture 2.8 was first formulated in 1989 by the authors [14] as an attempt to generalize Theorem 2.6. It requires some argument (cf. [14; §4]) to see that 2.6 and 2.8 give the same calculation for $\mathscr{P}_*(M)$ when M is as in 2.6.

§3. The Isomorphism Conjectures

The starting point of this section consists of Conjectures 1.8, 2.8. Simply stated these conjectures offer recipes for computing the values of the functors $\mathscr{L}_*^h(\ )$ and $\mathscr{P}_*(\ )$ on a space X in terms of their values on a collection of much simpler spaces (a point or a circle respectively). In this section we formulate a more general conjecture along these lines, where we drop the hypothesis of 1.8, 2.8 that X be aspherical with torsion free $\pi_1 X$, and we add to our list of Ω-spectra valued continuous functors $\mathscr{S}_*(\ )$ which we wish to analyze by such a conjectured recipe.

3.1 The Functors $\mathscr{P}_*(\ )$, $\mathscr{P}_*^{\text{diff}}(\ )$, $\mathscr{L}_*^{\infty}(\ )$, $\mathscr{K}_*(\ )$. We have already discussed $\mathscr{P}_*(\ )$ in §2; $\mathscr{P}_*^{\text{diff}}(\ )$ is the functor whose value on a space X is the Ω-spectrum defined by Hatcher [19] of stable smooth pseudoisotopies which have compact support in X; $\mathscr{L}_*^{\infty}(\ )$ is the functor whose value on a path connected space X is the Ω-spectrum (with four-fold periodicity) of surgery classifying spaces for the group $\pi_1 X$ with orientation data given by the zero homomorphism $w: \pi_1 X \to \mathbf{Z}_2$, having the surgery groups $L_*^{\infty}(\pi_1 X, w)$ defined by Ranicki [32] for homotopy groups; and $\mathscr{K}_*(\ )$ is the functor whose value on the path connected space X is the algebraic K-theory Ω-spectrum for the group ring $\mathbf{Z}\pi_1 X$ defined by Gersten [16] and by Wagoner [34].

3.2 The Generalized Siefert Fibration $\varrho(X): \mathscr{E}(X) \to \mathscr{B}(X)$ and the Map $f(X)$: $\mathscr{E}(X) \to X$. Recall that a group H is *cyclic by finite* (also called *virtually cyclic*) if either H is finite or if there is a short exact sequence of groups $\mathbf{Z} \to H \to F$ where F is finite and $\mathbf{Z}$ is the infinite cyclic group. Let X denote a path connected CW complex, and let $\mathscr{G}(X)$ denote the collection of all cyclic by finite subgroups of $\pi_1(X, x_0)$ where x_0 is a given base point of X. Define a category $\mathscr{C}(X)$ as follows. The objects of $\mathscr{C}(X)$ are equivalence classes of based connected covering space projections $p: (Y, y_0) \to (X, x_0)$ such that Image$(\pi_1 p) \in \mathscr{G}(X)$, where two such covering projections $p: (Y, y_0) \to (X, x_0)$ and $p': (Y', y_0') \to (X, x_0)$ are equivalent if there is a homeomorphism $h: (Y, y_0) \to (Y', y_0')$ satisfying $p = p' \circ h$. A map is $\mathscr{C}(X)$ from $p: (Y, y_0) \to (X, x_0)$ to $p': (Y', y_0') \to (X, x_0)$ consists of an (unbased) covering projection $q: Y \to Y'$ satisfying $p = p' \circ q$. We define $\mathscr{B}(X)$ to be the nerve of the category $\mathscr{C}(X)$. In more detail the vertices of $\mathscr{B}(X)$ are the objects of $\mathscr{C}(X)$, and a k-simplex in $\mathscr{B}(X)$ consists of a k-tuple $\langle p_1, p_2, \ldots, p_k\rangle$ of maps p_i in $\mathscr{C}(X)$ such that domain(p_i) = range(p_{i-1}) holds for all $1 < i \leq k$. The face operators are defined by $\partial_0\langle p_1,\ldots,p_k\rangle = \langle p_2,\ldots,p_k\rangle$, $\partial_k\langle p_1,\ldots,p_k\rangle = \langle p_1,\ldots,p_{k-1}\rangle$, $\partial_i\langle p_1,\ldots,p_k\rangle = \langle p_1,\ldots,p_{i-1},p_{i+1}\circ p_i,p_{i+2},\ldots,p_k\rangle$ for all $0<i<k$. We get $\varrho(X):\mathscr{E}(X)\to\mathscr{B}(X)$ as follows. For each k-simplex $\Delta=\langle p_1,\ldots,p_k\rangle$ in $\mathscr{B}(X)$ we set

$$\varrho(X):\varrho(X)^{-1}(\Delta-\partial_0\Delta)\to\Delta-\partial_0\Delta$$

equal to the standard projection

$$(\Delta-\partial_0\Delta)\times Y\to\Delta-\partial_0\Delta,$$

where Y is the domain of p_1 (or more precisely where the covering projection $(Y,y_0)\to(X,x_0)$ is the domain of p_1). We note that the union of all the composite maps

$$(\Delta-\partial_0\Delta)\times Y\xrightarrow{\text{proj.}}Y\xrightarrow{p_1}X$$

gives a well defined map $f(X):\mathscr{E}(X)\to X$.

3.3 Conjecture. *Let X denote any path connected CW complex and let $\mathscr{S}_*(\ )$ denote any of the Ω-spectra valued continuous functors $\mathscr{P}_*(\ )$, $\mathscr{P}_*^{\text{diff}}(\ )$, $\mathscr{L}_*^{\infty}(\ )$, $\mathscr{K}_*(\ )$. Then the composite map*

$$\mathbb{H}_*(\mathscr{B}(X),\mathscr{S}_*(\varrho(X)))\xrightarrow{\mathscr{S}_*(f(X))\circ A_*}\mathscr{S}_*(X)$$

is an equivalence of Ω-spectra, where A_* denotes the assembly map for the generalized Siefert fibration $\varrho(X):\mathscr{E}(X)\to\mathscr{B}(X)$ (cf. 2.4, 3.2) and where $\mathscr{S}_*(f(X))$ is the image of the map $f(X):\mathscr{E}(X)\to X$ under the functor $\mathscr{S}_*(\ )$.

There is a more general version of Conjecture 3.3 for Ω-spectra associated to surgery theory in which the orientation data $w:\pi_1X\to Z_2$ is not assumed to be the zero homomorphism. Space limitations prevent us from stating it here.

We can verify Conjecture 3.3 for many spaces X for the functor $\mathscr{L}_*^{\infty}(\ )$, but for these same spaces our proofs often break down for the functor $\mathscr{L}_*^h(\ )$ (or for the functor $\mathscr{L}_*^s(\ )$) if π_1X contains some elements of finite order. So at this time we do not know if 3.3 is a reasonable conjecture for $\mathscr{L}_*^h(\ )$ (or for $\mathscr{L}_*^s(\ )$). Is it possible that the truth of our 3.3 for $\mathscr{L}_*^h(\ )$ (or for $\mathscr{L}_*^s(\ )$) is just a formal consequence of the truth of 3.3 for both $\mathscr{L}_*^{\infty}(\ )$ and $\mathscr{K}_*(\ )$?

In the following remarks we consider some of the consequences of the truth of Conjecture 3.3 for each of the Ω-spectra valued functors in 3.1.

3.3.1 π_1X Torsion Free and X Aspherical. Conjecture 1.8 is equivalent to Conjecture 3.3 for $\mathscr{S}_*(\ )=\mathscr{L}_*^h(\ )$ in 3.3. Conjecture 2.8 is equivalent to Conjecture 3.3 for $\mathscr{S}_*(\ )=\mathscr{P}_*(\ )$ in 3.3, and each of these conjectures implies Conjecture 1.9. Conjecture 3.3 for $\mathscr{K}_*(\ )$ is equivalent to the usual assembly map $A_*:\mathbb{H}_*(X,\mathscr{K}_*(\text{pt.}))\to\mathscr{K}_*(X)$ being an equivalence of Ω-spectra, which in turn is equivalent to $\text{Wh}_i(\pi_1X)=0$ and $K_{-i}(\mathbf{Z}\pi_1X)=0$ for all $i\geq 1$ and $\tilde{K}_0(\mathbf{Z}\pi_1X)=0$.

3.3.2 The Novikov Conjectures. In general the truth of Conjecture 3.3 for the functor $\mathscr{L}_*^{\infty}(\ )$ implies the truth of Novikov's original conjecture, which stated in more modern terms claims that the assembly map $A_*:\mathbb{H}_*(X,\mathscr{L}_*^{\infty}(\text{pt.}))\to\mathscr{L}_*^{\infty}(X)$

is rationally injective on the homotopy group level. The truth of Conjecture 3.3 for either $\mathscr{K}_*(\ )$, $\mathscr{P}_*(\ )$, or for $\mathscr{P}_*^{\text{diff}}(\ )$ implies the truth of the K-theoretic Novikov conjecture, which claims that the assembly map $A_* : \mathbb{H}_*(X, \mathscr{K}_*(\text{pt.})) \to \mathscr{K}_*(X)$ is rationally injective on the homotopy group level.

3.3.3. On the Role of Finite Subgroups of $\pi_1 X$. Many authors (Quinn [31] and Yamasaki [38] were among the first) have recognized the role that finite subgroups should play in any calculation of the Ω-valued functors $\mathscr{S}_*(\ )$ of 3.1. However that the class of cyclic by finite subgroups of $\pi_1 X$ might be enough to detect all of $\mathscr{S}_*(X)$ first occured as a possibility in the authors' Theorem 2.6 and in their Conjecture 2.8.

The authors obtain the following partial verification of Conjecture 3.3 in an ongoing research project.

3.4 Theorem. *Suppose that in* 3.3 $\pi_1 X$ *is isomorphic to a cocompact discrete subgroup of a virtually connected Lie group. Then the map of* 3.3 *is an equivalence for* $\mathscr{S}_* = \mathscr{P}_*$, $\mathscr{P}_*^{\text{diff}}$, *or* $\mathscr{L}_*^{\infty}$, *and is a rational equivalence for* $\mathscr{S}_* = \mathscr{K}_*$.

References

1. Al'ber, S.I.: Spaces of mappings into manifolds of negative curvature. Dokl. Akad. Nauk. SSSR **178** (1968) 13–16
2. Anderson, D.R., Hsiang, W.-c.: The functors K_{-i} and pseudoisotopies of polyhedra. Ann. Math. **105** (1977) 201–223
3. Ballman, W., Gromov, M., Schroeder, V.: Manifolds of nonpositive curvature. Birkhäuser, 1985
4. Bass, H., Heller, A., Swan, R.: The Whitehead group of a polynomial extension. Inst. Hautes Etudes Sci. Publ. Math. **22** (1964) 64–79
5. Bieberbach, L.: Über die Bewegungsgruppen der Euklidischen Räume II: Die gruppen mit einem endlichen Fundamentalbereich. Math. Ann. **72** (1912) 400–412
6. Burghlea, D., Lashof, R.: Stability of concordances and the suspension homomorphism. Ann. Math. **105** (1977) 449–472
7. Eberlein, P.: Rigidity of lattices of nonpositive curvature. Erg. Theor. Dyn. Syst. **3** (1983) 47–85
8. Eberlein, P.: Euclidean de Rahm factor of a lattice of nonpositive curvature. J. Diff. Geom. **18** (1983) 209–220
9. Eells, J., Sampson, J.H.: Harmonic mappings of Riemannian manifolds. Amer. J. Math. **86** (1964) 109–160
10. Farrell, F.T., Hsiang, W.-c.: On Novikov's conjecture for nonpositively curved manifolds I. Ann. Math. **113** (1981) 199–209
11. Farrell, F.T., Jones, L.E.: Negatively curved manifolds with exotic smooth structures. J. Amer. Math. Soc. **2** (1989) 899–908
12. Farrell, F.T., Jones, L.E.: A topological analogue of Mostow's rigidity theorem. J. Amer. Math. Soc. **2** (1989) 257–370
13. Farrell, F.T., Jones, L.E.: Classical asperical manifolds. CBMS Regional Conference Series in Mathematics, vol. 75. Amer. Math. Soc. 1990
14. Farrell, F.T., Jones, L.E.: Computations of stable pseudoisotopy spaces of aspherical manifolds. Proc. of Intern. Conf. on Algebraic Topology, held in Poznan, Poland in June 1989 (to appear)

15. Farrell, F.T., Jones, L.E.: Rigidity and other topological aspects of compact nonpositively curved manifolds. Bull. Amer. Math. Soc. **22** (1990) 59–64
16. Gersten, S.: On spectrum of algebraic K-theory. Bull. Amer. Math. Soc. **78** (1972) 216–219
17. Gromoll, D., Wolf, J.: Some relations between the metric structure and the algebraic structure of the fundamental group in manifolds of nonpositive curvature. Bull. Amer. Math. Soc. **77** (1971) 545–552
18. Hartman, P.: On homotopic harmonic maps. Canad. J. Math. **19** (1967) 673–687
19. Hatcher, A.: Concordance spaces, higher simple homotopy theory, and applications. Proc. Symp. Pure Math. **32** (1978) 3–21
20. Hsiang, W.-c.: Geometric applications of algebraic K-theory. Proc. Int. Congress of Math., Warsaw 1983, pp. 99–118
21. Hu, B.: A PL geometric study of algebraic K-theory. Trans. Amer. Math. Soc., in press.
22. Igusa, K.: The stability theorem for pseudoisotopies. K-theory **2** (1988) 1–355
23. Igusa, K.: On the algebraic K-theory of A_∞-ring spaces. (Lecture Notes in Mathematics, vol. 967.) Springer, Berlin Heidelberg New York 1982, pp. 146–194
24. Kirby, R.C., Siebenmann, L.C.: Fundational essays on topological manifolds, smoothings, and triangulations. (Ann. Math. Stud., vol. 88.) Princeton Univ. Press, 1977
25. Mishchenko, A.S.: Infinite dimensional representations of discrete groups and higher signatures. Izv. Akad. Nauk. SSSR Ser. Mat. **38** (1974) 81–106
26. Mostow, G.D.: Quasi-conformal mappings in n-space and the rigidity of hyperbolic space forms. Inst. Hautes Etudes Sci. Publ. Math. **34** (1967) 53–104
27. Mostow, G.D.: Strong rigidity of locally symmetric spaces. Princeton Univ. Press, Princeton 1973
28. Nicas, A.: On higher Whitehead groups of a Bieberbach group. Trans. Amer. Math. Soc. **287** (1985) 853–859
29. Quinn, F.: Ends of maps II. Invent. math. **68** (1982) 353–424
30. Quinn, F.: A geometric formulation of surgery. Topology of manifolds. Proc. Georgia Topology Conf., Markham Press 1970. pp. 500–511
31. Quinn, F.: Applications of topology with control. Proc. Int. Congress of Math. Berkeley 1986, pp. 598–606
32. Ranicki, A.: Algebraic L-theory II. Proc. London Math. Soc. **27** (1973) 126–158
33. Schoen, R., Yau, S.T.: On univalent harmonic maps between surfaces. Invent. math. **44** (1978) 265–278
34. Wagoner, J.: Delooping classifying spaces in algebraic K-theory. Topology **11** (1972) 349–370
35. Waldhausen, F.: Algebraic K-theory of topological spaces I. Proc. Symp. Pure Math. **32** (1978) 35–60
36. Waldhausen, F.: Algebraic K-theory of generalized free products. Ann. Math. **108** (1978) 135–256
37. Wall, C.T.C.: Surgery on compact manifolds. Academic Press 1971
38. Yamasaki, M.: L-groups of crystallographic groups. Invent. math. **88** (1987) 571–602
39. Yau, S.T.: On the fundamental group of compact manifolds of nonpositive curvature. Ann. Math. **93** (1971) 579–585

Mapping Class Groups of Surfaces and Three-Dimensional Manifolds

Shigeyuki Morita

Department of Mathematics, Faculty of Science, Tokyo Institute of Technology
Ohokayama Tokyo 152, Japan

1. Introduction

Let Σ_g be a closed oriented surface of genus g and let $\mathcal{M}_g$ be its mapping class group. Namely it is the group of all isotopy classes of orientation preserving diffeomorphisms of Σ_g. It is also called the Teichmüller modular group because it acts on the Teichmüller space $\mathcal{T}_g$ properly discontinuously with the quotient space $\mathbf{M}_g$: the Riemann moduli space of compact Riemann surfaces of genus g. A classical theorem of Nielsen asserts that $\mathcal{M}_g$ can be naturally identified with the proper outer automorphism group of $\pi_1(\Sigma_g)$. Thus the mapping class group appears in diverse branches of mathematics and have been investigated from various points of view. In this article we would like to describe some of the recent progress in the topological aspects of the theory of the mapping class group. More precisely we will concern ourselves with the following three topics which are all of cohomological nature and mutually closely related: a brief review of some of the known results about the cohomology of $\mathcal{M}_g$, in particular various properties of some canonical cohomology classes of $\mathcal{M}_g$, called the characteristic classes of surface bundles (§2), the study of $\mathcal{M}_g$ via its action on the lower central series of $\pi_1(\Sigma_g)$ (§3) and the interplay between the structure of $\mathcal{M}_g$ and topological invariants of three dimensional manifolds (§4).

There has been a great deal of results which are relevant to the subjects of this article and our coverage here is necessarily limited. The reader is referred to survey articles [3, 10, 13, 14, 40] and their bibliographies for more informations.

2. The Cohomology of $\mathcal{M}_g$

By the cohomology of $\mathcal{M}_g$ we mean the Eilenberg-MacLane cohomology of the group $\mathcal{M}_g$ whose definition is given purely algebraically. However it has also geometrical meanings at least from the following two viewpoints. One is that, by virtue of the result of Earle-Eells [6], the classifying space for oriented Σ_g-bundles has the homotopy type of an Eilenberg-MacLane space $K(\mathcal{M}_g, 1)$ for $g \geq 2$ so that elements of the cohomology group of $\mathcal{M}_g$ can be naturally considered as *characteristic classes of surface bundles*. The other is the fact that the rational

Proceedings of the International Congress
of Mathematicians, Kyoto, Japan, 1990

cohomology of $\mathscr{M}_g$ is canonically isomorphic to that of the moduli space $\mathbf{M}_g$. For both of technical and theoretical reasons it is convenient to introduce the mapping class group $\mathscr{M}_{g,q}^p$ of a general compact oriented surface $\Sigma_{g,q}^p$ of genus g with p distinguished points and q boundary components (or equivalently q embedded discs). Here $\mathscr{M}_{g,q}^p$ is defined to be the group of path components of diffeomorphisms of $\Sigma_{g,q}^p$ which fix the p points and restrict to the identity on the boundary.

One of the most useful methods to compute (co)homology of a group is to find a space or a cell complex on which it acts naturally and then analyse the action. As for the mapping class group, besides the Teichmüller space it is natural to consider various cell complexes which are constructed out of isotopy invariants of surfaces, e.g. simple closed curves or arcs. By an extensive use of this method Harer obtained several fundamental results concerning the homology of the mapping class group (see [10] for a survey of his results). One of his results is about the stability of the (co)homology with respect to the genus and it may be stated roughly as

Theorem (Harer [8]). *The cohomology $H^k(\mathscr{M}_{g,q}^p;\mathbf{Z})$ is independent of g and q for $g \geq 3k+1$ and the same is true for the rational cohomology for $g \geq 3k$.*

In the case where $q \neq 0$ Ivanov has improved the stability range (see [14]). By virtue of the above theorem we can speak of the *stable cohomology* of $\mathscr{M}_g$.

Now we recall the definition of some canonical classes in the stable cohomology of $\mathscr{M}_{g,q}^p$. Thus let $\pi : E \to X$ be an oriented $\Sigma_{g,q}^p$-bundle. This means that it is an oriented differentiable fibre bundle with fibre Σ_g and there are given $(p+q)$ cross sections $s_i : X \to E$ $(i = 1, \cdots, p+q)$ such that their images are disjoint and the normal bundles of the last q sections are trivialized. Let ξ be the tangent bundle along the fibres, namely it is an oriented plane bundle over E consisting of all tangent vectors of E which are tangent to the fibres. We write $e \in H^2(E;\mathbf{Z})$ for the Euler class of ξ and consider the element $e_i = \pi_*(e^{i+1}) \in H^{2i}(X;\mathbf{Z})$ where $\pi_* : H^{2i+2}(E;\mathbf{Z}) \to H^{2i}(X;\mathbf{Z})$ is the Gysin homomorphism. We also set $\sigma_i = s_i^*(e) \in H^2(X;\mathbf{Z})$ $(i = 1, \cdots, p)$. These definitions are functorial in the obvious sense so that we have well defined cohomology classes $e_i \in H^{2i}(\mathscr{M}_{g,q}^p;\mathbf{Z})$ and $\sigma_i \in H^2(\mathscr{M}_{g,q}^p;\mathbf{Z})$. We may call e_i the i-th characteristic class of surface bundles. Mumford [31] defined, in the context of algebraic geometry, canonical classes κ_i in the Chow ring of the moduli space $\mathbf{M}_g$ (or in fact its compactification $\overline{\mathbf{M}}_g$) and it turned out that the class e_i is the topological version of κ_i (up to signs) although our work has been done independently. Miller [23] also considered the same class. According to a recent paper of Witten [39], these characteristic classes have physical meanings.

Theorem. *The homomorphism*

$$\Phi : \mathbf{Q}[e_1, e_2, \cdots, \sigma_1, \cdots, \sigma_p] \to H^*(\mathscr{M}_{g,q}^p;\mathbf{Q})$$

is injective up to degree $\frac{1}{3}g$.

The above theorem for the case $p = 0$ is due to Miller [23] and the author [24] independently and was proved by generalizing an earlier work by Atiyah [1]. Miller also observed that the stable cohomology $\lim H^*(\mathcal{M}_{g,1};\mathbf{Q})$,which exists by the stability theorem of Harer, is the tensor product of a polynomial algebra on even generators with an exterior algebra on odd generators. The general cases of the above theorem were proved by the author [24]. As a corollary to the above theorem, we have the following result which was one of our original motivation for the present work.

Corollary [24]. *The natural homomorphism* $\mathrm{Diff}_+\Sigma_g \to \mathcal{M}_g$ *does not have a right inverse for* $g \geq 18$.

We mention that the original Nielsen realization problem was solved affirmatively by Kerckhoff [20] and also that it is still unknown whether the homomorphism $\mathrm{Homeo}_+\Sigma_g \to \mathcal{M}_g$ has a right inverse or not. As Mumford says in [31] for the case $p = 0$ one of the main questions concerning the cohomology of the mapping class groups (or the moduli spaces of compact Riemann surfaces) is whether the stable cohomology $\lim H^*(\mathcal{M}^p_{g,q};\mathbf{Q})$ is isomorphic to $\mathbf{Q}[e_1, e_2, \cdots, \sigma_1, \cdots, \sigma_p]$ or not. In low dimensions it is known that $\mathcal{M}^p_{g,q}$ is perfect except for finite cases ([7, 33]) and also we have

Theorem (Harer [7]). $H^2(\mathcal{M}^p_{g,q};\mathbf{Z}) \cong \mathbf{Z}^{p+1}$ *with the generators* $\frac{1}{12}e_1, \sigma_1, \cdots, \sigma_p$ *for* $g \geq 5$.

On the other hand for a fixed g it is known that the homomorphism Φ : $\mathbf{Q}[e_1, e_2, \cdots, \sigma_1, \cdots, \sigma_p] \to H^*(\mathcal{M}^p_{g,q};\mathbf{Q})$ is far from being injective nor surjective. The latter statement is due to Harer-Zagier and is a consequence of the following result.

Theorem (Harer-Zagier [11], see also Penner [32]). *The orbifold Euler characteristic of* $\mathcal{M}^1_g$ *is equal to* $\zeta(1-2g)$ *where* ζ *is the Riemann zeta function.*

By making use of the above result, they have shown that the true Euler characteristic of $\mathcal{M}_g$ grows more than exponentially with respect to g and even takes negative values infinitely often.

Finally we consider the kernel of the homomorphism Φ, namely relations between the characteristic classes of surface bundles. It has certainly a big kernel simply because the rational cohomology of the mapping class groups are finite dimensional. More precisely Harer [9] has determined the virtual cohomological dimensions of the mapping class groups and in particular $H^*(\mathcal{M}_g;\mathbf{Q}) = 0$ for $k \geq 4g-4$. There are several known relations between the characteristic classes. For example Mumford proved that for a fixed g, the classes e_i are polynomials on $e_1, \cdots, e_{g-2}$ for all $i \geq g-1$ (see [24, 31] for the proof as well as other relations). Here we describe one particular relation which was proved by Harris [12] and the author [26] independently and by completely different methods. The following is along the lines of our papers [25, 26]. First of all it was proved that $H^1(\mathcal{M}^1_g;H)$, where H stands for the first homology of Σ_g, is an infinite cyclic group and a characterization was given for a crossed homomorphism $k : \mathcal{M}_g^1 \to H$ to represent its generator. It is amusing to point out here that there have been known various ways of describing such a crossed homomorphism reflecting the many-sided features of the mapping class group;complex analytic construction due to Earle [5], two definitions in the context of combinatorial group theory [25, 26], determinant of certain matrix representation [27], a very simple geometrical one due to Furuta (unpublished, see [30]) and a closely related one due to Trapp [36] in terms of the winding number and finally Wagoner's definition [37], a particular example of his pseudo isotopy invariants using algebraic K-theory, should be completed to give yet another one. By making use of the above crossed homomorphism, we proved in [26] a topological version of Earle's embedding theorem [5] which embedds any surface bundle into its associated family of Jacobian manifolds explicitly. We also proved that the crossed homomorphism k gives rise to a canonical cocycle for the first characteristic class $e_1 \in H^2(\mathcal{M}_{g,1};\mathbf{Z})$. Namely the 2-cochain c of $\mathcal{M}_{g,1}$ defined by $c(\varphi,\psi) = k(\varphi)\cdot k(\psi^{-1})$ $(\varphi,\psi \in \mathcal{M}_{g,1})$ is a cocycle representing the class e_1, where $\cdot$ denotes the intersection number. Then by pulling back a canonical cohomology class of the total space of the family of Jacobian manifolds to that of the given surface bundle, we obtain

Theorem (Harris [12], Morita [26]). *Let $e \in H^2(\mathcal{M}_g^1;\mathbf{Q})$ be the Euler class of the tangent bundle along the fibres of the universal Σ_g-bundle $\mathcal{M}_g^1 \to \mathcal{M}_g$ and let $e_1 \in H^2(\mathcal{M}_g^1;\mathbf{Q})$ be the first characteristic class. Then we have $\{2g(2-2g)e-e_1\}^{g+1} = 0$ in $H^{2g+2}(\mathcal{M}_g^1;\mathbf{Q})$.*

3. Representations of $\mathcal{M}_g$ via Its Action on $\pi_1(\Sigma_g)$

As was mentioned in the introduction, the mapping class group $\mathcal{M}_g$ is naturally isomorphic to the proper outer automorphism group of the fundamental group $\pi_1(\Sigma_g)$ and similarly $\mathcal{M}_{g,1}$ can be identified with a certain subgroup of the automorphism group of $\pi_1(\Sigma_{g,1})$ which is a free group of rank $2g$. These isomorphisms provide us with an algebraic method of investigating the structure of the mapping class groups in terms of their actions on the fundamental groups of surfaces. Apart from the general theory due to Magnus and his school on the structure of the (full) automorphism groups of free groups, it was Johnson who has first employed the above idea systematically in the study of the structure of (various subgroups of) the mapping class groups. By this method he obtained several fundamental results concerning the structure of the Torelli group $\mathcal{I}_g$ which is the subgroup of $\mathcal{M}_g$ acting trivially on the homology of Σ_g (see [18]). In this section we summarize the method and results of Johnson and their extensions by the author.

We denote Γ_1 for $\pi_1(\Sigma_{g,1})$ and inductively define $\Gamma_k = [\Gamma_{k-1},\Gamma_1]$. We may call the quotient group $N_k = \Gamma_1/\Gamma_k$ the k-th nilpotent quotient of Γ_1. We simply write H for $N_2 \cong H_1(\Sigma_g;\mathbf{Z})$ and let $\mathcal{L} = \oplus_{k\geq 1}\mathcal{L}_k$ be the free graded Lie algebra over $\mathbf{Z}$ generated by the elements of H. It is a classical result that we can identify $\mathcal{L}_k$

with Γ_k/Γ_{k+1} so that there exists a central extension $0 \to \mathcal{L}_k \to N_{k+1} \to N_k \to 1$. Now the action of $\mathcal{M}_{g,1}$ on Γ_1 induces that on N_k and we have a representation

$$\varrho_k : \mathcal{M}_{g,1} \longrightarrow \mathrm{Aut}N_k .$$

The image of the representation ϱ_2 is equal to the subgroup of $\mathrm{Aut}N_2$ which preserves the skew symmetric pairing on $N_2 = H$ defined by the intersection number so that ϱ_2 is essentially equal to the classical representation of $\mathcal{M}_g$ onto the Siegel modular group $\mathrm{Sp}(2g;\mathbf{Z})$. Now let us write $\mathcal{M}(k)$ for $\mathrm{Ker}\varrho_k$ so that in particular $\mathcal{M}(1) = \mathcal{M}_{g,1}$ and $\mathcal{M}(2)$ is the Torelli group $\mathcal{I}_{g,1}$. $\mathcal{M}(k)$ is the subgroup of $\mathcal{M}_{g,1}$ acting trivially on N_k. Now it can be shown that we have a central extension

$$0 \longrightarrow \mathrm{Hom}(H, \mathcal{L}_k) \longrightarrow \mathrm{Aut}N_{k+1} \longrightarrow \mathrm{Aut}N_k \longrightarrow 1$$

where the abelian group $\mathrm{Hom}(H, \mathcal{L}_k)$ acts on N_{k+1} by the formula $f(\gamma) = \gamma f(\bar{\gamma})(f \in \mathrm{Hom}(H, \mathcal{L}_k), \gamma \in N_{k+1}$ and $\bar{\gamma} \in H$ is the image of γ in $H_1(N_{k+1}) = H)$. Hence we obtain a homomorphism

$$\tau_k : \mathcal{M}(k) \longrightarrow \mathrm{Hom}(H, \mathcal{L}_k)$$

which is the restriction of the homomorphism $\varrho_{k+1} : \mathcal{M}_{g,1} \to \mathrm{Aut}N_{k+1}$ to the subgroup $\mathcal{M}(k)$. The canonical isomorphism $H^* \cong H$ makes it possible to write $\tau_k : \mathcal{M}(k) \to \mathcal{L}_k \otimes H$ and $\mathrm{Ker}\tau_k = \mathcal{M}(k+1)$. The homomorphisms τ_k were introduced by Johnson [16, 18] and we call them Johnson's homomorphisms.

Before going further, here we review Johnson's results on the structure of the Torelli group. There are two natural types of mapping classes which belong to the Torelli group. One is the Dehn twist along a bounding simple closed curve and is called a BSCC map. The other is the composition $\varphi\psi^{-1}$ of two (say right handed) Dehn twists φ and ψ along mutually homologous simple closed curves on Σ_g (or $\Sigma_{g,1}$) and is called a BP (bounding pair) map. At the time when Johnson began his study on the Torelli group, it was known that the Torelli group is generated by these two types of mapping classes (Powell [33]) and also it had been asked whether the subgroup $\mathcal{K}_g$ of $\mathcal{I}_g$ (resp. $\mathcal{K}_{g,1}$ of $\mathcal{I}_{g,1}$) generated by all BSCC maps has a finite index or not.

Theorem (Johnson [17]). *The Torelli group is generated by a finite number of BP maps for $g \geq 3$.*

Theorem (Johnson [16, 18]). *The image of the homomorphism $\tau_2 : \mathcal{I}_{g,1} \to \Lambda^2 H \otimes H$ is equal to $\Lambda^3 H \subset \Lambda^2 H \otimes H$ and* $\mathrm{Ker}\tau_2 = \mathcal{M}(3)$ *coincides with $\mathcal{K}_{g,1}$ so that we have a short exact sequence*

$$1 \longrightarrow \mathcal{K}_{g,1} \longrightarrow \mathcal{I}_{g,1} \xrightarrow{\tau_2} \Lambda^3 H \longrightarrow 1 .$$

In particular $\mathcal{K}_{g,1}$ has an infinite index in $\mathcal{I}_{g,1}$.

There are other abelian quotients of the Torelli group than the homomorphism τ_2. Namely Birman and Craggs [4] have produced many $\mathbf{Z}/2$-valued homomorphisms on the Torelli group by making use of the Rohlin invariant for homology 3-spheres. Combining these two types of abelian quotients, Johnson [19] succeeded to determine the abelianization of the Torelli group completely.

Now we go back to the general theory. We have Johnson's homomorphism $\tau_k : \mathcal{M}(k) \to \mathrm{Hom}(H, \mathcal{L}_k) = \mathcal{L}_k \otimes H$. Let us define a submodule $\mathcal{H}_k$ of $\mathcal{L}_k \otimes H$ to be the kernel of the natural surjection $\mathcal{L}_k \otimes H \to \mathcal{L}_{k+1}$ defined as $\xi \otimes u \mapsto [\xi, u]$ ($\xi \in \mathcal{L}_k, u \in H$). Then one can show that $\mathrm{Im}\tau_k \subset \mathcal{H}_k$ (see [28]). For $k = 2$, $\mathcal{H}_2 = \Lambda^3 H$ so that $\mathrm{Im}\tau_2 = \mathcal{H}_2$ and in [27] we have shown that $\mathrm{Im}\tau_3$ is a submodule of $\mathcal{H}_3$ of index a power of two. Now there is a natural structure of a Lie algebra over $\mathbf{Z}$ on the graded module $\oplus_{k\geq 2}\mathrm{Hom}(H, \mathcal{L}_k)$ such that the graded submodule $\oplus_{k\geq 2}\mathcal{H}_k$ is a Lie subalgebra (the degree should be shifted down by one). The graded module $\oplus_{k\geq 2}\mathcal{M}(k)/\mathcal{M}(k+1)$ also admits a similar structure and Johnson's homomorphisms $\{\tau_k\}_{k\geq 2}$ induce an injection

$$\oplus_{k\geq 2}\mathcal{M}(k)/\mathcal{M}(k+1) \longrightarrow \oplus_{k\geq 2}\mathcal{H}_k$$

of graded Lie algebras. It is a very important problem to identify the image of this homomorphism. Computations so far show that the rank of $\mathrm{Im}\tau_k$ is much smaller than that of $\mathcal{H}_k$ for even $k \geq 4$. Now $\mathrm{Aut}N_k$ is an extension of $\mathrm{GL}(2g; \mathbf{Z})$ by a nilpotent group whose associated graded module is $\oplus_{i=2}^{k-1}\mathrm{Hom}(H, \mathcal{L}_i)$ and the representation $\varrho_k : \mathcal{M}_{g,1} \to \mathrm{Aut}N_k$ induces an identification of the quotient group $\mathcal{M}_{g,1}/\mathcal{M}(k)$ as an extension of $\mathrm{Sp}(2g; \mathbf{Z})$ by a nilpotent group whose graded module is $\oplus_{i=2}^{k-1}\mathrm{Im}\tau_i$. Now we apply Sullivan's general theory [35]. Thus let $N_k \otimes \mathbf{Q}$ be the Malcev completion of the nilpotent group N_k. Then any automorphism of N_k extends uniquely to that of $N_k \otimes \mathbf{Q}$ so that there is an embedding $\mathrm{Aut}N_k \to \mathrm{Aut}N_k \otimes \mathbf{Q}$, where the latter group is an extension of $\mathrm{GL}(2g; \mathbf{Q})$ by a nilpotent Lie group $\mathcal{G}_k$ over $\mathbf{Q}$ whose associated graded vectorspace is $\oplus_{i=2}^{k-1}\mathrm{Hom}(H, \mathcal{L}_i) \otimes \mathbf{Q}$. Now it is easy to see that $\mathrm{Aut}N_k \otimes \mathbf{Q}$ is a linear algebraic group over $\mathbf{Q}$ and its maximal normal unipotent subgroup is precisely equal to $\mathcal{G}_k$. Hence, by virtue of the Levi-Chevalley decomposition theorem, we can conclude that $\mathrm{Aut}N_k \otimes \mathbf{Q}$ is isomorphic to a *split* extension $\mathcal{G}_k \rtimes \mathrm{GL}(2g; \mathbf{Q})$. Although there is no canonical splitting, we can make explicit calculations involving choices of $\mathrm{GL}(2g; \mathbf{Q})$-invariant 2-cocycles for the successive extension classes of the nilpotent group N_k for small k and we obtain

Theorem. *(i) The representation $\varrho_3 : \mathcal{M}_{g,1} \to \mathrm{Aut}N_3$ induces an embedding of $\mathcal{M}_{g,1}/\mathcal{K}_{g,1}$ into the split extension $\Lambda^3 H \rtimes \mathrm{Sp}(2g; \mathbf{Z})$ as a subgroup of finite index such that the associated crossed homomorphism $\tilde{k} : \mathcal{M}_{g,1} \to \Lambda^3 H$ has the property that its restriction to the Torelli group $\mathcal{I}_{g,1}$ is equal to $2\tau_2$.*

(ii) The representation $\varrho_4 : \mathcal{M}_{g,1} \to \mathrm{Aut}N_4$ induces an embedding of $\mathcal{M}_{g,1}/\mathcal{M}(4)$ into $(\mathcal{H}_3 \tilde{\times} \Lambda^3 H) \rtimes \mathrm{Sp}(2g; \mathbf{Z})$ as a subgroup of finite index where $\mathcal{H}_3 \tilde{\times} \Lambda^3 H$ denotes the central extension of $\Lambda^3 H$ by $\mathcal{H}_3$ defined by the bracket operation $\Lambda^3 H \otimes \Lambda^3 H \to \mathcal{H}_3$ and $\mathrm{Sp}(2g; \mathbf{Z})$ acts on it naturally.

In view of the fact that the crossed homomorphism $\tilde{k}$ in the above theorem followed by the contraction $\Lambda^3 H \to H$ is nothing but the crossed homomorphism $k : \mathcal{M}_{g,1} \to H$ of §2, it seems to be worthwhile to propose

Problem. Give a definition of the crossed homomorphism $\tilde{k} : \mathcal{M}_{g,1} \to \Lambda^3 H$ in the context of complex analysis and/or algebraic K-theory.

4. Algebraic Structures of $\mathcal{M}_g$ and Topological Invariants of 3-Manifolds

Let us recall the classical theorem on the existence and stable uniqueness of the Heegaard splittings of 3-manifolds. First of all it says that any closed oriented 3-manifold can be expressed as $M_\varphi = H_g \cup_\varphi -H_g$ for some $\varphi \in \mathcal{M}_{g,1}$ where H_g is an oriented handlebody of genus g with $\partial H_g = \Sigma_g$ and M_φ is the manifold obtained from the disjoint union of H_g and $-H_g$ by identifying the boundaries by the map φ. Secondly two such manifolds $M_\varphi(\varphi \in \mathcal{M}_{g,1})$ and $M_\psi(\psi \in \mathcal{M}_{g,1})$ are orientation preserving diffeomorphic each other if and only if the two elements φ and ψ are equivalent in the disjoint union $\amalg_g \mathcal{M}_{g,1}$ of the mapping class groups $\mathcal{M}_{g,1}$ generated by the following two types of moves. One is to replace $\varphi \in \mathcal{M}_{g,1}$ by any other element in its double coset with respect to the subgroup $\mathcal{N}_{g,1}$ consisting of all elements which extend to diffeomorphisms of H_g. The other is to replace $\varphi \in \mathcal{M}_{g,1}$ by $i_*(\varphi)\iota \in \mathcal{M}_{g+1,1}$ where $i : \mathcal{M}_{g,1} \to \mathcal{M}_{g+1,1}$ is a natural injection and ι is the right angled rotation of the "last handle" of Σ_{g+1}. Thus to define a topological invariant of 3-manifolds is equivalent to give a function defined on the mapping class groups $\mathcal{M}_{g,1}$ which is invariant under the above two moves. Although the situation here is much more subtle than that in the case of oriented links in S^3 and the braid groups, it seems to be still reasonable to expect that there should exist a deep connection between the algebraic structure of the mapping class group and topological invariants of 3-manifolds. As a supporting evidence for that, extending and completing earlier works by Birman-Craggs [4] and Johnson [15], we will describe below how the Casson invariant of homology 3-spheres influences the structure of the subgroup $\mathcal{K}_{g,1}$ of $\mathcal{M}_{g,1}$.

Let $\iota_g \in \mathcal{M}_{g,1}$ be the product of right angled rotation of each handle of (an appropriate model of) Σ_g so that M_{ι_g} is the 3-sphere. For each element $\varphi \in \mathcal{K}_{g,1}$ we consider the manifold $W_\varphi = H_g \cup_{\iota_g\varphi} -H_g$ which is easily seen to be an oriented homology 3-sphere. It was proved in [27] that any homology 3-sphere appears like this and two such manifolds $W_\varphi(\varphi \in \mathcal{K}_{g,1})$ and $W_\psi(\psi \in \mathcal{K}_{g,1})$ are orientation preserving diffeomorphic if and only if φ and ψ are equivalent in $\lim_{g\to\infty} \mathcal{K}_{g,1}$ generated by the move which replaces a given element $\varphi \in \mathcal{K}_{g,1}$ by $\iota_g^{-1}\eta\iota_g\varphi\eta' \in \mathcal{K}_{g,1}$ for any $\eta, \eta' \in \mathcal{N}_{g,1}$. Thus to define a topological invariant for oriented homology 3-spheres is equivalent to give a function on $\lim_{g\to\infty} \mathcal{K}_{g,1}$ which is invariant under the above move.

Now let $\lambda^* : \mathcal{K}_{g,1} \to \mathbf{Z}$ be the mapping defined by $\lambda^*(\varphi) = \lambda(W_\varphi)(\varphi \in \mathcal{K}_{g,1})$ where $\lambda(W_\varphi)$ is the Casson invariant of the homology 3-sphere W_φ. Recall that $\mathcal{K}_{g,1}$ is the subgroup of $\mathcal{M}_{g,1}$ generated by all BSCC maps. Suppose $\varphi \in \mathcal{K}_{g,1}$ is such a Dehn twist along a simple closed curve ω on $\Sigma_{g,1}$. Then since the surface Σ_g is now considered as a particular Heegaard surface of $S^3 = H_g \cup_{\iota_g} -H_g$, ω is naturally a knot in S^3. Moreover it is easy to see that the homology sphere W_φ is nothing but the one obtained by performing -1 surgery on S^3 along this knot. Hence $\lambda(W_\varphi)$ is equal to the negative of the Casson invariant of the knot ω and this in turn can be calculated at the level of the surface Σ_g (and not in S^3) because the compact surface which ω bounds on $\Sigma_{g,1}$ is naturally a Seifert surface of ω. Combining these considerations, it can be shown that the mapping $\lambda^* : \mathcal{K}_{g,1} \to \mathbf{Z}$ is a *homomorphism.* Here we recall from §3 that we have Johnson's homomorphism $\tau_3 : \mathcal{K}_{g,1} \to \mathcal{H}_3$ which gives a large abelian quotient of $\mathcal{K}_{g,1}$. However it turns out that the homomorphism λ^* cannot be covered by τ_3 and we need a deeper additive invariant for $\mathcal{K}_{g,1}$. Such an invariant can be obtained by making use of the first characteristic class $e_1 \in H^2(\mathcal{M}_{g,1};\mathbf{Z})$ of surface bundles introduced in §2 as follows. As was described in §2, the cohomology class e_1 can be represented by a cocycle $c \in Z^2(\mathcal{M}_{g,1};\mathbf{Z})$ defined as $c(\varphi,\psi) = k(\varphi)\cdot k(\psi^{-1})$ where $k : \mathcal{M}_{g,1} \to H$ is a crossed homomorphism representing a generator of $H^1(\mathcal{M}_{g,1};H)$. On the other hand Meyer [22] defined a 2-cocycle τ of the Siegel modular group $\mathrm{Sp}(2g;\mathbf{Z})$ which represents the signature class $\in H^2(\mathrm{Sp}(2g;\mathbf{Z});\mathbf{Z})$ and its pull back under the representation $\mathcal{M}_{g,1} \to \mathrm{Sp}(2g;\mathbf{Z})$ is $-\frac{1}{3}e_1$. Hence there exists a mapping $d : \mathcal{M}_{g,1} \to \mathbf{Z}$ such that $\delta d = c + 3\tau$. Although the mapping d does depend on the choice of the crossed homomorphism k, its restriction to the subgroup $\mathcal{K}_{g,1}$ does not and moreover it is a homomorphism there. This is the invariant we seek for. Namely there exists a certain homomorphism $q : \mathcal{H}_3 \to \mathbf{Q}$ and we have

Theorem [27]. *The function $\frac{1}{24}d + q\circ\tau_3$ defined on $\mathcal{K}_{g,1}$ is invariant under the equivalence relation on $\lim_{g\to\infty}\mathcal{K}_{g,1}$ so that it defines an invariant for oriented homology 3-spheres and this coincides with the Casson invariant.*

Remarks. 1. At present there are only a few known informations about the structure of the groups $\mathcal{K}_g$ and $\mathcal{K}_{g,1}$. For example it is not known whether they are finitely generated or not for $g \geq 3$. Also in view of the results in this section, it would be worthwhile to determine the abelianizations of them.

2. Recently the author obtained an interpretation of the invariant d as the signature defect of certain framed 3-manifolds [30]. It would be desirable to have further geometrical or analytical meaning of it, perhaps along the lines of Atiyah's paper [2] which treats the case of genus 1.

3. In some sense the invariant d can be understood to be the effect of the Casson invariant for knots on the structure of the group $\mathcal{K}_{g,1}$ (see [29]). It would be interesting to investigate how other invariants for knots influence the structure of $\mathcal{K}_{g,1}$.

4. It seems to be reasonable to conjecture that we can generalize the definition of the Casson invariant to possibly arbitrary 3-manifold by making use of the invariant d and the explicit description of the quotient $\mathcal{M}_{g,1}/\mathcal{M}(4)$ given in §3. In particular one can ask whether Walker's extension to rational homology 3-spheres [38] can be expressed like this or not.

5. Recently Reshetikhin and Turaev [34] have defined new topological invariants for 3-manifolds, motivated by Witten's ideas in quantum field theory, and there are certain representations of $\mathcal{M}_g$ associated to them. Kohno [21] has also made use of certain projective representations of $\mathcal{M}_g$, which arise naturally

in the conformal field theory, to derive topological invariants via the Heegaard splittings. There seems to be much work to be done here from the point of view of the structure of $\mathcal{M}_g$. For example we can ask whether there exists a natural number k such that the subgroup $\mathcal{M}(k)$ is contained in the kernel of these representations or not.

References

1. Atiyah, M.F.: The signature of fibre-bundles. In: Global analysis, papers in honor of K. Kodaira, Tokyo University Press, 1969, pp. 73–84
2. Atiyah, M.F.: The logarithm of the Dedekind η-function. Math. Ann. **278** (1987) 335–380
3. Birman, J.: Mapping class groups of surfaces. In: Braids, Contemp. Math. **78** (1988) 13–43
4. Birman, J., Craggs, R.: The μ-invariant of 3-manifolds and certain structural properties of the group of homeomorphisms of a closed oriented 2-manifold. Trans. Amer. Math. Soc. **237** (1978) 283–309
5. Earle, C.J.: Families of Riemann surfaces and Jacobi varieties. Ann. Math. **107** (1978) 255–286
6. Earle, C.J., Eells, J.: The diffeomorphism group of a compact Riemann surface. Bull. Amer. Math. Soc. **73** (1967) 557–559
7. Harer, J.: The second homology group of the mapping class group of an orientable surface. Invent. math. **72** (1983) 221–239
8. Harer, J.: Stability of the homology of the mapping class groups of orientable surfaces. Ann. Math. **121** (1985) 215–249
9. Harer, J.: The virtual cohomological dimension of the mapping class group of an orientable surface. Invent. math. **84** (1986) 157–176
10. Harer, J.: The cohomology of the moduli space of curves. In: Theory of moduli. (Lecture Notes in Mathematics, vol. 1337). Springer, Berlin Heidelberg New York 1988, pp. 138–221
11. Harer, J., Zagier, D.: The Euler characteristic of the moduli space of curves. Invent. math. **85** (1986) 457–485
12. Harris, J.: Families of smooth curves. Duke Math. J. **51** (1984) 409–419
13. Ivanov, N.V.: Algebraic properties of the mapping class groups of surfaces. Preprint, Leningrad 1985
14. Ivanov, N.V.: Complexes of curves and the Teichmüller modular group. Uspekhi Mat. Nauk **42** (1987) 49–91 [English transl.: Russ. Math. Surv. **42** (1987) 55–107]
15. Johnson, D.: Quadratic forms and the Birman-Craggs homomorphisms. Trans. Amer. Math. Soc. **261** (1980) 235–254
16. Johnson, D.: An abelian quotient of the mapping class group $\mathcal{I}_g$. Math. Ann. **249** (1980) 225–242
17. Johnson, D.: The structure of the Torelli group I: A finite set of generators for $\mathcal{I}_g$. Ann. Math. **118** (1983) 423–442
18. Johnson, D.: A survey of the Torelli group, Contemp. Math. **20** (1983) 165–179
19. Johnson, D.: The structure of the Torelli group II and III. Topology **24** (1985) 113–144
20. Kerckhoff, S.P.: The Nielsen realization problem. Ann. Math. **117** (1983) 235–265
21. Kohno, T.: Topological invariants for 3-manifolds using representations of mapping class groups I. Topology (to appear)
22. Meyer, W.: Die Signatur von Flächenbündeln. Math. Ann. **201** (1973) 239–264
23. Miller, E.Y.: The homology of the mapping class group. J. Diff. Geom. **24** (1986) 1–14
24. Morita, S.: Characteristic classes of surface bundles. Invent. math. **90** (1987) 551–577

25. Morita, S.: Families of Jacobian manifolds and characteristic classes of surface bundles I. Ann. Inst. Fourier **39** (1989) 777–810
26. Morita, S.: Families of Jacobian manifolds and characteristic classes of surface bundles II. Math. Proc. Camb. Phil. Soc. **105** (1989) 79–101
27. Morita, S.: Casson's invariant for homology 3-spheres and characteristic classes of surface bundles I. Topology **28** (1989) 305–323
28. Morita, S.: On the structure and the homology of the Torelli group. Proc. Japan Acad. **65** (1989) 147–150
29. Morita, S.: On the structure of the Torelli group and the Casson invariant. Topology (to appear)
30. Morita, S.: Signature defects of certain framed 3-manifolds and the Casson invariant. In preparation
31. Mumford, D.: Towards an enumerative geometry of the moduli space of curves. In: Arithmetic and geometry. Progr. Math. **36** (1983) 271–328
32. Penner, R.C.: Perturbative series and the moduli space of Riemann surfaces. J. Diff. Geom. **27** (1988) 35–53
33. Powell, J.: Two theorems on the mapping class group of a surface. Proc. Amer. Math. Soc. **68** (1978) 347–350
34. Reshetikhin, N., Turaev, V.G.: Invariants of 3-manifolds via link polynomials and quantum groups. Invent. math (to appear)
35. Sullivan, D.: Infinitesimal computations in topology. Publ. Math. I.H.E.S. **47** (1977) 269–331
36. Trapp, R.: A linear representation of the mapping class group $\mathcal{M}$ and the theory of winding numbers. Preprint
37. Wagoner, J.: K_2 and diffeomorphisms of two and three dimensional manifolds. In: Geometric topology. Academic Press 1979, pp. 557–577
38. Walker, K.: An extension of Casson's invariant to rational homology spheres. Preprint
39. Witten, E.: On the structure of the topological phase of two dimensional gravity. Nucl. Phys. **B 340** (1990) 281–332
40. Wolpert, S.: The topology and geometry of the moduli space of Riemann surfaces. In: Arbeitstagung Bonn 1984. (Lecture Notes in Mathematics, vol. 1111). Springer, Berlin Heidelberg New York 1985, pp. 431–451

Cyclic Cohomology and Invariants of Multiply Connected Manifolds

Henri Moscovici

Department of Mathematics, The Ohio State University
231 West 18th Ave., Columbus, OH 43210-1174, USA

To motivate the subject-matter of this report, let me begin by recalling two of the early applications of the Atiyah-Singer Index Theorem, concerning two important manifold invariants. The first result, which was in fact a precursor of the Index Theorem, is the Hirzebruch Signature Theorem. It asserts that the L-genus $\lambda(M) = L(M)[M]$ of a closed oriented smooth $4k$-dimensional manifold M is equal to the index of the signature operator $\not{B}^{+}$ attached to a Riemannian metric, which in turn coincides with the signature of the manifold. The second application requires M to be in addition a Spin-manifold and equates its $\widehat{A}$-genus $\alpha(M) = \widehat{A}(M)[M]$ with the index of the Dirac operator $\not{D}^{+}$ associated to a Spin-structure. These results not only explain the integrality of the above invariants (under the given conditions) but also have other, more subtle, implications. Thus, Hirzebruch's theorem implies that the L-genus is an invariant of the oriented homotopy type of M. On the other hand, in conjunction with Lichnerowicz's formula for the Dirac Laplacian, the index theorem for the Dirac operator implies that if $\alpha(M) \neq 0$, then M can admit no metric of (strictly) positive scalar curvature.

It is a remarkable fact that for simply connected manifolds both results are 'sharp'. That is, the L-genus is, up to a scalar multiple, the only rational oriented cobordism invariant which is also a homotopy type invariant. The parallel result for the $\widehat{A}$-genus, asserting that for a simply connected closed Spin-manifold M of dimension $4k$ ($k > 1$) the vanishing of $\alpha(M)$ is the only obstruction to the existence on M of a metric of positive scalar curvature, was recently proved by S. Stolz.

The uniqueness part of the preceding statements ceases to hold in the non-simply connected case. Indeed, let T be a torus and let $f : M \to T$ be a continuous map. For each cohomology class $\eta \in H^*(T, \mathbb{Q})$, define the corresponding higher L (resp. $\widehat{A}$)-genus by

$$\lambda_{f,\eta}(M) = (L(M) \cup f^*(\eta))[M] \quad (\text{resp. } \alpha_{f,\eta}(M) = (\widehat{A}(M) \cup f^*(\eta))[M]) .$$

Lusztig [Lu] found an ingenious index theoretical interpretation for the higher L-genera $\lambda_{f,\eta}(M)$, in terms of the index bundle of a family of twisted signature operators, and used it to prove that they are invariants of the oriented homotopy

Proceedings of the International Congress
of Mathematicians, Kyoto, Japan, 1990

type of M. The same method can be adapted to give the vanishing of the higher $\widehat{A}$-genera $\alpha_{f,\eta}(M)$ in the presence of a metric of positive scalar curvature (cf. [GL1]).

More generally, the torus in the above definition may be replaced by any $K(\Gamma, 1)$ space, where Γ is a finitely presented group. According to a conjecture of Novikov [N], the numbers $\lambda_{f,\eta}(M)$, $\eta \in H^*(\Gamma, \mathbb{Q})$ ought to be homotopy invariants. There is a parallel conjecture for the $\widehat{A}$-genera, made by Gromov and Lawson [GL2], implying that if the Spin-manifold M admits a metric of positive scalar curvature then $\alpha_{f,\eta}(M) = 0$, for all $\eta \in H^*(\Gamma, \mathbb{Q})$. For both conjectures one may assume without loss of generality that $\Gamma = \pi_1(M)$ and f is the classifying map of the universal cover $\widetilde{M} \to M$.

The preceding discussion was meant to underscore the following points. First, for non-simply connected manifolds, there are 'higher' versions of the classical genera, which are involved in interesting open problems. Secondly, an index theoretical interpretation of the higher genera provides an effective tool for handling such problems. It was also noted above that the Families Index Theorem supplies the desired interpretation in the case when Γ is free abelian, essentially because its dual, the torus $\widehat{\Gamma}$, is a nice space. In general, though, the 'dual object' of Γ is no longer an ordinary space and therefore beyond the scope of standard index theory. It is, however, a 'non-commutative space' in the sense of A. Connes [C1]. Moreover, Connes' theory of cyclic cohomology and its pairing with K-theory [C1, C2] offers precisely the right context for developing the appropriate generalization of index theory. The first two sections of this report will survey the 'higher' index theorems obtained along these lines in [CM2] (cf. also [CM1]) and [CGM], as well as their applications to the conjectures of Novikov and Gromov-Lawson.

Like the higher genera, the higher indices of elliptic operators are manufactured out of group cohomology classes, viewed as classes in the cyclic cohomology of the group ring $\mathbb{C}\Gamma$. Somewhat surprisingly, the unstable component of the cyclic cohomology of $\mathbb{C}\Gamma$ appears to be intimately related to non-local invariants when $K(\Gamma, 1)$ is a manifold. This point of view will be put forth in the last section, where I will interpret in this light the results in [MS1] and [MS2], relating η-invariants and analytic torsion to periodic geodesics.

1. Group Cocycles and Higher Indices

1.1 Covering Indices of Elliptic Operators

Let M be a closed smooth manifold, Γ a finitely presented group, $f : M \to B\Gamma$ a continuous map to the classifying space of Γ and $\widetilde{M}$ the pull-back of the universal cover $E\Gamma \to B\Gamma$. Let $C_r^*(\Gamma)$ denote the C^*-algebra obtained as the completion of the group ring $\mathbb{C}\Gamma$ acting on the Hilbert space $\ell^2(\Gamma)$. Then $\widetilde{M} \times_\Gamma C_r^*(\Gamma)$ is a $C_r^*(\Gamma)$-bundle over M. Using it, Mishchenko [Ms2] and Kasparov [K1] have associated to an elliptic pseudodifferential operator D on M a K-theory class $Index_{f,C_r^*(\Gamma)}D \in K_0(C_r^*(\Gamma))$. In [CM2] this assignment was refined to produce an element $Index_{f,\Gamma} D \in K_0(\mathbb{C}\Gamma \otimes \mathfrak{R})$, where $\mathfrak{R}$ stands for the ring of infinite complex matrices $(a_{ij})_{1\le i,j<\infty}$ satisfying the rapid decay condition $\sup((i+j)^k|a_{ij}|) < \infty$, for all $k \in \mathbb{R}$. We sketch the construction of this covering index.

Let $\Psi(M,E)$ denote the algebra of pseudodifferential operators of classical type and integral order acting on the space of smooth sections of a vector bundle E over M. Let $\Psi^{-\infty}(M,E)$ denote the ideal formed by all smoothing operators. The quotient algebra $CS(M,E)$ is referred to as the algebra of complete symbols with coefficients in E. When E is the trivial line bundle, we simplify the notation by omitting E and by setting $\mathfrak{A} = \Psi(M)$ and $\mathfrak{B} = \Psi^{-\infty}(M)$. Upstairs, on the covering manifold $\widetilde{M}$, we let $\widetilde{\mathfrak{A}}$ be the subalgebra of $\Psi(\widetilde{M})$ consisting of all Γ-invariant, properly supported pseudodifferential operators and let $\widetilde{\mathfrak{B}}$ be the subalgebra of $\widetilde{\mathfrak{A}}$ formed of smoothing operators. Clearly $\widetilde{\mathfrak{A}}/\widetilde{\mathfrak{B}}$ is isomorphic to $CS(M) = \mathfrak{A}/\mathfrak{B}$.

For E a $\mathbb{Z}_2$-graded vector bundle on M, an odd elliptic pseudodifferential operator $D \in \Psi(M,E)$ defines an $\mathfrak{A}$-linear map D^+ between two finitely generated projective right $\mathfrak{A}$-modules, which becomes an isomorphism, $CS(D^+)$, on tensoring with $CS(M) = \mathfrak{A}/\mathfrak{B}$. Note that D^+ can also be viewed as a quasi-isomorphism over $\widetilde{\mathfrak{B}}$. Now identify $CS(M)$ with $\widetilde{\mathfrak{A}}/\widetilde{\mathfrak{B}}$. A well-known construction in algebraic K-theory, employed in the definition of the connecting homomorphism ∂ : $K_1(\widetilde{\mathfrak{A}}/\widetilde{\mathfrak{B}}) \to K_0(\widetilde{\mathfrak{B}})$, associates to $CS(D^+)$ an element $\partial(D^+)$ in $K_0(\widetilde{\mathfrak{B}})$. We remark parenthetically that the same construction applied to $\mathfrak{A}/\mathfrak{B} = CS(M)$ gives rise to the ordinary index of D^+, on identifying $K_0(\mathfrak{B})$ with $\mathbb{Z}$.

The algebra $\widetilde{\mathfrak{B}}$ can also be regarded as the convolution algebra of smooth compactly supported functions on the differentiable groupoid $\widetilde{M} \times_\Gamma \widetilde{M}$. As such, it can be homomorphically mapped to $\mathbb{C}\Gamma \otimes \mathfrak{B}$ and further, after identifying $\mathfrak{B} = \Psi^{-\infty}(M)$ with $\mathfrak{R}$, to $\mathbb{C}\Gamma \otimes \mathfrak{R}$. Although these homomorphisms are non-canonical they induce however a well-defined map, independent of all choices made,

$$\theta : K_0(\widetilde{\mathfrak{B}}) \to K_0(\mathbb{C}\Gamma \otimes \mathfrak{R}) .$$

Definition 1. The *index of D^+ associated to the covering $\widetilde{M}$* is defined as

$$Index_{f,\Gamma} D^+ = \theta(\partial(D^+)) \in K_0(\mathbb{C}\Gamma \otimes \mathfrak{R}) .$$

Remarks. 1. The ring $\mathbb{C}\Gamma \otimes \mathfrak{R}$ is very likely the smallest group ring associated to Γ with the property that its K_0-group is nontrivial. Indeed, for Γ torsion-free, the reduced K_0-group of the group ring $\mathbb{C}\Gamma$ is believed to be 0. By contrast, the group $K_0(\mathbb{C}\Gamma \otimes \mathfrak{R})$ appears to be, in general, uncountable.

2. Under the map from $K_0(\mathbb{C}\Gamma \otimes \mathfrak{R})$ to $K_0(C_r^*(\Gamma) \otimes \mathfrak{K}) = K_0(C_r^*(\Gamma))$, induced by the inclusion of $\mathbb{C}\Gamma$ into $C_r^*(\Gamma)$ and of $\mathfrak{R}$ into the ring $\mathfrak{K}$ of compact operators on the infinite dimensional separable Hilbert space, the image of $Index_{f,\Gamma} D^+$ coincides with $Index_{f,C_r^*(\Gamma)} D^+$. However, our covering index seems to be a finer invariant, which depends on the *complete,* not just the principal, symbol of the given operator.

1.2 Higher Analytic Indices

Next, we will use group cocycles to produce numerical invariants out of the covering index. Let $c \in Z^k(\Gamma, \mathbb{C})$ be a homogeneous group cocycle; without loss of generality, we may assume that c is alternating. One associates to c a cyclic cocycle τ_c on the algebra $\mathbb{C}\Gamma \otimes \mathfrak{R}$ by

$$\tau_c(a_0 \otimes A_0, \ldots, a_k \otimes A_k) = \mathrm{Tr}(A_0 \ldots A_k) \sum_{g_0 g_1 g_2 \cdots g_k = 1} a_0(g_0) a_1(g_1) \ldots a_k(g_k) \cdot c(1, g_1, g_1 g_2, \ldots, g_1 g_2 \ldots g_k) ;$$

the cyclic cohomology class $\tau_{[c]} = [\tau_c] \in HC^k(\mathbb{C}\Gamma \otimes \mathfrak{R})$ depends only on the group cohomology class $[c] \in H^k(\Gamma, \mathbb{C})$.

One employs now Connes' pairing [C1] between cyclic cohomology and K-theory to define the *covering higher indices* of the elliptic operator D^+, as follows:

$$Index_{f,\eta} D^+ = \langle \tau_\eta, Index_{f,\Gamma} D^+ \rangle , \quad \eta = [c] \in H^k(\Gamma, \mathbb{C}) , \ k \text{ even} .$$

When $\eta = [1] \in H^0(\Gamma, \mathbb{C})$, the above index coincides with the Γ-index of Atiyah and Singer [A]. Their Index Theorem for Covering Spaces asserts that

$$Index_{f,[1]} D^+ = Index\, D^+ ,$$

the right hand side being the ordinary index of D^+.

In the same spirit, relying on the fact that the differentiable groupoids $\widetilde{M} \times_\Gamma \widetilde{M}$ and $M \times M$ are locally isomorphic, one can show that the pairing $\langle \tau_\eta, Index_{f,\Gamma} D^+ \rangle$ depends only on the Alexander-Spanier cohomology class $f^*(\eta) \in H^k(M, \mathbb{C})$ and the K-homology class $[D^+] \in K_0(M)$. More precisely, one has the equality

$$\langle \tau_\eta, Index_{f,\Gamma} D^+ \rangle = \langle f^*(\eta), [D^+] \rangle ,$$

where the right hand side is a certain natural pairing between $H^{\mathrm{ev}}(M, \mathbb{C})$, viewed as Alexander-Spanier cohomology, and the K-homology group $K_0(M)$. This pairing can also be used to attach *intrinsic higher indices* to the elliptic operator D^+ by setting

$$Index_\xi D^+ = \langle \xi, [D^+] \rangle ,$$

for each Alexander-Spanier cohomology class $\xi \in H^{\mathrm{ev}}(M, \mathbb{C})$. When $\xi = [1] \in H^0(M, \mathbb{C})$ one recovers the ordinary index of D^+. The above discussion can thus be summarized as follows.

Proposition 1 ([CM2], §5). $Index_{f,\eta} D^+ = Index_{f^*(\eta)} D^+$, $\forall \eta \in H^{\mathrm{ev}}(B\Gamma, \mathbb{C})$.

1.3 Cohomological Formulae

To relate the higher indices introduced above to higher genera, one needs to express them, just like in the classical case, in purely cohomological terms. The basic result in this direction is the following generalization of the Atiyah-Singer Index Theorem.

Theorem 2 ([CM2], §3). *Let $\xi \in H^{2q}(M, \mathbb{C})$. With $\sigma(D^+)$ = principal symbol of D^+ and $\mathfrak{I}(M)$ = the index class of M^n, one has:*

$$Index_{\xi} D^+ = \frac{1}{(2\pi i)^q} \frac{q!}{(2q)!} (-1)^n (ch\, \sigma(D^+) \cup \mathfrak{I}(M) \cup \xi)[T^*M].$$

Remarks. 3. In fact we obtained a more refined result, at the level of differential forms not just cohomology classes, which extends the Local Index Theorem. Its proof, based on the heat equation method, uses the full force of Getzler's symbolic calculus [G].

4. The above result, in a non-compact variant due to J. Roe [R], was also used, in a rather striking way, by S. Weinberger [W] to give a direct analytic proof of Novikov's celebrated theorem on the topological invariance of rational Pontrjagin classes.

Combining Proposition 1 and Theorem 2, one obtains the desired Higher Index Formula for Covering Spaces:

Theorem 3 [CM2], §5). *Let $\eta \in H^{2q}(B\Gamma, \mathbb{C})$. Then*

$$Index_{f,\eta} D^+ = \frac{1}{(2\pi i)^q} \frac{q!}{(2q)!} (-1)^n (ch\, \sigma(D^+) \cup \mathfrak{I}(M) \cup f^*(\eta))[T^*M].$$

1.4 Applications to the Novikov and Gromov-Lawson Conjectures

Theorem 3, applied to the signature operator $\not{B}^+$ on an even-dimensional manifold M^{2k}, gives the identity

$$\lambda_{f,\eta}(M) = C_{k,q} \langle \tau_\eta, Index_{f,\Gamma} \not{B}^+ \rangle, \quad \forall \eta \in H^{2q}(B\Gamma, \mathbb{C}),$$

where $C_{k,q}$ is an explicit (nonzero) constant depending only on the dimension of M and the degree of η. Thus, the Novikov conjecture would follow if one could prove that $Index_{f,\Gamma} \not{B}^+ \in K_0(\mathbb{C}\Gamma \otimes \mathfrak{R})$ is an invariant of the homotopy type of M. In view of Remark 2, this is unlikely to happen. However, its image $Index_{f,C_r^*(\Gamma)} \not{B}^+ \in K_0(C_r^*(\Gamma))$ *is* a homotopy invariant. Indeed, this follows from Mishchenko's fundamental construction of an a priori homotopy invariant $\sigma[M, f]$ in the Witt group $L(\mathbb{C}\Gamma)$ [Ms1], together with the fact that $Index_{f,C_r^*(\Gamma)} \not{B}^+$ coincides with the image of $\sigma[M, f]$ under the map sending $L(\mathbb{C}\Gamma)$ to $L(C_r^*(\Gamma)) = K_0(C_r^*(\Gamma))$ (cf. [MsS], [K2]).

The problem which emerges is therefore to extend the pairing of $H^{\mathrm{ev}}(\Gamma, \mathbb{C})$ with $K_0(\mathbb{C}\Gamma \otimes \mathfrak{R})$ to a pairing with $K_0(C_r^*(\Gamma))$. There is a strategy for that, devised by Connes in a similar context [C3]. It consists of finding an intermediate subalgebra, which is closed under holomorphic functional calculus and therefore has the same K_0-group, to which cyclic cocycles of the form τ_c extend by continuity. In our situation, the natural candidate for this subalgebra is $\mathfrak{C}\Gamma$ = the smallest subalgebra of $C_r^*(\Gamma) \otimes \mathfrak{R}$ (C^*-tensor product) which contains $\mathbb{C}\Gamma \otimes \mathfrak{R}$ and is closed under holomorphic functional calculus. The following two conditions, involving the choice of a word length function $g \mapsto |g|$, are sufficient for the successful application of the above strategy to Γ:

Polynomial Cohomology: $\forall \eta \in H^{2q}(\Gamma, \mathbb{C})$, $\exists c \in Z^{2q}(\Gamma, \mathbb{C})$ of polynomial growth, with $\eta = [c]$;

Rapid Decay (cf. [J]): $\exists s \in \mathbb{N}$ and $C > 0$ such that $\forall a \in \mathbb{C}\Gamma$

$$\|a\|_{C_r^*(\Gamma)} \leq C\left(\sum (1 + |g|)^{2s} |a(g)|^2\right)^{1/2}.$$

Theorem 4 ([CM2], §6). *Any finitely presented group Γ with the above two properties satisfies both the Novikov and the Gromov-Lawson conjectures.*

Indeed, for such a group one has, $\forall \eta \in H^{2q}(B\Gamma, \mathbb{C})$,

$$\lambda_{f,\eta}(M) = C_{k,q} \langle \tau_\eta \, , \; Index_{f, C_r^*(\Gamma)} B\!\!\!/^{+} \rangle \, ,$$

which proves the homotopy invariance of the higher L-genera. On the other hand, applying Theorem 3 to the Dirac operator $D\!\!\!/^{+}$ gives

$$\alpha_{f,\eta}(M) = (2\pi i)^q \frac{(2q)!}{q!} \langle \tau_\eta \, , \; Index_{f, C_r^*(\Gamma)} D\!\!\!/^{+} \rangle \, .$$

By an extension of Lichnerowicz' argument, Rosenberg [Ro] has shown that if M admits a metric of positive scalar curvature then $Index_{f, C_r^*(\Gamma)} D\!\!\!/^{+} = 0$. Therefore, all higher $\widehat{A}$-genera also vanish under this hypothesis.

Remark. 5. The conclusion of the above theorem can be strengthened to assert that the assembly map (see [K2], [K3]) $\mu : K_*(B\Gamma) \to K_*(C_r^*(\Gamma))$ is rationally injective.

The two properties stated above are satisfied by a large class of finitely presented groups, including the remarkable class of Gromov's hyperbolic groups [Gr]. The fact that a hyperbolic group satisfies the Polynomial Cohomology property is due to Gromov, while the Rapid Decay property was proved by P. de la Harpe [Ha], building on prior results by Haagerup [H] and Jolissaint [J].

Corollary 5 ([CM2], §6). *The hyperbolic groups satisfy the Novikov and the Gromov-Lawson conjectures.*

2. Asymptotic Cocycles and Higher Indices

We now turn to a different procedure for constructing numerical indices, based on Connes' pairing for entire cohomology [C2].

2.1 Almost Flat Bundles and Almost Representations

Let M be a closed connected Riemannian manifold, E a Hermitian vector bundle over M and ∇ a metric preserving connection on E. With $\Theta = \nabla^2$ being the curvature of ∇, we set

$$\|\Theta\| = \sup\{\|\Theta_x(X,Y)\| \; ; \; \|X \wedge Y\| \leqq 1, \; X, Y \in T_xM, \; x \in M\}.$$

Definition 2. (i) Given $\varepsilon > 0$, the connection ∇ is called *ε-flat* if $\|\Theta\| < \varepsilon$.

(ii) A vector bundle E is called *ε-flat* if it admits a metric preserving connection ∇ with $\|\Theta\| < \varepsilon$.

(iii) A class $\xi \in K^0(M)$ is called *almost flat* if for each $\varepsilon > 0$, there exist ε-flat bundles E^+, E^- such that $[E^+] - [E^-] = \xi$.

Note that the last notion is independent of the Riemannian structure on M. We denote by $K^0_{\mathrm{af}}(M)$ the subgroup of $K^0(M)$ formed of almost flat classes. If $h : N \to M$ is a continuous map, then $h^* : K^0(M) \to K^0(N)$ maps $K^0_{\mathrm{af}}(M)$ to $K^0_{\mathrm{af}}(N)$.

Remark. 6. If M is simply connected, there exists $\varepsilon_M > 0$ such that any ε_M-flat bundle is flat and therefore, trivial. On the other hand $K(\Gamma,1)$-manifolds with Γ 'large' tend to have 'many' almost flat bundles (see 2.4).

Let $\mathfrak{B}$ be a Banach algebra. Given a linear map ϱ of $\mathfrak{B}$ to the algebra $\mathfrak{L}(V)$ of all linear operators on a finite-dimensional Hilbert space V, we introduce the 'curvature' operators $\theta(a,b) = \varrho(ab) - \varrho(a)\varrho(b)$ and set, for any finite subset $F \subset \mathfrak{B}$,

$$\|\theta\|_F = \sup\{\|\theta(a,b)\| \; ; \; a, b \in F\}.$$

Definition 3. Given $\varepsilon > 0$ and a finite subset $F \subset \mathfrak{B}$, an *(ε, F)-representation* of $\mathfrak{B}$ is a linear map $\varrho : \mathfrak{B} \to \mathfrak{L}(V)$ such that: (i) $\|\varrho(a)\| \leq \|a\|$, $\forall a \in \mathfrak{B}$; (ii) $\varrho(1) = 1$; (iii) $\|\theta\|_F < \varepsilon$.

Let $p \in M$ be a fixed base point and let $\Gamma = \pi_1(M,p)$. We denote by $\Omega(M,p)$ the space of loops based at p. Choose a cross-section $\gamma : \Gamma \to \Omega(M,p)$ to the canonical projection $\pi : \Omega(M,p) \to \Gamma$ such that: $\gamma(1) = 1$ and $\forall g \in \Gamma$ with $g^2 \neq 1$, $\gamma(g^{-1}) = \gamma(g)^{-1}$. Given $F \subset \ell^1(\Gamma)$, one can associate to any ε-flat bundle E an $(\varepsilon c_F, F)$-representation $\varrho_E : \ell^1(\Gamma) \to \mathfrak{L}(E_p)$, by assigning to an element $g \in \Gamma$ the parallel transport operator $\varrho_E(g)$ along the loop $\gamma(g)$; if $g^2 = 1$, $\varrho_E(g)$ is replaced by $\frac{1}{2}(\varrho_E(g) + \varrho_E(g)^*)$. Indeed, since the effect of the parallel transport along loops can be estimated in terms of the curvature, one has

$$\|\theta_E\|_F \leq c_F \|\Theta_E\|,$$

where Θ_E denotes the curvature of E, θ_E is the curvature of ϱ_E and c_F is a constant depending on F.

2.2 Pairing with K-Theory and an Index Formula

Let $\varrho : \mathfrak{B} \to \mathfrak{L}(V)$ be an (ε, F)-representation of a Banach algebra $\mathfrak{B}$. Its character $\phi = (\phi_{2k})$ is a discrete version of the asymptotic cocycles introduced in [CM2, §4], and is given by the following formula:

$$\phi_{2k}(a_0, a_1, \dots, a_{2k}) = (-1)^k \frac{(2k)!}{k!} Tr(\varrho(a_0)\theta(a_1, a_2) \dots \theta(a_{2k-1}, a_{2k}))\,, \quad a_j \in \mathfrak{B}\,.$$

Recall that asymptotic cocycles pair with K-theory, by means of the pairing formula for entire cohomology ([C2], [GS]):

$$\langle \phi, e \rangle = \sum_{k=0}^{\infty} \frac{(-1)^k}{k!} \, \phi_{2k}(e - \frac{1}{2}, e, \dots, e)\,, \qquad e^2 = e \in \mathcal{M}_N(\mathfrak{B})\,.$$

Proposition 6 ([CGM], Prop. 9)**.** *Let $e^2 = e \in \mathcal{M}_N(\mathfrak{B})$. There exists a finite set $F \subset \mathfrak{B}$ and $\delta > 0$, such that for any (ε, F)-representation ϱ of $\mathfrak{B}$ with $0 < \varepsilon < \delta$, the above series converges. Moreover, with ϕ denoting the character of ϱ one has*

$$\langle \phi, e \rangle = Tr(\Pi - \frac{1}{2})\,,$$

where $\Pi \in \mathcal{M}_N(\mathfrak{L}(V)) = \mathfrak{L}(V \otimes \mathbb{C}^N)$ is the projection defined by

$$\Pi = \frac{1}{2\pi i} \int_{|z-1|=\frac{1}{2}} ((\varrho \otimes 1)(e) - z)^{-1} \, dz\,.$$

In the remainder of this section $\widetilde{M}$ will denote the universal cover of M^n and $f : M \to B\Gamma$ will be its classifying map. Consider an elliptic operator D^+ on M. From the preceding discussion, it follows that there exists a $\delta > 0$ such that, for any ε-flat bundle E on M with $0 < \varepsilon < \delta$, the pairing

$$Index_{E,\Gamma} D^+ := \langle \phi_E, Index_{f,\Gamma} D^+ \rangle\,,$$

between the character ϕ_E of the associated almost representation $\varrho_E : \ell^1(\Gamma) \to \mathfrak{L}(E_p)$ and the covering index $Index_{f,\Gamma} D^+ \in K_0(\mathbb{C}\Gamma \otimes \mathfrak{R})$ is well-defined.

Theorem 7 ([CGM], Thm. 10)**.** *With the above assumptions, one has*

$$Index_{E,\Gamma} D^+ = (-1)^n (ch\, \sigma(D^+) \cup \mathfrak{I}(M) \cup ch\, E)[T^* M] = Index\, D_E^+\,.$$

The proof is similar in spirit with that of Theorem 3 and involves two stages. The first, corresponding to Proposition 1, converts the above pairing into one between the 'asymptotic' Alexander-Spanier cocycle associated to the character ϕ_E and the K-homology class of D^+. The second consists in establishing the analogue of the index formula of Theorem 2.

2.3 Pairing with L-Theory and a Signature Formula

The extra virtue of the type of cocycles considered above is that they can be paired directly with the L-theory. Indeed, $\ell^1(\Gamma)$ is not merely a Banach algebra, but one equipped with a compatible involution. So, it is natural to work in the context of Banach $*$-algebras. An obvious modification of Definition 3 gives the notion of an $(\varepsilon, F) - *$*-representation* for a Banach $*$-algebra $\mathfrak{B}$.

Proposition 8 ([CGM], Prop. 5). *Let $\mathfrak{B}$ be a Banach $*$-algebra and let $H = H^*$ be a self-adjoint invertible element in $\mathcal{M}_N(\mathfrak{B})$.*

(i) If $\varrho : \mathfrak{B} \to \mathfrak{L}(V)$ is an $(\varepsilon, F) - *$-representation, with $F =$ the set of entries of H and H^{-1} and $0 < \varepsilon < 1/2N^2$, then $(\varrho \otimes 1)(H) \in \mathcal{M}_N(\mathfrak{L}(V)) = \mathfrak{L}(V \otimes \mathbb{C}^N)$ is self-adjoint and invertible.

(ii) If H' is another representative of the Witt class $[H] \in L(\mathfrak{B})$, there exists a finite set $F' \subset \mathfrak{B}$ and $\delta > 0$, such that for any (ε, F')-representation ϱ of $\mathfrak{B}$ with $0 < \varepsilon < \delta$ one has:

$$\textit{Signature}\,(\varrho \otimes 1)(H) = \textit{Signature}\,(\varrho \otimes 1)(H')\,.$$

Remark that the almost representation associated to an almost flat bundle is automatically a $*$-representation, therefore the above construction applies. From Proposition 6 and Theorem 7, one obtains:

Theorem 9 ([CGM], Thm. 6). *Let $H \in \mathcal{M}_N(\ell^1(\Gamma))$ be a representative of the Mishchenko class $\sigma[M,f] \in L(\ell^1(\Gamma))$. There exists a $\delta > 0$ such that, for any ε-flat bundle E on M^{2k} with $0 < \varepsilon < \delta$, one has:*

$$\textit{Signature}\,(\varrho_E \otimes 1)(H) = 2^k(\mathcal{L}(M) \cup ch\, E)[M]\,,$$

where $\mathcal{L}(M)$ denotes the stable L-class of M.

Corollary 10 [CGM]. *Let M, N be two closed oriented manifolds and let $h : N \to M$ be an oriented homotopy equivalence. Then, for any class $\xi \in K^0_{\mathrm{af}}(M)$, one has $(\mathcal{L}(N) \cup ch\, h^*\xi)[N] = (\mathcal{L}(M) \cup ch\xi)[M]$.*

Remark. 7. According to Theorem 7, the higher indices constructed in this section, unlike the higher indices of Section 1, are genuine indices of elliptic operators with suitable coefficients. Thus, it makes sense to try to obtain the above result by proving directly the homotopy invariance of the index of the signature operator with coefficients in an almost flat bundle. Such a direct proof was recently achieved by Hilsum and Skandalis [HS].

2.4 Application to the Novikov Conjecture

The notion of an almost flat bundle can be extended to the case when the base space is a simplicial complex and the bundle is infinite-dimensional (but endowed with a 'superconnection', in a suitable sense). Moreover, Theorem 9 continues to function in this extended context. With these modifications, one can show that $K^0(B\Gamma) = K^0_{\mathrm{af}}(B\Gamma)$, and therefore that the Novikov conjecture holds, for a 'very large' class of finitely presented groups Γ. This class includes all the cases for which the conjecture was previously established, in particular the following:

(1) $\Gamma = \pi_1(N)$ where N is a complete Riemannian manifold of non-positive sectional curvature (cf. [K3]) or, more generally, $\widetilde{N}$ is hyperspherical (see [GL2]);

(2) Γ is a discrete subgroup of a finitely connected Lie group (cf. [K3]) or of an algebraic group over a local field (cf. [KS]);

(3) Γ is hyperbolic in Gromov's sense (cf. [CM2]).

3. Unstable Cyclic Cohomology and Non-Local Invariants

3.1 Cyclic Cohomology of Group Rings

Let Γ be a finitely generated group. For simplicity, we will assume that it is torsion-free and of finite cohomological dimension (over $\mathbb{Q}$) n. This implies that the cyclic cohomology of $\mathbb{C}\Gamma$ stabilizes above n : $HC^{2k+n}(\mathbb{C}\Gamma) = HC^n(\mathbb{C}\Gamma)$ and $HC^{2k+1+n}(\mathbb{C}\Gamma) = HC^{n+1}(\mathbb{C}\Gamma)$, for $k = 0, 1, 2, \ldots$.

Burghelea [B] showed that $HC^*(\mathbb{C}\Gamma)$ decomposes into a direct sum parametrized by the conjugacy classes $[\gamma]$ in Γ :

$$HC^*(\mathbb{C}\Gamma) \cong H^*(B\Gamma, \mathbb{C}) \otimes HC^*(\mathbb{C}) \oplus \sum_{[\gamma]\neq 1} H^*(BN_\gamma, \mathbb{C}),$$

where $N_\gamma = \Gamma_\gamma / C_\gamma$, $\Gamma_\gamma =$ the centralizer of γ and $C_\gamma =$ the (free) group generated by γ. On the other hand, for certain classes of groups Γ, including discrete subgroups of general linear groups over fields of characteristic 0, Eckmann [E] showed that in the stable range, i.e. for $k \geq n$, $H^k(BN_\gamma, \mathbb{C}) = 0$, $\forall [\gamma] \neq 1$.

Thus, it was precisely the stable component of $HC^*(\mathbb{C}\Gamma)$ which was used in the preceding sections to construct higher indices of elliptic operators. The aim of this section is to suggest that the unstable component of $HC^*(\mathbb{C}\Gamma)$ plays a similar, albeit more subtle, rôle with respect to non-local invariants such as analytic torsion and η-invariants.

3.2 η-Invariants and Analytic Torsion via Cyclic Homology

Let M^{4k-1} be a closed connected Riemannian manifold. The η-invariant η of M (see [APS]), is meant to measure the 'signature' of the signature operator $\not{B}^{\mathrm{ev}}$ acting on the space $\Omega^{\mathrm{ev}} = \Omega^{\mathrm{ev}}(M)$ of exterior differential forms of even degree on M. We recall that η is defined as the value at $s = 0$ of the analytic continuation of the function

$$\eta(s) = \sum (\mathrm{sign}\lambda)|\lambda|^{-s}, \qquad \mathrm{Re}(s) \gg 0,$$

where λ runs over the set of non-zero eigenvalues of $\not{B}^{\mathrm{ev}}$, each eigenvalue being repeated according to its multiplicity. It can be easily shown that η involves only the 'middle' degree forms, more exactly than $\eta = \eta^{2k}(0)$, where

$$\eta^{2k}(s) = \sum (\text{sign}\,\lambda)|\lambda|^{-s}, \qquad \text{Re}(s) \gg 0,$$

with λ running this time over the set of non-zero eigenvalues of the operator $*d$ acting on $d^*\Omega^{2k}$.

This definition can be reformulated in terms of the cyclic homology of the algebra $\mathscr{A} = C^\infty(M)$. We recall ([C1]) that:

(i) $HH_*(\mathscr{A})$ is canonically isomorphic to $\Omega^*(M)$; under this isomorphism, the exterior derivative $d : \Omega^n(M) \to \Omega^{n+1}(M)$ corresponds to the operator $\beta = B \circ I$: $HH_n(\mathscr{A}) \to HH_{n+1}(\mathscr{A})$;

(ii) $HC_n(\mathscr{A}) \cong \Omega^n(M)/d\Omega^{n-1}(M) \oplus H^{n-2}(M) \oplus H^{n-4}(M) \oplus \cdots$.

By transport of structure, both $HH_*(\mathscr{A})$ and $HC_*(\mathscr{A})$ become (graded) pre-Hilbert spaces.

Thus, ignoring the zero modes, η can be viewed as the *regularized signature* of $*\beta : HC_{2k-1}(\mathscr{A}) \to HC_{2k-1}(\mathscr{A})$.

There is a similar interpretation for the analytic torsion. Namely, if τ denotes the analytic torsion of M (see [RS]), with the above notation one has:

$$\log \tau = \sum_{q \geq 0} (-1)^q \log \det{}_\zeta (\beta^* \beta : HC_q(\mathscr{A}) \to HC_q(\mathscr{A})).$$

Here $\deg_\zeta$ denotes the ζ-determinant of the corresponding positive elliptic operator, with the zero modes removed. Regarding $\log \deg_\zeta(\beta^*\beta : HC_q(\mathscr{A}) \to HC_q(\mathscr{A}))$ as a regularized dimension of $HC_q(\mathscr{A})$, the above formula identifies $\log \tau$ as the *regularized Euler characteristic* of $HC_*(\mathscr{A})$.

Let $\mathscr{E}_\varrho = \widetilde{M} \times_\Gamma E$ be a flat bundle associated to a unitary representation ϱ of $\Gamma = \pi_1(M)$ on E. Twisted invariants τ_ϱ and η_ϱ can be defined in similar terms. To this end, one needs a twisted version of the cyclic homology of the algebra $\mathscr{A} = C^\infty(M)$. This is obtained by replacing the entries $C(\mathscr{A})_{p,q} = \mathscr{A}^{\otimes q-p+1}$, if $q \geq p$ and 0 otherwise, in the (b, B)-bicomplex $C(\mathscr{A})$ of the algebra $\mathscr{A}$, with $C(\mathscr{A}, \mathscr{E}_\varrho)_{p,q} = (E \otimes \widetilde{\mathscr{A}}^{\otimes q-p+1})^\Gamma$, if $q \geq p$ and 0 otherwise. Here $\widetilde{\mathscr{A}} = C^\infty(\widetilde{M})$ and $(E \otimes \widetilde{\mathscr{A}}^{\otimes n})^\Gamma$ = the subspace of Γ-invariant elements of $E \otimes \widetilde{\mathscr{A}}^{\otimes n}$. Note that $1 \otimes b$ and $1 \otimes B$ commute with the (tensor product) action of Γ and therefore induce boundary operators for the bicomplex $C(\mathscr{A}, \mathscr{E}_\varrho)$, which we will continue to denote b and B. One can thus define the *cyclic homology of* $\mathscr{A}$ *with coefficients in* $\mathscr{E}_\varrho$ as $HC_*(\mathscr{A}, \mathscr{E}_\varrho)$ = the homology of the total complex associated to $C(\mathscr{A}, \mathscr{E}_\varrho)$ and the *Hochschild homology of* $\mathscr{A}$ *with coefficients in* $\mathscr{E}_\varrho$ as $HH_*(\mathscr{A}, \mathscr{E}_\varrho)$ = the homology of the 'vertical column complex' $\{(E \otimes \widetilde{\mathscr{A}}^{\otimes n})^\Gamma, b\}$.

3.3 Regularized Index Formulae

In [MS1] and [MS2], the η-invariants and the torsion invariants of locally symmetric manifolds of non-positive sectional curvature were expressed as special values of zeta-functions constructed out of the periodic geodesics. Our purpose here is to explain how to understand these results as some kind of 'regularized index formulae'.

Let $M = \Gamma\backslash\widetilde{M}$ with $\widetilde{M}$ a globally symmetric space of non-compact type and Γ a discrete, torsion-free, co-compact subgroup of orientation-preserving isometries. M inherits a (locally symmetric) Riemannian metric g, of non-positive sectional curvature. The connected components of the periodic set of the geodesic flow Φ, acting on the unit tangent bundle SM, are parametrized by the non-trivial conjugacy classes $[\gamma]$ in $\Gamma = \pi_1(M)$. Each connected component M_γ is itself a closed locally symmetric manifold of non-positive sectional curvature. Φ restricts to a periodic flow on each M_γ and the quotient $\widehat{M}_\gamma = M_\gamma/\Phi$ is an orbifold; we denote by $\chi(\widehat{M}_\gamma)$ its Euler characteristic. All orbits of Φ in M_γ have the same length ℓ_γ and m_γ will denote the multiplicity of a generic orbit.

Let ϱ be an *acyclic* unitary representation of Γ and denote by τ_ϱ the analytic torsion of M with coefficients in $\mathscr{E}_\varrho$. The main result in [MS2] reads as follows:

$$\log \tau_\varrho = - \sum_{[\gamma]\neq 1} \chi(\widehat{M}_\gamma)\mathrm{Tr}\varrho(\gamma)\frac{e^{-sl_\gamma}}{m_\gamma}\Bigg|_{s=0} ,$$

where the evaluation at 0 is made after analytic continuation.

We remark that, as a topological space, $\widehat{M}_\gamma = BN_\gamma$ (see 3.1). Moreover, the orientation of $\widehat{M}_\gamma$ defines a fundamental class $[\widehat{M}_\gamma]$ in the top-degree rational homology of BN_γ and one has:

$$\chi(\widehat{M}_\gamma) = e(\widehat{M}_\gamma)[\widehat{M}_\gamma],$$

where $e(\widehat{M}_\gamma)$ denotes the Euler class of the orbifold. The collection of classes $e(M,g) = \{e(\widehat{M}_\gamma) \in H^*(BN_\gamma,\mathbb{Q});\ [\gamma] \neq 1\}$ can be viewed as an element of a suitable completion $\widehat{HC}^*(\mathbb{Q}\Gamma)$ of $HC^*(\mathbb{Q}\Gamma)$, while the collection $[M,g] = \{[\widehat{M}_\gamma] \in H_*(BN_\gamma,\mathbb{Q});\ [\gamma] \neq 1\}$ defines an element in a similar completion of $HC_*(\mathbb{Q}\Gamma)$. Therefore, the right hand side of the above formula for torsion can be interpreted as a regularized pairing $\langle e(M,g),[M,g]\rangle_\varrho$, depending on the representation ϱ, the regularization being given by the geodesic spectrum. We recall that $\log\tau_\varrho$ is also defined by a regularization procedure, but involving the 'dual' data, namely the spectrum of the Laplace operator associated to the metric.

A similar interpretation can be given to the η-invariant formula proved in [MS1].

We conclude by observing that all the above expressions make sense for any closed $K(\Gamma,1)$-manifold which admits a metric of non-positive sectional curvature. Examples of such formulae, for the torsion of certain manifolds which are not locally symmetric, can be found in the work of D. Fried (cf. [F] and references therein). It is plausible that such formulae continue to hold on more general non-positively curved closed $K(\Gamma,1)$-manifolds.

References

[A] M. F. Atiyah: Elliptic operators, discrete groups and von Neumann algebras. Asterisque **32–33** (1976) 43–72

[APS] M. F. Atiyah, V. K. Patodi and I. M. Singer: Spectral asymmetry and Riemannian geometry. I, Math. Proc. Camb. Phil. Soc. **77** (1975) 43–69; II, **78** (1975) 405–432; III, **79** (1976) 71–99

[B] D. Burghelea: The cyclic homology of group rings. Comment. Math. Helv. **60** (1985) 354–365

[C1] A. Connes: Non-commutative differential geometry. Publ. Math. IHES **62** (1985) 41–144

[C2] A. Connes: Entire cohomology of Banach algebras and characters of θ-summable Fredholm modules. K-theory **1** (1988) 519–548

[C3] A. Connes: Cyclic cohomology and the transverse fundamental class of a foliation. In: H. Araki, E. G. Effros (eds.) Geometric Methods in Operator Algebras. Pitman, Research Notes in Mathematics Series **123** (1986) 52–144

[CGM] A. Connes, M. Gromov and H. Moscovici: Novikov's conjecture and almost flat vector bundles. C. R. Acad. Sci. Paris, Ser. I **310** (1990) 273–277

[CM1] A. Connes and H. Moscovici: Novikov's conjecture and hyperbolic groups. C. R. Acad. Sci. Paris, Ser. I **307** (1988) 475–480

[CM2] A. Connes and H. Moscovici: Cyclic cohomology, the Novikov conjecture and hyperbolic groups. Topology **29** (1990) 345–388

[E] B. Eckmann: Cyclic homology of groups and the Bass conjecture. Comment. Math. Helv. **61** (1986) 193–202

[F] D. Fried: Counting circles. In: Dynamical systems. (Lecture Notes in Mathematics, vol. 1342). Springer, Berlin Heidelberg New York, pp. 196–215

[G] E. Getzler: Pseudodifferential operators on supermanifolds and the Atiyah-Singer index theorem. Commun. Math. Phys. **92** (1983) 163–176

[GS] E. Getzler and A. Szenes: On the Chern character of a theta summable Fredholm module. J. Funct. Analysis **84** (1989) 343–357

[Gr] M. Gromov: Hyperbolic groups. In: S.M. Gersten (ed.) Essays in Group Theory. MSRI Publ. **8** (1987) 75–264

[GL1] M. Gromov and H. B. Lawson: Spin and scalar curvature in the presence of a fundamental group. Ann. Math. **111** (1980) 209–230

[GL2] M. Gromov and H. B. Lawson: Positive scalar curvature and the Dirac operator on complete Riemannian manifolds. Publ. Math. IHES **58** (1983) 83–196

[H] U. Haagerup: An example of a non-nuclear C^*-algebra which has the metric approximation property. Invent. math. **50** (1979) 279–293

[Ha] P. de la Harpe: Groupes hyperboliques, algèbres d'opérateurs et un théorème de Jolissaint. C. R. Acad. Sci. Paris, Ser. I **307** (1988) 771–774

[HS] M. Hilsum and G. Skandalis: Invariance par homotopie de la signature à coefficients dans un fibré presque plat. Preprint, Collège de France, 1990

[J] P. Jolissaint: Rapidly decreasing functions in reduced C^*-algebras of groups. Trans. Amer. Math. Soc. **317** (1990) 167–196

[K1] G. G. Kasparov: Topological invariants of elliptic operators, I: K-homology. Math. USSR Izv. **9** (1975) 751–792

[K2] G. G. Kasparov: K-theory, group C^*-algebras and higher signatures. Conspectus, Parts I, II. Chernogolovka Preprint 1981

[K3] G. G. Kasparov: Equivariant K-theory and the Novikov conjecture. Invent. math. **91** (1988) 147–201

[KS] G. G. Kasparov and G. Skandalis: Groupes agissant sur des immeubles de Bruhat-Tits, K-théorie opératorielle et conjecture de Novikov. C. R. Acad. Sci. Paris, Ser. I, **310** (1990) 171–174

[Lu] G. Lusztig: Novikov's higher signature and families of elliptic operators. J. Diff. Geom. **7** (1971) 229–256

[Ms1] A. S. Mishchenko: Homotopy invariants of non-simply connected manifolds. I, Math. USSR Izv. **4** (1970) 509–519

[Ms2] A. S. Mishchenko: C^*-algebras and K-theory. (Lecture Notes in Mathematics, vol. 763). Springer, Berlin Heidelberg New York, pp. 262–274

[MsS] A. S. Mishchenko and Yu. P. Solov'ev: Representations of Banach algebras and Hirzebruch type formulae. Math. USSR Sbornik **39** (1981) 189–205

[MS1] H. Moscovici and R. J. Stanton: Eta invariants of Dirac operators on locally symmetric manifolds. Invent. math. **95** (1989) 629–666

[MS2] H. Moscovici and R. J. Stanton: R-torsion and zeta functions for locally symmetric manifolds. Invent. math. (in press)

[N] S. P. Novikov: Analogues hermitiens de la K-théorie. Actes, Congrès Intern. Math. t. **2**, (1970) 39–45

[RS] D. B. Ray and I. M. Singer: R-torsion and the Laplacian on Riemannian manifolds. Adv. Math. **7** (1971) 145–210

[R] J. Roe: Exotic cohomology and index theory for complete Riemannian manifolds. Preprint 1990

[Ro] J. Rosenberg: C^*-algebras, positive scalar curvature and the Novikov conjecture. Publ. Math. IHES **58** (1983) 409–424

[W] S. Weinberger: An analytic proof of the topological invariance of rational Pontrjagin classes. Preprint 1990

State Sum Models in Low-Dimensional Topology

Vladimir G. Turaev

LOMI, Fontanka 27, Leningrad 191011, USSR and Louis Pasteur University
7, rue René Descartes, F-67084 Strasbourg, France

1. Introduction

In recent years certain fundamental ideas of statistical mechanics and quantum physics have penetrated through psychological barriers between physics and topology. The original impetus to this development was given in 1984 by the V. Jones's discovery of a new polynomial invariant of links in the 3-sphere S^3. His construction uses braids, Hecke algebras and von Neumann algebras. The later progress has involved 2-dimensional conformal field theory, theory of quantum groups and other subjects, creating a unified overwhelmingly rich area of study. Of especial importance have been the methods of statistical mechanics first applied by L. Kauffman to reproduce the Jones polynomial, and the ideas of quantum field theory brought into the subject by E. Witten.

From the viewpoint of topology the main new notion attained in the course of this development is the notion of topological quantum field theory (TQFT; see [At1]). There are several complementary approaches to constructing TQFTs in dimension 3. Here I present an approach based on the theory of quantum groups (see [Dr1, FRT1, Ji1]).

The state sum models appear in this context as a technical tool to define invariants of geometric objects. I give here an abstract description of the models; examples are abundant below. On the geometric part the models use certain "images" of the objects and "moves" relating these images. The moves should be local transformations of images following certain patterns. Any images of topologically equivalent objects should be related by a finite sequence of moves. One also distinguishes two sorts of local geometric blocks of the images called atoms and interactions. (This terminology is due to G. Kuperberg). Each interaction should give rise to a tuple of atoms, "involved in this interaction". On the statistical mechanics side one uses the notions of "algebraic initial data", "state", and "state sum". The initial data (in its simplest form) consists of a finite set I of "colors" and a weight function which is a numerical function on certain tuples of colors. A state on an image is an I-valued function on the set of atoms. Having such a state one associates with each interaction the tuple of colors of atoms involved in this interaction. One forms the product of weights of these tuples over all interactions and sums up these products over all states. To ensure topological invariance of the sum it suffices to check invariance under the moves. Their local nature usually enables one to reformulate the invariance

Proceedings of the International Congress
of Mathematicians, Kyoto, Japan, 1990

as a collection of algebraic equations on the weights. Solving these equations is fundamental to justify the model.

This scheme admits variations and extends to relative situations, producing linear operators (rather then numbers) forming a TQFT.

It seems that in the last decades of the 20th century characterized by the Bourbaki style in mathematics one may reject the methods outlined above as old-fashioned and non-elegant. A good excuse is that due mainly to E. Witten we have a global physical ideology behind the constructions. To justify this mathematically non-rigorous ideology we apply the down-to-earth technique of state sum models, moves et cetera. The models play here the role similar to that of simplicial chain complexes in homology theory.

The theory of state sum models has revived certain geometric ideas of K. Reidemeister, M. Newman, J. Alexander developed in the 1920–1930s; it involved more recent results of J. Cerf, R. Kirby, R. Fenn and C. Rourke, as well as the theory of special spines of 3-manifolds due to B. Casler, S. Matveev, and R. Piergallini, and the theory of categories of tangles due to D. Yetter and the author. Bringing these topological results in a row and relating them to the methods coming from physics and representation theory seem to be one of the most attractive features of the theory.

I will describe here 3 types of models: vertex models on link diagrams in $\mathbb{R}^2$, face models on link diagrams and link shadows, simplicial models on triangulations of 3-manifolds.

For an earlier discussion of state sum models in knot theory see [Jo2]. For other approaches see [Ja1, Ku1].

2. Vertex Models on Link Diagrams

The 2-dimensional vertex models of statistical mechanics are concerned with 4-valent graphs in $\mathbb{R}^2$, the edges and vertexes being respectively the atoms and interactions (see [Ba1]). These models have been adjusted to the topological setting of link diagrams in $\mathbb{R}^2$ where one has to take into account the over/undercrossing information and to treat the local extremums of the height function $(x, y) \mapsto y : \mathbb{R}^2 \to \mathbb{R}$ (reduced to the diagram at hand) as interactions. This produces isotopy invariants of colored framed oriented links in S^3 (see [AW1, Jo1, Tu1, Re1, RT1]).

Fix a set I of left modules over a commutative ring with unit K. Assume that each module $V \in I$ is equipped with a finite K-basis J_V.

By a colored link in S^3 we mean a finite collection of disjoint imbedded circles in S^3, each circle being equipped with a "color" $V \in I$. A link $L \subset S^3$ is framed (resp. oriented) if L is equipped with a non-singular normal vector field (resp. with orientation of L). Link diagrams are the images of links under pushing off into $\mathbb{R}^3 = S^3 \backslash \{\infty\}$ and projecting to $\mathbb{R}^2$, with an account taken of over/undercrossings. (To construct a diagram of a framed link L one first deforms L to make its framing orthogonal to $\mathbb{R}^2$).

We consider only the diagrams whose self-crossings are distinct from local extremums (of the height function). The classical Reidemeister moves relate any two diagrams of a link modulo ambient isotopy of the diagrams in $\mathbb{R}^2$. After a small modification the moves work also for diagrams of framed links. D. Yetter and independently the author gave a list of moves relating ambiently isotopic diagrams (see [Ye1, Tu2, Tu3]); the simplest of these moves introduces 1 point of local maxima and 1 point of local minima on a small vertical segment of a diagram.

The initial algebraic data of the model consists of 6 weight functions

$$R, \overline{R} : \coprod_{V,W \in I} J_V^2 \times J_W^2 \to K \, ; \quad \alpha, \beta, \mu, \nu : \coprod_{V \in I} J_V^2 \to K \, . \tag{1}$$

Let $\mathscr{D} \subset \mathbb{R}^2$ be a diagram of a colored framed oriented link L. The self-crossings of $\mathscr{D}$ and the local extremums of the height function split $\mathscr{D}$ into a finite number of arcs. A state s on $\mathscr{D}$ is a mapping, associating with each such arc e of $\mathscr{D}$ an element of the set $J_{V(e)}$ where $V(e) \in I$ is the color of the component of L projecting to the loop of $\mathscr{D}$ which contains e. For a state s on $\mathscr{D}$ and a positive (resp. negative) self-crossing a of $\mathscr{D}$ we put $\langle a \rangle_s = R(i, \ell, j, k)$ (resp. $\langle a \rangle_s = \overline{R}(i, \ell, j, k)$) where i, ℓ, j, k are the values of s on the 4 arcs incident to a; see Fig. 1. For a local extremum b, as in Fig. 2, we put $\langle b \rangle_s = f(i, \ell)$ where respectively $f = \alpha, \beta, \mu, \nu$. Put

$$\langle \mathscr{D} \rangle_s = \prod_a \langle a \rangle_s \prod_b \langle b \rangle_s \text{ and } \langle L \rangle = \langle \mathscr{D} \rangle = \sum_s \langle \mathscr{D} \rangle_s$$

where a, b, s run respectively over crossings, local extremums and states of $\mathscr{D}$.

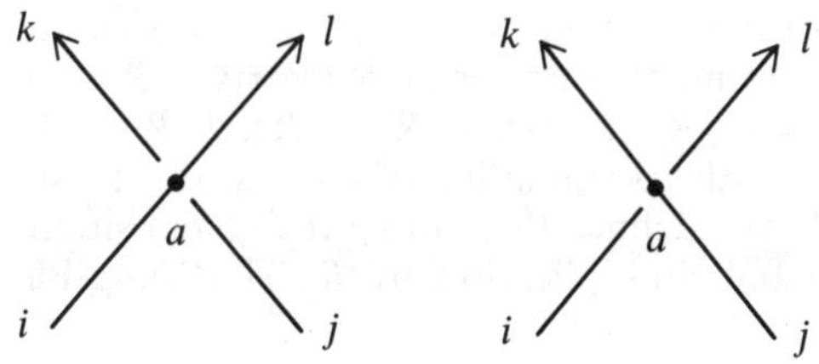

Fig. 1

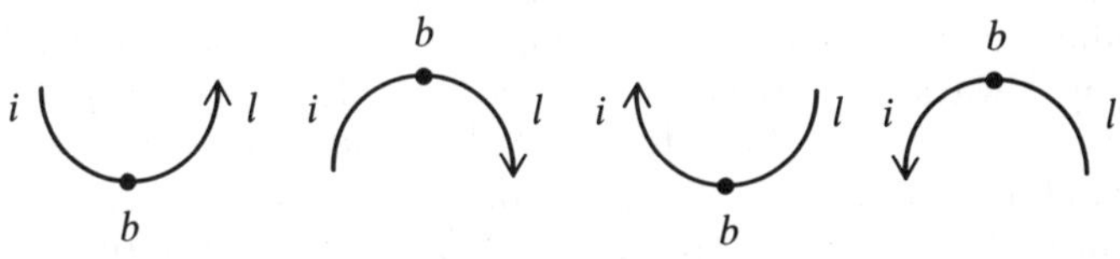

Fig. 2

To describe equations ensuring invariance of $\langle \mathscr{D} \rangle$ under the moves one treats (1) as matrices of linear operators

$$R_{V,W}, \overline{R}_{V,W} : V \otimes_K W \to W \otimes_K V \, ; \quad \alpha_V, \beta_V, \mu_V, \nu_V : V \to V. \tag{2}$$

The state sum $\langle \mathscr{D} \rangle$ depends solely on $\mathscr{D}$ and these operators. Here are equations corresponding to certain moves: $\beta_V = \alpha_V^{-1}$, $\nu_V = \mu_V^{-1}$, $\overline{R}_{V,W} = R_{W,V}^{-1}$,

$$(R_{V,W} \otimes 1_U) \circ (1_V \otimes R_{U,W}) \circ (R_{U,V} \otimes 1_W) \qquad (3)$$
$$= (1_W \otimes R_{U,V}) \circ (R_{U,W} \otimes 1_V) \circ (1_U \otimes R_{V,W}) .$$

The latter equation is the Yang-Baxter equation ; the operators satisfying the equation are called R-matrices. This equation ensures invariance of $\langle \mathcal{D} \rangle$ under the third Reidemeister move, known also as the braid relation. There are also a few equations relating R, α, μ (see [Tu3]). Note that without loss of generality one may assume that $\alpha_V = \beta_V = 1_V$ for all $V \in I$. Note also that one usually imposes an additional condition that a positive 360°-twist of the framing along a link component of color V leads to multiplication of the link invariant by an invertible constant depending only on V. This condition is fullfilled in the examples below.

According to [Re2], any system of invertible R-matrices $\{R_{V,W} : V \otimes W \to W \otimes V\}_{V,W \in I}$ satisfying a non-degeneracy condition may be enriched by $\{\alpha, \beta, \mu, \nu\}_V$ to satisfy all equations ensuring isotopy invariance of $\langle L \rangle$.

Using the operators (2) one may extend the invariant $L \mapsto \langle L \rangle$ to operator invariants of I-colored framed oriented tangles in $\mathbb{R}^2 \times [0,1]$ (see [Tu2, Tu3, Re2, RT1]). This produces a linear tensor representation of the category of tangles in the category of K-modules.

3. Quantum Groups as a Source of Invariants

A general method of constructing R-matrices starts with an algebra with unity A and an invertible element ("universal R-matrix") $R \in A \otimes A$ satisfying the equation $R_{12}R_{13}R_{23} = R_{23}R_{13}R_{12}$, where $R_{12} = R \otimes 1$, $R_{23} = 1 \otimes R \in A^{\otimes 3}$, and R_{13} is obtained from R_{23} by the permutation of the first two tensor multiples. For any left A-modules V, W one defines $R_{V,W}$ to be the composition of the permutation $V \otimes W \to W \otimes V$ and multiplication by R. The Yang-Baxter property of R implies (3).

The theory of quantum groups associates with each simple complex Lie algebra $\mathfrak{g}$ the quantized universal enveloping algebra $U_q(\mathfrak{g})$ over $\mathbb{C}$ possessing a canonical universal R-matrix (here $q \in \mathbb{C}$). All irreducible finite-dimensional $\mathfrak{g}$-modules are known to be deformable into $U_q(\mathfrak{g})$-modules. Thus, one may use these $\mathfrak{g}$-modules as the colors of link components. The system $\{R_{V,W}\}$ arising in this way may always be enriched by $\alpha_V = \beta_V = 1_V$, $\mu_V, \nu_V = \mu_V^{-1}$ to produce isotopy invariants of links (see [Re1], [We1]). Example: $\mathfrak{g} = sl_n(\mathbb{C})$, $V = \mathbb{C}^n$ with the standard action of $\mathfrak{g}$ and the canonical basis $e_1, \ldots, e_n$. Let E_i^j be the homomorphism $V \to V$ which transforms e_i in e_j and transforms e_k with $k \neq i$ into 0. Then

$$R_{V,V} = -q \sum_i E_i^i \otimes E_i^i + \sum_{i \neq j} E_i^j \otimes E_j^i + (q^{-1} - q) \sum_{i<j} E_i^i \otimes E_j^j ;$$

$$\mu_V = \sum_i q^{2i-m-1} E_i^i$$

(see [Ji1, Tu1]). One may color all components of an oriented framed link $L \subset S^3$ with this V and get an invariant $\langle L \rangle$ which is a Laurent polynomial over q. Multiplying $\langle L \rangle$ by a power of $-q^{-n}$ one makes it independent of the framing.

The resulting polynomial is (up to reparametrization) a reduction $P_L(q^n, q-q^{-1})$ of the Homfly polynomial $P_L(x, y)$ (see [Jo1, Tu1, Li1]). For $n=2$ this reduction is the Jones polynomial. In the case $n=2$ one may reformulate the model to produce the Kauffman [Ka1] bracket model for the Jones polynomial. Similarly, the fundamental representations of the Lie algebras of series B, C, D give rise to a series of 1-variable reductions of the 2-variable Kauffman polynomial (see [Tu1]). For invariants corresponding to $\mathfrak{g} = G_2$ see [Re1, Ku2].

For any m-component link in S^3 one may define a universal invariant sitting in $U_q(\mathfrak{g})^{\otimes m}$ and comprising all invariants corresponding to various colorings of components by $\mathfrak{g}$-modules (see [La1]).

4. Invariants of 3-Manifolds

The invariants of links in S^3 discussed above may be combined to produce topological invariants of closed oriented 3-manifolds and links in these manifolds. The relevant construction was introduced and applied to link invariants derived from $sl_2(\mathbb{C})$ in [RT2]. For the case of other classical semisimple Lie algebras see [TW1]. The construction is based on the classical theorem of V. Rochlin and W. Lickorish presenting any closed oriented (connected) 3-manifold as the result of surgery on S^3 along a framed link. R. Kirby introduced in 1978 moves on links which relate any presentations of homeomorphic manifolds. An equivalent system of local moves was introduced by R. Fenn and C. Rourke.

On the algebraic part the construction uses a modular Hopf algebra (see [RT2]). This is a quasitriangular Hopf algebra A with an R-matrix $R \in A \otimes A$ equipped with a finite collection $\{V_i\}_{i \in I}$ of irreducible finite dimensional A-modules satisfying certain axioms. Let, now, a closed oriented 3-manifold N be the result of surgery on a framed link $L \subset S^3$ with m components $L_1, \ldots, L_m$. Provide L with an arbitrary orientation. For each sequence $j = (i_1, \ldots, i_m) \in I^m$ denote by L^j the link L whose components $L_1, \ldots, L_m$ are colored respectively with $V_{i_1}, \ldots, V_{i_m}$. Here is the invariant of N:

$$\tau_A(N) = b^m c^{\sigma(L)} \sum_{j=(i_1,\ldots,i_m) \in I^m} \left(\langle L^j \rangle \prod_{r=1}^{m} \langle 0_{i_r} \rangle \right)$$

where: b, c are normalizing constants, associated with $A, R, \{V_i\}_i$; 0_i is the trivial knot in S^3 with zero framing and the color V_i; $\sigma(L)$ is the signature of the oriented 4-manifold W_L bounded by N and obtained by adding 2-handles to the 4-ball B along $L \subset \partial B$. This definition slightly differs from the one in [RT2] (cf. also [KM1]) ; equivalence of definitions follows in the case $A = U_q(sl_n)$ from computations of [RT2, TW1] and in general case from results of K. Walker (private communication). If L' is a colored framed oriented link in N then one pushes L' in $S^3 \backslash L$ and defines $\tau_A(N, L')$ using $L^j \coprod L'$ instead of L^j.

With a primitive complex root of unity q of degree $r \geq 3$ one associates a finite-dimensional modular Hopf algebra $U = U_q$ which is a quotient of $U_q(sl_2\mathbb{C})$ (see [RT2]). Here $I = \{0, 1/2, 1, \ldots, (r-2)/2\}$ and V_i is the q-deformation of the irreducible $sl_2\mathbb{C}$-module of dimension $2i+1$. Thus with the pair (N, L') we may associate the function $q \mapsto \tau_{U_q}(N, L')$ which is a topological invariant of this pair. For colored links in S^3 this function coincides with the value of the colored Jones polynomial in q. In general it is unknown if this function may be extended in a topologically invariant manner to the unit circle or, what is even more intriguing, to the open unit disc in $\mathbb{C}$. For a further study of $\tau_U(N)$ see [KM1, Li2].

For even r one may refine $\tau_U(N)$ to an invariant $\tau_U(N;\theta)$ where θ is an arbitrary spin structure on N, if $r = 0 \pmod 4$, and $\theta \in H^1(N;\mathbb{Z}/2\mathbb{Z})$, if $r = 2 \pmod 4$. Such θ defines a sublink $L(\theta)$ of the link $L \subset S^3$ presenting N: it consists of components of L such that θ does not extend across the corresponding 2-handles of W_L. One defines $\tau_U(N;\theta)$ in the same way as $\tau_U(N)$ but using only integral colors $0, 1, \ldots, [(r-2)/2]$ for the components of $L(\theta)$. Using [RT2, §7] and [KM1, Theorem 4.20] one may show that $\tau_U(N;\theta)$ is a topological invariant of $(N;\theta)$ and $\tau_U(N) = \sum_\theta \tau_U(N;\theta)$. The same refinement applies to (N, L'). In particular, for $q = \sqrt{-1}$, $r = 4$ the invariant $\tau_U(N;\theta)$ is equivalent to the Rochlin invariant $\mu(N;\theta) \in \mathbb{Z}/16\mathbb{Z}$: for any spin structure θ on N one has $\tau_U(N,\theta) = \exp(-3\pi\sqrt{-1}\mu(N;\theta)/8)$.

5. The Jones-Witten TQFT

Witten [Wi1] conjectured existence of a 3-dimensional TQFT extending the Jones polynomial of links in S^3 (or, more exactly, extending the value of this polynomial in a fixed root of unity q) to oriented 3-dimensional cobordisms and to 3-cobordisms with tangles sitting inside. It was shown in [RT2] that each modular Hopf algebra $(A, R, \{V_i\}_{i\in I})$ gives rise to a 3-dimensional TQFT. For $A = U_q$ one gets the Jones-Witten TQFT. (Though [RT2] treats only cobordisms with links, one may apply the same methods to tangles in cobordisms). The construction uses presentations of 3-cobordisms by tangles in $\mathbb{R}^3$ and operator invariants of tangles. One of the advantages of this approach is the relative simplicity of the definition of the linear space associated with a surface. For the closed oriented surface Σ_g of genus g this linear space is

$$\Psi_g = \bigoplus_{i_1,\ldots,i_g \in I} \left(V_{i_1} \otimes V_{i_1}^* \otimes \cdots \otimes V_{i_g} \otimes V_{i_g}^* \right)^0,$$

where for an A-module V one denotes by V^* the dual module and by V^0 the maximal splitting submodule of V with trivial action of A. (In [RT2] we involved a projector acting in Ψ_g which was later proven by K. Walker and the author, and independently N. Reshetikhin to be the identity). The Teichmüller modular group of Σ_g projectively acts in Ψ_g ; the corresponding 2-cocycle is closely related to the 2-cocycle of the projective Shale-Weil representation of the symplectic group, see [CLM1], [Tu5]. For $A = U_q$, q being a primitive root of even degree, one may refine the TQFT along the lines of Sect. 4.

6. Face Models on Link Shadows

Face models differ from vertex models mainly in that the role of atoms is played by 2-faces. A face model on link diagrams in $\mathbb{R}^2$ producing the value in $q \in \mathbb{C}$ ($q^r = 1$) of the Jones polynomial and its versions for colored links in S^3 was introduced in [KR1]. In [Tu4] this model was reformulated avoiding the use of coordinates, and generalized to diagrams on oriented surfaces. The natural topological setting for the model is the one of shadows (see [Tu4]). A shadow on an oriented closed surface Σ is a finite collection s of closed curves on Σ lying in general position (possibly with intersections) such that each component of $\Sigma\backslash s$ is equipped with a number, called the gleam of the component. Each link diagram $\mathscr{D}$ on Σ (presenting a link in $\Sigma \times \mathbb{R}$) canonically produces a shadow $s(\mathscr{D})$ on Σ; it has the same underlying loops as $\mathscr{D}$ whereas the gleam of a component X of $\Sigma\backslash s(\mathscr{D})$ is defined as the number of jumps up along ∂X diminished by the number of jumps down along ∂X and divided by 2 (the orientation in ∂X is specified by that of $X \subset \Sigma$). One defines Reidemeister-type moves on shadows and calls a class of shadows related by these moves a (framed) shadow link on Σ. If M is a closed oriented 3-manifold fibering over Σ with the fiber S^1 then each (framed) link in M considered up to isotopy canonically produces a (framed) shadow link on Σ. In particular, each link $L \subset S^3$ produces a shadow link on S^2 via the Hopf mapping of the diagrams. One may reconstruct L looking at its shadow on S^2.

The algebraic data used in the model consist of a finite set I, a set $\mathrm{adm} \subset I^3$ invariant under permutations, two functions $i \mapsto u_i$, $i \mapsto v_i : I \to \mathbb{C}\backslash 0$ and a function $I^6 \to \mathbb{C}$, called "the symbol".

A coloring of a shadow s is an I-valued function on the set of underlying loops of s. Let s be a colored shadow on Σ. An area-coloring η of s is an I-valued function on the set of components of $\Sigma\backslash s$ such that if 2 components X, Y of $\Sigma\backslash s$ meet along an arc of s of color i then $(i, \eta(X), \eta(Y)) \in \mathrm{adm}$. For each crossing point a of s one defines $|a|_\eta$ to be the symbol of the 6-tuple formed by the properly ordered colors of 2 loops of s traversing a and η-colors of 4 components of $\Sigma\backslash s$ incident to a. The symbol should be symmetric enough to make $|a|_\eta$ well-defined. For a component X of $\Sigma\backslash s$ with η-color i put $|X|_\eta = \exp(2u_i x)v_i^\chi$ where x and χ are the gleam and Euler characteristic of X. One forms the product of $|a|_\eta, |X|_\eta$ over all crossings and all components of $\Sigma\backslash s$ and sums up the products over all η. Under proper algebraic conditions, this model yields invariants of colored shadow links on Σ and colored framed links in M.

The model extends to colored shadow 3-valent graphs on Σ (and the graphs in M; a graph is colored if its edges are equipped with elements of I). The new ingredients are the additional multiples $|f|_\eta$ which are symbols of 6-tuples associated in the obvious way with all 3-valent vertices f of the graph. In fact, the setting of 3-valent graphs in 3-manifolds rather then links is natural for all models in this paper (cf. [RT1]).

The algebra $U_q(sl_2\mathbb{C})$ with $q = \exp(2\pi\sqrt{-1}h/r)$ gives rise to a suitable algebraic data. Here: $r \geq 3$; $(h, r) = 1$; $I = \{0, 1/2, 1, \ldots, (r-2)/2\}$; the symbol is the slightly normalized q-deformation of the Racah-Wigner $6j$-symbol (see [KR1]);

$$u_i = \pi\sqrt{-1}(i - i(i+1)hr^{-1}), v_i = (-1)^{2i}(q_0^{2i+1} - q_0^{-2i-1})(q_0 - q_0^{-1})^{-1},$$

where $q_0 = \exp(\pi\sqrt{-1}h/r)$; a triple (i, j, k) belongs to adm iff $r - 2 \geq i + j + k \in \mathbb{Z}$ and $|i - j| \leq k \leq i + j$. For colored links in S^3 the corresponding face model on S^2 produces the values in $\exp(\pi\sqrt{-1}h/2r)$ of the invariants considered in Sect. 3, the case $A = U_q(sl_2\mathbb{C})$.

Using shadow links on S^2 with complex gleams, and using the shadow version of the Kirby-Fenn-Rourke moves one may define shadow 3-manifolds (over $\mathbb{C}$) and the corresponding shadow version of the Jones-Witten TQFT. A remarkable property of shadow links and 3-manifolds over $\mathbb{C}$ is that one may continuously deform their topological types.

7. Simplicial Models

By a simplicial model I mean a statistical model on triangulations of polyhedra producing a topological invariant. Such a model for 3-manifolds was introduced in [TV1]. The geometric result underlying the model is the theorem of J. Alexander applicable in all dimensions: any two triangulations of a compact PL-manifold may be related by star subdivisions and their inverses. (Note that each 3-manifold admits a essentially unique PL-structure.) The algebraic data for the model is the same as in Sect. 6. The model gives rise to a 3-dimensional TQFT defined for non-oriented (possibly, non-orientable) 3-cobordisms (see [TV1]), and 3-cobordisms with links sitting inside (see [Tu6]). In particular for links in S^3 one gets a simplicial model on the link exteriors producing the values of the Jones polynomial in the roots of unity.

A state η on a triangulated compact 3-manifold N is an I-valued function on the set of edges of N. For a 3-simplex T of N denote by $|T|_\eta$ the symbol of the 6-tuple consisting of the η-colors of edges of a 2-face of T followed by the η-colors of the opposite edges of T. Put

$$|N, \partial N|_\eta = w^{-2\alpha} \prod_e v_{\eta(e)} \prod_T |T|_\eta$$

where e, T run respectively over all edges and 3-simplexes of N; α is the number of vertexes of N; w is a constant computed from the algebraic data. For a closed N the state sum $|N| = \Sigma_\eta |N|_\eta$ is a topological invariant of N if the algebraic data satisfy simple conditions which mimic the orthogonality relation and Biedenharn-Eliot identity for $6j$-symbols. Under the same algebraic conditions one similarly constructs for each 3-cobordism (N, Σ_1, Σ_2) a topologically invariant linear operator $Q(\Sigma_1) \to Q(\Sigma_2)$ where $Q(\Sigma)$ is a finite-dimensional Euclidean space associated with any closed surface Σ (see [TV1]). Elements of $Q(\Sigma)$ are geometrically represented by I-colored 3-valent graphs imbedded in Σ.

The invariant $N \mapsto |N|$ extends to I-colored framed links in N with oriented neighborhoods. Let $L = L_1 \cup \ldots \cup L_m \subset N$ be such a link with m components (here $\partial N = \emptyset$). Let U be a closed regular neighborhood of L. Put $M = \overline{N \backslash U}$ and triangulate M. Put $\Sigma = \partial M = m(S^1 \times S^1)$ and equip Σ with the triangulation induced by that of M and the orientation induced by that of U. Let $\psi_1, \ldots, \psi_m \subset \Sigma$ be meridians of $L_1, \ldots, L_m$.

For a sequence $j = (i_1, \ldots, i_m) \in I^m$ equip each ψ_r with the color i_r and denote by ψ_j the union of these m colored loops on Σ. Denote by L' the family of m colored loops on Σ obtained by shifting L along its framing. Denote by γ the 3-valent graph on Σ formed by the segments $\{e^*\}$ dual to edges $\{e\}$ of Σ. Here e^* crosses e once and connects the barycenters of the two 2-simplexes of Σ meeting along e. For a state η of M denote by γ^η the graph γ with the coloring $e^* \mapsto \eta(e)$.

Put

$$|N, L| = w^{2-2m} \sum_{j=(i_1,\ldots,i_m)} \left(\prod_{r=1}^{m} v_{i_r} \sum_\eta |M, \partial M|_\eta \langle \gamma^\eta | L' | \psi_j \rangle \right)$$

where the last multiple is the state sum invariant of the shadow graph $s = \gamma^\eta \cup L' \cup \psi_j$ on Σ. Here γ is assumed to lie in general position with respect to $L' \cup \psi_j$; the gleam of a component X of $\Sigma \backslash s$ is defined as follows: looking onto ∂X compute the number of passages from ψ_j to L' and from L' to γ, subtract the number of opposite passages, and divide by 2 (the orientation in ∂X is specified by that of $X \subset \Sigma$). Topological invariance of $|N, L|$ is ensured by the conditions mentioned above and the Racah identity for the symbol.

One may generalize this model to models derived from modular Hopf algebras or modular tensor categories. The definition of $|N|$, $|N, L|$ may be reformulated in terms of simple spines of 3-manifolds and link diagrams on the spines which makes computations more accessible.

The initial data associated with $q = \exp(2\pi\sqrt{-1}h/r)$ in Sect. 6 satisfy all necessary algebraic equations. The relationships between the corresponding invariants $|N| \in \mathbb{R}$ and $\tau_U(N) \in \mathbb{C}$ (defined for oriented N) are the following: $|N|$ equals $\tau_U(N)\overline{\tau_U(N)}$ (up to normalization). For links in S^3 the approaches of Sect. 2–3, Sect. 6, and Sect. 7 are equivalent. In particular, if all components of a framed link $L \subset S^3$ are provided with the color $1/2 \in I$ and the right-hand orientation of neighborhoods then $|S^3, L|$ is equal to the value of the Kauffman bracket polynomial $\langle L \rangle$ in $\exp(\pi\sqrt{-1}h/2r)$.

The invariant $|N|$ corresponding to q may be refined to an invariant $|N, \Delta|$ where $\Delta \in H^1(N; \mathbb{Z}/2)$ (see [TV1]). It is equal to the sum of $|N|_\eta$ over η such that the 1-cocycle $e \mapsto 2\eta(e)$ mod 2 presents Δ. Clearly, $|N| = \Sigma_\Delta |N, \Delta|$. For oriented closed N and even r one has $|N, \Delta| = \sum_\theta \tau_U(N; \theta)\overline{\tau_U(N; \theta + \Delta)}$.

Both $\tau_U(N)$ and $|N|$ have "physical" meaning: they are mathematical versions of invariants which come up respectively in the 3-dimensional quantum field theory with non-abelian Chern-Simons action (see [Wi1]) and (in the limit $q \to 1$) in 3-dimensional quantum gravity (see e.g. [HP1]).

References

[AW1] Akutsu, Y., Wadati, M.: Knot invariants and the critical statistical systems. J. Phys. Soc. Japan **56** (1987) 839–842

[At1] Atiyah, M.: Topological quantum field theories. Publ. Math. IHES **68** (1989) 175–186

[Ba1] Baxter, R.J.: Exactly solved models in statistical mechanics. Academic Press, London 1982

[CLM1] Cappell, S.E., Lee, R., Miller, E.Y.: Invariants of 3-manifolds from conformal field theory. Preprint 1990

[Dr1] Drinfeld, V.G.: Quantum groups. Proc. Int. Congress of Mathematicians. Academic Press, Berkeley, Cal., vol. 1, 1986, pp. 798–820

[FRT1] Faddeev, L.D., Reshetikhin, N.Y., Takhtajan, L.A.: Quantization of Lie groups and Lie algebras. Algebra i Analysis **1** (1989) 178–207 (Russian)

[HP1] Hasslacher, B., Perry, M.: Phys. Lett. **103**B (1981) 21

[Ja1] Jaeger, F.: Composition products and models for the Homfly polynomial. Preprint 1987

[Ji1] Jimbo, M.: Quantum *R*-matrix for the generalized Toda system. Commun. Math. Phys. **102** (1986) 537–547

[Jo1] Jones, V.F.R.: Notes on a talk in Atiyah's seminar, 1986

[Jo2] Jones, V.F.R.: On knot invariants related to some statistical mechanical models. Pacif. J. Math. **137** (1989) 311–334

[Ka1] Kauffmann, L.M.: State models and the Jones polynomial. Topology **26** (1987) 395–407

[Ku1] Kuperberg, G.: Involutory Hopf algebras and 3-manifold invariants. Preprint 1990

[Ku2] Kuperberg, G.: The quantum G_2 link invariant. Preprint 1990

[KM1] Kirby, R., Melvin, P.: On the 3-manifold invariants of Witten and Reshetikhin-Turaev for $sl(2, \mathbb{C})$. Preprint 1990

[KR1] Kirillov, A.N., Reshetikhin, N.Y.: Representations of the algebra $U_q(sl_2)$, q-orthogonal polynomials and invariants of links. In: Kac, V.G. (ed.), Infinite dimensional Lie algebras and groups. (Adv. Ser. In Math. Phys., vol. 7). World Scientific, Singapore 1988, pp. 285–339

[La1] Lawrence, R.J.: A universal link invariant using quantum groups. In: Differential geometric methods in theoretical physics, XVII International conference. World Scientific, Singapore 1988, pp. 55–63

[Li1] Lickorish, W.B.R.: Polynomials for links. Bull. London Math. Soc. **20** (1988) 558–588

[Li2] Lickorish, W.B.R.: Three-manifolds and the Temperley-Lieb algebra. Preprint 1990

[Re1] Reshetikhin, N.Y.: Quantized universal enveloping algebras, the Yang-Baxter equation and invariants of links, I, II. LOMI preprints E-4-87, E-17-87, Leningrad 1988

[Re2] Reshetikhin, N.Y.: Quasitriangular Hopf algebras and invariants of tangles. Algebra i Analysis **1** (1989) 169–188 (Russian)

[RT1] Reshetikhin, N.Y., Turaev, V.G.: Ribbon graphs and their invariants derived from quantum groups. Comm. Math. Phys. **127** (1990) 1–26

[RT2] Reshetikhin, N.Y., Turaev, V.G.: Invariants of 3-manifolds via link polynomials and quantum groups. Invent. math. **103** (1991) 547–597

[Tu1] Turaev, V.G.: The Yang-Baxter equations and invariants of links. Invent. math. **92** (1988) 527–553

[Tu2] Turaev, V.G.: The Conway and Kauffman modules of the solid torus with an appendix on the operator invariants of tangles. LOMI preprint E-6-88, Leningrad, 1988

[Tu3] Turaev, V.G.: Operator invariants of tangles and R-matrices. Izv. Akad. Nauk SSSR **53** (1989) 1073–1107 (Russian) [English transl.: Math. USSR-Izv. **35** (1990) 411–444]

[Tu4] Turaev, V.G.: Shadow links and IRF-models of statistical mechanics. Publ. Inst. Rech. Math. Av., Strasbourg (1990)

[Tu5] Turaev, V.G.: Quantum representations of modular groups and Maslov indices. To appear

[Tu6] Turaev, V.G.: Quantum invariants of links and 3-valent graphs in 3-manifolds. Preprint 1990

[TV1] Turaev, V.G., Viro, O.Y.: State sum invariants of 3-manifolds and quantum $6j$-symbols. Preprint 1990

[TW1] Turaev, V.G., Wenzl, H.: Quantum invariants of 3-manifolds associated with classical simple Lie algebras. Preprint 1991

[We1] Wenzl, H.: Representations of braid groups and the quantum Yang-Baxter equation. Pacif. J. Math. **145** (1990) 153–180

[Wi1] Witten, E.: Quantum field theory and Jones polynomial. Comm. Math Phys. **121** (1989) 351–399

[Ye1] Yetter, D.N.: Markov algebras. In:"Braids", AMS Contemporary Mathematics, vol. 78. AMS, Providence, RI, 1988

Canonical and Minimal Models of Algebraic Varieties

Yujiro Kawamata

Department of Mathematics, University of Tokyo, Hongo, Tokyo 113, Japan

In this talk we shall explain a minimal model program for arbitrary dimensional algebraic varieties over a field k of characteristic zero. Starting with the working hypothesis on the finite generatedness of the canonical ring, we shall discuss the process to obtain minimal and canonical models (we refer the reader to [KMM] for more general argument, or [Sh2] for more brief survey from the same point of view). We shall also review some recent results on the boundedness of indices of singularities in dimension 3.

We shall put emphasis on the idea of logarithmic formalism, which provides natural setup for the statements and proofs of new theorems such as vanishing or cone-contraction theorems, rather than just generalizations of previously known results. Many results will be obtained simultaneously for the log and non-log cases. There will also be applications to the reverse direction from the log case to the non-log case.

§ 1. Log-canonical Divisors

1.1 Let X be a nonsingular projective variety over k and K_X the canonical divisor, i.e., $\mathcal{O}_X(K_X) = \det \Omega_X^1$. The *canonical ring* $R(X) = \sum_{m=0}^{\infty} H^0(X, mK_X)$ is the fundamental birational invariant of X, and the *Kodaira dimension* $\kappa(X)$ is defined to be trans. $\deg_k R(X) - 1$ if $R(X) \neq k$ and $-\infty$ otherwise. Thus $\kappa(X) \leqq \dim X$, and if the equality holds, then X is said to be of *general type.* The basic question is the following:

Question 1. Is $R(X)$ a finitely generated graded k-algebra?

The answer to this question is known to be "yes" if $\dim X = 2$ by Zariski and Mumford [Z, Mu], and if $\dim X = 3$ by the combination of recent results of Mori, Kawamata, Benveniste, Shokurov and Fujita (for the case $\kappa(X) = 3$, the construction of the minimal model is by [Ka5, Ka8, Mo4, Sh1] and the rest by [Ka4] or [B], whereas the case $\kappa(X) < 3$ is by [Fu3]). The proof in dimension 2 works in

Proceedings of the International Congress
of Mathematicians, Kyoto, Japan, 1990

arbitrary characteristic, but our result in dimension 3 needs the assumption that the characteristic is zero because we use the vanishing theorem of Kodaira type.

1.2 The key word of our investigation is the $\mathbb{Q}$-*divisor*. A $\mathbb{Q}$-divisor $D = \sum_{j=1}^{t} d_j D_j$ is a formal sum of distinct prime divisors D_j on a normal variety X with rational coefficients d_j. The *integral part* $[D] = \sum [d_j] D_j$ is a usual Weil divisor. D is said to be $\mathbb{Q}$-*Cartier* if mD is a Cartier divisor for some positive integer m. X is called $\mathbb{Q}$-*factorial* if an arbitrary prime divisor on X is $\mathbb{Q}$-Cartier.

The *intersection number* $(D \cdot C) \in \mathbb{Q}$ of a $\mathbb{Q}$-Cartier $\mathbb{Q}$-divisor D and a complete curve C on X is defined by $(D \cdot C) = (mD \cdot C)/m$. D is said to be *nef* (or *numerically semipositive*) if $(D \cdot C) \geqq 0$ for all C.

1.3 Let us define the notion of the *log-canonical divisor*. Let X be a nonsingular variety and $B = \sum_{j=1}^{t} D_j$ a normal crossing divisor on X. The sheaf of logarithmic 1-forms $\Omega_X^1(\log B)$ is a locally free sheaf with a local basis $dx_1/x_1, \ldots, dx_r/x_r, dx_{r+1}, \ldots, dx_n$ if B is defined by $\prod_{j=1}^{r} x_j = 0$ for a local parameter system $\{x_1, \ldots, x_n\}$. Corresponding to its determinant sheaf, we define the log-canonical divisor to be $K_X + B$, which is related to questions concerning the open variety $X - B$.

But more important is its $\mathbb{Q}$-divisor version $K_X + D$ with $D = \sum_{j=1}^{t} d_j D_j$ for some rational numbers d_j which satisfy the inequalities $0 < d_j < 1$, i.e., D is effective and $[D] = 0$. Here, the experience shows that the conditions on the d_j are necessary to avoid counterexamples. This $\mathbb{Q}$-divisor version of the log-canonical divisor is related to the covering spaces. For example, if $f: X \to Y$ is a morphism of affine lines $X = Y = \mathbb{A}^1$ given by $f^*(y) = x^e$, then $K_X = f^*(K_Y + (e-1)/e \cdot B)$ for $B = \operatorname{div}(y)$.

In this paper a *smooth pair* (X, D) is meant to be a pair consisting of a nonsingular projective variety X and an effective $\mathbb{Q}$-divisor on it such that $[D] = 0$ and $\operatorname{Supp}(D)$ has only normal crossings. We can extend our question to the *log-canonical ring* of a smooth pair (X, D) defined by $R(X, D) = \sum_{m=0}^{\infty} H^0(X, [m(K_X + D)])$. (X, D) is said to be of *log-general type* if trans. $\deg_k R(X, D) = \dim X + 1$.

Question 2. Is $R(X, D)$ finitely generated?

The answer is known to be "yes" only in dimension 2 [Ka1, Fu2].

1.4 We shall explain Fujita's reduction ([Fu3], also [Mo3], [N]) which reduces Question 1 (*non-log-case*) for a variety X with $0 \leqq \kappa(X) < \dim X$ (or even Question 2 for a smooth pair (X, C), see [Mw]) to Question 2 (*log-case*) for another smooth pair (Y, D) of log-general type such that $\dim Y = \kappa(X)$.

After replacing X by a suitable birational model, one can find a proper surjective morphism $f: X \to Y$ with connected fibers to a nonsingular projective variety Y, a

$\mathbb{Q}$-divisor $D = D' + D''$ on Y, and a positive integer d which satisfy the following conditions:

(a) D' is nef,
(b) (Y, D'') is a smooth pair and f is smooth over $Y - \mathrm{Supp}(D'')$,
(c) $\dim Y = \kappa(X)$ and (Y, D) is of log-general type,
(d) $f^*: H^0(Y, [d \cdot m(K_Y + D)]) \simeq H^0(X, d \cdot m\, K_X)$ for all $m \in \mathbb{N}$.

The $\mathbb{Q}$-divisor D' comes from the moduli of fibers of f. The property (a) is proved by using the theory of variation of Hodge structures ([Fu1] if $\dim Y = 1$, and [Ka2] in general by using [G]). The degenerations of the fibers of f give rise to D''. The contribution of semistable degenerations to D'' is zero. So by the semistable reduction in codimension one, we understand that the coefficients of D'' are dominated by those of coverings, hence less than one. Then by using Kodaira's lemma (cf. [Ka3]), one can reduce Question 1 for X to Question 2 for some lower dimensional smooth pair.

§ 2. Log-Canonical Model

2.1 In higher dimensional algebraic geometry, varieties must be allowed to have certain mild singularities. Let X be a normal variety and $D = \sum_{j=1}^{t} d_j D_j$ an effective $\mathbb{Q}$-divisor such that $[D] = 0$. The pair (X, D) is said to have only *log-terminal singularities*, or simply (X, D) is called *log-terminal*, if the following conditions are satisfied:

(a) $K_X + D$ is $\mathbb{Q}$-Cartier,

(b) for a resolution of singularities ([H]) $\mu: Y \to X$ with a normal crossing divisor $\sum_{i=1}^{t+s} E_i$ on Y such that the E_i $(1 \leqq i \leqq t)$ are the strict transforms of the D_i and the E_i $(t + 1 \leqq i \leqq t + s)$ are all the prime divisors such that $\mathrm{codim}_X\, \mu(E_i) \geqq 2$, if we write

$$K_Y + \sum_{i=1}^{t+s} E_i = \mu^*(K_X + D) + \sum_{i=1}^{t+s} a_i E_i \tag{$*$}$$

then we have $a_i > 0$ for all i.

The rational numbers a_i are called *log-discrepancy coefficients*. Note that $a_i = 1 - d_i$ for $1 \leqq i \leqq t$. In the case $D = 0$, considering the formula

$$K_Y = \mu^* K_X + \sum_{i=t+1}^{t+s} b_i E_i, \tag{$**$}$$

we say that X has only *terminal* (resp. *canonical*) *singularities*, if $b_i > 0$ (resp. $b_i \geqq 0$) hold for all i. Since $b_i = a_i - 1$, we have the implications

$$X: \text{terminal} \Rightarrow X: \text{canonical} \Rightarrow (X, 0): \text{log-terminal}.$$

If (X, D) is log-terminal and $D^\#$ is an effective Cartier divisor on X, then the pair $(X, D + \varepsilon D^\#)$ is also log-terminal for a sufficiently small positive rational number ε.

For example, if dim $X = 2$, then

$$X : \text{terminal} \Leftrightarrow \text{nonsingular}$$

$$X : \text{canonical} \Leftrightarrow \text{rational double points } \mathbb{C}^2/\Gamma,\ \Gamma \subset \mathrm{SL}(2, \mathbb{C})$$

$$(X, 0) : \text{log-terminal} \Leftrightarrow \text{quotient singularities } \mathbb{C}^2/\Gamma,\ \Gamma \subset \mathrm{GL}(2, \mathbb{C}).$$

Though the above definitions are simple, it is hard to understand what these singularities really are. All quotient singularities of any dimension are log-terminal (with $D = 0$), but there are a lot more of them. In dimension 3, a detailed classification of terminal singularities are obtained by [MS], [D] and especially [Mo2]. The results in [Mo2] were fundamental for the success of Mori's flip theorem in dimension 3 ([Mo4]). But it seems too difficult to classify canonical or log-terminal singularities even in dimension 3. A structure theorem of three-dimensional canonical singularities is proved in [Ka8], and it was one of the key steps toward the flip theorem in [Mo4]. In general, we know only that if (X, D) is a log-terminal pair, then X has only rational singularities (Elkik [E], also [Fu4]).

2.2 Since $R(X)$ is determined by the function field $k(X)$, one might say that Question 1 is a purely algebraic one. But the solutions so far are geometric. In fact, the affirmative answer to Question 2 for a smooth pair (X, D) of log-general type is equivalent to the existence of the *log-canonical model* of (X, D); a pair (X', D') of a normal projective variety and a $\mathbb{Q}$-divisor is said to be the log-canonical model of (X, D) if the following conditions are satisfied:

(a) (X', D') has only log-terminal singularities,

(b) $K_{X'} + D'$ is an ample $\mathbb{Q}$-Cartier divisor,

(c) there exists a birational map $\gamma : X \dashrightarrow X'$ which is surjective in codimension one (i.e., every prime divisor on X' has a strict transform on X),

(d) $\gamma_* D = D'$ and $K_X + D \geqq \gamma^*(K_{X'} + D')$, where γ_* is the direct image homomorphism of codimension one cycles and $\gamma^* = \alpha_* \beta^*$ if $\beta = \gamma\alpha$ for birational morphisms α and β.

If $R(X, D)$ is finitely generated, then we set $X' = \mathrm{Proj}(R(X, D))$ (cf. [R1]). Conversely, the condition (d) implies that $R(X, D)$ is isomorphic to the log-canonical ring $R(X', D')$, which is similarly defined for the log-terminal pair. Hence the existence of the log-canonical model implies the finite generatedness.

§ 3. Minimal Model Program

3.1 We want to construct the log-canonical model via log-minimal models. A *log-minimal model* (X', D') of a smooth pair (X, D) is a pair which satisfies (a), (c), (d) of (2.2) and (b′) $K_{X'} + D'$ is nef.

A log-minimal model is defined just by a numerical property that $K + D$ is nef. But it has a good geometric property at least if it is of log-general type:

Theorem (Kawamata, Benveniste, Shokurov). *If (X, D) is a log-minimal model of log-general type, then there exists a positive integer m such that the linear system $|m(K_X + D)|$ is free. Therefore, the locus on X where $K_X + D$ is numerically trivial is contractible, and $R(X, D)$ is finitely generated.*

3.2 Starting from an arbitrary smooth pair (X, D) of log-general type, we should reduce the locus where the log-canonical divisor is *numerically negative* in order to obtain a log-minimal model. The key idea of the minimal model program is to look at the contraction morphisms associated with *extremal rays*, or the *extremal contractions* ([Mo1]). In the following, we combine cone and contraction theorems; in fact, they are proved at the same time ([Ka6]).

Cone-Contraction Theorem (Mori, Kawamata, Reid, Shokurov). *Let X be a normal projective $\mathbb{Q}$-factorial variety and D a $\mathbb{Q}$-divisor such that the pair (X, D) has only log-terminal singularities. Suppose that $K_X + D$ is not nef. Then there exists a projective surjective morphism $\varphi : X \to Y$ with connected fibers to a normal projective variety Y such that* (1) $-(K_X + D)$ *is φ-ample, and* (2) $\varrho(X/Y) = 1$ (*i.e., φ is not an isomorphism, and for any two curves C_i* ($i = 1, 2$) *with $\varphi(C_i)$ points, there exists a rational number q such that* $(B \cdot C_1) = q(B \cdot C_2)$ *for all prime divisors B on X*).

In the course of the proof of the above two theorems, the log-discrepancy coefficients a_i play a guiding role. The proof is by induction, and we need the relative version of the cone-contraction theorem over a base space S in order to have Property (2) (cf. [Ka6]).

3.3 φ is called an *extremal contraction*. We define its *degenerate locus* $E = \operatorname{Exc}(\varphi) = \{x \in X; \varphi \text{ is not an isomorphism at } x\}$. φ is a birational morphism if and only if $E \neq X$. There are three types of φ as follows.

(a) *Fibering type*: $\dim Y < \dim X$. Then $R(X, D) = k$.

(b) *Divisorial contraction*: E is a prime divisor. Then Y is again $\mathbb{Q}$-factorial and $(Y, \varphi_* D)$ is log-terminal. So we have a log-minimal model or apply the cone-contraction theorem again.

(c) *Small contraction*: $\operatorname{codim}_X E \geqq 2$. The *log-flip conjecture* asserts that there exists another projective birational morphism $\varphi^+ : X^+ \to Y$ such that $\operatorname{codim}_{X^+} \operatorname{Exc}(\varphi^+) \geqq 2$ and $K_{X^+} + D^+$ is φ^+-ample, where D^+ is the strict transform of D. Then (X^+, D^+) is again $\mathbb{Q}$-factorial and log-terminal. The birational transformation $X \dashrightarrow X^+$ is called the *log-flip* of φ.

Given a smooth pair, if the log-flip conjecture is true, and if there does not exist an infinite chain of successive log-flips (*termination conjecture*), then we obtain a log-minimal model after a finite number of birational transformations of divisorial contractions and log-flips.

Note that the log-flip conjecture is a special case of Question 2, because X^+ should coincide with $\operatorname{Proj}(\mathscr{R}(X/Y, D))$, where the *relative log-canonical ring* $\mathscr{R}(X/Y, D) = \sum_{m=0}^{\infty} \varphi_* \mathscr{O}_X([m(K_X + D)])$ is a graded $\mathscr{O}_Y$-algebra. In particular, the conjecture is local. It is easy to see that an affirmative answer to Question 2 implies the finite generatedness of $\mathscr{R}(X/Y, D)$.

A *flip* is a log-flip in the case where $D = 0$ and X has only terminal singularities. The existence of flips in dimension 3 is proved in [Mo4]. If X is a nonsingular and of dimension 4, the flip exists by [Ka9]. Shokurov has some idea toward the log-flip conjecture in dimension 3.

One can prove that $\mathrm{Exc}(\varphi)$ is covered by a family of rational curves ([Ka11, appendix] by using the vanishing theorem explained below and [MM]). As an application, Wilson [W] proved that there exists a rational curve on any Calabi-Yau 3-fold X whose Picard number is greater than 19. In fact, he proved the existence of an effective Cartier divisor D which is not nef. Since $(X, \varepsilon D)$ is log-terminal for $0 < \varepsilon \ll 1$, the cone-contraction theorem yields the result.

3.4 The following generalization [Ka3, V] of the Kodaira vanishing theorem is essential in the proof of the above theorems.

Vanishing Theorem (Kawamata, Viehweg). *Let (X, D) be a smooth pair and L a Cartier divisor on X. If $L - (K_X + D)$ is ample, then $H^i(X, L) = 0$ for $i > 0$.*

The idea of the proof is as follows. One constructs a $\mathbb{Q}$-divisor D' on X and a Galois covering $\pi : Y \to X$ from a nonsingular projective variety Y such that π^*D is a usual divisor, (X, D') is a smooth pair, $K_Y = \pi^*(K_X + D')$ and $D \leqq D'$ ([Ka2]). Then by [Kod], $H^i(Y, \pi^*(L + D' - D)) = 0$ for $i > 0$. Taking the $\mathrm{Gal}(Y/X)$-invariants, we obtain the theorem.

One can extend the theorem to the case where the pair (X, D) is log-terminal ([KMM]). As an application, if $L - K_X$ is nef and big (i.e., $(L - K_X)^{\dim X} > 0$), then $L - (K_X + \varepsilon B)$ is ample for some effective divisor B and $0 < \varepsilon \ll 1$ by Kodaira's lemma, hence the vanishing (cf. [Ka3]).

§ 4. Flip to Flop

4.1 In the notation of (3.3), a *flop* is a special kind of log-flip where X has only canonical singularities and K_X is numerically trivial along the fibers of φ. We can take D to be a small $\mathbb{Q}$-divisor such that (X, D) is log-terminal and $-D$ is φ-ample, since φ is an isomorphism in codimension one.

Given $\varphi : X \to Y$, the existence of flops in dimension 3 is proved by [R2] in the case where X has only terminal singularities and by [Ka8] in general. By [Ka8] and [Ko1], if X and X' are birationally equivalent minimal 3-folds (i.e., $\mathbb{Q}$-factorial projective 3-folds with terminal singularities and nef canonical divisors), there exists a sequence of flops which connects X and X'. In particular, a 3-fold of general type has only a finite number of minimal models ([KM]). Flops of 3-folds with terminal singularities are geometrically symmetric and preserves the analytic types of singularities ([Ko1]). But they are not symmetric in general even in dimension 3. In dimension > 3, the dimensions of $\mathrm{Exc}(\varphi)$ and $\mathrm{Exc}(\varphi^+)$ may also be different.

4.2 There are two approaches to make flips from flops. The first method uses double coverings. Given a small contraction $\varphi : X \to Y$ with φ-ample $-K_X$ as in

(3.2), we replace Y by its affine open subset, since flip is local. We take double covers $\tilde{X} \to X$ and $\tilde{Y} \to Y$ which ramify along a general element $B_X \in |-2K_X|$ and its image $B_Y \in |-2K_Y|$, respectively. The morphism φ induces a small contraction $\tilde{\varphi} : \tilde{X} \to \tilde{Y}$ such that $K_{\tilde{X}}$ is numerically trivial for $\tilde{\varphi}$. So if one can prove that $\tilde{X}$ has only canonical singularities, then the existence of flop $\tilde{X} \dashrightarrow \tilde{X}^+$ produces a flip $X \dashrightarrow X^+$ as the $\mathbb{Z}/(2)$-quotient. This approach is first found in [Ka8] to prove the existence of a flip for semistable degeneration of surfaces, and then applied to general 3-folds in [Mo4].

The second approach is the cone construction. Given $\varphi : X \to Y$ as before, we take the "total spaces" of K; we let $CX = \operatorname{Spec}\left(\sum_{m=0}^{\infty} \mathcal{O}_X(-mK_X)\right)$ and $CY = \operatorname{Spec}\left(\sum_{m=0}^{\infty} \mathcal{O}_Y(-mK_Y)\right)$. Here, both graded algebras are finitely generated, since $-K_X$ is a φ-ample $\mathbb{Q}$-Cartier divisor. One can prove that CX has only canonical singularities and K_{CX} is numerically trivial for the induced morphism $C\varphi : CX \to CY$. Thus if the flop $CX \dashrightarrow (CX)^+$ exists, the strict transform X^+ of X gives the flip in one dimension less.

§ 5. Bounding the Indices

5.1 The *index* r of a terminal singularity is the smallest positive integer such that rK is a Cartier divisor. For example, the cyclic quotient singularity $\mathbb{C}^3/\langle(\zeta^a, \zeta^{-a}, \zeta)\rangle$ for $\zeta = \exp(2\pi\sqrt{-1}/r)$ and $(r, a) = 1$ has index r. By [Mo2], an arbitrary terminal singularity of dimension 3 has a small deformation with only quotient singularities of the above type. Note that these quotient singularities are rigid [Sch].

We have to deal with singular varieties in dimension 3, and so results on nonsingular varieties must be modified. Let X be a projective 3-fold with only terminal singularities. Then we have

$$\chi(\mathcal{O}_X) = \frac{1}{24}\left\{c_1 c_2 + \sum_P (r_P - 1/r_P)\right\},$$

where $c_1 c_2 = -(\mu^* K_X \cdot c_2(Y))$ for a resolution $\mu : Y \to X$, the summation is taken over all virtual quotient singularities of X, where singular points of X are replaced by its generic small deformations, and r_P are their indices ([Ka7, R3]). More generally, Fletcher calculated the correction terms for $\chi(mK_X)$ ([Fl]).

As applications of the above formula, we have some boundedness results. By [Ka7] and [Ms], if K_X is numerically trivial, then $12K_X \sim 0$ if $\chi(\mathcal{O}_X) \geqq 2$ and $120K_X \sim 0$ if $\chi(\mathcal{O}_X) = 1$.

By [Ka10], there exist universal constants $m_1 \in \mathbb{N}$ and $m_2 \in \mathbb{Q}$ such that if X is a $\mathbb{Q}$-Fano 3-fold (a projective 3-fold having only terminal singularities and ample anti-canonical divisor $-K_X$) with Picard number $\varrho = 1$, then $m_1 K_X$ is a Cartier divisor and $(-K_X)^3 \leqq m_2$. Hence all the $\mathbb{Q}$-Fano 3-folds with $\varrho = 1$ form a bounded family.

Let S be a minimal nonsingular projective surface with $\kappa(S) = 1$, $f : X \to \Delta = \{t \in \mathbb{C}; |t| < 1\}$ a relative minimal model of a semistable degeneration of surfaces such that $f^{-1}(t)$ $(t \neq 0)$ are deformations of S, and m_i $(i = 1, \ldots, e)$ the multiplicities of multiple fibers of the elliptic fibration of S. If $m = \text{l.c.m.}\{12, m_i\}$, then mK_X is a Cartier divisor by [Ka11]. We note that a similar boundedness result for the case $\kappa(S) = 2$ would yield a geometric compactification of the moduli space of such surfaces (cf. [KSB]).

References

[B] Benveniste, X: Sur l'anneau canonique de certaines variétés de dimension 3. Invent. math. **73** (1983) 157–164

[D] Danilov, V.I.: Birational geometry of toric 3-folds. Math. USSR Izv. **21** (1983) 269–279

[E] Elkik, R.: Rationalité des singularités canoniques. Invent. math. **64** (1981) 1–6

[F1] Fletcher, A.R.: Contributions to Riemann-Roch on projective 3-folds with only canonical singularities and applications. Proc. Symp. Pure Math. **46**, no. 1 (1987) 221–231

[Fu1] Fujita, T.: On Kähler fiber spaces over curves. J. Math. Soc. Japan **30** (1978) 779–794

[Fu2] Fujita, T.: Fractionally logarithmic canonical rings of algebraic varieties. J. Fac. Sci. Univ. Tokyo IA-**30** (1984) 685–696

[Fu3] Fujita, T.: Zariski decomposition and canonical rings of elliptic threefolds. J. Math. Soc. Japan **38** (1986) 19–37

[G] Griffiths, P.: Periods of integrals on algebraic manifolds III. Publ. Math. IHES. **38** (1978) 779–794

[H] Hironaka, H.: Resolution of singularities of an algebraic variety over a field of characteristic zero. Ann. Math. **79** (1964) 109–326

[Ka1] Kawamata, Y.: On the classification of non-complete algebraic surfaces. (Lecture Notes in Mathematics, vol. 732.) Springer, Berlin Heidelberg New York 1979, pp. 251–232

[Ka2] Kawamata, Y.: Characterization of abelian varieties. Comp. Math. **43** (1981) 253–276

[Ka3] Kawamata, Y.: A generalization of Kodaira-Ramanujam's vanishing theorem. Math. Ann. **261** (1982) 43–46

[Ka4] Kawamata, Y.: On the finiteness of generators of a pluri-canonical ring for a 3-fold of general type. Amer. J. Math. **106** (1984) 1503–1512

[Ka5] Kawamata, Y.: Elementary contractions of algebraic 3-folds. Ann. Math. **119** (1984) 95–110

[Ka6] Kawamata, Y.: The cone of curves of algebraic varieties. Ann. Math. **119** (1984) 603–633

[Ka7] Kawamata, Y.: On the plurigenera of minimal algebraic 3-folds with $K \equiv 0$. Math. Ann. **275** (1986) 539–546

[Ka8] Kawamata, Y.: The crepant blowing-up of 3-dimensional canonical singularities and its application to degenerations of surfaces. Ann. Math. **127** (1988) 93–163

[Ka9] Kawamata, Y.: Small contractions of four dimensional algebraic manifolds. Math. Ann. **284** (1989) 595–600

[Ka10] Kawamata, Y.: Boundedness of $\mathbb{Q}$-Fano threefolds. To appear in Proc. Symp. Novosibirsk 1989

[Ka11] Kawamata, Y.: Moderate degenerations of algebraic surfaces. To appear in Proc. Symp. Bayreuth 1990

[KMM] Kawamata, Y., Matsuda, K., Matsuki, K.: Introduction to the minimal model problem. Adv. Stud. Pure Math. **10** (1987) 283–360

[KM] Kawamata, Y., Matsuki, K.: The number of minimal models for a 3-fold of general type is finite. Math. Ann. **276** (1987) 595–598

[Kod] Kodaira, K.: On a differential method in the theory of analytic stacks. Proc. Natl. Acad. Sci. USA **39** (1953) 1268–1273

[Ko1] Kollár, J.: Flops. Nagoya Math. J. **113** (1989) 15–36

[KSB] Kollár, J., Shepherd-Barron, N.I.: Threefolds and deformations of surface singularities. Invent. math. **91** (1988) 299–338

[MM] Miyaoka, Y., Mori, S.: A numerical criterion of uniruledness. Ann. Math. **124** (1986) 65–69

[Mo1] Mori, S.: Threefolds whose canonical bundles are not numerically effective. Ann. Math. **116** (1982) 133–176

[Mo2] Mori, S.: On 3-dimensional terminal singularities. Nagoya Math. J. **98** (1985) 43–66

[Mo3] Mori, S.: Classification of higher-dimensional varieties. Proc. Symp. Pure Math. **46**, no. 1 (1987) 269–331

[Mo4] Mori, S.: Flip theorem and the existence of minimal models for 3-folds. J. Amer. Math. Soc. **1** (1988) 117–253

[Mw] Moriwaki, A.: Semi-ampleness of the numerically effective part of Zariski decomposition. J. Math. Kyoto Univ. **26** (1986) 465–481

[Ms] Morrison, D.: A remark on Kawamata's paper [Ka7]. Math. Ann. **275** (1986) 547–553

[MS] Morrison, D., Stevens, G.: Terminal quotient singularities in dimensions three and four. Proc. Amer. Math. Soc. **90** (1984) 15–20

[Mu] Mumford, D.: The canonical ring of an algebraic surface (Appendix to [Z]). Ann. Math. **76** (1962) 612–615

[N] Nakayama, N.: The singularity of the canonical model of compact Kähler manifolds. Math. Ann. **280** (1988) 509–512

[R1] Reid, M.: Canonical 3-folds. In: Géométrie algébriques Angers 1979, pp. 273–310

[R2] Reid, M.: Minimal models of canonical 3-folds. Adv. Stud. Pure Math. **1** (1983) 131–180

[R3] Reid, M.: Young person's guide to canonical singularities. Proc. Symp. Pure Math. **46**, no. 1 (1987) 345–414

[Sch] Schlessinger, M.: Rigidity of quotient singularities. Invent. math. **14** (1971) 17–26

[Sh1] Shokurov, V.V.: The nonvanishing theorem. Math. USSR Izv. **26** (1986) 591–604

[Sh2] Shokurov, V.V.: Numerical geometry of algebraic varieties. Proc. Intl. Congress of Math. Berkeley 1986, pp. 672–681

[V] Viehweg, E.: Vanishing theorems. J. Reine Angew. Math. **335** (1982) 1–8

[W] Wilson, P.M.H.: Calabi-Yau manifolds with large Picard number. Invent. math. **98** (1989) 139–155

[Z] Zariski, O.: The theorem of Riemann-Roch for high multiples of an effective divisor on an algebraic surface. Ann. Math. **76** (1962) 560–615

Flip and Flop

János Kollár

Department of Mathematics, University of Utah, Salt Lake City, UT 84112, USA

Dedicated to my teacher
Teruhisa Matsusaka

The study of smooth projective varieties naturally breaks into two parts. One part is to study the relationship between birationally equivalent varieties and the other part is to study the different birational equivalence classes. The aim of this talk is to review some interesting features of birational equivalence for three dimensional varieties over $\mathbb{C}$.

For smooth projective curves birational equivalence implies isomorphism, hence the first part of the problem is not interesting.

For smooth projective surfaces the simplest birational transformation is the blowing up. One removes a point of a surface S and replaces it with a copy of $\mathbb{P}^1$ corresponding to the local complex directions at the removed point. Any birational equivalence is a composite of blowing ups and their inverses. There are no other kinds of birational transformations.

In dimension three one can blow up points and smooth curves. It is conjectured that any birational transformation is a composite of blowing ups and their inverses, though at the moment no one seems to know how to prove this. However it became clear in the last decade that even if the above factorisation is possible it may not be the best way of factoring birational transformations. A large theory has been developped by several people, I refer to the talk of (Mori 1990) in the same volume for a general overview. My aim in this talk is to concentrate on certain special birational transformations between threefolds, called flip and flop.

For technical reasons it is natural to investigate these transformations not only for smooth threefolds but also for certain singular ones. The smallest class of singularities that allows the theory to work well is given by the following definition:

Definition 1. Let X be a normal variety and let $f : X' \to X$ be a resolution of singularities with exceptional divisors $E_i \subset X'$. Assume that $\mathcal{O}(mK_X)$ is locally free for some $m > 0$. The smallest such m will be called the index of X. One can write

Proceedings of the International Congress
of Mathematicians, Kyoto, Japan, 1990

$$\mathcal{O}(mK_{X'}) \cong f^*\mathcal{O}(mK_X) \otimes \mathcal{O}(\sum a_i E_i),$$

for some integers a_i. We say that X has terminal (resp. canonical) singularities if $a_i > 0$ (resp. $a_i \geq 0$) for every i. The survey paper (Reid 1987) is the best source of information concerning terminal (and canonical) singularities.

The informal definition of flip or flop is the following. Let X be a threefold and let $C \subset X$ be a compact curve. Remove C from X and try to replace it with another compact curve C^+ in such a way that $C^+ \cup (X - C)$ becomes a threefold X^+. To avoid the unexciting possibility of $C = C^+$ we require that X and X^+ be nonisomorphic. Such an X^+ need not exist at all and it also need not be unique. It turns out that the nature of the transformation $X \dashrightarrow X^+$ depends on the sign of $C \cdot K_X$. The $C \cdot K_X = 0$ case is called flop, the $C \cdot K_X < 0$ case called flip. Very little is known about the case $C \cdot K_X > 0$ and I will not consider it at all.

The formal definitions are the following:

Definition 2. (2.1) A three dimensional curve neighborhood is a pair $C \subset X$ where C is a proper connected curve and X is the germ of a normal threefold along C. One can think of X as an analytic representative of the germ.

(2.2) A three dimensional curve neighborhood $C \subset X$ is called contractible if there is a morphism $f : C \subset X \to P \in Y$ satisfying the following properties:

(i) Y is the germ of a normal singularity around the point P;

(ii) $f(C) = P$;

(iii) $f : X - C \to Y - P$ is an isomorphism.

f and Y are uniquely determined by $C \subset X$. f is called the contraction morphism of $C \subset X$.

(2.3) Two three dimensional curve neighborhoods $C_i \subset X_i$ are called bimeromorphic if there is an isomorphism $(X_1 - C_1) \cong (X_2 - C_2)$.

(2.4) Let $C_1 \subset X_1$ be a three dimensional curve neighborhood. Let H_1 be a line bundle such that $H_1^{-1}|C_1$ is ample. A three dimensional curve neighborhood $C_2 \subset X_2$ bimeromorphic to $C_1 \subset X_1$ is called the opposite of $C_1 \subset X_1$ with respect to H_1 if there is a line bundle H_2 on X_2 such that $H_2|C_2$ is ample and

$$(X_2 - C_2, H_2^m|X_2 - C_2) \cong (X_1 - C_1, H_1^n|X_1 - C_1),$$

for some positive integers m, n. By (Kollár 1991, 2.1.6) the opposite is unique, though it need not exist.

Definition 3. Let $f : C \subset X \to P \in Y$ be a three dimensional contractible curve neighborhood. Assume that X has terminal singularities and that $K_X|C = 0$. Let H be a line bundle on X such that $H^{-1}|C$ is ample.

The opposite of $C \subset X$ (if it exists) will be called the flop of $C \subset X$. (It turns out that the flop is independent of the choice of H.)

Definition 4. Let $f : C \subset X \to P \in Y$ be a three dimensional contractible curve neighborhood. Assume that X has terminal singularities and that $-K_X|C$ is ample.

The opposite of $C \subset X$ with respect to K_X (if it exists) will be called the flip of $C \subset X$.

The simplest example of flops is the following:

Example 5. Let V be the total space of the line bundle $\mathcal{O}(-1,-1)$ over $\mathbb{P}^1 \times \mathbb{P}^1$. Both of the projections $\pi_i : \mathbb{P}^1 \times \mathbb{P}^1 \to \mathbb{P}^1$ can be extended to morphisms

$$p_i : (\mathbb{P}^1 \times \mathbb{P}^1 \subset V) \to (C_i \cong \mathbb{P}^1 \subset X_i).$$

It is easy to see that X_i is smooth and the normal bundle of $C_i \subset X_i$ is $\mathcal{O}(-1) + \mathcal{O}(-1)$. $p_2 \circ p_1^{-1} : X_1 \dashrightarrow X_2$ is bimeromorphic but it is not an isomorphism.

This example was first noticed by (Atiyah 1958). It was used in a systematic way by (Kulikov 1977) to study the birational transformations of threefolds that have a basepoint-free pencil of K3 surfaces.

Examples of flips are much more difficult to present since X cannot be smooth. A fairly exhaustive list is given in (Kollár-Mori 1991).

From the logical point of view (which is the reverse of the historical order) the theory of flips and flops has three large parts.

I. Prove that flops and flips exist.
II. Describe $C^+ \subset X^+$ in terms of $C \subset X$ as precisely as possible.
III. Applications.

The answer to part I is very simple to state:

Theorem 6 (Reid 1983). *Flops exist.*

Theorem 7 (Mori 1988). *Flips exist.*

The existence of flips is one of the most difficult results in three dimensional geometry; see (Kollár 1990a) for an introduction. Previously an important special case was settled by (Tsunoda 1987; Shokurov 1985; Mori 1985; Kawamata 1988) using different methods. (The order corresponds to the order in which the proofs were announced. The only complete published version is the last one.)

It is interesting to note that the above proofs provide very little information beyond the existence of X^+. Thus the answer to part II proceeds along very different lines.

For flops there is a detailed structure theory. This is based on the following observation:

Theorem 8 (Mori 1987). *Let $f : C \subset X \to P \in Y$ be a three dimensional contractible curve neighborhood such that $K_X|C = 0$. Assume in addition that X is smooth (or that it has only index one terminal singularities). Then Y embedds into $\mathbb{C}^4$ and in suitable local coordinates its equation can be written as $x^2 + h(y, z, t)$*

for some power series h. Let $\tau : Y \to Y$ *be the involution* $(x, y, z, t) \mapsto (-x, y, z, t)$. *Then*

$$(f^+ : X^+ \to Y) \cong (\tau \circ f : X \to Y) .$$

In particular, X^+ *is smooth iff* X *is smooth.*

If X has higher index points then the symmetry breaks down in some cases. However in these cases another explicit description is possible, see (Kollár 1989; Kollár 1991, 2.2.2). There are always many similarities between X and X^+:

Theorem 9. *Let* X_1 *and* X_2 *be projective threefolds with terminal singularities only. Assume that* X_2 *is obtained from* X_1 *by a sequence of flops. The following objects do not depend on j (up to isomorphism):*

(9.1) The intersection homology groups $IH^i(X_j, \mathbb{C})$ *together with their Hodge structures (Kollár 1989, 4.12).*

(9.2) The collection of analytic singularities of X_j *(Kollár 1989, 4.11).*

(9.3) The miniversal deformation space $\operatorname{Def} X_j$ *(Kollár-Mori 1991, 12.6).*

(9.4) The integral cohomology groups $H^i(X_j, \mathbb{Z})$ *(Kollár 1991, 3.2.2).*

(9.5) $\operatorname{Pic} X_j \subset \operatorname{Weil} X_j$ *(Kollár 1991, 3.2.2).*

(9.6) $h^0(X_j, \mathcal{O}(D))$ *for every Weil divisor D.*

(9.7) $h^i(X, \mathcal{O}(mK_{X_j}))$ *for every i and m.*

The isomorphisms are canonical except possibly for (9.2).

Describing $C^+ \subset X^+$ in terms of $C \subset X$ is harder for flips. The first breakthrough was a result of (Shokurov 1985) which showed that the singularities of X^+ are less "difficult" than the singularities of X in a complicated technical sense. For a given singularity the "difficulty" is not easy to compute but the definition is very well suited to inductive proofs.

It turns out that the situation is very complicated. The problem is that the singularities of X^+ depend not only on the singularities of X but also on the global structure of X. If X is given by coordinate patches and transition functions along $C \cong \mathbb{P}^1$ then the singularities of X^+ do not depend continuously on the transition functions. This makes any description very subtle. Several special cases are worked out in (Kollár-Mori 1991); many more – but not all – cases are treated in unpublished works of Mori. I would like to mention one patholgy that seems interesting:

Proposition 10 (Kollár-Mori 1991, 13.5, 13.7). *Let* $C^+ \subset X^+$ *be the flip of* $C \subset X$. *Then the number of irreducible components of* C^+ *is less than or equal to the number of irreducible components of C. For every* $m > 0$ *there are examples where* C^+ *is irreducible and C has m irreducible components.*

Let us now turn to some of the applications. The most important application of flops is the following result which shows that birational maps between certain threefolds can be factored into a sequence of flops. This factorisation is much more useful than a factorisation into blowing ups and downs would be. Its importance is clear only in light of the minimal model theory of threefolds (see e.g. (Mori 1990)).

Theorem 11 (Reid 1983; Kawamata 1988; Kollár 1989). *Let X_1 and X_2 be normal projective threefolds with terminal singularities only. Assume that they have $\mathbb{Q}$-factorial singularities. Assume furthermore that K_{X_i} has nonnegative intersection number with any curve $C \subset X_i$ for $i = 1, 2$. Then any birational map $f : X_1 --> X_2$ can be obtained as a composite of flops.*

Another application is to compact nonprojective threefolds. Under certain conditions there is a very economical way of finding a birationally equivalent model which is projective.

Theorem 12 (Kollár 1991, 5.2.3). *Let X be a proper algebraic threefold with $\mathbb{Q}$-factorial terminal singularities. Assume that K_X has nonnegative intersection number with any curve $C \subset X$. Then after finitely many flops one obtains a proper algebraic threefold X^+ with $\mathbb{Q}$-factorial terminal singularities such that X^+ admits a birational morphism $g : X^+ \to Y$ onto a projective variety Y. Y has terminal singularities and g contracts only finitely many curves.*

Flips are one step of Mori's program (also called minimal model program) for threefolds, in fact the most difficult step. Their main importance is thus derived from the importance and applications of the program. See (Kollár 1991) or (Mori 1990) for details.

There are many problems concerning flips even in dimension three. The most interesting open question about flips is the following.

Conjecture 13. Reid's Conjecture on General Elephants (Reid 1987; Kollár-Mori 1991). *The contraction map provides a one-to-one correspondence between the following two sets:*

Extremal neighborhoods:

$$EN := \left\{ \begin{array}{l} \text{Three dimensional contractible curve neighborhoods } C \subset X \text{ such that} \\ X \text{ has canonical singularities and } -K_X|C \text{ is ample.} \end{array} \right\}$$

Flipping singularities:

$$FS := \left\{ \begin{array}{l} \text{Three dimensional normal singularities } P \in Y \text{ such that } K_Y \text{ is not} \\ \mathbb{Q}\text{-Cartier and some } D \in |-K_Y| \text{ has a Du Val singularity at } P. \end{array} \right\}$$

Reid's original hope was that this equivalence can be used to obtain a proof of the existence of flips. To do this one needs to produce a member of $|-K_Y|$ with Du Val singularity and then to use this member to construct X^+. It is still to be seen whether either of these steps can be done in the spirit envisaged by Reid. A proof along these lines would considerably enhance our understanding of flips.

(Kollár-Mori 1991, 3.1) shows that for every flipping singularity there is a corresponding extremal neighborhood. The opposite direction is also proved for extremal neighborhoods with terminal singularities and irreducible C in (Kollár-Mori 1991, 1.7). The general case probably requires different methods. There is also the possibility that the correspondence is more complicated. The conjecture implies that every non-Gorenstein singularity on an extremal neighborhood is pseudo-terminal. I do not see any a priori reason why this should be so. On the other hand, even certain terminal singularities cannot occur on extremal neighborhoods, thus unexpected restrictions are possible.

The higher dimensional problems are discussed in the talk of (Kawamata 1990).

Acknowledgement. Partial financial support was provided by the NSF under grant numbers DMS-8707320 and DMS-8946082 and by an A. P. Sloan Research Fellowship.

References

Atiyah, M. (1958): On analytic surfaces with double points. Proc. Roy. Soc. **247**, 237–244

Kawamata, Y. (1988): The crepant blowing-up of 3-dimensional canonical singularities and its application to the degeneration of surfaces. Ann. Math **127**, 93–163

Kawamata, Y. (1990): Canonical singularities and minimal models of algebraic varieties. These Proceedings, p. 699

Kollár, J. (1989): Flops. Nagoya Math. J. **113**, 14–36

Kollár, J. (1990): Minimal models of algebraic threefolds: Mori's Program. Asterisque **177–178**, 303–326

Kollár, J. (1991): Flips, flops, minimal models etc. J. Diff. Geom. (to appear)

Kollár, J., Mori, S. (1991): Soon to be written up

Kulikov, V. (1977): Degenerations of K3 surfaces and Enriques surfaces. Math. USSR Izv. **11**, 957–989

Mori, S. (1985): Minimal models for semistable degenerations of surfaces. Lectures at Columbia University. Unpublished

Mori, S. (1987): Personal communication

Mori, S. (1988): Flip theorem and the existence of minimal models for 3-folds. Journal AMS **1**, 117–253

Mori, S. (1990): Birational Classification of algebraic threefolds. These Proceedings, p. 235

Reid, M. (1983): Minimal models of canonical threefolds, Algebraic Varieties and Analytic Varieties. Adv. Stud. Pure Math. vol. 1 (Iitaka, S., ed.). Kinokuniya and North-Holland, pp. 131–180

Reid, M. (1987): Young person's guide to canonical singularities. Algebraic Geometry Bowdoin 1985, Proc. Symp. Pure Math. vol. 46, 345–416

Shokurov, V. (1985a): The nonvanishing theorem. Izv. Akad. Nauk SSSR Ser. Mat. **49**, 635–651

Shokurov, V. (1985b): Letter to M. Reid

Tsunoda, S. (1987): Degenerations of surfaces. Algebraic Geometry, Sendai, Adv. Stud. Pure Math. vol. 10 (Oda, T., ed.). Kinokuniya and North-Holland, pp. 755–764

Linear Series on Algebraic Varieties

*Robert K. Lazarsfeld**

Department of Mathematics, University of California, Los Angeles, CA 90024, USA

In recent years, ideas and methods from the theory of vector bundles have been applied to study some classical sorts of questions concerning linear series on curves, surfaces and other algebraic varieties. Our purpose here is to survey some of the problems to which these and related techniques have been applied.

Many of the results we describe in one way or another involve the equations defining projective varieties. There are several motivations for studying questions along these lines. The first is historical—the equations defining varieties have been of interest to geometers at least since Noether, and it is natural to try to clarify and extend classical results as much as possible. Secondly, there is a "practical" motivation. It has become common for geometers to use computers to analyze explicit varieties, and for the most part the only way to describe a variety to a computer is by giving its equations. Finally, it seems likely that vector bundle methods will play an increasingly important role in the study of linear series, and the classical questions we consider here have proven to be good testing grounds for these techniques.

Limitations of space preclude more than a cursory discussion of methods of proof. We refer to [L3] for an overview of some of these. However we have attempted to survey some of the many interesting open problems in this area. We work throughout over the complex numbers $\mathbb{C}$.

§ 1. Castelnuovo-Mumford Regularity

Let $X \subset \mathbb{P}^r$ be an irreducible variety of dimension n, and denote by $\mathcal{I}_X$ the ideal sheaf of X in $\mathbb{P}^r$. Recall that one says that X is *k-regular* if $H^i(\mathbb{P}^r, \mathcal{I}_X(k-i)) = 0$ for $i \geq 1$ ([M2], [M3]); the *regularity* reg(X) of $X \subset \mathbb{P}^r$ is the least such k. The interest in this concept stems partly from the fact that reg(X) governs the complexity of computing the syzygies and other invariants of X ([BS1, BS2]). For example, a theorem of Mumford (cf. [EG]) states that X is k-regular if and only if for every $p \geq 0$, the minimal generators of the kth module of syzygies of the homogeneous

* Partially supported by N.S.F. grant DMS 89-02551.

Proceedings of the International Congress
of Mathematicians, Kyoto, Japan, 1990

ideal I_X occur in degrees $\leq k + p$. This has led to a substantial body of work aimed at bounding the regularity of X in terms of its invariants or (more recently) the degrees of its defining equations. In each case there is a fascinating tension between the situation for arbitrary schemes $X \subset \mathbb{P}^r$ — for which results and examples show that the regularity can be very large — and the geometric situation in which X is smooth or at least reduced, where one expects much stronger statements.

We start with the *Castelnuovo problem* of bounding the regularity of X in terms of its invariants. For an arbitrary scheme $X \subset \mathbb{P}^r$, Gotzmann [Gotz, G3] has given a best-possible bound in terms of the Hilbert polynomial of X. For instance, if $X \subset \mathbb{P}^r$ is a curve of degree d and arithmetic genus g, his result states that $\operatorname{reg}(X) \leq \binom{d-1}{2} + d - g$. The question now arises if one can do better assuming X is smooth or reduced.

In practice the main difficulty is to control $H^1(\mathbb{P}^r, \mathscr{I}_X(k))$, which is the classical question of whether hypersurfaces of degree k trace out a complete linear series on X. A theorem of Castelnuovo asserts that when $X \subset \mathbb{P}^r$ $(r \geq 3)$ is a smooth curve of degree d, then $H^1(\mathbb{P}^r, \mathscr{I}_X(k)) = 0$ for $k \geq d - 2$, and consequently X is $(d-1)$-regular. This is optimal for curves in $\mathbb{P}^3$, but not in general. In fact, Castelnuovo's result was completed in [GLP], where vector bundle techniques are used to show that if $X \subset \mathbb{P}^r$ is a reduced irreducible non-degenerate curve of degree d, then X is $(d + 2 - r)$-regular. Furthermore the borderline examples are classified: roughly speaking, X fails to be $(d + 1 - r)$-regular if and only if it has a $(d + 1 + r)$-secant line. This possibility of giving geometric explanations for extremal algebraic behavior is one of the interesting themes in this circle of ideas; another example appears in [G3].

These results for curves have led various people to hope that if $X \subset \mathbb{P}^r$ is a smooth non-degenerate complex projective variety of degree d and dimension n, then X is $(d + n + 1 - r)$-regular [EG]. At least when $r \geq 2n + 1$, this would be the best possible linear inequality. It is proved for surfaces in [P] and [L1], and for 3-folds in [R] when $r \geq 9$. In higher dimensions, the situation is much less clear. There is a rather weak bound of Mumford's (cf. [BEL, (2.1)]), but new ideas are apparently needed to treat the problem in general. It would already be very interesting to show that X is d-regular. There is also no clear understanding of what to expect if X is singular (but reduced and irreducible). For an arbitrary scheme $X \subset \mathbb{P}^r$, Ravi [Rav] asks whether one has the inequality $\operatorname{reg}(X_{\text{red}}) \leq \operatorname{reg}(X)$; this may be too much to wish for, but it would be nice to settle the matter.

As we have noted, attempts to bound the regularity of a projective variety $X \subset \mathbb{P}^r$ are motivated in part by the desire to bound the complexity of computing the syzygies and related invariants of X. Bayer has remarked that it is then natural to ask for bounds in terms of the degrees of the defining equations of X – which are usually primary data when one sets out to make an actual computation – rather than the degree of X, which may not be so obvious in a given situation. For an arbitrary scheme, this regularity can again be horrendously large: there are examples due to Mayr-Meyer-Bayer-Stillman [BS1] of schemes $X \subset \mathbb{P}^r$ defined by hypersurfaces of degree d with $\operatorname{reg}(X) \geq (d-2)^{2^{(r/10)}}$. However when X is smooth, the regularity grows much more slowly with d and r. In fact, it comes as a pleasant surprise to note that elementary arguments using the Kodaira vanishing theorem lead to the optimal bound in this case:

Theorem 1.1 [BEL]. *Assume that $X \subset \mathbb{P}^r$ is a smooth variety of dimension n and codimension e defined by hypersurfaces of degrees $d_1 \geq d_2 \geq \cdots \geq d_m$. Then X is $(d_1 + \cdots + d_e - e + 1)$-regular, and X fails to be $(d_1 + \cdots + d_e - e)$-regular if and only if it is the complete intersection of hypersurfaces of degrees $d_1, d_2, \ldots, d_e$.*

Note that only the degrees of the first $e = \operatorname{codim}(X, \mathbb{P}^r)$ defining equations come into play here. It would be very interesting to know to what extent one can allow X to be singular.

§ 2. Syzygies of Algebraic Varieties

It is useful to think of the results described in the previous section as being *extrinsic* in nature, in the sense that they refer to a given projective embedding $X \subset \mathbb{P}^r$, with the main difficulties coming from the fact that X might not be linearly normal. But one can also take a more *intrinsic* point of view, where one deals with the embedding $X \subset \mathbb{P} = \mathbb{P}(H^0(L))$ defined by the complete linear system associated to a very ample line bundle L on X. Here it is usually not so hard to compute $\operatorname{reg}(X)$, so one can ask for more precise information. A very interesting line of inquiry – inaugurated by Green in [G1] – is to study the syzygies of X.

We start with some notation. Let L be a very ample line bundle on a projective variety X, defining an embedding $X \subset \mathbb{P} = \mathbb{P}(H^0(L))$. Denote by $S = \operatorname{Sym}^{\cdot} H^0(L)$ the homogeneous coordinate ring of the projective space $\mathbb{P}$, and consider the graded S-module $R = R(L) = \oplus H^0(X, L^d)$. Let $E_{\cdot} = E_{\cdot}(L)$ be a minimal graded free resolution of R:

$$\begin{array}{ccccccccc} \cdots \longrightarrow & \oplus S(-a_{2,j}) & \longrightarrow & \oplus S(-a_{1,j}) & \longrightarrow & \begin{array}{c} S \\ \oplus \\ \oplus S(-a_{0,j}) \end{array} & \longrightarrow & R & \longrightarrow 0. \\ & \| & & \| & & \| & & & \\ & E_2 & & E_1 & & E_0 & & & \end{array}$$

We have indicated here the fact that R has a canonical generator in degree zero. It is easy to see that all $a_{0,j} \geq 2$ and all $a_{i,j} \geq i + 1$ when $i \geq 1$. To extend classical results on normal generation and presentation we ask when the first few E_i are as simple as possible:

Definition 2.1. The line bundle L saisfies *property* (N_p) if

$$E_0 = S \text{ when } p \geq 0$$

and

$$E_i = \oplus S(-i-1) \text{ [i.e. all } a_{i,j} = i+1] \qquad \text{for} \qquad 1 \leq i \leq p.$$

Note that if $E_0 = S$, then $E.$ determines a resolution of the homogeneous ideal $I = I_{X/\mathbb{P}}$ of X in $\mathbb{P}(H^0(L))$. Thus the definition may be summarized very concretely as follows:

L satisfies (N_0) $\Leftrightarrow$ X embeds in $\mathbb{P}(H^0(L))$ as a projectively normal variety;

L satisfies (N_1) $\Leftrightarrow$ (N_0) holds for L, and the homogeneous ideal I_X of X is generated by quadrics;

L satisfies (N_2) $\Leftrightarrow$ (N_0) and (N_1) hold for L, and the module of syzygies among quadratic generators $Q_i \in I$ is spanned by relations of the form

$$\sum L_i Q_i = 0,$$

where the L_i are *linear* polynomials;

and so on. Properties (N_0) and (N_1) are what Mumford [M3] calls respectively *normal generation* and *normal presentation*.

A classical theorem of Castelnuovo, Mattuck and Mumford states that if X is a smooth curve of genus g, and if $\deg(L) \geq 2g + 1$, then L is normally generated. Fujita and St. Donat proved that if $\deg(L) \geq 2g + 2$, then L satisfies (N_1). These results were extended and clarified by Green [G1], who showed that if $\deg(L) \geq 2g + 1 + p$, then L satisfies (N_p). Green's theorem was recovered as a consequence of an analogous statement of finite sets in [GL3], where it was also proved that if $\deg(L) = 2g + p$, then L fails to satisfy (N_p) if and only if either X is hyperelliptic or Φ_L embeds X with a $(p + 2)$-secant p-plane.

The result just quoted gives a first indication of the fact that (at least conjecturally) the syzygies of a smooth curve X are intimately connected with its geometry. The crucial invariant here is the *Clifford index* Cliff(X) of X. Referring for instance to [GL1] for the precise definition, suffice it to say that Cliff(X) is a non-negative integer which measures from the point of view of special divisors how general X is in moduli. For instance Cliff$(X) = 0$ if and only if X is hyperelliptic, and Cliff$(X) = 1$ if and only if X is trigonal or a smooth plane quintic. One has the inequality $0 \leq \text{Cliff}(X) \leq [(g - 1)/2]$, and $\text{Cliff}(X) = [(g - 1)/2]$ when X is a general curve of genus g.

The hope is that uniform results on the syzygies of a curve X can be strengthened to take into account its Clifford index. The clearest expression of this philosophy which has actually been proved to date is a theorem of [GL1] to the effect that if L is a very ample line bundle on X with $\deg(L) \geq 2g + 1 - 2 \cdot h^1(L) - \text{Cliff}(X)$, then L is normally generated. This includes the result of Castelnuovo et al. mentioned above, and for example taking L to be the canonical bundle Ω it yields Noether's celebrated theorem that Ω is normally generated unless X is hyperelliptic. One conjectures [GL1] that if L is a very ample line bundle with $\deg(L) \geq 2g + 1 + p - 2 \cdot h^1(L) - \text{Cliff}(X)$, then L satisfies (N_p) provided that Φ_L does not embed X with a $(p + 2)$-secant p-plane. This would imply all the known results on syzygies of curves. But its most interesting consequence would be a beautiful conjecture of Green's on the syzygies of canonical curves:

Conjecture 2.2 [G1]. Cliff(X) *is the least integer p for which* (N_p) *fails for the canonical bundle* Ω.

Green's conjecture has already sparked a considerable amount of work. For instance, drawing on some ideas of Chang and Ran as well as some computer computations by Bayer and Stillman, Ein [E] proves (2.2) on a general curve for $p \leq 3$. The most striking progress to date is due to Voisin [V1] and Schreyer [S], who verify the first non-classical case $p = 2$. Schreyer also gives examples to show that (2.2) fails in characteristic 2. A possible approach to this conjecture is suggested in [PR]. In a somewhat different direction, one has

Conjecture 2.3 ([GL1], Conj. 3.7). *One can read off the "gonality" of a curve X from the grading of the resolution* $E_{\bullet} = E_{\bullet}(L)$ *of any one line bundle of sufficiently large degree.*

In fact, one expects the gonality to be determined by the tail end of $E_{\bullet}$. One hopes that (2.3) should be easier to establish than (2.2).

Results on normal generation and presentation of curves have traditionally gone hand in hand with analogous results for abelian varieties. Suppose then that X is an abelian variety of dimension n, and let L be an ample line bundle on X. A theorem of Mumford [M1, M3], Koizumi [K] and Sekiguchi [Sek] states that $L^{\otimes k}$ is normally generated provided that $k \geq 3$, and Mumford [M3] proved that X is scheme-theoretically cut out by quadrics under the embedding defined by $L^{\otimes k}$ provided that $k \geq 4$. By analogy with the situation on curves, it is then natural to conjecture [L3] that if $k \geq p + 3$, then $L^{\otimes k}$ satisfies (N_p). Kempf [Kmf] has recently made considerable progress on this conjecture. Specifically, he proves if $p \geq 1$ then (N_p) holds for $L^{\otimes k}$ so long as $k \geq 2p + 2$. It would be wonderful to prove the conjecture in general.

Finally, what can one say in higher dimensions? Green [G2] proved that on an arbitrary smooth variety X of dimension n, any sufficiently positive line bundle L satisfies (N_p). But naturally one would like an explicit result. Mukai observed that the known theorems deal with embeddings defined by bundles of the type $K_X \otimes P$, where P is an explicit multiple of a suitably positive bundle. He suggested that in general one should aim for statements having this shape. In this direction one has

Theorem 2.4 [EL]. *Let A be a very ample line bundle on X, and let B be a numerically effective bundle on X. If* $k \geq n + 1 + p$, *then* $K_X \otimes A^{\otimes k} \otimes B$ *satisfies* (N_p).

So for example, $K_X \otimes A^{\otimes n+1}$ is normally generated (which has a quick proof: cf. [BEL], [AS] and [ABS]), and in the embedding defined by $K_X \otimes A^{\otimes n+2}$ the homogeneous ideal I_X is generated by quadrics. It would be very interesting to prove analagous statements assuming only that A is ample. Butler [But] shows that if X is a projective bundle over a curve, and if A is an ample line bundle on X, then $K_X \otimes A^{\otimes 2n+1}$ is normally generated. A rash optimist might be tempted to speculate that $K_X \otimes A^{\otimes n+2+p}$ satisfies (N_p) whenever A is an ample line bundle on an arbitrary smooth variety X. However even when $n = 3$ it is unknown whether $K_X \otimes A^{\otimes n+2}$ is very ample (this is a celebrated conjecture of Fujita), so that the moment it seems premature to hope for statements of this type for syzygies. However as Mukai points out, it might be reasonable to ask for sharp theorems for surfaces. When $X = \mathbb{P}^n$, results on syzygies and related issues amount to statements about multiplicative properties of subspaces of the polynomial ring. Green [G2] has obtained some general theorems along these lines, for which he has given interesting applications to the Hodge theory of hypersurfaces. We refer the reader to [G4] for a survey.

§ 3. Linear Series and Vector Bundles on Surfaces

As we remarked in the Introduction, most of the results described above are proved using vector bundle techniques. For questions involving regularity and syzygies, the arguments are largely cohomological in nature. However on surfaces, bundles have been used in a more geometrical fashion to study linear series. In this final section we survey two such applications. Henceforth X denotes a smooth complex projective surface.

Consider a curve C on X, and a line bundle A on C which is generated by its global sections. Thinking of A as a coherent sheaf on X, there is a natural surjective evaluation map $e_{C,A}: H^0(A) \otimes_{\mathbb{C}} \mathcal{O}_X \to A$. Set $F = F_{C,A} = \ker e_{C,A}$; thus F is a vector bundle of rank $h^0(C, A)$ on X. Philosophically, F encodes information about the pair (C, A) into a geometric object that lives globally on X. When X is a K3 surface, an analysis of these bundles together with a theorem of Mukai's [Mk1, Mk2] on the smoothness of the moduli space of simple bundles on X leads to the following

Theorem 3.1 [L2]. *Let X be a* K3 *surface, and let $C_0 \subset X$ be a smooth curve having the property that every curve in the linear series $|C_0|$ is reduced and irreducible. Then the general member $C \in |C_0|$ behaves generically in the sense of Brill-Noether theory, i.e. the varieties $W_d^r(C)$ of special linear series on C have the postulated dimensions and $W_d^r(C)$ is smooth away from $W_d^{r+1}(C)$.*

We refer for instance to [ACGH] for a fuller discussion of the Brill-Noether-Petri package. The hypothesis is satisfied for instance when $\mathrm{Pic}(X) = \mathbb{Z} \cdot [C_0]$, and the theorem then leads to a very quick proof of an important result of Gieseker's [Gies] concerning the behavior of the varieties $W_d^r(C)$ on a general curve of genus g. The bundles $F_{C,A}$ were studied on an arbitrary surface by Tjurin [T]. They also lead to the proof [GL2] of a conjecture of Harris, Mumford and Green to the effect that all curves in a given linear series on a K3 surface have the same Clifford index. Related results appear in [DM] and [Par]. Reid [Reid] has some intriguing ideas about how bundles of this type might be relevant to the study of canonical surfaces. More recently, Voisin [V2] has used an infinitesimal analogue of the $F_{C,A}$ to prove a very beautiful result about the Wahl map $\gamma_\Omega: \bigwedge^2 H^0(C, \Omega) \to H^0(C, \Omega^3)$ on a smooth curve C of genus g.

Finally, no survey of vector bundles and linear series would be complete without mentioning the very interesting work of Igor Reider [Rdr] on linear series on a smooth projective surface X. Suppose that L is a line bundle on X such that $K_X \otimes L$ fails to be globally generated at a point $x \in X$. Then $H^1(X, \mathcal{I}_x \otimes K_X \otimes L) \neq 0$, and so by Grothendieck duality and a well-known construction of Serre's one can construct a non-split extension of the form $0 \to \mathcal{O}_X \to E \to L \to 0$. Reider's idea — which has its antecedents in a proof by Mumford of the Ramanujam vanishing theorem — is that suitable numerical conditions on L will force E to be Bogomolov-unstable. Analyzing this instability geometrically then leads to the following

Theorem 3.2 [Rdr]. *Assume that L is a nef line bundle on X such that $K_X \otimes L$ is not globally generated. If $c_1(L^2) \geq 5$, then there is an effective divisor $D \subset X$ such that either $c_1(L) \cdot D = 0$ and $D^2 = -1$, or $c_1(L) \cdot D = 1$ and $D^2 = 0$.*

Reider proves an analogous statement for the failure of $K_X \otimes L$ to be very ample. This theorem has a surprising range of applications, among them a quick proof of Bombieri's results on pluricanonical maps of surfaces. Reider's theorem has attracted a lot of attention, and it has been extended in several directions. We refer to [Kot] and the conference proceedings [Som] for a sampling of some of the work in this direction.

References

[ABS] M. Andreatta, E. Ballico, A. Sommese: On the projective normality of the adjunction bundles II. Preprint

[AS] M. Andreatta, A. Sommese: On the projective normality of the adjunction bundles. Preprint

[ACGH] M. Arbarello, M. Cornalba, P. Griffiths, J. Harris: Geometry of algebraic curves. Springer, Berlin Heidelberg New York 1985

[BS1] D. Bayer, M. Stillman, On the complexity of computing syzygies. In: L. Robbianno (ed) Computational aspects of commutative algebra. Academic Press, 1989, pp. 1–13

[BS2] D. Bayer, M. Stillman: A criterion for detecting m-regularity. Inv. math. **87** (1987) 1–11

[BEL] A. Bertram, L. Ein, R. Lazarsfeld: Vanishing theorems, a theorem of Severi, and the equations defining projective varieties. To appear in J. of the AMS

[But] D. Butler: Tensor products of global sections for vector bundles over a curve with applications to linear series. To appear

[DM] R. Donagi, D. Morrison: Linear systems on K3 sections. J. Diff. Geom. **29** (1988) 49–64

[E] L. Ein: Some remarks on the syzygies of general canonical curves, J. Diff. Geom. **26** (1987) 361–366

[EL] L. Ein, R. Lazarsfeld: A theorem on the syzygies of smooth projective varieties of arbitrary dimension. To appear

[EG] D. Eisenbud, S. Goto: Linear free resolutions and minimal multiplicity. J. Alg. **88** (1984) 89–133

[Gies] D. Gieseker: Stable curves and special divisors: Petri's conjecture. Inv. math. **66** (1982) 251–275

[Gotz] G. Gotzmann: Eine Bedingung für die Flachheit und das Hilbertpolynom eines graduierten Ring. Math. Z. **158** (1978) 61–70

[G1, G2] M. Green, Koszul cohomology and the geometry of projective varieties I and II. J. Diff. Geom. **19** (1984) 125–171 and **20** (1984) 279–289

[G3] M. Green: Restrictions of linear series to hyperplanes, and some results of Macaulay and Gotzmann. In: E. Ballico, C. Ciliberto (eds), Algebraic curves and projective varieties (Lect. Notes in Mathematics, vol. 1389.) Springer, Berlin Heidelberg New York 1989, pp. 76–86

[G4] M. Green: Koszul cohomology and geometry. In: M. Cornalba et al. (eds) Lectures on riemann surfaces. World Scientific Press, 1989, pp. 177–200

[GL1] M. Green, R. Lazarsfeld: On the projective normality of complete linear series on an algebraic curve. Inv. math. **83** (1986) 73–90

[GL2] M. Green, R. Lazarsfeld: Special Divisors on curves on a K3 surface. Inv. math. **89** (1987) 357–370

[GL3] M. Green, R. Lazarsfeld: Some results on the syzygies of finite sets and algebraic curves. Comp. Math. **67** (1988) 301–314

[GLP] L. Gruson, R. Lazarsfeld, C. Peskine: On a theorem of Castelnuovo and the equations defining space curves. Inv. math. **72** (1983) 491–506

[Kmf] G. Kempf: Projective coordinate rings of abelian varieties. In: J.I. Igusa (ed.), Algebraic analysis geometry and number theory. Johns Hopkins Press 1989, pp. 423–438

[K] S. Koizumi: Theta relations and projective normality of abelian varieties. Am. J. Math. **98** (1976) 865–889

[Kot] D. Kotschick: Stable and unstable bundles on algebraic surfaces. To appear

[L1] R. Lazarsfeld: A sharp Castelnuovo bound for smooth surfaces. Duke Math. J. **55** (1987) 423–238

[L2] R. Lazarsfeld: Brill-Noether-Petri without degenerations. J. Diff. Geom. **23** (1986) 299–307

[L3] R. Lazarsfeld: A sampling of vector bundle techniques in the study of linear series. In: M. Cornalba et al. (eds) Lectures on Riemann surfaces. World Scientific Press, 1989, pp. 500–559

[Mk1] S. Mukai: Symplectic structure of the moduli space of sheaves on an abelian or K3 surface. Inv. math. **77** (1984) 101–116

[Mk2] S. Mukai, On the moduli space of bundles on K3 surfaces, I. In: Vector bundles on algebraic varieties. Oxford Univ. Press, 1987, pp. 341–413

[M1] D. Mumford: On the equations defining abelian varieties. Inv. math. **1** (1966) 287–354

[M2] D. Mumford: Lectures on curves on an algebraic surface. Ann. Math. Stud, vol. 59 (1966)

[M3] D. Mumford: Varieties defined by quadratic equations. Corso CIME 1969. In: Questions on algebraic varieties. Rome 1970, pp. 30–100

[P] H. Pinkham: A Castelnuovo bound for smooth surfaces. Inv. math. **83** (1906) 321–332

[Par] G. Pareschi: Exceptional linear systems on curves on del Pezzo surfaces. To appear

[PR] K. Paranjape, S. Ramanan: On the canonical ring of a curve. In: Algebraic geometry and commutative algebra. Kinokuniya, 1988

[R] Z. Ran: Local differential geometry and generic projections of threefolds. J. Diff. Geom. To appear

[Rav] M.S. Ravi: Regularity of ideals and their radicals. Manuscr. Math. **68** (1990) 77–87

[Reid] M. Reid: Quadrics through a canonical surface. In: A. Sommese et al. (eds) Algebraic geometry. (Lect. Notes in Mathematics, vol. 1417.) Springer, Berlin Heidelberg New York 1990, pp. 191–213

[Rdr] I. Reider: Vector bundles of rank 2 and linear systems on algebraic surfaces. Ann. Math. **127** (1988) 309–316

[S] F. Schreyer: A standard basis approach to the syzygies of canonical curves. To appear

[Sek] T. Sekiguchi: On the normal generation by a line bundle on an abelian variety. Proc. Jap. Acad. **54** (1978) 185–188

[Som] A. Sommese et al. (eds) Algebraic geometry. Proceedings, L'Aquila 1988. (Lecture Notes in Mathematics, vol. 1417.) Springer, Berlin Heidelberg New York 1990

[T] A. Tyurin: Cycles, curves and vector bundles on an algebraic surface. Duke Math. J. **55** (1987)

[V1] C. Voisin: Courbes tetragonales et cohomologie de Koszul. J. Reine Angew. Math. **387** (1988) 111–121

[V2] C. Voisin: Sur l'application de Wahl des courbes satisfaisant la condition de Brill-Noether-Petri. To appear

Mixed Hodge Modules and Applications

Morihiko Saito

Research Institute for Mathematical Sciences, Kyoto University, Kyoto 606, Japan

1. Mixed Hodge Module

Since the theory of mixed perverse sheaves was established in char $p > 0$ by Beilinson-Bernstein-Deligne-Gabber [3], it has been conjectured that there would exist objects in char 0 corresponding philosophically to mixed perverse sheaves. This conjecture was solved recently by introducing the notion of *mixed Hodge Modules* [25–28].

For a complex algebraic variety X, we have an abelian category $\mathrm{MHM}(X)$ of mixed Hodge Modules on X with a forgetful functor $\mathrm{rat} : \mathrm{MHM}(X) \to \mathrm{Perv}(\mathbb{Q}_X)$ which assigns the underlying perverse sheaf over $\mathbb{Q}$. By [2, 3], rat is naturally extended to a morphism of the bounded derived category $D^b\mathrm{MHM}(X)$ to $D_c^b(\mathbb{Q}_X)$, which is also denoted by rat. They satisfy the formalism of mixed sheaves, i.e.

Theorem 1.1. *The categories $D^b\mathrm{MHM}(X)$ are stable by the standard functors like f_*, $f_!$, f^*, $f^!$, ψ_g, φ_g, $\mathbb{D}$, $\boxtimes$, $\otimes$, $\mathcal{H}om$, and these functors are compatible with the corresponding functors on the underlying complexes of $\mathbb{Q}$-vector spaces via the functor* rat.

Proposition 1.2. *Every mixed Hodge Module $\mathcal{M}$ has a functorial increasing filtration W in $\mathrm{MHM}(X)$, called the weight filtration of $\mathcal{M}$, such that $\mathcal{M} \to \mathrm{Gr}_k^W \mathcal{M}$ is an exact functor and $\mathrm{Gr}_k^W \mathcal{M}$ is semisimple.*

If X is smooth, $\mathrm{MHM}(X)$ is a full subcategory of the category of (M, F, K, W) where M is a holonomic $\mathcal{D}_X$-Module with a good filtration F [14] and with rational structure given by an isomorphism $\alpha : \mathrm{DR}(M) \simeq \mathbb{C} \otimes K$ in $\mathrm{Perv}(\mathbb{C}_X)$ for a perverse sheaf K defined over $\mathbb{Q}$ [3], and W is a pair of filtrations on M and K compatible with α and gives the weight filtration in Proposition 1.2. Here DR is the de Rham functor (shifted to the left by $\dim X$), and $\mathrm{DR}(M) \in \mathrm{Perv}(\mathbb{C}_X)$ by [14]. (We assume that an analytic holonomic $\mathcal{D}$-Module on a smooth algebraic variety has an algebraic stratification. We can use also algebraic (filtered) $\mathcal{D}$-Modules [5] by GAGA and the extendability of mixed Hodge Modules.) For $X = \mathrm{pt}$, we have

Theorem 1.3. $\mathrm{MHM}(\mathrm{pt})$ *is the category of polarizable mixed $\mathbb{Q}$-Hodge structures* [8].

Proceedings of the International Congress
of Mathematicians, Kyoto, Japan, 1990

Here a mixed Hodge structure is called polarizable if its graded pieces are. This result can be generalized to the case of admissible variation of mixed Hodge structure in the sense of Steenbrink-Zucker [41] (in the one-dimensional case) and Kashiwara [16], cf. Theorem 1.5.

We say that a mixed Hodge Module $\mathscr{M}$ is *pure of weight n*, if $\mathrm{Gr}_k^W \mathscr{M} = 0$ for $k \neq n$. A pure Hodge Module is semi-simple, and a mixed Hodge Module is obtained by successive extensions of pure Hodge Modules. We say that a pure Hodge Module $\mathscr{M}$ has *strict support* if its support is irreducible and $\mathscr{M}$ has no sub or quotient with smaller support. By semisimplicity, a pure Hodge Module is a direct sum of pure Hodge Modules with different strict supports, called the *decomposition by strict support* (which is unique). A pure Hodge Module is also called a *polarizable Hodge Module* in [26], and we denote by $\mathrm{MH}_Z(X,n)^p$ the category of pure Hodge Modules of weight n with strict support Z. Then $\mathrm{MH}_Z(X,n)^p$ depends only on Z, and $\mathscr{M} \in \mathrm{MH}_Z(X,n)^p$ is generically a polarizable variation of Hodge structure in the sense of Griffiths [13]. An important fact is that its converse is true: any polarizable variation of Hodge structure can be extended uniquely to a pure Hodge Module, i.e.

Theorem 1.4. *We have an equivalence of categories*

$$\mathrm{MH}_Z(X,n)^p \simeq \mathrm{VHS}_{\mathrm{gen}}(Z, n - \dim Z)^p .$$

Here the right hand side is the category of polarizable variations of Hodge structures of weight $n - \dim Z$ defined on (nonempty) smooth subvarieties of Z. This result may be viewed as analogue of Brylinski's conjecture [6], cf. (5). Actually, the notion of polarizable Hodge Module is weaker than that of pure Hodge Module. But they coincide by Theorem 1.5 below, because a polarizable variation of Hodge structure is an admissible variation of Hodge structure by Schmid [37], and pure Hodge Modules are stable by intermediate direct images in the sense of [3] by Theorem 1.1 and the estimate of weight for direct images, cf. [27, 2.26].

Theorem 1.5. *Admissible variations of Hodge structures are mixed Hodge Modules,* cf. [27, 3.27].

Since the condition for mixed Hodge Modules is Zariski local, we can construct mixed Hodge Modules by induction on dimension, using Theorem 1.5 and the following:

Proposition 1.6. *Let g be a function on an algebraic variety X. Put $Y = g^{-1}(0)$, $U = X \setminus Y$. Let $\mathrm{MHM}(U,Y)_{\mathrm{ex}}$ be the category whose objects are pairs of mixed Hodge Modules $\mathscr{M}'$ on U and $\mathscr{M}''$ on Y endowed with morphisms $u : \psi_{g,1}\mathscr{M}' \to \mathscr{M}''$, $v : \mathscr{M}'' \to \psi_{g,1}M'(-1)$ such that $vu = N$, where $\psi_{g,1}$ is the unipotent monodromy part of ψ_g, and N is the logarithm of the unipotent part of the monodromy, tensored by $(2\pi i)^{-1}\mathbb{Z}$. Then we have an equivalence of categories*

$$\mathrm{MHM}(X) \simeq \mathrm{MHM}(U,Y)_{\mathrm{ex}}, \quad \text{cf. [27, 2.28]}.$$

We have also a generalization of the Kodaira vanishing:

Theorem 1.7. *Let Z be a projective variety with an ample line bundle L, and $\mathscr{M} \in \mathrm{MHM}(Z)$. Let X be a smooth variety containing Z so that $\mathscr{M}$ is represented by $(M,F,K,W) \in \mathrm{MHM}(X)$. Then $\mathrm{Gr}_p^F \mathrm{DR}\,\mathscr{M} := \mathrm{Gr}_p^F \mathrm{DR}\, M \in D(\mathscr{O}_Z)$ depends only on $\mathscr{M}$, and we have*

$$H^i(Z, \mathrm{Gr}_p^F \mathrm{DR}\,\mathscr{M} \otimes L) = 0 \quad \text{for} \quad i > 0,$$
$$H^i(Z, \mathrm{Gr}_p^F \mathrm{DR}\,\mathscr{M} \otimes L^{-1}) = 0 \quad \text{for} \quad i < 0.$$

Convention 1.8. In this paper we use left $\mathscr{D}$-Modules. In [26, 27], right Modules are used, because they are more convenient for the theory of direct image and duality. For the correspondence between left and right *filtered* $\mathscr{D}$-Modules, we use $\otimes_{\mathscr{O}_X}(\Omega_X^{\dim X}, F)$ with $\mathrm{Gr}_i^F \Omega_X^{\dim X} = 0$ for $i \neq -\dim X$, see [28, 34].

2. Idea of Construction

2.1 One-Dimensional Case [37]

Let S be an open disc with coordinate t, and $S^* = S \setminus \{0\}$. Let $\mathbb{H}$ be a polarized variation of Hodge structure of weight n on S^* with L the underlying local system defined over $\mathbb{Q}$. For simplicity, we assume the monodromy is *unipotent*. We denote by $\mathscr{O}_{S^*}(L)$ the corresponding locally free $\mathscr{O}_{S^*}$-Module with integrable connection, and F the Hodge filtration on it. Let $\mathscr{L}$ be Deligne's canonical extension [9] of $\mathscr{O}_{S^*}(L)$. By Schmid's nilpotent orbit theorem [37], the filtration F on $\mathscr{O}_{S^*}(L)$ is naturally extended to a filtration F of $\mathscr{L}$ whose graduation is free. Let $\mathscr{L}(0) = \mathscr{L}/t\mathscr{L}$ the fiber of the vector bundle $\mathscr{L}$ at the origin. Let $\pi : \tilde{S}^* \to S^*$ be a universal covering, and $L_\infty = \Gamma(\tilde{S}^*, \pi^* L)$ the multivalued sections of L. Then we have a canonical isomorphism

$$\mathscr{L}(0) \simeq \mathbb{C} \otimes L_\infty \,. \tag{1}$$

See [9]. The left hand side of (1) is equipped with the induced filtration F, called the limit Hodge filtration, and the right hand side with rational structure. The theorem of Schmid [37] (conjectured by Deligne) asserts that they form a mixed Hodge structure, called the limiting mixed Hodge structure, where the weight filtration is given by the monodromy filtration shifted by n. Here the monodromy filtration W is a finite increasing filtration which is uniquely characterized by the properties: $NW_j \subset W_{j-2}$ and $N^i : \mathrm{Gr}_i^W \xrightarrow{\sim} \mathrm{Gr}_{-i}^W$ $(i > 0)$, cf. [37, 11], and N is the logarithm of the unipotent part of the monodromy. The primitive part $P\,\mathrm{Gr}_{n+i}^W := \mathrm{Ker}\{N^{i+1} : \mathrm{Gr}_{n+i}^W \to \mathrm{Gr}_{n-i-2}^W\}$ is naturally polarized by the polarization induced by that of the variation of Hodge structure (combined with the isomorphism N^i). Since the above construction may be viewed as the definition of the nearby cycle functor ψ_t [10] for variations of Hodge structures, the result of Schmid can be interpreted as the stability theorem of polarized variations of Hodge structures under $P\,\mathrm{Gr}^W \psi_t$ in the one-dimensional case.

2.2 Geometric Case [38]

Assume there is a projective morphism $f : X \to S$ such that f is smooth over S^*, $f^{-1}(0)$ is a reduced divisor with normal crossings, and $L = R^n f_* \mathbb{Q}_X|_{S^*}$. Steenbrink [38] constructed a theory which gives a geometric proof of Schmid's theorem (see [39] for the nonreduced case). Although the notion of perverse sheaf and monodromy filtration on it was not yet known at that time, he constructed the weight filtration on the nearby cycle sheaf $\psi_f \mathbb{Q}_X$ [10], which represents the monodromy filtration in the abelian category of perverse sheaves (up to a shift of complex, cf. [3]). Moreover, he showed that the primitive part $P\,\mathrm{Gr}_i^W \psi_f \mathbb{Q}_X$ is a direct sum of constant sheaves supported on the intersections of $i+1$ irreducible components of the singular fiber (up to Tate twist and shift of complex), i.e. the stability of constant sheaves by $P\,\mathrm{Gr}_i^W \psi_f$ in the reduced normal crossing case. Using E_2-degeneration of the monodromy weight spectral sequence, this stability implies Schmid's theorem: the stability of the variations of Hodge structures by $P\,\mathrm{Gr}_{n+i}^W \psi_t$, because the direct image of the monodromy filtration on $\psi_f \mathbb{Q}_X$ induces the monodromy filtration on $\psi_t L$ (cf. also [26, 35] for further information).

The above construction can be extended to the vanishing cycle sheaf $\varphi_f \mathbb{Q}_X$ [10]. Using these, we can show the decomposition theorem for the direct image of filtered $\mathscr{D}$-Module $\int_f (\mathscr{O}_X, F)$ which implies that of $\mathbb{R} f_* \mathbb{Q}_X$, cf. [24, II]. The proof uses the V-filtration on the direct image, which was inspired by Varchenko's theory of asymptotic Hodge filtration [42], see also [24, I]. (Note that the definition of the direct image of filtered $\mathscr{D}$-Modules in [24] differs from the standard one by a shift of complex and filtration.)

2.3 Generalization

The first step to generalize the above proof of the decomposition theorem is the use of the filtration of Kashiwara [17] and Malgrange [22]. Let g be a holomorphic function on X, and $i_g : X \to X \times \mathbb{C}$ the embedding by graph. For a filtered holonomic $\mathscr{D}_X$-Module (M, F) (i.e. M is holonomic and F is a good filtration), we define $(\tilde{M}, F) = \int_{i_g} (M, F)$ (i.e. $\tilde{M} = M[\partial_t]$, cf. [28, (1.3)] for F). By [15], $\tilde{M}$ has the filtration of Kashiwara and Malgrange, denoted by V. We assume M regular and quasi-unipotent (i.e. V is indexed by $\mathbb{Q}$, and $t\partial_t - \alpha$ is nilpotent on Gr_V^α locally on X) and

$$\begin{aligned} t &: F_p V^\alpha \tilde{M} \xrightarrow{\sim} F_p V^{\alpha+1} \tilde{M} \quad \text{for} \quad \alpha > -1, \\ \partial_t &: F_p \mathrm{Gr}_V^\alpha \tilde{M} \xrightarrow{\sim} F_{p+1} \mathrm{Gr}_V^{\alpha-1} \tilde{M} \quad \text{for} \quad \alpha < 0. \end{aligned} \tag{2}$$

We define

$$\begin{aligned} \psi_g(M, F) &= \oplus_{-1<\alpha\le 0} \mathrm{Gr}_V^\alpha (\tilde{M}, F), \\ \varphi_g(M, F) &= \oplus_{-1<\alpha<0} \mathrm{Gr}_V^\alpha (\tilde{M}, F) \oplus \mathrm{Gr}_V^{-1} (\tilde{M}, F[-1]). \end{aligned} \tag{3}$$

(Same for $\psi_g M, \varphi_g M$, forgetting F.) Then we have canonical isomorphisms

$$\mathrm{DR}\, \psi_g M = \psi_g \,\mathrm{DR}\, M[-1], \quad \mathrm{DR}\, \varphi_g M = \varphi_g \,\mathrm{DR}\, M[-1], \tag{4}$$

generalizing (1). This was obtained by Malgrange [22] in the case $M = \mathscr{O}_X$, and the general regular holonomic case is due to Kashiwara [17]. In the case when the Hodge filtrations F of $\psi_g M, \varphi_g M$ are good and (2) is satisfied, these isomorphisms were also constructed in [26, 3.4.12] generalizing the construction of Malgrange (see also [33, 2.4] for further information), and used in the proof of the main theorems in an essential way.

For a smooth variety X, we define $\mathrm{MH}(X,n)$, the category of Hodge Modules of weight n, to be the largest full subcategory of the category of filtered holonomic $\mathscr{D}$-Modules with rational structure as in 1, whose objects admit decomposition by strict support and are stable by: (i) restrictions to Zariski-open subsets, (ii) strict support decomposition, (iii) $\mathrm{Gr}^W \psi_g$, $\mathrm{Gr}^W \varphi_g$ for any function g, where W is the monodromy filtration shifted by $n-1$ or n. Here (2) is always assumed when we take ψ_g, φ_g. We assume also that, if an object has point support, it is isomorphic to the direct image of Deligne's $\mathbb{Q}$-Hodge structure of weight n [8], cf. [26, 5.1.6]. The conditions are well-defined by induction on the dimension of the support. In fact, if $g^{-1}(0) \supset \operatorname{supp} M$, we have $\psi_g M = 0$, and $\varphi_g(M,F) = (M,F)$ is equivalent to $gF_pM \subset F_{p-1}M$ and follows from (2), cf. [26, 3.2.6]. So, in the stability by (iii), it is enough to consider the case when (M,F,K) has strict support and $g^{-1}(0) \not\supset \operatorname{supp} M$, using the stability by (ii). Moreover, in this case, it is enough to assume the stability by $\mathrm{Gr}^W \psi_g$ with condition (2) and the surjectivity of ∂_t in (2) for $\alpha = 0$, cf. [26, 5.1.15]. Note that the last conditions are equivalent to

$$F_p\tilde{M} = \sum_i \partial_t^i (V^{>-1}\tilde{M} \cap j_*j^{-1}F_{p-i}\tilde{M}) \tag{5}$$

where $j : X \times \mathbb{C}^* \to X \times \mathbb{C}$. This means that the Hodge filtration F is uniquely determined by its restriction to the complement of $g^{-1}(0)$ combined with the filtration V associated with g. The inverse functor of the equivalence of categories in Theorem 1.4 is given in this way, where we use also the regularity of the underlying $\mathscr{D}$-Modules [5, 18, 23]. (In the case of analytic $\mathscr{D}$-Modules, the regularity follows from the above inductive condition of Hodge Module, cf. [26, 5.1.18], but we have to assume it at infinity for algebraic $\mathscr{D}$-Modules. Note that the regularity [5, 18, 23] is used only for the uniqueness of the underlying $\mathscr{D}$-Modules of open direct images.) So Theorem 1.4 may be viewed as analogue of Brylinski conjecture [6], where Kashiwara-Kawai's order filtration is replaced by the filtration V. Here it is enough to take one g locally on X. But, for the stability by direct images, it is necessary to consider all g, because we have to deal with all f and the pull-backs of functions by f.

A polarization of a Hodge Module $\mathscr{M} = (M,F,K) \in \mathrm{MH}(X,n)$ is a duality $\mathbb{D}\mathscr{M} = \mathscr{M}(n)$, whose condition is defined inductively by the stability under $P\,\mathrm{Gr}^W \psi_g$, cf. [26, 5.2.10]. Here (n) is the Tate twist, which is defined essentially by the shift of filtration by n, cf. [27, (2.17.7)], and the dual $\mathbb{D}\mathscr{M} = (\mathbb{D}(M,F), \mathbb{D}K)$ is defined using the theory of dual of filtered induced $\mathscr{D}$-Modules and filtered differential complexes (e.g. its compatibility with de Rham functor), cf. [26, 2.4-5]. Note that $\mathbb{D}(M,F)$ is well-defined (e.g. $\mathbb{D}^2 = \mathrm{id}$ holds), because $\mathrm{Gr}^F M$ is Cohen-Macaulay over $\mathrm{Gr}^F \mathscr{D}_X$, cf. [26, 5.1.13], so that the filtration of the dual complex $\mathbb{D}(M,F)$ is strict and $\mathbb{D}(M,F)$ is filtered quasi-isomorphic to a filtered $\mathscr{D}_X$-Module. The category $\mathrm{MH}(X,n)^p$ is defined to be the category of polarizable Hodge Modules.

The merit of the inductive definition using nearby and vanishing cycles is that we can show by induction on dimension the stability of polarizable Hodge Modules by the cohomological direct image under a projective morphism and also the relative hard Lefschetz property. In fact, generalizing the proof of the decomposition theorem in 2.2, we can prove inductively the stability of the stability condition in the definition of polarizable Hodge Modules by the direct image, where the direct image of filtered $\mathscr{D}$-Modules is defined using a canonical resolution by filtered induced $\mathscr{D}$-Modules [26] (this definition coincides with that of Brylinski and Laumon using a factorization of a morphism f). A key point of the proof is the strictness of the Hodge filtration F on the direct image, which allows us to deal with filtered Modules instead of filtered complexes. When the image of the morphism is a point, we use Lefschetz pencil, and the assertion is reduced to the case of the projection of $\mathbb{P}^1$ to a point. Then we can apply Zucker's Hodge theory [44]. For the induced polarization on the direct image, we use also the duality of filtered $\mathscr{D}$-Modules for a proper morphism and its compatibility with Verdier duality [43], see [26, 2.5]. It should be noted that the definition of Hodge Module is obtained by axiomatizing some arguments which are repeatedly used in the proof of the stability by projective direct images. Combining with 1.4, we get a canonical pure Hodge structure on the intersection cohomology of a polarizable variation of Hodge structure. This Hodge structure coincides with that of Cattani-Kaplan-Schmid [7] and Kashiwara-Kawai [19, 20] where the singularity of the variation is a divisor with normal crossings.

2.4 Mixed Case

Once the pure objects are constructed, it is not difficult to define the mixed objects. We denote by $\mathrm{MHW}(X)^p$ the category of (M, F, K, W) as in 1 such that $\mathrm{Gr}_n^W(M, F, K) \in \mathrm{MH}(X, n)^p$. We can show by induction on dimension that $\mathrm{MHW}(X)^p$ is an abelian category such that every morphism is strictly compatible with the filtrations F and W in a strong sense, cf. [26, 5.1.14].

By definition, an object of $\mathrm{MHW}(X)^p$ is obtained by successive extension of pure Hodge Modules. Here arbitrary extensions are not allowed for mixed Hodge Modules, and we control the extensions by using mainly the stability by nearby cycle and vanishing cycle functors. The imposed conditions are natural generalizations of those of Steenbrink-Zucker [41]: the existence of the relative monodromy filtration and some condition on the filtrations F and W (i.e. the compatibility [26, 1.1.13] of the three filtrations F, W and V on $\tilde{M}$), cf. [27, 2.2-3]. We define the full subcategory of mixed Hodge Modules $\mathrm{MHM}(X)$ in $\mathrm{MHW}(X)^p$ using the stability by nearby cycle and vanishing cycle functors, smooth pull-backs and the existence of (and the stability by) the direct images by open embeddings whose complements are locally principal divisors, cf. [27, 4.2]. For singular X, $\mathrm{MHM}(X)$ is defined by using locally defined closed embeddings of X into smooth varieties. We can show the stability of the mixed Hodge Modules by the cohomological direct image under a projective morphism, generalizing the arguments of Steenbrink-Zucker [41] in the one-dimensional base space case. Using this, we can construct the functors in Theorem 1.1, where we use also Theorem 1.5 and Kashiwara's result [16] for the stability by external products $\boxtimes$, cf. [27]. Combining with 1.5, we get a canonical mixed Hodge structure on the cohomology of an admissible variation of mixed Hodge structure. We have also the decomposition theorem for the direct image of a pure Hodge Module by a proper morphism, using the semisimplicity of pure Hodge Modules, cf. [27, 4.5]. (See [30] for the last two functors in 1.1.)

3. Applications

3.1 Kollár's Conjecture [21, 32]

Let X be an irreducible algebraic variety, and $\mathbb{H}$ a polarizable variation of Hodge structure of weight $n - \dim X$ on a smooth open subvariety U of X. By Theorem 1.4, there exists uniquely a polarizable Hodge Module $\mathcal{M}$ on X extending $\mathbb{H}$. If X is a closed subvariety of smooth X', $\mathcal{M}$ is represented by $(M, F, K) \in \mathrm{MH}_X(X', n))^p$. Let $\mathcal{L}$ be the underlying $\mathcal{O}_U$-Module of $\mathbb{H}$ with Hodge filtration F. Put $q(\mathbb{H}) = \max\{p : F^p\mathcal{L} \neq 0\}$ and

$$S_X(\mathbb{H}) = \Omega_{X'}^{\dim X'} \otimes_{\mathcal{O}_{X'}} F_pM \quad \text{for} \quad p = \min\{p : F_pM \neq 0\}.$$

Then, $S_X(\mathbb{H})$ is an extension of $\Omega_U^{\dim U} \otimes_{\mathcal{O}_U} F^{q(\mathbb{H})}\mathcal{L}$, which is expressed using the filtration V, cf. (5). (It is more natural to use right $\mathcal{D}$-Modules, cf. 1.8.)

Let $f : X \to Y$ be a proper surjective morphism of irreducible algebraic varieties with relative dimension d. Then for a generic point $y \in Y$, the intersection cohomology $IH^i(X_y, \mathbb{H}|U_y)$ has a canonical pure Hodge structure, cf. 2.2, where $X_y = f^{-1}(y)$ and U_y is the smooth points of $U \cap X_y$. They form a variation of Hodge structure on a smooth open subvariety of Y, which we denote by $\mathbb{H}^i$. As a corollary of the decomposition theorem for the direct image of pure Hodge Modules, we get

$$\mathbf{R}f_*S_X(\mathbb{H}) \simeq \bigoplus_{q(\mathbb{H}^i)=q(\mathbb{H})+d} S_Y(\mathbb{H}^i)[-i]. \tag{6}$$

3.2 Illusie's Conjecture [35]

Let $f : X \to S$ be a projective morphism of a complex manifold to an open disc such that $X_0 := f^{-1}(0)$ is a divisor with normal crossings. Consider the (second) spectral sequence

$$E_2^{p,q} = H^p(X_0, \mathcal{H}^q\psi_f\mathbb{Q}_X) \Rightarrow H^{p+q}(X_0, \psi_f\mathbb{Q}_X)(\simeq H^{p+q}(X_t, \mathbb{Q})). \tag{7}$$

Here $X_t := f^{-1}(t)$ for $t \neq 0$ sufficiently near 0, and the last isomorphism depends on the lifting of t to the universal covering of $S \setminus \{0\}$. The spectral sequence degenerates at E_3, and the induced filtration coincides with the kernel filtration [41, 45]. This can be generalized to the theory of kernel spectral sequence, cf. [35].

3.3 Steenbrink's Conjecture [40, 29]

Let $f : (X, x) \to (\mathbb{C}, 0)$ be a germ of holomorphic function on a complex manifold. We define the *spectrum* $\mathrm{Sp}(f, x)$ using the Milnor monodromy and the mixed Hodge structure on the vanishing cohomology [40], (cf. also [39, 42]). Here we express the spectrum as a fractional Laurent polynomial (i.e., an element of $\mathbb{C}[t^{1/n}, t^{-1/n}]$ for some integer n), and multiply t on that of [40] to simplify the multiplication of spectra. Assume $\dim \mathrm{Sing} f = 1$, and let $\{Z_k\}$ be the irreducible components of $\mathrm{Sing} f$. Let $g : (X, x) \to (\mathbb{C}, 0)$ be a linear form (i.e., $dg(x) \neq 0$) such that $\{f = g = 0\}$ has isolated singularity. Let m_k be the mapping degree of the restriction of g to Z_k. Then, for $r \in \mathbb{N}$ sufficiently large, we have

$$\mathrm{Sp}(f+g^r,x)-\mathrm{Sp}(f,x)=\sum_{k,j} t^{\alpha_{k,j}+\beta_{k,j}/m_k r}(t-1)/(t^{1/m_k r}-1)\,. \tag{8}$$

Here $\alpha_{k,j}$ are the exponents of the spectrum of the restriction of f to $g^{-1}(t)(t\neq 0)$ near Z_k, and $\beta_{k,j}\in(0,1]$ are the logarithm (divided by $-2\pi i$) of the eigenvalues of the monodromy along Z_k. The range of r is determined by the discriminant of (f,g).

3.4 Intersection Cohomology of Link [12]

Let X be a complex algebraic variety, and Z a closed subvariety of X. Then the *link* of Z in X is defined to be a level set of a suitable real analytic function on X which vanishes exactly on Z. We can define a canonical mixed Hodge structure on the intersection cohomology of the link, and get estimates of weight, e.g. $IH^i_c(L)$ has weights $\le i$ for $i<\dim X-\dim Z$. Using these, we can show, for example, that a (real) torus cannot be a link of a subvariety of codimension >1.

3.5 Thom-Sebastiani Type Theorems

Let f and g be holomorphic functions on X and Y respectively. For $x\in X$ and $y\in Y$, put $z=(x,y)\in X\times Y$. Then

$$\mathrm{Sp}(f+g,z)=\mathrm{Sp}(f,x)\,\mathrm{Sp}(g,y)\,. \tag{9}$$

(See [36, 42] for the isolated singularity case.) This follows from the Thom-Sebastiani type theorems for regular holonomic filtered $\mathscr{D}$-Modules (with Steenbrink) and for mixed Hodge Modules (with Deligne).

3.6 Algebraic Cycles [31]

If the Hodge conjecture is true, the mixed motives over $\mathbb{C}$ and with $\mathbb{Q}$-coefficients, whose existence is still conjectural [1], might be quite close to the mixed Hodge structures of *geometric origin* (using mixed Hodge Modules instead of mixed perverse sheaves in [3]). In fact, let $\mathbb{Q}^H$ be the trivial Hodge structure of rank 1 and type $(0,0)$, and $\mathbb{Q}^H_X=a^*_X\mathbb{Q}^H$, $\mathbf{R}\Gamma(X,\mathbb{Q}^H)=(a_X)_*\mathbb{Q}^H_X$ with $a_X:X\to\mathrm{pt}$. Then we can define a cycle class map

$$\begin{aligned}\mathrm{cl}^{\mathrm{MH}}:\mathrm{CH}^p(X)_{\mathbb{Q}}&\to \mathrm{Ext}^{2p}_{D^b\mathrm{MHM}(X)^{\mathrm{go}}}(\mathbb{Q}^H_X,\mathbb{Q}^H_X(p)))\\ &=\mathrm{Ext}^{2p}_{\mathrm{MHS}(\mathbb{Q})^{\mathrm{go}}}(\mathbb{Q}^H,\mathbf{R}\Gamma(X,\mathbb{Q}^H)(p))\end{aligned} \tag{10}$$

where $\mathrm{CH}^p(X)_{\mathbb{Q}}$ is the Chow group tensored by $\mathbb{Q}$, $\mathrm{MHM}(X)^{\mathrm{go}}$ is the category of mixed Hodge Modules of geometric origin, and $\mathrm{MHS}(\mathbb{Q})^{\mathrm{go}}=\mathrm{MHM}(\mathrm{pt})^{\mathrm{go}}$. By the canonical filtration on $\mathbf{R}\Gamma(X,\mathbb{Q}^H)$, we get a filtration L on the target of the morphism such that

$$\mathrm{Gr}^i_L=\mathrm{Ext}^i_{\mathrm{MHS}(\mathbb{Q})^{\mathrm{go}}}(\mathbb{Q}^H,H^{2p-i}(X,\mathbb{Q})(p))\,, \tag{11}$$

and it induces a filtration L on $\mathrm{CH}^p(X)_{\mathbb{Q}}$ via $\mathrm{cl}^{\mathrm{MH}}$. The cycle map is surjective if and only if the Hodge conjecture is true for any variety. In this case $\mathrm{Gr}^i_L\,\mathrm{cl}^{\mathrm{MH}}$

are also surjective, and we get a filtration L on $\mathrm{CH}^p(X)_{\mathbb{Q}}$ such that (11) holds. Compare [4]. Moreover $\mathrm{Gr}_L^i \mathrm{CH}^p(X)_{\mathbb{Q}} = 0$ for $i > p$, and this filtration is separated if and only if $\mathrm{cl}^{\mathrm{MH}}$ is injective. The injectivity of $\mathrm{cl}^{\mathrm{MH}}$ seems to be related with the surjectivity of the cycle map of Bloch's higher Chow group $\mathrm{CH}^p(X,1)_{\mathbb{Q}}$ to $\mathrm{Ext}^{2p-1}_{D^b\mathrm{MHM}(X)^{\mathrm{go}}}(\mathbb{Q}_X^H, \mathbb{Q}_X^H(p))$.

References

1. Beilinson, A.: Height pairing between algebraic cycles. (Lecture Notes in Mathematics, vol. 1289.) Springer, Berlin Heidelberg New York 1987, pp. 1–26
2. Beilinson, A.: On the derived category of perverse sheaves. Ibid, pp. 27–41
3. Beilinson, A., Bernstein, J., Deligne, P.: Faisceaux pervers. Astérisque **100** (1982)
4. Bloch, S.: Lectures on algebraic cycles. Duke Univ. Math. Series 4, Durham 1980
5. Borel, A.: Algebraic $\mathscr{D}$-Modules. Academic Press, Boston 1987
6. Brylinski, J.-L.: Modules holonomes à singularités régulières et filtrations de Hodge II. Astérisque **101-102** (1983) 75-117
7. Cattani, E., Kaplan, A., Schmid, W.: L^2 and intersection cohomologies for a polarizable variation of Hodge structure. Inv. math. **87** (1987) 217–252
8. Deligne, P.: Théorie de Hodge I, II, III. Actes Congrès Intern. Math. (1970) 425–430; Publ. Math. IHES **40** (1971) 5–58; ibid. **44** (1974) 5–77
9. Deligne, P.: Equation différentielle à points singuliers réguliers. (Lecture Notes in Mathematics, vol. 163.) Springer, Berlin Heidelberg New York 1970
10. Deligne, P.: Le formalisme des cycles évanescents. In: SGA7 XIII and XIV. (Lecture Notes in Mathematics, vol. 340.) Springer, Berlin Heidelberg New York 1973, pp. 82–115 and 116–164
11. Deligne, P.: Conjecture de Weil II. Publ. Math. IHES **52** (1980) 137–252
12. Durfee, A., Saito, M.: Mixed Hodge structure on the intersection cohomology of links. Comp. Math. **76** (1990) 49–67
13. Griffiths, P.: Periods of integrals on algebraic manifolds I, II, III. Amer. J. Math. **90** (1968) 568–626; ibid. 805–865; Publ. Math. IHES **38** (1970) 125–180
14. Kashiwara, M.: On the maximally overdetermined system of linear differential equations I. Publ. RIMS, Kyoto Univ. **10** (1975) 563–579
15. Kashiwara, M.: On the holonomic systems of differential equations II. Inv. math. **49** (1978) 121–135
16. Kashiwara, M.: A study of variation of mixed Hodge structure. Publ. RIMS, Kyoto Univ. **22** (1986) 991–1024
17. Kashiwara, M.: Vanishing cycle sheaves and holonomic systems of differential equations. (Lecture Notes in Mathematics, vol. 1016.) Springer, Berlin Heidelberg New York 1983, pp. 136–142
18. Kashiwara, M., Kawai, T.: On the holonomic system of microdifferential equations III. Publ. RIMS, Kyoto Univ. **17** (1981) 813–979
19. Kashiwara, M., Kawai, T.: The Poincaré lemma for variations of polarized Hodge structures. Publ. RIMS, Kyoto Univ. **23** (1987) 345–407
20. Kashiwara, M., Kawai, T.: Hodge structure and holonomic systems. Proc. Japan Acad. **62** (1986) 1–4
21. Kollár, J.: Higher direct images of dualizing sheaves I, II. Ann. Math. **123** (1986) 11–42; **124** (1986) 171–202
22. Malgrange, B.: Polynôme de Bernstein-Sato et cohomologie évanescente, Astérisque **101-102** (1983) 243–267
23. Mebkhout, Z.: Une autre équivalence de catégories. Comp. Math. **51** (1984) 63–88

24. Saito, M.: Hodge filtration on Gauss-Manin systems I, II. J. Fac. Sci. Univ. Tokyo, Sect. I A **30** (1984) 489-498; Proc. Japan Acad., Ser. A **59** (1983) 37–40
25. Saito, M.: Hodge structure via filtered $\mathscr{D}$-Modules. Astérisque **130** (1985) 342-351
26. Saito, M.: Modules de Hodge polarisables. Publ. RIMS, Kyoto Univ. **24** (1988) 849–995
27. Saito, M.: Mixed Hodge modules. Publ. RIMS, Kyoto Univ. **26** (1990) 221–333
28. Saito, M.: Decomposition theorem for proper Kähler morphisms. Tohoku Math. J. **42** (1990) 127–148
29. Saito, M.: Vanishing cycles and mixed Hodge modules. Preprint IHES 1988
30. Saito, M.: Extension of mixed Hodge modules. Comp. Math. **74** (1990) 209–234
31. Saito, M.: Hodge conjecture and mixed motives I, II. Preprint RIMS-691 and 692, 1990
32. Saito, M.: On Kollár's conjecture. Preprint RIMS-693, 1990
33. Saito, M.: Duality for vanishing cycle functors. Publ. RIMS, Kyoto Univ. **25** (1989) 889–921
34. Saito, M.: Introduction to mixed Hodge Modules. Astérisque **179-180** (1989) 145–162
35. Saito, M., Zucker, S.: The kernel spectral sequence of vanishing cycles. Duke Math. J. **61** (1990) 329–339
36. Scherk, J., Steenbrink, J.: On the mixed Hodge structure of the Milnor fiber. Math. Ann. **271** (1985) 631–665
37. Schmid, W.: Variation of Hodge structure: the singularity of period mapping. Inv. math. **22** (1973) 211–319
38. Steenbrink, J.: Limits of Hodge structures. Inv. math. **31** (1976) 229-257
39. Steenbrink, J.: Mixed Hodge structure on the vanishing cohomology. In: Real and complex singularities, Oslo 1976. Sijthoff-Noordhoff, Alphen a/d Rijn 1977, pp. 525–563
40. Steenbrink, J.: The spectrum of hypersurface singularity. Astérisque **179-180** (1989) 163-184
41. Steenbrink, J., Zucker, S.: Variation of mixed Hodge structure I. Inv. math. **80** (1985) 489–542
42. Varchenko, A.: Asymptotic Hodge structure in the vanishing cohomology. Math. USSR Izv. **18** (1982) 469-512
43. Verdier, J.-L.: Dualité dans les espaces localement compacts. Séminaire Bourbaki no. 300 (1965/66)
44. Zucker, S.: Hodge theory with degenerating coefficients, L_2-cohomology in the Poincaré metric. Ann. Math. **109** (1979) 415–476
45. Zucker, S.: Variation of mixed Hodge structure II. Inv. math. **80** (1985) 543–565

L_2-Cohomology of Algebraic Varieties

*Leslie Saper**

Department of Mathematics, Duke University, Durham, NC 27706, USA

In this article I shall sketch the theory of L_2-cohomology and its applications in algebraic geometry. The basic principle is that for appropriate Riemannian metrics on a smooth, Zariski open subset of a complex projective variety X, the L_2-cohomology represents invariants of the full variety X, for example, the intersection cohomology and its Hodge structure. The primary obstacle to proving such results is to relate L_2-growth conditions to a suitable theory of "weights" for which one can establish a semi-purity theorem. Two examples will be presented whose proofs, although rather different, exhibit this pattern: the case of Kähler varieties with isolated singularities [37–39], and the case of locally symmetric varieties (Zucker's conjecture) [40, 41].

1. L_2-Cohomology

1.1 Definition and Basic Properties

References are [10] and [45]. Let M be a Riemannian manifold (assumed oriented for simplicity) and $\mathbb{E}$ a flat complex vector bundle with a (not necessarily flat) Hermitian metric. Let $L_2^{\bullet}(M;\mathbb{E})$ denote the Hilbert space of measurable $\mathbb{E}$-valued forms ϕ which satisfy the L_2-growth condition

$$\int_M |\phi|^2 dV < \infty .$$

Note that the dependence of this condition on the metric of M arises in two opposing fashions: in the metric on the exterior powers of the cotangent bundle, and in the volume form.

Definition 1. The *L_2-cohomology* $H_{(2)}^{\bullet}(M;\mathbb{E})$ is the cohomology of the complex

$$\mathrm{Dom}(d) = \{\, \phi \in L_2^{\bullet}(M;\mathbb{E}) \mid d\phi \in L_2^{\bullet+1}(M,\mathbb{E}) \,\} .$$

Here $d\phi$ is taken in the sense of currents.

* The author is an Alfred P. Sloan Research Fellow and is supported in part by NSF Grants DMS-8705849 and DMS-8957216.

Proceedings of the International Congress of Mathematicians, Kyoto, Japan, 1990

Similarly we can define the L_2-$\bar{\partial}$-cohomology $H^{p,\bullet}_{(2),\bar{\partial}}(M;E)$ if M is a Hermitian manifold and E is a Hermitian holomorphic vector bundle. By a regularization result of Cheeger [10], we obtain the same cohomology in either case if we require ϕ above to be smooth.

Quasi-isometry Invariance. Although in general the L_2-cohomology depends on the metrics involved, it is an invariant of their quasi-isometry class (two metrics g_1, g_2 are *quasi-isometric* if $Cg_2 \le g_1 \le C'g_2$ for constants C, $C' > 0$).

Hodge Theory. Clearly we have

$$\begin{aligned} H^{\bullet}_{(2)}(M;\mathbb{E}) &= \operatorname{Ker}(d)\,/\overline{\operatorname{Range}(d)} \bigoplus \overline{\operatorname{Range}(d)}/\operatorname{Range}(d) \\ &= \operatorname{Ker}(d) \cap \operatorname{Ker}(d^*) \bigoplus \overline{\operatorname{Range}(d)}/\operatorname{Range}(d)\,, \end{aligned} \tag{1}$$

where d^* is the Hilbert space adjoint. In general $\operatorname{Dom}(d^*)$ is smaller than $\operatorname{Dom}(\delta) = \{\,\phi \in L_2^{\bullet}(M;\mathbb{E}) \mid \delta\phi \in L_2^{\bullet-1}(M;\mathbb{E})\,\}$, where δ is the formal adjoint to d; by a result of Gaffney [15], they are equal if M is complete.

Since the second term of (1) is either 0 or infinite dimensional, we have the following result:

Theorem 2. *Say* $\operatorname{Range}(d)$ *is closed (e.g., if* $\dim H^{\bullet}_{(2)}(M;\mathbb{E}) < \infty$*). Then*

1) Hodge Theorem:

$$H^{\bullet}_{(2)}(M;\mathbb{E}) = \operatorname{Ker}(d) \cap \operatorname{Ker}(d^*) = \operatorname{Ker}(\Delta)\,,$$

where $\Delta = d^*d + dd^*$ *has domain in the sense of unbounded operators.*

2) Poincaré Duality: If furthermore, $d^* = \delta$ *(e.g., if M is complete), then the Hodge star operator induces conjugate linear isomorphisms*

$$H^{i}_{(2)}(M;\mathbb{E}) \xrightarrow{\sim} H^{m-i}_{(2)}(M;\mathbb{E}^*)\,,$$

where $m = \dim M$.

Kähler Package. Assume now that M is a complete Kähler manifold and $\mathbb{E} = \mathbb{C}$. In this case the formal identity $\Delta = 2\Box$ between the d- and $\bar{\partial}$-Laplacians holds in the sense of unbounded operators [47]. Consequently we have the

Theorem 3. *Let M be complete Kähler of dimension n and say* $\operatorname{Range}(d)$ *is closed. Then*

1) Hodge Decomposition:

$$H^{i}_{(2)}(M;\mathbb{C}) = \bigoplus_{p+q=i} H^{p,q}_{(2),\bar{\partial}}(M;\mathbb{C})\,.$$

2) Hard Lefschetz: If L denotes the operation of wedging with the Kähler class, there are isomorphisms

$$L^i : H^{n-i}_{(2)}(M;\mathbb{C}) \xrightarrow{\sim} H^{n+i}_{(2)}(M;\mathbb{C})\,.$$

Analogous statements hold when $\mathbb{E}$ underlies a polarizable variation of Hodge structure (see [44]).

1.2 Example: Conical Metrics

Let X be a closed triangulated pseudomanifold of dimension n, that is, a compact simplicial complex which equals the union of its n-simplices and in which every $(n-1)$-simplex is a boundary face of exactly two n-simplices. For example, it is well-known that any compact complex analytic variety admits such a structure. If $\Sigma \subset X$ is the codimension 2 skeleton, then $X^{\text{reg}} = X \setminus \Sigma$ is naturally a smooth manifold, dense in X.

Definition 4. A Riemannian metric on X^{reg} is called *conical* if it is quasi-isometric to a piecewise flat metric with respect to the triangulation of X.

Although the metric on X^{reg} is incomplete, Cheeger [10] has proved the

Theorem 5. *If X is an admissible triangulated pseudomanifold, the L_2-cohomology $H^{\bullet}_{(2)}(X^{\text{reg}};\mathbb{C})$ of a conical metric is finite dimensional and invariant under subdivision. It satisfies the Hodge theorem and (if X^{reg} is oriented) Poincaré duality.*

Here admissibility is an inductively defined condition on the links which is always satisfied if X admits a stratification with even codimensional strata (e.g., a complex analytic variety).

The proof of Theorem 5 is based on the following local calculation [10]:

Proposition 6. *Let N be a Riemannian manifold of dimension $m-1$; if m is even assume* Range(d) *is closed in $L_2^{m/2-1}(N)$. Let $c(N)$ be the closed cone on N and let its regular part $c(N)^{\text{reg}}$ have the conical metric $dx^2 + x^2 ds_N^2$. Then*

$$H^i_{(2)}(c(N)^{\text{reg}};\mathbb{C}) = \begin{cases} H^i_{(2)}(N;\mathbb{C}) & i < m/2\,, \\ 0 & i \geq m/2\,. \end{cases} \tag{2}$$

The truncation at middle degree arises from the interplay between the volume form and the metrics on the bundles of forms; see Sect. 4.1 for a further discussion in the case of a complete Kähler metric.

2. Intersection Cohomology

Independently of the preceding developments, and also partly motivated by the desire to find a homology theory on singular spaces satisfying Poincaré duality, Goresky and MacPherson had also encountered a local calculation such as (2), but in the topological context of their intersection homology. References are [5, 16, 17]; see also the historical survey [20].

2.1 Definition and Basic Properties

Let $X = X^0 \supset \Sigma = X^1 \supset X^2 \supset \cdots \supset X^n$ be a stratified topological pseudomanifold. In other words, the strata $S^k = X^k \setminus X^{k+1}$ are manifolds of even codimension $2k$ and S^0 is dense. (The restriction to even codimensional strata is not essential.) Furthermore, a point in S^k has a cofinal system of local neighborhoods of the form $B \times c(L)$, where B is a ball in S^k and L is a stratified topological pseudomanifold of dimension $2k-1$.

Let $\mathbb{E}$ be a flat complex vector bundle on $X^{\text{reg}} = X \setminus \Sigma$. The *(middle perversity) intersection cohomology* $IH^{\bullet}(X;\mathbb{E})$ was originally defined by imposing restrictions on the dimension of intersection of a geometric chain with the singular strata; it may also be defined as the hypercohomology $\mathbf{H}^{\bullet}(X;\mathcal{IC}^{\bullet}(\mathbb{E}))$, where

$$\mathcal{IC}^{\bullet}(\mathbb{E}) = \tau_{<n} Ri_{n*} \ldots \tau_{<1} Ri_{1*}\mathbb{E} \tag{3}$$

is Deligne's sheaf. Here $i_k : X \setminus X^k \hookrightarrow X \setminus X^{k+1}$ is the inclusion and $\tau_{<k}$ truncates sheaf cohomology in degrees $\geq k$.

Intersection cohomology is independent of the choice of stratification; when X is smooth and $\mathbb{E}$ is everywhere defined, it reduces to ordinary cohomology. True to its name, there is an intersection pairing effecting Poincaré duality. And when X is a projective algebraic variety and $\mathbb{E}$ underlies a polarizable variation of Hodge structure, Saito [35, 36] has given $IH^{\bullet}(X;\mathbb{E})$ a natural Hodge structure.

It is clear from (3) that intersection cohomology admits a local characterization. Namely, for $x \in S^k$, let $U = B \times c(L)$ be a local neighborhood of x as above. Then the local calculation

$$IH^i(U;\mathbb{E}) \cong IH^i(c(L);\mathbb{E}) \cong \begin{cases} IH^i(L;\mathbb{E}) & i < k, \\ 0 & i \geq k, \end{cases} \tag{4}$$

essentially characterizes intersection cohomology; here the isomorphism for $i < k$ is given by the natural "attaching" map.

2.2 Relations with L_2-Cohomology

In view of (2) and (4), Cheeger [10] proved:

Theorem 8. *Let X^{reg} have a conical metric. Then $H^{\bullet}_{(2)}(X^{\text{reg}};\mathbb{C}) \cong IH^{\bullet}(X;\mathbb{C})$.*

One is thus led to try extending the classical Hodge-de Rham theory to singular spaces by using metrics on X^{reg} which somehow reflect the singular compactification. However, for complex projective varieties, it would be more fruitful to have a Kähler metric. It is rare that there exist Kähler conical metrics, but there is a natural quasi-isometry class of Kähler metrics – namely, those induced from the Fubini-Study metric by a projective embedding. Cheeger, Goresky, and MacPherson have conjectured [12]:

Conjecture 9. *For the metric induced on X^{reg} by the Fubini-Study metric of any projective embedding, $H^{\bullet}_{(2)}(X^{\text{reg}};\mathbb{C}) \cong IH^{\bullet}(X;\mathbb{C})$. Furthermore, this induces a natural Hodge decomposition.*

Cheeger's work [10, 11] covers the case of varieties with analytically conical singularities, while Hsiang and Pati [18] and Nagase [26] have established the first part of the conjecture for surfaces with isolated singularities. Recently Ohsawa [31] has proven the first part for varieties with isolated singularities, using a result of Saper [39] (see Theorem 11 below). Since the metric is incomplete, the second part of the conjecture does not follow from Theorem 3. Nagase [25] has obtained partial results for surfaces; Pardon [32] has shown that a Hodge decomposition *cannot* in general be given by the L_2-$\bar{\partial}$-cohomology groups without conceivably imposing boundary conditions.

3. Complete Kähler Metrics

Let X be a projective algebraic variety of complex dimension n. In view of Theorem 3, we turn to *complete* Kähler metrics on X^{reg} in order to construct Hodge decompositions on intersection cohomology.

3.1 Degenerating Coefficients

Assume X is smooth and let $\mathbb{E}$ underlie a polarizable variation of Hodge structure on $X^{\text{reg}} = X \setminus D$, where D is a divisor with normal crossings. Endow X^{reg} with a complete Kähler metric which has Poincaré singularities normal to each component of D; in local coordinates, if $D = (z_1 \cdots z_k)$, the metric is quasi-isometric to

$$\sum_{i=1}^{k} \frac{dz_i d\bar{z}_i}{|z_i|^2(\log|z_i|^2)^2} + \sum_{i=k+1}^{n} dz_i d\bar{z}_i\,. \tag{5}$$

The isomorphism $H^{\bullet}_{(2)}(X^{\text{reg}};\mathbb{E}) \cong IH^{\bullet}(X;\mathbb{E})$ was proved by Zucker [44] when $n = 1$; the general case was proved independently by Cattani, Kaplan, and Schmid [9] and Kashiwara and Kawai [19]. Note that when $\mathbb{E}$ is trivial, the theorem is in fact quite simple; the main point here is the degeneration of coefficients.

3.2 Locally Symmetric Varieties

The prime examples of complete Kähler metrics naturally associated to *singular* varieties are those on locally symmetric varieties.

Let D be a bounded symmetric domain endowed with the Bergman metric, or equivalently, a Hermitian symmetric space of noncompact type G/K equipped with a group invariant metric. For Γ an arithmetically defined group of automorphisms of D, the quotient $\Gamma\backslash D$ is called a *locally symmetric variety*. It may have finite quotient singularities, but this is not important here; the main point is that it may be noncompact and that there is a natural, though singular, compactification $\Gamma\backslash D^*$ (due to Baily and Borel [1] and Satake [42]) as a normal projective variety. Briefly, $\Gamma\backslash D^*$ is obtained by adjoining arithmetic quotients $\Gamma_F\backslash F$ of rational boundary components F of D, which are themselves locally symmetric varieties of lower rank and dimension.

For example, when D is the complex n-ball, $\Gamma\backslash D^*$ is formed by adding singular cusp points. After resolving the singularities with a smooth exceptional divisor, the metric is quasi-isometric to (compare (5))

$$\frac{dz_1 d\bar{z}_1}{|z_1|^2(\log|z_1|^2)^2} + \frac{1}{|\log|z_1|^2|}\sum_{i=2}^{n} dz_i\, d\bar{z}_i\,. \tag{6}$$

The following theorem was conjectured by Zucker after verifying some cases with G having $\mathbb{Q}$-rank 1 [45]. It was established independently by Saper and Stern [40, 41], and by Looijenga [22]; Looijenga and Rapoport have also given a third proof [23]. Previously special cases were proven by Borel [4], Borel and Casselman [6] (see also [8]), and Zucker [45, 46]. An indication of Saper and Stern's proof will be given in Sect. 4.3.

Theorem 10. *Let E be a finite dimensional complex representation space of G with admissible metric, and let $\mathbb{E}$ be the corresponding metrized local system on $\Gamma\backslash D$. Then we have a natural isomorphism*

$$H^{\bullet}_{(2)}(\Gamma\backslash D;\mathbb{E}) \cong IH^{\bullet}(\Gamma\backslash D^*;\mathbb{E})\,.$$

3.3 Kähler Varieties

For a general Kähler variety X, one would like a complete Kähler metric on X^{reg} analogous to those on locally symmetric varieties. In particular, one would like a generalization of Theorem 10.

Note that one could always take a resolution $\tilde{X} \to X$ with exceptional divisor D having normal crossings, and consider a Poincaré metric on $\tilde{X}\backslash D$ as in Sect. 3.1. However this is not the metric we want, since its L_2-cohomology is strictly larger than $IH^{\bullet}(X;\mathbb{C})$, namely $H^{\bullet}(\tilde{X};\mathbb{C})$. Instead, Saper [37–39] constructed a metric which, locally on D, is quasi-isometric (modulo a perturbation) to the metrics on toroidal resolutions of locally symmetric models. Using this metric (which depends on $\tilde{X}$), he established the following theorem; a brief indication of the proof will be given in Sect. 4.2. The case of isolated analytically conical singularities was also proven later with a different metric by Ohsawa [30] (see also [28, 29]).

Theorem 11. *Let X be a Kähler variety with isolated singularities. There exists an infinite discrete set of complete Kähler metrics on X^{reg} for which we have*

$$H^{\bullet}_{(2)}(X^{\text{reg}};\mathbb{C}) \cong IH^{\bullet}(X;\mathbb{C}) \tag{7}$$

and

$$H^{0,\bullet}_{(2),\bar{\partial}}(X^{\text{reg}};\mathbb{C}) \cong H^{0,\bullet}_{\bar{\partial}}(\tilde{X};\mathbb{C})\,. \tag{8}$$

Here $\tilde{X}$ is any resolution of X.

A priori, the different metrics referred to in the theorem could induce different Hodge decompositions on $IH^{\bullet}(X;\mathbb{C})$ via (7); Zucker [47] has shown in fact that they all agree with Saito's Hodge structure. This is used in the proof of (8), which asserts that the L_2-$\bar{\partial}$-Betti numbers $h^{0,q}_{(2),\bar{\partial}}$ are birational invariants. The corresponding assertion for the L_2-$\bar{\partial}$-index $\sum_q(-1)^q h^{0,q}_{(2),\bar{\partial}}$ was conjectured in the contexts of Conjecture 9 and Theorem 10 by MacPherson [24]. Recently Pardon and Stern [33] have given a proof of MacPherson's conjecture (and (8)) for the metric induced from the Fubini-Study metric provided one uses L_2-$\bar{\partial}$-cohomology with Dirichlet boundary conditions.

4. Outline of Proofs

Let X be a projective variety of dimension n. To establish the isomorphism $H^{\bullet}_{(2)}(X^{\mathrm{reg}};\mathbb{E}) \cong IH^{\bullet}(X;\mathbb{E})$ for some metric on X^{reg}, the main point is to show that L_2-cohomology has the correct local calculation (see (4)). That is, for a neighborhood U of a point on a complex codimension k stratum with link L, we need

$$H^i_{(2)}(U^{\mathrm{reg}};\mathbb{E}) \cong H^i_{(2)}(c(L)^{\mathrm{reg}};\mathbb{E}) \cong \begin{cases} H^i_{(2)}(L^{\mathrm{reg}};\mathbb{E}) & i<k, \\ 0 & i \geq k. \end{cases} \tag{9}$$

In practise, proving just the vanishing condition is usually sufficient.

4.1 Heuristic Arguments

In order to develop a heuristic device for understanding why the vanishing result might hold, applicable to both Theorem 10 and Theorem 11, we will consider in this subsection the basic example of a space X with a single singular stratum of codimension k. For simplicity we will only treat constant coefficients, however this is not necessary.

Write $c(L)^{\mathrm{reg}} = \mathbb{R}^{\geq 0} \times L$ with r the coordinate on the first factor ($r \to \infty$ near the vertex) and π the projection on the second. Analogously to Borel's formula [3] for the metric on a cusp of a locally symmetric variety, consider the metric

$$dr^2 + \sum_{\gamma} e^{-\gamma r}\pi^* ds^2_{\gamma},$$

on $c(L)^{\mathrm{reg}}$, where $TL = \bigoplus_{\gamma} TL_{\gamma}$ is some decomposition into subbundles indexed by *weights* $\gamma \in \mathbb{R}^{\geq 0}$ and ds^2_{γ} is a metric on TL_{γ}. Then

$$dV = e^{-\delta r} dr\, dV_L \tag{10}$$

where $\delta = (1/2)\sum_{\gamma}(\dim TL_{\gamma})\gamma$ is the middle weight, while for a pure wedge product $\psi = \psi_1 \wedge \cdots \wedge \psi_i$ ($\psi_j \in \Gamma(TL^*_{\gamma_j})$) with weight $\mu = \sum_i \gamma_i$,

$$|\pi^*\psi|^2 = e^{\mu r}|\psi|^2_L. \tag{11}$$

By separation of variables, it is reasonable to assume that an L_2-cohomology class can be represented either as $\pi^*\psi$ or $f(r)dr \wedge \pi^*\psi$, where ψ is a closed form on L. From (10) and (11) we immediately see that $\pi^*\psi$ is L_2 if and only if $\mu < \delta$. On the other hand, if $f(r)dr \wedge \pi^*\psi$ is L_2, it may be expressed as a cocycle via $(\int_0^r f(t)dt)\pi^*\psi$ or $(-\int_r^\infty f(t)dt)\pi^*\psi$, which are L_2 precisely when $\mu < \delta$ or $\mu > \delta$ respectively [45]. This suggests that if the above notion of weight actually descends to a decomposition (or at least a filtration) of $H^i(L;\mathbb{C})$ and $H^i_{(2)}(c(L)^{\text{reg}};\mathbb{C})$, then

$$H^i_{(2)}(c(L)^{\text{reg}};\mathbb{C})_\mu = \begin{cases} H^i(L;\mathbb{C})_\mu & \mu < \delta\,, \\ H^1_{(2)}(\mathbb{R}^{\geq 0}) \otimes H^{i-1}(L;\mathbb{C})_\mu & \mu = \delta\,, \\ 0 & \mu > \delta\,, \end{cases} \tag{12}$$

where the subscript μ denotes the part of weight μ. The group $H^1_{(2)}(\mathbb{R}^{\geq 0})$ is infinite dimensional because, as is well known, Range(d) on $\mathbb{R}^{\geq 0}$ is not closed.

Even ignoring the infinite dimensional contribution, (12) does not achieve the desired truncation (9) at and above middle degree; instead we encounter a truncation at and above middle weight. We are thus forced to prove a semi-purity result, which that asserts these two truncations are the same, namely

$$H^i(L;\mathbb{C})_\mu = 0 \quad \text{for} \quad \begin{cases} i \geq k \text{ and } \mu \leq \delta, \text{ or} \\ i < k \text{ and } \mu \geq \delta\ . \end{cases} \tag{13}$$

4.2 Kähler Varieties with Isolated Singularities

Assume X has isolated singularities and let U be a contractible neighborhood of Sing X. We will realize the above heuristic argument in the proof of Theorem 11 by relating L_2-growth weights in U^{reg} to the weights of the mixed Hodge structure on $H^\bullet(U^{\text{reg}};\mathbb{C})$.

Let $\tilde{X} \to X$ be a resolution of X with normal crossing exceptional divisor D having smooth components. Let $D^{(p)}$ denote the disjoint union of p-fold intersections of components of D. Recall [13] that the q-th row of the E_1 term of the weight spectral sequence converging to $H^\bullet(U^{\text{reg}};\mathbb{C})$ is

$$\cdots \longrightarrow H^{q-4}(D^{(2)}) \longrightarrow H^{q-2}(D^{(1)}) \longrightarrow H^q(D^{(1)}) \longrightarrow H^q(D^{(2)}) \longrightarrow \cdots,$$

where the differentials are formed from alternating sums of Gysin maps and of restriction maps.

On the other hand, a principal technical result of [39] is the construction of an analogous spectral sequence $\{E_{(2),r}^{-p,q}\}$ converging to $H^\bullet_{(2)}(U^{\text{reg}};\mathbb{C})$. In this L_2 spectral sequence, the q-th row of $E_{(2),1}$ may be represented by forms with L_2-growth weight q, or by $dr\wedge$ forms with L_2-growth weight $q-1$. Consequently we see that

$$E_{(2),1}^{-p,q} = \begin{cases} E_1^{-p,q} & q < n\,, \\ H^1_{(2)}(\mathbb{R}^{\geq 0}) \otimes E_1^{-p,n} & q = n+1\,, \\ 0 & \text{otherwise}\,, \end{cases}$$

which is the rigorous version of (12) (note that $\delta = n$ here).

The required semi-purity (13) follows from the following stronger result. It was deduced as a consequence of the decomposition theorem [2, 35, 36] by Steenbrink [43]; an independent proof was given by Navarro Aznar [27].

Proposition 12. *The weight filtration W of the mixed Hodge structure on $H^\bullet(U^{\text{reg}};\mathbb{C})$ satisfies*

$$\operatorname{Gr}^W_\mu H^i(U^{\text{reg}};\mathbb{C}) = 0 \quad \textit{for} \quad \begin{cases} i \geq n \textit{ and } \mu \leq i, \textit{ or} \\ i < n \textit{ and } \mu > i\,. \end{cases}$$

Consequently, $E_2^{-p,q} = 0$ for $q - p \geq n$, $p \leq 0$ or $q - p < n$, $p > 0$.

For the intersection cohomology of the link of nonisolated singularities, the proposition has been generalized by Durfee and Saito [14].

4.3 Locally Symmetric Varieties

Let X now be the Baily-Borel-Satake compactification of a locally symmetric variety; we present some of the highlights in Saper and Stern's proof of Theorem 10, Zucker's conjecture. The technical approach used is to establish on $c(L)^{\text{reg}}$ the estimate

$$\|d\phi\|^2 + \|d^*\phi\|^2 \geq c\|\phi\|^2 \qquad (\phi \in \operatorname{Dom}(d) \cap \operatorname{Dom}(d^*),\ \deg\phi \geq k)\,, \tag{14}$$

where $c > 0$; it is well-known that this implies the desired vanishing of L_2-cohomology (9). The argument follows the heuristic reasoning of Sect. 4.1 by exploiting the fairly immediate connection between L_2-growth weights and Lie-theoretic weights. In contrast, the proof of Looijenga [22] (and even more so, Looijenga and Rapoport [23]) passes from L_2-growth weights to mixed Hodge weights via a local Hecke operator.

When H is a Lie subgroup of G, we will denote the Lie algebra of H by $\mathfrak{h}$, and when appropriate we will use the notation Γ_H (resp. K_H) to denote the arithmetic subgroup (resp. maximal compact subgroup) induced in H by Γ (resp. K).

For simplicity, we begin by assuming we are in the $\mathbb{Q}$-rank 1 case, with a single boundary component F. Let $Z = A \ltimes {}^\circ Z$ be the centralizer of F, where $A \cong \mathbb{R}^+$ is a one-dimensional $\mathbb{Q}$-split torus representing the dilations transverse to F, and ${}^\circ Z$ is a certain complement, representing movement in the link directions. The link L is $\Gamma_Z \backslash {}^\circ Z / K_Z$; thus since A commutes with K_Z, there is an adjoint action of A on the tangent bundle TL. We decompose TL into weight spaces $\bigoplus TL_\gamma$ under this action; after we have fixed a coordinate r on $\mathfrak{a}$ (with $r \to \infty$ near F), we may by abuse of notation consider the weights γ as nonnegative real numbers. Then Borel's formula for the metric [3] is

$$dr^2 + \sum_\gamma e^{-\gamma r} ds_\gamma^2\,,$$

where ds_γ^2 is a metric on TL_γ. In fact, after suitably rescaling r, we have $\gamma = 0$,

1, or 2. Thus we see the connection between L_2-growth weights and Lie-theoretic weights.

It turns out that it suffices to prove (14) assuming that ϕ is a weight vector for the induced action of $\mathfrak{a}$, say of weight μ. It also suffices to assume that ϕ is an eigenvector for $dr \wedge \iota_{\partial/\partial r}$, say with eigenvalue j; note that $j = 1$ (resp. 0) if ϕ belongs to (resp. is orthogonal to) the ideal generated by dr. In realizing the previous heuristic reasoning of Sect. 4.1, we never actually pass to cohomology; we merely seek to establish the estimate (14) for all possible values of μ and j. Thus the first step is accomplished by proving the estimate

$$\|d\phi\|^2 + \|d^*\phi\|^2 \geq (\mu-\delta)^2\|\phi\|^2 + (\mu-\delta)(1-j)\int_{\partial c(L)} |\phi|^2 \, dV_L, \qquad (15)$$

where δ is the middle weight (the weight of $\mathfrak{a}$ on $\left(\bigwedge^{\dim \mathfrak{z}} \mathfrak{z}\right)^{1/2}$). This proves (14) whenever

$$\mu \neq \delta \text{ and } j = 1, \quad \text{or} \\ \mu > \delta,$$

which corresponds to (12) of the heuristic argument.

For the second step, we concentrate on the link. Decompose ${}^\circ Z = M \ltimes N$, where M is reductive and N is the unipotent radical. Thus the link fibers over $\Gamma_M\backslash M/K_Z$ with nilmanifold fiber $\Gamma_N\backslash N$. By a result of Zucker [45], we may assume our forms are harmonic in the fibers. The Koszul Laplacian $\Delta_{\mathfrak{m},0}$ on $\bigwedge^\bullet(\mathfrak{m}/\mathfrak{k}_Z)^* \otimes H^\bullet(\mathfrak{n}; E) \otimes \delta$ extends to an action on forms and it suffices to assume that ϕ is also an eigenvector of $\Delta_{\mathfrak{m},0}$, say with eigenvalue ν. Similarly to (15), we may prove the estimate

$$\|d\phi\|^2 + \|d^*\phi\|^2 \geq \nu\|\phi\|^2 . \qquad (16)$$

The following proposition tells us when (16) yields a nontrivial estimate; it is the analogue of the heuristic semi-purity (13).

Proposition 13. *In the above context,*

$$\nu > 0 \quad \text{for} \quad \begin{cases} \deg\phi - j \geq k \text{ and } \mu \leq \delta, & \text{or} \\ \deg\phi - j < k \text{ and } \mu \geq \delta . \end{cases} \qquad (17)$$

The first line of (17) follows from [41, Prop. 10.2, Prop. 11.1, §12]; the second line follows by a dual argument. The proof uses a result of Raghunathan [34], valid for any coefficient sytem over $\Gamma_M\backslash M/K_Z$, to obtain a condition equivalent to $\nu = 0$; a Weyl group argument using Kostant's theorem [21] allows us to analyze this for the coefficients $H^\bullet(\mathfrak{n}; E) \otimes \delta$. In the $\mathbb{R}$-rank 1 case, the proposition reduces to a result of Casselman [7].

In the case of $\mathbb{Q}$-rank > 1, we form a distinguished cover of $c(L)^{\text{reg}}$, with each open set consisting of the points that are near a particular flag of boundary components. The above argument generalizes to prove the estimate for forms compactly supported in each such open set; one replaces Z by the $\mathbb{Q}$-parabolic subgroup Q which is the normalizer of the corresponding flag. In order to combine the separate estimates into one for all of $c(L)^{\text{reg}}$, we use a partition of unity $\{\eta_Q\}$; terms involving $|d\eta_Q|$ arise which can be controlled by choosing the elements of the distinguished cover to have sufficiently large overlap.

References

1. Baily W., Borel, W.: Compactification of arithmetic quotients of bounded symmetric domains. Ann. Math. **84** (1966) 442–528
2. Beilinson, A., Bernstein, J., Deligne, P.: Faisceaux pervers, Proceedings of the Conference "Analyse et topologie sur les espaces singuliers", Juillet 1981. Astérisque **100** (1982)
3. Borel, A.: Some properties of arithmetic quotients and an extension theorem. J. Diff. Geom. **6** (1972) 543–560
4. Borel, A.: L_2-cohomology and intersection cohomology of certain arithmetic varieties. In: Sally, J., Srinivasan, B. (eds.) Emmy Noether in Bryn Mawr. Springer, Berlin Heidelberg New York 1983
5. Borel, A., et. al.: Intersection Cohomology. Birkhäuser, Boston 1984
6. Borel, A., Casselman, W.: Cohomologie d'intersection et L^2-cohomologie de variétés arithmétiques de rang rationnel 2. C. R. Acad. Sci. Paris **301** (1985) 369–373
7. Casselman, W.: L^2-cohomology for groups of real rank one. In: Trombi, P. (ed.) Representation theory of reductive groups. Birkhäuser, Boston 1983
8. Casselman, W.: Introduction to the L^2-cohomology of arithmetic quotients of bounded symmetric domains. In: Suwa, T., Wagreich, P. (eds.) Complex analytic singularities. (Advanced Studies in Pure Mathematics, vol. 8.) North-Holland, Amsterdam 1986
9. Cattani, E., Kaplan, A., Schmid, W.: L^2 and intersection cohomologies for a polarizable variation of Hodge structure. Invent. math. **87** (1987) 217–252
10. Cheeger, J.: On the Hodge theory of Riemannian pseudomanifolds. In: Geometry of the Laplace operator. (Proc. Symposia Pure Math., vol. 36.) Amer. Math. Soc., Providence 1980
11. Cheeger, J.: Hodge theory of complex cones. Astérisque **101–102** (1983) 118–134
12. Cheeger, J., Goresky, M., MacPherson, R.: L^2-cohomology and intersection homology of singular algebraic varieties. In: Yau, S.-T. (ed.) Seminar on Differential Geometry. (Annals of Mathematics Studies, no. 102.) Princeton Univ. Press, Princeton 1982
13. Durfee, A.: Mixed Hodge structures on punctured neighborhoods. Duke Math. J. **50** (1983) 1017–1040
14. Durfee, A., Saito, M.: Mixed Hodge structures on the intersection cohomology of links. Comp. Math. To appear
15. Gaffney, M.: A special Stokes' theorem for complete Riemannian manifolds. Ann. Math. **60** (1954) 140–145
16. Goresky, M., MacPherson, R.: Intersection homology theory. Topology **19** (1980) 135–162
17. Goresky, M., MacPherson, R.: Intersection homology II. Invent. math. **72** (1983) 77–130
18. Hsiang, W.-C., Pati, V.: L^2-cohomology of normal algebraic surfaces I. Invent. math. **81** (1985) 395–412
19. Kashiwara, M., Kawai, T.: The Poincaré lemma for variations of polarized Hodge structure. Publ. R.I.M.S., Kyoto Univ. **23** (1987) 345–407
20. Kleiman, S.: The development of intersection homology theory. In: Duren, P. (ed.) A century of mathematics in America, part II. Amer. Math. Soc., Providence 1989
21. Kostant, B.: Lie algebra cohomology and the generalized Borel-Weil theorem. Ann. Math. **74** (1961) 329–387

22. Looijenga, E.: L^2-cohomology of locally symmetric varieties. Comp. Math. **67** (1988) 3–20
23 Looijenga, E., Rapoport, M.: Weights in the local cohomology of a Baily-Borel compactification. Preprint, January 1990
24. MacPherson, R.: Global questions in the topology of singular spaces. In: Proceedings of the International Congress of Mathematicians, Warsaw, 1983. Elsevier, Amsterdam 1984
25. Nagase, M.: Pure Hodge structure of the harmonic L^2-forms on singular algebraic surfaces. Publ. R.I.M.S., Kyoto Univ. **24** (1988) 1005–1023
26. Nagase, M.: Remarks on the L^2-cohomology of singular algebraic surfaces. J. Math. Soc. Japan **41** (1989) 97–116
27. Navarro Aznar, V.: Sur la théorie de Hodge des variétés algébriques à singularités isolées. Astérisque **130** (1985) 272–307
28. Ohsawa, T.: Hodge spectral sequence on compact Kähler spaces. Publ. R.I.M.S., Kyoto Univ. **23** (1987) 265–274. Also, Supplement to "Hodge spectral sequence on compact Kähler spaces" Preprint, August 1990
29. Ohsawa, T.: Hodge spectral sequence and symmetry on compact Kähler spaces. Publ. R.I.M.S., Kyoto Univ. **23** (1987) 613–625
30. Ohsawa, T.: An extension of Hodge theory to Kähler spaces with isolated singularities of restricted type. Publ. R.I.M.S., Kyoto Univ. **24** (1988) 253–263
31. Ohsawa, T.: Cheeger-Goresky-MacPherson's conjecture for the varieties with isolated singularities. Preprint, July 1990
32. Pardon, W.: The L_2-$\bar{\partial}$-cohomology of an algebraic surface. Topology **28** (1989) 171–195
33. Pardon, W., Stern, M.: L^2-$\bar{\partial}$-cohomology of complex projective varieties. Preprint, August 1990
34. Raghunathan, M.S.: Vanishing theorems for cohomology groups associated to discrete subgroups of semisimple Lie groups. Osaka J. Math. **3** (1966) 243–256. Also, Corrections to "Vanishing theorems ..." Osaka J. Math. **16** (1979) 295–299
35. Saito, M.: Modules de Hodge polarisables. Publ. R.I.M.S., Kyoto Univ. **24** (1988) 849–995
36. Saito, M.: Mixed Hodge modules. Publ. R.I.M.S., Kyoto Univ. **26** (1990) 221–333
37. Saper, L.: L_2-cohomology and intersection cohomology of certain varieties with isolated singularities. Invent. math. **82** (1985) 207–255
38. Saper, L.: L_2-cohomology of isolated singularities. Preprint, November 1985
39. Saper, L.: L_2-cohomology of Kähler varieties with isolated singularities. Preprint, July 1990
40. Saper, L., Stern, M.: L_2-cohomology of arithmetic varieties. Proc. Natl. Acad. Sci. USA **84** (August 1987) 5516–5519
41. Saper, L., Stern, M.: L_2-cohomology of arithmetic varieties. Ann. Math. **132** (1990) 1–69
42. Satake, I.: On compactifications of the quotient spaces for arithmetically defined discontinuous groups. Ann. Math. **72** (1960) 555–580
43. Steenbrink, J.H.M.: Mixed Hodge structures associated with isolated singularities. In: Singularities. (Proc. Symposia Pure Math., vol. 40, part 2.) Amer. Math. Soc., Providence 1983
44. Zucker, S.: Hodge theory with degenerating coefficients: L_2 cohomology in the Poincaré metric. Ann. Math. **109** (1979) 415–476
45. Zucker, S.: L_2 cohomology of warped products and arithmetic groups. Invent. math. **70** (1982) 169–218
46. Zucker, S.: L_2-cohomology and intersection homology of locally symmetric varieties, II. Comp. Math. **59** (1986) 339–398
47. Zucker, S.: The Hodge structures on the intersection homology of varieties with isolated singularities. Duke Math. J. **55** (1987) 603–616

Nonabelian Hodge Theory

Carlos T. Simpson

Laboratoire de Topologie et Geometrie, U.F.R.M.I.G.
Université Paul Sabatier, 118, route de Narbonne, F-31062 Toulouse-Cedex, France

Introduction

Classically, Hodge theory and related constructions provided extra structure to abelian topological invariants of the usual topological spaces associated to algebraic varieties over $\mathbb{C}$. I would like to explain how, in analogy with these abelian constructions, certain nonabelian topological invariants of complex algebraic varieties have extra structures.

We will be concerned with two related invariants, the space R of framed n-dimensional representations of $\pi_1(X,x)$, and M, its universal categorical quotient by the action of $G\ell(n)$. These are related to the nonabelian cohomology $H^1(X, G\ell(n))$, which is the set of isomorphism classes of representations, and which has a structure of non-Hausdorff space. The moduli space M is the set of Jordan equivalence classes of representations, the universal Hausdorff space to which the cohomology space maps. The representation space R is the cohomology of X relative to a choice of base point x. We will interpret R and M as nonabelian analogues of the abelian cohomology $H^1(X, \mathbb{C})$.

The Nonabelian Hodge Theorem

Suppose X is smooth and projective over $\mathbb{C}$. A *harmonic bundle* on X is a C^∞ vector bundle E with differential operators ∂ and $\overline{\partial}$, and algebraic operators θ and $\overline{\theta}$ (operators from E to one-forms with coefficients in E), such that the following hold. There exists a metric K so that $\partial + \overline{\partial}$ is a unitary connection and $\theta + \overline{\theta}$ is self adjoint. And if we set $D = \partial + \overline{\partial} + \theta + \overline{\theta}$ and $D'' = \overline{\partial} + \theta$, then $D^2 = 0$ and $(D'')^2 = 0$. With these conditions, (E, D) is a vector bundle with flat connection, and $(E, \overline{\partial}, \theta)$ is a *Higgs bundle*: a holomorphic vector bundle with holomorphic section $\theta \in H^0(\mathrm{End}(E) \otimes \Omega^1_X)$ such that $\theta \wedge \theta = 0$. Fix a class of Kähler metric for X. A Higgs bundle is *stable* (resp. *semistable*) if, for any coherent subsheaf $F \subset E$ preserved by θ with $0 < \mathrm{rk}(F) < \mathrm{rk}(E)$, we have

$$\deg(F)/\mathrm{rk}(F) < \text{ (resp. } \leq) \deg(E)/\mathrm{rk}(E) .$$

Proceedings of the International Congress
of Mathematicians, Kyoto, Japan, 1990

Theorem 1. *There is a natural equivalence between the categories of harmonic bundles on X and semisimple flat bundles (or representations of* $\pi_1(X)$*). There is a natural equivalence between the categories of harmonic bundles and direct sums of stable Higgs bundles with vanishing Chern classes. The resulting correspondence between representations and Higgs bundles can be extended to an equivalence between the categories of all representations of* $\pi_1(X)$*, and all semistable Higgs bundles with vanishing Chern classes.*

The first part of the theorem, relating harmonic bundles and representations, is due to Corlette [2] and Donaldson [6]. It is a generalization of the theorem of Eells and Sampson – a harmonic bundle is equivalent to a representation of the fundamental group in $G\ell(n)$ together with an equivariant harmonic map from the universal cover of X to $G\ell(n)/U(n)$. This equivalence uses the Bochner formula of Siu [18]. The part relating harmonic bundles and Higgs bundles is a generalization of the theorem of Narasimhan and Seshadri [11] (when $\theta = 0$ a Higgs bundle is just a vector bundle, and the corresponding representation is unitary). Part of this generalization involves the higher dimensional Hermitian-Yang-Mills theory for stable vector bundles developed by Donaldson [5] and Uhlenbeck, Yau [19]. Hitchin [9] gave the definition of stable Higgs bundles, and the proof of their correspondence with irreducible flat bundles, when X is a curve. These two directions are combined for the general statement in [14]. The last statement of the theorem comes from a formality result for the collection of complexes which control extensions [17].

A generalization to the case of noncompact curves is carried out in [16]. The noncompact higher dimensional case is still open (Corlette has some recent work on nonabelian L^2 cohomology in the higher dimensional case). For the remainder of the present discussion we will stick with the assumption that X is compact.

The set of flat bundles is analogous to the abelian de Rham cohomology, while the set of Higgs bundles (E, θ) is analogous to the abelian Dolbeault cohomology, $H^1(\mathcal{O}_X) \oplus H^0(\Omega^1_X)$. The first two parts of Theorem 1 may be interpreted as giving harmonic representatives for certain nonabelian de Rham and Dolbeault cohomology classes. The fact that the notion of harmonic representative (harmonic bundle) is the same in both cases, is the analogue of the abelian Kähler identity $\Delta_d = 2\Delta_{\overline{\partial}}$.

Variations of Hodge Structure. There is a natural $\mathbb{C}^*$ action on the category of semistable Higgs bundles with vanishing Chern classes, defined by $t : (E, \theta) \mapsto (E, t\theta)$.

Lemma 2. *The semisimple representations which are fixed by this action of* $\mathbb{C}^*$ *are exactly those which underly complex variations of Hodge structure.*

Proof. If (E, θ) is fixed, then there is an action of $\mathbb{C}^*$ on E, corresponding to a decomposition $E = \bigoplus E^p$ with $\theta : E^p \to E^{p-1} \otimes \Omega^1_X$. The spaces E^p will be mutually orthogonal with respect to the harmonic metric [14]. A harmonic bundle with orthogonal decomposition which is preserved by $\overline{\partial}$ and shifted once by θ is exactly a complex variation of Hodge structure. □

Rigid Representations. A representation ϱ of $\pi_1(X)$ is *rigid* if any nearby representation is conjugate to it. The correspondence given by Theorem 1 is continuous on the moduli space of semisimple representations (see below). Thus, if a semisimple representation is rigid, then it must be fixed by $\mathbb{C}^*$, and Lemma 2 implies that it comes from a complex variation of Hodge structure. This can be strengthened:

Proposition 3. *Suppose ϱ is a rigid semisimple representation of $\pi_1(X)$. Then there is a $\mathbb{Q}$-variation of Hodge structure $V_{\mathbb{Q}}$ such that ϱ is a direct factor of the monodromy representation of $V_{\mathbb{Q}} \otimes \overline{\mathbb{Q}}$ (that monodromy is a sum of conjugates of ϱ). If ϱ is two dimensional, then the variation of Hodge structure $V_{\mathbb{Q}}$ can be assumed to have Hodge types* $(1,0)$ *and* $(0,1)$ *only.*

For an ℓ-adic analogue, suppose X is defined over $K \subset \mathbb{C}$. If a representation ϱ is rigid, and integral at a prime ℓ, then the resulting representation into $PG\ell(n, \overline{\mathbb{Q}}_\ell)$ descends to a representation of the algebraic fundamental group $\pi_1^{\mathrm{alg}}(X_{K'})$ for some finite extension K'/K. These results suggest the following conjectures (which are actually special cases of the conjectures in [15]).

Conjecture 4 (rigid $\Rightarrow$ motivic). *Suppose ϱ is a rigid semisimple representation of $\pi_1(X)$. Then ϱ is a direct factor in the monodromy representation of a motive (i.e. family of varieties) over X.*

Conjecture 5 (rigid $\Rightarrow$ integral). *Suppose ϱ is a rigid semisimple representation of $\pi_1(X)$. Then ϱ is isomorphic to a representation with coefficients which are algebraic integers.*

Note that the first would imply the second. By Proposition 3, a rigid integral two dimensional representation comes from a family of abelian varieties, so in this case the second statement would imply the first. In the case of two dimensional representations, one can show that a Zariski dense representation into $S\ell(2)$ is either rigid, or else it is equal in $PS\ell(2)$ to a representation pulled back (via an algebraic map) from a curve with orbifold structure. So Conjecture 5 would imply that any such two dimensional representation is pulled back, either from a curve with orbifold structure, or a subspace of the moduli space of abelian varieties.

Restrictions on Fundamental Groups. The nonabelian Hodge theorem, and related considerations, have led to restrictions on which groups can be fundamental groups of smooth projective varieties (or really, of any compact Kähler manifolds). This subject begins with Siu's rigidity theorem [18]. Siu's method has been used in [1], [2], and [13]. A typical example is:

Theorem 6 (Carlson, Toledo). *Suppose Γ is a discrete cocompact subgroup of $SO(n,1)$ $(n > 2)$. Then Γ is not the fundamental group of a compact Kähler manifold.*

The monodromy group of a variation of Hodge structure must be a *Hodge group* [17]. For connected simple groups, the Hodge groups are all except the complex groups, and those isogenous to $S\ell(n,\mathbb{R})$ $(n \geq 3)$, $SU^*(2n)$ $(n \geq 3)$, $SO(p,q)$ (p, q odd), $E_{6(6)}$, or $E_{6(-26)}$. By work of many people, almost all lattices in real groups are known to be rigid, so some further nonexistence results can be derived from Proposition 3 (again, indirectly, from Siu's Bochner formula).

Theorem 7. *Suppose G is a real group which is not a Hodge group. Suppose $\Gamma \subset G$ is a rigid lattice. Then Γ is not the fundamental group of a compact Kähler manifold. If Γ is a group of the above form, and if Γ' is a semi-direct product of Γ with any other group (in which Γ is the quotient), then Γ' is not the fundamental group of a smooth projective variety.*

Siu's technique has been used to obtain other topological information. In a related direction, Corlette has used harmonic maps to prove some rigidity results about lattices in real groups. With a generalization of the Bochner technique applying to spaces with quaternionic structures, he proves that any cocompact lattice Γ in $Sp(n,1)$ $(n \geq 2)$, or F_4^{-20}, is superrigid over archimedean fields [3].

Moduli Spaces

Fix a number n. The *Betti* versions of our topological invariants are easy to define. $R_{\mathrm{B}}(X,x)$ is the space of representations of the fundamental group $\pi_1(X,x)$ on the fixed space $\mathbb{C}^n$. It is an affine variety defined by the generators and relations of the fundamental group. The group $G\ell(n,\mathbb{C})$ acts. There is a universal categorical quotient $M_{\mathrm{B}}(X)$, sometimes called the character variety. When X is a smooth projective variety, R and M have *algebraic de Rham* and *Dolbeault* realizations, as provided by the following theorem.

Theorem 8. *Suppose X is a smooth projective variety over $\mathbb{C}$, with a point $x \in X$. There exist algebraic varieties $R_{\mathrm{DR}}(X,x)$, $R_{\mathrm{Dol}}(X,x)$, $M_{\mathrm{DR}}(X)$, and $M_{\mathrm{Dol}}(X)$, quasiprojective over $\mathbb{C}$, which are moduli spaces as follows. R_{DR} is a fine moduli space for the set of (E,∇,β) where (E,∇) is a vector bundle with integrable algebraic connection and $\beta : E_x \cong \mathbb{C}^n$ is a frame. R_{Dol} is a fine moduli space for the set of (E,θ,β) where (E,θ) is a semistable Higgs bundle with vanishing Chern classes, and β is a frame for E_x. The group $G\ell(n,\mathbb{C})$ acts on R_{DR} and R_{Dol}, and M_{DR} and M_{Dol} are the universal categorical quotients. Their points parametrize semisimple objects.*

The construction of the moduli space of Higgs bundles was treated in Hitchin's paper [9], and later by Nitsure [12], in the case when X is a curve. The higher dimensional case is treated in a preprint by the author (containing some errors – particularly with regard to the next proposition – which have subsequently been corrected).

Proposition 9. *The Riemann-Hilbert correspondence between bundles with integrable connection and representations provides isomorphisms of the associated complex analytic spaces, $(R_{\mathrm{DR}})^{\mathrm{an}} \cong (R_{\mathrm{B}})^{\mathrm{an}}$ and $(M_{\mathrm{DR}})^{\mathrm{an}} \cong (M_{\mathrm{B}})^{\mathrm{an}}$.*

The Riemann-Hilbert correspondence is not algebraic. It provides an example of two different algebraic varieties whose associated complex analytic spaces are isomorphic, generalizing the example of Serre.

Proposition 10. *The correspondence of Theorem 1 provides a homeomorphism between the usual topological spaces underlying M_{Dol} and M_{B}. However, the extended correspondence between the points of R_{Dol} and R_{B} provided by the last part of Theorem 1, is not continuous.*

This proposition was treated by Hitchin in his case [9]. Hitchin also discussed several important properties of the moduli space M_{Dol}, all of which carry over to the case of higher dimensional X:

Theorem 11. *There is a natural algebraic action of $\mathbb{C}^*$ on the space M_{Dol}, representing the action defined previously. There is a proper map from M_{Dol} to a vector space, compatible with an action of $\mathbb{C}^*$ on the vector space having only positive weights. Thus if p is any point in M_{Dol}, then the limit $\lim_{t\to 0} tp$ exists in M_{Dol}.*

A corollary of this theorem, the continuity of Proposition 10, and the characterization of Lemma 2, is that any representation of $\pi_1(X)$ can be deformed to a complex variation of Hodge structure. This provides the mechanism for proving the restrictions stated in the second half of Theorem 7.

The Gauss-Manin Connection. Suppose $X \to S$ is a smooth projective family over a base scheme S, with section of base points $x : S \to X$. Then the above constructions give families of moduli spaces $R_{\mathrm{DR}}(X/S, x)$ and $M_{\mathrm{DR}}(X/S)$ over S. There is an analogue of the Gauss-Manin connection.

Theorem 12. *Suppose X/S is smooth and projective with base point $x(S)$. Let $(S \times S)^\wedge$ denote the formal neighborhood of the diagonal in $S\times S$, similarly in $S\times S\times S$. There is a canonical isomorphism*

$$\varphi : p_1^*(R_{\mathrm{DR}}(X/S))|_{(S\times S)^\wedge} \xrightarrow{\cong} p_2^*(R_{\mathrm{DR}}(X/S))|_{(S\times S)^\wedge},$$

such that on $(S\times S\times S)^\wedge$ the cocycle condition $p_{23}^*(\varphi)p_{12}^*(\varphi) = p_{13}^*(\varphi)$ is satisfied. The same holds for the universal categorical quotient $M_{\mathrm{DR}}(X/S)$. These isomorphisms are compatible with the local analytic trivializations which arise from the Riemann-Hilbert correspondence.

Proof. The category of vector bundles with connection on X/S can be interpreted as a category of crystals on X/S. If $S_0 \subset S$ is a closed subscheme defined by a nilpotent ideal, and X_0 is the inverse image of S_0, then the category of crystals on X/S is equivalent to the category of crystals on X_0/S. Hence the scheme $R_{\mathrm{DR}}(X/S)$ over S depends functorially only on the family $X_0 \to S_0$ and the inclusion $S_0 \subset S$, although the existence of $R_{\mathrm{DR}}(X/S)$ seems to depend on the existence of a family $X \to S$. This gives a crystal of schemes over the stratifying site of S, which is equivalent to the structure described in the theorem. □

At points where S and $R_{\text{DR}}(X/S)$ are smooth, the structure provided by this theorem may be described more simply as a splitting $\pi^*TS \to T(R_{\text{DR}}(X/S))$ of the quotient of tangent bundles, satisfying an integrability condition. If S is an artinian scheme, it gives a natural isomorphism $R_{\text{DR}}(X/S) \cong R_{\text{DR}}(X_0) \times S$.

The Hyperkähler Structure. The identification given by Theorem 1 is not complex analytic. Thus, the same space is given several complex structures.

Theorem 13 (Hitchin). *Denote by M^s, the set of smooth points of the moduli space. The isomorphism $M^s_{\text{Dol}} \cong M^s_{\text{B}}$ is smooth. Let I and J denote the complex structures on M^s obtained from M^s_{Dol} and M^s_{B} respectively. Then $K = IJ$ is a complex structure, and this triple gives a quaternionic structure in the tangent space at any point. The resulting quaternionic manifold M^s has a natural hyperkähler structure.*

This was pointed out by Hitchin [9] in the case when X is a curve; the general case has been treated in [4] and recently, in detail, by Fujiki [7]. We refer to Hitchin's paper for the definition of hyperkähler manifold.

The Space of λ-Connections. Here is a definition due to Deligne – the contents of this subsection were all described or outlined by him in correspondence [4]. Suppose $\lambda \in \mathbb{C}$. A *λ-connection* on a vector bundle E is an operator $\nabla : E \to E \otimes \Omega^1_X$, such that $\nabla(ae) = \lambda d(a)e + a\nabla(e)$. If $\lambda \neq 0$ then a λ-connection ∇ corresponds to a connection $\lambda^{-1}\nabla$. On the other hand, if $\lambda = 0$ then a bundle with integrable λ-connection is just a Higgs bundle (E, θ), $\theta = \nabla$. So this definition provides a deformation from the notion of connection to the notion of Higgs bundle. Its importance comes from the resulting moduli space.

Theorem 14. *Suppose X is smooth and projective over $\mathbb{C}$, with base point x. There exist quasiprojective algebraic varieties $W^+ \to \mathbf{A}^1$ and $Z^+ \to \mathbf{A}^1$ which are moduli spaces for λ-connections as follows. W^+ is a fine moduli space for $(\lambda, E, \nabla, \beta)$ where $\lambda \in \mathbf{A}^1$, (E, ∇) is a vector bundle with integrable λ-connection (semistable with vanishing Chern classes in the case $\lambda = 0$), and β is a frame for E_x. The group $G\ell(n, \mathbb{C})$ acts on W^+ with universal categorical quotient Z^+.*

The fibers W^+_1, W^+_0, Z^+_1, and Z^+_0 are respectively the moduli spaces R_{DR}, R_{Dol}, M_{DR}, and M_{Dol}. There are natural actions of $\mathbb{C}^*$ on W^+ and Z^+, defined by $t : (\lambda, E, \nabla, \beta) \mapsto (t\lambda, E, t\nabla, \beta)$, covering the standard action on $\mathbf{A}^1$.

Deligne thought of the space of λ-connections as a way of approaching the notion of quaternionic structure for the singular space M. There are complex analytic spaces W and Z over $\mathbb{P}^1$ characterized by the following properties. There are actions of $\mathbb{C}^*$ covering the standard action on $\mathbb{P}^1$, and antilinear involutions σ compatible with the action of $\mathbb{C}^*$ (and the map to $\mathbb{P}^1$) relative the involution $t \mapsto (\bar{t})^{-1}$. There are identifications

$$W|_{\mathbf{A}^1} \cong W^+ \quad Z|_{\mathbf{A}^1} \cong Z^+$$

compatible with the action of $\mathbb{C}^*$. On the fibers $W_1 \cong R_B$ and $Z_1 \cong M_B$, the involution σ takes a representation to the complex conjugate representation.

A harmonic bundle E yields a family of holomorphic bundles $(E, \overline{\partial} + \lambda\overline{\theta})$ with λ-connections $\nabla = \lambda\partial + \theta$, indexed by $\lambda \in \mathbf{A}^1$. The resulting section $\mathbf{A}^1 \to W^+$ is holomorphic and extends to a section $\mathbb{P}^1 \to W$. Refer to these sections (and their projections in Z) as *preferred sections.* The involution σ maps a preferred section to the preferred section corresponding to the complex conjugate representation. We get a trivialization $Z \cong M \times \mathbb{P}^1$, where M denotes the moduli space of harmonic bundles.

Proposition 15 (Deligne). *The trivialization $Z \cong M \times \mathbb{P}^1$ is a homeomorphism, and on the set of smooth points it identifies Z^s with the twistor space for the quaternionic structure of M^s.*

Proof. There is a $\mathbb{P}^1$ of complex structures q in the quaternions. The twistor space is $M^s \times \mathbb{P}^1$ endowed with a complex structure which makes the horizontal projective lines holomorphic, and which endows each fiber $M^s \times \{q\}$ with the complex structure q deduced from the quaternionic structure of M^s. One has to check that the identification $Z_q \cong M^s$ takes the complex structure of Z_q to the complex structure q of M^s. □

Deligne proposes a way of defining the notion of quaternionic structure for a singular space such as M by some axioms for a twistor space such as Z.

Compactifications. The space of λ-connections suggested by Deligne turns out to be very useful for compactifying the moduli spaces.

Lemma 16. *Suppose z is a point of Z^+. Then the limit $\lim_{t\to 0} tz$ exists in Z^+. Also, the set of fixed points of $\mathbb{C}^*$ in Z^+ is compact.*

Proposition 17. *Suppose Z^+ is any algebraic variety with an algebraic action of $\mathbb{C}^*$, such that the limits $\lim_{t\to 0} tz$ exist, and such that the set of fixed points is compact. Let $S^+ \subset Z^+$ denote the subset of points s such that $\lim_{t\to\infty} ts$ exists in Z^+. Then there is a universal geometric quotient $P = (Z^+ - S^+)/\mathbb{C}^*$, and P is a proper scheme (not necessarily projective).*

We can apply this construction to the moduli space of λ-connections Z^+, to obtain a compactification of M_{DR}.

Corollary 18. *Let Z^+ be the space of λ-connections. Let $P_{\mathrm{DR}} = (Z^+ - S^+)/\mathbb{C}^*$ be the proper scheme defined above. Then $M_{\mathrm{DR}} \subset P_{\mathrm{DR}}$ is an open subset, providing a compactification of the space of bundles with integrable connection.*

Proof. Note that the subset $S^+ \subset Z^+$ is contained in the fiber over $0 \in \mathbf{A}^1$. Let $D = (Z_0^+ - S^+)/\mathbb{C}^* \subset P_{\mathrm{DR}}$. It is a closed subset, and the complement is equal to $(Z^+ - Z_0^+)/\mathbb{C}^* \cong M_{\mathrm{DR}}$. □

There is a similar compactification $M_{\mathrm{Dol}} \subset P_{\mathrm{Dol}}$ where the complement is the same divisor D. In fact, these fit together into a relative compactification of Z^+ over $\mathbf{A}^1$.

The Hodge Filtration. Here is another use for the space of λ-connections. It can be thought of as the nonabelian analogue of the Hodge filtration on the de Rham moduli space. Suppose V is a finite dimensional vector space with an exhaustive decreasing filtration F^p. Define a locally free sheaf $\xi(V,F)$ over $\mathbf{A}^1$ as the submodule of $V \otimes \mathbb{C}[t,t^{-1}]$ generated by the $t^{-p}F^p$.

Lemma 19. *The construction ξ provides an equivalence between the category of vector spaces with exhaustive decreasing filtrations, and the category of locally free sheaves on $\mathbf{A}^1$ with action of $\mathbb{C}^*$ covering the standard action on $\mathbf{A}^1$.*

The vector space V is recovered as the fiber of $\xi(V,F)$ at 1, whereas the fiber at 0 is canonically the associated graded $\mathrm{Gr}_F^{\cdot}(V)$. Given a real structure, one can glue together $\xi(V,F)$ and $\xi(V,\overline{F})$ using the involution $t \mapsto (\overline{t})^{-1}$, to obtain a locally free sheaf $\xi(V,F,\overline{F})$ on $\mathbb{P}^1$ with action of $\mathbb{C}^*$ and antilinear involution.

In view of this lemma, we can define a *nonabelian filtration* to be a scheme over $\mathbf{A}^1$ with action of $\mathbb{C}^*$. The spaces W^+ and Z^+ over $\mathbf{A}^1$, together with their $\mathbb{C}^*$ actions, may then be interpreted as the *nonabelian Hodge filtrations* of the fibers $R_{\mathrm{DR}} = W_1^+$ and $M_{\mathrm{DR}} = Z_1^+$. The spaces $\sigma(W^+)$ and $\sigma(Z^+)$ (which are the restrictions of W and Z to the affine neighborhood of infinity) are the complex conjugates of the Hodge filtrations. The Dolbeault spaces R_{Dol} and M_{Dol} become the associated graded spaces of the nonabelian Hodge filtrations (just as for abelian cohomology). Lemma 16 is analogous to saying that the Hodge filtration is concentrated in positive degrees.

These definitions are compatible with the mixed Hodge structure defined by Morgan [10] and Hain [8] on the nilpotent completion of the fundamental group. Let $A = (\mathbb{C}\pi_1)^\wedge$ be the completion of the group algebra at the augmentation ideal. They show that this has a Hodge filtration F giving a mixed Hodge structure (the real structure being the completion of the real group algebra, and the weight filtration given by powers of the augmentation ideal). By Lemma 19, this data is equivalent to the sheaf of algebras $\xi(A,F,\overline{F})$ together with its $\mathbb{C}^*$ action and involution.

Given a complete augmented algebra such as A, one can define a formal scheme $R(A)$, the space of representations infinitesimally close to the trivial one. For $A = (\mathbb{C}\pi_1)^\wedge$ as above, the space $R(A)$ is naturally isomorphic to the completion of R_{B} at the point corresponding to the trivial representation.

Theorem 20. *The family of algebras $\xi(A,F,\overline{F})$ over $\mathbb{P}^1$ gives rise to a family of formal schemes $R(\xi(A,F,\overline{F}))$ over $\mathbb{P}^1$. This family of formal schemes (with $\mathbb{C}^*$ action and involution) is naturally isomorphic to the formal neighborhood of the trivial preferred section in W. The isomorphism is compatible with the one described above in the fiber over* 1.

Conversely, Tannaka duality allows the mixed Hodge structure $\xi(A,F,\overline{F})$ to be recovered from the data of the formal neighborhoods of the spaces W, together with some information on how they fit together for different ranks.

Speculations

It is natural to wonder how the results outlined above might fit into a notion of *nonabelian Hodge structure.* The action of $\mathbb{C}^*$ on the space of representations gives, by Tannaka duality, an action of $\mathbb{C}^*$ on the *proalgebraic completion* $\varpi_1(X)$ (the inverse limit of the Zariski closures of the images, over all finite dimensional representations of $\pi_1(X)$). The action of $S^1 \subset \mathbb{C}^*$ on the proreductive quotient ϖ_1^{red} satisfies some axioms which might be taken as a preliminary definition of pure Hodge structure on a group [17]. The map $S^1 \times \pi_1(X) \to \varpi_1^{\text{red}}$ is continuous when the right side is given the inverse limit of analytic topologies. And the action of $-1 \in S^1$ is a Cartan involution of the real form $(\varpi_1^{\text{red}})_{\mathbb{R}}$ (inverse limit of real Zariski closures). The restrictions of the first half of Theorem 7 may be derived from these axioms (one would like to interpret all those kinds of topological restrictions as following from the nonabelian Hodge structure).

This definition refers to the continuous structure of the space of representations only in the crudest way. Preferable would be a notion of nonabelian Hodge structure for the space of representations. We have defined the nonabelian Hodge filtration. What about weight filtrations and polarizations? Probably the hyperkähler metric will play a role. The mixed Hodge structure on $(\mathbb{C}\pi_1)^\wedge$ will provide the power series for the nonabelian Hodge structure at the trivial representation (Theorem 20).

With or without a precise definition, we are afforded the pleasant opportunity of revisiting all the old glories of Hodge theory, and trying to provide their nonabelian analogues. The subject started with Griffiths' transversality result. Preliminary heuristic calculations seem to show that an analogous transversality relates the Gauss-Manin connection defined in Theorem 12, to the Hodge filtration interpreted as the space Z^+. The lifts of vector fields from the base S, to the restriction of $Z^+(X/S)$ over $\mathbf{A}^1 - \{0\}$, will have poles of order ≤ 1 along Z_0^+ (and hence also along the divisor at infinity in the compactificaton).

This raises the question of whether there is a good notion of *variation of nonabelian Hodge structure* over S. One could try to replicate the works of Schmid and others about degenerating variations. Is there an appropriate sense in which the Gauss-Manin connection has *regular singularities?* And is there some type of limiting nonabelian Hodge structure? This type of theory would be in the service of devissage, trying to obtain information about the topology of a space X by fibering $X \to Y$ and studying the family of fundamental groups of the fibers. For example, the moduli space $M(X)$ (with all of its structures) becomes a constant subspace of the varying family $M(X_y)$. The wide array of strutures on $M(X_y)$ seems to make this an unlikely event.

Let me close with some wild speculation on how this picture might fit into an overall view of Hodge theory. We have treated two topological invariants, the space R of framed representations, and the quotient M. For which topological invariants will there be such a theory? Suppose F : Top $\to$ Sch is a functor from the category of topological spaces to the category of schemes. Are there some simple conditions which would guarantee that, for a smooth projective variety X, $F(X)$ has a "Hodge structure"? Perhaps (to say something entirely unsubstantiated) there might be an action of $\mathbb{C}^*$ on the set of such functors F, and the functors which are fixed are of Hodge type.

References

1. Carlson, J.A., Toledo, D.: Harmonic mappings of Kähler manifolds to locally symmetric spaces. Publ. Math. I.H.E.S. **69** (1989) 173–201
2. Corlette, K.: Flat *G*-bundles with canonical metrics. J. Diff. Geom. **28** (1988) 361–382
3. Corlette, K.: Archimedean superrigidity and hyperbolic geometry. Preprint
4. Deligne, P: Letter, March 20 (1989)
5. Donaldson, S.K.: Anti self dual Yang-Mills connections over complex algebraic surfaces and stable vector bundles. Proc. London Math. Soc. **50** (1985) 1–26
6. Donaldson, S.K.: Twisted harmonic maps and self-duality equations. Proc. London Math. Soc. **55** (1987) 127–131
7. Fujiki, A.: Hyperkähler structure on the moduli space of flat bundles. Preprint
8. Hain, R.: The de Rham homotopy theory of complex algebraic varieties, I. *K*-Theory **1** (1987) 271–324
9. Hitchin, N.J.: The self-duality equations on a Riemann surface. Proc. London Math. Soc. **55** (1987) 59–126
10. Morgan, J.: The algebraic topology of smooth algebraic varieties. Publ. Math. I.H.E.S. **48** (1978) 137–204
11. Narasimhan, M.S., Seshadri, C.S.: Stable and unitary bundles on a compact Riemann surface. Ann. Math. **82** (1965) 540–564
12. Nitsure, N.: Moduli space of semistable pairs on a curve. Proc. London Math. Soc. (3) **62** (1991) 275–300
13. Sampson, J.H.: Applications of harmonic maps to Kähler geometry. Contemp. Math. **49** (1986) 125–133
14. Simpson, C.T.: Constructing variations of Hodge structure using Yang-Mills theory and applications to uniformization. J.A.M.S. **1** (1988) 867–918
15. Simpson, C.T.: Transcendental aspects of the Riemann-Hilbert correspondence. Illinois J. Math. **34** (1990) 368–391
16. Simpson, C.T.: Harmonic bundles on noncompact curves. J.A.M.S. **3** (1990) 713–770
17. Simpson, C.T.: Higgs bundles and local systems. Publ. Math. I.H.E.S. (to appear)
18. Siu, Y.T.: Complex analyticity of harmonic maps and strong rigidity of complex Kähler manifolds. Ann. Math. **112** (1980) 73–110
19. Uhlenbeck, K.K., Yau, S.T.: On the existence of Hermitian-Yang-Mills connections in stable vector bundles. Comm. Pure and Appl. Math. **39-S** (1986) 257–293

Arithmetic and Hyperbolic Geometry

Paul Vojta

Department of Mathematics, University of California, Berkeley, CA 94720, USA

We begin by recalling Faltings' theorem (née Mordell's conjecture):

Theorem 0.1 (Faltings, 1983). *Let C be a curve of genus > 1 defined over a number field k. Then the set $C(k)$ of k-rational points on C is finite.*

Compact Riemann surfaces of genus > 1 are also characterized by the property that they are hyperbolic, in any of the senses discussed in the next section. Recent work has suggested that a similar link exists, formally at least, between arithmetic properties of algebraic varieties (of any dimension) and complex analytic properties of the corresponding complex manifold. The goal of this talk is to discuss generalizations of Theorem 0.1 which have been motivated by this analogy.

1. Hyperbolicity

We begin by defining several notions of hyperbolicity. Throughout this section let X be a connected complex manifold, not necessarily compact.

Definition 1.1. The manifold X is *negatively curved* if there exists a $(1, 1)$-form ω on X and a constant $\kappa > 0$ such that all holomorphic sectional curvatures of ω are $< -\kappa$.

Definition 1.2 (Kobayashi, 1970). The Poincaré distance d_{hyp} on the open unit disc $\mathbb{D} \subseteq \mathbb{C}$ is the distance given infinitesimally by the form

$$\frac{dz \wedge d\bar{z}}{(1 - |z|^2)^2} .$$

Then X is said to be *Kobayashi hyperbolic* if there exists a distance d_X on X (with $d_X(x, y) > 0$ whenever $x \neq y$) such that for all holomorphic maps $f : \mathbb{D} \to X$, $f^* d_X \leq d_{\text{hyp}}$.

Definition 1.3. The manifold X is *Brody hyperbolic* if all holomorphic maps from $\mathbb{C}$ to X are constant.

Proceedings of the International Congress
of Mathematicians, Kyoto, Japan, 1990

Actually, the latter two definitions are valid when X is a complex space. We have

$$\text{negatively curved} \implies \text{Kobayashi hyperbolic} \implies \text{Brody hyperbolic}.$$

The converse of the second arrow holds if X is compact, but not in general.

A general reference on hyperbolicity is Lang (1987); see also Kobayashi (1970).

2. Conjectures in Number Theory

The first conjecture relating complex analytic and arithmetic properties is the following.

Conjecture 2.1 (Lang, 1974). *A complete variety X/k has only finitely many rational points if the corresponding complex space is Kobayashi hyperbolic.*

More recently the above statement, which is qualitative in nature, has been replaced by a more quantitative version. To describe it, we first state a theorem.

Theorem 2.2. *Let D be an effective divisor on a complete irreducible nonsingular curve C, and assume that D has no multiple points. Let K be a canonical divisor on C, let A be an ample divisor on C, and fix $\varepsilon > 0$. Then for almost all symbols "?,"*

$$m(D, ?) + T_K(?) \le \varepsilon\, T_A(?) + O(1)\,. \tag{1}$$

Supplied with one set of definitions, this theorem is Nevanlinna's Second Main Theorem. Indeed, let C be a compact Riemann surface, and let $f : \mathbb{C} \to C$ be a holomorphic function, regarded as an infinite collection of maps $f_r : \overline{\mathbb{D}}_r \to C$ (here $\overline{\mathbb{D}}_r$ is the closed disc of radius r, $r > 0$). Replace the symbol "?" with r, and interpret "almost all" to mean all r outside a set of finite Lebesgue measure. Choose some distance function $\operatorname{dist}(P, Q)$ on C, and extend it to divisors by the formula

$$\operatorname{dist}(P, \Sigma n_Q \cdot Q) = \prod \operatorname{dist}(P, Q)^{n_Q}\,.$$

Then, for any divisor D on C let

$$m(D,r) = \int_0^{2\pi} -\log \operatorname{dist}(f(re^{i\theta}), D)\, \frac{d\theta}{2\pi}\,;$$

$$N(D,r) = \sum_{w \in \mathbb{D}_r} \operatorname{ord}_{z=w} f^*D \log \frac{r}{|w|}\,;$$

$$T_D(r) = m(D,r) + N(D,r)\,.$$

This gives a weak form of the Second Main Theorem for curves.

With a different set of definitions, we obtain an arithmetic result which goes by several names, depending on the genus of the curve. Let C be a curve defined over a number field k, and assume also that D is defined over k. We now replace "?" by points $P \in C(k) \setminus \operatorname{Supp} D$, and interpret "almost all" to mean all but finitely many points P. Fix a finite set S of places of k, and for all places v of k let $\operatorname{dist}_v()$ denote some fixed distance on C in the v-adic topology, extended to divisors as before. It is assumed that these are chosen uniformly in some sense, as in Lang (1983). Then for any divisor D on C defined over k, let

$$\begin{aligned} m(D,P) &= \frac{1}{[k:\mathbb{Q}]}\sum_{v\in S} -\log \operatorname{dist}_v(P,D)\,; \\ N(D,P) &= \frac{1}{[k:\mathbb{Q}]}\sum_{v\notin S} -\log \operatorname{dist}_v(P,D)\,; \\ T_D(P) &= m(D,P)+N(D,P) \\ &= \frac{1}{[k:\mathbb{Q}]}\sum_{v} -\log \operatorname{dist}_v(P,D)\,. \end{aligned}$$

The quantity T_D is thus the Weil height of P relative to D; in particular if D is ample then there are only finitely many points $P \in C(k)$ of bounded height. In the arithmetic case we will also write $h_D(P)$ for the height.

Proof of Theorem 2.2 (Algebraic Variant). Let g denote the genus of C.

Case 1. If $C(k)$ is finite then the inequality must hold; by Faltings' theorem 0.1 this must hold if $g > 1$. Conversely, if $g > 1$ then K is ample, so taking $D = 0$, $A = K$, and $\varepsilon < 1$, we see that Theorem 2.2 implies a bound on the height of rational points; hence it implies Mordell's conjecture.

Case 2. If $g = 0$ then Theorem 2.2 is equivalent to Roth's theorem:

Theorem (Roth, 1955). *For each $v \in S$ fix $\alpha_v \in \overline{\mathbb{Q}}$. Then for all $x \in k \setminus \{\alpha_v\}$,*

$$\frac{1}{[k:\mathbb{Q}]}\sum_{v\in S} -\log\min(\|x-\alpha_v\|_v, 1) < (2+\varepsilon)h(x)\,. \tag{2}$$

Here $h(x) = h_A(x)$ with $A = \mathcal{O}(1)$. For details, see Vojta (1987), Sect. 3.2.

Case 3. If $g = 1$, then Theorem 2.2 is equivalent to an approximation statement on elliptic curves, derived by Lang from Roth's theorem using methods of Siegel. See Lang (1960b) for details. □

Thus we find that the Second Main Theorem of Nevanlinna theory translates into the number field case, giving several known theorems. Generally speaking, any such theorem in Nevanlinna theory should translate into a true statement for number fields. Such statements often agree with diophantine conjectures previously made. See also Sect. 8, as well as Vojta (1987), Chap. 4.

Again, this framework is related to hyperbolicity because proofs of Nevanlinna theory exist using notions of curvature.

In Nevanlinna theory, the analogues of Roth's theorem and Mordell's conjecture are both proved by a common proof. This suggested that Mordell's conjecture should be provable by methods similar to those used in the proof of Roth's theorem. In fact, such a proof was found in Vojta (1989) and Vojta (1990).

3. Sketch of Roth's Proof

Roth's proof proceeds by contradiction. Assume that there are infinitely many $x \in k$ not satisfying (2). Then we choose a finite set with certain properties. These are used to construct an auxiliary polynomial, the properties of which lead to a contradiction.

To start, let $\alpha_1, \ldots, \alpha_m$ be all the conjugates of all α_v, $v \in S$. We construct a polynomial $P(x_1, \ldots, x_n)$, of degree d_i in x_i, with a specified type of zero at all points $(\alpha_i, \ldots, \alpha_i)$, $1 \le i \le m$. This is viewed as a linear algebra problem, with the coefficients of the polynomial as variables. The vanishing conditions are set up so as to use up almost all of the degrees of freedom in choosing P; by the Dirichlet Box Principle, we can find such a polynomial in $R[x_1, \ldots, x_n]$ with bounded coefficients, where R is the ring of integers of k. This step is often referred to as Siegel's lemma.

Next we choose a set of counterexamples to (2) such that

$$1 \ll h(x_1) \ll \cdots \ll h(x_n)$$

and such that, for all $v \in S$, the numbers

$$\frac{-\log \min(\|x_i - \alpha_v\|_v, 1)}{h(x_i)}, \qquad 1 \le i \le n$$

are close to each other. This involves a sphere packing argument.

Finally, for each $v \in S$, using the Taylor expansion of P at $(\alpha_v, \ldots, \alpha_v)$, the type of zero of P at that point, and the bound on the coefficients of P, we obtain a bound for $\|P(x_1, \ldots, x_n)\|_v$ and therefore for its norm. If the norm is too small, we obtain a contradiction unless $P(x_1, \ldots, x_n) = 0$. A similar argument applied to certain derivatives of P implies that they vanish as well. Hence P has a certain type of zero at $(x_1, \ldots, x_n)$. But this contradicts Roth's lemma: the type of zero of P is bounded from above, depending on the size of the coefficients of P and the heights of the x_i.

4. Vojta's Proof

A key difficulty in extending Roth's proof to the case of rational points is dealing with the absence of the α's, which are needed to (a) provide vanishing conditions which use up the degrees of freedom in choosing P, and (b) prove the vanishing of $P(x_1, \ldots, x_n)$, via the assumption that (2) fails to hold.

This difficulty is overcome by means of intersection theory; in particular, we make use of the fact that the diagonal divisor Δ on $C \times C$ is a divisor with negative self-intersection. Also let F_1 (resp. F_2) be a divisor on $C \times C$ coming from a divisor F of degree 1 on the first (resp. second) factor. Then let

$$\Delta' = \Delta - F_1 - F_2\,;$$
$$Y_r = \Delta' + a_1 F_1 + a_2 F_2\,,$$

where $a_i \in \mathbb{Q}$ are chosen so that $a_1/a_2 = r$ and Y_r^2 is positive but small. Then, by Riemann-Roch, high multiples of Y_r will have relatively few global sections; this is how we use up our degrees of freedom in this case. (Here r is a large rational number.)

The proof replaces many of the classical arguments involving polynomials and absolute values with the language of arithmetic intersection theory, as developed by Arakelov, Gillet, and Soulé. Therefore, we work on an arithmetic scheme W of dimension 3 corresponding to $C \times C$, and replace the Siegel's lemma argument with a use of the Riemann-Roch theorem of Gillet and Soulé, to construct a section y of a multiple of Y_r with certain arithmetical properties.

Next we choose points, only two this time, such that

$$1 \ll h(P_1) \ll h(P_2)$$

and satisfying a certain sphere packing condition again – this time in the Mordell-Weil group of the Jacobian of C. Let X be the arithmetic surface corresponding to C; for $i = 1, 2$ let E_i be the (Arakelov) divisor on X corresponding to P_i, and let E be the curve on W corresponding to (P_1, P_2). Then this sphere packing condition implies that $(E_1 \,.\, E_2)$ on X will be small, and therefore $(E \,.\, \Delta)$ on W will be small. Then $(E \,.\, Y_r)$ will be small and in fact negative. Thus y must be zero on E; similarly certain derivatives must vanish, giving a lower bound for the type of zero as before. Again, this produces a contradiction.

As is the case with Roth's theorem, this proof – as well as all proofs derived from it – is ineffective; i.e., it can give a bound on the *number* of rational points but not on their heights. Thus it is often true that one cannot prove that a given set of rational points is the set of *all* rational points.

5. Extensions Due to Faltings

In Faltings (1990), Vojta's methods were generalized, giving two new theorems.

Theorem 5.1 (Conjectured in Lang (1960a)). *Let X be a complete subvariety of an abelian variety A, and assume that X does not contain any translates of any nontrivial abelian subvarieties of A. Let k be a number field over which X and A are defined. Then $X(k)$ is finite.*

Theorem 5.2. *Let A be an abelian variety, and let $Y \subseteq A$ be a subvariety. Let k be a number field over which A and Y are defined, let v be a place of k, and let* $\mathrm{dist}_v(x, Y)$ *be a v-adic distance on A. Fix $\varepsilon > 0$, and fix a height function $h(x)$ on A relative to some ample divisor. Then the set of $x \in A(k) \setminus Y$ such that*

$$-\log \mathrm{dist}_v(x, Y) > \varepsilon h(x)$$

is finite.

Corollary 5.3 (Conjectured in Lang (1960a)). *Let H be an ample divisor on A. Then the set of integral points on A(k) relative to H is finite.*

In proving these results, Faltings adds the following ingredients to Vojta's work.

1. Use of the Poincaré divisor on $A \times A$ in place of Δ'.

2. Use of the product of A with itself n times instead of just two. The divisor he considers is again a combination of divisors from each factor, together with Poincaré divisors associated to the i-th and $(i+1)$st factors, $1 \le i < n$.

3. A new zero estimate (the "product theorem") is needed to deal with technical issues concerning the use of n factors.

4. Instead of proving separately an upper bound on the type of zero of the section y at $(P_1, \ldots, P_n)$, he builds this upper bound into the construction of y, using Minkowski's theorem on successive minima. Also, the use of the Gillet-Soulé Riemann-Roch theorem is replaced by more elementary manipulations.

6. Recent Work of Bombieri

A few months ago Bombieri reworked Vojta's proof, obtaining a more elementary proof using only the classical theory of heights, Siegel's lemma, and a few necessary results from algebraic geometry over a field; see Bombieri (1990). In particular, he also eliminated the use of the Gillet-Soulé Riemann-Roch theorem, but in a manner independent of the extensions due to Faltings.

Bombieri uses the same divisor Y as in Sect. 4, but he expresses it as a difference of two very ample divisors, as follows.

First choose a positive integer s such that

$$B := sF_1 + sF_2 - \Delta'$$

is very ample; write

$$\phi_B : C \times C \to \mathbb{P}^m$$

for the corresponding embedding. Likewise choose a divisor F of degree 1 on C and a positive integer N such that NF is very ample, giving an embedding

$$\phi_{NF} : C \to \mathbb{P}^n .$$

This gives another embedding

$$\psi : C \times C \to \mathbb{P}^n \times \mathbb{P}^n .$$

It then follows from the Enriques-Severi-Zariski lemma (Zariski 1952) that for all sufficiently large integers δ_1, δ_2, the map

$$\Gamma(\mathbb{P}^n \times \mathbb{P}^n, \mathcal{O}(\delta_1, \delta_2)) \to \Gamma(C \times C, \delta_1 NF_1 + \delta_2 NF_2)$$

is surjective. Thus, writing

$$dY = d_1 NF_1 + d_2 NF_2 - dB\,,$$

we find that for any sections s of dY and s' of dB, the product ss' is the restriction to $C \times C$ of a global section on $\mathbb{P}^n \times \mathbb{P}^n$. Thus, it is possible to revert to use of polynomials in this proof.

Since he is working with polynomials over a field, the arguments in Vojta's proof regarding analytic torsion and cohomology over finite fields are unnecessary.

Also, to obtain the upper bound on the type of zero of the section at (P_1, P_2), Bombieri uses suitably chosen projections from C to $\mathbb{P}^1$; thus he is able to use the same lemma as Roth uses.

This latter simplification, together with the ideas of Faltings concerning use of Siegel's lemma in place of the Gillet-Soulé Riemann-Roch theorem, remove the obstacles to combining this proof with Roth's original proof. Thus we have a unified proof of the number field version of Theorem 2.2.

Of course, combining the arithmetic proof with the proof in the case of Nevanlinna theory remains a distant goal.

7. Algebraic Points of Bounded Degree

The unified proof mentioned above can be further extended to give a generalization of Theorem 2.2 which contains Wirsing's generalization, Wirsing (1971), of Roth's theorem to algebraic points of bounded degree.

Before stating the theorem, however, let X be an arithmetic surface corresponding to C, and for all points $P \in C(\bar{k})$ let E_P denote the corresponding Arakelov divisor on X. Then let

$$d_a(P) = \frac{(E\,.\,E + \omega_{X/B})}{[k(P) : \mathbb{Q}]}\,,$$

where $\omega_{X/B}$ is the relative dualizing sheaf.

Theorem 7.1. *Let C, k, D, K, A, and ε be as in Theorem 2.2. Also fix $\nu \in \mathbb{Z}$, $\nu > 0$. Then for all points $P \in C(\bar{k})$ with $[k(P) : k] \le \nu$, we have*

$$m(D, P) + h_K(P) \le d_a(P) + \varepsilon\, h_A(P) + O(1)\,.$$

This is a weak form of the following conjecture.

Conjecture 7.2. *Under the same conditions as Theorem 7.1, we have*

$$m(D, P) + h_K(P) \le d(P) + \varepsilon\, h_A(P) + O(1)\,,$$

where

$$d(P) := \frac{\log |D_{k(P)/\mathbb{Q}}|}{[k(P) : \mathbb{Q}]}\,.$$

This conjecture is very strong; in particular it implies the *abc* conjecture:

Conjecture 7.3 (Masser-Oesterlé). *Given* $\varepsilon > 0$ *there exists a constant* $C = C(\varepsilon) > 0$ *such that for all relatively prime integers* a, b, *and* c *with* $a + b + c = 0$, *we have*

$$\max(|a|, |b|, |c|) < C \cdot \prod_{p|abc} p^{1+\varepsilon}.$$

This conjecture, in turn, implies the "asymptotic Fermat conjecture:"

Conjecture 7.4. *For all sufficiently large integers* n, *the only rational solutions to the equation*

$$X^n + Y^n + Z^n = 0$$

are the trivial ones, i.e., $XYZ = 0$.

8. Open Problems

In addition to the conjectures mentioned just above, the following problems are still unresolved at this time.

1. (Lang, 1965) Replace the error term $\varepsilon h_A(P)$ in the arithmetic variant of Theorem 2.2 with something sharper, e.g., $(1 + \varepsilon) \log h_A(P)$.

2. (Vojta, 1987) Prove a formula similar to (1) for rational points on varieties of higher dimension. In that case, however, one would first have to exclude a proper Zariski-closed subset Z from the set of points P. See also Schmidt (1975) and Vojta (1987), Example 3.5.1 for a discussion of a partial answer to this question, where the variety is projective space and the divisor D is a collection of hyperplanes.

3. (Lang, 1960b, p. 29) Generalize Theorem 5.1 to the case where X may contain the translate of an abelian subvariety of A. In this case the translated abelian subvarieties may contain infinitely many rational points, but it is conjectured that the rational points are not Zariski dense, unless X itself is a translated abelian subvariety of A. In Nevanlinna theory the analogue of this conjecture is called Bloch's conjecture; it was proved independently by Kawamata (1980) and Green and Griffiths (1980), both using work of Ochiai (1977). (In loc. cit., Lang actually makes the stronger conjecture that the rational points are contained in finitely many translated sub-abelian varieties contained in X; this conjecture is a consequence of the above conjecture, by the Kawamata structure theorem.)

References

Bombieri, E. (1990): The Mordell conjecture revisited. (To appear)

Faltings, G. (1983): Endlichkeitssätze für abelsche Varietäten über Zahlkörpern. Invent. math. **73**, 349–366; corrigendum Invent. math. **75**, 381

Faltings, G. (1990): Diophantine approximation on abelian varieties. Ann. Math. (to appear)

Green, M., Griffiths, P. (1980): Two applications of algebraic geometry to entire holomorphic mappings. The Chern Symposium 1979 (Proceedings of the International Symposium on Differential Geometry in Honor of S.-S. Chern, held in Berkeley, California, June, 1979). Springer, New York Heidelberg Berlin, pp. 41–74

Kawamata, Y. (1980): On Bloch's conjecture. Invent. math. **57**, 97–100

Kobayashi, S. (1970): Hyperbolic manifolds and holomorphic mappings. Marcel Dekker, New York

Lang, S. (1960a): Some theorems and conjectures in diophantine equations. Bull. AMS **66**, 240–249

Lang, S. (1960b): Integral points on curves. Publ. Math. IHES **6**, 27–43

Lang, S. (1965): Report on diophantine approximations. Bull. Soc. Math. France **93**, 177–192

Lang, S. (1974): Higher dimensional diophantine problems. Bull. AMS **80**, 779–787

Lang, S. (1983): Fundamentals of diophantine geometry. Springer, New York

Lang, S. (1987): Introduction to complex hyperbolic spaces. Springer, New York

Ochiai, T. (1977): On holomorphic curves in algebraic varieties with ample irregularity. Invent. math. **43**, 83–96

Roth, K. F. (1955): Rational approximations to algebraic numbers. Mathematika **2**, 1–20

Schmidt, W. M. (1975): Application of Thue's method in various branches of number theory. Proceedings of the International Congress of Mathematicians, R. P. James, ed., Canadian Mathematical Congress, Vancouver, pp. 177–186

Vojta, P. (1987): Diophantine approximations and value distribution theory. (Lecture Notes in Mathematics, vol. 1239). Springer, Berlin Heidelberg New York

Vojta, P. (1989): Mordell's conjecture over function fields. Invent. math. **98**, 115–138

Vojta, P. (1990): Siegel's theorem in the compact case. Ann. Math. (to appear)

Wirsing, E. (1971): On approximation of algebraic numbers by algebraic numbers of bounded degree. Proc. of Symp. in Pure Math. **XX** (1969 Number Theory Institute), pp. 213–247

Zariski, O. (1952): Complete linear systems on normal varieties and a generalization of a lemma of Enriques-Severi. Ann. Math. **55**, 552–592

References

Author Index

I indicates Volume I while II indicates Volume II.

Proceedings of the

International Congress of Mathematicians

August 21–29, 1990, Kyoto, Japan

II

Proceedings of the

International Congress of Mathematicians

ICM-90
International
Congress of
Mathematicians
KYOTO

August 21–29, 1990
Kyoto, Japan

Volume II

The Mathematical Society of Japan

Springer-Verlag

Tokyo Berlin Heidelberg New York
London Paris Hong Kong Barcelona
Budapest

International Congress of Mathematicians
August 21–29, 1990, Kyoto, Japan

Editor:
Ichiro Satake
Mathematical Institute
Faculty of Science
Tohoku University
Sendai 980, Japan

The logo for the ICM-90, designed by Kazuyoshi Aoki and Yuji Komai, symbolizes a Japanese stone lantern, the first character for Kyoto, as well as the character for 10^{16}.

With 96 figures, including 11 halftone illustrations

ISBN 4-431-70047-1 Set (2 volumes)
Springer-Verlag Tokyo Berlin Heidelberg New York

ISBN 3-540-70047-1 in 2 Bänden
Springer-Verlag Berlin Heidelberg New York Tokyo

ISBN 0-387-70047-1 Set (2 volumes)
Springer-Verlag New York Berlin Heidelberg Tokyo

Library of Congress Cataloging-in-Publication Data
International Congress of Mathematicians (1990: Kyoto, Japan)
Proceedings of the International Congress of Mathematicians, August 21-29, 1990, Kyoto / edited by Ichiro Satake.
p. cm. Includes bibliographical references. ISBN 0-387-70047-1
1. Mathematics – Congresses. I. Satake, Ichiro. 1927–. II. Title QA1.I82 1990 510–dc20
91-4972 CIP

Printed in Hong Kong

Typesetting: Asco Typesetting Ltd., Hong Kong; Springer TEX in-house system and typesetting output by Universitätsdruckerei H. Stürtz AG, Würzburg, Fed. Rep. of Germany
Printing and binding: Permanent Typesetting & Printing Co., Ltd., Hong Kong

543210

Contents

Volume I

Volume II

Scientific Program

Invited One-Hour Addresses at the Plenary Sessions

Invited Forty-Five Minute Addresses at the Section Meetings

Section 1: Mathematical Logic and Foundations

There were 15 Short Communications in this section.

Section 2: Algebra

There were 59 Short Communications in this section.

Section 3: Number Theory

There were 29 Short Communications in this section.

Section 4: Geometry

There were 38 Short Communications in this section.

Section 5: Topology

There were 43 Short Communications in this section.

Section 6: Algebraic Geometry

There were 25 Short Communications in this section.

Section 7: Lie Groups and Representations

There were 23 Short Communications in this section.

Section 8: Real and Complex Analysis

There were 73 Short Communications in this section.

Section 9: Operator Algebras and Functional Analysis

There were 52 Short Communications in this section.

Section 10: Probability and Mathematical Statistics

There were 37 Short Communications in this section.

Section 11: Partial Differential Equations

There were 48 Short Communications in this section.

Section 12: Ordinary Differential Equations and Dynamical Systems

There were 41 Short Communications in this section.

Section 13: Mathematical Physics

There were 39 Short Communications in this section.

Section 14: Combinatorics

There were 25 Short Communications in this section.

Section 15: Mathematical Aspects of Computer Science

There were 5 Short Communications in this section.

Section 16: Computational Methods

There were 26 Short Communications in this section.

Section 17: Applications of Mathematics to the Sciences

There were 18 Short Communications in this section.

Section 18: History, Teaching and the Nature of Mathematics

There were 22 Short Communications in this section.

In addition there were 2 Short Communications in the Post Deadline Session.

Unipotent Representations for Real Reductive Groups

Dan Barbasch

Cornell University, Ithaca, NY 14853 and
Rutgers University, New Brunswick, NJ 08903, USA

1. Introduction

The classification of the unitary dual of a real reductive group is an important problem in Representation theory. Typically one proceeds as follows. Given the Levi component of a parabolic subgroup, there are certain constructions such as unitary induction, derived functors and complementary series that preserve unitarity. Thus, given a group G it is reasonable to ask for a set $\mathscr{U}(G)$ such that the unitary dual is obtained by *unitarity preserving constructions* from all such sets on Levi components of the group. Such a set is provided in the case of integral infinitesimal character by the *special unipotent representations.* In particular the question of the unitarity of this set of representations arises.

The purpose of this lecture is to describe the proof of the unitarity of a particular subset of special unipotent representations. The precise result is as follows.

Theorem 1.1. *The representations in the Arthur L-packet in the special unipotent case (see Section 2) for the classical groups are unitary.*

The main idea of the proof of this theorem is the same as for the corresponding fact for the complex classical groups in Section 10 of [B]. The question of unitarity of complicated representations on a group is reduced to the same question for simpler representations on larger groups. The techniques are general, so that they apply to all special unipotent representations. For (serious) technical reasons, I can only carry out the proof for the aforementioned case.

Many of the more abstract results about unipotent representations can be phrased in a more general setting. This work, joint with J. Adams and D. Vogan will appear in [ABV].

In the case of a split classical group, [M] and [MW] have shown that the spherical unipotent representations actually occur in the residual spectrum.

In Section 2, I introduce the packet $\Pi({}^L\mathcal{O})$ of special unipotent reprsentations and give a definition of the Arthur L-packet. Section 3 is devoted to a computation of the size of $\Pi({}^L\mathcal{O})$. Combined with the constructions in Section 4, this gives all the special unipotent representations. Theorem 4.4 states that for the unitarity

Proceedings of the International Congress
of Mathematicians, Kyoto, Japan, 1990

of the representations in the Arthur L-packet it is enough to consider only *smoothly cuspidal* (definition at the start of Section 4) nilpotent orbits. The main result is contained in Theorem 5.3 where $\Pi({}^L\mathcal{O})$ is broken up according to WF-sets in the Lie algebras of the group and the dual group. In Section 6, I sketch the proof of the unitarity of the spherical representations in the Arthur L-packet for $Sp(2n,\mathbb{R})$. The idea is the following. We do an induction on how close π is to being obtained by *unitarity preserving functors* (unitary induction from a real parabolic subalgebra and derived functor construction in the appropriate range) from a strictly smaller parabolic subalgebra. Let $\mathfrak{g}(n)$ be the Lie algebra $sp(2n,\mathbb{R})$ of *rank* n and $\pi({}^L\mathcal{O})$ be the spherical representation in $\Pi({}^L\mathcal{O})$. Assume that it cannot be obtained by *unitarity preserving functors.* from a strictly smaller parabolic subalgebra. Then, for a well chosen value r, we embed $\mathfrak{g}(n)\times gl(r+1)$ as a Levi component of a real parabolic subalgebra $\mathfrak{p}(\mathbb{R})$ in $\mathfrak{g}(n+r+1)$. We then form $I(\pi)=\mathrm{Ind}_{P(\mathbb{R})}^{G(\mathbb{R})}[\Pi({}^L\mathcal{O})\otimes\mathrm{Triv}]$. Since $\Pi({}^L\mathcal{O})$ is hermitian, the induced representation inherits an *induced hermitian form.* On the other hand, it decomposes into a sum of two irreducible representations which can be seen to arise *earlier* than $\pi(\mathcal{O})$ in the induction. To finish the proof we need to see that the two factors have the same signature in the *induced form* of $I(\pi)$. This is a simple direct calculation of a signature on one K-type in $I(\pi)$ and $\Pi({}^L\mathcal{O})$.

Unless mentioned otherwise, I will use the following notation. The complex group $Sp(2n,\mathbb{C})$ and its Lie algebra will be denoted by G and $\mathfrak{g}$. The real form $Sp(2n,\mathbb{R})$ will be $G(\mathbb{R})$ and its real Lie algebra $\mathfrak{g}(\mathbb{R})$. The dual Lie algebra $so(2n+1)$ will be denoted ${}^L\mathfrak{g}$, its various real forms $so(p,q)$ by $\check{\mathfrak{g}}(\mathbb{R})$, and the corresponding groups by ${}^LG^0$ and $\check{G}(\mathbb{R})$.

Complex nilpotent orbits in $\mathfrak{g}$ will be denoted by $\mathcal{O}$, in the dual ${}^L\mathfrak{g}$ by ${}^L\mathcal{O}$.

θ will denote the complexification of the Cartan involution for $Sp(2n,\mathbb{R})$. Similarly $\check{\theta}$ will be the complexified Cartan involution of one of the real forms of ${}^L\mathfrak{g}$. The corresponding decompositions are $\mathfrak{g}=\mathfrak{k}+\mathfrak{s}$ and $\check{\mathfrak{g}}=\check{\mathfrak{k}}+\check{\mathfrak{s}}$.

Real forms of the orbit $\mathcal{O}$ will be denoted by $\mathcal{O}(\mathbb{R})$. Similarly for real forms of ${}^L\mathcal{O}$. By [S], (see also [V4] Section 4), real forms of $\mathcal{O}$ are in 1–1 correspondence with nilpotent orbits of K_c on $\mathfrak{s}$. We will use this identification without explicit mention.

Acknowledgements. This research was partially supported by NSF grant DMS–8803500. I also wish to thank David Vogan for several very valuable comments on the material of this talk.

2. Special Unipotent Representations

In [A1] and [A2], Arthur introduces a Langlands parameter attached to a homomorphism

$$\psi : W_{\mathbb{R}}\times SL(2,\mathbb{C})\longrightarrow {}^LG, \tag{2.1}$$

where $W_{\mathbb{R}}$ is the Weil group for $\mathbb{R}$. He conjectures that there should be an entire packet $\Pi(\psi)$ of representations attached to ψ, satisfying various properties relevant to the study of automorphic forms via the trace formula.

Because the group we are dealing with is split, ${}^LG$ is a direct product. Therefore maps as in (2.1) are determined by homomorphisms into ${}^LG^0$, the connected group. We will only consider the case of ψ's that are trivial on the identity component of $W_{\mathbb{R}}$ and such that $\psi|_{SL(2,\mathbb{C})}$ determines an even nilpotent orbit in the dual algebra. This case is called *special unipotent.*

Let ${}^L\mathcal{O}$ be the even nilpotent orbit determined by ψ and $\mathcal{O}$ be the dual orbit in $\mathfrak{g}$ in the sense of [L]. Let E, H, F, be the standard generators of $sl(2)$ and write

$$ {}^Le = d\psi(E)\,, \qquad {}^Lh = d\psi(H)\,, \qquad {}^Lf = d\psi(F)\,. \tag{2.2} $$

The infinitesimal character of the Arthur parameter is given by

$$ \lambda(\psi) = \lambda = \frac{1}{2}d\psi(H)\,. \tag{2.3} $$

This depends only on $\psi|_{SL(2,\mathbb{C})}$.

Definition 2.4. The *extended packet* $\Pi({}^L\mathcal{O})$ corresponding to ψ is the set of irreducible representations with annihilator in the universal enveloping algebra equal to the *special unipotent primitive ideal* attached to ${}^L\mathcal{O}$.

In particular, $\overline{\operatorname{Ad} G\cdot WF(\pi)} = \overline{\mathcal{O}}$, for $\pi\in\Pi({}^L\mathcal{O})$ as well as $\overline{\operatorname{Ad} G\cdot\mathcal{A}(\pi)} = \overline{\mathcal{O}}$, where $\mathcal{A}$ denotes the associated variety defined in [V4]. $\Pi({}^L\mathcal{O})$ corresponds to the union of the $\Pi(\psi)$ giving the same nilpotent orbit ${}^L\mathcal{O}$.

If $j\in W_{\mathbb{R}}$ is the element representing complex conjugation, then let $m=\psi(j)$. This is an element of order 2 in $C_{{}^LG}({}^Le, {}^Lh, {}^Lf)$. Then

$$ \check{\theta}_m = \operatorname{Ad}(m\cdot e^{i\pi\lambda}) \tag{2.5} $$

is an involution. Let $\check{G}_m(\mathbb{R})$ be the real form of ${}^LG$ with $\check{\theta} = \check{\theta}_m$. The parabolic subalgebra determined by λ is $\check{\theta}$-stable. Denote it by $\check{\mathfrak{p}}(\lambda)$ and fix a $\check{\theta}$-stable Cartan subalgebra $\check{\mathfrak{h}}$ containing λ and a Borel subalgebra $\check{\mathfrak{h}}\subset\check{\mathfrak{b}}$ for which λ is dominant. Then ${}^L\mathcal{O}$ meets $\check{\mathfrak{s}}\cap\check{\mathfrak{n}}(\lambda)$, so this also determines a real form ${}^L\mathcal{O}(\mathbb{R})$ of ${}^L\mathcal{O}$.

The representations in the Arthur L-packet attached to ψ can then be constructed as follows.

Let $\check{\mathcal{L}}(\psi) = \mathcal{A}_{\check{\mathfrak{p}}}$ be the irreducible representations (the Levi component may be disconnected) with infinitesimal character $\check{\varrho}$ which are obtained by derived functor construction from a character of $\check{\mathfrak{p}}(\lambda)$, as in [V1, VZ]. [V3] attaches to each such $\check{\mathcal{L}}(\psi)$ a standard modules $X_{\mathrm{reg}}(\psi)$ with regular integral infinitesimal character on a quasisplit real form $G_{\mathrm{qs}}(\mathbb{R})$ of G. Let $X(\psi)$ be the standard module obtained from $X_{\mathrm{reg}}(\psi)$ by applying the translation functor to infinitesimal character λ. Then the Arthur L-packet is the set of irreducible quotients of the $X(\psi)$. This coincides with the usual definition of L-packet used in [A1] and [A2].

Example. Let ${}^L\mathcal{O}$ be the nilpotent orbit with Jordan blocks 31^2 in $so(5)$. In the standard coordinates coming from embedding $sp(2n)$ in $gl(2n)$ the infinitesimal character is $(1,0)$ and $\mathcal{O}$ is the nilpotent with Jordan blocks 2^2. The centralizer of the Lie triple corresponding to ${}^L\mathcal{O}$ is isomorphic to $S[O(1)\times O(2)]$, so it has three conjugacy classes of elements of order 2. They correspond to two real orbits in $so(3,2)$ and one in $so(4,1)$. The most split Cartan subgroup of $Sp(4)$ is of the form MA where $M\cong\mathbb{Z}_2\times\mathbb{Z}_2$. If we assume that the first $\mathbb{Z}_2$ corresponds to the coordinate containing the 1 in the infinitesimal character, then the three principal series have M-characters Triv $\otimes$ Triv, Sgn $\otimes$ Sgn and Triv $\otimes$ Sgn. The Arthur L-packet of the first one contains one irreducible representation namely the spherical, the other ones two irreducible representations each.

3. The Coherent Continuation Representation

The results in this section hold for general reductive groups.

Fix a regular integral infinitesimal character χ_{reg}. Denote by $\mathcal{G}(\chi_{\text{reg}})$ the Grothendieck group of the category of $(\mathfrak{g}, K)$ modules with infinitesimal character χ_{reg}. Recall from [V1] that there is an action of the Weyl group on $\mathcal{G}(\chi_{\text{reg}})$, called the *coherent continuation action.* Then $\mathcal{G}(\chi_{\text{reg}})$ decomposes into a direct sum according to blocks $\mathcal{B}$,

$$\mathcal{G}(\chi_{\text{reg}})=\bigoplus\mathcal{G}_{\mathcal{B}}(\chi_{\text{reg}})\,. \tag{3.1}$$

Let $\mathfrak{h}_a\subset\mathfrak{g}$ be an abstract Cartan subalgebra and let Π_a be a set of (abstract) simple roots. For each irreducible representation $\mathcal{L}(\gamma)$, denote by $\tau(\gamma)$ the tau-invariant as defined in [V1]. Given a block $\mathcal{B}$ and disjoint orthogonal sets $S_1, S_2\subset\Pi_a$, define

$$\mathcal{B}(S_1,S_2)=\{\gamma\in\mathcal{B}|S_1\subset\tau(\gamma),\ S_2\cap\tau(\gamma)=\emptyset\}\,. \tag{3.2}$$

If in addition we are given a nilpotent orbit $\mathcal{O}\subset\mathfrak{g}$, we can also define

$$\mathcal{B}(S_1,S_2,\mathcal{O})=\{\gamma\in\mathcal{B}(S_1,S_2)|\ WF(\mathcal{L}(\gamma))\subset\overline{\mathcal{O}}\}\,. \tag{3.3}$$

Consider the case of $\mathfrak{g}$ viewed as a real Lie algebra. Then the case $S_1,\ S_2=\emptyset$ is called the double cone $\mathcal{C}(\mathcal{O})$. The double cell corresponding to $\mathcal{O}$ will be denoted $\overline{\mathcal{C}}(\mathcal{O})$.

Let $W_i=W(S_i)$, and define

$$\begin{aligned} m_S(\sigma)&=[\sigma:\operatorname{Ind}_{W_1\times W_2}^{W}(\text{Sgn}\otimes\text{Triv})]\,,\\ m_{\mathcal{B}}(\sigma)&=[\sigma:\mathcal{G}_{\mathcal{B}}(\chi_{\text{reg}})]\,. \end{aligned} \tag{3.4}$$

Theorem 3.5 (Vogan).

$$|\mathcal{B}(S_1,S_2,\mathcal{O})|=\sum_{\sigma\otimes\sigma\in\mathcal{C}(\mathcal{O})}m_{\mathcal{B}}(\sigma)m_S(\sigma)\,.$$

A sketch of the proof can be found in [BSS].

Recall $\lambda=\lambda(\psi)$ from Section 1. Then λ defines a set S_2 by

$$S_2=S(\psi)=\{\alpha\in\Pi_a|(\alpha,\lambda)=0\}\,. \tag{3.6}$$

Then

$$\Pi({}^L\mathcal{O}) = \bigcup_{\mathscr{B}} \mathscr{B}(\phi, S(\psi), \mathcal{O}). \tag{3.7}$$

In the classical groups case, $m_{\mathscr{B}}(\sigma)$ is explicitly computable. For the special unipotent case, $m_S(\sigma)$ equals 0 except for the representations occurring in the corresponding left cell $\overline{\mathscr{C}}^L(\mathcal{O})$ when it is 1 (see [BV]).

Theorem 3.8.

$$|\Pi({}^L\mathcal{O})| = \sum_{\mathscr{B}} \sum_{\sigma\otimes\sigma\in\overline{\mathscr{C}}^L(\mathcal{O})} m_{\mathscr{B}}(\sigma).$$

4. Induction from Parabolic Subgroups

Recall that a parabolic subalgebra $\mathfrak{p} = \mathfrak{m} + \mathfrak{n}$ is called *real for* $\mathfrak{g}(\mathbb{R})$ if $\theta(\mathfrak{p}) = \overline{\mathfrak{p}}$, *θ-stable for* $\mathfrak{g}(\mathbb{R})$, if $\theta(\mathfrak{p}) = \mathfrak{p}$.

A nilpotent $\mathcal{O}$ is called *induced from* $\mathcal{O}_{\mathfrak{m}}$ if $\mathcal{O} \cap (\mathcal{O}_{\mathfrak{m}} + \mathfrak{n})$ is dense in $\mathcal{O}_{\mathfrak{m}} + \mathfrak{n}$. It is called *smoothly induced* if it satisfies Hypotheses 8A and 8B in [BV].

Let $\mathfrak{p}(\xi) \subset \mathfrak{g}$ be be real or θ-stable. Let $\mathcal{O}_{\mathfrak{m}}(\mathbb{R}) \subset \mathfrak{m}(\mathbb{R})$ be a real form of $\mathcal{O}_{\mathfrak{m}}$. We define the *real induced set* from $\mathcal{O}_{\mathfrak{m}}(\mathbb{R})$ to be

$$\operatorname{ind}_{\mathfrak{p}}^{\mathfrak{g}}[\mathcal{O}_{\mathfrak{m}}(\mathbb{R})] = \overline{\bigcup_{t>0} \operatorname{Ad} G(\mathbb{R}) \cdot (t\xi + \mathcal{O}_{\mathfrak{m}}(\mathbb{R}))} \backslash \bigcup_{t>0} \operatorname{Ad} G(\mathbb{R}) \cdot (t\xi + \mathcal{O}_{\mathfrak{m}}(\mathbb{R})). \tag{4.1}$$

In the case of $\mathfrak{p}$ real, this set is the closure of $\operatorname{Ad} G[\mathcal{O}_{\mathfrak{m}}(\mathbb{R}) + \mathfrak{n}(\mathbb{R})]$.

In the case of θ-stable parabolic subalgebra in a classical group, this can be computed using unpublished results of D. Peterson. It is the closure of a single real orbit.

We give two constructions that yield special unipotent representations from special unipotent representations on Levi subgroups of proper parabolic subalgebras.

Induction. Let $\mathfrak{p}(\xi)$ be real or θ-stable. Assume that ${}^L\mathcal{O}_{\mathfrak{m}}$ and ${}^L\mathcal{O}$ are such that

$$\lambda|_{[\mathfrak{m},\mathfrak{m}]} = \lambda_{\mathfrak{m}}, \qquad \mathcal{O} = \operatorname{ind}_{\mathfrak{p}}^{\mathfrak{g}}[\mathcal{O}_{\mathfrak{m}}]. \tag{4.2}$$

Denote by λ_c the character of $\mathfrak{m}$ determined by λ on the center.

I. Let $\mathfrak{p}(\xi)$ be real. Then

$$\operatorname{Ind}_{P(\mathbb{R})}^{G(\mathbb{R})}[\pi_{\mathfrak{m}} \otimes \lambda_c]$$

has composition series formed of irreducible representations in $\Pi({}^L\mathcal{O})$ only.

II. Let $\mathfrak{p}(\xi)$ be θ-stable. Assume that $\pi \in \Pi(\mathcal{O}_{\mathfrak{m}})$ is such that $WF(\pi_{\mathfrak{m}})$ contains an orbit $\mathcal{O}_{\mathfrak{m}}(\mathbb{R})$ such that $\operatorname{ind}_{\mathfrak{p}(\mathbb{R})}^{\mathfrak{g}(\mathbb{R})}[\mathcal{O}_{\mathfrak{m}}]$ meets the orbit $\mathcal{O}$. Then

$$\mathscr{R}^i_{\mathfrak{p}}[\pi_{\mathfrak{m}} \otimes \lambda_c]$$

are not all zero and their composition series consists of irreducible representations in $\Pi({}^L\mathcal{O})$.

Coinduction. Let $\check{\mathfrak{p}} = \check{\mathfrak{m}} + \check{\mathfrak{n}}$ be a parabolic subalgebra, $\check{\theta}$-stable for some real form of $\check{\mathfrak{g}}$, such that $\check{\mathfrak{p}}(\lambda) \subset \check{\mathfrak{p}}$. Let $\mathfrak{p}$ be the dual real parabolic subalgebra. Let $\pi_{\mathfrak{m}} \in \Pi(\mathcal{O}_{\mathfrak{m}})$ be such that $\mathrm{ind}^{\check{\mathfrak{g}}}_{\check{\mathfrak{p}}}[{}^L\mathcal{O}_{\mathfrak{m}}(\mathbb{R})]$ meets the orbit ${}^L\mathcal{O}$. Then the irreducible representation

$$\check{\mathscr{R}}_{\mathfrak{p}} = P_\lambda([\mathscr{R}_{\check{\mathfrak{p}}}(\check{\pi}_{\mathfrak{m}})]^\vee) \tag{4.3}$$

is in $\Pi({}^L\mathcal{O})$. Recall that $\mathcal{O}$ is dual in the sense of [L] to the even nilpotent orbit ${}^L\mathcal{O}$. The same holds for $\mathcal{O}_{\mathfrak{m}}$ and ${}^L\mathcal{O}_{\mathfrak{m}}$.

Theorem 4.4. *Assume that $\mathcal{O}$ is smoothly induced from $\mathcal{O}_{\mathfrak{m}}$. Let π be in the Arthur L-packet determined by ${}^L\mathcal{O}$. Then there is $\pi_{\mathfrak{m}}$ in the Arthur L-packet determined by ${}^L\mathcal{O}_{\mathfrak{m}}$ such that π is the lowest K-type subquotient of* $\mathrm{Ind}^{G(\mathbb{R})}_{P(\mathbb{R})}[\pi_{\mathfrak{m}}]$.

In view of this, we say a nilpotent orbit is *smoothly cuspidal* if it is not smoothly induced from any proper parabolic subalgebra. They are as follows.

Type B. The largest Jordan block is odd size and occurs an odd number of times. All smaller odd sizes occur, and they each occur an even number of times.

Type C. The largest Jordan block is even size and occurs an even number of times. All smaller even sizes occur and they each occur an even number of times.

Type D. The largest Jordan block is odd size and occurs an even number of times. All smaller odd sizes occur, and they each occur an even number of times.

For the unitarity of representations in the Arthur L-packet, we can restrict to these cases.

5. The Main Result

The main result concerns the WF-set of the special unipotent representations. It is used to compute the decomposition of unitarily induced representations.

Recall the quotient of the component group $\overline{A}(\mathcal{O})$ from [L]. For a real form $\mathcal{O}(\mathbb{R})$, let $A(\mathcal{O}(\mathbb{R}))$ denote the component group of the correponding K_c-orbit in $\mathfrak{s}$.

Lemma 5.1. *If $\mathcal{O}$ is smoothly cuspidal, then $\overline{A}(\mathcal{O})$ can be identified with the full component group $A(\mathcal{O})$.*

For each finite dimensional representation F, recall the translation functors

$$T_F : X \mapsto P_\lambda[X \otimes F] \tag{5.2}$$

defined in [BV] Lemma 6.3.

Theorem 5.3. *Assume that $\mathcal{O}$ is smoothly cuspidal, and let $\mathcal{O}(\mathbb{R})$ be a real form.*

1. The WF-set of any $\pi \in \Pi(^L\mathcal{O})$ is the closure of a single orbit. Denote by $\Pi(\mathcal{O}(\mathbb{R}))$ the set of representations with WF-set $\overline{\mathcal{O}(\mathbb{R})}$, and by $\mathcal{G}(\mathcal{O}(\mathbb{R}))$ the corresponding (complexified) Grothendieck group.

2. The number of irreducible representations with WF-set $\mathcal{O}(\mathbb{R})$ is $|A(\mathcal{O}(\mathbb{R}))|$.

3. $A(\mathcal{O}(\mathbb{R}))$ maps onto $\overline{A}(\mathcal{O})$. $\mathcal{G}(\mathcal{O}(\mathbb{R}))$ decomposes under the (simultaneous) action of T_F in (5.1) into a direct sum of isomorphic subspaces, each of dimension $|A(\mathcal{O})|$.

Given $\pi \in \Pi(^L\mathcal{O})$, we can attach to it a pair of (unions of) real nilpotent orbits, $(WF(\pi), WF(\check{\pi}))$. This is an invariant for the action of T_F. The theorem is a consequence of the nontrivial computation of the behaviour of this invariant under induction and coinduction.

6. Unitarity

The proof of the unitarity of the distinguished representations attached to cuspidal nilpotent orbits proceeds by induction as in the complex case. Recall that the group is $Sp(2n, \mathbb{R})$. To conform to [B], we will write $\mathfrak{g}(n)$ for the Lie algebra $sp(2n, \mathbb{R})$ of rank n.

Recall that $^L\mathcal{O}$ is an even nilpotent and $\mathcal{O}$ is its dual. Assume that $\mathcal{O}$ is smoothly cuspidal. Let $\pi(\mathcal{O})$ be the *special unipotent spherical* representation in the Arthur L-packet attached to $^L\mathcal{O}$.

Proposition 6.1. *The WF-set of $\pi(\mathcal{O})$ is the closure of the unique $\mathcal{O}(\mathbb{R})$ for which there exists a Levi component of a real parabolic subalgebra $\mathfrak{m}$ such that $\mathcal{O}(\mathbb{R}) \cap \mathfrak{m}$ is a principal nilpotent in $\mathfrak{m}$.*

Proposition 6.2. *Assume that $\mathcal{O}$ is induced (not smoothly). Then the spherical representation in the Arthur L-packet is unitary. More precisely, it occurs as derived functor module from a unitary character on a θ-stable parabolic subalgebra satisfying the condition of Theorem 7.1(b) in* [V2]. *Its unitarity can also be seen from the fact that it occurs at the endpoint of a complementary series of an induced from a spherical distinguished representation on a Levi component of a real proper parabolic subalgebra.*

Proof of Theorem 1.1. Let $\mathcal{O}$ with Jordan blocks $1^{r_1}2^{r_2}\ldots(2k)^{r_{2k}}$ be a special cuspidal nilpotent as at the end of Section 4 (so that r_i are all even). We do an induction on the number of odd sized Jordan blocks. Assume that none of the $r_{2l-1} = 0$, for otherwise the representation is already unitary by Proposition 6.2. The case when $k = 1$, $r_1 = 2$ can be done by direct calculation.

Let $r = r_{2k}$. Denote the spherical distinguished representation attached to $\mathcal{O}$ by $\pi(^L\mathcal{O})$. Embed the Lie algebra $\mathfrak{m} = \mathfrak{g}(n) \times gl(r+1)$ as a Levi component in $\mathfrak{g}(n+r+1)$ and consider the unitarily induced reprsentation

$$\mathrm{Ind}_{\mathfrak{g}(n)\times gl(r+1)}^{\mathfrak{g}(n+r+1)} [\pi(\mathcal{O}) \otimes \mathrm{Triv}]. \tag{6.4}$$

The WF-set is the real form of

$$1^{r_1}2^{r_2}\dots(2k-2)^{r_{2k-2}}(2k-1)^{r_{2k-1}-2}(2k)^2(2k+2)^{r_{2k}}, \tag{6.5}$$

which is the support of the spherical distinguished representation corresponding to this nilpotent. The induced representation has two irreducible factors. They are both derived functor induced representations from a parabolic subalgebra with Levi factor $\mathfrak{g}(n+1)\times gl(r)$. The representation being induced is $\pi(\mathcal{O}')\otimes\chi$, where $\pi(\mathcal{O}')$ is the spherical distinguished representation correponding to the nilpotent orbit

$$1^{r_1}2^{r_2}\dots(2k-2)^{r_{2k-2}}(2k-1)^{r_{2k-1}-2}(2k)^{r_{2k}+2} \tag{6.6}$$

and χ is a unitary character so that the conditions of Theorem 7.1 in [V2] are satisfied. By induction (the occurence of r_{2k-1} has been reduced by 2 !) both factors are unitary. We only have to show that the signature on the lowest K-type of the other factor is the same as for the spherical K-type. If we write the highest weights of K-types as decreasing sequences of integers, this lowest K-type is

$$(\underbrace{1,\dots,1}_{r+1},0,\dots,0,\underbrace{-1,\dots,-1}_{r+1}). \tag{6.7}$$

The signature in the induced representation in (6.4) comes from K-types in $\pi(\mathcal{O})$ of the same type with the number of 1's $\le r+1$. Repeating the same argument using $\mathfrak{g}(n)\times gl(r+r_{2k-1}+r_{2k-2}+1)$ as in [B] Section 10, completes the proof. □

References

[A1] Arthur, J.: On some problems suggested by the trace formula. (Lecture Notes in Mathematics, vol. 1041). Springer, Berlin Heidelberg New York 1984, pp. 1–50

[A2] Arthur, J.: Unipotent automorphic representations: conjectures. Astérisque **171–172**, Orbites unipotentes et représentations II. Groupes p-adiques et réels, 1989, 13–71

[ABV] Adams J., Barbasch D., Vogan D.: Special unipotent representation for real reductive groups. Preprint in preparation

[B] Barbasch D.: The unitary dual for complex classical Lie groups. Invent. math. **96** (1989) 103–176

[BSS] Barbasch D., Sahi S., Speh B.: Degenerate series representations for $GL(n,\mathbb{R})$ and Fourier analysis. Symposia Mathematica **XXXI** (1990) 45–69

[BV] Barbasch D., Vogan D.: Unipotent representations of complex semisimple Lie groups. Ann. Math. **121** (1985) 41–110

[L] Lusztig G.: Characters of reductive groups over a finite field. Ann. Math. Studies **107**, Princeton University Press, Princeton NJ

[LV] Lusztig G., Vogan D.: Singularities of closures of K-orbits on flag manifolds Invent. math. **71** (1983) 365–379

[M] Moeglin C.: Orbites unipotentes et spectre discret non ramifie. Preprint 1990

[MW] Moeglin C., Waldspurger J.-L.: Le spectre residuel de $GL(n)$. Preprint 1989

[S] Sekiguchi J.: Remarks on nilpotent orbits of a symmetric pair. J. Math. Soc. Japan **39** (1987) 127–138

[V1] Vogan D.: Representations of real reductive Lie groups. Birkhäuser, Boston 1981

[V2] Vogan D.: Unitarizability of certain series of representations. Ann. Math. **120** (1984) 141–187

[V3] Vogan D.: Irreducible characters of semisimple Lie groups IV. Duke Math. J. **49** (1982) 943–1073

[V4] Vogan D.: Associated varieties and unipotent representations. Preprint 1989

[VZ] Vogan D., Zuckerman G.: Unitary representations with non-zero cohomology. Comp. Math. **53** (1984) 51–90

Eisenstein Cohomology of Arithmetic Groups and Its Applications to Number Theory

Günter Harder

Mathematisches Institut, Universität Bonn
Wegelerstraße 10, W-5300 Bonn, Fed. Rep. of Germany

In this note I want to discuss some questions concerning the cohomology of arithmetic groups. They concern the structure of the cohomology and especially the Eisenstein cohomology as a module under the Hecke algebra. My exposition here is rather vague and imprecise. But in a series of papers I investigated these questions in various special cases. Sometimes I proved theorems which give an answer to one of the questions in special cases, sometimes I discuss the conjectures in special situations and make them more precise. I will refer to these papers later on.

My main objective here is to show that a good understanding of these questions will have interesting applications in number theory, especially for the theory of special values of L-functions.

1. The General Setup

We start from a reductive group $G/\mathbb{Q}$, let $G^{(1)}/\mathbb{Q}$ be its derived group, let $Z/\mathbb{Q}$ be its centre. We denote the adele group of G by $G(\mathbb{A})$, and we decompose it into its finite and infinite part

$$G(\mathbb{A}) = G(\mathbb{R}) \times G(\mathbb{A}_f) = G_\infty \times G(\mathbb{A}_f)\,.$$

We choose a closed subgroup $K_\infty \subset G_\infty$ whose connected component of the identity is of the form

$$K_\infty^{(0)} = K_\infty^{(1)} \times Z(\mathbb{R})^{(0)}\,,$$

where $K_\infty^{(1)}$ is the connected component of the identity of a maximal compact subgroup of $G_\infty^{(1)}$. Moreover we choose a compact open subgroup $K_f \subset G(\mathbb{A}_f)$ (the level subgroup), and we consider the space

$$G_\infty/K_\infty \times G(\mathbb{A}_f)/K_f = X_\infty \times G(\mathbb{A}_f)/K_f\,.$$

The space X_∞ is a disjoint union of symmetric spaces (the centre contributes by a flat euclidean space). The group $G(\mathbb{Q})$ acts upon this space. We pass to the quotient and get a space

$$G(\mathbb{Q})\backslash X_\infty \times G(\mathbb{A}_f)/K_f = \mathscr{S}_K^G\,.$$

Proceedings of the International Congress
of Mathematicians, Kyoto, Japan, 1990

The quotient space $\mathcal{S}_K^G$ is a locally symmetric space. It is a finite disjoint union

$$\bigsqcup_{\underline{g}_f \in G(\mathbb{Q})\backslash G(\mathbb{A}_f)/K_f} \Gamma^{(\underline{g}_f)}\backslash X_\infty,$$

where $\underline{g}_f$ runs over a set of representatives in $G(\mathbb{A}_f)$ of the double coset space and where $\Gamma^{(\underline{g}_f)}$ is the arithmetic subgroup

$$\Gamma^{(\underline{g}_f)} = G(\mathbb{Q}) \cap \underline{g}_f K_f \underline{g}_f^{-1}.$$

We call this space the *locally symmetric space attached to* $G/\mathbb{Q}$ (of level K_f).

Finally we choose a rational representation

$$\varrho : G \longrightarrow GL(\mathcal{M})$$

which is defined over $\mathbb{Q}$ or over some number field (if we want it to be absolutely irreducible) or over $\overline{\mathbb{Q}}$ (see [Ha1], 1.3). This representation provides a sheaf $\tilde{\mathcal{M}}$ on $\mathcal{S}_K^G$ by a standard construction.

Our object of interest are the cohomology groups of $\mathcal{S}_K^G$ with coefficients in this sheaf

$$H^\bullet(\mathcal{S}_K^G, \tilde{\mathcal{M}}).$$

These cohomology groups are finite dimensional vector spaces over the field of definition of ϱ and they have some very important extra features (1.1 and 1.2).

1.1 The Borel-Serre Compactification

In general the quotient space $\mathcal{S}_K^G$ is not compact. There is a natural construction of a compactification $i : \mathcal{S}_K^G \longrightarrow \overline{\mathcal{S}}_K^G$, this is the Borel-Serre compactification (see [B-S]). It is known that $\overline{\mathcal{S}}_K^G$ is a manifold with corners (provided K_f is small enough). It is stratified by manifolds $\partial_P \mathcal{S}_K^G$ which are labelled by the conjugacy classes of parabolic subgroups over $\mathbb{Q}$. The closure of a stratum $\partial_P \mathcal{S}_K^G$ consists of the strata $\partial_Q \mathcal{S}_K^G$ where Q runs over the conjugacy classes of rational parabolic subgroups $Q \subset P$. One knows that the sheaves $\tilde{\mathcal{M}}$ extend nicely to $\overline{\mathcal{S}}_K^G$, the map $i : \mathcal{S}_K^G \to \overline{\mathcal{S}}_K^G$ is a homotopy equivalence and we get an isomorphism

$$H^\bullet(\mathcal{S}_K^G, \tilde{\mathcal{M}}) \xrightarrow{\sim} H^\bullet(\overline{\mathcal{S}}_K^G, \tilde{\mathcal{M}}).$$

We also consider the cohomology with compact supports $H_c^\bullet(\mathcal{S}_K^G, \tilde{\mathcal{M}})$, we get a long exact cohomology sequence

$$\to H_c^\bullet(\mathcal{S}_K^G, \tilde{\mathcal{M}}) \to H^\bullet(\overline{\mathcal{S}}_K^G, \tilde{\mathcal{M}}) \xrightarrow{r} H^\bullet(\partial\overline{\mathcal{S}}_K^G, \tilde{\mathcal{M}}) \to .$$

Let $M/\mathbb{Q}$ be the Levi quotient of the parabolic subgroup $P/\mathbb{Q}$, let $U_P/\mathbb{Q}$ be the unipotent radical of P and let $\mathfrak{u}_P$ be its Lie algebra. Then one knows that the cohomology groups $H^\bullet(\partial_P \mathcal{S}_K^G, \tilde{\mathcal{M}})$ can be expressed in terms of the cohomology of the locally symmetric space attached to M with coeffients in the the Lie algebra cohomology $H^\bullet(\mathfrak{u}_P, \mathcal{M})$ (which is a module for M). We write without further explanation

$$H^\bullet(\partial_P \mathcal{S}_K^G, \tilde{\mathcal{M}}) = \mathrm{Ind}_P^G H^\bullet(\mathcal{S}_{K^M}^M, \widetilde{H^\bullet(\mathfrak{u}_P, \mathcal{M})}).$$

(See for instance [Ha1], Theorem 1, [Ha5], 1.5–1.7) . The stratum $\partial_P \mathcal{S}_K^G$ has codimension $d(P)$ = corank of P this is the increase of the split rank of M against the split rank of G. The covering of $\partial \overline{\mathcal{S}}_K^G$ by the $\partial_P \mathcal{S}_K^G$ yields a spectral sequence with $E_1^{p,q}$-term

$$\bigoplus_{P: d(P)=p+1} H^q(\partial_P \mathcal{S}_K^G, \tilde{\mathcal{M}}) \Rightarrow H^{p+q}(\partial \overline{\mathcal{S}}_K^G, \tilde{\mathcal{M}}). \tag{Ss}$$

The exact sequence above induces a two-step filtration on the cohomology, but since the cohomology of the boundary will also be filtered, we get a many step filtration on the cohomology.

1.2 The Hecke Operators

Let $\mathcal{H}_{K_f} = \mathcal{C}_c(G(\mathbb{A}_f)//K_f)$ be the space of $\mathbb{Q}$ (or $\overline{\mathbb{Q}}$) valued functions on $G(\mathbb{A}_f)$ which have compact support and are biinvariant under K_f. These functions form an algebra under convolution; this is the so called Hecke algebra. It is a restricted tensor product of local Hecke algebras $\mathcal{H}_{K_p}$ if K_f is a product of local groups. We can construct an action of this algebra on all the groups

$$H_c^\bullet(\mathcal{S}_K^G, \tilde{\mathcal{M}}), H^\bullet(\mathcal{S}_K^G, \tilde{\mathcal{M}}), H^\bullet(\partial \overline{\mathcal{S}}_K^G, \tilde{\mathcal{M}})$$

which is compatible with the maps in the cohomolgy sequence. (To see this we pass to the limit for smaller and smaller level subgroups K_f', then the cohomology becomes a $G(\mathbb{A}_f)$-module and the $\varphi \in \mathcal{H}_{K_f}$ act by convolution on

$$H^\bullet(\mathcal{S}_K^G, \tilde{\mathcal{M}}) = \varinjlim_{K_f'} H^\bullet(\mathcal{S}_{K_f'}^G, \tilde{\mathcal{M}})^{K_f}).$$

The fundamental problem is to understand the structure of these cohomology groups as modules under the Hecke algebra.

A Digression. I will try to make a little bit more precise what it means to understand these modules. To do this I have to explain some ideas which go back to Eichler, Shimura, Deligne, Serre, Langlands and others. The picture that I will give is oversimplified and can only be true in simple cases.

We extend our coefficient system to $\tilde{\mathcal{M}}_\mathbb{C} = \tilde{\mathcal{M}} \otimes \mathbb{C}$. Then we can compute the cohomology groups $H^\bullet(\mathcal{S}_K^G, \tilde{\mathcal{M}}_\mathbb{C})$ by using differential forms. Within this space of differential forms we have the space of cusp forms which allows us to define a subspace

$$H^\bullet_{\mathrm{cusp}}(\mathcal{S}_K^G, \tilde{\mathcal{M}}_\mathbb{C})$$

in the cohomology (see [Bo] 5.5, [Schw]). To this space we can apply Hilbert space techniques and get an isotypical decomposition

$$H^{\bullet}_{\text{cusp}}(\mathscr{S}^G_K, \tilde{\mathscr{M}}_{\mathbb{C}}) = \bigoplus_{\pi} H^{\bullet}_{\text{cusp}}(\mathscr{S}^G_K, \tilde{\mathscr{M}}_{\mathbb{C}})[\pi]_{\mathbb{C}}$$

where $H^{\bullet}_{\text{cusp}}(\mathscr{S}^G_K, \tilde{\mathscr{M}}_{\mathbb{C}})[\pi] = H_{\pi}^{m(\pi)}$ is the sum of $m(\pi)$ copies of the irreducible $\mathscr{H}$-module H_π. Then π will be the finite component of certain cuspidal automorphic representation $\pi^* = \pi_\infty \otimes \pi$.

Let us assume that this subspace descends to a subspace $H^{\bullet}_{\text{cusp}}(\mathscr{S}^G_K, \tilde{\mathscr{M}})$ defined over $\overline{\mathbb{Q}}$ (or even $\mathbb{Q}$ if $\mathscr{M}$ is a $\mathbb{Q}$-vector space), this is true in special cases (see [Ha1], 3.2.5 and [Cl]). Then we get for this subspace

$$H^{\bullet}_{\text{cusp}}(\mathscr{S}^G_K, \tilde{\mathscr{M}}) \otimes \overline{\mathbb{Q}} = \bigoplus_{\pi} H^{\bullet}_{\text{cusp}}(\mathscr{S}^G_K, \tilde{\mathscr{M}}_{\mathbb{C}})[\pi],$$

where the isotypical components are now $\overline{\mathbb{Q}}$ vector spaces (if $\mathscr{M}$ is a $\mathbb{Q}$-vector space, we will get an action of $\mathrm{Gal}(\overline{\mathbb{Q}}/\mathbb{Q})$ on the set of the contributing π, it will permute the summands.)

There is some general belief that one can attach an arithmetical object $\mathbf{M}(\pi)$ (something like a motive or a motive with coefficients) to such an isotypical component.

It is not quite clear what such a thing is, but is should have different cohomological realizations, especially it should have λ-adic cohomology groups

$$H^{\bullet}(\mathbf{M}(\pi) \times \overline{\mathbb{Q}}, \overline{\mathbb{Q}}_\lambda)$$

which will be modules for $\mathrm{Gal}(\overline{\mathbb{Q}}/E)$, where E is some specific number field.

Then one expects some kind of reciprocity law: For almost all places p the local module H_{π_p} under the local component $\mathscr{H}_p$ of the Hecke algebra "is strongly related" to the structure of the $H^{\bullet}(\mathbf{M}(\pi) \times \overline{\mathbb{Q}}, \overline{\mathbb{Q}}_\lambda)$ as a module under $\prod_{\mathfrak{p}|p} \mathrm{Gal}(\overline{\mathbb{Q}}_{\mathfrak{p}}/E_{\mathfrak{p}})$. Another way of saying this is the following: Langlands attached to any cuspidal automorphic form π^* a series of L-functions

$$L(\pi^*, r, s)$$

where r is a parameter (a representation of the dual group) and where s is a complex variable (see [La]). These are the *automorphic* L-functions.

On the other hand a motive $\mathbf{M}$ is also a thing which yields an L-function

$$L(\mathbf{M}, s)$$

these are the *arithmetic* L-functions. One hopes that the above "strong relationship" says that for a suitable r_0 we have

$$L(\mathbf{M}(\pi), s) \sim L(\pi^*, r_0, s)$$

where $\sim$ means equality in simple cases.

For more precise information I refer to [Ko, Cl]; this problem has been investigated for Shimura varieties. For the classical case of $GL_2/\mathbb{Q}$ the existence of $\mathbf{M}(\pi)$ has been proved by Eichler, Shimura, Deligne and Scholl. In some other cases partial results are known ([Wi, Bl-Ro, Ta]). There exists also some numerical evidence for the truth of such an assertion in the case where $\mathcal{S}_K^G$ is not a Shimura variety ([Cr, E-G-M, A-P-T]).

Even if this is rather vague I think we should have this in the back of our mind if we say we want to "understand" the structure of the cohomology as a module under the Hecke algebra. I point out that for the case of the multiplicative group over a number field k i.e. for $G/\mathbb{Q} = R_{k/\mathbb{Q}}(G_m)$ the above is the content of class field theory.

1.3 Integral Cohomology

If we fix the level subgroup K_f, and if we choose a K_f-invariant lattice $\mathcal{M}_{\mathcal{O}} \subset \mathcal{M}$, then it is not difficult to define a sheaf $\tilde{\mathcal{M}}_{\mathcal{O}}$ on $\mathcal{S}_K^G$, and we can define the cohomology groups

$$H_c^\bullet(\mathcal{S}_K^G, \tilde{\mathcal{M}}_{\mathcal{O}}), H^\bullet(\mathcal{S}_K^G, \tilde{\mathcal{M}}_{\mathcal{O}}), H^\bullet(\partial\overline{\mathcal{S}}_K^G, \tilde{\mathcal{M}}_{\mathcal{O}}),$$

and we have the same exact sequence as above. Now we can consider the subalgebra $\mathcal{H}_{\mathbb{Z}}$ of $\mathbb{Z}$ valued functions in $\mathcal{H}$ and with some modification we can define an action of $\mathcal{H}_{\mathbb{Z}}$ on these integral cohomology groups (see [K-P-S], [Ha5], 1.3). It is certainly also of interest to study the integral cohomology groups as modules under the action of $\mathcal{H}_{\mathbb{Z}}$.

2. The Eisenstein Cohomology

The aim of the Eisenstein Cohomology is to provide an understanding of the maps

$$H^\bullet(\mathcal{S}_K^G, \tilde{\mathcal{M}}) \xrightarrow{r} H^\bullet(\partial\overline{\mathcal{S}}_K^G, \tilde{\mathcal{M}}) \xrightarrow{\delta} H_c^{\bullet+1}(\mathcal{S}_K^G, \tilde{\mathcal{M}})$$

as maps for modules under the Hecke algebra. Actually this of course also requires that we understand the cohomology $H^\bullet(\partial\overline{\mathcal{S}}_K^G, \tilde{\mathcal{M}})$ as a module under the Hecke algebra and hence at least a full understanding of the cohomology for lower dimensional groups.

We pick a parabolic subgroup P and look at the associate class $\mathcal{P} = \{P = P_1, P_2, \dots\}$. (Two parabolic subgroups are associate if and only if their Levi subgroups are conjugate.) Let M and $\mathfrak{u}_P$ be as in 1.1. Let us assume that $H^\bullet(\mathcal{S}_{K^M}^M, H^\bullet(\widetilde{\mathfrak{u}_P, \mathcal{M}}))[\pi_M]_{\mathbb{C}}$ is an isotypical contribution to the cuspidal cohomology $H_{\text{cusp}}^\bullet(\mathcal{S}_{K^M}^M, (\widetilde{\mathfrak{u}_P, \mathcal{M}})_{\mathbb{C}})$. We consider the group $W^{(M)} = N_G(M)/M$, this group gives us a collection of conjugate contributions

$$H^\bullet(\mathcal{S}_{K^M}^M, H^\bullet(\widetilde{\mathfrak{u}_P, \mathcal{M}}))[w \cdot \pi_M]_{\mathbb{C}}$$

(see [Ha1], IV, [Ha3], 1.1.2 for examples) and this gives us a contribution

$$\bigoplus_{P_\nu \in \mathcal{P}} \bigoplus_w \operatorname{Ind}_{P_\nu}^G H^\bullet(\mathcal{S}_{K^M}^M, H^\bullet(\widetilde{\mathfrak{u}_{P_\nu}, \mathcal{M}}))[w \cdot \pi_M]_{\mathbb{C}} \subset \bigoplus_{P_\nu \in \mathcal{P}} H^\bullet(\partial_{P_\nu} \mathcal{S}_K^G, \tilde{\mathcal{M}}_{\mathbb{C}}).$$

We now want to solve two problems:

The first problem is to attach to our orbit $[\pi_M]$ a $\mathscr{H}$-submodule

$$H^\bullet(\partial\overline{\mathscr{S}}^G_K,\tilde{\mathscr{M}}_{\mathbb{C}})[\pi_M]\subset H^\bullet(\partial\overline{\mathscr{S}}^G_K,\mathscr{M}_{\mathbb{C}})$$

which should be considered as the contribution of $[\pi_M]$ to the cohomology of the boundary. This is easily done in the case where $\mathscr{P}$ consists of maximal parabolic subgroups. In the general case it requires already the solution of the second problem below for groups of smaller semi simple rank. Sometimes this contribution satisfies an extra condition

$$H^\bullet(\partial\overline{\mathscr{S}}^G_K,\tilde{\mathscr{M}}_{\mathbb{C}})[\pi_M]\quad\begin{array}{l}\text{is a direct summand and does not weakly}\\ \text{intertwine with the complement}\end{array}\qquad\text{(bMD)}$$

where no weak intertwining means that the two summands do not have any Jordan-Hölder quotient in common.

If the condition (bMD) is satisfied, then it follows easily that the subspace

$$H^\bullet(\partial\overline{\mathscr{S}}^G_K,\tilde{\mathscr{M}}_{\mathbb{C}})[\pi_M]\subset H^\bullet(\partial\overline{\mathscr{S}}^G_K,\tilde{\mathscr{M}})\otimes\mathbb{C}$$

is rational, this means that it is defined over $\overline{\mathbb{Q}}$ and behaves nicely under the action of $\mathrm{Gal}(\overline{\mathbb{Q}}/\mathbb{Q})$ (see [Ha1], 4.3).

The second problem is to understand the intersection

$$\mathrm{Im}(r)(H^\bullet(\mathscr{S}^G_K,\tilde{\mathscr{M}}_{\mathbb{C}}))\cap H^\bullet(\partial\overline{\mathscr{S}}^G_K,\tilde{\mathscr{M}}_{\mathbb{C}})[\pi_M]=H^\bullet_{\mathrm{glob}}(\partial\overline{\mathscr{S}}^G_K,\tilde{\mathscr{M}}_{\mathbb{C}})[\pi_M]$$

I will call this the *image of* $H^\bullet(\mathscr{S}^G_K,\tilde{\mathscr{M}}_{\mathbb{C}})$ *in the subspace* $H^\bullet(\partial\overline{\mathscr{S}}^G_K,\tilde{\mathscr{M}}_{\mathbb{C}})[\pi_M]$.

To attack this second problem we use the theory of Eisenstein series.

As I explained earlier the datum $[\pi_M]$ yields a collection of spaces

$$\left\{\mathrm{Ind}^G_{P_\nu}H^\bullet_{\mathrm{cusp}}\big(\mathscr{S}^M_{K^M},H^\bullet(\mathfrak{u}_{P_\nu},\widetilde{\mathscr{M}_{\mathbb{C}}})\big)[w\cdot\pi_M]\right\}_{w\in W^{(M)},\nu=1,\dots,s}.$$

Attached to the $w\cdot\pi_M$ are isotypical spaces $\mathscr{X}(w\cdot\pi_M)$ of automorphic forms on M which "induce" certain spaces $\mathrm{Ind}^G_{P_\nu}(\mathscr{X}(w\cdot\pi_M))\subset\mathscr{C}_\infty(P_\nu(\mathbb{Q})\backslash G(\mathbb{A})/K_f)$ The classes in our spaces can be represented by differential forms $\omega_{\pi,P_\nu}\in\mathrm{Hom}_{K_\infty}(\Lambda^\bullet(\mathfrak{g}/\mathfrak{k}),\mathrm{Ind}^G_{P_\nu}(\mathscr{X}(w\cdot\pi_M))\otimes\mathscr{M}_{\mathbb{C}})$. These forms ω_{π,P_ν} are invariant under $P_\nu(\mathbb{Q})$. By a process of infinite summation over $P_\nu(\mathbb{Q})\backslash G(\mathbb{Q})$ we can try to make them invariant under $G(\mathbb{Q})$. But this summation may diverge. We introduce a complex parameter $\Lambda\in\mathbb{C}^{d(P)}$ and multiply our functions in $\mathrm{Ind}^G_{P_\nu}(\mathscr{X}(w\cdot\pi_M))$ by $\delta^\Lambda_{P_\nu}$. Then our summation converges for $\mathrm{Re}(\Lambda)$ in a certain positive cone and defines a holomorphic function in Λ which extends to a meromorphic function for all Λ (see [H-C]). These functions suitably evaluated at $\Lambda=0$ provide a space of automorphic forms

$$\mathrm{Eis}^*[\pi_M]\subset\mathscr{A}(G(\mathbb{Q})\backslash G(\mathbb{A})/K_f)$$

and a space (see [Ha3], 1.2.1)

$$\mathrm{Eis}^\bullet[\pi_M]_{\mathbb{C}} = \mathrm{Im}(H^\bullet(\mathfrak{g}, K, \mathrm{Eis}^*[\pi_M] \otimes \mathcal{M}_{\mathbb{C}}) \longrightarrow H^\bullet(\mathcal{S}_K^G, \tilde{\mathcal{M}}_{\mathbb{C}})).$$

We state the assertion

$$r : \mathrm{Eis}^\bullet[\pi_M]_{\mathbb{C}} \longrightarrow H^\bullet_{\mathrm{glob}}(\partial\overline{\mathcal{S}}_K^G, \tilde{\mathcal{M}}_{\mathbb{C}})[\pi_M] \quad \text{is a surjective map} \qquad \text{(Hope)}$$

The general philosophy expressed by (Hope) is that the entire cohomology $H^\bullet(\mathcal{S}_K^G, \tilde{\mathcal{M}}_{\mathbb{C}})$ is built up out of the cuspidal cohomology and the $\mathrm{Eis}[\pi_M]_{\mathbb{C}}$.

3. Arithmetical Applications

The structure of the module $\mathrm{Eis}[\pi_M]_{\mathbb{C}}$ as a module under the Hecke algebra will depend in a subtle way on the individual π_M. To be a little more precise: I explained earlier (Section 1) that we can attach various L-functions $L(\pi_M^*, r, s)$ to the automorphic form π_M^*. Certain products of some of these L-functions will occur in the computation of the constant term of the Eisenstein series (see [La]).

The module $\mathrm{Eis}[\pi_M]_{\mathbb{C}}$ resulting from the above construction will depend on the behavior of the occuring L-functions at $s = 0$ (vanishing, poles, special values, for a first subtle case see [Ha3], 3.3). It is my main goal to understand this behavior.

The same also applies to the contribution $H^\bullet(\partial\overline{\mathcal{S}}_{K_f}^G, \mathcal{M})[\pi_M]$, for instance it will be interesting to investigate the influence of the L-values on the (higher) differentials in (Ss).

I want to explain how such an understanding may have arithmetical consequences.

3.1 Special Values of L-Functions

We assume that our $[\pi_M]$ satisfies condition (bMD). As I explained earlier we know in this case that

$$H^\bullet(\partial\overline{\mathcal{S}}_K^G, \tilde{\mathcal{M}})[\pi_M]$$

is defined over $\overline{\mathbb{Q}}$, and it is even a rational subspace in the sense of [Ha1], 1.3. Let us assume we have proved that (Hope) is true, then we know that the space

$$r(\mathrm{Eis}^\bullet[\pi_M]_{\mathbb{C}}) \subset H^\bullet(\partial\overline{\mathcal{S}}_K^G, \tilde{\mathcal{M}})[\pi_M] \otimes \mathbb{C}$$

is a rational subspace. But this subspace can be described in terms of ratios of special values of L-functions and hence we get rationaltiy results for these values.

The typical result under good circumstances will be that

$$\frac{L(\pi_M^*, r, \nu - 1)}{L(\pi_M^*, r, \nu)} \times \pi \times \text{ discriminant factor } \in \overline{\mathbb{Q}}^*$$

where $\nu, \nu - 1$ are critical values on the sense of Deligne [De]. This gives us a tool to reduce the proof of Deligne's conjectures to the extreme critical values (for special examples see [Ha1], Theorem 2, (3), Corollary 4.2.2 and [Ha3], Corollary 3.5.1, it is clear that there will be many more examples where this works).

3.1.1 The Manin-Drinfeld Principle. Let us now fix the degree v of the cohomology, for a given $[\pi_M]$ we assume (bMD) and (Hope). If the map

$$\mathrm{Eis}^v[\pi_M]_{\mathbb{C}} \xrightarrow{\sim} \mathrm{Im}(r(\mathrm{Eis}^v[\pi_M]_{\mathbb{C}})) \subset H^\bullet(\partial\overline{\mathcal{S}}_K^G;\mathcal{M})[\pi_M]\otimes\mathbb{C},$$

is an isomorphism and if $\mathrm{Eis}^v[\pi_M]_{\mathbb{C}}$ does not weakly intertwine with its complement, then we say that the *Manin-Drinfeld Principle* (MD) holds (In the classical situations it just follows from the fact that the eigenvalues of the Hecke operators on cusp forms are different from the eigenvalues of Eisenstein classes (multiplicity one)). (The assertion (bMD) is something like (MD) for the cohomology of the boundary).

If we know (MD) then we can conclude that the space $\mathrm{Eis}^v[\pi_M]_{\mathbb{C}}$, which is constructed by transcendental means, descends to a subspace defined over $\overline{\mathbb{Q}}$ and it is actually rational. (See [Ha1], Corollary (4.2.1), (note the subtle point (c)), and [Ha3], Theorem II).

This is an interesting fact by itself, but in some cases we may evaluate these rational classes on certain cycles (modular symbols constructed from subgroups) and if we are lucky we may express the result in terms of special L-values. Hence we get rationality results for these L-values (see [Ha1], 5.7.1). As we explained in [H-S] this implies – combined with the results of Don Blasius [Bl] – the truth of Deligne's conjecture [De] in the case of algebraic Hecke characters. (See also [Ha3]).

3.1.2 Integrality. We go one step further. Let us assume for the given $[\pi_M]$ that (bMD) is true, then we can say: If we invert a finite number $S=\{p_1,p_2,\ldots,p_s\}$ of rational primes then we even get a decomposition

$$H^\bullet(\partial\overline{\mathcal{S}}_K^G,\tilde{\mathcal{M}}_{\mathcal{O}_S}) = \text{ complement } \oplus H^\bullet(\partial\overline{\mathcal{S}}_K^G,\tilde{\mathcal{M}}_{\mathcal{O}_S})[\pi_M]\,.$$

We assume moreover that (in a fixed degree v) (MD) holds. Then we may try to define an isotypical subspace

$$\mathrm{Eis}^v[\pi_M]_{\mathcal{O}_S} \subset H^v(\mathcal{S}_K^G,\mathcal{M}_{\mathcal{O}_S})$$

(There are some problems with torsion which I cannot discuss here), and we may ask for the image

$$r:\mathrm{Eis}^v[\pi_M]_{\mathcal{O}_S} \longrightarrow H^v_{\mathrm{glob}}(\partial\overline{\mathcal{S}}_K^G,\tilde{\mathcal{M}}_{\mathcal{O}_S})[\pi_M]\,.$$

If we have (Hope) then the cokernel of this map will be a finitely generated $\mathcal{O}_{\mathcal{S}}$-torsion module. It will be interesting to ask for the structure of this module. It will be nontrivial in many cases, this indicates a failure of an integral version of (MD). In general I hope that the structure of this cokernel may be related to the arithmetic of certain special values of L-functions of the form $L(\pi_M^*,r,v)$. (The conjectures of Deligne on special values say that these values divided by a suitable period Ω are rational numbers. But by construction Ω itself is only defined modulo $\mathbb{Q}^*$. To give sense to an assertion of the form above one has to define the periods modulo $\mathcal{O}_S^*$ which is already a problem in itself. Moreover we have to assume that for π_M the problem discussed in *Digression* has been settled.)

An example is discussed [Ha6] in detail: There I take for $G = GL_2/\mathbb{Q}$, $K_f = GL_2(\hat{\mathbb{Z}})$, the coefficient system is obtained from the module $\mathcal{M}_{n,\mathbb{Z}}$ of homogenous polynomials in two variables of degree n (n even) with coefficients in $\mathbb{Z}$. We invert the primes p dividing $n+2$ and those for which $p-1|n+2$. Then the structure of the above module is given by a value of the Riemann Zeta function: it is equal to $\mathbb{Z}_S/\zeta(-1-n)\mathbb{Z}_S$. (The value of the Zeta function is integral in $\mathbb{Z}_S$). A generalization of this result to a ramified situation will be discussed in the Bonn Diplom-thesis of Ch. Kaiser.

In [Ha2], IV, [Ha4], II, Beispiel $PGSp_2$ I discuss some other situations where the investigation of this question is of interest.

If our cokernel turns out to be cyclic, i. e. it is isomorphic to $\mathcal{O}_S/a(\pi_M)\mathcal{O}_S$, then we can interprete $a(\pi_M)$ as *the denominator of the Eisenstein class.* This number is of great interest for several reasons. If we pick a class $\omega \in H^{\nu}_{\text{glob}}(\partial\overline{\mathcal{S}}_K^G, \tilde{\mathcal{M}}_{\mathcal{O}_S})[\pi_M]$ and we look at

$$\text{Eis}(\omega) \in \text{Eis}^{\nu}[\pi_M] \subset H^{\nu}(\mathcal{S}_K^G, \tilde{\mathcal{M}}),$$

then $a(\pi_M)\text{Eis}(\omega)$ will be integral. This will give integrality results for special values of L-functions for instance those discussed in 3.1.1. In our example above this denominator is $\zeta(-1-n)$ we get integrality results for L-values of Dirichtlet series over real quadratic fields which are well known. The analogous problem for Gl_2/F for an imaginary quadratic extension of $\mathbb{Q}$ will be discussed in the Bonn dissertation of H. König.

3.2 The Mixed Motives

This is even more speculation. As I said already the structure of the space $\text{Eis}[\pi_M]_{\mathbb{C}}$ will depend on the behavior of certain L-functions $L(\pi_M^*, r, s)$ at $s = 0$, this argument $s = 0$ is the central point (for the functional equation). In [Ha3] I showed that in some cases (GL_3/F, F totally imaginary) this dependence may be quite subtle. The structure of $\text{Eis}[\pi_M]_{\mathbb{C}}$ depends on whether these L-values vanish or not and on the sign of the functional equation. I express the hope that for more complicated groups also the order of vanishing will influence the structure of the module (see [Ha3], 3.5.3). It would be exciting if we could read off the order of vanishing of L-functions in their central point from the structure of certain cohomology groups.

On the other hand one has the Beilinson-Deligne conjectures (originating from the Birch and Swinnerton-Dyer conjectures) which say that these special L-values should contain arithmetical information (see [R-S-S, Sch]). The order of vanishing should predict the rank of certain Ext-groups in the category of mixed motives. The arithmetic of the values (see remark above) should predict the order or structure of certain finite groups (K-groups, ideal class groups).

There is some hope to get information of this kind if we take the following detour:

We investigate the structure of $\mathrm{Eis}[\pi_M]_{\mathbb{C}}$ in dependence of the special value and let the module tell us something about the arithmetic.

Basically the following principle should be adopted in first approximation:

The vanishing of certain values $L(\pi_M^*, r, 0)$ may create a failure of the Manin-Drinfeld-*principle for* $\mathrm{Eis}[\pi_M]_{\mathbb{C}}$

This failure has the following effect. To our given π_M there should exist a module $\mathscr{E}[\pi_M] \subset H^\bullet(\mathscr{S}_K^G, \tilde{\mathscr{M}})$ which is filtered (see 1.1), which is a direct summand in the cohomology and for which no non trivial subquotient occurs in the Jordan-Hölder series of its complement and where finally the top quotient of the filtration maps isomorphically to $H^\bullet_{\mathrm{glob}}(\partial\overline{\mathscr{S}}_K^G, \tilde{\mathscr{M}})[\pi_M]$.

Now we try again to attach a motive $\mathbf{M}(\mathscr{E}[\pi_M])$ to this module as we did it in our *Digression* (for instance if $\mathscr{S}_K^G$ is a Shimura variety). But now this motive is also filtered because it inherits the filtration of the cohomology. In contrast to the cuspidal motive which is a pure motive this motive will be an extension of pure motives and we have (hopefully) constructed a mixed motive whose origin lies in the vanishing of an L-value.

This is discussed in greater detail and for special cases in [Ha4]. For instance in Section II, Beispiel $PGSp_2$ I show that such a failure of the Manin-Drinfeld principle is created by the Saito-Kurakawa lifting (see [PS]): We start with a modular cusp form for $Sl_2(\mathbb{Z})$ of weight 2mod4 which is an eigenform for the Hecke operators. In this case we see a mixed motive which is an extension of the Tate motive $\mathbb{Q}(-2)$ (which is obtained from the contribution of our modular form to the cohomology of the boundary) by a motive attached to the Saito-Kurakawa lifting of our form. This lifting contributes to the cuspidal cohomology of the group $PGSp_2$. It should be isomorphic to the motive attached to the original modular form (if we believe the Tate conjecture). In this example I am not able to compute any kind of extension class, i.e. I am not able to decide under which conditions this extension is nontrivial (for a further discussion see [Ha4]) but I can check certain neccessary consistencies, especially it is clear the the extension is predicted by the Beilinson-Deligne conjectures.

There is another such construction of a mixed motive for Hilbert modular surfaces (where I can prove nontriviality of the extension). I hope to include into a revised version of [Ha4].

In some sense the above principle has an integral analogue: If certain L-values are divisible by a prime p or a power of this prime then the Manin-Drinfeld principle should fail modulo p or a certain power p^δ of p. This failure is related to the fact that the Eisenstein class picks up a denominator (see 3.1.2). If we multiply it by its denominator to make it integral its restriction to the boundary becomes zero modulo the denominator and hence in a rather imprecise sense it becomes cuspidal modulo the denominator. This gives us congruences between Eisenstein classes and cuspidal cohomology classes which in the classical case of $Sl_2(\mathbb{Z})$ are related to the classical congruences.

Let us assume that p^δ is the exact divisor of $a(\pi_M)$. Then we get representations of Galois groups mod p^δ which contains the above class $a(\pi_M)\mathrm{Eis}(\omega)\mathrm{mod}\ p^\delta$ as an invariant vector by construction but which is bigger. Hence it is an extension of a rank one representation by another representation. This representation has controlled ramification (this requires deep results from the theory of p-adic representations of Galois groups) and hopefully it does not split. I discuss this in detail for the values $\zeta(-1-n)$ (for even positive n) in [Ha6], Chap VI. We get a different approach to some of the results of Mazur-Wiles [M-W] and of Ribet in [Ri].

References

[ASL] Automorphic forms, Shimura varieties and L-functions, I, II. (L. Clozel and J. S. Milne, eds.) Perspectives in Mathematics **10** (1990) 11

[MLG] Manifolds and Lie Groups. (Hano, J. et al., eds.) Progress in Math. **14** (1981)

[A-P-T] Ash, A., Pinch, R., Taylor, R.: An $\hat{A}_4$ extension of $\mathbb{Q}$ attached to a non-selfdual automorphic form on $GL(3)$. Preprint

[Bo] Borel, A.: Stable real cohomology of arithmetic groups, II. In: [MLG], 21–55

[Bl] Blasius, Don: On the critical values of Hecke L-series. Ann. Math. **124** (1986) 23–63

[Bl-Ro] Blasius, D., Rogawski, J.: Galois representations for Hilbert modular forms. Bulletin A.M.S., July 1989

[B-S] Borel, A., Serre, J.-P.: Corners and arithmetic groups. Comment. Math. Helv. **48** (1973) 436–491

[Cl] Clozel, L.: Motifs et formes automorphes. In: [ASL] I, 77–159

[Cr] Cremona, J. E.: Hyperbolic tessellations, modular symbols, and elliptic curves over complex quadratic fields. Comp. Math. **51** (1984) 275–323

[De] Deligne, P.: Valeurs de fonctions L et périodes d'intégrales. Proc. Symp. Pure Math. **33** (part. 2) (1979) 313–346

[E-G-M] Elstrodt, J., Grunewald, F., Mennicke, J.: $PSL(2)$ over imaginary quadratic fields. Proc. of thc Journée Arithmetique (Metz) (1982), Asterisque

[H-C] Harish-Chandra Automorphic forms on semisimple Lie groups. (Lecture Notes in Mathematics, vol. 62). Springer, Berlin Heidelberg New York 1968

[Ha1] Harder, G.: Eisenstein cohomology of arithmetic groups. The case GL_2. Inv. math. **89** (1987) 37–118

[Ha2] Harder, G.: Eisensteinkohomologie für Gruppen vom Typ $GU(2,1)$. Math. Ann. **278** (1987) 563–592

[Ha3] Harder, G.: Some results on the Eisenstein cohomology of arithmetic subgroups of GL_n. In: Cohomology of arithmetic groups. Proceedings of a conference held at CIRM (J.-P. Labesse, J. Schwermer, eds.). (Lecture Notes in Mathematics). Springer, Berlin Heidelberg New York (to appear)

[Ha4] Harder, G.: Arithmetische Eigenschaften von Eisensteinklassen, die modulare Konstruktion von gemischten Motiven und von Erweiterungen endlicher Galoismoduln. Preprint, Bonn 1989

[Ha5] Harder, G.: Eisenstein-Kohomologie arithmetischer Gruppen: Allgemeine Aspekte. Preprint

[Ha6] Harder, G.: Kohomologie arithmetischer Gruppen. (Textbook in preparation), Chap. VI (preprint)

[H-S] Harder, G., Schappacher, N.: Special values of Hecke L-functions and abelian integrals. (Lecture Notes in Mathematics, vol. 1111). Springer, Berlin Heidelberg New York 1985, pp. 17–49

[Ko] Kottwitz, R.: Shimura varieties and λ-adic representations. [ASL] I, 161–209

[K-P-S] Kuga, M., Parry, W., Sah: Group Cohomology and Hecke Operators. In: [MLG], 223–266

[La] Langlands, R.: Euler products. James K. Whittemore lectures, Yale University 1967

[M-W] Class fields of abelian extensions of $\mathbb{Q}$. Invent. math. **76** (1984) 179–330

[PS] Piateski-Shapiro, I.I.: On the Saito-Kurakawa lifting. Invent. math. **71** (1983) 309–338

[R-S-S] Beilinson's conjectures on special values of L-functions (Rapoport, M., Schappacher, N., Schneider, P., eds.). Perspectives in Mathematics, vol. 4, 1988

[Ra] Ramakrishnan, D.: Problems arising from the Tate and Beilinson conjectures in the context of Shimura varieties, [ASL] II, 227–252

[Ri] Ribet, K.: A modular construction of unramified p-extensions of $\mathbb{Q}(\mu_p)$. Invent. math. **34** (1976) 151–162

[Sch] Scholl, A.: Remarks on special values of L-functions. Preprint

[Schw] Schwermer, J.: Cohomology of arithmetic groups, automorphic forms and L-functions. Proceedings of a conference held at CIRM (J.-P. Labesse, J. Schwermer, eds.). (Lecture Notes in Mathematics, vol. 1447). Springer, Berlin Heidelberg New York 1990, pp. 1–29

[Ta] Taylor, R.: On Galois representations associated to Hilbert modular forms. Invent. math. **98** (1989) 265–280

[Wi] Wiles, A.: On ordinary λ-adic representatons associated to modular forms. Invent. math. **94** (1988) 503–514

Crystallizing the q-Analogue of Universal Enveloping Algebras

Masaki Kashiwara

Research Institute for Mathematical Sciences, Kyoto University, Kitashirakawa
Sakyo-ku, Kyoto 606, Japan

§ 0. Introduction

The notion of the q-analogue of universal enveloping algebras is introduced independently by Drinfeld and Jimbo in their study of exactly solvable models in statistical mechanics. This algebra $U_q(\mathfrak{g})$ contains a parameter q and it becomes the universal enveloping algebra when $q = 1$. This parameter is the one of temparature in the context of statistical mechanics and $q = 0$ corresponds to the absolute temparature zero. Therefore, we can expect that the theory of $U_q(\mathfrak{g})$ will be simplified at $q = 0$. We call the study of $U_q(\mathfrak{g})$ at $q = 0$ crystallization. Of course, we cannot deform $U_q(\mathfrak{g})$ at $q = 0$. However, we can construct the bases of representations of $U_q(\mathfrak{g})$ at "$q = 0$", and the $U_q(\mathfrak{g})$-module structure is described by combinatrics among them. This gives a purely combinatorial description of the tensor category of $U_q(\mathfrak{g})$-modules (and hence $U(\mathfrak{g})$-modules).

§ 1. Crystal Bases

1.1 Definition of $U_q(\mathfrak{g})$

Let us consider the following data:

(1.1) a finite-dimensional $\mathbf{Q}$-vector space $\mathfrak{t}$,
(1.2) an index set I,
(1.3) a linearly independent subset $\{\alpha_i; i \in I\}$ of $\mathfrak{t}^*$ and a subset $\{h_i; i \in I\}$ of $\mathfrak{t}$,
(1.4) an inner product $(\ ,\ )$ on $\mathfrak{t}^*$ and
(1.5) a lattice P of t^*.

We assume that they satisfy the following conditions:

(1.6) $\{\langle h_i, \alpha_j \rangle\}$ is a generalized Cartan matrix (i.e. $\langle h_i, \alpha_i \rangle = 2$, $\langle h_i, \alpha_j \rangle \in \mathbf{Z}_{\leq 0}$ for $i \neq j$ and $\langle h_i, \alpha_j \rangle = 0 \Longleftrightarrow \langle h_j, \alpha_i \rangle = 0$),
(1.7) $(\alpha_i, \alpha_i) \in \mathbf{Z}_{>0}$,
(1.8) $\langle h_i, \lambda \rangle = \frac{2(\alpha_i, \lambda)}{(\alpha_i, \alpha_i)}$,
(1.9) $\alpha_i \in P$ and $h_i \in P^* = \{h \in \mathfrak{t}; \langle h, P \rangle \subset \mathbf{Z}\}$.

Proceedings of the International Congress
of Mathematicians, Kyoto, Japan, 1990

The $\mathbf{Q}(q)$-algebra $U_q(\mathfrak{g})$ is then the algebra generated by the symbols e_i, f_i ($i \in I$) and q^h ($h \in P^*$) with the following fundamental relations:

(1.10) $q^h = 1$ for $h = 0$ and $q^{h+h'} = q^h q^{h'}$,
(1.11) $q^h e_i q^{-h} = q^{\langle h,\alpha_i\rangle} e_i$ and $q^h f_i q^{-h} = q^{-\langle h,\alpha_i\rangle} f_i$,
(1.12) $[e_i, f_j] = \delta_{ij}(t_i - t_i^{-1})/(q_i - q_i^{-1})$ where $q_i = q^{(\alpha_i,\alpha_i)}$ and $t_i = q^{(\alpha_i,\alpha_i)h_i}$,
(1.13) $\sum_n (-1)^n e_i^{(n)} e_j e_i^{(b-n)} = 0$ and $\sum_n (-1)^n f_i^{(n)} f_j f_i^{(b-n)} = 0$ for $i \neq j$ and $b = 1 - \langle h_i, \alpha_j \rangle$.

Here we used the notations $[n]_i = (q_i^n - q_i^{-n})/(q_i - q_i^{-1})$, $[n]_i! = \prod_{k=1}^n [k]_i$ and $e_i^{(n)} = e_i^n/[n]_i!$, $f_i^{(n)} = f_i^n/[n]_i!$. We understand $e_i^{(n)} = f_i^{(n)} = 0$ for $n < 0$.

1.2 Operators $\tilde{e}_i$ and $\tilde{f}_i$

For a $U_q(\mathfrak{g})$-module M and $\lambda \in P$, we set $M_\lambda = \{u \in M; t_i u = q^{\langle h_i,\lambda\rangle} u\}$ and call it the weight space of weight λ. We say that M is *integrable* if $M = \oplus_{\lambda\in P} M_\lambda$ and if M is a union of finitely dimensional sub-$U_q(\mathfrak{g}_i)$-modules for any i. Here $U_q(\mathfrak{g}_i)$ is the subalgebra of $U_q(\mathfrak{g})$ generated by e_i, f_i and t_i.

By the representation theory of $U_q(sl_2)$, any element u of M_λ is uniquely written in the form

(1.14) $u = \sum f_i^{(n)} u_n$ (resp. $= \sum e_i^{(n)} v_n$) where $u_n \in \ker e_i \cap M_{\lambda+n\alpha_i}$ and $u_n = 0$ except when $n + \langle h_i, \lambda \rangle \geq 0$ and $n \geq 0$ (resp. $v_n \in \ker f_i \cap M_{\lambda-n\alpha_i}$ and $v_n = 0$ except when $n \geq \langle h_i, \lambda \rangle$ and $n \geq 0$).

We define the endomorphisms $\tilde{e}_i$ and $\tilde{f}_i$ on M by

$$\tilde{e}_i u = \sum f_i^{(n-1)} u_n$$

and

$$\tilde{f}_i u = \sum f_i^{(n+1)} u_n .$$

Then $\tilde{e}_i$ and $\tilde{f}_i$ satisfy the relations symmetric to this: $\tilde{e}_i u = \sum e_i^{(n+1)} v_n$ and $\tilde{f}_i u = \sum e_i^{(n-1)} v_n$.

1.3 Crystal Base

Let A be the subring of $\mathbf{Q}(q)$ consisting of the rational functions regular at $q = 0$. Let M be an integrable $U_q(\mathfrak{g})$-module.

Definition 1.1. A *crystal base* of M is a pair (L, B) satisfying the following conditions.

(1.15) L is a free sub-A-module of M such that $M = \mathbf{Q}(q) \otimes_A L$.
(1.16) B is a base of the $\mathbf{Q}$-vector space L/qL.
(1.17) $L = \oplus_{\lambda\in P} L_\lambda$ and $B = \sqcup_{\lambda\in P} B_\lambda$ where $L_\lambda = L \cap M_\lambda$, $B_\lambda = B \cap (L_\lambda/qL_\lambda)$.
(1.18) $\tilde{e}_i L \subset L$ and $\tilde{f}_i L \subset L$. Hence $\tilde{e}_i$ and $\tilde{f}_i$ operate also on L/qL.
(1.19) $\tilde{e}_i B \subset B \cup \{0\}$ and $\tilde{f}_i B \subset B \cup \{0\}$.
(1.20) For $b, b' \in B$, $b' = \tilde{f}_i b$ if and only if $b = \tilde{e}_i b'$.

For a crystal base (L, B), the crystal graph is the oriented colored (by $i \in I$) graph with B as the set of vertices and $b \xrightarrow{i} b'$ if $b' = \tilde{f}_i b$.

The crystal graph describes completely the action of $\tilde{e}_i$ and $\tilde{f}_i$ on $B \sqcup \{0\}$.

For $b \in B$, we set

$$\varepsilon_i(b) = \max\{k \geq 0; \tilde{e}_i^k b \neq 0\}$$

and

$$\varphi_i(b) = \max\{k \geq 0; \tilde{f}_i^k b \neq 0\}.$$

For $b \in B_\lambda$, we have

$$\langle h_i, \lambda \rangle = \varphi_i(b) - \varepsilon_i(b).$$

Example 1.2. When $\mathfrak{g} = sl_2$, $U_q(sl_2)$ is the algebra generated by e, f, t, t^{-1} with the commutation relation $tet^{-1} = q^2 e, tft^{-1} = q^{-2} f$ and $[e, f] = (t - t^{-1})/(q - q^{-1})$. Then any irreducible $(l+1)$-dimensional representation is isomorphic to $V_l = \oplus_{k=0}^{l} \mathbf{Q}(q) u_k$ with $fu_k = [k+1]u_{k+1}$, $eu_k = [l+1-k]u_{k-1}$ and $tu_k = q^{l-2k}u_k$. Then $L = \oplus A u_k$ and $B = \{u_k; 0 \leq k \leq l\} \subset L/qL$. Then (L, B) is a crystal base of V_l. Its crystal graph is

$$u_0 \longrightarrow u_1 \longrightarrow \cdots \longrightarrow u_{l-1} \longrightarrow u_l.$$

1.4 Stability by Tensor Product

Let us define the comultiplication of $U_q(\mathfrak{g})$ by

$$\begin{aligned} \Delta(q^h) &= q^h \otimes q^h, \\ \Delta(e_i) &= e_i \otimes t_i^{-1} + 1 \otimes e_i, \\ \Delta(f_i) &= f_i \otimes 1 + t_i \otimes f_i. \end{aligned}$$

Then $U_q(\mathfrak{g})$ has a Hopf algebra structure with Δ as a comultiplication. By Δ, the tensor product of two $U_q(\mathfrak{g})$-modules has a structure of $U_q(\mathfrak{g})$-module.

Theorem 1 (Stability by $\otimes$). *Let M_1 and M_2 be two integrable $U_q(\mathfrak{g})$-modules and let (L_j, B_j) be a crystal base of M_j $(j = 1, 2)$. Set $L = L_1 \otimes_A L_2$ and $B = \{b_1 \otimes b_2 \in L/qL; b_j \in B_j\}$.*

(i) Then (L, B) is a crystal base of $M_1 \otimes_{\mathbf{Q}(q)} M_2$.
(ii) For $b_j \in B_j$ $(j = 1, 2)$, we have

$$\tilde{f}_i(b_1 \otimes b_2) = \begin{cases} \tilde{f}_i b_1 \otimes b_2 & \text{if } \varphi_i(b_1) > \varepsilon_i(b_2) \\ b_1 \otimes \tilde{f}_i b_2 & \text{if } \varphi_i(b_1) \leq \varepsilon_i(b_2), \end{cases}$$

$$\tilde{e}_i(b_1 \otimes b_2) = \begin{cases} b_1 \otimes \tilde{e}_i b_2 & \text{if } \varphi_i(b_1) < \varepsilon_i(b_2) \\ \tilde{e}_i b_1 \otimes b_2 & \text{if } \varphi_i(b_1) \geq \varepsilon_i(b_2). \end{cases}$$

1.5 Existence and Uniqueness

We set $P_+ = \{\lambda \in P; \langle h_i, \lambda\rangle \geq 0\}$. For $\lambda \in P_+$, let $V(\lambda)$ be the irreducible integrable $U_q(\mathfrak{g})$-module generated by a vector u_λ of weight λ satisfying $e_i u_\lambda = 0$. Then

$$V(\lambda) \cong U_q(\mathfrak{g}) \Big/ \Big(\sum_i (U_q(\mathfrak{g})e_i + U_q(\mathfrak{g})f_i^{1+\langle h_i,\lambda\rangle}) + \sum_{h\in P^*} U_q(\mathfrak{g})(q^h - q^{\langle h,\lambda\rangle}) \Big).$$

Let $L(\lambda)$ be the smallest sub-A-module of M such that $L(\lambda)$ contains u_λ and $L(\lambda)$ is stable by the $\tilde{f}_i$.

Let $B(\lambda)$ be the subset of $L(\lambda)/qL(\lambda)$ consisting of the non-zero vectors of the form $\tilde{f}_{i_1} \cdots \tilde{f}_{i_l} u_\lambda$.

Theorem 2 (Existence). *$(L(\lambda), B(\lambda))$ is a crystal base of $V(\lambda)$.*

Let $\mathcal{O}_{\rm int}$ be the category of integrable $U_q(\mathfrak{g})$-modules such that there exists a finite subset F of P such that $M = \oplus_{\lambda\in F+Q_-} M_\lambda$. Here, $Q_- = \sum \mathbf{Z}_{\leq 0}\alpha_j$. Then $\mathcal{O}_{\rm int}$ is a semi-simple abelian category and any irreducible object is isomorphic to $V(\lambda)$ for some $\lambda \in P_+$ ([L], [R]).

Theorem 3 (Uniqueness). *Let (L, B) be a crystal base of an object M in $\mathcal{O}_{\rm int}$. Then there exists an isomorphism $M \cong \oplus_j V(\lambda_j)$ by which (L, B) is isomorphic to $\oplus_j (L(\lambda_j), B(\lambda_j))$.*

Combining the Theorems 1, 2 and 3, we can describe completely the tensor category $\mathcal{O}_{\rm int}$.

First note that the crystal graph of $V(\lambda)$ is connected. Hence the irreducible decomposition of an object in $\mathcal{O}_{\rm int}$ is equivalent to the connected component decomposition of the crystal graph. Then Theorem 1 tells us the crystal graph of tensor products

Example. Take the case $\mathfrak{g} = sl_3$ (see §4 for the notation). Let $\{\Lambda_i\}$ be the dual base of $\{h_i\}$. Then the decompositions $V(\Lambda_1) \otimes V(\Lambda_1) = V(\Lambda_2) \oplus V(2\Lambda_1)$ and $V(\Lambda_1) \otimes V(\Lambda_2) = V(\Lambda_1 + \Lambda_2) \oplus V(0)$ are described as follows.

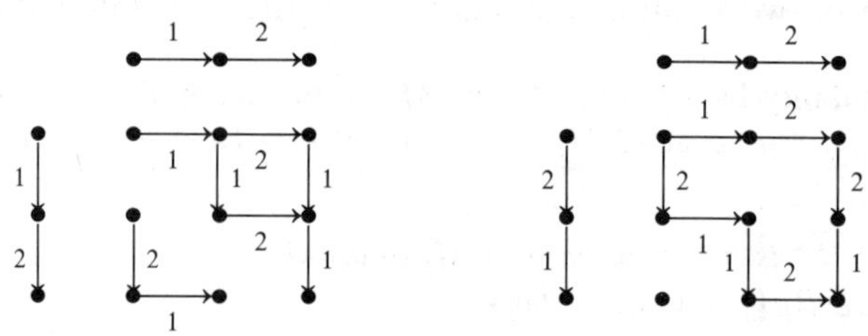

§ 2. Crystal Base of $U_q^-(\mathfrak{g})$

2.1 Operators $\tilde{e}_i$ and $\tilde{f}_i$ on $U_q^-(\mathfrak{g})$

Let $U_q^-(\mathfrak{g})$ be the subalgebra of $U_q(\mathfrak{g})$ generated by the f_i. Then $U_q^-(\mathfrak{g})$ has the unique endomorphisms e_i' and e_i'' such that

$$[e_i, P] = (t_i e_i''(P) - t_i^{-1} e_i'(P))/(q_i - q_i^{-1}) \quad \text{for any} \quad P \in U_q^-(\mathfrak{g}) .$$

Then e_i' and f_i satisfy the commutation relations:

$$e_i' f_j = q_i^{-\langle h_i, \alpha_j \rangle} f_j e_i' + \delta_{ij} . \tag{2.1}$$

Here we consider f_j as the left multiplication operator. Then any element u of $U_q^-(\mathfrak{g})$ can be uniquely written as

$$u = \sum_{n \geq 0} f_i^{(n)} u_n \quad \text{with} \quad e_i' u_n = 0 .$$

We define the endomorphisms $\tilde{e}_i$ and $\tilde{f}_i$ of $U_q^-(\mathfrak{g})$ by

$$\tilde{e}_i\Big(\sum f_i^{(n)} u_n\Big) = \sum f_i^{(n-1)} u_n$$
$$\tilde{f}_i\Big(\sum f_i^{(n)} u_n\Big) = \sum f_i^{(n+1)} u_n .$$

Then $\tilde{e}_i \tilde{f}_i = 1$ holds. Let $L(\infty)$ be the smallest sub-A-module of $U_q^-(\mathfrak{g})$ that contains 1 and that is stable by $\tilde{f}_i$. Let $B(\infty)$ be the subset of $L(\infty)/qL(\infty)$ consisting of the vectors of the form $\tilde{f}_{i_1} \cdots \tilde{f}_{i_l} \cdot 1$. Then $(B(\infty), L(\infty))$ has a similar property to crystal bases.

Theorem 4. *(i)* $\tilde{e}_i L(\infty) \subset L(\infty)$ *and* $\tilde{f}_i L(\infty) \subset L(\infty)$.
(ii) $\tilde{e}_i B(\infty) \subset B(\infty) \cup \{0\}$ *and* $\tilde{f}_i B(\infty) \subset B(\infty)$.
(iii) $B(\infty)$ *is a base of* $L(\infty)/qL(\infty)$.
(iv) If $b \in B(\infty)$ *satisfies* $\tilde{e}_i b \neq 0$, *then* $b = \tilde{f}_i \tilde{e}_i b$.

The relation of $(L(\infty), B(\infty))$ and $(L(\lambda), B(\lambda))$ is given by the following theorem.

Theorem 5. *For* $\lambda \in P_+$, *let* $\pi_\lambda : U_q^-(\mathfrak{g}) \to V(\lambda)$ *be the* $U_q^-(\mathfrak{g})$*-linear homomorphism sending 1 to* u_λ.
(i) $\pi_\lambda L(\infty) = L(\lambda)$.
Hence π_λ *induces the homomorphism* $\overline{\pi}_\lambda : L(\infty)/qL(\infty) \to L(\lambda)/qL(\lambda)$.
(ii) $\{b \in B(\infty); \overline{\pi}_\lambda(b) \neq 0\}$ *is isomorphic to* $B(\lambda)$ *by* $\overline{\pi}_\lambda$.
(iii) $\tilde{f}_i \circ \overline{\pi}_\lambda = \overline{\pi}_\lambda \circ \tilde{f}_i$.
(iv) For $b \in B(\infty)$ *such that* $\overline{\pi}_\lambda(b) \neq 0$, $\tilde{e}_i \overline{\pi}_\lambda(b) = \overline{\pi}_\lambda(\tilde{e}_i b)$.

§ 3. Global Crystal Base

Let $U_q^-(\mathfrak{g})_{\mathbf{Z}}$ be the sub-$\mathbf{Z}[q,q^{-1}]$-algebra of $U_q(\mathfrak{g})$ generated by the $f_i^{(n)}$. For $\lambda \in P_+$, we set $V_{\mathbf{Z}}(\lambda) = U_q^-(\mathfrak{g})_{\mathbf{Z}} \cdot u_\lambda$. Let $-$ be the ring automorphism of $U_q^-(\mathfrak{g})$ such that $\overline{q} = q^{-1}$ and $\overline{f}_i = f_i$. This induces the automorphism $-$ of $V(\lambda)$ by $\overline{Pu_\lambda} = \overline{P}u_\lambda$ for $P \in U_q^-(\mathfrak{g})$.

Theorem 6. *(i)* $(\mathbf{Q} \otimes U_q^-(\mathfrak{g})_{\mathbf{Z}}) \cap L(\infty) \cap L(\infty)^- \to L(\infty)/qL(\infty)$ *is an isomorphism.*

(ii) For any $\lambda \in P_+$, $(\mathbf{Q}\otimes V_{\mathbf{Z}}(\lambda))\cap L(\lambda)\cap L(\lambda)^- \to L(\lambda)/qL(\lambda)$ *is an isomorphism.*

Let G denote the inverse of these isomorphisms. Then we have $\overline{G(b)} = G(b)$ for $b \in L(\lambda)/qL(\lambda)$ with $\lambda \in P_+ \cup \{\infty\}$. Moreover, we have $G(b)u_\lambda = G(\overline{\pi}_\lambda b)$ for any $b \in L(\infty)/qL(\infty)$.

Theorem 7. *For any* $n \geq 0$ *and* i,

$$f_i^n U_q^-(\mathfrak{g}) \cap U_q^-(\mathfrak{g})_{\mathbf{Z}} = \bigoplus_{b \in \tilde{f}_i^n B(\infty)} \mathbf{Z}[q,q^{-1}]G(b),$$

$$f_i^n V(\lambda) \cap V_{\mathbf{Z}}(\lambda) = \bigoplus_{b \in \tilde{f}_i^n B(\lambda)\setminus\{0\}} \mathbf{Z}[q,q^{-1}]G(b).$$

We call $G(b)$ *global crystal base.*

It is proven by Lusztig ([L_3]) that the canonical bases introduced by himself in [L_2] in the case A_n, D_n and E_n coincide with the global canonical bases introduced here.

§ 4. Example

This example is a joint work with T. Nakashima. Let us take $\mathfrak{g} = sl_n$. Hence $I = \{1,\cdots,n-1\}$, $(\alpha_i,\alpha_j) = 1,-1/2,0$ according to $i = j$, $|i-j| = 1$, $|i-j| > 0$. Let $\Lambda_i \in \mathfrak{t}^*$ be the dual base of h_i and take $\oplus\mathbf{Z}\Lambda_i$ as P.

Then the crystal graph of $B(\Lambda_1)$ is

$$\boxed{1}\xrightarrow{\;1\;}\boxed{2}\longrightarrow\cdots\longrightarrow\boxed{n-1}\xrightarrow{\;n-1\;}\boxed{n}\;.$$

For $\lambda = \sum_{\nu=1}^{N}\Lambda_{i_\nu}$ $(1 \leq i_1 \leq \cdots \leq i_N)$ we embed $B(\lambda)$ into $B(\Lambda_1)^{\otimes i_1} \otimes B(\Lambda_1)^{\otimes i_2} \otimes \cdots$ by $u_\lambda \mapsto \left(\boxed{1}\otimes\cdots\otimes\boxed{i_1}\right)\otimes\left(\boxed{1}\otimes\cdots\otimes\boxed{i_2}\right)\otimes\cdots$. Then $B(\lambda)$ is parametrized by

$$\left(\boxed{m_{11}}\otimes\boxed{m_{12}}\otimes\cdots\otimes\boxed{m_{1i_1}}\right)\otimes\left(\boxed{m_{21}}\otimes\cdots\otimes\boxed{m_{2i_2}}\right)\otimes\cdots\;.$$

in $B(\Lambda_1)^{\otimes i_1} \otimes \cdots$. We associate to this base the Young diagram $Y(\lambda)$ with a positive integer in each box as follows

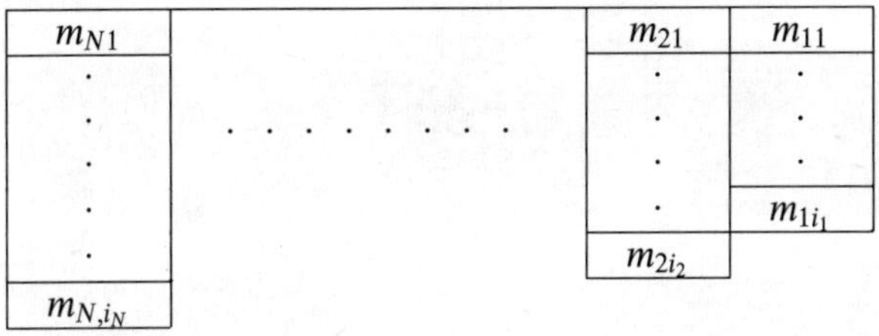

Here $Y(\lambda)$ is the Young diagram with the columns with length $i_1, \cdots, i_N$.

Theorem. *By this correspondence, $B(\lambda)$ is equal to the set of semi-standard tableaux with shape $Y(\lambda)$ (i.e. $\{m_{ij}\}$ satisfies $m_{ij} \leq m_{i'j}$ if $i > i'$ and $m_{ij} < m_{ij'}$ if $j < j'$).*

§ 5. Remarks

The notion of crystal base is introduced in $[K_1]$ under the form dual to the one given here. Theorem 2,3 in the case of A_n, B_n, C_n and D_n and Theorem 1 are proven there. In $[M]$, the crystal graph of basic representation of $U_q(\widehat{sl}_n)$ is given. The results here have been announced in $[K_2]$. Independently, Lusztig introduced the notion of canonical bases in the case A_n, D_n, E_n ($[L_2]$) and he showed that they coincide with global canonical bases ($[L_3]$).

References

[D] Drinfeld, V. G.: Hopf algebra and the Yang-Baxter equation. Sov. Math. Dokl. **32** (1985) 254–258

[J] Jimbo, M.: A q-difference analogue of $U(\mathfrak{g})$ and the Yang-Baxter equation. Lett. Math. Phys. **10** (1985) 63–69

[K] Kashiwara, M.: 1. Crystallizing the q-analogue of universal enveloping algebras. Commun. Math. Phys. **133** (1990) 249–260
2. Bases crystallines. C. R. Acad. Sci. Paris **311** (1990) 277–280

[L] Lusztig, G.: 1. On quantum groups. J. Algebra (1990)
2. Canonical bases arising from quantized enveloping algebra. J. AMS
3. Canonical bases arising from quantized enveloping algebra, II. Preprint

[M] Misra, K. C., Miwa, T.: Crystal bases for basic representation of $U_q(\widehat{sl}(n))$. Commun. Math. Phys. **134** (1990) 79–88

[R] Rosso, M.: Analogue de la forme de Killing et du théorème d'Harish-Chandra pour les groupes quantiques. Ann. Sci. Ec. Norm. Sup. **23** (1990) 445–467

Classification of Simple Graded Lie Algebras of Finite Growth

Olivier Mathieu *

D. M. I., ENS, 45, rue d'Ulm, F-75005 Paris, France, and
Rutgers University, Department of Mathematics, Hill Center
New Brunswick, NJ 08903, USA

Introduction

Conventions: The ground field is $\mathbb{C}$. By LA we mean Lie algebra.

Let us start with a few definitions.

- A LA $\mathscr{L}$ endowed with a decomposition

$$\mathscr{L} = \oplus_{n \in \mathbb{Z}} \mathscr{L}_n$$

is called a *graded LA* if we have $[\mathscr{L}_n, \mathscr{L}_m] \subseteq \mathscr{L}_{n+m}$. Moreover we will always assume that $\dim \mathscr{L}_n < \infty$ for any $n < \infty$. With our convention any graded LA is an *ordinary* LA and the notion should not be confused with super LA which are often called graded LA as well.

- A subspace V of $\mathscr{L}$ is called *homogeneous* if we have

$$V = \oplus_{n \in \mathbb{Z}} V_n$$

(where $V_n = V \cap \mathscr{L}_n$). The LA $\mathscr{L}$ is called *simple graded* if any homogenous ideal is trivial (i.e. O or $\mathscr{L}$) and if $\dim \mathscr{L} \geq 2$.

- Say that $\mathscr{L}$ has *finite growth* if

$$\dim \mathscr{L}_n \leq P(n)$$

for some polynomial P.

We have recently proved the following theorem [M2].

Theorem (1990). *Let $\mathscr{L}$ be a simple graded LA of* finite growth. *Then $\mathscr{L}$ is isomorphic to one of the following LA:*

1. A simple finite dimensional LA
2. An affine LA

---

* Work done under the hospitality of IAS at Princeton. I thank IAS for its support (DMS Grant 8610730).

Proceedings of the International Congress
of Mathematicians, Kyoto, Japan, 1990

3. *A LA of Cartan type*
4. **W** *(Virasoro-Witt LA).*

The previous theorem has been conjectured by V.G. Kac. Alltogether there are 14 infinite series and 13 exceptional LA. In part 1 (zoology) we will give precise definitions of the involved LA. Before we would like to make a few remarks. The origin of Kac conjecture comes from the following result **[K]**.

Theorem (V. G. Kac, 1967). *Let $\mathscr{L}$ be a simple graded LA of* finite growth. *Assume*
(*) $\mathscr{L}$ is generated by its "local part" $\mathscr{L}_{-1} \oplus \mathscr{L}_0 \oplus \mathscr{L}_1$
(**) the $\mathscr{L}_0$-module $\mathscr{L}_{-1}$ is irreducible.
Then $\mathscr{L}$ is isomorphic to

1. *a finite dimensional LA,*
2. *an affine LA or*
3. *a Cartan type LA.*

Moreover it follows from 1967 Kac paper the existence of "continous families" of simple graded LA. Thus there are *no hopes* for a classification *without the growth hypothesis.* Note also that in characteric $p \neq 0$ the classification of finite dimensional simple *LA* is still open. [1]

Part 1: Zoology

In the section we will describe some species i.e. the LA involved in the Theorem. Altough each of them admit infinitely many different gradings we can describe all of them. For the simplicity of the exposition we will describe theses gradings in one case only.

(1.1) Finite Dimensional Simple LA

Recall that finite dimensional simple Lie algebras have been classified around 1900 by Killing and Cartan. Four infinite series and five exceptional Lie algebras occur in their classification. The LA of the four infinite series are called classical LA. They are the following one.

A_n *or* $\mathfrak{sl}(n+1)$
B_n *or* $\mathfrak{so}(2n+1)$
C_n *or* $\mathfrak{sp}(2n)$
D_n *or* $\mathfrak{so}(2n)$

It is not easy to give a simple description of the five exceptional simple LA E_6, E_7, E_8, F_4 and G_2

[1] At ICM conference G. Seligman tells us that H. Strade and R. Wilson have rencently announced the classification of finite dimensional simple LA over field of characteristic $p > 7$.

(1.2) Affine (Kac-Moody) LA

Let $\mathfrak{g}$ be a finite dimensional simple LA, let ω be an automorphism of $\mathfrak{g}$ of finite order ℓ and let η be ℓ-root of unity. Set $L(\mathfrak{g}) = \mathfrak{g} \otimes \mathbb{C}[t, t^{-1}]$. Define the automorphism $\tilde{\omega}$ of $L(\mathfrak{g})$ by:

$$\tilde{\omega}(g \otimes t^n) = \eta^n \omega(g) \otimes t^n .$$

Let $L(\mathfrak{g}, \omega, \eta)$ be the LA of fixed points under $\tilde{\omega}$. A *LA* isomorphic to some $L(\mathfrak{g}, \omega, \eta)$ is called *affine*. The definition is not accurate because there are many non-trivial isomorphisms between various $L(\mathfrak{g}, \omega, \eta)$. Fortunately V.G. Kac found a one to one parametrization of these isomorphism classes [K]. Actually he proved that affine LA are exactely parametrized by automorphisms of Dynkin diagrams. All together there are 6 infinite series and 7 exceptional affine LA. With the usual notation affine algebras are

$$A_n^{(1)}, B_n^{(1)}, C_n^{(1)}, D_n^{(1)}, A_n^{(2)}, D_n^{(2)}, D_{(4)}^{(3)}, E_{(6)}^{(1)}, E_{(6)}^{(2)}, E_{(7)}^{(1)}, E_{(8)}^{(1)}, G_{(2)}^{(1)}, F_{(4)}^{(1)}.$$

Affine LA are also called loop algebras because any element of $L(\mathfrak{g})$ can be identified with a $\mathfrak{g}$-valued map on S^1 whose Fourrier decomposition is finite.

(1.3) Cartan Type LA

Let $\mathbf{W}_n$ be the LA of derivations of the ring of polynomials $\mathbb{C}[X_1, \cdots, X_n]$. Thus an element ∂ of $\mathbf{W}_n$ is a vector field with polynomial coefficients. Note $Lie(\partial)$ be the Lie derative action on spaces of differential forms.

Set $\mathbf{S}_n = \{\partial \in \mathbf{W}_n | Lie(\partial) \cdot v = 0\}$ where v is the usual volume form $dX_1 \wedge \cdots \wedge dX_n$.

For $n = 2m$ let $\omega = \sum_{1 \leq i \leq m} dX_i \wedge dX_{m+i}$ be the usual symplectic form and set $\mathbf{H}_n = \{\partial \in \mathbf{W}_n | Lie(\partial) \cdot \omega = 0\}$.

For $n = 2m + 1$ let $\alpha = dX_n + \sum_{1 \leq \leq i \leq m} X_i dX_{m+i} - X_{m+i} dX_i$ be the usual contact 1-form and set $\mathbf{K}_n = \{\partial \in \mathbf{W}_n | \alpha \wedge Lie(\partial) \cdot \alpha = 0\}$.

The LA $\mathbf{W}_n, \mathbf{S}_n, \mathbf{H}_n, \mathbf{K}_n$ are called *Cartan type LA*. These four infinite serie have been discovered by Cartan around 1910 [C].

(1.4) The Virasoro-Witt LA

Let $\mathbf{W}$ be the LA of derivations of $\mathbb{C}[T, T^{-1}]$.

Remarks: 1. Each of the previous LA admits *infinitely many* gradings but only *finitely many* of them satisfy the hypotheses (*) (**) of Kac theorem.

Example for $\mathcal{L} = \mathbf{W}_n$. Let $\underline{a} = a_1, \cdots, a_n$ be a sequence of non zero integers of same sign. There is a unique grading of $\mathcal{L}$ such that the element $X_1^{m_1} \cdots X_n^{m_n} \partial / \partial X_j$

is homogeneous of degree $a_1m_1 + \cdots + a_nm_n - a_j$. For any $\sigma \in S_n$ (where S_n is the symetric group) the gradings associated with $a_1, \cdots, a_n$ and with $a_{\sigma(1)}, \cdots, a_{\sigma(n)}$ are obviously isomorphic. It is easy to prove that the induced map from $(\mathbb{Z}_+^n \sqcup \mathbb{Z}_-^n)/S_n$ to the set of isomorphism classes of gradings of $\mathscr{L}$ is one to one and onto . But the grading associated with $(1, \ldots, 1)$ is the *only one* satisfying Kac's hypotheses (*) (**) $(n \geq 2)$.

Example: No grading of $\mathbf{W}, \mathbf{W}_1$ satisfies (*).

2. The terminology "affine LA", "Virasoro *LA*" is often used for central extension of LA considered here.

3. Affine LA are "simple graded" but not simple (because of evaluation maps $L(\mathfrak{g}) \to \mathfrak{g}$). Conversely a simple graded LA which is not simple is affine (it is a statement).

Table I. Simple graded LA of finite growth

dim $\mathscr{L} < \infty$	$A_n\ B_n\ C_n\ D_n$ $E_6\ E_7\ E_8\ F_4\ G_2$
Affine	$A_n^{(1)}\ B_n^{(1)}\ C_n^{(1)}\ D_n^{(1)}$ $A_n^{(2)}\ D_n^{(2)}$ $D_4^{(3)}\ E_6^{(1)}\ E_6^{(2)}\ E_7^{(1)}$ $E_8^{(1)}\ F_4^{(1)}\ G_2^{(1)}$
Cartan type	$\mathbf{W}_n \mathbf{S}_n \mathbf{H}_n \mathbf{K}_n$
Virasoro-Witt	$\mathbf{W}$

Part 2: About the Proofs

In this section we will describe the general plan of the proof. The proof divides into 3 Steps.

Step 1: Define 4 abstract classes of graded LA.

Step 2: Lemma: Any simple graded LA belongs to one of the 4 previous classes.

Step 3: Show 4 classification theorems (i.e. one for each class).

Step 1. *4 Definitions:* To meet step 1 we will define four abstract classes of graded Lie algebras.

Let h be a finite dimensional nilpotent LA and let M be a finite dimensional h-module. Recall that M decomposes as

$$M = \oplus M^\lambda$$

where λ runs over h^* and M^λ is the generalized eigenspace associated with λ (Engel Theorem). Let $\mathfrak{g}$ be a finite dimensional LA. A *Cartan subalgebra* (or CSA) is a nilpotent subalgebra equal to its normalizer. It is classical that CSA do exist. Moreover any two CSA are conjugated under a product of elementary automorphisms. Let $\mathscr{L}$ be a graded LA. Pick a CSA h of $\mathscr{L}_o$ and consider each $\mathscr{L}_n$ as an h-module. Thus we have:

$$\mathscr{L} = \bigoplus_{\substack{n\in\mathbb{Z}\\ \lambda\in h^*}} \mathscr{L}_n^\lambda .$$

Set: $\Delta = \{(n,\lambda)/\mathscr{L}_n^\lambda \neq 0\}$.
Let Q be the subgroup of $\mathbb{Z} \times h^*$ generated by Δ .

Definition 1. $\mathscr{L}$ is called *without roots* iff $\Delta \subseteq \mathbb{Z} \times 0$

Definition 2. $\mathscr{L}$ is called *weakly integrable* iff
1) $\Delta \nsubseteq \mathbb{Z} \times 0$

2) $\bigcap_{s\geq 1} \mathrm{ad}^s(\mathscr{L}_n^\lambda).\mathscr{L} = 0$

for any $(n,\lambda) \in \Delta, \lambda \neq 0$.
Set: $\mathscr{L}^+ = \oplus_{s>0}\mathscr{L}_s, \mathscr{L}^- = \oplus_{s<0}\mathscr{L}_s$.

Definition 3. Say that $\mathscr{L}$ is of *type* $\mathscr{C}$ iff dim $\mathscr{L}^+$ or dim $\mathscr{L}^-$ is finite but dim $\mathscr{L}$ is infinite.

A subset X of Q is called *quasi-order* iff

$\forall\tilde{\alpha} \in Q \exists N \geq 0 \forall m \geq N \forall \tilde{\beta}_1 \ldots \tilde{\beta}_m \in X$ we have $\tilde{\alpha} + \tilde{\beta}_1 + \cdots + \tilde{\beta}_m \in X$.

Let $\tilde{\alpha} \in Q$. The LA $\mathscr{L}$ is called $\tilde{\alpha}$-*deep* if we have $[\mathscr{L}_X, \mathscr{L}] = \mathscr{L}$ for any quasi-order X such that $X \cup \{\tilde{\alpha}\}$ is still a quasi-order (by definition we set $\mathscr{L}_X = \bigoplus_{(n,\lambda)\in X} \mathscr{L}_n^\lambda$).

Definition 4. $\mathscr{L}$ is called *deep* if $\mathscr{L}$ is $\tilde{\alpha}$-deep for some $\tilde{\alpha} = (n,\lambda) \in \Delta$ with $n \neq 0, \lambda \neq 0$.

Thus Definitions 1, 2, 3, 4 define 4 abstract classes of graded LA. Moreover any two CSA of $\mathscr{L}_O$ are conjugated under a degree 0 automorphism of $\mathscr{L}$. Hence the definitions do not depend on a choice for h.

Step 2

Lemma 1. Any simple graded LA $\mathscr{L}$ satisfies *exactely one* of the following assertions.
1) $\mathscr{L}$ is without roots.
2) $\mathscr{L}$ is weakly integrable.
3) $\mathscr{L}$ is of type $\mathscr{C}$.
4) $\mathscr{L}$ is deep.

Step 3. *4 Classification Theorems*

The previous lemma splits the category of simple graded LA into four subcategories. Each of the following four theorem is a classification theorem for each of the four classes. Thus The main theorem is an obvious corollary of these four theorems.

Theorem 1. *Let $\mathscr{L}$ be a simple graded LA* without roots. *Then $\mathscr{L}$ has infinite growth.*

Theorem 2. *Let $\mathscr{L}$ be a* weakly integrable *simple graded LA. Then $\dim \mathscr{L} < \infty$ or $\mathscr{L}$ is affine.*

Theorem 3. *Let $\mathscr{L}$ be a simple graded LA of* type C. *The $\mathscr{L}$ is of Cartan type.*

Theorem 4. *Let $\mathscr{L}$ be a* deep *simple graded LA of* finite growth. *Then $\mathscr{L}$ is isomorphic to W.*

Theorem 1 has the following consequence. Any simple graded LA $\mathscr{L}$ with $\mathscr{L}_0 = 0$ has infinite growth. The growth hypothesis in theorem 1 is crucial because there are simple graded LA $\mathscr{L}$ with $\mathscr{L}_0 = 0$. However there are no growth hypotheses for Theorems 2, 3.

Thus Theorems 1, 2, 3, 4 and the Lemma implies the classification of simple graded LA of finite growth.

Some References:
Say that $\mathscr{L}$ has *growth* ≤ 1 if we have $\dim \mathscr{L}_n \leq C$ for some constant C. In a previous paper [M1] we classify simple graded *LA* of growth ≤ 1.

1) The proof of Theorem 1 follows the same line as Theorem 1 in [M1].
2) 90 % of Theorem 2 was already proved in [M1].
3) The proof of Theorem 3 essentialy uses homological Kostant formula, Kac theorem and a calculation of characteric variety (following a nice trick of V. Guillemin).
4) The proof of Theorem 4 is the main difficulty. At some point we use Gabber-Kac theorem. Otherwise it is elementary.

Part 3: More About the Proofs

The proof of the theorem is quite long. The main "*tools*" are the following ones:

1) *Basic Tool*: We get informations from any *"formal construction"* of ideals. Obvious examples are centers, derived algebras... We can also use the notion of "quasi-order" for that purpose. Another typical example is the following. Let $\mathscr{L}$ be any graded LA.

Lemma 2. Assume that $\mathcal{L} = \mathcal{A} + \mathcal{B}$ where $[\mathcal{A}, \mathcal{A}] \subseteq \mathcal{A}$ and $[\mathcal{A}, \mathcal{B}] \subseteq \mathcal{B}$. Then $\mathcal{B} + [\mathcal{B}, \mathcal{B}]$ is an ideal.

The lemma is obvious but it is used many time to construct subalgebras which behaves like $sl(2)$ or like an Heisenberg algebra.

2) *Another Tool: Partial LA.* Let $a < 0$, $b > 0$ be integers. A *partial LA* is a graded vector space

$$\Gamma = \bigoplus_{a \leq i \leq b} \Gamma_i$$

endowed with *partial brackets* $\Gamma_i \times \Gamma_j \to \Gamma_{i+j}$ (for $a \leq i, j, i+j \leq b$) satisfying partial Jacobi identities . For a graded LA $\mathcal{L}$, its *partial part*

$$Part\ \mathcal{L} = \oplus_{a \leq i \leq b} \mathcal{L}_i$$

is a partial LA. Conversely any partial LA Γ is the partial part of some graded LA $\mathcal{L}$. Among such $\mathcal{L}$'s there is a *minimal model* $\mathcal{L}_{\min}(\Gamma)$.

Lemma 3. Let $\mathcal{L}$ be a graded LA and Γ be a partial LA. If Γ is a subquotient of Part $\mathcal{L}$ then $\mathcal{L}_{\min}(\Gamma)$ is a subquotient of $\mathcal{L}$.

Especially if $\mathcal{L}_{\min}(\Gamma)$ does have infinite growth, $\mathcal{L}$ does. It allows us to restrict the possible partial part of graded LA of finite growth because we prove that for particular 6 series of partial LA Γ their models $\mathcal{L}_{\min}(\Gamma)$ have infinite growth.

3) *Another Tool: Ranks.* The *rank* of a graded LA $\mathcal{L}$ is the dimension over $\mathbb{Q}$ of $\mathbb{Q} \otimes_{\mathbb{Z}} Q$. In the proof of Theorem 4 we study two cases :

1) rank $\mathcal{L} = 1$

2) rank $\mathcal{L} \geq 2$

Actually the rank ≥ 2 case is by far easier.

4) *Last Tool: Coadjoint Estimates.* Let $\mathcal{L}^* = \oplus \mathcal{L}_n^*$ be the *graded dual* of the graded LA $\mathcal{L}$. For a homogenous $\zeta \in \mathcal{L}^*$ the space $\mathcal{L} \cdot \zeta \subseteq \mathcal{L}^*$ is homogenous. For a simple graded LA $\mathcal{L}$ it is easy to show that the growth of $\mathcal{L} \cdot \zeta$ is *independant* of $\zeta \neq 0$. The following lemma is very crucial in proving Theorem 4.

Lemma 4. Assume $\mathcal{L}$ *deep,* simple graded and of *finite growth.* Then $\mathcal{L} \cdot \zeta$ has growth exactly 1.

Table II. Simple graded Lie algebras

Type	Finite growth	Infinite growth
Without root	$\emptyset$	A lot [Continous families]
Weakly integrable	Finite dim. or Affine	$\emptyset$
Type $\mathscr{C}$	Cartan type	$\emptyset$
Deep	Virasoro-Witt	A lot [Continous families]

References

[C] E. Cartan: Les groupes de transformations continus infinis simples. Ann. Sci. Ecole Norm. Sup. **26** (1909) 93–161

[K] V. G. Kac: Simple graded Lie algebras of finite growth. Math. USSR Izv. **2** (1968) 1271–1311

[M1] O. Mathieu: Classification des algebres de Lie graduées simples de croissance ≤ 1. Invent. math. **86** (1986) 371–426

[M2] O. Mathieu: Classification of simple graded Lie algebras of finite growth. Preprint 1990

Orbits on Flag Manifolds

Toshihiko Matsuki

College of Liberal Arts and Sciences, Kyoto University, Kyoto 606, Japan

1. H-Orbits on $X = G/P$

Let G be a connected real semisimple Lie group and X the flag manifold of G. X is a homogeneous space of G and the isotropy subgroup $P = P_x$ of each point x of X is called a minimal parabolic subgroup of G. Let σ be an involutive automorphism ($\sigma^2 = id.$) of G and H a subgroup of $G^\sigma = \{x \in G \mid \sigma x = x\}$ containing the identity component G_0^σ of G^σ. Irreducible pairs $(\mathfrak{g}, \mathfrak{h})$ of Lie algebras of G and H are classified by [Be].

The following are special cases of H-orbit decompositions of $X = G/P$.

(i) Let σ be a Cartan involution of G, $\mathfrak{g} = \mathfrak{k} \oplus \mathfrak{s}$ the Cartan decomposition of the Lie algebra $\mathfrak{g}$ of G for σ and $K = H = G^\sigma$. Then $P = MAN$ where $A = P \cap \exp \mathfrak{s}$, $M = Z_K(A)$ and N is the unipotent radical of P. The Iwasawa decomposition $G = KAN (\cong K \times A \times N)$ implies that $\#(K \setminus G/P) = 1$.

(ii) Let $G = G_1 \times G_1$, $P = P_1 \times P_1$ and $\sigma(x, y) = (y, x)$ for $(x, y) \in G_1 \times G_1$. Then $H = G^\sigma = \{(x, x) \in G \mid x \in G_1\}$. Since $H \setminus G \cong G_1$ by the map $H(x, y) \mapsto x^{-1}y$, the double coset decomposition $H \setminus G/P$ is identified with the Bruhat decomposition $P_1 \setminus G_1/P_1$.

(iii) When G is a complex semisimple Lie group and σ is a conjugation of G, H-orbits on X are studied in [A]. This study suggested the formulation for the following general cases.

Let θ be a Cartan involution of G such that $\sigma\theta = \theta\sigma$, $\mathfrak{g} = \mathfrak{k} \oplus \mathfrak{s}$ the Cartan decomposition of $\mathfrak{g}$ for θ and $K = G^\theta$.

Definition. An element x of X is called "special" when $A_x = P_x \cap \exp \mathfrak{s}$ is σ-stable. Put

$$U = \{x \in X \mid x \text{ is special }\} .$$

Theorem 1 [R, M1]. *$K \cap H \setminus U \cong H \setminus X$ by the inclusion map $U \hookrightarrow X$.*

There exists a unique subgroup H^a of G such that $G_0^{\sigma\theta} \subset H^a \subset G^{\sigma\theta}$ and that $K \cap H^a = K \cap H$. (Note $(H^a)^a = H$.)

Corollary [M1]. *There exists a one-to-one correspondence $D \mapsto D^a$ between H-orbits and H^a-orbits on X given by $K \cap H \setminus U \cong H \setminus X$ and $K \cap H \setminus U \cong H^a \setminus X$.*

Proceedings of the International Congress
of Mathematicians, Kyoto, Japan, 1990

Example 1. Let $G = SL(2, \mathbb{C})$. Then $X = P^1(\mathbb{C}) = \mathbb{C} \cup \{\infty\}$,

$$\text{where} \quad \begin{pmatrix} a & b \\ c & d \end{pmatrix} x = \frac{ax+b}{cx+d} \quad \text{for} \quad \begin{pmatrix} a & b \\ c & d \end{pmatrix} \in SL(2, \mathbb{C}), \ x \in X.$$

$$\text{Let} \quad \sigma \begin{pmatrix} a & b \\ c & d \end{pmatrix} = \begin{pmatrix} a & -b \\ -c & d \end{pmatrix}, \quad \text{and} \quad \theta g = {}^t\bar{g}^{-1}.$$

Then

$$K = SU(2), \quad H = G^\sigma = \left\{ \begin{pmatrix} a & 0 \\ 0 & a^{-1} \end{pmatrix} \mid a \in \mathbb{C}^\times \right\},$$

$$H^a = G^{\sigma\theta} = SU(1,1) = \left\{ \begin{pmatrix} a & b \\ \bar{b} & \bar{a} \end{pmatrix} \mid a\bar{a} - b\bar{b} = 1 \right\}$$

The H-orbits on X are $\{0\}$, $\mathbb{C}^\times$ and $\{\infty\}$ and the corresponding H^a-orbits are $\{|x| < 1\}$, $\{|x| = 1\}$ and $\{|x| > 1\}$, respectively. ($U = \{0\} \cup \{|x| = 1\} \cup \{\infty\}$.)

2. Expression by Symbols

Remark 1. If $H = G_0^\sigma$, then $H \setminus X$ depends only on the pair $(\mathfrak{g}, \sigma)$ because

$$X \cong \text{ the set of minimal parabolic subalgebras of } \mathfrak{g}$$

and

$$H \setminus X \cong \text{Ad}(H)\text{-conjugacy classes of minimal parabolic subalgebras of } \mathfrak{g} .$$

Theorem 2 [M-O]. *Let G and H be as in the following list (complex classical cases). Then we can express $H \backslash X$ (and $H^a \backslash X$) by symbols. ($p+q = n$, $[H : G_0^\sigma] = 1$ or 2.)*

Type	G	H	H^a
AI	$GL(n, \mathbb{C})$	$O(n, \mathbb{C})$	$GL(n, \mathbb{R})$
AII	$GL(n, \mathbb{C})$	$Sp(n/2, \mathbb{C})$ (n even)	$U^*(n)$
AIII	$GL(n, \mathbb{C})$	$GL(p, \mathbb{C}) \times GL(q, \mathbb{C})$	$U(p, q)$
BI	$SO(2n+1, \mathbb{C})$	$S(O(2p+1, \mathbb{C}) \times O(2q, \mathbb{C}))$	$SO(2p+1, 2q)$
CI	$Sp(n, \mathbb{C})$	$GL(n, \mathbb{C})$	$Sp(n, \mathbb{R})$
CII	$Sp(n, \mathbb{C})$	$Sp(p, \mathbb{C}) \times Sp(q, \mathbb{C})$	$Sp(p, q)$
DI	$SO(2n, \mathbb{C})$	$S(O(2p, \mathbb{C}) \times O(2q, \mathbb{C}))$	$SO(2p, 2q)$
DI'	$SO(2n, \mathbb{C})$	$S(O(2p+1, \mathbb{C}) \times O(2q-1, \mathbb{C}))$	$SO(2p+1, 2q-1)$
DIII	$SO(2n, \mathbb{C})$	$GL(n, \mathbb{C})$	$SO^*(2n)$

Note. In [M-O] p.155, we should read $GL(n, \mathbb{C})$ for $\mathbb{C}^\times \times PSL(n, \mathbb{C})$ on the line of DIII in Table 1.

Precise description of symbols and many examples are given in [M-O]. But we can explain shortly the essencial part as follows.

Let $x \in U \subset X$. Then $\mathfrak{a}_x = \mathrm{Lie}(P_x) \cap \mathfrak{s}$ is σ-stable by the definition of U. Let Σ_x be the root system of the pair $(\mathfrak{g}, \mathfrak{a}_x)$ and Σ_x^+ the positive system of Σ_x corresponding to P_x. Let Ψ_x denote the set of simple roots in Σ_x^+. Then we can take an orthogonal basis $\{e_1, \ldots, e_n\}$ of the dual $\mathfrak{a}_x^*$ of $\mathfrak{a}_x$ such that

$$\Psi_x = \begin{cases} \{\alpha_1, \ldots, \alpha_{n-1}\} & \text{if } G = GL(n, \mathbb{C}), \\ \{\alpha_1, \ldots, \alpha_n\} & \text{otherwise,} \end{cases}$$

where $\alpha_1 = e_1 - e_2, \ldots, \alpha_{n-1} = e_{n-1} - e_n$ and $\alpha_n = e_n$, $2e_n$ or $e_{n-1} + e_n$ if $G = SO(2n+1, \mathbb{C})$, $Sp(n, \mathbb{C})$ or $SO(2n, \mathbb{C})$, respectively.

To the left coset $(K \cap H)x$ in U, there corresponds a sequence $\varepsilon_1 \varepsilon_2 \ldots \varepsilon_n$ consisting of the following four kinds of letters.

($\pm$) If $\sigma e_i = e_i$, then $\varepsilon_i = +$ ("a boy") or $-$ ("a girl"). When $\varepsilon_i = \pm$ and $\varepsilon_j = \pm$ $(i \neq j)$,

$$\varepsilon_i = \varepsilon_j \iff \mathfrak{g}(\mathfrak{a}_x, e_i - e_j) \subset \mathrm{Lie}(H) .$$

(a) If $\sigma e_i = e_j$ with $i \neq j$, then we put a small letter ("a family name") to the couple $(\varepsilon_i, \varepsilon_j)$.

(A) If $\sigma e_i = -e_j$ with $i \neq j$, then we put a capital letter to the "old" couple $(\varepsilon_i, \varepsilon_j)$.

(O) If $\sigma e_i = -e_i$, then $\varepsilon_i = O$ ("the aged" or "dead"?).

Let w_i be the reflection with respect to the simple root α_i and $P_i = P \cup P w_i P$ $(P = P_x)$ the parabolic subgroup of G for α_i. Let π_i denote the projection of $X = G/P$ onto G/P_i.

Notation. For two H-orbits D_1 and D_2 on X, we write

$$D_1 \xrightarrow{i} D_2 \iff \pi_i(D_1) = \pi_i(D_2) \text{ and } \dim D_1 < \dim D_2 .$$

We put here two examples. (You can see 23 figures of examples in [M-O].)

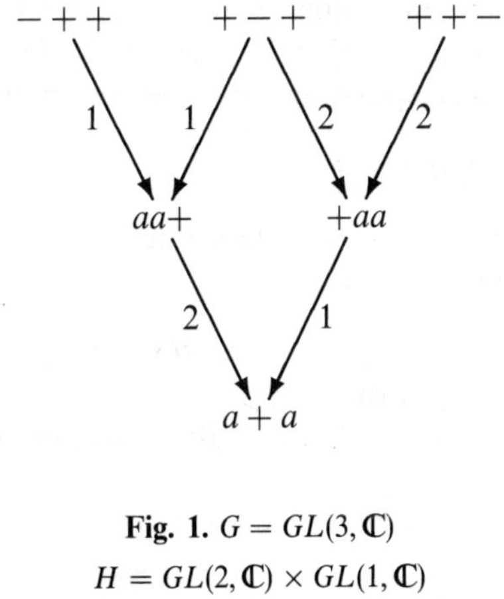

Fig. 1. $G = GL(3, \mathbb{C})$
$H = GL(2, \mathbb{C}) \times GL(1, \mathbb{C})$

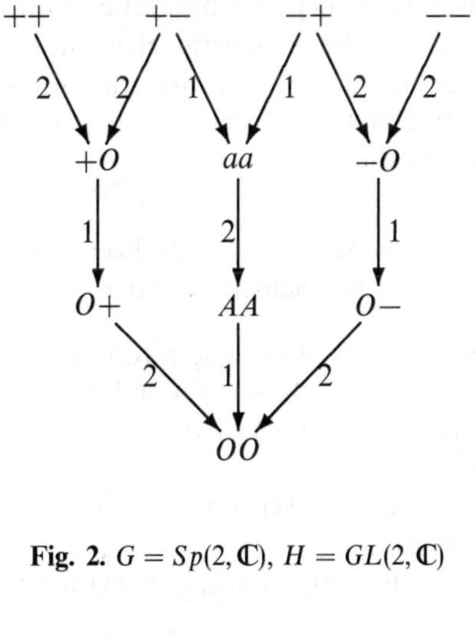

Fig. 2. $G = Sp(2, \mathbb{C})$, $H = GL(2, \mathbb{C})$

Remark 2 ([S], [M2]). In complex cases, we can find all the closure relations among H-orbits on X from the following two properties.

(a) $D_1 \xrightarrow{i} D_2 \Rightarrow D_1 \subset D_2^{cl}$.

(b) $D_1 \xrightarrow{i} D_2, D_3 \xrightarrow{i} D_4$ and $D_1 \subset D_3^{cl} \Rightarrow D_2 \subset D_4^{cl}$.

This is proved by the same argument as that of the Bruhat ordering since

$$D_1 \xrightarrow{i} D_2 \text{ and } D_1 \xrightarrow{i} D_3 \Rightarrow D_2 = D_3$$

in complex cases. To find all the closure relations in general real cases, we should follow a rather complicated procedure given in [M2].

Remark 3. These diagrams of orbits are useful to the study of the asymptotic behavior of spherical functions on semisimple symmetric spaces ([O]) and embeddings of Harish-Chandra modules into principal series ([M-O]).

Remark 4 (Problem). If $\Sigma = \Sigma(\mathfrak{g}, \mathfrak{a})$ is classical, then there exists (in principle) a similar (sometimes the same) expression of the H-orbits on X as that in a complex case. Give a complete list of such expressions by symbols. (For example, it is proved in [M2] that the diagram of $H^a \setminus X$ is upside-down to that of $H \setminus X$.)

Example 2 (= Exercise). When $G = GL(n, \mathbb{F})$ and $H = GL(p, \mathbb{F}) \times GL(n-p, \mathbb{F})$ for a division algebra $\mathbb{F}$ of characteristic $\neq 2$, the diagram of the H-orbits on X does not depend on $\mathbb{F}$.

Problem. Give good symbols for H-orbits on X when Σ is exceptional.

3. Uzawa's Function f and Vector Field v on X (Related to Intersections of H-Orbits and H^a-Orbits on X)

Recently, T. Uzawa discovered the following function f and vector field v on X which have very nice properties with respect to H-orbits and H^a-orbits.

Let Y_0 be a generic element of $\mathfrak{s}$. Then Y_0 defines a minimal parabolic subgroup P_0 of G such that $Y_0 \in \mathfrak{a}_0 = \text{Lie}(P_0) \cap \mathfrak{s}$ and that Y_0 is dominant for the positive system of the root system $\Sigma(\mathfrak{g}, \mathfrak{a}_0)$ corresponding to P_0. By the natural identification

$$G/P_0 \cong K/M_0 \cong \text{Ad}(K)Y_0$$

($K \cap P_0 = M_0 =$ the centralizer of Y_0 in K), $X = G/P_0$ is embedded into $\mathfrak{s}$. Let Y_x denote the element in $\text{Ad}(K)Y_0$ corresponding to $x \in X$.

Definition. (i) We define a function f on X by $f(x) = |Y_x^+|^2 = B(Y_x^+, Y_x^+)$ on X where $Y_x^+ = \frac{1}{2}(Y_x + \sigma Y_x)$ and $B(,)$ is the Killing form on $\mathfrak{g}$.

(ii) A vector field v on X is defined by $v_x =$ the (infinitesimal) Y_x^+-action at x for $x \in X$.

(iii) Φ_t ($t \in \mathbb{R}$) is the one-parameter group of transformations of X for the vector field v.

(iv) $\Phi_{\pm\infty}(x) = \lim_{t \to \pm\infty} \Phi_t(x)$ for $x \in X$.

Remark 5. The vector field v is the gradient of the function f with respect to the K-invariant Riemannian metric on $X = K/M_0$ induced from the inner product $(Z,Z') = B([Z,Y_0],Z'_{\mathfrak{s}})$ on $\mathfrak{k}^{\perp\mathfrak{m}_0}$ where $Z'_{\mathfrak{s}}$ is the element in $\mathfrak{s}$ such that $Z'_{\mathfrak{s}} - Z' \in \mathrm{Lie}(P_0)$ and $\mathfrak{k}^{\perp\mathfrak{m}_0}$ is the orthogonal compliment of $\mathfrak{m}_0 = \mathrm{Lie}(M_0)$ in $\mathfrak{k}$.

Remark 6. If the real rank of G is larger than one, then f and v depend essencially (not constant multiple) on the choice of Y_0.

Example 3. (continued to Example 1) Take

$$Y_0 = \begin{pmatrix} 1 & 0 \\ 0 & -1 \end{pmatrix} \in \mathfrak{s} = \left\{ \begin{pmatrix} z & x+iy \\ x-iy & -z \end{pmatrix} \mid x,y,z \in \mathbb{R} \right\} .$$

Since P_0 is the subgroup of G consisting of upper triangular matrices, eP_0 corresponds to ∞ in $P^1(\mathbb{C}) = \mathbb{C} \cup \{\infty\}$ and

$$kP_0 \mapsto \begin{pmatrix} a & b \\ -\bar{b} & \bar{a} \end{pmatrix} \infty = \frac{a}{-\bar{b}} \quad \text{for} \quad k = \begin{pmatrix} a & b \\ -\bar{b} & \bar{a} \end{pmatrix} \in K .$$

On the other hand,

$$\begin{aligned} \begin{pmatrix} a & b \\ -\bar{b} & \bar{a} \end{pmatrix} \begin{pmatrix} 1 & 0 \\ 0 & -1 \end{pmatrix} \begin{pmatrix} a & b \\ -\bar{b} & \bar{a} \end{pmatrix}^{-1} &= \begin{pmatrix} a & -b \\ -\bar{b} & -\bar{a} \end{pmatrix} \begin{pmatrix} \bar{a} & -b \\ \bar{b} & a \end{pmatrix} \\ &= \begin{pmatrix} a\bar{a} - b\bar{b} & -2ab \\ -2\bar{a}\bar{b} & -a\bar{a} + b\bar{b} \end{pmatrix} . \end{aligned}$$

So $\mathrm{Ad}(K)Y_0$ is the sphere given by $x^2+y^2+z^2 = 1$ and the function f is given by z^2. Two points $\{\infty\}$, $\{0\}$ and the unit circle in $P^1(\mathbb{C})$ correspond to $(0,0,1)$, $(0,0,-1)$ and the circle defined by $z = 0$, respectively, in $\mathrm{Ad}(K)Y_0$.

Theorem 3 [U]. (i) *v is tangent to H-orbits and H^a-orbits.*

(ii) $(df)_x = 0 \iff v_x = 0 \iff$ *x is special.*

(iii) *Let D be an H-orbit on X. Then there exists $m = \min_{x\in D} f(x)$ and for $x \in D$,*

$$f(x) = m \iff x \text{ is special} .$$

(iv) *$\Phi_{-\infty}(D) = D \cap U$ for H-orbits D on X.*

Corollary 1 [M3]. (a) *$D \cap D^a = (K \cap H)x$ for an $x \in U$.*

(b) *For two H-orbits D and E on X,*

$$D^{cl} \supset E \iff D \cap E^a \neq \phi \iff D^a \subset (E^a)^{cl} .$$

Proof ([U]). (a) Let $x \in D \cap D^a$. We have only to show that $x \in U$ by Theorem 1. Let m be the value of the function f at the points in $D \cap U$ $(= D^a \cap U)$. Suppose that $x \notin U$. Then $f(x) > m$ by (iii). Since the function for the H^a-orbit structure is $|Y_0|^2 - f(x)$, we have also $f(x) < m$ by (iii), a contradiction.

(b) Since $(H^a)^a = H$, we have only to prove the left $\iff$.

The assertion $D^{cl} \supset E \Rightarrow D \cap E^a \neq \phi$ is clear since

$$T_x(E) + T_x(E^a) = T_x(X)$$

for any $x \in E \cap E^a$ ([M3]).

Suppose that $D \cap E^a \neq \phi$ and let $x \in D \cap E^a$. Then

$$\Phi_\infty(x) = \lim_{t \to \infty} \Phi_t(x) \in D^{cl} \cap E^a \cap U = D^{cl} \cap E \cap U$$

by (i) and (iv). Hence $D^{cl} \cap E \neq \phi$ and therefore $D^{cl} \supset E$. □

Corollary 2. *Let D be an H-orbit on X and $x \in D \cap D^a$. Then*

(i) $$D \cong (K \cap H) \times_L \Phi_{-\infty}^{-1}(x)$$

where $L = K \cap H \cap P_x$ and

(ii) $$D \cap E^a \cong (K \cap H) \times_L (\Phi_{-\infty}^{-1}(x) \cap E^a)$$

for any H^a-orbit E^a on X. (Moreover it is clear that the fibers $\Phi_{-\infty}^{-1}(x)$ and $\Phi_{-\infty}^{-1}(x) \cap E^a$ are contractible to the point x.)

4. Remarks on Spherical Subgroups

Suppose that G is a complex semisimple Lie group. A complex Lie subgroup H of G is called "spherical" if there exists an open H-orbit on X. Such pairs (G, H) are classified by [K] when G is simple and H is reductive, and by [Br2] in general.

Theorem 4 [Br1, V]. *$H \subset G$ is spherical $\iff \#(H \setminus X)$ is finite. (Note that $\Leftarrow$ is clear.)*

There is a simple proof of $\Rightarrow$ using "rank-one sections" as follows.

Proof. We may assume that HP is open in G. Write $G = P_{\beta_1} P_{\beta_2} \cdots P_{\beta_m}$ where the β_i's are simple roots and $P_{\beta_i} = P \cup P w_{\beta_i} P$. Put $P^{(i)} = P_{\beta_1} P_{\beta_2} \cdots P_{\beta_i}$ $(P^{(0)} = P)$. We will show

$$\#(H \setminus HP^{(i)}/P) < \infty \text{ for } i = 0, 1, \ldots, m$$

by induction on i.

By the hypothesis of induction, we may assume that

$$HP^{(i-1)} = Hg_1 P \cup \cdots \cup Hg_k P \ .$$

Then we have

$$HP^{(i)} = Hg_1 P_{\beta_i} \cup \cdots \cup Hg_k P_{\beta_i} \ .$$

We have only to show that $\#(H \setminus Hg_j P_{\beta_i}/P) < \infty$ for $j = 1, \ldots, k$. Since $HP^{(i-1)}$ is open in G, $(g_j P_{\beta_i}/P) \cap (HP^{(i-1)}/P)$ is (Zariski) open in the one-dimensional subvariety $g_j P_{\beta_i}/P$ of the complex algebraic variety X. Hence the compliment of $(g_j P_{\beta_i}/P) \cap (HP^{(i-1)}/P)$ in $g_j P_{\beta_i}/P$ consists of finte points and therefore $\#(H \setminus Hg_j P_{\beta_i}/P) < \infty$. □

Let G be a real semisimple Lie group and H a Lie subgroup of G.

Conjecture 1. *If the real rank of* G *is one and there exists an open* H*-orbit on* $X = G/P$, *then* $\#(H \setminus X) < \infty$.

By the same argument as above for spherical subgroups, Conjecture 1 implies the following Conjecture 2.

Conjecture 2. *If there exists an open* H*-orbit on* X, *then* $\#(H \setminus X) < \infty$.

Remark 7. In general, $\#(H \setminus G/P) < \infty$ does not imply $\#(H_{\mathbb{C}} \setminus G_{\mathbb{C}}/P_{\mathbb{C}}) < \infty$. For example, if $G = SU(n, 1)$ $(n > 2)$ and $H = \theta N$ (where N is the unipotent radical of P), then $\#(H \setminus G/P) = 2$ and $\#(H_{\mathbb{C}} \setminus G_{\mathbb{C}}/P_{\mathbb{C}}) = \infty$.

References

[A] Aomoto, K.: On some double coset decompositions of complex semi-simple Lie groups. J. Math. Soc. Japan **18** (1966) 1–44

[Be] Berger, M.: Les éspace symmétriques non compacts. Ann. Sci. École Norm. Sup. **74** (1957) 85–177

[Br1] Brion, M.: Quelques propriétés des espaces homogènes sphériques. Manuscripta Math. **55** (1986) 191–198

[Br2] Brion, M.: Classification des espaces homogènes sphériques. Comp. Math. **63** (1987) 189–208

[K] Krämer, M.: Sphärische Untergruppen in Kompakten zusammenhängenden Liegruppen. Comp. Math. **38** (1979) 129-153

[M1] Matsuki, T.: The orbits of affine symmetric spaces under the action of minimal parabolic subgroups. J. Math. Soc. Japan **31** (1979) 331–357

[M2] Matsuki, T.: Closure relations for orbits on affine symmetric spaces under the action of minimal parabolic subgroups. Adv. Studies Pure Math. **14** (1988) 541–559

[M3] Matsuki, T.: Closure relations for orbits on affine symmetric spaces under the action of parabolic subgroups. Intersections of associated orbits, Hiroshima Math. J. **18** (1988) 59–67

[M-O] Matsuki, T., Oshima, T.: Embeddings of discrete series into principal series. In: The Orbit Method in Representation Theory. Birkhäuser, Boston 1990, pp. 147–175

[O] Oshima, T.: Asymptotic behavior of spherical functions on semisimple symmetric spaces. Adv. Studies Pure Math. **14** (1988) 561–601

[R] Rossmann, W.: The structure of semisimple symmetric spaces. Canad. J. Math. **31** (1979) 157–180

[S] Springer, T. A.: Some results on algebraic groups with involutions. Adv. Studies Pure Math. **6** (1984) 525–534

[U] Uzawa, T.: Invariant hyperfunction sections of line bundles. Preprint (1990)

[V] Vinberg, E. B.: Complexity of actions of reductive groups. Funct. Anal. Appl. **20** (1985) 1–11

Sur les Formes Automorphes de Carré Intégrable

Colette Mœglin

URA 748, UFR de Mathématiques, Université de Paris VII, 2, place de Jussieu
F-75251 Paris Cedex 05, France

1. Définitions et notations

Soit k un corps global, pour simplifier de caractéristique 0, i.e. une extension algébrique finie de $\mathbb{Q}$. On note $\mathbb{A}$ l'anneau des adèles de k c'est-à-dire la k-algèbre formée des éléments $\prod_v x_v$ où v parcourt l'ensemble des places de k et où x_v est un élément du complété k_v de k en la place v, entier pour tout v sauf un nombre fini.

Soit $\mathbf{G}$ un groupe algébrique affine connexe, défini sur k que l'on suppose réductif. Les exemples classiques s'obtiennent en considérant un espace vectoriel de dimension finie V défini sur k et une forme bilinéaire Φ sur V (définie sur k); on prend alors pour $\mathbf{G}$ le groupe des automorphismes linéaires de V qui respectent Φ. Pour toute k-algèbre, k', on note $\mathbf{G}(k')$ les points de $\mathbf{G}$ définis sur k'.

On appelle forme automorphe sur $\mathbf{G}$, une fonction ϕ à valeurs complexes sur $\mathbf{G}(k)\backslash\mathbf{G}(\mathbb{A})$ qui vérifie un certain nombre de propriétés pour lesquelles je renvoie à [3]. On appelle forme automorphe de carré intégrable modulo le centre, une forme automorphe, ϕ, pour laquelle il existe un caractère unitaire ω du centre $\mathbf{Z}(\mathbb{A})$ de $\mathbf{G}(\mathbb{A})$ tel que:

$$(i) \quad \phi(zg) = \omega(z)\phi(g), \qquad \forall z \in \mathbb{Z}(\mathbb{A}), \quad \forall g \in \mathbf{G}(\mathbb{A}),$$

$$(ii) \quad \int_{\mathbf{G}(k)\mathbf{Z}(\mathbb{A})\backslash\mathbf{G}(\mathbb{A})} \phi(g)\overline{\phi}(g)\, dg < \infty,$$

où dg est une mesure de Haar sur $\mathbf{G}(k)\mathbf{Z}(\mathbb{A})\backslash\mathbf{G}(\mathbb{A})$.

Dans cet exposé, on s'intéresse à la détermination des formes automorphes de carré intégrable. L'article de base pour cette question est [5] (cf. aussi [8]).

On appelle sous-groupe parabolique de $\mathbf{G}$, un sous-groupe fermé $\mathbf{P}$ de $\mathbf{G}$ tel que $\mathbf{G}(\bar{k})/\mathbf{P}(\bar{k})$ soit une variété projective ($\bar{k}$ est une clôture algébrique de k). Ces groupes sont (comme $\mathbf{G}$) connexes et jouent un rôle privilégié dans la théorie. Si $\mathbf{P} \neq \mathbf{G}$, $\mathbf{P}$ n'est plus un groupe réductif mais comme tout groupe algébrique affine connexe, il contient un unique sous-groupe fermé normal, noté ${}^u\mathbf{P}$ tel que $\mathbf{P}/{}^u\mathbf{P}$ soit réductif et ${}^u\mathbf{P}$ est minimal avec cette propriété. On pose

$$\mathbf{M}_P := \mathbf{P}/{}^u\mathbf{P}.$$

On note X_P le groupe des caractères de $\mathbf{M}_P(\mathbb{A})$, à valeurs complexes, engendré par les caractères du type $|\chi|^s$ où $|\,|$ est la valeur absolue adélique, où χ est

Proceedings of the International Congress
of Mathematicians, Kyoto, Japan, 1990

un caractère rationnel de $\mathbf{M}_P$ et où s est un nombre complexe. Tout caractère appartenant à X_P est évidemment trivial sur $\mathbf{M}_P(k)$. On vérifie que X_P est un espace vectoriel.

Par exemple si $\mathbf{G} = \mathrm{Aut}(V, \Phi)$ avec $\Phi \equiv 0$, i.e. $\mathbf{G} \simeq \mathrm{GL}(V)$ alors les sous-groupes paraboliques de $\mathbf{G}$ sont les stabilisateurs des drapeaux:

$$V_0 = 0 \subset V_1 \subset V_1 \subset \cdots \subset V_d = V$$

de V. Ici $\mathbf{M}_P \simeq \prod_{1 \le i \le d} \mathrm{GL}(V_i/V_{i-1})$ et $X_P \simeq \mathbb{C}^d$ isomorphisme donné par l'application qui à $(z_1, \cdots, z_d) \in \mathbb{C}^d$ associe le caractère $\chi\big((m_1, \cdots, m_d)\big) = |\det m_1|^{z_1} \cdots |\det m_d|^{z_d}$.

Soit $\mathbf{P}$ un sous-groupe parabolique de $\mathbf{G}$; on généralise la notion de formes automorphes en notant $A\big(\mathbf{P}(k)\,{}^u\mathbf{P}(\mathbb{A})\backslash\mathbf{G}(\mathbb{A})\big)$ l'ensemble des fonctions à valeurs complexes sur ce quotient qui vérifient les mêmes propriétés que les formes automorphes. En fait pour tout $\phi \in A\big(\mathbf{P}(k)\,{}^u\mathbf{P}(\mathbb{A})\backslash\mathbf{G}(\mathbb{A})\big)$ et pour tout $g \in \mathbf{G}(\mathbb{A})$ la fonction ϕ_g sur $\mathbf{M}_P(\mathbb{A})$ définie par:

$$\phi_g(m) = \phi(mg)\delta_P(m)^{-1/2}$$

est une forme automorphe pour $\mathbf{M}_P$ ($\delta_P(m)$ est le Jacobien de l'automorphisme de ${}^u\mathbf{P}(\mathbb{A})$ qui envoit $u \in {}^u\mathbf{P}(\mathbb{A})$ sur mum^{-1}.

On aurra aussi besoin de la notion de formes automorphes cuspidales; soit ϕ comme ci-dessus, on demande en plus que l'on ait pour tout sous-groupe parabolique propre $\mathbf{P}'$ de $\mathbf{M}_P$:

$$\forall g \in \mathbf{G}(\mathbb{A}), \qquad \int_{{}^u\mathbf{P}'(k)\backslash {}^u\mathbf{P}'(\mathbb{A})} \phi(ug)\, du = 0\,.$$

Cela revient à dire que la représentation de $\mathbf{M}_P(\mathbb{A})$ engendrée par ϕ_g (aux places à l'infini, c'est un $\mathfrak{g} - K$ module) est une représentation automorphe cuspidale.

2. Séries d'Eisenstein

Les séries d'Eisenstein fournisse un procédé pour transformer un élément $\phi \in A\big(\mathbf{P}(k)\,{}^u\mathbf{P}(\mathbb{A})\backslash\mathbf{G}(\mathbb{A})\big)$ en une famille, $E(\phi, \lambda)$ de formes automorphes dépendant méromorphiquement de $\lambda \in X_P$. Toutefois l'existence de $E(\phi, \lambda)$ n'a de démonstration écrite que si ϕ est de carrée intégrable modulo le centre (cf. [5], Chapitre 6 et 7) (Le résultat général est annoncé par Bernstein et dans le cas, que nous ne considérons pas d'un corps global de caractéristique > 0, résulte de résultats de [11] et [15]).

Dans ce qui suit, on suppose que $\phi \in A\big(\mathbf{P}(k)\,{}^u\mathbf{P}(\mathbb{A})\backslash\mathbf{G}(\mathbb{A})\big)$ vérifie la condition (i) ci-dessus (i.e. à un caractère central unitaire) pour l'action du centre de $\mathbf{M}_P(\mathbb{A})$ et est cuspidale. Alors ϕ est de carré intégrable modulo le centre de $\mathbf{M}_P(\mathbb{A})$. On sait que les pôles de $E(\phi, \lambda)$ sont de nature très simple: soit $\lambda_0 \in X_P$, alors il existe un ensemble fini E d'hyperplans de X_P passant par λ_0 et pour chacun de ces hyperplans $H \in E$ un entier n_H (E et n_H peuvent être choisi indépendamment de ϕ) tels que:

$$\prod_{H \in E} P_H^{n_H}\, E(\phi, \lambda)$$

soit holomorphe en λ_0 où P_H est l'équation de l'hyperplan H. On peut définir les résidus successsifs de $E(\phi, \lambda)$ le long d'un ensemble fini ordonné, E', d'hyperplans linéairement indépendants. On obtient une famille de formes automophes sur $\mathbf{G}(\mathbb{A})$ dépendant méromorphiquement de λ appartenant à $\cap_{H \in E'} H$.

Nature des pôles des Séries d'Eisenstein

Admettons que l'on sache normaliser les opérateurs d'entrelacements (cf. [5], appendice 2 et [14]); on sait en particulier le faire si $\mathbf{G} \simeq \mathrm{GL}(V)$. Les pôles des séries d'Eisenstein sont alors de 3 types différents:

- le premier type est de nature locale, il correspond aux pôles des opérateurs d'entrelacements normalisés locaux;
- les deux autres types sont de nature globale: les facteurs de normalisations des opérateurs d'entrelacemetns sont des quotients de fonctions L. Le deuxième type correspond aux zéros des fonctions L qui apparaissent aux dénominateurs et le troisième type correspond aux pôles des fonctions L qui apparaissent aux numérateurs.

Forme faible des résultats de Langlands

Théorème (Langlands). *Toute forme automorphe de carré intégrable modulo le centre est un résidu de séries d'Eisenstein.*

En fait la théorie de Langlands [5] donne beaucoup plus de renseignements puisque elle limite très sérieusement les hyperplans qui peuvent fournir des résidus de carré intégrable.

Conjecture. *Soient $\phi, \lambda, E(\phi, \lambda)$ comme ci-dessus. Les seuls hyperplans singuliers pour $E(\phi, \lambda)$ qui peuvent fournir des résidus de carré intégrable sont les hyperplans singuliers provenant des pôles des fonctions L des numérateurs des facteurs de normalisations.*

3. Le cas de GL(V)

La conjecture est vraie et on démontre alors aisément le théorème suivant (cf. [9], conjecturé et démontré dans des cas particulier par Jacquet [4] (la partie (i) avait été obtenue par Speh [13]):

Théorème. *Soit V un espace vectoriel de dimension n sur k; on suppose que* $\mathbf{G} = \mathrm{GL}(V)$.

(i) Soit $n = da$ une décomposition de n et soit

$$V_0 = 0 \subset \cdots \subset V_i \subset \cdots \subset V_a = V$$

un drapeau tel que $\dim V_i = id$, *pour* $0 \leq i \leq a$. *On note* $\mathbf{P}$ *le stabilisateur de ce drapeau et alors* $\mathbf{M}_P \simeq \prod_{1 \leq i \leq a} \mathrm{GL}(d)$. *Soit* $\phi \in A\big(\mathbf{P}(k)\,{}^u\mathbf{P}(\mathbb{A})\backslash \mathbf{G}(\mathbb{A})\big)$ *telle que pour tout* $g \in \mathbf{G}(\mathbb{A})$ *la représentation automorphe de* $\mathbf{M}_P(\mathbb{A})$ *engendrée par* ϕ_g *soit irréductible isomorphe à* $\varrho \otimes \cdots \otimes \varrho$ *où* ϱ *est une représentation automorphe cuspidale fixée de* $\mathrm{GL}(d)(\mathbb{A})$ *de caractère central unitaire. On rappelle que* $X_P \simeq \mathbb{C}^a$*. Alors l'ensemble des hyperplans siuguliers pour* $E(\phi, \lambda)$ *passant par le point:*

$$(a-1)/2, \ldots, (a-2i+1)/2, \ldots, -(a-1)/2$$

est l'ensemble des hyperplans $s_i - s_{i-1} = 1$ *pour* $1 \leq i \leq a$*. Le résidu obtenu successivement le long de ces hyperplans rangés dans n'importe quel ordre (l'ordre est indifférent) est une forme automorphe sur* $\mathbf{G}(\mathbb{A})$ *de carré intégrable modulo le centre.*

(ii) Toutes les formes automorphes de carré intégrable modulo le centre sont obtenues de cette façon.

Sur la preuve. (i) C'est un problème combinatoire peu profond (déjà résolu dans l'appendice 3 de [5] pour $n = 4$) que de montrer que les zéros des fonctions L des dénominateurs des facteurs de normalisations n'interviennent pas.

(ii) C'est un problème de décomposition de certaines induites (du moins il faut calculer leurs quotients irréductibles) qui permet de régler le cas des pôles des opérateurs d'entrelacements locaux.

4. Interprétation d'Arthur

Dans [6] Langlands suggère que les classes d'isomorphie de représentations automorphes tempérées se regroupent en paquets qui devraient être paramétrés par les classes de conjugaison d'homomorphismes "admissibles" d'un groupe (dont l'existence devrait provenir de la théorie des catégories tanakiennes), noté L_F, dans le L-groupe associé à G. Admettant cela, Arthur suggère que les classes d'isomorphie de représentations automorphes de carré intégrable se regroupent en paquets que devraient être paramétrés par les classes de conjugaison d'homomorphismes "admissibles" du produit direct de L_F par $\mathrm{SL}(2, \mathbb{C})$ dans le L-groupe. Arthur donne aussi une paramétrisation (conjecturale) des représentations intervenant dans un paquet donné ainsi que de la multiplicité avec laquelle elles devraient intervenir.

En admettant la conjecture de Ramanujan, i.e. que l'ensemble des formes automorphes cuspidales de $\mathrm{GL}(n)$, de caractère central unitaire, est exactement l'ensemble des formes automorphes tempérées, le résultat cité pour $\mathrm{GL}(n)$ prouve les conjectures d'Arthur dans ce cas (chaque paquet est ici réduit à un élément et la multiplicité est un).

Un cas particulier intéressant des conjectures d'Arthur est de considérer les homomorphismes "admissibles" triviaux sur L_F. Les paquets qu'ils devraient paramétriser sont appelés, par Arthur, unipotents. Supposons pour simplifier grandement que $\mathbf{G}$ est un groupe classique déployé. L'ensemble des paquets unipotents devrait alors être en bijection avec l'ensemble des orbites unipotentes du groupe complexe dual, G^*, associé à $\mathbf{G}$ (i.e. la composante neutre du L-groupe) ne rencontrant aucun sous-groupe de Levi propre de G^*. La méthode employée pour $\mathrm{GL}(n)$ devrait conduire aisément au résultat suivant:

soit $\mathbf{G}$ *un groupe classique déployé de centre fini; alors les résidus de carré intégrable des séries d'Eisenstein construites à partir des caractères non ramifiés d'un sous-groupe de Borel de* $\mathbf{G}$ *doivent réaliser avec multiplicité un les représentations automorphes de carré intégrable de* $\mathbf{G}(\mathbb{A})$ *paramétrées par les orbites unipotentes, O, de* G^* *ne rencontrant aucun sous-groupe de Levi propre de* G^* *et par les produits,*

$$\prod_v \chi_v,$$

indéxés par l'ensemble des places de k, de caractères du groupe, $\overline{A}(O)$*, introduit par Lusztig ([7], chapitre 14, où il est noté* $\overline{A}(u)$*), vérifiant:*

pour presque tout v, χ_v *est le caractère identité et le caractère de* $\overline{A}(O)$ *que l'on obtient en faisant le produit de tous les* χ_v *est le caractère identité.*

Pour un résultat dans ce sens, confer [10]. Ce qui me manque pour obtenir toutes les représentations d'un paquet unipotent est une construction générale des représentations automorphes cuspidales unipotentes; les séries theta en fournissent dans certains cas (cf. [12]).

Bibliographie

1. Arthur, J.: Unipotent automorphic representations: conjectures, in Orbites unipotentes et représentations II. Groupes p-adiques et réels. Astérisque **171-172** (1989) 13–72
2. Arthur, J.: Unipotent automorphic representations: global motivations. In: automorphic forms, Shimura varieties, and L-functions, vol. 1 (L. Clozel et J.S. Milne, eds.). Perspectives in Mathematics, 1990, pp. 1–75
3. Borel, A., Jacquet, H.: Automorphic forms and automorphic representaions. In: Proc. of Symp. in Pure Math., vol. 33, vol. 1. Amer. Math. Soc., Providence, RI, 1979, pp. 189–202
4. Jacquet, H.: On the residual spectrum of GL(n). In: Lie Group Representations II. (Lecture Notes in Mathematics, vol. 1041 (édité par R. Herb, S. Kudla, R. Lipsmann and J. Rosenberg.)) Springer, Berlin Heidelberg New York 1984, pp. 185–208
5. Langlands, R.P.: On the functional equations satisfied by Eisenstein series. (Lecture Notes in Mathematics, vol. 544.) Springer, Berlin Heidelberg New York 1976
6. Langlands, R.P.: Automorphic representations, Shimura varieties and motives – ein Märchen. In: Proc. of Symp. in Pure Math., vol. 33, vol. 2. Amer. Math. Soc., Providence, RI, 1979, pp. 205–246
7. Lusztig, G.: Characters of reductive groups over a finite field. Ann. Math. Studies, vol. 107. Princeton Univ. Press, 1984
8. Mœglin, C., Waldspurger, J.-L.: Décomposition spectrale et séries d'Eisenstein, paraphrase sur l'Ecriture. Prépublication, Paris 1989
9. Mœglin, C., Waldspurger, J.-L.: Le spectre résiduel de GL(n). Ann. de l'ENS, 1989
10. Mœglin, C.: Orbites unipotentes et spectre discret non ramifié, à paraitre à Compositio
11. Morris, L.E.: Eisenstein series for reductive groups over global function fields 1 et 2. Can. J. Math. **34** (1982) 1112–1182 et Can. J. Math. **35** (1983) 974–985
12. Piatetski-Shapiro, I.: On the Saito Kurokawa lifting. Math. inv. **71** (1983) 309–338
13. Speh, B.: Some results on principal series for GL(n, **R**). Ph. D. dissertation M.I.T., Juin 1977
14. Shahidi, F.: A proof of Langlands conjecture on Plancherel measures; complementary series for p-adic groups. Ann. Math., à paraitre
15. Waldspurger, J.-L.: Formes automorphes et series d'Eisenstein sur un corps de fonctions. Prépublication, Paris 1989

Semi-simple Groups and Arithmetic Subgroups

Gopal Prasad

Tata Institute of Fundamental Research, Homi Bhabha Road
Colaba, Bombay 400 005, India

To my teacher
M.S. Raghunathan

In this lecture I shall report on the recent results, and open questions, related to the congruence subgroup problem, computation of the covolume of S-arithmetic subgroups, bounds for the class-number of simply connected semi-simple groups and state the finiteness theorems of [3]. We shall also briefly mention the recent work on super-rigidity of *cocompact* discrete subgroups of $\mathrm{Sp}(n, 1)$ and the $\mathbf{R}$-rank 1 form of type F_4, which implies arithmeticity of these discrete subgroups.

Notation. Throughout this report k is a global field, that is either a number field (i.e. a finite extension of the field $\mathbf{Q}$ of rational numbers) or the function field of an algebraic curve over a finite field. Let V be the set of places of k, V_∞ (resp. V_f) be the set of archimedean (resp. nonarchimedean) places. For $v \in V$, k_v will denote the completion of k at v with the natural locally compact topology and $| \ |_v$ the normalized absolute value on k_v. For $v \in V_f$, $\mathfrak{o}_v$ will denote the ring of integers of k_v, $\mathfrak{f}_v$ the residue field, p_v the characteristic of $\mathfrak{f}_v$ and q_v its cardinality. In the sequel k_v is assumed to carry the "normalized" Haar measure, see [26, 0.1]. For any finite set S of places of k containing V_∞, $\mathfrak{o}_S$ will denote the ring of S-integers of k, i.e.

$$\mathfrak{o}_S = \{x \in k| \ \ |x|_v \leqq 1 \ \text{ for all } \ v \notin S\}.$$

A will denote the ring of adèles of k. For a finite set S of places of k, let A_S be the ring of S-adèles i.e. the restricted direct product of the k_v's for $v \notin S$.

Let G be a connected semi-simple algebraic group defined over k. We fix an embedding of G in SL_n defined over k and view G as a k-subgroup of SL_n in terms of this embedding. Let S be a fixed finite set of places of k containing V_∞. Let $G_S = \prod_{v \in S} G(k_v)$ with the locally compact topology induced by the topologies on $k_v, v \in S$. We shall let Γ denote the group $G(k) \cap \mathrm{SL}_n(\mathfrak{o}_S)$. Note that Γ depends on the embedding of G in SL_n fixed above. Embedded diagonally in G_S, Γ is a discrete subgroup of finite covolume. A subgroup of G_S is said to be *S-arithmetic* if it is commensurable with Γ.

Proceedings of the International Congress
of Mathematicians, Kyoto, Japan, 1990

1. The Congruence Subgroup Problem

For any non-zero ideal $\mathfrak{a}$ of $\mathfrak{o}_S$, we have the "reduction mod $\mathfrak{a}$"

$$\pi_{\mathfrak{a}} : \mathrm{SL}_n(\mathfrak{o}_S) \to \mathrm{SL}_n(\mathfrak{o}_S/\mathfrak{a}).$$

The kernel of $\pi_{\mathfrak{a}}|_\Gamma$ will be denoted by $\Gamma_{\mathfrak{a}}$, it is by definition the *principal S-congruence subgroup of Γ of level* $\mathfrak{a}$. Since $\mathfrak{o}_S/\mathfrak{a}$ is finite, $\mathrm{SL}_n(\mathfrak{o}_S/\mathfrak{a})$ is finite and hence $\Gamma_{\mathfrak{a}}$ is of finite index in Γ. An S-arithmetic subgroup is an *S-congruence subgroup* if it contains a principal S-congruence subgroup of Γ of some level. It is not difficult to see that this notion (of S-congruence subgroups) does not depend on the choice of the k-embedding of G in SL_n.

Henceforth, G will be assumed to be absolutely almost simple and simply connected. We shall assume further that G_S is noncompact or, equivalently, for some v in S, G is isotropic at v ([24]).

The *congruence subgroup problem* in its simplest form asks whether any S-arithmetic subgroup is an S-congruence subgroup. If the answer is in the affirmative, we say that G has the congruence subgroup property (for S-arithmetic subgroups). In general the answer to the above question is in the negative. For example, as has been known since 1880, the group $\mathrm{SL}_2/\mathbf{Q}$ does not have the congruence subgroup property for $S = V_\infty$ (but the group $\mathrm{SL}_n/\mathbf{Q}$ has the congruence subgroup property for all $n > 2$ – this was proved by Bass-Lazard-Serre and Mennicke independently in 1963). If k is a totally imaginary number field, the group SL_n/k fails to have the congruence subgroup property for any n ($S = V_\infty$); see [2]. To give a precise measure of the failure, J-P. Serre introduced "the S-congruence kernel" which is a profinite group defined as follows. On $G(k)$ we introduce the following two translation invariant topologies:

(1) The *S-congruence topology*: In this the S-congruence subgroups constitute a neighborhood base at the identity. It is obvious that this is the same topology as the one induced on $G(k)$ from $G(A_S)$. By strong approximation ([23], [14], [21]), the completion of $G(k)$ with respect to the S-congruence topology is $G(A_S)$.
(2) The *S-arithmetic topology*: In this the S-arithmetic subgroups contained in $G(k)$ constitute a neighborhood base at the identity. Completion of $G(k)$ with respect to this topology will be denoted by $\widehat{G}_S$.

As every S-congruence subgroup is S-arithmetic, the S-arithmetic topology on $G(k)$ is finer than the S-congruence topology and therefore there is a continuous homomorphism $\widehat{G}_S \to G(A_S)$. It is not difficult to show that this homomorphism is surjective and its kernel, denoted $C(S,G)$, is a profinite group. $C(S,G)$ is by definition the *S-congruence kernel*. It is clear that $C(S,G)$ is trivial if, and only if, G has the congruence subgroup property (for S-arithmetic subgroups). In the more precise formulation due to Serre, the congruence subgroup problem is the problem of determining the S-congruence kernel $C(S,G)$.

We have the following topological extension

$$(*) \qquad 1 \to C(S,G) \to \widehat{G}_S \to G(A_S) \to 1$$

of $G(A_S)$ by $C(S,G)$. The natural inclusion of $G(k)$ in $\widehat{G}_S$ provides a splitting of this extension over $G(k)(\hookrightarrow G(A_S))$. It has been conjectured that, under a fairly general hypothesis, (*) is a central extension i.e., $C(S,G)$ is central in $\widehat{G}_S$, see Section 4 below. We shall devote the next two sections to topological central extensions.

2. Topological Fundamental Group

A topological extension

$$(+) \qquad 1 \to \mathscr{C} \to \mathscr{E} \to \mathscr{G} \to 1 ,$$

of a locally compact and second countable topological group $\mathscr{G}$ by $\mathscr{C}$, with $\mathscr{E}$ locally compact and second countable, is a universal topological central extension (u.t.c.e.) of $\mathscr{G}$ if it is a central extension i.e., $\mathscr{C}$ is a closed central subgroup of $\mathscr{E}$, and given any topological central extension

$$1 \to C \to E \to \mathscr{G} \to 1 ,$$

with E locally compact and second countable, there is a *unique* continuous homomorphism $\varphi : \mathscr{E} \to E$ making the following diagram commutative

$$\begin{array}{ccccccccc} 1 & \to & \mathscr{C} & \to & \mathscr{E} & \to & \mathscr{G} & \to & 1 \\ & & \downarrow & & \downarrow \varphi & & \| & & \\ 1 & \to & C & \to & E & \to & \mathscr{G} & \to & 1 . \end{array}$$

It is clear that if $\mathscr{G}$ admits a u.t.c.e., the latter is unique upto natural equivalence. In case (+) is a u.t.c.e. of $\mathscr{G}$, $\mathscr{C}$ is by definition the *topological fundamental group* of $\mathscr{G}$ and it is denoted by $\pi_1(\mathscr{G})$. If $\mathscr{G}$ is a connected real semi-simple Lie group, then $\pi_1(\mathscr{G})$ coincides with the usual (algebraic topological) fundamental group of $\mathscr{G}$. It follows from certain results of Moore [20], that if $\mathscr{G}$ is *perfect* i.e. if it is its own commutator, and the cohomology group $H^2_m(\mathscr{G}, \mathbf{R}/\mathbf{Z})$, based on measurable cochains, is finite, then $\mathscr{G}$ admits a u.t.c.e. and $\pi_1(\mathscr{G})$ is isomorphic to the dual of $H^2_m(\mathscr{G}, \mathbf{R}/\mathbf{Z})$. It is also known that if $\mathscr{G}$ is totally disconnected, then the cohomology theory of $\mathscr{G}$ based on measurable cochains is identical with the theory based on continuous cochains [37].

If v is a nonarchimedean place where G is isotropic, then $G(k_v)$ is perfect (in fact any proper normal subgroup is central) and the cohomology group $H^2(G(k_v), \mathbf{R}/\mathbf{Z})$, defined in terms of continuous cochains, is essentially known:

Theorem 1. *Let v be a nonarchimedean place of k such that G is isotropic at v (or, equivalently, $G(k_v)$ is noncompact), then $H^2(G(k_v), \mathbf{R}/\mathbf{Z})$ is isomorphic to a subgroup, of index at most two, of the dual $\hat{\mu}(k_v)$ of the finite group $\mu(k_v)$ of roots of unity in k_v. Moreover, if at least one of the following three conditions holds, then it is isomorphic to $\hat{\mu}(k_v)$.*

(i) G is quasi-split over an odd degree extension of k_v;

(ii) k_v is not an extension of $\mathbf{Q}_2$;

(iii) k_v contains a primitive fourth root of unity.

As a consequence, if G is isotropic at v, then $G(k_v)$ admits a u.t.c.e. and $\pi_1(G(k_v))$ is isomorphic to the dual of $H^2(G(k_v), \mathbf{R}/\mathbf{Z})$.

It is expected that for any nonarchimedean place v where G is isotropic, $H^2(G(k_v), \mathbf{R}/\mathbf{Z})$ is isomorphic to $\hat{\mu}(k_v)$. For the spin group of a quadratic form over k which is of Witt index at least 2 at v, this is proved in [27, 1.9] and the same proof would take care of some other classical groups.

For the group SL_2 the above theorem is due to Moore [19]. For other Chevalley groups (i.e. groups which split over k) he proved that $H^2(G(k_v), \mathbf{R}/\mathbf{Z})$ is isomorphic to a subgroup of $\hat{\mu}(k_v)$, and about ten years later Deodhar [7] showed that this holds also when G is quasi-split over k_v (i.e. contains a Borel subgroup defined over k_v). Soon after Moore proved his result, Matsumoto showed, by constructing a suitable topological central extension of $G(k_v)$, that if G is a Chevalley group, the above cohomology group is actually isomorphic to $\hat{\mu}(k_v)$, and an observation of Deligne implies that this is also the case if G is quasi-split over k_v, see [28, §5]. Bak and Rehmann [1] have proved the above theorem, as well as Theorem 3 stated below, for groups of inner type A of relative rank $\geqq 2$ using K-theoretic methods.

In the generality stated above, the theorem is proved in [28] using the results of Moore, Matsumoto, Deodhar and Deligne and the Bruhat-Tits theory of reductive groups over nonarchimedean local fields. The complete proof of the above theorem is quite long and difficult and involves some case considerations. It is desirable to have a shorter and simpler proof.

The known results on $H^2(G(k_v), \mathbf{R}/\mathbf{Z})$ in case $G(k_v)$ is compact (or, equivalently, G is anisotropic at v) are summarised below.

Theorem 2. *Let v be a nonarchimedean place such that $G(k_v)$ is compact (then, as is well known, there is a central division algebra D_v over k_v such that $G(k_v)$ is isomorphic to the group $\mathrm{SL}_1(D_v)$ of elements of reduced norm 1 in D_v, and) the cohomology group $H^2(G(k_v), \mathbf{R}/\mathbf{Z})$, based on continuous cochains, is a finite group of order a power of p_v, where p_v is the characteristic of the residue field of k_v. It is cyclic if D_v is not the quaternion central division algebra over $\mathbf{Q}_2$ and is trivial if k_v does not contain a primitive p_v-th root of unity and D_v is not the quaternion central division algebra over $\mathbf{Q}_3$.*

This theorem is proved in [30]. The precise computation of $H^2(\mathrm{SL}_1(D_v), \mathbf{R}/\mathbf{Z})$ has not yet been done. We conjecture that it is isomorphic to $\mathbf{Z}/2\mathbf{Z} \oplus \mathbf{Z}/2\mathbf{Z}$ if D_v is the quaternion central division algebra over $\mathbf{Q}_2$, it is isomorphic to $\mathbf{Z}/3\mathbf{Z}$ if D_v is the quaternion central division algebra over $\mathbf{Q}_3$, and is isomorphic to the p_v-primary component of the dual $\hat{\mu}(k_v)$ of the group of roots of unity in k_v in all other cases.

Remark. Theorems 1 and 2 imply that, for any finite set S of places of k, $H^2(G(A_S), \mathbf{R}/\mathbf{Z})$ is the direct product of the $H^2(G(k_v), \mathbf{R}/\mathbf{Z}), v \notin S$. If moreover $G(A_S)$ is perfect, then it admits a u.t.c.e. and its topological fundamental group is the direct sum (with discrete topology) of the $\pi_1(G(k_v)), v \notin S$; see [27, Theorem 2.4]. This implies that if $C(S, G)$ is central in $\widehat{G}_S$, then it is actually finite [27, §2].

3. The S-Metaplectic Kernel

Let $\mathscr{H}$ be a subgroup of a locally compact second countable topological group $\mathscr{G}$. Assume that $\mathscr{H}$ is perfect and $\mathscr{G}$ admits a u.t.c.e. Then there is a topological central extension

$$1 \to C \to E \to \mathscr{G} \to 1,$$

with E locally compact and second countable, which splits over $\mathscr{H}$ and which is universal with respect to this property. The *relative topological fundamental group* $\pi_1(\mathscr{G}, \mathscr{H})$ is then by definition the group C.

The S-*metaplectic kernel* is the group

$$M(S, G) = \mathrm{Ker}\big(H^2(G(A_S), \mathbf{R}/\mathbf{Z}) \xrightarrow{\text{rest}} H^2(G(k), \mathbf{R}/\mathbf{Z})\big);$$

where $H^2(G(k), \mathbf{R}/\mathbf{Z})$ denotes the second cohomology of the abstract group $G(k)$ with coefficients $\mathbf{R}/\mathbf{Z}$. The topological central extensions of $G(A_S)$ by $\mathbf{R}/\mathbf{Z}$, which split over the subgroup $G(k)$, are classified by the S-metaplectic kernel. It is obvious that if $G(k)$ is perfect, then $M(S, G)$ is isomorphic to the Pontrjagin dual of the relative fundamental group $\pi_1(G(A_S), G(k))$.

We now come back to the congruence subgroup problem. Assume that $G(k)$ is perfect and $C(S, G)$ is central in $\widehat{G}_S$ (see Section 4 below). Then adapting an argument of [2, §15] and using Theorem 1, it can be proved that (∗) is the universal extension in the category of topological central extensions of $G(A_S)$ splitting over $G(k)$, see [27, §2]. In particular, $C(S, G)$ is isomorphic to the relative fundamental group $\pi_1(G(A_S), G(k))$ and so it is isomorphic to the Pontrjagin dual of the S-metaplectic kernel $M(S, G)$. Thus to determine the S-congruence kernel, it is enough to compute $M(S, G)$. Also, in the theory of automorphic forms (of fractional weights) it is of critical importance to know the topological central extensions of $G(A)$ which split over $G(k)$; these are determined by $M(\phi, G)$. Now we state the following theorem which "determines" $M(S, G)$ for all k-isotropic G.

Theorem 3. *Assume that G is isotropic over k. Let S be an arbitrary finite set of places of k. Then $M(S, G)$ is trivial if S contains either a nonarchimedean place, or a real place v such that the group $G(k_v)$ is not topologically simply connected. In general $M(S, G)$ is isomorphic to a subgroup of the dual $\hat{\mu}(k)$ of the group of roots of unity in k.*

For Chevalley groups this theorem was proved by Moore [19] and for groups which are quasi-split over k, it was proved by Deodhar [7]. For the group $G = \mathrm{SL}_2$, Moore in fact proved that $M(S, G)$ is trivial if S contains a noncomplex place and it is isomorphic to $\hat{\mu}(k)$ otherwise [1]. Soon after this, Matsumoto proved that for all Chevalley groups G, $M(S, G)$ is isomorphic to $\hat{\mu}(k)$ if S does not contain any noncomplex place. The same holds for any group which is quasi-split over k as was observed by Deligne. If either $S \supset V_\infty$ or k is a totally imaginary number field, the precise computation of $M(S, G)$ for the groups $\mathrm{SL}_n (n \geqq 3)$ and

[1] This result is equivalent to his theorem on the "uniqueness of the reciprocity law of global class field theory" – see [4] for an elegant proof of the latter.

$Sp_{2n}(n \geqq 2)$ is already in [2], and following the ideas of this paper, Vaserstein [35] computed the metaplectic kernel for many other classical groups. In 1981, Bak, in a Comptes Rendus note, outlined a proof of this theorem for all groups of classical type of relative rank at least two which uses the results of [1].

For arbitrary simply connected k-isotropic groups, the above theorem was proved by Prasad and Raghunathan in 1980 [27], and besides the results of Moore and Deodhar for split and quasi-split groups, the proof uses the results of [28] on topological central extensions of $G(k_v)$. Note that for a real place v, the condition that $G(k_v)$ is not topologically simply connected is equivalent to the condition imposed in [27, 3.4(ii)].

It is likely that if G is k-isotropic, $S \subset V_\infty$, and for every v in S, $G(k_v)$ is topologically simply connected, then $M(S, G)$ is isomorphic to $\hat{\mu}(k)$. This has been verified for many of the classical groups and some groups of exceptional types.

A variant of Moore's theorem on the "uniqueness of the reciprocity law", announced in [25], together with the results of [28, 30], can be used to compute $M(S, G)$, modulo 2-torsion, for all k-anisotropic G. For some results in this direction see [32].

4. Projective-Simplicity of $G(k)$ and Centrality of $C(S, G)$

It has been conjectured by Kneser, Platonov and Margulis that *if G is isotropic at each nonarchimedean place, then $G(k)$ is projectively-simple* i.e. it does not contain any proper noncentral normal subgroup, *and if it is anisotropic at some nonarchimedean place, then (as is well known, G is of type* A *and) any noncentral normal subgroup of $G(k)$ is the intersection of $G(k)$ with a normal subgroup of $\prod_{v \in \mathscr{S}} G(k_v)$, where $\mathscr{S}$ is the (finite) set of nonarchimedean places of k where G is anisotropic.* This conjecture is known to hold for all k-isotropic groups except possibly for certain *outer* forms of type E_6 of k-rank 1 which require division algebras of degree 3 for their construction. For anisotropic groups, the results are much less complete. In 1980, inspired by [22], Margulis [16] proved the above conjecture for groups of type A_1. This implies the projective-simplicity of the spin group of any quadratic form in 3 or 4 variables which is isotropic at all nonarchimedean places of k. Projective-simplicity of the spin group of any quadratic form in at least five variables was proved already in 1956 by Kneser [9] by an ingenious method. Borovoi and Chernousov have recently proved the projective-simplicity of $G(k)$ whenever G is of absolute rank at least two and it splits over a quadratic extension of k (this class includes all groups of type B, C, E_7, E_8, F_4 and G_2); and now Sury and Tomanov have independently established this for G of type A_3, which is isotropic at all nonarchimedean places – this implies the projective-simplicity of $G(k)$ for all groups G of type D (except the triality forms). But the anisotropic groups of (inner and outer type) A_n (n arbitrary) pose a serious challenge.

It may be of interest to note here that it follows from the well known result of Margulis [15] on normal subgroups of lattices [2] in semi-simple groups, and

[2] A *lattice* in a locally compact unimodular group is a discrete subgroup of finite covolume (with respect to any Haar measure).

the strong approximation property, that any noncentral normal subgroup of $G(k)$ is of finite index in $G(k)$ (see, for example, [23]). Moreover, it is easy to show that if G is isotropic at all the nonarchimedean places of k and has the congruence subgroup property for some S, then $G(k)$ is projectively-simple.

Based on the results of [2, 33] on SL_n and Sp_{2n}, and [18], where the centrality of the S-congruence kernel was proved for all Chevalley groups of rank $\geqq 2$, it has been conjectured that for arbitrary (simply connected) G, *$C(S,G)$ is central in $\widehat{G}_S$ if $\sum_{v \in S} k_v\text{-rank}(G) \geqq 2$ and G is isotropic at all nonarchimedean $v \in S$.* Using some of the ideas of [2, 33], Vaserstein [35] showed that this conjecture holds for all classical groups of k-rank at least two. Raghunathan has proved the above conjecture for all k-isotropic groups [31]; his proof does not require any case-by-case analysis.

For the spin group of an arbitrary (not necessarily isotropic) quadratic form in at least five variables the above conjecture on the centrality of $C(S,G)$ was proved by Kneser [10]. Refining and using his ideas, Rapinchuk and Tomanov have recently proved the conjecture for all anisotropic groups of type $\mathsf{B}_r (r \geqq 2)$, $\mathsf{C}_r (r \geqq 2)$, $\mathsf{D}_r (r \geqq 5)$, E_7, E_8, F_4, G_2, and the groups of type ${}^2\mathsf{A}_r (r \geqq 3)$ which split over a quadratic extension of k. The question of centrality of the S-congruence kernel for anisotropic groups of type A is a very interesting open problem–its solution may require new insight into the structure and geometry of central division algebras over global fields.

5. The Hasse Principle and Tamagawa Number

If k is a global function field, then the Galois cohomology $H^1(k,G)$ is trivial (this was proved by Harder). On the other hand, if k is a number field, it has been known for quite some time that the "Hasse principle" i.e., the assertion that the natural morphism

$$H^1(k,G) \to \prod_{v \in V_\infty} H^1(k_v,G)$$

is injective, holds for all (simply connected) G of type other than E_8. The Hasse principle has now been verified for groups of type E_8 by Chernousov [5].

If k is number field, let D_k be the absolute value of the discriminant of $k/\mathbf{Q}$ and if k is a global function field, let q_k be the cardinality of its field of constants, g_k the genus of k and $D_k = q_k^{2g_k-2}$.

Let ω be an invariant exterior form on G, of maximal degree, defined over k. Then for each place v, the form ω , and the normalized Haar measure on k_v, determine a Haar measure on $G(k_v)$ which we denote by ω_v.

Let $P = (P_v)_{v \in V_f}$ be a fixed *coherent* collection of parahoric subgroups: for each $v \in V_f, P_v$ is a parahoric subgroup of $G(k_v)$ such that the product $\prod_{v \in V_\infty} G(k_v) \cdot \prod_{v \in V_f} P_v$ is an open subgroup of $G(A)$. (Recall that a subgroup of $G(k_v)$ is said to be an *Iwahori subgroup* if it is the normalizer of a maximal pro-p_v subgroup of $G(k_v)$ or, equivalently, it is the stabilizer of a chamber (i.e. a simplex of maximal dimension) in the Bruhat-Tits building of $G(k_v)$. Any subgroup containing an Iwahori subgroup is called a *parahoric subgroup*.) It is known that, as G is semi-simple, the product $\prod \omega_v(P_v)$ is absolutely convergent and so there is a Haar measure μ on $G(A)$ which on the open subgroup $\prod_{v\in V_\infty} G(k_v)\cdot \prod_{v\in V_f} P_v$ coincides with the measure $D_k^{-\frac{1}{2}\dim G}\prod_{v\in V_\infty}\omega_v \cdot \prod_{v\in V_f}\omega_v|_{P_v}$. It is obvious from the product formula (i.e. $\prod_v |x|_v = 1$ for $x\in k^\times$) that the measure μ is independent of the k-form ω and it is called the *Tamagawa measure.* The *Tamagawa number* of G/k, to be denoted $\tau_k(G)$, is the positive real number $\mu(G(A)/G(k))$. It was conjectured by Weil that for all (simply connected absolutely almost simple) $G, \tau_k(G) = 1$. This conjecture has recently been proved by Kottwitz, over number fields, without any case-by-case considerations (see [11], and also [26, 3.3]). Using Arthur's trace formula and the Hasse principle, he has in fact shown that if k is a number field and $\mathscr{G}$ is the unique quasi-split *inner* k-form of G, $\tau_k(G) = \tau_k(\mathscr{G})$; this result was conjectured by Langlands. Now since the Tamagawa number of any simply connected quasi-split group is 1 [12, 13], Weil's conjecture follows.

Weil's conjecture remains unproven for groups defined over global function fields. It is still unknown, for example, if over such a k, the Tamagawa number of every outer k-form of type A is 1.

6. Covolumes of *S*-Arithmetic Subgroups, Bound for Class Numbers and the Finiteness Theorems

We shall now describe a formula for the covolume of S-arithmetic subgroups with respect to a natural Haar measure on G_S. We begin by describing a natural Haar measure μ_v on $G(k_v)$ for any place v of k. For a nonarchimedean place v of k, let μ_v be the Tits measure on $G(k_v)$ i.e. the Haar measure with respect to which the volume of any Iwahori subgroup of $G(k_v)$ is 1. If v is archimedean, then k_v is either $\mathbf{R}$ or $\mathbf{C}$ and μ_v is the Haar measure on $G(k_v)$ such that, in the induced measure, any maximal compact subgroup of $R_{k_v/\mathbf{R}}(G)(\mathbf{C})$ has volume 1. Now on $G_S = \prod_{v\in S} G(k_v)$ we take the product measure $\mu_S := \prod_{v\in S}\mu_v$.

Let $\mathscr{G}$ be the unique quasi-split *inner* k-form of G. For each nonarchimedean place v, we fix a parahoric subgroup $\mathscr{P}_v$ of $\mathscr{G}(k_v)$ of maximal volume such that $\prod_{v\in V_\infty}\mathscr{G}(k_v)\cdot\prod_{v\in V_f}\mathscr{P}_v$ is an open subgroup of $\mathscr{G}(A)$.

As in Section 5, let $P = (P_v)_{v\in V_f}$ be a fixed coherent collection of parahoric subgroups. Let S be a finite set of places containing V_∞ and let $\Lambda = G(k)\cap\prod_{v\notin S} P_v$. In its natural embedding in G_S, Λ is an S-arithmetic subgroup. Let G_v denote the smooth affine $\mathfrak{o}_v$-group scheme associated with the parahoric subgroup P_v by the Bruhat-Tits theory ([34, 3.4]). Let $\overline{G}_v := G_v\otimes_{\mathfrak{o}_v}\mathfrak{f}_v$ be the reduction mod $\mathfrak{p}_v$ of G_v. Let $\overline{T}_v$ be a maximal $\mathfrak{f}_v$-torus of $\overline{G}_v$ containing a maximal $\mathfrak{f}_v$-split torus and $\overline{M}_v$ be the maximal reductive $\mathfrak{f}_v$-subgroup containing $\overline{T}_v$. Note that $\overline{M}_v$ depends on the choice of P_v. Let $\overline{\mathscr{G}}_v, \overline{\mathscr{T}}_v$ and $\overline{\mathscr{M}}_v$ be similarly defined $\mathfrak{f}_v$-groups associated with $\mathscr{G}$ and the parahoric subgroup $\mathscr{P}_v$.

If $\mathscr{G}/k$ is not a triality form of type ${}^6\mathsf{D}_4$, let ℓ be the smallest extension of k over which $\mathscr{G}$ splits. If $\mathscr{G}/k$ is of type ${}^6\mathsf{D}_4$, let ℓ be a fixed extension of k of degree 3 contained in the Galois extension of degree 6 over which $\mathscr{G}$ splits. Let D_ℓ be the absolute value of the discriminant of $\ell/\mathbf{Q}$ if k is a number field and

$D_\ell = q_\ell^{2g_\ell - 2}$ if k is a global function field, where q_ℓ is the cardinality of the finite field of constants in ℓ and g_ℓ is the genus of ℓ.

The integer $\mathfrak{s}(\mathcal{G})$: If $\mathcal{G}$ splits over k, let $\mathfrak{s}(\mathcal{G}) = 0$. If $\mathcal{G}$ is a k-form of type ${}^2\mathsf{A}_r$, with r even, let $\mathfrak{s}(\mathcal{G}) = \frac{1}{2}r(r+3)$; if $\mathcal{G}$ is a k-form of type ${}^2\mathsf{A}_r$ (r odd), ${}^2\mathsf{D}_r$ (r arbitrary) or ${}^2\mathsf{E}_6$, let $\mathfrak{s}(\mathcal{G}) = \frac{1}{2}(r-1)(r+2), 2r-1$ or 26 respectively. If $\mathcal{G}$ is a triality form of type ${}^3\mathsf{D}_4$ or ${}^6\mathsf{D}_4$, then let $\mathfrak{s}(\mathcal{G}) = 7$.

The following theorem provides a "computable" formula for the volume of S-arithmetic quotients of G_S. It is proved in [26] .

Theorem 4. *Let $m_1 \leqq \cdots \leqq m_r$ be the exponents of the Weyl group of the absolute root system of G. Then*

$$\mu_S(G_S/\Lambda) = D_k^{\frac{1}{2}\dim G}(D_\ell/D_k^{[\ell:k]})^{\frac{1}{2}\mathfrak{s}(\mathcal{G})}\left(\prod_{v\in V_\infty}\left|\prod_{i=1}^{r}\frac{m_i!}{(2\pi)^{m_i+1}}\right|_v\right)\tau_k(G)\mathcal{E}(P) \;;$$

where

$$\mathcal{E}(P) = \prod_{v\in S_f}\frac{q_v^{(r_v+\dim\overline{\mathcal{M}}_v)/2}}{\#\overline{T}_v(\mathfrak{f}_v)}\cdot\prod_{v\notin S}\frac{q_v^{(\dim\overline{M}_v+\dim\overline{\mathcal{M}}_v)/2}}{\#\overline{M}_v(\mathfrak{f}_v)}\,,$$

$S_f = S\cap V_f$, and for $v\in V_f$, $r_v(=\dim\overline{T}_v)$ is the rank of G over the maximal unramified extension of k_v.

The results involved in the proof of this theorem provide the following lower bound for the class number of simply connected anisotropic groups (see [26, Theorem 4.3]).

Theorem 5. *Assume that G is anisotropic over k and moreover $G_\infty := \prod_{v\in V_\infty} G(k_v)$ is compact. Then the class number $\#\big(G_\infty\prod_{v\in V_f}P_v\backslash G(A)/G(k)\big)$ of G/k with respect to P is at least*

$$D_k^{\frac{1}{2}\dim G}(D_\ell/D_k^{[\ell:k]})^{\frac{1}{2}\mathfrak{s}(\mathcal{G})}\left(\prod_{v\in V_\infty}\left|\prod_{i=1}^{r}\frac{m_i!}{(2\pi)^{m_i+1}}\right|_v\right)\tau_k(G)\zeta(P) \;;$$

where

$$\zeta(P) = \prod_{v\in V_f}\frac{q_v^{(\dim\overline{M}_v+\dim\overline{\mathcal{M}}_v)/2}}{\#\overline{M}_v(\mathfrak{f}_v)}\,.$$

In [3, §7] this theorem is used to prove the following finiteness theorem.

Theorem 6. *Given a positive integer c, let $\mathcal{C}_c$ be the set of pairs (k,G) consisting of a number field k and a connected, simply connected absolutely almost simple group G such that G is anisotropic over k, $G_\infty := \prod_{v\in V_\infty} G(k_v)$ is compact, and the class number $\#\big(G_\infty\prod_{v\in V_f}P_v\backslash G(A)/G(k)\big)$ of G/k with respect to some coherent collection of parahoric subgroups $(P_v)_{v\in V_f}$ is less than c. Then (up to natural equivalence) $\mathcal{C}_c$ is finite.*

The formula for the volume of S-arithmetic quotients given above and certain number theoretic estimates have been used in [3] to prove that in characteristic zero, there are only finitely many distinct S-arithmetic subgroups Γ of covolume $\leqq c$, where c is a given positive number. Also, there are only finitely many S-arithmetic Γ with $0 \neq |\chi(\Gamma)| < c$, where $\chi(\Gamma)$ is the Euler-Poincaré characteristic of Γ in the sense of C. T. C. Wall. For precise results, see [3].

7. Super-Rigidity and Arithmeticity of Lattices

According to a celebrated theorem of Margulis (annonced at the ICM held in 1974), *irreducible* lattices [3] in real semi-simple groups of **R**-rank > 1 are super-rigid[4]. It follows rather easily from this that such lattices are arithmetic. On the other hand, it has been known for almost thirty years that the groups $\mathrm{SO}(n, 1)$ contain non-arithmetic lattices for $n \leqq 5$. In 1986, Gromov and Piatetski-Shapiro, employing a nice geometric construction, showed that for each n, $\mathrm{SO}(n, 1)$ contains plenty of non-arithmetic lattices([8]). Mostow has constructed non-arithmetic lattices in $\mathrm{SU}(2, 1)$ and $\mathrm{SU}(3, 1)$; however, whether lattices in $\mathrm{SU}(n, 1)$ are all arithmetic if n is sufficiently large is still an open question.

Corlette [6] has now established the super-rigidity of real representations of *cocompact* discrete subgroups in the remaining semi-simple groups of **R**-rank 1, namely the groups $\mathrm{Sp}(n, 1)$ and the **R**-rank 1 form of type F_4, using his basic theorem on the existence of a harmonic map in any given homotopy class of maps from a compact riemannian manifold into a locally symmetric space; and just a few weeks ago I have learnt that Gromov and Schoen have proved that any representation of such a discrete cocompact subgroup over a p-adic field is bounded by developing an analogue of the theory of harmonic maps for maps from a riemannian manifold into a Bruhat-Tits building. Now, as in the case of groups of **R**-rank > 1, arithmeticity of cocompact discrete subgroups of the groups $\mathrm{Sp}(n, 1)$ and the **R**-rank 1 form of type F_4 follows.

Acknowledgements. I thank the University of Michigan for inviting me to deliver the Fall 1989 Keeler lectures, and Mr. M.S. Keeler for instituting this lectureship. This address is in part based on my Keeler lectures. I also thank the Japan Association for Mathematical Sciences for financial support to attend the ICM.

References

1. Bak, A., Rehmann, U.: The congruence subgroup and metaplectic problems for $\mathrm{SL}_{n \geqq 2}$ of division algebras. J. Alg. **78** (1982) 475–547
2. Bass, H., Milnor, J., Serre, J–P.: Solution of the congruence subgroup problem for SL_n and Sp_{2n}. Publ. Math. IHES **33** (1967) 59-137

[3] A lattice in a semi-simple linear analytic group is said to be *irreducible* if no subgroup of it of finite index is a direct product of two infinite normal subgroups.

[4] For a detailed proof see [17]. Super-rigidity and arithmeticity in arbitrary characteristic has been proved in [36].

3. Borel, A., Prasad, G.: Finiteness theorems for discrete subgroups of bounded covolume in semi-simple groups. Publ. Math. IHES **69** (1989) 119–171; Addendum: *ibid* **71** (1990)
4. Chase, S., Waterhouse, W.: Moore's theorem on uniqueness of reciprocity laws. Invent. math. **16** (1972) 267–270
5. Chernousov, V.: On the Hasse principle for groups of type E_8. Soviet Math. Dokl.**39** (1989)
6. Corlette, K.: Archimedean super-rigidity and hyperbolic geometry (preprint)
7. Deodhar, V.: On central extensions of rational points of algrebaic groups. Amer. J. Math. **100** (1978) 303–386
8. Gromov, M., Piatetski-Shapiro, I.: Non-arithmetic groups in Lobachevsky spaces. Publ. Math. IHES **66** (1988) 93–103
9. Kneser, M.: Orthogonale Gruppen über algebraischen Zahlkörpern. J. Reine Angew. Math. **196** (1956) 213–220
10. Kneser, M.: Normalteiler ganzzahliger Spingruppen. J. Reine Angew. Math. **311/312** (1979) 191–214
11. Kottwitz, R.: Tamagawa numbers. Ann. Math. **127** (1988) 629–646
12. Langlands, R.: The volume of the fundamental domain. Proc. AMS Symp. Pure Math. **9** (1966) 143–148
13. Lai, K.: Tamagawa number of reductive algebraic groups. Compos. Math. **41** (1980) 153–188
14. Margulis, G.: Cobounded subgroups of algebraic groups over local fields. Funct. Anal. Appl. **11** (1977) 119–128
15. Margulis, G.: Finiteness of quotient groups of discrete subgroups. Funct. Anal. Appl. **13** (1979) 178–187
16. Margulis, G.: On the multiplicative group of a quaternion algebra over a global field. Soviet Math. Dokl. **21** (1980) 780–784
17. Margulis, G.: Arithmeticity of the irreducible lattices in the semi-simple groups of rank greater than 1. Invent. math. **76** (1984) 93–120
18. Matsumoto, H.: Sur les sous-groupes arithmétiques des groupes semi-simples déployés. Ann. Sci. École Norm. Sup., 4e sér. **2** (1969) 1–62
19. Moore, C.: Group extensions of *p*-adic and adelic linear groups. Publ. Math. IHES **35** (1968) 157–222
20. Moore, C.: Group extensions and cohomology for locally compact groups, III and IV. Transactions AMS **221** (1976) 1–58
21. Platonov, V.: The problem of strong approximation and the Kneser-Tits conjecture. Math. USSR Izv. **3** (1969) 1139–1147; Addendum: *ibid* **4** (1970) 784–786
22. Platonov, V., Rapinchuk, A.: On the group of rational points of three-dimensional groups. Soviet Math. Dokl. **20** (1979) 693–697
23. Prasad, G.: Strong approximation. Ann. Math. **105** (1977) 553–572
24. Prasad, G.: Elementary proof of a theorem of Bruhat-Tits and Rousseau. Bull. Soc. Math. France **110** (1982) 197–202
25. Prasad, G.: A variant of a theorem of Calvin Moore. C. R. Acad. Sci. (Paris) Sér. I **302** (1982) 405–408
26. Prasad, G.: Volumes of *S*-arithmetic quotients of semi-simple groups. Publ. Math. IHES **69** (1989) 91–117
27. Prasad, G., Raghunathan, M.S.: On the congruence subgroup problem: Determination of the "metaplectic kernel". Invent. math. **71** (1983) 21–42
28. Prasad, G., Raghunathan, M.S.: Topological central extensions of semi-simple groups over local fields. Ann. Math. **119** (1984) 143–268
29. Prasad, G., Raghunathan, M.S.: On the Kneser-Tits problem. Comment. Math. Helv. **60** (1985) 107–121

30. Prasad, G., Raghunathan, M.S.: Topological central extensions of $SL_1(D)$. Invent. math. **92** (1988) 645–689
31. Raghunathan, M.S.: On the congruence subgroup problem. Publ. Math. IHES **46** (1976) 107–161; second part: Invent. math. **85** (1986) 73–117
32. Rapinchuk, A.: Multiplicative arithmetic of division algebras over number fields and the metaplectic problem. Math. USSR Izv. **31** (1988) 349–379
33. Serre. J–P.: Le problème des groupes de congruence pour SL_2. Ann. Math. **92** (1970) 489-527
34. Tits, J.: Reductive groups over local fields. Proc. AMS Symp. Pure Math. **33** (1979), part 1, 29–69
35. Vaserstein, L.: The structure of classical arithmetic groups of rank greater than one. Math. USSR Sb. **20** (1973) 465–492
36. Venkataramana, T.: On super-rigidity and arithmeticity of lattices in semi-simple groups over local fields of arbitrary characteristic. Invent. math. **92** (1988) 255–306
37. Wigner, D.: Algebraic cohomology of topological groups. Transactions AMS **178** (1973) 83–93

Poles of Standard L Functions

Stephen Rallis

Department of Mathematics, The Ohio State University, Columbus, OH 43210, USA

Introduction

The Rankin-Selberg method in the theory of L functions gives explicit integral representations of certain L functions of automorphic representations of reductive algebraic groups. This allows one to determine the analytic continuation and functional equations of such L functions. Moreover the poles of these L functions can be determined explicitly. The major applications of this method include (i) the determination of how much of the complementary series contributes to cuspidal representations of GL_2 and (ii) the strong multiplicity one Theorem and the classification of automorphic representations of GL_n.

In recent years Rankin-Selberg integral representations have been found for several new classes of L functions [PS-R-II, PS-R-S]. We are concerned here with the standard L functions of the classical groups. The method of doubling a classical group (which is essentially a compactification of the group) is used in this case. The doubling method has the extra advantage that the special values of such L functions at integral points can be related to certain θ integrals arising from specific dual reductive pairs.

This leads to our second theme. Namely as a generalization of the classical Siegel-Weil identity, there exist, for dual reductive pairs, identities between certain regularized θ integrals and certain special values of Siegel type Eisenstein series. The point of such identities is to give criteria about the existence of a pole of the L functions mentioned above at a specific value in terms of the nonvanishing of a certain θ lift between groups of a given dual reductive pair.

The work in this lecture represents joint work (as specified below) with Ilya Piatetski Shapiro and Steve Kudla. For a survey of earlier related works, and a more extensive bibliography of the field, see [G-S].

§0. Notation

(1) Let G be a reductive group over k, a number field. Let G_v the associated group at the place v of k. Let $G(\mathbb{A})$ be the corresponding adelic group.

Let $\mathbb{K} = \prod_v K_v$ be a good maximal compact group where K_v is a special maximal compact subgroup of G_v (for all v).

Proceedings of the International Congress
of Mathematicians, Kyoto, Japan, 1990

Let $\Pi = \bigotimes_v \Pi_v$ be an irreducible automorphic cuspidal representation of $G(\mathbb{A})$. Define S_Π = the set of places of k consisting of the Archimedean places together with all finite places at which Π_v is not spherical. For $v \notin S_\Pi$ let $D_v(\Pi_v)$ be the conjugacy class in the L group of G_v associated to Π_v via the Satake isomorphism. Then for a representation ϱ of the L group ${}^LG$ let

$$L_v(s, \Pi_v, \varrho) = \det(1 - \varrho(D_v)q_v^{-s})^{-1}$$

be the usual Langlands Euler factor with $s \in \mathbb{C}$ and q_v = cardinality of residue field of k_v. Moreover let

$$L_S(s, \Pi, \varrho) = \prod_{v \notin S_\Pi} L_v(s, \Pi_v, \varrho)$$

be the corresponding global restricted L function.

We let $\zeta_v(s)$ be the usual zeta function associated to a local field k_v.

(2) We let $\mathrm{Sp}_n \times \mathrm{O}(Q)$ be a dual reductive pair, where Sp_n is the symplectic group of $2n \times 2n$ matrices and $\mathrm{O}(Q)$ is the orthogonal group of the quadratic form Q with $\dim Q$ even. The oscillator representation of this pair is denoted by $\omega_{Q,\psi}$(ψ an additive character) and is realized on the space $S(M_{mn}(k))(m = \dim Q)$ (See [W]). The construction works locally and globally (where the local field k is replaced by $\mathbb{A}_k = \mathbb{A}$ = adeles of the number field k). Following the work of [W] it is possible to define the space of θ kernels

$$\theta_\varphi(x, y) = \sum_{\xi \in M_{mn}(k)} \omega_{Q,\psi}(x, y)(\varphi)(\xi)$$

where $\varphi \in S(M_{mn}(\mathbb{A}))$ and $(x, y) \in \mathrm{Sp}_n \times \mathrm{O}(Q)(\mathbb{A})$. The kernel θ_φ is left invariant by $\mathrm{Sp}_n \times \mathrm{O}(Q)(k)$ and slowly increasing on $\mathrm{Sp}_n \times \mathrm{O}(Q)(k)\backslash\mathrm{Sp}_n \times \mathrm{O}(Q)(\mathbb{A})$. In particular this allows one to define for any f, which is rapidly decreasing on $\mathrm{O}(Q)(k)\backslash\mathrm{O}(Q)(\mathbb{A})$, the θ integral $\theta_\varphi(f)$ as

$$\int_{\mathrm{O}(Q)(k)\backslash\mathrm{O}(Q)(\mathbb{A})} \theta_\varphi(x, y) f(y)\, dy\,.$$

A similar $\theta_\varphi(F)$ can be defined for F rapidly decreasing on $\mathrm{Sp}_n(k)\backslash\mathrm{Sp}_n(\mathbb{A})$.

(3) Given a reductive group G we let $\mathscr{A}(G(\mathbb{A}))$ be the space of slowly increasing automorphic functions on $G(k)\backslash G(\mathbb{A})$ as given in [B-J]. Let $L^2_{\mathrm{cusp}}(G(\mathbb{A}))$ and $L^2_{\mathrm{res}}(G(\mathbb{A}))$ be the space of smooth cusp forms and smooth residual forms in $L^2(G(k)\backslash G(\mathbb{A}))$. A function f belonging to one of these spaces is smooth if f is fixed by a compact open subgroup K' of $G(\mathbb{A}_{\mathrm{fin}})$ and f is a C^∞ vector relative to G_∞ component of $G(\mathbb{A})$.

§1. The Doubling Method

We describe a setup generalizing the construction of [PS-R-II]. We let K be a 2 dimensional semisimple commutative algebra over a number field k. Then K is either a quadratic extension of k or $K = k \oplus k$. We let V be a finite dimensional k vector space provided with a nondegenerate ε symmetric bilinear form $\langle , \rangle (\varepsilon = \pm 1)$. We consider the extension of scalars functor $\sim$ applied to $V, \langle , \rangle$. Then $\langle , \rangle^{\sim} = \langle , \rangle \otimes_k K$ is a K valued bilinear form on the space $V \otimes_k K = W$. If $\mathrm{tr}_{K/k}$ is the canonical trace form on K then there exists an element $\xi \in K^{\times}$ so that the form $\langle\langle , \rangle\rangle = \mathrm{tr}_{K/k}(\xi \langle , \rangle^{\sim})$ is a totally split ε symmetric bilinear form on W over k (if K is a quadratic field $k(\sqrt{d})$ then we let $\xi = \sqrt{d}$ and if $K = k \oplus k$ we let $\xi = (1, -1)$).

We let $G = U(W, \langle\langle , \rangle\rangle)$ be the group of k linear isometries of $W, \langle\langle , \rangle\rangle$. If $2\ell = \dim_k(W)$, let $W = W_{2\ell}$. We let $H = U_K(W, \langle , \rangle^{\sim})$ be the group of K linear maps of W which preserve $\langle , \rangle^{\sim}$. Then $H \subseteq G$.

Moreover if $K = k \oplus k$ then $H \cong H_1 \times H_1$ where H_1 is $U(V, \langle , \rangle)$, the k isometry group of $V, \langle , \rangle$.

We let X_r = variety of r-dimensional $\langle\langle , \rangle\rangle$ isotropic subspaces of W (as a k space). Then there is a parabolic subgroup P_r of G so that $X_r \cong P_r \backslash G$. The set of double cosets $P_r \backslash G / H$ in general is not finite. However it is finite precisely when $\varepsilon = -1$ (for all r) and when $\varepsilon = 1$ with $r = \frac{1}{2} \dim_k(W) = \dim_k(V)$.

We are interested in determining the orbits of H in $P_r \backslash G$ which are negligible. Namely an H orbit $\mathcal{O}$ in X_r is negligible if the stablilizer in H of a subspace $T \in \mathcal{O}$ contains as a normal subgroup the unipotent radical of a proper parabolic subgroup of H. Then in the case $P_r \backslash G / H$ is finite the number of nonnegligible orbits is exactly one in the cases $r = 2, 3, \dim V - 1, \dim V (\varepsilon = -1)$ and $r = \dim V (\varepsilon = 1)$ ([Rab]).

Since the form $\langle\langle , \rangle\rangle$ is totally split the group $P_r = M_r U_r$ where U_r is the unipotent radical of P_r and $M_r \cong \mathrm{GL}_r(k) \times U(W_{2\ell-2r}, \langle\langle , \rangle\rangle_{2\ell-2r})$ with $\langle\langle , \rangle\rangle_{2\ell-2r}$ a totally split ε symmetric bilinear form on $W_{2\ell-2r}$ with $W_{2\ell-2r} \subseteq W_{2\ell}$. Then we consider the one dimensional character $\delta_{P_r} : P_r \to k^{\times}$ defined by $\delta_{P_r}(g) = \det(\mathrm{Ad}_{U_r}(m))$ with $g = mu$ relative to the decomposition above.

For each place v of k we define the quasi-character $\chi_{s,v}$ of $(M_r)_v$ by

$$\chi_{s,v}(m) = |\delta_{P_r}(m)|^{\kappa_r s}$$

with $\kappa_r^{-1} = 2\ell - r + 1 (2\ell - r - 1 \text{ resp.})$ if $\varepsilon = -1 (\varepsilon = 1 \text{ resp.})$

Then define the induced representation

$$I_v(s) = \mathrm{Ind}_{(P_r)_v}^{G_v}(\chi_{s,v})$$

where the induction is normalized. The global induced representation is given by

$$I(s) = \mathrm{Ind}_{P(\mathbb{A})}^{G(\mathbb{A})}(\bigotimes_v \chi_{s,v}) \simeq \bigotimes_v I_v(s) .$$

Starting with a section $\phi(\,, s) \in I(s)$ we form the Eisenstein series

$$E^r_\ell(g,s,\phi) = \sum_{\gamma \in P_r(k)\backslash G(k)} \phi(\gamma g, s).$$

For a fixed good maximal compact subgroup $\mathbb{K} = \prod_v K_v$ of $G(\mathbb{A})$ we say $\phi(\ ,s) \in I(s)$ is standard if the restriction of $\phi(\ ,s)$ to $\mathbb{K}$ is independent of s and $\phi(\ ,s)$ is K_∞ finite for each K_∞, ∞ an Archimedean prime.

We let $\prod = \bigotimes_v \Pi_v$ be an irreducible automorphic cuspidal representation of the group $H(\mathbb{A})$. We let $f_\Pi \in \Pi$.

Then we form the Rankin-Selberg integral

$$Z_r(f_\Pi, \phi(,s)) = \int_{H(k)\backslash H(\mathbb{A})} f_\Pi(h) E^r_\ell(h,s,\phi)\, dh$$

relative to some choice of suitable Tamagawa measure dh on $H(k)\backslash H(\mathbb{A})$ and ϕ standard.

Lemma 1.1. *In the case $P_r\backslash G/H$ admits one nonnegligible orbit (see conditions above)*

$$Z_r(f_\Pi, \phi(,s)) = \int_{H^0_\gamma(\mathbb{A})\backslash H(\mathbb{A})} \phi(\gamma g, s) \left(\int_{H_\gamma(k)\backslash H^0_\gamma(\mathbb{A})} f_\Pi(hg)\, dh \right) dg$$

where $H_\gamma(k)$ is the stabilizer of the point $P_r\gamma$ in the unique nonnegligible orbit of H in $P_r\backslash G$ and $H^0_\gamma(\mathbb{A})$ is the group of norm one elements of $H_\gamma(\mathbb{A})$.

Proof. The proof is the usual method of unwinding the series expansion of E^r relative to the $H(k)$ orbits in $P_r\backslash G$. We note that a negligible orbit contributes zero since (i) the stabilizer contains a normal subgroup which is the unipotent radical of a proper parabolic of H and (ii) the form f_Π is cuspidal. Thus we are left with one term as above (with the hypotheses of the Lemma). □

Remark. 1.1. We consider now the case where $\dim(V) = r$. In fact in these examples

$$H_\gamma(k) \cong \begin{cases} \{(g,g) | g \in H_1\} \cong H_1 & \text{if } K = k \oplus k \\ \text{the } k \text{ rational points of } H \simeq H_1 & \text{if } K \text{ is quadratic over } k. \end{cases}$$

The inner integral in Lemma 1.1 can be calculated. We discuss the two cases separately.

(i) If $K = k \oplus k$ and $\Pi = \Pi_1 \otimes \Pi_2, \Pi_i$ automorphic cuspidal irreducible representation of $H_1(\mathbb{A})$, the inner integral is

$$\int_{H_1(k)\backslash H_1(\mathbb{A})} f_{\Pi_1}(hg_1) f_{\Pi_2}(h)\, dh.$$

Such a term is zero unless $\Pi_1 \cong \Pi_2^{\sim}$ = the contragredient of Π_2. Moreover, the integral (with data defining f_{Π_1} and f_{Π_2} suitably normalized) then becomes a matrix coefficient of $\langle f_{\Pi_1} * g_1 | f_{\Pi_1^{\sim}} \rangle$ of Π_1.

(ii) If K is a field, the inner integral is

$$\int_{H_1(k)\backslash H_1(\mathbb{A})} f_\Pi(hg_1)\,dh.$$

This represents the period of f_Π over the subgroup $H_1 \subset H$. A fundamental question is to determine for which Π such an integral is nonvanishing. We note that a similar problem exists in the doubling for the GL_n case ([PS-R-II]). In fact in that case the set of automorphic representation Π which have nonvanishing period should be the set of Π which arise by quadratic base change from an appropriate unitary group (care must be taken in that case for the correct condition on central characters).

§2. Factorizability Properties

The point of expressing Z_r as an integral in Lemma 1.1 above is to determine whether the inner integral can be factorized as an infinite product (Euler Product) of local factors.

Indeed in the case $K = k \oplus k$ the general matrix coefficient of Π_1 factors as

$$\langle f_{\Pi_1} * g | f_{\Pi_1^{\sim}} \rangle = \prod_v \langle \xi_v * g_v | \xi_v^{\sim} \rangle_{(\Pi_1)_v}$$

provided there exists an embedding $\Pi \overset{i_\Pi}{\rightsquigarrow} L^2_{\mathrm{cusp}}(H_1(k)\backslash H_1(\mathbb{A}))$ (which is $H_1(\mathbb{A})$ intertwining) in such a way that $i_{\Pi_1}(\otimes \xi_v) = f_{\Pi_1}, i_{\Pi_1^{\sim}}(\otimes \xi_v^{\sim}) = f_{\Pi_1^{\sim}}$. The point here is that there is for each prime $v \in k$ *exactly one* $H_1 \cong H_\gamma(k_v)$ *invariant* form on the space $(\Pi_1)_v \otimes (\Pi_1^{\sim})_v$ for each irreducible admissible representation Π_1 (Schur's Lemma).

This example illustrates the very general principle of uniqueness that explains when a global Rankin integral (such as the general Z_r above) has an Euler product (independent of making the usual unwindings of the integral). Indeed in the example of Z_r we look at the following local problem. The global integral defining Z_r determines for each place v a bilinear functional on $(\Pi_1)_v \otimes I_v(s)$ which is H_v invariant. Thus we consider the space $\mathrm{Hom}_{H_v}((\Pi_1)_v \otimes I_v(s), 1)$ for all s.

If the dimension of such a space is at most 1 (for all s) and if this holds for all primes v, then $Z_r(\ )$ can be factorized as an infinite product of local factors. This is just the simple principle that local multiplicity one (for all primes v) implies global multiplicity one. The main problem here is that in order to express Z_r as such a factorizable product we need to know for some a priori reason that in fact the global Rankin integral is nonvanishing. However in the case Z_r given in Lemma 1.1 the nonvanishing property is determined by the nonvanishing of a certain period (see (ii) in Remark 1.1). What is known is that the principle of local uniqueness implies also the existence of the local functional equation and thus the determination of a local ε factor of a representation (see [PS-R-I] for the consideration of these matters in the doubling case).

Remark. 2.1. A typical Rankin-Selberg integral leads to the following type of uniqueness question of models. Namely let $M \subseteq N$ be reductive groups. Let Π_1 be an irreducible admissible representation of $M(k_v)$ and $\mathrm{Ind}_{P_{N_v}}^{N_v}(\sigma_v)$ an induced representation from a standard parabolic P_{N_v} of N_v with σ_v,a character on P_{N_v}/N_v. Then we consider the space $Y_v = \mathrm{Hom}_{M_v}(\Pi_1 \otimes \mathrm{Ind}_{P_{N_v}}^{N_v}(\sigma_v), 1)$. In the particular case $\Pi_1 = \mathrm{Ind}_{B_v}^{M_v}(\chi_v)$ (where B_v is a Borel subgroup of M_v and χ_v any quasicharacter on B_v), then the determination of Y_v reduces by the usual Bruhat theory to finding certain quasi-invariant distributions on N_v relative to the left action of B_v and the right action of P_{N_v}. The importance of using such a Π_1 is that "generically" Π_1 is an unramified spherical representation and thus will be a local constituent of a global representation of $M(\mathbb{A})$. In any case the relevant local analysis is concerned with the set of double cosets $B_v \backslash N_v / P_{N_v}$. If the set $B_v \backslash N_v / P_{N_v}$ is finite then we might expect that Y_v above is at most one dimensional (for "generic" values of σ_v and χ_v). We note the following analogy from the category of algebraic group representations that makes such a statement at least plausible. From [Kim] if the set $B(\mathbb{C}) \backslash N(\mathbb{C}) / P_N(\mathbb{C})$ is finite then every finite dimensional irreducible representation of $N(\mathbb{C})$, whose highest weight vector is fixed by $P_N(\mathbb{C})$ projectively, is multiplicity free when restricted to $M(\mathbb{C})$ (with the assumption that $N(\mathbb{C})$ and $M(\mathbb{C})$ are connected and semisimple). This is a multiplicity one Branching Rule. It is not clear at this point whether such a multiplicity one Branching Rule determines when Y_v is one dimensional but such a possibility bears future investigation.

Remark. 2.2. The Rankin-Selberg integral given in [PS-R-S] is a special case of the data in Remark 2.1. Namely $M = G_2$ and $N = \mathrm{SO}(7)$ and the embedding of G_2 into SO(7) is given by the standard action of G_2 on the space of trace zero elements of the 8 dimensional space of octonions. Here P_N is the parabolic subgroup of SO(7) which stabilizes a 2 dimensional isotropic flag. The L function represented in this example is the tensor product of $G_2 \times \mathrm{GL}_2$.

In the case $K = k \oplus k$ the zeta integral Z_ℓ in Lemma 1.1 (with $\ell = \dim V$) equals the product (with $\phi(,s) = \bigotimes_v \phi_v(,s)$)

$$\prod_v Z_{\ell,v}(\xi_v, \xi_v^\sim, s)$$

where $Z_{\ell,v}(\xi_v, \xi_v^\sim, s)$ is the local zeta integral given by

$$\int_{H_1(k_v)} \langle \xi_v * g_v | \xi_v^\sim \rangle \phi_v(\gamma_v(g_v, 1), s)\, dg_v \,.$$

Then from [PS-R-I, PS-R-II] we have

Proposition 2.1. *Let $K = k \oplus k$ and fix an arbitrary point $s_0 \in \mathbb{C}$. It is possible to choose data $\phi = \otimes\phi_v, f_{\Pi_1}$, and $f_{\Pi_1^\sim}$ so that*

$$Z_\ell(f_{\Pi_1} \otimes f_{\Pi_1^\sim}, \phi(,s)) = \frac{L_S(\frac{1}{2}+s, \Pi, r) Z_\infty(s)}{b_{\ell,S}(s)}$$

where

(i) $L_S(s, \Pi, r)$ is the restricted L function associated to the standard representation r of the L group of $U(V, \langle,\rangle)_0$ (= the connected component of $U(V, \langle,\rangle)$).

(ii) $b_{\ell,S}(s) = \prod_{v \notin S} \left\{\prod_{k=1}^{k=[\frac{\ell}{2}]} \zeta_v(2s+\ell+1-2k)\right\} \begin{cases} \zeta_v(s+\frac{\ell+1}{2}) & \textit{if } \varepsilon = -1 \\ 1 & \textit{if } \varepsilon = 1, \end{cases}$

(iii) $Z_\infty(s)$ is a meromorphic function in s which is nonvanishing at the arbitrarily chosen point s_0. Note that the choice of data above depends on s_0.

Remark. 2.3. The functional equation of $L_S(s, \Pi, r)$ is given in [PS-R-I]. In fact it is possible to define local L_v factors at the bad primes of Π (those v which are finite and Π_v is not spherical). Using then an extended definition of L we give the functional equation of the new L in [PS-R-III]. There we use the notion of local ε factors mentioned above.

Remark. 2.4. We note that if $L_S(s, \Pi, r)$ admits a pole at $s = s_0$ then $b_{\ell,S}(s) E_\ell^\ell(g, s, \phi)$ admits a nonzero pole at $s = 2s_0 - 1$ for some section ϕ. In fact by (iii) of Proposition 2.1 and for suitable choice of data $b_{\ell,S}(s) Z_\ell(f_{\Pi_1} \otimes f_{\Pi_1^\sim}, \phi(\ ,s))$ has a pole at $s = 2s_0 - 1$ and this pole comes from the pole of the normalized Eisenstein series.

§3. Poles of Eisenstein Series

To determine information about the possible poles of L functions associated to the doubling method we require knowledge about the poles of $E_n^n(\cdots)$. As we shall see the residues are explicitly describable and generically square integrable.

For the remaining 2 sections we assume that the number field k is totally real.

For the discussion below we adopt a slightly more general convention. Namely we let G_0 be the isometry group of the form

$$\begin{pmatrix} 0 & I_n \\ \varepsilon I_n & 0 \end{pmatrix}.$$

Thus

$$G_0 = \begin{cases} \mathrm{Sp}_n & \text{if } \varepsilon = -1 \\ \mathrm{O}(n,n) & \text{if } \varepsilon = +1. \end{cases}$$

We consider the parabolic $P_n \cong \mathrm{GL}_n \ltimes U_n$ of G_0 with U_n, the unipotent radical of P_n. We form the family of induced representations $\mathbb{I}_n(s) = \mathrm{Ind}_{P_n(\mathbb{A})}^{G_0(\mathbb{A})}(|\det g|_{\mathbb{A}^\times}^s \otimes 1_{U_n})$ (normalized induction) where $|\ |_{\mathbb{A}^\times}$ = the usual adelic norm on $\mathbb{A}^\times$. We form the family of Eisenstein series $E_n^n(g, s, \phi)$ where ϕ is a standard section.

Theorem 3.1 [K-R-IV]. *Let S be the set of primes v of k where v is Archimedean and all finite places v at which $\phi_v(\ ,s)$ is not K_v invariant.*

Then the normalized Eisenstein Series $b_{n,S}(s)E_n^n(g,s,\phi)$ has at most simple poles and these may occur only in the set $X_n = \{-\varrho_n, 1-\varrho_n, \cdots, \hat{0}, \cdots, \varrho_n - 1, \varrho_n\}$ where $\varrho_n = \frac{1}{2}(n+1)$ ($\frac{1}{2}(n-1)$ resp.) if $\varepsilon = -1$ ($\varepsilon = +1$ resp.) Here $\hat{0}$ means 0 is omitted in the case when n is odd.

Corollary. *We let Π be an automorphic irreducible cuspidal representation of $H_1(\mathbb{A})$ defined in §1. We let $G = G_0$ be the doubled group and $S(\Pi) = S$ as given in §0.*

The poles of $L_S(s,\Pi,r)$ are at most simple and these may occur only in the set $1/2 + X_\ell$ (X_ℓ defined relative to the doubled group $G = G_0$ with $\ell = \dim V$).

Thus to determine whether the poles of $b_{n,S}(s)E_n^n(\ )$ actually occur we must study what the residue representation is. That is we must determine the subspace

$$\operatorname*{Res}_{s=s_0} b_{n,S}(s)E_n^n(g,s,\phi) = (E_n^n)^*(g,s_0,\phi)$$

(with $s_0 \in X_n$) as ϕ varies.

We now consider the group $\mathrm{Sp}_n = G_0$. Much of the discussion remains valid for $\mathrm{O}(n,n)$ but since we have not checked all the details we restrict G_0 just to Sp_n. We consider the oscillator representation $\omega_{Q,\psi}$ (defined in §0) of $\mathrm{Sp}_n \times \mathrm{O}(Q)$ on the space $S(M_{mn}(k))(m = \dim Q)$. In the case where k_v is non Archimedean, we determine the space of $\mathrm{O}(Q)$ coinvariants in $S(M_{mn}(k))$, which is given by $S(M_{mn}(k))/\{\text{the span of } \varphi - \omega_{Q,\psi}((,y))\varphi \text{ as } \varphi \text{ varies in } S(M_{mn}(k))\} \cong S(M_{mn}(k))_{\mathrm{O}(Q)}$. Then we know that the functional $\varphi \overset{\delta}{\rightsquigarrow} \varphi(0)$ has the property that $f_\varphi(pg) = \langle \delta | \omega_{Q,\psi}(pg,y)\varphi \rangle = |\det m'|^{\frac{\dim Q}{2}} \chi_Q(\det m') \langle \delta | \varphi \rangle$ (where $p = m' \cdot v$ with $m' \in GL_n$). Thus $f_\varphi \in I_v(\frac{m}{2} - \frac{n+1}{2}, \chi_Q)$. Here $I_v(s,\chi_Q)$ denotes the local representation of Sp_n given by $\mathrm{Ind}_{P_n}^{\mathrm{Sp}_n}(|\det g|^s \otimes \chi_Q \otimes 1_{U_n})$ (normalized induction), where χ_Q is the quadratic character given by the Hilbert symbol $\langle\ |\Delta_Q\rangle$ with $\Delta_Q =$ discriminant of Q. In fact the structure of $S(M_{mn}(k))_{\mathrm{O}(Q)}$ as a Sp_n module is governed by the space of f_φ.

Proposition 3.1 [R-I, K-R-III]. *Every $\mathrm{O}(Q)$ invariant functional on the space $S(M_{mn}(k))$ factors through the map $S(M_{mn}(k)) \rightsquigarrow I_v(\frac{m}{2} - \frac{n+1}{2}, \chi_Q)$ given by $\varphi \rightsquigarrow f_\varphi$. In the case $k = k_v$ is non-Archimedean then $S(M_{mn}(k))_{\mathrm{O}(Q)}$ is equivalent as an Sp_n module to $\{\mathrm{Span}\, f_\varphi | \varphi \in S(M_{mn}(k))\} = R^n(Q)$. In case $k_v = \mathbb{R}$ we let $R^n(Q) = \{\mathrm{Span}\, f_\varphi | \varphi \in S(M_{mn}(\mathbb{R}))$ and φ is $\mathrm{U}(n)$ finite$\}$, $\mathrm{U}(n) =$ maximal compact subgroup of $\mathrm{Sp}_n(\mathbb{R})$. Moreover for each place v, $R^n(Q)$ is irreducible if $\dim Q \le n+1$.*

Remark. 3.1. This is the local version of the Siegel-Weil formula.

Then given a global quadratic form Q over k we define $R^n(Q) = \otimes_v R^n(Q_v)$, which is an admissible representation of $\mathrm{Sp}_n(\mathbb{A})$.

The representation $R^n(Q)$ has the following automorphic structure.

Theorem 3.2 [K-R-VI].
Let $\dim Q < n+1$. *Then*
(i) $\dim \mathrm{Hom}_{\mathrm{Sp}_n(\mathbb{A})}(R^n(Q), \mathscr{A}(\mathrm{Sp}_n(\mathbb{A}))) \le 1$; *In fact*
(ii) there exists a nonzero embedding r_Q^n *of* $R^n(Q)$ *into* $\mathscr{A}(\mathrm{Sp}_n(\mathbb{A}))$. *Moreover the image* $r_Q^n(R^n(Q)) \subseteq L^2_{\mathrm{res}}(\mathrm{Sp}_n(\mathbb{A}))$, *the residual spectrum of* $L^2(\mathrm{Sp}_n(\mathbb{A}))$, *except in the case when* Q *is the split 2 dimensional form.*

At this point we say two global quadratic forms Q and Q' are complementary if $\dim Q + \dim Q' = 2n+2$ and if $Q \cong A \oplus H_{2r}$, where A is anisotropic and H_{2r} is a direct sum of r hyperbolic planes, then $Q' \cong A \oplus H_{2r'}$. We note that Q uniquely determines Q' and vice-versa.

We note that if $\dim Q \ge n+1$ and Q admits a complementary form Q' then for each place v, $R^n(Q_v)$ admits a unique quotient representation $R^n(Q'_v)$ (this is basically just a simple case of the Howe duality conjecture for dual reductive pairs).

Thus we can describe the residue representation for $s_0 \in X_n$ and $s_0 > 0$. Indeed we note that $b_{n,S}(s)$ is analytic and non vanishing at such s_0. Thus we unambigously have that

$$\text{Subspace spanned by } (E_n^n)^*(g, s_0, \phi) =$$
$$\text{Subspace spanned by } \operatorname*{Res}_{s=s_0} E_n^n(g, s, \phi) = R_n(s_0)$$

This makes it possible to give a simple characterization of $R_n(s_0)$ in the following terms.

Theorem 3.3 [K-R-VI]. *Let* $s = s_0 \in X_n$ *and* $s_0 > 0$. *Then as a* $\mathrm{Sp}_n(\mathbb{A})$ *module* $R_n(s_0)$ *is isomorphic to the direct sum* $\oplus R^n(Q')$ *where the direct sum ranges over all classes of quadratic forms* Q' *with* $\dim Q' = (n+1) - 2s_0$ *and* $\chi_Q = 1$.

Remark. 3.2. We note that in case $s_0 = \frac{n+1}{2}$, $R_n(\frac{n+1}{2})$ is the identity representation if $\chi = 1$ and $\{0\}$ otherwise. In case $s_0 = \frac{n-1}{2}$, then $R_n(\frac{n-1}{2}) = R^n(H_2)$ if $\chi = 1$. Otherwise in the remaining cases $R_n(s_0)$ is a direct sum of infinitely many $R^n(\,)$.

Using Theorem 3.3 it is possible to show consequences concerning the existence of the poles of standard L function of the group Sp_n.

Theorem 3.4. *Suppose* $L_S(,\Pi,r)$ *admits a pole at* $s = \frac{1}{2} + s_0$ *where* $s_0 \in X_{2n}$ *and* $s_0 > 0$. *Then there exists a form* Q' *where* $\dim Q' = 2n+1-2s_0$ *and* $\chi_{Q'} = 1$ *with the property that relative to the dual pair* $\mathrm{Sp}_n \times \mathrm{O}(Q')$ *we can find* $\varphi \in S(M_{\dim Q',n}(\mathbb{A}))$ *and a function* $f_\Pi \in \Pi$ *so that the* θ *integral* $\theta_\varphi(f_\Pi) \ne 0$.

Remark. 3.3. We show in §4 that the actual set of possible poles of $L_S(s,\Pi,r)$ for $\mathrm{Re}(s) > 0$ is the set $s \in \{1, 2, \cdots, [\frac{n}{2}]+1\}$.

Remark. 3.4. If $L_S(s,\Pi,r)$ admits a pole at $s = \frac{1}{2} + s_0$, then there may be more than one Q' so that $\theta_\varphi(f_\Pi) \ne 0$.

§4. Siegel-Weil Formula

We consider the global version of Proposition 3.1. This constitutes a generalized Siegel-Weil formula, describing the residue of Eisenstein series in terms of θ series. First we recall the well known version of this formula established in [W] and extended in [K-R-I] and [K-R-II].

We define a particular type of global section built from f_φ for $\varphi \in S(M_{mn}(k))$. We consider $f_\varphi(g) = \prod_v f_{\varphi_v}(g_v)$ where $\varphi = \otimes \varphi_v$. Then $f_\varphi \in I(\frac{m}{2} - \frac{n+1}{2}, \chi_Q)$. Then we consider the character on P_n given by $p = m' \cdot u \rightsquigarrow |\det m'|^{s-s_0}$ $(s_0 = \frac{m}{2} - \frac{n+1}{2})$. We define the function on $\mathrm{Sp}_n(\mathbb{A})$

$$e^{sH_P(g)} = |\det m'|^{s-s_0}$$

where $g = m'uk$ relative to the Iwasawa decomposition. Then we form

$$F^n_\varphi(g, s) = f_\varphi(g) e^{sH_P(g)}$$

Now F^n_φ is a standard $\phi(,s)$ section defined in §1. Hence we may form Eisenstein series having certain remarkable properties as given by the Siegel-Weil formula.

Theorem 4.1 [W, K-R-I, K-R-II].
Let either Q be anisotropic or $\dim Q -$ *Witt index* $(Q) > n+1$. *Then the integral*

$$\int_{\mathrm{O}(Q)(k)\backslash \mathrm{O}(Q)(\mathbb{A})} \theta_\varphi(x, y)\, dy$$

is absolutely convergent. The Eisenstein series $E^n_n(g, s, F^n_\varphi)$ is holomorphic at $s = s_0 = \frac{m}{2} - \frac{n+1}{2}$ and defines an intertwining map of $R^n(Q)$ into $\mathcal{A}(\mathrm{Sp}_n(\mathbb{A}))$. Moreover we have the identity

$$E^n_n(x, s_0, F^n_\varphi) = c_Q \int_{\mathrm{O}(Q)(k)\backslash \mathrm{O}(Q)(\mathbb{A})} \theta_\varphi(x, y)\, dy$$

with $c_Q \neq 0$ depending only on Q.

The point in extending the Siegel-Weil formula is to have a procedure of regularizing the θ integral above.

At this point we look at the dual pair $\mathrm{Sp}_n \times \mathrm{O}(Q)$ with $\dim Q \leq 2n$ and $\chi_Q = 1$.

Let $v = \infty$ be an Archimedean real place. Then the local form Q_∞ is equivalent to $H_{2r} \oplus V_0$ where V_0 is anisotropic and H_{2r} = a direct sum of $2r$ hyperbolic planes. We consider the family of oscillator representations $\omega_{H_{2i} \oplus V_0, \psi}$ (with $i \leq r$).

We let $\mathcal{Z}_{\mathrm{Sp}_n} = \mathcal{Z}$ be the center of the universal enveloping algebra of $\mathrm{Sp}_n(\mathbb{R})$.

We define the *support ideal* I_{Q_∞} of ω_{Q_∞} as

$$\{\xi \in \mathcal{Z} | \omega_{H_{2i} \oplus V_0, \psi}(\xi)(\varphi_\infty) = 0 \quad \text{for all } \varphi_\infty \in S(M_{\dim(H_{2i} \oplus V_0), n}(\mathbb{R})) \quad \text{for } i = r-1\}.$$

Then it is straightforward to verify that $I_{Q_\infty} \neq \mathcal{Z}$ (since $\dim Q \leq 2n$). Moreover it is also easy to check that the space of θ kernels on the group $\mathrm{Sp}_n \times \mathrm{O}(Q)$

$$\{\theta_{\omega_{Q_\infty,\varphi_\infty}(\xi_\infty)(\varphi)}(x,y) | \xi_\infty \in I_{Q_\infty}\}$$

is *rapidly decreasing* in the $\mathrm{O}(Q)$ variable ($\dim Q \leq 2n$ and $Q \neq H_2$ split form).

Then as a substitute for the θ integral we use the regularized θ integral with an element $\xi_\infty \in I_{Q_\infty}$. That is,

$$i_{\xi,Q}(\varphi) = \int\limits_{\mathrm{O}(Q)(k)\backslash \mathrm{O}(Q)(\mathbb{A})} \theta_{\omega_{Q_\infty,\varphi_\infty}(\xi_\infty)(\varphi)}(x,y)\, dy\,.$$

Then $i_{\xi,Q} \in \mathrm{Hom}_{\mathrm{Sp}_n(\mathbb{A})}(R^n(Q), \mathscr{A}(\mathrm{Sp}_n(\mathbb{A}))$. But we also have

Lemma 4.1. *If* $\dim Q - $*Witt index* $(Q) \leq n+1$ *and* $2n \geq \dim Q > n+1$ *then* $i_{\xi,Q}$ *factors through the space* $R^n(Q')$*,* $Q' =$ *the complementary form to* Q*.*

Moreover if $\dim Q \leq n+1$ *and* $Q \neq H_2$ *, then the map* $s \rightsquigarrow E_n^n(g,s,F_\varphi^n)$ *is analytic at* $s = s_0 = \frac{m}{2} - \frac{n+1}{2}$*. Also* $\varphi \rightsquigarrow \operatorname*{val}_{s=s_0} E_n^n(g,s,F_\varphi^n)$ *defines an element in* $\mathrm{Hom}_{\mathrm{Sp}_n(\mathbb{A})}(R(Q), \mathscr{A}(\mathrm{Sp}_n(\mathbb{A}))$.

If $2n \geq \dim Q > n+1$ *and* $\dim Q - $*Witt index* $(Q) \leq n+1$ *then* $\varphi \rightsquigarrow \operatorname*{Res}_{s=s_0} E_n^n(g,s,F_\varphi^n)$ *(with* $s_0 = \frac{m}{2} - \frac{n+1}{2}$*) defines a intertwining map which factors through* $R^n(Q')$*. This determines a nonzero element in* $\mathrm{Hom}_{\mathrm{Sp}_n(\mathbb{A})}(R^n(Q'), \mathscr{A}(\mathrm{Sp}_n(\mathbb{A}))$.

By the Uniqueness Principle of Theorem 3.2 we have the following Siegel-Weil type identity.

Theorem 4.2. *Let* $s_0 = \frac{\dim Q}{2} - \frac{n+1}{2}$*.*

(1) *Let* $\dim Q \leq n+1$*, with* $Q \neq H_2$*. Then there is a constant* c_ξ *so that*

$$c_\xi E_n^n(g,s_0,F_\varphi^n) = i_{\xi,Q}(\omega_{Q,\psi}(g)\varphi)$$

(2) *Let* $2n \geq \dim Q > n+1$ *and* $\dim Q-$*Witt index* $(Q) \leq n+1$*. Then there is a constant* c'_ξ *so that*

$$c'_\xi \operatorname*{res}_{s=s_0} E_n^n(g,s,F_\varphi^n) = i_{\xi,Q}(\omega_{Q,\psi}(g)\varphi)$$

The goal then is to use Theorems 4.1 and 4.2 to glean information about special values of the L functions discussed above.

We indicate one type of result in this direction.

We fix a "Witt tower" of quadratic forms $Q_0, Q_0 \oplus H_2 = Q_2, \cdots, Q_0 \oplus H_{2r} = Q_{2r}$ where Q_0 is anisotropic (even dimensional) and H_{2r} is a direct sum of r hyperbolic planes.

We let $X_{Q_{2i}} = \{f \in L^2_{\mathrm{cusp}}(\mathrm{Sp}_n(\mathbb{A})| \theta_\varphi(f) = 0$ for all $\varphi \in S(M_{\dim Q_{2i},n}(\mathbb{A}))\}$. Then $X_{Q_{2r}}$ is a $\mathrm{Sp}_n(\mathbb{A})$ module and we define inductively the spaces $Y_{Q_{2i}} =$ the perpendicular complement of the space $X_{Q_{2i}} \cap X_{Q_{2i-2}} \cap \cdots \cap X_{Q_0}$ in $X_{Q_{2i-2}} \cap \cdots \cap X_{Q_0}$. Here perpendicular complement is taken relative to the Hermitian pairing given by the Petersson inner product on $L^2_{\mathrm{cusp}}(\mathrm{Sp}_n(\mathbb{A}))$. Then we have an orthogonal direct sum decomposition of $\mathrm{Sp}_n(\mathbb{A})$ modules of $L^2_{\mathrm{cusp}}(\mathrm{Sp}_n(\mathbb{A}) = Y_{Q_0} \oplus Y_{Q_2} \oplus$

$\cdots \oplus Y_{Q_{4n}}$. The space $Y_{Q_{2i}} = \{0\}$ if $\dim Q_{2i} < n$ ([PS-R-IV]). Also, if $f_\Pi \in \Pi$, an irreducible automorphic cuspidal representation which occurs in $Y_{Q_{2i}}$, then there exists $\varphi \in S(M_{mn}(\mathbb{A}))(m = \dim Q_{2i})$ ([R-I]) so that $\theta_\varphi(f_\Pi) \neq 0$. In fact, $\theta_\varphi(f_\Pi)$ is a cusp form on $\mathrm{O}(Q_{2i})(\mathbb{A})$.

On the other hand if $f \in Y_{Q_{2i}} \oplus \cdots \oplus Y_{Q_{4n}}$ then whether $\theta_\varphi(f)$ (for $\varphi \in S(M_{m,n}(\mathbb{A}))$ is zero or not is measured by the inner product formula developed in [R-II] and [R-III]. Indeed, using Theorems 4.1 and 4.2 we show the following identity:

Proposition 4.1. *Let* $f_\Pi \in \Pi$ *be chosen as in the beginning of* §2 *where* $\Pi \subset Y_{Q_{2i}} \oplus \cdots \oplus Y_{Q_{4n}}$.

(1) *Then if either* $\dim Q_{2i} \leq 2n+1$ *or* $\dim Q_{2i} - i > 2n+1$

$$\|\theta_{\varphi_1}(f_\Pi)\|^2_{\mathrm{O}(Q_{2i})(\mathbb{A})} = c'_{Q_{2i}} \operatorname*{val}_{s=s_0} Z_{2n}(f_{\Pi_1} \otimes \overline{f}_{\Pi_1}, F^{2n}_{\varphi_1^*}(\ ,s))$$

where φ_1^* *is a canonically determined function (coming from* φ_1*) in* $S(M_{\dim Q_{2i},2n}(\mathbb{A}))$ *(the doubled Weil representation space of* $\mathrm{Sp}_{2n} \times \mathrm{O}(Q_{2i})$*). Here* $s_0 = \frac{\dim Q_0}{2} + i - (n + \frac{1}{2})$. *Here* $\|\ \|^2_{\mathrm{O}(Q_{2i})(\mathbb{A})}$ *represents the Petersson norm of a function relative to a suitably normalized measure on* $\mathrm{O}(Q_{2i}(k)\backslash\mathrm{O}(Q_{2i})(\mathbb{A})$. *Also* $c'_{Q_{2i}}$ *is a nonzero constant independent of* φ *and* f_Π.

(2) *The inner product formula of (1) is also valid in the case where* $\dim Q_{2i} - i \leq 2n+1$ *and* $2n+1 < \dim Q_{2i} < 4n$, *but now* $\operatorname*{val}_{s=s_0}$ *is replaced by* $\operatorname*{Res}_{s=s_0}$.

Thus, using Proposition 2.1, it is possible to relate the Petersson norm of $\theta_{\varphi_1}(f_\Pi)$ to the special value of the L_S function defined in §2. The main problem that remains in getting an exact relation is to have control on the bad local factors, i.e. where the data is not spherical in $Z_v(\cdots)$. The point here is to be able to determine the local factor $Z_v(\cdots)$ when the local data has the form f_φ as given above.

However in any case using Proposition 2.1 and a much simpler version of the identity in Proposition 4.1 we deduce the statement about the poles of the L functions given in Theorem 3.4. Moreover Remark 3.3 follows from the fact that $Y_{Q_{2i}} = 0$ for $\dim Q_{2i} < n$.

References

[B-J] Borel, A., Jacquet, H.: Automorphic forms and automorphic representations. Proc. Symp. Pure Math. 33, vol. 1, 1979

[G-S] Gelbart, S., Shahidi, F.: Analytic properties of automorphic L functions. Perspectives in Math., Academic Press, 1988

[Kim] Kimelfeld, B.: Homogeneous domains on flag manifolds. J. Math. Anal. Appl. **121** (1987) 506–588

[K-R-I] Kudla, S., Rallis, S.: On the Weil-Siegel formula. J. Reine Angew. Math. **387** (1988) 1–68

[K-R-II] Kudla, S., Rallis, S.: On the Weil-Siegel formula II. J. Reine Angew. Math. **391** (1988) 65–84

[K-R-III] Kudla, S., Rallis, S.: Degenerate principal series and invariant distributions. Israel J. Math. **69** (1990) 25–45

[K-R-IV] Kudla, S., Rallis, S.: Poles of Eisenstein series and L functions. Preprint 1989

[K-R-V] Kudla, S., Rallis, S.: Ramified degenerate principal series. In preparation, 1990

[K-R-VI] Kudla, S., Rallis, S.: A regularized Siegel-Weil formula: the first term identity. In preparation, 1990

[PS-R-I] Piatetski-Shapiro, I., Rallis, S.: ε factor of representations of classical groups. Proc. Natl. Acad. Sci. **83** (1986) 4589–4593

[PS-R-II] Piatetski-Shapiro, I., Rallis, S.: L functions for classical groups. (Lecture Notes in Mathematics, vol. 1254). Springer, Berlin Heidelberg New York 1987

[PS-R-III] Piatetski-Shapiro, I., Rallis, S.: Rankin Triple L functions. Comp. Math. **64** (1987) 31–115

[PS-R-IV] Piatetski-Shapiro, I., Rallis, S.: A new way to get Euler products. J. Reine Angew. Math. **392** (1988) 110–124

[PS-R-S] Piatetski-Shapiro, I., Rallis, S., Schiffmann, G.: Rankin-Selberg integrals for G_2. To appear in Amer. J. Math.

[Rab] Rabau, P.: Preprint 1990

[R-I] Rallis, S.: On the Howe duality conjecture. Comp. Math. **51** (1984) 333–399

[R-II] Rallis, S.: Injectivity Properties of liftings associated to Weil representation. Comp. Math. **52** (1984) 139–169

[R-III] Rallis, S.: L functions and the oscillator representation. (Lecture Notes in Mathematics, vol. 1245). Springer, Berlin Heidelberg New York 1987

[W] Weil, A.: Sur la formule de Siegel dans la theorie des groupes classiques. Acta. Math. **113** (1965) 1–87

Iteration of Polynomial Automorphisms of $\mathbf{C}^2$

Eric Bedford

Department of Mathematics, Indiana University, Swain Hall East
Bloomington, IN 47405, USA

§ 1. Introduction

We will consider mappings $f = (f_1, f_2) : \mathbf{C}^2 \to \mathbf{C}^2$ such that f_1 and f_2 are holomorphic polynomials. A polynomial automorphism is a polynomial mapping which is 1-to-1 and onto; it follows that the inverse f^{-1} is also polynomial. We will discuss such mappings from the point of view of dynamical systems. That is, we will be primarily concerned with the forms of limiting and/or recurrent behavior that the iterates $\{f, f^2 = f \circ f, \ldots, f^{\circ n}, \ldots\}$ of such mappings can exhibit. Our purpose here is to present some work that we have done on these problems in collaboration with J. Smillie and which has been written up in [BS1, 2, 3]. Lack of time prevents us from discussing more recent work on entropy, Lyapunov exponents, and ergodicity (see [BS4]).

Analogous problems have been studied for polynomial mappings $p : \mathbf{C} \to \mathbf{C}$ in one complex variable. The starting point is the Julia set J_p, which is the complement of the set where the iterates $\{p^n : n = 1, 2, 3 \ldots\}$ are well behaved, i.e. $\mathbf{C} - J_p$ is the complement of the largest open set where $\{p^n\}$ forms a normal family. J_p turns out to be the boundary of $K_p^+ = \{z : p^n(z) \text{ is bounded for } n = 1, 2, 3, \ldots\}$ is a nonempty, compact, invariant set which carries all of the interesting dynamics of p. The fundamental properties of p on J_p were developed in the classical work of Fatou and Julia. One of the basic results of the theory is:

$$\text{The repelling periodic points for } p \text{ are dense in } J_p. \tag{1}$$

Here we will consider the potential theoretic approach to the dynamical properties of a polynomial p (cf. [Br, T]). This approach is well adapted to obtaining results of a more precise quantitative nature. The equilibrium measure μ of J_p, which is fundamental in potential theory, turns out to coincide with the unique measure of maximal entropy. Two results of Brolin [Br] show that μ arises as the limit of two sequences of sets of points. First, it is the limit of averages of point masses on the preimages of a point, i.e.

$$\mu = \lim_{n \to \infty} (\deg)^{-n} \sum_{a \in \{p^{-n}(z_0)\}} \delta_a , \tag{2}$$

where the limit is taken with respect to the weak topology of measures. Second, it is the limit of the average of point masses of periodic points, i.e.

Proceedings of the International Congress
of Mathematicians, Kyoto, Japan, 1990

$$\mu = \lim_{n\to\infty} (\#\,\mathrm{Per}(n))^{-1} \sum_{a\in \mathrm{Per}(n)} \delta_a, \tag{3}$$

where $\mathrm{Per}(n) = \{z \in \mathbf{C} : p^n(z) = z\}$ is the set of periodic points whose periods divide n.

While the potential theoretic approach gives useful and powerful tools, it applies only to polynomial mappings. Although much of the theory of Fatou and Julia applies also to rational mappings, potential-theoretic methods do not apply in this case. The equilibrium measure of J_p is not equal to the measure of maximal entropy unless p is a polynomial (see [Lo]). Although problems of iteration have been studied for some classes of entire functions, the theory is quite different; for instance, J_p is not the boundary of K_p^+.

A prototypical example of the polynomial automorphisms which we will consider is the mapping $f : \mathbf{C}^2 \to \mathbf{C}^2$ given by

$$f(x, y) = (y, p(y) - ax), \tag{4}$$

where $p(y)$ is a one-variable polynomial of degree $d > 1$. In fact, as will be shown in §2, there is no essential loss of generality in considering only these maps.

There are several sources of motivation for studying the iteration of automorphisms of $\mathbf{C}^2$. One is to generalize to the 2-dimensional case the results that have been obtained for complex polynomial maps. The two basic properties of a 1-variable map: that it is d-to-1 ($d > 1$) and that it is conformal, do not hold for automorphisms of $\mathbf{C}^2$. However, the fact that a 1-dimensional map $p : \mathbf{C} \to \mathbf{C}$ is a proper mapping of degree d generalizes to the observation that the mapping on homology given by (10) below is multiplication by d. Conformality of p is partially replaced by the fact that f is conformal when restricted to leaves of the stable/unstable foliations.

We define the *stable set* of a point p as

$$W^s(p, f) = \{z \in \mathbf{C}^2 : \lim_{n\to\infty} \mathrm{dist}(f^n(z), f^n(p)) = 0\}. \tag{5}$$

We say that p is a *periodic point* for f if $f^m(p) = p$ for some $m \geq 1$, and we call the smallest such m the *period* of p. For a periodic point p, we let λ_1, λ_2 denote the eigenvalues of $Df^m(p)$. If $|\lambda_1|, |\lambda_2| < 1$, then p is a *sink*, and it is well known that $W^s(p, f)$ is an open set containing p, which is called the *basin of attraction* of p. A proper open subset $\Omega \subset \mathbf{C}^2$ which is biholomorphically equivalent to $\mathbf{C}^2$ is called a *Fatou-Bieberbach domain.* It is classically known that if p is a sink, then the basin $B = W^s(p, f)$ is biholomorphically equivalent to $\mathbf{C}^2$, and in many cases, e.g. if f has more than one sink, B is a Fatou-Bieberbach domain which is not dense in $\mathbf{C}^2$. Basins of attraction were the original examples of Fatou-Bieberbach domains. The geometry of Fatou-Bieberbach domains in general is intriguing but not well understood, and the ones that arise as basins of attraction seem to be the most approachable. For instance, it was shown in [BS2] that a polynomial basin $B = W^s(p, f)$ cannot be "small" enough to avoid an algebraic variety, or "large" enough to contain one. In other words, if V is a 1-dimensional algebraic variety, then

$$B \cap V \neq \emptyset \quad \text{and} \quad V \not\subset \overline{B}.$$

Another motivation comes from polynomial automorphisms of $\mathbf{R}^2$. Hénon [Hn1, 2] showed that the automorphism $g : \mathbf{R}^2 \to \mathbf{R}^2$ given by

$$g(x, y) = (y, y^2 + c - ax) \tag{6}$$

can possess complicated dynamics. The stable set of a subset $\Lambda \subset \mathbf{C}^2$ is given by

$$W^s(\Lambda, g) = \{p \in \mathbf{C}^2 : \lim_{n\to\infty} \operatorname{dist}(g^n(p), \Lambda) = 0\}. \tag{7}$$

Let us call a set Λ an *attractor* for g if Λ is compact and invariant, and $W^s(\Lambda)$ contains a neighborhood of Λ. It was shown recently (see [BC]) that g has a "strange attractor" for certain values of the parameters a and c. "Strange" implies that the attractor Λ is not a union of sink orbits. One difference between the real and complex cases is worth pointing out here. Although this set Λ is an attractor for g in $\mathbf{R}^2$, it is not an attractor, when considered in $\mathbf{C}^2$. In fact a normal families argument shows that any attractor for g in $\mathbf{C}^2$ is a union of sink orbits. Although it is not clear exactly what relationship holds between the mapping g and the corresponding mapping extended to $\mathbf{C}^2$, it is felt that the study of the map in $\mathbf{C}^2$ should be analogous to approaching the study of polynomial maps $p : \mathbf{R} \to \mathbf{R}$ by the study of the same polynomial in the complex domain. When the theory is pursued in the complex domain, it seems to be more complete and show a more stable or continuous dependence on parameters.

§ 2. Elementary Mappings

The map f and a conjugate $h^{-1} \circ f \circ h$ will have the same properties under iteration. Friedland and Milnor [FM] have shown that the set of polynomial automorphisms naturally divides into two sets of equivalence classes under conjugation by polynomial automorphisms. One of these classes is called the "elementary" automorphisms, and the other consists of finite compositions of the form

$$f = f_1 \circ \ldots \circ f_m, \tag{8}$$

where

$$f_j(x, y) = (y, p_j(y) - a_j x) \tag{9}$$

and $p_j(y) = y^{d_j} + c_{d_j-2} y^{d_j-2} + \ldots$ is a monic polynomial of degree $d_j > 1$ in which the y^{d_j-1} coefficient vanishes.

The dynamics of the elementary maps has been studied in detail by Friedland and Milnor. They have shown that an elementary map f has rather simple dynamics; f has periodic points of only finitely many periods, the nonwandering set is quite simple, and f is conjugate to an isometry on the nonwandering set.

Thus the rest of this talk will be concerned with mappings of the form (8). If we define the sets

$$V^{\pm} = \{(x, y) \in \mathbf{C}^2 : |x| \gtrless |y|, \max |x|, |y| > \kappa\}$$

$$V = \{\max |x|, |y| < \kappa\}$$

then for κ large enough

$$f^{-1}(V^+) \subset V^+ \quad \text{and} \quad f^{-1}(V^+ \cup V) \subset V^+ \cup V$$

$$f(V^-) \subset V^- \quad \text{and} \quad f(V^- \cup V) \subset V^- \cup V.$$

It is evident, then that for a point p in V^+ (or V^-) $f^n(p)$ tends to infinity as n tends to $-\infty$ (or $+\infty$). Thus all recurrent behavior of f takes place in the bounded set V. (This is useful, since there does not seem to be a natural extension of f to a compactification of $\mathbf{C}^2$; the one point compactification of $\mathbf{C}^2$, for instance, cannot be given a complex structure.)

Although f is an automorphism, the behavior of p in the large is reflected by the fact that the mapping on homology

$$f_* : H_2(V^- \cup V, V^-) \to H_2(V^- \cup V, V^-) \tag{10}$$

corresponds to multiplication by d, i.e. $f_*(\tau) = (d)\tau$.

§ 3. The Sets $K^\pm$

The approach adopted by Hubbard and Oberste-Vorth [HO] is to consider the sets

$$K^\pm = \{q \in \mathbf{C}^2 : f^{\pm n}(q) \text{ is bounded for } n = 1, 2, 3, \ldots\}$$

where the iterates stay bounded in forward/backward time. Also of interest are the sets

$$J^\pm = \partial K^\pm$$

$$K = K^+ \cap K^- \quad \text{and} \quad J = J^+ \cap J^- .$$

It is evident that with $V^\pm$, V as in the previous section, then $K^\pm \subset V \cup V^\pm$.

A sequence $\{q_n\}$ is said to be an ε-orbit if $\operatorname{dist}(q_{n+1}, f(q_n)) < \varepsilon$ holds for all $n = 1, 2, 3, \ldots$. A point q is *chain recurrent* if for any $\varepsilon > 0$ there is an ε-orbit $\{q_n\}$ with $q = q_1 = q_{1+jN}$ for some N and all $j = 1, 2, 3, \ldots$. The set of chain recurrent points is denoted by $R(f)$. Since the iterates of any point outside of K tend to infinity in either positive or negative time, we see that $R(f) \subset K$.

In contrast to the case of mappings of one variable, where the Julia set can be either connected or a Cantor set, *the sets $K^\pm$ and $J^\pm$ are always connected* (cf. [BS1], Theorem 7.2.) The set $J = J^+ \cap J^-$, on the other hand, may be either connected or disconnected.

Hubbard and Oberste-Vorth [H, HO] have studied $U^\pm := \mathbf{C}^2 - K^\pm$ in some detail for mappings of the form (6) with a and c complex. For all a, c, the sets $U^\pm$ are homeomorphic to $(S^3 - \Sigma) \times \mathbf{R}$, where Σ is a solenoid, and the fundamental group $\pi_1(U^\pm) \cong \mathbf{Z}[\frac{1}{2}]$ is not finitely generated. [HO] also makes a detailed study of the topology of $K^\pm$ for mappings of the form (6) and certain values of $a, c \in \mathbf{C}$

If $(x, y) \notin K^+$, then the iterates $(x_n, y_n) := f^n(x, y)$ become unbounded with asymptotic behavior like $y_n \sim p(x_n)$. In the potential theoretic approach, we study $K^\pm$ and $\mathbf{C}^2 - K^\pm$ in terms of the rate at which the points (x_n, y_n) escape to infinity. From the specific polynomial form of f, we see that for fixed x

$$|f_n(x, y)| = |y|^{d^n} + O(|y|^{d^n - 2}) \tag{11}$$

as $|y| \to \infty$. This motivates the definition

$$G^\pm(x, y) = \limsup_{n \to +\infty} (d^{-n}) \log(|x_{\pm n}| + |y_{\pm n}|) .$$

Using (11) we may derive the facts

- $\{G^{\pm} = 0\} = K^{\pm}$.
- $G^{\pm}$ is pluriharmonic on $K^{\pm}$.
- $G^{+}(x, y) = \log^{+}|y| + o(1)$ for x fixed and $y \to \infty$.
- $G^{-}(x, y) = \log^{+}|x| - \log|a| + o(1)$ for y fixed and $x \to \infty$.

It is immediate from the definition that

- $G^{\pm} \circ f = (d^{\pm n})G^{\pm}$.

Further, $G^{\pm}$ is continuous on $\mathbf{C}^2$; and by a more subtle argument it may be shown to be Hölder continuous (see [FS]).

From these observations, we see that $G^{\pm}$ is the pluriharmonic Green function of $K^{\pm}$. In fact, if T denotes any complex line (i.e. a 1-dimensional subspace of $\mathbf{C}^2$), then the restriction $G^{\pm}|_T$ is the classical Green function of $K^{\pm} \cap T$ inside the Riemann surface T (which may be identified with $\mathbf{C}$).

§ 4. Potential Theory in C²

Although the Laplacian $\triangle$ is sometimes useful in problems of several complex variables, it is not an invariant operator. The defect of this lack of invariance shows up in problems of iteration, so we will use dd^c, which is a biholomorphically invariant version of $\triangle$. Similarly, the usual subharmonic functions, which are defined by the condition $\triangle v \geq 0$, are not preserved under holomorphic transformations. The class of subharmonic functions which remain subharmonic under holomorphic changes of coordinates is the class of *plurisubharmonic* or *psh* functions, and it is this class which will be useful to us.

By $\mathscr{D}_{(p,q)}$ we will denote the set of test (p, q)-forms, i.e. the forms which may be written as

$$\alpha = \sum_{1 \leq i_1 < \ldots < i_p \leq n} \sum_{1 \leq j_1 < \ldots < j_q \leq n} \alpha_{i_1,\ldots,i_p,j_1,\ldots,j_q} dz_{i_1} \wedge \ldots \wedge dz_{i_p} \wedge d\bar{z}_{j_1} \wedge \ldots \wedge d\bar{z}_{j_q}$$

where each $\alpha_{i_1,\ldots,i_p,j_1,\ldots,j_q}$ is a smooth function with compact support. The dual of this space is the space of (p, q) *currents*, which is denoted as $\mathscr{D}'_{(p,q)}$. For an example of a (1,1) current on $\mathbf{C}^2$ let us consider a 1-dimensional closed complex submanifold (or subvariety) M. The *current of integration* $[M]$ acts on a (1,1)-form φ by integration: $[M](\varphi) = \int_M \varphi$. A current may be thought of loosely as a (p, q)-form with distributions as coefficients. In case the current is positive, then the coefficients may be shown to be regular Borel measures. (See [L1] for further details.)

We consider the operator

$$dd^c : \mathscr{D}'_{(0,0)} \to \mathscr{D}'_{(1,1)},$$

which is given by

$$dd^c u = 2\sqrt{-1} \sum_{i,j=1}^{2} \frac{\partial^2 u}{\partial z_i \partial \bar{z}_j} dz_i \wedge d\bar{z}_j .$$

A function u is psh if it is upper semicontinuous and if $dd^c u$ is a positive (1,1) current. The functions $G^{\pm}$ defined above are psh, and $\mu^{\pm} := dd^c G^{\pm}$ are positive, closed currents which satisfy

$$f^* \mu^\pm = (d)^{\pm 1} \mu^\pm .$$

We will call $\mu^\pm$ the *stable/unstable currents.* It it easily seen that the support of $\mu^\pm$ is exactly $\partial K^\pm$.

Since $\mu^\pm$ are (1,1) currents, they, unlike measures, do not act directly on sets. But if T is a locally closed 1-dimensional complex manifold, then we may obtain a measure on T by restriction: $\mu^\pm|_T := (dd^c)_T(G^\pm|_T)$. That is, we let $G^\pm|_T$ denote the restriction of $G^\pm$ to T, and we let $(dd^c)_T$ denote the induced Laplacian on T, so $\mu^\pm|_T$ acts as a measure on subsets of T.

The first Theorem of Brolin mentioned above concerns the preimages of a point under iteration. Our way of adapting this to our situation is to work with (1,1) currents. We consider, instead of points, an algebraic variety V of codimension 1, so that $[V]$ is a (1,1) current. The pullback of the current of integration corresponds to the preimage under f and is given by $f^{n*}[V] = [f^{-n}V]$. Brolin's Theorem in this case becomes:

Theorem [BS1,2]. *If V is an algebraic variety of codimension 1 in $\mathbf{C}^2$, then*

$$\lim_{n\to\infty} (d^{-n}) f^{n*}[V] = c\mu^+ \tag{12}$$

holds for some constant $c > 0$, with the convergence being taken in the sense of currents.

(Another version of this Theorem was recently given in [FS].) The proof of this Theorem is based on the Poincaré-Lelong formula for currents

$$[\{x = 0\}] = \frac{1}{2\pi} dd^c \log |x| .$$

Similarly, if $V = \{h = 0\}$ is an algebraic variety defined by a polynomial h, then $dd^c \log |h|$ is a constant (positive) multiple of the current of integration on V. In vague outline, the proof of the theorem proceeds by showing that

$$\lim_{n\to\infty} \log |h(f^n(x, y))| = \text{const. } G^+(x, y) \tag{13}$$

holds locally in L^1. It follows, then, by applying the Poincaré-Lelong identity, that we obtain the convergence of the desired currents.

N. Sibony proposed working directly with invariant measures. If S is a positive, closed, (1,1) current on $\mathbf{C}^2$, and if u is a bounded, psh function, then we may form the wedge product $dd^c u \wedge S$, which will be a (2,2) current, and which acts on a test form φ according to the formula

$$(dd^c u \wedge S)(\varphi) := \int u dd^c \varphi \wedge S .$$

This formula is, formally, just an integration by parts, and the integral on the right hand side is justified by representing S as a form with measure coefficients and performing a (usual) wedge product with the smooth (1,1) form $dd^c\varphi$. In this way, we define

$$\mu := \mu^+ \wedge \mu^- , \tag{14}$$

and it is immediate that

$$f^*\mu = \mu .$$

The two sets of main importance from the dynamical point of view are the chain recurrent set and the nonwandering set. A point p is *wandering* if there is a neighborhood U of p such that $U \cap f^n U = \emptyset$ for all $n \neq 0$, and the *nonwandering set* $\Omega(f)$ is the set of all nonwandering points. It is generally true that $\Omega(f) \subset R(f)$. It is easily seen, too, that the support of an invariant, finite measure is contained in the nonwandering set, i.e. spt $\mu \subset \Omega(f)$.

It has been conjectured by N. Sibony that the analogue of (3) holds for the measure μ in (14) and mappings of the form (8), i.e.:

$$\lim_{n\to\infty} d^{-n} \sum_{z\in \mathrm{Per}(n)} \delta_z = \frac{1}{4\pi^2}\mu . \tag{15}$$

With N. Sibony, we have developed the following formal argument which makes (15) plausible. First $G := \max\{G^+, G^-\}$ is the psh Green function of the set K. (See, for instance [Be] for relevant information.) A calculation shows that $(dd^c G)^2 = \mu$. If we set

$$u_n = (d^{-n}) \log |f^n(x, y) - f^{-n}(x, y)| ,$$

then

$$\lim_{n\to\infty} u_n = G \tag{16}$$

holds a.e. Further, a standard calculation (cf. [BT1]) shows that

$$(dd^c u_n)^2 = d^{-2n}(2\pi)^2 \sum_{\{a : f^n(a) = f^{-n}(a)\}} \delta_a , \tag{17}$$

which may be interpreted as

$$(dd^c u_n)^2 = (\#\,\mathrm{Per}(2n))^{-1}(2\pi)^2 \sum_{\{a\in \mathrm{Per}(2n)\}} \delta_a . \tag{18}$$

This information, however, is not sufficient to obtain (15), because (16) does not in general imply that the measures $(dd^c u_n)^2$ converge to $(dd^c G)^2$. (See [C1,2] and [L2] for examples of bad behavior of $(dd^c)^2$.) For this approach to work, the limit (16) needs to be taken in a stronger sense.

§ 5. Iteration of Disks

We let M denote a 1-dimensional, locally closed, complex submanifold of $\mathbf{C}^2$, and we study with the current of integration $[M]$ under pull-backs by the iterates of f. In order to remove any technical problems that might be introduced by the boundary of M, we let χ be a test function such that spt $\chi \cap M$ is compact, and we work with $\chi[M]$. The following was obtained in [BS3, Theorem 3].

Theorem. *Let M be as above, and suppose that one of the following holds:*

(i) M is an open subset of an algebraic curve X.

(ii) $M \subset J^+$.

If we set $c = \int_M \chi\mu^-|_M$, then

$$\lim_{n\to\infty}(d^{-n})f^{*n}(\chi[M]) = \frac{c}{4\pi^2}\mu^+ .$$

This result has two interesting consequences for the study of K^+. The first is

Theorem. *Let Ω be a connected component of* int K^+. *If $\Omega \cap J^- \neq \emptyset$, then $\partial\Omega = J^+$.*

In particular, if p is a sink for f, then $p \in J^-$.

Corollary. *If p is a sink for f, and if B is the basin of attraction of p, then $\partial B = J^+$.*

The second application concerns hyperbolic periodic (saddle) points. Let us note that Smillie [Sm] has shown that the map f has positive topological entropy, so by a result of Katok [K], f always has saddle points. Such points have stable and unstable manifolds which are complex submanifolds of $\mathbf{C}^2$. The following result confirms a conjecture of J.H. Hubbard.

Theorem. *If p is a saddle point for f, then the stable manifold $W^s(p)$ is a dense subset of J^+.*

§ 6. Recurrence

Let us make some remarks on the behavior of f on int K^+. A connected component Ω of int K^+ is said to be a *wandering component* of int K^+ if $\Omega \cap f^n(\Omega) = \emptyset$ for $n \neq 0$. It is an interesting open question at this time whether there can be a wandering component for f (cf. the list of questions of Milnor in [Bi]). If there are no wandering components, then all components are periodic. The proof of Sullivan [Su2] that there are no wandering domains in dimension one uses quasiconformal methods, and analogous techniques are not available in the present case.

Here we consider a component Ω which is *recurrent*, i.e. there is a point $z \in \mathbf{C}^2$ such that Ω contains a point of the ω-limit set $\omega(z)$, which is the set of accumulation points of the forward iterates of z. It is easily shown that if Ω is recurrent, then $f^m(\Omega) = \Omega$ for some $m \geq 1$.

Let $A \subset \mathbf{C}$ denote either the unit disk or an annulus $\{r_1 < |\zeta| < r_2\}$. We say that an imbedding $\varphi : A \to \mathbf{C}^2$ is *rotation* if there is an irrational number a such that

$$f(\varphi(\zeta)) = \varphi(e^{i\pi a}\zeta)$$

holds for all $\zeta \in A$. In this case we call $\varphi(A)$ a *rotation domain.*

Theorem. *Suppose that Ω is a connected component of* int K^+ *and that f decreases volume. If Ω is recurrent, then either*

(i) Ω is the basin of attraction of a sink.

(ii) Ω is the basin of a rotation domain, i.e. $\Omega = W^s(\varphi(A)) = \bigcup_{\zeta\in A} W^s(\varphi(\zeta))$.

The possiblity of case (ii) was also shown in [FS]. In the second case, it is in fact possible to linearize f on Ω. That is, there is a biholomorphic mapping $h : A \times \mathbf{C} \to \Omega$ such that $h \circ L = f \circ h$, where $L(\zeta, w) = (e^{ia\pi}\zeta, \beta w)$.

We remark that in either case it is evident that $\Omega \cap J^- \neq \emptyset$, so by §5 we have $\partial\Omega = J^+$.

The remaining possibility is that a component of Ω is periodic but not recurrent. Not much is known about this case. T. Ueda [U1,2] has analyzed such domains which occur at a parabolic fixed point.

§ 7. Hyperbolic Mappings

Up to now our discussion has applied equally well to all choices of f. Now we see what extra information we can derive in the special situation of hyperbolicity. This should be an interesting special case, much in analogy with the case of uniformly expanding polynomial mappings in one variable. We say that a set $\Lambda \subset \mathbf{C}^2$ is a *hyperbolic set* for a mapping $f : \mathbf{C}^2 \to \mathbf{C}^2$ if there are constants $C < \infty$ and $\lambda < 1$ and for each $z \in \Lambda$ there is a splitting of the tangent bundle into stable and unstable directions, i.e. ${\mathbf{C}^2}_z = E^s_z \oplus E^u_z$, and

$$|Df^n(z)v| \leq C\lambda^n|v| \quad \text{for } z \in E^s_z$$

$$|Df^{-n}(z)v| \leq C\lambda^n|v| \quad \text{for } z \in E^u_z .$$

For our class of mappings satisfying (8), we will say that f is *hyperbolic* if J is a hyperbolic set for f. Not all of these mappings are hyperbolic, and it would be of interest to have criteria for hyperbolicity. Examples of hyperbolic maps are given by maps of the form (4) if p is uniformly expanding on the Julia set, and $|a|$ is small. These mappings are rather easily seen to be hyperbolic (see [HO, FS], and [DN] for the real case). The only examples known at the moment are for $|a|$ small or $|c|$ large. See [HO] and [FS] for further properties of these maps.

The choice of J as the set on which f should have a hyperblic splitting is justified by the following result [BS3, Theorem 6].

Theorem. *The following are equivalent:*
(i) f has a hyperbolic splitting over the chain recurrent set.
(ii) f has a hyperbolic splitting over the nonwandering set.
(iii) f has a hyperbolic splitting over J.

Using elementary properties of hyperbolicity and the fact that the iterates of f are a normal family on *int* K^+, we obtain

Theorem [BS1]. *If f is hyperbolic, then the interior of K^+ consists of the basins of finitely many hyperbolic sink orbits $\{s_1, \ldots, s_k\}$.*

One of the reasons for considering the case of hyperbolic mappings is that stable manifolds always exist. The following holds for a smooth (not necessarily holomorphic) hyperboic mapping.

Stable Manifold Theorem. *Let Λ be a compact hyperbolic set for a smooth diffeomorphism f. For every point $x \in \Lambda$, $W^s(x)$ is an immersed submanifold of dimension equal to that of E^s. Further, $T_xW^s(x) = E^s_x$, and the intersection of $W^s(x)$ and $W^u(x)$ at x is transverse.*

In the holomorphic case, we obtain more.

Theorem. *If f is hyperbolic, then the leaves of the stable and unstable foliations are complex submanifolds which are biholomorphically equivalent to* $\mathbf{C}$.

The relation between $J^{\pm}$ and the stable/unstable foliations is as follows. It is not hard to show that $W^s(J) = \bigcup_{x\in J} W^s(x) \subset J^+$, and similarly for J^-. We even have:

Theorem [BS1]. *If f is hyperbolic and $|\det Df| \leq 1$, then $W^s(J) = J^+$. If $s_1, s_2, \ldots s_k$ are the sinks of f then $W^u(J) = J^- - \{s_1 \ldots s_k\}$.*

The unstable foliation cannot be extended through the sinks. In fact, Fornaess and Sibony [FS] show that there is no germ of a variety V_j with $s_j \in V_j \subset J^-$.

We let $\mathcal{F}^s$ ($\mathcal{F}^u$) denote the stable (unstable) foliation, and we describe what the local situation looks like. We may cover J^+ with open sets $U \subset J^+$ such that there is a coordinate system (x, y) such that $U = J^+ \cap \{|x|, |y| < 1\}$, and each leaf of $\mathcal{F}^s \cap U$ is a graph $\{y = \varphi(x) : |x| < 1\}$ for an analytic function φ. If we let T denote a transversal to the leaves of $\mathcal{F}^s \cap U$, then the leaves are parametrized by the set $E := T \cap U$. We note that with this choice of transversal, the measure $\mu^+|_T$ gives a measure on the space of leaves of $\mathcal{F}^s \cap U$. If T' and T'' are two transversals, then we may describe the space of leaves as E' and E''; and we have a homemomorphism $\chi : E' \to E''$, which is given by following a leaf from T' to T''. Following Ruelle and Sullivan [RS], we say that the family of measures $\{\mu^+|_T\}$ defines a *transversal measure* on $\mathcal{F}^s \cap U$ if it is consistent with the homeomorphisms $\{\chi\}$, i.e. if $\chi_*\mu^+|_{T'} = \mu^+|_{T''}$. In [BS1, Theorem 6.5] it is shown that when f is hyperbolic, then the family $\{\mu^+|_T\}$ does define a transversal measure on $\mathcal{F}^s$. This yields the following structure for μ^+ restricted to U, which may be written as $\mu^+ \llcorner U$:

$$\mu^+ \llcorner U = \int_{t\in E} [M_t]\, \mu^+|_T(dt),$$

where $[M_t] = [\{y = \varphi_t(x)\}]$ denotes the current of integration over the leaf of $\mathcal{F}^s \cap U$ passing through the point $t \in E \subset T$.

Using this local representation for $\mu^{\pm}$ and the fact that the wedge product of currents of integration gives the current of integration over the intersection, i.e. $[M^s] \wedge [M^u] = [M^s \cap M^u]$, we obtain

$$\mu^+ \wedge \mu^- = \int_{t'\in E^s} \mu^+|_{T^s}(dt') \int_{t''\in E^u} \mu^-|_{T^u}(dt'')\, [M^s_{t'} \cap M^u_{t''}],$$

which gives a sort of product structure to the measure μ.

As a consequence, we obtain the result that *if f is hyperbolic, then the support of μ is J*. This property of the support of μ yields the existence of periodic points:

Corollary. *If f is hyperbolic, then the periodic points for f are dense in J.*

Thus the conjecture of J.H. Hubbard that the periodic points for f are dense in J is verified in the special case of hyperbolic mappings.

Further applications of the theory of hyperbolic mappings yield the following (see [BS1, Corollary 7.9]) *μ is mixing, μ is the unique measure of maximal entropy, and μ describes the distribution of periodic points.* In fact mixing has recently been shown to hold for all maps (see [BS4]). It would be interesting know whether the other two properties continue to hold for more general mappings.

References

[Be] Bedford, E.: Survey of pluri-potential theory. Several complex variables. Proceedings of the Mittag-Leffler-Institute, 1987–88, J.-E. Fornaess (ed.), Princeton Univ. Press, 1991

[BS1] Bedford, E., Smillie, J.: Polynomial diffeomorphisms of $\mathbf{C}^2$: Currents, equilibrium measure and hyperbolicity. Invent. math. **103** (1991) 69–99

[BS2] Bedford, E., Smillie, J.: Fatou-Bieberbach domains arising from polynomial automorphisms. Indiana Univ. Math. J. (to appear)

[BS3] Bedford, E., Smillie, J.: Polynomial diffeomorphisms of $\mathbf{C}^2$: Stable manifolds and recurrence. Preprint

[BS4] Bedford, E., Smillie, J.: Polynomial diffeomorphisms of $\mathbf{C}^2$: Ergodicity, exponents and entropy of the equilibrium measure. Preprint

[BT] Bedford, E., Taylor, B.A.: The Dirichlet problem for a complex Monge-Ampère equation. Invent. math. **50** (1976) 543–571

[BC] Benedicks, M., Carleson, L.: The dynamics of the Hénon map. Preprint

[Bi] Bielefeld, B.: Conformal dynamics problem list

[Br] Brolin, H.: Invariant sets under iteration of rational functions. Ark. Mat. **6** (1965) 103–144

[C1] Cegrell, U.: On the discontinuity of the complex Monge-Ampère operator. Proceedings, Toulouse 1983. (Lecture Notes in Mathematics, vol. 1094.) Springer, Berlin Heidelberg New York

[C2] Cegrell, U.: Discontinuité de l'operateur de Monge-Ampère complexe. C.R. Acad. Sci. Paris. Sér. I Math. **296** (1983) 869–871

[DN] Devaney, R., Nitecki, Z.: Shift automorphisms in the Hénon mapping. Comm. Math. Phys. **67** (1979) 137–146

[DE] Dixon, P.G., Esterle, J.: Michael's problem and the Poincaré-Fatou-Bieberbach phenomenon. Bull. AMS **15** (1986) 127–187

[FM] Friedland, S., Milnor, J.: Dynamical properties of plane polynomial automorphisms. Ergodic Theory Dyn. Syst. **9** (1989) 67–99

[FS] Fornæss, J.-E., Sibony, N.: Complex Hénon mappings in $\mathbf{C}^2$ and Fatou Bieberbach domains. Preprint

[Hn1] Hénon, M.: A two-dimensional mapping with a strange attractor. Comm. Math. Phys. **50** (1976) 69–77

[Hn2] Hénon, M.: Numerical study of quadratic area preserving mappings, Q. Appl. Math. **27** (1969) 291–312

[Hr] Herman, M.: Recent results and some open questions on Siegel's linearization theorem of germs of complex analytic diffeomorphisms of $\mathbf{C}^n$ near a fixed point. VIIIth International Congress on Mathematical Physics (Marseille, 1986). World Scientific Publishing, Singapore 1987, pp. 138–184

[Hu] Hubbard, J.: The Hénon mapping in the complex domain. In: M. Barnsley and S. Demko (eds.), Chaotic Dynamics and Fractals. Academic Press, New York 1986, pp. 101–111

[HO] Hubbard, J., Oberste-Vorth, W.: Hénon mappings in the complex domain. In preparation

[K] Katok, A.: Nonuniform hyperbolicity and structure of smooth dynamical systems. Proceedings of the International Congress of Mathematicians, Warsaw 1983

[L1] Lelong, P.: Fonctions plurisousharmoniques et formes différentielles positives. Gordon and Breach, New York 1968

[L2] Lelong, P.: Discontinuité et annulation de l'opérateur de Monge-Ampère complexe. Séminaire P. Lelong-P. Dolbeault-H. Skoda (1983). (Lecture Notes in Mathematics, vol. 1028.) Springer, Berlin Heidelberg New York 1983, pp. 219–224

[Lo] Lopes, A.: Equilibrium measures for rational maps. Ergod. Theor. Dyn. Syst. **6** (1986) 393–399

[M] Milnor, J.: Non-expansive Hénon maps. Adv. Math. **69** (1988) 109–114

[O] Oberste-Vorth, R.: Complex horseshoes and the dynamics of mappings of two complex variables. Thesis, Cornell University 1987

[RR] Rosay, J.-P., Rudin, W.: Holomorphic maps from $\mathbf{C}^n$ to $\mathbf{C}^n$. Trans. AMS **310** (1988) 47–86

[RS] Ruelle, D., Sullivan, D.: Currents, flows, and diffeomorphisms. Topology **14** (1975) 319–327

[S] Shub, M.: Global Stability of Mappings. Springer, Berlin Heidelberg New York 1987

[Sm] Smillie, J.: The entropy of polynomial diffeomorphisms of $\mathbf{C}^2$. Ergod. Theor. Dyn. Syst. **10** (1990) 823–827

[Su1] Sullivan, D.: Cycles for the dynamical study of foliated manifolds and complex manifolds. Invent. math. **36** (1976) 225–255

[Su2] Sullivan, D.: Quasiconformal homeomorphisms and dynamics II

[T] Tortrat, P.: Aspects potentialistes de l'itération des polynômes. Séminaire de Théorie du Potentiel, Paris, no. 8. (Lecture Notes in Mathematics, vol. 1235.) Springer, Berlin Heidelberg New York 1987

[U1] Ueda, T.: Local structure of analytic transformations of two complex variables. J. Math. Kyoto Univ. **26** (1986) 233–261

[U2] Ueda, T.: Local structure of analytic transformations of two complex variables. II. Preprint

[Z] Zehnder, E.: A simple proof of a generalization of a Theorem by C.L. Siegel, J. Palis & M. do Carmo (eds.) Geometry and Topology. (Lecture Notes of Mathematics, vol. 597.) Springer, Berlin Heidelberg New York 1977, pp. 855–866

Precise Analysis of $\bar{\partial}_b$ and $\bar{\partial}$ on Domains of Finite Type in $\mathbb{C}^2$

Michael Christ *

Department of Mathematics, University of California, 405 Hilgard Ave.
Los Angeles, CA 90024, USA

0. Introduction

In the past four years or so marked progress has been achieved in understanding precise regularity properties of solutions of the $\bar{\partial}$ equation on smoothly bounded two-dimensional complex manifolds, and of $\bar{\partial}_b$ on three-dimensional CR manifolds, under hypotheses of pseudoconvexity and finite type. Hölder and L^p Sobolev estimates, sharp nonisotropic estimates in L^2 norms, and pointwise bounds for Szegö and Bergman kernels have been obtained in partially overlapping works by a number of authors. This article represents a summary of those developments, and of two applications.

The approach taken is that of canonical solutions and L^2 theory. Important developments have also occurred in the method of explicit solution via integral kernels, but are neglected here because of the author's lack of expertise. See [Be, Fo, Ra].

Some notation: L^p_s denotes the Sobolev space of all functions having s derivatives in L^p, for $1 < p < \infty$ and $s \geq 0$. Λ_α denotes the space of functions Hölder continuous of order $\alpha > 0$, with the usual convention when α is an integer [St].

1. Definitions

The setting for our first group of results is a C^∞, compact (real) three-dimensional manifold M without boundary. A CR structure on such a manifold is a smooth, complex one-dimensional subbundle $T^{0,1}$ of the complexified tangent bundle of M, satisfying $T^{0,1}_x \cap T^{1,0}_x = \{0\}$ for all $x \in M$, where $T^{1,0} = \overline{T^{0,1}}$. Let $B^{0,1}$ be the bundle dual to $T^{0,1}$. Canonically associated to the CR structure is a first-order partial differential operator $\bar{\partial}_b$, mapping functions to sections of $B^{0,1}$, defined by

$$\bar{\partial}_b(f)(v) = v(f)$$

for $v \in T^{0,1}_x$ for some x. Thus $\bar{\partial}_b f$ is a portion of the usual differential df.

* Alfred P. Sloan fellow. Research also supported by the Institut des Hautes Etudes Scientifiques and National Science Foundation.

Proceedings of the International Congress
of Mathematicians, Kyoto, Japan, 1990

Although the problems to be considered here have also a global aspect, the bulk of the analysis will take place in a small coordinate patch, about an arbitrary point x_0. Then $B^{0,1}$ trivializes and $\bar{\partial}_b$ may be regarded as a complex vector field, which we write as $\bar{\partial}_b = X + iY$, where X, Y are smooth real vector fields, linearly independent at every point. Since M has dimension three, $\bar{\partial}_b$ is not elliptic.

Fix a real vector field T which is linearly independent of X, Y at X_0. Define the Levi form λ as a real-valued function by

$$[X, Y](x) = \lambda(x)T + O(X, Y).$$

Definition 1.1. *M is said to be pseudoconvex at x_0 if λ does not change sign in some neighborhood of x_0.*

The usual distinction between pseudoconvexity and pseudoconcavity is lost here; by replacing T by $-T$ if necessary, we may assume henceforth that $\lambda \geq 0$.

Definition 1.2. *M is said to be of finite type at x_0 if X, Y satisfy the Hörmander condition that they, together with all their Lie brackets of all orders, should span the tangent space to M at x_0.*

M is said to be pseudoconvex (respectively of finite type) if it is pseudoconvex (respectively of finite type) at every point. Both definitions are independent of various arbitrary choices involved – for instance the choice of T. x_0 is a point of type m if commutators of length m, but no smaller length, suffice to span the tangent space. ($[X, Y]$ is said to have length two, $[X, [X, Y]]$ length three, and so on.) M of of type $m = \max_{x \in M} \operatorname{type}(x)$. An equivalent condition is that some differential monomial D in X, Y should satisfy $D\lambda(x_0) \neq 0$.

The principal examples of CR manifolds are boundaries of open domains in $\mathbb{C}^2$; $T_x^{0,1}$ is taken to be the space of anti-holomorphic tangent vectors to $\mathbb{C}^2$, which are tangent to M at x. Globally real analytic boundaries are always of finite type. In this case $\bar{\partial}_b f$ may be computed by extending f smoothly to $\mathbb{C}^2$, applying $\bar{\partial}$ to the extension, and restricting the resulting $(0, 1)$ form to M. Thus $\bar{\partial}_b$ annihilates the restriction to M of any holomorphic function, hence has globally an infinite-dimensional kernel. Consequently $\bar{\partial}_b$ cannot satisfy any subelliptic estimate, despite the Hörmander condition.

It is necessary to choose a particular solution of $\bar{\partial}_b u = f$ in order to have any regularity theory; one candidate is the solution of minimal norm (after fixing a Hermitian metric on $B^{0,1}$ and a volume form on M) [K3]. Thus one studies

$$\bar{\partial}_b u = f \quad \text{with} \quad u \perp \operatorname{Kernel}(\bar{\partial}_b).$$

2. L^2 Theory

Having fixed a metric and volume form, we may speak of L^2 sections of $B^{0,1}$; henceforth we shall use the symbol L^2 to refer either to sections or to functions. In order to have any satisfactory existence theory, one requires

Hypothesis 2.1. *$\bar{\partial}_b$ is assumed to have closed range in L^2.*

More explicitly, if $f \in L^2$ is in the closure of $\bar{\partial}_b(L^2) \cap L^2$, then $f = \bar{\partial}_b u$ for some $u \in L^2$, and upon choosing the solution u orthogonal to the kernel, one has $\|u\| \leq C\|f\|$.[1] This hypothesis is violated generically; see §7. However:

Theorem 2.2 [K2]. *Suppose that M is pseudoconvex and may be realized as the boundary (in the C^∞ sense) of a complex variety in some $\mathbb{C}^n$. Then $\bar{\partial}_b$ has closed range in $L^2(M)$.*

For the case of the boundary of an open domain see [BS].

The principal result in the L^2 theory is then

Theorem 2.3 [K3]. *Suppose that M is a three-dimensional CR manifold, pseudoconvex and of finite type $\leq m$, on which $\bar{\partial}_b$ has closed range in L^2. Then for any C^∞ functions φ, ψ satisfying $\varphi\psi \equiv \psi$, for any $s \geq 0$ and any $f \in L^2$, the unique solution u of $\bar{\partial}_b u = f$ orthogonal to Kernel $(\bar{\partial}_b)$ satisfies*

$$\|\psi u\|_{L^2_{s+\delta}} \lesssim \|\varphi f\|_{L^2_s} + \|f\| .$$

In this theorem and the next, $\delta = m^{-1}$.

Indeed being orthogonal to the kernel, $u = \bar{\partial}_b^* v$ where $\|v\| \lesssim \|f\|$, so this follows from the fundamental local result:

Theorem 2.4. *Suppose that $U \Subset U'$ are open sets and that $\bar{\partial}_b \bar{\partial}_b^* v = f$ on U'. Then*

$$\|\bar{\partial}_b^* v\|_{L^2_{s+\delta}(U)} \lesssim \|f\|_{L^2_s(U')} + \|v\|_{L^2(U')} .$$

A well-known analogue is that $(X^2 + Y^2)v = f$ implies the same conclusion for Xv and Yv.

The fundamental innovation in the proofs of Theorems 2.2 and 2.3 was a rather simple microlocal analysis[2]. In brief: Choose coordinates (x, y, t) in which $x_0 = 0$, $X(0) = \partial/\partial x$, $Y(0) = \partial/\partial y$, and $T \equiv \partial/\partial t$. Let $(\xi', \tau) \in \mathbb{R}^2 \times \mathbb{R}$ be dual variables. Near 0, $\bar{\partial}_b$ is elliptic where $|\tau| \leq C|\xi'|$. Let P^+, P^- be classical pseudodifferential operators of order zero, with symbols supported where $\tau > 0$ and $|\tau| \geq C|\xi'|$, or where $\tau < 0$ and $|\tau| \geq C|\xi'|$, respectively.

---

[1] $\|\cdot\|$, with no subscript, will always denote the L^2 norm.

[2] The same idea is found in [HN], in the context of $\Box_b$.

$$\|\bar{\partial}_b P^- u\|^2 \gtrsim \|XP^- u\|^2 + \|YP^- u\|^2 - C\|u\|^2,$$

modulo some error terms, by Gårding's inequality. The right-hand side controls $\|u\|^2_{L^2_\delta} - C\|u\|^2$, by the result of [H] and [K1], so $\bar{\partial}_b$ is subelliptic in the support of the symbol of P^-. $\bar{\partial}_b$ is not subelliptic in the support of the symbol of P^+, but a similar application of Gårding's inequality demonstrates that instead, $\bar{\partial}_b\bar{\partial}_b^*$ is subelliptic there. Thus one studies P^+v, where $u = \bar{\partial}_b^* v$. One cannot eliminate the microlocalization by simply analyzing the equation $\bar{\partial}_b\bar{\partial}_b^* v = f$, as $\bar{\partial}_b\bar{\partial}_b^*$ fails to be subelliptic in the support of P^-.

3. Underlying Geometry

Of fundamental importance in the refined analysis of $\bar{\partial}_b$ is a second geometric structure induced on M by the CR structure [FS, RS, NSW, Sa, FP]. A curve $\gamma : [0, r] \mapsto M$ is said to be admissible if γ is absolutely continuous, $|\gamma'(t)| \le 1$ for almost every t, and if for almost every t, $\gamma'(t) \in T^{0,1} \oplus T^{1,0}$. Henceforth M is always assumed to be connected. Define:

$$\varrho(x, y) = \inf\{r : \exists \text{ an admissible curve } \gamma : [0, r] \mapsto M \text{ with } \gamma(0) = x \text{ and } \gamma(r) = y\}. \tag{3.1}$$

It is an easy theorem that under the hypothesis of finite type, any two points can be joined by an admissible curve. Then ϱ becomes a metric, in the sense of point-set topology. However it is not a Riemannian metric, and $\varrho(x, y)$ may be as large as $c[\text{dist}(x, y)]^{1/m}$, where dist is a Riemannian distance and M is of type m.

Let $B(x, r)$ denote the open ball centered at x, with respect to ϱ. Let B_0 denote the unit ball in $\mathbb{R}^3$. Given $x \in M$ and $r > 0$, define $\Phi : B_0 \mapsto M$ by

$$\Phi_{x,r}(u) = \exp(ru_1 X + ru_2 Y + \delta(x, r)u_3 T)(x)$$

where

$$\delta(x, r) = \sum_{0 \le |D| \le m-2} r^{|D|+2} |D\lambda(x)| \tag{3.2}$$

and D ranges over all differential monomials in X, Y, with $|D|$ defined to be the number of factors of X, Y. Then $\Phi_{x,r}$ is a diffeomorphism for all small r. Define $\tilde{B}(x, r) = \Phi_{x,r}(B_0) \subset M$. Then there exists c such that $B(x, cr) \subset \tilde{B}(x, r) \subset B(x, c^{-1}r)$ for all x, r.

Proposition 3.1 [NSW]. $|\tilde{B}(x, r)| \sim r^2\delta(x, r)$.

Corollary 3.2. *The ordered triple* (M, ϱ, *volume form*) *is a space of homogeneous type, in the sense of the theory of singular integral operators.*

See [CW] for the definition, and also [Ch7] for a recent exposition.

Fixing x, r, pull X, Y back to B_0 by

$$(\hat{X}f)(u) = r \cdot (X\check{f})(\Phi_{x,r}u)$$

where $\check{f} = f \circ \Phi_{x,r}^{-1}$, and similarly for $\hat{Y}$. Pull $\bar{\partial}_b$ back in the same way.

Theorem 3.3 [NSW].

- *The coefficients of* $\hat{X}, \hat{Y}$ *are* C^∞ *functions of* $u \in B_0$, *uniformly in* x, r.
- *The CR structure on* B_0 *associated to* $(\bar{\partial}_b)^\frown$ *is of finite type, uniformly in* x, r.

The last assertion means that $\hat{X}$, $\hat{Y}$ satisfy the Hörmander condition in a uniform way, hence satisfy a subelliptic estimate completely independent of x, r. Thus after pulling back to B_0, one has *both* upper and lower bounds on $(\bar{\partial}_b)^\frown$, uniformly in x, r; the "rescaling" maps $\Phi_{x,r}$ thus provide a way to analyze $\bar{\partial}_b$ uniformly at all points and all scales, and the geometry defined by ϱ is precisely adapted to the study of operators such as $\bar{\partial}_b$ and $X^2 + Y^2$.

Define $\tilde{\nabla} f = (Xf, Yf)$ and define $\tilde{\nabla}^\alpha$ analogously. Combining Theorems 2.4 and 3.3 yields

Corollary 3.4. *For any* N *there exists* N' *with the following property: for any* $x \in M$, *any* $0 < r \leq 1$, *and any* v, g *satisfying*

$$\|\tilde{\nabla}^\alpha g\|_{L^2(\tilde{B}(x,r))} \leq r^{-|\alpha|} \qquad \forall |\alpha| \leq N',$$
$$\bar{\partial}_b \bar{\partial}_b^* v = g \qquad \text{in } \tilde{B}(x,r),$$
$$\|\bar{\partial}_b^* v\|_{L^2(\tilde{B}(x,r))} \leq r, \text{ and}$$
$$\|v\|_{L^2(\tilde{B}(x,r))} \leq r^2,$$

one has $\bar{\partial}_b^* v \in C^\infty(\tilde{B}(x,r))$ *and*

$$\|\tilde{\nabla}^\alpha \bar{\partial}_b^* v\|_{L^\infty(\tilde{B}(x,r/2))} \leq C_N r^{1-|\alpha|} |B(x,r)|^{-1/2} \qquad \forall |\alpha| \leq N.$$

Let it be stressed that C_N is independent of x, r. This result is one of the principal ingredients in the proof of Theorem 4.1 below, and explains why that theorem should hold.

4. Main Results for CR Manifolds

Throughout this section M is assumed to be a smooth, compact three-dimensional CR manifold without boundary, pseudoconvex and of finite type not exceeding m, and $\bar{\partial}_b$ is assumed to have closed range. In addition to regularity results for $\bar{\partial}_b$, a goal has been to elucidate the nature of certain operators associated to $\bar{\partial}_b$, as had been done previously in the strictly pseudoconvex case [FS, Fe1, BSj]. These are principally the Szegö projection, that is, the orthogonal projection of $L^2(M)$ onto the kernel of $\bar{\partial}_b$, and the relative solving operator G, which is bounded and linear on L^2 and is uniquely defined by the relations $\bar{\partial}_b G f = \pi f$ and $Gf \perp$ Kernel $(\bar{\partial}_b)$, where πf denotes the orthogonal projection of f onto the range of $\bar{\partial}_b$. [3] Denote

[3] It can be shown that the kernel and cokernel of $\bar{\partial}_b$ both have infinite dimension.

by $S(x,y), G(x,y)$ their distribution-kernels, which are C^∞ off of the diagonal as a consequence of Theorem 2.3.

Define $V(x,y) = |B(x,\varrho(x,y))|$. Let D denote any differential monomial in X, Y, acting in either or both of the variables x, y, and let ℓ be the number of factors of X, Y.

Theorem 4.1 [Ch2, FK2, Ch5]. $|DG(x,y)| \lesssim \varrho(x,y)^{1-\ell} V(x,y)^{-1} \qquad \forall x \neq y.$

On the level of operators $S = I - G \circ \bar{\partial}_b$, so one has also:

Corollary 4.2 [Ch1, NRSW2, FK2]. $|DS(x,y)| \lesssim \varrho(x,y)^{-\ell} V(x,y)^{-1} \qquad \forall x \neq y.$

The estimates for S and its derivatives, together with its L^2 boundedness, imply that it is a singular integral operator in the sense of the Calderón-Zygmund theory, with respect to the geometric structure on M imposed by ϱ and the volume form. One of the fundamental results of that theory is that L^2 boundedness implies L^p boundedness.

In the same way, L^p boundedness of $X \circ G$, $Y \circ G$ would follow from L^2 boundedness. A second fundamental result of singular integral theory is the $T(1)$ theorem [DJ], which provides a necessary and sufficient condition for L^2 boundedness, and whose hypotheses are often checkable in practice. In the present instance the main condition to be verified is "weak boundedness", and this follows from some of the ingredients of the proof of the pointwise estimates for the kernels. Thus one has the case $s = 0$ of:

Theorem 4.3 [Ch1]. *Let $p \in (1,\infty)$ and $s \geq 0$. Then $X \circ G$ and $Y \circ G$ are bounded from L^p_s to L^p_s. Moreover $G : L^p_s \mapsto L^p_{s+m^{-1}}$, and $S : L^p_s \mapsto L^p_s$.*

For $s > 0$ one needs also to commute differential operators past G and S; this may be accomplished via microlocal arguments using standard pseudodifferential operators as in [FK1, Ch1, Ch5] or by the formalism of [CNS2]. Singular and fractional integral operators such as S, G are studied systematically in [NRSW2, CNS2].

Corollary 4.4 [Ch1, Ch5]. *If $U \Subset U' \subset M$ are open, $\bar{\partial}_b u = f$ and $u = \bar{\partial}_b^* v$ in U' for some $v \in L^2$, and if $f \in L^p_s(U')$, then $u \in L^p_{s+m^{-1}}(U)$.*

There are analogous results in the scale of Hölder spaces:

Theorem 4.5 [FK1]. *If $U \Subset U' \subset M$ are open, $\bar{\partial}_b u = f$ and $u = \bar{\partial}_b^* v$ in U' for some $v \in L^2$, if $\alpha > 0$ and if $f \in \Lambda_\alpha(U')$, then $u \in \Lambda_{\alpha+m^{-1}}(U)$* [4].

---

[4] This is proved with a nonstandard definition of Λ_α for $\alpha \in \mathbb{Z}$ in [FK1], but follows with the usual definition [St] from the machinery of [NRSW2, CNS2].

Another fundamental operator is $\Box_b^1 = \bar{\partial}_b\bar{\partial}_b^*$; associated to it is a relative solving operator whose distribution-kernel satisfies estimates like those for G, but with a power of $r^{2-\ell}$ [CNS2]. It has mapping properties like those of G, but gains two derivatives in the X, Y directions instead of one.

Basic difficulties in the proofs of these results are:

- There is no standard pseudodifferential calculus which permits inversion (modulo the kernel) of operators such as $\bar{\partial}_b$; one is led into classes $S^k_{\varrho,\delta}$ with $\varrho < \delta$.
- There is no single model by which one may approximate, as by the Heisenberg group in the strictly pseudoconvex case; and the natural model hypersurfaces $\{\text{Im}(z_2) = P(z_1)\}$ where P is a subharmonic, nonharmonic polynomial seem practically as difficult to analyze as the general case.
- The nonexistence of good holomorphic support functions [KN] renders the construction of explicit parametrices difficult.
- The lifting technique of [RS, NSW] and [Sa] is not (directly) applicable.
- The operators $P^{\pm}$ of the microlocal analysis sketched earlier do not preserve the type of estimates sought, thus introducing spurious singularities. [5]

The first proofs of these results were rather involved, but subsequently substantial simplification has occurred. See [Ch5].

5. The $\bar{\partial}$-Neumann Problem

Let $\Omega \Subset \mathbb{C}^2$ be pseudoconvex and smoothly bounded. Ω is said to be of finite type if $\partial\Omega$ is of finite type in the sense already defined; an equivalent notion of finite type (in $\mathbb{C}^2$) is that $\partial\Omega$ should have a bounded order of contact with all complex curves.

Fix a Riemannian metric on a neighborhood of $\bar{\Omega}$, let r be the signed geodesic distance to $\partial\Omega$, and set $\Box = \bar{\partial}\bar{\partial}^* + \bar{\partial}^*\bar{\partial}$, on $(0,1)$ forms. The $\bar{\partial}$–Neumann problem is the boundary value problem

$$\Box u = f \text{ on } \Omega \text{ with } \begin{cases} u \lrcorner \bar{\partial} r = 0 & \text{on } \partial\Omega \\ \bar{\partial} u \lrcorner \bar{\partial} r = 0 & \text{on } \partial\Omega \end{cases}$$

for $(0,1)$ forms on Ω, where $\lrcorner$ denotes the interior product of forms with respect to the Riemannian metric. $\Box$ is an elliptic second-order system, but the boundary conditions are non-coercive. Let N be the Neumann operator, which inverts $\Box$ on $(0,1)$ forms, modulo its finite-dimensional kernel. N sends $L^2_s(\Omega)$ to $L^2_{s+(2/m)}(\Omega)$ [K1, RS] for all $s \geq 0$, assuming $\partial\Omega$ to be of finite type m.

Precise results on the regularity properties of N in various function spaces have been obtained by [CNS2], generalizing previously known results for the strictly pseudoconvex case [GS]. One of the main steps is a computation of N in terms of simpler operators, as follows. Let $\bar{\omega}_1, \bar{\omega}_2$ be C^∞ and constitute an orthonormal basis for the space of $(0,1)$ forms at each point in a neighborhood

[5] This is particularly evident in [Ch1] and [FK1].

of $\partial\Omega$, with $\bar{\omega}_2 = \sqrt{2}\bar{\partial} r$. Let $\{\bar{L}_1, \bar{L}_2\}$ be a dual basis of anti-holomorphic vector fields. Let G solve for $(0,1)$ forms f:

$$\Box G f = f \text{ on } \Omega \qquad \text{with } Gf = 0 \text{ on } \partial\Omega\,;$$

write $Gf = G_1 f + G_2 f$ where $G_i f$ is a scalar multiple of $\bar{\omega}_i$. Let R denote restriction to $\partial\Omega$. Let P solve for $(0,1)$ forms f defined on $\partial\Omega$ (that is, f is a scalar multiple of $\bar{\omega}_1$)

$$\begin{cases} \Box P f = 0 & \text{on } \Omega \text{ to infinite order at } \partial\Omega \\ Pf \equiv f & \text{on } \partial\Omega. \end{cases}$$

Finally let K denote the relative solution operator for $\bar{\partial}_b \bar{\partial}_b^* = \Box_b^1$ on $\partial\Omega$. By a ψdo of order n we shall mean a classical pseudodifferential operator with symbol in $S^n_{1,0}$.

Theorem 5.1 [CNS2]. *$N = G + P \circ (\Box^- K \Gamma^+ + Q) \circ R \circ \bar{L}_2 G_1$ plus lower order terms.*

Here

- Q is a ψdo on $\partial\Omega$ of order -1
- Γ^+ is a ψdo of order 0 which microlocalizes to the region in phase space in which $\bar{\partial}_b$ is not subelliptic
- $\Box^-$ is a ψdo of order $+1$ which may be explicitly computed modulo a symbol of order -1.

The "lower order" terms are of a form similar to the principal one, so that in practice, any mapping property which one can establish for the main term, may also be proved for the lower-order part. The only ingredient which is not computable and quite well understood is K, and one must make do with properties established for it using the results of §4.

Combining Theorem 5.1 with results on K, plus commutation properties of K with vector fields, results in

Theorem 5.2 [CNS2]. *Let q be a polynomial of degree at most 2 in $L_1, \bar{L}_1$. Then*

$$q(L_1, \bar{L}_1) \circ N : L^p_s(\Omega) \mapsto L^p_s(\Omega) \qquad \forall\, p \in (1,\infty),\ s \geq 0$$
$$N : \Lambda_\alpha \mapsto \Lambda_{\alpha + \frac{2}{m}} \qquad \forall \alpha > 0\,.$$

Also $N : L^\infty \mapsto \Lambda_\alpha$ for all $\alpha < \frac{2}{m}$.

Let $B : L^2(\Omega) \mapsto L^2(\Omega)$ be the Bergman projection onto the holomorphic functions. If $\bar{\partial} f = 0$, then $\bar{\partial}(\bar{\partial}^* N f) = f$, so that $\bar{\partial}^* N$ solves the $\bar{\partial}$ equation and thus is of much interest.

Corollary 5.3 [CNS2].

$$\begin{aligned}
&\bar\partial^* N,\ L_1 \circ \bar\partial^* N,\ \bar L_1 \circ \bar\partial^* N : L^p_s(\Omega) \mapsto L^p_s(\Omega) \qquad \forall\, p \in (1,\infty),\ s \geq 0\\
&\bar\partial^* N : \Lambda_\alpha(\Omega) \mapsto \Lambda_{\alpha+m^{-1}}(\Omega) \qquad \forall \alpha > 0\\
&\bar\partial^* N : L^\infty(\Omega) \mapsto \Lambda_\beta(\Omega) \qquad \forall \beta < m^{-1}\\
&B : L^p_s(\Omega) \mapsto L^p_s(\Omega) \qquad \forall\, p \in (1,\infty),\ s \geq 0\,.
\end{aligned}$$

In order to formulate pointwise estimates on the Bergman kernel $B(z,w)$, extend ϱ to $\bar\Omega \times \bar\Omega$ by:

$$\varrho(z,w) = \max\Big(\varrho(\pi(z),\pi(w)),\ \mu(z),\ \mu(w)\Big),$$

where μ is defined by the relation

$$\delta(\pi(x),\mu(x)) = r(x)\,, \tag{5.1}$$

and where $\pi(x)$ denotes the point of $\partial\Omega$ closest to x, and δ is the invariant defined in (3.2). In the strictly pseudoconvex case, $\mu \sim r^{1/2}$, but on domains of finite type m one has only $c_1 r^{1/2} \leq \mu \leq c_2 r^{1/m}$. Let D be any differential monomial in $L_1, \bar L_1, L_2, \bar L_2$, with each factor permitted to act in either of the variables z, w, with k factors of $L_1, \bar L_1$, and ℓ factors of $L_2, \bar L_2$.

Theorem 5.4 [Mc, NRSW2]. $|DB(z,w)| \leq C_{k,\ell}\, \varrho(z,w)^{-2-k} \delta(\pi(z),\varrho(z,w))^{-2-\ell}$.

6. Higher Dimensions

The question of Hölder and L^p estimates and the nature of the Szegö and Bergman kernels is at present poorly understood in higher dimensions; what is clear is that everything is more subtle. The notion of finite type is itself more recondite. A boundary point x is said [D] to be of finite type m if no one-dimensional complex variety in the ambient space has order of contact greater than m with the boundary at x. [6] It is a theorem that the set of boundary points of finite type is open [D]. Given pseudoconvexity, finite type is a necessary [C1] and sufficient [C2] condition for the $\bar\partial$–Neumann problem to satisfy a subelliptic estimate. The proofs of these assertions are several orders of magnitude more difficult than in the two-dimensional case.

Question. *Let Ω be pseudoconvex and of finite type. Does there exist $\varepsilon > 0$ such that the Neumann operator maps $\Lambda_\alpha(\bar\Omega)$ to $\Lambda_{\alpha+\varepsilon}(\bar\Omega)$ and L^∞ to Λ_ε?*

The principal difficulty in higher dimensions stems from the nature of $\Box_b$ as a system, rather than a scalar equation. There has been definite progress in the simpler case where the Levi form of Ω is diagonalizable [M1, M2, FKM].

[6] In $\mathbb{C}^n$ for $n > 2$, this is not equivalent to the natural condition in terms of vector fields.

However certain negative as well as positive results have been obtained, so that even the diagonalizable case is more subtle than might have been anticipated.

In $\mathbb{C}^3$ consider a domain $\{\mathrm{Im}(z_3) > |z_1|^{2k} + |z_2|^2\}$, for $k = 2, 3, \ldots$. Identify the boundary with $\mathbb{C}^2 \times \mathbb{R}$ via the map $(z_1, z_2, t) \mapsto (z_1, z_2, t + i(|z_1|^{2k} + |z_2|^2))$. Set

$$L_1 = \frac{\partial}{\partial z_1} + ik\bar{z}_1|z_1|^{2k-2}\frac{\partial}{\partial t}, \qquad L_2 = \frac{\partial}{\partial z_2} + i\bar{z}_2\frac{\partial}{\partial t}, \qquad T = \frac{\partial}{\partial t}.$$

Then after being pulled back to $\mathbb{C}^2 \times \mathbb{R}$, $\Box_b := \bar{\partial}_b\bar{\partial}_b^* + \bar{\partial}_b^*\bar{\partial}_b$ on $(0,1)$ forms becomes

$$\Box_b = \begin{pmatrix} \mathscr{L} & 0 \\ 0 & \bar{\mathscr{L}} \end{pmatrix} \qquad \text{where } \mathscr{L} = -(\bar{L}_1 L_1 + L_2\bar{L}_2).$$

Thus one need only analyze $\mathscr{L}^{-1}$.

What geometric structure should underlie the analysis? Write $\bar{L}_j = X_j + iY_j$. Let ϱ denote the metric defined in terms of X_1, Y_1, X_2, Y_2 as in (3.1); one might hope that $\mathscr{L}^{-1}$ should behave as an operator smoothing of order two with respect to the balls associated to ϱ, as for the relative fundamental solution for $\Box_b$ in the two-dimensional case. This is false. Let $\lambda(z_1, z_2, t) = |z_1|^{2k-2}$ be the degenerate eigenvalue of the Levi form, up to a constant factor. Let $\tilde{\varrho}$ be the metric associated to the weaker, but still subelliptic, second-order operator $\tilde{\mathscr{L}} = -(X_1^2 + Y_1^2 + X_2\lambda X_2 + Y_2\lambda Y_2)$ according to the theory of [FP]. One has $\tilde{\varrho} \geq \varrho$, but no majorization $\tilde{\varrho} \leq C\varrho$ near where λ vanishes. Denote by B_ϱ and $B_{\tilde{\varrho}}$ the balls with respect to the two metrics. Define $V(z,w) = |B(z, \varrho(z,w))|$, $\tilde{V}(z,w) = |B(z, \tilde{\varrho}(z,w))|$. Let $K(z,w)$ denote the distribution-kernel for $\mathscr{L}^{-1}$, and S the Szegö kernel. Here $z, w \in \mathbb{C}^2 \times \mathbb{R}$.

Theorem 6.1 [M1].

- $|K(z,w)| \leq C\left(\tilde{\varrho}(z,w)^2/\tilde{V}(z,w)\right) \cdot \log\left(2 + \frac{\tilde{\varrho}(z,w)}{\varrho(z,w)}\right)$.
- *The last estimate becomes false if the logarithmic factor is removed.*
- $|S(z,w)| \leq C\tilde{V}(z,w)^{-1}$.
- $|L_2\bar{L}_2 K(z,w)| \leq CV(z,w)^{-1}$.

It is primarily $\tilde{\varrho}$, rather than ϱ, which figures in the first two estimates. Two further surprises: the extra singularity present in K disappears when appropriate derivatives are applied to it to yield S, while application of $L_2\bar{L}_2$ yields a kernel which satisfies natural estimates with respect to ϱ rather than $\tilde{\varrho}$.

Consider a pseudoconvex domain $\Omega \Subset \mathbb{C}^{n+1}$ of finite type, on which the Levi form has at most one degenerate eigenvalue at each boundary point. Fix smooth anti-holomorphic vector fields $\bar{L}_1, \ldots \bar{L}_n$ which form a basis for the anti-holomorphic tangent space to $\partial\Omega$ in some open set, with respect to which basis the Levi form is diagonalized, and such that the Levi form restricted to the span of $\bar{L}_2, \ldots \bar{L}_n$ is nondegenerate. Let $\tilde{\varrho}$ be the metric associated to the [FP]-type operator $\tilde{\mathscr{L}} = -(X_1^2 + Y_1^2 + \sum_{j\geq 2}[X_j\lambda X_j + Y_j\lambda Y_j])$, where λ denotes the degenerate eigenvalue. Define $\tilde{V}$ in terms of $\tilde{\varrho}$ as usual.

Theorem 6.2 [M2]. *Let Ω be pseudoconvex and of finite type, and suppose that its Levi form has at most one degenerate eigenvalue. Then the Szegö kernel on $\partial\Omega$ satisfies*

$$|S(z,w)| \leq C\tilde{V}(z,w)^{-1} \qquad \forall z \neq w \in \partial\Omega\,.$$

Moreover the appropriate bounds hold for derivatives, so that S is a singular integral operator with respect to the structure of a space of homogeneous type defined by $\tilde{\varrho}$. Thus it extends to an operator bounded on L^p for all $p \in (1,\infty)$; it also improves Hölder classes.

It is easy to see via microlocal analysis why two different geometric structures ought to come into play. Let $(\zeta,\tau) \in \mathbb{R}^4 \times \mathbb{R}$ be dual variables. Then microlocally where $\tau > 0$ and $|\tau| \geq C|\zeta|$, one has for any f:

$$\begin{aligned}\langle \mathscr{L}P^+f,\, P^+f\rangle &= \|L_1P^+f\|^2 + \|\bar{L}_2P^+f\|^2 \\ &\sim \|L_1P^+f\|^2 + \|\bar{L}_1P^+f\|^2 + \langle\lambda|T|P^+f,\, P^+f\rangle + \|\bar{L}_2P^+f\|^2\end{aligned}$$

modulo lower-order terms. One might also hope to estimate L_2P^+f, but

$$\|\bar{L}_2P^+f\|^2 - \|L_2P^+f\|^2 = \langle[\bar{L}_2, L_2]P^+f,\, P^+f\rangle \sim -\langle|T|P^+f,\, P^+f\rangle$$

(modulo a constant factor and zero-order terms) is potentially large and negative, and is not controlled by $\langle\mathscr{L}P^+f,\, P^+f\rangle$. The best that one can control is $\langle\lambda[\bar{L}_2, L_2]P^+f,\, P^+f\rangle$, so that

$$\begin{aligned}\langle \mathscr{L}P^+f,\, P^+f\rangle &\sim \|L_1P^+f\|^2 + \|\bar{L}_1P^+f\|^2 + \|\sqrt{\lambda}L_2P^+f\|^2 + \|\bar{L}_2P^+f\|^2 \\ &\sim \langle\tilde{\mathscr{L}}P^+f,\, P^+f\rangle + \|\bar{L}_2P^+f\|^2\,,\end{aligned}$$

again modulo lower-order terms. Thus $\tilde{\mathscr{L}}$ enters the picture. On the other hand, in any compact subset of $\mathbb{C}^2 \times \mathbb{R}$, λ is bounded above and one obtains

$$\langle\mathscr{L}P^-f,\, P^-f\rangle \sim \langle-(X_1^2+Y_1^2+X_2^2+Y_2^2)P^-f,\, P^-f\rangle\,.$$

At first the estimates of Theorem 6.2 for S might appear improbable, given the extra singularity exhibited by K. However, the Szegö projection is equal to the orthogonal projection onto the intersection of the kernels of $\bar{L}_1$ and $\sqrt{\lambda}\bar{L}_2$, and is thus *heuristically* expressible also as $I - \tilde{\bar\partial}_b^{\,*}\tilde{\Box}_b^{-1}\tilde{\bar\partial}_b$ where $\tilde{\bar\partial}_b f = (\bar{L}_1f)\bar\omega_1 + \sqrt{\lambda}(\bar{L}_2f)\bar\omega_2$ and $\tilde{\bar\partial}_b^{\,*}, \tilde{\Box}_b^{-1}$ are the associated operators. (The same holds with $\tilde{\bar\partial}_b$ defined to be $(\bar{L}_1f)\bar\omega_1 + \gamma(\bar{L}_2f)\bar\omega_2$, for any constant $\gamma \in \mathbb{R}$.) Thus the Szegö kernel can be described entirely in the "$\tilde{\varrho}$ – category", and Theorem 6.2 becomes less surprising; this is part of the idea in [M2].

Concerning Hölder estimates one has the following:

Theorem 6.3 [FKM]. *Let Ω be smoothly bounded, pseudoconvex and of finite type m. Assume that its Levi form is (smoothly) diagonalizable in a neighborhood of each boundary point. Then for any $0 < \alpha, \beta \in \mathbb{R}\backslash\mathbb{Z}$ satisfying $\beta < \alpha + 2m^{-1}$, for any $(0,1)$–form f on $\partial\Omega$,*

$$\|f\|_{\Lambda_\beta} \lesssim \|\Box_b f\|_{\Lambda_\alpha} + \|f\|_{L^2}.$$

Moreover for $\beta < 2/m$, $\|f\|_{\Lambda_\beta} \lesssim \|\Box_b f\|_{L^\infty}$.

The proof is far too involved to be summarized here. Corresponding results hold for the $\bar{\partial}$–Neumann problem, and for the Szegö projection. [7]

7. Two Applications

A. Characterization of Zero Varieties for the Nevanlinna Class

Let $\Omega \Subset \mathbb{C}^2$ be smooth, pseudoconvex and of finite type. Let r be a defining function, $\Omega = \{r < 0\}$. Let $\Omega_\varepsilon = \{r < -\varepsilon\}$. The Nevanlinna class is the set of all holomorphic functions on Ω satisfying $\sup_\varepsilon \int_{\partial\Omega_\varepsilon} \log^+ |f| < \infty$. The zero variety $Z \subset \Omega$ of a holomorphic function is said to satisfy the Blaschke condition if $\int_Z |r(z)|\, d\sigma(z) < \infty$, where σ is the induced volume element on Z.

Theorem 7.1 [CNS2]. *Let $Z \subset \Omega$ be a complex subvariety of codimension one, which satisfies the Blaschke condition. Then Z is the zero variety of a function in the Nevanlinna class.*

The converse was previously known and follows from Green's theorem. In the strictly pseudoconvex case this was proved earlier [He1, He2, Sk].[8]

The proof proceeds by the method of [L], in which one solves $i\partial\bar{\partial}\Theta = [Z]$ where $[Z]$ denotes the $(1,1)$ current defined by integration over Z. The desired Nevanlinna class function f will satisfy $\Theta = \log|f|$, so that one seeks a solution $\Theta \in L^1(\partial\Omega)$. Control of the L^1 norm at the boundary comes from:

Theorem 7.2 [CNS2]. $\|\bar{\partial}^* N f\|_{L^1(\partial\Omega)} \leq C\|f\|_{L^1(\Omega)} + \|\frac{\mu}{r} f \wedge \bar{\partial} r\|_{L^1(\Omega)}$.

The invariant $\mu \gg r$ is defined in (5.1).

In order to apply Theorem 7.2, one needs an estimate of [BC] to the effect that the Blaschke condition automatically implies a formally stronger estimate for $[Z]$, generalizing a result of [Mal] for the strictly pseudoconvex case. Moreover the higher is the type of a given boundary point, the *stronger* becomes this estimate near that point, in such a fashion that the second term on the right-hand side in the theorem is exactly under control. The full strength of the precise estimates for the Neumann operator is used; no loss of ε in regularity can be afforded.

B. Embeddability of Compact Three-Dimensional CR Manifolds.

Throughout this section M denotes a smooth, compact three-dimensional CR manifold without boundary. A CR embedding of M in $\mathbb{C}^n$ is a smooth embedding,

[7] It is not known to this author whether one may take $\beta = \alpha + 2m^{-1}$, nor whether similar L^p estimates hold.

[8] An additional hypothesis imposed on Z in [Sk] is satisfied in this case since $H^2(\Omega, \mathbb{C}) = 0$.

all of whose component functions are CR, that is, are annihilated by $\bar{\partial}_b$. There exist arbitrarily small, real analytic perturbations of the standard structure on S^3 which are not globally embeddable [Ro]. There exist strictly pseudoconvex CR structures, with the property that every function which is CR in a neighborhood of the origin is constant there [Ni].

There has been some interesting recent work in the negative direction. [BE] have proved that generic small perturbations of the standard structure on S^3 are non-embeddable, even in the real-analytic category. $\square_b^0 = \bar{\partial}_b^* \bar{\partial}_b$ is a self-adjoint operator which has (in the embeddable case) 0 as an eigenvalue of infinite multiplicity; the set of all positive eigenvalues is a locally discrete subset of the open interval $(0, \infty)$. Closed range fails if and only if there is a sequence of nonzero eigenvalues tending to 0, and [BE] show that generic perturbations create such eigenvalues.

[F] has given a simple construction of non-embeddable structures. One finds a simply connected, strictly pseudoconvex two-dimensional complex manifold Ω, whose boundary admits a nontrivial finite cover, M, and one pulls the CR structure from $\partial\Omega$ back to M. If embedded, M must bound a complex manifold, and the key is to prove that such a manifold would then be a nontrivial cover of Ω. [9]

A very simple construction of non-embeddable (global and local) examples is in [R]. Further pathologies are exhibited there.

A combination of results of [Bu] and [K3] yields embeddability provided $\bar{\partial}_b$ has closed range in L^2. Closed range would be a consequence of a subelliptic estimate for $\square_b^0$; but $\square_b^0$ is not subelliptic in dimension three. [10] In view of Theorem 2.2, closed range is equivalent to embeddability for strictly pseudoconvex three-dimensional CR manifolds. This result extends to the case of finite type:

Theorem 7.3 [Ch4]. *A compact, pseudoconvex three-dimensional manifold without boundary and of finite type is embeddable, provided $\bar{\partial}_b$ has closed range in L^2.*

To prove this one must show that there exist sufficiently many CR functions to provide local coordinates, and to separate points. In the strictly pseudoconvex case the latter is proved as follows: given $x \neq y \in M$, one first constructs a one-parameter family $\{g_t : 0 < t < 1\}$ of functions satisfying $g_t(x) = 1$, $g_t(y) = 0$, and $\|\bar{\partial}_b g_t\|_{C^N} \to 0$ for any N. This is easily done by a formal power series argument, which relies on the existence of holomorphic support functions. From [K3] it follows that for sufficiently large ℓ, one can solve $\bar{\partial}_b u_t = \bar{\partial}_b g_t$ with $\|u_t\|_{C^0} \leq C\|\bar{\partial}_b g_t\|_{C^\ell}$. Then $f_t = g_t - u_t$ is CR, and $f_t(x) \to 1$ while $f_t(y) \to 0$.

There exist [KN] real analytic pseudoconvex hypersurfaces of finite type in $\mathbb{C}^2$ for which any local holomorphic support function at a point x must vanish

[9] There do not exist complex manifolds of dimension greater than two with the required property.

[10] It is in higher dimensions, given strict pseudoconvexity, so that higher-dimensional compact, strictly pseudoconvex CR manifolds without boundary are always embeddable [Bt].

to infinite order; such functions cannot be used to construct adequate g_t. To circumvent this the strategy was to:

- show that on model hypersurfaces $\{\mathrm{Im}(z_2) = P(z_1)\}$, where P is any homogeneous subharmonic, nonharmonic real-valued polynomial, there exists a CR function φ in the Schwartz class such that $\varphi(0) \neq 0$,
- show that at any point of a pseudoconvex CR manifold of finite type, there exists a close approximation by one of the model domains,
- pull back a one-parameter family of dilates of φ from the model to $\{g_t\}$ on the given manifold, in such a way that $g_t(x) = 1$, $g_t(y) = 0$, and $\|\bar{\partial}_b g_t\|_{L^q} \to 0$ for some finite q, and
- solve $\bar{\partial}_b u_t = \bar{\partial}_b g_t$ with $\|u_t\|_{C^0} \leq C\|\bar{\partial}_b g_t\|_{L^q}$.

Then $g_t - u_t$ again does the job, for small t. In the last step one does not know how to make $\|\bar{\partial}_b g_t\|_{C^N}$ small for large N, and the L^p regularity theory is required to complete the proof. [11]

8. Analytic Hypoellipticity

In analogy with the C^∞ theory we say that $\bar{\partial}_b$ is (relatively) analytic hypoelliptic on M if whenever $\bar{\partial}_b u$ is real analytic in an open set V and $u = \bar{\partial}_b^* v$ for some $v \in L^2$ in V, u must be analytic in V.

Theorem 8.1 [G]. *Let $\Omega \Subset \mathbb{C}^2$ be strictly pseudoconvex with real analytic boundary. Then $\bar{\partial}_b$ is (relatively) analytic hypoelliptic, and the Szegö kernel is analytic off of the diagonal on $\partial\Omega \times \partial\Omega$.*

This issue is not currently well understood for domains of finite type, but there are some counterexamples: Let $\Omega = \{\mathrm{Im}(w) = [\mathrm{Re}(z)]^m\}$ for $m = 4, 6, 8, \ldots$.

Theorem 8.2 [CG]. *$\bar{\partial}_b$ is not (relatively) analytic hypoelliptic on $\partial\Omega$. Moreover the Szegö kernel is not analytic off the diagonal, nor is the Bergman kernel, restricted to $\partial\Omega \times \partial\Omega$ minus the diagonal.*

In fact one has only Gevrey regularity of order m for general u as above, and in particular for the Szegö kernel. The proof rests on an explicit, though not entirely transparent, formula for the Szegö kernel in [N].

In higher dimensions we retain the standard definition of analytic hypoellipticity. Let $n > 2$ and adopt coordinates $(z', \zeta, w) \in \mathbb{C}^{n-2} \times \mathbb{C} \times \mathbb{C}$.

Theorem 8.3 [CG]. *$\Box_b$ is not analytic hypoelliptic on $\{\mathrm{Im}(w) = |z'|^2 + [\mathrm{Re}(\zeta)]^m\}$ for $m = 4, 6, 8, \ldots$. Nor is the Szegö kernel analytic off the diagonal.*

---

[11] The existence of continuous peak functions may also be inferred from this argument; a stronger result was obtained some time ago in [BF], and another proof has been obtained in [FSi].

Various subelliptic second-order differential operators with analytic coefficients are known not to be analytic hypoelliptic, but these examples seem to be new. An equivalent formulation is that the second-order operator $\bar{\partial}_b \circ \bar{\partial}_b^*$ fails to be analytic hypoelliptic, in the standard sense, microlocally in a cone in which it is C^∞ hypoelliptic.

Conjecture 8.4 [CG]. *On a hypersurface $\{\mathrm{Im}(z_2) = P(z_1)\}$, where P is a subharmonic, nonharmonic polynomial, $\bar{\partial}_b$ is (relatively) analytic hypoelliptic if and only if all zeroes of $\triangle P$ are isolated in $\mathbb{C}^1$.*

9. Weighted Estimates for $\bar{\partial}$

The same philosophy as for the main results concerning $\bar{\partial}_b$ may be applied in other situations. Let $\varphi : \mathbb{C} \mapsto \mathbb{R}$ be subharmonic. Let $\mu = \triangle\varphi$, a positive, locally finite measure. Assume that

(1) μ is a doubling measure: $\mu(B(z,2r)) \le C\mu(B(z,r)) \quad \forall z \in \mathbb{C},\ r > 0.$
(2) There exists $\delta > 0$ such that $\mu(B(z,1)) \ge \delta > 0$ for all z.

(1) should be viewed as a finite type condition; (2) replaces the usual assumption that $\triangle\varphi \ge \delta$. Then for each $f \in L^2(\mathbb{C}, e^{-2\varphi})$ there exists $u \in L^2(\mathbb{C}, e^{-2\varphi})$ satisfying $\bar{\partial}u = f$ [Ch4, Ch6]. Let R be the bounded linear operator which assigns to each f the solution u with minimal norm in $L^2(\mathbb{C}, e^{-2\varphi})$, and denote also by R its distribution-kernel.

Let $\lambda : \mathbb{C} \mapsto \mathbb{R}^+$ be C^∞ and satisfy $\mu(B(z,\lambda(z))) \sim 1$ for all z. Define a Riemannian metric by $d\varrho^2 = \lambda^{-2}\, ds^2$ where ds^2 is the Euclidean metric.

Theorem 9.1 [Ch6]. *Assume that φ satisfies (1) and (2). Then there exist $C < \infty$, $\varepsilon > 0$ such that for all $z \ne \zeta \in \mathbb{C}$,*

$$|R(z,\zeta)| \le C|z-\zeta|^{-1} e^{-\varepsilon\varrho(z,\zeta)} e^{\varphi(\zeta)-\varphi(z)}\,.$$

Moreover C, ε depend only on the constant in (1).

Corollary 9.2 [Ch6]. *Let φ satisfy (1) and (2). Let $p \in [1,\infty]$ and assume that $f \in L^p(\mathbb{C}, e^{-\varphi})$. Then there exists u satisfying $\bar{\partial}u = f$, such that u, $\lambda^{-1}u \in L^p(\mathbb{C}, e^{-\varphi})$.*

These results are closely related to the analysis of $\bar{\partial}_b$ on the model hypersurfaces $\{\mathrm{Im}(z_2) = P(z_1)\}$; if P is a subharmonic but nonharmonic polynomial, then $\varphi = P$ satisfies our hypotheses. Taking a partial Fourier transform in $\mathrm{Re}(z_2)$ reduces $\bar{\partial}_b$ to $\bar{\partial} + \tau\cdot P_{\bar{z}}$ on $\mathbb{C}^1$, where τ is a real parameter; this last equals $e^{-\tau P} \circ \bar{\partial} \circ e^{\tau P}$, so that studying it on $L^2(\mathbb{C})$ with respect to Lebesgue measure is equivalent to studying the ordinary $\bar{\partial}$ operator with a weight. The counterexamples of the last section become entirely natural from this perspective.

L^p solvability of $\bar{\partial}$ on bounded domains in $\mathbb{C}^1$, with arbitrary subharmonic φ, has been studied in [FiS2, Ber] and [A].

References

[A] E. Amar: Manuscript

[BF] E. Bedford, J. E. Fornæss: A construction of peak functions on weakly pseudoconvex domains. Ann. Math. **107** (1978) 555–568

[Be] J. Belanger: Hölder estimates for $\bar{\partial}$. 1987 Ph.D. thesis, Princeton University

[Ber] B. Berndtsson: Weighted estimates for $\bar{\partial}$ in domains in $\mathbb{C}$. Duke Math. J. (to appear)

[BS] H. Boas, M. C. Shaw: Sobolev estimates for the Lewy operator on weakly pseudoconvex boundaries. Math. Ann. **274** (1986) 221–231

[BC] A. Bonami, P. Charpentier: Estimations des $(1,1)$ courants positifs fermés dans les domains de $\mathbb{C}^2$. In: Lectures Notes in Mathematics, vol. 1094. Springer, Berlin Heidelberg New York 1984, pp. 44–52

[BL] A. Bonami, N. Lohoué: Projecteurs de Bergman et Szegö pour une classe de domains faiblement pseudo-convexes et estimations L^p. Comp. Math. **46** (1982) 159–226

[Bt] L. Boutet de Monvel: Intégration des équations de Cauchy-Riemann induites formelles. Séminaire Goulaouic-Lions-Schwartz, Exposé IX

[BSj] L. Boutet de Monvel, J. Sjöstrand: Sur la singularité des noyaux de Bergman et de Szegö. Asterisque **34–5** (1976) 123–164

[Bu] D. Burns: Global behavior of some tangential Cauchy-Riemann equations. PDE and Geometry Conference, Park City, Utah, 1977. Dekker, New York 1979, pp. 51–56

[BE] D. Burns, C. Epstein: Embeddability for three dimensional CR-manifolds. Preprint

[C1] D. W. Catlin: Necessary conditions for subellipticity of the $\bar{\partial}_b$-Neumann problem. Ann. Math. **117** (1983) 147–171

[C2] D. W. Catlin: Subelliptic estimates for the $\bar{\partial}$-Neumann problem on pseudoconvex domains. Ann. Math. **126** (1987) 131–191

[C3] D. W. Catlin: Estimates of invariant metrics on pseudoconvex domains of domain two. Math. Z. **200** (1989) 429–466

[C4] D. W. Catlin: A Newlander-Nirenberg theorem for manifolds with boundary. Mich. Math. J. **35** (1988) 233–240

[CNS1] D. C. Chang, A. Nagel, E. M. Stein: Estimates for the $\bar{\partial}$-Neumann problem for pseudoconvex domains in $\mathbb{C}^2$ of finite type. Proc. Nat. Acad. Sci. USA **85** (1988) 8771–8774

[CNS2] D. C. Chang, A. Nagel, E. M. Stein: Estimates for the $\bar{\partial}$-Neumann problem in pseudoconvex domains of finite type in $\mathbb{C}^2$. Preprint

[Ch1] M. Christ: Regularity properties of the $\bar{\partial}_b$ equation on three-dimensional CR manifolds. J. AMS **1** (1988) 587–646

[Ch2] M. Christ: Pointwise bounds for the relative fundamental solution of $\bar{\partial}_b$. Proc. AMS **104** (1988) 787–792

[Ch3] M. Christ: Estimates for the $\bar{\partial}_b$ equation and Szegö projection on CR manifolds. In: Lecture Notes in Mathematics, vol. 1384, pp. 146–158

[Ch4] M. Christ: Embedding compact three-dimensional CR manifolds of finite type in $\mathbb{C}^n$. Ann. Math. **129** (1989) 195–213

[Ch5] M. Christ: On the $\bar{\partial}_b$ equation on three-dimensional CR manifolds. Proc. Symp. Pure Math. (to appear)

[Ch6] M. Christ: On the $\bar{\partial}$ equation in $\mathbb{C}^1$ with weights. J. Geom. Anal. (to appear)

[Ch7] M. Christ: Lectures on Singular Integral Operators. NSF-CBMS regional conference series, vol. 77. AMS, Providence 1990

[Ch8] M. Christ: Estimates for fundamental solutions of second-order subelliptic differential operators. Proc. AMS **105** (1989) 166–172

[CG] M. Christ, D. Geller: Counterexamples for analytic hypoellipticity on domains of finite type. Ann. Math. (to appear)

[CW] R. R. Coifman, G. Weiss: Analyse Harmonique Non-Commutative Sur Certains Espaces Homogènes. (Lecture Notes in Mathematics, vol. 242). Springer, Berlin Heidelberg New York 1971

[D] J. D'Angelo: Real hypersurfaces, orders of contact, and applications. Ann. Math. **115** (1982) 615–637

[DT] M. Derridj, D. S. Tartakoff: Local analyticity for $\Box_b$ and the $\bar{\partial}$-Neumann problem at certain weakly pseudoconvex points. Comm. PDE **13** (1988) 1521–1600

[DJ] G. David, J.-L. Journé: A boundedness criterion for generalized Calderón-Zygmund operators. Ann. Math. **120** (1984) 371–397

[F] E. Falbel: Non-embeddable CR-manifolds and surface singularities. Preprint

[Fe1] C. Fefferman: The Bergman kernel and biholomorphic mappings of pseudoconvex domains. Invent. math. **26** (1974) 1–66

[Fe2] C. Fefferman: The uncertainty principle. Bull. AMS **9** (1983)

[FK1] C. Fefferman, J. J. Kohn: Hölder estimates on domains of complex dimension two and on three dimensional CR manifolds. Adv. Math. **69** (1988) 223–303

[FK2] C. Fefferman, J. J. Kohn: Estimates of kernels on three dimensional CR manifolds. Rev. Mat. Iberoam. **4** (1988) 355–405

[FKM] C. Fefferman, J. J. Kohn, M. Machedon: Hölder estimates on CR manifolds with a diagonalizable Levi form. Rev. Mat. Iberoam. **4** (1988) 1–90

[FP] C. Fefferman, D. H. Phong: Subelliptic eigenvalue problems. In: Proceedings, Conference on Harmonic Analysis in honor of Antoni Zygmund. Wadsworth, Belmont, CA, 1981, pp. 590–606

[FSa] C. Fefferman, A. Sanchez-Callé: Fundamental solutions for second order subelliptic operators. Ann. Math. **124** (1986) 247–272

[FS] G. B. Folland, E. M. Stein: Estimates for the $\bar{\partial}_b$ complex and analysis on the Heisenberg group. Comm. Pure Appl. Math. **27** (1974) 429–522

[Fo] J. Fornæss: Sup-norm estimates for $\bar{\partial}$ in $\mathbb{C}^2$. Ann. Math. **123** (1986) 335–345

[FSi1] J. Fornæss, N. Sibony: Construction of p.s.h. functions on weakly pseudoconvex domains. Duke Math. J. **59** (1989) 633–655

[FSi2] J. Fornæss, N. Sibony: On L^p estimates for $\bar{\partial}$. Preprint, Université de Paris-Sud, 1989

[G] D. Geller: Analytic Pseudodifferential Operators for the Heisenberg Group and Local Solvability. Princeton University Press, Princeton, NJ 1990

[GS1] P. Greiner, E. M. Stein: On the solvability of some differential operators of type $\Box_b$. In: Several Complex Variables, Proceedings of International Conferences, Cortona, Italy 1976–7. Scuola Normale Superiore, Pisa

[GS2] P. Greiner, E. M. Stein: Estimates for the $\bar{\partial}$-Neumann problem. Princeton University Press, Princeton, NJ 1977

[HN] B. Helffer, J. Nourrigat: Hypoellipticité Maximale pour des Opérateurs Polynomes de Champs de Vecteurs. Birkhäuser, Boston 1985

[He1] G. M. Henkin: Lewy's equation and analysis on pseudoconvex manifolds. I. Russian Math. Surveys **32** (1977) 59–130

[He2] G. M. Henkin: Lewy's equation and analysis on pseudoconvex manifolds. II. Math. USSR Sb. **102** (1977) 63–64

[H] L. Hörmander: Hypoelliptic second order differential equations. Acta Math. **119** (1967) 147–171

[K1] J. J. Kohn: Boundary behavior of $\bar{\partial}$ on weakly pseudoconvex manifolds of dimension two. J. Diff. Geom **2** (1972) 523–542

[K2] J. J. Kohn: The range of the tangential Cauchy-Riemann operator. Duke Math. J. **53** (1986) 525–545

[K3] J. J. Kohn: Estimates for $\bar{\partial}_b$ on pseudoconvex CR manifolds. Proc. Symp. Pure Math. **43** (1985) 207–217

[KN] J. J. Kohn, L. Nirenberg: A pseudoconvex domain not admitting a holomorphic support function. Math. Ann. **201** (1973) 265–268

[L] P. Lelong: Fonctionelles analytiques et fonctions entières. Les Presses de L'Université de Montreal, Montreal, Canada 1968

[M1] M. Machedon: Estimates for the parametrix of the Kohn Laplacian on certain domains. Invent. math. **91** (1988) 339–364

[M2] M. Machedon: Szegö kernels on pseudoconvex domains with one degenerate eigenvalue. Ann. Math. **128** (1988) 619–640

[Ma] P. Malliavin: Fonctions de Green d'un ouvert strictement pseudoconvexe et de classe de Nevanlinna. C. R. Acad. Sci. Paris **278** (1974) 141–144

[Mc] J. D. McNeal: Boundary behavior of the Bergman kernel in $\mathbb{C}^2$. Duke Math. J. **58** (1989) 499–512

[N] A. Nagel: Vector fields and nonisotropic metrics. In: Beijing Lectures in Harmonic Analysis. Annals of Mathematics Studies 112. Princeton University Press, Princeton, NJ 1986, pp. 241–306

[NRSW1] A. Nagel, J.-P. Rosay, E. M. Stein, S. Wainger: Estimates for the Bergman and Szegö kernels in certain weakly pseudoconvex domains. Bull. AMS **18** (1988) 55–59

[NRSW2] A. Nagel, J.-P. Rosay, E. M. Stein, S. Wainger: Estimates for the Bergman and Szegö kernels in $\mathbb{C}^2$. Ann. Math. **129** (1989) 113–149

[NSW] A. Nagel, E. M. Stein, S. Wainger: Balls and metrics defined by vector fields, I: Basic properties. Acta Math. **155** (1985) 103–147

[Ni] L. Nirenberg: Lectures on linear partial differential equations. CBMS-NSF regional conference series, vol. 17, AMS 1973

[Ra] M. Range: Integral kernels and Hölder estimates for $\bar{\partial}$ on pseudoconvex domains of finite type in $\mathbb{C}^2$. Preprint

[R] J. P. Rosay: New examples of non-locally embeddable CR structures (with no non-constant CR distributions). Annales Institut Fourier **39** (1989) 811–823

[Ro] H. Rossi: Attaching analytic spaces to a space along a pseudo-convex boundary. In: Proceedings, Conference on Complex Analysis. Springer, New York Berlin 1965, pp. 242–256

[RS] L. P. Rothschild, E. M. Stein: Hypoelliptic differential operators and nilpotent groups. Acta Math. **137** (1976) 247–320

[Sa] A. Sanchez-Callé: Fundamental solutions and geometry of the sum of squares of vector fields. Invent. math. **78** (1984)

[S] M.-C. Shaw: Optimal Hölder and L^p estimates for $\bar{\partial}_b$ on the boundaries of real ellipsoids in $\mathbb{C}^n$. Trans. AMS **324** (1991) 213–234

[Si1] N. Sibony: Un exemple de domaine pseudoconvexe régulier où l'équation $\bar{\partial}u = f$ n'admet pas de solution bornée pour f bornée. Invent. math. **62** (1980) 235–242

[Si2] N. Sibony: On Hölder estimates for $\bar{\partial}$. Preprint, Université de Paris-Sud 1988

[Sk] H. Skoda: Valeurs au bord pour les solutions de l'opérateur d'', et caractérisation des zeros des fonctions de la classe de Nevanlinna. Bull. Soc. Math. France **104** (1976) 225–299

[Sm] H. Smith: A calculus for three-dimnnsional CR manifolds of finite type. Preprint

[St] E. M. Stein: Singular Integrals and Differentiability Properties of Functions. Princeton University Press, Princeton, NJ 1970

[W] S. Webster: On the proof of Kuranishi's embedding theorem. Ann. Inst. H. Poincaré Anal. Non Linéaire **6** (1989) 183–207

Adapted Multiresolution Analysis, Computation, Signal Processing and Operator Theory

Ronald R. Coifman

Department of Mathematics, Yale University, New Haven, CT 06520, USA

We would like to describe recent developments, relating signal processing, numerical analysis and harmonic analysis.

Some of the main tasks of these fields involve efficient description of various classes of functions or transformations on functions. The usual problems encountered require handling of smoothness, oscillation or scaling patterns, and obtaining efficient representations (or coding) in a small number of parameters.

As for transformations on functions, some of the main problems concerning harmonic analysts involve preservation of various function spaces, spectral theory and operator calculus. The numerical analyst is mostly concerned with efficient algorithms for computing the effect of an operator, or of its inverse.

Both activities require a detailed understanding of the action of an operator on functions. In order to prove estimates or compute efficiently a good knowledge of the geometry and cancellation effects are necessary.

We will start by describing various methods for efficient analysis or description of functions. In the second part this analysis will be used to obtain fast numerical algorithms linking ideas and problems from classical analysis, such as Littlewood-Paley theory, to the ability to compute.

In conclusion, it will become evident that efficient compression methods, i.e. methods in which an operator or functions are described by as small a number of parameters as is possible (for a given precision), are directly related to our ability to compute fast, as well as to our analytic understanding. Most of the results discussed in this talk were obtained in collaboration with Yves Meyer, Vladimir Rokhlin, Gregory Beylkin, and Victor Wickerhauser.

§ 1. Data Compression, Orthonormal Bases, and Best Basis

We describe various constructions of orthonormal bases in function spaces (or sample spaces) which will permit an adaptation of bases to given functions or signals.

Over the last fifty years we have seen within mathematics, the need to modify the Fourier transform by permitting regions or bands of frequencies to be lumped together. The most common analysis of this type, the so-called Littlewood-Paley theory, (as developped by Marcinkiewicz, Zygmund, Calderon, Stein and many others)in which frequencies are grouped in dyadic intervals, has proved to be a

Proceedings of the International Congress
of Mathematicians, Kyoto, Japan, 1990

powerful and flexible tool. This method permits a blend of analysis in space and frequency simultaneously.

The recent discoveries by Stromberg [S], Meyer [M], Mallat and Daubechies [D] of various orthonormal wavelet bases has opened the door for using these methods (Littlewood-Paley analysis) in signal processing and numerical computation and, stimulated a range of discoveries and constructions of other classes of orthonormal bases. We refer the reader to Meyer's talk in these proceedings for further detail.

Definitions of Modulated Waveform Libraries. We now introduce the concept of a "Library of orthonormal bases". For the sake of exposition we restrict our attention to two classes of numerically useful waveforms, introduced recently by Y. Meyer and the author.

We start with trigonometric waveform libraries. These are localized sine transforms associated to covering by intervals of **R** (more generally, of a manifold).

We consider a cover $\mathbf{R} = \bigcup_{-\infty}^{\infty} I_i$, $I_i = [\alpha_i \alpha_{i+1})$ $\alpha_i < \alpha_{i+1}$, write $\ell_i = \alpha_{i+1} - \alpha_i = |I_i|$ and let $p_i(x)$ be a window function supported in $[\alpha_i - \ell_{i-1/2}, \alpha_{i+1} + \ell_{i+1}/2]$ such that

$$\sum_{-\infty}^{\infty} p_i^2(x) = 1$$

and

$$p_i^2(x) = 1 - p_i^2(2\alpha_{i+1} - x) \quad \text{for } x \text{ near } \alpha_{i+1}$$

then the functions

$$S_{i,k}(x) = \frac{2}{\sqrt{2\ell i}} p_i(x) \sin\left[(2k+1)\frac{\pi}{2\ell_i}(x - \alpha_i)\right]$$

form an orghonormal basis of $L^2(\mathbf{R})$ subordinate to the partition p_i. The collection of such bases forms a library of orthonormal bases.

It is easy to check that if H_{I_i} denotes the space of functions spanned by $S_{i,k}$ $k = 0, 1, 2, \ldots$ then $H_{I_i} + H_{I_{i+1}}$ is spanned by the functions

$$P(x)\frac{1}{\sqrt{2(\ell_i + \ell_{i+1})}} \sin\left[(2k+1)\frac{\pi}{2(\ell_i + \ell_{i+1})}(x - \alpha_i)\right]$$

where

$$P = \left(p_i^2(x) + p_{i+1}^2(x)\right)^{1/2}$$

is a "window" function covering the interval $I_i \cup I_{i+1}$.

Another new library of orthonormal bases called the Wavelet packet library can be constructed. This collection of modulated wave forms, corresponds roughly to a covering of "frequency" space. This library contains the wavelet basis, Walsh functions, and smooth versions of Walsh functions called wavelet packets.

We'll use the notation and terminology of [D], whose results we shall assume.

We are given an exact quadrature mirror filter $h(n)$ satisfying the conditions of Theorem (3.6) in [D], p. 964, i.e.

$$\sum_n h(n-2k)h(n-2\ell) = \delta_{k,\ell}, \ \sum_n h(n) = \sqrt{2}.$$

We let $g_k = h_{l-k}(-1)^k$ and define the operations F_i on $\ell^2(\mathbf{Z})$ into "$\ell^2(2\mathbf{Z})$"

$$(1.0)\qquad \begin{aligned} F_0\{s_k\}(i) &= 2\sum s_k h_{k-2i} \\ F_1\{s_k\}(i) &= 2\sum s_k g_{k-2i} \,. \end{aligned}$$

The map $\mathbf{F(s_k)} = \mathbf{F_0(s_k)} \oplus \mathbf{F_1(s_k)} \in \ell^2(\mathbf{2Z}) \oplus \ell^2(\mathbf{2Z})$ is orthogonal and

$$(1.1)\qquad F_0^* F_0 + F_1^* F_1 = I \,.$$

We now define the following sequence of functions.

$$(1.2)\qquad \begin{cases} W_{2n}(x) = \sqrt{2}\sum h_k W_n(2x-k) \\ W_{2n+1}(x) = \sqrt{2}\sum g_k W_n(2x-k) \,. \end{cases}$$

Clearly the function $W_0(x)$ can be identified with the scaling function φ in [D] and W_1 with the basic wavelet ψ.

Let us define $m_0(\xi) = \frac{1}{\sqrt{2}}\sum h_k e^{-ik\xi}$ and

$$m_1(\xi) = -e^{i\xi}\bar{m}_0(\xi+\pi) = \frac{1}{\sqrt{2}}\sum g_k e^{ik\xi} \,.$$

Remark. The quadrature mirror condition on the operation $\mathbf{F} = (\mathbf{F_0}, \mathbf{F_1})$ is equivalent to the unitarity of the matrix

$$\mathcal{M} = \begin{bmatrix} m_0(\xi) & m_1(\xi) \\ m_0(\xi+\pi) & m_1(\xi+\pi) \end{bmatrix} .$$

Taking the Fourier transform of (1.2) when $n = 0$ we get

$$\hat{W}_0(\xi) = m_0(\xi/2)\hat{W}_0(\xi/2)$$

i.e.,

$$\hat{W}_0(\xi) = \prod_{j=1}^{\infty} m_0(\xi/2^j)$$

and

$$\hat{W}_1(\xi) = m_1(\xi/2)\hat{W}_0(\xi/2) = m_1(\xi/2)m_0(\xi/4)m_0(\xi/2^3)\cdots$$

More generally, the relations (1.2) are equivalent to

$$(1.3)\qquad \hat{W}_n(\xi) = \prod_{j=1}^{\infty} m_{\varepsilon_j}(\xi/2j)$$

and $n = \sum_{j=1}^{\infty} \varepsilon_j 2^{j-1} (\varepsilon_j = 0 \text{ or } 1)$.

The functions $W_n(x-k)$ form an orthonormal basis of $L^2(\mathbf{R^n})$.

We define a *library* of wavelet packets to be the collection of functions of the form $W_n(2^\ell x - k)$ where $\ell, k \in \mathbf{Z}, \mathbf{n} \in \mathbf{N}$. Here, each element of the library is determined by a scaling parameter ℓ, a localization parameter k and an oscillation parameter n. (The function $W_n(2^\ell x - k)$ is roughly centered at $2^{-\ell}k$, has support of size $\approx 2^{-\ell}$ and oscillates $\approx n$ times).

We have the following simple characterization of subsets forming orthonormal bases.

Proposition. *Any collection of indices (ℓ, n) such that the intervals $[2^\ell n, 2^\ell n + 1)$ form a disjoint cover of $[0, \infty)$ gives rise to an orthonormal basis of L^2.*

Motivated by ideas from signal processing and communication theory V. Wickerhauser and the author were led to measure the "distance" between a basis and a function in terms of the Shannon entropy of the expansion. More generally, let H be a Hilbert space.

Let $v \in H$, $\|v\| = 1$ and assume

$$H = \oplus \sum H_i$$

an orthogonal direct sum. We define

$$\varepsilon^2(v, \{H_i\}) = -\sum \|v_i\|^2 \ell n \|v_i\|^2$$

as a measure of distance between v and the orthogonal decomposition.

ε^2 is characterized by the Shannon equation which is a version of Pythagoras' theorem.

Let

$$H = \oplus(\sum H^i) \oplus (\sum H_j) = H_+ \oplus H_-$$

H^i and H_j give orthogonal decomposition $H_+ = \sum H^i, H_- = \sum H_j$. Then

$$\varepsilon^2(v; \{H^i, H_j\}) = \varepsilon^2(v, \{H+, H_-\} + \|v_+\|^2 \varepsilon^2 \left(\frac{v_+}{\|v_+\|}, \{H^i\} \right) + \|v_-\|^2 \varepsilon^2 \left(\frac{v_-}{\|v_-\|}, \{H_j\} \right).$$

This is Shannon's equation for entropy (if we interpret as in quantum mechanics $\|P_{H_+} v\|^2$ as the "probability" of v to be in the subspace H_+).

This equation enables us to search for a smallest entropy space decomposition of a given vector.

In fact, for the example of the first library restricted to covering by dyadic intervals we can start by calculating the entropy of an expansion relative to a local trigonometric basis for intervals of length one, then compare the entropy of an adjacent pair of intervals to the entropy of an expansion on their union. Pick the expansion of minimal entropy and continue until a minimum entropy expansion is achieved.

In practice, discrete versions of this scheme can be implemented in $CN \log N$ computations (where N is the number of discrete samples $N = 2^L$.)

For voice signals and images this procedure leads to remarkable compression algorithms (see [CMQW]).

Of course, while entropy is a good measure of concentration of an expansion, various other information cost functions are possible, permitting discrimination and choice between various special function expansion.

Other possible libraries can be constructed. The space of frequencies can be decomposed into pairs of symmetric windows around the origin, on which a smooth partition of unity is constructed. This and other constructions were obtained by one of our students E. Laeng [L].

Higher dimensional libraries can also be easily constructed (as well as libraries on manifolds), leading to new and direct analysis methods for linear transformations.

§ 2. Wavelets, Wavelet Packets and Numerical Algorithms

The usual way to analyze an integral operator relies on the Fourier transform. In principle, for convolution operators the problem is solved, although as we well know it is essentially impossible to rely solely on the Fourier transform to understand the effect of the operator on various classes of functions (such as Hölder, L^p etc.).

For operators which are not of convolution type even the problem of proving L^2 boundedness (for example), can be extremely difficult. We claim that good methods for fast computation of such operators shed light on their analysis and provide new approaches.

Initially the relations between computation and Calderón-Zygmund theory was pointed out to the author by V. Rokhlin who in his design of the Fast Multipole algorithm for computing potential interactions has essentially reinvented many of the ingredients of Calderón-Zygmund theory.

Rokhlin constructed a fast algorithm of order N to compute all sums

$$p_j = \sum_{i=1}^{N} \frac{g_i g_j}{|x_i - x_j|} \qquad \text{where} \quad x_i \in \mathbf{R}^3 \quad \mathbf{i = 1, \ldots, N}$$

although, naively it would seem to be impossible to do this calculation in less than N^2 computations, since this is the number of interactions. He observed that the effect of a cloud of charges located in a box can be described to any accuracy by the effect of a single multipole at the center of the box, requiring only a few numbers (Taylor coefficients of the field at the center of external boxes removed from the source). He organized all boxes in a dyadic hierarchy enabling an efficient $O(N)$ algorithm. (This algorithm is $O(N)$ independently of the configuration of the charges providing therefore a substantial improvement over FFT).

As we know, Littlewood-Paley and Calderón-Zygmund theory can achieve similar goals. Wavelet based algorithms providing an elegant reformulation and generalization of the multipole algorithms, were developed by Beylkin, Rokhlin and the author [BCR].

Before describing these methods in detail we return to the basic question of efficient computation of an integral operator.

$$Tf(x) = \int_0^1 k(x,y) f(y) dy \qquad x \in [0,1] .$$

Clearly, if we can write

$$k(x,y) = \sum a_{\alpha\beta} w_\alpha(x) w_\beta(y)$$

where the number of $a_{\alpha\beta}$ is small, we would reduce the computation to

$$\int f(y)w_\beta(y)dy = d_\beta$$

and to

$$\sum_\alpha \left(\sum_\beta a_{\alpha\beta} d_\beta \right) w_\alpha(x).$$

If we choose w_β to be the eigenvectors of T (say if T is self adjoint) at least the matrix $a_{\alpha\beta}$ would be diagonal. The price of course would be to compute d_β and the sum efficiently (say like FFT). Such algorithms of course are not known in general. Instead, we can compromise. View $k(x,y)$ as an image (i.e. $k(x,y)$ would represent the light intensity at pixel (x,y)) and try to compress the image. We are naturally led to consider a library such as the wavelet packets and to a selection of an orthonormal basis of $L^2((0,1)\times(0,1))$ such that the matrix $a_{\alpha\beta}$ has highest concentration (or lowest entropy).

Of course, when the kernel under consideration satisfies estimates invariant under translations and dilations. In particular, if $k(x,y)$ is a Calderón-Zygmund operator or a pseudo differential operator we expect the best basis to have a similar behavior i.e., we are forced to consider the wavelet basis. Remarkably, this algorithm (see [BCR]) corresponds on the one hand to the FMM methods of Rokhlin [R] and, on the other hand, to the so-called P_t, Q_t analysis of Calderón-Zygmund operators and the $T(1)$ theorem of David and Journé (see [DJ]).

Concretely these methods can be most simply described by using the Haar functions $h_I(x)$ where $h_I(x) = \frac{1}{|I|^{\frac{1}{2}}}$ on the left half of the dyadic interval I and $h_I(x) = -\frac{1}{|I|^{\frac{1}{2}}}$ on the right half, and zero elsewhere. We also let

$$\chi_I = \frac{1}{|I|^{\frac{1}{2}}} \qquad \text{on } I \quad \chi_I = 0 \quad x \notin I.$$

We expand $k(x,y)$ in terms of the two dimensional Haar functions $h_I(x)h_{I'}(y)$, $h_I(x)\chi_{I'}(y)$, $\chi_I(x)h_{I'}(y)$ as

$$k(x,y) = \sum \alpha_{II'} h_I(x)h_{I'}(y) + \sum \beta_{II'} h_I(x)\chi_{I'}(y) + \sum \gamma_{II'}\chi_I(x)h_{I'}(y)$$

$$\text{where} \quad \alpha_{II'} = \iint k(x,y)h_I(x)h_{I'}(y)dxdy$$

$$\beta_{II'} = \iint k(x,y)h_I(x)\chi_{I'}(y)dxdy$$

$$\gamma_{II'} = \iint k(x,y)\chi_I(x)h_{I'}(y)dxdy.$$

Introducing this representation of k we obtain

$$T(f)(x) = \sum_I h_I(x) \sum_{I'} \alpha_{II'} d_{I'} + \sum h_I(x) \sum_{I'} \beta_{II'} s_{I'} \sum_I \chi_I(x) \sum_{I'} \gamma_{II'} d_{I'} \tag{1.3}$$

where each sum in I' involves only dyadic intervals of length $|I|$ and where $s_I = \langle \chi_I, f\rangle, d_I = \langle h_I, f\rangle$. The matrix realisation of this computation (Fig. 1) does not correspond to the ordinary realization of the operator in terms of the Haar basis. In fact the s_I are dependant on the coordinates d_I; by doubling the size of the matrix we gain a block decomposition by scale (all interscale interactions occur through s_I). When applied to specific classes of operators such as C-Z or pseudo-differential operators this procedure yields banded matrices, with band width depending on the desired precision and the choice of wavelet . If we let P_j denote the orthogonal projection on the space of functions constant on dyadic intervals of length 2^{-j} and approximate T by $T_n = P_n T P_n$.

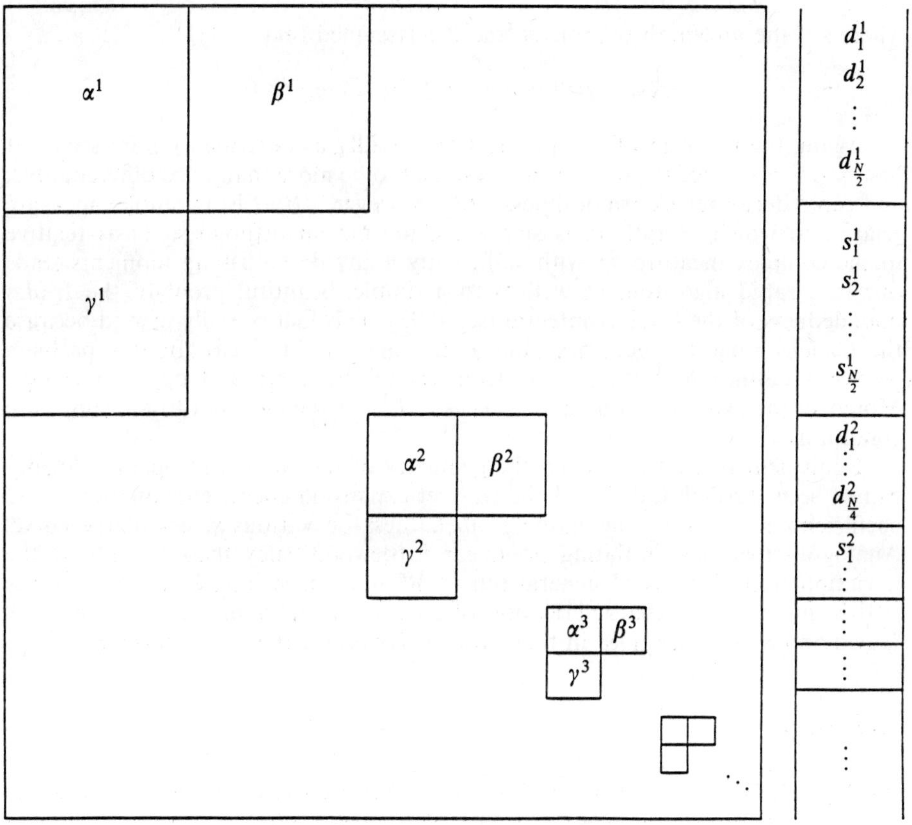

Fig. 1

We can rewrite

$$T_n = \sum_{j=1}^{n}(T_j - T_{j-1}) + T_0 = \sum_{j=1}^{n}(Q_jTQ_j + Q_jTP_j + P_jTQ_j) + T_0$$

where $Q_j = P_j - P_{j-1}$.

This decomposition relates the previous expansion to the Littlewood Paley approach and the proof of the "T of 1" theorem of David and Journe', see also [Se].

In general, we might hope for decompositions permitting us also to handle oscillatory integrals or higher singularities such as the one arising from wave propagation, Fourier integral operators, Radon transforms, and others.

It now seems clear that at least for some of these problems where previously microlocalization was the tool of choice for obtaining estimates, adapted orthonormal bases may become the tool for numerical computation.

Another class of examples relate to the Cauchy integral on chord arc curves. Here we study

$$C(f)(s) = p \cdot v \int \frac{f(t)}{z(s) - z(t)} dt$$

where s is the arclength parameter and it is assumed that

$$|z(s) - z(t)| > \delta |s - t| \quad \text{for some } \delta > 0 .$$

Again, the problem of computing $C(f)$ rapidly as described before seems at first impossible since the curve is not assumed to be more than once differentiable, and rapid decay reflects smoothness. Here however, $z'(t)dt$ is a complex measure relative to which everything is smooth. Choosing an orthogonal basis relative to the complex measure dz with sufficiently many dz vanishing moments leads one to a rapid algorithm as well as to a simple, beautiful proof of the $L^2(ds)$ boundedness of the Cauchy integral (see [CJS]. This fact is easily proved because the Cauchy operator becomes almost diagonal in this basis. In this problem we see a natural tie between the geometry of the curve and operator theory. Moreover the existence of a "good" basis in $L^2(ds)$ is equivalent to the chord-arc condition.

Equivalent relations between the geometry of the curve and operator theory can be seen through the size of the wavelet expansion coefficients of $z'(t)$. These coefficients measure the deviation from flatness on various scales of the curve. Analogous methods, initiating nonlinear Littlewood-Paley theory, studying the deviation from flatness of general sets in $\mathbf{R}^n$ have been introduced by P. Jones [PJ] in his beautiful characterization of subsets of rectifiable curves and by G. David and S. Semmes [DS] in their work on operator theory on surfaces.

References

[BCR] Beylkin, G., Coifman, R., Rokhlin, V.: Fast wavelet transforms and numerical algorithms I. To appear in Comm. Pure Appl. Math.

[CJS] Coifman, R., Jones, P.W., Semmes, S.: Two elementary proofs of the L^2 boundedness of Cauchy integrals on Lipschitz curves. J. AMS **2** (1989) 553–564

[CMQW] Coifman, R., Meyer, Y., Quake, S., Wickerhauser, V.: Signal processing and compression with wavelet packets. To appear

[D] Daubechies, I.: Orthonormal bases of compactly supported wavelets. Comm. Pure Appl. Math. **XL1** (1988)

[DJ] David, G., Journé, J.L.: A boundedness criterion for generalized Calderon-Zygmund operators. Ann. Math. **120** (2) (1984) 371–397

[DS] David, G., Semmes, S.: Audela des graphies Lepschitziens. To appear in Asterisque-Au-delá

[L] Laeng, E.: Une base orthonormale de $L^2(\mathbf{R})$...,. To appear in C. R. Acad. Sci. Paris (1990)

[M] Meyer, Y.: Principe d'incertitude, bases hilbertiennes et algébres d'opérateurs. Séminaire Bourbaki, 1985–85, 662. Astérisque (Société Mathématique de France)

[PJ] Jones, P.: Rectifiable sets and the traveling salesman problem. Invent. math. (1990)

[S] Stromberg, J.O.: A modified Haar system and higher order spline systems. Conference in Harmonic Analysis in Honor of Antoni Zygmund. Wadworth Math. Series, ed. by W. Beckner et al., vol. II, pp. 475–493

[Se] Semmes, S.: Nonlinear Fourier analysis. Bull. AMS **20** (1) (1989)

Rational Maps and Kleinian Groups

Curt McMullen

Princeton University, Department of Mathematics, Fine Hall, Washington Road
Princeton NJ 08544, USA

1. Introduction

There are many parallels between the theory of iterated rational maps f and that of Kleinian groups Γ, considered as dynamical systems on the Riemann sphere $\widehat{\mathbb{C}}$.[1] In this paper we will survey three chapters of this developing theory, and the Riemann surface techniques they employ:

1. The combinatorics of critically finite rational maps and the geometrization of Haken 3-manifolds via iteration on Teichmüller space.
2. Renormalization of quadratic polynomials and 3-manifolds which fiber over the circle.
3. Boundaries and laminations — Teichmüller space in Bers' embedding and the Mandelbrot set.

2. The Theme of Short Geodesics

What are the possible topological forms for a conformal dynamical system?

Part of the answer is provided by two theorems, due to Thurston, which employ *iteration on Teichmüller space* to construct rational maps and Kleinian groups of a given topological form. More precisely, the iteration either *finds a geometric model* or *reveals a topological obstruction* to its existence. This dichotomy stems from:

Theorem 1 [Mum]. *Let X_n be a sequence of points in the moduli space $\mathcal{M}_{g,k}$ of hyperbolic Riemann surfaces of genus g with k punctures. After passing to a subsequence, either*

- *X_n converges to X in $\mathcal{M}_{g,k}$, or*
- *there is a collection of disjoint simple closed geodesics S_n on X_n such that the hyperbolic length of S_n tends to zero.*

[1] See [Sul2] for part of the dictionary.

Proceedings of the International Congress
of Mathematicians, Kyoto, Japan, 1990

2.1 Critically Finite Rational Maps

Let $f : \mathbf{S}^2 \to \mathbf{S}^2$ be a branched covering of the sphere of degree greater than one, and let P denote the *post-critical set* of f, i.e.

$$P = \bigcup_{n=1}^{\infty} f^n(B),$$

where B denotes the branch points (at which f is locally many-to-one). If $|P| < \infty$ we say f is *critically finite.* Two such maps f and g are *combinatorially equivalent* if there is a homeomorphism $h : (\mathbf{S}^2, P_f) \to (\mathbf{S}^2, P_g)$ such that hfh^{-1} and g are isotopic rel P_g.

A critically finite map is a generalization to the complex domain of the *kneading sequence* for maps of the interval.

The following theorem provides a topological characterization of critically finite rational maps.

Theorem 2 [Th3, DH3]. *Let $f : \mathbf{S}^2 \to \mathbf{S}^2$ be critically finite with hyperbolic orbifold. Then either*

- *f is combinatorially equivalent to a rational map $g : \widehat{\mathbb{C}} \to \widehat{\mathbb{C}}$, unique up to automorphisms of $\widehat{\mathbb{C}}$, or*
- *there is an f-invariant system of disjoint simple closed curves Γ in $\mathbf{S}^2 - P$ providing a topological obstruction to such an equivalence.*

The technical condition "with hyperbolic orbifold" rules out certain elementary cases (which are also understood). It is satisfied, for example, if $|P| > 4$.

Sketch of the proof. The space of Riemann surface structures on $(\mathbf{S}^2, P)$, up to isotopy rel P is exactly the Teichmüller space of the sphere with $|P|$ distinguished points, denoted $\mathrm{Teich}(\mathbf{S}^2, P)$. Given such a structure, pull it back by f to obtain a new structure on the same space: this defines a map

$$T_f : \mathrm{Teich}(\mathbf{S}^2, P) \to \mathrm{Teich}(\mathbf{S}^2, P).$$

A fixed point for T_f gives an invariant complex structure and therefore a rational map g combinatorially equivalent to f.

This iteration has two fundamental features:

- T_f^k *contracts* the Teichmüller metric (for some fixed iterate k); and
- the contraction at a point X in $\mathrm{Teich}(\mathbf{S}^2, P)$ is less than $c[X] < 1$ where $c[X]$ is a continuous function depending only on the location of X in *moduli* space.

Now try to locate a fixed point of T_f by studying the sequence of iterates $X_n = T_f^n(X_0)$ of an arbitrary starting guess X_0. If $[X_n]$ returns infinitely often to a compact subset K of moduli space, then due to uniform contraction over K, the sequence converges to a fixed point and f is equivalent to a rational map.

Otherwise, by Mumford's theorem, the length of the shortest geodesic on X_n tends to zero. Set Γ = {isotopy classes of very short geodesics on X_n} for n sufficiently large. Since the Teichmüller distance from X_n to X_{n+1} is bounded, lengths change by only a bounded factor. Therefore Γ is f-invariant, in the sense that any geodesic representing a component of $f^{-1}(\gamma)$ is again in Γ.

Let $A : \mathbb{R}^\Gamma \to \mathbb{R}^\Gamma$ be defined by $A_{\delta\gamma} = \sum_\alpha \deg(f : \alpha \to \gamma)^{-1}$, where the sum is over components α of $f^{-1}(\gamma)$ homotopic to δ. By analyzing the geometry of short geodesics, one shows the leading eigenvalue of A is ≥ 1. This provides the desired topological obstruction.

Indeed, if f is equivalent to a rational map g, then one can thicken the curves in Γ to disjoint annuli, with conformal moduli $m_\gamma > 0$. By considering inverse images of these annuli under g, one finds the same curves can be represented by annuli with moduli $m'_\delta > \sum_\gamma A_{\delta\gamma} m_\gamma$. It follows that some curve can be thickened to an annulus of arbitrarily large conformal modulus, a contradiction. □

2.2 Haken 3-Manifolds

There is a parallel theory of iteration in Thurston's construction of hyperbolic structures on Haken manifolds. For simplicity we stick to closed manifolds.

To a Kleinian group Γ one associates the 3-dimensional *Kleinian manifold* $N = (\mathbb{H}^3 \cup \Omega)/\Gamma$, where $\Omega \subset \widehat{\mathbb{C}}$ is the domain of discontinuity. Then N has a hyperbolic structure on its interior and a conformal structure on its boundary.

Theorem 3 [Th1, Mor]. *Let M^3 be a closed Haken 3-manifold. Then either*

- *M^3 is diffeomorphic to a unique hyperbolic manifold $\mathbb{H}^3/\Gamma$, or*
- *there is a map of a torus into M^3, injective on π_1, providing a topological obstruction to a hyperbolic structure on M^3.*

Sketch of the proof. We combine Thurston's original approach with the Riemann surface techniques of [Mc2] and emphasize the parallel with the geometrization of rational maps.

A 3-manifold is *Haken* if it can be constructed by starting with 3-balls, and repeatedly gluing along incompressible submanifolds of the boundary. The idea of the proof is to carry out the construction geometrically, at each stage providing the pieces with hyperbolic structures. An orbifold technique [Mor, Fig. 14.6] reduces the problem to the case of gluing along the entire boundary.

Iteration enters at the inductive step: given a compact 3-manifold M^3 and gluing instructions encoded by an orientation-reversing involution $\tau : \partial M^3 \to \partial M^3$, we must construct a hyperbolic structure on M^3/τ. By induction M^3 is diffeomorphic to a Kleinian manifold N^3. Unlike the case of a closed manifold, which admits at most one hyperbolic structure by Mostow rigidity, the manifold N^3 is flexible. The set of possible shapes for N^3 is parameterized by the Teichmüller space of the boundary of M.

Which structure descends to M/τ? The answer can be formulated as a fixed point problem on Teichmüller space. Using the topology of M^3, Thurston defines the *skinning map*

$$\sigma : \mathrm{Teich}(\partial M) \to \mathrm{Teich}(\overline{\partial M})$$

by forming quasifuchsian covering spaces for each component of the boundary, and recording the conformal structure on the new ends which appear. The gluing instructions determine an isometry

$$\tau : \mathrm{Teich}(\overline{\partial M}) \to \mathrm{Teich}(\partial M),$$

and a fixed point for

$$T = \sigma \circ \tau$$

solves the gluing problem.

Here the parallel with the construction of critically finite maps emerges. The completion of the proof will follow [Mc2].

Assume M^3 is not an interval bundle over a surface (this special case is discussed in the next section). Then some fixed iterate T^k contracts the Teichmüller metric; in fact:

- The contraction of T^k at a point X in $\mathrm{Teich}(\partial M)$ is bounded by $c[X] < 1$ where $c[X]$ depends only on the location of X in *moduli* space.

As before, this reduces the proof to an analysis of short geodesics. Let $X_n = T^n(X_0)$ be the forward orbit of a starting guess X_0 in $\mathrm{Teich}(\partial M)$. If $[X_n]$ returns infinitely often to a compact subset of moduli space, the sequence converges and the gluing problem is solved.

Otherwise X_n develops short geodesics. With further analysis one finds these short geodesics bound cylinders in M^3, joined by τ to form an incompressible torus in M^3/τ. A closed hyperbolic manifold contains no such torus (it must correspond to a cusp), so we have located a topological obstruction to a hyperbolic structure. □

The bound on contraction $c[X]$ comes from a general result in the theory Riemann surfaces.

Let $Y \to X$ be a covering space of a hyperbolic Riemann surface X of finite area. Then there is a natural map $\theta : \mathrm{Teich}(X) \to \mathrm{Teich}(Y)$, defined by lifting complex structures from X to Y. Consider the case of the universal covering $\Delta \to X$ where Δ is the unit disk; G denotes the Fuchsian group of deck transformations.

Theorem 4 [Mc1]. *The map* $\theta : \mathrm{Teich}(X) \to \mathrm{Teich}(\Delta)$ *is a contraction for the Teichmüller metric. Moreover* $\|d\theta\| < c[X] < 1$ *where* c *depends continuously on the location of* X *in moduli space.*

This theorem is related to classical Poincaré series, as follows. For any Riemann surface R, let $Q(R)$ denote the Banach space of integrable holomorphic quadratic differentials $\phi(z)dz^2$ with $||\phi|| := \int_R |\phi| < \infty$. Starting with ϕ in $Q(\Delta)$, we can construct an automorphic form for G by the Poincaré series [Poin]:

$$\Theta(\phi) = \sum_G g^*(\phi).$$

Since $\Theta(\phi)$ is G-invariant, it determines an element of $Q(X)$.

In Teichmüller theory, $Q(X)$ is naturally identified with the cotangent space to Teich(X) at X, its norm is dual to the Teichmüller metric, and the operator $\Theta : Q(\Delta) \to Q(X)$ is the coderivative $d\theta^*$. This gives:

Corollary 5 (Kra's Theta Conjecture). $||\Theta|| < 1$ *for classical Poincaré series.*

On a global level the theorem says that lifts of Teichmüller mappings can be relaxed (isotoped to mappings of less dilatation):

Corollary 6. *Let $f : X_0 \to X_1$ be a Teichmüller mapping between distinct points in* Teich(X). *Then the map $\tilde{f} : \Delta \to \Delta$ obtain by lifting f to the universal covers of domain and range is not extremal among quasiconformal with the same boundary values on S^1.*

More generally, these contraction principles apply to a covering $Y \to X$ iff the covering is *nonamenable*; see [Mc1].

Now for a typical (acylindrical) M^3 the skinning map σ can be described as follows. Given a Riemann surface X in Teich(∂M),

(1) form countably many copies of its universal cover $\tilde{X}$, then
(2) glue them together in a pattern determined by the combinatorics of M^3 to obtain a new Riemann surface $\sigma(X)$.

The surface $\sigma(X)$ contains a dense full measure set of open disks each of which is canonically identified with the universal cover of X. By the results above, step (1) is a contraction for the Teichmüller metric. Step (2) is at worst an isometry, so $||d\sigma|| \leq ||d\theta|| < c[X] < 1$. (A more detailed expository account appears in [Mc5].)

3. Renormalization and 3-Manifolds Which Fiber over the Circle

For the special case of Haken manifolds presented as surface bundles over the circle, the construction of a hyperbolic structure is different in spirit and finds parallels with the construction of fixed points for renormalization.

3.1 Surface Bundles

Let S be a closed oriented surface of genus $g > 1$, and let $\phi : S \to S$ be a *mapping class*, that is a diffeomorphism determined up to isotopy. From this data one can construct a 3-manifold by starting with $M^3 = S \times [0, 1]$ and gluing the ends together by ϕ. Every 3-manifold which fibers over the circle admits such a description; ϕ is the *monodromy* of the fibration.

Theorem 7 [Th2]. *A 3-manifold M^3 which fibers over the circle admits a hyperbolic structure iff the monodromy ϕ is pseudo-Anosov.*

As in the preceding section, the construction of a hyperbolic structure can be formulated as a fixed point problem. There are two essential differences: (1) the desired fixed point lies on the boundary of Teichmüller space, rather than in its interior, and (2) it is dynamically hyperbolic rather than attracting.

Construction of the Hyperbolic Structure. We will work in the representation variety $\mathcal{V} = \mathrm{Hom}(\pi_1(S), \mathrm{Isom}(\mathbb{H}^3))/\text{conjugation}$. Let $AH(S) \subset \mathcal{V}$ denote the closed subset of discrete faithful representations. The idea is to find in $AH(S)$ the $\mathbb{Z}$-covering space of M^3 carrying the fundamental group of a fiber. The deck transformation acts by isometry on this covering space, so it is characterized as a fixed point in $AH(S)$ for the map

$$\Phi : \mathcal{V} \to \mathcal{V}$$

given by $\Phi(\varrho) = \varrho \circ \pi_1(\phi)$.

The construction of the fixed point can be organized into two steps. Let $QF(S) \subset AH(S)$ be the open subset of quasifuchsian groups; it is holomorphically parameterized by $\mathrm{Teich}(S) \times \mathrm{Teich}(\overline{S})$ (here $\overline{S}$ indicates reversal of orientation) and we denote by $\varrho(X, Y)$ the marked group corresponding to a pair of Riemann surfaces X and Y.

Step 1: Form the limit $\varrho_\infty = \lim_{n\to\infty} \varrho(X, \phi^{-n}(Y))$. Here X and Y are arbitrary Riemann surfaces and $\phi(Y)$ denotes the action of the mapping class on Teichmüller space. The representations $\varrho(X, \phi^{-n}(Y))$ range in a Bers' slice, which has compact closure in $AH(S)$, so the existence of some accumulation point is clear. Logically one can work with any accumulation point ϱ_∞; in fact, the sequence converges [CT, §7].

Step 2: Form the limit $\varrho = \lim \Phi^n(\varrho_\infty)$; this is a fixed point for Φ. Existence of this limit depends on

Compactness: The double limit theorem of [Th2], which assures there is some accumulation point ϱ; and

Rigidity: Sullivan's quasiconformal rigidity theorem [Sul1], which gives $\Phi(\varrho) = \varrho$ for any accumulation point.

The first step produces a point on the stable manifold of a fixed point of Φ, and the second iterates it to find the fixed point.

The limit in (Step 2) can be lifted to the level of marked groups $G_n \subset \mathrm{Aut}(\mathbb{H}^3)$ (rather than groups up to conjugacy) such that G_n tends algebraically to $G = \mathrm{Image}(\varrho)$. The groups G_n (conjugate to $\mathrm{Image}(\Phi^n(\varrho_\infty))$) are all isomorphic; we are viewing a single dynamical system from a changing perspective. As $n \to \infty$ the limit set of G_n becomes denser and denser, and the limit set of G is the full sphere.

The group G_{n+1} is obtained from G_n by a K-quasiconformal deformation with uniform K. By compactness of K-quasiconformal maps, one obtains a quasiconformal map ψ equivariant with respect to G and inducing the automorphism Φ. Since the limit set is the whole sphere, ψ is conformal by Sullivan's result. The group generated by G and ψ together is then a Kleinian group isomorphic to $\pi_1(M^3)$.

3.2 Quadratic-Like Maps

This discussion parallels the emerging complex viewpoint on *renormalization* of quadratic-like maps. For concreteness we will discuss the case of period doubling; see [Cvi, Milnor, Sul3] and [Sul4] for background and more details.

Consider the family of quadratic maps $z \mapsto z^2+c$ as the parameter c decreases along the real axis, starting at $c = 0$. One finds a sequence of parameter values $c(n)$ for which the attractor of $f_n(z) = z^2 + c(n)$ bifurcates from a cycle of order 2^n to 2^{n+1}; at $c(\infty) = \lim c(n)$ the attractor becomes a Cantor set. This *cascade of period doublings* was observed by Feigenbaum to have many universal features around $f_\infty(z) = z^2 + c(\infty)$. For example

$$\frac{c(n) - c(\infty)}{c(n+1) - c(\infty)} \to \lambda = 4.669201609\ldots$$

and this value of λ (as well as the fine structure of f_∞, such as the Hausdorff dimension of its attracting Cantor set) is the same for other families of smooth mappings with the same topological form as $z^2 + c$.

This universality is part of a larger renormalization picture proposed by Feigenbaum and established rigorously by Lanford and others. We present a version with the complex quadratic-like maps of Douady and Hubbard [DH2]; cf. [Sul3].

A *quadratic-like mapping* $f : U \to V$ is a proper degree two holomorphic map between open disks with $\overline{U} \subset V \subset \mathbb{C}$. Its *filled-in Julia set* $K(f)$ is $\bigcap_1^\infty f^{-n}(V)$. When $K(f)$ is connected, there is a unique quadratic polynomial $I(f)$ (the *inner class*) conjugate to f near $K(f)$ by a quasiconformal map which is conformal on $K(f)$. Thus I takes values in the Mandelbrot set M of polynomials $z^2 + c$ with connected Julia sets.

Let $\mathscr{Q}$ be the space of all analytic maps $f : \Omega_f \to \mathbb{C}$ defined on a region Ω_f containing the origin, such that $f'(0) = 0$ and f is quadratic-like on some neighborhood of zero. We identify $f(z)$ and $g(z)$ if some rescaling $\alpha f(z/\alpha)$ agrees with $g(z)$ on their common domain of definition. Finally $f_i \to f$ if there are representatives of f_i which converge to f uniformly on compact subsets of Ω_f.

The *renormalization operator* $\mathscr{R} : \mathscr{Q}' \to \mathscr{Q}$ is given by $\mathscr{R}(f) = f \circ f$. It is defined on an open set $\mathscr{Q}'$ such that $f \circ f$ is still quadratic-like near the origin.

Central to the picture is the existence of a unique fixed point g for $\mathscr{R}$. Since g and $\mathscr{R}(g)$ are equivalent, g satisfies the Cvitanović-Feigenbaum functional equation $\alpha g \circ g(z) = g(\alpha z)$. The universal constant λ above is the unique expanding eigenvalue for $\mathscr{R}$ at g.

We will sketch a construction of this fixed point g which parallels the geometrization of surface bundles.

Douady and Hubbard define a *tuning map* $\tau : M \to M$ such that $I(\mathcal{R}(f)) = \tau^{-1}(I(f))$ when defined. Thus τ describes the inverse of renormalization as it acts on the inner class. One finds that $\tau(f_n) = f_{n+1}$ and τ fixes the Feigenbaum polynomial f_∞. This accomplishes:

Step 1: Form the limit $f_\infty = \lim\ \tau^n(f_0)$.

Now let $\mathcal{Q}_\infty$ denote those f with inner class $I(f) = f_\infty$. Then $\mathcal{R}(\mathcal{Q}_\infty) \subset \mathcal{Q}_\infty$ and f and $\mathcal{R}(f)$ are quasiconformally conjugate near $K(f)$ for any f in $\mathcal{Q}_\infty$.

Step 2: Form the limit $g = \lim\ \mathcal{R}^n(f_\infty)$. This is a fixed point for $\mathcal{R}$, and in fact all f in $\mathcal{Q}_\infty$ are attracted to g.

The proof of (Step 2) again appeals to two principles.

Compactness: For any f in $\mathcal{Q}_\infty$, $< \mathcal{R}^n(f) >$ ranges in a compact subset of $\mathcal{Q}$. This is a fundamental result of Sullivan [Sul4]. Thus there is a subsequence of n such that $\mathcal{R}^n(f) \to g_0$, $\mathcal{R}^{n-1}(f) \to g_1$, $\ldots R^{n-k}(f) \to g_k$ and the tower $< g_0, g_1, \ldots >$ satisfies $\mathcal{R}(g_k) = g_{k-1}$. We can then apply:

Rigidity: Such a tower admits no quasiconformal deformations [Mc4].

Since $< g_0, g_1, \ldots >$ is conjugate to $< g_1, g_2, \ldots >$ by a suitable limit of a quasiconformal conjugacy between f and $\mathcal{R}(f)$, these towers are conformally identical, and in particular $\mathcal{R}(g_0) = g_0$. By rigidity of all limiting towers, the full sequence $\mathcal{R}^n(f) \to g_0$ and this fixed point is unique.

Geometric Limits. The fixed point of renormalization g is *not* itself rigid. Its universal structure is a result of being embedded deep in the dynamics of f. The tower $< g_0, g_1, \ldots >$ can be thought of as a geometric limit of the dynamical system generated by f as one rescales about its critical point. The limiting dynamic is *divisible* ($g_n = g_{n+1} \circ g_{n+1}$), and its Julia set fills the whole plane. If we set g_{-n} = the 2^nth iterate of g_0, then renormalization acts as a shift on the bi-infinite tower $< \ldots g_{-1}, g_0, g_1 \ldots >$ in a manner reminiscent of the deck transformation acting on the $\mathbb{Z}$-covering space of M^3 constructed before.

Self-Similarity in the Mandelbrot Set and in Bers' Slice. In the polynomial-like setting, one can actually define countably many renormalization and tuning operators $\mathcal{R}_c$, τ_c, one for each c such that the critical point of $z^2 + c$ is periodic. Milnor has made a detailed computer study of these operators, supporting many conjectures [Milnor]; among them, that τ_c has a unique fixed point and is differentiable there, with derivative given by the inverse of the expanding eigenvalue of $\mathcal{R}_c$ at *its* fixed point.

Similarly, we conjecture (in the case of one dimensional Teichmüller spaces) that Bers' boundary is self-similar about the point ϱ_∞ constructed in (Step 1), with similarity factor given by the expanding eigenvalue of the mapping class Φ at the fixed point of (Step 2). Dave Wright's computer study of the closely related Maskit boundary for the Teichmüller space of a punctured torus supports this conjecture [Wr]. In this case the mapping class group is $SL_2\mathbb{Z}$, and the expanding

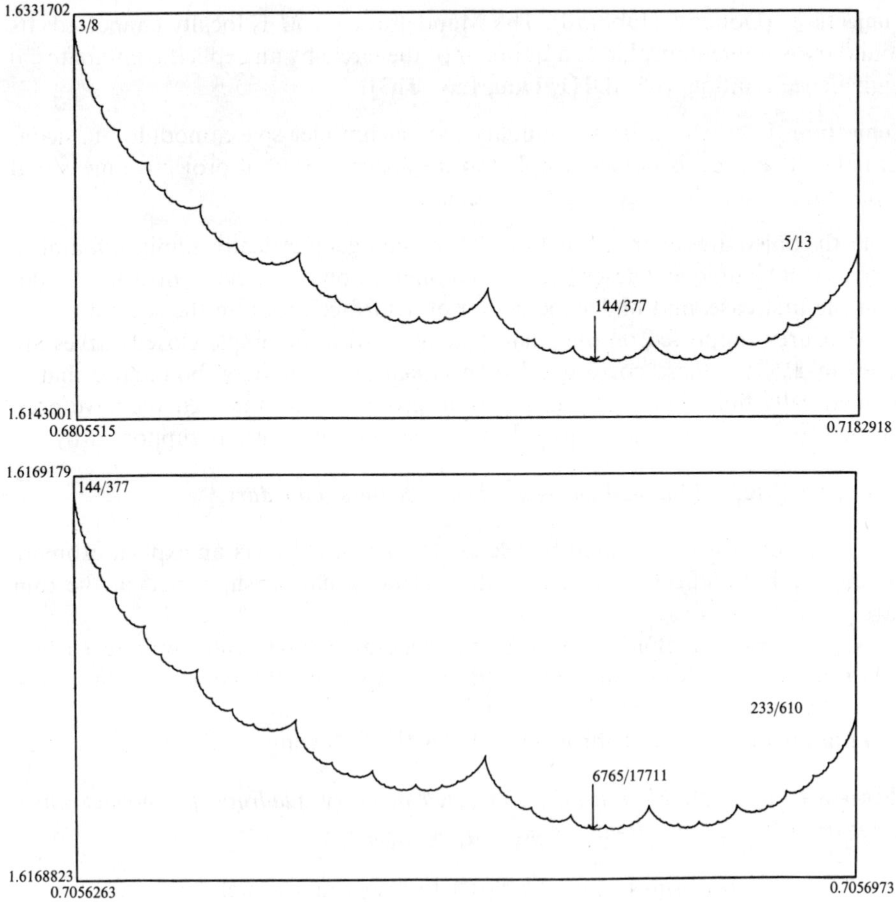

Fig. 1. Self-similarity at the edge of Teichmüller space

eigenvalue is algebraic (but different from the eigenvalues of the matrix). For example, when $\Phi = \begin{pmatrix} 2 & 1 \\ 1 & 1 \end{pmatrix}$, the boundary scales by $\lambda = 4.79129\ldots = \frac{5+\sqrt{21}}{2}$; see Fig. 1 for two blowups around ϱ_∞ computed by Wright.

Remark. Sullivan has established a compactness theorem for arbitrary compositions of a finite number of renormalization operators $\mathcal{R}_c$, with the condition that c is real. Our rigidity argument applies whenever such a compactness result is available. Much progress on a conceptual understanding of the full renormalization picture, including a different approach to rigidity, appears in [Sul4].

4. Boundaries and Laminations

We conclude with a very brief account of progress on the boundary of Teichmüller space and the boundary of the Mandelbrot set.

Conjecture (Douady–Hubbard). The Mandelbrot set M is locally connected. Its boundary is homeomorphic to a quotient of the circle by an explicit combinatorial equivalence relation. (Cf. [DH1, Dou, Lav, Th3])

Conjecture (Thurston). Bers' boundary for Teichmüller space, modulo quasiconformal equivalence, is homeomorphic to the space $\mathbb{P}\mathcal{ML}$ of projective measured laminations, modulo forgetting the measure.

Both conjectures express the hope that certain geometrically infinite dynamical systems can be uniquely described by a lamination on the circle — invariant under z^2 in the first case, and under the action of a surface group in the second.

Measures supported on maximal systems of disjoint simple closed curves are dense in $\mathbb{P}\mathcal{ML}$; these correspond to *maximal cusps* in Bers' boundary, that is geometrically finite limits of quasifuchsian groups where these curves have been pinched to form rank one cusps. Thus Thurston's conjecture is supported by:

Theorem 8 [Mc3]. *Maximal cusps are dense in Bers' boundary.*

This result was conjectured by [Bers]. The proof is by via an explicit estimate for the algebraic effect of a quasiconformal deformation supported in the thin part.

Also relevant is Bonahon's result: a general geometrically infinite surface group admits an ending lamination [Bon], supporting the conjecture that geometrically finite groups are dense in $AH(S)$.

Progress on the Mandelbrot set includes the following breakthrough:

Theorem 9 (Yoccoz). *M is locally connected at every quadratic polynomial which is not in the image of a tuning map.*

Yoccoz's result brings us a step closer to resolving the well-known:

Conjecture. Hyperbolic dynamics is open and dense in the space of complex quadratic polynomials.

It seems likely that Yoccoz's theorem generalizes to the case of polynomials lying in the image of only finitely many tuning maps. If so, by [MSS], the density of hyperbolic dynamics is equivalent to the quasiconformal rigidity of infinitely renormalizable polynomials.

References

[Bers] L. Bers: On boundaries of Teichmüller spaces and on kleinian groups: I. Ann. Math. **91** (1970) 570–600

[Bon] F. Bonahon: Bouts des variétés hyperboliques de dimension 3. Ann. Math. **124** (1986) 71–158

[CT] J. W. Cannon, W. P. Thurston: Group invariant Peano curves. Preprint.

[Cvi] P. Cvitanović: Universality in chaos. Adam Hilger Ltd, 1984.

[Dou] A. Douady: Algorithms for computing angles in the Mandelbrot set. In: M. F. Barnsley, S. G. Demko, (eds.) Chaotic dynamics and fractals. Academic Press, 1986, pp. 155–168

[DH1] A. Douady, J. Hubbard: Étude dynamique des polynômes complexes. Pub. Math. d'Orsay, 1984

[DH2] A. Douady, J. Hubbard: On the dynamics of polynomial-like mappings. Ann. Sci. Éc. Norm. Sup. **18** (1985) 287–344

[DH3] A. Douady, J. Hubbard: A proof of Thurston's topological characterization of rational maps. To appear in Acta Mathematica

[Lav] P. Lavaurs: Une description combinatoire de l'involution définie par M sur les rationnels à dénominateur impair. CRAS Paris **303** (1986) 143–146

[Mc1] C. McMullen: Amenability, Poincaré series and quasiconformal maps. Invent. math. **97** (1989) 95–127

[Mc2] C. McMullen: Iteration on Teichmüller space. Invent. math. **99** (1990) 425–454

[Mc3] C. McMullen: Cusps are dense. Ann. Math. **133** (1991) 217–247

[Mc4] C. McMullen: Renormalization and 3-manifolds which fiber over the circle. In preparation

[Mc5] C. McMullen: Riemann surfaces and the geometrization of 3-manifolds. Submitted to the Annals of the New York Academy of Sciences

[Milnor] J. Milnor: Self-similarity and hairiness in the Mandelbrot set. In M. C. Tangora (ed.) Computers in geometry and topology. Lect. Notes Pure Appl. Math. Dekker, 1989

[Mor] J. Morgan: On Thurston's uniformization theorem for three dimensional manifolds. In: The Smith conjecture. Academic Press, 1984, pp. 37–125

[Mum] D. Mumford: A remark on Mahler's compactness theorem. Proc. AMS **28** (1971) 289–294

[MSS] R. Mañe, P. Sad, D. Sullivan: On the dynamics of rational maps. Ann. Sci. Éc. Norm. Sup. **16** (1983) 193–217

[Poin] H. Poincaré: Mémoire sur les fonctions Fuchsiennes. Acta Math. **1** (1882/3), 193–294

[Sul1] D. Sullivan: On the ergodic theory at infinity of an arbitrary discrete group of hyperbolic motions. In: Riemann surfaces and related topics: Proceedings of the 1978 Stony Brook Conference. Ann. Math. Studies 97. Princeton, 1981

[Sul2] D. Sullivan: Quasiconformal homeomorphisms and dynamics I: Solution of the Fatou-Julia problem on wandering domains. Ann. Math. **122** (1985) 401–418

[Sul3] D. Sullivan: Quasiconformal homeomorphisms in dynamics, topology and geometry. In: Proceedings of the International Conference of Mathematicians. American Math. Soc., 1986, pp. 1216–1228

[Sul4] D. Sullivan: Bounds, quadratic differentials and renormalization conjectures. In: American Mathematical Society Centennial Publications, volume 2: Mathematics into the twenty-first century. Amer. Math. Soc., to appear

[Th1] W. P. Thurston: Hyperbolic structures on 3-manifolds I: Deformations of acylindrical manifolds. Ann. Math. **124** (1986) 203–246

[Th2] W. P. Thurston: Hyperbolic structures on 3-manifolds II: Surface groups and 3-manifolds which fiber over the circle. To appear in Ann. Math.

[Th3] W. P. Thurston: On the combinatorics and dynamics of iterated rational maps. Preprint.

[Wr] D. Wright: The shape of the boundary of Maskit's embedding of the Teichmüller space of once-punctured tori. Preprint.

Analytic Capacity for Arcs

Takafumi Murai

Department of Mathematics, School of Science, Nagoya University, Nagoya, 464-01 Japan

1. Introduction

For a domain Ω in the extended complex plane $\mathbb{C} \cup \{\infty\}$, $H^\infty(\Omega)$ denotes the Banach space of bounded analytic functions in Ω with supremum norm $\|\cdot\|_{H^\infty}$. For $\zeta \in \Omega$, we put $c(\zeta; \Omega) = \sup|f'(\zeta)|$, where the supremum is taken over all $f \in H^\infty(\Omega)$, $\|f\|_{H^\infty} \leq 1$, and $f'(\infty)$ is defined by $\lim_{z\to\infty} z\{f(\infty) - f(z)\}$. Given $f \in H^\infty(\Omega)$, $\|f\|_{H^\infty} \leq 1$, we take $g(z) = \{f(z) - f(\zeta)\}/\{1 - f(z)\overline{f(\zeta)}\}$. Then $\|g\|_{H^\infty} \leq 1$, $g(\zeta) = 0$ and $|g'(\zeta)| \geq |f'(\zeta)|$. Thus, for the computation of $c(\zeta; \Omega)$, we may restrict our attention to functions vanishing at ζ. For $p \geq 1$ and a domain Ω bounded by a finite number of analytic Jordan curves, $H^p(\Omega)$ denotes the H^p-space of analytic functions f in Ω with norm $\|f\|_{H^p} = \{(1/2\pi)\int_{\partial\Omega}|f|^p|dz|\}^{1/p}$, where $\partial\Omega$ is the boundary of Ω. The condition $\int_{\partial\Omega}(\mathcal{M}f)^p|dz| < \infty$ is required for each $f \in H^p(\Omega)$, where $\mathcal{M}f$ is a non-tangential maximal function which controls the behaviour of f near the boundary. The analytic capacity of a compact set E in $\mathbb{C}$ is defined by $\gamma(E) = c(\infty; \Omega_E)$, where Ω_E is the component of E^c containing ∞. Analytic capacity plays an important role in the theory of conformal mapping [SO], the 2-dimensional fluid dynamics [Mi], approximation theory [Ga, V2, Z] and singular integrals [Ch, J, Mu3]. Ahlfors [A] shows that a compact set E satisfying Riemann's theorem on removable singularities is characterized by $\gamma(E) = 0$, and Garabedian [G] studies $\gamma(\cdot)$ from the point of view of the dual extremum problem. Basic properties are mentioned in [AB, Ga, Z]. Here is the classical method of the computation of $\gamma(E)$ in the case where $E^c = \Omega_E$ and ∂E consists of a finite number of analytic Jordan curves.

The Ahlfors-Garabedian method [A, G]: *Construct a pair (f_0, ψ_0) of functions so that $f_0 \in H^\infty(E^c)$, $\|f_0\|_{H^\infty} \leq 1$, $f_0(\infty) = 0$, $\psi_0 \in H^1(E^c)$, $\psi_0(\infty) = 1$ and*

(1) $$\frac{1}{i} f_0 \psi_0 \, dz = |\psi_0||dz| \text{ almost everywhere (a.e.) on } \partial E,$$

where the orientation of dz is chosen so that E^c lies to the left. Once such a pair has been constructed, we have

(2) $$\gamma(E) = f_0'(\infty) = \|\psi_0\|_{H^1}.$$

Proceedings of the International Congress of Mathematicians, Kyoto, Japan, 1990

In fact,

$$|f_0'(\infty)| \le \gamma(E) = \sup{}_0 |f'(\infty)| = \sup{}_0 \left|\frac{1}{2\pi}\int_{\partial E} f\,dz\right|$$

$$= \sup{}_0 \left|\frac{1}{2\pi}\int_{\partial E} f\psi_0\,dz\right| \qquad (\text{by } \psi_0(\infty) = 1)$$

$$\le \|\psi_0\|_{H^1} = \frac{1}{2\pi i}\int_{\partial E} f_0\psi_0\,dz \qquad (\text{by } (1))$$

$$= \frac{1}{2\pi i}\int_{\partial E} f_0\,dz = f_0'(\infty),$$

which yields (2). Here $\sup_0$ is the supremum over all $f \in H^\infty(E^c)$, $\|f\|_{H^\infty} \le 1$, $f(\infty) = 0$. Thus it is essential to construct the pair (f_0, ψ_0). For any compact set F, there exists uniquely $f(\cdot; \Omega_F) \in H^\infty(\Omega_F)$ with norm 1 such that $f'(\infty; \Omega_F) = \gamma(F)$ [Ga, p. 24]. This is called the Ahlfors function of F. Let $\{\Omega_n\}_{n=1}^\infty$ be an increasing sequence of domains $\ni\infty$ with smooth boundaries such that $\bigcup_{n=1}^\infty \Omega_n = \Omega_F$. Then there exists uniquely $\psi_n \in H^1(\Omega_n)$ such that $\psi_n(\infty) = 1$ and $\|\psi_n\|_{H^1} = \gamma(\Omega_n^c)$. The sequence $\{\psi_n\}_{n=1}^\infty$ converges in Ω_F and the limit $\psi(\cdot; \Omega_F)$ is called the Garabedian function of F; $\psi(\cdot; \Omega_F)$ is determined independent of $\{\Omega_n\}$ [Sm, Su1, 2]. Note that $f_0 = f(\cdot; E^c)$ and $\psi_0 = \psi(\cdot; E^c)$. Our main theme is the study of $\gamma(\cdot)$, $f(\cdot; \cdot)$ and $\psi(\cdot; \cdot)$. The study from the point of view of Green's functions and harmonic measures is classical. In this talk, we focus on an approach based on the singular integral operator $\mathscr{H}$ defined by Calderón [C]. Let $\mathscr{A}$ denote the totality of sets E such that E consists of a finite number of mutually disjoint closed arcs $\{C_j\}_{j=1}^n$ and each arc C_j is a finite union of analytic arcs. For the computation of $\gamma(\cdot)$, we restrict our attention to sets in $\mathscr{A}$. Here are two motivations to deal with sets in $\mathscr{A}$. For a compact set F with a smooth boundary, we can discuss the Hadamard variation and the Schiffer variation [S2] of $\gamma(F)$. Then we can express $\gamma(F)$ as a perturbation from $\gamma(E)$ for a set E in $\mathscr{A}$. Thus, *in order to get global properties of* $\gamma(\cdot)$, it is necessary to study $\gamma(E)$, $E \in \mathscr{A}$. Another motivation is as follows. Given a compact set F, we can find a finite union G of closed disks such that $|\gamma(F) - \gamma(G)|$ is arbitrarily small. Note that $\gamma(G) = \gamma(\partial G)$. Removing some arcs on ∂G, we can find $E \in \mathscr{A}$ such that $|\gamma(G) - \gamma(E)|$ is arbitrarily small. (In this approximation, E can be chosen so that the connectivity of E is less than or equal to that of F.) Thus it is sufficient to study $\gamma(E)$, $E \in \mathscr{A}$. For $E \in \mathscr{A}$, ∂E denotes the boundary of E having two sides. The H^p-space $H^p(E^c)$ of analytic functions in E^c is analogously defined as above.

2. The Singular Integral Operator $\mathscr{H}$

For $E \in \mathscr{A}$, $L^2(E)$ denotes the L^2-space of functions on E with respect to the arc-length $|dz|$. The singular integral operator $\mathscr{H}_E$ from $L^2(E)$ to itself is defined by

$$\mathscr{H}_E h(z) = \frac{1}{\pi}\,\text{p.v.}\int_E \frac{1}{\zeta - z} h(\zeta)|d\zeta|,$$

where p.v. is the principal value. There are many articles about $\mathscr{H}$ [C, Ch, CJS, CMM, D, Me, Mu3]. The operator $\bar{\mathscr{H}}_E$ is defined by $\bar{\mathscr{H}}_E h = \overline{\mathscr{H}_E \bar{h}}$ and the inverse operator of $\mathrm{Id} - \mathscr{H}_E \bar{\mathscr{H}}_E$ is denoted by $\mathscr{T}_E$, where Id is the identity operator. Here are fundamental expressions of $\psi(\cdot; E^c)$, $f(\cdot; E^c)$ and $\gamma(E)$ in terms of $\mathscr{H}_E$ [Mu4]:

$$\psi(z; E^c) = \left\{1 + \frac{1}{\pi}\int_E \frac{1}{\zeta - z}\bar{\mathscr{H}}_E \mathscr{T}_E 1(\zeta)|d\zeta|\right\}^2,$$

$$f(z; E^c) = \frac{1}{\pi}\int_E \frac{1}{\zeta - z}\bar{\mathscr{T}}_E 1(\zeta)|d\zeta| / \sqrt{\psi(z; E^c)},$$

(3) $$\gamma(E) = \frac{1}{\pi}\int_E \mathscr{T}_E 1(\zeta)|d\zeta|.$$

The proof is founded on Garabedian's duality theorem [G]: $\gamma(E) = \inf\{\|\phi\|_{H^2}^2; \phi \in H^2(E^c), \phi(\infty) = 1\}$. This shows that

$$\gamma(E) = \inf\left\{\frac{1}{\pi}\int_E (|1 + \mathscr{H}_E h|^2 + |h|^2)|dz|; h \in L^2(E)\right\}.$$

Using the standard variational method, we obtain the required formulae. The following formula plays an important role to compute $\gamma(E)$ practically:

(4) $$\mathscr{H}_E\{u\mathscr{H}_E v + \mathscr{H}_E u \cdot v\} = \mathscr{H}_E u \cdot \mathscr{H}_E v - \varrho_E uv \quad (u, v \in L^2(E), \varrho_E = \overline{dz}/dz).$$

The other relations between $\gamma(E)$ and $\mathscr{H}_E$ are mentioned in [Mu3]; roughly speaking, $\gamma(E)$ is comparable with $1/\|\mathscr{H}_E\|_{1,w}$, where $\|\mathscr{H}_E\|_{1,w}$ denotes the norm of $\mathscr{H}_E$ as an operator from the L^1 space to the weak L^1 space. If E is contained in the real line $\mathbb{R}$, $\mathscr{H}_E$ is called the Hilbert transform and denoted by H_E. Formula (4) yields that

$$\mathscr{T}_E h = \tfrac{1}{2}h + \tfrac{1}{4}\{\tau_E H_E(h\tau_E^{-1}) - \tau_E^{-1}H_E(h\tau_E)\},$$

$$H_E\mathscr{T}_E h = \tfrac{1}{4}\{\tau_E H_E(h\tau_E^{-1}) + \tau_E^{-1}H_E(h\tau_E)\} \qquad (h \in L^2(E), E \subset \mathbb{R}),$$

where $\tau_E = \exp\{(\pi/4)H_E 1\}$. Thus $\mathscr{T}_E$ and $H_E\mathscr{T}_E$ handle easily in the case of $E \subset \mathbb{R}$. As application of our method, we obtain the following formulae:

$$\gamma(E) = |E|/4 \quad (E \subset \mathbb{R}) \quad [\mathrm{P}], \quad \gamma(E) = \sin(|E|/4) \quad (E \subset \mathbb{T}),$$

$$\delta(E; 0, \infty) = 2\tan(|E|/8) \quad (E \subset \mathbb{T}) \quad [\mathrm{Mu6}].$$

Here $\mathbb{T}$ is the unit circle, $|\cdot|$ is the 1-dimensional Lebesgue measure and $\delta(E; 0, \infty)$ denotes the supremum of $|f(0) - f(\infty)|$ over all $f \in H^\infty(E^c)$, $\|f\|_{H^\infty} \le 1$. Using (4), we can find the concrete forms of $(f(\cdot; \cdot), \psi(\cdot; \cdot))$ in various cases; once a pair has been found, the check is very easy as stated in the introduction.

3. Null Sets

The 1-dimensional Hausdorff measure is also denoted by $|\cdot|$ [F, p. 7]. We have $\gamma(E) \le |E|/\pi$ (Painlevé). This shows that $\gamma(E) = 0$ if the Hausdorff dimension

[F, p. 7] $\dim(E)$ is less than 1. On the other hand, $\gamma(E) > 0$ if $\dim(E) > 1$. Thus the case of $\dim(E) = 1$ is critical. Vitushkin [V1] constructed a planar Cantor set P_∞ such that $\gamma(P_\infty) = 0$ and $|P_\infty| > 0$ (cf. [Ga, p. 87]). We are interested in the geometric structure of sets of Vitushkin-Garnett type. Here is a deformation of P_∞ which handles easily: $Q_\infty = \bigcap_{n=0}^\infty \{\bigcup_{n=k}^\infty Q_k\}^{c\ell}$, where $Q_0 = [0, 1]$ and

$$Q_k = \left\{x + i \sum_{j=1}^{k} (-1)^j 2^{-j} \operatorname{sign}(\sin \pi 2^{j-1} x); 0 \leq x \leq 1\right\}^{c\ell} \qquad (k \geq 1).$$

(The notation $E^{c\ell}$ denotes the closure of E and sign $0 = 1$.) Then $\gamma(Q_\infty) = 0$ and $|Q_\infty| > 0$. There are two methods of the proof of $\gamma(Q_\infty) = 0$. The first method is as follows: *Supposing that the nontrivial Ahlfors function of Q_∞ exists, show a contradiction* [J, M1]. Mattila applies Besicovitch's set theory [F], and Jones uses the BMO norm $\{(i/2) \sup_{z \in \Omega} \iint_\Omega G(z, w)|f'(w)|^2\, dw \wedge \overline{dw}\}^{1/2}$ ($\Omega = Q_\infty^c$), where $G(z, w)$ is the Green's function of Ω. The second method is based on the construction of the approximate Garabedian function [Mu1]: *For $\varepsilon > 0$, construct a pair (ψ, R) of a function $\psi \in H^1(R)$ and a domain R so that $\infty \in R \subset Q_\infty^c$, $\psi(\infty) = 1$ and $\|\psi\|_{H^1} \leq \varepsilon$.* Once such a pair has been constructed, we have $\gamma(Q_\infty) \leq \gamma(R^c) \leq \|\psi\|_{H^1} \leq \varepsilon$, which gives $\gamma(Q_\infty) = 0$. The following dipole function plays an important role in the construction of such a pair:

$$p(z) = \exp\left\{\frac{\xi}{z - b} - \frac{\xi}{z - a}\right\} \qquad \left(a, b \in \mathbb{C}, \frac{\xi}{b - a} > 0\right).$$

We have $p(\infty) = 1$, $p(z) = 1 + O(|z|^{-2})$ $(z \to \infty)$ and $|p(z)| < 1$ in the strip with width $|b - a|$ which is perpendicular to the segment with endpoints a, b and contains it. The required function ψ is expressed as a product of dipole functions. This method works for non-homogeneous Cantor sets. Next we show some estimates of $\gamma(\cdot)$ from below. Calderón, Havin and Marshall [Ma] show that $\gamma(E) > 0$ if E is a compact set on a rectifiable curve satisfying $|E| > 0$; this theorem was formerly called the Denjoy conjecture [Ga, p. 36] and, in the proof, Calderón's theorem [C] on $\mathscr{H}$ plays an important role. Let $\mathrm{Bu}(\cdot)$ denote the Buffon needle probability (the Favard length). Since $\gamma(\cdot) \leq |\cdot|/\pi$ and $\mathrm{Bu}(\cdot) \leq \mathrm{Const}|\cdot|$, it is interesting to compare $\gamma(\cdot)$ with $\mathrm{Bu}(\cdot)$ (Vitushkin's problem [Ma, V2]). Jones and Murai [JM] show that there exists a compact set E such that $\mathrm{Bu}(E) = 0$ and $\gamma(E) > 0$ (cf. [M2, Mu2, 3]). This theorem suggests that $\mathrm{Bu}(\cdot) \leq \mathrm{Const}\, \gamma(\cdot)$. There are many problems about this topic [HHN, pp. 485–514]. We here note two problems: (I) *Suppose that $|E| < \infty$. Does* $\mathrm{Bu}(E) = 0$ *imply $\gamma(E) = 0$* [HHN, p. 491]? (II) *Construct a compact set E so that $\gamma_0(E) = 0$ and $\gamma(E) > 0$* [Ga, p. 55]. Here $\gamma_0(E)$ denotes the supremum of $|\int d\mu|$ over all Cauchy potentials $\mathscr{C}\mu(z) = \displaystyle\int \frac{1}{\zeta - z}\, d\mu$ of measures μ on E such that $\|\mathscr{C}\mu\|_{H^\infty} \leq 1$. Problem I originates in Besicovitch's set theory. A Borel set E satisfying $|E| < \infty$ is regular if $\underline{d}(z, E) = 1$ $|\cdot|$-a.e. on E, and E is irregular if $\underline{d}(z, E) < 1$ $|\cdot|$-a.e. on E, where $\underline{d}(z, E) = \liminf_{r \to 0} |\{\zeta \in E; |\zeta - z| \leq r\}|/(2r)$. If $|E| < \infty$, then $\mathrm{Bu}(E) = 0$ is equivalent to the irregularity of E [F, p. 89]. Problem II is posed to clarify the difference between {bounded Cauchy potentials} and $H^\infty(\cdot)$. Take a bad arc E (like a snowflake) such that $\dim(E) = 1$ and the diameter is equal to 1. Note that

$\gamma(E) \geq 1/4$ [Ga, p. 9]. *Is $\gamma_0(E)$ small? Study Ahlfors functions which cannot be expressed as Cauchy potentials of measures.*

4. Projection

Let $\mathscr{L}_\theta$ $(-\pi/2 < \theta \leq \pi/2)$ denote the straight line $x \sin\theta = y \cos\theta$ and let $\mathrm{pr}_\theta E$ denote the projection of E to $\mathscr{L}_\theta$. As is well known, a regular set is contained in a countable union of rectifiable graphs [F, p. 45]. Thus CHM's theorem [Ma] shows that $\gamma(E) > 0$ if E is a regular set satisfying $|E| > 0$. A set E satisfying $|E| < \infty$ is irregular if $|\mathrm{pr}_\theta E| = |\mathrm{pr}_{\theta'} E| = 0$ for two distinct numbers θ, θ' [F, p. 90]. From this point of view, it is interesting to estimate $\gamma(E)$ in terms of the projection of E to one direction. Recall that $\gamma(Q_\infty) = 0$ and $\mathrm{pr}\, Q_\infty = [0, 1]$, where $\mathrm{pr} = \mathrm{pr}_0$. To understand the geometric meaning of Q_∞, we begin with Q_1. Let $\Gamma(z) = [-1/2, 1/2] \cup (z + [-1/2, 1/2])$ and $\gamma(z) = \gamma(\Gamma(z))$. In hydrodynamics, $\Gamma(z)$ is regarded as a biplane wing section [Mi, Chap. VII] and there are many articles about $\Gamma(z)$ [Fe, Gar]. In order to practically compute $\gamma(z)$, we introduce the lift coefficient $\mathscr{L}(z)$ of $\Gamma(z)$ [Fe, Mi, p. 203]. There exists uniquely $f_z \in H^1(\Gamma(z)^c)$ such that f_z is real-valued continuous on $\partial\Gamma(z) - \{\pm 1/2, z \pm 1/2\}$, $f_z(\infty) = -i$ and f_z satisfies Joukowski's hypothesis [Mi, p. 199] "$|f_z(\zeta)| < \infty$ $(\zeta = 1/2, z + 1/2)$". Taking account of Blasius' formula [Mi, p. 173], we define the lift coefficient $\mathscr{L}(z)$ of $\Gamma(z)$ by

$$\mathscr{L}(z) = \frac{1}{4}\left|\frac{1}{2\pi}\int_{\partial\Gamma(z)} f_z(\zeta)^2\, d\zeta\right| \left(= \frac{1}{2}|f_z'(\infty)|\right).$$

It is sufficient to study $\gamma(z)$ in $\mathbb{P} = \{\mathrm{Re}\, z \geq 0, \mathrm{Im}\, z \geq 0\} - [0, 1]$. Then the following assertion [Mu5] holds:

$$\gamma(z) = \frac{1}{2} + \frac{\mathrm{Im}\, z}{2}\int_{\lambda(z)} \left\{\frac{\gamma(\zeta)}{\mathscr{L}(\zeta)} - 1\right\} \frac{d(\mathrm{Im}\,\zeta)}{(\mathrm{Im}\,\zeta)^2} \qquad (z \in \mathbb{P}), \tag{5}$$

where $\lambda(z)$ denotes the arc in $\mathbb{P}$ with endpoints z and a positive number such that (the modulus [SO, p. 199] *of $\Gamma(\zeta)^c$) is invariant on $\lambda(z)$ and, z is chosen as the starting point of the curvilinear integral. The inequality $\mathscr{L}(z) \leq \gamma(z)$ holds, and the equality $\mathscr{L}(z) = \gamma(z)$ holds if and only if* $\mathrm{Im}\, z = 0$. Using (5), we obtain the following equality [Mu5]:

$$\min_{x\geq 0, y \geq 0} \gamma(x + iy)/\gamma(x) = \min_{y\geq 0} \gamma(1 + iy)/\gamma(1)(0.9\ldots).$$

Note that $\gamma(x) = |\Gamma(x)|/4\Gamma(x)|/4$ $(x \in \mathbb{R})$ and the projection of $\Gamma(1 + iy)$ to $\mathbb{R}$ overlaps only at 1/2. This is a reason why we take Q_1 as the first step to construct a compact set Q_∞ of Vitushkin-Garnett type. Even $\Gamma(z)$, the behaviour of $\gamma(z)$ is not simple. If $0 < x_0 < 1$ is sufficiently near to 1, then $\gamma(x_0 + iy)$ has at least two local extrema in $(0, \infty)$ as a function of y. From this fact, we conjecture that there exists a compact set E such that $\gamma(E) = 0$ and $\gamma(T(E)) > 0$, where $T(x + iy) = x + i2y$ (or $= x + (iy/2)$). The following theorem gives the geometric information about compact sets E such that $\gamma(E)/|\mathrm{pr}\, E|$ is small.

Theorem 1 [Mu2-4]. *If E is a compact set on a graph Γ such that $|\Gamma| \leq 1$, then $\gamma(E) \geq C|\mathrm{pr}\ E|^{3/2}$, where C is an absolute constant. The power 3/2 is best possible.*

This is a generalization of CHM's theorem with a quantitative estimate and an interpretation, in terms of $\gamma(\cdot)$, of the optimal estimate of $\|C[\cdot]\|$ defined later. Dilating the coordinate axes, we obtain $\gamma(E) \geq C|\mathrm{pr}\ E|^{3/2}|\Gamma|^{-1/2}$ for any compact set E on a rectifiable graph Γ containing E. Thus, if $|\mathrm{pr}\ E| \geq 1$ and $|\Gamma| \leq M$, then $\gamma(E) \geq C/\sqrt{M}$. We have $\gamma(Q_n) \geq \mathrm{Const}/\sqrt{n}$, for example. The first half assertion in this theorem is rewritten in the following form also: *If $\gamma(E) \leq \varepsilon$, then a graph of length less than $C^2\delta^3\varepsilon^{-2}$ does not contain any subset F of E satisfying* $|\mathrm{pr}\ F| \geq \delta$. In order to prove the inequality in Theorem 1, it is necessary to investigate $\mathscr{H}$ in detail. Let BMO denote the Banach space of functions on $\mathbb{R}$, modulo constants, of bounded mean oscillation. For a real-valued function $a \in \mathrm{BMO}$, the singular integral operator $C[a]$ from the L^2-space of functions on $\mathbb{R}$ to itself is defined by

$$C[a]h(x) = \frac{1}{\pi}\mathrm{p.v.}\int_{-\infty}^{\infty} \frac{1}{y - x + i(A(y) - A(x))} h(y)\,dy,$$

where $A(x) = \int_0^x a(t)\,dt$. This is a version of $\mathscr{H}_\Gamma$, $\Gamma = \{x + iA(x); x \in \mathbb{R}\}$ by Calderón [C]. The following inequality [Mu3, p. 53] is established: $\|C[a]\| \leq \mathrm{Const}\{1 + \sqrt{\|a\|_{\mathrm{BMO}}}\}$. The proof in [Mu3] is not short, however, the method is founded on only one principle "the Calderón-Zygmund decomposition". Using the separation theorem and the Calderón-Zygmund decomposition, we can deduce the required inequality. In order to see the exactness of the power 3/2, it is sufficient to construct a compact set Q_n^* and a graph Γ_n^* containing Q_n^* so that $\gamma(Q_n^*) \leq \mathrm{Const}/\sqrt{n}$, $|\mathrm{pr}\ Q_n^*| = 1$ and $|\Gamma_n^*| \leq \mathrm{Const}\ n$. Our example is related to David's example [D]: *For any $M \geq 1$, there exists a real-valued function $a_M \in \mathrm{BMO}$ such that $\|a_M\|_{\mathrm{BMO}} \leq M$ and $\|C[a_M]\| \geq \mathrm{Const}\sqrt{M}$.* In the construction of (Q_n^*, Γ_n^*), the following fact [Mu4] is important: $\lim_{m\to\infty}\gamma(R(m)) < 1/4$, where $R(m) = \{x + i2^{-m}\mathrm{sign}(\sin \pi 2^m x); 0 \leq x \leq 1\}^{c\ell}$. (Note that $\{R(m)\}_{m=1}^{\infty}$ converges to $[0, 1]$ and $\gamma([0, 1]) = 1/4$.) Put

$$R(m_1, \ldots, m_n) = \left\{x + i\sum_{k=1}^{n} 2^{-m_1-\cdots-m_k}\,\mathrm{sign}(\sin \pi 2^{m_1+\cdots+m_k}x); 0 \leq x \leq 1\right\}^{c\ell}.$$

Choosing a sequence $\{m_k^*\}_{k=1}^{\infty}$ of positive integers so that $\{m_{k+1}^*/m_k^*\}_{k=1}^{\infty}$ is rapidly increasing, we put $Q_n^* = R(m_1^*, \ldots, m_n^*)$ $(n \geq 1)$. Then (3) yields that $\gamma(Q_{n+1}^*) \leq \gamma(Q_n^*) - \delta_0\gamma(Q_n^*)^3$ $(n \geq 1)$ with a small constant δ_0. Thus $\gamma(Q_n^*) \leq \mathrm{Const}/\sqrt{n}$ $(n \geq 1)$. Connecting endpoints of Q_n^* by segments parallel to the y-axis, we obtain an arc Γ_n^* of length less than Const. n, which we can regard as a graph. Thus the power 3/2 is best possible. It is interesting to try to deduce the exactness of 3/2 by the dipole functions.

5. The arc-Length Variation

Let $\mathscr{F}$ denote the totality of domains Ω such that $\Omega^c \in \mathscr{A}$. For $\Omega \in \mathscr{F}$, let $K(z, \overline{\zeta}; \Omega)$ denote the reproducing kernel of $H_0^2(\Omega) = \{f \in H^2(\Omega); f(\infty) = 0\}$ with respect to

$|dz|/(2\pi)$, i.e.

$$f(\zeta) = \frac{1}{2\pi}\int_{\partial\Omega} \overline{K(z, \bar{\zeta}; \Omega)f(z)}\,|dz| \qquad (f \in H_0^2(\Omega)), \qquad K(\infty, \bar{\zeta}; \Omega) = 0.$$

This is called the Szegö kernel of $H_0^2(\Omega)$ [B, Fa]. We have

$$\gamma(\Omega^c) = c(\infty; \Omega) = \frac{1}{(2\pi)^2}\int_{\partial\Omega}\int_{\partial\Omega} K(z, \bar{\zeta}; \Omega)\, dz\, d\bar{\zeta}.$$

Thus it is important to study the Szegö kernel of $H_0^2(\Omega)$. There are many articles about the variational approach to various Szegö kernels [GS, HS, S1, 2, Sm, SS]. We here show a variational formula for $\gamma(E)$, $E \in \mathscr{A}$ with respect to the arc-length. *This is the variation of degenerate boundaries and related to Löwner's differential equation.* There exists uniquely a pair $(g(\cdot; \Omega), \phi(\cdot; \Omega))$ of functions in $H^2(\Omega)$ such that $g(\infty; \Omega) = 0$, $\phi(\infty; \Omega) = 1$ and $\frac{1}{i}\phi(z; \Omega)\,dz = \overline{g(z; \Omega)}|dz|$ a.e. on $\partial\Omega$. We have $\psi(z; \Omega) = \phi(z; \Omega)^2$ and $f(z; \Omega) = g(z; \Omega)/\phi(z; \Omega)$ [G]. For $\zeta \in \Omega - \{\infty\}$, there exists uniquely a pair $(K(\cdot, \bar{\zeta}; \Omega), L(\cdot, \zeta; \Omega))$ of functions such that $K(\cdot, \bar{\zeta}; \Omega) \in H_0^2(\Omega)$, $(\cdot - \zeta)L(\cdot, \zeta; \Omega) \in H^2(\Omega)$, $L(z, \zeta; \Omega) = \dfrac{1}{z - \zeta}$ + (regular terms) near ζ and $\frac{1}{i}L(z, \zeta; \Omega)\,dz = \overline{K(z, \bar{\zeta}; \Omega)}|dz|$ a.e. on $\partial\Omega$ [B]. The function $K(z, \bar{\zeta}; \Omega)$ is none other than the Szegö kernel defined above. Here are two functions Dc and D^2c necessary for our variation. For three distinct numbers $w, z, \zeta \in \Omega$, we define

$$Dc(z, \zeta; \Omega) = |L(z, \zeta; \Omega)|^2 - |K(z, \bar{\zeta}; \Omega)|^2,$$

$$D^2c(w, z, \zeta; \Omega) = 2\,\mathrm{Re}\{DL(w, z, \zeta; \Omega)\overline{L(z, \zeta; \Omega)} - DK(w, z, \zeta; \Omega)\overline{K(z, \bar{\zeta}; \Omega)}\},$$

where

$$DK(w, z, \zeta; \Omega) = L(w, z; \Omega)\overline{L(w, \zeta; \Omega)} - \overline{K(w, \bar{z}; \Omega)}K(w, \bar{\zeta}; \Omega),$$

$$DL(w, z, \zeta; \Omega) = L(w, z; \Omega)\overline{K(w, \bar{\zeta}; \Omega)} - \overline{K(w, \bar{z}; \Omega)}L(w, \zeta; \Omega).$$

In the definition, we replace $K(\cdot, \bar{\infty}; \Omega) = \overline{K(\infty, \bar{\cdot}; \Omega)}$ by $-g(\cdot; \Omega)$, and replace $L(\cdot, \infty; \Omega) = -L(\infty, \cdot; \Omega)$ by $-\phi(\cdot; \Omega)$ if one of w, z, ζ is ∞. Let Γ be a closed analytic arc such that $\Gamma \subset \Omega^{cl}$, $\Omega - \Gamma \in \mathscr{F}$ and $\Gamma \cap \Omega^c$ is at most a singleton. A continuous function $w_t \in \Gamma$ on $[0, |\Gamma|]$ is called the arc-length representation of Γ if w_0, $w_{|\Gamma|}$ are endpoints of Γ and $|\Gamma_t| = t$ $(0 \le t \le |\Gamma|)$, where $\Gamma_t = \{w_s; 0 \le s \le t\}$; we define w_t so that $\Gamma \cap \Omega^c = \{w_0\}$ if $\Gamma \cap \Omega^c \ne \emptyset$. We write $\Omega_t = \Omega - \Gamma_t$ $(0 \le t \le |\Gamma|)$.

Theorem 2 [Mu7].

(6) *For any* $0 < t \le |\Gamma|$, *the derivative* $\partial c(\infty; \Omega_t)/\partial t$, *the limit* $\lim_{u\downarrow t} Dc(w_u, \infty; \Omega_t)$ $(= Dc(w_t, \infty; \Omega_t)$, *say) exist and* $\partial c(\infty; \Omega_t)/\partial t = Dc(w_t, \infty; \Omega_t)/4$. *The right-derivative* $\partial c(\infty; \Omega_0)/\partial t$ *at* $t = 0$ *exists and* $\partial c(\infty; \Omega_t)/\partial t$ *is continuous on* $[0, |\Gamma|]$.

(7) *For any* $0 < t \le |\Gamma|$ *and* $z \in \Omega - (\Gamma \cup \{\infty\})$, *the derivative* $\partial Dc(z, \infty; \Omega_t)/\partial t$, *the limit* $\lim_{u\downarrow t} D^2c(w_u, z, \infty; \Omega_t)$ $(= D^2c(w_t, z, \infty; \Omega_t)$, *say) exist and* $\partial Dc(z, \infty; \Omega_t)/\partial t =$

$D^2c(w_t, z, \infty; \Omega_t)/4$. *For any* $z \in \Omega - (\Gamma \cup \{\infty\})$, *the right-derivative* $\partial Dc(z, \infty; \Omega_0)/\partial t$ *at* $t = 0$ *exists and* $\partial Dc(z, \infty; \Omega_t)/\partial t$ *is continuous on* $[0, |\Gamma|]$.

In the case of $t = 0$ and $w_0 \in \Omega$, the above formulae correspond to the variation by cutting a hole [HS, Sm, SS, p. 283]. Our method is based on the comparison (by the aid of conformal mappings) with the variation of segments. This theorem is applied as follows. Given $E \in \mathscr{A}$, we can write $E = C_1 \cup \cdots \cup C_n$ with mutually disjoint closed arcs $\{C_j\}_{j=1}^n$. Using the arc-length representations of C_j $(j = 1, \ldots, n)$, we define a right-continuous arc-length representation W_t $(0 \le t \le |E|)$ of E. Then (6) shows that

$$\gamma(E) = c(\infty; E^c) = \frac{1}{4}\int_0^{|E|} Dc(W_t, \infty; E_t^c)\, dt, \tag{8}$$

where $E_t = \{W_s; 0 \le s \le t\}$. (The equality $\gamma(F \cup \{w\}) = \gamma(F)$ is important in our argument.) Recall that $Dc(W_t, \infty; E_t^c)$ is defined by the limit $\lim_{u \downarrow t} Dc(W_u, \infty; E_t^c)$. Given $0 < t < |E|$ and $z \in E_t^c - \{\infty\}$, we take a right-continuous arc-length representation W_s^* $(0 \le s \le t)$ of E_t; W_s^* may not be equal to W_s. Then (7) shows that

$$Dc(z, \infty; E_t^c) = Dc(z, \infty; F_s^c) + \frac{1}{4}\int_s^t D^2c(W_u^*, z, \infty; F_u^c)\, du, \tag{9}$$

where $F_u = \{W_x^*; 0 \le x \le u\}$. Using (8) and (9), we can study $\gamma(E)$. To investigate Dc and D^2c, we introduce a class $\mathscr{G}$ of domains. Let $\mathscr{G}$ denote the totality of domains Ω with the following property; Ω is expressed as $\Omega = \Omega^* - E$ with $E \in \mathscr{A}$ and a domain $\Omega^* \supset E$ bounded by a finite number of Jordan curves $\{C_j\}_{j=1}^n$ such that each C_j is a finite union of analytic arcs. The functions Dc and D^2c are defined for domains in $\mathscr{G}$. We see that $Dc(z, \zeta; \Omega)|dz||d\zeta|$ and $D^2c(w, z, \zeta; \Omega)|dw||dz||d\zeta|$ are conformally invariant. Thus we may discuss these differential forms in canonical domains. Now we show an application of Theorem 2. It is unknown whether $\gamma(\cdot)$ is subadditive [Da, DØ, V2]. Suita [Su3] shows that $\gamma(A \cup B) \le \gamma(A) + \gamma(B)$ if A and B are disjoint continua. Equalities (8) and (9) yield that $\gamma(\cdot)$ is subadditive if $D^2c \le 0$ for any domain.

Theorem 3 [Mu7]. *The inequality* $D^2c \le 0$ *holds for simply and doubly connected domains.*

Applying this theorem to simply-connected domains, we see that $\mathrm{Cap}(A \cup B) \le \mathrm{Cap}(A) + \mathrm{Cap}(B)$ if A and B are two continua with an intersection, where $\mathrm{Cap}(\cdot)$ is logarithmic capacity [Z, p. 134]. Note that $\mathrm{Cap}(\cdot)$ is not subadditive. The case of doubly-connected domains shows Suita's subadditivity and yields that $\gamma(A \cup B) \le \gamma(A) + \gamma(B)$ if A is a union of two continua and B is a continuum intersecting with A. If Ω is simply-connected, then we may assume that Ω is the open unit disk $\mathbb{D}$ and $w = 0$. We have

$$D^2c(0, z, \zeta; \mathbb{D}) = -\frac{(1 + |z\zeta|)^4 |1 + z\bar{\zeta}|^2}{(1 - |z|^2)(1 - |\zeta|^2)|z\zeta|^2|z - \zeta|^2} \times \left\{\left(\frac{|z| + |\zeta|}{1 + |z\zeta|}\right)^2 - \left|\frac{z + \zeta}{1 + z\bar{\zeta}}\right|^2\right\}\left\{\left(\frac{|z| + |\zeta|}{1 + |z\zeta|}\right)^2 - \left|\frac{z - \zeta}{1 - z\bar{\zeta}}\right|^2\right\} \quad (\le 0).$$

The hyperbolic distance in $\mathbb{D}$ is defined by $d(z, \zeta) = \operatorname{arctanh}(|z - \zeta|/|1 - z\bar{\zeta}|)$ $(z, \zeta \in \mathbb{D})$. We have $d(z, 0) + d(0, \zeta) = \operatorname{arctanh}\{(|z| + |\zeta|)/(1 + |z\zeta|)\}$. Thus the inequality $D^2c \leq 0$ is related to the triangle inequality with respect to $d(\cdot, \cdot)$. If Ω is doubly-connected, then we may assume that Ω is a ring $R_\varrho = \{\varrho < |z| < 1\}$. We have

$$D^2c(w, z, \zeta; R_\varrho) = \frac{2K^3}{\pi^3|wz\zeta|}$$
$$\times \operatorname{Im}\left[\left(\frac{dn(\xi - u)}{sn(\xi - u)}\frac{dn(\bar{\xi} - v)}{sn(\bar{\xi} - v)} - \frac{dn(\bar{\xi} - u)}{sn(\bar{\xi} - u)}\frac{dn(\xi - v)}{sn(\xi - v)}\right)\frac{dn(\bar{u} - \bar{v})}{sn(\bar{u} - \bar{v})}\right.$$
$$\left. - \left(\frac{dn(\xi - u)}{sn(\xi - u)}\frac{dn(\bar{\xi} - \bar{v})}{sn(\bar{\xi} - \bar{v})} - \frac{dn(\bar{\xi} - u)}{sn(\bar{\xi} - u)}\frac{dn(\xi - \bar{v})}{sn(\xi - \bar{v})}\right)\frac{dn(\bar{u} - v)}{sn(\bar{u} - v)}\right],$$
$$\xi = \frac{K}{i\pi}\log w, \quad u = \frac{K}{i\pi}\log z, \quad v = \frac{K}{i\pi}\log \zeta, \quad K = K(k),$$

where the modulus k is defined by $\log \varrho = -\pi K(\sqrt{1 - k^2})/K(k)$. The following expression is also applicable to compute D^2c: *Let* $\Omega^* = \{\bigcup_{k=1}^n [a_k, b_k]\}^c$ $(a_1 < b_1 < \cdots < a_n < b_n)$. *Then*

$$\text{(10)} \qquad D^2c(w, z, \zeta; \Omega^*) = \frac{1}{4|M(w)M(z)M(\zeta)|}$$
$$\times \operatorname{Re}\left[\left(\frac{M(w) + M(z)}{w - z}\frac{\overline{M(w)} - M(\zeta)}{\bar{w} - \zeta}\right.\right.$$
$$\left. - \frac{\overline{M(w)} - M(z)}{\bar{w} - z}\frac{M(w) + M(\zeta)}{w - \zeta}\right)\frac{\overline{M(z)} + M(\zeta)}{\bar{z} - \zeta}$$
$$- \left(\frac{M(w) + M(z)}{w - z}\frac{\overline{M(w)} + \overline{M(\zeta)}}{\bar{w} - \bar{\zeta}}\right.$$
$$\left.\left. - \frac{\overline{M(w)} - M(z)}{\bar{w} - z}\frac{M(w) - \overline{M(\zeta)}}{w - \bar{\zeta}}\right)\frac{\overline{M(z)} - M(\zeta)}{\bar{z} - \zeta}\right],$$
$$M(\xi) = \prod_{k=1}^n \sqrt{(b_k - \xi)/(a_k - \xi)} \quad (\xi = w, z, \zeta).$$

The ring R_ϱ is conformally mapped to a domain of this type with $n = 2$. To see $D^2c(w, z, \zeta; \Omega^*) \leq 0$ in the case of $n = 2$, we may assume that $w = x \in \mathbb{R}$, $z/i > 0$ and $\zeta/i < 0$. Then (10) shows that

$$D^2c(x, z, \zeta; \Omega^*) = \frac{Ax^2 + Bx + C}{|M(z)M(\zeta)||(x - z)(x - \zeta)|^2} \quad (A, B, C \in \mathbb{R}).$$

By an inequality of Möbius type in elementary geometry, we obtain $A < 0$ and $B^2 - 4AC < 0$.

References

[A] L. Ahlfors: Bounded analytic functions. Duke. Math. J. **14** (1947) 1–11

[AB] L. Ahlfors, A. Beurling: Conformal invariants and function-theoretic null-sets. Acta Math. **83** (1950) 101–129

[B] S. Bergman: The kernel function and conformal mapping. Math. Surv. V. Amer. Math. Soc., New York 1950

[C] A.P. Calderón: Commutators, singular integrals on Lipschitz curves and applications. ICM Helsinki 1978, pp. 85–96

[Ch] M. Christ: Lectures on singular integral operators. CBMS 77. Amer. Math. Soc., Providence 1990

[CJS] R.R. Coifman, P.W. Jones, S. Semmes: Two elementary proofs of the L^2 boundedness of Cauchy integrals on Lipschitz curves. J. Amer. Math. Soc. **2** (1989) 553–564

[CMM] R.R. Coifman, A. McIntosh, Y. Meyer: L'intégrale de Cauchy définit un opérateur borné sur L^2 pour les courbes lipschitziennes. Ann. Math. **116** (1982) 361–388

[D] G. David: Opérateurs de Calderón-Zygmund. ICM Berkeley 1986, pp. 890–899

[Da] A.M. Davie: Analytic capacity and approximation problems. Trans. Amer. Math. Soc. **171** (1972) 409–444

[DØ] A.M. Davie, B. Øksendal: Analytic capacity and differentiability properties of finely harmonic functions. Acta Math. **149** (1982) 127–152

[F] K.J. Falconer: The geometry of fractal sets. Cambridge Univ. Press, Cambridge 1985

[Fa] J.D. Fay: Theta functions on Riemann surfaces. (Lecture Notes in Mathematics, vol. 352.) Springer, Berlin Heidelberg New York 1973

[Fe] C. Ferrari: Sulla transformazione conforme di due cerchi in due profili alari. Memorie della R. Accad. delle Scienze di Torino Serie II **67** (1930) 1–15

[G] P. Garabedian: Schwarz's lemma and the Szegö kernel function. Trans. Amer. Math. Soc. **67** (1949) 1–35

[Ga] J. Garnett: Analytic capacity and measure. (Lecture Notes in Mathematics, vol. 297.) Springer, Berlin Heidelberg New York 1972

[Gar] I.E. Garrick: Potential flow about arbitrary biplane wing sections. Technical Report No. 542, NACA 1936, pp. 47–75

[GS] P. Garabedian, M. Schiffer: Identities in the theory of conformal mapping. Trans. Amer. Math. Soc. **65** (1949) 187–238

[H] S.Ya. Havinson: Analytic capacity of sets, joint nontriviality of various classes of analytic functions and the Schwarz lemma in arbitrary domains (Russian) Mat. Sb. **54** (1961) 3–50

[HS] N.S. Hawley, M. Schiffer: Half-order differentials on Riemann surfaces. Acta Math. **115** (1966) 199–236

[HHN] V.P. Havin, S.V. Hruščëv, N.K. Nikol'skii: Linear and complex analysis problem book. (Lecture Notes in Mathematics, vol. 1043.) Springer, Berlin Heidelberg New York 1984

[J] P.W. Jones: Square functions, Cauchy integrals, analytic capacity and harmonic measure. In: Harmonic analysis and partial differential equations, pp. 24–68 (Lecture Notes in Mathematics, vol. 1384.) Springer, Berlin Heidelberg New York 1989

[JM] P.W. Jones, T. Murai: Positive analytic capacity but zero Buffon needle probability. Pacific J. Math. **133** (1988) 99–114

[M1] P. Mattila: A class of sets with positive length and zero analytic capacity. Ann. Acad. Sci. Fenn. Ser. AI **10** (1985) 387–395

[M2] P. Mattila: Smooth maps, null-sets for integralgeometric measure and analytic capacity. Ann. Math. **123** (1986) 303–309

[Ma] D.E. Marshall: Removable sets for bounded analytic functions. In: Linear and complex analysis problem book, pp. 485–490 (Lecture Notes in Mathematics, vol. 1043.) Springer, Berlin Heidelberg New York 1984

[Me] Y. Meyer: Wavelets and operators. In: Analysis at Urbana 1 pp. 256–365 (Lecture Note Ser. 137.) London Math. Soc., Cambridge 1989

[Mi] L.M. Milne-Thomson: Theoretical hydrodynamics, Second edition. Macmillan, London 1949

[Mu1] T. Murai: Construction of H^1 functions concerning the estimate of analytic capacity. Bull. London Math. Soc. **19** (1987) 154–160

[Mu2] T. Murai: Comparison between analytic capacity and the Buffon needle probability. Trans. Amer. Math. Soc. **304** (1987) 501–514

[Mu3] T. Murai: A real variable method for the Cauchy transform, and analytic capacity. (Lecture Notes in Mathematics, vol. 1307.) Springer, Berlin Heidelberg New York 1988

[Mu4] T. Murai: The power 3/2 appearing in the estimate of analytic capacity. Pacific J. Math. **143** (1990) 313–340

[Mu5] T. Murai: Analytic capacity for two segments. Nagoya Math. J. **122** (1991), to appear

[Mu6] T. Murai: A formula for analytic separation capacity. Kōdai Math. J. **13** (1990) 265–288

[Mu7] T. Murai: The arc-length variation of analytic capacity and a conformal geometry. Submitted

[P] Ch. Pommerenke: Über die analytische Kapazität. Arch. Math. **11** (1960) 270–277

[S1] M. Schiffer: Variational methods in the theory of Riemann surfaces. In: Contributions to the theory of Riemann surfaces, pp. 15–30. (Ann. Math. Stud. 30.) Princeton Univ. Press, Princeton 1953

[S2] M. Schiffer: Some recent developments in the theory of conformal mapping. In: R. Courant, Dirichlet's principle. pp. 249–318. (Pure Appl. Math. III) Interscience, New York 1967

[Sm] E.P. Smith: The Garabedian function of an arbitrary compact set. Pacific J. Math. **51** (1974) 289–300

[Su1] N. Suita: On a metric induced by analytic capacity. Kōdai Math. Sem. Rep. **25** (1973) 215–218

[Su2] N. Suita: On a metric induced by analytic capacity II. Kōdai Math. Sem. Rep. **27** (1976) 159–162

[Su3] N. Suita: On subadditivity of analytic capacity for two continua. Kōdai Math. J. **7** (1984) 73–75

[SO] L. Sario, K. Oikawa: Capacity functions. Springer, Berlin Heidelberg New York 1969

[SS] M. Schiffer, D.C. Spencer, Functionals of finite Riemann surfaces. Princeton Univ. Press, Princeton 1954

[V1] A.G. Vitushkin: Example of a set of positive length but of zero analytic capacity (Russian). Dokl. Akad. Nauk SSSR **127** (1959) 246–249

[V2] A.G. Vitushkin: Analytic capacity of sets in problems of approximation theory (Russian). Uspehi Mat. Nauk **22** (1967) 141–199

[Z] L. Zalcman: Analytic capacity and rational approximation. (Lecture Notes in Mathematics, vol. 50.) Springer, Berlin Heidelberg New York 1968

Recent Applications of L^2 Estimates for the Operator $\bar{\partial}$

Takeo Ohsawa

Department of Mathematics, Nagoya University, Nagoya 464-01, Japan

1. In the theory of holomorphic functions of several variables, boundaries of complex manifolds arose as the singularities of analytic objects. Geometric structure of manifolds with boundaries of this sort is therefore of interest in function theory. As for the singularity of holomorphic functions, a general picture was given by the solution of the Levi problem given by Oka [39] over C^n and by Grauert [16] on complex manifolds, which characterized Stein manifolds by the existence of strictly plurisubharmonic exhaustion functions. The latter's work was based on the cohomology finiteness theorem and was generalized by Andreotti-Grauert [2] to noncompact complex spaces that admit certain exhaustion functions. As is well known, a method of partial differential equation is available to study the cohomology groups of Riemannian manifolds (cf. [7]). Importance of this method in function theory became apparent by the works of Andreotti-Vesentini [4] and Hörmander [19]. The latter is already a penetrating work that recovers main results of [39, 16] and [2] in a completely different way. The method consists in establishing an à priori L^2 estimate for a given $\bar{\partial}$-closed form, which is usually a direct consequence of commutator relations in a graded operator algebra generated by several covariant exterior differential operators like $\bar{\partial}$, ∂ and their adjoints. By and by it has turned out that this approach, namely the L^2 theory, has an advantage in obtaining more detailed information about the analytic cohomology groups and functions on noncompact complex manifolds (cf. [21, 25, 17, and 43]). The basic problems in this context are therefore to clarify the specific properties of the L^2 objects, like L^2 cohomology groups and harmonic forms, and relationship between the L^2 and the ordinary cohomology groups. We shall report below on recent results about the L^2 cohomology groups of noncompact manifolds with emphasis on the extension of the classical Hodge theory which turned out to have an application to intersection cohomology theory.

2. Basic technical devices are summarized here. Let (X, ds^2) be a connected Hermitian manifold of dimension n, and let (L, h) be a Hermitian line bundle over X. For any square integrable L-valued (p, q)-form u, $\|u\|_h$ will denote the L^2 norm of u with respect to ds^2 and h. The fundamental form of ds^2 will be denoted by ω and the curvature form of h by Θ_h.

Proceedings of the International Congress
of Mathematicians, Kyoto, Japan, 1990

Theorem 1. *If ds^2 is a complete Kähler metric on X and $i\Theta_h = \omega$, then for any L-valued (resp. L^{-1}-valued) (p, q)-form v with $\bar{\partial}v = 0$ $\deg v\,(:= p + q) > n$ (resp. $\deg v < n$) and $\|v\|_h < \infty$, there exists an L-valued (resp. L^{-1}-valued) $(p, q-1)$-form u satisfying $\bar{\partial}u = v$ and $\|u\|_h \leqq \|v\|_h$. Moreover the range of the maximal closed extension of $\bar{\partial}|C_0(X, L^{-1})$ is closed in the degrees $\leqq n$.*

The proof of Theorem 1 is based on the complex Weitzenböck's formula which we recall briefly. Let $C_0(X, L)$ denote the set of L-valued C^∞ differential forms with compact support on X, let $\bar{\partial}_h{}^*$ denote the Hilbert space adjoint of $\bar{\partial}|C_0(X, L)$ with respect to ds^2 and h, and let Λ be the (pointwise) adjoint of exterior multiplication by the fundamental form of ds^2. Then we have

$$[\bar{\partial}, \bar{\partial}_h{}^*]_{\mathrm{gr}} - *^{-1}[\bar{\partial}, \bar{\partial}_h{}^*]_{\mathrm{gr}}\, * = [i\Theta_h, \Lambda]_{\mathrm{gr}} \tag{1}$$

on $C_0(X, L)$, where $[a, b] := ab - (-1)^{\deg a \deg b} ba$, $*$ denotes the Hodge's star operator and $i\Theta_h$ is identified with the corresponding exterior multiplication. The à priori estimates needed for the above existence theorems follow directly from (1). Originally, Theorem 1 was stated only for compact manifolds for it was thought of as an analytic counterpart of Lefschetz's hyperplane section theorem on nonsingular projective varieties (cf. Akizuki-Nakano [1]). The above noncompact version is due to Andreotti-Vesentini [3] and recently it was used in an essential way in the proof of Cheeger-Goresky-MacPherson's conjecture for varieties with isolated singularities, as we shall see later. The following looks very likely to be a corollary of Theorem 1, but it was discovered much later by Donnelly-Fefferman [14] in the study of Schwartz kernels on strongly pseudoconvex domains, and the proof is actually independent of Theorem 1.

Theorem 2. *Under the situation of Theorem 1, assume particularly that L is the trivial bundle so that the connection form $\partial \log h$ is identified with a $(1, 0)$-form on X, and that $|\partial \log h|$ is bounded. Then for any $\bar{\partial}$-closed (p, q)-form v with $p + q \neq n$ which is square integrable with respect to the trivial fiber metric h_0, there exists a $(p, q-1)$-form u satisfying $\bar{\partial}u = v$ and $\|u\|_{h_0} \leqq \|v\|_{h_0}$. Moreover the range of the maximal closed extension of $\bar{\partial}$ with respect to the metric h_0 is closed.*

The proof of Theorem 2 is immediate from the equality

$$[\bar{\partial}, (\bar{\partial} \log h)^*]_{\mathrm{gr}} + *^{-1}[\bar{\partial}, (\bar{\partial} \log h)^*]_{\mathrm{gr}}\, * = [i\Theta_h, \Lambda]_{\mathrm{gr}}. \tag{2}$$

Another important machinery in the L^2 theory is the following.

Theorem 3. *Assume that $i\Theta_h = \omega$ for given Hermitian metrics ds^2 and h. If X admits a complete Kähler metric, then for any L-valued (n, q)-form v with $\bar{\partial}v = 0$, $q \geqq 1$ and $\|v\|_h < \infty$ (with respect to ds^2 and h) there exists an L-valued $(n, q-1)$-form u satisfying $\bar{\partial}u = v$ and $\|u\|_h \leqq \|v\|_h$*

Theorem 3 contains Hörmander's theorem (cf. [19]) whose applications are already widespread. The above formulation is given by Demailly [11] and Ohsawa [27, 31] independently, aiming at applying Hörmander's method to more geometric questions related to holomorphic n-forms and Bergman kernels (see also [30, 13, 15]). It is also regarded as a noncompact version of Kodaira vanishing theorem.

3. We are going to sketch how to apply the above mentioned tools to relate the ordinary and L^2 cohomology groups of Hermitian manifolds (X, ds^2). Let $C_0(X)$ be the set of C^∞ forms on X with compact support and let $\bar{\partial}_{\max}$ be the maximal closed extension of $\bar{\partial}|C_0(X)$ to the space of square integrable forms $L_{(2)}(X)$. Namely the domain of $\bar{\partial}_{\max}$, denoted by Dom $\bar{\partial}_{\max}$, consists of square integrable forms u for which $\bar{\partial}u$ is also square integrable. The (p, q)-component of $L_{(2)}(X)$ will be denoted by $L^{p,q}(X)$, and we call the space Ker $\bar{\partial}_{\max} \cap L^{p,q}(X)/\mathrm{Im}\ \bar{\partial}_{\max} \cap L^{p,q}(X)$ the L^2 $\bar{\partial}$-cohomology group of type (p, q), denoted by $H^{p,q}_{(2)}(X)$ or $H^{p,q}_{(2)}(X)_{ds^2}$. For any compact subset $K \subset X$ we set

$$L^{p,q}(X/K) := \{u \in L^{p,q}(X);\ \mathrm{supp}\ u \subset X \setminus K\}$$

and $H^{p,q}_{(2)}(X/K) := \mathrm{Ker}\ \bar{\partial}_{\max} \cap L^{p,q}(X/K)/\bar{\partial}(\mathrm{Dom}\ \bar{\partial}_{\max} \cap L^{p,q-1}(X))$. Then we have a long exact sequence

$$\cdots \to \varprojlim_{K \Subset X} H^{p,q}_{(2)}(X/K) \to H^{p,q}_{(2)}(X) \to H^{p,q}(X) \to \varprojlim_{K \Subset X} H^{p,q+1}_{(2)}(X/K) \to \cdots$$

where $H^{p,q}(X)$ denotes the ordinary $\bar{\partial}$-cohomology group of type (p, q). As for the $\bar{\partial}$-cohomology groups with compact support $H^{p,q}_0(X)$, we have

$$\cdots \to \varinjlim_{K \Subset X} H^{p,q-1}_{(2)}(X \setminus K) \to H^{p,q}_0(X) \to H^{p,q}_{(2)}(X) \to \varinjlim_{K \Subset X} H^{p,q}_{(2)}(X \setminus K) \to \cdots$$

Hence we have the following criterion.

Proposition 4. *The canonical homomorphism $\alpha^{p,q}$ (resp. $\beta^{p,q}$) is surjective if $\varprojlim H^{p,q+1}_{(2)}(X/K) = 0$ (resp. if $\varinjlim H^{p,q}_{(2)}(X \setminus K) = 0$) and injective if $\varprojlim H^{p,q}_{(2)}(X/K) = 0$ (resp. if $\varinjlim H^{p,q-1}_{(2)}(X \setminus K) = 0$).*

In case the boundary of X is "small", conditions of Proposition 4 are actually satisfied.

Proposition 5. *Suppose that X is the regular part of a projective variety Z such that $\dim(Z \setminus X) = 0$. Then there exists a complete Kähler metric on X for which $\varprojlim H^{p,q}_{(2)}(X/K) = 0$ if $p + q < n$ and $\varinjlim H^{p,q}_{(2)}(X \setminus K) = 0$ if $p + q > n$.*

We note that the à priori estimates for the (p, q)-forms on $X \setminus K$ imply the finite dimensionality of $H^{p,q}_{(2)}(X)$ and the separatedness of $H^{p,q+1}_{(2)}(X)$, so that we have the following straightforward consequence of Proposition 5.

Theorem 6 (cf [32]). *Let X be as in Proposition 5. Then there exists a complete Kähler metric on X such that*

1) *The canonical homomorphisms $\alpha^{p,q}$ (resp. $\beta^{p,q}$) are bijective if $p + q < n - 1$ (resp. if $p + q > n + 1$) and injective (resp. surjective) if $p + q = n - 1$ (resp. $p + q = n + 1$).*

2) *$H^{p,q}_{(2)}(X)$ are finite dimensional if $p + q \neq n$ and separated if $p + q = n$.*

The above metric is constructed as the sum of the Fubini-Study metric (restricted to X) and the complex Hessian of a C^∞ exhaustion function with values in $(-\infty, 0]$, say ϕ which behaves like $-\log(-\log \delta)$ near the points of $Z \setminus X$, where δ denotes the distance to $Z \setminus X$. Theorem 2 is then applied for a sufficiently small sublevel set of ϕ equipped with a complete Kähler metric of the form $\partial\bar{\partial}\lambda(\phi)$ for a suitable convex increasing function λ.

By weakening the assumptions of Theorems 1 and 2 one has a more general L^2 vanishing theorem that yields the following in a similar manner.

Theorem 7 (cf. [33]). *If X is the regular part of a projective variety Z such that $\dim(Z \setminus X) \leqq k$, there exists a complete Kähler metric on X such that*

1) *$\alpha^{p,q}$ (resp. $\beta^{p,q}$) are bijective if $p + q < n - k - 1$ (resp. $p + q > n + k + 1$) and injective (resp. surjective) if $p + q = n - k - 1$ (resp. $p + q = n + k + 1$)*

2) *$H^{p,q}_{(2)}(X)$ are finite dimensional if $|p + q - n| > k + 1$ and separated if $p + q = n - k - 1$.*

Let $H^r(X)$ and $H^r_0(X)$ denote respectively the r-th cohomology group of X and that with compact support. Then Theorem 7 implies an extension of Hodge theory to quasiprojective varieties. Namely, as a corollary of Theorem 7 we obtain

Theorem 8. *Under the situation of Theorem 7,*

1) *$H^{p,q}(X) \cong \overline{H^{q,p}(X)}$ for $p + q < n - k - 1$ and $H^{p,q}_0(X) \cong \overline{H^{q,p}_0(X)}$ for $p + q > n + k + 1$.*

2) *$H^r(X)$ (resp. $H^r{}_0(X)$) is canonically isomorphic to $\bigoplus_{p+q=r} H^{p,q}(X)$ for $r < n - k - 1$ (resp. isomorphic to $\bigoplus_{p+q=r} H^{p,q}_0(X)$ for $r > n + k + 1$).*

Remark. There are different proofs of Theorem 8. Namely Bauer and Kosarew [6] use characteristic p method and Arapura [5] the technique of logarithmic differential forms.

4. The preceeding discussion says nothing about the properties of $H^{p,q}{}_{(2)}(X)$ for $|p + q - n| \leqq k$. Since there is little hope to get any simple relation between the L^2 and the ordinary cohomology in this range, we must compromise at the moment to study the L^2 cohomology groups of (X, ds^2) for the exterior differential d instead of $\bar{\partial}$, which shall be denoted by $H^r_{(2)}(X)$ or $H^r_{(2)}(X)_{ds^2}$. Cheeger-Goreski-MacPherson [10] posed the following fundamental question for the L^2 cohomology with respect to the Fubini-Study metric. To fix our notation we set $H^r_{(2)}(U) := H^r_{(2)}(X \cap U)_{ds^2}$ for open subsets U of Z if ds^2 is the Fubini-Study metric.

Cheeger-Goresky-MacPherson's Conjecture. *Let $Z \subset P^N$ be any irreducible projective variety and let $X \subset Z$ be the set of regular points. Then $H^r_{(2)}(Z)$ are canonically isomorphic to the intersection cohomology groups $IH^r(Z)$ of the middle perversity (for the definition of $IH^r(Z)$, see [10]).*

In case $\dim(Z\setminus X) = 0$, the conjecture is equivalent to saying that

$$H^r_{(2)}(Z) \cong \begin{cases} H^r(X) & \text{if } r < n \\ \mathrm{Im}(H^n_0(X) \to H^n(X)) & \text{if } r = n \\ H^r_0(X) & \text{if } r > n, \end{cases} \tag{3}$$

which has been verified in this case by Cheeger [9] Hsiang-Pati [18] and Nagase [24] for $\dim Z \leqq 2$ and by the author in April of 1990, and we are going to sketch the idea of the proof. As we have seen in the case of the L^2 $\bar{\partial}$-cohomology, (3) is equivalent to saying that $\varprojlim H^r_{(2)}(Z/K) = 0$ for $r < n$ and $\varinjlim H^r_{(2)}(Z\setminus K) = 0$ for $r \geqq n$. It is more or less routine that these vanishing follow from Theorem 1 except for the case $r = n$ (cf. [32, 35]). Direct approaches to prove $\varinjlim H^n_{(2)}(Z\setminus K) = 0$ require subtle analysis as in [18] and [24]. Instead we take an indirect way. Namely we aim at proving the following.

Proposition 9. *There exists a Hermitian metric $d\sigma^2$ on X such that* $\dim H^r_{(2)}(X)_{d\sigma^2} = \dim H^r_{(2)}(Z)$ *and* $\dim H^r_{(2)}(X)_{d\sigma^2} = \dim IH^r(Z)$ *for $r = n \pm 1, n$.*

A candidate of $d\sigma^2$ is given by a result of Saper [41] (see also [23]):

Theorem 10. *If* $\dim(Z\setminus X) = 0$, *there exists a C^∞ exhaustion function $\phi : X \to (-\infty, 0]$ such that $d\sigma^2 := ds^2 + \partial\bar{\partial}\phi$ is a complete Kähler metric on X satisfying*

1) *The length of $\partial\phi$ is bounded*
2) $H^r_{(2)}(X)_{d\sigma^2} \cong IH^r(Z)$.

In order to establish the required equalities for the L^2 cohomology groups, we compare the spaces of harmonic forms by applying the approximation method of Runge-Hörmander. Namely let $d\sigma_\varepsilon^2 = ds^2 + \varepsilon\partial\bar{\partial}\phi$ for $\varepsilon \geqq 0$ and let $\mathscr{H}^r_\varepsilon$ be the set of harmonic r-forms with respect to $d\sigma_\varepsilon^2$. Then $\mathscr{H}^r_\varepsilon \cong \mathscr{H}^r_1$ for $\varepsilon > 0$ so that it suffices to show that $\dim \mathscr{H}^r_0 = \dim \mathscr{H}^r_1$, for the range of $d_{\max}$ (w.r.t. ds^2) is closed (cf. [35]). Let $\| \ \|_\varepsilon$ and $d_\varepsilon{}^*$ denote respectively the L^2 norm and the adjoint of d with respect to $d\sigma^2{}_\varepsilon$. Then the approximation argument proceeds as follows.

Lemma. *There exists a family of positive functions Φ_ε on X for $\varepsilon \searrow 0$ such that*

1) $\|\Phi_\varepsilon u\|_\varepsilon \leqq \text{const.}(\|u\|_\varepsilon + \|du\|_\varepsilon + \|d_\varepsilon{}^* u\|_\varepsilon)$ *uniformly in ε for all $u \in C_0(X)$ with* $\deg u \neq n$.

2) *From any sequence of locally square integrable forms f_ε with* $\deg f_\varepsilon \neq n$ *satisfying $\|\Phi_\varepsilon f_\varepsilon\|_\varepsilon = 1$ and $\|(d + d_\varepsilon{}^*)f_\varepsilon\|_\varepsilon < \infty$ one can choose a subsequence f_{ε_i} such that $\{f_{\varepsilon_i}\}$ converges to a nonzero element of* $\mathrm{Dom}(d_{\max} + d_{\max}{}^*)$ *strongly on compact subsets of X.*

As Φ_ε, we may take the length of $\varepsilon\partial\phi - \partial \log(\log(\delta^{-1} + 1))$ with respect to $d\sigma_\varepsilon^2$. Here δ is as before. The required property of Φ_ε follows from the non-integrability of $\delta^{-1} \log \delta$ on (0, 1/2) and an L^2 estimate

$$\|\Phi_\varepsilon u\|_\varepsilon \leqq \text{const.}(\|\chi_K u\|_\varepsilon + \|du\|_\varepsilon + \|d_\varepsilon{}^* u\|_\varepsilon),$$

which is valid uniformly in $\varepsilon \searrow 0$ for some compact set $K \subset X$ and all $u \in C_0(X)$ with deg $u \neq n$, where χ_K stands for the characteristic function of K.

Suppose that dim $\mathscr{H}_1{}^{n+1} >$ dim $\mathscr{H}_0{}^{n+1}$. Let $U \subset X$ be any relatively compact open subset which contains K. Then, by hypothesis there must exist $f_\varepsilon \in \mathscr{H}_\varepsilon^{n+1}$ for $\varepsilon > 0$ such that

1) $(\chi_U f_\varepsilon, g)_\varepsilon = 0$ for all $g \in \mathscr{H}_0{}^{n+1}$
2) $\|\Phi_\varepsilon f_\varepsilon\|_\varepsilon = 1$.

Here $(\ , \ )_\varepsilon$ denotes the inner product associated with $\| \ \|_\varepsilon$. But Lemma says then there exist $f \in \mathscr{H}_0{}^{n+1}$ with $f \neq 0$ satisfying $(\chi_U f, f)_0 = 0$, which contradicts the unique continuation theorem since f is harmonic. Since dim $H_0^{n+1}(X) \geqq$ dim $\mathscr{H}_0{}^{n+1}$, we may conclude that dim $\mathscr{H}_1{}^{n+1} =$ dim $\mathscr{H}_0{}^{n+1}$ in virtue of Saper's theorem. The proof for the case $r = n - 1$ is similar. If $r = n$, suppose that dim $\mathscr{H}_0{}^n >$ dim $\mathscr{H}_1{}^n$. Then there exists a finite dimensional subspace $\mathscr{H}' \subset \mathscr{H}_0{}^n$ from which we can choose g_ε ($\varepsilon \searrow 0$) satisfying $\|g_\varepsilon\|_\varepsilon = 1$ and $(\chi_U g_\varepsilon, g')_\varepsilon = 0$ for any $g' \in \mathscr{H}_\varepsilon^n$. Since $\mathscr{H}_\varepsilon^{n\pm 1}$ approximate $\mathscr{H}_0{}^{n\pm 1}$ in the above sense, it follows from (4) that there exist $(u_\varepsilon, v_\varepsilon)$ satisfying $du_\varepsilon + d_\varepsilon{}^* v_\varepsilon = \chi_U g_\varepsilon$ such that $\|\Phi_\varepsilon u_\varepsilon\|_\varepsilon$ and $\|\Phi_\varepsilon v_\varepsilon\|_\varepsilon$ are uniformly bounded in ε. Taking Lemma into account, this implies the existence of $g \neq 0$ in $\mathscr{H}'$ such that $\chi_U g$ belongs to the range of $d_{\max} + d_{\max}{}^*$, which is an absurdity. Thus we get dim $\mathscr{H}_0{}^n \leqq$ dim $\mathscr{H}_1{}^n$. Clearly dim $\mathscr{H}_0{}^n \geqq$ dim Im$(H_0^n(X) \to H^n(X))$. Therefore one must have dim $\mathscr{H}_0{}^n =$ dim $\mathscr{H}_1{}^n$ by Saper's theorem.

Concerning the L^2 cohomology with respect to the Fubini-Study metric, a more advanced question is whether there also exist geometric interpretation of the L^2 $\bar{\partial}$-cohomology groups $H_{(2)}^{p,q}(Z)$ $(:= H_{(2)}^{p,q}(X)_{ds^2})$. Recently Pardon-Stern [40] obtained the following.

Theorem 11. *Let Z be an irreducible projective variety of dimension n with isolated singularities. Then*

$$\chi^n{}_{(2)}(Z) := \Sigma_{q=0}^n (-1)^q \dim H_{(2)}^{n,q}(Z)$$

is a bimeromorphic invariant of Z.

Remark. There are variations of C–G–M conjecture arising from the theory of variations of Hodge structures. In this decade there have been fruitful works appearing on this topic (cf. [44, 42, 22, 8, 20]). As for the statement of the results, see Saper's exposition in this volume.

5. Let us mention other results on the L^2 cohomology groups, which are on pseudoconvex manifolds and therefore more directly related to function theory.

i) Let $(M, ds^2{}_M)$ be a Kähler manifold of dimension n and $D \subset M$ a relatively compact pseudoconvex domain with C^∞ smooth boundary. Then the Hodge theory extends to D as follows.

Theorem 12 (cf. [38]). *Let D be as above and let ϱ be any C^∞ defining function of $\partial\varrho$. If the Levi form of ϱ along the holomorphic tangent vectors of ∂D has everywhere at* *least $n-k$ positive eigenvalues, there exists a complete Kähler metric on D such that*

$$H_{(2)}^{p,q}(D) \cong \begin{cases} H_0^{p,q}(D) & \text{if } p+q \leqq n-k \\ H^{p,q}(D) & \text{if } p+q \geqq n+k. \end{cases}$$

Corollary. *Under the above situation,*

1) $H_0^r(D) \cong \bigoplus_{p+q=r} H_0^{p,q}(D) \quad \text{if } r \leqq n-k$

$H_0^{p,q}(D) \cong \overline{H_0^{q,p}(D)} \quad \text{if } p+q \leqq n-k.$

2) $H^r(D) \cong \bigoplus_{p+q=r} H^{p,q}(D) \quad \text{if } r \geqq n+k$

$H^{p,q}(D) \cong \overline{H^{q,p}(D)} \quad \text{if } p+q \geqq n+k.$

3) The restriction homomorphisms $H^{p,q}(D) \to \varinjlim H^{p,q}(D\backslash K)$ are surjective if $p+q<n-k$.

Remark. Under somewhat stronger assumptions, results in the above corollary can be proved by different methods. (cf. [28, 29, 26, 12 and 6]).

ii) Donnelly-Fefferman [14] determined the L^2 cohomology of strictly pseudoconvex domains in C^n with respect to the Bergman metrics.

Theorem 13. *Let $D \subset C^n$ be a bounded strictly pseudoconvex domain. Then, with respect to the Bergman metric,*

$$\dim H_{(2)}^{p,q}(D) = \begin{cases} 0 & \text{if } p+q \neq n \\ \infty & \text{if } p+q = n \end{cases}$$

We note that the infinite dimensionality of the L^2 cohomology can be proved by an elementary method (cf. [36]), but few things are known about the properties of the L^2 harmonic forms. In case $p=n$, the following is known.

Theorem 14 (cf. [37]). *Let X be a Stein manifold of dimension n, f a bounded holomorphic function on X such that $S := \{z \in X; f(z)=0\}$ has no singular points and $df|S \neq 0$, and let $\varrho_s : H^{n,0}(X\backslash S) \to H^{n-1,0}(S)$ be the residue homomorphism. Then there exists a bounded linear operator $I_s : H_{(2)}^{n-1,0}(S) \to H_{(2)}^{n,0}(X)$ such that* $\varrho_s \cdot (f^{-1} I_s) = \text{id}$.

Corollary. *Let $D \subset C^n$ be a bounded pseudoconvex domain and let $H \subset C^n$ be a complex hyperplane. Then, every L^2 holomorphic function f on $H \cap D$ has an L^2 holomorphic extension to D, say F that satisfies $\|F\| \leqq c_D \|f\|$. Here c_D is a positive number that depends only on the diameter of D.*

We leave two open questions related to the above results.

1. Under the situation of Theorem 12, let x and y be two distinct points in D. Does there exist an L^2 harmonic (p, q)-form that separates x and y if $p+q \neq n$?

2. Under the situation of Corollary to Theorem 14, let g be a holomorphic function on $H \cap D$ which admits a C^∞ extension to D, say G, such that $\|\bar{\partial}G\| < \infty$. Then does there exist an L^2 holomorphic extension of g to D?

References

1. Akizuki, Y., Nakano, S.: Note on Kodaira-Spencer's proof of Lefschetz theorems. Proc. Japan Acad. **30** (1954) 266–272
2. Andreotti, A., Grauert, H.: Théorème de finitude pour la cohomologie des espaces complexes. Bull. Soc. Math. France **90** (1962) 193–259
3. Andreotti, A., Vesentini, E.: Sopra un teorema di Kodaira. Ann. Sci. Norm. Sup. Pisa **15** (1961) 283–309
4. Andreotti, A., Vesentini, E.: Carleman estimates for the Laplace-Beltrami equation on complex manifolds. Publ. Math. IHES **25** (1965) 81–130
5. Arapura, D.: Local cohomology of sheaves of differential forms and Hodge theory. Preprint
6. Bauer, I., Kosarew, S.: On the Hodge spectral sequence for some classes of non-complete algebraic manifolds. Math. Ann. **284** (1989) 577–593
7. Bochner, S.: Curvature and Betti numbers, I, II. Ann. Math. **49** (1948) 379–390; **50** (1949) 77–93
8. Cattani, E., Kaplan, A., Schmidt, W.: Degeneration of Hodge structures. Ann. Math. **123** (1986) 457–535
9. Cheeger, J.: On the Hodge theory of Riemannian pseudomanifolds. Proc. Symp. Pure Math. **36** (1980) 91–146
10. Cheeger, J., Goresky, M., MacPherson, R.: L^2 cohomology and intersection homology for singular algebraic varieties. Seminar on differential geometry. (Ann. Math. Stud. 102.) Princeton Univ. Press 1982, pp. 303–340
11. Demailly, J.P.: Estimations L^2 pour l'opérateur $\bar{\partial}$ d'un fibré vectoriel holomorphe semi-positif au-dessus d'une variété kählérienne complète. Ann. Sci. Ec. Norm. Sup. **15** (1982) 457–512
12. Demailly, J.P.: Cohomology of q-convex spaces in top degrees. Math. Z. **204** (1990) 283–295
13. Diederich, K., Ohsawa, T.: On the parameter dependence of solutions to the $\bar{\partial}$-equation. To appear in Math. Ann.
14. Donnelly, H., Fefferman, C.: L^2 cohomology and index theorem for the Bergman metric. Ann. Math. **118** (1984) 593–619
15. Forster, O., Ohsawa, T.: Complete intersections with growth conditions. (Adv. Stud. Pure Math. 10.) Algebraic geometry, Sendai, 1987, pp. 91–104
16. Grauert, H.: On Levi's problem and the imbedding of real-analytic manifolds. Ann. Math. **68** (1958) 460–472
17. Grauert, H., Riemenschneider, O.: Kählersche Mannigfaltigkeiten mit hyper-q-konvexen Rand. Problems in Analysis. Symp. in Honor of S. Bochner. Princeton Univ. Press, 1970, pp. 61–79
18. Hsiang, W.C., Pati, V.: L^2-cohomology of normal algebraic surfaces. Invent. math. **81** (1985) 395–412
19. Hörmander, L.: L^2 estimates and existence theorems for the $\bar{\partial}$-operator. Acta Math. **113** (1965) 89–152
20. Kashiwara, M. Kawai, T.: The Poincaré lemma for variations of polarized Hodge structure. Publ. RIMS, Kyoto Univ. **23** (1987) 345–407
21. Kodaira, K.: On Kähler varieties of restricted type. Ann. Math. **60** (1954) 28–48

22. Looijenga, E.: L^2-cohomology of locally symmetric varieties. Comp. Math. **67** (1988) 3–20
23. Looijenga, E., Rappoport, M.: Weights in the local cohomology of a Baily-Borel compactification. Preprint
24. Nagase, M.: Remarks on the L^2-cohomology of singular algebraic surfaces. J. Math. Soc. Japan **41** (1989) 97–116
25. Nakano, S.: Vanishing theorems for weakly 1-complete manifolds, II. Publ. RIMS (1974) 101–110
26. Navarro-Aznar, V.: Sur la théorie de Hodge des variétés algébriques à singularités isolées. Astérisque **130** (1985) 272–397
27. Ohsawa, T.: On complete Kähler domains with C^1 boundary. Publ. RIMS **10** (1980) 929–940
28. Ohsawa, T.: A reduction theorem for cohomology groups of very strongly q-convex Kähler manifolds. Invent. math. **63** (1981) 335–354
29. Ohsawa, T.: Addendum to: A reduction theorem for cohomology groups of very strongly q-convex Kähler manifolds. Invent. math. **66** (1982) 391–393
30. Ohsawa, T.: Boundary behavior of the Bergman kernel function on pseudoconvex domains. Publ. RIMS **20** (1984) 897–902
31. Ohsawa, T.: Vanishing theorems on complete Kähler manifolds. Publ. RIMS **20** (1984) 21–38
32. Ohsawa, T.: Hodge spectral sequence on compact Kähler spaces. Publ. RIMS **23** (1987) 262–274
33. Ohsawa, T.: Hodge spectral sequence and symmetry on compact Kähler spaces. Publ. RIMS **23** (1987) 613–625
34. Ohsawa, T.: Cheeger-Goreski-MacPherson's conjecture for the varieties with isolated singularities. Math. Z. **206** (1991) 219–224
35. Ohsawa, T.: Supplement to "Hodge spectral sequence on compact Kähler spaces". To appear in Publ. RIMS
36. Ohsawa, T.: On the infinite dimensionality of the middle L^2 cohomology of complex domains. Publ. RIMS (1989) 499–502
37. Ohsawa, T. Takegoshi, K.: On the extension of L^2 holomorphic functions. Math. Z. **195** (1987) 197–204
38. Ohsawa, T. Takegoshi, K.: Hodge spectral sequence on pseudoconvex domains. Math. Z. **197** (1988) 1–12
39. Oka, K.: Domaines finis sans point critique intérieur, Jap. J. Math. **27** (1953) 97–155
40. Pardon, Stern, M.: $L^2 - \bar{\partial}$-cohomology of complex projective varieties. Preprint
41. Saper, L.: L_2-cohomology of Kähler varieties with isolated singularities. Preprint
42. Saper, L., Stern, M.: L_2-cohomology of arithmetic varieties. Ann. Math., to appear
43. Skoda, H.: Morphismes surjectifs de fibrés vectoriels semi-positif. Ann. Sci. Ec. Norm. Sup. 4e série **11** (1978) 577–611
44. Zucker, S.: L^2-cohomology of warped products and arithmetic groups. Invent. math. **70** (1982) 169–218

Differentiability and Measures in Banach Spaces

David Preiss

Department of Mathematics, University College London, London WC1E 6BT, UK

The purpose of this contribution is to give information about new results concerning natural questions about differentiability and measures in real Banach spaces (of infinite but also of finite dimension) and, possibly more importantly, to point out some of the many open problems we are still faced with in this area of research.

1. Differentiability

We recall two well known notions.

1. A real valued function f defined on an open subset G of a Banach space E is said to be Fréchet differentiable at a point $x \in G$ if there is $f'(x) \in E^\star$ such that

$$\lim_{u \to 0} \frac{|f(x+u) - f(x) - \langle f'(x), u \rangle|}{\|u\|} = 0,$$

$f'(x)$ is called the Fréchet derivative of f at x.

2. A real valued function f defined on an open subset G of a Banach space E is said to be Lipschitz on G if there is a constant C such that $|f(x) - f(y)| \le C\|x - y\|$ whenever $x, y \in G$. The smallest such constant C is denoted by $\mathrm{Lip}(f)$.

From the work of Lebesgue (in the one dimensional case) and of Rademacher (in the finite dimensional case) we know that Lipschitz functions on finite dimensional spaces are (Fréchet) differentiable almost everywhere with respect to the Lebesgue measure. Infinite dimensional results of similar nature are known for Gateaux differentiability. (See [1,3,5,6]). These extension are obtained by a linear approximation of the infinite dimensional situation by finite dimensional spaces. However, the question of Fréchet differentiability seems to need a different approach. This might be also seen from many examples of nowhere Fréchet differentiable Lipschitz mappings of a separable Hilbert space into itself, since for such mappings the Gateaux differentiability results mentioned above still hold.

Thus our first result answers a natural question.

Proceedings of the International Congress
of Mathematicians, Kyoto, Japan, 1990

Theorem 1. *Every Lipschitz function defined on a separable Hilbert space is Fréchet differentiable at least at one point.*

Hilbert spaces are, of course, not the most general spaces in which one would hope for such a result. Indeed, from the extensive investigations of differentiability questions for continuous convex functions (e.g., [13,14]) we know that the result may hold in all Asplund spaces. (A Banach space is said to be an Asplund space if the dual of every its separable subspace is separable.) This generalization of Theorem 1 is given in the following statement.

Theorem 2. *Every locally Lipschitz function defined on an open subset of an Asplund space is Fréchet differentiable on a dense subset of its domain.*

The method we use need not be confined to Fréchet differentiability. It applies also to so called $\mathscr{B}$ derivatives, in the definition of which we require the uniform convergence on the members of a given family $\mathscr{B}$ of bounded subsets of the Banach space satisfying some mild additional assumptions. (The details can be found in [10].) This gives the most general form of the above differentiability results. (However, a recent Haydon's example of an Asplund space without equivalent smooth norms shows that the deduction of Theorem 2 from Theorem 3 is not straightforward.)

Theorem 3. *Let E be a Banach space admitting an equivalent norm which is $\mathscr{B}$ differentiable away from the origin. Then every locally Lipschitz function defined on an open subset G of E is $\mathscr{B}$ differentiable on a dense subset of G.*

These statements, as given, are not satisfactory from the point of view of possible applications. For example, suppose that a Lipschitz function f on a separable Hilbert space has derivative zero at every point at which it is Fréchet differentiable. We would like to be able to deduce that f is constant. This can be done, since in all the above results the mean value theorem holds. For example, in case of Theorem 3 we prove that the increment of the function over any segment $[u,v] \subset G$ is majorized by the supremum of the derivatives in the direction $v-u$ at points at which the function is $\mathscr{B}$ differentiable.

The proof of the above results requires new information about Lipschitz functions in finite dimensional spaces. Thus, as a byproduct, we get the following curious statement.

Theorem 4. *There is a plane set N of Lebesgue measure zero such that every Lipschitz function defined on the plane is differentiable at some point of N.*

To describe a set having such a property is quite easy: Any G_δ plane set of Lebesgue measure zero containing all lines passing through two different points with rational coordinates will do. This particular example also suggests the reasons why our proof of Fréchet differentiability results is not straightforward. It combines in some way two notions of smallness of a set: First category (hence the G_δ part) and measure zero (hence the lines). It seems to be intuitively clear that a similar mixture is impossible on the line. That this is true has been shown in [2] and [15]: Theorem 4 is false on the line.

1.1 Construction of a Point of Fréchet Differentiability

The details of the proof can be found in [10]. Here we just point out the main observations. Because of that we restrict our attention to the proof of Theorem 1 only.

We introduce the directional derivatives of f by

$$f'(x,e) = \lim_{r\to 0} \frac{f(x+re) - f(x)}{r},$$

and we denote by M the set of all pairs $(x,e) \in E \times E^\star$ such that $\|e\| = 1$ and $f'(x,e)$ exists.

The first basic observation is that if $(x,e) \in M$ and $f'(x,e) = \mathrm{Lip}(f)$ then f is Fréchet differentiable at x. Even though such a pair need not exist, this suggests that we might attempt to use a maximizing procedure. Thus our plan is to construct inductively a sequence $(x_k, e_k) \in M$ so that:

1. The sequence x_k converges to some x.
2. The sequence e_k converges to some e.
3. The directional derivative $f'(x,e)$ exists.
4. For the pair (x,e) some variant of the above observation can be used.

To achieve 1, we simply choose x_{k+1} close to x_k. This is based on a local form of our observation, namely, that the equality of $f'(x,e)$ to the limit of the Lipschitz constants on balls around x suffices for Fréchet differentiability of f at x.

Unfortunately, to get 2 is not so simple. Since requirement 4 forces us to take $f'(x_k,e_k)$ as large as possible, we cannot at the same time prescribe how close should e_k be to e_{k-1}. There is also a different objection we should take into account: If our method worked, we would construct not only a point of differentiability, but also a point at which gradient vector exists. This causes no problem in Hilbert spaces, but is impossible in non-reflexive Asplund spaces. (Every liner functional not attaining its maximum on the unit ball gives an example.) Thus an idea suggests itself: We should change the norm (and the change should depend on f), at least in the general case. Recalling that we are constructing a sequence e_k of unit vectors, and observing that a small change of the norm can drastically change the set of pairs considered for the choice of (x_{k+1}, e_{k+1}), we find that 2 can be achieved by constructing, together with the sequence (x_k, e_k), a sequence of norms p_k, where p_{k+1} is the (e.g., l_2) sum of p_k and of a (small) multiple of the distance to the one dimensional subspace of H generated by e_k. Then the conditions $p_{k+1}(e_{k+1}) = 1$ $(= p(e_k))$ and $f'(x_{k+1}, e_{k+1}) > f'(x_k, e_k)$ already imply that e_{k+1} is close to e_k.

The requirement 3 seems to be the most difficult. To get it, we observe that the problem is essentially one dimensional and requires some method of interchange of limit and derivative. Since we cannot hope to be able to use anything like the uniform convergence of the derivatives, the only possibility seems to be to choose the points at which the increment of the function is approximated by the derivative globally. The following one dimensional lemma says that this can be done.

Lemma 5. *Suppose that $a < \xi < b$, $0 < \sigma < 1/4$, and $L > 0$ are real numbers, h is a Lipschitz function defined on $[a,b]$, $\mathrm{Lip}(h) \le L$, $h(a) = h(b) = 0$, and $h(\xi) \neq 0$. Then there is a measurable set $A \subset (a,b)$ such that*

1. *The Lebesgue measure of the set A is at least $\sigma|h(\xi)|/L$,*
2. *$h'(\tau) \ge \sigma|h(\xi)|/(b-a)$ for every $\tau \in A$, and*
3. *$|h(t) - h(\tau)| \le 4(1+2\sigma)\sqrt{h'(\tau)L}\,|t-\tau|$ for every $\tau \in A$ and every $t \in [a,b]$.*

The most important third statement of the lemma says that the approximation of the increment of the function by its derivative at the point τ is "globally good" in the whole interval $[a,b]$. The second statement just says that the derivative at τ increased as much as we could hope for. From the first statement we just use that τ can be chosen sufficiently far from the end points. This is needed in order to get a bilateral approximation.

Because in the first statement of Lemma 5 we do not have to speak about measure, the Lemma can be formulated without the notion of the Lebesgue measure. We can then try to prove it without any use of measure theory. This sounds difficult, since we also claim that h is differentiable at τ. But we can also replace the derivatives by lower derivatives and get a version of the lemma that really can be proved without any use of measure theory. Surprisingly enough, this statement then easily implies that Lipschitz functions on the real line have at least one point of differentiability. Though I did not follow this way, since to use the Lebesgue measure and maximal operator technique turned out to be much easier, these remarks suggest that the proof of differentiability discussed here is different from the usual measure theoretic proofs.

Having done this, we can already imagine how to construct the sequence (x_k, e_k) so that 3 holds: We will choose (x_{k+1}, e_{k+1}) so that the approximation of the the increment of the function by its directional derivative at the point x_{k+1} in the direction e_{k+1} is "globally good" on the whole line through x_{k+1} in the direction e_{k+1}.

However, the previous choice implies that our construction will lead to a pair (x,e) for which the equality $f'(x,e) = \mathrm{Lip}(f)$ is quite far from being true. Hence to achieve 4 we need to improve upon our main observation. We first reformulate this observation as:

A Lipschitz function f on a Hilbert space is Fréchet differentiable at x if there is a unit vector e such that $f'(x,e)$ exists and

$$\lim_{\delta \searrow 0} \sup\{f'(\tilde{x},\tilde{e}); (\tilde{x},\tilde{e}) \in M \text{ and } \|\tilde{x} - x\| < \delta\} \le f'(x,e).$$

A simple proof of this statement together with Lemma 5 gives the following differentiability criterion, which we formulate in the most general situation.

Theorem 6. *Suppose that E is a Banach space, $x_0 \in E$, $e_0 \in E$, $\|e_0\| = 1$, and that f is a Lipschitz function defined on E such that $f'(x_0, e_0)$ exists. Let M denote the set of all pairs $(x,e) \in E \times \{e \in E; \|e\| = 1\}$ such that $f'(x,e)$ exists, $f'(x,e) \ge f'(x_0,e_0)$, and*

$$|(f(x+te_0)-f(x))-(f(x_0+te_0)-f(x_0))| \leq 6|t|\sqrt{(f'(x,e)-f'(x_0,e_0))\mathrm{Lip}(f)}$$

for every $t \in \mathbb{R}$.

Then, if the norm is $\mathscr{B}$ *differentiable at* e_0*, and if*

$$\lim_{\delta \searrow 0} \sup\{f'(x,e); (x,e) \in M \text{ and } \|x-x_0\| \leq \delta\} \leq f'(x_0,e_0),$$

f *is* $\mathscr{B}$ *differentiable at* x_0.

Now, the way of constructing the sequences (x_k, e_k) and p_k is more or less clear. We always pick up the next pair from the set M described in the previous Theorem. The additional requirement is only that $f'(x_{k+1}, e_{k+1})$ is very close to the supremum of the directional derivatives $f'(x,e)$ for $(x,e) \in M$. Then we define the norm p_{k+1} and continue our construction. Though we still have to be quite careful and make some technical estimates, since, for example, the set M from the previous Theorem depends upon the choice of the norm, this construction leads to a sequence satisfying all our requirements.

1.2 Problems

From the previous discussion it is clear that the theory of differentiability still abounds with open problems. I would just like to point out the following two.

Problem 7. Does every pair of Lipschitz functions on a separable Hilbert space have a common point of differentiability?

Problem 8. For which finite Borel measures in separable Banach spaces is it true that every Lipschitz function is differentiable almost everywhere?

The second problem is purely finite dimensional since such measures do not exist in infinite dimensional spaces. (See [12].) The answer is not known in the plane (or in any higher dimensional space). In the one dimensional case the required measures are precisely those that are absolutely continuous with respect to the Lebesgue measure. In spite of Theorem 4 I do not know any example that would show that this is not true in all finite dimensional spaces.

2. Measures

The question whether measures on separable Banach spaces are determined by their values on balls has been around since R. O. Davies [4] published his beautiful example of two different probability measures on a compact metric space that agree on all balls. Together with J. Tišer [11] we recently answered it by proving:

Theorem 9. *Whenever two finite Borel measures in a separable Banach space agree on all balls, then they agree.*

To prove this statement, we first use the Fourier transform to reduce the problem to showing that the measures agree on all halfspaces. Then, by blowing up balls, we come to the situation when the halfspace contains a nonempty open cone C on every translate of which the measures agree. An approximation argument (or a differentiability result from [7]) reduces the problem further to the case when $\overline{C} \cap \overline{-C}$ is a subspace of finite codimension. Thus we can pass to the factor space and we have to solve the corresponding problem in finite dimensional spaces: Do we know the measure of a halfspace provided we know the measure of each translate of a nonempty open cone contained in it? Since this turned out to be true, our approach has been successful.

Instead of giving further details, it might be more interesting to point out some examples. The motivation for them comes from the Besicovitch-Morse differentiability theorem, which is a much stronger statement than that measures in finite dimensional normed spaces are determined by their values on balls:

For every (locally) finite Borel measure μ in a finite dimensional Banach space and for every μ integrable function f the limit

$$\lim_{r\searrow 0} \frac{1}{\mu(B(x,r))} \int_{B(x,r)} f(u)\, d\mu(u) \tag{1}$$

exists and equals $f(x)$ for μ almost every x.

As a corollary of this statement one can prove that, if μ and ν are two finite Borel measures in a finite dimensional Banach space satisfying $\mu(B) \geq \nu(B)$ for every ball B then $\mu \geq \nu$.

Example 1 ([9]). There is a Gaussian measure γ_1 in l_2 and a γ_1 integrable function f such that the limit in (1) is infinite uniformly for $x \in l_2$, i.e.,

$$\lim_{r\searrow 0} \inf_{x\in l_2} \frac{1}{\gamma_1(B(x,r))} \int_{B(x,r)} f(u)\, d\gamma_1(u) = \infty.$$

Example 2 ([8]). There is a Gaussian measure γ_2 in l_2 and a bounded γ_2 measurable function f such that, for μ almost every x, the limit in (1) does not exist.

Example 3 (J. Tišer). There is a non-degenerated Gaussian measure γ_3 in l_2 such that (1) holds for every $f \in L_p(\gamma_3)$, $p > 1$.

Example 4. In a separable Hilbert space the statement "$\mu(B) \geq \nu(B)$ for balls with radius less than one implies $\mu \geq \nu$" holds if and only if the dimension of the space is finite.

Example 5. In a separable Hilbert space the statement "$\mu(B) \geq \nu(B)$ for balls with radius greater than one implies $\mu \geq \nu$" holds if and only if the dimension of the space is infinite.

Example 6. In the l_∞ sum of a separable Hilbert space with the line there are measures μ and ν such that $\mu \not\geq \nu$ but $\mu(B) \geq \nu(B)$ for all balls.

However, in spite of the above result and examples, the investigation of the behaviour of measures on balls cannot be considered as finished. For example, the following question is still far from being answered.

Problem 10. Are finite Borel measures in separable Banach spaces determined by their values on balls with radii less than one?

References

1. Aronszajn, N.: Differentiability of Lipschitz functions in Banach spaces. Studia Math. **57** (1976) 147–160
2. Choquet, G.: Applications des propriétés descriptives de la fonction contingent à la théorie des fonctions de variable réelle et à la géometrie différentielle des variétés cartésiennes. J. Math. Pures Appl. **26** (1947) 115–226
3. Christensen, J. P. R.: Measure theoretic zero sets in infinitely dimensional spaces and application to differentiability of Lipschitz mappings. Actes du Deuxieme Colloque d'Analyse Fonctionele de Bordeaux **2** (1973) 29–39
4. Davies, R. O.: Measures not approximable or not specifiable by means of balls. Mathematika **18** (1971) 157–160
5. Mankiewicz, P.: On the differentiability of Lipschitz mappings in Fréchet spaces. Studia Math. **45** (1973) 15–29
6. Phelps, R. R.: Gaussian null sets and differentiability of Lipschitz maps on Banach spaces. Pacific J. Math. **77** (1978) 523–531
7. Preiss, D.: Almost differentiability of convex functions on Banach spaces and determination of measures by their values on balls. In: Proc. Conf. Geometry of Banach Spaces (Strobl 1989) (to appear)
8. Preiss, D.: Gaussian measures and the density theorem. Comment. Math. Univ. Carolinae **22** (1981) 181–193
9. Preiss, D.: Differentiation of measures in infinitely dimensional spaces. In: Proc. Conf. Topology and Measure III, pages 201–207, Greifswald, 1982
10. Preiss, D.: Differentiability of Lipschitz functions on Banach spaces. J. Functional Anal. **91** (1990) 312–345
11. Preiss, D., Tišer, J.: Measures on Banach spaces are determined by their values on balls (to appear)
12. Preiss, D., Zajíček, L.: Fréchet differentiation of convex functions in a Banach space with a separable dual. Proc. Amer. Math. Soc. **91** (1984) 202–204
13. Stegall, Ch.: The duality between Asplund spaces and spaces with the Radon-Nikodym property. Israel J. Math. **29** (1978) 408–412
14. Stegall, Ch.: The Radon-Nikodym property in conjugate Banach spaces II. Trans. Amer. Math. Soc. **264** (1981) 507–519
15. Zahorski, Z.: Sur l'ensemble des points de non-derivabilité d'une fonction continue. Bull. Soc. Math. France **74** (1946) 147–178

The Limit Element in the Configuration Algebra for a Discrete Group: A précis

Kyoji Saito

Research Institute for Mathematical Sciences, Kyoto University, Kyoto 606, Japan

§1. Introduction

(1.1) We construct certain infinitely generated Hopf algebra, which we call the *configuration algebra.* The algebra is generated by isomorphism classes of colored oriented finite graphs and is completed with respect to an adic topology filtered by the cardinality of graphs. Therefore certain limit process is admitted inside the algebra, which is absolutely necessary for the application explained in (1.2).

(1.2) The original attempt of the work, which is under investigation, is to apply the algebra for a construction of certain modular function on the moduli of discrete groups [8]. Let us explain this. Consider a finitely generated group Γ. By fixing a generator system, Γ gets naturally colored oriented graph structure, the Caylay graph. Let Γ_n be the subgraph of Γ consisting of elements of length $\leq n$ and let $\mathscr{A}(\Gamma_n)$ be the formal sum in the algebra of all non-void subgraphs of Γ_n. Then the following limit process is justified in the configuration algebra:

$$\omega_\Gamma := \lim_{n\to\infty} \log\left(\left(1+\mathscr{A}(\Gamma_n)\right)^{1/\sharp\Gamma_n}\right).$$

Namely, we show that these elements $\log\left(1+\mathscr{A}(\Gamma_n)\right)$ becomes Lielike in the algebra. By a use of certain basis $\{\varphi(S)\}_S$ (where S runs the set of isomorphism classes of connected graphs) for the space $\mathscr{L}_{\mathbb{R}}$ of Lielike elements (§8), the above limit element ω_Γ is developed as follows.

$$\omega_\Gamma = \sum_{S\in\mathrm{Conf}_0} \varphi(S)\cdot \lim_{t\to r} \frac{PM(S,t)}{PM(pt,t)}.$$

Here $PM(S,t) := \sum_{n=0}^{\infty} A(S,\Gamma_n)t^n$ is the generating function for $A(S,\Gamma_n) := \sharp\{\text{subgraphs of } \Gamma_n \text{ isomorphic to } S\}$, pt = the graph of one vertex, and $r > 0$ is the radius of convergence of $PM(pt,t)$. For a wide class of groups Γ, including hyperbolic groups and certain automatic groups (cf. [4, 2]), the ratio $PM(S,t)/PM(pt,t)$ extends to a rational function in t which is *regular* at $t = r$ (see §10). So one obtains a final formula,

Proceedings of the International Congress
of Mathematicians, Kyoto, Japan, 1990

$$\omega_\Gamma = \sum_{S \in \mathrm{Conf}_0} \varphi(S) \cdot \frac{PM(S,t)}{PM(pt,t)}\Big|_{t=r}$$

where $r := 1/\limsup_{n\to\infty} \sqrt[n]{\sharp\Gamma_n}$ is the first place of pole of $PM(pt,t)$ on the real axis and hence is a real algebraic number and $\omega_\Gamma \in \mathscr{L}_{\mathbb{Q}(r)}$.

To obtain the modular function, we need further study of representations of the Configuration algebra, which is a subject of another paper.

The §§2–4 treat generality on graphs and construct the configuration algebra. The §§5–9 treat Lielike elements and grouplike elements of the algebra. The §10 treats the limit elements in the agebra.

§2. Graphs and Covering Coefficients

Some basic combinatorial rules for graphs are discussed in this paragraph.

(2.1) Definition. 1. A pair (Γ, B) is called a *graph*, if Γ is a set of vertices and B (called the set of edges) is a subset of $\Gamma \times \Gamma \backslash \Delta$ with $\sigma(B) = B$, where σ is the involution $\sigma(\alpha,\beta) = (\beta,\alpha)$ and Δ is the diagonal. A graph is *connected*, if it is connected as a simplicial complex. A graph is *finite*, if $\sharp\Gamma < \infty$.

2. An *isomorphism* of graphs is a bijection of vertices inducing a bijection of edges. Any subset $\mathbb{S}$ of Γ is a *subgraph* by taking $B \cap (\mathbb{S} \times \mathbb{S})$ as edges for $\mathbb{S}$. The word "subgraph" is used only in this sense.

3. A graph is called *colored oriented*, if there exists a finite set G, called a coloring set, with an involution $\sigma_G : G \to G$ and a map $c : B \to G$ which is equivariant with the involutions: $c \circ \sigma = \sigma_G \circ c$. Isomorphisms and subgraphs of colored oriented graphs are defined as compatible with c.

Example. A group Γ with a finite generator system G with $G = G^{-1}$ and $e \notin G$ carries a colored oriented graph structure: $B := \{(\gamma,\delta) \in \Gamma^2 : \gamma^{-1}\delta \in G\}$, $c(\gamma,\delta) = \gamma^{-1}\delta$ and $\sigma_G(g) = g^{-1}$. The graph is called a *Caylay graph.*

(2.2) In all what follow, we fix an increasing sequence $\{G_p\}_{p\in\mathbb{N}}$ of coloring set. Associated to that the set of configurations is defined by

$$\begin{aligned}\mathrm{Conf}^{pq} &:= \{\mathbb{S} : \mathbb{S} \text{ is a finite colored oriented graph for } G_p \text{ such that} \\ &\qquad \text{number of edges at a vertex is at most } q.\}/\text{isomorphism}. \\ \mathrm{Conf} &:= \cup_{p,q} \mathrm{Conf}^{pq}, \\ \mathrm{Conf}_0 &:= \{S \in \mathrm{Conf} : S \text{ is connected}\}, \quad \mathrm{Conf}_+ := \mathrm{Conf} \backslash \{[\phi]\}.\end{aligned}$$

An isomorphisms class of a graph $\mathbb{S}$ is denoted by $[\mathbb{S}]$. The set of vertices of $\mathbb{S}$ is denoted by $|\mathbb{S}|$.

(2.3) The Conf has an abelian semigroup structure with a partial ordering:

$$[\mathbb{S}] \cdot [\mathbb{T}] := [\mathbb{S} \amalg \mathbb{T}] \qquad \text{for } [\mathbb{S}], [\mathbb{T}] \in \mathrm{Conf},$$

$S \leq T \overset{\text{def}}{\Longleftrightarrow}$ There are graphs $\mathbb{S}$ and $\mathbb{T}$ with $S = [\mathbb{S}]$, $T = [\mathbb{T}]$ and $\mathbb{S} \subset \mathbb{T}$. Here $1 = [\phi]$ is the minimal element in Conf. One has $\mathrm{Conf} \simeq \mathrm{Conf}_0^{\mathbb{Z}_{\geq 0}}$.

(2.4) For $S_1, \cdots, S_m$ and $S \in \mathrm{Conf}$, we define a numerical invariant, which we call the *covering coefficient*:

$$\binom{S_1, \cdots, S_m}{S} := \sharp\{(\mathbb{S}_1, \cdots, \mathbb{S}_m) : \mathbb{S}_i \subset \mathbb{S} \text{ s.t. } [\mathbb{S}_i] = S_i \ (i = 1, \cdots, m) \text{ and } \bigcup_{i=1}^{m} |\mathbb{S}_i| = |\mathbb{S}|.\}.$$

Here $\mathbb{S}$ is a graph s.t. $[\mathbb{S}] = S$. The definition does not depend on $\mathbb{S}$.

(2.5) We list some elementary properties of covering coefficients.

i) $\binom{S_1, \cdots, S_m}{S} = 0$ *unless* $S_i \leq S$ *for* $i = 1, \ldots, m$ *and* $\sum \sharp S_i \geq \sharp S$.

ii) $\binom{S_1, \cdots, S_m}{S}$ *is invariant by the permutation of* S_i*'s.*

iii) *For* $1 \leq {}^{\forall}i \leq m$, *one has elimination rule*:

$$\binom{S_1, \cdots, S_{i-1}, \phi, S_{i+1}, \cdots, S_m}{S} = \binom{S_1, \cdots, S_{i-1}, S_{i+1}, \cdots, S_m}{S}.$$

iv) *For the case* $m = 1$, $\binom{T}{S} = \begin{cases} 1 & \text{if } S = T, \\ 0 & \text{else.} \end{cases}$

v) *For the case* $S = \phi$, $\binom{S_1, \cdots, S_m}{\phi} = \begin{cases} 1 & \text{if } \cup S_i = \phi, \\ 0 & \text{else.} \end{cases}$

(2.6) **Composition Rule.** For $S_1, \cdots, S_m, T_1, \cdots, T_n$ and $S \in \mathrm{Conf}$, one has

$$\sum_{U \in \mathrm{Conf}} \binom{S_1, \cdots, S_m}{U} \binom{U, T_1, \cdots, T_n}{S} = \binom{S_1, \cdots, S_m, T_1, \cdots, T_n}{S}.$$

(2.7) **Decomposition Rule.** For $S_1, \cdots, S_m, U$ and $V \in \mathrm{Conf}$, one has

$$\binom{S_1, \cdots, S_m}{U \cdot V} = \sum_{S_1 = R_1 \cdot T_1} \cdots \sum_{S_m = R_m \cdot T_m} \binom{R_1, \cdots, R_1}{U} \binom{T_1, \cdots, T_m}{V}.$$

Here R_i and T_i run over Conf for all possible decompositions of S_i $(i = 1, \ldots, m)$.

§3. Configuration Algebra

(3.1) *Algebra Structure.* The free abelian group $\mathbb{Z} \cdot \mathrm{Conf}$ generated by Conf carries an algebra structure by a use of the semigroup structure on Conf (2.3) as the product (recall that $[\phi] = 1$). It is isomorphic to the free polynomial algebra generated by Conf_0. The algebra is graded by taking $\deg(S) := \sharp(S)$ for $S \in \mathrm{Conf}$, since the aditivity: $\sharp(S \cdot T) = \sharp(S) + \sharp(T)$.

(3.2) *Adic Topology.* For $n \geq 0$, let us define an ideal in $\mathbb{Z} \cdot \mathrm{Conf}^{pq}$

$$\mathscr{I}_n^{pq} := \text{the ideal generated by} \{S \in \mathrm{Conf}^{pq} : \sharp(S) \geq n\}.$$

Taking $\mathscr{I}_n^{pq}$ as a fundamental system of neighbourhoods of 0, we define the adic topology. The completion w.r.t. the adic topology is denoted by

$$\mathbb{Z}[[\mathrm{Conf}^{pq}]] := \varprojlim_n (\mathbb{Z} \cdot \mathrm{Conf}^{pq}) / \mathscr{I}_n^{pq}.$$

We put $\mathbb{Z}[[\mathrm{Conf}]] := \varinjlim \mathbb{Z}[[\mathrm{Conf}^{pq}]]$, even it is a confusing notation. For a commutative algebra $\mathbb{A}$ with a unit, we put $\mathbb{A}[[\mathrm{Conf}]] := \mathbb{A} \otimes_{\mathbb{Z}} \mathbb{Z}[[\mathrm{Conf}]]$ and call it the *configuration algebra* with coefficient in $\mathbb{A}$. Since $\mathbb{A}[[\mathrm{Conf}^{pq}]] \simeq \prod_{S \in \mathrm{Conf}^{pq}} \mathbb{A} \cdot S$, any element f of the algebra is expressed by a formal sum : $f = \sum_{S \in \mathrm{Conf}^{pq}} S \cdot f_S$ for a suitable $p, q > 0$ and constants $f_S \in \mathbb{A}$. Put, $\mathrm{Supp}(f) := \{S \in \mathrm{Conf} : f_S \neq 0\}$. A series f is said to be *of finite type,* if $\mathrm{Supp}(f)$ is contained in a finitely generated subsemigroup of Conf.

(3.3) A subset P of Conf is called saturated, if its *saturation* $\tilde{P} := \{T \in \mathrm{Conf} : {}^{\exists}S \in P \text{ s.t. } T \leq S\}$ coincides itself. A subalgebra generated by a saturated subset is called a *saturated subalgebra* of the configuration algebra. Let Γ be a Caylay graph of an infinite group and let P_Γ be the set of isomorphism classes of all finite subgraphs of Γ. Then P_Γ is a saturated semigroup contained in Conf^{pq} for some $p, q \geq 0$. $\mathbb{A}[[P_\Gamma]]$ is the configuration algebra for (Γ, G).

(3.4) *Exponential and Logarithmic Maps.* Assume $\mathbb{Q} \subset \mathbb{A}$. We define maps:

$$\exp(\mathscr{M}) := \sum_{n=0}^{\infty} \frac{1}{n!} \mathscr{M}^n \quad \text{and} \quad \log(1 + \mathscr{A}) := \sum_{n=1}^{\infty} \frac{(-1)^{n-1}}{n} \mathscr{A}^n$$

for $\mathscr{M}$ and $\mathscr{A} \in \mathbb{A}[[\mathrm{Conf}]]_+ :=$ the augmentation ideal generated by Conf_+.

§4. The Hopf Structure on the Configuration Algebra

(4.1) *Coproduct Φ_m.* For an $m \in \mathbb{N}$ and $U \in \mathrm{Conf}$, define a map

$$\Phi_m(U) := \sum_{S_1 \in \mathrm{Conf}} \cdots \sum_{S_m \in \mathrm{Conf}} \binom{S_1, \cdots, S_m}{U} S_1 \otimes \ldots \otimes S_m.$$

Owing to the decomposition rule (2.7) and (2.5) i), one has

$$\Phi_m(U \cdot V) = \Phi_m(U) \cdot \Phi_m(V) \qquad \text{for} \quad U, V \in \mathrm{Conf},$$
$$\Phi_m(\mathscr{I}_n^{pq}) \subset \sum_{n_1 + \cdots + n_m \geq n} \mathscr{I}_{n_1}^{pq} \otimes \cdots \otimes \mathscr{I}_{n_m}^{pq} \qquad \text{for} \quad p, q, n, m \in \mathbb{Z}_{\geq 0}.$$

Thus Φ_m extends to a homomorphism from the configuration algebra to its completed m-tensor product, denoted by Φ_m again, which we call the *coproduct.* The symmetric group $\mathfrak{S}_m$ acts on the range of Φ_m by permutation. The image of Φ_m is $\mathfrak{S}_m$-invariant, because of (2.5) ii). We shall call this property the *co-commutativity* of Φ_m.

The composition rule (2.6) implies, for $m, n \in \mathbb{N}$,

$$(1 \otimes \cdots \otimes 1 \otimes \Phi_m) \circ \Phi_{n+1} = \Phi_{m+n}\,.$$

Combining this fact with the cocommutativity, we see that any composits of Φ is expressed again by some Φ_m. We call this the *co-associativity* of Φ_m.

(4.2) The *augumentation map* for the algebra is defined as $\mathbb{Z}[[\mathrm{Conf}]] \to \mathbb{Z}$, $(S \in \mathrm{Conf}_+ \mapsto 0, [\phi] \mapsto 1)$. It is a *co-unit* in the sense: $(\mathrm{aug} \cdot \mathrm{id}) \circ \Phi_2 = \mathrm{id}$.

(4.3) *The involution map* ι. The following Lemma is nontrivial.

Lemma (Existense of Co-inverse). *There exists a unique algebra automorphism*

$$\iota : \mathbb{Z}[[\mathrm{Conf}]] \to \mathbb{Z}[[\mathrm{Conf}]]$$

such that

i) ι *is involutive. That is:* $\iota^2 = \mathrm{id}$.
ii) ι *is the co-inverse map w.r.t. the coproduct* Φ_2. *That is:* $(\iota \cdot \mathrm{id}) \circ \Phi_2 = \mathrm{aug}$.
iii) ι *leaves any saturated subalgebras of* $\mathbb{Z}[[\mathrm{Conf}]]$ *invariant.*
iv) ι *is continuous w.r.t. the adic topology. In particular,* $\mathrm{aug} \circ \iota = \mathrm{aug}$.

Remark. There is other coalgebra structure studied in combinatorics (Rota [5]).

§5. The Growth Functions for Configurations

In this and next paragraphs, we introduce some basic elements $1 + \mathscr{A}(T)$ and $\mathscr{M}(T)$ $(T \in \mathrm{Conf})$ of the configuration algebra, which turn out to be grouplike or Lielike respectively.

(5.1) *Growth Functions.* Let S and $T \in \mathrm{Conf}$ be given. Fix a graph $\mathbb{T}$ with $[\mathbb{T}] = T$ and put

$$\begin{aligned} \mathbb{A}(S, \mathbb{T}) &:= \sharp\{\mathbb{S} : \mathbb{S} \subset \mathbb{T} \quad \text{such that} \quad [\mathbb{S}] = S\}, \\ A(S, T) &:= \sharp\mathbb{A}(S, \mathbb{T})\,. \end{aligned}$$

We call $A(S, T)$ the growth function. By definition $A(S, T_1 \cdot T_2) = A(S, T_1) + A(S, T_2)$ for $S \in \mathrm{Conf}_0$ and $T_i \in \mathrm{Conf}$.

Let us introduce an element of $\mathbb{Z} \cdot \mathrm{Conf}$:

$$1 + \mathscr{A}(T) := 1 + \sum_{S \in \mathrm{Conf}_+} S \cdot A(S, T) = \sum_{\mathbb{S} \in 2^{\mathbb{T}}} [\mathbb{S}]\,,$$

where $2^{\mathbb{T}} := \{\text{subgraphs of } \mathbb{T}\}$. Obvious from definition, for T_1 and $T_2 \in \mathrm{Conf}$

$$(1 + \mathscr{A}(T_1 \cdot T_2)) = (1 + \mathscr{A}(T_1))(1 + \mathscr{A}(T_2))\,.$$

(5.2) *A numerical approximation* of the growth function.

Lemma. *For* $S \in \mathrm{Conf}$ *and for* $T \in \mathrm{Conf}^{pq}$, *one has*

$$\mathscr{A}(S,T) \leq \frac{1}{\sharp Aut(S)} \cdot (\sharp T)^{n(S)} \cdot q^{\sharp S - n(S)}.$$

(Here $n(S) := \sharp$ *of connected component of* S.*)*

(5.3) *Product expansion formula* for growth functions. The following basic formula can be shown by a combinatorial method.

Lemma. *Let* $S_1, \dots, S_m$ $(m \geq 1)$ *and* $T \in \mathrm{Conf}$ *be given. Then,*

$$\prod_{i=1}^{m} A(S_i, T) = \sum_{S \in \mathrm{Conf}} \binom{S_1, \dots, S_m}{S} A(S,T).$$

(5.4) *Grouplike property* of the growth function.

An element $g \in \mathbb{A}[[\mathrm{Conf}]] \backslash \{0\}$ is called grouplike [8], if it satisfies $\Phi_m(g) = g \otimes \cdots \otimes g$ for $^{\forall} m \in \mathbb{N}$. We put

$$G_{\mathbb{A}} := \{\text{all grouplike elements in } \mathbb{A}[[\mathrm{Conf}]]\},$$
$$G_{\mathbb{A},\text{finite}} := \{g \in G_{\mathbb{A}} : g \quad \text{is of finite type}\}.$$

The (5.3) Lemma implies the following Lemma.

Lemma (Grouplike Property). *For all* $T \in \mathrm{Conf}$, *one has* $1 + \mathscr{A}(T) \in G_{\mathbb{Z},\text{finite}}$. *That is: for any* $m \in \mathbb{N}$ *and* $T \in \mathrm{Conf}$, *one has*

$$(1 + \mathscr{A}(T)) \otimes \cdots \otimes (1 + \mathscr{A}(T)) = \Phi_m(1 + \mathscr{A}(T)).$$

Corollary.
$$\big(1 + \iota(\mathscr{A}(T))\big)\big(1 + \mathscr{A}(T)\big) = 1 \quad \textit{for} \quad T \in \mathrm{Conf},$$
$$\Phi_m \circ \iota = (\iota \otimes \cdots \otimes \iota) \circ \Phi_m \quad \textit{for} \quad m \in \mathbb{N}.$$

§6. The Logarithmic Growth Function

(6.1) For $S \in \mathrm{Conf}$, define the logarithmic growth function by:

$$\mathscr{M}(T) := \log(1 + \mathscr{A}(T))$$

(cf. (3.4). Develop $\mathscr{M}(T)$ in a series

$$\mathscr{M}(T) = \sum_{S \in \mathrm{Conf}} S \cdot M(S,T).$$

By definition $M(\phi, T) := 0$ for $T \in \mathrm{Conf}$ and $A(S,T) = M(S,T)$ for $S \in \mathrm{Conf}_0$. The multiplicativity of $\mathscr{A}(T)$ (5.1) implies the additivity:

$$\mathscr{M}(T_1 \cdot T_2) = \mathscr{M}(T_1) + \mathscr{M}(T_2)$$

for $T_i \in \mathrm{Conf}$ and hence the additivity:

$$M(S, T_1 \cdot T_2) = M(S, T_1) + M(S, T_2)$$

for S and $T_i \in \mathrm{Conf}$.

(6.2) *The linear dependence relations* on logarithmic functions.
The polynomial relation (5.4) Lemma implies a linear relation:

Lemma (Lielike Property). *For* $T \in \mathrm{Conf}$ *and* $m \geq 1$,

$$\sum_{i=1}^{m} 1 \otimes \cdots \otimes 1 \overset{i\text{-th}}{\mathscr{M}(T)} \otimes 1 \otimes \cdots \otimes 1 = \Phi_m(\mathscr{M}(T)).$$

Corollary. *Let* $S_1, \cdots, S_m \in \mathrm{Conf}_+$ *for* $m \geq 2$ *and* $T \in \mathrm{Conf}$ *be given. Then,*

$$\sum_{S \in \mathrm{Conf}} \binom{S_1, \cdots, S_m}{S} M(S,T) = 0.$$

Remark. The linear dependence relations among $M(S,T)$'s for $S \in \mathrm{Conf}$ is the key fact in all this paper. The Hopf algebra structure is introduced to deduce the relation. We shall solve this linear relation in (8.2) by a use of kabi coefficients introduced in §7.

(6.3) An element $\mathscr{M}$ of the configuration algebra satisfying the relation (6.2) is called *Lielike* ([7]). Let us fix notations:

$$\mathscr{L}_{\mathbb{A}}^{pq} := \{\text{all Lie-like elements in } \mathbb{A}[[\mathrm{Conf}^{pq}]]\},$$

and

$$\mathscr{L}_{\mathbb{A},\mathrm{finite}}^{pq} := \{M \in \mathscr{L}_{\mathbb{A}} : M \text{ is of finite type}\}.$$

We denote also: $\mathscr{L}_{\mathbb{A}} := \cup_{p,q} \mathscr{L}_{\mathbb{A}}^{pq}$ and $\mathscr{L}_{\mathbb{A},\mathrm{finite}} := \cup_{p,q} \mathscr{L}_{\mathbb{A},\mathrm{finite}}^{pq}$.

§7. Kabi Coefficients

(7.1) Definition. 1. A pair $(\mathbb{S}, \mathbb{U})$ of a graph $\mathbb{U}$ and its subgraph $\mathbb{S}$ is called a *kabi* over $\mathbb{S}$, if any vertex of $\mathbb{U} \backslash \mathbb{S}$ is connected to a vertx of $\mathbb{S}$ through an edge.
2. Let $U \in \mathrm{Conf}_0$ and let $\mathbb{U}$ be a graph with $[\mathbb{U}] \simeq U$. For $S \in \mathrm{Conf}_0$, put

$$\mathbb{K}(S, \mathbb{U}) := \{\mathbb{S} : \mathbb{S} \subset \mathbb{U} \ s.t.\ [\mathbb{S}] \simeq S \text{ and } (\mathbb{S}, \mathbb{U}) \text{ is a kabi}\},$$
$$K(S, U) := \sharp \mathbb{K}(S, \mathbb{U}).$$

We call $K(S,U)$ a *Kabi-coefficient*. Its definition does not depend on a choice of $\mathbb{U}$. From definition, $K(S,U)=0$ for $S \not\leq U$, and $K(S,S)=1$ for $S\in \mathrm{Conf}_0$. The word "kabi" means "mold" in Japanese.

(7.2) Lemma (The Inversion Formula of Kabi). *For $S,T\in \mathrm{Conf}_0$, one has*

(1) $$\sum_{U\in \mathrm{Conf}_0} (-1)^{\sharp U-\sharp S} K(S,U)\cdot A(U,T)=\delta(S,T).$$

(2) $$\sum_{U\in \mathrm{Conf}_0} (-1)^{\sharp U-\sharp T} A(S,U)\cdot K(U,T)=\delta(S,T).$$

Specializing (2), one gets: $\sum_{U\in \mathrm{Conf}_0}(-1)^{\sharp U}\sharp U\cdot K(U,T)=-\delta(pt,T)$.

(7.3) We remark the boundedness of non-zero entries of K.

Assertion. Let $T\in \mathrm{Conf}_0^{pq}$. Then $K(S,T)=0$, unless $\sharp T\leq \sharp S\cdot(q-1)+2$.

§8. Lielike Elements $\mathscr{L}_{\mathbb{A}}$

(8.1) *Basises of $\mathscr{L}_{\mathbb{A},\mathrm{finite}}$ and $\mathscr{L}_{\mathbb{A}}$.*

Lemma. *Let $\mathbb{A}$ be an algebra containing $\mathbb{Q}$. Then,*

i) The system $\{\mathscr{M}(T)\}_{T\in \mathrm{Conf}_0}$ give a $\mathbb{A}$-free basis for $\mathscr{L}_{\mathbb{A},\mathrm{finite}}$.

$$\mathscr{L}_{\mathbb{A},\mathrm{finite}} \simeq \oplus_{S\in \mathrm{Conf}_0}\mathbb{A}\cdot \mathscr{M}(S).$$

We introuduce another system $\{\varphi(S)\in \mathscr{L}_{\mathbb{Q},\mathrm{finite}}\cap \mathscr{I}_{\sharp S}\}_{S\in \mathrm{Conf}_0}$ of $\mathbb{A}$ basis of $\mathscr{L}_{\mathbb{A},\mathrm{finite}}$ (recall the inversion formula (7.2)):

$$\mathscr{M}(T)=\sum_{S\in \mathrm{Conf}_0}\varphi(S)\cdot A(S,T)$$

$$\varphi(S)=\sum_{T\in \mathrm{Conf}_0}\mathscr{M}(T)\cdot(-1)^{\sharp T-\sharp S}K(T,S).$$

ii) $\{\varphi(S)\}_{S\in \mathrm{Conf}_0}$ is a topological basis of $\mathscr{L}_{\mathbb{A}}$. That is:

$$\mathscr{L}_{\mathbb{A}}\simeq \varinjlim_{p,q}\Big(\prod_{S\in \mathrm{Conf}^{pq}}\mathbb{A}\cdot\varphi(S)\Big).$$

This means that any $\mathscr{M}\in\mathscr{L}_{\mathbb{A}}$ is express uniquely as an infinite sum

$$\mathscr{M}=\sum_{S\in \mathrm{Conf}_0^{pq}}\varphi(S)\cdot a_S$$

for some $p,q\in\mathbb{Z}_{\geq 0}$ and $a_S\in\mathbb{A}$ for $S\in \mathrm{Conf}_0^{pq}$.

(8.2) *An explicite formula* for $\varphi(S)$. For $S\in \mathrm{Conf}_0$, let us develop $\varphi(S)=\prod_{U\in \mathrm{Conf}}U\cdot\varphi(U,S)$ for $\varphi(U,S)\in\mathbb{Q}$. By a use of (3.4), (5.3), (7.2), one obtain:

$$\varphi(U,S) = \sum_{U=U_1^{k_1}\amalg\cdots\amalg U_m^{k_m}} \frac{(k_1+\cdots+k_m-1)!(-1)^{k_1+\cdots+k_m-1}}{k_1!\cdots k_m!}\binom{U_1,\dots,U_m}{S}.$$

Here $(U_1,\dots,U_m)$ means a sequence of U_i where each U_i appears with multiplicity k_i. Let $\mathscr{M} = \sum_{U\in\mathrm{Conf}} U \cdot M_U$ be an element of $\mathscr{L}_{\mathbb{A}}$. Then,

$$M_U = \sum_{S\in\mathrm{Conf}_0} \varphi(U,S) \cdot M_S .$$

So combining both formula, one is able to determine all coefficients of $\mathscr{M}$ from the knowledge of that for M_S ($S \in \mathrm{Conf}_0$).

Remark. In general, an element of $\mathscr{L}_{\mathbb{A}}$ cannot be expressed by an infinite sum of $\mathscr{M}(T)$ ($T \in \mathrm{Conf}_0$).

§9. Grouplike Elements $G_{\mathbb{A}}$

(9.1) Let $\mathbb{A}$ be an algebra with char=0 without an idempotent element and $\mathbb{Q} \subset \mathbb{A}$. Then one has isomorphisms: $\exp : \mathscr{L}_{\mathbb{A}} \simeq G_{\mathbb{A}}$ and $\mathscr{L}_{\mathbb{A},\mathrm{finite}} \simeq G_{\mathbb{A},\mathrm{finite}}$.

(9.2) *Generators for $G_{\mathbb{A},\mathrm{finite}}$ and $G_{\mathbb{A}}$.*

Lemma. *Let $\mathbb{A}$ be a commutative $\mathbb{Z}$-torsionfree algebra with a unit element and without an idempotent element.*

i) Any element g of $G_{\mathbb{A},\mathrm{finite}}$ is uniquely expressed as

$$g = \prod_{i\in I} (1 + \mathscr{A}(S_i))^{c_i}$$

for a finite index set I and $S_i \in \mathrm{Conf}_0$ and $c_i \in \mathbb{A}$ for $i \in I$. Therefore the correspondence: $S \mapsto 1 + \mathscr{A}(S)$ induces an isomorphism:

$$\ll \mathrm{Conf} \gg \otimes_{\mathbb{Z}} \mathbb{A} \simeq G_{\mathbb{A},\mathrm{finite}}$$

where $\ll \mathrm{Conf} \gg$ is the group generated by the semigroup Conf.

ii) The set $\{\exp(\varphi(s))\}_{S\in\mathrm{Conf}_0} \subset G_{\mathbb{Z},\mathrm{finite}}$ form a system of free topological generators of $G_{\mathbb{A}}$.

§10. Limit Elements in $\mathscr{L}_{\mathbb{R}}$

We finaly introduce the limit element ω_Γ for a finitely generated group Γ with fix generators. A further study of ω_Γ is beyond the scope of the present paper.

(10.1) We equip a *classical topology on the* $\mathbb{R}$*-vector space* $\mathscr{L}_{\mathbb{R}}^{pq}$ by:

$$\mathscr{L}_{\mathbb{R}}^{pq} := \varprojlim_n \mathscr{L}_{\mathbb{R}}^{pq} / \overline{\mathscr{J}_n^{pq}} \cap \mathscr{L}_{\mathbb{R}} = \prod_{S\in\mathrm{Conf}_0^{pq}} \mathbb{R}\cdot\varphi(S),$$

the projective limit topology of $\mathbb{R}$-vector spaces. $\mathscr{L}_{\mathbb{R},\mathrm{finite}}$ is dense in $\mathscr{L}_{\mathbb{R}}$. Samely, we put a classical topology on $\mathbb{R}[[\mathrm{Conf}]]$ by

$$\mathbb{R}[[\mathrm{Conf}^{pq}]] = \varprojlim_n \mathbb{R} \cdot \mathrm{Conf}^{pq} / \mathscr{J}_n^{pq} = \prod_{S \in \mathrm{Conf}^{pq}} \mathbb{R} \cdot S .$$

i) *The product and coproduct on* $\mathbb{R}[[\mathrm{Conf}^{pq}]]$ *are continuous w.r.t. the classical topology.*

ii) *The classical topology on* $\mathscr{L}_{\mathbb{R}}^{pq}$ *is homeomorphic to the topology induced from that on* $\mathbb{R}[[\mathrm{Conf}]]$.

iii) *Let us equip on* $G_{\mathbb{R}}$ *a classicial topology induced from* $\mathbb{R}[\mathrm{Conf}]]$ *as a subspace. Then* $\exp : \mathscr{L}_{\mathbb{R}}^{pq} \to G_{\mathbb{R}}^{pq}$ *is a homeomorphism.*

(10.2) *Limits of equally dividing points.*

The correspondence $1 + \mathscr{A}(T) \in G_{\mathbb{Z}} \mapsto \sharp T \in \mathbb{Z}$ extends continuously as an additive character from $G_{\mathbb{R}}$ to $\mathbb{R}$, which we denote by $\mathfrak{X}_{pt}$. For $S \in \mathrm{Conf}$, $(1 + \mathscr{A}(S))^{1/\sharp(S)}$ is called an equaly dividing point, where the exponent $1/\sharp(S)$ is chosen to get the equality: $\mathfrak{X}_{pt}((1 + \mathscr{A}(S))^{1/\sharp(S)}) = 1$. By a use of logarithm map, we define the set of log-equaly dividing points $\log(EDP^{pq}) := \{\mathscr{M}(T)/\sharp T : T \in \mathrm{Conf}_+^{pq}\}$ in $\mathscr{L}_{\mathbb{Q}}$ and its closure $\overline{\log(EDP^{pq})}$ w.r.t. the classical topology in $\mathscr{L}_{\mathbb{R}}$. The numerical approximation (5.2) will be repeatedly used to show the following Assertion and the next (10.3) Lemma.

Assertion. 1. The $\overline{\log(EDP^{pq})}$ *is a compact convex set.*

2. Let us develop any element ω *of* $\overline{\log(EDP^{pq})}$ *as* $\sum_{S \in \mathrm{Conf}_0} \varphi(S) \cdot a_S$. *Then i)* $0 \le a_S \le q^{\sharp S - 1}/\sharp\,\mathrm{Aut}(S)$ *for* $S \in \mathrm{Conf}_0$, *ii) if* $a_S \neq 0$ *then* $a_{S'} \neq 0$ *for* $S' \le S$.

(10.3) *Residual representation of the elements of* $\overline{\log(EDP^{pq})}$.

Let $\omega = \lim_{n\to\infty} M(T_n)/\sharp T_n$ be an element of $\overline{\log(EDP^{pq})}$ for a sequence $\{T_n\}_{n\ge 0}$ of Conf^{pq} for some p and q. We introduce formal power series

$$P(t) := \sum_{n=0}^{\infty} \sharp T_n \cdot t^n \in \mathbb{Z}[[t]] ,$$

$$P\mathscr{M}(t) := \sum_{n=0}^{\infty} \mathscr{M}(T_n) \cdot t^n \in \mathscr{L}_{\mathbb{Q}}[[t]] .$$

Lemma. *Suppose that the radius* r *of convergence of* $P(t)$ *is positive. Then,*

i) $PM(t)$ *converges in the same radius* r. *That is: for any* $S \in \mathrm{Conf}_0$,

$$PM(S,t) := \partial_S P\mathscr{M}(t) = \sum_{n=0}^{\infty} M(S, T_n) \cdot t^n \in \mathbb{Q}[[t]]$$

converges in the radius r *as for* $P(t)$.

ii) The value of the proportion $P\mathscr{M}(t)/P(t)$ *converges to* ω *in* $\mathscr{L}_{\mathbb{R}}$ *as* t *tends to the radius* r *along real axis from* 0. *That is:*

$$\omega = \lim_{t\to r} P\mathcal{M}(t)/P(t) = \sum_{S\in \mathrm{Conf}_0} \varphi(S)\cdot \lim_{t\to r} PM(S,t)/P(t)\,.$$

iii) If $P(t)$ and $PM(S,t)$ extend to meromorphic functions at $t=r$, then $PM(S,t)/P(t)$ is regular at $t=r$ and

$$\lim_{t\to\infty} PM(S,t)/P(t) = PM(S,t)/P(t)\big|_{t=r}\,.$$

(10.4) *The limit element ω_Γ for a discrete group Γ.*

Let Γ be a group and let G be a finite set of generators of Γ such that $G=G^{-1}$ and $e\notin G$. We regard Γ as a Caylay graph. For $\gamma\in\Gamma$, put $\ell(\gamma):=\inf\{n:\gamma=g_1\cdots g_n \text{ for some } g_i\in G,\ i=1,\cdots,n\}$. Define a sequence of increasing graphs $\Gamma_n:=\{\gamma\in\Gamma:\ell(\gamma)\le n\}$ for $n\in\mathbb{N}$ and associated generating functions,

$$P_\Gamma(t) := \sum_{n=0}^{\infty} \sharp\Gamma_n\cdot t^n\,, \tag{1}$$

$$P_\Gamma\mathcal{M}(t) := \sum_{n=0}^{\infty} \mathcal{M}(\Gamma_n)\cdot t^n\,. \tag{2}$$

Here $P_\Gamma(t)$ is well known in literature as growth power series for Γ relative to the generator system G. We have following facts i) and ii).

i) *For any finitely generated group Γ, the series $P_\Gamma(t)$ and $P_\Gamma\mathcal{M}(t)$ converge in a positive radius.*

ii) *For a wide class of group Γ with a generator system G, $PM(S,t)$ for $S\in \mathrm{Conf}_0$ becomes a rational function in t. This includes examples:*

a) *an automatic group Γ* ([2]) *such that the regular language $L(W)$ accepted by the word accepter contains at least one shortest word for every element of the group Γ.*

b) *Hyperbolic or negatively curved groups Γ* [5]. *Then $P_\Gamma(t)$ and $P_\Gamma\mathcal{M}(t)$ are rational functions with a common denominator.* (D. Epstein et al [3]).

Owe to i), we introduce the limit element for (Γ,G):

$$\begin{aligned}\omega_\Gamma &:= \lim_{n\to\infty}\log(1+\mathcal{A}(\Gamma_n))^{1/\sharp\Gamma_n}\\ &= \sum_{S\in\mathrm{Conf}_0}\varphi(S)\cdot\lim_{t\to r}\frac{P_\Gamma M(S,t)}{P_\Gamma M(pt,t)}\,.\end{aligned}$$

If Γ belongs to the case ii), owing to (10.3) Lemma iii), the limit process is replaced by a residue calculation of rational functions.

$$\omega_\Gamma = \sum_{S\in\mathrm{Conf}_0}\varphi(S)\cdot\frac{P_\Gamma M(S,t)}{P_\Gamma M(pt,t)}\bigg|_{t=0} \in \mathcal{L}_{\mathbb{Q}(r)}\,.$$

Here $r=1/\limsup_{n\to\infty}\sqrt[n]{\sharp\Gamma_n}$ is the first place of pole of $P_\Gamma(t)$ on the $\mathbb{R}_+$, and is a real algebraic number. We know little about ω_Γ except that it is defined.

(10.5) *A Kabi-kernel condition on ω_Γ.* The kabi coefficients induces a map:

$$\sum_{S\in \mathrm{Conf}_0^{pq}} \mathbb{A}\cdot\varphi(S) \simeq \sum_{T\in \mathrm{Conf}_0^{pq}} \mathbb{A}\cdot M(T), \varphi(S) \mapsto \sum_{T\in \mathrm{Conf}_0^{pq}} M(T)\cdot(-1)^{\sharp T-\sharp S}K(T,S)$$

for fix p and $q\in\mathbb{Z}_{\geq 0}$. By completing the map, we obtain a map:

$$K : \mathscr{L}_{\mathbb{A}}^{pq} \to \prod_{T\in \mathrm{Conf}_0^{pq}} \mathbb{A}\cdot M(T).$$

Since K is bijective on $\mathscr{L}_{\mathbb{A},\mathrm{finite}}^{p,q}$, one gets amazingly $\ker(K)\cap\log(EDP^{pq}) = \phi$. Let us give a characterization of an element of $\ker(K)\cap\overline{\log(EDP^{pq})}$:

Assertion. For an element $\omega = \lim_{n\to\infty}\mathscr{M}(T_n)/\sharp T_n$ *of* $\overline{\log(EDP^{pq})}$, *$\omega$ belongs to* $\ker(K)$, *if and only if* $\lim_{n\to\infty}\delta(S,T_n)/\sharp T_n = 0$ *for any* $S\in\mathrm{Conf}_0$, *where* $\delta(S,T) := \sharp\{$*connected components of* T, *which are isomorphic to* $S\}$.

Corollary. *The ω_Γ for an infinite group Γ belongs to* $\ker(K)$.

References

1. Baxter, R. J.: Exactly solved models in statistical Mechanics. Academic Press, New York 1982
2. Cannon, J. W., Epstein, D.B.A., Holt, D.F., Paterson, M.S., Thurston, W.P.: Word processing and group theory. Preprint 1988
3. Epstein, D.: A personal communication to the author. March 1990
4. Gromov, M.: Hyperbolic groups, essays in Group theory (edited by S.M. Gersten). MSRI Publications 8. Springer, Berlin Heidelberg New York 1987, pp. 75–263
5. Rota, G.-C.: Coalgebras and Bialgebras in Combinatrics. Lectures at the Umbral Calculus Conference, The University of Oklahoma, May 15–19, 1978. Notes by S.A. Joni
6. Saito, K.: Moduli space for Fuchsian groups. Algebraic Analysis, vol. II (edited by Kashiwara and Kawai). Academic Press, New York 1988, pp. 735–786
7. Saito, K.: The limit element in the configuration algebra for discrete groups. In preparation
8. Sweedler, M.E.: Hopf algebras. Benjamin, 1969

Some Recent Results on Weakly Pseudoconvex Domains

Nessim Sibony

CNRS URA D 0757, Université de Paris-Sud, Mathématiques, Bâtiment 425
F-91405 Orsay Cedex, France

We give here a survey of some recent results concerning analysis in weakly pseudoconvex domains in $\mathbb{C}^n$ with smooth boundary.

Let $\Omega \Subset \mathbb{C}^n$ be a smoothly bounded domain with defining function r, more precisely r is a smooth function defined in a neighborhood U of $\bar{\Omega}$ such that $\Omega = \{z \in U \; ; \; r(z) < 0\}$ and $dr \neq 0$ on $\partial\Omega$.

The domain Ω is pseudoconvex iff for $z \in \partial\Omega$ and $t \in \mathbb{C}^n$

$$\langle Lr(z)t, t\rangle = \sum_{j,k} \frac{\partial^2 r(z)}{\partial z_j \partial \bar{z}_k} t_j \bar{t}_k \geq 0 \quad \text{whenever} \quad \sum_{j=1}^{n} \frac{\partial r(z)}{\partial z_j} t_j = 0 .$$

If the inequality is strict for $t \neq 0$ the domain is said to be strictly pseudoconvex. Pseudoconvex domains with smooth boundary are just the domains of holomorphy with smooth boundary.

The analysis on strictly pseudoconvex domains received much attention in the late sixties and in the seventies, the explicit construction of kernels in order to solve $\bar{\partial}$-equation was one of the main tools to solve function theoretic questions in these domains, see [HeL]. The fact that a strictly pseudoconvex domain is locally biholomorphic to a strictly convex domain is crucial in this approach. Such a simple local model does not exist when the Levi form L is not positive definite. We give first few examples to show the type of difficulties we have to face.

1. Examples

a) The Kohn-Nirenberg example [KN1]. Let $\Omega = \{(z, w) \in \mathbb{C}^2 \; ; \; \mathrm{Re}\, w + | z |^{2k} + t \mid z \mid^2 \mathrm{Re}(z^{2k-2}) < 0\}$. If $| t | \leq \frac{k^2}{2k-1}$ then Ω is pseudoconvex. If $| t | > 1$ and $k \geq 3$ then there is no supporting analytic set to $\bar{\Omega}$ at the point 0. Hence Ω is not biholomorphically equivalent in a neighborhood of 0 to a convex domain.

b) Non embeddability into convex domains. There exists a smooth pseudoconvex domain $\Omega \Subset \mathbb{C}^3$ and $p \in \partial\Omega$, such that for any N and any convex domain $U \Subset \mathbb{C}^N$ there is no proper holomorphic map from $\Omega \cap B(p, r)$, $r > 0$, into U, [Si1].

c) The "worm" domain. Diederich and Fornaess [DF1] have exhibited a pseudoconvex domain $\Omega \Subset \mathbb{C}^2$ with smooth boundary such that $\bar{\Omega}$ does not have

Proceedings of the International Congress
of Mathematicians, Kyoto, Japan, 1990

a Stein neighborhood basis. Moreover given $\varepsilon > 0$ it is possible to construct a "worm" domain Ω_ε such that there is no plurisubharmonic (p.s.h.) function ϱ in Ω_ε satisfying $-Ad^\varepsilon \le \varrho \le -Bd^\varepsilon$ with positive constants A, B. Here d denote the distance to the boundary of Ω_ε, see [DF1] and [Ki]. This last property implies the non existence of a $\mathscr{C}^1$ plurisubharmonic defining function for Ω_ε, if $\varepsilon < 1$.

Despite the smoothness of $\partial\Omega$, the structure of the set $W(\partial\Omega) = \{z \in \partial\Omega$; there exists $t \in \mathbb{C}^n$, $t \neq 0$, $< \partial r(z), t >= 0$, $< Lr(z)t, t >= 0\}$ can be quite wild. Even when $W(\partial\Omega)$ is reduced to one point, most of the sharp results obtained in the strictly pseudoconvex case either do not generalize or require a more subtle analysis. Depending on the property under consideration one has to introduce specific classes of pseudoconvex domains.

2. *B*-Regular Domains

Let $U \Subset \mathbb{C}^n$ be a bounded domain. The boundary ∂U is B-regular iff for every $p \in \partial U$ there exists a function $\psi \in \mathscr{C}(\bar{U})$ p.s.h. in U with $\psi(p) = 0$ and $\psi < 0$ on $\bar{U}\backslash\{p\}$. See [Ca1] [Si2] [Si3] for this notion. The following result is proved in [Si2] [Si3].

Theorem 2.1. *Let $\Omega \Subset \mathbb{C}^n$ be a pseudoconvex domain with smooth boundary. Assume $\partial\Omega$ is B-regular. Fix $0 < \varepsilon < 1$. There exist two defining functions r_1, r_2 for Ω such that $\varrho_1 = -(-r_1)^{1-\varepsilon}$ is p.s.h. in Ω and $\varrho_2 = (r_2)^{1+\varepsilon}$ is p.s.h. in a neighborhood of $\bar{\Omega}$. Hence $\bar{\Omega}$ has a Stein neighborhood basis.*

The existence of a Stein neighborhood basis has been proved in many special cases by Diederich-Fornaess [DF2, DF3].

The assumption in the Theorem is the existence of a p.s.h. barrier at every point of the boundary. This is a stronger assumption than the non existence of analytic structure on $\partial\Omega$. It can be shown that if such barriers exist except on a "small" set then they exist everywhere [Si2].

When $W(\partial\Omega)$ is a countable union of closed sets where continuous functions can be approximated by p.s.h. ones, then $\partial\Omega$ is B-regular. This is the case when $\partial\Omega$ is real analytic [DF2] or of finite type [Ca1], in these cases $W(\partial\Omega)$ is of Hausdorff dimension $2n-2$. However B-regularity of $\partial\Omega$ allows the set $W(\partial\Omega)$ to be of positive Lebesgue measure in $\partial\Omega$.

3. $\bar{\partial}$-Problems

Many of the most natural problems that arise for weakly pseudoconvex domains are connected with the question of solving the $\bar{\partial}$-equation in the standard spaces L^p, H^s, Λ^s.

The most fundamental result in this context is probably the following theorem due to Hörmander [Ho1].

Theorem 3.1. *Let $\Omega \Subset \mathbb{C}^n$ be a pseudoconvex domain. For every $g \in L^2_{(0,1)}(\Omega)$ such that $\bar{\partial} g = 0$ there exists a function u such that*
i) $\bar{\partial} u = g$;
ii) $\int_\Omega |u|^2 \le C(\text{diam } \Omega) \int_\Omega |g|^2$.

The result holds without any assumption on the smoothness of $\partial\Omega$. The question of solving $\bar\partial$-equation is Sobolev spaces has been settled by Kohn, see [K1].

Theorem 3.2. *Let $\Omega \Subset \mathbb{C}^n$ be a pseudoconvex domain with smooth boundary. Given $s_0 > 0$ there exists a constant $C(s_0, \Omega)$ such that for every closed form $\alpha \in H^s_{(0,1)}(\Omega)$ with $s \le s_0$, there exists $u \in H^s(\Omega)$ such that*

$$\bar\partial u = \alpha$$

and

$$\| u \|_s \le C(s_0, \Omega) \| \alpha \|_s .$$

Similar results for the $\bar\partial_b$-equation have been proved by Kohn [K1], Shaw [S] and Boas-Shaw [BoS].

Theorem 3.3. *Let $\Omega \Subset \mathbb{C}^n$ be a pseudoconvex domain with smooth boundary. Then for any $s \ge 0$, $\bar\partial_b$ has closed range in $H^s_{(0,1)}(\partial\Omega)$.*

Let $\Lambda^s(\Omega)$ denote the space of Hölder continuous functions of order $s > 0$, with the usual convention when s is an integer, for $s = 0$ we identify $\Lambda^0(\Omega)$ and $\mathscr{C}^0(\bar\Omega)$. It is not in general possible to solve the $\bar\partial$-equation with Λ^s estimates as the following result shows, [Si3, Si4].

Theorem 3.4. *Let $m \ge 3$. There is an $\Omega \Subset \mathbb{C}^m$, pseudoconvex with smooth boundary, such that for every $0 \le s < \infty$, there exists a $\bar\partial$-closed form $\alpha \in \Lambda^s_{(0,1)}(\Omega) \cap H^{s+\frac{m-1}{2}}_{(0,1)}(\Omega)$ such that the equation $\bar\partial u = \alpha$ has no solution in $\Lambda^s(\Omega)$.*

Remark. For s fixed the domain can be made strictly pseudoconvex except at one point.

Concerning L^p estimates the problem was studied by Fornaess-Sibony [FSi1]. First of all there is no analogue of Hörmander's result for $p > 2$.

Theorem 3.5. *There exists a pseudoconvex domain $U \Subset \mathbb{C}^2$ and a $\bar\partial$-closed (0,1) form α, real analytic in a neighborhood of $\bar U$, such that the equation $\bar\partial u = \alpha$ has no solution in $L^p(U)$ for any $p > 2$.*

Observe that ∂U is not smooth. The example is based on the fact that the holomorphic functions in $L^p(U)$, $p > 2$, extend holomorphically to a domain $\tilde U$ which is not contained in $\bar U$. Such phenomenon cannot happen when U is Runge.

If one tries to solve the $\bar\partial$-equation in a Hartogs domain $U = \{(z, w) \in \mathbb{C}^2 ; z \in \Omega \subset \mathbb{C}, | w |< exp(-\varphi(z))\}$ where φ is subharmonic in an open set $\Omega \Subset \mathbb{C}$, one is led to the problem in one complex variable : solve $\partial u/\partial\bar z = f$ in $L^p(\Omega, \varphi)$ with the estimate

$$\left(\int_\Omega | u |^p e^{-\varphi}\right)^{1/p} \le C \left(\int_\Omega | f |^p e^{-\varphi}\right)^{1/p} .$$

When $1 < p \leq 2$ and $\Omega \Subset \mathbb{C}$ the above estimate holds. The proof is based on the fact that Hörmander's estimates work with the same constant on a domain in $\mathbb{C}^n$ and on any covering of this domain regardless of the number of sheets.

The situation is different for $p > 2$. There exists a subharmonic function φ in the unit disc Δ and $f \in \bigcap_{p>2} L^p(\Delta, \varphi)$ such that for any $p' > 2$ the equation $\partial u/\partial \bar{z} = f$ has no solution in $L^{p'}(\Delta, \varphi)$. Using these one variable results, we prove the following theorem.

Theorem 3.6. *There exists $\Omega \Subset \mathbb{C}^4$ pseudoconvex with smooth boundary, strictly pseudoconvex except at one point with the following property : for every $p > 2$ there exists a $\bar{\partial}$-closed form $\beta \in L^p_{(0,1)}(\Omega)$ such that the equation $\bar{\partial} u = \beta$ has no solution $u \in L^{p'}(\Omega)$ if $p' \in]\sqrt{8+9p} - 3, p]$.*

The result is probably not sharp. There is also a smooth pseudoconvex domain $\Omega \Subset \mathbb{C}^3$ such that $\bar{\partial} : L^p(\Omega) \to L^p_{(0,1)}(\Omega)$ does not have closed range.

The following questions are then quite natural.

i) Suppose $\Omega \Subset \mathbb{C}^n$ is pseudoconvex with smooth boundary. Let $s > 0$, for which $s' < s$ does the operator $\bar{\partial} : \Lambda^{s'}(\Omega) \to \Lambda^s_{(0,1)}(\Omega)$ has closed range ?

ii) Let $1 \leq p \leq \infty$, for which $p' < p$, does the operator $\bar{\partial} : L^{p'}(\Omega) \to L^p_{(0,1)}(\Omega)$ has closed range ?

The following partial answer due to Bonami-Sibony [BoSi] is a consequence of Kohn's result and of a Sobolev embedding theorem which basically says that if $\bar{\partial} u \in L^t(\Omega)$ for t large enough then the dimension $2n$ in the usual Sobolev embedding theorem may be replaced by $(n + 1)$.

Theorem 3.7. *Let $\Omega \Subset \mathbb{C}^n$ be a pseudoconvex domain with smooth boundary. Let $\alpha \in H^s_{(0,1)}(\Omega)$ a $\bar{\partial}$ -closed form.*

a) If $0 < s < \frac{n+1}{2}$ and $\alpha \in L^t_{(0,1)}(\Omega)$, with $\frac{1}{t} = \frac{1}{2} - \frac{s-1/2}{n+1}$, then the equation $\bar{\partial} u = \alpha$ has a solution in $L^r(\Omega)$, with $\frac{1}{r} = \frac{1}{2} - \frac{s}{n+1}$.

b) If $\frac{n+1}{2} < s \leq \frac{n+2}{2}$ and $\alpha \in L^t_{(0,1)}(\Omega)$, with $\frac{1}{t} = \frac{1}{2} - \frac{s-1/2}{n+1}$, then $\bar{\partial} u = \alpha$ has a solution in $\Lambda^\sigma(\Omega)$, with $\sigma = s - \frac{n+1}{2}$.

c) If $s > \frac{n+2}{2}$ and $\alpha \in \Lambda^{\sigma-1/2}_{(0,1)}(\Omega)$, then $\bar{\partial} u = \alpha$ has a solution in $\Lambda^\sigma(\Omega)$, with $\sigma = s - \frac{n+1}{2}$.

Remarks. i) In all cases the solution u belongs to $H^s(\Omega)$.

ii) The examples mentioned in Theorem 3.4 show that there is necessarily a loss in the regularity of the solution.

4. $\bar{\partial}$-Neumann Problem

Let $\Omega \Subset \mathbb{C}^n$ be a pseudoconvex domain with smooth boundary. Let α be a $\bar{\partial}$-closed (0,1) form with L^2 coefficients. An aspect of the $\bar{\partial}$-Neumann problem consists in studying the regularity of the solution u of the equation $\bar{\partial}u = \alpha$, where u is orthogonal to the holomorphic L^2 functions. This solution is called the canonical solution.

Let us say that Ω satisfies (R_k) if the canonical solution u is in $H^k(\Omega)$ whenever $\alpha \in H^k_{(0,1)}(\Omega)$; Ω satisfies (R) if the canonical solution u belongs to $\mathscr{C}^\infty(\bar{\Omega})$ whenever $\alpha \in \mathscr{C}C^\infty_{(0,1)}(\bar{\Omega})$.

It is known that property (R_k) holds for all k, whenever $\partial\Omega$ is B-regular, see Catlin [Ca1]. Very recently, it was proved by Boas-Straube [BoS] that (R_k) holds for all k if the domain Ω has a p.s.h. defining function.

On the other hand, inspired by ideas of Kiselman [Ki], Barrett has shown in [Ba] that worm domains do not have property (R_k) for k large enough, depending on the domain. It seems likely that property (R) fails in general.

Property (R) is connected with the boundary behavior of biholomorphic mappings. Indeed Bell-Ligocka have shown, in [BL], that a biholomorphic mapping between two smoothly bounded domains in $\mathbb{C}^n$, satisfying property (R), extends smoothly up to the boundary.

It is not known whether a biholomorphism between two smoothly bounded domains in $\mathbb{C}^n$ extends smoothly to the boundary.

5. Domains of Finite Type

A class of domains for which one can hope to generalize the results obtained for the strictly pseudoconvex domains is the class of domains of finite type.

Let M be a real hypersurface in $\mathbb{C}^n$ with defining function r. Let $p \in M$ and $\mathscr{V}_p$ the space of germs at p of irreducible one-dimensional analytic subvarieties. For $X \in \mathscr{V}_p$ define

$$\tau^*(X) = \sup\left\{\alpha > 0 \ : \ \limsup_{z\in X\setminus\{p\},z\to p} \frac{| r(z) |}{\| z - p \|^\alpha} < \infty\right\}.$$

The type of M at p is defined as $\tau(M,p) = \sup\{\tau^*(X), X \in \mathscr{V}_p\}$. M is of finite type if $\sup_{p\in M} \tau(M,p) < \infty$. This notion has been studied by D'Angelo who showed that $p \to \tau(M,p)$ is not uppersemicontinuous but that the set of points of finite type is open, [Dan1], [Dan2]. A point in M is a point of strict pseudoconvexity iff it is a point of type 2. Bounded pseudoconvex domains with real analytic boundary are of finite type, [DF2].

Let U be a neighborhood of $p \in \partial\Omega$. A subelliptic estimate of order ε holds for (0,1) forms supported in $U \cap \bar{\Omega}$ iff there exists a constant $C > 0$ such that

$$\| | u | \|_\varepsilon^2 \leq C(\| \bar{\partial}u \|^2 + \| \bar{\partial}^*u \|^2 + \| u \|^2)$$

for $u \in \mathscr{D}^{0,1}(U) = \{u \ ; u$ smooth (0,1)-form supported in $U \cap \bar{\Omega}, \bar{\partial}^*u \in \mathrm{Dom}(\bar{\partial}^*)\}$, where $\| | \ | \|_\varepsilon$ denotes the tangential Sobelev norm of order ε.

Following the work of Kohn [K2] on subelliptic estimates for domains with real analytic boundary, Catlin has proved the following result, [Ca2], [Ca3].

Theorem 5.1. *Let $\Omega \Subset \mathbb{C}^n$ be a pseudoconvex domain with smooth boundary. Subelliptic estimate for the $\bar\partial$-Neumann problem holds for some $\varepsilon > 0$ in a neighborhood of $p \in \partial\Omega$ iff $\partial\Omega$ is of finite type at p.*

The existence of a subelliptic estimate of order $\varepsilon > 0$ near a point p implies that the canonical solution to $\bar\partial u = g$ is smooth near p if g is smooth in a neighborhoof of p, [KN2].

The best ε in the subelliptic estimate is known only for domains of finite type in $\mathbb{C}^2$, [FeK, Ca4, FSi2] and for convex domains of finite type in $\mathbb{C}^n$ [FSi2].

Real progress has been made recently in the understanding of regularity properties of the canonical solution of the $\bar\partial$ and $\bar\partial_b$ equation for domains of finite type in $\mathbb{C}^2$, see the survey paper by Christ in these proceedings [Ch].

The understanding of the local geometry of domains of finite type in $\mathbb{C}^n$ is still incomplete. The following notion of type connected with p.s.h. barriers should be compared to the type τ.

For $p \in \partial\Omega$ let $\mathscr{P}_p = \{\varphi\ ;\ \varphi \text{ p.s.h. continuous in } \bar\Omega,\ \varphi \le 0,\ \varphi(p) = 0\}$. Define for $\varphi \in \mathscr{P}_p$

$$\tau^+(\varphi) = \limsup_{z \to p, z \in \Omega} \frac{\log \mid \varphi(z) \mid}{\log \parallel z - \varphi \parallel}, \qquad \tau^-(\varphi) = \liminf_{z \to p, z \in \Omega} \frac{\log \mid \varphi(z) \mid}{\log \parallel z - p \parallel}$$

and

$$P(p, \partial\Omega) = \inf_{\varphi \in \mathscr{P}_p} \frac{\tau^+(\varphi)}{\tau^-(\varphi)}.$$

It is shown in [FSi1] taht $\tau(p, \partial\Omega) \le P(p, \partial\Omega)$. The converse is proved in [FSi1] when Ω is pseudoconvex in $\mathbb{C}^2$ or convex in $\mathbb{C}^n$. The invariant P is connected with sharp subelliptic estimate.

It would be of interest to compare $P(p, \partial\Omega)$ and $\tau(p, \partial\Omega)$ for smooth pseudoconvex domains in $\mathbb{C}^n$.

6. Spectral Questions and Approximation Problems

For $0 \le k \le \infty$, let $A^k(\bar\Omega)$ denote the algebra of holomorphic functions in $\mathscr{C}^k(\bar\Omega)$. Let $H^\infty(\Omega)$ denote the algebra of bounded holomorphic functions in Ω with uniform norm.

When $\Omega \Subset \mathbb{C}^n$ is a pseudoconvex domain with smooth boundary, then the Gelfand spectrum of the algebra $A^k(\bar\Omega)$, $0 \le k \le \infty$, can be identified with $\bar\Omega$, [HaSi]. The case of $H^\infty(\Omega)$ is different. It is not known whether Ω is dense in the Gelfand spectrum of $H^\infty(\Omega)$ even when Ω is the unit ball or the unit polydisc in $\mathbb{C}^n$, $n > 1$. However there exists a pseudoconvex domain $\Omega \Subset \mathbb{C}^3$ with smooth boundary, strictly pseudoconvex except at one point of $\partial\Omega$, such that Ω is not dense in the spectrum of $H^\infty(\Omega)$, [Si5].

Very little is known about approximation problems of holomorphic functions defined on a smoothly bounded pseudoconvex domain $\Omega \Subset \mathbb{C}^n$. Because of the existence of domains Ω such that $\bar\Omega$ does not have a Stein neighborhood basis it is not possible, in general, to approximate uniformly functions in $A^\infty(\bar\Omega)$ by holomorphic functions in a neighborhood of $\bar\Omega$. If we assume that $\bar\Omega$ has a Stein neighborhoof basis, it is still not known if functions in $A^\infty(\bar\Omega)$ are uniform limits of functions holomorphic in a neighborhoof of $\bar{\Omega}$. This is however true if we assume that $\partial\Omega$ is B-regular [Si2].

When $\Omega \Subset \mathbb{C}^n$ is pseudoconvex and smoothly bounded, it is not known if $A^\infty(\bar{\Omega})$ is dense in $A^r(\bar{\Omega})$ for the $\mathscr{C}^r$ norm when $0 \leq r < \infty$.

For a more detailed discussion of these questions we refer to [Si3].

References

[Ba] Barrett, D.: Behavior of the Bergman projection on the Diederich-Fornaess worm. Preprint

[BL] Bell, S., Ligocka, E.: A simplification and extension of Fefferman's theorem on biholomorphic mappings. Invent. math. **57** (1980) 283–289

[BoS] Boas, H., Shaw, M.C.: Sobolev estimates for the Lewy operator on weakly pseudoconvex boundaries. Math. Ann. **274** (1986) 221–231

[BoSi] Bonami, A., Sibony, N.: Sobolev embedding theorem in $\mathbb{C}^n$ and the $\bar{\partial}$-equation. Preprint

[BoS] Boas, H., Straube, E.: Sobolev estimates for the $\bar{\partial}$-Neumann operator on domains in $\mathbb{C}^n$ admitting a defining function that is p.s.h. on the boundary. Preprint

[Ca1] Catlin, D.: Global regularity of $\bar{\partial}$-Neumann problem. Proc. Symp. Pure Math. **41** (1984) 39–49

[Ca2] Catlin, D.: Necessary conditions for subellipticity of the $\bar{\partial}$-Neumann problem. Ann. Math. **117** (1983) 147–171

[Ca3] Catlin, D.: Subelliptic estimates for the $\bar{\partial}$-Neumann problem on pseudoconvex domains. Ann. Math. **126** (1987) 131–191

[Ca4] Catlin, D.: Estimates of invariant metrics on pseudoconvex domains of dimension two. Math. Z. **200** (1989) 429–466

[Ch] Christ, M.: Analysis of $\bar{\partial}_b$ and $\bar{\partial}$ on domains of finite type in $\mathbb{C}^2$. These Proceedings, p. 859

[Dan1] D'Angelo, J.P.: Subelliptic estimates and failure of semi-continuity of orders of contact. Duke Math. J. **47** (1980) 955–957

[Dan2] D'Angelo, J.P.: Real hypersurfaces, order of contact and applications. Ann. Math. **115** (1982) 615–637

[DF1] Diederich, K., Fornaess, J.E.: An example with non trivial nebenhuelle. Math. Ann. **225** (1977) 275–292

[DF2] Diederich, K., Fornaess, J.E.: Pseudoconvex domains with real analytic boundary. Ann. Math. **107** (1978) 371–384

[DF3] Diederich, K., Fornaess, J.E.: Stein neighborhoods for finite preimage of regular domain. Manuscripta Math. **50** (1985) 11–27

[FeK] Fefferman, C.L., Kohn, J.J.: Hölder estimates on domains of complex dimension two and on three dimensional CR manifolds. Adv. Math. **62** (1988) 223–303

[FSi1] Fornaess, J.E., Sibony, N.: On L^p estimates for $\bar{\partial}$. To appear in Proc. Symp. Pure Math.

[HaSi] Hakim, M., Sibony, N.: Spectre de $A(\bar{\Omega})$ pour les domaines faiblement pseudoconvexes réguliers. J. Funct. Anal. **37** (1980) 127–135

[HeL] Henkin, G.M., Leiterer, J.: Theory of functions on complex manifolds. Birkhäuser, 1984

[Ho] Hormander, L.: L^2 estimates and existence theorems for the $\bar{\partial}$ operator. Acta Math. **113** (1965) 89–152

[Ki] Kiselman, Ch.: A study of the Bergman projection in certain Hartogs domains. Preprint

[K1] Kohn, J.J.: The range of the tangential Cauchy-Riemann operator. Duke Math. J. **53** (1986) 525–545

[K2] Kohn, J.J.: Subellipticity of the $\bar{\partial}$-Neumann problem on pseudoconvex domains: sufficient conditions. Acta Math. **142** (1979) 79–122

[S] Shaw, M.C.: L^2 estimates and existence theorems for the Cauchy Riemann complex. Invent. math. **82** (1985) 133–150

[Si1] Sibony, N.: Sur le plongement des domaines faiblement pseudoconvexes. Math. Ann. **273** (1986) 209–214

[Si2] Sibony, N.: Une classe de domaines pseudoconvexes. Duke Math. J. **55** (1987) 299–319

[Si3] Sibony, N.: Some aspects of weakly pseudoconvex domains. To appear in Proc. Symp. Pure Math.

[Si4] Sibony, N.: On Hölder estimates for $\bar{\partial}$. Ann. Math. Studies

[Si5] Sibony, N.: Problème de la couronne pour les domaines pseudoconvexes à bord lisse. Ann. Math. **126** (1987) 675–686

Analysis and Geometry on Groups

Nicholas Th. Varopoulos

Mathématiques, Université de Paris VI, 4 place Jussieu, F-75230 Paris Cedex 05, France

The results that I shall survey here can be seen from several different angles. There is a discrete point of view related to discrete finitely generated groups; there is also a C^∞ point of view related to connected Lie groups. One can, to a certain extent, unify the above two settings by considering general compactly generated locally compact groups but I shall not do so here. Both in the discrete and in the C^∞ case we can put forward either the Geometric formulation, such as Sobolev inequalities, or the analytical formulation that examines the behaviour of natural semigroups of operators on $L^2(G)$. What makes the theory hold together, in a final analysis, is that equivalence of all these different aspects. To explain how this comes about I have to start with some definitions.

1. Distance and Volume Growth

Let G be a discrete group generated by a finite number of generators $\gamma_1, \ldots, \gamma_k \in G$. One defines then a distance $d(.,.)$ on G by requiring that $d(gx, gy) = d(x, y)$ $(x, y, g \in G)$ and that $d(e, x)$, the distance of $x \in G$ from the neutral point $e \in G$ is, by definition, the smallest $n \geq 0$ for which we can write $x = \gamma_{i_1}^{\varepsilon_1} \ldots \gamma_{i_n}^{\varepsilon_n}$ $(i_1, \ldots, i_n = 1, \ldots, k;\ \varepsilon_j = 0, \pm 1)$.

Let G be a connected Lie group and let $X_1, \ldots, X_k \in \mathscr{L}(G)$ be a finite number of generators of the Lie algebra of G; in other words $X_1, \ldots, X_k$ are left invariant C^∞ fields on G that together with their successive brackets $[X_{i_1}[X_{i_2}, \ldots, X_{i_s}] \ldots]$ generate the tangent space. We say that an absolutely continuous path $l(t) \in G$ $(0 \leq t \leq T)$ is of length less or equal to T if its speed vector $\dot{l}(t) = dl\left(\dfrac{\partial}{\partial t}\right)$ (with respect to $X_1, \ldots, X_k$) is almost everywhere of length ≤ 1: This means that $\dot{l}(t) = \sum_{j=1}^k a_j X_j$ (p.p. $t \sum a_j^2 \leq 1$). We then say that $d(x, y) \leq T$ $(x, y \in G)$ if we can join x to y with a path of length $\leq T$.

The growth function $\gamma(t)$ $(t > 0)$ of G is in either of the above two cases defined to be $\gamma(t) =$ The Haar measure of a ball of radius t. For large t $(t \geq 1)$ the above function $\gamma(t)$ is essentially independent of the particular choice of the generators used: $\gamma(t)$ $(t \geq 1)$ is thus a group invariant. For Lie groups and $0 < t < 1$ the behaviour of $\gamma(t)$ does depend on the choice of $X_1, \ldots, X_k$ but we always have

Proceedings of the International Congress
of Mathematicians, Kyoto, Japan, 1990

$\gamma(t) \approx t^\delta$ with $\delta = \delta(G, X_1, X_2, \ldots, X_k) = 1, 2, \ldots$. (This is a theorem of Nagel-Stein-Wainger). For $t \geq 1$ and a Lie group we have either $\gamma(t) \approx t^D$ with $D = D(G) = 0$, $1, \ldots$ or $\gamma(t) \geq Ce^{ct}$ (This is a theorem of Guivarc'h). In the discrete case we have either $\gamma(t) \approx t^D$ ($D(G) = 0, 1, 2, \ldots$) if G is a finite extension of a nilpotent group or $\gamma(t)t^{-A} \to \infty$ for all $A \geq 1$ in all other cases (this is a theorem of Gromov).

2. The Diffusion and the Random Walks

Let G be a unimodular Lie group with $X_1, \ldots, X_k \in \mathscr{L}(G)$ as above, we can then consider $\Delta = -\sum X_j^2$ which can be identified with a self adjoint (positive) operator on $L^2(G)$ and we can also consider $T_t = \exp(-t\Delta)$ the corresponding submarkovian semigroup. The kernel of that semigroup will be denoted by $p_t(x, y) = p_t(x^{-1}y)$ ($t > 0$; $x, y \in G$). The discrete analogue of the above diffusion is of course the random walk defined on a discrete group by the transition matrix $M(x, y) = \mu(x^{-1}y)$ ($x, y \in G$) where $\mu \in \mathbb{P}(G)$ is a symmetric probability measure on G. We shall consider in what follows, essentially, only random walks that are defined by symmetric measures that have *generating* supports (: Gp (supp μ) $= G$). What we shall examine then is the convolution powers μ^n of that measure or equivalently $T_t = \exp(-t(\delta - \mu))$ the continuous time Markov semigroup that it generates.

3. Analytic and Probabilistic Formulation

One of the main accomplishments of the present methods is that it allows us to study the convolution powers of a finitely supported symmetric measure as considered in the previous section.

Theorem 1. *Let G be a discrete finitely generated group and let μ be a measure as above. Let us also assume that $\gamma(t) \geq ct^D$ for some c, $D > 0$. We then have $\mu^n(\{e\}) = O(n^{-D/2})$.*

The above theorem allows us in particular to classify the discrete groups for which the series $\sum \mu^n(\{e\}) = +\infty$. Such groups are called recurrent groups, the reason being that the random walk with transition matrix $M(x, y) = \mu(x^{-1}y)$ is a recurrent random walk (and this fact is independent of the particular choice of μ):

Corollary. *The only recurrent groups are the finite extensions of the following three groups*: $\{0\}$, $\mathbb{Z}$, $\mathbb{Z}^2$.

Theorem 1 easily generalizes to convolution products $\mu_1 * \cdots * \mu_n$ provided of course that the measures μ_j satisfy the appropriate conditions uniformly in j. Theorem 1 is a typical result of the discrete version of our theory. The continuous variant of the same result is the following.

Theorem 2. *Let G be a unimodular Lie group and let $X_1, \dots, X_k \in \mathscr{L}(G)$ be as before. Let us assume that the induced growth function satisfies $\gamma(t) \approx t^\delta\ (t \to 0)$ and $\gamma(t) \geq ct^D$ $(t \geq 1)$ for some $\delta, D = 0, 1, \dots$.*

We then have $\|p_t\|_\infty = O(t^{-\delta/2})\ (t \to 0)$ and $\|p_t\|_\infty = O(t^{-D/2})\ (t \to \infty)$.

The small time behaviour of $\|p_t\|_\infty$ described in the Theorem is contained in a previous more general result of A. Sanchez-Calle. The group structure is not essential for this small time behaviour of p_t. The above two theorems can of course be unified to a single result on locally compact groups and the methods of the proofs, as we shall see, have very little to do with "real analysis".

The metric $ds^2 = \varphi(y)(dx^2 + dy^2)$ on $\mathbb{R}^2$ where $\varphi(y) = y^{-2}$ for $|y| \geq 1$ gives an example of a Riemannian manifold that has exponential volume growth (since for $|y| \geq 1$ it is just the hyperbolic plane) but has "slow" decay for its canonical p_t as $t \to \infty$. Indeed the above metric is conformal with the Euclidean metric and therefore has no Green's function i.e. $\int_1^\infty p_t = +\infty$. This shows that the group structure in Theorem 2 is essential for the behaviour of p_t as $t \to \infty$.

4. Geometric Formulation

Let G be a unimodular Lie group and let $X_1, \dots, X_k \in \mathscr{L}(G)$ be as before. We shall denote the corresponding gradient by: $\nabla f = (X_1 f, \dots, X_k f) \in \mathbb{R}^k$ $(f \in C_0^\infty(G))$. The main Geometric Theorem is

Theorem 3. *Let G and $X_1, \dots, X_k$ be as before and let $\delta, D \geq 0$ be as in Theorem 2. Let also $n \geq 1$, $\delta \leq n \leq D$ we then have*

$$\|f\|_{n/(n-1)} \leq C\|\nabla f\|_1; \qquad f \in C_0^\infty.$$

Conversely if the above Sobolev inequality holds for some n then $n \geq \delta$ and $\gamma(t) \geq ct^n$ $(t \geq 1)$.

(All the $\|\ \|_p$ norms in what follows are taken in $L^p(G)$ for the Haar measure).

The above Sobolev type estimates are usually reformulated by the Geometers in terms of isoperimetric inequalities of the type $|A|_r^{(n-1)/n} \leq C|\partial A|_{r-1}$ $(A \subset G)$ where $|\ |_s$ refers to the appropriate s-dimensional Hausdorff measure and r is the topological dimension of G. The discrete analogue of the above theorem states:

Theorem 4. *Let G be a discrete finitely generated group, then the Sobolev inequality $\|f\|_{n/(n-1)} \leq C\|\nabla f\|_1$ $(f \in C_0(G))$ holds for some $1 \leq n \in \mathbb{R}$ if and only if $\gamma(t) \geq ct^n$ $(t \geq 1)$.*

In the above theorem the L_1-norm of the gradient is of course $\|\nabla f\|_1 = \sum_{d(x,y)=1} |f(x) - f(y)|$. Once more the above result can be stated in terms of discrete isoperimetric inequalities.

5. The Connection Between Analysis and Geometry

What unifies the Geometric and the Analytic point of view and what is, in a final analysis, the pivot of the proofs is the following general result of Functional Analysis:

Let A be the generator of an appropriate semigroup $T_t = e^{-tA}$ of operators on $L^p(X; dx)$ (the spaces of p-integrable functions on an abstract measure space $(X; dx)$). Let $n > 2$, the following two conditions are then equivalent:

(i) $\|f\|_{2n/(n-2)} \le C(Af, f)^{1/2}$; $f \in \mathrm{Dom}(A)$.
(ii) $\|T_t f\|_\infty \le ct^{-n/2}\|f\|_1$; $t > 0$, $f \in L^1$.

For the semigroups associated to our random walks on discrete groups the generator is: $A = \delta - \mu$ and the Dirichlet form satisfies $D_\mu(f) = (Af, f) \approx D_0(f)$ where we denote by $D_0(f) = \sum_{d(x,y)=1} |f(x) - f(y)|^2$ the "standard" Dirichlet form on G. This equivalence $D_\mu \approx D_0$ is trivial to see if μ has *finite* support but what is important is that it remains true for a more general class of measures; namely for all symmetric Probability measures on G with generating support and whose "variance" is finite:

$$E(\mu) = \sum_{x \in G} d^2(e, x)\mu(\{x\}) < +\infty.$$

This observation although not very difficult to prove is absolutely crucial for us.

In the case of a Lie group the Dirichlet form of our semigroup $T_t = e^{-t\Delta}$ is of course the familiar expression

$$(\Delta f, f) = \|\nabla f\|_2^2 = \int_G \sum |X_j f|^2.$$

Observe finally that the $L^1 \to L^\infty$ operator norm $\|e^{-t\Delta}\|_{1,\infty}$ on a Lie group is $p_t(e)$ and similarly $\|e^{-n(\delta-\mu)}\|_{1,\infty} \sim \mu^n(e)$ for a discrete group (This last $\sim$ has to be interpreted correctly but it certainly implies $\mu^n(e) = O(n^{-\alpha}) \Leftrightarrow \|e^{-t(\delta-\mu)}\|_{1,\infty} = O(t^{-\alpha})$).

With the above facts in mind the connection between the Geometric and the Analytic theory becomes obvious. Another thing that becomes apparent (and this is the single most important feature of all the proofs) is that *changing the measure* μ, say in Theorem 1, *makes no difference* as long as we restrict ourselves to measures of finite variance. Indeed such changes leave invariant (up to equivalence) the Dirichlet form $D_\mu(f)$. What remains to be done to complete the proof of, say Theorem 1, is to produce *one* symmetric probability measure with finite variance and with convolution powers that decay optimally: $\mu^n(\{e\}) = O(n^{-D/2})$.

This last step is done "by hand". We simply try out a measure of the form: $\mu = \sum \lambda_j \chi_j$ where $\lambda_j \ge 0, \sum \lambda_j = 1$ and where χ_j denotes the normalized characteristic function of the j-ball in G. The condition $E(\mu) < +\infty$ is easy to express in terms of the λ's and the convolution powers μ^n can be estimated by an elementary argument. The above construction does not seem to work if we restrict ourselves to measures of finite support and this is something that to this day I cannot really explain to myself in a satisfactory manner.

6. Further Development and Open Problems

In the interaction between the Geometric and Analytical results there is one point that remains obscure. Indeed what is natural to consider from the semigroup point of view is the norm $\|\Delta^{1/2}f\|_p$ ($f \in C_0^\infty$) (Δ is a self adjoint positive operator) and what occurs naturally in the Geometric formulation is the norm $\|\nabla f\|_p$. It is only for $p = 2$ that the two norms are obviously equivalent and it is an open problem whether we have in general $\|\nabla f\|_p \approx \|\Delta^{1/2}f\|_p$. That this is the case in the real variable situation $G = \mathbb{R}^n$ is the content of the classical M. Riesz theorem (for $p \neq 1, \infty$). This equivalence holds when G is a group of polynomial growth (This is a recent theorem of G. Alexopoulos). It also holds when G is non amenable e.g. a classical non compact semi-simple group (this is a result of N. Lohoué). The problem for a general unimodular group remains open and seems difficult. The above problem has an obvious discrete formulation that contains, no doubt, the essence of the difficulty.

When the group G is not unimodular then, as we already pointed out, the geometric aspect of our theory goes through in a very satisfactory fashion. What remains very much open is the analytical theory. Indeed the long time behaviour of the appropriate heat kernel remains untractable by the above methods. The problem is very much connected with the analysis of the canonical heat kernel on symmetric spaces. Indeed any symmetric space of non-compact type can be realized, by the Iwasawa decomposition KAN, as the *non-unimodular* group AN.

The last problem that I shall consider consists in obtaining a finer analysis of the behaviour of p_t as $t \to \infty$, for Lie group, or μ^n for a discrete group. Assume that G is a unimodular Lie group. If G is not amenable, and only then, we have $p_t(e) = O(e^{-\lambda t})$ where $\lambda > 0$ is the spectral gap of Δ and depends on the particular choice of the fields $X_1, \ldots, X_k$. There are good reasons to suspect that in fact $p_t(e) \sim t^{a/2}e^{-\lambda t}$ where a is some integer or possibly "$+\infty$" that only depends on the group and not on the choice of the fields (just as for amenable groups where we have $\lambda = 0$). The analogous conjecture for discrete groups is false (the counter example is due to D. Cartwright). If G is semi-simple this is, once more, related to the heat kernel on symmetric spaces (Ph. Bougerol has examined this case).

For a Lie group of polynomial growth G. Alexopoulos has proved a "local Central Limit" theorem: $p_t(e)t^{D/2} \xrightarrow[t\to\infty]{} \alpha_0 > 0$. The following asymptotic development $p_t(e) \sim t^{-D/2}[\alpha_0 + \alpha_1 t^{-1/2} + \cdots]$ should hold, but this is an open problem. Similarly for semi-simple groups and symmetric spaces C. Herz conjectures that $p_t(e) \sim e^{-\lambda t}t^{a/2}[\alpha_0 + \alpha_1 t^{-1/2} + \cdots]$ (as $t \to \infty$). Some logarithms could possibly appear in these asymptotic developments.

Let G be a discrete group and let $\mu, \nu \in \mathbb{P}(G)$ be two symmetric probability measure of finite variance. Let us also assume that $\mu^n(e) = O[\exp(-\alpha(n))]$ where $\alpha(t) \geq 0$ is an increasing positive function of ($t \geq 0$). By a slight variance of the previous methods (here we make essential use of E.B. Davies work in the subject) we can then show that: $\nu^n(e) = 0\ [\exp(-c\tilde{\alpha}(cn))]$ for some $0 < c$, where we denote by

$$\tilde{\alpha}(t) = \frac{1}{t}\int_0^t \alpha(t)\, dt.$$

The analogous result for unimodular Lie groups also holds. What makes this fact interesting is that for many natural functions e.g. $\alpha(t) = t^\alpha$, $t^\alpha \operatorname{Log}(1+t)$ e.c.t. we have $\alpha \approx \tilde{\alpha}$. This fact is used to analyse the groups that have *superpolynomial* growth:

Assume that G (discrete or Lie, amenable or not, but unimodular) satisfies the growth condition $\gamma(t) \geq \exp(ct^\alpha)$, $t \geq 1$, for some $0 < \alpha \leq 1$ [cf. R. Grigorchuk's paper in these proceedings]. Using our methods then we can easily establish that, say for a discrete group, we have $\gamma(n+1) - \gamma(n) \geq \exp(cn^\alpha)$ with possibly a different $c > 0$ but the same $0 < \alpha \leq 1$. Using this fact and refining our methods further (we use in particular here an idea of L. Saloff-Coste) we can then prove that (again for a discrete group) we have:

$$\mu^n(\{e\}) = O[\exp(-cn^{\alpha/(\alpha+2)})]$$

The analogous result when G is a Lie group and $\alpha = 1$ also holds. The above estimate is optimal. Indeed for any non virtuelly Nilpotent polycyclic group and every finitely supported symmetric $\mu \in \mathbb{P}(G)$ we have $\mu^{2n}(e) \geq C\exp[-cn^{1/3}]$ (this was shown by G. Alexopoulos) and for all these groups $\alpha = 1$. The details of the above result will appear elsewhere.

A decay of the type $\exp(-cn^\beta)$ for $p_t(e)$ gives rise of course to Orlicz type Sobolev inequalities of the form $\|f\|_{L\log^\gamma L} \leq C\|\nabla f\|_1$ where $\gamma = \gamma(\beta)$. In terms of isoperimetric inequalities for discrete groups for instance, we can say that if $\mu^n = O[\exp(-cn^{-\beta})]$ $(0 < \beta \leq 1)$ then we have:

$$|\partial\Omega| \geq C|\Omega|(\log|\Omega|)^{\frac{\beta-1}{\beta}}$$

for all finite $\Omega \subset G$ with $|\Omega| \geq 2$ where $|\ |$ denotes the cardinality of a finite set.

For exponential groups this gives:

$$|\partial\Omega| \geq C|\Omega|(\log|\Omega|)^{-2}.$$

A final result that I shall mention concerns $p_t(x, y)$ the canonical heat kernel on a Riemannian manifold that covers normally some compact manifold with deck transformation group G. With the present methods we can show that the behaviour of $\|p_t\|_\infty$ (as $t \to \infty$) is "identical" with the behaviour of $\mu^n(e)$ for $\mu \in \mathbb{P}(G)$ (as in Section 3). The term "identical" means for instance that $\mu^n(e) = O(n^{-a}) \Leftrightarrow \|p_t\|_\infty = O(t^{-a})$ or more generally that we have the:

$$O[\exp(-\alpha(.)] \rightleftarrows O[\exp(-c\tilde{\alpha}(c.)]$$

correspondence that we considered above. This is one of the very first results that I obtained in the subject and it is this that convinced me of the fundamental connection that existed between the discrete and the continuous theory.

In this survey I have said nothing about the Gaussian estimates of the heat kernels. It would take a different paper to do that. The interested reader could consult the literature below.

Literature

The theory that we surveyed in this paper is the subject matter of a forthcoming book [1]. A preliminary version of this book exists in the form of mimeographed notes: University Paris VI.

Most of the results that I presented were developed by the author in a series of papers the most significant being [2].

For the Functional Analytic tools of Section 5, and the work of E.B. Davies cf. [3, 4, 5, 6, 12].

For further developments in locally compact groups, cf. [7, 8, 9, 13]. Most of the work of G. Alexopoulos has not yet appeared in print, cf. [10, 14]. For the Symmetric space point of view, cf. [11, 15].

1. N. Th. Varopoulos, L. Saloff-Coste, Th. Coulhon: Analysis and geometry on groups. Cambridge University Press (to appear)
2. N. Th. Varopoulos: Analysis on Lie groups. J. Funct. Anal. **76**, no. 2 (1988) 346–410
3. N. Th. Varopoulos: Hardy-Littlewood theory for semigroups. J. Funct. Anal. **63**, no. 2 (1985) 240–260
4. P. Bénilan: Operateurs accrétifs et semi-groupes dans les espaces L^p $(1 \leq p \leq +\infty)$. In: Functional analysis and numerical analysis. Japan-France Seminar
5. A. Yoshikawa: Fractional powers of operators, interpolation theory and imbedding theorems. J. Fac. Sci. Univ. Tokyo, I.A. **18** (1971)
6. Th. Coulhon: Semigroup theory and evolution equations, Clement, Mitidien, De Pagter, ed. Marcel Dekker 1991
7. L. Saloff-Coste: Inégalité de Sobolev produite sur les groupes de Lie nilpotents. J. Funct. Anal. **79** (1) (1988) 44–56
8. L. Saloff-Coste: Sur la décroissance des puissances de convolution sur les groupes. Bull. Sci. Math., 2ᵉ série **113** (1989) 3–21
9. L. Saloff-Coste: Analyse sur les groupes de Lie à croissance polynomiale. Arkov for Mathematik **28** (2) 1990 315–331
10. G. Alexopoulos: C.R. Acad. Sci. Paris **309** (I) (1989) 661–662 and **305** (I) (1987) 777–779
11. N. Lohoué: Estimées de type Hardy pour l'opérateur $\Delta + \lambda$ d'un espace symétrie de type non compact. C.R. Acad. Sci. Paris **308** (I) (1989) 11–14
12. E.B. Davies: Heat kernels and spectral theory. Cambridge University Press, 1990
13. N.Th. Varopoulos: ICM-90 Satellite Conference Proceedings Harmonic Analysis (Sendai 1990)
14. G. Alexopoulos: ICM-90 Satellite Conference Proceedings Harmonic Analysis (Sendai 1990) and Canadian J. Math. (to appear)
15. P. Bougerol: Ann. Sci. Ec. Norm. Sup. 4ᵉ série **14** (1981) 403–432

Asymptotically Holomorphic Functions and Certain of Their Applications

Alexander L. Volberg

University of Kentucky, Lexington, KY 40506-0027, USA

1. Introduction

In one-dimensional complex analysis the essential part is played by approximation results and by dual uniqueness theorems. One of the general problems of rational approximation may be state as follows. Let X be a linear topological space of functions, Λ be a subset of $\mathbb{C}$ such that fractions $z \to (z-\lambda)^{-1}$, $\lambda \in \Lambda$, lies in X. How one can find out what subsets Λ', $\Lambda' \subset \Lambda$, will bring us a complete in X family of rational fractions $\{(z-\lambda)^{-1}\}_{z\in\Lambda'}$? Quite often it turns out that functions $f_{x^*}(\lambda) \stackrel{\text{def}}{=} \langle x^*, (z-\lambda)^{-1}\rangle$, $\lambda \in \Lambda$, $x^* \in X^*$, are "holomorphic" in Λ in a certain (sometimes pretty weak) sense. Thus we are dealing with a description of zero-sets of the class $\{f_{x^*}\}_{x^*\in X^*}$ of "holomorphic" functions. According to this it seems important to understand the nature of this "holomorphicity." The appropriate notion may be grasped by the conception of asymptotically holomorphic (AH) functions. These are (very roughly speaking) the functions defined in subdomains of $\mathbb{C}$ with vanishing $\bar{\partial}$-derivative on the boundary. More precisely (and we will see it below) only functions with some critical rate of decreasing of $\bar{\partial}$ deserved to be called AH.

To begin with we state here two typical results of AH function theory. Then we are going to show how to apply these results to

1. Problems in harmonic analysis (ideal description in the spirit of Domar);
2. A problem in dynamical systems (finiteness of number of limit cycles for QA vector fields);
3. Problems in approximation theory.

Perhaps, it is worthwhile to mention that 3 is the place, where the AH technique appeared. In what follows $\mathbb{D}$ is the unit disc, $\mathbb{C}^+$ is the right half-plane.

Theorem 1. *Let w be a positive increasing function on $(0,1)$ such that*

$$x \log \frac{1}{w(x)} \uparrow \infty, \ x \downarrow 0\,; \tag{R1}$$

$$\int_0^1 \log\log \frac{1}{w(x)} dx = \infty\,. \tag{1}$$

Proceedings of the International Congress of Mathematicians, Kyoto, Japan, 1990

And let $f \in C^1(\mathbb{D}) \cap L^\infty(\mathbb{D})$ *and*

$$|\bar\partial f(z)| \le w(1-|z|)\,. \tag{2}$$

Then either there exist constants c_1, c_2 *such that*

$$|f(x)| \le c_1 w(c_2(1-|z|),\ z \in \mathbb{D} \tag{S1}$$

or

$$\int_{\partial\mathbb{D}} \log\,|f|dm > -\infty\,. \tag{B1}$$

Theorem 1.2. *Let* $f \in C^1(\mathbb{C}^+) \cap L^\infty(\mathbb{C}^+)$ *be such that*

$$|\bar\partial f(z)| \le w\left(\frac{1}{Re\,z}\right), \tag{3}$$

where w *is a positive increasing function on* $(0,1)$ *which satisfies*

$$x^2 \frac{\log\,\frac{1}{w(x)}}{\log\,\frac{1}{w(2x)}} \uparrow \infty,\ x \downarrow 0\,. \tag{R2}$$

Then either there exist constants c_1, c_2 *such that*

$$|f(x)| \le c_1 w\left(\frac{c_2}{\sqrt{x}}\right) \tag{S2}$$

or there exists n *such that*

$$\lim_{x\to+\infty} |f(x)|e^{nx} > 0\,. \tag{B2}$$

Theorem 1.3. *Let* w *be a positive increasing function on* $(0,1)$ *such that for a positive* ε

$$x^\varepsilon \log\,\frac{1}{w(x)} \uparrow \infty,\ x \downarrow 0\,. \tag{R3}$$

And let $f \in C^1(\mathbb{D} \setminus \{0\})$ *and have estimates*

$$|f(z)| \le \left(\frac{1}{w(|z|)}\right)^{\frac{\varepsilon}{\log\,1/\varepsilon}} \tag{4}$$

and

$$|\bar\partial f(z)| \le w(|z|)\,. \tag{5}$$

Then either there exist constants c_1 *and* c_2 *such that*

$$|f(z)| \le c_1 w(c_2|z|) \tag{S3}$$

or there exists a function f^* *holomorphic in a punctured neighbourhood of the origin such that*

$$\forall c_2\, \exists c_1 \quad |f(z)| > |f^*(z)| - c_1 w(c_2|z|))\,.$$

Now let us discuss these results. These are theorems of "forced behaviour" of AH functions. They serve as a kind of the substitute for the uniqueness theorems for holomorphic function. In fact, a typical uniqueness theorem tells us that a holomorphic function either vanishes identically or is quite big. For AH functions we have a similar alternative: an AH function is either as small as its $\bar{\partial}$-derivative ((S1), (S2), (S3)) or is substantially big ((B1), (B2), (B3)). To obtain such a dichotomy one obviously has to have w in $|\bar{\partial} f| \leq w$ fast decreasing, which is indicated in conditions (1), (R2), (R3). The rate of this decreasing is sharp only in Theorem 1.1 and here we must impose an auxiliary conditions (R1) which is a regularity condition. (By the way, (R1) is indispensible in a sense.) (R2), (R3) are certainly too strong. But these actual versions of Theorems 1.2 and 1.3 have the advantage that regularity conditions and "rate-of-decay" conditions on w are united.

Assumption (4) deserves a separate comment. It singles out Theorem 1.3 from the row. Actually this is not the case. We could state Theorems 1.1, 1.2 in more general form, namely with f not bounded but with a mild growth (toward $\partial \mathbb{D}$ in Theorem 1.1 and towards ∞ in Theorem 1.2). The reason why we chosed these wordings is that we will use results only in this form. Also it is worthwhile to mention that Theorem 1.3 is absolutely trivial in the case of bounded f. As for Theorems 1.1, 1.2 all technical difficulties are here even for the case of bounded f.

2. Harmonic Analysis Applications

Our main object here is a weighted l^2 space, namely

$$l^2(p) \stackrel{\text{def}}{=} \{(a_n)_{n\in\mathbb{Z}} : \sum_{n\in\mathbb{Z}} |a_n|^2 e^{-p_n} < \infty\},$$

and the right-shift operator τ,

$$\tau : \{a_n\} \to \{a_{n-1}\}.$$

We will consider only weights with some regularity, and, for instance,

$$\lim_{n\to\infty} \frac{|p_n|}{\log |n|} = \infty. \tag{6}$$

Moreover we will say that p is of concave type if $n \to p_n$, $n > 0$, is concave and nonnegative, $n \to p_n$, $n < 0$, is convex negative. We will say that p is of convex type if $n \to p_n$, $n > 0$, is convex and nonnegative, $n \to p_n$, $n < 0$, is concave and negative. Our weights will be only either of concave or convex type. The description of τ-invariant subspaces in $l^2(p)$ is one of the model problems in harmonic analysis. Substantial contribution is made by Y. Domar [1], [2], [3] and A. Borichev [4], [5]. The idea of using AH function to these problems is due to A. Borichev. We follow his ideas here.This problem is still not fully understood in spite of considerable efforts undertaken since the classical Beurling description in the case

$$p_n = \begin{cases} 0, & n > 0 \\ -\infty, & n < 0\,. \end{cases} \tag{7}$$

It is natural to expect that if $\{p_n\}$ is pretty close to (7) then the structure of invariant subspace lattice is also classical. To state one positive result in this direction let us consider the sequence $p = (p_n)$ of concave type and such that

$$\left.\begin{aligned} & p_n = 0,\ n > 0\,; \\ & \frac{|p_{-n}|}{n^{1/2}} \uparrow \infty,\ n \to +\infty\,; \\ & \sum_{n\geq 1} \frac{|p_{-n}|}{n^2} = \infty\,. \end{aligned}\right\} \tag{8}$$

Let $c(p) \overset{\text{def}}{=} \bigcup\limits_{c>0} l^2(cp)$ be the projective limit of spaces $l^2(cp)$, $c > 0$.

Theorem 2.1. *Under the assumptions* (8) *all $c(p)$-closed invariant subspaces E of τ such that $\tau E \neq E$ are of the type IH^2, where I is a unimodular function, $I \in c(p)$.*

The Hilbert space setting is supposed to be much more difficult. For $l^2(p)$ with $p_n = 0,\ n \geq 0;\ p_n = e^{-cn}\ \ c > 0,\ n < 0$, the investigation of τ-invariant subspaces was started in [6] and finished in [7].

Remind that for a given space of sequences S is a subspace ($k \in \mathbb{Z}$)

$$S_k = \{\{s_m\} \in S :\ \ s_m = 0,\ m < k\}$$

is called standard. Any standard subspace is τ-invariant. When is the converse true? This seems to be interesting as operators with small amount of invariant subspaces naturally attract attention.

We need the result that follows for illustration. This result is weaker than similar results in [1], [2].

Let $p = (p_n)$, $p_{-n} = -p_n$, be of convex type and

$$\left.\begin{aligned} & \lim_{n\to\infty} \frac{p_n}{n \log n} = \infty \\ & \lim_{n\to\infty} (u_{\pm n} - u_{\pm 2^n}) > 0, \quad \text{where } u_n = p_n - p_{n+1} \end{aligned}\right\} \tag{9}$$

and $l(p) = \{(a_n) :\ \forall c_1 > 0 \sum\limits_{n<0} |a_n|^2 e^{-c_1 p_n} < \infty,\ \ \exists c_2 > 0 \sum |a_n|^2 e^{-c_2 p_n} < \infty\}$.

Theorem 2.2. *Under the assumption* (9) *all $l(p)$-closed τ-invariant subspaces E of $l(p)$ such that $\tau E \neq E$ are standard.*

The proofs are based on results of the first section. We outline briefly these proofs. In both theorems we need to solve convolution equation

$$(a * b)_n = 0, \quad n \leq 0,$$
$$a \in c(p) \text{ (or } l(p)), \; b \in c(p))^* (\text{or } l(p)^* = l(p)) \tag{10}$$

The usual Fourier transform $(a) \to \sum_{n\in\mathbb{Z}} a_n z^n$ turns out to be unappropriate in the case of $c(p)$ and even unapplicable in the second case ((a_n) grows too fast). Now we are going to define the generalized Fourier transform as follows.

Let

$$p^*(v) = \inf_{x<0} (p(x) - xv), \text{ if } p \text{ is of concave type;}$$
$$p^*(v) = \inf_{x>0} (p(x) - xv), \text{ if } p \text{ is of convex type,}$$

and we define a function $r(v)$ by the equality

$$p^*(v) = p(r(v)) - r(v)v.$$

Then for $a \in c(p)$

$$\tilde{\mathscr{F}} a(z) \stackrel{\text{def}}{=} \sum_{n=r(\log\frac{1}{|z|})}^{+\infty} a_n z^n, \quad |z| < 1\,;$$

for $a \in l(p)$

$$\tilde{\mathscr{F}} a(z) \stackrel{\text{def}}{=} \sum_{n=-\infty}^{r(\log\frac{1}{|z|})} a_n z^n, \quad |z| < 1\,.$$

Lemma 2.3. 1) *For* $a \in c(p)$

$$\forall c_2 \exists c_1 \; |\bar\partial(\tilde{\mathscr{F}} a)(z)| \leq c_1 e^{-p^*(c_2(1-|z|))}, \quad |z| \sim 1\,.$$

2) *For* $a \in l(p)$

$$\forall c_2 \exists c_1 \; |\bar\partial(\tilde{\mathscr{F}} a)(z)| \leq c_1 e^{-p^*(c_2 \log\frac{1}{|z|})}, \quad |z| \sim 0\,.$$

Let us remark that for concave-type p the function p^* grows near $v = 0$ and for convex-type p its p^* grows towards $v = \infty$. Conditions (8) and (9) guarantee that $\tilde{\mathscr{F}} a$ satisfies either conditions of Theorem 1.1 or conditions of Theorem 1.3. In other words $\tilde{\mathscr{F}} a$ is AH in the disc near the circle in the first case and AH near the origin in the second case.

Now suppose that

$$a, b \in l(p), \; (a * b)_n = 0, \; n < 0, \; (a * b)_0 = 1\,. \tag{10}$$

$\tilde{\mathscr{F}}$ is asymptotically multiplicative, which means that

$$|\tilde{\mathscr{F}}(a * b)(z) - \tilde{\mathscr{F}} a(z) \cdot \tilde{\mathscr{F}} b(z)| \to 0, \quad |z| \to 0\,. \tag{11}$$

Denoting $f \stackrel{\text{def}}{=} \tilde{\mathscr{F}} a$, $g \stackrel{\text{def}}{=} \tilde{\mathscr{F}} b$ and taking into account (10) and (11) we see that (S3) cannot occur neither for f nor for g.

Thus (B3) holds and so holomorphic f^* and g^* exist such that

$$\frac{1}{2} < |f^*(z)g^*(z)| < 2, \quad |z| \sim 0$$

$$\exists c_1, c_2 : \ |\log f^*(z)| + |\log g^*(z)| \leq c_1 \sum_{n \geq 0} e^{-c_2 p_n} |z|^{-n}, \quad |z| \sim 0 . \tag{13}$$

We need to prove that there exists $k \in \mathbb{Z}$ such that $a_m = 0$, $n < k$ and $b_n = 0$, $n < -k$. This a reformulation of the desirable standardness of all τ-invariant subspaces. Suppose this is not the case. Then f or g is not bounded. It is easy to see that f^* or g^* is not bounded either. Suppose this for f^*. Then denoting $\Phi(z) = \log f^*(z)$ we can assume that $|\Phi(z)| \geq c|z|^{-1}$ and taking (13) into account we see that

$$cr^{-1} \leq \sum_{n \geq 0} e^{-c_2 p_n} r^{-n}.$$

Now the second condition in (9) allows to show (see [4] e.g.) that

$$p_n \leq nC + \frac{1}{c_2} n \log n$$

which contradicts the first condition in (9).

The proof of Theorem 2.1 is based on the "inner-outer" factorization in $c(p)$. It turns out that every $f \in c(p)$, $f \neq 0$, can be represented in a form

$$f = Ih,$$

where $h \in H^2$ and I is a unimodular function in $c(p)$.

3. Dynamical System Application

Here we consider the system of differential equations

$$\left.\begin{aligned} \dot{x} &= \alpha(x, y) \\ \dot{y} &= \beta(x, y), \quad (x, y) \in R^2 \end{aligned}\right\} \tag{14}$$

where α, β are real functions belonging to a very smooth Carleman class $C\{M_n\}$. A limit cycle means that there are no other cycles in a neighbourhood. It is proved now by Yu. S. Il'yashenko [8], J. Ecalle and J. Martinet, R. Moussu, J.-P. Ramis [9], [10], [11], that for real analytic α, β in a domain Ω there is only a finite number of limit cycles of (14) in any compact part of Ω.

The AH technique allows to prove a similar result for quasianalytique vector fields (14), at least if all the singular points of (14) are non-degenerate. To state the result we need some notations.

For a given sequence $\{M_n\}$ the Carleman class $C\{M_n\}$ is defined by

$$C\{M_n\} = \{f \in C^\infty : \exists B_f, C_f : |f^{(n)}(x)| \leq B_f C_f^{|n|} M_{|n|}\}.$$

Here $n = (n_1, n_2)$, $|n| = n_1 + n_2$. We consider only Carleman classes with regular sequences $\{M_n\}$, namely if $m_n = M_n/n!$ then we assume that

1) $m_n^2 \le m_{n-1} m_{n+1}$,

2) $\sup \left(\frac{m_{n+1}}{m_n}\right)^{1/n} < \infty$,

3) $\lim m_n^{1/n} = \infty$.

Under these assumptions the properties of $C\{M_n\}$ can be reflected adequately by

$$w(x) \stackrel{\text{def}}{=} \inf_{n \ge 0} x^n \frac{M_n}{n!}.$$

For instance, $C\{M_n\}$ is quasianalytique (QA) iff

$$\int_0 \log \log \frac{1}{w(x)} dx = \infty \tag{15}$$

Adopting the ideas of Il'yashenko [12] to this new situation it is not hard to prove the following result [13].

Theorem 3.1. *Let* α, β *belong to* $C_\Omega\{M_n\}$ *and*

$$w(x) \le \exp\left(-\exp \frac{1}{x^{16}}\right), \tag{16}$$

then in any compact part of Ω *there is only a finite number of limit cycles providing that all singular points of* (14) *are non-degenerate.*

Proof repeats the one in [12] and proceeds by extending the monodromy transform to an AH function in a complex domain. Then Theorem 1.2 is applied. It is very likely that (16) can be weakened to (15).

4. Weighted Polynomial Approximation

Much interesting analysis has resulted from attempts to understand the structure of $P^2(\mu)$, the closure in $L^2(\mu)$ of the set $\mathcal{P}_A$ of all analytic polynomials, where μ is a positive finite Borel measure with compact support in the complex plane $\mathbb{C}$. The greatest achievement in this field due to J. Thomson [14] asserts that $P^2(\mu) \ne L^2(\mu)$ if and only if there is a point $c \in \mathbb{C}$ such that

$$|p(c)| \le k\|p\|_{L^2(\mu)}, \quad \forall p \in \mathcal{P}_A.$$

Such points are called points of bounded point evaluation. In other words Thomson has solved a problem of Mergelyan-Brennan. He even has achieved more: a description of $P^2(\mu)$ in terms of points of bounded evaluation.

But unfortunately this does not help much when one is interested in a description of $P^2(\mu)$ in terms of μ itself. Here are two examples. The first is a so-called splitting problem. In this problem

$$d\mu = w(1-r) r dr d\theta + h(\theta) dm(\theta) = \mu_{\mathbb{D}} + \mu_{\mathbb{T}}$$

is a sum of a radially symmetric measure $\mu_{\mathbb{D}}$ in the disc $\mathbb{D}$ and a measure $\mu_{\mathbf{T}}$ on $\mathbf{T} = \partial\mathbb{D}$. What are the conditions on the pair (w, h) necessary and sufficient for the splitting:

$$P^2(\mu) = P^2(\mu_{\mathbb{D}}) \oplus L^2(\mu_{\mathbf{T}})?$$

This problem was solved in [15], [16] and the next result is nothing more than another form of Theorem 1.1.

Theorem 4.1. 1) *Suppose that* (R1) *holds. Then the following conditions are sufficient for the splitting*

$$\int_{\mathbf{T}} \log h dm = -\infty, \quad \int_0 \log \log \frac{1}{w(x)} dx = \infty. \tag{17}$$

2) *Conditions* (17) *are also necessary if there exists an arc I, $I \in \mathbf{T}$, such that*

$$\int_I h^{-1} dm < \infty.$$

In our second example we deal with a problem of weighted polynomial approximation in an arbitrary simply connected domain. Let Ω be such a domain, u is a positive continuous function in Ω vanishing towards the boundary. It is always the case that

$$P^2(u dm_2) \subset L^2_a(u dm_2) \stackrel{\text{def}}{=} L^2(u dm_2) \cap \text{Hol}\,(\Omega).$$

Keldysh was the first who investigated when the equality

$$P^2(u dm_2) = L^2_a(u dm_2) \tag{18}$$

holds. He considered $u(z) = u(\text{dist}(z, \partial\Omega))$. This is not convenient as the conditions on U depend heavily on the smoothness of $\partial\Omega$ and vary from

$$u(x) \le \exp\left(-\exp\frac{1}{x^2}\right)$$

to

$$u(x) \le \exp\left(-\exp\frac{1}{x}\right).$$

J. Brennan considered the case $u(z) = w(G(z))$ where G is the Green function of Ω. The result of Brennan [17] is the following.

Theorem 4.2. 1) *Let Ω be an arbitrary simply connected domain, G be its Green function,*

$$u(z) = w(G(z)),$$

where w satisfies (R1). *Then the condition*

$$\int_0 \log \log \frac{1}{w(x)} dx = \infty$$

is sufficient for (18).

2) *If* $\partial\Omega$ *is smooth enough, then this condition is also necessary.*

The proof is based on Theorem 1.1. Under some auxiliary conditions on Ω this theorem was proved in [18].

References

[1] Domar, Y: A solution of the translation-invariant subspace problem for weighted L^p on $\mathbb{R}$, $\mathbb{R}_+$ or $\mathbb{Z}$. Lecture Notes in Mathematics, vol. 975. Springer, Berlin Heidelberg New York 1983, pp. 214–226

[2] Domar, Y: Extension of the Titchmarsh convolution theorem with applications in the theory of invariant subspaces. Proc. London Math. Soc. 1983, no. 46, pp. 288–300

[3] Domar, Y: Translation invariant subspaces of weighted L^p and l^p spaces. Math. Scand. **49** (1981) 133–144

[4] Borichev, A.A: A Titchmarsh-type convolution theorem on the group $\mathbb{Z}$. Arkiv for Matematik **27** (1989) 179–187

[5] Borichev, A.A: Generalized Fourier transform, the Titchmarsh theorem and almost analytic function. Algebra and Analysis, no. 4 (1989) 17–53 (in Russian; translation in Leningrad J. Math.)

[6] Sarason, D: The H^p spaces of an annulas. Mem. Amer. Math. Soc. **56** (1965) 1–78

[7] Hitt, D: Invariant subspaces of H^2 of an annulus. Pac. J. Math. **134** (1988) 101–133

[8] Il'yashenko, Yu. S: Finiteness theorems for limit cycles. Usp. Mat. Nauk **45**, no. 2 (1990) 143–200 (in Russian; translation in Russ. Math. Surv.)

[9] Ecalle, J., Martinet J., Moussu R., Ramis, J.-P: Non-accumulation des cycles-limites (I). C.R. Acad. Sci. Paris **304**, no. 13 (1987) 375–377

[10] Ecalle, J., Martinent, J., Moussu, Rl, Ramis, J.-P: Non-accumulation des cycles-limites (II). C.R. Acad. Sci. Paris **304**, no. 14 (1987) 431–434

[11] Ecalle, J: Finitude des cycles-limites et accelero-sommation de l'application de retour. Pré publications, Université de Paris-Sud, ORSAY, 1990, no. 90–36, pp. 1–86

[12] Il'ashenko, Yu.S: Limit cycles of polynomial vector fields with non-degenerate singular points on the real plane. Funk. Anal. Priloz. **18**, no. 3 (1984) 32–42 (in Russian; translation in Funct. Anal. Appl., January 1985, pp. 199–209)

[13] Volberg, A.L: Un théorème de Dulac-Ecalle-Il'yashenko-Martinet-Moussu-Ramis etendu aux fonction quasianalytique. Publ. Math. d'ORSAY, Seminaire d'Analyse Harmonique, Année 1989/90, pp. 152–171

[14] Thomson, J: Approximation in the mean by polynomials. To appear in Ann. Math.

[15] Volberg A.L: The logarithm of an almost analytic function is summable. Sov. Math. Dokl. **26** (1982) 238–243

[16] Volberg A.L., Jöricke, B: Summability of the logarithm of an almost analytic function and a generalizations of the Levinson-Cartwright theorem. Matem. Sbornik **130** (1986) 335–348 (in Russian; translation in Math. USSR Sbornik **58** (1987) 337–349)

[17] Brennan, J: Weighted polynomial approximation and quasianalyticity for general sets. Preprint, University of Kentucky, 1990

[18] Brennan, J: Weight polynomail approximation, quasianalyticity and analytic continuation. J. Reine Angew. Math. **357** (1985) 23–50

Cyclic Cohomology and K-Homology

Joachim Cuntz

Mathematisches Institut, Universität Heidelberg, Im Neuenheimer Feld 288
W-6900 Heidelberg, Fed. Rep. of Germany

Cyclic cohomology, as introduced by A. Connes (and independently by Tsygan), can be viewed as an algebraic version of K-homology in the sense of [BDF], [Kas]. In [Cu1], the author gave a description of K-homology for a complex algebra A using the free product algebra $QA = A * A$ and its ideal qA obtained as the kernel of the multiplication map $QA \to A$. The even K-homology group K^0A can be defined as the set $[qA, K]$ of homotopy classes of homomorphisms from qA into the elementary algebra K of compact operators (the completion of the algebra of complex matrices of arbitrary size). This set is an abelian group in a natural way. To describe the odd K-homology group K^1A one can use the universal extension $0 \to \varrho A \to RA \to A \to 0$, where RA is the free non-unital tensor algebra over A, described below in Section 1. In contrast with the situation for qA, the algebra ϱA admits several different C^*-algebra completions if A is a C^*-algebra. Two natural completions $\overline{\varrho A}^{\ell}$, $\overline{\varrho A}^{cp}$ can be defined by the condition that the maps $\overline{RA}^{\ell} \to A$, $\overline{RA}^{cp} \to A$ admit a linear, resp. completely positive splitting of norm 1. With these completions, $[\overline{\varrho A}^{\ell}, K]$ is a version of the Brown-Douglas-Fillmore group Ext$A$, while $[\overline{\varrho A}^{cp}, K]$ is the Kasparov group K^1A. The set $[\overline{\varrho A}^{\ell}, K]$ is only a semigroup in general. More generally, these algebras can be used to define the Kasparov groups $KK_*(A, B)$ as $KK_0(A, B) = [qA, K \otimes B]$, $KK_1(A, B) = [\overline{\varrho A}^{cp}, K \otimes B]$. The algebra $\overline{\varrho A}^{cp}$ is in fact isomorphic to the subalgebra of even elements in the natural C^*-completion of qA (it is a hereditary subalgebra of the algebra εA used by Zekri in [Zek]). The basic result of KK-theory – the existence of the Kasparov product – is in this setting the homotopy equivalence between $K \otimes qA$ and $K \otimes q(qA)$ [Cu1].

We now come to the connection with cyclic cohomology. It was shown in [Co-Cu] that cyclic cocycles, viewed as cocycles in the bicomplex of Tsygan-Loday-Quillen or in the $B - b$ bicomplex of Connes, correspond exactly to traces or supertraces on a power $(qA)^n$ of the ideal qA or to supertraces on QA that vanish on $(qA)^n$. In [Cu-Qu], among other things, it is shown that such a supertrace corresponds to a cyclic coboundary if and only if it is of the form $T' \circ d$ where $d : QA \to \Omega^1 QA$ is the universal derivation into the bimodule of abstract 1-forms over QA and T' is a supertrace on this bimodule. In fact, composition with d basically is the coboundary operator. Thus a cyclic cohomology class is roughly speaking a "homotopy class" of supertraces on QA or $(qA)^n$.

For this correspondence between supertraces and cocycles-coboundaries to be natural one has to introduce certain constants into the basic bicomplex (see

Proceedings of the International Congress
of Mathematicians, Kyoto, Japan, 1990

3.A and 3.B below). These constants become important when constructing cyclic cycles in the bicomplex. By what has been said above, such a cycle determines a "closed" element of QA, i.e. an element x, that varies under homotopies only by sums of commutators or, formally, for which dx is a sum of commutators in $\Omega^1 QA$. The natural cycles associated with idempotents or invertible elements in A are represented by certain power series in QA (whose coefficients are computed using the constants in the bicomplex) that have quite natural interpretations.

The exposition below is largely based on joint work with D. Quillen [Cu-Qu] and also on [Co-Cu, Cu2]. Needless to say, much of the material refines and continues ideas of A. Connes [Co1, Co2].

1. Universal Algebras Associated with an Algebra A

Let A, B be algebras and E a bimodule over B. We will see below that multilinear maps $\omega : A^n \to E$ satisfying

$$\begin{aligned} &p(x_0)\omega(x_1,\dots,x_n) - \omega(x_0x_1,\dots,x_n) + \omega(x_0, x_1x_2,\dots,x_n) \\ &- \dots + (-1)^n\omega(x_0,\dots,x_{n-1}x_n) + (-1)^{n+1}\omega(x_0,\dots,x_{n-1})p(x_n) = 0 \end{aligned} \tag{1.A}$$

where $p : A \to B$ is a linear map, together with a similar multilinear map ω' into the dual of E lead to cyclic cocycles on A. Let us see first how such maps may arise:

0. An example for $n = 0$ is given by a central element $z \in B$ and a linear map $p : A \to B$. One has $p(x)z - zp(x) = 0$.

1. In the one-variable case the main example is given by a quasihomomorphism, that is a pair of homomorphisms $\alpha, \bar{\alpha} : A \to B'$ into some algebra B' that contains B as an ideal such that $\alpha(x) - \bar{\alpha}(x) \in B$, $\forall x \in A$. Putting $p(x) = \frac{1}{2}\big(\alpha(x) + \bar{\alpha}(x)\big)$, $q(x) = \frac{1}{2}\big(\alpha(x) - \bar{\alpha}(x)\big)$ we have $q(xy) = p(x)q(y) + q(x)p(y)$. More generally, for $\alpha, \bar{\alpha}$ we may take two linear maps whose curvatures (see 2. below) coincide.

2. For the canonical example with 2 variables let $\varphi : A \to B$ be a linear map between two algebras. Let $\omega(x, y) = \varphi(xy) - \varphi(x)\varphi(y)$ be its "curvature", cf. [Qu]. Then $\varphi(x)\omega(y, z) - \omega(xy, z) + \omega(x, yz) - \omega(x, y)\varphi(z) = 0$.

Finally, if $E = B$ and ω_1, ω_2 are multilinear maps of n_1 resp. n_2 variables satisfying (1.A) then the product $\omega_1\omega_2$ is a function of $n_1 + n_2$ variables satisfying (1.A).

We now introduce, given an algebra A, algebras RA, QA with a linear map $p : A \to RA \subset QA$ in which these examples are universally realized. The algebra RA is simply the free tensor algebra over A, $RA \cong \oplus_{n\geq 1} A^{\otimes^n}$ or, in other words the universal algebra generated by symbols $p(x)$, $x \in A$ which are linear in x with no further relations ($p(x_1)\dots p(x_n)$ corresponding to $x_1 \otimes x_2 \otimes \dots \otimes x_n$ in the tensor algebra). Let

$$\omega(x, y) = p(xy) - p(x)p(y)$$

denote the curvature of the universal linear map p and ϱA the ideal in RA generated by $\omega(x, y)$, $x, y \in A$. There is a short exact sequence

$$0 \to \varrho A \to RA \to A \to 0.$$

This is a (uni)versal extension of A : if $0 \to J \to B \to A \to 0$ is any extension with a linear splitting $\varphi : A \to B$ then the map $p(x) \mapsto \varphi(x)$ induces a homomorphism $\widetilde{\varphi} : RA \to B$ that maps ϱA into J. Also, the algebra RA is contractible via the homotopy $\widetilde{\varphi}_t$, $\varphi_t : p(x) \mapsto tp(x)$ so that RA may be thought of as a noncommutative cone over A, the ideal ϱA corresponding to the suspension.

The algebra QA now is defined as the universal algebra generated by symbols $p(x)$, $q(x)$, $x \in A$ which are linear in x and satisfy

$$q(xy) = p(x)q(y) + q(x)p(y)$$
$$q(x)q(y) = p(xy) - p(x)p(y) .$$

There are two natural homomorphisms $\iota, \bar{\iota} : A \to QA$ mapping x to $p(x) \pm q(x)$ and a natural $\mathbb{Z}/2$-grading of QA with grading automorphism $^-$ defined by $\overline{(p(x) + q(y))} = p(x) - q(y)$. In fact, QA is isomorphic to the free product $A * A$ with the grading automorphism that exchanges the two copies of A.

Proposition 1.1. *The natural map $RA \to QA$ is injective and induces an isomorphism between RA and the algebra QA_+ of even elements in QA.*

In QA, let qA be the ideal generated by $q(x)$, $x \in A$. There is a short exact sequence

$$0 \to qA \to QA \to A \to 0$$

with two different splittings $A \to QA$. The map in 1.1 maps ϱA onto $\left((qA)^2\right)_+$. There is a simple argument showing that QA is homotopy equivalent in 2×2-matrices to the direct sum $A \oplus A$ (cf. [Cu1]). If we filter QA by the powers of the ideal qA, then the associated graded algebra is simply the algebra ΩA of abstract differential forms over A, i. e. the universal algebra generated by symbols $\varphi(x)$ and $d(x)$ where φ, d are linear in x and satisfy the relations

$$\varphi(xy) = \varphi(x)\varphi(y)$$
$$d(xy) = \varphi(x)d(y) + d(x)\varphi(y) .$$

Just as ϱA is the universal ideal in an extension of A, qA is universal for quasihomomorphisms as in 1. above: given a pair of homomorphisms $\alpha, \bar{\alpha} : A \to B'$ where B' contains B as an ideal, we get a homomorphism $QA \to B'$ mapping $p(x)$, $q(x)$ to $\frac{1}{2}\left(\alpha(x) + \bar{\alpha}(x)\right)$, $\frac{1}{2}\left(\alpha(x) - \bar{\alpha}(x)\right)$. This homomorphism sends qA into the ideal B. Quasihomomorphisms arise for instance from even Fredholm or Kasparov modules.

2. Transport of K-Theory and Cyclic Cohomology Classes by Certain Linear Maps

K-Theory elements may be transported not only by homomorphisms but also by certain linear (or even multilinear) maps between algebras. Let, for instance, $\varphi : A \to B$ be a linear map and $\omega(x, y) = \varphi(x, y) - \varphi(x)\varphi(y)$, $x, y \in A$ its curvature. If φ is a $*$-respecting map between C^*-algebras, $e = e^2 = e^*$ is a projection, and $\|\omega(e, e)\| < \frac{1}{2}$ then $\varphi(e)$ is close to a projection in B that can be obtained from $\varphi(e)$ by functional calculus, cf. also [CGM]. Similarly, if A, B are arbitrary algebras and ω is small algebraically e.g. nilpotent, then φ can be used to transport algebraic K-theory classes.

As a second example, consider a bounded linear map $\varphi : A \to B$ of two C^*-algebras for which $\varphi(x)\omega(y,z)$, $\omega(y,z)$ for $x,y,z \in A$ lie in some closed subalgebra $B_0 \subset B$. To show that φ induces a map $K_*A \to K_{*+1}(B_0)$ we use the algebras ϱA, RA completed with respect to the maximal C^*-norm for which p is of norm $\leq \|\varphi\|$. We obtain an exact sequence of C^*-algebras

$$0 \to \varrho A \to RA \to A \to 0$$

where RA is contractible. The long exact sequence of K-theory yields an isomorphism

$$K_*(A) \xrightarrow{\cong} K_{*+1}(\varrho A) .$$

Since φ induces a homomorphism $\widetilde{\varphi} : \varrho A \to B_0$, we obtain the desired map

$$K_*(A) \cong K_{*+1}(\varrho A) \to K_{*+1}(B_0) .$$

If φ is completely positive then it induces even a KK-element since then, for an appropriate completion, ϱA becomes a hereditary subalgebra of the C^*-algebra completion of $\varepsilon A = qA \times \mathbb{Z}/2$ on which $\widetilde{\varphi}$ is defined.

Consider further a quasihomomorphism, i.e. a pair α, $\overline{\alpha} : A \to B$ such that $\alpha(x) - \overline{\alpha}(x) \in J$, $x \in A$ for some ideal J of B. We assume that A, B, J are C^*-algebras, even though part of the discussion goes through more generally. This quasihomomorphism induces a homomorphism $QA \to B$ restricting to $\widetilde{\alpha} : qA \to J$. By split exactness of K_*, $K_*(QA) \cong K_*(qA) \oplus K_*(A)$ and the map $K_*(\imath) - K_*(\overline{\imath}) : K_*(A) \to K_*(QA)$, where $\imath, \overline{\imath}$ are as in Section 1, sends $K_*(A)$ into the first factor $K_*(qA)$. Composing with $K_*(\widetilde{\alpha})$ yields a map $K_*(A) \to K_*(J)$. We note that Kasparov or Fredholm modules usually give rise to quasihomomorphisms, thus can be used to transport K-theory classes.

We will see below that cyclic cocycles can be described as traces on RA or QA. Since quasihomomorphisms or linear maps from A into some algebra B induce homomorphisms QA, $RA \to B$ it will follow that traces on B can be transported under such maps to cyclic cocycles on A. The pairing between a trace and a transporte K-theory class corresponds to the pairing between the K-theory class and the transported trace.

3. Traces on RA, Supertraces on QA and Their Homotopy Classes

Let QA_+, QA_- denote the sets of even and odd elements with respect to the $\mathbb{Z}/2$-grading QA, respectively. We have seen above that $QA_+ \cong RA$ as an algebra. QA_+ and QA_- are RA-bimodules. Recall that a supertrace on a $\mathbb{Z}/2$-graded algebra or bimodule is a linear functional T satisfying $T(x\alpha) = (-1)^{\deg(x)\deg(\alpha)}\, T(\alpha x)$. Every even (resp. odd) supertrace on QA gives a bimodule trace T with $T(\alpha\omega) = T(\omega\alpha)$ for $\alpha \in RA$, $\omega \in QA_+$ (resp. $\omega \in QA_-$). Conversely, a bimodule trace T comes from a supertrace if and only if $T(qx_0 \dots qx_n) = T(qx_n qx_0 \dots qx_{n-1})$ for $x_0, \dots, x_n \in A$.

A linear functional on $\left((qA)^{2m}\right)_+ = (\varrho A)^m$ (an important special case is of course $(\varrho A)^0 = RA$) is given by its components $T^n, n \geq 2m$, defined by

$$T^n(x_0,\dots,x_k)=\begin{cases} T(qx_0qx_1\dots qx_n). & n \text{ odd}\\ T(px_0qx_1\dots qx_n) & n \text{ even}.\end{cases}$$

Proposition 3.1. *T represents a (bimodule)-trace if and only if for all even $n \geq 2m$*

$$b\,T^n=(1+\lambda)\,T^{n+1}$$
$$b'\,T^{n-1}=(1-\lambda)\,T^n$$

(here λ, b, b' are defined as in [Co1]*). Every bimodule trace can be modified canonically so as to extend to an even supertrace $\widetilde{T}$ on $(qA)^{2m}$.*

To prove Proposition 3.1, consider the map $\delta : QA_{\pm} \to QA_{\mp}$ defined by $\delta(\omega q(x)) = [p(x), \omega]$, $\omega \in QA$. The image of δ consists of all commutors in the bimodules QA_+, QA_- and one checks easily for $\delta : QA_- \to QA_+$

$$(T\circ\delta)^{2k+1}=-b\,T^{2k}+(1+\lambda)\,T^{2k+1}$$
$$(T\circ\delta)^{2k}=-b'\,T^{2k-1}-(1-\lambda)\,T^{2k}.$$

Since $(1-\lambda^2)T^k=(1-\lambda)bT^{k-1}=b'(1-\lambda)T^{k-1}=b'b'T^{k-2}$, the relations of Proposition 3.1 imply that $(1-\lambda^2)\ T^k=0$ for odd $k\geq 2n+1$, so that $(1+\lambda)\ T^k=\frac{2}{k+1}N\ T^k$ where $N=1+\lambda+\ldots+\lambda^k$. Thus T is a trace if and only if $(1-\lambda^2)\ T^k=0$ for such k and T is a cocycle in the total complex of the bicomplex

$$\begin{array}{ccccc}
C^{2n+1} & \xrightarrow{\frac{-2}{2n+2}N} & & & \\
\uparrow{\scriptstyle \frac{2n+2}{2}b'} & & \uparrow{\scriptstyle b} & & \\
\xrightarrow{N} & & C^{2n} & \xrightarrow{1-\lambda} & \\
& & \uparrow{\scriptstyle -nb} & & \uparrow{\scriptstyle -b'} \\
& & & \xrightarrow{-n(1-\lambda)} & C^{2n-1}
\end{array} \qquad (3.A)$$

It is obvious that every cocycle T in this bicomplex is cohomologous to one for which $(1-\lambda)\ T^k=0$ for all odd k. This procedure to obtain cyclic cocycles admits the following generalization:

Let A be an algebra and E, E' two RA-bimodules which are in duality, i.e. there is a bilinear pairing $(x,x')\mapsto\langle x|x'\rangle\in\mathbb{C}, x\in E,\ x'\in E'$ such that $\langle\alpha x|x'\rangle=\langle x|x'\alpha\rangle$ and $\langle x\alpha|x'\rangle=\langle x|\alpha x'\rangle$ for $\alpha\in RA$. Denote as above by $p : A \to RA$ the universal linear map and by ω its curvature. Suppose further that we are given two multilinear maps $c : A^r \to E$, $c' : A^s \to E'$ satisfying

$$p(x_0)c(x_1,\dots,x_r)-c(x_0x_1,\dots,x_r)+\dots$$
$$+(-1)^r c(x_0,x_1,\dots,x_{r-1}x_r)+(-1)^{r+1}c(x_0,\dots,x_{r-1})p(x_r)=0$$

and similarly for c' with r replaced by s. Then the following family $\{T^n\}_{n\geq r+s}$ of multilinear functions defines a cocycle in the bicomplex 3.A resp. 3.B:

$$T^n(x_0,..,x_n)=\begin{cases}\langle\,\omega(x_0,x_1)..\omega(x_{k-1},x_k)\,c(x_{k+1},..,x_{k+r})|c'(x_{k+r+1},..,x_n)\,\rangle & k \text{ odd}\\ \langle\,px_0\omega(x_1,x_2)..\omega(x_{k-1},x_k)\,c(x_{k+1},..,x_{k+r})|c'(x_{k+r+1},..,x_n)\,\rangle & k \text{ even}\end{cases}$$

where $k = n - r - s$. In fact, by the same computation as above we find that $b\,T^n = (1+\lambda)\,T^{n+1}$ for k even, and $b'\,T^n = (1-\lambda)\,T^{n+1}$ for k odd.

Applying this with $c(x_1,\ldots,x_{2n+1}) = q(x_1)q(x_2)\ldots q(x_{2n+1})$ we see that for bimodule traces on $\left((qA)^{2n+1}\right)_{-}$, Proposition 3.1 holds with odd and even interchanged. A linear functional on $\left((qA)^{2n+1}\right)_{-}$ is a trace if and only if $(1-\lambda^2)\,T^k = 0$ for even $k \geq 2n+1$ and if T is a cocycle in the bicomplex

$$\begin{array}{ccccc}
C^{2n+2} & \xrightarrow{\frac{-2}{2n+3}N} & & & \\
\uparrow{\scriptstyle \frac{2n+3}{2}b'} & & \uparrow{\scriptstyle b} & & \\
& \xrightarrow{N} & C^{2n+1} & \xrightarrow{1-\lambda} & \\
& & \uparrow{\scriptstyle \frac{-(2n+1)}{2}b} & & \uparrow{\scriptstyle -b'} \\
& & & \xrightarrow{\frac{-(2n+1)}{2}(1-\lambda)} & C^{2n}
\end{array} \qquad (3.B)$$

The question of which traces correspond to coboundaries now admits a very natural answer which is a basis for many different homotopy invariance results for cyclic cohomology. We form the algebra ΩQA of abstract differential forms over QA. An important fact which is easily checked is that the map $d\left(p(x)\right),\, d\left(q(x)\right) \mapsto p(dx),\, q(dx)$ induces an isomorphism (of bigraded algebras) between ΩQA and $Q\Omega A$. Let $\Omega^1 QA$ be the QA-bimodule of 1-forms over QA and $d : QA \to \Omega^1 QA$ the universal derivation. If T' is a (bimodule) supertrace on $\Omega^1 QA$ then $T' \circ d$ is a supertrace on QA.

We now determine the supertraces T' on $\Omega^1 QA$ and show that the cocycles corresponding to $T' \circ d$ are coboundaries. We restrict to the case of even supertraces. The odd case is quite similar but there are some slight complications in dimension 0 and 1 (an odd supertrace T corresponds to a cocycle in the bicomplex 3.B only if $T\left(q(x)\right) = 0, \forall x$) which we want to avoid.

Proposition 3.2. *There is a bijection between even supertraces T' on $\Omega^1 QA$ and cochains $\alpha = \{\alpha^k\}_{k\geq 0}$ given by*

$$\alpha^{2n}(x_0,\ldots,x_{2n}) = T'(dx_0qx_1\ldots qx_{2n})$$
$$\alpha^{2n+1}(x_0,\ldots,x_{2n+1}) = T'(x_0dx_1qx_2\ldots qx_{2n+1})$$

Let $\partial^{(1)}$, $\partial^{(2)}$ denote the two coboundary operators in the bicomplex 3.A, i.e.

$$\partial^{(1)}_{2n} = N + (n+1)b', \quad \partial^{(1)}_{2n-1} = -n\left(b + (1-\lambda)\right)$$
$$\partial^{(2)}_{2n} = b + 1 - \lambda, \quad \partial^{(2)}_{2n-1} = -(b' + \frac{1}{n}N)$$

One has $\partial^{(2)}\partial^{(1)} = \partial^{(1)}\partial^{(2)} = 0$ and we already have seen in 3.1 that a cochain T represents an even supertrace on QA if and only if $\partial^{(2)}\ T = 0$ and T is normalized, i.e. $(1-\lambda)\ T^{2n+1} = 0$, $\forall n$.

Every cochain α can be modified canonically to a cochain $P\alpha$ such that $\partial^{(1)}(P\alpha)$ is normalized. One has

$$(P\alpha)^{2n} = \alpha^{2n}$$

$$(P\alpha)^{2n-1} = \frac{1}{2n} N\alpha^{2n-1} + (1-\lambda)^{-1}(1 - \frac{1}{2n}N)b'\alpha^{2n-2}$$

If $\partial^{(1)}\alpha$ is already normalized then $P\alpha = \alpha$ so that $P^2 = P$. We can now formulate the main result of this section.

Theorem 3.3 [Cu-Qu]. *Let T' be an even supertrace on $\Omega^1 QA$ corresponding to the cochain α. Then the cocycle corresponding to $T' \circ d$ is given by $\partial^{(1)}(P\alpha)$.*

The analogous result holds for even supertraces on $(qA)^{2n}$ and for even supertraces on the $(qA)^{2n}$-bimodule generated in $\Omega^1 QA$ by $p(\Omega^1 A)$. Theorem 3.3 contains of course as a very special case the fact that cyclic cohomology is invariant under derivations of A [Go] and has many applications. We list here a few examples:

1. The cyclic cohomology classes associated with an n-summable Fredholm module or with a Θ-summable Fredholm module, [Co1], [Co2] are (in the first case upon application of S) invariant under homotopies of the homomorphisms and of the operators appearing in the module. In fact, these classes are obtained from homomorphisms of $(qA)^n$ or of QA into an algebra with a trace.

2. A normalized cyclic cocycle is a coboundary if and only if it comes from a supertrace of the form $T' \circ d$, T' a trace on $\Omega^1 QA$. Thus elements of cyclic cohomology can be described as homotopy classes of supertraces on QA or on $(qA)^n$. This applies to all different versions of cyclic cohomology (ordinary, periodic, entire). For instance, the periodic cyclic cohomology $HC_{\text{per}}^{\text{even/odd}}$ is given as

$$\{\text{even/odd supertraces on } \overline{QA}\}/\{T' \circ d \mid T' \text{ an even/odd supertrace on } \Omega^1\overline{QA}\}$$

where $\overline{QA} = \varprojlim_n QA/(qA)^n$ is the algebraic completion of QA.

3. Let $\varphi : A \to B$ be a linear map with nilpotent curvature ($\omega(x_0, x_1)\omega(x_2, x_3)\ldots \omega(x_{2k}, x_{2k+1}) = 0$ for $k \geq n$) and τ a trace on B. We obtain a homomorphism $\widetilde{\varphi} : RA \to B$ and a trace $T = \tau \circ \widetilde{\varphi}$ on $RA/(\varrho A)^{n+1}$ and thus an element $[T]$ of $HC^{2n}(A)$. Let $\varphi_t : A \to B$ be a differentiable homotopy of such maps and $\dot{\varphi}_t$ its derivative. Consider the linear map $\Psi_t : \Omega A \to B$ sending $x_0 dx_1 \ldots dx_n$ to $\varphi_t(x_0)\dot{\varphi}_t(x_1)\ldots\dot{\varphi}_t(x_n)$ and the corresponding homomorphism $\widetilde{\Psi}_t : R\Omega A \to B$. We obtain a trace T_t' on $R\Omega A$ which we can modify to an even supertrace still denoted by T_t' on $Q\Omega A \cong \Omega QA$ (see 3.1). $T_t' \circ d$ gives an even supertrace on $QA/(qA)^{2n+2}$ inducing a coboundary. This shows that the classes $[T_t]$ induced by φ_t remain constant.

4. Characteristic Classes as Elements of RA and QA

The idea of interpreting characteristic classes associated with idempotents or invertible elements as differential forms, elements of ΩA [Ka] or even as elements of QA [Co2] is of course not new. The description of these classes that we are now going to give is however at least partly new and will clarify their meaning.

Let e be an idempotent in A. With e we want to associate *odd* elements $ch_{2n}(e) \in (qA)^{2n+1}$ such that $d\left(ch_{2n}(e)\right)$ is a sum of supercommutators in $\Omega^1 QA$ and such that $ch_{2n+2k}(e)$ differs from $ch_{2n}(e)$ by a sum of supercommutators. Second, we associate an element $Ch(e)$ (an *even* element of $\overline{QA}$) of the algebraic completion $\overline{RA} = \lim\limits_{\leftarrow} RA/(\varrho A)^n$ such that $d\big(Ch(e)\big)$ is a sum of supercommutators. The elements $ch_{2n+2k}(e)$ represent cyclic cycles in $HC^-_{2n}(A)$ which may be paired with odd traces T on $(qA)^{2n}$. The identities $T\big(ch_{2n+2k}(e)\big) = T\big(ch_{2n}(e)\big)$, $(T'\circ d)\big(ch_{2n}(e)\big) = 0$ mean that we have a pairing with $\langle\, [T], ch_{2n}(e)\,\rangle = \langle\, S^k[T], ch_{2n-2k}(e)\,\rangle$. The element $Ch(e)$ represents a cycle for $HC^{\mathrm{per}}(A)$ (or for the entire cyclic homology $HC_{\mathrm{ev}}(A)$ in the sense of [Co2] or for other versions of cyclic cohomology for locally convex algebras – we only need $Ch(e)$ to converge in the completion of RA) and the pairing with a periodic cocycle represented by a trace T on RA is given by $T\big(Ch(e)\big)$. By functoriality, to construct $ch_k(e)$, $Ch(e)$, we may assume that $A = \mathbb{C}e \cong \mathbb{C}$.

The first part is easily achieved taking $ch_{2n}(e) = (2qe)^{2n+1}$. These are elements of $(qA)^{2n+1}$ differing from the class in dimension 0, given by $2q(e)$, only by sums of supercommutators by 3.1 (the difference is a boundary in the bicomplex dual to 3.A).

The element $Ch(e)$ is given by a series of the form

$$Ch(e) = \sum_{k\geq 0} c_{2k}\, p(e)\, q(e)^{2k} + \sum_{k\geq 1} c_{2k-1}\, q(e)^{2k}.$$

The fact that $\partial(\sum c_j e^{\otimes(j+1)}) = 0$ in the bicomplex dual to (3.A) leads, given $c_0 = 1$, immediately to

$$c_{2k} = -c_{2k-1} = \frac{(2k)!}{(k!)^2}$$

whence, formally,

$$Ch(e) = \frac{p(e)}{\sqrt{1-4(qe)^2}} - \frac{1}{2}\left(\frac{1}{\sqrt{1-4(qe)^2}} - 1\right)$$

which gives, using $(qe)^2 = p(e) - p(e)^2$

$$Ch(e) = \frac{1}{2}\left(1 - \frac{1-2p(e)}{\sqrt{(1-2p(e))^2}}\right).$$

This slightly surprising formula has a very natural interpretation: If we replace $p(e)$ by the complex variable z (or by an element of a Banach algebra) then $Ch(z)^2 = Ch(z)$ and $Ch(z)$ is the characteristic function of the half-plane $\mathrm{Re}\, z > \frac{1}{2}$, the power series representing $Ch(z)$ converging for $|z - z^2| < \frac{1}{2}$. This formula for $Ch(e)$ as an element of the algebra of differential forms on the algebra of differential forms on the cotangent bundle with modified multiplication) was used already by Fedosov [Fed] in his version of the index theorem. It was also known to A. Connes (private communication) and is more natural and useful than the one given in [Co2].

Let us turn now to the case of an invertible element u of the unital algebra A. To simplify the formulas for $ch_{2n+1}(u)$, $Ch(u)$ we work in the completions of the reduced algebras $QA' = A \underset{\mathbb{C}1}{*} A$, $(\Omega^1 QA)'$, RA', $(\Omega^1 QA)'$ with additional relations $q(1) = d(1) = \omega(1,x) = \omega(x,1) = 0$. Again we easily can define (this time even) elements $ch_{2n+1}(u) \in (\varrho A')^n$ as $\omega(u,u^{-1})^n - \omega(u^{-1},u)^n$ such that $d\big(ch_{2n+1}(u)\big)$ is a sum of supercommutators in $(\Omega' RA)'$ and such that $ch_{2n+1}(u)$ differs from $ch_1(u) = \omega(u,u^{-1}) - \omega(u^{-1},u)$ by a sum of supercommutators.

The construction of the odd element $Ch(u)$ is more involved and we have to pass to 2×2-matrices. Put

$$y = \begin{pmatrix} 0 & u^{-1}qu \\ \overline{u^{-1}qu} & 0 \end{pmatrix} \in M_2(\overline{QA'}) .$$

Using the identity $u^{-1}q(u)\overline{u^{-1}q(u)} = -q(u^{-1})q(u)$ one finds

$$y^{2n+1} = (-1)^n \begin{pmatrix} 0 & u^{-1}qu\big(qu^{-1}qu\big)^n \\ \overline{u^{-1}qu(qu^{-1}qu)^n} & 0 \end{pmatrix} .$$

Given $c_0 = 1$, there are unique coefficients c_k such that $\partial_{(1)}\varphi = 0$ in the bicomplex dual to 3.B for the chain

$$\varphi = \sum_{n\ge 0} c_{2n}\left(1 \otimes (u^{-1}\otimes u)^{\otimes n} - 1 \otimes (u \otimes u^{-1})^{\otimes n}\right) + \sum_{n \ge 0} c_{2n+1}(u^{-1}\otimes u)^{\otimes n+1}$$

and one computes

$$c_{2n} = \frac{2^{2n}(n!)^2}{(2n+1)!}, \qquad c_{2n+1} = c_{2n} .$$

Thus, putting

$$f(u) = \sum_{n\ge 0} c_{2n} q1\left(\big(qu^{-1}qu\big)^n - (qu\,qu^{-1})^n\right) + \sum_{n\ge 0} c_{2n+1} pu^{-1}qu\big(qu^{-1}qu\big)^n ,$$

$d\big(f(u)\big)$ is a sum of supercommutators and the first term is 0 if we impose $q(1) = 0$. To express $f(u)$ by a known power series one needs a square root for $q(u^{-1})q(u)$, which is given by y, and we put

$$Ch(u) = \sum_{n\ge 0} (-1)^n c_{2n+1}\, y^{2n+1} .$$

Given an odd trace T on $\overline{QA}'$ we extend T to $M_2(\overline{QA}')$ by

$$\widetilde{T}\left(\begin{pmatrix} a & b \\ c & d \end{pmatrix}\right) = T(b+c) .$$

We then have $\widetilde{T}\big(Ch(u)\big) = T\big(f(u)\big)$ where this number is the product $\langle\, [\varphi], [T]\, \rangle$ of the class of the cycle φ associated with u and the cyclic cohomology class $[T]$ associated with T.

Since the power series $\sum_{k\ge 0} (-1)^k \frac{2^{2k}(k!)^2}{(2k+1)!}\, z^{2k+1}$ represents the function $\frac{\ln(z+\sqrt{1+z^2})}{\sqrt{1+z^2}}$ we formally have

$$Ch(u) = \frac{\ln(y + \sqrt{1+y^2})}{\sqrt{1+y^2}} .$$

To understand the meaning of this formula, suppose that $\overline{u}$ differs from u by an exponential: $\overline{u} = ue^h$. Then

$$y^2 = \begin{pmatrix} \sinh^2(\frac{h}{2}) & 0 \\ 0 & \sinh^2(\frac{h}{2}) \end{pmatrix}$$

$$\ln(y + \sqrt{1+y^2}) = \begin{pmatrix} 0 & -\frac{h}{2}e^{\frac{h}{2}} \\ -\frac{h}{2}e^{-\frac{h}{2}} & 0 \end{pmatrix}.$$

This gives

$$\widetilde{T}\big(Ch(u)\big) = T\left(\frac{-\frac{h}{2}(e^{-\frac{h}{2}} + e^{\frac{h}{2}})}{\cosh(\frac{h}{2})}\right) = -T(h).$$

References

[BDF] Brown, L., Douglas, R., Fillmore, P.: Extensions of C^*-algebras and K-homology. Ann. Math. (2) **105** (1977) 265–324

[Co1] Connes, A.: Non commutative differential geometry. Publ. Math. IHES **62** (1985) 257–360

[Co2] Connes, A.: Entire cyclic cohomology of Banach algebras and characters of Θ-summable Fredholm modules. K-Theory **1** (1988) 519–548

[Co-Cu] Connes, A., Cuntz, J.: Quasi-homomorphismes, cohomologie cyclique et positivité. Commun. Math. Phys. **114** (1988) 515–526

[CGM] Connes, A., Gromov, M., Moscovici, H.: Novikov's conjecture and almost flat vector bundles. C.R. Acad. Sci. Paris, Sér. I **310** (1990) 273–277

[Cu1] Cuntz, J.: A new look at KK-theory. K-theory **1** (1987) 31–51

[Cu2] Cuntz, J.: Universal extensions and cyclic cohomology. C.R. Acad. Sci. Paris, Sér. I **309** (1989) 5–8

[Cu-Qu] Cuntz, J., Quillen, D. (In preparation)

[Fed] Fedosov: Analytic formulas for the index of an elliptic operator. Trudi Mosk. Math. **30** (1974) 159–240 (in Russian)

[Go] Goodwillie, R.: Cyclic homology, derivations and the free loopspace. Topology **24** (1985) 187–215

[Lo-Qu] Loday, J.-L., Quillen, D.: Cyclic homology and the Lie algebra homology of matrices. Comment. Math. Helv. **59** (1984) 565–591

[Ka] Karoubi, M.: Homologie cyclique et K-théorie. Astérisque **149** (1987)

[Kas] Kasparov, G.G.: The operator K-functor and extensions of C^*-algebras. Izv. Akad. Nauk SSSR, Ser. Mat. **44** (1980) 571–636

[Qu] Quillen, D.: Algebra cochains and cyclic cohomology. Publ. Math. IHES **68** (1989) 139–174

[Ts] Tsygan, B.L.: Homology of matrix algebras over rings and Hochschild homology. Uspekhi Mat. Nauk **38** 2 (1983) 217–218

[Zek] Zekri, R.: A new description of Kasparov's theory of C^*-algebra extensions. J. Funct. Anal. **84** (1989) 441–471

K-Theory for Groups Acting on Trees

Michael V. Pimsner

Mathematisches Institut der Universität Heidelberg, Im Neuenheimer Feld 288
W-6900 Heidelberg, Fed. Rep. of Germany

K-theory is by now such an established subject that it hardly needs any introduction. Recall however that if A is an associative ring with unit, Grothendieck defined $K_0(A)$ as the group generated by the semigroup of finitely generated, projective A-modules. Depending on the nature of the ring A one gets entirely different theories, one of the most vivid differences appearing for example when trying to define the higher K-groups.

One of the first achievements of K-theory, due to Atiyah, was in topology, in the case $A = C(X)$, the algebra of continuous complex valued functions on a compact topological space X. In this case the finitely generated projective A-modules are precisely the locally trivial vector bundles, so the whole theory admits a purely topological description. Due to Bott periodicity one gets a $\mathbb{Z}/2$ graded generalized cohomology theory $K_*(A)$ (Atiyah 1967). This is a consequence of the topology (and of course of the fact that the ground field are the complex numbers), so it is not really surprising that one gets more or less the same theory for complex Banach algebras (see e.g. Karoubi 1978).

However if $A = C(X)$ the theory is much richer, since along with K-theory comes another dual theory $K^*(A)$, called K-homology, that has deep connections with elliptic operators (Atiyah 1970), and that may be described either in terms of extensions as (Brown Douglas and Fillmore 1977) did it, or in terms of abstract elliptic operators as (Kasparov 1975) did it. The relevant fact for such a theory to exist is the fact that A is a C^*-algebra. Moreover in this case the two dual theories can be glued together. This was first done in (Pimsner, Popa, and Voiculescu 1979 and 1980) in the particular case when one variable is still commutative (and finite dimensional), motivated by the study of the homotopy invariance of extensions, and soon later (Kasparov 1980) came with what looks now to be the final theory. This is a functor of two (not necessarily commutative) C^*-algebras A and B, (which we will assume from now on for simplicity to be separable), with values in the $\mathbb{Z}/2$ graded abelian groups, denoted $KK_*(A, B)$, such that

$$K_*(A) = KK_*(\mathbb{C}, A)$$

and

$$K^*(A) = KK_*(A, \mathbb{C})$$

($\mathbb{C}$ stands as usual for the complex numbers). The most important feature of this theory is the product introduced by Kasparov, which is the analogue of the various cup, cap, slant... products in cohomology theory. In its simplest form it

Proceedings of the International Congress
of Mathematicians, Kyoto, Japan, 1990

may be described as a map

$$KK_*(A,B) \times KK_*(B,C) \longrightarrow KK_*(A,C)$$

denoted $(x,y) \to x \otimes_B y$.

The central theorem of KK-theory gives the construction of the product and asserts its associativity. For example the pairing between K-theory and K-homology is just the Kasparov product

$$KK(\mathbb{C},A) \otimes KK(A,\mathbb{C}) \longrightarrow KK(\mathbb{C},\mathbb{C})$$

since the latter is isomorphic to the additive group of integers $\mathbb{Z}$.

Another important consequence of the Kasparov product is the fact that one can regard the elements of $KK_*(A,B)$ as natural maps from $K_*(A)$ to $K_*(B)$,

$$KK_*(\mathbb{C},A) \times KK_*(A,B) \longrightarrow KK_*(\mathbb{C},B)$$

with composition given by the product. So even if one is interested only in K-theory, it is unnatural to discard KK-theory, since most of the interesting maps are not given by $*$-homomorphisms from A to B but by elements from $KK_*(A,B)$. An improvement of this aspect of KK-theory was recently proposed by Connes and Higson (1990).

Another consequence of the Kasparov product is the fact that $KK_*(A,A)$ becomes a (graded) ring with unit 1_A, given by the class of the identity map from A to A, and that $KK_*(A,B)$ has a left $KK_*(A,A)$ (and a right $KK_*(B,B)$) module structure.

It should be stressed at this point that to show that the product of two given elements equals a third one is often a deep theorem. It is enough to mention for example that the Atiyah-Singer theorem (Atiyah and Singer 1968) can be put into this form (Kasparov 1975, Connes and Skandalis 1984, Baum and Douglas 1982).

Of course KK-theory did not appear just for the sake of generalizing topological K-theory from spaces to general C^*-algebras. Already in the seventies K_* and K^* proved to be powerful invariants for certain classes of C^*-algebras (Elliott 1976, Pimsner and Popa 1978). One of the turning points in applying K-theory to C^*-algebras was probably the discovery (Pimsner and Voiculescu 1980), that the irrational rotation C^*-algebras A_θ are essentially nonisomorphic for different θ's. These are the C^*-algebras generated in L^2 of the unit circle by multiplication operators with continuous functions and by the rotation operator by the angle $2\pi\theta$. Their importance lies in the fact that they are the most simple nontrivial C^*-algebras of several classes of C^*-algebras. One can look at them as crossed products by $\mathbb{Z}$, as the group C^*-algebra of $\mathbb{Z}^2$ with a 2-cocycle (see e.g. Pedersen 1979) or as stably isomorphic to the C^*-algebra of the Kronecker foliation (Connes 1982). The computation of their K_*- and K^*-theories was done in (Pimsner and Voiculescu 1980) by proving a general six terms exact sequence for every crossed product by $\mathbb{Z}$. In the particular case of the irrational rotation algebras one gets:

Theorem (Pimsner and Voiculescu 1980).

$$K_0(A_\theta) \simeq \mathbb{Z}^2$$

with generators [1] *and* [R], *the class of the identity and respectively that of the Rieffel projection.*

$$K_1(A_\theta) \simeq \mathbb{Z}^2$$

with generators [u] *and* [v], *the unitaries corresponding to the rotation and respectively to multiplication by* z.

Surprisingly enough the same techniques could be used to solve the conjecture of Kaplansky-Kadison, that the reduced C^*-algebra of the free group on two generators (which was proved to be simple by (Powers 1975)) had no nontrivial (i.e. different from 0 and 1) projection. This was done (Pimsner Voiculescu 1982), again by proving a six terms exact sequence for reduced crossed products by free groups on any number of generators (not just one). In particular one gets:

Theorem (Pimsner and Voiculescu 1982).

$$K_0(C_r^*(\mathbb{F}_n)) \simeq \mathbb{Z},$$

with generator [1], *the class of* 1.

$$K_1(C_r^*(\mathbb{F}_n)) \simeq \mathbb{Z}^n$$

with generators $[g_1],\ldots,[g_n]$, *the classes of the generators of* $\mathbb{F}_n$.

As a corollary of this theorem one gets also that the reduced C^*-algebra of free groups are nonisomorphic if the number of generators are distinct. This may be just one hint of why the corresponding question for the von Neumann algebras of the free groups is so difficult.

Another motivation for studying the K-theory of crossed products by discrete groups came from topology. I won't pursue this any further, I will just mention that some knowledge of the K-theory of $\pi_1(M)$ (the fundamental group of the manifold M) gives a positive answer to the Novikov conjecture about the homotopy invariance of higher signatures for manifolds with prescribed homotopy group (Kasparov 1988).

It is quite clear that the K-theory of crossed products is important both for the study of C^*-algebras and for geometric applications. The basic tool for this problem is the equivariant KK^G-theory developed by Kasparov. Unlike the compact group case that can be obtained quite easily by taking only equivariant elements in the definition of the KK-groups, the general noncompact case is much more difficult and shows one of the advantages of KK-theory. Relevant for our purposes is the induction map

$$j : KK_*^G(A,B) \longrightarrow KK_*(A \rtimes G, B \rtimes G)$$

that commutes with the Kasparov product and sends the unit of $KK_*^G(A,A)$ to the unit of $KK_*(A \rtimes G, A \rtimes G)$. In particular it transforms KK^G-equivalence into KK-equivalence, where we have the following natural definition.

Definition. The C^*-algebras A and B are KK-equivalent if there exist $\alpha \in KK(A,B)$ and $\beta \in KK(B,A)$ such that $\alpha \otimes_B \beta = 1_A$ and $\beta \otimes_A \alpha = 1_B$.

(The KK^G-equivalence is defined in the same way using KK^G-theory). Up to now there are two ways of studying the K-theory of crossed products by discrete groups. One way, emphasized by Connes and Kasparov, takes the point of view that it is easier to deal with Lie groups, which then in turn will provide results for all their discrete subgroups. The second one treats the discrete group on its own right. Both methods are of geometric nature the main ingredient being the study of "spaces" on which the groups act naturally.

To show how this works let me describe, at least at a formal level, the Connes-Kasparov conjecture. Let G be a connected Lie group, K its maximal compact subgroup and $X = G/K$ the corresponding homogeneous space of dimension n. The class of the Dirac operator defines an element $\alpha \in KK_n^G(C_0(X), \mathbb{C})$. The needed $\beta \in KK_n^G(\mathbb{C}, C_0(X))$ is the element defined by Kasparov (1973) and Mishchenko (1973) out of the negative curvature of the homogeneous space X.

Theorem (Kasparov).

$$1. \quad \alpha \otimes_{\mathbb{C}} \beta = 1_{C_0(X)} \quad \in KK_0^G(C_0(X), C_0(X))$$

$$2. \quad \beta \otimes_{C_0(X)} \alpha = \gamma^G \quad \in KK_0^G(\mathbb{C}, \mathbb{C})$$

is an idempotent.

So if one can prove that $\gamma^G = 1_{\mathbb{C}}$ for the group G, then $C_0(X)$ is KK_n^G-equivalent to $\mathbb{C}$ and taking crossed products one gets that $C^*(G)$ is KK_n-equivalent to $C_0(X) \rtimes G$. Since this latter C^*-algebra is strong Morita equivalent (in the sense of Rieffel (1974); this is a particular, more easy, KK-equivalence) to $C^*(K)$ one gets in particular

$$K_*(C^*(G)) \simeq K_{*+n}(C^*(K)),$$

the isomorphism being natural.

Moreover if Γ is a discrete subgroup of the connected Lie group G, with $\gamma^G = 1_{\mathbb{C}}$, then the same holds: $C_0(X)$ being KK_n^Γ-equivalent to $\mathbb{C}$, one gets by taking crossed products by Γ that $C^*(\Gamma) \simeq_{KK_n} C_0(X) \rtimes \Gamma$ and again the latter C^*-algebra is strongly Morita equivalent to $C_0(X/\Gamma) \rtimes K$. Still better, we can start with $C_0(X) \otimes A \simeq_{KK_n^\Gamma} A$ to get the K-theory of $A \rtimes \Gamma$, for any C^*-algebra on which Γ acts. However this depends on $\gamma^G = 1_{\mathbb{C}}$ (or at least $\gamma^\Gamma = 1_{\mathbb{C}}$ for the discrete subgroup Γ). This is known for a large class of groups, including the amenable groups, $SO(n,1)$ (Kasparov 1984), $SU(n,1)$ (Julg and Kasparov 1990). It is also known to fail for groups having property T of Kazhdan. Note however that it would be enough to have $C^*(\Gamma) \simeq_{KK_n} C_0(X) \rtimes \Gamma$. In order to study this question Cuntz (1983) introduced the notion of K-amenability. But even this is not true in general as has been shown in a remarkable paper by Skandalis (1988).

Thus in general we can compute only what is called the "γ-part" of the K-groups, which is a direct summand of what we really want, since $\gamma^2 = \gamma$. If $\gamma^\Gamma = 1_{\mathbb{C}}$ (for the discrete subgroup of the Lie group G), we shall say that we know *the K-theory for the group Γ*. Note that this is really much more than knowing the K-groups of the crossed products.

Let us stop here this very incomplete presentation of the Lie group case and turn now to the case of groups that act on some oriented tree X, that is on a one-dimensional simply connected simplicial complex $X = (X^1, X^0)$, where X^1 will be the set of edges and X^0 the set of vertices (see Serre 1977). Since the tree is oriented there are two maps $t, o : X^1 \longrightarrow X^0$, which are thought to be the terminus and respectively the origin of the edge. An action of the group G on X is then an action of G on both X^1 and X^0, such that the maps t and o are G-equivariant.

The case of the discrete groups that act on some tree is well understood due to the Bass-Serre theory of graphs of groups (see Serre 1977). These consist of an oriented graph $\Sigma = (\Sigma^1, \Sigma^0)$ and of a collection $\{G_y\}_{y \in \Sigma^1}$ and $\{G_P\}_{P \in \Sigma^0}$ of groups, together with embeddings $t : G_y \longrightarrow G_{t(y)}$ and $o : G_y \longrightarrow G_{o(y)}$ for every $y \in \Sigma^1$. Out of a graph of groups one gets a group by constructing the so called "fundamental group of the graph of groups", that acts on the "universal cover of the graph of groups" which is a tree. The inverse construction is more straightforward: Σ is simply the orbit (homogeneous) space $G \backslash X$, while the G_y's (respectively the G_P's) are the stabilizer subgroups of the edges (resp. of the vertices) modulo some automorphism coming from the choice of a representative. The simplest graphs of groups, having only one edge, yield well known constructions in the theory of discrete groups, namely amalgamated products (if there are 2 vertices) and *HNN*-extensions (if there is only one vertex).

The nondiscrete groups that act on some tree are as interesting as the discrete ones (Serre 1977). The most interesting seem to be the reductive groups over local fields with one dimensional Bruhat-Tits building (see Tits 1979), e.g. $SL_2(\mathbb{Q}_p)$.

Since the computation of the K-groups of the reduced crossed products by free groups (Pimsner and Voiculescu 1982), partial results on the K-groups of free and amalgamated products and of *HNN*-extensions of groups have been obtained. Thus Lance (1983) introduced condition Λ, in order to use the methods of (Pimsner Voiculescu 1982) to compute the K-groups of the reduced C^*-algebra of certain free products of groups. This has been extended by Natsume (1985) to certain amalgamated products and finally Anderson and Paschke (1986) combined the above results with those of (Pimsner Voiculescu 1980) to get results for the K-groups of the reduced C^*-algebra of certain *HNN*-extensions. However the action of the group G on the tree X allows us to get much more, namely the knowledge of *the K-theory for the group G*. I will describe the original *Toeplitz Extension* approach of (Pimsner 1986), which is the generalisation of the methods of (Pimsner and Voiculescu 1980, 1982) to the general tree case. It is based on a rough analysis of actions of groups on trees due to Julg and Valette (1984). To this end one fixes a vertex O, called origin, and one denotes by χ_O the set of all edges that point to O. By adding to the continuous functions on the discrete space X^1 that vanish at infinity the characteristic function of χ_O, one gets one point at infinity that is fixed by the action of G. Denoting by X^1_+ the space thus constructed one gets the following G-equivariant exact sequence:

$$0 \longrightarrow C_0(X^1) \longrightarrow C_0(X^1_+) \longrightarrow \mathbb{C} \longrightarrow 0$$

which we call the (G-equivariant) Toeplitz extension. The name comes from the case of $\mathbb{Z}$ acting on its obvious tree, since after taking crossed products one gets the exact sequence

$$0 \longrightarrow K(L^2(\mathbb{T})) \longrightarrow T \longrightarrow C(\mathbb{T}) \longrightarrow 0$$

and where χ_O corresponds to the Hardy projection onto $H^2(\mathbb{T})$. With these notations in mind we have the following theorem.

Theorem. *The C^*-algebra $C_0(X^1_+)$ is KK^G_0-equivalent to the C^*-algebra $C_0(X^0)$.*

This really means that there are elements $\alpha \in KK_0^G(C_0(X^0), C_0(X_+^1))$ and $\beta \in KK_0^G(C_0(X_+^1), C_0(X^0))$ and that one is the inverse of the other in KK^G. This theorem together with the Toeplitz extension gives the K-theory for groups acting on trees. For example to compute the K-groups of crossed products, one tensors the Toeplitz extension with the C^*-algebra A, takes crossed products with G and applies the K-theory functor. Replacing in the periodic six-terms exact sequence the KK-equivalent terms one gets the following theorem

Theorem (Pimsner 1986). *Let G be a second countable group that acts on some oriented tree X and on the C^*-algebra A. The following six terms periodic sequence is exact:*

$$\begin{array}{ccccc} \oplus_{y\in\Sigma^1} K_0(A \rtimes G_y) & \xrightarrow{\sigma} & \oplus_{P\in\Sigma^0} K_0(A \rtimes G_P) & \xrightarrow{i} & K_0(A \rtimes G) \\ \uparrow\delta & & & & \downarrow\delta \\ K_1(A \rtimes G) & \xleftarrow{i} & \oplus_{P\in\Sigma^0} K_1(A \rtimes G_P) & \xleftarrow{\sigma} & \oplus_{y\in\Sigma^1} K_1(A \rtimes G_y) \end{array}$$

where $\sigma = t_* - o_*$ is the obvious map given by the difference of the terminus and origin maps, i is the map given by the inclusion maps $G_P \to G$ and δ are the maps induced by the boundary maps associated to the Toeplitz extension.

This is the generalisation of the six terms exact sequence for the free groups (Pimsner and Voiculescu 1982). Note that the above result is expressed in terms of the graph of groups associated to the action of G on X, so that one gets explicit exact sequences that express the K-groups of the group by the K-groups of the corresponding graph of groups. In particular one gets exact sequences for amalgamated products and for HNN-extensions of groups. Of course the same reasoning gives also the K^*-groups and more general the KK_*-groups of crossed products by groups acting on trees.

There is one technical point concerning crossed products which we did not mention at all, namely the difference between the reduced and the full ones. One reason is that the exact sequence exists for each of them, provided we fix one of the cross-norms. This has however an interesting corollary, which is a generalization of a result of Julg and Valette (1984)

Corollary. *If the countable discrete group G acts on some tree X, then G is K-amenable if and only if every stabilizer is K-amenable*

As we already mentioned, the Toeplitz extension together with the KK^G-equivalence of the Toeplitz algebra with $C_0(X^0)$, give us in fact more than just the above exact sequences. To illustrate this let us consider $\Gamma = \Gamma_1 \times \dots \Gamma_n$ the direct product of groups, each of them acting on some tree X_i. Then one still can express the KK-groups of the crossed product by Γ in terms of the the KK-groups of the crossed products by the stabilizer subgroups. This time one gets a *spectral sequence* associated to the simplicial decomposition of the n-dimensional simplicial complex $X = X_1 \times \dots X_n$. This is due to the following obvious consequence of the theorem

$$\oplus_{i_1,\dots,i_n} C_0(\dots \times X^1_{i_1,+} \times \dots \times X^1_{i_p,+} \times \dots) \simeq_{KK^\Gamma} C_0(X^{n-p})$$

where the sum in the left hand side of the formula is taken over all direct products that have on exactly p positions the edges with the point at infinity added and just the edges on the other ones.

This remark brings us to the last topic, namely the generalization to groups acting on buildings due to Kasparov and Skandalis (1989). Their approach makes the analogy with the Lie groups and with the Connes Kasparov conjecture even more transparent. Regarding the building X as a non Hausdorff manifold of negative sectional curvature they construct the C^*-algebra $\mathscr{A}_X$, which is the right analogue of the C^*-algebra of continuous functions on X and under the additional assumption that the building is locally finite, the "Dirac" element $\alpha \in KK_n^{\Gamma}(\mathscr{A}_X, \mathbb{C})$ and the "dual Dirac" element $\beta \in KK_n^{\Gamma}(\mathbb{C}, \mathscr{A}_X)$. Moreover they prove the following theorem.

Theorem (Kasparov Skandalis).

$$1. \quad \alpha \otimes_{\mathbb{C}} \beta = 1_{\mathscr{A}_X}$$
$$2. \quad \beta \otimes_{\mathscr{A}_X} \alpha = \gamma^{\Gamma}$$

is an idempotent.

The idempotent γ^{Γ} has been previously constructed by Julg and Valette (1988). They also showed (1984) that it equals $1_{\mathbb{C}}$ in the tree case. As in the case of Lie groups, if $\gamma^{\Gamma} = 1_{\mathbb{C}}$, we know the K-theory for Γ. In this case the crossed products by Γ are expressed in terms of a spectral sequence of the crossed products by the stabilizer subgroups. In general only the γ-*part* is known. This brings the study of the K-theory of groups acting on buildings at the same level as that of connected Lie groups. On one hand there is the γ-obstruction, on the other hand we know the positive answer to Novikov's conjecture for such groups (Kasparov 1988, Kasparov and Skandalis 1989).

However the one dimensional case, that is the case of a tree, is special. For first of all one does not need the locally finiteness of the tree, and second in this case $\gamma = 1$ for geometric reasons.

References

Anderson, J., Paschke, W. (1986): The K-theory of the reduced C^*-algebra of an *HNN*-extension. J. Operator Theory **16** (1986) 165–187

Atiyah, M.F. (1967): K-theory. Benjamin, New York Amsterdam

Atiyah, M.F. (1970): Global theory of elliptic operators. In: Proc. Intern. Conf. on Functional Analysis and Related Topics, Univ. of Tokyo Press, Tokyo, pp. 21–30

Atyiah, M.F., Singer, I.M. (1968): The index of elliptic operators III. Ann. Math. **87**, 546–604

Baum, P., Douglas, R.G. (1982): K-homology and index theory. In: Operator Algebras and Applications, Proc. Symp. Pure Math. Amer. Math. Soc. **38**, part 1, 117–173

Brown, L.G., Douglas, R.G., Fillmore, P.A. (1977) : Extensions of C^*-algebras and K-homology. Ann. Math. **105**, no.2, 265–324

Connes, A. (1982): A Survey of foliations and operator algebras. In: Operator Algebras and Applications, Proc. Symp. Pure Math. Amer. Math. Soc. **38**, part 1, 521–628

Connes, A., Higson, N. (1990): Déformations, morphismes asymptotiques et K-théorie bivariante. Preprint

Connes, A., Skandalis, G. (1984): The longitudinal index theorem for foliations. Publ. Res. Inst. Math. Sci. Kyoto Univ. **20**, 1139–1183

Cuntz, J. (1983): K-theoretic amenability for discrete groups. J. Reine Angew. Math. **344**, 180–195

Elliott, G.A. (1976): On the classification of inductive limits of sequences of semisimple finite dimensional algebras. J. Algebra **38**, 29–44

Julg, P., Valette, A. (1984): K-amenability for $SL_2(\mathbb{Q}_p)$ and the action on the associated tree. J. Funct. Anal. **58**, 194–215

Julg, P., Valette, A. (1988): Fredholm modules associated to Bruhat-Tits buildings. Preprint

Julg, P., Kasparov, G.G. (1990): L'anneau $KK_G(\mathbb{C}, \mathbb{C})$ pour $G = SU(n, 1)$. Preprint

Karoubi, M. (1978): K-theory. An Introduction. Springer, Berlin Heidelberg New York

Kasparov, G.G. (1973): The generalized index of elliptic operators. Funkts. Anal. Prilozhen. **7**, no. 3, 82–83 (in Russian)

Kasparov, G.G. (1975): Topological invariants of elliptic operators. In: K-homology. Izv. Akad. Nauk SSSR, Ser. Mat. **39**, 796–838

Kasparov, G.G. (1980): The operator K-functor and extensions of C^*-algebras. Izv. Akad. Nauk SSSR, Ser. Mat. **44**, 571–636

Kasparov, G.G. (1984): Lorentz groups: K-theory of unitary representations and crossed products. Dokl. Akad. Nauk USSR, **275**, 541–545 (in Russian)

Kasparov, G.G. (1988): Equivariant KK-theory and the Novikov conjecture. Invent. math. **91**, 147–201

Kasparov, G.G., Skandalis, G. (1989): Groups acting on buildings, operator K-theory, and Novikov's conjecture. Preprint

Lance, E.C. (1983): K-theory for certain group C^*-algebras. Acta. Math. **151**, 209–230

Mishchenko, A.S. (1973): Infinite dimensional representations of discrete groups and homotopic invariants of nonsimply connected manifolds. Usp. Mat. Nauk **28**, no. 2, 239–240 (in Russian)

Natsume, T. (1985): On $K_*(C^*(SL_2(\mathbb{Z})))$. J. Operator Theory **13**, 103–118

Pedersen, G.K. (1979): C^*-algebras and their automorphism groups. Academic Press, New York London

Pimsner, M.V. (1986): KK-groups of crossed products by groups acting on trees. Invent. math. **86**, 603–634

Pimsner, M.V. (1987): Cocycles on trees. J. Operator Theory **17**, 121–128

Pimsner, M.V., Popa, S.T. (1978): The Ext-groups of some C^*-algebras considered by J. Cuntz. Rev. Roum. de Math. Pures et Appl., Tome XXII **7**, 1069–1076

Pimsner, M.V., Popa, S.T., Voiculescu, D.V. (1979): Homogeneous C^*-extensions of $C(X) \otimes K(H)$, part 1. J. Operator Theory **1**, no. 1, 55–108

Pimsner, M.V., Popa, S.T., Voiculescu, D.V. (1980): Homogeneous C^*-extensions of $C(X) \otimes K(H)$, part 2. J. Operator Theory **4**, no. 2, 211–249

Pimsner, M.V., Voiculescu, D.V. (1980): Exact sequences for K-groups and Ext-groups of certain cross-product C^*-algebras. J. Operator Theory **4**, no. 1, 93–118

Pimsner, M.V., Voiculescu, D.V. (1982): K-groups of reduced crossed products by free groups. J. Operator Theory **8**, no. 1, 131–156

Powers, R.T. (1975): Simplicity of the C^*-algebra associated with the free group on two generators. Duke Math. J. **42**, 151–156

Rieffel, M.A. (1974): Induced representations of C^*-algebras. Adv. Math. **13**, 176–257

Serre, J.P. (1977): Arbres, amalgames, SL_2. Astérisque **46**, (Soc. Math. France)

Skandalis, G. (1988): Une Notion de nucléarité en K-théorie. K-Theory **1**, 549–573

Tits, J. (1979): Reductive groups over local fields. Proc. Symp. Pure Math. **33**, 29–69

Subfactors and Classification in von Neumann Algebras

Sorin Teodor Popa

University of California, Los Angeles, CA 90024, USA

0. Introduction

While initially introduced to study quantum mechanics and representations of groups, in recent years von Neumann algebras started to play a major role in many areas of mathematics. The class of von Neumann algebras that proved to be more important and of most physical significance are the ones that can be approximated by finite dimensional algebras, called hyperfinite. By 1985, as a result of 45 years of work initiated by Murray and von Neumann and culminating with the work of Connes and Connes-Haagerup, hyperfinite algebras were completely classified. The fundamental steps in accomplishing this proved to be the uniqueness of the hyperfinite type II_∞ factor (the hyperfinite factor with an infinite trace) and the classification of its automorphisms, both settled by Connes. Starting with his work on automorphisms, the problem of classifying actions of more general groups on the hyperfinite factors became an important trend of research in von Neumann algebras ([J1, Oc1, JT]).

In 1983, motivated by his study of actions of finite groups on hyperfinite von Neumann algebras, Jones initiated a Galois theory for von Neumann algebras by studying pairs of factors $N \subset M$ with finite index $[M : N] < \infty$, i.e. with M a finite N module. He proved the striking result that if the index of $N \subset M$ is less than 4, then it has to be equal to the square norm of a matrix with nonnegative integer entries, and must be of the form $4\cos^2\frac{\pi}{n}$ for some $n \geq 3$. Jones' results and ideas gave a new insight into the theory of operator algebras. It also brought together many other subjects and had an unexpected and far reaching impact in a number of fields of research such as statistical mechanics, quantum field theory, knot theory. Methods and results in either of these subjects proved to be inspiring for the others. Most of the intrinsic problems from the theory of subfactors, such as their classification, the construction of examples, the characterization of the values the index may take, seem to have physical significance. We will describe in this article the solution to some of these problems. We will present a classification result for a certain class of subfactors, called strongly amenable subfactors, a class that contains all subfactors of index ≤ 4. We will show that the index of an irreducible subfactor $N \subset M$ of the hyperfinite factor must always be the square norm of a (possibly infinite) matrix of nonnegative integers. In particular this

Proceedings of the International Congress
of Mathematicians, Kyoto, Japan, 1990

shows the existence of a gap in the set of indices, from 4 to 4.026.... Finally we will show that in striking contrast with the hyperfinite case, in the nonhyperfininte case any value > 4 may appear as the index of an irreducible subfactor. We will also explain how the theory of subfactors provides the natural framework and the necessary techniques for a unified approach to both subfactors problems and the classical operator algebra problems mentioned before: classification of hyperfinite von Neumann algebras and of their automorphisms.

1. Index for Subfactors

Let M be a type II_1 factor with a normalized trace τ and $N \subset M$ a subfactor. The Jones' index of N in M, denoted by $[M : N]$, is defined as $\dim_N L^2(M)$, the Murray-von Neumann coupling constant of N in its representation on $L^2(M)$, $L^2(M)$ being the completion of M in the Hilbert norm $\|x\|_2 = \tau(x^*x)^{1/2}$, $x \in M$. The definition of $[M : N]$ is in fact independent of the Hilbert space on which $N \subset M$ act, as $[M : N] = \dim_N \mathscr{H}/\dim_M \mathscr{H}$ for any M-Hilbert module $\mathscr{H}$. But, surprisingly enough, although each of $\dim_N \mathscr{H}$, $\dim_M \mathscr{H}$ may take any value from 0 to ∞ (a remarkable fact in itself!), Jones proved in [J2] that $[M : N]$ may only take the values $\{4\cos^2 \frac{\pi}{n} \mid n \geq 3\} \cup [4, \infty)$. The key idea in his proof is the so called basic construction. It later proved to be the fundamental construction of the theory. This construction associates to $N \subset M$ a new pair of factors $M \subset M_1$, with the same index $[M_1 : M] = [M : N]$, and with a projection $e_1 \in M_1$ that generates M_1 together with M and implements by compression the trace preserving conditional expectation of M onto N. By iteration one can thus produce a whole "tower" of factors $N \subset M \subset M_1 \subset \ldots$ together with a sequence of projections $e_i \in M_i$, $i \geq 1$, satisfying the remarkable axioms:

1.1.1 $[e_i, e_j] = 0$, $|i - j| \geq 2$.
1.1.2 $e_i e_{i\pm1} e_i = [M : N]^{-1} e_i, i \geq 1$.
1.1.3 $\tau(w e_{n+1}) = [M : N]^{-1}\tau(w), w \in \mathrm{Alg}\{1, e_1, \ldots, e_n\}$.

Conditions 1.1.1–1.1.3 is what forces $s = [M : N]$ to be in the set $\{4\cos^2 \frac{\pi}{n} \mid n \geq 3\}$, if less than 4. Moreover, for each $s \in \{4\cos^2 \frac{\pi}{n} \mid n \geq 3\} \cup [4, \infty)$ Jones proved the existence of such a sequence of projections with a trace satisfying 1.1.1–1.1.3. Then $R = \overline{\mathrm{Alg}\{e_i\}_{i\geq1}}^w$ is isomorphic to the hyperfinite type II_1 factor and $R^s = \overline{\mathrm{Alg}\{e_i\}_{i\geq2}}$ is a subfactor of R of index $[R : R^s] = s$. So one has:

1.2 Theorem [J2]. a) $[M : N] \in \{4\cos^2 \frac{\pi}{n} \mid n \geq 3\} \cup [4, \infty)$.
b) *Given any* $s \in \{4\cos^2 \frac{\pi}{n} \mid n \geq 3\} \cup [4, \infty)$ *there exists a pair of hyperfinite type* II_1 *factors* $R^s \subset R$ *with index* s.

Beside the above Jones' subfactors, we mention few other important classes of examples:

1.3.1 If $M = M_{n\times n}(N)$ is the algebra of n by n matrices over N, then $[M : N] = n^2$.

1.3.2 If σ is a properly outer action of a finite group G on a type II_1 factor N then $N \subset M = N \rtimes_\sigma G$ satisfies $[M : N] = |G|$. If μ is another action of G on N then $N \subset N \rtimes_\sigma G$ is isomorphic to $N \rtimes_\mu G$ if and only if σ is cocycle conjugate to μ. The classification of the actions σ is thus equivalent to the classification of the corresponding pairs of algebras, i.e. subfactors.

1.3.3 Let G be a finitely generated discrete group and fix $g_1, \ldots, g_n$, some generators. Let σ be an outer action of G on a type II_1 factor P or, more generally, an injective morphism of G into $\operatorname{Aut} P/\operatorname{Int} P$. Let $M = M_{(n+1)\times(n+1)}(P)$ and $N_{G,\sigma} = \{x \oplus (\oplus_i \sigma(g_i)(x)) \mid x \in P\} \subset M$ be the "diagonal" subfactor (in general $\sigma(g_i)$ are lifting automorphisms). Then $[M : N_{G,\sigma}] = (n+1)^2$. The isomorphism class of $N_{G,\sigma} \subset M$ doesn't depend on unitary perturbations of σ. In fact, $N_{G,\sigma} \subset M$ is isomorphic to $N_{G,\mu} \subset M$ iff σ is cocycle conjugate to μ, i.e. σ and μ are conjugate in $\operatorname{Aut} P/\operatorname{Int} P$ (for all this see [Po4]).

1.3.4 The Jones subfactors $R^s \subset R$ in 1.2 come from certain positive Markov traces on the infinite Hecke algebras $H_\infty(q)$ and are related to the Jones' polynomial invariant for knots. The general form of such a type of subfactor, obtained by investigating all possible positive Markov traces on $H_\infty(q)$ (found by Ocneanu, see [J3]), were constructed by Wenzl [We1], who computed their indices and proved that they are irreducible when q and the parameters describing the trace correspond to roots of unity.

1.3.5 Let G be a compact group acting minimally on a type II_1 factor M (i.e. such that the fixed point algebra M^G is irreducible in M). Let π be a unitary representation of G on a finite dimensional space V. Then $M^G \subset (M \otimes \operatorname{End} V)^G$ is an inclusion of type II_1 factors of index $(\dim V)^2$ which is irreducible if π is. Its basic construction is obtained by tensoring recursively with $\operatorname{End} V$ and by taking fixed point algebras. This example of subfactors is due to Wassermann [Wa].

1.3.6 Let $\{A_n \subset B_n\}_{n \geq 1}$ be an increasing sequence of inclusions of finite dimensional algebras, i.e. $A_n \subset A_{n+1}$, $B_n \subset B_{n+1}$, with a common trace τ. Assume the consecutive steps of these inclusions satisfy the so-called *commuting square condition*: $E_{A_{n+1}} E_{B_n} = E_{A_n}, n \geq 1$. Then, under suitable conditions on the trace, one has that $N = \overline{\cup A_n} \subset \overline{\cup B_n} = M$ are factors and $[M : N] = \lim [B_n : A_n]$, with $[B_n : A_n]$ appropriately defined ([PiPo1, PiPo2]). Moreover if T_n denotes the inclusion matrix for $A_n \subset B_n$, and τ is extremal in some sense then $[M : N] = \lim \|T_n\|^2$ ([PiPo2]).

2. Classification of Amenable Subfactors

The problem of classifying the subfactors of the hyperfinite type II_1 factor is of most importance for the theory of subfactors. Moreover, the example 1.3.3 shows that a classical problem such as the classification of actions of groups on factors can be translated into a classification problem for subfactors.

The classification of $N \subset M$ we are interested in is up to conjugacy of N by automorphisms of M. We will present now a result that classifies an important class of subfactors in terms of certain canonically associated combinatorial data.

2.1 Definition of the Invariant. Let $\{M_1' \cap M_k \subset M' \cap M_k\}_k$ be the sequence of inclusion of finite dimensional algebras of the higher relative commutants in the Jones' tower of factors. Consecutive steps of these inclusions satisfy the commuting square condition 1.3.6. The isomorphism class of this sequence depends only on the isomorphism class of $N \subset M$, so it is an invariant for $N \subset M$. We call it the *sequence of canonical (or standard) commuting squares* associated to $N \subset M$. The finite dimensional inclusions $M' \cap M_k \subset M' \cap M_{k+1}$, $k \geq 1$, are all described by a unique irreducible matrix of nonnegative integers A, called the *standard matrix* of $N \subset M$, and respectively by its transpose (for k even resp. odd), see [GHJ, Oc2, We1, Po2].

The standard matrix A can alternatively be regarded as a diagram or a graph. As first noted by Jones, for small values of the index (< 4) such a graph is necessarily a Coxeter graph of type A_n, D_n, E_6, E_7, E_8. Although the matrices (or graphs) A capture a great deal of information on the canonical sequence, in general it doesn't uniquely determine it. For the class of subfactors for which A is finite, a condition introduced by Ocneanu [Oc2] and that he calls *finite depth*, the canonical sequence of commuting squares is uniquely determined by just one of the commuting squares, involving two consecutive steps of the sequence

$$\begin{array}{ccc} M_1' \cap M_{k+1} & \subset & M' \cap M_{k+1} \\ \cup & & \cup \\ M_1' \cap M_k & \subset & M' \cap M_k \end{array}$$

with k large enough. Moreover k, the isomorphism class of the algebras, the trace and the inclusion matrices are all determined by the standard matrix of $N \subset M$ and $M \subset M_1$ ([Po2, Oc2]). An axiomatisation of such canonical commuting squares, in the finite depth case, describing them as paragroup type objects, was obtained in [Oc2]. In the case of the examples $N_{G,\sigma} \subset M$ in 1.3.3 with σ an action of the group G, the matrix A is just the Cayley matrix of G and it does describe completely the canonical invariant of the inclusion ([Po4]). In particular, if G is infinite, then A is infinite and so $N \subset M$ has infinite depth.

We mention that the basic construction can also be performed in a reverse way, this way obtaining a decreasing "tunnel" of subfactors $M \supset N \supset N_1 \supset \cdots$. The existence of each subfactor N_{k+1} in this tunnel is however unique only up to conjugacy by a unitary in N_k ([PiPo1]). But the isomorphism class of the sequence of finite dimensional inclusions $\{N_k' \cap N \subset N_k' \cap M\}_k$ as well as their resulting closures $R_0 = \overline{\cup(N_k' \cap N)} \subset \overline{\cup(N_k' \cap M)} = R$ depends only on the class of $N \subset M$ and is an obvious candidate for a model for $N \subset M$ ([Po2]).

Classifying $N \subset M$ by their canonical invariant amounts to proving the existence of a choice of a tunnel $M \supset N \supset N_1 \supset \cdots$ for which the higher relative commutants $N_k' \cap M$ generate M. We prove such an exhaustion result for subfactors for which the higher relative commutants satisfy a certain "amenable growth" condition.

2.2 Definition [Po3]. $N \subset M$ is *strongly amenable* if the k'th higher relative commutant for $R_0 \subset R$ is isomorphic to the k'th higher relative commutant for $N \subset M$, for each $k \geq 1$. This condition is also equivalent to the similar one obtained by considering the inclusion $M_1' \cap M_\infty \subset M' \cap M_\infty$ instead of $R_0 \subset R$, where $M_\infty = \overline{\cup M_k}$. We mention that in fact $\dim R_0' \cap R = \dim N' \cap M$ (or equivalently $\dim (M_1' \cap M_\infty)' \cap M' \cap M_\infty = \dim N' \cap M$) is sufficient for the strong amenability condition to hold true.

2.3 Theorem [Po3]. *Assume $N \subset M$ are hyperfinite type* II_1 *factors with finite index. The following conditions are equivalent:*
2.3.1 *$N \subset M$ is strongly amenable.*
2.3.2 *$N \subset M$ is isomorphic to $R_0 \subset R$.*
2.3.3 *The bicommutant of M in $M_\infty = \overline{\cup M_k}$ is equal to M.*
2.3.4 *$H(M \mid N) = H(R \mid R_0)$, where, for $A \subset B$, $H(B \mid A)$ denotes the Connes-Størmer relative entropy* [CS] *as used in* [PiPo1].
Moreover, the isomorphism class of a strongly amenable inclusion is completely determined by its canonical commuting squares.

Condition 2.3.4 in the above theorem can be interpreted as a Shanon-McMillan-Breiman type condition on the random walk with transition matrix $A^t A$ (A being the standard matrix for $N \subset M$) and transition probabilities proportional to the local indices of $N \subset M_k$: this random walk must have the largest possible entropy. The bicommutant property 2.3.3 is similar to the one stated for subfactors with finite depth and trivial relative commutant in [Oc2] and first pointed out in the case $[M : N] < 4$ by Skau [GHJ].

The proof of the theorem uses much of the subfactor techniques developed in [PiPo1], especially the probabilistic characterization of the index $[M : N]$ as the best constant s for which the inequality $E_N(x) \geq s^{-1}x$ holds true for all $x \in M_+$, E_N being the trace preserving conditional expectation of M onto N. The proof also uses in a crucial way the noncommutative local Rohlin theorem in [Po1].

The finite depth condition $\dim A < \infty$ (equivalently $\sup \dim \mathcal{Z}(M' \cap M_k) < \infty$, $\mathcal{Z}(B)$ being the center of B) is easily seen to imply the strong amenability condition. The particular case of Theorem 2.3 showing that subfactors with finite depth are classified by their canonical commuting squares was already settled in [Po2, Oc2].

All subfactors of index < 4 have finite depth. They are thus classified by their canonical invariants. A full list of the combinatorial objects arising this way from subfactors of index < 4 has been given in [Oc2] (for the case E_6 see also [BN]).

Condition 2.3.4 implies that all subfactors of index 4 are strongly amenable, although some do not have finite depth. A full list of the combinatorial objects that classify them, and thus of all subfactors of index 4 is given in [Po3].

2.4 Corollary. *Subfactors of index ≤ 4 are all strongly amenable and are thus completely classified by their canonical commuting squares. A full list of these subfactors can be given.*

Another important class of subfactors to which 2.3 applies are the Wenzl subfactors 1.3.4, which have finite depth and are thus strongly amenable, and the Wassermann subfactors 1.3.5, which are strongly amenable but have infinite depth. All these subfactors are thus uniquely determined by their combinatorial data.

3. Other Applications

Due to example 1.3.3 the classification Theorem 2.3 implies the classification, up to cocycle conjugacy, of actions σ of finitely generated discrete groups G on the hyperfinite factor, in the case when the associated subfactors $N_{G,\sigma} \subset M$ result strongly amenable. In fact, 2.3 translates into :

3.1 Theorem [Po4]. *Let P be the hyperfinite type II_1 factor and σ an outer action of a finitely generated discrete group G on P. With the notations of 1.3.3 the following are equivalent.*

3.1.1 *$N_{G,\sigma} \subset M$ is strongly amenable.*

3.1.2 *If g_i are some generators of G and $a = \frac{1}{2n+1}(\Sigma\, g_i + e + \Sigma\, g_i^{-1})$ then $\|g * a^n - a^n\|_1 \to 0$ for all $g \in G$.*

3.1.3 *The group G has 0 entropy, in the sense of* [KV].

Moreover, the class of $N_{G,\sigma} \subset M$ depends only on G and not on the cocycle conjugacy class of σ. Thus, there is a unique action, up to cocycle conjugacy, of the group G on the hyperfinite type II_1 factor.

Note that the groups G satisfying 3.1.2 or 3.1.3 are automatically amenable. The above theorem thus implies a particular case from Ocneanu's theorem on the uniqueness of actions of amenable groups (in fact, with a slight modification, for II_∞ factors as well). The class of groups with 0 entropy contains all groups with subexponential growth and thus most classes of important amenable goups. In particular 3.1 (and thus the general Theorem 2.3) does cover Connes' results on the classification of single automorphisms, which is essential for the classification of hyperfinite von Neumann algebras.

The relation between the ergodic properties of the inclusion of $N_G \subset M$ and of the random walk on G are investigated in [B].

Another application of 2.3 is the classification of minimal actions of compact groups.

3.2 Theorem [PoWa]. *Given any compact Lie group G there exists a unique outer minimal action of G on the hyperfinite factor.*

To prove this result one shows that, up to conjugacy, a minimal action of G is uniquely determined by the isomorphism class of the inclusion of factors $M^G \subset (M \otimes \mathrm{End}\, V)^G$ of 1.3.5, with π a finite dimensional representation of G containing a generating set of irreducible representations. This inclusion follows strongly amenable by the Wassermann's invariance principle ([Wa]) and so 2.3 applies.

The subfactor techniques can also be used to give a simple elementary proof to an old standing problem on cocycle actions:

3.3 Theorem [Po4]. *Let G be a discrete group with 0 entropy (e.g. with subexponential growth). Then any 2-cocyle action of G on an arbitrary type* II *factor can be perturbed to a genuine action.*

We mention that one can prove Theorem 2.3 by using only the injectivity property of the ambient factor M (i.e. for M having a certain invariant mean, called hypertrace). If one takes $N \subset M = M_{2\times 2}(N)$ then such a general form of 2.3 implies the hyperfiniteness of the injective factor M and thus the uniqueness of the hyperfinite type II_∞ factor, the other major result of Connes that led to the classification of hyperfinite algebras.

One can furthermore obtain an equivariant version of Theorem 2.3, involving an automorphism θ acting on M and leaving N globally invariant. The result shows that, aside from a commonly splitted part, θ acts only on the combinatorial part of $N \subset M$, resulting from the higher relative commutants. Due to the work in ([L]) this implies the classification of strongly amenable subfactors of type III_λ, $0 < \lambda < 1$, in terms of their combinatorial data.

4. Gaps for the Index of Hyperfinite Subfactors

Theorem 2.3 shows that not all hyperfinite type II_1 subfactors can be approximated (and thus classified) by finite dimensional subalgebras coming from higher relative commutants. It is in fact likely that not all hyperfinite type II_1 subfactors can be obtained by approximation with finite dimensional commuting squares like in example 1.3.6 (see 0.3 in [Po4]). For such subfactors though, one may hope that a classification can still be completed by establishing necessary and sufficient conditions for two increasing sequences of finite dimensional commuting squares (1.3.6) to give isomorphic subfactors, and then by classifying such sequences.

Approximation by finite dimensional commuting squares is also important for investigating the set of indices of irreducible subfactors, as subfactors with such approximation properties have index equal to square norms of matrices (cf. 1.3.6).

There is a rather general class of hyperfinite subfactors for which such approximations can be obtained. The key ingredient is the existence of certain special hypertraces ϕ for $N \subset M$, called *symmetric hypertraces*, for which both M, M' and the expectation of M onto N, e_N, are contained in the centralizer of ϕ. To prove existence of such hypertraces one needs $N \subset M$ to have certain ergodicity properties, much weaker though than the Shannon-Mc Millan-Breiman condition in 2.3.4.

4.1 Definition. $N \subset M$ is called *weakly amenable* if the random walk associated to it is ergodic, i.e. if the higher relative commutants generate a factor.

Note that all subfactors of sufficiently small indices ($< 2+\sqrt{5}$ will do) follow automatically weakly amenable.

4.2 Theorem [Po6]. *If $N \subset M$ is a weakly amenable inclusion of hyperfinite type* II_1 *factors and* $N' \cap M = \mathbb{C}$, *then* $N \subset M$ *has a symmetric hypertrace.*

Using the techniques in ([Po7]) one then proves:

4.3 Theorem [Po6]. *If $N \subset M$ is a weakly amenable inclusion of hyperfinite type* II_1 *factors with* $N' \cap M = \mathbb{C}$, *then* $N \subset M$ *can be locally approximated by finite dimensional commuting squares.*

By [PiPo2] (see 1.3.6) and taking into account that subfactors for which $[M : N] < 2+\sqrt{5}$ are all weakly amenable and that any number $\geq 2+\sqrt{5}$ is the square norm of an integral matrix ([GHJ]) one gets:

4.4 Corollary [Po6]. *If $N \subset M$ are hyperfinite* II_1 *factors with* $N' \cap M = \mathbb{C}$, *then* $[M : N]$ *is equal to the square of the norm of a (possibly infinite) matrix with nonnegative integer entries.*

The set of square norms of (possibly infinite) matrices with nonnegative integer entries is known (see e.g. [GHJ]): besides containing the half line $[2+\sqrt{5}, \infty]$, it has only countable accumulation points below $2+\sqrt{5}$, which converge to $2+\sqrt{5}$. So it has plenty of gaps below $2+\sqrt{5}$. The first gap arises between 4 and the square norm of the matrix corresponding to the Coxeter graph E_{10}, $\|E_{10}\|^2 = 4.026\ldots$, as this is the first matrix with square norm larger than 4. We mention that the existence of irreducible hyperfinite subfactors of such index was proved in [HOcS].

4.5 Corollary. *The set of indices of irreducible subfactors of the hyperfinite type* II_1 *factor has a gap between* 4 *and* 4.026.

In order to characterize completely the set of indices of irreducible subfactors of the hyperfinite II_1 factor one should be able to construct examples as well. Some examples were obtained by constructing commuting squares of finite dimensional algebras, like in 1.3.6, in [GHJ, We1, We2, HOcS, Ch, Su].

We mention that the inclusion matrices for the finite dimensional approximations of the weakly amenable hyperfinite type II_1 factors $N \subset M$ can be interpreted more "canonically", by introducing the *universal graph* (*or matrix*) of $N \subset M$, a general invariant that we will not explain here.

5. The Nonhyperfinite Case

Quite surprisingly, unlike the hyperfinite case when, as we have seen, the situation is quite rigid and the universal matrix of the subfactor is forced to have square norm equal to the index, for general irreducible pairs of type II_1 factors any index $s > 4$ may occur ([Po5]).

We will briefly discuss here a general method for constructing subfactors of finite index, which is quite different from the ones discussed before in 1.3 (e.g.

by producing finite dimensional commuting squares). The method consists in constructing Markov traces on certain universal algebras associated to an algebra Q and to the Jones' sequence of projections $\{e_i\}_i$, $\tau(e_i) = s^{-1}$. In particular, for certain extremal such traces, this produces one parameter families of irreducible inclusions of factors $N^s \subset M^s$ of index $[M^s : N^s] = s$, s ranging over the whole set $\{4\cos^2 \frac{\pi}{n} \mid n \geq 3\} \cup [4, \infty)$. For these extremal traces the resulting factors M^s are always non Γ (in the sense of [MvN]) and thus nonhyperfinite. The subfactors $N^s \subset M^s$ are not only irreducible, but also their higher relative commutants $(N^s)' \cap M_k^s$ are in some sense minimal, as they are generated by just $e_1, e_2, \cdots, e_k$.

5.1 Theorem [Po5]. *Given any $s > 4$ there are irreducible inclusions of nonhyperfinite factors of index s.*

The construction of $N^s \subset M^s$ can be briefly described as follows: Let Q be an algebra. Consider the universal algebra U^s generated by Q and by the Jones' projections $\{e_i\}_i$ of trace $\tau(e_i) = s^{-1}$, subject to the commutation relation $[Q, e_i] = 0$, for $i \geq 2$. Define on U^s a trace by letting $tr(w) = 0$ for all words with alternating letters $x_i \in Q$, $y_j \in \mathrm{Alg}\{1, e_1, e_2, \cdots\}$, for which $\tau(x_i) = 0$ and for which the projection of y_j onto $\mathrm{Alg}\{1, e_2, \cdots\}$ is 0 ([V]). Then M^s is defined as the smallest algebra of U^s/tr containing Q and on which e_1 implements by compression a conditional expectation and N^s is defined as the commutant of e_1 in M^s. The fact that $[M^s : N^s] = s$ results from a certain remarkable property of the above trace, called the Markov property.

In fact, any pair of factors arises this way, from some Q and some appropriate Markov trace on $U^s = U^s(Q)$. The free Markov trace defined before correspond to an extremal situation when no "information" is exchanged from Q to $\mathrm{Alg}\{e_1, e_2, \cdots\}$ through the "window" e_1. How to practically construct some others, especially hyperfinite ones, irreducible and with a prescribed index, is an open problem. This however may turn out to be a better method for constructing hyperfinte subfactors than the construction of finite dimensional commuting squares.

References

[B] D. Bisch: Entropy of groups and subfactors. To appear in Journal of Functional Analysis

[BN] J. Bion-Nadal: Subfactors associated to E_6. CNRS, Preprint 1989

[C] A. Connes: Classification des facteurs. Proc. Symp. Pure Math. **38** (1982) 43–109

[Ch] M. Choda: Index for factors generated by Jones' two sided sequence of projections. Pac. J. Math. **139** (1989) 1–16

[CS] A. Connes, E. Størmer: Entropy for automorphisms of II_1 von Neumann algebras. Acta Math. **134** (1975) 288–259

[GHJ] F. Goodman, P. de la Harpe, V.F.R. Jones: Coxeter graphs and tower of algebras. MSRI Publications 14. Springer 1989

[HOcS] U. Haagerup, A. Ocneanu, J. Schou: In preparation

[J1] V.F.R. Jones: Actions of finite groups on the hyperfinite II_1 factor. Mem. A.M.S. **28**, no. 237 (1980)

[J2] V.F.R. Jones: Index for subfactors. Invent. math. **72** (1983) 1–25

[J3] V.F.R. Jones: Hecke algebra representations of braid groups and link polynomials. Ann. Math. **126** (1987) 335–388

[JT] V.F.R. Jones, M. Takesaki: Actions of compact abelian groups on semifinite injective factors. Acta math. **153** (1984) 213–258

[KV] V. Kaimanovich, A. Vershik: Random walks on discrete groups: boundary and entropy. Ann. Probab. **11** (1983) 457–490

[L] P. Loi: On automorphisms of subfactors. UCLA, Preprint 1990

[MvN] F. Murray, J. von Neumann: Rings of operators IV. Ann. Math. **44** (1943) 716–808

[Oc1] A. Ocneanu: Actions of discrete amenable groups on von Neumann algebras. (Lecture Notes in Mathematics, vol. 1138.) Springer, Berlin Heidelberg New York 1985

[Oc2] A. Ocneanu: Quantized groups string algebras and Galois theory for algebras. In: Operator Algebras and Applications, 2. London Math. Soc. Lect. Notes Series **136** (1988) 119–172

[PiPo1] M. Pimsner, S. Popa: Entropy and index for subfactors. Ann. Sci. Ec. Norm. Sup. **19** (1986) 57–106

[PiPo2] M. Pimsner, S. Popa: Finite dimensional approximation of pairs of algebras and obstructions for the index. To appear in J. Funct. Anal.

[Po1] S. Popa: On a problem of R.V. Kadison on maximal abelian $*$-subalgebras in factors. Invent. math. **65** (1981) 269–281

[Po2] S. Popa: Classification of subfactors: reduction to commuting squares. Invent. math. **101** (1990) 19–43

[Po3] S. Popa: Sur la classification des sous-facteurs d'indice fini du facteur hyperfini. C. R. Acad. Sci. Paris, Série I **311** (1990) 95–100

[Po4] S. Popa: Sousfacteurs, actions des groupes et cohomologie. C. R. Acad. Sci. Paris, Série I **309** (1989) 771–776

[Po5] S. Popa: Markov traces on universal Jones algebras and subfactors of finite index. IHES Preprint 1990

[Po6] S. Popa: In preparation

[Po7] S. Popa: On amenability in type II_1 factors. In: Operator Algebras and Applications 2. London Math. Soc. Lect. Notes Series **136** (1988) 173–183

[PoWa] S. Popa, A. Wassermann: In preparation

[Su] S. Sunder: From hypergroups to subfactors. Preprint 1989

[V] D. Voiculescu: Circular and semicircular systems and free product factors. UCB Preprint 1989

[Wa] A. Wassermann: Coactions and Yang-Baxter equations for ergodic actions and subfactors. In: Operator Algebras and Applications 2. London Math. Soc. Lect. Notes Series **136** (1988) 202–236

[We1] H. Wenzl: Hecke algebras of type A_n and subfactors. Invent. math. **92** (1988) 349–383

[We2] H. Wenzl: Quantum groups and subfactors of type B, C and D. Commun. Math. Phys. **133** (1990) 383–432

Operator Algebras and Duality

Georges Skandalis

Report on joint work with *Saad Baaj*

Université Paris 7, UFR de Math., URA 212, Tour 45-55, 2, place Jussieu
F-75251 Paris Cedex 05, France

The Pontrjagyn duality associates to an abelian locally compact group a dual group and studies the properties of this correspondence. A natural idea is to try and generalize this duality to nonabelian groups, in particular to define an object dual to a group. Such dual objects were defined, first for compact groups [45, 19] then for locally compact groups ([38], [46]). Up to a large extent, it was the search of such a dual object for general locally compact groups that led to the theory of C^*-algebras. Then appeared the need of objects generalizing groups as well as their dual objects. These general objects can be called in a modern language "quantum groups". These "groups" can be studied as abstract groups, Lie groups, deformations of true groups .... It is certainly beyond our goals to review all aspects of the theory of "quantum groups" (see [3] and references therein). We will in fact concentrate our attention to the operator algebra approach, in other words to the study of the "locally compact quantum groups".

In terms of operator algebras Pontrjaygn duality takes the form of Takesaki-Takai duality [43, 40] based on the construction of W^*- and C^*-crossed-products. Along the years, under conjugated efforts of many specialists a set of axioms was built [11, 12, 42, 47, 16, 5] and duality was obtained [43, 20, 21, 30, 39, 4, 10, 6] for von Neumann algebras obeying these axioms called Kac von Neumann algebras. In a recent fundamental work [51–54], Woronowicz defined some objects that he called "compact matrix pseudogroups". Although they aren't Kac algebras, Woronowicz' "pseudo-groups" enjoy duality properties. Further examples with the same properties were given by Majid [26] and Podles-Woronowicz [32]. One of the motivations of this report is to describe a setting including both the Kac von Neumann algebras and these new examples, in which the duality results still hold.

Let H be a Hilbert space and $V \in L(H \otimes H)$ a unitary operator. Let us say V is multiplicative if it satisfies the pentagon equation $V_{12}V_{13}V_{23} = V_{23}V_{12}$. This relation appears in the framework of categories with associative tensor product (cf. [24, 25]); it is the one satisfied by the fusion operator (cf. [29]). It is also very similar to the Yang-Baxter equation and in some sense more primitive. In many papers concerned with operator algebras possessing duality properties, a multiplicative unitary plays a fundamental role (e.g. [12, 42, 16, 5, 18, 10]...); it is clear and more or less explicit in these papers that this unitary describes the whole situation.

Proceedings of the International Congress
of Mathematicians, Kyoto, Japan, 1990

It is therefore very natural to look for additional conditions that a mutliplicative unitary should satisfy in order to correspond to a "locally compact quantum group". Studying this problem, we were led to consider two conditions that we call *regularity* and *irreducibility*. We show how two pairwise dual Hopf C^*-algebras can be associated to a regular multiplicative unitary. When moreover this unitary is irreducible, we establish Takesaki-Takai duality results generalizing the previous ones.

An advantage of our approach is that a "quantum group" and its dual play completely symmetric roles. Also, it treats simultaneously the C^*- and W^*-algebra point of view. In fact, in our approach it is clear that for "locally compact quantum groups" the measure theory determines the topology. In some sense this can be thought of as a generalization of the famous theorem of Weil ([50], see also [23]): a "measurable quantum group" with an invariant (class of) measure(s) carries a unique structure of "locally compact quantum group".

Let us also mention that many algebraic constructions can be performed in our setting. In particular, we may associate a "quantum double" to any (irreducible) multiplicative unitary, and together with it, comes a solution of the quantum Yang Baxter equation (cf. [3]).

The question of the minimality of our axioms remains still unanswered: is a multiplicative unitary automatically regular? irreducible? Does one of these properties imply the other? Partial solutions to these questions were obtained: when the Hilbert space H is finite dimensional and when the unitary V satisfies a commutativity condition, regularity and irreducibility are both automatic; if the unitary V is of compact or discrete type (in other words if the associated quantum group is compact or discrete) its regularity implies its irreducibility.

In this report, we will first present some examples of occurrence of multiplicative unitaries, then explain the conditions of regularity and irreducibility and their consequences; we will finally construct the multiplicative unitaries associated with the examples of [26, 32] and discuss possible future developments. All the proofs, as well as more precise statements of the results given here can be found in [2].

1. Multiplicative Unitaries and Hopf Algebras

Let H be a Hilbert space. We will say that a unitary operator V acting on the tensor square $H \otimes H$ is *multiplicative* if it satisfies the pentagon equation:

$$V_{23}V_{12} = V_{12}V_{13}V_{23}$$

Here, by V_{12}, V_{23} and V_{13} we denote the operators $V \otimes 1_H$, $1_H \otimes V$ and $(1_H \otimes \Sigma)(V \otimes 1_H)(1_H \otimes \Sigma)$ acting on $H \otimes H \otimes H$, where Σ is the "flip" operator defined by $\Sigma(\xi \otimes \eta) = \eta \otimes \xi$ $(\xi, \eta \in H)$.

Note that the identity operator $1_{H \otimes H}$ is a multiplicative unitary. The importance of multiplicative unitaries in connection with operator algebras possessing duality properties was shown by many authors [12, 42, 16, 5, 10]. The multiplicative unitary associated with a locally compact group is constructed as follows:

Let G be a locally compact group and let m denote its right Haar measure. Let $H = L^2(G; m)$ be the Hilbert space of square integrable functions on G with respect with the measure m. Identify $H \otimes H$ with the space $L^2(G \times G; m \times m)$. Let then V be the operator acting on $H \otimes H$ by the formula $V(f)(s, t) = f(st, t)$ for every square integrable function f on $G \times G$ and $s, t \in G$. This operator is clearly unitary and its "multiplicativity" follows from the associativity of the composition law of G.

Operators satisfying this pentagon equation are naturally associated with Hopf algebras. Recall that a Hopf C^*-Algebra is a C^*-algebra A endowed with a coproduct which is a $*$-homomorphism $\delta : A \to A \otimes A^{(1)}$ satisfying the associativity condition: $(\delta \otimes \mathrm{id}) \circ \delta = (\mathrm{id} \otimes \delta) \circ \delta$. Let us describe three different ways for interpreting the pentagon relation:

a) Haar States and GNS Representations

A Haar state on a Hopf C^*-algebra is a state $\phi \in A^*$ such that for any form $\psi \in A^*$ we have $(\phi \otimes \psi) \circ \delta = (\psi \otimes \phi) \circ \delta = \psi(1)\phi$. Let then $(H_\phi, \pi_\phi, \xi_\phi)$ be the GNS construction associated with ϕ. Then the operator V_ϕ defined by $V_\phi(\pi_\phi(x)\xi_\phi \otimes \eta) = (\pi_\phi \otimes \pi_\phi)(\delta(x))(\xi_\phi \otimes \eta)$ is an isometry of $H_\phi \otimes H_\phi$ satisfying the pentagon equation. In particular, if V_ϕ is surjective, it is a multiplicative unitary.

b) Covariant Representations

Let (A, δ) be a Hopf C^*-algebra. A corepresentation of A in a Hilbert space H is a unitary $u \in L(H \otimes A)$ of the Hilbert A-module $H \otimes A$ satisfying the relation: $(\mathrm{id} \otimes \delta)(u) = u_{12}u_{13}$.

A coaction of A on some C^*-algebra B is a $*$-homomorphism $\delta_B : B \to B \otimes A$ satisfying the associativity condition: $(\delta_B \otimes \mathrm{id}) \circ \delta_B = (\mathrm{id} \otimes \delta) \circ \delta_B$. A covariant representation of A, B on a Hilbert space H is a pair (π, u) where $\pi : B \to L(H)$ is a $*$-representation and $u \in L(H \otimes A)$ is a corepresentation of A such that $\forall b \in B$, $(\pi \otimes \mathrm{id}) \circ \delta_B(b) = u(\pi(b) \otimes 1)u^*$.

The coproduct δ is a coaction of A on itself. Let (π, u) be a covariant representation on the Hilbert space H. Then $V = (\mathrm{id} \otimes \pi)(u)$ is a multiplicative unitary.

c) The Canonical Element

Let A be a finite dimensional Hopf algebra. Let E be the algebra of endomorphisms of the vector space A. Let us denote by v the canonical element of $A^* \otimes A$: through the identification of $A^* \otimes A$ with E, v is the identity of A. Denote by L the action of A on A by left multiplication. If $x \in A^*$ and $a \in A$, set $\varrho(x)a = (\mathrm{id} \otimes x)\delta(a)$. Consider L and ϱ as homomorphisms from A and A* into the algebra E. Simple computations then show:

[1] This is the C^*-algebraic "min" tensor product. If A has no unit, δ takes its values in the multiplier algebra $M(A \otimes A)$ of $A \otimes A$.

- For $a \in A$ we have in $E \otimes A$, $(\varrho \otimes \mathrm{id})(v)(L(a) \otimes 1) = (L \otimes \mathrm{id})(\delta(a))(\varrho \otimes \mathrm{id})(v)$.
- We have the equality $(\mathrm{id} \otimes \delta)(v) = v_{12}v_{13}$.

Therefore, the operator $V = (\varrho \otimes L)(v)$ satisfies the pentagon equation.

2. Algebras Associated with Multiplicative Unitaries

Let $V \in L(H \otimes H)$ be a multiplicative unitary. If ω is a continuous linear form on $L(H)$, we may form the operators $L(\omega) = (\omega \otimes \mathrm{id})(V) \in L(H)$ and $\varrho(\omega) = (\mathrm{id} \otimes \omega)(V) \in L(H)$. Let $L(H)_*$ denote the predual of $L(H)$ i.e. the set of linear mappings of the form $x \to \mathrm{Tr}(xT)$ where T spans the space of trace class operators.

2.1. Proposition. *The sets $A(V) = \{L(\omega)/\omega \in L(H)_*\}$ and $\hat{A}(V) = \{\varrho(\omega)/\omega \in L(H)_*\}$ are subalgebras of $L(H)$.*

Indeed, $L(\omega)L(\omega') = (\omega \otimes \omega' \otimes \mathrm{id})(V_{13}V_{23}) = L(\psi)$ where $\psi(x) = (\omega \otimes \omega') \times (V^*(1 \otimes x)V)$

since $V_{12}V_{13} = V_{23}V_{12}V_{23}^*$; in the same way, $\varrho(\omega)\varrho(\omega') = (\mathrm{id} \otimes \omega \otimes \omega')(V_{12}V_{13}) = \varrho(\psi')$ where $\psi'(x) = (\omega \otimes \omega')(V(x \otimes 1)V^*)$. In fact, all properties which may be proved for $A(V)$ are automatically proved for $\hat{A}(V)$ since $\Sigma V^* \Sigma$ is a multiplicative unitary.

There is a natural duality between $A(V)$ and $\hat{A}(V)$ expressed by the equalities $\langle L(\omega), \varrho(\omega')\rangle = \omega(\varrho(\omega')) = \omega'(L(\omega)) = (\omega \otimes \omega')(V)$.

It is also natural to consider the norm closures of the algebras $A(V)$ and $\hat{A}(V)$ that we denote by S_V and $\hat{S}_V$. It is not clear whether these are always C^*-algebras ie. if they are closed under the involution $x \to x^*$ of $L(H)$. For this reason, we are led to make, in the next sections, some extra assumptions.

In the case of the multiplicative unitary associated with a group, the algebras $A(V)$ and $\hat{A}(V)$ are respectively the Fourier algebra $A(G)$ acting by multiplication on $L^2(G)$ and $L^1(G)$ acting by (right) convolution on $L^2(G)$. Also S_V is the abelian C^*-algebra $C_0(G)$ of continuous functions vanishing at infinity and $\hat{S}_V$ the reduced C^*-algebra of the group G. In particular the Gelfand spectrum of S_V is G: we already have recovered G out of the associated multiplicative unitary. In fact, we get a converse to this statement:

2.2 Theorem. *If the associated algebra $A(V)$ is commutative, the multiplicative unitary V is (up to multiplicity) the multiplicative unitary associated with a locally compact group.*

This theorem is a generalization theorem of [50, 23, 12, 41, 42, 47, 5, 52]. Of course, this theorem also classifies the multiplicative unitaries for which $\hat{A}(V)$ is commutative, since this is equivalent to saying that $A(\Sigma V^* \Sigma)$ is commutative.

Let us mention another case where no extra assumptions are needed:

2.3 Theorem. *A multiplicative unitary acting on a finite dimensional Hilbert space is (up to multiplicity) the multiplicative unitary associated with a finite dimensional Kac von Neumann algebra.*

3. Regularity; the "Compact" Case

Let us begin with a rather easy fact:

3.1 Proposition. *The set* $\mathscr{C}(V) = \{(\mathrm{id} \otimes \omega)(\Sigma\omega)/\omega \in L(H)_*\}$ *is a subalgebra of* $L(H)$.

Studying this algebra in the case of locally compact groups and more generally in the examples to be discussed below, we find that this algebra is formed of compact operators and is norm dense in the algebra of compact operators. This leads to the following definition:

3.2 Definition. *We will say that the multiplicative unitary* V *is* regular *if the norm closure of* $\mathscr{C}(V)$ *coincides with the algebra* $K(H)$ *of compact operators of* H.

Regularity turns out to be extremely efficient in proving nice properties of the associated algebras:

3.3 Theorem. *Let* V *be a regular multiplicative unitary. Then the algebras* S *and* $\hat{S}$ *are Hopf* C^**-algebras with coproducts given by* $\delta(x) = V(x \otimes 1)V^*$ *and* $\hat{\delta}(y) = V^*(1 \otimes y)V$ $(x \in S,\ y \in \hat{S})$. *The operator* V *is a multiplier of the (spatial) tensor product* $\hat{S} \otimes S$.

This last property means that the closed subalgebra of $L(H \otimes H)$ generated by $y \otimes x$, $x \in S$, $y \in \hat{S}$ is closed under left and right multiplication by V. It is quite natural and helpful. In particular, it allows us to consider S and $\hat{S}$ as abstract C^*-algebras and still make sense of V in every representation.

We are also in position to form crossed products for algebras with coactions of the Hopf algebra S: if a C^*-algebra A is endowed with a coaction $\delta_A : A \to A \otimes S$ of S, the (reduced) crossed product $A \times \hat{S}$ is the C^*-algebra of operators acting on the Hilbert A-module $A \otimes H$ generated by the products of the form $\delta_A(a)(1 \otimes y)$, $a \in A$, $y \in \hat{S}$. For $a \in A$, $y \in \hat{S}$, $\delta_A(a)$ and $(1 \otimes y)$ are multipliers of $A \times \hat{S}$. We thus get homomorphisms π and $\hat{\theta}$ from A and $\hat{S}$ respectively into the multiplier algebra of $A \times \hat{S}$. Still our set of axioms is not complete in order to allow us to prove the suitable duality. On the other hand, this duality may now be proved in the "compact" case.

3.4 Definition. *A multiplicative unitary is said to be of compact type if the unit operator belongs to the algebra* $A(V)$.

If V is a multiplicative unitary associated with a compact group or with a Haar state of a unital Hopf algebra, it is of compact type. In a recent fundamental work

[51–54], S.L. Woronowicz introduced a set of axioms for "quantum groups" generated by a finite dimensional unitary representation. Woronowicz initially called his objects "compact matrix pseudogroups" but they are referred to as "compact quantum Lie groups". Nice and tractable examples were produced [51, 53]. It is natural to define "compact quantum groups" as projective limits (this corresponds to inductive limits for the Hopf algebra of functions) of "compact quantum Lie groups". It was shown in [52] that "compact quantum groups" possess a Haar state. It is quite clear that the corresponding operator is unitary. Its regularity is also easy. The converse to these facts is true, namely:

3.5 Theorem. *A regular multiplicative unitary of compact type is (up to multiplicity) the multiplicative unitary associated with a "compact quantum group" of Woronowicz.*

4. Irreducibility and Takesaki-Takai Duality

In order to introduce the last condition needed for the duality, let us examine again the case of locally compact groups: we have been able to produce out of the multiplicative unitary associated with a locally compact group, the multiplication operators and the right regular representation. On $L^2(G)$ acts moreover the left regular representation; moreover, left and right regular representations are equivalent and intertwined by a unitary operator U given by $(U\xi)(g) = \Delta(g)^{1/2}\xi(g^{-1})$, where Δ is the module of the group.

This leads us to assume the existence of an operator U satisfying some equations:

4.1 Definition. a) *A multiplicative unitary $V \in L(H \otimes H)$ is said to be irreducible if there exists a unitary $U \in L(H)$ such that $U^2 = 1_H$, $(V(U \otimes 1)\Sigma)^3 = 1_{H\otimes H}$ and such that the unitary $\hat{V} = \Sigma(U \otimes 1)V(U \otimes 1)\Sigma$ is multiplicative.*

b) *A Kac system is a triple (H, V, U) where H is a Hilbert space, $V \in L(H \otimes H)$ is a multiplicative unitary and $U \in L(H)$ satisfies the requirements of* a); *moreover, we require that V and $\hat{V}$ be regular.*

If (H, V, U) is a Kac system, $(H, \hat{V}, U)$ is also a Kac system and $\hat{\hat{V}} = (U \otimes U)V(U \otimes U)$. Taking the dual again, we find $\tilde{V} = \Sigma(1 \otimes U)V(1 \otimes U)\Sigma$; a fourth time will give us back V. This is the well noticed but still somewhat mysterious period 4 periodicity. (Note that as $\hat{\hat{V}} = (U \otimes U)V(U \otimes U)$ and $\tilde{V} = (U \otimes U)\hat{V}(U \otimes U)$ they are regular multiplicative unitaries).

We now have two representations of S and $\hat{S}$ in H: we will denote by $L : S \to L(H)$ and $\varrho : \hat{S} \to L(H)$ the inclusions considered up to now as identity representations; we will then set $R(x) = UL(x)U$ and $\lambda(y) = U\varrho(y)U$ $(x \in S,\ y \in \hat{S})$.

Replacing V by $\hat{V}$ we may now form crossed products for algebras with coactions of the Hopf algebra $\hat{S}$: if a C^*-algebra A is endowed with a coaction $\delta_A : A \to A \otimes \hat{S}$ of $\hat{S}$, the (reduced) crossed product $A \times S$ is the C^*-algebra of operators acting on the Hilbert A-module $A \otimes H$ generated by the products of the form

$$(\mathrm{id} \otimes \lambda)\delta_A(a)(1 \otimes L(x)), \qquad a \in A, \qquad x \in S.$$

We still have homomorphisms π and θ from A and S respectively into the multiplier algebra of $A \times S$, and $A \times S$ is spanned by products $\pi(a)\theta(x)$ $a \in A$, $x \in S$.

Let the C^*-algebra A be endowed with a coaction δ_A of S (resp. $\hat{S}$), then the crossed product $A \times \hat{S}$ (resp. $A \times S$) is endowed with a coaction $\hat{\delta}_A$ of $\hat{S}$ (resp. S) given by $\hat{\delta}_A(\pi(a)\hat{\theta}(y)) = (\pi(a) \otimes 1)(\hat{\theta} \otimes \mathrm{id})\hat{\delta}(y)$, $a \in A$, $y \in \hat{S}$ (resp. $\hat{\delta}_A(\pi(a)\theta(x)) = (\pi(a) \otimes 1)(\theta \otimes \mathrm{id})\delta(x)$, $a \in A$, $x \in S$).

4.2 Theorem (Takesaki-Takai Duality Theorem). *Let (H, V, U) be a Kac system. Then for any algebra A endowed with a coaction*[2] *δ_A of S the double crossed product $A \times \hat{S} \times S$ is naturally isomorphic with $A \otimes K(H)$.*

Remark. Replacing (H, V, U) by $(H, \hat{V}, U)$ we may exchange the roles of S and $\hat{S}$.

It is transparent in many papers (cf. eg. [21], [22], [39], [4], [10]) that Takesaki duality only relies on the "fundamental" operator; our proof is therefore just an adaption of methods used by these authors.

A first step to this duality is the case $A = \mathbf{C}$:

4.3 Lemma. *Let (H, V, U) be a Kac system. Then the closed vector span of $\{L(x)\varrho(y)/x \in S, y \in \hat{S}\}$ and the closed vector span of $\{L(x)\lambda(y)/x \in S, y \in \hat{S}\}$ are the algebra $K(H)$.*

This lemma, which can be thought of as a generalization of the famous Weyl-von Neumann theorem, explains the terminology of irreducibility.

Remark. It is also quite easy to prove a Takesaki duality theorem in the von Neumann algebra setting for Kac systems. In fact the regularity can be replaced by the weaker condition: the weak closure of $\mathscr{C}(V)$ is $L(H)$. The proof (if not the precise statement) of the main theorem of [10] applies in this context.

Also, one may generalize results of [1] and prove a Takesaki-Takai duality theorem for equivariant KK-theory with respect to the Hopf C^*-algebras S and $\hat{S}$.

5. Examples of Majid and Podles-Woronowicz

In [26] and [32] appeared a series of new constructions of interesting "quantum groups". These "quantum groups" are not in general Kac von Neumann algebras but they can still be expressed by a multiplicative unitary; in this way Takesaki-Takai duality is just an easy check.

The algebraic setting in the examples of [26, 27] and [32] is that of matched pairs of Hopf algebras (cf. [23, 28]). To such a matched pair are associated two new Hopf algebras: the one "generated by the matched pair" and the "bicrossproduct". These constructions were given in [37] and [28] in purely algebraic terms but may be performed in the multiplicative unitary setting. Examples of such matched pairs

[2] Provided a technical assumption called non-degeneracy in [21] is fulfilled by δ_A.

are given by any (locally compact) group G with two (closed) subgroups H and K such that every element g of G admits a unique decomposition $g = hk$ ($h \in H, k \in K$ – [44, 26]). Other examples are given by the "quantum double" construction of Drinfeld [3]. The examples of Podles and Woronowicz [32] are in fact based on this "quantum double" construction.

Let (A, δ_A) and (B, δ_B) be two Hopf C^*-algebras. Consider the *-homomorphism $\delta_A \otimes \delta_B : A \otimes B \to A \otimes A \otimes B \otimes B$. In order to put a Hopf C^*-algebra structure on $A \otimes B$, we use a *-isomorphism $\tau : A \otimes B \to B \otimes A$ and wish to put $\delta = (\mathrm{id}_A \otimes \tau \otimes \mathrm{id}_B)(\delta_A \otimes \delta_B)$. For δ to be coassociative it is enough that the following condition be satisfied:

$$(C)(\tau \otimes \mathrm{id}_A)(\mathrm{id}_A \otimes \tau)(\delta_A \otimes \mathrm{id}_B) = (\mathrm{id}_B \otimes \delta_A)\tau \qquad \text{and}$$

$$(\mathrm{id}_B \otimes \tau)(\tau \otimes \mathrm{id}_B)(\mathrm{id}_A \otimes \delta_B) = (\delta_B \otimes \mathrm{id}_A)\tau$$

Condition (C) is stated in [32] and, from a dual point of view, in [37] and [28] (in purely algebraic terms).

5.1 Definition. *Let (A, δ_A) and (B, δ_B) be two Hopf C^*-algebras. An inversion on A, B is a *-isomorphism $\tau : A \otimes B \to B \otimes A$ satisfying the conditions (C).*

Let (A, B, τ) be as in definition 5.1. Note that $b \to \tau(1 \otimes b)$ is a (right) coaction, called β, of the Hopf C^*-algebra A on the C^*-algebra B and $a \to \tau(a \otimes 1)$ is a (left) coaction, called α, of the Hopf C^*-algebra B on the C^*-algebra A.

Let $X \in L(H \otimes H)$ and $Y \in L(K \otimes K)$ be two regular multiplicative unitaries. Denote by S_X, $\hat{S}_X$, S_Y and $\hat{S}_Y$ the associated Hopf C^*-algebras associated with X and Y and by δ_X, $\hat{\delta}_Y$ the coproducts of S_X and $\hat{S}_Y$. Let $\tau : S_X \otimes \hat{S}_Y \to \hat{S}_Y \otimes S_X$ be an inversion on $(S_X, \hat{S}_Y)$.

5.2 Proposition. *The unitary operator $T = (\tau \otimes \mathrm{id})(Y_{23})(\mathrm{id} \otimes \tau)(X_{23})$ acting on $K \otimes H \otimes K \otimes H$ is multiplicative. It is called the bicrossproduct of X and Y with respect to τ.*

Let (H, X, u) and (K, Y, v) be two Kac systems. It is more natural to assume that τ is an inversion on (S_X, S_Y). Of course, in this case, we may form the bicrossproduct of X and $\hat{Y}$. In order to form the twisted and bicrossproducts of the Kac systems, we need the inversion τ to be suitably implemented.

5.3 Definition. *A matched pair of Kac systems is given by two Kac systems (H, X, u) and (K, Y, v) together with a unitary operator $Z \in L(H \otimes K)$ such that*

a) *There exists an inversion $\tau : S_X \otimes S_Y \to S_Y \otimes S_X$ such that for all $x \in S_X$, $y \in S_Y$ we have $Z\tau^{-1}(y \otimes x)Z^* = x \otimes y$.*

b) *$(H \otimes K, V, U)$ is a Kac system where $V = (Z_{12}^* X_{13} Z_{12}) Y_{24}$ and $U = (u \otimes v)Z$.*

5.4 Theorem. *Let $((H, X, u); (K, Y, v); Z)$ be a matched pair of Kac systems. Define the unitary operator $W = (Z_{34}{}^* \hat{Y}_{24} Z_{34})(Z_{12}{}^* X_{13} Z_{12})$ acting on $H \otimes K \otimes H \otimes K$.*

Then $(H \otimes K, W, U)$ is a Kac system. The Hopf algebra S associated with V is $(S_X \otimes S_Y, \delta_\tau)$. The algebras S and $\hat{S}$ associated with W are isomorphic respectively to $S_X \times_\alpha \hat{S}_Y$ and $S_Y \times_\beta S_X$.

Moreover, the multiplicative unitary operator T of proposition 5.2 is equivalent to W.

5.5 Definition. *With the notation of the above theorem, the Kac system $(H \otimes K, V, U)$ is called the product of (H, X, u) and (K, Y, v) twisted by Z; the Kac system $(H \otimes K, W, U)$ is called the bicrossproduct of (H, X, u) and (K, Y, v) relative to Z.*

5.6 Examples

a) Let (H, X, u) be a Kac system and G be a locally compact group acting by Hopf C^*-algebra automorphisms on S_X. Let $\tau: C_0(G) \otimes S_X \to S_X \otimes C_0(G)$ be given by $\tau(f)(x) = \alpha_x(f(x))$ $x \in G$, $f \in C_0(G; S_X)$ where we have identified $C_0(G) \otimes S_X$ and $S_X \otimes C_0(G)$ with the C^*-algebra $C_0(G; S_X)$ of continuous S_X-valued functions vanishing at ∞ on G. In this case, G acts naturally on the Hopf C^*-algebra $\hat{S}_X$ and the twisted and bicrossproducts are both obtained by the well known crossed-product constructions.

b) Let G_1 and G_2 be two locally compact groups. An inversion on $(C_0(G_1), C_0(G_2))$ is given by a homeomorphism $\tau: G_2 \times G_1 \to G_1 \times G_2$; then the product $(x_1, x_2)(y_1, y_2) = (x_1 z_1, z_2 x_2)$ where $(z_1, z_2) = \tau(x_2, y_1)$ is associative on $G_1 \times G_2$ and, endowed with this product, $G_1 \times G_2$ is a locally compact group G. Then, the twisted product of the associated Kac systems is the Kac system of the group G. The bicrossproduct construction gives new examples of Kac systems. In general, these examples are not associated with Kac von Neumann algebras [26, 27] and the antipode κ is unbounded. However, many computations may still be performed in this context.

Another way of understanding this example, is to start with a locally compact group G and assume that it has two closed subgroups G_1 and G_2 such that the map $(x_1, x_2) \to x_1 x_2$ is a homeomorphism from $G_1 \times G_2$ onto G. In this case, the actions of G_1 on G_2 and of G_2 on G_1 are the restrictions of the actions of G on $G_2 = G_1 \backslash G$ and on $G_1 = G/G_2$ and it is easy to compute the corresponding crossed products and thus the algebras S and $\hat{S}$ associated with W. Also, it is quite easy to construct groups with these properties:

- the Iwasawa decomposition $G = KP$ $(P = AN)$ of semisimple Lie groups;
- let G be a locally compact group acting by homeomorphisms on a locally compact group G_2 and containing the right translations of G_2; let then G_1 be the set of elements of G fixing the neutral element of G_2;
- in the above example, we may take G_2 to be any finite group and G be the group of all permutations of the set G_2 ...

A third way of interpreting this example (cf. [2] Appendix C) is the search of measure spaces X and transformations of $X \times X$ satisfying the pentagon relation.

It is then natural to add cocycles and form new multiplicative unitaries. In this way one recovers examples of Kac and Paliutkin [14, 15].

c) Let (H, X, u) be a Kac system and set $Y = sX^*s$ where $s \in L(H \otimes H)$ is the flip operator; then (H, Y, u) is a Kac system and $S_Y = \hat{S}_X$, $\hat{S}_Y = S_X$[3]. Let then $\tau : S_X \otimes S_Y \to S_Y \otimes S_X$ be given by $\tau(x) = XsxsX^*$. Since X is a multiplier of $S_Y \otimes S_X$ this is well defined. It turns out that τ is a non degenerate inversion and that $((H, X, u); (K, Y, v); Z)$ is a matched pair of Kac systems, where $Z = sX(u \otimes u)X^*(u \otimes u)s$. The corresponding twisted product $(H \otimes H, V, U)$ is the *quantum double* of the Kac system (H, X, u). Let S_V and $\hat{S}_V$ denote the corresponding Hopf C^*-algebras. There is a unitary operator R which is a multiplier of $\hat{S}_V \otimes \hat{S}_V$ which satisfies the algebraic properties of [3] and in particular R is a solution of the quantum Yang-Baxter equation.

Note that the construction of this twisted product was used by Podles and Woronowicz to build the "quantum Sl(2, C)" out of the "quantum SU(2)" ([32]).

In this case, the bicrossproduct is just a direct product.

6. Concluding Remarks

We developed here one point of view: find conditions easy to check on the "fundamental operator" that ensure Takesaki-Takai duality. However, we do not know if these conditions may turn out to be automatic.

Maybe one should look for a counterexample to regularity in transformations satisfying the pentagon equation.

The operator U defining irreducibility, is usually the product $\hat{J}J$ of the Tomita operators associated with Haar measures. Therefore, to prove irreducibility one would need to prove the existence of these Haar measures. Note that, in our context, this problem doesn't seem too difficult since we are given the regular representations and therefore the class of the Haar measures.

Once the Haar measures are found one needs to perform modular theory on them. Concerning this, we may formulate the following conjecture:

Call ϕ, ψ, $\hat{\phi}$ and $\hat{\psi}$ the left and right Haar measures of S and $\hat{S}$. Then there should exist positive unbounded operators F and $\hat{F}$ affiliated with the centralizers of ϕ and $\hat{\phi}$ such that for all $x \in S$ and $y \in \hat{S}$, $\psi(x) = \phi(FxF)\,\hat{\psi}(y) = \hat{\phi}(\hat{F}y\hat{F})$. The Hilbert spaces H_ϕ, H_ψ, $H_{\hat{\phi}}$ and $H_{\hat{\psi}}$ are naturally identified. The weights ϕ, ψ, $\hat{\phi}$ and $\hat{\psi}$ are faithful when extended to the bicommutants, therefore Tomita theory can be performed. Call J and $\hat{J}$ the Tomita operators associated with ϕ and $\hat{\phi}$, and put $U = \hat{J}J = J\hat{J}$. Let L and λ be the GNS representations associated with ϕ and $\hat{\phi}$; then form R and ϱ using L, λ and U. Since V is a multiplier of $\hat{S} \otimes S$, $(\varrho \otimes L)(V)$ acts naturally on $H_\phi \otimes H_\phi$. Then (H_ϕ, V, U) is a Kac system. Moreover, the operators F and $\hat{F}$ are representations of the Hopf algebras, therefore they are unbounded multipliers of S and $\hat{S}$. Moreover, the operators $L(F)$, $R(F)$, $\lambda(\hat{F})$ and $\varrho(\hat{F})$ commute pairwise. The modular operators are computed in terms of F and $\hat{F}$. We find:

---

[3] Note however that the coproducts of S_Y and $\hat{S}_Y$ differ from the ones of $\hat{S}_X$ and S_X by the flip.

$\Delta_\phi = L(F^{-1})R(F^{-1})\lambda(\hat{F})\varrho(\hat{F}^{-1})$, $\Delta_\psi = L(F)R(F)\lambda(\hat{F})\varrho(\hat{F}^{-1})$, $\Delta_{\hat{\phi}} = L(F^{-1})R(F)\lambda(\hat{F}^{-1})\varrho(\hat{F}^{-1})$ and $\Delta_{\hat{\psi}} = L(F^{-1})R(F)\lambda(\hat{F})\varrho(\hat{F})$. Also, the square of the antipode is implemented by the unbounded operator $K = L(F^{-1})R(F^{-1})\lambda(\hat{F})\varrho(\hat{F})$.

In the case of Kac von Neumann algebras occurs the simplification: $L(F)R(F) = 1 = \lambda(\hat{F})\varrho(\hat{F})$ and we recover the computation of the modular operators of [5, 35]. In the case of compact quantum groups, we have $F = 1$ and this conjecture agrees with the computation of Woronowicz [52]. This conjecture has been checked in the examples of bicrossproducts of groups (with cocycles) by Saad Baaj. Also, if this holds for a Kac system, it holds for its double (moreover, the double is unimodular i.e. $F = 1$).

Added in proof: It is quite easy to construct non regular multiplicative unitaries using a transformation satisfying the pentagon equation. In fact, one may associate to such an "irregular" transformation two non regular multiplicative unitaries one of which is irreducible but not the other.

Exactly the same situation occurs with the quantum deformation of the motion group constructed by S.L. Woronowicz.

References

1. Baaj, S., Skandalis, G.: C^*-algèbres de Hopf et théorie de Kasparov équivariante. *K*-theory **2** (1989) 683–721
2. Baaj, S., Skandalis, G.: Unitaries multiplicatifs et dualité pour les produits croisés de C^*-algèbres. Preprint
3. Drinfeld, V.G.: Quantum groups. Proc. ICM Berkeley 1986, pp. 798–820
4. Enock, M.: Produit croisé d'une algèbre de von Neumann par une algèbre de Kac. I.J.F.A. **26**, no. 1 (1977) 16–46
5. Enock, M., Schwartz, J.M.: Une dualité dans les algèbres de von Neumann. Bull. S.M.F. Suppl. mémoire **44** (1975) 1–144
6. Enock, M., Schwartz, J.M.: Produit croisé d'une algèbre de von Neumann par une algèbre de Kac. II. Publ. RIMS Kyoto Univ. **16**, no. 1 (1980) 189–232
7. Enock, M., Schwartz, J.M.: Extension de la catégorie des algèbres de Kac. Ann. de l'Inst. Fourier **36**, fasc. 1 (1986) 105–131
8. Enock, M., Schwartz, J.M.: Algèbres de Kac moyennables. Pacific J. of Math. **125**, no. 2 (1986) 363–379
9. Ernest, J.: Hopf von Neumann algebras. Proc. Conf. Funct. Anal. (Irvine, Calif.) Academic Press, 1967 pp. 195–215
10. van Heeswijck, L.: Duality in the theory of crossed products. Math. Scand **44** (1979) 313–329
11. Kac, G.I.: Ring groups and the duality principle. Trans. Moscow Math. Soc. (1963) 291–339. Translated from Trudy Moskov. Mat. Ob. **12** (1963) 259–301
12. Kac, G.I.: Ring groups and the duality principle II. Trans. Moscow Math. Soc. (1965) 94–126. Translated from Trudy Moskov. Mat. Ob. **13** (1965) 84–113
13. Kac, G.I.: Certain arithmetic properties of ring groups. Funkt. Anal. Prilozen. **6** (1972) 88–90
14. Kac, G.I., Paljutkin, V.G.: Finite group rings. Trans. Moscow Math. Soc. (1966) 251–294. Translated from Trudy Moskov. Mat. Obsc. **15** (1966) 224–261
15. Kac, G.I., Paljutkin, V.G.: Example of ring groups generated by Lie groups. (en russe) Ukr. Mat. J. **16**, no. 1 (1964) 99–105

16. Kac, G.I., Vainerman, L.I.: Nonunimodular ring-groups and Hopf-von Neumann algebras. Math. USSR Sb. **23** (1974) 185–214. Translated from Mat. Sb. **94** (136) (1974), 2 194–225
17. Katayama, Y.: Takesaki's duality for a non degenerate coaction. Math. Scand. **55** (1985) 141–151
18. Kirchberg, E.: Representation of coinvolutive Hopf-W^*-algebras and non abelian duality. Bull. Acad. Pol. Sc. **25** (1977) 117–122
19. Krein, M.G.: Hermitian-positive kernels in homogeneous spaces. Amer. Math. Soc. Transl. (2) **34** (1963) 109–164. Translated from Ukr. Mat. Z. **2**, no. 1 (1950) 10–59
20. Landstad, M.B.: Duality theory for covariant systems. Trans. AMS **248** (1979) 223–267
21. Landstad, M.B.: Duality for dual covariance algebras. Comm. Math. Phys. **52** (1977) 191–202
22. Landstad, M.B., Phillips, J., Raeburn, I., Sutherland, C.E.: Representations of crossed products by coactions and principal bundles. Trans. AMS **299**, no. 2 (1987) 747–784
23. Mackey, G.W.: Borel structures in groups and their duals. Trans. AMS **85** (1957) 134–165
24. Mac Lane, S.: Categories for the working mathematician. (Graduate Texts in Mathematics, vol. 5). Springer, New York Berlin Heidelberg 1974
25. Mac Lane, S.: Natural associativity and commutativity. Rice Univ. Studies **49** (1963) 4–28
26. Majid, S.H.: Non-commutative geometric groups by a bicrossproduct construction: Hopf algebras at the Planck scale. Thesis, Harvard Univ. 1988
27. Majid, S.H.: Hopf von Neumann algebra Bicrossproducts, Kac Algebras Bicrossproducts, and the classical Yang-Baxter equation. To appear in J. Funct. Anal.
28. Majid, S.H.: Physics for algebraists: Non-commutative and Non-cocommutative Hopf algebras by a Bicrossproduct construction. J. Algebra **130**, no. 1 (1990) 17–64
29. Moore, G., Seiberg, N.: Classical and quantum conformal field theory. Comm. Math. Phys. **123** (1989) 177–254
30. Nakagami, Y.: Dual action on a von Neumann algebra and Takesaki's duality for a locally compact group. Publ. RIMS Kyoto Univ. **12** (1977) 727–775
31. Nakagami, Y., Takesaki, M.: Duality for crossed products of von Neumann algebras. (Lecture Notes in Mathematics, vol. 731.) Springer, Berlin Heidelberg New York 1979
32. Podles, P., Woronowicz, S.L.: Quantum deformation of Lorentz group. Preprint
33. Rosso, M.: Comparaison des groupes SU(2) quantiques de Drinfeld et Woronowicz. Note aux C. R. Acad. Sci. **304** (1987) 323–326
34. Rosso, M.: Algèbres enveloppantes quantifiées, groupes quantiques compacts de matrices et calcul différentiel non commutatif. Prépublication
35. Schwartz, J.M.: Sur la structure des algèbres de Kac I. J. Funct. Anal. **34** (1979) 370–406
36. Schwartz, J.M.: Sur la structure des algèbres de Kac II. Proc. London Math. Soc. **41** (1980) 465–480
37. Singer, W.M.: Extension theory for connected Hopf algebras. J. Algebra **21** (1972) 1–16
38. Stinespring, W.F.: Integration theorems for gages and duality for unimodular groups. Trans. AMS **90** (1959) 15–56
39. Stratila, S., Voiculescu, D. and Zsido, L.: On crossed products. I and II. Rev. Roumaine Math. Pures Appl. **21** (1976) 1411–1449 and **22** (1977) 83–117
40. Takai, H.: On a duality for crossed products of C^*-algebras. J. Funct. Anal. **19** (1975) 25–39
41. Takesaki, M.: A characterization of group algebras as a converse of Tannaka-Stinespring-Tatsuuma duality theorem. Amer. J. Math. **91** (1969) 529–564
42. Takesaki, M.: Duality and von Neumann algebras. (Lecture Notes in Mathematics, vol. 427.) Springer, Berlin Heidelberg New York 1972, pp. 665–779
43. Takesaki, M.: Duality for crossed products and the structure of von Neumann algebras of type III. Acta Math. **131** (1973) 249–310

44. Takeuchi, M.: Matched pairs of groups and bismash products of Hopf algebras. Comm. Algebra **9**, no. 8 (1981) 841–882
45. Tannaka, T.: Über den Dualität der nicht-kommutativen topologischen Gruppen. Tôhoku Math. J. **45** (1938) 1–12
46. Tatsuuma, N.: A duality theorem for locally compact groups. J. Math. of Kyoto Univ. **6** (1967) 187–293
47. Vainerman, L.I.: Characterization of dual objects for locally compact groups. Funct. Anal. Appl. **8** (1974) 66–67. Translated from Funkt. Anal. Prilozen. **8**, no. 1 (1974) 75–76
48. Vallin, J.M.: C^*-algèbres de Hopf et C^*-algèbres de Kac. Proc. London Math. Soc. (3) **50** (1985) 131–174
49. Voiculescu, D.: Amenability and Katz algebras. Algèbres d'opérateurs et leurs applications en physique mathématique. Colloques Internationaux C.N.R.S. no. **274** (1977) 451–457
50. Weil, A.: L'intégration dans les groupes topologiques et ses applications. Act. Sc. Ind. no. **1145**. Hermann, Paris (1953)
51. Woronowicz, S.L.: Twisted SU(2) group. An example of a non commutative differential calculus. Publ. RIMS **23** (1987) 117–181
52. Woronowicz, S.L.: Compact matrix pseudogroups. Comm. Math. Phys. **111** (1987) 613–665
53. Woronowicz, S.L.: Tannaka-Krein duality for compact matrix pseudogroups. Twisted SU(N) group. Inv. math. **93** (1988)
54. Woronowicz, S.L.: Differential calculus on compact matrix pseudogroups (quantum groups). Comm. Math. Phys. **122** (1989) 125–170

Some Isoperimetric Inequalities and Their Applications

Michel Talagrand

Equipe d'Analyse, E.R.A. au C.N.R.S. no. 294, Université Paris VI, 4 Place Jussieu
F-75230 Paris Cedex 05, France

0. Introduction

Consider a product of measure spaces, provided with the product measure. Consider a subset A of this product, of measure at least one half. An important fact (the so-called concentration of measure phenomenon) is that even a small "enlargement" of A has measure very close to one. The inequalities we present describe this phenomenon for several notions of "enlargement".

1. The Isoperimetric Inequality for Gaussian Measure

We denote by S_n the Euclidean sphere of $\mathbb{R}^{n+1}$, equipped with the geodesic distance ϱ and a rotation invariant probability μ_n. For a (measurable) subset A of S_n, consider the set A_u of points of S_n within geodesic distance u of A. The isoperimetric inequality for the sphere, discovered by P. Lévy, is of fundamental importance. It states that $\mu_n(A_u) \geq \mu_n(C_u)$, where C is a cap (intersection of the sphere and of a half space) of the same measure as A.

We denote by γ_n the canonical Gaussian measure on $\mathbb{R}^n$, of density $(2\pi)^{-n/2}e^{-\|x\|^2/2}$ with respect to Lebesgue measure. Observe the simple, but essential fact that γ_n is the product measure on $\mathbb{R}^n$ when each factor is endowed with γ_1. It is an old observation, known as Poincaré lemma (although it does seem to be due to Maxwell) that, as $N \to \infty$, the projection of the normalized measure on $\sqrt{N}S_N$ onto $\mathbb{R}^n$ has γ_n as a limit. Therefore, it should not come as a surprise that Lévy's isoperimetric inequality on the sphere implies an isoperimetric inequality for γ_n. This was discovered independently by C. Borell [B1], and V. N. Sudakov and B. S. Tsirelson [S-T]. If we denote by A_u the set of points within Euclidean distance u of A, then $\gamma_n(A_u) \geq \gamma_n(H_u)$, where H_u is a half space with $\gamma_n(H) = \gamma_n(A)$. Taking this half space to be orthogonal to a coordinate axis, and remembering that γ_n is a product measure shows that if $\gamma_n(H) = \gamma_1((-\infty, a])$, then $\gamma_n(H_u) = \gamma_1((-\infty, a+u])$.

The author is also affiliated with the Ohio State University, Columbus, OH 43201, USA

Proceedings of the International Congress
of Mathematicians, Kyoto, Japan, 1990

For simplicity, we set $\Phi(u) = \gamma_1((-\infty, u]) = \frac{1}{\sqrt{2\pi}} \int_{-\infty}^{u} e^{-x^2/2} dx$. Thus we have

$$\text{if } \gamma_n(A) = \Phi(a), \text{ then } \gamma_n(A_u) \geq \Phi(a+u). \tag{1.1}$$

It is important to state this inequality, not only for the measure γ_n, but also for its infinite dimensional version γ, the product measure on $\mathbb{R}^{\mathbb{N}}$ when each factor is endowed with γ_1 (the result for γ follows from the result for γ_n and an obvious approximation). We denote by B the unit ball of ℓ^2, i.e. $B = \{x \in \mathbb{R}^{\mathbb{N}}, \sum_{k \geq 1} x_k^2 \leq 1\}$. The Gaussian isoperimetric inequality can then be stated as follows

$$\text{If } \gamma(A) = \Phi(a) \text{ then } \gamma_*(A + uB) \geq \Phi(a+u). \tag{1.2}$$

There $A + uB = \{x + uy; x \in A, y \in B\}$; the inner measure is needed as $A + uB$ might not be measurable. As became customary, we call (1.1) and (1.2) Borell's inequality. Lévy's inequality is usually proved using symmetrization (see e.g. the appendix of [F-L-M]). A. Ehrhard [E1] has developed a symmetrization method adapted to the measures γ_n that yields a direct proof of (1.2) as well as of the following remarkable inequality of Brunn-Minkowski's type: For two convex sets A, B of $\mathbb{R}^n$, and $0 \leq \lambda \leq 1$,

$$\Phi^{-1}(\gamma_n(\lambda A + (1-\lambda)B)) \geq \lambda \Phi^{-1}(\gamma_n(A)) + (1-\lambda)\Phi^{-1}(\gamma_n(B)). \tag{1.3}$$

(It is still open whether this inequality holds for non convex sets.)

Borell's inequality is a principle of remarkable power. It can be argued that, concerning applications, this inequality is used in two different forms.

The first type of use consist of rewriting (1.1) as $u^{-1}\gamma_n(A_u \backslash A) \geq u^{-1}\gamma_1([a, u+a])$ so that

$$\liminf_{u \to 0} u^{-1}\gamma_n(A_u \backslash A) \geq \frac{1}{\sqrt{2\pi}} \exp -\frac{a^2}{2}. \tag{1.4}$$

thereby recovering what is the more classical formulation of the isoperimetric inequality [O]. In this spirit (and using his symmetrization methods) A. Ehrhard has proved a number of interesting inequalities, that are versions for the Gauss measure of classical results [E2].

Inequality (1.4) for functions rather than sets [L] yields in particular that a function on $\mathbb{R}^n$ whose gradient belongs to $L^1(\gamma_n)$, belongs to the Orlicz space $L^1(\log L)^{1/2}$ of this measure, connecting with logarithmic Sobolev inequalities and hypercontractivity.

The second type of use of Borell's inequality is for "large" values of u (while Borell's inequality for large values of u follows from (1.4), the spirit of application is very different). It is mostly used in the following forms

$$\text{If } \gamma_n(A) \geq 1/2, \text{ then } \gamma_n(A_u) \geq \gamma_1((-\infty, u]) \tag{1.5}$$

$$\text{If } \gamma(A) \geq 1/2, \text{ then } \gamma_*(A + uB) \geq \gamma_1((-\infty, u]) \geq 1 - \frac{1}{2}\exp(-u^2/2). \tag{1.6}$$

In the terminology of V. Milman [M2] (1.5) is a "concentration of measure phenomenon". An immediate consequence of (1.5) is that if f is a Lipschitz function on $\mathbb{R}^n$, we have

$$\gamma_n(\{|f - M_f| \geq u\}) \leq 2\gamma_1\left(\left[\frac{u}{\|f\|_{\mathrm{Lip}}},\infty\right)\right) \leq \exp\left(-\frac{u^2}{2\|f\|_{\mathrm{Lip}}^2}\right) \tag{1.7}$$

where M_f is a median of f, i.e. $\gamma_n(\{f \geq M_f\}) = \gamma_n(\{f \leq M_f\}) = 1/2$, and where $\|f\|_{\mathrm{Lip}} = \sup_{x \neq y} |f(x) - f(y)|/\|x-y\|$.

It has been discovered by V. Milman [M1] that (1.7) (or, equivalently, a corresponding inequality on the sphere S_n) is at the root of the celebrated Dvoretzky's theorem. Actually, the following inequality is sufficient to prove Dvoretzky's theorem: There is a numerical constant K such that if f is a Lipschitz function on $\mathbb{R}^n$, then

$$\gamma_n(\{|f - \int f\, d\gamma_n| \geq u\}) \leq 2\exp\left(-\frac{u^2}{K\|f\|_{\mathrm{Lip}}^2}\right). \tag{1.8}$$

A very simple proof of this inequality (1.8) was discovered by B. Maurey and G. Pisier (cf. [P]; in that same reference is included a different proof due to Maurey using stochostic integrals which yields $K = 2$).

To understand better the relationship between (1.7) and (1.8) one should note that either of these inequalities imply the fact that $|M_f - \int f d\gamma_n| \leq K\|f\|_{\mathrm{Lip}}$. Here, as in the sequel, K denotes a universal constant, not necessarily the same at each occurence. It is not our purpose here to enter the topic of local theory of Banach spaces, that was covered by Milman's paper [M2], and we turn towards the application of (1.6) to probability theory. The importance of (1.6) stems from the fact that γ is the prototype for all Gaussian measures. To stress the point, we now outline the "canonical" way to look at Gaussian processes, that was put forward in [D] and that turned out to be of crucial importance. Given a point t in ℓ^2, the series $\sum_{k\geq 1} t_k x_k$ converges γ a.e. (since (x_k) is a sequence of independent r.v.) and thereby defines an element X_t of $L^2(\gamma)$, of law $N(0, \|t\|_2)$. Any subset T of ℓ^2 thus defines a Gaussian process $(X_t)_{t\in T}$. For many purposes all Gaussian processes can be reduced to this type. We say that the process is bounded if $\sup_{t\in T} X_t < \infty$ γ a.e. (to avoid technicalities, we assume from now on that T is countable).

A problem of historical importance was, given a Gaussian process (that is, in our setting a subset T of ℓ^2), to understand, under the conditions that G is bounded, what are the tails of $Y = \sup_{t\in T} |X_t|$, i.e. the behavior of the function $\gamma(\{Y \geq u\})$ as $u \to \infty$. It was proved by Landau and Shepp [L-S] and, independently by Fernique [F], that $E(e^{\alpha Y^2}) < \infty$ for some $\alpha > 0$, where for simplicity, we write Ef for $\int f d\gamma$. Interestingly, the proof of Landau and Shepp is isoperimetric in nature. In [B1], C. Borell use (1.5) as follows. Set $\sigma = \sup_{t\in T} \|t\|_2$. It is then clear that $Y(x) \leq Y(y) + \sigma u$ if $x \in y + uB$. Thus by (1.5)

$$\gamma(\{Y \geq M_Y + \sigma u\}) \leq \gamma_1([u,\infty)).$$

This implies that

$$\alpha > \sigma \Rightarrow E \exp \frac{Y^2}{2\alpha^2} < \infty . \tag{1.9}$$

C. Borell also used the same approach to obtain sharp integrability results for homogeneous chaos [B2, B3].

It turns out that, when more information is available on T (e.g. information about entropy numbers) results sharper that (1.10) can be obtained by specific methods. This has unfortunately lead some researchers to doubt the power of (1.6); the issue is that the usefulness of (1.6) is greatly enhanced by an appropriate use of A. This point was brought in particular to light in [T1], where the following is proved. Given a bounded process $T \subset \ell^2$, set

$$\tau = \inf\{u > 0; \gamma(\{\sup_{t\in T} |X_t| < u\}) > 0\} .$$

Then

$$\tau' > \tau \Rightarrow E \exp \frac{1}{2\sigma^2}(Y - \tau')^2 < \infty . \tag{1.10}$$

This result should be compared to (1.9). It can be interpreted as a tail estimate. It means that the function $f(u) = \Phi^{-1}(\gamma(\{y \le u\}))$ (that is concave by (1.3)) satisfies

$$0 \ge \lim_{u\to\infty} \left(f(u) - \frac{u}{\sigma} \right) \ge -\frac{\tau}{\sigma} . \tag{1.11}$$

Thus, $f(u)$ has an asymptote $(u/\sigma) + \ell$ with $-\tau/\sigma \le \ell \le 0$. This result is optimal in the sense that f can approach this asymptote arbitrarily slowly. We refer to [L-T2], Chapter 3, for an extension of this result to homogeneous chaos, and to [G-K] for further developments of the same idea.

While (1.10) is optimal for general processes, it can be improved when one has more information about T. In [T3] a method was introduced relying on (1.6) to improve the tail estimate (1.10) in the specific case where T is compact and there is a unique $t \in T$ with $\|t\| = \sigma$. The method has been developed further in [D-M-W]. It could also be used in many other situations, e.g. to simplify the results of [B-K].

While (1.10) uses in a rather precise form the information provided by (1.6), it is often sufficient (e.g. for the proof of Dvoretzky's theorem) to have a weaker information of the type

$$\gamma(A) \ge 1/2 \Rightarrow \gamma(A + uB) \ge 1 - K \, \exp\left(-\frac{u^2}{K} \right) \tag{1.12}$$

without precise information on the constant K. It is this principle, rather than (1.5) that we now on call the concentration of measure phenomenon (for the Gauss measure).

2. The Concentration of Measure Phenomenon

It seems rather unlikely that (1.6) could at all be improved, but it could come as a surprise that on the other hand (1.12) can be improved, in the sense that a similar inequality holds when the set $A + uB$ is replaced by a smaller (and, in some cases, much smaller) set. The central result of this section is that, in the class of product measures, the natural setting for the concentration of measure phenomenon is not the Gaussian measure γ but rather the product measure ν on $\mathbb{R}^{\mathbb{N}}$ obtained by providing each factor with the measure ν_1 of density $\frac{1}{2}e^{-|x|}$ with respect to Lebesgue measure. We set

$$B_1 = \left\{x \in \mathbb{R}^{\mathbb{N}}; \sum |x_k| \leq 1\right\}; B_2 = \left\{x \in \mathbb{R}^{\mathbb{N}}; \sum x_k^2 \leq 1\right\}.$$

Theorem 2.1 [T6]. *There exists a universal constant K such that for all subsets A of $\mathbb{R}^{\mathbb{N}}$, all $u \geq 0$, we have*

$$\nu(A) = \nu_1((-\infty, a]) \Rightarrow \nu_*(A + \sqrt{u}B_2 + uB_1) \geq \nu_1\left(\left(-\infty, a + \frac{u}{K}\right]\right). \tag{2.1}$$

In particular

$$\nu(A) \geq 1/2 \Rightarrow \nu_*(A + \sqrt{u}B_2 + uB_1) \geq \nu_1\left(\left(\infty, \frac{u}{K}\right]\right) = 1 - \frac{1}{2}\exp\left(-\frac{u}{K}\right). \tag{2.2}$$

A striking difference between this inequality and (1.6) is that the set A is enlarged by the mixture $\sqrt{u}B_2 + uB_1$ of the ℓ^2 and ℓ^1 balls, whose shape changes with the value of u. To understand the reason for the strange set $\sqrt{u}B_2 + uB_2$, it is instructive to derive from (2.1) the size of the tails $\nu(\{X_t \geq u\})$, where $X_t(x) = \sum t_k x_k$ and $t \in \ell^2$ (these can of course be obtained by a simple direct argument). The set $A = \{X_t \leq 0\}$ satisfies $\nu(A) \geq 1/2$ by symmetry. Thus by (2.1), we have $\nu_*(A + \sqrt{u}B_2 + uB_1) \geq 1 - \frac{1}{2}e^{-u/K}$. But obviously $X_t \leq \sqrt{u}\|t\|_2 + u\|t\|_\infty$ on $A + \sqrt{u}B_2 + uB_1$ (where $\|t\|_\infty = \sup_{k\geq 1}|t_k|$). Thus we get

$$\nu(\{X_t \geq \sqrt{u}\|t\|_2 + u\|t\|_\infty\}) \leq \frac{1}{2}e^{-u/K}$$

which can be formulated as

$$\nu(\{X_t \geq u\}) \leq \frac{1}{2}\exp\left(-\min\left(\frac{u^2}{K\|t\|_2^2}, \frac{u}{K\|t\|_\infty}\right)\right)$$

(and gives the correct order for $-\log \nu(\{X_t \geq u\})$.

Another difference between (2.1) and (1.6) is the unspecified constant K on the left side, that actually makes (2.1) closer to (1.12) than to (1.6). An interesting problem would be to find an "exact" version of (2.1). One could ask for example if there is a natural "smallest" set $W(u)$ (whose shape would depend on u in a possibly complex way) that could be used instead of $\sqrt{u}B_2 + uB_1$ in (2.1). The resulting inequality should give sharp estimates for $\nu(\{X_t \geq u\})$; the variety of competing estimates for this quantity [Hoe] might indicate the difficulty of the task.

The proof of Theorem 2.1 is made complicated by the fact that, in constrast with the Gauss measure or Lebesgue measure, the measure ν has less symmetries (in particular is not invariant by rotations) and thus that this restricts the use of rearrangements. The method of proof is to consider a statement similar to (2.1) (the set $\sqrt{u}B_2 + uB_1$ being replaced by a more amenable set $C(u)$ of comparable size) and prove it by induction over n, when the set A is assumed to depend on n coordinates only. The key observation is that the proof of the induction step can be deduced from a two-dimensional statement. While the proof in (2.1) is not simple, it is beyond doubt that the important part of (2.1) is (2.2) for large values of u ($u \geq K$). Fortunately, this is much simpler to prove. The idea is to prove, again by induction over the number of coordinates of which A depends, that, if one sets

$$h_A(x) = \inf\{u \geq 0; x \in A + C(u)\}$$

then $E\exp(h_A(x)/K) \leq 1/P(A)$, so that, by Chebyshev inequality,

$$\nu(A + C(u)) \geq 1 - \frac{1}{P(A)} \exp\left(-\frac{u}{K}\right),$$

that recovers (2.2) for u large enough.

We now explain why (2.2) is an improvement over (1.12). The argument that we will present will be referred to in the sequel as the "contraction argument". The precise form we use was introduced by G. Pisier [P, Ch. 2] and played an essential role in the discovery of the correct formulation of Theorem 2.1. (A similar idea occurs earlier in [G-M], Section 2-1).

Consider the increasing map ψ from $\mathbb{R}$ to $\mathbb{R}$ that transforms ν_1 into γ_1. It is a simple matter to see that

$$|\psi(x) - \psi(y)| \leq K \min(|x-y|, |x-y|^{1/2}). \tag{2.3}$$

Consider the map Ψ from $\mathbb{R}^{\mathbb{N}}$ to $\mathbb{R}^{\mathbb{N}}$, such that $\Psi((x_k)_{k\geq 1}) = (\psi(x_k))_{k\geq 1}$. Thus Ψ transforms ν into γ.

Consider a Borel set A of $\mathbb{R}^{\mathbb{N}}$ such that $\gamma(A) \geq 1/2$. Then

$$\begin{aligned}\gamma(\Psi(\Psi^{-1}(A) + \sqrt{u}B_2 + uB_1)) &= \nu(\Psi^{-1}(A) + \sqrt{u}B_2 + uB_1) \\ &\geq 1 - \frac{1}{2}\exp\left(-\frac{u}{K}\right).\end{aligned} \tag{2.4}$$

However, it follows from (2.3) that

$$A_u = \Psi(\Psi^{-1}(A) + \sqrt{u}B_2 + uB_1) \subset A + K\sqrt{u}B_2 \tag{2.5}$$

and thus (2.4) improves over (1.12). To illustrate the improvement of (2.4) over (1.12), consider the case where $A = \{x; \forall k \leq n, |x_k| \leq a_n\}$, where a_n is chosen so that $\gamma(A) = 1/2$ (and hence is of order $\sqrt{\log n}$). Then, for $u \ll \log n$, the set A_u is easily seen to be contained in

$$A+K\left(\sqrt{\frac{u}{\log n}}B_2+\frac{uB_1}{\sqrt{\log n}}\right)\subset A+\left(\frac{Ku}{\log n}\right)^{1/2}\sqrt{u}B_2,$$

where $u/\log n \ll 1$.

One intriguing aspect of Theorem 2.1, when used as an improvement over (1.12), is that it breaks the rotational invariance of the Gauss measure. Indeed, it not only tells us that $\gamma_n(A_u) \geq 1-\exp(-u/K)$ (where A_u is defined in (2.5)) but also that $\gamma_n((RA)_u) \geq 1-\exp(-u/K)$ for any rotation R of $\mathbb{R}^n$.

A natural question is whether (2.2) is the correct formulation of the concentration of measure phenomenon. This seems to be the case, at least in the setting of product measures. Indeed, consider a probability measure θ_1, on $\mathbb{R}$, and its product θ on $\mathbb{R}^{\mathbb{N}}$. Suppose that the following holds (that is much weaker than (1.2)). There exists $K>0$, such that

$$\theta(A)\geq 1/2 \Rightarrow \theta(A+KB_\infty)\geq 3/4$$

where $B_\infty = \{x\in\mathbb{R}^{\mathbb{N}}, \forall k\geq 1, |x_k|\leq 1\}$. Then the tails $f(u)=\theta(\{|x|\geq u\})$ must decay exponentially [T6]. Note that, if these tails decay exponentially in a smooth enough way, θ_1 is the image of ν_1 by a contraction, and Pisier's contraction argument presented before shows that (2.1) will also hold for θ.

Consider now $1\leq\alpha<\infty$ and the measure ν^α on $\mathbb{R}^{\mathbb{N}}$, obtained as the product measure when each factor is endowed with the probability measure $a_\alpha e^{-|x|^\alpha}dx$ (where a_α is a normalizing constant). The contraction argument presented above shows that

$$\nu^\alpha(A)\geq 1/2 \Rightarrow \nu_*^\alpha(A+U_\alpha(u))\geq 1-\exp\left(-\frac{u}{K(\alpha)}\right). \qquad (2.6)_\alpha$$

where $U_\alpha(u)=u^{1/2}B_2+u^{1/\alpha}B_\alpha$ for $\alpha\leq 2$, $U_\alpha(u)=u^{1/2}B_2\cap u^{1/\alpha}B_\alpha$ for $\alpha\geq 2$, and $B_\alpha=\{x\in\mathbb{R}^{\mathbb{N}}; \sum|x_k|^\alpha\leq 1\}$. For $\alpha>\beta$, $(2.6)_\alpha$ is a consequence of $(2.6)_\beta$ (by the contraction argument).

As in the Gaussian case, to each point $t\in\ell^2$ one can associate the random variable $X_t=\sum_{k\geq 1}t_kx_k$ on $(\mathbb{R}^{\mathbb{N}},\nu^\alpha)$; and each subset T of ℓ^2 thus defines a stochastic process. The main motivation for proving $(2.6)_\alpha$ was the discovery [T7, T8] of a new approach to the problem of finding lower bounds for $E\sup_{t\in T}X_t$ that makes $(2.6)_\alpha$ an essential step. This new approach eliminates the use of Slepian's lemma [S], which is a specific property of Gaussian processes. It replaces it by the use of $(2.6)_\alpha$, combined with a Sudakov-type minoration [Su]. It enables to describe $E\sup_{t\in T}X_t$ in terms of the geometry of T, thereby extending the results of [T2] for the Gaussian case $\alpha=2$. But due to limitations of space we cannot discuss this point further.

3. Concentration of Measure for Bernoulli Random Variables

Pisier used his contraction argument mentioned above to conclude from (1.2) that if λ_n denotes the product measure on $\mathbb{R}^n$ when $\mathbb{R}$ is equipped with the uniform measure λ_1 on $[-1, 1]$, then

$$\lambda_n(A) = \Phi(a) \Rightarrow \lambda_n(A_u) \geq \Phi\left(a + \frac{u}{K}\right). \tag{3.1}$$

Closely related to λ_n, but of somewhat greater importance in Probability (since it corresponds to random signs) is the probability μ_n on $\{-1, 1\}^n$ that gives mass 2^{-n} to each point. The problem arises whether a concentration of measure principle as strong as (3.1) holds for μ_n. This is not the case (as follows from the example given after (3.3)). The appropriate formulation for a substitute to (3.1) requires to think to $\{-1, 1\}^n$ as a subset of $\mathbb{R}^n$. For a non-empty subset A of $\{-1, 1\}^n$, we set $\varphi_A(x) = \inf\{\|x - y\|_2 ; y \in \text{ conv } A\}$, where conv A denotes the convex hull of A in $\mathbb{R}^n$.

Theorem 3.1 [T4]. $E \exp(\varphi_A^2/8) \leq 1/\mu_n(A)$.

Using Chebyshev inequality gives

$$\text{For } u \geq 0, \mu_n(\{\varphi_A \geq u\}) \leq \frac{1}{\mu_n(A)} e^{-u^2/8}. \tag{3.2}$$

We first explore the consequences of this result. Consider a *convex* function for $\mathbb{R}^n$. Then one can deduce from (3.2) that if M_f is median of f (for μ_n), we have

$$\mu_n(\{|f - M_f| > u\}) \leq 4 \exp\left(-\frac{u^2}{8\|f\|_{\text{Lip}}^2}\right). \tag{3.3}$$

This inequality should be compared to (1.7). A major difference with (1.7) is however that this result is really specific to *convex* functions. To see it, consider n even, and let $A = \{x \in \{-1, 1\}^n ; \sum_{i \leq n} x_i \leq 0\}$, so that $\mu_n(A) \geq 1/2$. Define $f(x) = \inf\{\|x - y\|_2 : y \in A\}$, so that $\|f\|_{\text{Lip}} = 1$, and $M_f = 0$. It is easy to see that for $y \in \{-1, 1\}^n$, we have $f(y) = \sqrt{2}((\sum_{i \leq n} y_i)^+)^{1/2}$. But the central limit theorem shows that $\mu_n(\{f \geq cn^{1/4}\}) \geq 1/4$ for some constant c independent of n. (Note then that $\mu_n(A_{cn^{1/4}}) \leq 3/4$ and that (3.1) fails.)

Consider now a Banach space E and vectors $(x_k)_{k \leq n}$ in E.

Set

$$\sigma = \sup\left\{\sum_{k \leq n} x^*(x_k)^2 ; x^* \in E^*, \|x^*\| \leq 1\right\}.$$

The function on $\mathbb{R}^n$ given by $f(y) = \|\sum_{k \leq n} y_k x_k\|_E$ is convex and satisfies $\|f\|_{\text{Lip}} = \sigma$. Consider a sequence $(\varepsilon_k)_{k \leq n}$ of (symmetric) Bernoulli random variables; that is, the sequence is independent identically distributed and $P(\varepsilon_i = 1) = P(\varepsilon_i = -1) = 1/2$. Then (3.3) implies

$$P\left(\left|\left\|\sum_{k\le n}\varepsilon_k x_k\right\| - M\right| \ge u\right) \le 4e^{-u^2/8\sigma^2} \tag{3.4}$$

where M is a median of $\|\sum_{k\le n}\varepsilon_k x_k\|$. From (3.4) and elementary computations follows that

$$\left\|\sum_{k\le n}\varepsilon_k x_k\right\|_p \le \left\|\sum_{k\le n}\varepsilon_k x_k\right\|_1 + K\sigma p^{1/2},$$

a precise form of the so called Khintchine-Kahane inequalities.

It is not known whether the exponent 1/8 in (3.2) can be improved; the best possible exponent would be 1/2. Another problem of interest would be the determination of $\min\{\mu_n(\{\varphi_A \ge u\}); \mu_n(A) = u\}$. It is likely that the sets which achieve this minimum depend on A, u; thus the problem might be difficult.

It is of interest to compare Theorem 3.1 with the classical results concerning Hamming distance. The Hamming distance $d(s,t)$ of two points s,t of a product of sets is the number of coordinates where s,t differ. For a subset A of $\{-1,1\}^n$, we set $d_A(x) = \inf\{d(x,y), y \in A\}$. It follows from an isoperimetric inequality of Harper [Ha] that for $\mu_n(A) \ge 1/2$, we have

$$\mu_n(\{d_A \ge u\sqrt{n}\}) \le \exp(-2u^2). \tag{3.5}$$

On the other hand, it is simple to see that $2d_A \le \sqrt{n}\varphi_A$. Thus $\{d_A \ge u\sqrt{n}\} \subset \{\varphi_A \ge 2u\}$. Now (3.2) provides the estimate

$$\mu_n(\{\varphi_A \ge 2u\}) \le 2\exp(-u^2/2).$$

Compared with (3.5), this provides a weaker bound (but of the same essential strength) for a larger set. The most important difference is however that (3.2), in contrast with (3.5), is independent of the dimension.

We now present an "abstract" extension of Theorem 3.1. Consider a sequence $(\Omega_k, \mu_k)_{k\le n}$ of probability spaces and denote by P the product measure on $\Omega = \prod_{k\le n}\Omega_k$. Consider a subset A of Ω. For $x \in \Omega$, consider the set

$$U_A(x) = \{t \in \{0,1\}^n; \exists y \in A, t_k = 0 \Rightarrow x_k = y_k\}.$$

We consider $U_A(x)$ as a subset of $\mathbb{R}^n$; we denote by $V_A(x)$ the convex hull of $U_A(x)$.

For $\alpha \ge 1$, $0 \le u \le 1$, we consider the function

$$\xi(\alpha,u) = \alpha(1-u)\log(1-u) - (\alpha+1-\alpha u)\log\left(1 - \frac{\alpha u}{1+\alpha}\right).$$

Elementary calculus show that this function increases in α, u, and is convex in u.

We set

$$\psi_{\alpha,A}(x) = \inf\left\{\sum_{i\le n}\xi(\alpha, y_i); y = (y_i)_{i\le n} \in V_A(x)\right\}.$$

Theorem 3.2 [T9]. $E\exp\psi_{\alpha,A} \le (1/P_*(A))^\alpha$.

Calculus shows that $\xi(1,u) \geq u^2/4$; thus Theorem 3.2 implies Theorem 3.1, but only with the worse exponent 1/16 instead of 1/8. An essential improvement of Theorem 3.2 over Theorem 3.1 is that for α large and u close to 1, $\xi(\alpha,u)$ is of order $\xi(\alpha,1) = \log(1+\alpha)$. The following bound seems to be of particular interest. For $t \geq 1$,

$$P_x(A) \geq 1/2 \Rightarrow P(\{\psi_{t,A} \geq t\}) \leq (2/e)^t .$$

It has been observed in [J-S] (using the method of [T4]) that if $0 < \eta < 1$, and if μ_n denotes now the measure $((1-\eta)\delta_0 + \eta\delta_1)^n$ on $\{0,1\}^n$, then for a set $A \subset \{0,1\}^n$, we have $E\exp\{\varphi_A^2/4\} \leq 1/\mu_n(A)$ (this also follows from Theorem 3.2). An interesting fact in that direction is that the tails of φ_A do *not* improve when η is small. This is somewhat unexpected. To see it, consider the case where $A = \{x \in \{0,1\}^n; \sum_{k\leq n} x_k \leq \eta n\}$, so that $\mu_n(A)$ is of order 1/2 by the law of large numbers. On the other hand, it is simple to see that (for ηn integer) $\varphi_A(y) \leq u$ if and only if $\sum y_k = p$ where $\sqrt{p}(1-\eta n/p) \leq u$, so that $p \leq \eta n + u\sqrt{p}$. For $u \leq (\eta n)^{1/2}$, this implies $p \leq 2\eta n$, so that $p \leq \eta n + u\sqrt{2\eta n}$. Thus for $p > \eta n + u\sqrt{2\eta n}$ we have $\varphi_A(y) > u$. It follows from the central limit theorem that if $0 < \eta \leq 1/2$, then for n large enough, we have $\mu_n(\{\varphi_A(y) > u\}) \geq \exp(-cu^2)$ for some c independent of n, η.

4. An Isoperimetric Inequality for Product Measure

An important concentration of measure phenomenon for product measures is as follows. Consider a sequence $(\Omega_k, \mu_k)_{k\leq n}$ of probability spaces. Denote by P the product measure on $\Omega = \prod_{k\leq n} \Omega_k$. Then

$$P(A) \geq 1/2 \Rightarrow P(\{d_A \geq u\}) \leq 2\exp(-u^2/Kn) . \tag{4.1}$$

where the Hamming distance d_A has been introduced in Section 3. This is an extension of (3.5) (with worse constants). It is easy to prove using the martingale approach introduced by B. Maurey and developed by G. Schechtman (see [M-S]). It also follows from Theorem 3.2 the way (3.5) follows from Theorem 3.1. (This approach gives a constant $K = 4$ in the exponent.)

For a set $A \subset \Omega$, and $k, q \geq 0$, consider

$$H(A,q,k) = \left\{ y \in \prod_{k\leq n} \Omega_k ; \exists x^1, \cdots, x^q \in A;\ \operatorname{card}\{i; \forall \ell \leq q, y_i \neq x_i^\ell\} \leq k \right\} .$$

For $q = 1$, this is exactly the set $\{d_A \leq k\}$. The set $H(A,q,k)$ can be thought of as an "enlargement" of A, although it does not seem possible to define it as a neighborhood of A for a distance.

Theorem 4.1 [T5]. *For some universal constant K, and all $k, q \geq 1$, we have*

$$P(A) \geq 1/2 \Rightarrow P_*(H(A,q,k)) \geq 1 - \left(\frac{K}{k} + \frac{K}{q\log q}\right)^k . \tag{4.2}$$

As stated, this theorem gives information only when k, q are large. However it is also possible to show that if $q \geq 2$, $k \geq k_0$, then

$$P(A) \geq 1/2 \Rightarrow P_*(H(A,q,k)) \geq 1 - \eta^k$$

where $\eta < 1$ is universal. In contrast with the case $q = 1$ (4.1), the estimate (4.2) is independent of the number of coordinates (and thus can be extended to the case of an infinity of factors.)

To gain some intuition about (4.2), it is useful to consider the case where $\Omega_i = \{0,1\}, \mu_i(\{0\}) = 1 - 1/n, \mu_i(\{1\}) = 1/n$, and

$$A = \left\{ (x_i); \sum_{i \leq n} x_i \leq 1 \right\}.$$

In that case, $P(A) \geq 1/2$ and

$$H(A,q,k) = \left\{ (x_i); \sum_{i \leq n} x_i \leq q + k \right\}.$$

For k of order $q \log q$, simple estimates show that

$$P(H(A,q,k)) \leq 1 - \left(\frac{1}{Kq \log q} \right)^k,$$

which should be compared to (4.2).

Theorem 4.1 has strong implications about the behavior of sums of independent random variables valued in a Banach space. Consider such variables $X_1, \cdots, X_n$ valued into a separable Banach space F. We now outline a method to obtain bounds on the tails of $\| \sum_{i \leq n} X_i \|$. (These bounds can now also be derived from Theorem 3.2, which has a considerably simpler proof than Theorem 4.1. Tail estimates are in particular at the root of classical theorems like strong laws of large numbers and laws of the iterated logarithm.) While this method might look complicated at first glance, it seems to capture the size of these tails in essentially all the situations; see e.g. [T5, L-T1]. Without essential loss of generality, one can assume that the variables are symmetric, i.e. X_i has the same distribution as $-X_i$. Consider a sequence $(\varepsilon_i)_{i \leq n}$ of Bernoulli random variables, that can be assumed to be independent of $(X_i)_{i \leq n}$. Thus $\sum_{i \leq n} \varepsilon_i X_i$ has the same distribution as $\sum_{i \leq n} X_i$. We then write

$$\begin{aligned} \left\| \sum_{i \leq n} \varepsilon_i X_i \right\| &= E_\varepsilon \left\| \sum_{i \leq n} \varepsilon_i X_i \right\| + \left(\left\| \sum_{i \leq n} \varepsilon_i X_i \right\| - E_\varepsilon \left\| \sum_{i \leq n} \varepsilon_i X_i \right\| \right) \\ &:= (\mathrm{I}) + (\mathrm{II}) \end{aligned} \tag{4.3}$$

where E_ε is the conditional expectation given $(X_i)_{i \leq n}$. Denote by μ_i the law of X_i on F; Consider a set $A \subset F^n$,and suppose that

$$A \subset \left\{ (x_1, \cdots, x_n) \in F^n, E_\varepsilon \left\| \sum_{i \leq n} \varepsilon_i x_i \right\| \leq M \right\}. \tag{4.4}$$

Then it is easy to see that

$$(x_1, \cdots, x_n) \in H(A, q, k) \Rightarrow \left\| \sum_{i \le n} \varepsilon_i x_i \right\| \le qM + \sum_{i \le k} \|x_i\|^*$$

where $\|x_i\|^*$ is the i-th largest term of the sequence $(\|x_i\|)_{i \le n}$. Suppose now that $P((X_1, \cdots, X_n) \in A) \ge 1/2$ (e.g. if $M = 2E\| \sum_{i \le n} \varepsilon_i X_i \|$ and there is equality in (4.4)). It then follows from (4.2) that, if $k \ge q$

$$P((\mathrm{I}) \ge qM + \sum_{i \le k} \|X_i\|^*) \le \left(\frac{K}{q}\right)^k . \tag{4.5}$$

On the other hand, if we set

$$\sigma_X^2 = \sup \left\{ \sum_{i \le n} x^*(X_i)^2 ; x^* \in E^*, \|x^*\| \le 1 \right\} ,$$

it follows from (3.4) that, conditionally on $X_1, \cdots, X_n$

$$P((\mathrm{II}) \ge K(1+u)\sigma_X) \le 4e^{-u^2} . \tag{4.6}$$

To make (4.5), (4.6) usable, it remains to control $\sum_{i \le n} \|X_i\|^*$ (which is a problem about real-valued random variables) and σ_X. Several methods have been developed for that purpose; adjusting the various parameters involved has allowed to get bounds of the right order in all the problems studied to date; cf. [L-T2], Chapters 6 to 8.

References

[A] R. J. Adler: An introduction to continuity, extrema, and related topics for general Gaussian processes. Forthcoming book

[B-K] S. Berman, N. Kôno: The maximum of a Gaussian process with non constant variance; a sharp bound for the distribution tail. Ann. Probab. **17**, 632–650 (1989)

[B1] C. Borell: The Brunn-Minkowski inequality in Gauss space. Invent. math. **30**, 207–216 (1975)

[B2] C. Borell: Tail probabilities in Gauss space. Vector Spaces Measures and Application, Dublin 1977. (Lecture Notes in Mathematics, vol. 644.) Springer, Berlin Heidelberg New York 1978, pp. 71–82

[B3] C. Borell: On polynomial chaos and integrability. Prob. Math. Statist. **3**, 191–203 (1984)

[D] R. M. Dudley: The sizes of compact subsets of Hilbert space and continuity of Gaussian processes. J. Funct. Anal. **1**, 290–330 (1987)

[D-M-W] V. Dobric, M. B. Marcus, M. Weber: The distribution of the large values of the supremum of a Gaussian process

[E1] A. Ehrhard: Symétrisation dans l'espace de Gauss. Math. Scand. **53**, 281–301 (1983)

[E2] A. Ehrhard: Inégalités isoperimétriques et intégrales de Dirichlet Gaussiennes. Ann. Sci. Ec. Norm. Sup. **17**, 317–332 (1984)

[E3] A. Ehrhard: Eléments extrémaux pour les inégalités de Brunn-Minkowski gaussiennes. Ann. Inst. H. Poincaré **22**, 149–168 (1986)

[F] X. Fernique: Intégrabilité des vecteurs gaussiens. C. R. Acad. Sci. Paris **270**, 1698–1699 (1970)

[F-L-M] T. Figiel, J. Lindenstrauss, V. Milman: The dimensions of almost spherical sections of convex bodies. Acta Math. **139**, 52–94 (1977)

[G-K] V. Goodman, J. Kuelbs: Rates of clustering of some self similar Gaussian processes. Manuscript, 1990

[G-M] M. Gromov, V. D. Milman: A topological application of the isoperimetric inequality. Amer. J. Math. **105**, 843–854 (1983)

[Ha] L. H. Harper: Optimal numbering and isoperimetric problems on graphs. J. Comb. Theory **1**, 385–393 (1966)

[Hoe] W. Hoeffding: Probability inequalities for sums of bounded random variables. J. Amer. Statist. Assoc. **58**, 13–30 (1963)

[Hof] J. Hoffmann-Jørgensen: Sums of independent Banach space valued random variables. Studia Math. **52**, 159–186 (1974)

[J-S] W. B. Johnson, G. Schechtman: Remarks on Talagrand's deviation inequality for Rademacher functions. Texas Functional Analysis Seminar 1988–89. The University of Texas, 1989

[K] J. P. Kahane: Some random series of functions. Cambridge University Press, 1985

[K-S] S. Kwapien, J. Szulga: Hypercontraction methods for comparison of moments of random series in normed spaces. Ann. Probab. (to appear)

[L-S] H. Landau, L. A. Shepp: On the supremum of a Gaussian process. Sankhyà **32**, 369–378 (1970)

[L] M. Ledoux: Isoperimétrie et inégalités de Sobolev logarithmiques gaussiennes. C. R. Acad. Sci. Paris **306**, 79–82 (1988)

[Lé] P. Lévy: Problèmes concrets d'analyse fonctionelle. Gauthier Villars, 1951

[L-T1] M. Ledoux, M. Talagrand: Some application of isoperimetric methods to strong limit theorems for sums of independent random variables, Ann. Probab. **18**, 754–789 (1990)

[L-T2] M. Ledoux, M. Talagrand: Probability in Banach spaces. (Ergebnisse der Mathematik, Bd. 23). Springer, Berlin Heidelberg New York 1991

[McK] H. P. McKean: Geometry of differential space. Ann. Probab. **1**, 197–206 (1973)

[M1] V. D. Milman: New proof of the theorem of Dvoretzky on sections of convex bodies. Funct. Anal. Appl. **5**, 28–37 (1971)

[M2] V. D. Milman: The Concentration Phenomenon and Linear Structure of Finite Dimensional Normed Spaces. Proceedings of the International Congress of Mathematicians, Berkeley 1986, pp. 961–975

[M3] V. D. Milman: The heritage of P. Lévy in functional analysis. Asterisque **157–158**, 273–302 (1988)

[M-S] V. D. Milman, G. Schechtman: Asymptotic theory of finite dimensional normed spaces. (Lecture Notes in Mathematics, vol. 1200.) Springer, Berlin Heidelberg New York 1986

[O] R. Osserman: The isoperimetric inequality. Bull. Amer. Math. Soc. **84**, 1182–1238 (1978)

[P] G. Pisier: Probabilistic methods in the geometry of Banach spaces. Probability and Analysis, Varenna (Italy) 1985. (Lecture Notes in Mathematics, vol. 1206.) Springer, Berlin Heidelberg New York 1986, pp. 167–241

[S] D. Slepian: The one sided barrier problem for Gaussian noise. Bell System Tech. J. **41**, 463–501 (1962)

[Su] V. N. Sudakov: Gaussian measures, Cauchy measures and ε-entropy. Sov. Math. Dokl. **10**, 310–315 (1969)

[S-T] V. N. Sudakov, B. S. Tsirelson: Extremal properties of half spaces for spherically invariant measures. J. Sov. Math. **9**, 9–18 (1978) [translated from Zap. Nauch. Sem. L.O.M.I. **41**, 14–24 (1974)]

[T1] M. Talagrand: Sur l'intégrabilité des vecteurs gaussiens. Z. Wahrsch. Verw. Gebiete **68**, 1–8 (1984)

[T2] M. Talagrand: Regularity of Gaussian processes. Acta Math. **159**, 99–149 (1987)

[T3] M. Talagrand: Small tails for the supremum of a Gaussian process. Ann. Inst. H. Poincaré **24**, 307–315 (1988)

[T4] M. Talagrand: An isoprimetric theorem on the cube and the Kintchine-Kahane inequalities. Proc. Amer. Math. Soc. **104**, 905–909 (1988)

[T5] M. Talagrand: Isoperimetry and integrability of the sum of independent Banach-space valued random variables. Ann. Probab. **17**, 1546–1570 (1989)

[T6] M. Talagrand: A new isoperimetric inequality and the concentration of measure phenomenon, to appear in Geometric aspects of functional analysis (Israel seminar). (Lecture Notes in Mathematics, vol. 1469.) Springer, Berlin Heidelberg New York 1991

[T7] M. Talagrand: Supremum of canonical processes. Manuscript, 1990

[T8] M. Talagrand: Regularity of infinitely divisible processes. Manuscript, 1990

[T9] M. Talagrand: A new isoperimetric inequality for product measure and the tails of sums of independent random variables. Manuscript, 1990

Random Walks and Diffusions on Fractals

Martin T. Barlow

Statistical Laboratory, University of Cambridge, 16 Mill Lane
Cambridge CB2 1SB, England

1. Introduction

The field this paper will survey could be called *analysis on fractals*: more specifically it is the study (by analytic or probabilistic means) of the fundamental second order PDEs on fractal spaces. The original motivation came from mathematical physicists working on the properties of disordered media, and interested in questions such as heat conduction, vibration modes etc. There is experimental evidence that fractals can provide a good model for certain kinds of disordered media; and it is hoped that the study of PDEs on fractal spaces would give at least some information about PDEs in disordered media. See [AO, RT] for early work by mathematical physicists, and [HBA] for a survey of the now very extensive physics literature.

Let $F \subseteq \mathbb{R}^d$ be a connected self-similar fractal with Hausdorff dimension $d_f = d_f(F)$, and let μ_F denote Hausdorff x^{d_f}-measure on F. (See Hutchinson [H] for a construction of such sets via families of linear maps.) The heat equation on F should take the form

$$\Delta_F u = \frac{\partial u}{\partial t}, \quad u(x,0) = u_0(x), \quad x \in F, \tag{1.1}$$

where $u : F \times \mathbb{R}_+ \to \mathbb{R}$, $u_0 \in C(F)$, and Δ_F is a 'Laplacian' operator (i.e. self-adjoint with respect to μ, local, non-positive, satisfying $\Delta_F 1 = 0$) acting on a subspace $\mathscr{D}(\Delta_F) \subset C(F)$. The following problems arise immediately:

(a) *Existence.* The construction of a suitable Δ_F which is F-isotropic, that is, locally invariant with respect to local isometries of F.
(b) *Uniqueness.* Is Δ_F characterised (up to a scale factor) by the property of being F-isotropic?
(c) *Properties.* The form of the functions in $\mathscr{D}(\Delta_F)$, and the properties of solutions to the Laplace, heat etc. equations associated with F.

Although the questions stated above are purely analytic, a natural approach (and historically the first one) is to approach them probabilistically, and seek to construct an F-isotropic diffusion process $X_t, t \geq 0$ on F. Then the infinitesimal generator of X is a natural candidate for the Laplacian Δ_F, and the transition density of X is a solution to the heat equation. Of course the process X is an interesting object, worthy of study in its own right.

Proceedings of the International Congress
of Mathematicians, Kyoto, Japan, 1990

In this survey I will concentrate on the probabilistic approach, on the construction and behaviour of diffusion processes on some specific classes of fractals. The subject is still quite young, and though some general patterns and unifying approaches are beginning to emerge, it still consists to a considerable extent of a collection of specific examples.

I will also concentrate on rigorous results in the mathematical literature. However, many of the key concepts, and in particular the indices d_f, d_w and d_s were introduced earlier in physics papers.

2. The Sierpinski Gasket

This is the simplest non-trivial connected fractal, and is the natural starting point for any investigation of fractal spaces. Hutchinson's theory ([H, F]) provides a convenient description. Let $a_1 = (0,0)$, $a_2 = (1,0)$, $a_3 = (\frac{1}{2}, \frac{1}{2}\sqrt{3})$, let $G_0 = \{a_1, a_2, a_3\}$, A_0 be the closed convex hull of G_0, and let ϕ_i, $1 \le i \le 3$, be the linear contractions defined by $\phi_i(x) = \frac{1}{2}(x + a_i)$. For any set $B \subseteq \mathbb{R}^2$ set

$$\Phi(B) = \bigcup_i \phi_i(B), \tag{2.1}$$

and let Φ^n be the n-fold convolution of Φ. The Sierpinski gasket G may be defined by

$$G = \bigcap_{n=0}^{\infty} \Phi^n(A_0) = cl\left(\bigcup_{n=0}^{\infty} \Phi^n(G_0)\right). \tag{2.2}$$

Note that $\Phi^n(A_0)$ consists of 3^n triangles each of side 2^{-n}. Following the terminology of Lindstrøm [L] we call any set of the form $B \cap G$ (respectively $B \cap \Phi^n(G_0)$) an *n-complex* (respectively *n-cell*), where B is any triangle of side 2^{-n} in $\Phi^n(A_0)$. Write $G_n = \Phi^n(G_0)$.

By [H] the Hausdorff dimension of G is given by $\dim G = \log 3/\log 2 = d_f(G) = d_f$. Let μ_G be the multiple of Hausdorff-x^{d_f} measure on G which assigns mass 1 to G. The key property of G used below is that it is finitely ramified:

(FR) if H_1 and H_2 are two adjacent n-complexes, with common point x, then any continuous path in $H_1 \cup H_2$ from H_1 to H_2 passes through x.

The natural approach to the construction of a diffusion X on G is to approximate X by a sequence of random walks on the discrete sets G_n. Call any pair of points $x, y \in G_m$ *neighbours* if x and y belong to the same m-cell, and write $N_m(x)$ for the set of neighbours of x: note that $\#N_m(x) = 4$ for $x \in G_m \setminus G_0$. Let $Y^m(r)$, $r \ge 0$ be a simple symmetric nearest-neighbour random walk on G_m. Initially we will take $Y^m(0) = 0$ for each m.

Define the sequences of stopping times

$$S_m = \min\{r \ge 0 : Y^m(r) \in G_0 \setminus \{0\}\},$$
$$T^m_{k+1} = \min\{r \ge T^m_k : Y^m(r) \in G_{m-1} \setminus \{Y^m(T^m_k)\}\}, \quad T^m_0 = 0.$$

Set

$$\tilde{Y}^{m-1}_k = Y^m(T^m_k), \quad k \ge 0;$$

we call $\tilde{Y}^{m-1}$ the *decimated random walk on* G_{m-1}. The following lemma, which is crucial in the analysis of the walks Y^m, is any easy consequence of the geometry of G, and in particular of (FR).

Lemma 1 ('Decimation invariance' [G, Ku1, BP]).
(a) *The random walks* Y^{m-1} *and* $\tilde{Y}^{m-1}$ *are equal in law.*
(b) *The r.v.* $(T^m_{k+1} - T^m_k, k \geq 0)$ *are i.i.d. and equal in law to* S_1.

A straightforward calculation gives $ES_1 = 5$, and using Lemma 1(b) it follows that $ES_m = 5^m$. This suggests that, to obtain a weak limit, one should consider the processes

$$X^{(m)}_t = Y^m([5^m t]) \quad t \geq 0\,. \tag{2.3}$$

Theorem 2 [G, Ku1, BP]. *The processes* $S^{(m)}$, $m \geq 0$, *converge weakly to a non-trivial continuous G-valued process* X.

The tightness of the sequence $X^{(m)}$ follows from the estimates on the crossing times S_m, and similar estimates on the crossing times of smaller n-complexes. The decimation invariance of the random walks Y^m provides the necessary connection between Y^m and Y^{m-1} to prove convergence of the entire sequence $X^{(m)}$, rather than just a subsequence.

A similar argument can be used to define X with $X_0 = x \in \bigcup G_n$. Let $X_t(w) = w(t)$ be defined on the canonical space $\Omega = C(\mathbb{R}_+, G)$, and let P^x be the law of X started at x. Then the law P^x for any $x \in G$ may be defined by considering the limits P^{x_n}, as $x_n \to x$. Set

$$P_t f(x) = E^x f(X_t)\,, \quad f \in C(G) \quad x \in G\,. \tag{2.4}$$

Theorem 3 [Ku1, BP]. (a) *The process* (P^x, X_t) *is a continuous Strong Markov process on* G.
(b) P_t *is a Feller semigroup.*
(c) P_t *is* μ*-symmetric, that is*

$$\int g(x)P_t f(x)\mu(dx) = \int P_t g(x) f(x)\mu(dx) \quad \textit{for all} \quad f,g \in C(G)\,.$$

The proofs of (a) and (b) use no really new ideas, but do involve a number of somewhat messy technicalities. It would be nice to see a really clean construction. (c) is an immediate consequence of the symmetry of the random walks Y^m.

Let $(\mathcal{L}, \mathcal{D}(\mathcal{L}))$ be the infinitesimal generator of the semigroup P_t: then $\mathcal{L}$ is a Laplacian operator on G, so taking $\Delta_G = \mathcal{L}$ gives a solution to the existence problem (2.1)(a). Write $\mathcal{E}(f,f)$ for the Dirichlet form associated with the semigroup P_t – see [Fu].

We now consider the second problem, that is, the uniqueness of X and $\mathcal{L}$. A *local isometry* of G is a triple (H,J,ϕ), where H,J are subsets of G, and $\phi : H \to J$ is an isometry in the intrinsic metric d_G. ($d_G(x,y)$ is the Euclidean length of the shortest path between x and y in G: it is easily seen that the intrinsic and Euclidean metrics are equivalent.) Let Π_1 be the set of local isometries of the form (H,J,ϕ) where H,J are n-complexes, and ϕ is the restriction to G of a linear isometry in $\mathbb{R}^2$. Let Π_2 be the set of isometries of the form $(H_1 \cup H_2, H_1 \cup H_2, \varphi)$, where H_1, H_2 are a pair of adjacent n-complexes, with common point x say, and φ fixes one of the H_i and reflects the other in the axis of symmetry which passes through x.

It is easy to see from the corresponding property of the Y^m that X is locally invariant under local isometries. By analogy with the case of $\mathbb{R}^d$ (where standard Brownian motion is the unique diffusion invariant with respect to translations and rotations) it is therefore reasonable to call X *Brownian motion* on F.

Theorem 4 [BP, Theorem 8.1]. *Let $(Y(t), \tilde{P}^x)$ be a diffusion process on G, and for $H \subset G$ write $T_H = \inf\{t \geq 0 : Y_t \notin H\}$. Suppose that, whenever $(H, J, \phi) \in \Pi_1 \cup \Pi_2$ is a local isometry of G, and $x \in H$, the $\tilde{P}^x$ law of $\phi(Y(T \wedge \cdot))$ and the $\tilde{P}^{\phi(x)}$ law of $Y(T_J \wedge \cdot)$ are equal. Then there exists a constant $a \in [0, \infty)$ such that the $\tilde{P}^x$ law of $(Y(t), t \geq 0)$ equals the P^x law of $(X_{at}, t \geq 0)$.*

Remark. The class of isometries required to ensure uniqueness is slightly larger than one would initially expect. The isometries in the class Π_2 are necessary to exclude the process (similar to some discussed in [G]) which moves along the edge of the largest m-complex it lies in.

We now turn to the properties of the process X. Set

$$d_w = d_w(G) = \log 5 / \log 2\,; \tag{2.5}$$

this number governs the space-time scaling of the process X. Set

$$\Phi(t, x) = t^{-d_f/d_w} \exp(-(x^{d_w} t^{-1})^{1/(d_w-1)})\,. \tag{2.6}$$

Theorem 5 [BP, Theorem 1.5]. *There exists a continuous symmetric function $p_t(x, y)$, $(t, x, y) \in (0, \infty) \times G \times G$ such that*
(a) *p_t is the transition density of X with respect to μ:*

$$P_t f(x) = \int f(y) p_t(x, y) \mu(dy)\,, \quad x \in G\,, \quad f \in bG\,,$$

(b) *there exist constants $c_1, \ldots, c_4$ such that*

$$c_1 \Phi(t, c_2|x - y|) \leq p_t(x, y) \leq c_3 \Phi(t, c_4|x - y|) \quad \textit{for} \quad x, y \in G, 0 < t \leq 1\,, \tag{2.7}$$

(c) *$u(t, x) = p_t(x_0, x)$ is the fundamental solution of the heat equation on G with pole at x_0:*

$$\frac{\partial u}{\partial t} = \mathcal{L} u\,; \quad u(0, \cdot) = \delta_{x_0}\,. \tag{2.8}$$

Remarks. 1. In [BP] this was actually proved for the process on the unbounded gasket $\tilde{G} = \bigcup_{n=0}^{\infty} 2^n G$, and in this case the estimate (2.7) holds for $t \in (0, \infty)$.
2. Setting $x = y$ in (2.7) one obtains

$$p_t(x, x) \approx t^{-d_f/d_w} \qquad 0 < t \leq 1\,. \tag{2.9}$$

(Here $\approx$ means 'bounded above and below by constants'.)

Let $-\lambda_1 \geq -\lambda_2 \geq \ldots$ be the eigenvalues of $\mathscr{L}$, let φ_i be the corresponding eigenfunctions, and set $N(\lambda) = \#\{\lambda_i : \lambda_i \leq \lambda\}$. Then the transition density $p_t(x, y)$ has an expansion

$$p_t(x, y) = \sum e^{-\lambda_i t} \varphi_i(x)\varphi_i(y) ;$$

setting $x = y$ and integrating over G one obtains for $t < 1$

$$t^{-d_f/d_w} \approx \int p_t(x, x)\mu(dx) = \sum e^{-\lambda_i t} = \int_0^\infty e^{-\lambda t} N(d\lambda),$$

and inverting the Laplace transform it follows that

$$N(\lambda) \approx \lambda^{d_f/d_w} \quad \text{as} \quad \lambda \to \infty . \tag{2.10}$$

The quantity $d_s = 2d_f/d_w$ is referred to as the 'spectral dimension' of G; from (2.9) and (2.10) one sees that it governs the short-time asymptotics of the transition function, and the frequency of small eigenvalues of $\mathscr{L}$.

There is now a well developed 'machine' for obtaining estimates on the transition functions of symmetric Markov processes from geometric properties of the underlying space – see [CKS, V] and the papers cited therein. However, this method does not appear to work well in the context of fractal spaces:

(a) It does not seem to be easy to obtain the correct L^1-Sobolev inequality as a starting point. In particular, using an isoperimetric inequality does not give the correct value of d_s – see [O, BBS and BB4].
(b) Even when one has on-diagonal estimates for p_t, there are problems in applying E.B. Davies' method (see [CKS]) for obtaining off-diagonal estimates. One major difficulty is that the measures $\nu_{[f,g]}$ associated with the Dirichlet form $\mathscr{E}(f, g)$ are singular with respect to μ_G – see [Ku2].

The proof of Theorem 6(b) proceeds as follows, and it is probable that a similar approach will work for other fractals with $d_s < 2$.

(1) Obtain estimates on the potential kernel density $u_A(x, y)$ for the process X killed on leaving the region $A \subset G$. In particular, one finds that if $B_m(x) = \{y \in G : |x - y| < 2^{-m}\}$ then

$$u_{B_m(x)}(x, x) \approx (\tfrac{3}{5})^m = (2^{-m})^{d_w - d_f} . \tag{2.11}$$

(2) Obtain estimates on
(a) the P^x-law of the times T_m to leave the region $B_m(x)$,
(b) the density $f_{xy}(s)ds = P^x(T_y \in ds)$.

(3) Combine (1) and (2)(a) to obtain bounds on $u_\lambda(x, y)$, the λ-potential density. If $\lambda = 2^{md_w}$ then the stopping times T_m and R_λ (an independent negative exponential r.v. with mean λ^{-1}) are of the same order of magnitude, and therefore so are $u_\lambda(x, x)$ and $u_{B_m(x)}(x, x)$. One has

$$u_\lambda(x, x) \approx \lambda^{\frac{1}{2}d_s - 1}, \quad \lambda \geq 1 . \tag{2.12}$$

(4) Applying a Tauberian theorem, and using the fact that since X is symmetric $t \to p_t(x, x)$ is decreasing, (2.12) implies that

$$p_t(x,x) \approx t^{-d_s/2}, \quad 0 < t \le 1. \tag{2.13}$$

(5) The off-diagonal bound follows by using the first entrance decomposition

$$p_t(x,y) = \int_0^t f_{xy}(s) p_{t-s}(y,y) ds, \tag{2.14}$$

and substituting the bounds given by (2.13) and (2)(b).

Many properties of the process X follow from Theorem 6, or the estimates used in the proof. For example, integrating (2.7) over G one obtains

$$E^x |X_t - x|^2 \approx t^{d_w}, \quad 0 < t < 1, \tag{2.15}$$

while the estimates on the tail of the distribution imply the Lévy Hölder law

$$c \le \limsup_{\delta \downarrow 0} \sup_{\substack{0 \le s \le t \le 1 \\ |t-s| < \delta}} |X_t - X_s| / h(t-s) \le c', \tag{2.16}$$

where $h(u) = u^{1/d_w} (\log(1/u))^{(d_w - 1)/d_w}$. The estimates on u_λ imply that X has a jointly continuous local time $(L_t^x, x \in G, t \ge 0)$ which is Hölder continuous of order $\frac{1}{2}(d_w - d_f) - \varepsilon$ in the space variable. From this it follows that X is space-filling:

$$\{X_t, 0 \le t \le T\} = G \quad \text{for all sufficiently large } T, \text{ a.s.}$$

I conclude this section with a brief account of further work on the Sierpinski gasket.

A purely analytic approach to the potential theory on G is given by Metz [M] and Kigami [Ki1]. Kusuoka [Ku2] gives a description of the Dirchlet form $\mathscr{E}(\cdot,\cdot)$ associated with X as an infinite matrix product. Two noteworthy consequences of this are, firstly, the result (mentioned above) that the measures $\nu_{[f,g]}$ are singular with respect to μ, and secondly, that the natural filtration of X is one-dimensional, in the sense of [DV].

Fukushima and Shima [FS], [S] give an explicit description of the eigenvalues λ_i of the operator $\mathscr{L}$: a particular consequence is that

$$\lambda^{-d_s/2} N(\lambda) \quad \text{oscillates boundedly as} \quad \lambda \to \infty. \tag{2.17}$$

Most attention has been given to the Brownian motion on G. However, a more complete analysis of the analytic behaviour of G would involve a study of diffusion processes with drift. Kumagai [Kum] has recently studied an interesting class of non-symmetric diffusions on G, the 'p-stream diffusions'.

3. Other Finitely Ramified Fractals

It is natural to ask whether the results of the previous section extend to other regular fractals. [G] discusses one other fractal, the Vicsek set, and Krebs [Kr] gave a fairly complete analysis. Lindstrøm [L] defined a large class of self-similar finitely ramified fractals, called 'nested fractals', and gave a construction of 'Brownian motion' on them. For a full description of this class, see [L]; to simplify the exposition here I will restrict myself to a subfamily which, however, seems large enough to capture the essential features.

Let $F_0 = \{a_1, \dots, a_k\}$ be the vertices of a regular k-sided polygon in $\mathbb{R}^2$, and let A_0 be the closed convex hull of F_0. Let $\lambda > 1$, $M \geq k$, $a_{k+1}, \dots, a_M$ be points in A_0 and let ϕ_i, $1 \leq i \leq M$ be linear contractions with scale factor λ such that

(a) $\phi_i(x) = a_i + \lambda^{-1}(x - a_i) \quad \text{for} \quad 1 \leq i \leq M$ (3.1)

(b) $\phi_i(A_0) \subset A_0$ for $1 \leq i \leq k$,

(c) the sets $\phi_i(A_0)$ are disjoint apart from their vertices,

(d) the set $A_1 = \bigcup_{i=1}^M \phi_i(A_0)$ is connected,

(e) A_1 satisfies the same symmetries as A_0.

The set

$$F = \bigcap_{n=0}^{\infty} \Phi^n(A_0), \quad \text{where} \quad \Phi(\cdot) = \bigcup_{i=1}^{M} \phi_i(\cdot),$$

is a nested fractal. Note that the condition (c) ensures that F is finitely ramified. The definition of the n-complexes and n-cells is the same as for the Sierpinski gasket; let also $F_n = \Phi^n(F_0)$.

An example is the Vicsek set, where $M = 5$, $\lambda = 3$, $a_1, \dots, a_4$ are the corners of the unit square, and $a_5 = \frac{1}{4}\Sigma a_i$ is its centre.

The first major difficulty in constructing a Brownian motion X on F is in finding a sequence of decimation invariant random walks. One's first thought might be to take Y^m to be a nearest neighbour random walk on F_m, where x and y are m-neighbours if x and y are adjacent vertices of a polygon in some m-cell. However, the example of the Vicsek set shows this will not work: if Y^m is chosen as above then $\tilde{Y}^{m-1}$ will make diagonal transitions, while Y^{m-1} can only make horizontal or vertical ones.

As the random walks Y^m on F_m are to be symmetric, it is easiest to define them by choosing a symmetric conductivity matrix $a_m(x, y)$, $x, y \in F_n$, and defining the transition matrix p_m by

$$p_m(x, y) = a_m(x, y)/a_m(x), \quad a_m(x) = \sum_y a_m(x, y). \tag{3.2}$$

Let $\alpha_1, \dots, \alpha_{k-1}$ satisfy $\Sigma\alpha_i = 1$, $\alpha_i \in [0, 1]$, $\alpha_i = \alpha_{k-i}$, and define

$$a_m(x, y) = \begin{cases} \alpha_i & \text{if } x, y \text{ belong to the same } m\text{-cell and } x \text{ and } y \\ & \text{are } i \text{ steps apart on the circumference;} \\ 0 & \text{otherwise.} \end{cases} \tag{3.3}$$

Let P_α^x be the law of the random walk Y^1 defined by (3.2) and (3.3) with $m = 1$. Let $T = \min\{r \geq 0 : Y \in F_0 - \{a_1\}\}$, and set $f_i(\alpha) = P_\alpha^{a_1}(Y_T = a_i)$, $f(\alpha) = (f_1(\alpha), \dots, f_{k-1}(\alpha))$. It is not hard to verify that if Y^m has transition probabilities given by $\alpha = (\alpha_1, \dots, \alpha_{k-1})$, then $\tilde{Y}^{m-1}$ has transition probabilities $f(\alpha)$, so that the search for a decimation invariant walk corresponds to the study of the fixed points of the map $\alpha \to f(\alpha)$. Let $\ell = [k/2]$.

Theorem 6 [L]. *Set $K = \{\alpha : \alpha_1 \geq \alpha_2 \geq \cdots \geq \alpha_\ell\}$. Then $f : K \to K$, and is continuous. Hence f has at least one fixed point in K.*

Let $\alpha \in K$ be a fixed point of f and Y^m, $m \geq 1$ be the associated random walks. Then there exists a number $\tau = \tau(\alpha, F)$ such that the processes $X^m(t) = Y^m(\tau^m t)$ converge weakly to a process X on F.

Theorem 7 [L]. (a) *X is a μ_F-symmetric Feller diffusion on F.*
(b) *X is invariant with respect to local isometries of F.*
(c) *Let $\Delta_{\alpha,F}$ be the infinitesimal generator of X, $-\lambda_1 \geq -\lambda_2 \geq \ldots$ be the eigenvalues of $\Delta_{\alpha,F}$ and $N(\lambda) = \#\{\lambda_i : \lambda_i < \lambda\}$. Then $N(\lambda) \approx \lambda^{\frac{1}{2}d_s(\alpha,F)}$, where*

$$d_s(\alpha, F) = 2 \log \tau / \log \lambda . \tag{3.4}$$

Remarks. 1. Lindstrøm actually uses nonstandard analysis to construct X, and the weak convergence is an immediate consequence.

2. If $\alpha \notin K$ is a fixed point, a limiting process can still be constructed, but it is no longer invariant.

Following Lindstrøm, let us call the numbers λ, M, τ the *length, mass* and *time scale factors* for the fractal F. Then

$$\dim(F) = d_f(F) = \log M / \log \lambda . \tag{3.5}$$

As in the case of the Sierpinski gasket, the number

$$d_w(F) = \log \tau / \log \lambda = 2d_f / d_s \tag{3.6}$$

governs the space/time scaling of X.

The three 'dimensions' d_f, d_s, d_w have been known to mathematical physicists for some time [RT, AO]. Together they appear to summarise fairly completely the behaviour of the process X. In view of the final relation in (3.6) (which seems to hold in great generality) there are really only two independent quantities, $d_f(F)$ (called the 'fractal dimension') and the spectral dimension $d_s(F)$. Of these d_f is a 'geometrical' quantity and for self-similar fractals can be calculated immediately from the length and mass scale factors. On the other hand τ and d_s are 'analytic': it appears necessary to solve some equation on F (or its approximations) to obtain their values.

The relation $d_s \leq d_f$ (or equivalently $d_w \geq 2$) holds for nested fractals [L], for graphs [Ku1] and for Sierpinski carpets [BB3], and it is probable that it holds in general. No other relation between these dimensions seems likely.

One problem left open by [L] is the uniqueness of the fixed point. That this cannot be altogether straightforward is shown by the case of the Vicsek set, where $\alpha = \left(\frac{1}{3}, \frac{1}{3}, \frac{1}{3}\right)$ and $\beta = (0, 1, 0)$ are both fixed points. However $\beta \notin K$ is unstable, and, as remarked above, does not give rise to an invariant process on the fractal. Theorem 7 shows that each fixed point in K gives rise to a Laplacian $\Delta_{\alpha,F}$, which in turn determines an 'analytic structure' on F. The uniqueness of fixed points therefore corresponds to the question as to whether a nested fractal carries only one natural analytic structure.

Conjecture 8. *The map $\alpha \to f(\alpha)$ has exactly one fixed point in K, and this fixed point is stable.*

Very recently I have proved this in the case $k = 4$ for the subfamily of nested fractals described above.

As far as the probabilistic properties of the process X are concerned, less is known than in the case of the Sierpinski gasket. However, there is little doubt that Theorem 5 holds for many nested fractals, with the single change of $d_.(F)$ for $d_.(G)$ in (2.6).

A much wider class of finitely ramified fractals (pcf self-similar sets) has been introduced in [Ki2]. Kigami uses analytic methods, and the question of the existence of a decimation invariant walk is replaced by the existence of a 'harmonic structure'. Kigami proves that, if a harmonic structure on F exists, then a Laplacian Δ_F and Dirichlet form $\mathscr{E}_F(\cdot,\cdot)$ can be constructed. However, there are at present neither existence nor uniqueness theorems for harmonic structures on general pcf self-similar sets.

4. Sierpinski Carpets

These are the simplest examples of infinitely ramified fractals, and have been studied in [BB1, BB2, BBS, BB3, and BB4]. I will restrict attention here to the basic Sierpinski carpet (SC) – the methods and results are the same for more general SCs. Let H_0 be the closed convex unit square, and $H_1 = H_0 \setminus (\frac{1}{3},\frac{2}{3})^2$. Repeating this operation (of removing the central square) one obtains a sequence of closed sets H_n (H_n consists of 8^n squares each of side 3^{-n}); the Sierpinski carpet H is defined by

$$H = \bigcap_{n=0}^{\infty} H_n \, .$$

It is fairly clear that the methods of Sections 2 and 3 will not work here. Instead, as each H_n has a connected interior, it is natural to consider continuous approximating processes. Let W^n be Brownian motion on H_n, with orthogonal reflection on ∂H_n, and let α_n be the maximum mean crossing time of H_n by W^n, defined by

$$\alpha_n = \sup_{x \in H_n} E^x(\inf\{t \geq 0 : W^n(t) \in J \cap H_n\}) \, , \tag{4.1}$$

where $J = \{z = (x,y) \in \mathbb{R}^2 : x = 1 \text{ or } y = 1\}$. Let $X_t^n = W^n(\alpha_n t)$; the processes X^n are tight, and so if (n_k) is a subsequence such that X^{n_k} converges weakly, the limiting process X is a continuous H-valued process. By using tightness of resolvents rather than processes (and, possibly, using a different subsequence), one can ensure that the limiting process is in fact a diffusion.

The length and mass scales of H are given by $\lambda_H = 3$, $M_H = 8$. There exists a number τ ($\tau \simeq 10.0118$) such that $\alpha_n \approx (\frac{1}{9}\tau)^n$: τ is the time scale for H. Let

$$d_f(H) = \frac{\log 8}{\log 3}\, , \quad d_s(H) = \frac{2 \log 8}{\log \tau}\, ;$$

$d_f(H)$ is the Hausdorff dimension of H, while the following result ensures that $d_s(H)$ is the spectral dimension of any Brownian motion constructed in this fashion.

Theorem 9 [BB4]. *Let (n_k) be any subsequence such that the processes X^{n_k} converge weakly to a diffusion process (X_t, P^x). Then X is Feller, and μ_H-symmetric, and has* *a transition density with respect to* μ_H *which satisfies the conditions (a)–(c) of Theorem* 5.

Remark. The main gap in the theory for the SC is the lack of any kind of uniqueness result: it is possible (though rather unlikely) that the laws of the X^n have more than one limit point.

The main steps in the construction of X, and the proof of Theorem 9, are as follows. Set $K_n = H_n \cap \left[0, \frac{1}{2}\right]^2$, and let L_i, $1 \le i \le 4$ denote the 4 edges of H_0.

(1) Using a reflection principle, and a path-crossing argument, one can prove the following Harnack inequality [BB1, Theorem 3.1]:

There exists a constant $c > 0$ such that if $u : H_n \to \mathbb{R}_+$ is harmonic in $H_n - J$ then

$$\sup_{u \in K_n} u(x) \le c \inf{}_{u \in K_n} u(x) \,. \tag{4.2}$$

(2) Let

$$R_n^{-1} = \inf\{\textstyle\int_{H_n} |\nabla u|^2 dx : u = 0 \text{ on } L_0, u = 1 \text{ on } L_2\} \,; \tag{4.3}$$

thus R_n is the resistance of H_n when two opposite edges are short-circuited, and a potential difference is applied between them. Using the Harnack inequality to construct a feasible function for the variational problem one obtains [BB2, Sect. 2]

$$\alpha_n \approx \left(\tfrac{8}{9}\right)^n R_n \,. \tag{4.4}$$

(3) Further comparison arguments, of H_n with associated wire networks, give [BBS, BB3]

$$\tfrac{1}{4} R_n R_m \le R_n R_m \le 4 R_n R_m \,. \tag{4.5}$$

Standard subadditivity results now imply that there exists a number ϱ such that $R_n \approx \varrho^n$. The time scale τ_H is defined by

$$\tau_H = 8\varrho = M_H \varrho \,. \tag{4.6}$$

(4) The proof of the analogue of Theorem 5 for H now proceeds in a fashion very similar to the proof in the case of the Sierpinski gasket.

The Equation (4.6), relating the resistance growth of the H_n with the time and mass scales, is called in the physics literature an Einstein relation, and has been known to physicists for some time – see [RT]. The use of ideas from electrical theory (or, equivalently, from a mathematical viewpoint, the theory of Dirichlet forms) seems to be a powerful tool in this area.

References

[AO] Alexander, S., Orbach, R.: Density of states on fractals: "fractons". J. Phys. (Paris) Lett. **43** (1982) L625–L631

[BB1] Barlow, M.T., Bass, R.F.: Construction of Brownian motion on the Sierpinski carpet. Ann. Inst. H. Poincaré **25** (1989) 225–257

[BB2] Barlow, M.T., Bass, R.F.: Local times for Brownian motion on the Sierpinski carpet. Prob. Theor. Rel. Fields **85** (1990) 91–104

[BB3] Barlow, M.T., Bass, R.F.: On the resistance of the Sierpinski carpet. Proc. R. Soc. Lond. A. **431** (1990) 345–360

[BB4] Barlow, M.T., Bass, R.F.: Transition densities for Brownian motion on the Sierpinski carpet. Preprint 1991

[BBS] Barlow, M.T., Bass, R.F., Sherwood, J.D.: Resistance and spectral dimension of Sierpinski carpets. J. Phys. A **23** (1990) L253–L258

[BP] Barlow, M.T., Perkins, E.A.: Brownian motion on the Sierpinski gasket. Prob. Theor. Rel. Fields **79** (1988) 543–623

[CKS] Carlen, E.A., Kusuoka, S., Stroock, D.W.: Upper bounds for symmetric Markov transition functions. Ann. Inst. H. Poincaré **23** (1987) 245–287

[DV] Davis, M.H.A., Varaiya, P.: The multiplicity of an increasing family of σ-fields. Ann. Prob. **2** (1987) 958–963

[F] Falconer, K.J.: Fractal geometry. Wiley, 1990

[Fu] Fukushima, M.: Dirichlet forms and Markov processes. North-Holland/Kodansha, Tokyo 1980

[FS] Fukushima, M., Shima, T.: On a spectral analysis for the Sierpinski gasket. Preprint, 1989

[G] Goldstein, S.: Random walks and diffusion on fractals. In: Kesten, H. (ed.) Percolation theory and ergodic theory of infinite particle systems (IMA Math. Appl., vol. 8.) Springer, Berlin Heidelberg New York 1987, pp. 121–129

[H] Hutchinson, J.E.: Fractals and self-similarity. Indiana Univ. Math. J. **30** (1981) 713–747

[HBA] Havlin, S., Ben-Avraham, D.: Diffusion in disordered media. Adv. Phys. **36** (1987) 675–798

[Ki1] Kigami, J.: A harmonic calculus on the Sierpinski space. Japan J. Appl. Math. **6** (1989) 259–290

[Ki2] Kigami, J.: Harmonic calculus on P.C.F. self-similar sets. Preprint, 1989

[Kr] Krebs, W.: A diffusion defined on a fractal. PhD thesis, University of California at Berkeley, 1988

[Ku1] Kusuoka, S.: A diffusion process on a fractal. In: Ito, K., N. Ikeda, N. (eds.) Symposium on Probabilistic Methods in Mathematical Physics, Taniguchi, Katata. Academic Press, Amsterdam 1987, pp. 251–274

[Ku2] Kusuoka, S.: Dirichlet forms on fractals and products of random matrices. Publ. RIMS Kyoto Univ. **25** (1989) 659-680

[Kum] Kumagai, T.: Construction and some properties of a class of non-symmetric diffusion processes on the Sierpinski gasket. Preprint, 1990

[L] Lindstrøm, T.: Brownian motion on nested fractals. Mem. Amer. Math. Soc. **420** (1990)

[M] Metz, V.: Brownshe Bewegung auf dem Sierpinski gasket aus potentialtheoretischer Sicht. Diplomarbeit, Bielefeld, 1988

[O] Osada, H.: Isoperimetric dimension and estimates on heat kernels of pre-Sierpinski carpets. Prob. Theor. Rel. Fields **86** (1990) 469–490

[RT] Rammal, R., Toulouse, G.: Random walks on fractal structures and percolation clusters. J. Phys. Lett. **44** (1983) L13–L22

[S] Shima, T.: On eigenvalue problems for the Sierpinski pre-gaskets. To appear Japan J. Appl. Math.

[V] Varopoulos, N.Th.: Isoperimetric inequalities and Markov chains. J. Funct. Anal. **63** (1985) 215–239

Applications of Group Representations to Statistical Problems

Persi Diaconis

Department of Mathematics, Harvard University, Science Center, 1 Oxford Street
Cambridge, MA 02138 USA

Abstract

Many problems in routine statistical analysis can be interpreted as the decomposition of a representation into irreducible components and the computation and interpretation of the projection of a given vector into these components. Examples include the usual spectral analysis of time series and the statistical analysis of variance. Recently, non-commutative representations have emerged as a practical tool. A variety of approaches have come together to give a unified theory.

1. Introduction

The study of a function through the size of its coefficients in an orthogonal expansion is a standard tool. This paper shows that expansions arising from the action of a group on a set occur naturally in a variety of statistical problems.

Example (Time Series Analysis). Let $f(0), f(1), \ldots f(N-1)$, be the observed value of a series of events. For example, the $f(k)$ might be the number of children born in New York City on successive days. Data collected in time often exhibit periodic behavior; New York City birth data looks like this:

$$411, 430, 418, 396, 401, 320, 322, \ldots,$$

the pattern of 5 high values followed by two low values persists. This seems surprising until one realizes that about 20% of all births are induced and physicians don't like to work on weekends. Izenman and Zabell (1978) discuss these data.

To find and interpret such periodicities, data $f(k)$ is often transformed as

$$\hat{f}(j) = \sum_{k=0}^{N-1} e^{2\pi ijk/N} f(k).$$

The data can be recovered by the inversion theorem

$$f(k) = \frac{1}{N} \sum_{j=0}^{N-1} \hat{f}(j) e^{-2\pi ijk/N}.$$

Proceedings of the International Congress
of Mathematicians, Kyoto, Japan, 1990

If the transform $\hat{f}(j)$ is relatively large for only a few values of j, the inversion theorem shows f is well approximated by these few periodic components. This gives a simple description of f and one can try to go back and understand why the few components are large.

This "bump hunting" part of spectral analysis is fully explained in books by Brillinger (1975) and Bloomfield (1976). It is only one part of the story – continuous spectra is the other part (see Tukey (1961) – but it, and its generalizations will dominate the present treatment.

The generalizations presented here involve a finite group G acting on a finite set X. Let $f: X \to \mathbb{R}$ be a given function. In the example above, G is a cyclic group of order N acting on itself by translation. The function $f(k)$ is the number of children born on day k. In the example of the next section, G is the symmetric group acting on itself and $f(\pi)$ is the number of people in an election who ranked candidates in the permutation π.

Let $L(X) = \{f: X \to \mathbb{C}\}$. This is a vector space on which G acts by $sf(x) = f(s^{-1}x)$. Mashke's theorem implies that $L(X)$ splits into a direct sum

$$L(X) = V_0 \oplus V_1 \oplus \cdots \oplus V_J$$

where each subspace is invariant under the group (so $g \in V_i$ implies $sg \in V_i$) and the pieces are irreducible, so no further splitting is possible. Clearly $f \in L(X)$ can be written as $f = \sum_{X=0}^{J} \hat{f}_i$ with $\hat{f}_i$ the projection into V_i.

The empirical finding, to be explored further, is that the subspaces often have simple interpretations and the decomposition of f into its projection into the V_i "makes sense".

Definition. *Spectral analysis consists of the computation and interpretation of $\hat{f}_i$ and the approximation of f by as few pieces as do a reasonable job.*

This necessarily vague definition encompasses a number of areas of classical statistics. In the next section an example is presented in some detail. Section 3 gives a group-theoretic version of the classical analysis of a variance as spectral analysis. Section 4 describes modern work on ANOVA of orthogonal designs as developed by Bailey, Nelder, Speed, Tjur, and their co-workers. That these two approaches lead to the same analysis in nice cases is an important recent result of Rosemary Bailey, Chris Rowley, and their co-workers. This is developed in Section 5. The final section gives pointers to the many topics which couldn't be covered in this brief review.

Spectral analysis as outlined here is a data analytic variation of ideas suggested earlier by Alan James and Ted Hannen. Hannen's (1965) monograph is filled with innovative ideas and treats continuous problems as well. Peter Fortini's (1977) thesis is also an important source of inspiration for the treatment presented here.

Only the rudiments of group representations are needed. The beginning of the books of Ledermann (1987) or Serre (1977) are ample background. I have tried to lay out the background in Diaconis (1988).

2. An Example

This section presents data on S_5 the symmetric group on 5 letters. The data arise from an election of the American Psychological Association. This organization asks its membership to rank order 5 candidates for president. Here $G = S_5$ and $f(\pi)$ is the number of voters choosing rank order π. For example, $f\begin{pmatrix} 1 & 2 & 3 & 4 & 5 \\ 5 & 4 & 3 & 2 & 1 \end{pmatrix} = 29$ so 29 voters ranked candidate one 5th, candidate two 4th and so on. The data is shown in Table 1.

Let $\varrho : S_5 \to GL_5(\mathbb{R})$ be the usual 5-dimensional permutation representation. Thus $\varrho(\pi)$ is a 5×5 matrix with (i, j) entry 1 if $\pi(i) = j$ and zero otherwise. The Fourier transform of f at ϱ is the matrix

Table 1. American Psychological Association election data

Ranking	No. of votes cast of this type	Ranking	No. of votes cast of this type	Ranking	No. of votes cast of this type	Ranking	No. of votes cast of this type
54321	29	43521	91	32541	41	21543	36
54312	67	43512	84	32514	64	21534	42
54231	37	43251	30	32451	34	21453	24
54213	24	43215	35	32415	75	21435	26
54132	43	43152	38	32154	82	21354	30
54123	28	43125	35	32145	74	21345	40
53421	57	42531	58	31542	30	15432	40
53412	49	42513	66	31524	34	15423	35
53241	22	42351	24	31452	40	15342	36
53214	22	42315	51	31425	42	15324	17
53142	34	42153	52	31254	30	15243	70
53124	26	42135	40	31245	34	15234	50
52431	54	41532	50	25431	35	14532	52
52413	44	41523	45	25413	34	14523	48
52341	26	41352	31	25341	40	14352	51
52314	24	41325	23	25314	21	14325	24
52143	35	41253	22	25143	106	14253	70
52134	50	41235	16	25134	79	14235	45
51432	50	35421	71	24531	63	13542	35
51423	46	35412	61	24513	53	13524	28
51342	25	35241	41	24351	44	13452	37
51324	19	35214	27	24315	28	13425	35
51243	11	35142	45	24153	162	13254	95
51234	29	35124	36	24135	96	13245	102
45321	31	34521	107	23541	45	12543	34
45312	54	34512	133	23514	52	12534	35
45231	34	34251	62	23451	53	12453	29
45213	24	34215	28	23415	52	12435	27
45132	38	34152	87	23154	186	12354	28
45123	30	34125	35	23145	172	12345	30

$$\hat{f}(\varrho) = \sum_{\pi} \varrho(\pi) f(\pi).$$

This has (i, j) entry the number of people ranking candidate i in position j. This natural summary is shown in Table 2 where entries are divided by the total number of voters to give proportions.

Table 2. Percentage of voters ranking candidate i in position j

	Rank				
Candidate	1	2	3	4	5
1	18	26	23	17	15
2	14	19	25	24	18
3	28	17	14	18	23
4	20	17	19	20	23
5	20	21	20	19	20

The largest number 28 in the (3, 1) position shows 28 percent of the voters ranked candidate 3 first. Candidate 3 also had some "hate vote": 23 percent ranked 3 last. This first order summary is the first thing anyone analyzing such data would try. It is natural to ask if it captures the essence of the voting pattern or if there is more to be learned.

The data is summarized by $f \in L(S_5)$. This last vector space splits into 7 invariant subspaces in its isotypic decomposition shown in Table 3.

Table 3. Decomposition of the regular representation

$M =$	V_1	$\oplus$	V_2	$\oplus$	V_3	$\oplus$	V_4	$\oplus$	V_5	$\oplus$	V_6	$\oplus$	V_7
Dim 120	1		16		25		36		25		16		1
SS/120	2286		298		459		78		27		7		0

Table 2 amounts to looking at the projection of f into $V_1 \oplus V_2$. If $L(S)$ is treated as an inner product space with $\langle f|g\rangle = \sum f(\pi)g(\pi)$ the function f decomposes into the pieces of its orthogonal projection. The norm square of f decomposes into the norm squared of its projections by Pythagoras's theorem. These squared lengths are shown in the last line of Table 3. As usual the largest contribution comes from the projection onto the constant functions. There is also a large projection onto the space V_3. This projection is not captured in the summary of Table 2.

The space V_3 is made up of "2nd order functions", a typical such being $\pi \mapsto \delta_{\{j,j'\}}\{\pi(i), \pi(i')\}$ which is 1 if the unordered pair $\{\pi(i), \pi(i')\} = \{j, j'\}$. The span of all 2nd order functions, orthogonal to $V_1 \oplus V_2$ make up V_3. In group representations language, V_3 is the isotypic subspace corresponding to the partition 3, 2. This V_3 is 25-dimensional. To understand the projection of f into V_3 a device of Colin Mallows was used. The function f corresponding to the data is projected onto V_3.

The inner product of this projection with functions $\delta_{\{j,j'\}}\{\pi(i), \pi(i')\}$ is then reported. The pairs $\{i, i'\}$, $\{j, j'\}$ can be chosen in 10 ways each. The 100 inner products are shown in Table 4.

Table 4. Second order, unordered effects

	Rank									
Candidate	1, 2	1, 3	1, 4	1, 5	2, 3	2, 4	2, 5	3, 4	3, 5	4, 5
1, 2	−137	−20	18	140	111	22	4	6	−97	−46
1, 3	476	−88	−179	−209	−147	−169	−160	107	128	241
1, 4	−189	51	113	24	−9	98	99	−65	23	−146
1, 5	−150	57	47	45	43	49	56	−48	−53	−48
2, 3	−42	84	19	−61	30	−16	82	−76	−39	72
2, 4	157	−20	−43	−25	−93	−76	−56	8	38	112
2, 5	22	−44	7	15	−117	69	25	62	99	−138
3, 4	−265	−7	72	199	39	140	85	19	−52	−233
3, 5	−169	10	88	70	78	44	47	−51	−36	−80
4, 5	296	−24	−142	−130	−5	−163	−128	38	−9	267

To explain, consider the $\{1, 3\}\{1, 2\}$ entry 476. This is the largest number in the table. It means that there is a large positive effect for ranking candidates one and three in the first two positions of the ballot. The last entry in row $\{1, 3\}$ shows that these candidates also had a lot of hate vote.

A very similar picture occurs for the last row of the table. With these observations, the tables pattern becomes apparent. The American Psychological Association consists of two groups, academicians and clinicians who are on uneasy terms. Candidates $\{1, 3\}$ are from one group, $\{4, 5\}$ from the other. Very few voters cross ranks so there is a large negative effect for ranking, say $\{1, 4\}$, first and second (or fourth and fifth). These observations are the main structure not revealed by the first order analysis.

In studying data as we have above it is natural to ask if the data were collected again, would the same patterns arise. I will not go into the details here, but a variety of stochastic analyses suggest that the natural scale of variability in Table 4 is ± 50, so the patterns observed are believable.

Further details of this analysis are given in Diaconis (1989). Diaconis and Smith (1989) give a different set of applications for these group-theoretic decompositions.

3. Analysis of Variance (ANOVA)

Consider data cross classified by I levels of one variable and J levels of a second variable. The observed data is then a function $f(i, j)$ from $X = \{(i, j) : 1 \le i \le I, 1 \le j \le J\}$ into $\mathbb{R}$. The product $S_I \times S_J$ acts on X and $L(X)$ splits into

$$\begin{array}{cccccccc} & L(X) = & V_0 \oplus & V_1 & \oplus & V_2 & \oplus & V_3 \\ \dim & I \times J & 1 & I-1 & & J-1 & & (I-1)(J-1) \end{array}$$

where V_0 is the space of constant functions, V_1 is the space of row functions $f(i, j) = f(i, j')$, V_2 is the space of column functions, and V_3 is the space orthogonal to $V_0 \oplus V_1 \oplus V_2$. The projection of f onto $V_0 \oplus V_1 \oplus V_2$ can be interpreted as the least squares approximation to f of form $f(i, j) = \alpha + \beta_i + \gamma_j$ for constants $\alpha, \beta_i, \gamma_j$.

This, and many more complex variants are known collectively in statistical literature as the Analysis of Variance. The classical book by Sheffe (1959) is still the best treatment of this widely used subject.

The group-theoretic treatment of ANOVA was pioneered by Alan James and Ted Hannen with important later work by Peter Fortini. Group theory is useful in analyzing more complex designs where the appropriate decomposition is not so easy to guess at. The dimensions and projections of various subspaces can be computed by character theory. Diaconis (1988) reviews these topics.

Even in the simple example given above, thinking group-theoretically has something to offer: Instead of $S_I \times S_J$ one can consider $S_I \times C_J$ or $C_I \times C_J$, with C_I a cyclic group of order I. These groups act transitively and their use would be appropriate if the order of the corresponding rows or columns matters. For example, if the rows of the table were birds, and the columns months of the year, with (i, j) entry the number of birds of type i cited in month j (all in a given location) the decomposition by $S_I \times C_{12}$ would be appropriate.

Carrying out the projections involves calculating the Fourier transform at many different irreducible representations. In Diaconis and Rockmore (1990) a non-commutative analog of the fast Fourier transform has been developed and used to make these computations efficient. Similar work is being developed by Beth (1984) and Clausen (1989a, b). Historically, the FFT on C_2^d was first developed by Yates (1937) to analyze multi way tables.

4. Modern ANOVA

Analysis of variance has developed along non group-theoretic lines. In this section a survey of works by Rosemary Bailey, Chris Rowley, John Nelder, Tue Tjur, Terry Speed, and their co-workers is given. The next section shows how the present treatment and group-theoretic treatment interact.

Begin with a finite set X. Let $L(X)$ be the real-valued functions on X. A design $\mathscr{D}$ is a set of partitions F of X. For example, in ANOVA with repeated observations in a cell $X = \{(i, j, k) : 1 \leq i \leq I, 1 \leq j \leq J, 1 \leq k \leq n_{ij}\}$ the design might be taken as $\mathscr{D} = \{U, R, C, R \wedge C, E\}$ where U is the universal partition with one block. R is the row partition $(i, j, k) \overset{R}{\sim} (i', j', k')$ iff $i = i'$, C is the column partition, $R \wedge C$, the minimum of R and C, has indices equivalent if they are in the same row and column, and E is the partition into singletons.

For each partition $F \in \mathscr{D}$, let $L_F = \{f \in L(X)$ which are F measurable$\}$. The projection $P_F : L(X) \to L_F$ is defined by the averaging matrix

$$(P_F)_{xy} = \begin{cases} 1/|f| & x, y \in f \text{ with } f \text{ a block in } F \\ 0 & \text{otherwise.} \end{cases}$$

Two partitions, $F, G \in \mathscr{D}$ are called orthogonal if their subspaces are geometrically orthogonal, or if $P_F P_G = P_G P_F$. For the ANOVA example, R and C are orthogonal provided $n_{ij} = |X| n_{i+} n_{+j}$ with $n_{i+} = \sum_j n_{ij}$. An orthogonal design has all factors orthogonal.

In recent years, orthogonal designs with the set of factors closed under maxima have come to be seen as a useful class with a unified theory. Adding maxima of orthogonal factors preserves orthogonality, so a design can always be completed in this way. Such designs are called Tjur designs because of the following basic result.

Theorem (Tjur (1984)). *A Tjur design admits a unique decomposition*

$$L(X) = \bigoplus_{G \in \mathscr{D}} V_G$$

with the property that

$$L_F = \bigoplus_{\substack{G \in \mathscr{D} \\ G \geq F}} V_G.$$

The projection of a given $f \in L(X)$ onto the various V_G constitute the analysis of variance for a Tjur design. The point of the decomposition is this: it is easy to complete the projection P_F of f onto L_F. The partially ordered set of factors then allows the computation onto the V_G by subtraction which amounts to Möbius inversion in the poset.

In the basic ANOVA example, the factors can be diagrammed as:

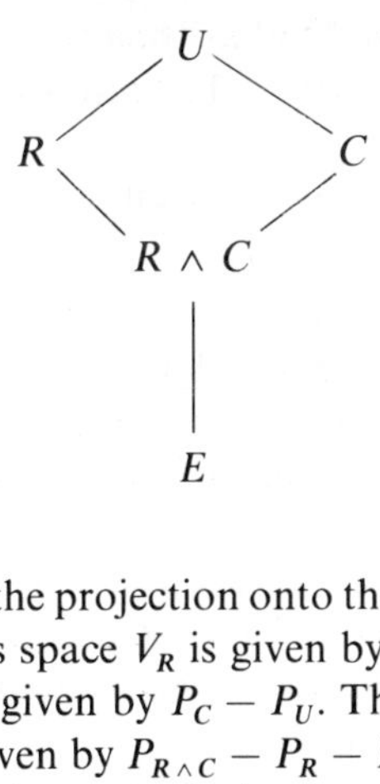

Given $f \in L(X)$ one computes the projection onto the constant functions by P_U. The projection onto the row effects space V_R is given by $P_R - P_U$. The projection onto the column effects space V_C is given by $P_C - P_U$. The projection onto the residual or interaction space $V_{R \wedge C}$ is given by $P_{R \wedge C} - P_R - P_C + P_U$.

More generally, the projection onto V_G is given by $\sum_{F \geq G} \mu(F, G) P_F$ with μ the Möbius function of the partially ordered set of factors.

This is an easy algorithm which is used by many large computer programs (e.g., Genstat) to analyze designed experiments. A splendid treatment of this point of view appears in Tjur (1984).

5. Groups and Modern ANOVA

Sections 3 and 4 above present two different approaches to the analysis of designed experiments. In both, data are represented as $f \in L(X)$. In the group case, a group G is found acting on X and the analysis consists of decomposing $L(X)$ and computing projections. In the second case, a collection of partitions is produced and one uses the splitting of $L(X)$ into parts indexed by these partitions.

It is natural to ask about the relation between these approaches. This problem has been solved for a large class of examples in recent work of Bailey, Prager, Rowley, and Speed (1983). Their work intertwines the two approaches. It is also important group-theoretically in providing examples where the Fourier transform can be computed using the simple averaging and difference algorithm outlined before.

To describe their result, let X be a finite set and $\mathscr{D} = \{F\}$ a design, or set of partitions of X. Assume that the blocks of each $F \in \mathscr{D}$ have uniform size, that all partitions F, G in $\mathscr{D}$ are orthogonal, that $\mathscr{D}$ is closed under max and min, and finally that $\mathscr{D}$ forms a distributive lattice under max and min.

These assumptions include many complex classical cases. However, adding the minimum of two partitions to an orthogonal design can destroy orthogonality.

As automorphism of a design $\mathscr{D}$ is a 1-1 map $\pi : X \to X$ such that for each $F \in \mathscr{D}$, if x and y are in the same block of F then $\pi(x)$ and $\pi(y)$ are in the same block of F. The set of all automorphisms of $\mathscr{D}$ is called the automorphism group of $\mathscr{D}$.

Bailey, Prager, Rowley, and Speed (1983) did three main things:

I) They identified the automorphism group of $\mathscr{D}$ as a generalized wreath product of symmetric groups indexed by a partially ordered set. These generalized wreath products have been extensively developed because of their role in the algebraic theory of semi-groups (Krohn-Rhodes theory). A marvelous introduction to this theory appears in Wells (1976). The result also builds on previous work by Silcock (1977).

II) They identify the characters of the automorphism group that appear in the representation $L(X)$.

III) They show that the group-theoretic and partition based analysis agree.

Much further work is not reported here. For example, they determine the commuting algebra of $L(X)$, give a natural language for describing the groups and decomposition, and finally they make the link with the large body of statistical work in a useful way.

Each approach has problems where it seems to be the superior mode of analysis. The approach by partitions works for some designs without enough symmetry to permit a useful group-theoretic analysis. For example, consider a 2-way array with n_{ij} entries per cell where n_{ij} are as shown:

1	2	3
2	4	6
3	6	9

Here it does not make sense to permute the rows or columns. However, the design is orthogonal and permits a straightforward analysis.

In the other direction, block designs are a widely used class of design which are not orthogonal. As an example, consider an experiment in which v levels of vanilla are to be compared to help decide how much to put into ice cream. If one asks people to taste many ice cream cones, they all taste like colored sugar water. Thus suppose people are asked to taste $k < v$ flavors. A complete block design involves $\binom{v}{k}$ people each of whom tastes k levels of vanilla. Suppose the response is a rating between 0 and 100. This yields $k\binom{v}{k}$ responses in total. The underlying set $X = \{(i, s): 1 \leq i \leq v, |s| = k, i \in s\}$ the responses give $f: X \to \mathbb{R}$.

The two natural partitions are for treatments and blocks. Thus $(i, s) \overset{T}{\sim} (i', s')$ if $i = i'$ and $(i, s) \overset{B}{\sim} (i', s')$ if $s = s'$. These two partitions do not yield an orthogonal design. Indeed, here $T \wedge B = U$ and the condition for orthogonality can be stated as $n_{is}|X| = n_i n_s$ where

- n_i is the number of elements in X receiving treatment i $\left(\text{so } n_i = \binom{v-1}{k-1}\right)$
- n_s is the number of elements in X in block s (so $n_s = k$)
- n_{is} is the number of elements in X with $x = (i, s)$ (so $n_s = 1$ if $i \in s$, 0 if $i \notin s$).

It follows that $n_{is}|X| \neq n_i \cdot n_s$ for $i \notin s$.

The group-theoretic analysis of this kind of data is straightforward but picks up aspects not developed in earlier analysis. The automorphism group can be identified with S_v. The representation $L(X)$ decomposes into a treatment space and a block space, but it also includes new pieces which may be interpreted as the effect of taster's rating by comparison. Fortini (1977) or Diaconis (1988) give further detail.

6. Other Topics

This section gives pointers to closely related research which cannot be adequately covered due to space limitations.

6.1 Stochastic Models

The approach to spectral analysis outlined above begins with data and a group. Almost all of the statistical literature begins with a probability model and presents the projections as estimates of parameters in a model. For example, for two way analysis of variance with one observation per cell the model would be written as

$$f(i, j) = \mu + \alpha_i + \beta_j + \varepsilon_{ij}$$

where μ, α_i, β_j are parameters to be estimated (and $\sum \alpha_i = \sum \beta_j = 0$ to yield identifiable parameters). The ε_{ij} are errors, or disturbance terms, which are usually assumed to be independent random variables with mean 0 and constant variance. The least squares estimates of these parameters are the projections described earlier.

Assuming a model leads to well understood ways to quantify standard errors for the estimates. It also allows analysis of data with no symmetry at all. Further, if more careful specifications are made on the distribution of the error terms, a variety of other estimates of the parameters become available.

One of the nice results of the past 10 years is a complete understanding of all possible covariance structures for the error terms which lead to the original projections being efficient estimators. This, and the closely related subject of general balance are treated by Speed (1987), or Bailey and Rowley (1990).

Rosemary Bailey has developed a more elaborate theory which allows incorporation of the randomization aspect of many designed experiments. Her treatment provides separate provision for treatment and design aspects. The theory makes extensive use of group theory and is well related with statistical practice. A recent survey with extensive pointers to other work is Bailey (1990).

There has also been an extensive development which assumes that the errors are Gaussian. The leading work here comes out of the Danish school. Andersson and Perlman (1989) is an important paper with pointers to other work.

6.2 Bayesian Methods and Shrinkage Estimators

Once a model is specified, the Bayesian approach to statistics proceeds by putting a prior distribution on the parameters and then using observations to get a posterior distribution. There has been very little work on analyzing the kind of data discussed above from a Bayesian perspective. Dawid (1988) presents some results as do Box and Tiao (1973) and Diaconis, Eaton, and Lauritzen (1991) but much remains to be done.

One of the exciting findings of recent statistical research has been the understanding that when many parameters must be estimated, the classical projection estimates can be uniformly improved. The improvement depends on the assumption of a model. Our current understanding of a reasonable way to go after the improvement involves a Bayesian (or empirical Bayesian) treatment of the problem. Again there has not been much work on shrinkage estimates for designed experiments but Bock (1975) and George (1986) are good starts.

6.3 Messy Data

Real data often contains a few stray or wild values that will foul up the classical linear estimates. There is a growing theory of robust statistics surveyed in Huber (1981) or Hoaglin, Mosteller and Tukey (1983, 1985). Much remains to be done in specializing the results available for robust regression to the demands of a complex designed experiment.

Tukey (1977) has developed robust analyses in much the same data analytic spirit as presented here. He supplements these with an extensive residual analysis for ferreting out non-linearities and wild values. He also gives techniques for fitting non-linear models such as

$$f(i, j) = \alpha + \beta_i + \gamma_j + \delta\beta_i\gamma_j + \varepsilon_{ij}.$$

There is active work on non-linear substitutes for classical least squares estimates. Projection pursuit, as developed by Friedman, Stutzle, and Schroeder (1984) or Huber (1985) is one of many varieties. Again, the adaption to analysis of designed experiments is largely open.

Finally, missing data is an annoying part of real statistical analysis. A neat design can become a nightmare with symmetry destroyed. The E.M. Algorithm is now a standard tool for beginning to deal with this problem. Dempster, Laird, and Rubin (1977) is a good reference and Little and Rubin (1987) is a comprehensive guide to the state of the art.

6.4 Final Words

The problem with a model based analysis is that usually the model is simply made up out of whole cloth, from linearity through stochastic assumptions. While in principle assumptions can be checked, in my experience they are wildly misused. See Freedman (1986, 1987) for an extensive discussion.

Complex models have the further disadvantage that their parameters are not simple-to-interpret averages but estimates of rather complex quantities whose interpretation depends crucially on the correctness of the model. The spectral estimate proposed in the first sections of this paper are relatively easy to understand averages.

Finally, as for block designs, group-theoretic considerations can lead to different analyses and new models. There is clearly much to be done in combining the best features of the various theories and then confronting them with reality.

References

Andersson, S., Perlman, M. (1988): Lattice models for conditional independence in a multivariate normal distribution. Technical report 155, Department of Statistics, University of Washington

Bailey, R. (1990): A unified approach to design of experiments. J. Roy. Statist. Soc. A **144**, 214–223

Bailey, R., Praeger, T., Rowley, C., Speed, T. (1983): Generalized wreath products of permutation groups. Proc. London Math. Soc. **47**

Bailey, R., Rowley, C. (1990): General Balance and Treatment Permutations. Lin. Alg. Appl. **127**, 183–225

Beth, T. (1984): Verfahren der Schnellen Fourier-Transform. Teubner, Stuttgart

Bloomfield, P. (1976): Fourier analysis of time series. An introduction. Wiley, New York

Bock, M.E. (1975): Minimax estimators of the mean of a multivariate normal distribution. Ann. Statist. **3**, 209–218

Box, G., Tiao, G. (1973): Bayesian inference in statistical analysis. Addison-Wesley, Reading, Mass

Brillinger, D. (1975): Time series, data analysis and theory. Holt, Rinehart, and Winston, New York

Clausen, M. (1989a): Fast Fourier transforms for meta-Abelian groups. SIAM J. Comput. **18**, 584–593

Clausen, M. (1989b): Fast generalized Fourier transforms. J. Theoret. Comput. Sci. **67**, 55–63

Dawid, P. (1988): Symmetry models and hypotheses for structured data layouts. J. Roy. Statist. Soc. B **50**, 1–26

Dempster, A., Laird, N., Rubin, D. (1977): Maximum likelihood from incomplete data via the EM algorithm. J. Roy. Statist. Soc. B **39**, 1–38

Diaconis, P. (1988): Group representations in probability and statistics. Institute of Math. Statistics, Hayward, CA

Diaconis, P. (1989): A generalization of spectral analysis with application to ranked data. Ann. Statist. **17**, 949–979

Diaconis, P. and Smith, L. (1989): Residual analysis for discrete longitudinal data. Technical report. Department of Statistics, Stanford University

Diaconis, P., Rockmore, D. (1990): Efficient computation of the Fourier Transform on finite groups. J. Amer. Math. Soc. **3**, 297–332

Diaconis, P., Eaton, M., Lauritzen, S. (1991): Finite de Finetti theorems in linear models and multivariate analysis. Technical report Dept. of Mathematics, Aalborg University. To appear in Scand. J. Statist.

Fortini, P. (1977): Representation of groups and the analysis of variance. Ph.D. Thesis, Department of Statistics, Harvard University

Friedman, J.H., Schroeder, A., Stutzle, W. (1984): Projection pursuit density estimation. J. Amer. Statist. Assoc. **79**, 599–608

Freedman, D. (1987): As others see us: A case study in path analysis. J. Educ. Statist. **12**, 101–206

Freedman, D., Navidi, W. (1986): Regression models for adjusting the 1980 census. Statist. Sci. **1**, 3–39

George, E. (1986): Minimax multiple shrinkage estimation. Ann. Statist. **14**, 188–205

Hannen, E.J. (1965): Group representations and applied probability. Methuen, New York

Hoaglin, D., Mosteller, F., Tukey, J. (1983): Understanding robust and exploratory analysis. Wiley, New York

Hoaglin, D., Mosteller, F., Tukey, J. (1985): Exploring data: Tables, trends, and shapes. Wiley, New York

Huber, P. (1981): Robust statistics, Wiley, New York

Huber, P. (1985): Projection pursuit. Ann. Statist. **13**, 435–525

Izenman, A., Zabell, S. (1978): Babies and the blackout: the genesis of a misconception. Technical report 38, Dept. of Statistics, University of Chicago

James, A. (1957): The relationship algebra of an experimental design. Ann. Math. Statist. **27**, 993–1002

James, A. (1982): Analysis of variance determined by symmetry and combinatorial properties of zonal polynomials. In G. Kallianpur et al. (eds.) Statistics and probability: essays in honor of C.R. Rao. North-Holland, New York

Ledermann, W. (1987): Introduction to group characters (2nd ed.). Cambridge University Press, Cambridge

Little, R., Rubin, D. (1987): Statistical analysis with missing data. Wiley. New York

Scheffe, H. (1959): Analysis of variance. Wesely, New York

Serre, J.P. (1977): Linear representations of finite groups. Springer, Berlin Heidelberg New York

Silcock, H.L. (1977): Generalized wreath products and the lattice of normal subgroups of a group. Algebra Universalis. **7**, 361–372

Speed, T. (1987): What is an analysis of variance? Ann. Statist. **15**, 885–941

Tjur, T. (1984): Analysis of variance models in orthogonal designs. Int. Statist. **52**, 33–82

Tukey, J. (1961): Discussion, emphasizing the connection between analysis of variance and spectrum analysis. Technometrics **3**, 191–219

Tukey, J. (1977): Exploratory data analysis. Addison-Wesley, Reading, Mass

Wells, C. (1976): Some applications of the wreath product construction. Amer. Math. Monthly **83**, 317–338

Yates, F. (1937): The design and analysis of factorial experiments. Imperial Bureau of Soil Science, Harpenden, England

Stochastic Models of Growth and Competition

Richard Durrett

Department of Mathematics, White Hall, Cornell University, Ithaca, NY 14853, USA

In this paper we will describe recent results on four interacting particle systems that model the growth and competition of plant species or the spread of an epidemic or forest fire. In each system there is a collection of sites, the d-dimensional integer lattice, that at each time $t \in [0, \infty)$ can be in one of a finite number of states, so the state of the process at time t is a function $\xi_t : \mathbb{Z}^d \to \{0, 1, \dots, k\}$. The time evolution is described by declaring that each site changes its state at a rate that depends upon the states of a finite number of neighboring sites. Here, we say that something happens at rate r if the probability of an occurrence between times t and $t + h$ is $rh + o(h)$.

1. The Basic Contact Process

In this model $\xi_t : \mathbb{Z}^d \to \{0, 1\}$, we think of 0 as vacant and 1 as occupied by a "particle," and the system evolves as follows:

(i) Particles die at rate one, give birth at rate β.
(ii) A particle born at x is sent to a y chosen at random from the $2d$ nearest neighbors $\{y : \|x - y\|_1 = 1\}$.
(iii) If y is occupied then the birth is suppressed.

Rule (iii) says that there can be at most one particle per site. This is a reasonable constraint if you are thinking of the spread of a plant species but this realism makes the model very difficult to analyze. Let ξ_t^A be the state at time t when initially $\xi_0^A(x) = 1$ if and only if $x \in A$, and let $\tau^A = \inf\{t : \xi_t^A \equiv 0\}$. If there are no particles then none can be born, so $\xi_t^A \equiv 0$ for all $t \geq \tau^A$. In words, the "all 0" state is an *absorbing state* and we say the system *dies out* at time τ^A.

The first question to be addressed is "When does the system have positive probability of not dying out starting from a single occupied site?" or "When is $P(\tau^{\{0\}} = \infty) > 0$?" It suffices to use a single occupied site as an initial configuration since $P(\tau^{\{0\}} = \infty) = 0$ implies $P(\tau^A = \infty) = 0$ for all finite A. Now, increasing β improves the chances for survival, so it should be clear that there is a critical value

$$\beta_c = \inf\{\beta : P(\xi_t^0 \not\equiv 0 \text{ for all } t) > 0\}.$$

Proceedings of the International Congress
of Mathematicians, Kyoto, Japan, 1990

If we delete rule (iii) from the definition, the resulting system is called a branching random walk and has $\beta_c = 1$. That is, in order for a branching random walk to survive it is sufficient to have a birth rate larger than the death rate. Since in the contact process some of the birth rate will be wasted on occupied sites, this proves the easy half of the following result.

Theorem 1A. $1 < \beta_c(\mathbb{Z}^d) \le 4$.

The lower bound is due to Harris (1974), the upper bound to Holley and Liggett (1978). Both bounds are reasonably accurate. Numerical results (see Brower, Furman, and Moshe (1978)) suggest that $\beta_c(\mathbb{Z}) \approx 3.299$ and $\beta_c(\mathbb{Z}^2) \approx 1.645$, and it has been shown (see Holley and Liggett (1981) or Griffeath (1983)) that $\beta_c(\mathbb{Z}^d) \to 1$ as $d \to \infty$.

Once it was established that $\beta_c \in (0, \infty)$, attention turned to "What does the process look like when it does not die out?" To answer this question we begin by introducing a special property of the contact process called *duality*

$$P(\xi_t^A(x) = 0 \text{ for } x \in B) = P(\xi_t^B(x) = 0 \text{ for } x \in A) .$$

An immediate consequence of duality is that if we start from $\xi_0^1(x) \equiv 1$ then $\xi_t^1 \Rightarrow \xi_\infty^1$. Here, $\Rightarrow$ is short for *converges weakly* and means that

$$P(\xi_t^1(x) = 0 \text{ for } x \in B) \to P(\xi_\infty^1(x) = 0 \text{ for } x \in B)$$

for all finite sets B. To prove the weak convergence we set $A = \mathbb{Z}^d$ in the duality equation to get

$$P(\xi_t^1(x) = 0 \text{ for } x \in B) = P(\xi_t^B(x) \equiv 0)$$

which increases to a limit as $t \to \infty$, since "all 0" is an absorbing state. It follows from standard results (see Chapter 1 of Liggett (1985)) that ξ_∞^1 is a stationary distribution for the contact process, i.e., if we start the process with this distribution it has this distribution for all time.

At the other extreme, the point mass on the "all 0" state, δ_0, is a trivial stationary distribution. Letting $B = \{y\}$ and $t \to \infty$ in the duality relation gives

$$P(\xi_\infty^1(y) = 0) = P(\tau_t^{\{y\}} < \infty) ,$$

so $\xi_\infty^1 = \delta_0$ if the contact process dies out, but is a nontrivial stationary distribution if the contact process survives. The next result, called the *complete convergence theorem* implies that ξ_∞^1 is the only nontrivial stationary distribution.

Theorem 1B. $\xi_t^A \Rightarrow P(\tau^A < \infty)\, \delta_0 + P(\tau^A = \infty)\, \xi_\infty^1$.

In words, when the process dies out it looks dead, but when it survives and t is large it looks like the system starting from all sites occupied.

The last result took fifteen years to evolve to its current form. Harris (1974), Griffeath (1978), Durrett (1980), Durrett and Griffeath (1982), and Durrett and

Schonmann (1987) proved increasingly more general results before Bezuidenhout and Grimmett (1990) finished the problem and in addition proved

Theorem 1C. *When* $\beta = \beta_c$, $P(\tau^{\{0\}} = \infty) = 0$.

In words, the contact process dies out at the critical value. For applications (including some we will make in this paper) it is worthwhile to note that all the results in this section hold if (ii) is replaced by

(ii) A particle born at x is sent to a y chosen at random from $x + \mathcal{N}$.

if we assume $\mathcal{N}$ is (a) *symmetric* with respect to reflection in any coordinate plane, and (b) *irreducible*, i.e., the group generated by $\mathcal{N}$ is $\mathbb{Z}^d$.

2. Multitype Contact Processes

It is well known, even to mathematicians, that there is more than one type of plant, so it is natural to generalize the contact process to have two (or more) types of particles. In this model, the state at time t $\xi_t : \mathbb{Z}^d \to \{0, 1, 2\}$ and we think of 0 as vacant and 1 and 2 as occupied by pine and maple trees respectively. With this in mind we formulate the evolution as follows:

(i) Particles of type i die at rate one, give birth at rate β_i.
(ii) A particle born at x is sent to a y chosen at random from $x + \mathcal{N}$ where $\mathcal{N}$ is symmetric and irreducible.
(iii) If y is occupied then the birth is suppressed.

When only one type of particle is present the system reduces to the basic contact process so if $\beta_1, \beta_2 > \beta_c(\mathbb{Z}^d)$ then there are three trivial equilibria: δ_0, μ_1 and μ_2, where μ_i is the limit starting from $\xi_t(x) \equiv i$. The main question to be answered about the new system is: "Is there a nontrivial stationary distribution?", i.e., one that concentrates on configurations that contain both 1's and 2's. The first result is a negative one.

Theorem 2A. *If* $\beta_1 > \beta_2$ *then there are no nontrivial translation invariant stationary distributions.*

Here translation invariant means that that the distribution is invariant under spatial shifts. This result and the others in this section are from Claudia Neuhauser's (1990) thesis. We conjecture that Theorem 2A holds without the assumption of translation invariance but that assumption is often difficult to remove. Note that Harris proved Theorem 1B for translation invariant initial distributions in 1974 but the general case was settled 15 years later.

Restricting our attention now to the special case $\beta_1 = \beta_2 > \beta_c(\mathbb{Z}^d)$, we have

Theorem 2B. *In dimensions* $d \leq 2$, *for any initial configuration, we have* $P(\xi_t(x) = 1, \xi_t(y) = 2) \to 0$ *for all* $x, y \in \mathbb{Z}^d$, *so all stationary distributions are trivial.*

Theorem 2C. *In dimensions $d \geq 3$, there is a one parameter family of stationary distributions ν_θ, $\theta \in [0,1]$, and all translation invariant stationary distributions are convex combinations of the ν_θ.*

As in the voter model, (see Liggett (1985) Chapter V or Durrett (1988) Chapter 2), the dichotomy between the behavior in $d \leq 2$ and $d \geq 3$ comes from the fact that random walks are recurrent in the first case and transient in the second. The stationary distributions are constructed by starting the system from an initial product measure in which 1's have density θ and 2's have density $1-\theta$, i.e., $\xi_0(x)$ are independent and take values 1 and 2 with probabilities θ and $1-\theta$. The reader should note that while the basic contact process has a single nontrivial stationary distribution, the two color version has a one parameter family in $d \geq 3$.

3. Successional Dynamics

In this model we again have $\xi_t : \mathbb{Z}^d \to \{0,1,2\}$ but this time we think of 0 as vacant and 1 and 2 as occupied by a bush or tree respectively. With this interpretation in mind the dynamics are formulated as follows:

(i) Particles of type i die at rate one, give birth at rate β_i.
(ii) A particle born at x is sent to a y chosen at random from $\{y : \|x-y\|_1 \leq M\}$, where M is an integer.
(iii) If $\xi_t(y) \geq \xi_t(x)$ then the birth is suppressed.

In words, trees can give birth onto sites occupied by bushes but not conversely. In biological terms the two species are part of a successional sequence. When only one type of particle is present, the system reduces, as in the last example, to a contact process so if $\beta_1, \beta_2 > \beta_c$ then there are three trivial equilibria: δ_0, μ_1, and μ_2, where μ_i is the limit starting from $\xi_t(x) \equiv i$.

Again, the main question to be answered is: "Are there nontrivial stationary distributions?" or more briefly "Is coexistence possible?" Our first answer is

Theorem 3A. *If $d = 1$ and $M = 1$ then for any initial configuration we have $P(\xi_t(x) = 1, \xi_t(y) = 2) \to 0$ as $t \to \infty$ for all $x, y \in \mathbb{Z}$ so there is no coexistence.*

This result can be proved by drawing a picture of a "typical" realization of the process starting with a single 2

0 0 1 0 1 0 2 0 2 2 0 2 0 0 0 2 1 0 1 0 0 1

and checking that since $M = 1$ there can never be a 1 between the leftmost and rightmost 2's. If the 2's do not die out, then the ends of the interval of 2's go to $-\infty$ and ∞ respectively (see Durrett (1980)) and the 1's get crowded out. In general either (a) all the 2's die out, or (b) some 2 starts an interval that grows forever. In either case $P(\xi_t(x) = 1, \xi_t(y) = 2) \to 0$ as $t \to \infty$.

We believe that coexistence is possible in all other cases

Conjecture 3A. *If $d > 1$ or $M > 1$ then coexistence is possible when $\beta_2 = \beta_c + \varepsilon$ and β_1 is large.*

The main trouble with proving this conjecture is that coexistence can only occur near the critical value. It is not hard to show that if $\beta_2 > \beta(d, M)$ then there is no coexistence for any $\beta_1 \le \infty$. Somewhat surprisingly, this problem which is difficult to solve when $d = 1$ and $M = 2$, or $d = 2$ and $M = 1$ turns out to be more tractable when M is large. In addition to proving Theorem 3A, Durrett and Swindle (1990) have shown

Theorem 3B. *If $\beta_1 > \beta_2^2 > 1$ then coexistence occurs for large M.*

To explain the last conclusion we need to introduce the long range contact process, a modification of the basic contact process in which (ii) is changed to:

(ii) A particle born at x is sent to a y chosen at random from $\{y : \|x - y\|_1 \le M\}$.

If we write $\beta_c(M)$ to indicate the dependence of the critical value on M and use ξ^1_∞ to denote the limit starting from all 1's then we have

Theorem 3C. *As $M \to \infty$, $\beta_c(M) \to 1$. Furthermore, if $\beta > 1$ then ξ^1_∞ converges weakly to a product measure with density $(\beta - 1)/\beta$.*

This result (for the neighborhood $\{y : \|x - y\|_\infty \le M\}$) was proved by Bramson, Durrett, and Swindle (1989) who identified the rate at which $\beta_c(M)$ approached 1. A simpler and more general proof, which does not give the right rate, can be found in Durrett (1989).

To explain the condition in Theorem 3B, observe that $\eta_t = \{x : \xi_t(x) = 2\}$ is a long range contact process, so if M is large and we are in equilibrium, η_t is approximately a product measure with density $(\beta_2 - 1)/\beta_2$. If the 2's were exactly that product measure, a 1 would die at rate $1 + \frac{\beta_2 - 1}{\beta_2}\beta_2$, (the second term representing births onto the site by 2's) and give birth at rate β_1/β_2 (the site must not be occupied by a 2 for a successful birth to occur). So for coexistence to occur we need $1 + \frac{\beta_2 - 1}{\beta_2}\beta_2 < \beta_1/\beta_2$ or $\beta_1 > \beta_2^2$. The careful reader will have noted that we have just argued the condition is necessary while Theorem 3B proves it is sufficient. Having faith in the heurisitc argument we make

Conjecture 3B. *If $\beta_1 < \beta_2^2$ then there is no coexistence for large M.*

Remark. The heuristic argument generalizes easily to show that if the two particles die at different rates then we need

$$\delta_1 + \frac{\beta_2 - \delta_2}{\beta_2}\beta_2 > \beta_1 \frac{\delta_2}{\beta_2}$$

and the proof of Theorem 3B generalizes to show that this condition is sufficient. It is natural to generalize the multitype contact process in this way but we do not know how to prove any results in that generality. The naive guess is that

$\beta_1/\delta_1 > \beta_2/\delta_2$ is right hypothesis for Theorem 2A. We believe this is correct but have no idea how to prove it.

Having discussed the existence of nontrivial stationary distributions, we turn to the question of uniqueness. Durrett and Møller (1991) have proved a "complete convergence theorem." To state their result let δ_0, μ_1, and μ_2 be the trivial stationary distributions mentioned at the beginning of this section. Let μ_{12} be the nontrivial stationary distribution constructed in Theorem 3B. Let $\eta_t = \{x : \xi_t(x) = 1\}$, $\zeta_t = \{x : \xi_t(x) = 2\}$, $\tau_1 = \inf\{t : \eta_t = \emptyset\}$, and $\tau_2 = \inf\{t : \zeta_t = \emptyset\}$.

Theorem 3C. *If $\beta_1 > \beta_2^2 > 1$ and M is large then*

$$\xi_t \Rightarrow P(\tau_1 < \infty, \tau_2 < \infty)\,\delta_0 + P(\tau_1 = \infty, \tau_2 < \infty)\,\mu_1 + P(\tau_1 < \infty, \tau_2 = \infty)\,\mu_2 + P(\tau_1 = \infty, \tau_2 = \infty)\,\mu_{12}\,.$$

In words if the 1's and/or 2's die out we end up with a trivial stationary distribution in which one or zero types of particles are present. If both the 1's and 2's survive and t is large, the system looks like μ_{12} so that is the only nontrivial stationary distribution. The value of M required for Theorem 3C is larger than that for Theorem 3B which is enormous. With more work this difference might be eliminated but the interesting problem is to show

Conjecture 3C. *The complete convergence theorem holds whenever coexistence occurs.*

4. An Epidemic Model

Our fourth system is a process $\xi_t : \mathbb{Z}^2 \to \{0, 1, 2\}$ that has been used to model the spread of epidemics and forest fires. In the epidemic interpretation 0 = healthy, 1 = infected, 2 = removed = immune or dead. In the forest fire interpretation, 0 = alive, 1 = on fire, and 2 = burnt. With these interpretations in mind, we formulate the dynamics as follows:

(i) A burning tree sends out sparks at rate β.
(ii) A spark emitted from x flies to one of the four nearest neighbors $\{y : \|y - x\|_1 = 1\}$ chosen at random. If the spark hits a live tree, the tree catches fire and begins immediately to emit sparks.
(iii) A tree remains on fire for an exponential amount of time with mean 1 then becomes burnt.
(iv) Burnt trees come back to life at rate α.

At first glance, the spontaneous re-appearance of trees may not seem reasonable. In the epidemic interpretation this is quite natural, however. Consider a disease like measles that upon recovery confers lifetime immunity. New susceptibles are born and immune individuals die. We combine the two transitions into the one in (iv) to keep a constant population size.

When $\alpha = \infty$, sites change instantaneously from 2 to 0 and the result is the contact process. At the other extreme, $\alpha = 0$, is the so-called "spatial epidemic with removal" in which regrowth is impossible. We begin by considering the behavior of our processes starting with a single burning tree at the origin in the midst of an otherwise virgin forest, i.e., $\xi_0^0(0) = 1$, $\xi_0^0(x) = 0$ for $x \neq 0$. Let $\eta_t^0 = \{x : \xi_t^0(x) = 1\}$, let $\zeta_t^0 = \{x : \xi_t^0(x) = 2\}$, and define a critical value by

$$\beta_c(\alpha) = \inf\{\beta : P(\zeta_t^0 \neq \emptyset \text{ for all } t) > 0\}.$$

Cox and Durrett (1988) considered the case $\alpha = 0$ and showed

Theorem 4A. *If $\beta > \beta_c(0)$ then there is a nonrandom convex set D so that on $\{\eta_t \neq \emptyset$ for all $t\}$ we have $\zeta_t^0 \approx \zeta_\infty^0 \cap tG$, and $\eta_t^0 \approx t\partial G$. To be precise, for any $\varepsilon > 0$ the following inequalities hold for large t*

$$\zeta_\infty^0 \cap (1-\varepsilon)tG \subset \zeta_t^0 \subset (1+\varepsilon)tG$$
$$\eta_t^0 \subset (1+\varepsilon)tG - (1-\varepsilon)tG.$$

In words, this result says that the fire expands linearly and has an asymptotic shape. The statement is made contorted by the fact that the set of trees that will ever burn, ζ_∞^0, is not all of $\mathbb{Z}^d$. Thus what we prove is that when t is large, ζ_t^0 is contained in $(1+\varepsilon)tG$ and (if nonempty) contains all the points of ζ_∞^0 in $(1-\varepsilon)tG$.

When $\alpha = 0$ the system cannot have a nontrivial stationary distribution but Durrett and Neuhauser (1991) have shown

Theorem 4B. *If $\beta > \beta_c(0)$ and $\alpha > 0$ then there is a nontrivial stationary distribution, i.e., one that assigns no mass to "all healthy" state.*

The last result illustrates some of the frustrations in "applied probability." The proof is intricate and required several months to put down on paper, but we have been repeatedly told by physicists and biologists that the conclusion is obvious. In view of our difficulties in proving existence the reader should not be surprised to learn that we have little to say about uniqueness.

Conjecture 4C. *If $\beta > \beta_c(\alpha)$ then there is a unique nontrivial stationary distribution.*

In the first three examples we have had varying degrees of success in identifying the set of stationary distributions. In each of those cases however there is a useful "duality equation" and we have not been able to find one here.

References

Bezuidenhout, C., Grimmett, G. (1990): The critical contact process dies out. Ann. Probab. (to appear)

Bramson, M., Durrett, R., Swindle, G. (1989): Statistical mechanics of crabgrass. Ann. Probab. **17**, 444–481

Brower, R.C., Furman, M.A., Moshe, M. (1978): Critical exponents for the Reggeon quantum spin model. Phys. Letters **76B**, 213–219

Cox, J.T., Durrett, R. (1988): Limit theorems for the spread of epidemics and forest fires. Stoch. Processes Appl. **30**, 171–191

Durrett, R. (1980): On the growth of one-dimensional contact processes. Ann. Probab. **8**, 890–907

Durrett, R. (1988): Lecture Notes on Particle Systems and Percolation. Wadsworth Pub. Co., Pacific Grove, CA

Durrett, R. (1989): A new method for proving the existence of phase transitions. To appear in the Proceedings of a Conference in Honor of Ted Harris. Birkhäuser, Boston

Durrett, R., Griffeath, D. (1982): Contact processes in several dimensions. Z. Wahrsch. Verw. Gebiete **59**, 535–552

Durrett, R., Møller, A.M. (1991): Complete convergence theorem for a competition model. Z. Wahrsch. Verw. Gebiete (to appear)

Durrett, R., Neuhauser, C. (1991): Epidemics with recovery in $d = 2$. Adv. Appl. Probab. (to appear)

Durrett, R., Schonmann, R.H. (1987): Stochastic growth models. In: Kesten (1987)

Durrett, R., Swindle, G. (1990): Are there bushes in a forest? Stoch. Processes Appl. (to appear)

Griffeath, D. (1978): Limit theorems for non-ergodic set-valued Markov processes. Ann. Probab. **6**, 379–387

Griffeath, D. (1983): The binary contact path process. Ann. Probab. **11**, 692–705

Harris, T.E. (1974): Contact interactions on a lattice. Ann. Probab. **2**, 969–988

Holley, R., Liggett, T.M. (1978): The survival of contact processes. Ann. Probab. **6**, 198–206

Holley, R., Liggett, T.M. (1981): Generalized potlatch and smoothing processes. Z. Wahrsch. Verw. Gebiete **55**, 165–195

Kesten, H. (ed.) (1987): Percolation theory and the ergodic theory of interacting particle systems. Springer, New York Berlin Heidelberg

Liggett, T.M. (1985): Interacting particle systems. Springer, New York Berlin Heidelberg

Neuhauser, C. (1990): Ergodic theorems for the multitype contact process. Ph.D. Thesis, Cornell University

Recurrent Ergodic Structures and Ramsey Theory

Hillel Furstenberg

Institute of Mathematics, Hebrew University, Givat Ram, Jerusalem, Israel

Introduction

Ramsey theory is best defined by example and the classic example of a Ramsey type theorem is the result of van der Waerden: if the integers $\mathbb{Z}$ are partitioned into finitely many sets, one of these contains arbitrarily long arithmetic progressions. An equivalent version states: given natural numbers r, l, $\exists N(r, l)$ so that if $N \geq N(r, l)$ and the integers $\{1, 2, \ldots, N\}$ are partitioned into r sets, one of these contains an l-term arithmetic progression. Erdös and Turán realized that this result would follow if it were the case that *any* subset of $\{1, 2, \ldots, N\}$ comprising δN elements contains an l-term arithmetic progression provided N is sufficiently large. This result, conjectured in 1936 [ET1] was proved by E. Szemerédi in 1973 [Sz1], and was reproved using ergodic theoretic methods in 1976 [Fu1]. The theorems of van den Waerden and Szemerédi illustrate the two sides of Ramsey theory: coloring – or partition – results, and density results. The latter are clearly stronger than the former, and the proofs are typically more recondite. The role of ergodic theory in density theorems of this type stems from the fact that in a number of situations theorems about patterns found in sets having "density" bounded from below are equivalent to theorems about the "return", or "recurrence", patterns for measure preserving transformations acting on a measure space. It would appear that the ubiquity of certain patterns in the combinatorics of large sets reflects the phenomenon of recurrence in ergodic theoretic contexts and the latter has to be studied to gain insight into the former.

We shall be examining three different contexts for recurrence results in ergodic theory. We shall mention the combinatorial (Ramsey theoretic) equivalents of these results, although we shall have to refer the reader elsewhere for the proof of the equivalence. Our purpose here is to display the common features of these recurrence phenomena. We shall find that in each setting there is a notion of *rigidity* and the notion of a special system constructed from scratch by a finite succession of rigid extension. For these systems (*distal* and *quasi-distal*) it will be possible to deduce recurrence properties directly, the key tool being Ramsey theory of the van der Waerden sort. For arbitrary systems we shall find a method to "factor out" the "independent" component, reducing the recurrence property

Proceedings of the International Congress
of Mathematicians, Kyoto, Japan, 1990

for a general system to that of a factor system, which will be one of the special systems for which the result has been established.

To motivate the ergodic theoretic discussion we list here some of the combinatorial consequences. For more details, particularly as regards the connection between the ergodic theoretic formulation and the Ramsey theoretic results, we refer the reader to [Fu2, FK2, FK4, and GRS1].

Theorem A. *There is a function $N(\delta, l)$ for $\delta > 0$ and $l \in \mathbb{N}$ so that if $N \geq N(\delta, l)$ and $S \subset \{1, 2, \ldots, N\}$ with $|S| \geq \delta N$ then S contains an l-term arithmetic progression. ($|S|$ denotes cardinality of S.)*

Theorem B. *Let P denote a finite subset of $\mathbb{R}^d$. There is a function $R(\delta, P)$ so that if $R \geq R(\delta, P)$ and B is a ball of radius R in $\mathbb{R}^d$ and $S \subset B$ is a measurable subset with $m(S) \geq \delta m(B)$, then S contains an integral dilation of P, i.e., a set of the form $u + nP$, $u \in \mathbb{R}^d, n \in \mathbb{N}$.*

Theorem C. *There exists a function $D(\delta, d, q)$ for $\delta > 0, d \in \mathbb{N}$, $q =$ a prime power, so that if V is a vector space over $\mathbb{E}_q$ of dimension $\geq D(\delta, d, q)$ and $S \subset V$ is a subset with $|S| \geq \delta |V|$, then S contains a d-flat (translate of d-dimensional subspace).*

Theorem D. *If Λ is a finite set (alphabet) let $W_N(\Lambda)$ denote words of length N in Λ. Let $W_N^*(\Lambda) = W_N(\Lambda \cup \{t\}) \setminus W_N(\Lambda)$ so that $\varphi(t) \in W_N^*(\Lambda)$ is a word in which the variable t appears. There exists a function $L(\delta, |\Lambda|)$ so that if $N \geq L(\delta, |\Lambda|)$ and $S \subset W_N(\Lambda)$ with $|S| \geq \delta |W_N(\Lambda)|$, then there exists $\varphi(t) \in W_N^*(\Lambda)$ so that the words $\varphi(a), \varphi(b), \ldots, \varphi(s)$, $a, b, \ldots, s \in \Lambda$ are all in S.*

Theorem A is Szemerédi's theorem. Theorem B is equivalent to a multidimensional version of Szemerédi's theorem. It was proved first in [FK1]. Theorem C is a density theoretic version of a coloring theorem first conjectured by Roth and proved by R.L. Graham and B. Rothschild [GR1]. The latter also follows from the theorem of A.W. Hales and R.I. Jewett [HJ1]. A special case of Theorem C was proved by T.C. Brown and J.P. Buhler [BB1], and in generality it was proved in [FK2]. Theorem D was conjectured by R.L. Graham and proved in [FK4].

By interpreting the letters of Λ as the digits for writing numbers to a base b, we can see that $D \Rightarrow A$. On the other hand, interpreting Λ as a vector space over a finite field, we can see that $D \Rightarrow C$. It is also not difficult to show that $D \Rightarrow B$, so that Theorem D holds the key to a number of results of Ramsey theory.

We remark that the major part of what is presented here is joint work with Y. Katznelson.

1. Classical Systems

In the classical setting for ergodic theory we speak of a *measure preserving system* (m.p.s.) $(X, \mathscr{B}, \mu, T)$ where $T : X \rightarrow X$ is a measurable, measure preserving map, and μ is a probability measure on the sets of $\mathscr{B}$. It is useful to distinguish between *ergodic* and *non-ergodic* systems. $(X, \mathscr{B}, \mu, T)$ is *ergodic* if $T^{-1}A = A$ implies $\mu(A) = 0$ or 1. There is a rather general decomposition theorem for measure preserving systems to ergodic components, and the recurrence phenomena we shall study here follow for arbitrary systems once they are established for ergodic systems. We also can assume without loss of generality that T is invertible.

If T is a measure preserving transformation on X then T induces an operator on functions on X: $T^{-1}f(x) = f(Tx)$. (The notation corresponds to the fact that $S^{-1}(T^{-1}f) = (TS)^{-1}f$.) T^{-1} defines a unitary operator on $L^2(X, \mathscr{B}, \mu)$ and we can study spectral properties of T^{-1}. In particular ergodicity is equivalent to 1 being a simple eigenvalue, and if, in addition, there are no other eigenvalues, we say that T, or $(X, \mathscr{B}, \mu, T)$, is *weak mixing*.

The mean ergodic theorem for L^2-functions is a special case of a general theorem for contraction operators in Hilbert space:

Theorem 1. *If $(X, \mathscr{B}, \mu, T)$ is an ergodic system and $f \in L^2(X, \mathscr{B}, \mu)$, then*

$$\frac{f + T^{-1}f + T^{-2}f + \cdots + T^{-N}f}{N+1} \longrightarrow \int f d\mu$$

in the norm of $L^2(X, \mathscr{B}, \mu)$.

V. Bergelson has given a far reaching generalization of this in the case of a weak mixing system [Be1].

Theorem 2. *Let $(X, \mathscr{B}, \mu, T)$ be a weak mixing system. Let $p_1(t), p_2(t), \ldots, p_k(t)$ be polynomials with integer coefficients with* $\deg p_i > 0$ *and* $\deg(p_i - p_j) > 0$ *for $i \neq j$. Then if $f_1, f_2, \ldots, f_k \in L^\infty(X, \mathscr{B}, \mu)$ we will have*

$$\frac{1}{N+1} \sum_{n=0}^{N} T^{-p_1(n)} f_1 T^{-p_2(n)} f_2 \cdots T^{-p_k(n)} f_k \to \int f_1 d\mu \int f_2 d\mu \cdots \int f_k d\mu \qquad (1)$$

in the norm of $L^2(X, \mathscr{B}, \mu)$.

One consequence of this theorem is that if $(X, \mathscr{B}, \mu, T)$ is weak mixing and $A_1, A_2, \ldots, A_k$ are any k subsets of positive measure, we will have for some (in fact, most) n:

$$\mu(A_1 \cap T^{-n} A_2 \cap \cdots \cap T^{-(k-1)n} A_k) > 0 \ , \qquad (2)$$

and, in particular, if $\mu(A) > 0$,

$$\mu(A \cap T^{-n} A \cap \cdots \cap T^{-(k-1)n} A) > 0 \ . \qquad (3)$$

The foregoing property, *that points of A return to A along arithmetic progressions,* is the recurrence phenomenon we have in mind in the case of a classical system. In the weak mixing case we have found that it follows from a generalized ergodic theorem.

The proof of Theorem 2 is based on the following lemma for convergence of averages in Hilbert space to 0.

Lemma 3. *Let $\{u_n\}$ be a bounded sequence of vectors in a Hilbert space. Assume for each m, that*

$$\gamma_m = \lim_{N\to\infty} \frac{1}{N}\sum_{n=1}^{N}\langle u_n, u_{n+m}\rangle$$

exists, and that

$$\frac{1}{M}\sum_{m=1}^{M}\gamma_m \longrightarrow 0 .$$

then

$$\left\|\frac{1}{N}\sum_{n=1}^{N}u_n\right\| \longrightarrow 0 .$$

Using the lemma and the ergodic theorem, one proves (1.1) for progressively more complex situations, beginning with

$$\frac{1}{N+1}\sum_{n=0}^{N}T^{-an}fT^{-bn}g,\ \frac{1}{N+1}\sum_{n=0}^{N}T^{-an}fT^{-bn}gT^{-cn}h,\cdots$$

We refer to [Be1] for details.

When $(X,\mathscr{B},\mu,T)$ is not weak mixing the result of Theorem 2 cannot be expected to be valid. For example assume φ is an eigenfunction, $T^{-1}\varphi = \lambda\varphi$, and set $f=\varphi^2, g=\varphi^{-1}$, then $T^{-n}fT^{-2n}g = \left(\lambda^{2n}\varphi^2\right)\left(\lambda^{-2n}\varphi^{-1}\right) = \varphi$ and

$$\frac{1}{N+1}\sum_{0}^{N}T^{-n}fT^{-2n}g \to \varphi$$

which is not constant. In the weak mixing case we would have obtained $\int f d\mu \int g d\mu$ as the limit.

2. Isometric Extensions and Distal Factors

If $(X,\mathscr{B},\mu,T)$ and $(Y,\mathscr{D},\nu,T)$ are two systems (it is convenient to use the same letter for the transformation), we say that a map $\pi : X \to Y$ defines $(Y,\mathscr{D},\nu,T)$ as a *factor* of $(X,\mathscr{B},\mu,T)$ if π is measurable, measure preserving, and $\pi(Tx) = T(\pi x)$. We lift functions from Y to X, $f \to f\circ\pi$ and since this imbeds $L^2(Y,\mathscr{D},\nu)$ isometrically into $L^2(X,\mathscr{B},\mu)$, we shall identify $f\in L^2(Y,\mathscr{D},\nu)$ with $f\circ\pi \in L^2(X,\mathscr{B},\mu)$. It will be convenient to denote the conditional expectation operator $f\mapsto E(f\mid\pi^{-1}\mathscr{D})$ by $f\mapsto E(f\mid Y)$ and we regard $E(f\mid Y)$ either as a function on Y or on X. Note that T^{-1} commutes with $E(\ \mid Y)$.

We call a m.p.s. *Kronecker* if it has the form $(Z,\mathscr{D},m,T)$ where $(Z,+)$ is a compact abelian group with a dense cyclic subgroup $\mathbb{Z}\alpha$, $\mathscr{D}$ = Borel sets, m = Haar measure, and $Tz = z+\alpha$. For a Kronecker system, $L^2(Z,\mathscr{D},m)$ is spanned by eigenfunctions (each character is an eigenfunction). It follows that if a system $(X,\mathscr{B},\mu,T)$ possesses a non-trivial Kronecker factor then it cannot be weak mixing. The converse is also true: if $(X, \mathscr{B}, \mu, T)$ is not weak mixing it has a non-trivial Kronecker factor.

We describe a relativized notion of a Kronecker factor, representing a certain relationship between a system and a factor.

Definition 1. $(X, \mathscr{B}, \mu, T)$ is an *isometric extension* of a factor $(Y, \mathscr{D}, \nu, T)$ if it can be represented – up to isomorphism – as follows: $X = Y \times M$, M is a homogeneous space of a compact group G, $\mu = \nu \times m$ where m is the G-invariant measure on M, and there exists a measurable function $\varrho : Y \to G$ so that for $y \in Y, u \in M$,

$$T(y, u) = (Ty, \varrho(y)u) .$$

An ergodic isometric extension of the trivial one-point system is a Kronecker system.

Definition 2. A m.p.s. $(X, \mathscr{B}, \mu, T)$ is (n-*step*) *distal* if it has a succession of factors

$$(X, \mathscr{B}, \mu, T) = (X_1, \mathscr{B}_1, \mu_1, T) \to \cdots \to (X_n, \mathscr{B}_n, \mu_n, T) \to \text{1-point} \tag{4}$$

where each extension is isometric.

The terminology comes from topological dynamics where distal systems are defined by the property that for $x \neq x'$, the distance $d(T^n x, T^n x')$ is bounded from below. It is easily seen that for metric spaces, a succession of isometric extensions leads to a distal system in this sense.

We now formulate the generalization of V. Bergelson's theorem 1.2 to the general ergodic situation. Note that weak mixing systems have only trivial distal factors.

Theorem 3. *Let $(X, \mathscr{B}, \mu, T)$ be an ergodic m.p.s. Let $p_1(t), p_2(t), \ldots, p_k(t)$ be polynomials with integer coefficients with* $\deg p_i > 0$ *and* $\deg(p_i - p_j) > 0$ *for $i \neq j$. There exists a distal factor $\pi : X \to Y$ so that for any $f_1, f_2, \ldots f_k \in L^\infty(X, \mathscr{B}, \mu)$*

$$\frac{1}{N+1}\sum_{n=0}^{N} T^{-p_1(n)} f_1 T^{-p_2(n)} f_2 \cdots T^{-p_k(n)} f_k$$

$$-\frac{1}{N+1}\sum_{n=0}^{N} T^{-p_1(n)} E(f_1 \mid Y) \cdots T^{-p_k(n)} E(f_k \mid Y) \to 0 \tag{5}$$

in $L^2(X, \mathscr{B}, \mu)$.

We call $(Y, \mathscr{D}, \nu, T)$ a *characteristic factor for* $(T^{-p_1(n)}, \ldots, T^{-p_k(n)})$. We use the indefinite article here because any extension of a characteristic factor is again a characteristic factor. We do not know if there exists a "smallest" one.

For special cases the information is more specific [Fu3]:

Theorem 4. *For any ergodic m.p.s.* $(X, \mathscr{B}, \mu, T)$ *there exists a "largest" Kronecker factor* $(Z, \mathscr{D}, m, T)$ *and for* $f, g \in L^\infty(X, \mathscr{B}, \mu)$, $a, b \in \mathbb{Z}$, $a \neq b$,

$$\lim_{N\to\infty} \frac{1}{N+1} \sum_{n=0}^{N} T^{-an} f T^{-bn} g = \lim_{N\to\infty} \frac{1}{N+1} \sum_{n=0}^{N} T^{-an} E(f \mid Z) T^{-bn} E(g \mid Z) \ . \quad (6)$$

Here we are also guaranteed existence of the limits in question (in $L^2(X, \mathscr{B}, \mu)$). If we write $\bar{f} = E(f \mid Z), \bar{g} = E(g \mid Z)$ and regard the right hand side of (2.3) as a function $\psi(\zeta)$ on Z, then it can be rewritten as

$$\psi(\zeta) = \int \bar{f}(\zeta + a\theta)\bar{g}(\zeta + b\theta) d\theta$$

We can also be more explicit regarding characteristic factors for (T^{-an}, T^{-bn}, T^{-cn}). To formulate this result we need the notion of a 2-*step nilpotent system.* This is a system for which the underlying space has the form N/Γ where N is a 2-step nilpotent group, Γ is a closed cocompact subgroup and N possesses a sequence of closed subgroups N_i so that N/N_i is locally compact, $\Gamma/\Gamma \cap N_i$ is discrete and cocompact in N/N_i, the measure on N/Γ is the (unique) N-invariant probability measure (lifted from $(N/N_i) \ / \ (\Gamma/\Gamma \cap N_i)$) and the measure preserving transformation is $T(g\Gamma) = (\tau g)\Gamma$ for some $\tau \in N$.

Theorem 5. *For any ergodic* $(X, \mathscr{B}, \mu, T)$ *there exists a 2-step nilpotent factor* $(Y, \mathscr{D}, \nu, T)$ *so that for* $f, g, h \in L^\infty(X, \mathscr{B}, \mu)$ *and* a, b, c *distinct integers, we have*

$$\lim_{N\to\infty} \frac{1}{N+1} \sum_{n=0}^{N} T^{-an} f T^{-bn} g T^{-cn} h =$$

$$\lim_{N\to\infty} \frac{1}{N+1} \sum_{n=0}^{N} T^{-an} E(f \mid Y) T^{-bn} E(g \mid Y) T^{-cn} E(h \mid Y)$$

Theorem 5 was obtained jointly with B. Weiss and overlaps results of Conze and Lesigne [CL1]. It may be conjectured that similar results are valid for more general expressions with 2-step nilpotent systems replaced by arbitrary nilpotent systems.

3. Multiple Recurrence

We begin this section by formulating a version of (1.3) for distal systems.

Theorem 1. *Let* $(Y, \mathscr{D}, \nu, T)$ *be a distal system, let* $f \in L^\infty(Y, \mathscr{D}, \nu)$ *be non-negative but not vanishing a.e. Then for any* $l \in \mathbb{N}$,

$$\liminf_{N\to\infty} \frac{1}{N+1} \sum_{n=0}^{N} \int f \, T^{-n} f \, T^{-2n} f \cdots T^{-(l-1)n} f \, d\mu > 0 \ . \quad (7)$$

To explain the role of distality in this result we shall prove here by way of illustration that for 1-step distal systems one has the weaker statement: With $(Y,\mathscr{D},\nu,T)$ and f,l as above, $\exists n$ such that

$$\int fT^{-n}fT^{-2n}f\cdots T^{-(l-1)n}f d\mu > 0 \ . \tag{8}$$

A refinement of the argument will show that (7) is valid. Moreover a relativized version of this argument can be reapplied successively to handle the general distal case.

If M is a homogeneous space of the compact group G, then G acts continuously on $L^2(M,m) : \psi(u) \longrightarrow \psi^g(u) = \psi(gu)$. In particular, if $\psi \in L^\infty(M,m)$, its orbit under G, $\Psi = \{\psi^g \mid g \in G\}$, is compact in $L^2(M,m)$. Suppose $\psi \geq 0$ and that ψ is not 0 a.e. For any l, $\exists\varepsilon > 0$ so that if $\psi_1,\psi_2,\ldots,\psi_l \in L^\infty(M,m)$ satisfy $\|\psi_j - \psi\| < \varepsilon$ then $\int \psi_1\psi_2\ldots\psi_l dm > 0$. Since Ψ is compact we can partition Ψ into finitely many subsets of diameter $< \varepsilon$. So we can write $\Psi = \Psi_1 \cup\cdots\cup \Psi_r$ so that if $\psi_1,\psi_2,\ldots,\psi_l \in \Psi_j$ then $\int \psi_1\psi_2\cdots\psi_l dm > 0$.

Now consider a 1-step distal system, i.e., a Kronecker system, $(Z,\mathscr{D},\nu,T)$. Z plays the role of G as well as M. Take f as in the theorem and set $\psi = f$. By van der Waerden's theorem there is an l-term arithmetic progression, $a, a+n, a+2n,\ldots,a+(l-1)n$ such that each $T^{-(a+in)}f, i = 0,1,\ldots,l-1$, belongs to the same Ψ_j. Then

$$\int T^{-a}fT^{-(a+n)}f\cdots T^{-(a+(l-1)n)}f\, d\mu > 0$$

or

$$\int f\, T^{-n}f\cdots T^{-(l-1)n}f\, d\mu > 0$$

as claimed.

We now combine Theorem 1 with Theorem 2.1 to obtain

Theorem 2. *If $(X,\mathscr{B},\mu,T)$ is any ergodic m.p.s., $f \in L^\infty(X,\mathscr{B},\mu)$ is a non-negative function not vanishing a.e., then*

$$\liminf_{N\to\infty} \frac{1}{N+1}\sum_{n=0}^{N}\int fT^{-n}fT^{-2n}f\cdots T^{-(l-1)n}f\, d\mu > 0 \ .$$

Decomposition to ergodic components gives the result for arbitrary measure preserving systems. Finally taking $f = 1_A$ we obtain the

Theorem (Multiple Recurrence). *Let $(X,\mathscr{B},\mu,T)$ be any m.p.s., let $A \in \mathscr{B}$ with $\mu(A) > 0$. Then for any l there exists n with*

$$\mu(A\cap T^{-n}A\cap T^{-2n}A\cap\cdots\cap T^{-(l-1)n}A) > 0 \ . \tag{9}$$

As we show in [Fu1] and [Fu2] this is equivalent to Theorem A of the Introduction.

4. IP-Systems

We denote by $\mathscr{F}$ the collection of all finite subsets of $\mathbb{N} = \{1, 2, 3, \ldots\}$. If $\alpha, \beta \in \mathscr{F}$ we write $\alpha < \beta$ if $\max \alpha < \min \beta$.

Definition 1. An *IP-subset* of $\mathscr{F}$ consists of a sequence $\alpha_1 < \alpha_2 < \ldots$ in $\mathscr{F}$ together with all finite unions of these.

The elements of an IP-subset of $\mathscr{F}$ may be placed in one-one correspondence with $\mathscr{F}$: $\beta = \{i_1, i_2, \ldots i_k\} \leftrightarrow \alpha_{i_1} \cup \alpha_{i_2} \cup \cdots \alpha_{i_k} = \alpha_\beta$.

Definition 2. An *$\mathscr{F}$-sequence* is any sequence $\{x_\alpha\}$ labelled by $\alpha \in \mathscr{F}$. An *$\mathscr{F}$-subsequence* of $\{x_\alpha\}$ is the subsequence corresponding to an IP-subset of $\mathscr{F}$: $x'_\beta = x_{\alpha_\beta}$.

Definition 3. If X is a topological space, $\{x_\alpha\}$ an $\mathscr{F}$-sequence in X, we say that $x_\alpha \to x_0 \in X$, or $\lim_{\alpha \to \infty} x_\alpha = x_0$, if for any neighborhood V of x_0, $\exists \alpha_0$ so that for $\alpha > \alpha_0$ we have $x_\alpha \in V$.

The following theorem is equivalent to N. Hindman's theorem on finite sum sets [Hi1]:

Theorem 4. *If X is a compact metric space and $\{x_\alpha\}$ is an $\mathscr{F}$-sequence in X, then $\{x_\alpha\}$ has a convergent $\mathscr{F}$-subsequence.*

If X is finite this can be reformulated as: if $\mathscr{F} = \mathscr{C}_1 \cup \cdots \mathscr{C}_r$ then some $\mathscr{C}_j$ contains an IP-subset of $\mathscr{F}$. This latter formulation is due to Hindman.

The next definition describes the principal concept of this section.

Definition 5. An $\mathscr{F}$-sequence $\{\sigma_\alpha\}$ with values in a semigroup is an *IP-system* if $\alpha < \beta$ implies that

$$\sigma_{\alpha \cup \beta} = \sigma_\alpha \sigma_\beta \ . \tag{10}$$

Writing σ_i for $\sigma_{\{i\}}$ we see that $\sigma_{\{i_1, i_2, \ldots, i_k\}} = \sigma_{i_1} \sigma_{i_2} \cdots \sigma_{i_k}$ when $i_1 < i_2 < \cdots < i_k$. We note that in the commutative case (4.1) is valid whenever $\alpha \cap \beta = \emptyset$.

Lemma 6. *An $\mathscr{F}$-subsequence of an IP-system is an IP-system.*

We call this an *IP-subsystem.*

The classical notion of a m.p.s. will now be replaced by that of an IP-system of measure preserving transformations (m.p.t.). Inside a classical system we can form numerous IP-systems, taking $T_\alpha = T^{n_\alpha}$ where $n_\alpha = \sum_{i \in \alpha} n_i$ is an IP-system in $(\mathbb{Z}, +)$. In addition to providing more information regarding classical systems, the analysis of IP-systems provides recurrence results for situations where the classical setup doesn't prevail – or is trivial, because as might be the case T^p could equal the identity for some p and all T under consideration.

The main theorem of this section is

Theorem (IP-Recurrence). *Let $(X,\mathscr{B},\mu)$ be a measure space and let $\{T^{(1)}_\alpha\},\cdots,\{T^{(k)}_\alpha\}$ be k IP-systems of m.p.t. of $(X\ell,\mathscr{B},\mu)$ where all the T^i_α commute with one another. Then if $A \in \mathscr{B}, \mu(A) > 0$, $\exists \alpha \in \mathscr{F}$ so that*

$$\mu(T_\alpha^{(1)-1} A \cap T_\alpha^{(2)-1} A \cap \cdots \cap T_\alpha^{(k)-1} A) > 0 \ . \tag{11}$$

We remark that this theorem implies Theorems B and C of the Introduction. We refer the reader to [FK2] for details.

The proof of the IP-recurrence theorem depends on a development of ergodic theory for IP-systems analogous to classical theory. We proceed to outline the salient points of such a theory. One important point should be made at the outset. Since in the recurrence theorem we are looking for one index α for which we have a multiple intersection, we may pass from the IP-systems to IP-subsystems that may be more convenient. Since on account of Theorem 4, an $\mathscr{F}$-subsequence of an $\mathscr{F}$-sequence in a compact metric space converges, we shall often find it convenient to deal with an appropriately chosen $\mathscr{F}$-subsequence, which means we are really dealing with IP-subsystems of the initially given systems.

If $\{T_\alpha\}$ is an IP-system of m.p.t. of $(X,\mathscr{B},\mu)$ then writing as before $T_\alpha^{-1} f(x) = f(T_\alpha x)$ we obtain, in the commutative case, an IP-system of isometry operators $\{T_\alpha^{-1}\}$ on $L^2(X,\mathscr{B},\mu)$. Let $\mathscr{K}$ be the semigroup of linear operators on L^2 with norm ≤ 1, endowed with the weak operator topology. $\mathscr{K}$ is a compact metrizable space. Using Theorem 4 we can find a subsystem so that T_α^{-1} converges in $\mathscr{K}$. We now have

Proposition 7. *If $\{K_\alpha\} \subset \mathscr{K}$ is an IP-system and $K_\alpha \to Q$ then Q is a self-adjoint projection operator. When $K_\alpha = T_\alpha^{-1}$ for m.p.t. $\{T_\alpha\}$, then Q is the conditional expectation operator relative to a subalgebra $\mathscr{D} \subset \mathscr{B}$.*

Proof. If $Q = \lim\limits_{\alpha\to\infty} K_\alpha$ then $Q^2 = \lim\limits_{\alpha\to\infty}\lim\limits_{\beta\to\infty} K_\alpha K_\beta = \lim\limits_{\alpha\to\infty}\lim\limits_{\beta\to\infty} K_{\alpha\cup\beta} = Q$. The functions in the range of Q are characterized by: $K_\alpha f \to f$ weakly. Since $\|f\| \geq \|K_\alpha f\|$, weak convergence implies strong convergence. For strong convergence $K_\alpha f \to f, K_\alpha g \to g \Rightarrow K_\alpha f \vee K_\alpha g \to f \vee g$. Now $T_\alpha^{-1} f \vee T_\alpha^{-1} g = T_\alpha^{-1}(f \vee g)$ and so if $f,g \in$ range Q also $f \vee g \in$ range Q. This characterizes conditional expectation operators. □

For an IP-system $\{T_\alpha\}$ we now suppose having passed to a subsystem for which $\lim T_\alpha^{-1} = Q_T$ exists. For IP-systems this operator Q_T plays the role of the ergodic average. Since Q_T is self-adjoint and the T_α^{-1} are isometric this convergence will generally not be in the strong topology, and certainly not pointwise. In the extreme case when $Q_T =$ identity then the convergence is strong and we speak of $\{T_\alpha\}$ as a *rigid* system. An example of a rigid system is one obtained from the Kronecker system $(Z,\mathscr{B},\mu,T)$ setting $T_\alpha = T^{n_\alpha}$ for $\{n_\alpha\}$ an IP-system in $\mathbb{Z}$. if $T_\alpha \to Q_T$ it is not hard to show, applying T_α to eigenfunctions of T, that Q_T =identity. In IP-theory rigid systems play the role of Kronecker systems.

The opposite extreme to rigidity occurs when $Q_T f - E(f)$. This means that $\int f T_\alpha^{-1} g\, d\mu \to \int f\, d\mu \int g\, d\mu$ so that f and $T_\alpha^{-1} g$ are asymptotically independent. We call this case *mixing*.

Let $(X, \mathscr{B}, \mu)$ be a measure space, $(Y, \mathscr{D}, \nu)$ a factor of $(X, \mathscr{B}, \mu)$ such that a system $\{T_\alpha\}$ is defined both on X and on Y, that is to say, we have $\pi : X \to Y$ with $\pi T_\alpha = T_\alpha \pi$. A subspace $\mathscr{M} \subset L^2(X, \mathscr{B}, \mu)$ is called a *Y-module* if it is closed under multiplication by $L^\infty(Y, \mathscr{D}, \nu)$. A Y-module is of *finite rank* if it is spanned by finitely many functions over $L^\infty(Y, \mathscr{D}, \nu)$. Finally $\mathscr{M}$ is called $\{T_\alpha\}$ *quasi-invariant* if for each $g \in \mathscr{M}$ and $\varepsilon > 0$, $\exists \alpha_0$ so that for $\alpha > \alpha_0$ the L^2 distance satisfies

$$\operatorname{dist}(T_\alpha^{-1} g, \mathscr{M}) < \varepsilon \ .$$

Definition 8. $(X, \mathscr{B}, \mu, \{T_\alpha\})$ is a *rigid extension* of $(Y, \mathscr{D}, \nu, \{T_\alpha\})$ if $L^2(X, \mathscr{B}, \mu)$ is spanned by $\{T_\alpha\}$ quasi-invariant Y-modules of finite rank.

Now consider commuting IP-systems $\{T_\alpha^{(1)}\}, \cdots \{T_\alpha^{(k)}\}$ of m.p.t. on $(X, \mathscr{B}, \mu)$. We call $(X, \mathscr{B}, \mu, \{T_\alpha^{(1)}\}, \cdots \{T_\alpha^{(k)}\})$ a *k-fold IP-system.*

Definition 9. $(X, \mathscr{B}, \mu, \{T_\alpha^{(1)}\}, \cdots \{T_\alpha^{(k)}\})$ is *quasi-distal* if there is a sequence of k-fold IP-systems $(X_l, \mathscr{B}_l, \mu_l, \{T_\alpha^{(1)}\}, \cdots \{T_\alpha^{(k)}\})$, $l = 0, 1, 2, \ldots, n$, with $(X_0, \mathscr{B}_0, \mu_0) = (X, \mathscr{B}, \mu)$, and with X_n =point, and with maps $\pi_l : X_l \to X_{l+1}$ which are measure preserving and satisfy $\pi_l T_\alpha = T_\alpha \pi_l$ and such that for each l, $(X_l, \mathscr{B}_l, \mu_l, \{T_\alpha^{(i)} T_\alpha^{(j)-1}\})$ is a rigid extension of $(X_{l+1}, \mathscr{B}_{l+1}, \mu_{l+1}, \{T_\alpha^{(i)} T_\alpha^{(j)-1}\})$ for some $i \neq j$.

We now have

Theorem 10. *Given a k-fold IP-system $(X, \mathscr{B}, \mu, \{T_\alpha^{(1)}\}, \cdots \{T_\alpha^{(k)}\})$ there is a subsystem and a factor $(Y, \mathscr{D}, \nu, \{T_\alpha^{(1)}\}, \cdots \{T_\alpha^{(k)}\})$ such that the latter is quasi-distal and for any $f_1, f_2, \ldots, f_k \in L^\infty(X, \mathscr{B}, \mu)$ we have*

$$\lim_\alpha \int T_\alpha^{(1)-1} f_1 \cdots T_\alpha^{(k)-1} f_k d\mu = \lim_\alpha \int T_\alpha^{(1)-1} E(f_1 \mid Y) \cdots T_\alpha^{(k)-1} E(f_k \mid Y) d\nu \ .$$

The implication of Theorem 10 for us is that the IP-recurrence theorem will follow in general if we can show that for quasi-distal systems and $f \geq 0$ with f not vanishing a.e.,

$$\lim_{\alpha \to \infty} \int T_\alpha^{(1)^{-1}} f T_\alpha^{(2)^{-1}} f \cdots T_\alpha^{(k)^{-1}} f d\mu > 0 \ .$$

This can indeed be shown in a manner analogous to the proof of Theorem 3.1, by successively lifting the desired property via the rigid extensions that constitute the links of the quasi-distal system. This outlines the proof of the IP-recurrence theorem.

5. $W(\Lambda)$-Systems

In this section we describe the ergodic structure whose recurrence properties lead to Theorem D, the density version of the Hales-Jewett theorem. The structure will be similar to that of the previous section except that now the operators are no longer assumed to commute.

We take $\Lambda = \{1, 2, \ldots, k\}$. Suppose we are given k sequences of m.p.t. $\{T_n^{(1)}, \{T_n^{(2)}\}, \ldots, \{T_n^{(k)}\}$ of a measure space $(X, \mathscr{B}, \mu)$. Given a word $w \in W_l(\Lambda)$ we can form the transformation

$$T(w) = T_1^{w(1)} T_2^{w(2)} \cdots T_l^{w(l)} . \tag{12}$$

Definition 1. A $W(\Lambda)$-system consists of a family of m.p.t. of a measure space $(X, \mathscr{B}, \mu)$ corresponding to $w \in W(\Lambda)$ and formed in accordance with (12). We denote the system $(X, \mathscr{B}, \mu, \{T(w)\}_{w \in W(\Lambda)})$.

We can now state the

Theorem ($W(\Lambda)$-Recurrence). *Let $(X, \mathscr{B}, \mu, \{T(w)\}_{w \in W(\Lambda)})$ be a $W(\Lambda)$-system. If $f \in L^\infty(X, \mathscr{B}, \mu)$ is non-negative and not 0 a.e. then $\exists \varphi(t) \in W^*(\Lambda)$ so that*

$$\int T(\varphi(1))^{-1} f\, T(\varphi(2))^{-1} f \cdots T(\varphi(k))^{-1} f d\mu > 0 . \tag{13}$$

The proof of this theorem is patterned after the proof of the IP-recurrence theorem of §4. There is a new feature which arises on account of the non-commutativity. This feature also reflects an aspect of Ramsey theory for $W^*(\Lambda)$ as opposed to $\mathscr{F}$.

We use the following notation. If $w \in W_N(\Lambda)$ and $\alpha \subset \{1, 2, \ldots, N\}$, we designate by w_α the word of $W^*(\Lambda)$ in which the letters at positions of α are replaced by the variable t. If we write $\varphi(t)$ for w_α then we denote $\varphi(i)$ by w_α^i, $i \in \Lambda$. Finally set $\Omega = \Lambda^{\mathbb{N}}$. We regard Ω as part of the "boundary" of $W^*(\Lambda)$; namely for sequences $w^{(n)} \in W(\Lambda)$ of increasing length, and $\alpha^{(n)} \in \mathscr{F}$, we say $w_{\alpha_n}^{(n)} \to \omega \in \Omega$ if $\alpha_n \to \infty$ and $w^{(n)}(p) = \omega(p)$ for each $p \in \mathbb{N}$ and $n > n(p)$.

We next define *subspaces* of $W(\Lambda)$ and $W^*(\Lambda)$. Let $\Sigma = \{\varphi_1(t), \varphi_2(t), \ldots\}$ be a sequence in $W^*(\Lambda)$. We define $W_\Sigma^*(\Lambda)$ as consisting of words of the form $\varphi_1(u_1)\varphi_2(u_2) \cdots \varphi_h(u_h)$, h arbitrary, with $u_i \in \Lambda$ or $u_i \in \Lambda \cup \{t\}$ respectively. It is clear how to define Ω_Σ.

By analogy with $\mathscr{F}$-sequences we have

Definition 2. A function $x(w_\alpha)$ is called a $W^*(\Lambda)$-*sequence.* Its restriction to a subspace is a $W^*(\Lambda)$-subsequence.

Definition 3. If $x(w_\alpha)$ takes values in a metric space M. Then we say $x(w_\alpha)$ is *coherent* if there exists a function $x^* : \Omega \to M$ such that $x(w_\alpha) \to x^*(\omega)$ uniformly as $w_\alpha \to \omega \in \Omega$.

The following theorem is a far reaching extension of the Hales-Jewett theorem. It includes Hindman's theorem. See [Ca1, FK3].

Theorem 4. *A $W^*(\Lambda)$-sequence with values in compact metric space has a coherent $W^*(\Lambda)$-subsequence.*

Given a $W(\Lambda)$-system we will restrict to appropriate subspaces in order to achieve coherence of certain expressions. Note that the expression in (5.2) is a $W^*(\Lambda)$-sequence.

The next lemma shows how IP-systems occur within $W(\Lambda)$-systems.

Lemma 5. *Let $w \in W_N(\Lambda), \alpha, \beta \subset \{1, 2, \cdots, N\}, \alpha < \beta$ and suppose that for $n \in \alpha \cup \beta, w(n) = i$. Then for any $W(\Lambda)$-system $\{T(w)\}$*

$$T(w^j_\alpha)T(w^i_\alpha)^{-1}T(w^j_\beta)T(w^i_\beta)^{-1} = T(w^j_{\alpha\cup\beta})T(w^i_{\alpha\cup\beta})^{-1} \ . \tag{14}$$

Using Theorem 4 we pass to a subsystem for which all

$$Q_{ij}(\omega) = \lim_{\substack{\alpha\to\infty \\ w\to\omega}} T(w^i_\alpha)T(w^j_\alpha)^{-1}$$

exist uniformly. On account of Lemma 5 and Proposition 4.7 these define conditional expectation operators $E(\ |\ \mathscr{D}_{ij}(\omega))$ provided each $i \in \Lambda$ occurs infinitely often in ω. These $Q_{ij}(\omega)$ represent the "ergodic averaging" operator in the $W(\Lambda)$ context. Note that we obtain a "field" of operators. In a similar way we can form rigid extensions, and quasi-distal factor systems, except that the entire structure will vary (continuously) with $\omega \in \Omega$. The σ-algebras would not be invariant but for α far out the operators

$$T^i_j(w_\alpha) = T(w^i_\alpha)T(w^j_\alpha)^{-1}$$

will move $\mathscr{D}_{ij}(\omega)$ to $\mathscr{D}_{ij}(\omega')$ with ω' close to ω. With this apparatus we can proceed as in Sect. 4.

References

[BB1] Brown, T.C., Buhler, J.P.: A density version of a geometric Ramsey theorem. J. Comb. Theory Ser. A **32** (1982) 20–34

[Be1] Bergelson, V.: Weakly mixing PET. Ergodic Theory Dyn. Systems **7** (1987) 337–349

[Ca1] Carlson, T.J.: Some unifying principles in Ramsey theory. Discrete Math. **68** (1988) 117–169

[CL1] Conze, J.P., Lesigne, E.: Sur un théorème ergodique pour les mesure diagonales. Preprint

[ET1] Erdös, P., Turán, P.: On some sequences of integers. J. London Math. Soc. **11** (1936) 261–264

[Fu1] Furstenberg, H.: Ergodic behavior of diagonal measures and a theorem of Szemerédi on arithmetic progressions. J. d'Analyse Math. **31** (1977) 204–256

[Fu2] Furstenberg, H.: Recurrence in Ergodic Theory and Combinatorial Number Theory. Princeton University Press (1981)

[Fu3] Furstenberg, H.: Nonconventional ergodic theorems. In: Symposium on Legacy of von Neumann. Proceedings of Symposia in Pure Mathematics, Amer. Math. Soc. **50** (1990) 43–56

[FK1] Furstenberg, H., Katznelson, Y.: An ergodic Szemerédi theorem for commuting transformations. J. d'Analyse Math. **34** (1978) 275–291

[FK2] Furstenberg, H., Katznelson, Y.: An ergodic Szemerédi theorem for IP-systems and combinatorial theory. J. d'Analyse Math. **45** (1985) 117–168

[FK3] Furstenberg, H., Katznelson, Y.: Idempotents in compact semigroups and Ramsey theory. Israel J. Math. **68** (1989) 257–270

[FK4] Furstenberg, H., Katznelson, Y.: A density version of the Hales-Jewett theorem. J. D'Analyse Math. (to appear)

[GR1] Graham, R.L., Rothschild, B.L.: Ramsey's theorem for *n*-parameter sets. Trans. Amer. Math. Soc. **159** (1971) 257–292

[GRS1] Graham, R.L., Rothschild, B.L., Spencer, J.H.: Ramsey theory. Wiley, New York 1980

[HJ1] Hales, A.W., Jewett, R.I.: Regularity and positional games. Trans. Amer. Math. Soc. **106** (1963) 222–229

[Hi1] Hindman, N.: Ultrafilters and combinatorial number theory. In: Number theory Carbondale 1979. Lecture Notes in Mathematics, vol. 751. Springer, Berlin Heidelberg New York 1979, pp. 119–184

[Sz1] Szemerédi, E.: On sets of integers containing no *k* elements in arithmetic progression. Acta Arith. **27** (1975) 199–245

Random Schrödinger Operators

Shinichi Kotani

Department of Mathematics, University of Tokyo, Hongo, Tokyo 113, Japan

1. Introduction

As for the present subject of random Schrödinger operators, two articles had already appeared in the proceedings of the last ICM at Berkeley presenting various viewpoints (Pastur [13], Spencer [15]). Since then several important works have been done for Schrödinger operators with certain quasi periodic potentials in the one dimensional case. There does not seem to be much to add in this note to what already exists in the literature; however, we would like to mention several open problems in the case where random potentials are nearly almost periodic. We focus our attention only on the one dimensional case.

2. Setup of Random Schrödinger Operators

Let

$$\Omega = L^2_{\text{real}}\left(\mathbb{R}, \frac{dx}{1+|x|^3}\right).$$

For $q \in \Omega$ set

$$L(q) = -\frac{d^2}{dx^2} + q.$$

It is not difficult to see that $L(q)$ defines a selfadjoint operator in $L^2(\mathbb{R}, dx)$. We consider a family of operators $\{L(q); q \in \Omega\}$ under a probability measure P on Ω. We impose the following conditions on P:

(1) Shift invariance: For any $A \in \mathfrak{B}(\Omega)$ and $x \in \mathbb{R}$ it holds that

$$P(T_x A) = P(A)$$

where T_x is the shift operator on Ω.

(2) Ergodicity: $T_x A = A$ a.e. with respect to P for every $x \in \mathbb{R}$ implies $P(A) = 0$ or 1.

Proceedings of the International Congress
of Mathematicians, Kyoto, Japan, 1990

(3) Integrability:

$$\int_{\Omega}\left\{\int_0^1 q(x)^2\,dx\right\}P(dq) < \infty.$$

There are many examples satisfying these conditions such as periodic potentials, almost periodic potentials and any stationary ergodic process with finite second moment.

Now we introduce quantities related to the operator $L(q)$, which are necessary for subsequent discussions. For any $\lambda \in \mathbb{C}\setminus\mathbb{R}$, there exist unique solutions $f_{\pm}(x, \lambda, q)$ of

$$L(q)u = \lambda u, \text{ with } u(0) = 1 \text{ and } u \in L^2(\mathbb{R}\pm, dx),$$

respectively. Define

$$h_{\pm}(\lambda, q) = \pm\frac{d}{dx} f_{\pm}(x, \lambda, q)|_{x=0}.$$

It is of some interest to see

$$f_{\pm}(x, \lambda, q) = \exp\left(\pm\int_0^x h_{\pm}(\lambda, T_y q)\,dy\right). \tag{2.1}$$

The Green function of $L(q)$ is given by

$$g_{\lambda}(x, y, q) = -(h_{+}(\lambda, q) + h_{-}(\lambda, q))^{-1} f_{+}(x, \lambda, q) f_{-}(y, \lambda, q)$$

if $x \geq y$. The importance of $h_{\pm}$ in the study of almost periodic Schrödinger operators was first recognized by R. Johnson and J. Moser [3].

3. Fundamental Theorems and Open Problems

Define a holomorphic function $w(\lambda)$ on $\mathbb{C}\setminus\mathbb{R}$ by

$$w(\lambda) = -\frac{1}{2}\int_{\Omega} g_{\lambda}(0, 0, q)^{-1} P(dq)$$

for $P \in \mathscr{P}(\Omega)$. This function was first introduced by R. Johnson and J. Moser [3] in the almost periodic case. Without difficulty we find

$$w(\lambda) = \int_{\Omega} h_{+}(\lambda, q) P(dq) = \int_{\Omega} h_{-}(\lambda, q) P(dq).$$

The function w has a positive imaginary part on the upper half plane $\mathbb{C}_{+}$ and such a holomorphic function is called a Herglotz function, which has appeared often in the spectral theory of differential operators and the classical theory of moment problems. Our w has more special properties by which we see the existence of the finite limit $w(\xi + i0)$ for any $\xi \in \mathbb{R}$. We set for $\xi \in \mathbb{R}$

$$w(\xi + i0) = -\gamma(\xi) + i\pi n(\xi).$$

Then it is known that $\gamma(\xi)$ is non-negative and $n(\xi)$ is continuous, non-negative and non-decreasing. $\gamma(\xi)$ is called the Lyapunov exponent and $n(\xi)$ is called the integrated density of states. $\pi n(\xi)$ is called the rotation number. The readers will find a suitable explanation of γ, n in R. Johnson and J. Moser [3] or S. Kotani [5]. Now we can state

Theorem (S. Kotani [4]). *Set*

$$N = \{\xi \in \mathbb{R}; \gamma(\xi) = 0\}.$$

Then for almost all $q \in \Omega$ with respect to P

$$h_+(\xi + i0) = -\overline{h_-(\xi + i0)}$$

holds for almost all $\xi \in N$.

This theorem has several interesting implications, among which we see that, by the reflection principle, $h_\pm(\lambda, q)$ have analytic continuation from $\mathbb{C}_+$ to the lower half plane $\mathbb{C}_-$ through the interior $\mathring{N}$. Therefore this combined with the formula (2.1) implies that $L(q)$ has generalized Bloch solutions if $\lambda \in \mathring{N}$, and in particular $L(q)$ has pure absolutely continuous spectrum on $\mathring{N}$. A more careful study shows

Theorem (K. Ishii [2], L.A. Pastur [12], S. Kotani [4]). *With probability one the absolutely continuous spectrum of $L(q)$ coincides with $\bar{N}^\mu := \{\xi \in \mathbb{R}; \mu(U \cap N) > 0$ for any neighbourhood U of $\xi\}$, where μ denotes Lebesgue measure.*

Therefore we can tell everything about the absolutely continuous spectrum by looking at the function w alone. Then the following question naturally arises:

P1. *Can we determine the pure absolute continuity of the spectrum from w alone?*

The above argument shows that if $(dn)(\mathbb{R} \setminus \check{N}) = 0$, then $L(q)$ has only absolutely continuous spectrum with probability one. However this condition is far from the necessity. In this respect, S. Kotani, M. Krishna [6] gives more recent information. It gives a more delicate sufficient condition which was essentially introduced by B.M. Levitan [7]

Once this problem was solved, the next question would be

P2. *Does the absolute continuity of the spectrum imply the almost periodicity of the random potentials?*

It is well known that if the spectrum of $L(q)$ coincides with the set N and N consists of only finitely many closed intervals, then q must be a quasi periodic potential (S.P. Novikov [11], H.P. McKean and P. van Moerbecke [9]). We have also several papers treating potentials having infinitely many gaps of certain types in the spectrum (H.P. McKean and E. Trubowitz [10], V.A. Marchenko and I.V. Ostrovskii [8], L.A. Pastur and B.A. Tkachenko [14]). Among them B.M. Levitan [7] is very closely related to our purpose and is restated in our framework in S. Kotani and M. Krishna [6]. W. Craig [1] provides a non-probabilistic treatment of this problem and gives a simpler proof of the trace formula. Although much effort has been made to clarify the relationship between the spectrum and the potentials, it seems that we have not grasped a key point yet.

References

1. Craig, W.: The trace formula for Schrödinger operators on the line. Preprint
2. Ishii, K.: Localization of eigenstates and transport phenomena in one dimensional disordered systems. Supp. Prog. Theor. Phys. **53** (1973) 77–138
3. Johnson, R., Moser, J.: The rotation number of almost periodic potentials. Comm. Math. Phys. **84** (1982) 403–438
4. Kotani, S.: Ljapounov indices determine absolutely continuous spectrum of stationary random one-dimensional Schrödinger operators. Proc. of Taniguchi Symp. SA. Katata 1982, pp. 225–247
5. Kotani, S.: One dimensional Schrödinger operators and Herglotz functions. Proc. Taniguchi Symp. PMMP. Katata 1985, pp. 219–250
6. Kotani, S., Krishna, M.: Almost periodicity of some random potentials. J. Funct. Anal. **78** (1988) 390–405
7. Levitan, B.M.: On the closure of the set of finite-zone potentials. Math. USSR Sb. **51** (1985) 67–89
8. Marchenko, V.A., Ostrovskii, I.V.: A characterization of the spectrum of Hill's operator. Math. USSR Sb. **26** (1975) 493–554
9. McKean, H.P., van Moerbecke, P.: The spectrum of Hill's equation. Invent. math. **30** (1975) 217–274
10. McKean, H.P., Trubowitz, E.: Hill's operator and hyperelliptic function theory in the presence of infinitely many branch points. Comm. Pure Appl. Math. **29** (1976) 143–226
11. Novikov, S.P.: The periodic problem for the *KdV* equation I. Funct. Anal. Appl. **8** (1974) 236–246
12. Pastur, L.A.: Spectral properties of disordered systems in one-body approximation. Comm. Math. Phys. **88** (1980) 167–196
13. Pastur, L.A.: Spectral properties of metrically transitive operators and related problems. Proc. of ICM Berkeley 1986, pp. 1296–1311
14. Pastur, L.A., Tkachenko, B.A.: On the spectral theory of the one-dimensional Schrödinger operator with a limit periodic potential. Dokl. Acad. Nauk USSR **279** (1984) 1050–1054
15. Spencer, T.: Random and quasiperiodic Schrödinger operators. Proc. of ICM Berkeley 1986, pp. 1312–1318

De Rham Cohomology of Wiener-Riemannian Manifolds

Shigeo Kusuoka

Research Institute for Mathematical Sciences, Kyoto University Kyoto 606, Japan

0. Introduction

In the study of Riemannian manifolds, we often observe operators in L^2-spaces. The study of infinite dimensional manifolds had started a long time ago, but we did not think of such L^2-spaces because of a lack of good Riemannian volumes. However, some people (e.g. Gross, Kuo, Eells, Elworthy, Ramer [7]) had tried to study infinite dimensional manifolds which possess a good Hilbert-Riemannian metric and a good measure which is modeled after the Wiener measure. Let us call such an infinite dimensional manifold a Wiener-Riemannian manifold.

After Malliavin introduced stochastic calculus of variation, which is called Malliavin calculus now, there was a great progress on stochastic analysis. (If we made a list of contributors, it might be very long). And this progress enables us to challenge the study of Wiener-Riemannian manifolds again. Such challenge started already by some people (e.g. Malliavin, Airault, Biesen [1, 2], Watanabe, Kazumi, Aida).

In this paper, we focus on the de Rham cohomology of Wiener-Riemannian manifolds and explain what the purpose of this topic is and what we can prove at the moment.

1. Preliminary Facts and Motivation

We say that a triple (μ, H, B) is an abstract Wiener space, if

(1) B is a separable real Banach (or Fréchet) space,

(2) H is a separable real Hilbert space continuously and densely embedded in the Banach space B,

and

(3) μ is a probability measure in B such that

$$\int_B \exp(\sqrt{-1}\,{}_B\langle z, u\rangle_{B^*})\mu(dz) = \exp\left(-\frac{1}{2}\|u\|_{H^*}{}^2\right)$$

for any $u \in B^* \subset H^*$.

Proceedings of the International Congress of Mathematicians, Kyoto, Japan, 1990

Here B^* and H^* are the dual space of B and H respectively. Often the dual space H^* is identified with the Hilbert space H itself. However, we will not take this convention here.

An important example of abstract Wiener spaces is an ordinary Wiener space (μ_0, H_0, W^d), where $W^d = \{w \in C([0, 1]; \mathbb{R}^d); w(0) = 0\}$, $H_0 = \{h \in W^d; h(t)$ is absolutely continuous in t and $\int_0^1 |\dot{h}(t)|^2\, dt < \infty\}$, and μ_0 is the Wiener measure on W^d.

Now let E be a separable Hilbert space. For any bounded measurable map $f: B \to E$, we define a measurable map $P_t f: B \to E$, $t \geqq 0$, by

$$P_t f(z) = \int_B f(e^{-t/2} z + (I_H - e^{-t})^{1/2} w)\mu(dw), \qquad z \in B.$$

Then it is easy to see that $\{P_t\}_{t \geqq 0}$ can be regarded as a strongly continuous contraction semigroup in $L^p(B; E, d\mu)$, $p \in (1, \infty)$. Let $\mathscr{L} = \mathscr{L}_{p,E}$ denote the infinitesimal generator of the semigroup $\{P_t\}_{t \geqq 0}$ in $L^p(B; E, d\mu)$. For any separable real Hilbert space E, $s \geqq 0$ and $p \in (1, \infty)$, we define a Banach space $\mathbf{D}_p^s(E)$ by

$$\mathbf{D}_p^s(E) = \mathrm{Image}(I - \mathscr{L}_{p,E})^{-s/2} \text{ in } L^p(B; E, d\mu),$$

and

$$\|u\|_{\mathbf{D}_p^s} = \|(I - \mathscr{L}_{p,E})^{s/2} u\|_{L^p(B; E, d\mu)}, \qquad u \in \mathbf{D}_p^s(E).$$

Also, we define $\mathbf{D}_p^s(E)$, $s < 0$, $p \in (1, \infty)$, to be the dual Banach space of $\mathbf{D}_q^{-s}(E^*)$, where $\dfrac{1}{p} + \dfrac{1}{q} = 1$. As usual, we identify the Banach space $L^p(B; E, d\mu)$ with the dual space of $L^q(B; E^*, d\mu)$. Then we see that

$$\mathbf{D}_p^s(E) \supset \mathbf{D}_{p'}^{s'}(E) \text{ if } 1 < p \leqq p' < \infty \quad \text{and} \quad -\infty < s \leqq s' < \infty,$$

and

$$\|u\|_{\mathbf{D}_p^s(E)} = \|(I - \mathscr{L})^{s/2} u\|_{L^p(B; E, d\mu)}, \qquad u \in L^p(B; E, d\mu),$$

if $s \in (-\infty, 0]$ and $p \in (1, \infty)$.

We define $\mathbf{D}_p^\infty(E)$, $p \in (1, \infty)$ and $\mathbf{D}_{p-}^\infty(E)$, $p \in (1, \infty]$, by

$$\mathbf{D}_p^\infty(E) = \bigcap_{s \in (0,\infty)} \mathbf{D}_p^s(E),$$

and

$$\mathbf{D}_{p-}^\infty(E) = \bigcap_{q \in (0,p)} \mathbf{D}_q^\infty(E).$$

Then they become Fréchet spaces naturally.

Also, we have the following.

Proposition 1. *For any separable real Hilbert space E, if $u \in \mathbf{D}_2^1(E)$, there is a $Du \in \mathbf{D}_2^0(H^* \otimes E)$ such that*

$$\mu\left(\left\{z \in B; \left\|\frac{1}{t}(u(z+th) - u(z)) - Du(z)h\right\|_E > \varepsilon\right\}\right) \to 0, \qquad t \to 0,$$

for any $h \in H$ *and* $\varepsilon > 0$.

The following is due to Meyer.

Theorem 2. *For any separable real Hilbert space E and* $p \in (1, \infty)$, *the linear operator* D *from* $\mathbf{D}_2^1(E)$ *into* $\mathbf{D}_2^0(H^* \otimes E)$ *can be extended (or restricted) to a bounded linear operator from* $\mathbf{D}_p^s(E)$ *into* $\mathbf{D}_p^{s-1}(H^* \otimes E)$ *for any* $s \in \mathbb{R}$ *and* $p \in (1, \infty)$.

Therefore the dual operator D^* of D is a bounded linear operator from $\mathbf{D}_p^s(H \otimes E)$ into $\mathbf{D}_p^{s-1}(E)$ for any $s \in \mathbb{R}$ and $p \in (1, \infty)$. Let $\iota_H : H^* \to H$ be an isometry given by $(\iota_H u, h)_H = {}_{H^*}\langle u, h\rangle_H$, $h \in H$. Then we have

$$\mathscr{L} = -\tfrac{1}{2}D^*\iota_H D.$$

The following is due to Malliavin and Watanabe. By virtue of it we can consider hypersurfaces in abstract Wiener spaces.

Theorem 3. *Let* $F = (F_1, \ldots, F_n) \in \mathbf{D}_{\infty-}^{\infty}(\mathbb{R}^n)$ *and assume that*

$$\det(\{(DF_i, DF_j)_{H^*}\}_{i,j=1,\ldots,n})^{-1} \in \bigcap_{p \in (1,\infty)} L^p(B; d\mu).$$

Then the linear map $S_F : \mathscr{S}(\mathbb{R}^n) \to \mathbf{D}_{\infty-}^{\infty}(\mathbb{R})$ *given by* $S_F f(z) = f(F(z))$, $z \in B$, *is extensible to a continuous linear map from* $\mathscr{S}'(\mathbb{R}^n)$ *into* $\bigcup_{\substack{s<0 \\ p \in (1,\infty)}} \mathbf{D}_p^s(\mathbb{R})$.

In particular, we can think of a generalized Wiener functional $\delta(F)$ if F satisfies the assumption of Theorem 3.

For any separable real Hilbert space E, we denote by $E^{\otimes n}$ the completion of the algebraic tensor product $E \otimes E \otimes \cdots \otimes E$, $n \geqq 1$. Then $E^{\otimes n}$ is a real Hilbert space and $\{e_{k_1} \otimes e_{k_2} \otimes \cdots \otimes e_{k_n}\}_{k_1,\ldots,k_n=1}^{\infty}$ is a complete orthonormal basis if $\{e_k\}_{k=1}^{\infty}$ is a complete orthonormal basis of E. We also use the convention that $E^{\otimes 0} = \mathbb{R}$. For each $n \geqq 1$, let $A_n : E^{\otimes n} \to E^{\otimes n}$ be a bounded linear map given by $A_n(h_1 \otimes h_2 \otimes \cdots \otimes h_n) = \dfrac{1}{n!}\sum_{\sigma \in S_n} \mathrm{sgn}(\sigma) h_{\sigma(1)} \otimes h_{\sigma(2)} \otimes \cdots \otimes h_{\sigma(n)}$, $h_1, \ldots, h_n \in E$. Here S_n is a set of permutations of $\{1, \ldots, n\}$. We define a closed subspace $\bigwedge^n E$ of $E^{\otimes n}$ by $\bigwedge^n E = \{x \in E^{\otimes n}; A_n(x) = x\}$.

Now we define a continuous linear operator $\tilde{d}_n : \mathbf{D}_{\infty-}^{\infty}(\bigwedge^n H^*) \to \mathbf{D}_{\infty-}^{\infty}(\bigwedge^{n+1} H^*)$ by $\tilde{d}_n u = (n+1)A_{n+1} Du$. Then we have $d_{n+1} d_n = 0$. We also define an inner product $(\cdot, *)_n$ on $\mathbf{D}_{\infty-}^{\infty}(\bigwedge^n H^*)$ by

$$(u, v)_n = (n!)^{-1} \cdot \int_B (u(z), v(z))_{H^{*\otimes n}} \mu(dz), \; u, v \in \mathbf{D}_{\infty-}^{\infty}(\textstyle\bigwedge^n H^*).$$

Then we see that the formal adjoint operator $\tilde{d}_n^*$ is a continuous linear operator from $\mathbf{D}_{\infty-}^{\infty}(\bigwedge^{n+1} H^*)$ into $\mathbf{D}_{\infty-}^{\infty}(\bigwedge^n H^*)$.

The following is due to Shigekawa [8].

Theorem 4. (1) $L_n = \tilde{d}_n{}^*\tilde{d}_n + \tilde{d}_{n-1}\tilde{d}_{n-1}{}^*$ *has a natural self-adjoint extension in* $L^2(B; \bigwedge^n H^*, d\mu)$, $n = 0, 1, \ldots$.

(2) $$\dim(\ker L_n) = \dim(\{u \in \mathbf{D}^\infty_{\infty-}(\textstyle\bigwedge^n H^*); \tilde{d}_n u = 0\}/\{\tilde{d}_{n-1}v; v \in \mathbf{D}^\infty_{\infty-}(\bigwedge^{n-1}H^*)\}) = \begin{cases} 1 & n = 0 \\ 0 & n \geqq 1 \end{cases}.$$

Our main purpose is to show similar results for curved spaces.

Hypersurfaces in Wiener Spaces

Let M be a compact Riemannian manifold of finite dimension embedded in the Euclidean space $\mathbb{R}^d$. Then for each $x \in M$, the tangent space $T_x(M)$ is regarded as a subvector space in $\mathbb{R}^d$. Let P_x denote the orthogonal projection in $\mathbb{R}^d$ onto $T_x(M)$ for each $x \in M$. Let us think of the following stochastic differential equation.

$$\begin{cases} dX(t, x; w) = P_{X(t,x;w)} \circ dw(t), & t \in [0, 1] \\ X(0, x; w) = x \in M \end{cases}.$$

Here $\{w(t); t \in [0, 1]\}$ is a d-dimensional Brownian motion. This is the Stroock's representation of Brownian motion on the manifold M.

Now let us take $x, y \in M$ and fix them throughout. Let us think of a subset $\tilde{M}_{xy} = \{w \in W^d; X(1, x; w) = y\}$. This is not a closed set in W_d in general. Let $\nu = \nu_{xy}$ be a measure on $\tilde{M}_{xy}$ given by

$$\nu(dw) = \delta_y(X(1, x; w))\mu(dw).$$

Let $N(w) = \iota_H DX(1, x; w)^*(DX(1, x; w)DX(1, x; w)^*)^{-1}DX(1, x; w)$, $w \in W^d$. Then $N \in \mathbf{D}^\infty_{\infty-}(H^* \otimes H)$. If $w \in \tilde{M}_{xy}$, then $N(w)$ gives normal direction of $\tilde{M}_{xy}$ in principle, and so $I_H - N(w)$ is the orthogonal projection in H on to $T_w(\tilde{M}_{xy})$ in principle.

Now we introduce an equivalence relation $\sim$ on $\mathbf{D}^\infty_{\infty-}(\bigwedge^n H^*)$ by the following. $u \sim v$ if $u(w)(h_1 - N(w)h_1, \ldots, h_n - N(w)h_n) = v(w)(h_1 - N(w)h_1, \ldots, h_n - N(w)h_n)$ ν-a.e.w for all $h_1, \ldots, h_n \in H$. Also, let $\tilde{\mathscr{A}}^n = \mathbf{D}^\infty_{\infty-}(\bigwedge^n H^*)/\sim$. We can define a map $d_n : \tilde{\mathscr{A}}_n \to \tilde{\mathscr{A}}_{n+1}$ by $d_n(u/\sim) = (\tilde{d}_{n+1}u)/\sim$. Then obviously $d_{n+1}d_n = 0$.

We can introduce an inner product on $\tilde{\mathscr{A}}_n$ by $(u/\sim, v/\sim)_n = (n!)^{-1}\int_{M_{xy}}((\Lambda^n((I_H - N(w))^*)u(w), (\bigwedge^n((I_H - N(w))^*)v(w))_{H_0{}^{*\otimes n}}\nu(dw)$, $u, v \in \tilde{\mathscr{A}}^n$. Then we can see that the formal self adjoint map $d_n{}^*$ is a map from $\tilde{\mathscr{A}}_{n+1}$ into $\tilde{\mathscr{A}}_n$. So we have an operator $L_n = d_n{}^*d_n + d_{n-1}d_{n-1}{}^*$ in the Hilbert space which is the completion of $\tilde{\mathscr{A}}_n$. It is not difficult to see that L_n is closable. So let us denote by the same L_n the Friedlichs extension of L_n.

Our main interest is to study ker L_n. It is a quite natural guess that there is a relation between ker L_n and the topological cohomology of $\tilde{M}_{xy}$. However, the topology of $\tilde{M}_{xy}$ is not well-defined. So we have to modify this guess as follows.

Let us think of the following ordinary differential equation.

$$\begin{cases} \dfrac{d}{dt}Y(t, x; h) = P_{Y(t,x;h)} \cdot \dot{h}(t), & t \in [0, 1], \\ Y(0, x; h) = x \in M \end{cases}$$

for each $h \in H_0$, and let

$$\overline{M}_{xy} = \{h \in H_0; Y(1, x; h) = y\}.$$

Then $\overline{M}_{xy}$ is a closed Hilbertian submanifold in H_0. A natural guess is that there is a natural isomorphism between ker L_n and $H^n(\overline{M}_{xy}; \mathbb{R})$, nth-topological cohomology of the space $\overline{M}_{xy}$.

The results which we can prove at the moment is the following.

Theorem. (1) *There is a natural map* $j_n : \ker L_n \to H^n(\overline{M}_{xy}; \mathbb{R})$, $n \geqq 0$.
(2) j_n, $n = 0, 1, 2, \ldots$, *is surjective.*
(3) j_n, $n = 0$ or 1, *is injective.*

Let us give some remarks on this theorem. First, it is well-known that $\nu_{xy}(\overline{M}_{xy}) = 0$. So even the construction of the map j_n is not trivial. Also, let

$$\text{Path}_{xy} = \left\{\gamma : [0, 1] \to M; \gamma(t) \text{ is absolutely continuous in } t \text{ and } \int_0^1 |\dot{\gamma}(t)|_{T_{\gamma(t)}}(M)^2\, dt < \infty\right\}.$$

Then we have a natural map $\pi : \overline{M}_{xy} \to \text{Path}_{xy}$ given by $\pi(h)(t) = Y(t, x; h)$. It is easy to see that $(\overline{M}_{xy}, \pi, \text{Path}_{xy})$ is a vector bundle. So we have a natural isomorphism between $H^n(\overline{M}_{xy}; \mathbb{R})$ and $H^n(\text{Path}_{xy}; \mathbb{R})$. Since the topological cohomology of path spaces have been studied by topologists, we can use their results to study $H^n(\overline{M}_{xy}; \mathbb{R})$.

The strategy to show our theorem is the following. If we think of a finite dimensional compact manifold M, it is well known by the name of de Rham-Hodge-Kodaira theory that there is a natural isomorphism between $\ker(d_n{}^* d_n + d_{n-1} d_{n-1}{}^*)$ and $Hn(M; \mathbb{R})$, and this isomorphism is usually constructed via de Rham cohomology. So we have to define de Rham cohomology for Wiener-Riemannian manifolds. To compare $\ker(d_n{}^* d_n + d_{n-1} d_{n-1}{}^*)$ and de Rham cohomology, the hypoellipticity of the operator $d_n{}^* d_n + d_{n-1} d_{n-1}{}^*$ plays a key role. Also, to compare de Rham cohomology and $H^n(M; \mathbb{R})$, we use Cech-de Rham's argument which requires the existence of partition of unity and the cohomology vanishing theorem of Poincare type.

In the rest of this paper, we explain in what sense we have the definition of de Rham cohomology, the existence of partition of unity, the cohomology vanishing theorem and the hypoellipticity.

3. Tools

Let X be a separable metric space and let $\mathscr{P}(X)$ denote the set of all subsets of X.

Definition 1. We call a function $\alpha : \mathscr{P}(X) \to [0, \infty)$ a *finite capacity on* X if the following are satisfied.

(1) $\alpha(\phi) = 0$.
(2) $\alpha(A) \leqq \alpha(B)$ for any $A, B \in \mathscr{P}(X)$ with $A \subset B$.
(3) $\alpha(A_1 \cup A_2) \leqq \alpha(A_1) + \alpha(A_2)$ for any $A_1, A_2 \in \mathscr{P}(X)$.
(4) For any $A \in \mathscr{P}(X)$, $\alpha(A) = \inf\{\alpha(G); A \subset G, G \text{ is an open set in } X\}$.
(5) There is a sequence $\{K_n\}_{n=1}^{\infty}$ of compact sets in X such that $\alpha(X \setminus K_n) \to 0$ as $n \to \infty$.

We denote by $\mathscr{CAP}(X)$ the set of finite capacities on X.

Definition 2. Let $\alpha \in \mathscr{CAP}(X)$.

(1) We say that an element $A \in \mathscr{P}(X)$ is *α-quasi-closed*, if there is a sequence $\{K_n\}_{n=1}^{\infty}$ of compact sets in X such that $K_n \subset A$, $n \geqq 1$, and $\alpha(A \setminus K_n) \to 0$, $n \to \infty$.

(2) We say that an elememt $A \in \mathscr{P}(X)$ is *α-quasi-open* if $X \setminus A$ is α-quasi-closed.

Definition 3. We say that an $\alpha \in Cap(X)$ is *countably dominated* if α satisfies the following.

$$\alpha\left(\bigcup_{n=1}^{\infty} A_n\right) \leqq \sum_{n=1}^{\infty} \alpha(A_n) \text{ for any } \{A_n\}_{n=1}^{\infty} \subset \mathscr{P}(X).$$

Example 1. For each $s \in (0, \infty)$ and $p \in (1, \infty)$, let

$$C_{s,p}(G) = \inf\{\|u\|_{\mathbf{D}_p^s(\mathbb{R})}{}^p; u \in \mathbf{D}_s^p(\mathbb{R}), u(z) \geqq 1 \ \mu\text{-a.e. } z \in G\}$$

for any open set G in B, and

$$C_{s,p}(C) = \inf\{C_{s,p}(G); G \text{ is an open set in } B \text{ and } C \subset G\}$$

for any subset C in B. Also, let us define $C_\infty : \mathscr{P}(B) \to [0, 1]$ by

$$C_\infty(C) \stackrel{\text{def}}{=} \sum_{n=1}^{\infty} 2^{-n} \cdot C_{n,n}(C) \text{ for any subset } C \text{ in } B.$$

Then $C_{s,p}$, C_∞ are countably dominated finite capacities in B. These capacities were first introduced by Malliavin.

Example 2. For any $s \in (0, \infty)$, $p \in (1, \infty)$, we define

$$\text{Cap}_{s,p}^{\infty}(G) = \inf\{\|u\|_{\mathbf{D}_p^s(\mathbb{R})}{}^p; u \in \mathbf{D}_{\infty-}^{\infty}(\mathbb{R}), u(z) \geqq 1, \mu\text{-a.e.} z \in G\}$$

for each open set G in B, and

$$\text{Cap}_{p,s}^{\infty}(C) = \inf\{\text{Cap}_{p,s}^{\infty}(G); C \subset G, G \text{ is open}\}$$

for any subset C in B. Also, let us define $\text{Cap}^\infty : \mathscr{P}(B) \to [0, 1]$ by

$$\text{Cap}^\infty(C) \stackrel{\text{def}}{=} \sum_{n=1}^{\infty} \text{Cap}_{n,n}^{\infty}(C) \text{ for any subset } C \text{ in } B.$$

Then $\text{Cap}_{p,s}^{\infty}$ and Cap^∞ are finite capacities on B.

Definition 4. Let U be a C_∞-quasi-open set. For any $p \in (1, \infty]$ and a separable Hilbert space E, we say that $f: U \to E$ belongs to $\mathscr{D}_{p,\text{loc}}^{w}(U; E)$ if for any $n \geqq 2 +$

$[1/(p+2)]$ and a compact set K with $K \subset U$, there is $\varphi \in \mathbf{D}_n^n(\mathbb{R})$ such that

(1) $0 \leqq \varphi \leqq 1$, $\varphi(z) = 0$, $z \in B \backslash U$,

(2) $\varphi f \in \mathbf{D}^n_{(n \wedge p) - 1/n}(E)$,

and

(3) $C_\infty(K \backslash \{\varphi = 1\}) < 1/n$.

Then we have the following ([4, Lemma (5.24)]).

Theorem 5. *Let U_n, $n = 1, 2, \ldots$, and U are* Cap$^\infty$*-quasi-open sets, and assume that*

(i) $\bigcup_{n=1}^{\infty} U_n \subset U$,

and

(ii) *for any compact set K with $K \subset U$,* $\mathrm{Cap}^\infty\left(K \backslash \left(\bigcup_{k=1}^{n} U_k\right)\right) \to 0$ *as $n \to \infty$.*

Then there are $\varphi_n \in \mathscr{D}^w_{\infty,\mathrm{loc}}(U; \mathbb{R})$ such that

(1) $\varphi_n(z) = 0$, $U \backslash U_n$, $n \geqq 1$,

(2) $0 \leqq \varphi_n \leqq 1$,

(3) $\varphi_n f \in \mathscr{D}^w_{p,\mathrm{loc}}(U; E)$ *for any* $f \in \mathscr{D}^w_{p,\mathrm{loc}}(U_n; E)$, $n \geqq 1$,

and

(4) *for any $\varepsilon > 0$ and compact set K with $K \subset U$, there are $m \geqq 1$ and a compact set $K' \subset U$ such that*

(i) $\varphi_n(z) = 0$, $z \in K'$, $n \geqq m + 1$,

(ii) $\sum_{n=1}^{m} \varphi_n(z) = 1$, $z \in K'$,

and

(iii) $\mathrm{Cap}^\infty(K \backslash K') < \varepsilon$.

Now let U be a C_∞-quasi-open set in W^d. We give an equivalence relation $\sim$ in $\mathscr{D}^w_{p,\mathrm{loc}}(U; \bigwedge^n H_0{}^*)$ $n \geqq 0$, $p \in (1, \infty]$, by the following:
$\alpha \sim \beta$ if

$$\alpha(w)(h_1 - N(w)h_1, \ldots, h_n - N(w)h_n) = \beta(w)(h_1 - N(w), \ldots, h_n - N(w)h_n)$$

ν-a.e. $w \in U$ for any $h_1, \ldots, h_n \in H_0$. Now we define a vector space $\mathscr{A}_p^n(U)$, $n \geqq 0$, $p \in (1, \infty]$, by $\mathscr{A}_p^n(U) \overset{\text{def}}{=} \mathscr{D}^w_{p,\mathrm{loc}}(U; \bigwedge^n H_0{}^*)/\sim$. Then $d_n : \mathscr{A}_p^n(U) \to \mathscr{A}_p^{n+1}(U)$ is well-defined and $d_{n+1} d_n = 0$.

For any $\gamma \in C^\infty([0, 1]; M)$ and $\varepsilon > 0$, let $U(\gamma, \varepsilon)$ be a Cap$^\infty$-quasi-open set given by

$$U(\gamma, \varepsilon) = \left\{ w \in W^d;\ \max_{t \in [0,1]} \mathrm{dis}_M(X(t, x; w), \gamma(t)) < \varepsilon \right\}.$$

Then we can prove the following.

Theorem 6. *There is an $\varepsilon_0 > 0$ satisfying the following. If $\gamma_k \in C^\infty([0, 1]; M)$, $\varepsilon_k \in (0, \varepsilon_0)$, $k = 1, \ldots, n$, and $\nu_{xy}\left(\bigcap_{k=1}^{n} U(\gamma_k, \varepsilon_k)\right) > 0$, then for any $\alpha \in \mathscr{A}_p^m\left(\bigcap_{k=1}^{n} U(\gamma_k, \varepsilon_k)\right)$,*

$m \geqq 1$, *with* $d_m \alpha = 0$, *there is a* $\beta \in \mathscr{A}_p^{m-1}\left(\bigcap_{k=1}^{n} U(\gamma_k, \varepsilon_k)\right)$ *such that* $d_{m-1}\beta = \alpha$.

For any $\gamma \in C^\infty([0, 1]; M)$ and $\varepsilon > 0$, let $\bar{U}(\gamma, \varepsilon)$ be an open set in H_0 given by

$$\bar{U}(\gamma, \varepsilon) = \left\{h \in H_0;\ \max_{t \in [0,1]} \operatorname{dis}_M(Y(t, x; h), \gamma(t)) < \varepsilon\right\}.$$

Then the support theorem tells us the following

Proposition 7. *Let* $\gamma_k \in C^\infty([0, 1]; M)$ *and* $\varepsilon_k \in (0, \varepsilon_0)$, $k = 1, \ldots, n$. *Then* $\nu_{xy}\left(\bigcap_{k=1}^{n} U(\gamma_k, \varepsilon_k)\right) > 0$ *if and only if* $\overline{M}_{xy} \cap \left(\bigcap_{k=1}^{n} \bar{U}(\gamma_k, \varepsilon_k)\right) \neq \phi$.

Then by using Cech-de Rham's argument, we have the following.

Corollary 8. $\{\alpha \in \mathscr{A}_p^n(W^d); d_n\alpha = 0\}/\{d_{n-1}\beta; b \in \mathscr{A}_p^{n-1}(W^d)\} \simeq H^n(\overline{M}_{xy}; \mathbb{R})$ *for any* $n \geqq 0$ *and* $p \in (1, \infty]$.

Also, we have the following hypoelliptic result

Theorem 9. Ker $L_n \subset \mathscr{A}_2^n(W^d)$, $n \geqq 0$.

References

1. Airault, H., Biesen J.: Géométrie riemannienne en codimension finie sur l'espace de Wiener. C.R. Acad. Sci. Paris Série I, **311** (1990) 125–130
2. Airault, H., Malliavin, P.: Intégration géométrique sur l'espace de Wiener. Bull Sci. Math. **112** (1988) 13–55
3. Kusuoka, S.: On the foundation of Wiener-Riemannian manifolds. In: Elworthy, K.D., Zambrini, J-C. (eds.) Stochastic analysis, path integration and dynamics. (Pitman Research Notes in Math. Series vol. 200). Longman Scientific & Technical, Essex 1989, pp. 130–164
4. Kusuoka, S.: Analysis on Wiener spaces I, Nonlinear maps. (To appear in J. Funct. Anal.)
5. Kusuoka, S.: Analysis on Wiener spaces II, Differential forms. Preprint RIMS Kyoto Univ
6. Kusuoka, S.: Analysis on Wiener spaces III. In preparation
7. Ramer, R.: On the de Rham complex of finite codimensional differential forms on infinite dimensional manifolds. Preprint 1974
8. Shigekawa, I.: De Rham-Hodge-Kodaira's decomposition on an abstract Wiener space. J. Math. Kyoto Univ. **26** (1986) 191–202

Some Recent Results in the Asymptotic Theory of Statistical Estimation

Lucien M. Le Cam *

Department of Statistics, University of California, Berkeley, CA94720, USA

1. Introduction

One of the simplest results in asymptotic theory of estimation is the Hájek-Le Cam asymptotic minimax theorem. Besides being simple, it has many applications. We review the theorem and give brief indications on some applications.

The theorem is called Hájek-Le Cam because it was proved by Hájek (1972) for the asymptotically normal (more precisely LAN) case. There was a previous theorem by Le Cam (1953). Hájek's result was substantially extended in Le Cam (1979).

Section 2 below gives a summary of definitions and notation. Section 3 reviews the asymptotic minimax theorem. Section 4 indicates how the theorem can be applied to problems recently studied by Donoho and Liu (1990), by M. Low (1989) and by Golubev and Nussbaum (1990). For further applications of the asymptotic minimax theorem, see Millar (1983).

2. Definitions and Notation

We shall use the definitions of Le Cam (1986) with indication of conditions under which these definitions reduce to the more usual ones.

An *experiment* $\mathscr{E} = \{P_\theta; \theta \in \Theta\}$ will be given by a σ-field $\mathscr{A}$ carried by a set $\mathscr{X}$ and a family $\{P_\theta; \theta \in \Theta\}$ of probability measures on $\mathscr{A}$. The set Θ is usually called the parameter space. The L-space $L(\mathscr{E})$ of an experiment $\mathscr{E}$ is the set of all finite signed measures defined on $\mathscr{A}$ and dominated by some convergent sum $\sum_\theta c_\theta P_\theta$, $c_\theta \geq 0$, $\sum_\theta c_\theta < \infty$. Let $\mathscr{E}$ and $\mathscr{F}$ be two experiments, with $\mathscr{E} = \{P_\theta; \theta \in \Theta\}$ on a σ-field $\mathscr{A}$ and $\mathscr{F} = \{Q_\theta; \theta \in \Theta\}$ on some other σ-field $\mathscr{B}$. A *transition* T from $L(\mathscr{E})$ to $L(\mathscr{F})$ is a positive linear map from $L(\mathscr{E})$ to $L(\mathscr{F})$ such that $\|T\mu\| = \|\mu\|$ if $\mu \geq 0$. Here $\|\mu\|$ is the L_1-norm $\|\mu\| = \sup_f\{|\int f d\mu|; |f| \leq 1, f \text{ measurable}\}$. The deficiency $\delta(\mathscr{E}, \mathscr{F})$ is the number $\delta(\mathscr{E}, \mathscr{F}) = \inf_T \sup_\theta \|Q_\theta - TP_\theta\|$ where the inf is over all transitions. The distance $\Delta(\mathscr{E}, \mathscr{F})$ is $\max\{\delta(\mathscr{E}, \mathscr{F}), \delta(\mathscr{F}, \mathscr{E})\}$. Two experiments $\mathscr{E}$ and $\mathscr{F}$ are equivalent if $\Delta(\mathscr{E}, \mathscr{F}) = 0$.

* Research supported by NSF grant DMS-8701426.

Proceedings of the International Congress of Mathematicians, Kyoto, Japan, 1990

The reader who would prefer working only with transitions given by Markov kernels can satisfy himself or herself that all transitions from $L(\mathscr{E})$ to $L(\mathscr{F})$ are given by Markov kernels if 1) the family $\{P_\theta\}$ is dominated and 2) the Q_θ are Borel measures on a Borel subset of a complete separable metric space.

An estimation problem consists of an experiment $\mathscr{E} = \{P_\theta; \theta \in \Theta\}$ together with a set Z and a loss function W defined on $\Theta \times Z$ to $(-\infty, +\infty]$ such that $\inf_z W_\theta(z) > -\infty$. The set Z will also be assumed to carry a vector lattice Γ of bounded numerical functions, complete for the sup norm and such that $1 \in \Gamma$.

A decision procedure ϱ is then a transition ϱ from $L(\mathscr{E})$ to the dual Γ' of Γ (for the sup norm). Such a transition has a value $\gamma\varrho P$ for $\gamma \in \Gamma$ and $P \in L(\mathscr{E})$. (This is a contraction of $\int[\int \gamma(z)K(dz,x)]P(dx)$.) The risk of ϱ at θ is $R(\theta,\varrho) = W_\theta\varrho P_\theta = \sup\{\gamma\varrho P_\theta; \gamma \in \Gamma, \gamma \leq W_\theta\}$.

Here again the reader who prefers to work with Markov kernels (K, as above) can assume that 1) the $\{P_\theta\}$ are dominated 2) Z is compact, $\Gamma = C(Z)$ and each W_θ is lower semicontinuous.

An estimation problem given by an experiment $\mathscr{E} = \{P_\theta; \theta \in \Theta\}$ and a loss function W has a set $\mathscr{R}(\mathscr{E}, W)$ of possible risk functions, the set of functions f from Θ to $(-\infty, +\infty]$ such that there is a decision procedure ϱ for which $W_\theta\varrho P_\theta \leq f(\theta)$ for all $\theta \in \Theta$.

Often we shall need to work with subsets $F \subset \Theta$. Then $\mathscr{E}_F$ will be $\mathscr{E}_F = \{P_\theta; \theta \in F\}$.

3. The Asymptotic Minimax Theorem

The distance defined in Sect. 2 gives a topology on the set of (equivalence classes) of experiments indexed by a set Θ. Another topology is the weak topology: A directed set $\{\mathscr{E}_\nu\}; \mathscr{E}_\nu = \{P_{\theta,\nu} : \theta \in \Theta\}$ converges weakly to $\mathscr{F}$ if for every *finite* set $F \subset \Theta$, the distances $\Delta(\mathscr{E}_{\nu,F}, \mathscr{F}_F)$ tend to zero. This is equivalent to convergence in distribution of the vector of likelihood ratios $\left\{\frac{dP_{t,\nu}}{dP_{s,\nu}}, t \in F\right\}$ for all $s \in F$.

To state the theorem call a loss function V *special* if $V_\theta \in \Gamma$ for each $\theta \in \Theta$.

Theorem 1. *Let f be a function that does not belong to $\mathscr{R}(\mathscr{F}, W)$. Then there is a special $V \leq W$, a number $\alpha > 0$, a finite set F and an $\varepsilon > 0$ such that if $\Delta(\mathscr{E}_F, \mathscr{F}_F) < \varepsilon$ then $f + \alpha$ restricted to F does not belong to $\mathscr{R}(\mathscr{E}_F, V)$.*

For a proof, see Le Cam (1979) or Le Cam (1986) pp. 109–110.

Remark 1. There is a weaker version of the theorem that might be easier to visualize. Let $\{\mathscr{E}_\nu\}$ be a directed family of experiments $\mathscr{E}_\nu = \{P_{\theta,\nu}; \theta \in \Theta\}$. Assume that the $\mathscr{E}_\nu$ converge weakly to $\mathscr{F}$ and that for each ν the function f_ν belongs to $\mathscr{R}(\mathscr{E}_\nu, W)$. If f_ν converges pointwise to f then $f \in \mathscr{R}(\mathscr{F}, W)$. This easy version is not sufficient for applications where one wants to truncate W. The fact that the finite set F the special V and the ε of Theorem 1 depends only on the triplet $(\mathscr{F}, W, f)$ is also lost in the weaker version.

Remark 2. Theorem 1 has been stated in the general framework of Sect. 2 with procedures that are "transitions". If one wants to restrict oneself to transitions representable by Markov kernels it is sufficient to put restrictions on the limit $\mathscr{F}$ and the loss W. Call $\mathscr{R}(\mathscr{F}, W$, Markov) the set of functions defined as in Sect. 2 for $\mathscr{R}(\mathscr{F}, W)$ but for transitions that are Markov kernels. It is enough to assume that $\mathscr{R}(\mathscr{F}, W) = \mathscr{R}(\mathscr{F}, W$, Markov) for the limit experiment $\mathscr{F}$. Assumptions that insure this are given in Sect. 2 and in Le Cam (1986) pp. 11–14. No assumptions need to be placed on the experiments $\mathscr{E}$ such that $\Delta(\mathscr{E}, \mathscr{F}) < \varepsilon$.

Theorem 1 uses only weak convergence to $\mathscr{F}$ of the experiments $\mathscr{E}$. There is another mode of convergence that is usually available at very little cost. It is as follows.

Take a fixed $\mathscr{F} = \{Q_\theta; \theta \in \Theta\}$ and call a set $S \subset \Theta$ compact if the set $\{Q_\theta; \theta \in S\}$ is compact in $L(\mathscr{F})$ for the L_1-norm. Let $\{\mathscr{E}_\nu\}$ be a directed family of experiments, $\mathscr{E}_\nu = \{P_{\theta,\nu}; \theta \in \Theta\}$. It is said to converge to $\mathscr{F}$ on compacts if for each compact set S the restrictions $\mathscr{E}_{\nu,S}$ are such that $\Delta(\mathscr{E}_{\nu,S}, \mathscr{F}_S)$ tends to zero.

The standard LAN conditions of Le Cam (1960) imply convergence on compacts. (Hájek's 1972 do not.) According to Lindae (1972) convergence on compacts follows from pointwise convergence plus some tail equicontinuity of differences $\|P_{s,\nu} - P_{t,\nu}\|$, $s, t \in S$ compact. In many cases one would wish to consider convergence on *precompact* sets instead of compacts. The precompact convergence can be reduced to the compact one by completing $\mathscr{F}$. This can be achieved without any difficulty.

Now if $\mathscr{E}_\nu$ converges on compacts to $\mathscr{F}$, Theorem 1 is certainly applicable, but can one say more? In the direction of lower bounds for the risk, perhaps very little can be said. However here are two results, that are of some interest.

Theorem 2. *Assume that, for compacts defined as above, $\mathscr{E}_\nu$ converges to $\mathscr{F}$ on compacts and that W is bounded (that is $\sup\{|W_\theta(z)|; \theta \in \Theta, z \in Z\} < \infty$.) Then if $f \in \mathscr{R}(\mathscr{F}, W)$ there is for each ν an $f_\nu \in \mathscr{R}(\mathscr{E}_\nu, W)$ such that $f_\nu \to f$ uniformly on the compact subsets of Θ.*

This is easy to see. It tends to indicate that some results that can be achieved on the limit $\mathscr{F}$ can also be achieved asymptotically on the directed set $\{\mathscr{E}_\nu\}$.

Another result extends the lower bound of Theorem 1. To state it, let $W_\theta^c = c \wedge W_\theta$ for $c \geq 0$. For risk functions $W_\theta \sigma_\nu P_{\theta,\nu}$ that might not be measurable, let $\int_* W_\theta \sigma_\nu P_{\theta,\nu}\ \mu(d\theta)$ be the lower integral, supremum of integrals of measurable functions not exceeding $W_\theta \sigma_\nu P_{\theta,\nu}$. Consider an experiment $\mathscr{F} = \{Q_\theta; \theta \in \Theta\}$ and loss functions satisfying the following assumption:

(A) If Θ is pseudometrized by the distance $d(s,t) = \|Q_s - Q_t\|$ then the risk functions $W_\theta^c \varrho Q_\theta$ are Borel measurable in θ for all c and all procedures ϱ available in $\mathscr{F}$.

We shall state our next theorem assuming that $d(s,t) = \|Q_s - Q_t\|$ is in fact a metric on Θ. Modifications for a more general case are easy.

Theorem 3. *Suppose that condition* (A) *above is satisfied for the experiment* $\mathscr{F}$ *and that* $d(s,t)$ *defined above is a metric.*

Let μ *be a finite Radon measure on* Θ *(metrized by* d*). Assume that* $W \geq 0$, *and let* $A = \inf_\varrho \int W_\theta \varrho Q_\theta \mu(d\theta)$ *be the Bayes risk for* μ *and* $\mathscr{F}$.

Then for each $b < A$ *there is a* $c < \infty$, *a compact* $K \subset \Theta$ *and an* $\alpha > 0$ *such that if* $\Delta(\mathscr{E}_K, \mathscr{F}_K) < \alpha$ *then* $\inf_\sigma \int_* I_K(\theta) W_\theta^c \sigma P_\theta \mu(d\theta) \geq b$, *the infimum being over all procedures* σ *available for* $\mathscr{E} = \{P_\theta; \theta \in \Theta\}$

Proof. Let ϱ_0 be a Bayes procedure for $\mathscr{F}$, W and μ. Let $\mathscr{V}$ the class of special loss functions V with $V \leq W$. Then, by definition, $W_\theta \varrho_0 Q_\theta = \sup_{V \in \mathscr{V}} V_\theta \varrho_0 Q_\theta = \sup_c \sup_{\mathscr{V}} V_\theta^c \varrho_0 Q_\theta = \sup_c W_\theta^c \varrho_0 Q_\theta$. Thus if the $W_\theta^c \varrho_0 Q_\theta$ are measurable $\int W_\theta \varrho_0 Q_\theta \mu(d\theta) = \sup_c \int W_\theta^c \varrho_0 Q_\theta \mu(d\theta)$. Since $W \geq 0$, this implies that for any number b', $b < b' < A$ there is a finite c and a compact $K \subset \Theta$ such that $\int_K W_\theta^c \varrho_0 Q_\theta \mu(d\theta) > b'$. Let $\alpha > 0$ be such that $b + \|\mu\| c\alpha < b'$. If $\Delta(\mathscr{E}_K, \mathscr{F}_K) < \alpha/2$ there is a transition T from $L(\mathscr{F}_K)$ to $L(\mathscr{E}_K)$ such that $\|P_\theta - TQ_\theta\| < \alpha$ for all $\theta \in K$. This T extends to a transition from $L(\mathscr{F})$ to $L(\mathscr{E})$. Thus, if σ is any procedure on $\mathscr{E}$, the procedure $\varrho = \sigma T$ defined for $\mathscr{F}$ is such that $|W_\theta^c(\sigma T)Q_\theta - W_\theta^c \sigma P_\theta| < c\alpha$ for all $\theta \in K$. This implies

$$\begin{aligned}\int_* (W_\theta \sigma P_\theta) \mu d(\theta) &\geq \int_* I_K(\theta) W_\theta \sigma P_\theta \mu(d\theta) \\ &\geq \int_* I_K(\theta) W_\theta^c \sigma P_\theta \mu(d\theta) \\ &\geq \int_K W_\theta^c (\sigma T) Q_\theta \mu(d\theta) - \|\mu\| c\alpha \geq b.\end{aligned}$$

Hence the result. □

Remark 1. It should be noted that the measurability requirement (A) is imposed only on the limit experiment $\mathscr{F}$, not on the approximating experiments $\mathscr{E}$. In the cases considered in the literature the functions $\theta \rightsquigarrow W_\theta^c \varrho Q_\theta$ are in fact continuous. Thus measurability is not a serious problem. However it seems to be needed for the validity of Theorem 3.

Remark 2. Let $\mathscr{M}$ be a class of Radon probability measures on Θ. The conclusion of the theorem can be replaced by: Let b denote any number strictly inferior to $\sup_\mu \inf_\varrho \int W_\theta \varrho Q_\theta \mu(d\theta)$. Then there is a compact $K \subset \Theta$ and numbers $\alpha > 0$ and $c < \infty$ such that if $\Delta(\mathscr{E}_K, \mathscr{F}_K) < \alpha$ one has

$$\sup_\mu \inf_\sigma \int_* W_\theta^c \sigma P_\theta \mu(d\theta) \geq b.$$

This can be seen as in Theorem 3 taking a Bayes procedure ϱ_0 for a μ that almost achieves the $\sup_\mu$ for procedures on $\mathscr{F}$.

Remark 3. One might ask whether the conclusion of Theorem 3 would remain valid under only weak convergence of the experiments instead of compact convergence. This is perhaps not so. The difficulty arises from the fact that pointwise convergence of a bounded directed set of functions does not imply convergence of their integrals.

4. Some Applications

A) Let us start by an example of M. Low (1989) since it is very simple. Consider, on the line $\mathbb{R}$, a fixed probability density f_0 (with respect to Lebesgue measure) such that $f_0(0) > 0$, $\sup_x f(x) < \infty$ and such that f_0 be continuous at zero. Let $\{\alpha_n\}$ and $\{\beta_n\}$ be nondecreasing sequences of positive numbers such that $\alpha_n \to \infty$ and $(\alpha_n^2 \beta_n)(f_0(0)n)^{-1} \to 1$. Consider the class H of functions from $\mathbb{R}$ to $\mathbb{R}$ such that $\int h^2 < \infty$, $\int |h| < \infty$ and $\sup_x |h(x)| < \infty$. Let h_n be the number $h_n = \int \alpha_n^{-1} h(\beta_n x) f_0(x) dx$. Define $f_n(h,x) = [1 + \alpha_n^{-1} h(\beta_n x) - h_n] f_0$ if $1 + \alpha_n^{-1} h(\beta_n x) - h_n \geq 0$. Let $f_n(h,x) = f_0(x)$ otherwise. The standard Gaussian shift experiment $\mathscr{G}$ of H is one where one takes under $\theta = 0$ the distribution G_0 of a Gaussian linear process Z indexed by H and such that $E\langle Z,h\rangle = 0$ and $E|\langle Z,h\rangle|^2 = \|h\|^2 = \int h^2(x)dx$. For another value $h \in H$ one takes for G_h the measure $dG_h = \exp\{\langle Z,h\rangle - \frac{1}{2}\|h\|^2\} dG_0$.

Now let $\mathscr{E}_n = \{P_h^n; h \in H\}$ be defined by taking for P_h^n the joint distribution of n independent observations from the density $f_n(h,x)$. Low shows that $\mathscr{E}_n$ converges weakly to the Gaussian $\mathscr{G}$ as $n \to \infty$.

By restricting oneself to subsets of H one can obtain a variety of results from Theorem 1 (or 2). For instance Low considers a set of densities subject to a condition $\sup_x |f^k(x)| \leq M$ and estimates of $f(0)$. By selecting $\alpha_n = c_1 n^{k(2k+1)^{-1}}$ he shows that the appropriate rate of convergence of the estimate is in $n^{k(2k+1)^{-1}}$. This was known otherwise but Low obtains the exact limit of the risk for several loss functions.

The technique of rescaling through coefficients α_n and β_n had been previously used by Has'minskii (1979) to study estimation of a mode. For $\beta_n \equiv 1$, it has been used extensively.

B) A more complicated example appears in a paper by Golubev and Nussbaum (1990). They consider the problem of estimating a signal $t \rightsquigarrow f(t)$, $t \in [0,1]$ when the observations are of the form $Y_i = f(x_{i,n}) + \xi_i$, $i = 1,\ldots,n$ with, for instance $x_{i,n} = i/n$ and with noise ξ where the ξ_i are independent, mean zero, fixed variance σ^2 and fourth moment $E\xi_i^4$ less than a fixed constant c. The problem has been studied by many authors. A major breakthrough is due to Pinsker (1980) who considered the case where the ξ_i are Gaussian. Pinsker and subsequent authors consider the Sobolev class $W_2^m = \{f \in L_2; D^m f \in L_2\}$ where L_2 is the Hilbert space of the Lebesgue measure on $[0,1]$. For the subset $W_2^m(B) = \{f \in W_2^m; \|D^m f\|^2 \leq B\}$ let $\Delta = \lim_n \inf_{\hat{f}} \sup_f n^{(2m)(2m+1)^{-1}} E_{f,n}\|\hat{f} - f\|^2$ where the sup is on $f \in W_2^m(B)$ and the inf is over all estimators depending on n observations. The papers of Pinsker (1980) and Nussbaum (1985) give the result

$$\Delta = \gamma(m) B^r \sigma^{4mr} \quad \text{for } r = (2m+1)^{-1}$$

and $\gamma(m) = (2m+1)^r [m/\pi(m+1)]^{2mr}$. The fact that the ξ_i were Gaussian was essential in the proofs. Golubev and Nussbaum use only the restrictions $E\xi_i = 0$, $E\xi_i = \sigma^2$, $E\xi_i^4 \leq c$ and obtain a similar result.

The proof is full of ingenious devices. The relation with Theorems 1, 2 and 3 is obtainable through a series of arguments that go about as follows. Consider a particular f_0, for instance $f_0 \equiv 0$ and deviations from it. Let $W_2^{m,0}$ on $[0, 1]$ be that part of W_2^m formed by functions whose derivatives of order $0, 1, \ldots, m$ vanish at 0 and 1. For $f \in W_2^{m,0}$ one can obtain an orthogonal expansion $f = \sum_j c_j \varphi_j$ with $\|\varphi_j\| = 1$ and $\|D^m \varphi_j\|^2 = \lambda_j$ increasing in j. Now take an integer q and for $k = 1, 2, \ldots, n$ let $I_{k,q} = ((k-1)/q, k/q]$. Transport $W_2^{m,0}$ to $I_{k,q}$, by proper scaling. Look at deviations of the type $\sum_k \sum_{j=1}^{s} \varphi_{j,k,q}(x) f_{j,k}$ where $\varphi_{j,k,q}$ is φ_j transported to $I_{k,q}$ and put equal to zero outside $I_{k,q}$. Take only deviations that remain in $W_2^m(B)$. This allows to separate the observations by classes, the k-th class yielding a model $y_i = \sum_{j=1}^{s} \varphi_{j,k,q}(x_{i,n}) f_{j,k} + \xi_i$ for those $x_{i,n}$ that fall in $I_{k,q}$.

Golubev and Nussbaum let q depend on n, so it becomes $q(n)$ of the order of n^r. They then proceed to show that the part of the regression model restricted to one of the intervals $I_{k,q}$ converges to a Gaussian shift one.

Selecting the parameters $f_{j,k}$ independently according to some measure v one can try to find a lower bound on the Bayes risk.

The bound in the limit is given by Theorem 1 or 3 for each of the subintervals $I_{k,q}$; $k = 1, 2, \ldots, q$. Since the Bayes risk for the entire problem is $q(n)$ times the risk on each $I_{k,q}$ the global lower bound can be computed for each fixed s. Then one will let s tend to infinity. Of course this is only a brief sketch of the method of proof. There are many other difficult steps on the way. One of them is to make sure that the product measure v^{sq} on $\mathbb{R}^{sq}$ concentrates on the Sobolev ball $W_2^m(B)$. This was also crucial in Pinsker (1980).

In Low (1989) or Golubev and Nussbaum (1990) Theorems such as Theorems 1, 2 and 3 are used to reduce a complex problem to one in which the distributions are Gaussian and where one can often get more precise information.

C) The estimation problem treated by Donoho and Liu (1990) differs considerably from the one described in (B) above. Yet the two are closely connected. Let $\mathscr{F}$ be a class of probability densities with respect to Lebesgue measure λ on an interval $[-a, +a]$ of the line. Assume that $\mathscr{F}$ is convex, closed and bounded for the L_2-norm, $\|f\|^2 = \int f^2 d\lambda$. Donoho and Liu study the problem of estimating the value $T(f)$ of a real valued *linear* function T defined on $\mathscr{F}$ when one takes n independent observations $X_1, \ldots, X_n$ from some $f \in \mathscr{F}$. For example one may want to estimate the value at zero of the k-th derivative of f subject to a local constraint on the m-th derivative, with $k \leq m$.

Let v_n be the empirical measure of the first n observations. One can either limit oneself to estimates $\hat{T}$ that are *linear affine* in v_n (with risk R_A indicated by a suffix A) or use any arbitrary measurable function $\hat{T}$ of v_n (with risk R_M, indicated by a suffix M). A first remark is that, for *affine* estimates and square loss the problem of estimation of T is not more difficult than the estimation problem for a certain Gaussian shift experiment where one observes $Y = f + \sigma_n W$, $f \in \mathscr{F}$, W a white noise or a Gaussian process defined on subsets of $[-\alpha, +\alpha]$, with expectations zero and a given covariance function. This is quite analogous to (B) above, but now we need to estimate only the value of $T(f)$ instead of the whole f as in (B).

Let $\mathscr{G}_n$ be the Gaussian experiment with observations $Y = f + \sigma_n W$, $f \in \mathscr{F}$. Donoho and Liu proceed as follows

1) $\mathscr{F}$ being as described, there is a worst pair $(f_{0,n}, f_{1,n})$ of elements of $\mathscr{F}$ such that the minimax risk for *affine* estimates and for the one dimensional system $S_n = \{f_{\theta,n} = (1-\theta)f_{0,n} + \theta f_{1,n};\ \theta \in [0,1]\}$ is the same as the minimax risk for affine estimates for the entire $\mathscr{G}_n$. Furthermore the estimate for the worst pair is minimax for $\mathscr{G}_n$ among affine estimates. It is given by an explicit formula.

2) Consider the problem of estimating θ for the segment S_n described above and observations

$$\int u(t)Y(dt) \text{ where } u = (f_{1,n} - f_{0,n})\|f_{1,n} - f_{0,n}\|^{-1}.$$

By sufficiency, this is equivalent to the problem where all of Y would be observed.

For the problem the risk R_A for affine estimates is a certain function $\sigma \rightsquigarrow R_A(\sigma)$ of the standard deviation σ of $\int u(t)Y(dt)$. Similarly for the minimax risk $R_M(\sigma)$ for all measurable estimates. From Ibragimov-Has'minskii (1984) one knows that $\sup_\sigma R_A(\sigma)/R_M(\sigma)$ is bounded by a constant μ^*. From Donoho, Liu and McGibbon (1989) one knows that $\mu^* \leq 5/4$. This essentially solves the problem for the Gaussian case, at least if one considers a 25% margin acceptable.

The method "almost" solves the initial problem of estimation of T defined on $\mathscr{F}$ for the independent observations $X_1, \ldots, X_n$ at least if one selects σ_n and the white noise W properly, since for *affine* estimates the two problems are essentially asymptotically equivalent. (Asymptotically only because to get exact equivalence one has to select the Gaussian set function W with a covariance that depends on the true f_0). However that is for *affine* estimates. Would there be a possibility of doing much better for estimation of $T(f)$ by general measurable functions of the $X_1, \ldots, X_n$?

Donoho and Liu resolve the difficulty, at least for usual cases, by an appeal to a theorem similar to Theorem 2, Sect. 3 above.

Let $P_{\theta,n}$ be the joint distribution of $X_1, \ldots, X_n$ for the densities $f_{\theta,n} = (1-\theta)f_{0,n} + \theta f_{1,n}$, $\theta \in [0,1]$. Consider the experiments $\mathscr{E}_n = \{P_{\theta,n}; \theta \in [0,1]\}$. Consider also a Gaussian experiment

$$\mathscr{F}_n = \{Q_{\theta,n}; \theta \in [0,1]\}$$

where $Q_{\theta,n}$ is $\mathscr{N}(\theta, \sigma_n^2)$ on the line. One can prove the following

Proposition 1. *Assume that the Lévy distance between the distribution under $P_{0,n}$ of $\sum_{j=1}^n \left[\frac{f_{1,n}(X_j)}{f_{0,n}(X_j)} - 1\right]$ and a normal distribution $\mathscr{N}(0, \tau_n^2)$ tends to zero as $n \to \infty$. Assume that τ_n stays bounded. Then if $\tau_n^2\sigma_n^2 \to 1$ the distance $\Delta(\mathscr{E}_n, \mathscr{F}_n)$ between the experiments $\mathscr{E}_n = \{P_{\theta,n}; \theta \in [0,1]\}$ and the Gaussian $\mathscr{F}_n$ tends to zero.*

This is easy to see. It follows then that the difference between the minimax risk $R_M(\mathscr{E}_n)$ for $\mathscr{E}_n$ and $R_M(\mathscr{F}_n)$ for $\mathscr{F}_n$ tends to zero.

Of course, the bulk of the argumentation of Donoho and Liu takes place on the Gaussian experiment. Donoho and Nussbaum have now extended these arguments to the estimation of certain *quadratic* functionals of the density f instead of linear ones. That the problem can be very different can be seen from an article of Bickel and Ritov (1990). The subject is still progressing.

References

[1] Bickel, P.J., Ritov, Y. (1990): Achieving information bounds in non and semiparametric models. Ann. Statist. **18**, no. 2, 925-938

[2] Donoho, D.L., Liu, R.C. (1990): Geometrizing rates of convergence, III. Tech. Report no. 138, Department of Statistics, Berkeley

[3] Donoho, D.L., Liu, R.C., MacGibbon, B. (1988). Minimax risk for hyperrectangles. Tech. Report no. 123, Department of Statistics, Berkeley

[4] Donoho, D.L., Nussbaum, M. (1990): Minimax quadratic estimation of a quadratic functional. Tech. Report no. 236, Department of Statistics, Berkeley

[5] Golubev, G.K., Nussbaum, M. (1990): A risk bound in Sobolev class regression. Ann. Statist. **18**, no. 2, 758–778

[6] Hájek, J. (1972): Local asymptotic minimax and admissibility in estimation. Proc. 6th Berkeley Symp. Math. Stat. Proba., vol. 1, pp. 175–194

[7] Has'minskii, R.Z. (1979): Lower bound for the risk of nonparametric estimates of the mode. Contribution to Statistics: J. Hájek memorial volume (Jurečková, J., ed.). Akademia, Praha, pp. 90–97

[8] Ibragimov, I.A., Has'minskii, R.Z. (1984): On nonparametric estimation of the value of a functional in Gaussian white noise. Theory Probab. Appl. **24**, 18–32

[9] Le Cam, L. (1953): On some asymptotic properties of maximum likelihood estimates and related Bayes estimates. Univ. California Pub. Statist. **1**, 125–142

[10] Le Cam, L. (1960): Locally asymptotically normal families of distributions. Univ. California Pub. Statist. **3**, 37–98

[11] Le Cam, L. (1979): On a theorem of J. Hájek. Contributions to statistics: J. Hájek memorial volume (Jurečková, J., ed.) Akademia, Praha, pp. 119–137

[12] Le Cam, L. (1986): Asymptotic methods in statistical decision theory. Springer, New York Berlin Heidelberg

[13] Lindae, D. (1972): Distributions of likelihood ratios and convergence of experiments. Unpublished Ph.D Thesis, Univ. California, Berkeley

[14] Low, M.G. (1989): Local convergence of nonparametric density estimation problems to Gaussian shift experiments on a Hilbert space. Tech. Report no. 225, Department of Statistics, Berkeley

[15] Millar, P.W. (1983): The minimax principle in asymptotic statistical theory. Ecole d'Eté de Probabilités, Saint Flour, 1981. (Lecture Notes in Mathematics, vol. 976.) Springer, Berlin Heidelberg New York

[16] Nussbaum, M. (1985): Spline smoothing in regression models and asymptotic efficiency in L_2. Ann. Statist. **13**, 984–997

[17] Pinsker, M.S. (1980): Optimal filtering of square integrable signals in Gaussian white noise. Probl. Inform. Transmission **16**, 120–133

Localization and Intermittency: New Results

Stanislav A. Molchanov

Department of Mathematics and Mechanics, Moscow State University,
Moscow, 119808, USSR

The theory of the Anderson localization was developed for a long time in the initial probabilistic framework for the operators of the form

$$H = \Delta + V(x, w), \qquad x \in R^d(\mathbb{Z}^d), \qquad w \in (\Omega, F, P). \tag{1}$$

Here Δ is the Laplacian (continuous or on lattice) and $V(x, w)$ is the random homogeneous field (or process) on a probability space (Ω, F, P). The central achievement of one-dimensional theory was a series of S. Kotani's articles [1]–[3], where he discovered deep connections between Ljapunov's exponents $\gamma(\lambda)$, the structure of the prediction of the potential $V(\cdot)$ and the spectral localization of H.

Techniques of the cluster resolvent expansions, developed by Fröhlich and Spencer [4], together with generalization of Kotani's idea [2], proposed almost simultaneously (in different forms by Souillard et al. [5] and Simon-Wolff [6]) made it possibile to solve the point spectrum problem for Anderson's tight-binding model (1) in the multidimensional lattice case $\mathbb{Z}^d$, $d > 1$.

These results are summed up in the review article of Martinelli-Scoppola [7]. Simplifications and generalizations were proposed recently in this field by Dreifus and Klein [8].

Probabilistic approach, nevertheless, did not clear up the central physical idea of localization—the absence of resonance between a quantum particle with a given admissible energy λ and some (rich enough) family of the blocks of the potential. It's natural to attempt to formulate the direct geometric conditions for the individual potential $V(\cdot)$, which will lead to the localization of the spectrum. The first and a very important step in this direction was made in the paper Simon-Spencer [9]. They proved (in the one-dimensional case) that unboundness of potential V gives us the singular spectrum.

Moreover, in a multidimensional situation ($d > 1$), even in the lattice case, there are many open problems on the thin structure of the spectrum H, and on the existence of bifurcations with respect to some parameters of the model, etc. It's related, for example, to the simplest functional of $\sigma(H)$, namely the integral density of states $N(\lambda)$ (the so-called problem of Lifshitz tails).

In this lecture I'll give an account of some recent results in this direction. It is based on the papers, which I wrote together with my friends and colleagues, especially J. Görtner, A. Gordon, L. Pastur and B. Simon.

This lecture was prepared at the time of my visit to Caltech (spring 1990) and I am very grateful to Barry Simon for his hospitality, to A. Klein, R. Carmona,

Proceedings of the International Congress
of Mathematicians, Kyoto, Japan, 1990

B. Simon and T. Wolff for their useful discussions and to M. D'Elia, V. Jaksic, S. Katok and J. Madow for their support and assistance.

§ 1. One-Dimensional Localization. Lattice Case

The next group of results can be obtained by the combination of the ideas of Simon-Spencer [9] (appearance of the family of non-resonant blocks), Fröhlich-Spencer [4, 7] (cluster expansions of resolvent) and some new considerations, the principal of which is a randomization of energy.

Theorem 1. *Consider in* $l^2(\mathbb{Z}^1_+)$, $\mathbb{Z}^1_+ = \{0,1,\ldots\}$ *the Schrödinger operator* $H^\theta = \Delta + V(x)$, $x \in \mathbb{Z}^1_+$. Δ *is a discrete Laplacian,* $\Delta\psi(x) = \psi(x+1) + \psi(x-1)$ *and the parameter* θ *is a boundary phase connected with the boundary condition* $\psi(-1)\cos\theta + \psi(0)\sin\theta = 0$. *Assume, that for some energy interval* $I \subset R^1$ *there exists the sequence of the "nonresonant blocks"* $[x_n, y_n]$, $n = 1,2,\ldots$, *that is a sequence of points* $0 < x_1 \le y_1 < x_2 \le y_2 \ldots$ *such that for every* $\lambda \in I$

$$|R^{(n)}_\lambda(x_n, y_n)| \le \delta_n\,; \quad \delta_n \to 0\,, \quad n \to \infty\,. \tag{2}$$

Here $R^{(n)}_\lambda = (H^{(n)} - \lambda)^{-1}$ *is the resolvent of the operators on the block* $[x_n, y_n]$: $H^{(n)}_\psi = \Delta\psi + V(x)\psi(x)$, $x_n \le x \le y_n$, $\psi(x_n - 1) = \psi(y_n + 1) = 0$.

Suppose that there exists a nondecreasing sequence of constants $A_n \ge 1$ *and constant* $c \ge 1$ *such that*

$$\sum_n A_n\delta_n < \infty\,, \quad L_n = |y_{n+1} - x_n| + 1 \le A_1 \ldots A_n \cdot c^n\,. \tag{3}$$

Then $\sigma(H^\theta) \cap I = \sigma_{pp}(H^\theta) \cap I$ *almost everywhere (a.e.) in* θ.

In the particular case, which is important for many applications, when $x_n \le c^n$ (coordinates of blocks do not increase faster than exponential) conditions (2) and (3) become simple. It is enough to require $\sum_n \delta_n < \infty$.

Effective verification of (2) and (3) gives the next result. It is based on the physical idea of the absence of resonance.

Theorem 2. *Suppose (under the conditions of Theorem 1) that for every* $\lambda \in R^1$ *there exist a positive constant* $\varrho = \varrho(\lambda)$ *and a sequence of non-resonant blocks* $[x_n, y_n](\lambda)$, *such that* $\operatorname{dist}\{\lambda, \sigma(H^{(n)})\} \ge \varrho(\lambda)$ *and* $l_n = |x_n - y_n| + 1 \ge A(\varrho)\ln n$ *where the function* $A = A(\varrho)$ *could be specified. Then, if* $x_n < c^n$, $c > 0$,

$$\sigma(H^\theta) = \sigma_{pp}(H^\theta) \qquad \text{a.e. in } \theta\,.$$

Theorems 1 and 2 include all well-known mechanisms of localization: high barriers, long bumps, gaps in periodic potentials, etc. (See for comparison [9].)

The typical examples where Theorems 1 and 2 could (and will) be applied are random (usually nonhomogeneous) potentials $V(x,w)$, $x \in \mathbb{Z}^1$.

Example 1 (Independent Random Variables). Let $V(x) = \xi_x(w)$, $x \ge 0, w \in (\Omega, F, P)$ be i.r.v and $\xi_x(w) = a(x) + \eta_x(w), x > 0$, where $a(x)$ is an arbitrary nonrandom function and $\{\eta_x(w),\ x \geq 0\}$ is uniformly non-degenerate in the following sense: There exist positive constants ε_0, δ_0 such that

$$\underline{P}\{\eta_x > 1 + \delta_0\} \geq \varepsilon_0, \quad P\{\eta_x < -1 - \delta_0\} \geq \varepsilon_0, \quad x = 0, 1, \ldots .$$

Then $\sigma(H^\theta = \Delta + \xi_x) = \sigma_{pp}(H^\theta)$ a.s. $\underline{P}$ and a.e. in θ.

Example 2 (Gaussian Potentials). Let $V(x) = \xi_x$, $x \geq 0$, be a nonstationary gaussian sequence, $\langle \xi_x \rangle = 0$, $0 < c_1 < \langle \xi_x^2 \rangle < c_2 < \infty$, $\mathrm{cov}(\xi_x, \xi_y) = \langle \xi_x \xi_y \rangle \leq \frac{c_3}{\ln^{1+\varepsilon}(|x-y|)}$, $|x - y| \geq c_4$. The last condition is well known in the theory of gaussian fields and processes: if correlations decay only logarithmically, then the structure of the high peaks suffers from bifurcations. It's possible to prove that in our case, P-a.s. in any interval $[2^n, 2^{n+1})$, $n > n_0(w)$, there exists a "triplet" $|\xi_x| > \delta\sqrt{n}$, $|\xi_{x+1}| > \delta\sqrt{n}$, $|\xi_{x+2}| > \delta\sqrt{n}$ of the high peaks. It's enough to allow us to apply Theorem 2 and so $\sigma(H^\theta) = \sigma'_{pp}(H^\theta)$ a.e. in θ and P-a.s.

Example 3 (Unbounded Quasiperiodic Potentials). Let $f(t) = f(t+1)$ be periodic function on the unit circle S^1 for which there exists at least one point t_0 of logarithmic singularity:

$$|f(t)| \geq c\left(\ln|t - t_0|^{-1}\right)^{1+\varepsilon}, \quad t \in S^1, \quad \varepsilon > 0. \tag{4}$$

Then for quasirandom potentials of the form

$$V(x) = f(\alpha x), \qquad V(x) = f(\alpha x^2 + \beta) \tag{5}$$

for almost all values of parameters α, (α, β) we do have a pure point spectrum (a.e. in θ). The second of these potentials (for the case $f(t) = \mathrm{ctg}\,\pi t$) is the popular model of "quantum chaos" (see [18]).

For the case of entire axis $\mathbb{Z}^1$, the treatment for the operator

$$H = \Delta + V(x), \qquad -\infty < x < \infty$$

is more subtle. Of course, if conditions of Theorems 1 and 2 are satisfied for $x > 0$ and $x < 0$, then it's possible to prove (as in Theorems 1 and 2) that for a.e. $\lambda \in I$ (or $\lambda \in R^1$)

$$R_\lambda(0, x) = (H - \lambda)^{-1}(0, x) \in l^2(\mathbb{Z}^1).$$

In this spectral problem there is no exterior random parameter (such as θ in the case $\mathbb{Z}^1_+$) and (following [5] and [6]) the randomness must be included in the potential. The next lemma generalizes the one-dimensional results of [5] and [6].

Lemma 1. *Let $H^a = \Delta + V_0(x) + a\varphi(x)$, $x \in \mathbb{Z}^1$, for* a.e. *$\lambda < R^1$ and denote by*

$$R^0_\lambda(0, x) = (\Delta + V_0 - \lambda)^{-1}(0, x) = (H^0 - \lambda)^{-1}(0, x) \in l^2(\mathbb{Z}^1).$$

If perturbation φ decays "sufficiently fast", then for a.e. λ any two Weyl's solutions $\psi^\pm_\lambda(x)$ of the equation $(H^0 - \lambda)\psi = 0$ satisfy

$$\sum_{x \in \mathbb{Z}'} |\psi^-| \, |\psi^+_\lambda(x)| \, |\varphi(x)| < \infty$$

Then, a.e. in a, $\sigma(H^a) = \sigma_{pp}(H^a)$. *The typical* ψ *is the one which satisfies* $|\varphi(x)| \leq \exp(-\varepsilon|x|)$.

In [5] and [6] similar results were proved for functions φ with finite support. We will give two examples of the above lemma.

Example 4. Let $V(x) = \xi_x$, $x \in \mathbb{Z}^1$ be i.r.v. which satisfy the conditions of Example 1. If, in addition, one of them, say ξ_0, has absolutely continuous distribution, then

$$\sigma(\Delta + \xi_x) = \sigma_{pp}(\Delta + \xi_x) \qquad P\text{—a.s.}$$

Example 5. Let ξ_x, $x \in \mathbb{Z}^1$ be i.i.d.r.v. with common absolutely continuous distribution $\langle \xi_x \rangle < \infty$ and $\varphi(x)$, such that $|\varphi(x)| \leq \exp(-\varepsilon|x|)$, be the elementary potential. Consider a homogeneous random potential of alloy type

$$V(x) = \sum_{n=\infty}^{+\infty} \xi_n \varphi(x-n) .$$

It's not very difficult to verify the application of Theorem 2 in this case (see [17], where similar problems were analyzed in a more complicated situation). Application of Lemma 1 to the "partition"

$$V(x) = \sum_{n \neq 0} \xi_n \varphi(x-n) + \xi_0 \varphi(x) = V_0(x) + \xi_0 \varphi(x)$$

shows that the Schrödinger operator $H = \Delta + \sum_n \xi_n \varphi(x-n)$ under the above conditions has P-a.s. p.p. spectrum in $l^2(\mathbb{Z}^1)$. Earlier (see [17]) it was known only that $\sigma(H) = \sigma_{\text{sing}}(H)$ P-a.s.

§ 2. Some Generalizations

The main idea of Theorems 1 and 2 can be extended to more general one-dimensional lattice systems: Jacobi operators of the form $H\psi(x) = l(x-1)\psi(x-1) + l(x)\psi(x+1)$, operators where the Laplacian Δ is replaced by nonlocal convolution $\sum_y l(x-y)\psi(y) = A\psi(x)$ and kernel $l(x-y)$ decays fast enough off the diagonal and so on.

Let us formulate two theorems of this type. The first result about the "random string" has a clear mechanical meaning.

Theorem 3. *Let* $H^\theta\psi(x) = l(x-1)\psi(x-1) + l(x)\psi(x+1)$, $l(x) > 0$, $x \geq 0$ *be the operator of the lattice string with boundary condition* $\psi(-1)\cos\theta + \psi(0)\sin\theta = 0$, $\theta \in [0, \pi)$. *Assume that for some sequence of blocks* $[x_n, y_n]$: $l_n = |x_n - y_n| + 1 > c \ln^{1+\varepsilon} n, \varepsilon$, $c > 0$; $x_n < c_1^n$, $c_1 > 1$ *we have* $l(x) \leq \lambda_0$, $x \in [x_n, y_n]$, $n = 1, 2, \ldots$. *Then a.e.* $\theta \in [0, \pi)$

$$\sigma(H^\theta) \bigcap (|\lambda| > \lambda_0) = \sigma_{pp}(H^\theta) \bigcap (|\lambda > \lambda_0) .$$

Physically this means that "long, not very elastic inclusions in an elastic medium" lead to localization of short waves. In some cases it is possible to prove that for $|\lambda| < |\lambda_0|$ there is no p.p. spectrum.

The following is connected with nonlocal Laplacian.

Theorem 4. *Let* $H = \widetilde{\Delta} + V(x)$, *where* $\widetilde{\Delta}\psi(x) = \sum_{y\in\mathbb{Z}^1} l(x-y)\psi(y)$, $|l(z)| \leq \frac{c}{1+|Z|^{\beta}}$, $\beta > 8$ *and* $V(x)$ *is i.i.d.z.v. with common a.c. distribution. Then P-a.s.*

$$\sigma(H) = \sigma_{pp}(H)\,.$$

Other generalizations referred to increasing the dimension.

Schrödinger Operator in the Strip. Let $D = \mathbb{Z}^1_+ + \mathbb{Z}_N$, where $\mathbb{Z}_N = (0, 1, \ldots, N-1)$, $N \equiv 0$, is a group of residue mod N and hamiltonian H in $l^2(D)$ has a form

$$\begin{aligned}
&H^{\theta} = \Delta + V(x,z)\,, \quad (x,z) \in D\,, \quad \Delta\psi(x,z) = \psi(x+1,z) + \psi(x-1,z)\\
&+\psi(x,z+1) + \psi(x,z-1)\,, \quad x \geq 1\,, \quad z \pm 1 = z \pm 1(\mathrm{mod} N)\,,\\
&\cos\,\theta_z\psi(-1,z) + \sin\,\theta_z\psi(0,z) = 0\,, \quad z \in \mathbb{Z}_N\,,\\
&\theta = (\theta_0, \ldots, \theta_{N-1}) \in [0,\pi)^N\,.
\end{aligned} \tag{6}$$

It is well known, that for homogeneous fields $V(x,z)$ (with group of shifts $x \to x+h$, $x, h \in \mathbb{Z}^1$) the localization theorems may be proven by the classical method of Ljapunov exponents. This method in the case of the strip is more complicated than for $\mathbb{Z}^1$. On the contrary, the cluster method of Section 1 doesn't feel the difference between $\mathbb{Z}^1$ and $\mathbb{Z}^1 \times \mathbb{Z}_N$.

Theorem 5. *Assume that for given operator* H^{θ} *and energy interval* $I \subset R^1$ *there exists a system of blocks* $B_n = \{(x,z) : (x_n \leq x \leq y_n,\ z \in \mathbb{Z}_n),\ n = 1, 2, \ldots$ *and constants* h_n, ϱ_n, *such that*

a) $V(x,z)| \geq h_n$, $(x,z) \in B_n$

b) $\mathrm{dist}(h_n, I) = 4 + \varrho_n$, $\varrho_n > 0$.

If $x_n < c^n$, $c > 1$, $l_n = |x_n - y_n| + 1$ *and* $\sum_n (1+\varrho_n)^{-l_n} < \infty$, *then*

$$\sigma(H^{\theta}) \bigcap I = \sigma_{pp}(H^{\theta}) \bigcap I \qquad \text{a.e. } \theta \in [0,\pi)^N\,.$$

Note that the central conditions

$$\mathrm{dist}(h_n, I) \geq 4 + \varrho_n\,, \quad \varrho_n > 0\,, \quad n = 1, 2, \ldots$$

have a slightly different form with respect to "pure one-dimensional" theory. Early on we used $\mathrm{dist}(h_n, I) > 2 + \varrho_n$. This is because $\sigma(\Delta) = [-2, +2]$ in $l^2(\mathbb{Z}^1)$ and $\sigma(\Delta) = [-4, 4]$ in $l^2(\mathbb{Z}^1 \times \mathbb{Z}_N)$.

In virtue of Theorem 5 and its generalization to the case of $(\mathbb{Z}^1 \times \mathbb{Z}_N)$, which uses simple variants of Lemma 1, all previous examples (1-5) automatically transfer to the corresponding examples in the strip (half-strip). (The number of a.c. conditions now equals N. For example, in the analog of Lemma 1 we must consider N perturbations: $\xi_1\varphi_1(x,z_1), \xi_2\varphi_2(x,z_2), \ldots, \xi_N\varphi_N(x,z_N), (\xi_1,\ldots,\xi_N)$ has a.c. distribution.)

Multidimensional generalizations ($d > 1$) of Theorems 1 and 2 of Section 1 exist, but are noneffective in the case of homogeneous potentials. They act as follows: For any $n = 1,2,\ldots$ in $\mathbb{Z}^d$ let there exist two 1-connected (in the sense of percolation theory) surfaces Γ_n^+, Γ_n^-, $D_n = \operatorname{diam}\Gamma_n^-$, $d_n = \min|x-y|_{x\in\Gamma^+, y\in\Gamma^-}$, $\operatorname{Int}\Gamma_1^- \subset \operatorname{Int}\Gamma_1^+ \subset \operatorname{Int}\Gamma_2^- \subset \operatorname{Int}\Gamma_2^+ \subset \ldots$. Let $|\Gamma_n^\pm| = \underline{\underline{0}}(D_n^{d-1})$, $|\operatorname{Int}\Gamma_n^\pm| = \underline{\underline{0}}(D_n^d)$, $n \to \infty$ and for potential $V(x)$, $x \in \mathbb{Z}^d$ the next estimation takes place:

$$|V(x)| \geq c\left(D_n^{d-1}n^{1+\varepsilon}\right)^{1/d_n}, \quad x \in B_n = \operatorname{Int}\Gamma_n^+ \backslash \operatorname{Int}\Gamma_n^- .$$

Under these conditions the cluster expansion of resolvent $R_\lambda^0(0,x) = (\Delta + V(x))^{-1}(0,x)$, with respect to non-resonant families of blocks $B_n(\partial B_n = \Gamma_n^+, \Gamma_n^-)$, shows that

$$R_\lambda(0,x) \in l^2(\mathbb{Z}^d) \quad \text{a.e. } \lambda.$$

Using the Simon-Wolff theorem under some additional technical conditions it is possible to prove a few concrete results about the p.p. spectrum.

Example 6. Let $V(x) = \xi_x|x|^\alpha$, $x \in \mathbb{Z}^d$, $\alpha > 0$ and ξ_x is i.i.d.r.v. with some moments properties and a.c. distribution. For example, $\xi_x \in [-1,1]$ or $[0,1]$ and is uniformly distributed. Another case is where ξ_x is standard $N(0,1)$ gaussian r.v. Here P-a.s. for all $\alpha > 0$

$$\sigma(\Delta + V(x)) = \sigma_{pp}(\Delta + V(x)) .$$

The corresponding eigenfunctions decay superexponentially. The structure of the σ_{ess} in the case $\xi_x \in [0,1]$ is not trivial and dependent on α. Note that $\sigma(\Delta + V(x)) = \sigma_{\text{discr}}(\Delta + V(x))$ if $\alpha > d$.

Example 7. There exist potentials $V(x,w)$, $x \in \mathbb{Z}^d$, $d > 1$ such that

a) $V(x,w)$ is homogeneous and ergodic with “good” mixing properties. For example it has strong mixing condition with respect to the family of bounded subsets $\mathbb{Z}^d$.

b) $\langle|V(x,w)|^p\rangle < \infty$ for every $p > 0$.

c) Operator $H = \Delta + \sigma V(x)$ has p.p. spectrum for all positive coupling constants σ.

Note, however, that in this “counterexample” to the Anderson’s hypothesis the correlations of $V(x,w)$ decay (in any sense) very slowly. Potentials $V(x,w)$ of this example do not percolate from above. That is, for any $\lambda > 0$ the set $A^-(\lambda) = \{x : |V(x,w)| < \lambda\}$ is the union of finite components. Typical potentials (in physical applications) have a finite level of percolation [16].

§ 3. Continuous One-Dimensional Case

In the transition from the results of Section 1 for lattice Laplacians to their analogs for the operator

$$H^\theta \psi(x) = -\frac{d^2\psi}{dx^2} + V(x), \quad x \ge 0; \quad \psi(0)\cos\theta + \psi'(0)\sin\theta = 0$$

there are a few technical obstacles connected with unboundness of $-\Delta = -d^2/dx^2$ in $l^2(R^1_+)$. If $V(x) \ge 0$ the methods and results are similar to Theorems 1–3 but seem stronger.

Theorem 6. *Let potential $V(x) \ge 0$ for some $\delta_0 > 0$ have a next estimation from below. For some sequence $x_n \uparrow \infty$, $x_{n+1} - x_n \uparrow \infty$*

$$V(x) > h_n, \quad x \in [x_n, x_n + \delta_0]. \tag{8}$$

If

$$\overline{\lim}_{n\to\infty} \frac{x_{n+1}}{x_n} \exp(-\delta_0\sqrt{h_n}) < 1, \quad \sum_n \exp(-\delta_0\sqrt{h_n}) < \infty \tag{8'}$$

then

$$\sigma(H^\theta) = \sigma_{pp}(H^\theta) \quad \text{a.e. } \theta \in [0,\pi).$$

A. Gordon has analyzed in detail the particular case $V(x) \equiv 0$, $x \notin \cup_n [x_n, x_n + \delta_0]$, $V(x) = h_n \uparrow \infty$, $x \in [x_n, x_n + \delta_n]$ (rear high scatterers). He proved that in this case under condition

$$\underline{\lim}_{n\to\infty} \frac{x_{n+1}}{h_n \cdot x_n} \exp(-\delta_0\sqrt{h_n}) > 1 \tag{9}$$

$$\operatorname{Sp} H^\theta = \operatorname{Sp}_{sc} H^\theta \quad \text{for a.e. in } \theta.$$

The next result is a variant of Kotani's theorem [3], but for p.p. spectrum (but not singular as in [3]).

Theorem 7. *Let $F_0(x) = F_0(x+T)$ be a continuous periodic function and $[x_n, y_n]$ be a sequence of blocks such that $\sup_{x\in[x_n,y_n]} |V(x) - F_0(x)| = \varepsilon_n \to 0$ $_{n\to\infty}$, $x_n < c^n$, $c > 1$, $l_n = |x_n - y_n| \ge \ln^{1+\varepsilon} n$, $\varepsilon > 0$, $n \ge n_0$. If Δ is one of the gaps in the $\sigma_{ess}(-d^2/dx^2 + F_0(x))$, then for a.e. in θ*

$$\sigma(H^\theta) \bigcap \Delta = \sigma_{pp}(H^\theta) \bigcap \Delta.$$

If potential $V(x)$ is unbounded from below, but has a logarithmic estimation of the form $V(x) \ge -c_1 \ln(|x|+1) + c_2$, then it is possible to prove variants of Theorems 5 and 6 under stronger assumptions on the increase of $\{h_n\}$ in Theorem 5 or $\{l_n\}$ in Theorem 6.

In the case of the entire line R^1 (that is in $l^2(R^1)$) the continuous analog of Lemma 1 (Sect. 1) can be used. It makes it possible to prove the theorems on the point spectrum for many physically interesting models.

Example 8. Let $V(x) = \sum_{n=-\infty}^{+\infty} \xi_n \varphi\left(\frac{x-x_n}{\theta_n}\right)$ be a "shot noise" potential. Here $\{x_n\}$ is the Poissonian points flow, $|\varphi(x)| \le \exp(-\varepsilon|x|)$ the elementary potential, $\{\theta_n, \xi_n\}$ i.i.d.r. vectors with finite exponential moments: $\langle \exp(z\xi)\rangle < \infty$, $\langle \exp(\theta z)\rangle < \infty$, $\langle \exp(z\frac{1}{\theta}\rangle < \infty$, $|z| < z_0$, $z_0 > 0$ and the distribution of ξ_n is a.c. Then P-a.s. in $l^2(R^1)$

$$\sigma(H) = \sigma\left(-d^2/dx^2 + V(x,w)\right) = \sigma_{pp}(H).$$

This example is closely connected with the paper [17] in which the authors proved that in the same situation $\sigma(H) = \sigma_{\text{sing}}(H)$ P-a.s., but under additional restrictions, the elementary potential φ is not the *soliton*. Although this condition is not essential and the spectrum is p.p., the appearance of the soliton in this context is not accidental.

Theorem 8. *There exists potential* $V(x,w)$, $x \in R^1$ *and large parameter* $L > 0$ *for which*

$$\text{a)} \quad \left| V(x,w) - \sum_{A=-\infty}^{+\infty} - \left(\frac{2\xi_n^2}{\text{ch}^2 \xi_n (x - l_n)} \right) \right| \le e^{-\delta L}, \quad \delta > 0, \quad x \in R^1. \tag{10}$$

Here $\{\xi_n\}$ *is i.i.d.r.v. (for instance uniformly distributed in* $[0,a]$, $a > 0$*),* $\{l_n\}$ *is a homogeneous random point process in* R^1 *(dependent on* $\{\xi_n\}$*) with good mixing properties. Note that* $\varphi(x) = -2/\text{ch}^2 x$ *is the simplest soliton (1-soliton).*

$$\text{b)} \quad \sigma(H) \bigcap [0,\infty) = \sigma_{\text{ac}}(H) \bigcap [0,\infty), \quad \sigma(H) \bigcap [-\infty, 0] = \sigma_{pp}(H) \bigcap [-\infty, 0].$$

This potential is one of the realizations of "soliton's gase". It is closely connected with the problem of statistical solutions of the KdV-equation.

§ 4. Parabolic Problems for the Anderson Model. Intermittency and Related Topics

Evolution problems for the physical fields in the random medium (chemistry kinetics, hydrodynamics, etc.) very often have the form of a parabolic equation with random coefficients, in particular, with random potential. The simplest example is

$$\begin{aligned} \partial c/\partial t &= D\Delta c + \xi(x)c \\ c(0,x) &\equiv 1. \end{aligned} \tag{11}$$

Function $c(t,x)$, $t \ge 0$ has the meaning of the concentration of the particles at moment t in the point x. The kinematic part of the hamiltonian $D\Delta$ describes its diffusion (D–diffusion coefficient) and potential $\xi(x)$ its transformation. If $\xi(x) > 0$, then $\xi(x)dt$ is the probability that in the time interval $(t, t+dt)$ any particle in the point x will split (birth of a particle). If $\xi(x) < 0$, then $\xi(x)$ is the intensity of the death process in the point x. We will consider problem (11) in the discrete case where $x \in \mathbb{Z}^d$, $D\Delta\psi(x) = D\sum_{|x'-x|=1}(\psi(x') - \psi(x))$ is the lattice

Laplacian (which is the generator of the symmetrical random walk $x_t, t \geq 0$ in continuous time) and $\xi(x, w)$ is the homogeneous ergodic random field.

If $V(x) = \xi_x$, $x \in \mathbb{Z}^d$ are i.i.d.r.v., then $H - D\Delta + \xi$ is the hamiltonian of the tight-binding Anderson model. The diffusion coefficient D is the inverse coupling constant $D = 1/\sigma$. It's well known [7], [8], that for $\sigma \gg 1$ (strong disorder) or for small σ, but $\lambda \gg 1$ (fluctuation part of the spectrum) under some technical restrictions P-a.s. $\sigma(H) = \sigma_{pp}(H)$.

Asymptotic Properties. The solution $c = c(t, x)$, $t \to \infty$ can be represented in the spectral terms. It allows investigation by direct probabilistic methods. This gives us essential information about the structure $\sigma(H)$, $\lambda \gg 1$.

The central qualitative property of the field $c(t, x)$, $t \to \infty$ is its intermittency – that is, informally, the existence of strongly pronounced spatial structures (in this case sharp and high peaks). The definition of intermittency is given in terms of statistical moments $c(t, x)$.

Remember that $c(0, x) \equiv 1$. This means that $c(t, x)$ is a homogeneous ergodic field for every $t > 0$. It is not very difficult to prove that the condition $\langle \exp(t\xi) \rangle < \infty$, $t > 0$ is necessary and sufficient for the existence of $\langle c^p(t, x) \rangle$, $t > 0$, $p = 1, 2, \ldots$.

Definition 1. We will say that the family of the fields $c(t, x)$, $x \in \mathbb{Z}^d$, $t > 0$ is the asymptotic intermittency parameter of the family when $t \to \infty$, if for the functions

$$\wedge_p(t) = \ln\langle c^p(t, \cdot)\rangle$$

the following relations take place as $t \to \infty$

$$\wedge_t(t) \ll \frac{\wedge_2(t)}{2} \ll \frac{\wedge_B}{B}(t) \ll \ldots$$

Here $A(t) \ll B(t)$ means $B(t) - A(t) \to_{t\to\infty} +\infty$.

Theorem 9. *If potential $\xi(x)$ is unbounded from above and $G(t) = \langle \exp(t\xi(\cdot))\rangle < \infty$, $t > 0$, then the solution $c(t, x)$ is asymptotically intermitted in the sense of Definition 1. The logarithmic asymptotics of the statistical moment has a form*

$$\frac{\ln\langle c^p(t, 0)\rangle}{p} = \frac{\wedge_p(t)}{p} \sim_{t\to\infty} \frac{G(pt)}{p}. \tag{12}$$

The more exact asymptotics depends upon the structure of the tails of the one-dimensional distributions of potential $\xi(\cdot)$.

Theorem 10. *Let $\xi(x)$, $x \in \mathbb{Z}^d$ be i.i.d.r.v. (Anderson model) and $P\{\xi(1) > t\} \sim_{t\to\infty} \exp(-ct^\beta)$, $\beta > 1$. Then*

$$\frac{\wedge_p(t)}{p} = \frac{G(pt)}{p} - 2d\, Dpt + \bar{\bar{0}}(t) = \frac{c(\beta)t^{\beta/\beta - 1}}{p} - 2d\, Dpt + \bar{\bar{0}}(t), \quad t \to \infty. \tag{13}$$

But if $P\{\xi(\cdot) > t\} \sim \exp(-c \exp(c, t^\beta))$, $\beta > 1$, then

$$\frac{\wedge_p(t)}{p} = \frac{G(pt)}{p} + \bar{\bar{0}}(t) = c(\beta, p) \ln^{\frac{1}{\beta}} t + \bar{\bar{0}}(t). \tag{14}$$

The difference between these two formulas are due to physical reasons. In the first case "strong centers" of the potential $\xi(\cdot)$, which contribute mainly to the growth of the number of particles, have the form of a single high peak. In the second case it is wide but not very high islands. The second term of asymptotics of $\wedge_p(t)/p$ describes the probability of "keeping" particles by "strong centers".

It is possible to observe the same effect in the results about the almost sures (a.s.) behavior of $c(t,x), \quad t \to \infty$.

Theorem 11. *Let* $\ln\ln(P\{\xi > t\})^{-1} < c_1 t^\beta, \quad \beta < 1, c_1 > 0$ *(roughly speaking,* $P\{\xi > t\} \geq \exp\{-c\exp(c_1 t^\beta)\}, \beta < 1$. *Then P-a.s.* $t \to \infty$

$$\frac{\ln c(t,x)}{t} = \left(\log\left(\frac{1}{H(0)}\right)\right)^{-1}(d \ln t) - 2dD + \bar{\bar{0}}(1)\,. \tag{15}$$

(Here, $H(t) = P\{\xi > t\}$, $(\ )^{-1}$ *means inverse function.)*

If, however,

$$\ln\ln H(t) > c_2 t^\beta\,, \quad \beta > 1\,, \quad c_2 > 0 \tag{16}$$

then P-a.s.

$$\frac{\ln c(t,x)}{t} = \left(\log\frac{1}{H(\cdot)}\right)^{-1}(d \ln t)\,. \tag{17}$$

Consider now the initial parabolic Anderson problem (11) for the localized initial condition $c(0,x) = \delta_0(x)$. Assume, that $\xi(x) > 0$ is an i.i.d.r.v. with "exponential tails". For simplicity let $P\{\xi > t\} = \exp(-ct^\beta), \quad \beta > 1$. In the beginning we have only one particle but birth and diffusion processes lead to the "occupation of the space". This problem of the quantitative description of this phenomena is similar to the famous problem KPP (Kolmogorov-Petrovski-Piskunov).

It is not very difficult to show that the boundary of "occupied" region in moment t is given by a sphere

$$S_t = \{x : |x| \leq t \ln^{1/\beta} t\}$$

in the following way: if $|x| > t \ln^{1/\beta+\varepsilon} t$, then $c(t,x) \xrightarrow{P} 0$ and if $|x| < t \ln^{1/\beta-\varepsilon} t$, then $c(t,x) \xrightarrow{P} \infty$.

However, the set σ_t is extremely nonuniformly occupied, as is a typical effect of intermittency.

Theorem 12. *For every* $t > 0$, $\varepsilon > 0$ *it's possible to find (random) points* $x_1(t,w), \ldots, x_k(t,w)$, $k < \ln t$, *such that*

$$\sum_{x \in \mathbb{Z}^d} c(t,x) \geq (1-\varepsilon)\sum_{i=1}^{k} c(t,x_1)\,. \tag{18}$$

It is likely that $k = k(t,w)$ is bounded in probability. This theorem shows that it is necessary to be very careful in the applications of the results on the averaging description of the propagation front of concentration in random media. The field of concentration inside the front has an extremly non-uniform structure.

§ 5. On the Basic States in the Anderson Model. Precision of the Asymptotical Formulas for "Lifshitz Tails"

The limit theorems for the boundary part of the spectrum, that is for the basic states will be considered in finite, but big volume V when $V \to \infty$. The integral density of states $N(\lambda)$ can be studied in the framework of the same procedure.

Let $S_N^d = [-N, N]^d$, where points $N, -N$ are identified, be the d dimensional lattice torus of the volume $V_N = (2N)^d$ and $H = \Delta + \xi(x)$ be the operator of the tight-binding Anderson model. This means that $\xi(x)$, $x \in S_N^d$ is an i.i.d.r.v. We will consider a typical one for the theory of "Lifshitz tails" case

$$\underline{P}\{\xi_x > t\} = \exp\{-ct^\beta\}, \quad t > 0, \quad \beta > 0.$$

We are interested in the structure of higher eigenvalues and corresponding eigenfunctions. It's easy to understand that they are closely connected with the higher peaks of the potential $\xi(x)$, $x \in S_N^d$. Consider the two variational series:

$$\xi_N^{(1)} > \xi_N^{(2)} > \ldots > \xi_N^{(V)}$$
$$\lambda_N^{(1)} > \lambda_N^{(2)} > \ldots > \lambda_N^{(V)}$$

The limit distribution of any fixed number k of the first r.v. in the ξ-series is described by the Weibull's type low:

$$P\left\{\frac{\xi_N^{(1)} - \xi_N^{(2)}}{\frac{1}{p}\ln^{(1/p-1)} V_N} \in (x_1 + dx_1), \frac{\xi_N^{(2)} - \xi_N^{(3)}}{\frac{1}{p}\ln^{(1/p-1)} V_N} \in (x_2 + dx_2), \ldots \right.$$
$$\left. \frac{\xi_N^{(k)} - \ln^{1/p} V_N}{\frac{1}{p}\ln^{(1/p-1)} V_N} \in (x_k + dx_k)\right\} \tag{19}$$
$$\to_{N\to\infty} P_k(x_1, \ldots, x_k) = \exp(-x_1 - 2x_2 \ldots - kx_k - e^{-x_k}), \quad x_1, \ldots x_k \geq 0.$$

It seems very reasonable, that the corresponding (or near) formulas are valid for $\lambda_N^{(i)}$, $i = 1, 2, \ldots, k$. In some sense it is true.

Theorem 13. *For some positive* $A_N^{(1)}, \ldots, A_N^{(k)}, B_N$,

$$P\left\{\frac{\lambda_N^{(1)} - \lambda_N^{(2)} - A_N^{(1)}}{B_N} > x_1, \frac{\lambda_N^{(2)} - \lambda_N^{(3)} - A_N^{(2)}}{B_N} > x_2, \ldots, \frac{\lambda_N^{(k)} - A_N^{(k)}}{B_N} > x_k\right\} \tag{20}$$
$$\to_{N\to\infty} \int_{x_1}^{\infty} \cdots \int_{x_k}^{\infty} \exp(-y_1 - 2y_2 - \ldots - ky_k - \exp(-y_k)) dy_1 \ldots dy_k,$$
$$x_1, \ldots, x_k > 0.$$

The structure of normalizing constants, however, depends on β. There exist many bifurcations with respect to β. The nature of these bifurcations is very simple. It is obvious that only ξ-peaks which have an order $\underline{\underline{0}}(\ln^{1/\beta} V)$, for example, bigger then $(1 - \varepsilon)\ln^{1/\beta} V$ can contribute to the initial part of the λ-series. But $\#\{x \in S_N^d : \xi(x) > (1 - \varepsilon)\ln^{1/\beta} V\} = \underline{\underline{0}}(V^{\delta(\varepsilon)})$; $\delta(\varepsilon) \to 0$, $\varepsilon \to 0$ and the distances between these ξ-peaks have an order $N^{1-\delta_1(\varepsilon)}$; $\delta_1(\varepsilon) \to 0$, $\varepsilon \to 0$. The interaction between peaks is very small and any of them as shown by standard perturbation calculations gives an eigenvalue

$$\lambda(x_0) = \xi(x_0) - \frac{c_0}{\xi(x_0)} + c_1 \frac{\sum_{|x'-x|=1} \xi(x')}{\xi^2(x_0)} + \bar{\bar{0}}\left(\frac{1}{\xi^3(x_0)}\right). \tag{21}$$

If $p < 2$, then the second and all other terms are small enough with respect to the "gaps" between neighboring $\xi^{(i)}$ and we can use formula (19), changing $\xi^{(i)}$ by $\lambda^{(i)}$. If $2 \le p < 3$, then the second term of expansion (21) is essential. It is necessary to slightly change constant $A_N^{(k)}$. However, as in the case $p < 2$, we have the correspondence $\xi^{(1)} \leftrightarrow \lambda^{(1)}$, $\xi^{(2)} \leftrightarrow \lambda^{(2)}, \ldots$, $\xi^{(k)} \leftrightarrow \lambda^{(k)}$. If $p \ge 3$, then the gaps between $\xi^{(1)} - \xi^{(2)}, \ldots$ are smaller than $\frac{1}{\xi^2(x_i)}$, this correspondence is destroyed and all normalized constants are new. It is not possibile to write out explicit formulas for $A_N^{(i)}, B_N$ as functions of β and dimension d.

The same analysis applies to the problem of "Lifshitz tails" as to the high energy asymptotic of $N^*(\lambda) = 1 - N(\lambda) = \lim_{V\to\infty} \frac{1}{V} \sum_{\lambda_N^{(i)} > \lambda} 1$. It is well known that

$$P\{\xi > \lambda + 2d\} < N^*(\lambda) < P\{\xi > \lambda - 2d\}.$$

(Of course, $2d = ||\Delta||_{l_2}$). It follows from these estimations that

$$-\ln N^*(\lambda) \sim_{\lambda\to+\infty} \lambda^\beta .$$

What is the more precise form of this asymptotics? The answer depends primarily on β and contains information on the structure of corresponding eigenfunctions.

Theorem 14.

a) *If $p < 2$, then*

$$-\ln N^*(\lambda) = \lambda^\beta + \bar{\bar{0}}(1).$$

b) *If $p = 2$ (gaussian case), then*

$$-\ln N^*(\lambda) = \lambda^\beta + c_1(d) + \bar{\bar{0}}(1). \tag{22}$$

c) *If $2 < p < 3$*

$$-\ln N^*(\lambda) = \lambda^\beta + c_2\lambda^{\beta-2} + \bar{\bar{0}}(1). \tag{23}$$

d) *If $3 < p < 4$*

$$-\ln N^*(\lambda) = \lambda^\beta + c_2\lambda^{\beta-2} + c_3\lambda^{(\beta-2)-\frac{2}{\beta-1}}(1 + \bar{\bar{0}}(1)) \tag{24}$$

and so on.

There is no room here to discuss other types of bifurcations of $N(\lambda)$ in the case of "double exponential" tails of the distribution $\xi(\cdot)$. The situation here is similar to the results of Theorems 10 and 11.

References

[1] Kotani, S.: (1983), Ljapunov indices determine absolute continuous spectra of stationary one-dimensional Schrödinger operators. In: Proc. Taniguchi Intern. Symp. on Stochastic Analysis, Katata and Kyoto (1982), ed. K. Ito. North-Holland, pp. 225–247

[2] Kotani, S. (1986): Ljapunov exponent and spectra for one dimensional random Schrödinger operators. Proc. Conf. on Random Matrices and their Applications. Contemp. Math. **50**, Providence R.I., pp. 277–286

[3] Kotani, S. (1985): Support theorems for random Schrödinger operators. Comm. Math. Phys. **97**, 443–452

[4] Fröhlich, G., Spencer, T. (1983): Absence of diffusion in the Anderson tight-binding model for large disorder or low energy. Comm. Math. Phys. **88**, 151–189

[5] Delyon, F., Levy, Y., Souillard, B. (1985): Anderson localization for multidimensional systems at large disorder or low energy. Comm. Math. Phys. **100**, 4670–470

[6] Simon, B., Wolff, T. (1986): Singular continuous spectrum under rank one perturbations and localization for random Hamiltonian. Commun. Pure Appl. Math. **39**, 75–90

[7] Martinelli, F., Scoppola, E. (1987): Introduction to the mathematical theory of Anderson localization. La Rivesta del Nuovo Cimento **10**, 3

[8] Dreifus, H. von, Klein, A. (1989): A new proof of localization in the Anderson tight-binding model. Comm. Math. Phys. **124**, 285–299

[9] Simon, B., Spencer, T. (1989): Trace class perturbations and the absence of absolutely continuous spectra. Comm. Math. Phys. **125**, 113–125

[10] Kirsch, W., Molchanov, S., Pastur, L. (1989): The point spectrum of the one-dimensional Schrödinger operator with an unbounded potential. Preprint Ruhr-Universität, Bochum

[11] Kirsch, W., Molchanov, S., Pastur, L. (1990): One-dimensional Schrödinger operator with unbounded potential: pure point spectrum, I. Funct. Anal. Appl. **2** (in Russian)

[12] Kirsch, W., Molchanov, S., Pastur, L. (1990): One-dimensional Schrödinger operator with unbounded potential: continuous case, II. Funct. Anal. Appl. (to appear) (in Russian)

[13] Kirsch, W., Molchanov, S. (1989): Schrödinger operator with the potential of "solitons gas" type. Preprint Ruhr-Universität, Bochum

[14] Gordon, A., Molchanov, S. (1990): One-dimensional Schrödinger operators with a strongly fluctuating random potential. Funct. Anal. Appl. (to appear) (in Russian)

[15] Görtner, J., Molchanov, S. (1990): Parabolic problems for the Anderson model. Intermittency and related topics, I. Comm. Math. Phys. (to appear)

[16] Men'shikov, M., Molchanov, S., Sidorenko, A. (1986): Percolation theory and some applications. Itogi Nauki, Ser. Teor. Veroyatn **25**, 53–110 (in Russian)

[17] Kirsch, W., Kotani, S., Simon, B. (1985): Absence of absolutely continuous spectrum for one-dimensional random but deterministic Schrödinger operators. Ann. Inst. H. Poincare **42**, 383

[18] Casati, G., Chirikov, B., et al. (1979): Stochastic behavior in classical and quantum Hamiltonian systems. Lecture Notes in Physics, vol. 93. Springer, Berlin Heidelberg New York

[19] Molchanov, S., Simon, B.: Localization theorem for non-local Schrödinger operators in one-dimensional lattice case. Comm. Math. Phys. (to appear)

[20] Jacksic, V., Molchanov, S., Simon, B.: Spectral theory of multidimensional lattice Schrödinger operator with random potential, which increase in probability. Comm. Math. Phys. (to appear)

The Laws of Some Brownian Functionals

Marc Yor

Laboratoire de Probabilités, Université P. et M. Curie, Tour 56, 3ème étage
4, place Jussieu, F-75252 Paris Cedex 05, France

Thanks mainly to the relationship between the heat equation, newtonian potential theory and Brownian motion, the laws of a large number of Brownian functionals have been obtained during the last fifty years, at least via explicit expressions of their Laplace and Fourier transforms. Much pioneering work in this area was done by Paul Lévy.

Gradually, with the development of Itô's stochastic calculus, excursion theory, path decompositions and the technique of enlargement of filtrations, these studies of individual distributions on $\mathbb{R}$, sometimes exhibiting identities between two laws, which looked a priori to be mere "coincidences", have been understood in a deeper way, in fact often by showing that two *processes* are identical in law; see Biane [3], for a recent survey in that spirit.

The most elementary examples of Brownian functionals are linear functionals: if $f \in L^2(\mathbb{R}_+, dt)$, and $(B_t, t \geq 0)$ is a real-valued BM, the Wiener integral $\int_0^\infty f(t)\, dB_t$ is a centered Gaussian variable, with variance $\int_0^\infty f^2(t)\, dt$. Quadratic functionals of BM represent the next level of complexity; those functionals are of great interest as, somewhat surprisingly, they occur in a number of very different studies of Brownian motion, such as the Ray-Knight theorems for Brownian local times, the Ciesielski-Taylor identities, some limiting laws of planar BM, and principal values of Brownian local times.

We shall take here, as a prototype of a quadratic Brownian functional, the stochastic area of planar BM, and it will be shown how Paul Lévy's formula for this stochastic area appears again and again in most of the above mentioned studies of Brownian motion.

1. On Lévy's Area Formula

Consider a two-dimensional Brownian motion $Z_t = X_t + iY_t$, $t \geq 0$, starting from 0, and the stochastic area process

$$S_t = \int_0^t (X_s\, dY_s - Y_s\, dX_s), \qquad t \geq 0.$$

Proceedings of the International Congress
of Mathematicians, Kyoto, Japan, 1990

Lévy's formula:

$$E[\exp(i\lambda S_1)|Z_1 = z] = \left(\frac{\lambda}{\sinh \lambda}\right)\exp -\frac{|z|^2}{2}(\lambda \coth \lambda - 1) \tag{1}$$

has played some important rôle in recent years, for example in the Bismut approach [7, 8] to the Atiyah-Singer theorems.

To prove formula (1), Lévy [14] used a diagonalization procedure. A different approach (Williams [25], Yor [28]) is to use a change of probability method, which reduces the computation of the law of the quadratic functional S_1 to that of the variance of a Gaussian variable.

First, by the rotational invariance of the law of BM, and independence properties, we have, for any $\lambda \in \mathbb{R}$:

$$E[\exp(i\lambda S_1)|Z_1 = z] = E\left[\exp -\frac{\lambda^2}{2}\int_0^1 ds|Z_s|^2 \Big| |Z_1| = |z|\right]. \tag{2}$$

Next, we introduce the new probability P^λ:

$$P^\lambda|_{\mathscr{F}_t} = \exp\left\{\frac{\lambda}{2}(|Z_t|^2 - 2t) - \frac{\lambda^2}{2}\int_0^t ds|Z_s|^2\right\}\cdot P|_{\mathscr{F}_t} \tag{3}$$

under which $(Z_t, t \le 1)$ is a Gaussian process (more precisely: an Ornstein-Uhlenbeck process) for which the variance at time 1 is easily computed. Formula (1) now follows from formulae (2) and (3).

We note a simple consequence of formulae (1) and (2):

$$E\left[\exp -\left(a|Z_1|^2 + \frac{\lambda^2}{2}\int_0^1 ds|Z_s|^2\right)\right] = \left(\cosh \lambda + 2a\frac{\sinh \lambda}{\lambda}\right)^{-1}. \tag{4}$$

Many variants of Lévy's formula (1) have now been developed; in particular, Biane-Yor [5] obtained a sequence of extensions of Lévy's formula by decomposing $(Z_u, u \le 1)$ into the sum of the Brownian bridge $(Z_u - uZ_1, u \le 1)$ and $(uZ_1, u \le 1)$, then developing the left-hand side of (1) with respect to this decomposition, and finally iterating this procedure. The Lévy formulae obtained in this way are closely connected, on the one hand, with the linear decomposition of BM along the orthogonal basis of the Legendre polynomials and, on the other hand, with the continued fraction:

$$\lambda \coth \lambda - 1 = \frac{\lambda^2}{3+}\,\frac{\lambda^2}{5+}\,\frac{\lambda^2}{7+}\cdots.$$

2. Squares of Bessel Processes and Ray-Knight Theorems

Let $\delta \ge 1$ be an integer, and $(q_t, t \ge 0)$ be the square of a BES(δ) process, that is the square of the euclidean norm of a δ-dimensional BM $(B_t, t \ge 0)$; then, q satisfies the

SDE:

$$q_t = q_0 + 2\int_0^t \sqrt{q_s}\, d\beta_s + \delta t, \tag{5}$$

where $(\beta_s, s \geq 0)$ is a real-valued BM.

It is well-known that this equation, with $q_0 = x \geq 0$, and δ any positive real number, has a unique pathwise solution in $\mathbb{R}_+$, hence a unique law on $C(\mathbb{R}_+, \mathbb{R}_+)$, which we shall denote by Q_x^δ.

Shiga and Watanabe [23] remarked that

$$Q_x^\delta * Q_{x'}^{\delta'} = Q_{x+x'}^{\delta+\delta'}, \text{ for any } \delta, \delta', x, x' \geq 0 \tag{6}$$

(where $P * Q$ is the convolution of P and Q), thus extending to all starting points and dimensions the obvious additivity property for integer dimensions.

From (6), Q_x^δ is an infinitely divisible probability distribution on $C(\mathbb{R}_+, \mathbb{R}_+)$, which admits the following Lévy-Khintchine representation (Pitman-Yor [18]): there exist two σ-finite ≥ 0 measures M and N on $C(\mathbb{R}_+, \mathbb{R}_+)$ such that:

$$Q_x^\delta(\exp - \langle \omega, f \rangle) = \exp - (xM + \delta N)(1 - \exp - \langle \omega, f \rangle) \tag{7}$$

(we use the notation: $\langle \omega, f \rangle = \int_0^\infty dt\, \omega(t) f(t)$, for $f \geq 0$).

This representation may be obtained, including an explicit description of M and N in terms of the Itô characteristic measure $n(d\omega)$ of the Poisson point process of Brownian excursions, with the help of the Ray-Knight theorems on Brownian local times $(L_t^a;\, a \in \mathbb{R},\, t \geq 0)$, which are now presented:

(RK$_1$) if $T_1 \equiv \inf\{t : B_t = 1\}$, the law of $(L_{T_1}^{1-a};\, 0 \leq a \leq 1)$ is Q_0^2

(RK$_2$) if $\tau_x \equiv \inf\{t : L_t^0 = x\}$, the law of $(L_{\tau_x}^b;\, b \geq 0)$ is Q_x^0

(after the original proofs of Ray [22] and Knight [12], several different derivations of (RK$_1$) and (RK$_2$) have been given; see, for example, Jeulin [11] for a recent survey based on Tanaka's formula).

Conversely, we may now deduce from the Lévy-Khintchine representation (7) some extensions of the Ray-Knight theorems; here is one (Le Gall-Yor [13]): for $\delta > 0$, the law of the process $(C_a^{(\delta)};\, a \geq 0)$ of the local times of $\left(|B_t| + \frac{2}{\delta} L_t^0;\, t \geq 0\right)$ is Q_0^δ.

In the particular case $\delta = 2$, we recover (RK$_1$), using the representation of BES(3) as $(|B_t| + L_t^0;\, t \geq 0)$, due to Pitman ([17]), jointly with a well-known time-reversal relation between BES(3) and Brownian motion (see Williams [26]).

3. An Explanation of the Ciesielski-Taylor Identities

Ciesielski-Taylor [9] obtained the following puzzling identity in law:

$$\int_0^\infty ds\, 1_{(R_{\delta+2}(s) \leq 1)} \overset{\text{(law)}}{=} T_1(R_\delta), \tag{8}$$

where R_δ resp. $R_{\delta+2}$, is a BES process, with dimension δ, resp. $\delta + 2$, starting from 0, and $T_1(R_\delta) \equiv \inf\{t : R_\delta(t) = 1\}$.

In the case $\delta = 1$, the identity (8) can be understood by time-reversal, but no such pathwise explanation has been obtained for other dimensions. Below, a spectral-type explanation and extensions are presented, following [29].

Writing both sides of (8) as integrals of the local times of the two BES processes, and using the Ray-Knight theorems, (8) then appears as a particular case of the identity in law:

$$-\int_0^1 da\, f'(a) B^2_{g(a)} \overset{\text{(law)}}{=} \int_0^1 da\, g'(a) B^2_{f(a)} \tag{9}$$

where $f, g : [0, 1] \to \mathbb{R}_+$, are C^1, with f decreasing, g increasing, and $f(1) = g(0) = 0$.

In fact, a more general identity in law holds:

$$-\int_a^b dx\, f'(x) B^2_{g(x)} + f(b) B^2_{g(b)} \overset{\text{(law)}}{=} g(a) B^2_{f(a)} + \int_a^b dx\, g'(x) B^2_{f(x)} \tag{10}$$

where $f, g : [a, b] \to \mathbb{R}_+$, are C^1, with f decreasing, and g increasing. In turn, this implies some extensions of the C-T identities.

Now, it is easily shown that (10) is a particular case of the following Fubini-type identity in law:

$$\int_0^\infty ds \left(\int_0^\infty dB_u \varphi(s, u) \right)^2 \overset{\text{(law)}}{=} \int_0^\infty ds \left(\int_0^\infty dB_u \varphi(u, s) \right)^2 \tag{11}$$

where $\varphi \in L^2([0, \infty[^2; \mathbb{R})$

A striking application of (11) is the following identity in law, obtained with C. Donati-Martin:

$$\int_0^1 ds\, (B_s - G)^2 \overset{\text{(law)}}{=} \int_0^1 ds\, (B_s - sB_1)^2, \tag{11'}$$

where $G = \int_0^1 du\, B_u$.

The general identity (11) is an infinite dimensional extension of the elementary identity in law: if $\underline{X}_n = (X_1, \ldots, X_n)$ is an n-dimensional sample of the $N(0, \sigma^2)$ law, then, for any $n \times n$ matrix:

$$\|A\underline{X}_n\| \overset{\text{(law)}}{=} \|A^*\underline{X}_n\|, \qquad \text{where } A^* \text{ is the transpose of } A. \tag{12}$$

The above arguments may be developed to give an explanation of the large class of extensions of the C-T identities obtained by Biane [4], between functionals of pairs of diffusions which satisfy a certain duality property.

4. Some Limiting Laws of Planar BM

Let $(Z_t, t \geq 0)$ be a planar BM starting from z_0, and consider $(\theta_t, t \geq 0)$ a continuous determination of the argument of $(Z_u, u \leq t)$ around $z_1 \neq z_0$.

Spitzer [24] showed that:

$$\frac{2}{\log t}\theta_t \xrightarrow[t\to\infty]{\text{(law)}} C_1 \tag{13}$$

where C_1 is a standard Cauchy variable.

This may be refined by decomposing θ_t into: $\theta_t^- + \theta_t^+$, where:

$$\theta_t^- = \int_0^t d\theta_s 1_{(|Z_s - z_1| \le R)} \quad \text{and} \quad \theta_t^+ = \int_0^t d\theta_s 1_{(|Z_s - z_1| \ge R)}.$$

Then, considering moreover $(A_t, t \ge 0)$ an integrable additive functional, we have (Messulam-Yor [16]):

$$\frac{2}{\log t}(\theta_t^-, \theta_t^+, A_t) \xrightarrow[t\to\infty]{\text{(law)}} (W^-, W^+, c_A\Lambda)$$

where: $W^- = \int_0^\sigma d\gamma_s 1_{(\beta_s \le 0)}$, $W^+ = \int_0^\sigma d\gamma_s 1_{(\beta_s \ge 0)}$, $\Lambda = l_\sigma^0$, c_A is a constant depending only on A, β and γ are two independent linear BM's starting from 0, $\sigma = \inf\{t : \beta_t = 1\}$, and $(l_u^0, u \ge 0)$ is the local time of β at 0. The law of (W^-, W^+, Λ) is characterized by:

$$E[\exp(-a\Lambda + ibW^- + icW^+)] = \left(\cosh c + \frac{2a + |b|}{c}\sinh c\right)^{-1} \tag{14}$$

a formula which is very similar to (4), this being easily explained thanks mainly to the Ray-Knight theorem (RK_1).

The convergence in law (13) may be further extended by considering $(\theta_t^1, \dots, \theta_t^n; t \ge 0)$, the winding numbers of $(Z_u, u \le t)$ around n distinct points $z_1, \dots, z_n$.

One obtains (Pitman-Yor [19, 20]):

$$\frac{2}{\log t}(\theta_t^1, \dots, \theta_t^n) \xrightarrow[t\to\infty]{\text{(law)}} (W_1, \dots, W_n)$$

and the characteristic function of $(W_1, \dots, W_n)$ is:

$$E\left[\exp i\left(\sum_{j=1}^n \lambda_j W_j\right)\right] = \left(\cosh\left(\sum_{j=1}^n \lambda_j\right) + \frac{\sum|\lambda_j|}{\sum \lambda_j}\sinh\left(\sum_{j=1}^n \lambda_j\right)\right)^{-1}. \tag{15}$$

5. Arc sine Laws for Linear Brownian Motion

Let $(B_t, t \ge 0)$ be the linear BM, starting from 0. Lévy [15] showed that: $\Gamma_+ \equiv \int_0^1 ds\, 1_{(B_s > 0)}$ follows the arc sine distribution, that is:

$$P(\Gamma_+ \in dt) = \frac{dt}{\pi\sqrt{t(1-t)}}. \tag{16}$$

(Notice that, in paragraph 4, we considered $\int_0^\sigma ds\, 1_{(B_s \ge 0)}$, where $\sigma \equiv \inf\{t : B_t = 1\}$; now, σ is replaced by 1).

The r.v. $g_1 \equiv \sup\{t < 1 : B_t = 0\}$ is also arc sine distributed, but this is easier to prove.

Here is a recent proof of (16), obtained with M. Barlow and J. Pitman [1], using excursion theory; one can show:

$$\frac{1}{l_t^2}(\Gamma_+(t), \Gamma_-(t)) \overset{\text{(law)}}{=} (T_+, T_-), \tag{17}$$

where T_+ and T_- are two independent stable $(\frac{1}{2})$ variables.

It now follows that:

$$\Gamma_+ \equiv \Gamma_+(1) \overset{\text{(law)}}{=} \frac{T_+}{T_+ + T_-}, \qquad \text{which proves (16).}$$

Several infinite dimensional extensions of (17) have now been obtained, jointly with J. Pitman [21]. Here is one: let $V(t)$ be the infinite sequence of lengths of excursions of B away from 0, during the time interval $(0, t)$, including the last unfinished excursion, arranged in decreasing order, so that:

$$V(t) = (V_1(t), \ldots, V_n(t), \ldots), \qquad \text{with } V_1(t) \geq V_2(t) \geq \cdots \geq V_n(t) \geq \cdots.$$

Then, for every $t > 0$, and $s > 0$,

$$\frac{V(t)}{t} \overset{\text{(law)}}{=} \frac{V(\tau_s)}{\tau_s}, \tag{18}$$

where $(\tau_s, s \geq 0)$ is the inverse of the local time $(l_t, t \geq 0)$.

6. Cauchy's Principal Value of Brownian Local Times

Consider again $(B_t, t \geq 0)$ a linear BM starting from 0, $(l_t, t \geq 0)$ its local time at 0, and $(\tau_t, t \geq 0)$ the inverse of l. As a consequence of the regularity of Brownian local times,

$$H_t \overset{\text{def}}{=} \lim_{\varepsilon \to 0} \int_0^t \frac{ds}{B_s} 1_{(|B_s| \geq \varepsilon)}$$

exists a.s., uniformly in t in a compact set $\left(\text{whereas: } \int_0^t \frac{ds}{|B_s|} = \infty \text{ a.s.}\right)$. From the scaling property and the Markov property of BM, it follows that: $(H_{\tau_t}, t \geq 0)$ is a symmetric Cauchy process (with parameter π!).

Independently, it was remarked by Spitzer [24] that if $(X_u, u \geq 0)$ is a linear BM independent of $(\tau_t, t \geq 0)$, then $(X_{\tau_t}, t \geq 0)$ is a symmetric Cauchy process.

However, this identity in law does not extend to the two 2-dimensional processes:

$$\left(\frac{1}{\pi} H_{\tau_t}, \tau_t; t \geq 0\right) \qquad \text{and} \qquad (X_{\tau_t}, \tau_t; t \geq 0)$$

since we have the formula:

$$E\left[\exp\left(i\frac{\lambda}{\pi} H_{\tau_t} - \frac{\theta^2}{2}\tau_t\right)\right] = \exp\left(-t\lambda \coth\left(\frac{\lambda}{\theta}\right)\right). \tag{19}$$

From this formula, we deduce:

$$E\left[\exp\left(i\frac{\lambda}{\pi}H_T\right)|l_T = x\right] = \frac{\lambda}{\sinh\lambda}\exp - x(\lambda\coth\lambda - 1) \tag{20}$$

where T is an exponential variable (with parameter $(\frac{1}{2})$) independent of B. From Lévy's formula (1), we deduce:

$$\left(\frac{1}{\pi}H_T; l_T\right) \overset{\text{(law)}}{=} \left(S_1; \frac{1}{2}|Z_1|^2\right) \tag{21}$$

which has not yet received a simple direct explanation.

As a consequence, we obtain, for fixed $t > 0$:

$$P(H_t \in dx) = \left(\frac{2}{\pi^3 t}\right)^{1/2} \sum_{n=0}^{\infty} (-1)^n \exp\left(-\left(n + \frac{1}{2}\right)^2 \frac{x^2}{2t}\right) dx. \tag{22}$$

The fact that some relation between the laws of the processes $(H_t, t \geq 0)$ and $(S_t, t \geq 0)$ exists may be understood, at least in some sense, via easier identities in law, such as:

$$\int_0^1 \frac{ds}{R_s} \overset{\text{(law)}}{=} 2\left(\int_0^1 ds\, \tilde{R}_s^2\right)^{-1/2}, \tag{23}$$

where $(R_t, t \geq 0)$, resp: $(\tilde{R}_t, t \geq 0)$ is a BES process with dimension $\delta \geq 2$, resp: $\tilde{\delta} \equiv 2\delta - 2$.

All the results presented in this paragraph are taken from Biane-Yor [6]. Principal values of Brownian local times have been studied in depth by Yamada (see, in particular, [27]) and Bertoin [2]; they have also been investigated, for physical purposes, by Ezawa et al [10].

Note added in proof: Further results about Cauchy's principal value of local times have now been obtained by Bertoin [30] and Fitzsimmons and Getoor [31].

References

1. M.T. Barlow, J.W. Pitman, M. Yor: Une extension multidimensionnelle de la loi de l'arc sinus. Sém. Probas. XXIII. (Lecture Notes in Mathematics, vol. 1372.) Springer, Berlin Heidelberg New York 1989, pp. 294–314
2. J. Bertoin: Applications de la théorie spectrale des cordes vibrantes aux fonctionnelles additives principales d'un brownien réfléchi. Ann. I.H.P., **25**, 3 (1989) 307–323
3. Ph. Biane: Decompositions of Brownian trajectories and some applications. Notes from lectures given at the Probability Winter-school of Wuhan, China, Fall 1990
4. Ph. Biane: Comparaison entre temps d'atteinte et temps de séjour de certaines diffusions réelles. Sém. Probas. XIX. (Lecture Notes in Mathematics, vol. 1123.) Springer, Berlin Heidelberg New York 1985, pp. 291–296
5. Ph. Biane, M. Yor: Variations sur une formule de Paul Lévy. Ann. I.H.P. **23** (1987) 359–377
6. Ph. Biane, M. Yor: Valeurs principales associées aux temps locaux browniens. Bull. Sci. Maths. **111** (1987) 23–101
7. J.M. Bismut: The Atiyah-Singer theorems. J. Funct. Anal. **57** (1984) 56–99 and 329–348
8. J.M. Bismut: Formules de localisation et formules de Paul Lévy. Astérisque **157–158**. Colloque Paul Lévy sur les processus stochastiques 1988, pp. 37–58

9. Z. Ciesielski, S.J. Taylor: First passage time and sojourn density for Brownian motion in space and the exact Hausdorff measure of the sample path. Trans. Amer. Math. Soc. **103** (1962) 434–450
10. H. Ezawa, J.R. Klauder, L.A. Shepp: Vestigial effects of singular potentials in diffusion theory and quantum mechanics. J. Math. Phys. **16**, 4 (1975) 783–799
11. T. Jeulin: Ray-Knight's theorems on Brownian local times and Tanaka's formula. Seminar on Stochastic Processes, eds. E. Cinlar, K.L. Chung, R.K. Getoor. Birkhäuser, Basel 1983, pp. 131–142
12. F.B. Knight: Random walks and a sojourn density process of Brownian motion. Trans. Amer. Math. Soc. **107** (1963) 56–86
13. J.F. Le Gall, M. Yor: Excursions browniennes et carrés de processus de Bessel. C. R. Acad. Sci. Paris, Série I **303** (1986) 73–76
14. P. Lévy: Wiener random functions and other Laplacian random functions. Proc. Second Berkeley Symp., vol. II, 1950, pp. 171–186
15. P. Lévy: Sur certains processus stochastiques homogènes. Comp. Math. **7** (1939) 283–339.
16. P. Messulam, M. Yor: On D. Williams' "pinching method" and some applications. J. London Math. Soc. **26** (1982) 348–364
17. J.W. Pitman: One dimensional Brownian motion and the three-dimensional Bessel process. Adv. Appl. Prob. **7** (1975) 511–526
18. J.W. Pitman, M. Yor: A decomposition of Bessel Bridges. Z. Wahrscheinlichkeitstheorie **59** (1982) 425–457
19. J.W. Pitman, M. Yor: Asymptotic laws of planar Brownian motion. Ann. Prob. **14** (1986) 733–779
20. J.W. Pitman, M. Yor: Further asymptotic laws of planar Brownian motion. Ann. Prob. **17** (1989) 965–1011
21. J.W. Pitman, M. Yor: Arc sine laws and interval partitions derived from a stable subordinator. Submitted to Proc. London Math. Soc. (November 1990)
22. D.B. Ray: Sojourn times of a diffusion process. Ill. J. Math. **7** (1963) 615–630
23. T. Shiga, S. Watanabe: Bessel diffusions as a one-parameter family of diffusion processes. Z. Wahrscheinlichkeitstheorie **27** (1973) 37–46
24. F. Spitzer: Some theorems concerning 2-dimensional Brownian motion. Trans. Amer. Math. Soc. **87** (1958) 187–197
25. D. Williams: On a stopped Brownian motion formula of H.M. Taylor. Sém. Proba. X. (Lecture Notes in Mathematics, vol. 511.) Springer, Berlin Heidelberg New York 1976, pp. 235–239
26. D. Williams: Path decomposition and continuity of local time for one-dimensional diffusions I. Proc. London Math. Soc. **28** (3) (1974) 738–768
27. T. Yamada: On some limit theorems for occupation times of one-dimensional Brownian motion and its continuous additive functionals locally of zero energy. J. Math. Kyoto Univ. **26** (2) (1986) 309–322
28. M. Yor: Remarques sur une formule de Paul Lévy. Sém. Proba. XIV. (Lecture Notes in Mathematics, vol. 784.) Springer, Berlin Heidelberg New York 1980, pp. 343–346
29. M. Yor: Une explication du théorème de Ciesielski-Taylor. To appear in Ann. I.H.P. **27** (2) (1991)
30. J. Bertoin: Complements on the Hilbert transform and the fractional derivatives of Brownian local times. J. Math. Kyoto **30** (4) (1990) 651–670
31. P.J. Fitzsimmons and R.K. Getoor: On the distribution of the Hilbert transform of the local time of a symmetric Lévy process. To appear in Ann. Prob.

The Stability of Minkowski Spacetime

Demetrios Christodoulou

Courant Institute of Mathematical Sciences, New York University, 251 Mercer Street
New York, NY 10012, USA

According to Einstein, spacetime is a 4-dimensional manifold M with a metric $g_{\mu\nu}$ of signature (3, 1). General relativity is a unified theory of space, time and gravitation in which the connection of $g_{\mu\nu}$ is identified with the gravitational force and the Einstein equations

$$R_{\mu\nu} - \tfrac{1}{2} g_{\mu\nu} R = 8\pi T_{\mu\nu}$$

hold, where $R_{\mu\nu}$ and R are respectively the Ricci curvature and scalar curvature of $g_{\mu\nu}$ and $T_{\mu\nu}$ is the energy tensor of matter. In the absence of matter the equations reduce to the Einstein vacuum equations for the spacetime manifold:

$$R_{\mu\nu} = 0.$$

These equations are at first sight a degenerate differential system. That is, the null space of the symbol σ_ξ at a given covector ξ is nonzero for all covectors ξ. This is due to the fact that the equations are generally covariant; proper account must be taken of the geometric equivalence of metrics related by diffeomorphism. This is done by considering σ_ξ not on the space of 2-covariant symmetric tensors at a point but rather on the quotient of this space by the equivalence relation induced by the symbol of the diffeomorphisms; two such tensors are equivalent if they differ by $\xi X + X \xi$ for some covector X. The null space of σ_ξ is then found to be nonzero if and only if ξ belongs to the null cone defined by the metric g. The Einstein equations are therefore of hyperbolic character.

The central mathematical problem of the theory is the initial value problem. An initial data set is a 3-dimensional manifold Σ with a positive definite metric $\bar{g}_{ij}$ and a 2-covariant symmetric tensorfield k_{ij}. The problem is to find a 4-dimensional manifold M with a metric $g_{\mu\nu}$ of signature (3, 1) satisfying the Einstein vacuum equations and an imbedding of Σ into M such that $\bar{g}_{ij}$ and k_{ij} are respectively the first and second fundamental forms induced on Σ by the imbedding. The Einstein vacuum equations impose on the initial data set the constraint equations:

$$\bar{\nabla}^i k_{ij} - \bar{\nabla}_j \operatorname{tr} k = 0$$

$$\bar{R} - |k|^2 + (\operatorname{tr} k)^2 = 0$$

where $\bar{R}$ is the scalar curvature of $\bar{g}_{ij}$. These are respectively the Codazzi and Gauss equations of the imbedding of Σ in M.

Proceedings of the International Congress
of Mathematicians, Kyoto, Japan, 1990

The local existence of solutions of the initial value problem was proven by Choquet-Bruhat [1] by introducing a harmonic system of coordinates, thereby reducing the equations to a differential system for the metric components in these coordinates which is hyperbolic in the standard sense. By a global solution of the initial value problem we mean a spacetime $(M, g_{\mu\nu})$ which is geodesically complete. Penrose [2] found a basic obstruction to global existence in his singularity theorem: the spacetime cannot be geodesically complete if Σ is non-compact and contains a trapped sphere, namely a surface S diffeomorphic to S^2 such that the divergence of the outgoing future directed null normals to S in M is everywhere negative.

An initial data set is called asymptotically flat if the complement of a compact set in Σ is diffeomorphic to the complement of a ball in $\mathbf{R}^3$, $\bar{g}_{ij}$ is a complete metric on Σ and the curvature of $\bar{g}_{ij}$ as well as k_{ij} tend to zero at infinity in an appropriate way. Since the formulation of general relativity, the Minkowski spacetime has been the only known global solution of the vacuum Einstein equations arising from asymptotically flat initial data. A basic problem in the theory is the question of stability of Minkowski spacetime, that is, whether any asymptotically flat initial data set which is sufficiently close to the trivial one gives rise to a global solution of the vacuum Einstein equations. In a recent joint work Sergiu Klainerman and myself have answered this question in the affirmative. We have also studied in detail the asymptotic behaviour of the solutions.

Before giving an outline of our proof I wish to give an indication of the difficulty involved. The naive approach to the problem would be to try to extend the solution obtained by Choquet-Bruhat of the system of equations for the metric components in harmonic coordinates, to a global one. It turns out that this approach would work if the space dimension were greater than 3. However, it fails in 3 space dimensions, a fact already recognized by Choquet-Bruhat.

Our proof uses two main ideas. The first is the relationship between conserved quantities and symmetry and the second is the relationship between symmetry and the causal structure of spacetime. The first idea goes back to Noether. Consider a field theory in a given spacetime whose field equations are derivable from an action S. The energy tensor is then defined by:

$$T_{\mu\nu} = \frac{\delta S}{\delta g^{\mu\nu}}$$

and the invariance of S under diffeomorphisms implies that $T_{\mu\nu}$ is divergence-free:

$$\nabla^{\mu} T_{\mu\nu} = 0.$$

Now suppose that X is a vectorfield generating a 1-parameter group of isometries of (M, g) (killing vectorfield). Then the 1-form

$$P_{\mu} = -T_{\mu\nu} X^{\nu}$$

is divergence-free:

$$\nabla^{\mu} P_{\mu} = 0.$$

It follows that the integral

$$\int_{\Sigma} {*P}$$

on a Cauchy hypersurface Σ is a conserved quantity, that is, its value is the same for all Cauchy hypersurfaces. If the action is invariant under conformal transformations of the metric then the energy tensor is trace-free $\operatorname{tr} T = 0$ and the above considerations extend to the case where X generates a 1-parameter group of conformal isometries of (M, g) (conformal killing vectorfield). An important requirement on a physical theory is that the energy tensor should satisfy the positivity condition

$$T(X_1, X_2) \geq 0$$

for any two future directed timelike vectors X_1, X_2 at a point. If the vectorfield X above is timelike and future directed then the quantity

$$\int_\Sigma *P = \int_\Sigma T(X, N)\, d\mu_{\bar{g}}$$

is nonnegative, N being the future directed unit normal to Σ. As its value is the same as that on the Cauchy hypersurface on which the initial data is given, it provides an estimate for the solution in terms of the initial data.

In the case of gravitation the energy tensor, defined as above,

$$G_{\mu\nu} = \frac{\delta S}{\delta g^{\mu\nu}}$$

vanishes, as this expresses the Euler-Lagrange equations of gravitation, namely the Einstein equations. Thus the above considerations fail at first sight. The way out of this impasse is to consider the Bianchi identities

$$\nabla_{[\alpha} R_{\beta\gamma]\delta\varepsilon} = 0$$

(where [ ] stands for cyclic permutation). We define a Weyl field $W_{\alpha\beta\gamma\delta}$, in a given spacetime, to be a 4-covariant tensorfield possessing the algebraic symmetries of the Weyl or conformal curvature tensor. The natural field equations for a Weyl field are the Bianchi equations

$$\nabla_{[\alpha} W_{\beta\gamma]\delta\varepsilon} = 0$$

which we write simply as

$$DW = 0.$$

In a spacetime satisfying the vacuum Einstein equations the curvature is an example of a Weyl field satisfying the Bianchi equations. In a 4-dimensional spacetime the dual $*W$ of a Weyl field W is also a Weyl field and if W satisfies the Bianchi equations so does $*W$. The operator D although formally identical to the exterior derivative, is not an exterior differential operator and $D^2 \neq 0$. As a consequence, the Bianchi equations imply an algebraic condition:

$$R_\mu{}^{\alpha\beta\gamma} * W_{\nu\alpha\beta\gamma} - R_\nu{}^{\alpha\beta\gamma} * W_{\mu\alpha\beta\gamma} = 0.$$

The Bianchi equations are conformally covariant. If f is a conformal isometry of

(M, g), that is $f * g = \Omega^2 g$ for some positive function Ω, and W is a solution of the Bianchi equations then so is $\Omega^{-1} f * W$.

Associated to a Weyl field W is a 4-covariant tensorfield Q, quadratic in W, called the Bel-Robinson tensor [4]:

$$Q_{\alpha\beta\gamma\delta} = W_{\alpha\beta\gamma\sigma} W_{\beta}{}^{\varrho}{}_{\delta}{}^{\sigma} + *W_{\alpha\varrho\gamma\sigma} * W_{\beta}{}^{\varrho}{}_{\delta}{}^{\sigma}.$$

It is totally symmetric and trace-free and satisfies the following positivity condition

$$Q(X_1, X_2, X_3, X_4) \geq 0$$

for any four future directed timelike vectors X_1, X_2, X_3, X_4 at a point, with equality if and only if W vanishes at that point. Furthermore, if W satisfies the Bianchi equations then Q is divergence free:

$$\nabla^{\alpha} Q_{\alpha\beta\gamma\delta} = 0.$$

It follows that if X_1, X_2, X_3 are three vectorfields each generating a 1-parameter group of conformal isometries of (M, g), then the 1-form

$$P = -Q(\cdot, X_1, X_2, X_3)$$

is divergence-free, consequently the integral

$$\int_{\Sigma} *P$$

on a Cauchy hypersurface Σ is a conserved quantity, which is positive definite in the case that X_1, X_2, X_3 are all timelike and future directed.

Given a Weyl field W and a vectorfield X the Lie derivative $\mathscr{L}_X W$ of W with respect to X is not in general a Weyl field. However we can define a modified Lie derivative $\hat{\mathscr{L}}_X W$ which is a Weyl field:

$$\begin{aligned}\hat{\mathscr{L}}_X W_{\alpha\beta\gamma\delta} = \mathscr{L}_X W_{\alpha\beta\gamma\delta} &- \tfrac{1}{8} \operatorname{tr} \pi W_{\alpha\beta\gamma\delta} \\ &- \tfrac{1}{2}(\hat{\pi}_{\mu\alpha} W_{\mu\beta\gamma\delta} + \hat{\pi}_{\mu\beta} W_{\alpha\mu\gamma\delta} \\ &+ \hat{\pi}_{\mu\gamma} W_{\alpha\beta\mu\delta} + \hat{\pi}_{\mu\delta} W_{\alpha\beta\gamma\mu})\end{aligned}$$

where $\pi_{\mu\nu} = \mathscr{L}_X g_{\mu\nu}$ and $\hat{\pi}$ is the deformation tensor of X, namely the trace-free part of π. The modified Lie derivative commutes with the Hodge dual:

$$\hat{\mathscr{L}}_X * W = * \hat{\mathscr{L}}_X W.$$

As a consequence of the linearity and the conformal covariance of the Bianchi equations, if W is a solution of these equations and X is a vectorfield generating a 1-parameter group of conformal isometries f_t, then

$$\hat{\mathscr{L}}_X W = \frac{d}{dt} (\Omega_t^{-1} f_t * W)|_{t=0}$$

is also a solution of the same equations. Therefore the above considerations regarding conserved quantities can be applied to the Weyl field $\hat{\mathscr{L}}_X W$ as well.

Minkowski spacetime has a 15 dimensional conformal group 0(4, 2) consisting of the abelian subgroups of translations and inverted translations, the scalings and the Lorentz group 0(3, 1). The generators of time translations and inverted time translations are the only timelike conformal killing vectorfields. A general spacetime will not have a nontrivial conformal group. If the condition that the vectorfields in the above considerations are conformal killing is dropped, then, although the quantities

$$\int_{\Sigma} *P$$

will not be conserved, their growth shall be determined by the spacetime integrals of expressions which are quadratic in W and linear in $\hat{\pi}$, the deformation tensors of the vectorfields. The idea is to find a subgroup of 0(4, 2) and an action of this subgroup on the spacetime manifold having the following property: the deformation tensors of the vectorfields generating the action should decay at infinity in such a way that the growth of the corresponding quantities is bounded in terms of the quantities themselves. It turns out that the subgroup consisting of time translations, scalings, inverted time translations and the rotation group 0(3), leaving the total energy vector invariant, suffices.

The problem is then how to define the action of these groups on a general spacetime arising from asymptotically flat initial data in such a way as to satisfy the above requirement. Now the group of time translations is the simplest to define and is related to the choice of a time function t. As our argument is one of continuity starting from the initial Cauchy hypersurface, a time function seems to enter the problem naturally. A canonical choice is that of a maximal time function, namely one whose level sets Σ_t are maximal spacelike hypersurfaces. The second fundamental form k_{ij} of Σ_t is then trace-free: $\operatorname{tr} k = 0$. There is a unique such function t with the property that the total momentum vector relative to Σ_t, that is, the projection to Σ_t of the total energy vector, vanishes. To define the action of the other groups we consider the fact that the spacetime is expected to be asymptotically flat. Since these groups act canonically on Minkowski spacetime, there is a canonical action defined at infinity. What we need is a way to extend this action to the spacetime. This is provided by the causal structure. The causal structure on a manifold M is the fundamental structure defined on M by a metric of signature $(\dim M - 1, 1)$. The causal future $J^+(S)$ of a set $S \subset M$ is the set of points q which can be reached by a future directed causal curve initiating at S. Similarly $J^-(S)$ is the set of points q which can be reached, from S, by a past directed causal curve. The specification of $J^+(p)$ and $J^-(p)$ for every $p \in M$ defines the causal structure, which is equivalent to the conformal geometry of M. In our problem we consider the boundaries of the causal pasts of a 1-parameter family of surfaces at "infinity" which are related to each other by a time translation. However we cannot quite start from "infinity" since the existence of a global solution is precisely what we wish to establish. Nevertheless, following our continuity argument we can assume that a spacetime slab has been constructed. The role of "infinity" is then played by the final maximal hypersurface Σ_{t*}. We construct a 1-parameter family of surfaces diffeomorphic to S^2 on Σ_{t*} by solving a certain equation of motion for 2-surfaces on a 3-dimensional manifold. We then consider the inner boundaries of the causal pasts of these surfaces in the spacetime slab. These null hypersurfaces C_u are the level sets of what we call the optical function u. Let $S_{t,u}$ be the surfaces of intersection of the C_u with the Σ_t. Let l and $\underline{l}$ be respectively the outgoing and incoming null normals to $S_{t,u}$ whose component along T, the generator of time translations, is equal to T. Then we have

$$T = \tfrac{1}{2}(l + \underline{l})$$

and we define the generator of scalings by

$$S = \tfrac{1}{2}(\underline{u}l + u\underline{l})$$

and the generator of inverted time translations by

$$K = \tfrac{1}{2}(\underline{u}^2 l + u^2 \underline{l})$$

where

$$\underline{u} = u + 2r, \qquad r = \sqrt{\frac{A}{4\pi}}$$

and A is the area of $S_{t,u}$. An action of the rotation group $O(3)$ on Σ_{t*} is defined starting from the standard action on the sphere at infinity in such a way that the group orbits are the level surfaces of u on Σ_{t*}. The action is then extended to the spacetime slab by conjugation: Given an element $O \in O(3)$ and a point $p \in S_{t,u}$, to obtain the point Op we follow the generator of C_u through p toward the future until p_*, the point of intersection with Σ_{t*}. We then move to Op_* and follow the generator of C_u through that point toward the past until the point of intersection with Σ_t. This point, which again lies on $S_{t,u}$, is the sought for point Op. The three rotation vectorfields ${}^{(a)}\Omega$, $a = 1, 2, 3$, generating this action satisfy:

$$[{}^{(a)}\Omega, l] = 0,$$
$$g({}^{(a)}\Omega, l) = g({}^{(a)}\Omega, T) = 0$$

and, of course, the commutation relations of the Lie algebra of $O(3)$:

$$[{}^{(a)}\Omega, {}^{(b)}\Omega] = \epsilon_{abc}{}^{(c)}\Omega.$$

The group orbits are the surfaces $S_{t,u}$.

By the above construction, the deformation tensors of the generating vectorfields depend entirely on the geometric properties of the hypersurfaces C_u and Σ_t. These properties differ significantly from those in the case of Minkowski spacetime. Consider for example, on a given C_u, the second fundamental form θ of $S_{t,u}$ relative to Σ_t, in particular the ratio

$$f = \frac{|\hat{\theta}|^2}{(\mathrm{tr}\,\theta)^2}$$

where $\hat{\theta}$ is the trace-free part of θ. This ratio, in contrast to the case in Minkowski spacetime, does not tend to zero as $t \to \infty$. In fact, $\lim_{t\to\infty} f$ measures the flux of energy radiated to infinity. Another example is the area of $S_{t,u}$ on a given C_u, which in Minkowski spacetime verifies $r - t = O(1)$ as $t \to \infty$, while in general we find

$$r - t = -2M_0 \log t + 0(1),$$

where M_0 is the initial total mass. These differences, which are absent in the case of space dimension greater thean 3, are the reason for the failure of the naive approach I mentioned earlier.

In carrying out the plan outlined above we encounter a certain difficulty. The positive mass theorem of Schoen-Yau [5] and Witten [6] implies that the curvature of any nontrivial initial data set decays not faster than d^{-3} where d is the distance from a given point. However, general initial data with d^{-3} decay lead to a logarithmic divergence in the construction of the optical function. The difficulty is overcome by observing that the leading terms at infinity in the initial data correspond to the Schwarzschild solution which is both spherically symmetric and static and is therefore annihilated by both the rotations and the time translations. We therefore take as our basic Weyl fields not the spacetime curvature R but rather $\hat{\mathscr{L}}_T R$ and $\hat{\mathscr{L}}_0 R$, where 0 stands for the collection $\{{}^{(a)}\Omega : a = 1, 2, 3\}$. Due to the fact that there are no nontrivial solutions of the vacuum Einstein equations which have a Cauchy hypersurface diffeomorphic to $\mathbf{R}^3$ and are either spherically symmetric or stationary, we are able to control R itself in terms of $\hat{\mathscr{L}}_0 R$ and $\hat{\mathscr{L}}_T R$.

The proof of the theorem is by the method of continuity and it involves a bootstrap argument. Using an appropriate version of the local existence theorem we can assume that the spacetime is maximally extended up to a value t_* of the time function. This value is defined to be the maximal one such that certain geometric quantities defined by the level sets of the time function and the optical function remain bounded by a small positive number ε_0. These quantities control in particular the isoperimetric constant of the surfaces $S_{t,u}$, on which the Sobolev inequalities depend. It then follows that a certain norm of the deformation tensors of the vectorfields T, S, K and 0 in the spacetime slap bounded by Σ_0 and Σ_{t^*} is less than another small positive number ε_1. We then consider the 1-form P, where

$$\begin{aligned} P &= P_1 + P_2, \\ P_1 &= -Q(\hat{\mathscr{L}}_0 R)(\cdot, \bar{K}, \bar{K}, T) - Q(\hat{\mathscr{L}}_T R)(\cdot, \bar{K}, \bar{K}, \bar{K}), \\ P_2 &= -Q(\hat{\mathscr{L}}_0^2 R)(\cdot, \bar{K}, \bar{K}, T) - Q(\hat{\mathscr{L}}_0 \hat{\mathscr{L}}_T R)(\cdot, \bar{K}, \bar{K}, \bar{K}) \\ &\quad - Q(\hat{\mathscr{L}}_S \hat{\mathscr{L}}_T R)(\cdot, \bar{K}, \bar{K}, \bar{K}) - Q(\hat{\mathscr{L}}_T^2 R)(\cdot, \bar{K}, \bar{K}, \bar{K}) \end{aligned}$$

and

$$\bar{K} = K + T.$$

We define the quantity $E = \max\{E_1, E_2\}$ where

$$E_1 = \sup_t \int_{\Sigma_t} *P, \qquad E_2 = \sup_u \int_{C_u} *P$$

and everything is restricted to the spacetime slab under consideration. The crucial point is the estimate of the error terms which control the growth of $\int_{\Sigma_t} *P$ and $\int_{C_u} *P$. Using the bound for the deformation tensors we are able to estimate the integral in the spacetime slab of these error terms by $c\varepsilon_1 E$ and thus arrive at an inequality of the form

$$E \leq c(D + \varepsilon_1 E)$$

where D stands for initial data. When ε_1 is chosen sufficiently small, which is achieved by choosing ε_0 suitably small, this implies $E \leq cD$. On the other hand we are able to show that the aforementioned geometric quantities associated to the level sets of t and u are bounded by cE. Thus if D is suitably small this bound does not exceed $\varepsilon_0/2$, which by continuity contradicts the maximality of t_*, unless of course $t_* = \infty$, in which case the theorem is proved.

The estimate of the error terms would fail if it were not for the fact that the worst error terms vanish due to a simple algebraic identity: if A, B, C are any three symmetric trace-free 2-dimensional matrices then $\operatorname{tr}(ABC) = 0$. The reason why such matrices appear here can be traced back to the symbol of the Einstein equations. As I mentioned at the beginning the null space of σ_ξ is nontrivial if and only if ξ is a null convector. But we can say more; when ξ is a (nonzero) null covector, the null space of σ_ξ is in fact isomorphic to the space of symmetric trace-free 2-dimensional matrices. This is therefore the space of dynamical degrees of freedom of the gravitational field at a point. There is no product in this space because for any two such matrices A, B we have $AB + BA - I\operatorname{tr}(AB) = 0$. This implies the identity mentioned above.

It remains for me to state the smallness condition on the initial data which is required. Take a point $p \in \Sigma_0 = \Sigma$ and a positive real number a. Let d_p be the distance function on Σ from p. Set:

$$\begin{aligned} D(p, a) = {} & \sup_{\Sigma} \{a^{-2}(d_p^2 + a^2)^3 |\overline{Ric}|^2\} \\ & + a^{-3}\left\{\int_\Sigma \sum_{l=0}^{3} (d_p^2 + a^2)^{l+1} |\bar{\nabla}^l k|^2 \, d\mu_{\bar{g}} \right. \\ & \left. + \int_\Sigma \sum_{l=0} (d_p^2 + a^2)^{l+3} |\bar{\nabla}^l B|^2 \, d\mu_{\bar{g}} \right\} \end{aligned}$$

here $\overline{Ric}$ is the Ricci curvature and B the Bach tensor or conformal curvature of $(\Sigma, \bar{g})$:

$$B = \operatorname{curl} \hat{\overline{Ric}},$$

where $\hat{\overline{Ric}}$ is the trace-free part of $\overline{Ric}$. Then it is the dimensionless invariant

$$\inf_{p \in \Sigma, a > 0} D(p, a)$$

which must be sufficiently small.

In concluding I would like to emphasize that many deeper and more difficult mathematical problems remain in general relativity. Among these are the formation of trapped surfaces, the nature of the singularities and the so called cosmic censorship question: whether or not singularities are generically proceeded by a region of trapped surfaces. All these are aspects of the initial value problem in the large.

References

1. Y. Choquet-Bruhat: Theoreme d'existence pour certain systems d'equations aux derivées partielles nonlinaires. Acta Math. **88** (1952) 141–225
2. R. Penrose: Gravitational collapse and spacetime singularities. Phys. Rev. Lett. **14** (1965) 57–59
3. D. Christodoulou, S. Klainerman: The global nonlinear stability of the Minkowski space. Ann. Math Studies (to appear)
4. L. Bel: Introduction d'un tenseur du quatrième ordre. C. R. Acad. Sci. Paris **248** (1959) 1094–1096
5. R. Schoen, S.T. Yau: On the proof of the positive mass conjecture in general relativity. Commun. Math. Phys. **65** (1979) 45–76
6. E. Witten: A new proof of the positive energy theorem. Commun. Math. Phys. **80** (1981) 381–402

Harmonic Maps with Values into Spheres

Jean-Michel Coron

Université Paris-Sud, Département de Mathématiques
Bâtiment 425, F-91405 Orsay, France

0. Introduction

Let (M, g) and (N, h) be two compact Riemannian manifolds of dimension m and n; the manifold M may have a boundary but not N. Without any loss of generality we may assume, using Nash's theorem, that N is isometrically embedded in $\mathbb{R}^k$. Associated with a map u from M into N is the Dirichlet energy

$$E(u) = \int_M e(u)(x) dM(x) \tag{0.1}$$

where $e(u)(x)$ is the square of the Hilbert-Schmidt norm of $u'(x) : T_xM \to T_{u(x)}N$. Let us note that E is defined on the Sobolev space $H^1(M; N) = \{u \in H^1(M; \mathbb{R}^k);\ u(x) \in N \text{ for a.e. } x\}$. The critical point of E on $H^1(M; N)$ are called harmonic maps. The Euler-Lagrange equation satisfied by a harmonic map is

$$\Delta_M u^i = g^{\alpha\beta} A^i_{u(x)} \left(\frac{\partial u}{\partial x^\alpha}, \frac{\partial u}{\partial x^\beta} \right) \quad \text{for } i \in \{1, \ldots, m\} \tag{0.2}$$

where A_u is the second fundamental form of N. This generalization of the usual harmonic functions is due to J. Eells and J. Sampson [ES]. Since their pioneer work, the subject of harmonic maps between manifolds has drawn the attention of many analysts, geometers and physicists. For quite complete surveys on that subject we refer to two papers by J. Eells and L. Lemaire: [EL1] and [EL2].

When N is the unit sphere S^n of $\mathbb{R}^{n+1}$ and $M = \bar{\Omega}$, where Ω is a smooth bounded open set of $\mathbb{R}^m$, the energy of u is (with standard notations)

$$E(u) = \int_\Omega |\nabla u|^2 = \int_\Omega u_i^\alpha u_i^\alpha \tag{0.3}$$

and $u : M \to N$ is harmonic if and only if $u \in H^1(M; N)$ and satisfies, in the sense of distributions,

$$-\Delta u = u |\nabla u|^2 . \tag{0.4}$$

A first natural question is the regularity of harmonic maps in the interior of M. One easily checks that if $m = 1$ any harmonic map is smooth. This is an open question for $m = 2$ but quite interesting partial results are known (Section 1). For $m \geq 3$ a harmonic map needs not to be smooth on the interior of M

Proceedings of the International Congress
of Mathematicians, Kyoto, Japan, 1990

(e.g. $u : B^m = \{x \in \mathbb{R}^m;\ |x| < 1\} \to S^{m-1}$, $u(x) = x/|x|$). On the other hand S. Hildebrandt, H. Kaul and K.O. Widman have proved in [HKW] that if the image of u is included in a "small" ball then u is smooth; for $N = S^n$ a ball included in an open hemisphere is "small"; in that case their result is optimal since $u : B^3 \to S^3$ defined by $u(x) = (x/|x|, 0)$ is harmonic. Unfortunately there is no general result on the singular set $S(u)$ of a harmonic map u. Much more is known on $S(u)$ if u is a *minimizing* harmonic map. A map $u : M \to N$ is a minimizing harmonic map if for any $v : M \to N$ such that $v = u$ on ∂M then $E(u) \leq E(v)$. In that case R. Schoen and K. Uhlenbeck have proved in [SU1] and [SU2] (see also M. Giaquinta and E. Giusti [GG] if the image of u is included in a chart) that $S(u)$ is a compact set of dimension less or equal to $(m-3)$ and that, near a singular point, u behaves like a singular homogeneous minimizing harmonic map from $\bar{B}^m$ into N. Such maps are called minimizing tangent maps (MTM). R. Schoen and K. Uhlenbeck have proved in [SU3] that there are no MTM from B^m into S^n if $m \leq d(n)$ where $d(3) := 3$ and $d(n) := 1 + min\{n/2, 5\}$ otherwise. A classification of the MTM for $m = 3$ and $N = S^2$ is given in [BCL] – see Section 2 –. Moreover in [BCL] a sharp lower bound of $E(u)$ for $u : \Omega \subset \mathbb{R}^3 \to S^2$ with prescribed singularities is given (Section 3). In Section 4 we give examples of MTM due to F.H. Lin [L] and [CG] for $N = S^n$.

When $n = m - 1$ and $N = S^m$ then if the degree of u restricted to ∂M is not zero, $S(u)$ cannot be empty. On the other hand R. Hardt and F.H. Lin have proved in [HL1] that $S(u)$ may be not empty even if u restricted to $\partial\Omega$ is of degree zero. More precisely they have in particular constructed smooth maps $\gamma : \partial B^3 \to S^2$ of degree zero such that (gap phenomenon)

$$\operatorname{Min}\{E(v)\,;\ v \in H^1_\gamma(B^3;S^2)\} < \operatorname{Inf}\{E(v)\,;\ v \in H^1_\gamma(B^3;S^2) \cap C^1(\bar{B}^3;S^2)\} \tag{0.5}$$

where $H^1_\gamma(B^3;S^2) = \{u \in H^1(B^3;S^2);\ u = \gamma \text{ on } \partial B^3\}$. Inequality (0.5) implies that $H^1_\gamma(B^3;S^2) \cap C^\infty(B^3;S^2)$ is not dense in $H^1_\gamma(B^3;S^2)$. More generally one may ask under which condition $C^\infty(M;N)$ is dense in the Sobolev space $W^{1,p}(M;N)$. In Section 5 we will describe recent results concerning this problem.

Inequality (0.5) leads also to the question: is the infimum in the right hand side of (0.5) achieved ? The answer is not known but it is proved in [BBC] that a positive answer to that question is equivalent to the smoothness of a minimizer for a relaxed energy associated to E. Moreover, using various relaxed energies, it is possible to prove – see [BBC] – that, if the degree of γ is not zero or if (0.5) holds, then the Dirichlet problem with boundary data γ has infinitely many solutions.

Our paper is organised as follows:

1. Regularity of Harmonic Maps from a Surface.
2. Classification of MTM for $m = 3$ and $N = S^2$.
3. Lower Bound of $E(u)$ for $\Omega \subset \mathbb{R}^3$, $N = S^2$ when $S(u)$ is Known.
4. Example of MTM for $N = S^n$.
5. Gap Phenomenon and the Problem of the Density of Smooth Maps.
6. Relaxed Energies.

1. Regularity of Harmonic Maps from a Surface

In this section we assume $m = 2$. Let us first mention that the following question is still open

Question 1.1. *Let u be a harmonic map from M into N. Is u smooth in the interior of M?*

A positive answer has been given under some extra assumptions. One has

Theorem 1.2. *Assume $u : M \to N$ is harmonic; then u is smooth in the following cases:*

(i) *u is a minimizing harmonic map (C.B. Morrey* [M]*);*
(ii) *there exists x_0 in the interior of M such that u is smooth on $M \backslash \{x_0\}$ (J. Sacks and K. Uhlenbeck* [SaU]*);*
(iii) $\phi := (|u_1|^2 - |u_2|^2 - 2i < u_1, u_2 >)\,(dx_1 + i dx_2)^2$ *is holomorphic (M. Grüter* [G] *if $\phi = 0$, R. Schoen* [Sc] *in the general case);*
(iv) *N is a sphere (F. Hélein* [H2]*).*

Case (iv) is a quite recent result. Let us briefly sketch the proof of (iv). F. Hélein first notes that

$$\nabla u^i = u^j (u^j \nabla u^i - u^i \nabla u^j), \quad \text{for } 1 \le i \le 3. \tag{1.1}$$

Next, using (0.4), one has (see [C, KRS and Sh]):

$$\operatorname{div}(u^j \nabla u^i - u^i \nabla u^j) = 0 \quad \text{for } 1 \le i \le 3 \text{ and } 1 \le j \le 3\,; \tag{1.2}$$

hence there exists B^{ij} such that (at least locally)

$$\operatorname{curl}\ B^{ij} = u^j \nabla u^i - u^i \nabla u^j \quad \text{for } 1 \le i \le 3 \text{ and } 1 \le j \le 3\,. \tag{1.3}$$

From (1.2) and (1.3) F. Hélein gets

$$\Delta u^i = \det(\nabla u^j, \nabla B^{ij}) \quad \text{for } 1 \le i \le 3\,. \tag{1.4}$$

Finally the continuity of u^i – hence the smoothness of u (see [LU] Chap. 8) – follows from (1.4) and a result due to H. Wente [We].

We end up this section by mentionning two recent results slightly related to Question 1.1 or Theorem 1.2.

Theorem 1.3 (F. Hélein [H1]). *Let u be a quasi-conformal homeomorphism between two Riemannian surfaces. Then, if $(|u_1|^2 - |u_2|^2 - 2i < u_1, u_2 >)\,(dx_1 + i dx_2)^2$ is holomorphic, u is a smooth harmonic map.*

Theorem 1.4 [CH]. *Let u be a smooth harmonic map from M into N and let $\varphi : M \to M$ be a continuous map homotopic to the identity map, then $E(u) \le E(u \circ \varphi)$.*

2. Classification of the MTM for $m = 3$ and $N = S^2$

In this section $m = 3$ and $N = S^2$. Then a MTM u is a non-constant minimizing harmonic map from $\bar{B}^3$ into S^2 such that $u(x) = \omega(x/\mid x\mid)$ for some harmonic map $\omega : S^2 \to S^2$. Let us recall that these maps are important since they give the behavior of a minimizing harmonic map from a 3-dimensional manifold into S^2 (see [SU1, Si and GW]).

The classification of MTM is given by the following theorem proved in [BCL].

Theorem 2.1. *The map $u : \bar{B}^3 \to S^2$ is a MTM if there exists R in $SO(3)$ such that $u(x) = \pm Rx/\mid x\mid$.*

Sketch of the proof of Theorem 2.1. Let u be a MTM from $\bar{B}^3$ into S^2. We have $u(x) = \omega(x/\mid x\mid)$ where ω is a harmonic map from S^2 into S^2 and therefore there exist two polynomials P and Q prime together such that, if we denote by π the stereographic projection $\pi : S^2 \to (\mathbb{R}^2 \cup \{\infty\} \times \{0\} \simeq \mathbb{C} \cup \{\infty\}$ with pole (0,0,1)

$$\text{either} \quad \pi\omega\pi^{-1}(z) = P(z)/Q(z) \quad \text{or} \quad \pi\omega\pi^{-1}(z) = P(\bar{z})/Q(\bar{z}) \tag{2.1}$$

– see [EL1 (10.6)] –. One first notices that, since u is a minimizing harmonic map, then for any smooth vector field X on B^3 with compact support $\frac{d}{d\varepsilon}E(u(x+\varepsilon X(x))|_{\varepsilon=0} = 0$ which leads to

$$\mathbb{R}^3 \ni \int_{S^2} S|\nabla\omega|^2 dS = 0\,. \tag{2.2}$$

Then one checks that, if $d :=$ Max(deg P, deg Q) = 1, (2.2) implies that $\omega = \pm R$ for some R in $SO(3)$. In the case $|d| \geq 2$ it is proved in [BCL] that one can decrease the energy of u by splitting the singularity $\{0\}$ into $|d|$ distinct points.

It remains only to verify that $Rx/|x|$ is a MTM. There are now a lot of proofs available for that – see [BCL, L and CG] –. The shortest method is perhaps to say that since there are singular harmonic maps $\bar{B}^3 \to S^2$ there exists at least a MTM from $\bar{B}^3$ into S^2; such a MTM needs to be of the form $\pm R_0 x/|x|$ for some R_0 in $SO(3)$; but clearly, if $\pm R_0 x/|x|$ is a MTM, then $\pm Rx/|x|$ is also a MTM for any R in $SO(3)$.

Remark 2.2. In an earlier paper [HKL] R. Hardt, D. Kinderlehrer and F.H. Lin had proved that the degree d of a MTM is bounded by a universal constant. Moreover, R. Cohen, R. Hardt, D. Kinderlehrer, S.Y. Lin and M. Luskin had proved numerically that if $\pi\omega\pi^{-1}(z) = z^2$, u is not a MTM ([HKLL]; see also some recent numerical studies by F. Alouges [A]).

3. Lower Bound of $E(u)$ for $\Omega \subset \mathbb{R}^3$ and $N = S^2$ when $S(u)$ is Known

We start with the special case $\Omega = \mathbb{R}^3$. Let $(a_i)_{1\le i\le \ell}$ be ℓ points in $\mathbb{R}^3$ and $(d_i)_{1\le i\le \ell}$ be ℓ integers. We are interested in the value of

$$I = \inf\left\{\int_{\mathbb{R}^3} |\nabla u|^2 \; ; \; u \in C^1(\mathbb{R}^3\backslash\{a_1,\dots,a_\ell\};S^2) \right. \\ \left. \text{and} \quad \deg(u,a_i) = d_i \quad \text{for } 1 \le i \le \ell \right\}. \tag{3.1}$$

In (3.1) $\deg(u, a_i)$ denotes the degree of u restricted to a small ball centered at a_i. One easily checks that, if $m < \infty$, then

$$\sum_{i=1}^{\ell} d_i = 0\,. \tag{3.2}$$

So we will assume that (3.2) holds true. In order to give an explicit formula for I, we construct with the points a_i such that $d_i > 0$ a family of points $(P_j)_{1\le j\le p}$ (called "positive" points) in the following way : each P_j belongs to the set $\{a_i; d_i > 0\}$ and each a_i with $d_i > 0$ is repeated exactly d_i times in the family (P_j) ; we do the same with the set $\{a_i; d_i < 0\}$ and get the family of "negative" points $(N_j)_{1\le j\le n}$. It follows from (3.2) that $p = n$. Let L be defined by

$$L = \mathrm{Min}\left\{\sum_{j=1}^{p} |P_j - N_{\sigma(j)}| \; ; \sigma \text{ is a permutation of } \{1,2,\dots,p\}\right\}. \tag{3.3}$$

Then we have [BCL]

Theorem 3.1. *The infimum in* (3.1) *is never achieved and* $I = 8\pi L$.

Sketch of the proof of $I = 8\pi L$. a) $I \le 8\pi L$. This part relies on the following lemma (dipole construction).

Lemma 3.2. *Let P and N be two points in $\mathbb{R}^3$ and let $\varepsilon > 0$. There exists a sequence of maps $(u_n)_n$ in $C^1(\mathbb{R}^3\backslash\{P,N\};S^2)$ such that*

$$E(u_n) \to 8\pi|P - N| \text{ as } n \to \infty$$

$$\deg(u_n, P) = 1\,,\ \deg(u_n, N) = -1$$

$$u_n = \text{North pole on } \left\{Q \in \mathbb{R}^3;\ \mathrm{dist}(Q,[P,N]) \ge \varepsilon\right\}.$$

Lemma 3.2 gives $I \le 8\pi L$ if $n = p = 1$. The general case is obtained by gluing together dipoles for the pairs $(P_i, N_{\sigma(i)})$.

b) $I \ge 8\pi L$. Two different proofs are known.

α) See [BCL]. Let, for $u \in \{v \in C^1(\mathbb{R}^3\backslash\{a_1,\dots,a_\ell\};S^2);\ \deg(v,a_i) = d_i\}$,

$$D(u) = (u\cdot(u_2 \times u_3),\ u\cdot(u_3 \times u_1),\ u\cdot(u_1 \times u_3))\,.$$

Then one has

$$\operatorname{div} D(u) = 4\pi\left(\sum_{i=1}^{n}\left(\delta_{P_i} - \delta_{N_i}\right)\right) \quad \text{and } |\nabla u|^2 \geq 2|D(u)|\,. \tag{3.4}$$

From (3.4) one gets

$$\begin{aligned} E(u) \geq 8\pi \ \operatorname{Min} & \left\{\sum_{i=1}^{n}(\xi(P_i) - \xi(N_i));\ \xi : \mathbb{R}^3 \to \mathbb{R} \quad \text{such that} \right. \\ & \left. |\xi(x) - \xi(y)| \leq |x - y| \ \forall x \in \mathbb{R}^3,\ \forall y \in \mathbb{R}^3 \right\}. \end{aligned} \tag{3.5}$$

Finally, using Kantorovitch's theorem on min-max and the characterization due to G. Birkhoff of the extremal points of the set of doubly stochastics matrices, one can prove that the right-hand side in (3.5) is $8\pi L$.

β) This proof is due to F. Almgren, W. Browder and E.H. Lieb [ABL]. It relies on the co-area formula and goes as follows. Let u be in $C^1(\mathbb{R}^3\backslash\{a_1,\ldots,a_\ell\};S^2)$ with $\deg(u,a_i) = d_i$ for any i in $\{1,\ldots,\ell\}$. One has

$$E(u) \geq 2\int_{\mathbb{R}^3} J_2(u)dx = 2\int_{S^2} \mathscr{H}^1(u^{-1}(S))d\sigma(S)\,, \tag{3.6}$$

$$\partial(u^{-1}(S)) = \sum_{i=1}^{P}\left(\delta_{P_i} - \delta_{N_i}\right)\,. \tag{3.7}$$

Finally from (3.6) and (3.7) one gets $E(u) \geq 8\pi L$.

Two entensions of Theorem 3.1 to the case $\Omega \neq \mathbb{R}^3$ are possible. We assume that the points a_i are in Ω. Let

$$\begin{gathered} I_1 = \operatorname{Inf}\{E(u);\ u \in C^1(\Omega\backslash\{a_1,\ldots,a_\ell\};S^2)\,, \\ \deg(u,a_i) = d_i \ for\ 1 \leq i \leq \ell \text{ and u is constant on } \partial\Omega\} \end{gathered}$$

$$I_2 = \operatorname{Inf}\{E(u);\ u \in C^1(\Omega\backslash\{a_1,\ldots,a_\ell\};S^2),\ \deg(u,a_i) = d_i \quad \text{for } 1 \leq i \leq \ell\}\,.$$

Let us denote by D the geodesic distance in Ω. Let

$$L_2 = \operatorname{Min}\left\{\sum_{i=1}^{p} D(P_i, N_{\sigma(i)});\ \sigma \text{ is a permutation of } \{1,\ldots,p\}\right\}$$

if $p = n$; $L_2 = +\infty$ if $p \neq n$. Let us define L_1 for $p \leq n$ by

$$\begin{aligned} L_1 = \operatorname{Min} & \left\{\sum_{i=1}^{p} D(P_i, N_{\sigma(i)}) + \sum_{i=p+1}^{n} \operatorname{dist}(N_{\sigma(i)}, \partial\Omega)\ ;\right. \\ & \left. \sigma \text{ is a permutation of } \{1,\ldots,n\}\right\}. \end{aligned} \tag{3.8}$$

For $p \geq n$ L_1 is defined by replacing in (3.8) n by p, p by n, N by P and P by N. Then one has (see [BCL]) $I_1 = 8\pi L_1$ and $I_2 = 8\pi L_2$.

Remark 3.3. a) The analogous of Theorems 3.1 for the liquid crystals energy has been obtained by M. Giaquinta, G. Modica and J. Souček in [GMS3]. b) F. Almgren, E.H. Lieb in [AL] and R. Hardt, F.H. Lin in [HL2] have obtained estimates on the number of singularities of minimizing harmonic maps from $\Omega \subset B^3$ into S^2.

4. Examples of MTM

The following maps have been proved to be MTM:

a) $u_0 : B^{n+m} \subset \mathbb{R}^{n+m} = \mathbb{R}^{n+1} \times \mathbb{R}^{m-1} \to S^n$, $u_0(x', x'') = x'/|x'|$.

b) $u_0 : B^{2n} \to S^n$, $u_0(x) = H(x/|x|)$ where $H : S^{2n-1} \to S^n$ are the Hopf maps related to the multiplication of complex numbers ($n = 2$), quaternions ($n = 4$) and Cayley numbers ($n = 8$). Example a. for $m = 1$ is due to F.H. Lin [L]. Example a) for $m \geq 2$ and Example b) are proved in [CG].

Lin's proof relies on the null Lagrangian method. It can be divided into three steps.

Step 1. Let $u : \mathbb{R}^{n+1} \to S^n$ then

$$|\nabla u|^2 \geq (n-1)^{-1}\{\mathrm{tr}(\nabla u)^2 - (\mathrm{div}\ u)^2\} \text{ with equality if } u = u_0\,. \tag{4.1}$$

Step 2. Let $u : B^{n+1} \to \mathbb{R}^{n+1}$ then

$$\mathrm{tr}(\nabla u)^2 - (\mathrm{div}\ u)^2 = \mathrm{div}\{(\mathrm{div}\ u)u - u.\nabla u\}\,. \tag{4.2}$$

Step 3. Let $u \in H^1(B^{n+1}; S^n)$ with $u = u_0$ on ∂B^{n+1}. Inequality (4.1) gives

$$\int_{B^{n+1}} |\nabla u|^2 \geq (n-1)^{-1} \int_{B^{n+1}} (\mathrm{tr}(\nabla u)^2 - (\mathrm{div}\ u)^2)\,. \tag{4.3}$$

Using (4.2) one gets

$$\int_{B^{n+1}} (\mathrm{tr}\ u)^2 - (\mathrm{div}\ u)^2 = \int_{B^{n+1}} (\mathrm{tr}\ u_0)^2 - (\mathrm{div}\ u_0)^2\,. \tag{4.4}$$

Finally using (4.3), (4.4) and (4.1) for $u = u_0$ one has $\int_{B^{n+1}} |\nabla u|^2 \geq \int_{B^{n+1}} |\nabla u_0|^2$.

The proof in [CG] relies on the co-area formula as in [ABL] and on a projection-averaging procedure. Let us explain the method for Example b) with the quaternionic Hopf map. Let $u : B^8 \to S^4$ such that $u = u_0$ on ∂B^8. We want to prove that

$$E(u) \geq E(u_0)\,. \tag{4.5}$$

For ψ in $G_3(\mathbb{R}^5)$, write $\pi_\psi : S^4 \to S^2$ for the nearest point projection. One first checks that

$$E(u) = \frac{2}{3} \fint_{G_3(\mathbb{R}^5)} E(\pi_\psi \circ u)\,. \tag{4.6}$$

Let $v = \pi_\psi \circ u$ and $v_0 = \pi_\psi \circ u_0$. Inequality (4.5) is a consequence of (4.6) and

$$E(v) \geq E(v_0)\,. \tag{4.7}$$

The proof of (4.7) goes as follow

$$E(v) \geq 2\int_{B^8} J_2(v) = 2\int_{S^2} \mathcal{H}^6(v^{-1}(S)) = \int_{S^2} \mathcal{H}^6(v^{-1}(-S) \cup v^{-1}(S))\,; \tag{4.8}$$

Note that, at least if v is not too singular – see [CG] for more details –,

$$\begin{aligned}\partial(v^{-1}(-S) \cup v^{-1}(S)) &= (\pi_\psi \circ H)^{-1}(S) - (\pi_\psi \circ H)^{-1}(-S) \\ &= \partial(v_0^{-1}(-S) \cup v_0^{-1}(S))\,.\end{aligned} \tag{4.9}$$

One can prove that, for some complex structure on $\mathbb{R}^8$, $v_0^{-1}(-S) \cup v_0^{-1}(S)$ is a complex variety; hence, by (4.9) and [F, p. 435 and 652],

$$\mathcal{H}^6(v^{-1}(-S) \cup v^{-1}(S)) \geq \mathcal{H}^6(v_0^{-1}(-S) \cup v_0^{-1}(S))\,. \tag{4.10}$$

Inequality (4.7) follows from (4.8), (4.10) and

$$E(v_0) = 2\int_{B^8} J_2(v_0) = \int_{S^2} \mathcal{H}^6(v_0^{-1}(-S) \cup v_0^{-1}(S))\,. \tag{4.11}$$

5. The Gap Phenomenon and the Problem of the Density of Smooth Maps

Let us start with a quite interesting theorem proved by R. Hardt and F.H. Lin in [HL1].

Theorem 5.1 (Gap Phenomenon). *The exist smooth maps $\gamma : \partial B^3 \to S^2$ of degree zero such that*

$$\begin{aligned}&\mathrm{Min}\{E(u);\ u \in H^1(B^3; S^2) \quad \text{and} \quad u = \gamma \text{ on } \partial B^3\} \\ &< \mathrm{Inf}\{E(u);\ u \in C^1(\bar{B}^3; S^2) \quad \text{and} \quad u = \gamma \text{ on } \partial B^3\}\,.\end{aligned} \tag{5.1}$$

Sketch of the construction of γ. We follow a method proposed by H. Brezis in [Br]. Let ε be a small positive number. Let $P_1 = (0, 0, 1-\varepsilon)$, $N_1 = (0, 0, 1+\varepsilon)$, $N_2 = (0, 0, -1+\varepsilon)$, $P_2 = (0, 0, -1-\varepsilon)$. By Lemma 3.2 there exists a map $u_\varepsilon : \mathbb{R}^3 \to S^2$ smooth on $\mathbb{R}^3 \backslash \{N_1, N_2, P_1, P_2\}$ such that the degree of u_ε at the N_i (resp. P_i) is -1 (resp. $+1$) and

$$\int_{\mathbb{R}^3} |\nabla u_\varepsilon|^2 \leq 8\pi(2\varepsilon + 2\varepsilon) + \varepsilon\,. \tag{5.2}$$

One takes for γ the restriction of u_ε to ∂B^3. The degree of γ is zero. For v in $C^1(\bar{B}^3; S^2)$ with $v = \gamma$ on ∂B^3 let $w : \mathbb{R}^3 \to S^2$ be defined by $w = v$ on B^3 and $w = u_\varepsilon$ on $\mathbb{R}^3 \backslash B^3$. That map is continuous on $\mathbb{R}^3 \backslash \{N_1, P_2\}$ and has degree $+1$ at P_2, -1 at N_1 hence by Theorem 3.1

$$\int_{\mathbb{R}^3} |\nabla w|^2 \geq 8\pi(2 + 2\varepsilon) \tag{5.3}$$

which implies, with (5.2),

$$\int_{B^3} |\nabla v|^2 \geq 16\pi(1-\varepsilon) - \varepsilon . \tag{5.4}$$

The gap phenomenon follows from (5.2) and (5.4).

Note that the gap phenomenon implies that even if γ is of degree zero $C^1_\gamma(B^3;S^2) = \{v;\ v \in C^1(B^3;S^2) \text{ and } v = \gamma \text{ on } \partial B^3\}$ may not be dense in $H^1_\gamma(B^3;S^2) = \{v \in H^1(B^3;S^2);\ v = \gamma \text{ on } \partial B^3\}$. Similar non-density had been observed previously by R. Schoen and K. Uhlenbeck; they had proved in [SU2]

Theorem 5.2. *The map u_0 in $H^1(B^3;S^2)$ defined by $u_0(x) = x/|x|$ cannot be approximated in the H^1-norm by maps in $C^1(\bar{B}^3;S^2)$.*

Proof of Theorem 5.2. Assume $u^n \in C^1(\bar{B}^3;S^2)$ satisfy $u^n \to u_0$ in $H^1(B^3;\mathbb{R}^3)$. This implies that $D(u^n) \to D(u_0)$ in $L^1(B^3;\mathbb{R}^3)$ but div $D(u^n) = 0$ and div $D(u_0) = 4\pi\delta_0$. A contradiction.

This leads to the question raised by J. Eells and L. Lemaire in [EL2]: let p be in $[1,\infty)$, is $C^\infty(M,N)$ dense in the Sobolev space $W^{1,p}(M;N)$? The answer is now known. One has

Theorem 5.3. *If $p \geq m$, then $C^\infty(M;N)$ is dense in $W^{1,p}(M;N)$. If $p < m$ then $C^\infty(M;N)$ is dense in $W^{1,p}(M;N)$ if and only if the homotopy group $\pi_{[p]}(N)$ is trivial* ($[p]$ *is the largest integer less or equal to* p).

Theorem 5.3 is easy if $p > m$ since in that case, by the Sobolev embeddings, $W^{1,p}(M;N) \subset C(M;N)$. The case $p = m$ has been proved by R. Schoen and K. Uhlenbeck in [SU2] and [SU3]. When $p < m$ and $\pi_{[p]}(N) \neq 0$, F. Bethuel and X. Zheng have constructed in [BZ] a map in $W^{1,p}(M;N)$ which cannot be approximated in the $W^{1,p}$-norm by maps in $C^\infty(M;N)$; their proof uses previous arguments due to B. White [Wh]. The fact that $\pi_{[p]}(N)$ is trivial implies density is a difficult theorem due to F. Bethuel [B2].

When one does not have density we may ask for a characterization of the closure $\widehat{W^{1,p}}(M,N)$ of $C^\infty(M,N)$ in $W^{1,p}(M;N)$. Only partial results are known. One has

Theorem 5.4. *Assume that N is $([p]-1)$ connected, $H_{[p]}(N)$ is torsion free and that $\pi_1(N)$ is abelian if $[p] = 1$. Then the two following conditions are equivalent*

$$u \in \widehat{W^{1,p}}(M,N) \tag{5.5}$$

the pullback by u of any closed $[p]$-form on N is (weakly) closed. (5.6)

The implication (5.5) $\Rightarrow$ (5.6) has been noticed by R. Schoen and K. Uhlenbeck in [SU3] – it holds without any topological assumption on N –. The converse has been proved by F. Bethuel in [B1] for $p = 2$, $M = B^3$ and $N = S^2$, by F. Demengel in [D] for $1 \leq p < 2$, $M = B^m$ and $N = S^1$; the general case is proved in [BCDH].

Theorem 5.3 leads also to the question: which size of singularities one has to allow in order to have density ? F. Bethuel has proved in [B2] – see also [BZ] for $M = B^m$, $N = S^{m-1}$, $2 \leq p < m$ –,

Theorem 5.5. *Assume $p < m$. The maps from M into N which are smooth except on a manifold of dimension $m - [p] - 1$ are dense in $W^{1,p}(M;N)$.*

6. Relaxed Energies

The gap phenomenon leads naturally to the following question

Question 6.1. *Let $\gamma : \partial B^3 \to S^2$ be a smooth map of degree zero. Is the infimum $Inf\{E(u);\ u \in C^1(\bar{B}^3;S^2)$ and $u = \gamma$ on $\partial B^3\}$ achieved?*

The answer to this question is still not known but in [BBC] it has been proved that a positive answer to that question is equivalent to the regularity of a minimizer for a relaxed energy E_1 associated to E. In order to define E_1 we need some notations. Let $H^1_\gamma(B^3;S^2) = \{u \in H^1(B^3;S^2);\ u = \gamma$ on $\partial B^3\}$, $C^1_\gamma(B^3;S^2) = \{u \in C^1(\bar{B}^3;S^2);\ u = \gamma$ on $\partial B^3\}$, $R_\gamma(B^3;S^2) = \{u \in H^1_\gamma(B^3;S^2);\ u$ is C^1 except at a finite number of points$\}$. Recall – see Theorem 5.4 – ([B2] or [BZ]) that R_γ is dense in $H^1_\gamma(B^3;S^2)$ for the H^1—norm. For u in R_γ let

$$L(u) = \text{Min}\left\{\sum_{i=1}^{n} |P_i - N_{\sigma(i)}|\ ;\quad \sigma \text{ is a permutation of } \{1,\ldots,n\}\right\} \tag{6.1}$$

where the P_i (resp. N_i) are the singularities of u of positive (resp. negative) degree counted according to their degree. By a result of [BCL], for any u in R_γ,

$$L(u) = \frac{1}{4\pi}\text{Sup}\left\{\int_{B^3} D(u)\cdot\nabla\theta - \int_{\partial B^3}\theta \text{Jac}(\gamma); |\theta(x) - \theta(y)| \leq |x - y|\ \forall x, \forall y\right\}$$

from which it follows that $L(u)$ makes sense for u in $H^1_\gamma(B^3;S^2)$. Let now $E_1 : H^1_\gamma(B^3;S^2) \to [0,+\infty)$ be defined by

$$E_1(u) = E(u) + 8\pi L(u)\,.$$

Note $E_1 = E$ on $C^1_\gamma(B^3;S^2)$. The following properties of E_1 are proved in [BBC].

$$E_1 \text{ is weakly lower semicontinuous} \tag{6.2}$$

$$\text{Min}\left\{E_1(u);\ u \in H^1_\gamma(B^3;S^2)\right\} = \text{Inf}\left\{E(u);\ u \in C^1_\gamma(B^3;S^2)\right\} \tag{6.3}$$

$$E_1(u) = \text{Inf}\left\{\underline{\lim}\ E(u^n);\ u^n \in C^1_\gamma(B^3;S^2) \quad \text{and} \quad u^n \to u \text{ a.e.}\right\}. \tag{6.4}$$

In particular a positive answer to question: is any minimizer of E_1 smooth ? will give a positive answer to Question 6.1. Recently M. Giaquinta, G. Modica and J. Souček have proved in [GMS2] that the (eventual) singular set of any minimizer of E_1 has Hausdorff dimension less or equal to 1.

Even if this approach does not give, for the moment being, the answer to Question 6.1 it allows to prove

Theorem 6.2. *Let $\gamma : \partial B^3 \to S^2$ be a smooth map. Assume that either the degree of γ is not zero or the degree of γ is zero and (5.1) holds. Then there are infinitely many harmonic maps in $H^1_\gamma(B^3; S^2)$.*

Let us give the main steps of the proof of Theorem 6.2 when the degree of γ is zero and (5.1) holds. Let for λ in $[0,1]$, $E_\lambda(u) = E(u) + 8\pi\lambda L(u)$. From (6.2) we get

$$E_\lambda \text{ is weakly lower semicontinuous ;} \tag{6.5}$$

hence

$$\text{the infimum } m_\lambda \text{ of } E_\lambda \text{ on } H^1_\gamma(B^3; S^2) \text{ is achieved}. \tag{6.6}$$

Using the gap phenomenon one can prove, see [BBC],

$$m_\lambda > m_0 \quad \text{for any } \lambda > 0. \tag{6.7}$$

Note also that clearly

$$\lim_{\lambda \to 0} m_\lambda = m_0 . \tag{6.8}$$

The infinitely many harmonic maps in $H^1_\gamma(B^3; S^2)$ follows from (6.6), (6.7) and (6.8) and

$$\text{any minimizer of } E_\lambda \text{ is a harmonic map}. \tag{6.9}$$

It remains to prove (6.9). Let u be in R_γ and let φ be in $C^1_0(B^3; \mathbb{R}^3)$; then the singularities of $(u+\varepsilon\varphi)/|u+\varepsilon\varphi|$ are the same as those of $(u+\varepsilon\varphi)/|u+\varepsilon\varphi|$ if $|\varepsilon||\varphi|_{L^\infty} < 1$ hence, see (6.1),

$$L((u+\varepsilon\varphi)/|u+\varepsilon\varphi|) = L(u) \quad \text{if } |\varepsilon||\varphi|_{L^\infty} < 1 . \tag{6.10}$$

By the density of R_γ in $H^1_\gamma(B^3; S^2)$, (6.10) also holds for u in $H^1_\gamma(B^3; S^2)$; it gives (6.9).

Remark 6.3. a) F. Bethuel and H. Brezis have proved in [BB] that any minimizer of E_λ for $\lambda \in [0,1)$ is smooth on $\bar{B}^3$ except at a finite number of points. b) A quite interesting different approach to relaxed energies have been given by M. Giaquinta, G. Modica and J. Souček in [GMS1] and [GMS2].

Note added in proof. In a very interesting paper (to appear in C. R. Acad. Sci. Paris) F. Hélein has given a positive answer to Question 1.1.

References

[A] Alouges, F.: Un schéma numérique pour le calcul d'applications harmoniques de $\mathbb{R}^3$ dans la sphère. C.R. Acad. Sci. Paris, 1990

[ABL] Almgren, F., Browder, W., Lieb, E.H.: Co-area, liquid crystals and minimal surfaces. (Lecture Notes in Mathematics, vol. 1306.) Springer, Berlin Heidelberg New York 1988, pp. 1–12

[AL] Almgren, F., Lieb, E.H.: Singularities of energy minimizing maps from the ball to the sphere: examples, counter-examples and bounds. Ann. Math. **128** (1988) 483–530

[B1] Bethuel, F.: A characterization of maps in $H^1(B^3, S^2)$ which can be approximated by smooth maps. Ann. IHP Analyse non Linéaire (to appear)

[B2] Bethuel, F.: Approximations dans des espaces de Sobolev et groupes d'homotopie. C.R. Acad. Sci. Paris **307** (1988) 293–296, and The approximation problem for Sobolev maps between two manifolds. Acta Math. (to appear)

[BB] Bethuel, F., Brezis, H.: Régularité de minima pour des énergies relaxées. C.R. Acad. Sci. Paris (1990), and Regularity of minimizers of relaxed problems for harmonic maps. Preprint

[BBC] Bethuel, F, Brezis, H., Coron, J.M.: Relaxed energies for harmonic maps. In: Variational problems, H. Berestycki, J.M. Coron, I. Ekeland (eds.). Birkhäuser, 1990

[BCDH] Bethuel, F., Coron, J.M., Demengel, F., Hélein, F.: A cohomological cirterion for density of smooth maps in Sobolev spaces between two manifolds. In: Defects, singularities and patterns in nematic liquid crystals, J.M. Coron, J.M. Ghidaglia, F. Hélein (eds.). Kluwer Academic Press, 1991

[Br] Brezis, H.: S^k-valued maps with singularities. In: Topics in calculus of variations, M. Giaquinta (ed.). (Lecture Notes in Mathematics, vol. 1365.) Springer, Berlin Heidelberg New York 1989, pp. 1–30

[BCL] Brezis, H., Coron, J.M., Lieb, E.H.: Harmonic maps with defects. Comm. Math. Phys. **107** (1986) 649–705

[BZ] Bethuel, F., Zheng, X.: Density of smooth functions between two manifolds in Sobolev spaces. J. Funct. Anal. **80** (1988) 60–75

[C] Chen, Y.: Weak solutions to the evolutions problems of harmonic maps. Math. Z. **201** (1989) 69–74

[CG] Coron, J.M., Gulliver, R.: Minimizing p-harmonic maps into spheres. J. Reine Angew. Math. **401** (1989) 82–100

[CH] Coron, J.M., Hélein, F.: Minimizing harmonic diffeomorphisms. Compositio Math. **69** (1989) 175–228

[D] Demengel, F.: Une caractérisation des applications de $W^{1,p}(B^N, S^1)$ qui ne peuvent être approchées par des fonctions C^∞. C.R. Acad. Sci. Paris **310** (1990) 553–557

[EL1] Eells, J., Lemaire, L.: Report on harmonic maps. Bull. London Math. Soc. **10** (1978) 1–68

[EL2] Eells, J., Lemaire, L.: Another report on harmonic maps. Bull. London Math. Soc. **20** (1988) 385–524

[ES] Eells, J., Sampson, J.H.: Harmonic mappings of Riemannian manifolds. Amer. J. Math. **86** (1964) 109–160

[F] Federer, H.: Geometric measure theory. Springer, Berlin Heidelberg New-York, 1969

[G] Grüter, M.: Regularity of weak H-surfaces. J. Reine Angew. Math. **329** (1981) 1–15

[GG] Giaquinta, M., Giusti, E.: The singular set of the minima of certain quadratic functionals. Ann. S.N.S. Pisa **11** (1984) 45–55

[GMS1] Giaquinta, M., Modica, G., Souček, G.: Cartesian currents, weak diffeomorphisms and existence theorems in nonlinear elasticity. Arch. Rational Mech. Anal. **106** (1989) 97–159

[GMS2] Giaquinta, M., Modica, G., Souček, G.: The Dirichlet energy of mappings with values into the sphere. Manuscripta Math. **65** (1989) 489–507

[GMS3] Giaquinta, M., Modica, G., Souček, G.: Liquid crystals: relaxed energies, dipoles, singular lines and singular points. Annali S.N.S. Pisa (to appear)

[GW] Gulliver, R., White, B.: The rate of a convergence of a harmonic map at a singular point. Math. Ann. **283** (1989) 539–549

[H1] Hélein, F.: Homéomorphismes quasi conformes entre surfaces riemanniennes. C.R. Acad. Sci. Paris **307** (1988) 725–730

[H2] Hélein, F.: Régularité des applications faiblement harmoniques entre une surface et une sphère. C.R. Acad. Sci. Paris **311** (1990) 519–524

[HKL] Hardt, R., Kinderlehrer, D., Lin, F.H.: Existence and partial regularity of static liquid crystal configurations. Comm. Math. Phys. **105** (1986) 547–570

[HKLL] Hardt, R., Kinderlehrer, D., Lin, S.Y., Luskin, M.: Minimum energy configurations for liquid crystals: computational results. In: Theory and applications of liquid crystals, J.L. Ericksen, D. Kinderlehrer (eds.), IMA **5**, 1987

[HKW] Hildebrandt, S., Kaul, H., Widman, K.O.: An existence theorem for harmonic mappings of Riemannian manifolds. Acta Math. **138** (1977) 1–16

[HL1] Hardt, R., Lin, F.H.: A remark on H^1 mappings. Manuscripta Math. **56** (1986) 1–10

[HL2] Hardt, R., Lin, F.H.: Stabilities of singularities of minimizing harmonic maps. J. Diff. Geom. **29** (1989) 113–123

[KRS] Keller, J., Rubinstein, J., Sternberg, P.: Reaction-diffusion processes and evolution to harmonic maps. Preprint

[L] Lin, F.H.: Une remarque sur l'application $x/|x|$. C.R. Acad. Sci. Paris **305** (1987) 529–531

[LU] Ladyshenskaya, O.A., Vral'tseva: Linear and quasilinear elliptic equations. Academic Press, New York 1968

[M] Morrey Jr., C.B.: The problem of plateau on a Riemannian manifold. Ann. Math. **49** (1948) 807–851

[SaU] Sacks, J., Uhlenbeck, K.: The existence of minimal immersions of 2-spheres. Ann. Math. **113** (1981) 1–24

[Sc] Schoen, R.: Analytic aspects of the harmonic maps problem. M.S.R.I. **2** (1984) 321–358

[Sh] Shatah, J.: Weak solutions and development of singularities of the $SU(2)$ σ-model. Comm. Pure Appl. Math. **41** (1988) 459–469

[Si] Simon, L.: Asymptotics for a class of nonlinear evolution equations with applications to geometric problems. Ann. Math. **118** (1983) 525–571

[SU1] Schoen, R., Uhlenbeck, K.: A regularity theory for harmonic maps. J. Diff. Geom. **17** (1982) 307–335

[SU2] Schoen, R., Uhlenbeck, K.: Boundary regularity and the Dirichlet problem for harmonic maps. J. Diff. Geom. (1983) 253–268

[SU3] Schoen, R., Uhlenbeck, K.: Regularity of minimizing harmonic maps into the sphere. Invent. math. **78** (1984) 89–100

[SU4] Schoen, R., Uhlenbeck, K.: Approximation theorems for Sobolev mappings. Preprint

[We] Wente, H.: An existence theorem for surfaces of constant mean curvature. J. Math. Anal. Appl. **26** (1969) 318–344

[Wh] White, B.: Infima of energy functionals in homotopy classes. J. Diff. Geom. **23** (1986) 127–142

Isometric Embeddings of Riemannian Manifolds

Matthias Günther

Universität Leipzig, Sektion Mathematik, Augustusplatz 10
O-7010 Leipzig, Fed. Rep. of Germany

1. Explanation of the Problem

Let M be an n-dimensional manifold of class C^∞ and g any given Riemannian metric on M. We will consider the following classical problem motivated by differential geometry. Does there exist an embedding $u = (u^1, \ldots, u^q) : M \to \mathbb{R}^q$ such that the usual euclidian metric of $\mathbb{R}^q$ induces on the submanifold $u(M)$ the given metric g? In other words, u must satisfy

$$F(u) := du \cdot du = g, \tag{1}$$

or in local coordinates

$$\sum_{l=1}^{q} \frac{\partial u^l}{\partial x^i} \frac{\partial u^l}{\partial x^j} = g_{ij}.$$

The dot in (1) denotes the usual scalar product of $\mathbb{R}^q$. The notion embedding means, that u is locally an immersion and globally a homeomorphism of M onto the subspace $u(M)$ of $\mathbb{R}^q$. If an embedding $u : M \to \mathbb{R}^q$ satisfies (1) on the whole M, we speak of an *isometric embedding.* If u is an immersion and a solution of (1) in a (possibly small) neighbourhood of any point of M, we speak of a *local isometric embedding.* A further question is the regularity of the embedding in dependence on the regularity of the metric. And finally, what can be said about the minimal value of q?

We will give some historical remarks. There exists a great number of beautiful papers which handle the isometric embedding problem (local or global) under further assumptions on the manifold M or the metric g (e.g. special values of dimension n, positivity assumptions on the curvature), but we are interested only in the general problem. Janet (1926), Cartan (1927) and Burstin (1931) proved the existence of a local isometric embedding with $q = n(n + 1)/2$ in the analytical case; the essentials of their proofs are suitable applications of the Cauchy-Kowalewski theorem. Up to date a corresponding result (with the same value of q and without further assumptions) is unknown in the nonanalytical case, even for the dimension $n = 2$. Nash (1954) and Kuiper (1955) proved the existence of a (global) isometric embedding of class C^1 with $n < q \leq 2n + 1$ (in fact their results are more subtle),

Proceedings of the International Congress
of Mathematicians, Kyoto, Japan, 1990

provided that the metric g is continuous. In an outstanding paper Nash (1956) showed the existence of a (global) isometric embedding $u \in C^s(M, \mathbb{R}^q)$ $(s \geq 3, s = \infty)$ if $g \in C^s$ and $q = 3n(n+1)/2 + 4n$ in the compact case, $q = (n+1)q_{\text{com}}$ in the noncompact case. Note the surprising difference of the value of q in the case C^1 and C^s, $s \geq 3$, respectively. Nash's paper (1956) is fundamental not only for the problem under consideration, but also because of the applied method, which gives the foundation of the so-called hard implicit function theorems or Nash-Moser technique. The latter plays an important role in the modern theory of nonlinear partial differential equations. There are many papers concerning this technique, see e.g. Moser (1961, 1966), Jacobowitz (1972), Zehnder (1975), Hamilton (1982) and Hörmander (1985, 1988). Finally we mention the book "Partial Differential Relations" by Gromov (1986), which contains many material and references belonging to our problem. Gromov gives $q = (n+2)(n+3)/2$ as best value in the smooth case.

One of the main steps in Nash's demonstration (1956) is the solution of the perturbation problem associated to (1), i.e. the determination of $v : M \to \mathbb{R}^q$ such that $F(u + v) = g + f$, if a solution u of $F(u) = g$ is known and f small in some sense. The linearized equations $F'(u)v = h$ of our nonlinear problem (1) are $2du \cdot dv = h$ or locally written

$$\partial_i u \cdot \partial_j v + \partial_j u \cdot \partial_i v = h_{ij}.$$

Here the embedding $u : M \to \mathbb{R}^q$ and a symmetric covariant 2-tensor field h on M is given and $v : M \to \mathbb{R}^q$ is unknown. This is a linear first order system of partial differential equations for v. The system seems to be very simple, but it does not settle down in a standard class. Now the idea is as follows. Additionally we demand $\partial_i u \cdot v = 0$. Then we find the relations

$$\partial_i u \cdot v = 0, \qquad \partial_i \partial_j u \cdot v = -h_{ij}/2. \tag{2}$$

which form an algebraic system of $n(n+3)/2$ equations for v. Following Gromov and Rohlin (1970) a C^2-mapping $u : M \to \mathbb{R}^q$ is called *free*, if for each $x \in M$ the $n(n+3)/2$ vectors $\partial_i u(x)$, $\partial_i \partial_j u(x)$ of $\mathbb{R}^q$ are linearly independent. Now let u be free, then we have a unique solution v of (2) with minimal pointwise $\mathbb{R}^q$-norm, which we will denote by

$$v = -E(u)(0, h)/2.$$

We have the following mapping properties

$$F : C^s \to C^{s-1}, \qquad E(u) : C^s \to C^s \quad \text{if} \quad u \in C^{s+2}.$$

This shows the loss of differentiability; therefore we cannot use simple successive approximations in order to solve the nonlinear equations. One way out is given by the above mentioned method of hard implicit function theorems consisting in a complicated combination of successive approximations and smoothing operators.

2. A New Method to Solve the Perturbation Problem

In this section we give another and short way to handle the perturbation problem, which gives moreover somewhat better results. Let M be compact. We use certain

Hölder spaces $C^{s,\lambda}$ of functions or sections of vector bundles over M with non-negative integers s and Hölder exponent λ, $0 < \lambda < 1$; the latter is fixed once for all. $S_*^{(2)}M$ denotes the bundle of symmetric covariant 2-tensors on M. Let $\|E(u)\|_{2,\lambda}$ be the norm of the linear mapping

$$E(u): C^{2,\lambda}(M, T_*M) \times C^{2,\lambda}(M, S_*^{(2)}M) \to C^{2,\lambda}(M, \mathbb{R}^q),$$

if u is a free and smooth mapping. Now we can formulate the following

Theorem 1. *Let $u \in C^\infty(M, \mathbb{R}^q)$ be a free mapping and $f \in C^{s,\lambda}(M, S_*^{(2)}M)$ with $s \geq 2$ or $s = \infty$. There is a positive number θ (independent of u, s and f) with the property: If*

$$\|E(u)\|_{2,\lambda} \|E(u)(0, f)\|_{2,\lambda} \leq \theta, \tag{3}$$

then there exists a $v \in C^{s,\lambda}(M, \mathbb{R}^q)$ such that one has

$$d(u + v) \cdot d(u + v) = du \cdot du + f \quad \text{on } M.$$

Remark. Our solubility condition (3) contains two factors coming from the linearized problem taken at the given initial mapping u. The first depends only on the coefficients of the unknowns of the linearized problem, the second depends only on the solution of the linearized problem with the given perturbation term f. These two influences must keep a certain balance.

We only sketch the *proof* here. In order to explain the idea, let M be the n-dimensional torus. For this special case see also a recent paper of Hörmander (1988), where a much more complicated technique is used. We write our nonlinear equations as

$$\partial_i u \cdot \partial_j v + \partial_j u \cdot \partial_i v + \partial_i v \cdot \partial_j v - f_{ij} = 0. \tag{4}$$

It is well known, that

$$(\Delta - 1): C^{s+2,\lambda}(M) \to C^{s,\lambda}(M)$$

is an isomorphism. Therefore we can apply the operator $(\Delta - 1)$ to (4) and obtain after some rearrangement the crucial relations

$$\begin{aligned} &\partial_i\{(\Delta - 1)(\partial_j u \cdot v) + \Delta v \cdot \partial_j v\} + \partial_j\{(\Delta - 1)(\partial_i u \cdot v) + \Delta v \cdot \partial_i v\} \\ &\quad - 2\{(\Delta - 1)(\partial_i \partial_j u \cdot v) + \tfrac{1}{2}\partial_i v \cdot \partial_j v - \partial_i \partial_l v \cdot \partial_j \partial_l v \\ &\quad + \Delta v \cdot \partial_i \partial_j v + \tfrac{1}{2}(\Delta - 1) f_{ij}\} = 0. \end{aligned} \tag{5}$$

Hence it suffices to solve the new system

$$\partial_i u \cdot v = -(\Delta - 1)^{-1}(\Delta v \cdot \partial_i v),$$

$$\partial_i \partial_j u \cdot v = -\tfrac{1}{2} f_{ij} + (\Delta - 1)^{-1}(\partial_i \partial_l v \cdot \partial_j \partial_l v - \Delta v \cdot \partial_i \partial_j v - \tfrac{1}{2}\partial_i v \cdot \partial_j v).$$

The properties of $(\Delta - 1)^{-1}$ guarantee that there is no loss of differentiability, hence we can solve the system by simple successive approximations if f fulfils the smallness condition (3). The only difficulty is to treat the C^∞-case. We must take care of the constants arising in the inequalities and their dependence on the number s.

Now some words to the general case. We equip M with an auxiliary Riemannian metric g_0 of class C^∞. The covariant derivatives, the Laplacians as well as the occuring curvature tensors are meant always with respect to g_0. We have to use the Lichnerowicz-Laplacians for vector fields and symmetric 2-tensor fields

$$\Delta_{(1)} t_i = \Delta t_i - R_{i\cdot}^{\;l} t_l, \; \Delta = \nabla^l \nabla_l,$$

$$\Delta_{(2)} t_{ij} = \Delta t_{ij} - 2R_{i\cdot j\cdot}^{\;k\;l} t_{kl} - R_{i\cdot}^{\;l} t_{lj} - R_{j\cdot}^{\;l} t_{il}.$$

If α is a sufficiently large real number, then the mappings

$$(\Delta_{(1)} - \alpha): C^{s+2,\lambda}(M, T_* M) \to C^{s,\lambda}(M, T_* M),$$

$$(\Delta_{(2)} - \alpha): C^{s+2,\lambda}(M, S_*^{(2)} M) \to C^{s,\lambda}(M, S_*^{(2)} M)$$

are isomorphisms. We define a vector field $N(v)$ and a symmetric 2-tensor field $L(v)$ by the formulas

$$N_i(v) = -\Delta v \cdot \nabla_i v,$$

$$L_{ij}(v) = \nabla^l \nabla_i v \cdot \nabla_l \nabla_j v - \Delta v \cdot \nabla_i \nabla_j v - R_{i\cdot j\cdot}^{\;k\;l} \nabla_k v \cdot \nabla_l v - \frac{\alpha}{2} \nabla_i v \cdot \nabla_j v.$$

One has instead of (5) the identity

$$\begin{aligned}(\Delta_{(2)} - \alpha)\{\nabla_i u \cdot \nabla_j v + \nabla_j u \cdot \nabla_i v + \nabla_i v \cdot \nabla_j v\} \\ = \nabla_i\{(\Delta_{(1)} - \alpha)(\nabla_j u \cdot v) - N_j(v)\} + \nabla_j\{(\Delta_{(1)} - \alpha)(\nabla_i u \cdot v) - N_i(v)\} \\ - 2\{(\Delta_{(2)} - \alpha)(\nabla_i \nabla_j u \cdot v) - L_{ij}(v) - (\nabla_i R_{j\cdot}^{\;l} + \nabla_j R_{i\cdot}^{\;l} - \nabla^l R_{ij})(\nabla_l u \cdot v)\}.\end{aligned}$$

These equations are shown by a straightforward calculation with repeated use of the Ricci identity. This procedure leads to the final system

$$\nabla_i u \cdot v = ((\Delta_{(1)} - \alpha)^{-1} N(v))_i,$$

$$\nabla_i \nabla_j u \cdot v = -\tfrac{1}{2} f_{ij} + ((\Delta_{(2)} - \alpha)^{-1} M(v))_{ij} \tag{6}$$

with

$$M_{ij}(v) = L_{ij}(v) + (\nabla_i R_{j\cdot}^{\;l} + \nabla_j R_{i\cdot}^{\;l} - \nabla^l R_{ij})((\Delta_{(1)} - \alpha)^1 N(v))_l.$$

Again we can solve (6) with the help of simple successive approximations. For details see Günther (1989a). □

We give another variant of the perturbation result, which is important for applications in Sect. 3.

Theorem 2. *Let $B \subseteq \mathbb{R}^n$ be the open unit ball and $B_1, B_2 \subseteq \mathbb{R}^n$ open sets with $\bar{B}_1 \subseteq B_2$, $\bar{B}_2 \subseteq B$. Further, let $u \in C^\infty(\bar{B}, \mathbb{R}^q)$ be a free mapping and $f = (f_{ij}) \in C^{s,\lambda}(B, \mathbb{R}^{n(n+1)/2})$ with $s \geq 2$ or $s = \infty$. There exists a positive number θ (independent of u, s and f) with* *the property*: *If*

$$\operatorname{supp} f \subseteq B_1 \quad \text{and} \quad \|E(u)\|_{2,\lambda}\|E(u)(0,f)\|_{2,\lambda} \le \theta,$$

then there exists a $v \in C^{s,\lambda}(M, \mathbb{R}^q)$ *with*

$$\operatorname{supp} v \subseteq B_2 \quad \text{and} \quad \partial_i(u+v)\cdot\partial_j(u+v) = \partial_i u\cdot\partial_j u + f_{ij} \quad \text{in} \quad B.$$

We indicate the *proof* of Theorem 2. We choose a cut-off function $a \in C^\infty(B)$ with $a(x) = 1$ for all $x \in B_1$ and $a(x) = 0$ for all $x \in \bar{B}\backslash B_2$. Further, the Dirichlet problem

$$\Delta w = h \quad \text{in } \bar{B}, \qquad w = 0 \quad \text{on } \partial B$$

possesses a unique solution $w =: \Delta^{-1}h \in C^{s+2,\lambda}(\bar{B})$ if $h \in C^{s,\lambda}(\bar{B})$. The definition of $N_i(v)$, $M_{ij}(v)$ is now modified as follows

$$N_i(v) = 2\partial_i a(\Delta v\cdot v) + a(\Delta v\cdot\partial_i v),$$

$$M_{ij}(v) = \Delta(4\partial_i a\partial_j a(v\cdot v) + 2a\partial_i a(\partial_j v\cdot v) + 2a\partial_j a(\partial_i v\cdot v) + a^2(\partial_i v\cdot\partial_j v))$$
$$- \Delta(a\partial_i\Delta^{-1}N_j(v) + a\partial_j\Delta^{-1}N_i(v) + 3\partial_i a\Delta^{-1}N_j(v) + 3\partial_j a\Delta^{-1}N_i(v)).$$

Then $M_{ij}(v)$ is a linear combination of terms

$$\partial^{\alpha_1}a\partial^{\alpha_2}\Delta^{-1}N_l(v) \quad \text{with} \quad |\alpha_1| + |\alpha_2| = 3,\ |\alpha_2| \le 2,$$

$$\partial^{\alpha_1}a\partial^{\alpha_2}a(\partial^{\alpha_3}v\cdot\partial^{\alpha_4}v) \quad \text{with} \quad |\alpha_1| + \cdots + |\alpha_4| = 4,\ |\alpha_3|, |\alpha_4| \le 2$$

and therefore we have $N_i(v), M_{ij}(v) \in C^{s,\lambda}$ if $v \in C^{s+2,\lambda}$. Let v be a solution of the system

$$\partial_i u\cdot v = -a\Delta^{-1}N_i(v),$$
$$\partial_i\partial_j u\cdot v = \tfrac{1}{2}(-f_{ij} + \Delta^{-1}M_{ij}(v)) \tag{7}$$

then after some easy calculations it follows that

$$\partial_i(u + a^2v)\cdot\partial_j(u + a^2v) = \partial_i u\cdot\partial_j u + a^2 f_{ij} \quad \text{in } B.$$

The properties of $N_i(v)$, $M_{ij}(v)$ and Δ^{-1} guarantee that (7) can be solved by simple approximations too.

3. A Result Concerning the Value of *q*

Now lt M be any C^∞-manifold, not necessarily compact. We have the following

Theorem 3. *Let g be of class C^∞, let $u_0 \in C^\infty(M, \mathbb{R}^q)$ be a free embedding with $q \ge n(n+3)/2 + 5$ and $du_0\cdot du_0 < g$. Further, let δ be any positive continuous function on M. Then there exists an embedding $u \in C^\infty(M, \mathbb{R}^q)$ such that*

$$du\cdot du = g \quad \text{and} \quad |u(x) - u_0(x)| \le \delta(x) \quad \text{for every } x \in M.$$

Concerning the existence of free embeddings u_0 we have the following proposition. Its proof is based on the well known theorem of Sard, see e.g. Gromov and Rohlin (1970, Sect. 2.5). The condition $du_0 \cdot du_0 < g$ can be reached by easy manipulations.

Proposition. *If g is a continuous metric on M, then there exists a free embedding $u_0 \in C^\infty(M, \mathbb{R}^q)$ with $q = n(n+5)/2$ and $du_0 \cdot du_0 < g$ on M.*

Combining Theorem 3 and the Proposition, we obtain the

Corollary. *Every smooth Riemannian manifold possesses a smooth isometric embedding into $\mathbb{R}^q$ with*

$$q = \max\{n(n+5)/2,\ n(n+3)/2 + 5\}.$$

Remark. In general, we have for our q, that $q = q_G - 3$ and in special cases, i.e. $M = S^n$ (n-dimensional sphere), that $q = q_G - (n-2)$. Here q_G means that value of q, which was given by Gromov (1986, Sect. 3.1.7). The main restriction for our q comes from the existence of a free embedding and not from our method to construct the isometric embedding.

We will give a brief sketch for the *proof* of Theorem 3. Firstly, one can write

$$g = du_0 \cdot du_0 + \sum_{l \geq 1} h^{(l)}.$$

Thereby $h^{(l)}$ are symmetric covariant 2-tensor fields on M, such that in suitable local coordinate systems $x : u^{(l)} \to \mathbb{R}^n$

$$h_{11}^{(l)}(x) = a^4(x), \qquad h_{ij}^{(l)}(x) = 0 \qquad (1 \leq i \leq j \leq n, j \neq 1) \tag{8}$$

with $a \in C_0^\infty(u^{(l)})$; hence supp $h^{(l)} \subseteq U^{(l)}$. Only the one-one-coordinate is different from zero! The family $\{U^{(l)}\}$ are to be a locally finite covering for M. Hence it suffices to show: u_0 can be changed into a free embedding $\tilde{u}_0$ with

$$d\tilde{u}_0 \cdot d\tilde{u}_0 = du_0 \cdot du_0 + h^{(1)} \quad \text{in } U^{(1)},$$

$$\tilde{u}_0 = u_0 \quad \text{in } M \setminus U^{(1)},$$

$$|\tilde{u}_0(x) - u_0(x)| < \delta(x) \quad \text{for all} \quad x \in M.$$

The construction of such a $\tilde{u}_0$ will be done in the steps two and three.

Secondly, identifying $U^{(1)}$ with the open unit ball B of $\mathbb{R}^n$, we choose to every integer $k \geq 2$ and $\varepsilon \in (0, 1)$ a free mapping $u_{\varepsilon,k} \in C^\infty(\bar{B}, \mathbb{R}^q)$ with

$$du_{\varepsilon,k} \cdot du_{\varepsilon,k} = du_0 \cdot du_0 + h^{(1)} + 0(\varepsilon^{k+1}),$$

$$u_{\varepsilon,k} = u_0 \quad \text{in } B \setminus B_1.$$

$B_1 \subseteq \mathbb{R}^n$ is an open set with $\bar{B}_1 \subseteq B$. The construction of $u_{\varepsilon,k}$ is made in form of a series

$$u_{\varepsilon,k}(x) = u_0(x) + \varepsilon u_1(\varepsilon, x) + \cdots + \varepsilon^k u_k(\varepsilon, x).$$

(9) means equality up to powers ε^k. The proof offers certain technical difficulties; one must take advantage of the special form of $h^{(1)}$ in (8) as well as the assumption

$q \geq n(n + 3)/2 + 5$. We note, that $u_{\varepsilon,k}$ has in some sense a singular behaviour as $\varepsilon \to 0$. For details see Günther (1989b, Sects. 3, 4).

Thirdly we apply Theorem 2 with the initial mapping $u = u_{\varepsilon,k}$; the tensor field $f = f_{\varepsilon,k}$ is determined by the remainder term $0(\varepsilon^{k+1})$ in (9). From the properties of the $u_{\varepsilon,k}$ we obtain

$$\|E(u_{\varepsilon,k})\|_{2,\lambda} = 0(\varepsilon^{-k_0}), \qquad \|f_{\varepsilon,k}\|_{2,\lambda} = 0(\varepsilon^{k-2})$$

with an integer k_0 independent of k. If we choose k sufficiently large and ε sufficiently small, then we can satisfy the solubility condition of Theorem 2. This finishes the proof of Theorem 3.

References

Burstin, C. (1931): Ein Beitrag zum Problem der Einbettung der Riemannschen Räume in euklidischen Räumen. Mat. Sb. USSR **38**, 74–85

Cartan, E. (1927): Sur la possibilité de plonger un espace riemannien donné dans un espace euclidien. Ann. Soc. Polon. Math. **6**, 1–7

Gromov, M.L., Rohlin, V.A. (1970): Embeddings and immersions in Riemannian geometry. Usp. Mat. Nauk **25**, 3–62 (Russian). [English transl.: Russ. Math. Surv. **25** (1970) 1–57]

Gromov, M.L. (1986): Partial differential relations. Springer, New York Berlin Heidelberg

Günther, M. (1989a): On the perturbation problem associated to isometric embeddings of Riemannian manifolds. Ann. Global Anal. Geom. **7**, 69–77

Günther, M. (1989b): Zum Einbettungssatz von J. Nash. Math. Nachr. **144**, 165–187

Hamilton, R.S. (1982): The inverse function theorem of Nash and Moser. Bull. Amer. Math. Soc. **7**, 65–222

Hörmander, L. (1985): On the Nash-Moser implicit function theorem. Ann. Acad. Sci. Fenn. **10**, 255–259

Hörmander, L. (1988): The Nash-Moser theorem and para-differential operators. Preprint

Jacobowitz, H. (1972): Implicit function theroems and isometric embeddings. Ann. Math. **95**, 191–225

Janet, M. (1926): Sur la possibilité de plonger un espace riemannien donné dans un espace euclidien. Ann. Soc. Polon. Math. **5**, 38–43

Kuiper, N. (1955): On C^1 isometric imbeddings I, II. Proc. Kon. Acad. Wet. Amsterdam A **58**, Indigationes Mathematicae **17**, 545–556, 683–689

Moser, J. (1961): A new technique for the construction of solutions of non-linear differential equations. Proc. Nat. Acad. Sci. USA **47**, 1824–1831

Moser, J. (1966): A rapidly convergent iteration method and non-linear partial differential equations I, II. Ann. Scuola Norm. Sup. Pisa **20**, 265–315, 499–535

Nash, J. (1954): C^1 isometric imbeddings. Ann. Math. **60**, 545–556

Nash, J. (1956): The imbedding problem for Riemannian manifolds. Ann. Math. **63**, 20–63

Nash, J. (1966): Analyticity of the solutions of implicit function problems with analytic data. Ann. Math. **84**, 345–355

Zehnder, E. (1975): Generalized implicit function theorems with applications to some small divisor problems I. Comm. Pure Appl. Math. **28**, 91–140

On Scattering by Obstacles

Mitsuru Ikawa

Department of Mathematics, Faculty of Science, Osaka University
Toyonaka, Osaka 560, Japan

1. Introduction

Let n be an odd integer ≥ 3, and let $\mathcal{O}$ be an open bounded set in $\mathbb{R}^n$ with smooth boundary Γ. We assume that

$$\Omega = \mathbb{R}^n - \overline{\mathcal{O}} \quad \text{is connected}.$$

Consider the following acoustic problem

$$\begin{cases} \Box u = \dfrac{\partial^2 u}{\partial t^2} - \Delta u = 0 & \text{in } \Omega \times (-\infty, \infty) \\ u = 0 & \text{on } \Gamma \times (-\infty, \infty) \\ u(x,0) = f_1(x), \quad \dfrac{\partial u}{\partial t}(x,0) = f_2(x). \end{cases} \tag{1.1}$$

It is known that every solution $u(x,t)$ of (1.1) with finite energy approaches to a solution $u_0^+(x,t)(u_0^-(x,t))$ of the wave equation in the free space as $t \to \infty (t \to -\infty)$. The mapping from the initial data of $u_0^-(x,t)$ to those of $u_0^+(x,t)$ is called scattering operator, and the scattering matrix $\mathcal{S}(\sigma)$ $(\sigma \in \mathbb{R})$ is a representation of this mapping (for the definition, see for example Lax-Phillips [10, p. 170]). $\mathcal{S}(\sigma)$ is a unitary operator in $L^2(S^{n-1})$ for all $\sigma \in \mathbb{R}$ where $S^{n-1} = \{\omega \in \mathbb{R}^n; |\omega| = 1\}$, and is of the form

$$\mathcal{S}(\sigma) = I + \mathcal{K}(\sigma) \tag{1.2}$$

where I denotes the identity operator and $\mathcal{K}(\sigma)$ is an integral operator. The kernel $K(\omega, \theta; \sigma)$ of $\mathcal{K}(\sigma)$ is given in the following way:

Let $v_-(x, \omega, \sigma)$ be the solution of

$$\begin{aligned} &\Delta v + \sigma^2 v = 0 && \text{in } \Omega, \\ &v = -\exp(-i\sigma\omega) && \text{on } \Gamma, \end{aligned}$$

v satisfies the incoming radiation condition.

Then v_- has an asymptotic expansion

$$v_-(r\theta, \omega, \sigma) \sim \frac{e^{i\sigma r}}{r^{(n-1)/2}} s(\theta, \omega; \sigma), \quad \text{as } r \to \infty.$$

Proceedings of the International Congress
of Mathematicians, Kyoto, Japan, 1990

Then the kernel $K(\omega,\theta;\sigma)$ is given by

$$K(\omega,\theta;\sigma)=\left(\frac{\sigma}{2\pi i}\right)^{(n-1)/2}\overline{s(-\theta,\omega;\sigma)}\,.$$

Roughly speaking, the kernel $K(\omega,\theta;\sigma)$ represents the rate of the wave reflected by $\mathcal{O}$ in the direction θ for the incident plane wave of frequency σ propagating in the direction ω.

Concerning $\mathcal{S}(\sigma)$, the following fact is known:

$\mathcal{S}(\sigma)$ is the restriction to the real axis of an $\mathcal{L}(L^2(S^{n-1}))$-valued function $\mathcal{S}(z)$ that is analytic for $\operatorname{Im} z \le 0$ *and meromorphic in the whole complex plane* $\mathbb{C}$ ([10, p. 166]).

It is an intrinsic subject of scattering theory to consider relationships between the geometry of the obstacle and the analytic property of the scattering matrix. Concerning this subject, the following theorem is fundamental:

Theorem 5.6 of Chapter V of [10]. *The scattering matrix uniquely determines the scatterer.*

The above theorem shows that all the geometric informations of the obstacle are contained in the scattering matrix. Indeed, all the informations of $\mathcal{O}$ are in $\mathcal{S}(z)$, but how do we extract the geometric informations from the scattering matrix? This is one of the most important and interesting problems of scattering theory.

In this note we would like to consider scattering matrices in connection with the above problem. In Section 2 we present several results on scattering matrices. In Section 3 we propose a conjecture which was given originally by Lax and Phillips in [10]. In Section 4 the validity of this conjecture will be considered for obstacles consisting of several strictly convex bodies.

2. Several Results on Scattering Matrices

As to the geometry of obstacles, we introduce the following notion.

Definition 1. Suppose that $\mathcal{O}$ is contained in $\{x;|x|<\varrho\}$. We say that $\mathcal{O}$ is nontrapping if there exists $T>0$ such that every broken ray in Ω according to the geometric optics starting from a point in $\Omega(\varrho)=\Omega\cap\{x;|x|<\varrho\}$ necessarily goes out from $\Omega(\varrho)$ within time T. We say that $\mathcal{O}$ is trapping if, for any $T>0$, there is a broken ray staying in $\Omega(\varrho)$ for a period longer than T.

2.1 Asymptotic Behavior of the Scattering Phase

Since $\mathscr{K}(\sigma)$ is of trace class, $\det \mathscr{S}(\sigma)$ is well defined, and the unitarity of $\mathscr{S}(\sigma)$ implies that $|\det \mathscr{S}(\sigma)| = 1$. The scattering phase $s(\sigma)$ is defined by

$$s(\sigma) = -i \log \det \mathscr{S}(\sigma) \qquad \text{for } \sigma \in \mathbb{R}. \tag{2.1}$$

Melrose [13] proved that

$$s(\sigma) = c_n \mathrm{Vol}(\mathcal{O})\, \sigma^n + O(\sigma^{n-1}) \qquad \text{as } \sigma \to \pm\infty. \tag{2.2}$$

Analoguous to the Weyl formula which is an asymptotics of the distribution of eigenvalues of the interior problem, the volume of the obstacle can be got from the asymptotic behavior of the scattering phase. For nontrapping obstacles Petkov-Popov [17] obtained the full asymptotic expansion of $s(\sigma)$.

2.2 On the Poles of the Scattering Matrix

As mentioned in the Introduction, the scattering matrix is meromorphic in the whole complex plane. Thus, it is very natural to pose the question:

How does the geometry of $\mathcal{O}$ relate to the distribution of poles of $\mathscr{S}(z)$?

Purely Imaginary Poles. Lax and Phillips showed in [11] that there are an infinite number of purely imaginary poles.

For Nontrapping Obstacles. By combining the general result in [12] with the results on the propagation of singularities for the problem (1.1) due to [15], we have that, if $\mathcal{O}$ is nontrapping, there are positive constants a and b such that

$$\{z; \mathrm{Im}\, z \le a \log(|z|+1) + b\} \text{ is free from poles of } \mathscr{S}(z). \tag{2.3}$$

Obstacles Consisting of Two Convex Bodies. (i) The existence of non-purely imaginary poles was proved for the first time by Bardos-Guillot-Ralston [1]. They considered the case that

$$\mathcal{O} = \mathcal{O}_1 \cup \mathcal{O}_2, \quad \overline{\mathcal{O}_1} \cap \overline{\mathcal{O}_2} = \phi, \tag{2.4}$$

$$\mathcal{O}_1 \text{ and } \mathcal{O}_2 \text{ are strictly convex}, \tag{2.5}$$

and showed that, for any $\varepsilon > 0$, the logarithmic domain $\{z; \mathrm{Im}\, z \le \varepsilon \log |z|\}$ contains an infinite number of poles of $\mathscr{S}(z)$.

(ii) This result was improved by Ikawa [4] as follows: Let $a_j \in \Gamma_j = \partial \mathcal{O}_j$, $j = 1, 2$, be the points such that $|a_1 - a_2| = \mathrm{dis}(\mathcal{O}_1, \mathcal{O}_2) = d$. Then, $\mathscr{S}(z)$ has poles at approximately the points

$$\frac{\pi k}{d} + ic, \qquad k \in \mathbb{Z} \tag{2.6}$$

where c is a positive constant determined by d and the curvatures of Γ_j at a_j, $j = 1, 2$.

Later, Gérard [3] gave more precise descriptions of poles.

(iii) Ikawa [6] considered $\mathcal{O} \subset \mathbb{R}^3$ of the form (2.4) such that the principal curvatures $\kappa_{j,k}(x)$, $j = 1, 2$, of Γ_j at x vanish only at a_j of order $2l$, $l \geq 1$, that is, for $j = 1, 2$

$$C^{-1}|x - a_j|^{2l} \leq \kappa_{j,k}(x) \leq C|x - a_j|^{2l} \quad \text{for } x \in \Gamma_j,\ k = 1, 2 .$$

In this case there is a sequence of poles $\{z_j\}_{j=1}^{\infty}$ such that

$$\operatorname{Im} z_j \to 0 \quad \text{as } j \to \infty .$$

Poles for Convex Obstacles. For nontrapping obstacles, as (2.3) shows, all the poles of the scattering matrix are over a logarithmic curve. Bardos-Lebeau-Rauch [2] considered strictly convex obstacles with analytic boundary, and showed that, under a certain additional condition, there is a positive constant $c > 0$ such that $\{z; \operatorname{Im} z \leq c|z|^{1/3}\}$ contains only a finite number of poles and $\{z; \operatorname{Im} z \leq (c + \varepsilon)|z|^{1/3}\}$ contains an infinite number of poles for any $\varepsilon > 0$.

Upper Bound of Distribution of Poles. Melrose showed in [13] the following estimate:

$$\#\{z; \text{poles of } \mathscr{S}(z) \text{ such that } |z| \leq \lambda\} \leq C\lambda^n \quad \text{for all } \lambda > 0. \tag{2.7}$$

Remark. Zworski considered in [18, 19] the same problem for scattering by potentials, and showed that (2.7) holds also in this case. Moreover, he showed the estimate (2.7) is optimal in general.

3. Modified Lax-Phillips Conjecture

Consider the boundary value problem with parameter $\mu \in \mathbb{C}$

$$\begin{cases} (\triangle - \mu^2)w(x) = 0 & \text{in } \Omega \\ w(x) = g(x) & \text{on } \Gamma \end{cases} \tag{3.1}$$

for $g(x) \in C^\infty(\Gamma)$. It is well known that for $\operatorname{Re}\mu > 0$ (3.1) has a unique solution $w(x)$ in $H^2(\Omega)$. Denote by $R(\mu)$ the operator defined by

$$w(x) = (R(\mu)g(\cdot))(x) .$$

Then $R(\mu)$ is an $\mathscr{L}(L^2(\Gamma), L^2(\Omega))$-valued holomorphic function in $\operatorname{Re}\mu > 0$. By the regularity theorem for elliptic operators, $R(\mu)$ can be regarded as an operator in $\mathscr{L}(C^\infty(\Gamma), C^\infty(\overline{\Omega}))$. As an $\mathscr{L}(C^\infty(\Gamma), C^\infty(\overline{\Omega}))$-valued function $R(\mu)$ can be prolonged analytically to the whole complex plane as a meromorphic function. Concerning the poles of $\mathscr{S}(z)$ and $R(\mu)$ we have from Theorem 5.1 of Chapter V of [10]

$$z \text{ is a pole of } \mathscr{S}(z) \text{ if and only if } \mu = iz \text{ is a pole of } R(\mu) .$$

Let z be a pole of $\mathscr{S}(z)$. Then $\mu = iz$ is a pole of $R(\mu)$. The above Theorem 5.1 asserts also the existence of a non-trivial μ-outgoing solution $w(x; \mu)$ of (3.1) for

$g \equiv 0$. Therefore $u(x,t;\mu) = w(x;\mu)\,e^{\mu t}$ satisfies (1.1) for $f_1(x) = w(x;\mu)$, $f_2(x) = \mu w(x;\mu)$. Thus, if $\operatorname{Im} z = -\operatorname{Re}\mu$ is very small, $u(x,t;\mu)$ decays very slowly as $t \to \infty$. This suggests us that, if poles of $\mathscr{S}(z)$ appear near the real axis, we have probably solutions of (1.1) with finite energy decaying very slowly as $t \to \infty$. Thus, we suspect that the stronger solutions to (1.1) are trapped by obstacle $\mathcal{O}$, the nearer to the real axis the poles of $\mathscr{S}(z)$ will appear, and we suspect also that the stronger rays of geometric optics in Ω are trapped, the stronger solutions to (1.1) will be trapped by $\mathcal{O}$.

Thus, we would like to propose the following conjecture:

Modified Lax-Phillips Conjecture. *If $\mathcal{O}$ is trapping, there exists $\alpha > 0$ such that $\mathscr{S}(z)$ has an infinite number of poles in $\{z; \operatorname{Im} z \leq \alpha\}$.*

Recall that, for nontrapping obstacle $\mathcal{O}$, all the poles of $\mathscr{S}(z)$ are over a logarithmic curve $\operatorname{Im} z = a\log(|z|+1)+b$. Then, for any $\alpha > 0$ there are only a finite number of poles in $\{z; \operatorname{Im} z \leq \alpha\}$. Thus, if the above conjecture is true, the existence of such α becomes a characterization of trapping obstacles by means of distribution of poles of scattering matrices.

Hereafter, we say that MLPC (abbreviation of the Modified Lax-Phillips Conjecture) is valid for obstacle $\mathcal{O}$ when there is $\alpha > 0$ such that the scattering matrix $\mathscr{S}(z)$ corresponding to $\mathcal{O}$ has an infinite number of poles in $\{z; \operatorname{Im} z \leq \alpha\}$.

Even though it is more than 20 years since the original conjecture was given, the examples of obstacles for which the validity of MLPC is proved are few. As presented in Section 2, obstacles consisting of two convex bodies were known as these examples.

4. Obstacles Consisting of Several Strictly Convex Bodies

As mentioned in the previous section, MLPC is valid for $\mathcal{O}$ consisting of two strictly convex bodies. In this section we consider an extension of this result to obstacles consisting of several strictly convex bodies. Here we would like to mention about the geometrical difference between $\mathcal{O}$ consisting of two strictly convex bodies and $\mathcal{O}$ consisting of more than two bodies. For $\mathcal{O}$ consisting of two strictly convex bodies, there is only one primitive periodic rays in Ω. On the other hand, when $\mathcal{O}$ is consisted of three convex bodies for example, there are infinitely many primitive periodic rays in Ω in general. The infiniteness of the number of primitive periodic rays in Ω makes the problem difficult to treat. In this case we have to control the complexity caused by the infiniteness of the number of primitive periodic rays in Ω. It seems to us that the asymptotic behavior of periodic rays is closely related to the ergodic property of rays in Ω. Actually we can control the complexity of periodic rays only for obstacles consisting of several small balls.

Now we shall state our theorem. Let P_j, $j = 1,2,\dots,L$ $(L \geq 3)$, be points in $\mathbb{R}^3$. For $\varepsilon > 0$ we set

$$\mathcal{O}_\varepsilon = \cup_{j=1}^{L} \mathcal{O}_{j,\varepsilon}, \quad \mathcal{O}_{j,\varepsilon} = \{x; |x - P_j| < \varepsilon\}.$$

Theorem 4.1. *Assume that P_j, $j = 1, 2, \dots, L\,(L \geq 3)$, satisfy the condition*

$$\textit{any triple of } P_j\textit{'s does not lie on a straight line.} \tag{A}$$

Then, there is a positive constant ε_0 such that, for all $0 < \varepsilon \leq \varepsilon_0$, MLPC is valid for $\mathcal{O}_\varepsilon$.

The proof of Theorem 4.1 will be devided into several steps.

4.2 General Theorem

We present a theorem given in [8]. Let $\mathcal{O}_j$, $j = 1, 2, \dots, L\,(L \geq 3)$, be bounded open sets $\mathbb{R}^3$ with smooth boundary Γ_j satisfying

$$\text{every } \mathcal{O}_j \text{ is strictly convex}, \tag{H.1}$$

$$\begin{gathered}\text{for every } \{j_1, j_2, j_3\} \in \{1, 2, \dots, L\}^3 \text{ such that } j_l \neq j_{l'} \text{ if } l \neq l', \\ (\text{convex hull of } \overline{\mathcal{O}_{j_1}} \text{ and } \overline{\mathcal{O}_{j_2}}) \cap \overline{\mathcal{O}_{j_3}} = \phi.\end{gathered} \tag{H.2}$$

We set

$$\mathcal{O} = \cup_{j=1}^{L} \mathcal{O}_j, \quad \Gamma_j = \partial \mathcal{O}_j. \tag{4.1}$$

Denote by γ an oriented periodic ray in $\Omega = \mathbb{R}^3 - \overline{\mathcal{O}}$, and we shall use the following notations:

d_γ: the length of γ,
T_γ: the primitive period of γ,
i_γ: the number of the reflecting points of γ,
P_γ: the linearized Poincaré map of γ.

We define a function $F_D(s)$ ($s \in \mathbb{C}$) by

$$F_D(s) = \sum_\gamma \frac{(-1)^{i_\gamma} T_\gamma}{|I - P_\gamma|^{1/2}} e^{-sd_\gamma} \tag{4.2}$$

where the summation is taken over all the oriented periodic rays in Ω and $|I - P_\gamma|$ denotes the determinant of $I - P_\gamma$.

Concerning the periodic rays in Ω we have

$$\#\{\gamma; \text{periodic ray in } \Omega \text{ such that } d_\gamma < r\} < e^{a_0 r} \tag{4.3}$$

and

$$|I - P_\gamma| \geq e^{2a_1 d_\gamma}, \tag{4.4}$$

where a_0 and a_1 are positive constants depending on $\mathcal{O}$. The estimates (4.3) and (4.4) imply that the right hand side of (4.2) converges absolutely in $\{s \in \mathbb{C};\ \mathrm{Re}\, s > a_0 - a_1\}$. Thus $F_D(s)$ is well defined in $\{s \in \mathbb{C};\ \mathrm{Re}\, s > a_0 - a_1\}$, and holomorphic in this domain.

Now we have

Theorem 4.2. *Let $\mathcal{O}$ be an obstacle given by* (4.1) *satisfying* (H.1) *and* (H.2). *If $F_D(s)$ cannot be prolonged analytically to an entire function, then $MLPC$ is valid for $\mathcal{O}$.*

The proof is based on the trace formula due to Bardos-Guillot-Ralston [1]. The essential part of the proof is given in [8, Section 2].

4.3 Zeta Functions of Symbolic Flows

In order to consider singularities of $F_D(s)$ we shall use the fact that $F_D(s)$ has a close relationship to a zeta function of a symbolic flow. Denote by v_0 the abscissa of the convergence of the right hand side of (4.2), that is,

$$v_0 = \inf\{v;\ \text{the right hand side of (4.2) converges absolutely for } \operatorname{Re} s > v\}.$$

We shall show the following

Proposition 4.3. *Let $\zeta(s)$ be the zeta function defined by (4.6). Then, we have that*

$$F_D(s) - (-\frac{d}{ds}\log\zeta(s)) \quad \textit{is holomorphic in} \quad \operatorname{Re} s > v_0 - a_2$$

where a_2 is a positive constant depending on $\mathcal{O}$.

Proposition 4.3 implies that singularities in $\operatorname{Re} s > v_0 - a_2$ of $F_D(s)$ coincide with those of $\zeta(s)$. Therefore, in order to apply Theorem 4.2 it suffices to show the existence of singularities of $\zeta(s)$ in $\operatorname{Re} s > v_0 - a_2$.

We introduce some notations of symbolic flows. Let $A = (A(i,j))_{i,j=1,2,\cdots,L}$ be the $L \times L$ matrix defined by $A(i,j) = 1 (i \neq j)$ and $A(j,j) = 0$. Following Parry-Pollicott [16] we set

$$\Sigma_A = \{\xi = (\cdots,\xi_{-1},\xi_0,\xi_1,\cdots);\ \xi_j \in \{1,2,\cdots,L\} \text{ and } A(\xi_j,\xi_{j+1}) = 1 \text{ for all } j\}.$$

Denote by σ_A the shift transformation defined by

$$(\sigma_A\xi)_j = \xi_{j+1}\,.$$

We consider relationships between Σ_A and bounded broken rays in the outside of $\mathcal{O}$. Let $(\cdots,l_{-1},l_0,l_1,\cdots)$ be the reflection order of a bounded broken ray in Ω. Then, as was shown in [5], it is an element of Σ_A. Conversely, for each element of $\xi \in \Sigma_A$ there exists a unique broken ray with the reflection order ξ. Note that a periodic ray in Ω corresponds to a periodic element $\xi \in \Sigma_A$, that is, $\sigma_A^n\xi = \xi$ for some n.

We shall define real valued functions f and g on Σ_A. We set

$$f(\xi) = |X_0X_1|$$

where X_j denote the j-th reflection point of the broken ray corresponding to ξ. Suppose that $\xi \in \Sigma_A$ satisfies $\sigma_A^n\xi = \xi$ for some n. Set $\mathbf{i} = (\xi_0,\ldots,\xi_{n-1})$, and let

$\varphi_{i,0}^{\infty}$ be the phase function defined in [5, Section 5]. Denote by $\lambda_1(\xi)$ and $\lambda_2(\xi)$ the eigenvalues of P_γ greater than 1, and by $\kappa_l(\xi), l = 1, 2$, the principal curvatures at X_0 of the wave front of the phase function $\varphi_{i,0}^{\infty}$. Then we have

$$\lambda_1(\xi)\lambda_2(\xi) = \prod_{j=1}^{n}(1 + f(\sigma_A^j\xi)\kappa_1(\sigma_A^j\xi))(1 + f(\sigma_A^j\xi)\kappa_2(\sigma_A^j\xi)). \tag{4.5}$$

Define $g(\xi)$ for an periodic element ξ by

$$g(\xi) = -\frac{1}{2}\log(1 + f(\xi)\kappa_1(\xi))(1 + f(\xi)\kappa_2(\xi)).$$

By using the fact that the periodic elements are dense in Σ_A, we can extend $g(\xi)$ for all $\xi \in \Sigma_A$ by the continuity.

Define $\zeta(s)$ by

$$\zeta(s) = \exp\left(\sum_{n=1}^{\infty}\frac{1}{n}\sum_{\sigma_A^n\xi=\xi}\exp S_n(-sf(\xi) + g(\xi) + \pi i)\right) \tag{4.6}$$

where we set

$$S_n(-sf(\xi) + g(\xi) + \pi i) = \sum_{k=0}^{n-1}(-sf(\sigma_A^k\xi) + g(\sigma_A^k\xi) + \pi i).$$

It is easy to see that the right hand side of (4.6) converges for $\mathrm{Re}\, s$ large. We call $\zeta(s)$ a zeta function of the symbolic flow (Σ_A, σ_A). Then, we have the following relation for $\mathrm{Re}\, s > v_0$:

$$\begin{aligned} &F_D(s) - \left(-\frac{d}{ds}\log\zeta(s)\right) \\ &= \sum_\gamma T_\gamma(-1)^{i_\gamma}\{|I - P_\gamma|^{-1/2} - (\lambda_1\lambda_2)^{-1/2}\}\exp(-sd_\gamma). \end{aligned} \tag{4.7}$$

Note that the following estimate holds:

$$\big|\, |I - P_\gamma|^{-1/2} - (\lambda_1\lambda_2)^{-1/2} \big| \le C(\lambda_1\lambda_2)^{-1/2}\, e^{-a_2 d_\gamma} \tag{4.8}$$

where a_2 is a positive constant depending on $\mathcal{O}$. Then, we have immediately Proposition 4.3 by substituting (4.8) in the right hand side of (4.7).

4.4 Singular Perturbations of Symbolic Flows

It is likewise difficult in general to show the existence of singularities of the zeta function $\zeta(s)$, because there is no $s \in \mathbb{C}$ such that $-sf(\xi) + g(\xi) + \pi i$ is real for all $\xi \in \Sigma_A$. Namely, this fact makes impossible to apply the Ruelle-Perron-Frobenious theorem to get the existence of poles of $\zeta(s)$.

When the bodies composing $\mathcal{O}$ are small in comparison with the distances between each other, $\zeta(s)$ can be approximated by a zeta function of a graph, which is much easier to treat. To this end, we use theorems on singular perturbations of symbolic flows proved in [7, 9]. Denote by ζ_ε, f_ε, g_ε the ζ, f, g attached to $\mathcal{O}_\varepsilon$ respectively. If we set

$$f_0(\xi) = |P_{\xi_0}P_{\xi_1}|, \quad \tilde{g}_0(\xi) = \frac{1}{2}\log\left(\frac{1}{4}\cos\frac{\Theta(\xi)}{2}\right)$$

where $\Theta(\xi)$ denotes the angle $P_{\xi_{-1}}P_{\xi_0}P_{\xi_1}$, it holds that

$$|\log\varepsilon|\,|||f_\varepsilon - f_0|||_\theta, \quad |||g_\varepsilon - (\log\varepsilon + \tilde{g}_0)|||_\theta \to 0 \quad \text{as } \varepsilon \to 0$$

for every fixed $\theta > 0$ (For the definition of the norm $|||\cdot|||_\theta$, see [16]). By using the above relations we have the following expression of $\zeta_\varepsilon(s)$:

$$\zeta_\varepsilon(s) = Z_\varepsilon(s - (\log\varepsilon + \pi i)/d_{\max}).$$

Here $d_{\max} = \max_{j\neq k}|P_jP_k|$ and $Z_\varepsilon(\tilde{s})$ is a function defined by

$$Z_\varepsilon(\tilde{s}) = \exp\left(\sum_{n=1}^{\infty}\frac{1}{n}\sum_{\sigma_A^n\xi=\xi}\exp S_n r_\varepsilon(\xi,\tilde{s})\right),$$
$$r_\varepsilon(\xi,\tilde{s}) = -\tilde{s}f_\varepsilon(\xi) + h_\varepsilon(\xi) + k(\xi)\log\varepsilon,$$
$$k(\xi) = 1 - f_0(\xi)/d_{\max},$$

where $h_\varepsilon(\varepsilon \geq 0)$ satisfies

$$|||h_\varepsilon - h_0|||_\theta \to 0 \quad \text{as } \varepsilon \to 0,$$
$$h_0(\xi) \text{ is real if } k(\xi) = 0.$$

Then Theorem 1 of [9] guarantees that $Z_\varepsilon(\tilde{s})$ has a pole near $s_0 \in \mathbb{R}$ when ε is small.[1] Thus, we have the existence of pole of $\zeta_\varepsilon(s)$. Thus Theorem 4.1 is proved.

References

1. Bardos C., Guillot J.C., Ralston J.: La relation de Poisson pour l'équation des ondes dans un ouvert non borné. Application à la théorie de la diffusion. Comm. Partial Diff. Equations **7** (1982) 905–958
2. Bardos C., Lebeau G., Rauch J.: Scattering frequencies and Gevrey 3 singularities. Invent. math. **90** (1987) 77–114
3. Gérard C.: Asymptotique des poles de la matrice de scattering pour deux obstacles strictement convexes. Bull. Soc. Math. France **116** (31) (1989)
4. Ikawa M.: On the poles of the scattering matrix for two strictly convex obstacles. J. Math. Kyoto Univ. **23** (1983) 127–194
5. Ikawa M.: Decay of solutions of the wave equation in the exterior of several convex bodies. Ann. Inst. Fourier **38** (1988) 113–146
6. Ikawa M.: Trapping obstacles with a sequence of poles of the scattering matrix converging to the real axis. Osaka J. Math. **22** (1985) 657–689

[1] In [7] the existence of poles is proved under some additional assumptions.

7. Ikawa M.: Singular perturbation of symbolic flows and poles of the zeta functions. Osaka J. Math. **27** (1990) 281–300
8. Ikawa M.: On the distribution of poles of the scattering matrix for several convex bodies. Proc. of the Conference in Honor of Prof. T. Kato "Functional Analytic Methods for Partial Differential Equations". Lecture Notes in Mathematics, vol. 1450. Springer, Berlin Heidelberg New York 1990, pp. 210–225
9. Ikawa M.: On the existence of poles of zeta functions for certain symbolic dynamics. In preparation
10. Lax P.D., Phillips R.S.: Scattering theory. Revised edition. Academic Press, New York 1989
11. Lax P.D., Phillips R.S.: Decaying modes for the wave equation in the exterior of an obstacle. Comm. Pure Appl. Math. **22** (1969) 737–787
12. Lax P.D., Phillips R.S.: A logarithmic bound on the location of the poles of the scattering operator. Arch. Rat. Mech. **40** (1971) 268–280
13. Melrose R.: Polynomial bound on the distribution of poles in scattering by an obstacle. Journées Equations aux Dérivées Partielles, St. Jean de Monts, 1984
14. Melrose R.: Weyl asymptotics for the phase in obstacle scattering. Comm. Partial Diff. Equations **13** (1988) 1431–1439
15. Melrose R., Sjöstrand J.: Singularities of boundary problems, I and II. Comm. Pure Appl. Math. **31** (1978) 593–617; **35** (1982) 129–168
16. Parry W., Pollicott M.: An analogue of the prime number theorem for closed orbits of Axiom A flows. Ann. Math. **118** (1983) 573-591
17. Petkov V., Popov G.: Asymptotic behavior of the scattering phase for nontrapping obstacles. Ann. Inst. Fourier **32** (1982) 111–149
18. Zworski M.; Sharp polynomial bounds on the number of scattering poles of radial potentials. J. Funct. Anal. **82** (1989) 370–403
19. Zworski M.: Sharp polynomial bounds on the number of scattering poles. Duke Math. J. **59** (1989) 311–323

Interaction des Singularités Faibles Pour les Équations d'Ondes Semi-linéaires

Gilles Lebeau

Université de Paris-Sud, Centre d'Orsay, Département de Mathématiques, Bâtiment 425
F-91405 Orsay Cedex, France

I. Introduction

Soit $\Box = \partial_t^2 - \Delta_x$ l'opérateur des ondes, où $t \in \mathbb{R}$, $x \in \mathbb{R}^d$, et Ω un ouvert de $\mathbb{R}^{1+d}$ qui est un domaine de détermination pour $\omega = \Omega \cap \{t = 0\}$.

Soit $u(t, x)$ une fonction réelle continue sur Ω appartenant localement à l'espace

$$C^0(\mathbb{R}_t, H^{s+1}(\mathbb{R}^d)) \cap C^1(\mathbb{R}_t, H^s(\mathbb{R}^d))$$

où H^s est l'espace de Sobolev usuel et vérifiant dans Ω l'équation des ondes semi-linéaires

$$(1)\qquad \begin{cases} \Box u = F(t, x, u, \nabla u) & \nabla u = (\partial_t u, \nabla_x u) \\ u\Big|_{t=0} = u_0 \in H^{s+1}_{\mathrm{loc}}(\omega) & \dfrac{\partial u}{\partial t}\Big|_{t=0} = u_1 \in H^s_{\mathrm{loc}}(\omega) \end{cases}$$

où F est une fonction C^∞ de ses arguments et où $s > d/2$ (ou $s > (d/2) - 1$ si F ne dépend pas de ∇u).

On s'intéresse à déterminer les singularités de la solution u de (1) dans $\Omega_+ = \Omega \cap \{t > 0\}$ en fonction de ses données de Cauchy u_0 et u_1 [ou encore en fonction de $u|_{\Omega_-}$, $\Omega_- = \Omega \cap \{t < 0\}$]. Plus précisément si $p \in T^*\Omega_+ \setminus \Omega_+$, on cherche à déterminer pour quelles valeurs de σ a-t-on

$$(2)\qquad u \in H^\sigma_p \quad \text{(espace de Sobolev microlocal)}.$$

Depuis les travaux de pionnier de J.-M. Bony [8-13] ce type de problème a été intensivement étudié, et son contenu s'est avéré très riche.

Rappelons les résultats généraux sur les solutions du (1).

(3) **Théorème 1** (J.-M. Bony [13]). *Soit u solution de* (1) *vérifiant $u \in H^{s_0}_{\mathrm{loc}}(\Omega)$, $s_0 = (1 + d)/2 + 1 + \varrho$, $\varrho > 0$. Si p est non caractéristique, on a $u \in H_p^{s_0+\varrho+1}$. Si p est caractéristique et si la bicaractéristique de $\Box$ passant par p est issue de $q = (x_0, \xi_0) \in T^*\omega \setminus \omega$, on a $u \in H^\sigma_p$ pour $\sigma \le s_0 + \varrho$ si les données vérifient $u_j \in H_q^{\sigma-j}$.*

En d'autres termes, il n'y a pas d'effets non linéaires juqu'à la régularité $s_0 + \varrho$.

Proceedings of the International Congress
of Mathematicians, Kyoto, Japan, 1990

(4) **Théorème 2** (M. Beals [4], J.-Y. Chemin [17]). *Sous les mêmes hypothèses, si γ est une bicaractéristique de $\Box$, p_1, p_2 deux points de γ, on a $u \in H^\sigma_{p_1}$ ssi $u \in H^\sigma_{p_2}$ pour $\sigma \leq 3s_0 - (1+d) - 2 = s_0 + 2\varrho$.*

Les Théorèmes 1 et 2 ont leur analogue pour les équations totalement non linéaires d'ordre quelconque à linéarisé strictement hyperbolique (voir [8, 17]).

Dans ce cadre, le Théorème 2 exprime que jusqu'à la régularité $s_0 + 2\varrho$ les effets non linéaires se réduisent à une opération de somme fibre à fibre sur le front d'onde.

Pour les régularités plus élevées, il n'y a pas de théorème de nature purement géométrique sur le front d'onde. Dans [4], M. Beals a construit une solution $u \in H^s$ de $\Box u + \beta u^3 = 0$ ($\beta \in C^\infty$), dont les données de Cauchy sont C^∞ hors de l'origine et $u \notin H^{3s-d+2+\varepsilon}$ à l'intérieur du cône d'onde, avec $d \geq 2$. [Le cas de la dimension 1 d'espace a été élucidé par J. Rauch et M. Reed [36], et J.-Y. Chemin [16] qui donnent un résultat complet jusqu'à $\sigma = \infty$.]

Il est toutefois possible d'obtenir des résultats de localisation du front d'onde de u jusqu'à $\sigma = \infty$ en faisant des hypothèses de conormalité sur les données de Cauchy ou sur la solution u dans le passé Ω_-. Dans ce cadre, le théorème d'interaction de trois ondes progressives a été obtenu simultanément par J.-M. Bony [11] et R. Melrose et N. Ritter [32] et [33], puis généralisé et amélioré par J.-Y. Chemin [18] et Sa-Baretto [41].

La difficulté principale de ce type de problème provient du fait que la géométrie qui porte les singularités de la solution est elle-même singulière. Les techniques utilisées font intervenir soit des microlocalisations d'ordre supérieur (J.-M. Bony) soit des espaces de distributions conormales définis par éclatement (R. Melrose).

On se propose ici de décrire certains résultats qu'on peut obtenir en faisant des hypothèses d'analyticité sur la géométrie qui porte les singularités. Dans cette direction, on a:

(5) **Théorème 3** [26] ($d = 3$). *Soit $u \in H^s(\Omega)$, $s > 2$ vérifiant* (1) *où $F(t, x, u)$ est polynomial en u et où les données u_0, u_1 sont des distributions intégrales de Fourier C^∞ sur Λ, lagrangienne analytique réelle lisse de $T^*\omega$. Pour tout réel σ, il existe un ensemble sous-analytique, homogène isotrope L_σ de $T^*\Omega$ tel que $WF^\sigma(u) \subset L^\sigma$. En particulier, pour tout k, u est de classe C^k sur un ouvert dense de Ω.*

La preuve du Théorème 3, qui utilise le théorème de désingularisation de Hironaka, ne fournit pas d'estimation sur L^σ. Les résultats qui suivent fournissent pour les solutions de (1) une majoration de L^σ et comme corollaire, la preuve d'une conjecture de J.-M. Bony sur le pincement d'une onde progressive en dimension 2 d'espace. (Démontré dans [27] pour F polynômiale en u.)

(6) **Théorème 4** ($d = 2$). *Soit u solution de* (1) *avec u_0 et u_1 conormales classiques sur $\Lambda = T^*_V\omega$, où V est une courbe analytique lisse possédant un minimum non dégénéré de son rayon de courbure en $A \in \mathbb{R}^2$. Soit S la queue d'aronde issue de V et Q_+ le demi-cône d'onde d'avenir issu du point de pincement B de S. Alors près de B, on a $WF(u) \subset T^*_{S \cup Q_+}$.*

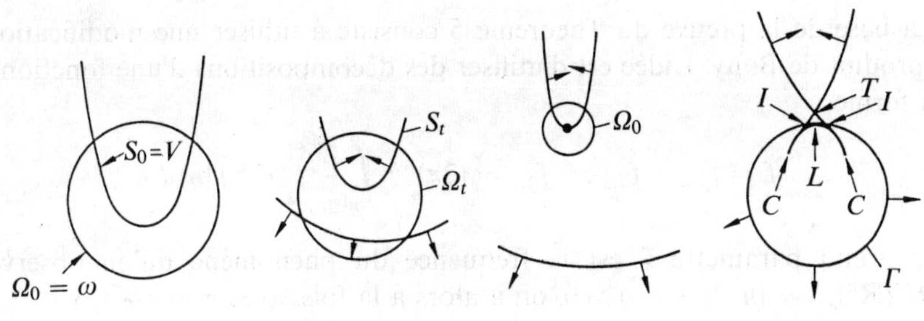

$t = -t_0$ $\quad$ $|t - t_0|$ petit $\quad$ $t = 0$ $\quad$ $t > 0$

Fig. 1. Pincement d'une onde progressive

Ce résultat a été amélioré par J.-M. Delort [23] qui prouve en particulier que u est conormale aux points lisses de $T^*_{S \cup Q_+}$.

On n'abordera pas ici les travaux relatifs aux équations totalement non linéaires (voir Alinhac [1, 2, 3]), ni les problèmes aux limites ([6, 7, 21, 30, 43, 44, 45, 46]). Signalons à cet égard que R. Melrose, Zworski, Sa-Baretto ont obtenu le théorème de diffraction non-linéaire d'une onde conormale par un obstacle convexe.

II. Calcul Multilinéaire

La première étape pour obtenir par exemple le Théorème 4 est de majorer le front d'onde de u par le front d'onde de distributions construites explicitement à partir des données u_0, u_1, et qui décrivent les phénomènes de propagation-interaction.

On appelle diagramme D la donnée d'un ensemble fini $I = \{1, \ldots, N\}$ muni d'une partition et d'une application $f : I \to I \cup \{0\}$ telle que $f(I_k) \subset I_{k-1}$, $f(I_1) = \{0\}$. On définit le degré de D par $\deg D = \operatorname{card} J$, $J = \{i \in I, f^{-1}(i) = \phi\}$. Soit $e_+(z)$, $(z = (t, x))$ la solution élémentaire de $\Box$ à support dans l'avenir. Pour $(z_0, z_1, \ldots, z_N) \in \mathbb{R}^{(1+d)(N+1)}$ on pose:

$$\text{(8)} \qquad \begin{cases} [D] = \prod_{i \notin J} \nabla^{\beta_i} e_+(z_{f(i)} - z_i) \prod_{i \in J} e_+(z_{f(i)} - z_i) \\ \{D\} = \prod_{i \in J} (v_i^0(x_i)\, \delta'_{t_i=0} + v_i^1(x_i)\, \delta_{t_i=0}) \, . \end{cases}$$

Ici β_i vérifie $|\beta_i| \leq 1$, $v_i^0 \in \operatorname{vect}\{u_1, \nabla^\beta u_0, |\beta| \leq 1\}$ et $v_i^1 \in \operatorname{vect}\{\nabla^\beta u_1, \nabla^\gamma u_0, |\beta| \leq 1, |\gamma| \leq 2\}$, et on omet dans la notation la dépendance en β_i, v_i^0. Avec les techniques de [27, 29] on prouve:

(9) **Théorème 5.** *Soit u vérifiant* (1) *avec $s = (d/2) + \varrho$, $\varrho > 0$. Pour $M \geq 1$ et $\sigma \in [s + 1 + (M-1)\varrho, s + 1 + M\varrho[$ on a:*

$$\text{(10)} \qquad \begin{aligned} & WF^\sigma(u)\big|_{t>0} \subset \\ & \Big\{(z_0, \zeta_0)\, ; \text{il existe un diagramme } D \text{ avec } \deg(D) \leq M \\ & \quad \text{et } (z_0, z_1, \ldots, z_N, \zeta_0, 0, \ldots, 0) \in WF\big[[D] \cdot \{D\}\big]\Big\} . \end{aligned}$$

La base de la preuve du Théorème 5 consiste à utiliser une modification du paraproduit de Bony. L'idée est d'utiliser des décompositions d'une fonction $f(y)$ de la forme

$$f = f_{1,\lambda} + f_{2,\lambda}\,; \quad f_{1,\lambda} = (2\pi)^{-n} \int_{|\eta| \leq c_0\,\lambda^{\delta}} e^{iy\eta} \widehat{f}(\eta)\, d\eta \tag{11}$$

où le grand paramètre λ est la fréquence du phénomène qu'on observe. Si $f \in H^s(\mathbb{R}^n)$, $s = (n/2) + \varrho$, $\varrho > 0$, on a alors à la fois, avec $\sigma = s - \nu$, $\nu \in]0, \varrho[$

$$\|\partial_x^\alpha f_1\|_s \leq C_\alpha \lambda^{|\alpha|\,\delta} \qquad \|f_2\|_\sigma \leq C\lambda^{-\nu\delta}\,. \tag{12}$$

(Le paraproduit correspondrait à $\delta = 1$ et C_0 petit.)

Pour une distribution $f(y, \lambda)$ dépendant d'un paramètre $\lambda \geq 1$, telle que $|\widehat{\varphi f}(\eta, \lambda)| \leq$ polynôme(η, λ) pour tout $\varphi \in C_0^\infty$, on définit le λ-WF de f par

$$(13)\qquad \begin{aligned} &(y_0, \eta_0) \notin \lambda\text{-}WF(f) \Leftrightarrow \exists\, \varphi \text{ localisant près de } y_0\,, \\ &V \text{ voisinage de } \eta_0 \text{ tels que } \int_{\eta \in \lambda V_0} |\widehat{\varphi f}(\eta, \lambda)|^2 d\eta \in \mathcal{O}(\lambda^{-\infty})\,. \end{aligned}$$

Si $C_p = \{2^{p-1} \leq |\eta| < 2^{p+1}\}$, et $\mu \in \mathbb{R}$ on pose

$$a_p^2(f) = \int_{C_p} |\widehat{f}(\eta, 2^p)|^2 d\eta \qquad |f|_\mu^2 = \sum_{p \geq 0} 2^{2p\mu} a_p^2(f)\,. \tag{14}$$

Alors si $t > n/2$ et $a(x, \lambda)$ vérifie pour un $\delta \in]0, 1[$ $\|\partial_x^\alpha a\|_t \leq C_\alpha \lambda^{|\alpha|\delta}$, on a λ-$WF(af) \subset \lambda$-$WF(f)$ et pour $v(x) \in H^\mu$, $|av|_\mu \leq \text{cte}\, \|v\|_\mu$ pour tout μ, et on a:

(15) **Lemme 1.** *Supposons donné pour tout λ une décomposition $g(y) = g_1(y, \lambda) + g_2(y, \lambda)$ avec $(y_0, t\eta_0) \notin \lambda$-$WF(g_1)$, $\forall t > 0$ et $|g_2|_\mu < \infty$. Alors $g \in H^\mu_{(y_0, \eta_0)}$.*

D'où on déduit par exemple par la formule de Taylor en choisissant $\delta = 1 - 0$ et $\sigma = (n/2) + 0$ dans (12).

(16) **Proposition 1** [29]. *Soit $u \in H^s(\mathbb{R}^n)$, $s = (n/2) + \varrho$, $\varrho > 0$ et $F \in C^\infty$. Pour $N \geq 1$ et $\mu \in [(d/2) + N\varrho, d/2 + (N+1)\varrho[$ on a*

$$WF^{(\mu)}(F(u)) \subset \bigcup_{j=1}^{N} WF^\mu(u^j)\,.$$

III. Deuxième Microlocalisation

La difficulté à pouvoir tirer de l'information à partir du Théorème 5 est que le produit $[D] \cdot \{D\}$ est caractéristique, c'est-à-dire le front d'onde du produit tensoriel $[D] \otimes \{D\}$ rencontre le fibré conormal à la diagonale. C'est à ce stade qu'intervient la deuxième microlocalisation. C'est M. Kashiwara qui, dans les années 70, a l'idée d'introduire les 2-microfonctions associées à une sous-variété involutive du fibré cotangent. Dans ce cadre, le calcul symbolique des opérateurs 2-microdifférentiels a été développé par Y. Laurent [25]. Toujours dans la théorie analytique, mais en utilisant ses propres outils, plutôt que la théorie cohomologique de Kashiwara, les définitions de seconde microlocalisation et de microlocalisation d'ordre supérieur dans le cas lagrangien sont dues à J. Sjöstrand [42]. Un peut plus tard, et pour traiter le problème d'interaction non linéaire de trois ondes progressives, J-M. Bony introduit la seconde microlocalisation C^∞ sur une variété lagrangienne. Ce type de théorie a par ailleurs eu d'autres extensions: théorie isotrope [28], microlocalisations C^∞ d'ordre supérieur associées à des métriques de J.-M. Bony et N. Lerner [15], deuxième microlocalisation simultanée de J.-M. Delort [22]. Le champ d'application de la seconde microlocalisation ne se limite d'ailleurs pas à la compréhension des phénomènes non-linéaires: cet outil s'est révélé très performant pour l'étude des phénomènes de diffraction d'ondes linéaires.

On se limitera ici au cas de la deuxième microlocalisation analytique à croissance sur la lagrangienne T_Δ^*, où Δ est une sous-variété de $\mathbb{R}^n$, en théorie de J. Sjöstrand.

Soit $x = (x', x'') \in \mathbb{R}^n$, $x' \in \mathbb{R}^{n'}$, $x'' \in \mathbb{R}^{n''}$, $n' + n'' = n$ et Δ la sous-variété d'équation $x'' = 0$. On note $(x, \xi) = (x', x'', \xi', \xi'')$ les points du fibré cotangent $T^*\mathbb{R}^n$. Le fibré conormal à Δ, a pour équation $T_\Delta^* = \{x'' = 0, \xi' = 0\}$. On note $(x', \xi'', x'^*, \xi''^*)$ les points du fibré cotangent $T^*(T_\Delta^*)$. Le fibré cotangent à Δ, $T^*\Delta$ s'identifie à un sous-espace de $T^*(T_\Delta^*)$ par l'injection

$$T^*\Delta \ni (x', x'^*) \mapsto (x', 0; x'^*, 0) \in T^*(T_\Delta^*)\,. \tag{17}$$

Soit f une distribution à support compact dans $\mathbb{R}^n$. Pour $w \in \mathbb{C}^n$, $\lambda \in [1, \infty[$, $\mu \in]0, \mu_0]$, $\mu_0 < 1$ on pose

$$T^2 f(w, \lambda, \mu) = \int_{\mathbb{R}^n} e^{-(\lambda/2)\bar{\mu}^2 (w-y)^2}\, T^1 f(y, \lambda)\, dy\,, \quad \bar{\mu}^2 = \mu^2/1 - \mu^2 \tag{18}$$

où $T^1 f(y, \lambda)$ est défini pour $y \in \mathbb{C}^n$, $\lambda \in [1, \infty[$ par

$$T^1 f(y, \lambda) = \int e^{-(\lambda/2)\, y''^2 - \lambda/2 (y'-x')^2 - \lambda/2 (iy''-x'')^2} f(x)\, dx\,. \tag{19}$$

La transformation $f \mapsto T^1 f$ est une transformation de Fourier-Bros-Iagolnitzer usuelle [42] qui envoie le complexifié $(T_\Delta^*)^{\mathbb{C}}$ sur la section nulle. Les transformées $T^1 f$ et $T^2 f$ vérifient des estimations uniformes d'espaces de Sjöstrand

$$\begin{aligned} &|T^1 f(y, \lambda)| \le \mathrm{cte}\, \lambda^A e^{\lambda/2 (\mathrm{Im}\, y)^2}\,; \\ &|T^2 f(w, \lambda, \mu)| \le \mathrm{cte}\, \lambda^A \Big(\frac{2\pi}{\lambda \bar{\mu}^2}\Big)^{n/2} e^{(\lambda/2)\, \mu^2 (\mathrm{Im}\, w)^2}\,. \end{aligned} \tag{20}$$

Pour $\alpha = (x', \xi''; x'^*, \xi''^*)$ on pose

$$\chi(\alpha) = (w', w'') \quad w' = x' - i x'^*\,, \ \ w'' = -\xi'' + i \xi''^*\,. \tag{21}$$

(22) **Définition 1.** Soit $\alpha_0 \in T^*(T^*_\Delta)$ et $f \in \mathscr{E}'(\mathbb{R}^n)$. On dit que α_0 n'appartient pas au deuxième micro-support à croissance de f le long de T^*_Δ, et on note $\alpha_0 \notin WF^2_{T^*_\Delta}(f)$ ssi il existe W voisinage de $w_0 = \chi(\alpha_0)$, A, B, C, $\mu_0 > 0$ tels que

$$(23)\qquad \begin{aligned} &\forall \mu \in]0,\mu_0]\,,\ \forall w \in W\,,\ \forall \lambda, \lambda\mu^2 \geq 1 \\ &\Rightarrow |T^2 f(w,\lambda,\mu)| \leq A\lambda^B e^{\lambda(\mu^2/2)[(\operatorname{Im} w)^2 - c]}\,. \end{aligned}$$

Alors $WF^2_{T^*_\Delta}(f)$ est un fermé homogène de $T^*(T^*_\Delta)$ et si P est un opérateur différentiel sur $\mathbb{R}^n$, on a $WF^2_{T^*_\Delta}(Pf) \subset WF^2_{T^*_\Delta}(f)$.

(24) **Proposition 2** [27]. *Soit $f \in \mathscr{E}'(\mathbb{R}^n)$, dont la transformée de Fourier vérifie*

$$(25)\qquad \exists M, \delta > 0 \ t.q \quad \int (1+|\xi''|)^\delta |\widehat{f}(\xi',\xi'')|\, d\xi'' \leq cte(1+|\xi'|)^M\,.$$

La trace $f|_\Delta$ est bien définie et vérifie

$$(26)\qquad SS(f|_\Delta) \subset WF^2_{T^*_\Delta}(f) \cap T^*\Delta\,.$$

Soit à présent u solution de (1), telle que ses données de Cauchy u_0 et u_1 soient conormales classiques sur $\Lambda = T^*_V(\mathbb{R}^d)$ où V est une hypersurface analytique réelle. Si D est un diagramme, on définit $\Lambda_{[D]}$ et $\Lambda_{\{D\}}$ comme les lagrangiennes complexes naturellement associées aux distributions $[D]$ et $\{D\}$, à savoir

$$(27)\qquad \begin{aligned} \Lambda_{\{D\}} = \big\{ &(z_0, z_1, \ldots, z_N, \zeta_0, \zeta_1, \ldots, \zeta_N), \zeta_0 = 0, \zeta_i = 0 \text{ si } i \notin J \\ &t_i = 0,\ (x_i,\xi_i) \in T^*_V \mathbb{C}^d \text{ ou } \xi_i = 0 \text{ si } i \in J \big\}\,. \end{aligned}$$

$$(28)\qquad \begin{aligned} &\Lambda_{[D]} = \big\{ (z_0, z_1, \ldots, z_N, \zeta_0, \zeta_1, \ldots, \zeta_N), \text{ tel qu'il existe} \\ &\Xi_1, \ldots, \Xi_N \text{ avec } (z_i - z_{f(i)}, \Xi_i) \in \Lambda_\square,\ \zeta_0 = \sum_{f(i)=0} \Xi_i\,, \\ &\zeta_i = -\Xi_i + \sum_{f(j)=i} \Xi_j \text{ si } i \neq 0 \big\} \end{aligned}$$

où $\Lambda_\square = T^*_{\Gamma^{\mathbb{C}}} \cup \{\zeta = 0\}$, $\Gamma^{\mathbb{C}}$ étant le complexifié du cône d'onde ($T^*_0 \subset T^*_{\Gamma^{\mathbb{C}}}$).

Si Δ est la diagonale de $M \times M$, $M = \mathbb{R}^{(1+d)(N+1)}$, on montre alors (en utilisant des résultats de finitude en géométrie sous-analytique réelle):

(29) **Proposition 3** [27]. $WF^2_{T^*_\Delta}([D] \otimes \{D\}) \subset C_{(T^*_\Delta)^{\mathbb{C}}}(\Lambda_{[D]} \otimes \Lambda_{\{D\}})$.

Ici $C_\Lambda(A)$ désigne le cône réel de A le long de Λ, sous ensemble du fibré normal à Λ, identifié au fibré conormal à Λ via la structure symplectique du fibré cotangent ambiant.

On déduit alors des Propositions 2 et 3 le résultat suivant, qui permet de rendre effectif le Théorème 5.

(30) **Proposition 4.** *$WF([D]\text{-}\{D\}) \subset (\Lambda_{[D]}\widehat{+}\Lambda_{\{D\}}) \cap T^*\mathbb{R}^{(1+d)(N+1)}$ où $\widehat{+}$ est l'opérateur de M. Kashiwara et P. Schapira* [24]

(31) $$(x,\xi) \in S_1\widehat{+}S_2 \text{ ssi il existe des suites } (x_n^i, \xi_n^i) \in S_i \text{ telles que } x_n^i \to x,\ \xi_n^1 + \xi_n^2 \to \xi,\ |x_n^1 - x_n^2|\,|\xi_n^i| \to 0\,.$$

IV. Estimations Géométriques des Singularités d'Ondes Non Linéaires

Soit u une onde semi-linéaire solution de (1) qui vérifie:

(32) les données de Cauchy u_0, u_1 sont conormales classiques sur $\Lambda = T_V^*\mathbb{R}^d$ où V est une hypersurface analytique réelle.

Le Théorème 5 et la Proposition 4 fournissent des estimations géométriques des singularités de u à partir de l'ensemble des points limites d'ensembles de suites tracées dans le fibré cotangent complexe $T^*\mathbb{C}^{(1+d)}$, comme suit.

On note $\mathscr{E}$ des ensembles de suites $(z_n, \zeta_n) \in T^*\mathbb{C}^{1+d}$, $z = (t,x)$, $n \in \mathbb{N}$, qui vérifient:

(33) la suite z_n converge vers un point de Ω,

(34) il existe une suite convergente $\eta_n \in \mathbb{C}^{1+d}$, $|\eta_n| = 1$ et une suite $\lambda_n \in \mathbb{C}^*$ telles que $\zeta_n = \lambda_n \eta_n$,

(35) la suite (z_n, ζ_n) est caractéristique, i.e. $\zeta_n = (\tau_n, \xi_n)$, $\tau_n^2 = \xi_n^2$.

On suppose en outre que $\mathscr{E}$ est stable par extraction de sous-suites, et vérifie:

(36) $\mathscr{E}$ contient toute suite (z_n, ζ_n) satisfaisant (33), (34), (35) et $\lim \zeta_n = 0$,

(37) si $(z_n, \zeta_n) \in \mathscr{E}$ et z_n' vérifie $\lim z_n' = \lim z_n$ et $\lim |z_n - z_n'|\,|\zeta_n| = 0$ alors $(*)(z_n', \zeta_n) \in \mathscr{E}$,

(38) si $(z_n, \zeta_n) \in \mathscr{E}$ et si z_n' est telle que $\lim z_n' \in \Omega$ appartient au demi-cône d'onde de sommet $\lim z_n$ qui ne rencontre pas l'hyperplan $t = 0$ et si (z_n, ζ_n) et (z_n', ζ_n) sont sur la même bicaractéristique complexe de $\square$, alors $(*)(z_n', \zeta_n) \in \mathscr{E}$.

[(*) signifie: il existe une suite extraite telle que.]

Soit à présent

$$\underline{\mathscr{E}} = \{\mathscr{E}_1, \mathscr{E}_2, \dots, \mathscr{E}_M, \dots\} \tag{39}$$

une suite (croissante) de tels ensembles de suites telles que

$$\text{si } (z_n, \zeta_n^j) \in \mathscr{E}_{k_j} \text{ sont } N \text{ suites } (j = 1, \dots, N) \tag{40}$$

possédant le même point de base z_n et si ζ_n est une suite telle que (z_n, ζ_n) vérifie (33), (34), (35) et $\lim(\zeta_n^1 + \dots + \zeta_n^N - \zeta_n) = 0$ alors $(*)(z_n, \zeta_n) \in \mathscr{E}_k$ avec $k = k_1 + \dots + k_N$.

On définit alors $Z_M(\underline{\mathscr{E}})$ par:

$$\begin{aligned} Z_M(\underline{\mathscr{E}}) = \big\{ &(z,\zeta) \in T^*\mathbb{C}^{1+d}|_\Omega, \text{ il existe } N \text{ suites} \\ &(z_n, \zeta_n^j) \in \mathscr{E}_{k_j}, \ z = \lim z_n, \ \zeta = \lim \zeta_n^1 + \dots + \zeta_n^N \\ &\text{et } k_1 + \dots + k_N \le M \big\} \end{aligned} \tag{41}$$

[le point (z, ζ) n'est pas caractéristique en général.]

Enfin on note $\mathscr{A}_V$ l'ensemble des suites (z_n, ζ_n) vérifiant (33), (34) et (35), $z_n = (0, x_n)$, $\zeta_n = (\tau_n, \xi_n)$, $(x_n, \xi_n) \in T^*_{V^\mathbb{C}}$ où $V^\mathbb{C}$ est un complexifié de V.

Théorème 6. *Soit u une onde semi-linéaire solution de* (1) *avec $s = (d/2) + \varrho$, $\varrho > 0$, vérifiant* (32). *Pour $M \ge 1$ et $\sigma \in [s + 1 + (M-1)\varrho, s + 1 + M\varrho[$ on a $WF^\sigma(u)|_{t>0} \subset Z_M(\underline{\mathscr{E}}) \cap T^*\Omega_+$ dès que $\underline{\mathscr{E}} = (\mathscr{E}_1, \mathscr{E}_2, \dots)$ vérifie $\mathscr{A}_V \subset \mathscr{E}_1$.*

On obtient alors le Théorème 4 comme conséquence du Théorème 6 dans [27] en construisant explicitement un $\underline{\mathscr{E}}$ de la forme $(\mathscr{E}_1, \mathscr{E}_1, \mathscr{E}_1, \dots)$ tel que $Z(\underline{\mathscr{E}}) \cap T^*\Omega_+ = T^*_{s \cup Q_+}$ près de B.

Le Théorème 6 a été amélioré depuis par J.-M. Delort [22], [23].

Références

[1] S. Alinhac: Paracomposition et opérateurs paradifférentiels. Comm. in P.D.E. **11-4** (1986) 87–121

[2] S. Alinhac: Evolution d'une onde simple pour des équations non-linéaires générales. Current Topics in P.D.E. Kinokuniya Tokyo 1986, pp. 63–90

[3] S. Alinhac: Interaction d'ondes simples pour des équations complètement non-linéaires. Ann. Sci. Ec. Norm. Sup. 4ème série **21** (1988) 91–132

[4] M. Beals: Self-spreading and strength of singularities for solutions to semi-linear wave equations. Ann. Math. **118** (1983) 187–214

[5] M. Beals: Propagation of smoothness for nonlinear second order strictly hyperbolic equations. Proc. of Symposia in Pure Math. **43** (1985) 21–45

[6] M. Beals et G. Métivier: Progressing wave solutions to certain nonlinear mixed problems. Duke Math. J. **53** (1986) 125–137

[7] M. Beals et G. Métivier: Reflection of transversal progressing waves in nonlinear strictly hyperbolic mixed problems. Am. J. Math. **109** (1987) 335–360

[8] J.-M. Bony: Calcul symbolique et propagation des singularités pour les équations aux dérivées partielles non linéaires. Ann. Sci. Ec. Norm. Sup., 4ème série **14** (1981) 209–246

[9] J.-M. Bony: Propagation des singularités .... Sém. Goulaouic-Schwartz, Ec. Polytechnique, 1979–80, no. 22

[10] J.-M. Bony: Interaction des singularités .... Sém. Goulaouic-Schwartz, Ec. Polytechnique, 1981–82, no. 2

[11] J.-M. Bony: Interaction des singularités pour les équations de Klein-Gordon non linéaires. Sém. Goulaouic-Meyer-Schwartz, Ec. Polytechnique, 1983–84, no. 10

[12] J.-M. Bony: Second microlocalization and interaction of singularities for non linear P.D.E.. Hyperbolic equations and related topics (Mizohata, ed.). Kinokuniya, 1986, pp. 11–49

[13] J.-M. Bony: Singularités des solutions de problèmes hyperboliques non linéaires. Advances in Microlocal Analysis (Garnir, ed.), NATO ASI Series, vol. 168, Reidel, 1985, pp. 15–39

[14] J.-M. Bony et N. Lerner: Quantification asymptotique .... Sém. E.D.P., Ec. Polytechnique, 1986–87, no. 2 et 3

[15] J.-M. Bony et N. Lerner: Quantification asymptotique et microlocalisations d'ordre supérieur. Ann. Sci. Ec. Norm. Sup., 4ème série **22** (1989) 1–57

[16] J.-Y. Chemin: Calcul paradifférentiel précisé et applications aux équations aux dérivées partielles non semilinéaires. Duke Math. J. **56**, no. 3 (1988) 431–469

[17] J.-Y. Chemin: Interaction contrôlée dans les E.D.P. non linéaires strictement hyperboliques. Bull. Soc. Math. France **116** (1988) 341–383

[18] J.-Y. Chemin: Interaction de trois ondes dans les équations semilinéaires strictement hyperboliques d'ordre 2. Comm. in P.D.E. **12** (11) (1987) 1203–1225

[19] J.-Y. Chemin: Régularité de la solution d'un problème de Cauchy fortement non linéaire à données singulières en un point. Ann. Inst. Fourier **39** (1989)

[20] J.-Y. Chemin: Evolution d'une singularité ponctuelle dans des équations strictement hyperboliques non linéaires. à paraître

[21] F. David et M. Williams: Singularities of solutions to semilinear boundary value problems. Amer. J. Math. **109** (1987) 1087–1109

[22] J.-M. Delort: Deuxième microlocalisation simultanée et front d'onde de produits. Ann. Sci. Ec. Norm. Sup., 4ème série **23** (1990) 257–310

[23] J.-M. Delort: Conormalité des ondes semi-linéaires le long des caustiques. Séminaire E.D.P., Ec. Polytechnique, 1988–89, no. 15

[24] M. Kashiwara, P. Shapira: Microlocal study of sheaves. Astérisque **128** (1985)

[25] Y. Laurent: Théorie de la deuxième microlocalisation dans le domaine complexe. Progress in Mathematics, vol. 53. Birkhäuser, 1985

[26] G. Lebeau: Problème de Cauchy semi-linéaire en 3 dimensions d'espace. Un résultat de finitude. J. Funct. Anal. **77** (1988)

[27] G. Lebeau: Equations des ondes semi-linéaires II. Contrôle des singularités et caustiques non linéaires. Invent. math. (1989)

[28] G. Lebeau: Deuxième microlocalisation sur les sous-variétés isotropes. Ann. Inst. Fourier Grenoble **35**, no. 2 (1985) 145–216

[29] G. Lebeau: Front d'onde des fonctions non linéaires et polynômes. Séminaire E.D.P., Ec. Polytechnique, 1988–89, no. 10

[30] E. Leichtnam: Régularité microlocale pour des problèmes de Dirichlet non linéaires non caractéristiques d'ordre deux à bord peu régulier. Bull. Soc. Math. France **115** (1987) 457–489

[31] R. Melrose: Semi-linear waves with cusp singularities. Actes Journées E.D.P., St Jean de Monts 1987, no. 10

[32] R. Melrose et N. Ritter: Interaction of progressing waves for semi-linear wave equation I. Ann. Math. **121** (1985) 149–236

[33] R. Melrose et N. Ritter: Interaction of progressing waves for semi-linear wave equation II. Arkiv för Math. **25** (1987) 91–114

[34] Y. Meyer: Remarques sur un théorème de J.-M. Bony. Suppl. di Rend. del Circolo Mat. di Palermo, 1981, pp. 1–20

[35] Y. Meyer: Régularité des solutions des équations aux dérivées partielles non linéaires. Sém. Bourbaki. Lecture Notes in Mathematics, vol. 842. Springer, Berlin Heidelberg New York 1980, pp. 293–302

[36] A. Piriou: Calcul symbolique non linéaire pour une onde conormale simple. Ann. Inst. Fourier **38**, no. 4 (1988) 173–186

[37] J. Rauch et M. Reed: Nonlinear microlocal analysis of semi-linear hyperbolic systems in one space dimension. Duke Math. J. **49** (1982) 397–475

[38] J. Rauch et M. Reed: Singularities produced by the nonlinear interaction of three progressing waves; examples. Comm. in P.D.E. **7** (1982) 1117–1133

[39] J. Rauch et M. Reed: Classical, conormal, semilinear waves. Séminaire E.D.P., Ec. Polytechnique, 1985–86, no. 5

[40] N. Ritter: Progressing wave solutions to non-linear hyperbolic Cauchy problems. Ph. D. Thesis M.I.T., 1984

[41] A. Sa Barreto: Interaction of conormal waves for fully semilinear wave equations. (à paraître)

[42] J. Sjöstrand: Singularités analytiques microlocales. Astérisque **95** (1982)

[43] M. Sablé-tougeron: Régularité microlocale pour des problèmes aux limites non linéaires. Ann. Inst. Fourier **36**, no. 1 (1986) 39–82

[44] M. Williams: Spreading of singularities at the boundary in semilinear hyperbolic mixed problems I: microlocal $H^{s,s'}$ regularity. Duke Math. J. **56** (1988) 17–40

[45] M. Williams: Spreading of singularities at the boundary in semilinear hyperbolic mixed problems II: crossing and self-spreading. Trans. Amer. Math. Soc. **310** (1988)

[46] C.J. Xu: Propagation au bord des singularités pour des problèmes de Dirichlet non linéaires d'ordre deux. Actes Journées E.D.P., St Jean de Monts, 1989, no. 20

Static and Moving Defects in Liquid Crystals

Fang Hua Lin

Courant Institute of Mathematical Sciences, New York University
New York, NY 10012, USA

1. Introduction

There have been many recent activities in the analysis of defects (or singularities) of solutions of partial differential equations. In these solutions defects often reveal crucial facets of certain nonlinear problems which they model. Here I shall survey some recent studies on static and moving defects in liquid crystals. One may find references [1], [2] and [3] useful for the discussion below.

Liquid crystals are optically anisotropic, even when they are at rest. Scientifcally, defects of the optical director in liquid crystals have long been of interest. One can resolve, in experiment, individual defects and details of configurations near them, with relatively simple optical (polarizing) microscopes. Often, such configurations are not static but, in many cases, they change very slowly with time. Thus it is useful to first consider defects in static liquid crystals.

In the classical Oseen-Frank model, energy minimizing static configurations of liquid crystals can be described by unit vector fields on a 3-dimensional domain. They are closely related to the theory of harmonic maps into the sphere. The latter also provides us with precise information concerning those isolated point defects such as bounds on the number of point defects and behavior of configurations near each such isolated point.

It was observed, at least experimentally [4], that line and surface defects do occur in liquid crystals. Following the general order-parameter theory of Ginzburg-Landau, J. Ericksen posed a new mathematical model to tackle these phenomena. It turns out the study of Ericksen's model is related to the study of harmonic maps to singular spaces (in this case the singular spaces are circular cones in $\mathbb{R}^4$ or Minkowski space $\mathbb{R}^{3,1}$). Problems of harmonic maps into singular spaces arise also in the study of super rigidity and other geometrical or topological problems for which we refer to a recent work of M. Gromov and R. Schoen [5].

In studying line and surface defects, we introduced a new mathematical device which was based on H. Federer's dimension reduction principle (see [6] and [3]). It is a useful tool also to study level sets of solutions to elliptic and parabolic equations [7]. Defect sets can be characterized as preimages of the vertex of the circular cone under these maps, or equivalently, the vanishing sets of the orientational order of liquid crystals. We can use this device to estimate the Hausdorff dimension, as well as the Hausdorff measure of defects of static and moving liquid crystals. Moreover, local behavior of liquid crystal configurations near these defects may also be described rather precisely.

Proceedings of the International Congress
of Mathematicians, Kyoto, Japan, 1990

2. Oseen-Frank Model and Point Defects

2.1 In the Oseen-Frank model, the energy minimizing static configurations of liquid crystals can be described by maps $n : \Omega \subseteq \mathbb{R}^3 \to \mathbb{S}^2$ which minimize the energy functional:

$$\int_\Omega W(n, \nabla n)\, dx \,, \tag{2.1}$$

where

$$\begin{aligned} W(n, \nabla n) = {} & k_1|\text{div } n|^2 + k_2(q + n \cdot \text{curl } n)^2 + k_3|n \wedge \text{ curl } n|^2 \\ & + (k_2 + k_4)\, [\text{tr}(\nabla n)^2 - (\text{div } n)^2] \,, \end{aligned} \tag{2.2}$$

the k_i's and q are material constants. The defect set is defined to be the discontinuity set of the map n.

It was shown by Hardt, Kinderlehrer and myself [8] that *for any bounded lipschitz domain* Ω in $\mathbb{R}^3$ *and any* $n_0 \in H^{1/2}(\partial\Omega, \mathbb{S}^2)$, *there is a minimizer n of* (2.1) *with* $n = n_0$ *on* $\partial\Omega$ *provided that* $k_1, k_2, k_3 > 0$. *Moreover, n satisfies the following:*

(i) *for any compact* $K \subset \Omega$,

$$\int_K |\nabla n|^2\, dx \leq c(K, \Omega, k_1, k_2, k_3) \;; \tag{2.3}$$

(ii) $\nabla n \in L^q_{loc}(\Omega)$ *for some* $q > 2$;

(iii) *the defect set* Σ *of n has Hausdorff dimension strictly less than one, and n is analytic in* $\Omega \sim \Sigma$. (See [9].)

In the special case $k_1 = k_2 = k_3 = 1$ and $k_4 = q = 0$,

$$W(n, \nabla n) = |\nabla n|^2 \tag{2.4}$$

which is the integrand for harmonic maps from Ω into $\mathbb{S}^2$. Schoen and Uhlenbeck [10] have shown, in this case, that the defect set of n consists of isolated points. Moreover, when $\partial\Omega$ and $n_0 : \partial\Omega \to \mathbb{S}^2$ are smooth, n is also smooth near $\partial\Omega$. This is unknown for minimizers of (2.1).

2.2 Much has been learned recently about defects of an energy minimizing harmonic map from a 3-dimensional domain to the sphere $\mathbb{S}^2$. First of all, at each such isolated defect, there is a unique tangent map, which follows from a general theorem of L. Simon [11]. Second, these energy minimizing tangent maps are classified by Brezis-Coron and Lieb [12]. They are of the form $\pm R \circ \frac{x}{|x|}$ for some rotation R of $\mathbb{R}^3$. Moreover, by a theorem of L. Simon [13] and Gulliver-White [14], there are two positive constants C and α (independent of maps) such that if n is energy minimizing from $\mathbb{B}^3$ to $\mathbb{S}^2$ and $\underline{0}$ is a defect of n, then

$$|n(x) - \phi\left(\frac{x}{|x|}\right)| \leq c|x|^\alpha \tag{2.5}$$

for some $\phi(\cdot) = \pm R(\cdot)$.

An interesting application of the classification of energy minimizing tangent maps is the following a priori estimate on the number of defect points which is shown by Almgren-Lieb [15] and, independently, by Hardt and myself [16]:

For each compact $K \subset B^3$ there is a universal constant $C(K)$ such that

$$\begin{aligned} &\#\text{ of defects of } n \text{ in } K \le C(K) \\ &\text{for any energy minimizing map } n \text{ from } \mathbb{B}^3 \text{ to } \mathbb{S}^2 . \end{aligned} \tag{2.6}$$

Related to (2.6) is the following stability theorem [16]:

If $g : \mathbb{S}^2 \to \mathbb{S}^2$ is such that $\|g - id\|_{C^1(\mathbb{S}^2)} \le \varepsilon_0$ (for some $\varepsilon_0 > 0$), then any energy minimizing map $n : B^3 \to \mathbb{S}^2$ with $n = g$ on $\mathbb{S}^2$ has a unique defect point a so that

$$\|n(x) - R_a \circ \frac{(x-a)}{|x-a|}\|_{C^\alpha(B^3)} \le C\varepsilon_0^{1/4} \tag{2.7}$$

and

$$|a| \le C\varepsilon_0^{1/2} \ , \quad \|R_a - id\| \le C\varepsilon_0^{1/4} \tag{2.8}$$

for some rotation R_a of $\mathbb{R}^3$ and for some positive constants C and α.

The classification of energy minimizing tangent maps in higher dimensions or target spheres with nonstandard metrics remains as a difficult open problem.

When the domain is four-dimensional and the target is $\mathbb{S}^2$, one deduces from a recent theorem of L. Simon [17] and Hardt and myself [18] the following:

Theorem A. *Let $u : B^4 \to \mathbb{S}^2$ be an energy minimizing map, then the defect set of u is locally a finite union of a finite set and a finite family of $C^{1,\alpha}$ curves with finitely many crossings. Moreover, the 1-dimensoinal Hausdorff measure of the defect set is locally finite.*

2.3 Continuous non energy minimizing harmonic maps are of interest: from both analysis and differential geometry aspects. For example a classical problem is to represent a homotopy class of maps between compact Riemannian manifolds by harmonic maps (see [19], [20]). Some partial results related to the theory of liquid crystals have been found by Bethuel, Brezis and Coron [21] and Giaquinta, Monica and Soucèk [22].

It should be noted also that solutions to harmonic map systems with the Dirichlet boundary condition are not unique (see [21], [23] and [24]), and the defect sets of these maps can be much more complicated [24].

3. Line and Surface Defects

3.1 To explain line and surface defects in liquid crystals, one uses Ericksen's model. In this new models, the energy minimizing configuration of liquid crystals is described by a pair, (s, n), where $s : \Omega \to [-1/2, 1]$ is a real function which denotes the variable degree of orientation and $n : \Omega \to \mathbb{S}^2$ dentoes the axis of optical director. Thus (s, n) minimizes a bulk-energy functional which, for particular choices of material constants involved, reduces to

$$\int_\Omega [k|\nabla s|^2 + s^2|\nabla n|^2 + \psi(s)]\, dx \tag{3.1}$$

with $k > 0$, where the potential function ψ is a positive $\mathbb{C}^2$-function defined on $(-1/2, 1)$ (cf. [25] and [3]).

Since s may be zero somewhere in Ω, (3.1) is a rather degenerate variational integral. However, the change of variables $u = sn$ reduces (3.1) to

$$\int_\Omega [(k-1)|\nabla s|^2 + |\nabla u|^2 + \psi(s)]\, dx\,. \tag{3.2}$$

The variational integral (3.2) is, essentially, the energy of the map $(s, u) : \Omega \to \mathbb{C}_k$ where $\mathbb{C}_k$ is the circular cone $\{(t, u) \in \mathbb{R} \times \mathbb{R}^3 : |t| = \sqrt{k-1}|u|\}$, for $k \geq 1$, in the Euclidian space $\mathbb{R}^4$, or, the circular cone $\{(t, u) \times \mathbb{R} \times \mathbb{R}^3 : |t| = \sqrt{1-k}|u|\}$, for $k < 1$, in the Minkowski space $\mathbb{R}^{3,1}$. Now the direct method in the calculus of variations implies the existence of a minimizer of (3.2) under the Dirichlet boundary condition that $(s, t)\big|_{\partial\Omega} \in H^{1/2}(\partial\Omega, \mathbb{C}_k)$. Moreover, when $0 < k \leq 1$, and $\psi \equiv 0$, the minimizer of (3.2) is unique (see [25]).

Regarding the regularity of the map (s, u), one has the following result, see [25]:

Theorem B. *Let Ω be a bounded $C^{1,\alpha}$ domain in $\mathbb{R}^3$ and let $(s_0, u_0) \in C^{1,\alpha}(\partial\Omega, \mathbb{R}^4)$, with $|s_0| = |u_0|$ on $\partial\Omega$. Suppose (s, u) is a minimizer of (3.2) which satisfies the constraint $|s| = |u|$ a.e. in Ω. Then $(s, u) \in C^\beta(\overline{\Omega})$. Moreover, when $0 < k < 1$, both s and u are lipschitz continuous in $\overline{\Omega}$.*

It should be pointed out that (s, u) is analytic (if ψ is) or smooth (if ψ is) on the open set $\{x \in \Omega : s(x) > 0\}$. This follows from standard elliptic regularity theory.

We also note that the existence and partial regularity of minimizers of (3.2) were also established by L. Ambrosio in [26] and [27].

3.2 Having seen the regularity of the minimizers of (3.2), one is then interested in the defect sets of the optical director n. It is shown in [3] and [25] that defect sets are precisely the nodal set of the orientational order s, i.e., the set $\{x \in \Omega : s(x) = 0\}$.

Theorem C. *Let (s, u) be a minimizer of (3.2). Then the set $\{s = 0\}$ is either all of Ω or is of Hausdorff dimension ≤ 2. If, in addition, $k > 1$ and $s \gneq 0$, then the set $\{s = 0\}$ has the Hausdorff dimension ≤ 1.*

One notices that, for any $0 < k < 1$, examples of minimizers (s, u) of (3.2) with $s \geq 0$ and with $\{s = 0\}$ being 2-dimensional were explicitly constructed in [28].

The proof of *Theorem C* is based on the dimension reduction principle of H. Federer [6] and the monotonicity of the function $N(r)$ defined below (see also [3], [7] and [25]).

For $a \in \Omega$, $r \in (0, d_a)$, $d_a = \mathrm{dist}(a, \partial\Omega)$, $B_r(a) = \{x \in \Omega : |x - a| < r\}$, we define

$$\begin{aligned} D(r) &= \int_{B_r(a)} [(k-1)|\nabla s|^2 + |\nabla u|^2 + s \cdot \psi'(s)]\, dx\,, \\ H(r) &= \int_{\partial B_r(a)} (k-1)|s|^2 + |u|^2 \end{aligned} \tag{3.3}$$

and

$$N(r) = \frac{r\,D(r)}{H(r)} \tag{3.4}$$

provided that $H(r) \neq 0$. Then one has the following

Lemma D. *There are two positive constants r_0 and C depending only on ψ such that the function*

$$N(r)\,e^{Cr} \tag{3.5}$$

is a monotone increasing function of $r \in (0, r_a)$, $r_a = \min[d_a, r_0]$.

The monotonicity of $N(r)$ is also a useful fact in the work [5]. In fact, suppose u is a harmonic map from $\mathbb{R}^n$ to N with curvature of $N \leq 0$. Then the function

$$N(r) = \frac{r \int_{B_r(a)} |\nabla u|^2\,dx}{\int_{\partial B_r(a)} d^2(\underline{0}, u)} \tag{3.6}$$

is a monotone increasing function of r. Where $d(\underline{0}, u)$ is the intrinsic distance from a fixed point $\underline{0}$ to u in N. The proof of (3.6) is based on the monotonicity formula for energy and the fact that

$$\Delta d^2(\underline{0}, u) \geq 2|\nabla u|^2 \tag{3.7}$$

for any harmonic maps $u : \mathbb{R}^n \to N$.

As a consequence of (3.6) u is locally uniformly lipschitz continuous (independent on N).

Finally we also note that, by combining Lemma D and [18], one can estimate 2-dimensional (or 1-dimensional in some special cases) Hausdorff measure of $\{s = 0\}$ and describe the structure of $\{s = 0\}$.

4. Moving Defects

4.1 Equations adequate for the treatment of both static and dynamic phenomena in liquid crystals, called the Leslie-Ericksen theory, were developed during the 1960's [1]. In [29], Ericksen derived a full set of dynamic equations for nematic liquid crystals with variable degree of orientation. Since motions of liquid crystals are generally slow, the motion of the optical directors is our main concern. After neglecting the small velocity of the fluid, the motion of the optical director can be described as evolution of harmonic maps from Ω to $\mathbb{C}_k$ (a circular cone in $\mathbb{R}^4$ or $\mathbb{R}^{3,1}$). More precisely, we let $v = (s, u) : \Omega \to \mathbb{C}_k$ and let

$$\tau(v) = 0 \quad \text{in } \Omega \tag{4.1}$$

be the equations for harmonic maps from Ω to $\mathbb{C}_k$. Then the evolution of the optical director satisfies

$$\frac{\partial}{\partial t} v = \tau(v)\,, \qquad \text{for } (x, t) \in \Omega \times (0, \infty) \tag{4.2}$$

with initial Cauchy data

$$v(x,0) = v_0(x)\ , \quad x \in \Omega \tag{4.3}$$

and Dirichlet boundary condition

$$v(x,t) = v_0(x) \quad \text{for } x \in \partial\Omega\ . \tag{4.4}$$

The problem (4.2), (4.3), (4.4) was first studied by Eells-Sampson [20] and more recently by Struwe and Chen [30] and many others. The main difference between our problem and those studied earlier is that the target manifold contains singularities. Nevertheless one has the following

Theorem E. *If v_0 is uniformly lipschitz continuous on $\overline{\Omega}$ and $v_0\big|_{\partial\Omega} \in C^{1,\alpha}(\partial\Omega)$, then* (4.2), (4.3), (4.4) *has a global weak solution $v \in L^\infty((0,\infty), H^1(\Omega))$. Moreover, for $k \le 1$, v is unique and satisfies*

$$\sup_\Omega |\nabla_x v|(\cdot,t) \le c \sup_\Omega |\nabla_x v_0(x)|\ , \tag{4.5}$$

and for $k > 1$,

$$\|v\|_{C^\beta(\overline{\Omega})}(\cdot,t) \le c \sup_\Omega |\nabla_x v_0(x)|\ . \tag{4.6}$$

Here c and β are constants depending only on Ω and k.

The proof of (4.5) is based on the observation that $\mathbb{C}_k$ in $\mathbb{R}^{3,1}$ is a negatively curved Riemannian submanifold. Even when $k > 1$, s^2 is a strictly convex function on $\mathbb{C}_k$. This geometrical fact can be used to show (4.6).

4.2 Since the moving defects are precisely the set $\{(x,t) : s(x,t) = 0\}$ in $\Omega \times \mathbb{R}^+$, the problem reduces to studying the nodal set of a solution to certain parabolic systems. In general it is still an interesting subject for future studies. However, there are several recent works by C. Fefferman and H. Donnelly [31] and by Hardt and Simon [32] concerning nodal sets of solutions to elliptic equations. In [7] we studied the corresponding question for a class of heat equations. Interestingly enough, the function $N(r)$ introduced in (3.4) can also be used to control not only the local behavior of solutions near nodal sets but also global Hausdorff measure of nodal sets. Generalizing those arguments in [7] to the problem (4.2) will be an interesting problem for future researchers.

References

1. Ericksen, J. L., Kinderlehrer, D.: Theory and applications of liquid crystals. IMA vol. 5. Springer, Berlin Heidelberg New York 1986
2. Brezis, H.: $\mathbb{S}^k$-valued maps with singularities. (Lecture Notes in Mathematics, vol. 1365.) Springer, Berlin Heidelberg New York 1989
3. Lin, F.-H.: Nonlinear theory of defects in nematic liquid crystals; phase transition and flow phenomena. CPAM, vol. XLII, 1989, pp. 789–8914
4. Kléman, M.: Points, lines and walls. Wiley, New York 1983
5. Gromov, M., Schoen, R.: Harmonic maps into singular spaces and superrigidity. Preprint
6. Federer, H.: The singular sets of area-minimizing rectifiable currents with codimension one and of area-minimizing flat chain modulo two with arbitrary codimension. Bull. Amer. Math. Soc. **76** (1970) 667–711

7. Lin, F.-H.: Nodal sets of solutions of elliptic and parabolic equations. To appear in CPAM
8. Hardt, R., Kinderlehrer, D., Lin, F. H.: Existence and partial regularity of static liquid crystal configurations. Comm. Math. Physics **105** (1986) 547–570
9. Hardt, R., Kinderlehrer, D., Lin, F. H.: Stable defects of minimizers of constrained variational principles. Ann. Inst. H. Poincaré, Anal. Nonlineaire **5**, no. 4 (1988) 297–322
10. Schoen, R., Uhlenbeck, K.: A regularity theory of harmonic maps. J. Diff. Geom. **18** (1983) 253–268
11. Simon, L.: Asymptotics for a class of nonlinear evolution equations. Anal. Math. **118** (2) (1983) 525–571
12. Brezis, H., Coron, J. M., Lieb, E.: Harmonic maps with defects. Comm. Math. Phys. **107** (1986) 649–705
13. Simon, L.: Isolated singularities for extreme of geometrical variational problems. (Lecture Notes in Mathematics, vol. 1161.) Springer, Berlin Heidelberg New York 1986
14. Gulliver, R., White, B.: The rate of convergence of a harmonic map at a singular point. Math. Ann. **283** (1989) 539–550
15. Almgren, F. Jr., Lieb, E.: Singularities of energy minimizing maps from the ball to the sphere: examples, counterexamples, and bounds. Anal. Math. **128** (1988) 483–530
16. Hardt, R., Lin, F. H.: Stability of singularities of minimmizing harmonic maps. J. Diff. Geom. **29** (1988) 113–123
17. Simon, L.: On the singularities of harmonic maps. Preprint
18. Hardt, R., Lin, F. H.: The singular set of an energy minimizing map from $\mathbb{B}^4$ to $\mathbb{S}^2$. Preprint
19. Eells, J., Lemaire, L.: Another report on harmonic maps. Bull. London Soc. **20** (8) (1988) Part 5
20. Eells, J., Sampson, J. H.: Harmonic mappings of Riemannian manifolds. Amer. J. Math. **86** (1964) 109–160
21. Bethuel, F., Brezis, H., Coron, J. M.: Relaxed energies for harmonic maps. Preprint
22. Giaquinta, M., Monica, G., Soucèk, J.: Cartesian currents and liquid crystals, dipoles, singular lines and singular points. Preprint
23. Hardt, R., Kinderlehrer, D., Lin, F. H.: The variety of configurations of static liquid crystals. To appear
24. Hardt, R., Poon, C. C., Lin, F. H.: Axially symmetric harmonic maps minimizing a relaxed energy. Preprint
25. Lin, F. H.: On nematic liquid crystals with variable degree of orientations. To appear in CPAM
26. Ambrosio, L.: Existence of minimal energy configurations of nematic liquid crystals with variable degree of orientation. Preprint
27. Ambrosio, L.: Regularity of solutions of a degenerate elliptic variational problem. Preprint
28. Ambrosio, L., Virga, E.: A boundary value problem for nematic liquid crystals with variable degree of orientations. Preprint
29. Ericksen, J. L.: Liquid crystals with variable degree of orientation. IMA preprint # 559 (1989)
30. Struwe, M.: The evolution of harmonic maps. In these proceedings, pp. 1197–1203
31. Donnelley, H., Fefferman, C.: Nodal sets of eigenfunctions on Riemannian manifolds. Inv. Math. **93** (1988) 161–183
32. Hardt, R., Simon, L.: Nodals sets for solutions of elliptic equations. J. Diff. Geom. **30** (1989) 505–522

On Kinetic Equations *

Pierre-Louis Lions

CEREMADE, Université Paris-Dauphine, Place de Lattre de Tassigny
F-75775 Paris Cedex 16, France

I. Introduction

We will review some recent progress on various *kinetic equations* which include the well-known *Boltzmann equation* and also *Vlasov models* like Vlasov-Poisson or Vlasov Maxwell systems. Before describing more precisely these mathematical results, we would like to recall first a few of the basic physical notions underlying these models.

First of all, kinetic models are a branch of *Statistical Physics*. And they arise in a large number of different physical contexts like, for instance, in the study of the dynamics of electrons or ions in plasmas, in the study of the dynamics of nucleons in Nuclear Physics, in the modelling of semi-conductors, in the modelling of the reentry of various aircrafts in a rarefied atmosphere, in the study of the formation and stability of planetary rings or even in the study of the formation of galaxies. It is worth noting that this list, by no means exhaustive, includes different physical interactions on very different scales. Of course, each of these applications reveals specific mathematical questions that we will not address here. Instead, we will concentrate here on the main mathematical issues raised by all these applications.

In spite of these extremely different physical backgrounds, the main principle underlying these models can be summarized as follows. Let us suppose we want to study the evolution of a large number of particles that we take to be identical in order to simplify the presentation. The reader should be aware that the word particle above is used here in a vague sense and might be rather misleading since in the applications listed above the "particles" may be electrons, ions, nucleons, molecules, rocks or stars ! Therefore, in order to be more precise, we consider these objects as classical point-particles.

Next, we observe that when we deal with a large number of particles, it is impossible to study the evolution of each particle and we wish to look instead for a statistical description of this evolution. In other words, the unknowns which would have been otherwise the positions and velocities of each particle "reduce" to a function f of (x, v, t), $(x, v \in \mathbb{R}^N,\ t \geq 0)$ which is the density of particles

* Dedicated to the memory of R. J. DiPerna.

Proceedings of the International Congress
of Mathematicians, Kyoto, Japan, 1990

at position x, time t and with velocity v. Of course, f is nonnegative. Next, we simply indicate that many kinetic models, but not all of them, take the following form

$$\frac{\partial f}{\partial t} + v \cdot \nabla_x f + F \cdot \nabla_v f = C \quad \text{in } \mathbb{R}^N \times \mathbb{R}^N \times (0, \infty). \tag{1}$$

Here and everywhere below, $\nabla_x f$ and $\nabla_v f$ denote respectively the gradient of f with respect to x and v. Variants involve different velocities than v or convolution terms instead of $F \cdot \nabla_v f$ ... In (1), F stands for a *force* acting on the particles while C is a term which takes into account the possible *collisions* between particles.

Let us now give a few specific examples: we begin with collision-less models that is we set $C = 0$. These models are called *Vlasov models*. In the case of charged particles thus obeying the electromagnetic interaction, one uses two models namely the *Vlasov-Maxwell system* and the *Vlasov-Poisson system*. The Vlasov-Maxwell system consists of (1) (with $C = 0$ and $N = 3$) and of

$$F(x, v, t) = E(x, t) + v \times B(x, t) \quad \text{in} \quad \mathbb{R}_x^3 \times \mathbb{R}_v^3 \times (0, \infty) \tag{2}$$

$$\begin{gathered} \frac{\partial E}{\partial t} - \frac{1}{c} \operatorname{curl} B = -j, \quad \frac{\partial B}{\partial t} + \frac{1}{c} \operatorname{curl} E = 0, \\ \operatorname{div} B = 0, \quad \operatorname{div} E = \varrho \quad \text{in } \mathbb{R}_x^3 \times (0, \infty) \end{gathered} \tag{3}$$

$$j_k = \int_{\mathbb{R}^3} f\, v_k\, dv, \quad \varrho = \int_{\mathbb{R}^3} f\, dv. \tag{4}$$

Observe that (2) means that F is the *Lorentz force* associated to the *electromagnetic field* (E, B) which of course satisfies the *Maxwell equations* (3). Finally, (4) means that the electromagnetic field is created by the particles. And this field creates a force (F) which acts on the particles – this is why one speaks of self-consistent forces for Vlasov models.

The parameter c is the speed of light and one sees that, in the limit $c \to \infty$, (2)–(4) becomes

$$F(x, t) = E(x, t) = -\nabla_x \Phi(x, t) \quad \text{in} \quad \mathbb{R}_x^3 \times (0, \infty) \tag{5}$$

$$-\Delta \Phi = \varrho \quad \text{in} \quad \mathbb{R}_x^3 \times (0, \infty) \tag{6}$$

$$\varrho(x, t) = \int_{\mathbb{R}^3} f\, dv. \tag{7}$$

This system coupled with (1) (with $C = 0$, $N = 3$) is called the *Vlasov-Poisson* system.

It is important for the analysis we will present to observe two facts. First, in Vlasov models or more specifically in collisionless models ($C = 0$), the equation (1) simply means that f *is constant along particle paths.* Of course, these paths are given by Newton's law ($\dot{x} = v$, $\dot{v} = F$). Next, a general feature of Vlasov models illustrated by the two examples above is the particular dependence of the force upon f: indeed, F depends in fact only on *macroscopic quantities* that is averages of f in v like

$$\int_{\mathbb{R}^3} f(x,v,t)\,\psi(v)\,dv$$

for some given ψ.

We now give one example of *collision models* or in other words one example for the term C. This is a famous nonlinear term introduced by J.C. Maxwell [35] and L. Boltzmann [6] which is given by

$$C = Q(f,f) = \int_{\mathbb{R}^3} dv_* \int_{S^2} d\omega\, B(v-v_*,\omega)\,\{f'f'_* - ff_*\} \tag{8}$$

where $f_* = f(x,v_*,t)$, $f' = f(x,v',t)$, $f'_* = f(x,v'_*,t)$ and we have

$$v' = v - (v-v_*,\omega)\omega\,, \quad v'_* = v_* + (v-v_*,\omega)\omega\,. \tag{9}$$

The collision operator C determines the rate at which the density f is modified by collisions taking place at (x,t), between two particles with velocities v and v_* – in other words, one makes a statistical balance of all possible collisions affecting the density $f(x,v,t)$. Now, if we assume these collisions to be elastic, the velocities v', v'_* after the collision satisfy

$$v' + v'_* = v + v_*\,, \quad |v'|^2 + |v'_*|^2 = |v|^2 + |v_*|^2\,. \tag{10}$$

All possible solutions of (10) are given by (9), where ω is a parameter allowing a simple description of the set of solutions of (10).

Finally, B is always assumed to be a nonnegative function of $|v-v_*|$ and $(v-v_*,\omega)$ only. It depends on the interactions between the particles and the simplest (and most famous) example is the so-called *hard-spheres* case where $B(z,\omega) = |(z,\omega)|$.

Let us mention that the *Boltzmann equation* is the equation (1) with $F=0$ and C given as above.

Other models for the collision term C exist: some are modifications of the Boltzmann model like the so-called Boltzmann-Enskog or Boltzmann-Dirac models while others are, in some vague sense, simplifications like the Fokker-Planck or the Landau models. We refer to C. Cercignani [8], S. Chapman and T.G. Cowling [10], C. Truesdell and R.G. Muncaster [40] for more details and also for derivations of the Boltzmann model from first principles.

After this brief description of three famous examples of kinetic models, we now turn to a quick presentation of the numerous mathematical problems raised by these models. Let us mention immediately that we shall concentrate on the main mathematical issues forgetting many specific questions of interest for one or several of the physical applications listed above.

The first category of problems is the study of the Cauchy problems for these models, prescribing (f,E,B) (or f depending on the model) at time $t=0$ with the usual questions regarding existence, uniqueness, regularity, approximation and numerical analysis, special solutions, steady states, stability, long-time behavior, boundary conditions ...

But there is much more at hand. In fact, it seems fair to say that Boltzmann's equation is not only famous because of its fascinating mathematical structure but also because of its formal relations with other famous mathematical physics models like hydrodynamical models (compressible Euler equations i.e. gaz dynamics systems, incompressible Navier-Stokes or Stokes or Euler equations ...). It is worth recalling Hilbert's goals to solve Boltzmann equation and to recover from it the Fluid Dynamics equations. Therefore, an important category of mathematical problems concerns the systematic study of the numerous links between kinetic models and other models in Physics. We just mentioned hydrodynamical limits and the link with Fluid Dynamics but one has to add to that theme the derivation of MHD equations, combustion models, reaction-diffusion systems and in fact hyperbolic systems of conservation laws. Indeed, arbitrary (symmetric) systems of conservation laws can be formally approximated by ad hoc kinetic models.

Other limits involve the derivation of kinetic models from "large number of particles limits" and the study of statistical solutions or hierarchies of equations and the propagation of chaos. Another example of a connection with another physical regime is the derivation of kinetic models from Quantum models via Wigner transforms and semi-classical limits ...

We will concentrate here on the first category of problems namely the analysis of the Cauchy problem for kinetic models even if it is quite clear that progress on this theme should and has already yielded progress on the second theme as well. And even if we restrict our attention to the basic existence and uniqueness questions, three types of results have been obtained:

- smooth and unique solutions in the small that is locally in time or globally with smallness restrictions: for the Boltzmann equation alone, many important works have been given in that direction and we can quote only a few of them (complete lists of references can be found in the references given in the bibliography here) like H. Grad [27], C. Cercignani [8], R. Illner and M. Shinbrot [30], T. Nishida and K. Imai [36], S. Ukaï [41], K. Hamdache [28] ...
- existence and uniqueness of special solutions: a famous example is given by the study of space-homogeneous solutions, i.e. x-independent solutions of Boltzmann equation, study initiated by the work of T. Carleman [7] ...
- global existence of weak solutions.

This last category of results has been obtained in a series of works for all kinetic models by R.J. DiPerna and the author [14–18] (see also the survey [19]). And we are going to give one sample of these results by specializing our attention to the Vlasov-Poisson-Boltzmann model in the next section. Finally, in the last section, we will indicate what are the main tools of independent interest that are being used in the proofs of these results.

II. Global Existence of Weak Solutions

As we said before, we concentrate on the Vlasov-Poisson-Boltzmann (VPB in short) system:

$$\frac{\partial f}{\partial t} + v \cdot \nabla_x f - \lambda \nabla_x \Phi \cdot \nabla_v f = Q(f,f) \quad \text{in} \quad \mathbb{R}^3_x \times \mathbb{R}^3_v \times (0,\infty) \tag{11}$$

$$-\Delta \Phi = \varrho \quad \text{in} \quad \mathbb{R}^3_x \times (0,\infty) \tag{12}$$

and in order to make Φ well-defined by (12) we prescribe $\Phi = 0$ at infinity in a weak sense (like for instance $\Phi \in L^\infty(0,\infty; L^6(\mathbb{R}^3_x))$. And we recall

$$\begin{aligned} &Q(f,f) = Q^+(f,f) - Q^-(f,f), \\ &Q^-(f,f) = fL(f), \quad L(f) = f \underset{v}{*} A \end{aligned} \tag{13}$$

$$\begin{aligned} Q^+(f,f) &= \int_{\mathbb{R}^3} dv_* \int_{S^2} d\omega \, B(v - v_*, \omega) f' f'_*, \\ Q^-(f,f) &= \int_{\mathbb{R}^3} dv_* \int_{S^2} d\omega \, B(v - v_*, \omega) f f_*, \end{aligned} \tag{14}$$

$$v' = v - (v - v_*, \omega)\omega, \quad v'_* = v + (v - v_*, \omega)\omega. \tag{16}$$

Finally, λ is a nonnegative parameter: when $\lambda = 0$, the (VPB) system reduces to the standard Boltzmann equation, while when $\lambda > 0$ and $B \equiv 0$ the (VPB) system becomes the Vlasov-Poisson system.

We will make the following assumptions

$$B(z,\omega) \text{ is a function of } z \text{ and } (z,\omega) \text{ only, } B \geq 0 \tag{17}$$

$$\begin{gathered} \int_{|z| \leq R} \int_{S^2} B \, dz \, d\omega < \infty, \quad \frac{1}{1+|z|^2} \int_{|v-z| \leq R} \int_{S^2} B \, dv \, d\omega \to 0, \\ \text{as } |z| \to \infty, \text{ (for all } R < \infty). \end{gathered} \tag{18}$$

These assumptions are clearly satisfied in the case of the hard-spheres model $B(v,\omega) = |(v,\omega)|$. Let us also mention that (18) corresponds to the classical angular cut-off assumption.

Let us now recall the known a priori estimates (when $B \not\equiv 0$) which have all a physical origin (conserved quantities or decay of entropy): formally, a solution of (11)–(12) satisfies

$$\begin{gathered} \iint_{\mathbb{R}^3 \times \mathbb{R}^3} f \psi \, dv \, dx \quad \text{is independent of} \quad t \geq 0, \\ \text{where} \quad \psi = 1, v_\alpha \quad (\alpha = 1,2,3), \end{gathered} \tag{19}$$

$$\iint_{\mathbb{R}^3\times\mathbb{R}^3} f|v|^2\,dx\,dv + \lambda\int_{\mathbb{R}^3} |\nabla_x\Phi|^2\,dx \quad \text{is independent of} \quad t \geq 0, \tag{20}$$

$$\frac{d}{dt}\iint_{\mathbb{R}^3\times\mathbb{R}^3} f|x-vt|^2 = -\lambda t\int_{\mathbb{R}^3} |\nabla_x\Phi|^2\,dx. \tag{21}$$

Finally, the decay of (the mathematical) entropy which is a crucial part of the famous H-theorem is expressed by the following formal identity:

$$\begin{aligned} &\frac{d}{dt}\iint_{\mathbb{R}^3\times\mathbb{R}^3} f\,\log f\,dx\,dv + \frac{1}{4}\int_{\mathbb{R}^3} dx \iint_{\mathbb{R}^3\times\mathbb{R}^3} dv\,dv_* \\ &\qquad \cdot \int_{S^2} B(f'f'_* - ff_*)\,\log\frac{f'f'_*}{ff_*}\,d\omega = 0. \end{aligned} \tag{22}$$

Notice that the second term is clearly nonnegative.

From this, one deduces easily that, if the initial condition f^0 satisfies for some $R \in (0,\infty)$

$$\iint_{\mathbb{R}^3\times\mathbb{R}^3} f^0(1+|x|^2+|v|^2+|\log f^0|)\,dx\,dv + \lambda\int_{\mathbb{R}^3} |\nabla_x\Phi^0|^2\,dx \leq R, \tag{23}$$

then, formally at least, a solution of (11)–(12) satisfies for all $t \geq 0$

$$\begin{cases} \displaystyle\iint_{\mathbb{R}^3\times\mathbb{R}^3} f(1+|v|^2)\,dx\,dv + \lambda\int_{\mathbb{R}^3} |\nabla_x\Phi|^2\,dx \leq R \\ \displaystyle\iint_{\mathbb{R}^3\times\mathbb{R}^3} f(|x|^2 + \log|f|) \leq K(1+t)^2 \\ \displaystyle\int_0^t ds \int_{\mathbb{R}^3} dx \iint_{\mathbb{R}^3\times\mathbb{R}^3} dv\,dv_* \\ \qquad\displaystyle\int_{S^2} B(f'f'_* - ff_*)\,\log\frac{f'f'_*}{ff_*}\,d\omega \leq K(1+t)^2, \end{cases} \tag{24}$$

for some positive constant K which depends only on R. In fact, if $\lambda = 0$ (the case of the pure Boltzmann equation), the factor $(1+t)^2$ can be omitted in the last inequality. In the collisionless case (Vlasov-Poisson system) i.e. when $B \equiv 0$, the Liouville conservation is translated by the following a priori estimate (formally)

$$\iint_{\mathbb{R}^3\times\mathbb{R}^3} \beta(f)\,dx\,dv \quad \text{is independent of } t \tag{25}$$

whenever $\beta \in C([0,\infty);[0,\infty))$ and $\beta(f^0) \in L^1(\mathbb{R}^3_x \times \mathbb{R}^3_v)$.

However, all these bounds are not enough to allow to define $Q(f,f)$ in a meaningful way. It only allows to define $Q^{\pm}(f,f)$ as an a.e. finite function on $(0,\infty)\times\mathbb{R}^3_x\times\mathbb{R}^3_v$ ($L^{1/2}$ in fact). In particular, it is not clear how one should define a solution of (11)–(12).

To circumvent this difficulty, we introduced in [14, 15] the notion of renormalized solutions. The first step is to prove that an additionnal bound can be obtained:

$$\int_0^M dt \int_{\mathbb{R}^3} dx \int_{|v|\le M} \{Q^+(f,f)+Q^-(f,f)\}\, \frac{1}{1+f}\, dv \le C \quad \text{for all } M>0 \tag{26}$$

where C is a positive constant depending only on M and on R (in (23)).

In particular, $Q^{\pm}(f,f)(1+f)^{-1}$ makes sense in L^1_{loc}. We may now give the

Definition. $f \in C([0,\infty); L^1(\mathbb{R}^3_x \times \mathbb{R}^3_x))$ *is a renormalized solution of* (11)–(12) *if it satisfies* (24), (26), (12) *in the sense of distributions, and*

$$\frac{\partial}{\partial t}\beta(f) - \operatorname{div}_x(v\beta(f)) - \lambda \operatorname{div}_v(\Phi\beta(f)) = Q(f,f)\beta'(f) \quad \text{in } \mathscr{D}' \tag{27}$$

for all $\beta \in C([0,\infty);\mathbb{R})$ *such that* $\beta'(t)(1+t)$ *is bounded on* $[0,\infty)$.

Remarks. 1) (27) is nothing else than writing formally the chain rule using (11).

2) It is enough to consider a single function β provided it is one to one, like for instance $\beta(t) = \log(1+t)$.

3) One can show (see [15]) that the notion of renormalized solution for Boltzmann equations ($\lambda = 0$) is stronger than the more standard notion of mild solution (an integrated form of the equation along almost all particle paths).

4) It is worth recalling that in the collision-less case ($B = 0$), the physical law behind the equation is that f should be constant along particle paths, a fact which is clearly equivalent to the fact that $\beta(f)$ is constant along particle paths (for all β or for a single 1-1 β …). This is precisely what we request in the above definition.

5) In the collision-less case ($B = 0$) i.e. in the case of the Vlasov-Poisson system, renormalized solutions are stronger than the classical weak solutions as built by Arsenev [4]. In particular (see [18]) they satisfy the Liouville conservation (25) and are continuous in time (with values in $L^p_{x,v}$ …).

Our main existence result is the

Theorem 1. *Let* (17)–(18) *hold and let* f^0 *satisfy (23), then there exists a renormalized solution* f *of* (11)–(12) *in* $C([0,\infty)\,;\, L^1(\mathbb{R}^3_x \times \mathbb{R}^3_v))$ *which satisfies* $f|_{t=0} = f^0$ *on* $\mathbb{R}^3_x \times \mathbb{R}^3_v$ *and*

$$\begin{aligned} &\frac{d}{dt}\iint_{\mathbb{R}^3_x\times\mathbb{R}^3_v} f\log f\,dx\,dv + \frac{1}{4}\int_{\mathbb{R}^3} dx \iint_{\mathbb{R}^3\times\mathbb{R}^3} dv\,dv_* \\ &\cdot \int_{S^2} B(f'f'_* - ff_*)\log\frac{f'f'_*}{ff_*}\,d\omega \\ &\le \iint_{\mathbb{R}^3_x\times\mathbb{R}^3_v} f^0 \log f^0\,dx\,dv. \end{aligned} \tag{28}$$

Remarks. 1) Uniqueness and regularity of solutions are major open problems. Also, the equality in (28) is another important open question. In some vague sense, this result is analogous to the existing results (essentially due to J. Leray [31]) on three-dimensional incompressible Navier-Stokes equations.

2) Theorem 1 and its proof has been applied, adapted or extended to various other kinetic equations and related questions: K. Hamdache [29] (boundary conditions); B. Perthame [37] (BGK model); J. Polewczak [39], L. Arkeryd and C. Cercignani [3] (Boltzmann-Enskog model); M.J. Esteban and B. Perthame [23] (inelastic collisions); J.M. Dolbeault [13] (Boltzmann-Dirac model); L. Arkeryd [2], L. Desvillettes [11], C. Cercignani [9] (long time behavior); L. Desvillettes [12] (convergence of splitting methods); C. Bardos, F. Golse and D. Levermore [5] (convergence to some incompressible Fluid Dynamics models when the mean free path goes to 0).

The main ingredient in the proof of Theorem 1 is the following "stability under weak convergence" result taken from [15, 16].

Theorem 2. *Let f_n^0 satisfy* (23) *and let* (17)–(18) *hold. Let $f_n \in C([0,\infty); L^1(\mathbb{R}_x^3 \times \mathbb{R}_v^3))$ be a renormalized solution of* (11)–(12) *satisfying $f_n|_{t=0} = f_n^0$. Without loss of generality, we may assume that f_n^0, f_n converge weakly in L^1 to f^0, f respectively. Then, we have*

1) $\int_{\mathbb{R}_v^3} f_n \psi(v)\, dv \overset{n}{\to} \int_{\mathbb{R}_v^3} f \psi(v)\, dv$ *in $L^1((0,T) \times \mathbb{R}_x^3))$ for all $T \in (0,\infty)$ and for all $\psi \in L^\infty(\mathbb{R}_v^3)$.*

2) $\int_{\mathbb{R}_v^3} Q^\pm(f_n, f_n)\, \psi(v)\, dv \overset{n}{\to} \int_{\mathbb{R}_v^3} Q^\pm(f,f)\, \psi(v)\, dv$ *locally in measure on $(0,\infty) \times \mathbb{R}_x^3$ for all $\psi \in L^\infty(\mathbb{R}_v^3)$ with compact support.*

3) $\liminf_{t\to\infty} \iint_{\mathbb{R}^3\times\mathbb{R}^3} dv\, dv_* \int_{S^2} B\, d\omega (f'_n f'_{n*} - f_n f_{n*}) \log \frac{f'_n f'_{n*}}{f_n f_{n*}} \geq \iint_{\mathbb{R}^3\times\mathbb{R}^3} dv\, dv_* \int_{S^2} B\, d\omega (f' f'_* - f f_*) \log \frac{f' f'_*}{f f_*}$, a.e. t, x.

4) $f \in C([0,\infty); L^1(\mathbb{R}_x^3 \times \mathbb{R}_v^3))$ *is a renormalized solution of* (11)–(12) *which satisfies $f|_{t=0} = f^0$ on $\mathbb{R}_x^3 \times \mathbb{R}_v^3$.*

As we said in the above remarks, uniqueness and regularity of solutions are not known in general. However, in the special case of the Vlasov-Poisson system that is when $B \equiv 0$, these questions are solved in [33] and follow from the following new a priori estimates.

Theorem 3. *Let $B \equiv 0$ (Vlasov-Poisson system) and let $f^0 \in L^1 \cap L^\infty(\mathbb{R}_x^2 \times \mathbb{R}_v^3)$ satisfy for some $k > 3$*

$$\iint_{\mathbb{R}_x^3\times\mathbb{R}_v^3} f^0 |v|^m \, dv\, dv < \infty \quad \textit{for all} \quad m \in [0,k)\,. \tag{29}$$

Then, there exists a renormalized solution $f \in C([0,\infty); L^1(\mathbb{R}_x^3 \times \mathbb{R}_v^3))$ satisfying

$$\sup_{t\in[0,T]} \iint_{\mathbb{R}_x^3\times\mathbb{R}_v^3} f(x,v,t) |v|^m \, dv\, dx < \infty\,, \qquad \textit{for all} \quad m \in [0,k)\,,\ T \in (0,\infty)\,. \tag{30}$$

Remarks. 1) We prove in fact in [33] a bound on $\iint_{\mathbb{R}^3_x \times \mathbb{R}^3_v} f(x,v,t) \cdot |v|^m \, dv \, dx$ which depends only on the bounds on f^0.

2) Some uniqueness and regularity results have been independently proven by K. Pfaffelmoser [38].

III. Mathematical Tools for the Global Existence of Weak Solutions in Kinetic Models

In this last section, we briefly present the three tools of independent interest that we use in the proofs of the global existence results (like Theorem 1) for weak solutions of kinetic models. These tools are 1) the notion of renormalized solutions and their properties, 2) a theory of a.e. flow-solutions of ordinary differential equations, 3) velocity-averaging lemmata.

We already described above the first tool (and its physical interpretation): we just want to mention here that this notion can be useful even for nonlinear equations involving second-order terms like Fokker-Planck-Boltzmann equations [14] or Landau equations [19] or like nonlinear elliptic equations (see P.L. Lions and F. Murat [32]). Let us emphasize the elementary fact that this notion consists essentially in writing down the equation formally satisfied by a nonlinear change of unknown.

The second tool is a theory of a.e. flows developed by R.J. DiPerna and the author [20, 21] that can be illustrated by the following example. Let Ω be a bounded smooth domain of $\mathbb{R}^N$, $(N \geq 1)$, let $B(x)$ be a vectorfield on Ω whose regularity will be discussed below. We want to study the flow associated to the following ordinary differential equation

$$\frac{dX}{dt} = B(X) \quad \text{for } t \in \mathbb{R}, \quad X(0) = x \in \overline{\Omega}. \tag{31}$$

We consider here only a homogeneous equation to simplify the presentation. This is also why we assume that $\overline{\Omega}$ is an invariant region, i.e.

$$B(x) \cdot \nu(x) = 0 \quad \text{on} \quad \partial\Omega \tag{32}$$

where ν is the unit exterior normal on $\partial\Omega$. Of course, X is a function of t and $x \in \overline{\Omega}$ and we expect $X(t,x)$ to belong to $\overline{\Omega}$ for all $t \in \mathbb{R}$, $x \in \overline{\Omega}$. We shall say that X is an a.e. flow associated to (31) if $X \in C(\mathbb{R}; L^1(\Omega))$ and if X satisfies

$$X(t,x) \in \overline{\Omega} \quad \text{a.e.} \quad x \in \Omega\,, \ \forall\, t \in \mathbb{R} \tag{33}$$

$$X(t+s,x) = X(t, X(s,x)) \quad \text{a.e.} \quad x \in \Omega\,, \ \forall\, t,s \in \mathbb{R} \tag{34}$$

$$\lambda(\{x \in \Omega \ / \ X(t,x) \in N\}) = 0 \quad \text{if} \quad \lambda(N) = 0\,, \ \forall\, t \in \mathbb{R} \tag{35}$$

$$\frac{\partial X}{\partial t} = B(X) \quad \text{in} \quad \mathscr{D}'(\mathbb{R} \times \Omega) \tag{36}$$

where λ denotes the Lebesgue measure restricted to Ω.

We also denote by $Y \circ \lambda$ the image measure of λ by an application Y from $\overline{\Omega}$ into $\overline{\Omega}$ i.e. the measure on $\overline{\Omega}$ defined by

$$\forall\, \varphi \in C(\overline{\Omega}), \quad \int \varphi\, d(Y \circ \lambda) = \int \varphi(Y(x))\, dx. \tag{37}$$

We also introduce the following condition

$$X_t \circ \lambda \leq e^{C_0|t|}\, \lambda, \quad \forall\, t \in \mathbb{R} \tag{38}$$

where $C_0 \geq 0$ and $X_t(x) = X(t,x)$. Of course, (38) implies (35). Then, we have the

Theorem 4. *Let* $B \in W^{1,1}(\Omega)$. *We assume* div $B \in L^\infty(\Omega)$ *and* (32). *We denote by* $C_0 = \|\operatorname{div} B\|_{L^\infty}$.

1) *There exists a unique a.e. flow associated to* (31) *satisfying* (38).

2) *In addition,* $X \in C^1(\mathbb{R}; L^{\frac{N}{N-1}}(\Omega)) \cap L^{\frac{N}{N-1}}(\Omega; W^{1,\frac{N}{N-1}}_{\text{loc}}(\mathbb{R}))$ *and, for almost all* $x \in \Omega$, $X \in C^1(\mathbb{R})$, $B(X) \in C(\mathbb{R})$, (31), (33) *and* (34) *hold.*

3) *The a.e. flow* X *satisfies also*

$$\frac{\partial X}{\partial t} = \operatorname{div}_x(BX) - (\operatorname{div} B)X \quad \text{in} \quad \mathcal{D}'(\mathbb{R} \times \Omega) \quad , \quad X|_{t=0} = x \quad \text{a.e. in } \Omega \tag{39}$$

and for all $u_0 \in L^1(\Omega)$, $u(X(t,x))$ *is the unique renormalized solution in* $C(\mathbb{R}; L^1(\Omega))$ *of*

$$\frac{\partial u}{\partial t} = \operatorname{div}_x(Bu) - (\operatorname{div} B)u, \quad u|_{t=0} = u_0 \quad \text{a.e. in } \Omega. \tag{40}$$

If $u^0 \in L^\infty(\Omega)$, $u(X(t,x))$ *is also the unique solution of* (40) *in* $L^\infty(\mathbb{R} \times \Omega)$ *of* (40) *in the senese of distributions.*

4) *Let* $B_n \in W^{1,1}(\Omega)$ *satisfy* (32) *and* div $B_n \in L^\infty(\Omega)$ *for all* $n \geq 1$. *We assume that* B_n *converges to* B *in* $L^1(\Omega)$ *and that* div B_n *converges to* div B *in* $L^1(\Omega)$. *We denote by* X_n *the a.e. flow satisfying* (38) *corresponding to* B_n. *Then,* X_n *converges to* X *in* $C([-T,+T]; L^q(\Omega))$, $(\forall\, q < \infty)$; $\frac{\partial X_n}{\partial t}$ *converges to* $\frac{\partial X}{\partial t}$ *in* $C([-T,+T]; L^1(\Omega))$ *for all* $T < \infty$ *and* $X_n(\cdot,x)$ *converges to* $X(\cdot,x)$ *uniformly on compact sets of* $\mathbb{R}$ *for almost all* $x \in \Omega$. *Finally, if* u_n^0 *converges in* $L^p(\Omega)$ *to* u^0 $(1 \leq p < \infty)$, *then* $u_n(t,x) = u_n^0(X_n(t,x))$ *converges to* $u(t,x) = u^0(X(t,x))$ *in* $C([-T,+T]; L^p(\Omega))$, $(\forall\, T < \infty)$.

Remarks. 1) Analogous results hold for time-dependent vector fields $B(t,x) \in L^1((-T,+T); W^{1,1}(\Omega))$ with div $B \in L^1((-T,+T); L^\infty(\Omega))$, $(\forall\, T < \infty)$ or for flows in the whole space.

2) In [20], we show by a counterexample that the $W^{1,1}(\Omega)$ regularity is in general optimal.

The very definition of renormalized solutions of (11)–(12) requires to be able to build an a.e. flow as above for the vector field $B(t,x,v) = (v, -\nabla_x \Phi(x,t))$. Of course, we have $\operatorname{div}_{(x,v)} B = 0$. Therefore, in order to prove Theorem 1 and in view of the sharpness of the $W^{1,1}$ regularity, we have to show that B is in (say) $L^1_{\text{loc}}(W^{1,1}_{\text{loc}})$ or in other words that $\Phi \in L^1_{\text{loc}}(W^{2,1}_{\text{loc}})$. But, this is not clear in view of the estimates on f and ϱ which are basically L^1 estimates. However, classical results from Harmonic analysis indicate that a rather sharp condition for this regularity to hold is to check that $\varrho \in L^1_{\text{loc}}(\mathscr{H}^1_{\text{loc}})$ where $\mathscr{H}^1$ denotes the multi-dimensional Hardy space. But, since f and ϱ are nonnegative, this is known to be equivalent to $\varrho \in L^1_{\text{loc}}(L^1 \log L^1_{\text{loc}})$ where we denote by $L^1 \log L^1_{\text{loc}}$ the set of functions φ such that $|\varphi| \log |\varphi|$ is in L^1_{loc}. Again this integrability of ϱ is not clear and requires some careful analysis. In fact, this integrability of ϱ is a consequence of the estimates (24) and again this is a sharp statement making thus a rather remarkable chain of sharp statements in order to produce the desired a.e. flow (that is the particle paths). Indeed, one can show that if $f|v|^2 \in L^1_{x,v}$, $f \log f \in L^1_{x,v}$ then $\varrho \log \varrho \in L^1_x$, a striking fact in view of the following classical interpolation result: if $f|v|^2 \in L^1_{x,v}$, $f \in L^p_{x,v}$, $(1 \le p \le \infty)$ then $\varrho \in L^q_{x,v}$ with $q = \frac{Np+2p}{Np+2}$. Observe that $q < p$ if $p > 1$ and that, in some sense, this loss of integrability vanishes "before L^1 at the $L^1 \log L^1$ level".

The final tool that we use in the proof of Theorem 1 is the improved regularity of macroscopic quantities that is velocity averages. The first indications of such a phenomenon were some results by V.I. Agoshkov [1] and by F. Golse, B. Perthame and R. Sentis [26]. The first general and sharp results were obtained in F. Golse, P.L. Lions, B. Perthame and R. Sentis [25] and partially extended in R.J. DiPerna and P.L. Lions [17], P. Gérard [24]. A complete theory is now available (R.J. DiPerna, P.L. Lions and Y. Meyer [22]) and we now present one example of such results.

Theorem 5. *Let $1 < p \le 2$, let $\tau \in [0,1)$, let $m \ge 0$, let $\psi \in C_0^\infty(\mathbb{R}^N)$ and let $f(x,v)$ (resp. $f(x,v,t)$) $\in L^p(\mathbb{R}^N_x \times \mathbb{R}^N_v)$ (resp. $L^p(\mathbb{R}^N_x \times \mathbb{R}^N_v \times \mathbb{R}_t)$) satisfy*

$$(-\Delta_x + 1)^{-\tau/2} (-\Delta_v + 1)^{-m/2} \{v \cdot \nabla_x f\} \in L^p_{x,v} \tag{41}$$

(resp.

$$(-\Delta_{x,t} + 1)^{-\tau/2} (-\Delta_v + 1)^{-m/2} \left\{ \frac{\partial f}{\partial t} + v \cdot \nabla_x f \right\} \in L^p_{x,v,t} \text{ .)} \tag{42}$$

Then, $\int_{\mathbb{R}^N} f(x,v)\psi(v)\,dv$ (resp. $\int_{\mathbb{R}^N} f(x,v,t)\,\psi(v)\,dv$) $\in B_2^{s,p}(\mathbb{R}^N_x)$ (resp. $B_2^{s,p}(\mathbb{R}^N_x \times \mathbb{R}_t)$) where $s = (1-\tau)\,\frac{1}{(1+m)}\,\frac{1}{p'}$.

This result proved by careful Fourier analysis admits various variants or extensions (see [22]).

We would like to conclude by indicating that these tools and the methods of proofs we used for the construction of global weak solutions admit various applications not only to the study of kinetic models and to the two categories of problems listed in the Introduction but also to other fields of Nonlinear Partial Differential Equations. One example of this fact is a recent work by P.L. Lions, B. Perthame and E. Tadmor [34] on conservation laws.

References

1. V.I. Agoshkov: Spaces of functions with differential-difference characteristics and smoothness of solutions of the transport equation. Sov. Math. Dokl. **29** (1984) 662–666
2. L. Arkeryd: On the long time behaviour of the Boltzmann equation in a periodic box. Preprint
3. L. Arkeryd, C. Cercignani: On the convergence of solutions of the Enskog equation to solutions of the Boltzmann equation. Preprint
4. A.A. Arsenev: Global existence of a weak solution of Vlasov's systems of equations. USSR Comp. Math. Math. Phys. **15** (1975) 131–143
5. C. Bardos, F. Golse, D. Levermore: Work in preparation
6. L. Boltzmann: Weitere Studien über das Wärmegleichgewicht unter Gasmolekülen. Sitzungsberichte der Akademie der Wissenschaften Wien **66** (1872) 275–370 [Translation: Further studies on the thermal equilibrium of gaz molecules. In: Kinetic Theory 2, ed. S.G. Brush. Pergamon Press, Oxford 1966]
7. T. Carleman: Problèmes mathématiques dans la théorie cinétique des gaz. Notes written by L. Carleson and O. Frostman. Publications Mathématiques de l'Institut Mittag-Leffler. Almquist and Wiksell, Uppsala 1957
8. C. Cercignani: The Boltzmann equation and its applications. Springer, Berlin Heidelberg New York 1988
9. C. Cercignani: Work in preparation
10. S. Chapman, T.G. Cowling: The mathematical theory of non-uniform gases. 2nd edn. Cambridge Univ. Press, Cambridge 1952
11. L. Desvillettes: Convergence to equilibrium in large time for Boltzmann and B.G.K. equations. Arch. Rat. Mech. Anal. **110** (1990) 73–91
12. L. Desvillettes: On the convergence of splitting algorithms for some kinetic equations. Preprint
13. J.M. Dolbeault: Work in preparation
14. R.J. DiPerna, P.L. Lions: On the Fokker-Planck-Boltzmann equation. Comm. Math. Phys. **120** (1988) 1–23
15. R.J. DiPerna, P.L. Lions: On the Cauchy problem for Boltzmann equations: Global existence and weak stability. C.R. Acad. Sci. Paris **306** (1988) 343–346; Ann. Math. **130** (1989) 321–366
16. R.J. DiPerna, P.L. Lions: Global solutions of Boltzmann equations and the entropy inequality. Arch. Rat. Mech. Anal. (1990)
17. R.J. DiPerna, P.L. Lions: Global weak solutions of Vlasov-Maxwell systems. Comm. Pure Appl. Math. **XLII** (1989) 729–757
18. R.J. DiPerna, P.L. Lions: Solutions globales d'équations du type Vlasov-Poisson. C.R. Acad. Sci. Paris **307** (1988) 655–658; and work in preparation
19. R.J. DiPerna, P.L. Lions: Global weak solutions of kinetic equations. Sem. Matematico Torino (1990)
20. R.J. DiPerna, P.L. Lions: Ordinary differential equations, Sobolev spaces and transport theory. Invent. math. **98** (1989) 511–547
21. R.J. DiPerna, P.L. Lions: Equations différentielles ordinaires et équations de transport avec des coefficients irréguliers. In: Séminaire EDP 1988-1989, Ecole Polytechnique, Palaiseau 1989
22. R.J. DiPerna, P.L. Lions, Y. Meyer: L^p regularity of velocity averages. Ann. I.H.P. Anal. Non-Lin. (1991)
23. M.J. Esteban, B. Perthame: On the modified Enskog equation with elastic or inelastic collisions; Models with spin. Ann. I.H.P. Anal. Non-Lin. (1991)

24. P. Gérard: Moyennes de solutions d'équations aux dérivées partielles. In: Séminaire EDP 1986–1867, Ecole Polytechnique, Palaiseau 1987
25. F. Golse, P.L. Lions, B. Perthame, R. Sentis: Regularity of the moments of the solution of a transport equation. J. Funct. Anal. **76** (1988) 110–125
26. F. Golse, B. Perthame, R. Sentis: Un résultat de compacité pour les équations de transport et applications au calcul de la limite de la valeur propre principale d'un opérateur de transport. C.R. Acad. Sci. Paris **301** (1985) 341–344
27. H. Grad: Principles of the kinetic theory of gases. In: Flügge's Handbuch der Physik XII. Springer, Berlin Heidelberg New York 1958
28. K. Hamdache: Existence in the large and asymptotic behaviour for the Boltzmann equation. Japan J. Appl. Math. **2** (1985) 1–15
29. K. Hamdache: Global existence for weak solutions for the initial boundary value problems of Boltzmann equation. Arch. Rat. Mech. Anal. (1991)
30. R. Illner, M. Shinbrot: The Boltzmann equation. Global existence for a rare gas in an infinite vacuum. Comm. Math. Phys. **95** (1984) 217–226
31. J. Leray: Etude de diverses équations intégrales nonlinéaires et de quelques problèmes que pose l'hydrodynamique. J. Math. Pures Appl. **12** (1933) 1–82
32. P.L. Lions, F. Murat: Work in preparation
33. P.L. Lions, B. Perthame: Régularité des solutions du système de Vlasov-Poisson en dimension 3. C.R. Acad. Sci. Paris, **311** (1990) 205–210; and work in preparation
34. P.L. Lions, B. Perthame, E. Tadmor: Formulation cinétique des lois de conservation scalaires multidimensionnelles. C.R. Acad. Sci. Paris 1990; and work in preparation
35. J.C. Maxwell: On the dynamical theory of gases. Phil. Trans. Roy. Soc. London **157** (1966) 49–88
36. T. Nishida, K. Imai: Global solutions to the initial-value problem for the nonlinear Boltzmann equation. Publ. R.I.M.S. Kyoto Univ. **12** (1976) 229–239
37. B. Perthame: Global existence of solutions to the BGK model of Boltzmann equations. J. Diff. Eq. **82** (1989) 191–205
38. K. Pfaffelmoser: Globale klassische Lösungen des dreidimensionalen Vlasov-Poisson systems. Preprint
39. J. Polewczak: Global existence in L^1 for the modified nonlinear Enskog equation in $\mathbb{R}^3$. Preprint
40. C. Truesdell, R.G. Muncaster: Fundamentals of Maxwell's kinetic theory of a simple monoatomic gas. Academic Press, New York 1980
41. S. Ukai: Solutions of the Boltzmann equation. In: Patterns and Waves. Qualitative analysis of differential equations. North-Holland, Amsterdam 1986, pp. 37–96

Sheaf Theory for Partial Differential Equations

Pierre Schapira

Département de Mathématiques, Université Paris-Nord, F-93430 Villetaneuse, France

0. Introduction

In order to analyze the singularities of hyperfunction solutions of systems of partial differential equations, M. Sato introduced in 1969 the microlocalization functor and, more fundamentally, the microlocal point of view. Then began an intense activity, in what is now called "microlocal analysis", and in the field of analytical partial differential equations the main tools, microdifferential operators and quantized contact transformations, were developed in Sato-Kawai-Kashiwara's paper [S-K-K]. However, after their study of micro-hyperbolic systems [K-S1], M. Kashiwara and the author realized that for many problems these analytical tools were not really necessary on the condition to work with the complex of holomorphic solutions of the system, $R\mathcal{H}om_{\mathcal{D}_X}(\mathcal{M}, \mathcal{O}_X)$, and only to keep in mind the codirections of non-propagation of this complex, here the characteristic variety of $\mathcal{M}$. In other words one simply works with a complex of sheaves F on a real manifold X, and what we defined as its micro-support, $SS(F)$, a closed conic involutive subset of T^*X.

This was the starting point of the "microlocal study of sheaves", developed in [K-S3, K-S4].

It is not our purpose to discuss this theory here, but we need to recall a few basic facts in order to introduce the new notion of an elliptic pair (obtained in collaboration with J.-P. Schneiders), a generalization of that of an elliptic system, the real manifold M on which the system is elliptic being replaced by an **R**-constructible sheaf. We construct a characteristic class associated with an elliptic pair, and prove that when the pair has compact support, the complex of its holomorphic solutions has finite dimensional cohomology and the index is calculated as the integral of the characteristic class.

1. Microlocal Study of Sheaves

In this section we fix some notations and recall a few results of [K-S3, K-S4].

Let X be a real C^∞-manifold. One denotes by $\tau : TX \to X$ and by $\pi : T^*X \to X$ its tangent and cotangent bundles, respectively. If M is a submanifold,

Proceedings of the International Congress
of Mathematicians, Kyoto, Japan, 1990

one denotes by $T_M X$ and $T^*_M X$ the normal and conormal bundles to M in X, respectively. In particular $T^*_X X$ is the zero-section of T^*X.

One denotes by $\delta : \Delta \hookrightarrow X \times X$ the diagonal embedding and we identify X with Δ and T^*X with $T^*_\Delta(X \times X)$ by the first projection defined on $X \times X$ and on $T^*(X \times X) \simeq T^*X \times T^*X$, respectively. If Λ is a subset of T^*X, Λ^a will denote its image by the antipodal map on T^*X.

If X and Y are two manifolds, one denotes by q_1 and q_2 the first and second projection, defined on $X \times Y$.

Let A be a commutative unitary ring with finite global homological dimension (e.g. $A = \mathbf{Z}$). One denotes by $D(X)$ the derived category of the category of sheaves of A-modules on X, and by $D^b(X)$ the full subcategory consisting of objects with bounded cohomology. If Z is a locally closed subset of X, one denotes by A_Z the sheaf on X which is constant with stalk A on Z and zero on $X \setminus Z$. One denotes by or_X the orientation sheaf on X, and by ω_X the dualizing complex on X. Hence:

$$\omega_X \simeq or_X[\dim X],$$

where $\dim X$ is the dimension of X.

The "six operations" (as says Grothendieck), that is, the operations $\otimes^L$, $R\mathcal{H}om$, Rf_*, $Rf_!$, f^{-1}, $f^!$, are now classical tools that we shall not recall. We simply introduce some notations. For $F \in \mathrm{Ob}(D^b(X))$ and $G \in \mathrm{Ob}(D^b(Y))$, one sets:

$$F \boxtimes^L G = q_1^{-1}F \otimes^L q_2^{-1}G,$$

$$D'F = R\mathcal{H}om(F, A_X),$$

$$DF = R\mathcal{H}om(F, \omega_X).$$

There are other operations of interest on sheaves. If M is a closed submanifold of X and $F \in \mathrm{Ob}(D^b(X))$, the specialization of F along M, $\nu_M(F)$, is an object of $D^b(T_M X)$ and the microlocalization of F along M, $\mu_M(F)$, an object of $D^b(T^*_M X)$. Sato's functor μ_M has been generalized in [K-S3] as follows. For F and G in $D^b(X)$ on sets:

$$\mu\hom(G, F) = \mu_\Delta R\mathcal{H}om(q_2^{-1}G, q_1^!F).$$

Then:

$$R\pi_*\mu\hom(G, F) \simeq R\mathcal{H}om(G, F),$$

$$\mu\hom(A_M, F) \simeq \mu_M(F).$$

After the introduction of the functor μ_M it became natural to work in T^*X, and M. Kashiwara and the author introduced in 1982 (cf. [K-S2]) the micro-support $SS(F)$ of an object F of $D^b(X)$. This is a closed conic subset of T^*X which roughly speaking describes the set of codirections of non-propagation of F. More precisely:

Definition 1.1. We say that an open subset U of T^*X does not meet $SS(F)$ if for any real C^1-function φ on X and any $x_0 \in X$ such that $(x_0; d\varphi(x_0)) \in U$, one has:

$$\left(R\Gamma_{\{x;\varphi(x)\geq\varphi(x_0)\}}(F)\right)_{x_0} = 0\,.$$

An important property of the micro-support is given by:

Theorem 1.2. *Let $F \in \mathrm{Ob}(D^b(X))$. Then $SS(F)$ is an involutive subset of T^*X.*

(For the precise definition of "involutive", cf. [K-S4, Ch. VI].)

One can evaluate the micro-support of sheaves after the main operations described above. For example one proves that for F and G in $D^b(X)$:

$$SS(\mu\hom(G,F)) \subset C(SS(F), SS(G))\,, \tag{1.1}$$

where $C(\Lambda_1,\Lambda_2)$ is the normal cone of Λ_1 along Λ_2, a closed subset of TT^*X that we identify with a subset of T^*T^*X by the Hamiltonian isomorphism. In particular:

$$\mathrm{supp}(\mu\hom(G,F)) \subset SS(G) \cap SS(F)\,. \tag{1.2}$$

Let $f : Y \to X$ be a morphism of manifolds. One associates the maps:

$$T^*Y \xleftarrow[{}^tf']{} Y \underset{X}{\times} T^*X \xrightarrow[f_\pi]{} T^*X \tag{1.3}$$

and one sets:

$$T^*_YX = {}^tf'^{-1}(T^*_YY)\,. \tag{1.4}$$

Using (1.1), one can evaluate the micro-support of $f^{-1}F$ or $f^!F$ (cf. [K-S3]). In particular if f is non-characteristic for F, that is

$$T^*_YX \cap f_\pi^{-1}(SS(F)) \subset Y \underset{X}{\times} T^*_XX\,, \tag{1.5}$$

one gets:

$$SS(f^{-1}F) \subset {}^tf' f_\pi^{-1}(SS(F))\,. \tag{1.6}$$

Similarly, if $G \in \mathrm{Ob}(D^b(Y))$ and f is proper on supp(G), one proves:

$$SS(Rf_*G) \subset f_\pi\, {}^tf'^{-1}(SS(G))\,. \tag{1.7}$$

Remark that formulas (1.6) and (1.7) are similar to classical formulas obtained when calculating the wave front set of distributions or hyperfunctions or when calculating the characteristic variety of $\mathscr{D}$-modules, after non characteristic inverse images of proper direct images.

2. Constructible Sheaves (cf. [K-S3, K-S4])

In this section we assume all manifolds are real analytic and the base ring A is noetherian. An object F of $D^b(X)$ is called weakly **R**-constructible(w-**R**-

constructible for short) if there exists a subanalytic stratification $X = \sqcup_\alpha X_\alpha$ such that for all α, all $j \in \mathbf{Z}$, the sheaves $H^j(F)|_{X_\alpha}$ are locally constant. If moreover for each $x \in X$, each $j \in \mathbf{Z}$, the stalk $H^j(F)_x$ is finitely generated, one says F is **R**-constructible. One denotes by $D^b_{\text{w}-\mathbf{R}-c}(X)$ (resp. $D^b_{\mathbf{R}-c}(X)$) the full subcategory of $D^b(X)$ consisting of w-**R**-constructible (resp. **R**-constructible) objects.

The involutivity Theorem 1.2 allows us to characterize microlocally w-**R**-constructible objects.

Theorem 2.1. *Let $F \in \mathrm{Ob}(D^b(X))$. The following conditions are equivalent.*

*(a) F is w-**R**-constructible.*

(b) $SS(F)$ is contained in a closed conic subanalytic isotropic subset of T^*X.

(c) $SS(F)$ is a closed conic subanalytic Lagrangian subset of T^*X.

By this result one proves easily that the category of w-**R**-constructible (resp. **R**-constructible) sheaves is stable by the main operations on sheaves (Rf_* when f is proper, f^{-1}, $f^!$, $\otimes^L$, $R\mathcal{H}om$, μ_M, ν_M, $\mu\mathrm{hom}$).

If X is a complex manifold one defines similarly the notions of w-**C**-constructible and **C**-constructible sheaves, by assuming that the stratas of the subanalytic stratification $X = \sqcup_\alpha X_\alpha$ are complex analytic submanifolds. Then the link between **R**- and **C**-constructibility is given by:

Theorem 2.2. *Let $F \in \mathrm{Ob}(D^b_{\text{w}-\mathbf{R}-c}(X))$. Then F is w-**C**-constructible if and only if $SS(F)$ is conic for the action of $\mathbf{C}^\times$ on T^*X.*

Remark 2.3. Note that the microlocal study of constructible sheaves was initiated by Kashiwara [K1].

3. $\mathcal{D}$-Modules

We shall not review this theory here and refer to [S-K-K, K1, S1] for detailed expositions. We shall only fix a few notations and make the link with the micro-support.

From now on the base ring A is the field **C** of complex numbers.

Let $(X, \mathcal{O}_X)$ be a complex manifold of complex dimension n. One denotes by $\mathcal{D}_X$ (resp. $\mathcal{D}_X^\infty$) the sheaf on X of finite order (resp. infinite order) holomorphic differential operators and one sets $\Omega_X = \mathcal{O}_X^{(n)} \otimes or_X$, where $\mathcal{O}_X^{(n)}$ is the sheaf of holomorphic n-forms. One denotes by $D(\mathcal{D}_X)$ (resp. $D(\mathcal{D}_X^{op})$) the derived category of the abelian category of left (resp. right) $\mathcal{D}_X$-modules, and by $D^b(\mathcal{D}_X)$ (resp. $D^b_{\text{coh}}(\mathcal{D}_X)$) the full triangulated subcategory of $D(\mathcal{D}_X)$ consisting of objects with bounded (resp. bounded and coherent) cohomology. One defines similarly $D^b(\mathcal{D}_X^{op})$ and $D^b_{\text{coh}}(\mathcal{D}_X^{op})$.

If $\mathcal{M}$ is an object of $D^b_{\text{coh}}(\mathcal{D}_X)$, its characteristic variety denoted $\mathrm{char}(\mathcal{M})$, is a closed conic analytic subset of T^*X, which is involutive ([S-K-K]). In fact one has:

Theorem 3.1 ([K-S2]). *Let* $\mathcal{M} \in Ob(D^b_{\text{coh}}(\mathcal{D}_X))$. *Then:*

$$SS(R\,\mathcal{H}om_{\mathcal{D}_X}(\mathcal{M}, \mathcal{O}_X)) = \text{char}(\mathcal{M}).$$

Note that the inclusion $\subset$ in Theorem 3.1, which is the most useful for applications, is easily deduced from the Cauchy-Kowalevski theorem, in its precised form due to Leray [L]. The converse inclusion makes use of the sheaf of rings $\mathcal{E}_X^{\mathbf{R}}$ of [S-K-K]. Also note that this result, combined with Theorem 1.2, gives a new proof of the involutivity of the characteristic variety of $\mathcal{D}$-modules.

Let $f : Y \to X$ be a morphism of complex manifolds. One denotes by $\mathcal{D}_{Y\to X}$ the sheaf $\mathcal{O}_Y \otimes_{f^{-1}\mathcal{O}_X} f^{-1}\mathcal{D}_X$ endowed with its natural structure of a $(\mathcal{D}_Y, f^{-1}\mathcal{D}_X)$-bimodule.

Let $\mathcal{M} \in \text{Ob}(D^b(\mathcal{D}_X))$. One sets:

$$\underline{f}^{-1}\mathcal{M} = \mathcal{D}_{Y\to X} \otimes^L_{f^{-1}\mathcal{D}_X} f^{-1}\mathcal{M}.$$

Let $\mathcal{N} \in \text{Ob}(D^b(\mathcal{D}_Y^{op}))$. One sets

$$\underline{f}_*\mathcal{N} = Rf_*(\mathcal{N} \otimes^L_{\mathcal{D}_Y} \mathcal{D}_{Y\to X}).$$

If $\mathcal{M} \in \text{Ob}(D^b(\mathcal{D}_X))$ and $\mathcal{N} \in \text{Ob}(D^b(\mathcal{D}_Y))$ one sets:

$$\mathcal{M}\underline{\boxtimes}\mathcal{N} = \mathcal{D}_{X\times Y} \otimes_{\mathcal{D}_X \boxtimes \mathcal{D}_Y} (\mathcal{M} \boxtimes \mathcal{N}),$$

and there is a similar formula for right modules.

Finally one sets:

$$\underline{D}'\mathcal{M} = R\,\mathcal{H}om_{\mathcal{D}_X}(\mathcal{M}, \mathcal{D}_X),$$
$$\underline{D}\mathcal{M} = R\,\mathcal{H}om_{\mathcal{D}_X}(\mathcal{M}, \Omega_X \otimes_{\mathcal{O}_X} \mathcal{D}_X[n]).$$

(In this last formula, $\mathcal{M}$ is a right module.)

4. Microfunction Solutions of $\mathcal{D}$-Modules

Let M be a real analytic manifold, X a complexification of M, $\mathcal{M}$ a left coherent $\mathcal{D}_X$-module. By considering the complex $R\,\mathcal{H}om_{\mathcal{D}_X}(\mathcal{M}, \mathcal{O}_X)$ and using Theorem 3.1, one may recover many classical results. For example, applying (1.2) with $G = D'(\mathbf{C}_M)$ one gets:

$$\text{supp}(R\,\mathcal{H}om_{\mathcal{D}_X}(\mathcal{M}, \mathcal{C}_M)) \subset T^*_M X \cap \text{char}(\mathcal{M}),$$

where $\mathcal{C}_M$ is the sheaf of Sato microfunctions. In particular this shows that the analytic wave front set of a hyperfunction solution of a system of linear differential equations is contained in the characteristic variety of the system ("Sato's principle"). More generally, the inclusion (1.1) immediately implies that the microfunction solutions of the system $\mathcal{M}$ extend in the micro-hyperbolic directions, and one recovers the results of [K-S1] in the differential case. Microdifferential systems can be treated similarly, once the microlocal action of $\mathscr{E}_X^{\mathbf{R}}$ on $\mathscr{O}_X$ is defined as in [K-S3].

These techniques can also be applied to the study of boundary value problems and diffraction, including the case of non smooth obstacles. It is then useful to introduce new sheaves of microfunctions (using the functor μhom) and new wave front sets. We refer to [S2] for details.

5. Elliptic Pairs

In this section we expose new results obtained in collaboration with J.-P. Schneiders. Let X be a complex manifold of complex dimension n. If there is no risk of confusion, we identify X with the real underlying manifold.

Definition 5.1. An elliptic pair $(\mathscr{M}, F)$ on X is the data of $\mathscr{M} \in \mathrm{Ob}(D^b_{\mathrm{coh}}(\mathscr{D}_X))$ and $F \in \mathrm{Ob}(D^b_{\mathbf{R}-c}(X))$ satisfying: $\mathrm{char}(\mathscr{M}) \cap SS(F) \subset T^*_X X$.

We use the same terminology for objects of $D^b_{\mathrm{coh}}(\mathscr{D}_X^{op})$.

Theorem 5.2 (cf. [S-Sc]). *Let $(\mathscr{M}, F)$ be an elliptic pair.*

(i) Regularity. The natural morphism:

$$R\mathscr{H}om_{\mathscr{D}_X}(\mathscr{M}, D'F \otimes \mathscr{O}_X) \to R\mathscr{H}om_{\mathscr{D}_X}(\mathscr{M} \otimes F, \mathscr{O}_X)$$

is an isomorphism.

(ii) Assume $\mathrm{supp}(\mathscr{M}) \cap \mathrm{supp}(F)$ *is compact.*

(a) Finiteness. For all $j \in \mathbf{Z}$, the **C***-vector spaces*

$$H^j R\Gamma(X; R\mathscr{H}om_{\mathscr{D}_X}(\mathscr{M} \otimes F, \mathscr{O}_X))$$

are finite dimensional.

(b) Duality. The pairing

$$R\mathscr{H}om_{\mathscr{D}_X}(\mathscr{M} \otimes F, \mathscr{O}_X) \otimes (\Omega_X \otimes^L_{\mathscr{D}_X} \mathscr{M} \otimes F) \longrightarrow \Omega_X \otimes^L_{\mathscr{D}_X} \mathscr{O}_X$$

and the integration morphism $H^n_c(X; \Omega_X \otimes^L_{\mathscr{D}_X} \mathscr{O}_X) \simeq H^{2n}_c(X; or_X) \longrightarrow \mathbf{C}$ induce a perfect duality on the spaces $H^j R\Gamma(X; R\mathscr{H}om_{\mathscr{D}_X}(\mathscr{M} \otimes F, \mathscr{O}_X))$ and $H^{n-j} R\Gamma(X; \Omega_X \otimes^L_{\mathscr{D}_X} \mathscr{M} \otimes F)$.

(c) Parameters. Let Y be another complex manifold. Then the natural morphism

$$(Rq_{2*} q_1^{-1} R\mathscr{H}om_{\mathscr{D}_X}(\mathscr{M} \otimes F, \mathscr{O}_X)) \otimes \mathscr{O}_Y \longrightarrow Rq_{2*} R\mathscr{H}om_{q_1^{-1}\mathscr{D}_X}(q_1^{-1}(\mathscr{M} \otimes F), \mathscr{O}_{X \times Y})$$

is an isomorphism.

Sketch of Proof. (i) follows from a general result of [K-S4] which asserts that the natural morphism

$$R\mathscr{H}om(F, A_X) \otimes^L G \to R\mathscr{H}om(F, G)$$

is an isomorphism as soon as $SS(F) \cap SS(G)$ is contained in $T^*_X X$.

(ii) Using techniques of [Sc] one can reduce the problem to the case where $\mathcal{M}$ admits a free presentation. Next by adding the Cauchy-Riemann system to $\mathcal{M}$, one can assume F is supported by a real analytic manifold M whose complexification is X. Then one represents F by a bounded complex whose components are direct sums of sheaves $\mathbf{C}_U$, U open, relatively compact, subanalytic in M and such that $D'_M\mathbf{C}_U = \mathbf{C}_{\overline{U}}$, (here D'_M is the duality functor on M). Then (ii) is proved by similar arguments as those in [B-S] or [R-R]. □

By adapting Kashiwara's construction of the characteristic cycle of **R**-constructible sheaves (cf. [K-S4, Ch. IX]) we shall now construct a characteristic class associated with an elliptic pair.

Let $(\mathcal{M}, F)$ be an elliptic pair and assume $\mathcal{M}$ is a right module. Sato's isomorphism $\mathcal{D}_X^\infty \simeq \delta^! \mathcal{O}_{X\times X}^{(0,n)}[n]$ induces a morphism:

$$R\mathcal{H}om_{\mathcal{D}_X}(\mathcal{M}\otimes F, \mathcal{M}\otimes F) \to \delta^!(\underline{D}(\mathcal{M}\otimes F)\boxtimes(\mathcal{M}\otimes F)\otimes^L_{\mathcal{D}_{X\times X}}\mathcal{O}_{X\times X}). \tag{5.1}$$

Moreover the natural morphism:

$$\Omega_X \otimes_{\mathcal{O}_X} \mathcal{D}_X[n] \otimes^L_{\mathcal{D}_X\otimes\mathcal{D}_X} \delta^{-1}\mathcal{O}_{X\times X} \to \Omega_X \otimes^L_{\mathcal{D}_X} \mathcal{O}_X[n]$$

induces a morphism:

$$\delta^{-1}(\underline{D}(\mathcal{M}\otimes F)\boxtimes(\mathcal{M}\otimes F)\otimes^L_{\mathcal{D}_{X\times X}}\mathcal{O}_{X\times X}) \to \omega_X. \tag{5.2}$$

Set for short:

$$H = \underline{D}(\mathcal{M}\otimes F)\boxtimes(\mathcal{M}\otimes F)\otimes^L_{\mathcal{D}_{X\times X}}\mathcal{O}_{X\times X}. \tag{5.3}$$

We get the chain of morphisms:

$$\begin{aligned} R\mathcal{H}om_{\mathcal{D}_X}(\mathcal{M}\otimes F, \mathcal{M}\otimes F) &\to \delta^! H \\ &\to \delta^!\delta_*\delta^{-1}H \simeq \delta^{-1}H \\ &\to \omega_X. \end{aligned}$$

Hence setting:

$$S = \operatorname{char}(\mathcal{M}) \cap \operatorname{supp}(F), \tag{5.4}$$

we get a morphism:

$$\operatorname{Hom}_{\mathcal{D}_X}(\mathcal{M}\otimes F, \mathcal{M}\otimes F) \to H^0_S(X;\omega_X). \tag{5.5}$$

In fact this construction can be made "microlocal". Set:

$$\Lambda = \operatorname{char}(\mathcal{M}) + SS(F)^a. \tag{5.6}$$

Since the micro-support of H is contained in $\Lambda \times \Lambda^a$, we have the isomorphisms:

$$\begin{aligned} \delta^! H &\simeq R\pi_*\mu_\Delta H \\ &\xleftarrow{\sim} R\pi_* R\Gamma_\Lambda \mu_\Delta H, \end{aligned}$$

and we get the morphism:

$$\mathrm{Hom}_{\mathscr{D}_X}(\mathscr{M} \otimes F, \mathscr{M} \otimes F) \to H^0_\Lambda(T^*X; \pi^{-1}\omega_X). \tag{5.7}$$

Definition 5.3. We call the image of 1 in $H^0_S(X;\omega_X)$ (resp. in $H^0_\Lambda(T^*X;\pi^{-1}\omega_X)$) by the morphism (5.5) (resp. (5.7)) the Euler class (resp. the microlocal Euler class) of the elliptic pair $(\mathscr{M}, F)$ and we denote it by $\mathrm{eu}(\mathscr{M}, F)$ (resp. $\mu\mathrm{eu}(\mathscr{M}, F)$).

Of course $\mathrm{eu}(\mathscr{M}, F)$ is the restriction of $\mu\mathrm{eu}(\mathscr{M}, F)$ to the zero-section of T^*X.

Definition 5.4. Let $(\mathscr{M}, F)$ be an elliptic pair such that $\mathrm{supp}(\mathscr{M}) \cap \mathrm{supp}(F)$ is compact and assume $\mathscr{M}$ is a right module. One sets:

$$\chi(X;\mathscr{M},F) = \sum_j (-1)^j \dim(H^j(R\Gamma(X; F \otimes \mathscr{M} \otimes^L_{\mathscr{D}_X} \mathscr{O}_X))).$$

By adapting to the case of $\mathscr{D}$-modules Kashiwara's proof of the index theorem for constructible sheaves (cf. [K-S4, Ch. IX]), we can prove:

Theorem 5.5. *In the situation of Definition 5.4, one has:*

$$\chi(X;\mathscr{M},F) = \int_X \mathrm{eu}(\mathscr{M}, F). \tag{5.8}$$

Examples and Comments 5.6. (a) Assume X is the complexification of a real analytic manifold M, and let $\mathscr{M}$ be a coherent $\mathscr{D}_X$-module. Then $\mathscr{M}$ is elliptic on M in the classical sense if an only if $(\mathscr{M}, \mathbf{C}_M)$ is an elliptic pair, which simply means that:

$$\mathrm{char}(\mathscr{M}) \cap T^*_M X \subset T^*_X X.$$

In this case the isomorphism of Theorem 5.2 applied to the elliptic pair $(\mathscr{M}, D'\mathbf{C}_M)$ gives the isomorphism:

$$R\mathscr{H}om_{\mathscr{D}_X}(\mathscr{M}, \mathscr{A}_M) \simeq R\mathscr{H}om_{\mathscr{D}_X}(\mathscr{M}, \mathscr{B}_M)$$

where $\mathscr{A}_M$ (resp. $\mathscr{B}_M$) denotes the sheaf of real analytic functions (resp. hyperfunctions) on M. If M is compact, one recovers the classical finiteness theorem for elliptic systems, and the index is calculated by the Atiyah-Singer theorem [A-S].

(b) Let Ω be an open subset of X with real analytic boundary. Then $(\mathscr{M}, \mathbf{C}_\Omega)$ is an elliptic pair if and only if $\partial\Omega$ is non-characteristic for $\mathscr{M}$, that is:

$$\mathrm{char}(\mathscr{M}) \cap T^*_{\partial\Omega} X \subset T^*_X X.$$

If Ω is relatively compact, one gets the finiteness of the spaces $\mathrm{Ext}^j_{\mathscr{D}_X}(\Omega, \mathscr{M}, \mathscr{O}_X)$, a result of Bony and Schapira [B-S] (in case $\mathscr{M} = \mathscr{D}_X/\mathscr{D}_X.P$, cf. Kashiwara [K1] and Kawai [Ka] for various generalizations), extended to the relative case by

Houzel and Schapira [H-S], the index being calculated by Boutet de Monvel and Malgrange [B-M].

(c) For any $\mathcal{M} \in \mathrm{Ob}(D^b_{\mathrm{coh}}(\mathcal{D}_X))$, $(\mathcal{M}, \mathbf{C}_X)$ is an elliptic pair. In this case the duality theorem is due to Mebkhout [M]. If $\mathcal{M}$ has compact support one recovers many classical results. In particular if $\mathcal{G}$ is a coherent $\mathcal{O}_X$-module with compact support, one can apply the theorem with $\mathcal{M} = \mathcal{D}_X \otimes_{\mathcal{O}_X} \mathcal{G}$ and recover theorems of Cartan and Serre (cf. [C-S, Se]). Concerning the index, let us recall that O'Brian, Toledo and Tong [O-T-T], generalizing the Hirzebruch-Riemann-Roch formula [H], constructed the Chern class of coherent $\mathcal{O}_X$-modules with compact support, and proved an index theorem in this case. For the case of $\mathcal{D}_X$-modules with compact support, cf. Angéniol-Lejeune [A-L].

(d) For any $F \in \mathrm{Ob}(D^b_{\mathbf{R}-c}(X))$, $(\mathcal{O}_X, F)$ is an elliptic pair and its microlocal Euler class coincides with the Lagrangian cycle of F defined by Kashiwara in [K2]. Remark that if G is an **R**-constructible object on a real manifold M, one can associate to it an elliptic pair, namely $(\mathcal{O}_X, i_* G)$ where $i : M \hookrightarrow X$ is a complexification of M. In this case the index is calculated by Kashiwara (loc. cit.) (cf. also Dubson [D] and Ginsburg [G] in the complex case).

(e) Let $B(x_0; \varepsilon)$ denote the open ball (in a local chart at x_0) with center x_0 and radius ε and let $\mathcal{M}$ be an holonomic $\mathcal{D}_X$-module. Then for ε small enough, $(\mathcal{M}, \mathbf{C}_{B(x_0;\varepsilon)})$ is an elliptic pair (cf. [K1]).

References

[A-L] B. Angéniol, M. Lejeune-Jalabert: Le théorème de Riemann-Roch singulier pour les D-modules. In: Systèmes différentiels et singularités. Astérisque **130** (1985) 130–160

[A-S] M. Atiyah, I.M. Singer: The index of elliptic operators on compact manifolds. Bull. Am. Math. Soc. **69** (1963) 422–433

[B-M] M. Boutet de Monvel, B. Malgrange: Le théorème de l'indice relatif. Ann. Sc. Ec. Norm. Sup. **23** (1990) 151–192

[B-S] J.-M. Bony, P. Schapira: Existence et prolongement des solutions holomorphes des équations aux dérivées partielles. Invent. math. **17** (1972) 95–105

[C-S] H. Cartan, J.-P. Serre: Un théorème de finitude concernant les variétés analytiques compactes. C. R. Acad. Sci. **237** (1953) 128–130

[D] A. Dubson: Formules pour l'indice des faisceaux constructibles et $\mathcal{D}$-modules holonomes. C. R. Acad. Sci. **298** (1984) 113–116

[G] V. Ginsburg: A theorem on the index of differential systems and the geometry of manifolds with singularities. Sov. Math. Dokl. **31** (1985) 309–313

[H] F. Hirzebruch: Topological methods in algebraic geometry. (Grundlehren der mathematischen Wissenschaften, vol. 131.) Springer, Berlin Heidelberg New York 1966

[K1] M. Kashiwara: Systems of microdifferential equations. Progress in Math., Birkhäuser, 1983

[K2] M. Kashiwara: Index theorem for constructible sheaves. In: Systèmes différentiels et singularités. Astérisque **130** (1985) 193–209

[K-S1] M. Kashiwara, P. Schapira: Micro-hyperbolic systems. Acta Math. **142** (1979) 1–55

[K-S2] M. Kashiwara, P. Schapira: Micro-support des faisceaux. C. R. Acad. Sci. **295** (1982) 487–490

[K-S3] M. Kashiwara, P. Schapira: Microlocal study of sheaves. Astérisque **128** (1985)

[K-S4] M. Kashiwara, P. Schapira: Sheaves on manifolds. (Grundlehren der mathematischen Wissenschaften, vol. 292.) Springer, Berlin Heidelberg New York 1990

[Ka] T. Kawai: Finite dimensionality of cohomology groups attached to systems of linear differential equations. J. Math. Kyoto Univ. **13** (1973) 73–95

[L] J. Leray: Problème de Cauchy I. Bull. Soc. Math. France **85** (1957) 389–430

[M] Z. Mebkhout: Théorèmes de dualité globale pour les D-modules cohérents. Math. Scand. **50** (1982) 25–43

[O-T-T] N. O'Brian, D. Toledo, Y.L. Tong: Hirzebruch-Riemann-Roch for coherent sheaves. Amer. J. Math. **103** (1981) 253–271

[R-R] J.-P. Ramis, G. Ruget: Complexe dualisant et théorèmes de dualité en géométrie analytique complexe. Publ. Math. I.H.E.S. **38** (1971) 77–91

[S-K-K] M. Sato, T. Kawai, M. Kashiwara: Hyperfunctions and pseudo-differential equations. (Lecture Notes in Mathematics, vol. 287.) Springer, Berlin Heidelberg New York 1973

[S1] P. Schapira: Microdifferential systems in the complex domain. (Grundlehren der mathematischen Wissenschaften, vol. 269.) Springer, Berlin Heidelberg New York 1985

[S2] P. Schapira: Microfunctions for boundary value problems. In: Algebraic Analysis, papers dedicated to Prof. Sato (M. Kashiwara and T. Kawai, eds.). Academic Press, New York 1988, pp. 809–819

[S-Sc] P. Schapira, J.-P. Schneiders: Paires elliptiques I – Finitude et dualité. C. R. Acad. Sci. **311** (1990) 83–86; II – Classe d'Euler et indice. C. R. Acad. Sci. **312** (1991) 81–84

[Sc] J.-P. Schneiders: Un théorème de dualité relative pour les modules différentiels. C. R. Acad. Sci. **303** (1986) 235–238. Thèse, Univ. Liège 1986 et article à paraître.

[Se] J.-P. Serre: Un théorème de dualité. Comm. Math. Helv. **29** (1954) 9–26

The Evolution of Harmonic Maps

Michael Struwe

Mathematatik, ETH-Zentrum, CH-8092 Zürich, Switzerland

1. Let M and N be compact Riemannian manifolds of dimensions m and ℓ and with metrics γ and g, respectively. We may assume that N is isometrically embedded in some Euclidean $\mathbb{R}^n$.

For C^1-maps $u : M \to N \subset \mathbb{R}^n$ let

$$e(u) = \frac{1}{2}|\nabla u|^2 = \frac{1}{2}\sum_{\alpha,\beta=1}^{m}\sum_{i=1}^{n}\gamma^{\alpha\beta}\frac{\partial}{\partial x_\alpha}u^i\frac{\partial}{\partial x_\beta}u^i \tag{1.1}$$

be the energy density and – with $dM = \sqrt{\det(\gamma_{\alpha\beta})}dx$ – let

$$E(u) = \int_M e(u)dM$$

be the energy of u. If $M \subset \mathbb{R}^m$ carries the Eulidean metric, E is nothing but the standard Dirichlet integral

$$E(u) = \frac{1}{2}\int_M |\nabla u|^2 dx\,.$$

A map u is harmonic if E is stationary at u, which for $u \in C^2$ is equivalent to the condition that the vector $\big(\Delta_M u\big)(x)$ at all points $x \in M$ is orthogonal to the tangent space $T_{u(x)}N$ to N at $u(x) \in \mathbb{R}^n$; that is,

$$\Delta_M u \perp T_u N\,. \tag{1.2}$$

(Δ_M is the Laplace-Beltrami operator on M.) In local coordinates on N, (1.2) takes the form

$$\Delta_M u = \Gamma(u)(\nabla u, \nabla u) \tag{1.3}$$

with a bilinear form Γ involving the Christoffel symbols of the metric g on N. Again, if $M \subset \mathbb{R}^m$ and if, in particular, $N = S^\ell \subset \mathbb{R}^n$ $(n = \ell + 1)$, (1.3) takes the following simple form

$$\Delta u = u|\nabla u|^2\,.$$

The notion of harmonic map generalizes the concepts of closed geodesic (for $M = S^1$) and harmonic function (for $N = \mathbb{R}$).

Proceedings of the International Congress of Mathematicians, Kyoto, Japan, 1990

The study of harmonic maps was initiated by Fuller, Nash, and Sampson; see Eells-Lemaire [12] for further background material and references. The first general existence result is due to Eells and Sampson. Since standard variational techniques fail, in their pioneering work Eells and Sampson [13] introduce the evolution problem

$$\partial_t u - \Delta_M u \perp T_u N \quad \text{on } M \times]0, \infty[\tag{1.4}$$

with initial condition

$$u = u_0 \quad \text{at } t = 0. \tag{1.5}$$

Upon multiplying (1.4) by $\partial_t u \in T_u N$ and integrating by parts, we at once obtain the energy inequality

$$E\big(u(T)\big) + \int_0^T \int_M |\partial_t u|^2 dM dt \leq E(u_0), \quad \forall T > 0 \tag{1.6}$$

for any solution u of (1.4), (1.5). That is, (1.4) is the L^2-gradient flow for E. In local coordinates again, (1.4) takes the form

$$\partial_t u - \Delta_M u = \Gamma(u)(\nabla u, \nabla u). \tag{1.7}$$

Upon differentiation the latter expression and multiplying by ∇u, we moreover obtain the following Bochner-type differential inequality for the energy density

$$(\partial_t - \Delta_M)e(u) + c|\nabla^2 u|^2 \leq K_N e(u)^2 + C_M e(u), \tag{1.8}$$

where K_N denotes an upper bound for the sectional curvature of $N, c > 0$, and C_M depends on (the Ricci curvature of) the metric γ. (1.6), (1.8) and Moser's weak Harnack inequality for parabolic equations now lead to the following result.

Theorem 1 (Eells-Sampson [13]). *Suppose $K_N \leq 0$. Then for any smooth map $u_0 : M \to N$ problem (1.4), (1.5) admits a unique, smooth solution $u(t)$ which, as $t \to \infty$ suitably, converges to a smooth harmonic map u_∞ homotopic to u_0.*

Theorem 1 was extended to manifolds with boundary by Hamilton [17]. Moreover, in this case the curvature restriction $K_N \leq 0$ can be weakened if the initial and boundary data have small range (Jost [21]). However, for general data, the curvature restriction $K_N \leq 0$ is in some sense optimal as Eells and Wood [14] show that Theorem 1 ceases to be true for $M = T^2, N = S^2$ and an initial map u_0 of topological degree 1.

2. Nevertheless, in two dimensions ($m = 2$) Lemaire [23] and Sacks-Uhlenbeck [29] independently showed that also the topological condition $\pi_2(N) = 0$ suffices to find harmonic representatives for all homotopy classes of maps $u_0 : M \to N$. In fact, a new proof of this result, using the evolution problem (1.4), can be given [35]. (Since by the result of Eells and Wood we must expect singularities, we consider also maps in the Sobolev space

$$H^{1,2}(M; N) = \left\{ u \in H^{1,2}(M; \mathbb{R}^n);\ u(M) \subset N \right\}$$

of measurable, finite energy maps $u : M \to N$; that is, with distributional derivative in L^2.) Then we have the following generalization of Theorem 1.

Theorem 2 ([33; Theorem 4.2]). *Suppose $m = 2$. For any $u_0 \in H^{1,2}(M;N)$ there exists a global weak solution $u : M\times]0,\infty[\to N$ of* (1.4), (1.5) *which satisfies* (1.6) *and is* (C^∞–) *regular away from finitely many points* $(\bar{x}_k, \bar{t}_k)$, $1 \le k \le K$. *The solution u is unique in this class.*

At a singularity $(\bar{x}, \bar{t})$ a "harmonic sphere" $\bar{u} : S^2 \cong \overline{\mathbb{R}^2} \to N$ separates in the sense that for suitable $x_m \to \bar{x}$, $R_m \searrow 0$, $t_m \nearrow \bar{t}$ we have

$$u_m(x) := u\big(\exp_{x_m}(R_m x), t_m\big) \longrightarrow \tilde{u} \quad \text{in } H^{2,2}_{\text{loc}}(\mathbb{R}^2;N), \tag{2.1}$$

where $\tilde{u} \not\equiv$ const. is harmonic, has finite energy and extends to a smooth harmonic map $\bar{u} : S^2 \cong \overline{\mathbb{R}^2} \to N$.

Finally, as $t \to \infty$ suitably, $u(t) \rightharpoonup u_\infty$ weakly in $H^{1,2}(M;N)$, where $u_\infty : M \to N$ is smooth and harmonic. Convergence is strong away from finitely many points $(\bar{x}_\ell, \bar{t}_\ell = \infty), 1 \le \ell \le L$, where harmonic spheres separate in the sense (2.1).

Remark. Let $\varepsilon_0 = \inf\{E(u); u : S^2 \to N \text{ is non-constant and harmonic}\} > 0$. Then by (2.1) we can bound $K + L \le \varepsilon_0^{-1} E(u_0)$. In particular, for initial data such that $E(u_0) < \varepsilon_0$ the solution u constructed in Theorem 2 is smooth and converges uniformly to a harmonic limit.

Theorem 2 was extended to 2-manifolds M with boundary $\partial M \neq \emptyset$ by Chang [3] with applications to results by Brezis-Coron and Jost on the Dirichlet problem for harmonic maps into the sphere, and by Chen-Musina [7] to the case of target manifolds with boundary.

Moreover, variants of (1.4) arise if one attempts to solve free boundary problems for minimal surfaces by a deformation method and results completely analogous to Theorem 2 hold; see [34]. M. Li [24] has recently generalized these results to free boundary problems for harmonic maps.

To this day it is not known whether in general (in two dimensions) the flow (1.4) will encounter singularities in finite time. (Of course, the result of Eells and Wood provides us with an example where either such singularities exist or the flow fails to converge asymptotically.) However, some recent results of Chang and Ding [4], respectively work by Grayson and Hamilton [16] lend support to the conjecture that for $m = 2$ the flow (1.4) does not develop singularities in finite time.

3. The situation is quite different in higher dimensions, as there are energy minimizing harmonic maps with singularities. An example is given by the well-known map $u(x) = \frac{x}{|x|} : B_1(0) \subset \mathbb{R}^m \to S^{m-1} \subset \mathbb{R}^m$ (Brezis-Coron-Lieb [2], Lin [26]); if $m \ge 7$, u is minimizing even if we regard u as a map $u : B_1(0) \to S^m \subset \mathbb{R}^{m+1}$ (Jäger-Kaul [20]). For energy-minimizing maps therefore only partial regularity results can be expected. Such a regularity theory was developed by Schoen and Uhlenbeck [30] – and independently by Giaquinta-Giusti [15] in the case that the image is covered by a single chart. A crucial role is played by a subtle monotonicity estimate.

Similarly, progress on the evolution problem (1.4) for $m \ge 3$ and general targets came as a consequence of a peculiar monotonicity formula for (1.4), discovered in [36; Lemma 3.2] for maps $u : \mathbb{R}^m \times [0,T] \to N$ and extended to curved domains by Chen-Struwe [8]. Let

$$G_{z_0}(x,t) = \frac{1}{(4\pi(t_0-t))^{m/2}} \exp\left(-\frac{|x-x_0|^2}{4(t_0-t)}\right), \quad \text{if } t < t_0 \tag{3.1}$$

be the fundamental solution to the heat equation on $\mathbb{R}^m \times \mathbb{R}$ with singularity at $z_0 = (x_0, t_0)$, and let $\varphi \in C_0^\infty(\mathbb{R}^m)$ be a smooth cut-off function such that $\varphi \equiv 1$ in a neighbourhood of 0 and such that the support of φ is contained in a ball of radius ϱ less than the injectivity radius ϱ_M of M. For a solution $u : M \times [0, T[\to N$ and $R^2 < t_0 < T$ let

$$\Phi(R; z_0) = \frac{1}{2} R^2 \int_{B_\varrho(x_0)} |\nabla u|^2 \varphi^2 G_{z_0} \, dx_{|t=t_0-R^2},$$

in local normal coordinates around x_0 on M.

Theorem 3 (Chen-Struwe [8; Lemma 4.2]). *There exists a constant C depending only on M and N such that for any $T > 0$, any $0 < R^2 < R_0^2 \le t_0 < T$ and any regular solution $u : M \times]0, T[\to N$ of* (1.4) *there holds*

$$\Phi(R; z_0) \le \exp\big(C(R_0 - R)\big) \Phi(R_0; z_0) + CE(u_0)(R_0 - R). \tag{3.2}$$

A particular consequence of the monotonicity formula (3.2) is the following.

Theorem 4 ([36; Theorem 5.1]; Chen-Struwe [8; Lemma 4.4]). *There exists a constant $\varepsilon_0 > 0$ depending only on M and N such that for any solution $u : M \times [0, T[\to N$ of* (1.4) *the following is true:*

If $\Phi(R; z_0) < \varepsilon_0$ for some $z_0 = (x_0, t_0)$, $0 < R^2 \le t_0 < T, R < \iota_M$, then $|\nabla u(z_0)| \le C$, with a constant $C = C(M, N, R)$.

Theorem 4 at once leads to a new proof of Mitteau's [27] global existence result for initial data with small energy density. More important, Theorem 4 together with a penalization device to obtain approximate solutions for (1.4), due to Chen [5], Keller-Rubinstein-Sternberg [22], and Shatah [31] led to a global existence and partial regularity result for (1.4) in higher dimensions ($m \ge 3$).

Theorem 5 (Chen-Struwe [8; Theorem 1.5]). *For any (smooth) map $u_0 : M \to N$ there exists a global weak solution $u : M \times]0, \infty[\to N$ of* (1.4), (1.5) *satisfying* (1.6) *and regular off a set of co-dimension ≥ 2. (In fact, this dimension estimate holds for all $t > 0$; see Cheng* [9].*) As $t \to \infty$ suitably, $u(t) \to u_\infty$ weakly in $H^{1,2}(M, N)$ where $u_\infty : M \to N$ is harmonic and regular off a set of co-dimension ≥ 2.*

Moreover, u satisfies a variant of the monotonicity formula (3.2). This fact was used by Coron [10] to prove that the solution obtained in Theorem 5 is in general not unique – even among partially regular solutions satisfying (1.6).

The estimate on the co-dimension of the singular set very likely can be improved to 3, as for energy-minimizing (stationary) harmonic maps; see Giaquinta-Giusti [15] and Schoen-Uhlenbeck [30]. However, as was first observed by Coron-Ghidaglia [11], in higher dimensions singularities may appear in finite time. Subsequently, Chen and Ding [6] gave an argument relating singularities to the fact that in higher dimensions the infimum of the energy in certain non-trivial homotopy classes of maps may be 0, an observation due to B. White [37].

In fact, their reasoning can be considerably simplified by combining Theorem 4 with Moser's weak Harnack inequality for parabolic equations. First note:

Theorem 6. *For any $T > 0$ there exists $\varepsilon_1 > 0$ depending only on T, M and N such that any smooth solution $u : M \times [0,T] \to N$ of* (1.4), (1.5) *with $E(u_0) < \varepsilon_1$ can be extended to a global, smooth solution $u : M \times [0,\infty[\to N$, converging, as $t \to \infty$ suitably, to a constant harmonic map.*

Proof. Let $R_0^2 = \inf\{\iota_M^2, T\}$. For $0 < R_0^2 \le t_0 \le T$, $x_0 \in M$ we can estimate with constants C depending on M, N, and T only

$$\Phi(R_0; z_0) \le C R_0^{2-m} E\big(u(t_0 - R_0^2)\big) \le C E(u_0) < \varepsilon_0 , \tag{3.3}$$

if $\varepsilon_1 > 0$ is sufficiently small. Hence by Theorem 4 we have

$$|\nabla u(x,t)| \le C \ \text{ uniformly for } \ t \ge R_0^2, x \in M ,$$

and u can be extended as a smooth solution of (1.4), (1.5) on $M \times [0,\infty[$.

By (1.8) and Moser's [28] supremum estimate for weak sub-solutions of linear parabolic equations, moreover we obtain

$$|\nabla u(x,t)|^2 \le C\, E\big(u(t - R_0^2)\big) \le C\, E(u_0) \quad \text{for } \ t \ge 2R_0^2, x \in M . \tag{3.4}$$

From this uniform estimate, asymptotic convergence follows as in Eells-Sampson [13] or Jost [21]. Finally, if $\varepsilon_1 > 0$ is sufficiently small, by (3.4) the image of any map $u(t), t \ge 2R_0^2$, and hence also of the limiting harmonic map u_∞ is contained in a strictly convex geodesic ball on N. It follows that $u_\infty \equiv$ const.; see Jäger-Kaul [19]. □

By Theorem 6, of course, for homotopically non-trivial initial data u_0 with $E(u_0) < \varepsilon_1(T)$ the flow (1.4), (1.5) must blow up before time T. In fact, blow-up time approaches 0 as the initial energy decreases to 0.

Finally, we remark that in dimensions $m \ge 3$ singularities – as in the case $m = 2$ – seem to be related to harmonic spheres or to self-similar solutions $u(x,t) = w\left(\frac{x-x_0}{\sqrt{t_0-t}}\right)$ of (1.4); see Struwe [36; Theorem 8.1]. (The work of Coron-Ghidaglia strongly suggests that solutions of the latter kind in dimensions $m \ge 3$ actually exist.)

The approach of [8] in general cannot be extended to initial maps belonging to $H^{1,2}(M;N)$, only. A different approach via time-discrete minimization was proposed by Horihata-Kikuchi [18]. Based on their ideas, Bethuel et al. [1] recently established global existence of distribution solutions to (1.4) for finite energy maps into spheres.

Further directions of present research into (1.4) include the study of (1.4) on complete, non-compact manifolds; see Li-Tam [25] for some recent work in this regard.

A subject related to (1.4) is the Cauchy problem for harmonic maps into Minkowski space. See Shatah [31], Sideris [32].

References

1. Bethuel, F., Coron, J.M., Ghidaglia, J.M., Soyeur, A.: Heat flows and relaxed energies for harmonic maps. Preprint 1990
2. Brezis, H., Coron, J.M., Lieb, E.: Harmonic maps with defects. Comm. Math. Phys. **107** (1986) 649–705
3. Chang, K.-C.: Heat flow and boundary value problem for harmonic maps. Preprint 1988
4. Chang, K.-C., Ding, W.-Y.: A result on global existence for heat flows of harmonic maps from D^2 into S^2. Preprint 1989
5. Chen, Y.: Weak solutions to the evolution problems of harmonic maps. Math. Z. **201** (1989) 69–74
6. Chen, Y., Ding, W.-Y.: Blow-up and global existence for heat flows of harmonic maps. Preprint
7. Chen, Y., Musina, R.: Harmonic mappings into manifolds with boundary. Preprint, Trieste 1989
8. Chen, Y., Struwe, M.: Existence and partial regularity results for the heat flow for harmonic maps. Math. Z. **201** (1989) 83–103
9. Cheng, X.: Estimate of singular set of the evolution problem for harmonic maps. Preprint, Rice Univ. 1990
10. Coron, J.-M.: Nonuniqueness for the heat flow of harmonic maps. Preprint 1988
11. Coron, J.-M., Ghidaglia, J.-M.: Explosion en temps fini pour le flot des applications harmoniques. Preprint 1988
12. Eells, J., Lemaire, L.: A report on harmonic maps. Bull. London Math. Soc. **10** (1978) 1–68
13. Eells, J., Sampson, J.H.: Harmonic mappings of Riemannian manifolds. Amer. J. Math. **86** (1964) 109–160
14. Eells, J., Wood, J.C.: Restrictions on harmonic maps of surfaces. Topology **15** (1976) 263–266
15. Giaquinta, M., Giusti, E.: On the regularity of the minima of variational integrals. Acta Math. **148** (1982) 31–46
16. Grayson, M., Hamilton, R.S.: The formation of singularities in the harmonic map heat flow. Preprint
17. Hamilton, R.S.: Harmonic maps of manifolds with boundary. (Lecture Notes in Mathematics, vol. 471.) Springer, Berlin Heidelberg New York 1975
18. Horihata, H., Kikuchi, N.: A construction of solutions satisfying a Caccioppoli inequality for nonlinear parabolic equations associated to a variational functional of harmonic type. Preprint
19. Jäger, W., Kaul, H.: Uniqueness and stability of harmonic maps, and their Jacobi fields. Manuscr. math. **28** (1979) 269–291
20. Jäger, W., Kaul, H.: Rotationally symmetric harmonic maps from a ball into a sphere and the regularity problem for weak solutions of elliptic systems. J. Reine Angew. Math. **343** (1983) 146–161
21. Jost, J.: Ein Existenzbeweis für harmonische Abbildungen, die ein Dirichletproblem lösen, mittels der Methode des Wärmeflusses. Manuscr. math. **34** (1981) 17–25
22. Keller, J., Rubinstein, J., Sternberg, P.: Reaction – diffusion processes and evolution to harmonic maps. Preprint
23. Lemaire, L.: Applications harmoniques de surfaces riemanniennes. J. Diff. Geom. **13** (1978) 51–78
24. Li, M.: Harmonic map heat flow with free boundary. Preprint, Trieste 1990
25. Li, P., Tam, L.-F.: The heat equation and harmonic maps of complete manifolds. Preprint 1989

26. Lin, F.H.: Une remarque sur l'application $x/|x|$. C.R. Acad. Sc. Paris **305** (1987) 529–531
27. Mitteau, J.C.: Sur les applications harmoniques. J. Diff. Geom. **9** (1974) 41–54
28. Moser, J.: A Harnack inequality for parabolic differential equations. Comm. Pure Appl. Math. **17** (1964) 101–134
29. Sacks, J., Uhlenbeck, K.: The existence of minimal immersions of 2-spheres. Ann. Math. **113** (1981) 1–24
30. Schoen, R.S., Uhlenbeck, K.: A regularity theory for harmonic maps. J. Diff Geom. **17** (1982) 307–335, and **18** (1983) 329
31. Shatah, J.: Weak solutions and the development of singularities of the $SU(2)$ σ-model. Comm. Pure Appl. Math. **41** (1988) 459–469
32. Sideris, T.: Global existence of harmonic maps in Minkowski space. Comm. Pure Appl. Math. **42** (1989) 1–13
33. Struwe, M.: On the evolution of harmonic maps of Riemannian surfaces. Comment. Math. Helv. **60** (1985) 558–581
34. Struwe, M.: The existence of surfaces of constant mean curvature with free boundaries. Acta Math. **160** (1988) 19–64
35. Struwe, M.: Heat flow methods for harmonic maps of surfaces and applications to free boundary problems. (Lecture Notes in Mathematics, vol. 1324.) Springer, Berlin Heidelberg New York 1988, pp. 293–319
36. Struwe, M.: On the evolution of harmonic maps in higher dimensions. J. Diff. Geom. **28** (1988) 485–502
37. White, B.: Infima of energy functionals in homotopy classes of mappings. J. Diff. Geom. **23** (1986) 127–142

Integrable Systems in Gauge Theory, Kähler Geometry and Super KP Hierarchy – Symmetries and Algebraic Point of View

Kanehisa Takasaki

Research Institute for Mathematical Sciences, Kyoto University
Sakyo-ku, Kyoto 606, Japan

1. Introduction

The following nonlinear systems all provide valuable material to search for new "nonlinear integrable systems".

- self-duality equation in Yang-Mills theory
- self-duality equation in Kähler geometry
- super Kadomtsev-Petviashvili (KP) hierarchy.

From these equations, one will be able to imagine several types of extensions of so called "soliton equations" such as the celebrated Korteweg-de Vries (KdV) equation etc. The first two cases are in a sense "higher dimensional" (or "multi-dimensional") nonlinear integrable systems; the last case will be interesting as an extension of M. Sato's work on the KP hierarchy [SS] and background ideas [S] referred to under the key words "algebraic analysis."

This lecture is a summary of my recent work on these equations, in particular, the self-duality equations, with focus on their symmetry properties. It is nowadays widely recognized that symmetries of soliton equations can be described by representation theory of Kac-Moody Lie algebras [DJKM]. A similar observation to the self-duality equation of Yang-Mills theory has been known for years [UN, CGW, D, T1]. The case of the self-duality equation in Kähler geometry seems to have remained less obscure [BP]. Recently I obtained an explicit description of infinitesimal symmetries, which exhibits a Poisson algebra structure [T2]. Very recently, inspired by a work of Leznov et al. [LMS], I noticed that these infinitesimal symmetries can be "exponentiated" by a simple method [T3]. This leads to a kind of "perturbative" construction of a class of general (local) solutions. To stress underlying symplectic structures, I will illustrate these results for a $4N$-dimensional generalization of the self-duality equations rather than in the original form.

The basic standpoint of my work largely relies on the philosophy of "algebraic analysis," which understands differential equations as a differential algebra, i.e., a set of abstract symbols and differential-algebraic relations among them. This language has turned out to be particularly useful [T4] in the case of the super KP hierarchy of Manin and Radul [MR] as well as the original KP hierarchy. For the treatment of the self-duality equations, we shall not specify such a differential-algebraic interpretation; however, its spirit is included therein.

Proceedings of the International Congress
of Mathematicians, Kyoto, Japan, 1990

2. Generalized Self-Duality Equations

2.1 The Case of Yang-Mills Theory

We consider a $4N$-dimensional space-time with coordinates

$$(x, p) = (x^1, \dots, x^{2N}, p^1, \dots, p^{2N}) \tag{1}$$

and a generalized self-duality equation of Yang-Mills theory on this space-time. This equation, as in the four dimensional case, has two equivalent expressions [C]. As we shall see later on, these two expressions have analogues in Kähler geometry. The first expression is given by the equations

$$\frac{\partial^2 K}{\partial x^i \partial p^j} - \frac{\partial^2 K}{\partial x^j \partial p^i} + \left[\frac{\partial K}{\partial x^i}, \frac{\partial K}{\partial x^j}\right] = 0, \tag{2}$$

where the unknown function $K = K(x, p)$ takes values in the Lie algebra Lie$G$ of the structure group G. The second one is given by

$$\frac{\partial}{\partial x^i}\left(\frac{\partial J}{\partial p^j} J^{-1}\right) - \frac{\partial}{\partial x^j}\left(\frac{\partial J}{\partial p^i} J^{-1}\right) = 0, \tag{3}$$

where the unknown function $J = J(x, p)$ now takes values in G.

As well known, these equations are the integrability condition (in the sense of Frobenius) of the linear system

$$\left(\frac{\partial}{\partial p^i} - \lambda \frac{\partial}{\partial x^i} + A_i\right) \Psi(\lambda) = 0. \tag{4}$$

The gauge potentials A_i are combined with the previous unknown functions J and K as:

$$A_i = -\frac{\partial K}{\partial x^i} = -\frac{\partial J}{\partial p^i} J^{-1}. \tag{5}$$

We consider, in particular, a special pair of solutions

$$\begin{aligned} \Psi(\lambda) &= W(\lambda), \quad W(\lambda) = 1 + \sum_{n \le -1} W_n \lambda^n, \\ \Psi(\lambda) &= V(\lambda), \quad V(\lambda) = \sum_{n \ge 0} V_n \lambda^n \end{aligned} \tag{6}$$

connected with J and K by the relation

$$K = -W_{-1}, \quad J = V_0. \tag{7}$$

The linear system, with these expressions inserted, gives rise to a *nonlinear* system with the new unknown functions W_n and V_n. Symmetries are to be constructed for this nonlinear system rather than the original equation.

2.2 The Case of Kähler Geometry

We now turn to Kähler geometry. Our notational conventions are as follows. Let $i, j, \ldots$ be symplectic indices with values in integers $1, \ldots, 2N$. ϵ^{ij} and ϵ_{ij} denote the standard symplectic ϵ-symbols normalized as $\epsilon_{2i-1,2i} = -\epsilon_{2i,2i-1} = 1$ and $\epsilon^{2i-1,2i} = -\epsilon^{2i,2i-1} = 1$. The Einstein summation convention is understood only for symplectic indices. Symplectic indices are raised and lowered as $\xi_i = \epsilon_{ij}\,\xi^j$ and $\eta^j = \eta_i\,\epsilon_{ij}$.

A $4N$-dimensional generalization of the self-duality equation in Kähler geometry is provided by hyper-Kähler geometry. As pointed out (or re-discovered) by Plebanski [P] in the four dimensional (self-dual) case, hyper-Kähler geometry (also called "$\mathscr{H}$-space") has two equivalent local pictures based upon the first and second "heavenly equations." The "second" picture consists of a $4N$-dimensional coordinate system $(x, p) = (x^1, \ldots, x^{2N}, p^1, \ldots, p^{2N})$, a scalar unknown function $\Theta = \Theta(x, p)$, and the "second heavenly equation"

$$\frac{\partial^2 \Theta}{\partial x^i \partial p^j} - \frac{\partial^2 \Theta}{\partial x^j \partial p^i} + \left\{ \frac{\partial \Theta}{\partial x^i}, \frac{\partial \Theta}{\partial x^j} \right\}_{(x)} = 0, \tag{8}$$

where $\{\ ,\ \}_{(x)}$ stands for the Poisson bracket in x,

$$\{F, G\}_{(x)} = \epsilon^{ij} \frac{\partial F}{\partial x^i} \frac{\partial G}{\partial x^j}. \tag{9}$$

In the "first" picture, one has a $4N$-dimensional coordinate system $(p, \hat{p}) = (p^1, \ldots, p^{2N}, \hat{p}^1, \ldots, \hat{p}^{2N})$, a scalar unknown function $\Omega = \Omega(p, \hat{p})$, and the "first heavenly equation"

$$\left\{ \frac{\partial \Omega}{\partial p^i}, \frac{\partial \Omega}{\partial p^j} \right\}_{(\hat{p})} = \epsilon_{ij}, \tag{10}$$

where we use another Poisson bracket,

$$\{F, G\}_{(\hat{p})} = \epsilon^{ij} \frac{\partial F}{\partial \hat{p}^i} \frac{\partial G}{\partial \hat{p}^j}. \tag{11}$$

Geometrically, Ω represents a Kähler potential, and p^i and $\hat{p}_i$ correspond to complex coordinates and their complex conjugates. In the following, however, we understand $(p, \hat{p})$ or (x, p) as $4N$ *independent* complex variables and consider formal aspects of the above differential equations.

The role of $W(\lambda)$ and $V(\lambda)$ is now to be played by two sets of functions (or formal Laurent series)

$$\begin{aligned} u^i(\lambda) &= \sum_{n \le -1} u^i_n \lambda^n + x^i + p^i \lambda \quad (1 \le i \le 2N), \\ \hat{u}^i(\lambda) &= \hat{p}^i + \sum_{n \ge 1} \hat{u}^i_n \lambda^n, \quad (1 \le i \le 2N) \end{aligned} \tag{12}$$

subject to the exterior differential equations

$$\epsilon_{ij}\, du^i(\lambda) \wedge du^j(\lambda) = \epsilon_{ij}\, d\hat{u}^i(\lambda) \wedge d\hat{u}^j(\lambda), \tag{13}$$

and

$$d\Theta = \in_{ij} u^i_{-2} dp^j + \in_{ij} u^i_{-1} dx^i,$$
$$d\Omega = - \in_{ij} u^i_0 dp^j + \in_{ij} \hat{u}^i_1 d\hat{p}^j. \tag{14}$$

Here u^i_n and $\hat{u}^i_n$ are understood as unknown functions of (x, p) (in the second heavenly picture) or of $(p, \hat{p})$ (in the first heavenly picture); λ is considered a constant under the total differential d, i.e., $d\lambda = 0$. Symmetries are to be constructed for this nonlinear system.

3. Infinitesimal Symmetries

3.1 The Case of Yang-Mills Theory [T1]

For the $(W(\lambda), V(\lambda))$-system, a one-parameter family of transformations

$$\big(W(\lambda), V(\lambda)\big) \mapsto \big(W(\varepsilon, \lambda), V(\varepsilon, \lambda)\big) \tag{15}$$

of solutions is defined by the Riemann-Hilbert factorization

$$W(\varepsilon, \lambda) e^{-\varepsilon X(\lambda)} W(\lambda)^{-1} = V(\varepsilon, \lambda) e^{-\varepsilon Y(\lambda)} V(\lambda)^{-1}. \tag{16}$$

Here $X(\lambda) = X(\lambda, x, p)$ and $Y(\lambda) = Y(\lambda, x, p)$, the data of transformations, are Lie$G$-valued functions of $4N + 1$ variables of the form

$$X(\lambda) = \mathbf{X}(\lambda, x^1 + p^1\lambda, \ldots, x^{2N} + p^{2N}\lambda),$$
$$Y(\lambda) = \mathbf{Y}(\lambda, x^1 + p^1\lambda, \ldots, x^{2N} + p^{2N}\lambda), \tag{17}$$

where $\mathbf{X}$ and $\mathbf{Y}$ are arbitrary Lie$G$-valued functions of $2N + 1$ variables with Laurent expansion

$$\mathbf{X}(\lambda, u) = \sum_{n=-\infty}^{\infty} \mathbf{X}_n(u)\lambda^n, \quad \mathbf{Y}(\lambda, u) = \sum_{n=-\infty}^{\infty} \mathbf{Y}_n(u)\lambda^n. \tag{18}$$

[In fact, some restriction on these data is required for the Riemann-Hilbert factorization to work well; a prescription is to put upper and lower bounds to the range of n as $-\infty < n \le n_X$ for $\mathbf{X}(\lambda)$ and $-n_Y \le n < \infty$ for $\mathbf{Y}(\lambda)$. A similar remark also applies to the hyper-Kähler case. This is a somewhat technical issue.] The associated infinitesimal transformations

$$\delta_{X,Y} W(\lambda) = \left.\frac{\partial W(\varepsilon, \lambda)}{\partial \varepsilon}\right|_{\varepsilon=0},$$
$$\delta_{X,Y} V(\lambda) = \left.\frac{\partial V(\varepsilon, \lambda)}{\partial \varepsilon}\right|_{\varepsilon=0} \tag{19}$$

have the following structure.

Proposition 1. *The infinitesimal symmetries act on $W(\lambda)$ and $V(\lambda)$ as follows.*

$$\delta_{X,Y} W(\lambda) \cdot W(\lambda)^{-1} = \big(W(\lambda)X(\lambda)W(\lambda)^{-1} - V(\lambda)Y(\lambda)V(\lambda)^{-1}\big)_{\le -1},$$
$$\delta_{X,Y} V(\lambda) \cdot V(\lambda)^{-1} = \big(V(\lambda)Y(\lambda)V(\lambda)^{-1} - W(\lambda)X(\lambda)W(\lambda)^{-1}\big)_{\ge 0}, \tag{20}$$

where $(\ )_{\geq 0}$ *and* $(\ )_{\leq -1}$ *are linear maps on the space of Laurent series of* λ *defined by*

$$\begin{aligned}(\ )_{\geq 0} &: \lambda^n \mapsto \theta(n \geq 0)\lambda^n,\\ (\ )_{\leq -1} &: \lambda^n \mapsto \theta(n \leq -1)\lambda^n.\end{aligned} \tag{21}$$

Further, these infinitesimal symmetries obey the commutation relations

$$\left[\delta_{X_1,Y_1}, \delta_{X_2,Y_2}\right] = \delta_{[X_1,X_2],[Y_1,Y_2]}. \tag{22}$$

Thus, in particular, the infinitesimal symmetries give rise to a nonlinear realization of a direct sum of two loop algebras (with extra $2N$ variables $u^1, \ldots, u^{2N}$). The associated infinitesimal transformations of $J = V_0$ and $K = -W_{-1}$ can be readily derived from the above result.

3.2 The Case of Kähler Geometry [T2]

The case of $(u(\lambda), \hat{u}(\lambda))$-system requires a more involved factorization, i.e., a factorization with respect to composition of maps. Let us consider this issue within the (x, p)-coordinate system. [A fully parallel treatment is possible with the $(p, \hat{p})$-coordinate system.] $u(\lambda)$ and $\hat{u}(\lambda)$ are now interpreted as maps

$$\begin{aligned}u(\lambda) &: x \mapsto u(\lambda, x, p),\\ \hat{u}(\lambda) &: x \mapsto \hat{u}(\lambda, x, p)\end{aligned} \tag{23}$$

from the x-space into the u-space or $\hat{u}$-space respectively. A one-parameter family of transformations

$$\big(u(\lambda), \hat{u}(\lambda)\big) \mapsto \big(u(\varepsilon, \lambda), \hat{u}(\varepsilon, \lambda)\big) \tag{24}$$

of solutions can be defined by the Riemann-Hilbert factorization

$$u(\varepsilon, \lambda)^{-1} \circ e^{-\varepsilon \xi_F(\lambda)} \circ u(\lambda) = \hat{u}(\varepsilon, \lambda)^{-1} \circ e^{-\varepsilon \xi_{\hat{F}}(\lambda)} \circ \hat{u}(\lambda), \tag{25}$$

where $\xi_F(\lambda)$ and $\xi_{\hat{F}}(\lambda)$ are Hamiltonian vector fields of the form

$$\begin{aligned}\xi_F(\lambda) &= \in^{ij} \frac{\partial F(\lambda)}{\partial u^i} \frac{\partial}{\partial u^j},\\ \xi_{\hat{F}}(\lambda) &= \in^{ij} \frac{\partial \hat{F}(\lambda)}{\partial \hat{u}^i} \frac{\partial}{\partial \hat{u}^j}.\end{aligned} \tag{26}$$

The generating functions $F(\lambda) = F(\lambda, u)$ and $\hat{F}(\lambda) = \hat{F}(\lambda, \hat{u})$ are arbitrary functions of $2N + 1$ variables with Laurent expansion

$$F(\lambda) = \sum_{n=-\infty}^{\infty} F_n(u)\lambda^n, \quad \hat{F}(\lambda) = \sum_{n=-\infty}^{\infty} \hat{F}(\hat{u})\lambda^n. \tag{27}$$

The infinitesimal transformations

$$\delta_{F,\hat{F}} u^i(\lambda) = \left.\frac{\partial u^i(\varepsilon,\lambda)}{\partial \varepsilon}\right|_{\varepsilon=0},$$

$$\delta_{F,\hat{F}} \hat{u}^i(\lambda) = \left.\frac{\partial \hat{u}^i(\varepsilon,\lambda)}{\partial \varepsilon}\right|_{\varepsilon=0} \tag{28}$$

have the following structure.

Proposition 2. *The infinitesimal symmetries act on $u(\lambda)$ and $\hat{u}(\lambda)$ as:*

$$\delta_{F,\hat{F}} u^i(\lambda) = \left\{ \left[F\left(\lambda, u(\lambda)\right) - \hat{F}\left(\lambda, \hat{u}(\lambda)\right) \right]_{\le -1}, u^i(\lambda) \right\}_{(x)},$$

$$\delta_{F,\hat{F}} \hat{u}^i(\lambda) = \left\{ \left[\hat{F}\left(\lambda, \hat{u}(\lambda)\right) - F\left(\lambda, u(\lambda)\right) \right]_{\ge 0}, \hat{u}^i(\lambda) \right\}_{(x)}. \tag{29}$$

Further, the infinitesimal symmetries obey the commutation relations

$$\left[\delta_{F_1,\hat{F}_1}, \delta_{F_2,\hat{F}_2} \right] = \delta_{\{F_1,F_2\}_{(u)},\{\hat{F}_1,\hat{F}_2\}_{(\hat{u})}}. \tag{30}$$

Thus the infinitesimal symmetries give a nonlinear realization of a direct sum of two Poisson (loop) algebras.

Remarkably, the above infinitesimal symmetries can be extended to Θ and Ω without modifying the Poisson algebra structure.

Proposition 3. *The infinitesimal symmetries can be consistently extended to Θ and Ω by the following rule.*

$$\delta_{F,\hat{F}} \Theta = \operatorname*{res}_{\lambda=\infty} F\left(\lambda, u(\lambda)\right) + \operatorname*{res}_{\lambda=0} \hat{F}\left(\lambda, \hat{u}(\lambda)\right),$$

$$\delta_{F,\hat{F}} \Omega = -\operatorname*{res}_{\lambda=\infty} \lambda^{-2} F\left(\lambda, u(\lambda)\right) - \operatorname*{res}_{\lambda=0} \lambda^{-2} \hat{F}\left(\lambda, \hat{u}(\lambda)\right), \tag{31}$$

where the residues are normalized as

$$\operatorname*{res}_{\lambda=\infty} \lambda^n = -\delta_{n,-1}, \quad \operatorname*{res}_{\lambda=0} \lambda^n = \delta_{n,-1}. \tag{32}$$

These extended infinitesimal symmetries obey the same commutation relations as in Proposition 2.

4. Perturbative Method [T3]

The infinitesimal symmetries, as we have seen, have a considerably simple and beautiful structure. The Riemann-Hilbert factorization problems in general are hard to solve explicitly. For the case of Yang-Mills fields, several solution methods are developed; for the hyper-Kähler case, only existence theorems are known (except for a few very special families of solutions). The method presented here, so to speak, "exponentiate" the infinitesimal symmetries by expanding everything in powers of ε. As Leznov et al. [LMS] pointed out, the parameter ε plays the role of "coupling constants" in field theory; therefore we call the following method "perturbative."

4.1 The Case of Yang-Mills Theory

Let us consider the previous Riemann-Hilbert factorization in case where

$$W(\lambda) = V(\lambda) = 1 \text{ (trivial solution)}, \quad Y(\lambda) = 0. \tag{33}$$

Let us define

$$\mathscr{X}(\varepsilon, \lambda) = W(\varepsilon, \lambda) X(\lambda) W(\varepsilon, \lambda)^{-1} \tag{34}$$

Since $\partial/\partial\varepsilon$ corresponds to the action of $\delta_{X,0}$ on $\big(W(\varepsilon, \lambda), V(\varepsilon, \lambda)\big)$, one can readily find a closed differential equation satisfied by $\mathscr{X}(\varepsilon, \lambda)$ with respect to ε.

Proposition 4. *$\mathscr{X}(\varepsilon, \lambda)$ satisfies the differential equation*

$$\frac{\partial \mathscr{X}(\varepsilon, \lambda)}{\partial \varepsilon} = \left[\big(\mathscr{X}(\varepsilon, \lambda)\big)_{\leq -1}, \mathscr{X}(\varepsilon, \lambda)\right] \tag{35}$$

and the initial condition

$$\mathscr{X}(\varepsilon = 0, \lambda) = \mathbf{X}(\lambda, x + p\lambda). \tag{36}$$

Further,

Proposition 5. $K(\varepsilon) = -W_{-1}(\varepsilon)$ *and* $J(\varepsilon) = V_0(\varepsilon)$ *obey the differential equations*

$$\begin{aligned} \frac{\partial K(\varepsilon)}{\partial \varepsilon} &= \operatorname*{res}_{\lambda=\infty} \mathscr{X}(\varepsilon, \lambda), \\ \frac{\partial J(\varepsilon)}{\partial \varepsilon} J(\varepsilon)^{-1} &= \operatorname*{res}_{\lambda=\infty} \lambda^{-1} \mathscr{X}(\varepsilon, \lambda). \end{aligned} \tag{37}$$

Substitution of the Taylor expansion ("perturbation series")

$$\mathscr{X}(\varepsilon, \lambda) = \sum_{k=0}^{\infty} \mathscr{X}^{(k)}(\lambda) \varepsilon^k / k! \tag{38}$$

into the above equation yields a set of recursive relations

$$\begin{aligned} \mathscr{X}^{(0)}(\lambda) &= X(\lambda) = \mathbf{X}(\lambda, x + p\lambda), \\ \mathscr{X}^{(k+1)} &= \sum_{\ell=0}^{k} \binom{k}{\ell} \left[\big(\mathscr{X}^{(k-\ell)}(\lambda)\big)_{\leq -1}, \mathscr{X}^{(\ell)}(\lambda)\right]. \end{aligned} \tag{39}$$

The unknown functions $K(\varepsilon)$ and $J(\varepsilon)$ of the generalized self-duality equations, too, can be determined by expansion into powers of ε.

In the original formulation of Leznov et al. [LMS], the projection $(\ )_{\leq -1}$ is represented by an integral operator; they exploit its algebraic properties to check, by brute force, the validity of their formula.

3.2 The Case of Kähler Geometry

We now start from the Riemann-Hilbert factorization with

$$u^i(\lambda) = \hat{u}^i(\lambda) = x^i + p^i\lambda \text{ (trivial solution)}, \quad \hat{F}(\lambda) = 0, \tag{40}$$

and derive differential equations satisfied by

$$\mathscr{F}(\varepsilon, \lambda) = F\big(\lambda, u(\varepsilon, \lambda)\big) \tag{41}$$

and $\Theta(\varepsilon)$ with respect to ε.

Proposition 6. *$\mathscr{F}(\varepsilon, \lambda)$ satisfies the differential equation*

$$\frac{\partial \mathscr{F}(\varepsilon, \lambda)}{\partial \varepsilon} = \{[\mathscr{F}(\varepsilon, \lambda)]_{\le -1}, \mathscr{F}(\varepsilon, \lambda)\}_{(x)} \tag{42}$$

and the initial condition

$$\mathscr{F}(\varepsilon = 0, \lambda) = F(\lambda, x + p\lambda). \tag{43}$$

Proposition 7. *One can obtain $\Theta(\varepsilon)$ by solving the differential equation*

$$\frac{\partial \Theta(\varepsilon)}{\partial \varepsilon} = \operatorname*{res}_{\lambda=\infty} \mathscr{F}(\varepsilon, \lambda) \tag{44}$$

under the initial condition

$$\Theta(\varepsilon = 0) = 0. \tag{45}$$

These equations can be solved by the same "perturbative method" as illustrated in the case of Yang-Mills fields.

The above construction is not suited for the first heavenly picture based upon $(p, \hat{p}, \Omega)$. To give a similar construction for the first heavenly picture, we just have to restart from the situation where

$$u^i(\lambda) = \hat{u}^i(\lambda) = \hat{p}^i + p^i\lambda, \quad F(\lambda) = 0, \tag{46}$$

and consider equations satisfied by $\Omega(\varepsilon)$ and

$$\hat{\mathscr{F}}(\varepsilon, \lambda) = \hat{F}\big(\lambda, \hat{u}(\varepsilon, \lambda)\big). \tag{47}$$

5. KP and Super KP Hierarchies [T4]

In the differential-algebraic approach mentioned in the introduction, a nonlinear system is represented by a commutative algebra $\mathscr{A}$ with a set of derivations $\partial_1, \partial_2, \ldots$. If one is not interested in a particular choice of such derivations, it is convenient to understand a differential algebra as a pair $(\mathscr{A}, \Delta)$ of a commutative algebra and an $\mathscr{A}$-module Δ of derivations in $\mathscr{A}$. Infinitesimal symmetries are then, by definition, derivations $\delta : \mathscr{A} \to \mathscr{A}$ that satisfy the condition

$$[\delta, \partial] \in \Delta \quad (\forall \partial \in \Delta). \tag{48}$$

In most applications, the derivations $\partial_1, \partial_2, \ldots$, are chosen to be commutative, and symmetries are characterized as extra derivations of $\mathscr{A}$ that commute with these derivations. (The super KP hierarchy is somewhat distinct; not only $\mathscr{A}$ being a supercommutative algebra, the set of derivations are neither commutative nor supercommutative. One can however see that its basic structure is almost parallel to the case of the KP hierarchy.)

The generalized self-duality equations, too, can be formulated as such an abstract differential algebra. Its algebraic part $\mathscr{A}$ should be a commutative algebra (over a suitable differential subalgebra that specifies in which domain to seek for solutions) generated by the Laurent coefficients of $W(\lambda)$ and $V(\lambda)$, or of $u(\lambda)$ and $\hat{u}(\lambda)$. In the latter case, one may also add Θ or Ω. This certainly provides an umbiguous framework for the notion of infinitesimal symmetries; however, one will gain nothing practically new from this reinterpretation.

The situation is considerably different for the case of the KP and super KP hierarchies. For these equations, the differential-algebraic language seems to have a substantial meaning. Of particular importance is a $\mathscr{D}$-module structure hidden in the formulation of the KP and super KP hierarchy. (This observation for the case of the KP hierarchy is due to Sato, who stresses the relevance of the notion of $\mathscr{D}$-modules even in more general perspectives [S].) With the aid of this $\mathscr{D}$-module structure, one can find a new set of generators w_{ij} $(i \geq 0, j \leq -1)$ in $\mathscr{A}$. This is the most direct way to see a connection with the geometry of infinite dimensional (super) Grassmannian manifolds; w_{ij}'s can be identified with affine coordinates on an open subset therein. This also leads to: an explicit description of infinitesimal symmetries δ_A parametrized by elements A of an infinite matrix Lie algebra $gl(\infty)$ (for the KP hierarchy) or of its super-version $gl(\infty|\infty)$ (for the super KP hierarchy), a differential-algebraic characterization of the τ function, its symmetry contents related to central extensions of $gl(\infty)$ and $gl(\infty|\infty)$, etc.

In fact, Θ and Ω may be in a sense considered an analogue of the τ function. This analogy becomes quite reasonable if we consider a hierarchy of the generalized self-duality equations discussed here. Their representation-theoretic and geometric properties are however considerably different from the τ function.

References

[BP] Boyer, C.P., Plebanski, J.F.: An infinite hierarchy of conservation laws and nonlinear superposition principles for self-dual Einstein spaces. J. Math. Phys. **26** (1985) 229–234

[C] Chau, L.-L.: Chiral fields, self-dual Yang-Mills fields as integrable systems, and the role of the Kac-Moody algebra. In: Nonlinear phenomena, K.B. Wolf (ed.). (Lecture Notes in Physics, vol. 189.) Springer, Berlin Heidelberg New York 1983

[CGW] Chau, L.-L., Ge, M.-L., Wu, Y.-S.: Kac-Moody algebra in the self-dual Yang-Mills equation. Phys. Rev. **D25** (1982) 1086–1094

[D] Dolan, L.: A new symmetry group of real self-dual Yang-Mills theory. Phys. Lett. **113B** (1982) 387–390

[DJKM] Date, E., Jimbo, M., Kashiwara, M., Miwa, T.: Transformation theory for soliton equations III-VI. J. Phys. Soc. Japan **50** (1982) 3806–3812, 3813–3818; Physica **4D** (1982) 343–365. Publ. RIMS., Kyoto Univ. **18** (1982) 1077–1110

[LMS] Leznov, A.N., Mukhtarov, M.A.: Deformation of algebras and solutions of self-duality equation. J. Math. Phys. **28** (1987) 2574–2578. Leznov, A.N., Saveliev, V.M.: Exactly and completely integrable nonlinear dynamical sysmtems. Acta Appl. Math. **16** (1989) 1–74

[MR] Manin, Yu.I., Radul, A.O.: A supersymmetric extension of the Kadomtsev-Petviashvili hierarchy. Commun. Math. Phys. **98** (1985) 65–77

[P] Plebanski, J.F.: Some solutions of complex Einstein equations. J. Math. Phys. **16** (1975) 2395–2402

[SS] Sato, M., Sato, Y.: Soliton equations as dynamical systems in an infinite dimensional Grassmann manifold. In: Nonlinear Partial Differential Equations in Applied Sciences, P.D. Lax, H. Fujita, G. Strang (eds.). North-Holland, Amsterdam, and Kinokuniya, Tokyo, 1982

[S] Sato, M.: $\mathscr{D}$-modules and nonlinear systems. In: Integrable Systems in Quantum Theory and Statistical Mechanics, M. Jimbo, T. Miwa, A. Tshuchiya (eds.). Adv. Stud. Pure Math., vol. 19, pp. 417-434. Kinokuniya, Tokyo 1989

[T1] Takasaki, K.: A new approach to the self-dual Yang-Mills equations. Commun. Math. Phys. **94** (1984) 35–59. Hierarchy structure in integrable systems of gauge fields and underlying Lie algebras. Commun. Math. Phys. **127** (1990) 225–238

[T2] Takasaki, K.: An infinite number of hidden variables in hyper-Kähler metrics. J. Math. Phys. **30** (1989) 1515-1521. Symmetries of hyper-Kähler (or Poisson gauge field) hierarchy. J. Math. Phys. **31** (1990) 1877–1888

[T3] Takasaki, K.: Perturbative approach to self-duality equations. In preparation

[T4] Takasaki, K.: Symmetries of the super KP hierarchy. Lett. Math. Phys. **17** (1989) 351-357. Differential algebras and $\mathscr{D}$-modules in super Toda lattice hierarchy. Lett. Math. Phys. **19** (1990) 229-236

[UN] Ueno, K., Nakamura, Y.: Transformation theory for anti-self-dual equations and the Riemann-Hilbert problem. Phys. Lett. **109B** (1982) 273–278

H-Measures and Applications

Luc Tartar

Department of Mathematics, Carnegie Mellon University, Pittsburgh, PA 15213, USA

Introduction

Historians interested in the evolution of Science will probably be very surprised when they will analyse all the strange fashions that have struck the scientific community in the second part of this century. They will certainly be aware of the political orientation of those who had launched many of these fake new ideas and wonder why mathematicians had not been more rational in their behaviour. Will they find another example of political censorship than the one that had suppressed two pages of the introduction of an article where I had described my scientific ideas [1]? Will everyone agree that it was indeed slanderous that I had thanked there two of my teachers, Laurent Schwartz and Jacques-Louis Lions and that this part should definitely be cut? Had the censors thought that they could suppress the mention of other political facts and avoid me describing them elsewhere [2]? Historians may wonder at the stupidity of these censors and ponder if they had even understood the meaning of the few lines that they had spared at the beginning: "Il y a une différence énorme entre l'étude des singularités d'équations aux dérivées partielles (linéaires ou non) et celle de leurs oscillations: c'est la différence entre la physique classique et la physique quantique".

In this article I had described my point of view that the study of oscillating solutions of partial differential equations was the key mathematical question to investigate in order to shed some light on the strange rules invented by physicists for explaining natural phenomena. Every specialist of differential equations is aware of the distinction between finite and infinite dimensional effects and it is only the result of an intensive propaganda that so many have adopted a point of view about mechanics which was adequate in the eighteenth century when partial differential equations had not yet found their place and that continuum mechanics and electromagnetism were not even thought of. However, even if all the extensive knowledge about linear partial differential equations contained in the treatise of L. Hörmander [3] had been available at the beginning of the century, it would not have helped so much the physicists puzzled as they were by the spectroscopic measurements of light absorbed and emitted in some gases. One cannot blame then those who have invented the strange rules of quantum

[1] There is a huge difference between the study of singularities of partial differential equations (be them linear or not) and that of their oscillations: it is the difference between classical and quantum physics.

Proceedings of the International Congress of Mathematicians, Kyoto, Japan, 1990

mechanics for their lack of knowledge of partial differential equations but one should blame those who have transformed these rules into dogma. As R. Penrose once wrote "Quantum theory, it may be said, has two things in its favour and only one against. First, it agrees with all the experiments. Second, it is a theory of astonishing and profound mathematical beauty. The only thing to be said against the theory is that it makes absolutely no sense" . In order to give a rational explanation of the puzzling measurements made in these experiments one needs an increased knowledge of partial differential equations and there are a few new mathematical questions that should be understood for that purpose.

As light is involved we know that there will be some hyperbolic equations, the wave equation or Maxwell's system or some even larger system, and we expect that the linearised system will only have the velocity of light as characteristic speed so that we may reasonably restrict our attention to semilinear systems. Even standard questions as the relation between the wave equation and geometrical optics needs to be thought again. Fifteen years ago it was already clear why the mathematical results now found in [3] were not adapted to the goal that I was pointing at and a first reason was that one cannot expect to understand the partial differential equations of continuum mechanics without accepting discontinuous coefficients; even if one was ready to make smoothness assumptions and stay away from interfaces one definitely had to avoid assuming the coefficients to be infinitely differentiable or analytic. There is however a more important drawback of the linear theory of propagation of singularities which became more apparent once I had obtained my personal version of propagation using the tool of H-measures [4] which I will describe in a moment. What the linear theory is really interested in is the propagation of regularity and this leads to a quite negative concept of a singularity which is not defined as a quantitative object; of course, measuring the propagation of H^s regularity instead of C^∞ regularity does not correct this defect in any way. The physically intuitive idea of a beam of light is then absolutely not described by the theory of "propagation of singularities" for partial differential equations and the inadequacy is hidden by the fact that the bicharacteristic rays which have appeared in the linear theory are precisely those which physicists had thought important in their formal computations. One should then criticise this approach of propagation of singularities for describing the properties of light as not making more sense than some physicists' rules; a better test for a mathematical tool than making the bicharacteristic rays appear is to be able to measure what is transported along them and tell what happens along the bicharacteristic rays to important quantities for physics like energy and momentum.

As matter is also involved we face much more trouble because the question of what matter could be is at stake anyway and, even if the rules of quantum mechanics are indeed wrong, one cannot forget about the real defects of the classical concepts of light and matter. A probably good mathematical model to understand is the coupled Maxwell-Dirac system where matter is described by a complex four dimensional vector field and light is described by the electromagnetic field, the coupling involving quadratic terms with the famous Planck constant h appearing as a coupling parameter between light and matter and not as this mysterious parameter that the dogma wants to attach to every hamiltonian.

[2] This is the first paragraph of a review by R. Penrose of a book by J.C. Polkinghorne "The Quantum World" in The Times Higher Education Supplement, March 23, 1984.

If we were to follow this indication we would be interested in understanding some mathematical properties of semilinear hyperbolic systems with quadratic interaction and the special role played by the four dimensional space-time where we apparently live would probably be linked to Sobolev's embedding theorem, but before playing such a game which accepts too much of the physicists' dogmatic postulates, we should look at a more rational explanation of the mathematical difficulties encountered in the spectroscopic measurements.

Even if we do not know what atoms really are they do appear as tiny obstacles and we have then to face the difficulty of working with at least two scales, a microscopic one and a macroscopic one. In some way the practical goal of quantum physics is to compute corrections in the effective equations satisfied by the macroscopic quantities from a fine and possibly wrong description of what equation the microscopic quantities do satisfy. As mathematicians we should describe a general framework in order to understand more of this question and there are indeed choices to make and difficulties to overcome.

The first obvious choice that we have already made is to work with partial differential equations and not with ordinary differential equations; this can be considered a lesson learned from A. Einstein about the defects of I. Newton's classical approach. There has been much propaganda in recent years for works emphasising finite dimensional effects in partial differential equations and one may indeed be attracted by some of the difficult and interesting mathematical questions which had led H. Poincaré to introduce so many tools and ideas before the development of quantum physics. It was another great mathematician who formalised some of the rules followed by physicists in their quantic games but it is surprising that J. Von Neuman would show that no ordinary differential equation could produce the same results as the rules of quantum mechanics and forget to question the very nature of that set of rules. Certainly if one believed that one should create a game that will generate a sequence of numbers one might be tempted by the mathematical properties of spectra of linear operators. Was then the dogma already well accepted before it became obvious that the spectroscopic experiments were not generating mere lists of numbers? Were mathematicians so impressed by this apparent success of functional analysis? Had there been an intentional effort of propaganda around functional analysis in order to avoid that mathematicians study more relevant partial differential equations of continuum physics in the spirit of what S. Sobolev and J. Leray had already been doing in the 1930s? Certainly, and L. Hörmander [3] is right in pointing at some misconceptions created by L. Schwartz's approach, but some other misconceptions have been propagated by his own approach to partial differential equations. Is there indeed a classical treatise on partial differential equations which does mention these properties of partial differential equations related to the strange effects observed in spectroscopy or more simply which does quote the relation between microscopic and macroscopic levels which is such a crucial question in physics?

The mathematical tool of H-measures which I have introduced [4, 5] is a new step toward a better understanding of these questions and I have chosen the prefix H as a reminder that these objects had arisen naturally in the theory of homogenisation, a term to which I attribute a more general meaning than which is usually given in the rare books related to the subject like those of A. Bensoussan, J.-L. Lions and G. Papanicolaou [6] or of E. Sanchez-Palencia [7] where periodicity assumptions often obscure the methods which I had developed for more general situations [8], partly in collaboration with F. Murat [9] and as an extension of earlier results of S. Spagnolo [10, 11]. The H-measures added to my previous description of the role of oscillations in partial differential equations [1] that of concentration effects whose importance in continuum physics I had not foreseen before the work of P.-L. Lions, R. DiPerna and A. Majda [12-16]. My initial purpose for introducing H-measures was to derive small amplitude homogenisation theorems [4, 5, 17] in order to explain why some particular formula obtained by physicists [18] was indeed accurate despite the fact that the arguments used in its derivation did not make any sense. I can trace back my intuitive understanding about these objects to some formula for computing an exact quadratic correction term [19] appearing in a model which I had introduced earlier in order to understand some averaging question in hydrodynamics. It was only later that I found a way to use the same H-measures for describing the propagation of oscillations and concentration effects in some partial differential equations [4, 5] obtaining then a quantitative transport property in the form of partial differential equations in x and ξ satisfied by the H-measures. I wanted to avoid the standard theory of pseudo-differential operators [3] and construct what I needed for my quadratic microlocal tool of H-measures in order to be able to study partial differential equations of continuum mechanics without making spurious hypotheses of smoothness for the coefficients. However even for those who have devoted a long time reading [3] H-measures may still appear to be natural as they have been introduced independently by P. Gérard [20, 21] although the name of microlocal defect measures which he has chosen for them may reflect a negative attitude inherent in [3].

Of course H-measures are only a step toward the mathematical understanding of these questions of physics which I had sketched at the beginning and there are other pieces of that scientific puzzle which should not be left aside like the apparition of memory effects by homogenisation, which seems the mathematical explanation of what physicists attribute to their strange rules of spontaneous absorption and emission; it must be emphasised that these homogenisation results are obtained without any postulate of a probabilistic nature. There are some more or less classical cases of memory effects induced by homogenisation like viscoelasticity which can be found in Sanchez-Palencia [7] but the effects which I was mentioning are related to hyperbolic situations and have not received much attention apart from my own tentatives [22, 23] and that of Y. Amirat, K. Hamdache and A. Ziani [24, 25] and so a lot remains to be done.

H-Measures

Contrary to wave front sets which can be attached to general distributions but are mere geometric sets endowed with a negative property of lack of smoothness, H-measures are only defined for sequences of functions converging weakly to zero in $L^2(R^N)$ and express in a quantitative way the limit of quadratic quantities, the H-measure being zero in the case of strong convergence in $L^2(R^N)$ and, because they are measures on $R^N \times S^{N-1}$, they can see the action of a class of pseudo-differential operators of order zero.

Definition 1. An *admissible symbol* s is a continuous function on $R^N \times S^{N-1}$ admitting a decomposition $s(x, \xi) = \Sigma_n a_n(\xi) b_n(x)$ with the functions a_n being continuous on S^{N-1}, the functions b_n being continuous on R^N and converging to zero at infinity, and such that $\Sigma_n ||a_n||.||b_n|| < \infty$ where the norms are sup norms. The *standard operator* S with symbol s is the continuous operator on $L^2(R^N)$ defined by $\mathbf{F}(Su)(\xi) = \Sigma_n a_n(\xi/|\xi|)\mathbf{F}(b_n u)(\xi)$ where $\mathbf{F}$ denotes the Fourier transform. A continuous operator L on $L^2(R^N)$ is said to have symbol s if $L - S$ is a compact operator on $L^2(R^N)$.

The only technical point to check is that the commutator $L_1L_2 - L_2L_1$ of two such operators is a compact operator on $L^2(R^N)$.

Proposition 2. *If U^n is a sequence converging weakly to zero in $(L^2(R^N))^p$, then there is a subsequence and measures $\mu^{i,j}$ on $R^N \times S^{N-1}$, i, j = 1,..,p, such that for every operators L_1, L_2 with symbols s_1, s_2 the limit of $L_1(U_i^n)L_2(U_j^n)*$ is a measure ν on R^N defined by $< \nu, \phi > = < \mu^{i,j}, \phi s_1 s_2 * >$ for every test function ϕ continuous with compact support in R^N.*

One immediately finds that μ is hermitian nonnegative and has a few other obvious properties, one of them being the following localization principle for H-measures, which is analogous to the information on the wave front sets derived from application of the stationary phase method.

Proposition 3. *If a sequence U^n converges weakly to zero in $(L^2(R^N))^p$, corresponds to a H-measure μ, and is such that $\Sigma_{ij}\partial_i(b_{ij}U_j^n)$ converges strongly to zero in $H^{-1}_{\text{loc}}(R^N)$ where the functions b_{ij} are continuous, then one has $\Sigma_{ij}\xi_i b_{ij}\mu^{jk} = 0$ for k = 1,..,p.*

Before describing the more technical property of propagation let us give a few examples of what was just mentioned.

Example 4. Let $u^n(x) = v(x, x/\varepsilon)$ where ε is a sequence converging to zero with v defined on $R^N \times R^N$ and $v(x, y)$ having period 1 in each component y_j, j = 1,..,N; denoting by Y the unit cube, we assume that v is continous in x with values in $L^2(Y)$ and decompose v in Fourier series in y, $v(x, y) = \Sigma_m v_m(x)e^{2i\pi(m.y)}$, assuming moreover that v_0 is zero. Under these hypotheses, without extraction of a subsequence, the H-measure μ associated to u^n is defined by $< \mu, \Phi > = \Sigma_m \int |v_m(x)|^2\Phi(x, m/|m|)dx$ for every continuous function Φ on $R^N \times S^{N-1}$ with compact support in x.

Example 5. Let $u^n(x) = \varepsilon^{-N/2}v(x/\varepsilon)$ where ε is a sequence converging to zero and v belongs to $L^2(R^N)$. Without extraction of a subsequence the H-measure μ associated to u^n is defined by $< \mu, \Phi > = \int |\mathbf{F}v(\xi)|^2\Phi(0, \xi/|\xi|)d\xi$ for every continuous function Φ on $R^N \times S^{N-1}$ with compact support in x.

Example 6. Let u^n be a sequence converging weakly to zero in $L^2(R^N)$ and corresponding to a H-measure μ; assume moreover that for some continuous

---

[3] I use L. Schwartz's notations so that $\mathbf{F}(Su)(\xi) = \int s(x, \xi/|\xi|)e^{-2i\pi(x.\xi)}u(x)dx$ for u smooth with compact support.

[4] z^* denotes the complex conjugate of z.

functions b_j, j = 1,..,N, $\Sigma_j\partial_j(b_ju^n)$ converges to zero in $H^{-1}_{loc}(R^N)$ strong. Then μ satisfies $P(x,\xi)\mu = 0$ with P defined by $P(x,\xi) = \Sigma_j b_j(x)\xi_j$.

Example 7. Assume that the sequence U^n converges weakly to zero in $(L^2(R^N))^N$ corresponds to a H-measure μ and satisfies $\partial_i U^n_j = \partial_j U^n_i$ for i, j = 1,..,N. Then there exists a scalar nonnegative measure ν on $R^N \times S^{N-1}$ such that $\mu^{i,j} = \xi_i\xi_j\nu$ for i, j = 1,..,N.

Example 8. Let u^n be a sequence converging weakly to zero in $H^1(R^{N+1})$ and assume that for some continuous functions ϱ and a_{ij}, i, j = 1,..,N independent of x_0, $\varrho\partial_0^2u^n$ - $\Sigma_{ij}\partial_i(a_{ij}\partial_ju^n)$ converges to zero in $H^{-1}_{loc}(R^{N+1})$ strong; assume moreover that U^n defined by $U^n_i = \partial_iu^n$ for i = 0,..,N, corresponds to a H-measure μ. Then $\mu^{i,j} = \xi_i\xi_j\nu$ for i, j = 0,..,N and ν satisfies $Q(x,\xi)\nu = 0$ with Q defined by $Q(x,\xi) = \varrho(x)\xi_0^2$ - $\Sigma_{ij}a_{ij}(x)\xi_i\xi_j$.

The propagation effects for H-measures are related to the existence of quadratic balance laws and they take the form of partial differential equations in (x,ξ) satisfied by the H-measures. This is more quantitative than what can be said for wave front sets where it is the complementary property of regularity which is actually propagated. A precise commutation lemma is needed.

Proposition 9. *Under additional regularity hypotheses, if S_1 and S_2 are the standard operators of symbols s_1 and s_2 then $\partial_j(S_1S_2 - S_2S_1)$ is a continuous operator on $L^2(R^N)$ with symbol $\xi_j\{s_1,s_2\}$ where $\{,\}$ denotes the usual Poisson bracket.*

In particular if $s_1 = a(\xi)$ and $s_2 = b(x)$ then the formula is valid for a smooth and b merely of class C^1, thanks to a result of A. Calderon [26]. In the case of the scalar equation of Example 6 one can obtain then a propagation result under some natural regularity hypotheses.

Proposition 10. *Let u^n be a sequence converging weakly to zero in $L^2(R^N)$ and assume that $\Sigma_jb_j\partial_ju^n + cu^n = f^n$ with f^n converging weakly to zero in $L^2(R^N)$, the coefficients b_j being assumed to be real and of class C^1 while c is only assumed to be continuous. Assume moreover that (u^n, f^n) corresponds to a H-measure μ. Then $\mu^{1,1}$ satisfies the following transport equation $< \mu^{1,1}, \{\Phi, P\} - \Phi\,\mathrm{div}\,b + 2\Phi\,\mathrm{Re}\,c >$ $= < 2\,\mathrm{Re}\,\mu^{1,2}, \Phi >$ for every function Φ of class C^1 in (x,ξ) with compact support in x, with P defined by $P(x,\xi) = \Sigma_jb_j(x)\xi_j$.*

In the case of the wave equation of Example 8 one finds a similar result.

Proposition 11. *Let u^n be a sequence converging weakly to zero in $H^1(R^{N+1})$ and assume that $\varrho\partial_0^2u^n$ - $\Sigma_{ij}\partial_i(a_{ij}\partial_ju^n) = f^n$ with f^n converging weakly to zero in $L^2(R^N)$, the coefficient ϱ being real positive independent of x_0 and of class C^1, the matrix a with entries a_{ij}, i, j = 1,..,N being hermitian positive independent of x_0 and of class C^1. Let U^n be defined by $U^n_i = \partial_iu^n$ for i = 0,..,N, and assume that (U^n, f^n) corresponds to a H-measure μ. Then $\mu^{i,j} = \xi_i\xi_j\nu^{1,1}$ for i, j = 0,..,N, and $\mu^{i,N+1} = \xi_i\nu^{1,2}$ for i = 0,..,N and $\nu^{1,1}$ satisfies the following transport equation $< \nu^{1,1}, \{\Phi, Q\} > =$ $< 2\,\mathrm{Re}\,\nu^{1,2}, \Phi >$ for every function Φ of class C^1 in (x,ξ) with compact support in x, with Q defined by $Q(x,\xi) = \varrho(x)\xi_0^2$ - $\Sigma_{ij}a_{ij}(x)\xi_i\xi_j$.*

One can complete Proposition 10 as in [4] by proving a trace theorem on a noncharacteristic hyperplane and from that deduce a result of change of variables for H-measures under local C^1 diffeomorphism, so that a theory on manifolds could be developed. A more interesting question lies in understanding the effect of semilinearity, that is when f^n does depend upon u^n or U^n in the framework of Proposition 10 or 11. The difficulty lies in the fact that H-measures are intrinsically quadratic objects and do not then predict anything about the limits of trilinear quantities for instance. H-measures do provide an improvement on the method of compensated compactness that I had developed with F. Murat [27, 28] but so far have not provided an alternative approach for quasilinear hyperbolic systems of conservation laws in order to replace the method that I had introduced [28] based on Young measures and compensated compactness, a method which had been successfully applied by R. DiPerna [29, 30, 31, 32]. Ron DiPerna had pointed out many years ago the defects of that old method and the need for a dynamic way of describing oscillations; H-measures is still the best answer to that quest and it is obviously not sufficient. It is important to point out that results like Proposition 11 correspond to the possibility of preparing initial data as a beam concentrated at a point and pointing in some direction and then follow where the energy goes; in the spirit of Example 4 one can also prepare initial data that correspond to a H-measure concentrated at a point in space and charging a countable number of points of the unit sphere and still follow the energy along each of these countably many small beams of light; as wave front sets are closed they cannot even see only a countable dense set of the sphere.

Other Results

It would be unfair not to point out that P. Gérard has introduced quite interesting variants which I cannot cover here [20, 21]. I cannot either discuss of other applications like the relation with homogenisation [4, 17].

Conclusion

In conclusion I was quite wrong in that small paragraph spared by the political censors [1] where I associated the study of "propagation of singularities" with classical physics as much better mathematical results connected to the propagation of light are like the above Proposition 11. I was also partly wrong in my previous ideas on quantum physics and oscillations as I had forgotten to include concentration effects in my description. Both these conclusions were learned from the possibilities created by the new mathematical tool of H-measures but a lot remains to be done on the way to a better understanding of physics through increased knowledge of some precise aspects of partial differential equations.

Obviously none of these new results are difficult and they could have been proved a long time ago by any of the best specialists of partial differential equations had they been interested in Science.

Acknowledgements. I would not have come here to preach one more time to induce mathematicians to be really interested in Science without the convincing arguments of Michael Crandall; I want to thank him for his continuous support along all these years.

This work is supported by a grant from the National Science Foundation DMS-8803317.

References

1. Tartar, L.: Etude des oscillations dans les équations aux dérivées partielles non linéaires. In: Trends and Applications of Pure Mathematics to Mechanics. (Lecture Notes in Physics, vol. 195.) Springer, Berlin Heidelberg New York 1984, pp. 384–412
2. Tartar, L.: Moscou sur Yvette, Souvenirs d'un Mathématicien Exclus. 90pp. 1986. Unpublished
3. Hörmander, L.: The analysis of linear partial differential operators I-IV. Springer, Berlin Heidelberg New York 1983–1985
4. Tartar, L.: H-measures, a new approach for studying homogenisation, oscillations and concentration effects in partial differential equations. Proc. Roy. Soc. Edinburgh **115A** (1990) 193–230
5. Tartar, L.: How to describe oscillations of solutions of nonlinear partial differential equations. In: Transactions of the Sixth Army Conference on Applied Mathematics and Computing, ARO Report 89-1, pp. 1133–1141
6. Bensoussan, A., Lions, J.-L., Papanicolaou, G.: Asymptotic analysis for periodic structures. Studies in Mathematics and its Applications, vol. 5. North-Holland, Amsterdam 1978
7. Sanchez-Palencia, E.: Non homogeneous media and vibration theory. (Lecture Notes in Physics, vol. 127.) Springer, Berlin Heidelberg New York 1980
8. Tartar, L.: Cours Peccot. Collège de France, Paris, 1977. Unpublished
9. Murat, F.: H-convergence. Séminaire d'Analyse Fonctionnelle et Numérique, Université d'Alger, 1977–1978
10. Spagnolo, S.: Sul limite delle soluzioni di problemi di Cauchy relativi all' equazione del calore. Ann. Scuola Norm. Sup. Pisa (3) **21** (1967) 657–699
11. Spagnolo, S.: Sulla convergenzia di soluzioni di equazioni paraboliche ed ellitiche. Ann. Scuola Norm. Sup. Pisa (3) **22** (1968) 571–597
12. Lions, P.-L.: The concentration compactness principle in the calculus of variations: the locally compact case. Part 1 and 2. Ann. Inst. H. Poincaré, Anal. non Linéaire (1984) 109–145 and no. 4, pp. 223–283
13. Lions, P.-L.: The concentration compactness principle in the calculus of variations: the limit case. Part 1 and 2. Rev. Mat. Iberoamericana **1** (1985) (1) 145–201; (2) 45–121
14. DiPerna, R.J.: Oscillations and concentration in solutions to the equations of mechanics. In: Directions in partial differential equations, M.G. Crandall, P.H. Rabinowitz, R.E.L. Turner (eds.), pp. 43–53. Academic Press, New York 1987
15. DiPerna, R.J., Majda, A.J.: Oscillations and concentration in weak solutions of the incompressible fluid equations. Comm. Math. Phys. **108** (1987) 667–689
16. DiPerna, R.J., Majda, A.J.: Concentration in regularizations for 2-D incompressible flow. Comm. Pure Appl. Math. **40** (1987) 301–345
17. Tartar, L.: H-measures and small amplitude homogenization. In: Random Media and Composites, R.V. Kohn, G. Milton (eds.), pp. 89–99. SIAM, Philadelphia 1989
18. Landau, L.D., Lifschitz, E. M.: Electrodynamics of continuous media. Pergamon Press, 1984
19. Tartar, L.: Remarks on homogenization. In: Homogenization and effective moduli of materials and media, J.L. Ericksen, D. Kinderlehrer, R.V. Kohn, J.-L. Lions (eds.), pp. 228-246. The IMA Volumes in Math. and its Applic., vol. 1. Springer, Berlin Heidelberg New York 1986
20. Gérard, P.: Compacité par compensation et régularité 2-microlocale. Séminaire Equations aux Dérivées Partielles 1988–1989 (Ecole Polytechnique, Palaiseau, exp VI)
21. Gérard, P.: Microlocal defect measures. To appear

22. Tartar, L.: Nonlocal effects induced by homogenization. In: Partial differential equations and the calculus of variations, vol. 2. Essays in Honor of E. DeGiorgi, pp. 925–938. Birkhäuser, 1989
23. Tartar, L.: Memory effects and homogenization. To appear in Arch. Rational Mech. Anal.
24. Amirat, Y., Hamdache, K., Ziani A.: Homogénéisation d'équations hyperboliques du premier ordre et application aux écoulements miscibles en milieu poreux. Ann. Inst. H. Poincaré, Anal. non Linéaire (5) **6** (1989) 397–417
25. Amirat, Y., Hamdache, K., Ziani A.: Etude d'une équation de transport à mémoire. C. R. Acad. Sci. Paris, Sér. I Math. **311** (1990) 685–688
26. Calderon, A.P.: Commutators of singular integral operators. Proc. Nat. Acad. Sci. **53** (1978) 1092–1099
27. Murat, F.: Compacité par compensation. Ann. Scuola Norm. Sup. Pisa Cl. Sci. (4) **5** (1978) 489–507
28. Tartar, L.: Compensated compactness and applications to partial differential equations. In: Nonlinear analysis and mechanics. Heriot-Watt Symposium IV, R.J. Knops (ed.). Research Notes in Mathematics, vol. 39, pp. 136–212. Pitman, London 1979
29. DiPerna, R.J.: Convergence of approximate solutions to conservation laws. Arch. Rat. Mech. Anal. **82** (1983) 27–70
30. DiPerna, R.J.: Convergence of the viscosity method for isentropic gas dynamics. Comm. Math. Phys. **91** (1983) 1–30
31. Tartar, L.: The compensated compactness method applied to systems of conservation laws. In: Systems of Nonlinear Partial Differential Equations, J.M. Ball (ed.). NATO ASI Series C 111, pp. 263–285. Reidel, New York 1983
32. Tartar, L.: Discontinuities and oscillations. In: Directions in partial differential equations, M.G. Crandall, P.H. Rabinowitz, R.E.L. Turner (eds.), pp. 211–233. Academic Press 1987

Microlocal Analysis in Spectral and Scattering Theory and Index Theory

Michael E. Taylor

Department of Mathematics, University of North Carolina
Chapel Hill, NC 27599-3250, USA

1. Introduction

Microlocal analysis has the dual foundation of the theory of singular integral operators, which arose originally to treat elliptic PDE, and geometrical optics, which arose to describe solutions to wave equations. It was forged into a powerful general tool for linear PDE a little over 20 years ago, through the efforts of at least five groups, including Calderon and Zygmund and their students; Kohn and Nirenberg; Sato and his collaborators; Maslov and Egorov and their collaborators; and Hörmander.

The basic object of microlocal analysis is the Fourier integral operator, which can be written as an oscillatory integral

$$\mathrm{Au}(x) = \int a(x, y, \theta) e^{i\psi(x,y,\theta)} u(y) dy d\theta \,, \tag{1.1}$$

where ψ is a real valued phase function, typically homogeneous of degree 1 in θ, and the amplitude $a(x, y, \theta)$ belongs to a symbol class, typically characterized by estimates on its derivatives, e.g.,

$$|D_y^\gamma D_x^\beta D_\theta^\alpha a(x, y, \theta)| \le C_{\alpha\beta\gamma} \langle\theta\rangle^{m-\varrho|\alpha|+\delta|\beta|+\delta|\gamma|} \tag{1.2}$$

to define $S^m_{\varrho,\delta}$. If a is asymptotic to a sum of terms homogeneous of degree $m - j$ in θ, $j = 0, 1, 2, \ldots$, we say $a \in S^m$. An important special case of (1.1) is

$$\mathrm{Au}(x) = \int a(x, \xi) e^{i\varphi(x,\xi)} \hat{u}(\xi) d\xi \,, \tag{1.3}$$

where $\hat{u}$ is the Fourier transform of u; the phase function is $\psi = \varphi(x, \xi) - y \cdot \xi$. One imposes a nondegeneracy condition implying that $(\nabla_\xi \varphi, \xi) \mapsto (x, \nabla_x \varphi)$ is locally well defined; this is the canonical transformation associated to (1.3). Under appropriate conditions on ψ in (1.1), one obtains a Fourier integral operator associated to a more general sort of canonical relation. The identity canonical transformation arises from the phase function $x \cdot \xi - y \cdot \xi$, for which (1.3) specializes to the formula

$$\mathrm{Au}(x) = \int a(x, \xi) e^{ix\cdot\xi} \hat{u}(\xi) d\xi \tag{1.4}$$

Proceedings of the International Congress
of Mathematicians, Kyoto, Japan, 1990

for a pseudodifferential operator. If $a \in S^m_{\varrho,\delta}$, we say the operator A in (1.4) belongs to $OPS^m_{\varrho,\delta}$.

Other sorts of pseudodifferential operators and Fourier integral operators have also arisen over the past 20 years, as tools in the study of various problems in PDE. Some of them, useful for certain problems in spectral theory and in scattering theory, will be described below.

2. Diffraction Effects

Operators of the form (1.3) describe propagation of waves away from a boundary, and also transversal reflection of waves off a boundary. Following earlier formal developments, Melrose [Me1] and Taylor [T1] made rigorous analyses of propagation of singularities along rays hitting a boundary at grazing incidence, producing Fourier-Airy operators of the form

$$L(F) = \int [gA(\zeta) + ihA'(\zeta)]A(\zeta_0)^{-1}e^{i\theta}\hat{F}(\xi)d\xi\,, \tag{2.1}$$

in which $A = A_{\pm}$ is an Airy function, $A_{\pm}(z) = Ai(e^{\mp 2\pi i/3}z)$, $\theta(x,\xi), \zeta(x,\xi)$ is a pair of phase functions, and $g(x,\xi), h(x,\xi)$ is a pair of amplitudes. The phase functions satisfy a system of eikonal equations and the amplitudes satisfy a system of transport equations. The parametrix (2.1) applies for example to solutions to the wave equation $u_{tt} - \Delta u = 0$ on $\Omega = \mathbb{R} \times \mathcal{O}$ where $\mathcal{O} = \mathbb{R}^n \setminus K$, K smooth and strictly convex, with lateral boundary $\partial\Omega = \mathbb{R} \times \partial K$. The Dirichlet condition can be formulated as $u = f$ on $\partial\Omega$ with $f \in \mathcal{E}'(\partial\Omega)$, and we require $u = 0$ for $t << 0$. Then $L(F)|_{\partial\Omega}$ is given by an elliptic Fourier integral operator J, acting on F, and the parametrix is given by (2.1) with $F = J^{-1}(f)$.

An important associated operator is the Neumann operator N, defined by $Nf = \partial u/\partial v|_{\partial\Omega}$ where u solves the Dirichlet problem with data f. From (2.1) one obtains the fundamental formula

$$N = J(A_1\Phi + B_1)J^{-1} \tag{2.2}$$

for a certain elliptic $A_1 \in OPS^{\frac{2}{3}}, B_1 \in OPS^0$. Here Φ is defined by $(\Phi f)\hat{}\,(\xi) = \Phi(\zeta_0)\hat{f}(\xi)$, where $\zeta_0 = |\xi|^{-\frac{1}{3}}\xi_n$ appears in (2.1), and $\Phi(z) = A'(z)/A(z)$ is an Airy quotient. The operator $A_1\Phi + B_1$ belongs to $OPS^1_{1/3,0}$; some of its operator properties are discussed in §3 below.

In [MeT1]–[MeT2] this parametrix construction was applied to produce some detailed results on scattered high frequency waves, relating to the behavior of the outgoing solution $v(x,\lambda)$ to

$$(\Delta + \lambda^2)v = 0 \text{ on } \mathbb{R}^n \setminus K, \quad v = e^{-i\lambda x\cdot\omega} \text{ on } \partial K\,. \tag{2.3}$$

This included a corrected Kirchhoff approximation, describing the behavior of $\partial_v v(x,\lambda)$ for $x \in \partial K$. We obtained an expansion of the form $K(\omega,x,\lambda)e^{-i\lambda x\cdot\omega}$ with

$$K(\omega,x,\lambda) \sim \sum_{j,k\geq 0} \lambda^{\frac{2}{3}-k-\frac{2}{3}j} b_{jk}(\omega,x)\Psi^{(j)}(\lambda^{\frac{1}{3}}Z)\,, \tag{2.4}$$

Z being a smooth function vanishing at the shadow boundary, determined in a subtle fashion by the symplectic geometry (in particular, *not* generally equal to $\nu \cdot \omega$). The function Ψ is given by

$$\Psi(\tau) = e^{-i\tau^3/3} \int A(s)^{-1} e^{-is\tau} ds \, ; \tag{2.5}$$

it is rapidly decreasing as $\tau \to -\infty$ and satisfies

$$\Psi(\tau) \sim \sum_{j \geq 0} \alpha_j \tau^{1-3j}, \quad \tau \to +\infty. \tag{2.6}$$

This appears to sharpen and extend some formulas of Fok. It demonstrates rigorously the often divined transitional layer of width $\sim \lambda^{-\frac{1}{3}}$ about the shadow boundary. Work on the corrected Kirchhoff approximation for solutions to Maxwell's equations was done in Yingst [Yi].

The asymptotic behavior of $v(x, \lambda)$ away from ∂K exhibits *two* transitional regions near the shadow boundary, one of width $\sim \lambda^{-\frac{1}{3}}$ and another of width $\sim \lambda^{-\frac{1}{2}}$. Near the shadow boundary, it was shown in [MeT2] that

$$v(x, \lambda) \sim \int b(x, \zeta, \lambda) \Big(\frac{A_+}{A_-}\Big)(\lambda^{\frac{2}{3}}\zeta) e^{i\lambda\varphi(x,\zeta)} d\zeta \tag{2.7}$$

where b has compact support in ζ and an asymptotic expansion in powers of λ. This has a further expansion of the form

$$\lambda^\alpha \left[e^{i\lambda\psi_0} p_0(x, \lambda) + e^{i\lambda\psi_1} p_1(x, \lambda) + e^{i\lambda\psi_2} p_2(x, \lambda)\right]. \tag{2.8}$$

The first term captures the direct wave, the third term the reflected wave, and the middle term represents a diffraction effect confined very close to the shadow boundary. The amplitudes $p_j(x, \lambda)$ have the following form:

$$p_0(x, \lambda) = a_0(x, \lambda^{\frac{1}{2}} s_0(x), \lambda^{\frac{1}{2}}) \tag{2.9}$$

where here and below $s_j(x)$ are smooth functions vanishing simply on the shadow boundary. $a_0(x, \tau, \mu)$ is a symbol of product type in τ, μ.

$$p_2(x, \lambda) = a_2(x, \lambda^{\frac{1}{3}} s_2(x), \lambda^{\frac{1}{3}}), \tag{2.10}$$

and $a_2(x, \tau, \mu)$ is also a symbol of product type. The most subtle term is

$$p_1(x, \lambda) = a_1(x, \lambda^{\frac{1}{3}} s_1(x), \lambda^{-\frac{1}{6}}) \tag{2.11}$$

where $a_1(x, \tau, \mu)$ is C^∞ away from $(\tau, \mu) = (0, 0)$, rapidly decreasing as $\tau \to \infty$, and has a conormal singularity at $(0, 0)$, which can be approached as $\lambda \to \infty$ due to the negative exponent on λ in the last argument of a_1.

Zworski [Zw1] has extended the analysis of the asymptotic behavior of $v(x, \lambda)$ to be uniformly valid as x approaches ∂K. In [Zw2] a rigorous study of the shift in the shadow boundary predicted by Keller and Rubinow is made; the shift is asymptotic to $C_0\lambda^{-\frac{2}{3}}$, well inside the smaller transitional region, hence in a region where the asymptotic expansion (2.8) simplifies.

In [MeT1] there is also a uniform analysis of the near peak scattering amplitude, extending earlier work of [MjT] and [Me3] on the on peak behavior. The expansion is somewhat like (2.8). Again there are two transitional regions and a subtle term connecting them. This time the widths of the transition regions are $\sim \lambda^{-\frac{1}{3}}$ and $\sim \lambda^{-1}$.

A key ingredient in these analyses was a normal form for Fourier integral operators with folding canonical relations, namely

$$P_1 \mathscr{A}i + P_2 \mathscr{A}i', \quad P_1 \in OPS^m, \ P_2 \in OPS^{m-\frac{1}{3}}, \tag{2.12}$$

where $\mathscr{A}i$ is Fourier multiplication by $Ai(\zeta_0)$ and $\mathscr{A}i'$ is similarly defined. Using this, Farris [Fa] found a microlocal model for solutions to boundary problems with grazing rays, namely

$$K_1(\mathscr{A}_+/\mathscr{A}_-)K_2 \tag{2.13}$$

where K_j are elliptic Fourier integral operators and $(\mathscr{A}_+/\mathscr{A}_-)$ is Fourier multiplication by $A_+(\zeta_0)/A_-(\zeta_0)$.

Gliding ray problems arise for example if one considers the wave equation $u_{tt} - \Delta u = 0$ on $\mathbb{R} \times K$ rather than $\mathbb{R} \times \mathcal{O}$, where as above $K \subset \mathbb{R}^n$ has smooth strictly convex boundary. In such a case, one has a parametrix like (2.1) with $A_\pm(\zeta)$ replaced by $Ai(\zeta)$. Since $Ai(z)$ has real zeros, it is convenient to make an almost analytic continuation of $\zeta(x,\xi)$ and evaluate it at $\xi_n + iT$ for some $T \neq 0$ rather than at ξ_n. Then one has instead of (2.2) a formula for the Neumann operator involving Φi, Fourier multiplication by $Ai'(\zeta_0)/Ai(\zeta_0)$. It is of great help that, thanks to Melrose's work on equivalence of glancing hypersurfaces, one can arrange that $\zeta = \zeta_0$ on the boundary, $\zeta_0(\xi) = |\xi|^{-\frac{1}{3}}\xi_n$. Not having this introduces complications, which however were tackled in [Es1]. One needs to resort to energy estimates rather than use symbol calculus, and hence produces a less explicit sort of parametrix.

Having the replacement for (2.1)–(2.2) described above for gliding ray problems does not end one's job here. The operator Φi is a rather complicated operator, a singular sort of Fourier integral operator with an infinite number of canonical transformations accumulating along the leaf relation of the characteristic variety for the boundary. One key method for taming Φi is described in the next section.

3. Airy Operator Calculus

Solving boundary problems other than the Dirichlet problem for wave equations with grazing rays involves, in addition to (2.1), the solution to an equation of the form

$$(A\Phi + B)f = g, \tag{3.1}$$

with $A \in OPS^{\frac{2}{3}}, B \in OPS^1$. If B is elliptic, one has an elliptic operator in $OPS^1_{\frac{1}{3},0}$, while if A is elliptic and the symbol of B vanishes on $\{\zeta_0 = 0\}$, one has a hypoelliptic operator. There is a special calculus of Airy operators, extending that of [Me2], sketched in [T5] and treated in detail in [MeT3], which provides more analytical detail on compositions and parametrices of such operators than the $S^m_{1/3,0}$ calculus, based on some remarkable identities which follow from constructing the Neumann operator (2.2) using different choices of solutions g, h to the transport equations for the amplitudes in (2.1).

The full advantages of this approach are apparent when one treats the analogues that arise in problems with gliding rays, i.e., composition and parametrices for operators like

$$A\Phi i + B, \tag{3.2}$$

with A and B as in (3.1). These are highly singular variants of Fourier integral operators, with a rather subtle operator calculus. Nevertheless, if B is elliptic, the operator (3.2) is shown to have a parametrix in the class $\mathscr{A}i_\sigma^{-1,\pm}$, where by definition $\mathscr{A}i_\sigma^{m,\pm}$ consists of operators with asymptotic expansions of the form

$$T \sim B + \sum_{j\geq 0} A_j \Phi i C_j, \quad B \in OPS^m, \ A_j \in OPS^{m_j}, \ C_j \in OPS^0, \tag{3.3}$$

where $m_j + \frac{1}{3} = m - \ell_j, \ell_j \geq 0$ is an integer, $\ell_j \to \infty$ as $j \to \infty$. The sign + or − reflects the choice of sign of T, with ζ_0 evaluated at $\xi_n + iT$. If A is elliptic and B has vanishing symbol on $\{\zeta_0 = 0\}$, the operator (3.2) has a parametrix of the form

$$(C\Phi i^{-1} + D)R(1 + E\Phi i^{-1}R)^{-1} \tag{3.4}$$

with $C \in OPS^{-1}, D \in OPS^{-\frac{4}{3}}, R \in OPS^{\frac{1}{3}}$ elliptic, $E \in OPS^{-1}$. Then

$$E\Phi i^{-1}R : H^s \longrightarrow H^{s+\frac{1}{3}}, \tag{3.5}$$

so the Neumann series for the last factor is asymptotic. This affords a symbolic construction of parametrices for numerous boundary problems involving gliding rays.

There are alternative approaches to many boundary problems, both in grazing and gliding cases, that involve applying energy estimates to such equations as (3.1) or its analogue using the operator (3.2). Information so obtained is less explicit than by a symbolic parametrix construction, but can be useful all the same. Energy estimate approaches also have a flexibility to apply to cases not amenable to explicit constructions. Eskin [Es2] has proposed a number of techniques along those lines. In [T6] the Fefferman-Phong inequality was brought to bear on a number of equations of the form (3.1).

4. Functional Calculus

Let A be a self adjoint elliptic operator in OPS^1. It is convenient to analyze many functions of A via

$$f(A) = \int \hat{f}(t)\, e^{itA}\, dt, \tag{4.1}$$

since e^{itA} is a group of Fourier integral operators, of the form (1.3) for $|t|$ small. If f belongs to a symbol space, $\hat{f}(t)$ is singular only at the origin, and is rapidly decreasing as $|t| \to \infty$. One can write $\hat{f}(t) = \hat{f}_1(t) + \hat{f}_2(t)$ where $\hat{f}_1$ has support near

$t = 0$ and $\hat{f}_2 \in \mathcal{S}(\mathbb{R})$. We can use (1.3) to analyze $f_1(A)$ as a pseudodifferential operator. If A acts on functions on a compact manifold M, then $f_2(A) \in OPS^{-\infty}$ is for many purposes negligible. If M is not compact, $A = (c - \Delta)^{\frac{1}{2}}$, (with $c - \Delta \geq 0$), and f is even, one can replace (4.1) by

$$f(A) = \int \hat{f}(t) \cos tA \, dt, \tag{4.2}$$

and exploit finite propagation speed to get useful information on $f_2(A)$.

Using (4.1) to evaluate the trace of $f(A)$ and hence give sharp results on the spectral asymptotics of A was one of the early spectacular applications of Fourier integral operators. In Hörmander's paper [Ho1], a key point is to analyze the trace of (4.1) for $f(\lambda) = f_\tau(\lambda) = f(\lambda - \tau)$ as $\tau \to \infty$, given $\hat{f} \in C_0^\infty(\mathbb{R})$ having small support. [DG] extended this to any $\hat{f} \in C_0^\infty(\mathbb{R})$. In [T2] and Chapter 12 of [T3] a number of applications of the fact that $f(A) \in OPS^m_{1,0}$ when $f \in S^m_1(\mathbb{R})$ were discussed. It was also shown how functional calculus applied to the Laplace operator on S^2 led to a clean treatment of the problem of scattering of waves in $\mathbb{R}^3$ by a sphere, a special case of the general problem discussed in §1 which has been treated by many authors, generally with a heavier dependence on special function theory. About the same time, Colin de Verdiere [CV] discussed a similar theory of functions of pseudodifferential operators, emphasizing applications to semiclassical asymptotics for spectra of Schrödinger operators.

The method of separation of variables, combined with a harmonic analysis on the base using the techniques described above, were used by [CT] to give a detailed analysis of diffraction of waves by a cone.

The formula (4.2) was exploited in [CGT] to produce fine estimates on the heat kernel on a variety of complete Riemannian manifolds. Of particular interest were those for which the Ricci tensor was bounded from below; also stronger hypotheses, such as C^∞-bounded geometry, yielded stronger results, such as L^p-boundedness of (4.2) for $f \in S^0_1(\mathbb{R})$ holomorphic in a strip of width related to the volume growth of M. In [CGT] we used this to recover known results on L^p-boundedness when M is a symmetric space of rank 1. In [DST] a study was made of the L^p-spectrum of the Laplace operator on geometrically finite quotients of hyperbolic space. In [T8] the argument of [CGT] was honed and yielded results on general symmetric spaces of noncompact type, sharp enough to establish conjectures on what is precisely the L^p-spectrum of the Laplace operator. In such a case, as is well known, the L^2-spectrum of $-\Delta$ is $[|\varrho|^2, \infty)$. Then the L^p-spectrum is shown to be precisely $\{|\varrho|^2 + z^2 : |\mathrm{Im} z| \leq |\frac{2}{p} - 1| \cdot |\varrho|\}$. Furthermore, with $H = -\Delta - |\varrho|^2 \geq 0, L = H^{\frac{1}{2}}, f(L)$ is bounded on L^p provided $1 < p < \infty$ and f is holomorphic on a strip $|\mathrm{Im} z| \leq |\frac{2}{p} - 1| \cdot |\varrho|$ and is a symbol of order 0 there. Recently, J.Anker [An], [An2] has further extended these arguments.

5. Semiclassical Asymptotics and Gauge Fields

Let M be a Riemannian manifold, G a compact Lie group, $P \to M$ a principal G-bundle. A gauge field is defined by a connection on P. Corresponding to each irreducible unitary representation λ of G is a Hermitian vector bundle $E_\lambda \to M$ with connection, ∇_λ, and a Hamiltonian operator $H_\lambda^0 = \nabla_\lambda^* \nabla_\lambda$. If a scalar potential V is also given, a semiclassical analysis of the resulting quantum system involves study of the spectrum of

$$H_\lambda = \hbar^2 H_\lambda^0 + V, \quad \hbar = |\lambda + \delta|^{-1}, \tag{5.1}$$

as $\hbar \to 0$. Thus $\lambda \to \infty$ in a Weyl chamber. This problem was studied in [ST1]–[ST2]. In particular, given $f \in \mathscr{S}(\mathbb{R})$, an asymptotic expansion was made of $\operatorname{Tr} f(H_\lambda)$; it was shown that

$$\operatorname{Tr} f(H_\lambda) = \langle \kappa, \chi_\lambda \rangle = d_\lambda \beta(\lambda + \delta) \tag{5.2}$$

for a Weyl group invariant symbol β, defined by the spectrum of a certain central distribution κ on G arising from the operator $f(-A^{-1}L) \in OPS^0(P)$, where

$$L = \Delta + (V-1)\Delta_G^P - |\delta|^2 V, \quad A = -\Delta_G^P + |\delta|^2, \tag{5.3}$$

Δ being the Laplace operator on P and Δ_G^P the vertical Laplacian. One arranges that $V > 1$. Methods of §4 were used to analyze $f(-A^{-1}L)$. We have $\kappa = \operatorname{Tr}_G f(-A^{-1}L)$, where the '$G$-trace' of an operator on $C^\infty(P)$ with Schwartz kernel $K(p,q)$ is defined to be

$$\kappa(g) = \int_P K(p \cdot g, p)\, dV(p). \tag{5.4}$$

In [GU], using different techniques, a different sort of asymptotic study of the spectrum of (5.1) was made, as $\lambda \to \infty$ along a ray. In [TU] a synthesis of these results has been achieved. We study the asymptotic behavior of

$$\operatorname{Tr} f(\hbar^{-1} H_\lambda^{-\frac{1}{2}}(H_\lambda - c)), \quad \hbar = |\lambda + \delta|^{-1} \tag{5.5}$$

as $\hbar \to 0$, for a given $c \in \mathbb{R}$, a regular value of V, given $\hat{f} \in C_0^\infty(\mathbb{R})$. This is analyzed in terms of the G-trace of $f(Q)$, where

$$Q = (-L)^{-\frac{1}{2}}(-L - cA) \in OPS^1(P). \tag{5.6}$$

Now Q is typically not elliptic, but it is an operator of principal type, with real principal symbol. A modification of the method of §4 shows that, for $\hat{f} \in C_0^\infty(\mathbb{R})$, $f(Q)$ is a Fourier integral operator on $C^\infty(P)$ with a canonical relation given by the leaf relation on the characteristic variety of Q. Then $\operatorname{Tr}_G f(Q)$ is under reasonable geometrical conditions a central Lagrangian distribution on G, whose nature reflects the classical dynamics of the gauge and scalar field on the Wong bundle. Harmonic analysis of $\operatorname{Tr}_G f(Q)$ involves restriction to a maximal torus $\mathbf{T}$, an operation for which clean intersection hypotheses often fail, if G has rank ≥ 2. In the simplest cases there can arise distributions on $\mathbf{T}$ associated to transversally intersecting Lagrangians, thus giving rise to nonclassical asymptotics for (5.5).

6. Operator K-Theory

Atiyah proposed K-homology as an abstract setting for index theory, and this was developed by Brown, Douglas, and Fillmore and by Kasparov and others; see [Bl]. In [BDT] an investigation was made of cycles defined by elliptic differential operators. It was shown that every closed extension of a first order elliptic differential operator D, acting on sections of a vector bundle over a smooth Riemannian manifold M, not necessarily complete, defines a class $[D]$ in $KK(C_0(M), \mathbb{C})$, independent of the choice of closed extension. The proof involves a use of (4.1) and finite propagation speed. In case M is contained in $\overline{M}$, compact with boundary, consideration of the boundary map $\partial : KK(C_0(M), \mathbb{C}) \to K_1(\partial M)$ applied to cycles coming from different closed extensions leads to identities in K-homology.

For example, if $M = \Omega$ is a pseudoconvex manifold with the property that the $\overline{\partial}$-Neumann problem has compact resolvent on $(0,p)$-forms, for $p \neq 0$, then taking two natural closed extensions of $D = \overline{\partial} + \overline{\partial}^*$ and applying the boundary map produces the identity

$$[D_{\partial\Omega}] = [\tau_\Omega] \text{ in } K_1(\partial\Omega) \tag{6.1}$$

where $D_{\partial\Omega}$ is the Dirac operator on $\partial\Omega$, with its natural spin^c-structure, and $[\tau_\Omega]$ is the Toeplitz extension. This identity refines and generalizes Boutet de Monvel's index theorem for elliptic Toeplitz operators. It also leads to a number of other identities in K-homology.

Identities in $K_1(M)$, including (6.1), can be used together with the Bott map to produce identities in $K_0(M)$.

In [T9] an examination was made of ways in which differential operators and pseudodifferential operators define elements of $K^j(\Psi^0(M))$, where $\Psi^0(M)$ is the C^*-algebra which is the L^2-operator norm closure of $OPS^0(M)$, in case M is compact. It was shown that in some cases (6.1) has a further refinement as an identity in $K^1(\Psi^0(\partial\Omega))$. I suspect that in general the two elements differ by a quantity which can be described in terms of

$$\tau : K^0(M) \to K^1(\Psi^0(M)), \tag{6.2}$$

which arose in [T9] to describe an obstruction for an elliptic pseudodifferential operator acting on sections of a vector bundle to define an element of $K^0(\Psi^0(M))$.

The K-theory of $\Psi^0(M)$ can be thought of as a microlocal version of the K-theory of M. Extra structure arises, partly due to the richer structure of ideals; in particular, if $\Lambda \subset T^*M \setminus 0$ is a closed conic set, one can form $\Psi^0_\Lambda(M)$, the closure of the set of elements of $OPS^0(M)$ whose principal symbols vanish on Λ. For example, if M has a contact structure, one can let Λ be the contact line bundle $(\setminus 0)$; then $\Psi^0/\Psi^0_\Lambda \approx C(M) \oplus C(M)$. To give one example of a natural cycle arising in this context, if $M = \partial\Omega$ is the boundary of a strongly pseudoconvex domain in $\mathbb{C}^2$, $\overline{\partial}_b$ defines an element of $KK(\Psi^0_\Lambda(M), \mathbb{C})$. It is of interest to compute the (co)boundary map $\delta : KK(\Psi^0_\Lambda, \mathbb{C}) \to K^1(\Psi^0/\Psi^0_\Lambda)$, isomorphic to $K_1(M) \oplus K_1(M)$ in this last case. It is noted in [T9] that, with D_M as in (6.1),

$$\delta[\overline{\partial}_b] = ([D_M], -[D_M]) . \tag{6.3}$$

References

[An] Anker, J.: L^p Fourier multipliers on Riemannian symmetric spaces of the non-compact type. Preprint

[An2] Anker, J.: Handling the inverse spherical Fourier transform. Preprint

[BDT] Baum, P., Douglas, R., Taylor, M.: Cycles and relative cycles in analytic K-homology. J. Diff. Geom. **30** (1989) 761–804

[Bl] Blackadar, B.: K-theory for operator algebras. Springer, Berlin Heidelberg New York 1986

[BdM] Boutet de Monvel, L.: On the index of Toeplitz operators of several complex variables. Invent. math. **50** (1979) 249–272

[CGT] Cheeger, J., Gromov, M., Taylor, M.: Finite propagation speed, kernel estimates for functions of the Laplace operator, and the geometry of complete Riemannian manifolds. J. Diff. Geom. **17** (1982) 15–53

[ChT] Cheeger, J., Taylor, M.: Diffraction of waves by conical singularities. Comm. Pure Appl. Math. **35** (1982) 275–331, 487–529

[CV] Colin de Verdiere, Y.: Spectre conjoint d'operateurs pseudodifferentiels qui commutent, I: Le cas non integrable. Duke Math. J. **46** (1979) 169–182. II: Le cas integrable. Math. Z. **171** (1980) 51–73

[Cor] Cordes, H.O.: Elliptic pseudo-differential operators – An abstract theory. (Lecture Notes in Mathematics, vol. 756.) Springer, Berlin Heidelberg New York 1979

[DST] Davies, E.B., Simon, B., Taylor, M.: L^p spectral theory of Kleinian groups. J. Funct. Anal. **78** (1988) 116–136

[DG] Duistermaat, J.J., Guillemin, V.: The spectrum of positive elliptic operators and periodic bicharacteristics. Invent. math. **29** (1975) 39–79

[Es1] Eskin, G.: Parametrix and propagation of singularities for the interior mixed hyperbolic problem. J. Anal. Math. **32** (1977) 17–62

[Es2] Eskin, G.: General initial-boundary problems for second order hyperbolic equations. In: Singularities in boundary value problems. D. Reidel, Boston 1981, pp. 19–54

[Fa] Farris, M.: Egorov's theorem on a manifold with diffractive boundar., Comm. PDE **6** (1981) 651–688

[GU] Guillemin, V., Uribe, A.: Reduction, the trace formula, and semiclassical asymptotics. Proc. Nat. Acad. Sci. USA **84** (1987) 7799–7801

[HR] Hellfer, B., Robert, D.: Comportement semi-classique du spectre des Hamiltoniens quantiques periodiques. Ann. Inst. Fourier **31** (1981) 169–223

[Ho1] Hörmander, L.: The spectral function of an elliptic operator. Acta Math. **121** (1968) 193–218

[Ho2] Hörmander, L.: The analysis of linear partial differential operators, vols. 3 and 4. Springer, Berlin Heidelberg New York 1985

[Ki] Kirchhoff, G.: Vorlesungen über Math. Physik (Optik). Leipzig, 1891.

[MjT] Majda, A., Taylor, M.: The asymptotic behavior of the diffraction peak in classical scattering. Comm. Pure Appl. Math. **30** (1977) 639–669

[Me1] Melrose, R.: Microlocal parametrices for diffractive boundary value problems. Duke. Math. J. **42** (1975) 605–635

[Me2] Melrose, R.: Airy operators. Comm. PDE **3** (1978) 1–76

[Me3] Melrose, R.: Forward scattering by a convex obstacle. Comm. Pure Appl. Math. **33** (1980) 461–499

[MeS] Melrose, R., Sjöstrand, J.: Singularities of boundary value problems. I.: Comm. Pure Appl. Math. **31** (1978) 593–617, II: Comm. Pure Appl. Math. **35** (1982) 129–168

[MeT1] Melrose, R., Taylor, M.: Near peak scattering and the corrected Kirchhoff approximation for convex bodies. Adv. Math. **55** (1985) 242–315

[MeT2] Melrose, R., Taylor, M.: The radiation pattern of a diffracted wave near the shadow boundary. Comm. PDE **11** (1986) 599–672

[MeT3] Melrose, R., Taylor, M.: Boundary problems for wave equations with grazing and gliding rays. To appear

[ST1] Schrader, R., Taylor, M.: Small $\hbar$ asymptotics for quantum partition functions associated to particles in external Yang-Mills potentials. Commun. Math. Phys. **92** (1984) 555–594

[ST2] Schrader, R., Taylor, M.: Semiclassical asymptotics, gauge fields, and quantum chaos. J. Funct. Anal. **83** (1989) 258–316

[Str] Strichartz, R.: A functional calculus for elliptic pseudodifferential operators. Amer. J. Math. **94** (1972) 711–722

[T1] Taylor, M.: Grazing rays and reflection of singularities of solutions to wave equations. Comm. Pure Appl. Math. **29** (1976) 1–38, 463–481

[T2] Taylor, M.: Fourier integral operators and harmonic analysis on compact manifolds. AMS Proc. Symp. Pure Math. **35** (2) (1979) 115–136

[T3] Taylor, M.: Pseudodifferential operators. Princeton Univ. Press, 1981

[T4] Taylor, M.: Diffraction efects in the scattering of waves. In: Singularities in boundary value problems, D. Reidel (ed.). Boston 1981, pp. 271–316

[T5] Taylor, M.: Airy operator calculus. Contemp. Math. AMS (1984) 169–192

[T6] Taylor, M.: Fefferman-Phong inequalities in diffraction theory. AMS Proc. Symp. Pure Math. **43** (1984) 261–300

[T7] Taylor, M.: Noncommutative microlocal analysis. Memoirs AMS **313** (1984)

[T8] Taylor, M.: L^p-estimates on functions of the Laplace operator. Duke Math. J. **58** (1989) 773–793

[T9] Taylor, M.: Pseudodifferential operators and K-homology. I: AMS Proc. Symp. Pure Math. **51** (1) 561–583; II: Contemp. Math. **106** (1990) 245–269

[T10] Taylor, M.: Pseudodifferential operators and nonlinear PDE. Progress in Math. Birkhäuser, 1991 (to appear)

[TU] Taylor, M., Uribe, A.: Semiclassical spectra of gauge fields. To appear

[Yi] Yingst, D.: The Kirchhoff approximation for Maxwell's equations. Indiana Math. J. **32** (1983) 543–562

[Zw1] Zworski, M.: High frequency scattering by a convex obstacle. Duke Math. J. (to appear)

[Zw2] Zworski, M.: Shift of the shadow boundary in high frequency scattering. Commun. Math. Phys. (to appear)

Problems on Limit Sets of Foliations on Complex Projective Spaces

César Camacho

Instituto de Matemática Pura e Aplicada, Estrada Dona Castorina 110
CEP-22460, Rio de Janeiro, Brasil

We consider differential equations

$$\eta = P(x,y)dy - Q(x,y)dx = 0$$

where $P(x,y)$ and $Q(x,y)$ are complex polynomials in the complex variables $(x,y) \in \mathbf{C}^2$.

The integrals of this equation are either open Riemann surfaces, passing through points where not both P and Q vanish, or singular points $(x_0, y_0) \in \mathbf{C}^2$ where $P(x_0, y_0) = Q(x_0, y_0) = 0$. These integrals define a foliation of $\mathbf{C}^2$ that extends naturally to a foliation with singularities, $\mathscr{F}$, on $\mathbf{C}P(2)$ the complex projective 2-space. We assume from now on that singularities are isolated.

The first to study these foliations from the local point of view around singularities were C.A.Briot and J.C.Bouquet in 1856 and later by H.Poincaré, P.Painlevé, H.Dulac. Nowadays this local theory is very much developed specially due to the contributions from the Soviet, French and Brazilian schools on this subject, despite the existence of several interesting problems, for instance the developing of a bifurcation theory.

On the other hand, from the global point of view the dynamics of these foliations is far from being understood and it is really in the beginning. Here we wish to pose three problems concerning the most elementary concept of the dynamics of a foliation, that is, its limit set.

As any leaf L of $\mathscr{F}$ is open, we define the *limit set of* L as

$$\lim(L) := \bigcap_{n\geq 1} \overline{L \setminus K_n}$$

where $K_n \subset K_{n+1} \subset L$ is a sequence of compact subsets of L such that $\bigcup_{n\geq 1} K_n = L$. Then define

$$\lim \mathscr{F} = \overline{\bigcup_L \lim(L)}.$$

So $\lim \mathscr{F}$ is a closed, invariant subset of $\mathbf{C}P(2)$.

Problem 1. Classify all foliations whose limit set is analytic.

When $\lim \mathscr{F}$ is analytic and has dimension zero we have the following

Proceedings of the International Congress
of Mathematicians, Kyoto, Japan, 1990

Theorem (G. Darboux). *If* $\lim \mathcal{F}$ *is finite then* $\mathcal{F}$ *admits a meromorphic first integral, i.e. there are polynomials* f *and* g *such that the leaves of* $\mathcal{F}$ *are given by* $f - cg = 0$, $c \in \mathbf{C}$.

Notice that by Remmert-Stein theorem $\lim(L)$ is finite if and only if $\overline{L}$ is analytic.

A simple example where the limit set has dimension one is the linear foliation given by

$$\mathcal{L} : \lambda x dy - y dx = 0, \quad \lambda \notin \mathbf{R}.$$

Then $\lim \mathcal{L}$ is a union of three projective lines: the x and y-axes, and the line at infinity. The holonomy of each one of these lines is hyperbolic, i.e. it contains an element whose linear part is of the form $z \mapsto \lambda z$ with $|\lambda| \neq 1$. The following theorem shows that foliations with hyperbolic limit set of dimension one are essentially linear.

Theorem (C. Camacho, A. Lins N., P. Sad). *Let* $\Lambda = \lim \mathcal{F}$ *be an analytic subset of dimension one such that:*

(i) The holonomy of each irreducible component of Λ *is hyperbolic.*

(ii) The number of separatrices at each singularity is finite.

Then there is a rational map F *of* $\mathbf{C}P(2)$ *and a linear flow* $\mathcal{L}$ *such that* $\mathcal{F} = F^* \mathcal{L}$.

Sketch of Proof. The proof consists of four parts. The first one contains all the dynamics of the problem.

I. Lemma. *Let* $0 \in V \subset \mathbf{C}$ *be a neighborhood and* $f, g : V, 0 \to \mathbf{C}, 0$ *holomorphic local diffeomorphisms such that* $|f'(0)| < 1$. *Suppose that for any* $p \in V$ *the orbit* $0(p)$ *of the pseudogroup generated by* f *and* g *satisfies* $\overline{0(p)} \setminus 0(p) \subset \{0\}$. *Then* $f \circ g = g \circ f$.

Thus the holonomy group of each irreducible component of Λ is abelian and linearizable.

II. The Resolution of $\mathcal{F}$. *Suppose that the singular set of* $\mathcal{F}$ *is,* $\operatorname{sing} \mathcal{F} = \{p_1, \ldots, p_m\}$. *The resolution of* $\mathcal{F}$ *(Theorem of Bendixson-Seidenberg [1, 9]) consists of a proper holomorphic map* $\pi : M \to \mathbf{C}P(2)$ *which is obtained as a certain composition of finitely many quadratic blow up's at the points of* $\operatorname{sing} \mathcal{F}$, *such that if* $D = \pi^{-1} \operatorname{sing} \mathcal{F}$ *is the divisor, then:*

(i) $\pi|_{M \setminus D}$ *is a diffeomorphism onto its image.*

(ii) $\tilde{\mathcal{F}} = \pi^* \mathcal{F}$ *is a foliation leaving* D *invariant with isolated elementary singularities which in local charts* (x, y) *have one of the following forms*

a) simple: $\lambda_1 x dy - (\lambda_2 y + \ldots) dx = 0$, $\frac{\lambda_1}{\lambda_2} \notin \mathbf{Q}_+$

b) saddle-nodes: $(x + \ldots) dy - y^p dx = 0$, $p \geq 2$.

Proposition. *All components of* D *(and so of* $\tilde{\Lambda} = \Lambda \cup D$*) are hyperbolic and there are no saddle nodes in* $\operatorname{sing} \tilde{\mathcal{F}}$.

This follows from the index theorem proved in [2] applied to the resolution of $\mathcal{F}$.

III. Proposition. *There is a closed meromorphic one form ω in $\mathbf{C}P(2)$ such that:*
(i) The polar divisor of ω, $(\omega)_\infty$, has order one and is contained in Λ.
(ii) In $\mathbf{C}P(2) \setminus (\omega)_\infty$, ω induces $\mathscr{F}$.

The proof of this proposition consists in showing that once all singularities and holonomies of $\tilde{\Lambda}$ are linearizable it is posssible to cover $\tilde{\Lambda}$ by local chart neighborhoods $(x_\alpha, y_\alpha) \in U_\alpha$, $\alpha \in A$, such that if *sing* $\tilde{\mathscr{F}} \cap U_\alpha = \phi$ then $\tilde{\mathscr{F}}|_{U_\alpha}$ is defined by $dy_\alpha = 0$ and if *sing* $\tilde{\mathscr{F}} \cap U_\alpha \neq \phi$ then this intersection is a point $p \in U_\alpha$ where $x(p) = y(p) = 0$ and $\tilde{\mathscr{F}}|_{U_\alpha}$ is given by $x_\alpha dy_\alpha - \lambda_\alpha y_\alpha dx_\alpha = 0$. We assume also that if $U_\alpha \cap U_\beta \neq \phi$ then it is simply connected. A form $\tilde{\omega}$ is then defined in $U_\alpha U_\alpha$ as $\tilde{\omega}_\alpha := \frac{dy_\alpha}{y_\alpha}$ in regular neighborhoods U_α and as $\tilde{\omega}_\alpha := \frac{dy_\alpha}{y_\alpha} - \lambda_\alpha \frac{dx_\alpha}{x_\alpha}$ in singular neighborhoods U_α. If $U_\alpha \cap U_\beta \neq \phi$ are two regular neighborhoods it is proven using the existence of a hyperbolic element of the holonomy that the change of coordinates $y_\beta = L_{\beta\alpha}(y_\alpha)$ is linear, thus $\tilde{\omega}_\alpha = \tilde{\omega}_\beta$ in $U_\alpha \cap U_\beta \neq \phi$. Similarly if U_α is singular, U_β regular and $U_\alpha \cap U_\beta \neq \phi$ one can show that $\tilde{\omega}_\beta = g_\alpha \cdot \tilde{\omega}_\alpha$ in $U_\alpha \cap U_\beta$, where g_α is holomorphic independent of β.

The form $\tilde{\omega}$ induces $\omega = \pi_* \tilde{\omega}$ in a neighborhood of $\Lambda \subset \mathbf{C}P(2)$. Now, since $\mathbf{C}P(2) \setminus \Lambda$ is a Stein manifold this form ω extends to $\mathbf{C}P(2)$ by a generalization of Levi's theorem.

IV. Construction of $\mathscr{L}$ and F. Here we follow Cerveau-Mattei [4]. The differential form ω restricted to $\mathbf{C}^2$ has the following polar divisor:

$$\Gamma = (\omega)_\infty \cap \mathbf{C}^2 = \bigcup_{j=0}^{m} \Gamma_j \quad \text{where } \Gamma_j = (f_j = 0)$$

and each f_j is a polynomial.

Since

$$\mathscr{H}^1_{DR}\left(\mathbf{C}^2 \setminus \bigcup_{j=0}^{m} \Gamma_j\right) = \left[\frac{df_0}{f_0}, \ldots, \frac{df_m}{f_m}\right].$$

Then we can show that

$$\omega|_{\mathbf{C}^2} = \sum_{j=0}^{m} \lambda_j \frac{df_j}{f_j}$$

where

$$\lambda_j = 12\pi i \int_{\gamma_j} \omega$$

and γ_j is a closed path making a simple tour around Γ_j in a small transverse cross section to Γ_j. The holonomy group of Γ_0 is generated by $\{\mu_1^{\ell_1}, \ldots, \mu_m^{\ell_m}\} \subset \mathbf{C}^*$ where the ℓ_j's are integers and $\mu_j = \exp 2\pi i \lambda_j / \lambda_0$. Since this group is discrete it is possible to find integers $u_1, \ldots, u_m$; $v_0, \ldots, v_m$ such that for $\Phi = f_1^{u_1} \ldots f_m^{u_m}$ and $\Psi = f_0^{v_0} \ldots f_m^{v_m}$ we have

$$\frac{d\Psi}{\Psi} + \lambda \frac{d\Phi}{\Phi} = \delta \omega|_{\mathbf{C}^2}, \quad \lambda, \delta \in \mathbf{C}^*.$$

Then the map is $F = (\Phi, \Psi)$ and $\mathscr{L} : \frac{dy}{y} + \lambda \frac{dx}{x} = 0$. □

It was shown by Il'iashenko and his students [6, 8] that there are open sets of foliations exhibiting dense leaves, see also Cerveau [5].

The following problem concerns only those foliations which do not admit algebraic leaves.

Theorem (A. Lins N. [7]). *For an open and dense set of foliations of* $\mathbf{C}P(2)$ *there are no algebraic leaves.*

When no algebraic leaves exist questions about limit sets are more subtle. Along this direction we have.

Problem 2. Do nontrivial minimal sets exist?

A minimal set of $\mathscr{F}$ is a closed, invariant, nonempty subsets of $\mathbf{C}P(2)$ which is minimal with these three properties and nontrivial means it is not a singularity. Thus the problem above refers to the existence of a leaf of $\mathscr{F}$ which does not accumulate on *sing* $\mathscr{F}$. What is known about this problem is the following

Theorem (C. Camacho, A. Lins N., P. Sad [3]). *If* $\mathscr{M}$ *is a nontrivial minimal set of* $\mathscr{F}$. *Then*

1. $\mathscr{M}$ *is unique.*
2. *Any leaf accumulates in* $\mathscr{M}$.
3. *Any leaf in* $\mathscr{M}$ *has exponential growth, is a hyperbolic Riemann surface and has no parabolic ends.*
4. *There is no* $\mathscr{F}$*-transverse invariant measure with support in* $\mathscr{M}$.

Recently Bonatti, Langevin, Moussu, Tischler proved also that there is a leaf in $\mathscr{M}$ with nontrivial linear holonomy.

Finally we have the following question which intends to establish a relation between these foliations and the iteration of endomorphisms of the Riemann sphere.

Problem 3 (Sad's Conjecture). Assuming all singularities of $\mathscr{F}$ generic then lim $\mathscr{F}$ is the closure of the separatrices of $\mathscr{F}$.

References

1. Bendixson, I.: Sur les points singuliers d'une équation differentielle linéaire. Ofv. Kongl. Ventenskaps, Akademiens Forhandlinger **148** (1895) 81–89
2. Camacho, C., Sad, P.: Invariant varieties through singularities of holomorphic vector fields. Ann. Math. **115** (1982) 579–595
3. Camacho, C., Lins Neto, A., Sad, P.: Minimal sets of foliations in complex projective spaces. Publ. Math. IHES **68** (1988) 187–203
4. Cerveau, D., Mattei, J.F.: Formes intégrables holomorphes singulières. Astérisque **97** (1982)
5. Cerveau, D.: Densité des feuilles de certaines équations de Pffaf à 2 variables. Ann. Inst. Fourier **1** (1983) 185–194
6. Il'iashenko, J.: Global and local aspects of the theory of complex differential equations. Proc. of Int. Cong. of Math. Helsinki 1978, pp. 821–826

7. Lins Neto, A.: Algebraic solutions of polynomial differential equations and foliations in dimension two. Holomorphic Dynamics Proceedings, Mexico 1986. (Lecture Notes in Mathematics, vol. 1345.) Springer, Berlin Heidelberg New York 1988
8. Mjuller, B.: On the density of solutions of an equation in $\mathbf{C}P^2$. Mat. Sbornik **27** (1975) 325–338
9. Seidenberg, A.: Reduction of singularities of the differential equation $AdY = BdX$. Amer. J. Math. (1968) 248–269

The Dynamics of Non-uniformly Hyperbolic Systems in Two Variables

Lennart Carleson

Department of Mathematics, Royal Institute of Technology, S-10044 Stockholm, Sweden and University of California, Los Angeles, CA 90024, USA

One of the most interesting aspects of dynamical systems is to try to understand the nature of 'chaotic' behavior and in particular to describe the 'strange' attractors. This problem is in a natural way related to the nature of turbulence and there is an overwhelming mass of numerical information but the rigorous mathematical results only apply so far in very special situations.

Nevertheless, there exists today a solid foundation and basic definitions. Let me first recall part of this. We consider a smooth mapping $x \to T(x)$ of a compact domain $\mathcal{D}$ into itself where $\mathcal{D} \subset \mathbb{R}^n$ or $\mathbb{T}^n$. As first realized by Pesin (see e.g. [12]) the starting point should be an invariant measure μ, i.e. $\mu \geq 0$, $\mu(T^{-1}(E)) = \mu(E)$, $\mu(\mathcal{D}) = 1$. The existence of such a measure is obvious, we can e.g. start from a point mass at a point, push this forward, take an average and a weak limit. The basic fact is now that Liapounov exponents exist a.e. with respect to such a measure. This means that we can split the tangent space $\oplus E_j(x)$ so that for $v \in E_j$

$$\lim \frac{\log |DT^n(x)v|}{n} = \lambda_j(x)$$

exists a.e.(μ). In the case of Anosov-mapping this splitting is continuous and $\lambda_j(x) \neq 0$, so that expanding and contracting directions are uniformly separated. This may be called the uniformly hyperbolic case. In this case the metric entropy h_μ of the pair (T, μ) can be written

$$h_\mu = \int \left(\sum \lambda_i(x)^+\right) d\mu(x)$$

and the conditional measures along the unstable foliation $W^u(x)$ are absolutely continuous with respect to Lebesgue measure in these directions.

From the point of view of numerical studies a third property of (T, μ) is relevant. Let's call μ a physical measure if there exists A of positive n-dimensional Lebesgue measure so that for Lebesgue a.a. $x \in A$ and all $\varphi \in C_0^\infty(\mathcal{D})$

$$\frac{1}{n} \sum_1^n \varphi(T^j(x)) \to \int \varphi \, d\mu .$$

This holds for every ergodic component of μ in the Anosov case.

Proceedings of the International Congress of Mathematicians, Kyoto, Japan, 1990

It turns out that all of the above can be generalized to non-uniformly hyperbolic systems. For simplicity assume that $\lambda_i(x) \neq 0$ a.e. (μ). Briefly stated: the validity of the entropy formula and the absolute continuity of μ in unstable directions are equivalent (Ledrappier, Young) [7]. μ is in this case a physical measure (if it is ergodic) [13]. We call such a μ a Sinai-Bowen-Ruelle-measure.

In spite of the generality and completeness, the main question remains open: do (SBR)-measures exist? This question has to be addressed from the beginning and seems to mean that we must build unstable and stable manifolds by hand so that the constructions work Lebesgue almost everywhere. When this has been done one can then read off the Pesin properties of the implied measures.

Let us first discuss the 1-dimensional case. At the previous congress Jakobson [6] gave a report and he also constructed an absolutely continuous invariant measure for the mapping $T = 1 - a\varphi(x)$ on $(-1, 1)$ into itself, $\varphi(t) = t^2$, for $a \in$ set of positive Lebesgue measure [5]. The result applies to more general $\varphi(t)$, e.g. $\varphi = |t|^p$. It is well understood that once we know the increase of the derivative at the critical value

$$D_n = |T^{n'}(1)|$$

then the existence of a (SBR)-measure follows. In fact, Nowicki-van Strien [10], have the following beautiful result: if

$$\sum \varphi^{-1}(D_n^{-1}) < \infty$$

then μ exists. (This result should be optimal).

Let us consider somewhat further how one can obtain the increase of $D_n(a)$, now considered as a function of a for, say, $T = 1 - ax^2$. Setting $x_j(a) = T^j(1; a)$ we have $D_n = \prod_0^{n-1} |-2ax_j(a)|$. If a_0 corresponds to a situation where 0 is strictly pre-periodic and the resulting cycle is repelling, $D_n(a_0)$ is obviously exponentially increasing (the simplest case is $a_0 = 2$; $1 \to -1 \to -1 \to \cdots$). It is natural to study parameters a close to a_0 and use the parameter to avoid very small values for $x_n(a)$. It is easy to prove that $|x_n'(a)| \sim D_n(a)$ provided

$$\sum \frac{1}{D_n} < \infty$$

and the larger D_n the more effectively we can avoid very small values of x_n. It can be proved that Jakobson's result holds around every a_0 (see [2] for $a_0 = 2$).

The basic property of a dynamical system — as opposed to purely random motions — is the quasi-periodicity. A return of x_n to $(-\varepsilon, \varepsilon)$ forces an especially strong repetition of orbits since $|x_{n+1} - 1| \leq C \cdot \varepsilon^2$ and x_{n+j} behaves as x_j as long as $|x_{n+j} - x_j| < |x_j|/j^2$ (say). Observe also that during this period $|x_{n+j} - x_j| \sim D_j x_n^2$. If we slightly weaken the sufficient condition $D_j x_n^2 < 1/j^2$ to < 1 we obtain the following model (see [4]) for the amount of randomness in the sequence $D_n(a)$.

Let $\{\Delta_n(\omega)\}_0^\infty$ be the following stochastic process.

(a) $\Delta_0(\omega) \equiv 1$;
(b) there exist stopping times $\{n_j(\omega)\}_1^\infty$, $n_1(\omega) \equiv 1$;

(c) If n is a stopping time, choose a number t_n at random with uniform distribution in $(0, A)$ and independent of the past and define

$$\Delta_{n+j}(\omega) = \Delta_{n-1}(\omega) t \Delta_j(\omega)$$

for $j = 0, 1, \cdots, m-1$ as long as

$$t^2 \Delta_j(\omega) \leq 1 .$$

If $m < \infty$ the next stopping time is $n + m$.

It is an easy fact that if $A > e$ then

$$\lim \frac{\log \Delta_n}{n} > 0$$

with probability 1.

Interpreting parameter space as probability space and $\Delta_n = D_n$ this statement contains all the essentials of the chaotic behavior for these one-dimensional systems.

Let us now turn to the two-variable case. J. Palis [11] has proposed the following scenario for the creation of strange attractors (i.e. an invariant set Λ, $W^s(\Lambda) \supset$ a neighborhood of Λ, with a dense orbit with positive Liaponov exponent). Let $T(x; a)$ depend on a parameter and let 0 be a fixed point which is hyperbolic (one eigenvalue > 1, one < 1, the product < 1). Let $W^s(a)$, $W^u(a)$ be the stable and unstable manifolds through 0 and suppose $W^s(a_0)$ is tangent to $W^u(a_0)$ at some point p_0. Then, for a close to a_0, $T(x; a)$ has, for a set of positive measure of the parameter, a strange attractor. This has in fact been proved recently by Mora-Viana [8], see Fig. 1.

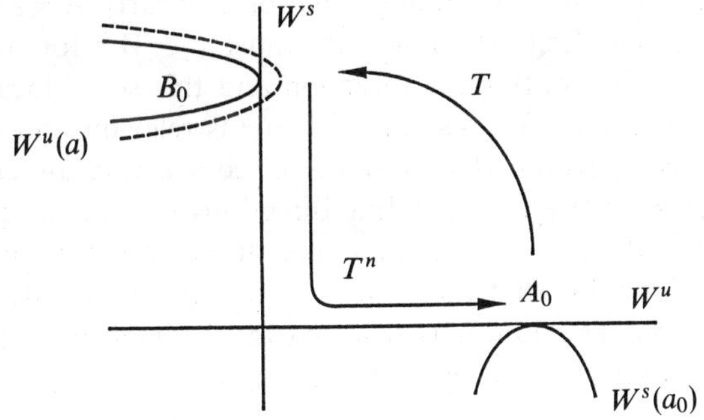

Fig. 1

It turns out that the basic properties of T show up already for the special (Hénon) map

$$T_0 : \quad \begin{cases} 1 - ax^2 + y & 0 < b < b_0 \\ bx & a_0 < a < 2 \end{cases}$$

with b_0 arbitrarily small and a_0 close to 2. The reason is that some suitable high iterate T^n of T maps a small box close to p_0 into itself and the basic behavior of T^n is obtained from its second order terms and the eigenvalues at 0. Let us therefore try to understand the special case T_0, see Fig. 2.

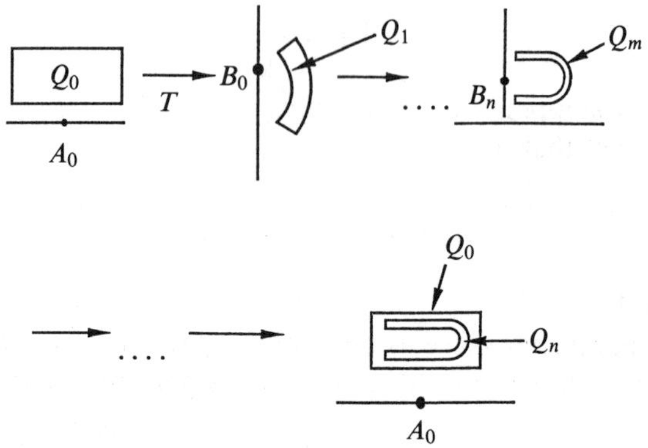

Fig. 2

$T_0(x, y)$ has a fixed point F at $\sim (\frac{1}{2}, \frac{b}{2})$ and the eigenvalues are $\sim (-2, \frac{b}{2})$. The unstable manifold W^u looks like Figure 3 (see next page), and it is easy to see that if an attractor exists it is $\subset \overline{W^u}$. We would expect $\overline{W^u}$ to be the strange attractor. Since b is small we expect $\|DT_0^n\| \sim D_n$ corresponding to the case $b = 0$. Most directions would then expand for T_0^n but since $\det(T_0^n) = (-b)^n$ some very strong contracting direction must exist. We need to find this and to make a foliation along these directions. The expanding foliation is clearly $\overline{W^u}$ and we need most leaves to be long. The 'critical' values are those $c \in W^u$ for which the tangent direction of $W^u =$ the contracting direction and this set is located in $(-b, b)$ as indicated below (Only the intersection with $\overline{W^u}$ is relevant and this is formally a Cantor set). It now turns out that existence of contracting directions depends on the simultaneous existence of expanding directions but the accuracy is increased by powers in b and the situation therefore allows for a boot-strap-argument. We prove in the end [3] that all critical points are uniformly expanding after parameter exclusion. This amounts to using the 1-dimensional argument and to verifying, for the Cantor set of possible starting points c,

$$\|DT_0^n(c)\| \geq e^{\gamma n}.$$

We now have the setup for a foliation and for the understanding of W^u. A (SBR)-measure can be constructed [1] starting from arc length on a sheet of W^u and pushing this forward. Most mass will be located on rather long arcs $\subset W^u$.

Let me elaborate some more on this constructive aspect of the Pesin theory.

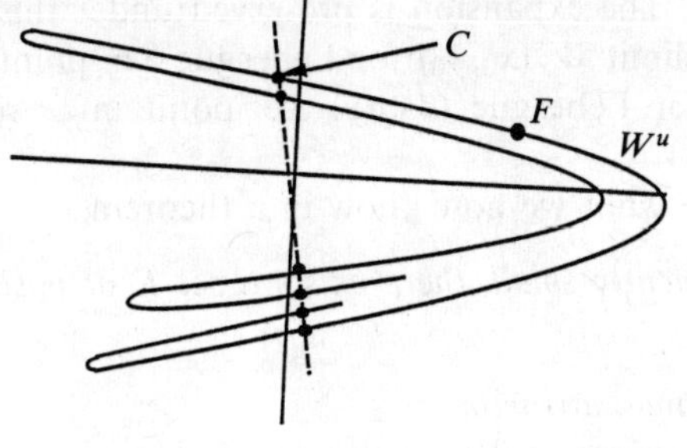

Fig. 3

$$DT_0 = \begin{pmatrix} -2ax_0 & 1 \\ b & 0 \end{pmatrix}$$

transfers a vector $(1, q)$ of slope q at (x_0, y_0) to a vector of slope q_1 where

$$q = 2ax_0 + \frac{b}{q_1}$$

and

$$\|DT(1,q)\|^2 = (-2ax_0 + q)^2 + b^2 .$$

For $q = q_0 = 2ax_0$ we have length b so this direction is almost maximally contracting. If we choose

$$q_1 = 2ax_1$$

we again contract and get in the general case

$$q = 2ax_0 + \cfrac{b}{2ax_1 + \cfrac{b}{q_2}} .$$

It is natural to define

$$\bar{q}_n = 2ax_0 + \cfrac{b}{2ax_1 + \cfrac{b}{\ddots + 2ax_{n-1}}}$$

and call this the contracting direction and

$$V_n : \quad y' = \bar{q}_n(x, y)$$

the n-th order contracting vector field. The $\bar{q}_n$ are very unstable but if we have expansion along $(x_0, \cdots, x_{\nu-1})$

$$|\bar{q}_\nu(x, y) - \bar{q}_{\nu-1}(x, y)| \leq b^{\nu-1} .$$

We now obtain a stable foliation by integrating along the vector field V_n where we get contraction by b^n. The expansion is preserved under this small perturbation and we obtain in the limit $W^s(x_0, y_0)$ for Lebesgue a.a. points (x_0, y_0) along W^u. Whether this holds for Lebesgue $(dx\,dy)$ a.e. point in $\mathscr{D}$ remains open but is probably true.

Let me summarize what we now know in a theorem.

Given $b > 0$, sufficiently small, there exists a set E of positive measure so that for $a \in E$

(i) *$T_0(x, y)$ has a strange attractor*
(ii) *$T_0(x, y)$ admits a unique (SBR)-measure*
(iii) *for any $a \in E$, there exists $a' \notin E$, so that any $\varepsilon > 0$ $|a - a'| < \varepsilon$ and $T(x, y, a')$ has an attractive cycle.*

It is still not known if for $a \in E$ almost all (Lebesgue) points in $\mathscr{D}$ generate the (SBR)-measure but this seems very likely.

Let me end by some comments.

The original Hénon simulation concerned the parameter values (1.4, 0.3). It is most likely that the Mora-Viana-result applies and we have strange attractors arbitrarily nearby. However, the most natural conjecture here is that the problem of proving existence of a strange attractor for a particular parameter value is in some rigorous sense undecidable. To develop this aspect of (even one-dimensional) dynamics seems an interesting task.

The approach can be generalized to n dimensions when $(n - 1)$ eigenvalues are very small and one eigenvalue is > 1, so that the unstable manifold is 1-dimensional [9]. The case of > 1-dimensional W^u has not been studied. One can see possibilities of progress also then.

It was perhaps not so clear from the presentation but it is of central importance for the argument that the Cantor set of critical points is very thin. This made it possible to consider every critical point independently and obtain uniform expansion. We could then disregard the distribution of the orbit in the "vertical" direction. However, for an interesting case such as the standard map on $\mathbb{T}^2$:

$$\begin{cases} x' = k \sin x - 2x - y \\ y' = x \end{cases}$$

the critical set basically is two 'curves' ($x \sim \frac{\pi}{2}, \frac{3\pi}{2}$) and the distribution along these 'curves' must be understood. The problem here remains open even if k is very large.

References

1. Benedicks, M., Carleson, L.: On iterations of $1 - ax^2$ on $(-1, 1)$. Ann. Math. **122** (1985) 1–25
2. Benedicks, M., Carleson, L.: The dynamics of the Hénon map. Ann. Math. **132** (1990) 629–725
3. Benedicks, M., Young, L.-S.: To appear

4. Carleson, L.: A model for chaotic dynamics. To appear in Lecture Notes in Mathematics. Springer, Berlin Heidelberg New York
5. Jakobson, M.: Absolutely continuous invariant measures for one-parameter families of one-dimensional maps. Comm. Math. Phys. **81** (1981) 39–88
6. Jakobson, M.: Families of one-dimensional maps and nearby diffeomorphisms. Proc. Int. Congr. ICM-86 Berkeley 1986, vols. 1, 2. Amer. Math. Soc., pp. 1150–1160
7. Ledrappier, F., Young, L.-S.: The metric entropy of diffeomorphisms, I, II. Ann. Math. **122** (1985) 509–574
8. Mora, L., Viana, M.: Abundance of strange attractors. To appear in Acta Math.
9. Mora, L., Viana, M.: To appear
10. Nowicki, J., van Strien, S.: Invariant measures exist under a summability condition for unimodal maps. Preprint Delft
11. Palis, J., Takens, F.: Homoclinic bifurcations. Cambridge Univ. Press. To appear
12. Pesin, Ya. B.: Characteristic Liapunov exponents and smooth ergodic theory. Russ. Math. Surv. **32** (4) (1977) 55–114
13. Pugh, C., Shub, M.: Ergodic attractors. Trans. Amer. Math. Soc. **312** (1989) 1–54

The Acceleration Operators and Their Applications to Differential Equations, Quasianalytic Functions, and the Constructive Proof of Dulac's Conjecture

Jean P. Ecalle

Mathématiques, Bâtiment 425, Université de Paris-Sud, Centre d'Orsay
F-91405 Orsay, France

We introduce (§§1,2) a general apparatus (resurgence, alien derivations, acceleration, etc.) that enables one to study and resum most divergent expansions of natural origin. We then proceed to give three select applications (§§3–5).

§1. Resurgent Functions, Alien Derivations and Medianization

The Singularity Algebras $\mathscr{S}(S_\theta)$ and $\mathscr{S}^{\text{int}}(S_\theta)$

Let $\mathbb{C}$ be the Riemann surface of $\log\zeta$ and S_θ (resp. S_{θ_1,θ_2}) the semi-axis $\arg\zeta = \theta$ (resp. the sector $\theta_1 \le \arg\zeta \le \theta_2$). A *major* $\check{\varphi}$ on S_θ is an analytic germ defined on $S_{\theta-2\pi,\theta}$ close to 0 ("at the root of $S_{\theta-2\pi,\theta}$"). The *minor* $\hat{\varphi}$ of $\check{\varphi}$ is defined at the root of S_θ by $\hat{\varphi}(\zeta) = \check{\varphi}(\zeta) - \check{\varphi}(\zeta \cdot e^{-2\pi i})$. A *singularity* of direction θ is a *class* $\overset{\triangledown}{\varphi}$ of majors $\check{\varphi}$ modulo the space of regular (i.e. holomorphic) functions at 0. A singularity $\overset{\triangledown}{\varphi}$ is said to be *integrable* iff:

$$(1.1)\qquad \begin{cases} \zeta\check{\varphi}(\zeta) \longrightarrow 0 \text{ as } \zeta \to 0 & \text{on } S_{\theta-2\pi,\theta} \quad \text{and} \\ \displaystyle\int_0^{\zeta_0} |\hat{\varphi}(\zeta)|\,|d\zeta| < +\infty & \text{for } \zeta_0 \in S_\theta \text{ close to } 0\,. \end{cases}$$

Integrable singularities are fully determined by their minor.

For any two classes $\overset{\triangledown}{\varphi}_1, \overset{\triangledown}{\varphi}_2 \in \mathscr{S}(S_\theta)$ and $u \in S_{\theta-2\pi}$ close to 0, the class $\overset{\triangledown}{\varphi}_3$ of the *major* $\check{\varphi}_{3,u}$ defined by

$$(1.2)\qquad \check{\varphi}_{3,u}(\zeta) = \int_u^{\zeta-u} \check{\varphi}_1(\zeta_1)\check{\varphi}_2(\zeta-\zeta_1)\,d\zeta\,,$$
$$\big(u \in S_{\theta-2\pi}\,;\zeta \quad \text{and} \quad \zeta - u \in S_{\theta-2\pi,\theta-\pi}\big)$$

depends neither on u nor on the choice of $\check{\varphi}_i$ in $\overset{\triangledown}{\varphi}_i$. The convolution $\overset{\triangledown}{\varphi}_1 * \overset{\triangledown}{\varphi}_2 = \overset{\triangledown}{\varphi}_3$ thus defined turns $\mathscr{S}(S_\theta)$ into a commutative algebra. The space $\mathscr{S}^{\text{int}}(S_\theta)$ of integrable singularities is a subalgebra and its convolution reduces to:

$$(1.3)\quad \hat{\varphi}_1 * \hat{\varphi}_2(\zeta) = \int_0^\zeta \hat{\varphi}_1(\zeta_1)\hat{\varphi}_2(\zeta-\zeta_1)\,d\zeta_1 \quad (\zeta, \zeta_1, \zeta-\zeta_1 \text{ on } S_\theta \text{ and close to } 0)\,.$$

Proceedings of the International Congress
of Mathematicians, Kyoto, Japan, 1990

The Resurgence Algebras $\mathscr{R}(S_\theta)$ and $\mathscr{R}^{\text{int}}(S_\theta)$

The subspace $\mathscr{R}(S_\theta) \subset \mathscr{S}(S_\theta)$ of all $\overset{\triangledown}{\varphi}$ whose minor $\hat{\varphi}$ can be analytically continued along any path that keeps close to S_θ (without going back) and bypasses to the right or to the left all intervening singular points $\omega_i \in S_\theta$, is closed under convolution. For any sequence ε_i of $\pm$ signs and any ζ in $]\omega_r, \omega_{r+1}[$, denote by $\varphi^{\varepsilon_1,\dots,\varepsilon_r}_{\omega_1,\dots,\omega_r}(\zeta)$ the determination of $\hat{\varphi}(\zeta)$ obtained by starting from 0 and bypassing each ω_i to the right (resp. left) if $\varepsilon_i = +$ (resp. $-$). The space of all $\overset{\triangledown}{\varphi}$ in $\mathscr{R}(S_\theta)$ whose minors $\hat{\varphi}$ have all their determinations $\hat{\varphi}^{\varepsilon_i}_{\omega_i}$ integrable on their segment of definition $]\omega_r, \omega_{r+1}[$, constitutes a subalgebra $\mathscr{R}^{\text{int}}(S_\theta)$. We call $\mathscr{R}(S_\theta)$ (resp. $\mathscr{R}^{\text{int}}(S_\theta)$) the algebra of resurgent functions (resp. integrable resurgent functions) of direction θ.

Alien Derivations and Medianization

For any finite sequence $\varepsilon_i = \pm$ denote by p (resp. q) the number of $+$ (resp. $-$) signs and consider the weights:

$$\delta^{\varepsilon_1,\dots,\varepsilon_{r-1}} = \delta_{p,q} = \frac{p!\,q!}{(p+q+1)!}\,; \quad \lambda^{\varepsilon_1,\dots,\varepsilon_r} = \lambda_{p,q} = \frac{(2p)!\,(2q)!}{4^{p+q}\,p!\,q!\,(p+q)!} \tag{1.4}$$

For any $\omega \in S_\theta$ the operator Δ_ω of $\mathscr{R}(S_\theta)$ onto itself defined by:

$$\Delta_\omega : \overset{\triangledown}{\varphi} \longmapsto \overset{\triangledown}{\varphi}_\omega \qquad \text{with} \qquad \check{\varphi}_\omega(\zeta) = \sum_{\varepsilon_i} \delta^{\varepsilon_1,\dots,\varepsilon_{r-1}}\,\hat{\varphi}^{\varepsilon_1,\dots,\varepsilon_{r-1},+}_{\omega_1,\dots,\omega_{r-1},\omega}(\zeta+\omega) \tag{1.5}$$

(for ζ on S_θ and close to 0) is a *derivation* of the algebra $\mathscr{R}(S_\theta)$:

$$\Delta_\omega(\overset{\triangledown}{\varphi}_1 * \overset{\triangledown}{\varphi}_2) = (\Delta_\omega \overset{\triangledown}{\varphi}_1) * \overset{\triangledown}{\varphi}_2 + \overset{\triangledown}{\varphi}_1 * (\Delta_\omega \overset{\triangledown}{\varphi}_2)\,. \tag{1.6}$$

We call Δ_ω the *alien derivation* with index ω. On $\mathscr{R}^{\text{int}}(S_\omega)$ it reduces to:

$$\Delta_\omega : \hat{\varphi} \longmapsto \hat{\varphi}_\omega \tag{1.7}$$

$$\text{with} \qquad \hat{\varphi}_\omega(\zeta) = \sum_{\varepsilon_i} \delta^{\varepsilon_1,\dots,\varepsilon_{r-1}} \left\{ \hat{\varphi}^{\varepsilon_1,\dots,\varepsilon_{r-1},+}_{\omega_1,\dots,\omega_{r-1},\omega}(\zeta+\omega) - \hat{\varphi}^{\varepsilon_1,\dots,\varepsilon_{r-1},-}_{\omega_1,\dots,\omega_{r-1},\omega}(\zeta+\omega) \right\}.$$

Along with the natural derivation $\hat{\partial} : \check{\varphi} \mapsto -\zeta\check{\varphi}$, the Δ_ω finitely generate all continuous derivations of $\mathscr{R}(S_\theta)$.

Similarly, the operator med ("medianization"):

$$\text{med} : \hat{\varphi}(\zeta) \longmapsto \text{med}\,\hat{\varphi}(\zeta) = \sum_{\varepsilon_i} \lambda^{\varepsilon_1,\dots,\varepsilon_r}_{\omega_1,\dots,\omega_r}(\zeta) \qquad (\text{if } \omega_r < \zeta < \omega_{r+1}) \tag{1.8}$$

is an homomorphism of the algebra $\mathscr{R}^{\text{int}}(S_\theta)$ into the algebra $\mathbb{L}^{\text{int}}(S_\theta)$ of univalued, locally integrable functions on S_θ:

$$\text{med}\,(\hat{\varphi}_1 * \hat{\varphi}_2) = (\text{med}\,\hat{\varphi}_1) * (\text{med}\,\hat{\varphi}_2) \qquad \text{with } * \text{ as in (1.3)}\,. \tag{1.9}$$

Medianization has the added advantage of preserving realness: if the *germ* $\hat{\varphi}$ is real, so is the *function* med $\hat{\varphi}$.

The Resurgence Algebras $\mathscr{R}$ and $\mathscr{R}^{\text{int}}$

The convolution algebra of all classes $\overset{\triangledown}{\varphi}$ which, along with their successive alien derivatives $\Delta_{\omega_r} \dots \Delta_{\omega_1} \overset{\triangledown}{\varphi}$ ($\forall \omega_i \in \mathbb{C}$) belong to all $\mathscr{R}(S_\theta)$ (resp. all $\mathscr{R}^{\text{int}}(S_\theta)$) is known as the general algebra $\mathscr{R}$ (resp. $\mathscr{R}^{\text{int}}$) of resurgent functions (resp. integrable r.f.). Their *majors* $\check{\varphi}$ are defined in spiral-like neighbourhoods of $\underset{\cdot}{0}$ on $\mathbb{C}$ and their *minors* $\hat{\varphi}$ can be continued (starting from 0 and bypassing intervening singular points) along any split line on $\mathbb{C}$.

Resurgent functions of natural origin tend to reproduce themselves at their singular points. This self-reproduction is exactly described by *resurgence equations* linking $\overset{\triangledown}{\varphi}$ to its *alien derivatives* $\Delta_\omega \overset{\triangledown}{\varphi}$. By a slight abuse, we often extend the label of "resurgent function" to those power series whose Borel transforms (see §2) belong to $\mathscr{R}$. See [1–3].

§2. The Acceleration Operators

The Laplace Transform $\mathscr{L}$ and the Borel Transforms $\mathscr{B}$ and $\widetilde{\mathscr{B}}$

$$\mathscr{L} : \hat{\varphi}(\zeta) \longmapsto \varphi(z) = \int_0^{+\infty} e^{-\zeta z} \hat{\varphi}(\zeta)\, d\zeta \tag{2.1}$$

$$\mathscr{B} : \varphi(z) \longmapsto \hat{\varphi}(\zeta) = \frac{1}{2\pi i} \int_{c-i\infty}^{c+i\infty} e^{z\zeta} \varphi(z)\, dz \tag{2.2}$$

The classical *Laplace transform* $\mathscr{L}$ is a homomorphism of the convolution algebra $\mathbb{L}^{\text{int}}_{\text{exp}}(\mathbb{R}^+)$ of all univalued, locally integrable functions on $\mathbb{R}^+$ with (at most) exponential growth at $+\infty$, into the multiplicative algebra $\mathbb{B}$ of holomorphic germs bounded in half planes $\operatorname{Re} z \geq x_0$. Its inverse $\mathscr{B}$ is known as the Borel transform. For each formal series $\tilde{\varphi}(z) = \sum \varepsilon_n(z)$ whose general term $\varepsilon_n \in \mathbb{B}$ has a Borel transform $\hat{\varepsilon}_n$, we have a notion of *formal* (or term-wise) Borel transform:

$$\widetilde{\mathscr{B}} : \tilde{\varphi}(z) = \textstyle\sum \varepsilon_n(z) \longmapsto \hat{\varphi}(\zeta) = \sum \hat{\varepsilon}_n(\zeta) \quad \left(\text{e.g. } \sum a_n z^{-n} \longmapsto \sum \frac{a_n}{\Gamma(n)} \zeta^{n-1}\right). \tag{2.3}$$

The Acceleration Operators and Their Kernels

An acceleratrix is a function F holomorphic in a neighbourhood of $\underset{\cdot}{\infty} \in \mathbb{C}$, real positive on $\mathbb{R}^+ = S_0$ and such that for $z \to \infty$:

$$x^{-1} F(x) \to 0\,;\ \delta F(z) \sim \delta F(x)\,;\ \delta^2 F(z) \sim \delta^2 F(x) \tag{2.4}$$

with $0 < x \to \infty$; $z = xe^{i\theta}$ (θ fixed in $\mathbb{R}$) and:

$$\delta F(z) = zF'(z)F(z)\ ;\ \delta^2 \varphi(z) = 1 + zF''(z)/F'(z) - zF'(z)/F(z)\,. \tag{2.4bis}$$

The co-acceleratrix G of F is the germ G defined on $[+0, \dots]$ by:

$$G\big(f(z)\big) \equiv F(z) - zf(z) \text{ with } f(z) = F'(z) \qquad \big(G(\zeta) \to +\infty \text{ as } \zeta \to 0\big)\,. \tag{2.5}$$

Borel-Laplace takes the multiplicative endomorphism $\mathscr{C}_F : \varphi_1(z_1) \mapsto \varphi_2(z_2) \equiv \varphi_1\big(F(z_2)\big)$ of $\mathbb{B}$ into a convolution endomorphism $\widehat{\mathscr{C}}_F$ of $\mathbb{L}^{\text{int}}_{\text{exp}}(\mathbb{R}^+)$:

$$\widehat{\mathscr{C}}_F = \mathscr{B}.\mathscr{C}_F.\mathscr{L} : \hat{\varphi}_1(\zeta_1) \longmapsto \hat{\varphi}_2(\zeta_2) = \int_0^{+\infty} C_F(\zeta_2,\zeta_1)\hat{\varphi}_1(\zeta_1)\, d\zeta_1 . \tag{2.6}$$

$\widehat{\mathscr{C}}_F$ is known as the *acceleration* $z_1 \to z_2$. Its integral kernel is given by:

$$C_F(\zeta_2,\zeta_1) = \frac{1}{2\pi i}\int_{c-i\infty}^{c+i\infty} \exp\big(\zeta_2 z_2 - \zeta_1 F(z_2)\big)\, dz_2 \tag{2.7}$$

and it has faster-than-exponential decrease in ζ_1:

$$\log C_F(\zeta_2,\zeta_1) \sim -\zeta_1 G\big(\zeta_2/\zeta_1\big) \quad \text{for } 0 < \zeta_2 \text{ fixed and } \zeta_1 \to +\infty . \tag{2.8}$$

Therefore, the natural domain of definition of $\widehat{\mathscr{C}}_F$ is much larger than $\mathbb{L}^{\text{int}}_{\text{exp}}(\mathbb{R}^+)$; it is the convolution algbera $\mathbb{L}^{\text{int}}_{F\text{-acc}}(\mathbb{R}^+)$ of all functions $\hat{\varphi}_1(\zeta_1)$ with F-accelerable growth, i.e. for which there exists $c > 0$ such that:

$$\big|\hat{\varphi}_1(\zeta_1)\big| \leq \text{Cst.}/C_F(c,\zeta_1) \quad \text{or} \quad \log|\hat{\varphi}_1(\zeta_1)| \leq \zeta_1 G(c/\zeta_1) \text{ as } \zeta_1 \to +\infty . \tag{2.9}$$

The largest such c is the *acceleration abscissa* of $\hat{\varphi}_1$.

Strong, Moderate, Weak Accelerations

Strong accelerations ($\log z_2/\log z_1 \to +\infty$) have kernels with *slightly* over–exponential decrease in ζ_1 and yield germs $\hat{\varphi}_2(\zeta_2)$ which are defined in a spiral-like neighbourhood of $\underset{\cdot}{0} \in \underset{\cdot}{\mathbb{C}}$ with infinite aperture. *Moderate accelerations* ($\log z_2/\log z_1 \to 1/\alpha$ with $0 < \alpha < 1$) have kernels decreasing like $\exp\big(-c_\alpha \zeta_1^{1/\beta}\zeta_2^{-\alpha/\beta}\big)$ with $\beta = 1-\alpha$ and they yield germs $\hat{\varphi}_2(\zeta_2)$ which are defined in a sector of aperture $\pi\beta/\alpha$. *Weak accelerations* ($\log z_2/\log z_1 \to 1$ but $z_2/z_1 \to 1$) have very fast decreasing kernels but they yield germs $\hat{\varphi}_2(\zeta_2)$ which are defined only at the root of $\mathbb{R}^+ = S_0$ and are usually non-analytic, but only Denjoy-quasianalytic (cf. §4). Of great practical importance are the elementary accelerations $z_1 \to z_2 = \exp(\sigma z_1)$ and $z_1 \to z_2 = {z_1}^{1/\alpha}$ with their respective kernels:

$$C_F(\zeta_2,\zeta_1) = (\zeta_2)^{\zeta_1/\sigma - 1}\Big/\Gamma(\zeta_1/\sigma) \tag{2.10}$$

$$C_F(\zeta_2,\zeta_1) = {\zeta_2}^{-1}C_\alpha(X) \tag{2.11}$$
$$\text{with} \quad X = {\zeta_1}^{1/\beta}{\zeta_2}^{-\alpha/\beta} \qquad \big(0<\alpha,\ \beta<1,\ \alpha+\beta=1\big)$$

$$C_\alpha(X) \sim (c/2\pi)^{1/2}.X^{1/2}.\exp(-cX) \tag{2.11bis}$$
$$\text{when} \quad X \to \infty \quad \text{with} \quad \operatorname{Re} X > 0 \ (c = \alpha^{\alpha/\beta}\beta) .$$

Accelero-summability

A formal series $\tilde{\varphi}(z) = \sum \varepsilon_n(z)$ is said to be accelero-summable with sum $\varphi(z)$ and critical times $z_1, z_2, \ldots, z_r$ if it can be subjected to the following operations (algebra homomorphisms):

$$\begin{array}{ccc}
\tilde{\varphi}_1(z_1) = \tilde{\varphi}(z) & \dashrightarrow & \varphi(z) = \varphi_r(z_r) \quad \text{multiplicative algebras} \\
\tilde{\mathscr{B}} \downarrow & & \uparrow \mathscr{L} \\
\hat{\varphi}_1(\zeta_1) \underset{12}{\longrightarrow} \hat{\varphi}_2(\zeta_2) \underset{23}{\longrightarrow} \hat{\varphi}_3(\zeta_3) \to \cdots & \to & \hat{\varphi}_r(\zeta_r) \quad \text{convolution algebras}
\end{array}$$

with arrows $(i, i+1)$ denoting the acceleration $z_i \to z_{i+1}$. See [4, 5, 7].

§3. Acceleration Applied to Many-Levelled Differential Systems

Resurgent functions are truly ubiquitous. They arise as formal solutions of differential equations (or difference equations, or general functional equations) with analytic coefficients, or of systems of such equations. They occur in the study (normalization, conjugacy, iteration) of local analytic objects, chiefly: local singular vector fields and local diffeomorphisms. Again, most expansions in a "singular parameter" (such as the Planck constant in the Schrödinger equation) turn out to be divergent and resurgent. Indeed, it is no exaggeration to claim that most divergent expansions met with in actual life are not only resurgent but also summable by $\mathscr{L}\tilde{\mathscr{B}}$ (one critical time) or by $\mathscr{L}\widehat{\mathscr{C}}_{F_{r-1}} \ldots \widehat{\mathscr{C}}_{F_1}\tilde{\mathscr{B}}$ (r critical times; $r \geq 2$). The former case ($r = 1$) is by far the more common. Criticity $r \geq 2$ occurs only in connection with objects of a certain complexity. Thus it never arises with vector fields (resp. diffeomorphisms) in less than 3 (resp. 2) dimensions.

We shall describe that phenomenon (namely, $r \geq 2$) in the case of a *many-levelled* but formally separable differential system, because it illustrates all the relevant analysis while keeping formal complications down to a minimum. So, consider a local analytic system (3.1) that is formally conjugate to the normal system (3.2) under transformation (3.3).

$$\text{(3.1)} \qquad \frac{1}{p_i} t^{1+p_i} \dot{x}_i + \lambda_i x_i = b_i(t, x_1, \ldots, x_v) \in \mathbb{C}\{t, x_1, \ldots, x_v\} \quad (i = 1, \ldots, v)$$

$$\text{(3.2)} \qquad \frac{1}{p_i} t^{1+p_i} \dot{y}_i + \lambda_i y_i = 0 \qquad \left(i = 1, \ldots, v \; ; \; \lambda_i \in \mathbb{C}^* \; ; \; p_i \in \mathbb{N}^*\right)$$

$$\text{(3.3)} \qquad x_i = h_i(t, y_1, \ldots, y_v) \in \mathbb{C}[[t, x_1, \ldots, x_v]] \qquad (i = 1, \ldots, v) \, .$$

Let $q_1 < q_2 < \ldots < q_r$ be the distinct values taken by the *levels* p_i and assume for simplicity's sake that the various λ_i attached to any given level display neither resonance nor quasiresonance (i.e. the combinations $\sum n_i \lambda_i$ neither vanish nor do they get abnormally close to 0). Under those mild genericity assumptions, the formal integral $x(t,u)$ of system (3.1):

$$(3.4) \qquad x(t,u) = \sum u^n E^n(t)\varphi^n(t)$$

$$\left(n \in \mathbb{N}^\nu \; ; \; u^n = \prod u_i^{n_i} \; ; \; E^n(t) = \exp \sum n_i \lambda_i t^{-p_i} \; ; \; \varphi^n \in \left(\mathbb{C}[[t]]\right)^\nu\right)$$

obtained by plugging into (3.3) the elementary solution $y_i = u_i \exp(\lambda_i t^{-p_i})$ of (3.2), can be shown to be convergent in u and divergent in t, but resurgent and accelero-summable with critical times $z_i = t^{-q_i}$ $(i = 1,\ldots,r)$. This compact statement translates into the following. Let θ be a multipolarization, i.e. a choice of angles $\theta_1,\ldots,\theta_r$ satisfying the self-compatibility condition:

$$(3.5) \qquad \left|\frac{\theta_i}{q_i} - \frac{\theta_{i+1}}{q_{i+1}}\right| < \frac{\pi}{2}\left(\frac{1}{q_i} - \frac{1}{q_{i+1}}\right) \qquad (1 \le i \le r-1 \; ; \; \theta_i \in \mathbb{R}).$$

Further, let Ω_i be the set of all $\omega \in \mathbb{C}$ whose projection $\dot\omega$ on $\mathbb{C}$ is of the form $\sum n_j \lambda_j$ with $p_j = q_i$ and $n_j \in \mathbb{N}$ (or $n_j = -1$ for one j at most). Then each component $\varphi^n(t) = \varphi_1^n(z_1)$ of $x(t,u)$ has a Borel transform $\hat\varphi_1(\zeta_1)$ with only isolated singularities and a growth rate not exceeding $\exp(\mathrm{cst.}|\zeta_1|^{q_2/q_2-q_1})$. Thus it has accelerable growth for the acceleration $z_1 \to z_2$ taken along any axis $\arg\zeta_1 = \theta_1$ that avoids singularities. The corresponding accelerate $\hat\varphi_2(\zeta_2 \parallel \theta)$ can be analytically continued within a sector $\widehat{S}_2$ containing (at least) all directions θ_2 linked to θ_1 by (3.5). In that sector it possesses only isolated singularities and has accelerable growth for the acceleration $z_2 \to z_3$. Thus we get a succession of accelerates $\hat\varphi_i(\zeta_i \parallel \theta)$, the last of which (for $i = r$) has exponential growth and can be laplaced along any semi-axis θ_r compatible with θ_{r-1}, yielding the sought-after sum $\varphi_r(z_r \parallel \theta) = \varphi(t \parallel \theta)$.

Moreover, each i-th Borel transform $\hat x(\zeta_i, u \parallel \theta)$ satisfies the so-called Bridge Equation, which reads:

$$(3.6) \qquad \dot\Delta_\omega \, \hat x(\zeta_i, u \parallel \theta) = \mathbb{A}_{\omega \parallel q_i,\theta} . \hat x(\zeta_i, u \parallel \theta) \quad \left(\dot\Delta_\omega = e^{-\omega z}\Delta_\omega ; \omega \in \Omega_i \cap \hat S_i\right)$$

$$(3.6\text{bis}) \qquad \mathbb{A}_{\omega \parallel q_i,\theta} = u^{n(\omega)} \left\{ \sum_{p_j \ge q_i} \mathbb{A}^j_{\omega \parallel q_i,\theta}(u) . u_j \frac{\partial}{\partial u_j} + \sum_{p_j < q_i} \mathbb{A}^j_{\omega \parallel q_i,\theta}(u) . \frac{\partial}{\partial u_j} \right\}$$

$$(3.6\text{ter}) \qquad \begin{cases} \dot\omega = \sum_{p_j = q_i} n_j(\omega)\lambda_j \; ; \quad u^{n(\omega)} = \prod u_j^{n_j(\omega)} \; ; \\ \mathbb{A}^j_{\omega \parallel q_i,\theta} \in \mathbb{C}[[u_k \text{ with } p_k < q_i]] \end{cases}$$

and which describes in compact form the resurgence properties of all the $\hat\varphi_i^n(\zeta_i \parallel \theta)$. The Bridge Equation holds, in some form or other, for all local objects. It says in effect that alien derivations $\dot\Delta_\omega$ act on formal integrals like ordinary differential operators $\mathbb{A}_\omega$, while at the same time enabling one to calculate those $\mathbb{A}_\omega$.

Here, the components of the $\mathbb{A}_{\omega \parallel q_i,\theta}$ relative to each level q_i are formal power series in all the parameters u_k attached to the lower levels $p_k < q_i$. For $i = 1$, they are scalar-valued. Lastly, and crucially, these differential operators, taken together, constitute a complete set of *analytic invariants* of the system (3.1).

The proof [6, 8] relies heavily on the study of the operators:

$$D_{\varpi} = e^{-\varpi(z)}.\partial^{-1}.e^{+\varpi(z)} = \Big(\varpi'(z) + \partial\Big)^{-1} \tag{3.7}$$

$$\Big(\partial = \partial/\partial z \ ; \ z = 1/t \ ; \ \varpi(z) = \omega_0 + \omega_1 z + \cdots + \omega_q z^q\Big)$$

and their equivalents in the various ζ_i planes, for all three cases: *precritical* $(q_i < q)$, *critical* $(q_i = q)$ and *postcritical* $(q_i > q)$.

§4. Acceleration and Quasianalyticity. Cohesive Functions

Transfinite Denjoy Classes of Quasianalytical Functions. Cohesive Functions

Let $\mathscr{L}$ be a C^∞ automorphism of $]\ldots,+\infty]$ with $\mathscr{L}(x) \ll x$. An *iterator* $\mathscr{L}^*$ of $\mathscr{L}$ is *any* C^∞ automorphism of $]\ldots,+\infty]$ such that $\mathscr{L}^* \circ \mathscr{L} = -1 + \mathscr{L}^*$. For any transfinite ordinal $\alpha = \omega^r.n_r + \cdots + \omega.n_1 + n_0 < \omega^\omega$ $(n_i \in \mathbb{N})$ we put:

$$L_\alpha = (L)^{\circ n_0} \circ (L^*)^{\circ n_1} \circ (L^{**})^{\circ n_2} \circ \cdots \circ (L^{*\cdots *})^{\circ n_r} = \alpha\text{-th iterate of } L = \log\ . \tag{4.1}$$

The function L_α is not uniquely determined (unless $\alpha < \omega$) but the algebra ${}^\alpha D$ of all C^∞ functions on $I = [x_1, x_2]$ with derivatives bounded by:

$$\left|\varphi^{(n)}(x)\right| < c^n.\left(\frac{L_\alpha(n)}{L'_\alpha(n)}\right)^n \qquad \big(c = c(\varphi) = \text{cst}\ ;\ \forall x \in I\big) \tag{4.2}$$

depends only on α. We call it the Denjoy class of order α. Its elements φ are *quasianalytic*, i.e. they vanish if all their derivatives at a given point vanish. The classes ${}^\alpha D$ increase with α and their union for all $\alpha < \omega^\omega$ is the algebra COHES of *cohesive functions*.

Weak Accelerates are Cohesive and Cohesive Functions are Weak Accelerates

It can be shown [4, 5] that any cohesive function $\hat{\varphi}_2(\zeta_2)$ on an interval $[0,\sigma]$ is a *weak accelerate* (i.e. the result of a weak acceleration $z_1 \to z_2$ with $\log z_2/\log z_1 \to 1$) and, under very mild assumptions on the acceleration, the converse holds: each weak accelerate $\hat{\varphi}_2(\zeta_2)$ is cohesive on some interval $]0,\sigma]$, i.e. on each $[\varepsilon,\sigma]$ with $\varepsilon > 0$.

The *cohesiveness* of weak accelerates (just as the *analyticity* of moderate or strong accelerates) is truly providential, for each acceleration $z_1 \to z_2$ is actually a two-stepped process: *first*, we calculate $\hat{\varphi}_2(\zeta_2)$ as a *germ* (for small ζ_2) by means of integral (2.6); and *then* we must take its continuation (analytic or quasianalytic) to get $\hat{\varphi}_2$ as a global, multivalued *function* on $\mathbb{R}^+$. Of course, when singular points ω_i stand in the way of quasianalytic continuation, their "circumvention" (right or left) calls for a special construction [4, 5] since $\hat{\varphi}_2(\zeta_2)$ is not defined outside $\mathbb{R}^+$.

The direct statement (cohesive functions are weak accelerates) is also highly meaningful, as it leads to a new and fairly elementary procedure for quasianalytic continuation [4–6].

§5. The Finiteness of Limit-Cycles. Analysable Functions

I am indebted to J. Martinet, R. Moussu and J.-P. Ramis for drawing my attention to a conjecture by Dulac (long known as Dulac's theorem, but unproven by him) to the effect that the limit-cycles of a vector field on $\mathbb{R}^2$ with polynomial or real coefficients, cannot possibly accumulate anywhere. Since accumulation could take place only close to a polycycle (or a point) and since polycyles, under repeated blowing-ups, can be brought down to a simple form, the problem may be rephrased as follows. Let $\mathscr{C}$ be a closed curve on $\mathbb{R}^2$ consisting of r analytic arcs $\mathscr{C}_i$ intersecting at points $S_i = \mathscr{C}_i \cap \mathscr{C}_{i+1}$. Let X be a real analytic vector field defined on a neighbourhood of $\mathscr{C}$, with the $\mathscr{C}_i$ as integral curves and with a non-vanishing linear part at each summit S_i. Next, draw an analytic curve Γ_i across each $\mathscr{C}_i$ and endow it with an analytic abscissa $x_i = 1/z_i$ ($x_i \sim +0\,; z_i \sim +\infty$) positive towards the "interior" of $\mathscr{C}$. The integral curve crossing Γ_i at the point with inverse abscissa z_i crosses Γ_{i+1} at the point z_{i+1}. The germ $G_i : z_i \mapsto z_{i+1}$ is the *local map* of summit S_i and the germ $F = G_r \circ \cdots \circ G_1$ is the *return map* of X. Limit-cycles close to $\mathscr{C}$ clearly correspond to large fixed points of F. Thus it is all a matter of establishing the trichotomy:

$$F(z) \equiv z \quad \textit{or} \quad F(z) > z \quad \textit{or} \quad F(z) < z \qquad (z \gg 1)\,. \tag{5.1}$$

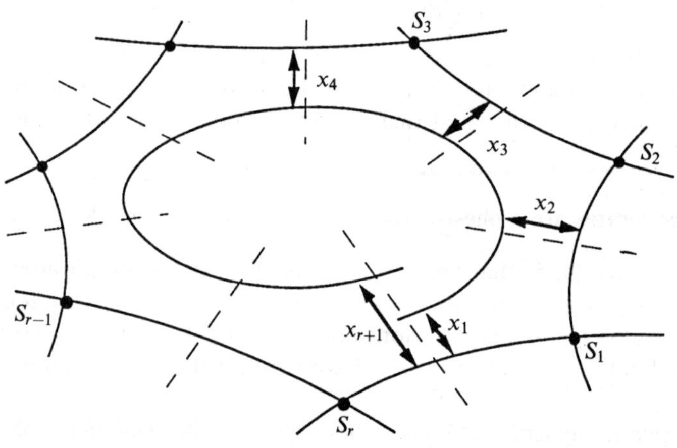

Due to reduction, the field X has, at each summit S_i, either two non-zero eigenvalues of negative ratio $-\lambda_i \notin \mathbb{Q}$ (type I) or $-\lambda_i \in \mathbb{Q}$ (type II) or only one non-zero eigenvalue (type III). For all three types, the local map G_i has a formal counterpart $\widetilde{G}_i$ which is an asymptotic or transasymptotic series of the form:

$$\widetilde{G}_i = \widetilde{K}_i \circ P_{\lambda_i} \circ \widetilde{H}_i \qquad \text{with } P_{\lambda_i}(z) \equiv z^{\lambda_i} \qquad \text{(type I)} \tag{5.2}$$

$$\widetilde{G}_i = {}^*\widetilde{V}_i \circ \widetilde{U}_i^* \qquad \text{(type II)} \tag{5.3}$$

$$\widetilde{G}_i = \widetilde{K}_i \circ E \circ \widetilde{U}_i^* \qquad \text{with } E(z) \equiv \exp z \qquad \text{(type III}^+\text{)} \tag{5.4}$$

$$\widetilde{G}_i = {}^*\widetilde{V}_i \circ L \circ \widetilde{H}_i \qquad \text{with } L(z) \equiv \log z \qquad \text{(type III}^-\text{)} \tag{5.5}$$

with the $\widetilde{H}_i$ and $\widetilde{K}_i$ denoting ordinary real power series of the form $az.\{1 + \sum a_n.z^{-n}\}$ ($a > 0$, $a_n \in \mathbb{R}$) and with $\widetilde{U}_i^*$ and ${}^*\widetilde{U}_i$ standing for the *formal* iterators (direct and inverse) of the "unitary" maps U_i which describe the holonomy of X at S_i:

$$\overline{U} \circ U(z) \equiv z \; ; \; \widetilde{U}^* \circ {}^*\widetilde{U}(z) \equiv z \; ; \; \widetilde{U}^* \circ \widetilde{U}(z) \equiv 2\pi i + \widetilde{U}^*(z) . \tag{5.6}$$

The $\widetilde{G}_i$ are usually divergent (except for type I and diophantine λ_i) but can always be resummed by $\mathscr{LB}$ with respect to a single critical time $z_i = h_i(z)$ of the form:

$$\begin{cases} h_i(z) = \log z - c \log\log z \; ; & c \text{ large (type I or II)} \\ h_i(z) \sim \log G_i(z) \text{ (type III}^+) & h_i(z) \sim \log z \text{ (type III}^-) . \end{cases} \tag{5.7}$$

As for the return map $F = G_r \circ \cdots \circ G_1$, its formal counterpart $\widetilde{F} = \widetilde{G}_r \circ \cdots \circ \widetilde{G}_1$ is a transseries with a unique "pulled-down" expansion,

$$\widetilde{F}(z) = z + \sum a_{\underline{n}} \widetilde{A}_{\underline{n}}(z) \qquad (0 \leq \underline{n} \leq \underline{n}_0 < \omega^\omega \; ; \; a_{\underline{n}} \in \mathbb{R}) \tag{5.8}$$

with finite or transfinite ordinals $\underline{n}$ and decreasing *transmonomials* $\widetilde{A}_{\underline{n}}$ that are irreducible concatenations of real coefficients and symbols $+, \times, \log, \exp$.

Like its factors $\widetilde{G}_i$, the transseries $\widetilde{F}$ is usually divergent, but always accelero-summable with at most r critical times z_i associated with its irregular summits S_i (actually, the intrinsic notion is that of *critical class* $\{z_i\}$ regrouping all times equivalent to z_i). For each z_i, only those transseries $\tilde{\varphi}(z) = \tilde{\varphi}_i(z_i)$ that are *carried* by $\widetilde{F}(z)$ and formally *subexponential* in z_i possess a Borel transform $\hat{\varphi}_i(\zeta_i)$ in the ζ_i plane. The rest must provisionnally retain their status as symbols and bide their "times" to be actualized as true functions! Of course, in order to preserve realness, it is the *median functions* med $\hat{\varphi}_i(\zeta_i)$ which are being accelerated or Laplaced.

Accelero-summability is proven by induction on q for $\widetilde{F}_q = \widetilde{G}_q \circ \cdots \circ \widetilde{G}_1$. Crucial to the argument is the comparability of non-equivalent critical times z_i and z_j (one is either faster or slower than the other). Since the function F_{ji} taking z_i to z_j and its transseries $\widetilde{F}_{ji}$ have the same factorization structure as F and $\widetilde{F}$, but with a lesser number of factors, one and the same induction takes care of the accelero-summability of $\widetilde{F}$ and $\widetilde{F}_{ji}$. Despite descriptive and notational hurdles, the proof [4, 5] is amazingly simple*. It reduces to showing, by induction on q, that accelero-summing the factors $\widetilde{G}_q$ and $\widetilde{F}_{q-1}$ to G_q and F_{q-1} and composing them to get $F_q = G_q \circ F_{q-1}$, yields the same result as composing the transseries $\widetilde{G}_q$ and $\widetilde{F}_{q-1}$ to get $\widetilde{F}_q$ and then accelero-summing it to get F_q. This, in turn, follows from a permutability of type $\int \sum = \sum \int$, due to absolute summability, with $\int$ representing an acceleration or Laplace integral and $\sum$ standing for the infinite sum which translates, in each model ζ_i, the operation of composition $\circ$.

* Another proof, apparently quite different and non-constructive in nature, has been announced by Y.S. Ilyashenko.

We end up with a formal trichotomy:

$$(5.9)\qquad \begin{cases} \widetilde{F}(z) \equiv z \quad \textbf{\textit{or}} \quad \widetilde{F}(z) = z + a_0\widetilde{A}_0(z) + o\big(\widetilde{A}_0(z)\big) \quad \textbf{\textit{or}} \\ \widetilde{F}(z) = z - a_0\widetilde{A}_0(z) + o\big(\widetilde{A}_0(z)\big) \qquad \big(a_0 >, \ \widetilde{A}_0 > 0\big) \end{cases}$$

which, after accelero-summation, translates into the wanted trichotomy (5.1).

Analysable Functions

The return map F is only a special instance of *analysable functions*. Those are real-analytic germs on $]\ldots,+\infty]$ that can be represented by an accelero-summable transseries $\widetilde{F}$ with critical times z_i which are themselves linked by analysable functions F_{ji}, etc..., with a *finite critical tree* $z_{i_1,\ldots,i_r}$. Unlike analytic functions, the class of analysable functions enjoys extreme stability (under all common operations) while retaining the two essential properties of real-analytic functions (*i*) being locally comparable (*ii*) being totally reducible to a formal object, viz. an infinite set of coefficients. Analysable functions are of very frequent occurence. See [6, 8]. There, we also introduce an even more comprehensive notion of analysable function, which subsumes both complex-analytic and cohesive functions and seems to stretch the Analytic Principle to its farthest possible limit.

Bibliography

[1–3] J. Ecalle: Les fonctions résurgentes (vols. 1, 2, 3). Pub. Math. Orsay (1981, 1981, 1985)

[4] J. Ecalle: Finitude des cycles-limites et accéléro-sommation de l'application de retour. Dans les Actes du Colloque sur les cycles-limites, Luminy 1989. Lecture Notes in Mathematics, vol. 1455. Springer, Berlin Heidelberg New York 1990

[5] J. Ecalle: La conjecture de Dulac: une preuve constructive. (à paraître à Travaux en Cours, Décembre 1990)

[6] J. Ecalle: Calcul accélératoire et applications. (à paraître à Travaux en Cours, Décembre 1990)

[7] J. Ecalle: Calcul compensatoire et linéarisation quasianalytique des champs de vecteurs locaux. (à paraître à Travaux en Cours, 1991)

[8] J. Ecalle: Fonctions résurgentes, calcul étranger, calcul accélératoire et applications. (cours général – en préparation)

[9] J. Ecalle, J. Martinet, R. Moussu, J.P. Ramis: Non-accumulation des cycles-limites. C. R. Acad. Sci., série I **304**, n° 14 (1298), (I) pp. 375–378, (II) pp. 431–434

Finiteness Theorems for Limit Cycles

Ju. S. Il'yashenko

Department of Mathematics and Mechanics, Moscow State University
117234 Moscow, USSR

1. Statement of Results

Theorem 1. *Polynomial vector field in the real plane has only finitely many limit cycles.*

Theorem 2. *Analytic vector field in the closed two dimensional surface has only finitely many limit cycles.*

Theorem 3. *A singular point of any analytic vector field has a neighbourhood free of limit cycles.*

Theorem 4 (Nonaccumulation Theorem). *A polycycle of an analytic vector field in the closed two dimensional surface has a neighbourhood free of limit cycles.*

It is known from the time of Poincaré that the Theorem 4 implies Theorems 1–3. The Theorem 3 is the direct consequence of the Theorem 4: The singular point is a particular case of a polycycle. It is distinguished because the wrong opinion that it was proved long ago is widely spread in mathematical literature and folklore.

In the case when the polycycle is a cycle (contains no singular points) Theorem 4 is an immediate consequence of the analytic dependence of solutions on the initial conditions and a uniqueness theorem for analytic functions, applied to the Poincaré map. In the case of the polycycle (separatrix polygone) an analog of the Poincaré map may be defined (Fig. 1); it is a germ of transformation of the semiinterval onto itself. Its natural form is $(\mathbb{R}^+, 0) \to (\mathbb{R}^+, 0)$

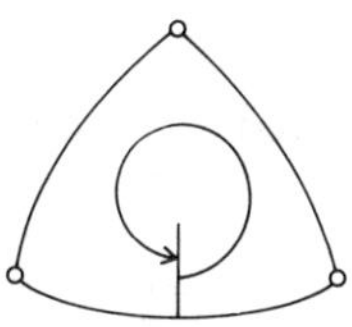

Fig. 1

Proceedings of the International Congress
of Mathematicians, Kyoto, Japan, 1990

Theorem 5 (Identity Theorem). *If the monodromy map of the polycycle of an analytic vector field has an infinite set of fixpoints, then it is identical.*

Theorem 4 is a trivial consequence of the Theorem 5.

The same results are independently and quite in a different way obtained by J. Ecalle (France).

2. Problems Related to the Hilbert's 16th

The classical problem is "... the question on the maximal number and situation of the Poincaré limit cycles for the equation of the form

$$dy/dx = Y/X \tag{1}$$

where X, Y are polynomials on x, y of degree n ..."

The *Hilbert number* for the family of equations (in particular, for a single equation) is the maximum number of limit cycles for the equations of the family. By definition, the Hilbert number is finite or does not exist. Mark some versions of the Hilbert problem.

Algebraic Versions

1. *The Dulac's problem. Prove the existence of the Hilbert number for the equation* (1).
2. *Prove the existence of the Hilbert number for the whole family* (1) (*it is noted as* $H(n)$).
3. *Give the* (*explict*) *upper estimate on* $H(n)$.

Analytic Versions

4. *Prove the existence of the Hilbert number for the analytic vector field in the sphere* S^2.
5. *Prove the existence of the Hilbert number for any family of analytic vector fields in* S^2 *with the finite dimensional compact base.*

Smooth Version (Hilbert-Arnold Problem)

6. *Prove the existence of the Hilbert number for a "typical" family of smooth vector fields in* S^2 *with the finite dimensional compact base. Here and below "smooth" is "infinitely smooth".*

Only the Problems 1 and 4 are solved up to now. There are the following natural implications:

$$\begin{array}{ccc} 5 & \Rightarrow & 4 \\ \Downarrow & & \Downarrow \\ 3 \Rightarrow 2 & \Rightarrow & 1 \end{array}$$

Remark, that even if the Problem 3 will be solved, it will not imply the solution of the Problems 4–6. The exact meaning of the word "typical" in Problem 6 is the part of the solution. The Hilbert-Arnold problem is the illustration of the heuristical principle: the smooth function behaves like an analytic one, when it is met in the finite dimensional typical family.

3. Sources of the Proof

The Dulac's problem concentrates many branches of the theory of differential equations, and the tools, used in its solution, have many other applications. These tools are the following: *desingularization, Dulac's asymptotic expansion, complex extension in sense of Petrovskii-Landis, functional cochains, super exact asymptotic series.* The main goal of this talk is to describe these tools. Other applications and further developments are briefly expressed in four Appendices.

4. Desingularization

The simplest variant of the desingularization is the polar change of coordinates with the subsequent division by the proper power of the distance to the pasted circle instead of the origin. In more details, let v be an analytic vector field with an isolated singular point 0. The map $(r, \varphi) \mapsto (r + 1, \varphi)$, (r, φ) being polar coordinates, brings the initial vector field, defined in the neighbourhood of 0 to the vector field, defined in the exterior part of the neighbourhood of the unit circle. This new field may be analytically extended in the interior part of this neighbourhood. After the division by some proper power of $r - 1$ it becomes a field with a finite number of singular points, located on the unit circle pasted instead of the singular point. These new singular points are in some sense "simpler" than the previous one. It may be repeated several times. The simplest form of singular points obtained by this way is given in the

Definition. A singular point of the vector field is called *elementary*, if the linearisation of the field at this point has at least one nonzero eigenvalue.

These eigenvalues are called those of the singular point.

Bendixson-Seidenberg-Dumortier Theorem. *After a finite number of steps of the blowing-up process an isolated singular point of an analytic vector field can be split to a finite number of elementary ones.*

Using this theorem one may consider the polycycle in the Identity theorem as to be elementary, that is to say, with only elementary singular points. Other applications and developments of the desingularization method are described in the Appendices I and II.

5. The Dulac's Theorem

Dulac (1923) found an asymptotic expansion of the monodromy of the polycycle up to any power of the distance from the polycycle. The more convenient chart is the logarithmic one: if x is a "natural" chart on the semiinterval $(R^+, 0)$, then $\xi = -\ln x$ is a logarithmic chart.

Dulac's Theorem. *Let γ be an elementary polycycle of an analytic vector field in the plane. Then the semitransversal to the polycycle may be so chosen, that the correspondent monodromy map Δ_γ will be either flat, or inverse to flat, or admits an asymptotic expansion $\hat{\Delta}_\gamma$ having the following form in the logarithmic chart*

$$\hat{\Delta}_\gamma = \alpha\xi + \beta + \sum P_j(\xi)\exp(-\nu_j\xi), \tag{2}$$

$\alpha > 0$, $\beta \in R$, *P_j are real polynomials*, $0 < \nu_j \to \infty$.

Denote by Fix_∞ the set of germs $(\mathbb{R}^+, \infty) \to (\mathbb{R}^+, \infty)$ with the infinite number of fixpoints.

Dulac's Lemma. *If $\Delta_\gamma \in \mathrm{Fix}_\infty$ then $\Delta_\gamma = \mathrm{id}$.*

The proof, given in (Dulac 1923) uses only the properties of Δ_γ, listed in the above theorem. For the maps only with such properties the lemma is wrong; counterexample: $x \mapsto x + (\exp(-1/x)) \sin 1/x$. Note, by means of the theorem 5 that such a map cannot appear as a monodromy in the analytic case.

In fact Dulac proves, that the conditions of the lemma imply $\hat{\Delta}_\gamma = \mathrm{id}$. It is trivial, because the terms of the summation in the right hand side of (2) are nonoscillating, and each term tends to zero faster than the previous one.

The Dulac's Theorem is proved by means of smooth, not analytic, theory (Il'yashenko, 1985). The recent development of this theory is described in the Appendix III.

6. Complex Extension: The Hyperbolic Case

Let the polycycle γ have only hyperbolic singular points (having no eigenvalues in the imaginary axis). In this case the monodromy map in the logarithmic chart may be extended in the domain, which is "like the right halfplane" $\mathbb{C}^+ = \{\operatorname{Re}\zeta > 0\}$ and has the form

$$\Omega_C = \Phi_C C^+, \qquad \Phi_C : \zeta \mapsto \zeta + C\sqrt{\zeta + 1}$$

On the other hand Δ_γ may be decomposed in this domain in the asymptotic series (2). The germs with these two properties are called "*almost regular*".

In the hyperbolic case the Identity Theorem may be easily proved (Il'yashenko, 1984). Indeed, if $\Delta_\gamma \in \mathrm{Fix}_\infty$, then $\hat{\Delta}_\gamma = \mathrm{id}$ by the Dulac's lemma. Then $\Delta_\gamma - \mathrm{id} = o(\exp(-\nu\xi))$ for any $\nu > 0$ on (R^+, ∞). At the same time $\Delta_\gamma - \mathrm{id}$ is holomorphic and bounded in Ω_C. The theorem of Phragmen-Lindelöf type (namely, Watson's Theorem) implies then, that $\Delta_\gamma - \mathrm{id} \equiv 0$. This proves the Identity Theorem in the hyperbolic case. Pass now to the general case.

7. Semihyperbolic Singular Points, Functional Cochains and Reduction to Complex Analysis

In the general case the polycycle in the Identity Theorem, assumed to be elementary without loss of generality (see Section 4) contains either hyperbolic, or *semihyperbolic* singular points. These last points have, by definition, one zero and one nonzero eigenvalue. The monodromy of an elementary polycycle may be decomposed in the product

$$\Delta_\gamma = \Delta_N \circ \cdots \circ \Delta_1 \tag{3}$$

of the so called "*correspondence maps*" (Fig. 2) related to the vertices of the polycycle (Fig. 3). Semihyperbolic points bring in the composition (3) the maps of the three following kinds: exponents, logarithms and functional cochains. The appearance of exponents is easily seen in the example $\dot{x} = x^2$, $\dot{y} = -y$. The correspondence map of the semiinterval $x > 0$, $y = e$ onto the semiinterval $y > 0$, $x = 1$ is $x \mapsto y(x) = \exp(-1/x)$. In the logarithmic chart it is $\xi \mapsto \exp \xi$.

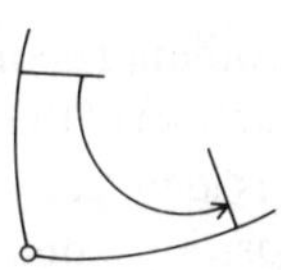

Fig. 2

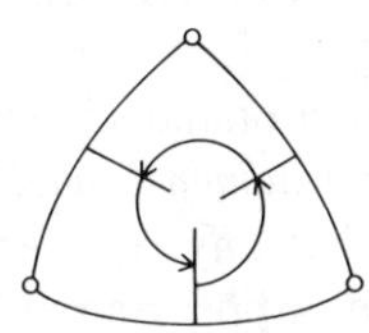

Fig. 3

The functional cochains appear in the description of the correspondence maps for the semihyperbolic singular points of real analytic vector fields. A field of this type has a holomorphic invariant curve, tangent in the semihyperbolic singular point to the eigenvector of the linearization with the nonzero eigenvalue. A monodromy transformation (named the monodromy of the singular point) corresponding to the loop on this curve, surrounding this singular point, is tangent to identity in zero. Such a transformation is formally equivalent to the time one shift along the orbits of the holomorphic vector field with the zero linearization in the singular point. The conjugating formal Teylor series (called also normalizing) is divergent in general. Yet it is asymptotic to some actual function. Namely, there is a covering of a punctured neighbourhood of the fixpoint 0 by sectors of angle $\alpha \in \left(\frac{\pi}{p}, \frac{2\pi}{p}\right)$ invariant under the rotation by π/p with the vertex at 0. Here $p + 1$ is the multiplicity of the fixpoint. In each sector a chart is defined, conjugating the initial map with its formal normal form. This chart has an asymptotic expansion at the vertex, given by the normalizing series. The collection of these charts form a normalizing atlas or, in other words, a normalizing cochain. The transition functions of this atlas are called coboundary of the cochain and contain all the information on the geometric properties of the initial germ. They give also the complete invariant of the analytic classification of the germs $(\mathbb{C}, 0) \mapsto (\mathbb{C}, 0)$ tangent to identity. The difference between the maps forming the coboundary of the cochain, and identity, is decreasing exponentially and faster, than the nonzero holomorphic function in the sector with the angle, larger than π/p, can decrease.

The correspondence map for the semihyperbolic singular point of a real analytic vector field in the natural map is given by the formula (Il'yashenko, 1986)

$$\Delta = g \circ \exp \circ h_{p,\lambda} \circ H, \tag{4}$$

Here H is a normalizing cochain for the monodromy of a singular point, $g : (\mathbb{C}, 0) \mapsto (\mathbb{C}, 0)$ is a holomorphic germ, Δ is real on $(\mathbb{R}^+, 0)$, $h = pz^p/(1 - apz^p \ln z)$

The Identity Theorem is now reduced to the following problem of the complex analysis: prove, that a composition of almost regular germs and of germs, obtained from (4) by transition to the logarithmic map, and also of germs inverse to the previous ones, is either identical, or has no fixpoints near infinity. The ideas for getting a solution of this problem will be described below.

8. Phragmen-Lindelöf Property of the Functional Cochains

The class of the functional cochains defined in some domain Ω containing $(\mathbb{R}^+, \infty)$ is called having a *Phragmen-Lindelöf property*, if any cochain of this class, decreasing on $(\mathbb{R}^+, \infty)$ faster than any exponent: $\exp(-\nu\xi)$, $\nu > 0$, is identically zero on $(\mathbb{R}^+, \infty)$.

All the cochains, appearing in study of the compositions, defined in the end of the Section 7, have the Phragmen-Lindelöf property.

9. Super Exact Asymptotic Series

The plan of the further proof is the following: decompose the monodromy map in the asymptotic series with nonoscillating terms; using the Phragmen-Lindelöf Theorem prove, that if this series is equal to identity, then the map itself is identical. The following problem arises in this way. It is not difficult to construct a polycycle of an analytic vector field having the nonidentical monodromy map with the identical Dulac series. Thus the Dulac series does not uniquely determine the monodromy map of the elementary polycycle (though in the hyperbolic case it does). So, we have to construct series, describing both exponential and transexponential decreasing. At first glance it is impossible: any remainder term of the Dulac series may be larger than the transexponential terms. This difficulty can be overcome by the way, shown by the following example. Take two classes of germs of decreasing functions at infinity: M_0 and M_1, both containing 0 and identity. Let the germs of these classes have the Phragmen-Lindelöf type property: the nonzero germ of the class M_0 (M_1) cannot decrease faster than $\exp(-\nu\xi)$ (respectively, $\exp(-\nu \exp \xi)$) for any $\nu > 0$. Let the germs of class M_0 be decomposed in asymptotic Dulac series and the germs of the class M_1 be decomposed in the asymptotic series of the type

$$a_0 + \sum a_j \exp(-\nu_j \exp \xi) \tag{5}$$

$0 < \nu_j \to \infty$, $a_j \in M_0$, $j = 0, 1, \ldots$. The series (5) is called "*super exact asymptotic series*" (the Russian abbreviation is STAR). Its free term contains all the information of the exponential asymptotics of the germ, and the higher terms give the trans-exponential one. The terms of this series do not oscillate and the germs of the class M_1 are uniquely determined by the corresponding STAR's. Thus for such germs the Identity Theorem holds.

10. Additional Decomposition Theorem

In order to apply the previous ideas to the monodromy map, one must decompose it in terms each one of which will be uniquely determined by the corresponding STAR; this STAR is like (5), but much more complicated. The Decomposition Theorem will be stated here for the so called alternant case, when the maps exp and ln in the composition (3) arise in turn.

Definition. Two germs f, g of functions $(\mathbb{R}^+, \infty) \mapsto (\mathbb{R}, 0)$ belong to the same class of Archimedean equivalence iff such positive a and b exist, that $|f|^a < |g|$ and $|g|^b < |f|$ near infinity.

Additional Decomposition Theorem. *In the alternant case the monodromy map of the polycycle may be decomposed in the sum*

$$\Delta_\gamma = \alpha\xi + \beta + \varphi + \sum \psi_j \tag{6}$$

with $|\varphi| < C \exp(-\nu\xi)$, $|\psi_j| < C_1 \exp(-C_2 \exp \nu_j \xi)$ *for some positive* ν, ν_j, C, C_1, $C_2 (\nu_j < \nu_{j+1})$. *If* $\alpha = 1$, $\beta = 0$, $\varphi \neq 0$, then $|\varphi|$ *belongs to the same Archimedean class as* $\exp(-\xi)$; *if* $\alpha = 1$, $\beta = 0$, $\varphi \equiv 0$, *then* $|\psi_1|$ *belongs to the same Archimedean class as* $\exp(-\exp \nu_1 \xi)$.

This theorem implies the Identity Theorem in the alternant case.

Remark. In fact, in the decomposition (6) φ has an asymptotic Dulac series, and possesses the Phragmen-Lindelöf property. The germs $\psi_j \circ \nu_j^{-1} \circ \ln$ have also the same property; if $\alpha = 1$, $\beta = 0$, $\varphi \equiv 0$, then the germ $\psi_1 \circ \nu_1^{-1}$ may be decomposed in STAR with nonoscillating terms.

11. On the Publications and References

The Identity Theorem for the alternant polycycles is proved in full detail in (Il'yashenko, 1990). The complete proof is exposed in a book (Il'yashenko, to appear), which is four times larger than the previous paper forming the first part of the publication. The book contains a complete exposition of a proof with the large introductory chapter, planned for nonspecialists, and is independent of the first part. The first part contains almost all the ideas, used in the book, but is much simpler; it is a kind of digest of the general proof.

The reference list below in relation to Section 7 and Appendices is incomplete for the lack of space. I should mention here that an important contribution in this topic was done by Bogdanov, Bryuno, Denkowska, Dumortier, Ecalle, Elisarov, Malgrange, Martinet, Moussu, Ramis, Roussari, Trifonov Van den Essen, Voronin and others.

Appendix I. Topological Classification of Nonmonodromic Singular Points in the Plane; the Order of Topologically Sufficient Jet

Main Alternative. *A germ of a smooth vector field in the singular point, which is nonflat, is either monodromic (that's to say, admits the Poincaré map, Fig.* 4a) *or has a characteristic orbit (Fig.* 4b). *The characteristic orbit is the phase curve, which enters into the singular point after infinite time, in positive or negative sense, tangent to some ray at this point.*

(a) (b)

Fig. 4

The Bendixson-Dumortier Theorem guarantees that a finite number of blowing up steps allows to distinguish the two kinds of germs in the Main Alternative, if the singular point has finite multiplicity. By definition, the multiplicity of the singular point of a smooth germ of vector field v is the dimension of the local ring $Q = \mathcal{O}/(v)$, where $\mathcal{O}$ is the ring of all germs of the smooth functions at the origin, and (v) is an ideal, generated in this ring by the components of v. The jet of the vector field is Main-Alternative-sufficient, iff all its representators are simultaneously monodromic or have a characteristic orbit. It is said to be topologically sufficient if all its representators are orbitally topologically equivalent. This means, that the phase curves of any representator may be transformed by the fixpoint preserving homeomorphism into the phase curves of any other representator.

Dumortier proved, that the Main-Alternative-sufficient jet for a smooth germ with finite multiplicity always exists, and for the germ having moreover a characteristic orbit, a topologically sufficient jet also exists.

Theorem (Kleban, to appear). *The order of the Main-Alternative- and topologically sufficient jet, defined above, is not larger, than the triple multiplicity of the singular point.*

Appendix II. Desingularization in the Families of Vector Fields

An important step in understanding of the Hilbert-Arnold's problem is the desingularization theorem for the families of vector (or line) fields. Explain the main statement, omitting the details. A family of complex analytic surfaces is a triple $M \xrightarrow{\pi} B$ with holomorphic $n+2$ and n-manifolds M and B, and a holomorphic map π with the constant rank n. A line field α on M is a section of the projective tangent bundle to M extended in the maximal possible domain Ω_α with the following property: the set $\Sigma = M \backslash \Omega_\alpha$ is analytic and $\dim \Sigma \leq \dim M - 2$. By definition, the singular points of the field α are all the points of Σ. The family of the line fields, corresponding to the family $M \xrightarrow{\pi} B$ is the line on the total space M, tangent to the fibers of π. It may be proved, that the restriction of such a family on the fiber, having at least one nonsingular point (such a fiber is called noncritical), may be extended in all the points of the fiber, except some discrete set, called essential singular points. A blowing up of the family of surfaces $M \xrightarrow{\pi} B$ is the commutative diagram

$$\begin{array}{ccc} \tilde{M} & \xrightarrow{H} & M \\ {\scriptstyle \tilde{\pi}}\downarrow & & \downarrow{\scriptstyle \pi} \\ \tilde{B} & \xrightarrow{\varrho} & B \end{array}$$

where the left column is the family of surfaces with the same dimension of the base, H and ϱ are holomorphic, and the restriction of H on each fiber of $\tilde{\pi}$ is the finite number of inverse blowing ups. If α is the family of the vector fields, corresponding to the right column, then the blown up field α^*, corresponding to the left column, is: $\alpha = H_* \alpha^*$.

Theorem (Trifonov, 1990). *For any analytic family of line fields, corresponding to the family of surfaces $M \xrightarrow{\pi} B$ with the compact total space M with the boundary, having noncritical fibers only, a blowing up exists, giving a new family of line fields, for which all the essential singular points are elementary* (*that's to say, the extended restriction of the family of fields on each fiber is locally generated by the holomorphic vector field, having only elementary singular points*).

Appendix III. Smooth Normal Forms in Local Dynamics

The formal equivalence of smooth vector fields is necessary for their smooth equivalence. The inverse implication

$$\textbf{Formal equivalence} \Rightarrow \textbf{smooth equivalence} \tag{7}$$

takes place in the following cases:

1°. Hyperbolic singular points (Sternberg, Chen).

2°. Singular points with one zero eigenvalue (a saddlenode) or an imaginary pair; all other eigenvalues lie outside the imaginary axis (Belitsky, 1986).

For the singular points with the degeneration of the codimension two in the linear part the implication (7) is not proved; in the case of codimension three (for instance, three imaginary pairs) it is wrong (Takens).

Similar results are valid for the local families. In the following cases the families may be brought to the integrable normal forms by a finitely smooth change of coordinates and time (the formulas below give normal forms)

1°.	Perturbation of the nonresonant hyperbolic singular point	$\dot{x} = A(\varepsilon)x$
2°.	Perturbation of the one-resonant hyperbolic singular point (all the resonances are consequences of a single one $(\lambda, r) = 0$	$\dot{x} = (\operatorname{diag} x)(\lambda + P(u))$ $u = x^r$ – a resonant monomial, $x \in \mathbb{R}^n$, P – a vector polynomial.
3°.	Perturbation of a saddle-node of finite multiplicity $p + 1$ with all the resonances being the consequences of $\lambda_1 = 0$	$\dot{x} = x^{p+1}(1 + ax^p)^{-1}$ $\dot{y} = (\operatorname{diag} a(x, \varepsilon))y$ $x \in \mathbb{R}^1$, $y \in \mathbb{R}^{n-1}$

(Kostov, 1984; Il'yashenko and Yakovenko, 1990)

The finitely smooth classification of the perturbations of the germs with an imaginary pair of eigenvalues has functional moduli (Il'yashenko and Yakovenko, to appear). Parallel theory is developed for diffeomorphisms.

Appendix IV. Nonlinear Stokes Phenomena

In the general case local dynamics near a stable point gives a local chart near this point, defined up to a finite number of parameters. The genericity assumption is the Siegel condition, sufficient to the analytic equivalence of the germ of vector field or diffeomorphism to its linear part. The normalizing map is uniquely determined by its linear part. In the resonant case the normalizing chart is replaced by the normalizing atlas, formed by the sector-like domains, covering the punctured neighbourhood of the fixpoint with the point itself on the boundary. The map, conjugating the resonant vector field or the diffeomorphism with its formal normal form is defined and biholomorphic in each sector. The transition functions give the complete invariant of the analytic classification of such germs and contain all the information on the geometric properties of the germ. This program is brought up for germs of maps $(\mathbb{C}, 0) \to (\mathbb{C}, 0)$ with the resonant linear part (the multiplicator is the root of unity) and for germs of vector fields in $(\mathbb{C}^2, 0)$: the resonant saddles and semihyperbolic singular points (giving rise to the Ecalle-Voronin and Martinet-Ramis moduli). All this material with different extensions and applications is discussed in the forthcoming book (Elisarov et al.)

References

Belitsky, G.R.: Smooth equivalence of germs of vector fields with one zero and one nonzero or one pair of pure imaginary eigenvalues. Funct. Anal. Appl. (Russ.) **20**, no. 4 (1986) 1–8

Dulac, H.: Sur les cycles limites. Bull. Soc. Math. France **51** (1923) 45–188

Elisarov, P.M., Il'yashenko, Ju.S., Scherbakov, A.A., Voronin, S.M., Yakovenko, S.Yu.: Nonlinear stokes phenomena. Collection of papers. To appear

Il'yashenko, Ju.S.: Limit cycles of polynomial vector fields in the real plane with the non-degenerated singular points. Funct. Anal. Appl. (Russ.) **18**, no. 3 (1984) 32–42

Il'yashenko, Ju.S.: The Dulac's memoir "On the limit cycles" and related topics of the local theory of differential equations. Russ. Math. Sur. **40**, no. 6 (1985) 41–78

Il'yashenko, Ju.S.: Separatrix biangles of analytic vector fields in the plane. Vestnik MGU, Ser. Math. no. 4 (1986) 25–31

Il'yashenko, Ju.S.: Finiteness theorems for limit cycles. Russ. Math. Surv. **45**, no. 2 (1990) 143–200

Il'yashenko, Ju.S.: Finiteness theorems for limit cycles. A book, submitted to AMS in spring 1990

Il'yashenko, Ju.S., Yakovenko, S.Yu.: Normal forms of local families and nonlocal bifurcations. Russ. Math. Surv. **46**, no. 1 (1991)

Il'yashenko, Ju.S., Yakovenko, S.Yu.: Paper V in the book of Elisarov

Kostov, V.P., Versal deformations of the differential forms of degree α in the line. Funct. Anal. Appl. (Russ.) **18**, no. 4 (1984) 81–82

Kleban, O.A.: Order of topologically sufficient jet of the vector field in the plane. To appear

Trifonov, S.I.: Desingularization in the families of analytic vector fields. VINITI, 1990

Averaging and Passage Through Resonances

Anatoly I. Neishtadt

Space Research Institute, USSR Academy of Sciences, Profsoyuznaya 84/32
Moscow 117810, USSR

The problems, concerning averaging of perturbations in systems passing through resonances, are dealt with in this report. Such interesting phenomena as capture into resonance, probabilistic scattering of trajectories, destruction of adiabatic invariants, and delay of stability loss, appear to be connected with influence of resonances.

1. Slow-Fast Systems, Systems with Rotating Phases, Averaging Method

Slow-fast systems (systems having fast and slow variables) are systems of differential equations of the form

$$\dot{x} = f(x, y, \varepsilon), \qquad \dot{y} = \varepsilon g(x, y, \varepsilon), \qquad 0 < \varepsilon \ll 1. \tag{1}$$

Variables x are called fast variables, and variables y are called slow ones. If $\varepsilon = 0$, the system is unperturbed. The equation for x with $y = \text{const}$, $\varepsilon = 0$ is called the fast equation (system). The system (1) is investigated using various methods of the perturbation theory corresponding to various properties of the fast equation.

The following case is of great importance for many applications: the fast variables are angle variables (phases) on the m-torus $\mathbb{T}^m = \mathbb{R}^m/2\pi\mathbb{Z}^m$, and a trajectory of the fast system winds round the torus with frequencies ω which depend on slow variables. In this case the system (1) is called a system with rotating phases; it has the form

$$\dot{\varphi} = \omega(I) + \varepsilon f(I, \varphi, \varepsilon), \qquad \dot{I} = \varepsilon g(I, \varphi, \varepsilon). \tag{2}$$

Below we suppose that right-hand sides of the system are smooth enough (infinitely differentiable for simplicity) functions of all the arguments when $(\varphi, I, \varepsilon) \in \mathbb{T}^m \times D \times [0, \varepsilon_0]$, where D is a compact domain in $\mathbb{R}^n$. The classical averaging method gives a recipe for approximate description of the slow variable I evolution in a time interval of order $1/\varepsilon$. According to this method one should replace the system (2) with the averaged system

Proceedings of the International Congress
of Mathematicians, Kyoto, Japan, 1990

$$\dot{J} = \varepsilon G(J), \qquad G(J) = (2\pi)^{-m} \oint_{\mathbb{T}^m} g(J, \varphi, 0)\, d\varphi. \tag{3}$$

It is supposed that solutions of (3) are good approximations of solutions of (2).

"This principle is neither a theorem, nor an axiom, nor definition, but only a physical suggestion, in other words a vaguely formulated and, to put it strictly, wrong statement. Such statements often appear to be fruitful sources for mathematical theorems" [4].

We denote by $J(t)$ the solution of the averaged system (3) with initial condition $I_0 \in D_0 \subset D$ at $t = 0$ and by $(I(t), \varphi(t))$ the solution of the slow-fast system (2) (the "precise system") with initial condition $(I_0, \varphi_0) \in D_0 \times T^m$. So $I(0) = J(0) = I_0$. We suppose that the solution $J(t)$ is defined and kept at the positive distance from the bound of the domain D for $0 \le t \le 1/\varepsilon$. The problem of estimating the difference between $I(t)$ and $J(t)$ when $0 \le t \le 1/\varepsilon$ is traditionally called the averaging method justification problem [11].

If the system possesses only one frequence ($m = 1$) and this frequence ω does not vanish, the following estimate is valid: $|I(t) - J(t)| < c\varepsilon$ for $0 \le t \le 1/\varepsilon$, $c =$ const > 0 (P. Fatou; L.I. Mandelshtam and N.D. Papaleksi; this estimate is the first result in the averaging method justification, see [21]). An analogous result holds for multifrequency systems with constant ($\omega =$ const) nonresonant frequencies [11].

For Hamiltonian systems close to integrable ones, the averaging method justification was promoted in frames of the Kolmogorov-Arnold-Moser (KAM) theory [4] and estimates were obtained for an infinite or exponentially large (Nekhoroshev's theorem [4]) time interval. The averaged system in this case is of the form $\dot{J} = 0$; there is no evolution of variables I and frequencies $\omega(I)$ in the averaging method approximation.

2. Averaging in Multifrequency Systems

In a general case, slow variables I and, consequently, frequencies $\omega(I)$ of the system with rotating phases (2) may change in time. In multifrequency ($m \ge 2$) systems at some moments of time a linear dependence of frequencies with integer coefficients, i.e. resonance, occurs. At the resonance a trajectory of the fast system fills a torus of lower dimension and there is no reason to expect that the averaging over the whole torus $\mathbb{T}^m$ in (3) describes the motion correctly.

The resonance condition $(k, \omega(I)) = 0$ for each $k \in \mathbb{Z}^m \setminus \{0\}$ defines a surface, called the resonant surface, in the space of slow variables (here $(\ ,\ )$ denotes the standard scalar product in $\mathbb{R}^m$). The union of resonant surfaces for all $k \in \mathbb{Z}^m \setminus \{0\}$ is called the resonant set. The averaging method estimates can be obtained basing on the following idea: if the set of points which are close to resonant surfaces is of small measure, the influence of motion within this set must be small for the majority of initial conditions.

General results were obtained in this direction by D.V. Anosov [1] and T. Kasuga [18]. Let $E(\varrho, \varepsilon)$ be the set of initial conditions for which the error due to the averaging method exceeds ϱ for time $1/\varepsilon$:

$$E(\varrho, \varepsilon) = \left\{ (I_0, \varphi_0) \in D_0 \times \mathbb{T}^m : \sup_{0 \le t \le 1/\varepsilon} |I(t) - J(t)| > \varrho \quad \text{for} \quad I(0) = J(0) = I_0 \right\} \tag{4}$$

(if $I(t)$ is not defined on the whole time interval, we formally assume that supremum is equal to 1). According to the theorem by D.V. Anosov [1], if the resonant set is of measure zero, then mes $E(\varrho, \varepsilon) \to 0$ as $\varepsilon \to 0$. Really, the D.V. Anosov theorem has been proved for averaging in general slow-fast systems (1). The following estimate is obtained for system (2).

Theorem 1 [27]. *Let at least one of the following two conditions be satisfied:*

$$\operatorname{rank}(\partial\omega/\partial I) = m \quad \text{or} \quad \omega \neq 0 \quad \text{and} \quad \operatorname{rank}(\partial(\omega/\|\omega\|)/\partial I) = m - 1. \tag{5}$$

Then the mean (over initial conditions) error of the averaging method does not exceed a quantity of order $\sqrt{\varepsilon}$:

$$\int_{D_0 \times \mathbb{T}^m} \sup_{0 \le t \le 1/\varepsilon} |I(t) - J(t)| \, dI_0 \, d\varphi_0 < c_1 \sqrt{\varepsilon}. \tag{6}$$

Here c_1 and furthermore c_i are positive constants.

Corollary. *Under the hypotheses of Theorem* 1

$$\text{mes } E(\varrho, \varepsilon) < c_1 \sqrt{\varepsilon}/\varrho. \tag{7}$$

An equivalent formulation: outside of the set of measure κ the following estimate of the averaging method error is valid:

$$|I(t) - J(t)| < c_1 \sqrt{\varepsilon}/\kappa. \tag{8}$$

The estimate (6) is unimprovable. The estimates (7), (8) are unimprovable within the class of power estimates. It looks reasonable, that these latter estimates can be improved if we confine ourselves to generic perturbations (see below Sect. 3).

If $m > n + 1$, the condition (5) cannot be satisfied. In this case the frequency-mapping $I \mapsto \omega(I)$ defines the submanifold $M = \omega(D)$ in $\mathbb{R}^m$, and to obtain some averaging method estimates one should use Diophantine approximations on this submanifold. The following results have been obtained in this direction.

It is shown in [7] that for arbitrary m, n the estimates (6)–(8) hold for almost all members of a typical family of frequencies which depends on a sufficiently large number of parameters.

The problem was considered in [16] under the following restriction on the curvature of the manifold $M = \omega(D)$. Let $x \in M$, vector $\gamma \in T_x M^\perp$ (the normal space to M at x) and let h_γ be the second fundamental quadratic form of M with respect to γ. It is assumed, that for every x, γ there exists a two-dimensional subspace in $T_x M$ where the form h_γ is defined positively or negatively. For $m < 2 +$

$(n-1)(n-2)/2$ this condition is satisfied for generic mappings ω, and it is proved that the estimates (6)–(8) are valid.

The estimates for any m, n and generic mappings ω have been obtained.

Theorem 2 (V.I. Bakhtin [7]). *For systems with m fast and n slow variables and with a generic frequency mapping ω the mean (over initial conditions) error of the averaging method, does not exceed a quantity of order $\varepsilon^{1/p+1}$, provided that* $\binom{n+p}{p} \geq n+m$.

Consequently, for such systems

$$\text{mes}\, E(\varrho, \varepsilon) < c_1 \varepsilon^{1/p+1}/\varrho.$$

The genericity condition is presented in the explicit form. The mappings ω not subject to this condition belong to a set of codimension 1 in the functional space.

3. Passage Through and Capture into Resonances in Two-Frequency Systems

In two-frequency ($m = 2$, $\varphi = (\varphi_1, \varphi_2)$, $\omega = (\omega_1, \omega_2)$) systems resonant surfaces of different resonances do not cross each other if $\omega \neq 0$. So the influence of each resonance can be investigated apart from others.

Definition [3]. The system (2) meets the condition A ($\bar{A}$, respectively) if the ratio of frequencies ω_1/ω_2 changes with a nonzero speed along its trajectories (along the trajectories of the averaged system, respectively):

$$A: (\omega_1 \partial\omega_2/\partial I - \omega_2 \partial\omega_1/\partial I) g > c_1^{-1},$$

$$\bar{A}: (\omega_1 \partial\omega_2/\partial I - \omega_2 \partial\omega_1/\partial I) G > c_1^{-1},$$

Theorem 3 (V.I. Arnold [3, 5]). *If the condition A is satisfied, then*

$$|I(t) - J(t)| < c_2\sqrt{\varepsilon}, \qquad 0 \leq t \leq 1/\varepsilon.$$

Theorem 4 [24, 26]. *If the system meets both the condition $\bar{A}$ and some other condition B (which is actually almost always satisfied), then for all initial points (I_0, φ_0) with the exception of a set U_c of measure not greater than $c_2\sqrt{\varepsilon}$, the following estimate holds*:

$$|I(t) - J(t)| < c_3\sqrt{\varepsilon}\, |\ln \varepsilon|, \qquad 0 \leq t \leq 1/\varepsilon.$$

For any $\kappa \geq c_2\sqrt{\varepsilon}$ outside the initial point set of measure not greater than κ the following estimate is valid (*compare with* (8))

$$|I(t) - J(t)| < c_4\sqrt{\varepsilon}\, |\ln c_5^{-1}\kappa|.$$

The estimates are unimprovable. For the proofs of Theorems 3, 4 see also [20].

For initial conditions belonging to the set U_c (c is for "capture") the capture into resonance takes place. The essence of this phenomenon is that the phase point reaches a resonant surface and begins drifting along this surface, resonance conditions being kept approximately. Therefore, solutions of the precise system and the averaged one diverge by a quantity of order 1 for a time interval $1/\varepsilon$. The initial conditions for captured trajectories tend to fill the phase space densely, as $\varepsilon \to 0$. So, it is expedient to consider capture into resonance as a random event and to estimate its probability. At first these phenomena were studied in connection with problems of celestial mechanics [17, 22].

To describe both the capture phenomenon and the behaviour of the captured trajectories, we have to make some additional constructions. It can be shown that for a fixed resonance $k_1\omega_1 + k_2\omega_2 = 0$ variation of the resonant phase $\gamma = k_1\varphi_1 + k_2\varphi_2$ close to the resonant surface is described by the following equations:

$$\gamma'' = -\partial V(\gamma, \sigma)/\partial\gamma + L(\sigma) + O(\sqrt{\varepsilon}), \qquad \gamma' = O(\sqrt{\varepsilon}). \tag{9}$$

Here $\sigma \in \mathbb{R}^{n-1}$ are coordinates of the phase point projection onto the resonant surface, and prime denotes the derivative with respect to $\tau = \sqrt{\varepsilon}t$ [5]. This reduction of the problem was used in [14, 22–24]. Putting in (9) $\varepsilon = 0$ we obtain the pendulum-like system (shortly, the pendulum) describing the one-dimensional motion in a periodic potential under constant torque L, where $L \neq 0$ because of condition $\bar{A}$:

$$\gamma'' = -\partial V(\gamma, \sigma)/\partial\gamma + L(\sigma), \qquad \sigma = \text{const}. \tag{10}$$

Two possible types of this pendulum's phase portraits are shown in Fig. 1a, b. In Fig. 1a the capture into resonance is impossible. In Fig. 1b there is a domain of oscillation corresponding to phase points captured into resonance. Under the influence of terms $O(\sqrt{\varepsilon})$ in (9) the phase point can cross the separatrix in Fig. 1b and transit from the rotational domain to the oscillational one. This is a capture into resonance. The backward transition is also possible. This is an escape from resonance.

The oscillational domain of the portrait exists for a finite number of resonances only (because as the order of resonance $|k| = |k_1| + |k_2|$ grows, the purely periodical term $\partial V/\partial\gamma$ in (10) tends to 0 and the torque constant $|L|$ is separated from 0); such resonances are said to be strong. For every oscillational domain of a strong resonance the capture probability can be calculated [31] (this probability can be equal to 0 as well). To describe approximately the motion of captured points within a

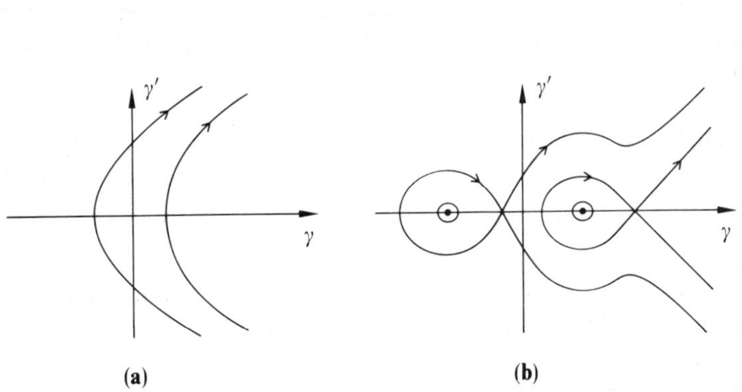

Fig. 1

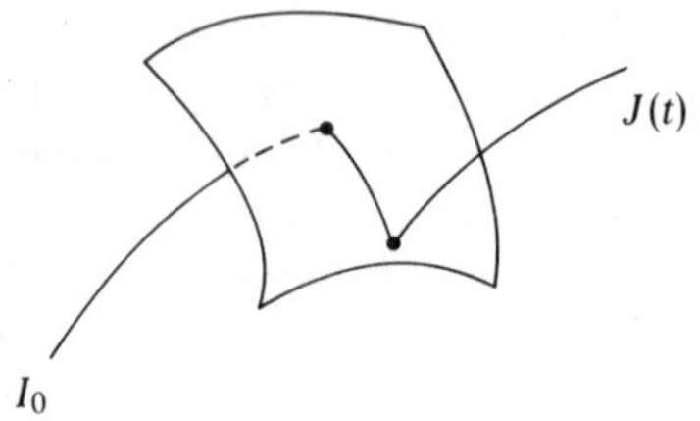

Fig. 2

given oscillational domain we average the velocities of changing of σ and that of the pendulum's energy h over unperturbed oscillations. So we obtain the set of equations for γ, h, which can be called the internal averaged system for a given oscillational domain of the given resonance. Solving this system one can determine the moment of escape from the resonance if the moment of the capture is known (as the saying is, one can construct the in-out function). To describe approximately trajectories with capture into resonance we glue solutions of the averaged system (3) and the σ-component of solutions of the internal averaged system as it is shown in Fig. 2. It turns out that a finite number of trajectories of the internal averaged system is glued to a trajectory of the averaged system; each of them corresponds to a capture into some oscillational domain of one of strong resonances. It can be shown that under some conditions (satisfied almost always) for the majority of initial conditions in the set U_c the behaviour of $I(t)$ can be described by one of the trajectories glued in this way with an accuracy $O(\sqrt{\varepsilon}\ln\varepsilon)$. The only exception is a subset which measure tends to 0 as $\varepsilon \to 0$ faster than any given power of ε.

The condition B of Theorem 4 is that unstable singular points of the phase portraits are nondegenerate. If the condition $\bar{A}$ is satisfied (and B is not), outside a set of initial conditions with measure $\kappa \geq c_2\sqrt{\varepsilon}$ the following unimprovable estimate is valid

$$|I(t) - J(t)| < c_3\sqrt{\varepsilon}/\sqrt{\kappa}, \qquad 0 \leq t \leq 1/\varepsilon$$

(the case $n = 1$ is considered in [24], $n \geq 2$, in [34]).

The estimates in the case where the condition $\bar{A}$ is not satisfied seem to be unknown. Yet the model problem of estimates has been entirely solved in the case of a single resonance only or, which appears to be the same, for a one-frequency system with the vanishing frequency.

Theorem 5 (V.I. Bakhtin [8, 9]). *In a generic one-frequency ($m = 1$) system outside a set of initial conditions of measure $\kappa \geq c_2\sqrt{\varepsilon}$ the following estimate holds*

$$|I(t) - J(t)| < c_3\sqrt{\varepsilon}/\sqrt{\kappa}, \qquad 0 \leq t \leq 1/\varepsilon.$$

If $n \geq 2$ this estimate is unimprovable. The genericity conditions are imposed on the functions ω, $G(I)$ near the surface $\{\omega = 0\}$ and on the function $\partial\omega/\partial I \cdot g(I, \varphi, 0)$ considered at $I \in \{\omega = 0\}$.

In [35] the procedure is proposed, which allows to describe the evolution of slow variables with accuracy better than $\sqrt{\varepsilon}$, for the case of a single resonance.

Captures into resonance can play the key role in the system's motion for time interval $\sim 1/\varepsilon^{3/2}$. Let the averaged system have a periodic solution crossing the resonant surface. Then for time $1/\varepsilon^{3/2}$ phase points belonging to a set of measure of order 1 will be captured (not to pay attention to changing of the phase volume we deal with systems preserving phase volume). It seems reasonable, that quasi-random captures into resonance and subsequent escapes from resonance can give rise to chaotic dynamics. For examples of such systems see [33, 40].

These phenomena result in destruction of adiabatic invariants in multi-frequency systems (the problem which can be traced back to the paper by P.A.M. Dirac [15]). Let a Hamiltonian system depending on the parameter λ be completely integrable and possess action-angle variables I, φ and frequencies $\omega = \omega(I, \lambda)$ for each fixed λ. Let λ be a slow periodic function of time: $\lambda = \lambda(\varepsilon t)$. Changing of I, φ is described by a system of the form (2), and the averaged system is of the form $\dot{J} = 0$, so in the averaging method approximation $I = \text{const}$. In one-frequency systems, according to the theorem by V.I. Arnold [2], the action I remains eternally near its initial value or as the saying goes, appears to be a perpetual adiabatic invariant (provided that some nondegeneracy conditions are satisfied). In [32] an example of a two-frequency system is constructed, where due to captures into resonance I changes by a quantity of order 1 for a set of initial conditions of measure of order 1 and for time interval $1/\varepsilon^{3/2}$. In systems passing through resonance without being captured (for example, under the condition A) adiabatic invariants seem to be also destructed, though slowly, for times of order $1/\varepsilon^2$.

The approach based on the analysis of joint statistical properties of different passages seems to be fruitful to study multiple passages through resonances. To the author's knowledge, there are no strict results obtained on this way yet.

4. Passage Through a Separatrix

Let us consider a slow-fast system (1) such that the corresponding fast system is a Hamiltonian system with one degree of freedom (on the plane for simplicity). In other words, we consider a system of the form

$$\dot{p} = -\partial E/\partial q + \varepsilon f_1, \qquad \dot{q} = \partial E/\partial p + \varepsilon f_2, \qquad \dot{z} = \varepsilon f_3, \tag{11}$$

$$E = E(p, q, z), \qquad f_i = f_i(p, q, z, \varepsilon), \qquad 0 < \varepsilon \ll 1, \qquad (p, q) \in \mathbb{R}^2, \qquad z \in \mathbb{R}^m$$

with fast variables p, q and slow variables z. Let us assume that for all values of z the phase portrait of the fast system is divided into regions G_i by separatrices l_i of a saddle point C (as, for example, in Fig. 3). A perturbation causes the evolution, and the projection of a phase point onto the (p, q)-plane may cross the separatrix of the fast system. Far from separatrices the system (11) can be transformed into the form of the system with one rotating phase and a nonzero frequency. To achieve this we have to choose as new variables E, z and the phase of the unperturbed motion. When phase point approaches the separatrix, the frequency tends to zero. Separatrix may be considered as a special type of resonance.

Passage through a separatrix leads to probability phenomena discovered by I.M. Lifshitz, A.A. Slutskin and V.M. Nabutovskii [19]. Phase points, being initially

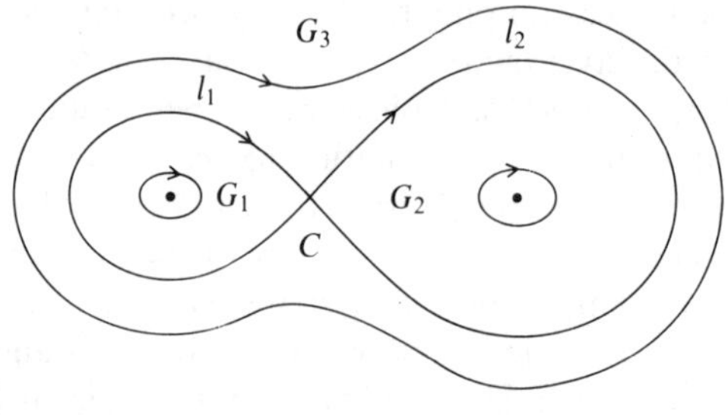

Fig. 3

at the distance of order ε from each other, can be captured after separatrix crossing into different regions, and their motions after crossing will be entirely different. Since initial conditions are known always with some finite accuracy, the deterministic approach to the problem fails when $\varepsilon \to 0$. But it is possible to consider the captures into different regions as random events and calculate their probabilities.

Let $(p_0, q_0) \in G_3$ for $z = z_0$. Let U_δ be the δ-ball around $M_0 = (p_0, q_0, z_0)$ in $\mathbb{R}^{n+2}$. Let $U^{(i)}_{\delta,\varepsilon}$ be the subset of U_δ containing initial points, which will be captured into G_i, $i = 1, 2$. By definition due to V.I. Arnold [2], the probability of capture into G_i, $i = 1, 2$, of a point M_0 is

$$Q_i(M_0) = \lim_{\delta\to 0} \lim_{\varepsilon\to 0} \frac{\text{mes } U^{(i)}_{\delta,\varepsilon}}{\text{mes } U_\delta}$$

(of course, it must be proved that this limit exists).

Let Hamiltonian $E = 0$ at the saddle point C and, therefore, at the separatrices. We will consider the problem under the assumption that the following quantities are different from zero:

$$\Theta_i(z) = -\oint_{l_i} \left(\frac{\partial E}{\partial p} f_1^0 + \frac{\partial E}{\partial q} f_2^0 + \frac{\partial E}{\partial z} f_3^0 \right) dt, \qquad i = 1, 2,$$

$$\Theta_3(z) = \Theta_1(z) + \Theta_2(z),$$

the integrals being calculated along the unperturbed separatrices, $f_j^0 = f_j(p, q, z, 0)$. These integrals are improper, because the motion along a separatrix takes infinite time. Our normalization of E ensures the convergence of the integrals. The value $(-\varepsilon\Theta_j)$ approximates the change of energy E along the connected part of the perturbed trajectory, which lies near the unperturbed separatrix l_j. The condition $\Theta_j \neq 0$ ensures sufficiently fast passage through the separatrix. Let, for certainty, $\Theta_j > 0$ and $E > 0$ into G_3. In this case phase points from G_3 will be captured either into G_1 or into G_2.

For an approximate description of the evolution we can glue at the separatrices the solutions of the averaged system from different regions. A solution terminating on the separatrix in region G_3 must be glued with solutions beginning at the separatrix in the regions G_1 and G_2 (in Fig. 4 the behaviour of E for such glued solutions is shown). As is shown in [26], for the majority of initial conditions one

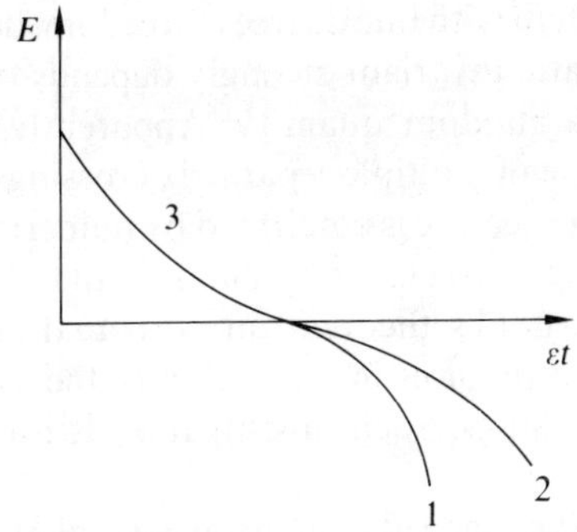

Fig. 4

of these glued solutions describes the evolution of E, z with accuracy $O(\varepsilon \ln \varepsilon)$ over time interval of order $1/\varepsilon$ (an unimprovable estimate). The measure of the set of "bad" initial points, for which such estimate of accuracy is not valid, tends to zero faster than ε^r for every given $r \geq 1$. The probability of capture into G_i is calculated by the formula

$$Q_i(M_0) = \frac{\Theta_i(z_*)}{\Theta_1(z_*) + \Theta_2(z_*)}, \qquad i = 1, 2 \tag{12}$$

where z_* is the value of variable z at the moment of separatrix crossing, calculated by the averaging method. It is easy to understand this formula for probability: the probability of capture into G_i is equal to ratio of the flux of the phase points in G_i to the whole flux in $G_1 \cup G_2$ for $|z - z_*| < \delta$ in the limit, when $\varepsilon \to 0$ and then $\delta \to 0$. Various particular cases of formula (12) were considered in [2, 10, 17, 25].

Analogous phenomena take place and analogous description of the evolution is applicable in the following more general situation. The unperturbed system of differential equations in $\mathbb{R}^{n+2}$ possesses $n+1$ first integrals. A joint level of n integrals is a smooth 2-dimensional manifold. On this manifold the picture of the level lines of the $(n+1)$-th integral contains saddle points and separatrices. A perturbation causes the evolution and a phase point crosses these separatrices.

Passages through separatrices destroy adiabatic invariants. Let the system (11) be a Hamiltonian system, depending periodically on slow time $z = \varepsilon t$, $f_{1,2} = 0$. The averaged system possesses an integral (adiabatic invariant) in each region G_i. This integral is "the action" of the fast system $I = I(E, z)$ [4]. In the case of separatrix crossing the value of I along a trajectory changes. Through the period of the slow time this value in general does not return to its initial value even in the averaging method approximation (which is called here the adiabatic approximation). This is an important mechanism of the transport in the phase space of such systems. In addition to this "big" (~ 1) change of I there exists the "small" ($\sim \varepsilon\,|\ln \varepsilon|$) change, caused by the difference between exact and averaged solutions. The change of the adiabatic invariant due to the separatrix crossing was calculated through order ε first in [38] for the specific case of the simple pendulum in a slowly varying gravity field (the problem considered by P. Ehrenfest) and then in [12, 28] in the general case of a Hamiltonian system with one degree of freedom, depending slowly on time. The change of the adiabatic invariant strongly depends on initial conditions and must be considered as a random quantity. Apparently, accumulation of these random changes in the case of multiple separatrix crossings leads to diffusion of the adiabatic invariant (this conjecture is confirmed by numerical calculations for many cases). In absence of the separatrix crossing diffusion of the adiabatic invariant does not exist according to Arnold's theorem on perpetual adiabatic invariance [2]. Study of these phenomena requires investigation of the joint statistical properties of different separatrix crossings. Such investigation is now only at the beginning [13].

Similar phenomena take place also in more general Hamiltonian systems with fast and slow motions (the Hamiltonian is $E = E(p, q, z_1, z_2)$, where (p, q) and $(z_1, \varepsilon^{-1}z_2)$ are pairs of conjugated canonical variables). These phenomena, for example, play an important role in J. Wisdom's theory of the origin of some gaps in the asteroid belt [39, 29].

5. Delay of Stability Loss

Let the fast equation of a slow-fast system (1) have an equilibrium position or a limit cycle. Let us suppose that drift of the slow variables leads to a dynamical bifurcation: the equilibrium position (cycle) loses its stability but remains non-degenerate. For an equilibrium a pair of conjugated eigenvalues leaves the left half-plane without passing through point 0, for a cycle multipliers leaves the unit circle without passing through point 1. In analytic systems the stability loss is inevitably delayed [30]: the phase points which moved near the stable equilibrium (cycle) for a time interval of order $1/\varepsilon$ before the moment of bifurcation, remain near the unstable equilibrium (cycle) for a time of order $1/\varepsilon$ after the bifurcation. For this time slow variables change by quantity of order 1. Such delay of stability loss is not in general found in nonanalytic systems: receding from the unstable equilibrium takes place near the bifurcation value of the slow variables.

The delay of stability loss for an equilibrium was first discovered by L.S. Pontrjagin and M.A. Shishkova [36] for some model equation system. For this system the asymptotic behaviour of the time of delay was calculated using analytic continuation of solutions in the plane of complex time. The existence of delay itself may be also derived from an earlier statement of Y. Sibuya [37].

The delay of the stability loss for an equilibrium is shown graphically in Fig. 5. It is assumed that the fast motion loses its stability softly, i.e. for $\tau = \tau_* + 0$ the stable limit cycle is emanating from the equilibrium, and the size of the cycle (the amplitude of selfoscillations) increases like $\sqrt{\tau - \tau_*}$ (here $\tau = \varepsilon t$). In the precise slow-fast system receding from this equilibrium and transition to the cycle occurs for $\tau > \tau_* + \text{const}$, when the cycle is already of size of order 1. The oscillation amplitude increases with a jump (for a time t of order $\ln \varepsilon$) from the order ε to order 1, i.e. the stability loss appears to be produced abruptly.

This phenomenon is in a somewhat unexpected way connected with averaging and passage through resonances. It is convenient to explain this connection using

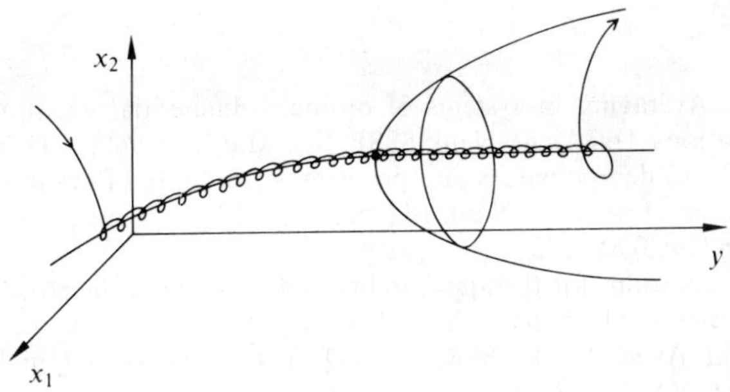

Fig. 5

an example. Let us consider the slow-fast system which is written in a complex form:

$$\dot{z} = (\tau - i)(z - \tau), \qquad z = x_1 + ix_2, \qquad \tau = \varepsilon t \tag{13}$$

(the equation for the slow variable τ has been already integrated, the example in [36] differs from (13) in nonlinear terms). The fast equation has an equilibrium position $z = \tau$, which is stable for $\tau < 0$ and is unstable for $\tau > 0$. Let us introduce the function $\Psi = (\tau - i)^2/2$ and consider, following [36], level lines $\operatorname{Re} \Psi = \text{const}$ in the plane of the complex slow time τ (Fig. 6). If time t is considered along the path $\operatorname{Re} \Psi = \text{const}$, then the equilibrium has purely imaginary eigenvalues. In the polar coordinates $z = \tau + \varrho e^{i\varphi}$ the system with the rotating phase φ is obtained. The value ϱ is an integral of the averaged system and an adiabatic invariant of the precise system. At the level lines $\operatorname{Re} \Psi = 0$, connecting the points $-1, i, +1$ (Fig. 6) the resonance occurs: the frequency of φ is equal to zero at the point i. The passage through this resonance changes the adiabatic invariant by the value of order $\sqrt{\varepsilon}$. So it follows that the phase points which were attracted to the equilibrium at the moment of time $\tau < -1$, will be ejected at $\tau \simeq +1$. The phase points, which were attracted at $\tau = \tau_0 \in [-1, 0]$, will be ejected at $\tau \simeq |\tau_0|$.

Analogous phenomena lead to the delay of stability loss of equilibria in the general case. For the motion along some paths in the plane of complex time the system has an adiabatic invariant, which prevents the departure from the equilibrium position, and some critical path with passage through resonance determines the maximal time of delay.

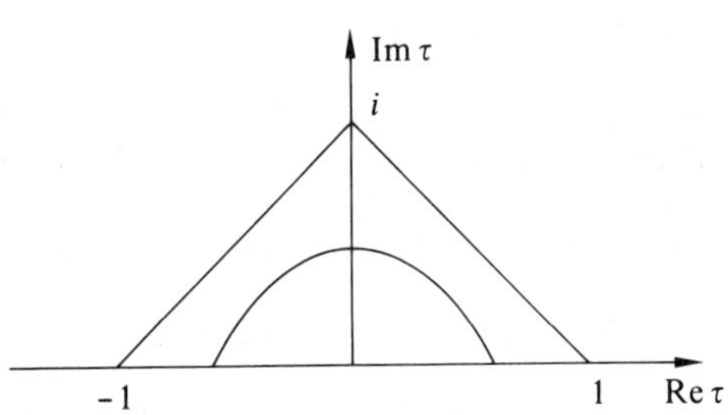

Fig. 6

References

1. Anosov, D.V.: Averaging in systems of ordinary differential equations with rapidly oscillating solutions. Izv. Akad. Nauk SSSR, Ser. Mat. **24** (1960) 721–742 (Russian)
2. Arnold, V.I.: Small denominators and problems of stability of motion in classical and celestial mechanics. Usp. Mat. Nauk **18** (1963) 91–192 (Russian). [English transl.: Russ. Math. Surv. **18** (1963) 85–192]
3. Arnold, V.I.: Conditions for the applicability and estimate of the error of an averaging method for systems which pass through states of resonance in the course of their evolution. Dokl. Akad. Nauk SSSR 161, 9–12 (1965) (Russian). [English transl.: Sov. Math. Dokl. **6** (1965) 331–334]
4. Arnold, V.I.: Mathematical methods of classical mechanics. Nauka, Moscow 1974 (Russian). [English transl.: Springer, New York 1978]
5. Arnold, V.I.: Geometrical methods in the theory of ordinary differential equations. Nauka, Moscow 1978 (Russian). [English transl.: Springer, New York, 1983]
6. Arnold, V.I., Kozlov, V.V., Neishtadt, A.I.: Mathematical aspects of classical and celestial mechanics (Contemporary problems of mathematics. Dynamical systems-3). VINITI, Moscow 1985 (Russian). [English transl.: Encyclopaedia of Mathematical Sciences, vol. 3. Springer, Berlin 1988]
7. Bakhtin, V.I.: Averaging in multifrequency systems. Funkts. Anal. Prilozh. **20** (1986) 1–7 (Russian). [English transl.: Funct. Anal. Appl. **20** (1986) 83–88]
8. Bakhtin, V.I.: On averaging method in multifrequency systems. Ph.D. Thesis. Moscow State Univ., Moscow, 1986 (Russian)
9. Bakhtin, V.I.: Averaging in generic one-frequency system. Differ. Uravn. (to appear) (Russian)
10. Bakai, A.S.: On coefficient of capture of particles into accelerator. Atomnaya energiya **21** (1966) 503–504 (Russian)
11. Bogolyubov, N.N., Mitropol'skii Yu. A.: Asymptotic methods in the theory of nonlinear oscillations. 2nd ed. Nauka, Moscow 1958 (Russian) [English transl.: Gordon and Breach, New York 1961]
12. Cary, J.R., Escande, D.F., Tennyson, J.L.: Adiabatic invariant change due to separatrix crossing. Phys. Rev. A **34** (1986) 4526–4275
13. Cary J.R., Scodje, R.T.: Phase change between separatrix crossings. Physica D **36** (1989) 287–316
14. Chirikov, B.V.: Passage of a nonlinear oscillatory system through resonance. Dokl. Akad. Nauk SSSR **125** (1959) 1015–1018 (Russian). [English transl.: Sov. Phys. Dokl. **4** (1959) 390–394]
15. Dirac, P.A.M.: The adiabatic invariance of the quantum integral. Proc. R. Soc. A **107** (1925) 725–734
16. Dodson, M.M., Rynne, B.P., Vickers, J.A.G.: Averaging in multifrequency systems. Nonlinearity **2** (1989) 137–148
17. Goldreich, P., Peale, S.: Spin-orbit coupling in the Solar system. Astron. J. **71** (1966) 425–438
18. Kasuga, T.: On the adiabatic theorem for the Hamiltonian system of differential equations in classical mechanics. I, II, III. Proc. Japan. Acad. **37** (1961) 366–382
19. Lifshitz, I.M., Slutskin, A.A., Nabutovskii, V.M.: On phenomenon of scattering of charged quasiparticles at singular points in *p*-space. Dokl. Akad. Nauk SSSR **137** (1961) 553–556 (Russian)
20. Lochak, P., Meunier, C.: Multiphase averaging for classical systems. Springer, New York 1988
21. Mitropol'skii, Yu.A.: Averaging method in nonlinear mechanics. Naukova Dumka, Kiev 1971 (Russian)

22. Moltchanov, A.M.: The resonant structure of the Solar system. Icarus **8** (1968) 203–215
23. Morozov, A.D.: A complete qualitative investigation of Duffing's equation. Differ. Uravn. **12** (1976) 241–255 (Russian) [English transl.: Differ. Equations **12** (1976) 164–174]
24. Neishtadt, A.I.: Passage through a resonance in the two-frequency problem. Dokl. Akad. Nauk SSSR **221** (1975) 301–374 (Russian). [English transl.: Sov. Phys. Dokl. **20** (1975) 189–191]
25. Neishtadt, A.I.: Passage through a separatrix in a resonance problem with a slowly varying parameter. Prikl. Mat. Mekh. **39** (1975) 621–632 (Russian) [English transl.: J. Appl. Math. Mech. **39** (1975) 594–605]
26. Neishtadt, A.I.: Some resonance problems in nonlinear systems. Ph.D. Thesis. Moscow State Univ., Moscow 1975 (Russian)
27. Neishtadt, A.I.: Averaging in multifrequency systems. II. Dokl. Akad. Nauk SSSR **226** (1976) 1295–1298 (Russian). [English transl.: Sov. Phys., Dokl. **21** (1976) 80–82]
28. Neishtadt, A.I.: Change of an adiabatic invariant at a separatrix. Fiz. Plazmy **12** (1986) 992–1001 (Russian). [English transl.: Sov. J. Plasma Phys. **12** (1986) 568–573]
29. Neishtadt, A.I.: Change of an adiabatic invariant at a separatrix in the systems with two degrees of freedom. Prikl. Mat. Mekh. **51** (1987) 750–757 (Russian). [English translation: PMM USSR **51** (1987) 586–592]
30. Neishtadt, A.I.: Persistence of stability loss for dynamical bifurcations. I, II. Differ. Uravn. **23** (1987) 2060–2067; **24** (1988) 226–233 (Russian). [English transl.: Differ. Equations. **23** (1987) 1385–1390; **24** (1988) 171–176]
31. Neishtadt, A.I.: Averaging, capture into resonances and chaos in nonlinear systems. In.: Campbell, D. (ed.) Chaos. AIP, New York 1990, pp. 261–275
32. Neishtadt, A.I.: Probability phenomena in perturbed systems. In.: Bazikin A., Zarkhin, Yu. (eds.) Mathematics and modelling. Puschino, 1990, pp. 141–155 (Russian)
33. Neishtadt, A.I., Petrovichev, B.A., Chernikov, A.A.: On capture of the particles into regime of unlimited acceleration. Fiz. Plazmy **15** (1989) 1026–1029 (Russian). [English translation: Sov. J. Plasma Phys. **15** (1989) 593–594]
34. Pronchatov, V.E.: On estimate of the error of the averaging method in two-frequency problem. Mat. Sb. **134** (1987) 28–41 (Russian)
35. Sanders, J.: On the passage through resonance. SIAM J. Math. Anal. **10** (1979) 1220–1243
36. Shishkova, M.A.: Examination of one system of differential equations with a small parameter in the highest derivatives. Dokl. Akad. Nauk SSSR **209** (1973) 576–579 (Russian)
37. Sibuya, Y.: Sur reduction analytique d'un systeme d'equations differentiales ordinares lineaires content un parametre. J. Univ. Tokio, sect. 1, **10** (1958) 527–540
38. Timofeev, A.V.: On the constancy of an adiabatic invariant when the nature of the motion changes. Zh. Eksp. Teor. Fiz. **75** (1978) 1303–1306 (Russian). [English translation: Sov. Phys. JETP **48** (1978) 656–659]
39. Widsom, J.: A perturbative treatment of motion near the 3/1 commensurability. Icarus **63** (1985) 272–289
40. Zaslavsky, G.M., Neishtadt, A.I., Petrovichev, B.A., Sagdeev, R.Z.: Intensification of diffusion due to interaction wave-particle in a weak magnetic field. Fiz. Plazmy **15** (1989) 638–641 (Russian). [English translation: Sov. J. Plasma Phys. **15** (1989) 368–370]

Entropy in Smooth Dynamical Systems

Sheldon E. Newhouse

Department of Mathematics, University of North Carolina
Chapel Hill, NC 27599-3250, USA

1. Introduction

The topological entropy of a continuous dynamical system is now well established as an important invariant of the system. It was first defined by Adler, Konheim, and McAndrew [AKM] in 1965 using open coverings [1]. Nowadays it is convenient to think of the topological entropy as a limit of the number of the distinct orbits of a given length which can be obtained with a fixed small precision. Thus, it measures the *orbit growth* of the system.

More precisely, suppose that $f : M \to M$ is a continuous self-mapping of the compact metric space M with metric d. Given a positive integer n, and a small real number $\delta > 0$, we say that a set E is an (n, δ)-separated set if, for any $x \neq y \in E$ there is a $j \in [0, n)$ such that $d(f^j x, f^j y) > \delta$. Let $r(n, \delta, f)$ be the maximum cardinality of an (n, δ)-separated set in M. Let

$$h(\delta, f) = \limsup_{n\to\infty} \frac{1}{n} \log r(n, \delta, f)$$

and

$$h(f) = \lim_{\delta\to 0} h(\delta, f) = \sup_{\delta>0} h(\delta, f) .$$

This is the topological entropy of f. The most interesting properties of $h(f)$ arise from its relation to set of invariant probability measures of f. Let us denote this set by $\mathcal{M}(f)$, and recall that it is a compact metrizable set. Given $\mu \in \mathcal{M}(f)$, the measure-theoretic or metric entropy, $h_\mu(f)$, is defined as follows. For a finite Borel measurable partition α, let

$$H_\mu(\alpha) = -\sum_{A\in\alpha} \mu(A) \log \mu(A) .$$

Given two finite partitions α, β, set $\alpha \vee \beta = \{A \bigcap B : A \in \alpha, B \in \beta\}$. Then, set

$$h_\mu(\alpha, f) = \lim \frac{1}{n} H_\mu(\bigvee_{i=0}^{n-1} f^{-i}(\alpha))$$

[1] Adler recently pointed out to us that topological entropy generalizes Shannon's notion of *channel capacity* (see [SW, p.7])

Proceedings of the International Congress
of Mathematicians, Kyoto, Japan, 1990

and

$$h_\mu(f) = \sup_\alpha h_\mu(\alpha, f) = \lim_{\text{diam } \alpha \to 0} h_\mu(\alpha, f).$$

The following properties hold:

1. $h(f^n) = nh(f), \quad h_\mu(f^n) = nh_\mu(f)$ for $n \in \mathbf{Z}^+$
2. $h(f^t) = |\, t\,| h(f^1)$ if $\{f^t\}_{t\in\mathbf{R}}$ is a continuous flow.
3. $h(f) = \sup_{\mu \in \mathcal{M}(f)} h_\mu(f)$
4. $h(f) = h(g)$ if f is topologically conjugate to g; i.e., there is a homeomorphism ϕ with $\phi f \phi^{-1} = g$.
5. $h(f)$ can be computed using (n, δ)—separated subsets of sets of large measure for various $\mu \in \mathcal{M}(f)$(see [N1]). That is,

$$h(f) = \sup_{\mu \in \mathcal{M}(f)} \lim_{\sigma \to 1} \inf_{\mu(\Lambda) > \sigma} \lim_{\delta \to 0} \bar{r}(\delta, \Lambda)$$

where

$$\bar{r}(\delta, \Lambda) = \limsup_{n \to \infty} \frac{1}{n} \log \bar{r}(\delta, n, \Lambda)$$

and $\bar{r}(\delta, n, \Lambda)$ is the maximal cardinality of an (n, δ)-separated subset of Λ.

The relationship between topological and metric entropies (statement 3 above) is a combination of the work of Goodman, Goodwyn, and Dinaburg. It is referred to as the Variational Principle for Topological Entropy. An elegant proof has been given by Misiurewicz (see [DGS], pp. 140–146 for a more general result). Statement 4 above states that $h(f)$ is a topological conjugacy invariant of f. It is generally not a complete invariant of topological conjugacy. However, for certain important systems it is close to being complete. For instance, Adler and Marcus [AM] have shown that two mixing subshifts of finite type with the same topological entropy are almost topologically conjugate in the sense that each is a boundedly finite to one factor of a third mixing subshift of finite type. Also, Milnor and Thurston [MT] have shown that a piecewise monotone continuous map f of an interval with positive topological entropy is semi-conjugate in a simple way to a piecewise linear map g with the same number of turning points as f and such that the slope of each monotone piece of g has absolute value equal to $e^{h(f)}$.

It was known quite early that some form of regularity affected the finiteness of the topological entropy. Embedding a sequence of larger and larger shift automorphims topologically in a homeomorphism of the two-sphere shows that homeomorphisms need not have finite topological entropy. However, Kushnirenko proved that every C^1 self-map of a compact manifold has finite topological entropy. A natural question arises: When are there measures μ for which $h_\mu(f) = h(f)$?. Misiurewicz was the first to construct examples of C^k diffeomorphisms of compact manifolds with no measure of maximal entropy. His first examples were on manifolds of dimension greater than three, but now it is known that such examples arise in small perturbations of diffeomorphisms having a degenerate homoclinic orbit (an intersection of stable and unstable manifolds of a hyperbolic saddle point with infinite order contact) even on the two dimensional sphere. His construction worked for every finite k, but it failed in the C^∞ case. The question of the existence of measures of maximal entropy was quite important. One consequence of the results described here is that for every C^∞ self-map f of a compact smooth manifold, the function $\mu \to h_\mu(f)$ is uppersemicontinuous on $\mathcal{M}(f)$. Hence, f does indeed possess measures of maximal entropy.

In this article we shall survey several recent results about topological and metric entropy, particularly as they relate to smooth systems. We view the results described here as part of the natural evolution of certain *topological* aspects of the qualitative theory of dynamical systems following the rich development in the sixties due, in large part, to Anosov, Sinai, and Smale. The recent developments on quadratic mappings due to Carleson and Benedicts may be viewed as part of the evolution of certain *quantitative* aspects of this theory. In the case of uniformly hyperbolic dynamics, these two types of aspects come together beautifully in the theory of *equilibrium states* as described, for example, in [B2]. It is natural to search for a generalization of the Equilibrium State Theory which encompasses all of these results.

2. Entropy and Volume

To motivate the results, we first consider the case of mappings of the interval. Let M be the unit interval and let $f : M \to M$ be a continuous map with finitely many turning points. Let $\log^+(x) = \max(\log x, 0)$.

Let $\ell(f)$ = length of image of f with multiplicities:

$$\ell(f) = \int_M | f'(x) | \, dx$$

Theorem 1 (Misiurewicz-Szlenk [MS]). *With f as above, the following results hold.*

1. $h(f) = \lim_{n\to\infty} \frac{1}{n} \log^+ \ell(f^n)$.
2. *$\mu \to h_\mu(f)$ is uppersemicontinuous on $\mathcal{M}(f)$.*
3. *$f \to h(f)$ is uppersemicontinuous for f C^1 where one perturbs in the C^1 topology keeping the same number of turning points.*
4. *$f \to h(f)$ is lowersemicontinuous for certain f.*

Theorem 2 (Misiurewicz [M2]). *The map $f \to h(f)$ is lowersemicontinuous for all C^0 f.*

Corollary 3. *The map $f \to h(f)$ is continuous for f in the set of C^1 maps with a uniformly bounded number of turning points.*

This last corollary was also proved by Milnor and Thurston [MT] .

We consider a direct generalization of the preceding results. A clue as to how to proceed comes from work of Margulis in the sixties on the geodesic flow on compact negatively curved manifolds. Following his work, it became known that the topological entropy of the time-one map equals the maximum volume growth rate of compact disks in the unstable manifolds.

Let D^k be the unit k-disk in $\mathbf{R}^k$, and let M be a C^∞ manifold. Let $C^r(M,M)$ be the space of C^r self maps of M , and let $\mathscr{D}^r(M,M)$ be the space of C^r diffeomorphisms of M where r is an integer greater than 1.

A C^r k-disk in M is a C^r map $\gamma : D^k \to M$.
Define the k-volume of γ by

$$| \gamma |_k = \int_{D^k} | \Lambda^k T\gamma | d\lambda$$

where $d\lambda$ is Lebesgue measure on D^k, and $\Lambda^k T\gamma$ is the k—th exterior power of the derivative $T\gamma$

For $f \in C^r(M,M)$, let Λ be a compact f—invariant set, and let U be a compact neighborhood of Λ. Given a positive integer n, set $W^s(n,U) = \bigcap_{0\leq j<n} f^{-j}(U)$. This is just the set of points whose iterates from time 0 through $n-1$ remain in U. For a k-disk γ in U, set

$$G_k(\gamma, f, U) = \limsup_{n\to\infty} \frac{1}{n} \log^+(| f^{n-1} \circ \gamma \mid \gamma^{-1}(W^s(n,U)) |_k) .$$

This is the volume growth of the f^{n-1}-st iterate of the part of γ which remains in U from time 0 through time $n-1$.

A collection $\mathscr{A}$ of k-disks in U with $1 \leq k \leq$ dim M, is called *ample* for Λ if it contains a subcollection $\mathscr{A}_1$ for which $\exists K > 0$ such that

1. $K^{-1} \leq | D_x\gamma(v) | \leq K$ and $| D_x^2\gamma(v,v) | \leq K$
$\forall x \in$ *domain* $\gamma, | v | = 1$ and $\gamma \in \mathscr{A}_1$

2. For $x \in \Lambda$, and for each k-dimensional subspace H of T_xM, there exists a sequence of k-disks $\gamma_1, \gamma_2, \ldots, \in \mathscr{A}_1$ whose tangent spaces at $\gamma_i(0)$ approach H in the Grassmann sense as $i \to \infty$.

If M is complex analytic with a hermitian metric, Λ and U are as above, and $\mathscr{A}$ is a collection of holomorphic disks in U, then $\mathscr{A}$ is *holomorphically ample* or *h-ample* if, in condition 2 above, H is assumed to be a complex subspace of the holomorphic tangent space of T_xM.

Clearly the family of all disks through points in Λ is ample for Λ. If $M = \mathbf{R}^N$, with the usual metric, then the collection $\mathscr{A}$ of affine disks through points in Λ is ample.

Theorem 4 gives an upper bound for entropy in terms of volume growth rates of smooth disks. An earlier upper bound in terms of the average (relative to Lebesgue measure) of the maximum growth of the norms of the exterior powers of the derivative had been obtained by Przytycki [P] for diffeomorphisms.

Theorem 4 [N1]. *Suppose $f \in C^r(M,M)$, Λ is a compact invariant set, U is a compact neighborhood of Λ, and $\mathscr{A}$ is an ample family of C^r disks for Λ. If f and M are complex analytic assume that $\mathscr{A}$ is h-ample. Then,*

$$h(f \mid \Lambda) \leq \sup_{\gamma \in \mathscr{A}} G(\gamma, f, U).$$

If $f \in \mathscr{D}^r(M,M)$, then

$$h(f \mid \Lambda) \leq \sup_{\substack{\gamma \in \mathscr{A} \\ \dim \gamma < \dim M}} G(\gamma, f, U).$$

In particular, for $f \in \mathscr{D}^r(M^2)$, $h(f \mid \Lambda) \leq \sup_{\text{curves } \gamma} G(\gamma, f, U)$.

Using Theorem 4, one can give simple proofs of the following results.

Theorem 5. 1. *Let $f : \mathbf{R}^N \to \mathbf{R}^N$ be a polynomial map with coordinate functions of degree $\leq d$; i.e., for $x \in \mathbf{R}^N, f(x) = (f_1(x), f_2(x), \ldots, f_N(x))$ with $f_i(x)$ a polynomial in x of degree $\leq d$. If Λ is a compact invariant set for f, then*

$$h(f \mid \Lambda) \leq N \log d \quad (\textit{Gromov}).$$

2. *If $f : S^2 \to S^2$ is a rational map of degree d, then*

$$h(f) \leq \log d \quad (\textit{Gromov} - \textit{Ljubich}).$$

3. *If $f : \mathscr{P}^N(\mathbf{C}) \to \mathscr{P}^N(\mathbf{C})$ is globally defined and holomorphic, then $h(f) \leq$ $\log$(topological degree) (Gromov)*

S. Friedland [F] has obtained results on volume growth in quasi-projective varieties which generalize Theorem 5.1 .

It is well-known (Misiurewicz-Przytycki [MP]) that if M is compact and $f : M \to M$ is C^1, then $h(f) \geq \log(\text{topological degree})$.

So, Theorems 5.2 and 5.3 above are equalities.

From now on, we assume that M is a compact C^∞ manifold.

The above theorems give an upper estimate of $h(f)$ in terms of volume growth rates of disks. For the lower estimate we have

Theorem 6 (Yomdin [Y1]). *For $f \in C^\infty(M,M)$, and any C^∞ disk γ in M,*

$$h(f) \geq G(\gamma, f).$$

Yomdin's results can be used to give a proof of a generalization of the Shub entropy conjecture in the C^∞ case. To recall this conjecture, first define the *homology growth* of a map f, $HG(f)$, to be $\limsup_{n\to\infty} \frac{1}{n} \log |f_\star^n|$ where $f_\star : H_\star(M,\mathbf{R}) \to H_\star(M,\mathbf{R})$ is the induced map on the direct sum of the real homology groups of f (given any norm). The entropy conjecture states that for a C^1 diffeomorphism f of the compact manifold M, one has $h(f) \geq HG(f)$. Of course, $HG(f)$ is the same as the maximum logarithm of the absolute values of the eigenvalues of $f_\star : H_\star(M,\mathbf{R}) \to H_\star(M,\mathbf{R})$. Yomdin's results show that this holds for arbitrary C^∞ maps.

Corollary 7 (Yomdin) (C^∞ Entropy Conjecture). *If $f \in C^\infty(M,M)$, then $h(f) \geq HG(f)$.*

We note that the Entropy Conjecture fails in general for piece-wise linear homeomorphisms although it is true for "typical" piecewise linear maps. There is a large literature on various cases of the entropy conjecture (see [FS]). The general conjecture is still unproved for $f \in \mathscr{D}^r(M,M)$ with $1 \leq r < \infty$.

The next result states that, for positive $h(f)$, there always exist disks γ for which $G(\gamma, f)$ assumes the maximum value. In addition, it can be shown that there are such disks for which $G(\gamma, f)$ is actually a limit and not just a lim sup.

Theorem 8. *For $f \in C^\infty(M,M)$, $\mathscr{A}$, an ample family of C^∞ disks, we have*

$$h(f) = \sup_{\gamma \in \mathscr{A}} G(\gamma, f) = \max_{\gamma \in \mathscr{A}} G(\gamma, f).$$

In general, the disks γ for which $G(\gamma, f) = h(f)$ are not easily identifiable. However, for an area decreasing self-embedding of a surface with boundary, the entropy is just the growth rate of the length of the boundary.

Theorem 9 [N2]. *For $f \in D^\infty(M^2), \partial M^2 \neq \emptyset, f$ area decreasing, we have*

$$h(f) = G(\partial M^2, f).$$

3. Continuity Properties of Entropy

The methods used in the proofs of the above results concerning volume growth and entropy have local analogs by which we mean that one considers the growth rates of the cardinalities of (n,δ)—separated sets or of the volumes of disks which remain in small neighborhoods of the orbits of given points. These can be combined with general results estimating the defect in uppersemicontinuity of both topological and metric entropy to obtain various continuity properties of entropy for C^∞ systems.

We begin with a description of the so-called local entropy of a dynamical system $f : M \to M$.

Given $\Lambda \subset M, x \in \Lambda, n \in \mathbf{Z}^+, \varepsilon > 0$, set

$$W_x(n, \varepsilon, \Lambda) = \{y \in \Lambda : d(f^j y, f^j x) < \varepsilon \; \forall j \in [0, n)\},$$

$$r(n, \delta, \varepsilon, \Lambda) = \sup_{x \in \Lambda} \max\{card\ E : E \subset W_x(n, \varepsilon, \Lambda), E \text{ is } (n, \delta) - \text{separated}\},$$

$$r(\varepsilon, \Lambda) = \lim_{\delta \to 0} \limsup_{n} \frac{1}{n} \log r(n, \delta, \varepsilon, \Lambda).$$

If $\mu \in \mathscr{M}(f)$, let

$$h_{\mu \mathrm{loc}}(\varepsilon, f) = \lim_{\sigma \to 1} \inf_{\mu(\Lambda) > \sigma} r(\varepsilon, \Lambda),$$

and set,

$$h_{\mathrm{loc}}(\varepsilon, f) = \sup_{\mu} h_{\mu\,\mathrm{loc}}(\varepsilon, f).$$

The quantity $h_{\mathrm{loc}}(\varepsilon, f)$ is called the ε– local entropy of f, and $h_{\mu\mathrm{loc}}(\varepsilon, f)$ is called the ε–local entropy of f relative to μ.

The next theorem states that $h_{\mathrm{loc}}(\varepsilon, f)$ gives an upper bound for the difference $h(f) - h(\varepsilon, f)$ while $h_{\mu\mathrm{loc}}(\varepsilon, f)$ gives and upper bound for the difference $h_\mu(f) - h_\mu(\beta, f)$ for any partition β with diameter less than ε. Earlier estimates of these differences were given by Bowen in [B].

Theorem 10 [N2]. *For any continuous self map f of the compact metric space M, and $\varepsilon > 0$,*

1. $h(f) \leq h(\varepsilon, f) + h_{\mathrm{loc}}(\varepsilon, f)$
2. *If $\mu \in \mathcal{M}(f)$ and β is a finite Borel partition with diam $\beta < \varepsilon$, then* $h_\mu(f) \leq h_\mu(\beta, f) + h_{\mu\,\mathrm{loc}}(\varepsilon, f)$

Next we consider local volume growth.

Let $W_x(n, \varepsilon) = W_x(n, \varepsilon, M)$.

For a disk γ, set

$$G_{\mathrm{loc}}(\varepsilon, \gamma) = \limsup_{n\to\infty} \frac{1}{n} \log^+ \sup_{x\in M} | f^{n-1} \circ \gamma \mid \gamma^{-1}(W_x(n, \varepsilon)) |$$

and

$$G_{\mathrm{loc}}(\varepsilon, f) = \sup_{\gamma} G_{\mathrm{loc}}(\varepsilon, \gamma)$$

Theorem 11 [N2]. *For $f \in C^r(M, M)$, $r > 1$,*

$$h_{\mathrm{loc}}(\varepsilon, f) \leq G_{\mathrm{loc}}(2\varepsilon, f)$$

Theorem 12 (Yomdin[Y1]). *For $f \in C^\infty(M, M)$,*

$$\lim_{\varepsilon\to 0} G_{\mathrm{loc}}(\varepsilon, f) = 0$$

From Theorems 10, 11, 12 with some elementary arguments (see [N2]) we have

Theorem 13. 1. *For $f \in C^\infty(M, M), \mu \to h_\mu(f)$ is uppersemicontinuous*
2. *$f \to h(f)$ is uppersemicontinuous on $C^\infty(M, M)$.*

Yomdin [Y1] has an independent proof of Theorem 13.2.

In general, $f \to h(f)$ is not lowersemicontinuous, but Katok has proved that this does hold on surfaces.

Theorem 14 (Katok). *$f \to h(f)$ is lowersemicontinuous on $\mathcal{D}^2(M^2)$*

Corollary 15. *$f \to h(f)$ is continuous on $\mathcal{D}^\infty(M^2)$.*

Question 16. Is the preceding map Holder continuous?

Answer: No
Yomdin [Y2] has examples of curves $\{f_t\}$ of real analytic maps on S^2, $t \in [-\varepsilon, \varepsilon]$ with $h(f_t) = 0$ for $t \in [-\varepsilon, 0]$, and for $t \in (0, \varepsilon)$,

$$h(f_t) - h(f_0) > C\frac{\log|\log t|}{|\log t|}.$$

In view of Corollary 15, we wish to point out some analogies between the entropy map $f \to h(f)$ on $\mathscr{D}^\infty(M^2)$ and the rotation number map $f \to \varrho(f) \in \mathbf{R}/\mathbf{Z}$ on $Homeo^+(S^1)$, the set of orientation preserving homeomorphisms of the circle. Both $h(f)$ and $\varrho(f)$ are topological invariants which depend continuously on f. Moreover, $\varrho(f)$ is rational iff f has periodic points for $f \in Homeo^+(S^1)$ while (as proved by Katok) $h(f)$ is positive iff f has transverse homoclinic points for $f \in \mathscr{D}^\infty(M^2)$.

Problems

1. (Monotonicity of entropy) For $f_r(x) = r - x^2$, the function $r \to h(f_r)$ is monotone increasing (Douady and Hubbard). What about $r \to h(f_{r,b})$ for fixed b with $f_{r,b}(x, y) = (r - x^2 + by, x)$?.

2. Let $\mathscr{M}_{\max}(f)$ denote the set of measures of maximal entropy for a mapping f. As a consequence of Theorem 13, for any C^∞ f, $\mathscr{M}_{\max}(f) \neq \emptyset$. For $f \in \mathscr{D}^\infty(M^2)$ with $h(f) > 0$, is $\mathscr{M}_{\max}(f)$ a finite dimensional simplex?

Related to this problem, Hofbauer ([H]) has developed an interesting theory concerning piecewise monotone mappings of an interval with finitely many monotone continuous pieces. His theory can be described by introducing the notion of a zero-entropy set (0-entropy set). Hofbauer calls these *small* sets.

Let $f : X \to X$ be a Borel automorphism of a standard Borel space. A *0-entropy* set is an f-invariant subset $X_1 \subset X$ such that for any ergodic $\mu \in \mathscr{M}(f)$, with $\mu(X_1) = 1$, we have $h_\mu(f) = 0$. By convention, if $\mathscr{M}(f) = \emptyset$, then, every invariant subset of X is a 0-entropy set. Periodic orbits are simple 0-entropy sets as are stable manifolds of hyperbolic periodic orbits in the smooth setting. There is a natural notion of isomorphism mod 0-entropy: Borel automorphisms $(f, X), (g, Y)$ are isomorphic mod 0-entropy if there are 0-entropy sets $X_1 \subset X$, $Y_1 \subset Y$, and a Borel isomorphism $\phi : X \setminus X_1 \to Y \setminus Y_1$, with $g\phi = \phi f$. We say two Borel endomorphisms $f : X \to X, g : Y \to Y$ are isomorphic mod 0-entropy if their natural extensions are isomorphic mod 0-entropy. Finally, we say that the Borel endomorphism (f, X) is Markov mod 0-entropy if it is isomorphic mod 0-entropy to a finite or countable state topological Markov chain (σ, Σ_A). We will say that a measure preserving endomorphism (X, f, μ) is *essentially Markov* if its natural extension $(\hat{X}, \hat{f}, \hat{\mu})$ is isomorphic to a Markov process. In this case we also say that μ is *essentially Markov.*

Theorem 17 (Hofbauer). *Let $f : I \to I$ be a piecewise monotone map of the interval. Then, (f, I) is Markov mod 0-entropy. Moreover, there are only finitely many ergodic measures of maximal entropy and each is essentially Markov.*

Moving to general C^∞ maps of an interval with positive topological entropy, we can prove that there are at most a countable number of ergodic measures of maximal entropy, and that each is essentially Markov.

In dimension greater than 1, it is not always true that positive entropy implies that $\mathcal{M}_{max}(f)$ is a finite dimensional simplex: take the direct product of the identity transformation and any transformation with positive entropy. However, it is possible that, generically, i.e. for elements of a residual set of diffeomorphisms, $\mathcal{M}_{max}(f)$ is a finite dimensional simplex.

3. For $f \in \mathcal{D}^r(M^2), h(f) = 0$, can f be C^r perturbed to be Morse-Smale? This is true if the limit set of f is finite and hyperbolic [MaP], but not even known if the non-wandering set is finite.

4. For $f \in \mathcal{D}^\infty(M^2)$, let $\chi^+(x)$ denote the positive characteristic exponent of x (defined for a total probability set of x). Let $\phi(x) = -\chi^+(x)$. Then, ϕ is bounded and Borel measurable. Define

$$P(\phi) = \sup_{\mu \in \mathcal{M}(f)} h_\mu(f) + \int \phi d\mu .$$

Is there always a μ_0 such that

$$P(\phi) = h_{\mu_0}(f) + \int \phi d\mu_0 .$$

This is true for continuous ϕ. Such $\mu_0's$ are called ϕ-equilibrium states. For a hyperbolic attractor Λ and any μ supported in the basin of Λ and absolutely continuous with respect to Lebesgue measure, it is known that

$$\lim_{n\to\infty} \frac{1}{n} \sum_{k=0}^{n-1} f_\star^k(\mu)$$

is the unique ϕ-equilibrium state on Λ (Ruelle). In general, one would expect ϕ-equilibrium states to be related to weak limits of the measures $\{\frac{1}{n}\sum_{k=0}^{n-1} f_\star^k(\mu)\}$. In this connection, Pesin and Sinai have shown that the weak limits of the iterates of Lebesgue measure have absolutely continuous densities along unstable leaves for partially hyperbolic attractors [PS].

References

[AKM] R. Adler, A. Konheim, M. McAndrew: Topological entropy. Trans. Amer. Math. Soc. **114** (1965) 309–319

[AM] R. Adler, B. Marcus: Topological entropy and equivalence of dynamical systems. Memoirs Amer. Math. Soc. 219 (1979)

[B] R. Bowen: Entropy-expansive maps. Trans. Amer. Math. Soc. **164** (1972) 323–331

[B2] R. Bowen: Equilibrium states and the ergodic theory of Anosov diffeomorphisms. (Lecture Notes in Mathematics, vol. 470.) Springer, Berlin Heidelberg New York 1975

[DGS] M. Denker, C. Grillenberger, and K. Sigmund: Ergodic Theory on Compact Spaces. (Lecture Notes in Mathematics, vol. 527.) Springer, Berlin Heidelberg New York 1976

[FS] D. Fried and M. Shub: Entropy, linearity, and chain recurrence. Publ. Math. IHES **50** (1979) 203–214

[F] Friedland: Entropy of polynomial and rational maps. Preprint, Dept. of Math., Univ. of Illinois at Chicago

[H] F. Hofbauer: On intrinsic ergodicity of piecewise monotonic transformations with positive entropy II. Israel J. Math. **38** (1981) no. 1–2, 107–115, and The structure of piecewise monotonic transformations. Erg. Theory Dyn. Syst. **1** (1981) no. 2, 159–178

[M1] M. Misiurewicz: Diffeomorphisms without any measure with maximum entropy. Bull Acad. Polon. Sci., Ser. Math. Astron. Phys. **21** (1973) 903–910

[M2] M. Misiurewicz: Horseshoes for mappings of the interval. Bull. Acad. Polon. Sci., Ser. Math. Astron. Phys. **27** (1979) 167–169

[MaP] I. Malta and M. Pacifico: Breaking cycles on surfaces. Invent. math. **74** (1983) no. 1, 43-62.

[MP] M. Misiurewicz and F. Przytycki: Topological entropy and degree of smooth mappings. Bull Acad. Polon. Sci., Ser. Math. Astron. Phys. **25** (1977) no. 6, 573–574

[MS] M. Misiurewicz and W. Szlenk: Entropy of piecewise monotone mappings. Asterisque **50** (1977) 299–310

[MT] J. Milnor and W. Thurston: Iterated maps of the interval. Dynamical Systems (Maryland 1986-87). (Lecture Notes in Mathematics, vol. 1342) Springer, Berlin Heidelberg New York 1988, pp. 465–563

[N1] S. Newhouse: Entropy and Volume. Erg. Theory Dyn. Syst. **8** (1988) 283–299

[N2] S. Newhouse: Continuity properties of entropy. Ann. Math. **129** (1989) 215–235 and Corrections, Ann. Math. **131** (1990) 409–410

[P] F. Przytycki: An upper estimate for topological entropy of smooth diffeomorphisms. Invent. math. **59** (1980) 205–213

[PS] Pesin-Sinai: Gibbs measures for partially hyperbolic attractors. Erg. Theory Dyn. Syst. **2** (1982), no. 3-4, 417–438

[R] D. Ruelle: A measure associated with Axiom A attractors. Amer. J. Math. **98** (1976) 619–654

[SW] C. Shannon, W. Weaver: The mathematical theory of communication. Univ. of Illinois Press, Urbana, Ill. 1949

[Y1] Y. Yomdin: Volume growth and entropy. Israel J. Math. **57**, no. 3 (1987) 285–301, and C^k-resolution of semialgebraic mappings – Addendum to 'Volume growth and entropy'. Israel J. Math. **57**, no. 3 (1987) 301–318

[Y2] Y. Yomdin: Preprint

Combinatorial Models Illustrating Variation of Dynamics in Families of Rational Maps

Mary Rees

Department of Pure Mathematics, University of Liverpool, P.O. Box 147
Liverpool L69 3BX, UK

I have been interested in a particular case of the following question:

If f_t $(t \in T)$ is a family of (rational) maps, how do dynamics vary with t?

There are a number of advantages in considering this question for families of rational maps. Variation of dynamics in such a family is usually extremely rich, and even the topological nature of the Julia set of a map – the invariant set on which all the interesting dynamics occur – often varies widely. There are usually many hyperbolic maps within such a family. A rational map f is *hyperbolic* if all critical orbits converge to attractive periodic orbits. In this case, the *Julia set* $J(f)$ can be defined as the set of all z such that $f^n(z)$ does *not* converge to an attractive periodic orbit, and f is expanding on $J(f)$. Such a map f is relatively easy to analyse, and is dynamically stable in a neighbourhood $U \supset f^{-1}(U)$ of its Julia set, that is, for all g sufficiently near f, there exists a homeomorphism $\varphi_g : U \to \overline{\mathbf{C}}$ with $\varphi_g(J(f)) = J(g)$ and $\varphi_g \circ f = g \circ \varphi_g$ on $f^{-1}(U)$. Thus, most families of rational maps contain both great variation in dynamics and open subsets – stable components – on which dynamics are constant. (It is still not known, however, whether all stable maps are hyperbolic.) Furthermore, a hyperbolic component usually contains a unique *critically finite* map, that is, one for which the critical forward orbits are finite. So dynamics on a hyperbolic component can be examined by examining this map. Finally, perhaps the most important advantage in considering rational families the classical work of Fatou and Julia ([F], [J]) implies that, in general, the dynamics of a rational map is much influenced by the behaviour of its finitely many critical points. This feature, that the dynamics should be rather dependent on the dynamics of finitely many distinguished points, is not unique to families of rational maps.

A Particular Family

Thanks to the work of Douady and Hubbard [DH1, DH2] and a reinterpretation of some of this by Thurston [T], the best understood family of rational maps is

$$f_a : z \mapsto z^2 + a (a \in \mathbf{C}).$$

The critical point ∞ is fixed throughout this family (as for any polynomial family, of course), and the most significant orbit is that of the other critical point, 0, or of its image, the critical *value* $a = f_a(0)$. Since $f_a''(\infty) \neq 0$, there is a holomorphic

Proceedings of the International Congress
of Mathematicians, Kyoto, Japan, 1990

map φ_a of some set $\{z : z > r_a\}$ onto a neighbourhood of ∞ and satisfying $\varphi_a(z^2) = f_a \circ \varphi_a(z)$. Moreover, a is in the image of φ_a for large a. This very classical result is the key to the very detailed description of the global dynamics of the family f_a. Douady and Hubbard showed [DH1], [DH2] that the map

$$\Phi : a \mapsto \varphi_a^{-1}(a)$$

is a holomorphic bijection of $\{a : f_a^n(0) \to \infty\}$ onto $\{z : z > 1\}$, which, significantly, can be rewritten as $\{z : f_0^n(z) \to \infty\}$. Thus, the critical value a and $\Phi(a)$ have isomorphic dynamics under the maps f_a, f_0 respectively. The map Φ extends continuously and with similar isomorphic-dynamics properties to map the Mandelbrot set (see Fig. 1)

$$\{a : f_a^n(0) \not\to \infty\},$$

not onto the closed unit disc $\{z : f_0^n(z) \not\to \infty\}$, but onto a *quotient* space of this. This quotient space is interesting, but highly computable (as Thurston's work [T] makes particularly clear). If $\Phi(a) \leq 1$, $\Phi(a)$ does not always take on under f_0 exactly the dynamics of a under f_a (although it often does), partly because $J(f_a)$ moves with a. For instance, points on the unit circle (which is the Julia set of f_0) come together, and that changes the dynamics of points inside the unit disc. It is not known if Φ is a homeomorphism, but recent progress has been made on this outstanding open problem, by Yoccoz.

Other Families

Local dynamics near ∞ are also significant in the study of other polynomial families, as in the work of Branner and Hubbard, for example ([BH1, BH2]). For other families of rational maps, it makes sense to take slices of one complex dimension by keeping the orbits of all but one critical point constant and finite – such slices are conjectured dense (since hyperbolic components are conjectured dense), and one can always examine later the question of how the slices intersect and fit together. A rational map of degree d has $2d-2$ critical points (up to multiplicity), all of multiplicity $\leq d-1$. Thus, a degree two rational map f has exactly two critical points c_1, c_2, which vary continuously with f. I have considered, for $m \geq 1$, the variety

$$\{f : f \text{ is degree two, } c_1 \text{ has period } m\}/\text{Möbius conjugation}$$

If $m = 1$, this gives the family $f_a : z \mapsto z^2 + a$. For the purposes of discussion, I wish to consider the case $m = 3$, which gives

$$g_a : z \mapsto \frac{(z-a)(z-1)}{z^2} (a \neq 0)$$

in which one critical point 0 has orbit $0 \mapsto \infty \mapsto 1 \mapsto 0$ of period 3, and the other critical point $\dfrac{2a}{a+1}$ has image (or critical value) $\dfrac{-(a-1)^2}{4a}$. There are obviously three values of a for which this second critical point is fixed, giving, up to Möbius conjugation, the three polynomials from the family $\{f_b\}$ with 0 of period 3. Again for the purposes of discussion, we fix one of these polynomials, p,

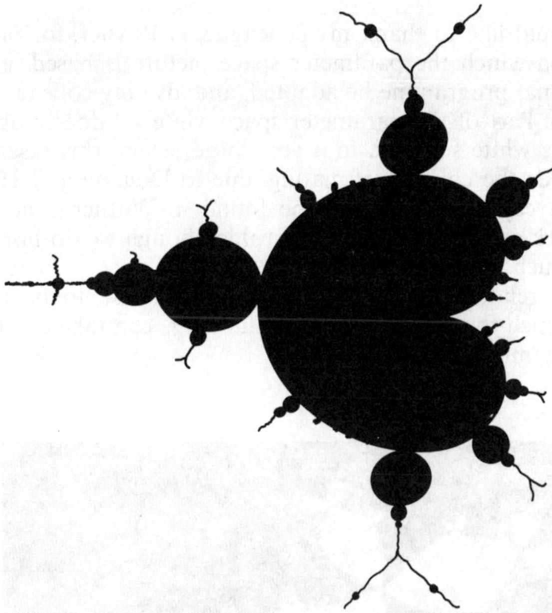

Fig. 1. The Mandelbrot set

the anticlockwise rabbit polynomial for which the periodic attractive basins are rotated anticlockwise by p around the common fixed boundary point. To avoid confusion, (since p is identified with its Möbius conjugate within the family $\{g_a\}$), we refer to the fixed critical point of p as c_2, and the period 3 critical point as c_1.

Comparison of a Dynamical Plane and Parameter Space

A crude, inaccurate, but extremely useful idea in analysing such a parameter space is that one should be able to describe the dynamics of any map in the family by movement of one critical value across the dynamical plane of one fixed map in the family, so that the critical value takes in turn the dynamics of all points of the fixed map. We have already indicated how this idea comes into the analysis of the family $f_a : z \mapsto z^2 + a$. For the purposes of discussion, we wish, now, to compare the parameter space $\{g_a : a \neq 0\}$ and the dynamical plane of the anticlockwise rabbit polynomial p. Within these spaces, the sets

$$\{z : p^n(z) \to \{c_1, p(c_1), p^2(c_1)\}$$

and

$$\{a : g_a^n\left(\frac{2a}{a+1}\right) \to \{0, 1\infty\}\}$$

should obviously be compared, remembering that 0 and $\dfrac{2a}{a+1}$ are the critical points of g_a. We refer to these sets as the *white sets*, for obvious reasons. See

Figs. 2 and 3. I should like to thank my colleague, F. Rayner, for producing the computer picture on which the parameter space picture is based, and Tan Lei, who wrote the original programme he adapted, and also my colleague R. Morris, for his contribution. Part of the parameter space white set does indeed resemble the dynamical plane white set, and, to a very large extent, this resemblance can be proved. One needs the concept of mating, due to Douady and Hubbard [D], and the concept of capture, which can be found in Wittner's thesis [W]. The complementary black sets are not homeomorphic, though we do not expect that, since there is no such homeomorphism for the family $\{f_a\}$. Neither are they clearly dynamically related. So we need, at the very least, to be more precise about the way in which the moving critical value of g_a can take on the dynamics of points in the dynamical plane of p.

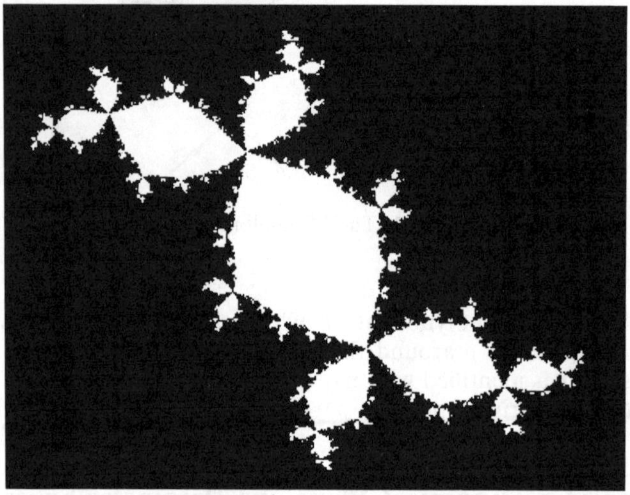

Fig. 2. The dynamical plane of p

How to Take on the Dynamics of *p*

If we do not try to produce rational maps, it is easy to produce maps taking on the dynamics of various points in the dynamical plane of p, and this turns out to be worthwhile. We restrict to critically finite branched coverings. If f is a branched covering, then we define

$$X(f) = \{f^n(c) : c \text{ is critical}, n > 0\}.$$

Then f is *critically finite* if $\#(X(f)) < \infty$. We are going to give some examples based on the critically finite polynomial p with critical points c_1, c_2. Let $\beta : [0,1] \to \overline{\mathbf{C}}$ be a path, which for simplicity, we assume is simple, and let $\beta(0) = c_2$. Let σ_β be a homeomorphism which is the identity outside a (small) disc neighbourhood of $Image(\beta)$, and maps $\beta(0)$ to $\beta(1)$. See Fig. 4. Obviously, $\sigma_\beta \circ p$, like p, has critical points c_1, c_2 (provided $\beta(1) \neq p(c_1)$). Provided β avoids the

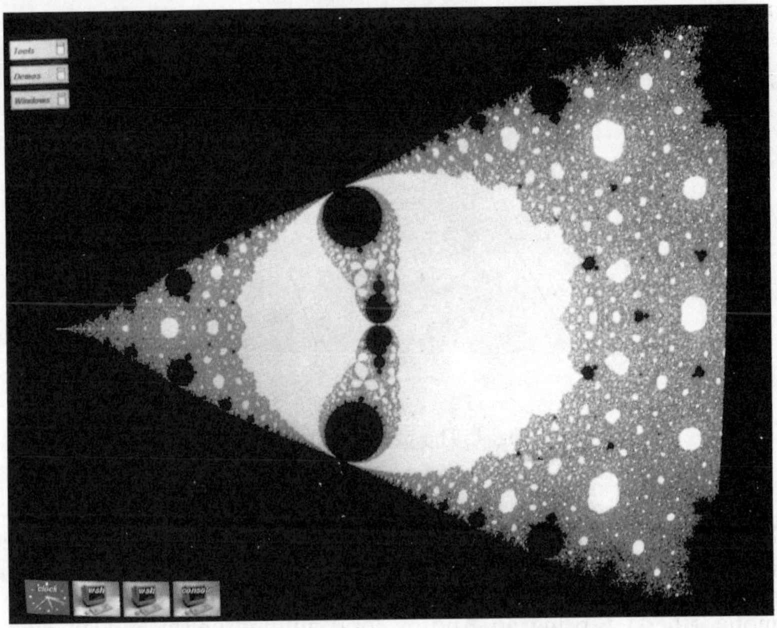

Fig. 3. The parameter space of the g_a

forward orbits of $\beta(1)$, c_1, $\sigma_\beta \circ p$ is a critically finite branched covering, whenever $\beta(1)$ is a strictly preperiodic point under p. If $\beta(1)$ has period $q > 1$ under p, then $\zeta : [0,1] \to \overline{\mathbf{C}}$ is uniquely defined by $p \circ \zeta = \beta$, $\zeta(1) = p^{q-1}(\beta(1))$, and $\sigma_\zeta^{-1} \circ \sigma_\beta \circ p$ is a critically finite branched covering, with two periodic critical points.

Obviously, the maps $\sigma_\beta \circ p$ and $\sigma_\zeta^{-1} \circ \sigma_\beta \circ p$ are not rational, but they are *equivalent* to rational maps, in a sense yet to be explained. Indeed, we have the following theorem, which, for simplicity, is stated only for the set $\{g_a : a \neq 0\}$ and the polynomial p. See [R1].

The Polynomial-and-Path Theorem. *Any hyperbolic critically finite rational map g_a is equivalent to $\sigma_\beta \circ p$ or $\sigma_\zeta^{-1} \circ \sigma_\beta \circ p$ for some β*

In the proof, the path β is obtained from a path g_{a_t} from g_a to p.

Equivalence of Branched Coverings

This is a homotopy-type equivalence, due to Thurston [T]. The concept occurs in Thurston's theorem [T] about which critically finite branched coverings are equivalent to rational maps. I am avoiding stating this theorem, although it is crucial to the work I am describing.

Let f g be critically finite branched coverings of $\overline{\mathbf{C}}$. Then f is *equivalent* to g, written $f \simeq g$, if there exists an orientation-preserving homeomorphism φ, and a path g_t through critically finite branched coverings, such that $\varphi \circ f \circ \varphi^{-1} = g_0$, $g = g_1$, $X(g_t) = X(g)$ for all t.

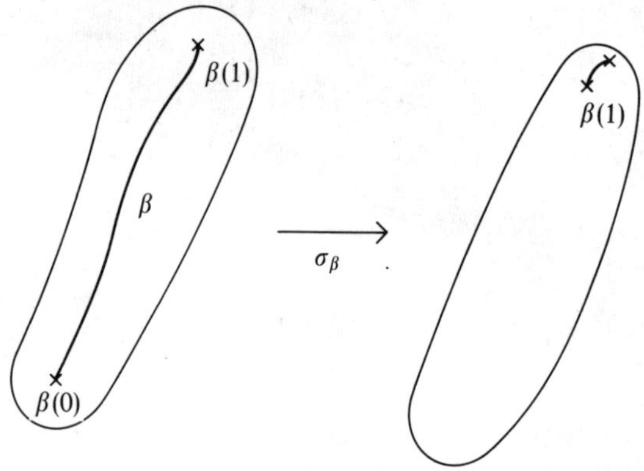

Fig. 4. The homeomorphism

There are many situations in dynamics when some sort of homotopy equivalence between maps f and g implies semiconjugacy. Always, some sort of hyperbolicity is required for f. See, for example, Franks [Fr], where the case of (among others) f being an Anosov toral automorphism is dealt with. The theorem in the present context is as follows. (See, for example, [R1].)

The Semiconjugacy Proposition. *If f is a critically finite hyperbolic rational map and g is a critically finite branched covering with $f \simeq g$, and f and g are conjugate in neighbourhoods of any periodic critical orbits, then there is a continuous map $\varphi : \overline{\mathbf{C}} \to \overline{\mathbf{C}}$ such that $\varphi \circ g = f \circ \varphi$.*

So a description of a hyperbolic critically finite rational map up to equivalence, as in the Polynomial-and-Path Theorem, is useful. In fact, when the equivalence is to a map of the form $\sigma_\beta \circ p$ or $\sigma_\zeta^{-1} \circ \sigma_\beta \circ p$, the techniques can be extended to describe completely the topological dynamics of the hyperbolic map, and hence of all maps within that hyperbolic component [R1].

The Information Obtained, and Problems with It

In summary, the Polynomial-and-Path Theorem gives a multivalued map from the set of hyperbolic components within the family $\{g_a\}$ (the example we are discussing) into a set of paths in the dynamical plane of p. There are two large problems with this, which we shall refer to as the *Existence Problem* and the *Uniqueness Problem.*

The Existence Problem. For what paths in the dynamical plane does there exist a corresponding critically finite rational map? The path space is certainly too large, as it stands (although, of course, no very precise definition has been given).

The Uniqueness Problem. How can one decide when two paths give the same rational map?

In order to address these problems, one needs to work with a space obtained by modifying the space of paths. This can be done. See [R1]. (The space is obtained by adapting the ideas of Thurston's combinatorial description of the Mandelbrot set [T], where he also worked out the theory to a considerable extent for other polynomial families. I suspect that similar adaptations may have been made by others, since very substantial work has been done on polynomial families. An adaptation that I am aware of, for a different family, is due to my student D. Ahmadi [A]. Shishikura and Tan Lei [ST] work with a partial combinatorial description in their work on matings of degree 3 polynomials, but following the original Douady-Hubbard approach to the combinatorics.) The modified path space is a *tree with balls*. See Fig. 5 (which is not intended to be accurate). A *tree with balls* is a simply-connected space which is a connected union of a tree and pairwise disjoint balls. In this space, one considers paths from a base point in a ball, and one identifies these paths with their second endpoints.

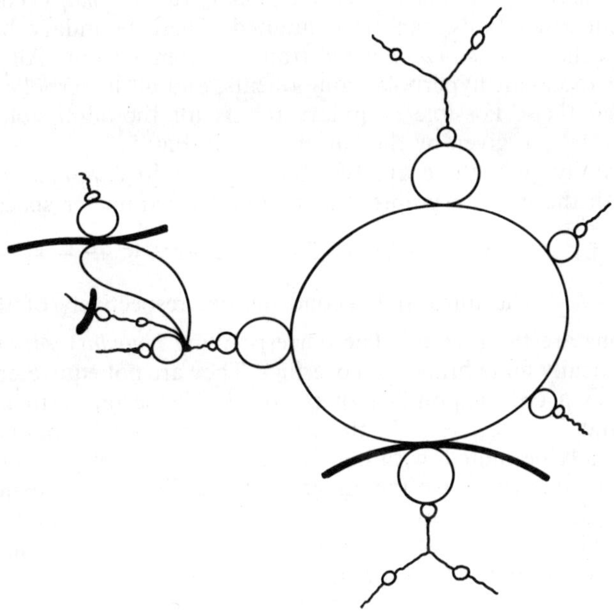

Fig. 5. The tree with balls

All balls in the tree with balls represent critically finite degree two branched converings (with one critical point of period 3, for the example under discussion). Among them are represented all the hyperbolic critically finite maps with both critical points periodic (in the family $\{g_a\}$), and it makes sense to identify these balls with the corresponding hyperbolic components. Since the tree with balls is a model for the black set in parameter space, boundaries of hyperbolic components from the white set – with one critical point in the backward orbit of the other, periodic, one – are represented simply by points in the tree with balls. It should be stressed that the tree with balls, and the nature of the balls in it, can be computed entirely from the dynamics of p, which, in turn, can be computed from the dynamics of $z \mapsto z^2$. This is not surprising, since the tree with balls is derived from paths in the dynamical plane of p. Any hyperbolic component is represented in the tree with balls countably many times, giving countably many paths from the base hyperbolic component. This is rather curious, when there are uncountably many paths between any two hyperbolic components in the true parameter space. Nevertheless, I would hope that the paths in the tree with balls represent *some* of the real paths in parameter space. But little work has been done, as yet, on the relation between the topologies of the two spaces.

Solution of the Existence Problem

The *Admissable Boundary Theorem* ([R1], [R2]) can be interpreted as follows. Some balls in the tree with balls represent *boundary rational maps*. Their positions, like that of all other balls, can be computed. These boundary balls separate an open set called the *admissible set* from its complement. All balls in the admissible set represent hyperbolic components, and all hyperbolic components are represented there. Possible boundary points for the admissible set (one of which is accurate) are given by the cut-off lines in Fig. 5.

For the family $\{g_a\}$, there are two boundary rational maps, which can be associated with the missing points 0 and ∞ in the parameter space. Note that, in fact, $g_0(z) = \dfrac{z-1}{z}$, and after conjugation by $z \mapsto (-a)^{\frac{1}{2}} z$, g_a converges, as $a \to \infty$, to $z \mapsto \dfrac{1}{z}$. The third and second iterates respectively of these Möbius transformations are the identity. The corresponding *boundary rational maps* are degree two critically finite branched coverings. They are not equivalent to rational maps. Each has a critical point c_1 of period 3. The second critical point c_2 is of period 3 and 2 respectively. In the first case, three circles bounding disjoint discs are cyclically permuted, with one of the discs containing the critical points, one the critical values, and one the images of these. The complement of the discs is invariant under the map. In the second case, there is a simple loop which is mapped to itself, with orientation reversed. It separates the critical values from the other 3 points in the critical forward orbits. See Fig. 6.

The Uniqueness Problem

One can define an equivalence relation on paths with eventually periodic end-points in the dynamical plane of p simply by $\beta_1 \simeq \beta_2$ if $\sigma_{\beta_1} \circ p \simeq \sigma_{\beta_2} \circ p$ or $\sigma_{\zeta_1}^{-1} \circ \sigma_{\beta_1} \circ p \simeq \sigma_{\zeta_2}^{-1} \circ \sigma_{\beta_2} \circ p$ (whichever is appropriate). This equivalence can be transferred to the tree with balls. The equivalence relation on the tree with balls is rather close to being a group orbit equivalence relation.

Open Question. Is the equivalence relation sufficiently close to a group orbit equivalence relation for it to be possible to construct a fundamental domain?

Such a fundamental domain would presumably be rather like a Dirichlet fundamental domain.

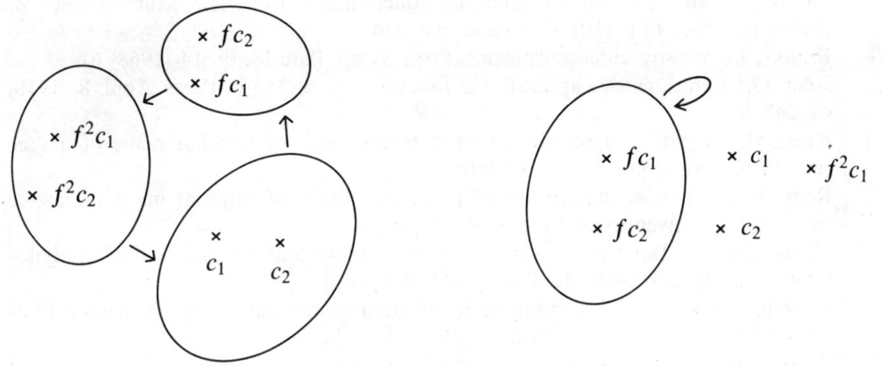

Fig. 6. Boundary rational maps

If $V = \{g_a : a \neq 0\}$, then, for $m \geq 0$,

$$V_m = \{g_a : g_a^m\left(\frac{-(a-1)^2}{4a}\right) \notin \{0, 1, \infty\}\}.$$

The techniques of the Polynomial-and-Path Theorem give a natural isnomorphism of $\pi_1(V_m)$ into a certain subgroup $G(m+1)$ of the group of homeomorphisms of $\overline{\mathbf{C}}$ preserving $p^{-m-1}(X(p))$, modulo the appropriate isotopies, namely, those which are constant on $p^{-m-1}(X(p))$. Modulo these, the group $G(m+1)$ consists of all homeomorphisms φ for which there exists a closed loop α based at c_2 such that

$$(p, p^{-m-1}(X(p))) \simeq_\varphi (\sigma_\alpha \circ p, p^{-m-1}(X(p))).$$

This notation means that there is a path g_t through critically finite branched coverings such that $g_0 = \varphi \circ p \circ \varphi^{-1}$, $g_1 = \sigma_\alpha \circ p$, $g_t^{-m-1}(X(g_t))$ is independent of t. (The relation $\simeq_\varphi$ is only an equivalence relation when φ is allowed to vary.) The isomorphism is not surjective for $m \geq 1$. This seems to be because V sits in a natural larger space, which includes nonrational maps.

References

[A] Ahmadi, D.: Thesis in preparation.

[BH1] Branner, B., Hubbard, J.H.: The iteration of cubic polynomials. Part I: The global topology of parameter space. Acta Math. **160** (1988) 143–206

[BH2] Branner, B., Hubbard, J.H.: The iteration of cubic polynomials. Part II: Patterns and parapatterns. Danmarks Tekniske Hojskole, 1989 (preprint)

[D] Douady, A.: Systèmes dynamiques holomorphes. Séminaire Bourbaki 1983, Astérisque **105/6** (1983) 39–63

[DH1] Douady, A., Hubbard, J.H.: Etudes dynamiques des polynômes complexes, avec la collaboration de P. Lavaurs, Tan Lei, P. Sentenac. Parts I and II. Publications Mathématiques d'Orsay, 1985

[DH2] Douady, A., Hubbard, J.H.: Itérations des polynômes quadratiques complexes. C.R. Acad. Sci. Paris, Série I **294** (1982) 123–126

[F] Fatou, P. : Mémoire sur les équations fonctionelles. Bull. Soc. Math. France **47** (1919) 161–271, **48** (1920) 33–96 and 208–314

[Fr] Franks, J.: Anosov diffeomorphisms. Proc. Symp. Pure Math. **14** (1968) 61–93

[J] Julia, G.: Itérations des applications fonctionelles. J. Math. Pures Appl. **8** (1918) 47–245

[R1] Rees, M.: A partial description of parameter space of rational maps of degree two: Part 1. To appear in Acta Math.

[R2] Rees, M.: A partial description of parameter space of rational maps of degree two: Part 2. University of Liverpool (preprint)

[ST] Shishikura, M., Tan Lei: A family of rational maps and matings of cubic polynomials. Max Planck Institute, Bonn, 1988 (preprint)

[T] Thurston, W.P.: On the combinatorics of iterated rational maps. Princteton University and I.A.S., 1985 (preprint)

[W] Wittner, B.: On the bifurcation loci of rational maps of degree two. Thesis, Cornell University, 1988

Hyperelliptic Function Parametrization for the Chiral Potts Model

R. J. Baxter

Mathematics and Theoretical Physics, I.A.S., The Australian National University
G.P.O. Box 4, Canberra, A.C.T. 2601, Australia

Abstract. The chiral Potts model in statistical mechanics is characterized by sets of variables a, b, c, d , lying on a high-degree algebraic curve. (Each such set being a "rapidity" vector.) Here we show how they can be parametrized by the hyperelliptic theta functions used by Sonya Kowalevski.

Introduction

Considerable progress has been made in the last three years in understanding the integrable chiral Potts model [1]–[27]. This statistical mechanical model was first formulated [4, 10] in terms of homogeneous sets of variables (a, b, c, d) satisfying the relations (only two of which are independent)

$$a^N + k'b^N = kd^N , \quad k'a^N + b^N = kc^N , \tag{1}$$
$$ka^N + k'c^N = d^N , \quad kb^N + k'd^N = c^N .$$

Here N is an integer greater than one and k, k' are real constants satisfying $k^2 + k'^2 = 1$: it is convenient to introduce a parameter θ such that

$$k = \sin\theta , \quad k' = \cos\theta , \quad 0 < \theta < \pi/2 . \tag{2}$$

For $N = 2$, the relations (1) are of genus 1 and can be uniformized using Jacobi's elliptic functions. For $N > 2$ the relations are of higher genus and one needs hyperelliptic functions. Here we show how this can be done.

It has been found convenient to introduce [5, 20] other variables v, u, x, y, μ , related to $a : b : c : d$ and satisfying

$$k \sin u = \sin v , \tag{3a}$$

$$x = e^{i(u-v)/N} , \quad y = e^{i(\pi+u+v)/N} , \tag{3b}$$

$$\mu^N = k'/(1 - k x^N) = (1 - k y^N)/k' , \tag{3c}$$

$$a : b : c : d = x\mu : y : 1 : \mu . \tag{3d}$$

Proceedings of the International Congress
of Mathematicians, Kyoto, Japan, 1990

Once N, θ and any one of the variables v, u, x, y, μ is given, the rest are determined (to within a discrete set of choices). Here we regard v as an independent complex variable; u, x, y, μ, and the ratios $a : b : c : d$, as functions thereof.

We have shown [27] that low-temperature corner transfer matrix calculations introduce integrals of the form

$$\int^{v} \frac{e^{i(N-2\alpha)v'/N}\, dv'}{\Delta(v')} , \tag{4}$$

where $\alpha = 1, \ldots, N-1$ and

$$\Delta(v) = -ik \cos u = \surd\{\sin^2 v - \sin^2 \theta\} . \tag{5}$$

Although these calculations were only for k' small, these are *precisely* the hyperelliptic integrals that occur if we specialize the classic work of Kowalevski [28] to the function $\Delta(v)$, considering it as the square root of a Laurent polynomial in $e^{2iv/N}$ (or equivalently in $\cot[(v + \pi - \theta)/N]$).

Here we make this specialization and find, rather remarkably, that it provides a uniformizing parametrization, not just of Δ and $e^{2iv/N}$, but of all the variables $v, u, \ldots, d$. Indeed, our main result (48) is that a, b, c, d can be normalized so that (to within elementary factors) each is an hyperelliptic theta function.

We hope this parametrization will be useful in pursuing the corner transfer matrix calculations.

Kowalevski's Notation

In §6 (pp. 217–221) of [28], take $\varrho = N - 1$, $A_0 < 0$, and define $a_0, \ldots, a_{2\varrho}$ by

$$a_{2j-1} = -\cot(\pi j/N) , \quad a_{2j} = -\cot[(\pi j + \pi - 2\theta)/N] . \tag{6}$$

Then $a_0 < a_1 < \cdots < a_{2\varrho}$. Changing the variable x in [28] to v, where $x = -\cot[(v + \pi - \theta)/N]$, the expression $\surd R(x)$ therein becomes

$$\surd R(x) = (-1)^{N-1} c_0 \Delta(v) / \{\sin[(v + \pi - \theta)/N]\}^N , \tag{7}$$

where $c_0 = [-A_0/(N \sin 2\theta)]^{1/2}$ is a positive real constant. The values of v corresponding to $x = a_0, \ldots, a_{2\varrho}$ are $v = b_0, \ldots, b_{2\varrho}$, where

$$b_{2j-1} = \pi j - \pi + \theta , \quad b_{2j} = \pi j - \theta .$$

The sign conventions in [28] are such that $i^{2N-j} \surd R(x)$ is real and positive for $a_{j-1} < x < a_j$. It follows that $\Delta(v)$ is negative imaginary for $-\theta < v < \theta$, positive real for $\theta < v < \pi - \theta$, and satisfies the anti-periodicity relation

$$\Delta(v + \pi) = -\Delta(v) . \tag{8}$$

The continuation of $\Delta(v)$ is therefore analytic in the lower half of the complex v-plane. When $\mathrm{Im}(v) \to -\infty$, then $\Delta(v) \to -i\, e^{iv}/2$.

The functions $F_1(x),\ldots,F_{N-1}(x)$ of [28] are linearly independent polynomials of degree $N-2$. Allowing complex coefficients, we can choose them so that, for $\alpha = 1,\ldots,N-1$

$$\frac{F_\alpha(x)}{\sqrt{R(x)}}dx = \frac{e^{i(N-2\alpha)v/N}\,dv}{\Delta(v)} \tag{9}$$

which is proportional to the integrand that occurs in (4).

Hyperelliptic Integrals

With these notations, the definite integrals $K_{\alpha\beta}$, $\overline{K}_{\alpha\beta}$ of [28] are (for $\alpha,\beta = 1,\ldots,N-1$)

$$K_{\alpha\beta} = \int_{\pi\beta-\pi+\theta}^{\pi\beta-\theta} \frac{e^{i(N-2\alpha)v/N}}{\Delta(v)}\,dv\,,$$
$$\overline{K}_{\alpha\beta} = -i\int_{\pi\beta-\pi-\theta}^{\pi\beta-\pi+\theta} \frac{e^{i(N-2\alpha)v/N}}{\Delta(v)}\,dv\,. \tag{10}$$

In [27] we defined the function

$$I(\theta,\alpha) = \int_{-\theta}^{\theta} \frac{e^{i(1-2\alpha)y}\,dy}{\sqrt{\{\sin^2\theta\ -\sin^2 y\}}} \tag{11}$$

and showed that $I(\theta,\alpha) = \pi\,F(\alpha,1-\alpha\,;1;\,\sin^2\theta)$, where $F(\alpha,\beta;\gamma;z)$ is the usual hypergeometric function. Define

$$L_\alpha = I\left(\frac{\pi}{2}-\theta,\alpha/N\right)\,,\quad L'_\alpha = I(\theta,\alpha/N)\,. \tag{12}$$

Then L_α and L'_α are positive real and

$$L_{N-\alpha} = L_\alpha\,,\quad L'_{N-\alpha} = L'_\alpha \tag{13}$$

$$K_{\alpha\beta} = i\omega^{-\alpha\beta}e^{\pi i\alpha/N}L_\alpha\,,\quad \overline{K}_{\alpha\beta} = \omega^{\alpha-\alpha\beta}L'_\alpha\,, \tag{14}$$

where $\omega = e^{2\pi i/N}$.

The $G_{\alpha\beta}$ defined in [28] are the elements of the transposed inverse of the matrix $(2K_{\alpha\beta})$, and are given by

$$G_{\alpha\beta} = i(1-\omega^{\alpha\beta})e^{-\pi i\alpha/N}/(2NL_\alpha)\,,$$

while the $K'_{\alpha\beta}$ are

$$K'_{\alpha\beta} = \sum_{\gamma=1}^{\beta}\overline{K}_{\alpha\gamma} = \frac{ie^{\pi i\alpha/N}(\omega^{-\alpha\beta}-1)}{2\sin(\pi\alpha/N)}\,L'_\alpha\,. \tag{15}$$

From now on take sums and products over α, β, γ to be from 1 to $N-1$ unless otherwise indicated. Inserting some omitted primes in the relevant equation in [28], define

$$\tau_{\alpha\beta} = 2i \sum_{\gamma} G_{\gamma\alpha} K'_{\gamma\beta} \,. \tag{16}$$

Using (13), this gives

$$\tau_{\alpha\beta} = \frac{2i}{N} \sum_{\gamma} \frac{\cos \frac{\pi\gamma(\alpha-\beta)}{N} \, \sin \frac{\pi\alpha\gamma}{N} \, \sin \frac{\pi\beta\gamma}{N} \, L'_{\gamma}}{\sin \frac{\pi\gamma}{N} \, L_{\gamma}} \tag{17}$$

for $\alpha, \beta = 1, \ldots, N-1$. The matrix with elements $-i\tau_{\alpha\beta}$ is real, symmetric and positive-definite.

Hyperelliptic Functions

Write the variables $u_1, \ldots, u_{\varrho}; v_1, \ldots, v_{\varrho}$ of [28] as $w_1, \ldots, w_{\varrho}; s_1, \ldots, s_{\varrho}$. Define new variables $v_1, \ldots, v_{\varrho}$ by $x_{\alpha} = -\cot[(v_{\alpha} + \theta)/N]$, $x_1, \ldots, x_{\varrho}$ being as in [28] . Then for $\alpha = 1, \ldots, N-1$,

$$w_{\alpha} = \sum_{\beta} \int_{\pi\beta-\pi+\theta}^{v_{\beta}} \frac{e^{i(N-2\alpha)v/N} \, dv}{\Delta(v)} \; ; \tag{18}$$

$$w_{\alpha} = 2 \sum_{\beta} K_{\alpha\beta} s_{\beta} \; , \quad s_{\alpha} = \sum_{\beta} G_{\beta\alpha} w_{\beta} \,. \tag{19}$$

Writing the ordered set of variables $\{s_1, \ldots, s_{N-1}\}$ simply as s , the hyperelliptic theta function is

$$\Theta\{s\} = \sum_{m} \exp\{2\pi i \textstyle\sum_{\alpha} m_{\alpha} s_{\alpha} + \pi i \sum_{\alpha} \sum_{\beta} m_{\alpha} \tau_{\alpha\beta} m_{\beta}\} \,, \tag{20}$$

the outer sum being over all values of the integers $m = \{m_1, \ldots, m_{N-1}\}$.

More generally, let $n = \{n_1, \ldots, n_{N-1}\}$ be a given set of rational numbers, and define

$$\Theta\{s|n\} = \sum_{m} \exp\{2\pi i \textstyle\sum_{\alpha} (m_{\alpha} + n_{\alpha}) s_{\alpha} + \pi i \sum_{\alpha} \sum_{\beta} (m_{\alpha} + n_{\alpha}) \tau_{\alpha\beta} (m_{\beta} + n_{\beta})\} \,. \tag{21}$$

If $n_1, \ldots, n_{N-1}$ are all integers, then obviously $\Theta\{s|n\} = \Theta\{s\}$. Allow one of them to be half an odd integer and define $\Theta\{s\}_0, \ldots, \Theta\{s\}_{2N-2}$ by

$$\Theta\{s\}_{2\lambda} = \Theta(s_1, \ldots, s_{\lambda}, s_{\lambda+1} - \tfrac{1}{2}, \ldots, s_{N-1} - \tfrac{1}{2} \mid 0, \ldots, 0, \tfrac{1}{2}, 0, \ldots, 0) \tag{22}$$

$$\Theta\{s\}_{2\lambda-1} = \Theta(s_1, \ldots, s_{\lambda-1}, s_{\lambda} - \tfrac{1}{2}, \ldots, s_{N-1} - \tfrac{1}{2} \mid 0, \ldots, 0, \tfrac{1}{2}, 0, \ldots, 0) \,,$$

the arguments $n_1, \ldots, n_{N-1}$ on the RHS being zero except in position λ, i.e. $n_{\lambda} = \frac{1}{2}$, provided $1 \leq \lambda < N$. (For $\Theta\{s\}_0$ all the $n_1, \ldots, n_{N-1}$ are zero.)

The Case $N = 3$

To fix our ideas, for $N = 3$, $\Theta\{s|n\} = \Theta(s_1, s_2 \mid n_1, n_2)$ and these last definitions become

$$\begin{aligned} \Theta(s_1,s_2)_0 &= \Theta(s_1-\tfrac{1}{2}, s_2-\tfrac{1}{2} \mid 0,0)\,, & \Theta(s_1,s_2)_1 &= \Theta(s_1-\tfrac{1}{2}, s_2-\tfrac{1}{2} \mid \tfrac{1}{2},0) \\ \Theta(s_1,s_2)_2 &= \Theta(s_1, s_2-\tfrac{1}{2} \mid \tfrac{1}{2},0)\,, & \Theta(s_1,s_2)_3 &= \Theta(s_1, s_2-\tfrac{1}{2} \mid 0,\tfrac{1}{2}) \\ \Theta(s_1,s_2)_4 &= \Theta(s_1, s_2 \mid 0,\tfrac{1}{2})\,. \end{aligned}$$

In this case, setting $\varrho = \tau_{12} = \tau_{21}$, we have $\tau_{11} = \tau_{22} = 2\varrho$ and (20) simplifies to

$$\Theta\{s\} = \sum_{m_1,m_2} \exp\{2\pi i(m_1 s_1 + m_2 s_2) + 2\pi i\varrho(m_1^2 + m_1 m_2 + m_2^2)\}\,. \tag{23}$$

Changing the summation indices to $m_1 + m_2$ and $m_1 - m_2$ (both odd or both even), it follows that

$$\Theta\{s\} = \theta_2(\pi s_1 + \pi s_2, q^3)\theta_2(\pi s_1 - \pi s_2, q) \;+\; \theta_3(\pi s_1 + \pi s_2, q^3)\theta_3(\pi s_1 - \pi s_2, q) \tag{24}$$

where $q = e^{2\pi i\varrho}$ and $\theta_2(u,q)$, $\theta_3(u,q)$ are the ordinary single-variable Jacobi elliptic theta functions:

$$\theta_2(u,q) = \sum_{n=-\infty}^{\infty} e^{(2n-1)iu} q^{(2n-1)^2/4}$$

$$\theta_3(u,q) = \sum_{n=-\infty}^{\infty} e^{2niu} q^{n^2}\,.$$

(A similar decomposition of $\Theta\{s\}$ into sums of products of θ_2, θ_3 functions occurs for $N = 4$, but not apparently for higher N.)

The case $N = 3$ is further discussed in the Appendix.

Identities

Once more allowing N to be an arbitrary integer greater than one, we can now write down the first two of the three identities given by Kowalevski at the end of §6 of [28], specialized to this case. They are, for all values (real or complex) of $v_1, \ldots, v_{N-1}$,

$$\prod_\beta \frac{\sin[(v_\beta + \theta - \pi\alpha)/N]}{\sin[(v_\beta - \theta + \pi)/N]} = (-1)^\alpha [\Theta\{s\}_{2\alpha}/\Theta\{s\}]^2\,,$$

$$\prod_\beta \frac{\sin[(v_\beta - \theta + \pi - \pi\alpha)/N]}{\sin[(v_\beta - \theta + \pi)/N]} = (-1)^{\alpha-1} [\Theta\{s\}_{2\alpha-1}/\Theta\{s\}]^2\,, \tag{25}$$

where $\alpha = 0, \ldots, N-1$ in the first equation; $1, \ldots, N-1$ in the second.

There is a wealth of information in just these identities. For instance, taking $v_\beta = \pi\beta - \pi + \theta$ (all β), we see from (18) and (19) that $s_\beta = 0$, so the first identity gives

$$\frac{\sin 2\theta}{\sin[(\pi + \pi\alpha - 2\theta)/N]} = N[\Theta\{\mathbf{0}\}_{2\alpha}/\Theta\{\mathbf{0}\}]^2 \tag{26}$$

for $\alpha = 0,\ldots,N-1$, writing $\mathbf{0}$ for $\{0,\ldots,0\}$. Also, taking $v_\beta = \pi\beta - \theta$, we obtain $w_\alpha = \sum_\beta K_{\alpha\beta}$, $s_\beta = 1/2$, and the second identity gives

$$\frac{\sin[(\pi - 2\theta)/N]}{\sin[(\pi\alpha - \pi + 2\theta)/N]} = [\Theta\{\tfrac{1}{2}\}_{2\alpha-1}/\Theta\{\tfrac{1}{2}\}]^2 \tag{27}$$

for $\alpha = 1,\ldots,N-1$, writing $\frac{1}{2}$ for $\{\frac{1}{2},\ldots,\frac{1}{2}\}$.

Symmetries of the Θ Functions

Define

$$\varrho_\alpha = \frac{i}{N}\sum_\gamma \frac{\sin^2(\pi\alpha\gamma/N)L'_\gamma}{\sin(\pi\gamma/N)L_\gamma} . \tag{28}$$

Then $\varrho_\alpha = \varrho_{\alpha+N} = \varrho_{N-\alpha} = \varrho_{-\alpha}$; $\varrho_0 = \varrho_N = 0$; and from (17)

$$\tau_{\alpha\beta} = \varrho_\alpha + \varrho_\beta - \varrho_{\alpha-\beta} . \tag{29}$$

Given θ, we can regard either (26) or (27) as defining $\varrho_1,\ldots,\varrho_{N-1}$.

The definition (20) can be written more symmetrically by setting $s_\alpha = r_\alpha - r_N$ (r_N arbitrary). Then

$$\Theta\{s\} = \sum_m \exp\{2\pi i\textstyle\sum_{\alpha=1}^N m_\alpha r_\alpha - \pi i\sum_{\alpha=1}^N\sum_{\beta=1}^N \varrho_{\alpha-\beta} m_\alpha m_\beta\} , \tag{30}$$

the outer sum now being over all values of the N integers $m = \{m_1,\ldots,m_N\}$, subject to the condition

$$m_1 + \cdots + m_N = 0 .$$

For α an integer other than $1,\ldots,N-1$, define s_α by $s_0 = s_N = 0$, $s_\alpha = s_{\alpha+N} = s_{\alpha-N}$. (This is consistent with (19) .) Write $\Theta\{s\}$ alternatively as $\Theta\{s_\alpha\}$, the index α being understood to range over all integers, in particular over $0,\ldots,N-1$. Define $\delta_{\alpha\beta}$ to be one if $\alpha = \beta$ (mod N), zero otherwise. Then from (20) we readily see that $\Theta\{s_\alpha\}$ satisfies the quasi-periodicity and evenness relations

$$\Theta\{s_\alpha + \delta_{\alpha j} - \delta_{0j}\} = \Theta\{s_\alpha\} , \tag{31a}$$

$$\Theta\{s_\alpha + \tau_{\alpha j}\} = \exp\{-2\pi i(s_j + \varrho_j)\}\ \Theta\{s_\alpha\} \tag{31b}$$

$$\Theta\{-s_\alpha\} = \Theta\{s_\alpha\} , \tag{31c}$$

for $j = 0,\ldots,N-1$.

The RHS of (30) is clearly unchanged by cyclic permutations of $r_1,\ldots,r_N$ (similarly permuting $m_1,\ldots,m_N$), in particular by $r_\alpha \to r_{\alpha-1}$, so

$$\Theta\{s_\alpha\} = \Theta\{s_{\alpha-1} - s_{N-1}\} . \tag{31d}$$

Also, replacing m_α, r_α in (30) by $m_{N+1-\alpha}, r_{N+1-\alpha}$, we see that

$$\Theta\{s_\alpha\} = \Theta\{s_{N+1-\alpha} - s_1\}\,. \tag{31e}$$

The other functions $\Theta\{s\}_0, \ldots, \Theta\{s\}_{2N-2}$ can be expressed in terms of $\Theta\{s\}$ and satisfy similar symmetry relations. In particular,

$$\Theta\{s_\alpha\}_1 = -i \exp\left\{\pi i \left(s_1 + \tfrac{1}{2}\varrho_1\right)\right\}\ \Theta\left\{s_\alpha + \tfrac{1}{2}(\delta_{\alpha 0} - 1 + \tau_{\alpha 1})\right\} \tag{32}$$

and is an odd function:

$$\Theta\{-s_\alpha\}_1 \ = \ -\Theta\{s_\alpha\}_1\,, \quad \Theta\{\mathbf{0}\}_1 \ = \ 0\,. \tag{33}$$

Specialization to a Single Variable v

Now we focus on the case which seems to be relevant to the chiral Potts model, namely when

$$\begin{aligned} v_1 &= v \\ v_\beta &= \pi\beta - \pi + \theta\,, \quad \beta = 2, \ldots, N-1\,. \end{aligned} \tag{34}$$

Thus there is only one variable v, and the expression (18) simplifies to

$$w_\alpha = \int_\theta^v \frac{e^{i(N-2\alpha)v/N}}{\Delta(v)}\,, \tag{35}$$

where $\alpha = 1, \ldots, N-1$. Obviously $w_1, \ldots, w_{N-1}$ are not independent: once one (and N, k) is specified, the rest are determined to within a discrete set of choices. Similarly, $s_1, \ldots, s_{N-1}$ are not independent. In fact, from (25),

$$\Theta\{s\}_3 \ = \ \Theta\{s\}_5 \ = \ \cdots \ = \ \Theta\{s\}_{2N-3} \ = \ 0\,, \tag{36}$$

which relations can be regarded as defining $s_2, \ldots, s_{N-1}$ in terms of s_1 .

The function $\Theta\{s\}_1$ does not vanish identically, but is zero when $v = 0$, i.e. when $s_1 = s_2 = \cdots = s_{N-1} = 0$.

Let $\mathscr{D}$ be the region of the complex v-plane consisting of the lower half plane, together with the real axis except for the points $v = \pi\alpha \pm \theta$ (for all integers α). If we use the above sign conventions for $\Delta(v)$, and take the path of integration in (35) to be a straight line, then the s_α are analytic in $\mathscr{D}$. So are u and μ , but they are not uniquely defined by (3a)–(3c).

We can fix the choice of u and μ by specifying them when v is real, between $-\theta$ and θ . Then they can be chosen so that $-\pi/2 < u < \pi/2$ and $\mu e^{-iv/N}$ is real and positive; each s_α is negative imaginary, between $-\tau_{\alpha 1}/2$ and 0 .

These conventions define the s_α, u, μ uniqely for $v \in \mathscr{D}$. We extend these definitions by analytic continuation beyond $\mathscr{D}$. The points $v = \pi\alpha \pm \theta$ are branch points, and s_α, u, μ become multi-valued functions of v . A convenient tool for handling this multi-valuedness is the following set of automorphisms.

Automorphisms

There are five mappings (or sets of mappings) that change u, v, μ (and hence x, y, a, b, c, d), but leave the relations (3a)–(3d) unchanged. Including the corresponding transformation of $s_1, \dots, s_{N-1}$, they are

$$\begin{aligned}
M_j^{(1)} &: v, u; \mu, s_\alpha \;\to\; v,\ u;\ \omega\mu,\ s_\alpha + \delta_{\alpha j} - \delta_{0j} \\
M_j^{(2)} &: v, u; \mu, s_\alpha \;\to\; v,\ u + 2j\pi;\ \mu,\ s_\alpha + \tau_{\alpha j} \\
M^{(3)} &: v, u; \mu, s_\alpha \;\to\; v,\ \pi - u;\ e^{2iv/N}\mu^{-1},\ -s_\alpha \\
M^{(4)} &: v, u; \mu, s_\alpha \;\to v + \pi, u + \pi;\ \omega\mu,\ s_{\alpha-1} - s_{N-1} + \tfrac{1}{2}(\delta_{\alpha 1} + \tau_{\alpha 2} - \tau_{\alpha 1}) \\
M^{(5)} &: v, u; \mu, s_\alpha \;\to -v,\ -u;\ e^{-2iv/N}\mu,\ s_{N+1-\alpha} - s_1 - \tfrac{1}{2}\tau_{\alpha 1}\,.
\end{aligned} \tag{37}$$

The mappings $M_j^{(1)}$, $M_j^{(2)}$, $M^{(3)}$ manifest the fact that $u, \mu, s_1, \dots, s_{N-1}$ are multi-valued functions of v. By deforming the path of integration in (35) to make an extra loop round the line segment $(\pi j - \pi + \theta, \pi j - \theta)$, we increase w_α by $2K_{\alpha j}$, resulting in the mapping $M_j^{(1)}$ of μ and s_α . Here j can be any integer, but only the values $0, 1, \dots, N-1$ give distinct mappings.

Similarly, including one extra loop round each of the j line segments $(-\theta, \theta)$, $(\pi - \theta, \pi + \theta)$, $(\pi j - \pi - \theta, \pi j - \pi + \theta)$ increases w_α by $2iK'_{\alpha j}$, and results in the mapping $M_j^{(2)}$. The mapping $M^{(3)}$ is obtained by negating $\Delta(v)$ in (35), i.e. moving onto the other Riemann sheet for $\Delta(v)$.

Incrementing v by π replaces w_α by $\omega^{-\alpha}(w_\alpha + K_{\alpha 0} + iK'_{\alpha 1})$ and gives $M^{(4)}$; negating v replaces w_α by $-w_{N-\alpha} - iK'_{\alpha 1}$ and gives $M^{(5)}$. Once v, u, μ are known, then x, y and the ratios $a : b : c : d$ are uniquely determined by (3b) and (3d). Hence we can normalize a, b, c, d so that

$$\begin{aligned}
M_j^{(1)} &: x, y;\ a, b, c, d \to x,\ y;\ \omega a, b, c, \omega d \\
M_j^{(2)} &: \qquad\qquad\quad\ \to \omega^j x,\ \omega^j y;\ \omega^j a, \omega^j b, c, d \\
M^{(3)} &: \qquad\qquad\quad\ \to \omega/y,\ \omega/x;\ c, \omega^{1/2} d, \omega^{-1/2} a, \omega^{-1} b \\
M^{(4)} &: \qquad\qquad\quad\ \to x,\ \omega y,\ a, b, \omega^{-1} c, d \\
M^{(5)} &: \qquad\qquad\quad\ \to x^{-1},\ \omega/y;\ d, \omega^{1/2} c, \omega^{-1/2} b, a\,.
\end{aligned} \tag{38}$$

This makes it clear that the five mappings merely permute (and possibly negate) a^N, b^N, c^N, d^N : they leave the set of relations (1) unchanged. These automorphisms are related to the R, S, T, U of previous papers [4], [10], [20] ; in particular

$$R = M^{(3)} M^{(5)}\,, \quad S = M^{(3)} M_1^{(2)}\,. \tag{39}$$

Zeros of *a*, *b*, *c*, *d*

Take $v \in \mathscr{D}$ and choose u , μ as specified above. By applying Cauchy's theorem to the rectangle with corners $-\theta, \pi - \theta, \pi - \theta - iR, -\theta - iR$ (R large and positive), we can verify that in the limit $v \to -i\infty$,

$$w_\alpha = (\omega^\alpha K_{\alpha 1} + iK'_{\alpha 1})/(\omega^\alpha - 1)\,. \tag{40}$$

It follows that s_α is then

$$s_\alpha = \frac{N-\alpha}{2N} + \tfrac{1}{2}(\varrho_{\alpha-1} - \varrho_1) \,, \quad \alpha = 1, \ldots, N \,, \tag{41}$$

and $u - v = i \ln k$.

Applying the mappings $M^{(3)}$ and $M^{(5)}$, we deduce that if (for $\alpha = 1, \ldots, N$)

$$s_\alpha = \lambda \frac{\alpha - N}{2N} + \frac{\lambda\nu}{2}(\varrho_{\alpha-1} - \varrho_1) \,, \tag{42}$$

where $\lambda, \nu = \pm 1$, then $u \to i\lambda\infty$, $v \to i\nu\infty$. Writing the RHS of (42) as $\zeta_{\lambda\nu\alpha}$, the function $\Theta\{s_\alpha - \zeta_{\lambda\nu\alpha}\}_1$ then vanishes. From (32), so therefore does

$$\Theta\{s_\alpha - \tfrac{1}{2}\lambda g_\alpha - \tfrac{1}{2}\lambda\nu\varrho_\alpha\} \,,$$

where

$$\begin{aligned} g_\alpha &= \alpha/N \,, \quad \alpha = 0, \ldots, N-1 \,, \\ &= g_{\alpha+N} \,, \quad \forall\alpha \,. \end{aligned} \tag{43}$$

Define

$$\begin{aligned} \hat{x} &= C\,\Theta\{s_\alpha - \tfrac{1}{2}g_\alpha + \tfrac{1}{2}\varrho_\alpha\}/\Theta\{s_\alpha + \tfrac{1}{2}g_\alpha + \tfrac{1}{2}\varrho_\alpha\} \,, \\ \hat{y} &= C'\,\Theta\{s_\alpha - \tfrac{1}{2}g_\alpha - \tfrac{1}{2}\varrho_\alpha\}/\Theta\{s_\alpha + \tfrac{1}{2}\lambda g_\alpha - \tfrac{1}{2}\varrho_\alpha\} \,, \\ \hat{\mu} &= C''\,e^{2\pi i\langle s\rangle}\,\Theta\{s_\alpha + \tfrac{1}{2}g_\alpha + \tfrac{1}{2}\varrho_\alpha\}/\Theta\{s_\alpha + \tfrac{1}{2}g_\alpha - \tfrac{1}{2}\varrho_\alpha\} \,, \end{aligned} \tag{44}$$

where

$$\langle s\rangle = N^{-1} \sum_{\alpha=1}^{N-1} s_\alpha$$

and C, C', C'' are some constants. Then from (3b), (3c) these functions $\hat{x}, \hat{y}, \hat{\mu}$ have the same zeros and poles at $v = \pm i\infty$ as do x, y, μ, respectively.

Further, using (31a) and (31b), we find that $\hat{x}, \hat{y}, \hat{\mu}$ transform under $M_j^{(1)}$ and $M_j^{(2)}$ in the same way as x, y, μ; the ratios $\hat{x}/x$, $\hat{y}/y$, $\hat{\mu}/\mu$ being invariant under these automorphisms. It follows at once that the ratios are only two-valued functions of v, being uniqely determined by v and $\Delta(v)$. They can be regarded as defined on two Riemann sheets of the complex v-plane.

By itself, (31d) does not appear sufficient to establish the invariance of the ratios under $M^{(4)}$: one presumably needs to also use the relations (36). However, the s_α are invariant under ${M^{(4)}}^N$, so the ratios are periodic functions of v, of period $N\pi$.

The following statements have not yet been rigorously proved, but (guided by the small-k' limit) they appear to be correct: (i) The ratios $\hat{x}/x$, $\hat{y}/y$, $\hat{\mu}/\mu$ remain finite and non-zero as $v \to \pm i\infty$ on either sheet. (ii) The Θ functions in (44) are non-zero for all finite v.

Assuming these statements to be true, it follows that $\hat{x}/x$ is bounded on both Riemann sheets; analytic except possibly at the branch points, where it is finite. By a straightforward extension of Liouville's theorem in complex variable theory, it follows that $\hat{x}/x$ is a constant. We can choose C so that this ratio is one. Arguing similarly for y and μ, it follows that

$$x = \hat{x}\,, \quad y = \hat{y}\,, \quad \mu = \hat{\mu}\,. \tag{45}$$

From the mapping $M^{(3)}$, the constants C, C', C'' must satisfy

$$CC' = \omega\,, \quad C' = e^{i\pi/N} CC''^{2}\,. \tag{46}$$

Also, when $v = \theta$ and $s_1, \ldots, s_{N-1} = 0$, $x\mu = CC' = e^{i\pi/2N}$. It follows that there exists a single constant χ such that

$$C = e^{i\pi/N}\chi^{-1}\,, \quad C' = e^{i\pi/N}\chi\,,$$
$$C'' = e^{-i\pi/2N}\chi\,, \tag{47}$$

and

$$a:b:c:d = e^{i\pi/2N} e^{2\pi i\langle s\rangle}\,\Theta\left\{s_\alpha - \tfrac{1}{2}g_\alpha + \tfrac{1}{2}\varrho_\alpha\right\} : \chi\, e^{i\pi/N}\,\Theta\left\{s_\alpha - \tfrac{1}{2}g_\alpha - \tfrac{1}{2}\varrho_\alpha\right\} :$$
$$\Theta\left\{s_\alpha + \tfrac{1}{2}g_\alpha - \tfrac{1}{2}\varrho_\alpha\right\} : \chi\, e^{-i\pi/2N} e^{2\pi i\langle s\rangle}\,\Theta\left\{s_\alpha + \tfrac{1}{2}g_\alpha + \tfrac{1}{2}\varrho_\alpha\right\}\,. \tag{48}$$

Thus, in this normalization, and to within factors that are constant or exponential in $s_1, \ldots, s_{N-1}$, the variables a, b, c, d are hyperelliptic theta functions. (For the case $N = 2$, which corresponds to the Ising model, they are proportional to the four Jacobi theta functions $\theta_1, \theta_4, \theta_3, \theta_2$, all with the same argument.)

Expressions for k, k', χ

We can evaluate χ by taking $s_\alpha = \pm\frac{1}{2}(\varrho_{\alpha-1} - \varrho_1) \pm \frac{1}{2}(N - \alpha)/N$, in which case one of a, b, c, d vanishes. Substituting the forms (48) into (1), we obtain three distinct relations:

$$(k'/k)^{1/N} = i\, e^{-\pi i(\varrho_1 + \cdots + \varrho_{N-1})/N}\,\Theta\{g_\alpha\}_1/\Theta\{\varrho_\alpha\}_1\,, \tag{49}$$

$$k^{1/N} = \chi\,\Theta\{\varrho_\alpha\}_1/\Theta\{\varrho_\alpha - g_\alpha\}_1\,,$$
$$= \chi^{-1}\,\Theta\{\varrho_\alpha\}_1/\Theta\{\varrho_\alpha + g_\alpha\}_1\,. \tag{50}$$

We can regard these equations as defining χ, and providing alternative expressions for k, k' . In fact, for N = 2, 3 and 4 we have observed that 13-digit numerical calculations are fitted by (48)–(50), with

$$\chi = i\, e^{i(2-N)\theta/N}\,, \tag{51}$$

As explained in the Appendix, we have also verified this formula to 201 terms in a series expansion for $N = 3$. It seems likely that it is exactly correct, for all N. Remembering that g_α is real and ϱ_α is pure imaginary, (51) is equivalent to

$$\arg\Theta\{g_\alpha + \varrho_\alpha\}_1 = (N-2)\theta/N \tag{52}$$

an intriguingly simple conjecture.

Appendix: The case $N = 3$

Here we consider the case $N = 3$, when the hyperelliptic $\Theta\{s\} = \Theta(s_1, s_2)$ functions can be expressed, using (24), in terms of ordinary single-variable Jacobi theta functions of nomes q and q^3, where $q = e^{2\pi i \varrho}$.

Setting $\phi = \pi/2 - \theta$, the relation (26), for $\alpha = 0$, becomes

$$4\sin^2(2\phi/3) = \frac{3[\Theta(0,0)^2 - \Theta(\frac{1}{2},\frac{1}{2})^2]}{\Theta(0,0)^2}\,, \tag{A1}$$

and hence, using (2),

$$k'^2 \;=\; 1 - k^2 \;=\; \sin^2\phi \;=\; 27q\{1 - 15q + 171q^2 + \cdots\}\,. \tag{A2}$$

Here we regard these equations as defining ϕ, k, k' as functions of q , for $|q| < 1$.

From these definitions, we should like to directly verify the above results (49), (50) and the conjecture (51). These can be written, using (32), as

$$(k'/k)^{1/3} = -i\, e^{\pi i(1+\varrho)/3}\, \Theta\left(-\tfrac{1}{6}+\varrho, \tfrac{1}{6}+\tfrac{1}{2}\varrho\right)/\Theta\left(\tfrac{1}{2},\tfrac{1}{2}+\tfrac{1}{2}\varrho\right)\,, \tag{A3}$$

$$\begin{aligned} k^{1/3} &= \chi\, e^{\pi i/3}\, \Theta\left(\tfrac{1}{2},\tfrac{1}{2}+\tfrac{1}{2}\varrho\right)/\Theta\left(\tfrac{1}{6},-\tfrac{1}{6}-\tfrac{1}{2}\varrho\right)\,, \\ &= \chi^{-1}\, e^{-\pi i/3}\, \Theta\left(\tfrac{1}{2},\tfrac{1}{2}+\tfrac{1}{2}\varrho\right)/\Theta\left(\tfrac{1}{6},-\tfrac{1}{6}+\tfrac{1}{2}\varrho\right)\,, \end{aligned} \tag{A4}$$

$$\chi = e^{i(\pi+\phi)/3}\,. \tag{A5}$$

Define

$$Q(q) \;=\; \prod_{n=1}^{\infty}(1 - q^n) \;=\; \sum_{n=-\infty}^{\infty}(-1)^n q^{n(3n-1)/2}\,, \tag{A6}$$

and write $Q(q^r)$ simply as Q_r . Then from (24) we can prove that

$$\begin{aligned} \Theta(0,0) \;+\; \Theta\left(\tfrac{1}{2},\tfrac{1}{2}\right) \;&=\; 2(Q_2Q_6)^5/(Q_1Q_3Q_4Q_{12})^2\,, \\ \Theta(0,0) \;-\; \Theta\left(\tfrac{1}{2},\tfrac{1}{2}\right) \;&=\; 8\,q\,(Q_4Q_{12})^2/(Q_2Q_6)\,, \\ \Theta\left(\tfrac{1}{2},\tfrac{1}{2}+\tfrac{1}{2}\varrho\right) \;&=\; \Theta\left(\tfrac{1}{2},\tfrac{1}{2}-\tfrac{1}{2}\varrho\right) \;=\; Q(q)^2\,. \end{aligned} \tag{A7}$$

We also conjecture the following five identities

$$\Theta\left(\tfrac{1}{2},\tfrac{1}{2}\right) = (Q_1Q_3)^2/(Q_2Q_6)\,, \tag{A8a}$$

$$\Theta(0,0)^3 Q_1^3 Q_3^3 = Q_1^{12} \;+\; 27q\, Q_3^{12}\,, \tag{A8b}$$

$$\Theta\left(-\tfrac{1}{6}+\varrho,\tfrac{1}{6}+\tfrac{1}{2}\varrho\right) = (1 - e^{-2\pi i/3})\, Q_3^2\,, \tag{A8c}$$

$$\Theta\left(\tfrac{1}{6},-\tfrac{1}{6}\pm\tfrac{1}{2}\varrho\right) = [Q_2^3Q_3/(Q_1Q_6)] \;\pm\; i\,(3q)^{1/2}[Q_1Q_6^3/(Q_2Q_3)]\,, \tag{A8d}$$

$$\Theta\left(\tfrac{1}{6},-\tfrac{1}{6}+\tfrac{1}{2}\varrho\right)\,\Theta\left(\tfrac{1}{6},-\tfrac{1}{6}-\tfrac{1}{2}\varrho\right) = \Theta(0,0)\, Q_1\, Q_3\,. \tag{A8e}$$

We have verified these conjectures to order q^{200} in series expansions in powers of q, using Fortran with integer arithmetic. The hyperelliptic Θ functions herein, as well as $Q(q)$, have q-expansions that are sparse, with the few non-zero coefficients being of order one. By writing the identities solely in terms of sums of products of such functions (with *no* divisions), we avoided integer overflows: for instance, to this order the largest coefficient in the expansion of $Q(q)^{12}$ is only 5,187,456.

The desired formulae (A3), (A4), (A5) follow from (A1), (A2) and (A6)–(A8e), so we have verified them to order q^{200}. The equation (A3) becomes very simple:

$$(k'/k)^{1/3} = 3^{1/2}\, q^{1/6}\, \{Q(q^3)/Q(q)\}^2 . \tag{A9}$$

Note

A number of colleagues have pointed out that the conjectures (A8a)–(A8b) can be proved, either by the general theory of modular functions, or from specific identities already in the literature. In particular, the author is indebted to G.E. Andrews for showing their connection with the theory of generalized Frobenius partitions [29].

References

1. H. Au-Yang, B.M. McCoy, J.H.H. Perk, S.Tang, M.L. Yan: Phys. Lett. A **123** (1987) 219–223
2. B.M. McCoy, J.H.H. Perk, S. Tang, C.H. Sah: Phys.Lett A **125** (1987) 9–14
3. H. Au-Yang, B.M. McCoy, J.H.H. Perk, S. Tang: In: M. Kashiwara and T. Kawai (eds.) Algebraic Analysis vol. 1. Academic Press, New York 1988, pp. 29–39
4. R.J. Baxter, J.H.H.Perk, H.Au-Yang: Phys.Lett. A **128** (1988) 138–142
5. R.J. Baxter: J.Stat.Phys. **52** (1988) 639–667
6. R.J. Baxter: Phys. Lett. A **133** (1988) 185–189
7. J.H.H. Perk Proc. of Symposia in Pure Mathematics, vol. **49**, part 1 (1989) 341–354
8. G. Albertini, B.M. McCoy, J.H.H. Perk, S. Tang: Nucl. Phys. B **314** (1989) 741–763
9. G. Albertini, B.M. McCoy, J.H.H. Perk: Adv. Stud. Pure Math. **19** (1989) 1–55
10. H. Au-Yang, J.H.H.Perk: Adv. Stud. Pure Math. **19** (1989) 57–94
11. R.J. Baxter: Adv. Stud. Pure Math. **19** (1989) 95–116
12. G. Albertini, B.M. McCoy, J.H.H. Perk: Phys. Lett. A **135** (1989) 159–166
13. G. Albertini, B.M. McCoy, J.H.H. Perk: Phys. Lett. A **139** (1989) 204–212
14. R.J. Baxter: Phys. Lett. A **140** (1989) 155–157
15. R.J. Baxter: J. Stat. Phys. **57** (1989) 1–39
16. V.B. Mateev and A.O. Smirnov: Some comments on the solvable chiral Potts model. Preprint 1989
17. H. Itoyama, B.M. McCoy, J.H.H. Perk: Intnl. J. Mod. Phys. B **4** (1990) 995–1001
18. D. Bernard and V. Pasquier: Intnl. J. Mod. Phys. B **4** (1990) 913–927
19. V.V. Bazhanov, Yu. G. Stroganov: J. Stat. Phys. **59** (1990) 799–817
20. R.J. Baxter, V.V. Bazhanov, J.H.H. Perk: Intnl. J. Mod. Phys. B **4** (1990) 803–870
21. G. Albertini, B.M. McCoy: Nucl. Phys. B**350** (1991) 745–788
22. R.J. Baxter: Phys. Lett. A **146** (1990) 110–114
23. B.M. McCoy, S. Roan: Phys. Lett. A **150** (1990) 347–354
24. R.J. Baxter: Calculation of the eigenvalues of the transfer matrix of the chiral Potts model. For the Proceedings of the Fourth Asia Pacific Conference 1990 (World Scientific)

25. V.V. Bazhanov, R.M. Kashaev: Cyclic L-operator related with a 3-state R-matrix, preprint RIMS-702 (1990). To appear in Comm. Math. Phys.
26. V.V. Bazhanov, R.M. Kashaev, V.V. Mangazeev, Yu.G. Strogarov: $(Z_N)^{n-1}$ generalization of the chiral Potts model. To appear in Comm. Math. Phys.
27. R.J. Baxter: J. Stat. Phys. **63** (1991) 509–529
28. S. Kowalevski: Acta Mathematica **12** (1889) 177–232
29. G.E. Andrews: Memoirs of the American Math. Soc. **49** (1984) no. 301, iv.

Abstract Compact Group Duals, Operator Algebras and Quantum Field Theory

Sergio Doplicher

Dipartimento di Matematica, Università di Roma "La Sapienza", I-00185 Rome, Italy

1. Introduction

In this report we will review joint work with J. E. Roberts on the existence of a unique compact group of internal symmetries, and of field operators, on which the group acts as global gauge transformations, which commute or anticommute in the normal way at spacelike separations [16].

These results are deduced from first principles (essentially, the postulate of locality) involving solely the observable quantities, for local Quantum Theories without massless particles on the four dimensional Minkowski space.

On the mathematical side, these results stem out of a new duality theory for compact groups which has been developed for this purpose. While the classical theory of Tannaka and Krein (cf. e.g. [22]) characterizes the dual of a compact group within the category of finite dimensional vector spaces, the new theory characterizes such a dual within abstract categories.

More specifically, our abstract compact group duals will be strict symmetric monoidal C^*-categories with conjugates, having subobjects and direct sums, where the selfintertwiners of the monoidal unit reduce to $\mathbb{C}$ [15].

Each such category can be "locally" realized as a full subcategory of End $\mathfrak{A}$ (cf. Sect. 2), i.e. as an action on a C^*-algebra.

Now if a category $\mathcal{T}$ with the mentioned properties acts as a full subcategory of End $\mathfrak{A}$ on a C^*-algebra $\mathfrak{A}$ with centre $\mathbb{C} \cdot I$, a key construction [14] provides a cross product $\mathfrak{A} \times \mathcal{T}$ containing $\mathfrak{A}$ as a C^*-subalgebra with trivial relative commutant, and a compact group G of automorphisms of $\mathfrak{A} \times \mathcal{T}$ having $\mathfrak{A}$ as the fixed point subalgebra, such that the objects of $\mathcal{T}$ are the restrictions to $\mathfrak{A}$ of inner endomorphisms of $\mathfrak{A} \times \mathcal{T}$, induced by Hilbert spaces with support I in $\mathfrak{A} \times \mathcal{T}$; these Hilbert spaces together with $\mathfrak{A}$ generate $\mathfrak{A} \times \mathcal{T}$ and are G-spaces.

Assigning to each object of $\mathcal{T}$ the representation of G on the corresponding Hilbert space in $\mathfrak{A} \times \mathcal{T}$ and to each arrow the G-module map given by multiplication in $\mathfrak{A} \times \mathcal{T}$ defines a functor of $\mathcal{T}$ into the category of G-Hilbert spaces in $\mathfrak{A} \times \mathcal{T}$ which is an isomorphism of strict symmetric monoidal categories of $\mathcal{T}$ with a category of continuous unitary finite dimensional representations of G.

These central results, when applied "locally" to an abstract category $\mathcal{T}$ in the earlier mentioned class, yield at the same time the existence of a unique compact

Proceedings of the International Congress of Mathematicians, Kyoto, Japan, 1990

group G and of an isomorphism of $\mathcal{T}$ with a representation category of G as strict symmetric monoidal categories [15, Theorem 6.1].

By completely different methods, a parallel result has been independently established by Pierre Deligne in a recent paper [7] characterizing as abstract categories the categories of representations of an algebraic group on finite dimensional vector spaces over a field of characteristic zero.

In the Quantum Field Theory context, we let $\mathfrak{A}$ be the C^*-algebra generated by all local observables (Sect. 3). The superselection structure associated to the given vacuum state ω_0 on $\mathfrak{A}$ is determined by $\mathfrak{A}$ itself and can be described by a category $\mathcal{T}$ fulfilling exactly the axioms mentioned above. In the important case of "localizable charges", $\mathcal{T}$ is actually a full subcategory of End $\mathfrak{A}$. The monoidal structure induced from End $\mathfrak{A}$ is related to composition of superselection charges; the symmetry to the intrinsic notion of statistics of superselection sectors; the existence of conjugates is a consequence of particle-antiparticle symmetry of superselection quantum numbers. All these pieces of structure have been deduced, essentially from the locality principle, in earlier works [9, 10], and can be summarized saying that superselection structure is described by a full subcategory $\mathcal{T}$ of End $\mathfrak{A}$ with the above properties. The corresponding cross product $\mathfrak{A}\times\mathcal{T}$ describes *field operators* and the dual action G on $\mathfrak{A}\times\mathcal{T}$ provides the compact group of internal symmetries. With some technical modifications needed to take care of "quantum topological charges" this construction yields the more general results described in Section 3.

2. Abstract Compact Group Duals and Operator Algebras

Which *abstract* categories can appear as the dual object of a compact group? The general answer can be summarized in the following theorem [15]:

2.1. Theorem. *Let $\mathcal{T}$ be a strict symmetric monoidal C^*-category with conjugates, having subobjects and direct sums, such that the selfinterwiners of the monoidal unit are $\mathbb{C}$. There is a unique compact group G and an isomorphism of symmetric monoidal C^*-categories of $\mathcal{T}$ with a category of finite dimensional continuous unitary representations of G.*

Rather than giving the formal definitions involved in this statement, it is easier to see how they arise naturally in an important model.

Let $\mathfrak{A}$ be a C^*-algebra with centre $\mathbb{C}\cdot I$. We will denote by End $\mathfrak{A}$ the category whose objects are the unital endomorphism of $\mathfrak{A}$ and whose arrows (ϱ,ϱ') for object ϱ,ϱ' are their intertwiners in $\mathfrak{A}$:

$$(\varrho,\varrho') = \{T\in\mathfrak{A} \mid T\varrho(A) = \varrho'(A)T\ ,\ A\in\mathfrak{A}\}\,. \tag{1}$$

End $\mathfrak{A}$ is a C^**-category* with the linear structure, norm and * induced by $\mathfrak{A}$ and composition of arrows $T\circ S$, when defined, given by product in $\mathfrak{A}$. Further, it is a *monoidal* C^*-category, where the monoidal operations are composition in the semigroup with identity ι (the identity automorphism) of unital endomorphisms, and on arrows

$$\begin{aligned} &T \in (\varrho, \varrho') \, , \; S \in (\sigma, \sigma') \to T \times S \in (\varrho\sigma, \varrho'\sigma') \, ; \\ &T \times S = T\varrho(S) = \varrho'(S)T \, . \end{aligned} \tag{2}$$

These operations are always defined, *strictly* associative, and fulfill obvious compatibility relations with composition, so that End $\mathfrak{A}$ is a *strict* monoidal C^*-category. The selfinterwiners of the monoidal unit ι are the elements of the centre of $\mathfrak{A}$, i.e. $(\iota, \iota) = \mathbb{C} \cdot I$.

We are interested in *full monoidal subcategories* $\mathcal{T}$ of End $\mathfrak{A}$ (i.e. $\mathcal{T}$ is specified by a semigroup of unital endomorphisms with identity ι and all arrows given by (1)) which are the model for a representation category of a group. The monoidal operations on $\mathcal{T}$ should behave like the *tensor product* in a subcategory of Hilbert spaces where $\otimes$ is *strictly associative*. This suggests the following

2.2. Definition. *A* symmetry ε *for a strict monoidal C^*-category $\mathcal{T}$ is an assignment of a unitary $\varepsilon(\varrho, \sigma) \in (\varrho\sigma, \sigma\varrho)$ for each pair of objects ϱ, σ, so that if $R \in (\varrho, \sigma)$, $R' \in (\varrho', \sigma')$,*

$$\varepsilon(\sigma, \sigma') \circ R \times R' = R' \times R \circ \varepsilon(\varrho, \varrho') \tag{3}$$

and

$$\varepsilon(\sigma, \varrho) \circ \varepsilon(\varrho, \sigma) = I_{\varrho\sigma} \tag{4}$$

$$\varepsilon(\iota, \varrho) = \varepsilon(\varrho, \iota) = I_\varrho \tag{5}$$

$$\varepsilon(\varrho\sigma, \tau) = \varepsilon(\varrho, \tau) \times I_\sigma \circ I_\varrho \times \varepsilon(\sigma, \tau) \, . \tag{6}$$

The symmetry ε defines canonical unitary representations $\varepsilon_\varrho^{(n)}$ of the permutation groups $\mathbb{P}(n)$ of n objects with values in (ϱ^n, ϱ^n), $n = 2, 3, \cdots$, obtained assigning $I_{\varrho^{r-1}} \times \varepsilon(\varrho, \varrho) \times I_{\varrho^{n-r-1}}$ to the exchange of r with $r+1$ (where $I_\sigma \in (\sigma, \sigma)$ is the identity interwiner).

Crucial for *compactness* is the existence of conjugates. The following definition is modelled on the properties of complex conjugates of finite dimensional group representations.

2.3. Definition. *The strict symmetric monoidal C^*-category $(\mathcal{T}, \varepsilon)$ has conjugates if to each object ϱ of $\mathcal{T}$ there is an object $\overline{\varrho}$ and arrows $R, \overline{R}$ fulfilling*

$$R \in (\iota, \overline{\varrho}\varrho) \, , \; \overline{R} \equiv \varepsilon(\overline{\varrho}, \varrho) \circ R \, , \tag{7}$$

$$\overline{R}^* \times I_\varrho \circ I_\varrho \times R = I_\varrho \, , \tag{8}$$

$$R^* \times I_{\overline{\varrho}} \circ I_{\overline{\varrho}} \times \overline{R} = I_{\overline{\varrho}} \, . \tag{9}$$

Eventually we will say that $\mathcal{T}$ has *subobjects* if each non zero selfadjoint projection $E \in (\varrho, \varrho)$ is the range of an isometry $W \in (\sigma, \varrho)$ for some object σ; $\mathcal{T}$ has *direct sums* if any pair of objects σ, σ' are subobjects of some object ϱ, associated to complementary projections in (ϱ, ϱ).

Now the statement of Theorem 2.1 is explained. When $\mathscr{T}$ is actually a full subcategory of End $\mathfrak{A}$, and $\mathfrak{A}$ has centre $\mathbb{C}\cdot I$, slightly more general axioms suffice, replacing existence of conjugates with abundance of objects with "determinant one". We can define determinants in an abstract category $\mathscr{T}$ as above along the following lines.

It can be shown [15, Sect. 2] that each object ϱ has an *integer dimension* $d(\varrho)$ given by

$$R^* \circ R = d(\varrho) \cdot I \tag{10}$$

where R obeys the conjugate equations (7,8,9). The objects with dimension one are shown to form a *group* and their equivalence classes form an *abelian group* $\mathscr{G}_0$. The *determinant map*

$$\text{Objects } \mathscr{T} \to \mathscr{G}_0$$

assigns to ϱ the class of the subobject (automatically of dimension one) of $\varrho^{d(\varrho)}$ associated to the projection $\varepsilon_\varrho^{d(\varrho)}(A_{d(\varrho)}) \in (\varrho^{d(\varrho)}, \varrho^{d(\varrho)})$, where $A_{d(\varrho)}$ is the totally antisymmetric projection in the group algebra of $\mathbb{P}(d(\varrho))$ [15, Sect. 3].

The determinant map has the properties

$$\det \varrho_1 \oplus \varrho_2 = \det \varrho_1 \cdot \det \varrho_2 \tag{11}$$

$$\det \overline{\varrho} = (\det \varrho)^{-1} \tag{12}$$

$$\det \varrho_1\varrho_2 = 1 \quad \text{if} \det \varrho_1 = \det \varrho_2 = 1\,. \tag{13}$$

Objects with determinant one are called *special.* An object ϱ of dimension d is special if and only if there is an *isometry* $R \in (\iota, \varrho^d)$ such that

$$R \circ R^* = \varepsilon_\varrho^{(d)}(A_d) \tag{14}$$

$$R^* \times I_\varrho \circ I_\varrho \times R = (-1)^{d-1}\frac{1}{d} I_\varrho\,.$$

We conclude that $(\mathscr{T}, \varepsilon)$ is *specially directed* in the sense that each finite set of objects $\varrho_1, \cdots, \varrho_n$ are dominated by a special object ϱ: it suffices to choose

$$\varrho = \varrho_1 \oplus \cdots \oplus \varrho_n \oplus \overline{\varrho}_n \oplus \cdots \oplus \overline{\varrho}_1\,, \tag{15}$$

which will be special by (11) and (12).

Now we can state precisely in which sense $\mathscr{T}$ is locally described by a full subcategory of End $\mathfrak{A}$ [15, Sect. 4].

2.4. Theorem. *Let $\mathscr{T}$ be as in Theorem 2.1, ϱ a special object and $\mathscr{T}_\varrho$ the full subcategory with objects $\{\iota, \varrho, \varrho^2, \cdots\}$. There is a C^*-algebra $\mathscr{O}_\varrho$ with centre $\mathbb{C}\cdot I$, a strict symmetric monoidal full subcategory $\mathscr{T}_{\hat{\varrho}}$ of* End $\mathscr{O}_\varrho$ *with objects the powers of a special object $\hat{\varrho}$, and a functor of $\mathscr{T}_\varrho$ onto $\mathscr{T}_{\hat{\varrho}}$ which is an isomorphism of symmetric monoidal categories.*

To construct $\mathscr{O}_\varrho$ one considers, for each $k \in \mathbb{Z}$, the inductive limit $\mathscr{O}_\varrho^k$ of the Banach spaces $(\varrho^r, \varrho^{r+k})$, $r, r+k \in \mathbb{N}_0$, under the maps

$$T \in (\varrho^r, \varrho^{r+k}) \to T \times I_\varrho \in (\varrho^{r+1}, \varrho^{r+k+1})\,.$$

The composition of maps and the adjoint in $\mathscr{T}$ make $\mathcal{O}^k$, $k \in \mathbb{Z}$, into a $\mathbb{Z}$-graded *-algebra which can be completed in a unique way into a C^*-algebra $\mathcal{O}_\varrho$ such that the grading is given by a continuous action of $\mathbb{T}$.

Denoting by $i: T \to i(T)$ the embedding of (ϱ^r, ϱ^s) into $\mathcal{O}_\varrho$ we have

$$\begin{aligned} i(T \times I_\varrho) &= i(T)\,, \\ i(I_\varrho \times T) &= \hat{\varrho} \circ i(T)\,, \end{aligned} \tag{16}$$

which defines the endomorphism $\hat{\varrho}$.

Theorem 2.4 brings the abstract duality problem into its natural context, non commutative operator algebras.

The main result, crucial both for the proof of Theorem 2.1 and for the application to QFT, deals with a monoidal full subcategory $\mathscr{T}$ of End $\mathfrak{A}$ (hence $\mathscr{T}$ is specified by its objects, a semigroup Δ of unital endomorphisms with identity ι) which has a symmetry ε (but $\mathscr{T}$ does not need to have subobjects and direct sums: if it did, $\mathscr{T}$ would have conjugates if and only if it were specially directed, cf. the comments above and [12, 9, 10]).

Such a category arises naturally if $\mathfrak{A}$ is embedded in a C^*-algebra $\mathfrak{B}$ as the fixed points under a compact group G of automorphisms of $\mathfrak{B}$ and $\mathfrak{A}' \cap \mathfrak{B} = \mathbb{C} \cdot I$. In this case a finite dimensional *Hilbert space* H in $\mathfrak{B}$ (i.e. $\psi^* \psi' \in \mathbb{C} \cdot I$ if $\psi, \psi' \in H$) with support I (i.e. the left annihilator of H is zero) induces an endomorphism ϱ_H of $\mathfrak{A}$ if and only if H is G stable; then

$$H = \{\psi \in \mathfrak{B} \mid \psi A = \varrho_H(A)\psi\,, \quad A \in \mathfrak{A}\}\,. \tag{17}$$

Then H carries a unitary representation U_H of G given by $U_H(g)\psi = g(\psi), \psi \in H$. The representations so obtained and their intertwiners form a category denoted $\mathscr{U}(\mathfrak{B}, G)$. The linear operators (H, H') between Hilbert spaces H, H' as above form a linear subspace of $\mathfrak{B}$ (spanned by $H'H^*$) and their G-invariant elements $(H, H')_G$ lie in $\mathfrak{A}$ and are the arrows of $\mathscr{U}(\mathfrak{B}, G)$. Also $(H, H')_G \subset (\varrho_H, \varrho_{H'})$ and $U_H \to \varrho_H$, $T \in (U_H, U_{H'}) \to T \in (\varrho_H, \varrho_{H'})$ sets up an isomorphism of monoidal categories of $\mathscr{U}(\mathfrak{B}, G)$ with *the full subcategory* $\mathscr{S}(\mathfrak{B}, G)$ *of* End $\mathfrak{A}$ *with objects* $\{\varrho_H, H$ *a finite dimensional G-Hilbert space in* $\mathfrak{B}$ *with support* $I\}$. The monoidal structure in $\mathscr{U}(\mathfrak{B}, G)$ is induced by the operator product in $\mathfrak{B}$ which defines a *strictly associative* tensor product of Hilbert spaces in $\mathfrak{B}$.

This monoidal structure has a natural symmetry θ defined by

$$\begin{aligned} &\theta_{H,H'} \in (HH', H'H)\,, \\ &\theta_{H,H'}\psi\psi' = \psi'\psi \;;\; \psi \in H\;,\; \psi' \in H'\,, \end{aligned} \tag{18}$$

where actually $\theta_{H,H'} \in \mathfrak{A}$. Then $(\mathscr{U}(\mathfrak{B}, G), \theta)$ and $(\mathscr{S}(\mathfrak{B}, G), \theta)$ are isomorphic as strict symmetric monoidal categories [13]. Representations with determinant one correspond to special objects.

Every category of finite dimensional continuous representations of a compact group G can be embedded in a category $\mathscr{U}(\mathfrak{B}, G)$ as above [13, Sect. 7].

2.5. Definition. *We will say that* $(\mathscr{T}, \varepsilon)$ *is a* **symmetric monoidal subcategory** *of* $(\mathscr{S}(\mathfrak{B}, G), \theta)$ *if* $\mathscr{T}$ *is a monoidal subcategory of* $\mathscr{S}(\mathfrak{B}, G)$ *and* ε *is just the restriction* *of the symmetry θ to objects in $\mathcal{T}$. In this case we will denote by $H(\mathcal{T})$ the Hilbert spaces in $\mathfrak{B}$ inducing the objects of $\mathcal{T}$.*

The main result of [14] provides a converse construction given $\mathfrak{A}$ and $\mathcal{T}$.

2.6. Theorem. *Let $\mathfrak{A}$ be a C^*-algebra with centre $\mathbb{C} \cdot I$ and $(\mathcal{T}, \varepsilon)$ a symmetric, specially directed full monoidal subcategory of* End $\mathfrak{A}$.

There is a C^*-algebra $\mathfrak{B}$ and a strongly compact group G of automorphisms of $\mathfrak{B}$ such that

1) $\mathfrak{A} = \mathfrak{B}^G$;
2) $\mathfrak{A}' \cap \mathfrak{B} = \mathbb{C} \cdot I$;
3) *$(\mathcal{T}, \varepsilon)$ is a symmetric monoidal subcategory of $(\mathcal{S}(\mathfrak{B}, G), \theta)$;*
4) *$\mathfrak{B}$ is generated as a C^*-algebra by $\mathfrak{A}$ and $H(\mathcal{T})$.*

As a consequence, G is the stabilizer of $\mathfrak{A}$ in Aut $\mathfrak{B}$ *and $(\mathfrak{B}, G)$ is unique up to isomorphisms that leave $\mathfrak{A}$ pointwise fixed.*

The C^*-algebra $\mathfrak{B}$ may be called the crossed product $\mathfrak{A} \times \mathcal{T}$ and G the dual action. The group G would be automatically compact provided we replace it by its strong closure if needed [13].

Note that the Theorem provides at the same time G and a representation category of G isomorphic to $\mathcal{T}$, namely the image in $\mathcal{U}(\mathfrak{B}, G)$ of the subcategory $\mathcal{T}$ of $\mathcal{S}(\mathfrak{B}, G)$.

The special case $\mathcal{T} = \mathcal{T}_\varrho$, ϱ a special object yields a compact Lie group. Combined with Theorem 2.4 this yields a compact Lie group G_ϱ for each special object of a category $\mathcal{T}$ as in Theorem 2.1; the group G of Theorem 2.1 is obtained as the projective limit of the compact Lie groups G_ϱ.

The proof of Theorem 2.6 is reduced to the case $\mathcal{T} = \mathcal{T}_\varrho$. Indeed, if $\hat{\mathcal{T}}_\varrho$ is the full subcategory of End $\mathfrak{A}$ with objects the objects of $\mathcal{T}$ dominated by a special object ϱ, then $\mathfrak{A} \times \hat{\mathcal{T}}_\varrho$ is easily constructed from $\mathfrak{A} \times \mathcal{T}_\varrho$ and $\mathfrak{A} \times \mathcal{T}$ can be obtained as the inductive limit of the C^*-algebras $\mathfrak{A} \times \hat{\mathcal{T}}_\varrho$.

The construction of $\mathfrak{A} \times \mathcal{T}_\varrho$ is then central, and we outline the main ideas involved. Being generated by a special G-Hilbert module H and by $\mathfrak{A}$, $\mathfrak{A} \times \mathcal{T}_\varrho$ will then contain both $\mathfrak{A}$ and the Cuntz algebra [6] $\mathcal{O}_d$ generated by H, where $d = d(\varrho) = \dim H$, with intersection the fixed points $\mathcal{O}_G$ in $\mathcal{O}_d$ under the action of G. This action is canonical in the sense that it is induced by the unitary representation of G on H, $U(g)\psi = g(\psi)$, $\psi \in H$, $g \in G$. Thus, if the system $(\mathfrak{B}, G)$ is given, we have monomorphisms $\mu : \mathcal{O}_G \to \mathfrak{A}$, $\pi : \mathfrak{A} \to \mathfrak{B}$, $\zeta : \mathcal{O}_d \to \mathfrak{B}$ such that

$$\zeta(\psi)\pi(A) = \pi\varrho(A)\zeta(\psi), \quad A \in \mathfrak{A}, \quad \psi \in H \tag{19}$$

$$\mu \circ \sigma = \varrho \circ \mu \tag{20}$$

(where σ is the endomorphism of $\mathcal{O}_d$ induced by the generating Hilbert space H) such that the following diagram is commutative

$$\begin{array}{ccc} \mathfrak{A} & \xrightarrow{\pi} & \mathfrak{B} \\ \uparrow{\scriptstyle \mu} & & \uparrow{\scriptstyle \zeta} \\ \mathcal{O}_G & \longrightarrow & \mathcal{O}_d\,. \end{array} \tag{21}$$

As a consequence of $\mathfrak{A}' \cap \mathfrak{B} = \mathbb{C} \cdot I$, we have

$$\mu((\sigma^r, \sigma^s)) = (\varrho^r, \varrho^s)\ ;\ r, s \in \mathbb{N}_0\,. \tag{22}$$

But G and μ are *not* known a priori. The solution of the problem involves three steps.

a) Since ϱ is special, identifying G with its faithful representations by $d \times d$ unitary matrices given by the action on H, we have $G \subset SU(d)$. Then $\mathcal{O}_{SU(d)} \subset \mathcal{O}_G$. Now $\mu_0 \equiv \mu|_{\mathcal{O}_{SU(d)}}$ is actually *determined* by the data, $\varepsilon_\varrho^{(n)}, n = 2, 3, \cdots$ and R fulfilling (14) for ϱ.

b) Assuming that G and μ are given fulfilling (20) and $\mu((\sigma^r, \sigma^s)) \subset (\varrho^r, \varrho^s)\ ;\ r, s \in \mathbb{N}_0$ (cf. (22)), find a *universal solution* to the diagram (21) with condition (19).

c) Apply this construction to $SU(d)$, μ_0. We get a C^*-algebra $\tilde{\mathfrak{B}}$ with an action of $SU(d)$ with $\mathfrak{A}$ as fixed points but, since (22) does not hold for μ_0 in general, we will not have $\mathfrak{A}' \cap \tilde{\mathfrak{B}} = \mathbb{C} \cdot I$.
However $\mathfrak{A}' \cap \tilde{\mathfrak{B}}$ is *commutative*, and the system $(\tilde{\mathfrak{B}}, SU(d))$ can be obtained as the induced C^* system from a unique system $(\mathfrak{B}, G)$ where now $\mathfrak{A} = \mathfrak{B}^G$ fulfills $\mathfrak{A}' \cap \mathfrak{B} = \mathbb{C} \cdot I$.

Then G and μ fulfilling (20),(22) are obtained and we can identify $(\mathfrak{B}, G)$ with the universal solution of step b).

Step a) involves the C^*-algebraic version of the theory of invariants for $SU(d)$ [12].

If $S \in H^d$ is a normalized totally antisymmetric vector, $S \in \mathcal{O}_{SU(d)}$ and, together with $\theta^{(n)}(p), p \in \mathbb{P}(n)$, $n = 2, 3, \cdots$ (where $\theta^{(n)}(p) \in (H^n, H^n)$ permutes the factors in H^n by p), generates $\mathcal{O}_{SU(d)}$ as a C^*-algebra, and this C^*-algebra is *simple*.

Now if ϱ is special symmetric, $\varepsilon_\varrho^{(n)}$ and R (cf. eq. (14)) will generate a C^*-algebra isomorphic to $\mathcal{O}_{SU(d)}$, and we can define a unique monomorphism μ_0 of $\mathcal{O}_{SU(d)}$ into $\mathfrak{A}$, fulfilling (20), by

$$\mu_0(\theta^{(n)}(p)) = \varepsilon_\varrho^{(n)}(p)\,,\ p \in \mathbb{P}(n)\,, \quad n = 2, 3, \ldots$$
$$\mu_0(S) = R\,.$$

Step b) involves first an algebraic construction [14, Sect. 2]. With ${}^\circ\mathcal{O}_d$ the dense * subalgebra of $\mathcal{O}_d$ generated by H, and ${}^\circ\mathcal{O}_G$ its G invariant part, we can regard ${}^\circ\mathcal{O}_d$ and $\mathfrak{A}$ as ${}^\circ\mathcal{O}_G$-bimodules letting $X \in {}^\circ\mathcal{O}_G$ act by multiplication in ${}^\circ\mathcal{O}_d$ and by multiplication with $\mu(X)$ in $\mathfrak{A}$. The ${}^\circ\mathcal{O}_G$-module tensor product $\mathfrak{A} \otimes_{\mu_0} {}^\circ\mathcal{O}_d$ can be made into an algebra by requiring that $I \otimes_{\mu_0} H$ induces ϱ on $\mathfrak{A} \otimes I$, i.e. setting

$$\begin{gathered} A \otimes_\mu X \cdot B \otimes_\mu Y = A\varrho^n(B) \otimes_\mu XY\,, \\ A,\ B \in \mathfrak{A}\ ,\ X \in H^n,\ Y \in {}^\circ\mathcal{O}_d\,. \end{gathered} \tag{23}$$

To define the *-operation it is essential to use the fact that $S \in {}^\circ\mathcal{O}_G$ since $G \subset SU(d)$. Since $\psi^* = S^*S\psi^* = S^*j(\psi)$, where j is the antilinear map of H into $H^{d-1}, j(\psi) = d(-1)^{d-1}\psi^*S$ (as $S^*\sigma(S) = (-1)^{d-1}\frac{1}{d}I$, cf. also (14)), we can define the adjoint map on the generators $\mathfrak{A}\otimes_\mu H$ by

$$\begin{aligned}(A \otimes_\mu \psi)^* &= (A \otimes_\mu I \cdot I \otimes_\mu \psi)^* = I \otimes_\mu \psi^* \cdot A^* \otimes_\mu I = \\ &= I \otimes_\mu S^* j(\psi) \cdot A^* \otimes_\mu I = \mu(S^*)\varrho^{d-1}(A^*) \otimes_\mu j(\psi)\,.\end{aligned} \tag{24}$$

Along these lines it is possible to make $\mathfrak{A}\otimes_\mu {}^\circ\mathcal{O}_d$ into a *-algebra and there is a unique C^*-norm extending the norm of $\mathfrak{A}$ which is continuous for the action of G. The completion $\mathfrak{A}\otimes_\mu\mathcal{O}_d$ in this norm is the desired universal solution to (19), (21) [14, Sect. 3].

Taking now $\mu = \mu_0$ given by step a) in this construction we find a C^*-system $(\tilde{\mathfrak{B}}, SU(d))$ with fixed points $\mathfrak{A}$ such that, as a consequence of the conditions $\mu_0(\theta(p)) = \varepsilon_\varrho(p)$ we have [13, Theorem 5.2]

$$\mathfrak{A}' \cap \tilde{\mathfrak{B}} = \tilde{\mathfrak{B}}' \cap \tilde{\mathfrak{B}}\,. \tag{25}$$

The desired system $(\mathfrak{B}, G)$ is now constructed in step c) noting that by (25) and $\mathfrak{A}' \cap \mathfrak{A} = \mathbb{C} \cdot I$, $SU(d)$ acts ergodically on $\tilde{\mathfrak{B}}' \cap \tilde{\mathfrak{B}}$, hence, by compactness, transitively on its spectrum. Picking ϕ in that spectrum, G will be the stabilizer of ϕ; the smallest closed two sided ideal J in $\tilde{\mathfrak{B}}$ containing $\ker\phi$ will be G stable hence G will act on $\mathfrak{B} = \tilde{\mathfrak{B}}/J$ providing the desired solution [11].

Now G is known and the identification of $\mathcal{O}_G$ with $\mathfrak{A} \cap \mathcal{O}_d$ in $\mathfrak{B}$ defines μ, fulfilling (22) since $\mathfrak{A}' \cap \mathfrak{B}$ is now $\mathbb{C} \cdot I$. The corresponding universal solution of (19), (21) $(\mathfrak{A} \otimes_\mu \mathcal{O}_d, G)$ is now isomorphic to $(\mathfrak{B}, G)$ [14].

Theorem 2.6 admits spatial versions of direct use in QFT. If m denotes the conditional expectation $\mathfrak{A} \times \mathcal{T} \to \mathfrak{A}$ given by integration of the action of G over the normalized Haar measure, a faithful representation π_0 of $\mathfrak{A}$ induces a representation π of $\mathfrak{B}$ via m [14, Sect. 6].

If $\pi_0(\mathfrak{A})$ is a factor M, $N \equiv \pi(\mathfrak{B})''$ will be a factor with $M' \cap N = \mathbb{C} \cdot I$; conversely if N is a factor with separable predual and $G \subset \operatorname{Aut} N$ is compact for the strong topology of the action on the predual, such that $M = N^G$ is infinite with trivial relative commutant, then N arises from M as above [14, Theorem 7.1].

We could generalize End $\mathfrak{A}$ to Bimod $\mathfrak{A}$, a strict monoidal C^*-category whose objects are homomorphisms ϱ of $\mathfrak{A}$ into square matrix algebras over $\mathfrak{A}$ and where an arrow in (ϱ, σ) is a matrix over $\mathfrak{A}$ intertwining ϱ and σ [15, Sect. 1].

Bimod $\mathfrak{A}$ can be locally identified with End $\tilde{\mathfrak{A}}$ for a suitable C^*-algebra $\tilde{\mathfrak{A}}$. If the object ϱ is a unital mapping of $\mathfrak{A}$ into $M_n(\mathfrak{A})$, let $\tilde{\mathfrak{A}}$ be the C^*-tensor product of $\mathfrak{A}$ with the UHF algebra $M_n(\mathbb{C}) \otimes M_n(\mathbb{C}) \otimes \ldots$, and define an endomorphism $\tilde{\varrho}$ of $\tilde{\mathfrak{A}}$ by

$$\begin{gathered}\tilde{\varrho}(A \otimes B_1 \otimes B_2 \otimes B_3 \otimes \ldots) = \varrho(A) \otimes B_1 \otimes B_2 \otimes \ldots\,, \\ B_i \in M_n(\mathbb{C})\,, \quad A \in \mathfrak{A}\,.\end{gathered} \tag{26}$$

It can be proved that $\mathcal{T}_\varrho$ is isomorphic to $\mathcal{T}_{\tilde{\varrho}}$ [15, Sect. 5]. Actions described by Bimod $\mathfrak{A}$ arise, e.g. if the C^*-system $(\mathfrak{B}, G)$ does not have G-Hilbert spaces. A unitary matrix $X \in \mathfrak{B} \otimes M_n(\mathbb{C})$ such that

$$g \otimes \imath(X) = X \cdot I \otimes v(g), \quad g \in G, \tag{27}$$

where v is a continuous unitary n-dimensional representation of G, defines an object of Bimod $\mathfrak{A}$, $\mathfrak{A} = \mathfrak{B}^G$, by

$$\varrho(A) = XA \otimes IX^*, \quad A \in \mathfrak{A}. \tag{28}$$

By the foregoing device this action could be equally well described by an action in End $\tilde{\mathfrak{A}}$.

On the other side, every category of continuous unitary finite dimensional representations of a compact group G can be embedded in End $\mathfrak{A}$ for a *simple* C^*-algebra $\mathfrak{A}$, which is obtained as the fixed point subalgebra under a canonical action of G on an infinite tensor product of Cuntz algebras [13, Sect. 7].

As a variant of that construction one can consider the fixed points $\mathcal{O}_{U(G)}$ in the Cuntz algebra $\mathcal{O}_\infty$ under the action α of a group G defined by a unitary representation U of G on the infinite dimensional generating Hilbert space H, by

$$\alpha_g(\psi) = U(g)\psi\,; \quad \psi \in H, \quad g \in G.$$

It can be shown [5] that this action is *ergodic* if and only if U admits no non zero finite dimensional subrepresentation, and is *prime* if $U(G)$ has compact strong closure. If furthermore U is the direct sum of finite dimensional subrepresentations with determinant one, $\mathcal{O}_{U(G)}$ is *simple* [5], generalizing to $\mathcal{O}_\infty$ the analogous result for $\mathcal{O}_d$ [12].

By taking, for each metrizable compact group G, the representation U to be the doubled left regular representation λ, we get a *simple* C^*-algebra $\mathcal{O}_{\lambda(G)}$, and each continuous isomorphism $G_1 \to G_2$ induces a canonical isomorphism $\mathcal{O}_{\lambda(G_2)} \to \mathcal{O}_{\lambda(G_1)}$.

Current research deals with the question: when are the $\mathcal{O}_{\lambda(G)}$ non isomorphic?

3. Operator Algebras and Quantum Field Theory

Quantum Mechanics says that observables generate a C^*-algebra $\mathfrak{A}$ [31]; local quantum theory, providing a general basis to Quantum Field Theory, requires that $\mathfrak{A}$ is generated by *local observables* [21]. Local observables are specified by an *inclusion preserving* map $\mathcal{O} \to \mathfrak{A}(\mathcal{O})$ from the set $\mathcal{K}$ of double cones (bounded set obtained intersecting past with future open light cones) in Minkowski space to von Neumann algebras acting on a separable Hilbert space $\mathcal{H}_0$. This map is extended to any $S \subset \mathbb{R}^4$ defining $\mathfrak{A}(S)$ as the C^*-algebra generated by $\mathfrak{A}(\mathcal{O})$, $\mathcal{O} \subset S$, $\mathcal{O} \in \mathcal{K}$; so that $\mathfrak{A} = \mathfrak{A}(\mathbb{R}^4)$.

The C^*-algebra $\mathfrak{A}$ is assumed to be *irreducible*, i.e. the defining representation π_0 of $\mathfrak{A}$ describes a single *superselection sector*, which should contain the *vacuum state* ω_0 induced by a unit vector $\Omega \in \mathcal{H}_0$.

Other superselection sectors will be described by other unitary equivalence classes of irreducible representations of $\mathfrak{A}$.

As a manifestation of Einstein causality, if the double cones $\mathcal{O}_1$, $\mathcal{O}_2$ cannot be joined by any signal travelling at most with the speed of light, i.e. $\mathcal{O}_1$ is included in the spacelike complement $\mathcal{O}_2'$ of $\mathcal{O}_2$, observables localized in $\mathcal{O}_1$ and $\mathcal{O}_2$ should not interfere with one another. Quantum Mechanics says that they will commute; hence, for each $\mathcal{O} \in \mathcal{K}$, *the locality principle* says that

$$\mathfrak{A}(\mathcal{O}) \subset \mathfrak{A}(\mathcal{O}')'. \tag{29}$$

In QFT, fields are basically associated with points or, possibly, with loops or infinite strings. While a double cone is an appropriate neighbourhood of a point, an appropriate neighbourhood of a string is the cone joining a point to a double cone at spacelike infinity. The set of such *spacelike cones* is denoted by $\mathcal{J}$.

Locality is often sharpened to *duality*

$$\mathfrak{A}(\mathcal{S})^- = \mathfrak{A}(\mathcal{S}')' \tag{30}$$

where $\mathcal{S}$ is a double cone (so that weak closure on the left-hand side is unnecessary) or a spacelike cone.

Duality is related to the absence of spontaneously broken gauge symmetries [29]. Otherwise, *essential duality* [30], requiring that the net $\mathcal{O} \in \mathcal{K} \to \mathfrak{A}^d(\mathcal{O}) \equiv \mathfrak{A}(\mathcal{O}')'$ fulfills (30) for each double cone, would hold quite generally, as a consequence of a result of Bisognano and Wichmann [1], whenever there are underlying Wightman fields.

Our last assumption is Property B: if $\mathcal{S}$, $\tilde{\mathcal{S}}$ are either double cones or spacelike cones such that $\mathcal{S} + \mathcal{N} \subset \tilde{\mathcal{S}}$ for some neighborhood $\mathcal{N}$ of the origin, non zero self adjoint projections in $\mathfrak{A}(\mathcal{S})^-$ are equivalent to I mod $\mathfrak{A}(\tilde{\mathcal{S}})^-$, i.e.

$$\mathcal{O} \neq E = E^*E \in \mathfrak{A}(\mathcal{S})^- \Rightarrow E = WW^*,\ W^*W = I,\ W \in \mathfrak{A}(\tilde{\mathcal{S}})^-. \tag{31}$$

If we would assume the basic requirements of translation covariance and spectrum condition in the vacuum sector, Property B would be a consequence, as discovered by Borchers [2].

These few axioms on the inclusion preserving map $\mathcal{O} \to \mathfrak{A}(\mathcal{O})$ (irreducibility, duality and Property B), though too general on their own to characterize physically the vacuum sector, turn out to be sufficient for a discussion of superselection structure tailored to theories without massless particles.

In such a theory one particle excitations of the vacuum would be associated to (factorial) representations π of $\mathfrak{A}$ which are translationally covariant, with energy momentum spectrum (the spectrum of the representation of the translation group) in the forward light cone starting with an isolated hyperboloid of positive mass (massive particle representations). If π describes an excitation of the vacuum ω_0, ω_0 can be obtained as a mean of π over translations.

However, Buchholz and Fredenhagen [4] derived, from the above assumptions, a much stronger relation expressing that π can be localized in spacelike cones:

$$\pi|_{\mathfrak{A}(\mathcal{C}')} \cong \pi_0|_{\mathfrak{A}(\mathcal{C}')},\ \mathcal{C} \in \mathcal{J}. \tag{32}$$

The general definition and classification of *statistics* [9, 10] (cf. also J. E. Roberts contribution to [26], and comments below) can be extended to representations fulfilling (32), ([4], cf. also [16, Sect. 4]) and moreover massive particle representations have necessarily *finite statistics* [17] (so that in particular π is of type I).

These basic facts motivate, in our general frame where covariance and spectrum condition are not assumed, the following

3.1. Definition. *In a theory specified by the inclusion preserving irreducible map $\mathcal{O} \to \mathfrak{A}(\mathcal{O})$ fulfilling duality and Property* B, *the superselection sectors are the unitary equivalence classes of irreducible representations π of $\mathfrak{A}$ fulfilling* (32) *with finite statistics.*

Superselection sectors are usually associated with charged field operators which connect the vacuum sector to the other sectors. These fields might be attached to points (Wightman fields) or to strings from points to spacelike infinity, in presence of "quantum topological charges". We can formalize these notions in the following definition [16].

3.2. Definition. *An* extended field system with gauge symmetry $(\pi, G, \mathfrak{F})$ *consists of*

- *a representation π of $\mathfrak{A}$ on a Hilbert space $\mathscr{H}$ containing π_0 as a subrepresentation on $\mathscr{H}_0 \subset \mathscr{H}$;*
- *a strongly compact group G of unitaries on $\mathscr{H}$ leaving $\mathscr{H}_0$ pointwise fixed;*
- *an inclusion preserving map $\mathfrak{F}$ from spacelike cones to von Neumann algebras on $\mathscr{H}$ such that for each $\mathscr{C} \in \mathscr{J}$, the $g \in G$ induce automorphisms α_g of $\mathfrak{F}(\mathscr{C})$, with fixed points $\pi(\mathfrak{A}(\mathscr{C}))^-$; moreover $\mathscr{H}_0$ is cyclic for $\mathfrak{F}(\mathscr{C})$ and the union of $\mathfrak{F}(\mathscr{C} + a)$ for all $a \in \mathbb{R}^4$ is irreducible; $\mathfrak{F}(\mathscr{C})$ commutes with $\pi(\mathfrak{A}(\mathcal{O}))$ if $\mathcal{O}$ is a double cone spacelike to $\mathscr{C}$.*

The system $(\pi, G, \mathfrak{F})$ is normal *if the map $\mathfrak{F}$ obeys graded local commutativity for the $\mathbb{Z}_2$-grading defined by a central element $k \in G$ with square the identity.*

The system is complete *if every irreducible representation $\tilde{\pi}$ of $\mathfrak{A}$ fulfilling (32) with finite statistics is a subrepresentation of π, i.e. π describes all superselection sectors.*

The main result of [16] states that such a system can be canonically constructed from the net $\mathcal{O} \to \mathfrak{A}(\mathcal{O})$ of local observables.

3.3. Theorem. *Let the irreducible, inclusion preserving map $\mathcal{O} \to \mathfrak{A}(\mathcal{O})$ fulfill duality and Property* B*; there exists a complete, normal extended field system with gauge symmetry, and this system is unique up to unitary equivalence.*

Unitary equivalence of two systems $(\pi_i, G_i, \mathfrak{F}_i)$ acting on $\mathscr{H}_i$; $i = 1, 2$, is expressed by a unitary operator W of $\mathscr{H}_1$, onto $\mathscr{H}_2$ intertwining π_1 and π_2, the maps $\mathfrak{F}_1$ and $\mathfrak{F}_2$, and taking G_1 onto G_2:

$$\begin{aligned} W\pi_1(A) &= \pi_2(A)W, \quad A \in \mathfrak{A}\,; \\ W\mathfrak{F}_1(\mathscr{C}) &= \mathfrak{F}_2(\mathscr{C})W, \quad \mathscr{C} \in \mathscr{J}\,; \\ WG_1 &= G_2W\,. \end{aligned}$$

An important subclass of superselection sectors, carrying "localizable charges", is obtained by requiring (32) to hold for *double cones*:

$$\pi_\xi|_{\mathfrak{A}(\mathscr{O}')} \cong \pi_0|_{\mathfrak{A}(\mathscr{O}')}\,, \quad \mathscr{O} \in \mathscr{K}\,. \tag{33}$$

This subclass is described by a unique complete normal *local* field system with gauge symmetry $(\pi_l, G_l, \mathfrak{F}_l)$, where now $\mathfrak{F}_l$ is an inclusion preserving map from *double cones* to von Neumann algebras on a Hilbert space $\mathscr{H}_l$, with irreducible range and each $\mathfrak{F}_l(\mathscr{O})$ is cyclic on $\mathscr{H}_0$.

There is a closed normal subgroup N of G identifying $\mathscr{H}_l$ with the N-fixed vectors in $\mathscr{H}$ and π_l with $\pi|_{\mathscr{H}_l}$; the von Neumann algebra generated by the $\mathfrak{F}_l(\mathscr{O})$ with $\mathscr{O} \subset \mathscr{C}$ is identified with the restriction to $\mathscr{H}_l$ of the N-fixed points in $\mathfrak{F}(\mathscr{C})$, $\mathscr{C} \in \mathscr{J}$ [16].

Superselection sectors described by representations fulfilling (32) but not (33) are often said to carry "quantum topological charges".

Superselection sectors are in 1–1 correspondence with the spectrum $\hat{G}$ of G. Quantum topological charges are carried by sectors corresponding to representations which are non trivial on N.

The superselection structure of localizable charges is described by a full subcategory $\mathscr{T}$ of End $\mathfrak{A}$ [9, 10].

The objects of $\mathscr{T}$ are *localized morphisms* of $\mathfrak{A}$ (i.e. they act trivially in the spacelike complement of some double cone which, by choosing an equivalent morphism, can be arbitrarily placed) with *finite statistics* in the following sense.

The full subcategory of End $\mathfrak{A}$ with objects the localized morphisms has a *unique symmetry* ε which is I for spacelike separated morphisms [9]. The canonically associated representations $\varepsilon_\varrho^{(n)}$ of $\mathbb{P}(n)$ describe the *statistics* of ϱ: they permute the factors in the *product state vectors* in the representation ϱ^n which carry the "charges" of ϱ localized in n mutually spacelike double cones, whilst the induced *product states* are totally symmetric.

If ϱ is irreducible, the irreducible representations of $\mathbb{P}(n)$ occurring in $\varepsilon_\varrho^{(n)}$ are precisely those with at most $d(\varrho) \in \mathbb{N}$ antisymmetrizations (resp. symmetrizations) and ϱ is *paraBose* (resp. *paraFermi*) of order $d(\varrho)$, or everyone and ϱ has infinite statistics ($d(\varrho) = \infty$) [9].

Every representation π_ξ of $\mathfrak{A}$ fulfilling (33) is unitarily equivalent to a localized morphism composed with π_0. In view of applications to massive theories on the four-dimensional spacetime we can disregard infinite statistics ([17]; see also [10; Appendix]).

We can change ε to a "Bosonized" symmetry $\hat{\varepsilon}$ (differing from ε on irreducible elements only by sign) so that $(\mathscr{T}, \hat{\varepsilon})$ is a strict symmetric monoidal C^*-category *with conjugates* [10, 15]. The complete normal field system with gauge symmetry can be constructed from the covariant representation of the C^*-system $(\mathfrak{A} \times \mathscr{T},\, G)$ induced from π_0 via m [16, Sect. 3].

If only essential duality holds, we can construct the complete normal local field system with gauge symmetry from the dual net $\mathfrak{A}^d$. The stabilizer of $\pi(\mathfrak{A}^d)$ in Aut $\mathfrak{F}$ is G, the unbroken part of the gauge group. The full gauge group, possibly including *broken symmetries*, can be defined as the stabilizer $\mathscr{G}$ of $\pi(\mathfrak{A})$ in Aut $\mathfrak{F}$ [16]. It can be shown that each $\gamma \in \mathscr{G}$ leaves the local field algebras $\mathfrak{F}(\mathscr{O})$ globally stable and $\mathscr{G}$ can be in part analyzed in terms of degeneracy of the vacuum, arising from different extensions of ω_0 to $\mathfrak{A}^d$ [3].

The duality theory reviewed in Sect. 2 suggests a natural generalization. If we drop the equation (4) in Definition 2.2 of a symmetry and we symmetrize the condition (6) adding

$$\varepsilon(\varrho, \sigma\tau) = I_\sigma \times \varepsilon(\varrho, \tau) \circ \varepsilon(\varrho, \sigma) \times I_\tau \tag{6'}$$

we obtain a *strict braided* monoidal category [25].

Problem [15, 16]. *Which class of mathematical objects has as abstract duals the strict braided monoidal C^*-categories with conjugates, with subobjects and direct sums, such that $(\iota, \iota) = \mathbb{C}$? If such a category $\mathscr{T}$ acts on a C^*-algebra $\mathfrak{A}$ with centre $\mathbb{C} \cdot I$ as a full subcategory of* End $\mathfrak{A}$, *can we generalize Theorem 2.6 to a construction of a cross product $\mathfrak{A} \times \mathscr{T}$?*

This natural route to define a kind of "quantum groups" is required in order to extend the previous results to theories on low dimensional spacetimes.

To describe statistics in this case we have to replace the permutation groups by braid groups [18, 19, 20]. The category $\mathscr{T}$ describing superselection structure is strict monoidal with a unitary braiding. Solution to the problem above would clarify which "quantum groups" may appear as internal symmetries of low dimensional models [28] (cf. also most contributions to [26]).

There are remarkable relations between the analysis of braid statistics of two dimensional theories and the Jones theory of invariant polynomials for knots and of index of W^*-inclusions [23, 24]. These relations have been discovered by R. Longo who, extending the notion of Jones index to inclusions of type III factors, showed that [27]

$$d(\varrho)^2 = \operatorname{ind}(\varrho(\mathfrak{A}(\mathscr{O})), \mathfrak{A}(\mathscr{O})) \tag{34}$$

when ϱ is localized in $\mathscr{O}$; in two dimensional theories $d(\varrho)$ need not be an integer but its values are limited by the equation (39). The same equation shows that $d(\varrho)$ is related to a quantum version of the Fredholm index of an endomorphism (cf. also the comments in [8]).

References

1. Bisognano, J.J., Wichmann, E.H.: On the duality condition for quantum fields. J. Math. Phys. **17** (1976) 303–321
2. Borchers, H.J.: A Remark on a theorem of B. Misra. Comm. Math. Phys. **4** (1967) 315–323
3. Buchholz, D., Doplicher, S., Longo, R., Roberts, J.E.: Broken symmetries and degeneracy of the vacuum in quantum field theory. Preprint in preparation

4. Buchholz, D., Fredenhagen, K.: Locality and the structure of particle states, Comm. Math. Phys. **84** (1982) 1–54
5. Ceccherini, T., Pinzari, C.: Canonical actions on $\mathcal{O}_\infty$. J. Funct. Anal. (to appear)
6. Cuntz, J.: Simple C^*-algebras generated by isometries. Comm. Math. Phys. **57** (1977) 173–185
7. Deligne, P.: Categories Tannakiennes. Grothendieck Festschrift. Birkhäuser, Basel 1990 (to appear)
8. Doplicher, S.: Problems in quantum field theory and operator algebras. In: Araki, H., Moore, C.C., Stratila, S., Voiculescu, D. (eds.) Operator Algebras and their connections with Ergodic Theory and Topology. (Lecture Notes in Mathematics, vol. 1132). Springer, Berlin Heidelberg New York 1985
9. Doplicher, S., Haag, R., Roberts, J.E.: Local observable and particle statistics I. Comm. Math. Phys. **23** (1971) 199–230
10. Doplicher, S., Haag, R., Roberts, J.E.: Local observables and particle statistics II. Comm. Math. Phys. **35** (1974) 49–85
11. Doplicher, S., Roberts, J.E.: A remark on compact automorphism groups of C^*-algebras. J. Funct. Anal. **66** (1986) 67–72
12. Doplicher, S., Roberts, J.E.: Duals of compact Lie groups realized in the Cuntz algebras and their actions on C^*-algebras. J. Funct. Anal. **74** (1987) 96–120
13. Doplicher, S., Roberts, J.E.: Compact group actions on C^*-algebras. J. Operator Theory **19** (1988) 283–305
14. Doplicher, S., Roberts, J.E.: Endomorphism of C^*-algebras, cross products and duality for compact groups. Ann. Math. **130** (1989) 75–119
15. Doplicher, S., Roberts, J.E.: A new duality theory for compact groups. Invent. math. **89** (1989) 157–218
16. Doplicher, S., Roberts, J.E.: Why there is a field algebra with a compact gauge group describing the superselection structure in particle physics. Comm. Math. Phys. **131** (1990) 51–107
17. Fredenhagen, K.: On the existence of antiparticles. Comm. Math. Phys. **79** (1981) 141–151
18. Fredenhagen, K., Rehren, K.H., Schroer, B.: Superselection sectors with braid group statistics and exchange algebras. Comm. Math. Phys. **125** (1989) 201–226
19. Fröhlich, J., in Non perturbative quantum field theory. G. t'Hooft et al. (eds.) Plenum Press, New York 1988
20. Fröhlich, J., Marchetti, P.A.: Quantum field theory of vortices and anyons. Comm. Math. Phys. **121** (1989) 177–224
21. Haag, R., Kastler, D.: An algebraic approach to quantum field theory. J. Math. Phys. **5** (1964) 848–861
22. Hewitt, E., Ross, K.A.: Abstract harmonic analysis II. Springer, Berlin Heidelberg New York 1970
23. Jones, V.F.R.: Index for subfactors. Invent. math. **72** (1983) 1–25
24. Jones, V.F.R.: Hecke algebra representations of Braid groups and link polynomials. Ann. Math. **126** (1987) 335–388
25. Joyal, A., Street, R.: Braided monoidal categories. Mcquarie Mathematics Reports No. 860081 (1986)
26. Kastler, D. (ed.): Algebraic theory of superselection sectors. World Science Publ. Co., Singapore 1990
27. Longo, R.: Index of subfactors and statistics of quantum fields I. Comm. Math. Phys. **126** (1989) 217–247; II. Comm. Math. Phys. **130** (1990) 285–310
28. Mack, G., Schomerus, V.: Conformal field algebras with quantum symmetry from the theory of superselection sectors. Comm. Math. Phys **134** (1990) 139

29. Roberts, J.E.: Spontaneously broken gauge symmetries and superselection rules. Proceedings of the International School of Mathematical Physics, Camerino 1974, Gallavotti, G. (ed.). Università di Camerino (1976)
30. Roberts, J.E.: Net cohomology and its applications to field theory. In: L. Streit (ed.) Quantum Fields – Algebras, Processes. Springer, Berlin Heidelberg New York 1980
31. Segal, I.E.: Postulates for general quantum mechanics. Ann. Math. **48** (2) (1947) 930–948

Introduction to Constructive Quantum Field Theory

Joel Feldman

University of British Columbia, Vancouver, BC, Canada, V6T 1Y4

1. Introduction

Quantum field theories (QFT) are used to model physical systems that share two common features. Firstly, they are of atomic or subatomic scale, so that they exhibit quantum mechanical behaviour. Secondly, they involve fields. To a physicist a field is an observable (i.e. something you measure) that is a function on space-time. Two examples are the electric and magnetic fields. Indeed the best known QFT is Quantum Electrodynamics (QED), which models the interaction of electrons with the electromagnetic field.

In addition to the obvious fields, it is often wise to associate fields with particles. For example, at the high energies typical of QED electrons and positrons are continually being created and destroyed. So, rather than attempt to explicitly keep track of how many electrons there are at all times, one invents a field, called the electron field from which one may calculate the charge and current densities. The former tells you the positions of all electrons and the latter tells you their velocities.

An electron field is also used in modelling another physical system that will appear later in this talk. The system is a crystal consisting of a large population of electrons interacting with each other and with a lattice of stationary or almost stationary ions. As there is a fixed number of electrons per ion and as the lattice is effectively of infinite size the population is really huge. Fortunately, almost all of these electrons do nothing of interest. They just sit there forming what is called the Fermi sea. However, there are small numbers of electrons continually jumping out of and back into the Fermi sea. Once again an electron field is introduced to keep track of them all.

I will use the Gårding-Wightman axioms [SW] to tell you what a quantum field theory is from a mathematical point of view. These axioms are really designed for relativistic quantum field theories so some of them are not appropriate for our current purposes and I will just mention those in passing.

The first axiom sets up the state space.

1. *State Space.* There exists a separable Hilbert space F.

The vectors (actually rays) of F are the states of our system. Next come the field operators. There are a couple of technical complications. First, fields are like distributions. They need not be well-defined at points, but instead have to

Proceedings of the International Congress of Mathematicians, Kyoto, Japan, 1990

be smeared against nice test functions. Secondly, even after smearing, fields are often unbounded operators. So there are domain questions. Gårding-Wightman handles these by postulating the existence of a common dense domain.

2. *Field Operators.* There exists a dense linear subspace D of F and for each $f \in S(R^4), j \in \{1, ..., n\}$, there exists an operator $\phi_j(f)$ with domain containing D such that

$$\phi_j(f)D \subset D$$

$$\phi_j(f)^* D \subset D$$

$$f \to (\phi_j(f)v, w) \in S'(R^4) \text{ for each } v, w \in D\,.$$

As with ordinary tempered distributions one writes

$$\phi_j(f) = \int dt d^3\mathbf{x} \phi_j(t, \mathbf{x}) f(t, \mathbf{x})\,.$$

Time evolution is determined by a special self-adjoint operator, the Hamiltonian.

3. *Hamiltonian.* There exists a self-adjoint operator H such that

$$e^{isH} D \subset D \quad \forall s \in R$$

$$e^{isH} \phi_j(t, \mathbf{x}) e^{-isH} = \phi_j(t + s, \mathbf{x})$$

$$H \geq 0$$

$$0 \text{ is a simple eigenvalue of } H \text{ with eigenvector } \Omega \in D\,.$$

In practice the vector Ω plays an important role. For example, in Quantum Electrodynamics, Ω is the vacuum. That is, the state corresponding to a completely empty world. All physically interesting states are constructed by adding small numbers of particles to the vacuum.

In a relativistic QFT one wants covariance not just under time translations but also under the full Poincaré group. So there are axioms postulating the existence and basic properties of a unitary representation of (the covering group, inhomogeneous $\mathrm{SL}(2, C), of$) the Poincaré group. One also wants causality, i.e. measurements made at space-like separated points better not influence each other, so there is an axiom imposing this.

A constructive quantum field theorist is someone who tries, at least, to rigorously construct and determine properties of quantum field theories.

Just as in classical mechanics the Hamiltonian $H = H_\lambda$ of a physically interesting system invariably contains a parameter (or several parameters) λ called the coupling constant. For a special value of λ, normally 0, the Hamiltonian is trivial in the sense that one can determine its properties and, in particular, the time evolution it generates, explicitly. So it is natural to view H_λ as a perturbation of H_0. However, for a QFT, this perturbation is extremely singular. I will discuss three symptoms of this singularity and how they are treated.

2. Change of Hilbert Space

The first appears even in the models that are easiest to deal with. The construction of the, so called, weakly coupled $\lambda\phi_2^4$ model can now be reasonably presented in an advanced undergraduate analysis course. This model is completely characterized by a measure $d\mu_\lambda$ on $S'(R^2)$. Even though not true, you might want to interpret $S'(R^2)$ as the space of possible values of the field $\phi(t, x)$ and the measure as the probability density for the field configurations when the coupling constant takes the value λ.

Theorem [Fr]. *If $0 \leq \lambda \neq \lambda'$ then $d\mu_\lambda$ and $d\mu_{\lambda'}$ are mutually singular.*

This theorem is not as much of an impediment as you might think. It suffices to deal with the measure weakly. In terms of the field operators, you should attempt to construct inner products $(\Pi\phi_j(f_j)\Omega, \Omega)$ rather than to construct the field operators directly.

3. Renormalization

The second symptom is much more serious. It is the problem of renormalization [BS, H, FHRW]. I will illustrate it using the crystal model mentioned earlier [BG, FT]. For simplicity I will even discard the lattice, leaving only the "electrons". Consider a system of N particles each of mass m living in a periodic box of side L and interacting with each other through an even two body potential $\lambda V \in S(R^3)$. The Hamiltonian for this system is

$$H_{L,N,\lambda} = \frac{1}{2m}\sum_{i=1}^{N}(-\Delta_{\mathbf{x}_i}) + \lambda\sum_{i<j} V(\mathbf{x}_i - \mathbf{x}_j)$$

and acts on the Hilbert space

$$F_N = \{v \in L^2((R^3/LZ^3)^N) \mid v(\mathbf{x}_{\sigma(1)}, ..., \mathbf{x}_{\sigma(N)}) = (-1)^{\mathrm{sgn}(\sigma)} v(\mathbf{x}_1, ..., \mathbf{x}_N) \text{ for all } \sigma \in S_N\}.$$

The antisymmetry of the vectors $v \in F_N$ reflects the hypothesized Fermionic character of the particles. We would now like to take the thermodynamic limit $L, N \to \infty$ with the density $\varrho = N/L^3$ fixed. In order to avoid having to introduce the electron field I will concentrate on the ground state energy density

$$p(\lambda) = \lim_{\substack{L,N\to\infty \\ N/L^3=\varrho}} \frac{\inf \operatorname{spec} H_{L,N,\lambda}}{L^3}.$$

Of course it is not easy to determine $\inf \operatorname{spec} H_{L,N,\lambda}$ except when $\lambda = 0$. So we do some perturbation theory. Assume $p(\lambda)$ exists (even this is not obvious), is C^∞ and that we can move derivatives inside the limit sign. Then it is not so difficult to develop an explicit formula for $\frac{d^n p}{d\lambda^n}(0)$. This formula is conventionally stated in terms of Feynman diagrams

$$\frac{d^n p}{d\lambda^n}(0) = \sum_{\substack{\text{connected vacuum} \\ \text{graphs of order } n}} \mathrm{Val}(G)\,.$$

The sum is over roughly $(n!)^2$ Feynman diagrams each of which is a mnemonic device for a specific integral, called $\mathrm{Val}(G)$.

For example one diagram with $n = 2$ is given in Fig. 1.

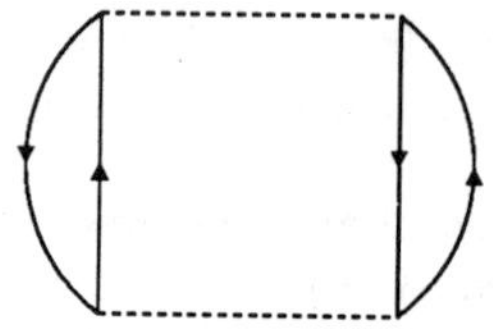

Fig. 1

It has the value

$$\int \frac{d^4p}{(2\pi)^4}\frac{d^4q}{(2\pi)^4}\frac{d^4r}{(2\pi)^4}\widehat{V}(\mathbf{q})^2 \frac{1}{ip_0 - e(\mathbf{p})}\frac{1}{i(p_0-q_0)-e(\mathbf{p}-\mathbf{q})}\frac{1}{i(r_0-q_0)-e(\mathbf{r}-\mathbf{q})}$$
$$\times \frac{1}{ir_0 - e(\mathbf{r})} \quad \text{where } p = (p_0, \mathbf{p})\,,\ q = (q_0, \mathbf{q})\,,\ r = (r_0, \mathbf{r})\,,$$

$$e(\mathbf{p}) = \frac{1}{2m}\mathbf{p}^2 - \mu_0$$

$$\mu_0 = \frac{\pi}{4m}\left(\frac{3}{2}\sqrt{\pi}\varrho\right)^{3/2}.$$

It is not clear whether the integral giving the value of this (or for that matter any other) graph converges. There are two potential obstructions to convergence. Firstly the integrand decays rather slowly at infinity. This turns out to be harmless. All graphs turn out to be convergent (though not necessarily absolutely convergent) at infinity. Secondly the integrand has singularities. The factor $ip_0 - e(\mathbf{p})$ is zero when $p_0 = 0$ and $|\mathbf{p}| = \sqrt{2m\mu_0}$. More precisely

$$\frac{\text{const}'}{\sqrt{p_0^2 + R^2}} \le \left|\frac{1}{ip_0 - e(\mathbf{p})}\right| \le \frac{\text{const}}{\sqrt{p_0^2+R^2}}\,, \qquad R = |\mathbf{p}| - \sqrt{2m\mu_0}$$

on $|p| \le$ const. Hence $[ip_0 - e(\mathbf{p})]^{-1}$ is locally L^1 but not locally L^n for any $n \ge 2$.

While the graph in the example converges there are tons of them, in fact infinitely many, that diverge. It turns out that, in this model, every single divergent graph has

$$\mathrm{Val}(G) = \int \frac{d^4k}{(2\pi)^4}\frac{f(k)}{[ik_0 - e(\mathbf{k})]^n} \qquad n \ge 2$$

with $n \ge 2$. I have lumped all the other integrals into $f(k)$. These integrals may also diverge if they have this same form.

I claim that all of these divergences arise, not because of any pathology in the model, but merely because we are trying to do the perturbation expansion in a stupid way. To see what I mean let's pretend that the exact

$$p(\lambda) = \int \frac{d^4k}{(2\pi)^4} \frac{f(k,\lambda)}{ik_0 - \frac{1}{2m}\mathbf{k}^2 + \mu_0 + \delta\mu(\mu_0,\lambda)}$$

for some reasonable functions $f(k,\lambda)$ and $\delta\mu(\mu_0,\lambda)$ with $\delta\mu(\mu_0,0) = 0$. The integral converges. Yet when we apply $(d^n/d\lambda^n)|_{\lambda=0}$ we get large powers of $[ik_0 - e(\mathbf{k})]^{-1}$ and hence divergence. It is not wise to try to expand a function, $[ik_0 - e(\mathbf{k}) + \delta\mu]^{-1}$, with a singularity somewhere, in powers of a function, $[ik_0 - e(\mathbf{k})]^{-1}$ with a singularity somewhere else.

Once we know that this is the source of the divergences, the treatment is simple. Just parametrize our models, not by μ_0 and λ, but rather by $\mu = \mu_0 + \delta\mu(\mu_0,\lambda)$ and λ. Then the position of the singularity in

$$\frac{1}{ip_0 - e(\mathbf{p}) + \delta\mu(\mu_0,\lambda)} = \frac{1}{ip_0 - \frac{1}{2m}\mathbf{p}^2 + \mu}$$

becomes independent of λ and the source of divergence mentioned above disappears. Of course to implement this you still have to find the right $\delta\mu(\mu_0,\lambda)$. But there is a renormalization algorithm which tells you how to choose $\delta\mu(\mu_0,\lambda)$, inductively in powers of λ.

Theorem [FT]. *There exists a formal power series (with finite coefficients),* $\mu_0(\lambda,\mu) \sim \mu + \sum_{i=1}^{\infty} m_i(\mu)\lambda^i$, *such that every coefficient in the composite f.p.s.*

$$p(\lambda, \mu_0(\lambda,\mu), L) \sim \sum_{n=0}^{\infty} p_n^{\mathrm{ren}}(L)\lambda^n$$

converges as $L \to \infty$. *Furthermore*

$$|\lim_{L\to\infty} p_n^{\mathrm{ren}}(L)| \leq \mathrm{const}^n n!.$$

This phenomenon – the divergence of coefficients in a perturbation expansion unless parameters are carefully adjusted so as to hold some physical quantity fixed – occurs commonly. For example, in classical mechanics, when you perturb a periodic orbit in a system of one degree of freedom you must be careful to keep the period fixed. That is, it is not generally possible to find a symplectic map converting $\frac{1}{2}(p^2+q^2) + \lambda v(p,q)$ to $\frac{1}{2}(P^2+Q^2)$. One must use $f_\lambda(\frac{1}{2}(P^2+Q^2))$ with a very carefully chosen f_λ as the target.

4. Symmetry Breaking

The final symptom of the singularity of perturbations in QFT that I will discuss is the phenomenon of symmetry breaking. It is possible for all the input data used to specify a model to have a given symmetry without the resulting model having the symmetry. At first this may seem like a shocking, even inflammatory, statement. It is possible to rephrase it in an innocuous form but I feel that the above one gives a more realistic reflection of the impact of symmetry breaking.

Once again my illustration uses the many-electron model [FW, S]. Recall that the space of N-electron states in our periodic box is

$$F_N = \{v \varepsilon L^2((R^3/LZ^3)^N) | v(\mathbf{x}_{\sigma(1)}, ..., \mathbf{x}_{\sigma(N)}) = (-1)^\sigma v(\mathbf{x}_1, ..., \mathbf{x}_N) \quad \forall \sigma \in S_N\}.$$

To avoid having to change Hilbert spaces every time we change N we consider

$$F = \bigoplus_N F_N$$

with the modified "Hamiltonian"

$$\tilde{H}_L = H_{L,\lambda} - \mu N - c(L)1$$

($c(L)$ is chosen to make $\inf \operatorname{spec} \tilde{H}_L = 0$) defined by

$$\tilde{H}_L | F_N = \frac{1}{2m} \sum_{i=1}^{N} -\Delta_{\mathbf{x}_i} + \sum_{i<j} \lambda V(\mathbf{x}_i - \mathbf{x}_j) + \text{interaction with ions} - \mu_0 N - c(L)1 .$$

As $L \to \infty$, states with the wrong density end up with "$\tilde{H} = \infty$" and get pushed out of the domain of $\tilde{H}$.

Now

$$V(\alpha) = e^{i\alpha(N - d(L))}$$

($d(L)$ is a normalization constant) is a unitary representation of R, the covering group of $U(1)$, which commutes with $\tilde{H}_L$ for every L since $\tilde{H}_L : F_N \to F_N$. Suppose that $\tilde{H} = \lim_{L \to \infty} \tilde{H}_L$ were some ordinary (e.g. strong) limit. If $\tilde{H}$ has a simple eigenvalue at the bottom of its spectrum, then the corresponding eigenvector Ω must also be an eigenvector of $V(\alpha)$ and in fact

$$V(\alpha)\Omega = \Omega$$

if the normalization constant $d(L)$ is chosen appropriately.

The value $(O\Omega, \Omega)$ of the observable (= operator) O measured in the state Ω obeys $(O\Omega, \Omega) = (V(\alpha)^* O V(\alpha)\Omega, \Omega)$. There are whole orbits of observables that give the same answer when measured in the state Ω. The symmetry $U(1)$ is not broken.

On the other hand if there is an eigenspace of dimension greater than one at the bottom of the spectrum of $\tilde{H}$, then each individual eigenvector Ω need not be an eigenvector of $V(\alpha)$, though of course the eigenspace is still invariant under $V(\alpha)$. The value of O measured in the state Ω need no longer be constant as O moves over the orbit.

In a QFT the limit $\tilde{H} = \lim_{L \to \infty} \tilde{H}_L$ is not an ordinary (e.g. strong) limit. Instead of getting one model with a degenerate ground state one gets a whole family [I, R] of independent models $F^{(\beta)}$ each with a unique ground state. Each independent model no longer carries a represenation of $U(1)$ because $V(\alpha)$ tries to map $\Omega_\beta \in F^{(\beta)}$ to $\Omega_{\alpha+\beta} \in F^{(\alpha+\beta)} \neq F^{(\beta)}$.

This is symmetry breaking. It has important physical consequences. In the example we have been discussing superconductivity arises. It also has a big impact on any attempted construction. In practice one must always attempt to construct the interacting broken symmetry model as a perturbation of a free model which also has the same broken symmetry.

References

[BS] Bogoliubov, N.N., Shirkov, D.V.: Introduction to the theory of quantized fields. Wiley, New York 1980

[BG] Benfatto, G., Gallavotti, G.: Perturbation theory of the Fermi surface in a quantum liquid. A general quasi particle formalism and one dimensional systems. J. Stat. Phys. (1990) (to appear)

[FHRW] Feldman, J., Hurd, T., Rosen, L., Wright, J.: QED: A proof of renormalizability. (Lecture Notes in Physics, vol. 312.) Springer, Berlin Heidelberg New York 1988

[FT] Feldman, J., Trubowitz, E.: Perturbation theory for many fermion systems. Helv. Phys. Acta **63** (1990) 157–260

[FW] Fetter, A.L., Walecka J.D.: Quantum theory of many-particle systems. McGraw Hill, New York 1971

[Fr] Froehlich, J.: Verification of axioms for Euclidean and relativistic fields and Haag's theorem in a class of $P(\phi)_2$ models. Ann. L'Inst. H. Poincaré **21** (1974) 271–317

[H] Hepp, K.: Théorie de la renormalisation. (Lecture Notes in Physics, vol. 2.) Springer, Berlin Heidelberg New York 1969

[I] Israel, R.B.: Convexity in the theory of lattice gases. Princeton University Press, Princeton 1979

[R] Ruelle, D.: Statistical mechanics, rigorous results. Benjamin, New York Amsterdam 1969

[S] Schrieffer, J.R.: Theory of superconductivity. (Frontiers in Physics, vol. 20.) Addison-Wesley, Redwood City 1964

[SW] Streater, R., Wightman, A.: PCT, spin and statistics and all that. Benjamin, New York Amsterdam 1964

Solvable Lattice Models and Quantum Groups

Michio Jimbo

Department of Mathematics, Faculty of Science, Kyoto University, Kyoto 606, Japan

Our main concern in this survey is the following question. What is the rôle of quantum groups in solvable lattice models; more specifically, what is their rôle in building models and in solving models respectively?

Building models amounts to constructing solutions to the Yang-Baxter equation (YBE). Through the development of the last 4–5 years, the situation has been clarified relatively well. In the context of quantum groups, the theory of YBE is the theory of intertwiners of their representations. In the first half of this article we shall elucidate this statement on three constructions.

In contrast, the rôle of quantum groups in solving models is very obscure. By 'solving' the model thus constructed we mean to calculate physically important quantities in closed form. The one point functions are one of the few quantities which can be evaluated exactly, thanks to Baxter's corner transfer matrix method. In many cases they are expressible in terms of certain modular forms arising from representations of affine Lie algebras. In the latter half of this paper we shall explain these results in the light of Kashiwara's theory of crystal base for quantum groups.

1. Yang-Baxter Equation

Let V be a finite dimensional vector space over $\mathbf{C}$, and let $R(\xi,\eta)$ be a function of some variables ξ, η with values in $\mathrm{End}(V \otimes V)$. Then YBE is the following functional equation for $R(\xi,\eta)$, written in $\mathrm{End}(V \otimes V \otimes V)$:

$$R_2(\xi,\eta)R_1(\xi,\zeta)R_2(\eta,\zeta) = R_1(\eta,\zeta)R_2(\xi,\zeta)R_1(\xi,\eta) \tag{1}$$

where

$$R_1(\xi,\eta) = R(\xi,\eta) \otimes 1, \quad R_2(\xi,\eta) = 1 \otimes R(\xi,\eta).$$

Following the common terminology we call a solution of YBE an R matrix, and $\xi,\eta,\cdots$ spectral parameters.

Throughout this article, by quantum groups we mean the quantized enveloping algebras $U_q(\mathfrak{g})$. For us the most important case is when $\mathfrak{g}$ is an *affine* (rather than finite dimensional) Lie algebra. Let us fix notations. $(a_{ij})_{1\le i,j\le l}$ signifies the Cartan matrix of $\mathfrak{g}$, $d_i \in \mathbf{Q}^\times$ are such that $d_i a_{ij} = d_j a_{ji}$, and q is a complex

Proceedings of the International Congress
of Mathematicians, Kyoto, Japan, 1990

number satisfying $q_i^2 \neq 0$ with $q_i = q^{d_i}$. Then $U_q(\mathfrak{g})$ is the $\mathbf{C}$-algebra with the generators $e_i, f_i, t_i^{\pm 1}$ $(1 \le i \le l)$ subject to the following defining relations.

$$t_i t_j = t_j t_i,\ t_i t_i^{-1} = 1 = t_i^{-1} t_i,$$

$$t_i e_j t_i^{-1} = q_i^{a_{ij}} e_j,\ t_i f_j t_i^{-1} = q_i^{-a_{ij}} f_j,\ [e_i, f_j] = \delta_{ij} \frac{t_i - t_i^{-1}}{q_i - q_i^{-1}},$$

$$\sum_{l=0}^{1-a_{ij}} (-1)^l e_i^{(1-a_{ij}-l)} e_j e_i^{(l)} = 0,\ \sum_{l=0}^{1-a_{ij}} (-1)^l f_i^{(1-a_{ij}-l)} f_j f_i^{(l)} = 0,\ (i \neq j)$$

where $e_i^{(l)} = e_i^l/[l]_{q_i}!$, $f_i^{(l)} = f_i^l/[l]_{q_i}!$, $[l]_t! = \prod_{j=1}^l (t^j - t^{-j})/(t - t^{-1})$. The comultiplication is given by

$$\Delta(e_i) = e_i \otimes 1 + t_i \otimes e_i,\ \Delta(f_i) = f_i \otimes t_i^{-1} + 1 \otimes f_i,\ \Delta(t_i) = t_i \otimes t_i .$$

1.1 Trigonometric R Matrices

The first construction associates an R matrix with each pair $(U_q(\mathfrak{g}), \pi)$, consisting of a qauntum group and its finite dimensional irreducible representation. Here q is assumed to be 'generic', i. e. $q^n \neq 1$ for all positive integers n.

Let us take the simplest example of $U_q(\widehat{\mathfrak{sl}}(2, \mathbf{C}))$ corresponding to $(a_{ij}) = \begin{pmatrix} 2 & -2 \\ -2 & 2 \end{pmatrix}$, and its two dimensional representation $(\pi_\xi, V = \mathbf{C}^2)$ depending on $\xi \in \mathbf{C}^\times$ given as follows.

$$\pi_\xi(e_0) = \begin{pmatrix} 0 & 0 \\ \xi & 0 \end{pmatrix},\ \pi_\xi(f_0) = \begin{pmatrix} 0 & \xi^{-1} \\ 0 & 0 \end{pmatrix},\ \pi_\xi(t_0) = \begin{pmatrix} q^{-1} & 0 \\ 0 & q \end{pmatrix},$$

$$\pi_\xi(e_1) = \begin{pmatrix} 0 & 1 \\ 0 & 0 \end{pmatrix},\ \pi_\xi(f_1) = \begin{pmatrix} 0 & 0 \\ 1 & 0 \end{pmatrix},\ \pi_\xi(t_1) = \begin{pmatrix} q & 0 \\ 0 & q^{-1} \end{pmatrix}.$$

An intertwiner $R(\xi, \eta)$ between $\pi_\xi \otimes \pi_\eta$ and $\pi_\eta \otimes \pi_\xi$ is a linear isomorphism $V \otimes V \to V \otimes V$ such that

$$R(\xi,\eta) \left(\pi_\xi \otimes \pi_\eta\right) \circ \Delta(X) = \left(\pi_\eta \otimes \pi_\xi\right) \circ \Delta(X) R(\xi,\eta) \quad \forall X \in U_q(\widehat{\mathfrak{sl}}(2, \mathbf{C})).$$

Solving these linear equations for R one finds that up to a scalar

$$R(\xi,\eta) = \begin{pmatrix} a & 0 & 0 & 0 \\ 0 & c & b & 0 \\ 0 & b & \bar{c} & 0 \\ 0 & 0 & 0 & a \end{pmatrix},$$

with $a = \xi q - \eta q^{-1}$, $b = \xi - \eta$, $c = \xi(q - q^{-1})$ and $\bar{c} = \eta(q - q^{-1})$. This is the well known R matrix of the 6 vertex model [1].

The left and right hand sides of YBE (1) are both intertwiners of the threefold tensor products $\pi_\xi \otimes \pi_\eta \otimes \pi_\zeta \to \pi_\zeta \otimes \pi_\eta \otimes \pi_\xi$. It can be checked that, if the spectral parameters are generic, any intertwiner of $\pi_\xi \otimes \pi_\eta \otimes \pi_\zeta$ into itself must be a scalar. From this one deduces that $R(\xi, \eta)$ solves YBE.

The explicit form of the R matrices are available for the vector representation of $U_q(\mathfrak{g})$ of classical types [2, 3]; further results are given in [4–8]. In the general case the existence of the intertwiners is guaranteed by Drinfeld's universal R matrix [9]. Irreducible finite dimensional representations of Yangians, which are certain degeneration of $U_q(\mathfrak{g})$, are classified also by Drinfeld [10] (the results for $U_q(\mathfrak{g})$ are almost the same; cf. [11]). In order to describe the corresponding R matrices in full generality, further work remains to be done.

1.2 Chiral Potts-Type Models

The second construction is related to R matrices whose spectral parameters live on certain algebraic curves. Here we take q to be a primitive N-th root of unity: $q^N = 1$ (we assume N is odd). For our purposes we enlarge the algebra $U_q(\mathfrak{g})$ slightly by adding central elements $z_i^{\pm 1}$, which modify the comultiplication as

$$\Delta(e_i) = e_i \otimes 1 + z_i t_i \otimes e_i, \quad \Delta(f_i) = f_i \otimes t_i^{-1} + z_i^{-1} \otimes e_i,$$
$$\Delta(t_i) = t_i \otimes t_i, \quad \Delta(z_i) = z_i \otimes z_i .$$

The resulting algebra $\tilde{U}_q(\mathfrak{g})$ is known as the quantum double of a Borel subalgebra of $U_q(\mathfrak{g})$. A characteristic feature of q being a root of 1 is that the elements e_i^N, f_i^N, t_i^N lie in the center $\mathscr{Z}$ of $\tilde{U}_q(\mathfrak{g})$.

Again we take the example of $\mathfrak{g} = \widehat{\mathfrak{sl}}\,(2, \mathbf{C})$. We consider an N dimensional irreducible representation of $\tilde{U}_q(\mathfrak{g})$ containing 5 parameters $\xi = (a_0, a_1, x_0, x_1, c) \in (\mathbf{C}^\times)^5$.

$$\pi_\xi(e_0) = x_0 F, \ \pi_\xi(f_0) = x_0^{-1} E, \ \pi_\xi(t_0) = T^{-1},$$
$$\pi_\xi(e_1) = x_1 E, \ \pi_\xi(f_1) = x_1^{-1} F, \ \pi_\xi(t_1) = T,$$
$$\pi_\xi(z_0) = c_0, \ \pi_\xi(z_1) = c_0^{-1} .$$

Here

$$E = \frac{a_1^2 Z - 1}{q - q^{-1}} X, \ F = (a_0 a_1 X)^{-1} \frac{a_0^2 Z^{-1} - 1}{q - q^{-1}}, \ T = q \frac{a_1}{a_0} Z,$$

and

$$X = \begin{pmatrix} 1 & & & \\ & q^2 & & \\ & & \ddots & \\ & & & q^{2N-2} \end{pmatrix}, \quad Z = \begin{pmatrix} 0 & 1 & & \\ & 0 & 1 & \\ & & \ddots & \ddots \\ & & & \ddots & 1 \\ 1 & & & & 0 \end{pmatrix}.$$

Suppose that $\pi_\xi \otimes \pi_\eta$ is equivalent to $\pi_\eta \otimes \pi_\xi$. Since $\Delta(e_i^N) = e_i^N \otimes 1 + (z_i t_i)^N \otimes e_i^N \in \mathscr{Z} \otimes \mathscr{Z}$, one must have

$$\frac{\pi_\xi(e_i)^N}{1 - \pi_\xi(z_i t_i)^N} = \frac{\pi_\eta(e_i)^N}{1 - \pi_\eta(z_i t_i)^N} (\equiv \Gamma_i)$$

and likewise for the f_i^N. Hence ξ, η are forced to lie on a common algebraic variety $\mathscr{S}_\Gamma$ with $\Gamma = (\Gamma_i)$ being the moduli parameters. It turns out that $\mathscr{S}_\Gamma$ is essentially a product $\mathscr{C}_k \times \mathscr{C}_k$ of curves

$$\mathscr{C}_k = \left\{(x, y, \mu) \mid x^N + y^N = k(1 + x^N y^N), \quad \mu^N = \frac{\sqrt{1-k^2}}{1 - kx^N}\right\}.$$

Conversely, if $\xi = (r, r')$, $\eta = (\tilde{r}, \tilde{r}')$ ($r = (x, y, \mu) \in \mathscr{C}_k$ etc.) lie on $\mathscr{S}_\Gamma$, then there exists an intertwiner and is given by the following formulas.

$$R(\xi, \eta) = R'(\tilde{r}, r') \left(R''(r', \tilde{r}') \otimes R''(r, \tilde{r})\right) R'(r, \tilde{r}'),$$

$$R'(r, \tilde{r}) = \sum_{a=0}^{N-1} \widehat{W}_{r\tilde{r}}(a) \left(X^{-1} \otimes X\right)^a, \quad R''(r, \tilde{r}) = \sum_{a=0}^{N-1} \overline{W}_{r\tilde{r}}(a) Z^a,$$

$$\widehat{W}_{r\tilde{r}}(a) = \prod_{l=1}^{a} \frac{\tilde{\mu} y - \mu \tilde{y} q^{2l-2}}{\tilde{\mu}\tilde{x} - \mu x q^{2l}}, \quad \overline{W}_{r\tilde{r}}(a) = \prod_{l=1}^{a} \mu\tilde{\mu} \frac{x q^2 - \tilde{x} q^{2l}}{\tilde{y} - y q^{2l}}.$$

The same reasoning as before shows that $R(\xi, \eta)$ solves YBE. The $\widehat{W}_{r\tilde{r}}$, $\overline{W}_{r\tilde{r}}$ are the Boltzmann weights (the former being Fourier transformed) of the chiral Potts model found by Au-Yang, Baxter, McCoy, Perk and others [12–13]. The connection with the quantum groups at roots of 1 was first noticed by Bazhanov-Stroganov [14] in a different language. The formulation above follows [15].

The study of intertwiners at roots of 1 has begun only recently. The chiral Potts model has been generalized to the case corresponding to a class of representations of $U_q(\widehat{\mathfrak{sl}}(n, \mathbf{C}))$ [16, 17]. In the general case the structure of irreducible representations at $q^N = 1$ and the existence of intertwiners are the major open questions. We remark that, for simple finite dimensional $\mathfrak{g}$, irreducible representations have been classified by De Concini and Kac [18].

1.3 Face Models

The third construction deals with solutions of YBE (1) which are not of the form $R(\xi, \eta) \in \mathrm{End}(V \otimes V)$.

Let us return to the example of Sect. 1.1., q being generic again. Let V_a denote the a dimensional irreducible representation of the subalgebra $U_q(\mathfrak{sl}(2, \mathbf{C})) \subset U_q(\widehat{\mathfrak{sl}}(2, \mathbf{C}))$. We consider the decomposition with respect to $U_q(\mathfrak{sl}(2, \mathbf{C}))$ of the tensor product representation

$$V_a \otimes V_2^{\otimes \mathscr{N}} = \bigoplus_c \Omega_{ac}^{\mathscr{N}} \otimes V_c,$$

$$\Omega_{ac}^{\mathscr{N}} = \{v \in V_a \otimes V_2^{\otimes \mathscr{N}} \mid e_1 v = 0, \ t_1 v = q^{c-1} v\}.$$

The $R = R(\xi, \eta) \in \mathrm{End}(V_2 \otimes V_2)$ commutes with the action of $U_q(\mathfrak{sl}(2, \mathbf{C}))$, hence

$$W_i = \mathrm{id} \otimes \mathrm{id} \otimes \cdots \otimes \overset{i\, i+1}{R} \otimes \cdots \otimes \mathrm{id} \ \in \mathrm{End}(V_a \otimes V_2^{\otimes \mathscr{N}})$$

are well-defined operators on $\Omega_{ac}^{\mathscr{N}}$ and satisfy YBE.

Repeated use of the Clebsch-Gordan decomposition gives rise to a basis of $\Omega_{ac}^{\mathcal{N}}$ indexed by a sequence $\mu = (\mu_0, \mu_1, \dots, \mu_{\mathcal{N}})$ ($\mu_0 = a, \mu_{\mathcal{N}} = c$) such that $\mu_i = \mu_{i+1} \pm 1$ for each i. In this basis the matrix elements of W_i look as follows.

$$(W_i)_{\mu\nu} = \delta_{\mu_0,\nu_0} \cdots \delta_{\mu_{i-1},\nu_{i-1}} W \begin{pmatrix} \mu_{i-1} & \nu_i \\ \mu_i & \mu_{i+1} \end{pmatrix} \delta_{\mu_{i+1},\nu_{i+1}} \cdots,$$

$$W \begin{pmatrix} a & a \pm 1 \\ a \pm 1 & a \pm 2 \end{pmatrix} = \frac{[1+u]}{[1]}, \quad W \begin{pmatrix} a & a \pm 1 \\ a \pm 1 & a \end{pmatrix} = \frac{[a \mp u]}{[a]},$$

$$W \begin{pmatrix} a & a \mp 1 \\ a \pm 1 & a \end{pmatrix} = \frac{[u]}{[1]} \sqrt{\frac{[a+1][a-1]}{[a]^2}},$$

where $q^u = \xi/\eta$ and $[u] = (q^u - q^{-u})/(q - q^{-1})$. Starting from the Verma module instead of V_a, we obtain the same formulas with a being any complex number.

In fact if one replaces the symbol $[u]$ by the elliptic theta function

$$\theta_1\left(\frac{\pi u}{L}, p\right) = 2p^{1/8} \sin \frac{\pi u}{L} \prod_{k=1}^{\infty} \left(1 - 2p^k \cos \frac{2\pi u}{L} + p^{2k}\right)(1 - p^k),$$

these formulas still solve YBE ($L \in \mathbf{C}^\times$ is an arbitrary parameter). Such elliptic solutions are known in correspondence to many of the representations of $U_q(\mathfrak{g})$ [19–23]. Pasquier [24] pointed out that their trigonometric degenerations ($p \to 0$) can be described as above in terms of quantum groups. It remains an open problem to explain the existence of elliptic solutions.

2. One-Point Functions

2.1 Formulation

Let us now come to our second topic about the calculation of physical quantities called one point functions. The problem is set up as follows. We consider a square lattice $\mathscr{L}$ on the plane $\mathbf{Z}^2$. Each site $i = (i_1, i_2)$ of the lattice has a random variable s_i taking values in some set $\mathscr{S}$. A configuration is an assignment $s = (s_i)$ of the values for each i. One also imposes the condition that the s_i for the boundary sites $i \in \partial\mathscr{L}$ are fixed to a given configuration $\bar{s}$ (which is to be chosen a 'ground state'; see below).

A model is specified by giving a Boltzmann weight $W \begin{pmatrix} a & b \\ d & c \end{pmatrix}$ for each quadruple (a, b, c, d) of elements of $\mathscr{S}$. Given a configuration s, to each 'face' (= an elementary square) of the lattice we attach a number $W \begin{pmatrix} s_i & s_j \\ s_l & s_k \end{pmatrix}$ where i, j, k, l are the NW, NE, SE, SW corner sites of the face. The probability of a given configuration s to occur is defined to be

$$p(s) = Z^{-1} \prod_{\text{faces}} W \begin{pmatrix} s_i & s_j \\ s_l & s_k \end{pmatrix}$$

where the product ranges over the faces of $\mathscr{L}$, and Z is a normalization constant to make the total probability $= 1$. Let us now fix $\bar{s}$, a site $0 \in \mathscr{L}$ and $a \in \mathscr{S}$. By definition the one point function is the expected value of $\delta_{s_0,a}$:

$$P_a(\bar{s}) = \mathrm{Prob}(s_0 = a) = \lim \sum_s \delta_{s_0\,a} p(s)\,.$$

Here the lim signifies the limit of lattice size tending to ∞.

Let us take again the R matrix of Sect. 1.1. Taking base vectors $\{v_\lambda\}_{\lambda=\pm1}$ ($\{\pm1\}$=the set of weights of V_2), we set $R(\xi,\eta)v_\lambda \otimes v_\mu = \sum v_\kappa \otimes v_\nu R(\xi,\eta)_{\kappa\nu,\lambda\mu}$. Let $\mathscr{S} \simeq \mathbf{Z}$ be the set of level 1 integral weights of $\widehat{\mathfrak{sl}}\,(2,\mathbf{C})$. For $a,b,c,d \in \mathscr{S}$ we set

$$\begin{aligned} W\begin{pmatrix} a & b \\ d & c \end{pmatrix} &= 0 \quad \text{unless } a-b, b-c, a-d, d-c \text{ all belong to } \{\pm1\}, \\ &= R(\xi,\eta)_{\kappa\nu,\lambda\mu} \quad \text{if } a+\lambda = b, b+\mu = c, a+\kappa = d, d+\nu = c\,. \end{aligned}$$

Hereafter we consider the region $0 < q < 1 < x = \xi/\eta$.

The boundary configuration $\bar{s}$ is so chosen as to maximize $p(s)$. In the present case, $\bar{s}$ turns out to be in one-to-one correspondence with *dominant* integral weights of fixed level ($= 1$ here). For example we take $\bar{s}_i = 0$ or 1 according as $i_1 + i_2$ is even or odd. This corresponds to the fundamental weight Λ_0.

Baxter's corner transfer matrix (CTM) is the only known tool to evaluate the one point function exactly. It is an infinite dimensional matrix whose trace gives the one point function up to a simple factor. In the sequel we shall focus on this trace. According to Baxter's argument based on YBE [1], it is completely determined by the trace at $q = 0$, where CTM becomes diagonal. We quote here the conclusion of the method. The CTM eigenvectors at $q = 0$ are labeled by an object called *path*. In the present example, it is a sequence $\mu = (\mu_0, \mu_1, \mu_2, \cdots)$ ($\mu_i \in \mathscr{S}$) such that

$$\mu_i = \mu_{i-1} \pm 1 \ \text{ for all } i, \quad \mu_i = \bar{\mu}_i \ \text{ for } i \gg 0,$$

where $\bar{\mu}_i = (0,1,0,1,0,1,\ldots)$ corresponds to the ground state $\bar{s}$. The trace of CTM becomes the combinatorial sum (called the one dimensional configuration sum)

$$\begin{aligned} c_a^{\Lambda_0}(t) &= \sum_\mu \delta_{\mu_0\,a} t^{\omega(\mu)}, \\ \omega(\mu) &= \sum_{j=1}^{\infty} j\left(H(\mu_{j-1},\mu_j,\mu_{j+1}) - H(\bar{\mu}_{j-1},\bar{\mu}_j,\bar{\mu}_{j+1})\right). \end{aligned} \tag{2}$$

Here $H(\lambda,\mu,\nu) = 0$ if $\lambda + 1 = \mu = \nu + 1$, $= 0$ otherwise, and the sum ranges over the paths whose 'initial point' μ_0 is fixed to $a \in \mathscr{S}$. This is a counting problem – counting the number of paths such that μ_0 and $\omega(\mu)$ are fixed. It is not hard to show that

$$c_a^{\Lambda_0}(t) = \frac{t^{a^2/4}}{\prod_{j=1}^{\infty}(1-t^j)}\,.$$

The quantity on the right hand side coincides with the string function of the irreducible highest weight module $M(\Lambda_0)$ of the affine Lie algebra $\widehat{\mathfrak{sl}}(2,\mathbf{C})$ in the sense of [25].

2.2 Crystal Base and One-Point Functions

In the same manner each R matrix of type $(U_q(\mathfrak{g}), \pi)$ gives rise to a similar model. The theory of crystal base [26] provides a natural framework to deal with the sum (2) in general. According to [26], for any integrable highest weight representation $M(\Lambda)$ of $U_q(\mathfrak{g})$ with highest weight Λ, there exits a unique canonical base $B(\Lambda)$ (crystal base) 'at $q = 0$'. Moreover they behave extremely simply with respect to taking tensor products. For details we refer the reader to the exposition of Kashiwara [27] in these proceedings.

Let us formulate the results in a generalized setting. We take

$U_q(\mathfrak{g}) = U_q(\widehat{\mathfrak{sl}}(n,\mathbf{C}))$,

$V =$ l-th symmetric tensors $S^l(\mathbf{C}^n)$ of n dimensional representation of $U_q(\mathfrak{g})$,

$\mathscr{S} =$ the set of integral weights of level l.

Let Λ_i denote the fundamental weights of $\widehat{\mathfrak{sl}}(n,\mathbf{C})$ and let $\varepsilon_i = \Lambda_{i+1} - \Lambda_i$. The set of weights of V is identified with $\mathscr{A}_l = \{\varepsilon_{i_1} + \cdots + \varepsilon_{i_l} \mid 0 \le i_1, \cdots, i_l \le n-1\}$. Let P_l^+ denote the set of dominant integral weights of level l. For $\Lambda \in P_l^+$, a Λ-path is defined to be a sequence $\mu = (\mu_0, \mu_1, \mu_2, \cdots)$ $(\mu_i \in \mathscr{S})$ such that

$$\mu_i - \mu_{i-1} \in \mathscr{A}_l \text{ for all } i, \quad \mu_i = \sigma^i(\Lambda) \text{ for } i \gg 0,$$

where σ signifies the linear automorphism such that $\sigma(\Lambda_i) = \Lambda_{i+1}$. Finally if $\mu - \lambda = \varepsilon_{i_1} + \cdots + \varepsilon_{i_l}$ and $\nu - \mu = \varepsilon_{j_1} + \cdots + \varepsilon_{j_l}$, then

$$H(\lambda, \mu, \nu) = \max_{\tau \in \mathfrak{S}_l} \sum_{k=1}^{l} \theta\left(i_{\tau(k)} - j_k\right) \tag{3}$$

with $\theta(x) = 0$ if $x < 0$, $= 1$ if $x \ge 0$. With these notations we have

Theorem 1 [28, 29]. *There exists a bijective correspondence between the set of Λ-paths and the crystal base $B(\Lambda)$ of $M(\Lambda)$, such that each path μ has weight $\mu_0 - \omega(\mu)\delta$ in the latter. Here δ signifies the null root.*

Corollary 2. *The one dimensional sum* (2) *is the string function*

$$c_a^\Lambda(t) = \sum_{n \in \mathbf{Z}} \dim M(\Lambda)_{a-n\delta} t^n .$$

This result has been found first by a purely combinatorial method in [30]. There is also a relative version of these results. Here one takes

$$\mathcal{S} = P^+_{l'+l},$$

$$W\begin{pmatrix} a & b \\ d & c \end{pmatrix} = \text{elliptic solutions in Sect.1.3. corresponding to } S^l(\mathbf{C}^n).$$

In this case $\bar{s}$ corresponds to a pair $(\Lambda', \Lambda) \in P^+_{l'} \times P^+_l$. A sequence $\mu = (\mu_0, \mu_1, \mu_2, \cdots)$ is called a (Λ', Λ)-path if

a) for any i there exists a pair $(\tilde{\Lambda}', \tilde{\Lambda}) \in P^+_{l'} \times P^+_l$ such that $\mu_i = \tilde{\Lambda}' + \tilde{\Lambda}$ and $\mu_{i+1} = \tilde{\Lambda}' + \sigma(\tilde{\Lambda})$,

b) $\mu_i = \Lambda' + \sigma^i(\Lambda)$ for $i \gg 0$.

Theorem 3 [29]. *A vector $u \otimes v \in B(\Lambda') \otimes B(\Lambda)$ is a highest weight vector if and only if u is highest and v corresponds to a (Λ', Λ)-path.*

Let

$$M(\Lambda') \otimes M(\Lambda) = \bigoplus_{a \in P^+_{l'+l}} \Omega_{\Lambda' \Lambda a} \otimes M(a)$$

be the decomposition of $U_q(\widehat{\mathfrak{sl}}(n, \mathbf{C}))$ modules where $\Omega_{\Lambda' \Lambda a}$ denotes the space of highest weight vectors of weight a. Let $c_a^{\Lambda' \Lambda}$ denote the one dimensional configuration sum (3) where μ ranges over (Λ', Λ)-paths such that $\mu_0 = a$, and $H(\lambda, \mu, \nu)$ is the same as (3). Then we have

Corollary 4.

$$c_a^{\Lambda' \Lambda}(t) = \sum_n \dim(\Omega_{\Lambda' \Lambda a})_{a - n\delta} t^n.$$

The left hand side is called the branching function studied in [25, 31]. Similar results for other classical type algebras (for level 1) can be found in [6, 7, 32].

3. Summary

We have seen two aspects of the interplay between solvable lattice models and quantum groups. Our conclusions are summarized as follows:

1) The theory of intertwiners lead to solutions of YBE, and hence to construction of solvable lattice models.

2) The one point functions are reduced to one dimensional configuration sums by CTM method. The theory of crystal base enables one to identify the latter with modular forms arising in the representation theory of affine Lie algebras.

Although crystal base provides a powerful combinatorial tool, there is no direct explanation why these modular forms should arise at all. We feel the intrinsic understanding is still missing.

References

1. Baxter, R. J.: Exactly Solved Models in Statistical Mechanics. Academic, London 1982
2. Bazhanov, V.V.: Integrable quantum systems and classical Lie algebras. Comm. Math. Phys. **113** (1987) 471–503
3. Jimbo, M.: Quantum R matrix for the generalized Toda system. Comm. Math. Phys. **102** (1986) 537–547
4. Reshetikhin, N. Yu.: The spectrum of the transfer matrices connected with Kac-Moody algebras. Lett. Math. Phys. **14** (1987) 235-246
5. Reshetikhin, N. Yu.: Zapiski Nauchnyi Seminarov LOMI **169** (1988) 122
6. Okado, M.: Quantum R matrices related to the spin representations of B_n and D_n. Comm. Math. Phys. **134** (1990) 467–486
7. Kuniba, A.: Quantum R matrix for G_2 and a solvable 175 vertex models. J. Phys. A **23** (1990) 1349–1362
8. Chung, H. J., Koh, I. G.: Solutions to the quantum Yang-Baxter equation for the exceptional Lie Algebras with a spectral parameter. Preprint 1990
9. Drinfel'd, V. G.: Quantum groups. Proceedings of the International Congress of Mathematicians, Berkeley, California, USA 1986, pp. 798–820
10. Drinfel'd, V. G.: A new realization of Yangians and quantized affine algebras. Sov. Math. Dokl. **36** (1988) 212–216
11. Chari, V., Presseley, A.: Quantum affine algebras. Preprint 1990
12. Au-Yang, H., McCoy, B. M., Perk, J. H. H., Tang, S.: Solvable models in statistical mechanics and Riemann surfaces of genus greater than one. In: Algebraic Analysis, vol. 1, eds. M. Kashiwara and T. Kawai. Academic, 1989, pp. 29–39; and references therein
13. Baxter, R. J., Perk, J. H. H., Au-Yang, H.: New solutions of the star-triangle relations for the chiral Potts model. Phys. Lett. A **128** (1988) 138–142
14. Bazhanov, V. V., Stroganov, Yu. G.: Chiral Potts model as a descendant of the six vertex models. J. Stat. Phys. **51** (1990) 799–817
15. Date, E., Jimbo, M., Miki, K., Miwa, M.: New R matrices associated with cyclic representations of $U_q(A_2^{(2)})$. Preprint RIMS **706** (1990)
16. Bazhanov, V. V., Kashaev, R. M., Mangazeev, V. V., Stroganov, Yu. G.: $(\mathbf{Z}_N \times)^{n-1}$ generalization of the chiral Potts model. Preprint 1990
17. Date, E., Jimbo, M., Miki, K., Miwa, M.: Generalized chiral Potts models and minimal cyclic representations of $U_q(\widehat{\mathfrak{gl}}\,(n, \mathbf{C}))$. Preprint RIMS **715** (1990), Comm. Math. Phys. (to appear)
18. De Concini, C., Kac, V. G.: Representations of quantum groups at roots of 1. Preprint 1990
19. Baxter, R. J.: Eight-vertex model in lattice statistics and one-dimensional anisotropic Heisenberg chain II. Equivalence to a generalized ice-type model. Ann. Phys. **76** (1973) 25-47
20. Andrews, G. E., Baxter, R. J. and Forrester, P. J.: Eight-vertex SOS model and generalized Rogers-Ramanujan-type identities. J. Stat. Phys. **35** (1984) 193–266
21. Date, E., Jimbo, M., Kuniba, A., Miwa, T., Okado, M.: Exactly solvable SOS models II : Proof of the star-triangle relation and combinatorial identities. Adv. Studies Pure Math. **16** (1988) 17-122
22. Jimbo, M., Miwa. T., Okado, M.: Solvable lattice models related to the vector representation of classical simple Lie algebras. Comm. Math. Phys. **116** (1988) 507-525
23. Jimbo, M., Kuniba, A., Miwa. T., Okado, M.: The $A_n^{(1)}$ face models. Comm. Math. Phys. **119** (1988) 543-565
24. Pasquier, V.: Etiology of IRF models. Comm. Math. Phys. **118** (1988) 335–364

25. Kac, V. G., Peterson, D. H.: Infinite-dimensional Lie algebras, theta functions and modular forms. Adv. Math. **53** (1984) 125–264
26. Kashiwara, M.: Bases cristallines. C. R. Acad. Sci. Paris **311** (1990) 277–280
27. Kashiwara, M.: Crystalizing the q analogue of universal enveloping algebras. In these Proceedings, pp. 791–797
28. Misra, K. C., Miwa, T.: Crystal bases for the basic representations of $U_q(\widehat{\mathfrak{sl}}(n, \mathbf{C}))$. Comm. Math. Phys. **134** (1990) 79–88
29. Jimbo, M., Misra, K. C., Miwa, T., Okado, T.: Combinatorics of representations of $U_q(\widehat{\mathfrak{sl}}(n, \mathbf{C}))$ at $q = 0$. Preprint RIMS **709** (1990). Comm. Math. Phys. (to appear)
30. Date, E., Jimbo, M., Kuniba, A., Miwa, T., Okado, M.: Paths, Maya diagrams and representations of $\widehat{\mathfrak{sl}}(2, r)$. Adv. Studies Pure Math. **19** (1989) 149–191
31. Kac, V. G., Wakimoto, M.: Modular and conformal invariance constraints in representation theory of affine Lie algebras. Adv. Math. **70** (1988) 156–236
32. Date, E., Jimbo, M., Kuniba, A., Miwa, T., Okadao, M.: One-dimensional configuration sums in vertex models and affine Lie algebra characters. Lett. Math. Phys. **17** (1989) 69–77

The Periodic Problems for Two-Dimensional Integrable Systems

Igor Krichever

Landau Institute for Theoretical Physics, Academy of Sciences of the USSR
GSP-1, 117940 ul. Kosygina 2, Moscow, USSR

1. Introduction

Since the middle of the seventies algebraic geometry has become a very powerful tool in various problems of mathematical and theoretical physics. In the theory of integrable equations the algebraic geometrical methods provide a construction of the periodic and quasi-periodic solutions which can be written exactly in terms of theta functions of the auxiliary Riemann surfaces.

All the integrable equations which are considered in the soliton theory can be represented as compatibility conditions of the auxiliary linear problems. One of the most general types of such representations has the form:

$$[\partial_y - L, \partial_t - A] = 0\,, \tag{1.1}$$

where L, A are differential operators of the form

$$L = \sum_{i=0}^{n} u_i(x, y, t)\partial_x^i\,, \quad A = \sum_{i=0}^{m} v_i(x, y, t)\partial_x^i \tag{1.2}$$

with scalar or matrix coefficients.

The most important example of these equations is the Kadomtsev-Petviashvilii (KP) equation

$$\frac{3}{4}\sigma^2 u_{yy} + \left(u_t - \frac{3}{2}uu_x + \frac{1}{4}u_{xxx}\right)_x = 0 \tag{1.3}$$

which is equivalent to (1.1), where

$$L = \sigma(-\partial_x^2 + u(x, y, t))\,, \quad A = \partial_x^3 - \frac{3}{2}u\partial_x - w(x, y, t)\,. \tag{1.4}$$

The algebraic geometrical construction of the solutions of integrable equations is based on the concept of the Baker-Akhiezer functions which are definded by their very specific analytical properties on the auxiliary Riemann surfaces. For example, the Baker-Akhiezer functions in the case of the KP equation are defined for each smooth algebraic curve Γ (Riemann surface of finite genus g) with the fixed point P_0 on it, and the local parameter $k^{-1}(P)$ in the neighbourhood of this point, $k^{-1}(P_0) = 0$. For any set of generic points γ_j, $j = 1, \ldots, g$, there exists a unique function $\Psi(x, y, t, P)$, $P \in \Gamma$, such that:

Proceedings of the International Congress
of Mathematicians, Kyoto, Japan, 1990

1^0. It is meromorphic on Γ outside the point P_0 and has no more than simple poles at the points γ_j (if they are distinct);

2^0. The function Ψ has the form:

$$\Psi(x,y,t,P) = \left(1 + \sum_{s=1}^{\infty} \xi_s^i(x,y,t)k^{-1}\right) \exp(ikx + \sigma^{-1}k^2y + ik^3t) \tag{1.5}$$

$k = k(P)$, near the point P_0.

For any formal series of the form (1.5) there exist unique operators L and A of the form (1.4) such that the following relations

$$\begin{aligned}(\partial_y - L)\Psi &= O(k^{-1})\exp(ikx + \sigma^{-1}k^2y + ik^3t)\\ (\partial_t - A)\Psi &= O(k^{-1})\exp(ikx + \sigma^{-1}k^2y + ik^3t)\end{aligned} \tag{1.6}$$

are valid. From (1.6) it follows that the coefficient $u(x,y,t)$ of these operators is equal to

$$u(x,y,t) = 2i\xi_{1,x}(x,y,t)\,. \tag{1.7}$$

The left hand sides of (1.6) define the functions which have the same analytical properties outside P_0 as Ψ, and have the form (1.6) near this point. From the uniqueness of the Baker-Akhiezer function Ψ, it follows that they are equal to zero. Hence,

$$(\partial_y - L)\Psi = 0\,, \quad (\partial_t - A)\Psi = 0 \tag{1.8}$$

and $u(x,y,t)$, which is given by (1.7) is a solution of the KP equation.

The Baker-Akhiezer function $\Psi(x,y,t,P)$ can be exactly written in terms of the Abelian differentials and Riemann theta-function. From the corresponding formulae it follows that the above constructed solutions of the KP equation have the form

$$u(x,y,t) = 2\partial_x^2 \ln \theta(Ux + Vy + Wt + \Phi/\tau) + \text{const}\,. \tag{1.9}$$

Here, $\theta(z_1, \ldots, z_g)$ is the Riemann theta-function which is defined by the matrix τ_{ij} of the b-periods of the normalized holomorphic differentials on Γ. The vectors $2\pi iU$, $2\pi iV$, $2\pi iW$ are the vectors of b-periods of the normalized Abelian differentials of the second kind with the only poles at P_0 of orders 2, 3, 4, respectively. The vector Φ corresponds to the set of the points γ_j and can be considered in (1.9) as an arbitrary vector.

The construction was proposed in [1, 2] and was developed in different ways for various types of integrable equations (see, for example, the reviews [3, 4, 5, 6]. The analytical properties of the Baker-Akhiezer functions are the natural generalization of the analytical properties of Bloch functions of the ordinary periodic finite-gap differential operators which were obtained in the remarkable works by Novikov, Dubrovin, Matveev, Its in which the algebraic geometrical solutions of the KdV equation, sine-Gordon equation and some other Lax-type equations were constructed.

In this report we shall present a brief review of the latest results obtained in the theory of periodic problems for the two-dimensional integrable systems. First of all, why is it algebraic geometry ? What is the meaning of the algebraic geometrical solutions for the general periodic (in x and y) initial value problem for such equations ? For the one-dimensional evolution integrable equations, the algebraic geometrical solutions are dense in the space of all periodic solutions (though this statement has not been proved rigorously for all such equations). In the case of the two-dimensional integrable equations the situation is much more complicated.

There are two real forms of the KP-1 ($\sigma^2 = -1$) and KP-2 ($\sigma^2 = 1$). It turns out, that the periodic problems for these equations differ dramatically from each other.

The formal non-integrability of the periodic problem for the KP-1 equation was proved in [7]. The proof of the integrability of such problem for the KP-2 equation was obtained by the author [8] and is based on the spectral theory of the operator

$$M = \sigma \partial_y - \partial_x^2 + u(x, y) \tag{1.10}$$

with the periodic potential.

The second problem which will be considered in this talk is the perturbation theory for two-dimensional integrable equations. We shall concentrate our attention on the so-called Whithem equation which is in our case a system of equations on bundles over the Teichmüller spaces. Finally, we shall demonstrate how the Whithem theory and other aspects of the perturbation theory of integrable equations will be married to each other in attemps to solve the Heisenberg relations

$$[L_n, A_m] = 1 , \tag{1.11}$$

for the ordinary differential linear operators

$$L_n = \sum_{i=0}^{n} u_i(x)\partial_x^i , \quad A_m = \sum_{i=0}^{m} v_i(x)\partial_x^i , \quad u_n = v_m = 1 . \tag{1.12}$$

The latter are the most popular subject in the field of string theory.

2. The Spectral Theory of Two-Dimensional Periodic Linear Differential Operators

The solutions $\Phi(x, y, w_1, w_2)$ of the nonstationary Schrödinger equation

$$\left(\sigma \partial_y - \partial_x^2 + u(x, y)\right) \Phi(x, y, w_1, w_2) = 0 \tag{2.1}$$

with the periodic potential are called Bloch solutions, if they are eigenfunctions of the monodromy operators, i.e.

$$\begin{aligned} \Psi(x + a_1, y, w_1, w_2) &= w_1 \Psi(x, y, w_1, w_2) \\ \Psi(x, y + a_2, w_1, w_2) &= w_2 \Psi(x, y, w_1, w_2) . \end{aligned} \tag{2.2}$$

The set of pairs $Q = (w_1, w_2)$, for which there exists such a solution is called the Floque set, and will be denoted by Γ. The multivalued functions $p(Q)$, $E(Q)$ such that

$$w_1 = \exp(ipa_1) , \quad w_2 = \exp(iEa_2)$$

are called quasi-momentum and quasi-energy, respectively.

For the "free" operator with zero potential $u_0 = 0$, the Floque set is parametrized by the points of the complex k-plane

$$w_1^0 = \exp(ika_1)\,, \quad w_2^0 = \exp(-\sigma^{-1}k^2a_2) \tag{2.3}$$

and the Bloch solutions have the form

$$\Psi_0(x, y, k) = \exp(ikx - \sigma^{-1}k^2y)\,. \tag{2.4}$$

It turns out that if Re $\sigma \neq 0$, then the Floque set of the operator (2.1) with the smooth potential $u(x, y)$ is isomorphic to the Riemann surface Γ (which has in a generic case infinite genus). The corresponding Riemann surface has such a specific structure that the theory of abelian differentials, theta-functions and so on can be constructed for it as well as for the finite genus case.

The source of the difference between the two cases Re $\sigma = 0$ and Re $\sigma \neq 0$ is the difference between the structure of the "resonant" points for the free operators. The resonant points are the points on the complex k-plane which are the pre-images of the self-intersection points of the imbedding $C \to C^2$, which is defined by (2.3). The points k and k' are resonant, if

$$w_i^0(k) = w_i^0(k')\,, \quad i = 1, 2\,. \tag{2.5}$$

From (2.3) it follows that such points are parametrized be integers $(N > 0, M)$ and have the form:

$$k = k_{N,M}\,, \quad k' = k_{-N,-M}\,, \tag{2.6}$$

where

$$k_{N,M} = \frac{\pi N}{a_1} + i\sigma\frac{Ma_1}{Na_2}\,. \tag{2.7}$$

In case Re $\sigma \neq 0$, the resonant points tend to infinity and, hence, have no limiting points outside infinity. In case Re $\sigma = 0$, the resonant points are dense on the real axis which makes it impossible (at least by means of our methods) to construct the global Riemann surface of the Bloch functions.

For the real smooth potential u the Floque set can be described in the following form. Let us call the set of pairs of the complex numbers $\pi = \{p_{s,1}, p_{s,2}\}$ (where s belongs to any finite or infinite subset of integer pairs $(N > 0, M)$) "adimissible", if

$$\text{Re } p_{s,i} = \frac{\pi N}{a_1}\,, \quad |p_{s,i} - k_s| = O\left(\frac{1}{|k_s|}\right)\,, \quad i = 1, 2$$

and the intervals $[p_{s,1}, p_{s,2}]$ do not intersect each other. Let us define the Riemann surface $\Gamma(\pi)$ for any admissible set π. It is obtained from the complex k-plane by cutting it along the intervals $[p_{s,1}, p_{s,2}]$ and $[-\overline{p_{s,1}}, -\overline{p_{s,2}}]$ and by sewing after that the left side of the first cut with the right side of the second cut and vise versa.

Theorem 1. *For any real periodic potentials $u(x, y)$, which can be analytically extended in some neighbourhood of the real values x, y, the Bloch solutions of the Equation (2.1) with $\sigma = 1$ are parametrized by the points Q of the Riemann surface $\Gamma(\pi)$ corresponding to some admissible set π. The function $\Psi(x, y, Q)$ which is normalized by the condition $\Psi(0, 0, Q) = 1$, is meromorphic on Γ and has a simple pole γ_s on each cycle a_s which corresponds to the cut $[p_{s,1}, p_{s,2}]$. If the admissible set π contains only a finite number of pairs, then $\Gamma(\pi)$ has finite genus and is compactified by only one point $P_0(k = \infty)$, in the neighbourhood of which the Bloch function Ψ has the form (1.5).*

The potentials u for which $\Gamma(\pi)$ has finite genus, are called finite-gap potentials and as it follows from the last statement of the theorem that they coincide with the algebraic geometrical potentials.

Theorem 2. *Any smooth periodic potential u of the Equation* (2.1) *(with* Re $\sigma \neq 0$*), which can be analytically extended in the complex neighbourhood of the real x, y, can be approximated uniformly with any number of the derivatives by means of the finite-gap (algebraic geometrical) potentials.*

The Floque set is the "integral" of the KP equation. From the previous theorems we have:

Theorem 3. *For any smooth periodic function $v(x, y)$ there exists a unique solution of the* KP-2 *equation $u(x, y, t)$, such that $u(x, y, 0) = v(x, y)$. This solution is regular for all t and quasi-periodic in t. Any smooth periodic solutions of the* KP-2 *equation can be approximated by means of the finite-gap solutions.*

3. The Perturbation Theory of the Finite-Gap Solutions. Whithem Equations

The non-linear WKB (or Whithem) method can be applied to any non-linear equation which has the set of the exact solutions of the form

$$u_0(x, y, t) = u_0(Ux + Vy + Wt + \Phi|I_1, \ldots, I_N), \tag{3.1}$$

where $u_0(z_1, \ldots, z_g|I_k)$ is a periodic function of the variable z_i depending on the parameters I_k. The vectors U, V, W are also functions of the same parameters: $U = U(I)$, $V = V(I)$, $W = W(I)$.

In the framework of the non-linear WKB-method the asymptotic solutions of the form

$$u(x, y, t) = u_0(\varepsilon^{-1}S(X, Y, T)|I_k) + \varepsilon u_1 + \ldots \tag{3.2}$$

are constructed for the perturbed or non-perturbed initial equation. Here $X = \varepsilon x$, $Y = \varepsilon y$, $T = \varepsilon t$ are the "slow variables". If the vector $S(X, Y, T)$ is defined from the relations

$$\begin{gathered} \partial_X S = U(I(X, Y, T) = U(X, Y, T) \\ \partial_Z S = V(X, Y, T), \ \partial_T S = W(X, Y, T) \end{gathered} \tag{3.3}$$

the main term u_0 in the expansion (3.2) satisfies the initial equation up to the first order in ε. After that all the other terms of the series (3.2) are defined from the non-homogeneous linear equations. The construction of such asymptotic solutions even for integrable equations is very important, because when using the slow modulation of the parameters of their exact solutions one can sometimes solve the integrable equation with "non-integrable boundary conditions".

For the KdV equation and for some other Lax-type equations, the Whithem method was developed and applied to various problems in [9, 10, 11]. For the two-dimensional integrable systems the Whithem method was proposed in [12]. We shall present here only a part of the corresponding results.

The asymptotic solutions of the form (3.2) can be constructed with an arbitrary dependence of the parameters I_k on slow variables. In this case the expansion

(3.2) is valid on the scales of order 1. The right hand side of the non-homogeneous linear equation which defines the first order term u_1 contains the first derivatives of the parameters I_k. Therefore, the choice of the dependence of I_k on slow variables can be used for the cancellation of the "secular" term in u_1. The corresponding equations on I_k are usually called the Whithem equations.

Let us consider again the KP equation as an example of the two-dimensional integrable systems. Its finite-gap solutions have the form (3.1). The set of their parameters are the system of local coordinates of the manifold M_g which has dimension $N = 3g + 1$.

$$M_g\big((\Gamma, P_0[k^{-1}]_2\big) . \tag{3.4}$$

(Two local parameters are m-equivalent if $k_1 = k + O(k^{-m})$; the corresponding equivalence class of the local parameter is denoted by $[k^{-1}]_m$.)

Let us consider the second kind differentials on Γ with the only poles at the point P_0 of the form

$$dp = dk(1 + O(k^{-2}), \quad dE = i\sigma^{-1}dk^2(1 + O(k^{-3}), \quad d\Omega = dk^3(1 + O(k^{-4}) \tag{3.5}$$

which have the real periods for any cycle on Γ. Their integrals $p(Q)$, $E(Q)$, $\Omega(Q)$ are multivalued functions on the manifold M_g^* which is a bundle over M_g

$$M_g^* = \big(\Gamma, P_0, [k^{-1}]_2, Q \in \Gamma\big) . \tag{3.6}$$

If $(\lambda, I_1, \ldots, I_{3g+1})$ is a system of local coordinates on M_g^* and I_k are functions of the variables X, Y, T then $p = p(\lambda, X, Y, T)$, $E = E(\lambda, X, Y, T)$, $\Omega = \Omega(\lambda, X, Y, T)$ become functions of these variables.

Theorem 4. *The necessary conditions for the existence of the asymptotic solutions of the equation*

$$\frac{3}{4}\sigma^2 u_{yy} + (u_t - \frac{3}{2}uu_x + \frac{1}{4}u_{xxx})_x + \varepsilon K[u] = 0 \tag{3.7}$$

which has the form (3.2) *with uniformly bounded first-order term are equivalent to the equation*

$$\frac{\partial p}{\partial \lambda}\left(\frac{\partial E}{\partial T} - \frac{\partial \Omega}{\partial T}\right) - \frac{\partial E}{\partial \lambda}\left(\frac{\partial p}{\partial T} - \frac{\partial \Omega}{\partial X}\right) + \frac{\partial \Omega}{\partial \lambda}\left(\frac{\partial p}{\partial Y} - \frac{\partial E}{\partial X}\right) = \frac{\langle \Psi K \Psi_+ \rangle_x}{\langle \Psi \Psi_+ \rangle_x}\frac{\partial p}{\partial \lambda} . \tag{3.8}$$

Here $K[u]$ is an arbitrary differential polynomial; Ψ, Ψ_+ are the corresponding Baker-Akhiezer function and its dual, respectively.

Remark. It turns out that there are only $3g + 1$ independent equations among the Equation (3.8) which should be fulfilled for any point Q of the curve Γ.

For the KdV equation and $K = 0$ the Equation (3.8) have the form

$$\partial_T p = \partial_X \Omega \tag{3.9}$$

which was obtained in [11]. The construction of the exact solutions of the Equation (3.8) with $K = 0$ was proposed in the work [12]. We shall present the particular case of this scheme in the next section where the Heisenberg relations would be considered.

4. The Heisenberg Relations for the Ordinary Linear Differential Operators

Great progress has been made recently in non-perturbative two-dimensional gravity coupled to various matter fields. It was shown that the dependence of physical quantities (such as specific heat) on scaled coefficients of the models is described by the KP-hierarchy on the space of the ordinary linear differential operators L_n, A_m such that the relations (1.11) are fulfilled. For pure two-dimensional gravity $n = 2$, $m = 3$ the Equation (1.11) is equivalent to the Painlevé 1 equation

$$\frac{1}{4}u_{xxx} - \frac{3}{2}uu_x = 1 . \tag{4.1}$$

The Equation (1.12) has a simple scaling transformation

$$u_1(x) = \varepsilon^{(i-n)\beta}\widetilde{u_i}\left(\varepsilon^{-\beta}x\right), \quad v_i = \varepsilon^{(i-m)\beta}\widetilde{v_i}\left(\varepsilon^{-\beta}x\right) \tag{4.2}$$

$\beta = (n+m)^{-1}$. For the operators $\widetilde{L}_n, \widetilde{A}_m$ with the coefficients $\widetilde{u}_i, \widetilde{v}_i$ we have

$$[\widetilde{L}_n, \widetilde{A}_m] = \varepsilon . \tag{4.3}$$

The formal asymptotic solutions of the equation (4.3) can be constructed using any commuting operators $[L_{n,0}, A_{m,0}] = 0$

$$\widetilde{L}_n = L_{n,0} + \varepsilon L_{n,1} + \dots , \; \widetilde{A}_m = A_{m,0} + \varepsilon A_{m,1} + \dots . \tag{4.4}$$

Unfortunately, these asymptotic solutions are well-defined only in the interval $x \sim 1$. For our purposes it is necessary to have the solutions for $x \sim \varepsilon^{-1/(n+m)}$. It can be done in framework of the Whithem theory.

The commuting operators of co-prime orders $(n, m) = 1$ are parametrized by the coefficients of the polynomial

$$w^n + E^m + \sum_{in+jm\leq nm-2} \alpha_{ij} w^i E^j = 0 \tag{4.5}$$

and by the points of the Jacobian of the corresponding algebraic curve [13]. In [1, 2] the exact formulae for the coefficients of the generic commuting operators in terms of the Riemann theta-function were found. For example,

$$u_{n-2} = -n\partial_x^2 \ln \theta(Ux + \Phi/\tau) + \text{const} . \tag{4.6}$$

Here the matrix τ of b-periods of Γ depends on the values α_{ij}. The vector U is also a function of the variables α_{ij}. The phase vector Φ is arbitrary. All the other coefficients have the same structure

$$u_i = u_{i,0}(Ux + \Phi/\tau), \quad v_i = v_{i,0}(Ux + \Phi/\tau) . \tag{4.7}$$

Let me consider the operators $L_n^\#, A_m^\#$ with the coefficients

$$u_i^\# = u_{i,0}\left(\frac{1}{\varepsilon}S(X)/\tau(X)\right), \quad v_i^\# = v_{i,0}\left(\frac{1}{\varepsilon}S(X)/\tau(X)\right) . \tag{4.8}$$

If the vector $S(X)$ is defined by the relation $\partial_X S = U(\alpha_{ij}(X))$ then the operators $L_n^\#, A_m^\#$ commute up to the order ε. As was shown in [12] in more general situation the requirement that the first order terms in the expansion (4.4) should be uniformly bounded for all x leads to the equations on the variables α_{ij}. They are particular cases of (3.8) and have the form

$$\frac{\partial w(E,X)}{\partial X} = \frac{\partial p(E,X)}{\partial E}. \tag{4.9}$$

It turns out that they are integrable and we present the construction of their solutions below. Our conjecture (which is partly proved now for $n = 2$, $m = 3$) is that all the other terms of the asymptotic solutions (4.4) are also bounded and the series (4.4) are convergent. If this is true, it is possible to make the inverse rescaling and find the limit for $\varepsilon \to 0$. To begin with we shall give the final answer for the KdV equation with the "string" boundary conditions (1.12) $n = 2$, $m = 2k + 1$.

Let us consider an arbitrary hyperelliptic curve Γ

$$y^2 = \prod_{i=1}^{2k+1}(E - E_i) = E^{2k+1} + \prod_{i=1}^{2k} c_i E^i = R(E). \tag{4.10}$$

As is well-known, this curve defines the solutions of the KdV equation which have the form (1.9) (with $V = 0$).

For any given set of the parameters: the complex constants $c_{k,0}, c_{k+1,0}, \ldots, c_{2k,0}$, the real constants h_i, h_i', $i = 1, \ldots, k$, we shall consider the hyperelliptic curve which is defined by the polynomial R with the coefficients

$$c_i = c_{i,0}\,,\ i = k+2, \ldots, 2k\,; \quad c_k = x + c_{k,0}\,; \quad c_{k+1} = t + c_{k+1,0} \tag{4.11}$$

and such that

$$\operatorname{Im} \int_{E_{2i}}^{E_{2i+1}} \sqrt{R}\, dE = h_i\,, \quad \operatorname{Im} \int_{E_1}^{E_{2i}} \sqrt{R}\, dE = h_i'\,, \qquad i = 1, \ldots, k. \tag{4.12}$$

The Equation (4.12) are the set of $2k$ real equations which define k unknown complex coefficients c_i, $i = 0, \ldots, k-1$ of the polynomial $R(E)$. They become functions of the variable x, t. The τ matrix of the corresponding curve becomes a (known) function of the variables x, t. Let us define the vector

$$S_i(x,t) = \frac{1}{\pi}\left(\int_{E_1}^{E_{2i}} \sqrt{R}\, dE - \sum_{j=1}^{k} \tau_{ij} \int_{E_{2j}}^{E_{2j+1}} \sqrt{R}\, dE\right). \tag{4.13}$$

The Main Conjecture. *The functions*

$$u(x,t) = -2\partial_x^2 \ln\, \theta(S(x,t) + \Phi/\tau(x,t)) - 2r_1(x,t) \tag{4.14}$$

are the exact solutions of the KdV equation with the "boundary conditions" (1.12).

Here $r_1(x,t)$ is the coefficient of the differential

$$d\Omega_1 = \frac{E^k + \sum_{i=0}^{k-1} r_i E^i}{2\sqrt{R}}\, dE\,, \quad \int_{E_{2i}}^{E_{2i+1}} d\Omega_1 = 0\,.$$

Thus the Equations (4.12) are the only transcendental equations in the definition of $u(x,t)$.

Let us consider now the general Heisenberg relations. Any equation of the form (4.5) has the formal solution

$$w = k^m + \sum_{i=-m+2}^{\infty} a_i k^{-i}\,, \quad k^n = E\,. \tag{4.15}$$

This means that the affine curve (4.5) is compactified by a single point P_0. Let us fix a first few coefficients of the expansion (4.15) and denote them by

$$a_{n-j} = \frac{j}{n} t_j\,, \quad j = 1, \ldots, m+n-2\,. \tag{4.16}$$

They uniquely define the following coefficients of (4.5)

$$\alpha_{ij}, im + jn \geq (m-1)(n-1) = 2g\,. \tag{4.17}$$

For any given real numbers h_i, h_i', $i = 1, \ldots, g$, all the other coefficients of the polynomial (4.5) can be defined (at least locally) as functions of the parameters t_j with the help of the relations

$$\operatorname{Im} \int_{a_i} w\, dE = h_i\,, \quad \operatorname{Im} \int_{b_i} w\, dE = h_i'\,. \tag{4.18}$$

They give $2g$ real equations on g complex variables $\alpha_{ij}, in + jm < (n-1)(m-1)$. Therefore, the curve Γ and the algebraic function $w(E)$ become functions of the variables t_j.

Theorem 5. *The function* $w(E, t_1, \ldots)$ *satisfies the Whithem equations* (4.9) *if* $t_1 = x$.

Let us define the differentials $d\Omega_j$, $j = 1, \ldots, m+n-2$, whose only poles at infinity have the form

$$d\Omega_j = dk^j \left(1 + O(k^{-j-1})\right) \tag{4.19}$$

and such that

$$\operatorname{Im} \int_\gamma d\Omega_j = 0\,, \quad \gamma \in H_1(\Gamma)\,. \tag{4.20}$$

Corollary. *If the relations* (4.18) *are fulfilled, then*

$$\frac{\partial p}{\partial t_j} = \frac{\partial \Omega_j}{\partial x} \quad \frac{\partial \Omega_j}{\partial t_j} = \frac{\partial \Omega_j}{\partial t_i}\,.$$

Remark. It can be shown that the conjecture which was proposed recently in [14] leads to one particular solution of the Painlevé 1 which belongs to our set of solutions.

References

1. Krichever, I.M.: An algebraic-geometrical construction of the Zakharov-Shabat equation and their periodic solutions. Dokl. Akad. Nauk SSSR **227**, 291–294
2. Krichever, I.M.: The integration of non-linear equations by methods of algebraic-geometry. Funkt. Anal. Priloz. **11** (1) (1977) 15–31
3. Krichever, I.M.: Methods of algebraic geometry in the theory of non-linear equations. Uspekhi Matem. Nauk **32** (6) (1977) 183–208
4. Dubrovin, B.A.: Theta-functions and non-linear equations. Uspekhi Matem. Nauk **36** (6) (1981) 11–80
5. Krichever, I.M., Novikov, S.P.: Holomorphic bundles over algebraic curves and non-linear equations. Uspekhi Matem. Nauk **35** (6) (1980) 47–68
6. Dubrovin, B.A., Krichever, I.M., Novikov, S.P.: Integrable systems. Dynamical systems **4**, Itogi Nauki i Tekhniki, Fund. invest., VINITI Akad. Nauk SSSR, 1985
7. Zakharov, V.E., Schulmann, E.I.: On problems of the integrability of two-dimensional systems. Dokl. Akad. Nauk SSSR **283** (6) (1985) 1325–1329
8. Krichever, I.M.: Spectral thoery of two-dimensional periodic operators and its applications. Uspekhi Matem. Nauk **44** (2) (1989) 121–184
9. Gurevich, A.V., Pitaevskii, L.P.: Nonstationary structure of the interactionless shock-wave. JETP **65** (3) (1973) 590–604
10. Dobrokhotov, S.Yu., Maslov, V.P.: Multiphase asymptotics of non-linear partial differential equations with a small parameter. Soviet Scientific Reviews, Math. Phys. Rev. **3** (1982) 221–280, OPA Amsterdam
11. Flashka, H., Forest, M., McLaughlin, L.: The multiphase averaging and the inverse spectral solution of the Korteweg-de Vries equation. Comm. Pure Appl. Math. **33** (6) (1980) 739–784
12. Krichever, I.M.: The averaging method for two-dimensional integrable equations. Funkt. Anal. Priloz. **22** (3) (1988) 37–52
13. Burchnall, J.L., Chaundy, T.W.: Commuting ordinary differential operators, II. Proc. Roy. Soc. London **118** (1928) 557–583
14. Novikov, S.P.: On relations $[L, A] = 1$. To appear in Funkt. Anal. Priloz. 1990

Renormalization Group and Random Systems

Antti Kupiainen

Rutgers University, Department of Mathematics, New Brunswick, NJ 08903, USA

1. Introduction

Some of the most interesting and challenging problems in theoretical and mathematical physics have been the ones involving in an essential way several distance scales. Such problems include the theory of critical phenomena and phase transitions, the problem of the existence of relativistic quantum field theories, the turbulent behaviour in hydrodynamic systems and many others. While many of these problems date back several decades (or, indeed, to the previous century), significant progress was made only in the 60s, mainly in the context of critical phenomena, using ideas stressing the role of scale invariance in such problems. This line of thought culminated with the creation of the theory of the Renormalization Group (RG) by K.Wilson (see [1] for the history) that provided a unified method of attack for multiscale problems. The 70s saw the RG-theory vindicated in approximative analytic treatments of a multitude of problems in critical phenomena, hydrodynamics and other many-body systems. With the increasing power of computers, the ideas entered numerical analysis of such problems in the 80s.

On the mathematical side, RG ideas lead to the rigorous theory of so-called renormalizable quantum field theories (and a non-renormalizable one too) and some critical statistical mechanics problems [2] (these works by no means exhaust the mathematical progress in these fields, nor even the RG inspired works, for a rather recent set of references, see [3]. In this talk I would like to review some recent attempts to develope a mathematical RG theory for the study of so-called random (or disordered) systems. Again, this field has seen plenty of beautiful mathematical work, some of it RG inspired too, but no attempt is being made here for a comprehensive review.

2. Disordered Spin Systems

The mathematical framework of the problems we are going to study is that of *path space measures.* This is a unified framework for classical statistical mechanics, Euclidean quantum field theory, diffusion processes and many other systems. One

Proceedings of the International Congress
of Mathematicians, Kyoto, Japan, 1990

considers measures on some space of maps

$$\phi : \mathcal{S} \to \mathcal{M}$$

where $\mathcal{S}$ is the "space" or "time" or "space-time" and $\mathcal{M}$ the "field-space". For QFT, $\mathcal{S}$ is a manifold ($\mathbf{R}^n$, a Riemann surface...), for statistical mechanics a lattice (e.g. $\mathbf{Z}^n$) and $\mathcal{M}$ can be a manifold, a group or just some set, depending on the concrete model.

We will concentrate in this talk mainly on two examples of the above, which occur in a wide variety of physical contexts. The first is the *Ising Model* , where $\mathcal{S}$ is $\mathbf{Z}^d$ and $\mathcal{M}$ is $\mathbf{Z}_2 = \{1, -1\}$. To define the measure on the set of such maps $\mathcal{F} = (\mathbf{Z}_2)^{\mathbf{Z}^d}$, one first considers a subset of $\mathcal{F}$, $\mathcal{F}^{\pm}_{\Lambda} = \{\phi \mid \phi(x) = \pm 1, x \in \Lambda^c\}$ and on this finite set (take Λ finite) the probability measure

$$\mu_{\Lambda}^{\beta,\pm}(\{\phi\}) = \mathcal{Z}^{-1} e^{\beta \sum (\phi(x)\phi(y)-1)} \tag{1}$$

where x and y run through nearest neighbors on the lattice. The measures $\mu^{\beta,\pm}$ are constructed as limits of (1) as $\Lambda \to \mathbf{Z}^d$. They describe *symmetry breaking*: for β large and $d > 1$,

$$\lim_{\Lambda \to \mathbf{Z}^d} \int \phi(x) d\mu_{\Lambda}^{\beta,\pm}(\phi) = \int \phi(x) d\mu^{\beta,\pm}(\phi) = \pm m \tag{2}$$

with $m \neq 0$. The boundary conditions $\pm$ in (1) select non-zero *magnetization* $\pm m$ at the infinite volume limit.

Let us now present a disordered version of (1). The physical idea behind the disorder is that real materials, which are modelled by (1), have impurities, such as atoms of different kind randomly placed in the crystal structure. Mathematically, this translates into spatially varying randomness in the various parameters of the model. It turns out that a large class of disorder is modelled by replacing the exponent in (1) by

$$\beta(\sum_{x,y \in \mathbf{Z}^d} J(x,y)(\phi(x)\phi(y) - 1) + \sum_{x \in \mathbf{Z}^d} (\phi(x) - (\pm 1))h(x)) \,. \tag{3}$$

The functions J and h are *random* with some prescribed probability distribution. Thus, as $\Lambda \to \mathbf{Z}^d$, we get *random measures* $\mu^{J,h}(\phi)$ on $\mathcal{F}$ and can ask whether the m in (2) is nonzero almost everywhere in J and h (this is, of course, only interesting if the randomness preserves the $\phi \to -\phi$ symmetry of the system).

An example of (3) is the *Spin Glass*, where $h = 0$ and e.g. $J(x,y) = 0$, unless $\mid x - y \mid = 1$ and the non-zero J's are taken independent and identically distributed, with a $J \to -J$ symmetric distribution.

The example we will discuss below is the *Random Field Ising Model*, given by (3) with J non-random as in (1) and the $h(x)$, $x \in \mathbf{Z}^d$ independent, identically distributed, with a distribution invariant under $h(x) \to -h(x)$ and with variance $Eh(x)^2 = \varepsilon^2$.

Theorem 1 [4, 5, 6]. *Let $d > 2$. Then, there exists a $\beta_0 < \infty$ and an $\varepsilon_0 > 0$ such that for $\beta > \beta_0, \varepsilon < \varepsilon_0$ the limit* (2) *exists and $m \neq 0$ with probability* 1. *For $d \leq 2$ the magnetization m vanishes for all β and ε.*

Remark. Let us define the *lower critical dimension* d_L as the largest d where the magnetization vanishes for all β. Then, the Theorem states that $d_L = 2$. The corresponding dimension for the non-random system is 1. In the physics literature there was a long controversy on whether $d_L = 2$ or $d_L = 3$. Original arguments due to Imry and Ma [7] suggested that $d_L = 2$. These were later challenged by arguments originating in quantum field theory, coined as "dimensional reduction". This was a rule, partially justified, stating that d_L for the random system is two more than the one of the non-random one. Since the latter is 1, it was predicted that $d_L = 3$. The controversy was solved, at $\beta = \infty$, i.e. for the ground state of the Hamiltonian (3), in [4], and for large β in [5]. The $d = 2$ result is due to [6].

We will outline below the RG proof of the $d > 2$ part of Theorem 1.

3. Diffusion in Random Media

Ordinary diffusion provides another example, where randomness brings interesting effects. Consider random walk on the lattice $\mathbf{Z}^d$ described by *transition probabilities* $p(x, y)$ from $x \in \mathbf{Z}^d$ to $y \in \mathbf{Z}^d$:

$$p : \mathbf{Z}^d \times \mathbf{Z}^d \to [0, 1] \tag{4}$$

satisfying

$$\sum_{y \in \mathbf{Z}^d} p(x, y) = 1\,. \tag{5}$$

p allows us to define measures μ_T, $T \in \mathbf{N}$ on the space Ω_T of walks $\omega : \{0, 1, \ldots, T\} \to \mathbf{Z}^d$ starting from $\omega(0) = 0$:

$$\mu_T(\{\omega\}) = \prod_{i=1}^{T} p(\omega(i-1), \omega(i))\,. \tag{6}$$

Diffusion is a property of the large T limit of such measures. It will be convenient to realize them as measures ν_T on $C([0, 1])$, the space of continous paths $\omega : [0, 1] \to \mathbf{R}^d$, by a simple rescaling. Thus, given an $\omega \in \Omega_T$, we obtain a piecewise linear path

$$\tilde{\omega}(t) = T^{-1/2}(\omega(i-1) + (Tt - i + 1)(\omega(i) - \omega(i-1)) \tag{7}$$

where $i - 1 = [Tt]$ and $[\ ]$ denotes the integral part. ν_T is the measure induced by (7) on $C([0, 1])$ with its standard σ-algebra, and we will study the limit

$$\lim_{T\to\infty} \nu_T \tag{8}$$

also called the *scaling limit*, and its properties.

The walk is diffusive, if the *diffusion constant*

$$D(p) = \lim_{T\to\infty} D(T,p) \equiv \lim_{T\to\infty} T^{-1} \sum_{\omega} \omega(T)^2 \mu_T(\{\omega\}) \tag{9}$$

exists and is non-zero. In terms of the scaling limit

$$D(p) = \int d\nu(\omega)\omega(1)^2 . \tag{10}$$

For random walks in *homogenous* environments, the matrix p is translationally invariant $p(x,y) = p(x-y)$. If p has, say, exponential falloff, the scaling limit is given by the Wiener process. In the disordered system p is taken as a *random matrix* from some ensemble $\mathcal{P}$. One then asks whether the scaling limit and the diffusion constant exist for almost all p.

Let us now describe the ensemble $\mathcal{P}$. We take p a small random perturbation of the simple random walk:

$$p(x,y) = \frac{1}{2d} + b(x,y) \tag{11}$$

for $|x-y| = 1$ and $p(x,y) = 0$ otherwise. Here b is taken as a random matrix such that

a) $b(x,\cdot)$ and $b(x',\cdot)$ are independent if $x \neq x'$, and identically distributed.

b) The distribution of $b(x,\cdot)$ is invariant under the natural action of the subgroup of $O(d)$ fixing the lattice $\mathbf{Z}^d$.

c) The generating function of b satisfies the bound

$$Ee^{tb(x,y)} < e^{\varepsilon^2 t^2}$$

where ε is taken small. So, in particular, the variance of b is small.

d) $\mathrm{Prob}(p(x,y) < e^{-N}) < e^{-\Gamma N}$, *for $N > 1$. Γ is taken large.*

Under these conditions we have the

Theorem 2 [8]. *Let $d > 2$. Then there exist $\varepsilon_0 > 0$ and $\Gamma_0 < \infty$ such that whenever $\varepsilon < \varepsilon_0$ and $\Gamma > \Gamma_0$ the diffusion constant $D(p)$ exists and takes a constant value $D \neq 0$ a.s. in $\mathcal{P}$. The measures ν_T converge weakly to the Wiener measure with diffusion constant D, a.s. in $\mathcal{P}$.*

Remarks. 1. Theorem 2 holds for a much wider class of p's with exponential falloff in $|x-y|$ [8]. It also holds for a continous time version of the problem, where the transition probability $P(x,t)$ in time t from 0 to x satisfies the equation

$$\partial_t P = \Delta P + \nabla \cdot (\mathbf{b}P) \tag{12}$$

where $\mathbf{b} : \mathbf{Z}^d \to \mathbf{R}^d$ is random [8].

2. We expect Theorem 2 also to hold for $d = 2$. If $d = 1$, the situation is completely different [9]. The walk is subdiffusive ; indeed, the mean square distance is proportional to $(\log t)^4$ with probability tending to 1 as $t \to \infty$. On the other hand, if b in (11) is taken symmetric i.e. $b(x, y) = b(y, x)$, or, if the vector field $\mathbf{b}$ in (12) is a gradient (of white noise), then under quite general conditions the walk is diffusive in all d [10].

4. The Renormalization Group

We sketch now the method of proof of Theorems 1 and 2 using the RG. Originally, the RG was developed for the study of (1) (and other similar measures) in a neighbourhood of the *critical point* $\beta = \beta_c$ at which $\mu^{\beta,\pm}$ has *long range correlations* (at $\beta = \beta_c$ the magnetization m vanishes):

$$\int \phi(x)\phi(y)d\mu \sim C \mid x - y \mid^{-\alpha} \tag{13}$$

as $\mid x - y \mid \to \infty$. Here $\alpha > 0$. Such long-range correlations make the analysis hard by conventional methods, such as the Taylor expansion in β at $\beta = 0$ or at $\beta = \infty$, since β_c is a point of nonanalyticity of the correlation functions. In a nutshell, the RG approach as applied to such critical situations consists of three steps:

Coarse Graining. Choose an integer L and smear ϕ suitably on scale L to get a ϕ_L, defined on $L\mathbf{Z}^n$.

Scaling. Set $\phi'(x) = L^\gamma \phi_L(Lx)$. Then $\phi \to \phi' \equiv r_L\phi$ defines a map in $\mathscr{F}$ (often one needs to enlarge $\mathscr{F}$, i.e. allow for more general random fields than the ϕ one started with), which induces a map for the measures, $\mu \to \mu' \equiv \mathscr{R}_L\mu$. $\mathscr{R}_L$ is the RG map.

Iteration. For L fixed, the control of $\mathscr{R}_L$ is a problem involving short-range correlations and thus often managable. The solution of the full problem is now translated to the iteration of $\mathscr{R}_L$: if Λ is, say, a cube of side L^N in $\mathbf{Z}^n$, the integral (2) is done after N iteration steps. The scale-invariance of the problem is in this approach seen as stabilization of $\mathscr{R}_L^k\mu$: this is expected to tend to a *fixed point* of the map $\mathscr{R}_L$ in a space of probability measures on $\mathscr{F}$, provided the scale-parameter γ is chosen properly.

The disordered systems we are considering are not critical in the sense of (13). However, as we will see, they are problems involving several distance scales and the RG approach turns out to be the natural one here too. We first illustrate these steps for the random walks, where the RG will be especially simple.

For coarse graining, given an $\omega \in \Omega_T$, let ω_L be ω restricted to $L\mathbf{N}$. Then, for scaling, set

$$\omega'(t) = L^{-1/2}\omega_L(Lt) \equiv (r_L\omega)(t). \tag{14}$$

Note that ω' takes values in $L^{-1/2}\mathbf{Z}^d$. We get the new measure

$$\mu'(\{\omega'\}) = \mu(\{\omega \mid r_L\omega = \omega'\}) \tag{15}$$

which is given in terms of new transition probabilities

$$\mu'(\{\omega'\}) = \prod_{i=1}^{L^{-1}T} L^{-\frac{d}{2}} p'(\omega'(i-1), \omega'(i)) \tag{16}$$

with

$$p'(x,y) = L^{-d/2} p^L(L^{1/2}x, L^{1/2}y) \tag{17}$$

where p^L is the L-th power of the matrix p. Note that, since $\omega(i)$ are summed over $L^{-1/2}\mathbf{Z}^d$, the powers of L in (16) are natural. Thus the renormalized measure μ' is described in terms of the new matrix p', obtained from p via the non-linear map (17), which we shall call the RG map $\mathscr{R}$. In particular we have the identity for the diffusion constants

$$D(t,p) = D(L^{-1}t, \mathscr{R}p) = D(1, \mathscr{R}^n p) \tag{18}$$

if $t = L^n$, which means that we have translated the study of the long time behaviour of the walks to the iteration of the RG map.

In (18), $\mathscr{R}^n p$ are transition probability densities for a walk on $L^{-n/2}\mathbf{Z}^d$, and $\mathscr{R}$ maps such densities to ones for a walk on the finer lattice $L^{-(n+1)/2}\mathbf{Z}^d$. In the limit $n \to \infty$, we have walks on $\mathbf{R}^d$, and then $\mathscr{R}$ has a 1-parameter family of Gaussian fixed points

$$p^*_D(x,y) = (2\pi\frac{D}{d})^{-d/2} e^{-d|x-y|^2/2D} \,. \tag{19}$$

i.e. the transition probabilities of the Wiener process with diffusion constant D. For example, an exponentially decaying homogenous p is driven to this fixed point upon iteration of $\mathscr{R}$, since in that case $\mathscr{R}$ is just convolution composed with scaling:

$$\widehat{R^n p}(k) = \hat{p}(L^{-n/2}k)^{L^n} \to e^{-Dk^2/2d}$$

as $L \to \infty$, where $\hat{p}$ is the Fourier transform and D is determined by $\hat{p}(k) = 1 - Dk^2/2d + \mathscr{O}(|k|^4)$. This is of course nothing but a reformulation of the central limit theorem.

The main part of the proof of Theorem 2 consists of showing that $\mathscr{R}^n p \to p^*_D$ for $\mathscr{P}$-almost all p in a sufficiently strong sense as $n \to \infty$. There are two aspects in this convergence. First, the derivative $(D\mathscr{R})_{p^*}$ turns out to be contractive if $d > 2$, with largest eigenvalue $L^{(2-d)/2}$. Hence, it turns out that writing $\mathscr{R}^n p = p_n = Ep_n + b_n$, b_n has variance proportional to

$$\varepsilon_n^2 = L^{(2-d)/2}\varepsilon^2 . \tag{20}$$

The second aspect concerns the condition *d)* above. This is a condition for "traps" in the "environment": it turns out that, if the bound is violated for Γ small enough, the walk is likely not to be diffusive. The renormalized p should also satisfy a similar bound (which describes the "traps" in longer distance scales), and indeed, one finds

$$\text{Prob}\left(\int_{\Box^c} p_n(x, y) dy < e^{-N} \right) < L^{-n\Gamma} e^{-\Gamma N} \tag{21}$$

where $\Box$ is a unit cube centered at x. Thus the iteration of the RG "wipes out" the randomness in the matrix p.

It should be mentioned that, if one starts with p having correlations with sufficiently slow decay in $|x - y|$, the bound (21) may be violated for some n and thus traps may occur in large scales. Indeed, it is possible to construct $\mathcal{P}$ where p have correlations with exponential decay (but slow enough), such that the walk is subdiffusive in all dimensions [11, 12].

Now we turn to the Ising model. It is convenient to represent $\phi \in \mathcal{F}$ as a (in general disconnected) closed surface S in the dual lattice: an $(d-1)$-cell c_{xy} dual to a bond $\{xy\}$ with $|x - y| = 1$ is in S if $\phi(x) \neq \phi(y)$. Thus, to ϕ there corresponds a surface S and an assignement of signs $\pm$ to the components of the complement of S. The measure $\mu^{\beta,h}$ (we drop the $\pm$) corresponding to (3) can then be viewed as a measure on the set $\mathcal{S}$ of all such surfaces

$$\mu^{\beta,h}(\{S\}) = \mathcal{Z}^{-1} e^{-\beta(\text{Area}(S) + \sum_{V_+} h - \sum_{V_-} h)} \tag{22}$$

where $V_\pm$ are the $\pm$ regions determined by S.

If $h = 0$ and β is large enough, the S's are suppressed in the typical configurations of (22). Thus, the probability that the point x in (2) is in a different component of the complement of S than ∞ is small and the boundary condition determines the sign at x. If h is non-zero, let us consider the case of a connected S. Then, e.g. for $+$ boundary conditions,

$$\frac{\mu(\{S\})}{\mu(\{\emptyset\})} = e^{-\beta(\text{Area}(S) - 2\sum_{x \in V_S} h(x))}$$

with V_S the interior of S. Such an S will be probable if

$$h_S = 2 \sum_{x \in V_S} h(x) > \text{Area}(S) . \tag{23}$$

h_S, being a sum of independent random variables, has variance $Eh_S^2 = |V_S|\varepsilon^2$, and thus (23) is unprobable if $d > 2$ since $\varepsilon|V_S|^{1/2} < \text{Area}(S)$ for all S. However, we see that disordering configurations, where (23) holds, occur in all scales of the problem. We translate now these observations to the RG language.

For the coarse graining, let $S \in \mathcal{S}$ be $S = \cup_i S_i$ with S_i connected. We set

$$S_L = [\bigcup_{d(S_i)\geq L} S_i] \tag{24}$$

where $d(S)$ is the diameter of the set and $[S]$ denotes the smallest union of L-sided cubes, centered at $L\mathbf{Z}^d$, and covering S.

For scaling, set $S' = r_L S = L^{-1} S_L$. Note that S' is not any more in $\mathcal{S}$; it is a "thick" surface. We set again

$$\mu'(\{S'\}) = \mu^{\beta,h}(\{S \mid r_L S = S'\}). \tag{25}$$

It turns out that μ' is approximately of the form (22) with, however, a new h' having variance $\sim L^{2-d}$ times the variance of h, and a new $\beta' = L^{d-1}\beta$. These powers are easy to understand: the L^{d-1} comes from the scaling of the area of surfaces of dimension $d-1$, whereas the L^{2-d} comes from the fact that

$$h'(x) \sim L^{1-d} \sum h(x)$$

with the sum having L^d independent random variables. The L^{1-d} is due to the β in (22) multiplying h.

Thus, approximatively

$$\mathcal{R}^n \mu^{\beta,h} \sim \mu^{\beta_n,h_n} \tag{26}$$

with β_n and the variance ε_n^2 of h_n given by

$$\beta_n = L^{n(d-1)}\beta \ , \ \varepsilon_n^2 = L^{n(2-d)}\varepsilon^2 . \tag{27}$$

The proof of Theorem 1 thus consists in showing that, upon renormalizing, the measure (22) is driven to the trivial $\beta = \infty$ fixed point with no randomness.

In both cases discussed above, the reason why the disorder gets weaker under $\mathcal{R}$ is simple: the linearized RG is contractive at the fixed point. However, it is contractive only in the probabilistic sense: e.g. the variance is contracted. There are regions in the space of disorder (in $\mathcal{P}$ or in the space of h's) where $\mathcal{R}$ is expanding and the main mathematical problem is to show that such regions become more and more unprobable upon the iteration of $\mathcal{R}$. These regions correspond to the "traps" in long distance scales or configurations of h that dominate in (22) the area term.

5. Conclusions

There are several problems in the theory of random systems where the RG approach is likely to be the natural one. Among such problems are the classical motion of a particle in the presence of randomly located scatterers (the Lorentz gas), the problem of extended states and diffusion for a Schrödinger operator with random potential and the Spin Glass mentioned above.

For the first two problems it is conceivable that the methods used in [8] for the random walks are sufficient to prove the existence of diffusive behaviour.

In the case of the Spin Glass, there presumably is a non-trivial $\beta = \infty$ fixed point for the RG. However, here even setting up a useful RG-scheme constitutes a challenge.

References

1. K. Wilson: The renormalization group and critical phenomena. Rev. Mod. Phys. **55** (1983) 583-600
2. K. Gawędzki, A. Kupiainen: Massless lattice ϕ_4^4 theory: Rigorous control of a renormalizable asymptotically free model. Commun. Math. Phys. **99** (1985) 197–252
   - Gross-Neveu model through convergent perturbation expansions. Commun. Math. Phys. **102** (1985) 1–30
   - Renormalization of a non-renormalizable quantum field theory. Nucl. Phys. **B262** (1985) 33–48
   - J. Feldman, J. Magnen, V. Rivasseau, R. Seneor: A renormalizable field theory: the massive Gross-Neveu model in two dimensions. Commun. Math. Phys. **103** (1986) 67–103
3. J. Glimm, A. Jaffe: Quantum physics. A functional integral point of view. Springer, New York 1987
4. J. Imbrie: The ground state of the three dimensional random field ising model. Commun. Math. Phys. **98** (1985) 145-176
5. J. Bricmont, A. Kupiainen: Phase transition in the 3d random field ising model. Commun. Math. Phys. **116** (1988) 539–572
6. M. Aizenmann, J. Wehr: Rounding effects of quenched randomness on first order phase transitions. Commun. Math. Phys., to appear (1990)
7. Y. Imry, S. K. Ma: Random field instability of the ordered state of continous symmetry. Phys. Rev. Lett. **35** (1975) 1399–1401
8. J. Bricmont, A. Kupiainen: Random walks in asymmetric random environments. Preprint, Rutgers University 1990
9. Y. G. Sinai: Theor. Prob. Appl. **27** (1982) 256
10. G. Papanicolaou, S. R. S. Varadhan: In: J. Fritz, J. Lebowitz, D. Szasz (eds.) Random fields. North-Holland, 1981, p. 835
11. R. Durrett: Commun. Math. Phys. **104** (1986) 87
12. M. Bramson, R. Durrett: Commun. Math. Phys. **119** (1988) 119

References

Invariants of Links and 3-Manifolds Related to Quantum Groups

Nicolai Reshetikhin *

Department of Mathematics, Harvard University, Cambridge, MA 02138, USA

Recent years were signified by "strong interaction" between various ideas coming from physics and mathematics. Operator algebra, representation theory, low-dimensional topology, q-analysis and others are some of the fields involved in this interesting area. Algebraic constructions distilled from the theory of quantum integrable systems are quantum groups.

In the theory of quantum integrable systems it was found in [Ba, Ya] that a certain equation known today as the Yang-Baxter equation is playing the fundamental role in integrability. The essential role of it becomes clear after works by Zamolodchikov's [Z] concerned with factorizable scattering and integrability and by Faddeev and Sklyanin [FS] where quantum inverse transformation methods were developed as a method for studying quantum integrable systems [FT]. On the basis of these developments Drinfeld [Dr1] and Jimbo [J] introduced the Hopf algebras $U_q(\hat{\mathscr{G}})$ which can be considered as deformations of universal enveloping algebras of the Kac-Moody algebras. The concept of quantization of Lie groups and Lie algebras was presented by Drinfeld in his address to Berkeley ICM [Dr1]. The construction of quantum groups based on a given solution of the Yang-Baxter equation was presented in [FRT]. The general algebraic framework of quantum groups in terms of algebras with quadratic relations [S] was developed by Manin [M].

Applications of quantum groups to low dimensional topology are my subject today. The first results in this direction were obtained in the pioneering work by Jones [Jo1] where he found a new invariant of links using certain constructions from operator algebras. Shortly after this invariant was generalized by a group of authors [HOMFLY] which is abbreviated now as HOMFLY. Another invariant similar to this was proposed by Kauffman [Ka]. The HOMFLY invariant is related to Hecke algebra, the Kauffman invariant to Birman-Wenzl algebra [BW]. Some results were generalized for invariants of graphs in $\mathbb{R}^3$ ([Mi]).

The relation of these invariants to solutions of the Yang-Baxter equation was explained by Jones [Jo] and Turaev [Tu1]. The invariant of links related to quantum groups $U_q(\mathscr{G})$ for any simple $\mathscr{G}$ were studied in [Re1], where several new invariants were found. The relation between quantum groups and Hecke and Birman-Wenzl algebra was studied in [J1, Re1]. Finally it becomes clear that all these invariants can be generalized to be considered as invariants of framed graphs in $\mathbb{R}^3$ ([RT1]).

---

* On leave of absence, LOMI, Fontanka 27, Leningrad, 191011, USSR

Proceedings of the International Congress
of Mathematicians, Kyoto, Japan, 1990

Quantum groups produce interesting examples of tensor categories with non-trivial square of commutativity morphism (quasitensor categories). Algebras $U_q(\mathcal{G})$ when q is a root of 1 give examples of semisimple tensor categories without fiber functor. The categorial explanation of invariants of links related to quantum group were given in [Tu2, TR1, Re2]. It is based on the notion of category of tangles [JS]. The general fact is that for any given tensor category [McL, DM] (with some special required properties which plays the role of ribbon element in ribbon Hopf algebra [RT1]) one can define invariants of framed links associated with these categories.

The important fact about invariants of framed links associated with quantum groups (or with tensor categories) is [RT2] that one can formulate simple conditions under which these invariants produce the invariant of framed links which is invariant under Kirby moves [Ki]. The nontrivial fact is that the algebras $U_q(\mathcal{G})$ satisfy these conditions. It was proven for $\mathcal{G} = sl_2$ in [RT2].

The conditions mentioned above can also be formulated as conditions on corresponding tensor category. The categorial language seems most natural for the explanation of relations between these invariants and those which arise from topological field theory. For simplest values of $q : q^3 = 1, q^4 = 1, q^6 = 1$ invariants of 3-manifolds related to sl_2-algebra were computed in [KiM]. It was shown there that for these values of q the invariant can be reduced to known invariants. It seems that $q^5 = 1$ should be the first value when it is not so.

Topological field theory is another important branch of theoretical physics, close to the theory of integrable systems [A] [Sh] [Wi]. The most impressive results in the conception of topological field theory were the quantum field theoretical description of Donaldson's invariants, the description of new invariants of 3-manifolds, and links in 3-manifolds obtained by Witten [Wi, Wi1].

The relation between these invariants and those mentioned above is very natural in the language of category theory and essentially based on the concept of modular category of G. Segal [Se].

References

[A] Atiyah, M.: Topological field theories. Publ. Math. IHES **68** (1989) 175–186

[Ba] Baxter, R.J.: Ann. Phys. **70**, no. 1 (1972) 193–228

[DM] Delinge, P., Milne, J.: Tannakian categories. Lecture Notes in Mathematics, vol. 900. Springer, Berlin Heidelberg New York 1981

[Dr1] Drinfeld, V.G.: Quantum groups. Proc. Intern. Congr. of Math., vol. 1, pp. 798–820. Amer. Math. Soc., Berkeley 1986

[FRT] Faddeev, L.D., Reshetikhin, N.Yu., Takhtajan, L.A.: Quantization of Lie groups and Lie algebras. Algebraic Anal., vol. 1, Academic Press, 1988, p. 129; Algebra and Analysis **1** (1989) (in Russian)

[FS] Faddeev, L.D., Sklyanin, E.K.: Sov. Phys. Doklady **23** (1978) 902

[FT] Faddeev, L.D., Takhtajan, L.A.: Russ. Math. Surv. **34** (1979) 11

[HOMFLY]Freyd, P., Yetter, D., Hoste, J., Lickorish, W.B.R., Millett, K., Ocneanu, A.: A new polynomial invariant of knots and links. Bull. Amer. Math. Soc. **12** (1985) 239–246

[J] Jimbo, M.: Lett. Math. Phys. **10** (1985) 63–70

[J1] Jimbo, M.: Lett. Math. Phys. **11** (1986) 247–252

[Jo] Jones, V.F.R.: Bull. Amer. Math. Soc. **12** (1985) 103–111

[Jo1] Jones, V.F.R.: Talk at Atiyah Seminar, fall 1987 (unpublished)

[JS] Joyel, A., Street, R.: Braided tensor categories. Preprint 1989

[Ka] Kauffman, L.M.: An invariant of regular isotopy. Trans. Amer. Math. Soc. **130**, no. 2 (1990) 417–471

[Ki] Kirby, R.: Invent. math. **45** (1978) 35–56

[KiM] Kirby, R., Melvin, P.: On the invariant of Witten and Reshetikhin-Turaev for $sl(2, \mathbb{C})$. Preprint 1990

[Ko] Kohno, T.: Invariants of 3-manifolds and representations of mapping class groups. Preprint 1990

[KW] Kac, V.G., Wakimoto, M.: Adv. Math. **70** (1988) 156–236

[LaS] Larson, R.G., Seedler, M.E.: Amer. J. Math. **141**, no. 1 (1969) 75–94

[Lic] Lickorish, W.B.R.: Ann. Math. **76** (1962) 531–540

[Lic1] Lickorish, W.B.R.: Variants of 3-manifold invariants. Preprint 1990

[Lu] Lusztig, G.: Quantum groups at roots of 1. MIT preprint, Oct. 1989

[M] Manin, Yu.I.: Quantum groups and non-commutative geometry. Université de Montréal preprint, CRM-1561, 1988

[McL] MacLane, S.: Categories for the working mathematicians. Springer, Berlin Heidelberg New York 1971

[Mi] Millett, K.C.: An invariant of 3-valent spatial graphs. UCSB preprint 1989

[Mu] Murakami, J.: Osaka J. Math. **26** (1989) 1–55

[R] Radford, D.: Amer. J. Math. **98**, no. 2 (1976) 333-355

[Re1] Reshetikhin, N.Yu.: Quantized universal enveloping algebras. The Yang-Baxter equation and invariants of links. I. LOMI preprint E-4-87, 1988; II. LOMI preprint E-17-87, 1988

[Re2] Reshetikhin, N.Yu.: Algebra and Analysis **1**, no. 2 (1989) (in Russian)

[RT1] Reshetikhin, N.Yu., Turaev, V.G.: Comm. Math. Phys. **127** (1990) 1–26

[RT2] Reshetikhin, N.Yu., Turaev, V.G.: Invariants of 3-manifolds via link polynomials and quantum groups. M.S.R.I. preprint (April, 1989); Invent. math. **103** (1991) 547–597

[S] Skyanin, E.K.: Funct. Anal. Appl. **17** (1983) 273

[Se] Segal, G.: The definition of conformal field theory. St. Catherine's College preprint, Oxford 1989

[Sh] Shwartz, A.: Lett. Math. Phys. **2** (1978) 247–252

[Tu1] Turaev, V.G.: Invent. math. **92** (1988) 527–553

[Tu2] Turaev, V.G.: Izv. ANSSSR **53** (1989) 1073–1107 (in Russian)

[Wi] Witten, E.: Comm. Math. Phys. **117** (1988) 353–386

[Wi1] Witten, E.: Comm. Math. Phys. **121** (1989) 351–399

[Ya] Yang, C.N.: Phys. Rev. Lett. **19**, no. 23 (1967) 1312–1314

[Z] Zamolodchikov, A., Zamolodchikov, A/B: Ann. Phys. **120** (1979) 253–290

Geometry of Fermionic String

Albert Schwarz

Department of Mathematics, University of California at Davis, Davis, CA 95616, USA

Physicists hope that the Green-Schwarz superstring theory describes all interactions existing in the Nature. However the verification of this conjecture is connected with very difficult and very interesting mathematical problems. We consider here only some problems arising in the Polyakov approach to the fermionic string. (Fermionic string is closely related with the Green-Schwarz superstring.) We explain the connection between string theory and superconformal geometry, the origin of string measure on superconformal moduli space and analytic properties of this measure, the construction of universal moduli space and the expression of string measure in terms of super τ-function etc. The lecture is based on the papers [1–10]. The results concerning the measure on the moduli space of $N = 2$ superconformal manifolds are new.

Superconformal Manifolds and Strings

Let us consider a domain U in $(1|N)$-dimensional complex superspace $C^{1|N}$. One can define N-superconformal transformation of this domain as a complex analytic transformation preserving up to multiplier the 1-form

$$\alpha = dz + \sum_i \theta_i d\theta_i \,. \tag{1}$$

Here $(z, \theta_1, \ldots, \theta_N)$ denote complex coordinates in U (z is even, $\theta_1, \ldots, \theta_N$ are odd). N-superconformal manifold can be defined as a manifold pasted together from $(1|N)$-dimensional complex superdomains by means of N-superconformal transformations. The (super)space of classes of all compact N-superconformal manifolds having genus p (N-superconformal moduli space) will be denoted by $\mathcal{M}_p^N$. The most important cases are $N = 0, N = 1$ and $N = 2$. In the case $N = 0$ we obtain the moduli space of conformal manifolds (or moduli space of complex curves). This moduli space arises in bosonic string theory. $N = 1$ superconformal manifolds (or simply superconformal manifolds) arise in the superstring theory. Analogously $N = 2$ superconformal manifolds are connected with $N = 2$ superstrings.

Let us explain the connection of string theory with moduli spaces following the Polyakov approach. The action functional of the bosonic string can be

Proceedings of the International Congress
of Mathematicians, Kyoto, Japan, 1990

represented in the form:

$$S(X,g) = \langle dX, dX \rangle = \int_M g^{\alpha\beta} \partial_\alpha X^a \partial_\beta X^a dV \tag{2}$$

where M is a 2-dimensional manifold, X denotes a string field (i.e. a map of M into R^D), $g = (g_{\alpha\beta})$ is a riemannian metric in M, dV denotes the corresponding volume element and $\langle dX, dX \rangle$ denotes the scalar square of 1-form dX with respect to the metric g. The functional (2) is invariant with respect to reparametrizations and to Weyl transformations (i.e. it remains intact if we make a diffeomorphism of M or replace the metric $g_{\alpha\beta}$ by the metric $\varrho g_{\alpha\beta}$ where ϱ denotes a non-vanishing real function on M). To calculate the partition function of bosonic string we must integrate $\exp(-S)$ over all string fields and over all riemannian metrics on compact two-dimensional surfaces. The contribution of surfaces having genus p is known as the p-loop contribution to the partition function. The action functional S is quadratic with respect to string field X and therefore the calculation of the integral over X can be reduced to the calculation of $\det \Delta_0$ where $\Delta_0 = d^+ d$ is the Laplace operator on the scalars. Taking into account the reparametrization invariance by means of the Faddeev-Popov trick we reduce the calculation of the p-loop contribution to the partition function to the integration of

$$(\det \Delta_0)^{-D/2} \det \Delta_{gh} \tag{3}$$

over the space $\tilde{\mathscr{M}}_p$. Here $\tilde{\mathscr{M}}_p$ denotes the space of orbits of the group of diffeomorphisms in the space of riemannian metrics on two-dimensional compact surface M of genus p, $\det \Delta_{gh}$ denotes so called ghost determinant. The action functional S is Weyl invariant; but this is not true for the expression (3) due conformal anomaly. However for $D = 26$ (critical dimension) conformal anomaly vanishes; this permits us to represent the p-loop contribution to the partition function for critical string in the form:

$$\int_{\mathscr{M}_p} (\det \Delta_0)^{-13} \det \Delta_{gh} dv \tag{4}$$

where $\mathscr{M}_p$ denotes the space obtained from $\tilde{\mathscr{M}}_p$ by means of factorization with respect to Weyl transformations. In such a way $\mathscr{M}_p$ consists of classes of riemannian metrics; metrics connected by diffeomorphisms and by Weyl transformations are identified. By the construction of the integrand in (4) we have chosen one of metrics in every class. The $\det \Delta_0$, $\det \Delta_{gh}$ and the volume element dv depend on this choice but the integrand in (4) is well defined on $\mathscr{M}_p$. The integrand

$$d\mu = (\det \Delta_0)^{-13} \det \Delta_{gh} dv \tag{5}$$

is known as the string measure on the moduli space $\mathscr{M}_p$. It is evident that the space $\mathscr{M}_p$ coincides with the moduli space of conformal manifolds $\mathscr{M}_p^{N=0}$. The constructions above can be generalized to the supercase. In this case one has to replace the riemannian metrics on 2-dimensional surfaces by supperriemannian metrics on (2|2)-dimensional supermanifolds. By definition superriemannian metrics on a (2|2)-dimensional superdomain U is an odd vector field e on U

satisfying the conditions a) (anti)commutator of e and the complex conjugate field $\bar{e}$ is a linear combination of e and $\bar{e}$ (i.e. $[e,\bar{e}]_+ = \alpha e + \bar{e}\bar{\alpha}$ where α is an odd function), b) the vectors $e, \bar{e}, E = [e,e]_+, \bar{E} = [\bar{e},\bar{e}]_+$ form a basis in the tangent space. If $e' = \exp(i\lambda)e$ where λ is a real function then e and e' determine the same superriemannian metric; if $e' = \exp(i\lambda)e$, where λ is an arbitrary function, one says that e' and e are connected by Weyl transformations. The action functional of a fermionic string can be written in the form

$$S(X,e) = \langle \hat{e}X, \hat{e}X \rangle \tag{6}$$

where X denotes the string field (a map of a (2|2)-dimensional supermanifold M into R^D) and $\hat{e}$ denotes the first order differential operator corresponding to the supperriemannian metric e. This functional is invariant under reparametrizations and Weyl transformations. Slight modification of the considerations above permits us to express the p-loop contribution to the partition function of critical fermionic string ($D = 10$) in terms of an integral over the space of classes of superriemannian metrics on (2|2)-dimensional supermanifolds of genus p. (We identify two metrics connected with reparametrization or Weyl transformations). One can check that this space coincides with the moduli space $\mathscr{M}_p^{N=1}$ of superconformal manifolds. This follows from the assertion that for every superriemannian metrics on (2|2)-dimensional manifolds M one can find a covering of M by charts U_i with complex coordinates $(z^{(i)}, \theta^{(i)})$ in such a way that in every chart superriemannian metrics takes the form $\hat{e}^{(i)} = \Phi^{(i)}(\partial/\partial\theta^{(i)} + \theta^{(i)}\partial/\partial z^{(i)})$ (here $\Phi^{(i)}$ denotes a non-vanishing function). We obtain a measure $d\mu$ on the supermoduli space $\mathscr{M}_p^{N=1}$ (one can give an expression of $d\mu$ in terms of determinants of superLaplacians; this expression is similar to (5)).

Superconformal Geometry

Let us return to the consideration of N-superconformal transformations of a $(1|N)$-dimensional complex superdomain U with coordinates $Z = (z, \theta_1, \ldots, \theta_N)$. The operators

$$D_i = \frac{\partial}{\partial\theta_i} + \theta_i \frac{\partial}{\partial z}$$

are called covariant derivatives. Let us suppose that by the transformation $\tilde{Z} = f(Z)$ the operators D_i transform into operators $\tilde{D}_i$. The transformation $\tilde{Z} = f(Z)$ is N-superconformal if and only if

$$\tilde{D}_i = F_{ij}(Z)D_j \tag{7}$$

(the operators $\tilde{D}_1, \ldots, \tilde{D}_N$ are linear combinations of $D_1, \ldots, D_N$). This follows immediately from the remark that the vector fields corresponding to the operators D_i are orthogonal to the 1-form (1). It is easy to check that the matrices $F_{ij}(Z)$ in (7) satisfy

$$F_{ij}(Z)F_{kj}(Z) = \Phi(Z)\delta_{ik}\,. \tag{8}$$

In other words the matrix function $F_{ij}(Z)$ takes on values in the group $G = O(N, C) \times \mathbf{C}^*$ where $O(N, C)$ denotes the complex orthogonal group, $\mathbf{C}^*$ denotes the group of non-zero complex numbers. In the case $N = 1$ we have $G = \mathbf{C}^*$; in the case $N = 2$ the group G is disconnected and its connected part is isomorphic to $\mathbf{C}^* \times \mathbf{C}^*$. The $N = 2$ superconformal transformation is called untwisted if the matrix F_{ij} in (7) belongs to the connected component of G. Recall that N-superconformal manifold is pasted together from superdomains by means of N-superconformal transformations; if in the case $N = 2$ all these transformations are untwisted one says that $N = 2$ superconformal manifold is untwisted. In the case $N = 2$ it is convenient to introduce linear combinations of covariant derivatives: $D_\pm = (2)^{-1/2}(D_1 \pm iD_2)$; the behavior of $D_\pm$ by untwisted superconformal transformations is given by

$$\tilde{D}_+ = F_+ D_+, \quad \tilde{D}_- = F_- D_- . \tag{9}$$

The transformation law of the form (1) by N-superconformal transformations can be written as

$$\tilde{\alpha} = \Phi(Z)\alpha \tag{10}$$

where $\Phi(Z) = (\det F_{ij})^{2/N}$ for $N \geq 1$. (For $N = 0$ the matrix $F_{ij}(Z)$ has no sense, but the function $\Phi(Z)$ is well defined.) One says that a field ϱ on conformal manifold M has type k (or that ϱ is a k-differential) if the transformation law of this field by conformal transformation connecting two charts in M is given by $\tilde{\varrho} = \Phi^{-k}\varrho$. The field ϱ on $N = 1$ superconformal manifold M has type k if the transformation law of ϱ by superconformal transformations is $\tilde{\varrho} = F^{-k}\varrho$ (in the case $N = 1$ we have only one covariant derivative D and $\tilde{D} = FD$, i.e. the matrix F is simply a number.) The field ϱ on untwisted $N = 2$ superconformal manifold has type (k, l) if its transformation law is $\tilde{\varrho} = F_+^{-k} F_-^{-l}\varrho$. In other words the field of type k in the case $N = 0$ is a section of a line bunde K^k, where K is a holomophic line bundle with transition functions $\Phi(z)$ and in the case $N = 1$ it is a section of a line bundle ω^k where holomorphic line bundle ω is defined by means of transition functions $F(Z)$. In $N = 2$ case the field of type (k, l) is a section of $\omega_+^k \omega_-^l$ where ω_+ and ω_- are defined by means of transition functions F_+ and F_- correspondingly. Note that the bundles K and ω can be interpreted also as canonical line bundles. (Recall that for an arbitrary supermanifold one can define the canonical line bundle using as transition functions (super)Jacobians of transformations connecting different charts.)

Let us denote by $\mathscr{A}(L)$ the space of a holomorphic sections of holomorphic line bundle L over a supermanifold M. In the case $N = 0$ we will use the notation $\mathscr{A}_k$ for $\mathscr{A}(K^k)$ and in the case $N = 1$ we will use the same notation for $\mathscr{A}(\omega^k)$. If M is a conformal manifold (a superconformal manifold) then the cotangent space at the corresponding point of $\mathscr{M}_p^{N=0}$ (of $\mathscr{M}_p^{N=1}$) can be identified with $\mathscr{A}_2$ (with $\Pi\mathscr{A}_3$). Here Π denotes the parity reversion. Note that $\mathscr{A}_2$ and $\Pi\mathscr{A}_3$ are complex linear spaces; this permits us to introduce complex structures in $\mathscr{M}_p^{N=0}$ and $\mathscr{M}_p^{N=1}$. It is well known that for $p \geq 1$ the complex dimension of $\mathscr{M}_p^{N=0}$ is equal to $3p - 3$. One can prove that $\mathscr{M}_p^{N=1}$ is a $(3p - 3|2p - 2)$ dimensional complex supermanifold for $p \geq 1$. Let us show that one can construct measures on moduli spaces using the so called Mumford form and its generalizations. First of all we will introduce the notion of a function of weight k on bases in linear space E by means of the condition $f(\tilde{e}) = (\det P)^k f(e)$. Here $e = \{e_i\}$ and $\tilde{e} = \{\tilde{e}_i\}$ are two bases in E, and P denotes the matrix connecting e and $\tilde{e}$; i.e. $\tilde{e}_i = P_i^j e_j$. If E is a linear superspace we use the same definition, but instead of determinants one has to considers superdeterminants (Berezinians). If E is infinite-dimensional one has to consider only admissible bases (bases connected with a standard basis in E by a matrix P having well defined determinant in some sense.) The measure in linear space E can be defined as a function of weight 1 on the bases of E. We will denote the one-dimensional linear space consisting of measures in E by $m(E)$. In the definitions above one can take E as a complex or real linear space; correspondingly $m(E)$ will be complex or real too. However if E is a real linear space it is natural to modify the definition of a function of bases having weight k. Namely one has to replace $\det P$ by $|\det P|$ in this definition. Then the measure in E can be defined as a positive function of bases having weight 1. If E is a complex linear space and E^{real} corresponding real linear space then a hermitian metric on $m(E)$ generates positive measures in E^{real} and in $(E^{\text{real}})^*$. To determine a measure in a (super)manifold $\mathcal{M}$ one has to fix the measures in all tangent spaces (or functions of weight -1 on the bases of all cotangent spaces.) Let us consider a conformal manifold M and complex linear spaces

$$\Sigma_k(M) = \mathscr{A}_k(M) \dotplus \mathscr{A}_{1-k}(M)^* \tag{11}$$

$$\lambda_k(M) = m(\Sigma_k(M)) = m(\mathscr{A}_k(M)) \otimes m(\mathscr{A}_{1-k}(M))^* . \tag{12}$$

Mumford proved that there is a canonical isomorphism

$$\lambda_k(M) \approx \lambda_1(M)^{6k^2-6k+1} , \tag{13}$$

i.e. the one-dimensional linear space $\lambda_k(M)$ is isomorphic to the $(6k^2 - 6k + 1)$-th tensor power of $\lambda_1(M)$. (One can consider the spaces $\lambda_k(M)$ as fibres of a holomorphic line bundle λ_k over $\mathcal{M}_p^{N=0}$; then (13) can be interpreted as holomorphic equivalence of line bundles λ_k and $\lambda_1^{6k^2-6k+1}$.) Using (13) for $k = 2$ we obtain an isomorphism

$$\lambda_2(M) = \lambda_1(M)^{13} ; \tag{14}$$

this isomorphism is known as the Mumford form. In other words, the Mumford form can be interpreted as a function $\mathcal{M}(e, f|M)$ of bases in $\Sigma_2(M)$ and $\Sigma_1(M)$, having weight -1 with respect to the basis e of $\Sigma_2(M)$ and weight 13 with respect to the basis f of $\Sigma_1(M)$. There exists a natural scalar product in $\mathscr{A}_1(M)$ (in the space of holomorphic abelian differentials). This scalar product generates a hermitian metric in $\lambda_1(M) = m(\mathscr{A}_1(M))$; using this metric and the isomorphism (14) we obtain a hermitian metric in $\lambda_2(M)$. If a conformal manifold M has genus greater than 1, $\lambda_2(M)$ can be identified with $m(\mathscr{A}_2(M))$. Remembering that $\mathscr{A}_2(M)$ can be considered as a cotangent space to $\mathcal{M}_p^{N=0}$ we obtain a measure in $\mathcal{M}_p^{N=0}, p > 1$. Using the Mumford form $\mathcal{M}(e, f)$ we can describe this measure as follows. Let $t_1, \ldots, t_s, s = 3p - 3$, denote a basis in the complex tangent space at the point of $\mathcal{M}_p^{N=0}$ corresponding to the conformal manifold M and let $e_1, \cdots, e_s$ be the dual basis in the cotangent space $\mathcal{A}_2(M) = \Sigma_2(M)$. Then the real measure in $\mathcal{M}_p^{N=0}$ can be defined by the formula

$$\mu(t_1, \ldots, t_q, \bar{t}_1, \ldots, \bar{t}_s) = |\mathcal{M}(e, f)|^2 \tag{15}$$

where f is an arbitrary orthonormal basis in $\mathcal{A}_1(M)$.

A. Voronov [11] generalized the Mumford construction to the case when M is a superconformal manifold. In this case one has to modify the definitions of $\Sigma_k(M)$ and $\lambda_k(M)$ as follows:

$$\Sigma_k(M) = \mathcal{A}_k(M) + \Pi \mathcal{A}_{1-k}(M)^*, \tag{16}$$

$$\lambda_k(M) = m(\Sigma_k(M)) = m(\mathcal{A}_k(M)) \otimes m(\mathcal{A}_{1-k}(M)). \tag{17}$$

Voronov proved that $\lambda_k(M)$ is canonically isomorphic to $\lambda_1(M)^{2k-1}$; in particular

$$\lambda_3(M) \approx \lambda_1(M)^5. \tag{18}$$

This isomorphism is known as a super-Mumford form; it can be interpreted as a function $\mathcal{M}(e, f)$ having weight -1 with respect to the basis e in $\Sigma_3(M)$ and weight 5 with respect to the basis f in $\Sigma_1(M)$. We will say that M is a normal superconformal manifold if $\mathcal{A}_0(M) \equiv \mathbf{C}$ (all holomorphic functions on M are constant). Then $\Sigma_1(M) = \mathcal{A}_1(M)$ is provided with a natural scalar product. If the genus of M is greater than 1, we have $\Sigma_3(M) = \mathcal{A}_3(M)$ and remembering the description of the cotangent space to $\mathcal{M}_p^{N=1}$ we obtain a measure on $\mathcal{M}_p^{N=1}$, $p > 1$. Belavin and Knizhnik [12] proved that the measure on $\mathcal{M}_p^{N=0}$ constructed by means of the Mumford form coincides with the string measure. A corresponding result for the measure on $\mathcal{M}_p^{N=1}$ was proved in [3].

Let us describe similar constructions for the moduli space of untwisted $N = 2$ superconformal manifolds. First of all one has to mention that this moduli space is isomorphic to the moduli space $\mathcal{M}_p^{1|1}$ of all complex $(1|1)$-dimensional supermanifolds. (For definiteness we consider compact supermanifolds of genus p.) To explain the coincidence of these two moduli spaces it is useful to note that a N-superconformal manifold can be considered as a $(1|N)$-dimensional complex contact supermanifold; the converse assertion is also true. (Recall that the contact structure can be specified by means a of 1-form satisfying non-degeneracy condition; the form (1) entering in the definition of N-superconformal transformations specifies contact structure in a $(1|N)$-dimensional complex superdomain.) For an $(m|n)$-dimensional contact supermanifold one can construct a $(2m-1|2n)$-dimensional contact supermanifold PT^*X by means of projectivization of the cotangent bundle T^*X. If X is a $(1|1)$-dimensional complex supermanifold we obtain $(1|2)$-dimensional contact complex supermanifold, i.e. $N = 2$ superconformal manifold. One can check that this construction gives all untwisted $N = 2$ superconformal manifolds. The fields on the $(1|1)$-dimensional manifold X can be identified with chiral fields on the $N = 2$ superconformal manifold PT^*X (i.e. with fields satisfying the condition $D_+\Phi = 0$). Let L denote a holomorphic vector bundle over a $(1|N)$-dimensional supermanifold R. We define a one-dimensional complex linear space $\lambda(L)$ by the formula

$$\lambda(L) = m(H^0(L)) \otimes m(H^1(L))^* . \tag{19}$$

Here $H^k(L)$ are Čech cohomology groups. It is easy to check that the spaces $\lambda_k(M)$ defined above can be considered as $\lambda(\omega^k)$ where ω denotes the canonical bundle over M. To define a measure on $\mathcal{M}_p^{1|1} = \mathcal{M}_p^{N=2}$ (untwisted) we must introduce hermitian metric in $\lambda(\omega_+^{-1}\omega_-^{-1})$ for every untwisted $N = 2$ superconformal manifold R. (The tangent space to $\mathcal{M}^{1|1}$ can be interpreted as $H^1(\omega_+^{-1}\omega_-^{-1})$.) However $\lambda(\omega_+^{-1}\omega_-^{-1})$ is canonically isomorphic to $\lambda(\mathcal{O}_R)$ where $\mathcal{O}_R$ is a trivial line bundle over R. (There exists a canonical isomorphism $\lambda(L_1) = \lambda(L_2)$ for every two holomorphic line bundles L_1, L_2 over a $(1|N)$-dimensional manifold R where $N \geq 2$.) Let us consider the $(1|1)$-dimensional manifold R' corresponding to R. If $x \in R'$ we denote by $V(x)$ a $(1|1)$-dimensional linear space of holomorphic functions on the $(0|1)$-dimensional manifold $\pi^{-1}(x)$. (Here π is a natural projection of R onto R'.) The spaces $V(x)$ can be considered as fibres of a $(1|1)$-dimensional vector bundle V over R'; one can check that $H^i(V) = H^i(\mathcal{O}_R)$ and therefore $\lambda(V) = \lambda(\mathcal{O}_R)$. From the other side one can identify $\lambda(V)$ with $\lambda(\mathcal{O}_{R'}) \otimes \lambda(\Pi\omega_{R'}) = \lambda(\mathcal{O}_{R'})^2$. (The bundle $\mathcal{O}_{R'}$ can be considered as a subbundle of V and the corresponding quotient bundle is $\Pi\omega_{R'}$.) In the normal case ($H^0(\mathcal{O}_{R'}) = C$) the natural scalar product in $\Pi H^1(\mathcal{O}_{R'}) = H^0(\omega_{R'})$ induces a hermitian metric in $\lambda(\mathcal{O}_{R'})$ and therefore in $\lambda(\omega_+^{-1}\omega_-^{-1}) = \lambda(\mathcal{O}_{R'})^2$.

Universal Moduli Space

Let us consider the space H of square integrable functions on the supercircle (i.e. H consists of functions $F(z,\theta) = f(z) + \varphi(z)\theta$, where z is a complex number, $|z| = 1$ and θ is an odd variable). We introduce an odd bilinear scalar product (bilinear pairing) in H by the formula

$$\langle F, \tilde{F}\rangle = \oint F(z,\theta)\tilde{F}(z,\theta)dzd\theta = \oint f(z)\tilde{\varphi}(z)dz + \oint \varphi(z)\tilde{f}(z)dz . \tag{20}$$

The functions $z^n, z^n\theta$ form the standard basis of H. (Here n is an integer.) The subspace of H spanned by the vectors corresponding to $n < 0$ (to $n \geq 0$) will be denoted by H_- (H_+). The natural projection of H onto H_- (H_+) will be denoted by π_- (π_+). We will say that a linear subspace W of H belongs to the super-Grassmannian Gr, if the projection π_- of W into H_- is a Fredholm operator and the projection π_+ of W into H_+ is a compact operator. (Recall that in the bosonic case the Grassmannian Gr can be defined in a similar way but the role of H is played by the space of functions on the circle $|z| = 1$.) Let us denote by Γ the supergroup of even invertible functions on the supercircle. The group Γ acts in Gr by means of multiplication operators. The subspace of Gr consisting of elements $W \in \text{Gr}$ satisfying $W^\perp = FW$, $F \in \Gamma$ will be denoted by UMS. (Here $W^\perp$ denotes the orthogonal complement to W with respect to the bilinear pairing (20).) All moduli spaces can be embedded in UMS by means of the Krichever construction. More precisely, let us consider a $(1|1)$-dimensional complex manifold N, a point $n \in N$ and a coordinate system (z, θ), $|z| \leq 1$, in the neighbourhood U of n. The space of triples $(N, n, (z, \theta))$ will be denoted by $\mathscr{P}$. For every point $P \in \mathscr{P}$ we define a space $W(P)$ as a space of functions on the supercircle admitting holomorphic extension to $N \backslash U$ (to the exterior of supercircle). One can prove that $W(P) \in$ UMS for every $P \in \mathscr{P}$ and therefore $\mathscr{P}$ is embedded in UMS. Let us consider for every element $W \in$ Gr finite-dimensional spaces $\mathscr{A}(W), \mathscr{A}(W^\perp)$ and $\Sigma(W)$ defined by the formulas

$$\mathscr{A}(W) = W \cap H_+ = \operatorname{Ker} \pi_-^W \tag{21}$$

$$\mathscr{A}(W^\perp) = W^\perp \cap H_+ = \Pi(H_-/\operatorname{Im} \pi_-^W) \tag{22}$$

$$\Sigma(W) = \mathscr{A}(W) + \Pi\mathscr{A}(W^\perp)\ . \tag{23}$$

Here π_-^W denotes the projection π_- considered as a map from W into H_- and Π denotes the parity reversion. If $W = W(P, L)$, $P = (N, n, (z, \theta)) \in \mathscr{P}$, L is a holomorphic line bundle over N, we can identify $\mathscr{A}(W)$ with the space $H^0(L)$ of holomorphic sections of L over N. In particular if N is a superconformal manifold, $L = \omega^k$, the space $\mathscr{A}(W, P, \omega^k)$ coincides with the space $\mathscr{A}_k(N)$ of holomorphic sections of the bundle ω^k over N (i.e. with the space of holomorphic fields of type k). The space $\Sigma(W(P, \omega^k))$ can be identified with the space $\Sigma_k(N)$. We will construct the extension of the super-Mumford form to UMS as a function $\mathscr{M}(w, w', W)$, where w denotes a basis in $\Sigma(F^3W)$ and w' denotes a basis in $\Sigma'(FW) = \Sigma(W^\perp) = \Sigma(W)$. (Recall that for $W \in$ UMS we have $W^\perp = FW$, $F \in \Gamma$.) The construction of this extension is based on the notion of super τ-function. Let us first define the τ-function $\tau(W, P)$, where $W \in$ Gr, $F \in \Gamma$ in the case when the projections π_-^W and π_-^{FW} are isomorphisms. In this case we can consider the counterimages $(\pi_-^W)^{-1}e$ and $(\pi_-^{FW})^{-1}e$ of the standard basis $e = \{z^n, z^n\theta, n < 0\}$ in H_-. The τ-function $\tau(W, F)$ can be defined as the determinant of the matrix connecting two bases $F(\pi_-^W)^{-1}e$ and $(\pi_-^{FW})^{-1}e$ in FW. One can prove that this determinant is well-defined. (Note that the corresponding determinant in the bosonic case is ill-defined; therefore the definition of the super τ-function is simpler than the definition of Sato's τ-function.) In the general case, one can assign to every basis w of $\Sigma(W)$ a basis $\hat{w}$ of $W \in$ Gr, determined up to the unimodular transformation. For example, if $\mathscr{A}(W^\perp) = 0$, $\Sigma(W) = \mathscr{A}(W)$ to construct the basis $\hat{w}$ we add to the basis w a set u of the vectors in W satisfying the condition $\pi_-^W(u) = e$. (In other words the set u is transformed into the standard basis e of H_- by the projrction π_-.) We define the τ- function $\tau(w, w', W, F)$ where w is the basis in W, w' is a basis in FW as a determinant of the matrix connecting the bases $F\hat{w}$ and $\hat{w}'$ in FW.

$$\tau(w, \hat{w}, W, F) = \det(F\hat{w}|\hat{w}')\ . \tag{24}$$

This infinite-dimensional determinant is well-defined. We can now construct the extension of the super-Mumford form to UMS by the formula

$$\mathscr{M}(w, \hat{w}, W) = \frac{\tau(\Pi w', w, W, F^3)}{\tau(\Pi w', w', W, F)^3}\ . \tag{25}$$

Here $W \in$UMS, $F \in \Gamma$, w is a basis in $\Sigma(F^3 W)$, w' is a basis in $\Sigma(FW)$ and $\Pi w'$ is a basis in $\Sigma(W) = \Pi\Sigma(FW)$. It is easy to check that the weights of $\mathcal{M}(w, w', W)$ with respect to w and w' are correct (-1 and 5 correspondingly). One can prove that for the case $W = W(P)$, $P = (N, n, (z, \theta)) \in \mathcal{P}$, N is a superconformal manifold the function $\mathcal{M}(w, w', W)$ coincides with the usual super-Mumford form $M(w, w', N)$. Using the expression of the super τ-function through the Reidemeister torsion we can express $\mathcal{M}(w, w', W(P))$ through holomorphic fields on N and their zeroes. The expression obtained in such a way coincides with the similar expression for $M(w, w', N)$ given in [11]. This remark gives the simplest proof of the relation

$$\mathcal{M}(w, w', W(P)) = M(w, w', N) . \tag{26}$$

Note that the function $\mathcal{M}(w, w', W(P))$ gives an extension of the super-Mumford form to $\mathcal{P}$. Moreover, it is easy to check that this function depends only on the complex manifold N (i.e. does not depend on the point $n \in N$ and on the coordinate system (z, θ)). Therefore the super-Mumford form can be considered as a function on the moduli space $\mathcal{M}^{1|1}$ of compact $(1|1)$-dimensional complex manifolds. (More rigorously this function depends on the point $N \in \mathcal{M}^{1|1}$ and on the bases w, w' in $\Sigma_3(N) = \Gamma(\omega^3)$ and in $\Sigma_1(N) = \Gamma(\omega)$.) The same assertion can be obtained from [11]. The moduli space $\mathcal{M}^{1|1}$ coincides with the moduli space of untwisted $N = 2$ superconformal manifolds and therefore the statement above permits us to assert that the super-Mumford form admits hidden $N = 2$ superconformal symmetry; see [10]. It is well known that the $N = 2$ world-sheet superconformal symmetry is related with the $N = 1$ space-time symmetry in the string theory. The remarks above show that the $N = 2$ superconformal symmetry may play a fundamental role in the string theory.

References

1. Baranov, M., Schwarz, A.: Multiloop contribution to string theory. Pis'ma ZhETF 42 (1985) 340 (Russian). [English transl.: JETP Lett. **42** (1986) 419]
2. Baranov, M., Frolov, I., Schwarz, A.: Geometry of two-dimensional superconformal field theories. Teor. Mat. Fiz. **70** (1987) 92 (Russian). [English transl.: Teor. Math. Fiz. **70** (1987)]
3. Baranov, M., Schwarz, A.: On the multiloop contribution to the string theory. Int. J. Mod. Phys. A**3** (1987) 28
4. Schwarz, A.: Fermionic string and universal moduli space. Pis'ma ZhETF 46 (1987) 340 (Russian). [English transl.: JETP Lett. **46** (1988) 428]
5. Rosly, A., Schwarz, A., Voronov, A.: Geometry of superconformal manifolds. CMP **119** (1988) 129
6. Rosly, A., Schwarz, A., Voronov, A.: Superconformal geometry and string theory. CMP **120** (1989) 437
7. Schwarz, A.: Fermionic string and universal moduli space. Nucl. Phys. B**317** (1989) 323
8. Dolgikh, S., Schwarz, A.: SuperGrassmannians, super-τ-functions and strings. In: V. Knizhnik memorial volume (Brink, L., Polyakov, A. eds.). 1990
9. Dolgikh, S., Rosly, A., Schwarz, A.: Supermoduli spaces. ICTP preprint (1989), CMP **135** (1990) 91

10. Baranov, M., Frolov, I., Schwarz, A.: Geometry of the superconformal moduli space. Teor. Mat. Fiz. **72** (1989) 241 (Russian)
11. Voronov, A.: A formula for the Mumford measure in superstring theory. Funk. Anal. i Prilozhen. **22** (2) (1988) 67 (Russian). [English transl.: Funct. Anal. Appl. **22** (1988) 139]
12. Belavin, A., Knizhnik, V.: Algebraic geometry and the geometry of quantum strings. Phys. Lett. **168B** (1986) 201

Geometric Aspects of Quantum Field Theory

Graeme Segal

Mathematical Institute, 24-29 St. Giles, Oxford OX1 3LB, England

Quantum field theory has been the basic tool of particle physics for more than half a century, but unlike earlier such tools it has not been accompanied by a satisfying mathematical theory. Recently this has begun to change. One reason is that the ideas of quantum field theory have turned out to shed light on purely mathematical questions. These applications are my subject today. So far, nevertheless, the field theory has played either a heuristic or an explanatory role in the mathematics, and the actual theorems can, and often must, be proved by other means. I hope that this will be less true as the mathematics of field theory becomes better developed. Meanwhile I shall just indicate some areas where field theory and geometry have come together, trying to illustrate the point of view rather than formulate theorems. For the most part I shall be summarizing other people's work, predominantly Witten's.

§ 1. The Framework

In his address to the Berkeley ICM Witten described $d + 1$ dimensional quantum field theory as follows. One considers "fields" defined on some class of oriented $d + 1$ dimensional manifolds M. A "field" might mean a map from M to some auxiliary manifold X, or a section of some natural fibre bundle on M, or even an equivalence class of such sections. In any case one has a space $F(M)$ of fields for each compact manifold M with boundary. A field $f \in F(M)$ has a boundary value $f|\partial M$ which belongs to some space $F_0(\partial M)$ of fields on the boundary. We also suppose given an "action" functional $S: F(M) \to \mathbb{R}$, defined uniformly for all M. Then field theory is the study of the functions Ψ_M on $F_0(\partial M)$ of the form

$$\Psi_M(f_0) = \int_{F(M;f_0)} e^{-S(f)}\mathscr{D}f, \tag{1.1}$$

where $F(M;f_0) = \{f \in F(M): f|\partial M = f_0\}$. More generally, if the boundary $\partial M = \Sigma_0 \amalg \Sigma_1$ consists of an incoming part Σ_0 and an outgoing part Σ_1 we are interested in operators $\Psi_M: H_{\Sigma_0} \to H_{\Sigma_1}$, where H_{Σ_i} is a space of functions on $F_0(\Sigma_i)$, on the form

Proceedings of the International Congress of Mathematicians, Kyoto, Japan, 1990

$$(\Psi_M \phi)(f_1) = \int_{F_0(\Sigma_0)} K_M(f_0, f_1)\phi(f_0)\mathscr{D}f_0,$$

where

$$K_M(f_0, f_1) = \int_{F(M; f_0, f_1)} e^{-S(f)}\mathscr{D}f. \tag{1.2}$$

(A boundary component is 'outgoing' or 'incoming' according as its orientation agrees or not with that of M; and $\bar{\Sigma}_0$ denotes Σ_0 with reversed orientation.)

The preceding formulae are only schematic, and so far it has proved impossible to develop an integration theory of the type needed. But let us at least try to abstract the essential structure. It comprises

(i) a vector space H_Σ for each closed oriented d-dimensional manifold with whatever structure is appropriate;

(ii) a bilinear pairing $H_{\bar{\Sigma}} \times H_\Sigma \to \mathbb{C}$;

(iii) an element $\Psi_M \in H_{\partial M}$ for each $d + 1$ dimensional manifold M with appropriate structure.

The most obvious properties these data should have are

(a)
$$H_{\Sigma_1 \amalg \Sigma_2} = H_{\Sigma_1} \otimes H_{\Sigma_2}$$

and

$$\Psi_{M_1 \amalg M_2} = \Psi_{M_1} \otimes \Psi_{M_2};$$

(in particular $H_\Sigma = \mathbb{C}$ when $\Sigma = \varnothing$, and so $\Psi_M \in \mathbb{C}$ when M is closed.)

(b) if two components Σ_1 and Σ_2 of the boundary of M are sewn together by an orientation-reversing diffeomorphism to form a new manifold $\check{M}$ such that $\partial M = \Sigma_1 \amalg \Sigma_2 \amalg \partial\check{M}$ then the map $H_{\partial M} \to H_{\partial\check{M}}$ defined by the bilinear pairing takes Ψ_M to $\Psi_{\check{M}}$.

In particular, when $\partial\check{M} = \varnothing$ and Ψ_M is regarded as an operator $H_{\Sigma_1} \to H_{\Sigma_2}$, property (b) asserts that

$$\operatorname{trace}(\Psi_M) = \Psi_{\check{M}} \in \mathbb{C}.$$

I do not know how far an axiomatization of this kind is appropriate or helpful in traditional quantum field theory, but with some especially simple kinds of theory it works well and is a useful tool in geometry, rather like a new kind of cohomology theory. I shall mention some limitations of the framework in §§ 5 and 6 below.

A feature of each of the examples I shall describe is that either the phase space is finite dimensional because of the presence of a large group of gauge symmetries, or else the path integral (1.1) reduces to a finite dimensional integral because the integrand is an exact differential form outside a finite dimensional submanifold of the space of fields (i.e. "the stationary phase calculation is exact"). One might take this to mean that genuine quantum field theory is not involved. A more optimistic moral, however, is that one can sometimes best study finite dimensional problems by the infinite dimensional methods of field theory.

§ 2. Index Theory and the Elliptic Genus

A particle moving in a Riemannian manifold X affords the simplest example of the path integral idea, and can be regarded as a $0 + 1$ dimensional field theory. The action for a path $\gamma : [0, T] \to X$ is $S(\gamma) = \frac{1}{2}\int_0^T \|\gamma'(t)\|^2\, dt$. For a point P the vector space H_P is $L^2(X)$, and to the 1-manifold $[0, T]$ is associated the heat operator $e^{-T\Delta}$ in H_P, where Δ is the Laplacian of X. The formula (1.2) is then the usual path integral representation of the heat kernel; and if we replace $[0, T]$ by a circle S_T of length T we have a formula for trace$(e^{-T\Delta})$ as an integral over the loop space $\mathscr{L}_T X = \mathrm{Map}(S_T; X)$. This path integral does not reduce to a finite dimensional integral.

The position is different and more relevant if we replace the action $S : \mathscr{L}_T X \to \mathbb{R}$ with the inhomogeneous differential form $\hat{S} = S + \omega$, where ω is the 2-form on $\mathscr{L}_T X$ which to two deformations ξ, η of a loop γ assigns the number

$$\omega(\gamma; \xi, \eta) = \int_0^T \langle \xi(t), \eta'(t) \rangle\, dt.$$

(Here $\eta'(t)$ is the covariant derivative.) (*) Witten observed (see [2], [15]) that the action $\hat{S}$ corresponds to the $0 + 1$ dimensional field theory for which H_P is the mod 2 graded space of L^2 spinor fields on X, while the operator associated to $[0, T]$ is the spinorial heat operator $e^{-T\Delta}$. The graded trace (or "supertrace") tr$(e^{-T\Delta})$ is now independent of T, and is the index of the Dirac operator on X. This is a topological invariant of X called its $\hat{A}$-genus. On the other hand the top degree component of the differential form $e^{-\hat{S}}$ on $\mathscr{L}_T X$ is exact outside the finite dimensional manifold of point loops, so by Stokes's theorem the path integral can be reduced to an integral over X (identified with the point loops). The outcome is the Atiyah-Singer formula for the index of the Dirac operator. The elaboration of this idea was described by Bismut [8] at the Berkeley ICM.

So far we have been dealing with well-known material. But we can go on to consider a $1 + 1$ dimensional theory whose action is a differential form on the space $F(\Sigma)$ of maps from a surface Σ to X. Then the vector space H_S associated to a circle S will be the space of L^2 spinors on the loop space $\mathscr{L}X$. When Σ is a torus the path integral over $F(\Sigma)$ is called the *elliptic genus* $e_\Sigma(X)$ of X. (In fact there are a number of variants, applying in slightly different situations.) The $0 + 1$ dimensional result that the supertrace of $e^{-T\Delta}$ was independent of T has the analogue that $e_\Sigma(X)$ depends only on the *conformal* structure of Σ, i.e. for each X it is a modular function on the upper $\frac{1}{2}$-plane. The elliptic genus can be interpreted formally as the equivariant index of a version of the Dirac operator on the manifold X. As with the $\hat{A}$-genus the path integral defining $e_\Sigma(X)$ collapses to an integral over X, and this expresses it in terms of the characteristic numbers of X already familiar in algebraic topology. Nevertheless the elliptic genus has striking and unexpected properties, especially in connection with the topology of circle actions, and it stimulated the discovery of elliptic cohomology, a new theory whose true nature remains obscure. (An account of this subject can be found in [22]. Cf. also [1], [26]).

---

* More accurately, we replace $e^{-S(\gamma)}\mathscr{D}\gamma$ by $e^{-\hat{S}}$. The integral of an inhomogeneous form means the integral of its component of top degree.

§3. Topological Field Theories

A field theory is *topological* if it is defined for smooth manifolds with no additional structure (apart from a question of fixing projective multipliers which I shall suppress in this talk.) The vector spaces H_Σ must then be finite dimensional. A discussion of the formal properties can be found in [4].

(a) $1 + 1$ *dimensional Theories.* These are completely described by giving a commutative ring A with a 1 together with a linear map $\theta : A \to \mathbb{C}$ such that the bilinear form $(a, b) \mapsto \theta(ab)$ is non-degenerate. In fact $A = H_{S^1}$, and the product $A \otimes A \to A$ is Ψ_M, where M is a disc with two holes.

(b) $2 + 1$ *dimensional Theories.* These are by far the most studied, and the structure is much richer: it appears to be roughly equivalent to a quantum group. A theory gives us an invariant for each closed 3-manifold, and a representation of the mapping class group of each closed surface. If we choose an element $\xi \in H_{S^1 \times S^1}$ we get an invariant $\phi_\xi(K, M) \in \mathbb{C}$ for each knot K in a 3-manifold M by defining

$$\phi_\xi(K, M) = \langle \xi, \Psi_{M-U} \rangle,$$

where U is a tubular neighbourhood of K. In Reshetikhin's, Turaev's, and Feigin's talks at this Congress we heard how the same output arises from a quantum group. The relation between the two approaches does not seem completely understood, but I shall say a little more about it in §4 below.

$2 + 1$ dimensional theories are important because there is a supply of natural examples which lead to the knot invariants of Vaughan Jones and others. There is a theory for each compact Lie group G and choice of "level" k. The level is an element $k \in H^4(BG; \mathbb{Z})$, i.e. an integer if G is simple and simply connected. Regarding k as a characteristic class for G-bundles there corresponds to it a secondary Chern-Simons characteristic class S_k with values in $\mathbb{R}/2\pi\mathbb{Z}$ which is defined on the space $F(M)$ of isomorphism classes of G-bundles with connection on a 3-manifold M. This, or rather iS_k, is the action defining the theory [33]. But the theory can be constructed without mentioning path-integrals in the following way.

The vector space H_Σ associated to a surface Σ is the "quantization" of the symplectic manifold $\mathcal{M}_\Sigma$ of flat G-bundles on Σ. (This is the symplectic quotient [5] of the space of *all* connections on Σ by the action of the gauge group.) The symplectic structure of $\mathcal{M}_\Sigma$ depends on the level: its class is the image of k under the transgression $H^4(BG) \to H^2(\mathcal{M}_\Sigma)$. To obtain a definite quantization one method is to

(i) choose a complex structure on Σ,
(ii) identify $\mathcal{M}_\Sigma$ with the moduli space of stable holomorphic G-bundles on Σ by the Narasimhan-Seshadri theorem [23], thereby giving $\mathcal{M}_\Sigma$ a Kähler structure,
(iii) represent the symplectic form as the curvature of a holomorphic line bundle L on $\mathcal{M}_\Sigma$, and
(iv) define H_Σ as the space of holomorphic sections of L.

One must show that H_Σ is essentially independent of the complex structure chosen. Even after that one must construct the vectors Ψ_M associated to 3-manifolds. No natural way of doing this is known, though in general terms one can say that if

$\Sigma = \partial M$ then the boundaries of flat G-bundles on M form a Lagrangian submanifold in $\mathcal{M}_\Sigma$, and this should define a vector in the quantization H_Σ. But the connection of the spaces $\mathcal{M}_\Sigma$ with 3-manifolds was a great surprise, for they arise more obviously from 1 + 1 dimensional conformal theories, as we shall see below.

The application of this theory to the study of knots and 3-manifolds is discussed elsewhere at this Congress, so here I shall just emphasize that it has led to many new results about the geometry of the spaces $\mathcal{M}_\Sigma$, notably Verlinde's beautiful formula [30] for the dimension of the space H_Σ. By applying quantum field theory to $\mathcal{M}_\Sigma$ in a slightly different way Witten has recently been led to conjecture a formula for the volume of $\mathcal{M}_\Sigma$ in terms of $\zeta_G(2g-2)$, where g is the genus of Σ, and

$$\zeta_G(s) = \Sigma(\dim V)^{-s},$$

the sum being over the irreducible representations V of G.

(c) 3 + 1 *dimensional Theories.* For each compact group G there is an important 3 + 1 dimensional theory [3, 13, 17] which assigns to a closed 4-manifold W its Donaldson invariant, i.e. (roughly) the number of "instantons" on W. (An instanton is a solution of the self-dual Yang-Mills equations.) This theory was described in Floer's talk at this Congress. The vector space H_M for a 3-manifold M is the Floer cohomology group defined by applying infinite dimensional Morse theory to the space $F(M)$ of isomorphism classes of G-connections on M, the Morse function being the circle-valued Chern-Simons form already mentioned. A 4-manifold W with boundary M has a relative Donaldson invariant in H_M. Unfortunately field theory, although strikingly exemplified here, has not so far helped much with the geometry, except insofar as it is a field-theoretic idea to study the instanton moduli spaces in terms of the space of all connections.

I should say a word about Floer cohomology. The infinite dimensional manifolds F which arise in field theory are usually *polarized*, in the sense that their tangent spaces are roughly decomposed into positive- and negative-energy halves. (Cf. [26] § 4.) Floer's Morse function defines a decomposition of this kind, into the positive and negative eigenspaces of the Hessian. For such a manifold F one expects to be able to define "middle dimensional cohomology", by considering infinite dimensional cycles whose tangent spaces roughly fill the negative half of the tangent spaces to F. This idea goes back, of course, to Dirac's treatment of electrodynamics in terms of a sea of negative energy electrons. The same idea has been formalized by Feigin in his "semi-infinite" cohomology of Lie algebras [14]. Apart from the space of connections above, Floer cohomology has also been applied to the loop space of a symplectic manifold [12], and there too it arises as the state space of a field theory, the 1 + 1 dimensional topological σ-model of [32].

§ 4. 1 + 1 Dimensional Conformal Field Theory

Conformal field theory is akin to topological field theory in the sense that, up to isomorphism, a compact surface has only a finite dimensional space of conformal structures. Conformal theories can be axiomatized in the same way as topological ones [27, 28]. They have been much studied since the influential paper [7], partly for their relevance to string theory, but also because of their role in at least three areas of mathematics:

(i) the representation theory of loop groups and of Diff(S^1),
(ii) the study of the moduli spaces of Riemann surfaces and holomorphic bundles, and
(iii) the construction of the monster simple group and its representations.

For the third of these areas I refer to [15]. A slightly more conventional approach to conformal field theory is summarized in [18].

A conformal theory consists of a vector space H naturally associated to the standard circle S^1, together with an operator $\Psi_\Sigma : H^{\otimes m} \to H^{\otimes n}$ for each Riemann surface Σ with m incoming and n outgoing parametrized boundary circles. Thus Diff(S^1) acts (projectively) on H, and so does the semigroup $\mathscr{A}$ of surfaces which are topologically cylinders. (The composition-law is sewing end-to-end.) The semigroup $\mathscr{A}$ has twice the dimension of Diff(S^1), and is a complex manifold. One of the important ideas of the theory is that $\mathscr{A}$ plays the role of a complexification of the group Diff(S^1): more precisely, the relation between them is the same as that between the unitary group U_n and the semigroup $\{g \in GL_n(\mathbb{C}) : \|g\| < 1\}$ of contraction operators. (Cf. [28] and also Neretin [24].)

Let us recall that the loop group $\mathscr{L}G$ of a compact group G has an interesting class of irreducible projective representations $\{H_{k,V}\}$ – the positive energy representations – which are parametrized by their level $k \in H^4(BG; \mathbb{Z})$, which describes the projective multiplier of the representation, and an irreducible representation V of G. (The image of k in $H^2(\mathscr{L}G)$ is the class of the circle bundle defined by the central extension.) For a given level k only a finite set of representations V can occur. An important fact about the representations $H_{k,V}$ is that they possess a canonical projective action of Diff(S^1) intertwining with that of $\mathscr{L}G$. This action extends to – more accurately, is the boundary value of – an action of $\mathscr{A}$ by trace-class operators $\Psi_A : H \to H$. In fact Ψ_A is characterized by intertwining with the group G_A of holomorphic maps $A \to G_{\mathbb{C}}$, which acts on the source and target of Ψ_A via restriction to the two ends of A. The remarkable fact is that the irreducible representations of a given level constitute something very close to a conformal field theory. To state this precisely one needs the concept of a modular functor [27, 28]. (In the literature modular functors are usually referred to as "conformal blocks" [7] or "solutions of the Knizhnik-Zamolodchikov equations". For the latter, see Varchenko's talk at this Congress.)

A *modular functor* has a finite set Φ of labels. It assigns a finite dimensional vector space E_Σ to each Riemann surface with boundary where each boundary circle is labelled with an element of Φ. The axioms are

(i) $E_\Sigma = \mathbb{C}$ when Σ is the Riemann sphere,
(ii) $E_{\Sigma^1 \amalg \Sigma_2} \cong E_{\Sigma_1} \otimes E_{\Sigma_2}$,
(iii) $E_{\check{\Sigma}} \cong \bigoplus_{\phi \in \Phi} E_{\Sigma, \phi}$ where (Σ, ϕ) is obtained from $\check{\Sigma}$ by cutting it along a simple closed curve and giving both new boundary circles the label ϕ.

For the application Φ is the set of irreducible representations of $\mathscr{L}G$ of a given level. There is a modular functor E such that when Σ is a Riemann surface with m

incoming and n outgoing boundary circles labelled with representations $H_{\alpha_1}, \ldots, H_{\alpha_m}$ and $H_{\beta_1}, \ldots, H_{\beta_n}$ there is an operator

$$\Psi_{\Sigma,\xi} : H_{\alpha_1} \otimes \cdots \otimes H_{\alpha_m} \to H_{\beta_1} \otimes \cdots \otimes H_{\beta_n}$$

for each $\xi \in E_\Sigma$ which intertwines with the action of the group G_Σ of holomorphic maps $\Sigma \to G_{\mathbb{C}}$. (In fact E_Σ can be *defined* as the space of such intertwining operators: then the point to establish is property (iii) above, which amounts to a version of the Peter-Weyl theorem for loop groups.) The first complete proof of this result is in [29]. (Cf. Tsuchiya's talk at this Congress.)

One of the advantages of the field-theoretic viewpoint in the representation theory of loop groups is to make plain the otherwise mysterious modularity properties of the characters: in field theory the values of the characters are naturally associated to complex tori.

Witten realized [33] that the modular functor E_Σ just described is essentially independent of the complex structure of Σ, and is the state space of the corresponding $2 + 1$ dimensional topological theory based on the Chern-Simons action. More recently Kontsevich [21] has sketched an argument to show, still more surprisingly, that the concepts of modular functor and $2 + 1$ dimensional topological theory are exactly equivalent.

A "topological" modular functor is closely related to a quantum group, for the quantum deformation of G amounts essentially to a way of defining an exotic tensor product on the category of representations of G. For a modular functor E we can define

$$V_1 \otimes_E V_2 = \bigoplus_W E_{\Sigma, V_1, V_2, W} \otimes W,$$

where W runs through the irreducible representations of G, and Σ is a disc with two holes whose boundary components are labelled V_1, V_2 (incoming) and W (outgoing).

It is easy to relate the modular functor E_Σ to the space $H_\Sigma = \Gamma(\mathscr{M}_\Sigma; L)$ of holomorphic sections described in § 3. Let us decompose the closed surface Σ as $\Sigma_1 \cup \Sigma_2$ by a simple closed curve S. A holomorphic bundle on Σ is automatically trivial on Σ_1 and Σ_2, so it can be described by a clutching function on S, i.e. by an element of $\mathscr{L}G_{\mathbb{C}}$. The set of isomorphism classes of bundles on Σ – essentially the same as $\mathscr{M}_\Sigma$ – is therefore the double coset space $G_{\Sigma_1} \backslash \mathscr{L}G_{\mathbb{C}}/G_{\Sigma_2}$, and the space $\Gamma(\mathscr{M}_\Sigma; L)$ is the G_{Σ_1}-invariant part of $H = \Gamma(\mathscr{L}G_{\mathbb{C}}/G_{\Sigma_2}; \pi^* L)$, where $\pi : \mathscr{L}G_{\mathbb{C}}/G_{E_2} \to \mathscr{M}_\Sigma$. If we now take Σ_2 to be a standard disc then H is the basic representation of $\mathscr{L}G$ of level k, constructed by the Borel-Weil method [25]. Finally, it is easy to see that when the boundary of Σ_1 is labelled with H we have $E_\Sigma = E_{\Sigma_1}$, and so

$$E_\Sigma \cong H^{G_{\Sigma_1}} \cong \Gamma(\mathscr{M}_\Sigma; L).$$

The preceding argument, which shows how representations of $\mathscr{L}G$ define functions on the moduli space of G-bundles, also shows how representations of $\mathrm{Diff}(S^1)$ give functions on the moduli space $\mathscr{C}_\Sigma$ of complex structures on a smooth surface Σ. For $\mathscr{C}_\Sigma$ behaves like a double coset space of the semigroup $\mathscr{A}$: if we write $\Sigma = \Sigma_1 \cup \Sigma_2$, and choose fixed complex structures on Σ_1 and Σ_2, then $\Sigma_1 \cup A \cup \Sigma_2$ runs through an open set of $\mathscr{C}_\Sigma$ as A runs through $\mathscr{A}$. The representation theory of

Diff(S^1) allows us, for example, to identify and classify holomorphic line bundles on $\mathscr{C}_\Sigma$ much more simply than does conventional algebraic geometry. (Cf. [6, 28].)

§ 5. Zamolodchikov's *c*-Theorem

The point of view of this talk is successful with topological and conformal field theories, but so far it has never been taken seriously in a wider context. At present the only general definition of a field theory is the classical one in terms of the vacuum expectation values of a class of operators varying from theory to theory. This is not sufficiently manageable for one to be able to speak, for instance, of the "space $\mathscr{T}$ of all 1 + 1 dimensional theories". Nevertheless one of the most interesting recent developments has been the following result of Zamolodchikov [36], which is framed in terms of the space $\mathscr{T}$.

Whatever may be the definition of a theory, it will presumably be true that from any theory T one can derive a 1-parameter family of theories T_λ (for $\lambda \in \mathbb{R}^\times$) simply by multiplying all lengths by λ. The resulting flow on $\mathscr{T}$ is the *renormalization group flow*. Conformal theories are fixed points of this flow. Zamolodchikov's idea is to define a Riemannian metric on $\mathscr{T}$ and a smooth function $c : \mathscr{T} \to \mathbb{R}$ such that

(i) the renormalization group flow is the gradient flow of c, and
(ii) $c(T)$ is equal to the central charge if T is a conformal theory.

The *central charge* of a conformal theory is the number describing the central extension of Diff(S^1) which acts on the state space of the theory.

It is fairly straightforward to calculate the possible non-conformal infinitesimal deformations of a conformal theory, so Zamolodchikov's theorem suggests that one could in principle discover the global topology of the space $\mathscr{T}$ by Morse theory. Vafa and others have made some steps in this direction.

Zamolodchikov's argument is based on perturbation theory. I think it is a fascinating challenge to put it in a better mathematical setting.

§ 6. 1 + 1 Dimensional Quantum Gravity

The most dramatic recent development in quantum field theory has been a breakthrough in 2-dimensional quantum gravity. It has suddenly appeared possible to perform integrals over the space of all metrics on a surface and get very explicit answers. The main success has come from the technique of random matrices [9, 11, 19], and I cannot discuss it here. One very unexpected outcome is to link the theory with the classical completely integrable systems of non-linear partial differential equations such as the KdV equation. So far the situation has not really been assimilated mathematically, but the results seem to describe the algebraic topology of the space of metrics on a surface, or – equivalently – the moduli spaces of complex structures.

From ordinary algebraic geometry one knows (see Morita's talk at this Congress) a ring of stable cohomology classes on the moduli spaces $\mathscr{M}_g$ of surfaces of genus g ("stable" means that they are defined independently of g), and Witten has claimed that the field-theoretic results are the integrals of these classes over the spaces $\mathcal{M}_g$, i.e. the *characteristic numbers* of $\mathcal{M}_g$. This has led him to the striking conjecture [34] that the generating function for the characteristic numbers is a certain specific solution of the KdV hierarchy. His lectures [35] give an excellent account of the present state of the subject.

From the point of view of this talk it is interesting that quantum gravity does not fit directly into the framework of § 2 above, but nevertheless seems likely to be axiomatizable along related but more subtle lines. The crucial point that distinguishes the gauge-theory situations to which § 2 applies from the gravitational ones to which it does not is simply that an automorphism of a bundle P on $M_1 \cup M_2$ is just a pair of automorphisms of $P|M_1$ and $P|M_2$, but a diffeomorphism of $M_1 \cup M_2$ cannot be broken into two diffeomorphisms.

References

1. O. Alvarez, T. Killingback, M. Mangano, P. Windey: String theory and loop space index theorems. Comm. Math. Phys. **111** (1987) 1–10
2. M.F. Atiyah: Circular symmetry and stationary phase approximation. Proc. Conf. in honour of L. Schwartz, Astérisque **131** (1985) 43–59
3. M.F. Atiyah: New invariants of 3- and 4-dimensional manifolds. Amer. Math. Soc., Proc. Symp. Pure Math. **48** (1988) 285–99
4. M.F. Atiyah: Topological quantum field theories, Publ. Math. I.H.E.S. Paris **68** (1989) 175–86
5. S. Axelrod, S. Della Pietra, E. Witten: Geometric quantization of Chern-Simons gauge theory. J. Diff. Geom. **33** (1991) 787–902
6. A.A. Beilinson, V.V. Schechtman: Determinant line bundles and Virasoro algebras. Comm. Math. Phys. **118** (1990) 651–701
7. A.A. Belavin, A.M. Polyakov, A.B. Zamolodchikov: Infinite conformal symmetry in two-dimensional quantum field theory. Nucl. Phys. **B241** (1984) 333–380
8. J.-M. Bismut: Index theorem and the heat equation, Proc. Internat. Cong. Math. Berkeley 1986, pp. 491–504
9. E. Brezin, V.A. Kazakov: Exactly solvable field theories of closed strings. Phys. Lett. **236B** (1990) 144–150
10. S.K. Donaldson, P. Kronheimer: Geometry of 4-manifolds. Oxford U.P. 1990
11. M. Douglas, S. Shenker: Strings in less than one dimension. Nucl. Phys. B **335** (1990) 635–654
12. A Floer: Morse theory for fixed points of symplectic diffeomorphisms. Bull. Amer. Math. Soc. **16** (1987) 279–281
13. A. Floer: An instanton invariant for 3-manifolds. Comm. Math. Phys. **118** (1988) 215–240
14. I.B. Frenkel, H. Garland, G. Zuckerman: Semi-infinite cohomology and string theory, Proc. Nat. Acad. Sci. USA. **83** (1986) 8442–6
15. I.B. Frenkel, J. Lepowsky, A. Meurman: Vertex operator algebras and the monster. Academic Press, 1988
16. D. Friedan, P. Windey: Supersymmetric derivation of the Atiyah-Singer index theorem and the chiral anomaly. Nucl. Phys. **B235** (1984) 395–416.
17. M. Furuta, D. Kotschick, S. Donaldson: Floer homology groups in Yang-Mills theory. (To appear)
18. K. Gawedzki: Conformal field theory. Seminaire Bourbaki **704** (1988) in Astérisque.

19. D.J. Gross, A.A. Migdal: A nonperturbative treatment of two dimensional quantum gravity. Phys. Rev. Lett. **64** (1990) 127–130
20. N.J. Hitchin: Flat connections and geometric quantization. Comm. Math. Phys. **131** (1990) 347–380
21. M. Kontsevich: Rational conformal field theory and invariants of 3-dimensional manifolds (Marseilles preprint, to appear)
22. P.S. Landweber (ed.): Elliptic curves and modular forms in algebraic topology, Proceedings, Princeton 1986. (Lecture Notes in Mathematics, vol. 1326.) Springer, Berlin Heidelberg New York 1986
23. M.S. Narasimhan, C.S. Seshadri: Stable and unitary bundles on a compact Riemann surface. Ann. Math. **82** (1965) 540–67
24. Yu. A. Neretin: A complex semigroup containing the group of diffeomorphisms of a circle. Funkts. Anal. Prilozh. **21** (1987) 82–83
25. A. Pressley, G. Segal: Loop groups. Oxford, U.P. 1986
26. G. Segal: Elliptic cohomology, Seminaire Bourbaki **695** (1988). In Astérisque **161**–**162** (1988) 187–201
27. G. Segal: Two dimensional conformal field theories and modular functors. IXth Proc. Internat. Cong. Math. Phys. Swansea 1988 (B. Simon, A. Truman, I.M. Davies, (eds).) Adam Hilger, 1989, pp. 22–37
28. G. Segal: The definition of conformal field theory. (To appear)
29. A. Tsuchiya, K. Ueno, Y. Yamada: Conformal field theory on the universal family of stable curves with gauge symmetry. In Conformal field theory and solvable lattice models. Adv. Stud. Pure Math. **16** (1988) 297–372
30. E. Verlinde: Fusion rules and modular transformations in two dimensional conformal field theory. Nucl. Phys. B **300** (1988) 360–376
31. E. Witten: Topological quantum field theory. Comm. Math. Phys. **117** (1988) 353–86
32. E. Witten: Topological σ-models. Comm. Math. Phys. **118** (1988) 411–449
33. E. Witten: Quantum field theory and the Jones polynomial. Comm. Math. Phys. **121** (1989) 351–99
34. E. Witten: On the structure of the topological phase of two dimensional gravity. IAS Preprint IAS SNS HEP 89/66
35. E. Witten: Two-dimensional gravity and intersection theory on moduli space. Lectures, Harvard 1990 (To appear)
36. A.B. Zamolodchikov: Renormalization group and perturbation theory near fixed points in two-dimensional field theory. Yad. Fiz. **46** (1987) 1819–1831

Quantum Mechanics of Many-Particle Systems*

I.M. Sigal[1]

Department of Mathematics, 100 St. George Street, University of Toronto
Toronto, Ontario, M5S 1A1

1. Introduction

In this talk I will discuss some mathematical questions arising in the theory of quantum many-body systems. Examples of such systems are atoms, molecules, nuclei, solids and, to some extent, stars. The remarkable fact is that such diverse objects are described within a single mathematical framework given in terms of the Schrödinger operator:

$$H = -\Delta + V(x) \quad \text{on } L^2(X).$$

Here X is the configuration space of a system in question, it is either $\mathbb{R}^{3N}$ or a linear subspace thereof, where N is the number of particles, Δ is the Laplacian on X and $V(x)$ is the total potential energy, e.g.

$$V(x) = \sum V_{ij}(x_i - x_j),$$

the sum of pair potentials, with x_i being the coordinate of the ith particle and $x = (x_1, \ldots, x_N) \in X$. Whenever the Pauli principle is taken into account, $L^2(X)$ should be replaced by its appropriate subspace (see the paragraph after Equation (12)).

Physical properties of quantum systems are associated with spectral characteristics of H. For instance, existence and uniqueness of unitary dynamics, i.e. solution of the Cauchy problem

$$i\frac{\partial \psi_t}{\partial t} = H\psi_t \quad \text{and} \quad \psi_0 = \psi, \tag{1}$$

satisfying $\|\psi_t\| = \|\psi\|$, is equivalent to the statement that H is self-adjoint. Once this is established, the next problem is classification of orbits ψ_t. Though ψ_t is an orbit in the Hilbert space $L^2(X)$, of fundamental interest is its behaviour in the configuration space X. According to the latter, orbits are classified as bounded, wandering and escaping (to ∞). This rough decomposition turns out to be equivalent to splitting the spectrum into pure point, singular continuous and absolutely continuous parts. Strange, recurrent orbits coming from the singular continuous spectrum subspace are ruled out by

* Supported by NSERC under Grant NA7901.

[1] I.K. Killam Research Fellow.

Proceedings of the International Congress
of Mathematicians, Kyoto, Japan, 1990

Theorem [Mo 1981, PSS 1981]. *Assume that*

$$|y|^{|\alpha|}\partial^{\alpha} V_{ij}(y) \quad \text{are} \quad \Delta_y\text{-compact for } |\alpha| \leq 2. \tag{2}$$

Then

$$\textit{sing. cont. spec.} \;=\; \phi .$$

The next basic questions are the asymptotic behaviour of unbounded orbits and stability of bounded ones under perturbations of H. They lead to the three main problems of the theory of quantum many-particle systems: scattering, binding and resonances. We consider here the first two problems, referring the interested reader to [Sig 1989] for a review of the last one.

2. Scattering Theory

The basic problem here is to show that as $|t| \to \infty$, every orbit starting in the continuous spectrum subspace obeys

$$\left\|\psi_t - \sum \text{ simple orbits}\right\| \;\to\; 0. \tag{3}$$

Here the sum is taken over all possible break-ups of the system into subsystems and over all stable states (i.e. eigenfunctions of the corresponding Schrödinger operators) of the resulting subsystems. Given a break-up and a specification of stable states, the corresponding simple motion can be written as

$$\text{simple motion} \;=\; \phi \otimes u_t ,$$

where ϕ is the product of the eigenfunctions of the subsystems (depending on the internal coordinates) and u_t, a free (no potentials) orbit of the centers-of-mass of the subsystems (see Fig. 1). The time factors corresponding to the eigenvalues of the subsystems (the internal energies) are included into u_t.

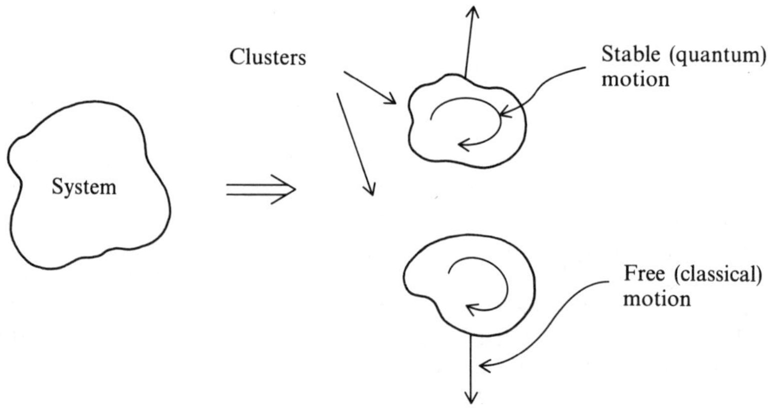

Fig. 1. Break up of a system into independent stable clusters

This problem is called *asymptotic completeness*. Over the last 40 years it was a focus of attention of many mathematicians and mathematical physicists. Important contributions were made by T. Kato, S. Kuroda, T. Ikebe, M. S. Birman, L. D. Faddeev, V. Enss, B. Simon, S. Agmon and L. Hörmander among others (see references in [SigSof 1987, 1990ab]). The problem was solved for the short-range potentials in [SigSof 1987] (see [Graf 1990] for a beautiful and concise proof, different proofs are given in [Kit 1990, Tam 1990]):

Theorem [SigSof 1987]. *Assume* (2) *and that*

$$\langle y\rangle^{\mu} V_{ij}(y) \quad \textit{are } \Delta_y\textit{-bounded} \tag{4}$$

with $\mu > 1$. *Then asymptotic completeness holds.*

Precise restrictions on smoothness of V_{ij} are not essential and can be relaxed if some extra care is exercised. What is important is their decay at infinity. If $\mu > 1$, then the potentials are called *short-range*, and if $\mu \leq 1$, *long-range*. Choices of the free evolutions are different for short-range and long-range cases. In particular, in the short-range case they are generated by the Hamiltonians

$$H^{\text{asympt}} = E^{\text{intern}} - \Delta^{\text{CM}}, \tag{5}$$

where E^{intern} is the sum of the eigenvalues of the subsystems and Δ^{CM} is the Laplacian in the center-of-mass coordinates, and in the long-range case, by more complicated, time-dependent operators. For the long-range potentials only partial results are, presently, obtained. Namely, for $N \leq 3$ and $\mu > \sqrt{3} - 1$ asymptotic completeness is proven in [Enss 1985] and for $N \leq 4$ and $\mu = 1$, in [SigSof 1990a,b].

Open Problem. Prove asymptotic completeness for all N and long-range (especially, Coulomb-type: $\mu = 1$) potentials.

We make a few comments about the proofs.

Propagation Set. The first step one makes in the modern scattering theory is changing the viewpoint. Instead of studying the evolution of the system in the physical space $\mathbb{R}^3$ one investigates the behaviour of orbits ψ_t in the phase-space T^*X. For $\Omega \subset T^*X$ we define (modulo boundary terms) the probability that the system in question is in Ω at time t:

$$(\text{Prob. syst. in } \Omega \text{ at } t) \equiv \|\phi\psi_t\|^2, \tag{6}$$

where ϕ is a pseudodifferential operator whose symbol is supported in Ω and is equal to 1 in a slightly smaller set. If $\Omega \subset X$ and ϕ is the multiplication operator by the characteristic function of Ω, then (6) is a standard definition for a quantum probability for the system to be in the region Ω of the configuration space. One expects that as $|t| \to \infty$, ψ_t concentrates on the set

$$\text{PS} = \cup\,(\text{Class. phase-space traj. of quant. stable subsystems}),$$

called the *propagation set* (see Fig. 1). Indeed, we have

Theorem (SigSof 1987). *Assume* (2) *and* (4) *with* $\mu > 0$. *Then as* $|t| \to \infty$

$$\text{(Prob. syst. in } T^*X\backslash PS \text{ at } t) \to 0 \tag{7}$$

in the mean sense explained below.

Note that unlike the wavefront set in the propagation of singularities the propagation set consists of *partially quantized* bicharacteristics (classical trajectories). A peculiarity of quantum propagation is that the motion along these bicharacteristics is unstable: the system can tunnel from one bicharacteristic to another. One of the consequences of this is that the convergence in (7) is in the mean sense:

$$\int_{|t|\geq 1} (\text{l.h.s. of (7)}) \frac{dt}{|t|} < \infty .$$

Thresholds. A key restriction, omitted from the statement of the above theorem, says that initial conditions for the orbits ψ_t must be from Range $E_\Delta(H)$, where $E_\Delta(H)$ is the spectral projection of H and Δ is some interval away from the thresholds. The *thresholds* or threshold energies are critical values of Hamiltonians driving partially quantized bicharacteristics, (i.e. E^{intern} of (5)). Understanding the evolution at threshold energies is a delicate problem, particularly, because of the tunneling between critical and non-critical bicharacteristics (i.e. bicharacteristics for which a given energy is a critical value and those for which it is not).

One of the principal differences between treatments of short-range and long-range scattering is that while in the short-range case one can avoid considering threshold energies, in the long-range case, one cannot.

Long-Range Scattering. In the long-range case as a system breaks apart, the interaction between departing subsystems cannot be neglected entirely. It is replaced by a potential depending only on the internal coordinates of the subsystems and time. As a result the induction in the number of particles, which is used in the proof, forces one to estimate orbits ψ_t generated by time-dependent Schrödinger operators of the form

$$H(t) = H + W(x,t),$$

where the potential W obeys

$$|\partial^\alpha_{(x,t)} W(x,t)| \leq C_\alpha (1 + |x| + |t|)^{-\mu-|\alpha|} .$$

Since the energy is not conserved anymore, one cannot separate and remove threshold energies by cut-off functions in H, as it is done in the short-range case. To study ψ_t we use a version of microlocal analysis with several fine time scales. This allows us to separate critical bicharacteristics from non-critical ones. More precisely, we split the energy axis as

$$\text{dist}(E, \text{thresholds}) \geq t^{-\beta} \tag{8}$$

$$\text{dist}(E, \text{threshold}) \leq t^{-\beta} \tag{9}$$

with some $\beta < \mu$ $(= 1)$. For $E_{\Delta(t)}(H)\psi_t$ with $\Delta(t)$ from (8) we still manage to prove that subsystems separate (though with relative velocities vanishing in time). For energies in (9), we consider separately regions

$$|x| \le t^{\alpha} \tag{10}$$

and

$$|x| \ge \frac{1}{2}t^{\alpha} \tag{11}$$

with α depending on the directions in X and obeying $(1 =)\ \mu > \alpha > 1 - \beta/2 > 0$. In the first of these regions we show that the system in question is localized. To this end we use that non-critical bicharacteristics lie outside (10), while the critical ones stay inside. Indeed, their velocities, due to (9), are bounded by

$$\begin{aligned} \text{critic. vel.} &\le C\big(\text{dist}(E, \text{thresholds})\big)^{1/2} \\ &= Ct^{-\beta/2}. \end{aligned}$$

As a result we can set $x = 0$ in $W(x,t)$ which reduces the problem to the time-independent one. Since the critical bicharacteristics do not reach the second region, the system propagates there along non-critical bicharacteristics. This means it breaks up as $|t| \to \infty$. Subtlety here is that we can conduct the analysis above only in fixed directions, not globally, reducing the problem to a simple one step by step. Thence stems our limitation on the number of particles.

3. Binding

Binding is the property of matter to form stable compounds, such as atoms, molecules, nuclei, etc. It is measured by a gap between inf spec H and cont spec H. This gap is called the *binding energy*,

$$\text{BE} = \inf \text{cont spec}\, H - \inf \text{spec}\, H.$$

In fact, this is the energy needed to destroy the most stable bounded orbit, the one corresponding to inf spec H.

Example: Atoms. Consider a system consisting of N electrons and a nucleus of charge z. Let $z = N$. For simplicity we assume the nucleus to be infinitely heavy. Then the Schrödinger operator of such a system is

$$H = \sum_{i=1}^{N} \big(-\Delta_i + V(x_i)\big) + \frac{1}{2}\sum_{i \ne j} |x_i - x_j|^{-1}, \tag{12}$$

where $V(x) = -z/|x|$ and the units used are such that the electron mass is $1/2$, the electron charge is 1 and the Planck constant is 2π. It acts on $\bigwedge_{i=1}^{N} L^2(\mathbb{R}^3 \times \mathbb{Z}_q)$, where $q = 2$, the number of spin states of electron. It was shown in [LiebSim 1977] that

$$-\inf \text{spec} = O(z^{7/3}).$$

On the other hand it is believed that

$$\text{BE} = O(1).$$

Thus the existence of matter as we know it is due to a subtle phenomenon, indeed.

Theorem [SecoSigSol 1990].

$$\mathrm{BE} \leq \mathrm{const}\, z^{20/21}. \tag{13}$$

The proof of this theorem is rather instructive. Some of the tools used are described in the following sections.

Open Problem. Show that $\mathrm{BE} = O(1)$.

Another way to measure binding is by the maximal number of electrons a nucleus of charge z can hold together, $N(z)$. It was shown in [Rus 1982, Sig 1982] that $N(z)$ is finite.

Theorem. $$N(z) = z + O(z^{5/7}).$$

It was shown in [BengLieb 1983, Solov 1990] that if the electrons were bosons, then

$$N^{\mathrm{boson}}(z) = 1.21z + O(z)$$

(the numerical value of the coefficient in front of z was computed in [Baum 1984]). [Bach 1990] proves that

$$\mathrm{BE}^{\mathrm{boson}} = O(z^2).$$

Thus Pauli principle plays a crucial role in the formation of atoms.

The asymptotic behaviour of $N(z)$ was established in [LSST 1988] and the remainder estimate was derived in [FeffSeco 1990a]. A simpler proof is given in [SecoSigSol 1990].

Ground State Energy. One of the basic characteristics of a quantum system is its ground state energy, i.e. the lowest eigenvalue of the corresponding Schrödinger operator. As an example we consider a molecule with N electrons and M nuclei of charges $z_1, \ldots, z_M$. Let $\sum z_i = N$. We assume the nuclei to be infinitely heavy and located at positions $r_1, \ldots, r_M$ (Born-Oppenheimer model). Let $Z = (z_1, \ldots, z_M)$, $|Z| = \sum_{j=1}^{M} z_j$ and $R = (r_1, \ldots, r_M)$. The Schrödinger operator of such a molecule is (12) with $V(x)$ given by

$$V(x) = -\sum_{j=1}^{M} \frac{z_j}{|x - r_j|}, \tag{14}$$

the potential of the interaction of an electron with the nuclei. Denote by $E(Z, R)$ the ground state energy of such a molecule. One of the most elementary questions here is to understand asymptotic behaviour of $E(Z, R)$ as $\sum z_j$ tends to ∞. To make this problem physically and mathematically interesting we have also to scale r_j. Knowing such asymptotic behaviour is one of the key inputs in the proof of the previous theorem.

TF Gas. L. H. Thomas and E. Fermi have suggested in 1927 that a large Coulomb system (atom or molecule) in the ground state (= eigenfunction corresponding to the lowest eigenvalue) looks like a classical gas but with Pauli principle, namely, there could be at most 2 electrons per volume $(2\pi)^3$ in the phase space. Such an object is called now the *Thomas-Fermi gas.* Its states are described by the *electron density* $\varrho \geq 0$ on $\mathbb{R}^3$ normalized as

$$\int \varrho \;=\; N \;=\; \#\text{ of electrons}\,.$$

The energy of the Thomas-Fermi gas is given by a simple non-linear functional

$$\mathscr{E}^{\mathrm{TF}}(\varrho) \;=\; \gamma \int \varrho^{5/3} + \int V\varrho + \frac{1}{2}\int \varrho(|x|^{-1} * \varrho)\,, \tag{15}$$

where $\gamma = (3/5)(3\pi^2)^{2/3}$ and $V(x)$ is given by (14). The ground state energy, $E^{\mathrm{TF}}(Z,R)$, is the infimum (in fact, minimum if $N \leq \sum z_j$) of this functional. It has the following scaling property

$$E^{\mathrm{TF}}(Z,R) \;=\; a^{7/3}E^{\mathrm{TF}}(a^{-1}Z, a^{1/3}R)\,.$$

Hence

$$E^{\mathrm{TF}}(Z,R) \;=\; O(|Z|^{7/3})\,.$$

Asymptotics. The next theorem shows that the Thomas-Fermi theory is asymptotically correct for large Z systems but only to the leading order.

Theorem. *Let $Z \to \infty$ along a given direction and let the mutual distances between r_j be bounded from below by* const $|Z|^{-2/3+\varepsilon}$. *Then*

$$E(Z,R) \;=\; E^{\mathrm{TF}}(Z,R) + \frac{1}{4}\sum z_j^2 + o(|Z|^2)\,. \tag{16}$$

It will follow from the analysis below that the leading term on the r.h.s. represents the quasiclassical energy of the bulk of electrons and the second term, the quantum spectrum of Coulomb singularities.

The leading term in (16) was obtained in [LiebSim 1977]. The second term of asymptotics was conjectured by J. M. C. Scott in 1952 as a contribution of those electrons which move very close to the nuclei (see [LiebSim 1977 and Lieb 1981] for a discussion). For atoms the Scott conjecture was proven in [Hughes 1990, SiedWeik 1987, 1989] and for molecules, in [IvSig 1990]. A proof of the next, $z^{5/3}$ term for atoms is announced in [FeffSeco 1990b]. The approach in [Hughes 1990, SiedWiek 1987, 1989, FeffSeco 1990b] is based on an expansion in angular momentum channels. This is possible since the electron interaction with the nucleus $V(x)$, is spherically symmetric in atoms. The problem is then reduced to a one-dimensional one which is treated by the standard WKB method. The proof in [IvSig 1990] is rather general. I will make a few comments about it.

Mean Field Theory. The main utility of the Thomas-Fermi theory for us is to provide a sufficiently simple Schrödinger operator approximating the rather complex H. Let ψ be the ground state of H, i.e. the eigenfunction corresponding to the smallest eigenvalue. Consider the random variable of electron density $\sum_{i=1}^{N} \delta(x - x_j)$ with the probability distribution $|\psi(x_1,\ldots,x_N)|^2 dx_1,\ldots,dx_N$. The ideas of Thomas and Fermi suggest that for large N this random variable is close to some mean electron density, namely the one, $\varrho^{\mathrm{TF}}(x)$, which minimizes $\mathcal{E}^{\mathrm{TF}}(\varrho)$. Hence the potential experienced by any one electron is approximately

$$\phi(x) = V(x) + |x|^{-1} * \varrho^{\mathrm{TF}}(x),$$

i.e. the one produced by the nuclei screened by this mean electron density ϱ^{TF}. Thus we introduce the mean-field (or quasiparticle) Schrödinger operator

$$H^{\mathrm{ind}} = \sum_{i=1}^{N} (-\Delta_i + \phi(x_i)) - D$$

acting on $\bigwedge_{i=1}^{N} L^2(\mathbb{R}^3 \times \mathbb{Z}_2)$, where D is a number compensating for overcounting the electron–electron interaction in ϕ,

$$D = \frac{1}{2}\int\int \frac{\varrho^{\mathrm{TF}}(x)\varrho^{\mathrm{TF}}(y)}{|x-y|} dx\, dy.$$

This operator describes independent (quasi-) electrons moving in an external potential $\phi(x)$. It was realized for some time that in many calculations H can be replaced by H^{ind}.

Some potential theory, Lieb-Thirring inequality, which combines uncertainty and Pauli principles, and a simple version of quasiclassical estimates discussed below yield the following result (cf. Lieb 1981, Hughes 1990, SiedWeik 1987, 1990, FeffSeco 1990b).

Theorem. *Let $E^{\mathrm{ind}}(Z,R)$ be the ground state energy of H^{ind}. Then*

$$E(Z,R) = E^{\mathrm{ind}}(Z,R) + O(|Z|^{5/3}).$$

The One-Body Problem. The separation of variables on $\bigwedge_{i=1}^{N} L^2(\mathbb{R}^3 \times \mathbb{Z}_2)$ gives

$$E^{\mathrm{ind}}(Z,R) = \sum_{i=1}^{N} E_i - D_{\mathrm{TF}},$$

where $E_1, E_2, \ldots$ are the eigenvalues of the one-particle operator $P = -\Delta + \phi(x)$ acting on $L^2(\mathbb{R}^3 \times \mathbb{Z}_2)$, labeled in order of their increase and counting their multiplicities. This is a much used relation in Quantum Physics and is a consequence of the Pauli principle: at most two electrons (the double degeneracy corresponding to $\mathbb{Z}_2$) per a quantum state. We show

Theorem. *Let $Z \to \infty$ along a given direction and let the mutual distances between the r_j's $\geq$ const$|Z|^{-2/3+\varepsilon}$. Then*

$$\sum_{i=1}^{N} E_i = \mathrm{Weyl} + \mathrm{Scott} + o(|Z|^2), \tag{17}$$

where, with $p(x,\xi) = |\xi|^2 + \phi(x)$, the symbol of P,

$$\text{Weyl} = \iint_{p \le 0} p\, dx\, d\xi$$

and

$$\text{Scott} = \frac{1}{8}\Sigma z_j^2 .$$

The last two theorems yield the main result.

Quasiclassical Asymptotics. Now we discuss the proof of (17). Let λ be a fixed vector in $\mathbb{R}^M$ with non-negative components and let $Z = \beta^{-3}\lambda$. The scaling $x \to \beta x$ and properties of $\phi(x)$ show that the large Z problem is, in fact, the quasiclassical problem with $\beta = O(|Z|^{-1/3})$ playing the role of the Planck constant. Thus we have to find the quasiclassical asymptotic (as $\beta \to 0$) for the sum of the first $N = |Z|$ eigenvalues of the *one-particle* Schrödinger operator P. The problem is that $\phi(x)$ is singular. Thus standard quasiclassical methods based on pseudodifferential calculus do not work here. Our method originates in ideas of [Ivr 1986].

There are two ingredients in our proof. First of all we estimate global quantities through local ones. For instance, we study

$$\operatorname{tr}\big(\psi(x) g(P)\big), \tag{18}$$

where $g(\lambda) = \lambda$ for $\lambda \le 0$ and $= 0$ for $\lambda \ge 0$. If $\psi \equiv 1$, then the trace above is just the sum of negative eigenvalues of P. We take for ψ smooth functions localized outside of the singularities of the potentials. Then it is not difficult to obtain asymptotic expansion in the quasiclassical parameter β of the trace (18). Pseudodifferential calculus provides convenient tools for such a purpose. Adapting a standard technique, one represents $g(P)$ as

$$g(P) = \int \hat{g}(t) e^{-iPt} dt,$$

where $\hat{g}(t)$ is the Fourier transform of g. The evolution operator e^{-iPt} is then approximated on a small time interval to any power in β by Fourier integral operators in the spirit of the geometrical optics. Such an approximation is possible because of finite speed of propagation of singularities for the Schrödinger equation: for sufficiently small times and for bounded energies the singularities of $\phi(x)$ do not reach $\operatorname{supp}\psi$. The appropriate Fourier integral operators are then expanded by the method of stationary phase. The information about (18) is recovered using the Tauberian technique. However, the remainder estimates here depend on $\operatorname{supp}\psi$ and on estimates of $\partial^\nu \phi$ on $\operatorname{supp}\psi$.

The second ingredient is a multiscale analysis. There are three scales in the problem: momentum scale determined by the quasiclassical parameter $\beta = O(|Z|^{-1/3})$, space scale, $l(x)$, determined by how the potential differentiates and the energy scale, $f(x)$, determined by the size of the potential. The first scale is constant while the other two depend on x. In our problem

$$l(x) - \text{dist of } x \text{ to the singularities}$$

and $f(x) = l(x)^{-1}$. At each point outside of the singularities we rescale the problem using the scales at this point in such a way that the problem is mapped into a model one, i.e. the one with a potential U obeying $|\partial^\alpha U(x)| \leq C_\alpha$ on a unit ball, with the effective quasiclassical parameter

$$\alpha_{\text{eff}}(x) = \frac{\beta}{l(x)f(x)^{1/2}},$$

which depends on all the scales. The new problem admits a quasiclassical expansion discussed above with a remainder bound independent of the singular structure of $\phi(x)$. This implies an expansion for the original problem outside of small balls around the singularities of $\phi(x)$. Inside each of those balls we analyze the problem differently. Namely, we replace the potential by its leading term near the singularity and solve the quasiclassical problem for this truncated operator more accurately. Leading terms near and outside the singularities combine into a single Weyl term over $\mathbb{R}^3$ which gives the Thomas-Fermi energy. Precise quasiclassical expansion of low-lying eigenvalues of the truncated problem near the singularities yields the Scott correction.

4. Conclusion

The rigorous Quantum Mechanics of many-particle systems is a fast developing branch of Mathematical Physics. This paper is not a comprehensive review of the subject. A rather versatile account of the Schrödinger operators as well as many references can be found in [CFKS]. What we attempted here is a brief glimpse into some of the recent developments and trends, just a small sample of many fresh and exciting problems with which this field abounds.

Acknowledgements. This paper was written while the author was visiting the Institut für Theoretische Physik, ETH-Zürich; the author is grateful to J. Fröhlich and W. Hunziker for the hospitality. It is a pleasure to thank V. Bach, J. Fröhlich, W. Hunziker, B. Simon and J.-P. Solovej for reading the manuscript and making many useful remarks. This paper sums up results of the joyful collaboration with A. Soffer, V. Ivrii, J.-P. Solovej and L. Seco.

References

Bach, V. (1991): Ionization energies of bosonic Coulomb systems. Lett. Math. Phys. **21**, 139–149

Baumgartner, B. (1984): On the Thomas-Fermi-von-Weizsäcker and Hartree energies as functions of the degree of ionization. J. Phys. **A17**, 1593–1602

Benguria, R., Lieb, E.H. (1983): Proof of the stability of highly negative ions in the absence of the Pauli principle. Phys. Rev. Lett. **50**, 1771–1774

Cycon, H.L., Froese, R.G., Kirsch, W., Simon, B. (1987): Schrödinger operators. Springer, Berlin Heidelberg New York

Enss, V. (1985): Quantum scattering theory for two- and three-body systems with potentials of short- and long-range. In: Schrödinger Operators (S. Graffi, ed.). (Lecture Notes in Mathematics, vol. 1159.) Springer, Berlin Heidelberg New York

Fefferman, C., Seco, L. (1990a): Asymptotic neutrality of large ions. Comm. Math. Phys. **128**, 109–130

Fefferman, C., Seco, L. (1990b): On the energy of a large atom. Bulletin AMS **23**, 525–530

Graf, G.-M. (1990): Asymptotic completeness for N-body short-range quantum systems: a new proof. Comm. Math. Phys. **132**, 73–101

Hughes, W. (1990): An atomic energy lower bound that gives Scott's correction. Adv. Math. **79**, 213

Ivrii, V. (1986): Weyl's asymptotic for the Laplace-Beltrami operator in Riemann polyhedra. Dokl. Akad. Nauk SSSR **38**, 35–38

Ivrii, V., Sigal, I.M. (1990): Asymptotics on the ground state energies of large Coulomb systems. Ann. Math. (submitted)

Kitada, H. (1990): On the completeness of N-body wave operators. Preprint, Tokyo

Lieb, E. (1981): Thomas-Fermi and related theories of atoms and molecules. Rev. Modern Phys. **53**

Lieb, E., Sigal, I.M., Simon, B., Thirring, W. (1988): Approximate neutrality of large Z-ions. Comm. Math. Phys. **116**, 635–644

Lieb, E., Simon, B. (1977): Thomas-Fermi theory of atoms, molecules and solids. Adv. Math. **23**, 22–116

Mourre, E. (1981): Absence of singular continuous spectrum of certain self-adjoint operators. Comm. Math. Phys. **78**, 391–408

Perry, P., Sigal, I.M., Simon, B. (1981): Spectral analysis of N-body Schrödinger operators. Ann. Math. **114**, 519–567

Ruskai, M.B. (1982): Absence of discrete spectrum of highly negative ions. Comm. Math. Phys. **82**, 457–469

Seco, L., Sigal, I.M., Solovej, J.-P. (1990): Bound on the ionization energy of large atoms. Comm. Math. Phys. **131**, 307–315

Siedentop, H., Weikard, R. (1987): On the leading energy correction for the statistical model of the atom: interacting case. Comm. Math. Phys. **112**, 471–490

Siedentop, H., Weikard, R. (1989): On the leading correction of the Thomas-Fermi model: lower bound. Invent. math. **97**, 159–193

Sigal, I.M. (1982): Geometric methods in the quantum many-body problem. Nonexistence of very negative ions. Comm. Math. Phys. **85**, 309–324

Sigal, I.M. (1989): Geometrical theory of resonances in multi-particle systems. In: Proceedings IX Congress Math. Phys. (B. Simon et al., eds.). Adam Hilger

Sigal, I.M., Soffer, A. (1987): The N-particle scattering problem: asymptotic completeness for short-range systems. Ann. Math. **125**, 35–108

Sigal, I.M., Soffer, A. (1990a): Long-range many-body scattering. Asymptotic clustering for Coulomb-type potential. Invent. math. **99**, 115–143

Sigal, I.M., Soffer, A. (1990b): Asymptotic completeness for 4 and 5 particle systems with the Coulomb-type potentials. University of Toronto. Preprint

Solovej, J.-P. (1990): Asymptotics for bosonic atoms. Lett. Math. Phys. **20**, 165–172

Tamura, H. (1990): Asymptotic completeness for N-body Schrödinger operators ... Ibaraki. Preprint

Moduli of Stable Curves, Conformal Field Theory and Affine Lie Algebras

Akihiro Tsuchiya

Department of Mathematics, School of Science, Nagoya University, Nagoya 464-01, Japan

§1. Introduction

Conformal field theory was initiated by Belavin, Polyakov and Zamolodchikov [1] as 2-dimensional quantum field theory describing the 2-dimensional critical phenomena. Conformal field theory is characterized by infinite-dimensional symmetries such as Virasoro algebra, and its correlation functions are characterized by differential equations arising from representations of infinite-dimensional Lie algebras.

In this article, we report our works with Y. Kanie [12], K. Ueno and Y. Yamada [13] concerning the construction of conformal field theory on the universal family of stable curves under the gauge symmetries associated with integrable representations of Lie algebras.

We realize it by constructing a coherent $\mathscr{O}_{M^{(1)}_{g,N}}$-module $\mathscr{V}_\Lambda$ over the modular stack $M^{(1)}_{g,N}$ of N-pointed stable curves of genus g with first order infinitesimal structure, the sheaf of twisted first order differential operators $\mathscr{D}^1_{M^{(1)}_{g,N}}(-\log D^{(1)}_{g,N} : c_v)$, which is the geometric counterpart of the Virasoro algebra, and the action of $\mathscr{D}^1_{M^{(1)}_{g,N}}(-\log D^{(1)}_{g,N} : c_v)$ on $\mathscr{V}_\Lambda$. The solution sheaf of $\mathscr{V}_\Lambda$ gives what physicists call the current-conformal block which is the most fundamental object in conformal field theory. Moreover it turns out to be locally free and the factorization property at the normal crossing divisors $D^{(1)}_{g,N}$ holds.

The monodromy of these solution sheaves give the representations of the central extension $\widehat{\Gamma}_{g,N}$ of the mapping class group $\Gamma_{g,N}$, by the multiplicative subgroup K of $\mathbf{C}^*$ generated by $e^{2\pi\sqrt{-1}c_v/24}$.

The construction of conformal blocks was also done using a quite different method (topological point of view) by G. Moore and N. Seiberg [9].

§2. Integrable Modules of Affine Lie Algebras and the Sugawara Form

Let $\mathfrak{g}$ be a simple Lie algebra over $\mathbf{C}$. We fix a Cartan subalgebra $\mathfrak{h}$, a simple root system Σ, and the invariant bilinear form $(\ ,\ )$ on $\mathfrak{g}$ normalized by $(\theta,\theta) = 2$

Proceedings of the International Congress of Mathematicians, Kyoto, Japan, 1990

for the highest root θ. Let P_+ denote the set of dominant integral weights of $\mathfrak{g}$ for $\lambda \in P_\ell$. We denote by V_λ the irreducible $\mathfrak{g}$-module with highest weight λ. And V_0 is the one dimensional trivial $\mathfrak{g}$-module, $V_{\lambda^\dagger}$ denotes the contragradient $\mathfrak{g}$-module. For each positive integer ℓ, we define the finite subset P_ℓ of P_+ by $\{\lambda \in P_\ell; 0 \le (\theta, \lambda) \le \ell\}$. We fix an orthonormal basis $\{J_a\}$ of $\mathfrak{g}$.

Let $\widehat{\mathfrak{g}} = \mathfrak{g} \otimes \mathbf{C}((\xi)) \oplus \mathbf{C}c$ denote the associated affine Lie algebra of $\mathfrak{g}$. For $X \in \mathfrak{g}$, $f \in \mathbf{C}((\xi))$ and $n \in Z$, we set $X[f] = X \otimes f$ and $X(n) = X \otimes \xi^n$.

In the sequel we fix a positive integer ℓ. The set P_ℓ parametrizes the integrable highest weight $\widehat{\mathfrak{g}}$-modules of level ℓ, and for each $\lambda \in P_\ell$ we denote the associated integrable $\widehat{\mathfrak{g}}$-module by $\mathscr{H}_\lambda$. For each $\lambda \in P_\ell$, put $\Delta_\lambda = (\lambda, \lambda + 2\varrho)/2(\ell + g^*)$, where ϱ is half the sum of the positive roots of $\mathfrak{g}$.

For each $n \in \mathbf{Z}$, we define the Sugawara operator $T(n)$ acting on $\mathscr{H}_\lambda$ by

$$T(n) = \frac{1}{2(\ell + g*)} \sum_a \sum_{k \in Z} {}^{\circ}_{\circ} J_a(k) J_a(n-k) {}^{\circ}_{\circ} \tag{2.1}$$

where the symbol ${}^{\circ}_{\circ} \quad {}^{\circ}_{\circ}$ denotes the so-called normal ordering and $g*$ denotes the dual Coxeter number of $\mathfrak{g}$. For an element $\ell = \sum_n b_n \xi^{n+1} d/d\xi \in \mathbf{C}((\xi))d/d\xi$, define an operator $T[\ell]$ on $\mathscr{H}_\lambda$ by $T[\ell] = -\sum_n b_n T(n)$. Then we have the following fundamental relations of the operators on $\mathscr{H}_\lambda$.

$$\begin{aligned} [X[f], Y[g]] &= [X, Y][fg] + \ell(X, Y)\,\mathrm{Res}(df \cdot g)\,\mathrm{id} \\ [T[\ell], X[f]] &= X[\ell(f)] \\ [T[\ell_1], T[\ell_2]] &= T[[\ell_1, \ell_2]] + \frac{c_v}{12} \mathrm{Res} \frac{d^3 \ell_1}{d\xi^3} \ell_2 d\xi\, \mathrm{id} \end{aligned} \tag{2.2}$$

where $c_v = \ell \dim \mathfrak{g}/(\ell + g*)$.

Consider the automorphism group $\mathscr{D}$ of the $\mathbf{C}$-algebra $\mathbf{C}[[\xi]]$. Then an element h of $\mathscr{D}$ is represented by a formal power series $h(\xi) = a_0\xi + a_1\xi^2 + \cdots$, $a_0 \neq 0$. For each $n \ge 0$, define the normal subgroup $\mathscr{D}_n$ of $\mathscr{D}$ by $\mathscr{D}_n = \{h(\xi) = \xi + a_n\xi^{n+1} + \cdots\}$. Since each element h of $\mathscr{D}_1$ is uniquely written as $h = \exp(\ell)$, $\ell \in \mathbf{C}[[\xi]]\xi^2 d/d\xi$, we define the operator $G[h]$ of $\mathscr{H}_\lambda$ by $G[h] = \exp(T[\ell])$. These operators $G[h]$ define a representation of $\mathscr{D}_1$ on $\mathscr{H}_\lambda$.

§3. Local Universal Family of N-Pointed Stable Curves of Genus g

Let (g, N) be a pair of non-negative integers with $2g - 2 + N \ge 1$. Consider a local universal family of N-pointed connected stable curves of genus g.

$$\mathscr{F} = (\pi : C \to B : s_1, \cdots, s_N) \tag{3.1}$$

where $s_j : B \to C$ are cross-sections of π, and set $S_j = s_j(B)$ and $S = \bigcup_j S_j$.

Let D be the discriminant locus of $\pi : C \to B$. Then D is a normal crossing divisor of B, and B is a $3g - 3 + N$ dimensional complex manifold.

For each non-negative integer n or $n = \infty$ we consider the associated local universal family of N-pointed stable curves of genus g with n-th infinitesimal neighborhoods.

$$\mathcal{F}^{(n)} = (\pi^{(n)} : C^{(n)} \longrightarrow B^{(n)} : s_1^{(n)}, \cdots, s_N^{(n)}; t_1^{(n)}, \cdots, t_N^{(n)}) \tag{3.2}$$

where $t_j^{(n)} : \mathcal{O}_{B^{(n)}}[[\xi]]/(\xi^{n+1}) \to \mathcal{O}_{C^{(n)}}/I_{S_j^{(n)}}^{n+1}$ are $\mathcal{O}_{B^{(n)}}$-algebra isomorphisms.

We get the sequence of fibering

$$B = B^{(0)} \longleftarrow B^{(1)} \longleftarrow \cdots \longleftarrow B^{(\infty)} \tag{3.3}$$

where $p_m^n = p : B^{(n)} \to B^{(m)}$ is a principal fibering with structure group $\mathcal{D}_m^{n \oplus N}$, where $\mathcal{D}_m^n = \mathcal{D}_m/\mathcal{D}_n$.

By the local universality of the system $\mathcal{F}^{(n)}$, we have the following isomorphism of $\mathcal{O}_{B^{(n)}}$-modules.

$$\varrho_n : \Theta_{B^{(n)}}(-\log D^{(n)}) \xrightarrow{\sim} R^1\pi_*^{(n)}(\Theta_{C^{(n)}/B^{(n)}}(-(n+1)S^{(n)})) \tag{3.4}$$

where $D^{(n)} = p^{-1}(D)$, $S_j^{(n)} = s_j^{(n)}(B^{(n)})$, $S^{(n)} = \cup S_j^{(n)}$.

With each local universal family $\mathcal{F} = (\pi : C \to B : s_1, \cdots, s_N)$, we associate the following $\mathcal{O}_{B^{(n)}}$-modules

$$\begin{aligned} K_{\widehat{S}_j^{(n)}} &= \varprojlim_m \mathcal{O}_{C^{(n)}}(*S_j^{(n)})/I_{S_j^{(n)}}^{m+1} \\ \Theta_{\widehat{S}_j^{(n)}/B^{(n)}}(*) &= \mathcal{D}er_{\mathcal{O}_{B^{(n)}}}(K_{\widehat{S}_j^{(n)}}, K_{\widehat{S}_j^{(n)}}). \end{aligned} \tag{3.5}$$

The $\mathcal{O}_{B^{(n)}}$-module $\Theta_{\widehat{S}_j^{(n)}/B^{(n)}}(*)$ has the canonical Lie algebra structure over $\mathcal{O}_{B^{(n)}}$. We denote its Lie bracket by $[\ ,\ ]_0$.

For $n = \infty$ we have the following canonical $\mathcal{O}_{B^{(\infty)}}$-module isomorphisms.

$$\begin{aligned} \widetilde{t}_j^{(\infty)} : \mathcal{O}_{B^{(\infty)}}((\xi_j)) &\xrightarrow{\sim} K_{\widehat{S}_j^{(\infty)}} \\ \widetilde{t}_j^{(\infty)} : \mathcal{O}_{B^{(\infty)}}((\xi_j))\frac{d}{d\xi_j} &\xrightarrow{\sim} \Theta_{\widehat{S}_j^{(\infty)}/B^{(\infty)}}(*). \end{aligned} \tag{3.6}$$

We have the following canonical inclusion mappings.

$$\begin{aligned} \pi_*^{(n)}\mathcal{O}_{C^{(n)}}(*S^{(n)}) &\hookrightarrow \sum_{j=1}^N K_{\widehat{S}_j^{(n)}} \\ \pi_*^{(n)}\Theta_{C^{(n)}/B^{(n)}}(*S^{(n)}) &\hookrightarrow \sum_{j=1}^N \Theta_{\widehat{S}_j^{(n)}/B^{(n)}}(*). \end{aligned} \tag{3.7}$$

Now consider the following condition (Q) for the system $\mathcal{F} = (\pi : C \to B : s_1, \cdots, s_N)$.

(Q) For each point $b \in B$, and for each irreducible component C of $C_b = \pi^{-1}(b)$, there exists j with $s_j(b) \in C$.

Proposition 3.1. *Under the condition* (Q) *for* $\mathscr{F}$, *we have the following surjective* $\mathcal{O}_{B^{(n)}}$*-module homomorphism*

$$\theta : \sum_{j=1}^{N} \Theta_{\widehat{S}_j^{(n)}/B^{(n)}}(*) \longrightarrow \Theta_{B^{(n)}}(-\log D^{(n)}). \tag{3.8}$$

Furthermore, we introduce the Lie bracket [,] *on* $\sum_{j=1}^{N} \Theta_{S_j^{(n)}/B^{(n)}}(*)$ *by*

$$[v_1, v_2] = [v_1, v_2]_0 + \theta(v_1)(v_2) - \theta(v_2)(v_1). \tag{3.9}$$

Then the map θ *is a homomorphism of Lie algebras.*

As the group $\mathscr{D}^{\oplus N}$ acts on $B^{(\infty)}$, the induced action of $h = (h^1, \cdots, h^N) \in \mathscr{D}^{\oplus N}$ on $f = \sum_{j=1}^{N} \sum_n a_n^j \xi_j^n \in \sum_{j=1}^{N} \mathcal{O}_{B^{(\infty)}}((\xi_j))$ and $\ell = \sum_{j=1}^{N} \sum_n b_n^j \xi_j^{n+1} d/d\xi_j \in \sum_{j=1}^{N} \mathcal{O}_{B^{(\infty)}}((\xi_j)) d/d\xi_j$ are defined by $\pi(h)(f) = \sum_j \sum_n \varrho(h)(a_n^j) h^j(\xi_j^{n+1})$ and $\pi(h)(\ell) = \sum_j \sum_n \varrho(h)(b_n^j) \operatorname{Ad}(h^j)(\xi_j^{n+1} d/d\xi_j)$ respectively.

Proposition 3.4. *The action of* $\mathscr{D}^{\oplus N}$ *preserves the subspace* $\pi_*^{(\infty)} \mathcal{O}_{B^{(\infty)}}(*S^{(\infty)})$ *of* $\sum_{j=1}^{N} \mathcal{O}_{B^{(\infty)}}((\xi_j))$.

We remark that the invariant part of the action of $\mathscr{D}_n^{\oplus N}$ on $\sum_{j=1}^{N} \mathcal{O}_{B^{(\infty)}}((\xi_j))$ and $\sum_{j=1}^{N} \mathcal{O}_{B^{(\infty)}}((\xi_j)) \frac{d}{d\xi_j}$ are $\sum_{j=1}^{N} K_{\widehat{S}_j^{(n)}}$ and $\sum_{j=1}^{N} \Theta_{\widehat{S}_j^{(n)}/B^{(n)}}(*)$ respectively.

§4. Sheafification and the Module $\mathscr{V}_{\vec{\lambda}}$

In this section, we assume the condition (Q) for any local universal family $\mathscr{F}$.

With each system $\mathscr{F} = (\pi : C \to B : s_1, \cdots, s_N)$, we associate the $\mathcal{O}_{B^{(\infty)}}$-Lie algebra sheaves

$$\begin{aligned} \widehat{\mathfrak{g}}(\mathscr{F}^{(n)}) &= \sum_{j=1}^{n} \mathfrak{g} \otimes K_{\widehat{S}_j^{(n)}} \oplus \mathcal{O}_{B^{(n)}} c \\ \widehat{\mathfrak{g}}_-(\mathscr{F}^{(n)}) &= \mathfrak{g} \otimes \pi_*^{(n)} \mathcal{O}_{C^{(n)}}(*S^{(n)}) \subseteq \widehat{\mathfrak{g}}(\mathscr{F}^{(n)}). \end{aligned} \tag{4.1}$$

The Lie algebra structure is given by

$$\left[\sum_{j=1}^{n} X_j \otimes f_j, \sum_{j=1}^{n} Y_j \otimes g_j \right] = \sum_{j=1}^{n} [X_j, Y_j] \otimes f_j g_j + \sum_{j=1}^{n} \operatorname{Res}(g_j df_j)(X_j, Y_j) c. \tag{4.2}$$

Then $\widehat{\mathfrak{g}}_-(\mathscr{F}^{(n)})$ is a Lie subalgebra of $\widehat{\mathfrak{g}}(\mathscr{F}^{(n)})$. In the case of $n = \infty$, we have

$$\widehat{\mathfrak{g}}(\mathscr{F}^{(\infty)}) = \sum_{j=1}^{N} \mathfrak{g} \otimes \mathcal{O}_{B^{(\infty)}}((\xi_j)) \oplus \mathcal{O}_{B^{(\infty)}} c. \tag{4.3}$$

The Lie group $\mathscr{D}^{\oplus N}$ acts on $\widehat{\mathfrak{g}}(\mathscr{F}^{(\infty)})$ as Lie algebra automorphisms, and the sub-algebra $\widehat{\mathfrak{g}}_{-}(\mathscr{F}^{(\infty)})$ is preserved by the action of $\mathscr{D}^{\oplus N}$.

For each $\vec{\lambda} = (\lambda_1, \cdots, \lambda_N)$, a quasi-coherent $\mathscr{O}_{B^{(\infty)}}$-module $\mathscr{H}_{\vec{\lambda}}(\mathscr{F}^{(\infty)})$ is defined by

$$\mathscr{H}_{\vec{\lambda}}(\mathscr{F}^{(\infty)}) = \mathscr{O}_{B^{(\infty)}} \otimes \mathscr{H}_{\vec{\lambda}}, \quad \mathscr{H}_{\vec{\lambda}} = \mathscr{H}_{\lambda_1} \otimes \cdots \otimes \mathscr{H}_{\lambda_N}. \tag{4.4}$$

Then the Lie algebra sheaf $\widehat{\mathfrak{g}}(\mathscr{F}^{(\infty)})$ acts on $\mathscr{H}_{\vec{\lambda}}(\mathscr{F}^{(\infty)})$ $\mathscr{O}_{B^{(\infty)}}$-linearly. Define the $\mathscr{O}_{B^{(\infty)}}$-modules $\mathscr{H}'_{\vec{\lambda}}(\mathscr{F}^{(\infty)})$ and $\mathscr{V}_{\vec{\lambda}}(\mathscr{F}^{(\infty)})$ by

$$\begin{aligned} \mathscr{H}'_{\vec{\lambda}}(\mathscr{F}^{(\infty)}) &= \widehat{\mathfrak{g}}_{-}(\mathscr{F}^{(\infty)}) \cdot \mathscr{H}_{\vec{\lambda}}(\mathscr{F}^{(\infty)}) \\ \mathscr{V}_{\vec{\lambda}}(\mathscr{F}^{(\infty)}) &= \mathscr{H}_{\vec{\lambda}}(\mathscr{F}^{(\infty)}) / \mathscr{H}'_{\vec{\lambda}}(\mathscr{F}^{(\infty)}). \end{aligned} \tag{4.5}$$

Since the Lie group $\mathscr{D}_1^{\oplus N}$ acts on $\mathscr{H}_{\vec{\lambda}}(\mathscr{F}^{(\infty)})$, preserving $\mathscr{H}'_{\vec{\lambda}}(\mathscr{F}^{(\infty)})$, the $\mathscr{O}_{B^{(\infty)}}$-module $\mathscr{V}_{\vec{\lambda}}(\mathscr{F}^{(\infty)})$ has the structure of a $\mathscr{D}_1^{\oplus N}$-module.

Define the $\mathscr{O}_{B^{(1)}}$-modules $\mathscr{H}_{\vec{\lambda}}(\mathscr{F}^{(1)})$, $\mathscr{H}'_{\vec{\lambda}}(\mathscr{F}^{(1)})$ and $\mathscr{V}_{\vec{\lambda}}(\mathscr{F}^{(1)})$ as the invariant part of the $\mathscr{D}_1^{\oplus N}$-actions on $\mathscr{H}_{\vec{\lambda}}(\mathscr{F}^{(\infty)})$, $\mathscr{H}'_{\vec{\lambda}}(\mathscr{F}^{(\infty)})$ and $\mathscr{V}_{\vec{\lambda}}(\mathscr{F}^{(\infty)})$ respectively. Then the $\mathscr{O}_{B^{(1)}}$-module $\mathscr{H}_{\vec{\lambda}}(\mathscr{F}^{(1)})$ has a canonical structure of $\widehat{\mathfrak{g}}(\mathscr{F}^{(1)})$-module and we have the following canonical $\mathscr{O}_{B^{(1)}}$-module isomorphism.

$$\begin{aligned} \mathscr{H}'_{\vec{\lambda}}(\mathscr{F}^{(1)}) &\simeq \widehat{\mathfrak{g}}_{-}(\mathscr{F}^{(1)}) \mathscr{H}_{\vec{\lambda}}(\mathscr{F}^{(1)}) \\ \mathscr{V}_{\vec{\lambda}}(\mathscr{F}^{(1)}) &\simeq \mathscr{H}_{\vec{\lambda}}(\mathscr{F}^{(1)}) / \mathscr{H}'_{\vec{\lambda}}(\mathscr{F}^{(1)}). \end{aligned} \tag{4.6}$$

The $\mathscr{O}_{B^{(1)}}$-module $\mathscr{V}_{\vec{\lambda}}(\mathscr{F}^{(1)})$ is the most fundamental object in conformal field theory. Our first main theorem is the following one.

Theorem I. *The module $\mathscr{V}_{\vec{\lambda}}(\mathscr{F}^{(1)})$ is a coherent $\mathscr{O}_{B^{(1)}}$-module.*

For the proof of this theorem, we use Gabber's theorem on the involutiveness of the characteristic variety [Gabber 5].

§5. Differential Equations Defining Conformal Blocks

In this section we also assume the condition (Q), unless otherwise stated. Now to each $\mathscr{F}$, we associate the following sheaf version of the Virasoro algebra.

$$\mathrm{Vir}(\mathscr{F}^{(\infty)} : c_v) = \sum_{j=1}^{N} \mathscr{O}_{B^{(\infty)}}((\xi_j)) \frac{d}{d\xi_j} \oplus \mathscr{O}_{B^{(\infty)}} \tag{5.1}$$

with the Lie bracket

$$[(\ell, r), (m, s)] = \left([\ell, m], \frac{c_v}{12} \sum_{j=1}^{N} \mathrm{Res}_{\xi_j = 0} \left(\frac{d^3 \ell_j}{d\xi_j^3} m_j d\xi_j \right) + \theta(\ell)(s) - \theta(m)(r) \right) \tag{5.2}$$

where

$$\ell = \left(\ell_1 \frac{d}{d\xi_1}, \cdots, \ell_N \frac{d}{d\xi_N}\right), \quad m = \left(m_1 \frac{d}{d\xi_1}, \cdots, m_N \frac{d}{d\xi_N}\right)$$
$$\in \sum_{j=1}^{N} \mathcal{O}_{B^{(\infty)}}((\xi_j)) \frac{d}{d\xi_j} \text{ and } r, s \in \mathcal{O}_{B^{(\infty)}} .$$

The group $\mathcal{D}^{\oplus N}$ acts on $\mathrm{Vir}(\mathcal{F}^{(\infty)} : c_v)$ as Lie algebra automorphisms as follows. For $h = (h_1, \cdots, h_N) \in \mathcal{D}^{\oplus N}$, $v = (\ell, r) \in \mathrm{Vir}(\mathcal{F}^{\infty)}, c_v)$, $\ell = (\ell_1 d/d\xi_1, \cdots, \ell_N d/d\xi_N)$,

$$\pi(h)(\ell, r) = (\pi(h)(l), \varrho(h)(r)$$
$$+ \frac{c_v}{12} \sum_{j=1}^{N} \mathrm{Res}_{\xi_j=0}(\pi(h)(\ell_j)\{h_j(\xi_j), \xi_j\} \left(\frac{dh_j(\xi_j)}{d\xi_j}\right)^{-1} d\xi_j)),$$

where $\{h_j(\xi_j), \xi_j\}$ denotes the Schwarzian derivative. Taking the $\mathcal{D}_1^{\oplus N}$ invariant part, we get the following exact sequence of Lie algebras.

$$0 \longrightarrow \mathcal{O}_{B^{(1)}} \longrightarrow \mathrm{Vir}(\mathcal{F}^{(1)} : c_v) \longrightarrow \sum_{j=1}^{N} \Theta_{\widehat{S}_j^{(1)}/B^{(1)}}(*) \longrightarrow 0 . \tag{5.3}$$

The Lie algebra sheaf $\mathrm{Vir}(\mathcal{F}^{(\infty)}; c_v)$ acts on $\mathcal{H}_{\vec{\lambda}}(\mathcal{F}^{(\infty)})$ by the following formula

$$D(v)(F \otimes |\Phi\rangle) = \theta(\ell)(F) \otimes \Phi + F \otimes \sum_{j=1}^{N} \varrho_j(T[\ell_j])|\Phi\rangle + rF \otimes |\Phi\rangle \tag{5.4}$$

where $F \in \mathcal{O}_{B^{(\infty)}}$, $|\Phi\rangle \in \mathcal{H}_{\vec{\lambda}}$, $v = (\ell, r) \in \mathrm{Vir}(\mathcal{F}^{(\infty)}, c_v)$.

Proposition 5.1. *The action of* $\mathrm{Vir}(\mathcal{F}^{(\infty)} : c_v)$ *on* $\mathcal{H}_{\vec{\lambda}}(\mathcal{F}^{(\infty)})$ *preserves* $\mathcal{H}'_{\vec{\lambda}}(\mathcal{F}^{(\infty)})$, *so* $\mathrm{Vir}(\mathcal{F}^{(\infty)} : c_v)$ *acts on* $\mathcal{V}_{\vec{\lambda}}(\mathcal{F}^{(\infty)})$.

Finally taking the invariant part of the action $\mathcal{D}_1^{\oplus N}$, we get an action of the Lie algebra sheaf $\mathrm{Vir}(\mathcal{F}^{(1)} : c_v)$ on $\mathcal{H}_{\vec{\lambda}}(\mathcal{F}^{(1)})$, $\mathcal{H}'_{\vec{\lambda}}(\mathcal{F}^{(1)})$ and $\mathcal{V}_{\vec{\lambda}}(\mathcal{F}^{(1)})$.

Proposition 5.2. *For any system* $\mathcal{F}$, *there exists a unique* $\mathcal{O}_{B^{(\infty)}}$*-module homomorphism,*

$$a^{(\infty)} : \pi_*^{(\infty)} \Theta_{C^{(\infty)}/B^{(\infty)}}(*) \oplus \mathcal{O}_{B^{(\infty)}} \longrightarrow \mathcal{O}_{B^{(\infty)}} \tag{5.5}$$

such that for any $v \in \pi_*^{(\infty)} \Theta_{C^{(\infty)}/B^{(\infty)}}(*) \oplus \mathcal{O}_{B^{(\infty)}}$ *and* $|\Phi\rangle \in \mathcal{V}_{\vec{\lambda}}(\mathcal{F}^{(\infty)})$ *we have,*

$$D(v)|\Phi\rangle = a^{(\infty)}(v)|\Phi\rangle . \tag{5.6}$$

This map $a^{(\infty)}$ is expressed, locally on B, in the following way. For $\mathcal{F} = (\pi : C \to B; s_1, \cdots, s_N)$, taking B small enough, there exists an element $\omega \in H^0(C \times_B C, \omega_{C/B}^{\boxtimes 2}(2\Delta))$ such that near the diagonal $\Delta \subseteq C \times_B C$

$$\omega = \frac{dw\,dz}{(w-z)^2} + \text{regular at } \Delta, \tag{5.7}$$

where w and z represent fiber coordinates of $\pi : C \to B$ and $\omega_{C/B}$ denotes the relative dualizing sheaf of π. Define, then, the associated projective connection by $S_\omega(z)dz^2 = -6\lim_{w\to z}\{\omega - dw\,dz/(w-z)^2\}$. Then for $v = ((\ell_1 d/d\xi_1, \cdots, \ell_N d/d\xi_N), r)$ we have

$$a^{(\infty)}(v) = \frac{c_v}{12}\sum_{j=1}^{N} \operatorname{Res}_{\xi_j=0}(\ell_j(\xi_j)S_\omega(\xi_j)d\xi_j) + r\,. \tag{5.8}$$

Taking the $\mathscr{D}_1^{\oplus N}$ invariant part we get the $\mathscr{O}_{B^{(1)}}$-module homomorphism,

$$a : \pi_*^{(1)}\Theta_{C^{(1)}/B^{(1)}}(*S^{(1)}) \oplus \mathscr{O}_{B^{(1)}} \longrightarrow \mathscr{O}_{B^{(1)}}\,. \tag{5.9}$$

The sheaf of twisted first-order differential operators on $B^{(1)}$ is defined by

$$\mathscr{D}^1_{B^{(1)}}(-\log D^{(1)} : c_0) = \operatorname{Vir}(\mathscr{F}^{(1)} : c_v) \Big/ \left(\ker a \oplus \sum_{j=1}^{N} \Theta_{\widehat{S}_j^{(1)}/B^{(1)}}(-2S_j^{(1)})\right). \tag{5.10}$$

Then we have the following exact sequence of Lie algebra sheaves on $B^{(1)}$

$$0 \longrightarrow \mathscr{O}_{B^{(1)}} \longrightarrow \mathscr{D}^1_{B^{(1)}}(-\log D^{(1)} : c_v) \longrightarrow \Theta_{B^{(1)}}(-\log D^{(1)}) \longrightarrow 0\,. \tag{5.11}$$

By Proposition 5.2, we get the following second main theorem of this paper.

Theorem II. *On $\mathscr{V}_{\vec{\lambda}}^{\dagger}(\mathscr{F}^{(1)})$, the sheaf of Lie algebras $\mathscr{D}^1_{B^{(1)}}(-\log D^{(1)} : c_v)$ acts as twisted first order differential operators.*

As a corollary of this theorem we have

Corollary 5.3. *On $B^{(1)} - D^{(1)}$, the coherent $\mathscr{O}_{B^{(1)}}$-module $\mathscr{V}_{\vec{\lambda}}^{\dagger}(\mathscr{F}^{(1)})$ is locally free.*

The sheaf of twisted first order differential operators $\mathscr{D}^1_{B^{(1)}}(-\log D : c_v)$ can be trivialized locally on B as follows. Take $\omega \in H^0(C \times_B C, \omega_{C/B}^{\boxtimes}(2\Delta))$ satisfying the property (5.7). Then we can associate canonically the $\mathscr{O}_{B^{(1)}}$-module isomorphism

$$A_\omega = A_\omega(\mathscr{F}) : \Theta_{B^{(1)}}(-\log D^{(1)} : c_v) \oplus \mathscr{O}_{B^{(1)}} \longrightarrow \mathscr{D}^1_{B^{(1)}}(-\log D^{(1)} : c_v) \tag{5.12}$$

which is indeed a Lie algebra isomorphism.

By this isomorphism $A_\omega(\mathscr{F})$, the sheaf $\mathscr{O}_{B^{(1)}}$ has a structure of a $\mathscr{D}^1_{B^{(1)}}(-\log D^{(1)} : c_v)$ module. We denote this $\mathscr{D}^1_{B^{(1)}}(-\log D^{(1)} : c_v)$ module by $\mathscr{L}_\omega(\mathscr{F}^{(1)})$. Then we can consider the solution sheaf,

$$\mathscr{H}om_{\mathscr{D}^1_{B^{(1)}}(-\log D^{(1)}:c_v)}(\mathscr{V}_{\vec{\lambda}}^{\dagger}(\mathscr{F}^{(1)}), \mathscr{L}_\omega(\mathscr{F}^{(1)}))\,. \tag{5.13}$$

A fiber of this sheaf is nothing but what physicists call the space of conformal blocks.

Finally in this section, we remark how to define the sheaves $\mathscr{V}_{\vec{\lambda}}(\mathscr{F}^{(1)})$ and $\mathscr{D}^1_{B^{(1)}}(-\log D^{(1)} : c_v)$ in the case when the condition (Q) is not satisfied. For any local universal family $\mathscr{F} = (\pi : C \to B; s_1, \cdots, s_N)$, not necessarily satisfying the condition (Q), we can take a local universal family of $(N+k)$ pointed stable curves $\mathscr{F}' = (\pi' : C' \to B'; s'_1, \cdots, s'_{N+k})$ satisfying the condition (Q) and surjective smooth maps $F : C' \to \mathbf{C}$ and $f : B' \to B$ with following properties: 1) $f \circ \pi' = \pi \circ F$ 2) $Fs'_j = s_j f, j = 1, \cdots, N$ 3) for any $b' \in B'$, put $b = f(b')$, $C_{b'} = \pi'^{-1}(b')$, $C_b = \pi^{-1}(b)$. Then, the map $F_{b'} : (C_{b'} : s'_1(b'), \cdots, s'_N(b')) \to (C_b; s_1(b), \cdots, s_N(b))$ is an isomorphism of N-pointed stable curves. Note that $D' = f^{-1}(D)$.

For $\vec{\lambda} \in P_\ell^N$, put $\vec{\lambda} = (\vec{\lambda}, 0, \cdots, 0) \in P_\ell^{N+k}$. Then it can be shown that there exists a Lie algebra sheaf $\mathscr{D}^1_{B^{(1)}}(-\log D^{(1)} : c_v)$ on $B^{(1)}$, and a coherent $\mathscr{O}_{B^{(1)}}$-module $\mathscr{V}_{\vec{\lambda}}(\mathscr{F}^{(1)})$ with the action of $\mathscr{D}^1_{B^{(1)}}(-\log D^{(1)} : c_v)$ and canonical isomorphisms $f^*\mathscr{D}^1_{B^{(1)}}(-\log D^{(1)} : c_v) \to \mathscr{D}^1_{B'^{(1)}}(-\log D'^{(1)} : c_v)$ and $f^*\mathscr{V}_{\vec{\lambda}}(\mathscr{F}^{(1)}) \to \mathscr{V}_{\vec{\lambda}'}(\mathscr{F}'^{(1)})$. The objects $\mathscr{D}^1_{B^{(1)}}(-\log D^{(1)} : c_v)$ and $\mathscr{V}_{\vec{\lambda}}(\mathscr{F}^{(1)})$ do not depend on the choice of $\mathscr{F}'$, and these are the objects we wanted to define.

§6. Local Freeness and Factorization

Here we study the behavior of the $\mathscr{D}^1_{B^{(1)}}(-\log D^{(1)} : c_v)$ module $\mathscr{V}_{\vec{\lambda}}(\mathscr{F}^{(1)})$ near the discriminant locus $D^{(1)}$. Since the problem is local, we can take $\mathscr{F} = (\pi : C \to B; s_1, \cdots, s_N)$ with condition (Q), with coordinate $(\tau_1, \cdots, \tau_M)$ on B, $M = 3g - 3 + N$, so that the discriminant locus is of the form $D = D_1 \cup \cdots \cup D_k$, $D_i = \{(\tau) \in B; \tau_i = 0\}, i = 1, \cdots, k$. Set $E = \bigcap_{j=1}^k D_j$, $E^{(1)} = \bigcap_{j=1}^k D_j^{(1)}$, and denote by $\pi_E : C_E \to E$ the restriction of $\pi : C \to B$ on E. Let $\widetilde{\pi}_E : \widetilde{C}_E \to E$ be the simultaneous normalization of $\pi_E : C_E \to E$, and $\sigma'_p, \sigma''_p : E \to \widetilde{C}_E$ be the cross-sections corresponding to the normalized double points, $p = 1, \cdots, k$. Then the family $\widetilde{\mathscr{F}}_E = \widetilde{\pi} : \widetilde{C}_E \to E; \sigma'_p, \sigma''_p, (p = 1, \cdots, k), s_1, \cdots, s_N)$ is a local universal family of $(N + 2k)$-pointed (not necessary connected fiber) nonsingular curves satisfying the condition (Q). Put $\widetilde{\mathscr{F}}_E^{(1)} = \widetilde{\pi}_E^{(1)} : \widetilde{C}_E^{(1)} \to \widetilde{E}^{(1)})$, the associated family of 1-structure. Then the canonical map $\widetilde{E}^{(1)} \to E^{(1)}$ is a $(\mathbf{C}^*)^{2k}$-principal fibering.

Take $\omega \in H^0(C \times_B C, \omega_{C/B}^{\boxtimes 2}(2\Delta))$ with the condition (5.7) as well as the property $i_E^*(\omega) \in H^0(\widetilde{C}_E \times_E \widetilde{C}_E, \omega_{\widetilde{C}_{E/E}}^{\boxtimes 2}(2\Delta))$. Using this element ω, we fix the trivializations, $\Theta_{B^{(1)}}(-\log D^{(1)}) \oplus \mathscr{O}_{B^{(1)}} \to \mathscr{D}^1_{B^{(1)}}(-\log D^{(1)} : c_v)$ and $\Theta_{\widetilde{E}^{(1)}} \oplus \mathscr{O}_{\widetilde{E}^{(1)}} \to D^1_{\widetilde{E}^{(1)}}(c_v)$.

Now we have the following fundamental theorem, which we call the sewing process.

Theorem III. *Fix $\vec{\lambda} = (\lambda_1, \cdots, \lambda_N)$. Then for each $\vec{\mu} = (\mu_1, \cdots, \mu_k) \in P_\ell^k$, and for each*

$$\Phi \in Hom_{\mathscr{O}_{\widetilde{E}^{(1)}}(c_v)}(\mathscr{V}_{(\vec{\mu}, \vec{\mu}^\dagger, \vec{\lambda})}(\widetilde{\mathscr{F}}_E^{(1)}), \mathscr{O}_{\widetilde{E}^{(1)}}) \tag{6.1}$$

we can associate the formal solution having the initial term Φ,

$$\widetilde{\Phi} \in Hom_{\mathscr{O}_{B^{(1)}}}(\mathscr{V}_{\vec{\lambda}}(\mathscr{F}^{(1)}), \mathscr{O}_{E^{(1)}}[[\tau_1, \cdots, \tau_k]]\tau^\Delta \vec{\mu}) \tag{6.2}$$

where $\tau^{\Delta\vec{\mu}} = \tau_1^{\Delta\mu_1}, \cdots, \tau_k^{\Delta\mu_k}$, $\vec{\mu}^\dagger = (\mu_1^\dagger, \cdots, \mu_k^\dagger)$ *and* $\mathcal{O}_{E^{(1)}}[[\tau_1, \cdots, \tau_k]] = \varprojlim_m \mathcal{O}_{B^{(1)}}/I_{E^{(1)}}^{m+1}$.

And we can show that the formal power series $\widetilde{\Phi}$ *converge with respect to the variables* $\tau_1, \cdots, \tau_k$.

Then we have

Theorem IV. *The coherent* $\mathcal{O}_{B^{(1)}}$*-module* $\mathcal{V}_{\vec{\lambda}}^{\dagger}(\mathcal{F}^{(1)})$ *is locally free.*

For each $\mathcal{O}_{B^{(1)}}$-module $\mathcal{N}$ such as $\mathcal{O}_{B^{(1)}}$, $\Theta_{B^{(1)}}(-\log D^{(1)})$, $\mathcal{V}_{\vec{\lambda}}^{\dagger}(\mathcal{F}^{(1)})$, the V-filtration along $E^{(1)}$ is defined as $V_{\mathbf{p}}\mathcal{N} = I_1^{-p_1} \cdots I_k^{-p_k}\mathcal{N}$, for $\mathbf{p} = (p_1, \cdots, p_k) \in \mathbf{Z}^k$ where $I_j = \mathcal{O}_{B^{(1)}}\tau_j$, $j = 1, \cdots, k$.

The associated graded module is defined by $\mathrm{Gr}_*^V \mathcal{N} = \sum_{p\in\mathbf{Z}^k} \mathrm{Gr}_p^V \mathcal{N}$, $\mathrm{Gr}_p^V \mathcal{N} = V_p\mathcal{N} / \sum_{j=1}^k V_{p-e_j}\mathcal{N}$ where $e_j = (0, \cdots, 1, 0, \cdots, 0) \in \mathbf{Z}^k$. Then we have $\mathrm{Gr}_*^V \mathcal{O}_{B^{1)}} = \mathcal{O}_{E^{(1)}}[\tau_1, \cdots, \tau_k]$, $\mathrm{Gr}_*^V \Theta_{B^{(1)}}(-\log D^{(1)}) = (\sum_{p=1}^k \mathcal{O}_{E^{(1)}}\tau_p\partial/\partial\tau_p + \Theta_{E^{(1)}}) \otimes \mathbf{C}[\tau_1, \cdots, \tau_k]$ where $\deg\tau_j = -e_j$, and $\deg \tau_j\partial/\partial\tau_j = 0$, $j = 1, \cdots, k$.

Theorem V. 1) *There exists a canonical* $\mathcal{O}_{\widetilde{E}^{(1)}}$*-module isomorphism*

$$\mathrm{Gr}_0^V \mathcal{V}_{\vec{\lambda}}^{\dagger}(\mathcal{F}^{(1)}) \otimes_{\mathcal{O}_{E^{(1)}}} \mathcal{O}_{\widetilde{E}^{(1)}} \simeq \sum_{\vec{\mu}\in P_\ell^k} \mathcal{V}_{(\vec{\mu},\vec{\mu}^\dagger,\vec{\lambda})}^{\dagger}(\widetilde{\mathcal{F}}_E^{(1)}) .$$

2) *There exists a canonical isomorphism of* $\mathrm{Gr}_*^V \Theta_{B^{(1)}}(-\log D^{(1)}) \oplus \mathrm{Gr}_*^V \mathcal{O}_{B^{(1)}}$ *modules*

$$\mathrm{Gr}_0^V \mathcal{V}_{\vec{\lambda}}^{\dagger}(\mathcal{F}^{(1)}) \otimes \mathbf{C}[\tau_1, \cdots, \tau_k] \simeq \mathrm{Gr}_*^V \mathcal{V}_{\vec{\lambda}}^{\dagger}(\mathcal{F}^{(1)}) .$$

Let $k_\theta = \mathbf{C}E_\theta \oplus \mathbf{C}E_{-\theta} \oplus \mathbf{C}H_\theta$ denote the principal 3-dimensional subalgebra of $\mathfrak{g}$, and for each $\lambda \in P_+$, let $V_\lambda = \sum_{j\in\frac{1}{2}\mathbf{Z}_{\geq 0}} V_{\lambda,j}$ denote the homogeneous decomposition of the k_θ module V_λ into $(2j+1)$-dimensional irreducible components.

For each $(\lambda, \mu, \nu) \in P_\ell^3$, define

$$\begin{aligned} W_{\lambda,\mu,\nu} =&\{\varphi \in Hom_{\mathfrak{g}}(V_\lambda \otimes V_\mu \otimes V_\nu, \mathbf{C}), \\ &\varphi|_{V_{\lambda,j_1}\otimes V_{\mu,j_2}\otimes V_{\nu,j_3}} = 0 \quad \text{if} \quad j_1 + j_2 + j_3 > \ell\} \end{aligned} \tag{6.3}$$

and set $N_{\lambda,\mu,\nu} = \dim_{\mathbf{C}} W_{\lambda,\mu,\nu}$. Then we have

Proposition 6.1. *For each* $\vec{\lambda} \in P_\ell^N$, *the rank of* $\mathcal{V}_{\vec{\lambda}}^{\dagger}(\mathcal{F}^{(1)})$ *is computed combinatorially in terms of* $N_{\lambda,\mu,\nu}$, $(\lambda, \mu, \nu) \in P_\ell^3$ *only.*

§7. Monodromy Representations of the Mapping Class Group

Let $\Gamma_{g,N}$ be the mapping class group of N-pointed oriented surfaces of genus g with first-order infinitesimal structure, and put $\Gamma_g = \Gamma_{g,0}$. Then there exist surjective homomorphisms, $\Gamma_{g,N} \to S_N \to 1$ and $\Gamma_{g,N} \to \Gamma_g \to 1$, and also an exact sequence $1 \to \mathbf{Z}^N \to \Gamma_{g,N} \to \Gamma_g^N \to 1$, where Γ_g^N is the mapping class group of N-pointed oriented surfaces of genus g.

Next consider the modular stack $M_{g,N}^{(1)}$ of the N-pointed (unordered) stable curves of genus g with first-order infinitesimal structure, and the normal crossing divisor $D_{g,N}^{(1)} \subset M_{g,N}^{(1)}$ representing singular curves. Set $\overset{\circ}{M}{}_{g,N}^{(1)} = M_{g,N}^{(1)} - D_{g,N}^{(1)}$. Then the fundamental group $\pi_1(\overset{\circ}{M}{}_{g,N}^{(1)})$ is isomorphic to $\Gamma_{g,N}$. For an S_N-invariant subset Λ of P_ℓ^N, we denote $\mathscr{V}_\Lambda(\mathscr{F}^{(1)}) = \bigoplus_{\vec{\lambda}\in\Lambda} \mathscr{V}_{\vec{\lambda}}^{\dagger}(\mathscr{F}^{(1)})$. Then the functors $\mathscr{F} \to \mathscr{D}^1_{B^{(1)}}(-\log D^{(1)} : c_v)$ and $\mathscr{F} \to \mathscr{V}_\Lambda(\mathscr{F}^{(1)})$ define a sheaf of twisted first-order differential operators $\mathscr{D}^1_{M_{g,N}^{(1)}}(-\log D_{g,N}^{(1)} : c_v)$ on $M_{g,N}^{(1)}$, and a locally-free coherent $\mathscr{O}_{M_{g,N}^{(1)}}$-module $\mathscr{V}_\Lambda(M_{g,N}^{(1)})$ on which $\mathscr{D}^1_{M_{g,N}^{(1)}}(-\log D_{g,N}^{(1)} : c_v)$ acts. The functor $(\mathscr{F}, \omega) \to \mathscr{L}_\omega(\mathscr{F}^{(1)})$ define a system of invertible $\mathscr{O}_{U_i}$-modules $\mathscr{L}_i$ on which $\mathscr{D}^1_{U_i}(-\log D^{(1)} : c_v)$ acts for some open covering $\{U_i\}$ of $M_{g,N}^{(1)}$. Now restrict the system $\{\mathscr{L}_i, U_i\}$ on $\overset{\circ}{M}{}_{g,N}^{(1)}$. Then the twisting system $\{\mathscr{L}_{ij}, U_i \cap U_j\}$ on $\overset{\circ}{M}{}_{g,N}^{(1)}$ is defined by $\mathscr{L}_{ij} = \mathscr{H}om_{\mathscr{D}^1_{\overset{\circ}{M}{}_{g,N}^{(1)}}(c_v)}(\mathscr{L}_{i|U_i\cap U_j}, \mathscr{L}_{j|U_i\cap U_j})$, cf. Kashiwara [7]. Then using some results of Oda [10], about the homotopy type of $\overset{\circ}{M}{}_{g,N}^{(1)}$, we have

Proposition 7.1. *The twisting data $\{\mathscr{L}_{ij}\}$ on $\overset{\circ}{M}{}_{g,N}^{(1)}$ define central extensions $\widehat{\Gamma}_{g,N}$ and $\widehat{\Gamma}_g$ of $\Gamma_{g,N}$ and Γ_g by the multiplicative subgroup K of $\mathbf{C}^*$ generated by $e^{2\pi\sqrt{-1}c_v/24}$, with the following commutative diagram.*

$$\begin{array}{ccccccccc} 1 & \longrightarrow & K & \longrightarrow & \widehat{M}_{g,N} & \longrightarrow & M_{g,N} & \longrightarrow & 1 \\ & & \downarrow = & & \downarrow & & \downarrow & & \\ 1 & \longrightarrow & K & \longrightarrow & \widehat{M}_g & \longrightarrow & M_g & \longrightarrow & 1. \end{array} \tag{7.1}$$

Theorem VI. *For each Λ, the system of solution sheaves on $\overset{\circ}{M}{}_{g,N}^{(1)}$,*

$$\mathscr{H}om_{D^1_{\overset{\circ}{M}{}_{g,N}^{(1)}}(c_v)}(\mathscr{V}_{\Lambda|U_i}, \mathscr{L}_i) \tag{7.2}$$

defines a monodromy representation ϱ of the group $\widehat{\Gamma}_{g,N}$ such that an element k of K acts as $\varrho(k) = k \cdot \mathrm{id}$.

References

1. Belavin, A.A., Polyakov, A.M., Zamolodchikov, A.B.: Infinite conformal symmetry in two dimensional quantum field theory. Nucl. Phys. **B241** (1984) 333–380
2. Beilinson, A.A., Schechtman, Y.A.: Determinant bundles and Virasoro algebras. Commun. Math. Phys. **118** (1988) 651–701
3. Deligne, P., Mumford, D.: The irreducibility of the space of curves of given genus. Publ. Math. I.H.E.S. **36** (1969) 75–110

4. Friedan, D., Shenker, S.: The analytic geometry of two dimensional conformal field theory. Nucl. Phys. **B281** (1987) 509–545
5. Gabber, O.: The integrability of the characteristic variety. Amer. J. Math. **103** (1981) 445–468
6. Kac, V.G.: Infinite dimensional Lie algebras. 2nd ed. Cambridge University Press, 1985
7. Kashiwara, M.: Representation theory and $\mathscr{D}$-modules on flag varieties. Astérisque **173-174** (1989) 55–109
8. Knizhnik, V.G., Zamolodchikov, A.B.: Current algebra and Wess-Zumino model in two dimensions. Nucl. Phys. **B247** (1984) 83–103
9. Moore, G., Seiberg, N.: Classical and quantum conformal field theory. Commun. Math. Phys. **123** (1989) 177–254
10. Oda, T.: Etale homotopy type of the moduli space of algebraic curves. Preprint, R.I.M.S., 1990
11. Segal, G.: Conformal field theory. In these Proceedings, pp. 1387–1396
12. Tsuchiya, A., Kanie, Y.: Vertex operators in conformal field theory on P^1 and monodromy representations of the braid group. In: Conformal field theory and solvable lattice models. Adv. Stud. Pure Math. **16** (1988) 297–372
13. Tsuchiya, A., Ueno, K., Yamada, Y.: Conformal field theory on universal family of stable curves with gauge symmetries. In: Integrable systems in quantum field theory and statistical mechanics. Adv. Stud. Pure Math. **19** (1989) 459–566
14. Verlinde, E.: Fusion rules and modular transformations in $2d$ conformal field theory. Nuclear Phys. **B300** (1988) 360–376
15. Witten, E.: Quantum field theory and the Jones polynomial. Commun. Math. Phys. **121** (1989) 351–399

Non-Constructive Proofs in Combinatorics

Noga Alon

Department of Mathematics, Raymond and Beverly Sackler Faculty of Exact Sciences
Tel Aviv University, Tel Aviv, Israel and
IBM Almaden Research Center, San Jose, CA 95120, USA

One of the main reasons for the fast development of Combinatorics during the recent years is certainly the widely used application of combinatorial methods in the study and the development of efficient algorithms. It is therefore somewhat surprising that many results proved by applying some of the modern combinatorial techniques, including Topological methods, Algebraic methods, and Probabilistic methods, merely supply existence proofs and do not yield efficient (deterministic or randomized) algorithms for the corresponding problems.

We describe some representing non-constructive proofs of this type, demonstrating the applications of Topological, Algebraic and Probabilistic methods in Combinatorics, and discuss the related algorithmic problems.

1. Topological Methods

The application of topological methods in the study of combinatorial objects like partially ordered sets, graphs, hypergraphs and their coloring have become in the last ten years part of the mathematical machinery commonly used in combinatorics. Many interesting examples appear in [12]. Some of the more recent results of this type deal with problems that are closely related to certain algorithmic problems. While the topological tools provide a powerful technique for proving the required results, they give us no clue on an efficient way for solving the corresponding algorithmic questions.

A typical result of this type is the following theorem, proved in [2].

Theorem 1.1. *Let N be an open necklace with ka_i beads of color i, $1 \le i \le t$. Then one can cut N in $(k-1)t$ places and partition the resulting intervals into k collections, each containing precisely a_i beads of color i for all $1 \le i \le t$.*

The bound $(k-1)t$, conjectured in [17] (where it is proved for $k = 2$) is sharp. This can be seen by considering the necklace in which the beads of each type appear contiguously. The proof supplies no efficient procedure, which finds, given a necklace as above, a partition of it with the desired properties. By an efficient procedure

Proceedings of the International Congress
of Mathematicians, Kyoto, Japan, 1990

we mean here, and in what follows, either a deterministic algorithm whose running time is polynomial (in the length of the input) or a randomized algorithm whose expected running time (on the worst-case input) is polynomial.

Here is a sketch of the proof of the above theorem. A similar method is used in [6]. First we need a continuous version of it. Let $I = [0, 1]$ be the (closed) unit interval. An *interval t-coloring* is a coloring of the points of I by t colors, such that for each i, $1 \leq i \leq t$, the set of points colored i is (Lebesgue) measurable. Given such a coloring, a *k-splitting of size r* is a sequence of numbers $0 = y_0 \leq y_1 \leq \cdots \leq y_r \leq y_{r+1} = 1$ and a partition of the family of $r + 1$ intervals $F = \{[y_i, y_{i+1}]: 0 \leq i \leq r\}$ into k pairwise disjoint subfamilies $F_1, \ldots, F_k$ whose union is F, such that for each j, $1 \leq j \leq k$, the union of all intervals in F_j captures precisely $1/k$ of the total measure of each of the t colors.

The following result is the continuous analogue of Theorem 1.1.

Proposition 1.2. *Every interval t-coloring has a k-splitting of size* $(k - 1)t$.

We note that a similar statement can be proved for general continuous probability measures instead of those defined by the colors. This generalizes the Hobby-Rice Theorem on L_1-approximation [18]. It is also related to one of the cake-splitting problems of Steinhaus. It is easy to see that the classical theorem of Liapounoff [20] implies the existence of an even splitting in this more general setting, but unlike the above result does not supply any finite bound on the number of cuts required to form the splitting. For more details see [2].

It is not difficult to see that Proposition 1.2 implies Theorem 1.1. This is because any open necklace with $\sum_{i=1}^{t} ka_i = kn$ beads as in the theorem can be converted into an interval t-coloring by partitioning the interval I into kn segments of equal size and by coloring the j-th part by the color of the j-th bead of the necklace. By Proposition 1.2 there is a k splitting with $(k - 1)t$ cuts. Of course, these cuts need not occur at the endpoints of the segments, but a simple induction argument can be used to show that the cuts may be shifted until they form a partition of the discrete necklace satisfying the assertion of Theorem 1.1. We omit the details.

Another simple observation, whose details we omit, is the fact that the validity of Proposition 1.2 for (t, k) and for (t, k') implies its validity for (t, kk'). Therefore, it suffices to prove the proposition for prime values of k. To do so we define, following [11], a CW-complex $Y = Y(k, m)$ as follows.

For two integers k and m, put $N = N(k, m) = (k - 1)(m + 1)$ and let $\Delta = \Delta^N$ denote the N-dimensional simplex; i.e., $\Delta = \{(x_0, \ldots, x_N): x_i \geq 0, \sum_{i=0}^{N} x_i = 1\}$. The *support* of a point $x \in \Delta$, denoted by Supp(x), is the minimal face of Δ that contains x. Define

$$Y = Y(k, m) = \{(y_1, \ldots, y_k): y_1, \ldots, y_k \in \Delta, \operatorname{Supp}(y_i) \cap \operatorname{Supp}(y_j) = \emptyset \text{ for all } 1 \leq i < j \leq k\}.$$

The cyclic group Z_k acts freely on Y by letting its generator ω cyclically shift the coordinates of each point $y \in Y$, i.e., $\omega(y_1, \ldots, y_k) = (y_2, \ldots, y_k, y_1)$.

The following lemma is proved in [11].

Lemma 1.3. *If k is a prime, $m \geq 1$, $N = N(k, m) = (k-1)(m+1)$ and $Y = Y(k, m)$ and ω are as in the preceding paragraph, then Y is $N-k$ connected and hence for every continuous mapping $h: Y \mapsto R^m$ there is a whole orbit of the Z_k action on Y that is mapped by h into one point. I.e., there is a $y \in Y$ such that $h(y) = h(\omega(y)) = \cdots = h(\omega^{k-1}(y))$.*

We can now prove Proposition 1.2 for primes k. Let c be an interval t-coloring. Define $N = N(k, t-1) = (k-1)t$, $Y = Y(k, t-1)$ and consider the continuous function $h: Y \mapsto R^{t-1}$ defined as follows.

Suppose $y = (y_1, \ldots, y_k) \in Y$. By the definition of Y, each y_i is a point of Δ^N, i.e., a real vector of length N with nonnegative coordinates whose sum is 1. Moreover, the supports of the points y_i are pairwise disjoint. Put $x = (x_0, \ldots, x_N) = \frac{1}{k}\sum_{i=1}^{k} y_i$, and define a partition of the [0, 1]-interval I into $N+1$ intervals $I_0, \ldots, I_N$ by

$$I_0 = [0, x_0] \qquad \textit{and} \qquad I_j = \left[\sum_{l=0}^{j-1} x_l, \sum_{l=0}^{j} x_l\right], \qquad (1 \leq j \leq N).$$

Observe that since the supports of the points y_i are pairwise disjoint, then for each interval I_j with a positive length there is a unique l such that the j-th coordinate of y_l is positive.

For each l, $1 \leq l \leq k$, let F_l be the family of all the intervals I_j such that the j-th coordinate of y_l is positive. Note that the sum of lengths of the intervals in each F_l is precisely $1/k$, and that $F_1, \ldots, F_l$ form a partition of all the intervals I_j whose lengths are positive. For each i, $1 \leq i \leq t-1$, define $h_i(y)$ to be the measure of the i-th color in the union of the intervals of F_1. The function $h(y)$ is now defined by $h(y) = (h_1(y), \ldots, h_{t-1}(y))$.

This function is clearly continuous. Also, for every $1 \leq l \leq k$ and $1 \leq i \leq t-1$, $h_i(\omega^{l-1}(y))$ is precisely the measure of the i-th color in the union of the intervals of F_l. By Lemma 1.3 there is a $y \in Y$ such that $h(y) = h(\omega(y)) = \cdots = h(\omega^{k-1}(y))$. This means that each of the k families F_l corresponding to this point y captures precisely $1/k$ of the measure of each of the first $t-1$ colors. Since the total measure of each F_l is $1/k$, it follows that the last color is evenly distributed between the families as well. This completes the proof for the case of prime k, and hence implies the validity of Proposition 1.2 and Theorem 1.1. □

The main topological tool in the above proof is the Borsuk-type theorem stated in Lemma 1.3. This proof does not seem to supply an efficient way of producing a partition whose existence is guaranteed by the theorem.

In the classification of algorithmic problems according to their complexity, it is customary to try and identify the problems that can be solved efficiently, and those that *probably* cannot be solved efficiently. A class of problems that can be solved efficiently is the class P of all problems for which there are deterministic algorithms whose running time is polynomial in the length of the input. A class of problems that probably cannot be solved efficiently are all the NP-complete problems. An extensive list of such problems appears in [16]. It is well known that if any of them can be solved efficiently, then so can all of them, since this would imply that the two complexity classes P and NP are equal.

It is not too difficult to show that the following problem is NP-complete: Given a necklace satisfying the assumptions of Theorem 1.1, decide if one can form an even k-splitting of it by using less than b cuts. On the other hand, we know that $(k - 1)t$ cuts always suffice, so although the problem of finding the minimum possible number of cuts cannot be solved efficiently, unless $P = NP$, it is considerable and seems likely that the problem of finding an even k-splitting using $(k - 1)t$ cuts is much easier. We do not know any efficient algorithm for this problem.

Another result whose (simple) proof applies the Borsuk-Ulam theorem is the following fact, proved in [1]:

Theorem 1.4. *Let $A_1, \ldots, A_d$ be d pairwise disjoint subsets of R^d, each containing precisely n points, and suppose that no hyperplane contains $d + 1$ of the points in the union of all the sets A_j. Then there is a partition of $\bigcup A_j$ into n pairwise disjoint sets $S_1, \ldots, S_n$, each containing precisely one point from each A_j, such that the n simplices $\mathrm{conv}(S_1), \ldots, \mathrm{conv}(S_n)$ are pairwise disjoint.*

Here, again, the proof does not supply an efficient way of finding the sets S_i if the sets A_j are given, (although the proof does provide an efficient way of doing it for each *fixed* dimension d.)

2. Algebraic Methods

Many combinatorial proofs rely on methods from linear and multilinear algebra. Extensive survey of results of this type is given in [9]. These proofs rarely supply constructive procedures for the corresponding algorithmic problems. Here is a simple example, which is a special case of one of the results in [5].

Proposition 2.1. *Every (not necessarily simple) graph with maximum degree 5 and average degree greater than 4, contains a 3-regular subgraph.*

The proof relies on the classical theorem of Chevalley and Warning (see, e.g., [10]). This theorem, that deals with the number of solutions of a system of multivariable polynomials over a finite field, is the following.

Theorem 2.2. *Let $P_j(x_1, \ldots, x_m)$, $(1 \leq j \leq n)$ be n polynomials over a finite field F of characteristic p. If the number of variables, m, is greater than the sum of the degrees of the polynomials then the number of common zeros of the polynomials (in F^m) is divisible by p. In particular, if there is one common zero then there is another one.*

The proof is extremely simple; If F has q elements, then the number N of common zeros satisfies

$$N \equiv \sum_{x_1,\ldots,x_m \in F} \prod_{j=1}^{n} (1 - P_j(x_1,\ldots,x_m)^{q-1}) (\mathrm{mod}\ p).$$

By expanding the right hand side we get a linear combination of monomials of the form $\prod_{i=1}^{m} x_i^{k_i}$ and for each such monomial at least one of the exponents k_i is strictly smaller than $q-1$. This implies that in F, $\sum_{x_i \in F} x_i^{k_i} = 0$, showing that the contribution of each monomial to the sum expressing N is $0 (\mathrm{mod}\ p)$ and completing the proof. □

We can now prove Proposition 2.1. Given a graph $G = (V, E)$ satisfying the assumptions of the proposition, let n denote the number of its vertices. For each edge $e \in E$ and for each vertex $v \in V$, let $a(v, e)$ be 0 if e is not incident with v, 1 if e is a non-loop incident with v, and 2 if e is a loop incident with v. For each $e \in E$ let x_e be a variable and consider the following system of polynomial equations over $GF(3)$:

$$\sum_{e \in E} a(v, e) x_e^2 = 0 \qquad (v \in V).$$

This is a system of n degree-2 polynomial equations with $|E| > 2n$ variables. Moreover, it clearly has the trivial solution $x_e = 0$ for all e. Hence there is, by Theorem 2.2, a non-trivial solution $(y_e : e \in E)$. Let H be the subgraph of G consisting of all edges e for which $y_e \neq 0$. By the equations above, the degree of every vertex of H is divisible by 3, and since the maximum degree in G is 5 it follows that H is 3-regular, completing the proof. □

It is known that the decision problem: "Given a graph G, decide if it contains a 3-regular subgraph", is NP-complete. By the proposition above in certain cases we know that the answer to the decision problem is "yes" and yet the proof does not yield an efficient procedure for finding such a subgraph.

Another result proved by applying some extension of the Chevalley Warning Theorem is the following statement, proved in [7]. Recall that a *hypergraph* is a pair $(V, \mathscr{F})$ (sometimes denoted only by $\mathscr{F}$), where V is a finite set of vertices, and $\mathscr{F}$ is a finite set of subsets of V. The *degree* of a vertex is the number of edges that contain it.

Theorem 2.3. *Let q be a prime power, and let $\mathscr{F} = \{F_1, \ldots, F_{d(q-1)+1}\}$ be a hypergraph whose maximal degree is d. Then there exists $\emptyset \neq \mathscr{F}_0 \subset \mathscr{F}$, such that $|\bigcup_{F \in \mathscr{F}_0} F| \equiv 0$ (mod q).*

Here, again, we do not know how to quickly find such a subset $\mathscr{F}_0$. Moreover, it can be shown that the problem of finding such an $\mathscr{F}_0$ is equivalent to the following problem: Given a polynomial h of degree at most d with $d(q-1)+1$ variables over $GF(q)$, suppose that $h(0) = 0$. Find another zero of $h(x)$ in which each variable is either 0 or 1.

3. Probabilistic Methods

Probabilistic methods have been useful in combinatorics for almost fifty years. Many examples can be found in [14] and in [21].

In a typical application of the probabilistic method we try to prove the existence of a combinatorial structure (or a substructure of a given structure) with certain prescribed properties. To do so, we show that a randomly chosen element from an appropriately defined sample space satisfies all the required properties with positive probability. In most applications, this probability is not only positive, but is actually high and frequently tends to 1 as the parameters of the problem tend to ∞. In such cases, the proof usually supplies an efficient randomized algorithm for producing a structure of the desired type, and in many cases this algorithm can be derandomized and converted into an efficient deterministic one.

There are, however, certain examples, where one can prove the existence of the required combinatorial structure by probabilistic arguments that deal with rare events; events that hold with positive probability which is exponentially small in the size of the input. Such proofs usually yield neither randomized nor deterministic efficient procedures for the corresponding algorithmic problems.

A class of examples demonstrating this phenomenon is the class of results proved by applying the Local Lemma. This result, proved in [13] (see also, e.g., [21]), supplies a way of showing that certain events hold with positive probability, although this probability may be extremely small. The exact statement (for the symmetric case) is the following.

Lemma 3.1. *Let $A_1, \ldots, A_n$ be events in an arbitrary probability space. Suppose that the probability of each of the n events is at most p, and suppose that each event A_i is mutually independent of all but at most b of the other events A_j. If $ep(b + 1) < 1$ then with positive probability none of the events A_i holds.*

One of the applications of this lemma, given already in the original paper [13], deals with *hypergraph coloring.* A hypergraph is *k-uniform* if each if its edges contains precisely k vertices. It is *k-regular* if each of its vertices is contained in precisely k edges. A hypergraph is 2-*colorable* if there is a two-coloring of the set of its vertices so that none of its edges is monochromatic. Erdös and Lovász proved the following result.

Proposition 3.2. *For each $k \geq 9$, every k-regular, k-uniform hypergraph is two colorable.*

The proof follows almost immediately from Lemma 3.1. Let (V, E) be a k-uniform, k-regular hypergraph, and let $f: V \mapsto \{0, 1\}$ be a random 2-coloring obtained by choosing, for each $v \in V$ randomly and independently, $f(v) \in \{0, 1\}$ according to a uniform distribution. For each $e \in E$ let A_e denote the event that f restricted to e is a constant, i.e., that e is monochromatic. It is obvious that $\text{Prob}(A_e) = 2^{-(k-1)}$ for every e, and that each event A_e is mutually independent of all the events A_f but those for which $f \cap e \neq \emptyset$. Since there are at most $k(k - 1)$

edges f that intersect e we can substitute $b = k(k-1)$ and $p = 2^{-(k-1)}$ in Lemma 3.1 and conclude that for $k \geq 9$ with positive probability none of the events A_e holds, completing the proof. □

We note that a different, algebraic proof of the statement of the last proposition (that works for all $k \geq 8$) is given in [4]. Both proofs do not supply an efficient way of finding a proper two-coloring for a given hypergraph satisfying the assumptions of the proposition. Note that, in general, the problem of deciding whether a hypergraph is 2-colorable is NP-complete.

Another application of the Local Lemma, which appears in [8], is the following.

Proposition 3.3. *Every directed simple graph $D = (V, E)$ with minimum outdegree δ and maximum indegree Δ contains a directed (simple) cycle of length* 0(mod k), *provided* $e(\Delta\delta + 1)\left(1 - \frac{1}{k}\right)^{\delta} < 1$.

The proof here first applies the Local Lemma to show that there exists a function $f: V \mapsto \{0, 1, \ldots, k\}$ such that for every $v \in V$ there is a vertex $u \in V$ such that (v, u) is a directed edge of D and $f(u) \equiv f(v) + 1 \pmod{k}$.

Given such an f, the rest of the proof is very simple. We just choose, for every vertex $v \in V$, some vertex $p(v)$ such that $(v, p(v))$ is a directed edge and $f(p(v)) \equiv f(v) + 1 \pmod{k}$. Suppose $v \in V$ and consider the sequence

$$v_0 = v, \qquad v_1 = p(v_0), \qquad v_2 = p(v_1), \ldots$$

Let j be the minimum index such that there is an $i < j$ with $v_i = v_j$. The cycle $v_i v_{i+1} \ldots v_{j-1} v_j = v_i$ is a directed cycle of length 0(mod k), as needed.

Here, again, the proof is not constructive in the sense that it does not provide an efficient way of finding such a cycle in a directed graph satisfying the assumptions. This is because the proof that a function f as above exists is non-constructive.

We note that it is not known if the related decision problem "Given a directed graph, decide if it contains a directed even cycle" is polynomial, but it is easy to deduce from the results of [15] that the similar problem "Given a directed graph and an edge e in it, decide if there is an even cycle containing e" is NP-complete.

The proof of the next result also relies on the Local Lemma, but contains several additional ingredients as well. The details appear in [3].

Theorem 3.4. *There is an absolute constant c with the following property: For any two graphs $G_1 = (V, E_1)$ and $G_2 = (V, E_2)$ on the same set of vertices, where G_1 has maximum degree at most d and G_2 is a vertex disjoint union of cliques of size cd each, the chromatic number of the graph $G = (V, E_1 \cup E_2)$ is precisely cd.*

The proof, again, does not supply an efficient (deterministic or randomized) algorithm for producing a proper cd-vertex coloring of G.

We close this section mentioning the following result of J. Spencer, whose proof, given in [22], which combines the probabilistic method with a counting argument, also fails to supply an efficient procedure for the corresponding algorithmic problem.

Theorem 3.5. *Let $v_1, \ldots, v_n$ be n real vectors of length n each, and suppose that the l_∞-norm of each v_i is at most 1. Then there are $\varepsilon_1, \ldots, \varepsilon_n \in \{-1, 1\}$, such that the l_∞-norm of the sum $\sum_{i=1}^{n} \varepsilon_i v_i$ is at most $6\sqrt{n}$.*

4. Concluding Remarks

We have seen several examples of combinatorial results proved by topological, algebraic or probabilistic methods. One natural question that arises is whether these methods are necessary. After all, we may tend to believe that simply stated combinatorial results should have simple combinatorial proofs. Although this sounds plausible, there are no known natural combinatorial proofs for any of the results mentioned here (as well as for various other known similar examples).

Another question that should be addressed is whether the proofs given here are really inherently non-constructive. Is it possible to modify them so that they yield efficient ways of solving the corresponding algorithmic problems? There are no known efficient algorithms for any of the problems mentioned here. However, it seems very likely that such algorithms do exist. This is related to questions regarding the complexity of search problems that have been studied by several researchers. See, e.g., [19].

In the study of complexity classes like P and NP one usually considers only decision problems, i.e., problems for which the only two possible answers are "yes" or "no." However, the definitions extend easily to the so called "search" problems, which are problems where a more elaborate output is sought. The search problems corresponding to the complexity classes P and NP are sometimes denoted by FP and FNP.

Consider, for example, the obvious algorithmic problem suggested by Theorem 1.1, namely, given a necklace satisfying the assumptions of the theorem, find a partition of it satisfying the conclusions of the theorem. This problem is in FNP, since it is a search problem, and given a proposed solution for it we can check in polynomial time that it is indeed a solution.

Notice that this problem always has a solution, by Theorem 1.1, and hence it seems plausible that finding one should not be a very difficult task. The situation is similar with all the other algorithmic problems corresponding to the various results mentioned here. Still, the problem of solving efficiently the corresponding search problems remains an intriguing open question.

References

1. J. Akiyama, N. Alon: Disjoint simplices and geometric hypergraphs. In: Combinatorial Mathematics; Proc. of the Third International Conference, New York, NY 1985 (G.S. Blum, R.L. Graham and J. Malkevitch, eds.). Ann. New York Acad. Sci. **555** (1989) 1–3
2. N. Alon: Splitting necklaces. Adv. Math. **63** (1987) 247–253

3. N. Alon: The strong chromatic number of a graph. To appear
4. N. Alon, Z. Bregman: Every 8-uniform, 8-regular hypergraph is 2-colorable. Graphs and Combinatorics **4** (1988) 303–305
5. N. Alon, S. Friedland, G. Kalai: Regular subgraphs of amost regular graphs. J. Combin. Theory Ser. B **37** (1984) 79–91
6. N. Alon, P. Frankl, L. Lovász: The chromatic number of Kneser hypergraphs. Trans. Amer. Math. Soc. **298** (1986) 359–370
7. N. Alon, D.J. Kleitman, R. Lipton, R. Meshulam, M.O. Rabin, J. Spencer: Set systems with no union of cardinality 0 modulo *m*. To appear
8. N. Alon, N. Linial: Cycles of length 0 modulo *k* in directed graphs. J. Combin. Theory Ser. B **47** (1989) 114–119
9. L. Babai, P. Frankl: Linear algebra methods in combinatorics I. Preliminary version. Department of Computer Science, University of Chicago, 1988
10. Z.I. Borevich, I.R. Shafarevich: Number theory. Academic Press, New York 1966
11. I. Bárány, S.B. Shlosman, A. Szücz: On a topological generalization of a theorem of Tverberg. J. London Math. Soc. (2), **23** (1981) 158–164
12. A. Bjorner: Topological methods. To appear in: Handbook of Combinatorics (R.L. Graham, M. Grötschel and L. Lovász, eds.). North-Holland, Amsterdam 1991
13. P. Erdös, L. Lovász: Problems and results on 3-chromatic hypergraphs and some related questions. In: Infinite and finite sets (A. Hajnal et. al., eds.). Colloq. Math. Soc. J. Bolyai, vol. 11. North-Holland, Amsterdam 1975, pp. 609–627
14. P. Erdös, J. Spencer: Probabilistic methods in combinatorics. Academic Press/Akademiai Kiado, New York Budapest 1974
15. S. Fortune, J.E. Hopcroft, J. Wyllie: The directed subgraph homeomorphism problem. Theor. Comp. Sci. **10** (1980) 111–120
16. M.R. Garey, D.S. Johnson: Computers and intractability: A guide to the theory of *NP*-completeness. Freeman, 1979
17. C.H. Goldberg, D.B. West: Bisection of circle colorings. SIAM J. Algeb. Discrete Methods **6** (1985) 93–106
18. C.R. Hobby, J.R. Rice: A moment problem in L_1 approximation. Proc. Amer. Math. Soc. **16** (1965) 665–670
19. D.S. Johnson, C.H. Papadimitriou, M. Yannakakis: How easy is local search? JCSS **37** (1988) 79–100
20. A. Liapounoff: Sur les fonctions vecteurs completement additives. Izv. Akad. Nauk SSSR **4** (1940) 465–478
21. J. Spencer: Ten lectures on the probabilistic method. SIAM, Philadelphia, 1987
22. J. Spencer: Six standard deviations suffice. Trans. Amer. Math. Soc. **289** (1985) 679–706

Infinite Permutation Groups in Enumeration and Model Theory

Peter J. Cameron

School of Mathematical Sciences, QMW, University of London, Mile End Road
London E1 4NS, UK

1. Introduction

A permutation group G on a set Ω has a natural action on Ω^n for each natural number n. The group is called *oligomorphic* if it has only finitely many orbits on Ω^n for all $n \in \mathbb{N}$. (The term means "few shapes". Typically our permutation groups are groups of automorphisms of structures of some kind; oligomorphy implies that the structure has only finitely many non-isomorphic n-element substructures for each n.)

Oligomorphic permutation groups have close connections with both model theory and combinatorial enumeration. For the former, a basic result is the theorem of Engeler, Ryll-Nardzewski and Svenonius characterizing $\aleph_0$-categorical countable structures as those whose automorphism groups are oligomorphic. The connection with enumeration is via homogeneous structures, those for which orbits on n-sets are isomorphism types of induced substructures. A theorem of Fraïssé gives us a rich supply of homogeneous structures. These matters are described in Section 2. Section 3 develops some tools of enumeration theory (cycle index) in this context, with a few applications. In Section 4, the famous countable "random graph" of Erdős and Rényi is used to introduce the ideas of measure and Baire category. In the fifth section, some results and problems on the rate of growth of orbit-counting sequences are presented. Finally, the search for cyclic automorphisms of certain interesting graphs leads to sum-free sets, which have a fascinating theory.

For an oligomorphic permutation group G, I let $f_n(G)$, $F_n(G)$ and $F_n^*(G)$ be the numbers of orbits of G on n-sets, n-tuples of distinct elements, and all n-tuples respectively. By convention, $f_0(G) = F_0(G) = F_0^*(G) = 1$. There are some interesting relations among these sequences, notably

$$F_n^*(G) = \sum_{k=1}^{n} S(n,k) F_k(G),$$

where $S(n,k)$ is the Stirling number of the second kind; this fact has a number of combinatorial consequences (Cameron and Taylor 1985).

I will always assume that the set Ω on which a group acts is finite or countable. (Little is lost here; the downward Löwenheim–Skolem theorem of model theory guarantees that any sequences $(f_n(G))$, $(F_n(G))$, etc. realized by an oligomorphic group can be realized by a group of countable degree.)

Proceedings of the International Congress
of Mathematicians, Kyoto, Japan, 1990

Notation: $G \times H$ and $G \operatorname{Wr} H$ denote the direct and wreath products of the permutation groups G and H (acting on the disjoint union and cartesian product respectively of the sets admitting G and H); G_α is the stabilizer of the point $\alpha \in \Omega$.

S_∞ is the symmetric group of countable degree; C_n, the cyclic group of order n; and A, the group of order-preserving permutations of $\mathbb{Q}$. Note that $f_n(A) = 1$ for all n: any order-preserving bijection between finite subsets of $\mathbb{Q}$ can be extended to a (piecewise-linear) order-preserving permutation of $\mathbb{Q}$.

Three properties of $\mathbb{Q}$ will be important. First of course is Cantor's (1895) characterization of $\mathbb{Q}$ as countable dense ordered set without endpoints. Second, as just mentioned, any order-preserving bijection between finite subsets of $\mathbb{Q}$ extends to an automorphism. Third is the fact that, if X, Y are finite ordered sets with $X \subseteq Y$, then any embedding of X in $\mathbb{Q}$ can be extended to an embedding of Y. These observations are the starting point for the next section.

For a fuller and more leisurely discussion of oligomorphic permutation groups, see my lecture notes (Cameron 1990).

2. $\aleph_0$-Categoricity and Homogeneity

A countable structure M over a first-order language is *$\aleph_0$-categorical* if it is the unique countable model of its theory, i.e. determined up to isomorphism by countability and first-order sentences. (The prototype is $\mathbb{Q}$, characterized by Cantor's theorem, as we saw.) In the spirit of geometry (where, since Klein's Erlanger Programm, we have known of a connection between axiomatizability and symmetry), the following remarkable result was found independently by Engeler (1959), Ryll-Nardzewski (1959) and Svenonius (1959):

Theorem 2.1. *The countable structure M is $\aleph_0$-categorical if and only if* $\operatorname{Aut}(M)$ *is oligomorphic.*

The proof involves the well-known tool of "back-and-forth". Without going into details (familiar to the experts), I note that back-and-forth is used, not just to show that two countable structures are isomorphic, but also to describe orbits on n-tuples structurally. The pre-requisite for back-and-forth is the ability to extend finite isomorphisms one point at a time.

A countable relational structure M is *homogeneous* if every isomorphism between finite substructures of M can be extended to an automorphism of M. The *age* of M, written $\operatorname{Age}(M)$, is the class of finite structures embeddable in M. Now back-and-forth shows that M is homogeneous if and only if, for any $X, Y \in \operatorname{Age}(M)$ with $X \subseteq Y$, any embedding of X into M can be extended to Y. It suffices to require this when $|Y| = |X| + 1$. Using these ideas, Fraïssé (1953) showed:

Theorem 2.2. *A class $\mathcal{A}$ of finite structures is the age of a countable homogeneous structure M if and only if $\mathcal{A}$ is closed under isomorphism and under taking substructures, has only countably many non-isomorphic members, and has the amalgamation property. If these conditions hold, then M is unique up to isomorphism.*

We say that M is the *Fraïssé limit* of the class $\mathscr{A}$. This result is extremely useful for constructing examples. For instance, the class of finite triangle-free graphs has Fraïssé's properties; so there is a unique countable homogeneous universal triangle-free graph T (Henson 1971).

There is a natural topology on the symmetric group S_∞, that of pointwise convergence. In this topology, the full automorphism group of any first-order structure is closed in S_∞. Moreover, for $G \leq H$, G is a dense subgroup of H if and only if G and H have the same orbits on n-tuples for all n. Now, given any permutation group G, there is a "canonical relational structure" M such that G is a dense subgroup of Aut(M), and M is homogeneous.

The automorphism group of a countable homogenous structure over a finite relational language is thus a closed oligomorphic permutation group. The converse is false, however; I do not know any nice characterization of this class of permutation groups. (The closed oligomorphic groups are precisely the automorphism groups of countable $\aleph_0$-categorical structures; we may add the word "homogeneous" to the right-hand side of this equivalence.)

If M is a homogeneous structure and $G = \mathrm{Aut}(M)$ is oligomorphic, then $f_n(G)$ is equal to the number of unlabelled n-element structures in the class Age(M) (i.e. up to isomorphism), and $F_n(G)$ to the number of labelled structures (i.e. on the point set $\{0,\ldots,n-1\}$). Thus, enumeration of unlabelled or labelled structures in classes satisfying Fraïssé's hypotheses is equivalent to description of the appropriate orbit-counting sequence for an oligomorphic group G. As hinted above, many interesting combinatorial enumeration problems are of this type.

3. Generating Functions

It is common practice in combinatorial enumeration to use different forms of generating function in labelled and unlabelled enumeration problems (Goulden and Jackson 1983). Thus, we define the *ordinary generating function*

$$f_G(t) = \sum_{n\geq 0} f_n(G)t^n$$

for $(f_n(G)$, and the *exponential generating function*

$$F_G(t) = \sum_{n\geq 0} F_n(G)t^n/n!$$

for $(F_n(G))$. Convergence properties of these functions connect with growth rates (e.g. finite non-zero radius of convergence of f_G is equivalent to exponential growth of $(f_n(G))$), but usually we treat the series formally. Both turn out to be specializations of a series in infinitely many variables, as follows.

If H is a finite permutation group of degree n, its *cycle index* is the polynomial

$$Z(H;s_1,\ldots,s_n) = \frac{1}{|H|}\sum_{h\in H} s_1^{c_1(h)}\ldots s_n^{c_n(h)},$$

where $c_d(h)$ is the number of d-cycles in the cycle decomposition of h. Its rôle in enumeration is well known. Now, if G is a finite or oligomorphic permutation group, we define the *modified cycle index* of G to be

$$\tilde{Z}(G; s_1, s_2, \ldots) = \sum_i Z(H_i; s_1, s_2, \ldots),$$

where the summation is over representatives of the G-orbits on finite sets, and H_i is the group induced on the i-th set by its setwise stabilizer. (By convention, the empty set contributes a term 1.) For finite G, it can be shown that

$$\tilde{Z}(G; s_1, s_2, \ldots) = Z(G; s_1 + 1, s_2 + 1, \ldots).$$

A couple of examples for infinite groups are

$$\tilde{Z}(S_\infty) = \exp\Big(-\sum_{n\geq 1} \frac{s_n}{n}\Big)$$

and

$$\tilde{Z}(A) = \frac{1}{1 - s_1}.$$

These series specialize as follows:

Proposition 3.1. *For any oligomorphic permutation group G,*
(a) $f_G(t) = \tilde{Z}(G; t, t^2, t^3, \ldots)$;
(b) $F_G(t) = \tilde{Z}(G; t, 0, 0, \ldots)$.

The behaviour under direct and wreath products and point stabilizers can be described:

Proposition 3.2. *(a) $\tilde{Z}(G \times H) = \tilde{Z}(G)\tilde{Z}(H)$;*
(b) $\tilde{Z}(G \operatorname{Wr} H) = \tilde{Z}(H; \tilde{Z}(G) - 1)$;
(c) $\sum_i \tilde{Z}(G_{\alpha_i}) = \frac{\partial}{\partial s_1}\tilde{Z}(G)$.

(The substitution in (b) is defined by

$$A(B) = A(B(s_1, s_2, s_3, \ldots), B(s_2, s_4, s_6, \ldots), B(s_3, s_6, s_9, \ldots), \ldots).$$

In (c), the summation is over a set of orbit representatives α_i; G_{α_i} is the stabilizer of α_i, acting on the remaining points.)

Thus, $f_{G \operatorname{Wr} H}$ can be determined from f_G and $\tilde{Z}(H)$. For example,

$$f_{G \operatorname{Wr} S_\infty}(t) = \prod_{n\geq 1}(1 - t^n)^{-f_n(G)},$$

$$f_{G \operatorname{Wr} A}(t) = \frac{1}{2 - f_G(t)}.$$

I close this section with a few simple examples.

Example 1. $f_{C_2}(t) = 1 + t + t^2$; so $f_{C_2 \operatorname{Wr} A}(t) = 1/(1 - t - t^2)$. This is the generating function for the Fibonacci numbers.

Example 2. $f_n(S_\infty \operatorname{Wr} S_\infty)$ is the partition function $p(n)$, whose generating function is $\prod_{n\geq 1}(1 - t^n)^{-1}$. Moreover, the generating function for $f_n(S_\infty \operatorname{Wr} S_\infty \operatorname{Wr} S_\infty)$ is $\prod_{n\geq 1}(1 - t^n)^{-p(n)}$, which converges for all t; so the growth rate is slower than exponential, though faster than $\exp(n^{1-\varepsilon})$ for any $\varepsilon > 0$. this sequence arises in work of Cayley (1889) counting canonical forms.

Example 3. We have $F_{G\,\mathrm{Wr}\,H}(t) = F_H(F_G(t) - 1)$, where the usual substitution is intended. In particular,

$$F_{S_\infty \mathrm{Wr}\, G}(t) = F_G(e^t - 1).$$

Let $F^*(t)$ be the exponential generating function for $F_n^*(G)$. Since $F_n^*(G) = F_n(S_\infty \text{ Wr } G)$, we have

$$F_G^*(t) = F_G(e^t - 1),$$

which is thus equivalent to the relation involving Stirling numbers given in Section 1.

Example 4. Let C be the group preserving the cyclic order on the roots of unity. Then

$$\tilde{Z}(C) = 1 + \sum_{n\geq 1} \frac{1}{n} \sum_{d|n} \phi(d) s_d^{n/d}.$$

From the fact that $f_C(t) = 1/(1-t)$, it can be deduced that

$$\exp(t/(1-t)) = \prod_{n\geq 1} (1-t^n)^{-\phi(n)/n}.$$

For several further amusing examples, including groups realizing several familiar sequences, see Cameron (1987b), (1989). There are also close connections with Joyal's (1981) combinatorial formal power series. (Joyal regards an age as a category whose morphisms are the embeddings; the objects of cardinality n occur as the coefficients in formal power series.)

4. The Random Graph

It follows from Fraïssé's theorem (2.2) that there is a unique countable homogeneous graph R in which every finite graph is embedded. R is characterized by the property that, if U and V are finite disjoint sets of vertices, there is a vertex z joined to every vertex in U and to no vertex in V. (This property translates the equivalent of homogeneity given just before the statement of (2.2).)

The graph R first appeared in the literature in a paper by Erdős and Rényi (1963), who showed that, with probability 1, a countable random graph is isomorphic to R. (The probability measure is defined by the rule "choose edges independently with probability $\frac{1}{2}$"; but in fact the same graph is obtained if we take any fixed edge probability p with $0 < p < 1$, or even if we let p vary a bit, e.g. tend to infinity not too slowly.) The proof of this paradoxical assertion is remarkably easy. First, given fixed finite disjoint sets U and V, the probability that the required vertex exists is 1. Now, since there are only countably many choices for U and V, and a countable intersection of sets of measure 1 has measure 1, the characteristic property of R holds with probability 1.

Although an explicit description of R is unnecessary for this d iscussion (the probabilistic argument shows its existence, while back-and-forth gives uniqueness), there are a couple of simple constructions for it:

Construction 1. The vertices are the natural numbers; x is joined to y if $x < y$ and 2^x occurs in the (binary) expression for y as a sum of distinct powers of 2 (or vice versa).

Construction 2. The vertices are the primes congruent to 1 (mod 4); p and q are joined if p is a quadratic residue mod q. (This is symmetric, by quadratic reciprocity.)

In Construction 1, the asymmetric form of the relation gives a model for the Zermelo-Fraenkel axioms of set theory excluding the axiom of infinity. The proof of the characteristic property of R in Construction 2 is a pleasant exercise using the Chinese Remainder Theorem and Dirichlet's Theorem.

As noted, the back-and-forth argument shows not only the uniqueness of R, but the homogeneous action of its automorphism group. If we use the first explicit construction of R, we find a group of primitive recursive automorphisms acting homogeneously on R. It would be interesting to investigate this group; for example, to see how the "recursive presentation" of R affects the structure of the group.

Truss (1985) showed that the full automorphism group of R is simple, and described all the cycle types of its elements. We know that $F_n(\mathrm{Aut}(R))$ and $f_n(\mathrm{Aut}(R))$ are equal to the numbers of labelled and unlabelled graphs on n vertices respectively; the former is $2^{\frac{1}{2}n(n-1)}$, the latter asymptotically $2^{\frac{1}{2}n(n-1)}/n!$.

R is not the only countable homogeneous graph. All such graphs were found by Lachlan and Woodrow (1980) (the finite ones had been found earlier by Gardiner (1976)).

Theorem 4.1. *A countably infinite homogeneous graph is one of the following:*

(i) the disjoint union of m complete graphs of size n, where at least one of m and n is infinite;

(ii) complement of (i) (complete multipartite);

(iii) the Fraïssé limit of the class of finite graphs containing no complete subgraph of size n ($n \geq 3$)*;*

(iv) complement of (iii);

(v) the random graph R.

The universal K_n-free graphs in (iii) were constructed by Henson (1971). We met Henson's graph for $n = 3$ earlier, where it was called T.

I turn now to more general relational structures. Suppose that we have some class $\mathscr{X}$ of objects, each of which is described by a countable sequence of choices. There are two ways of assigning structure to the set $\mathscr{X}$:

Method 1. By assigning non-negative numbers summing to 1 to the outcomes of each choice, $\mathscr{X}$ becomes a measure space. For example, a countable graph can be determined by choosing "edge" or "non-edge" for each pair of vertices; if we give each choice the value $\frac{1}{2}$, we obtain the above model. Equivalently, we could assign $1/2^n$ to each possible extension of an n-vertex graph to an $(n+1)$-vertex graph. This technique can in principle be extended to many other classes of relational structures; but it is not clear what values to assign in general.

Method 2. We can make $\mathscr{X}$ into a complete metric space by defining the distance between two objects to be a suitable (decreasing) functions of the length of the common initial sequence of choices defining them. (The longer we have to wait to distinguish two objects, the closer they are.) Although the metric involves a choice of function, the topology does not. A set $\mathscr{Y} \subseteq \mathscr{X}$ is open if, for any $Y \in \mathscr{Y}$, there is an initial subsequence of the choice sequence defining Y which forces membership in $\mathscr{Y}$; and a set $\mathscr{Y} \subseteq \mathscr{X}$ is dense if, following any finite number of choices, there is a continuation defining a member of $\mathscr{Y}$.

A set is *residual* if it contains a countable intersection of open dense sets. The *Baire category theorem* asserts that, in a complete metric space, a residual set is non-empty (and, perforce, dense). Residual sets are regarded as "large", comparable to sets of measure 1 in a probability space.

Analogously to the Erdős-Rényi theorem, it is possible to show that the set of graphs isomorphic to R is residual in the set of countable graphs. We now generalize this observation.

Let $\mathscr{A}$ be any age. We define $\mathscr{X}(\mathscr{A})$ to be the class of all structures on the set $\mathbb{N}$ whose age is contained in $\mathscr{A}$. (Thus, for example, if $\mathscr{A} = \text{Age}(R)$, then $\mathscr{X}(\mathscr{A})$ is the set of all graphs on the vertex set $\mathbb{N}$.) An element of $\mathscr{A}$ is determined by countably many choices, the nth choice describing how to extend an element of $\mathscr{A}$ on the set $\{0,\ldots,n-1\}$ to one on the set $\{0,\ldots,n\}$. So $\mathscr{X}(\mathscr{A})$ is a complete metric space. Now we have:

Proposition 4.2. *Let M be a countable homogeneous structure. Then the set of structures isomorphic to M is residual in $\mathscr{X}(\text{Age}(M))$.*

No such result holds for measure, which seems much harder to deal with. However, there are some specific results. For example, $\mathbb{Q}$ is the random total order (where the measure is defined by making all possible orderings of any finite set equally likely).

5. Growth Rates

Quite a bit is known about possible growth rates of the sequence $(f_n(G))$. (Of course, these general results apply to the numbers of unlabelled structures in a class satisfying Fraïssé's conditions. However, some of them are known to hold for any age.) A basic fact is that this sequence is non-decreasing: see Cameron (1976). Most of the results are due to Pouzet (1981) and Macpherson (1985a), (1985b).

Pouzet showed that the rate of growth is either polynomial ($an^d \leq f_n(G) \leq bn^d$, where $n \in \mathbb{N}$ and $a, b > 0$) or faster than any polynomial. In the latter case, Macpherson found a fractional exponential lower bound $\exp(n^{\frac{1}{2}-\varepsilon})$, comparable to the partition function. For primitive groups, Macpherson's result is much more striking:

Theorem 5.1. *There is an absolute constant $c > 1$ such that, if G is primitive, then either $f_n(G) = 1$ for all n, or $f_n(G) \geq c^n$ for all sufficiently large n.*

Macpherson gave $c = 2^{\frac{1}{5}} - \varepsilon$; it is conjectured that the result holds with $c = 2 - \varepsilon$ (this would be best possible, see below).

Polynomial growth of degree $k-1$ is realized by, among others, S_∞^k (acting with k orbits) and $S_\infty \operatorname{Wr} S_k$ (acting imprimitively). $S_\infty \operatorname{Wr} S_\infty$ realizes the partition function. Other fractional exponential growth rates, roughly $\exp\left(n^{\frac{p+1}{p+2}}\right)$ for $p \in N$, can also be realized. We saw that the growth rate for $S_\infty \operatorname{Wr} S_\infty \operatorname{Wr} S_\infty$ is faster than fractional exponential but slower than exponential. There are many imprimitive examples with exponential growth; we saw the Fibonacci numbers realized by $C_2 \operatorname{Wr} A$.

Primitive groups exhibiting exponential growth are fairly rare. Most of them are automorphism groups of "treelike objects" (Cameron 1987b) related to the $\mathbb{Q}$-trees of combinatorial group theory (Alperin and Bass 1987). There are also structures related to circular orders, including Lachlan's (1984) "circular tournament" L. The group of order preserving and reversing permutations of L has the slowest known growth rate of any primitive group ($f_n(\operatorname{Aut}(L)) \sim 2^{n-2}/n$).

For growth rates faster than exponential, we see in nature a gap between factorial growth (for the homogeneous pair of linear orders, we have $f_n(G) = n!$), and growth like $\exp(cn^2)$ (realized by the random graph, and projective and affine spaces over finite fields). A result of Macpherson (1987) throws some light on this. The *independence property* (Shelah 1978) forces growth at least $\exp(cn^2)$; for homogeneous structures over finite languages, negating the independence property bounds the growth by $\exp(n^{1+\varepsilon})$. Also, for ω-*stable* structures, the same gap occurs, the criterion being the types of *strictly minimal* sets around which the structure is built. (See Cherlin–Harrington–Lachlan (1985) for the theory of ω-stable, $\aleph_0$-categorical structures.)

In general, there is no upper bound for the rate of growth of $(f_n(G))$: take a homogeneous structure over a language where the number of n-ary relation symbols grows as fast as you please with n. On the other hand, for a homogeneous structure over a finite language, $f_n(G) \leq \exp(P(n))$ for some polynomial P.

Many of the most interesting open questions about growth rate concern its smoothness. For example, do the limits

$\lim_{n\to\infty} f_n(G)/n^d$ (for polynomial growth of degree d),
$\lim_{n\to\infty} \log\log f_n(G)/\log n$ (for fractional exponential growth),
$\lim_{n\to\infty} f_n(G)^{1/n}$ (for exponential growth),

exist? If so, what possible values can these limits take? Apart from Macpherson's gap in values of the third limit for primitive groups, almost nothing is known.

It is known that $f_n(G) = f_{n+1}(G) = f_{n+2}(G)$ can only hold in the trivial case where G fixes a set of size at most n and is transitive on $(n+2)$-subsets of the complement. The situation $f_n(G) = f_{n+1}(G)$ has been studied; a few examples are known, and some have been characterized, but a general result seems difficult.

6. Cyclic Automorphisms of Graphs

Measure and Baire category have been used for constructing subgroups of oligomorphic groups. Though much more general results are available, I will consider here only a special case, regular cyclic subgroups.

Let g be a cyclic automorphism of a countable graph Γ. Then the vertices of Γ can be indexed by the integers in such a way that g acts as a shift $\alpha_i \mapsto \alpha_{i+1}$. Let $S = S(\Gamma, g) = \{n > 0 : \alpha_0 \sim \alpha_n\}$. Then S determines Γ up to isomorphism ($\alpha_i \sim \alpha_j$ if and only if $|i - j| \in S$) , and g up to conjugacy in $\mathrm{Aut}(\Gamma)$. We write $\Gamma = \Gamma_S$, $g = g_S$. Now we can ask: for which sets S is Γ_S isomorphic to some interesting graph?

A subset S of $\mathbb{N}$ is determined in an obvious way by infinitely many choices; so the methods of measure and Baire category apply.

S is called *universal* if every finite sequence of zeros and ones occurs as a consecutive subsequence of the characteristic function of S. It is easily checked that $\Gamma_S \cong R$ if and only if S is universal. A weak form of the law of large numbers says that the set of universal sequences has measure 1; it can also be shown to be residual. Since residual sets and measure-1 sets have cardinality $2^{\aleph_0}$, we conclude that the random graph has $2^{\aleph_0}$ non-conjugate cyclic automorphisms !

Now consider the homogeneous universal triangle-free graph T. First note that, for $S \subseteq \mathbb{N}$, Γ_S is triangle-free if and only if S is *sum-free*, i.e. $x, y \in S \Rightarrow x + y \notin S$. Now sum-free sets can be determined by countably many binary choices, in an obvious way: considering natural numbers in turn, if $n = x + y$ where x, y have already been put into S, then $n \notin S$; otherwise we are free to choose.

Let S be a sum-free set. The only obvious necessary condition for a finite zero-one sequence ε to be a subsequence of the characteristic function of S is that, if $j - i \in S$, then ε_i and ε_j cannot both be 1. Call a sum-free set *sf-universal* if every finite zero-one sequence satisfying this condition is a subsequence of the characteristic function of S. Now $\Gamma_S \cong T$ if and only if S is sf-universal.

Henson (1971) showed that T has cyclic automorphisms, i.e. that sf-universal sets exist. Can we prove this using category or measure? It is easily seen that a residual subset of all sum-free sets are sf-universal, so we do indeed get $2^{\aleph_0}$ non-conjugate cyclic automorphisms of T. (Incidentally, I *conjecture* that an sf-universal set has density 0. This would, if true, give a "density version" of Schur's theorem (1916) that $\mathbb{N}$ cannot be covered by finitely many sum-free sets.)

However, when we turn to measure, the situation is different. Any set of odd numbers is obviously sum-free, and no such set is sf-universal. It came as a surprise to me to find that the probability that a random sum-free set consists entirely of odd numbers is non-zero. (This probability is about 0.218... .) Furthermore, there are infinitely many "periodic" sum-free sets whose subsets have positive probability. (After the odd numbers, the next two are the congruence classes 2 and 3 (mod 5) and the congruence classes 1 and 4 (mod 5).) I do not know whether or not almost all sum-free sets are contained in periodic ones. (For details, see Cameron (1987a).)

Experiment suggests the possibility of quasi-periodic behaviour: a periodicity is established for a few cycles and then disrupted, to return with its phase shifted as part of a longer period. Maybe this process can continue infinitely and yield non-periodic sets whose subsets have positive probability.

What of the corresponding graphs Γ_S? For each periodic set S whose subsets have positive probability, there is an "almost homogeneous" graph Γ^* such that $\Gamma_{S'} \cong \Gamma^*$ for almost all subsets S' of S. (For the set of odd numbers, Γ^* is the "universal bipartite graph".) It is not known whether any other triangle-free graphs occur with positive probability. (In particular, this is not known for T.)

Many other questions about random sum-free sets remain open. For example, what is the average density, and how is the density distributed? (There are "spectral lines" corresponding to the periodic sets described above, e.g. a delta-function of weight 0.218... at density $\frac{1}{4}$ (from the sets of odd numbers). Does almost every sum-free set have a density? Is the density almost surely in the interval $(0, \frac{1}{4}]$? Does the spectrum have a continuous part?)

I conclude with an example with very different behaviour. Covington (1989) calls a graph *N-free* if it contains no induced path of length 3. She showed that there is an "almost homogeneous" universal countable N-free graph C, unique up to isomorphism, but admitting no cyclic automorphisms. One can write down a condition on sets S equivalent to Γ_S being N-free. There are $2^{\aleph_0}$ sets satisfying this condition, but the corresponding graphs Γ_S are pairwise non-isomorphic! In other words, an N-free graph has at most one conjugacy class of cyclic automorphisms.

References

Alperin, R., Bass, H. (1987): Length functions of group actions on Λ-trees. In: Combinatorial group theory and topology (Alta, Utah, 1984) (Ann. Math. Stud. 111). Princeton Univ. Press, Princeton, NJ, pp. 265–378

Cameron, P.J. (1976): Transitivity of permutation groups on unordered sets. Math. Z. **148**, 127–139

Cameron, P.J. (1987a): Portrait of a typical sum-free set. In: Whitehead, C. (ed), Surveys in combinatorics. London Math. Soc. (Lecture Notes, vol. 123). Cambridge Univ. Press, Cambridge, pp. 13–42

Cameron, P.J. (1987b): Some treelike objects. Quart. J. Math. Oxford (2) **38**, 155–183

Cameron, P.J. (1989): Some sequences of integers. Discrete Math. **75**, 85–102

Cameron, P.J. (1990): Oligomorphic Permutation Groups. London Math. Soc. (Lecture Notes, vol. 152). Cambridge Univ. Press, Cambridge

Cameron, P.J., Taylor, D.E. (1985): Stirling numbers and affine equivalence. Ars Combinatoria **20B**, 2–14

Cantor, G. (1895): Beiträge zur Begründung der transfiniten Menge. Math. Ann. **46**, 481–512

Cayley, A. (1889): Collected mathematical papers. Cambridge Univ. Press, London

Cherlin, G.L., Harrington, L.A., Lachlan, A.H. (1985): $\aleph_0$-categorical, $\aleph_0$-stable structures. Ann. Pure Appl. Logic **28**, 103–135

Covington, J. (1989): A universal structure for N-free graphs. Proc. London Math. Soc. (3) **58**, 1–16

Engeler, E. (1959): Äquivalenzklassen von n-Tupeln. Z. Math. Logik Grundl. Math. **5**, 340–345

Erdős, P., Rényi, A. (1963): Asymmetric graphs. Acta Math. Acad. Sci. Hungar. **14**, 295–315

Fraïssé, R. (1953): Sur certains relations qui généralisent l'ordre des nombres rationnels. C. R. Acad. Sci. Paris **237**, 540–542

Gardiner, A.D. (1976): Homogeneous graphs. J. Comb. Theory (B) **20**, 94–102

Goulden, I.P., Jackson, D.M. (1983): Combinatorial enumeration. Wiley, New York
Henson, C.W. (1971): A family of countable homogeneous graphs. Pacific J. Math. **38**, 69–83
Joyal, A. (1981): Une théorie combinatoire des séries formelles. Adv. Math. **42**, 1–82
Lachlan, A.H. (1984): Countable homogeneous tournaments. Trans. Amer. Math. Soc. **284**, 431–461
Lachlan, A.H., Woodrow, R.E. (1980): Countable ultrahomogeneous undirected graphs. Trans. Amer. Math. Soc. **262**, 51–94
Macpherson, H.D. (1985a): Orbits of infinite permutation groups. Proc. London Math. Soc. (3) **51**, 246–284
Macpherson, H.D. (1985b): Growth rates in infinite graphs and permutation groups. Proc. London Math. Soc. (3) **51**, 285–294
Macpherson, H.D. (1987): Permutation groups of rapid growth. J. London Math. Soc. (2) **35**, 276–286
Pouzet, M. (1981): Application de la notion de relation presque-enchaïnable au dénombrement des restrictions finis d'une rélation. Z. Math. Logik Grundl. Math. **27**, 289–332
Ryll-Nardzewski, C. (1959): On category in power $\leq \aleph_0$. Bull. Acad. Pol. Sér. Math. Astr. Phys. **7**, 545–548
Schur, I. (1916): Über die Kongruenz $x^m + y^m \equiv z^m (\mathrm{mod} p)$. Jber. Deutsch. Math.-Verein. **25**, 114–117
Shelah, S. (1978): Classification theory and the number of non-isomorphic models. North-Holland, Amsterdam
Svenonius, L. (1959): $\aleph_0$-categoricity in first-order predicate calculus. Theoria **25**, 82–94
Truss, J.K. (1985): The group of the countable universal graph. Math. Proc. Camb. Philos. Soc. **98**, 213–245

Geometric Presentations of Groups with an Application to the Monster

Alexander A. Ivanov

Institute for System Studies, Academy of Sciences of the USSR, 9, Prospect 60 Let Oktyabrya, 117312, Moscow, USSR

1. Introduction

Let $\mathscr{G} = (\mathscr{G}, I, \Delta, t)$ be a *geometry*, that is a set $\mathscr{G}$ of elements together with a symmetric reflexive *incidence relation* I and a *type function* $t : \mathscr{G} \to \Delta$. A geometry is supposed to satisfy the following axiom: *the restriction of t to any maximal set of pairwise incident elements is a bijection onto* Δ.

Let Φ be a *flag* of $\mathscr{G}$, i.e. a set of pairwise incident elements. Let $\mathscr{G}_\Phi = (\mathscr{G}_\Phi, I_\Phi, \Delta_\Phi, t_\Phi)$ where $\mathscr{G}_\Phi$ is the set of elements which are not contained in Φ and are incident to all elements in Φ; $\Delta_\Phi = \Delta - t(\Phi)$; I_Φ and t_Φ are the restrictions on $\mathscr{G}_\Phi$ of I and t, respectively. Then $\mathscr{G}_\Phi$ is a geometry in the above sense and is called the *residual geometry* of $\mathscr{G}$ with respect to Φ.

Let $\mathscr{G}$ and $\mathscr{G}'$ be geometries over the same set of types. A mapping $\phi : \mathscr{G}' \to \mathscr{G}$ is a *morphism of geometries* if ϕ preserves the incidence relation and the type function. A morphism is called a *covering* if its restriction to any proper residual geometry is an isomorphism. A morphism from a geometry onto itself is an *automorphism*. Let $G \leq \mathrm{Aut}(\mathscr{G})$ be an automorphism group of $\mathscr{G}$. G is said to be *flag-transitive* if it acts transitively on the set of maximal flags of $\mathscr{G}$.

We usually assume that $\Delta = \{0, 1, \ldots, r-1\}$ where r is the *rank* of the geometry. A geometry $\mathscr{G}$ is *connected* if the graph with $\mathscr{G}$ as vertices and I as edges is connected. All geometries we will consider are supposed to be connected. A considerable amount of information about geometry is carried by its *diagram*. The latter is a graph on Δ where the edge joining i and j symbolizes the rank 2 residual geometries of type $\{i, j\}$. The empty edge stands for the generalized digons (any two elements of different types are incident), an ordinary edge stands for projective planes, etc.

Let $\mathscr{G}$ be a geometry, G be a flag-transitive group of automorphisms of $\mathscr{G}$ and $\Phi = \{\alpha_0, \alpha_1, \ldots, \alpha_{r-1}\}$ be a maximal flag of $\mathscr{G}$. Let $G_i = G(\alpha_i)$ be the stabilizer of α_i in G and $\mathscr{A}$ be the amalgam of these stabilizers. The *members* G_i of $\mathscr{A}$ are the *maximal parabolics*. In the flag-transitive case $\mathscr{G}$ can be reconstructed from G and $\mathscr{A}$, namely its elements of type i are the (right) cosets of G_i in G, $0 \leq i \leq r-1$; two cosets are incident if they have a nonempty intersection. In this situation we write $\mathscr{G} \cong \mathscr{G}(G, \mathscr{A})$.

A geometry $\mathscr{G}$ possesses a *universal covering* $\phi_u : \widetilde{\mathscr{G}} \to \mathscr{G}$ such that for any other covering $\phi : \mathscr{G}' \to \mathscr{G}$ there is a covering $\psi : \widetilde{\mathscr{G}} \to \mathscr{G}'$ such that $\phi_u = \phi\psi$. Let G act flag-transitively on $\mathscr{G}$. Then automorphisms from G can be lifted to automorphisms of

Proceedings of the International Congress
of Mathematicians, Kyoto, Japan, 1990

$\tilde{\mathscr{G}}$ and all these liftings form a group $\tilde{G}$ which acts flag-transitively on $\tilde{\mathscr{G}}$. Let $\mathscr{A}$ be an amalgam. By definition an *$\mathscr{A}$-group* is a group which contains $\mathscr{A}$ and is generated by the elements of $\mathscr{A}$. An *$\mathscr{A}$-homomorphism* is a homomorphism of $\mathscr{A}$-groups whose restriction on $\mathscr{A}$ is the identity mapping. If the class of $\mathscr{A}$-groups is nonempty then there exists a *universal $\mathscr{A}$-group $U(\mathscr{A})$* such that any $\mathscr{A}$-group is an image of $U(\mathscr{A})$ under an $\mathscr{A}$-homomorphism. One can define $U(\mathscr{A})$ as a group having all elements of $\mathscr{A}$ as generators and all equalities valid in the members of $\mathscr{A}$ as relations.

The following fundamental result was proved almost simultaneously in [Pas1, Tit2] and in an unpublished manuscript by S.V. Shpectorov.

Theorem 1.1. *Let $\mathscr{G}$ be a geometry, G be a flag-transitive automorphism group of $\mathscr{G}$ and $\mathscr{A}$ be the amalgam of maximal parabolics. Let $\phi_u : \tilde{\mathscr{G}} \to \mathscr{G}$ be the universal covering, $\tilde{G}$ be the group of all liftings of the automorphisms from G and $U(\mathscr{A})$ be the universal $\mathscr{A}$-group. Then $\tilde{\mathscr{G}} \cong \mathscr{G}(U(\mathscr{A}), \mathscr{A})$ and $\tilde{G} \cong U(\mathscr{A})$.*

A geometry whose universal covering is an isomorphism is said to be *simply connected.* By Theorem 1.1 $\mathscr{G}$ is simply connected if and only if G is the universal $\mathscr{A}$-group.

In terms of generators and relations Theorem 1.1 gives a practical method for the determination of universal coverings of flag-transitive geometries. In this method one writes down a presentation for $U(\mathscr{A})$ starting with $\mathscr{A}$ and constructs $U(\mathscr{A})$ by a coset enumeration. Nowadays this method is rather popular and on its base a number of very interesting results on classification of flag-transitive geometries have been obtained (cf. [Hei, KTs, Pas2, WY] and many other papers). In some papers generators with relations are constructed just from conditions on the geometry. The language of generators and relations is rather convenient for concrete calculations. On the other hand the language of amalgams is more abstract. The correlation between these languages looks like the correlation between the languages of matrices and modules in the theory of linear operators.

Let $\mathscr{G}$, G and $\mathscr{A}$ be as above. A *geometric presentation* associated with the action of G on $\mathscr{G}$ is a set of elements of $\mathscr{A}$ and a set of relations on these elements valid in the members of $\mathscr{A}$, giving a presentation of $U(\mathscr{A})$. Notice that a particular geometric presentation is involved in the definition of $U(\mathscr{A})$. If the geometry $\mathscr{G}$ is simply connected then $G \cong U(\mathscr{A})$ and the geometric presentation is said to be *faithful*, so it is a presentation of G.

Construction of faithful geometric presentations for large sporadic groups form the content of the proposed lecture. In the next section we start with a simple illustrative example. We consider a presentation for Janko's group J_1 from [ATLAS]. This presentation can be treated as a geometric one and it is faithful since the corresponding geometry is simply connected. In Sect. 3 we discuss practical methods for proving simple connectedness. Here we present two new ideas. One of them relies on consideration of simply connected subgeometries, the second one deals with the fixed points of involutions from the automorphism group. Section 4 is devoted to simple connectedness of geometries of the sporadic simple groups J_4 (the Janko's group), F_2 (the Baby Monster) and F_1 (the Monster). The result for J_4 implies that a presentation from [SW] is faithful. In the case of F_1 our result enabled S. Norton to prove that the famous Y-presentation (c.f. [ATLAS], [CNS]) is faithful. In Sect. 5 we discuss some uniqueness aspects of sporadic groups.

2. An Illustrative Example: The Group J_1

In [ATLAS] Janko's group J_1 is presented as a Coxeter group having diagram

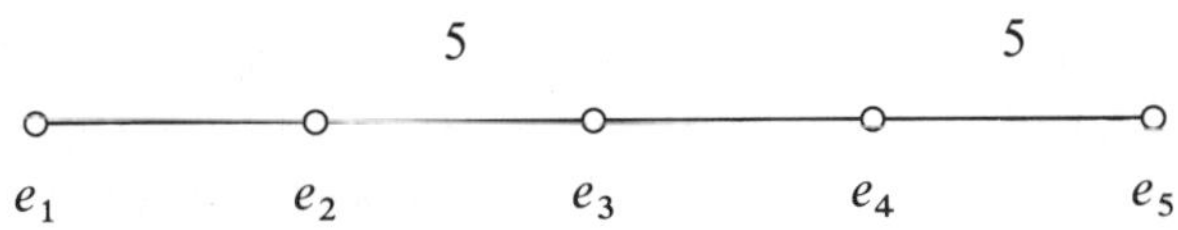

with the additional relations: $z_1 = z_2 = z_3 e_1 = 1$. Here z_i is the nontrivial element in the center of $\langle e_i, e_{i+1}, e_{i+2}\rangle \cong A_5 \times Z_2$, $1 \le i \le 3$. Let $H_i = \langle e_j | 1 \le j \le 5, j \ne i\rangle$, $2 \le i \le 5$ be subgroups of $H \cong J_1$. Then the relations for H which involve only generators of H_i give a presentation for H_i and $H_2 \cong H_4 \cong A_5 \times Z_2$, $H_3 \cong D_6 \times D_{10}$, $H_5 \cong L_2(11)$. Let $\mathscr{H}$ be the amalgam of the subgroups H_i, $2 \le i \le 5$ and $\mathscr{G}(J_1) = \mathscr{G}(H, \mathscr{H})$. The geometry $\mathscr{G}(J_1)$ possesses a nice combinatorial interpretation. The group J_1 acts distance-transitively on a graph $\Gamma(J_1)$ of valency 11 with the following intersection diagram:

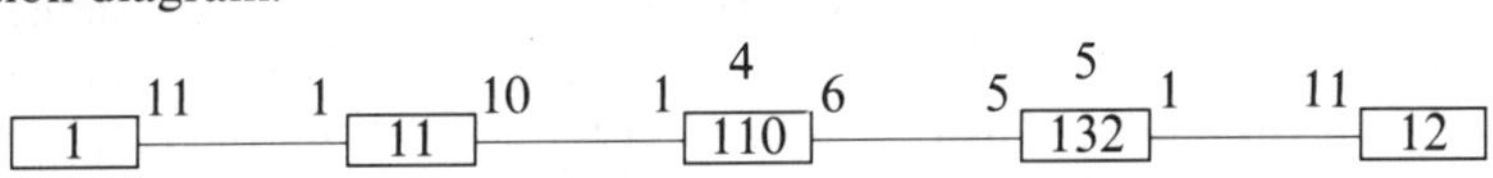

Then the elements of type 2, 3, 4 and 5 in $\mathscr{G}(J_1)$ are the Petersen subgraphs, the length 5 cycles fixed by Z_3-subgroups, edges and vertices, respectively. The incidence corresponds to the symmetrized inclusion.

The above presentation is a geometric one with respect to the action of J_1 on $\mathscr{G}(J_1)$. So it is faithful if and only if $\mathscr{G}(J_1)$ is simply connected. Let $\phi : \tilde{\mathscr{G}} \to \mathscr{G}(J_1)$ be the universal covering. Then ϕ induces a covering $\hat{\phi} : \tilde{\Gamma} \to \Gamma(J_1)$ of graphs. Each 5-cycle in $\Gamma(J_1)$ is contained in a Petersen subgraph. Thus to prove that $\hat{\phi}$ is an isomorphism one should split all cycles of $\Gamma(J_1)$ into cycles of length 5. The existence of these splittings is proved in [Ivn2, Lemma 6.9] by induction on the length of cycles.

The above presentation was independently obtained in [Ivn1] just as a consequence of the simple connectedness of $\mathscr{G}(J_1)$.

3. Proving Simple Connectedness of Geometries

A direct combinatorial proof of simple connectedness of a geometry basically consists of the following two steps (cf. [IS2, Pas2, Ron2]).

Step 1. With the geometry $\mathscr{G}$ in question one associates a graph $\Gamma = \Gamma(\mathscr{G})$ with the property that the universal covering $\phi : \tilde{\mathscr{G}} \to \mathscr{G}$ induces a covering $\hat{\phi} : \tilde{\Gamma} \to \Gamma$ of graphs. From the condition that ϕ is a covering of $\mathscr{G}$ one makes a conclusion that cycles of Γ from a certain class $\mathscr{K}$ are *contractible* with respect to $\hat{\phi}$. This means that each cycle from $\mathscr{K}$ can be lifted to an isomorphic cycle in $\tilde{\Gamma}$.

Step 2. Now to prove that $\hat{\phi}$ (and hence ϕ) is an isomorphism it is sufficient to show that the normal closure of the cycles from $\mathscr{K}$ generates the fundamental group of Γ. This is equivalent to the claim that an arbitrary cycle in Γ can be split into cycles from $\mathscr{K}$.

There is a number of ways to associate a graph with a geometry. The typical strategy is the following. One chooses two types in $\mathscr{G}$ (points and lines) and considers the *point graph* of $\mathscr{G}$, that is, the graph on the set of points where adjacency corresponds to collinearity. Here we consider a special situation when $\mathscr{G}$ contains a simply connected subgeometry.

Let $\mathscr{G}$, G and $\mathscr{A}$ be as above and let $\mathscr{F}$ be a simply connected subgeometry in $\mathscr{G}$ over the set of types $0 \in \Delta_{\mathscr{F}} \subseteq \Delta$. Suppose that the stabilizer F of $\mathscr{F}$ in G acts flag-transitively on $\mathscr{F}$. Let $\mathscr{B}$ be the subamalgam of $\mathscr{A}$ consisting of the subgroups $F_i = G_i \cap F$, $i \in \Delta_{\mathscr{F}}$. Then by Theorem 1.1 the simple connectedness of $\mathscr{F}$ implies that $\mathscr{B}$ generates in $\tilde{G} \cong U(\mathscr{A})$ a subgroup isomorphic to F and it is easy to see that this subgroup stabilizes in $\tilde{\mathscr{G}}$ a subgeometry which is an isomorphic lifting of $\mathscr{F}$. Let $\mathscr{C}$ be the set of all images of $\mathscr{F}$ under G. We assume that distinct subgeometries from $\mathscr{C}$ have distinct sets of elements of type 0. Let $\mathscr{C}(\alpha_0)$ be the set of subgeometries from $\mathscr{C}$ which contain α_0. Then G_0 acts on $\mathscr{C}(\alpha_0)$ as it acts on the cosets of F_0. Let $R(\alpha_0)$ be a nontrivial symmetric 2-orbit in this action (if any such exists). Let Σ be a graph on $\mathscr{C}$ where two subgeometries are adjacent if they have an element β of type 0 in common and are in the relation $R(\beta)$. Σ will be called an *intersection graph of subgeometries*. Let $\mathscr{F}$ and $\mathscr{F}'$ be adjacent in Σ. Let $m = m(\mathscr{G})$ be the number of elements β of type 0 in $\mathscr{F} \cap \mathscr{F}'$ such that $\mathscr{F}$ and $\mathscr{F}'$ are in the relation $R(\beta)$. Then the valency of Σ is equal to $|\mathscr{F}^0| \cdot v/m$, where $\mathscr{F}^0$ is the set of elements of type 0 in $\mathscr{F}$, and we have the following

Lemma 3.1. *If $m(\tilde{\mathscr{G}}) = m(\mathscr{G})$ then ϕ induces a covering $\hat{\phi} : \tilde{\Sigma} \to \Sigma$ where $\tilde{\Sigma}$ is an intersection graph of subgeometries in $\tilde{\mathscr{G}}$.*

Corollary 3.2. *Let $v = 1$ and suppose that F_0 is maximal in F. Then the valency of Σ is equal to $|\mathscr{F}^0|$ and ϕ induces a covering $\hat{\phi} : \tilde{\Sigma} \to \Sigma$.*

Let $\Xi(\alpha_0)$ be a subgraph in Σ having $\mathscr{C}(\alpha_0)$ as vertices and $R(\alpha_0)$ as edges. This subgraph can be lifted to an isomorphic subgraph in $\tilde{\Sigma}$. This implies that a cycle from $\Xi(\alpha_0)$ as well as its images under G are contractible. An additional set of contractible cycles can give other classes of simply connected subgeometries in $\mathscr{G}$ (if any).

The intersection graphs are crucial in our simple connectedness proof for the geometries in Sect. 4. The point is that for $G \cong J_4$, F_2 and F_1 the corresponding geometry contains a simply connected subgeometry whose stabilizer is a maximal order subgroup of G. It turns out that in each case an intersection graph of subgeometries corresponds to the minimal antireflexive 2-orbit of G acting on the subgeometries.

Now let us turn to the second step. In all examples we meet in the lecture, $\mathscr{K}$ contains all shortest cycles of Γ. These are pentagons in the J_1-geometry and triangles in the geometries of J_4, F_2 and F_1. So we come to a *standard problem*: to present the cycles of Γ as sums of the shortest ones. If the shortest cycles are triangles then the standard problem is the problem of *triangulation* of graphs.

Let us give a sufficient condition for a graph to be triangulable (compare [Ron 2, Lemma 5]). As usual $\Gamma_j(x)$ denotes the set of vertices of Γ which are at distance j from a vertex x, $0 \le j \le d$, where d is the *diameter* of Γ.

Lemma 3.3. *If for each* $2 \le j \le d$ *the conditions* (i) *and* (ii) *are satisfied then* Γ *is triangulable*

(i) *For* $y \in \Gamma_j(x)$ *the subgraph induced by* $\Gamma_1(x) \cap \Gamma_{j-1}(y)$ *is connected.*

(ii) *For* $y, z \in \Gamma_j(x)$, y *and* z *adjacent, either* $\Gamma_1(x) \cap \Gamma_{j-1}(y) \cap \Gamma_{j-1}(z) \neq \emptyset$ *or* $\Gamma_1(x) \cap \Gamma_{j-1}(y)$ *and* $\Gamma_1(x) \cap \Gamma_{j-1}(z)$ *are joined by an edge.*

Let G act on Γ and let τ be an involution in G. Let $\Theta(\tau)$ be the subgraph of Γ induced by the vertices fixed by τ and $\mathscr{K}(\tau)$ be the set of cycles from $\mathscr{K}$ which lie in $\Theta(\tau)$. Then we can try to prove that normal closure of $\mathscr{K}(\tau)$ generates the fundamental group of $\Theta(\tau)$. If we would succeed doing this for all representatives of conjugacy classes of involutions in G then we will get the following nice criterion: if a cycle is fixed by an involution then it is contractible.

4. Geometric Presentations of Large Sporadic Groups

4.1 Generic Properties

Let G be one of the groups J_4, F_2 or F_1. Then G contains an elementary abelian subgroup K of order 2^r ($r = 4$ for J_4 and $r = 5$ otherwise) such that $N_G(K)/C_G(K) \cong L_r(2)$. Let $1 < K_0 < K_1 < \cdots < K_{r-1} = K$ be a chain of subgroups in K. Let $\mathscr{G}(G)$ be a geometry whose elements of type i are the subgroups of G conjugate to K_i, $0 \le i \le r-1$ and the incidence be defined by inclusion. Let $G_i = N_G(K_i)$, $0 \le i \le r-1$ and $\mathscr{A}$ be the amalgam of these subgroups. Then $\mathscr{A}$ is the amalgam of maximal parabolics and $\mathscr{G}(G) = \mathscr{G}(G, \mathscr{A})$. The diagram of $\mathscr{G}(G)$ is a *string*. Namely the edge $\{i, j\}$ is empty for $|i-j| > 1$. The residue of type $\{i, i+1\}$, $0 \le i \le r-3$ is the projective plane over $GF(2)$. For $\mathscr{G}(J_4)$ and $\mathscr{G}(F_2)$ the residue of type $\{r-2, r-1\}$ is the geometry of edges and vertices of the Petersen graph while for $\mathscr{G}(F_1)$ it is the famous triple cover of the generalized quadrangle of order (2, 2) denoted by $\circ\!\!=\!\!\overset{\sim}{=}\!\!=\!\!\circ$.

The parabolic G_0 is of the form $G_0 \cong A.\,B$ where $A = O_2(G_0)$ is an extraspecial group 2_+^{1+n}, $n = 12$, 22 and 24; $B \cong 3 \cdot M_{22}.2$, Co_2 and Co_1 for $G \cong J_4$, F_2 and F_1, respectively. The subgroup A coincides with the kernel of G_0 acting on the residual geometry $\mathscr{G}(\alpha_0)$ of $\mathscr{G}$ with respect to α_0. The geometry $\mathscr{G}(\alpha_0)$ possesses a *natural representation* in the module $\bar{A} = A/Z(A)$. This means that elements of type i in

$\mathscr{G}(\alpha_0)$ correspond to certain i-dimensional subspaces of $\bar{A}$, $1 \leq i \leq r - 1$ and the incidence in $\mathscr{G}(\alpha_0)$ corresponds to inclusion of the subspaces. In particular the image of K in $\bar{A}$ corresponds to an element of type $r - 1$. This property can be taken as a definition of K.

The geometries $\mathscr{G}(J_4)$ and $\mathscr{G}(F_2)$ are so-called *P-geometries*. A characterization of a nice class of girth 5 graphs is reduced to the classification of P-geometries [Inv3]. The simple connectedness of $\mathscr{G}(J_4)$ and $\mathscr{G}(F_2)$ was proved just within this classification. Other results in this direction can be found in [IS1, IS2, Shp1, Shp2, Shp3]. The geometry $\mathscr{G}(F_1)$ and the truncations of $\mathscr{G}(J_4)$ and $\mathscr{G}(F_2)$ over type $r - 1$ are the minimal 2-local parabolic geometries of these groups described in [RSt].

4.2 Janko's Group J_4

The action of J_4 on $\mathscr{G}(J_4)$ is described by the following *diagram of parabolics* where under the node of type i the structure of G_i is given ($[2^n]$ stands for an arbitrary group of order 2^n).

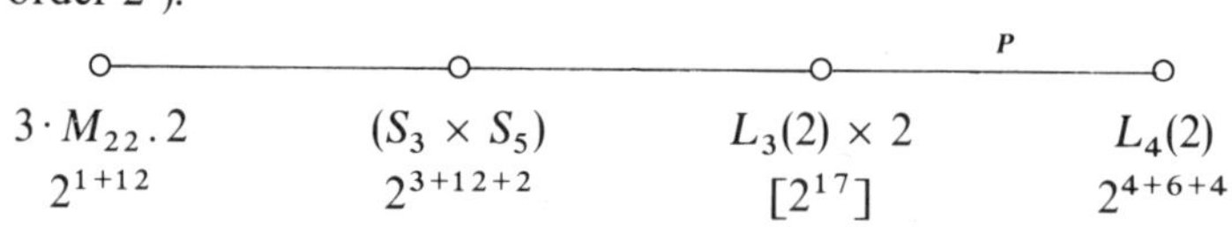

The parabolic $G_2 = N_G(K_2)$ is contained in a subgroup $L \cong 2^{11} : M_{24}$, and $K_2 \leq O_2(L)$. Let us consider the set of subgroups of $O_2(L)$ conjugated in L to nontrivial subgroups of K_2. Then we obtain a subgeometry $\mathscr{F}$ in $\mathscr{G}(J_4)$ over the type set $\{0, 1, 2\}$. This subgeometry is isomorphic to the minimal parabolic geometry $\mathscr{G}(M_{24})$ of M_{24}. Notice that $\mathscr{G}(M_{24})$ arises as a residue of type $\{2, 3, 4\}$ in $\mathscr{G}(F_1)$.

The following result was proved in [Hei] and independently checked on a computer by S.V. Shpectorov and the author.

Lemma 4.1. *The geometry $\mathscr{G}(M_{24})$ is simply connected.*

Thus we obtain a family $\mathscr{C}$ of simply connected subgeometries in $\mathscr{G}(J_4)$. G_0 acts on $\mathscr{C}(\alpha_0)$ as $M_{22}.2$ acts on the set of 77 blocks of $S(3, 6, 22)$. So there are two symmetric antireflexive 2-orbits in this action, having valencies 16 and 60. Let $R(\alpha_0)$ be the 2-orbit of valency 60.

Lemma 4.2. $m(\tilde{\mathscr{G}}) = m(\mathscr{G}(J_4)) = 7$.

Hence by Lemmas 3.1 and 4.2 the universal covering of $\mathscr{G}(J_4)$ induces a covering of an intersection graph of subgeometries $\Sigma(J_4)$ of valency 15, $180 = 1771 \cdot 60/7$. Notice that the elements of type 0 in $\mathscr{G}(M_{24})$ are the 1771 sextets of $S(5, 8, 24)$.

The subgraph $\Xi(\alpha_0)$ contains representatives of all classes of triangles. So for simple connectedness of $\mathscr{G}(J_4)$ it is sufficient to prove that $\Sigma(J_4)$ is triangulable. The latter fact was proved by application of Lemma 3.3 and an analysis of the subgraphs induced by the fixed points of involutions. A description of the 2-point stabilizers of J_4 acting on the cosets of $L \cong 2^{11} : M_{24}$ given in [Nor1] was used in a crucial way. A detailed proof is presented in [Ivn4].

The parabolic G_3 is contained in a subgroup $H \cong 2^{10}:L_5(2)$. It is shown in [Ivn2] that the permutational action of G on the cosets of H preserves a graph Π of valency 31 and girth 5. Let $\mathscr{H}(J_4)$ be a rank 3 geometry whose elements are vertices, edges and pentagons of Π with the natural incidence relation. In [SW] an explicit geometric presentation associated with the action of J_4 on $\mathscr{H}(J_4)$ is given. This is a compact and nice presentation which is similar to the Steinberg presentation for the group $D_5(2)$. Notice that H is isomorphic to a maximal parabolic in $D_5(2)$. As a corollary of the simple connectedness of $\mathscr{G}(J_4)$ it was proved in [Ivn4] that $\mathscr{H}(J_4)$ is simply connected.

Theorem 4.3. *The geometries $\mathscr{G}(J_4)$ and $\mathscr{H}(J_4)$ are simply connected and the presentation by G. Stroth and R. Weiss is faithful.*

The above result implies also the simple connectedness of a rank 3 2-local geometry of J_4 from [Bue] where maximal parabolics are L, G_0 and a subgroup $2^{3+12}.(S_5 \times L_3(2))$.

It is mentioned in [AS] that an independent proof of the triangulability of $\Sigma(J_4)$ is obtained by the authors of that paper.

4.3 The Baby Monster F_2

The action of F_2 on $\mathscr{G}(F_2)$ is described by the following diagram of parabolics.

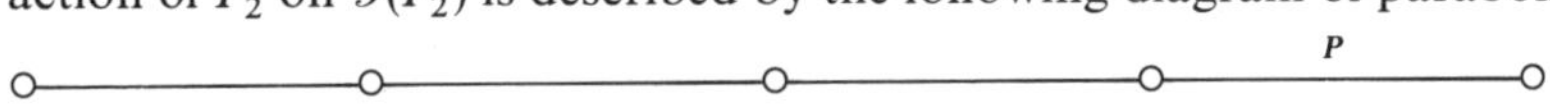

Co_2	$S_3 \times M_{22}.2$	$L_3(2) \times S_5$	$L_4(2) \times 2$	$L_5(2)$
2^{1+22}	$[2^{32}]$	$[2^{35}]$	$[2^{34}]$	$2^{5+10+10+5}$

Let $S \cong 2^{9+16}.Sp_8(2)$ and $E \cong 2 \cdot {}^2E_6(2).2$ be subgroups of $G \cong F_2$ and let $\bar{S} = S/O_2(S)$, $\bar{E} = E/O_2(E)$. Then up to a suitable choice of S with respect to $K \cong 2^5$, the intersections $S \cap G_i$, $0 \le i \le 3$ contain $O_2(S)$ and their images in $\bar{S}$ are the maximal parabolics of the natural geometry of $\bar{S}$. Up to a suitable choice of E with respect to K and S the intersections $E \cap G_i, 0 \le i \le 2, E \cap S$ contain $O_2(E)$ and their images in $\bar{E}$ are the maximal parabolics of the natural geometry of $\bar{E}$.

So S gives us a subgeometry $\mathscr{S}$ of type C_4 while E gives a subgeometry $\mathscr{E}$ which is a truncation of a geometry of type F_4. Both $\mathscr{S}$ and $\mathscr{E}$ are simply connected by [Tit1], [Tit2]. Let $\mathscr{C}$ be the family of subgeometries which contains $\mathscr{E}$. Put $E_0 = E \cap G_0$. Then E_0 is maximal in E and $[O_2(G_0):O_2(G_0) \cap E_0] = 2$. So we can apply Corollary 3.2 and come to a covering of an intersection graph $\Sigma(F_2)$ of subgeometries. Then $\Sigma(F_2)$ is a graph on the class of $\{3,4\}$-transpositions in F_2 where two transpositions are adjacent if their product is a central involution of F_2. The subgroup S acting on $\Sigma(F_2)$ has an orbit of length 120 which induces a complete subgraph Θ. Since $\mathscr{S}$ is simply connected one can show that Θ is contractible. On the other hand Θ contains representatives of all triangles in $\Sigma(F_2)$. So we come to the triangulation problem. Here we are in a position to prove the triangulability just by Lemma 3.3 using the structure constants of $\Sigma(F_2)$ calculated in [Hig]. Eventually we come to the following result whose proof will be given in [Ivn5].

Theorem 4.4. *The geometry $\mathcal{G}(F_2)$ is simply connected.*

It would be interesting to find a nice explicit presentation of F_2 as a corollary of Theorem 4.4.

Some independent results on triangulability of $\Sigma(F_2)$ were obtained by T. Meixner [Mei] in his study of a c-extension of the natural geometry of ${}^2E_6(2)$ related to F_2 (cf. [Bue]).

4.4 The Monster F_1

In the case of $\mathcal{G}(F_1)$ we have the following diagram

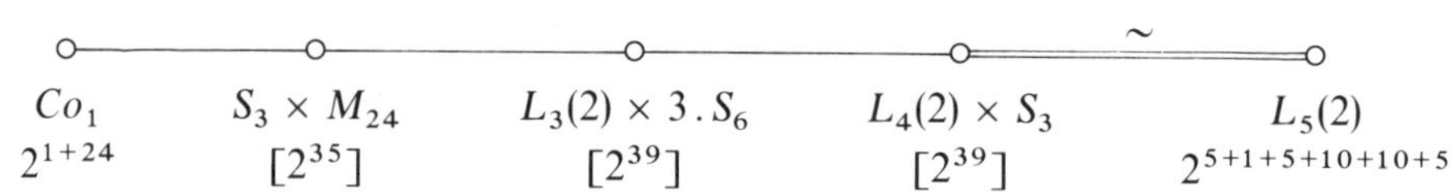

$\mathcal{G}(F_1)$ contains $\mathcal{G}(F_2)$ as a subgeometry. This embedding can be described as follows. $G \cong F_1$ contains exactly two conjugacy classes ($2A$ and $2B$) of involutions with centralizers $2 \cdot F_2$ and $2_+^{1+24} . \mathrm{Co}_1$, respectively. Let K be the subgroup involved in the definition of $\mathcal{G}(F_1)$ and $\tau \in K^\#$. Then τ is of type $2B$ and $O_2(C_G(\tau))$ contains an involution σ of type $2A$ such that $C_G(\sigma) \cap N_G(K)$ induces $L_5(2)$ on K. Let $\mathcal{F}$ consist of all images of the subgroups of K under conjugation by $C_G(\sigma)$. Then $\mathcal{F}$ is isomorphic to $\mathcal{G}(F_2)$ and its stabilizer F is $C_G(\sigma) \cong 2 \cdot F_2$. So by Theorem 4.4 we obtain a family $\mathcal{C}$ of simply connected subgeometries in $\mathcal{G}(F_1)$ whose members are in a one-to-one correspondence with the $2A$-involutions in G.

The subgroup $F_0 = F \cap G_0$ is maximal in F and $[O_2(G_0) : O_2(G_0) \cap F_0] = 2$. So by Corollary 3.2 the universal covering of $\mathcal{G}(F_1)$ induces a covering $\hat{\phi}$ of a graph $\Sigma = \Sigma(F_1)$. The latter is a graph on the set of $2A$-involutions in G where two involutions are adjacent if their product is a $2B$-involution. So to prove that $\mathcal{G}(F_1)$ is simply connected it is sufficient to prove that the normal closure of the cycles from $\Xi(\alpha_0)$ and their conjugacies under G generates the fundamental group of Σ. Let $\Sigma^* = \Sigma^*(F_1)$ be the *Monster graph*, that is a graph on the same set as Σ where two involutions are adjacent if their product is an involution of type $2A$. Detailed information about Σ^* is contained in [Nor2], [GMS]. Let us show that $\hat{\phi}$ induces a covering of Σ^*. Let $x, y \in 2A$ and $x \cdot y \in 2A$. Then $y \in \Sigma_2(x)$. Let Ω be a graph on the set $\Sigma_1(x) \cap \Sigma_1(y)$ where u and v are adjacent if the triangle $\{x, u, v\}$ is conjugated to a triangle from $\Xi(\alpha_0)$. Then Ω is connected and hence $\hat{\phi}$ induces a covering of Σ^*. Now, Σ^* is also an intersection graph of subgeometries. This means that two subgeometries from $\mathcal{C}$ which are adjacent in Σ^* have a common element of type 0. Let $\Xi^*(\alpha_0)$ be the subgraph in Σ^* on the set of subgeometries passing through α_0. Then $\Xi^*(\alpha_0)$ contains representatives of all triangles in Σ^*. The triangulability of Σ^* was proved by application of Lemma 3.3 with an analysis of the fixed points of involutions and using some results from [Nor2] and [GMS]. So we come to the following result whose proof will appear in [Ivn6].

Theorem 4.5. *The geometry $\mathcal{G}(F_1)$ is simply connected.*

An independent triangulation proof for the Monster graph is given in [AS, Sect. 8].

Let $\mathscr{A}$ be the amalgam of maximal parabolics associated with the action of F_1 on $\mathscr{G}(F_1)$. Then by Theorems 1.1 and 4.5, $U(\mathscr{A}) \cong F_1$. Let $\mathscr{B}$ be the subamalgam of $\mathscr{A}$ consisting of the parabolics G_0, G_1 and G_2. The residues of types $\{0, 1, 2\}$ and $\{0, 1, 2, 3\}$ are classical and simply connected. Hence, if $i = 4$ or 5 then G_i coincides with the universal group of the amalgam consisting of the subgroups $G_i \cap G_j$ for $0 \leq j \leq i - 1$. So we have the following

Corollary 4.6. $U(\mathscr{B}) \simeq F_1$.

By arguments similar to the above ones it is possible to prove simple connectedness of a geometry $\mathscr{G}(Co_1)$ isomorphic to the residue $\mathscr{G}(\alpha_0)$ in the Monster geometry. Here the crucial subgeometry is $\mathscr{G}(Co_2)$ isomorphic to the analogous residue in the Baby Monster geometry. It is proved in [Shp3] that $\mathscr{G}(Co_2)$ is simply connected. So in view of Lemma 4.1 and Theorem 4.5 the following proposition holds.

Theorem 4.7. *The geometry $\mathscr{G}(F_1)$ is 2-simply connected.*

A geometry is 2-simply connected if any of its covering whose restrictions on rank 2 residues are isomorphisms is an isomorphism itself. To describe the universal 2-cover of the Baby Monster geometry $\mathscr{G}(F_2)$ is an interesting open problem. I conjecture that the automorphism group of this 2-cover is a nonsplit extension of the elementary abelian group of order 3^{4371} by F_2.

The above results were announced at the Durham Symposium on Groups and Combinatorics in July 1990. During this symposium S. Norton [Nor4], using his recent results [Nor3] and Corollary 4.6, proved that the famous Y-presentation (cf. [ATLAS, CNS]) is faithful. Namely he proved that the group Y_{555} in the notation of [ATLAS] is isomorphic to the Bimonster, i.e. to the wreath product of the Monster and a group of order 2. This result gives an explicit presentation for the Monster and for other related groups.

5. On Uniqueness of Sporadic Groups

Here we indicate some relationship between the above results and the uniqueness problem for sporadic simple groups.

We consider a geometric approach to the uniqueness problem which involves the following steps. (1) Starting with some general properties of a group G in question, prove that it should act flag-transitively on a geometry $\mathscr{G}(G)$. (2) Using geometrical properties of $\mathscr{G}(G)$, prove uniqueness of the corresponding amalgam of maximal parabolics. (3) Prove that $\mathscr{G}(G)$ is simply connected.

Notice that this scheme is very close to the approach based on so-called *uniqueness systems* proposed in [AS].

First of all the existence of the geometries $\mathscr{G}(J_4)$, $\mathscr{H}(J_4)$, $\mathscr{G}(F_2)$ and $\mathscr{G}(F_1)$ follows from some general properties of the groups such as the structure of the centralizers of involutions. In the case of J_4 the second step was realized in [SW] for $\mathscr{H}(J_4)$ in terms of generators and relations and in [Shp] for $\mathscr{G}(J_4)$ in terms of amalgams. So Theorem 4.3 implies the following

Corollary 4.8. *There is a unique group of type* J_4.

For F_2 and F_1 the step (2) is not done yet but we belive that [GMS] and [Sev] contain enough information for this purpose.

Notes added in proof. (1) The simple connectedness of the geometry $\mathscr{G}(\mathrm{Co}_1)$ which is a residue in the Monster geometry is proved in [Ivanov, A.A.: The minimal parabolic geometry of the Conway group Co_1 is simply connected. In: Proc. Int. Conf. "Combinatorics 90", Gaeta, Italy 1990. (To appear)].

(2) The geometric approach to the uniqueness problem (cf. Section 5) is now completed for the groups F_1 and F_2 [Ivanov, A.A.: A geometric approach to the uniqueness problem for the sporadic simple groups. Dokl. Akad. Nauk SSSR **316** (1991) 1043–1046 (Russian)].

(3) Using the simple connectedness of the Baby Monster P-geometry the author has proved that Y_{433} is isomorphic to $2 \times 2.\, F_2$. This gives an explicit presentation for F_2 and hence answers the problem posed in the paragraph after Theorem 4.4 [Ivanov A.A.: Presenting the Baby Monster. Preprint, 1991].

References

[AS] Aschbacher, M., Segev, Y.: Extending morphisms of groups and graphs. (To appear)

[Bue] Buekenhout, F.: Diagram geometries for sporadic groups. Contemp. Math. **45** (1985) 1–31

[ATLAS] Conway, J.H., et al.: Atlas of Finite Groups. Oxford Univ. Press, 1985

[CNS] Conway, J.H., Norton, S.P., Soicher, L.H.: The Bimonster, the group Y_{555} and the projective plane of order 3. In: Tangora, M.C. (ed.) Computers in algebra. Marcel Dekker, 1988, pp. 27–50

[GMS] Griess, R.J., Jr., Meierfrankenfeld, U., Segev, Y.: A uniqueness proof for the Monster. Ann. Math. **130** (1989) 567–602

[Hei] Heiss, S.: On a parabolic system of type M_{24}. J. Algebra (to appear)

[Hig] Higman, D.G.: A monomial character of Fischer's Baby Monster. In: Gross, F. (ed.) Proc. of the Conference on Finite Groups. Acad Press, New York, 1976, pp. 277–283

[Ivn1] Ivanov, A.A.: A geometric characterization of the group J_1. In: Proc. X All-Union Conf. on Group Theory. Gomel', 1986, p. 91 (Russian)

[Ivn2] Ivanov, A.A.: On 2-transitive graphs of girth 5. Europ. J. Combin. **8** (1987) 393–420

[Ivn3] Ivanov, A.A.: Graphs of girth 5 and diagram geometries related to the Petersen graph. Dokl. Akad. Nauk. **295** (1987) 529–533 (Russian) [English transl. Sov. Math. Dokl. **36** (1988) 83–87]

[Ivn4] Ivanov, A.A.: A presentation for J_4. Proc. London Math. Soc. (To appear)

[Ivn5] Ivanov, A.A.: A geometric characterization of Fischer's Baby Monster. University of Western Australia, Preprint, May 1991.

[Ivn6] Ivanov, A.A.: A geometric characterization of the Monster. Proc. Symp. "Groups and Combinatorics", Durham 1990. (To appear)

[IS1] Ivanov, A.A., Shpectorov, S.V.: Geometries for sporadic groups related to the Petersen graph. II. Europ. J. Combin. **10** (1989) 347–361

[IS2] Ivanov, A.A., Shpectorov, S.V.: The P-geometry for M_{23} has no nontrivial 2-coverings. Europ. J. Combin. **11** (1990) 373–379

[KTs] Komissartschik, E.A., Tsaranov, S.V.: Construction of finite group amalgams and geometries. Geometries of the group $U_4(2)$. Comm. Algebra **18** (1990) 1071–1118

[Mei] Meixner, T.: Personal communication, 1990

[Nor1] Norton, S.: The construction of J_4. In: Cooperstein, B. and Mason, G. (eds.) The Santa Cruz Conference on Finite Groups. Amer. Math. Soc., 1980, pp. 271–278

[Nor2] Norton, S.: The uniqueness of the Fischer-Griess Monster. Contemp. Math. **45** (1985) 271–285

[Nor3] Norton, S.: Presenting the monster? Bull. Math. Soc. of Belgium (A) **42** (1990) 595–605

[Nor4] Norton, S.: Constructing the Monster. Proc. Symp. "Groups and Combinatorics", Durham 1990. (To appear)

[Pas1] Pasini, A.: Some remarks on covers and apartments. In: Baker, C.A. and Batten, L.M. (eds.) Finite geometries. Marcel Dekker Inc., New York Basel 1985, pp. 233–250

[Pas2] Pasini, A.: A classification of a class of Buekenhout geometries exploiting amalgams and simple connectedness. Preprint 1989

[Ron1] Ronan, M.: Covers and automorphisms of chamber systems. Europ. J. Combin. **1** (1980) 259–269

[Ron2] Ronan, M.: Coverings of certain finite geometries. In: Cameron, P., Hirschfeld, J., and Hughes, D. (eds.) Finite geometries and designs. Cambridge Univ. Press, 1981, pp. 316–331

[RSt] Ronan, M.A., Stroth, G.: Minimal parabolic geometries for the sporadic groups. Europ. J. Combin. **5** (1984) 59–91

[Seg] Segev, Y.: On the uniqueness of Fischer's baby monster (to appear)

[Shp1] Shpectorov, S.V.: A geometric characterization of the group M_{22}. In: Klin, M.H., Faradzev, I.A. (eds.) Investigations in algebraic theory of combinatorial objects. Moscow, VNIISI, 1985, pp. 112–123 (Russian)

[Shp2] Shpectorov, S.V.: On geometries with the diagram P^n. Preprint, 1988 (Russian)

[Shp3] Shpectorov, S.V.: The universal 2-cover of the P-geometry $\mathcal{G}(Co_2)$ (To appear)

[SW] Stroth, G., Weiss, R.: Modified Steinberg relations for the group J_4. Geom. Dedic. **25** (1988) 513–525

[Tit1] Tits, J.: A local approach to buildings. In: The Geometric Vein. The Coxeter Festschrift. Springer, Berlin 1981, pp. 519–547

[Tit2] Tits, J.: Ensembles ordonnés, immeubles et sommes amalgameés. Bull. Soc. Math. Belg. **A38** (1986) 367–387

[WY] Weiss, R., Yoshiara, S.: A geometric characterization of the groups Suz and HS. J. Algebra **133** (1990) 182–196

Some Developments in Ramsey Theory

Vojtech Rödl

Department of Mathematics, Emory University, Atlanta, GA 30322, USA

Introduction

Ramsey Theory is a part of combinatorial mathematics that studies the behaviour of structures under partitions. Many Ramsey type results assert that complete disorder is impossible – they find some regular substructures in general combinatorial structures.

Generic results are due to F.P. Ramsey [R3] and B.L. van der Waerden [V]. The systematic study of these and related statements was initiated by the work of P. Erdös and R. Rado. This includes extending van der Waerden's theorem by giving a complete description of those systems of linear equations that are preserved under partitions [R1, R2], developing the partition calculus – a theory which deals with transfinite extensions of Ramsey's theorem [EHR, EHMR] – and setting many further directions for systematic study [Er1, E4]. Another step was done by Hales and Jewett [HJ], who found a new Ramsey type combinatorial principle, the theorem for partitions of words over finite alphabets. This result was further generalized by R.L. Graham and B.L. Rothschild [GR2] for partitions of multidimensional words (parameter words). This subsequently led R.L. Graham, K. Leeb and B.L. Rothschild [GLR] to the proof of the conjecture of G. C. Rota on finite vector spaces. Moreover, the Hales-Jewett Theorem played an important role in studying induced subgraphs, distributive lattices and (m, p, c)-sets. A survey of the work in this area can be found in the book "Ramsey Theory" by R.L. Graham, B.L. Rothschild and J. Spencer [GRS].

Most of the work in Ramsey theory has been focused toward two directions: structural Ramsey type theorems (finding new statements of Ramsey type) and quantitative Ramsey type theorems (evaluating the size and number of the corresponding objects). Results were found that deal with vector spaces (and their combinatorial analogues – parameter words), set and relational-systems, groups and lattices, and other structures. Various generalizations were found for partitions restricted by topological concepts. (cf. [CS])

Ramsey type statements have a number of striking applications and connections to other parts of mathematics (e.g. [PH, CW, AH, KV] cf. also [NR11]). On the other hand some deep theorems were discovered using non-combinatorial means (e.g. [Fu, FK1, and FK2]).

Proceedings of the International Congress
of Mathematicians, Kyoto, Japan, 1990

In this account, we will outline the development in a few areas of this subject.

The Ramsey Theorem and Ramsey Numbers

We will adopt the usual convention of identifying the positive integer m with the set of its predecessors $\{0, 1, 2, \dots, m-1\}$. For a set X, $[X]^k$ denotes the set of all k-element subsets of X.

The Ramsey Theorem (1930). *For all positive integers $n_1, n_2, \dots, n_l$ and k there exists an integer m such that for every partition*

$$[m]^k = C_1 \cup C_2 \cup \cdots \cup C_l$$

there exists i and a set $Y_i \subseteq m = \{0, 1, 2, \dots, m-1\}$ $|Y_i| = n_i$ such that $[Y_i]^k \subseteq C_i$.

The Ramsey number $R_k(n_1, n_2, \dots, n_l)$ is the smallest integer m such that the Ramsey Theorem is valid. Clearly $R_1(n_1, n_2, \dots, n_l) = \sum_{i=1}^{l} (n_i - 1) + 1$; to determine the Ramsey numbers for $k \geq 2$ is much harder. Even in the next simplest case, i.e., $k = 2, l = 2$ progress on determining the corresponding Ramsey numbers $R(n_1, n_2) = R_2(n_1, n_2)$ has been very slow. For example, the lower bound for $R(n, n)$ due to P. Erdös [E1] was already known in 1947:

$$R(n, n) > (1 + o(1)) \frac{1}{e\sqrt{2}} n 2^{n/2}. \tag{1}$$

In more than forty years since its proof, this bound has been improved only by a factor of 2. This was done by J. Spencer [S4] implementing the Lovász Local Lemma [EL]. The upper bound

$$R(n_1, n_2) \leq \binom{n_1 + n_2 - 2}{n_1 - 1} \tag{2}$$

given by Erdös and Szekeres [ES] in 1935 has been improved for n_1 fixed and $n_2 \to \infty$ by Graver and Yackel [GY] and Ajtai, Komlós and Szemerédi [AKS]. For other values of n_1, n_2 there have been improvements on (1) only by a constant factor ([Fr1]). The bound in (2) was improved in [R6]. (see also [GR1])

$$R(n_1, n_2) \leq C_1 \binom{n_1 + n_2}{n_1} \Big/ [\log(n_1 + n_2)]^{c_2}. \tag{3}$$

Subsequently, for values of n_1 and n_2 which are close to each other $\left(\frac{n_2}{\sqrt{\log n_2}} \ll n_1 \leq n_2\right)$, (3) was improved by A. Thomason [T2] who established

$$R(n_1, n_2) \leq C_3 \binom{n_1 + n_2}{n_1} \exp\left[-\frac{n_1}{2n_2} \log n_2 + C\sqrt{\log n_2} \right]. \tag{4}$$

The proof of both inequalities is based on the idea of counting triangles in graphs and their complements (cf. [Go, L2]). Both (3) and (4) can probably be significantly improved.

Of all asymptotic bounds for $R(n_1, n_2)$ the only satisfactory ones are known for $n_1 = 3$ and $n_2 = n$ large:

$$C_1\left(\frac{n}{\log n}\right)^2 \leq R(3, n) \leq C_2\frac{n^2}{\log n}.$$

The lower bound was established by P. Erdös [E2] and the upper bound by Ajtai, Komlós, Szemerédi [AKS] (cf. Shearer [S2]).

For $n_1 \geq 4$ and $n_2 = n$ large, the situation is much less satisfactory. For instance,

$$C_3\left(\frac{n}{\log n}\right)^{5/2} \leq R(4, n) \leq C_4\frac{n^3}{\log^2 n}$$

(cf. [S5] [AKS]) are the best bounds currently known. A substantial effort has gone into finding the exact values $R(n_1, n_2)$ for small values of n_1, n_2. It is unlikely that $R(n_1, n_2)$ will be determined for many new values of n_1 and n_2 (cf. table below). For example, it would appear that determination of $R(5, 5)$ will require some essentially new ideas. The reader is referred to [GRS], [RK] and [R10] for further discussion.

n_1/n_2	3	4	5	6	7	8	9
3	6	9	14	18	23	28	36
4		18	25–27	34–43	47–66		
5			43–52	51–94	76–160		
6				102–169			

Let us also mention that various extensions of the Ramsey numbers have been investigated quite extensively in the last 15 years. For a survey on this work, see [GR1]. We will close this section by mentioning three old problems.

1) From the bound on Ramsey numbers $R(n, n)$ one can deduce that $\sqrt{2} \leq R(n, n)^{1/n} \leq 4$.

So, the first problem is to improve either of these bounds for $n \to \infty$. (i.e. improve the base of the exponent in either case) It is, for example, not even known whether the $\lim_{n\to\infty} R(n, n)^{1/n}$ exists.

2) Another old problem is to decide the growth of $R_2(3, 3, \ldots, 3)$ (n-times). It is not even known whether $\lim_{n\to\infty} [R_2(3, 3, \ldots, 3)]^{1/n}$ is finite or infinite. (The existence of the limit here is easy to prove). The known bounds are

$$(315)^{1/5} \leq [R_2(3, 3, \ldots, 3)]^{1/n} \leq \left(\frac{n}{e}\right)(1 + o(1)). \tag{5}$$

Note that the upper bound in (5) has not been improved in the 74 years since its discovery [S1].

3) Perhaps the most important problem concerning Ramsey numbers is to decide about the behavior of $R_3(n, n)$. The following is known (cf. [EHMR]) and it is believed that the upper bound is correct

$$\frac{n^2}{6}(1 + o(1)) \leq \log_2 R_3(n, n) \leq 2^{cn}.$$

Structural Induced and Restricted Ramsey-Type Theorems

Most of my contributions to Ramsey Theory have been in this area, much of this is joint work with J. Nesetril. The area was opened by work of J. Folkman [Fo]. Answering a question of Erdös and Hajnal, he proved that for every n there exists a graph H not containing K_{n+1}, but for which every 2-coloring of the edges results in a monochromatic K_n. A related result was obtained by W. Deuber [D1], P. Erdös, A. Hajnal, L. Posa [EHP] and myself [R4]. These results, together with work of Graham, Leeb, Rothschild and others, (cf. [D2, DR, L1, NR1]) layed the ground work for further systematic investigation of Ramsey classes.

Let **K** be a class of objects (e.g. graphs, parameter sets, ...) endowed with isomorphism and subobjects. For A, $B \in \mathbf{K}$, consider the set $\binom{B}{A}$ consisting of all subobjects of B which are isomorphic to A (A-subobjects of B). Let $A \in \mathbf{K}$. The class **K** has the *A-Ramsey Property* if for every $B \in \mathbf{K}$ there exists $C \in \mathbf{K}$ with $C \to (B)_2^A$, i.e. for every 2-coloring of the A-subobjects of C there exists a B-subobject of C with all its A-subobjects colored the same. If **K** has the A-Ramsey property for each $A \in \mathbf{K}$ we say that **K** is a *Ramsey class.*

In this language, the classical Ramsey theorem means that the class of all finite sets together with inclusion forms a Ramsey class. Also the class of all finite dimensional vector spaces over finite fields is a Ramsey class – this is the statement of Graham-Leeb-Rothschild Theorem [GLR].

Effort has been focused toward finding new Ramsey classes and, for the other classes, one has tried to characterize $A \in \mathbf{K}$ for which **K** has the A-Ramsey property.

So, for example, the class of all k-uniform hypergraphs with linearly ordered vertex sets, Gra(k) (more general, the class Soc(Δ) of all set systems of type Δ) is a Ramsey class. This was proved in [NR3] (see also [NR5]) and independently by Abramson and Harrington [AH]. On the other hand, the same classes with vertex sets not ordered fail to be Ramsey. Here the A-Ramsey property holds only for highly symmetric hypergraphs A, i.e., those which are either complete or empty. Subsequently, analogous statements were proved for the class of partially ordered sets [NR6] and the classes of parameter sets and vector space systems, [P1, P2].

Another direction was to consider the same type of questions for certain sub classes **K** of the class of all k-uniform hypergraphs, partially ordered sets, and others. We will explain here the result of [NR3].

A type $\Delta = (m_\delta; \delta \in \Delta)$ is an indexed collection of positive integers. A system A of type Δ is a pair $(X, \mathbf{M})$ where X is a finite linearly ordered set, $\mathbf{M} = (\mathbf{M}_\delta; \delta \in \Delta)$ and $\mathbf{M}_\delta \in [X]^{m_\delta}$. The elements of **M** are called *edges*. The system A is a *subsystem* (subobject) of $B = (Y, \mathbf{N})$ if X is a subset of Y with the induced order, and $\mathbf{M}_\delta = \mathbf{N}_\delta \cap P(X)$ for every $\delta \in \Delta$. Two systems $(X, \mathbf{M})$ and $(Y, \mathbf{N})$ are isomorphic if there is a monotone bijection $f: X \to Y$ taking $\mathbf{M}_\delta$ onto $\mathbf{N}_\delta$ for every $\delta \in \Delta$. Let $\binom{B}{A}$ be the set of all subsystems of B which are isomorphic to A. Let $\mathrm{Soc}(\Delta)$ be the class of all set systems of type Δ. For example, if $\Delta = (2)$ the class $\mathrm{Soc}(\Delta)$ is the class of graphs with ordered vertices and induced embedding.

A system $F \in \mathrm{Soc}(\Delta)$ is called irreducible if every pair of points of F is contained in an edge of F. For $\mathbf{F} \subseteq \mathrm{Soc}(\Delta)$ ($\mathbf{F}$ may be infinite), denote by Forb($\mathbf{F}$) the set of all $A \in \mathrm{Soc}(\Delta)$ with $\binom{A}{F} = \emptyset$ for every $F \in \mathbf{F}$.

The following theorem was proved in [NR3], (see also [NR5]).

Let $\mathbf{F} \subseteq \mathrm{Soc}(\Delta)$ be a (possibly infinite) set of irreducible set systems. Then $\mathbf{K} = \mathrm{Forb}(\mathbf{F})$ is a Ramsey class. Moreover, if $\mathbf{K}$ satisfies certain additional assumptions, the implication can be reversed, i.e., if $\mathbf{K} \subseteq \mathrm{Soc}(\Delta)$ is a Ramsey class satisfying additional assumptions then $\mathbf{K}$ is of the form $\mathbf{K} = \mathrm{Forb}(\mathbf{F})$, where $\mathbf{F}$ is the class of irreducible set systems (cf. also [N1])

As there are many naturally defined classes that are not described in terms of forbidden irreducible set systems (note for example that for type $\Delta = (2)$ the only irreducible systems are complete graphs) the following general problem (asked in many particular forms by P. Erdös) arises:

Given $A \in \mathrm{Soc}(\Delta)$ determine which classes $\mathbf{K} \subseteq \mathrm{Soc}(\Delta)$ have the A-Ramsey property.

The answer to this question is hard even in the simplest cases, i.e. for $|A| = 1$ and $\Delta = (2)$. However, quite general satisfactory answers can be found.

In [NR4] and [NR6], partite amalgamation was introduced. This method was successful in [NR7] [NR8] and moreover allowed simplification of (difficult) proofs of some earlier statements. It was furthermore observed that it is possible to apply this proof technique to Hales Jewett cubes and to arithmetic progression. This motivated further research with P. Frankl and R.L. Graham, [FRG] where extending the earlier results of H.J. Prömel [P1, P2], we succeeded in applying partite amalgamation to vector spaces and parameter words. This paper was followed by work of J. Nesétril, H.J. Prömel, B. Voigt and myself ([NR9, NR10, PV]) which describes the general form of results which can be obtained by partite amalgamation.

Euclidean Ramsey Theory

Hadwiger and Nelson [H1] posed the problem of determining the chromatic number $\chi(n)$ of Euclidean space $\mathbf{R}^n$ – the minimum integer $r + 1$ such that for every decomposition $\mathbf{R}^n = X_1 \cup X_2 \cup \cdots \cup X_r$ and every positive real α there are two points in X_i for some i with distance precisely α. It was proven in [H1], [MM], [W] that $4 \leq \chi(2) \leq 7$ and it follows moreover from [Fa] that under the assumption that the sets X_i are measurable, the lower bound can be replaced by 5. The best current bounds for $\chi(n)$ are

$$(1.2)^n \le \chi(n) \le (3 + o(1))^n.$$

The lower bound is due to Frankl and Wilson [FW], the upper bound to Larman and Rogers [LR].

Generalizing the problem of Hadwiger and Nelson, the following concept was investigated in [EG]. A finite subset $A \subset \mathbf{R}^d$ is called *Ramsey* if for every r there exists $n = n(r, A)$ such that for every partition $\mathbf{R}^n = X_1 \cup X_2 \cup \cdots \cup X_r$ there is some i and $A' \subset X_i$ with A' congruent to A.

In a series of papers, Erdös et al. [EG] have investigated this property. They have shown that all Ramsey sets A are *spherical*, that is, A has a circumcenter, a point equidistant to all points of A. On the other hand, it is proved in [EG] that the vertex sets of rectangular parallelepipeds (and therefore all their subsets) are Ramsey.

The simplest sets that are spherical but can not be embedded into rectangular parallelepipeds are obtuse triangles.

In [FR3], (cf. also [FR2]) it is shown that all vertex sets of nondegenerate simplicies are Ramsey. Moreover, for both simplices and rectangular parallelepipeds one can in fact choose $n(r, A) = C(A) \log r$, where $C(A)$ is an appropriate positive constant. This can be further strengthened as follows:

Let $S^m(\varrho) = \{(x_0, x_1, \ldots, x_m) \in R^{m+1} : x_0^2 + x_1^2 + \cdots + x_m^2 = \varrho^2\}$ be the m-dimensional sphere of radius ϱ. In [Gr1], a configuration A is called *sphere-Ramsey* if for every r there exists $m = m(A, r)$ and $\varrho = \varrho(A)$ such that for every partition $S_\varrho^m = X_1 \cup \cdots \cup X_r$ there exists $A' \subset X_i$ with A' congruent to A. R.L. Graham [Gr1] proved that rectangular parallelepipeds are sphere-Ramsey and asked whether for a parallelepiped A with circumradius α one can choose $\varrho(A) = \alpha + \delta$, where $\delta > 0$ is arbitrarily small. A positive answer to this question for 2-point configurations, follows from a result of Lovász [L3]. In [FR3], a positive answer to Graham's equation is given in the following stronger form:

Call a set A *hyper-Ramsey* if for all $\delta > 0$ there exist positive constants $c = c(A, \delta)$, $\varepsilon = \varepsilon(\varepsilon, \delta)$ and subsets $X = X(m) \subset S^m(\varrho(A) + \delta)$ for $m > m_0(\delta)$ such that

i) $|X(m)| < c^m$ but if
ii) $|Y| \ge (1 - \varepsilon)^m |X(m)|$ and $Y \subset X(m)$

then Y contains a congruent copy of A.

It has been stated in [FR3] that if $A \subset R^n$, $B \subset R^m$ are hyper-Ramsey, then, so is their product $A * B = \{x * y | x \in A, y \in B\}$ where for $x = (x_1, x_2, \ldots, x_n)$ and $y = (y_1, \ldots, y_m)$ one defines $x * y = (x_1, \ldots, x_n, y_1, \ldots, y_m)$. From [FW], it follows that every 2-element set is hyper-Ramsey (cf. also [R5]). Thus, also the products of 2-element sets, i.e. all rectangular parallelepipeds are hyper-Ramsey (which answers Graham's conjecture affirmatively).

Our current knowledge does not exclude the possibility that all spherical sets are hyper-Ramsey, but we can not prove this even for non-degenerate simplices. The first step towards settling this problem would be to decide whether all obtuse triangles are hyper-Ramsey. We can, however, prove that non-degenerate simplices (and also their products) are exponentially sphere-Ramsey. This means that they are sphere-Ramsey and there exists $\varepsilon = \varepsilon(A) > 0$ and constant $C = C(A)$ such that $m = m(A, r)$ can be chosen to satisfy $m \le C(1 + \varepsilon)^r$. The proof of this fact is given in [FR3] and the main tool in proving it is a rather difficult theorem about intersection properties of partitions [FR1].

Concerning the original problem of characterizing Ramsey configurations, the first problem left in [FR3] was to decide whether the vertices of the regular pentagon form a Ramsey configuration. Recently, I. Kriz [K] proved that the vertices of the regular $k-gon$ A_k form a Ramsey configuration for any positive integer k. This supports the hope that Ramsey configurations might coincide with spherical ones. The bounds on $n(A_k, r)$ which follow from the proof given in [*K*], seem to be rather weak.

Finally, we will state here one more problem. In [F5], P. Erdös asked whether $n(T, 2) = 2$ for any configuration of three vertices of a non-regular triangle.

Van der Waerden's Theorem

In 1927, B.L. van der Waerden [V] published the proof of the following result:

Theorem (Van der Waerden's Theorem). *For all positive integers k and r, there exists an integer W(k, r) so that if the set of integers* $\{1, 2, \ldots, W(k, r)]$ *is partitioned into r classes, then at least one class contains a k-term arithmetic progression.*

Set $W(k) = W(k, 2)$, known values are $W(2) = 3$, $W(3) = 9$, $W(4) = 35$, $W(5) = 178$ (see [Gr2]). The original proof of van der Waerden, as well as the later proof (cf. [GR3], [D3], [T1]), give very weak upper bounds on the function $W(k)$. In 1987, S. Shelah [S2] found a fundamentally new proof that yields a much better upper bound.

Set $f_1(n) = 2n$, $f_2(n) = 2^n$, $\ldots$, $f_{i+1}(n) = f_i^n(1)$, where f_i^n denotes the n-times iterated function f_i. So, $f_3(n)$ is the n-times iterated exponential consisting of a tower of n twos. The function $f_4(n)$ grows much faster: $f_4(1) = f_3(1) = 2$, $f_4(2) = f_3 \circ f_3(1) = 2^2$, $f_4(3) = f_3 \circ [f_3 \circ f_3(1)] = f_3[f_4(2)] = 65536$ which is a tower of 4 twos, and $f_4(4) = f_3[f_4(3)]$ is a tower of 65536 twos. While the earlier best upper bounds on $W(k)$ were of the order of the Ackerman function, i.e. $f_k(ck)$ for $c > 0$, the bound given by S. Shelah [S2] is "only" $W(k) \leq f_4(k + 1)$ (cf. also [GRS]). The problem of finding reasonable estimates for $W(k)$ is one of the main problems of the area and the result of Shelah is a vast improvement of the previous bound. The best known lower bounds are exponential: $W(k + 1) > k \cdot 2^k$ for k prime was proved in [B2], while $W(k) > k^{1-\varepsilon}2^k$ for $k > k_0(\varepsilon)$ was recently established in [S6].

In [ET], P. Erdös and P. Turán considered the quantity $r_k(n)$ defined as the maximum cardinality $|Z|$ of a set $Z \subset \{1, 2, \ldots, n\}$ that does not contain an arithmetic progression of k terms. One of the reasons for investigating $r_k(n)$ is that $r_k(n) < n/2$ implies $W(k) < n$ and bounds on $r_k(n)$ could improve the upper bound for $W(k)$. Erdös and Turán noted that $r_k(n + m) \leq r_k(n) + r_k(n)$ which implies that $\lim_{n \to \infty} = c_k$ exists. Erdös and Turán conjectured that $c_k = 0$ for all k. They also conjectured that $r_k(n) < n^{1-\varepsilon_k}$ which was disproved by Salem and Spencer [SS] who proved $r_3(n) > n \exp\left(-\dfrac{c \ln n}{\ln \ln n}\right)$. This was improved in 1946 by F.A. Behrend who established $r_3(n) > n \exp\left(-\frac{c \ln n}{\sqrt{\ln n}}\right)$, the best current upper bound for $r_3(n)$ is due to Heath-Brown [H2] and Szemerédi. In 1952, Roth proved that $r_3(n) < c\, n/\log\log n$. In 1967, E. Szemerédi [S7] proved $r_4(n) = o(n)$. His proof used the general statement of van der Waerden's theorem. Later, Roth [R8, R9] gave a different proof that $r_4(n) = o(n)$ (and his proof probably yields $r_4(n) < n/\log_l n$ where l is a large fixed integer and $\log_l n$ denotes the l-times iterated logarithm cf [S8]). In 1973, Szemerédi [S9] gave a proof of the Erdös-Turán conjecture, establishing that $c_k = 0$ for every k. A different proof based on ergodic theory was found in 1977 by H. Fürstenberg [Fu]. Since then, the methods of ergodic theory have proven to be a powerful tool for investigating related questions; proofs of several extensions of Szemerédi's theorem were found [FK1]. Recently, Fürstenberg and Katznelson, establishing a conjecture of R.L. Graham [Gr1], proved the density form of Hales-Jewett's theorem. The methods of ergodic theory as well as the proof of Szemerédi do not yield bounds on $r_k(n)$. Therefore, finding alternative proof techniques is still of interest. Recently, P. Frankl and the author considered the following question. Let $\{a_i, b_i\}$, $1 \le i \le k$ be pairwise disjoint 2-element sets. Define $F_i = \{a_1, a_2, \ldots, a_k, b_i\} - \{a_i\}$ and $\mathbf{F} = \mathbf{F}(k) = \{F_1, F_2, \ldots, F_k\}$. Let $\mathscr{E} \subset [V]^k$ be such that i) $|V| = n$, ii) $|E_1 \cap E_2| \le k-2$, for any two distinct $E_1, E_2, \in \mathscr{E}$ and iii) $\mathscr{E}$ does not contain $\mathbf{F}(k)$ as a subconfiguration.

Set $e\tilde{x}(n, \mathbf{F}(k)) = \mathrm{Max}\{|\mathscr{E}|; \mathscr{E}$ satisfies i), ii), and iii)$\}$. With P. Frankl, we recently noticed that $e\tilde{x}(n, \mathbf{F}(k)) \ge c_k\, n^{k-2} r_k(n)$, and thus,

$$e\tilde{x}(n, \mathbf{F}(k)) = o(n^{k-1}) \tag{6}$$

would imply $r_k(n) = o(n)$. (In fact, this implies a density version of the Graham-Leeb-Rothschild Theorem as well.) This is known to be true for $k = 3$ by [RS] and with P. Frankl, we recently gave the proof for $k = 4$. We conjecture that (6) holds for every k.

References

[AH] F.G. Abramson, L.A. Harrington: Models without indiscernibles. J. Symbolic Logic **43** (1978) 572–600

[AKS] M. Ajtai, J. Komlós, E. Szemerédi: A note on Ramsey numbers. J. Comb. Th. (A) **29** (1980) 354–360

[B1] F.A. Behrend: On sets of integers which contain no three in arithmetic progression. Proc. Nat. Acad. Sci. **23** (1946) 331–332

[B2] E.R. Berlekamp: A Construction for partitions which avoid long arithmetic progressions. Can. Math. Bull. **11** (1968) 409–414

[CS] T.J. Carlson, S.G. Simpson: Topological Ramsey theory. To appear in: Mathematics of Ramsey theory (J. Nesetril and V. Rödl, eds.). Springer, Berlin

[CW] D. Coppersmith, S. Winograd: Matrix multiplication via arithmetic progressions. 19th Annual ACM Symposium on Theory of (1987)

[D1] W. Deuber: Generalization of Ramsey's theorem. In: Infinite and Finite Sets (A. Hajnal, R. Rado and V.T. Sós, eds.) Colloq. Math. Soc. János Bolyai 10. North-Holland, Amsterdam 1975, pp. 323–332

[D2] W. Deuber: Partitionen und lineare Gleichungssysteme. Math. Z. **133** (1973) 109–123

[D3] W. Deuber: On van der Waerden's theorem on arithmetic progressions. J. Comb. Th. **A32** (1982) 115–118

[DR] W. Deuber, B.L. Rothschild: Categories without the Ramsey property. In: Combinatorics (A. Hajnal, V.T. Sós, eds.) Coll. Math. Sco. János Bolyai. **18** (1976) 225–249

[E1] P. Erdös: Some remarks on the theory of graphs. Bull. Amer. Math. Soc. **53** (1947) 292–294

[E2] P. Erdös: Graph theory and probability. Canad. J. Math **11** (1959) 34–38

[E3] P. Erdös: The art of counting. Selected Writings. The MIT Press, Cambridge, Massachusets and London, England (1973)

[E4] P. Erdös: Problems and results on finite and infinite graphs. In: Recent advances in graph theory. Czchoslovak Academy of Sciences 1975, pp. 183–192

[E5] P. Erdös: Combinatorial problems in geometry and number theory. In: Proceedings of the Symposium in Pure Math, vol. 34 (1979) AMS, pp. 149–162

[EG] P. Erdös, R.L. Graham, P. Montgomery, B.L. Rothschild, J. Spencer, E.G. Straus: Euclidean Ramsey Theorems. I. J. Combin. Theory Ser. A **14** (1973) 341–363

[EHP] P. Erdös, A. Hajnal, L. Pósa: Strong embeddings of graphs into colored graphs. In: Infinite and finite sets (A. Hajnal, R. Rado and V.T. Sós, eds.). Colloq. Math. Soc. János Bolyai 10. North-Holland, Amsterdam 1975, pp. 585–595

[EHMR] P. Erdös, P. Hajnal, A. Máté, R. Rado: Combinatorial set theory. Partition relations for cardinals. North-Holland, Amsterdam 1984

[EHR] P. Erdös, A. Hajnal, R. Rado: Partition relations for cardinal numbers. Acta Math. Acad. Sci. Hungar. **16** (1965) 93–196

[EL] P. Erdös, L. Lovász: Problems and results on 3-chromatic hypergraphs and some related questions. In: Infinite and finite sets (A. Hajnal, R. Rado and V.T. Sós, eds.). North-Holland, Amsterdam 1975, pp. 609–628

[ES] P. Erdös, G. Szekeres: A Combinatorial problem in geometry. Comp. Math. **2** (1935) 463–470

[ET] P. Erdös, and P. Turán: On some sequences of integers. J. London Math. Soc. **11** (1936) 261–264

[Fa] K.J. Falconer: The realization of disances in measurable subsets covering R^n. J. Combin. Theory, Ser A **31** (1981) 184–189

[Fo] J. Folkman: Graph with monochromatic complete subgraphs in every edge coloring. SIAM J. Appl. Math. **18** (1970) 19–24

[FGR] P. Frankl, R.L. Graham, V. Rödl: Induced restricted Ramsey theorems for spaces. J. Comb. Theory, Ser. A **44** (1987) 120–128

[FR1] P. Frankl, V. Rödl: Forbidden Intersections. Trans. Amer. Math. Soc. **299** (1987)

[FR2] P. Frankl, V. Rödl: All triangles are Ramsey. Trans. Amer. Math. **297** (1986) 777–779

[FR3] P. Frankl, V. Rödl: A partition property of simplices in Euclidean space. J. Amer. Math. **3**(1) (1990) 1–7

[FR4] P. Frankl, V. Rödl: The uniformity lemma for hypergraphs. Graphs and Combinatorics. (To appear)

[FW] P. Frankl, R.M. Wilson: Intersection theorems with geometric consequences. Combinatorica **1** (1981) 357–368

[Fr] Frasnay: Sur des functions d'entiers se rapportant an Ramsey. C. R. Acad. Sci. Paris **256** (1963) 2507–2510

[Fu] H. Fürstenberg: Ergodic behavior of diagonal measures and a theorem of Szemerédi on arithmetic progressions. J. Anal. Math. **31** (1977) 204–256

[FK1] H. Fürstenberg, Y. Katznelson: An ergodic Szemerédi theorem. IP Systems and Combinatorial Theory. J. Anal. Math. **45** (1985) 117–168

[FK2] H. Fürstenberg, Y. Katznelson: A density version of the Hales-Jewett theorem. (1990)

[FW] P. Frankl, R.M. Wilson: Intersection theorems with geometric consequences. Combinatorica **1** (1981) 357–368

[Go] Goodman: On sets of acquaintances and strangers at any party. Amer. Math. Monthly **66** (1959) 778–783

[Gr1] R.L. Graham: Old and new euclidean Ramsey theorems. Discrete geometry and convexity. Ann. New York Acad. Sci. **440** (1985) 20–30

[Gr2] R.L. Graham: Recent developments in Ramsey theory. Proceedings of the International Congress of Mathematicians, Warszawa 1983, pp. 1555–1567

[GLR] R.L. Graham, K. Leeb, B.L. Rothschild: Ramsey's theorem for a class of categories. Adv. Math. **8** (1972) 417–433

[GR1] R.L. Graham, V. Rödl: Numbers in Ramsey theory. In: Surveys in Combinatorics (ed. C. Whitehead). (LMS Lecture Note Series 123.) Cambridge University Press, pp. 111–153

[GR2] R.L. Graham, B.L. Rothschild: Ramsey's theorem for *n*-parameter sets. Trans. Amer. Math. Soc. **159** (1971) 257–292

[GR3] R.L. Graham, B.L. Rothschild: A short proof of van der Waerden's theorem on arithmetic progressions. Proc. Amer. Math. Soc. **42** (1974) 356–386

[GRS] R.L. Graham, B.L. Rothschild, J.H. Spencer: Ramsey Theory. John Wiley and Sons, New York 1980

[GY] J.E. Graver, J. Yackel: Some graph theoretic results associated with Ramsey's theorem. J. Comb. Theory Ser. A. **4** (1968) 125–175

[H1] H. Hadwiger: Ungelöste Probleme. No. 40. Elemente der Math. **16** (1961) 103–104

[H2] D.R. Heath-Brown: Integer sets containing no arithmetic progressions. 1986

[HJ] A.W. Hales, R.I. Jewett: Regularity and positional games. Trans. Amer. Math. Soc. **106** (1963) 222–229

[K] I. Kriz: Permutation groups in Euclidean Ramsey theory. (To appear)

[KV] P.G. Kolaitis, M. Vardi: The decision problem for the probabilities of higher-order properties. In: 19th 'Annual ACM Symposium (STOC)', pp. 425–435

[L1] K. Leeb: Vorlesungen über Pascaltheorie. Erlangen 1973

[L2] G. Lorden: Blue-empty chromatic graphs. Amer. Math. Monthly **69** (1962) 114–120

[L3] L. Lovász: Self-dual polylopes and the chromatic number of distance graphs on the sphere. Acta Sci. Math. **45** (1983) 317–323

[LR] D.G. Larman, C.A. Rogers: The realization of distances within sets in Euclidean space. Mathematika **19** (1972) 1–24

[MM] L. Moser, W. Moser: Solution to Problem 10. Cand. Math. Bull. **16** (1961) 187–189

[N1] J. Nesetril: There are 4 types of Ramsey classes of graphs. J. Comb. Theorems B **46**(2) (1989) 127–132

[N2] J. Nesetril: Ramsey theory. (To appear in Handbook of Combinatorics, R.L. Graham, M. Grötschel, L. Lovász, ed.)

[NR1] J. Nesetril, V. Rödl: Type theory of partition properties of graphs. In: Recent Advances in Graph Theory (ed. M. Fiedler). Academia, Prague 1975, pp. 405–412

[NR2] J. Nesetril, V. Rödl: A structural generalization of the Ramsey theorem. Bull. Amer. Math. Soc. **83**(1) (1977) 127–128

[NR3] J. Nesetril, V. Rödl: Partition of relational and set-systems. J. Comb. Theorems **A22** (1977) 289–312

[NR4] J. Nesetril, V. Rödl: A short proof of the existence of restricted Ramsey graphs by means of a partite construction. Combinatorica **1**(2) (1981) 199–202

[NR5] J. Nesetril, V. Rödl: Ramsey classes of set systems. J. Comb. Theory **A34** (1983) 183–201

[NR6] J. Nesetril, V. Rödl: Combinatorial partitions of finite cosets and Lattices – Ramsey lattices. Algebra Universalis **19** (1984) 106–119

[NR7] J. Nesetril, V. Rödl: Sparse Ramsey graphs. Combinatorica **4**(1) (1984) 71–78

[NR8] J. Nesetril, V. Rödl: Strong Ramsey theorems for Steiner systems. Trans. Amer. Math. Soc. **303**(1) (1987) 183–197

[NR9] J. Nesetril, V. Rödl: Partite construction and Ramseyian theorems for sets, numbers and spaces. Comment. Math. Univ. Carol. **28** (1987) 569–580

[NR10] J. Nesetril, V. Rödl: The partite construction and Ramsey set systems. Discr. Math. **75** (1989) 327–334

[NR11] Mathematics of Ramsey theory (ed. J. Nesetril and V. Rödl). (Algorithms and Combinatorics, vol. 5). Springer, Berlin Heidelberg New York 1990

[NR12] J. Nesetril, V. Rödl: The partite construction and Ramsey space systems. In: Mathematics of Ramsey theory. Springer, Berlin Heidelberg New York 1990

[P1] H. J. Prömel: Induced partition properties of combinatorial cubes. J. Comb. Theory Ser. A **39** (1985) 177–208

[P2] H. J. Prömel: Partition properties of q-hypergraphs. J. Comb. Theory Ser. B **41** (1986) 356–385

[PH] J. Paris, L. Harrington: A mathematical incompleteness in Peano arithmetic. In: Handbook of Mathematical Logic (ed. J. Bairwise). North-Holland, Amsterdam 1977, pp. 1133–1142

[PV] H.J. Prömel, B. Voigt: A sparse Graham-Rothschild theorem. Trans. Amer. Math. Soc. **309**(1) (1988) 113–137

[RK] S.P. Radziszowski, D.L. Kreher: Search algorithm for Ramsey graphs by union of group orbits. J. Graph Theory **12**(1) (1988) 59–72

[R1] R. Rado: Studien zur Kombinatorik. Math. Z. **36** (1933) 425–480

[R2] R. Rado: Note on combinatorial analysis. Proc. London Math. Soc. **48** (1943) 122–160

[R3] F.P. Ramsey: On a Problem for Formal Logic. Proc. London Math. Soc. **30** (1930) 264–286

[R4] V. Rödl: A generalization of the Ramsey theorem. In: Graphs, hypergraphs and block systems, Zielona Gora 1976, pp. 211–219

[R5] V. Rödl: On a problem in combinatorial geometry. Discr. Math. **45** (1983) 129–131

[R6] V. Rödl: Upper bounds on Ramsey numbers $R(k, l)$. To appear

[R7] K. Roth: On Certain Sets of Integers. J. London Math. **28** (1953) 104–109

[R8] K. Roth: Irregularities of Sequences Relative to Arithmetic Progressions, III. J. Number Theory MR **41** (1970) 125–142

[R9] K. Roth: Irregularities of sequences relative to arithmetic progressions, IV. Per. Math. Hungar. (1972) 301–326

[R10] S.P. Radziszowski: Small Ramsey numbers (manuscript 1991)

[RS] I.Z. Ruzsa, E. Szemerédi: Triple systems with no six points carrying three triangles. Colloq. Math. J. Bolyai **18** (1978) 939–945

[SS] R. Salem, D.C. Spencer: On sets of integers which contain no three terms in arithmetic progression. Proc. Nat. Acad. Sci. **28** (1942) 561–563

[S1] I. Schur: Uber die Kongruenz $x^m + y^m \equiv z^m \pmod{\varrho}$. Jber. Deutsch. Math.-Verein. **25** (1916) 114–117

[S2] J.B. Shearer: A note on the independence number of a triangle-free graph. Discr. Math. **46** (1983) 83–87

[S3] S. Shelah: Primitive recursive bounds for van der Waerden numbers. J. Amer. Math. Soc. **1** (1988) 683–697

[S4] J. Spencer: Ramsey's theorem – A new lower bound. J. Comb. Theory (A) **18** (1975) 108–115

[S5] J. Spencer: Asymptotic lower bounds for Ramsey functions. Discr. Math. **20** (1977) 69–76

[S6] Z. Szabó: An application of Lovász local lemma – A new lower bound for the van der Waerden number. Random structures and algorithms **1**(3) (1990) 343–360

[S7] E. Szemerédi: On sets of integers containing no four elements in arithmetic progression. Acta Math. Acad. Sci. Hung. **20** (1969) 89–104

[S8] E. Szemerédi: On sets of integers containing no *k*-elements in arithmetic progression. Proceedings of the International Congress of Mathematicians, Vancouver 1974, pp. 503–505

[S9] E. Szemerédi: On sets of integers containing no *k* elements in arithmetic progression. Acta Arith. **27** (1975) 199–245

[T1] A.D. Taylor: A note on van der Waerden theorem. J. Comb. Theory Ser. A **33** (1982) 215–219

[T2] A. Thomasson. Upper bounds for Ramsey numbers (to appear)

[V] B.L. van der Waerden: Beweis einer Baudetschen Vermutung. Nieuw Arch. Wisk. **15** (1927) 212–216

[W] D.R. Woodal: Distances realized by sets covering plane. J. Comb. Theory A **14** (1973) 187–200

Strongly Polynomial and Combinatorial Algorithms in Optimization

Éva Tardos

School of Operations Research and Industrial Engineering
Cornell University, Ithaca, NY 14853, USA

1. Introduction

The input to a combinatorial optimization problem usually consists of two parts: the first part describes the combinatorial structure; the second part is a list of numerical data. For example, the input to the maximum-flow and the shortest path problems consists of a network (the combinatorial structure) and numbers that define the capacity and the length of each arc, respectively. An algorithm for these problems is *polynomial* if its running time can be bounded by a polynomial in the size of the underlying combinatorial structure (the number of nodes and edges) and the number of digits needed to write the numerical data. Most of the early polynomial algorithm satisfied, in fact, a stronger notion of efficiency; these algorithms are not only polynomial, but the number of arithmetic operations performed can be bounded by a polynomial in the size of the underlying combinatorial structure alone, independent of the size of the numbers involved. Such algorithms are called *strongly polynomial.*

Strongly polynomial algorithms are more appealing theoretically and the additional insight needed to develop such algorithms can lead to practical improvements. It is an important theoretical question to understand which problems can be solved in strongly polynomial time. In this paper we shall survey partial results in this direction. We discuss some techniques to turn polynomial algorithms into strongly polynomial ones.

Khachiyan's [15] proof that the linear programming problem, $\max(cx : Ax \leq b)$, can be solved in polynomial time has been one of the major breakthroughs in the design and analysis of algorithms in the last ten years. The first polynomial linear programming algorithm was developed using the ellipsoid method of Yudin and Nemirovskii [33] from non-linear programming. Grötschel, Lovász and Schrijver [11] have used the ellipsoid method to develop a powerful technique for proving that many combinatorial optimization problems can be solved in polynomial time (see also [14] and [25]). The polynomial algorithms developed using the ellipsoid method are generally not strongly polynomial. There have been several other polynomial linear programming algorithms developed in the recent years (e.g., Karmarkar's algorithm [13]), however no strongly polynomial

Proceedings of the International Congress
of Mathematicians, Kyoto, Japan, 1990

is known. It is an important open problem if there exists a strongly polynomial linear programming algorithm.

A somewhat related question is if the combinatorial optimization problems that were proved to be solvable in polynomial time using the ellipsoid method can be solved more efficiently using combinatorial techniques, rather than employing the analytic techniques from non-linear programming.

In this paper we shall survey two techniques for converting polynomial algorithms into strongly polynomial ones. The first one uses Diophantine approximation. It replaces the numbers occurring in the problem description by small integers that define an equivalent problem. Any polynomial algorithm can be used to solve the resulting problem in strongly polynomial time. The second technique is iterative. After some preprocessing each iteration finds an approximately optimal solution to the problem. From such an approximate solution the algorithm either finds an optimal solution, or it concludes that some of the defining constraint are not necessary, and these are deleted before the next iteration starts.

There are further techniques known for converting some polynomial algorithms into strongly polynomial ones. The first such technique has been developed by Megiddo [20] for solving combinatorial ratio minimization problems in strongly polynomial time. This technique can turn algorithms that use binary search (which is not strongly polynomial) into strongly polynomial algorithms. One of the most powerful application of this technique, due to Megiddo [21], is testing the feasibility of linear programs with at most two variables in each inequality. A more recent application by Norton, Plotkin and Tardos [22] gives a further extension of the class of linear programs solvable in strongly polynomial time: If a linear program is known to be solvable in strongly polynomial time, then so is its extension by a constant number of additional variables and side constraints. A nice example in this class, which we shall discuss later in more detail, is the concurrent flow problem.

2. Using Diophantine Approximation

One of the most powerful techniques for making combinatorial optimization algorithms strongly polynomial, due to Frank and Tardos [7], uses simultaneous Diophantine approximation. Many combinatorial optimization problems can be defined as a pair of a (highly structured) system of subsets $\mathscr{I}$ of a finite set E and weights $w(e)$ for each $e \in E$ (e.g., perfect matchings in a graph form a system of the subsets of the edges). The weight of a set $I \in \mathscr{I}$ is $w(I) = \sum_{e \in I} w(e)$. The problem is to find a set in $\mathscr{I}$ with maximum weight. Let n denote the size of the set E. The size of such the combinatorial structure $\mathscr{I}$ depends on the way $\mathscr{I}$ is specified, but generally it is at least n.

The idea of converting polynomial algorithms for problems in the above form into strongly polynomial ones is as follows. Consider first a weight function $\hat{w}$ whose coordinates are integers with size no more than a polynomial in n. For such inputs there is no distinction between polynomial and strongly polynomial algorithms. The key lemma is that for every weight function w there is an equivalent weight function $\hat{w}$ that consists of small integers. Furthermore, the new weight function $\hat{w}$ can be found in strongly polynomial time. The strongly polynomial algorithm first computes such an equivalent weight function $\hat{w}$, and then uses the polynomial algorithm to find the maximum weight set subject to the new weights.

Two weight functions are *equivalent* for a problem $\mathcal{I}$, if for every two sets $I, J \in \mathcal{I}$ we have that $w(I) < w(J)$ if and only if $\hat{w}(I) < \hat{w}(J)$. Clearly, the maximum weight set in $\mathcal{I}$ is the same for any two equivalent weight functions. The existence of an equivalent weight function whose coordinates are small integers was first observed by Orlin [23]. Orlin's proof is not algorithmic. Frank and Tardos [7] gave an algorithmic proof using the simultaneous Diophantine approximation technique of Lovász (see in [18]).

Throughout the paper we shall use different norms. For a vector x we use $||x||_\infty$ to denote the maximum absolute value of a coordinate of x, and $||x||_1$ to denote the sum of the absolute values of the coordinates of x.

Theorem 1. *For a given n-dimensional vector w, and an integer N one can find an integer vector $\hat{w}$ in time polynomial in n and $\log N$, such that $||\hat{w}||_\infty \leq 2^{O(n^3)} N^{O(n^2)}$ and for every integer vector a such that $||a||_1 \leq N$, $aw < 0$ if and only if $a\hat{w} < 0$.*

Let w be the weight function in a combinatorial optimization problem. If Theorem 1 is applied with w and $N = n + 1$, we obtain an equivalent weight function $\hat{w}$.

Next we try to give an idea of the proof of Theorem 1. Consider a vector w and an integer N. Lovász's [18] algorithm finds a positive integer q and an integer vector w' such that (1) $||qw - w'||_\infty \leq 1/N$, and (2) $q \leq 2^{O(n^2)} N^n$. This vector w' satisfies a property similar to that required by Theorem 1.

Lemma 2. *Let w' be the vector obtained from w and N by the Diophantine approximation algorithm, and let a be an integer vector such that $||a||_1 \leq N$. If $aw \leq 0$ then also $aw' \leq 0$.*

Proof. Assume $aw \leq 0$. Consider aw'. The sum of the coordinates of a is at most N. Using this fact, and the first property of Diophantine approximation we obtain that $a(w' - qw) < 1$. We also have that $aw \leq 0$ and $q \geq 0$, therefore $aw' < 1$. Since both a and w' are integral, it follows that $aw' \leq 0$. □

The lemma implies that the vector w' satisfies all the relevant non-strict inequalities satisfied by w. The vector $\hat{w}$ in Theorem 1 must also satisfy all of the strict inequalities. Theorem 1 can be proved by a procedure that uses Diophantine approximation repeatedly.

A nice application of Theorem 1 shows that a maximum weight clique in a perfect graph can be found in strongly polynomial time. The first polynomial algorithm for the problem was developed by Grötschel, Lovász and Schrijver [11] using the ellipsoid method. Theorem 1 can be used to convert this algorithm into a strongly polynomial one.

Many combinatorial optimization problems are related to linear programming. Combinatorial special cases of linear programming include several network flow problems. These problems generally have constraint matrix with small integer entries. In Section 4 we shall discuss the maximum-flow, the transshipment and the multi-commodity flow problems in more detail. The constraint matrix in these problems describes the combinatorial structure (e.g., it is the incidence matrix of the network). The simultaneous Diophantine approximation technique can be used to convert polynomial algorithms for these problems into strongly polynomial ones.

Consider the linear program $\max(cx : Ax \leq b)$ where A is an n by m matrix. Assume that the matrix A is integral, $m \geq n$, and let $\Delta(A)$ denote the maximum absolute value of a subdeterminant of A. We say that two objective functions c and $\hat{c}$, and the right-hand sides b and $\hat{b}$ are equivalent for the constraint matrix A if the same set of inequalities are satisfied as equations at the optimal solution for the two linear programs $\max(cx : Ax \leq b)$ and $\max(\hat{c}x : Ax \leq \hat{b})$. Using linear programming duality and Theorem 1 we can prove the following theorem.

Theorem 3. *For any linear program* $\max(cx : Ax \leq b)$ *with an n by m integer matrix A, there is an equivalent integral right-hand side* $\hat{b}$ *and objective function* $\hat{c}$ *such that* $\log ||b||_\infty$ *and* $\log ||c||_\infty$ *are polynomially bounded in n, m and* $\log \Delta(A)$.

Corollary 4. *Linear programs where the constraint matrix A is integral and has coefficients whose size is at most polynomial in n and m, can be solved in strongly polynomial time.*

3. Iterative Approach

In this section we discuss an iterative method for converting polynomial algorithms into strongly polynomial ones. When using the iterative technique it is possible to take advantage of additional insight into the combinatorial structure of the problem in question. In some cases this leads to very efficient algorithms. For example, highly efficient algorithms for the maximum-flow and transshipment problems have been obtained this way.

Consider a linear program $\max(cx : Ax \leq b)$. The *dual* of this problem is the problem $\min(by : A^T y = c, y \geq 0)$. We say that x' is *feasible* if $Ax' \leq b$, and y' is *dual feasible* if $y' \geq 0$ and $A^T y' = c$. It is well-known that the linear program and its dual have the same optimal value. *Complementary slackness* gives a localized condition that helps recognize a pair of primal and dual optimal solutions. A feasible solution x' and a feasible dual solution y' are optimal if and only if for every row $a_i x \leq \beta_i$ of $Ax \leq b$ and the corresponding coordinate y_i of y, we have

$$a_i x' < \beta_i \Longrightarrow y'_i = 0 \ . \qquad (1)$$

The key concept of the iterative method is ε-optimality. This is a relaxation of the complementary slackness conditions. A feasible solution x' and feasible dual solution y' are ε-optimal if for every row i we have that

$$a_i x' < \beta_i - \varepsilon \Longrightarrow y'_i = 0 \ . \qquad (2)$$

Note that ε-optimality with $\varepsilon = 0$ is the same as the complementarity slackness conditions. It is not too difficult to show that an ε-optimal pair (x', y') is close to being optimal both in terms of its objective function value and in terms of the distance to an optimal solution. This was first proved by Mangasarian [19]. For our purposes it will be important that the bound on the distance of the ε-optimal x' to an optimal solution is independent of the size of the numbers in the objective function. Such a bound was proven by Cook, Gerards, Schrijver and Tardos [4].

Theorem 5. *Suppose x' and y' form a pair of ε-optimal primal and dual solutions to the linear program* $\max(cx : Ax \leq b)$. *Then there exists an optimal x^* such that* $||x' - x^*||_\infty \leq n\varepsilon\Delta(A)$.

Assume that A is an integer matrix, and let α denote $||A||_\infty$, the largest absolute value of a coefficient of A.

Corollary 6. *Suppose that x' and y' are a pair of ε-optimal solutions to the linear program* $\max(cx : Ax \leq b)$ *and its dual. Consider a constraint $ax \leq \beta$. Suppose x' satisfies $\beta - ax' \geq n\varepsilon\alpha\Delta(A)$. The coefficient corresponding to the inequality $ax \leq \beta$ is zero in every optimal dual solution, and every optimal solution is also optimal if the constraint $ax \leq \beta$ is deleted.*

This theorem and corollary are the key of the iterative strongly polynomial algorithm. The algorithm repeatedly finds an ε-optimal primal-dual solution pair, and deletes all constraints that are known (by Corollary 6) not to be tight at an optimal solution. This procedure is repeated until only the tight constraints are left. The algorithm terminates in at most as many iterations as the number of defining inequalities. There are two remaining issues that need to be discussed: how can one guarantee that at least one inequality will be deleted each iteration, and how can one find an ε-optimal solution in strongly polynomial time.

In order to guarantee that at least one inequality will be deleted one has to do some preprocessing before the iteration, and choose an appropriate ε. In most special cases the preprocessing is quite simple, in the general case it involves a projection. See [30, 27] for more details for the general case. After such preprocessing the appropriate choice of ε turns out to be $||b||_\infty/(n^2\Delta(A))$.

Next consider the issue of finding an ε-optimal solution. This can be accomplished in strongly polynomial time if $\Delta(A)$ is "small". Round every coordinate of b to an integer multiple of ε. A pair of optimal primal and dual solutions for the rounded problem is ε-optimal for the original problem. If $\Delta(A)$ is small then the coefficients of the rounded b are small multiples of ε. If a polynomial, but not strongly polynomial, algorithm is used to solve the rounded problem, then the resulting running time will be independent of the size of the numbers in the original vector b. However, it will depend on the size of the numbers in the objective function, c. In order to get rid of the dependence on c we need to solve the rounded problem by applying the same iterative scheme to its dual. The running time of the resulting algorithm is independent of the size of the numbers in the objective function as well as in the right-hand side.

4. Network Flow Problems

The most well-known linear programs with small integral constraint matrices are network flow problems. These problems can be used to illustrate some of the ideas from the previous sections. By taking advantage of the special structure one can make the general algorithms simpler and more efficient.

A network is a directed graph $G = (V, E)$. We shall use n and m to denote the number of nodes and edges, respectively. The *maximum-flow problem* is defined by a network, a source $s \in V$ and a sink $t \in V$ and nonnegative capacities $u(v, w)$ associated with the edges $(v, w) \in E$. A vector $f(v, w)$ for $(v, w) \in E$ is a *preflow* if $0 \leq f(v, w) \leq u(v, w)$ for every $e \in E$. Its *excess* at a node $v \in V$ is defined to be

$$e_f(v) = \sum_w f(w, v) - \sum_w f(v, w) \ . \tag{3}$$

A preflow f is a *flow* if $e_f(v) = 0$ for every node $v \neq s, t$. The *value* of the flow is $e_f(t)$. The problem is to find a flow with maximum value.

In the *transshipment problem* there are costs $c(v, w)$ associated with the edges instead of capacities, and there are demands $b(v)$ on the nodes $v \in V$. A preflow f is a *transshipment* if $e_f(v) = b(v)$ for every node $v \in V$. The *cost of a preflow* f is $\sum_{vw \in E} f(v, w)c(v, w)$. The problem is to find a transshipment of minimum cost.

Both the maximum-flow and the transshipment problems are linear programs with $0, \pm 1$ constraint matrix. The capacities and the demands form the right-hand side, the costs are the coefficients of the objective function. The strongly polynomial solvability of these problems follows, for example, from Corollary 4.

The first polynomial algorithm for the maximum-flow problem is due independently to Dinic [5] and Edmonds and Karp [6]. Both algorithms are strongly polynomial. The first polynomial algorithm for the transshipment problem is due to Edmonds and Karp [6]. This algorithm is *not* strongly polynomial. The first strongly polynomial algorithm was developed much later [29] using the iterative method of Section 3.

The *multi-commodity flow problem* is defined similarly. There are k pairs of sources and sinks $s_i, t_i \in V$ and each has an associated demand, d_i. The problem is to find flows f_i from s_i to t_i of value d_i such that a joint capacity constraint, $\sum_{(v,w) \in E} f_i(v, w) \leq u(v, w)$, is satisfied for each edge $(v, w) \in E$. The multi-commodity flow problem is also a linear program with $0, \pm 1$ constraint matrix, and therefore it can be solved in strongly polynomial time.

The multi-commodity flow problem is fairly a general linear program. Dinic (see in [1]) and Itai [12] have independently proved that any linear program can be reduced to a 2-commodity flow problem in polynomial time.

4.1 Transshipment Problem

The first strongly polynomial algorithm for the transshipment problem [29] is rather slow and complicated. It is more appropriate to refer to it as a proof of solvability in strongly polynomial time, rather than as an efficient algorithm. Since then, many strongly polynomial algorithms have been discovered and some of them are surprisingly simple and efficient, such as the recent algorithms due to Goldberg and Tarjan [9] and Orlin [24]. Here we shall focus Orlin's algorithm.

The dual variables in the transshipment problem are prices $p(v)$ at the nodes $v \in V$. The dual constraints require that the *reduced* cost $c_p(v,w) = p(v)+c(v,w)-p(w)$ is nonnegative for every arc. The complementary slackness conditions require that the reduced cost is zero for every arc with positive flow.

Orlin's algorithm uses a variant of the ε-optimality conditions that relax the flow conservation constraints rather than the complementary slackness constraints. The algorithm maintains a preflow f and a dual feasible price function p that satisfy the complementary slackness conditions. We say that such a pair is Δ-optimal if $|b(v) - e_f(v)| \leq \Delta$ for every node v.

First consider a simple variant of the Edmonds and Karp algorithm. One step of the algorithm selects a node v with large excess (with $e_f(v) >> b(v)$) and a node w with large deficit ($e_f(w) << b(w)$). It sends flow from v to w along the cheapest path between them. Such a step decreases the excess or the deficit of a node. Roughly speaking the maximum excess or deficit decreases by a constant factor after every n shortest path computations. Therefore, if $B = ||b||_\infty$, then this algorithm finds an εB-optimal transshipment in $O(n \log(1/\varepsilon))$ shortest path computations. If all demands are integral, then throughout the flow is integral, and hence after at most $O(n \log B)$ shortest path computations an optimal transshipment is found.

The constraint matrix, A, of the transshipment problem is totally unimodular, *i.e.* $\Delta(A)$ is 1. Therefore, the nonnegativity requirement for an edge (v,w), can be deleted if the value of a Δ-optimal flow f on an arc (v,w) is at least $n\Delta$. Deleting the nonnegativity requirement, $f(v,w) \geq 0$, corresponds, in graph theoretic terms, to contracting the arc (v,w).

The general iterative method starts each iteration by some preprocessing to guarantee that at least one inequality will be deleted each iteration. There is no need for preprocessing in the case of the transshipment problem. It is easy to see that a $\Delta = B/n^2$-optimal flow must have an arcs that can be contracted. Indeed, let v be the node with largest demand in absolute value. In a Δ-optimal flow all excess or deficit is at most Δ, hence most of the demand of v must be satisfied, and at least one arc entering or leaving v must carry sufficient flow that it can be contracted.

The above ideas give the sketch of a strongly polynomial algorithm for the transshipment problem that consists of $O(n^2 \log n)$ shortest path computations. One iteration of the algorithm computes and ε-optimal solution for some ε. This consists of $O(n \log n)$ shortest path computations. Then some edges are contracted (at least one each iteration), and a new iteration starts. One reason why this algorithm is slow is, that it restarts the flow computation after every iteration. Orlin [24] gave a more sophisticated version of this algorithm that does not have separate iterations, but instead contracts edges and then continues the same flow computation. Amortized analysis is used to show that a minimum cost transshipment is found in $O(m \log n)$ shortest path computations.

Let us compare running times of the resulting strongly polynomial algorithms with the algorithm of Edmonds & Karp. Note that while the iterative algorithm is already strongly polynomial, its running time does not compare favorably with the Edmonds & Karp algorithm. In order to make the algorithm strongly polynomial, the $O(\log B)$ in the running time had to be replaced by $O(m \log n)$. However, in the running time of Orlin's algorithm the $O(\log B)$ is replaced by only an $O(\log n)$, which is much more attractive.

4.2 The Maximum-Flow Problem

The first polynomial algorithms for the maximum-flow problem, due independently to Dinic [5] and Edmonds and Karp [6], were strongly polynomial. The preflow-push algorithm of Goldberg [10] is one of the biggest recent breakthroughs in network algorithms. Using sophisticated data-structures Goldberg and Tarjan [10] obtained an $O(nm \log(n^2/m))$ implementation of this algorithm. This was the fastest known strongly polynomial algorithm for all values of n and m until very recently. Ahuja and Orlin [2] have developed a simple data-structure free version of the algorithm, that is more efficient for all but very large values of the capacities. Its running time is $O(mn + n^2 \log U)$, if we assume that the capacities are integral and at most U. This algorithm was the starting point of the strongly polynomial algorithm of Cheriyan and Hagerup [3]. Combining the ideas for converting polynomial algorithms into strongly polynomial ones, the data-structures used by Goldberg and Tarjan and a very sophisticated amortized analysis, they give a maximum-flow algorithm that runs in $O(mn + n^2 \log^3 n)$ time. This time bound is better than the Goldberg and Tarjan bound if the graph is not too sparse.

For practical problems the size of the numbers involved often compares favorably to the size of the network. Therefore, polynomial algorithms that are not strongly polynomial might be more efficient in practice. To compare polynomial algorithms with strongly polynomial ones Gabow has suggested the *similarity assumption*, that is, to assume that all numbers involved are integers that have size at most $O(\log n)$. Unfortunately, algorithms whose running time depends exponentially on the size of the numbers involved appear polynomial under this assumption. A slightly weaker assumption that avoids this problem, is to assume that all numbers involved are integral of size is at most $\log^{O(1)} n$. Under this assumption both Orlin's transshipment algorithm and the Cheriyan & Hagerup algorithm compares favorably to its polynomial counterpart. It is an interesting open problem to see what other polynomial algorithms be converted to strongly polynomial ones by replacing the size of the numbers in the running time by $O(\log^c n)$ for some constant c where n is the combinatorial size of the problem.

4.3 Multi-Commodity Flows

There are several important special cases of the linear programming problem which have a combinatorial structure similar to the transshipment problem and also play a fundamental role in applications, but are not nearly as well understood. For many of these problems, the only known polynomial algorithm is obtained by using polynomial linear programming algorithms. Such algorithms use continuous methods and do not take advantage of the combinatorial structure. It is an interesting question whether some of these problems can be solved more efficiently using combinatorial techniques. The multi-commodity flow problem is an example of such a problem.

Many polynomial algorithms, if stated with an appropriate initial point, find an approximately optimal solution in time polynomial in n, m and $\log(1/\varepsilon)$. If ε is large this is strongly polynomial. In the definition of ε-optimality used for designing strongly polynomial algorithms, the complementary slackness conditions have been relaxed in a one-sided way. A symmetric version would require only $y_i \leq \varepsilon$ instead of $y_i = 0$. Primal-dual potential reduction versions of Karmarkar's method find "ε-optimal" solutions in this symmetric sense in time polynomial in n, m and $\log(1/\varepsilon)$. See Todd [31] for a survey of such methods.

Recently, Klein, Stein and Tardos [16] have obtained a combinatorial algorithm that significantly outperforms the general purpose techniques for a special case of the multi-commodity flow problem where ε is large. Consider a multi-commodity flow that satisfies the demands, but not necessarily the capacity constraints. Let $f(v, w) = \sum_i f_i(v, w)$ denote the sum of the different commodities on an arc (v, w). We consider the maximum ratio $\lambda = \max_{(v,w)\in E} f(v, w)/u(v, w)$ as the value of the flow. The objective of the *concurrent flow problem* is to minimize this ratio λ.

The concurrent flow problem has one variable, λ whose coefficients are not $0, \pm 1$ in the corresponding linear program. The coefficients of λ are the capacities. The problem is an extension with one new variable of a linear program that is known to be solvable in strongly polynomial time. This implies that the problem can be solved in strongly polynomial time [22].

Shahrokhi and Matula [28] gave a fully polynomial approximation algorithm for the special case of this problem with unit capacities, that is, an algorithm that finds a concurrent flow with value no worse than $(1 + \varepsilon)$ times the optimal, in time polynomial in n, m and $1/\varepsilon$. This algorithm is quite slow, both in terms of its dependence on ε and in its dependence on the other parameters. Klein, Stein and Tardos [16] gave a much faster algorithm extending the Shahrokhi & Matula technique. The algorithm uses a notion of ε-optimality similar to the symmetric notion used by the general linear programming algorithms. The algorithm runs roughly (ignoring log's) in $O(\varepsilon^{-2}m^2k)$ time. This compares quite favorably to known algorithms as long as ε is large (e.g., a constant).

Approximation algorithms for the concurrent flow problem are especially interesting because several of its applications require only approximately optimal solutions. Examples of such applications are graph partitioning by Leighton and Rao [17] and VLSI routing by Raghavan and Thompson [26]. It remains a significant open problem to design algorithms for this problem that outperform general purpose linear programming techniques for the case when ε is small.

The key dual variables for the concurrent flow problem are nonnegative lengths $\ell(v, w)$ associated with the edges. Let λ denote the value of the multi-commodity flow f. The complementary slackness conditions stated in terms of these variables require that (1) all arcs with nonzero length are fully utilized, that is, for each arc (v, w) either the total flow on it is exactly λ times its capacity, or the arc has zero length; and (2) for each commodity the flow can be decomposed into flows along simple paths each of which is shortest according to the length function ℓ.

The variant of ε-optimality used by the algorithm is defined by relaxing both of the above constraints. We require that (1) each arc is either almost fully utilized or has close to zero length; and (2) almost all the flow is carried on close to shortest paths. It is not too difficult to prove that these conditions imply that the value of multi-commodity flow is close to optimal.

The idea of the approximation scheme is to maintain a length function defined as a function of the flow, $\ell(v, w) = \exp(\alpha f(v, w))$ for an appropriate constant α, such that the first part of the ε-optimality conditions is satisfied. The second assumption is gradually enforced as flow is repeatedly rerouted from long paths to the corresponding shortest path.

5. Generalized Flow: An Open Problem

The most intriguing special case of linear programming for which no strongly polynomial algorithm is known is the generalized flow problem, which is essentially the problem of an arbitrager who wants to maximize his profit by converting currencies at different exchanges. The problem is defined by a network, a source $s \in V$, capacities and positive gainfactors $\gamma(v, w)$ on the arcs. When x units of flow enter an arc (v, w) then $x\gamma(v, w)$ units leave the arc. The excess of a preflow f is defined to be $\sum_w f(w, v)\gamma(w, v) - \sum_w f(v, w)$. A preflow is a *generalized flow* if all nodes, except the source, have zero excess. The problem is to find a generalized flow with maximum excess at the source.

The generalized flow problem is a linear programming problem. The constraint matrix is not $0, \pm 1$; its nonzero entries are the gainfactors. The problem can also be considered as a slight generalization of the class of feasibility problems with two variables per inequality considered by Megiddo [21]. Consider the (equivalent) version of the problem with demands at the nodes instead of the capacities. The linear programming dual of this version has two variables in each inequality. However, it does not fit into Megiddo's framework, because it has an objective function with more than two variables; the demands at the nodes are the coefficients of the objective function in the dual.

The fastest known algorithm for the generalized flow problem is due to Vaidya [32]. It combines general linear programming techniques and fast matrix inversion techniques. It makes some use of the network structure by speeding up the matrix inversions involved. Goldberg, Plotkin and Tardos [8] gave two combinatorial algorithms for this problem. One of the algorithms uses a transshipment algorithm as a subroutine, the other one combines techniques from maximum-flow and transshipment algorithms. Both run in time comparable to the fastest known algorithms for the problem. It remains an interesting open problem to find algorithms that significantly outperform general linear programming techniques.

Finally, we point out that a combination of the multi-commodity flow and the generalized flow problem covers the full generality of linear programming.

We have mentioned in the first part of of Section 4. that every program can be reduced to an equivalent 2-commodity flow problem [1,12]. However, the reduction is not strongly polynomial. The following theorem is an easy extension.

Theorem 7. *Every linear program can be reduced in strongly polynomial time to an equivalent 2-commodity generalized flow problem.*

References

1. G. M. Adelson-Velskii, E. A. Dinic, A. V. Karzanov: Flow algorithms. Izdatelstvo Nauka, Moskwa, 1975 (in Russian)
2. R. K. Ahuja, J. B. Orlin: A fast and simple algorithm for the maximum flow problem. Operations Res. **37**, 748–759 (1989)
3. J. Cheriyan, T. Hagerup: A randomized maximum-flow algorithm. In: Proceedings of the 30st Annual Symposium on Foundation of Computer Science, 1989.
4. W. Cook, A. M. H. Gerards, A. Schrijver, É. Tardos: Sensitivity theorems in integer and linear programming. Math. Programming **34**, 251–264 (1986)
5. E. A. Dinic: Algorithm for solution of a problem of maximum flow in networks with power estimation. Sov. Math. Dokl. **11**, 1277–1280 (1970)
6. J. Edmonds, R. M. Karp: Theoretical improvements in algorithmic efficiency for network flow problems. J. Assoc. Comput. Mach. **19**, 248–264 (1972)
7. A. Frank, É. Tardos: An application of simultaneous Diophantine approximation in combinatorial optimization. Combinatorica **7** (1) 49–65 (1987)
8. A. V. Goldberg, S. A. Plotkin, É. Tardos: Combinatorial algorithms for the generalized circulation problem. In: Proceedings of the 29th FOCS Symposium, 1988, pp. 432–443 (to appear in Mathematics of Operations Research)
9. A. V. Goldberg, R. E. Tarjan: Finding minimum-cost circulations by canceling negative cycles. J. Assoc. Comput. Mach. **36**, 873–886 (1989)
10. A. V. Goldberg, R. E. Tarjan: A new approach to the maximum flow problem. J. Assoc. Comput. Mach. **35**, 921–940 (1988)
11. M. Grötschel, L. Lovász, A. Schrijver: The ellipsoid method and its consequences in combinatorial optimization. Combinatorica **1**, 169–197 (1981)
12. A. Itai: Two-commodity flow. JACM **25**, 596–611 (1978)
13. N. Karmarkar: A new polynomial-time algorithm for linear programming. Combinatorica **4**, 373–395 (1984)
14. R. M. Karp, C. H. Papadimitriou: On linear characterization of combinatorial optimization problems. SIAM J. Comput. **11**, 620–632 (1982)
15. L. G. Khachyian: A polynomial algorithm in linear programming. Dokl. Akad. Nauk SSSR **244**, 1093–1096 (1979) [English translation: Sov. Math. Dokl. **20** (1979) 191–194]
16. P. Klein, C. Stein, É. Tardos: Leighton-Rao might be practical: Faster approximation algorithms for concurrent flow with uniform capacities. In: Proceedings of the ACM Symposium on the Theory of Computing, 1990, pp. 310–321
17. T. Leighton, S. Rao: An approximate max-flow min-cut theorem for uniform multicommodity flow problems with applications to approximation algorithms. In: Proceedings of the 29th Annual Symposium on Foundations of Computer Science, 1988, pp. 422–431
18. A. K. Lenstra, Jr. H. W. Lenstra, L. Lovász: Factoring polynomials with rational coefficients. Math. Ann. **261**, 515–548 (1982)
19. O. L. Mangasarian: Condition number for linear inequalities and linear programs. Methods of Operations Research **43**, 3–15 (1981)

20. N. Megiddo: Combinatorial optimization with rational objective functions. Mathematics of Operations Research **4**, 414–424 (1979)
21. N. Megiddo: Towards a genuinely polynomial algorithm for linear programming. SIAM J. Comput. **12**, 347–353 (1983)
22. C. H. Norton, S. Plotkin, É. Tardos: Using separation algorithms in fixed dimension. In: Proceedings of the ACM-SIAM Symposium on Discrete Algorithms 1990, pp. 377–387
23. J. B. Orlin: On the simplex algorithm for networks and generalized networks. Math. Prog. Studies **24**, 166–178 (1985)
24. J. B. Orlin: A faster strongly polynomial minimum cost flow algorithm. In: Proc. 20th ACM Symp. on Theory of Computing 1988, pp. 377–387
25. M. W. Padberg, M. R. Rao: The Russian method for linear programming III: Bounded integer programming. Technical Report 81-38, New York University, Graduate School of Business Administration, 1981
26. P. Raghavan, C. D. Thompson: Randomized rounding: a technique for provably good algorithms and algorithmic proofs. Combinatorica **7**, 365–374 (1987)
27. A. Schrijver: Theory of integer and linear programming. Wiley and Sons, 1986
28. F. Shahrokhi, D. W. Matula: The maximum concurrent flow problem. J. ACM **37** (2), 318–334 (1990)
29. É. Tardos: A strongly polynomial minimum cost circulation algorithm. Combinatorica **5**(3) 247–255 (1985)
30. É. Tardos: A strongly polynomial algorithm for solving combinatorial linear programs. Operations Research, pp. 250–256 (1986)
31. M. J. Todd: Recent developments and new directions in linear programming. In: M. Iri, K. Tanabe (eds.) Mathematical Programming: Recent Developments and Applications. KKT Scientific Publishers, Tokyo 1989, pp. 109–158
32. P. M. Vaidya: Speeding up linear programming using fast matrix multiplication. In: Proc. 30th IEEE Annual Symposium on Foundations of Computer Science, 1989
33. D. B. Yudin, A. S. Nemirovskii: Evaluation of the informational complexity of mathematical programming problems. Ekonomika i Matematicheskie Metody **12**, 128–142 (1976) (in Russian)

Computational Complexity in Finite Groups

László Babai *

Department of Computer Science, University of Chicago, Chicago, IL 60637, USA
and Department of Algebra, Eötvös University, Budapest, Hungary H-1088

We survey recent results on the *asymptotic* complexity of some of the fundamental computational tasks in finite groups in a variety of computational models. A striking recent feature is that techniques motivated by the problems of the more abstract models (nondeterminism, extreme parallelization) have turned out to provide powerful tools in the design of surprisingly efficient algorithms on realistic models (e.g. *a nearly linear time* membership test for permutation groups with a small base).

The techniques involve a combination of *elementary combinatorial results* on finite groups, some classical elementary group theory, and the extensive use of certain consequences of the *classification of finite simple groups* (CFSG).

Most of the recent work surveyed is due to E. M. Luks, G. Cooperman, L. Finkelstein, Á. Seress, E. Szemerédi, and the author.

1. Group Models and Measures of Complexity

Rubik's Cube illustrates some of the basic problems of computational group theory. We may want to decide whether or not a particular configuration is feasible (accessible without pulling the cube apart); determine the total number of feasible configurations; or construct "typical" configurations. In group theoretical terms, we are *given a group G by a list S of generators* (the "legal moves"), and we wish to determine whether or not a particular element of a larger group belongs to *G (membership testing)*; determine the *order* of *G*; generate uniformly distributed *random members* of *G*. The gourmet will ask more sophisticated questions such as deciding solvability, nilpotence, constructing normal closures, the center, composition factors, Sylow subgroups, etc.

The *cost* of answering these questions depends on two factors: the way group operations are performed, and the measure of cost.

For greatest generality, we consider *black box groups*, a model where no restriction is made on the way group operations are performed. In this model, elements of an unknown group *B* (the "group in the box") are encoded by

* Research supported in part by NSF Grant CCR-8710078 and Hungarian National Foundation for Scientific Research Grant 1812.

Proceedings of the International Congress
of Mathematicians, Kyoto, Japan, 1990

binary strings of uniform length n. (In particular, $|B| \leq 2^n$.) Group operations are performed by the "black box" at unit cost. A *black box group* is a subgroup of B given by a list of generators. (Generators of B, or a recognition method of strings in B, are *not* assumed to be known.)

Two implementations of the "black box" are of particular interest: *permutation groups* (subgroups of the symmetric group $B = \mathrm{Sym}(\Omega)$), and *matrix groups* over finite fields (subgroups of the linear group $B = GL(d, q)$).

The *models of computation* to be studied include deterministic as well as Monte Carlo (randomized) computation, both sequential and parallel (several processors). Timing estimates refer to the logarithmic cost RAM model [AHU].

A *Monte Carlo algorithm* uses randomization, hence its outcome may be in error. However, on any input, the probability of error is required to be $\leq 1/4$. By repeating the algorithm m times and taking majority vote, the chance of error is reduced to $< e^{-m/8}$ (Chernoff's bound).

The *cost* (in terms of a specific resource such as time, space, number of processors, length of proofs) is measured as a function of the *length of the input*, i.e. the number of input bits. An algorithm is said to have cost $O(f(n))$ (or cost $O^{\sim}(f(n))$) if for $n \geq n_0$ and on *all* inputs of length n, the cost is at most $cf(n)$ (at most $f(n)(\log n)^c$, resp.) (n_0, c will denote various constants throughout).

A function $f(n)$ is *polynomially bounded* (or *"short"*, *"small"*) if $f(n) \leq n^c$ for some c and all $n \geq n_0$. A *polylog bound* means $f(n) \leq (\log n)^c$. When used as technical terms, *"short"*, *"small"* will be Italicized.

NC ("Nick's Class") denotes (somewhat informally) the class of functions computable in *polylog* time, using a *small* number of parallel processors.

NP ("nondeterministic polynomial time") stands for the class of decision problems where the "yes" answers have *short* proofs. More precisely, a predicate $A(x)$ belongs to NP if there exists a polynomial time computable predicate $B(x, w)$ such that for every input string x, $A(x) \leftrightarrow (\exists^P w) B(x, w)$, where $\exists^P w$ refers to *short* strings w. The string w is called a *witness* of the statement $A(x)$. The negations of NP-predicates form the class $coNP$. (Cf. [GJ].)

We shall also consider the class AM ("Arthur–Merlin"), a randomized extension of NP, defined as follows: the predicate $A(x)$ belongs to AM if there exists an NP-predicate $B(x, r)$ such that for every input string x, $A(x)$ is equivalent to $B(x, r)$ for most *short* strings r. (The definition of "most" is flexible; asking more than 51% will define the same complexity class as asking, say, a $1 - 2^{-n}$ fraction, where n is the length of x.) Informally, $A(x)$ has *short* "interactive proofs" in the sense that if the all-knowing but untrusted Merlin is able to present a *short* "witness" in response to a random question r of polynomial time bounded Arthur, this should convince skeptical Arthur by way of overwhelming statistical evidence that Merlin's claim $A(x)$ is true. (Cf. [Ba2, BM, GMR, Go].)

2. General Methods: Black Box Groups

In this section we demonstrate the somewhat unexpected fact that nontrivial computational tasks, such as constructing *random elements* and deciding *solvability*, can be accomplished in Monte Carlo polynomial time in the extremely general model of *black box groups*. Here and throughout the paper, G will denote a finite group and S a set of generators of G.

2.1 Two Combinatorial Lemmas

We begin with two elementary results which will play a key role. The first one concerns the number of group operations required in order to construct an element from a given set of generators.

A *straight line program* from $S \subseteq G$ to $g \in G$ is a sequence $u_1, \ldots, u_m$ of elements of the group G such that each u_i either belongs to S or is obtained from one or two previous elements by a group operation; and $g = u_m$. The *straight line cost* $\text{cost}(g|S)$ is the smallest m such that such a straight line program exists.

Lemma 1 (Reachability Lemma [BSz]**).** *Given any set S of generators of a group G and any $g \in G$, we have* $\text{cost}(g|S) \leq (1 + \log|G|)^2$.

(All log's in this paper are to the base 2.) A *subproduct* of the elements $h_1, \ldots, h_k \in G$ is a product of the form $h_1^{e_1} \cdots h_k^{e_k}$, where $e_i \in \{0, 1\}$. The k-dimensional *cube* $C = C(h_1, \ldots, h_k)$ is the set of all the 2^k subproducts of the h_i. The cube C is *nondegenerate* if $|C| = 2^k$. The basic structure established in the proof of Lemma 1 is a *chain of nondegenerate cubes:* we prove the existence of a sequence of elements $h_1, \ldots, h_t$ which generate a nondegenerate cube C such that $G = C^{-1}C$; and for every i, $\text{cost}(h_i|h_1, \ldots, h_{i-1}, S) \leq 2i - 1$. The h_i are found inductively; we can continue as long as $C^{-1}C \neq G$: the element outside $C^{-1}C$ of lowest straight line cost will do. We shall refer to this procedure as *doubling the cube*. Clearly, we must stop at some $t \leq \log|G|$.

The proof just sketched is non-constructive; it does not tell how to find an element that will double the current cube. The following lemma provides the key to an efficient Monte Carlo procedure.

A *graph* is a pair $X = (V, E)$ where V is the set of *vertices*, and E is a set of unordered pairs of vertices, called *edges*. Two vertices v, v' are *adjacent* if $\{v, v'\} \in E$. An *isomorphism* of two graphs is a bijection of the vertex sets preserving adjacency. The group of self-isomorphisms of X is the *automorphism group* $\text{Aut}(X)$. We say that X is *vertex-transitive* if $\text{Aut}(X)$ is a transitive subgroup of $\text{Sym}(V)$. The number of vertices adjacent to $v \in V$ is the *degree* of v. X is *locally finite* if its vertices have finite degrees. The boundary ∂W of a subset $W \subseteq V$ consists of all vertices in $V \setminus W$ adjacent to some vertex in W. X is *connected* if $\partial W \neq \emptyset$ for any nonempty proper subset W. A walk of length ℓ in X is a chain of $\ell + 1$ vertices, each adjacent to its predecessor. The *distance* of $v, v' \in V$ is the length of the shortest walk between them. Let $X^t(v)$ denote the set of vertices at distance $\leq t$ from v.

Lemma 2 (Local Expansion Lemma [Ba4,5]**).** *Let $X = (V, E)$ be a locally finite connected vertex-transitive graph and $v \in V$. If $W \subseteq X^t(v)$ and $|W| \leq |V|/2$ then $|\partial W| \geq |W|/(4t)$.*

This lemma has the interesting consequence that random walks on a vertex-transitive graph "don't get stuck" in a corner.

Theorem 3 [Ba5]. *Let X be a connected vertex-transitive graph of finite degree d. Assume that $|X^{4t}(v)| \leq |V|/2$. Let τ be a random integer chosen uniformly from $\{t, t+1, \ldots, \ell\}$ where $\ell = Ct^2 d \log|G|$ ($C = 500$). Then with probability $\geq 1/16$, a random walk of length τ, starting at v, will end outside $X^t(v)$.*

Among the ingredients of the proof is a Cheeger-type [Ch] eigenvalue estimate for graphs, derived from the local expansion property, following the lines of [Alo].

The graphs this theorem will be applied to are *Cayley graphs.* The vertex set of the Cayley graph $X(G,S)$ is G; and $g \in G$ is adjacent to gh for $h \in S \cup S^{-1}$.

2.2 Membership and Random Generation

We begin with a nondeterministic result.

Theorem 4 [BSz]. *Membership in black box groups belongs to NP.*

Indeed, a *short* straight line program qualifies as a *witness* of membership.

There is little hope for making this proof constructive, even in the very special case when the "group in the box" is the multiplicative group of $GF(q)$. Indeed in this case, finding a straight line program to generate g from $S = \{h\}$ is equivalent to solving the equation $h^x = g$. This is the *discrete logarithm* problem, not believed to be solvable in polynomial time. (Known algorithms require time $\exp(c\sqrt{q \log q})$; whereas polynomial time would mean polylog(q) steps.)

Yet, part of the proof *can* be turned into an efficient Monte Carlo algorithm.

Theorem 5 [Ba5]. *Nearly uniformly distributed random elements of a black box group can be constructed in Monte Carlo polynomial time.*

Nearly uniform distribution means each element has probability $(1 \pm \varepsilon)/|G|$ to be selected; and the reliability of the algorithm is $\geq 1 - \delta$, where ε, δ are input parameters, and the number of operations is polynomially bounded in $k = \log|G| + \log(1/\varepsilon) + \log(1/\delta)$. The idea is to construct a set S' of generators such that the diameter of the Cayley graph $X(G,S')$ is *small.* Once this is accomplished, *short* random walks are known to produce nearly uniformly distributed elements [Ald, Alo].

To reduce the diameter, we should like to adapt the "doubling the cube" trick. The difficulty is, how to obtain the next h_i, which must be outside the set $C^{-1}C$, where C is the current cube. The solution is a *short* random walk of random length over the Cayley graph. By Theorem 3, such a walk has a fixed positive chance of reaching a desired element.

2.3 Nonmembership, Order

These two problems are known to belong to the class AM [Ba4]. Indeed, to prove this was the original motivation behind inventing the Local Expansion Lemma.

For black box groups, the nonmembership and order verification problems are provably *not* in NP [BSz]. (To be precise, here we are talking about a *relativized* version of NP: computations refer to the black box, an *"oracle"* [GJ].)

We conjecture, however, that for matrix groups over finite fields, these problems belong to NP. This will follow from the conjecture below. The *length* of a *presentation* of a group (in terms of generators and relations) is the number of bits required to write down the presentation. E. g., the presentation $\langle a | a^N = 1\rangle$ of the cyclic group has length $\log N + O(1)$ (we write exponents in binary).

Short Presentation Conjecture. (i) Every finite simple group G has a presentation R of length polylog($|G|$). (ii) If G is of Lie type over $GF(q)$ then such R is computable from the standard name of G in time polylog($|G|$), assuming $GF(q)$ and a primitive root in it are explicitly given.

One can prove, using Lemma 1, that part (i) of the Conjecture, if true, automatically extends to all finite groups [BKLP]. The conjecture itself has been verified for all G except those of rank one twisted Lie type [BKLP], cf. [Ka4]. Note that for a Lie-type simple group of rank d over $GF(q)$, the Conjecture requires presentations of length $\leq (d \log q)^c$; whereas the Steinberg presentations [St, Car] require an exponentially greater number, about $d^2 q$ generators.

Verification of the *order* is a central problem. If it belongs to NP (as we expect for matrix groups), this brings a number of other verification problems into NP, including composition factors, homomorphisms, isomorphisms, kernels, minimal normal subgroups. On the other hand, problems known to be in AM (for black box groups in general) but not expected to belong to NP (even for permutation groups) include verification of the intersection of two subgroups, the centralizer of an element, non-conjugacy of two elements [Ba4].

2.4 Random Subproducts

Algorithms often depend on access to random elements of G (e.g. [CFS, NP]). Theorem 5 constructs such elements in reasonable polynomial time, but not efficiently enough for some applications. *Random subproducts*, however, often emulate truly random elements very efficiently.

Let $S = \{g_1, \ldots, g_s\}$. A *random subproduct* is a subproduct $\varphi = g_1^{e_1} \cdots g_s^{e_s}$, where the e_i are independent uniform $(0, 1)$-variables (coin flips).

Lemma 6 [BLS2]. *Let $H < G$ be a proper subgroup. Then* $\text{Prob}(\varphi \notin H) \geq 1/2$.

Let L be the maximum length of subgroup chains in G. ($L \leq \log |G|$.) Lemma 6 implies that the probability that $2L(1+\alpha)$ random subproducts do not generate G is less than $\exp(-\alpha^2 L/(1+\alpha))$. A refined argument yields:

Lemma 7 [BCFLS]. *If G is given by a list of s generators, then a Monte Carlo procedure, using $O(s \log L)$ group operations, produces (with large but fixed probability) a set of $O(L)$ generators for G.*

This keeps the number of generators down when constructing subgroups. Some additional combinatorics yields a particularly efficient *normal closure* algorithm. Recall that the notation $O^{\sim}(f(n))$ refers to an upper bound $f(n)(\log n)^{O(1)}$.

Theorem 8 [BCFLS]. *Let $H \leq G$ be black box groups, each given by a list of $O(L)$ generators. Then the normal closure of H in G can be constructed (in the form of $O(L)$ generators) in Monte Carlo $O^{\sim}(L^2)$ operations.*

As an immediate application, we obtain polynomial time Monte Carlo algorithms to decide *solvability* and *nilpotence* of G (both require $O^{\sim}(L^3)$ operations).

We should stress that the results of this section apply to all black box groups, including their implementations as *matrix groups* or *permutation groups*. In spite of their generality, the results are strong enough to yield asymptotic savings even in the well-studied area of permutation groups (see Sect. 4.2).

3. Permutation Groups: Survey of Complexity Status

We consider groups $G \leq \mathrm{Sym}(\Omega)$ where $|\Omega| = n$. The *stabilizer* of $x \in \Omega$ is the subgroup $G_x = \{g \in G : x^g = x\}$. The *pointwise set stabilizer* of $\Delta \subseteq \Omega$ is the subgroup $G_\Delta = \bigcap_{x \in \Delta} G_x$. We call Δ a *base* if $G_\Delta = \{1\}$.

Let $G = G^{(0)} \geq G^{(1)} \geq \ldots \geq G^{(m)} = \{1\}$ be a chain of subgroups. A *strong generating set* (SGS) w.r. to this chain is a set $S \subseteq G$ such that $G^{(i)} = \langle S \cap G^{(i)} \rangle$ for every i. A *transversal system* (TS) is a family $\{T_i : 1 \leq i \leq m\}$, where T_i is a (right) transversal (set of coset representatives) of $G^{(i)}$ in $G^{(i-1)}$. A *partial transversal system* is a family of partial transversals $T_i' \subseteq T_i$.

The *stabilizer chain* w.r. to a given ordered base $\Delta = \{x_1, \ldots, x_m\}$ is defined as $G^{(i)} = G_{x_1,\ldots,x_i}$. The concept of an SGS w.r. to an ordered base was introduced by C. C. Sims in the early 60's as a central tool in computational group theory ([Sim1, 2]). Given an SGS, a TS is readily constructed, solving the *membership* and *order* problems and the construction of truly uniformly distributed *random elements*. A slight modification yields *normal closures*, clearing the way for more advanced applications. The central problem of constructing an SGS was efficiently solved by Sims [Sim1, 2].

The *asymptotic complexity* of these algorithms was not analyzed until 1980 when the complexity of many of these algorithms was recognized to be polynomial time in [FHL]. In particular, an $O(n^6 + sn^2)$ variant of Sims's SGS algorithm was constructed, where s is the number of input generators. As a consequence, membership, order, normal closures, solvability were shown to be computable in polynomial time [FHL]. E. M. Luks has subsequently added an array of elegant polynomial time algorithms which, for the first time, required deeper group theoretic analysis. The list includes the center, a composition chain [Lu2], and subcases of the *coset intersection* problem, i.e. determining $G \cap Hh$ where $G, H \leq \mathrm{Sym}(\Omega)$, $h \in \mathrm{Sym}(\Omega)$. The subcases solved in polynomial time in Luks's seminal paper [Lu1] include the case when G is solvable, or more generally, the nonabelian composition factors of G are restricted to the set $\mathcal{G}(c)$ consisting of the alternating groups of degree $\leq c$, the groups of Lie type of rank $\leq c$, and the sporadic groups. The algorithm uses classical *divide and conquer* algorithmic techniques, splitting the domain into orbits and then into domains of imprimitivity ([Wi]). When G is primitive, some of the algorithms use exhaustive search. In such cases, the polynomial time claim depends on the following result.

Theorem 9 [BCP]. *If $G \leq \mathrm{Sym}(\Omega)$ is primitive and $G \in \mathcal{G}(c)$ then G is small.*

(Small means $|G| \leq n^{c'}$ for some constant c', depending on c.) For primitive solvable groups, the precise bound is $|G| \leq 24^{-1/3} n^c$ where $c = 1+\log_9(48 \cdot 24^{1/3}) = 3.24399\ldots$ [Pá, Wo].

Sylow subgroups and Sylow normalizers were added to the polynomial time library by Kantor [Ka1, 2, 3]. For a long list of additional results see [KL].

Another important observation of [Lu1] was that a number of problems, including coset intersection, setwise stabilizer of a subset, centralizer of an element, and centralizer of a subgroup, are *equivalent* (polynomial time reducible to one another), and the *graph isomorphism* problem (to decide whether or not two given graphs are isomorphic) is reducible to each. In particular, as long as graph isomorphism is not solved in polynomial time (the best current algorithm requires $\exp(O^{\sim}(\sqrt{n}))$ for graphs on n vertices, cf. [BL]), coset intersection, etc., are not expected to be efficiently solvable. On the other hand, the decision versions of these problems ("is $G \cap Hh \neq \emptyset$?") are *not* NP-complete, unless the so-called *polynomial time hierarchy* of complexity classes collapses [BM, GMW]. (The conjecture that the polynomial time hierarchy does not collapse is a stronger version of the famous $NP \neq coNP$ conjecture [Sto].)

Some related problems *are* NP-complete; the nicest is A. Lubiw's result: the predicate "G has a fixed-point-free element" is NP-complete, even for elementary abelian 2-groups [Lub]. An even harder problem is to determine minimum generating sequences; the length of the shortest word in S representing $g \in G$ is $PSPACE$-complete [Je1]. (For related problems, see [BHKLS].)

On the other end of the spectrum, some of the basic problems were shown to admit ultra-fast *parallel algorithms.* Most notably, *membership, order*, and even a *composition chain* are computable in NC [BLS1]. A striking feature of the algorithm is that even for the rudimentary tasks of membership testing, we are forced to determine the composition factors first, using several facts of asymptotic group theory currently derivable only via the *classification of finite simple groups* (CFSG) (cf. Sect. 5.). – Coset intersection is *not* known to be in NC; if it is in NC, then so is the isomorphism of graphs of degree 3 [LM].

4. Efficient Construction of Strong Generators

The $O(n^6 + sn^2)$ analysis of the SGS algorithm of [FHL] was soon replaced by $O(n^5 + sn^2)$ [Kn, Je2]. Knuth's is the closest to Sims's original approach and is quite efficient in practice, but there exist large collections of examples where its typical behavior is as bad as the n^5 worst case bound (while $s = n - 1$) [Kn].

The n^5 bottleneck was broken, using machinery developed for the NC result, in [BLS2] ($O^{\sim}(n^4 + sn^2)$). Further improvements yield $O^{\sim}(sn^3)$ [BLS3], the best deterministic bound to date. These results heavily depend on the CFSG.

4.1 A Fast and Elementary Monte Carlo SGS Algorithm

We outline a new $O^{\sim}(n^3 + sn)$ *time* Monte Carlo SGS algorithm with a perfectly *elementary* analysis [BCFLS].

Let $H \leq G$ be a subgroup of index k; and $T = \{t_1, \ldots, t_k\}$ a right transversal. For $g \in G$, let $\bar{g} = t_i$ where $Hg = Ht_i$. Schreier's lemma asserts that the sk elements $t_i h_j \overline{t_i h_j}^{-1}$ (*"Schreier generators"*) generate H, where $S = \{h_1, \ldots, h_s\}$ [Ha]. Noting that a transversal for G_x in G is easily constructed, a natural approach to constructing an SGS would be to consider the Schreir generators for G_x and repeat. The difficulty is that the number of generators grows rapidly.

Random subproducts are the new tool. (When using the results of Sect. 2.4, the following bound comes handy: *Every subgroup chain in S_n has length $< 2n$* [Ba3, CST].) Lemma 7 alone saves nearly an order of magnitude over [Kn, Je2]: we reduce the number of generators of G to $O(n)$, then construct a transversal and the Schreier generators of G_x; repeat. The cost of this naive approach is $O^{\sim}(n^4 + sn)$. When $G \geq A_n$, we have a particularly speedy variant, based on the following consequence of Lemma 6.

Proposition 10 [BLS2]. *If $G = \langle S \rangle \leq S_n$ and S' is a set of $c \log n$ random subproducts of S then with large probability, the orbits of $\langle S' \rangle$ and G agree.*

Applying this to the action of $G \geq A_n$ on the set of ordered 6-tuples we find that $O(\log n)$ random subproducts are likely to generate all of G. So the previous argument, using only $O(\log n)$ random subproducts in each round, constructs an SGS for $G \geq A_n$ in time $O^{\sim}(n^3)$. This process works also when G induces A_k or S_k on some orbit $\Delta \subseteq \Omega$ ($|\Delta| = k$). However, now we don't get generators of G_Δ. (The procedure preserves the action of G on Δ only.) Instead, we use a set of $O(k)$ defining relations of A_k or S_k to construct *normal generators* of G_Δ; and then use our *normal closure* algorithm (Theorem 8) to obtain G_Δ. Another ingredient of the $O^{\sim}(n^3 + sn)$ algorithm is *computation of G_x in $O^{\sim}(snk)$*, where k is the length of the orbit of x (apply Lemma 7 to the Schreier generators). This bound is exploited through a "smallest orbit first" strategy. By adding action on maximal blocks to Ω we ensure that the next x to be stabilized is from an orbit with primitive action. The timing depends on a combinatorial observation:

Lemma 11 [BCFLS]. *If $G \leq S_n$ is primitive and G_x has a nontrivial orbit of length $k < n/2$ then every subgroup $H \neq \{1\}$ of G_x has a nontrivial orbit of length $\leq k$.*

4.2 Small Base Groups in Nearly Linear Time

Groups with a small base are of particular importance; e.g. linear groups, treated as permutation groups on a vector space, always have a base of size $\leq \log n$. Let us say that a family of groups has small bases if they have base size polylog(n). The SGS methods of [Si1,2], [Kn], [Je2], [BCFLS] require $> n^2$ time for such groups. Combinatorial techniques based on Lemmas 1 and 2 have recently led to a *Monte Carlo SGS algorithm in time $O^{\sim}(n)$ for small base groups* [BCFS].

The basic ideas are (i) a very efficient implementation of Sims's "Schreier vector" data structure to store transversals, based on the "doubling the cube"trick (Sect. 2.1); and (ii) the use of Lemma 2 to rapidly locate elements not yet reached by the current partial transversal system. A key new feature of these methods is that rather than operating with the coarse *subgroup* structure, we are able to handle *chains of certain subsets,* such as cubes and their generalizations.

5. CFSG vs. Elementary

We mention some of the consequences of the simple groups classification (CFSG) used in the analysis of the algorithms quoted. *Schreier's conjecture* that the outer automorphism groups of simple groups are solvable, is used in Luks's composition chain algorithm [Lu2] and the algorithms building on it [BLS1, 2, 3, Ka1, 2, 3]. In Sect. 4.1 we used that the degree of transitivity of $G \leq S_n$ is $t \leq 6$ (unless $G \geq A_n$) [CKS]. At the cost of some extra log factors (swallowed by the $O^\sim$ notation) this can be replaced by the 19th century bound $t = o(\log^2 n)$ [Jo]. Using the CFSG, Cameron has shown that if G is a primitive group of order $> n^{2\log n}$ then $n = \binom{k}{\ell}^m$ and G is a subgroup of $S_k wr S_m$ with socle A_k^m acting on the ordered m-tuples of ℓ-subsets of a k-set [Cam]. This result helps reduce the case of "large" primitive groups to S_n; the remaining primitive groups have small bases. This is indispensable for [BLS1, 2, 3], even if all we need is to *test membership!* Kantor uses detailed knowledge of the CFSG even just to find an element of order p.

Other elementary estimates that may help avoid CFSG references (at a cost of some extra log's) include the bound $|G| < \exp(4\sqrt{n}\log^2 n)$ for G primitive but not doubly transitive [Ba1] and $|G| < n^{c\log^2 n}$ for $G \not\geq A_n$ doubly transitive [Py]. Bochert's 1892 estimate [Bo] that a doubly transitive group $G \not\geq A_n$ has *minimal degree* $\geq n/4$ (cf. [Wi]) is used in [BCFS]. *Combinatorial proofs* may directly suggest *efficient algorithms.* A case in point is the algorithm derived from a simple proof of Jordan's $o(\log^2 n)$ bound on the degree of transitivity [BS], allowing ultra-parallelized (NC) management of S_n [BLS1].

Conclusion. During the past decade, the asymptotic complexity of computation in finite groups has been analyzed in a variety of models of computation. New combinatorial and algebraic tools have been developed. Structural insights gained from the study of models ranging from the *unrealistic* (extreme parallelization: NC) to the *absurd* (nondeterminism: NP, AM) have contributed to the design of new efficient algorithms with a reasonable expectation of competitive implementations.

Acknowledgment. I feel privileged to have had an exciting ongoing collaboration with Gene Luks, now for over a decade. I have also greatly benefited from joint work with Gene Cooperman, Larry Finkelstein, and Ákos Seress.

References

[AHU] A. Aho, J. Hopcroft, J. Ullman: The design and analysis of computer algorithms. Addison-Wesley 1974

[Ald] D. Aldous: On the Markov chain simulation method for uniform combinatorial distributions *etc.* Probab. Eng. Info. Sci. **1** (1987) 33–46

[Alo] N. Alon: Eigenvalues and expanders. Combinatorica **6** (1986) 83–96

[Ba1] L. Babai: On the order of uniprimitive permutation groups. Ann. Math. **113** (1981) 553–568

[Ba2] L. Babai: Trading group theory for randomness. Proc. 17th ACM STOC, Providence RI, 1985, pp. 421–429

[Ba3] L. Babai: On the length of chains of subgroups in the symmetric group. Comm. Algebra **14** (1986) 1729–1736

[Ba4] L. Babai: Bounded-round interactive proofs in finite groups. SIAM J. Discr. Math. (to appear)

[Ba5] L. Babai: Local expansion of vertex-transitive graphs and random generation in finite groups. Proc. 23rd ACM STOC 1991, pp. 164–174

[BCFLS] L. Babai, G. Cooperman, L. Finkelstein, E.M. Luks, Á.Seress: Fast Monte Carlo algorithms for permutation groups. 23rd ACM STOC 1991, pp. 90–100

[BCFS] L. Babai, G. Cooperman, L. Finkelstein, Á. Seress: Permutation groups with small base in almost linear time. In: Proc. ISSAC '91, Bonn (to appear)

[BCP] L. Babai, P.J. Cameron, P.P. Pálfy: On the orders of primitive groups with restricted nonabelian composition factors. J. Algebra **79** (1982) 161–168

[BHKLS] L. Babai, G. Hetyei, W. M. Kantor, A. Lubotzky, Á. Seress: On the diameter of finite groups. Proc. 31st IEEE FOCS 1990, pp. 857–865

[BKLP] L. Babai, W. M. Kantor, E. M. Luks, P. P. Pálfy: Short presentations for finite groups. In preparation

[BL] L. Babai, E.M. Luks: Canonical labeling of graphs. Proc. 15th ACM STOC 1983, pp. 171–183

[BLS1] L. Babai, E. Luks, Á. Seress: Permutation Groups in NC. Proc. 19th ACM STOC 1987, pp. 409–420

[BLS2] L. Babai, E. Luks, Á. Seress: Fast management of permutation groups. Proc. 28th IEEE FOCS 1988, pp. 272–282

[BLS3] L. Babai, E. Luks, Á. Seress: Fast deterministic management of permutation groups. In preparation

[BM] L. Babai, S. Moran: Arthur–Merlin games: a randomized proof system, and a hierarchy of complexity classes. J. Comp. Sys. Sci. **36** (1988) 254–276

[Bo] A. Bochert: Ueber die Classe der Transitiven Substitutionsgruppen. Math. Ann. **40** (1892) 176–193

[BS] L. Babai, Á. Seress: On the degree of transitivity of permutation groups: A short proof. J. Comb. Theory A **45** (1987) 310–315

[BSz] L. Babai, E. Szemerédi: On the complexity of matrix group problems I. Proc. 25th IEEE FOCS, Palm Beach, FL, 1984, pp. 229–240

[Cam] P.J. Cameron: Finite permutation groups and finite simple groups. Bull. London Math Soc. **13** (1981) 1–22

[Car] R. Carter: Simple groups of Lie type. Wiley 1972, 1989

[Ch] J. Cheeger: A lower bound for the smallest eigenvalue of the Laplacian. In: Problems in Analysis. Princeton Univ. Press 1970, pp. 195–199

[CFS] G. Cooperman, L. Finkelstein, N. Sarawagi: A random base change algorithm for permutation groups. In: Proc. ISSAC 1990, Tokyo, ACM Press and Addison-Wesley 1990, pp. 161–168

[CKS] C.W. Curtis, W.M. Kantor, G. Seitz: The 2-transitive permutation representations of the finite Chevalley groups. Trans. A.M.S. **218** (1976) 1–57

[CST] P.J. Cameron, R. Solomon, A. Turull: Chains of subgroups in symmetric groups. J. Algebra **127** (1989) 340–352

[FHL] M. L. Furst, J. Hopcroft, E. M. Luks: Polynomial-time algorithms for permutation groups. 21st IEEE FOCS 1980, pp. 36–41

[GJ] M. Garey, D.S. Johnson: Computers and Intractability: A guide to the theory of NP-completeness. Freeman, New York 1979

[GMR] S. Goldwasser, S. Micali, C. Rackoff: The knowledge complexity of interactive proofs. SIAM J. Comp. **18** (1989) 186–208

[GMW] O. Goldreich, S. Micali, A. Wigderson: Proofs that yield nothing but their validity and a methodology of cryptographic protocol design. Proc. 27th IEEE FOCS 1986, pp. 174–187

[Go] S. Goldwasser: Interactive proofs and applications. These proceedings, pp. 1521–1535

[Ha] M. Hall, Jr.: The theory of groups. Macmillan, New York 1959

[Je1] M. R. Jerrum: The complexity of finding minimum length generator sequences. Theor. Comp. Sci. **36** (1985) 265–289

[Je2] M.R. Jerrum: A compact representation for permutation groups. J. Algorithms **7** (1986) 60–78

[Jo] C. Jordan: Nouvelles recherches sur la limite de transitivité des groupes non alternés. Bull. Soc. Math. France **1** (1873) 35–60

[Ka1] W.M. Kantor: Polynomial-time algorithms for finding elements of prime order and Sylow subgroups. J. Algorithms **6** (1985) 478–514

[Ka2] W.M. Kantor: Sylow's theorem in polynomial time. J. Comp. Sys. Sci. **30** (1985) 359–394

[Ka3] W.M. Kantor: Finding Sylow normalizers in polynomial time. J. Algorithms **11** (1990) 523–563

[Ka4] W.M. Kantor: Some topics in asymptotic group theory. Proc. Durham Group Theory Conf., 1990. (To appear)

[KL] W.M. Kantor, E. M. Luks: Computing in quotient groups. In: Proc. 22nd ACM STOC, Baltimore, 1990, pp. 524–534

[Kn] D.E. Knuth: Efficient representation of perm groups. Combinatorica **11** (1991) 57–68 (preliminary version circulated since 1981)

[Lub] A. Lubiw: Some *NP*-complete problems similar to graph isomorphism. SIAM J. Comp. **10** (1981) 11–21

[Lu1] E.M. Luks: Isomorphism of graphs of bounded valence can be tested in polynomial time. J. Comp. Sys. Sci. **25** (1982) 42–65

[Lu2] E.M. Luks: Computing the composition factors of a permutation group in polynomial time. Combinatorica **7** (1987) 87–99

[LM] E.M. Luks, P. McKenzie: Fast parallel computation with permutation groups. Proc. 27th IEEE FOCS 1985, pp. 505–514

[NP] P.M. Neumann, Cheryl E. Praeger: A recognition algorithm for the special linear groups. Manuscript, 1990

[Pá] P.P. Pálfy: A polynomial bound on the orders of primitive solvable groups. J. Algebra **77** (1982) 127–137

[Py] L. Pyber: The orders of doubly transitive groups: elementary estimates. (To appear)

[Sim1] C.C. Sims: Computation with permutation groups. In: Proc. Second Symp. Symb. Algeb. Manipulation. ACM, New York 1971, pp. 23–28

[Sim2] C.C. Sims: Some group theoretic algorithms. (Lecture Notes in Mathematics 697.) Springer, Berlin Heidelberg New York 1978, pp. 108–124

[St] R. Steinberg: Generators for simple groups. Can. J. Math. **14** (1962) 277–283

[Sto] L. Stockmeyer: The polynomial-time hierarchy. Theoret. Comp. Sci. **3** (1976) 1–22

[Wi] H. Wielandt: Finite permutation groups. Acad. Press, New York 1964

[Wo] T.R. Wolf: Solvable and nilpotent subgroups of $GL(n, q^m)$. Can. J. Math. **34** (1982) 1097–1111

A Theory of Computation and Complexity over the Real Numbers

Lenore Blum [1]

International Computer Science Institute, 1947 Center Street, Berkeley, CA 94704; Mills College, Oakland, CA 94613; and University of California, Berkeley, CA 94720, USA

1. Introduction

Classically, the theories of computation and computational complexity deal with discrete problems, for example over the integers, about graphs, etc. On the other hand, most computational problems that arise in numerical analysis and scientific computation, in optimization theory and more recently in robotics and computational geometry, have as natural domains the reals **R**, or complex numbers **C**. A variety of ad hoc methods and models have been employed to analyze complexity issues in this realm, but unlike the classical case, a natural and invariant theory has not yet emerged. One would like to develop theoretical foundations for a theory of computational complexity for numerical analysis and scientific computation that might embody some of the naturalness and strengths of the classical theory.

Toward this goal, we have been developing a new theory of computation and complexity which attempts to integrate key ideas from the classical theory in a setting more amenable to problems defined over continuous domains. Our approach is both algebraic and concrete; the underlying space is an arbitrary commutative ring (or field) and the basic operations are polynomial (or rational) maps and tests.

The theory yields results in the continuous setting analogous to the pivotal classical results of *undecidability* and *NP-completeness* over the integers, yet reflecting the special mathematical character of the underlying space. For example, over the reals we have that (1) the Mandelbrot set as well as most Julia sets are undecidable[2] and (2) the problem of deciding if an algebraic variety has a real point is NP-complete. While there are many subtle differences between the new and classical results, the ability to employ mathematical tools of more mainstream mathematics (such as from algebra, analysis, geometry and topology) in the domain of the reals may suggest new approaches for tackling the classical, as well as new, "$P = NP$?" questions.

The material covered here is based in large part on (Blum, Shub and Smale 1989) denoted in this paper by BSS, (Blum and Smale 1990) and (Blum 1990).

[1] This work was partially supported by National Science Foundation grants CCR-8712121 and CCR-8907663 and the Letts-Villard Chair at Mills College.

[2] Indeed, the complements of these sets provide examples of *semi-decidable* sets that are undecidable over the reals.

Proceedings of the International Congress of Mathematicians, Kyoto, Japan, 1990

Discussions of related work and references are contained in those papers. See also (Shub 1990a, 1990b) and (Smale 1990). Additional relevant literature is listed in the References.

2. Computable Functions and Decidable Sets

The classical theory of computation had its origins in work of logicians – of Gödel, Turing, Church, Kleene, Post – in the 1930s. Of course there were no computers at the time; this work, in particular Turing's (1937), clearly anticipated the development of the modern digital computer. But even more, a primary motivation for the logicians was to formulate and understand the concept of *decidability*, or of a *decidable* set, thus to make sense of such questions: "Is the set of theorems of arithmetic decidable?" or "Is the set of polynomials with integer coefficients and integer solutions decidable?"[3]

Intuitively, a set $S \subset U$ is *decidable* if there is an "effective procedure" that given any element u of U (some natural universe) will decide in a finite number of steps whether or not u is in S, i.e. if the characteristic function of S (with respect to U) is "effectively computable." The models of computation designed by these logicians were intended to capture the essence of this concept of effective procedure/computation. The idea was to design formal "machines" with operations, and finitely described rules for proceeding step by step from one operation to the next, so simple and constructive that it would be self-evident that the resulting computations were effective.

In each formalism (e.g. Turing's), a function f from the natural numbers $\mathbf{N}$ to $\mathbf{N}$ is defined to be *computable* if it is the *input-output* function of some such machine (e.g. a Turing machine). It is quite remarkable that even though the formalisms were often markedly different, in each case, the resulting class of computable functions (and hence decidable sets) was exactly the same. Thus, the class of computable functions appears to be a natural class, independent of any specific model of computation.[4] This gives one a great deal of confidence in the theoretical foundations of the theory of computation. Indeed, what is known as *Church's thesis* is an assertion of belief that the classical formalisms completely capture our intuitive notion of computable function. Compelling motivation clearly would be required to justify yet a new paradigm.

3. Examples

In order to motivate our theory, we briefly discuss three examples, one from complex analytic dynamics, one from numerical analysis and one from classical complexity theory.

---

[3] It was originally taken for granted (by mathematicians in general, and Hilbert (1901–1902) in particular) that the answers to these questions were both affirmative. The queries were actually posed as tasks: "Produce decision procedures for the given sets." The incompleteness/undecidabilty results of Gödel (1931) in the first place, and of Matijasevich (1971) on the unsolvabilty of Hilbert's Tenth Problem in the second, show such tasks cannot be carried out in full generality.

[4] These functions are often called the *(partial) recursive functions*.

3.1 Is the Mandelbrot Set Decidable?

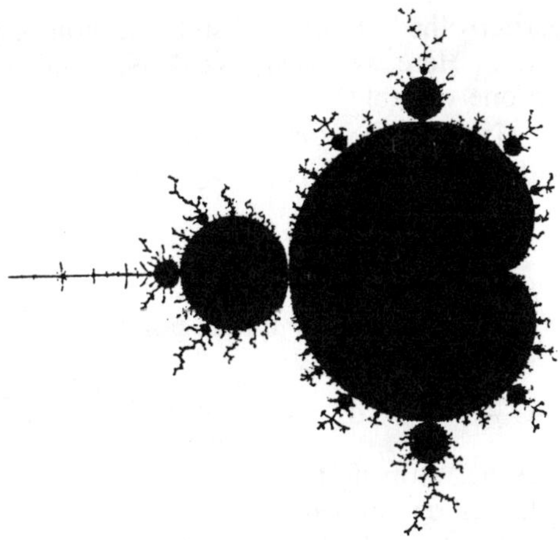

Fig. 1. The Mandelbrot set[5]

This question was asked by Penrose (1989) in his book *The Emperor's New Mind.* Recall the Mandelbrot set **M** can be defined:

$$\mathbf{M} = \{c \in \mathbf{C} | p_c^n(0) \nrightarrow \infty\},$$

where $p_c(z) = z^2 + c$ and p_c^n is the n-th *iterate of* p_c.

It is well known (see e.g. (Branner 1989)) that the boundary of **M** has a rich and extraordinarily complex structure. Hence, the reasonableness of Penrose's query.

However, the classical theory presupposes all underlying spaces are countable and hence ipso facto cannot handle such questions about arbitrary sets of real or complex numbers. One way to deal with this might be to consider the rational or algebraic skeletons of the sets in question. Problems quickly arise with this approach (e.g. consider the rational skeleton of the points on the curve $x^3 + y^3 = 1$ in the positive orthant). Another way might be to take a recursive analysis approach. For example, we might imagine a Turing machine being input a real number bit by bit by oracle. Using its internal instructions, the machine operates on what it sees, possibly every so often outputing a bit. The resulting sequence, if any, would be considered in the limit the (binary expansion of the) real output. Problems arise here when one wants to decide if two numbers are equal.

Penrose speculates on various such approaches and concludes (p. 129) "One is left with the strong feeling that the correct viewpoint has not yet been arrived at."

---

[5] This illustration is from (Penrose 1989) and is reproduced with permission by the publisher.

3.2 The Newton Machine

Newton's method is perhaps the "algorithm" sine qua non of numerical analysis and scientific computation. Here we briefly recall Newton's method for finding zeros of polynomials in one variable.

Given a polynomial $f(z)$ over the complex numbers $\mathbf{C}$, define the *Newton map* $N_f : S \to S$ of the Riemann sphere $S = C \cup \{\infty\}$ into itself by

$$N_f(z) = z - (f(z)/f'(z)).$$

Now for Newton's method: Pick an initial point $z_0 \in \mathbf{C}$ and generate the *orbit*

$$z_0, z_1 = N_f(z_0), z_2 = N_f(z_1), \ldots, z_{k+1} = N_f(z_k) = N_f^{k+1}(z_0), \ldots$$

Some stopping rule such as "stop if $|f(z_k)| < \varepsilon$ and output z_k (else pick a new initial point if k is too large)" is implied.[6]

We can represent Newton's method schematically as in Fig. 2.

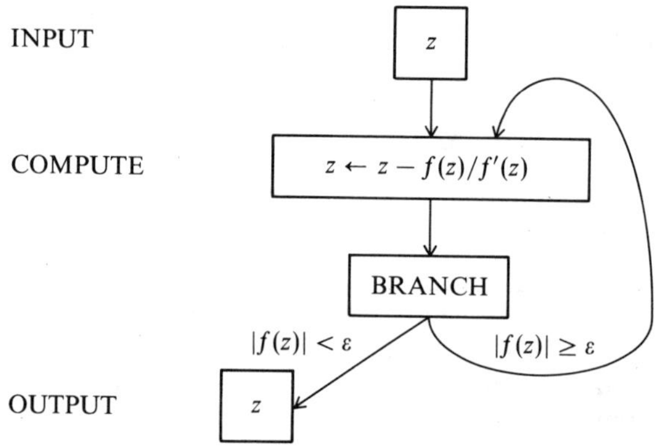

Fig. 2. The Newton machine for f

A Turing machine for implementing Newton's method, by reducing all operations to bit operations, would wipe out its basic underlying structure. We would like to have a model of computation in which Newton could be represented as naturally as in the Newton machine, and in which its salient features would be as apparent.

---

[6] It is well known, however, that Newton's method is not *generally convergent*. The main obstruction to general convergence is the existence of attracting periodic points of period at least 2. (See (Smale 1985) and (Friedman 1989) for estimates on measures for the basin of attraction of the Newton map.)

3.3 Does $P = NP$?

If a problem has a solution that can be easily verified, can such a solution be found quickly? This question is formalized by means of the fundamental open problem of classical (discrete) complexity theory, namely does $P = NP$? We would like to pose this question within a more general setting, thus perhaps increasing the mathematical tools and perspectives available to tackle it.

4. Finite Dimensional Machines over a Ring R

Now we describe our formal model of computation over a ring. Let R be an arbitrary ordered commutative ring (or field).

Definition. A finite dimensional *machine M over R* consists of three spaces: *input space $\overline{I}$*, *state space $\overline{S}$*, and *output space $\overline{O}$* of the form R^l, R^n, R^m respectively, together with a finite directed connected graph with four types of nodes: *input, computation, branch* and *output.*

The *unique* input node has no incoming edges and only one outgoing edge. All other nodes have (possibly several) incoming edges. Computation nodes have only one outgoing edge, branch nodes exactly two (left and right), and output nodes none. Each node has associated maps:

At the *input node,* there is a linear map I taking points from the input space to the state space.

Each *computation node* has an associated polynomial or rational map $g : R^n \to R^n$ of the state space to itself.

Each *branch node* has an associated polynomial function $h : R^n \to R$ from the state space R^n to the ring R. For a given *state* z in R^n at such a node, branching left or right will depend upon whether or not $h(z) < 0$.

Finally each *output node* has an associated linear map from the state space to the output space.

If R is a field and g is a rational map associated with some computation node, we will assume that previous nodes have tested for the vanishing of the denominators occurring in g and branched away as necessary. (Thus we are assuming that a map associated with a computation node is defined at every input to the node.)

Thus the Newton machine is an example of a machine over $\mathbf{R}$. (Here we are viewing $\mathbf{C}$ as $\mathbf{R}^2$ and the Newton map N_f as a rational map $g = (g_1, g_2) : \mathbf{R}^2 \to \mathbf{R}^2$. By expressing the stopping rule as $|f(z)|^2 < \varepsilon^2$, we get an equivalent real polynomial condition $h(x, y) < 0$.)

It is quite natural to view M as a discrete dynamical system. Here it is convenient to assume there is only one output node with associated map O. Thus we may let $\overline{N} = \{1, \ldots N\}$ be the set of nodes of M , where 1 is the input node and N the output node. We call the space of node/state pairs $\overline{N} \times \overline{S}$ the *full state space* of the machine.

Implicitly associated to M is the *computing endomorphism*

$$H : \overline{N} \times \overline{S} \to \overline{N} \times \overline{S}$$

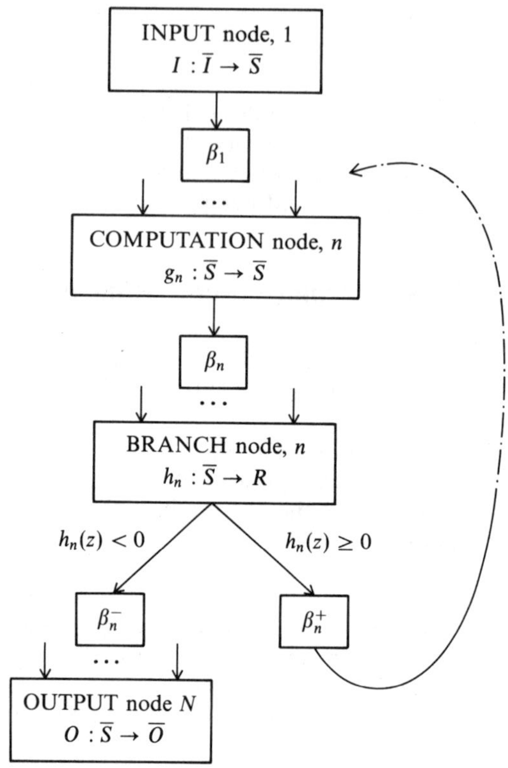

Fig. 3. A finite dimensional machine M: I and O are the maps associated with the input and output nodes respectively. For n a computation node, g_n is the associated "computation map"; and for n a branch node, h_n is the associated "branching function."

For n an input node or a computation node, β_n is the unique *next node* following n. For n a branch node, β_n^- is the next node along the left outgoing edge and β_n^+ the next node along the right outgoing edge

of the full state space to itself. That is, H maps each node/state pair (n, x) to the unique *next node/next state* pair determined by the directed graph of the machine and its associated maps (see Fig. 3) as follows:

$$H(1,x) = (\beta_1, x)\,; \quad H(N,x) = (N,x)\,; \quad H(n,x) = (\beta_n, g_n(x))$$

if n is a computation node; and if n is a branch node,

$$H(n,x) = (\beta_n^-, x) \text{ if } h_n(x) < 0, \quad \text{else } (\beta_n^+, x) \text{ if } h_n(x) \geq 0.$$

The computing endomorphism is our main technical as well as conceptual tool. For example, we can use it to define the *input-output map* φ_M of a machine M as follows:

With *input* y in $\overline{I}$, let $x = I(y)$. Then with *initial point* $z_0 = (1, x)$ of the full state space $\overline{N} \times \overline{S}$ generate the *computation* (i.e. the orbit under iterates of H)

$$z_0 = (1,x), z_1 = H(z_0), z_2 = H(z_1), \ldots, z_k = H(z_{k-1}) = (n_k, x_k), \ldots$$

Halt when (if ever) the first point z_T is produced which has the form $z_T = (N, w)$. If this is the case, the resulting finite sequence is called a *halting computation*; we say M *halts* on input y in *(halting) time* T with *output* $O(w)$ and define $\varphi_M(y) = O(w)$. If there is no such T, then M *does not halt* on input y (i.e. the halting time is infinite) and φ_M is not defined.

The *halting set* of M, Ω_M, is the set of all points in $\overline{I}$ on which M halts. Thus, $\varphi_M : \Omega_M \to \overline{O}$.

The conditions describing halting computations are essentially (semi-)algebraic; they serve as the key technical tool in the proof of the NP-Completeness Theorem, as well as in an algebraic proof of Gödel's Theorem (see (Blum and Smale, 1990)). The basic idea is that the relevant sets can be defined in terms of these conditions. For example, the *time T halting set* of M can be defined as the set of all points y in $\overline{I}$ for which there are solutions $z_0, \ldots, z_T$ and w to the (time T) *register equations* (of M):

$$z_0 = (1, I(y)), \quad z_T = (N, w) \quad \text{and } z_k = H(z_{k-1}) \text{ for } k = 1, \ldots, T .$$

Now having defined our formal notion of machine over R, we can easily formalize all related concepts including those in Sects. 2 and 3. For example, we define a map

$$\varphi : Y \to R^m, \quad Y \subset R^l$$

to be *computable over* R if it is the input-output map of some machine M over R, i.e. if $\varphi = \varphi_M$ and $Y = \Omega_M$. We say M *computes* φ.[7] A set $S \subset R^l$ is *decidable over* R if its characteristic function is computable over R. Otherwise it is *undecidable over* R.

In this setting, Penrose's question may thus be posed quite formally: Is the Mandelbrot set $\mathbf{M}$ decidable over $\mathbf{R}$? (Again we are viewing $\mathbf{C}$ as $\mathbf{R}^2$.)

But before addressing this, it is worth noting that $\mathbf{M}'$, the complement of $\mathbf{M}$, is *semi-decidable* over $\mathbf{R}$. That is, there is a machine over $\mathbf{R}$ that on input $x \in \mathbf{R}^2$ outputs 1 if $x \in \mathbf{M}'$ and otherwise outputs 0 or is undefined. A semi-decidable machine for $\mathbf{M}'$ can be constructed (see Fig. 4) using the fact that $\mathbf{M}$ is also characterized as $\{c \in C \mid |p_c^n(0)| \leq 2\}$.

Now as in the classical theory, it is easy to see that a set is decidable just in case both it and its complement are semi-decidable, and that the semi-decidable sets are exactly the halting sets.[8]

Thus we are now ready to take a closer look at halting sets. Here it is convenient to return to the directed graph picture of a machine M over R. To each point y in the halting set Ω_M we associate its *halting path*, i.e. the finite sequence of nodes

---

[7] We remark that the new theory reduces to the classical when $R = \mathbf{Z}$. That is, the computable functions over $\mathbf{Z}$ are exactly the recursive functions (see BSS). Therefore, our model of computation is sufficiently powerful to develop the classical theory. (By Church's Thesis, we would have cause for concern had we produced more functions computable over $\mathbf{Z}$.)

[8] In the classical theory, the halting sets are exactly the *output sets* of machines. This is true also for machines over real closed fields (BSS), but not in general over arbitrary ordered rings or fields (Michaux 1990). *Open Problem.* Which results of the classical theory of computation generalize and which do not? See (Blum, Smale 1990) and (Friedman, Mansfield 1988) for additional examples.

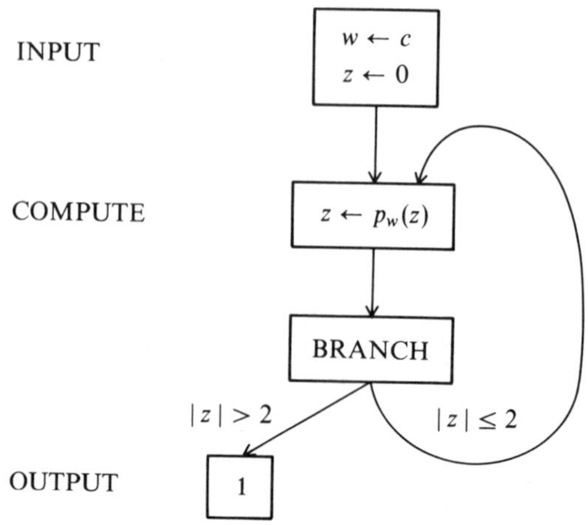

Fig. 4. A semi-decision machine for the complement of **M**

$$n_0 = 1, n_1, \ldots, n_T = N$$

traversed from input to output in the *computation* of $\varphi_M(y)$.

There are only a countable number of halting paths. For each halting path γ let $\bar{I}\gamma$ be the set of all points in the halting set Ω_M that have γ as their halting path. It is easy to see that for distinct γ's the $\bar{I}\gamma$'s are disjoint. Also, each $\bar{I}\gamma$ is a *semi-algebraic* set.[9] Note also that M acts like a "straight line program" on $\bar{I}\gamma$. Indeed, by concatenating the input, computation, and output maps that occur along the path γ, we see that φ_M restricted to $\bar{I}\gamma$ is just a polynomial (or rational) map φ_γ. Thus we have the following:

Proposition 1. *The halting set of a machine M over R is a (disjoint) countable union of semi-algebraic sets (over R); the input-output map φ_M is a piecewise polynomial (or rational) map.*

So, for example, halting sets have integral Hausdorff dimension.

Proposition 2 (Sullivan 1990). *The Mandelbrot set is not the countable union of semi-algebraic sets over* **R**.[10]

Corollary. *The Mandelbrot set is not decidable over* **R**.

The same holds for most Julia sets since, from the theory of complex analytic dynamical systems, we know that most Julia sets have fractional Hausdorff dimension. Indeed, for hyperbolic rational maps of the Riemann sphere, we have the following

[9] A set $S \subset R^l$ is *basic semi-algebraic* (over R) if it is the set of elements in R^l that satisfy a (fixed) finite set of polynomial equalities and inequalities (over R). A *semi-algebraic* set is a finite union of basic semi-algebraic sets.

[10] Here I would like to acknowledge helpful discussions with Michel Herman and Adrien Douady and also with John Hubbard who provided an independent proof of this result.

Theorem (BSS). *A Julia set is decidable if and only if it is*

1. *a round circle,*
2. *an arc of a round circle, or*
3. *the whole sphere.*

5. Infinite Dimensional Machines over R

The classical construction of a universal machine assumes an effective coding of machines by (natural) numbers. In effect, the coding is a collapsing of sequences of numbers into a single number. Over the integers this can be done by a Gödel coding. However, in general over a ring R (e.g. over the reals), such an invertible collapsing cannot be done by a computable map. This suggests that a universal machine over R should have the facility to take as input finite sequences of unbounded length.

In addition, if we wish to have a natural framework for dealing with uniform procedures for solving problem instances of arbitrary dimension (say for the Travelling Salesman Problem), we are also led to consider machines that handle unbounded sequences.

With these considerations in mind we are motivated to extend our notions to *infinite dimensional machines* over R:

The underlying spaces $\overline{I}$, $\overline{S}$, and $\overline{O}$ for an infinite dimensional machine over R each will be R^∞, the *infinite direct sum* space over R. A point $y = (y_1, y_2, \ldots)$ in R^∞ satisfies $y_k = 0$ for k sufficiently large. The *length* of y is the largest n such that $y_n \neq 0$. Polynomial (or rational) maps in this context are still defined by a fixed finite number of polynomials (rational functions) that and depend only on a fixed finite number of variables.

The machine will consist of a finite connected directed graph now containing five types of nodes, four as before, with associated maps. If the machine had only the previous type nodes it would essentially be a finite dimensional machine. The increased power comes from the addition of *fifth nodes* that allow accessing of coordinates of arbitrarily high dimension.

A *fifth node* may have several incoming edges but only one outgoing edge. The associated map transforms state $x = (i, j, x_1, \ldots, x_j, \ldots, x_k, \ldots, x_i, \ldots)$ to state $x' = (i, j, x_1, \ldots, x_i, \ldots, x_k, \ldots, x_i, \ldots)$, assuming i and j are positive integers. That is, the fifth node map writes x_i in the "j-th place" of x and leaves everything else alone.

Thus, the first two coordinates of the state space play a special role which require some minor modification of the other maps. For $y = (y_1, y_2, \ldots)$ in $\overline{I}$ we let $I(y) = (1, 1, \text{length}(y), y_1, 0, y_2, 0, y_3 \ldots)$. This initializes the indices i and j and leaves room for workspace. Information about the length of y is often useful and so is also included. For $x = (i, j, x_1 x_2, \ldots)$ in $\overline{S}$, we suppose a computation node map can alter the first two coordinates only by adding 1 or by setting to 1. Finally we let $O(x) = (x_2, x_4, \ldots)$.

Computing endomorphisms, input-output maps, halting sets and all such related notions are defined exactly as before. Note that a finite dimensional machine can be considered as a special case of infinite dimensional machine.

See BSS for an explicit construction of a universal machine over **R**.

6. Complexity Theory over a Ring R

A goal of computational complexity theory is to quantify the intrinsic difficulty of solving problems. This theory had its origins in the 1960s.[11] It was developed primarily by researchers, originally trained in mathematics and logic, but who found more hospitable environments for these interests in the newly emerging computer science departments. Here, in the realm of the solvable (decidable), they discovered a rich and natural hierarchy, with the dichotomy of tractability/intractability mirroring the dichotomy of decidability/undecidability studied by the logicians. (For the seminal work in the theory, see (Rabin 1960), (M. Blum 1967), (Winograd 1970), (Cook 1971), (Karp 1972) and (Levin 1972).)

Classical complexity theory deals primarily with combinatorial (discrete, integer) problems. We extend the theory in order to consider a wider class of problems. As has been traditional however, we focus on *decision problems*. These are problems with "yes/no" answers (to questions generally of the form "Does there exist a solution to ...?") and are classified as to their difficulty into *classes* P, NP or as being NP-*complete*.

Definition. A *decision problem* over R is a pair (Y, Y_{yes}) with

$$Y_{\text{yes}} \subset Y \subset R^{\infty}.$$

Y is the set of *problem instances*, and Y_{yes} is the set of yes-*instances*.

For example the Travelling Salesman Problem, stated over an ordered ring R, can be put in this form by letting:

$$\begin{aligned} Y &= \{(n, A, k) \mid n \text{ is a positive integer, } k > 0 \text{ and} \\ &\qquad A = (a_{ij}) \text{ is an } n \times n \text{ matrix over } R\} \\ Y_{\text{yes}} &= \{(n, A, k) \text{ in } Y \mid \text{there is a tour } \tau(n) \text{ with Distance } (A, \tau(n)) \leq k\}. \end{aligned}$$

Here a tour $\tau(n) = (\tau_1, \tau_2, \ldots, \tau_n)$ is a cycle on the entire set $\{1, 2, \ldots, n\}$ and Distance $(A, \tau(n)) = \left(\sum_{i=1}^{n-1} a_{\tau_i \tau_{i+1}}\right) + a_{\tau_n \tau_1}$. By representing A by the sequence of its rows one after the other, we have $Y \subset R^{\infty}$.

A second example, which will be prominent in out theory, is the *4-Feasibility Problem (4-FEAS)* over R. Here,

$$\begin{aligned} Y &= \{\text{multivariable polynomials } f \text{ over } R \mid \text{degree } f \leq 4\} \\ Y_{\text{yes}} &= \{f \text{ in } f(\xi) = 0 \text{ for some } \xi = (\xi_1, \ldots, \xi_k) \text{ in } R^k\} \end{aligned}$$

We are supposing that polynomials are represented as elements of R^{∞} via the *standard representation* (see BSS).

Thus the 4-FEAS problem is: Given a multivariable polynomial f of degree 4 with coefficients from R, does $f(x) = 0$ have a solution over R? While it may not at all be obvious how to decide if such a solution exists, it is a straightforward procedure to verify one that may be presented to us. Just plug the purported solution into the equation and check it out. Is this verification tractable in our model of computation? The answer will depend on the underlying mathematical

[11] However, as early as 1948, von Neumann (1963) had articulated the need for such a theory.

properties of the ring or field, as well as our measure of complexity. But first we must formalize the basic concepts of size and cost.

We first suppose we have a function *height* defined on R with values in the non-negative reals, e.g. for $R = \mathbf{Z}$ or $\mathbf{R}$, and $y \in R$ we might choose *height* (y) to be *logarithmic* height, $log_2(|y|+1)$, or *unit* height, 1. Then for $y = (y_1, y_2, \ldots, y_n, 0, 0, \dot{)} \in R^\infty$, we define

$$\text{size}(y) = \text{length}(y) + \text{height}(y)$$

where $\text{height}(y) = \max \text{height}(y_i)$. Thus with unit height, *size* reflects the "dimension" of input, whereas over the integers with logarithmic height size reflects the traditional bit length. For the remainder of this paper, unless otherwise stated, we will suppose unit height for $\mathbf{R}$ and logarithmic height for $\mathbf{Z}$.

Now suppose M is a machine over R with a height function defined. Then

$$\text{cost}_M(y) = T_M(y) \times h_{\max}(y)$$

where $T_M(y)$ is the halting time of M on input y (which may be finite or infinite depending on whether or not y is in the halting set of M) and $h_{\max}(y)$ is the maximum height of any element occurring in the computation of M on input y. Over the reals, the cost reflects the number of basic algebraic operations, whereas over the integers, the cost reflects the number of bit operations.

The following definitions make sense only in case height has been defined over R.

Definition. A map φ on *(admissible)* inputs $Y \subset R^\infty$ is *polynomial time computable over* R if there is a machine M over R that computes φ and

$$\text{cost}_M(y) \leq \text{poly}(\text{size}(y)), \quad \text{for all } y \text{ in } Y .$$

Here poly is some polynomial with nonnegative integer coefficients. Polynomial time is meant to formalize our notion of tractability.

Now we are in a position to formally define class P and class NP over R. While the first definition is straightforward, the second is considerably more subtile.

Definition. A decision problem (Y, Y_{yes}) is in *class* P (*polynomial time*) over R if the characteristic function of Y_{yes} in Y is polynomial time computable over R.

Definition. (Y, Y_{yes}) is in *class* NP (*non-deterministic polynomial time*) if there is a machine M which takes as input pairs (y, w) (where y in Y is a problem instance and w in R^∞ is thought of as a "guess" or "witness" for a solution to y), outputs 1 or 0 (yes or no) and satisfies:

1. If y is a yes-instance then there exists some (guess for a solution) w such that $\varphi_{M'}(y, w) = 1$ and
$$\text{cost}_M(y, w) \leq \text{poly}(\text{size}(y)) .$$
2. If y is a *no-instance* (i.e. not a yes-instance) then there is *no* (guess) w such that $\varphi_M(y, w) = 1$.

We remark that we need only consider guesses w for which size(w)poly(size(y)).

M is called an *NP-decision machine* for the *NP-problem* (Y, Y_{yes}). Property 1 reflects the non-deterministic aspect of this notion, i.e. for each yes-instance, we just require that some polynomial time verifiable solution exists, not necessarily that one can be found. Property 2 requires that the verification process have some integrity, i.e. it can never output yes for a no-instance input.

In this general setting it is natural to ask (analogous to the classical question over $\mathbf{Z}$): Does $P = NP$ over R? For $R = \mathbf{R}$, we have a new *open problem.*

Now let us return to 4-FEAS. The halting time for verifying a purported solution to a polynomial equation $f(x) = 0$ using a straightforward evaluation process can easily be seen to be bounded above by a polynomial function of the length of f. (Recall we are supposing f is standardly represented as an element of R^∞.) Thus over the reals, since size is length and cost is halting time, this verification is polynomial time.

On the other hand, over the integers we know (by the undecidability of Hilbert's Tenth Problem) that even the *smallest* size of integer solutions to polynomial equations $f(x) = 0$ (solvable over $\mathbf{Z}$) cannot be bounded above by any polynomial (in size(f)). Thus, even if we only consider solutions of smallest size, there is no polynomial that will bound the cost of verification; in general it would just take too long to even read in purported solutions.

The above arguments show that 4-FEAS is in class NP over $\mathbf{R}$ but not over $\mathbf{Z}$.

A key impetus for the development of classical complexity theory was the discovery (by Cook (1971) and Levin (1973)) of the existence of NP problems (over $\mathbf{Z}$) that efficiently encode all NP problems.

Definition. $\left(\widehat{Y}, \widehat{Y}_{\text{yes}}\right)$ is *NP-complete* if it is in class NP and *universal* in the following sense:

For every (Y, Y_{yes}) in NP there is a polynomial time map $\varphi : Y \to \widehat{Y}$ such that for all y in Y

$$y \text{ is in } Y_{\text{yes}} \text{ if and only if } \varphi(y) \text{ is in } \widehat{Y}_{\text{yes}}\,.$$

Here φ is the efficient (i.e. polynomial-time) *coding* function. Thus any decision procedure for $\left(\widehat{Y}, \widehat{Y}_{\text{yes}}\right)$ can be easily converted (in polynomial time) into one for (Y, Y_{yes}) of not worse complexity (up to a polynomial): To decide if y is in Y_{yes}, simply encode y into $\widehat{Y}$ using a polynomial time machine for φ and then decide if $\varphi(y)$ is in $\widehat{Y}_{\text{yes}}$.

Thus an NP-complete problem is the "hardest" problem in the class NP; any NP problem can be efficiently "reduced" to it.

We have the following analogue over $\mathbf{R}$, to the pivotal Cook Theorem (3–SAT is NP-complete)[12] over $\mathbf{Z}$:

Main Theorem (BSS). *The 4-Feasibility Problem (4-FEAS) is NP-complete over the reals.*

[12] Cook's *Satisfiability Problem* is: Given a Boolean formula $\varphi(u_1, \ldots, u_k)$ is there an assignment to the variables $u_1, \ldots, u_k$ that makes the formula true? For 3-SAT the Boolean formulas considered are conjunctions of clauses of the form "U or V or W". Here each of U, V or W is either a variable or the negation of a variable.

Remarks. This theorem has a number of immediate consequences which point to the subtle differences between the theory of NP over the integers and over the reals.

For example, over the integers it is easy to see, using a simple counting argument, that NP problems are decidable in exponential time (in the size of the instance). This is because, as noted earlier, for problem instances y, we need only consider guesses of size at most poly(size(y)). Over **Z**, there are at most $2^{\text{poly(size}(y))}$ such guesses, and so a perfectly good decision procedure is to check out each one in turn using an NP-decision machine for the problem.

On the other hand, over **R** there are a continuum number of such guesses, and so it is not even clear that NP problems are decidable over **R**, no less decidable in exponential time. However, by Tarski (1951), 4-FEAS is decidable over **R**. So by the NP-completeness of 4-FEAS we see that all NP problems are decidable over **R**. Moreover, (by Canny (1988) and Renegar (1988)) 4-FEAS is decidable in exponential time (over **R**),[13] and so all NP problems must be decidable in exponential time. Thus we have the same result here over the reals as over the integers but now for much deeper reasons.

The Main Theorem implies that the $P = NP$? problem over **R** is equivalent to the new *open problem*: Is 4-FEAS in class P over **R**? (thus focusing our attention on an intrinsic algebraic-geometric problem new to complexity theory.) In contrast, recall over the integers, 4-FEAS is not *even* decidable over **Z**.

The analogous NP-complete problem over the complex numbers **C** is related to an effective version of Hilbert's Nullstellensatz.[14] Thus, as in the case of the reals, the NP-complete problem here is of a fundamental nature. However, for the moment, these are essentially the only NP-complete problems known over the reals or complex numbers.

To contrast, what makes the classical theory of NP-completeness so compelling has been the discovery (indicated first by the work of (Karp 1972)) of a large number of seemingly unrelated NP-complete problems. A polynomial time decision method for one would yield polynomial time decision methods for all.

Open Problem. Find other (seemingly unrelated) NP-complete problems over the reals or complex numbers.

The TSP is NP-complete over **Z** and, as remarked earlier, in class NP over **R**.

Open problem: Is TSP NP-complete over **R**?

Proof of Main Theorem (Idea). The task at hand is to show, for each NP-problem (Y, Y_{yes}) over **R**, how to encode in polynomial time any problem instance $\mathbf{y}$ as a degree 4 polynomial $\widehat{\mathbf{y}}$ over **R** such that:

$$\mathbf{y} \text{ is a yes-instance if and only if } \widehat{\mathbf{y}} = 0 \text{ has a solution over } \mathbf{R}.$$

---

[13] For related results with respect to bit complexity see (Grigor'ev and Vorobjov 1988).

[14] A machine over **C** is similar to one over **R** except at branching nodes; over **C** branching left of right will depend on whether or not a polynomial h evaluated at the current state x is equal to 0. The NP-complete problem over **C** is: Given a system $f_1, \ldots, f_k$ of polynomials in n variables $x_1, \ldots, x_n$ over **C**, decide if there is a common solution over **C**. By Hilbert's Nullstellensatz, $f_1, \ldots, f_k$ has no common solution just in case 1 is in the ideal generated by $f_1, \ldots, f_k$, i.e. if and only if $1 = a_1 f_1 + \ldots + a_k f_k$ for some polynomials $a_1, \ldots, a_k$ over **C** in the variables $x_1, \ldots, x_n$.

The basic idea is to utilize the register equations for an NP-decision machine for (Y, Y_{yes}).

First suppose M is any machine over R. Note that the assertion "M with input $\mathbf{y}$ outputs $\mathbf{x}$ in time $\mathbf{T}$" is equivalent to asserting the system of equations

$$z_0 = (1, I(\mathbf{y})), \quad z_{\mathbf{T}} = (N, x_{\mathbf{T}}), \quad O(x_{\mathbf{T}}) = \mathbf{x} \quad \text{and } z_k = H(z_{k-1}) \text{ for } k = 1, \ldots, T.$$

is solvable over R. We can convert (in polynomial time) this essentially semi-algebraic system over $\mathbf{R}$ to a single polynomial equation of degree 4

$$f(\mathbf{y}, \mathbf{x}, u_1, \ldots, u_{T'}) = 0$$

such that the original system is solvable over $\mathbf{R}$ if and only if the single equation is. Here $\mathbf{T}' = p(\mathbf{T})$ where p is some polynomial dependent only on M. See BSS for details.

Now suppose M is an NP-decision machine for (Y, Y_{yes}) with time bound a polynomial q. For $\mathbf{y}$ in Y let $\mathbf{T} = q(\text{size}\mathbf{y})$. Suppose $\mathbf{w}$ is in $\mathbf{R}^\infty$ and size$(\mathbf{w}) = \mathbf{T}$. By the above we have:

"M with input $(\mathbf{y}, \mathbf{w})$ halts with output 1 in time $\mathbf{T}$" if and only if there is a solution to $f((\mathbf{y}, \mathbf{w}), 1, u_1, \ldots, u_{T'}) = 0$. Here $\mathbf{T}' = p(\mathbf{T}) = p(q(\text{size } \mathbf{y}))$.

Now we are ready to encode: For each $\mathbf{y}$ in Y let $\widehat{\mathbf{y}}$ be the degree 4 polynomial $f((\mathbf{y}, w), 1, u_1, \ldots, u_{T'})$ as above (having constant $\mathbf{y}$ and $\mathbf{T} + \mathbf{T}'$ variables $w = (w_1, \ldots, w_T)$ and $u_1, \ldots, u_{T'}$). This is a polynomial time encoding over $\mathbf{R}$.

Now by the definition of NP-problems and NP-decision machines, $\mathbf{y}$ is in Y_{yes} if and only if: there is a w in $\mathbf{R}^\infty$ with size$(w) = \mathbf{T}$ such that M with input $(\mathbf{y}, w)$ halts with output 1 in time T. By the above, this holds if and only if: there is a solution over $\mathbf{R}$ to $f((\mathbf{y}, w), 1, u_1, \ldots, u_{T'}) = 0$, i.e. to $\widehat{y} = 0$.

7. Conclusion and Directions

Since this framework is new there are a number of open problems, some of which have already been indicated and new directions to take. Many questions naturally arise concerning the relationship between the classical and new fundamental "$P = NP$?" questions, and even whether the various possible extensions of the classical notion of NP (e.g. via guesses or via non-deterministic computation) are equivalent. (This is related to the question of whether or not the TSP is NP-complete over $\mathbf{R}$.) Here it is proving fruitful to investigate the fundamental question under varying assumptions on height, computing power and branching criteria. (See (Shub 1990b).) In the general setting, algebraic topology is providing useful tools for lower bound arguments on the topological complexity (i.e. branching complexity) of problems (Smale 1987; Vasiliev 1988; Levine 1989; Hirsch 1990). Related questions are also being pursued over other fields e.g. the p-adics (Bishop 1990).

Another direction is to study questions of parallel and distributed computation, as well as probabilistic algorithms, in this context. For the latter it would be natural to add "coin tossing" nodes to machines. Related to distributed computation, Luo and Tsitstklis (1990) have recently given tight lower bounds for the communication complexity of several algebraic problems.

To bring the theory closer to numerical analysis and scientific computation one must extend the new model of computation to incorporate notions of round-off errors, condition numbers and approximate solutions. See (Renegar 1990) and (Priest 1990). Here questions of the relationship between the complexity and the condition of a problem arise. In this direction it would also seem natural to adjoin nodes to compute limits of (rapidly) converging sequences, as well as other reasonable functions.

Finally there are interconnections between logic (and computation/complexity theory) and the theory of complex analytic dynamical systems to pursue. This is an intriguing direction. For example, inspired by the "degree theory" of classical recursive function theory, one is led to study the hierarchy of Julia sets imposed by various notions of relative decidability. Roughly we say a set A is *decidable relative to* a set B ($A \leq B$) if a machine with an additional node for deciding B (i.e. an "oracle" for B) can be used to decide A. The question then is: what is the resulting hierarchy? Classically, it was an open problem for a number of years (a variant of Post's problem) to find two semi-decidable sets of integers that were incomparable with respect to relative decidability. (See (Rogers 1967).) Over **R**, Chong (1990) has shown that the situation appears to be quite the opposite, at least for undecidable Julia sets of quadratic maps.[15] Thus we ask: are there two comparable undecidable Julia sets? Alternatively, is there a natural way to increase the power of machines so that the resulting hierarchy is meaningful?

In the opposite direction we have exploited the analog between computing machines and dynamical systems in our NP-completeness proofs over the reals (Blum, Shub and Smale 1989) and for a new proof of Gödel's Theorem (Blum and Smale 1990). Here the computing endomorphism is our key technical tool. Can we exploit this analogy further and use techniques of dynamical systems to better understand the nature of complexity of computations and of formal computing machines? In concrete form, this approach has been successfully used by Batterson (1990) (following Shub (1983) and Smale (1985)) for the global analysis of classical algorithms of numerical linear algebra.

References

Batterson, S. (1990): Dynamics of eigenvalue computation. To appear in: Proceedings of the Smalefest

Bishop, E. (1990): Computation and complexity over $\mathbf{Q}_p$. Work in progress

Blum, L. (1990): Lectures on a theory of computation and complexity over the reals (or an arbitrary ring). Lectures in the Sciences of Complexity II (ed. E. Jen). Addison-Wesley, pp. 1–47

Blum, L., Shub, M. (1986): Evaluating rational functions: Infinite precision is finite cost and tractable on average. SIAM J. Computing **15** (2) 384–398

Blum, L., Shub, M., Smale, S. (1989): On a theory of computation and complexity over the real numbers: NP-completeness, recursive functions and universal machines. Bull. Amer. Math. Soc. **21**, no. 1, 1–46

Blum, L., Smale, S. (1990): The Gödel incompleteness theorem and decidability over a ring. To appear in: Proceedings of the Smalefest

Blum, M. (1967): A machine-independent theory of the complexity of recursive functions. J. Assoc. Comp. Mach. **14**, 322–336

Branner, B. (1989): The Mandelbrot set. Chaos and Fractals, Proceedings of Symposia in Applied Mathematics (AMS) **39**, 75–105

[15] Technically, we are talking about the complements of Julia sets which are semi-decidable over **R**.

Canny, J. (1988): Some algebraic and geometric computations in PSPACE. Proceedings of the 20th Annual ACM Symposium on the Theory of Computing, pp. 460–467

Chong, C.T. (1990): Reducibility of Julia sets in complex analytic dynamics. Preprint

Cook, S.A. (1971): The complexity of theorem proving procedures. In: Proceedings 3rd ACM STOC, pp. 151–158

Friedman, H., Ko, K. (1982): Computational complexity of real functions. J. Theoret. Comput. Sci. **20**, 323–352

Friedman, H., Mansfield, R. (1988): Algorithmic procedures. Preprint, Penn State

Friedman, J. (1989): On the convergence of Newton's method. J. Complexity **5**, 12–33

Garey, M.R., Johnson, D.S. (1979): Computers and intractability. W.H. Freeman and Co., San Francisco

Gödel, K. (1931): Über formal unentscheidbare Sätze der Principia Mathematica und verwandter Systeme I. Monatshefte für Mathematik und Physik **38**, 173–198

Grigor'ev, D.Yu., Vorobjov, N.N. (1988): Solving systems of polynomial inequalities in subexponential time. J. Symb. Comput. **5**, 37–74

Heintz, J., Roy, M.-F., Solerno, P. (1989): Sur la complexite du principe de Tarski-Seidenberg. Preprint

Hilbert, D. (1901–1902): Mathematical problems. Bull. Amer. Math. Soc. **8**, 437–479

Hirsch, M. (1990): Applications of topology to lower bound estimates in computer science. PhD thesis, UC Berkeley

Karp, R. (1972): Reducibility among combinatorial problems. Complexity of Computer Computations, edited by R. Miller and J. Thatcher. Plenum Press, New York pp. 85–10

Levin, L.A. (1973): Universal sorting problems. Problemy Peredaci Informacii **9**, 115–116 (in Russian). [English transl.: Problems of Information Transmission **9**, 265–266]

Levine, H. (1989): A lower bound for the topological complexity of poly (d,n). J. Complexity **5**, 34–44

Luo, Z.-Q., Tsitsiklis, J.N. (1990): Communication complexity of algebraic computation (extended abstract.). Proceedings of the 31st Annual Symposium on Foundations of Computer Science, IEEE Computer Society Press, pp. 758–765

Matijasevich, Y. (1970): Enumerable sets are Diophantine. Sov. Math. Dokl. **11**, 354–357

Michaux, C. (1990): Ordered rings over which output sets are recursively enumerable sets. Preprint, Universite de Mons, Belgium

Penrose, R. (1989): The Emperor's New Mind. Oxford University Press

Pour-El, M.B., Richards, I. (1983): Computability and noncomputability in classical analysis. Trans. Amer. Math. Soc **275**, 539–560

Priest, D. (1990): Precision analysis: An approach to designing accurate computations in floating point arithmetic. Preprint

Rabin, M.O. (1969): Degree of difficulty of computing a function and a partial ordering of the recursive sets. Technical Report No. 2. Hebrew University, Jerusalem

Renegar, J. (1988): A faster PSPACE algorithm for deciding the existential theory of the reals. Proceedings of the 29th Annual Symposium of Computer Science. IEEE Computer Society Press, pp. 291–295

Renegar, J. (1989): On the computational complexity and geometry of the first-order theory of the reals, Parts I, II and III. Technical Report nos. 853, 854, 856, School of Operations Research and Industrial Engineering, Cornell University, Ithaca

Renegar, J. (1990): Some foundations for a general theory of condition numbers. To appear in: Proceedings of the Smalefest

Rogers, Jr., H. (1967): Theory of Recursive Functions and Effective Computability. McGraw Hill, New York

Shub, M. (1983): The geometry and topology of dynamical systems and algorithms for numerical problems. Lectures at Beijing University, Beijing, China

Shub, M. (1990a): Some remarks on Bezout's theorem and complexity theory. IBM Research Report # 71048, 1990 and to appear in: Proceedings of the Smalefest

Shub, M. (1990b): On the work of Steve Smale on the theory of computation. IBM Research Report # 71048, 1990 and to appear in: Proceedings of the Smalefest

Smale, S. (1985): On the efficiency of algorithms of analysis. Bull. Amer. Math. Soc. **13** (2) 87–121

Smale, S. (1987): On the topology of algorithms I. J. Complexity **3**, 81–89

Smale, S. (1990): Some remarks on the foundations of numerical analysis. SIAM Review **32**, no. 2, 211–220

Sullivan, D. (1990): Personal communication

Turing, A.M. (1937): On computable numbers, with an application to the Entscheidungsproblem. Proc. Lond. Math. Soc. (Ser. 2) **42**, 230–265

Vasiliev, V. (1988): Cohomology of the braid group and the complexity of algorithms. Preprint and to appear in: Proceedings of the Smalefest

Von Neumann, J. (1963): The general and logical theory of automata. Collected Works, vol. V (A. Taub, ed.). MacMillan, New York, pp. 288–328

Von zur Gathen, J. (1988): Algebraic complexity theory. Technical Report No. 207/88, Department of Computer Science, University of Toronto, 37 pp.

Winograd, S. (1970): On the number of multiplications necessary to compute certain functions. Comm. Pure Appl. Math. **23**, 65–79

Efficient Factoring Polynomials over Local Fields and Its Applications

Alexandre L. Chistov

Leningrad Institute for Informatics and Automation of the Academy of Sciences of the USSR (LIIAN), 14th line 39, Leningrad 199178, USSR

Introduction

In this paper an algorithm is described for factoring multivariable polynomials over local fields. The complexity of the algorithm is polynomial in the size of input and the characteristic p of the residue field of the local field. As an application a polynomial equivalence is ascertained for the problem of constructing a basis of the ring of all integers of a given number field and the problem of finding square free part of an integer. It means that the first (respectively second) of these problems can be solved within polynomial time if there is an oracle for solving the second (respectively first) problem within polynomial time. Even the more general result is proved which is also valid in the case of non-zero characteristic, see Theorem 2.

In proofs of the last results we use on the one hand the factorization of polynomials over local fields and on the other hand an idea which is applied for obtaining efficient bounds for sizes of coefficients in the Newton-Puiseux expansion, see [8, 7] and also Lemma 1 below.

The present results solve in particular problems posed by H.W. Lenstra, Jr. in [9]. In the general case, even for one variable, earlier known algorithms required for factoring polynomials over local fields an enumeration exponential in the size of input data before applying Hensel's lemma, see [1]. Elements of local fields are represented as sums of infinite series. Here and below we regard a series as computable in time polynomial in $A_1, \dots, A_m$ iff its ith partial sum S_i is computed in time polynomial in $A_1, \dots, A_m$ and i for all i. Besides that, if computation of S_i involves other infinite series, then it should involve a number of initial terms polynomial in i and $A_1, \dots, A_m$.

Our algorithm of the factorization uses the method of Newton's polygons for constructing roots of polynomials in one variable. However, in its classical form, as in the case when the residue field is of zero characteristic, this method does not succeed because of the presence of higher ramification for extention of local fields, when one cannot choose in advance a uniformizing element in the extension. For solving the problem we use additionally expansions of a special type in the factor algebra modulo the polynomial under the factorization. Our algorithm is of the greatest interest in the case when the characteristic of the local field is zero. In the case of non-zero characteristic an analog of this algorithm is the classical algorithm

Proceedings of the International Congress of Mathematicians, Kyoto, Japan, 1990

for resolution of singularities of algebraic curves over finite fields (it is also of polynomial complexity in the size of input and the characteristic of the finite field). Thus, our algorithm can be considered as a new efficient method for local resolution of singularities in rings which are finite over $\mathbb{Z}$ of $\mathbb{F}_p[t]$, where t is an element algebraically independent over the field.

Now we proceed with a more detailed statement of results. Let k, o, $\hat{k}$, $\hat{o}$ and π denote $\mathbb{Q}$, $\mathbb{Z}$, $\mathbb{Q}_p$, $\mathbb{Z}_p$ and p respectively in the case of zero characteristic, and $\mathbb{F}_p(t)$, $\mathbb{F}_p[t]$, $\mathbb{F}_p((t))$, $\mathbb{F}_p[[t]]$ and t in the case of non-zero characteristic (concerning standard notations see [2]). Let $\bar{\hat{k}}$ be the algebraic closure of $\hat{k}$ and $\mathrm{ord}: \bar{\hat{k}} \to \mathbb{Q} \cup \{\infty\}$ the order function with respect to π-adic metric on $\hat{k}$, and $\mathrm{ord}(\pi) = 1$, see [1]. Let K' denote a finite separable extension of k, $\hat{K}$ a composite of K' and $\hat{k}$ in $\bar{\hat{k}}$ over k, $\hat{O}$ the subring of $\hat{K}$ of all the integral elements over $\hat{o}$. The field K' is given over k by a primitive element θ with its minimal polynomial $\varphi \in k[Z]$. Let $\varphi_1 \in \hat{k}[Z]$ be an irreducible factor of φ such that $\hat{K} \simeq \hat{k}[Z]/(\varphi_1(Z))$, i.e. $\varphi_1(\theta) = 0$ in $\hat{K}$. Without loss of generality we suppose that $\varphi \in o[Z]$, $\varphi_1 \in \hat{o}[Z]$ and the leading coefficients $\mathrm{lc}_Z\, \varphi = \mathrm{lc}_Z\, \varphi_1 = 1$. The polynomial φ_1 is uniquely defined by its reduction $\bar{\varphi}_1 = \varphi_1 \bmod \pi^{r+1} \in \hat{o}/\pi^{r+1}\hat{o}[Z]$ where $r = \mathrm{ord}\, \mathrm{Res}_Z(\varphi, \varphi')$ is the order of the discriminant of φ; see [1], chap. 4, § 3, th. 1. The polynomial φ_1 and consequently the field $\hat{K}$ are given accordingly by $\bar{\varphi}_1$.

Furthermore (see Remark 1 below), let π_1 be a uniformizing element of $\hat{K}$, and $\eta \in \hat{K}$ such that $\hat{k}[\eta]$ is a maximal unramified extension of $\hat{k}$ contained in $\hat{K}$. Let the minimal polynomial h for η over $\hat{k}$ be given, and the minimal polynomial g for π_1 over $\hat{k}[\eta]$. Besides that, let $h \in o[Z]$, $\mathrm{lc}_Z\, h = 1$, $g \in k[\eta][Z]$, $\mathrm{lc}_Z\, g = 1$. Set the field $K = k[\eta, \pi_1]$, $O = o[\eta, \pi_1]$.

Each element $\lambda \in \hat{K}$ can be decomposed with respect to either the $\hat{k}$-basis $\{\theta^i\}_{0 \le i < \deg \varphi_1}$, or the $\hat{k}$-basis $\{\eta^i \pi_1^j\}_{0 \le i < \deg h,\, 0 \le j < \deg g}$, or it can be expanded into the series $\sum_{j \ge j_{0,i}} a_{ij} \eta^i \pi_1^j$, $a_{ij} \in \{0, 1, \ldots, p-1\}$. And we can go from any of these three representations to any other within polynomial time (as it was previously defined).

Let $f \in K'[X_1, \ldots, X_n]$ be an arbitrary polynomial and

$$f = \frac{1}{a} \sum_{i_1, \ldots, i_n} \left(\sum_{0 \le j < \deg \varphi} a_{i_1, \ldots, i_n, j} \theta^j \right) X_1^{i_1} \ldots X_n^{i_n}$$

where $a, a_{i_1, \ldots, i_n, j} \in o$. We define the length $l(b)$ of the element $0 \ne b \in o$ as $l(b) = (1 + \deg_t b)([\log_2 p] + 1)$ if $k = \mathbb{F}_p(t)$, and $l(b) = \min\{s \in \mathbb{Z} : |b| < 2^{s-1}\}$ if $k = \mathbb{Q}$. Set $l(0) = 1$. If $b \in k$, $b = b_1/b_2$, $\mathrm{GCD}(b_1, b_2) = 1$ then $l(b) = l(b_1) + l(b_2)$. The size $L(f)$ of f is set to be

$$l(a) + \max_{i, i_1, \ldots, i_{n,j}} (1 + \deg_{X_i} f)^n (1 + \deg \varphi)\, l(a_{i_1, \ldots, i_n, j})$$

Similarly we define the size $L(\varphi)$ and the sizes of other polynomials.

Now we can formulate the main result.

Theorem 1. (i) *In time polynomial in* $L(f)$, $L(\varphi)$ *and* p, *the decomposition* $f = \lambda \prod_{i \in I} f_i^{r_i}$ *is constructed where* $f_i \in \hat{K}[X_1, \ldots, X_n]$ *are pairwise distinct polynomials irreducible over the field* $\hat{K}$, $1 \le r_i \in \mathbb{Z}$, *I is a finite set, and* $0 \ne \lambda \in K$.

(ii) *When* $n = 1$ *set* $X = X_1$. *Then additionally within the same time for each field* $\hat{K}_i = \hat{K}[X]/(f_i)$, $i \in I$, *one can construct an element* $\eta_i \in \hat{K}_i$ *with its minimal polynomial* $h_i(Z)$ *over the field* $\hat{K}$ *such that* $\hat{K}[\eta_i]$ *is a maximal unramified extension of* $\hat{K}$ *contained in* $\hat{K}_i$, *and a uniformizing element* π_i *of the field* $\hat{K}_i$ *with its minimal polynomial* $g_i(Z)$ *over the field* $\hat{K}[\eta_i]$. *Herewith the leading coefficients* $\mathrm{lc}_Z\, h_i = \mathrm{lc}_Z\, g_i = 1$ *and* $h_i \in K[Z]$, $g_i \in K[\eta_i][Z]$

Remark 1. Note that $\varphi_1, \eta, \pi_1, h, g$ can be constructed in time polynomial in $L(\varphi)$ and p by applying Theorem 1 to the polynomial $\varphi \in k[Z]$ instead of $f \in K[X]$.

Considering derivates of polynomials Theorem 1 can be reduced to the case when f is separable, i.e. when all $r_i = 1$. Then for the case of one variable we obtain the following result.

Theorem 2. *Let* $f \in \hat{O}[X]$, $\varphi \in \hat{o}[Z]$, $\mathrm{lc}_X\, f = \mathrm{lc}_Z\, \varphi = 1$ *and the order of discriminants* $\mathrm{ord}\ \mathrm{Res}_x(f, f') \le \delta$, $\mathrm{ord}\ \mathrm{Res}_Z(\varphi, \varphi') \le \delta_1$. *Then one can construct* $f_i, \eta_i, \pi_i, g_i, h_i$; $i \in I$, *from Theorem* 1 *in time polynomial in* $\deg_X f, \deg_Z \varphi, \delta, \delta_1, p$.

The general plan of the proof of Theorem 1 is given in Section 1.

Now we go to the applications. Let k, o be the same as above, $f \in k[X]$ a separable polynomial, $K' = k[X]/(f)$ the factor algebra, and O' the integral closure of o in K'. The square free part of an element $0 \neq a \in o$ is defined as an element $a_1 \in o$ which is equal to the product of all the irreducible divisors of a in the first power, i.e. $a_1 = \prod_{q|a} q$ where $q \in o$ are irreducible.

Theorem 3. *The following problems are polynomial equivalent.*

(1) *For a given element* $a \in o$ *find the square free part* a_1 *of* a.

(2) *For a given separable polynomial* $f \in k[X]$ *construct an* o*-basis of the integral closure* O' *of the ring* o *in the algebra* $K' = k[X]/(f)$.

In the case of non-zero characteristic of k the square free part of an element $a \in \mathbb{F}_p[t]$ can be found in polynomial time by considering the derivate $\frac{d}{dt}$.

Thus, the following statement is valid

Theorem 4. *The integral closure of the ring* $\mathbb{F}_p[t]$ *in a separable algebra* $\mathbb{F}_p(t)[X]/(f)$ *can be constructed within the time polynomial in* $\deg_X f$ *and* $\log p$.

Note that the earlier known algorithms for constructing the integral closure from the statement of Theorem 4 required the time polynomial in p instead of $\log p$.

In the case when the characteristic of the field k is 0 and f is an irreducible polynomial we obtain from Theorem 4 the next

Theorem 5. *The following problems are polynomial equivalent*

(1) *For a given integer* a *find the square free part* a_1 *of* a.

(2) *For a given irreducible polynomial* $f \in \mathbb{Q}[X]$ *construct a* $\mathbb{Z}$*-basis of the ring of all integers of the field* $\mathbb{Q}[X]/(f)$.

The polynomial reducibility of the problem (1) to the problem (2) is obtained easily by considering fields $k(\sqrt{b})$ where $b \in o$. The inverse reducibility is proved much more difficult. Here we have the following

Theorem 6. *Let* $f \in o[X]$ *be a separable polynomial with the leading coefficient* $\mathrm{lc}_X f = 1$, *and let be known the square free part D of the discriminant of the polynomial f. Then within the time polynomial in* $L(f)$, *one can construct an o-basis of the integral closure of o in the algebra* $k[X]/(f)$.

The description of the algorithms from Theorems 3–6 is given in Section 2. The whole construction is based on the algorithm of the factorization from Theorem 1 and formulated below Lemma 1.

Let $q \in o$ be irreducible, k_q be q-adic completion of the field k, $\bar{k}_q$ an algebraic closure of k_q, and $\mathrm{ord}_q : \bar{k}_q \to \mathbb{Q} \cup \{\infty\}$ the order function with $\mathrm{ord}_q(q) = 1$. Further, let $\mathrm{Res}_X(f, f')$ be the discriminant of the polynomial f, $\mathrm{char}(o/qo)$ the characteristic of the residue field.

Lemma 1. *Let f be a polynomial from the statement of Theorem* 6 *and the degree* $\deg_X f < \mathrm{char}(o/qo)$. *Let* $x_1, x_2 \in k_q$ *be two distinct roots of the polynomial f. Then there exists an integer* $1 \le s < \deg_X f$ *for which* $\mathrm{ord}_q\left(\frac{d^s f}{dX^s}(x_i) - b_i q^{\mu}\right) > \mu$, $i = 1, 2$, *where* $\mu \in (1/v)\mathbb{Z}$, $v \in \mathbb{Z}$, $\mathrm{GCD}(v, \mathrm{char}(o/qo)) = 1$, $1 \le v \le \deg_X f$, $0 \le \mu \le (\mathrm{ord}_q(\mathrm{Res}_X(f, f')))/2$; $b_1, b_2 \in \bar{k}_q$; $\mathrm{ord}_q b_1, \mathrm{ord}_q b_2 \ge 0$; $\mathrm{ord}_q(b_1 - b_2) = 0$.

An analog to this lemma (but without a restriction to $\deg_X f$) is valid also in the case of zero characteristic residue fields. It was proved by the author earlier and it was central for obtaining of efficient bounds for sizes of coefficients in the Newton-Puiseux expansion, see [8]. Namely, the ith coefficient in the Newton-Puiseux expansion can be constructed within the time polynomial in the size of input and i^l where l is the transcendence degree of the field of constants over the primitive subfield. The main problem here is to estimate the first coefficients till the stabilization of the process of the expansion, i.e. the distinction of roots. Note that the direct methods give here, even in the case when there are no extensions of the constant field, an exponential or subexponential growth of sizes of coefficients. After the stabilization of the expansion one can apply Hensel's lemma in the ordinary way.

The last result has many important applications. These are algorithms with polynomial complexity for factoring multivariable polynomials over fields of power and fraction-power series and also algorithms in the theory of algebraic curves: computing the genus of a curve, indices of ramification, uniformizing elements, the smooth model of a curve, etc. within polynomial time, see [7].

1. Description of the Algorithm for Factoring Polynomials over Local Fields

At first we reduce the problem of factoring a polynomial $f \in K'[X_1, \dots, X_n]$ over the field $\hat{K}$ to the principle case when $n = 1$. Using the algorithm, for example from [4], one can assume f to be irreducible over K'. Applying the algorithm from [4] we can find an absolutely irreducible (i.e. irreducible over $\bar{K'}$) factor f_1 of f. Herewith some coefficient of f_1 is equal to 1. Besides that, see [4], we construct also the field $K'[Z]/(\psi)$ generated over K' by the coefficients of the polynomial f_1. Here $\psi \in K'[Z]$ is an irreducible polynomial in one variable. Factor ψ over $\hat{K}$. Let $\psi = \prod_{i \in I} \psi_i$ be the obtained decomposition, $\psi_i \in \hat{K}[Z]$, and $\hat{K}_i = \hat{K}[X]/(\psi_i)$ the fields. Compute the norms $f_i = N_{\hat{K}_i(X_1,\dots,X_n)/\hat{K}(X_1,\dots,X_n)}(\hat{f}_i)$ of the polynomials $\hat{f}_i = f_1 \bmod \psi_i \in \hat{K}_i[X_1, \dots, X_n]$ for all $i \in I$. Then $f = \lambda \prod_{i \in I} f_i$ is a decomposition of f into irreducible factors over the field $\hat{K}$, where $0 \neq \lambda \in K'$. Thus, everything is reduced to the factorization of ψ over $\hat{K}$, i.e. to the case when $n = 1$.

Let now $n = 1$, and f be a separable polynomial. Without loss of generality we can assume that $f \in O[X]$ and $\mathrm{lc}_X f = 1$. Construction of some imbedding which we are going to define is central in the proof of Theorem 1.

It is an embedding of $\hat{K}$-algebras

$$\hat{K}[X]/(f) \hookrightarrow \prod_{j \in J} \hat{K}_j[X]/(f_j) \tag{1}$$

where the fields $\hat{K}_j$ are weakly ramified extensions of $\hat{K}$. More precisely, $\hat{K}_j = \hat{K} \otimes_K K_j$ where $K_j = K[\eta_j, \pi_j]$, $\operatorname{ord} \eta_j = 0$, $h_j \in K[Z]$ is a minimal polynomial for the element η_j over $\hat{K}$, $\mathrm{lc}_Z h_j = 1$, $g_j = Z^{v(j)} - \pi_1$ is a minimal polynomial for the element π_j over the field $\hat{K}[\eta_j]$, $\mathrm{GCD}(v(j), p) = 1$. Further, $f_j \in \hat{K}_j[X]$ is an Eisenstein's polynomial with respect to the field $\hat{K}_j$, and $\deg_X f_j = p^{\varepsilon(j)}$, $0 \leq \varepsilon(j) \in \mathbb{Z}$. Therefore, f_j is irreducible over the field $\hat{K}_j$. Finally, for discriminants we have $\operatorname{ord} \operatorname{Res}(f_j, f_j') \leq \operatorname{ord} \operatorname{Res}(f, f')$ for all $j \in J$. J is a finite set.

To construct the embedding we need the Newton broken line of the polynomial and other notions connected with it. Now we go to their definitions. Let $\Sigma = \{0, 1, \dots, p-1\}$ and $\Sigma_1 = \{\sum_{0 \leq i < \deg h} a_i \eta^i : a_i \in \Sigma\}$ be systems of representatives of the residue fields of $\hat{k}$ and $\hat{K}$ respectively. Let $\psi \in O[X]$, $\mathrm{lc}_X \psi = 1$. We write

$$\psi = \sum_{0 \leq i \in \mathbb{Z}} \sum_{0 \leq u \leq \deg \psi} a_{i,u} \pi_1^i X^u, \qquad a_{i,u} \in \Sigma. \tag{2}$$

We define

$$\psi_{\#}(N, \hat{K}) = \sum_{0 \leq i \leq N} \sum_{0 \leq u \leq \deg \psi} a_{i,u} \pi_1^i X^u \in K[X]$$

to be an approximation to the polynomial ψ in the ring $K[X]$;

$$P(\psi) = \{(i, u) : a_{i,u} \neq 0\}.$$

If $\alpha, \beta \in \mathbb{Q}$, $\alpha > 0$ $\beta \geq 0$ then we set $P(\psi, \alpha, \beta) = \{(i, u) \in P(\psi) : \forall (i_1, u_1) \in P(\psi)$ $[\alpha i_1 + \beta u_1 \geq \alpha i + \beta u]\}$,

$$\psi^*(\alpha, \beta) = \sum_{(i,u) \in P(\psi, \alpha, \beta)} (a_{i,u} \bmod \pi_1) X^u \in \hat{O}/\pi_1 \hat{O}[X],$$

$$\psi(\alpha, \beta) = \sum_{(i,u) \in P(\psi, \alpha, \beta)} a_{i,u} \pi_1^i X^u \in K[X];$$

$$V(\psi) = \{(0, \deg \psi)\} \cup \{(i, u) : \exists \alpha, \beta [\{(i, u)\} = P(\psi, \alpha, \beta)]\}$$

to be the set of vertices of the Newton broken line of ψ;

$$E(\psi) = \{((i_1, u_1), (i_2, u_2)) \in V(\psi)^2 : \exists (\alpha, \beta) [(i_1, u_1), (i_2, u_2) \in P(\psi, \alpha, \beta) \,\&\, u_1 > u_2]\}$$

to be the set of edges of the Newton broken line of f. If $w = ((i_1, u_1), (i_2, u_2)) \in E(\psi)$, $\{(i_1, u_1), (i_2, u_2)\} \subset P(\psi, \alpha, \beta)$ then we define

$$\psi_w^* = \psi^*(\alpha, \beta), \qquad \lambda(w) = (u_1 - u_2)^{-1}(i_1 - i_2) = \beta/\alpha$$

to be the coefficient of the slope of the edge w, and $0 \leq \alpha(w), \beta(w) \in \mathbb{Z}$ for which $\beta(w)/\alpha(w) = \lambda(w)$, $\mathrm{GCD}(\beta(w), \alpha(w)) = 1$.

Finally, if ψ is not a monomial then $E(\psi) \neq \emptyset$, and let $w_0 \in E(\psi)$ be an edge for which $\lambda(w_0) = \min\{\lambda(w) : w \in E(\psi)\}$, $\psi^* = \psi_{w_0}^*$, $\alpha_0 = \alpha_0(\psi, \hat{K}) = \alpha(w_0)$, $\beta_0 = \beta_0(\psi, \hat{K}) = \beta(w_0)$, $\alpha_1 = \alpha_1(\psi, \hat{K})$ be the greatest divisor of α_0 relatively prime with p.

These notions are defined (and we shall use them) for polynomials with coefficients from an arbitrary other local field $\tilde{K}$ if $\tilde{\Sigma}$ and $\tilde{\pi}$ are fixed analogous to Σ_1 and π_1 in (2).

The embedding (1) is constructed recursively with the iteration of two cases. The first case is when f^* has two or more different roots in $\overline{\mathbb{F}}_p$, the second one is when f^* is a power of a linear polynomial. In the first case we construct an intermediate embedding of $\hat{K}$-algebras

$$\hat{K}[X]/(f) \hookrightarrow \prod_{l \in L} \hat{K}_l[X]/(f_l) \tag{3}$$

where the fields $\hat{K}_l$ have the same properties as $\hat{K}_j$ above. In particular, η_l, π_l, h_l, g_l, $\hat{O}_l = \hat{O}[\eta_l, \pi_l]$ are defined. The polynomial f_l^* is a power of a linear one for every $l \in L$ (in the definition of f_l^* we change $\hat{K}$, π_1, Σ_1 for $\hat{K}_l$, π_l, $\Sigma_l = \{\sum_{0 \leq i < \deg h_l} a_i \eta_l^i : a_i \in \Sigma_1\}$ in (2)).

Now we go the construction of the embedding (3). We suppose f not to be a monomial, and besides that, $\lambda(f, K) > 0$ changing if it is necessary f for $\pi_1^{\deg f} f(X/\pi_1)$. We factor f^* over $\hat{O}/(\pi_1)$. Namely, $f^* = \prod_{\gamma \in \Gamma} \bar{h}_\gamma^{e_\gamma}$, $\bar{h}_\gamma \in O/(\pi_1)[X]$ are irreducible, $\mathrm{lc}_X \bar{h}_\gamma = 1$, $1 \leq e_\gamma \in \mathbb{Z}$. Let h_γ be a polynomial with coefficients from Σ_1 for which $h_\gamma \bmod \pi_1 = \bar{h}_\gamma$, η_γ a root of h_γ in $\bar{K}$, $\pi_\gamma^{\alpha_1(\hat{K}, f)} = \pi_1$ if $h_\gamma \neq X$ and $\pi_\gamma = \pi_1$ if $h_\gamma = X$, $K_\gamma = K[\eta_\gamma, \pi_\gamma]$, $\hat{K}_\gamma = \hat{K} \otimes_K K_\gamma$, $\pi_0^{\alpha_0(\hat{K}, f)} = \pi_1$, $\hat{K}_\gamma' = \hat{K}[\eta_\gamma, \pi_0]$. Set $F = f(X\pi_0^{\beta_0})/\pi_0^{\beta_0 \deg f} \in \hat{K}_\gamma'[X]$. Then $F \in \hat{O}[\eta_\gamma, \pi_0]$, $\mathrm{lc}_X F = 1$, $F^* = f^*$, $\lambda(F, \hat{K}_\gamma') = 0$. The decomposition $F^* = (X - \bar{\eta}_\gamma)^{e_\gamma} \bar{\psi}_\gamma$ over the residue field of $\hat{K}_\gamma'$ is lifted by Hensel's lemma till $F = F_\gamma \psi_\gamma$ over $\hat{K}_\gamma'$ where $F_\gamma \bmod \pi_0 = (X - \bar{\eta}_\gamma)^{e_\gamma}$. Set $f_\gamma = F_\gamma(X/\pi_0^{\beta_0})\pi_0^{\beta_0 \deg f}$. Then $f_\gamma \in \hat{K}_\gamma[X]$ is a divisor of f, $\mathrm{lc}_X f_\gamma = 1$, $f_\gamma^* = (X - \bar{\eta}_\gamma)^{e_\gamma}$ if $\bar{\eta}_\gamma \neq 0$. Compute all the polynomials $\hat{f}_\gamma = (f_\gamma)_\#(\mathrm{ord}\,\mathrm{Res}(f, f')/\mathrm{ord}(\pi_\gamma), \hat{K}_\gamma)$. Then there exists an imbedding $\hat{K}[X]/(f) \hookrightarrow \prod_{\gamma \in \Gamma} \hat{K}_\gamma[X]/(\hat{f}_\gamma)$, and $\hat{f}_\gamma^* = f_\gamma^*$ if $\bar{h}_\gamma \neq X$. To construct (3) it is sufficient now to do it recursively for $\hat{f}_{\gamma_0} \in K[X]$ with $\bar{h}_{\gamma_0} = X$ (if such an element γ_0 exists at all). Thus, we have reduced the first case to the second one.

Consider the second case which is central. Using some double recursion for each $l \in L$ from (3) we construct a polynomial $\tilde{f}_l \in K_l[X]$ such that $\hat{K}_l[X]/(f_l) \simeq \hat{K}_l[X]/(\tilde{f}_l)$ and either $\tilde{f}_l^*$ is not a power of a linear polynomial, or $\tilde{f}_l$ is an Eisenstein's polynomial with $\deg \tilde{f}_l = p^{\varepsilon(l)}$, $\varepsilon(l) \in \mathbb{Z}$, and it coincides with some f_j from (2). Therefore, the second case will be reduced to the first one.

To construct $\tilde{f}_l$ we carry out a recursive procedure within a finite number of steps beginning from step 0. At step a if it is not final we construct a polynomial $H_{a+1} \in K_l[X]$ such that $\mathrm{lc}_X H_{a+1} = 1$, $H_{a+1} \in \hat{O}_l[X]$, $\hat{K}_l[X]/(f_l) \simeq \hat{K}_l[X]/(H_{a+1})$, $\lambda(H_{a+1}, \hat{K}_l) = p^{-\varepsilon(a+1)}, 0 \le \varepsilon(a+1) \in \mathbb{Z}$; $\varepsilon(a+1) > \varepsilon(a)$ if $a > 0$; and H_{a+1}^* is a power of a linear polynomial, $\operatorname{ord} \operatorname{Res}(H_{a+1}, H'_{a+1}) \le \operatorname{ord} \operatorname{Res}(f_l, f_l')$.

Now we describe step 0. Denote $x = X \bmod f_l \in K_l[X]/(f_l) = \Lambda$, $\lambda(f_l, K_l) = \beta_0/p^{\varepsilon} > 0$. We can assume without loss of generality that

$$\operatorname{card} \Sigma_l > \deg f(\deg f + 1)/2 + 1.$$

Choose $0 \le v_1, v_2 \in \mathbb{Z}$, $c \in \Sigma_l$, such that v_2 is minimal, $x^{v_2}/\pi_l^{v_1} + cx$ is a primitive element of Λ over K_l with its minimal polynomial ϕ, and ϕ^* is a power of a linear polynomial, $\lambda(\phi, \hat{K}_l) = p^{-\varepsilon}$. Set $H_1 = \phi$. Step 0 is not final.

We describe step a with $a \ge 1$. Let $H = H_a$, $\Lambda = K_l[X]/(H_a)$, $x = X \bmod H_a$, $H^*(\bar{\xi}) = 0$, $\xi \in \Sigma_l$, $\xi \bmod \pi_l = \bar{\xi}$, $N_0 = p^{\varepsilon(a)}$, $Z_{N_0-1} = x^{N_0} \in \Lambda$, ψ_{N_0-1} be the minimal polynomial for Z_{N_0-1} over K_l, $\mathrm{lc}\, \psi_{N_0-1} = 1$. Using recursion on $N \ge N_0 - 1$ we construct an element $Z_N \in \Lambda$ with its minimal polynomial $\psi_N \in K_l[X]$, $\mathrm{lc}_X \psi_N = 1$, $\psi_N \in \hat{O}_l[X]$, such that if $N \ge N_0$ then $Z_N = Z_{N-1} + \gamma_N x^b \pi_l^u$ where $0 \le b < p^{\varepsilon(a)}$, $b + up^{\varepsilon(a)} = N$, $\gamma_N \in \Sigma_l$, $u, b \in \mathbb{Z}$. Further, $\lambda_N = \lambda(\psi_N, \hat{K}_l) > N/p^{\varepsilon(a)}$; if z_N is not final then $\lambda_N = N_1/p^{\varepsilon(a)}$, $N_1 \in \mathbb{Z}$ and ψ_N^* is a power of a linear polynomial.

We show how to find $\gamma_N \in \Sigma_l$. If $\lambda_{N-1} > N/p^{\varepsilon(a)}$ then $\gamma_N = 0$. If $\lambda_N = N/p^{\varepsilon(a)}$ we find the root $\bar{\xi}_{N-1}$ of ψ_{N-1}^*. Then $\bar{\xi}_{N-1} \in \hat{O}_l/(\pi_l)$. Set $\bar{\gamma}_N = -\bar{\xi}_{N-1}/\bar{\xi}^b$, $\gamma_N \in \Sigma_l$, $\gamma_N \bmod \pi_l = \gamma_N$.

The process of constructing z_N is finished at the element z_{N_2} for which one of two conditions is valid:

(i) $\lambda(\psi_{N_2}, \hat{K}_l) = N_3/p^{e(a)}$, where $e(a) > \varepsilon(a)$, $\mathrm{GCD}(N_3, p) = 1$, $0 \le e(a)$, $N_3 \in \mathbb{Z}$ and $\psi_{N_2}^*$ is a power of a linear polynomial;

(ii) $\psi_{N_2}^*$ is not a power of a linear polynomial. We have $N_2 < \deg f \operatorname{ord} \operatorname{Res}(f, f')/(2 \operatorname{ord} \pi_l)$.

In both cases when (i) or (ii) is fulfilled choose $0 \le v_1, v_2 \in \mathbb{Z}$, $c \in \Sigma_l$ such that v_2 is minimal, $z_{N_2}^{v_2}/x^{v_1} + cx$ is a primitive element of Λ over K_l with its minimal polynomial $\phi \in K_l[X]$, $\mathrm{lc}_X \phi = 1$, and $\lambda(\phi, \hat{K}_l) = 1/\mathrm{LCM}(p^{\varepsilon(a)}, \alpha_0(\psi_{N_2}, \hat{K}_l))$. Besides that in the case (i) ϕ^* is a power of a linear polynomial and in the case (ii) ϕ^* is not a power of a linear polynomial. Set $H_{a+1} = \phi_{\#}(\operatorname{ord} \operatorname{Res}(f_l, f_l')/\operatorname{ord}(\pi_l), \tilde{K}_l)$.

Now let condition (i) be fulfilled. Then we set $\varepsilon(a+1) = e(a)$. Step a is not final if $\deg H_{a+1} > p^{\varepsilon(a+1)}$. If $\deg H_{a+1} = p^{\varepsilon(a+1)}$ then step a is final, and we set $\tilde{f}_l = H_{a+1}$. It is Eisenstein's polynomial.

Let condition (ii) be fulfilled. Then step a is final. Set $\tilde{f}_l = H_{a+1}$. The polynomial $\tilde{f}_l^*$ has at least two different roots in $\bar{\mathbb{F}}_p$. Thus, we have finished the description of the algorithm for constructing the embedding (1).

Having constructed (1) we can find all the roots of f contained in fields $\hat{K}_j[X]/(f_j), j \in J$, using the standard method of Newton's polygons, since we know uniformizing elements and systems of representatives of residue fields in $\hat{K}_j[X]/(f_j)$. After that we can find all the irreducible factors of f over $\hat{K}$ computing, for example, normes of linear factors. Further π_i, η_i, h_i, g_i are constructed for the fields $\hat{K}[X]/(f_i)$, $i \in I$. These fields are subfields of $\hat{K}_j[X]/(f_j), j \in J$. So here arise only some technical difficulties. We have completed the description of the algorithm from Theorem 1.

2. Constructing a Basis of the Ring of Integral Elements of a Global Field

For the proof of Theorems 3–6 it is sufficient to reduce problem (1) to problem (2) in Theorem 3 and to prove Theorem 6.

So let $0 \neq b \in o$. We find the square free part b_1 of b. In the case when $\mathrm{char}(k) > 0$ it is possible to do this even without an algorithm for (2) considering $\frac{d}{dt}$. Let $\mathrm{char}(k) = 0$. Set $\theta = \sqrt{b}$, $K' = k(\sqrt{b})$, $\omega_1, \omega_2 \in k(\sqrt{b})$ to be $\mathbb{Z}$-basis of O'. We represent $\sqrt{b} = c_1\omega_1 + c_2\omega_2$; $c_1, c_2 \in \mathbb{Z}$ and compute $b_3 = \mathrm{GCD}(c_1, c_2)$, $e_1 = c_1/b_3, e_2 = c_2/b_3$. Then $\sqrt{b/b_3^2} \in O'$ and $b_2 = b/b_3^2$ is square free. Recursively we find the square free part b_4 of b_3. Then $b_1 = \mathrm{LCM}(b_2, b_4)$.

Now we go to the proof of Theorem 6. We denote by B the set of all the irreducible divisors $q \in o$ of D such that $2 + (\deg_X f)^2 \geq \mathrm{char}(o/qo)$ and find B by factoring D over $\mathbb{F}_p$ or the enumeration. Set $\delta_0 = D/\prod_{q \in B} q$.

For each $\delta \in o$ we consider the localizations $O'_{(\delta)} = S^{-1}O'$ and $o_{(\delta)} = S^{-1}o$ where $S = o \setminus \bigcup_{q|\delta} qo$. We define a δ-integral basis of O' as a family of elements of O' which is an $o_{(\delta)}$-basis of $O'_{(\delta)}$. Consider also completions in δ-adic topology $o_\delta, O'_\delta, k_\delta, K'_\delta$ of o, O', k, K' respectively. Then $\omega_1, \ldots, \omega_m \in O'$ is a δ-integral basis iff it is an o_δ-basis of O'_δ. Each polynomial $0 \neq \psi \in k_\delta[X]$ can be represented in the form

$$\psi = \sum_{i_0 \leq i \in \mathbb{Z}} \sum_{0 \leq u \leq \deg \psi} a_{i,u}\delta^i X^u, \qquad a_{i,u} \in \Sigma_\delta \tag{4}$$

where $\Sigma_\delta = \{z \in \mathbb{Z} : 0 \leq z < |\delta|\}$ if $\mathrm{char}(k) = 0$ and $\Sigma_\delta = \{z \in \mathbb{F}_p[t] : \deg z < \deg \delta\}$ if $\mathrm{char}(k) = p > 0$; there exists $a_{i_0,u} \neq 0$. Set $\mathrm{ord}_\delta \psi = i_0$. Let now $\psi \in o_\delta[X], \mathrm{lc}_X \psi = 1$. Then we define $P(\psi)$, $P(\psi, \alpha, \beta)$, $\psi^*(\alpha, \beta)$, $\psi(\alpha, \beta)$, $V(\psi)$, $E(\psi)$, ψ^*_w, $\psi^* = \psi^*(\delta, X)$, $\lambda = \lambda(\psi, \delta)$, $\alpha_0 = \alpha_0(\psi, \delta)$, $\beta_0 = \beta_0(\psi, \delta)$ for (4) analogous to that as it was above for (2) with changing π_1, Σ_1 for δ, Σ_δ. Also we define $\psi_\#(N, \delta)$ analogous to $\psi_\#(N, K)$. Further, if $\deg \psi = 0$, $\psi \in k_\delta$ we set $\psi^* = a_{i_0,0} \bmod \delta o \in o/\delta o$

Lemma 2. *Let $\psi, \psi_i \in o[X]$, $\mathrm{lc}\, \psi = \mathrm{lc}\, \psi_i = 1$, be separable polynomials for all $i \in I$. Let be given $\omega_{i,1}, \ldots, \omega_{i,m_i}$ a δ-integral basis of the integral closure O_i of o in $k[X]/(\psi_i)$ for every $i \in I$. Further, let $\mathrm{ord}_\delta(\psi - \prod_{i \in I} \psi_i) > (2 \deg \psi + 1) \mathrm{ord}_\delta \mathrm{Res}(\psi, \psi')$ where $\mathrm{Res}(\psi, \psi')$ is the discriminant of ψ. Then in time polynomial in $L(\psi)$, $L(\psi_i)$, $L(\omega_{i,j})$, $1 \leq j \leq m_i$, $i \in I$, one can either construct a δ-integral basis $\omega_1, \ldots, \omega_m$ of the integral closure O of o in $k[X]/(\psi)$ or find a decomposition $\delta = \delta_1\delta_2$ where $\delta_1, \delta_2 \in o$ are* *non-inversible in* o. *Besides that,* $L(\omega_i) \le P(L(\psi), L(\delta))$, $1 \le i \le m$, *where the polynomial* P *does not depend on* ψ_i, ω_{ij}.

The following lemma is an analog of Hensel's lemma.

Lemma 3. *Let* ψ *be a polynomial from the statement of Lemma 2,* $\lambda = \lambda(\psi, \delta)$; $g_0, h_0 \in o[X]$, $\mathrm{lc}\, g_0 = \mathrm{lc}\, h_0 = 1$, $g_0 = g_0(1, \lambda)$, $h_0 = h_0(1, \lambda)$, $\psi^* = g_0(1, \lambda) h_0(1, \lambda)$, $\mathrm{Res}(g_0, h_0) \bmod \delta o$ *be inversible in* $o/\delta o$. *Then for every* $0 \le i \in \mathbb{Z}$ *there exist polynomials* $g^{(i)}, h^{(i)} \in o[X]$ *such that* (i) $\mathrm{lc}\, g^{(i)} = \mathrm{lc}\, h^{(i)} = 1$; (ii) $\lambda(g^{(i)}, \delta) > \lambda$, $\lambda(h^{(i)}, \delta) \ge \lambda$ *and the equality takes place for* $g^{(i)}$ *if* g_0 *is not a monomial and analogously for* $h^{(i)}$; (iii) $g_0^*(1, \lambda) = (g^{(i)})^*(1, \lambda)$, $h_0^*(1, \lambda) = (h^{(i)})^*(1, \lambda)$; (iv) $\mathrm{ord}_\delta(\psi - g^{(i)} h^{(i)}) > i/\alpha_0(\psi, \delta)$; (v) $g^{(i)}, h^{(i)}$ *can be found in time* $P(L(\psi), \delta, i)$.

We need one more auxiliary algorithm which uses Euclid's algorithm for finding GCD and derivates. This algorithm has at input a polynomial $f_0 \in o/\delta o[X]$ with $\mathrm{lc}_X f_0 = 1, \deg(f_0) < \mathrm{char}(o/qo)$ for every irreducible $q|\delta$. It has one of three outputs:

1) The decomposition $\delta = \delta_1 \delta_2$ where $\delta_1, \delta_2 \in o$ are noninversible.

2) The decomposition $f_0 = f_1 f_2$ where $f_1, f_2 \in o/\delta o[X]$ are polynomials of non-zero degree $\mathrm{lc}_X f_1 = \mathrm{lc}_X f_2 = 1$ and the resultant $\mathrm{Res}(f_1, f_2)$ is inversible in $o/\delta o$.

3) The representation $f = f_3^\varepsilon$ where $f_3 \in o/\delta o[X]$, $\mathrm{lc}_X f_3 = 1$, $1 \le \varepsilon \in \mathbb{Z}$, $f_3 \bmod q \in o/qo[X]$ are separable and X does not divide $f_3 \bmod q$ for all $q|\delta$.

Now we return to the description of the algorithm from Theorem 6. We suppose without loss of generality that f is not a monomial, i.e. $f \ne X$. Then changing f for $D^{\deg f} f(X/D)$ if it is necessary we assume $\lambda(f, D) > 0$.

For every $q \in B$ using algorithms from section 1 we find $f_i \in o[X]$ irreducible over k_q such that $\mathrm{ord}_q(f - \prod_{i \in I} f_i) > (2 \deg f + 1) \mathrm{ord}_q \mathrm{Res}(f, f')$. We find also $\eta_i^{(1)}, \pi_i^{(1)} \in k[X]/(f_i)$ $h_i^{(1)}, g_i^{(1)}$ analogous to η_i, π_i, h_i, g_i from Section 1 with changing $\hat{K}$ for k_q. Then $(\eta_i^{(1)})^{m_1} (\pi_i^{(1)})^{m_2}$, $0 \le m_1 < \deg h_i^{(1)}$, $0 \le m_2 < \deg g_i^{(1)}$, is a q-integral basis of the integral closure of o in $k[X]/(f_i)$, $i \in I$. Applying Lemma 2 we construct a q-integral basis E_q of O' for every $q \in B$.

Recall that $\delta_0 = D/\prod_{q|B} q$. We describe an algorithm which constructs some decomposition $\delta_0 = \prod_{j \in J} \delta_j$ where $\delta_j \in o$ are non-inversible and for $j \in J$ constructs a δ_j-integral basis E_j of O'. Then the union of families E_q, $q \in B$; E_j, $j \in J$; $X^i \bmod f$, $0 \le i < \deg f$, is a system of generators of the o-module O'. Using the result from [3] about the construction of a basis of a lattice over $\mathbb{Z}$ by its system of generators in polynomial time or the similar result for the case of non-zero characteristic we find the required o-basis of O'.

Let a non-inversible divisor δ of δ_0 be given. Now it is sufficient for us to describe an algorithm which finds either a δ-integral basis of O' or a non-trivial decomposition $\delta = \delta_1 \delta_2$. Indeed, if δ_0 is non-inversible then applying this algorithm to $\delta = \delta_0$ and again to the obtained divisors δ_1 and δ_2 (if they arise) etc. we shall construct the required decomposition $\delta = \prod_{j \in J} \delta_j$ and δ_j-integral basises of O'.

Our algorithm is recursive on $\deg(f)$ and under fixed degree on the maximal multiplicity of roots from $\overline{o/qo}$ of polynomials $f^*(\delta, X) \bmod q \in o/qo[X]$ herewith the maximum is taken over all the roots and all irreducible $q|\delta$. In this recursive algorithm a case may occur when $f = X$. Then $f^*(\delta, X)$ is not determined but in this case $O' = o$ and there are no difficulties.

So let $\delta | \delta_0$. At first we apply the auxiliary algorithm (see above) to the polynomial $f^*(\delta, X)$. If we have input 1) then the algorithm finishes its work. If we have input 2) then there exist g_0, h_0 satisfying the conditions of Lemma 3 such that $g_0^*(1, \lambda) = f_1$, $h_0^*(1, \lambda) = f_2$, $\lambda = \lambda(f, \delta)$. We use Lemma 3 for $\psi = f$, $i = (2 \deg f + 1) \operatorname{ord}_\delta \operatorname{Res}(f, f')\alpha_0(\delta, f)$ and construct $g^{(i)}$, $h^{(i)}$. We apply separately the algorithm described to $g^{(i)}$ and $h^{(i)}$ instead of f and further use Lemma 2.

It remains to consider input 3), i.e. $f^*(\delta, X) = h^\varepsilon$. At first suppose that $\varepsilon > 1$ or $\lambda(f, \delta)^{-1} \notin \mathbb{Z}$. Set $N_1 = \deg f(\deg f + 1)/2 + 1$, $N_2 = (\deg f)^2 + 2$ and choose subsets $C_1, C_2 \subset \Sigma_\delta$ of N_1 and N_2 elements such that for them the natural through mappings $C_i \subset \Sigma_\delta \to o/\delta o \to o/qo$ are injective for all $q | \delta$, $i = 1, 2$. It is possible since $q \notin B$. Let $C_3 = \{\gamma \in \mathbb{Z} : 1 \le \gamma \le \deg f\}$, $C_4 = C_3 \times C_2 \times C_1$, $C_4' = \{m\} \times C_2 \times C_1$. We shall enumerate elements of $C_4 \setminus C_4'$ if $\varepsilon > 1$ and C_4' if $\varepsilon = 1$.

Let $(\gamma, c_2, c_1) \in C_4$ be an arbitrary but fixed element. We compute $\phi_1 \in k[X]$, $\operatorname{lc}_X \phi_1 = 1$ the minimal polynomial over k of the element $f^{(\gamma)}(x) = \dfrac{d^\gamma f}{dX^\gamma} \bmod f \in k[X]/(f) = K'$ where $x = X \bmod f \in K'$. We have $\phi_1 \in o_{(\delta)}[X]$ and if $\gamma < m$ then $\lambda(\phi_1, \delta) > 0$. On the other hand $f^{(m)}(x) = m!$ and $\lambda(\phi_1, \delta) = 0$ for $\gamma = m$.

We compute $\mu_3 = \operatorname{LCM}(\alpha_0(f, \delta), \alpha_0(\phi_1, \delta)) > 0$ and $v_1, v_2, v_3 \in \mathbb{Z}$ such that $v_1 \lambda(f, \delta) + v_2 \lambda(\phi_1, \delta) + v_3 = 1/\mu_3; |v_1|, |v_2|, |v_3| \le \mu_3 \max\{\lambda(f, \delta), \lambda(\phi_1, \delta)\}; v_3 = 0$ if $\lambda(\phi_1, \delta) \ne 0$. Set $\mu_1 = \mu_3 \lambda(f, \delta) \in \mathbb{Z}$, $\mu_2 = \mu_3 \lambda(\phi_1, \delta)$.

When $\gamma \ne m$ it is fulfilled $\lambda(\phi_1, \delta) > 0$, $v_3 = 0$. We choose $0 < v_2 \le \mu_1$ and define $v_4 = v_1 - \mu_2$, $v_5 = v_2 + \mu_1$, $v_6 = 0$.

When $\gamma = m$ it is fulfilled $\lambda(\phi_1, \delta) = 0$, $v_3 = 0$. We choose $0 < v_1 \le \mu_3$, $v_2 = 0$ and define $v_4 = v_1 + \mu_3$, $v_5 = 0$, $v_6 = v_3 - \mu_1$.

Return to the case of an arbitrary γ. We compute $\phi \in k[X]$, $\operatorname{lc}_X \phi = 1$, the minimal polynomial over k of the element $z = x^{v_1}(f^{(\gamma)}(x))^{v_2}\delta^{v_3} + c_1 x^{v_4}(f^{(\gamma)}(x))^{v_5}\delta^{v_6} + c_2 x \in K'$. The following lemma is based on Lemma 1 from Introduction.

Lemma 4. *Fix an irreducible divisor q of δ. Then there exists $(\gamma, c_2, c_1) \in C_4$, among those which we enumerate, such that* $\operatorname{ord}_\delta \operatorname{Res}(\phi, \phi') \le r = \operatorname{ord}_\delta \operatorname{Res}(f, f')$, $\deg \phi = m = \deg f$, $\lambda(\phi, q) \ne 0$, $\lambda(\phi, q)^{-1} \in \mathbb{Z}$ *and*

(i) *if $\varepsilon > 1$ then either X divides $\phi^*(q, X)$ or there exist more than* $\deg h = m/\varepsilon$ *different roots of $\phi^*(q, X)$* in $\overline{o/qo}$.

(ii) *if $\varepsilon = 1$ then X does not divide $\phi^*(q, X)$* and *$\phi^*(q, X)$ is separable.*

Note that if $\phi_2 = \phi^*(\delta, X) \bmod q \ne X^m$ then $\phi^*(q, X) = \zeta^{-m}\phi_2(X\zeta)$, $\zeta \in \overline{o/qo}$. Lemma 4 permits either to change f for $\tilde{\phi} = \phi_\#((2m + 1)r, \delta)$ for some (γ, c_2, c_1) and continue our construction recursively or obtain some decomposition $\delta = \delta_1 \delta_2$.

It remains to consider the case $f = h^\varepsilon$, $\varepsilon = 1$ and $\lambda(f, \delta)^{-1} \in \mathbb{Z}$. But here a δ-integral basis can be obtained immediately as it follows from

Lemma 5. *Let* $\lambda(f, \delta)^{-1} = e \in \mathbb{Z}$, $f^*(\delta, X) \bmod q$ *be separable polynomial and X does not divide* $f^*(\delta, X) \bmod q$ *for all $q | \delta$. Then the family $x^i(x^e/\delta)^j$, $0 \le i < e$, $0 \le j < m/e$, is a δ-integral basis of O'.*

References

1. Borevich Z.I., Shafarevich I.R.: Number theory, 2nd edn. Nauka, Moskow, 1972 (Russian) [English transl. of 1st edn.: Academic Press, 1966]
2. Bourbaki, N.: Algèbre commutative, Chap. 1–7. Actualités Sci. Indust., nos. 1290, 1293, 1308, 1314. Hermann, Paris 1961, 1964, 1965
3. Frumkin, M.A.: Application of modular arithmetics to the construction of algorithms for solving systems of linear equation. Dokl. Akad. Nauk SSSR **229** (5) (1976) 1067–1070 (Russian)
4. Chistov, A.L.: Polynomial complexity algorithm for factoring polynomials and constructing components of a variety in subexponential time. Zap. Nauchn. Semin. Leningrad. Otdel. Mat. Inst. Steklov (LOMI) **137** (1984) 124–188 (Russian) [English transl.: J. Sov. Math. **34** (4) (1986)]
5. Chistov, A.L.: Efficient factorization of polynomials over local fields. Dokl. Akad. Nauk SSSR **293** (5) (1987) 1073–1077 (Russian) [English transl.: Sov. Math. Dokl. **35** (2) (1987) 430–433]
6. Chistov, A.L.: Complexity of constructing the ring of integral elements of a global field. Dokl. Akad. Nauk SSSR **306** (5) (1989) 1063–1067 (Russian)
7. Chistov, A.L.: Polynomial complexity algorithms for computational problems in the theory of algebraic curves. Zap. Nauchn. Semin. Leningrad. Otdel. Mat. Inst. Steklov (LOMI) **176** (1989) 127–150 (Russian)
8. Chistov, A.L.: Polynomial complexity of the Newton-Puiseux algorithm. (Lecture Notes in Computer Sciences, vol. 233.) Springer, New York Berlin Heidelberg 1986, pp. 247–255
9. Open Problems from FCT'83. Preprint, Borgholm Univ., Borgholm 1983

Interactive Proofs and Applications

Shafi Goldwasser *

Laboratory for Computer Science, Electrical Engineering and Computer Science Department, Massachusetts Institute of Technology, Cambridge, MA, USA

1. Introduction

Proofs whose correctness can be verified efficiently play a central role in complexity theory. The famous complexity class NP consists of those sets for which "short" proofs of membership exist. For example, the set of all satisfiable Boolean formulas is in NP. A short proof that a boolean formula ϕ is satisfiable would be a truth assignment to the boolean variables which makes ϕ true. Formally, NP $=$ $\{L \subseteq \{0,1\}^*$ s.t. $\exists$ a polynomial time computable function f_L and constant $c > 0$ such that $x \in \{0,1\}^n$ is in L if and only if $\exists y \in \{0,1\}^{n^c}$ such that $f_L(x,y) = 1\}$.

How about proving that there is *no* assignment which makes ϕ true? It is generally believed that no short proof exists. Still, by some very recent work, we now know of "procedures" which can convince us quickly and beyond a shadow of a doubt that a formula ϕ is not satisfiable. Such procedures, introduced by Goldwasser, Micali, and Rackoff [GMR], and in somewhat different form by Babai [Ba] are called *interactive proofs* .

Informally, an interactive proof-system is a method by which one algorithm of unlimited resources, called the *prover*, convinces another algorithm which runs in polynomial time, called the *verifier*, of the truth of a proposition. The verifier may toss coins, ask repeated questions of the prover, and run efficient tests upon the prover's responses before deciding whether to be convinced or not. Interactive proofs do not yield proofs in the strict mathematical sense: the verifier may be incorrectly convinced with an exponentially small, though non-zero probability.

Formally, a set L is said to have an *interactive proof-system* if for all x in L, there exists a prover that can convince the verifier that x is in L with high probability, and for all x not in L, no prover can convince the verifier that x is in L with better than negligible probability. The class IP denotes the sets for which interactive proofs of membership exist.

Essentially, this procedure adds two new ingredients to the classical notion of proof (which can be written down and does not require active participation of the verifier): *randomness* and *interaction* with the prover.

* Supported in part by NSF Grant 865727-CCR and ARO grant DAAL03-86-k-017. Article was partially written while on a sabbatical in Princeton University, Computer Science Department.

Proceedings of the International Congress of Mathematicians, Kyoto, Japan, 1990

Lund, Fortnow, Karloff, Nisan [LFKN], and Shamir [Sh] show the exact unexpected power of adding randomness and interaction to classical proofs: a polynomial time verifier can be convinced of membership in L if and only if membership in L can be computed in polynomial space (PSPACE)[2]. Examples of problems that can be solved in polynomial space are showing that a boolean formula is non-satisfiable, computing the permanant of a matrix, or showing that a generalized geography game has a forced win. This particular list is of increasingly difficult "complete" problems, each as hard as a class of problems (Co-NP, FP, PSPACE).

What exactly gives interactive proofs their enhanced power? Both randomization and interaction are necessary: if the verifier did not toss coins then the prover could simply simulate all the moves of the verifier on his own and interactive proofs would be the same as the NP short proofs; in the absence of interaction, IP equals probabilistic polynomial time computation. The amount of interaction and randomness necessary is studied in [Ba, AGH, GS]. It seems that polynomial number of message exchanges are necessary to obtain the full power of interactive proofs.

Randomness is used in a different way in interactive proofs as defined in [GMR] than in the Arthur Merlin games as defined in [Ba]. In Arthur Merlin games the verifier moves are restricted to only tossing coins and sending their outcome to the prover, while in interactive proofs the verifier can toss coins in secrecy and send the prover messages computed based on the hidden coins. It is this secrecy which made possible some of the early examples of interactive proofs (see Section 4.). Still, Goldwasser and Sipser [GS] showed that the Arthur Merlin games and interactive proof systems are equal in power.

How about efficiently verifying membership in even harder sets? Ben-Or, Goldwasser, Kilian and Wigderson [BGKW] introduce procedures called *two prover interactive proofs*. Instead of one prover, it is two provers who jointly attempt to convince the verifier of the truth of a proposition. The two provers can decide on a common strategy before the interaction with the verifier starts, but once they start interacting with the verifier they can no longer communicate or see the messages exchanged between the verifier and the "other prover". (This is reminiscent of the police practice to interrogate two suspects in a crime separately. The consistency of the alibi is what assures the police of its correctness). The class IP_2 denote the sets for which membership has a two-prover interactive proof. Babai, Fortnow, and Lund [BFL] show that IP_2 equals exactly nondeterministic exponential time (which contains PSPACE and is known to strictly contain NP).

The notion of interactive proof generalizes in the right way to attack a novel problem: how to convince a verifier of the truth of a proposition without giving him any extra "knowledge". For example, convince a verifier that a Boolean formula is satisfiable without revealing a truth assignment (or *anything* he can not compute in polynomial time). This is made precise with the introduction of *zero-knowledge interactive proofs* by Goldwasser, Micali and Rackoff in [GMR]. Goldreich, Micali and Wigderson [GMW] showed that every set in NP has a zero-

[2] Note that polynomial space computation may take exponential time.

knowledge interactive proof if one way functions exist. Their result was extended [BGG, IY] to every set in IP. In contrast, [BGKW] show that zero-knowledge two-prover interactive proof exist for every set in IP_2 without resorting to the unproven assumption of one-way functions.

The notion of zero-knowledge has important applications to the design of cryptographic protocols and fault tolerant computation.

2. Background

2.1 Notation

We use the symbol $|x|$ to denote the binary length of a string $x \in \{0,1\}^*$. Whenever we refer to picking an element out of a set at random we mean with uniform probability distribution. A *language* L is a subset of $\{0,1\}^*$. We use the term algorithm and Turing machine interchangably throughout the paper.

2.2 Complexity Classes

Traditionally, efficient computation in theoretical computer science has been associated with polynomial time computation. The complexity class **P** is defined to be the set of languages for which membership can be computed by a polynomial time algorithm. (Intuitively, these languages correspond to easy to solve problems.)

In recent years probabilistic polynomial time computation has emerged as an alternative accepted formalism of efficient computation. A probabilistic algorithm is an algorithm which can toss coins as an additional primitive operation.

A language L is said to be accepted by a probabilistic polynomial time algorithm M if for all $x \in L$, the prob(M accepts x) $> \frac{2}{3}$; and for all x not in L, the prob(M rejects x) $> \frac{2}{3}$. The class of languages accepted by probabilistic polynomial time algorithms is called **BPP**. (Intuitively, these languages correspond to problems that are easy to solve by probabilistic algorithms.

A notable example of a problem which is in BPP but not known to be in P is integer primality testing.

The class of languages in which membership can be verified by a polynomial time algorithm is called **NP**. (Intuitively, **NP** corresponds to the set of problems for which, once solved, it is easy to verify that the solution is correct.) A language $L \in$ **NP** iff there exists a deterministic polynomial time algorithm M_L and a polynomial p_L such that $x \in L$ if and only if there exists a y such that $|y| \leq p_L(|x|)$ and $M_L(x,y)$ accepts.

It is generally believed that **P** $\neq$ **NP**. An NP-complete problem is a language L such that $L \in$ **NP** and if $L \in$ **P** then **P** $=$ **NP**. An example of an NP-complete language is the set of satisfiable Boolean formulas.

The *Polynomial Time Hierarchy* (**PH**) was introduced to classify computational problems with a more complex logical structure than **NP**. It is defined inductively as follows: (in the definition below the $|y_i| \leq$ polynomial($|x|$) and R is a polynomial time computable relation.)

$$\textstyle\sum_0^P = \Pi_0^P = P$$

$$\textstyle\sum_{k+1}^P = \{L |\text{ for some } R \in \Pi_k^P, L = \{x | \exists_y R(x, y)\}\}$$

$$\Pi_{k+1}^P = \{L |\text{ for some } R \in \sum_k^P, L = \{x | \forall_y R(x, y)\}\}$$

$$\mathbf{PH} = \bigcup_{k>0} \textstyle\sum_k^P$$

A common conjecture made in complexity theory is that the hierarchy defined is strict and $\sum_k^P \neq \sum_{k+1}^P$.

So far, we concentrated on the time taken by an algorithm to solve a problem as the "resource" in question. Alternatively, the amount of memory used by an algorithm can be considered. We call **PSPACE** the set of languages accepted by deterministic Turing machines which are restricted to use a polynomial in the length of the input amount of space. The polynomial time hierarchy is contained in **PSPACE**, as it requires only a polynomial amount of space to go through all possible values of the universal and existential quantifiers of a logical formula to evaluate if it is satisfiable even though it requires exponential time.

3. Definitions

3.1 Interactive Proof Systems

Before defining notion of interactive proof-systems, we define the notion of interactive Turing machine.

Definition. An *Interactive Turing machine (ITM)* is a probabilistic Turing machine which in addition to its input tape, random tape, work tape, and output tape, has a one read only communication tape, and one write only communication tape. The contents of the write-only communication tape can be thought of as *messages sent* by the machine; while the contents of the read-only communication tape can be thought of as *messages received* by the machine.

Definition. An *interactive protocol* is an ordered pair of ITMs (A, B) which share the same input tape; B's write-only communication tape is A's read-only communication tape and vice versa. The machines take turns in being active with B being active first. During its active stage, the machine first performs some internal computation based on the contents of its tapes, and second writes a string on its write-only communication tape. The i-th *message* of $A(B)$ is the string $A(B)$ writes on its write-only communication tape in i-th stage. The protocol is terminated by machine B which *accepts* (or rejects) the input by entering an accept (or reject) state. The first member of the pair, A, is a computationally unbounded Turing machine. The *computation time* of machine B is defined as the sum of B's computation time during its active stages, and is bounded by a polynomial in the length of the input string.

A few parameters of an interactive protocol are of interest. The $value_{A,B}(x)$ of an interactive protocol (A, B) on input x is the probability (taken over A and B's coin tosses) that B accepts x. The number of *rounds* of an interactive protocol (A, B) on input x is defined to be the number of message exchanges between A and B. The *size* of an interactive protocol (A, B) on input x be the total number of bits exchanged between A and B.

Definition. We say that a language $L \subset \{0, 1\}^*$ has an *interactive proof-system* if $\exists ITM\ V$ s.t.

1. $\exists\ ITM\ P$ s.t (P, V) is an interactive protocol and $\forall x \in L$ the $value_{(P,V)}(x) > \frac{2}{3}$.
2. $\forall\ ITM\ P$ s.t (P, V) is an interactive protocol $\forall x \notin L$ the $value_{(P,V)}(x) < \frac{1}{3}$. Note that it does not suffice to require that the verifier cannot be fooled by the predetermined prover (such a mild condition would have presupposed that the "prover" is a trusted oracle).

We say that (P, V) (P for which condition 1 holds) accepts L or is an interactive proof-system for L. Let **IP** denote all languages accepted by some interactive proof-system. Note that **NP** is a special case of interactive proofs, where the interaction is trivial and the verifier tosses no coins.

We say that (P, V) is a *$t(n)$-round interactive proof-system* if $\forall x \in L$, the number of rounds of (P, V) on input x is bounded by $t(|x|)$, and let $\mathbf{IP}[t(n)]$ denote all languages accepted by some $t(n)$-round interactive proof-system.

We say that (P, V) has *error probability* ε if $\forall x \in L, value_{P,V}(x) > 1 - \varepsilon$, and $\forall x \notin L, \forall P' value_{P',V}(x) < \varepsilon$.

The error probability of an interactive protocol (P, V) can be decreased to be exponentially small by a simulating[3] protocol (P', V') which runs the (P, V) protocol independently several times in parallel. V' accepts the input if and only if V accepts the input in a majority of the runs.

Amplification Lemma 1. *Let (P, V) be an interactive proof-system for L. On inputs of length n to (P, V), let $m(n)$ = length of messages exchanged, and $g(n)$ = number of rounds, Then $\forall$ polynomials $k(n)$, $\exists$ interactive proof system (P', V') for L with error probability $< \frac{1}{2^{k(n)}}$, $g(n)$ rounds and length of messages exchanged $O(m(n)k(n))$.*

3.2 Arthur Merlin Games (AM)

Another, seemingly more restricted, proof-system was defined by Babai [Ba1]. Babai called his proof-system an Arthur-Merlin game where Arthur corresponds to the verifier and Merlin corresponds to the prover. The difference with interactive proof-system of (GMR) is that Arthur messages consist only of the outcome of this coin tosses.

[3] The notion of one interactive-protocol simulating another is used throughout this paper. We say that protocol (P', V') *simulates* (P, V) if V' can use V as an oracle and P' can use P as an oracle and the same language is received by (P', V') as by (P, V). The *cost* of the simulation is the ratio of the sizes of (P', V') and (P, V).

An Arthur Merlin protocol (game) (M, A) on common input x is an interactive-protocol in which A's i-th message consist of the next block of $m(|x|)$ bits on its random tape (where $m()$ is a polynomial). The protocol is terminated by A, who then evaluates a polynomial time function defined over the contents of its tapes to decide whether to accept or reject the input x.

Thus, Arthur can be thought of as a verifier who flips public coins which Merlin, the prover, can look at.

Similarly to [GMR], Babai [Ba1] defined a language L to have an *Arthur-Merlin proof-system* if $\exists ITM\ A$ s.t

1. $\exists\ ITM\ M\ s,t$ (M, A) is an Arthur Merlin Protocol and $\forall x \in L$, the $value_{(M,A)}(x) > \frac{2}{3}$.
2. $\forall ITM\ M$ s.t (M, A) is an Arthur Merlin Protocol, $\forall x \notin L$, the $value_{(M,A)}(x) < \frac{1}{3}$.

We say that an Arthur Merlin game (M, A) accepts L or is an Arthur Merlin proof-system for L if it obeys condition 1. **AM** will denote the set of all languages accepted by some Arthur Merlin proof-system, and **AM**$[t(n)]$ denote the set of all languages accepted by a $t(n)$-round Arthur Merlin proof-system.

4. Examples of Interactive Proofs

Notation. Whenever an interactive protocol is demonstrated, we let $B \longrightarrow A$: denote an active stage of machine B, in the end of which B sends A a message. Similarily, $A \longrightarrow B$: denotes an active stage of machine A.

Example 1: Quadratic Residuosity

Let $Z_n^* = \{x < n,\ (x, n) = 1\}$
$QR = \{(x, n) \mid x < n, (x, n) \text{ and } \exists y \text{ s.t } y^2 \equiv x \bmod n\}$
$QNR = \{(x, n) \mid x < n, (x, n) \text{ and } \nexists y \text{ s.t } y^2 \equiv x \bmod n\}$
We demonstrate an interactive proof-system for QNR.
On input (x, n) to interactive protocol (A, B):
$B \longrightarrow A$: B sends to A the list $w_1 \cdots w_k$ where $k = |n|$ and

$$w_i = \begin{cases} z_i^2 \bmod n & \text{if } b_i = 1 \\ x \cdot z_i^2 \bmod n & \text{if } b_i = 0 \end{cases}$$

where B selected $z_i \in Z_n^*, b_i \in \{0, 1\}$ at random.
$A \longrightarrow B$: A sends to B the list $c_1 \cdots c_k$ s.t.

$$c_i = \begin{cases} 1 \text{ if } w_i \text{ is a quadratic residue mod } n \\ 0 \text{ otherwise} \end{cases}$$

B accepts iff $\forall_{1 \le i \le k},\ c_i = b_i$
B interprets $b_i = c_i$ as evidence that $(x, n) \in QRN$; while $b_i \neq c_i$ leads him to reject.

We claim that (A, B) is an interactive proof-system for QNR. If $(x, n) \in QNR$, then w_i is a quadratic residue mod n iff $b_i = 1$. Thus, the all powerful A can easily compute whether w_i is a quadratic residue mod n or not, compute c_i correctly and make B accept with probability 1. If $(x, n) \notin QNR$ and $(x, n) \in QR$ then w_i is a random quadratic residue mod n regardless of whether $b_i = 0$ or 1. Thus, the probability that A (no matter how powerful he is) can send c_i *s.t* $c_i = b_i$, is bounded by $\frac{1}{2}$ for each i and probability that B accepts is at most $(\frac{1}{2})^k$.

Recognizing quadratic non-residues is in **NP**, but is not known to be solvable by probabilistic polynomial time algorithms (BPP). A short (non-interactive) proof that x is a quadratic non-residue modulo n is a prime factor p of n such that the Legendre symbol of x mod p is -1.

The interest of the interactive proof for QNR described above is that none of the prime factors of n are disclosed to the verifier during the interactive proof. In fact, this example was the first zero-knowledge interactive proof (see [GMR] for definition) known for a language not known to be solvable in polynomial time. This was shown by Goldwasser Micali and Rackoff in [GMR].

Example 2: Graph Non-Isomorphism

The most famous interactive proof for a problem not known to be in **NP** is for the *graph non-isomorphism* problem. This was shown by Goldreich, Micali, and Wigderson [GMW]. The input is a pair of graphs G_1 and G_2, and one is required to prove that there exists no 1-1 edge-invariant mapping of the vertices of the first graph to the vertices of the second graph. (A mapping π from the vertices of G_1 to the vertices G_2 is *edge-invariant* if the nodes v and u are adjacent in G_1 iff the nodes $\pi(v)$ and $\pi(u)$ are adjacent in G_2.)

The length of the shortest (non-interactive) proof of non-isomorphism is no better than the best deterministic isomorphism test, i.e $e^{O(\sqrt{n \log n})}$.

The interactive proof (A, B) on input (G_1, G_2) proceeds as follows:

$B \longrightarrow A$: B chooses at random one of the two input graphs, G_{α_i}

where $\alpha_i \in \{1, 2\}$. B creates a random isomorphic copy of G_{α_i} and sends it to A. (This is repeated for $1 \le i \le k$, with independent random choices, where $k =$ number of vertices in G_{α_i}.)

$A \longrightarrow B$: A sends B $\beta_i \in \{1, 2\}$ for all $1 \le i \le k$.

B accepts iff $\beta_i = \alpha_i$ for all $1 \le i \le k$.

B interprets $\beta_i = \alpha_i$ as evidence that the graphs are not isomorphic; while $\beta_i \ne \alpha_i$ leads him to reject.

If the two graphs are not isomorphic, the prover has no difficulty to always answer correctly (i.e., a β equal to α), and the verifier will accept. If the two graphs are isomorphic, it is impossible to distinguish a random isomorphic copy of the first from a random isomorphic copy of the second, and the probability that the prover answers correctly to one "query" is at most $\frac{1}{2}$. The probability that the prover answers correctly all k queries is $\le (\frac{1}{2})^k$.

5. Equivalence of Interactive Proof Systems and Arthur Merlin Proof Systems

If we review the examples of interactive-proof systems for quadratic non-residuosity and graph non-isomorphism, we see that the ability of the verifier to keep his coin flips secret, and thus unable the prover to predict the correct answer to his questions unless the statement in question was true, was the *crucial* ingredient which made these proofs go through. Thus, the interactive proofs described in the example are not Arthur Merlin games. This leads to our next question: as Arthur is a special case of a general verifier who can not hide the results of his coin flips, **AM** is contained in **IP**, but is it a strict containment?

Surprisingly, it turns out that **AM** = **IP**. Goldwasser and Sipser [GS] show a transformation from a general interactive proof-system to an Arthur Merlin protocol accepting the same language. Moreover, the transformation preserves the number of rounds.

Theorem 2 [GS]. *∀ polynomially bounded* $t(n) \geq 2$ $AM[t(n)] = IP[t(n)]$.

In particular, the theorem implies that the graph non-isomorphism language is accepted by a constant round Arthur Merlin protocol. This has certain implication as to the complexity of graph non-isomorphism (see Section 7.2).

The equivalence between **AM** and **IP** is highly convenient. It is usually easier to prove membership of languages using the interactive proof formulation. The elegant simplicity of the Arthur-Merlin games definition facilitates proving complexity results.

For example, we have defined **IP** to allow two sided error: the verifier can err both when x is in the language and when x is not in the language. Goldreich, Sipser and Mansour [GSM] show that every $L \in \mathbf{AM}[t(n)]$ has a one sided error Arthur Merlin protocol (M, A) of $t(n)$-rounds. Namely, when $x \in L$, the Pr(A accepts x) = 1.

6. Interaction: Essential Ingredient?

When the verifier is deterministic it is clear that interaction does not add power with respect to language recognition, since the all powerful prover can anticipate in advance all messages of the verifier and answer them with no need for interaction. However, when the verifier is a probabilistic algorithm, it is an interesting and not fully resolved question whether more rounds of interaction enable a proof-system to recognize more languages. It has been shown by Babai [Ba1], that any language which has a constant round interactive proof-system, has an interactive proof-system that requires only two rounds.

Theorem 3 [Ba]. *For constant* $c \geq 2$ $AM[c] = AM[2]$.

Babai and Moran [BM] obtained a stronger result showing how the number of rounds can be halved .

Theorem 4 [BM]. *For polynomially bounded* $t(n) \geq 2$ $AM\ [2t(n)] = AM[t(n)+1]$.

The simulation technique used to prove the collapse and speed-up theorem result in a $O(t(n))$ cost. Thus, if the number of rounds is halved a logarithmic number of times, the size of game grows faster than a polynomial in n. Thus, the speed-up theorem can not resolve the question whether for unbounded $t(n)$, $\mathbf{AM}[t(n)] = \mathbf{AM}[2]$.

One piece of evidence known in regard to this question was obtained by Aiello, Goldwasser, and Hastad [AGH] as follows.

Theorem 5 [AGH].

$$\forall f, g \text{ s.t } g(n) = o(f(n)), \exists \text{ oracle set } B \text{ s.t } AM^B[f(n)] \neq AM^B[g(n)]$$

This result does not imply any unrelativized conclusion, but implies that resolving the above question will involve proof techniques which do not relativize (e.g., diagonalization type arguments will not suffice). As a word of warning we note that quite recently a few relativized results in this field have shown not to hold in a non-relativized setting. See Section 8..

7. The Complexity of Interactive Proofs

An interesting question is what is the complexity of language in **IP** in terms of more traditional complexity classes. For example, can a verifier be convinced that a Boolean formula ϕ has *no* satisfying assignments, or that the *shortest* traveling salesman tour in a graph is bounded by integer k?

The answer to this question is different for interactive proofs with bounded (constant independent of the input) number of rounds and interactive proofs with polynomial number of rounds.

We start with the case of bounded rounds.

7.1 Bounded Rounds Interactive Proofs

The *Polynomial Time Hierarchy* (PH) (see Section 2.2) was introduced to classify computational problems with a more complex logical structure than **NP**. We can extend the definition of $\sum_k^P$ and $\prod_k^P$ from a finite number of alternation of quantifiers, to a number of alternations which is a function of the input length. Namely, $\sum_{t(n)}^P = \{L|L = \{x|\exists y_{t(n)} \forall y_{t(n)-1} \cdots R(x, y_1 \cdots, y_{t(n)})\}, R \in P\}$ and $\prod_k^P = \{L|L = \{x|\forall_{y_{t(n)}} \exists_{y_{t(n)-1}} \cdots R(x, y_1 \cdots y_{t(n)})\}, R \in P\}$.

It can be shown (using ideas from a proof of Lautman [L]) that membership in any language accepted by an interactive proof of $t(n)$ rounds can be expressed as the question of satisfying a logical formula with $t(n)$ alternation of universal and existential quantifiers starting with a universal quantifiers.

Theorem 6 [Ba, AGH]. $AM[t(n)] \subseteq \Pi_{t(n)}^P$ *for all* $t(n) \geq 2$.

This was shown by Babai [Ba1] for $t(n) = 2$, and Aielo-Goldwasser-Hastad for unbounded $t(n)$. The converse is not known to be true.

An especially interesting case in question is whether languages in **Co-NP** $= \prod_1^P$ are in **IP**[2]. A result which may be taken as supporting a separation between **Co-NP** and **IP**[2] is by Boppana-Hastad-Zachos [BHZ].

Theorem 7 [BHZ]. If $Co\text{-}NP \subseteq AM[2] \Longrightarrow PH = \prod_2^P$.

We conclude that showing that $Co\text{-}NP$ has bounded round interactive proofs would be quite hard if true. The proof of Theorem 7, however, does not carry over to to **AM**$[t(n)]$ for the case of unbounded number of rounds $t(n)$.

A language is said to have non-uniform polynomial time deterministic (non-deterministic) algorithms if for every size of input n, there exists a polynomial in n time deterministic (non-deterministic) algorithm which decides membership of strings of length n in the language. Non-uniform complexity has generated much interest in complexity theory in the last years.

Theorem 8. *Languages in $IP[2]$ have non-uniform polynomial-time non-deterministic algorithms.*

7.2 The Complexity of Graph Non-Isomorphism

The graph isomorphism problem is one of a select group of problems in **NP** which have not been classified as solvable in polynomial time, nor have been proven NP-complete.

The developments regarding interactive proof-system have shed light on the complexity of the graph non-isomorphism problem.

Let $ISO = \{\text{graphs } (G_1, G_2) \ s.t \ G_1 \text{ is isomorphic to } G_2\}$ and $Non\text{-}ISO$ as its complement.

In Section 4. we showed that $Non\text{-}ISO$ has a bounded round interactive proof an thus is in **IP**[2]. Combining the results above we can show the following conditional implications.

Corollary 9. *If ISO is NP-complete then*

1. Polynomial Time Hierarchy (PH) collapses to $\prod_2^P$.
2. Co-NP has non-uniform polynomial time non-deterministic algorithms.

This is usually taken as evidence that ISO is *not* NP-complete. In general, showing that both a language L and its complement $\bar{L}$ are in **IP**[2] imply that either L is not NP-complete or the polynomial time hierarchy collapses and Co-NP has non-uniform polynomial size non-deterministic algorithms. This can help when trying to classify the complexity of an **NP** problem. An intersting **NP** problem which is not known to be NP-complete is the problem of the existence of a short vector in an integer lattice. It is open question whether the complement problem (all vectors in a lattice are bigger than a given value) is in **IP**[2].

8. IP = PSPACE

We saw in the previous section that bounded round interactive proofs are quite low in the polynomial time hierarchy. In this section, we investigate how powerful are general interactive proofs.

We first bound the complexity of languages accepted by interactive proofs. Membership in every $L \in \mathbf{IP}$ can be decided by some Turing machine which uses a polynomial amount of space and $\mathbf{IP} \subseteq \mathbf{PSPACE}$ as follows.

Let L be in **IP** accepted by an interactive proof (P, V) with error probability ε. On input x a Turing machine can in polynomial space go through all possible histories of communication between P and V and compute the probability that the verifier accepts. If this probability is greater than ε, then x is in L.

Until recently, it was not known whether the containment of $\mathbf{IP} \subseteq \mathbf{PSPACE}$ is strict. In fact, only a handful of languages not known to be in **NP** were shown to be in **IP**.

In December 1989, Lund Fortnow Karloff and Nisan showed that every language in the polynomial time hierarchy has an interactive proof system. Their proof used novel techniques showing how to reduce proving to the verifier the number of accepting computations of a non-deterministic Turing machine, to proving the value of polynomials of low degree over a finite field. These algebraic techniques are of general interest to complexity theory at large as they do not relativize.

Theorem 10 [LFKN]. *$PH \subseteq IP$.*

The result was announced to a few colleagues using the international electronic mail system. A few days later A. Shamir [Sh] closed the gap and showed that in fact languages accepted by interactive proofs with polynomial number of rounds of interaction are exactly those languages accepted in polynomial space.

Theorem 11 [Sh]. *$IP = PSPACE$.*

Shamir's proof shows an interactive proof for the *quantified Boolean formula*(QBF) language defined as follows:

$$QBF = \{F = (Q_1 y_1, Q_2 y_2, Q_n y_n, R(y_1, ..., y_n)\},$$

where R is an unquantified Boolean formula and $Q_i \in \{\exists, \forall\}\}$. Membership in any language in **PSPACE** can be reduced to membership in QBF.

We briefly sketch the ideas in the proof. The idea of [LFKN] can be described as follows. Let $L \in \Pi_k^P$. Then, $x \in L$ if and only if $F(x) = \forall y_1 \exists y_2 \ldots \exists y_k R(x, y_1, \ldots, y_k)$ is true. (Here k is *finite*, and R an unquantified Boolean formula). We can write an arithmetic expression $\tilde{F} = \prod_{y_1=0}^{1} \sum_{y_2=0}^{1} \cdots \sum_{y_k=0}^{1} \tilde{R}(x, \tilde{y}_1, ..., \tilde{y}_k)$, replacing every Boolean variable y_i by integer variable $\tilde{y}_i$, every logical $\wedge$ to arithmetic $\times$, every logical $\neg E$ to $1 - E$, every $\forall y$ to $\prod_{y \in \{0,1\}}$, and every $\exists y$ to $\sum_{y \in \{0,1\}}$. $\tilde{F}$ evaluates to 0 if and only if the input x does not satisfy F. Thus in order to prove that x satisfies F the prover has to prove that $\tilde{F} \neq 0$. Define $g_{y_1}(z) = \tilde{F}_{y_1=z}$

(evaluated $\tilde{F}$ with y_1 replaced everywhere by variable z). Note that since k is finite, $g_{y_1}(z)$ is a polynomial of polynomial degree in variable z. The prover sends the coefficients of $g_{y_1}(z)$ to the verifier. The verifier checks that the value of g received on $z = 0, 1$ matches the assertion that $\tilde{F} \neq 0$. If this assertion was false, then it must be that the g received is not the correct polynomial, and since both g received and the correct polynomial are of low degree they can agree on only few points. Now, the verifier chooses a random value z' from some sufficiently large range of integers, and asks the prover to prove that the value of $g_{y_1}(z')$ equals $\tilde{F}_{y_1/z'}$ ($\tilde{F}$ when y_1 is assigned value z'.) Since z' was chosen randomly among integers from a sufficiently large range, with high probability this new assertion is false too. The verifier and prover iterate on that until they exhaust all the variables y_i, and are left with a fully instanciated expression, at which point the verifier can verify whether the value the prover claims for it is true.

When extending this proof to **PSPACE**, the number of quantifiers k is not finite, and the degree of the polynomial $\tilde{F}$ and the intermediate g's in the variables y_i can be exponential! Shamir's [Sh] solution is to replace expressions of the form $(\forall y_i R)$ by $(\forall y_i \exists y'_1, ..., y'_{i-1}$, such that $(y_1 = y'_1, y_2 = y'_2, ..., y_{i-1} = y'_{i-1}, y_i = y'_i) \wedge R_{y'=y})$ (i.e replace each y_l in R by y'_l , $l \leq i$ in R). This way, no variable in $\tilde{F}$ will have higher than constant degree and the number of variables is squared. When k is not finite the value of $\tilde{F}$ may be doubly exponential in size, thus the prover will prove to the verifier that $\tilde{F} \neq 0$ modulo a prime of smaller size. The above procedure is modified accordingly.

A few remarks are in order. First, the equivalence with **PSPACE** implies that **IP** is closed under complementation. Second, a polynomial number of rounds of interaction seem to be crucial for the argument used in the proofs of Theorems 10 and 11. Showing that **IP** = **IP[2]** would imply that **PSPACE** = $\Pi_{\mathbf{2}}^{\mathbf{P}}$. We conclude that resolving the question of whether bounded round interactive proofs accept all of **IP** would settle questions complexity theorists have been struggling with for years.

The current proof technique of Theorem 11 requires the prover to be **PSPACE** powerful. Possible conjectures are (1) that the ability to decide whether or not a Boolean formula is satisfiable does not enable a prover to convince a polynomial time verifier that a formula is not satisfiable (i.e membership in **Co-NP** complete languages), and (2) that the ability to *count* the number of satisfying assignments to a Boolean formula is sufficient to enable a prover to convince a verifier of membership in **PSPACE** complete languages. It seems likely progress can be made on these questions in near future.

9. Multi-Prover Interactive Proofs

Is membership in a **PSPACE** language the limit of what a polynomial time verifier can hope to be convinced of?

Ben-Or, Goldwasser, Kilian and Wigderson [BGKW] proposed an extended definition of interactive proof-systems which they called multi-prover interactive proofs. Instead of one prover attempting to convince a verifier that x, the input string, is in L, the prover consists of k separate agents (or rather k provers) who jointly attempt to convince a verifier that x is in L. The k provers can cooperate and communicate between them to decide on a common optimal strategy before the interaction with the verifier starts. But, once they start to interact with the verifier, they can no longer send each other messages or see the messages exchanged between the verifier and the "other prover".

As in single prover interactive proofs, the verifier is a probabilistic polynomial time algorithm, and can exchange upto a polynomial number of messages with each one of the provers (with no restriction on interleaving the exchanged messages) before deciding to accept or reject the input x.

Definition. Let P_1, P_2,..., P_k be Turing machines which are computationally unbounded and V be a probabilistic polynomial time Turing machine. All machines have a read-only input tape, a work tape and a random tape. All machines share the same input tape. The P_i's share an infinite read-only random tape of 0's and 1's. Every P_i has one write-only communication tape and one write-only communication tape. V has k write-only communication tapes and k read only communication tapes. The ith write-only communication tape of V is P_i's read-only communication tape, and vice versa. Namely, on its ith communication tape, V writes messages to P_i. We call $(P_1, P_2, ..., P_k, V)$ a *k-prover interactive protocol.*

Definition. We say that $L \subset \{0,1\}^*$ has a *k-prover interactive proof-system* (IPS) if there exists an interactive probabilistic Turing machine V such that:

1 $\exists P_1, P_2, ..., P_k$ such that $(P_1, P_2, ..., P_k, V)$ is a k-prover interactive protocol and $\forall x \in L$, prob(V accepts input x) $\geq \frac{2}{3}$.
2 $\forall P_1, P_2, ..., P_k$ such that $(P_1, P_2, ..., P_k, V)$ is a k-prover interactive protocol, prob(V accepts input x) $\leq \frac{1}{3}$.

We let $\mathbf{IP}_k$ denote the class of languages which are accepted by k-prover interactive proof-systems and $\mathbf{IP}_k[t(n)]$ denote the class of languages which are accepted by k-prover interactive proof-system using $t(n)$ rounds.

It can be shown that as in the single-prover case, for any k, if $L \in \mathbf{IP}_k$ then there exists an an interactive proof $(P_1, P_2, ...P_k, V)$ for L such that for all $x \in L$, the Pr(V accepts x) $= 1$. The number of rounds, size, and error probability of a multi-prover interactive proof is defined similarly to interactive proofs.

10. The Complexity of Multi-Prover Interactive Proofs

Is $\mathbf{IP}_k$ be different than $\mathbf{IP}$? The hope is that the inability of the provers to communicate with each other will allow the verifier to check the consistency of the provers responses against each other. Intuitively, this additional check makes the verifier trust the answers of the prover more than in the single-prover case. This potentially may allow the verifier to be convinced of membership in harder languages.

It was shown in the original paper of [BGKW] that languages accepted by interactive proofs with a polynomial number of provers are equal to languages accepted by interactive proofs with two provers.

Theorem 12 [BGKW]. *For all* $k \geq 2$, *if* $L \in IP_k$ *then* $L \in IP_2$.

Shortly after the results showing **IP** = **PSPACE**, Babai, Fortnow and Lund [BFL] proved that the class of languages having two-prover interactive proof systems is exactly nondeterministic exponential time (NEXPTIME).

Theorem 13 [BFL]. $IP_2 = NEXPTIME$.

This represents a further step demonstrating the unexpected power of randomization and interaction in efficient provability. It follows that multiple provers with coins are strictly stronger than without, since $NEXPTIME \neq NP$. In particular, for the first time, provably polynomial time intractable languages turn out to have "efficient proof systems" since $NEXPTIME \neq P$.

The first part of the proof extends the techniques of [LFKN] and [Sh] for the single prover case. The second part is a verification scheme for the multilinearity of an n-variable function held by the two provers.

10.1 Interaction in Multi-Prover Interactive Proofs

Unbounded number of rounds of interaction seemed necessary to realize the full power of interactive proofs. In contrast, work of Fortnow, Rompel and Sipser [FRS], Cai, Condon and Lipton [CCL], and Kilian shows that a two prover proof system can accept every language in IP_2 using bounded number of rounds and achieving a constant error probability.

Theorem 14 [FRS, CCL], kilian. *For any constant* ε, *any* $L \in IP_2$ *has a bounded round two-prover interactive proof with error probability* ε.

11. Approximating the Clique Problem

A new connection between approximation of combinatorial graph problems and multi-prover interactive proofs was recently discovered.

It is well known that the Clique problem (i.e., finding the size of the largest complete subgraph in a given graph G) is **NP**-complete. However, the related problem of *approximating* this size (say within a constant factor) is not known to have an efficient algorithm, nor is it known to be **NP**-complete. The best known polynomial time approximation algorithm for clique is a factor $O(\frac{n}{\log^2 n})$ below the optimal [Boppana et al.]

Using the characterization of $\mathbf{IP_2} = \mathbf{NEXPTIME}$, Feige, Goldwasser, Lovasz, and Safra [FGLS] showed the following.

Theorem 15 [FGLS]. *If approximating the clique problem within factor* $2^{\log^c n}$ *for some* $c < 1$ *can be done in time* $2^{log^k n}$ *for some* $k > 1$ *then*

1. $EXPTIME = NEXPTIME$
2. $NP \subseteq \bigcup_{k>0} DTIME(2^{log^k n})$

References

[AGH] W.Aiello, S. Goldwasser, J. Hastad: On the power of interaction. Proc. 27th IEEE Foundations of Computer Science Conf. 1986 pp. 368–379

[BFL] L. Babai, L. Fortnow, C. Lund: Non-deterministic exponential time has two-prover interactive protocols. Proceedings, 31st Annual IEEE Symp. on Foundations of Computer Science, 1990, pp. 16–25.

[Ba] L. Babai: Trading group theory for randomness. Proc. 17th ACM Symp. on Theory of Computing 1985 pp. 421–429

[Ba2] L. Babai: E-mail and the unexpected power of interaction. University of Chicago Tech Report CS90–15, April 24, 1990

[BHZ] R. Boppana, J. Hastad, S. Zachos: Does co-NP have short interactive proofs? Information Processing Letters **25** (1987) 127–132

[BM] L. Babai, S. Moran: Arthur-Merlin Games: a randomized proof system, and a hierarchy of complexity classes. To appear in J. Computer Science and Systems.

[C] S. Cook, The Complexity of Theorem-Proving Procedures. Proc. of 3rd Symposium of Theory of Computation, 1971 pp. 151–158

[BGG] M. Ben-Or, O. Goldreich, S. Goldwasser, J. Håstad, J. Kilian, S. Micali, P. Rogaway: IP is in zero-knowledge. Proceedings, Advances in Cryptology, Crypto 1988

[BGKW] M. Ben-Or, S. Goldwasser, J. Kilian, A. Wigderson: Multi-prover interactive proof systems: Removing intractability assumptions. Proceedings of the 20th STOC, ACM, 1988

[CCL] J. Cai, A. Condon, R. Lipton: On bounded round multi-prover interactive proof systems. STACS90

[FRS] L. Fortnow, J. Rompel, M.Sipser: On the power of multi-prover interactive protocols. Proceedings, Structures **88**, pp. 156–161

[FGLS] L. Lovasz, S. Goldwasser, U. Feige, S. Safra: On the difficulty approximating the clique function. In preparation

[GMR] S. Goldwasser, S. Micali, and C. Rackoff: The knowledge complexity of interactive proof-systems. Proceedings of the 17th STOC, ACM, 1985, pp. 291–304

[GMS] O. Goldreich, Y. Mansour, and M. Sipser: Interactive proof systems: Provers that never fail and random selection. Proc. FOCS87

[GMW] Goldreich, Oded, Silvio Micali, Avi Wigderson: Proofs that yield nothing but the validity of the assertion, and a methodology of cryptographic protocol design. Proceedings of the 27th FOCS, IEEE, 1986, pp. 174–187

[GS] S. Goldwasser and M. Sipser: On public versus private coins in interactive proof systems. Proc. 18th Symp. on Theory of Computing (1986) pp. 59–68

[IY] R. Impagliazzo, M. Yung: Direct Minimum Knowledge Computations. Proceedings, Advances in Cryptology, Crypto 1987

[LFKN] C. Lund, L. Fortnow, H. Karloff, N. Nisan: Algebraic methods for interactive proof systems. Proceedings, 31st Annual IEEE Symp. on Foundations of Computer Science, 1990, pp. 2–10

[L] C. Lautemann: BPP and the polynomial hierarchy. Info. Proc. Letters **17**, no. 4 (1983) 215–217

[Sh] A. Shamir: IP=PSPACE. Proceedings, 31st Annual IEEE Symp. on Foundations of Computer Science, 1990, pp. 11–15

Information Theoretic Reasons for Computational Difficulty

Avi Wigderson *

The Hebrew University, Jerusalem, Israel and
Princeton University, Princeton, NJ 08544, USA

1. Introduction

We give an intuitive account of the concepts in the title, by considering the following simple number-theoretic example. Imagine two distant players who communicate by exchanging binary messages (bits). One player is given a *prime* number x, and the second a *composite* number y, where $x, y < 2^n$. The players' task is to find a prime number p, with $p < 2n$, such that $x \not\equiv y \pmod{p}$. The existence of such a small prime p is guaranteed by the prime number theorem and the Chinese remainder theorem.

The players agree beforehand on a "protocol" for exchanging messages. The protocol dictates to each player what message to send at each point, based on his input and the messages he received so far. It also dictates when to stop, and how do determine the answer from the information received. There is no limit on the computational complexity of these decisions, which are free of charge. The cost of the protocol is the number of bits they have to exchange on the *worst case* choice of inputs. We shall be interested in the cost of the best protocol under this measure, which we denote by $t(n)$.

There is a trivial protocol in which one player sends his input to the second (n bits), who computes the answer and sends it ($\log n$ bits) back to the first. This shows that $t(n) \leq n + \log n$.

How small can $t(n)$ be? Is it possible that [1] $t(n) = O(\log n)$, which is (essentially) the trivial lower bound? At present, these trivial upper and lower bounds are the best known!

Why should anyone take the time to think about this problem, besides its innocently simple statement and the challenge of the exponential gap in our knowledge? The reason is that this *information theoretic* problem encodes the *computational difficulty* of primality testing! Answering it is extremely important for computational number theory and theoretical computer science as follows:

---

* This research was partially supported by the American-Israeli Binational Science Foundation grant number 87-00082.

[1] For two functions $f(n), g(n)$ we write $f(n) = O(g(n))$ (resp. $f(n) = \Omega(g(n))$) if there is a fixed positive constant that bounds the ratio $f(n)/g(n)$ from above (resp. from below) for every integer n.

Proceedings of the International Congress
of Mathematicians, Kyoto, Japan, 1990

If $t(n) = O(\log n)$, then for every n there is a polynomial (in n) size logical formula of n boolean variables, that evaluates to *'true'* if and only if the values of these variables represent (in binary) a prime integer! This means that primality can be tested highly efficiently, both sequentially and in parallel, which has far reaching practical consequences (e.g. in cryptography and coding).

If $t(n) \neq O(\log n)$, the primality function has no such formulæ, and it would be the first explicit example of such a function. This would resolve a close cousin of the P vs. NP question, which is perhaps the single most important problem of theoretical computer science. (In a nutshell, this question asks if there exist mathematical theorems whose proofs are much harder to find than to verify.) Essentially no progress was made on this problem since it was first posed informally in a letter of Gödel to von Neumann in the 50s, and formulated by Edmonds, Cook and Levin in the 70s.

There is nothing special about choosing the primality function. If, instead, the number x given to the first player is a perfect square, and y is a non-square, then the communication complexity $t(n)$ of the same communication problem would correspond to the formula complexity of testing if an integer (given in binary) is a perfect square. For this problem it turns out that $t(n) = O(\log n)$.

It is not hard to see, by a simple counting argument, that for almost all Boolean functions on the integers, the communication complexity is as bad as can be, i.e. $t(n) = \Omega(n)$ (again, for input integers less than 2^n), and the associated Boolean formulae computing them must be of exponential size. But no explicit example of a function requiring superpolynomial formulae size is known.

The computational complexity of Boolean functions, in the model of Boolean circuits and formulæ, called *Circuit Complexity*, was initiated by Shanon and others in the 40s. Its importance stems from the direct connection between resources in this model, circuit size and circuit depth ($\approx$ the logarithm of formula size), and the "standard" resources time and space (respectively) in Turing machines. Non-trivial results exist mainly for restricted circuits, especially monotone (no negations allowed) circuits. Comprehensive texts on circuit complexity are the books of Wegener [W] and Dunne [D]. [BS] is a beautiful and concise account of the most important issues.

The information theoretic model of *Communication Complexity* was introduced by Yao [Y] in '79. In his model the two players have to compute a Boolean function of their two inputs (in contrast to the "search" problem defined above). This model was extensively studied in the last ten years, and many significant questions about it were resolved. No complete survey of this area exists, but many of the issues and results are discussed and referenced in [BFS].

In 1988 Karchmer and Wigderson [KW] suggested to study the communication complexity of relations (search problems), and showed how to associate with an arbitrary Boolean function a relation (in essentially the way described above), such that the circuit depth of the given function is exactly the communication complexity of the associated relation. They also showed how to capture in a similar fashion monotone computation. This connection established that computational difficulty (at least of the "depth" resource), stems from purely information theoretic limitations.

The elegance of the communication model, and the existence of non-trivial machinary to handle it, enabled the sequence of recent lower bounds on the monotone circuit depth of several important functions [KW, RW2, RW1]. We have beleive that this approach will be instrumental in breaking the ice and finally obtaining a non-trivial lower bound for some explicit function on general (non-monotone) circuits!

In Section 2 we give the neccesary background from circuit complexity and communication complexity. In Section 3 we describe the connection between the two, and in Section 4 we state the lower bounds obtained via this connection. In Section 5 we give some concluding remarks.

2. Preliminaries

2.1 Boolean Functions and Circuit Complexity

For an integer r let $[r]$ denote the set $\{1, 2, ..., r\}$. The set $\{0, 1\}^n$ will denote all binary sequences of length n. We interpret such a sequence as truth assignment to n Boolean variables $x_1, x_2, ..., x_n$. These variables will encode some object, e.g. a number, a graph, a matrix in a natural way. Any function $f : \{0, 1\}^n \to \{0, 1\}$ is a **Boolean function**. We will be interested in asymptotic complexity of functions, so n should be thought of as a parameter describing "input size", and the function f as a sequence of functions, $\{f_n\}$, one for each value of n. Here are some functions we will deal with in this paper. Let $x = (x_1, x_2, ..., x_n) \in \{0, 1\}^n$.

- *parity*$(x) = 1$ iff $\sum_{i=1}^{n} x_i$ is odd.
- *majority*$(x) = 1$ iff $\sum_{i=1}^{n} > n/2$.
- *prime*$(x) = 1$ iff x is the binary representation of a prime integer.
- *stconn*, the st-connectivity function. Here x represents a graph (binary relation) on a set V, with $|V| = m$, $n = m^2$. So x can be thought of as an $m * m$ matrix A with $A_{ij} = 1$ $(i, j \in V)$ iff there is a direct connection (an *edge*) from i to j. For fixed $s, t \in V$, *stconn*$(x) = 1$ iff there is a path from s to t in this graph (i.e. if s and t are related in the reflexive transitive closure of the relation A).
- *clique*. Here x represents an undirected graph on m vertices, so $n = m^2$ and $x_{ij} = x_{ji}$. A clique is a subset U of the vertices such that every pair of distinct vertices $i, j \in U$ is connected by an edge. Now *clique*$(x) = 1$ iff the graph represented by x has a clique of size (say) $m/10$.
- *matching*, the perfect matching function. Again x represents an $m * m$ matrix A with rows B and columns G with $A_{bg} = 1$ if boy $b \in B$ and girl $g \in G$ are willing to be matched. Now *matching*$(x) = 1$ iff all boys and girls can be simultanously matched. Equivalently, *matching*$(x) = 1$ iff the permanent of the matrix A is positive.

There is a natural partial order on the set of sequences $\{0, 1\}^n$, namely for sequences x, y we say that $x \leq y$ if for all $i \in [n]$ $x_i \leq y_i$. A Boolean function

f is called **monotone** if $f(x) \leq f(y)$ whenever $x \leq y$. For example, the functions *majority, stconn, clique* and *matching* above are monotone, but *parity* and *prime* are not.

A **Boolean circuit** over $\{x_1, x_2, ..., x_n\}$ is a sequence of functions $g_1, g_2, ..., g_s$ such that for every $k \in [s]$ either:

1. $g_k = x_i$ for some $i \in [n]$.
2. $g_k = \bar{x}_i$ (the negation of x_i) for some $i \in [n]$.
3. $g_k = g_i \vee g_j$ for some $i, j < k$.
4. $g_k = g_i \wedge g_j$ for some $i, j < k$.

Functions of type 1 and 2 are called inputs of the circuit, those of type 3 and 4 are called gates, and the last function g_s is called the output of the circuit. Thus, a circuit computes a Boolean function of the variables in a natural way — its output. The **size** of the circuit is s, the number of functions. The **depth** is the length of the longest "path" from an input to the output. More precisely, the depth is defined to be $d(g_s)$, where $d(g_k) = 0$ if g_k is an input, and $d(g_k) = 1 + \max(d(g_i), d(g_j))$ if g_k is a gate.

The **size** of a function f, denoted $s(f)$, is the size of the smallest circuit computing f. The **depth** of f, denoted $d(f)$, is the depth of the shallowest circuit computing f. In a vauge sense which can be formalized, the size corresponds to the sequential time to compute f, and the depth to the "parallel" time, as well as space, needed to compute f. The relationship between these two fundamental complexity measures is far from understood. It is trivial that $\log s(f) \leq d(f) \leq s(f)$ for every function f. A nontrivial construction of Paterson and Valiant [PV] slightly improves the right inequality to $d(f) \leq s(f)/\log s(f)$.

If we disallow negated inputs, i.e. functions of type 2 in our circuits, it is clear that they can now compute only monotone functions, and indeed we call such circuits **monotone circuits**. Every monotone function can be computed by a monotone circuit. We let $s^m(f)$ denote the smallest size of a monotone circuit for f, and similarly with $d^m(f)$ for monotone depth. The same weak relationship between monotone size and depth as above is the best that is known. To summerize:

Theorem 1 [PV].

$$\log s(f) \leq d(f) \leq s(f)/\log s(f) \tag{1}$$

$$\log s^m(f) \leq d^m(f) \leq s^m(f)/\log s^m(f)\,. \tag{2}$$

Finally, a **Boolean formula** is a circuit in which every function g_i can appear at most once on the right hand side of type 3 or 4. In other words, in a formula subfunctions need to be recomputed, and thus it looks like a tree. The **formula size** of a function f, denoted $L(f)$, is the size of the smallest formula for f, and similarly $L^m(f)$ stands for the monotone formula size. The depth of a circuit and a formula is clearly the same. Formula size and (circuit) depth are closely related:

Theorem 2 [Sp].

$$d(f) = \Theta(\log L(f)) \tag{3}$$

$$d^m(f) = \Theta(\log L^m(f))\,. \tag{4}$$

2.2 Communication Complexity

Let X, Y, Z be finite sets, and let $R \subseteq X \times Y \times Z$ be a relation. A *deterministic communication protocol* A over (X, Y, Z) specifies the exchange of information bits by two players, I and II, that initially receive as inputs $x \in X$ and $y \in Y$, respectively. The protocol dictates who sends the first message, and how a player computes his next message from his input and the previously received messages. It also dictates when they terminate, and how they compute a value $A(x, y) \in Z$. There is no limit on the complexity of these computations.

Denote by $c_A(x, y)$ the number of bits exchanged by I and II on the input pair (x, y) when using protocol A. Let $c_A(R) = \max_{(x,y)\in X\times Y} c_A(x, y)$.

We say that A *computes* R if for all $(x, y) \in X \times Y$ we have $(x, y, A(x, y)) \in R$. Then the *deterministic communication complexity* of the relation R is $c(R) = \min\{c_A(R)|A \text{ computes } R\}$. When R is clear from the context, we sometimes use $c(X', Y')$ for the communication complexity of R restricted to the subdomain $X' \times Y'$, where $X' \subseteq X$ and $Y' \subseteq Y$.

The original model for communication complexity, as introduced by Yao [Y], dealt only with the computation of functions by the two players (i.e. for every pair of inputs there is a unique output). We accomodate this common notion of computing functions (rather than relations) in a natural way. If R has the property that for every (x, y) there is a unique $z(x, y) \in Z$ with $(x, y, z) \in R$, then we identify the relation R with the function $R : X \times Y \to Z$ where $R(x, y) = z(x, y)$.

The main advantage in generalising the model to compute relations, is that now we can relate it to the circuit model.

3. Boolean Relations and Circuit Depth

3.1 Unrestricted Computation

The study of relations (search problems) was initiated by [KW], who showed that the circuit depth of a boolean function is exactly captured by the communication complexity of a related relation.

Specifically, for $f : \{0,1\}^n \to \{0,1\}$ define the relation

$$R_f \subset f^{-1}(1) \times f^{-1}(0) \times [n]$$

by $(x, y, i) \in R_f$ iff $x_i \neq y_i$. In words, player I gets an input $x \in \{0,1\}^n$ with $f(x) = 1$, player II gets $y \in \{0,1\}^n$ with $f(y) = 0$, and their task is to find a coordinate where their inputs differ. Note that this search problem is somewhat different than the one defined in the introduction, where the players' task was to find a small prime rather than a coordinate. However, this is just a matter of representation of integers (either binary notation or modular notation), and since there is an efficient conversion between the two, they are essentially equivalent. The heart of this research is the following theorem.

Theorem 3 [KW]. *For every function* $f : \{0,1\}^n \to \{0,1\}$ *we have*

$$d(f) = c(R_f)\,. \tag{5}$$

The simple proof of this theorem follows from the fact that there is a one-to-one correspondance between formulæ for f and communication protocols for the relation R_f. We illustrate this by an example. Consider the parity function *parity* on n variables. We will assume that n is a power of two, and denote the complement of this function by $\overline{parity}$. Also, for $x \in \{0,1\}^n$, let $x_L, x_R \in \{0,1\}^{n/2}$ denote respectively the left and right half-portions of the sequence x.

A simple recursive construction of a formula for $parity(x)$ (and $\overline{parity}(x)$ by de-Morgan's rules) is as follows:

If $n > 1$ then $parity(x) = (parity(x_L) \wedge \overline{parity}(x_R)) \vee (\overline{parity}(x_L) \wedge parity(x_R))$.
If $n = 1$ then x is a single variable x_i, and $parity(x_i) = x_i$.

Reducing problem size by a factor of two costs depth two ($\vee$ of $\wedge$'s) so this formula has depth $2\log n$ (and size n^2).

A protocol for R_{parity} can also be described recursively. Recall that by definition player I has an odd sequence x ($parity(x) = 1$), and player II has an even sequence y, and they want to find a coordinate where x and y differ.

If $n > 1$ then player I sends the bit $parity(x_L)$ and player II responds by sending to player I the bit $parity(y_L)$. If these two bits are different, then the players can discard the right portion, and recursively find a coordinate in which x_L and y_L differ. If the two bits are the same, this implies $parity(x_R) \neq parity(y_R)$, and they can recursively continue with the right portions.
If $n = 1$, then the index of this bit is the required coordinate.

Again, with two bits the problem size is reduced by a factor of two, so the communication complexity is $2\log n$.

Both the formula and the protocol can be viewed as binary trees. In the formula the boolean values propagate from the inputs to the output via the Boolean gates at the nodes. In the protocol the players move along the tree from output to leaves according to their input values. These two object are essentially the same, if we syntactically identify player I with tree nodes labeled by an $\vee$ gate, and player II with the $\wedge$ gate. The formal proof of Theorem 3 below (in two lemmas) follows this simple idea.

Lemma 4. *For all functions f and all $B_0, B_1 \subseteq \{0,1\}^n$ such that $B_0 \subseteq f^{-1}(0)$ and $B_1 \subseteq f^{-1}(1)$ we have*

$$C(B_1, B_0) \le d(f)\,. \tag{6}$$

Proof. By induction on $d(f)$.

If $d(f) = 0$ then f is either x_i or $\bar{x}_i$. In either case, we have that for all $x \in B_1$ and $y \in B_0$, $x_i \neq y_i$ so that i is always an answer and $C(B_1, B_0) = 0$.

For the induction step we suppose that $f = f_1 \wedge f_2$ (the case $f = f_1 \vee f_2$ is treated similarly) so that $d(f) = \max(d(f_1), d(f_2)) + 1$. Let $B_0^j = B_0 \cap f_j^{-1}(0)$ for $j = 1, 2$. By induction we have that $C(B_1, B_0^j) \le d(f_j)$ for $j = 1, 2$. Consider the following protocol for B_1 and B_0: II sends a 0 if $y \in B_0^1$, otherwise he sends a 1; the players then follow the best protocol for each of the subcases. We have

$$C(B_1, B_0) \le 1 + \max_{j=1,2}(C(B_1, B_0^j)) \le 1 + \max_{j=1,2}(d(f_j)) = d(f)\,. \tag{7}$$

The converse is as follows, and its similar proof is omitted.

Lemma 5. *Let $B_0, B_1 \subseteq B_n$ such that $B_0 \cap B_1 = \emptyset$. Then, there exists a function f with $B_0 \subseteq f(0)$ and $B_1 \subseteq f(1)$ such that*

$$d(f) \leq C(B_1, B_0)\,. \tag{8}$$

3.2 Monotone Computation

We conclude this section with the analog of Theorem 3 for monotone computation. Let $f : \{0,1\}^n \to \{0,1\}$ be a monotone function, and define the relation

$$R_f^m \subset f^{-1}(1) \times f^{-1}(0) \times [n]$$

by $(x, y, i) \in R_f$ iff $x_i > y_i$. In words, player I gets an input $x \in \{0,1\}^n$ with $f(x) = 1$, player II gets $y \in \{0,1\}^n$ with $f(y) = 0$, and their task is to find a coordinate i where $x_i = 1$ and $y_i = 0$. The existance of such a coordinate is guaranteed by the monotonicity of f. This relation captures exactly the depth of monotone circuits for f.

Theorem 6 [KW]. *For every monotone function $f : \{0,1\}^n \to \{0,1\}$ we have*

$$d^m(f) = c(R_f^m)\,. \tag{9}$$

The proof is essentially the same as the previous one, with a difference in the base case of the induction — the leaves of the tree — which is the only place monotone and nonmonotone computations differ.

It is very convenient to reformulate Theorem 6 in the following way. A **minterm** (or **minimal 1-witness**) of a monotone function f is a minimal subset $S \subseteq [n]$ such that any input x with $x_i = 1$ for all $i \in S$ satisfies $f(x) = 1$. Let MIN(f) be the set of all minterms of f. It is easy to see that MIN(f) characterizes f.

Similarly, one can define a **maxterm** (or **minimal 0-witness**) as any minimal subset of the variables which, if set to 0, force the function value to 0. Let MAX(f) denote the set of all maxterms of f.

It is easy to see that every minterm must intersect every maxterm (otherwise f can be simultaniously be forced to be 0 and 1 on the same input). Also, monotone computation of f does not become easier if we restrict ourselves to inputs which are defined only by minterms and maxterms. Thus the task of the players in R_f^m turns out to be: Player I gets a minterm $S \in$ MIN(f), player II gets a maxterm $T \in$ MAX(f), and they must find an element of the intersection $S \cap T$. And Theorem 6 becomes:

Theorem 7. *For every monotone function f we have*

$$d^m(f) = c(\mathrm{MIN}(f), \mathrm{MAX}(f))\,. \tag{10}$$

As an example, consider the *majority* function on n bits, say with $n = 2k - 1$, an odd number. Then

$$\text{MIN}(majority) = \text{MAX}(majority) = \{S \in [n] : |S| = k\}.$$

Clearly, by the pigeonhole principle every minterm and maxterm intersect, and the players should find a member of the intersection. A recursive procedure similar to the one describes above yields only an upper bound of $O((\log n)^2)$ bits (the reader may wish to find this protocol). It is a remarkable result, which was given two different beautiful proofs, that one can do much better!

Theorem 8 [AKS, V].

$$d^m(majority) = O(\log n).$$

Even though the same upper bound holds for the communication complexity of the above relation, the two proofs were given in terms of circuits. They are, in some inherent way, construct the circuit in a bottom-up fashion (from inputs to output), and it is hard to translate them into a simple protocol. In the remaining sections we will see that in other cases it is much easier to handle the communication problem, working top-down on the protocol, especially for lower bound purposes.

4. Applications

4.1 Size vs. Depth

Shanon, who initiated circuit complexity over 40 years ago, observed that simple counting arguments show that most functions are difficult to compute.

Theorem 9 [Sh]. *Almost all Boolean functions on n variables f satisfy*

$$s(f) = \Omega(2^n/n) \tag{11}$$

$$d(f) = \Omega(n). \tag{12}$$

On the other hand, there is no known sequence of functions (that can be explicitely described) f for which $s(f) \neq O(n)$ or $d(f) \neq O(\log n)$. Finding such a function is a major problem of theoretical computer science.

With this problem seeming too difficult for present techniques, the monotone analoge of this problem was attacked. Nevertheless, explicit monotone functions f for which $s^m(f)$ is super linear or $d^m(f)$ is super logarithmic were not discovered till 1985. That year, in a breakthrough paper Razborov [R1] proved a **super-polynomial** monotone size lower bound on the clique function above, which is known to be in the class NP.

Theorem 10 [R1].

$$s^m(clique) = n^{\Omega(\log n)}.$$

Soon after, exponential lower bounds were given for functions in NP by Andreev [A] and by Alon and Boppana [AB]. In particular, the later improved the clique lower bound to

Theorem 11 [AB].

$$s^m(clique) = \exp(n^{1/6}).$$

These results, together with theorem 1 immediately imply a super logarithmic lower bound on monotone depth.

Corollary 12.

$$d^m(clique) = \Omega(n^{1/6}).$$

As mentioned in the discussion preceding theorem 1 , it is not known if the logarithmic relashionship between size and depth is tight. This means that we are looking for a depth lower bound which is superlogarithmic in the size. The first application of the communication complexity viewpoint enabled to find such a result for the *st*-connectivity function, *stconn*.

There is a natural monotone circuit for this function, that is based on raising the matrix describing the graph to the mth power, by repeatedly squaring it $\log m$ times. Each such step takes size $O(m^3) = O(n^{3/2})$ and depth $O(\log m) = O(\log n)$, so we get $s^m(stconn) = O(n^{3/2} \log n)$, and $d^m(stconn) = O((\log n)^2)$ Thus, in this circuit, depth is quadratic in the logarithm of the size. Karchmer and Wigderson [KW] proved that this depth bound is optimal regardless of size.

Theorem 13 [KW].

$$d^m(stconn) = \Omega((\log n)^2).$$

The proof of this theorem draws intuition from the communication protocol for R^m_{stconn} which corresponds to the above circuit. It uses probabilistic arguments to show that after $O(\log n)$ bits of communication sent by the players in any protocol, one can interpret the remaining steps of the protocol as solving an *st*-connectivity problem of roughly half the size. For more detail, we refer the reader to the original paper. We just note here, that this direction, working from the output of the circuit down to the inputs is natural in the communication complexity framework, but rare in existing circuit lower bounds.

Together with (4) of Theorem 2, we conclude that there is a superpolynomial gap between circuit and formula size in the monotone model.

Corollary 14.

$$L^m(stconn) = n^{\Omega(\log n)}.$$

We end this section by noting that it is not known if this superpolynomial gap can be improved to exponential (for any monotone function, not neccessarily explicitely given).

4.2 Monotone vs. Nonmonotone

Once we are able to prove monotone lower bounds, it is interesting to ask if negations can help when computing a monotone function. The *clique* function is unlikely to provide the answer, as being NP-complete, we expect an exponential size lower bound even for nonmonotone circuits.

The perfect matching function *match*, on the other hand, has a simple polynomial time algorithm and therefore a polynomial size (nonmonotone) circuit. Razborov [R2] was able to apply his methods to this function as well, and prove the first superpolynomial gap between monotone and nonmonotone computation.

Theorem 15 [R2].

$$s^m(match) = n^{\Omega(\log n)} .$$

The methods of Alon and Boppana [AB] which improved the clique lower bound to exponential failed to improve the perfect matching bound. However, E. Tardos [T] found that their arguments would work for another function, which while being very similar to *clique*, has a polynomial time algorithm, and thus established an exponential gap between monotone and nonmonotone size.

The analogous question for depth was resolved last year by Raz and Wigderson [RW2]. Again, the function that examplifies the gap is the perfect matching function *match*. A sequence of results on this problem and parallel algorithms for it led Borodin, von zur Gathen and Hopcroft [BGH] to the following depth upper bound.

Theorem 16 [BGH].

$$d(match) = O((\log n)^2) .$$

Note that the monotone depth lower bound that follows from taking the logarithm in Theorem 15 only matches this upper bound. However, the communication complexity approach provided a depth bound which is independent of the size:

Theorem 17 [RW2].

$$d^m(match) = \Omega(m) = \Omega(\sqrt{n}) .$$

The proof of this theorem utilizes the communication complexity approach in a different way. We show that the communication problem R^m_{match} encodes another communication problem, called set disjointness. In this problem, each of the two players is given a subset of $\{1, 2, ..., m\}$, and their task is to find if these two subsets are disjoint or not.

More precisely, any protocol for R^m_{match} using c communication bits would lead to a **probabilistic** protocol for set disjointness which uses only $O(c)$ bits on average. However, a difficult result of Kalyanasundaram and Schnitger [KS] shows that this is possible only if $c = \Omega(m)$, from which our theorem follows.

For the exact definitions of probabilistic communication complexity, as well as the probabilistic *reduction* which establish the relathionship above we refer the reader to [RW2].

We note that a straightforward reduction from the *clique* to *match* establishes a better depth lower bound for *clique* than the one which followes from the size lower bound of Alon and Boppana in Corollary 12.

Corollary 18 [RW2].

$$d^m(clique) = \Omega(m) = \Omega(\sqrt{n}).$$

5. Conclusions

We feel that the communication complexity approach to circuit lower bounds was fruitful for a few reasons. The lack of computation in that model. The ability to view computations top-down; this simplifies several knwon upper and lower bounds (see [K]). And last, but not least, the existing techniques and results for this model that can be used via reductions.

This paper was not intended as a complete survey, and indeed some new results were recently obtained. However, there are still no nontrivial nonmonotone lower bounds, and we believe that the communication complexity approach will play a role in changing this state of affairs!

Acknowledgements. The work described in this paper was done in an exciting and enlightening collaboration with Mauricio Karchmer and Ran Raz.

References

[A] Andreev, A.E.: On a method for obtaining lower bounds on the complexity of individual monotone functions. Dokl. Ak. Nauk. SSSR **282** (1985) 1033–1037 (in Russian). [English translation in: Sov. Math. Dokl. **31** (1985) 530–534]

[AB] Alon, A., Boppana, R.: The monotone circuit complexity of Boolean functions. Combinatorica **7**, no. 1 (1987) 1–22

[AKS] Ajtai, M., Komlós, J., Szemerédi, E.: An $O(n \log n)$ sorting network. Proceedings 15th STOC 1983, pp. 1–9

[BFS] Babai, L., Frankel, P., Simon, J.: Complexity classes in communication complexity theory. Proceedings of the 27th FOCS 1988, pp. 337–347

[BGH] Borodin, A., von zur Gathen, J., Hopcroft, J.: Fast parallel matrix and GCD Computations. Proceedings of the 23rd STOC 1982, pp. 65–71

[BS] Boppana, R., Sipser, M.: The complexity of finite functions. MIT/LCS/TM-405 (1989)

[D] Dunne, P.E.: The complexity of Boolean networks. Academic Press, 1988

[K] Karchmer, M.: Communication complexity — A new approach to circuit depth. MIT Press, 1989

[KS] Kalyanasundaram, B., Snitger, G.: The probabilistic communication complexity of set intersection. Proceedings Structure in Complexity Theory 1987, pp. 41–49

[KW] Karchmer, M., Wigderson, A.: Monotone circuits for connectivity require super-logarithmic depth. Proceedings of the 20th STOC 1988, pp. 539–550

[PV] Paterson, M., Valiant, L.G.: Circuit size is non-linear in depth. TCS **2** (1976) 397–400

[R1] Razborov, A.A.: Lower bounds for the monotone complexity of some boolean functions. Dokl. Ak. Nauk. SSSR **281** (1985) 798–801 (in Russian) [English transl. in: Sov. Math. Dokl. **31** (1985) 354–357]

[R2] Razborov, A.A.: Lower bounds on the monotone network complexity of the logical permanent. Mat. Zametki **37** (1985) 887–900 (in Russian) [English transl. in: Math. Notes of the Academy of Sciences of USSR **37** (1985) 485–493]

[RW1] Raz, R., Wigderson, A.: Probabilistic communication complexity of boolean relations. Proc. of the 30th FOCS 1989

[RW2] Raz, R., Wigderson, A.: Monotone circuits for matching require linear depth. Proccedings of the 22th STOC 1990

[Sh] Shannon, C.: The synthesis of two-terminal switching circuits. Bell Syst. Techn. J. **28** (1949) 59–98

[Sp] Spira, P.M.: On time-hardware complexity tradeoffs for Boolean functions. Proceedings of 4th Hawaii Symp. on System Sciences 1971, pp. 525–527

[T] Tardos, E.: The gap between monotone and non-monotone circuit complexity is exponential. Combinatorica **8** (1988) 141–142

[V] Valiant, L.G.: Short monotone formulæ for the majority function. J. Algorithms **5** (1984) 363–366

[W] Wegner, I.: The complexity of boolean functions. John Wiley, 1988.

[Y] Yao, A.C.-C.: Some complexity questions related to distributive computing. Proceedings of 11th STOC 1979, pp. 209–213

Recent Developments in Shock-Capturing Schemes

Ami Harten

School of Mathematical Sciences, Tel Aviv University, Ramat Aviv, Tel Aviv, Israel, and Department of Mathematics, University of California, Los Angeles, CA 90024-1555, USA

Abstract. In this paper we review the development of the shock-capturing methodology, paying special attention to the increasing nonlinearity in its design and its relation to interpolation. It is well-known that high-order approximations to a discontinuous function generate spurious oscillations near the discontinuity (Gibbs phenomenon). Unlike standard finite-difference methods which use a fixed stencil, modern shock-capturing schemes use an adaptive stencil which is selected according to the local smoothness of the solution. Near discontinuities this technique automatically switches to one-sided approximations, thus avoiding the use of discontinuous data which brings about spurious oscillations.

1. Introduction

In this paper, we describe and analyze numerical techniques that are designed to approximate weak solutions of hyperbolic systems of conservation laws in several space dimensions. For sake of exposition, we shall describe these methods as they apply to the pure initial value problem (IVP) for a one-dimensional scalar conservation law

$$u_t + f(u)_x = 0, \quad u(x,0) = u_0(x)\,. \tag{1.1}$$

To further simplify our presentation, we assume that the flux $f(u)$ is a convex function, i.e., $f''(u) > 0$ and that the initial data $u_0(x)$ are piecewise smooth functions which are either periodic or of compact support. Under these assumptions, no matter how smooth u_0 is, the solution $u(x,t)$ of the IVP (1.1) becomes discontinuous at some finite time $t = t_c$. In order to extend the solution for $t > t_c$, we introduce the notion of weak solutions, which satisfy

$$\frac{d}{dt}\int_a^b u\,dx + f(u(b,t)) - f(u(a,t)) = 0 \tag{1.2a}$$

for all $b \geq a$ and $t \geq 0$. Relation (1.2a) implies that $u(x,t)$ satisfies the PDE in (1.1) wherever it is smooth, and the Rankine-Hugoniot jump relation

$$f(u(y+0,t)) - f(u(y-0,t)) = [u(y+0,t) - u(y-0,t)]\frac{dy}{dt} \tag{1.2b}$$

across curves $x = y(t)$ of discontinuity.

Proceedings of the International Congress of Mathematicians, Kyoto, Japan, 1990

It is well-known that weak solutions are not uniquely determined by their initial data. To overcome this difficulty, we consider the IVP (1.1) to be the vanishing viscosity limit $\varepsilon \downarrow 0$ of the parabolic problem

$$(u^\varepsilon)_t + f(u^\varepsilon)_x = \varepsilon u^\varepsilon_{xx} \quad u^\varepsilon(x,0) = u_0(x) \ , \tag{1.3a}$$

and identify the unique "physically relevant" weak solution of (1.1) by

$$u = \lim_{\varepsilon \downarrow 0} u^\varepsilon \ . \tag{1.3b}$$

The limit solution (1.3) can be characterized by an inequality that the values $u_L = u(y-0,t), u_R = u(y+0,t)$ and $s = dy/dt$ have to satisfy; this inequality is called an entropy condition; admissible discontinuities are called shocks. When $f(u)$ is convex, this inequality is equivalent to Lax's shock condition

$$a(u_L) > s > a(u_R) \tag{1.4}$$

where $a(u) = f'(u)$ is the characteristic speed (see [8] for more details).

We turn now to describe finite difference approximations for the numerical solution of the IVP (1.1). Let v_j^n denote the numerical approximation to $u(x_j, t_n)$ where $x_j = jh, t_n = n\tau$; let $v_h(x,t)$ be a globally defined numerical approximation associated with the discrete values $\{v_j^n\}, -\infty < j < \infty, n \geq 0$.

The classical approach to the design of numerical methods for partial differential equations is to obtain a solvable set of equations for $\{v_j^n\}$ by replacing derivatives in the PDE by appropriate discrete approximations. Therefore, there is a conceptual difficulty in applying classical methods to compute solutions which may become discontinuous. Lax and Wendroff [9] overcame this difficulty by considering numerical approximations to the *weak formulation* (1.2a) rather than to the PDE (1.1). For this purpose, they have introduced the notion of schemes in conservation form:

$$v_j^{n+1} = v_j^n - \lambda(\bar{f}_{j+\frac{1}{2}} - \bar{f}_{j-\frac{1}{2}}) \equiv (E_h \cdot v^n)_j \ ; \tag{1.5a}$$

here $\lambda = \tau/h$ and $\bar{f}_{i+\frac{1}{2}}$ denotes

$$\bar{f}_{i+\frac{1}{2}} = f(v_{i-k+1}^n, \ldots, v_{i+k}^n) \ ; \tag{1.5b}$$

$\bar{f}(w_1, \ldots, w_{2k})$ is a numerical flux function which is consistent with the flux $f(u)$, in the sense that

$$\bar{f}(u, u, \ldots u) = f(u) \ ; \tag{1.5c}$$

E_h denotes the numerical solution operator. Lax and Wendroff proved that if the numerical approximation converges boundedly almost everywhere to some function u, then u is a weak solution of (1.1), i.e., it satisfies the weak formulation (1.2a). Consequently discontinuities in the limit solution automatically satisfy the Rankine-Hugoniot relation (1.2b). We refer to this methodology as shock-capturing (a phrase coined by H. Lomax).

In the following, we list the numerical flux function of various 3-point schemes ($k = 1$ in (1.5b)):

(i) The Lax-Friedrichs scheme [7]

$$\bar{f}(w_1, w_2) = \frac{1}{2}[f(w_1) + f(w_2) - \frac{1}{\lambda}(w_2 - w_1)] \tag{1.6}$$

(ii) Godunov's scheme [1]

$$\bar{f}(w_1, w_2) = f(V(0; w_1, w_2)) \ ; \tag{1.7a}$$

here $V(x/t; w_1, w_2)$ denotes the self-similar solution of the IVP (1.1) with the initial data

$$u_0(x) = \begin{cases} w_1 & x < 0 \\ w_2 & x > 0 \end{cases} . \tag{1.7b}$$

(iii) The Cole-Murman scheme [12]:

$$\bar{f}(w_1, w_2) = \frac{1}{2}[f(w_1) + f(w_2) - |\bar{a}(w_1, w_2)|(w_2 - w_1)] \tag{1.8a}$$

where

$$\bar{a}(w_1, w_2) = \begin{cases} \frac{f(w_2) - f(w_1)}{w_2 - w_1} & \text{if } w_1 \neq w_2 \\ a(w_1) & \text{if } w_1 = w_2 \end{cases} . \tag{1.8b}$$

(iv) The Lax-Wendroff scheme [9]:

$$\bar{f}(w_1, w_2) = \frac{1}{2}\{f(w_1) + f(w_2) - \lambda a\left(\frac{w_1 + w_2}{2}\right)[f(w_2) - f(w_1)]\} . \tag{1.9}$$

Let $E(t)$ denote the evolution operator of the exact solution of (1.1) and let E_h denote the numerical solution operator defined by the RHS of (1.5a). We say that the numerical scheme is r-th order accurate (in a pointwise sense) if its local truncation error satisfies

$$E(\tau) \cdot u - E_h \cdot u = 0(h^{r+1}) \tag{1.10}$$

for all sufficiently smooth u; here $\tau = 0(h)$. If $r > 0$, we say that the scheme is consistent.

The schemes of Lax-Friedrichs (1.6), Godunov (1.7), and Cole-Murman (1.8) are first order accurate; the scheme of Lax-Wendroff (1.9) is second order accurate.

We remark that the Lax-Wendroff theorem states that if the scheme is convergent, then the limit solution satisfies the weak formulation (1.2b); however, it need not be the entropy solution of the problem (see [4]). It is easy to see that the schemes of Cole-Murman (1.8) and Lax-Wendroff (1.9) admit a stationary "expansion shock" (i.e., $f(u_L) = f(u_R)$ with $a(u_L) < a(u_R)$) as a steady solution. This problem can be easily rectified by adding sufficient numerical dissipation to the scheme (see [11] and [3]).

2. Interpolatory Schemes and Linear Discontinuities

Let us consider the constant coefficient case $f(u) = au, a = \text{const.}$ in (1.1), i.e.,

$$u_t + au_x = 0, \quad u(x,0) = u_0(x)\,, \tag{2.1a}$$

the solution to which is

$$u(x,t) = u_0(x - at)\,. \tag{2.1b}$$

In this case the schemes (1.6) – (1.9) take the form

$$v_j^{n+1} = \sum_{\ell=-K^-}^{K^+} C_\ell(v) v_{j+\ell}^n \equiv (E_h \cdot v^n)_j \tag{2.2}$$

where $v = \lambda a$ is the CFL number. The coefficients $C_\ell(v)$ are independent of the numerical solution v^n; this makes E_h a linear operator.

We say that the numerical scheme E_h is (linearly) stable if

$$\|(E_h)^n\| \leq C \quad \text{for} \quad 0 \leq n\tau \leq T, \quad \tau = O(h)\,. \tag{2.3a}$$

In the constant coefficient case the scheme is stable if and only if it satisfies von Neumann's condition

$$\left| \sum_{\ell=-K^-}^{K^+} C_\ell(v) \ell^{i\ell\xi} \right| \leq 1 \quad \text{for all} \quad 0 \leq \xi \leq \pi\,. \tag{2.3b}$$

It is easy to see that all the 3 point schemes (1.6) – (1.9) are stable under the CFL condition

$$|v| = |\lambda a| \leq 1\,. \tag{2.3c}$$

The notion of stability (2.3a) is related to convergence through Lax's equivalence theorem, which states that a consistent linear scheme is convergent if and only if it is stable (see [13] for more details).

Let us denote by S_i^r the stencil of $(r+1)$ successive points starting with x_i

$$S_i^r = \{x_i, x_{i+1}, \ldots, x_{i+r}\}\,, \tag{2.4a}$$

let $P(x; S_i^r, u)$ denote the unique polynomial of degree r interpolating the $(r+1)$ values of u on this stencil and let $Q(x;u)$ denote the piecewise polynomial interpolation of u

$$Q(x;u) = P(x; S_{i(j)}^r; u) \quad x_{j-1} \leq x \leq x_j\,. \tag{2.4b}$$

We refer to the numerical scheme

$$v_j^{n+1} = Q(x_j - a\tau; v^n) \tag{2.4c}$$

as interpolatory scheme. Clearly, the interpolatory scheme (2.4) is r-th order accurate. When $Q(x;v^n)$ is the piecewise linear interpolation of v^n (i.e., $r = 1, i(j) = j-1$ in (2.4b)) then (2.4c) is the first-order accurate upwind scheme; in the constant coefficient case this scheme is identical to those of Godunov (1.7) and Cole-Murman (1.8).

Next let us assume $a > 0$ and consider the second order case $r = 2$ in which $Q(x; v^n)$ is a piecewise-parabolic interpolation of v^n. There are two different choices of stencil in (2.4): Taking Q in $[x_{j-1}, x_j]$ to be the parabola through $S^2_{j-1} = \{x_{j-1}, x_j, x_{j+1}\}$ (i.e., $i(j) = j - 1$) results in the Lax-Wendroff scheme (1.9); taking Q in $[x_{j-1}, x_j]$ to be the parabola through $S^2_{j-2} = \{x_{j-2}, x_{j-1}, x_j\}$ (i.e., $i(j) = j - 2$) results in the second-order upwind scheme.

We turn now to consider the application of these schemes to the step function $H(x)$

$$H(x) = \begin{cases} 0 & x \le 0 \\ 1 & x > 0 \end{cases}, H_j = \begin{cases} 0 & j \le 0 \\ 1 & j \ge 1 \end{cases}. \tag{2.5a}$$

For the first order upwind scheme we get that

$$Q(x; H) = \begin{cases} 0 & x \le 0 \\ x/h & 0 \le x \le h \\ 1 & h \le x \end{cases}; \tag{2.5b}$$

for the Lax-Wendroff scheme

$$Q(x; H) = \begin{cases} 0 & x \le -h \\ \frac{1}{2}\frac{x}{h}(1 + \frac{x}{h}) & -h \le x \le 0 \\ 1 - \frac{1}{2}(1 - \frac{x}{h})(2 - \frac{x}{h}) & 0 \le x \le h \\ 1 & h \le x \end{cases}; \tag{2.5c}$$

for the second order upwind scheme we get that

$$Q(x; H) = \begin{cases} 0 & x \le 0 \\ \frac{1}{2}\frac{x}{h}(1 + \frac{x}{h}) & 0 \le x \le h \\ 1 + \frac{1}{2}(\frac{x}{h} - 1)(2 - \frac{x}{h}) & h \le x \le 2h \\ 1 & 2h \le x \end{cases}. \tag{2.5d}$$

We observe that Q in (2.5b) is a monotone function of x; consequently the numerical solution by Godunov's scheme to these data is also monotone. On the other hand Q for the second order schemes (2.5c) – (2.5d) is not a monotone function. For the Lax-Wendroff scheme Q is negative in $-h \le x \le 0$ and has a minimum of -0.125; similarly for the second order upwind scheme Q is larger than 1 in $h \le x \le 2h$ with a maximum of 1.125. This observation explains the Gibbs-like phenomenon of generating spurious oscillations in calculating discontinuous data with these second order schemes.

We say that the scheme E_h is monotonicity preserving if

$$v \text{ monotone } \Rightarrow E_h \cdot v \text{ monotone}. \tag{2.6}$$

Clearly the numerical solution of a monotonicity preserving scheme to initial data of a step-function is always monotone and therefore the discontinuity propagates without generating spurious oscillations.

Godunov has shown that the *linear* scheme (2.2) is monotonicity preserving if and only if

$$C_\ell(v) \geq 0, \quad -K_- \leq \ell \leq K_+; \tag{2.7}$$

this implies that a monotonicity-preserving scheme which is linear is necessarily only first-order accurate. It took some time to realize the Godunov's monotonicity theorem does not mean that there are no high-order accurate monotonicity preserving schemes; it only means that there are no such linear ones. Hence high-order accurate monotonicity-preserving schemes are nonlinear in an essential way.

The second-order accurate schemes mentioned above are linear because the choice of the stencil (2.4) is fixed. Let us consider now a piecewise-quadratic interpolation which is made nonlinear by an adaptive selection of the stencil in (2.4b). For the interval $[x_{j-1}, x_j]$ let us consider the two stencils $S^2_{j-2} = \{x_{j-2}, x_{j-1}, x_j\}$ and $S^2_{j-1} = \{x_{j-1}, x_j, x_{j+1}\}$, and select the one in which the interpolant is smoother. If we measure the smoothness of u by the second derivative of the corresponding parabola we select

$$i(j) = \begin{cases} j-2 & \text{if } |\frac{d^2}{dx^2} P(x; S^2_{j-2}, u)| \leq |\frac{d^2}{dx^2} P(x; S^2_{j-1}, u)| \\ j-1 & \text{otherwise} \end{cases} . \tag{2.8a}$$

When we apply this selection of stencil to the step-function $H(x)$ (2.5a) we get that for $[x_{-1}, x_0]$ we choose the stencil $S^2_{-2} = \{x_{-2}, x_{-1}, x_0\}$ for which $P(x; S^2_{-2}, H) \equiv 0$; for the interval $[x_1, x_2]$ we choose the stencil $S^2_1 = \{x_1, x_2, x_3\}$ for which $P(x; S^2_1, H) \equiv 1$. As is evident from comparing (2.5c) and (2.5d) it does not matter which stencil we assign to $[x_0, x_1]$ since both parabolae are monotone there; with (2.8a) we select S^2_{-1} for $[x_0, x_1]$. Thus we get in (2.4)

$$Q(x; H) = \begin{cases} 0 & x \leq 0 \\ \frac{1}{2}\frac{x}{h}(1 + \frac{x}{h}) & 0 \leq x \leq h \\ 1 & h \leq x \end{cases} \tag{2.8b}$$

which is a monotone function of x although it is actually a piecewise-quadratic polynomial.

The use of an adaptive stencil is the main idea behind the Essentially Non-Oscillatory (ENO) schemes to be described later in this paper. It extends to high order of accuracy in a straightforward manner: For r-th order accuracy we consider for $[x_{j-1}, x_j]$ the r stencils $S^r_{j-r}, S^r_{j-r+1}, \ldots, S^r_{j-1}$. We choose $i(j)$ in (2.4b) to be the one which minimizes

$$\left| \frac{d^r}{dx^r} P(x; S^r_i, u) \right| \quad \text{for } i = j-r, \ldots, j-1 . \tag{2.9}$$

3. Total Variation Stability and TVD Schemes

An immense body of work has been done to find out whether stability of constant coefficient scheme with respect to all "frozen coefficients" associated with the problem, implies convergence in the variable coefficient case and in the nonlinear case.

In the variable coefficient case, where the numerical solution operator is linear and Lax's equivalence theorem holds, it comes out that the stability of the variable coefficient scheme depends strongly on the dissipativity of the constant coefficient one, i.e., on the particular way it damps the high-frequency components in the Fourier representation of the numerical solution.

In the nonlinear case, under assumptions of sufficient smoothness of the PDE, its solution and the functional definition of the numerical scheme, Strang proved that linear stability of the first variation of the scheme implies its convergence; we refer the reader to [13] for more details.

In the case of discontinuous solutions of nonlinear problems, linearly stable schemes are not necessarily convergent; when such a scheme fails to converge, we refer to this case as "nonlinear instability." The occurrence of a nonlinear instability is usually associated with insufficient numerical dissipation which triggers exponential growth of the high-frequency components of the numerical solution.

The following theorem states that a stronger sense of stability, namely uniform boundedness of the total variation of the numerical solution, does imply convergence to a weak solution.

Theorem 3.1. *Let v_h be a numerical solution of a conservative scheme* (1.5).

(i) If

$$\mathrm{TV}(v_h(\cdot, t)) \leq C \cdot \mathrm{TV}(u_0) \tag{3.1}$$

where TV( ) *denotes the total variation in x and C is a constant independent of h for $0 \leq t \leq T$, then any refinement sequence $h \to 0$ with $\tau = 0(h)$ has a convergent subsequence $h_j \to 0$ that converges in L_1^{loc} to a weak solution of* (1.1).

(ii) If v_h is consistent with an entropy inequality which implies uniqueness of the IVP (1.1), *then the scheme is convergent (i.e., all subsequences have the same limit, which is the unique entropy solution of the IVP* (1.1)).

We say that the scheme E_h is Total Variation Diminishing (TVD) if

$$\mathrm{TV}(E_h \cdot v) \leq \mathrm{TV}(v) \tag{3.2}$$

where

$$\mathrm{TV}(w) = \sum_j |w_{j+1} - w_j|. \tag{3.3}$$

Clearly TVD schemes satisfy (3.1) with $C = 1$ and therefore are TV stable.

In [2] we have shown that if the scheme can be written in the form

$$v_j^{n+1} = v_j^n + C_{j+\frac{1}{2}}^{+} \Delta_{j+\frac{1}{2}} v^n - C_{j-\frac{1}{2}}^{-} \Delta_{j-\frac{1}{2}} v^n \tag{3.4a}$$

where $C^{\pm}_{j+\frac{1}{2}}$ satisfy for all j

$$C^{\pm}_{j+\frac{1}{2}} \geq 0, \quad C^{+}_{j+\frac{1}{2}} + C^{-}_{j+\frac{1}{2}} \leq 1 \tag{3.4b}$$

then the scheme is TVD; here $\Delta_{i+\frac{1}{2}} v^n = v^n_{i+1} - v^n_i$. Applying this lemma to the general scheme

$$v_j^{n+1} = v_j^n - \lambda(\bar{f}_{j+\frac{1}{2}} - \bar{f}_{j-\frac{1}{2}}) \tag{3.5a}$$

$$\bar{f}_{j+\frac{1}{2}} = \frac{1}{2}(f_j + f_{j+1} - q_{j+\frac{1}{2}} \Delta_{j+\frac{1}{2}} v^n) \tag{3.5b}$$

we get that if λq satisfies

$$\lambda|\bar{a}_{j+\frac{1}{2}}| \leq \lambda q_{j+\frac{1}{2}} \leq 1 \tag{3.6a}$$

then the scheme (3.5) is TVD; here

$$\bar{a}_{j+\frac{1}{2}} = \frac{f_{j+1} - f_j}{\Delta_{j+\frac{1}{2}} v}. \tag{3.6b}$$

This shows that the Cole-Murman scheme (1.8) for which $q = |\bar{a}|$ is TVD subject to the CFL restriction $\lambda|\bar{a}_{j+\frac{1}{2}}| \leq 1$.

Using conditions (3.4b) it is possible to construct TVD schemes which are second-order accurate in the L_1-sense (see [2] and [14]). However, TVD schemes are at most second-order accurate (see [5]). In order to design higher-order accurate shock capturing schemes we introduce the notion of Essentially Non-Oscillatory (ENO) schemes.

4. ENO Schemes

In this section we describe high-order accurate Godunov-type schemes which are a generalization of Godunov's scheme (1.7) and van Leer's MUSCL scheme [10].

We start with some notations: Let $\{I_j\}$ be a partition of the real line; let $A(I)$ denote the interval-averaging (or "cell-averaging") operator

$$A(I) \cdot w = \frac{1}{|I|} \int_I w(y)dy; \tag{4.1}$$

let $\bar{w}_j = A(I_j) \cdot w$ and denote $\bar{w} = \{\bar{w}_j\}$. We denote the approximate reconstruction of $w(x)$ from its given cell-averages $\{\bar{w}_j\}$ by $R(x; \bar{w})$. To be precise, $R(x; \bar{w})$ is a piecewise-polynomial function of degree $(r-1)$, which satisfies

(i) $\quad R(x; \bar{w}) = w(x) + O(h^r)$ wherever w is smooth (4.2a)

(ii) $\quad A(I_j) \cdot R(\cdot; \bar{w}) = \bar{w}_j$ (conservation). (4.2b)

Finally, we define Godunov-type schemes by

$$v_j^{n+1} = A(I_j) \cdot E(\tau) \cdot R(\cdot\,; v^n) \equiv (\overline{E}_h \cdot v^n)_j \tag{4.3a}$$

$$v_j^0 = A(I_j) u_0\,; \tag{4.3b}$$

here $E(t)$ is the evolution operator of (1.1).

In the scalar case, both the cell-averaging operator $A(I_j)$ and the solution operator $E(\tau)$ are order-preserving, and consequently also total-variation diminishing (TVD); hence

$$\mathrm{TV}(\overline{E}_h \cdot \overline{w}) \leq \mathrm{TV}(R(\cdot\,; \overline{w}))\,. \tag{4.4}$$

This shows that the total variation of the numerical solution of Godunov-type schemes is dominated by that of the reconstruction step.

We turn now to describe the recently developed essentially non-oscillatory (ENO) schemes of [5, 6], which can be made accurate to any finite order r. These are Godunov-type schemes (4.3) in which the reconstruction $R(x; \overline{w})$, in addition to relations (4.2), also satisfies

$$\mathrm{TV}(R(\cdot\,; \overline{w})) \leq \mathrm{TV}(\overline{w}) + O(h^{1+p})\,, \quad p > 0 \tag{4.5}$$

for any piecewise-smooth function $w(x)$. Such a reconstruction is essentially non-oscillatory in the sense that it may not have a Gibbs-like phenomenon at jump-discontinuities of $w(x)$, which involves the generation of 0(1) spurious oscillations (that are proportional to the size of the jump); it can, however, have *small* spurious oscillations which are produced in the smooth part of $w(x)$, and are usually of the size $O(h^r)$ of the reconstruction error (4.2a).

When we use an essentially non-oscillatory reconstruction in a Godunov-type scheme, it follows form (4.4) and (4.5) that the resulting scheme (4.3) is likewise essentially non-oscillatory (ENO) in the sense that for all piecewise-smooth function $w(x)$

$$\mathrm{TV}(\overline{E}_h \cdot \overline{w}) \leq \mathrm{TV}(\overline{w}) + O(h^{1+p})\,, \quad p > 0; \tag{4.6}$$

i.e., it is "almost TVD." Property (4.6) makes it reasonable to believe that the total variation of the numerical solution is uniformly bounded. We recall that by Theorem 3.1, this would imply that the scheme is convergent (at least in the sense of having convergent subsequences). This hope is supported by a very large number of numerical experiments.

Next we describe one of the techniques to obtain an ENO reconstruction. To simplify our presentation we assume that $\{I_j\}$ is a uniform partition

$$I_j = (x_{j-1}, x_j)\,, \quad x_j = jh\,.$$

Given cell averages $\{\overline{w}_j\}$ of piecewise-smooth function $w(x)$, we observe that

$$h\overline{w}_j = \int_{x_{j-1}}^{x_j} w(y)dy = W(x_j) - W(x_{j-1}) \tag{4.7a}$$

where

$$W(x) = \int_{x_0}^{x} w(y)dy \tag{4.7b}$$

is the primitive function of $w(x)$. Hence we can easily compute the point values $\{W(x_j)\}$ by summation

$$W(x_i) = h \sum_{j=i_0}^{i} \overline{w}_j . \tag{4.7c}$$

Once we have computed the point values of the primitive function we use the ENO interpolation technique (2.4), (2.9) to obtain $Q(x; W)$, an r-th order piecewise-polynomial interpolation of W, i.e.,

$$Q(x; W) = P(x; S^r_{i(j)}, W) \quad \text{for} \quad x_{j-1} \le x \le x_j \tag{4.8a}$$

where $P(x; S^r_i, W)$ is the unique r-th degree polynomial which interpolates W over the stencil $S^r_i = \{x_i, x_{i+1}, \ldots, x_{i+r}\}$, and $i(j)$ is chosen so that

$$\left| \frac{d^r}{dx^r} P(x; S^r_{i(j)}, W) \right| = \min_{j-r \le i \le j-1} \left| \frac{d^r}{dx^r} P(x; S^r_i, W) \right| . \tag{4.8b}$$

We define $R(x; \overline{w})$ by

$$R(x; \overline{w}) = \frac{d}{dx} Q(x; W) . \tag{4.9}$$

We observe that if $w(x)$ is smooth in (x_{j-1}, x_j) then for h sufficiently small the algorithm (4.8b) will select a stencil $S^r_{i(j)}$ in which $w(x)$ is smooth. It follows then from standard interpolation theorems that

$$R(x; \overline{w}) = \frac{d}{dx} P(x; S^r_{i(j)}, W) = \frac{d}{dx} W + O(h^r) = w(x) + O(h^r) \tag{4.10}$$

which is property (4.2a). Furthermore (4.10) holds in every interval except for those in which $w(x)$ has a discontinuity. As we have seen in the examples (2.5) and (2.8b) the Gibbs-phenomenon is associated with intervals near the discontinuity and not with the interval that contains the discontinuity. This is why the reconstruction (4.8) – (4.9) satisfies the ENO property (4.5); in [2] we show that the second-order accurate ENO scheme is actually TVD. The conservation property (4.2b) follows directly from the definition (4.9):

$$\begin{aligned} A(I_j)R(\cdot; \overline{w}) &= \frac{1}{h} \int_{x_{j-1}}^{x_j} \frac{d}{dx} Q(x; W) dx = \frac{1}{h} [Q(x_j; W) - Q(x_{j-1}; W)] \\ &= \frac{1}{h} [W(x_j) - W(x_{j-1})] = \overline{w}_j . \end{aligned} \tag{4.11}$$

The abstract scheme (4.3) can be written in the standard conservation form (1.5). To do so let us denote by $\tilde{v}(x, t)$ the solution in the small of the IVP

$$\begin{cases} (\tilde{v})_t + f(\tilde{v})_x = 0 \\ \tilde{v}(x, 0) = R(x; v^n) \end{cases}, \quad 0 \le t \le \tau \tag{4.12}$$

and integrate this PDE over $I_j \times [0, \tau]$; using the divergence theorem and (4.2b) we get that v^{n+1} in (4.3) can be expressed by

$$v^{n+1}_j = v^n_j - \lambda [\hat{f}_{j+\frac{1}{2}} - \hat{f}_{j-\frac{1}{2}}] \tag{4.13a}$$

where

$$\hat{f}_{j+\frac{1}{2}} = \frac{1}{\tau}\int_0^\tau f(\tilde{v}(x_j,t))dt \tag{4.13b}$$

In the first-order case the scheme (4.13) is identical to Godunov's scheme and the numerical flux (4.13b) can be expressed in a closed form by (1.7b). For higher order schemes we use a numerical flux which is an appropriate approximation to (4.13b) (see [6] for more details).

We remark that the ENO schemes are related to the interpolatory schemes of Sect. 2 as follows: In the constant coefficient case a fixed choice of stencil (i.e., $i(j)-j=$ constant in (4.8a)) results in the interpolatory scheme (2.4) corresponding to the same choice of stencil.

References

1. Godunov, S. K.: A difference scheme for numerical computation of discontinuous solutions of equations of fluid dynamics. Math. Sbornik **47**, 271–306 (1959) (in Russian)
2. Harten, A.: High resolution schemes for hyperbolic conservation laws. J. Comp. Phys. **49**, 357–393 (1983)
3. Harten, A., Hyman, J. M.: A self-adjusting grid for the computation of weak solutions of hyperbolic conservation laws. J. Comp. Phys. **50**, 235–269 (1983)
4. Harten, A., Hyman, J. M., Lax, P. D.: On finite-difference approximations and entropy conditions for shocks. Comm. Pure Appl. Math. **29**, 297–322 (1976)
5. Harten, A., Osher, S.: Uniformly high-order accurate non-oscillatory schemes, I. SIAM J. Numer. Anal. **24**, 279 (1987)
6. Harten, A., Engquist, B., Osher, S., Chakravarthy, S. R.: Uniformly high-order accurate non-oscillatory schemes, III. J. Comp. Phys. **71**, 231 (1987)
7. Lax, P. D.: Weak solutions of nonlinear hyperbolic equations and their numerical computation. Comm. Pure Appl. Math. **7**, 159–193 (1954)
8. Lax, P. D.: Hyperbolic systems of conservation laws and the mathematical theory of shock waves. Society for Industrial and Applied Mathematics, Philadelphia, 1972
9. Lax, P. D., Wendroff, B.: Systems of conservation laws. Comm. Pure Appl. Math. **13**, 217–237 (1960)
10. van Leer, B.: Towards the ultimate conservative difference schemes V. A second order sequel to Godunov's method. J. Comp. Phys. **32**, 101–136 (1979)
11. Majda, A., Osher, S.: Numerical viscosity and entropy condition. Comm. Pure Appl. Math. **32**, 797–838 (1979)
12. Murman, E. M.: Analysis of embedded shock waves calculated by relaxation methods. AIAA J. **12**, 626–633 (1974)
13. Richtmyer, R. D., Morton, K. W.: Difference methods for initial value problems. 2nd edn. Interscience-Wiley, New York 1967
14. Sweby, P. K.: High resolution schemes using flux limiters for hyperbolic conservation laws. SIAM J. Numer. Anal. **21**, 995–1011 (1984)

Undirected Multiflow Problems and Related Topics – Some Recent Developments and Results

Alexander V. Karzanov

Institute for System Studies, Academy of Sciences of USSR, 9, Prospect 60 Let Oktyabrya
117312 Moscow, USSR

Abstract. A multiflow (multicommodity flow), arising as a natural extension of the well-known notion of a network flow, is a popular object studied in linear programming and combinatorial optimization. We discuss some recent results on undirected multiflow problems and ideas behind them, concerning combinatorial and computational aspects, such as: (i) the existence of feasible and optimal solutions with small denominators, (ii) special solvabilility criteria and minimax relations, (iii) efficient solution algorithms, (iv) a relationship between multiflows and packings of cuts and metrics, and some others.

1. Preliminaries

Fractionality. We start with some basic notion. Suppose that $\mathscr{K}$ is a collection (a *class*) of linear programs P of the form: (i) find $x \in \mathbf{Q}^n$ satisfying $Ax \le b$ (the *feasibility problem*), or (ii) maximize (or minimize) $c^T x$ subject to $Ax \le b$, $x \in \mathbf{Q}^n$ (the *optimization problem*), where A is an integral $m \times n$-matrix and b (c) is an integral m-vector (n-vector). Define the *fractionality* $\varphi(P)$ of P to be the minimum positive integer k for which P has a rational feasible (optimal) solution x such that kx is integral; if P has no feasible (optimal) solution, we put $\varphi(P) := 1$. The fractionality $\varphi(\mathscr{K})$ of $\mathscr{K}$ is defined as $\sup\{\varphi(P) | P \in \mathscr{K}\}$; if $\varphi(\mathscr{K}) = \infty$ we say that $\mathscr{K}$ has unbounded fractionality.

In combinatorial applications, a class $\mathscr{K}$ usually describes the linear programming relaxation of a certain combinatorial problem, and determining $\varphi(\mathscr{K})$ can be rather difficult. For example, let A_G be the $\{0, 1\}$-matrix whose rows correspond to the edges $e \in E$ of a graph $G = (V, E)$ and the columns are the incidence vectors of the perfect matchings $M \subseteq E$ in G, that is, each vertex of G is covered by exactly one edge of M. Then the well-known conjecture of Berge-Fulkerson [Fu2] (cf. [Se3]) that "any bridgless cubic graph G can be covered by six perfect matchings so that each edge is covered twice" is equivalent to the assertion that $\varphi(\mathscr{K}) = 2$ where $\mathscr{K}$ is the set of programs $\max\{1^T x | A_G x \le 1, x \ge 0\}$ for such Gs (actually it is even unknown whether $\varphi(\mathscr{K})$ is finite).

Flows and Multiflows. We shall deal only with undirected flows and multiflows. By a *graph* we mean a finite undirected graph without multiple edges and loops; an edge $\{u, v\}$ will be denoted by uv.

Consider a *network* $N = (G, H, c)$ consisting of a (*supply*) graph $G = (V, E)$, a (*demand*) graph $H = (T, U)$ with $T \subseteq V$ and a nonnegative integer-valued

Proceedings of the International Congress
of Mathematicians, Kyoto, Japan, 1990

vector $c \in \mathbf{Z}_+^E$ (of edge *capacities*); we suppose that each vertex of H is covered by an edge. An $s-t$ *chain* in G is a subgraph $L = (V_L, E_L)$ of G with $V_L = \{s = v_0, v_1, \ldots, v_m = t\}$ and $E_L = \{v_i v_{i+1} | i = 1, \ldots, m\}$. Define $\mathscr{L} := \mathscr{L}(G, H)$ to be $\cup_{st \in U} \mathscr{L}(G, st)$ where $\mathscr{L}(G, st)$ is the set of $s-t$ chains in G. A *multiflow* in N is a nonnegative rational-valued function f on $\mathscr{L}$ satisfying the capacity constraints:

$$\zeta^f(e) := \sum (f(L) | L \in \mathscr{L}, e \in E_L) \leq c(e) \quad \text{for all } e \in E\,. \tag{1}$$

For $st \in U$ the restriction f_{st} of f to $\mathscr{L}(G, st)$ is called a *flow* between s and t of *value* $v(f_{st}) := \sum (f(L) | L \in \mathscr{L}(G, st))$. The *total* value $v(f)$ of f is $\sum_{st \in U} v(f_{st})$.

Three types of problems on multiflows are well-known.

(2) *Demand problem*, $D(G, H, c, d)$: given a vector $d \in \mathbf{Z}_+^U$ (of *demands*), find f satisfying $v(f_{st}) = d(st)$ for all $st \in U$ (or establish that such an f does not exist).

(3) *Maximization problem*, $M(G, H, c)$: find f with $v(f)$ maximum (a *maximum multiflow*).

(4) *Minimum cost multiflow problem*, $M(G, H, c, a)$: given a vector $a \in \mathbf{Z}_+^E$ (of edge *costs*), find a *maximum* multiflow f whose *total cost* $\sum_{e \in E} a(e)\zeta^f(e)$ is minimum (clearly this is reduced to a linear program).

An obvious necessary condition of solvability of (2) (that is, the existence of a required f) is the *cut condition*

$$\Delta(c, d, X) := c(\delta_G(X)) - d(\delta_H(X \cap T)) \geq 0 \quad \text{for all } X \subseteq V\,, \tag{5}$$

where $\delta(X') = \delta_{G'}(X')$ denotes the set of edges of a graph $G' = (V', E')$ with one endvertex in $X' \subseteq V'$ and the other in $V' \setminus X'$ (a "cut" in G'); and for $c' \in \mathbf{Q}^{E'}$ and $E'' \subseteq E'$, $c'(E'')$ stands for $\sum (c'(e) | e \in E'')$. When $|U| = 1$ (2)–(4) are specified to be the classical (single-commodity) demand, maximum and minimum cost maximum flow problems. Well-known results on flows are that if $|U| = 1$ then: (i) (5) is sufficient for solvability of (2), and (ii) (2) has an integral (feasible) solution whenever it is solvable [Me, FF]. Similarly, if $|U| = 1$ then (3) and (4) have integral optimal solutions. This can be written as $\varphi(D(H)) = \varphi(M(H)) = \varphi(C(H)) = 1$ for $|U| = 1$, where $D(H)$ is the set of problems $D(G, H, c, d)$ with fixed H and arbitrary G, c and d; $M(H)$ and $C(H)$ are defined in a similar way.

When G and H vary, determining $\varphi(P)$ for an "individual" problem P in (2), (3) or (4) seems to be rather difficult. However, as we shall see in Sects. 2–5, this task has been successfully accomplished for many classes $D(H)$, $M(H)$ and $C(H)$. In particular, the values $\varphi(C(H))$ have been found for all H, and there is only one H for which $\varphi(D(H))$ is still unknown. In Sect. 5 results on the fractionality for planar supply graphs G are discussed. For some other results on multiflows see [Fr].

The cut condition (5) gives a special solvability criterion for (2) in the case $|U| = 1$. This is equivalent to the minimax relation "the maximum value of a flow between vertices s and t is equal to the minimum capacity $c(\delta(X))$ of a cut $\delta(X)$ separating s and t" ($\delta(X)$ *separates* s and t if $|X \cap \{s, t\}| = 1$). We shall see that the bounded fractionality behaviour for $D(H)$ or $M(H)$ with other H's is also accompanied by appearance of certain special solvability criteria or minimax relations.

As to computational aspects, we are not going to present here a survey of various approaches developed to solve multiflow problems. The purpose is only to outline the idea of one general approach, the so-called "splitting-off" method. By applying this method one can prove constructively the existence of an integral or half-integral solutions for certain "difficult" problems of type (2) and (3), and develope efficient algorithms to solve them. Apparently splitting-off techniques appeared originally in [RW] in connection with the two-commodity flow problem.

Complexity. We briefly recall the notions of polynomial and strongly polynomial algorithms. For details, see, e.g., [GJ]. An algorithm is said to be *polynomial* if its running time (the number of bit operations) is bounded by a polynomial in the input size of the problem (the *input size* is the number of bits occurring in the input). Speaking of a strongly polynomial algorithm, one means that the input consists of two parts. The first part describes a "combinatorial structure", and the second part is a list of "numerical data". The *dimension* of the input is the amount of bits in description of this structure plus the amount of numerical data. For instance, the input size of $D(G, H, c, d)$ is $O(|V|^2 + \sum_{e \in E} \log_2(c(e) + 1) + \sum_{u \in U} \log_2(d(u) + 1))$ while its dimension is $O(|V|^2)$. A polynomial algorithm is *strongly polynomial* if, roughly speaking, it consists of elementary arithmetic operations and data transfers, the number of these operations is bounded by a polynomial in the dimension of the input, and the size of data arised in the algorithm is polynomially bounded in the input size.

Remarks. (i) The above definition of a flow in the "edge-chain" form is known to be equivalent to the usual "node-edge" definition of a flow as a function on directed edges satisfying conservation conditions [FF]. The first form is preferable for us to explain some combinatorial ideas and methods. Though this form leads to appearance of constraint matrices for (2)–(4) with possibly exponential in $|V|$ number of columns we support explicitly only chains ("columns") with non-zero values of a multiflow; this provides efficiency of algorithms mentioned below.

(ii) The problems (2)–(4), being stated in the node-edge form, can be solved by general-purpose polynomial algorithms like the ellipsoid method [Kh]. Moreover, there are polynomial algorithms of finding a basis solution [GLS] and even, by a method of Tardos [Ta], strongly polynomial algorithms (since the constraint matrix has entries $0, 1, -1$, and hence its size is polynomially bounded in $|V|$). However these general methods do not guarantee (at least for the problems (3) and (4)) that the denominators of the obtained (optimal) solution will not exceed the fractionality of a problem; this will be explained in Sect. 3.

Related Problems. There is a special kind of duality between multiflows and packings of cuts or metrics. This can be illustrated by the following example. Let $l \in \mathbf{Z}_+^E$ be a vector (of edge *lengths*) associated with a connected graph $G = (V, E)$. A simple fact is that for $s, t \in V$ there exist cuts $\delta(X_1), \ldots, \delta(X_k)$ in G such that: (i) $|\{i | e \in \delta(X_i)\}| \leq l(e)$ for all $e \in E$, and (ii) $|\{i | \delta(X_i)$ separates s and $t\}| = \text{dist}_l(s, t)$. Here $\text{dist}_l(u, v)$ is the minimum l-length $\sum(l(e) | e \in E_L)$ of an $s - t$ chain L in G. The problem of determining cuts satisfying (i)–(ii) is dual, in a sense, to (2) with $U = \{st\}$. In Sect. 6 we discuss similar problems on cuts and metrics and their relation to multiflows.

2. Demand Problem

Applying Farkas' lemma to (2) and making simple transformations one obtains the following *metrical criterion* [Lo2]: $D(G,H,c,d)$ is solvable if and only if

$$\sum_{e\in E} c(e)m(e) - \sum_{u\in U} d(u)m(u) \geq 0 \tag{6}$$

holds for each metric m on V such that

$$m \text{ is primitive and has an extremal graph } \Gamma \text{ with } E_\Gamma \subseteq U\,. \tag{7}$$

Here by a *metric* on V we mean a nonnegative rational-valued function m on the set of unordered pairs in V satisfying $m(xx) = 0$ and $m(xy) + m(yz) \geq m(xz)$ for any $x, y, z \in V$ (we use the term "metric" rather than "semimetric"); m is called *primitive* if $m' + m'' = m$, where m' and m'' are metrics, implies $m' = \lambda m$ for some λ; an *extremal graph* of m is a minimal graph $\Gamma = (V_\Gamma, E_\Gamma)$ with $V_\Gamma \subseteq V$ such that for any distinct $u, v \in V$ there is $st \in E_\Gamma$ for which $m(st) = m(su) + m(uv) + m(vt)$.

We say that a metric m on V *is induced* by a graph $Q = (V_Q, E_Q)$ if there is a mapping σ from V *onto* V_Q such that for $u, v \in V$, $m(uv)$ is equal to the distance in Q between $\sigma(u)$ and $\sigma(v)$. A simplest example of a primitive metric gives that induced by the graph K_2, called the *cut metric* (K_p is the complete graph on p vertices). Clearly, for a cut metric (6) turns into the inequality in (5).

Papernov found the complete list of demand graphs H such that for all G any metric m as in (7) is a cut metric. By the argument above, this describes the set of H for which the cut condition (5) gives a criterion of solvability of (2).

Theorem 1 [Pa]. *Suppose that H is either K_4 or C_5 or a 2-star. Then* (2) *is solvable if and only if* (5) *holds. For any other H there are G, c and d such that* (5) *holds but* (2) *has no solution.*

(Here C_5 is the circuit on 5 vertices, and a graph is called a *p-star* if there are at most p vertices in it covering all edges.) Theorem 1 says nothing about numbers $\varphi(D(H))$ for H occurring in it. This number gives another theorem, due to Lomonosov, which extendes results in [Hu, RW, ADK, Se4]. We say that a vector (c,d) is *Eulerian* if $c(\delta_G(X)) + d(\delta_H(X \cap T))$ is even for every $X \subseteq V$. Clearly $\varphi(D(H)) \leq 2\varphi(D^e(H))$ where $D^e(H)$ is the set of problems in $D(H)$ with Eulerian (c,d).

Theorem 2. [Lo1, Lo3]. *If H is as in Theorem 1, (c,d) is Eulerian and* (5) *holds then $P = D(G,H,c,d))$ has an integral solution, that is, $\varphi(D^e(H)) = 1$.*

Theorem 2 implies that $\varphi(D(H)) \leq 2$ for H as above (actually $\varphi(D(H)) = 2$ unless H is a 1-star). There are strongly polynomial algorithms to find a half-integral or integral (in the Eulerian case) solution when H is a 2-star [Ch1, Se2] or $H \in \{K_4, C_5\}$ [Ka1]. These algorithms exploit, in essence, augmentation techniques that construct a solution starting with zero multiflow. We now explain how to derive Theorem 2 directly from Theorem 1 by use of splitting-off operations, as shown in [Ka6, Sc3].

Suppose that (5) holds. Then, by Theorem 1, P has a solution f. One may assume that $d \neq 0$, $c(e) > 0$ for all $e \in E$, and $st \notin E$ whenever $st \in U$. For a pair $\pi = \{uv, vw\}$ of edges of G let $\alpha = \alpha(\pi) \le \min\{c(uv), c(vw)\}$ be the maximum rational for which $D(G', H, c', d)$ is still solvable, where G' and c' arise from G and c by decreasing $c(uv)$ and $c(vw)$ and increasing $c(uw)$ by α (if $uw \notin E$ we add the edge uw of capacity α); this operation is called *splitting off* π by α. Let π be chosen so that α is maximum. If $\alpha \ge 1$ we split off π by 1, obtaining a "simpler" solvable problem $D(G', H, c'', d)$ for which (c'', d) is again Eulerian, and the result follows by induction. But if $\alpha < 1$ then, by Theorem 1, there is $X \subseteq V$ such that $\Delta'' := \Delta(c'', d, X) < 0$ ($\Delta(\cdot,\cdot,X)$ is defined in (5)). Since $\Delta := \Delta(c, d, X)$ and Δ'' are even, $\Delta \ge 0$ and, obviously, $c(\delta(X)) - c''(\delta(X)) \le 2$, we have $\Delta = 0$, whence $\alpha = 0$. This implies $f = 0$, and hence $d = 0$; a contradiction.

The above proof can be transformed to a strongly polynomial algorithm as follows. Choose a vertex $v \in V$ of the current network (G, H, c), consider pairs $\pi = \{uv, vw\}$, one by one, and split off π by $\lfloor \alpha(\pi) \rfloor$. If $v \in V \setminus T$, remove v from G. Repeat the same for a new vertex v', and so on. As a result, one eventually gets $\tilde{G} = (\tilde{V}, \tilde{E})$ and $\tilde{c}$ such that $\tilde{V} = T$ and $\tilde{c}(e) \ge d(e)$ for $e \in U$. Now a required multiflow in the original network is constructed in a natural way by using the obtained numbers α. One shows that calculation of $\alpha(\pi)$ can be reduced to solving $O(1)$ minimum cut problems. This provides a strongly polynomial algorithm.

Now let $\Gamma_1 + \ldots + \Gamma_p$ denote the graph that is the union of disjoint graphs $\Gamma_1, \ldots, \Gamma_p$. For $H = K_2 + K_2 + K_2$ and arbitrary $k \in \mathbf{Z}_+$ one can construct a solvable problem (2) of fractionality at least k [Lo1, Lo3]. This and simple observation that if H' is a subgraph of H then $\varphi(D(H')) \le \varphi(D(H))$ imply the following result.

Theorem 3. *If H contains a matching of 3 edges then $\varphi(D(H)) = \infty$.*

The only graphs different from those in Theorems 2 and 3 are: (i) certain subgraphs of K_5, (ii) the union of K_3 and a 1-star, (iii) $K_3 + K_3$. The case (ii) is easily reduced to (i). The following theorem generalizes Theorem 2.

Theorem 4 [Ka6]. *If $H = K_5$ then $\varphi(D^e(H)) = 1$.*

The proof of this theorem given in [Ka6] and based on splitting-off techniques is rather complicated. First, one shows that a metric satisfying (7) for $H = K_5$ is either a cut metric or a metric induced by $K_{2,3}$, called a 2,3-*metric* ($K_{p,q}$ is the complete bipartite graph with parts of p and q vertices). Thus (2) is solvable if and only if (6) holds for all cut metrics and 2,3-metrics on V. Second, unlike the proof above (when $\alpha = \alpha(\pi)$ is always an integer), in our case α can take half-integer values; in particular, $\alpha = 1/2$ is possible (in which case the "obstacle" m violating (6) after splitting off π by 1 is a 2,3-metric). The core of the proof is to show, using combinatorial properties of 2,3-metrics, that if $\alpha(\pi) < 1$ for all π then α is 0 everywhere.

This proof also can be turned into a strongly polynomial algorithm (however, using the ellipsoid method). To get such an algorithm, one shows that determining $\alpha(\pi)$ is reduced to solving $O(1)$ problems P: given $c' \in \mathbf{Z}_+^E$ and a metric ϱ on T that is either a cut metric or a 2,3-metric, find a metric m on V such that m coincides with ϱ on T and $\sum(c'(e)m(e) | e \in E)$ is minimum. The size of the constraint matrix for P is a polynomial in $|V|$, hence P is solvable in a strongly polynomial time, by [Ta].

Theorems 2–4 yield $\varphi(D(H))$ for all H except $H = K_3 + K_3$. A simple example shows that $\varphi(D^e(K_3 + K_3))$ is at least 2, and a conjecture is that it is exactly 2.

Theorems 2 and 4 have some consequence in lattice theory. A set $S = \{a_1, \dots, a_k\} \subseteq \mathbf{Q}^n$ is said to form a *Hilbert basis* if the intersection of the lattice $a_1\mathbf{Z} + \dots + a_k\mathbf{Z}$ with the cone in $\mathbf{Q}^n$ generated by S coincides with $\{\lambda_1 a_1 + \dots + \lambda_k a_k | \lambda_1, \dots, \lambda_k \in \mathbf{Z}_+\}$ (cf. [GP]). Let $S(G, H)$ be the set of the following vectors in $\mathbf{Q}^E \times \mathbf{Q}^U$: (i) $(\chi_L, \varepsilon_{st})$, L is an $s-t$ chain in G, $st \in U$; (ii) $(\chi_C, 0)$, C is a circuit in G; (iii) $(2\varepsilon_e, 0)$, $e \in E$; here χ_L (χ_C) is the incidence vector of E_L (E_C) in $\mathbf{Q}^E$, and ε_e (ε_{st}) is the e-th (st-th) unit basis vector in $\mathbf{Q}^E$ ($\mathbf{Q}^U$). Then $S(G, H)$ forms a Hilbert basis when H is a subgraph of K_5, or a 2-star, or the union of K_3 and a 1-star.

3. Maximization Problem

The set of H's for which $M(H)$ has bounded fractionality turns out to be larger than that for $D(H)$. The complete list of such H's is unknown, but the values $\varphi(M^*(H))$ have been determined for all H. Here $M^*(H)$ is the set of programs $P = M^*(G, H, c)$ dual to those in $M(H)$; P can be written as: minimize $c^T l$ subject to $l \in \mathbf{Q}_+^E$ and $\mathrm{dist}_l(s, t) \geq 1$ for each $st \in U$.

We say that c is *inner Eulerian* if $c(\delta(\{v\}))$ is even for any $v \in V \setminus T$. Clearly $\varphi(M(H)) \leq 2\varphi(M^e(H))$ where $M^e(H)$ is the set of problems in $M(H)$ with inner Eulerian c. The following theorem, due to Lomonosov and the author, generalizes results in [Hu, RW, Lov, Ch2]. Let $\mathscr{A}(H)$ denote the collection of maximal independent sets (*anticliques*) of H ($A \subset T$ is *independent* if $st \notin U$ for all $s, t \in A$).

Theorem 5. [Ka3, Lo3]. *Let $\mathscr{A}(H)$ have a partition $\{\mathscr{A}_1, \mathscr{A}_2\}$ such that each $\mathscr{A}_i$ consists of pairwise disjoint sets (in other words, H is the complement of the line graph of a bipartite graph). Then $\varphi(M^e(H)) = 1$.*

(A weaker statement that $\varphi(M(H)) \leq 2$ occurred in [KL].) Again, as for Theorems 2 and 4, splitting-off techniques can be applied to prove Theorem 5 [Ka3]. A sketch is as follows. It turns out that the dual problem for H in question can be solved separately from the primal one. More precisely, one shows that: (i) the dual problem has an optimal solution l of a special form [KL], namely,

$$l = \frac{1}{2} \sum_{A \in \mathscr{A}(H)} \chi(\delta(X_A)) \tag{8}$$

where X_A, $A \in \mathscr{A}$, are subsets of V such that $X_A \cap T \subseteq A$, and each $s \in T$ is contained in just one X_A ($\chi(E')$ is the incidence vector of $E' \subseteq E$); and (ii) these X_A's can be found by solving the minimum cut problem in a certain network of dimension $O(|E||V|)$. One can see that $c^T l'$ is an integer when c is inner Eulerian. This observation implies that $\alpha(\pi)$ is a multiple of $1/2$ for any $\pi = \{uv, vw\} \subseteq E$, where $\alpha(\pi)$ is the maximum of $\lambda \leq \min\{c(uv), c(vw)\}$ for which splitting off π by λ preserves the maximum value of a multiflow. The central point of the proof is to show that if uv, wv, zv are three edges with $\alpha(\{uv, vw\}) = \alpha(\{wv, vz\}) = 1/2$ then $\alpha(uv, vz) \geq 1$; this enables us to apply induction. The proof can be turned into a "pure combinatorial" strongly polynomial algorithm.

The expression (8) shows that $\varphi(M^*(H)) \leq 2$ for H as in Theorem 5. The exact values of $\varphi(M^*(H))$ were pointed out in [Ka7] for all H. In particular, the following result was stated there. Specify the set of H's with the property:

(9) for any three pairwise intersecting anticliques A, B, C in H, $A \cap B = B \cap C = C \cap A$.

Theorem 6. *$\varphi(M^*(H)) \in \{1, 2, 4\}$ if H satisfies (9) and $\varphi(M^*(H)) = \infty$ otherwise.*

The first part can be reformulated in polyhedral terms as follows. Let $P(G, H)$ be the polyhedron $\{l \in \mathbf{Q}^E | l \geq 0,\ \mathrm{dist}_l(s, t) \geq 1$ for any $st \in U\}$. If H is as in (9) then $P(G, H)$ is 1/4-integral, that is, each face in it contains a 1/4-integral point.

Another consequence of Theorem 6 is that if H is not as in (9) then $\varphi(M(H)) = \infty$ because $\varphi(M(H')) \geq \varphi(M^*(H'))$ for all H'. The latter follows from a general statement [Ka7] (extending a result on totally dual integral systems [Fu1, EG]): let A be a nonnegative $m \times n$-matrix, b be an integral m-vector, and let the program $D(c) := \max\{y^T b | y \geq 0,\ y^T A \leq c\}$ have a $1/k$-integral optimal solution for every nonnegative integral n-vector c; then the polyhedron $\{x \in \mathbf{Q}^n | x \geq 0, Ax \geq b\}$ is $1/k$-integral.

It is unknown whether $\varphi(M(H))$ is finite for each H as in (9); a conjecture is that $\varphi(M(H)) \leq 4$.

Finally, we show that $M(G, H, c)$ can have an optimal basis solution f such that the denominator of some component of f exceeds $\varphi(M(H))$, as it was mentioned in Sect. 1. Take a demand problem $D = D(G', H', c, d)$ with $H' = (T', U') = K_2 + K_2 + K_2$ such that $\varphi(D) > 2$ (existing by Theorem 3). Let $U = \{s_i t_i | i = 1, 2, 3\}$. Add new vertices p_i, q_i and edges $p_i s_i$, $q_i t_i$, $i = 1, 2, 3$, to G', forming the graph G, and put $c(p_i s_i) := c(q_i t_i) := d(s_i t_i)$. Let $H = (T, U)$ be the complete graph on T, and $B := \{p_i q_i | i = 1, 2, 3\}$. Then $\varphi(M(H)) \leq 2$, by Theorem 5. Now take the objective function $h(f) := \sum(v(f_u) | u \in B)$. Then any optimal basis solution of $\max\{h(f) | f$ a multiflow in $(G, H, c)\}$ is an optimal basis solution of $M(G, H, c)$ and it determines a solution of $D(G'H'c, d)$ in a natural way, whence the vector $2f$ is not integral.

4. Minimum Cost Problem

According to the multi-terminal version of the minimum-cost maximum-flow theorem [FF], $\varphi(C(H)) = 1$ when H is a complete bipartite graph. The following result was stated in [Ka2].

Theorem 7. *If H is a complete p-partite graph with $p \geq 3$ (that is, $\mathcal{A}(H)$ consists of p pairwise disjoint sets) then $\varphi(C(H)) = 2$.*

Theorem 7 is a consequence of a pseudo-polynomial algorithm (an algorithm of complexity $O(c(E)Q(|V|))$, $Q(n)$ is a polynomial in n) which finds a half-integral optimal primal solution. This algorithm extends the minimum-cost augmenting path method in [FF] based on ideas of the primal-dual method in linear programming. Recently the author found a strongly polynomial algorithm using a general method in [Ta].

On the other hand, it was shown in [Ka5] that if H is not a complete p-partite $(p \geq 2)$ then $\varphi(C(H)) = \infty$.

5. Demand Problem for Planar Graphs

Speaking of a planar graph G, we mean that G is explicitly embedded in the plane without intersecting edges. There are known several cases of the demand problem (2) when it is solvable provided that the cut condition (5) holds. The most interesting of them are the following:

(C1) the graph $(V, E \cup U)$ is planar [Se5];

(C2) there is a set $\mathscr{H}$ of two faces in G such that each edge of H connects vertices in the boundary of some face in $\mathscr{H}$ [Ok];

(C3) $U = \{s_1t_1, \ldots, s_kt_k\}$ and there are two inner faces I and J in G so that $s_1, \ldots, s_k$ occur in clockwise order in I and $t_1, \ldots, t_k$ do so in J [Sc2].

Moreover, in (C1)–(C3), if (c, d) is Eulerian and (5) holds then (2) has an integral solution.

Now we consider the case similar to (C2) for $|\mathscr{H}| \geq 3$. A simple example with $G = K_{2,3}$ shows that (5) is, in general, not sufficient for solvability of (2). However, the above result is extended, in a sense, as follows.

Theorem 8 [Ka10]. *Let* $|\mathscr{H}| = 3$.

(i) (2) *is solvable if and only if the metrical condition* (6) *holds for all* m *such that* m *is a cut metric or a* 2,3*-metric on* V.

(ii) *If* (c, d) *is Eulerian and* (2) *is solvable then* (2) *has an integral solution.*

It was shown in [Ka10] that if $|\mathscr{H}| = 4$ (or more) then (ii) is, in general, not true, and there are infinitely many "types" of metrics m necessary for cheking solvability of (2) for all corresponding G and H.

The statement (i) follows from a result on packing of cuts and 2,3-metrics (Theorem 10(ii)(b) below). To prove (ii) we use (i) and the splitting-off method as in the proof of Theorem 4. There are certain difficulties when applying this method, because in order to keep planarity we should take only those pairs π of edges of G which are contained in the boundary of a face of G. The core of the proof is to show that if $\alpha(\pi) < 1$ for all such π's then there are three edges of capacity 1 in G such that the graph G' obtained by removing these edges consists of three components, each containing just one face in $\mathscr{H}$. Now (ii) is proved by using Okamura's theorem for (C2).

6. Packings of Cuts and Metrics

There is a kind of duality that connects solvability conditions for the demand problem (2) with a certain packing problem on metrics. It can be expressed in a general form as follows.

Proposition 9. *Given* $G = (V, E)$, $H = (T, U)$ *and a set* M *of metrics on* V, *the following statements are equivalent:*

(i) *for any* c *and* d, (2) *is solvable if and only if* (6) *holds for all* $m \in M$;

(ii) *for any* $l \in \mathbf{Z}_+^E$, *there exist* $m_1, \ldots, m_k \in M$ *and* $\lambda_1, \ldots, \lambda_k \in \mathbf{Q}_+$ *so that:*

$$\lambda_1 m_1(e) + \ldots \lambda_k m_k(e) \leq l(e) \quad \textit{for all } e \in E\,; \tag{10}$$

and

$$\lambda_1 m_1(st) + \ldots + \lambda_k m_k(st) = \mathrm{dist}_l(st) \quad \textit{for all } st \in U\,. \tag{11}$$

This is easily proved by applying Farkas' lemma or the cone polarity. Proposition 9 enables us to derive results on packing of metrics directly from corresponding solvability theorems for (2) (like Theorems 1, 4, 8), and vice versa. Note that this relationship gives only theorems on the existence of *rational* λ. There are stronger, integral, version for some of these theorems; as a rule, their proofs are based on special, sometimes complicated, combinatorial approaches. Now we present some results in this area. We say that a vector $l \in \mathbf{Z}_+^E$ is *bipartite-like* if the l-length of every circuit in G is even.

Theorem 10. *Let l be bipartite-like.*

(i) (10) *and* (11) *hold for some cut metrics m_i's and integral λ_i's in the following cases: (a) H is K_4 or C_5 or a 2-star* [Ka4] *(cf.* [Se1] *for $H = K_2 + K_2$); (b) G and H are as in* (C1) *in Sect.* 5 [Se5]*; (c) G and H are as in* (C2) *in Sect.* 5 [Sc1] *(see* [Ka8] *for a strongly polynomial algorithm).*

(ii) (10) *and* (11) *hold for $m_1, \ldots, m_k$, where m_i is a cut metric or a 2,3-metric, and integers $\lambda_1, \ldots, \lambda_k$ in the following cases: (a) H is K_5 or the union of K_3 and a 1-star* [Ka9]*; (b) G and H are as in Theorem* 8 [Ka10].

There is a connection of the problem (10)–(11) and the problem (P): given a metric m, decide whether m is contained in the conic hull of metrics from a certain collection M. Such a connection was demonstrated in [Ka4] in terms of an extremal graph of m for M consisting of the set of cut-metrics. It was also shown there that for this M the problem (P) (or, equivalent, the problem "whether m is embeddable isometrically in the space L^1" [De]) is NP-hard.

References

[ADK] Adelson-Velsky, G.M., Dinits, E.A., Karzanov, A.V.: Flow Algorithms. Nauka, Moscow, 1975 (Russian)

[Ch1] Cherkassky, B.V.: A finite algorithm for solving the two-commodity flow problem. Ekon. Mat. Metody **9** (1973) 1147–1149 (Russian)

[Ch2] Cherkassky, B.V.: A solution of a problem of multicommodity flows in a network. Ekon. Mat. Metody **13** (1) (1977) 143–151 (Russian)

[De] Deza, M.: On the Hamming geometry of unitary cubes. Dokl. Akad. Nauk SSSR **134** (1960) 1037–1040 (Russian)

[EG] Edmonds, J., Giles, R.: A min-max relation for submodular function on graphs. In: Hammer, P.L., Johnson, E.L., Korte, B., Nemhauser, G.L. (eds.), Studies in integer programming. (Ann. Discr. Math., vol. 10.) North-Holland, Amsterdam 1977, pp. 185–204

[Fr] Frank, A.: Packing paths, circuits and cuts, a survey. Report 88532-OR, Inst. für Diskrete Mathematik, Bonn 1988

[Fu1] Fulkerson, D.R.: Blocking polyhedra. In: Harris, B. (ed.), Graph theory and its applications. Academic Press, New York 1970, pp. 93–112

[Fu2] Fulkerson, D.R.: Blocking and antiblocking pairs of polyhedra. Math. Progr. **1** (1971) 168–194

[FF] Ford, L.R., Fulkerson, D.R.: Flows in networks. Princeton Univ. Press, Princeton, NJ, 1962

[GJ] Garey, M.R., Johnson, D.S.: Computers and intractability: A guide to the theory of NP-completeness. W.H. Freeman, San Francisco, CA, 1979

[GP] Giles, R., Pulleyblank, W.: Total dual integrality and integral polyhedra. Linear Algebra and Appl. **25** (1979) 191–196

[GLS] Grötschel, M., Lovász, L., Schrijver, A.: Geometric algorithms and combinatorial optimization. (Algorithms and Combinatorics, vol. 2.) Springer, Berlin Heidelberg New York, 1988

[Hu] Hu, T.C. Multi-commodity network flows. J. ORSA **11** (1963) 344–360

[Ka1] Karzanov, A.V.: Combinatorial methods to solve cut-determined multiflow problems. In: Combinatorial methods for flow problems. Inst. for System Studies, Moscow, 1979, iss. 3, pp. 6–69 (Russian)

[Ka2] Karzanov, A.V.: A minimum cost maximum multiflow problem. In: Combinatorial methods for flow problems. Inst. for System Studies, Moscow, 1979, iss. 3, pp. 138–156 (Russian)

[Ka3] Karzanov, A.V.: On multicommodity flow problems with integer-valued optimal solutions. Dokl. Akad. Nauk SSSR **280** (4) (1985) 789–792 (Russian) (English transl.: Soviet Math. Dokl. **31** (1) (1985) 151–154)

[Ka4] Karzanov, A.V.: Metrics and undirected cuts. Math. Programming **35** (1985) 183–198

[Ka5] Karzanov, A.V.: Unbounded fractionality in maximum multiflow and minimum cost multiflow problems. In: Fridman, A.A. (ed.), Problems of discrete optimization and methods to solve them. Centr. Econom. and Math. Inst., Moscow, 1987, pp. 123–135 (Russian)

[Ka6] Karzanov, A.V.: Half-integral five-terminus flows. Discr. Appl. Math. **18** (3) (1987) 263–278

[Ka7] Karzanov, A.V.: Polyhedra related to undirected multicommodity flows. Linear Algebra and Appl. **114–115** (1989) 293–328

[Ka8] Karzanov, A.V.: Packing of cuts realizing distances between certain vertices of a planar graph. Discr. Math. **85** (1990) 73–87

[Ka9] Karzanov, A.V.: Sums of cuts and bipartite metrics. Europ. J. Comb. **11** (1990) 473–484

[Ka10] Karzanov, A.V.: Paths and metrics in a planar graph with three and more distinguished faces, In two parts. Res. Reps. RR 816-M and RR 817-M, IMAG ARTEMIS, Grenoble, 1990 (to appear in J. Comb. Theory (B))

[KL] Karzanov, A.V., Lomonosov, M.V.: Systems of flows in undirected networks. In: Mathematical programming etc. Inst. for System Studies, Moscow, 1978, iss. 1, pp. 59–66 (Russian)

[Kh] Khachiyan, L.G.: Polynomial algorithms in linear programming. Zhurnal Vychislitelnoj Matematiki i Matematicheskoi Fiziki **20** (1980) 53–72 (Russian)

[Lo1] Lomonosov M.V.: Solutions of two problems on flows in networks. Unpublished manuscript, 1976 (Russian)

[Lo2] Lomonosov M.V.: On a system of flows in a network. Problemy Peredatchi Informacii **14** (1978) 60–73 (Russian)

[Lo3] Lomonosov M.V.: Combinatorial approaches to multiflow problems. Discr. Appl. Math. **11** (1) (1985) 1–94

[Lov] Lovász, L.: On some connectivity properties of Eulerian graphs. Acta Math. Akad. Sci. Hung. **28** (1976) 129–138

[Me] Menger, K.: Zur allgemeinen Kurventheorie. Fundam. Math. **10** (1927) 96–115

[Ok] Okamura, H.: Multicommodity flows in graphs. Discr. Appl. Math. **6** (1983) 55–62

[Pa] Papernov, B.A.: On existence of multicommodity flows. In: Fridman, A.A. (ed.), Studies in discrete optimizations. Nauka, Moscow, 1976, pp. 230–261 (Russian)

[RW] Rothschild, B., Whinston, A.: Fealibility of two-commodity network flows. Oper. Res. **14** (1966) 1121–1129

[Sc1] Distances and cuts in planar graphs. Report OS-R8610, Mathematical Centre, Amsterdam, 1986 (to appear in J. Comb. Th. (B))

[Sc2] Schrijver, A.: The Klein bottle and multicommodity flows. Report OS-R8810, Mathematical Centre, Amsterdam, 1986 (to appear in Combinatorica)

[Sc3] Schrijver, A.: Short proofs on multicommodity flows and cuts. Preprint, 1988 (to appear in J. Comb. Th. (B))

[Se1] Seymour, P.D.: A two-commodity cut theorem. Discr. Math. **23** (1978) 177–181

[Se2] Seymour, P.D.: A short proof of the two-commodity flow theorem. J. Comb. Th. (B) **26** (1979) 370–371

[Se3] Seymour, P.D.: On multi-colouring of cubic graphs, and conjectures of Fulkerson and Tutte. Proc. Lond. Math. Soc. (3) **38** (1979) 423–460

[Se4] Seymour, P.D.: Four-terminus flows. Networks **10** (1980) 79–86

[Se5] Seymour, P.D.: On odd cuts and planar multicommodity flows. Proc. Lond. Math. Soc., Ser. III. **42** (1981) 178–192

[Ta] Tardos, E.: A strongly polynomial algorithm to solve combinatorial linear programs. Oper. Res. (1986) 250–256

Computing Vortex Sheet Motion

Robert Krasny

Department of Mathematics, University of Michigan, Ann Arbor, MI 48109, USA

1. Introduction

Coherent vortex structures occur in many types of fluid flow including mixing layers, jets and wakes. A vortex sheet is a mathematical model for such structures, in which the shear layer is approximated by a surface across which the tangential fluid velocity has a jump discontinuity. Vortex sheet motion belongs to the field of vortex dynamics, one of the main approaches to understanding fluid turbulence.

Careful numerical experiments have helped advance the mathematical study of vortex sheets. Difficulties arise in computing vortex sheet motion due to short wavelength instability, singularity formation, and spiral roll-up. This paper reviews the problem of computing vortex sheet motion and presents several applications. See [2] for a sample of other vortex models and numerical methods.

2. Analytic Evolution and Singularity Formation

A vortex sheet is defined by a curve $z(\Gamma, t)$ in the complex plane, where Γ is the circulation parameter and t is time. The evolution equation is [4,32],

$$\overline{\frac{\partial z}{\partial t}}(\Gamma, t) = \int_{-\infty}^{\infty} K(z(\Gamma, t) - z(\tilde{\Gamma}, t))d\tilde{\Gamma} \quad , \quad K(z) = \frac{1}{2\pi i z} \; . \tag{1}$$

The Cauchy principal value of the integral is taken. Equation (1) says that a point on the vortex sheet moves with the average of the two limiting velocities, as the curve is approached from either side.

A flat vortex sheet of constant strength $z(\Gamma, t) = \Gamma$ is an equilibrium solution of (1). Linear stability analysis shows that short wavelength perturbations can grow arbitrarily fast (Kelvin-Helmholtz instability). This means that the linearized initial value problem is ill-posed in the sense of Hadamard. However, Sulem et al. [35] have proven that if the initial perturbation is an analytic function of Γ, then the solution of (1) remains analytic for a positive time interval.

Birkhoff conjectured that instability and nonlinearity would cause a singularity to form during the vortex sheet's evolution [4, 5]. An asymptotic analysis by Moore [24, 26] supports this conjecture, indicating that with initial perturbation amplitude ε, a $\Gamma^{3/2}$ branch point forms in the vortex sheet at a finite critical time $t = t_c(\varepsilon)$. Meiron et al. [23] analyzed the Taylor series coefficients of $z(\Gamma, t)$

Proceedings of the International Congress of Mathematicians, Kyoto, Japan, 1990

with respect to the time variable and obtained results consistent with Moore's. The validity of Moore's approximation for $t < t_c$ has been proven [6] and special solutions have been studied [7, 14], but proving that a singularity forms for general initial data is an open problem.

3. The Point Vortex Approximation

Rosenhead performed the first vortex sheet computation in 1931 [31], using the periodic Cauchy kernel in (1). The sheet was discretized by a finite number of point vortices per period $z_j(t) \sim z(\Gamma_j, t), j = 1, ..., N$, leading to the ordinary differential equations,

$$\overline{\frac{dz_j}{dt}} = \sum_{k \neq j} K(z_j - z_k)N^{-1}, \quad K(z) = \frac{1}{2i}\cot \pi z\,. \tag{2}$$

The sum omits the singular term $k = j$, but if the vortex sheet has a bounded 2nd Γ-derivative, then the discretization error is $O(N^{-1})$ [25]. If the vortex sheet is analytic, then infinite order accuracy may be obtained by applying one step of Richardson extrapolation [34, 16].

Rosenhead used $N \sim 10$ points and the 1st order Euler method with time step $\Delta t \sim 0.05$ to integrate in time. He drew a smooth interpolating curve through the point vortices, suggesting that a perturbed vortex sheet rolls up into a smooth spiral. In the 1950's, Birkhoff performed computations using a larger number of point vortices and more accurate time integration [4, 5]. In contrast to Rosenhead's results, the points' computed motion was irregular, leading Birkhoff to question whether the vortex sheet rolls up into a spiral. Later workers sought to obtain convergent numerical results by using higher order accurate quadrature rules for the principal value integral, e.g. [15, 38]. Another approach was to stabilize the problem by adding surface tension [28]. In spite of much effort, the computations failed to converge as the number of points increased.

The key to obtaining convergent numerical results for $t < t_c$ lies in Fourier analysis of the computed solution [19]. Sulem et al. [36] showed that the singularity structure of nonlinear evolution equations could be obtained from spectral computations, by analyzing the rate of decay of the discrete Fourier coefficients. For vortex sheet computations, discrete Fourier coefficients of the perturbation quantities $p_j(t) = z_j(j) - \Gamma_j$ can be similarly analyzed.

Figure 1 shows computations with $N = 50$ in single and double precision arithmetic. Irregular small scale motion develops in single precision, but the double precision results are smooth. The corresponding spectral amplitudes are plotted in Fig. 2. The initial spectrum has a spike at wavenumber $k = 1$ (an explicit perturbation of amplitude $\varepsilon = 0.01$), as well as broad band noise in the higher modes. In Fig. 2a, the noise is amplified by the system's instability, leading to the irregular motion in Fig. 1a for $t \geq 0.3$. In Fig. 2b, the spectrum spreads smoothly to higher wavenumbers, due to genuine nonlinear effects [19].

A stable physical process is modeled by a well-posed initial value problem, and if the difference scheme is consistent and stable, then the solution converges as the mesh is refined [30]. Shear flows however are physically unstable and this appears as ill-posedness in the vortex sheet initial value problem. The point vortex approximation for an analytic vortex sheet defines a consistent but unstable

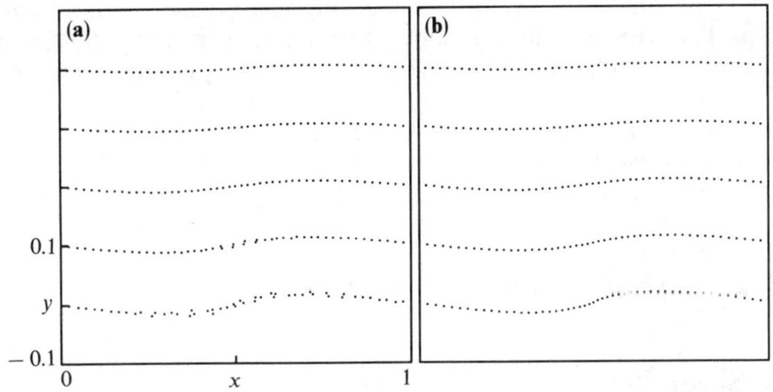

Fig. 1. Point vortex computations at times $t = 0, 0.1, 0.2, 0.3, 0.4$. **(a)** single precision. **(b)** double precision

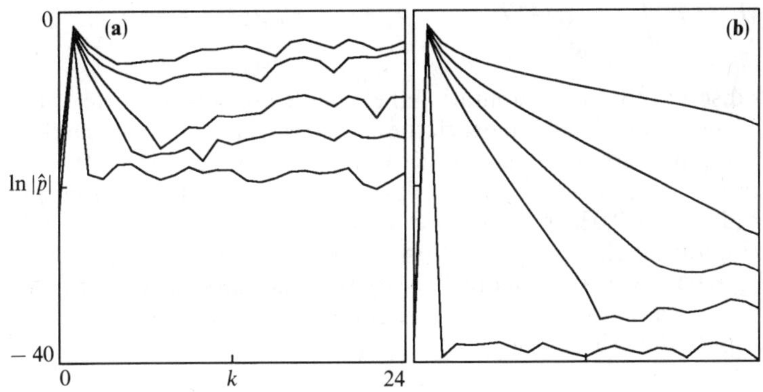

Fig. 2. Discrete Fourier coefficient amplitudes corresponding to Fig. 1. **(a)** single precision. **(b)** double precision

difference scheme. Fritz John has observed [18], "Instability of a difference scheme under small perturbations does not exclude the possibility that in special cases the scheme converges towards the correct function, if no errors are permitted in the data or the computation." This refers to roundoff error, due to the computer's finite precision arithmetic, as opposed to discretization error, due to replacing a continuous operator by a discrete approximation. Using higher precision arithmetic is one way to see convergence as the mesh is refined, but for vortex sheet computations, a more practical remedy is to filter out the spurious roundoff error perturbations [19]. Computations and theory [8] now show that the point vortex approximation converges as $N \to \infty$ for $t < t_c$. A consistent picture of singularity formation in a vortex sheet has been obtained: infinite curvature forms at an isolated point, but the vortex sheet remains continuously differentiable at $t = t_c$, showing no sign of roll-up [19, 23, 24, 26, 34].

An obvious question is whether the vortex sheet continues to exist past the critical time. Note that in other problems, a physically valid weak solution can be defined past a critical time, e.g. shock formation in a nonlinear hyperbolic equation. Computations show that the point vortex approximation does not converge for $t > t_c$ as $N \to \infty$ [19]. A different type of small scale motion occurs in the point vortex system (2) for $t > t_c$, but it is not relevant to vortex sheet evolution. Based on work with self-similar vortex sheets [27], Pullin conjectured that a periodically perturbed sheet rolls up into a spiral for $t > t_c$, the spiral vanishes in size as $t \to t_c^+$, and for any $t > t_c$ it has an infinite number of turns [29]. As described in the next section, numerical experiments using Chorin's vortex blob method support this conjecture [1, 9, 10, 11, 20, 21, 22].

4. Vortex Sheet Roll-Up

Let $\delta > 0$ be a smoothing parameter and consider a regularized approximation to (1),

$$\overline{\frac{\partial z}{\partial t}}(\Gamma,t) = \int_{-\infty}^{\infty} K_\delta(z(\Gamma,t) - z(\tilde{\Gamma},t))d\tilde{\Gamma}\,, \quad K_\delta(z) = K(z)\frac{|z|^2}{|z|^2+\delta^2}\,. \tag{3}$$

When (3) is discretized, the computational elements are called "vortex blobs". For fixed $\delta > 0$, short wavelength perturbations no longer have unbounded growth rates and computed solutions converge as the number of blobs $N \to \infty$, even for $t > t_c$ [20]. Figure 3a shows the evolution for $0 \le t \le 1$, with the smoothing parameter value $\delta = 0.03$, in a case for which the vortex sheet's critical time is $t_c \sim 0.375$. Figure 3b shows the solution at time $t = 1$ with decreasing amounts of smoothing $0.05 \le \delta \le 0.2$. Figure 4 shows that the smoothed solutions at $t = 1$ converge to a spiral as $\delta \to 0$. The limit spiral is a candidate extension for the vortex sheet past the critical time.

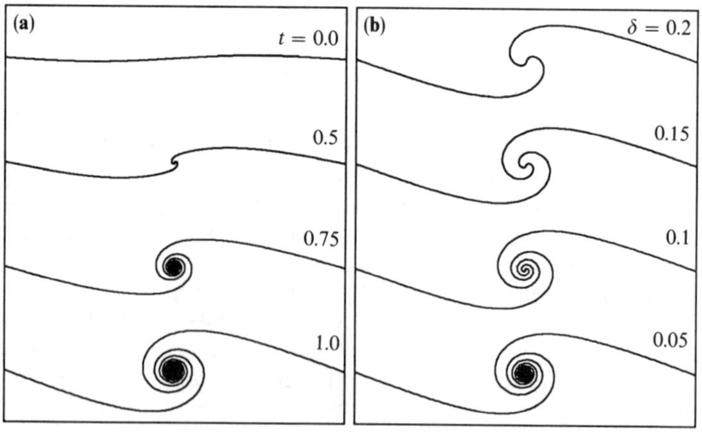

Fig. 3. Regularized vortex sheet roll-up past the critical time $t_c \sim 0.375$. **(a)** $\delta = 0.03$, increasing time. **(b)** $t = 1$, decreasing δ

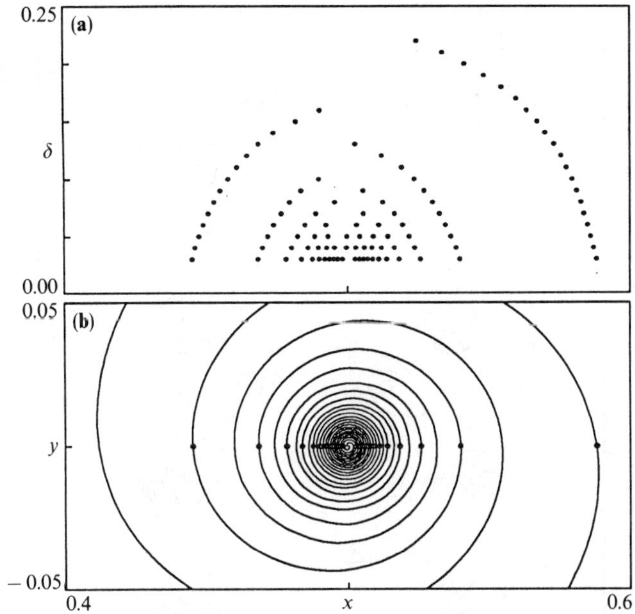

Fig. 4. Convergence as $\delta \to 0$ for $t = 1 > t_c \sim 0.375$. **(a)** x-axis intercepts of one spiral branch plotted against δ. **(b)** closeup of the solution for $\delta = 0.03$

The numerical experiments suggest that the vortex blob method provides a convergent discretization of vortex sheet motion for $t > 0$. This has been proven for $t < t_c$ [8], but proving convergence for the physically important roll-up regime $t \geq t_c$ is an outstanding problem. Other interesting issues concern uniqueness of the limit for different regularizations [3, 37], existence of a weak solution to the incompressible Euler equations with general vortex sheet initial data and the possible presence of concentrations in the limit $\delta \to 0$ [12, 13].

5. Applications

The vortex blob method has advantageous mathematical and numerical properties, but the smoothing parameter δ has no precise physical meaning. One would like to know whether computations performed with a value $\delta > 0$ approximate real fluid motion. Some applications presented below demonstrate the vortex blob method's potential for simulating shear layer dynamics.

Aircraft Trailing Vortices. On takeoff and landing, an aircraft sheds vortices at the wing's trailing edge. Figures 5 and 6 show a free-space vortex sheet simulation of this process, including the effects of the wing tips and deployed flaps [21]. The computation illustrates different types of vortex interactions: rotation of like-sign vortex pairs, translation of opposite-sign vortex pairs, core deformation due to collision, and vortex sheet folding.

Separation at a Sharp Edge. Vortices are shed from the edges of a flat plate that is moving in a viscous fluid. As the viscosity is reduced, an ideal flow emerges

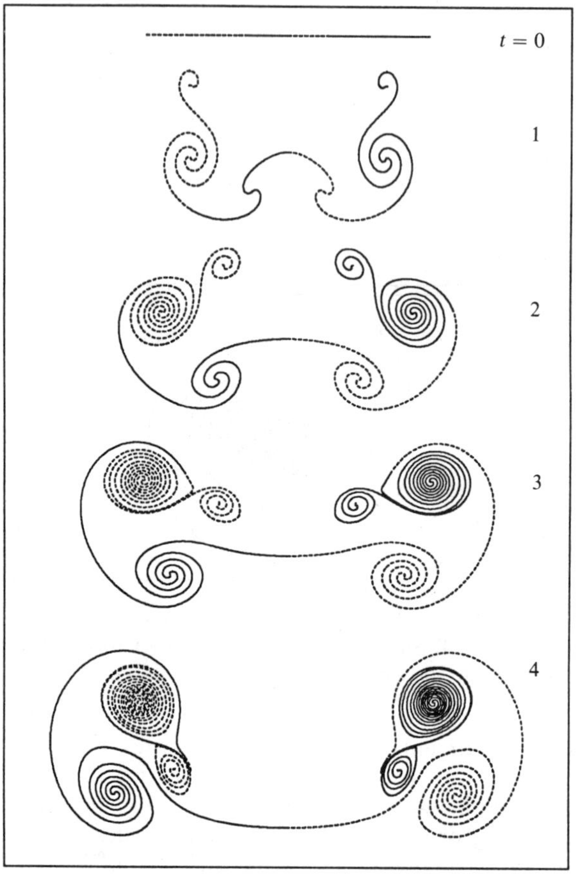

Fig. 5. Roll-up of an aircraft trailing vortex sheet, including tip and flap vortices. The solid and dotted lines indicate opposite senses of rotation

having embedded vortex sheets that emanate from the edges. This problem is more difficult to compute than the periodic and free space problems considered above. New issues arise, in satisfying the flow tangency condition on the plate, and shedding the correct amount of circulation at the edges. Previous numerical studies did not obtain smooth spiral roll-up, e.g. [17, 33].

A new implementation of the vortex blob method has been developed. Figure 7a is a computation of the vortex sheets that separate from an impulsively started flat plate. The velocity field plotted in Fig. 7b shows that the sheets form a recirculating region behind the plate.

To validate the algorithm, a comparison with Pullin's computation of self-similar vortex sheet roll-up [27] has been performed. The similarity assumption circumvents the difficulty of solving the initial value problem. Figure 8 compares a time dependent vortex blob computation with Pullin's self-similar result. The two plots may be superimposed to verify that the spiral shapes are in good agreement. Further details are given in [22] and a more complete validation is in preparation.

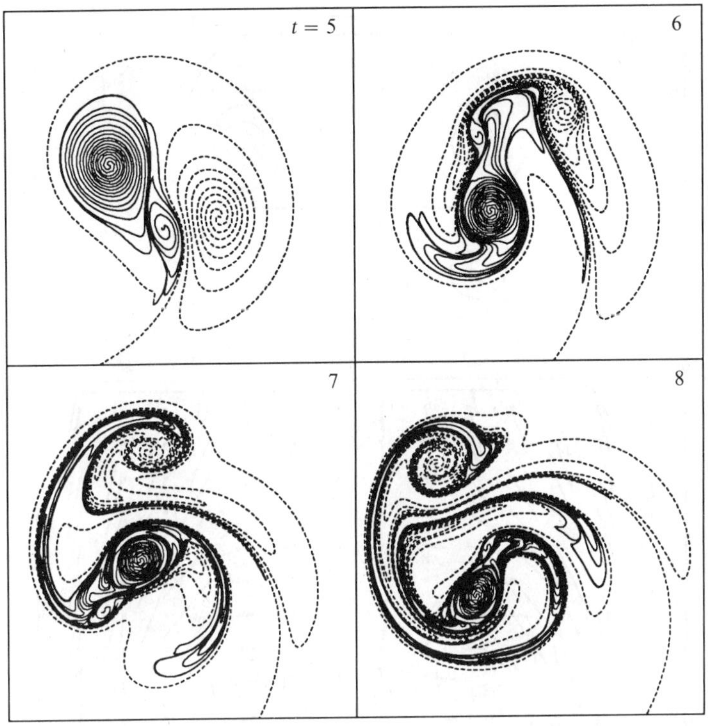

Fig. 6. Continuation of Fig. 5, showing details of core deformation and vortex sheet folding

Instability of a Jet. Figure 9 shows the evolution of a jet being expelled from a box. The jet is driven by two point sources in the lower corners of the box, which are turned on at time $t = 0$. A starting vortex forms and propagates away from the outlet, leaving behind a thin straight jet. Waves form along the jet, rolling up into a small vortex which propagates through the large starting vortex.

6. Final Remarks

Vortex sheet motion poses interesting mathematical problems concerning singular integrals, weak limits, and nonlinear dynamics. Vortex blob computations may provide a useful tool for clarifying the role of coherent vortex structures in shear flow. Future computational work will focus on improved treatment of boundary conditions, the effects of parametric forcing, and three dimensionality.

Acknowledgements. This work was supported in part by GRI Contract #5088-260-1692, NSF Grant DMS-#8801991, and ONR URI#N000184-86-K-0684. The computations were performed at the NSF San Diego Supercomputer Center and the University of Michigan.

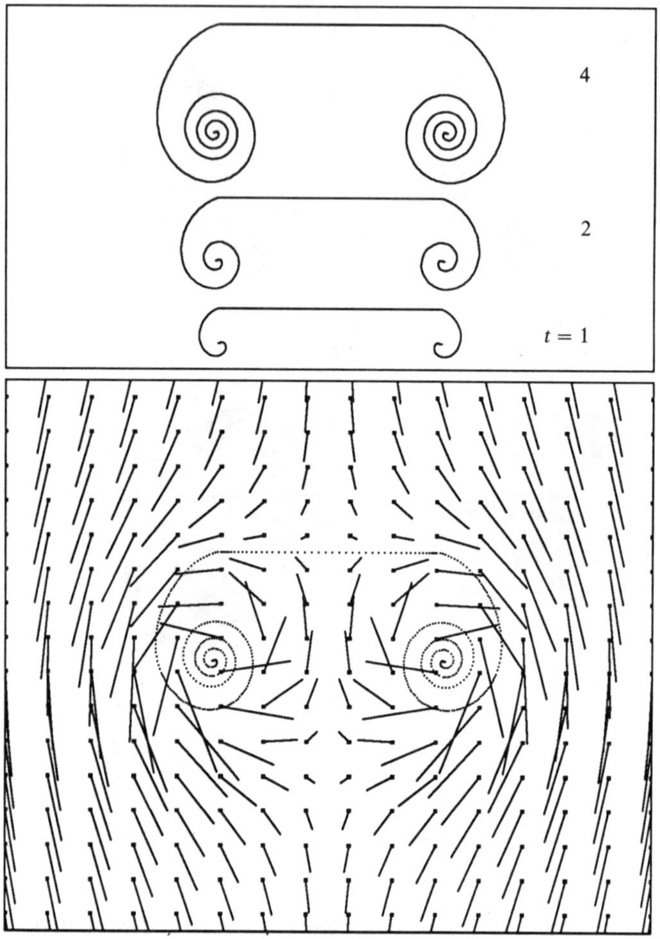

Fig. 7. **(a)** Vortex sheet roll-up due to the impulsively started upward motion of a flat plate. **(b)** Velocity field at time $t = 4$

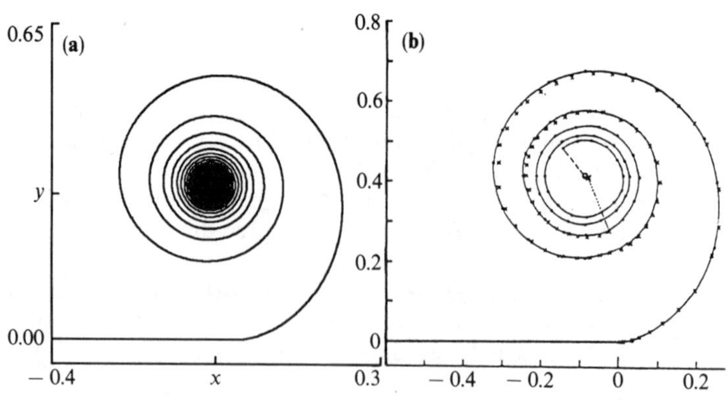

Fig. 8. **(a)** Time dependent vortex blob computation, $\delta = 0.025, t = 1$. **(b)** Self-similar vortex sheet roll-up past a semi-infinite flat plate, reproduced from [27]

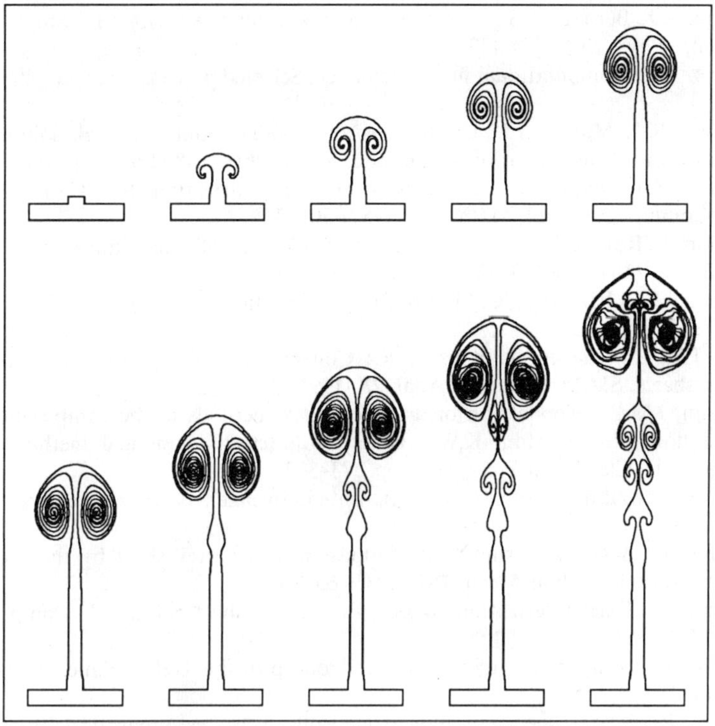

Fig. 9. Computation of a jet being expelled from a box

References

1. Anderson, C.: A vortex method for flows with slight density variations. J. Comp. Phys. **61** (1985) 417
2. Anderson, C., Greengard, C. (eds.): Vortex Dynamics and Vortex Methods. (Lectures in Applied Mathematics, AMS). Proc. 1990 AMS-SIAM Summer Seminar in Appl. Math.
3. Baker, G.R., Shelley, M.J.: On the connection between thin vortex layers and vortex sheets. J. Fluid Mech. **215** (1990) 161–194
4. Birkhoff, G.: Helmholtz and Taylor instability. Proc. Symp. Appl. Math. **XIII** Amer. Math. Soc., Providence, R. I. (1962) 55–76
5. Birkhoff, G., Fisher, J.: Do vortex sheets roll-up? Rend. Circ. Mat. Palermo Ser. 2, **8** (1959) 77–90
6. Caflisch, R., Orellana, O.: Long time existence for a slightly perturbed vortex sheet. Comm. Pure Appl. Math. **39** (1986) 807–838
7. Caflisch, R., Orellana, O.: Singular solutions and ill-posedness for the evolution of vortex sheets. SIAM J. Math. Anal. **20** (1989) 293–307
8. Caflisch, R., Lowengrub, J.: Convergence of the vortex method for vortex sheets. SIAM J. Numer. Anal. **26** (1989) 1060–1080
9. Chorin, A.J.: Numerical study of slightly viscous flow. J. Fluid Mech. **57** (1973) 785–796

10. Chorin, A.J., Bernard, P.S.: Discretization of a vortex sheet with an example of roll-up. J. Comp. Phys. **13** (1973) 423–429
11. Chorin, A.J.: Computational fluid mechanics: Selected papers. Academic Press, Boston 1989
12. DiPerna, R.J., Majda, A.J.: Oscillations and concentrations in weak solutions of the incompressible fluid equations. Commun. Math. Phys. **108** (1987) 667–689
13. DiPerna, R.J., Majda, A.J.: Concentrations in regularizations for 2-D incompressible flow. Comm. Pure Appl. Math. **XL** (1987) 301–345
14. Duchon, J., Robert, R.: Global vortex sheet solutions of Euler equations in the plane. J. Diff. Eq. **73** (1988) 215–224
15. Higdon, J.J.L., Pozrikidis, C.: The self-induced motion of vortex sheets. J. Fluid Mech. **150** (1985) 203–231
16. Hou, T.Y., Lowengrub, J., Krasny, R.: Convergence of a point vortex method for vortex sheets. SIAM J. Numer. Anal. **28** (1991) 308–320
17. Graham, J.M.R.: Application of discrete vortex methods to the computation of separated flows. In: Morton, K.W., Baines, M.J. (eds.), Numerical methods for fluid dynamics II. Clarendon Press 1985, pp. 273–302
18. John, F.: Partial differential equations, 4th ed. Springer, Berlin Heidelberg New York 1982, p. 8
19. Krasny, R.: A study of singularity formation in a vortex sheet by the point vortex approximation. J. Fluid Mech. **167** (1986) 65–93
20. Krasny, R.: Desingularization of periodic vortex sheet roll-up. J. Comp. Phys. **65** (1986) 292–313
21. Krasny, R.: Computation of vortex sheet roll-up in the Trefftz plane. J. Fluid Mech. **184** (1987) 123–155
22. Krasny, R.: Vortex sheet computations: Roll-up, wakes, separation. In: Anderson, C., Greengard, C. (eds.), Vortex dynamics and vortex methods. (Lectures in Applied Mathematics, AMS). Proc. 1990 AMS-SIAM Summer Seminar in Appl. Math.
23. Meiron, D.I., Baker, G.R., Orszag, S. A.: Analytic structure of vortex sheet dynamics. 1. Kelvin-Helmholtz instability. J. Fluid Mech. **114** (1982) 283–298
24. Moore, D.W.: The spontaneous appearance of a singularity in the shape of an evolving vortex sheet. Proc. Roy. Soc. Lond. A **365** (1979) 105–119
25. Moore, D.W.: On the point vortex Method. SIAM J. Sci. Stat. Comp. **2** (1981) 65–84
26. Moore, D.W.: Numerical and analytical aspects of Helmholtz instability. In: F.I. Niordson, N. Olhoff (eds.), Theoretical and applied mechanics. Proc. XVI IUTAM Conf. North-Holland, Amsterdam 1984, pp. 629–633
27. Pullin, D.I.: The large-scale structure of unsteady self-similar rolled-up vortex sheets. J. Fluid Mech. **88** (1978) 401–430
28. Pullin, D.I.: Numerical studies of surface-tension effects in nonlinear Kelvin-Helmholtz and Rayleigh-Taylor instability. J. Fluid Mech. **119** (1982) 507–532
29. Pullin, D.I.: Personal communication, 1983
30. Richtmyer, R.D., Morton, K.W.: Difference methods for initial-value problems. Interscience, New York London Sydney 1967
31. Rosenhead, L.: The formation of vortices from a surface of discontinuity. Proc. R. Soc. Lond. A **134**, 170–192
32. Rott, N.: Diffraction of a weak shock with vortex generation. J. Fluid Mech. **1** (1956) 111–128
33. Sarpkaya, T.: An inviscid model of two-dimensional vortex shedding for transient and asymptotically steady separated flow over an inclined plate. J. Fluid Mech. **68** (1975) 109–128
34. Shelley, M.: A study of singularity formation in vortex sheet motion by a spectrally accurate vortex method. J. Fluid Mech. (to appear 1991)

35. Sulem, C., Sulem, P.L., Bardos, C., Frisch, U.: Finite time analyticity for the two- and three-dimensional Kelvin-Helmholtz instability. Comm. Math. Phys. **80** (1981) 485–516
36. Sulem, C., Sulem, P.L., Frisch, U.: Tracing complex singularities with spectral methods. J. Comp. Phys. **50** (1983) 138
37. Tryggvason, G.: Simulations of vortex sheet roll-up by vortex methods. J. Comp. Phys. **75** (1988) 253
38. van de Vooren, A.I.: A numerical investigation of the rolling up of vortex sheets. Proc. R. Soc. Lond. A **373** (1980) 67–91

Developments in the Double Exponential Formulas for Numerical Integration

Masatake Mori

Department of Applied Physics, Faculty of Engineering,
University of Tokyo, 7-3-1 Hongo, Bunkyo-ku, Tokyo 113, Japan

1. Optimality of the Trapezoidal Rule

The double exponential formula, abbreviated as the DE-formula, was first presented by Takahasi and Mori [18] in 1974 as an efficient and robust quadrature formula to compute integrals with end point singularity, e.g.

$$I = \int_{-1}^{1} \frac{dx}{(x-2)(1-x)^{1/4}(1+x)^{3/4}} , \tag{1}$$

or over the half infinite interval, e.g.

$$I = \int_0^\infty e^{-x} \log x \sin x dx . \tag{2}$$

The DE-formula is based on the optimality of the trapezoidal rule over $(-\infty, \infty)$ in the following sense. Consider the integral

$$I = \int_{-\infty}^{\infty} g(u) du , \tag{3}$$

where $g(u)$ is analytic over $(-\infty, \infty)$ and $|g(u)|$ is integrable. We apply the trapezoidal rule, or equivalently the midpoint rule, to (3) with an equal mesh size h:

$$I_h = h \sum_{k=-\infty}^{\infty} g(kh) . \tag{4}$$

Then the error of (4) is expressed in terms of a contour integral [16]

$$\Delta I_h = \frac{1}{2\pi i} \int_{\hat{C}} \hat{\Phi}_h(w) g(w) dw , \tag{5}$$

where $\hat{\Phi}_h(w)$ is called the characteristic function of the error and defined by

$$\hat{\Phi}_h(w) = \begin{cases} \dfrac{+2\pi i}{1 - \exp(-\dfrac{2\pi i}{h} w)} ; & \operatorname{Im} w > 0 \\[2ex] \dfrac{-2\pi i}{1 - \exp(+\dfrac{2\pi i}{h} w)} ; & \operatorname{Im} w < 0 \end{cases} \tag{6}$$

Proceedings of the International Congress
of Mathematicians, Kyoto, Japan, 1990

and $\hat{C}$ consists of two infinite curves one of which runs to the left above the real axis and the other to the right below the real axis in such a way that there exists no singularity of $g(w)$ between these two curves. Since ΔI_h is a linear functional over the family of analytic functions on $(-\infty, \infty)$, it can be regarded as a Sato's hyperfunction [10] and $-\hat{\Phi}_h(w)/(2\pi i)$ is nothing but its defining function.

Although there are infinite number of quadrature formulas for the integral (3), the trapezoidal rule (4) is proved to be optimal in the following sense. Let an arbitrary quadrature formula for (3) be

$$I_A = \sum_{k=-\infty}^{\infty} A_k g(a_k) \ . \tag{7}$$

Then its error is expressed also in terms of the contour integral

$$\Delta I_A = I - I_A = \frac{1}{2\pi i} \int_{\hat{C}} \hat{\Phi}_A(w) g(w) dw \ . \tag{8}$$

Since $|\hat{\Phi}_A(w)|$ usually decays exponentially as $|\mathrm{Im}\, w|$ becomes large for quadrature formulas of practical use we define the average decay rate r of $|\hat{\Phi}_A(w)|$ for large $|\mathrm{Im}\, w|$ as follows:

$$r = \lim_{d\to\infty} \left(\lim_{R\to\infty} \frac{1}{2R} \int_{-R+id}^{R+id} \left\{ -\frac{\partial}{\partial v} \log |\hat{\Phi}_{\mathrm{A}}(w)| \right\} dw \right), \quad w = u + iv \ . \tag{9}$$

It is easy to see that the error of numerical integration is smaller if the decay rate r is larger. Then we have

Theorem (Takahasi-Mori, 1970). *Suppose that A_k and a_k in I_A for I satisfy*

$$\sum_{k=-\infty}^{\infty} \frac{|A_k|}{|a_k|^2} \leq M < \infty. \tag{10}$$

Then, among quadrature formulas I_A whose average density of a_k's per unit length is equal ($= \nu_{\mathrm{P}}$) to each other, the trapezoidal rule I_h with equal mesh size $h = 1/\nu_{\mathrm{P}}$ is optimal in the sense that r attains its maximum

$$r_{\max} = 2\pi\nu_{\mathrm{P}} = \frac{2\pi}{h} \tag{11}$$

by the trapezoidal rule.

The decay rate in case of the Simpson's rule is $r = \pi\nu_{\mathrm{P}} = \pi/h$, so that the Simpson's rule is as twice inefficient as the trapezoidal rule.

2. The Double Exponential Formula

Now that the trapezoidal rule over $(-\infty, \infty)$ is optimal, we can get a new efficient quadrature formula by means of a variable transformation. Let the given integral be

$$I = \int_a^b f(x)dx \ . \tag{12}$$

The variable transformation

$$x = \phi(u), \quad \phi(-\infty) = a, \ \phi(+\infty) = b \tag{13}$$

leads to

$$I = \int_{-\infty}^{\infty} f(\phi(u))\phi'(u)du \ . \tag{14}$$

Since this is an integral over $(-\infty, \infty)$ we apply the trapezoidal rule with equal mesh size h, which results in a quadrature formula

$$I_h = h \sum_{k=-\infty}^{\infty} f(\phi(kh))\phi'(kh) \ . \tag{15}$$

This is an infinite summation and in actual computation we need to truncate the sum

$$I_h^{(N)} = h \sum_{k=-N_-}^{N_+} f(\phi(kh))\phi'(kh) \ , \tag{16}$$

where $N = N_- + N_+ + 1$ is the number of function evaluations. Therefore the overall error of (16) is

$$\Delta I_h^{(N)} = I - I_h^{(N)} = I - I_h + I_h - I_h^{(N)} = \Delta I_h + \varepsilon_t \ , \tag{17}$$

where ΔI_h is the discretization error defined by

$$\Delta I_h = I - I_h = \int_{-\infty}^{\infty} f(\phi(u))\phi'(u)du - h \sum_{k=-\infty}^{\infty} f(\phi(kh))\phi'(kh)$$

$$= \frac{1}{2\pi i} \int_{\hat{C}} \hat{\Phi}_h(w) f(\phi(w))\phi'(w)dw \ , \tag{18}$$

and ε_t is the truncation error defined by

$$\varepsilon_t = I_h - I_h^{(N)} = h \sum_{k=-\infty}^{-N_-} f(\phi(kh))\phi'(kh) + h \sum_{k=N_+}^{\infty} f(\phi(kh))\phi'(kh) \ . \tag{19}$$

In general if an analytic function $g(w)$ decays rapidly as $\mathrm{Re}\, w \to \pm\infty$, then it grows rapidly as $\mathrm{Im}\, w \to \pm\infty$, and vice versa. Therefore $|\Delta I_h|$ and $|\varepsilon_t|$ cannot be made small at the same time and there should be an optimal decay rate of $|g(w)|$ as $\mathrm{Re}\, w \to \pm\infty$.

In order to get an optimal quadrature formula Takahasi and Mori [18] investigated the efficiency of the formulas based on the three kinds of variable transformations $x = \phi(u)$ which have the following asymptotic behaviors under the condition that ΔI_h and ε_t are of the same order of magnitude:

$$(a)\quad |\phi'(u)| \approx \exp(-|u|^m),\ m = 1, 2, \ldots \tag{20}$$

$$(b)\quad |\phi'(u)| \approx \exp(-c \exp |u|) \tag{21}$$

$$(c)\quad |\phi'(u)| \approx \exp(-c \exp |u|^m),\ m = 3, 5, \ldots. \tag{22}$$

They found that the optimal decay of $|g(u)|$ or $|f(\phi(u))\phi'(u)|$ is double exponential, i.e.

$$|f(\phi(u))\phi'(u)| \sim \exp(-c \exp |u|), \quad |u| \to \infty\ , \tag{23}$$

and the quadrature formula obtained based on this optimal transformation is called a double exponential formula, abbreviated as DE-formula.

Specifically, for the integral over $(-1, 1)$

$$I = \int_{-1}^{1} f(x)dx \tag{24}$$

the transformation

$$x = \tanh\left(\frac{\pi}{2} \sinh u\right) \tag{25}$$

gives a DE-formula, and for the integral

$$I = \int_{0}^{\infty} f_1(x) \exp(-x)dx \tag{26}$$

the transformation

$$x = \exp(u - \exp(-u)) \tag{27}$$

gives a DE-formula over $(0, \infty)$. It is also shown that the asymptotic error of the formula in terms of the mesh size h of the trapezoidal rule is expressed as

$$|\Delta I_h| \approx \exp\left(-\frac{C}{h}\right)\ , \tag{28}$$

and that the asymptotic error in terms of the number N of the function evaluations is

$$|\Delta I_h^{(N)}| \approx \exp\left(-c\frac{N}{\log N}\right)\ . \tag{29}$$

Before the DE-formula was developed a quadrature formula also based on variable transformation called the IMT-formula had been proposed by Iri, Moriguti and Takasawa in 1969 [3], which was characterized by the fact that the original finite interval of integration $(0, 1)$ was transformed onto itself. The asymptotic error behavior of the formula was shown to be

$$\Delta I^{(N)} \approx \exp(-c\sqrt{N})\ , \tag{30}$$

from which we see that asymptotically the efficiency of the DE-formula is superior to that of the IMT-formula. Mori [6] presented a formula based on the transformation from $(-1, 1)$ onto itself having an asymptotic error behavior

$$\Delta I^{(N)} \approx \exp\left(-c\frac{N}{(\log N)^2}\right) \tag{31}$$

which is slightly inferior to the DE-formula. In 1982 Murota and Iri [8] tried to improve the IMT-formula by means of parameter tuning and repeated application of the IMT-transformation and it turned out that, although the efficiency is improved by the repeated application of the IMT-transformation step by step, its limit does not attain the efficiency of the DE-formula as shown below:

$$\text{IMT-single}: \ \Delta I^{(N)} \approx \exp(-c\sqrt{N}) \tag{32}$$

$$\text{IMT-double}: \ \Delta I^{(N)} \approx \exp\left(-c\frac{N}{(\log N)^2}\right) \tag{33}$$

$$\text{IMT-triple}: \ \Delta I^{(N)} \approx \exp\left(-c\frac{N}{(\log N)(\log\log N)^2}\right) \tag{34}$$

$$\cdots$$

$$\text{DE-formula}: \ \Delta I^{(N)} \approx \exp\left(-c\frac{N}{\log N}\right) . \tag{35}$$

3. Analysis of the DE-Formula on Function Spaces

At a research meeting held at the Research Institute for Mathematical Sciences of Kyoto University in 1985 M.Sugihara presented a detailed theoretical analysis on the optimality of the DE-formula introducing function spaces for integrands and his hand-written note on the analysis appeared in [12] in Japanese. Although he is now preparing a full paper about the details of the analysis, fragrance of his analysis will be worth while to be given here.

Basically he extended the analysis by Stenger [11] on H^p space to the analysis on spaces of functions defined not on the unit circle but directly on the real axis $u \in (-\infty, \infty)$ in the w-plane, where $w = u + iv$. These spaces are characterized by the decay of their elements at large $|\mathrm{Re}\, w|$. First denote the strip domain in the w-plane

$$D(d) \overset{d}{\equiv} \left\{ w \in \mathbb{C} \,\middle|\, |\mathrm{Im}\, w| < \frac{\pi}{2} d \right\} \tag{36}$$

and define

$$\mathscr{A}(D(d)) \equiv \{\text{analytic functions on } \mathrm{D}(d)\} . \tag{37}$$

He introdeuced a function space

$$H_{\text{double}}(D(d); A, B) \ (B < 1/d)$$
$$\equiv \left\{ g \in \mathscr{A}(D(d)) \,\middle|\, \sup_{w \in D(d)} \{|g(w)| \cdot |\exp(A\cosh Bw)|\} < +\infty \right\} \tag{38}$$

with an appropriate norm $\| g \|_{\text{double}}$. In short this space is characterized by the double exponential decay of its elements at large $|\text{Re}\, w|$. Consider the integral (3) and an approximation (7) to it in H_{double}. As to the norm of the error

$$E_N = \int_{-\infty}^{\infty} g(u)du - \sum_{k=1}^{N} A_k g(a_k) \tag{39}$$

Sugihara proved

Theorem 1. *In* $H_{\text{double}}(D(d); A, B)$

$$\inf_{a_k, A_k \in \mathbb{R}} \| E_N(a_1, \ldots, a_N; A_1, \ldots, A_N) \| \geq C \exp\left(-c\frac{N}{\log N}\right) . \tag{40}$$

On the other hand, the following theorem holds for the trapezoidal rule.

Theorem 2. *In* $H_{\text{double}}(D(d); A, B)$

$$\| E_N(\text{trapezoidal rule}) \| \leq C' \exp\left(-c'\frac{N}{\log N}\right) . \tag{41}$$

From these two theorems we immediately see that the trapezoidal rule is optimal in H_{double}.

Next, in a similar way as H_{double}, Sugihara introduced another function space H_{single} characterized by the single exponential decay of its elements at large $|\text{Re}\, w|$, and showed again the optimality of the trapezoidal rule in H_{single}. However, the inequality for the error in H_{single} corresponding to (41) is

$$\| E_N(\text{trapezoidal rule}) \| \leq C' \exp\left(-\sqrt{\pi^2 d A}\sqrt{N + \frac{1}{2}}\right) , \tag{42}$$

where A is some constant, so that it is clear that the double exponential transformation asymptotically leads to a more efficient quadrature formula than the single exponential one. Then it is quite natural to raise a question: is the trapezoidal rule more efficient in a space whose elements decay more rapidly than those in H_{double}? Sugihara answered the question negatively by proving that there exists no element except zero function in such a space. Consequently we conclude that the DE-formula is asymptotically optimal.

4. The DE-Formula for Slowly Decaying Oscillatory Integrals

Consider the integral

$$I = \int_0^{\infty} f(x)dx , \tag{43}$$

where

$$f(x) = f_1(x)\cos x, \quad f_1(x) = \text{algebraic function} . \tag{44}$$

In this case $f(\phi(w))\phi'(w)$ does not belong to H_{double}, so that the DE-formula does not work well as seen from the analysis in the previous section. Toda and Ono

[19] applied the DE-formula followed by the Richardson's extrapolation method successfully to

$$I = \lim_{z \to 0} \int_0^\infty f(x) \exp(-zx) dx \ . \tag{45}$$

Afterwards Sugihara showed both theoretically and experimentally [13] that the Richardson's method is more efficiently applied to

$$I = \lim_{z \to 0} \int_0^\infty f(x) \exp(\ zx^2) dx \ . \tag{46}$$

Recently Ooura and Mori [9] presented an interesting transformation which gives an efficient formula for integrals such as

$$I = \int_0^\infty f_1(x) \sin \omega x dx \tag{47}$$

$$I = \int_0^\infty f_1(x) \cos \omega x dx. \tag{48}$$

Consider the variable transformation

$$x = M\phi(t) = \frac{Mu}{1 - \exp(-K \sinh u)}, \quad M, K = \text{constant} \ . \tag{49}$$

Then $\phi(u)$ satisfies $\phi(-\infty) = 0$ and $\phi(+\infty) = \infty$. Moreover,

$$\lim_{u \to +\infty} \phi(u) = u \quad \text{double exponentially} \tag{50}$$

$$\lim_{u \to -\infty} \phi'(u) = 0 \quad \text{double exponentially} \tag{51}$$

hold. If we apply the variable transformation $x = M\phi(u)$ to (47) and compute it by the trapezoidal rule, we have

$$I_h = Mh \sum_{k=-\infty}^{\infty} f_1(M\phi(kh)) \sin(\omega M \phi(kh)) \phi'(kh) \ . \tag{52}$$

Choose h such that $\omega Mh = 2\pi$, then

$$\sin(\omega M \phi(kh)) \sim \sin \omega Mkh = \sin 2\pi k = 0 \ , \tag{53}$$

so that we can truncate the sum (52) at some moderate value of k.

5. Problems Arising in Coding Automatic Integrator

When you write an automatic integrator based on the DE-formula you must be careful about the loss of significant digits which may occur when computing, say, $(1 + x)^{-3/4}$ in (1) in the close neighborhood of $x = -1$, and about the overflow. A device to avoide the loss of significant digits is given in [18].

Recently a useful method to avoide the overflow which may occur when computing the weights of the DE-formula was presented by Watanabe [21]. Consider again the integral

$$I = \int_{-1}^{1} f(x)dx \ . \tag{54}$$

The weights of the DE-formula obtained by

$$x = \phi(u) = \tanh(\frac{\pi}{2} \sinh u) \tag{55}$$

are

$$A_k = \frac{\cosh kh}{\cosh^2(\frac{\pi}{2} \sinh kh)}, \quad k = 0, \pm 1, \pm 2, \ldots \ , \tag{56}$$

and a careless coding often gives rise to the overflow when computing the denominator in (56) because it grows double exponentially as k becomes large. Watanabe found a recurrence relation

$$A_{k+1} = A_k \times r_k \ , \tag{57}$$

where

$$r_k = \frac{\cosh h + \sinh h \tanh kh}{(\cosh s_k + \phi(kh) \sinh s_k)^2} \tag{58}$$

and

$$s_k = \pi \sinh \frac{h}{2} \cosh((k + \frac{1}{2})h) \ , \tag{59}$$

and showed that the integral (54) can be computed by the following small code:

```
      I=0
      DO 10 k = N, 1, −1
         I = (I + f(a_k) + f(−a_k)) × r_{k−1}
   10 CONTINUE
      I = (πh/2)(I + f(a_0)) .
```

Although the denominator of r_k has a double exponential factor $\sinh s_k$, its inner exponential factor has a small coefficient $\sinh(h/2)$, so that the overflow in r_k will not occur until k becomes much larger than such k for which the overflow occurs in A_k.

6. Applications of the DE-Formula

The DE-formula is used for multiple integration. See [14] and [1], and the references therein.

The DE-formula is installed in many computer centers in Japan and is easily found in subroutine packages in the Japanese market. It is actually used in various fields of science and technology. In the references one paper is listed from each of the fields, the boundary element method [2], the suface charge method [20], filter analysis[4], and molecular chemistry [5]. Very recently also in the field of statistics it is proved to be quite efficient for numerical evaluation of risk of improved estimation [15]. For further reference see these papers and the references therein.

References

1. V.U. Aihie, G.A. Evans: A comparison of the error function and the tanh transformation as progressive rules for double and triple singular integrals. J. Comput. Appl. Math. **30** (1990) 145–154
2. T. Higashimachi, N. Okamoto, Y. Ezawa, T. Aizawa, A. Ito: Interactive structural anaysis system using the advanced boundary element method. *Boundary Elements* — Proceedings of the Fifth International Conference, Hiroshima, Japan, November 1983, eds. Brebbia, C. A., Futagami, T., Tanaka, M., A Computational Mechanics Center Publication. Springer, Berlin Heidelberg New York, pp. 847–856
3. M. Iri, S. Moriguti, Y. Takasawa: On a certain quadrature formula. J. Comput. Appl. Math. **17** (1987) 3–20 — translation of the original paper in Japanese that appeared in Kokyuroku, RIMS. Kyoto Univ. no. 91 (1970) 82–118
4. T. Kida, T. Kurogochi: A simple numerical derivation of group delay and impulse response from the prescribed characteristic function of a reactance low-pass filter (in Japanese). The Transactions of the Institute of Electronics and Communication Engineers **J63** (1980) 421–428
5. T. Momose, T.Shida: Efficient formulas for molecular integrals over the Hiller-Sucher-Feinberg identity using Cartesian Gaussian functions: Towards the improvement of spin density calculation. J. Chem. Phys. **87** (1987) 2832–2846
6. M. Mori: An IMT-type double exponential formula for numerical integration. Publ. RIMS, Kyoto Univ. **14** (1978) 713–729
7. M. Mori: The double exponential formulas for numerical integration over the half infinite interval. Numerical Mathematics Singapore 1988, International Series of Numerical Mathematics, vol. 86. Birkhäuser, Boston 1988, pp. 367–379
8. K. Murota, M. Iri: Parameter tuning and repeated application of the IMT-type transformtaion in numerical quadrature. Numer. Math. **38** (1982) 327–363
9. T. Ooura, M. Mori: A quadrature formula for oscillatory integrals over the half infinite interval based on variable transformation (in Japanese). Kokyuroku, RIMS Kyoto Univ. no. 717 (1990) 68–75
10. M. Sato: Theory of hyperfunctions, I. J. Fac. Sci. Univ. Tokyo **8** (1959) 139–193
11. F. Stenger: Optimal convergence of minimum norm approximation on H^p. Numer. Math. **29** (1978) 345–362
12. M. Sugihara: On the optimality of the quadrature formula based on the DE-type variable transformation (in Japanese). Kokyuroku, RIMS Kyoto Univ. no. 585 (1986) 150–175

13. M. Sugihara: Methods of numerical integration of oscillatory functions by the DE-formula with the Richardson extrapolation. J. Comput. Appl. Math. **17** (1987) 47–68
14. M. Sugihara: Method of good matrices for multi-dimensional numerical integrations — An extension of the method of good lattice points. J. Comput. Appl. Math. **17** (1987) 197–213
15. N. Sugiura: Private communication
16. H. Takahasi, M. Mori: Error estimation in the numerical integration of analytic functions. Rep. Comput. Centre Univ. Tokyo **3** (1970) 41–108
17. H. Takahasi, M. Mori: Quadrature formulas obtained by variable transformation. Numer. Math. **21** (1973) 206-219
18. H. Takahasi, M. Mori: Double exponential formulas for numerical integration. Publ. RIMS Kyoto Univ. **9** (1974) 721–741
19. H. Toda, H. Ono: Some remarks for efficient usage of the double exponential formulas (in Japanese). Kokyuroku, RIMS, Kyoto Univ. no. 339 (1978) 74–109
20. Y. Uchikawa, T. Ohye, K. Gotoh: Improved surface charge method (in Japanese). The Transactions of the Institute of Electrical Engineers of Japan **A 101** (1981) 263–270
21. T. Watanabe: On the double exponential formula for numerical integration (in Japanese). Kakuyugokenkyu **63** (1990) no. 5, 397–411

Computational Complexity of Solving Real Algebraic Formulae

James Renegar

School of Operations Research and Industrial Engineering, Cornell University
Ithaca, NY 14853-7501, USA

1. Introduction

We briefly survey recent computational complexity results for certain algebraic problems that are relevant to numerical analysis and mathematical programming. Topics include (i) linear programming, (ii) decision methods and quantifier elimination methods for the first order theory of the reals, (iii) solving real algebraic formulae approximately, and (iv) ill-posed problem instances.

2. Linear Programming

In this section we discuss complexity results for the linear programming problem

$$\begin{aligned} &\text{maximize } c^T x \\ &\text{subject to } Ax \geq b \end{aligned} \tag{2.1}$$

where $c \in \mathbb{R}^n$, $b \in \mathbb{R}^m$ and A is an $m \times n$ matrix.

In the last four years there has been a vast amount of work on "interior point" algorithms, motivated by Karmarkar's algorithm [16]. Unlike the traditional simplex method which moves from vertex to vertex around the feasible region $\{x; Ax \geq b\}$, interior point methods proceed through the interior $\{x; Ax > b\}$ of the feasible region.

Karmarkar's algorithm is a "projective" interior point algorithm, the basic computation for each iteration being a projective transformation. In the last four years another breed of interior point algorithms has received a lot of attention, "path-following" algorithms. These are closer to traditional numerical analysis than are projective algorithms, having Newton's method at their heart. The best upper bounds known for the complexity of linear programming are based on the analysis of particular path-following algorithms.

Following are the simple ideas behind the first path-following algorithm proven to have a polynomial-time bound. Let α_i^T denote the i-th row of the constraint matrix A.

The *center* of the system of linear inequalities $Ax \geq b$ is the point z which maximizes $\prod_i(\alpha_i^T x - b_i)$, viewed as a function restricted to the feasible region. The center exists and is unique if the feasible region is bounded and has non-empty interior, as we assume in what follows. Equivalently, then, the center is

Proceedings of the International Congress
of Mathematicians, Kyoto, Japan, 1990

the maximizer of the strictly concave function $f(x) := \sum_i \ln(\alpha_i^T x - b_i)$, viewed as being defined only on the interior of the feasible region.

The center has a natural physical interpretation which arises from the equations $\nabla f(z) = 0$. Consider each hyperplane $\{x;\, \alpha_i^T x = b_i\}$ as emitting a force which acts on an arbitrary point x not in the hyperplane. The direction of the force is orthogonal to, and away from, the hyperplane, and its magnitude at x equals the reciprocal of the distance from x to the hyperplane, i.e., the force at x is simply $\alpha_i/(\alpha_i T_x - b_i)$. Then the center is the unique equilibrium point in the interior of the feasible region.

Now assume $k^{(0)}$ is a known strict lower bound on the optimal objective value for (2.1). Let $z^{(0)}$ denote the center of the extended system

$$\begin{aligned} Ax &\geq b \\ c^T x &\geq k^{(0)} \end{aligned} \tag{2.2}$$

and assume that $x^{(0)}$ is known to be a feasible "good" approximation to $z^{(0)}$. We know $x^{(0)}$ but not necessarily $z^{(0)}$. We want to move from $x^{(0)}$ towards an optimal solution for (2.1). A natural way to proceed is to create a new system with center $z^{(1)}$ closer to an optimal solution than $z^{(0)}$, and then move from $x^{(0)}$ to a feasible "good" approximation $x^{(1)}$ for $z^{(1)}$.

To create a new system we simply increase $k^{(0)}$. Of course we must be careful that $k^{(0)}$ not be increased above the optimal objective value for (2.1). Hence, it is natural to replace $k^{(0)}$ by $k^{(1)} = \delta c^T x^{(0)} + (1 - \delta)k^{(0)}$ where $0 < \delta \leq 1$. This corresponds to bringing the hyperplane $\{x;\, c^T x = k^{(0)}\}$ for (2.2) towards $x^{(0)}$, thus causing the equilibrium point to move to another point with better objective value.

To compute an approximation $x^{(1)}$ for $z^{(1)}$ we apply one iteration of Newton's method, beginning at $x^{(0)}$, to the equations $\nabla f^{(1)}(x) = 0$ where $f^{(1)}(x) := \ln(c^T x - k^{(1)}) + \sum_i \ln(\alpha_i^T x - b_i)$. (The special structure of $f^{(1)}$ makes the gradient and Hessian very easy to compute).

It is easily proven that if δ is sufficiently small and $x^{(0)}$ is a "good" approximation to $z^{(0)}$ then $x^{(1)}$ obtained in this manner will be a "good" approximation to $z^{(1)}$. However, to establish noteworthy complexity bounds we need to prove something to the effect that δ need not be "too" small. In [23], the author proved that any δ satisfying $0 < \delta \leq 1/13$ is allowable, i.e., proceeding iteratively the algorithm will then generate points $x^{(j)}$, $j = 0, 1, \ldots$, converging to an optimal solution. In practice, much larger δ are acceptable.

The algorithm terminates with standard procedures for computing an exact optimal solution from sufficiently close approximations.

The number "1/13" arises from an analysis of Newton's method for a carefully chosen coordinate system. The analysis can be greatly simplified by relying on work of Smale [32] as was shown by Renegar and Shub [28]. The latter paper presents a unified complexity analysis for several path-following interior point algorithms.

The best complexity bound known for linear programming is due to Vaidya [38]. He established the bound for a path-following algorithm not suggested to be practical, as it relies on fast matrix multiplication. Vaidya's bound is $O((m+n)^{3/2} nL^2 \log(L) \log\log(L))$ bit operations where (very roughly) L is the number of bits required to specify the particular problem instance, i.e., required to specify A, b and c. This bound is an improvement on the earlier record bound of Gonzaga [11] and Vaidya [37] (derived from a modification of the algorithm described above), which in turn was an improvement on the record bound of Karmarkar [16], which in turn was an improvement on the original polynomial-time bound of Khachiyan [17].

The complexity bounds for interior point algorithms are generally obtained by bounding the number of iterations required by an algorithm and multiplying the bound by the amount of work required per average iteration. The best iteration bound known for an interior point method is $O(\sqrt{m+n}\,L)$ and was established by the author in [23]. The recent record complexity bounds arise from clever ways to (theoretically) reduce the amount of work required per average iteration. Much effort has been expended by researchers (including the author) to decrease the iteration bound, but to no avail.

Other seminal papers in the complexity theory of interior point methods include work by Kojima, Mizuno and Yoshise [18] (motivated by Megiddo [19]), Monteiro and Adler [21], and Ye [40]. The amount of recent literature on interior point algorithms is staggering. Relevant surveys have been written by Gonzaga [12], Goldfarb and Todd [10], and Megiddo [20].

All polynomial time algorithms for linear programming require polynomial time in the Turing machine sense, i.e., the number of bit operations is bounded by a polynomial function in the bit length L of the input. The number of arithmetic operations (over the rationals Q) for all of the algorithms tends to infinity as L does, even when m and n are fixed. (By contrast, the number of arithmetic operations required by the simplex method can be bounded above by a function of m and n alone).

A major open question in the complexity theory of linear programming is whether or not there exist polynomial-time real number machine algorithms in the sense of Blum, Shub and Smale [2], i.e., is "uniform," accepts arbitrary real number coefficients as inputs, and has arithmetic operation count bounded by a polynomial in m and n. Tardos [35] has made some progress on this question by devising an algorithm with arithmetic operation bound independent of the coordinates of the data vectors b and c, but dependent on the coefficients of A which she assumes to be integers.

"Experts" are divided in their opinions as to the answer of the question. If the answer was affirmative then most likely there would be practically important linear programming algorithms yet to be discovered. If the answer was negative then the complexity heirarchy for real number machines would be very different from that for Turing machines.

3. Decision Methods and Quantifier Elimination Methods

Now we move to a very general setting which includes many problems from numerical analysis and mathematical programming, e.g., eigenvalue problems, non-linear programming with multi-variate polynomial objective and constraint functions, sensitivity analysis problems, etc. All of these problems arise in the *classical* (by computational complexity standards) setting of the decision problem for the first order theory of the reals. We begin with a quick introduction for readers unfamiliar with this setting.

A *sentence* is an expression composed from certain ingredients. Letting $\mathbb{R}$ denote a real-closed field, the following is an example of a sentence:

$$(\exists x_1 \in \mathbb{R}^{n_1})(\forall x_2 \in \mathbb{R}^{n_2}) \left[(g_1(x_1,x_2) > 0) \bigvee (g_2(x_1,x_2) = 0)\right] \bigwedge (g_3(x_1,x_2) \neq 0)\,. \tag{3.1}$$

The ingredients are: vectors of variables (x_1 and x_2); the quantifiers $\exists$ and $\forall$; atomic predicates (e.g. $g_1(x_1,x_2) > 0$) which are polynomial inequalities ($>, \geq, =, \neq, <,$); and a Boolean function holding the atomic predicates ($[B_1 \vee B_2] \wedge B_3$).

A sentence asserts something. The above sentence asserts that there exists $x_1 \in \mathbb{R}^{n_1}$ such that for all $x_2 \in \mathbb{R}^{n_2}$, (i) either $g_1(x_1,x_2) > 0$ or $g_2(x_1,x_2) = 0$, and (ii) $g_3(x_1,x_2) \neq 0$. Depending on the specific coefficients of the atomic predicate polynomials this assertion is true or it is false.

The set of all true sentences constitutes the first order theory of the reals. A *decision method* for the first order theory of the reals is an algorithm which, given any sentence, correctly determines if the sentence is true. Decision methods for the reals were first proven to exist by Tarski [36] who constructed one.

A sentence is a special case of a more general expression, called a *formula.* Here is an example of a formula:

$$(\exists x_1 \in \mathbb{R}^{n_1})(\forall x_2 \in \mathbb{R}^{n_2}) \left[(g_1(z,x_1,x_2) > 0) \bigvee (g_2(z,x_1,x_2) = 0)\right] \bigwedge (g_3(z,x_1,x_2) \neq 0)\,. \tag{3.2}$$

A formula has one thing that a sentence does not, namely, a vector $z \in \mathbb{R}^{n_0}$ of *free variables.* When specific values are substituted for the free variables, the formula becomes a sentence.

A vector $\bar{z} \in \mathbb{R}^{n_0}$ is a *solution* for the formula if the sentence obtained by substituting $\bar{z}$ is true.

Two formulae are *equivalent* if they have the same solutions.

A *quantifier elimination method* is an algorithm which, given any formula, computes an equivalent quantifier-free formula, i.e., for the above formula $(\exists x_1 \in \mathbb{R}^{n_1})(\forall x_2 \in \mathbb{R}^{n_2})P(z,x_1,x_2)$ such a method would compute an equivalent formula $Q(z)$ containing no quantified variables.

When a quantifier elimination method is applied to a sentence, it becomes a decision method. Thus, a quantifier elimination method is in some sense more general than a decision method.

Tarski [36] actually constructed a quantifier elimination method.

Many problems in numerical analysis and mathematical programming can be cast as the problem of computing a solution for a particular formula. The reader will easily verify that this can be done for the problems mentioned earlier. It can be done for many other problems as well. Of course determining if a solution for a formula exists can be done with a decision method. In the next section we discuss the complexity of approximating solutions for formulae.

Both (3.1) and (3.2) are said to be in prenex form, i.e., all quantifiers occur in front. More generally, a formula can be constructed from other formulae just as (3.1) was constructed from the atomic predicates.

We now present a brief survey of some complexity highlights for quantifier elimination methods, considering only formulae in prenex form. General bounds follow inductively. (If a formula is constructed from other formulae, first apply quantifier elimination to the innermost formulae, then to the innermost formulae of the resulting formula, etc.)

We consider the general formula

$$(Q_1 x_1 \in \mathbb{R}^{n_1}) \ldots (Q_\omega x_\omega \in \mathbb{R}^{n_\omega}) P(z, x_1, \ldots, x_\omega), \tag{3.3}$$

where $Q_1, \ldots, Q_\omega$ are quantifiers, assumed without loss of generality to alternate, i.e., Q_i is not the same as Q_{i+1}. Let m denote the number of distinct polynomials occurring among the atomic predicates and let $d \geq 2$ be an upper bound on their degrees.

In the case of Turing machine computations where all polynomial coefficients are restricted to be integers, we let L denote the maximal bit length of the coefficients. In this context we refer to the number of "bit operations" required by a quantifier elimination method. In the general and idealized case that the coefficients are not integers we rely on the computational model of Blum, Shub and Smale [2], and refer to "arithmetic operations", these essentially being field operations, including comparisons.

The sequential bit operation bounds that have appeared in the literature are all basically of the form

$$(md)^E \left[L^{O(1)} + \text{Cost}\right] \tag{3.4}$$

where E is some exponent and "Cost" is the worst-case cost of evaluating the Boolean function holding the atomic predicates, i.e., worst-case over 0–1 vectors.

The first reasonable upper bound for a quantifier elimination method was proven by Collins [5]. He obtained $E = 2^{O(n)}$ where $n := n_0 + \ldots + n_\omega$. Collins' bound is thus "doubly exponential" in the number of variables. His method requires the formula coefficients to be integers, the number of arithmetic operations (not just bit operations) growing with the size of the integers. This is reminiscent of the polynomial time algorithms for linear programming discussed earlier. Also, Collins' algorithm was not shown to parallelize, although enough is now known that a parallel version probably could be developed. Collins' work has been enormously influential in the area.

The next major complexity breakthrough was made by Grigor'ev [13] who developed a decision method for which $E \approx [O(n)]^{4\omega}$. Grigor'ev's bound is doubly exponential only in the number of quantifier alternations. Many interesting problems can be cast as sentences with only a few quantifier alternations. For these, Grigor'ev's result is obviously significant. Like Collins' quantifier elimination method, Girgor'ev's decision method requires integer coefficients and was not proven to completely parallelize.

Slightly incomplete ideas of Ben-Or, Kozen and Reif [1] were completed by Fitchas, Galligo and Morgenstern [9] to construct a quantifier elimination method with arithmetic operation bound

$$(md)^E \text{Cost} \tag{3.5}$$

where $E = 2^{O(n)}$. This provides an arithmetic operation analog of Collins' bit operation bound. When restricted to integer coefficients, the method also yields the Collins' bound if the arithmetic operations are carried out bit by bit. Moreover, the algorithm parallelizes. Assuming each arithmetic operation requires one time unit, the resulting time bound is

$$[E \log(md)]^{O(1)} + \text{Time}(N) \tag{3.6}$$

if $(md)^E N$ parallel processors are used, where Time(N) is the worst-case time required to evaluate the Boolean function holding the atomic predicates using N parallel processors. The analogous time bound for bit operations is also valid, namely, $[E \log(Lmd)]^{O(1)} + \text{Time}(N)$ if $(md)^E[L^{O(1)} + N]$ parallel processors are used.

In [25], the author introduced a new quantifier elimination method for which $E = \prod_{k=0}^{\omega} O(n_k)$. This was established for arithmetic operations and bit operations, i.e., (3.4) and (3.5). Regarding bit operations, the dependence of the bounds on L was shown to be very low, $L^{O(1)}$ in (3.4) being replaced by $L(\log L)(\log\log L)$, i.e., the best bound known for multiplying two L-bit integers. Moreover, the method was shown to parallelize, resulting in the arithmetic operation time bound (3.6) if $(md)^E N$ parallel processors are used, and the bit operation time bound $\log(L)[E \log(md)]^{O(1)} + \text{Time}(N)$ if $(md)^E[L^2 + N]$ parallel processors are used.

Independently and simultaneously, Heintz, Roy and Solerno [15] developed a quantifier elimination method for which $E = O(n)^{O(\omega)}$, both for arithmetic and bit operations. Their method also completely parallelizes.

The various bounds are best understood by realizing that quantifier elimination methods typically work by passing through a formula from back to front. First the vector x_w is focused on, then the vector $x_{\omega-1}$, and so on. Some methods ([5, 25]) make a second pass, from front to back. The work arising from each vector results in a factor for E. For Collins' quantifier elimination method, the factor corresponding to x_k is $2^{O(n_k)}(2^{O(n)} = 2^{O(n_0)} \dots 2^{O(n_\omega)})$. For the method introduced by the author, the factor is $O(n_k)$. The factor corresponding to Grigor'ev's decision method is $\approx O(n)^4$ independently of the number of variables in x_k. In that method a vector with few variables can potentially create as much work as one with many variables. Similarly, the factor corresponding to the quantifier elimination method of Heintz, Roy and Solerno is $O(n)^{O(1)}$ independently of x_k.

For the record, the quantifier elimination method in [25] produces a quantifier-free formula of the form

$$\bigvee_{i=1}^{I} \bigwedge_{j=1}^{J_i} (h_{ij}(z) \Delta_{ij} 0)$$

where $I \leq (md)^E$, $J_i \leq (md)^{E/n_0}$, $E = \prod_{k=0}^{\omega} O(n_k)$, the degree of h_{ij} is at most $(md)^{E/n_0}$ and the Δ_{ij} are standard relations $(\geq, >, =, \neq, <, \leq)$. If the coefficients of the original formula are integers of bit length at most L, the coefficients of the polynomials h_{ij} will be integers of bit length at most $(L + n_0)(md)^{E/n_0}$.

Results of Weispfenning [39], and Davenport and Heintz [6], show the double exponential dependence on ω of the above bound on the degrees of the polynomials h_{ij} cannot be improved in the worst case.

Fisher and Rabin [8] proved an exponential worst-case lower bound for decision methods. However, the lower bound is exponential only in the number of quantifier alternations, and is only singly exponential in that. A tremendous gap remains between the known upper and lower bounds for decision methods.

In closing this section we mention that work of Canny ([3], [4]) has been especially influential in this area in recent years, both for the techniques he has developed and employed and for the connections he has established between the area and robotics. Work of Vorobjov [14] has also been very influential.

4. Solving Formulae Approximately

In this section we discuss the complexity of approximating solutions for formulae, restricting attention to the field $\mathbb{R}$ of real numbers.

The most basic problem in this vein is that of approximating roots of univariate polynomials. Sequential bounds for this problem are numerous, for various models of computation, and have been proven over many years. Discussions have been provided by Smale ([31, 33]), and Schonhage [29].

Until very recently significant time bounds for the parallel computation of roots of univariate polynomials have been missing. However, Neff [22] has proven, in the parlance of computer science, that the problem is in NC. He has shown that all roots of a univariate polynomial can be approximated to within Euclidean distance $\varepsilon > 0$ in time $O[\log(Ld)+\log\log(4+\frac{1}{\varepsilon})]^3$ using $[Ld \log(2+\frac{1}{\varepsilon})]^{O(1)}$ parallel processors, where $d \geq 3$ is the degree of the univariate polynomial and L is the maximal bit length of the coefficients, assumed to be integers. Although Neff does not present an arithmetic operation time-bound for arbitrary real number coefficients, his ideas can be extended to do so. Assuming we desire to approximate all roots lying within distance r of the origin, the resulting time bound is of the form $[\log(d)\log\log(4+\frac{r}{\varepsilon})]^{O(1)}$ if $[d \log(2+\frac{r}{\varepsilon})]^{O(1)}$ parallel processors are used, assuming one time unit is required per processor per arithmetic operation.

Neff's result and techniques have implications beyond the univariate setting. For example, the principal bottleneck in parallelizing Collins' quantifier elimination method has been its reliance on univariate polynomial root approximation. Neff's result removes that bottleneck.

In [26], the author reduces the problem of approximating solutions for formulae (3.3) to the problem of approximating roots for univariate polynomials. Both sequential and parallel complexity bounds for the reduction are provided. Using Neff's algorithm and ignoring the cost of evaluating the Boolean function holding the atomic predicates (which generally is a relatively negligible cost), the resulting arithmetic operation time bound is $[E \log(md)\log\log(4+\frac{r}{\varepsilon})]^{O(1)}$ if $[(md)^E \log(2+\frac{r}{\varepsilon})]^{O(1)}$ parallel processors are used, where as in the last section, $E = \prod_k O(n_k)$. For each connected component of the solution set that intersects $\{z; \|z\| \leq r\}$, a point within Euclidean distance ε of the component is computed within this time, assuming that either (i) both r and ε are input to the algorithm or (ii) only ε is input and r is defined to be the infimum of distances of all solutions from the origin.

The resulting bit operation time bound is $[E \log(Lmd)\log\log(4+\frac{1}{\varepsilon})]^{O(1)}$ if $[L \log(2+\frac{1}{\varepsilon})]^{O(1)}(md)^E$ parallel processors are used. For each connected component of the solution set, a point within distance ε of the component is computed within this time.

(The solution set consists of at most $(md)^E$ connected components. If the coefficients of the formula are all integers of bit length at most L then each connected component of the solution set has a point within distance $2^{\overline{L}}$ of the origin, where $\overline{L} = L(md)^E$.)

Sequential bounds established in [26], the best presently available, are $(md)^E \log\log(4+\frac{r}{\varepsilon})$ arithmetic operations and $\hat{L}^2\log(\hat{L})\log\log(\hat{L})(md)^E$ bit operations where $\hat{L} = L + \log(2+\frac{1}{\varepsilon})$. The dependence of this arithmetic operation bound on r and ε cannot be improved, as was proven by the author in [24].

The occurrence of the ratio r/ε in both the sequential and parallel arithmetic operation bounds naturally leads one to suspect that analogous bounds hold for relative error, that is, for the problem of computing a point within distance $\varepsilon/\|\bar{x}\|$ of an actual solution $\bar{x}$. However, it is easily proven that even for the problem of computing relative approximations to roots of quadratic polynomials, no real number model algorithm has a uniform arithmetic operation bound independent of the quadratic polynomial, depending only on the degree and relative error desired. Uniform bounds for computing points within specified relative error of a solution require that the basic algorithmic operations include more than just arithmetic operations and comparison operations, e.g., radicals.

5. Ill-Posed Problem Instances

Complexity theory has been developed almost exclusively for problems for which exact data is used in computations. A theory which fully incorporates the use of approximate data has yet to be developed. Central to such a development will be the notion of an ill-posed problem instance. In this section we relate an answer to the question, "Is it possible to know a problem instance is ill-posed?" A complete development can be found in [27]. Here we can only sketch a few of the ideas, rather vaguely.

Define a problem to be a formula $P(x, y)$ with two vectors $x \in \mathbb{R}^n$, $y \in \mathbb{R}^m$ of free variables, $\mathbb{R}$ denoting the real numbers. A *problem instance* corresponds to specifying values for y. The values are the *data* for the instance. We say "*x is a solution for instance y*" if the pair (x, y) is a solution for the formula.

Many problems can be cast in this format. For example, linear programming with a fixed number of constraints and variables fits this format. The vector y then specifies A, b and c in (2.1).

We assume a formula $P(x, y)$ encoding the problem of interest is known and we assume arbitrarily accurate approximate data for the actual instance is available through an oracle. Input to the oracle is $\delta > 0$ and output is $\bar{y}$ strictly within error δ of the data for the actual instance. (Very general functions are allowed in measuring solution and data errors, but for this brief synopsis assume errors are measured by norms.)

The goals:

1) Determine if the actual instance has a solution.
2) If it has a solution, compute a ε-approximate solution, i.e., $\bar{x} \in \mathbb{R}^n$ guaranteed to be within error ε of a solution for the actual instance.

The goals are to be achieved using approximate data and any other information available about the actual instance, e.g., it might be known the actual instance has infinitely many solutions, even if the exact data for the instance is not known. There are no restrictions on the known information, including its form, except it is required to be consistent, i.e., a contradiction cannot be deduced from it.

Depending on the known information and the exact data for the actual instance, the goals may not be achievable. Roughly, the actual instance is *ill-posed* if the goals cannot be achieved regardless of how accurate the approximate data is.

To be more definite, we introduce three definitions, which are discussed at length in [27].

Definition. A problem instance $\tilde{y}$ is *indistinguishable from the actual instance* if the known information regarding the actual instance does not exclude the possibility that $\tilde{y}$ is the actual instance.

Definition. An *acceptable algorithm* for the problem:

1) Accepts as input any tuple $(\bar{y}, \delta, \varepsilon)$ where $\delta > 0$, $\varepsilon > 0$ and $\bar{y}$ might be provided by the oracle upon input δ.
2) Replies one of the following three statements:
   (a) "All instances which
      (i) arc indistinguishable from the actual instance and
      (ii) are strictly within error δ of $\bar{y}$,
      have solutions, and $\bar{x}$ is strictly within error ε of a solution for all such instances," (where $\bar{x}$ is computed by the algorithm).
   (b) "All instances which
      (i) are indistinguishable from the actual instance and
      (ii) are strictly within error δ of $\bar{y}$,
      do not have solutions."
   (c) "Please provide better data accuracy."
3) Can be proven correct, i.e., correct in the sense that whenever it replies with statement (2a) or (2b) and the input tuple satisfies the condition that $\bar{y}$ is strictly within error δ of data for an instance which is indistinguishable from the actual instance then the statement replied is indeed true.

The motivation for requiring that the algorithm can be proven correct in the sense of (3) is that if the algorithm does, say, reply with statement (2a) upon some input for which $\bar{y}$ is strictly within error δ of the actual instance then one can prove, in terms of one's knowledge regarding the actual instance, that the point $\bar{x}$ is indeed within error ε of a solution for the actual instance. In other words, one can be certain in terms of what one knows about the actual instance that the algorithm will not erroneously claim a certain point to be within error ε of a solution for the actual instance when in fact it is not. *Indeed, the reader should regard the definition of an acceptable algorithm to simply formalize the requirement that one be able to trust the algorithm not to reply with an incorrect answer for the actual instance.*

Definition. The actual instance is *definitely well-posed* if some acceptable algorithm replies (2a) or (2b) whenever $\bar{y}$ is strictly within error δ of the actual instance and $\delta > 0$ is sufficiently small, where what constitutes sufficiently small δ may depend on ε.

We do not provide a precise definition of an ill-posed problem instance. We only assume that whatever definition is chosen, it excludes 'definitely well-posed' instances.

In [27], the author argues that if the information known about the actual instance can be used to deduce the actual instance is not 'definitely well-posed' then it can be used to deduce the actual instance is 'definitely well-posed', a contradiction. Consequently, if the known information is consistent, it cannot be known the actual instance is ill-posed. (Somewhat inaccurately, the result amounts to saying that if one knows arbitrarily accurate approximate data is insufficient for the goals, then that knowledge provides sufficient information to design acceptable algorithms for which accurate approximate data is sufficient, a contradiction.) Consequently, for problems corresponding to real formulae it is impossible to know if for the instance of interest it is pointless to collect better approximate data and try new algorithms (although it is certainly possible to sometimes know it is not pointless).

The impossibility of being able to know the actual instance is ill-posed is primarily a consequence of the existence of decision methods for the first order theory of the reals. If instead, for example, solutions are required to be rational vectors, examples can be easily constructed showing it is possible to deduce from the known information that the actual instance is not 'definitely well-posed' without arriving at a contradiction. This is discussed at greater length in [27].

Again, much work needs to be done to develop a complexity theory which incorporates the use of approximate data. Some reflections on this have been provided by Demmel [7], Smale [34] and Shub [30]. In [27], basics for a very general theory of condition numbers are developed.

References

1. Ben-Or, M., Kozen, D., Reif, J.: The complexity of elementary algebra and geometry. J. Comp. System Sci. **32** (1986) 251–264
2. Blum, L., Shub, M., Smale, S.: On a theory of computation and complexity over the real numbers: NP-completeness, recursive functions and universal machines. Bull. Amer. Math. Soc. **21** (1989) 1–46
3. Canny, J.: The Complexity of Robot Motion Planning. MIT Press, Cambridge, Mass., 1989
4. Canny, J.: Some algebraic and geometric computations in PSPACE. Proceedings of the 20th Annual ACM Symposium on the Theory of Computing 1988, pp. 460–467
5. Collins, G.: Quantifier elimination for real closed fields by cylindrical algebraic decomposition. (Lectures Notes in Computer Science, vol. 33). Springer, Berlin Heidelberg New York 1975, pp. 515–532
6. Davenport, J., Heintz, J.: Real quantifier elimination is doubly exponential. J. Symb. Comp. **5** (1988) 29–35
7. Demmel, J.: On condition numbers and the distance to the nearest ill-posed problem. Num. Math. **51** (1987) 251–289
8. Fischer, M., Rabin, M.: Super-exponential complexity of Presburger arithmetic. In: Complexity of Computations (SIAM-AMS Proc., 7), 1974, pp. 27–41
9. Fitchas, N., Galligo, A., Morgenstern, J.: Algorithmes rapides en sequentiel et en parallel pour l'élimination de quantificateurs en geométrie élémtaire. Séminaire Structures Ordonnées, U.E.R. de Math. Univ. Paris VII, 1987
10. Goldfarb, D., Todd, M.: Linear programming. In: Nemhauser, G., Rinnooy Kan, A., Todd, M. (eds.) Handbooks in Operations Research and Management Science, vol. I: Optimization, Chapter 2. North-Holland, Amsterdam 1989
11. Gonzaga, C.: An algorithm for solving linear programming problems in $O(n^3L)$ operations. In: Megiddo, N. (ed.) Progress in Mathematical Programming – Interior Point and Related Methods, Chapter 1. Springer, Berlin Heidelberg New York 1989
12. Gonzaga, C.: Path following methods for linear programming. To appear in SIAM Review
13. Grigor'ev, D.Yu.: The complexity of deciding Tarski algebra. J. Symb. Comp. **5** (1988) 65–108

14. Grigor'ev, D.Yu., Vorobjov, N.N. Jr.: Solving systems of polynomial inequalities in subexponential time. J. Symb. Comp. **5** (1988) 37–64
15. Heintz, J., Roy, M.-F., Solernó, P.: Sur la complexite du principe de Tarski-Seidenberg. Bull. Soc. Math. France **118** (1990) 101–126
16. Karmarkar, N.: A new polynomial time algorithm for linear programming. Combinatorica **4** (1984) 373–395
17. Khachiyan, L.G.: Polynomial algorithms in linear programming. USSR Computational Mathematics and Mathematical Physics **20** (1980) 53–72
18. Kojima, M., Mizuno, S., Yoshise, A.: A polynomial-time algorithm for a class of linear complementarity problems. Mathematical Programming **44** (1989) 1–26
19. Megiddo, N.: Pathways to the optimal set in linear programming. In: Megiddo, N. (ed.) Progress in Mathematical Programming – Interior Point and Related Methods, Chapter 8. Springer, Berlin Heidelberg New York 1989
20. Megiddo, N.: On the complexity of linear programming. In: Bewley, T. (ed.) Advances in Economic Theory. Cambridge University Press, 1987, pp. 225–268
21. Monteiro, R.C., Adler, I.: Interior path-following primal-dual algorithms. Mathematical Programming **44** (1989) 27–66
22. Neff, A.C.: Specified precision polynomial root isolation is in NC. To appear in J. Comp. System Sci.
23. Renegar, J.: A polynomial-time algorithm, based on Newton's method, for linear programming. Mathematical Programming **40** (1988) 59-94
24. Renegar, J.: On the worst-case arithmetic complexity of approximating zeros of polynomials. J. Comp. **3** (1987) 90–113
25. Renegar, J.: On the computational complexity and geometry of the first-order theory of the reals. To appear in J. Symb. Comp.
26. Renegar, J.: On the computational complexity of approximating solutions for real algebraic formulae. To appear in SIAM J. Computing
27. Renegar, J.: Is it possible to know a problem instance is ill-posed? To appear in: From Topology to Computation, Proceedings of the Smalefest
28. Renegar, J., Shub, M.: Unified complexity analysis for Newton LP methods. To appear in Mathematical Programming
29. Schonhage, A.: Equation solving in terms of computational complexity. Proceedings of the International Congress of Mathematicians, Berkeley, CA, 1986. American Mathematical Society, 1987, pp. 131–153
30. Shub, M.: On the work of Steve Smale on the theory of computation. To appear in: From Topology to Computation, Proceedings of the Smalefest
31. Smale, S.: On the efficiency of algorithms of analysis. Bull. Amer. Math. Soc. **13** (1985) 87–121
32. Smale, S.: Newton's method estimates from data at one point. The Merging of Disciplines in Pure, Applied and Computational Mathematics. Springer, Berlin Heidelberg New York 1986, pp. 185–196
33. Smale, S.: Algorithms for solving equations. Proceedings of the International Congress of Mathematicians, Berkeley, CA, 1986. American Mathematical Society, 1987, pp. 87–121
34. Smale, S.: Some remarks on the foundations of numerical analysis. SIAM Review **32** (1990) 211–220
35. Tardos, E.: A strongly polynomial algorithm for solving combinatorial linear programs. Operations Research **34** (1986) 250-256
36. Tarski, A.: A Decision Method for Elementary Algebra and Geometry. University of California Press, 1951
37. Vaidya, P.: An algorithm for linear programming which requires $O(((m+n)n^2+(m+n)^{1.5}n)L)$ arithmetic operations. Mathematical Programming **47** (1991) 175–202

38. Vaidya, P.: Speeding-up linear programming using fast matrix multiplication. Proceedings of the 30th IEEE Symposium on the Foundations of Computer Science, pp. 332–337
39. Weispfenning, V.: The complexity of linear problems in fields. J. Symb. Comp. **5** (1988) 3–27
40. Ye, Y.: An $O(n^3L)$ potential reduction algorithm for linear programming. To appear in Mathematical Programming

Turbulence, Dynamical Systems and the Unreasonable Effectiveness of Empirical Eigenfunctions

Philip Holmes[1], *Gal Berkooz*[2], *and John L. Lumley*[3]

[1] Departments of Theoretical and Applied Mechanics and Mathematics and Center for Applied Mathematics, Cornell University, Ithaca, NY 14853, USA
[2] Center for Applied Mathematics, Cornell University, Ithaca, NY 14853, USA
[3] Sibley School of Mechanical and Aerospace Engineering and Center for Applied Mathematics, Cornell University, Ithaca, NY 14853, USA

1. Introduction

Paul Halmos [1981] once claimed that applied mathematics is bad mathematics. Naturally, we do not share this view (nor did Halmos, at least not entirely, in the article cited above). With the analysis of turbulence as an example, in this short paper we hope to show that, while its concerns are neither as clean nor as circumscribed, applied mathematics can be rich and fascinating and that it often interacts deeply with the best "pure" mathematics.

Two threads can be detected in the application of dynamical systems theory to turbulence. The first and better known originates with Landau [1944] and Hopf [1948], who proposed a "soluble" model equation which shared features of Navier-Stokes and could be shown to exhibit a sequence of *bifurcations* to flows of increasing complexity as a parameter ($\sim$ Reynolds number) increases. Hopf also suggested that there should be a finite dimensional attracting manifold in the infinite dimensional phase space of the evolution equations. Ruelle and Takens [1971], following an idea also proposed by Arnold, introduced the notions of generic properties and structural stability to the discussion and argued that *strange attractors,* characteristic of low dimensional systems, would more likely provide an explanation for the complex, apparently statistical motions of systems ostensibly governed by the deterministic Navier-Stokes equations than the quasi-periodic flows of Landau and Hopf. This idea led to enormous activity – experimental, analytical and numerical – especially in studies of closed fluid systems such as Bénard convection and the Taylor-Couette problem (cf. Swinney and Gollub [1981]).

Center manifold theory and the *unfolding* of *degenerate bifurcations* (cf. Guckenheimer and Holmes [1983], Golubitsky and Guckenheimer [1986]) have been major tools in the study of interacting instability modes. However, this local approach seems best suited to problems in hydrodynamic instability and the transition to turbulence rather than the fully developed turbulence characteristic of "open" flows. More recently, proofs of finite Hausdorff dimension for attractors of various partial differential equations (PDEs), including Navier-Stokes, and of *inertial manifolds* which globally attract all initial data exponentially fast, have helped connect these finite dimensional ideas with infinite dimensional evolution equations, cf. Constantin et al. [1989], Temam [1988].

Proceedings of the International Congress of Mathematicians, Kyoto, Japan, 1990

The second thread, also traceable to a paper of Hopf [1952], cf. Foias-Prodi [1967], Foias [1973], takes a statistical viewpoint and addresses the notions of *invariant measures* and other probabilistic descriptions of turbulent fields. Of course, since the original work of Reynolds [1895], statistical descriptions of turbulence have been widely used in engineering and physics.

This paper outlines aspects of current work in which we attempt to bring these statistical and deterministic approaches together. Taking the near wall region of a fully developed turbulent boundary layer as a specific case and using a basis of empirical eigenfunctions, arrived at by statistical means, the Navier-Stokes equations are projected into a low dimensional subspace and the resulting ordinary differential equations (ODEs) studied by dynamical systems techniques. See Aubry et al. [1988, 1989, 1990], Holmes [1990], Berkooz et al. [1991] for original material and background. Here we concentrate on mathematical issues involved in the projection, truncation and modelling processes and indicate how some of the "intuitive" simplifications made in the work cited above can be justified.

In Sect. 2 we discuss the proper orthogonal decomposition, by which an "optimal" basis is generated from data ensembles. Section 3 addresses the averaging implicit in representations of boundary layer flows lacking streamwise variation and shows that the many spatial scales and modes neglected in such truncations can be rationally modelled. In Sect. 4 we draw conclusions.

We hope that this brief paper provides at least a taste of the fascinating interplay between physical foundations, experimental work, modelling and diverse types of mathematical analysis characteristic of "good" applied mathematics.

2. The Proper Orthogonal Decomposition and "Optimality"

Lumley [1967, 1970] first suggested the use of the *proper orthogonal* or *Karhunen-Loève decomposition* in turbulence studies (cf. Loève [1955]). Motivated by experimental observation of coherent structures (cf. Cantwell [1981]) in open flows such as fully developed jets, wakes, shear layers and boundary layers, he sought an unbiased method for the recognition and "extraction" of such structures as objects in space-time.

We describe the method for scalar fields; the vectorial generalization is not difficult. Suppose that $U = \{u^i(x) | i \in I\}$ is an ensemble of realizations of turbulent field on some region $\Omega \subseteq \mathbb{R}^n$; each u^i belonging to a suitable Hilbert space $\mathcal{H}$ with inner product $(\cdot,\cdot)$ and norm $\|\cdot\|$. Here I is an index set for the realizations, $\langle\cdot\rangle$ denotes the ensemble average. For simplicity one can think of time averages in a statistically stationary flow. We seek a basis $\Phi = \{\phi_j(x)\}_1^\infty$ for $\mathcal{H}$ such that the ensemble averaged normalized projections $\langle P_j(u)\rangle = \langle(u, \phi_j)\rangle/\|\phi_j\|$ onto each element in turn are maximized among all bases Ψ. The desired basis is produced by solution of the Fredholm integral equation

$$\int_\Omega R(x, x')\phi_j(x')\,dx' = \lambda_j \phi_j(x), \tag{2.1}$$

where $R(x, x') = \langle u(x)u(x')\rangle$ is the ensemble averaged two point autocorrelation function. The basis elements ϕ_j are called *empirical eigenfunctions*, since they derive from $R(x, x')$, itself the result of experimental observation or numerical simulation; they are orthogonal (henceforth assumed *orthonormal*) and the eigenvalues λ_j correspond to the ensemble averaged kinetic energy in each "mode", ϕ_j, via the expressions

$$u = \sum_1^\infty a_j \phi_j, \quad \|u\|^2 = \sum_1^\infty |a_j|^2. \tag{2.2}$$

Also, since the a_j are uncorrelated, we have

$$\left\langle \sum_1^\infty |a_j|^2 \right\rangle = \sum_1^\infty \langle |a_j|^2 \rangle = \sum_1^\infty \lambda_j. \tag{2.3}$$

Φ may have other desirable properties; for example, when the tensor $R(x, x')$ derives from measurements on an incompressible fluid, the ϕ_j are divergence-free vectors: an advantage in Galerkin projection of the Navier-Stokes equations, since the projected pressure term $(\nabla p, \phi_j)$ can be removed by integration by parts.

The basis Φ is optimal in the following sense. Let $\Psi = \{\psi\}_1^\infty$ be any other orthonormal basis, so that any field $u \in U$ can be approximated by the n'th order truncations $u \approx \sum_1^n a_j \phi_j$ and $u \approx \sum_1^n b_j \psi_j$:

Proposition 1. *For each $n \geq 1$ and any Ψ, $\sum_1^n \langle a_j \rangle^2 \geq \sum_1^n \langle b_j^2 \rangle$.*

This may be proved by a manipulation involving the correlation matrix $R(x, x')$ and the fact that, if K is a self-adjoint operator and Q an orthogonal projector onto span $\{\phi_1, \ldots, \phi_n\}$, then

$$\mathrm{Tr}(K \cdot Q) = \sum_1^n (K\phi_j, \phi_j) \leq \sum_1^n \kappa_j, \tag{2.4}$$

where $\kappa_1, \ldots, \kappa_n$ are the n largest eigenvalues of K (cf. Temam [1988, p. 260]). Proposition 1 guarantees that use of Φ minimizes the error, in a mean square sense, among all possible truncations of any fixed order. In fact Φ spans the subspace which contains almost all realizations in a measure theoretic sense of the flow from which $R(x, x')$ was computed.

Several groups have recently been using empirical eigenfunctions for the representation of turbulent fields: see Moin et al. [1984, 1989] and Sirovich et al. [1987, 1988, 1989, 1990] for examples based on computer simulation. However, in spite of all the numerical activity and studies of convergence of averages (cf. Foias et al. [1990]), there has been little study of the way in which the mean square optimality of Φ relates to the dynamics of ordinary differential systems produced via projection of the governing PDEs onto (low dimensional) subspaces spanned by finite sets $\Phi^N = \{\phi_j\}^N$. The original work of our group (Aubry et al. [1988]) indicated that the statistical optimality of Φ^N led to systems which exhibited *instantaneous dynamical behavior* representative of the full system, in spite of the low order truncations employed (only 5 complex modes!). Sirovich and his colleagues have made similar observations for other dissipative PDEs. The ideas introduced in the rest of this paper are directed toward a rational justification of these observations.

3. Dynamics in Subspaces Lacking Streamwise Variation

In any finite dimensional representation of an infinite dimensional evolution equation, such as Navier-Stokes, the phase space $\mathscr{H}$ is divided into a finite subset of resolved modes, R, and its (orthogonal) complement, S, elements of which are modelled in terms of elements of R, or simply neglected. For example, if an inertial manifold, M, exists, it is expressible as a function $h : R \to S$ and one relies on the attractivity of M to guarantee that all states asymptotically approach points of M with coordinates $(r, h(r))$. The reduced dynamical system or *inertial form* (Sell [1989]) is then a closed set of ODEs for r: in its simplest form it is merely the projection of the full flow onto R with $h \equiv 0$.

In measure theoretic terms, there is a conditional measure $\mu_r(s)$ on each unresolved fiber $r + S$ over the base $R \ni r$ and an associated measure $\mu_R(r)$ on R itself, forming a measurable partition, so that for a set $B \subset \mathscr{H}$:

$$\mu(B) = \int_R \mu_r(s \in B)\, d\mu_R(r)\,. \tag{3.1}$$

In this context, the physical notion of small scale or local isotropy (Tennekes and Lumley [1974]) is the assumption that μ_r is independent of the base point r. More generally, modelling of activity in small spatial scales in terms of the large scales can be seen as an attempt to estimate μ_r.

Throughout the analysis to follow two notions of ensemble average are implicit. The first is the usual one of averaging over many separate (experimental) realizations. The second is based on the observation that, when length scales in the homogeneous directions (those lacking a distinguished origin) are long compared to those of typical turbulent phenomenon, integration over those directions will yield a characteristic measure of all the dynamical phenomena. An implicit equivalence or ergodicity assumption is thus invoked.

We now focus on a turbulent boundary layer over a flat plate, the domain $\Omega = [0, L_1] \times [0, L_2] \times [0, L_3]$ being of streamwise extent L_1, spanwise L_3 and normal L_2, with periodic boundary conditions in x_1 and x_3. In Aubry et al. [1988, 1989, 1990] Ω is the "wall region".

Flow visualizations of the boundary layer by Kline et al. [1967] (one of which is reproduced in Aubry et al. [1988]) demonstrate the presence of coherent structures (streaks) with long streamwise spatial scales and relatively small spanwise spacing. This, together with explicit evaluation of empirical eigenfunctions, prompts the split of $\mathscr{H}$ into R and S developed below.

In this section we identify R with (a finite dimensional subset of) the subspace of flows $u(x,t) = u(x_2, x_3, t)$ having no streamwise dependence. Thus, if P is the orthogonal projector $P : \mathscr{H} \to R$, its application is identical to averaging in the streamwise direction, as can be seen by appeal to the representation

$$\mathbf{u}(\mathbf{x}, t) = \sum_{k,l,n} a_{k,l,n}(t) \Phi^n_{k,l}(x)\,, \tag{3.2}$$

where

$$\Phi^n_{k,l}(x) = e^{2\pi i\left((kx_1/L_1)+(lx_3/L_3)\right)}\, \phi^n_{k,l}(x_2)/\sqrt{L_1 L_3}\,,$$

and

$$P(u) = \sum_{l,n} a^n_{0,l}(t)\Phi^n_{0,l}(x) \equiv \frac{1}{L_1}\sum a^n_{k,l}(t)\int_0^{L_1} \Phi^n_{k,l}(x)\,dx_1\,. \tag{3.3}$$

Here $u(x,t)$ is the fluctuating field riding on the mean flow $U = \big(u_1(x_2),0,0\big)$.

Let

$$\frac{du}{dt} = N(u) \tag{3.4}$$

denote the Navier-Stokes equations. We wish to determine the evolution of the resolved state $r \in R$. Ideally, we should solve (3.4) for an ensemble of initial conditions (r,s_i) to find μ_r and determine the vector field in R at each such r by integration with respect to this measure:

$$\left(\frac{dr}{dt}\right)_a = \int_{S_r} P\left(\frac{du}{dt}\bigg|_{u=r+s}\right) d\mu_r(s)\,. \tag{3.5}$$

A simpler alternative is to project (3.4) onto R and solve the resulting reduced equation

$$\left(\frac{dr}{dt}\right)_b = P\big(N(r+s)\big)\,, \tag{3.6}$$

with $s = h(r)$ modelled in some way. This latter is computationally accessible and fortunately we have

Proposition 2. *For statistically stationary flows, as $L_1 \to \infty$ so $\left(\frac{dr}{dt}\right)_b \to \left(\frac{dr}{dt}\right)_a$.*

Proof. The right hand side of (3.5) is the conditional ensemble average over S_r such that $P(u) = r$. However, if L_1 is large enough Ω may be divided into $M = L_1/d \gg 1$ regions of length d in each of which the flow is statistically independent. (Thus d is assumed to be much greater than the streamwise length scale.) The ensemble average may then be written

$$\begin{aligned}\int_{S_r} P\left(\frac{du}{dt}\bigg|_{u=r+s}\right) d\mu_r(s) &= \frac{1}{M}\sum_1^M \frac{dr}{dt}\bigg|_{r+s} \to \frac{1}{L_1}\int_0^{L_1} \frac{dr}{dt}\bigg|_{r+s} dx_1\\ &= P\big(N(r+s)\big)\,.\end{aligned}$$

□

Physically, we assume that Ω is long enough to contain "something of everything" at any instant, and so effectively to yield the measure μ_r. (It is also related to the small scale isotropy assumption referred to earlier). Thus the projected evolution equations (3.6) differ from the "ideal" case (3.5) only in that the unresolved modes, s, must be modelled. We now turn to this aspect.

The evolution equation for the fluctuations u, including Reynolds stress terms due to the fact that the mean flow $U(x) = \big(U_1(x_2),0,0\big)$ only solves the Navier-Stokes in a suitably averaged sense, may be written as in Aubry et al. [1988], using the Einstein summation convention:

$$\frac{\partial u}{\partial t} + u_{i,1}U_1 + U_{1,2}u_2\delta_{i1} + u_j u_{i,j} - \langle u_{i,j}u_j\rangle = -\frac{1}{\varrho}\pi_{,j} + \nu u_{i,jj}\,. \tag{3.7}$$

Here $\langle\cdot\rangle$ denotes the streamwise-spanwise spatial average:

$$\langle\cdot\rangle = \frac{1}{L_1 L_3}\int_0^{L_1}\int_0^{L_3} (\cdot)\, dx_1\, dx_3 ,$$

π the fluctuating pressure and ν is the viscosity. Letting $u_i = r_i + s_i$ and applying P, this yields for (3.6)

$$\begin{aligned}\frac{\partial r_i}{\partial t} &+ r_{i,1}U_1 + U_{1,2}r_2\delta_{i1} + P(s_j s_{i,j}) - \langle s_j s_{i,j}\rangle \\ &+ r_j r_{i,j} - \langle r_j, r_{i,j}\rangle = -\frac{1}{\varrho} p_{,i} + \nu r_{i,jj} ,\end{aligned} \tag{3.8}$$

where $\pi = p + q$ and p corresponds to the resolved pressure modes. Note that, since r_i is independent of x_1 and s_i has zero mean in x_1, several "mixed" terms vanish: specifically, the Leonard stresses are zero.

$$P\Big((r_j s_{i,j}) + s_j r_{i,j} - \big\langle (r_j s_{i,j} + (s_j r_{i,j})\big\rangle\Big) = 0 , \tag{3.9}$$

as Aubry et al. [1988] assumed.

Now (3.8) is "closed" apart from the term $P(s_j, s_{i,j}) - \langle s_j s_{i,j}\rangle$ which represents interaction between modes in R and S. Normally one might expect this to result in transfer of energy from R to S *and* from S to R, as well as modulate its transport among the R modes. However, when S represents a subspace of no streamwise dependence, and is spanned by the leading empirical eigenfunctions, we have

Proposition 3. (1) *$P(s_j s_{i,j}) - \langle s_j s_{i,j}\rangle$ on the average can only transfer energy from R to S or move it around in R. No energy on the average enters R from S.*

(2) *The ratio $\langle r_1 r_2\rangle/\|r\|^2$ of Reynolds stress to turbulent kinetic energy is restricted to an interval appropriate for the ensemble U. For the truncations of Aubry et al. [1988] this interval is bounded away from zero.*

(3) *Provided only low wavenumbers are retained in R, the ratio of energy loss from R to S to turbulent kinetic energy in R is compatible with simple Heisenberg or eddy viscosity modelling.*

Proof. Multiplying (3.8) by r_i and averaging by $\langle\cdot\rangle$ yields the evolution equation for the resolved turbulent kinetic energy $\frac{\langle r_i r_i\rangle}{2}$:

$$\begin{aligned}\frac{D}{Dt}\Big\langle\frac{r_i r_i}{2}\Big\rangle = &- U_{1,2}\langle r_1 r_2\rangle - \Big\langle\Big(\frac{r_i r_i}{2} + \frac{p}{\varrho}\Big) r_j - \nu\Big(\frac{r_i r_i}{2}\Big)_{,j} - \nu(r_i r_j)_{,i}\Big\rangle_{,j} \\ &- \big\langle P(s_j s_{i,j}) r_i\big\rangle - 2\nu\langle\varrho_{ij}\varrho_{ij}\rangle ,\end{aligned} \tag{3.10}$$

where $\varrho_{ij} = (r_{i,j} + r_{j,i})/2$ and $D/Dt = \frac{\partial}{\partial t} + r_i\frac{\partial}{\partial x_i}$ denotes the convective deriative. In (3.10) the term $\big\langle P(s_j, s_{i,j}) r_i\big\rangle$ alone represents interactions between R and S. Using incompressibilty and the fact that r_i is independent of x_1 it may be rewritten explicitly as

$$\left\langle \frac{1}{L_1}\int_0^{L_1} s_j s_{i,j}\, dx_1\, r_i \right\rangle = \left\langle \frac{1}{L_1}\int_0^{L_1} (s_i s_j)_{,j}\, dx_1 r_i \right\rangle$$
$$= \left\langle \frac{1}{L_1} s_i s_1 \Big|_0^{L_1} r_i + \frac{1}{L_1}\int_0^{L_1} (s_i s_2)_{,2} r_i\, dx_1 + \frac{1}{L_1}\int_0^{L_1} (s_i s_3)_{,3} r_i\, dx_1 \right\rangle$$
$$= \left\langle \Big(\frac{1}{L_1}\int_0^{L_1} s_i s_2 r_i\, dx_1\Big)_{,2} + \Big(\frac{1}{L_1}\int_0^{L_1} s_i s_3 r_i\, dx_1\Big)_{,3} \right.$$
$$\left. - \frac{1}{L_1}\int_0^{L_1} s_i s_2\, dx_1 r_{i,2} - \frac{1}{L_1}\int_0^{L_1} s_i s_3\, dx_1 r_{i,3} \right\rangle$$
$$= \langle P(s_i s_j r_i)_{,j}\rangle - \langle P(s_i s_j) r_{i,j}\rangle = \langle s_i s_j r_i\rangle_{,j} - \langle P(s_i s_j) r_{i,j}\rangle . \tag{3.11}$$

The first term in (3.11) represents transport of energy among modes in R while the second represents straining of modes in S by those in R – i.e. losses from R to S. No term exists for transfer of energy from S to R on the average. This establishes (1).

To establish (2) we consider the ratio $\tau = \langle u_1 u_2\rangle / \langle u_i u_i\rangle$. Since this argument has already appeared (Berkooz et al. [1991]), we merely sketch it. Expanding τ by (3.2)-(3.2)with $k = 0$ and maximizing and minimizing over the available modes in any specified truncation $|l| \le L$, $1 \le n \le N$ leads to a study of expressions of the form

$$\frac{(\phi^n_{0,l})_1 (\phi^{n*}_{0,l})_2}{(\phi^n_{0,l})_i (\phi^n_{0,l})_i} \tag{3.12}$$

and thus the relative signs and magnitudes of the streamwise $(\cdot)_1$ and spanwise $(\cdot)_2$ components of the empirical basis vectors $\phi^n_{0,l}$ determine the upper and lower bounds for the range of ratios which can be represented by velocity fields belonging to R. Reference to Fig. 4 of Aubry et al. [1988] shows that the lower bound is strictly positive, establishing (2).

Having shown in (1) that energy can be lost from R to S we now wish to model this effect by expressing the loss as a function purely of the resolved modes r_i. To do this we estimate the ratio

$$\frac{\langle P(s_i s_j) r_{i,j}\rangle}{\langle \varrho_{ij}\varrho_{ij}\rangle} \sim \frac{\langle r_{i,j} r_{i,j}\rangle^{1/2} \langle P(s_i s_j) P(s_i s_j)\rangle^{1/2}}{\langle (r_{i,j}+r_{j,i})(r_{i,j}+r_{j,i})\rangle}$$
$$\sim \frac{\langle r_{i,j} r_{i,j}\rangle^{1/2} \langle (s_i, s_i)\rangle}{\langle r_{i,j} r_{i,j}\rangle} \sim \frac{\langle (s_i, s_i)\rangle}{\langle r_{i,j} r_{i,j}\rangle^{1/2}} . \tag{3.13}$$

Here $\sim$ means "equal within an order 1 number"(the correlation coefficient).

In the conventional Heisenberg type model as used by Aubry et al. [1988] (cf. Tennekes and Lumley [1972]) an effective (eddy) viscosity

$$\nu_T = \frac{\int_0^{L_2} \langle s_i s_i\rangle\, dx_2}{\left(L_2 \int_0^{L_2} \langle r_{i,j} r_{i,j}\rangle\, dx_2\right)^{1/2}} \tag{3.14}$$

is introduced by averaging the numerator and square of the denominator of (3.13) over the wall region normal to the wall. From (3.13) a more natural choice might seem to be

$$\tilde{\nu}_T = \frac{1}{L_2}\int_0^{L_2}\left(\frac{\langle s_i s_j\rangle}{\langle r_{i,j} r_{i,j}\rangle^{1/2}}\right) dx_2, \tag{3.15}$$

but using the facts that the velocities s_i, r_i and their derivatives are represented by the empirical eigenfunctions, which may in turn be well fitted by (low) order polynomials in the wall region we find that (3.14) and (3.15) are of the same order. Specifically, let

$$\langle s_i s_i\rangle \sim x_2^{2p}, \quad \langle r_{i,j} r_{i,j}\rangle \sim x_2^{2q}, \tag{3.16}$$

so that a typical scale in the unresolved mode is represented by a polynomial of order p and in the resolved modes by order q ($p > q$). Then (3.14) and (3.15) yield, respectively

$$\nu_T \sim \frac{\sqrt{2q+1}}{2p+1} L_2^{2p-q}, \quad \tilde{\nu}_T \sim \frac{L_2^{2p-q}}{2p-q+1}, \tag{3.17}$$

so that

$$\frac{\tilde{\nu}_T}{\nu_T} \sim \frac{2p+1}{\sqrt{2q+1}(2p-q+1)}. \tag{3.18}$$

If, as in Aubry et al. [1988], the energy is assumed to be lost to the next higher wavenumbers (the lowest in S), then $p = q+1$ and

$$\frac{\tilde{\nu}_T}{\nu_T} \sim \frac{2q+3}{(q+3)\sqrt{2q+1}} \in (c/\sqrt{q}, 1), \tag{3.19}$$

while if we assume that energy is transferred to high wavenumbers in S, $p \gg q$ and

$$\frac{\tilde{\nu}_T}{\nu_T} \sim \frac{1}{\sqrt{q}}. \tag{3.20}$$

In either case, if a small range of wavenumbers $0 \le q \le Q$ are retained in R, as in Aubry et al. [1988] the ratio $\tilde{\nu}_T/\nu_T$ does not vary radically and (3) is established. □

As Berkooz et al. [1991] point out, the restriction, by the projection, of the ratio τ of Reynolds stress to kinetic energy to a range appropriate to the experimental observations is crucial to the success of low dimensional representations lacking streamwise variation in producing relevant dynamics. If τ could drop to zero, decoupling spanwise from streamwise motions, then one expects turbulent kinetic energy to decay (cf. Moffatt [1990]). When τ is strictly positive the turbulent fluctuations can (indeed must) extract energy from the mean velocity gradient $U_{1,2}$ via the third term in (3.7) (the second is absent if there is no streamwise dependence).

The first conclusion of Proposition 3 is striking in that it shows that motions contained high spanwise and wall-normal wavenumbers but lacking streamwise variations (in R) cannot on the average extract energy even from *low* streamwise wavenumber modes (in S). This is distinct, of course, from their ability to extract energy from the mean velocity gradient referred to above.

4. Conclusions

We conclude that projections on the subspace S, lacking streamwise variation (and finite subspaces of it) have some special properties. These become apparent due to the convenient interpretation of the projection as an average in the streamwise direction (Proposition 2). The notion of equivalence of ensemble averages is used in an implicit way and must be considered an assumption. We then can make the following observations. The projected system is in effect a spatially averaged system, constraining the reduced dynamics to what will be physically observable. Assumptions previously made neglecting the Leonard stresses prove to be exact in an average sense. The expression used for the effective (Heisenberg) viscosity, which was based on physical intuition, is proven to be correct (within an order 1 number) in an average sense. We also observe that on the average energy does not pass from S to its complement, R, justifying our intuitive feeling that R is a fundamentally important subspace. (It was previously referred to as a "backbone" for the analysis Holmes [1990].) We also recalled a previous observation that the turbulent energy production is held in an experimentally appropriate range, again confirming our physical intuition.

In work of this type we use a wide range of mathematical tools. Statistical methods are used to extract key features from experimental data or simulations. Recent ideas from PDE, analysis and dynamical systems theory motivate our derivation of reduced (projected) dynamical systems. This is not a bag of unrelated tricks which happen to work; there are deep relations among the different pieces of mathematics and the physical problems which have prompted their development and use. Although we do not describe them here, ideas from the global theory of dynamical systems permit us to give fairly complete analyses of large $\left(O(10-50)\right)$ sets of ODEs, to understand the effects of symmetries on them (cf. Armbruster et al. [1988, 1989]) and the influence of noise and other perturbations on the heteroclinic attractors they possess (Stone and Holmes [1989, 1990]). Physical intuition and reasoning come to our aid when rigorous mathematical arguments are inadequate. Perhaps more significantly for mathematics, attempts to analyze problems such as turbulence continue to provide a wealth of challenging mathematical problems and even to suggest whole new fields of study.

Acknowledgements. The work described in this paper was supported by the Air Force Office of Scientific Research and the National Science Foundation under AFOSR 89 0226A (Wall Layers).

References

Armbruster, D., Guckenheimer, J., Holmes, P. (1988): Heteroclinic cycles and modulated travelling waves in systems with $O(2)$ symmetry. Physica **29D**, 257–282

Armbruster, D., Guckenheimer, J., Holmes, P. (1989): Kuramoto-Sivashinsky dynamics on the center-unstable manifold. S.I.A.M. J. Appl. Math. **49**, 676–691

Aubry, N., Holmes, P., Lumley, J., Stone, E. (1988): The dynamics of coherent structures in the wall region of a turbulent boundary layer. J. Fluid Mech. **192**, 115–173

Aubry, N., Lumley, J., Holmes, P.J. (1990): The effect of modelled drag reduction on the wall region. Theor. Comp. Fluid Dyn. **1**, 229–248

Aubry, N., Sanghi, S. (1989): Streamwise and spanwise dynamics of the turbulent wall layer. In: Forum on Chaotic Flow, Ghia (ed.). Proc. ASME, New York

Berkooz, G., Holmes, P.J., Lumley, J.L. (1991): Intermittent dynamics in simple models of the turbulent wall layer. J. Fluid Mech. (in press)

Blackwelder, R.F. (1989): Some ideas on the control of near wall eddies. AIAA paper #89–1009, AIAA 2nd Shear Flow Conference

Cantwell, B.J. (1981): Organized motion in turbulent flow. Ann. Rev. Fluid Mech. **13**, 457–517

Constantin, P., Foias, C., Temam, R., Nicolaenko, B. (1989): Integral manifolds and inertial manifolds for dissipative partial differential equations. Springer, New York

Foias, C. (1973): Statistical study of Navier-Stokes equations I and II. Rend. Sem. Mat. Univ. Padova **48**, 219–348 and **49** 9–123

Foias, C., Manley, O., Sirovich, L. (1990): Empirical and Stokes eigenfunctions and the far-dissipative turbulent spectrum. Phys. Fluids **A 2**, 464–467

Foias, C., Prodi, G. (1967): Sur le comportement global des solutions non-stationnaires des équations de Navier-Stokes en dimension 2. Rend Sem. Mat. Univ. Padova **39**, 1–34

Golubitsky, M., Guckenheimer, J. (1986) (eds.): Multiparameter bifurcation theory. Contemp. Math. **56**. AMS, Providence, R.I.

Guckenheimer, J., Holmes, P. (1983): Nonlinear oscillations, dynamical systems and bifurcations of vector fields. (Applied Mathematical Sciences, vol. 42.) Springer, New York, 1986 (Second printing)

Halmos, P. (1981): Applied mathematics is bad mathematics. In: Mathematics tomorrow (ed. L.A. Steen). Springer, New York, pp. 9–20

Holmes, P.J. (1990): Can dynamical systems approach turbulence? In: Whither turbulence: Turbulence at the crossroads (ed. J.L. Lumley). (Lecture Notes in Physics, vol. 357). Springer, New York, pp. 195–249, 306–309

Hopf, E. (1948): A mathematical example displaying the features of turbulence. Commun. Pure Appl. Math. **1**, 303–322

Hopf, E. (1952): Statistical hydromechanics and functional calculas. J. Rat. Mech. Anal. **1**, 87–123

Landau, L. (1944): On the problem of turbulence. Dokl. Akad. Nauk. SSSR. **44**, 339-342

Loève, M. (1955): Probablity theory. Van Nostrand

Lorenz, E.N. (1963): Deterministic nonperiodic flow. J. Atmos. Sci. **20**, 130–141

Lumley, J.L. (1967): The structure of inhomogeneous turbulent flows. In: Atmospheric turbulence and radio wave propagation (eds. A.M. Yaglom and V.I. Tatarski). Nauka, Moscow, pp. 166–178

Lumley, J.L. (1970): Stochastic tools in turbulence. Academic Press, New York

Moffatt, H.K. (1990): Fixed points of turbulent dynamical systems and suppression of nonlinearity. In: Whither turbulence (ed. J.L. Lumley). (Lecture Notes in Physics, vol. 357). Springer, New York, pp. 250–257

Moin, P. (1984): Probing turbulence via large eddy simulation. Proceedings AIAA Aerospace Sciences Meeeting

Moin, P., Moser, R.D. (1989): Characteristic-eddy decomposition of turbulence in a channel. J. Fluid Mech. **200**, 471–509

Ruelle, D., Takens, F. (1971): On the nature of turbulence. Commun. Math. Phys. **20**, 167–192 and **23**, 343–344

Sell, G.R. (1989): Approximation dynamics: hyperbolic sets and intertial manifolds. (Submitted for publication)

Sirovich, L. (1987a): Turbulence and the dynamics of coherent structures: I. Q. Appl. Math. **45**, 561–571

Sirovich, L. (1987b): Turbulence and the dynamics of coherent structures: II. Q. Appl. Math. **45**, 561–571

Sirovich, L. (1987c): Turbulence and the dynamics of coherent structures: III. Q. Appl. Math. **45**, 561–571

Sirovich, L. (1989): Chaotic dynamics of coherent structures. Physica **D37**, 126–145

Sirovich, L., Kirby, M., Winter, M. (1990): An eigenfunction approach to large scale transitional structures in jet flow. Phys. Fluids **A 2**, 127–136

Sirovich, L., Maxey, M., Tarman, H. (1987): Analysis of turbulent thermal convection. In: Turbulent shear flows, vol. 6 (eds. F. Durst et al.). Springer, New York

Sirovich, L., Rodriguez, J.D. (1987): Coherent structures and chaos: a model problem. Phys. Lett. **A 120**, 211–214

Stone, E., Holmes, P. (1989): Noise induced intermittency in a model of a turbulent boundary layer. Physica **D37**, 20–32

Stone, E., Holmes, P. (1990): Random perturbations of heteroclinic attractors. SIAM J. Appl. Math. **50**, 726–743

Swinney, H.L., Gollub, J.P. (1981) (eds.): Hydrodynamic instabilities and the transition to turbulence. (Topics in Applied Physics, vol. 45). Springer, New York

Temam, R. (1988): Infinite-dimensional dynamical systems in mechanics and physics. Springer, New York

Tennekes, H., Lumley, J.L. (1972): A first course in turbulence. MIT Press, Boston, MA

Wavelets and Applications

Yves F. Meyer

CEREMADE, Université Paris Dauphine, F-75775 Paris Cedex 16, France

1. Introduction

Some investigations conducted by (1) D. Marr in psycho-physiology of human vision, (2) J.S. Lienard in speech signal processing and (3) J. Morlet in seismic signal processing led these scientists to switch from short-time Fourier analysis to some more specific algorithms better suited to detect and analyze abrupt changes in images or signals.

These algorithms are strikingly similar and in the three of them the functions $e^{i\omega t}$, which have a given frequency ω, and are the building blocks of the standard Fourier analysis, are replaced by "wavelets" which are time and frequency items and are the building blocks of "wavelets analysis". Wavelets have a finite duration (which can be arbitrarily small) but nevertheless, should also possess a well defined average frequency.

The success of the wavelets theory is due to the remarkable formulation by A. Grossmann of J. Morlet's ideas. Today this theory has applications in various branches of science whenever complicated interactions between events occuring at different scales appear. This happens in astrophysics [7] or in turbulence [3, 13, 14].

Independently of the above mentioned research, heavy constraints imposed by digital speech processing have led to the discovery of the so-called quadrature mirror filters. These filters also have some applications in image processing where they improve pyramidal algorithms [1].

During the fall of 1986, S. Mallat discovered that some quadrature mirror filters were the key to the construction of orthonormal wavelet bases generalizing the Haar system.

This program was completed by I. Daubechies (1987) and A. Cohen (1990) and culminated with the discovery (1989) by G. Beylkin, R. Coifman and V. Rokhlin of striking new algorithms in numerical analysis [6].

Working on submarine passive detection, J.M. Nicolas set up a new hierarchical organization of quadrature mirror filters, distinct from the one proposed by S. Mallat.

R. Coifman and the speaker proved the convergence of these schemes to new "libraries of orthonormal bases" resembling the waveforms used by J.S. Lienard.

Proceedings of the International Congress
of Mathematicians, Kyoto, Japan, 1990

R. Coifman and V. Wickerhauser are planning to improve speech signal compression substantially through new algorithms to select a most efficient representation from such libraries. This line of research might end the dispute between the advocates of Gabor wavelets and those of Grossmann since both types belong to the library [9].

2. The Windowed Fourier Transform

In signal analysis one often encounters the problem of extracting the frequency content of a signal $f(t)$ for which one has only local information. The so-called short-time Fourier transform, or windowed Fourier transform uses cut-off functions $g_k(t)$ which vanish when the information on $f(t)$ is missing. The frequency content of each block $g_k(t)f(t)$ is given by its Fourier coefficients. Let, $g_k(t)$, for instance, be the indicator function of the interval $I_k = [2k\pi, 2(k+1)\pi[$ and let us expand each block $g_k(t)f(t)$ into its Fourier series $\sum \alpha(k,l)\, e^{ilt}$ on the interval I_k. It would amount to the same thing if (1), we define "trivial wavelets" by $\sqrt{2\pi}\, w_{(k,l)}(t) = e^{ilt}\,\chi(t-2k\pi)$, $k \in \mathbb{Z}$, $l \in \mathbb{Z}$, $\chi(t)$ being the indicator function of $[0, 2\pi[$, (2) observe that these trivial wavelets form an orthonormal basis of $L^2(\mathbb{R})$, (3) use this basis for expanding our signal.

As everyone knows, this way of splitting into "hard blocks" produces numerical artifacts and the coefficients $\alpha(k,l) = (2\pi)^{-1/2}\langle f, w_{k,l}\rangle$ do not give the frequency content of the signal f around I_k. To suppress these artifacts, D. Gabor decided in 1945 to replace $\chi(t)$ by a smoother window function $g(t)$. Gabor wavelets are

$$w_{k,l}(t) = e^{ilt}\, g(t-2k\pi)\,, \quad l \in \mathbb{Z}\,,\ k \in \mathbb{Z} \tag{2.1}$$

where $g(t) = \pi^{-1/4}\exp(-t^2/2)$.

But we would like the l^2 norm of the wavelets coefficients to provide an energy (L^2) estimate on the signal. This happens when $g = \chi$. But if $g(t)$ satisfies the two condition $\int_{-\infty}^{\infty} t^2|g(t)|^2\, dt < \infty$ and $\int_{-\infty}^{\infty} \xi^2\, |\hat{g}(\xi)|^2\, d\xi < \infty$, it is never the case (R. Balian, G. Battle, R. R. Coifman and S. Semmes [11]). For that reason, the definition of Gabor wavelets has been modified to

$$w_{k,l}(t) = e^{ilt}\, g(t-ak)\,, \quad 0 < a < 2\pi\,, \tag{2.2}$$

and this over complete system is currently used in signal analysis [11].

3. Wavelets with Constant Shape

A function $\psi(x)$ belonging to $L^2(\mathbb{R}^n)$ is an *analyzing wavelet* if its Fourier transform $\hat{\psi}(\xi)$ vanishes at 0 in a precise manner, given by the *convergence* of the following integral

$$c(\psi) = \int_{\mathbb{R}^n} |\hat{\psi}(\xi)|^2\, |\xi|^{-n}\, d\xi\,. \tag{3.1}$$

If $\int_{\mathbb{R}^n} |\psi(x)|^2 (1+|x|)^{n+\alpha}\, dx < \infty$ for some $\alpha > 0$, then $\psi \in L^1(\mathbb{R}^n)$ and $c(\psi)$ is finite iff the integral of ψ vanishes. More generally $c(\psi)$ is finite for any ψ in the Hardy space $H^1(\mathbb{R}^n)$.

We now pick an analyzing wavelet ψ, the mother wavelet, and decide that the other wavelets ψ_g of the family (which are used in the wavelet analysis) will only differ from ψ by their orientation, their position and their size. This can be formulated in terms of the group G with elements $g(x) = a\varrho(x)+b$, $a > 0$, $b \in \mathbb{R}^n$, $\varrho \in SO(n)$. In other words, $\psi_g(x) = a^{-n/2}\,\psi(g^{-1}(x))$. The wavelet coefficients $\alpha(g)$, $g \in G$, of $f \in L^2(\mathbb{R}^n)$ are $\langle f, \psi_g\rangle$ and satisfy

$$\|f\|_2^2 = (c(\psi))^{-1} \int_G |\alpha(g)|^2\, dg\,, \tag{3.2}$$

where $dg = a^{-n-1}\, da\, db\, d\varrho$ is the left-invariant Haar measure of the group G. This implies [16, 23]

$$f(x) = (c(\psi))^{-1} \int_G \alpha(g)\, \psi_g(x)\, dg\,. \tag{3.3}$$

Let φ be a radial function in the Schwartz class such that $\hat{\varphi}(0) = 1$ and $\hat{\varphi}(\xi) = 0$ when $|\xi| \geq 1$. Then J. Morlet's analyzing wavelet is

$$\psi(x) = \exp(i5x_1)\, \varphi(x) \tag{3.4}$$

and looks like a Gabor wavelet. But the other wavelets ψ of the Morlet family differ from Gabor wavelets by the fact that both their size and their average frequency are modified. Indeed we have

$$\psi_g(x) = a^{-n/2}\, e^{i5a^{-1}v\cdot x}\, \varphi\left(\frac{x-b}{a}\right) \tag{3.5}$$

where v is a unit vector, $a > 0$ and $b \in \mathbb{R}^n$. The Fourier transform of ψ_g is contained in the ball $|\xi - 5a^{-1}v| \leq a^{-1}$ which amounts to saying that the average frequency of ψ_g is $5a^{-1}v$.

On the other hand, the size of the support of ψ_g is $0(a)$ and Morlet wavelets unlike Gabor wavelets have the sharpest spatial resolution at high frequencies, which is consistent with the Heisenberg uncertainty principle. In other words, when ψ is the Morlet analyzing wavelet, the multiscale analysis given by (3.3) also provides a multichannel analysis.

Some scientists, like D. Marr [21], have accepted a looser spectral resolution and they impose on the Fourier transform of the analyzing wavelet the following conditions: $\hat{\psi}(\xi) \in C^\infty(\mathbb{R}^n)$ and

$$(\partial^\alpha \hat{\psi})(0) = 0 \qquad \text{for} \quad |\alpha| \leq m \tag{3.6}$$

$$\hat{\psi}(\xi) = 0(|\xi|^{-N}) \qquad \text{as} \quad |\xi| \to \infty\,. \tag{3.7}$$

The larger m and N, the better the spectral resolution.

Scientists who apply wavelets to signal or image processing believe that the diagnostic given by the wavelet coefficients will not depend too much on the choice of ψ as long as ψ satisfies (3.6) and (3.7). This belief is supported by functional analysts who have been obtaining the same results when using either the standard Littlewood-Paley theory or one of its variants. Today it is easy to recognize that the standard Littlewood-Paley theory is a wavelet analysis based on Morlet wavelets. A. Calderón, E. Stein and G. Weiss developed a program in which a function f on $\mathbb{R}^n$ is analyzed by the normal derivative $\frac{\partial u}{\partial t}(x,t)$ of its harmonic extension $u(x,t)$ to the upper half space $(x \in \mathbb{R}^n,\ t > 0)$. This also amounts to a wavelet analysis in which the mother wavelet is $\psi(x) = (|x|^2 - n^2)\,(|x|^2+1)^{-(n+3)/2}$.

Grossmann's simple and elegant formalism has been directly used in astrophysics [7], in experiments on turbulence [3, 13, 14] and in many other fields of science or technology. In all these applications, the role played by wavelet analysis is to provide a better localization of small scale structures. These fine details are enhanced after being extracted from a background which is either cancelled or strongly attenuated by a correctly tuned wavelet.

Discrete versions of (3.3) would be $f(x) = \sum_{\lambda\in\Lambda} c(\lambda)\psi_\lambda(x)$ where Λ is a suitable discrete subset of G. J. Morlet proposed $\lambda(x) = 2^{-\alpha j}(r(x) + \beta k)$, $j \in \mathbb{Z}$, $k \in \mathbb{Z}^n$, $r \in F$, where $\alpha > 0$, $\beta > 0$ are small enough and F is a finite set of rotations. I. Daubechies proved in [11] that

$$\left\| \sum_{\lambda\in\Lambda} c(\lambda)\psi_\lambda(x) \right\|_{L^2(\mathbb{R}^n)} \le C \left(\sum_{\lambda\in\Lambda} |c_\lambda|^2 \right)^{1/2}$$

and that the corresponding mapping from $l^2(\Lambda)$ into $L^2(\mathbb{R}^n)$ is onto when α and β are small enough. The values of α and β depend strongly on the choice of ψ which should satisfy (3.6) and (3.7) [22].

Daubechies' theorem implies that a canonical decomposition $f(x) = \sum_{\lambda\in\Lambda} c(\lambda)\psi_\lambda(x)$ exists in which the coefficients are given by $c(\lambda) = \langle f, \tilde{\psi}_\lambda\rangle$. Unfortunately these "dual wavelets" $\tilde{\psi}_\lambda$ might be badly behaved in terms of size and regularity. When this happens, it prevents us from using (in G. Weiss terminology) the "atomic decomposition" $f(x) = \sum c(\lambda)\psi_\lambda(x)$ in situations other than the "trivial L^2 setting". But if one is limited to the L^2 theory, the Haar system (1909) will provide the most efficient wavelet analysis, since it uses the simplest analyzing wavelet and since it is an orthonormal basis.

All these difficulties would disappear if ψ_λ, $\lambda \in \Lambda$, forms an orthonormal basis of $L^2(\mathbb{R}^n)$.

4. Pyramidal Algorithms, Quadrature Mirror Filters and Orthonormal Wavelets

We fix an integer $N \ge 1$ and assume we are given two trigonometric polynomials $m_0(\theta) = h_0 + h_1 e^{i\theta} + \cdots + h_{2N-1}e^{i(2N-1)\theta}$, $m_1(\theta) = g_0 + g_1 e^{i\theta} + \cdots + g_{2N-1}e^{i(2N-1)\theta}$ such that $m_0(0) = 1$, $|m_0(\theta)|^2 + |m_0(\theta+\pi)|^2 = 1$ and $m_1(\theta) = e^{i(2N-1)\theta}\,\overline{m_0(\theta+\pi)}$.

We fix these polynomials and for every $\delta > 0$ we consider the corresponding operators $F_0 : l^2(\delta\mathbb{Z}) \to l^2(2\delta\mathbb{Z})$, $F_1 : l^2(\delta\mathbb{Z}) \to l^2(2\delta\mathbb{Z})$ which are defined by

$$F_0\{x_{(k\delta)}\}(2l\delta) = \sqrt{2}\sum_{-\infty}^{\infty} h_k\, x_{(k+2l)\delta}\,, \tag{4.1}$$

$$F_1\{x_{(k\delta)}\}(2l\delta) = \sqrt{2}\sum_{-\infty}^{\infty} g_k\, x_{(k+2l)\delta}\,. \tag{4.2}$$

These two operators are called "quadrature mirror filters" and have the following remarkable properties. The operator $F = (F_0, F_1) : l^2(\delta\mathbb{Z}) \to l^2(2\delta\mathbb{Z}) \times l^2(2\delta\mathbb{Z})$ is a unitary isomorphism and both adjoints $F_0^* : l^2(2\delta\mathbb{Z}) \to l^2(\delta\mathbb{Z})$, $F_1^* : l^2(2\delta\mathbb{Z}) \to l^2(\delta\mathbb{Z})$ are partial isometries. The ranges of F_0^* and F_1^* are orthogonal in $l^2(\delta\mathbb{Z})$ and, finally, one has

$$I = F_0^* F_0 + F_1^* F_1\,. \tag{4.3}$$

We now consider the increasing sequence $\Gamma_j = 2^{-j}\mathbb{Z}$ of lattices $(0 \le j)$ together with the corresponding partial isometries $F_0^* : l^2(\Gamma_j) \to l^2(\Gamma_{j+1})$. It makes sense to compose these operators and we are led to study the asymptotic behavior of $(F_0^*)^j : l^2(\Gamma_0) \to l^2(\Gamma_j)$. By construction, this operator commutes with integral translations τ_k, $k \in \mathbb{Z}$. Let $\varepsilon_k \in l^2(\Gamma_0)$ be defined by $\varepsilon_k(l) = 0$ if $l \neq k$, 1 if $l = k$. Then $(F_0^*)^j(\varepsilon_k)$, $k \in \mathbb{Z}$, is an orthonormal sequence in $l^2(\Gamma_j)$ and we would like to know if, in some sense, this sequence converges to an orthonormal sequence of the form $\varphi(x-k)$, $k \in \mathbb{Z}$, where φ belongs to $L^2(\mathbb{R})$. The convergence procedure is defined in the following way. To each element $f_j(k2^{-j})$ in $l^2(\Gamma_j)$ we attach the corresponding atomic measure $\sigma_j = 2^{-j/2}\sum f_j(k2^{-j})\delta_{(k2^{-j})}$ where δ_a is the Dirac mass at the point a. We next define the convergence of the sequence f_j by the weak convergence of the corresponding sequence σ_j.

Theorem 2 (A. Cohen, 1989). *The two following conditions are equivalent*

(4.4) *the discrete orthonormal sequences* $(F_0^*)^j(\varepsilon_k)$, $k \in \mathbb{Z}$, *converge to* $\varphi(x-k)$, $k \in \mathbb{Z}$ *and this sequence is orthonormal in* $L^2(\mathbb{R})$.
(4.5) *for* $0 < \theta < 2\pi$, $\lim_{j\to\infty} m_0(\theta)m_0(2\theta)\dots m_0(2^j\theta) = 0$.

If (4.5) is satisfied, the function $\varphi \in L^2(\mathbb{R})$ which is called the *scaling function* can be characterized by an other property. Consider the functional equation

$$\varphi(x) = 2\sum_{0}^{2N-1} h_k\, \varphi(2x-k)\,, \tag{4.6}$$

where we impose the condition $\varphi \in L^1(\mathbb{R})$ and $\int \varphi(x)\,dx = 1$. Then (4.6) admits a unique solution which is precisely the function φ of (4.4). Moreover the Fourier transform $\hat{\varphi}(\xi)$ is given by

$$\hat{\varphi}(\xi) = m_0(\xi/2)\dots m_0(\xi/2^j)\dots \tag{4.7}$$

We next define $\psi \in L^2(\mathbb{R})$ by $\psi(x) = 2\sum_0^{2N-1} g_k\,\varphi(2x-k)$ or equivalently by

(4.8) $$\hat{\psi}(\xi) = m_1(\xi/2)\,m_0(\xi/4)\ldots m_0(\xi/2^j)\ldots$$

If (4.5) is satisfied, then $2^{j/2}\psi(2^jx-k)$, $j \in \mathbb{Z}$, $k \in \mathbb{Z}$, is an orthonormal basis of $L^2(\mathbb{R})$.

It remains, for a given N, to choose $m_0(\theta)$ and $m_1(\theta)$ carefully in order to obtain a smooth wavelet $\psi(x)$. One way to proceed is the following. There exists (F. Riesz) $m_0(\theta) = h_0 + \cdots + h_{2n-1}\,e^{i(2N-1)\theta}$ such that $m_0(0) = 1$ and $|m_0(\theta)|^2 = c_N \int_\theta^\pi (\sin x)^{2N-1}\,dx$ ($c_N > 0$ will permit $m_0(0) = 1$). We then have [12].

Theorem 3 (I. Daubechies, 1987). *There exists a constant $\alpha > 0$ such that for each $N \geq 2$, the corresponding functions φ and ψ will belong to $C^{\alpha N}$.*

This construction also gives orthonormal wavelet bases in several dimensions. For the sake of simplicity, we stick to $n = 2$. We then consider the three wavelets $\psi_1(x,y) = \psi(x)\,\varphi(y)$, $\psi_2(x,y) = \varphi(x)\,\psi(y)$ and $\psi_3(x,y) = \psi(x)\,\psi(y)$. The full collection $2^j\psi_q(2^jx-k, 2^jy-l)$, $q = 1, 2$ or 3, $j \in \mathbb{Z}$, $k \in \mathbb{Z}$, $l \in \mathbb{Z}$ is an orthonormal basis of $L^2(\mathbb{R}^2)$.

When compared to Fourier series expansions, orthonormal wavelets expansions provide a much deeper insight into the local or global properties of the function to be analyzed [17, 18].

For example, J. O. Strömberg proved in 1980 that orthonormal wavelets form an unconditional basis for the Hardy space $H^1(\mathbb{R}^n)$. A suitable regrouping of the wavelet expansion of a function f in $H^1(\mathbb{R}^n)$ yields its atomic decomposition [22].

5. Wavelets Packets

Wavelet analysis gives its best performance when it is applied to signals with abrupt changes or to functions which have simple discontinuities on smooth surfaces and which are smooth elsewhere. On the other hand, wavelet analysis gives its worst performance on stationary signals.

The speech signal obviously contains these two components and one would like to switch freely from wavelet analysis to windowed Fourier analysis in speech signal processing. R. Coifman and his co-workers have constructed a library of orthonormal bases in which can be found I. Daubechies orthonormal wavelets, a second orthonormal basis resembling the Gabor wavelets as well as many other bases.

Let us fix some notations to describe this library. We start with two quadrature mirror filters $F_0 : l^2(\mathbb{Z}) \to l^2(2\mathbb{Z})$ and $F_1 : l^2(\mathbb{Z}) \to l^2(2\mathbb{Z})$ as in Theorem 3. Let $\mathscr{I}$ be the collection of all dyadic intervals $I = [l2^j, (l+1)2^j)$, $l \in \mathbb{N}$, $j \in \mathbb{Z}$, which are contained in $[0, \infty)$. To each $I \in \mathscr{I}$ we attach a closed subspace $W_I \subset L^2(\mathbb{R})$ and a function $w_I(x)$ which are uniquely defined by the four following properties

(5.1) $w_I(x - k2^{-j})$, $k \in \mathbb{Z}$, is an orthonormal basis of W_I;

(5.2) if I_0 is the left half of I and I_1 the right half, then W_I is the direct orthogonal sum of W_{I_0} and W_{I_1};

(5.3) if $f(x) = \sum_{-\infty}^{\infty} \alpha_I(k)\, w_I(x - k2^{-j})$
$= \sum_{-\infty}^{\infty} \beta_{I_0}(2k)\, w_{I_0}(x - 2k2^{-j}) + \sum_{-\infty}^{\infty} \gamma_{I_1}(2k)\, w_{I_1}(x - 2k2^{-j})$,
then $\beta_{I_0} = F_0(\alpha_I)$ and $\gamma_{I_1} = F_1(\alpha_I)$;

(5.4) if $I = [0, 2^j)$, then $W_I = V_j$ and $w_I(x) = 2^{j/2}\varphi(2^j x)$.

It is easy to compute $w_I(x)$. We write $I = [l2^j, (l+1)2^j)$, $l = \varepsilon_0 + \varepsilon_1 2 + \cdots + \varepsilon_q 2^q + \ldots$ where $\varepsilon_q = 0$ or 1 and we have

$$\hat{w}_I(\xi) = 2^{-j/2}\, m_{\varepsilon_0}(\xi/2^{j+1})\, m_{\varepsilon_1}(\xi/2^{j+2}) \ldots \tag{5.5}$$

With these notations, the library of orthonormal bases generated by the quadrature mirror filters (F_0, F_1) is described by the following theorem.

Theorem 4. *Let $\mathcal{I}_* \subset \mathcal{I}$ be any collection of dyadic intervals $I \subset [0, \infty)$ with the property that, excepting a set D which is either finite or denumerable, each $x \in [0, \infty[$, $x \notin D$, belongs to one (and only one) interval $I \in \mathcal{I}_*$.*

Then the corresponding family $w_I(x - k2^{-j})$, $k \in \mathbb{Z}$, $I = [l2^j, (l+1)2^j) \in \mathcal{I}_*$, is an orthonormal basis of $L^2(\mathbb{R})$.

When $\mathcal{I}_*$ is the obvious collection of the dyadic intervals $[2^j, 2^{j+1})$, $j \in \mathbb{Z}$, this basis happens to be the one described in theorem 3 and when $\mathcal{I}_*$ is the collection of $[l, l+1)$, $l \in \mathbb{N}$, the corresponding $w_I(x)$ will be denoted by $w_l(x)$ and resemble Gabor wavelets. Moreover, if $m_0(\xi) = \frac{1+e^{i\xi}}{2}$ and $m_1(\xi) = \frac{1-e^{i\xi}}{2}$, the orthonormal basis $w_l(x-k)$, $l \in \mathbb{N}$, $k \in \mathbb{Z}$, is the well known Walsh system which is widely used in signal processing. When $N \geq 2$ and $m_0(\xi)$ is chosen following I. Daubechies, the corresponding orthonormal basis $w_l(x-k)$, $l \in \mathbb{N}$, $k \in \mathbb{Z}$, should be compared to a smooth version of the Walsh system.

In their work in speech signal compression, R. R. Coifman and V. Wickerhauser are using this full library of bases together with an entropy criterion for selecting the specific basis among the library which provides the "best" expansion [9].

References

1. E.H. Adelson, R. Hingorani, E. Simoncelli: Orthogonal pyramid transforms for image coding. SPIE Visual Communications and Image Processing II. **845** (1987)
2. M. Antonini, M. Barlaud, I. Daubechies, P. Mathieu: Image coding using wavelet transform. Lassy, Nice-Sophia Antipolis, 06560 Valbonne, France
3. F. Argoul, A. Arnéodo, Y. Gagne, G. Grasseau, U. Frisch, E.J. Hopfinger: Wavelet analysis of turbulence reveals the multifractal nature of the Richardson cascade. Nature **338**, no. 6210 (1989)
4. G. Battle: Heisenberg proof of the Balian-Low theorem. Math. Phys. Letters (1990)
5. G. Battle, P. Federbush: Ondelettes and phase cluster expansions: a vindication. Comm. Math. Phys. **109** (1987) 417–419

6. G. Beylkin, R. Coifman, V. Rokhlin, Fast Wavelet Transforms and Numerical Algorithms, I. Research Report YALEU/DCS/RR-696, 1989
7. A. Bijaoui, G. Mars, E. Slezak: Identification of structures from galaxy counts: use of the wavelet transform. Astron. Astrophys. **227** (1990) 301–316
8. A. Cohen: Ph.D. Dissertation, CEREMADE, Université Paris-Dauphine, 75775 Paris, Cedex 16
9. R. Coifman, V. Wickerhauser: Best-adpated wave packet bases. Numerical Algorithms Research Group, Dpt. Mathematics, Yale, New Haven, CT 06520, USA
10. J.M. Combes, A. Grossmann Ph. Tchamitchian: Time frequency methods and phase space. Proceedings of the International conference, Marseille, France, Dec. 14–18, 1987
11. I. Daubechies: The wavelet transform, time-frequency localization and signal analysis. IEEE Trans. Inf. Theory (1990)
12. I. Daubechies: Orthogonal bases of compactly supported wavelets. Comm. Pure Appl. Math. **41** (1988) 909–996
13. M. Farge, M. Holschneider, J.F. Colonna: Wavelet analysis of coherent structures in two-dimensional turbulence. Topological Fluid Mechanics, ed. K. Moffat. Cambridge University Press, 1989, pp. 765–776
14. M. Farge, G. Rabreau: Wavelet transform to detect and analyze coherent structures in two-dimensional turbulent flows. C. R. Acad. Sci. Paris **307** Ser. II (1988) 1479–1486
15. D. Gabor: Theory of communication. J. Inst. Elec. Eng. (London) **93** III (1946) 429–457
16. A. Grossmann, J. Morlet: Decomposition of Hardy functions into square integrable wavelets of constant shape. SIAM J. Math. Anal. **15** (1984) 723–736
17. M. Holschneider, Ph. Tchamitchian: Pointwise Analysis of Riemann's "nondifferentiable" function. Invent. math. (to appear)
18. S. Jaffard: Exposants de Hölder en des points donnés et coefficients d'ondelettes. C. R. Acad. Sci. Paris **308** Ser. I (1989) 79–81
19. P.G. Lemarié: Les ondelettes en 1989. (Lecture Notes in Mathematics, vol. 1438.) Springer, Berlin Heidelberg New York 1990
20. S. Mallat: Review of multifrequency channel decompositions of images and wavelet models. IEEE Trans. Acoustics, Speech and Signal Processing **37**, no. 12 (1989)
21. D. Marr: Vision: A computational investigation into the human representation and processing of visual information. Freeman, New York 1982
22. Y. Meyer: Ondelettes et opérateurs, tomes I, II et III. Hermann, Paris 1990
23. R. Murenzi: Wavelet transforms associated to the n-dimensional euclidean group with dilations. Preprint Institut de Physique Théorique, Université Catholique de Louvain, B-1348 Louvain-la-Neuve, Belgium

Pattern Formation in Reaction-Diffusion Systems

Masayasu Mimura

Department of Mathematics, Faculty of Science, Hiroshima University
Hiroshima, 730 Japan

1. Introduction

Reaction-diffusion approach has been used to explain pattern formation arising in neurobiology, chemical physics, population ecology, developmental biology and other fields. Despite its simple structure, a class of reaction-diffusion systems exhibit a lot of spatial and spatio-temporal patterns. Some of these patterns in a reacting and diffusing medium can be often observed by internal layers or interfaces which are boundaries between qualitatively different states in the system. Such interfaces exhibit a variety of geometrical patterns such as rotating patterns in the Belousov-Zhabotinsky reagent [Wi], dendritic patterns in solidifications [Ca], pigmentation patterns on shells [MK] and animal coat marking [Mu], for instance.

The term "reaction-diffusion equations" is usually taken to mean the following semilinear parabolic equations:

$$\frac{\partial u}{\partial t} = D\Delta u + F(u), \tag{1.1}$$

where $u(t, x) = (u_1, u_2, \ldots, u_n)(t, x)$ means the concentration, density and other physical component with time and space variables t and x, D is an $n \times n$ nonnegative matrix and in most cases, it is a diagonal one and F is the reaction term.

Among so many reaction-diffusion equation models, we are concerned with activator-inhibitor systems which arise in modelling of morphogenesis ([GM]). The most simple and suggestive system is the following two-component model:

$$\begin{cases} \dfrac{\partial u}{\partial t} = d_1 \Delta u + f(u, v) \\ \dfrac{\partial v}{\partial t} = d_2 \Delta v + \delta g(u, v), \end{cases} \tag{1.2}$$

where u and v are called respectively the activator and its inhibitor in morphogenesis [GM] or the propagator and its controller in excitable media [Fi1]. Here d_1 and d_2 are the diffusion rates of u and v. δ is the ratio of reaction rates. The nonlinearities of f and g in which we are interested in this paper are restricted to two types in

Proceedings of the International Congress
of Mathematicians, Kyoto, Japan, 1990

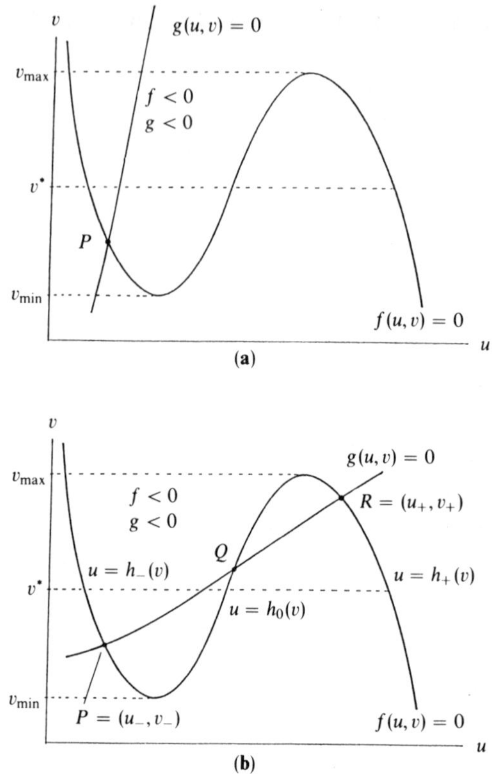

Fig. 1

Figs. 1a and b. A prototype of f and g is

$$\begin{cases} f(u, v) = u(1 - u)(u - a) - v \\ g(u, v) = u - \gamma v \end{cases} \tag{1.3}$$

with constants $0 < a < 1$ and $\gamma > 0$. When v is totally absent in (1.2) with (1.3), it is called the time dependent Ginzburg-Landau equation and when v does not diffuse, it reduces to the well-known FitzHugh-Nagumo equation which models a nerve impulse propagating along the axon([NAY]).

The kinetic system of (1.2) is given by

$$\begin{cases} \dfrac{du}{dt} = f(u, v) \\ \\ \dfrac{dv}{dt} = \delta g(u, v), \end{cases} \tag{1.4}$$

which is called the Bonhoeffer-Van der Pol equation when f and g take (1.3).

For the case in Fig. 1a, (1.4) has the only one equilibrium state P, which is globally stable. In this case, a small disturbance from this state is rapidly damped, while a large disturbance wanders far from the state but then eventually returns to the state. Because of this feature, the state P is called a rest state in excitable media. On the other hand, for the case in Fig. 1b. (1.4) has three equilibria P, Q and R where P and R are stable and Q is unstable. That is, (1.4) is called a monostable system for the former case, while it is a bistable one for the latter case.

We consider the situation where diffusion terms are present in (1.2) and one component u diffuses slower than the other v, moreover, u reacts much faster than v. More specifically, we introduce the new parameters ε, τ and D instead of d_1, d_2 and δ through

$$\varepsilon = \sqrt{d_1}, \qquad \tau = \delta/\sqrt{d_1} \qquad \text{and} \qquad D = d_2/\delta.$$

Moreover, we specify the nonlinearities of f and g as (1.8) for simplicity only, though the results which will be stated later are valid for more general nonlinearities under appropriate conditions. We thus rewrite (1.2) as

$$\begin{cases} \varepsilon\tau \dfrac{\partial u}{\partial t} = \varepsilon^2 \Delta u + f(u) - v \\ \\ \dfrac{\partial v}{\partial t} = D\Delta v + u - \gamma v \end{cases} \tag{1.5}$$

with $f(u) = u(1-u)(u-a)$, where ε is sufficiently small and τ and D are of the order $0(1)$ compared with ε.

Assuming that ε is sufficiently small, we use singular perturbation techniques to study the existence and stability of stationary pulse solutions in a monostable system and traveling front solutions in a bistable one in Sections 2 and 3, respectively. Both solutions exhibit internal layers with width $0(\varepsilon)$. In Section 4, in order to study the dynamics of such layers, we derive the equation of motion for interfaces in the limit $\varepsilon \downarrow 0$. Finally, we would like to give some remarks on our system in Section 5.

Acknowledgement. I have benefitted from the discussions with my colleagues, H. Fujii, Y. Nishiura, H. Ikeda, T. Tsujikawa, R. Kobayashi and T. Ohta.

2. Localized Patterns in a Monostable Medium

In this section, we consider (1.5) under the situation where there is only one trivial (constant) equilibrium state $P = (0, 0)$ as in Fig. 1a. In addition, we assume

$$\int_0^1 f(u)\, du > 0$$

under which the equilibrium state P is asymptotically stable. This situation is realizable when a and γ are chosen to satisfy $0 < a < 1/2$ and $0 < \gamma < \max(u-a)(1-u)$.

We treat the system (1.5) in the whole domain R^N. The boundary condition at infinity is

$$\lim_{|x|\to\infty} (u, v) = (0, 0). \tag{2.1}$$

For this problem, there arises naturally a question of interest: Are there any nontrivial equilibrium states? Intuitively we can imagine the following situation: Suppose that there is a local disturbance in u. If it is not so small, then it possibly forms into a large peak and expands, as an activator, but its expanding may be stopped due to the relatively faster diffusion of the inhibitor. This argument suggests the possibility of the localization of two components, if there exists a suitable balance between them.

Motivated by this suggestion, we are interested in the existence problem of (hopefully stable) nonconstant equilibrium states of (1.5) under the boundary condition (2.1).

For one dimensional case, singular perturbation techniques [Fi2] or shooting arguments [EHT] can be applied to (1.5), (1.6) if ε is sufficiently small.

Theorem 1. *There is* ε_0 *such that the stationary problem of* (1.5), (2.1) *has two different types of* 1D*-pulse solutions* $(\bar{u}_\varepsilon, \bar{v}_\varepsilon)$ *and* $(\underline{u}_\varepsilon, \underline{v}_\varepsilon)$, *which are symmetric with* $x = 0$, *for* $0 < \varepsilon < \varepsilon_0$ *(the profiles of these solutions are Fig.* 2).

We now address the following questions: (1) Is there any other pulse-solution for small ε? (2) Is there any pulse solution for not small ε? (3) How is the stability of the equilibrium solutions $(\bar{u}_\varepsilon, \bar{v}_\varepsilon)$ and $(\underline{u}_\varepsilon, \underline{v}_\varepsilon)$? Unfortunately, the first two equations (1) and (2) have not yet been answered except for the special case when f is piecewise-linear

$$f(u, v) = -1 + H(u - a) - v$$

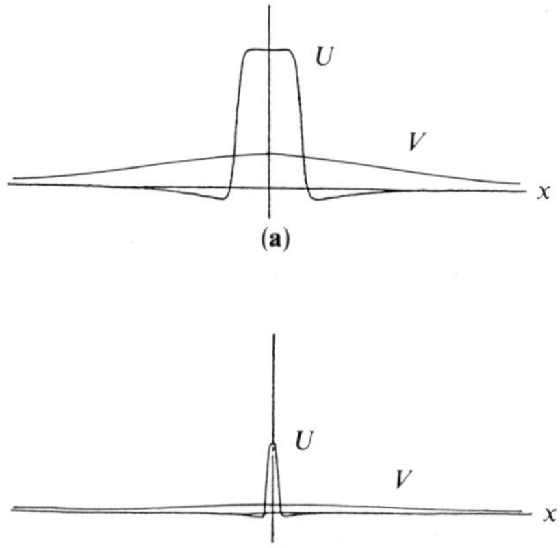

Fig. 2

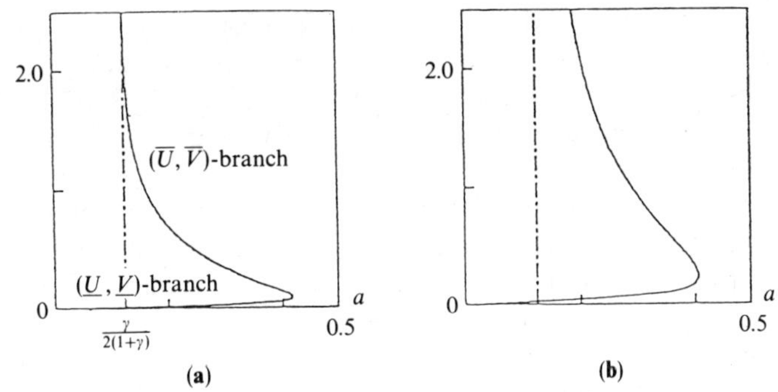

Fig. 3

with constants a and γ satisfying $0 < a < 1/2$ and $0 < \gamma < 2a/(1-2a)$, where $H(x)$ is the Heaviside step function. For this case, the global structure of symmetric pulse solutions can be completely understood ([OMK]). For instance, Figs. 3a and b show respectively 1D- and 3D-radially symmetric solutions. That is, for suitably fixed a and γ, there is the critical value ε_c such that there are exactly two different solutions for $0 < \varepsilon < \varepsilon_c$, where the upper and lower branches correspond to $(\overline{u}_\varepsilon, \overline{v}_\varepsilon)$ and $(\underline{u}_\varepsilon, \underline{v}_\varepsilon)$, respectively, while there are no solutions for $\varepsilon_c < \varepsilon$.

For the third question (3), we have recently obtained the following:

Theorem 2 [NM]. *Let* $(\overline{u}_\varepsilon, \overline{v}_\varepsilon)$ *and* $(\underline{u}_\varepsilon, \underline{v}_\varepsilon)$ *be the equilibrium solutions of* (1.5), (2.1) *which are given in Theorem* 1. *There is the critical value* τ_c *such that* $(\overline{u}_\varepsilon, \overline{v}_\varepsilon)$ *is asymptotically stable (except for translation free) for* $\tau_c < \tau$, *while it is unstable through Hopf bifurcation for* $0 < \tau < \tau_c$. *On the other hand,* $(\underline{u}_\varepsilon, \underline{v}_\varepsilon)$ *is unstable for any* $\tau > 0$.

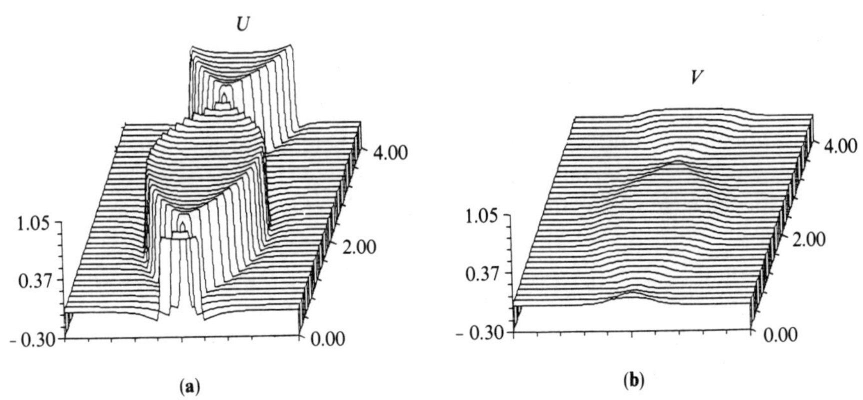

Fig. 4

Remark [[NM]). When τ decreases, the destabilization of $(\bar{u}_\varepsilon, \bar{v}_\varepsilon)$ occurs and there appears a bifurcating solution which exhibits oscillating internal layers as in Fig. 4.

It is quite interesting to consider the stability of pulse solutions in higher dimensional space. For the piecewise nonlinearity, it has been recently shown in [OMK] that even if τ is fixed to satisfy $\tau_c < \tau$, the stability of 2D- and 3D-solutions is different from the 1D-one. For higher dimensional cases, the stability crucially depends on parameters a and γ. For instance, when γ is suitably fixed, there is the critical value a_c such that the solution is stable for $a_c < a$, while it is unstable through radially symmetric breaking. For more general nonlinearities, numerical simulations show that such destabilization also occurs ([OMK]). Its rigorous treatment is in progress.

3. Traveling Fronts in a Bistable Medium

In this section, we consider (1.5) in the case when there are two equilibrium states which are stable as in Fig. 1b. We simply write them as $P:(u_-, v_-)$ and $R:(u_+, v_+)$. This situation is realizable when $0 < a < 1/2$ and

$$\gamma > \gamma_0 = q/\{(2 - a - \sqrt{a^2 - a + 1})(1 - 2a + \sqrt{a^2 - a + 1})\}$$

Under this situation, we consider one dimensional traveling front solutions which correspond to propagating transition from one state P to the other R. This kind of problem occurs in the study of diffusive waves in population genetics, combustion theory, chemical reaction and population dynamics.

These solutions can be represented by $(u, v)(z)$ with $z = x + ct$ where c is the velocity. Thus, (1.5) is written as

$$\begin{cases} \varepsilon^2 u_{zz} - \varepsilon c\tau u_z + f(u) - v = 0 \\ v_{zz} - cv_z + u - \gamma v = 0 \end{cases} \qquad z \in R, \tag{3.1}$$

where we may take $D = 1$ without loss of generality. The corresponding boundary conditions are

$$(u, v)(\pm\infty) = (u_\pm, v_\pm). \tag{3.2}$$

Our problem is to find c such that (3.1) has a heteroclinic orbit $u(z)$ connecting (u_+, v_+) and (u_-, v_-) at the infinities $z = \pm\infty$.

For a special case when γ tends to ∞, (3.1), (3.2) formally reduces to the scalar problem for u

$$\begin{cases} \varepsilon^2 u_{zz} - \varepsilon c\tau u_z + f(u) = 0, \qquad z \in R \\ u(\pm\infty) = u_\pm \end{cases} \tag{3.3}$$

because $v \equiv 0$. For this problem, it is already known that there is only one traveling front solution $u(z)$ with the unique velocity c and it is stable (except for translation free) ([FM], for instance). We easily find that when ε is sufficiently small, this solution exhibits a single internal layer with width $O(\varepsilon)$.

We consider the problem as to how many traveling fronts exist and how their stability are for (3.1), (3.2). Before stating the result, we introduce the following two critical values in addition to γ_0:

$$\gamma_1 = 9/\{(2-a)(1-2a)\}$$

and

$$\gamma_2 = 9/(1 + a + \sqrt{3(a^2 - a + 1)}/\{(1+a)(2-a)(1-2a)\}$$

When ε is sufficiently small, singular perturbation techniques can apply to (3.1), (3.2) so that the global bifurcation pictures with respect to τ are drawn for some values of γ, as in Fig. 5 ([IMN]). We find that the number of traveling front solutions depend upon the parameters τ and γ, which is totally different from the scalar version. Fig. 5a is the structure for $\gamma_0 < \gamma < \gamma_1$, Fig. 5b is for $\gamma = \gamma_1$, Fig. 5c is for $\gamma_1 < \gamma < \gamma_2$ and Fig. 5d is for $\gamma_2 < \gamma$. For the special case when $\gamma = \gamma_1$ so that the kinetics possess odd symmetry, there exist an odd symmetric standing front solution (which has zero velocity) for any τ and a pitchfork bifurcation occurs at $\tau = \tau_c$ such that the trivial solution is stable for $\tau_c < \tau$, while it is unstable for $0 < \tau < \tau_c$. When γ is slightly different from γ_1, this symmetric structure is deformed into the imperfection structures in which one limit point appears.

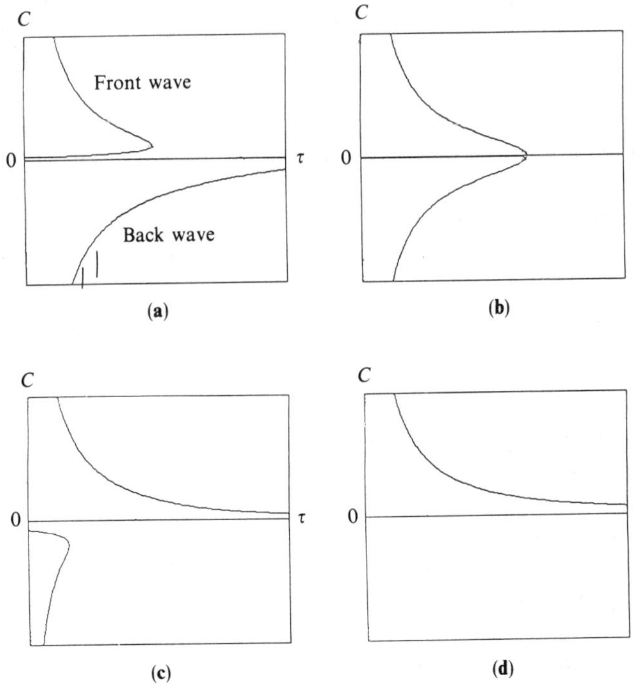

Fig. 5

We call the front with positive velocity a front wave, while the one with negative velocity a back wave. For the stability of these solutions in one dimensional case, we have

Theorem 3 (NMIF]). *Consider the solution branches in Fig. 5.*

(1) (*Fig.* 5a) *The fast front wave and the back wave are stable, while the slow front wave is unstable;*

(2) (*Fig.* 5b) *The front and back waves are both stable;*

(3) (*Fig.* 5c) *The front wave and the fast back wave are stable, while the slow back wave is unstable;*

(4) (*Fig.* 5d) *The front wave is stable.*

The fourth case corresponds to the scalar problem which was already noted.

A natural question arises whether these planar traveling front solutions are stable or not in higher dimensional spaces? Recently we have found that the stability crucially depends on the value of τ. The result will be stated in a forthcoming paper.

4. Interfacial Dynamics

We have shown the stability as well as existence of stationary pulse solutions and traveling front solutions which possess internal layers with width $0(\varepsilon)$. From an application view point, we address the following question: When ε is sufficiently small, how does a solution of the initial-boundary value problem of (1.5), (2.1) or (1.5), (3.2) evolve into these patterns? In order to answer it, we derive the approximating evolution equation from the original reaction-diffusion system (1.5) when ε tends to zero. In this case, one could expect that an internal layer becomes an interface in the limit $\varepsilon \downarrow 0$. First, we consider a bistable scalar equation of (1.5) when v is assumed to be constant, say $v = q$ such that $f(u) = q$ has three zeros $h_-(q)$, $h_0(q)$ and $h_+(q)$.

$$\varepsilon\tau \frac{\partial u}{\partial t} = \varepsilon^2 \Delta u + f(u) - q. \tag{4.1}$$

We consider (4.1) in the whole plane R^2 for simplicity. When ε is sufficiently small, one could expect that the process consists of two different stages. The first stage is that even if the initial data is smooth, the solution $u(t, x)$ tends, in a short time, to one stable equilibrium state $u = h_+(q)$ in a region where $u(0, x) > h_0(q)$ or to the other stable state $u = h_-(q)$ in a region where $u(0, x) < h_0(q)$, since the system is a bistable one. This indicates that there appear internal layers with width $0(\varepsilon)$ which separate the plane into two different subplanes ([FH]). In the limit $\varepsilon \downarrow 0$, it implies the appearance of interfaces in the plane. The next stage is that such interfaces propagate. Suppose that the interface is described by a gentle curve $\Gamma(t)$ in a way that $R^2 \backslash \Gamma(t) = \Omega_+(t) \cup \Omega_-(t)$ where the relations $u = h_\pm(q)$ hold in the regions $\Omega_\pm(t)$. The motion of $\Gamma(t)$ is described by the following evolution equation ([KT], for instance):

$$\tau V = \{-c(q) + \varepsilon K\} n \tag{4.2}$$

where V is the normal velocity of the interface, K is the mean curvature at $\Gamma(t)$, n is the unit normal vector of $\Gamma(t)$ pointing from $\Omega_+(t)$ to $\Omega_-(t)$ and $c(q)$ is the velocity of the 1D-traveling front solution of the bistable equation

$$\begin{cases} \dfrac{\partial u}{\partial t} = \dfrac{\partial^2 u}{\partial x^2} + f(u) - q, \qquad t > 0, \qquad x \in R \\ \lim\limits_{x \to \pm\infty} u = h_{\pm}(q) \end{cases}$$

It is known that $c(q)$ is explicitly represented as

$$c(q) = \frac{h_+(q) - 2h_0(q) + h_-(q)}{\sqrt{2}(h_+(q) - h_-(q))}$$

Recently, for the study of the curvature effect on the motion of interfaces, (4.2) has been investigated in the mathematical community ([CGG], [ES], [Gr] and the references therein).

Come back to our system of equations (1.5) in R^2. Keeping the scalar equation (4.1) in mind, one could expect that the process also consists of two stages: The first stage is the appearance of internal layers or interfaces in the limit $\varepsilon \downarrow 0$. In the region where is away from interfaces, we may put $\varepsilon = 0$ in (1.5) so that it becomes

$$\frac{\partial v}{\partial t} = D\Delta v + g_{\pm}(v), \tag{4.3}$$

where $g_{\pm}(v) = h_{\pm}(v) - \gamma v$. The functions $u = h_{\pm}(v)$ stand for the two branches of $f(u) = v$, as shown in Fig. 1. Thus, the whole plane R is decomposed into two subplanes $\Omega_{\pm}(t)$ where $u = h_{\pm}(v)$ hold. The boundary between $\Omega_+(t)$ and $\Omega_-(t)$ is an interface, say $\Gamma(t)$. The second stage is the motion of interfaces. As in the similar way to the scalar version, we obtain

$$\tau V = \{-c(q) + \varepsilon K\} n \tag{4.4}$$

where q is the value of v on the interface. We thus have the evolution equation for (v, Γ)

$$\begin{cases} \dfrac{\partial v}{\partial t} = D\Delta v + g_{\pm}(v), \qquad (t, x) \in \Omega_{\pm}(t) \\ \tau V = -\{c(q) + \varepsilon K\} n \quad \text{on } P(t) \end{cases} \tag{4.5}$$

with suitable continuity conditions for v on the interfaces. For the one dimensional case, the global existence of a smooth solution is proved in [HNM]. On the other hand, for higher dimensional cases, complicated singularities may appear so that we can not expect the global existence of smooth solutions. Recently, the local existence of a smooth solution has been shown in [Ch].

For the dynamics of interfaces, there is a great difference between the scalar case (4.1) and the system (4.4). For the former case, if the configuration of an initial-interface is convex, it is still convex for any time, while it possibly becomes non-convex for the latter case ([OMK]).

The analysis of this system is in the beginning stage.

5. Concluding Remarks

In the previous sections, we have two different systems. One, system (1.5). is the so-called reaction diffusion system with a small parameter ε and the other (4.5) is the interface equation associated with (1.4) in the limit as $\varepsilon \downarrow 0$. The relation between (1.5) and (4.5) can be studied by intuitive and formal asymptotic analysis. The rigorous understanding has been achieved for the scalar case (3.1) ([MS]) and the numerical study ([IK]). This is one of the important problems which we should study in future, from pattern formation view point.

Note after submission. After completing this work, the author learned the paper by X. Chen: Generation and propagation of interfaces in reaction diffusion systems (IMA, preprint series #708), which discuss the relation of (1.5) and (4.5).

References

[Ca] Caginalp, G.: An analysis of a phase field model of a free boundary. Arch. Rat. Mech. Anal. **92** (1986) 205–245

[Ch] Chen, X.Y.: Dynamics of interfaces in reaction diffusion systems. Hiroshima Math. J. (to appear)

[CGG] Chen, Y.G., Giga, Y., Goto, S.: Uniqueness and existence of viscosity solutions of generalized mean curvature equations. Preprint

[EHT] Ermentrout, G.B., Hastings, S.P., Troy, W.C.: Large amplitude stationary waves in an excitable lateral-inhibitory medium. SIAM J. Appl. Math. **44** (1984) 1133–1149

[ES] Evans, L.C., Spruck, J.: Motion of level sets by mean curvature I. Preprint

[Fi1] Fife, P.C.: Understanding the patterns in the BZ reagent. J. Statist. Phys. **39** (1985) 687–703

[Fi2] Fife, P.C.: Dynamics of internal layers and diffusive interfaces. CBMS-NSF Regional Conference Series in Applied Mathematics, vol. 53 (1988)

[FH] Fife, P.C., Hsiao, L.: The generation and propagation of internal layers. Nonlinear Analysis **12** (1988) 19–41

[FM] Fife, P., McLeod, J.B.: The approach of solutions of nonlinear diffusion equations to travelling front solutions. Arch. Rat. Mech. Anal. **65** (1977) 335–361

[GM] Gierer, A., Meinhardt, H.: A theory of biological pattern formation. Kybernetika **12** (1972) 30–39

[Gr] Grayson, M.A.: The heat equation shrinks embedded plane curves to round points. J. Diff. Geom. **26** (1987) 285–314

[HNM] Hilhorst, D., Nishiura, Y., Mimura, M.: A free boundary problem arising from some reaction-diffusion system. Preprint

[IMN] Ikeda, H., Mimura, M., Nishiura, Y.: Global bifurcation phenomena of traveling wave solutions for some bistable reaction-diffusion systems. Nonlinear Analysis **13** (1989) 507–526

[IK] Ikeda, T., Kobayashi, R.: Numerical simulations to interfacial dynamics. 1989 (videotape)

[KT] Keener, J.P., Tyson, J.J.: Spiral waves in the Belousov-Zhabotinsky reaction. Physica D **21** (1986) 307–324

[MK] Meinhardt, H., Klinger, M.: Pattern formation by a coupled oscillations: The pigmentation patterns on the shell of molluscs. (Lecture Notes in Biomathematics, vol. 71.) Springer, Berlin Heidelberg New York 1987, pp. 184–198

[MS] de Mottoni, P., Schatzman, M.: Development of interfaces in N-dimensional space. 1989 (preprint)

[Mu] Murray, J.D.: A prepattern formation for animal coat markings. J. Theor. Biol. **88** (1981) 161–199

[NAY] Nagumo, J., Arimoto, S., Yoshizawa, S.: An active pulse transmission line simulating nerve axon. Proc. Inst. Radio Engrs., vol. 50, 1962, pp. 2061–2070

[NM] Nishiura, Y., Mimura, M.: Layer oscillations in reaction-diffusion systems. SIAM J. Appl. Math. **49** (1989) 481–514

[NMIF] Nishiura, Y., Mimura, M., Ikeda, H., Fujii, H.: Singular limit analysis of stability of traveling wave solutions in bistable reaction-diffusion systems. SIAM J. Math. Anal. **52** (1990) 142–164

[OMK] Ohta, T., Mimura, M., Kobayashi, R.: Higher-dimensional localized patterns in excitable media. Physica D **34** (1989) 115–144

[Wi] Winfree, A.T.: The geometry of biological time. Springer, Berlin Heidelberg New York 1980

The Development of Algebraic Methods of Problem-Solving in Japan in the Late Seventeenth and the Early Eighteenth Centuries

Annick M. Horiuchi

REHSEIS (CNRS), 49 rue Mirabeau, F-75016 Paris, France

I. Introduction

The rapid growth of mathematical knowledge during the Edo period (1600–1868) is one of the most remarkable features of the history of science in Japan before the modernization of the Meiji era. The chief outcomes of this long-standing tradition were:

- the accumulation of a significant body of high-standard mathematical works jealously kept within private academies where they were communicated to small numbers of selected disciples,
- the wide diffusion of mathematical practice through the publication of popular textbooks and the activity of schoolmasters and itinerent teachers.

To appreciate the extent of this development, one must recall that, at the turn of the 17th century, mathematics meant little more than elementary computations performed with the abacus. The situation was modified to a considerable extent after the Japanese turned their attention to ancient Chinese scientific works. In less than half a century, Japanese mathematics developed from the primary art of computation which served the practical needs of merchants, craftsmen and low-grade warriors into a discipline appealing to a scholarly audience.

Before dealing with the original contribution of Japanese mathematicians, it is to be noted that the assimilation of Chinese mathematics was quite effectively prompted by the novel practice of leaving several problems unsolved at the end of the books as a challenge to other mathematicians. Almost all the difficult problems discussed in Chinese works were integrated in this way into the Japanese corpus.

A further impetus was added in 1658 by the discovery and the subsequent reprint of a 13th century Chinese treatise [Zhu Shijie 1299], the level of which far surpassed those of previously available works. The 13th century in the history of Chinese mathematics was a very productive period (often described as its golden age) when significant achievements were made, most particularly in the area of algebraic devices [Li Yan and Du Shiran 1987, chap. 5]. The Japanese scholars' attention soon focused on the *tengen* (or *tianyuan* in Chinese) method which Zhu Shijie used to solve a range of tricky problems.

Proceedings of the International Congress
of Mathematicians, Kyoto, Japan, 1990

The *tengen/tianyuan* method of solving problems was basically similar to the one which is called algebra in the West. The Japanese scholars spent many years before getting at the meaning of the word *tengen* (literally "celestial origin"), the name given by the Chinese to the "unknown". The level of the difficulties involved was such that the contemporary Chinese mathematicians who were unaware of the past achievements had to wait for the introduction of the western algebra in the 18th century to rediscover the meaning of the method. [Martzloff 1987, pp. 105–106]

Japanese mathematicians worked on problems which were modelled on the older Chinese problems. These problems dealt with concrete situations and were expressed in numerical terms; their solution included both the numerical result and the procedure (*jutsu* in Japanese), the sequence of the operations to perform on the abacus or with counting-rods to get to the result.

The proliferation of small problems of this kind throughout the Edo period and the increasingly artificial character of most of them led some historians to stress the artistic and recreational character of Japanese mathematics and the mathematicians' indifference to the 'utility' of their art [Mikami, 1921].

This point of view relied chiefly on the examination of one side of Japanese mathematics: the problems. But one cannot ignore the fact that the "utility" of mathematics in the past very often depended on the general and efficient *methods* and *tools* which were obtained through solving particular problems. Therefore, the issue of the utility of Japanese mathematics cannot be settled before having examined the other side of the mathematicians' work, that of elaborating *methods* of solving problems.

The aim of this paper is to discuss some prominent features of the development of these methods in the late 17th and the early 18th centuries. I will focus on the achievements of Seki Takakazu (?–1708) and Takebe Katahiro (1664–1739), two major mathematicians of this period. I will discuss the way Seki improved the Chinese algebraic methods and stress the importance of the Chinese root-extraction procedure in the course of his research. I will then turn to one of Takebe's main contributions, the introduction of the infinite power series in the scope of algebraic calculation.

II. SEKI's Study of Algebraic Devices

Let us begin with Seki's synthesis of earlier methods of problem-solving. The synthesis, achieved by 1683, took the form of a trilogy. Each treatise was devoted to a particular method of problem-solving [Hirayama et al. 1974, sects. 6, 7 and 8]. I will examine only the last treatise where Seki expounds an original method of problem-solving which can be understood as an extension of the Chinese *tengen/tianyuan* method.

Let us consider first the main features of this Chinese method which had a considerable influence on the development of algebraic devices in Japan. Here is an example of a problem requiring the *tengen* method for its solution:

> "Given a rectangular field of 8 mu 5 fen and 5 li [1 mu = 240 square bu; 1 fen = 0.1 mu; 1 li = 0.1 fen]. We only say that the sum of the length and the width is 92 bu. Find the length and the width". [Zhu Shijie 1299, chapter kaifang shisuo]

The solution followed a regular pattern: first, a quantity was set up as the unknown. Then two different algebraic expressions for a certain quantity were built up (in the problem above, the area was expressed as $x(92 - x)$ and 2052). The equation was derived by subtracting one of the expressions from the other. Finally, the numerical value of the unknown was determined by extracting the root of the equation, digit by digit.

One basic feature of the *tengen/tianyuan* method was the use of counting rods to carry out the polynomial calculation as well as the extraction of the root. The solution itself, as I have said in the introduction, consisted of the sequence of operations to be performed with this instrument. The instructions given by the author were like: "Put one rod for the unknown", "Multiply by itself four times", "Add it to the area", etc.

The way the counting-rods were to be handled could be reconstructed from specific symbols representing the polynomials obtained on the counting-board (a large sheet of paper with horizontal and vertical lines drawn on it). These symbols were inserted in the solution after each instruction. Polynomial expressions as well as equations of one unknown were represented by the column of their coefficients arranged in the order of increasing powers of x (see Fig. 1):

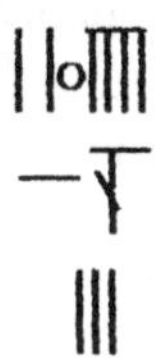

209 (constant term)

-16 (coefficient of x)

3 (coefficient of x^2)

Ex: $3x^2 - 16x + 209$

Fig. 1. The Chinese notation for polynomials

Let us now turn to Seki's improvement of the Chinese method. The need for an improvement originated in the publication in 1671 of fifteen unsolved problems by Sawaguchi Kazuyuki, a Kyoto mathematician. Sawaguchi's problems were so intricate that none of them could be handled with the Chinese method. This is an example of Sawaguchi's problems:

> "We have now *A*, *B* and *C*, such that each is a cube.
> We say first that the volumes of *A* and *B* altogether make 137,340 tsubo and also that the volumes of *B* and *C* altogether make 121,750 tsubo.
> We say in addition that the square root of the edge of *A*, the cube root of the edge of *B* and the fourth root of the edge of *C* altogether make 1 shaku 2 sun. Find the sides of *A*, *B* and *C*." [Sawaguchi 1671, Problem 4].

The third treatise of Seki's trilogy gave a global solution to two questions implicitly raised by Sawaguchi's problems which can be formulated as follows: How can one proceed to the calculation when the quantities involved cannot be

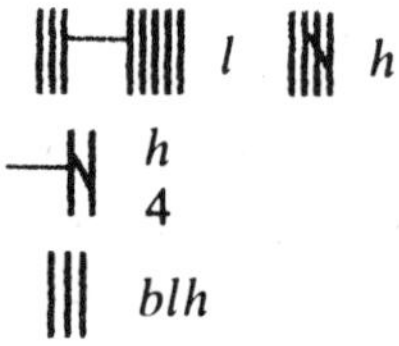

$(315l - 4h)$	(constant term)
$(-12h^5)$	(coefficient of x)
$3bhl$	(coefficient of x^2)

Ex: $3lbhx^2 - 12h^5x + (315l - 4h)$

Fig. 2. Seki's notation for polynomials with literal coefficients.

The literal and the numerical parts were dealt with separately. The latter part was transcribed by using the rod-numerals; the former was written beside these numerals using Chinese characters.

In the example above, the Chinese characters meaning length, breadth, height and number four have been replaced by the letters l, b and h and the arabic numeral 4. h^5 is represented by $\overset{h}{4}$ because Japanese mathematicians used to consider the number of times a quantity was multiplied by itself, that is the power minus one unit.

expressed in terms of one unknown? How can one eliminate the unknown within two equations?[1]

Seki answered by adding supplementary steps to the *tengen* pattern. In the course of these steps, additional unknowns were introduced and eliminated. Seki's improvement can be characterized by two main features. First, the general pattern of the *tengen* was maintained by considering one unknown at each stage and by integrating the other unknown quantities into the data. Second, the calculation was reduced to a single process of eliminating one unknown within two given algebraic equations.

Seki's ability to cope with such general contexts and questions was closely related to his use of adequate notations to represent polynomials and equations with literal coefficients (see Fig. 2). Seki's notations were obtained by extending the traditional representation of polynomials with numerical coefficients. The rules of calculation with the new notations remained unchanged. [Hirayama et al. 1974, Sect. 29].

This step has been described by many historians as a shift from an "instrumental" algebra into a "written" algebra. But this description is misleading in at least one respect. In fact, the solution maintained its algorithmic character and consisted of a rhetorical description of the operational instructions to be performed on an imaginary counting-board. In the solution, the notation was only used to represent the polynomial expressions obtained at each main step of the calculation. The calculation itself was performed somewhere else and was not made explicit. We do not even know if Seki performed the calculation on the paper or went on using the counting-board in some manner. In this respect, his calculation still preserved an instrumental character.

---

[1] Regarding this point, an extension to the case of four unknowns was achieved in China as early as the beginning of the 14th century [Zhu Shijie, 1303]. Japanese mathematicians of Edo period did not know about this extension.

Additional features should be noted about Seki's use of this notation. Seki described the procedure of elimination in very general terms, by considering sets of two equations with arbitrary coefficients.

The coefficients were represented by means of Chinese characters taken from a series of twelve characters (the ten "stems" and the twelve "branches"; *kanshi*) in the same way as alphabet letters symbolise numbers in western algebra.

Many studies have been devoted to Seki's method of elimination by which he solved problems through cancelling a certain determinant [Mikami 1913]. Leaving aside detailed analysis of this method, we note that Seki's interest in the general rules of calculating the determinant was closely connected to his systematic use of Chinese characters as symbols in place of particular numbers.

To conclude the first part of this study, I will add that the whole trilogy of Seki's reveals a clear shift of his concern from particular problems to general methods of solution. In all the treatises composing the trilogy, Seki displays both his intention and his ability to reduce any particular problem to general processes of calculation.

This tendency towards generalization, however, was not entirely new in the Sino-Japanese tradition. The search for general procedures of problem-solving had always been part of the Chinese tradition, as is well exemplified by the classification adopted in one of the oldest mathematical treatises in China, the *Nine Chapters of Mathematics* [Chemla 1988; Wu 1986]. Seki's originality lay rather in departing from the general tendency of his time to favour particular problems, and in his introduction of a range of efficient tools to describe the methods in general terms. As a consequence, Seki's attention gradually concentrated on general objects like equations and polynomials.

III. 'Defective' Problems and Seki's Reflection on Root-Extraction Procedures

Let us now examine more closely Seki's study of equations. This part of Seki's achievement, which took place shortly after the trilogy, is particularly interesting in that it brings forward a quite different way of discussing properties of equations and roots from the one so far better-known to modern mathematicians. A particular computational device plays a central role in Seki's research: the root-extraction procedure.

Seki's interest in equations stemmed from a particular, quite pragmatic preoccupation: the search for methods to correct what he called 'defective' problems (*byôdai*). According to Seki's definition [Hirayama et al. 1976, Sect. 9], problems were "defective" or wrongly stated if they led either to more than one acceptable solution or to none. Defective problems had to be corrected by changing the terms of the problem.

In the course of his research, Seki's interest shifted to the equations themselves and he thought out a general method of transforming an equation with more than one root into one equation with a single root [Hirayama et al. 1974, Sect. 8]. This was done by changing the value of one of the coefficients in the initial equation.

Before discussing Seki's method, we must note that the notion of "equation' did not have a strict equivalent in traditional Chinese methematics. Instead, we find the concept of "configuration for extraction" (*kaifangshi*), referring to the numbers set up on the counting-board to perform the extraction. The configuration itself was similar to the polynomial expression (Fig. 1). Likewise, instead of the concept of "root", we find that of "quotient", which referred to the location on the counting-board of the result of the extraction. The coefficients of the equation were similarly called by the names of their respective locations on the counting-board.

The extraction procedure which Seki used in this context was a general procedure which allowed him to compute successively all the real roots of a given equation.

This procedure, in fact, was an outcome of a long tradition of research in China [Li Yan and Du Shiran 1987, Sect. 5.2] but let us concentrate here only on the procedure as was set forth in Seki's treatise [Hirayama et al. 1976, Sect. 8]. The whole extraction was built on one basic pattern of computation involving the coefficients of the equation (see Table 1), in which one can easily recognize the so-called Horner-Ruffini process. The successive configurations can be interpreted as a gradual alteration of the coefficients of (1) into the coefficients of (2) where (2) is satisfied by y with $y = x - a$.

$$f(x) = m + nx + lx^2 + px^3 = 0 \tag{1}$$

$$\varphi(y) = f(y + a) = 0 \tag{2}$$

The above interpretation of the root-extraction pattern as a substitution of the unknown was not explicitly stated by Seki in this context. But all the improvements he introduced suggest that Seki did have a similar explanation of it.

Let us examine more closely how Seki used this procedure to find all the roots of a given equation. To begin, the first root of the equation, let us call it a, was sought by carrying out successively this basic pattern with several quotients

Table 1. The basic pattern of Seki's root extraction procedure

	(1)	(2)
Quotient	a	a
Constant term	m	$m + (n + la + pa^2).a = m + na + la^2 + pa^3$
Coefficient of x	n	$n + (l + pa).a = n + la + pa^2$
Coefficient of x^2	l	$l + p.a$
Coefficient of x^3	p	p

(3)	(4)
a	a
$m + na + la^2 + pa^3$	$m + na + la^2 + pa^3$
$n + la + pa^2 + (l + 2pa).a = n + 2la + 3pa^2$	$n + 2la + 3pa^2$
$l + pa + p.a = l + 2pa$	$l + 2pa + p.a = l + 3pa$
p	p

The calculation is carried out for each configuration from the bottom up.

(positive or negative), suitably chosen so that the number in the top line gets closer and finally becomes equal to zero. The first root was then derived by adding the list of quotients.

The equation $\varphi(y) = 0$ corresponding to this last configuration would therefore be inferior by one unit compared to the initial equation.

Seki's originality lay in his idea of iterating the previous process to the new equation. Assuming for example that b is a root of the new equation $\varphi(y) = 0$, which means that at the end of the iterated process with b as quotient, the top line of the last configuration gets to zero, then the number $a + b$ would be the second root of the initial equation.

The central role played by this root-extraction procedure in Seki's reflection on equations and roots is particularly obvious in the way he tried to solve the aforementioned question of removing the "excess" roots.

Seki's method was based on a general criterion which was to be fulfilled by the coefficients of any equation having a single root. This criterion which is actually a condition of existence of a double root was the outcome of a computation in which the extraction procedure played a central role.

The fact that a is the single root of (1) meant for Seki that at the end of the extraction procedure, the top two lines of the last configuration (the configuration (4) in Table 1) were equal to zero.

$$\begin{cases} m + na + la^2 + pa^3 = 0 \\ n + 2la + 3pa^2 = 0. \end{cases}$$

The criterion was then derived by eliminating a within these two equations:

$$27m^2p^2 + 4ml^3 + 4n^3p - 18mnlp - n^2l^2 = 0.$$

As shown by this example, the root-extraction procedure in Seki's mathematics was clearly something more than a device for computing the roots of an equation. This procedure was automatically involved in any discussion pertaining to equations, coefficients or roots. No hypothesis on the roots or any property of the equation could be stated without referring to this procedure. "Theory of extraction" would thus be the most suitable name for this part of Seki's research.

IV. Takebe's Method of Computing the Length of An Arc

Let us turn now to another significant response given by the Japanese mathematicians to a very ancient problem of trigonometry, the problem of finding a general procedure for calculating the length of an arc a in terms of the sagitta s and the diameter of the circle d (see Fig. 3).

We shall focus here on the work of Seki's disciple, Takebe Katahiro, whose contribution provides the crowning step of a long tradition of research in China and Japan [Li Yan and Du Shuran, § 3.2]. In 1720, Takebe expressed for the first time the square of the length of the arc of the circle as an infinite power series of the sagitta. Takebe's work displays an unprecedented intensity in the application of algebraic methods and especially of the root-extraction procedure.

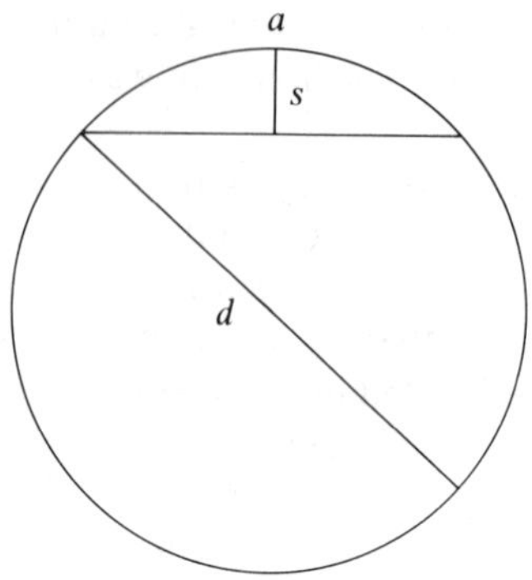

Fig. 3

To begin with, let us look at the infinite procedure of calculating the quantity $(a/2)^2$ as formulated in Takebe's treatise [Takebe 1722, sect. 12]. a, s and d are respectively the length of the arc, the sagitta and the diameter of the circle.

> "The fundamental procedure runs as follows:
>
> Sagitta and diameter are multiplied. This gives the approximate value of the square of half the arc. [sd]
>
> Divide by three the square of the sagitta. This gives the first discrepancy. [$X_1 = s^2/3$]
>
> Put down the first discrepancy and multiply by the sagitta. Divide by the diameter. Then, multiply by 8 and divide by 15. This gives the second discrepancy. [$X_2 = X_1 . (s/d). (8/15)$]
>
> Put down the second discrepancy. Multiply by the sagitta. Divide by the diameter. Multiply by 9. Divide by 14. This gives the third discrepancy. [$X_3 = X_2 (s/d). (9/14)$]
>
> Put down the third discrepancy. Multiply by the sagitta. Divide by the diameter. Multiply by 32. Divide by 45. This gives the fourth discrepancy. [$X_4 = X_3 (s/d). (32/45)$]
>
> [...]
>
> The successive discrepancies are added to the approximate value of the square of half the arc. This gives the fixed value of the square of half the arc." [Takebe 1722, Chap. 12]

Leaving aside the highly empirical method through which Takebe found his way to this procedure, we shall focus on his extension of the scope of algebraic calculation.

The first thing to examine is the original commentary he added to the explanation of his method. In this commentary, Takebe took as his starting point a common distinction between 'exhaustible' numbers (*tsukuru kazu*) and 'inexhaustible' ones (*tsukizaru kazu*). Numbers were exhaustible (respectively inexhaustible) if they had a limited (respectively unlimited) decimal expression. He then extended this distinction to procedures and formed the following programme:

> "Numbers like 1/4 and 1/5 are exhaustible numbers. 1/3 and 1/7 are examples of inexhaustible numbers. Addition, subtraction, and multiplication are exhaustible procedures. Division and extraction are inexhaustible procedures. The perimeter of the square or the areas of rectangles have an exhaustible form. The circumference and the segment of the circle have an inexhaustible form. Thus, just as the forms of the arc and the circle are inexhaustible, the procedure involved is also inexhaustible. Since the procedure is inexhaustible, so are the resulting numbers."
> [Takebe 1722, Chap. 12]

Takebe's commentary aimed at conferring a legitimate status to the infinite procedure, a thing which must have been quite unfamiliar and uncomfortable to most of his contemporaries. The link he established between the infinite procedures and the numbers with an infinite decimal expression may be considered as his primary argument for integrating the new procedure into the existing mathematical corpus.

The role played by this link was not only pedagogical. As can be seen in Takebe's later works, this analogy between the decimal numbers and the infinite series allowed him to venture into still more new fields of research.

This last point is exemplified by the second method of calculating the length of arc which Takebe devised shortly after the first. The method rested on the idea of deriving an infinite series by performing the root-extraction on an equation with literal coefficients.

We shall limit our discussion to the first step of the process which contains the main idea. Takebe resorted to a classical device consisting in inscribing polygonal lines inside the arc and getting approximate values of the arc by calculating the length of the polygonal lines (see Fig. 4).

Now, let us call a_0 the length of the arc to be calculated and s_0, the sagitta in terms of which the arc is to be expressed.

As can easily be seen, the sagitta s_1 of the arc $a_1 = a_0/2$ satisfies the following equation (1).

$$-s_0 d + 4dx - 4x^2 = 0 \tag{1}$$

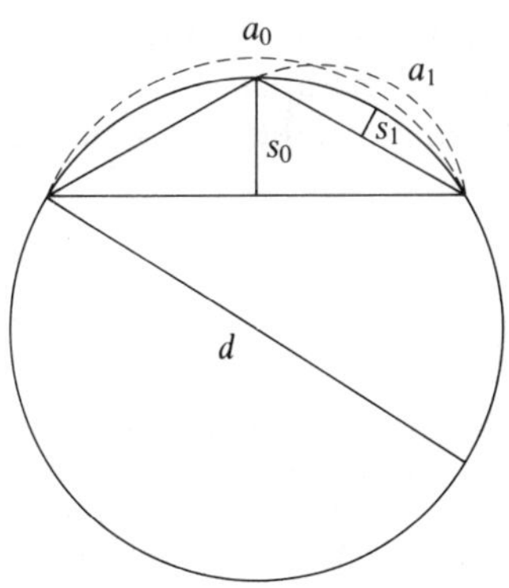

Fig. 4

The extraction procedure is performed on (1) exactly as if the coefficients were numerical. As a result of this extraction s_1 is expressed as an infinite power series of s_0/d.

$$s_1 = s_0/4 + s_0{}^2/16d + s_0{}^3/32d^2 + 5s_0{}^4/256d^3 + 7s_0{}^5/512d^4 + 21s_0{}^6/2048d^5 + \cdots$$

The same process is then performed on (2) in which the constant term $-s_1 d$ is given as an infinite series of s_0/d.

$$-s_1 d + 4dx - 4x^2 = 0 \tag{2}$$

s_2 could then be expressed in terms of s_0/d by performing the root-extraction procedure on (2). We note that in the course of the calculation successive terms of the infinite series are handled as if they were successive digits of a decimal number.

This example shows clearly the central role played by the aforedescribed analogy in Takebe's second method of calculating the arc. His extension of the extraction procedure to cases where coefficients are literal and even infinite series, as well as his way of handling the infinite series, were direct outcomes of this analogy.

From a historical point of view, this later aspect of Takebe's research had a fundamental significance in offering a very effective method of constructing infinite series. Moreover, in his explanation of the second method, Takebe clearly displayed his belief that the previously described extraction procedure could be applied to an equation of any degree. He also suggested that the general law which governs the successive terms of the series was resulting from the extraction procedure [Takebe s.d, p. 24b]. Both points bring Takebe's method very close to the idea of the binomial theorem.

Conclusion

Even though the analyses above do not cover the whole scope of mathematical research in the Edo period, they show well enough the growing interest in algebraic devices during the 17th and the 18th centuries in Japan. We have seen how the Chinese algebraic devices, including the procedure of extraction, have been extended and enriched by Seki and Takebe. By studying these devices, they not only sought to solve the largest amount of problems, but also strove to clarify and make explicit the general patterns of methods of problem-solving. In my view, this outstanding feature of the Japanese tradition has more claim to the utilitarian cause than has usually been recognized.

References

Chemla, Karine (1988): La pertinence du concept de classification pour l'analyse des textes mathématiques chinois. Extrême-orient, Extrême-occident, 10

Hirayama, Akira, Shimodaira, Kazuo, Hirose, Hideo (eds.) (1974): Seki Takakazu zenshû (Seki Takakazu's Complete Works). Osaka kyôiku tosho, Tokyo

Lam, Lay Yong (1982): Chinese polynomial equations in the thirteenth century. In: Explorations in the history of science and technology in China. Shanghai

Li, Yan, Du, Shiran (1987): Chinese mathematics – A concise history. Clarendon Press, Oxford
Martzloff, Jean-Claude (1987): Histoire des mathématiques chinoises. Masson, Paris
Mikami, Yoshio (1913): On the Japanese theory of determinants. Isis **1**, 9–36
Mikami, Yoshio (1921): Bunkashijô yori mitaru nihon no sûgaku (Japanese mathematics considered from the point of view of cultural history). Kôseisha Kôseikaku, Tokyo (reprint, 1984)
Qin, Jiushao (1247): Shushu jiuzhang (Mathematical treatise in nine chapters)
Sawaguchi, Kazuyuki (1671): Kokon sanpôki (Treatise of ancient and modern mathematics)
Takebe, Katahiro (1722): Tetsujutsu sankyô (The mathematical classic of *tetsujutsu*)
Takebe, Katahiro: Enri kohai jutsu (The procedure for calculating the arc based on the principle of the circle)
Wu, Wentsun (1986): Recent studies of the history of Chinese mathematics. Proceedings of the International Congress of Mathematicians, Berkeley, California, USA. Amer. Math. Soc., Berkeley 1987
Zhu, Shijie (1299): Suanxue qimeng (Introduction to mathematical studies)
Zhu, Shijie (1303): Siyuan Yujian (The jade mirror of four unknowns)

The Birth of Spectral Theory – Joseph Liouville's Contributions

Jesper Lützen

Mathematics Department, University of Copenhagen, Universitetsparken 5
DK-2100 Copenhagen Ø, Denmark

1. Introduction

The history of spectral theory is the history of a beautiful and important area of mathematics with close links to physics and with a strong influence on the development of functional analysis. Its roots lies in three areas: 1) discrete systems described by matrices (or quadratic forms) and continuous systems described by 2) differential equations or 3) integral equations. These different appearences of spectral theory were not formally connected until around 1900. The history of spectral theory of matrices has been studied in detail by Hawkins (e.g. 1975), so here I shall concentrate on the last two areas. In both of these, Joseph Liouville (1809–1882) played a role in the early period. His and his friend Charles Sturm's (1803–1855) work from the 1830s on spectral theory of self adjoint second order differential equations is emphasized in all histories of mathematics of the 19th century (e.g. Dieudonné 1981). The bulk of his work on spectral theory of integral operators, on the other hand, remained unpublished in his notebooks, until I recontructed and published it in a scientific biography of Liouville that has recently appeared (Lützen 1990 Chap. XV). Therefore Fredholm's and Hilbert's works from the beginning of the 19th century are the earliest published works on spectral theory of integral operators, but Liouville's notes from the mid 1840s show that he anticipated many of the basic ideas with more than half a century. Moreover they reveal that Liouville used a variational technique, named after Rayleigh and Ritz (1877 and 1909 respectively), to determine the eigenfunctions[1], and that he questioned the naive version of the Dirichlet principle more than 20 years before Weierstrass' famous criticism.

Liouville's "integral operator"originated in a study of potential theory of charged surfaces, and Liouville's approach has turned out to be of interest even to modern potential theorists (Berg and Lützen 1990), (Berg and Fuglede 1990), (Bang and Fuglede 1990), (Berg 1990).

[1] In the 1810s continuous spectra had implicitly been considered by Fourier and Cauchy in their work on the Fourier Integral. Otherwise continuous spectra do not occur until 1897 when Wirtinger was led to them in his study of Hill's equation (Dieudonné 1981 p. 149). Therefore, in this paper, we shall only consider discrete spectra of eigenvalues.

Proceedings of the International Congress
of Mathematicians, Kyoto, Japan, 1990

2. Separation of Variables

Sturm and Liouville considered the following differential equation

$$(k(x)V'(x))' + (g(x)r - l(x))V(x) = 0 \quad \text{for} \quad x \in (\alpha, \beta) \tag{1}$$

with the boundary conditions

$$k(x)V'(x) - hV(x) = 0 \quad \text{for} \quad x = \alpha \tag{2}$$

and

$$k(x)V'(x) + HV(x) = 0 \quad \text{for} \quad x = \beta \tag{3}$$

where k, g and l are positive functions, h and H are positive constants and r is a parameter. They both arrived at this problem by separating variables (i.e. by setting $u(x,t) = V(x)e^{-rt}$) in the partial differential equation

$$g\frac{\partial u}{\partial t} = \frac{\partial(k\frac{\partial u}{\partial x})}{\partial x} - lu \tag{4}$$

governing the heat conduction in a heterogeneous, unequally polished bar.[2] Here u represents the temperature, and the boundary conditions (2) and (3) reflect that the ends of the bar are maintained at zero degrees.

Sturm and Liouville saw that there were only non-trivial solutions to (1)–(3) when r belongs to a countable set of eigenvalues[3] $r_1, r_2, r_3, \cdots$, and they studied the following questions:

1) properties of the eigenvalues;
2) behaviour of the corresponding eigenfunctions $V_1, V_2, V_3, \cdots$;
3) expansion of arbitrary functions in a series of eigenfunctions.

The expansion problem arose in attempting to fit the solution $\sum A_n V_n e^{-r_n t}$, found by superposing simple solutions of (4), to a given initial temperature distribution $u(x,0) = f(x)$. Indeed this poses the problem of finding A_n's so that

$$\sum A_n V_n(x) = f(x). \tag{5}$$

Separation of variables was at the origin of all the early eigenvalue problems of differential operators (Lützen 1987). It was used in its full generality by Fourier in his influential work on heat conduction in homogeneous equally polished materials (1822). Stationary heat conduction in a rectangular plate and non-stationary heat conduction in a rod led Fourier to the following simple special case of (1):

$$V''(x) = -m^2 V(x) \tag{6}$$

with suitable boundary conditions.

---

[2] In his earliest work, Liouville (1830) had g and k constant.

[3] I shall use Hilbert's terminology "eigenvalue" "spectrum" etc. Moreover, I shall use the word "orthogonal", although Liouville had no such geometric ideas.

In this case the series (5) is the usual trigonometric Fourier series. When studying stationary heat conduction in spheres and cylinders Fourier used spherical and cylindrical coordinates in order to have one of the coordinates constant at the boundary. In these coordinates, the Laplacian has variable coefficients and so, separation of variables led to more complicated differential operators with variable coefficients, for which the eigenfunctions could only be found as infinite series.

Similar ideas had been used already in 1759 by Euler, to study vibrations of a circular membrane (eigenfunctions are Bessel functions) and implicitly in the 1780s by Legendre and Laplace in their study of attraction from a spheroid (eigenfunctions are spherical harmonics). In the 1830s Lamé developped such applications of curvilinear coordinate systems into a flexible method. In particular he studied stationary heat conduction in an ellipsoid and was thus led to the so-called Lamé functions or ellipsoidal harmonics. Later in the century the method of separation of variables applied to the partial differential equations of physically interesting phenomena gave rise to many other special functions.

3. Sturm-Liouville Theory

In their study of the eigenvalue problem (1)–(3) with variable, and not even explicitly given coefficients, Sturm and Liouville hit on problems of an entirely new nature. Indeed, since no workable explicit expression of the solutions can be found, they had to work directly with the equation, and the results they found were necessarily qualitative. Except for Cauchy's theorem on the existence of a solution to the Cauchy problem of a first order differential equation, (1824–1981) all previous works on differential equations had asked the question: given a differential equation, find its solutions. Sturm and Liouville broadened the question to: given a differential equation, find some property of its solutions. Conceptually this is a great step toward a qualitative theory, taken up by Poincare in the 1880s in the case of non-linear equations.

Sturm mainly studied point 1) and 2) above. By a variational technique, he followed the behaviour of the roots of a solution V_r of (1) and (2). He showed that these roots are decreasing functions of r. From this observation he concluded that there is a countable infinity of eigenvalues of the system (1)–(3) and he deduced his famous comparison and oscillation theorems. He developped most of his remarkable theory in the period from 1829 to 1833, but it did not appear until Liouville invited him to publish it in the first volume of his journal (Sturm 1836a,b).

3.1 Liouville's 1830 Paper

Liouville primarily studied the expansion problem 3), finding, in the process, additional results related to 1) and 2). His first paper of 1830 was a peculiar mixture of ingenious ideas and unrigorous methods. Among the ingenious ideas was his use of the method of successive approximations to express the solution of (1) and (2). He proved the convergence of the series and had thus published the first theorem of existence of a differential equation (1830, 1836).[4] Moreover he established the orthogonality relation

$$\int_{\alpha}^{\beta} g(x)V_m(x)V_n(x)\,dx = 0 \quad \text{for} \quad m \neq n. \tag{7}$$

This is the only general result in Sturm-Liouville theory, that had been shown earlier, namely by Poisson (1826).

Multiplying (5) by $g(x)V_m(x)$ and integrating from α to β, Liouville found by (7), that if $f(x)$ can be expressed in a series eigenfunctions, the "Fourier" coefficients must be of the form

$$A_n = \frac{\int_{\alpha}^{\beta} g(x)V_n(x)f(x)\,dx}{\int_{\alpha}^{\beta} g(x)V_n^2(x)\,dx}. \tag{8}$$

In this argument Liouville integrated term by term in an infinite series. We know that this is problematic, but he never questioned this exchange of limit procedures.

Finally Liouville determined the asymptotic behaviour of V_n:

$$V_n(x) \sim \frac{\sin\sqrt{n}\,x}{\sqrt{n}} \quad \text{for} \quad n \quad \text{large}, \tag{9}$$

and used this to prove the convergence of the "Fourier" series of f:

$$\sum_n A_n V_n = \sum_n \frac{\int_{\alpha}^{\beta} g(x)V_n(x)f(x)\,dx}{\int_{\alpha}^{\beta} g(x)V_n^2(x)\,dx} V_n. \tag{10}$$

In 1830 he used the expression for V_n, found by successive approximations, to establish the latter two theorems, and his proofs are highly questionable even with the standards of rigour of that time.

3.2 Liouville's Mature Papers

Liouville returned to these questions in a series of papers from the period 1836–1837. Now he did much better, for two reasons; 1) he could use Sturms's investigations of the oscillatory behaviour of V_n and 2) he could refer to Dirichlet's proof (1829) of the convergence of ordinary trigonometric Fourier series. With these tools Liouville gave a beautiful and rigorous proof (in fact two of increasing generality) (1837a,b) of the convergence of the Fourier series (10) for a large class of functions f. This may be considered his most important contribution to Sturm-Liouville theory.

However the problem remains: what is the limit of the Fourier series? In his first paper in the series of 1836–37 (1836), Liouville claimed that the Fourier series converges to the function f, i.e. if

[4] Cauchy's first existence proof was not published until 1981. The method of successive approximations was later attributed to Picard (1890).

$$F(x) = \sum_{n=1}^{\infty} \frac{\int_\alpha^\beta g(x)V_n(x)f(x)\,dx}{\int_\alpha^\beta g(x)V_n^2(x)\,dx} V_n(x) \tag{11}$$

then $F = f$. In fact he showed that

$$\int_\alpha^\beta g(x)(F(x) - f(x))V_n(x)\,dx = 0 \quad \text{for all} \quad n \tag{12}$$

and inferred that $F(x) - f(x) = 0$. As Liouville himself later realized, the last inference is wrong. One can only conclude that $F - f$ oscillates "infinitely fast". In a note in his notebooks he attributed this insight to a certain Mr. D, probably his good friend Dirichlet, who had explicitly excluded such functions from his convergence proof. Dispite repeated efforts by Liouville, partly in colaboration with Sturm, this major lacuna in Sturm-Liouville theory was left open untill arround 1900.

3.3 Higher Order Sturm-Liouville Theory

In a course at the College de France, Liouville (1838) generalized Sturm-Liouville theory to certain non-self adjoint higher order equations. Already Lagrange had introduced the adjoint equation and now Liouville introduced the adjoint boundary values. With this new concept he could prove a biorthogonality relation, that takes the place of (7) and he succeeded in generalizing most of Sturm's results. However, he was not able to generalize his own main theorem on the convergence of Fourier series, in spite of the importance he attached to this question. There were good reasons for this failure: Even in simple cases the Fourier series does not converge to the expanded function (see Haagerup's example in Lützen 1984).

Why did Liouville's intuition fail in this case? I think the reason is that Liouville (as well as many of his French contemporaries) were often inspired by physics and borrowed their intuition from this science. However the generalizations proposed by Liouville did not correspond to any physical problem. Thus ended the first epoque in the history of Sturm-Liouville theory with an only partially successfull generalization for the sake of generality. When the theory was revived about 40 years later, it was again a physically interesting problem that lay behind, and this was also the case when Liouville in the 1840s took up spectral theory of integral operators.

4 Liouville on Integral Operators

Fourier and Laplace transforms had given rise to integral equations already in the 18th century, but if we disregard these implicit occurrences, Abel (1823) was the first to study such an equation. Liouville broadened the approach in the 1830s by emphasizing the great physical interest of integral equations (1832) and by offering his theory of differentiation of arbitrary order, as a means to solve them. Moreover integral equations played a central role in his work on Sturm-Liouville theory (Lützen 1982).

4.1 The Published Note

At the end of 1845 Liouville published a $1\frac{1}{2}$ page note on spectral theory for a general class of integral operators. He considered the eigenvalue equation[5]

$$\int_D l(x')T(x,x')\zeta(x')\,dx' = m\zeta(x) \tag{13}$$

where l is a real valued function defined in a subset D of $\mathbb{R}^n$ (or perhaps D is an n dimensional manifold) and T is real and symmetric on $D \times D$. His note contained two theorems:

Theorem 1. *Let ζ_1, ζ_2 be solutions (eigenfunctions) corresponding to two different (eigen) values m_1 and m_2. Then the following (orthogonality) relation holds:*

$$\int_D l(x)\zeta_1(x)\zeta_2(x)\,dx = 0 \quad \textit{for} \quad m_1 \neq m_2 \,. \tag{14}$$

Theorem 2. *If l is always positive all the eigenvalues m are real.*

Liouville even concluded his brief note with the following prophetic remark:

> Moreover, one can easily see that instead of the left-hand side of the equation A[(13)] one can substitute more complicated integrations or even operations of another kind without the theorem (conveniently modified, if necessary, to suit these new operations) and even the proof ceasing to be correct; for the statements above rely essentially on a certain symmetry which it will be sufficient to retain (Liouville 1845b).

With our present day knowledge of symmetric operators we must admire Liouville's foresight.

4.2 The Unpublished Notes

My admiration grew considerably when, two years ago, I discovered that this note contained but a small fraction of a great spectral theory of an integral operator in potential theory that Liouville confided to his notebooks arround 1846.

Liouville's notes begin as follows:

1°. Render $\iiiint \frac{\lambda\lambda' d\omega d\omega'}{\Delta}$ a minimum with $\iint \lambda\, d\omega =$ const.; and you have a function l for which $\iint \frac{l' d\omega'}{\Delta} =$ const. $= 1$, say; that corresponds to the equilibrium distribution of electricity.

2°. Render <u>maximum</u>
Let $\iint \lambda\, d\omega = 0$ and $\iint \frac{\lambda^2 d\omega}{l} =$ const. and you find a second function $l\zeta_1$ such that

$$\iint \frac{l'\zeta_1' d\omega'}{\Delta} = m_1\zeta_1 \,; \quad m_1 < 1 \quad \text{if} \quad \iint \frac{l' d\omega'}{\Delta} = 1 \,.$$

[5] Of course Liouville did not use vector notation. He wrote $x, y, \cdots, z$ instead of x.

3°. Let $\iint \lambda\, d\omega = 0$, $\iint \lambda\zeta_1 d\omega = 0$ and $\iint \frac{\lambda^2 d\omega}{l}$ = const. You find the function $l\zeta_2$ such that

$$\iint \frac{l'\zeta_2' d\omega'}{\Delta} = m_2\zeta_2\,, \quad m_2 < m_1$$

and so on.

And a function Q can be expressed as

$$Q = l(A_0 + A_1\zeta_1 + A_2\zeta_2 + \cdots + A_n\zeta_n + \cdots)\,,$$
$$\iint l\zeta_m\zeta_n d\omega = 0\,, \quad A_n = \frac{\iint Q\zeta_n d\omega}{\iint [l]\zeta_n^2 d\omega}\,.$$

That is what we found a long time ago (Liouville Ms 3618 (1), pp. 15v–16r).

What goes on here? From the rest of Liouville's approximately 100 pages of messy notes on this subject it appears that he considers a closed surface S with a single layer charge distribution of which the charge density at a point x is given by the function $\lambda = \lambda(x)$. If this function is considered a function of another point x' on S Liouville writes $\lambda' = \lambda(x')$. Moreover Δ denotes the distance $|x - x'|$ between the two points and $d\omega$ is the surface measure on S.

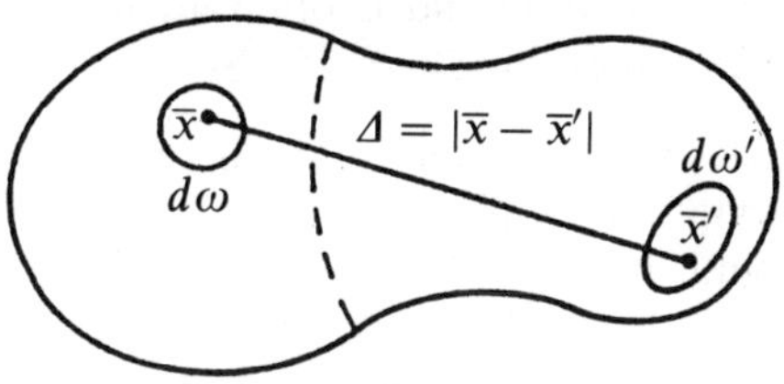

In the first step Liouville minimizes the potential energy of the charge distribution under the assumption that the total charge is kept constant. This is similar to the way Gauss (1840) had shown the existence of the equilibrium distribution, and Liouville concludes that the minimizing function l is indeed the equilibrium distribution. This means that its potential on S is a constant that we may assume to be equal to 1:

$$\iint_S \frac{l(x')}{|x - x'|}\, d\omega(x') = 1\,. \tag{15}$$

In order to abbreviate the rest of the argument, and to see the connection with modern spectral theory I shall introduce some notation, that cannot be found by Liouville. Consider the Hilbert space $L^2(S, l d\omega)$ with the inner product

$$\langle \zeta, \eta \rangle = \iint_S \zeta(x)\eta(x)l(x)\, d\omega(x)\,. \tag{16}$$

Moreover define the integral operator A on $L^2(S, l d\omega)$ by

$$A\zeta = \iint_S \frac{\zeta(x')}{|x - x'|} l(x')\, d\omega(x')\,. \tag{17}$$

In this terminology (15) can be written as

$$A1 = 1\,, \tag{18}$$

i.e. the function 1 is an eigenfunction of A with eigenvalue 1. In the second step Liouville finds the next eigenfunction and eigenvalue by maximizing $\langle A\zeta, \zeta\rangle$ under the condition that $\langle 1, \zeta\rangle = 0$ and $\|\zeta\|^2 =$ constant. The maximizing function ζ_1 has

$$A\zeta_1 = m_1\zeta_1 \quad \text{where} \quad m_1 \leq 1\,. \tag{19}$$

Generally Liouville's n'th step calls for the maximization of $\langle A\zeta, \zeta\rangle$ in the orthogonal complement of the subspace spanned by the eigenfunctions found in the $n-1$ previous steps. The maximizing function ζ_n has

$$A\zeta_n = m_n\zeta_n \quad \text{where} \quad m_n \leq m_{n-1} \leq \cdots \leq m_1 \leq 1 \tag{20}$$

It is clear that these eigenfunctions are mutually orthogonal, and in the last lines of the quote Liouville claims that any function on S can be expanded in a Fourier series of the eigenfunctions.

The above variational technique for finding eigenfunctions is now the standard method for compact operators and is often attributed to Rayleigh (1877). Liouville's note raises many questions:

1) What motivated Liouville to develop this method?
2) How much of a proof did he give?
3) Why did he not publish his results?
4) Are Liouville's results correct?
5) What are the historical and mathematical connections to later works?

4.3 Motivation

Liouville was motivated mainly by Gauss' great paper on potential theory (1840) in which the following problem was formulated:

Gauss' Problem. *Given a function $\overline{V}$ on a surface S. Determine a single layer distribution λ on S whose potential has the value $\overline{V}$ on S.*

In 1842, in connection with his (also unpublished) research on stability of rotating masses of fluid Liouville introduced the Lamé functions S_{iB} of the second kind, in addition to Lame's R_{iB}, M_{iB} and N_{iB}. The only thing we need to know about these functions is that they are defined in $\mathbb{R}^3$ in such a way that R_{iB} and S_{iB}, for each value of i and B, are constant on a certain ellipsoid S, whereas M_{iB} and N_{iB} can be considered as functions on this surface. In 1845 and 1846 Liouville published a series of results about the Lamé functions. The main theorem states that:

$$\iint_S \frac{l(x')M_{iB}(x')N_{iB}(x')}{|x-x'|}\, d\omega(x') = \frac{4\pi R_{iB}(x)S_{iB}(x)}{2i+1} M_{iB}(x)N_{iB}(x)\,. \tag{21}$$

Here l is a function on S that Liouville later interpreted as the equilibrium charge distribution. Liouville remarked, that if ζ_n denotes the product $M_{iB}N_{iB}$, ζ_n will satisfy the eigenvalue equation

$$\iint_S \frac{l(x')\zeta_n(x')}{|x-x'|}\, d\omega(x') = m_n\zeta_n(x)\,. \tag{22}$$

It is now easy to solve Gauss' problem for the ellipse. Indeed if we write $\overline{V}$ as

$$\overline{V} = \sum_n A_n\zeta_n \tag{23}$$

(Liouville showed that this could "always" be done), then

$$\lambda = l\sum_n \frac{A_n}{m_n}\zeta_n \tag{24}$$

is the desired charge distribution. The proof is an easy application of (22) and the orthogonality of the ζ_n's.

At this point Liouville got the great idea of generalizing this method to an arbitrary surface S, i.e. to find its equilibrium distribution l, and define the eigenfunctions ζ_n by (22), such that any function $\overline{V}$ can be expanded in a series (23). Then λ defined by (24) is a solution of Gauss' problem. He sketched this idea briefly in his published paper (1845a) and even emphasized its importance:

> After having studied the matter I do not hesitate to regard the functions ζ as being of the utmost importance in analysis (Liouville 1845a).

The problem raised in Liouville's notebooks is the following: How can the ζ_n's be found? As we saw, the variational method provided the answer. His notebooks even reveal how he arrived at the method: First he used the idea *a posteriori*: Asuming that the ζ_n's exist with $1 > m_1 > m_2 > \cdots$ and supposing that any function can be expanded as $\sum A_n\zeta_n$, it is easy to see that m_n and ζ_n can be found by the variational method. Some 40 pages later he returned to this method and remarked that "considered *a priori* it demonstrates the existence of these functions".

4.4 Proofs

In my Liouville biography (Lützen 1990 Chap. XV) one can find as much of Liouville's proofs as I have been able to reconstruct from Liouville's rather disorganized notes. These proofs are really beautiful up to a certain point and are strikingly similar to the arguments found in a modern book on spectral theory of compact operators. Liouville succeeded in proving Bessel's inequality, Parseval's equality and showed that if a function is orthogonal to all the ζ_n's it must vanish identically. From this he concluded that if the Fourier series of a function Q converges, it has the sum Q. However, since he did not have the Lebesgue integral at his disposal, and since he was interested in pointwise or uniform convergence rather than L^2 convergence, he could not prove that the

Fourier series converges. Thus the situation was the oposite of the situation in Sturm-Liouville theory, where he could prove the convergence of the Fourier series but could not show that it had the desired sum.

4.5 Why Were Liouville's Results Never Published?

Liouville developed several interesting theories and results that he never published. In general the main reason for not publishing was lack of time, but in the case we consider here, there were also inner mathematical reasons. Indeed there were holes in Liouville's argument. First, he does not seem to have been able to prove that the decreasing series of eigenvalues m_n tend to zero, and this is of great importance for the whole theory. Liouville may not have been aware of this shortcoming but toward the end of his notes he began to sense another problem, in connection with the variational technique for finding l and the eigenfunctions ζ_n. Indeed he discovered that though the upper (lower) bound exists the max (min) may not always be attained; instead there might be what he called a "tendence indéfinie vers le but".

It is very remarkable that Liouville here pointed to the weakness of the naive version of the Dirichlet-principle, and similar principles, about 25 years before Weierstrass presented his famous counterexample (1870–1895). Of course, once the variational method was called into question the existence of the ζ_n's was not secured, and this may explain why Liouville did not publish his ideas.

4.6 Are Liouville's Results Correct?

This question is urgent, once we have realized that Liouville's arguments are not entirely satisfactory. In modernized language it is enough to show that Liouville's operator A is compact on $L^2(S, l d\omega)$ and, indeed, Christian Berg has succeeded in proving this theorem under rather weak assumptions on the surface S (Berg and Lützen 1990).

4.7 Connections to Later Works

In fact it was not Rayleigh but Heinrich Weber (1869) who first published a variational method for finding eigenfunctions. He applied the method to the two dimensional Laplacian, but as Liouville, he did not show that the infima were attained by minimizing functions. It was left to Poincaré (1894) to establish Weber's results rigorously. In connection with Liouville's notes a later paper by Poincaré (1896) is even more interesting. Here Poincaré introduced what he called fundamental functions of a surface S. They were defined by a variational method, and are in fact identical to Liouville's ζ_n. Poincaré was unable to prove rigorously that these functions exist, and so he only used them as a heuristic tool. However his student Le Roy (1898) succeeded in giving a very long proof of their existence.

Weber's, Rayleigh's, and Poincaré's work, together with works by Kirchhoff, Klein and others from arround 1880 mark the first continuation of Sturm-Liouville theory, in the sence that they dealt with differential equations. Integral operators (or equations) implicitly arose out of the Beer-Neumann method for solving the Dirichlet problem. They were studied by Volterra in the 1890s but their spectral theory was not studied until 1900 in an elegant paper by Fredholm.

Fredholm's ideas were carried much further in Hilbert's monumental work (1904–1910), that in a sense marks the conclusion of the development we have considered. Hilbert introduced the so-called "vollstätige" operators, and showed that their eigenvalues can be found by the variational technique used by Liouville. Moreover he joined the two types of spectral theory that Liouville had studied. Indeed he showed how one can use a Green's function to reduce problems in Sturm-Liouville theory to (easier) problems of integral operators. He even connected both these theories to spectral theory of discrete systems, by showing how the spectral theorem of quadratic forms in the limit would lead to the spectral theorem of integral operators, or more generally of quadratic forms in infinitely many variables. This idea had already been suggested by Poincaré but Hilbert was the fist to carry out the limit procedure rigorously. In the case of infinite quadratic forms he made an ingenious use of the Stieltjes integral in order to deal with the continuous spectrum.

About the same time, the structural movement made another unification of the three branches of spectral theory possible. It became clear that they all dealt with one and the same question for different *operators* between different *spaces.* Liouville clearly did not have such a unifying concept, but there is no doubt that he was aware of the formal similarity between his work on Sturm-Liouville theory and his unpublished work on integral operators. Indeed many of the theorems are the same and their proofs often go along the same lines.

5. Conclusion

One may wonder if the development of spectral theory would have been speeded up if Liouville had published his research on his integral operator in 1846. This is not at all certain. In fact, Sturm-Liouville theory was not developped further until half a century after Sturm and Liouville had published their work, and Liouville's small published note on symmetric integral operators went completely unnoticed. How can this lack of interest be explained?

The reason is probably the very qualitative nature of the results. I have stressed this in connection with Sturm-Liouville theory and it is equally true of the results concerning Liouville's operator A. In fact, except for simple surfaces, even the equilibrium distribution l is impossible to determine. Most mathematicians in the middle of the 19th century were not excited by existence theorems, unless they could find the objects. Therefore they did not continue Sturm-Liouville theory and would probably have neglected Liouville's spectral theory of his integral operator even if it had been published. Only at the end of the 19th century, the mathematical community was ripe for such questions.

This shows Liouville's farsightedness.

Liouville is usually mentioned in histories of spectral theory in connection with his work on Sturm-Liouville theory. I hope this paper has demonstrated, that he was an even greater pioneer in this field than it is usually acknowledged.

References

Abel, N.H. (1823): Oplösning af et Par Opgaver ved Hjelp af bestemte Integraler. Magazin for Naturvidenskaberne, ser. 1, **I** (1923), 11–27. [French translation in Oeuvres]

Bang, T., Fuglede, B. (1990): No two quotients of Normalized Binomial Mid-coefficients are equal. J. Number Theory **35**, 345–349

Berg, C. (1990): Integrals involving Gegenbauer and Hermite polynomials on the imaginary axis. Equationes Math. (to appear)

Berg, C., Fuglede, B. (1990): Liouville's operator for a disc in space. Manuscr. math. **67**, 165–185

Berg, C., Lützen, J. (1990): J. Liouville's Unpublished Work on an Integral Operator in Potential Theory. A Historical and Mathematical Analysis. Expositiones Mathematicae **8**, 97–136

Cauchy, A.L. (1824–1981): Equations Différentielles Ordinaires. Etudes Vivantes, Paris. New York 1981. Fragment of lecture notes for the second year at the École Polytechnique. Introduction by C. Gilain

Dieudonné, J. (1981): History of Functional Analysis. North-Holland, Amsterdam 1981

Dirichlet, P.G. Lejeune (1829): Sur la convergence des séries trigonométriques qui servent à représenter une fonction arbitraire entre des limites données. J. Reine Angew. Math. **4**, 157–169; Werke **1**, 117–132

Fourier, J. (1822): Thèorie analytique de la chaleur. Paris 1922

Fredholm, I. (1900): Sur une nouvelle mèthode pour la résolution du problème de Dirichlet. Öfversigt af Kungliga Svenska Vetenskabs-Akademiens Förhandlinger, Stockholm **57**, 39–46; Oeuvres Complètes, 61–68

Gauss, C.F. (1840): Allgemeine Lehrsätze in Beziehung auf die im verkehrten Verhältnisse des Quadrats der Entfernung wirkenden Anziehungs- und Abstossungs-Kräfte. Resultate aus den Beobachtungen des magnetischen Vereins im Jahre 1839 (4) Leipzig 1840; Werke **5**, 197–242; Ostwald's Klassiker Nr. 2, Leipzig 1889. [French transl. in J. Math. Pures Appl. **7** (1842) 273–324]

Hawkins, T. (1975): Cauchy and the Spectral Theory of Matrices. Historia Mathematica **2**, 1–29

Hilbert, D. (1904–1910): Grundzüge einer allgemeinen Theorie der linearen Integralgleichungen. Nachrichten Königl. Ges. Wiss. Göttingen Math.-Phys. Kl. (1904) 49–91; 213–259; (1905) 307–338; (1906) 157–227; 439–480; (1910) 355–417; 595–618. Collected as monograph Leipzig 1912. References refer to monograph

Le Roy E. (1898): Sur l'intégration de l'équation de la chaleur (2e partie). Ann. Ec. Norm. Sup. (3) **15**, 9–178

Liouville, J. (1830): Mèmoire sur la thèorie analytique de la chaleur. Ann. Math. Pures Appl. **21** (1830–1831) 131–181; Summary by Sturm in Bull. Sci. Math. Phys. et Chim.

Liouville, J. (1832): Sur quelques questions de géometrie et de mécanique, et sur un nouveau genre de calcul pour résoudre ces questions. J. Ec. Polyt. **13** (21. cahier) 1–69

Liouville, J. (1836): Mémoire sur le développement des fonctions ou parties de fonctions en séries, dont les divers termes sont assujettis à satisfaire à une même équation différentielle du second ordre, contenant un paramètre variable. J. Math. Pures Appl. **1**, 253–265; C. R. Acad. Sci. Paris. **1** (1835) 418

Liouville, J. (1837a): Second Mémoire sur le développement des fonctions ou parties de fonctions en séries, dont les divers termes sont assujettis à satisfaire à une même équation différentielle du second ordre, contenant un paramètre variable. J. Math. Pures Appl. **2**, 16–35

Liouville, J. (1837b): Troisième Mémoire sur le développement des fonctions ou parties de fonctions en séries, dont les divers termes sont assujettis à satisfaire à une même équation différentielle du second ordre, contenant un paramètre variable. J. Math. Pures Appl. **2**, 418–437; C. R. Acad. Sci. Paris. **5**, 205–207

Liouville, J. (1838): Premier mémoire sur la théorie des équations différentielles linéaires, et sur le développement des fonctions en séries. J. Math. Pures Appl. **3**, 561–614; C. R. Acad. Sci. Paris. **7**, 1112–1116

Liouville, J. (1845a): Sur diverses questions d'analyse et de physique mathématique. J. Math. Pures Appl. **10**, 222–228

Liouville, J. (1845b): Sur une propriété génerale d'une classe de fonctions. J. Math. Pures Appl. **10**, 327–328

Liouville, J. (1846): Lettres sur diverses questions d'analyse et de physique mathématique, concernant l'ellipsoide, adressées à M.P.H. Blanchet – Première Lettre. J. Math. Pures Appl. **11**, 217–236. Deuxième Lettre. J. Math. Pures Appl. **11**, 261–290

Liouville, J. (Ms 3615–3640) The Liouville Nachlass at the Institut de France

Lützen, J. (1982): Joseph Liouville's Contribution to the Theory of Integral Equations. Historia Mathematica **9**, 373–391

Lützen, J. (1984): Sturm and Liouville's Work on Ordinary Linear Differential Equations. The Emergence of Sturm-Liouville Theory. Arch. Hist. Exact. Sci. **29**, 309–376f

Lützen, J. (1987): The Solution of Partial Differential Equations by Separation of Variables: A Historical Survey. Studies in Mathematics **26** (ed. E. Phillips) 242–277. Math. Ass. of America

Lützen, J. (1990): Joseph Liouville (1809–1882). Master of Pure and Applied Mathematics. Springer, New York

Poincaré, H. (1894): Sur les équations de la physique mathématique. Rend. Circ. Mat. Palermo **8**, 57–155; Oeuvres **9**, 123–196

Poincaré, H. (1896): La méthode de Neumann et le problème de Dirichlet. Acta Math. **20**, 59–142; Oeuvres **9**, 202–272

Poisson, S.D. (1826): Note sur les racines des équations transcendantes. Bull. Soc. Philomatique 145–148

Rayleigh, Lord (1877): The Theory of Sound. London

Sturm, C. (1836a): Mémoire sur les Équations différentielles linéaires du second ordre. J. Math. Pures Appl. **1**, 106–186

Sturm, C. (1836b): Mémoire sur une classe d'Équations à différences partielles. J. Math. Pures Appl. **1**, 373–444

Weber, H. (1869): Ueber die Integration der partiellen Differentialgleichung $\frac{\partial^2 u}{\partial x^2} + \frac{\partial^2 u}{\partial y^2} + k^2 u = 0$. Math. Ann. **1**, 1–36

Weierstrass, K. (1870–1895): Über das sogenannte Dirichletsche Princip. (Read in 1870); Werke, Berlin 1895, vol. 2, pp. 49–54

Mathematics as Metaphor

*Yuri Ivanovich Manin**

Steklov Mathematical Institute, 42 Vavilova, 117966 GSP-1, Moscow, USSR

あかあかやあかあかあかやあかあかや
あかやあかあかあかあかや月

Myōo (1173–1232)

Ordre. [...] Je sais un peu ce que c'est et combien peu de gens l'entendent. Nulle science humaine ne le peut garder. Saint Thomas ne l'a pas gardé. La mathématique le garde, mais elle est inutile en sa profondeur.

Pascal, Pensées

Introduction

When H. Poincaré first published in 1902 his book *La Science et l'hypothèse,* it became a bestseller. The first chapter of this book was devoted to the nature of the mathematical reasoning. Poincaré discussed an old philosophical controversy whether the mathematical knowledge could be reduced to long chains of tautological transformations of some basic ("synthetic") truths or it contained something more. He argued that the creative power of mathematics was due to a free choice of the initial hypotheses-definitions which were later on constrained by the comparison of deductions with the observable world.

The society of our days seems to be much less interested in the philosophical subtleties than Poincaré's contemporaries. I do not want to say that science itself became less popular. Such books as S. Weinberg's *The first three minutes* and S.W. Hawking's *A brief history of time* are sold by hundreds of thousands and favorably reviewed in widely distributed newspapers. What has changed is the general mood. The paradoxality of the new physical theories is perceived less dramatically and more pragmatically. (We can note that the perception of visual arts evolved in much the same way: if the first exhibitions of Impressionists were a kind of spiritual revolution, each new wave of the post-war avant-garde immediately acquired family traits of academism).

* Delivered by Barry Mazur.

Proceedings of the International Congress of Mathematicians, Kyoto, Japan, 1990

In this atmosphere, the heated discussions of the bygone days on the foundational crisis of mathematics and the nature of infinity seem almost irrelevant and certainly inappropriate. The audience responds much livelier to the opinions about school education or a new generation of computers.

This is why I have decided to present at this section an unpretentious essay in which our science is considered as a specialized dialect of the natural language, and its functioning as a special case of speech. This implies certain suggestions about the high school and University training.

Metaphorism

The word "metaphor" is used here in a non-technical sense, which is best rendered by the following quotations from James P. Carse's book *Finite and Infinite Games*:

"Metaphor is the joining of like to unlike such that one can never become the other."

"At its root all language has the character of metaphor, because no matter what it intends to do, it remains language, and remains absolutely unlike whatever it is about".

"The unspeakability of nature is the very possibility of language".

Considering mathematics as a metaphor, I want to stress that the interpretation of the mathematical knowledge is a highly creative act. In a way, mathematics is a novel about Nature and Humankind. One cannot tell precisely what mathematics teaches us, in much the same way as one cannot tell what exactly we are taught by "War and Peace". The teaching itself is submerged in the act of re-thinking this teaching.

This opinion seemingly disagrees with the time-honored tradition of applied mathematics in scientific and technological calculations.

In fact, I want only to restore a certain balance between the technological and the humanitarian sides of mathematics.

Two Examples

Let me try to illustrate the metaphoric potential of mathematics by discussing two disjoint subjects: the Kolmogorov complexity and the "Dictator Theorem" due to K. Arrow.

i) Kolmogorov's complexity of a natural number N is the length of a shortest program P that can generate N, or the length of a shortest code of N. A reader should imagine a way of coding integers which is a partial recursive function $f(P)$ taking natural values. Kolmogorov's theorem states that among all such functions there exist the most economical ones in the following sense: if $C_f(N)$ is the minimal value of P such that $f(P) = N$, then $C_f(N) \leq \text{const}\,.\,C_g(N)$, where const. depends only on f, g but not N.

Since P can be reconstructed from its binary notation, the length $K_f(N)$ of the shortest program generating N is bounded by $\log_2 C_f(N)$. This function, or rather the class of all such functions defined up to a bounded summand, is the Kolmogorov complexity.

First of all, $K(N) \leq \log N + \text{const}$. Of course, this conforms nicely with the historical successes of the positional notation systems which provided us with the number generating programs of logarithmic length. However, there are arbitrary large integers whose Kolmogorov's complexity is much smaller than the length of their notation, e.g., $K(10^N) \leq K(N) + \text{const}$. In general, when we use large numbers at all, we seemingly use only those which have a relatively small Kolmogorov's complexity. Even decimal decompositions of π which are, probably, the longest well-defined numbers ever produced by mathematicians, are Kolmogorov simple, because $K([10^N\pi]) \leq \log N + \text{const}$. In general, small Kolmogorov complexity = high degree of organization.

On the other hand, almost all integers N have the complexity close to $\log N$. For example, if $f(P) = N$ for an optimal f, then $K(P)$ is equivalent to $\log P$. Such integers have many remarkable properties which we usually connect with "randomness".

Second, Kolmogorov's complexity can be easily defined for discrete objects which are not numbers, for example, Russian or English texts. Therefore, "War and Peace" has a pretty well defined measure of its complexity; the indeterminacy is connected with the choice of an optimal coding and seems to be pretty small if one chooses one of the small number of reasonable codings.

From this viewpoint, is "War and Peace" a highly organized or an almost random combinatorial object?

Third, Kolmogorov's complexity is a non-computable function. More precisely, if f is optimal, there is no recursive function $G(N)$ which would differ from $C_f(N)$ by $\exp(O(1))$. One can only bound complexity by computable functions.

I feel that Kolmogorov's complexity is a notion that is very essential to keep in mind in any discussion of the nature of human knowledge.

As long as the content of our knowledge is expressed symbolically (verbally, digitally, ...) there are physical restrictions on the volume of information that can be kept and handled. We always rely upon various methods of information compressing. Kolmogorov complexity puts absolute restrictions on the efficiency of such a compression. When we speak, say, of physical laws, expressed by the equations of motion, we mean that a precise description of the behaviour of a physical system can be obtained by translating these laws into a computer program. But the complexity of laws we can discover and use is clearly bounded. Can we be sure that there are no laws of arbitrary high complexity, even governing the "elementary" systems?

At this point, our discussion becomes totally un-mathematical, and before a mathematically-minded audience I must stop here. But such is the fate of any metaphor.

ii) Arrow's Dictator Theorem was discovered around 1950. Mathematically, it is a combinatorial statement describing certain functions with values in binary relations. Intuitively, it is a formalized discussion of the problem of Social Choice. Suppose that a lawmaker has to establish a law which governs the processing of individual wills of voters into a collective decision. If the problem is to choose one of the two alternatives, the standard solution is do it by the majority of votes. However, usually there are more than two alternatives (imagine the funds allocation problems), and voters may be asked to order them according to their preferences. What should be the algorithm extracting the collective preference from any set of individual preferences? Arrow considered algorithms satisfying some natural and democratic axioms (e.g., when everybody prefers A to B, the society prefers A to B). Nevertheless, he discovered that when there are more than two alternatives, the only way to achieve a solution is to nominate a member of the society ("the Dictator") and to equate his personal preference order to the social one. (Actually, this is one of the versions of Arrow's theorem discovered later. Also, it refers to the case of a finite society; in the infinite case, the social decisions can be made by ultrafilters, appropriately called "the ruling hierarchies".)

In a way, this theorem illustrates the content of Jean-Jacque Rousseau's idea of a Contrat Social.

The fundamental inconsistency of the image of the ideal democratic choice can be illustrated by the following story referring to three voters and three alternatives. It is the story of three knights errant at the cross-roads with a stone before them. The inscription on the stone prophesies only losses: who goes to the left will lose his sword; who goes to the right will lose his horse; who goes straight will lose his head. The knights dismount and start taking council. In a Russian version of this story, the knights have names and personalities: the youngest and ardent Alyosha Popovich, the eldest and wisest Dobrynya Nikitich, and the slow peasant Ilya Muromets. So Alyosha values sword more than horse, and horse more than his head; Dobrynya values most his head, then his sword, then his horse; and Ilya prefers his horse to head to sword.

A reader will note that the three individual preference orders constitute one and the same cyclic order on the set of alternatives. As a result, one can decide by majority the choice between any two of the alternatives, but the union of these decisions will be inconsistent: the democratic procedure cannot provide us with a well-ordered list. The knights sigh and delegate the decision-making power to Dobrynya.

Does the Arrow theorem tell us something that we did not know beforehand? Yes, I think it does if we are ready to discuss it seriously, that is, to look closely at the combinatorial proof, to imagine the possible real life content of various assumptions and elementary logical steps made on the way, in general, to enhance our imprecise imagination by the rigid logic of a mathematical reasoning. We can understand better, for example, some tricks of policy-making and some pitfalls the society can leap whole-heartedly into (like accepting without questioning a list of alternatives imposed by a ruling hierarchy, although precisely the compilation of this list can be the central issue of the social decision making).

At this stage, we come to the main topic of our discussion: what distinguishes a mathematical discourse from a natural language discourse, why the Pascalian "ordre" came to reign over our specialized symbolic activities, and is it truly so "useless in its profundity"?

Language and Mathematics

A very interesting chapter of the interaction between mathematics and humanities started about thirty years ago when the first serious attempts of automatic translation were made. These first attempts were a painful failure, at least so for many optimists who believed that in this domain, there are no fundamental obstacles, and it remains only to overcome technical difficulties connected with the sheer amount of information to be processed. In other words, they took for granted that the translation is in principle performed by a not very complex algorithm which only must be made explicit and then implemented as a computer program.

This assumption is a nice example of a mathematical metaphor (actually, a specialization of the general "computer metaphor" used in the brain sciences).

This metaphor proved to be extremely fruitful for the theoretical linguistics in general because it forced linguists to start describing vocabulary, semantics, accidence, and syntax of human languages with unprecedented degree of explicitness and completeness. Some totally new notions and tools were discovered thanks to this program.

However, the successes of the automatic translation itself were (and still are) scanty. It became clear that written human speech is an extremely bad input data for any algorithmic processing planned as translation or even as a logical deduction. (I add this proviso because there is nothing special in human speech considered as a material for, say, statistical studies).

This fact can be considered as a universal property of human languages, and it deserves some attention. One must first of all reject as too naive a usual explanation that the universe of meanings of a human language is too vast and poorly structured to admit a well organized metalanguage describing this world. The point is that even if we severely restrict this universe to the subset of arithmetic of small integer quantities, we shall still have to face the same difficulty. In fact, this difficulty was a decisive reason for the crystallization of the whole system of arithmetical notation and the basic algorithms of calculation, and later on of symbolic algebra. Even the vocabulary of elementary arithmetic in human language is basically archaic: the finite natural series of primitive societies "one, two, three, indefinitely many" is reproduced in the exponential scale in our "thousand, million, billion, zillion". The expressions for relatively small numbers like "1989" are actually names of the decimal notation and not of the numbers themselves.

The advantage of F. Viete's algebra over the semi-verbal algebra of Diophantus was due not to the fact that it could express new meanings but to the incomparably greater susceptibility to the algorithmic processing ("identical transformations" of our high school algebra).

The rupture of the intuitive and emotional ties between a text and its producer/user so characteristic of the language of science was compensated by the new computational automatisms. In their (albeit restricted) domain they proved to be infinitely more efficient that the traditional Platonian and Aristotelian culture of everyday language discourse. Why then our scientific papers are still written as a disorganized mixture of words and formulas? Partly because we still need those emotional ties; partly becaue some meanings (like human values) are best rendered in human language. But even as a medium of scientific speech, human language has some inherent advantages: appealing to the spatial and qualitative imagination, it helps to understand "structurally stable" properties like the number of free parameters (dimension), existence of extrema, symmetries. To put it bluntly, it makes possible the metaphorical use of science.

Metaphor and Proof

The views professed here can be considered in relation to the high school and graduate curriculum.

The general mathematical education of the first half of this century was application oriented. It provided the basic minimum for the practical life problems and a smooth transition to the study of engineering and scientific calculations at the college level. The break of this curriculum with the activity of professional mathematicians became more and more pronounced. As is well known, this brought the reaction in the form of NewMath in USA and similar programs in other countries. These programs introduced into the high school mathematics the notions and principles borrowed from professionals: set theory, axiomatic presentation of proofs, strict culture of definitions.

NewMath became widely accepted but its expansion was accompanied by the protesting voices which in the 70s and 80s merged into a loud chorus. The critics disagreed with the basic arguments of the NewMath proponents. Leaving aside the objections based upon the data from cognitive sciences and learning psychology I shall only recall those concerning the general evaluation of the role of the proof in mathematics.

The one pole is represented by the well known statement due to Nicolas Bourbaki: "Dépuis les Grécs, qui dit Mathématiques, dit démonstration". According to this perception, the rigorous proof was made a matter of principle in the NewMath programs. It was argued that: a) a proof helps to understand a mathematical fact; b) a rigorous proof is the most essential component of the modern professional mathematics; c) mathematics possesses the universally recognized criteria of rigour.

These views were extensively criticised, e.g. by Gila Hanna in the book "Rigorous Proof in Mathematics Education", OISE Press, Ontario 1983. In particular, Gila Hanna pointed out that the mathematicians are far from unanimous in accepting the criteria of rigour (referring to quarrels between logicists, formalists and intuitionists), and that working mathematicians constantly break all rules in the book.

In my opinion, this is irrelevant.

What is relevant, is the imbalance between various basic values which is produced by the emphasis on proof. Proof itself is a derivate of the notion of "truth". There are a lot of values besides truth, among them "activities", "beauty" and "understanding", which are essential in the high school teaching and later. Neglecting precisely these values, a teacher (or a university professor) tragically fails. Unfortunately, this also is not universally recognized. A sociological analysis of the controversies around the Catastroph Theory of René Thom shows that exactly the shift of orientation from the formal truth to understanding provoked such a sharp criticism. But of course, the Catastroph Theory is one of the developed mathematical metaphors and should only be judged as such.

Pedagogically, a proof is just one of the genres of a mathematical text. There are many different genres: a calculation, a commented sketch, a computer program, a description of an algorithmic language, or such a neglected kind as a discussion of the connections between a formal definition and intuitive notions. Every genre has its own laws, in particular, laws of rigour, which only are not codified because they were not payed a special attention.

A central problem of a teacher is to demonstrate at the restricted area of his or her course the variety of types of mathematical activities and underlying value orientations. Of course, this variety is hierarchically organized. The goals may vary from achieving an elementary arithmetical and logical literacy to programming skills, and from the simplest everyday problems to the principles of modern scientific thinking. In the spectrum of these goals, the emphasis on the norms of "rigorous proof" can safely occupy a peripheral position.

But having said all this, I must stress that my argumentation by no means undermines the ideal of a rigorous mathematical reasoning. This ideal is a fundamental constituting principle of mathematics, and in this sense Bourbaki is certainly right. Having no external object of study, being based on a consensus of a restricted circle of devotees, the mathematics could not develop without the permanent control of rigid rules of game. Applicability of mathematics in the strict sense of this word (like its indispensability in the Apollo project) is due to our ability to control series of symbolic manipulations of fantastic length.

The existence of this ideal is far more essential than its unattainability. The freedom of mathematics (G. Cantor) can only develop in the limits of iron necessity. The hardware of modern computers is an incarnation of this necessity.

Metaphor helps a human being to breathe in this rarefied atmosphere of Gods.

Teaching Mathematics to Students Not Majoring in Mathematics

– Present Situation and Future Prospects –

Haruo Murakami

Department of Applied Mathematics, Faculty of Engineering, and
Division of Intelligence Science, Graduate School of Science and Technology
Kobe University, Kobe, 657 Japan

The present era, described variously as a Highly Industrialised Society, the Information Era or even a Postindustrial Society, places greater emphasis than before on the importance of teaching mathematics. Teaching mathematics to students who select it as their major subject is of course a problem of vital importance, but compared with all the students studying mathematics, those actually majoring in it would be a drop in the ocean; the problem of teaching mathematics to those majoring in other subjects is a much larger problem.

University education has become more available and more widely spread. In Japan, for instance, there are three times as many university students as there were middle-school pupils before the 2nd World War, and the present university teaching staff is five times more numerous than pre-war middle-school teachers. Universities have rid themselves of their ivory tower image and have adapted to the need for practical sciences. At present, there is probably more demand for this, a more practically oriented area.

Most of those students who require mathematical education at university level specialise in engineering, the rest in natural sciences such as physics, chemistry, and biology, as well as those who specialise in economics, business studies, or some areas of medicine, etc. In fact, in Japan, there are only 3,200 students specialising in pure mathematics, whereas almost all the students in engineering, who number well over 65,000, receive lectures in mathematics.

1. *What kind of mathematics and mathematical teaching is required?*

Mathematics is indispensable for the study of engineering, no matter what the field of engineering and no matter what the area of technology the student selects for his or her future professional skill. Students of engineering cannot therefore avoid learning mathematics. At present, although the extent varies according to the field, all students of engineering get some mathematical tuition. What passes for mathematics here is mathematical skills, i.e. mathematical knowledge and the ability to apply this knowledge to engineering, ability to calculate and compute, etc. The subjects normally taught to them are: calculus, linear algebra, theory of complex functions, vector analysis, special functions such as Bessel functions,

Proceedings of the International Congress
of Mathematicians, Kyoto, Japan, 1990

differential equations, probability and statistics, etc. They would be quite unable to come to grips with their field of engineering without knowing these subjects. It is therefore unlikely that the need for teaching these subjects should decrease in the future.

There are naturally some differences depending on the specialisation. For example, those who enter fields where fluid dynamics is important, such as civil and mechanical engineering, may find vector analysis important. But for those who aim at signal theory, it would be essential to study abstract algebra and in particular Galois fields. There are some areas of engineering where topology and graph theory may be more important, or it may be essential to study Boolean algebra. How would one satisfy such diverse demands of engineering students?

One might say that except for engineering, a basic knowledge of mathematics would suffice for any discipline. For an ordinary student in economics, elementary calculus and linear algebra used to be sufficient. However, econometrics, which is increasingly becoming an essential part of economics, demands some knowledge of probability and statistics, and further, depending on one's specialty, a student in economics may have to know such subjects as fixed point theorems and the theory of differential equations. Similarly, the students of physics, chemistry and many other fields all require some degree of specialised knowledge in mathematics.

There are also some subjects in mathematics which seem irrelevant or unimportant to one's own field at first, and yet turn out to be important or even essential later. Since it is impossible to teach all potential areas of mathematics covering many different subjects, we ought to lay a certain amount of groundwork in order to equip students with the ability to handle new subjects as they arise. For example, there was a time when Boolean algebra was not considered a part of applied mathematics. Now, however, Boolean algebra is even more common in applied than in pure mathematics. We shall probably see this sort of thing happening in the future too. We have no idea which techniques will be used and no idea of what to provide in the way of basic education for these new techniques. But to go back to the above example, it is not difficult to deal with Boolean algebra as long as one has some grounding in abstract algebra. This shows how important it is to provide training in the basics of mathematics as a whole. In the information age, beset as it is by violent changes, it is difficult, and even dangerous, to make prior judgments and predictions about the sort of mathematics that will be needed. Surely our aim should be to give our students the ability to learn new topics, and, to enable them to confront new things without fear, although this is easier said than done.

2. *Should one confine oneself to teaching students only the knowledge and skills immediately required for their study, or should they be taught some different aspects of mathematics as well? If so, what should these be?*

If we consider the future development of applied mathematical techniques on the assumption that computers will be more widely used, then we should probably place the emphasis on *intelligence* rather than *skill*, on *thought* rather than *computation*, and on *concepts* rather than *formulae*.

Applied mathematicians of the old school were concerned solely with the memorization and use of formulae. They were, so to speak, *walking reference books* of mathematical formulae. Nowadays, however, many more scientific fields have a very high mathematical content, and mathematics is now applied to a much wider range of subjects. Thanks to modern computer techniques, computational skills will be replaced by the use of computers. Take the example of teaching techniques of integration. The basics such as integration by parts and integration by substitution must be taught as a part of differential and integral calculus, but there is no need to go into finer technical points, or give the students hours of computational practice. It is more efficient to train them to use computer algebra systems or formula manipulation systems such as REDUCE or MATHEMATICA, just as students in the past were trained to use formula books. In the same way, we can delegate the specialised techniques of solving differential equations to computer formula manipulation systems.

In the past, engineering departments tended to view mathematics as something required for physics. The mathematics used in physics generally deals with phenomena which are quantified from the start, and in many cases, it is possible to manage with just classical mathematics or statistics. In recent years, however, even in engineering, there has been a rapid expansion in subjects dealing with social phenomena, such as environmental engineering, urban engineering, and management engineering, most of which deal with unquantifiable things. This has led to a greater need for abstract mathematics. Taking this into account, the emphasis of mathematical education from now on should be placed on training students to become able to grasp abstraction and logical thought.

Many engineers retain the old-fashioned "walking reference book" or "cook-book-method" approach to mathematics. This tends to filter through to engineering students who then attempt to try to find applicable formula from existing lists and apply the result of the computation to an engineering problem. I would like to stress the enormous importance of logical thought through the teaching of mathematics. We should highlight mathematical way of thinking, and concentrate on basic mathematical concepts. Thus, when teaching integration, we should stress the *meaning* of the integral more than how to *evaluate* a given integral. Even when explaining a theorem, we should keep the proof to a minimum. It is more important to stress premises and offer counter-examples where these premises do not hold true.

3. *How do computers change the teaching of mathematics?*

In the face of changing times and especially with the advent of the Information Era, is it still correct to teach the same things as we have done so far, or should we reconsider the contents of our syllabus? How can we introduce the computer more effectively in education? Which new ways of teaching, making full use of computers, can be devised?

The contents of the mathematics to be taught in the computer era must, of necessity, have changed. Discrete mathematics and finite mathematics have become more important than before. Although more syllabus time should be devoted to these areas, this does not mean that traditional continuous mathematics should lose importance. Moreover, to gain deeper understanding, one also needs these skills for concrete computation and calculation. What exactly are these skills, then? As the expertise required for modern warfare are different from that used during the age of bows and arrows, so the computational skills required in the age of computers are vastly different from those required in the days of pencils and paper. The skill of handling computers has already become a part of a mathematician's skill. There may be other new kinds of skills required in the future.

Originally, mathematics possessed an experimental side. Euler speaks of pure mathematics thus: "The properties of numbers which everybody knows have been discovered through ordinary observation. They were discovered long before their correctness could be verified by proof. It is by observation that more and more properties are discovered, and afterwards people spend all their efforts on proving them." The emergence of the computer expands the possibilities of observation and experimentation in mathematics.

Since it is easy to draw graphs of functions and the solution curves of differential equations on a personal computer, this can be used to enhance learning and research. For instance, it is now easy to draw the solution graphs of a differential equation for as many cases as one wishes, extract those properties which are in common, and make a conjecture on the nature of the solutions. One can then attempt to prove the conjecture, but before reaching this stage, the skill of handling computer graphics is required. There are still some pure mathematicians who have an aversion to computers, but surely this aversion is merely backwardness.

Once upon a time, man used only his own strength, but he then learned to use the strength of domestic animals such as horses to do his labour. Later, engines were developed which were hundreds or thousands of times more powerful than the animals. These engines now routinely do work that would be almost impossible by manpower alone. Life without these machines is virtually unthinkable.

In the same way, man has also learned to use technology to perform his intellectual work. Computers have been developed which can process information hundreds or thousands of times faster than the human brain, and can routinely do work that would be almost impossible by brainpower alone. There are still many areas where computers cannot replace the power of human thought, but they are now becoming essential tools, and it would be difficult to contemplate life without them. Mathematics research and teaching must adapt accordingly in order to benefit from the progress of computers.

At the very least, we could install a microcomputer in the classroom with a big screen to demonstrate effectively the locus of functions, solution curves of differential equations and so on. We could then ask students to do their assignments either by using their own personal computers or by using computing facilities provided by the university. As I mentioned above, a computer algebra system could be employed as a very powerful tool in the mathematics education.

4. *Which subjects should be taught? One example*

One topic proposed as a practical example of mathematical over-abstraction was that old chestnut, the existence theorem of solutions to differential equations. This can be taught with the use of Picard's successive approximation. Of course, instead of carrying out a rigorous proof of the Picard successive approximation, one could just settle it by arm waving. There may be people who think it unnecessary to teach the existence theorem. I believe that it would be better to teach the theorem. One reason is that successive approximation is a concrete way of showing the existence of a solution by construction and this is useful in engineering. Another reason is that the mathematics which students have learned at high school does not prepare them for the possibility that there may be no solution to a problem. When they were given a problem to solve, they could always take it for granted that it had a solution. It would be very useful, for several purposes, to give the students a little culture shock by teaching them the existence theorem. However, those teaching first-year university students should be aware of the fact that their students are completely new to mathematics involving things like existence theorem. Also, topics like this should not take up too much time in the syllabus.

However, it may be a little hard on the students if there are too few subjects of computation, like calculation of indefinite integrals for instance, which follow on from the high-school curriculum, even though it should not be covered in great depth. It is by no means easy for students to overcome the gap between high-school mathematics and university mathematics. Should we change high-school mathematics, then? It would be quite inappropriate to cram subjects like existence theorems wholesale into the high-school curriculum. In fact, mathematics is not only an abstract science, it also is a technique. If it is taught under the sole auspices of mathematicians, they might overimpose their taste for the abstract. The useful aspects may be obscured, and worse still, there is the danger that even the deep understanding of mathematics may be lost. This is because there can be no deep understanding without a fair amount of skill in concrete calculation, etc. I was told that some students of Tokyo University, while knowing all sorts of abstract things, were quite unable to expand $\sin x$ into Taylor series, and that primary school pupils in France who were asked what "$1 + 2 =?$ " is, answered "$1 + 2 = 2 + 1$," adding that the operation of addition is commutative. Such anecdotes are in fact quite embarrassing.

5. *In which direction should the teaching be aimed? Is it appropriate to choose the same method as applied to teaching mathematics specialists, or is a different approach called for altogether?*

Although mathematics for engineering students should not be very different from that for pure mathematics students, one should be aware of the fact that pure mathematics students are those who are already attracted or enchanted by the beauty of mathematics. These students are willing to study mathematics as their major subject, whereas students of other subjects think of mathematics as merely a tool to use for their own studies. Thus we need to give the students motivation to study. How can this be achieved? We can begin by showing them how one can construct mathematical models and how one can solve them. What I have been doing in my ODE course for engineering students at Kobe University is this. At the first lecture, they are given a mathematical model of population dynamics. The equation given first is

$$\frac{dx}{dt} = a\,x\,.$$

As is well known, this simple Malthus model is not very good at all. They are then introduced to the famous Verhulst model:

$$\frac{dx}{dt} = a\,x\,\left(1 - \frac{x}{b}\right).$$

By giving this model, I can talk about the asymptotic behaviour of a solution, equilibrium points, and stability of a solution. Then, a small parameter c is added.

$$\frac{dx}{dt} = a\,x\,\left(1 - \frac{x}{b}\right) - c\,.$$

This parameter c may correspond to capturing animals in Africa, or to deforestation by human beings in South America. The parameter c is then increased a little. Here I am in a position to be able to talk about the structural stability. The parameter can then be increased a little more. If the parameter c is increased still further, then, all of sudden, the structural stability is lost, and students realise that all the solution curves drop away from the top left to the bottom right. I can then explain what environmental capacity means to us and how easily we can destroy our world.

I have also found it very effective to demonstrate this example of population dynamics by installing a micro-computer in the classroom displaying families of logistic curves on the screen.

Concerning the way of teaching, one can say that there are two ways of teaching, namely, the *top down method* in which general theory is taught first and examples are explained by applying the general theory, and the *bottom up method* in which examples are taught first and general theory is introduced by extracting properties common to these examples. I think the most effective way is perhaps to give a few typical examples like the ones above in order to motivate the students and then formulate a general theory by extracting properties common to these examples. After proving theorems, more examples of application should be given to show how powerful the theory is. This combination of bottom up and top down methods would be the most efficient.

6. *What qualifications should the instructors ideally possess in order to offer the most effective instruction? Should the teaching be undertaken by those who have specialised in pure mathematics and have carried out some research, or should it*

be undertaken by experienced users of mathematics? Or, indeed, is the ideal tutor for this kind of education someone who is primarily a teaching specialist, research being of secondary concern?

On this matter, Dr. Satsuma of Tokyo University carried out an interesting survey some years ago. According to his report, the students expressed the following views in response to the question: "Who would be an ideal tutor of mathematics to engineering students?" "Those who once worked hard to learn mathematics because they understand how we feel and how much we suffer from an allergy to mathematics;" "Get rid of those who lecture from inside their own world, while ignoring students;" "Mathematics professionals are likely to stay wrapped up in their cocoon, so give us a non-mathematician;" "An expert on teaching mathematics, who understand those who do not understand."

These comments provoke a little awakening. Why should it be all right for middle-school or high-school teachers to be teaching experts, but when it comes to university, those who do no research at all are suddenly able to teach well? Surely, someone who has never done any research cannot fully understand its importance and, therefore, cannot teach it properly. This would surely halve the value of university education. This is why university teachers must be researchers. Even if this researcher were to lack eloquence, his pervading dedication to research is bound to come across and inspire students. Of course, if this researcher also were to strive to "understand those who do not understand," he or she would become quite outstanding. There are those who would leave research to academies and research institutions and have universities solely as organs of education. This, for the above-stated reasons, would not seem appropriate.

7. *It is safe to assume that the importance of teaching mathematics to students of non-mathematics majors will further increase in the future. What sort of policies and methods have we to cope with this increased importance?*

Mathematics itself, to be sure, has undergone many changes. Not only has there been an expansion in the way mathematics provides the basic means of expression for other sciences but, at the same time, it has itself undergone internal changes. Some famous long standing problems, such as the four colour problem or the continuum hypothesis, have already been solved. At the same time, mathematics has become extremely complex and greatly more diversified. It is now almost impossible to understand a paper on a subject completely unrelated to one's area of study. Small wonder that non-mathematicians regard mathematics as a secret sect of mystics, with some few high priests allowed to carry out secret rituals after long years of specialised study, totally incomprehensible to ordinary mortals. Abstraction has advanced to this degree, particularly since the beginning of this century. A good example of this is the Bourbaki group of French mathematicians in the 1940s, who advanced the rigorousness of expressions. Thus mathematics proceeds to greater refinement, creating a beautifully elegant logical structure – the Temple of Pure Mathematics. This results in weakening the link with the other sciences and engineering, for which mathematics ought to provide the basic mode of expression. Even the partition walls between mathematics and physics, and also between mathematics and engineering have grown wider.

This trend, however, has changed a little in the last few years. For example, the latest developments in mathematical physics, particularly the advancements concerning nonlinear problems, are an indication of the fact that the two disciplines are again beginning to come closer together. An involvement of algebra and geometry becomes also noticeable, as well as the traditional analysis.

Since mathematics is different from engineering, it is not desirable that only engineers are to be in charge of teaching mathematics at the faculty of engineering. On the other hand, as transpired from the survey mentioned earlier, if somebody who studied mathematics and does research in pure mathematics is put in charge of teaching students of applied sciences, such a tutor has the tendency to become blinkard and teach pure mathematics, as would be taught to students of mathematics. This occasionally confuses the students.

What, then, is the best way of instructing? Those who teach mathematics to students majoring in other subjects, such as engineering, should ideally understand engineering to some extent, so that they may sufficiently comprehend the attitudes and the enthusiasm which students of engineering have towards their subject. At the same time they should have a sound grasp of mathematics as such and should do research in mathematics or in a very closely related subject. Of course, it is almost impossible for one person to know all branches of engineering in detail, to know what is easy and what is difficult, to know what is important and what is trivial, and at the same time have a top-class grasp of pure mathematics. Nonetheless, mathematics should be taught to engineering students by someone who is not entirely ignorant of what engineering is, who has an adequate understanding of the role mathematics plays within engineering, and who knows how one should teach mathematics to them. Unfortunately, it is difficult to see how people of such profile might easily come out of the present educational establishment in any significant numbers.

In consequence, we should consider some effective strategies for producing teachers of that kind. As I mentioned before, both in engineering and in other applied sciences, there has emerged an increasing number of fields which positively utilise mathematics and entirely depend on it. This now covers a very wide area and what is common to this area is that it no longer relies solely on recalling theorems or formulae, but relies more on a thorough understanding of the underlying mathematical structure. This is applied mathematics in the contemporary sense. It lies between pure mathematics and engineering, bridging the gap between the two. This is why applied mathematics has suddenly acquired much more importance. Reflecting this new importance, Japan SIAM (Japan Society of Industrial and Applied Mathematics) was established earlier this year.

I would now like to make a proposal, which I address firstly to the government of Japan. In each faculty of engineering, create a department of applied mathematics or a department of mathematical engineering that would be responsible for the teaching of mathematics in the faculty and would also create a base for research in applied mathematics and mathematical engineering. Since this department would be involved with teaching mathematics to students of other departments, the staff-student ratio in the department should be appropriately weighted. Those who graduate from such a department are likely to have a foot both in engineering and in mathematics, and should therefore understand both ways of thinking. They would therefore be the ideal graduates for our job. Such people may exceptionally come from computer science or from informatics, but not as a rule. Preferably, they should be inclined more toward mathematics. However, as the role of the computer in those disciplines is quite extensive, the tutors should not suffer from computer allergy.

I should furthermore like to add a few words concerning the staff composition of this department. First, it is likely that a well balanced team could be appointed containing the right number of pure mathematicians in proportion to others. As the number of graduates of the department increased, these graduates may begin occupying a larger part of the staff. At that time there may arise the danger that this would lead to a weakening of the feeling for the mathematical spirit. Unlike other departments within engineering, this department would depend on retaining the mathematical spirit, i.e. mathematics as such. There should always be a reasonable number of pure mathematicians on the staff of these departments if we are to avoid a deterioration. Sufficient attention must be paid to this point.

The situation in other countries is probably similar. It is not a bad idea to establish a faculty of mathematical sciences consisting of a department of pure mathematics, applied mathematics, statistics, and possibly mathematical informatics. Also, it may be thought, for instance, that in developing countries, where the task of developing national resources and industrialisation is pre-eminent, the relevance of applied mathematics far outstrips that of pure mathematics. In these countries, it might be better to begin with application-oriented mathematics and hope that a move towards pure mathematics will take place.

Naturally, we would not be creating this department solely for the purpose of training teachers of mathematics. Most of the graduates from this department would seek employment in industry. It is said that the recent trend in industry is from heavy-thick-long-big to light-thin-short-small. It is impossible to know how long this will continue, but it has resulted in mathematics being directly, rather than indirectly, involved in industry. The mathematics involved is, of course, mainly applied mathematics. Thus the graduates of the departments of applied mathematics or of mathematical engineering that I am proposing here would be the very people now required in industry and would play an important role there. In fact, those countries and industries which do not have such people may well be jeopardizing their future, just as the structure of industries is rapidly upgraded. I should therefore like to emphasise to the industrialists the importance of creating such departments.

Author Index

I indicates Volume I while II indicates Volume II.